Active Control of Noise and Vibration

Also available from E & FN Spon

Engineering Noise Control
Theory and practice
2nd Edition
David A. Bies and Colin H. Hansen

Underwater Acoustic Modelling
Principles, techniques and applications
2nd Edition
Paul C. Etter

Sound Intensity
2nd Edition
Frank J. Fahy

Active Control of Noise and Vibration

Colin H. Hansen and Scott D. Snyder
Department of Mechanical Engineering
University of Adelaide
South Australia

E & FN SPON
An Imprint of Chapman & Hall
London · Weinheim · New York · Tokyo · Melbourne · Madras

Published by E & FN Spon, an imprint of Chapman & Hall, 2–6 Boundary Row, London SE1 8HN, UK

Chapman & Hall, 2–6 Boundary Row, London SE1 8HN, UK

Chapman & Hall, GmbH, Pappelallee 3, 69469 Weinheim, Germany

Chapman & Hall USA, 115 Fifth Avenue, New York, NY 10003, USA

Chapman & Hall Japan, ITP-Japan, Kyowa Building 3F, 2-2-1 Hirakawacho, Chiyoda-ku, Tokyo 102, Japan

Chapman & Hall Australia, 102 Dodds Street, South Melbourne, Victoria 3205, Australia

Chapman & Hall India, R. Seshadri, 32 Second Main Road, CIT East, Madras 600 035

First edition 1997

Typeset in 10/12 Times by Amy Boyle, Walderslade, Kent
Printed in Great Britain by Cambridge University Press

ISBN 0 419 19390 1

A catalogue record for this book is available from the British Library

∞ Printed on acid-free paper, manufactured in accordance with ANSI/NISO Z39.48-1992 (Permanence of Paper)

Contents

This book is dedicated
to Susan
to Gillian
to Thomas
to Kristy
and to Laura

Preface

Active control of sound and vibration is a relatively new and fast growing field of research and application. The number of papers published on the subject have been more than doubling every year for the past ten years and each year more researchers are becoming involved in this fascinating subject. Because of this rapid growth and continuing new developments, it is difficult for any book to claim to cover the subject completely. However, we have attempted to include the most recent theoretical and practical developments, while at the same time devoting considerable space to fundamental principles which will not become outdated with time. We have also devoted space to explaining how active control systems may be designed and implemented in practice, and the practical pitfalls which must be avoided to ensure a reliable and stable system.

We have treated the active control of noise and the active control of vibration in a unified way, even though later on in the book some noise and vibration control topics are treated separately. The reason for the unified treatment is that it is becoming increasingly difficult to keep the two disciplines separate, as one depends so much on the other. For example, the treatment of the active control of sound radiated by vibrating structures would be incomplete if either active control of the radiated acoustic field or the active control of the structural vibration were omitted. Thus, in the first part of the book, which is concerned entirely with fundamental concepts of relevance to active noise and vibration control, an attempt has been made to combine the two subjects so that it can be seen how they are related and how they share many common concepts.

One interesting topic which we have omitted from this book is a discussion of patents. Not only is the subject a large one (with hundreds of patents already granted), it is rapidly growing and it is difficult to do justice to it in a book of this type. One gem of wisdom which we would like to share with our readers is that some of the patent holders are only too willing to sue for patent infringement, even though it can be shown that prior knowledge existed before many of the patents. Thus, any company preparing to market any products containing active noise or vibration control should be prepared to fight a legal battle for their right to do so unless they have obtained a licence to use the technology from the patent owners. So far, it seems that lawyers and judges have made more money from active control than any engineering company, and they don't even own any patents! It is also of interest to ponder upon the number of patents which are granted which closely describe an aspect of active noise and vibration control which has been patented previously. All we can do here is try to appeal to some sense of reason as too much litigation of this type will stifle research, slow down new product development and enrich the legal profession – all of which we would be better off without.

Acknowledgements

The authors would like to thank Gill Snyder and Myriam Piccinino for their expert help in word processing a substantial amount of the first draft. The second author would also like to thank his daughter, Kristy, for her assistance with some of the figures. Many of the graduate students supervised by the authors spent much of their time proof reading the manuscript and offering constructive criticisms and for this we thank them. We would like to thank Mike Brennan from the UK for providing the information on electrorheological actuators which we used in Chapter 15. The first author would also like to thank the University of Adelaide for the opportunity to spend a substantial amount of two six-month sabbaticals working on the manuscript. Finally we would like to thank our wives Sue and Gill and daughter Kristy for their patience and support during the last four years when it has seemed that 'the book' has dominated our lives.

1

Background

1.1 INTRODUCTION AND POTENTIAL APPLICATIONS

The control of low frequency noise and vibration has traditionally been difficult and expensive and in many cases not feasible, because of the long acoustic wavelengths involved. If only passive control techniques are considered, these long wavelengths make it necessary to use large mufflers and heavy enclosures for noise control, and very soft isolation systems and/or extensive structural damping treatment (including application of vibration absorbers) for vibration control. As long ago as the 1930s the idea of using active sound cancellation as an alternative to passive control for low frequency sound was proposed by Paul Lueg (1933, 1936). The idea was to use a transducer (control source) to introduce a secondary (control) disturbance into the system to cancel the existing (primary) disturbance, thus resulting in an attenuation of the original sound. The cancelling disturbance was to be derived electronically based upon a measurement of the primary disturbance. It is this alternative means of noise and vibration control which emerged from such modest beginnings to which this book is dedicated.

Since the original idea was conceived in those very early days, the active control of sound and vibration as a technology has been characterized by transition: transition from a dream to practical implementation and from a laboratory experiment to mass production. This transition has taken a long time, partly because of the time it took to develop sufficiently powerful signal processing electronics, partly because of a lack of understanding of the physical principles involved and partly because of the multidisciplinary nature of the technology which combines a wide range of technical disciplines including physics, electrical engineering, materials science and mechanical engineering. Being a collection of pieces, in which the strength of the chain is only as strong as its weakest link, it is little wonder that the technology has been characterized by advances which have come in a series of spurts rather than in a continuous flow.

After the exposition of the original idea of active control of noise in ducts in the 1930s, it was not until the 1950s that the idea was rekindled, this time by a man named Olson (1953, 1956), who investigated possibilities for active sound cancellation in rooms, in ducts and in headsets and earmuffs. Again, limitations in the available electronic control hardware as well as limitations in control theory prevented this technology from being commercially realized.

In the late 1970s and 1980s there was a resurgence of interest in active sound cancellation. Advances in control theory and, perhaps more importantly, advances in microelectronics meant that commercial systems were technically achievable. The result was the installation of a number of 'prototype' systems in industrial facilities to control low frequency noise in situations where existing passive control techniques were exorbitantly expensive or impractical. In spite of these advances, there were still a number of technical problems which prohibited widespread implementation of the technology. Perhaps the most limiting of these was associated mainly with the available of transducers and actuators: stable, high power, low-frequency-response sound and vibration sources and rugged sensors capable of continuous operation for long periods of time in harsh industrial environments were simply not available. Other factors which slowed the development of commercial active control systems in the 1980s include:

- insufficient experience of practical installations;
- complexity and cost of systems;
- lack of education of designers and potential users;
- insufficient evidence of cost savings, long-term performance and reliability; and
- lack of sufficient marketing effort.

In the 1990s, many of these problems will be overcome; inexpensive multichannel electronic controllers and inexpensive, robust transducers and actuators are being developed and books such as this will help to educate potential system designers. 'Chipsets' aimed specifically at mass market implementations of active sound and vibration control exist in the prototype stage and are being aggressively marketed by their manufacturers.

Research in active sound and vibration control is also expanding: the number of technical papers published on the topic since Lueg's work in the 1930s is increasing exponentially, from approximately 240 before 1970 to 850 in the 1970s and to 2200 in the 1980s, a trend which is continuing in the 1990s.

Modern active sound or vibration control systems consist of one or more control sources used to introduce a secondary (or controlling) disturbance into the structural/acoustic system. This disturbance suppresses the unwanted noise or vibration originating from one or more primary sources. The (control) signals which drive the control actuators are generated by an electronic controller which uses as inputs measurements of the residual field (remaining after introduction of the control disturbance) and in the case of feedforward systems, a measure of the incoming primary disturbance. Active noise and vibration control systems are ideally suited for use in the low frequency range, below approximately 500 Hz. Although higher frequency active control systems have been built, a number of technical difficulties, both structural/acoustic (for example, more complex vibration and radiated sound fields) and electronic (where higher sampling rates are required), limit their efficiency. At higher frequencies passive systems also become more cost effective. A 'complete' noise or vibration control system would generally consist of active control for low frequencies and passive control for higher frequencies.

An important property of many modern active sound and vibration control systems (particularly feedforward systems) is that they are self-tuning (adaptive) so that they can adapt to small changes in the system being controlled. Non-adaptive controllers are generally confined to the feedback type in cases where slight changes in the environmental conditions will not be reflected in significant degradation in controller performance. This book will concentrate heavily on adaptive control systems, although many of the principles outlined apply equally well to non-adaptive systems. Characteristics of feedforward and feedback control systems are discussed in the next section.

As integrated microprocessors dedicated to signal processing become less expensive and faster (the speed having doubled every 18 months for the last 10 years), potential active control applications increase in number. However, it should not be assumed that more processing power will extend the applications endlessly. There are some supposedly potential applications (for example, control of traffic noise in living rooms) which will remain impractical, no matter how much processing power is available, because the limitations are a result of the structural/acoustic characteristics of the problem. Although more powerful signal processing electronics help to alleviate the electronic problems associated with extending the application of active control to higher frequencies and to more complex multichannel problems, the structural/acoustic limitations mentioned remain. For the example cited above, to provide significant (or any) attenuation of the unwanted disturbance, a vast array of sensors and actuators would be required: it would be cheaper to build a thicker wall! These limitations will become more apparent later in this book.

The efficiency of active noise and vibration control systems depends upon the design of and harmony of operation between, two major subsystems; the 'physical' system, and the electronic control system. The physical system encompasses the required transducers; the 'control sources' for inducing the secondary disturbance, and the 'error sensors' which monitor the performance of the active control system by providing some measure of the residual noise and/or vibration field. Thus, the physical system provides the structural/acoustic interface for the active control systems, and the electronic control system drives the physical system in such a way that the unwanted primary source noise and/or vibration field is attenuated.

The quality of the design of these two major subsystems is the critical factor in determining the ability of the active control system to produce the desired results. The design of the physical system, comprising the arrangement of control sources and error sensors, limits the maximum noise or vibration control that can be achieved by an ideal active controller. The control electronics limit the ability of the active control system to reach this maximum achievable result. Thus, although the influence of the quality (or lack thereof) of the two major subsystems manifest themselves in different ways, no active control system can function efficiently with an inefficient physical or electronic subsystem.

The design requirements for the electronic and physical subsystems are very different from, although not completely independent of, one another. Similarly, the design of these subsystems varies from application to application in that the appropriate control strategy is dependent upon the control objective, whether it be vibration control, radiated sound power control, sound transmission control or some other objective. For example, the physical control system for reducing aircraft interior noise is not the same as the physical control system for reducing noise transmission in an air handling duct or the system for vibration isolation of an electron microscope. Similarly, the electronic controller for an adaptive feedforward system is not the same as the electronic controller for a feedback system. However, the underlying principles of efficient design for each subsystem are the same. The purpose of this book is to outline these principles; to explain the physical mechanisms of active noise and vibration control; to provide general methodologies for the design of both the physical and electronic control system; to apply these to a variety of common low frequency noise and vibration problems; and to describe the means of physically realizing active noise and vibration control systems. As time goes on it is becoming more and more difficult to keep the disciplines of active control of sound and active control of vibration separate because noise control is often dependent on structural vibration control and because the control of acoustic fields can result in structural vibration control. This is the reason for the integrated treatment in this book; even though some noise and vibration control topics are treated separately later on, the fundamental principles governing the physical system behaviour and controller design are treated in a unified way, as is the control of structural sound radiation.

Active control of sound and vibration is an exciting field of research which is relatively new. Practical systems are now being realized and installed, and efforts directed at understanding the behaviour of experimental systems are providing us with new insights and a better overall understanding of the fascinating subjects of acoustics, vibration and structural acoustics. Some typical applications for active noise and vibration control are listed below.

1. Control of aircraft interior noise by use of lightweight vibration sources on the fuselage and acoustic sources inside the fuselage.
2. Reduction of helicopter cabin noise by active vibration isolation of the rotor and gearbox from the cabin.

3. Reduction of noise radiated by ships and submarines by active vibration isolation of interior mounted machinery (using active elements in parallel with passive elements) and active reduction of vibratory power transmission along the hull, using vibration actuators on the hull.
4. Reduction of internal combustion engine exhaust noise by use of acoustic control sources at the exhaust outlet or by use of high intensity acoustic sources mounted on the exhaust pipe and radiating into the pipe at some distance from the exhaust outlet.
5. Reduction of low frequency noise radiated by industrial noise sources such as vacuum pumps, forced air blowers, cooling towers and gas turbine exhausts, by use of acoustic control sources.
6. Lightweight machinery enclosures with active control for low frequency noise reduction.
7. Reduction of low frequency vibration of structures (including structures for future space stations) by use of lightweight vibration actuators such as piezoelectric ceramic (often referred to as piezoceramic) crystals.
8. Reduction of sway in tall buildings.
9. Reduction of low frequency noise propagating in air conditioning systems by use of acoustic sources radiating into the duct airway.
10. Reduction of electrical transformer noise either by using a secondary, perforated lightweight skin surrounding the transformer and driven by vibration sources or by attaching vibration sources directly to the transformer tank. Use of acoustic control sources for this purpose is also being investigated but a large number of sources are required to obtain global control.
11. Reduction of noise inside automobiles using acoustic sources inside the cabin and lightweight vibration actuators on the body panels.
12. Active suspension systems for all vehicle and machinery types. These systems will usually involve an active system in parallel with a passive system.
13. Active headsets and earmuffs.

The first commercially available active control systems were those for controlling low frequency, plane wave sound propagation in air handling ducts. Systems for other noise and vibration control problems are currently under development; however, each type of application presents its own peculiar problems which must be solved on an individual basis. It would be rare for success to be obtained by an arbitrary application of a single-channel controller or even a multichannel controller to a system for which the physical arrangement of the control sources and error sensors has not been properly optimized by detailed analysis. One of the aims of this book is to explain how such a detailed analysis and optimization may be undertaken for any system of interest and how a multichannel control system may be designed and implemented.

As a good understanding of the underlying physics of a particular noise or vibration problem is an essential part of designing an optimum control system, a significant part of this book will be devoted to the development of this understanding by discussing in depth the underlying theoretical concepts. Although it may not seem so at times, the content of this book is directed solely at active control of noise and vibration. The basic principles and underlying theoretical concepts of the disciplines relevant to this topic are covered at a level and depth necessary for their application to active control. Some may argue that the coverage of some topics is superficial and best left to dedicated texts on the particular subject. We believe that the coverage given here is adequate for active control applications, and that there is a considerable advantage in presenting the material in a form and in the detail suitable just for active noise and vibration control, thus saving the reader from searching various other books for the material needed to understand the remainder of this book.

This book is divided into 15 chapters followed by an appendix. To begin, essential fundamentals of acoustics, vibration, signal processing and control are covered. This is followed by a discussion of the physical principles of active noise control, and the principles of active vibration control. Next the design and practical implementation of active noise and vibration control systems, including a description of sound and vibration transducers and electronic hardware is considered.

The remainder of Chapter 1 is devoted to providing a brief overview of how active noise and vibration control systems work in principle. Chapter 2 is concerned with the fundamental physical principles of structural acoustics, which includes a review of the basics of acoustics and vibration which are a necessary foundation for understanding the work in the remainder of the book. Chapter 3 is concerned with the principles of frequency analysis, Chapter 4 covers both theoretical and experimental modal analysis, Chapter 5 covers the fundamentals of modern feedback control and Chapter 6 covers the principles of feedforward control. The physical principles of active control of sound in ducts (plane wave and higher order mode transmission), sound transmission into enclosed spaces and sound radiation from vibrating surfaces are discussed in Chapters 7, 8 and 9 respectively. Feedforward control of vibration in structures such as beams and plates is discussed in Chapter 10 followed by a treatment of feedback control of structural vibration in Chapter 11; and in Chapter 12 both feedforward and feedback isolation of vibration of one system from another is considered. Finally, the purpose of Chapters 13, 14 and 15 is to tell the reader how to construct a physical control system, as well as to describe the hardware which may be used for the signal processing, the control actuators and the sensors.

1.2 OVERVIEW OF ACTIVE CONTROL SYSTEMS

One of the simplest applications of active noise control is associated with the attenuation of plane waves propagating in air ducts. The simplicity of this application is a result of the problem being one dimensional; the sound field is restricted from spreading out in three dimensions by the walls of the duct and

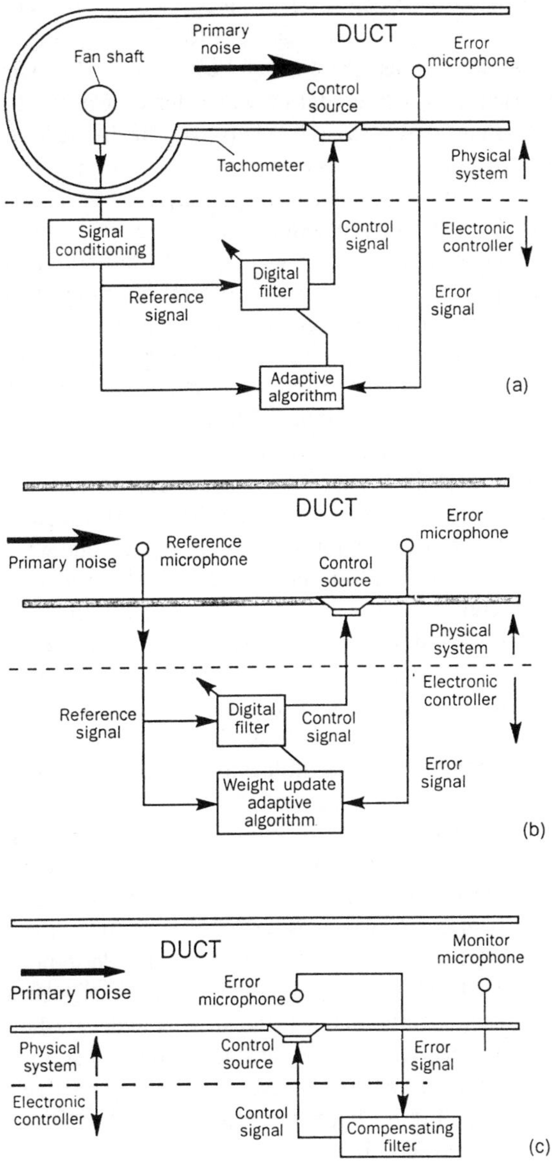

Fig. 1.1 Various schemes for active control of plane waves propagating in ducts: (a) feedforward with tachometer signal for the controller reference input; (b) feedforward with a microphone signal for the controller reference input; (c) feedback.

can only propagate in one direction along the duct. Commercial active systems to control duct noise were the first to be applied in practice by a number of years, with the first systems being installed in the early 1980s. It seems appropriate, therefore, to use the duct application to explain some of the various different approaches to active control which have been used in the past. The active control systems described in this book may be divided into two categories: feedforward and feedback. A schematic diagram of a typical implementation of each of these controller types is shown in Fig. 1.1. Each acts to suppress the noise generated by some source, referred to here as the primary source.

Feedforward controllers (which are invariably digital) rely on the availability of a reference signal which is a measure of the incoming disturbance (noise or vibration). This signal must be received by the controller in sufficient time for the required control signal to be generated and output to the control source when the disturbance (from which the reference signal was generated) arrives. For stationary or slowly varying periodic disturbances this time constraint need not be satisfied, as the assumption can be made that the signal during one period will be very similar to that during the previous period. Thus it is relatively easy to obtain a reference signal for a periodic disturbance. Although it is more difficult to obtain a satisfactory reference signal for a random or non-periodic disturbance, a predictive measure can usually be obtained if the disturbance is travelling in a confined space such as a duct. In this case, an upstream measurement can be used to predict the disturbance at some downstream location at a later time. Systems such as these, for which the active control system produces the control signal at the downstream location at the same time that the primary signal arrives, are referred to as 'causal'.

Causality is a condition which all feedforward controller designs must satisfy if the noise or vibration to be controlled is not periodic. This means that for random noise, the time it takes for the primary acoustic or vibration signal to travel from the reference sensor location to the error sensor location must be greater than the processing time of the controller, plus the time delay associated with the control source electroacoustics, plus the time taken for the signal to travel from the control sources to the error sensors. Periodic signals which are slowly varying need not satisfy this condition as it may be assumed that the characteristics of one period are sufficiently similar to those of the period preceding it. Causality also affects feedback controllers to a certain extent, depending on the type of control law used. For example, if velocity feedback is used, the system damping is effectively increased, and no control is possible for the first cycle of a transient disturbance.

An important point to note is that if the reference signal is supplied by a microphone in the duct, the signal will be subject to contamination by the upstream travelling control disturbance, an effect which must be taken into account in system design.

The digital feedforward control system depicted in Fig. 1.1(a) specifically targets active attenuation of the discrete tones in the duct, such as those generated by a fan (in which case the tonal frequencies would be equal to the blade passage frequency and its harmonics). In this arrangement, the tachometer output is synchronized with the rotating shaft of the fan responsible for the periodic primary noise. The signal conditioning electronics convert the tachometer signal into a combination of sinusoids which provide a predictive measure of the fundamental rotational frequency of the fan and the harmonics which are to be controlled.

To generate the appropriate signal to drive the control source, the reference signal (see Fig. 1.1(a)) is passed through a digital filter to generate the resulting control signal which is fed to a control source which introduces the control disturbance into the duct. Non-adaptive controllers are characterized by fixed digital filters, the parameters of which are determined by using acoustic system analysis or by using trial and error in such a way that the signal at the 'error' microphone is minimized. Unfortunately, the physical acoustic or vibration system to be controlled rarely remains the same for very long (as even small changes in temperature or flow speed change the speed of sound significantly), resulting in large phase errors between the desired and actual control signals. At first, attempts were made to overcome this problem by updating the filter weights iteratively, based on a measurement of the root mean square (r.m.s.) signal at the error microphone. Current practice involves the use of an adaptive algorithm, as shown in Figs. 1.1(a) and (b), (discussed in Chapter 6) to adjust the characteristics of the adaptive filter to minimize the downstream residual disturbance, the measure of which is the instantaneous value of the squared signal detected by an 'error' microphone. In this case the filter weights are updated in a time frame of the order of the digital sampling rate, and much better results are obtained. Various other aspects which need to be taken into account in the filter weight update algorithm, such as the electroacoustic transfer functions of the loudspeaker and error microphone are discussed in detail in Chapter 6.

If broadband random noise is to be controlled, a reference signal correlated with all components of the primary signal must be obtained. For the duct noise control case being considered here, this could be done by substituting a microphone in the duct for the tachometer, as shown in Fig. 1.1(b). Unfortunately, this arrangement often leads to the reference signal being contaminated with the component of the signal from the control source which is transmitted upstream, possibly leading to system instability, which is why this arrangement is not preferred if only periodic or tonal noise is to be controlled. Ways of overcoming this stability problem are discussed in Chapter 6.

Feedback control systems differ from feedforward systems in the manner in which the control signal is derived. Whereas feedforward systems rely on some predictive measure of the incoming disturbance to generate an appropriate 'cancelling' disturbance, feedback systems aim to attenuate the residual effects of the disturbance after it has passed. Feedback systems are thus better at reducing the transient response of systems, while feedforward systems are best at reducing the steady state response. In structures and acoustic spaces, feedback controllers effectively add modal damping and in the duct system shown in Fig.

1.1(c) the feedback controller also reflects incoming waves by modifying the duct wall impedance at the control loudspeaker. Thus, unlike feedforward systems for which the physical system and controller can be optimized separately, feedback systems must be designed by considering the physical system and controller as a coupled system.

Referring to Fig. 1.1(c), it can be seen that a feedback controller derives a control signal by filtering an error signal, not by filtering a reference signal as is done by a feedforward controller. In active noise and vibration control systems, the characteristics of the feedback control system are chosen so as to return the system (as measured at an error sensor) to its unperturbed state as quickly as possible, subject to some system stability constraints.

It is of interest to explore the physical mechanisms which are involved in active noise and vibration control. In systems characterized by propagating waves, such as the duct system shown in Fig. 1.1, feedback systems function primarily by reflecting incoming waves (which are subsequently dissipated by internal losses in the vibroacoustic system) as they attempt to force a pressure or vibration null at the control source location, resulting an impedance mismatch. In structures or acoustic spaces which contain standing waves such that they can be described modally, feedback controllers result in a change in system resonance frequencies and damping.

Feedforward control systems for periodic noise, which are directed at achieving global noise reductions such that the total sound power radiated or total system potential energy is reduced, function by affecting the source radiation impedance and in some cases the control sources can absorb sound or vibratory power. For example, loudspeakers can absorb sound power if their cone motion is appropriate. Maintaining the appropriate cone motion, however, takes considerably more power than is available for absorption. Thus, all that will be noticed is a minuscule reduction in the power required to drive the loudspeaker. However, it can be shown (see Chapter 8) that this absorption phenomenon only occurs for a suboptimally adjusted controller. For an optimally adjusted controller, no absorption occurs and the attenuation of the disturbance is achieved entirely by modification of the radiation impedance presented to the source of the primary disturbance. When the noise or vibration is completely random the control mechanisms associated with feedforward control are a little more restricted. As the disturbance is random, causality constraints prevent the control source from optimally affecting the primary source radiation impedance and for sound propagating along a duct (or vibration propagating along a structure), the principal mechanism is one of reflection and subsequent dissipation of the energy by internal system losses, although absorption of energy by the control sources can often play an important role.

For free field, random sound sources, there are no known physical mechanisms that would allow global control (either feedback or feedforward) using acoustic sources, although it is possible to achieve local zones of cancellation which are generally at the expense of increased levels elsewhere. Global control is sometimes possible by controlling the vibration of the structure generating the noise, provided that a sample of the disturbance driving the structure can be obtained sufficiently far in advance. Zones of cancellation can be achieved using either feedforward or feedback control and the reduced local sound field results from interference between the primary sound field and the field produced by the control sources. A local minimum can be achieved by placing an error sensor in the region to be controlled. Unfortunately, the region in which the noise level can be reduced by 10 dB or more is only about one-tenth of a wavelength in radius, which makes this technique less than useful for anything but very low frequency noise.

It is important to remember that the physical arrangement of control sources and error sensors plays a very important role in determining the effectiveness of an active control system. Moving the locations of the control sources and sensors affects both system controllability and stability. For feedforward systems, the physical system arrangement can be optimized independently of the controller, but for feedback systems, the physical system arrangement is an important part of the controller design. Also of importance is the size of the source to be controlled compared to an acoustic (for sound control) or structural (for vibration control) wavelength at the lowest frequency to be controlled. Clearly, one small control source will be ineffective in achieving global control of a primary source which is many wavelengths in dimensions because of its inability to significantly change the primary source radiation impedance. These concepts will be demonstrated in later chapters.

Some of the other problems which must be addressed in the design of feedforward electronic controllers include acoustic or vibration feedback from the control source to the reference sensor and non-linear control sources (that is, when the acoustic or vibration output of the control source contains frequencies not present in the electrical input; a condition often caused by harmonic distortion in the control actuators or driving amplifiers). In addition it is necessary to either provide on-line system identification of the electroacoustic transfer functions of the control sources and error microphones and the acoustic delay between them (cancellation path) or else design a complex controller which does not need this information. This latter alternative could require the use of complex filter structures such as neural networks or non-linear adaptive algorithms such as genetic algorithms.

One question which may be asked is how one decides whether to use a feedforward or a feedback controller for a particular application. The answer is that a feedforward system should be implemented whenever it is possible to obtain a suitable reference signal, because the performance of a feedforward system is, in general, superior to a feedback system. Unfortunately, in many instances it is not possible to obtain a suitable reference signal, as when attempting to reduce the resonant response of an impulsively excited structure. In cases such as these, feedback control systems are especially suitable and, indeed, are the only alternative.

As has been mentioned previously, active control can be applied effectively to both noise and vibration problems. For noise problems, feedforward control has been applied successfully to ducts and on an experimental basis to noise in aircraft cabins and motor vehicle interiors and exteriors. Feedback control has been applied successfully to ear defenders where it is not easy to sample in advance the incoming signal, making it difficult to generate an appropriate reference signal for a feedforward controller.

In the area of structural vibration, the use of feedback control has been popular because of its ability to damp structural vibrations without the need to be able to measure in advance a reference signal. However, if it is possible to obtain a reference signal sufficiently far ahead of the required control source action, feedforward control is almost invariably preferred over feedback control because of its inherent stability characteristics and usually superior performance.

As already mentioned in the preface, it is important to realize that many aspects of active noise and vibration control are the subject of numerous US and worldwide patents (numbering approximately 1600 at the end of 1994); the reader is warned that a patent search is an essential prerequisite to the commercialization of any active noise or vibration control system. Patent infringement is likely to be costly, as many patents are owned by companies with sufficient resources to sue offenders successfully.

REFERENCES

Lueg, P. (1933). Process of Silencing Sound Oscillations. German Patent DRP No. 655 508.

Lueg, P. (1936). Process of Silencing Sound Oscillations. US Patent No. 2043 416.

Olson, H.F. (1953). Electronic sound absorber. *Journal of the Acoustical Society of America*, **25**, 1130–1136.

Olson, H.F. (1956). Electronic control of noise, vibration and reverberation. *Journal of the Acoustical Society of America*, **28**, 966–972.

2

Fundamentals of acoustics and vibration

In this chapter we cover many of the fundamental concepts which form the basis of the theoretical analyses undertaken throughout the rest of the book. Both fundamentals of acoustics and fundamentals of vibrations are discussed in what we hope is a unified way. We begin with the derivation of the acoustical wave equation, followed by a discussion of the fundamentals of structural mechanics. These fundamentals are used as a basis to derive the wave equation for various wave types in beams, plates and cylinders. The concepts of acoustic waves in fluid media are extended to waves in structures. Next is discussed the concept and application of Green's functions to the determination of the pressure response in an acoustic medium to an arbitrary excitation force distribution (which could be a structural surface), and the determination of the displacement response of a structure to an arbitrary point or distributed excitation force. Finally, the concepts of acoustic impedance, structural impedance, acoustic intensity, structural intensity and power transmission are discussed with reference to a number of examples.

2.1 ACOUSTIC WAVE EQUATION

Before we derive the acoustic wave equation, we will discuss some underlying fundamental concepts. Acoustic disturbances travel through fluid media in the form of longitudinal waves and are generally regarded as small amplitude perturbations to an ambient state. For a fluid such as air or water, the ambient state is characterized by the values of the physical variables (pressure P, velocity U and density ρ_0) which exist in the absence of the disturbance. The ambient state defines the medium through which the sound propagates. A homogeneous medium is one in which all ambient quantities are independent of position. A quiescent medium is one in which they are independent of time and in which $U = 0$. The idealization of homogeneity and quiescence will be assumed in the following derivation of the wave equation, as this generally provides a satisfactory quantitative description of acoustic phenomena.

The previously mentioned ambient field variables satisfy the fluid dynamic equations, and when an acoustic disturbance is present, the effect this has on the variables must be included. Thus in the presence of an acoustic disturbance we have

$$\begin{aligned} &\text{Pressure:} && P_{tot} = P + p(\boldsymbol{r}, t) \\ &\text{Velocity:} && \boldsymbol{U}_{tot} = \boldsymbol{U} + \boldsymbol{u}(\boldsymbol{r}, t) \\ &\text{Temperature:} && T_{tot} = T + \tau(\boldsymbol{r}, t) \\ &\text{Density:} && \rho_{tot} = \rho_0 + \sigma(\boldsymbol{r}, t) \end{aligned} \qquad \text{(2.1.1a,b,c,d)}$$

Note that vector quantities are in bold typeface.

The derivation of the acoustic wave equation is based on three fundamental fluid dynamic equations; the continuity (or conservation of mass) equation, Euler's equation (or the equation of motion) and the equation of state. Each of these equations will be discussed separately.

2.1.1 Conservation of mass

Consider an arbitrary volume V as shown in Fig. 2.1. The total mass contained in this volume is $\int_V \rho_{tot}\,\mathrm{d}V$. The law of conservation of mass states that the rate of mass leaving the volume V must equal the rate of change of mass in the volume. That is,

$$\int_A \rho_{tot}\boldsymbol{U}_{tot}\cdot\boldsymbol{n}\,\mathrm{d}A = -\frac{\mathrm{d}}{\mathrm{d}t}\int_V \rho_{tot}\,\mathrm{d}V \qquad (2.1.2)$$

where A is the area of surface enclosing the volume V, and $\boldsymbol{n}$ is the unit vector normal to the surface A at location $\mathrm{d}A$.

At this stage it is convenient to transform the area integral on the left-hand side of (2.1.2) to a volume integral by use of Gauss's integral theorem, which is written as follows:

$$\int_A \psi\cdot\boldsymbol{n}\,\mathrm{d}A = \int_V \nabla\cdot\psi\,\mathrm{d}V \qquad (2.1.3)$$

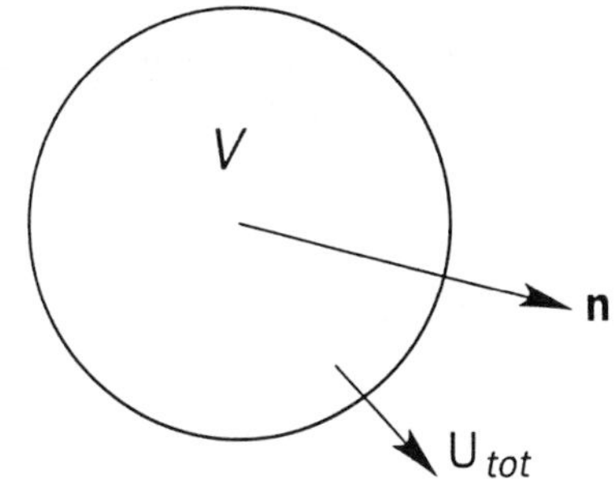

Fig 2.1 Arbitrary volume for illustrating conservation of mass.

where ψ is an arbitrary vector and the operator ∇ is the scalar divergence of the vector ψ. Thus, in cartesian coordinates,

$$\nabla \cdot \psi = \frac{\partial \psi}{\partial x} + \frac{\partial \psi}{\partial y} + \frac{\partial \psi}{\partial z} \tag{2.1.4}$$

and (2.1.2) becomes

$$\int_V \nabla \cdot (\rho_{tot} \boldsymbol{U}_{tot}) \, \mathrm{d}V = -\frac{\mathrm{d}}{\mathrm{d}t} \int_V \rho_{tot} \, \mathrm{d}V = -\int_V \frac{\partial \rho_{tot}}{\partial t} \mathrm{d}V \tag{2.1.5a,b}$$

Rearranging gives

$$\int_V \left[\nabla \cdot (\rho_{tot} \boldsymbol{U}_{tot}) + \frac{\partial \rho_{tot}}{\partial t} \right] \mathrm{d}V = 0 \tag{2.1.6}$$

or,

$$\nabla \cdot (\rho_{tot} \boldsymbol{U}_{tot}) = -\frac{\partial \rho_{tot}}{\partial t} \tag{2.1.7}$$

Equation (2.1.7) is the continuity equation.

2.1.2 Euler's equation

In 1775 Euler derived his well-known equation of motion for a fluid, based on Newton's first law of motion. That is, the mass of a fluid particle multiplied by its acceleration is equal to the sum of the external forces acting upon it.

Consider the fluid particle of dimensions Δx, Δy and Δz shown in Fig. 2.2. The external forces F acting on this particle are equal to the sum of the pressure differentials across each of the three pairs of parallel forces. Thus,

$$F = \boldsymbol{i} \cdot \frac{\partial P_{tot}}{\partial x} + \boldsymbol{j} \cdot \frac{\partial P_{tot}}{\partial y} + \boldsymbol{k} \cdot \frac{\partial P_{tot}}{\partial z} = \nabla P_{tot} \tag{2.1.8a,b}$$

where ***i***, ***j*** and ***k*** are the unit vectors in the x, y and z directions and where the operator ∇ is the gradient operator and is the vector gradient of a scalar quantity.

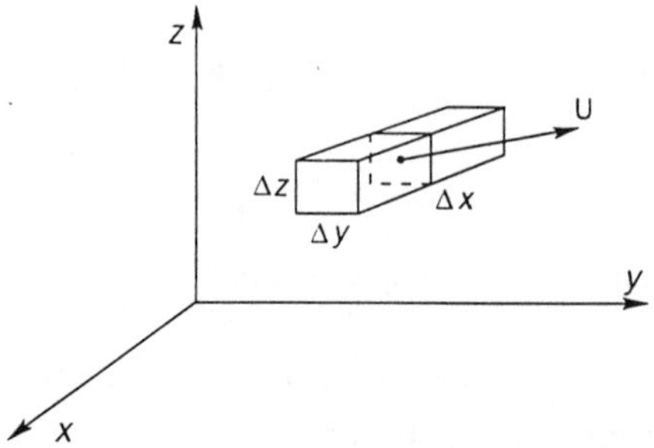

Fig. 2.2 Particle of fluid.

The inertia force of the fluid particle is its mass multiplied by its acceleration and is equal to

$$m\dot{\boldsymbol{U}}_{tot} = m\frac{\mathrm{d}\boldsymbol{U}_{tot}}{\mathrm{d}t} = \rho_{tot}V\frac{\mathrm{d}\boldsymbol{U}_{tot}}{\mathrm{d}t} \tag{2.1.9a,b}$$

At this stage it is important to check that the signs of the terms are correct. Assume that the fluid particle is accelerating in the positive x, y and z directions. Then the pressure across the particle must be decreasing as x, y and z increase, and the external force must be negative. Thus,

$$\boldsymbol{F} = -\nabla P_{tot} = \rho_{tot}V\frac{\mathrm{d}\boldsymbol{U}_{tot}}{\mathrm{d}t} \tag{2.1.10a,b}$$

This is the Euler equation of motion for a fluid.

If we were interested in considering sound propagation through porous acoustic media, then it would be necessary to add the term $R\boldsymbol{U}_{tot}$ to the right-hand side of (2.1.10), where R is a constant dependent upon the properties of the fluid.

The term $\mathrm{d}\boldsymbol{U}_{tot}/\mathrm{d}t$ on the right-hand side of (2.1.10) can be expressed in partial derivative form as follows:

$$\frac{\mathrm{d}\boldsymbol{U}_{tot}}{\mathrm{d}t} = \frac{\partial\boldsymbol{U}_{tot}}{\partial t} + (\boldsymbol{U}_{tot}\cdot\nabla)\boldsymbol{U}_{tot} \tag{2.1.11}$$

where

$$(\boldsymbol{U}_{tot}\cdot\nabla)\boldsymbol{U}_{tot} = \frac{\partial\boldsymbol{U}_{tot}}{\partial x}\cdot\frac{\partial x}{\partial t} + \frac{\partial\boldsymbol{U}_{tot}}{\partial y}\cdot\frac{\partial y}{\partial t} + \frac{\partial\boldsymbol{U}_{tot}}{\partial z}\cdot\frac{\partial z}{\partial t} \tag{2.1.12}$$

2.1.3 Equation of state

As sound propagation is associated with only very small perturbations to the ambient state of a fluid, it may be regarded as adiabatic. Thus the total pressure P will be functionally related to the total density ρ_{tot} as follows:

$$P_{tot} = f(\rho_{tot}) \tag{2.1.13}$$

Since the acoustic perturbations are small, and P and ρ_0 are constant, $\mathrm{d}p = \mathrm{d}P_{tot}$, $\mathrm{d}\sigma = \mathrm{d}\rho_0$ and (2.1.13) can be expanded into a Taylor series:

$$\mathrm{d}p = \frac{\partial f}{\partial \rho_0}\mathrm{d}\sigma + \frac{1}{2}\frac{\partial f}{\partial \rho_0}(\mathrm{d}\sigma)^2 + \text{ higher order terms} \tag{2.1.14}$$

The equation of state is derived by using (2.1.14) and ignoring all of the higher order terms on the right-hand side of (2.1.14). This approximation is adequate for moderate sound pressure levels, but becomes less and less satisfactory as the sound pressure level exceeds 130 dB (60 Pa). Thus, for moderate sound pressure levels we have

$$\mathrm{d}p = c_0^2\,\mathrm{d}\sigma \tag{2.1.15}$$

where $c_0^2 = \dfrac{\partial f}{\partial \rho_0}$ is assumed to be a constant. Integrating (2.1.15) gives

$$p = c_0^2\sigma + \text{const} \tag{2.1.16}$$

which is the linearized equation of state.

Thus the curve $f(\rho_0)$ of (2.1.13) has been replaced by its tangent at P_{tot}, ρ_{tot}. The constant may be eliminated by differentiating (2.1.16) with respect to time. Thus,

$$\frac{\partial p}{\partial t} = c_0^2\frac{\partial \sigma}{\partial t} \tag{2.1.17}$$

Equation (2.1.17) will be used to eliminate $\dfrac{\partial \sigma}{\partial t}$ in the derivation of the wave equation to follow.

2.1.4 Wave equation (linearized)

The wave equation may be derived from (2.1.7), (2.1.10) and (2.1.17) by making the linearizing approximations listed below. These assume that the acoustic pressure p is small compared with the ambient pressure P and that P is constant over time and space. It is also assumed that the mean velocity $U = 0$. Thus,

$$P_{tot} = P + p \approx P \tag{2.1.18a,b}$$

$$\rho_{tot} = \rho_0 + \sigma \approx \rho_0 \tag{2.1.19a,b}$$

$$U_{tot} = \boldsymbol{u} \tag{2.1.20}$$

$$\frac{\partial P_{tot}}{\partial t} = \frac{\partial p}{\partial t} \tag{2.1.21}$$

$$\frac{\partial \rho_{tot}}{\partial t} = \frac{\partial \sigma}{\partial t} \tag{2.1.22}$$

$$\nabla P_{tot} = \nabla p \tag{2.1.23}$$

Using (2.1.11), the Euler equation, (2.1.10), may be written as

$$-\nabla P_{tot} = \rho_{tot}\left[\frac{\partial \boldsymbol{U}_{tot}}{\partial t} + (\boldsymbol{U}_{tot}\cdot\nabla)\boldsymbol{U}_{tot}\right] \tag{2.1.24}$$

Using (2.1.18) (2.1.19) and (2.1.20), (2.1.24) may be written as

$$-\nabla p = \rho_0\left[\frac{\partial \boldsymbol{u}}{\partial t} + \boldsymbol{u}\cdot\nabla\boldsymbol{u}\right] \tag{2.1.25}$$

As $\boldsymbol{u}$ is small and $\nabla\boldsymbol{u}$ is approximately the same order of magnitude as $\boldsymbol{u}$, the quantity $\boldsymbol{u}\cdot\nabla\boldsymbol{u}$ may be neglected and (2.1.25) is written as

$$-\nabla p = \rho_0\frac{\partial \boldsymbol{u}}{\partial t} \tag{2.1.26}$$

Using (2.1.19), (2.1.20) and (2.1.22), the continuity equation (2.1.7) may be written as

$$\nabla\cdot(\rho_0\boldsymbol{u} + \sigma\boldsymbol{u}) = -\frac{\partial \sigma}{\partial t} \tag{2.1.27}$$

As $\sigma\boldsymbol{u}$ is so much smaller than $\rho_0\boldsymbol{u}$, the equality in (2.1.27) can be approximated as

$$\nabla\cdot(\rho_0\boldsymbol{u}) = -\frac{\partial \sigma}{\partial t} \tag{2.1.28}$$

Using (2.1.17), (2.1.28) may be written as

$$\nabla\cdot(\rho_0\boldsymbol{u}) = -\frac{1}{c_0^2}\frac{\partial p}{\partial t} \tag{2.1.29}$$

Taking the time derivative of (2.1.29) gives

$$\nabla \cdot \rho_0 \frac{\partial \boldsymbol{u}}{\partial t} = -\frac{1}{c_0^2}\frac{\partial^2 p}{\partial t^2} \tag{2.1.30}$$

Substituting (2.1.26) into the left-hand side of (2.1.30) gives

$$-\nabla \cdot \nabla p = -\frac{1}{c_0^2}\frac{\partial^2 p}{\partial t^2} \tag{2.1.31}$$

or

$$\nabla^2 p = \frac{1}{c_0^2}\frac{\partial^2 p}{\partial t^2} \tag{2.1.32}$$

The operator ∇^2 is the (div grad) or the Laplacian operator, and (2.1.32) is known as the linearized wave equation or the Helmholtz equation.

The wave equation can be expressed in terms of the particle velocity by taking the gradient of the linearized continuity equation (2.1.29). Thus,

$$\nabla(\nabla \cdot \rho_0 \boldsymbol{u}) = -\nabla\left[\frac{1}{c_0^2}\frac{\partial p}{\partial t}\right] \tag{2.1.33}$$

Differentiating the Euler equation (2.1.26) with respect to time gives

$$-\nabla \frac{\partial p}{\partial t} = \rho_0 \frac{\partial^2 \boldsymbol{u}}{\partial t^2} \tag{2.1.34}$$

Substituting (2.1.34) into (2.1.33) gives

$$\nabla(\nabla \cdot \boldsymbol{u}) = \frac{1}{c_0^2}\frac{\partial^2 \boldsymbol{u}}{\partial t^2} \tag{2.1.35}$$

However, it may be shown that grad div = div grad + curl curl, or

$$\nabla(\nabla \cdot \boldsymbol{u}) = \nabla^2 \boldsymbol{u} + \nabla \times (\nabla \times \boldsymbol{u}) \tag{2.1.36}$$

Thus (2.1.35) may be written as

$$\nabla^2 \boldsymbol{u} + \nabla \times (\nabla \times \boldsymbol{u}) = \frac{1}{c_0^2}\frac{\partial^2 \boldsymbol{u}}{\partial t^2} \tag{2.1.37}$$

which is the wave equation for the acoustic particle velocity.

2.1.5 Velocity potential

The velocity potential is a scalar quantity which has no particular physical significance. It is introduced as a convenience to avoid having to solve the vector differential equation (2.1.35) to find the acoustic particle velocity. Thus to avoid this, the velocity is expressed as a scalar potential ϕ as follows:

$$\boldsymbol{u} = -\nabla\phi \tag{2.1.38}$$

However, if this is substituted into (2.1.37) we find by Stokes theorem (curl grad = 0 or $\nabla \times \nabla\phi = 0$) that the second term on the left-hand side of the equation vanishes. This effectively means that postulating a velocity potential solution to the wave equation causes some loss of generality and restricts the solutions to those which do not involve fluid rotation. Fortunately, acoustic motion in liquids and gases is nearly always rotationless.

Introducing (2.1.38) for the velocity potential into Euler's equation, (2.1.26) becomes

$$-\nabla p = -\rho_0 \frac{\partial \nabla\phi}{\partial t} = -\rho_0 \nabla \frac{\partial \phi}{\partial t} \tag{2.1.39a,b}$$

Integrating gives

$$p = \rho_0 \frac{\partial \phi}{\partial t} + \text{const} \tag{2.1.40}$$

Introducing (2.1.40) into the wave equation (2.1.32) for acoustic pressure, integrating with respect to time and dropping the integration constant gives

$$\nabla^2\phi = \frac{1}{c_0^2}\frac{\partial^2\phi}{\partial t^2} \tag{2.1.41}$$

This is the preferred form of the Helmholtz equation as both acoustic pressure and particle velocity can be derived from the velocity potential solution by simple differentiation.

2.1.6 Inhomogeneous wave equation (medium containing acoustic sources)

Assume that the acoustic medium contains acoustic sources with a net volume velocity output of q units per unit volume per unit time (note that by this definition, the quantity q has the dimensions of T^{-1}). The mass introduced by these sources must be added to the left-hand side of the continuity equation (2.1.29):

$$\rho_0 q - \nabla \cdot (\rho_0 \boldsymbol{u}) = \frac{1}{c_0^2}\frac{\partial p}{\partial t} \tag{2.1.42}$$

which reduces to

$$\nabla \cdot \boldsymbol{u} = -\frac{1}{\rho_0 c_0^2}\frac{\partial p}{\partial t} + q \tag{2.1.43}$$

Substituting (2.1.38) and (2.1.40) into (2.1.43) gives

$$\nabla^2\phi = \frac{1}{c_0^2}\frac{\partial^2\phi}{\partial t^2} - q \tag{2.1.44}$$

which is often referred to as the inhomogeneous wave equation. Note that the inhomogeneous wave equation describes the response of a forced acoustic system, whereas the homogeneous wave equation describes the response of an acoustic medium which contains no sources or sinks.

The quantity q is referred to as a source distribution or source strength per unit volume. For a point source of strength q located at r_0 in the acoustic medium, the quantity q in (2.1.44) would be replaced by $q\delta(r - r_0)$. For a source of unit strength at $\boldsymbol{r}_0$, the quantity q in (2.1.44) would be replaced by $\delta(\boldsymbol{r} - \boldsymbol{r}_0)$. Note that $\boldsymbol{r}$ is the location at which the velocity potential, ϕ, is to be evaluated. The use of the Dirac delta function, $\delta(\boldsymbol{r} - \boldsymbol{r}_0)$, is discussed in more detail in Section 2.4. An identical equation to (2.1.44) applies for the acoustic pressure, p. It is obtained simply by replacing ϕ with p and replacing q with $j\omega\rho_0 q$.

2.1.7 Wave equation for one-dimensional mean flow

Consider a medium with a mean flow of U_z along the z-axis in the positive direction. Consider also a reference frame X, Y, Z and T moving along with the fluid. Then the wave equation in this moving reference frame is

$$\frac{\partial^2\phi}{\partial X^2} + \frac{\partial^2\phi}{\partial Y^2} + \frac{\partial^2\phi}{\partial Z^2} = \frac{1}{c^2}\frac{\partial^2\phi}{\partial T^2} \tag{2.1.45}$$

Introducing a second, stationary reference frame x, y, z and t we have the following relationships between the two sets of coordinates:

$$X = x \tag{2.1.46}$$

$$Y = y \tag{2.1.47}$$

$$Z = z - U_z t \tag{2.1.48}$$

$$T = t \tag{2.1.49}$$

In terms of the stationary reference frame the quantity $\dfrac{\partial\phi}{\partial T}$ can be written as

$$\frac{\partial\phi}{\partial T} = \frac{\partial\phi}{\partial t}\frac{\partial t}{\partial T} + \frac{\partial\phi}{\partial z}\frac{\partial z}{\partial T} \tag{2.1.50}$$

From (2.1.49) $\dfrac{\partial T}{\partial t} = 1$ and from (2.1.48)

$$\frac{\partial z}{\partial T} = \frac{\partial(z + U_z t)}{\partial T} = U_z \tag{2.1.51a,b}$$

Thus (2.1.50) becomes

$$\frac{\partial \phi}{\partial T} = \frac{\partial \phi}{\partial t} + U_z \frac{\partial \phi}{\partial z} \tag{2.1.52}$$

Differentiating a second time

$$\frac{\partial}{\partial T}\left[\frac{\partial \phi}{\partial T}\right] = \frac{\partial}{\partial T}\left[\frac{\partial \phi}{\partial t} + U_z \frac{\partial \phi}{\partial z}\right] \tag{2.1.53}$$

Expanding gives

$$\begin{aligned}\frac{\partial^2 \phi}{\partial T^2} &= \frac{\partial^2 \phi}{\partial t^2} \cdot \frac{\partial t}{\partial T} + \frac{\partial^2 \phi}{\partial z \partial t} \cdot \frac{\partial z}{\partial T} + U_z \frac{\partial^2 \phi}{\partial t \partial z} \cdot \frac{\partial t}{\partial T} + U_z \frac{\partial^2 \phi}{\partial z^2} \cdot \frac{\partial z}{\partial t} \\ &= \frac{\partial^2 \phi}{\partial t^2} + 2U_z \frac{\partial^2 \phi}{\partial z \partial t} + U_z^2 \frac{\partial^2 \phi}{\partial z^2}\end{aligned} \tag{2.1.54a,b}$$

Thus,

$$\frac{\partial^2 \phi}{\partial T^2} = \left[\frac{\partial}{\partial t} + U_z \frac{\partial}{\partial z}\right]^2 \phi \tag{2.1.55}$$

Expressing the left-hand side of (2.1.45) in terms of the stationary reference frame gives

$$\frac{\partial^2 \phi}{\partial X^2} + \frac{\partial^2 \phi}{\partial Y^2} + \frac{\partial^2 \phi}{\partial Z^2} = \frac{\partial^2 \phi}{\partial x^2} + \frac{\partial^2 \phi}{\partial y^2} + \frac{\partial^2 \phi}{\partial z^2} \tag{2.1.56}$$

Substituting (2.1.55) and (2.1.56) into (2.1.45) gives the following wave equation for a fluid with a mean velocity of U_z along the positive z direction in terms of a stationary reference frame x, y, z and t :

$$\nabla^2 \phi = \frac{1}{c_0^2}\left[\frac{\partial}{\partial t} + U_z \frac{\partial}{\partial z}\right]^2 \phi \tag{2.1.57}$$

2.1.8 Wave equation in cartesian, cylindrical and spherical coordinates

2.1.8.1 Cartesian coordinates

$$\nabla^2 \phi = \frac{\partial^2 \phi}{\partial x^2} + \frac{\partial^2 \phi}{\partial y^2} + \frac{\partial^2 \phi}{\partial z^2} \tag{2.1.58}$$

Thus the wave equation in cartesian coordinates is

$$\frac{\partial^2\phi}{\partial x^2} + \frac{\partial^2\phi}{\partial y^2} + \frac{\partial^2\phi}{\partial z^2} = \frac{1}{c_0^2}\frac{\partial^2\phi}{\partial t^2} \tag{2.1.59}$$

2.1.8.2 Cylindrical coordinates (see Fig. 2.3)

$$\nabla\phi = \frac{\partial\phi}{\partial r} + \frac{1}{r}\frac{\partial\phi}{\partial\theta} + \frac{\partial\phi}{\partial z} \tag{2.1.60}$$

$$\nabla^2\phi = \frac{\partial^2\phi}{\partial r^2} + \frac{1}{r}\frac{\partial\phi}{\partial r} + \frac{1}{r^2}\frac{\partial^2\phi}{\partial\theta^2} + \frac{\partial^2\phi}{\partial z^2} \tag{2.1.61}$$

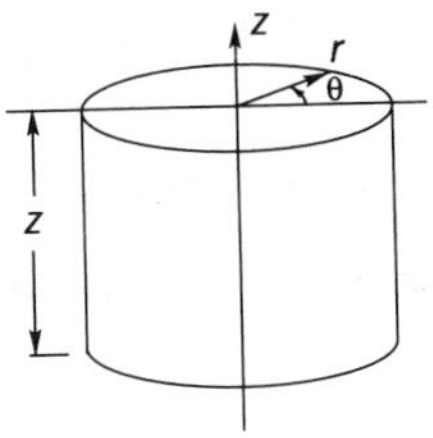

Fig 2.3 Cylindrical coordinate system.

Thus the wave equation in cylindrical coordinates is

$$\frac{\partial^2\phi}{\partial r^2} + \frac{1}{r}\frac{\partial\phi}{\partial r} + \frac{1}{r^2}\frac{\partial^2\phi}{\partial\theta^2} + \frac{\partial^2\phi}{\partial z^2} = \frac{1}{c_0^2}\frac{\partial^2\phi}{\partial t^2} \tag{2.1.62}$$

2.1.8.3 Spherical coordinates (see Fig. 2.4)

$$\nabla\phi = \frac{\partial\phi}{\partial r} + \frac{1}{r}\frac{\partial\phi}{\partial\theta} + \frac{1}{r\sin\theta}\frac{\partial\phi}{\partial\vartheta} \tag{2.1.63}$$

$$\nabla^2\phi = \frac{\partial^2\phi}{\partial r^2} + \frac{2}{r}\frac{\partial\phi}{\partial r} + \frac{1}{r^2}\frac{\partial^2\phi}{\partial\theta^2} + \frac{1}{r^2\sin\theta}\frac{\partial\phi}{\partial\theta} + \left(\frac{1}{r\sin\theta}\right)^2\frac{\partial^2\phi}{\partial\vartheta^2} \tag{2.1.64}$$

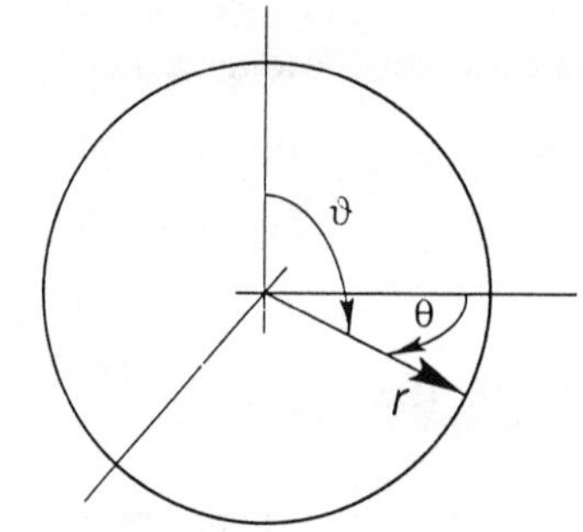

Fig 2.4 Spherical coordinate system

Thus the wave equation in spherical coordinates is

$$\frac{\partial^2\phi}{\partial r^2} + \frac{2}{r}\frac{\partial\phi}{\partial r} + \frac{1}{r^2}\frac{\partial^2\phi}{\partial\theta^2} + \frac{1}{r^2\sin\theta}\frac{\partial\phi}{\partial\theta} + \left(\frac{1}{r\sin\theta}\right)^2\frac{\partial^2\phi}{\partial\vartheta^2} = \frac{1}{c_0^2}\frac{\partial^2\phi}{\partial t^2} \tag{2.1.65}$$

2.1.9 Speed of sound, wavenumber, frequency and period

The one-dimensional wave equation in cartesian coordinates for a wave travelling along the x-axis is

$$\frac{\partial^2\phi}{\partial x^2} = \frac{1}{c_0^2}\frac{\partial^2\phi}{\partial t^2} \tag{2.1.66}$$

It can be shown by substituting the following equation back into (2.1.66) that a solution of (2.1.66) is

$$\phi = f(c_0 t \pm x) \tag{2.1.67}$$

where f is any continuous function. A special case of (2.1.67) for single frequency sound is

$$\phi = A\cos(k(c_0 t \pm x) + \beta) \tag{2.1.68}$$

where A, k and β are constants. The quantity c_0 is the constant from the wave equation.

The general solutions to the wave equation for plane and spherical waves are discussed more fully elsewhere (Bies and Hansen, 1996, Chapter 1). Equation (2.1.68) can be plotted as shown in Fig. 2.5.

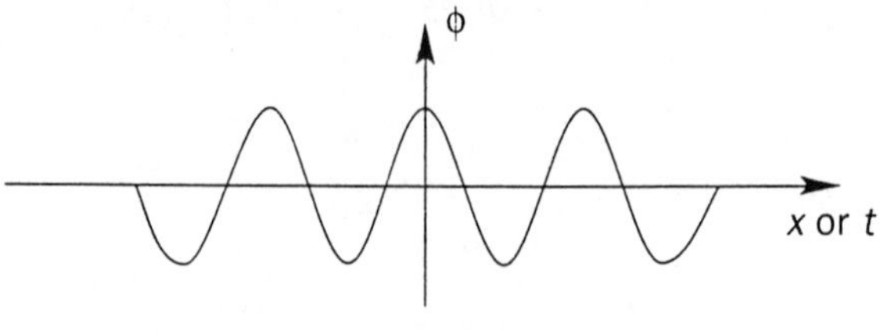

Fig. 2.5 Harmonic one-dimensional potential function.

The constant k of (2.1.68) can be evaluated by fixing the time t so that $kc_0t + \beta = 0$. Thus,

$$\phi = A\cos kx = A\cos 2\pi \frac{x}{\lambda} \tag{2.1.69}$$

This defines the constant k in terms of wavelength λ. Thus $k = 2\pi/\lambda$ is referred to as the wavenumber.

The period of the wave can be defined by fixing the spatial location x in such a way that $\beta + kx = 0$. Thus,

$$\phi = A\cos kc_0t \tag{2.1.70}$$

which defines a period $T = 2\pi/kc_0$ and frequency $1/T = f$. Note that angular frequency is defined as,

$$\omega = 2\pi f = kc_0 \quad (\text{rad s}^{-1}) \tag{2.1.71}$$

The speed of the wave can be determined by fixing $c_0t \pm x = 0$. Thus $c_0 = x/t$.

But x/t is the wave speed; thus, c_0 must be the speed of the wave, or the speed of sound. This speed is often referred to as the phase speed of the wave. In some cases (for example sound propagating in ducts or bending waves propagating in structures), the speed of sound is frequency dependent and the wave field is described as dispersive. When this occurs, wave phase speed is not very meaningful if a band of frequencies are involved. A more useful quantity is then group velocity which is a measure of the speed of the energy propagation of the disturbance. This quantity is discussed in more detail by Fahy (1985) and is defined as

$$c_g = \frac{d\omega}{dk} = \frac{d(c_0k)}{dk} = c_0 + k\frac{dc_0}{dk} \tag{2.1.72a,b,c}$$

2.1.10 Speed of sound in gases, liquids and solids

For an ideal adiabatic gas, the adiabatic state equation relating the total pressure P and the volume V is,

$$PV^\gamma = P_0V_0^\gamma = \text{constant} \tag{2.1.73a,b}$$

Taking natural logs of all terms gives

$$\gamma \ln V + \ln P = 0 \tag{2.1.74}$$

Differentiating (2.1.74) gives

$$\frac{\gamma dV}{V} + \frac{dP}{P} = 0 \tag{2.1.75}$$

Hence,

$$\frac{dP}{dV} = -\frac{\gamma P}{V} = \frac{dP}{d(m/\rho)} = -\frac{\rho^2}{m}\frac{dP}{d\rho} \qquad (2.1.76a,b,c)$$

where m is the mass of fluid of density ρ_0 contained in volume V. But (2.1.15) shows

$$dp = c_0^2 d\sigma \qquad (2.1.77)$$

which is the same as

$$\frac{dP}{d\rho_0} = c_0^2 \qquad (2.1.78)$$

Substituting (2.1.78) into (2.1.76) gives

$$c_0^2 = \frac{\gamma P}{\rho_0} \qquad (2.1.79)$$

Making use of the equation of state for ideal gases

$$PV = \frac{m}{M}RT \qquad (2.1.80)$$

where T is the gas temperature, M is the molecular weight of the gas in kilograms per mole and R is the universal gas constant (8.314 J K^{-1} mol^{-1}), we have for the speed of sound

$$c_0 = \sqrt{\gamma RT/M} \qquad (2.1.81)$$

The speed of sound (longitudinal wave) in a three-dimensional solid is given by (Fahy, 1985):

$$c_L = \sqrt{\frac{E(1-\nu)}{\rho(1+\nu)(1-2\nu)}} \qquad (2.1.82)$$

where E is Young's modulus of elasticity and ρ is the density of the material.

For a two-dimensional solid such as a thin plate the speed of sound is given by (Fahy, 1985):

$$c_L = \sqrt{\frac{E}{\rho(1-\nu^2)}} \qquad (2.1.83)$$

where ν is Poisson's ratio for the solid.

For a one-dimensional solid such as a long slender rod the speed of sound is given by (Cremer *et al.*, 1973)

$$c_L = \sqrt{\frac{E}{\rho}} \qquad (2.1.84)$$

For a liquid, Young's modulus of (2.1.84) is replaced by the bulk modulus B of the liquid, which is defined as

$$B = -V\left[\frac{\partial V}{\partial P}\right]^{-1} \tag{2.1.85}$$

and the speed of sound is

$$c_0 = \sqrt{\frac{B}{\rho}} \tag{2.1.86}$$

2.1.11 Sound propagation in porous media

As mentioned at the end of Section 2.1.2, viscous effects due to internal friction in an acoustic medium such as a porous acoustic material (e.g. rockwool) can be taken into account by adding the term RU_{tot} to the right-hand side of the Euler equation of motion. Alternatively, these effects could be taken into account by replacing the speed of sound c with a complex quantity, $c_0(1+j\eta)^{1/2}$ as outlined elsewhere (Skudrzyk, 1971, p. 283). Information of a more practical nature on sound propagation in porous media is given by Bies and Hansen (1996, 2nd edn., Appendix 3).

2.2 STRUCTURAL MECHANICS: FUNDAMENTALS

In developing analytical models for active vibration control systems in later chapters, a knowledge of the fundamentals of structural mechanics will be assumed. Rather than considering the subject of structural mechanics in an exhaustive way, we will concern ourselves here with a discussion of the fundamental principles which will be directly relevant to the analyses in later chapters.

One way of analysing a mechanical system is to extend Newton's laws for a single particle to systems of particles and use the concepts of force and momentum, both of which are vector quantities. An approach such as this is referred to as Newtonian mechanics or vectorial mechanics and will be discussed briefly in the next section. For more detailed treatment of this subject the reader is advised to consult Meirovitch (1970) or McCuskey (1959).

A second approach known as analytical mechanics is attributed principally to Lagrange, and later Hamilton, and considers the system as a whole rather than as a number of individual components. Using this approach problems are formulated in terms of two scalar quantities; energy and work. This second approach will be discussed in detail in Section 2.2.2.

2.2.1 Summary of Newtonian mechanics

The three fundamental physical laws describing the dynamics of particles are those enunciated by Isaac Newton in the *Principia* (1686). The first law states that a particle of constant mass will remain at rest or move in a straight line unless acted upon by a force. The most useful concept used in the solution of dynamics problems is embodied in Newton's second law, which effectively states that the time rate of change of the linear momentum vector for a particle is equal to the force vector acting on the particle. That is,

$$\boldsymbol{F} = \frac{\mathrm{d}(m\boldsymbol{u})}{\mathrm{d}t} \tag{2.2.1}$$

For the systems which will be discussed in this book, the mass will remain constant and thus (2.2.1) may be written as

$$\boldsymbol{F} = m\frac{\mathrm{d}\boldsymbol{u}}{\mathrm{d}t} \tag{2.2.2}$$

Newton's third law of motion states that the force exerted on one particle by a second particle is equal and opposite to the force exerted by the first particle on the second. That is,

$$\boldsymbol{F}_{12} = -\boldsymbol{F}_{21} \tag{2.2.3}$$

A particle is an idealization of a body whose dimensions are very small in comparison with the distance to other bodies and whose internal motion does not affect the motion of the body as a whole. Mathematically, it is represented by a mass point of infinitesimally small size.

All three of Newton's laws apply equally well to particles experiencing angular motion. In this case, the force vector is replaced by a torque vector and the linear momentum vector is replaced by an angular momentum vector equal to $J\dot{\boldsymbol{\theta}}$, where J is the moment of inertia of the particle and $\dot{\boldsymbol{\theta}}$ is the angular velocity vector.

It can be shown (McCuskey, 1959) that Newton's laws lead to three very important conservation laws.

1. If the vector sum of the linear forces acting on a particle are zero, then the linear momentum of the particle will remain unchanged.
2. If the vector sum of the torques acting on a particle are zero, then the angular momentum of the particle will remain unchanged.
3. If a particle is acted on by only conservative forces, then its total energy (kinetic plus potential) will remain unchanged.

A conservative force is one which is not associated with any dissipation. Thus, a friction force is a non-conservative force. If a particle is acted on by a non-conservative force $\boldsymbol{F}_1$, then the rate of change of energy of the particle moving with velocity $\boldsymbol{u}$ is:

$$\frac{\mathrm{d}E}{\mathrm{d}t} = \boldsymbol{F}_1 \cdot \boldsymbol{u} \tag{2.2.4}$$

The kinetic energy K of a particle is given by

$$K = \frac{1}{2}mu^2 \tag{2.2.5}$$

where m is its mass and $u = |\mathbf{u}|$ is its speed. The potential energy of a particle is given by

$$V = mgh \tag{2.2.6}$$

where g is the acceleration due to gravity and h is the height above some reference datum.

For a particle attached to a horizontal massless spring, the potential energy is given by

$$V = -kx \tag{2.2.7}$$

where k is the spring stiffness and x is its extension.

2.2.1.1 Systems of particles

Newton's laws and the conservation laws also apply to bodies and systems of particles as well as to single particles. Thus, as shown by Meirovitch (1970), for a system of particles acted on by no external forces, the linear and angular momentum of the system will remain unchanged and will be equal to the vector sum of the linear and angular momentum of each particle. In addition the total system energy will remain unchanged, provided that all of the external forces acting on the system are conservative.

Thus, the motion of the centre of mass of a system of particles or a body is the same as if all the system mass were concentrated at that point and were acted upon by the resultant of all the external forces.

In other words, for a linear conservative system, the mass of a body multiplied by its acceleration is equal to the sum of the external forces acting upon it. This principle will be used in later sections of this text to derive the equations of motion for both discrete and continuous systems.

2.2.2 Summary of analytical mechanics

The following topics, which are relevant to the analysis in later parts of this book, will be considered briefly: generalized coordinates; principle of virtual work; d'Alembert's principle; Hamilton's principle; and Lagrange's equations.

2.2.2.1 Generalized coordinates

In many physical problems, the bodies of interest are not completely free but are subject to some kinematic constraints. As an example, consider two point masses m_1 and m_2 connected by a rigid link of length L, as shown in Fig. 2.6.

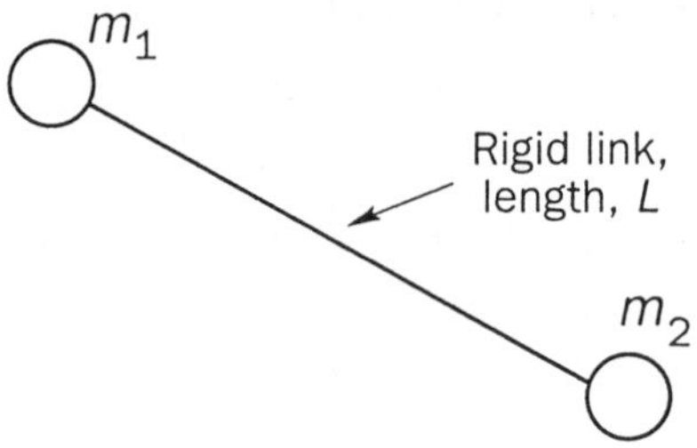

Fig. 2.6 Rigid link.

The motions of m_1 and m_2 are completely defined in cartesian coordinates using x_1, y_1, z_1, x_2, y_2 and z_2. However, the six coordinates are not independent of each other. In fact,

$$(x_1 - x_2)^2 + (y_1 - y_2)^2 + (z_1 - z_2)^2 = L \tag{2.2.8}$$

Thus, in this case, the problem is solved if only five of the coordinates are known, as the sixth one may be determined using (2.2.8). The motion of this system may be completely defined by the three coordinates of either m_1 or m_2 and two angles which define the orientation of the rigid link.

In general, the motion of N particles subject to c kinematic restraints can be described uniquely by n independent coordinates q_k ($k = 1, 2, \ldots n$), where

$$n = 3N - c \tag{2.2.9}$$

The n coordinates $q_1 \ldots q_n$ are referred to as generalized coordinates – each is independent of all of the others. The generalized coordinates may not be physically identifiable and they are not unique. Thus, there may be several sets of generalized coordinates which can describe a particular physical system. They must be finite, single-valued, differentiable with respect to time and continuous.

The generalized coordinates for a given system represent the least number of variables required to specify the positions of the elements of the system at any particular time. A system which can be described in terms of generalized independent coordinates is referred to as holonomic.

2.2.2.2 Principle of virtual work

This principle was first enunciated by Johann Bernoulli in 1717, and is essentially a statement of the static equilibrium of a mechanical system.

Consider a single particle at a vector location $\boldsymbol{r}$ in space acted upon by a number of forces, N. Defining $\delta\boldsymbol{r}$ as any small virtual (or imaginary) displacement arbitrarily imposed on the particle, the virtual work done by the forces is

$$\delta W = \boldsymbol{F}_1 \cdot \delta\boldsymbol{r} + \boldsymbol{F}_2 \cdot \delta\boldsymbol{r} + \ldots\ldots \boldsymbol{F}_N \cdot \delta\boldsymbol{r} \tag{2.2.10}$$

or

$$\delta W = \sum_{i=1}^{N} \boldsymbol{F}_i \cdot \delta \boldsymbol{r} \tag{2.2.11}$$

or

$$\delta W = \boldsymbol{F} \cdot \delta \boldsymbol{r} \tag{2.2.12}$$

where $\boldsymbol{F}$ is the resultant of all the forces acting on the particle.

The principle of virtual work states that if and only if, for any arbitrary virtual displacement $\delta \boldsymbol{r}$ the virtual work $\delta w = 0$ under the action of the forces $\boldsymbol{F}_i$, the particle is in equilibrium. For non-zero $\delta \boldsymbol{r}$, (2.2.11) shows either that $\delta \boldsymbol{r}$ is perpendicular to $\sum \boldsymbol{F}_i$ or that $\sum \boldsymbol{F}_i = 0$. Since (2.2.11) must hold for any $\delta \boldsymbol{r}$, the first possibility is ruled out and $\sum \boldsymbol{F}_i = 0$. Thus, for a particle to be in equilibrium,

$$\delta W = \boldsymbol{F} \cdot \delta \boldsymbol{r} = 0 \tag{2.2.13}$$

where $\boldsymbol{F}$ is the resultant of all forces acting on the particle.

For a system subject to constraint forces, and distinguishing between the applied forces $\boldsymbol{F}_i$ and the constraint forces $\boldsymbol{F}_i'$ we obtain, using (2.2.13), the following

$$\delta W = \sum_{i=1}^{n} \boldsymbol{F}_i \cdot \delta \boldsymbol{r}_i + \sum_{i=1}^{m} \boldsymbol{F}_i' \cdot \delta \boldsymbol{r}_i = 0 \tag{2.2.14}$$

An example of a constraint force would be the reaction force on a particle resting on a smooth surface.

As the work of the constraint forces through virtual displacements compatible with the system constraints is zero, (2.2.14) becomes

$$\delta W = \sum_{i=1}^{n} \boldsymbol{F}_i \cdot \delta \boldsymbol{r}_i = 0 \tag{2.2.15}$$

However, for systems with constraints, all of the virtual displacements $\delta \boldsymbol{r}_i$ are not independent and so (2.2.15) cannot be interpreted simply to imply that $\boldsymbol{F}_i = 0$ $(i = 1,2 \ldots n)$.

If the problem is described by a set of independent generalized coordinates, (2.2.15) may be written as

$$\delta W = \sum_{k=1}^{m} \boldsymbol{Q}_k \delta \boldsymbol{q}_k = 0 \tag{2.2.16}$$

where $\boldsymbol{Q}_k$ $(k = 1 \ldots m)$ are known as generalized forces. Since the number of generalized coordinates is now the same as the number of degrees of freedom of the system, the virtual displacements $\delta \boldsymbol{q}_k$ are all independent so that $\boldsymbol{Q}_k = 0$ $(k = 1 \ldots m)$.

Example 2.1

Use the principle of virtual work to calculate the angle θ for the link of length L shown in Fig. 2.7 to be in static equilibrium. Here, x = elongation of the spring, and y = lowering of one end of the link from the horizontal position.

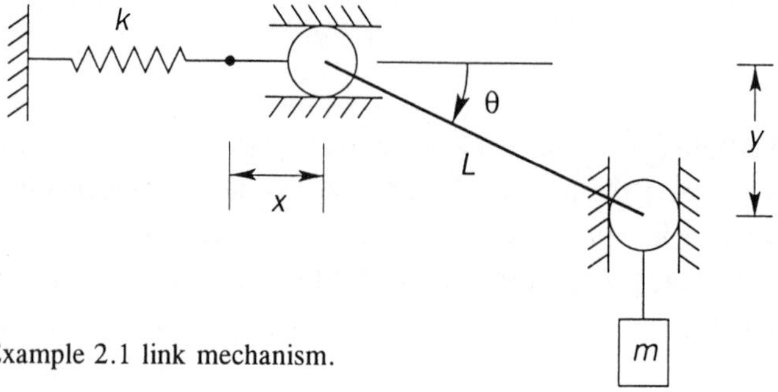

Fig. 2.7 Example 2.1 link mechanism.

The position of the ends of the link at equilibrium may be described by

$$x = L(1 - \cos\theta), \qquad y = L\sin\theta \tag{a}$$

The virtual work can be written as

$$\delta W = -kx\delta x + mg\delta y = 0 \tag{b}$$

where the virtual displacements δx and δy are obtained from (a) and are equal to

$$\delta x = L\sin\theta\,\delta\theta, \qquad \delta y = L\cos\theta\,\delta\theta \tag{c}$$

Introducing the first of the equations (a) and both equations (c) into (b) we obtain

$$\delta W = -kL(1 - \cos\theta)L\sin\theta\,\delta\theta + mgL\cos\theta\,\delta\theta = 0 \tag{d}$$

from which we obtain

$$(1 - \cos\theta)\tan\theta = \frac{mg}{kL} \tag{e}$$

Solving equation (e) gives the value of θ corresponding to the equilibrium position of the system shown in the figure.

2.2.2.3 D'Alembert's principle

In 1743 in his *Traité de Dynamique*, D'Alembert proposed a principle to reduce a dynamics problem to an equivalent statics problem. By introducing 'inertial forces' he applied Bernoulli's principle of virtual work to systems in which motion resulted from the applied forces. The inertial force acting upon the ith particle of mass m_i in a system is $-m_i\ddot{r}$ or $-m_i\ddot{\boldsymbol{u}}_i$. If the resultant force acting on the ith particle of a system of n particles is $\boldsymbol{F}_i$, then D'Alembert's principle states that the system is in equilibrium if the total virtual work performed by the inertia forces and applied forces is zero. That is,

$$\sum_{i=1}^{n} (\boldsymbol{F}_i - m_i \ddot{\boldsymbol{r}}_i) \cdot \delta \boldsymbol{r}_i = 0 \tag{2.2.17}$$

D'Alembert's principle represents the most general formulation of dynamics problems and all the various principles of mechanics (including Hamilton's principle to follow) are derived from it.

The problem with D'Alembert's principle is that it is not very convenient for deriving equations of motion as problems are formulated in terms of position coordinates which, in contrast with generalized coordinates, may not all be independent. A different formulation, Hamilton's principle, which avoids this difficulty will now be discussed.

2.2.2.4 Hamilton's principle

Hamilton's principle is a consideration of the motion of an entire system between two times t_1 and t_2. It is an integral principle and reduces dynamics problems to a scalar definite integral.

Consider a system of n particles of masses m_i located at points $\boldsymbol{r}_i$ and acted upon by resultant external forces $\boldsymbol{F}_i$. By D'Alembert's principle we can write

$$\sum_{i=1}^{n} (m_i \ddot{\boldsymbol{r}}_i - \boldsymbol{F}_i) \cdot \delta \boldsymbol{r}_i = 0 \tag{2.2.18}$$

This is the dynamic condition to be satisfied by the applied forces and inertia forces for arbitrary $\delta \boldsymbol{r}_i$ consistent with the constraints on the system.

The virtual work done by the applied forces is

$$\delta W = \sum_{i=1}^{n} \boldsymbol{F}_i \cdot \delta \boldsymbol{r}_i \tag{2.2.19}$$

Furthermore,

$$\frac{\mathrm{d}}{\mathrm{d}t}(\dot{\boldsymbol{r}}_i \cdot \delta \boldsymbol{r}_i) = \dot{\boldsymbol{r}}_i \frac{\mathrm{d}}{\mathrm{d}t}(\delta \boldsymbol{r}_i) + \ddot{\boldsymbol{r}}_i \cdot \delta \boldsymbol{r}_i \tag{2.2.20}$$

Interchanging the derivative and variational operators gives

$$\frac{\mathrm{d}}{\mathrm{d}t} \delta \boldsymbol{r}_i = \delta \dot{\boldsymbol{r}}_i \tag{2.2.21}$$

Thus,

$$\dot{\boldsymbol{r}}_i \cdot \delta \dot{\boldsymbol{r}}_i = \delta \left[\frac{1}{2} \dot{\boldsymbol{r}}_i^2 \right] = \delta \left[\frac{1}{2} u_i^2 \right] \tag{2.2.22}$$

where u_i is the speed of the ith particle. When the mass m_i is included as a factor, then (2.2.22) represents the variation in kinetic energy, δT, of the particle.

Substituting (2.2.21) and (2.2.22) into (2.2.20) and rearranging gives

$$\ddot{r}_i \cdot \delta r_i = \frac{\mathrm{d}}{\mathrm{d}t}(\dot{r}_i \cdot \delta r_i) - \delta\left[\frac{1}{2}u_i^2\right] \tag{2.2.23}$$

Multiplying (2.2.23) by m_i and summing over all particles in the system gives

$$\sum_{i=1}^{n} m_i \ddot{r}_i \cdot \delta r_i = \frac{\mathrm{d}}{\mathrm{d}t}(\dot{r}_i \cdot \delta r_i) - \delta T \tag{2.2.24}$$

Substituting (2.2.24) and (2.2.19) into (2.2.18) gives

$$\delta T + \delta W = \sum_{i=1}^{n} m_i \frac{\mathrm{d}}{\mathrm{d}t}(\dot{r}_i \cdot \delta r_i) \tag{2.2.25}$$

The instantaneous configuration of a system is given by the values of the n generalized coordinates defining a representative point in the n-dimensional configuration space.

The system configuration changes with time tracing a true path (or dynamic path) in the configuration space. In addition to this true path there will be an infinite number of imagined variations of this path. Consider two times t_1 and t_2 at which we assume that $\delta r_i = 0$; that is, two times at which the dynamic path and all the imagined variations of this path coincide (see Fig. 2.8).

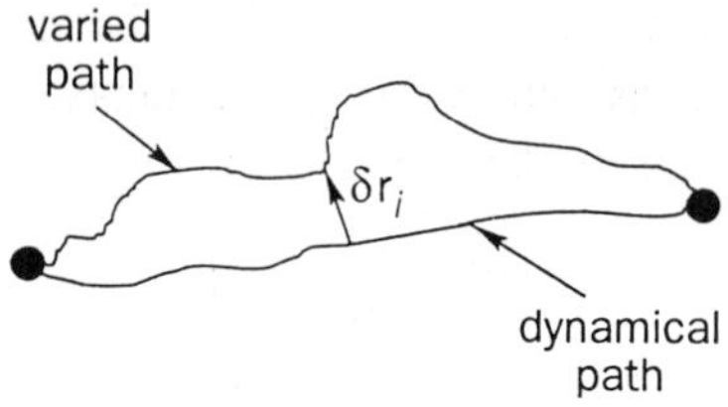

Fig. 2.8 Dynamic path.

$$\delta r_i(t_1) = \delta r_i(t_2) = 0 \tag{2.2.26}$$

Integrating (2.2.25) between t_1 and t_2 gives

$$\int_{t_1}^{t_2} (\delta T + \delta W)\mathrm{d}t = \sum_{i=1}^{n} m_i(\dot{r}_i \cdot \delta r_i) \Big|_{t_1}^{t_2} \tag{2.2.27}$$

which, given the conditions of (2.2.26), reduces to

$$\int_{t_1}^{t_2} (\delta T + \delta W)\mathrm{d}t = 0 \tag{2.2.28}$$

If we assume that δW is a work function arising from the potential energy so that $\delta W = -\delta V$, then (2.2.28) may be written for a holonomic system as

$$\delta \int_{t_1}^{t_2} (T - V)\,\mathrm{d}t = 0 \tag{2.2.29}$$

The function $T - V$ is called the Lagrangian L of the system.

Equation (2.2.29) is a mathematical statement of Hamilton's principle, which may be stated as follows: 'of all paths in time that the coordinates r_i may be imagined to take between two instants t_1 and t_2, the dynamic (or true) path actually taken by the system will be that for which $\int_{t_1}^{t_2} (T - V)\,\mathrm{d}t$ will have a stationary value, provided that the path variations vanish at the end points t_1 and t_2.'

Hamilton's principle provides a formulation rather than a solution of dynamic problems and as will be shown shortly, it is better used to derive equations of motion which can be used for the solution of dynamic problems.

Example 2.2

Consider the system shown in Fig. 2.7 and use Hamilton's principle to derive the equation of motion. Using equation (a) of Example 2.1 we can write the Lagrangian

$$\begin{aligned} L &= T - V = \frac{1}{2}m\dot{y}^2 - \frac{1}{2}kx^2 + mgy \\ &= \frac{1}{2}mL^2\left[\dot{\theta}^2\cos^2\theta - \frac{k}{m}(1 - \cos\theta)^2 + \frac{2g}{L}\sin\theta\right] \end{aligned} \tag{a}$$

Substituting equation (a) into (2.2.29) gives

$$\delta \int_{t_1}^{t_2} L\,\mathrm{d}t = \int_{t_1}^{t_2} \left(\frac{\partial L}{\partial \theta}\delta\theta + \frac{\partial L}{\partial \dot{\theta}}\delta\dot{\theta}\right)\mathrm{d}t$$

$$= -mL^2\int_{t_1}^{t_2}\left\{\left[\dot{\theta}^2\sin\theta\cos\theta + \frac{k}{m}(1 - \cos\theta)\sin\theta - \frac{g}{L}\cos\theta\right]\delta\theta - \dot{\theta}\cos^2\theta\,\delta\dot{\theta}\right\}\mathrm{d}t = 0 \tag{b}$$

The second term in the integrand contains $\delta\dot{\theta}$, making it incompatible with the remaining terms, which are all multiplying $\delta\theta$. But $\delta\dot{\theta} = \mathrm{d}(\delta\theta)/\mathrm{d}t$, so that after an integration by parts of the term containing $\delta\dot{\theta}$ we arrive at

$$\begin{aligned} &\int_{t_1}^{t_2}\left[\frac{\mathrm{d}}{\mathrm{d}t}(\dot{\theta}\cos^2\theta) + \dot{\theta}^2\sin\theta\cos\theta + \frac{k}{m}(1 - \cos\theta)\sin\theta - \frac{g}{L}\cos\theta\right] \\ &\quad \times \delta\theta\,\mathrm{d}t - \dot{\theta}\cos^2\theta\,\delta\theta\,\Big|_{t_1}^{t_2} = 0 \end{aligned} \tag{c}$$

where the constant $-mL^2$ has been ignored.

Invoking the requirement that the variation $\delta\theta$ vanish at the two instants t_1 and t_2, the second term in equation (c) reduces to zero. Moreover, $\delta\theta$ is arbitrary in the time interval between t_1 and t_2, so that the only way for the integral to be zero is for the coefficient of $\delta\theta$ to vanish for any time t. Hence, we must set

$$\frac{\mathrm{d}}{\mathrm{d}t}(\dot{\theta}\cos^2\theta) + \dot{\theta}^2\sin\theta\cos\theta + \frac{k}{m}(1 - \cos\theta)\sin\theta - \frac{g}{L}\cos\theta = 0 \tag{d}$$

which is the desired equation of motion. We note that by letting $\dot{\theta} = \ddot{\theta} = 0$ we obtain the same equation for the system equilibrium position as the one derived in Example 2.1.

In general, it is not necessary to use Hamilton's principle directly for the solution of dynamic problems. Instead we use Hamilton's principle to derive Lagrange's equations of motion which can then be used to solve dynamic problems.

2.2.2.5 Lagrange's equations of motion

It was mentioned earlier that the generalized D'Alembert's principle (2.2.17) is not convenient for the derivation of equations of motion. It is more advantageous to express (2.2.17) in terms of a set of generalized coordinates q_k ($k = 1, 2, \ldots . n$) in such a way that the virtual displacements δq_k are independent and arbitrary. Under these circumstances, the coefficients of δq_k ($k = 1, 2, \ldots . n$) can be set equal to zero separately, thus obtaining a set of differential equations in terms of generalized coordinates, known as Lagrange's equations of motion.

Instead of using D'Alembert's principle, Lagrange's equations of motion can be derived using Hamilton's principle which is simply an integrated form of D'Alembert's principle. The derivation of Lagrange's equations of motion using Hamilton's principle is discussed in detail elsewhere (McCuskey, 1959; Meirovitch, 1970; and Tse *et al.*, 1978) and only the results will be given here.

For each generalized coordinate q_k, the Lagrange equation of motion is

$$\frac{\mathrm{d}}{\mathrm{d}t}\left[\frac{\partial T}{\partial \dot{q}_k}\right] - \frac{\partial T}{\partial q_k} = Q_k' \qquad (k=1,\ 2,\ \ldots .\ n) \tag{2.2.30}$$

where Q_k' is the kth generalized force, q_k is the kth generalized coordinate, the system has n degrees of freedom, and T is the kinetic energy of the total system.

The forces Q_k' may be made up of potential forces Q_v, damping forces Q_s and the applied forces Q_k. The potential forces can be either due to changes in height, spring forces or both. The damping forces are either viscous or hysteretic and are associated with the internal system damping.

The potential energy is a function of the generalized coordinates. Thus,

$$V = V(q_1, q_2, \ldots .. q_n)$$

Expanding this about the stable equilibrium position of the system gives

$$V = V_0 + \sum_{k=1}^{n} \left(\frac{\partial V}{\partial q_k} \right)_0 q_k + \frac{1}{2} \sum_{k=1}^{n} \sum_{i=1}^{n} \left(\frac{\partial^2 V}{\partial q_k \partial q_i} \right)_0 q_k q_i + \ldots . \tag{2.2.31}$$

where the subscript 0 denotes the value at the equilibrium position. V_0 can be defined as zero if we measure the potential energy with respect to this datum. Since V is a minimum at V_0 its first derivative must vanish. If also the third and higher order terms are neglected, (2.2.31) becomes

$$V = \frac{1}{2} \sum_{k=1}^{n} \sum_{i=1}^{n} \left(\frac{\partial^2 V}{\partial q_k \partial q_i} \right)_0 q_k q_i \tag{2.2.32}$$

For small variations, the second partial derivative term may be assumed constant. Denoting this as an equivalent spring constant k_{ki} we obtain

$$V = \frac{1}{2} \sum_{k=1}^{n} \sum_{i=1}^{n} k_{ki} q_k q_i \tag{2.2.33}$$

where $k_{ki} = k_{ik}$. Note that k_{ki} are the elements of an $n \times n$ stiffness matrix.

The spring force (or potential force) associated with coordinate q_k is

$$Q_v = \frac{\partial V}{\partial q_k} = -\sum_{k=1}^{n} k_{ki} q_k \tag{2.2.34}$$

By analogy with the potential energy, a dissipation function may be defined as

$$D = \frac{1}{2} \sum_{k=1}^{n} \sum_{i=1}^{n} C_{ki} \dot{q}_k \dot{q}_i \tag{2.2.35}$$

and the damping force associated with the velocity $\dot{q}_k$ is

$$Q_D = \frac{\partial D}{\partial \dot{q}_k} = -\sum_{k=1}^{n} C_{ki} \dot{q}_k \tag{2.2.36}$$

Including the spring force and the damping force in the Lagrangian, (2.2.30) becomes

$$\frac{\mathrm{d}}{\mathrm{d}t} \left(\frac{\partial T}{\partial \dot{q}_k} \right) - \frac{\partial T}{\partial q_k} + \frac{\partial D}{\partial \dot{q}_k} + \frac{\partial V}{\partial q_k} = Q_k \tag{2.2.37}$$

For the free vibration of a conservative system we obtain

$$\frac{\mathrm{d}}{\mathrm{d}t}\left(\frac{\partial T}{\partial \dot{q}_k}\right) - \frac{\partial T}{\partial q_k} + \frac{\partial V}{\partial q_k} = 0 \tag{2.2.38}$$

where the kinetic energy may be defined as

$$T = \frac{1}{2}\sum_{k=1}^{n}\sum_{i=1}^{n} m_{ki}\,\dot{q}_k\,\dot{q}_i \tag{2.2.39}$$

Since V is a function of the coordinates only and $\dfrac{\partial V}{\partial \dot{q}_k} = 0$, (2.2.38) may be written in terms of the Lagrangian $L = T - V$ as

$$\frac{\mathrm{d}}{\mathrm{d}t}\left(\frac{\partial L}{\partial \dot{q}_k}\right) - \frac{\partial L}{\partial q_k} = 0 \quad (k=1, 2, \ldots n) \tag{2.2.40}$$

Returning to (2.2.33) and (2.2.39) we may express these using matrix notation respectively (where matrices and vectors are indicated in bold typeface) as

$$V = \frac{1}{2}\boldsymbol{q}_k^T \boldsymbol{K}\,\boldsymbol{q}_k \tag{2.2.41}$$

and

$$T = \frac{1}{2}\dot{\boldsymbol{q}}_k^T \boldsymbol{M}\,\dot{\boldsymbol{q}}_k \tag{2.2.42}$$

where $\boldsymbol{K}$ and $\boldsymbol{M}$ are referred to respectively as the system stiffness and mass matrices.

Substituting (2.2.41) and (2.2.42) into (2.2.38) gives the following equation of motion for the free vibration of a conservative system.

$$\boldsymbol{M}\,\ddot{\boldsymbol{q}}_k + \boldsymbol{K}\,\boldsymbol{q}_k = 0 \tag{2.2.43}$$

$$\boldsymbol{M} = \begin{bmatrix} m_{11} & m_{12} & \cdots & m_{1n} \\ m_{21} & m_{22} & \cdots & \vdots \\ \vdots & \vdots & \vdots & \vdots \\ m_{n1} & \cdots & \cdots & m_{nn} \end{bmatrix} \tag{2.2.44}$$

$$\boldsymbol{K} = \begin{bmatrix} k_{11} & k_{12} & \cdots & k_{1n} \\ k_{21} & k_{22} & \cdots & \vdots \\ \vdots & \vdots & \vdots & \vdots \\ k_{n1} & \cdots & \cdots & k_{nn} \end{bmatrix} \tag{2.2.45}$$

Example 2.3

Use Lagrange's equations to derive the equations of motion for the double pendulum shown in Fig. 2.9. This type of system is known as a two-degree-of-freedom system as it possesses two point masses, each of which is capable of movement in only one coordinate direction (θ_1 and θ_2 respectively).

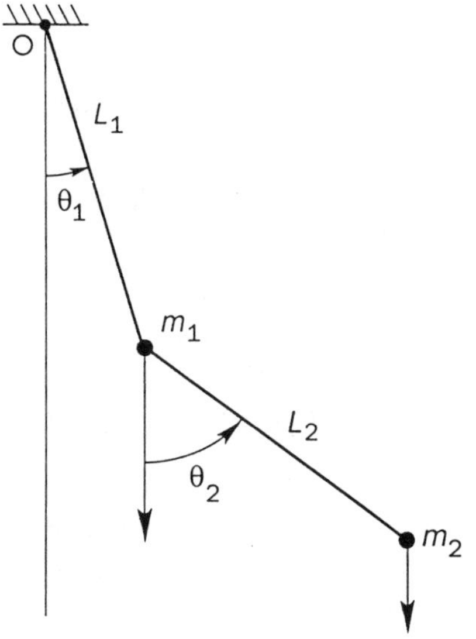

Fig. 2.9 Double pendulum (Example 2.3).

Because the system is conservative and the constraints of the system are taken into account by choosing θ_1 and θ_2 as generalized coordinates, we can use (2.2.40) to derive the required equations.

The kinetic energy of the system is

$$T = \frac{1}{2}(m_1 + m_2)L_1^2\dot{\theta}_1^2 + \frac{1}{2}m_2 L_2^2\dot{\theta}_2^2 + m_2 L_1 L_2 \dot{\theta}_1 \dot{\theta}_2 \cos(\theta_1 - \theta_2) \tag{a}$$

and the potential energy referred to the point of support is

$$V = -(m_1 + m_2)gL_1\cos\theta_1 - m_2 g L_2 \cos\theta_2 \tag{b}$$

From (a) and (b) the Lagrangian L is

$$L = T - V = \text{(a)} - \text{(b)} \tag{c}$$

From (a), (b) and (c) we have

$$\frac{\partial L}{\partial \dot{\theta}_1} = (m_1 + m_2)L_1^2\dot{\theta}_1 + m_2 L_1 L_2 \dot{\theta}_2 \cos(\theta_1 - \theta_2) \tag{d}$$

$$\frac{\partial L}{\partial \dot{\theta}_2} = m_2 L_2^2 \dot{\theta}_2 + m_2 L_1 L_2 \dot{\theta}_1 \cos(\theta_1 - \theta_2) \tag{e}$$

$$\frac{\partial L}{\partial \theta_1} = -m_2 L_1 L_2 \dot{\theta}_1 \dot{\theta}_2 \sin(\theta_1 - \theta_2) - (m_1 + m_2) g L_1 \sin\theta_1 \tag{f}$$

$$\frac{\partial L}{\partial \theta_2} = m_2 L_1 L_2 \dot{\theta}_1 \dot{\theta}_2 \sin(\theta_1 - \theta_2) - m_2 g L_2 \sin\theta_2 \tag{g}$$

Substituting (a), (b) and (c) into (2.2.40) gives for each of the generalized coordinates θ_1 and θ_2

$$\begin{aligned}(m_1 + m_2) L_1 \ddot{\theta}_1 + m_2 L_2 \ddot{\theta}_2 \cos(\theta_1 - \theta_2)& \\ + m_2 L_2 \dot{\theta}_1 \dot{\theta}_2 \sin(\theta_1 - \theta_2) + (m_1 + m_2) g \sin\theta_1 &= 0\end{aligned} \tag{h}$$

$$\begin{aligned}L_2 \ddot{\theta}_2 + L_1 \ddot{\theta}_1 \cos(\theta_1 - \theta_2) + L_1 \ddot{\theta}_1 \cos(\theta_1 - \theta_2)& \\ - L_1 \dot{\theta}_1 \dot{\theta}_2 \sin(\theta_1 - \theta_2) + g \sin\theta_2 &= 0\end{aligned} \tag{i}$$

Equations (h) and (i) are two simultaneous differential equations which may be solved for the coordinates θ_1 and θ_2 as a function of time. These equations can also be derived with some patience and difficulty using Newton's equations of motion. However, the advantage of the Lagrange approach will soon become obvious to any readers who wish to attempt the derivation using Newton's formulation.

Example 2.4

Use Lagrange's equations to derive the equation of motion for the system shown in Fig. 2.10.

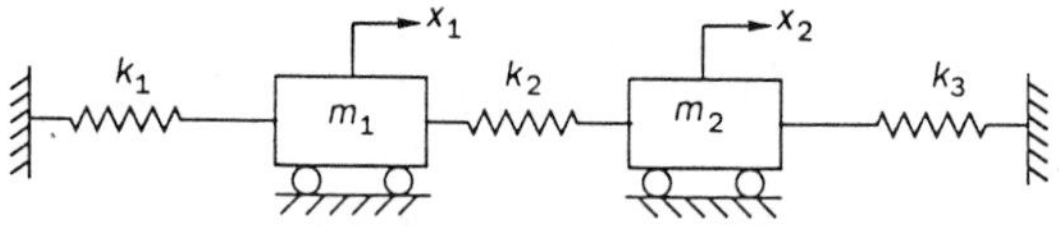

Fig. 2.10 Simple two-degree-of-freedom undamped system.

From the figure, we can see that x_1 and x_2 are acceptable generalized coordinates q_1 and q_2. The spring force acting on x_1 due to displacement x_1 is $-(k_1 + k_2)x_1$. Thus, $k_{11} = k_1 + k_2$. Similarly it can be seen that $k_{22} = k_2 + k_3$. The force acting on m_1 due to a displacement x_2 is $k_2 x_2$. Thus, $k_{12} = -k_2$ and

$$\begin{aligned}V &= \frac{1}{2}\left[k_{11}x_1^2 + k_{22}x_2^2 + k_{12}x_2^2 x_1 + k_{21}x_1 x_2\right] \\ &= \frac{1}{2}\left[(k_1 + k_2)x_1^2 + (k_2 + k_3)x_2^2 - 2k_2 x_1 x_2\right]\end{aligned} \tag{a}$$

The kinetic energy is calculated by first determining the elements of the mass matrix $\boldsymbol{M}$. In this case, $m_{11} = m_1$, $m_{21} = m_{12} = 0$ and $m_{22} = m_2$ (as x_1 and x_2 refer to the motion of m_1 and m_2 directly). Thus, the kinetic energy T is given by

$$T = \frac{1}{2}\left(m_1\dot{x}_1^2 + m_2\dot{x}_2^2\right) \tag{b}$$

Substituting (a) and (b) into (2.2.40) gives the following two equations of motion:

$$\left.\begin{aligned} m_1\ddot{x}_1 + (k_1 + k_2)x_1 - k_2x_2 &= 0 \\ m_2\ddot{x}_2 + (k_2 + k_3)x_2 - k_2x_1 &= 0 \end{aligned}\right\} \tag{c}$$

Alternatively, the equations of motion could have been derived directly from (2.2.43) by substituting the appropriate quantities for the elements of the mass and stiffness matrices.

2.2.3 Influence coefficients

In addition to the use of Newton's laws (described in Section 2.2.1) or variational techniques leading to Lagrange's equations (described in Section 2.2.2) for the derivation of the equations of motion of a vibrating system, there is a third option referred to as the influence coefficient method. There are two types of influence coefficient, both of which define the static elastic property of a system, in common use: the stiffness influence coefficient, k_{ij}, which is the force acting on mass j (or location j) due to a unit static displacement at mass i (with all other masses stationary); and the flexibility influence coefficient d_{ij} which is the static displacement at mass i due to a unit force at mass j with no other forces acting. For convenience we will only consider here the stiffness influence coefficients k_{ij}.

Consider the two-degree-of-freedom system shown in Fig. 2.10. The force acting on n_1 due to a unit displacement at m_1 is $k_{11} = k_1 + k_2$. The force acting on m_2 due to a unit displacement of m_1 is $k_{12} = -k_2$. It can be seen easily that the force acting on m_1 due to a unit displacement at m_2 is $k_{21} = -k_2$. Thus, $k_{12} = k_{21}$ which is Maxwell's theory of reciprocity, which can be stated in general as: the force produced at any location j due to a unit displacement at location i in a system is the same as the force produced at location i due to a unit displacement at location j.

Example 2.5

Use the influence coefficient method to derive the equations of motion for the two-degree-of-freedom system shown in Fig. 2.10.

It has been shown that the equation of motion for a multi-degree-of-freedom undamped system can be written as

$$\boldsymbol{M}\ddot{\boldsymbol{q}} + \boldsymbol{K}\boldsymbol{q} = 0 \tag{a}$$

For the two-degree-of-freedom system illustrated in Fig. 2.10

$$\begin{bmatrix} m_{11} & m_{12} \\ m_{21} & m_{22} \end{bmatrix} \begin{Bmatrix} \ddot{q}_1 \\ \ddot{q}_2 \end{Bmatrix} + \begin{bmatrix} k_{11} & k_{12} \\ k_{21} & k_{22} \end{bmatrix} \begin{Bmatrix} q_1 \\ q_2 \end{Bmatrix} = 0 \tag{b}$$

In this case, the generalized coordinates used are x_1 and x_2; thus, $m_{12} = m_{21} = 0$ and $m_{11} = m_1$, $m_{22} = m_2$ and $x_1 = \bar{x}_1 e^{j\omega t}$, $x_2 = \bar{x}_2 e^{j\omega t}$. Thus, (b) becomes

$$-\omega^2 \begin{bmatrix} m_1 & 0 \\ 0 & m_2 \end{bmatrix} \begin{bmatrix} \bar{x}_1 \\ \bar{x}_2 \end{bmatrix} + \begin{bmatrix} k_{11} & k_{12} \\ k_{21} & k_{22} \end{bmatrix} \begin{bmatrix} \bar{x}_1 \\ \bar{x}_2 \end{bmatrix} = 0 \tag{c}$$

By inspection of Fig. 2.10, the influence coefficients are given by

$$k_{11} = k_1 + k_2$$
$$k_{22} = k_2 + k_3$$
$$k_{12} = k_{21} = -k_2$$

Thus, the equation of motion becomes

$$\begin{bmatrix} -m_1\omega^2 & 0 \\ 0 & -m_2\omega^2 \end{bmatrix} \begin{bmatrix} x_1 \\ x_2 \end{bmatrix} + \begin{bmatrix} k_1+k_2 & -k_2 \\ -k_2 & k_2+k_3 \end{bmatrix} \begin{bmatrix} x_1 \\ x_2 \end{bmatrix} = 0 \tag{d}$$

2.3 VIBRATION OF CONTINUOUS SYSTEMS

Vibrating structures are generally characterized by the propagation of waves, although if analyses of the motion are undertaken by use of the discretization techniques discussed in the previous section, this may not always be obvious. The motion of simple structures such as beams, plates and cylinders can be analysed from first principles using wave analysis in much the same way as was done for fluid media in Section 2.1. In structures, there will be two wave types in addition to the longitudinal waves present in fluid media and analysed in Section 2.1. These are flexural waves and shear (or torsional) waves, and all of these wave types must be considered when active control of structural vibration and noise radiation is undertaken. Even though shear and longitudinal waves do not significantly contribute directly to sound radiation, the phenomenon of energy conversion from one wave type to another at structural discontinuities means that these wave types must be controlled if structural sound radiation is to be controlled.

The motion of simple beams, plates and cylinders, analysed using a wave approach, is the subject of Sections 2.3.1, 2.3.2 and 2.3.3. More complex structures must be analysed by dividing them up into discrete elements and then using finite element analysis and the techniques of Chapter 4.1 to determine their motion.

In this section, the equations of motion (or wave equations) for beams, plates and thin cylinders will be derived and summarized. More detailed treatments are available in books devoted entirely to the topic (Skudrzyk, 1968; Leissa, 1969, 1973; and Soedel, 1993). The results will be used in later sections where the active control of vibration and noise radiation from these structural types is considered.

2.3.1 Nomenclature and sign conventions

The literature on the vibrations of beams, plates and shells is characterized by a wide range of nomenclature and sign conventions. The sign conventions adopted in the past for moments, forces and rotations show little consistency between various authors and are rarely even discussed. Inconsistency in the use of sign conventions can cause problems in the derivation of the equations of motion, generally resulting in one or more sign errors in the final equation. In this book, the right-hand rule for axis labelling in the cartesian coordinate system and positive external moments and rotations will be followed consistently. This convention is illustrated in Fig. 2.11.

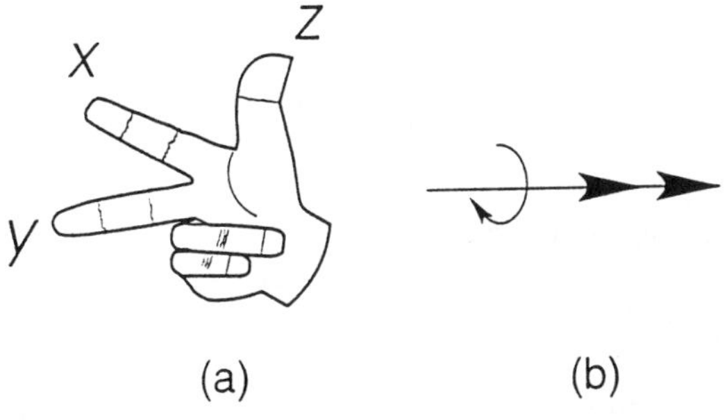

Fig. 2.11 Sign conventions: (a) right-hand axis labelling convention for the cartesian coordinate system; (b) general positive rotation convention.

Positive external forces and displacements are in the positive directions of the corresponding axes. Thus, in this book, the sign convention adopted for displacements, shear forces, bending moments and twisting moments follows that of Leissa (1969, 1973) for plates and shells, and the convention adopted for beams is the same as that used by Fahy (1985) and Cremer *et al.* (1973). Displacements and applied forces are generally positive upwards and angular deflections and applied moments are positive in the counter-clockwise direction when looking along an axis towards the origin.

For plates and shells, the quantities Q, N and M represent forces and moments per unit width, whereas in beams they represent simply forces and moments. It should be noted that the final equations of motion for each type of structure are independent of the sign conventions used in their derivation. The nomenclature used here is identical to that used by Leissa except that here, in-plane displacements are denoted by ξ_x, ξ_y and ξ_θ rather than by u and v to avoid confusion with the use of these symbols for particle velocity and volume velocity respectively, in other parts of this book. Double subscripts are also used here to denote stresses and strains and the same symbols with different subscripts are used to denote shear stresses and shear strains. For example σ_{xy} is used to denote shear stress in the $x-y$ plane and σ_{xx} is used for normal stress in the x-direction. Leissa (1969) used τ_{xy} and σ_x to denote these same quantities. Use of the same symbols for normal and shear stresses is preferred as it allows expressions relating stresses and strains to be more easily generalized. As it will be useful in the derivation of the equations of motion of beams plates and shells, the general three-dimensional relationship between stresses and strains will now be discussed. It is derived using the generalized three-dimensional Hooke's law (Cremer *et al.*, 1973, p.133; Heckl, 1990; Pavic, 1988) and may be written as follows:

$$\sigma_{ik} = \frac{E}{1+\nu}\left[\frac{\nu}{1+2\nu}(\epsilon_{11}+\epsilon_{22}+\epsilon_{33})\,\delta_{ik}+\epsilon_{ik}\right] \tag{2.3.1}$$

where $i = 1, 2, 3$ and $k = 1, 2, 3$. For normal stresses $i = k$ and for shear stresses $i \neq k$.

The strains are defined as

$$\epsilon_{ik} = \frac{1}{\delta_{ik}+1}\left[\frac{\partial \xi_i}{\partial k}+\frac{\partial \xi_k}{\partial i}\right] \tag{2.3.2}$$

and the coefficient δ_{ik} is defined as

$$\begin{aligned}\delta_{ik} &= 1 \text{ if } i = k\\ &= 0 \text{ if } i \neq k\end{aligned} \tag{2.3.3}$$

For the cartesian coordinate system $x = 1$, $y = 2$, $z = 3$. For the cylindrical coordinate system $r = 1$, $a\vartheta = 2$, $x = 3$ where a is the radius at which the stresses and strains are evaluated, and ϑ is the cylindrical angular coordinate.

In this section, the nomenclature and sign conventions used are illustrated at the beginning of each discussion and will remain unchanged throughout the remainder of the book.

One inconsistency in nomenclature between that used in this section and that used by Leissa (1969, 1973) is in the definition of angular rotation. Here, angular rotations are given a subscript to represent the axis about which the rotation is made, that is, θ_x means rotation about the x-axis. This is consistent with generally accepted elasticity conventions. However, in Leissa's work (1969, 1973), the subscript refers to the axis normal to the plane which is rotating.

Thus, θ_x means rotation of a plane perpendicular to the x-axis or rotation about a normal perpendicular to a surface parallel with the x-axis. In both cases, angular displacement is positive in the counter-clockwise direction when viewed along the axis of rotation towards the origin of the coordinate system (right-hand rule – see Fig. 2.11). The convention used here is illustrated in Fig. 2.12.

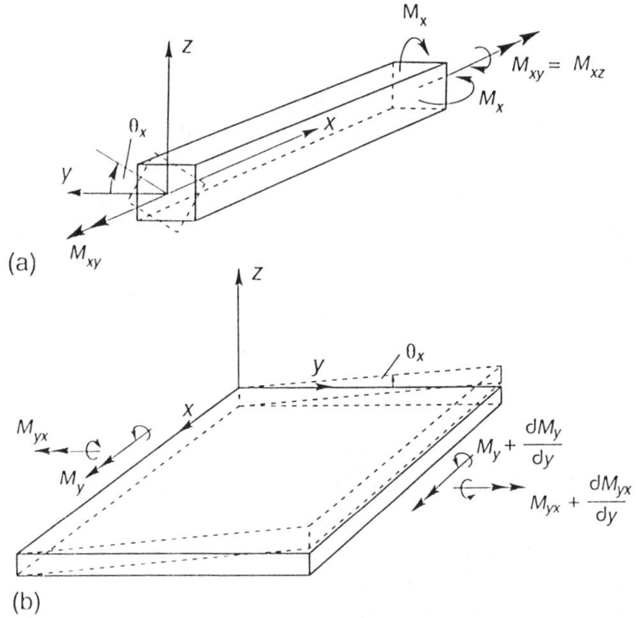

Fig. 2.12 Sign conventions for beams, plates and shells: (a) moment and angle convention used for beams; (b) moment and angle convention used for plates and shells.

The convention used here to define internal moments is consistent with that used by Leissa and is the same for beams, plates and shells. The subscript on a moment refers to a moment acting to rotate the plane perpendicular to the axis denoted by the subscript. In a beam, for example, M_x could refer to a moment about either the y or z axes, depending upon the type of flexural wave it is characterizing. However, in a plate or a shell, M_x refers only to a moment about the y-axis or θ-axis respectively and not about the axis normal to the surface. A double subscript on the moment refers to a twisting moment about the axis denoted by the first subscript, acting to twist the plane perpendicular to the axis defined by the second subscript. Thus, M_{yx} represents the twisting of a plane perpendicular to the x-axis about the y-axis.

The convention used for externally applied moments is different to that used for internal moments, and again this is consistent with Leissa (1969, 1973). For externally applied moments, the first subscript is e to denote an external moment and the second subscript denotes the axis about which the moment is acting.

Thus, M_{ex} refers to an external moment acting about the x-axis. For beams,

M_{ex} has the dimensions of moment per unit length and for plates and shells it has the dimensions of moment per unit area. Similarly, for externally applied forces q, the dimensions are force per unit length for beams and force per unit area for plates and shells. For applied point forces and line or point moments, it is necessary to use the Dirac delta function to express them in terms of force or moment per unit length or area, to allow them to be used in the equations of motion.

2.3.2 Damping

Damping may be included in any of the following analyses by replacing Young's modulus E with the complex modulus $E(1+j\eta)$ and the shear modulus G with the complex shear modulus $G(1+j\eta)$ where η is the structural loss factor. Unfortunately the complex elastic model is not strictly valid in the time dependent forms of the equations of motion as it can lead to non-causal solutions. However, it is valid if restricted to steady state and simple harmonic (or multiple harmonic) vibration.

2.3.3 Waves in beams

By definition, beams are long in comparison with their width and depth and are sufficiently thin that the cross sectional dimensions are only a small fraction (less than 10%) of a wavelength at the frequency of interest. Beams that do not satisfy the latter criterion must be analysed by making a correction for the lateral inertia of the section as will be discussed later.

In a simple rectangular section beam it is possible for four different waves to coexist; one longitudinal, one torsional and two flexural. In this section we will derive the wave equations which describe each wave type of interest. The results will be of use in Chapter 10 where active control of beam vibration is considered.

The coordinates x, y and z, displacements w_z, w_y and ξ_x and angular rotation θ of a small beam segment of length δx in the axial direction are shown in Fig. 2.13.

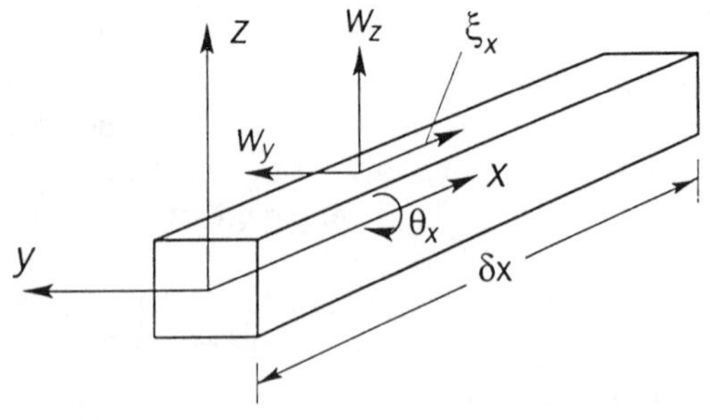

Fig. 2.13 Coordinates and displacements for wave motion in a beam.

2.3.3.1 Longitudinal waves

Longitudinal waves in solids which extend many wavelengths in all directions are very similar to acoustic waves in fluid media. However, when the solid medium is a bar (often referred to as a beam) or thin plate and only extends many wavelengths in one (or two) directions, we are left with a situation where pure longitudinal wave motion cannot occur and the term 'quasi-longitudinal' is used. This is because the lateral surfaces of the beam or the top and bottom surfaces of the plate are free from constraints, allowing the presence of longitudinal stress to produce lateral strains due to the Poisson contraction phenomenon.

The ratio of longitudinal stress to longitudinal strain in a beam is by definition equal to Young's modulus, E. Consider a segment of a beam as shown in Fig. 2.14. The x-coordinate is along the axis of the beam and the cross-sectional area is S.

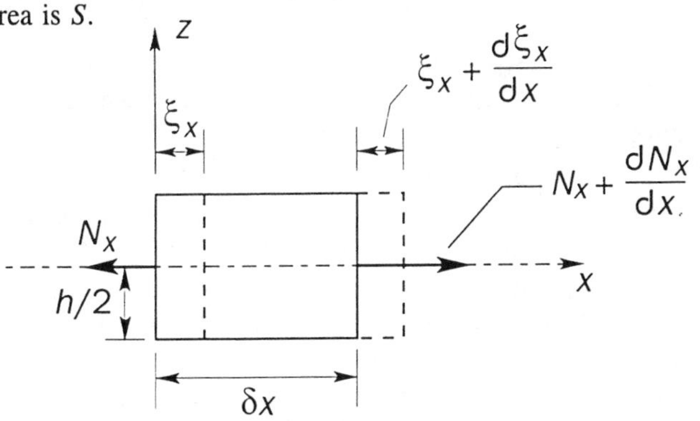

Fig. 2.14 Axial beam element of infinitely short length δx.

The strain which the element in Fig. 2.14 undergoes in the x-direction along the axis of the beam is given by

$$\epsilon_{xx} = \frac{\partial \xi_x}{\partial x} \tag{2.3.4}$$

By definition the stress necessary to cause this strain is

$$\sigma_{xx} = E\epsilon_{xx} = E\frac{\partial \xi_x}{\partial x} \tag{2.3.5}$$

The force acting in the x-direction is given by,

$$N_x = b\int_{-h/2}^{h/2} \sigma_{xx}\,\mathrm{d}z = bh\sigma_{xx} = S\sigma_{xx} \tag{2.3.6}$$

where S is the cross-sectional area of the beam of thickness h and width b.

The equation of motion for the element in Fig. 2.14 is obtained by equating the force acting on the element with the mass of the element multiplied by its acceleration as follows:

$$(\rho S \delta x) \frac{\partial^2 \xi_x}{\partial t^2} = \left[\sigma_{xx} + \frac{\partial \sigma_{xx}}{\partial x} \delta x - \sigma_{xx} \right] S = \frac{\partial \sigma_{xx}}{\partial x} \delta x\, S \tag{2.3.7}$$

Combining (2.3.5) and (2.3.7) results in the wave equation for longitudinal waves propagating in a beam:

$$\frac{\partial^2 \xi_x}{\partial x^2} = \frac{\rho}{E} \frac{\partial^2 \xi_x}{\partial t^2} \tag{2.3.8}$$

This is analogous to the one-dimensional wave equation in an acoustic medium given in Section 2.1.

For the solid beam or bar the longitudinal wave speed (phase speed) is:

$$c_L = (E/\rho)^{1/2} \tag{2.3.9}$$

As this expression is independent of frequency, longitudinal waves in a beam may be described as non-dispersive, a property also characterizing longitudinal waves in any other structure or medium.

The equation for longitudinal waves in a two dimensional solid such as a thin plate is obtained from (2.3.8) by replacing E with $E/(1-\nu^2)$ (see Cremer *et al.* 1973). Similarly for waves in a three-dimensional solid, E in (2.3.8) is replaced by $E(1-\nu)/(1+\nu)(1-2\nu)$ where ν is Poisson's ratio for the material.

Equation (2.3.8) is based on assumptions which are accurate provided that the cross-sectional dimensions of the beam are less than one-tenth of a wavelength at the frequency of interest. The assumptions are:

1. The lateral motion due to Poisson's contraction has no effect on the kinetic energy of the beam.
2. The axial displacement is independent of the location on the beam cross section (that is, plane surfaces remain plane).

At higher frequencies when these assumptions are no longer accurate, it is necessary to add correction terms to the equation of motion to account for them (lateral inertia and lateral shear corrections). The most important correction is that of lateral inertia and this can be accounted for by increasing the density ρ of the beam in the equation of motion by the factor (Skudrzyk, 1968, p.161).

$$1 + \frac{\Delta\rho}{\rho} = 1 + \frac{1}{2}\nu^2 k^2 r^2 \tag{2.3.10}$$

where ν is Poisson's ratio, r the effective radius of the beam (the actual radius for a circular section beam) and k is the wave number ($=2\pi/\lambda$) of the longitudinal wave. This correction provides accurate results for beams with cross-sectional dimensions less than two tenths of a wavelength, as shown in Fig. 2.15.

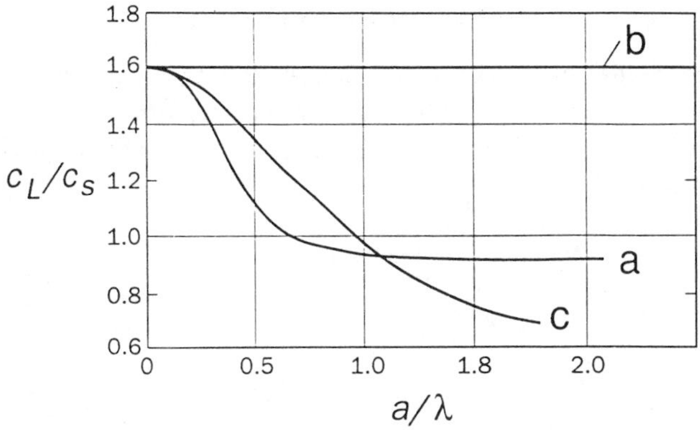

Fig. 2.15 Longitudinal wave velocities in a beam: (a) exact solution from elasticity theory; (b) classical theory; (c) classical theory with correction for lateral inertia.

Beams with larger cross-sectional dimensions must be analysed using a lateral shear correction term as well as the lateral inertia term. This is discussed in detail elsewhere (Skudrzyk, 1968, Chapter 14).

Solutions to the equation of motion (2.3.8) are dependent upon the boundary conditions of the beam. The general solution to (2.3.8) is:

$$\xi_x(x,t) = (A_1 e^{-jkx} + A_2 e^{jkx}) e^{j\omega t} \tag{2.3.11}$$

or alternatively,

$$\xi_x(x,t) = A_3 \cos(kx + \alpha) e^{j\omega t} \tag{2.3.12}$$

Only two boundary condition types are meaningful for longitudinal vibrations. They are fixed or free. At a fixed boundary $\xi = 0$ and at a free boundary $\frac{\partial \xi}{\partial x} = 0$. Thus, the solution to (2.3.11) or (2.3.12) for clamped (fixed) ends is:

$$\xi_x(x,t) = \left[A \sin \frac{n\pi x}{L} \right] e^{j\omega t}, \qquad n = 1,2 \ldots \tag{2.3.13}$$

where L is the length of the beam and n is the mode number.

For free ends, the solution is:

$$\xi_x(x,t) = \left[A \cos \frac{n\pi x}{L} \right] e^{j\omega t}, \qquad n = 1,2 \ldots \tag{2.3.14}$$

Equations (2.3.13) and (2.3.14) are effectively the beam mode shape functions (if the constant A is set = 1).

2.3.3.2 Torsional waves (transverse shear waves)

Solids, unlike fluids, can resist shear deformation and thus are capable of transmitting shear type waves. In a beam, these waves appear as torsional waves where the motion is characterized by a twisting of the cross-section about the longitudinal axis of the beam.

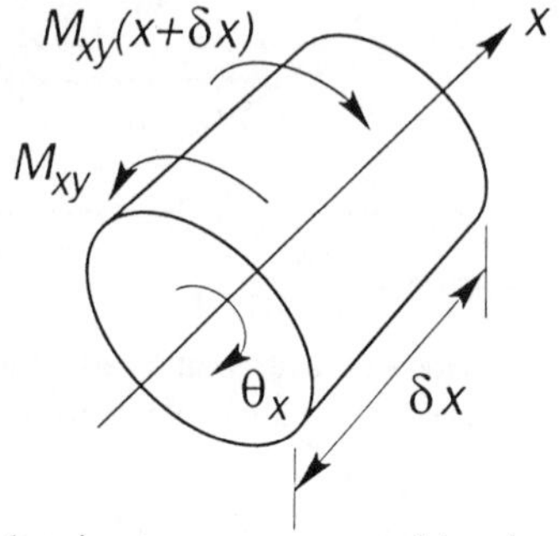

Fig. 2.16 Beam element showing torques generated by shear waves.

Consider a section of infinitesimal length δx of a uniform beam as shown in Fig. 2.16. When a beam is twisted, the torque changes with distance along it due to torsional vibration. The difference in torque between the two ends of an element will be equal to the polar moment of inertia multiplied by the angular acceleration. Thus,

$$M_{xy}(x + \delta x) - M_{xy}(x) = \rho J \delta x \frac{\partial^2 \theta_x}{dt^2} \tag{2.3.15}$$

where M_{xy} is the twisting moment (or torque) acting on the beam, J is the polar second moment of area of the beam cross-section, ρ is the density of the beam material and θ_x is the angle of rotation about the longitudinal x-axis.

It is also clear that

$$M_{xy}(x + \delta x) - M_{xy}(x) = \frac{\partial M_{xy}}{\partial x} \delta x \tag{2.3.16a,b}$$

The angular deflection of a shaft is related to the torque (twisting moment M_{xy}) by (see Fig. 2.16)

$$M_{xy} = C \frac{\partial \theta_x}{\partial x} \tag{2.3.17}$$

where C is the torsional stiffness of the beam. Differentiating (2.3.17) gives

$$\frac{\partial M_{xy}}{\partial x} = C \frac{\partial^2 \theta_x}{\partial x^2} \tag{2.3.18}$$

Equating (2.3.15) and (2.3.16b), and substituting (2.3.18) into the result gives

$$C\frac{\partial^2\theta_x}{\partial x^2}\delta x = \rho J\frac{\partial^2\theta_x}{\partial t^2}\delta x \tag{2.3.19}$$

or

$$\frac{\partial^2\theta_x}{\partial x^2} = \frac{\rho J}{C}\frac{\partial^2\theta_x}{\partial t^2} \tag{2.3.20}$$

which is the wave equation for torsional waves.

Values for the torsional stiffness, C, of rectangular section bars are given in Table 2.1 for various ratios of the thickness h to width b, where h is smaller than b.

Table 2.1 Torsional stiffness of solid rectangular bars

b/h	$\frac{C}{Gh^3b}$
1	0.141
1.5	0.196
2	0.230
3	0.263
4	0.283
5	0.293
10	0.312
20	0.333

From (2.3.20) it can be seen that the phase speed of a torsional wave in the beam is given by

$$c_s^2 = \frac{\rho J}{C} \tag{2.3.21}$$

For a circular section beam, $C = GJ$ and (2.3.21) becomes

$$c_s^2 = G/\rho \tag{2.3.22}$$

where G is the shear modulus for the material. The same boundary conditions and solutions apply as for longitudinal waves, with the displacement ξ_x replaced by the angular displacement θ_x and c_L replaced with c_s.

2.3.3.3 Flexural waves

Flexural waves are by far the most important in terms of sound radiation as they result in a displacement, normal to the surface of the beam, which couples well with any adjacent fluid.

Consider a beam subject to bending as shown in Fig. 2.17. The forces acting on a small segment of δx are also shown in this figure and the displacements and rotations of this small segment are shown in Fig. 2.18.

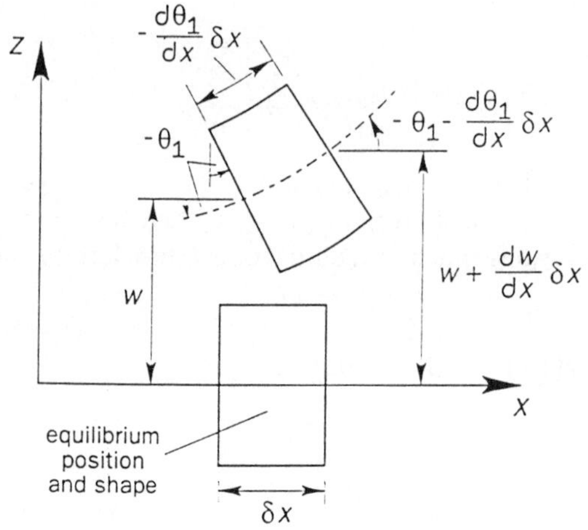

Fig. 2.17 Forces on an element of a beam.

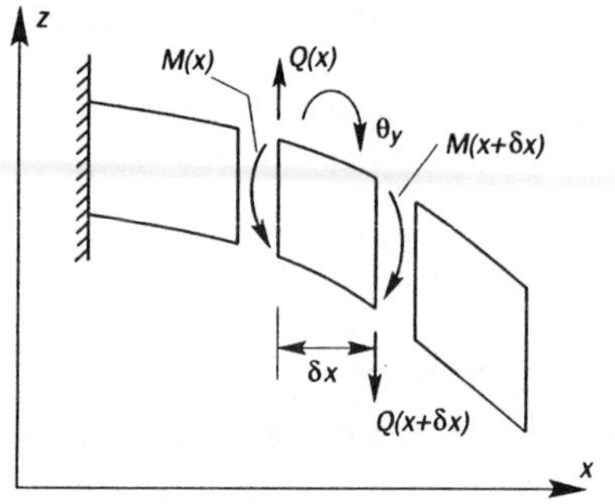

Fig. 2.18 Displacements of a beam segment of length δx.

To simplify the notation to follow, the displacement w_z in the z-direction will be denoted w. From Fig. 2.18, it can be seen that the vertical displacement w and the angle of rotation of the element about the y-axis are related by

$$\theta_y = \theta_1 + \theta_2 = -\frac{\partial w}{\partial x} \tag{2.3.23}$$

where θ_1 is the slope of the beam due to rotation of the section and θ_2 is the additional slope due to shear.

The resultant shear force acting on the small segment is given by

$$-Q_x(x + \delta x) + Q_x(x) = -\frac{\partial Q_x(x)}{\partial x}\delta x \tag{2.3.24}$$

and the motion of the segment obeys Newton's second law, thus,

$$-\frac{\partial Q_x}{\partial x}\delta x = m\delta x\frac{\partial^2 w}{\partial t^2} \qquad (2.3.25)$$

where w is the vertical displacement of the beam at location x and m is the mass per unit length. The total deflection w of the beam element consists of two components; w_1 due to the bending of the beam and w_2 due to the shearing of the beam. The shear motion changes the slope of the beam by an additional angle θ_2 which does not contribute to the rotation of the beam section as a result of the bending slope θ_1.

By examination of Fig. 2.18, it can be seen that the resultant bending moment, which tends to rotate the element in a clockwise direction, is

$$M_x(x+\delta x) - M_x(x) = \frac{\partial M_x}{\partial x}\delta x \qquad (2.3.26)$$

The shear forces $Q_x(x+\delta x)$ and $Q_x(x)$ exert an additional moment on the element given approximately by $Q_x(x)\delta x$. These two moments may be equated to the rotary inertia of the element using Newton's second law of motion. Thus,

$$\frac{\partial M_x}{\partial x}\delta x + Q_x\delta x = I'\ddot{\theta}_1 = \rho I\delta x\ddot{\theta}_1 \qquad (2.3.27a,b)$$

where I' is the moment of inertia of the element δx about the y-axis and I is the second moment of area of the cross-section about the transverse axis in the neutral plane. Equation (2.3.27) simplifies to

$$\frac{\partial M_x}{\partial x} + Q_x = \rho I\ddot{\theta}_1 \qquad (2.3.28)$$

Classical beam theory would assume that $\rho I = 0$. Indeed, for the static case this is also true and the shear force is related to the bending moment by

$$Q_x = -\frac{\partial M_x}{\partial x} \qquad (2.3.29)$$

The relationship between the bending moment M and the deflection w_1 that is produced as a result will now be developed. Consider the same element as shown in Fig. 2.19. Now consider a segment δz of this beam element. The strain in this segment due to bending deformation is given by

$$\epsilon = 2z\frac{\partial \theta_1}{\partial x}\frac{\delta x}{2}\cdot\frac{1}{\delta x} \qquad (2.3.30)$$

Using (2.3.23) we obtain

$$\epsilon = -z\frac{\partial^2 w_1}{\partial x^2} \qquad (2.3.31)$$

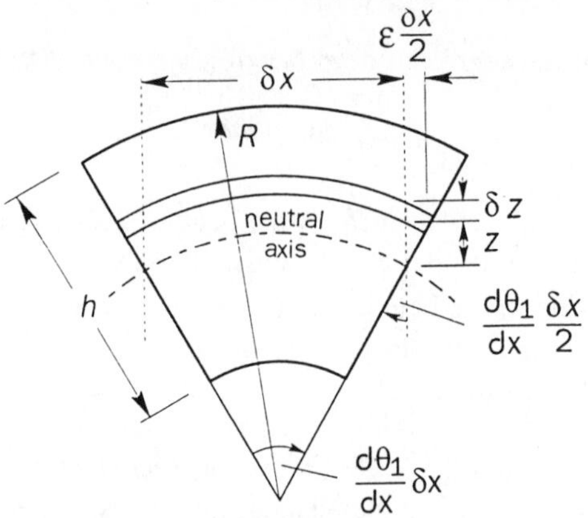

Fig. 2.19 Beam element undergoing pure bending.

Assuming that the relationship between stress and strain in the shaded element of Fig. 2.19 is the same as that for a longitudinal beam, we obtain

$$\sigma = E\epsilon \tag{2.3.32}$$

where E is Young's modulus of elasticity.

The bending moment generated by the shaded volume of Fig. 2.19 is given by

$$M_x = b\int_{-h/2}^{h/2} \sigma z \, dz \tag{2.3.33}$$

where b is the width of the beam and h is its thickness.

Substituting (2.3.31) and (2.3.32) into (2.3.33) gives

$$M_x = -b\int_{-h/2}^{h/2} zE\frac{\partial^2 w_1}{\partial x^2} z \, dz = -bE\frac{\partial^2 w_1}{\partial x^2}\int_{-h/2}^{h/2} z^2 \, dz \tag{2.3.34}$$

Thus,

$$M_x = -EI\frac{\partial^2 w_1}{\partial x^2} \tag{2.3.35}$$

and from (2.3.29), assuming that $\rho I = 0$, the shear force may be written as

$$Q_x = -\frac{\partial M_x}{\partial x} = EI\frac{\partial^3 w_1}{\partial x^3} \tag{2.3.36}$$

However, the actual deflection is greater than w_1 due to the shearing of the beam. Timoshenko (1921) postulated that these additional shearing forces produced an additional deflection w_2 of the beam, given by

$$\frac{\partial w_2}{\partial x} = -\frac{\gamma Q_x}{SG} \tag{2.3.37}$$

where S is the beam cross-sectional area and γ is a constant dependent on the shape of the beam cross-section. For circular beams $\gamma = 1.11$, for rectangular section beams $\gamma = 1.20$ and for I-beams, γ varies between 2.00 and 2.40 (see Skudrzyk, 1968, p. 200). In (2.3.37) G is the shear modulus or modulus of rigidity of the beam material. The deflection increases the slope of the beam by $\theta_2 = -\frac{\partial w_2}{\partial x}$, but does not cause a rotation of the element.

Thus, in summary we have the following equations which describe the motion of the beam and include the effects of transverse shear and rotary inertia:

$$\frac{\partial Q_x}{\partial x} = -m\frac{\partial^2 w}{\partial t^2} \tag{2.3.38}$$

where m is the mass per unit length of the beam ($=\rho S$)

$$\frac{\partial M_x}{\partial x} + Q_x = \rho I \ddot{\theta}_1 \tag{2.3.39}$$

$$M_x = -EI\frac{\partial^2 w_1}{\partial x^2} \tag{2.3.40}$$

$$\frac{\partial w_2}{\partial x} = -\frac{\gamma Q_x}{SG} \tag{2.3.41}$$

$$w = w_1 + w_2 \tag{2.3.42}$$

$$\theta_1 = -\frac{\partial w_1}{\partial x} \tag{2.3.43}$$

The quantities w_1, w_2, θ_1, Q_x and M_x must be eliminated to give an equation in w, the total displacement. This is done by first using (2.3.40) to (2.3.42) to write M_x in terms of Q_x and w. This result for M_x is substituted into (2.3.39) and θ_1 is eliminated from that equation using (2.3.41), (2.3.42) and (2.3.43). Q_x is then eliminated from this result using (2.3.38). When this is done we obtain:

$$EI\frac{\partial^4 w}{\partial x^4} + m\frac{\partial^2 w}{\partial t^2} - \left[\rho I + \frac{\rho\gamma IE}{G}\right]\frac{\partial^4 w}{\partial x^2 \partial t^2} + \frac{\rho^2\gamma I}{G}\frac{\partial^4 w}{\partial t^4} = 0 \tag{2.3.44}$$

which is the beam equation of motion for flexural waves.

The first two terms in (2.3.44) represent the classical beam equation for flexural waves. In this case (with the remaining terms ignored), the bending wave speed is given by $c_b^4 = \omega^2 EI/m$. The third term is the correction for transverse shear and the fourth term is the correction for rotary inertia. The above equation, although not exact (as it assumes that the effects of rotary inertia and shear can be treated separately), gives excellent results even for very thick beams. On the other hand, the classical wave equation (first two terms in (2.3.44)) results in significant errors if the ratio of the wavelength to the equivalent section radius exceeds 0.1. That is, good results may be expected with classical beam theory provided that the bending wavelength is 10 times greater than the equivalent section radius. The bending wave speed in the beam is related to the bending wave number k_b by $c_b = \omega/k_b$ (see Fig. 2.20).

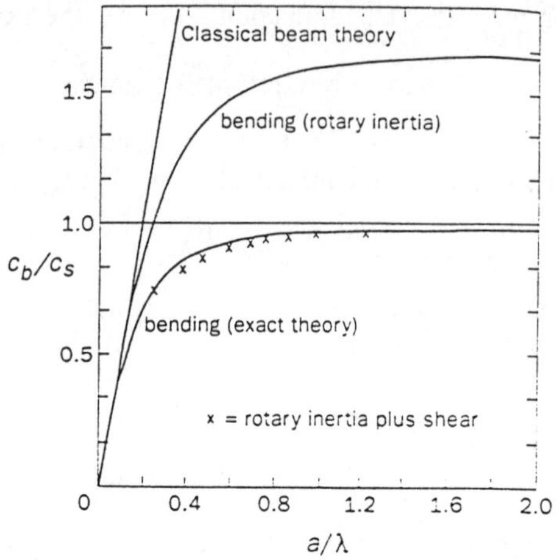

Fig. 2.20 Effect of rotary inertia and shear on the calculated bending wave speed in a beam.

If an external force (of $q(x)$ force units per unit length) is applied to the beam in the positive z-direction, (2.3.38) becomes

$$-\frac{\partial Q_x}{\partial x} + q(x) = m\frac{\partial^2 w}{\partial t^2} \tag{2.3.45}$$

and (2.3.44) becomes

$$EI\frac{\partial^4 w}{\partial x^4} + m\frac{\partial^2 w}{\partial t^2} - \left[\rho I + \frac{\rho\gamma IE}{G}\right]\frac{\partial^4 w}{\partial x^2 \partial t^2} + \frac{\rho^2\gamma I}{G}\frac{\partial^4 w}{\partial t^4}$$
$$= -\frac{\gamma IE}{SG}\frac{\partial^2 q(x)}{\partial x^2} + q(x) + \frac{\rho\gamma I}{SG}\frac{\partial^2 q(x)}{\partial t^2} \tag{2.3.46}$$

If classical beam theory is used then we would be left with the first two terms on the left-hand side of the equation and only the second term on the right.

Note that a point force F applied to the beam at $x = x_o$ may be represented by

$$q(x) = F\delta(x - x_o) \tag{2.3.47}$$

where $\delta(\ )$ is the Dirac delta function.

If we apply an external moment to the beam (of $M_e(x)$ units of moment per unit length) (2.3.39) becomes

$$M_e + \frac{\partial M}{\partial x} + Q_x = \rho I \ddot{\theta}_1 \tag{2.3.48}$$

and the equation of motion is

$$EI\frac{\partial^4 w}{\partial x^4} + m\frac{\partial^2 w}{\partial t^2} - \left[\rho I + \frac{\rho\gamma IE}{G}\right]\frac{\partial^4 w}{\partial x^2 \partial t^2} + \frac{\rho^2\gamma I}{G}\frac{\partial^4 w}{\partial t^4} = \frac{\partial M_e(x)}{\partial x} \tag{2.3.49}$$

A line moment M_L applied across the beam at $x = x_o$ may be represented by

$$M_e(x) = M_L\delta(x - x_o) \tag{2.3.50}$$

For a beam which is very thin compared to its width, the quantity E in the preceding equations must be replaced by $E/(1 - \nu^2)$, to give the one-dimensional plate equation.

It is interesting to note that if the y- and z-axes were interchanged, so that the beam displacement along the y-axis was of interest, the right-hand sign convention would change (2.3.23) to

$$\theta_z = \frac{\partial w}{\partial x}$$

(2.3.39) to

$$\frac{\partial M_x}{\partial x} + Q_x = -\rho I \ddot{\theta}_1$$

(2.3.43) to

and (2.3.48) to

$$M_e - \frac{\partial M_x}{\partial x} - Q_x = \rho I \ddot{\theta}_1$$

where it is assumed that the conventions for positive internal shear force and bending moment remain the same.

Note that the equations of motion remain unchanged except for (2.3.49) involving an externally applied moment M_e where the positive sign on the right-hand side of the equation is now replaced with a negative sign, and in all equations, the moment of inertia, I, will be about the z-axis rather than the y-axis.

The bending wave speed c_b in a beam is related to the bending wavenumber k_b by $c_b = \omega/k_b$. An expression for c_b can be determined by substituting the expression for a simple harmonic progressive wave:

$$w = Ae^{j(\omega t - k_b x)} \tag{2.3.51}$$

into (2.3.44) to give:

$$\alpha^4 k_b^4 = \omega^2 + \beta k_b^2 \omega^2 - \delta\omega^4 \tag{2.3.52}$$

where

$$\alpha^4 = \frac{EI}{m}, \; \beta = \left[\frac{I}{S} + \frac{\gamma IE}{SG}\right] \text{ and } \delta = \frac{\rho\gamma I}{SG}$$

As (2.3.52) is invalid when the corrections for shear and rotary inertia are large, we may replace k_b^2 on the right-hand side with ω/α^2 (its zero order approximation). Equation (2.3.52) then can be written as

$$k_b^4 = \frac{\omega^2 + \beta\omega^3/\alpha^2 - \delta\omega^4}{\alpha^4} \tag{2.3.53}$$

Hence

$$k_b = \pm j\Delta \text{ and } \pm\Delta \tag{2.3.54}$$

where

$$\Delta = (\omega^2 + \beta\omega^3/\alpha^2 - \delta\omega^4)^{1/4}/\alpha \tag{2.3.55}$$

Thus, the wave speed c_b for bending waves is given by:

$$c_b = \frac{\omega}{k_b} = \frac{\alpha\omega^{1/2}}{(1 + \beta\frac{\omega}{\alpha^2} - \delta\omega^2)^{1/4}} \tag{2.3.56a,b}$$

At low frequencies, when the shear or rotary inertia corrections can be ignored,

$$k_b^4 = \frac{\omega^2 m}{EI}; \qquad c_b = \frac{\omega}{k_b} = \alpha\sqrt{\omega} \tag{2.3.57a,b}$$

and the group wavespeed is:

$$c_g = \left[\frac{\partial\omega}{\partial k}\right]^{-1} = 2\omega^{1/2}\left[\frac{m}{EI}\right]^{-1/4} \tag{2.3.58}$$

Note that for a rectangular section beam there will be two flexural waves with displacements perpendicular to one another. The two equations describing the motion are identical to (2.3.44), but each type of motion will be characterized by a different second moment of area, I. For motion in the z-direction, I_{yy} is used and vice versa.

From (2.3.54) and (2.3.51) it can be seen that the complete solution for bending waves in a beam is given by:

$$w(x,t) = \left[A_1 e^{-jk_b x} + A_2 e^{jk_b x} + A_3 e^{-k_b x} + A_4 e^{k_b x}\right] e^{j\omega t} \tag{2.3.59}$$

or, in terms of transcendental and hyperbolic functions,

$$w(x,t) = \left[A\cos kx + B\sin kx + C\cosh kx + D\sinh kx\right]e^{j\omega t} \tag{2.3.60}$$

In the above equations the constants A_1 to A_4 and A to D may all be complex, depending on the boundary conditions.

Solutions to (2.3.59) and (2.3.60) for various beam boundary conditions are summarized below

$$\text{Clamped:}\quad \left.\begin{aligned} w(x,t) &= 0 \quad \text{at } x = 0,\ L \\ \frac{\partial w(x,t)}{\partial x} &= 0 \quad \text{at } x = 0,\ L \end{aligned}\right\} \tag{2.3.61}$$

$$\text{Free:}\quad \left.\begin{aligned} \frac{\partial^2 w(x,t)}{\partial x^2} &= 0 \text{ at } x = 0,\ L \\ \frac{\partial^3 w(x,t)}{\partial x^3} &= 0 \text{ at } x = 0,\ L \end{aligned}\right\} \tag{2.3.62}$$

$$\text{Simply supported:}\quad \left.\begin{aligned} w(x,t) &= 0 \quad \text{at } x = 0,\ L \\ \frac{\partial^2 w(x,t)}{\partial x^2} &= 0 \quad \text{at } x = 0,\ L \end{aligned}\right\} \tag{2.3.63}$$

Of course both ends of the beam do not have to have the same boundary conditions. Any combination of the above is possible.

If damping is included by replacing E with $E(1 + j\eta)$ in the beam equation of motion, the solution may still be expressed in the form of (2.3.59), but with k_b replaced by a complex bending wave number k_b' defined as (Fahy, 1988):

$$k_b' = k_b(1 - j\,\eta/4) \tag{2.3.64}$$

and the solution for a single travelling wave is written as:

$$\xi(x,t) = Ae^{j(\omega t - k_b x)}\,e^{-k_b \eta\, x/4} \tag{2.3.65}$$

showing that the travelling wave attenuates exponentially with increasing distance x from its source. Note that the constant A may be complex ($A = \bar{A}e^{j\phi}$), where $\bar{A}$ is a real constant.

2.3.4 Waves in thin plates

The coordinate system and notation convention for displacements, stresses, forces and moments for waves in thin plates is shown in Fig. 2.21. The sign convention followed is consistent with that of Leissa (1969). As mentioned above, internal forces and stresses are defined in terms of quantities per unit plate width (or unit length) and the externally applied force is defined as force per unit area. The three well-known equilibrium equations (Timoshenko and Woinowsky-Krieger, 1959), $\tau_{xy} = \tau_{yx}$, $\tau_{zx} = \tau_{xz}$ and $\tau_{yz} = \tau_{zy}$ have been included in Fig. 2.21(d). Twisting moments are defined such that $M_{xy} = M_{yx}$. They are equal because the shear stresses causing them are equal.

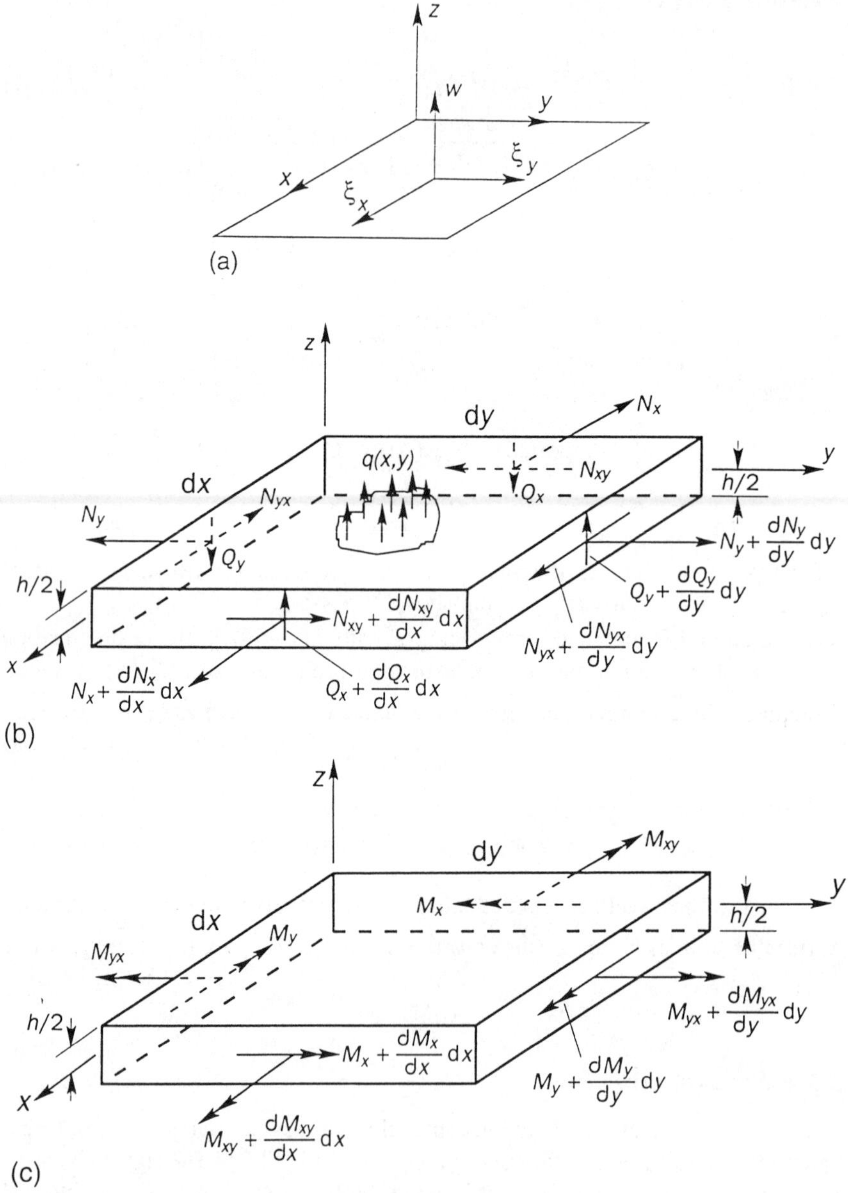

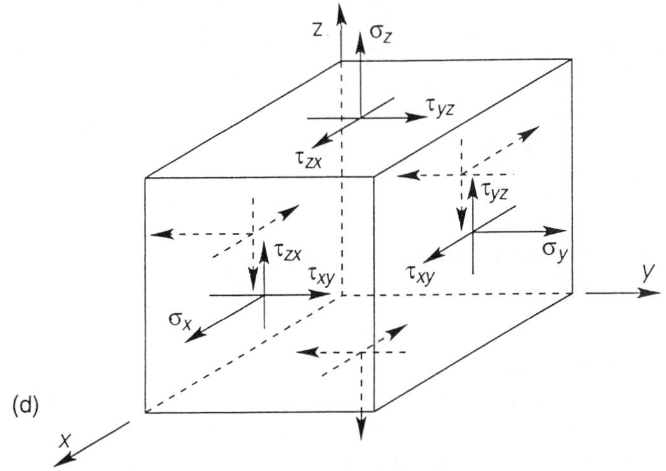

Fig. 2.21(a) Coordinate system and displacements for a thin plate. w is the displacement for a flexural wave, ξ_y and ξ_x are the in-plane displacements in the y and x directions respectively; (b) forces (intensities) acting on a plate element; (c) moments (intensities) acting on a plate element; (d) notation and positive directions of stress.

In the following sections it will be assumed that the plates are isotropic. Orthotropic plates are also of interest as stiffened plates can often be modelled as such. These are adequately treated by Leissa (1969) and will not be considered further here.

2.3.4.1 *Longitudinal waves*

The derivation of the wave equation for longitudinal wave propagation in one direction in a thin plate is similar to the derivation for a beam. Figure 2.14 and (2.3.8) also apply to longitudinal propagation in the x-direction in a plate. The only difference is that the relationship between stress and strain is

$$\sigma_{xx} = \frac{E}{(1-\nu^2)}(\epsilon_{xx} + \nu\epsilon_{yy}) \tag{2.3.66}$$

For wave propagation in the x-direction, ϵ_{yy} is assumed to be negligible. Thus, the wave equation for longitudinal wave propagation in the x-direction in a thin plate is

$$\frac{\partial^2 \xi_x}{\partial x^2} = \frac{\rho(1-\nu^2)}{E}\frac{\partial^2 \xi_x}{\partial t^2} \tag{2.3.67}$$

and the corresponding phase speed is

$$c_L = \sqrt{E/\rho(1-\nu^2)} \tag{2.3.68}$$

For propagation in the y-direction, the subscript x is replaced with y. Note that (2.3.67) contains no correction for shear or rotary inertia and is only valid for thin plates, less than 0.1 wavelengths thick.

Boundary conditions and solutions are identical to those found for longitudinal waves in a beam (for classical beam theory).

2.3.4.2 *Transverse shear waves*

In-plane shear waves travelling in thin plates are influenced very little by the free surfaces of the plate, thus, the shear wave speed and the wave equation is very similar to that for a large volume. Although in-plane shear waves in thin plates are difficult to generate by applied forces, they can play a significant role in the transportation of vibrational energy through a plate structure such as a ship. This is because some flexural wave energy is converted to shear wave energy at structural junctions and discontinuities, some of which in turn is converted back to flexural wave energy at other joints.

For a shear wave propagating in the x direction, the corresponding displacement is in the y-direction. We will denote this as ξ_y and the resulting situation for the plate element is shown in Fig. 2.22, where (x, y) is in the plane of the plate.

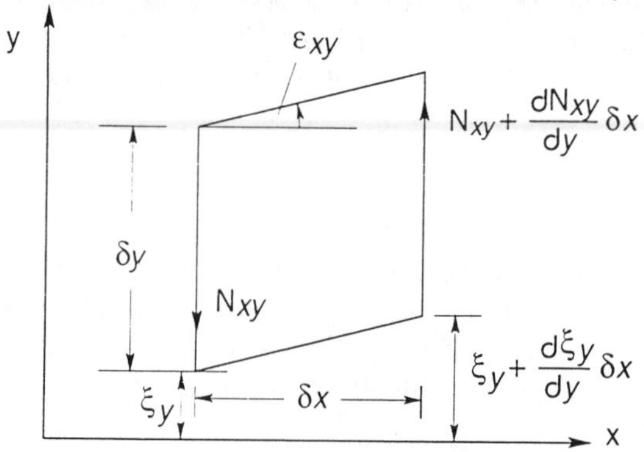

Fig. 2.22 Shear stresses and displacement associated with motion only in the y-direction.

From Fig. 2.22 it can be seen that the net shear force per unit width acting on the element (of thickness δ_z) in the y-direction is

$$\left[N_{xy}(x + \delta x) - N_{xy}\right] = \frac{\partial N_{xy}}{\partial x}\delta x \tag{2.3.69}$$

The equation of motion for an element of thickness h can then be written as

$$\rho\delta_x\delta_y h\frac{\partial^2\xi_y}{\partial t^2} = \frac{\partial N_{xy}}{\partial x}\delta_x\delta_y \tag{2.3.70}$$

where N_{xy} is the shear force per unit plate width (or length in the y-direction), ξ_y is the transverse displacement of the element in the y-direction and ρ is the density of the plate material.

The shear force N_{xy} is obtained by integrating the shear stress over the plate thickness h.

Thus,

$$N_{xy} = \int_{-h/2}^{h/2} \sigma_{xy} \, dz \tag{2.3.71}$$

For shear wave propagation only, this simplifies to

$$N_{xy} = h\sigma_{xy} \tag{2.3.72}$$

Thus, (2.3.70) may be written as

$$\rho \frac{\partial^2 \xi_y}{\partial t^2} = \frac{\partial \sigma_{xy}}{\partial x} \tag{2.3.73}$$

The shear strain is related to the shear stress by the shear modulus G as follows:

$$\sigma_{xy} = G\epsilon_{xy} \tag{2.3.74}$$

From (2.3.2) we obtain

$$\epsilon_{xy} = \frac{\partial \xi_y}{\partial x} \tag{2.3.75}$$

as the quantity $\frac{\partial \xi_x}{\partial y}$ is zero in this case. Substituting (2.3.74) and (2.3.75) into (2.3.73) gives the following wave equation for a transverse in-plane shear wave propagating in a thin plate in the x-direction:

$$\frac{\partial^2 \xi_y}{\partial x^2} = \frac{\rho}{G} \frac{\partial^2 \xi_y}{\partial t^2} \tag{2.3.76}$$

The frequency independent phase speed is then

$$c_s = \sqrt{\frac{G}{\rho}} \tag{2.3.77}$$

Boundary conditions and solutions are identical with those found for longitudinal waves, except that E is replaced with G, and the displacement is normal (but in the plane of the plate) to the direction of wave propagation.

2.3.4.3 Flexural waves

Referring to Fig. 2.22, we can write the following relationships between stress and forces and bending moments acting on the plate element:

$$N_x = \int_{-h/2}^{h/2} \sigma_{xx} \, dz \tag{2.3.78}$$

$$N_y = \int_{-h/2}^{h/2} \sigma_{yy}\, dz \tag{2.3.79}$$

$$N_{xy} = N_{yx} = \int_{-h/2}^{h/2} \sigma_{xy}\, dz \tag{2.3.80}$$

$$M_x = \int_{-h/2}^{h/2} \sigma_{xx}\, z\, dz \tag{2.3.81}$$

$$M_y = \int_{-h/2}^{h/2} \sigma_{yy}\, z\, dz \tag{2.3.82}$$

$$M_{xy} = M_{yx} = \int_{-h/2}^{h/2} \sigma_{xy}\, z\, dz \tag{2.3.83}$$

The above quantities have the units of force or moment per unit length.

Before proceeding further, it is necessary to relate these quantities to the plate displacements w, ξ_x, ξ_y. Consider the edge view of a portion of plate shown in Fig. 2.23. The centre line of the plate at point P is displaced longitudinally by a distance ξ_{x0} and laterally by w.

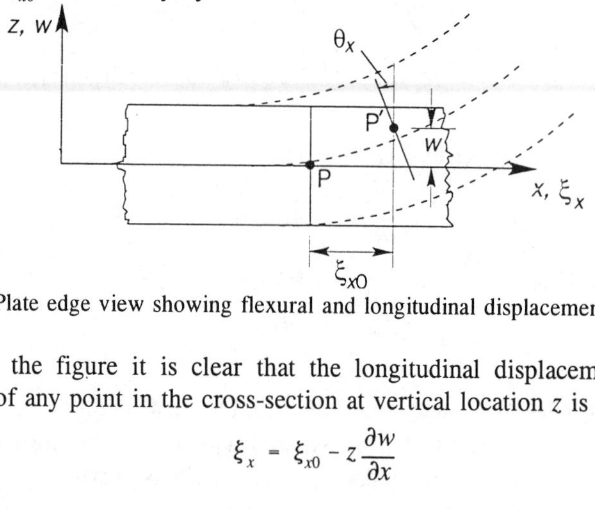

Fig. 2.23 Plate edge view showing flexural and longitudinal displacement.

From the figure it is clear that the longitudinal displacement in the x direction of any point in the cross-section at vertical location z is given by

$$\xi_x = \xi_{x0} - z\frac{\partial w}{\partial x} \tag{2.3.84}$$

and the longitudinal displacement in the y-direction is

$$\xi_y = \xi_{y0} - z\frac{\partial w}{\partial y} \tag{2.3.85}$$

The in-plane strains and shear strains can be related to the quantities in the preceding equations using (2.3.2), which gives

$$\epsilon_{xx} = \frac{\partial \xi_x}{\partial x} \tag{2.3.86}$$

$$\epsilon_{yy} = \frac{\partial \xi_y}{\partial y} \tag{2.3.87}$$

$$\epsilon_{xy} = \frac{\partial \xi_y}{\partial x} + \frac{\partial \xi_x}{\partial y} \tag{2.3.88}$$

The stresses for an isotropic plate are related to the strains as follows:

$$\sigma_{xx} = \frac{E}{1-\nu^2}\left(\epsilon_{xx} + \nu\epsilon_{yy}\right) \tag{2.3.89}$$

$$\sigma_{yy} = \frac{E}{1-\nu^2}\left(\epsilon_{yy} + \nu\epsilon_{xx}\right) \tag{2.3.90}$$

$$\sigma_{xy} = G\epsilon_{xy} \tag{2.3.91}$$

Substituting (2.3.84) and (2.3.85) into (2.3.86) to (2.3.88) and the results into (2.3.89) to (2.3.91) gives expressions for the stresses in terms of the plate mid-plane displacements ξ_{x0}, ξ_{y0} and w. These results can then be substituted into (2.3.78) to (2.3.83) to obtain expressions for N_x, N_y, N_{xy}, M_x, M_y and M_{xy} in terms of plate displacements.

For flexural wave propagation, $\xi_{x0} = \xi_{y0} = 0$, and the resulting integrations give $N_x = N_y = N_{xy} = 0$. The integration results for the moments are:

$$M_x = -D\left[\frac{\partial^2 w}{\partial x^2} + \nu\frac{\partial^2 w}{\partial y^2}\right] \tag{2.3.92}$$

$$M_y = -D\left[\frac{\partial^2 w}{\partial y^2} + \nu\frac{\partial^2 w}{\partial x^2}\right] \tag{2.3.93}$$

$$M_{xy} = M_{yx} = -D(1-\nu)\frac{\partial^2 w}{\partial x \partial y} \tag{2.3.94}$$

where the bending stiffness, D, is defined as,

$$D = \frac{Eh^3}{12(1-\nu^2)} \tag{2.3.95}$$

and ν = Poisson's ratio.

Referring to Figs. 2.21(b) and (c), and summing moments about the x and y axes (remembering that for flexural waves, $N_x = N_y = N_{xy} = 0$), we obtain

$$Q_x - \frac{\partial M_x}{\partial x} - \frac{\partial M_{xy}}{\partial y} = \frac{\rho h^3}{12}\frac{\partial^3 w}{\partial x \partial t^2} \tag{2.3.96}$$

$$Q_y - \frac{\partial M_{xy}}{\partial x} - \frac{\partial M_y}{\partial y} = \frac{\rho h^3}{12}\frac{\partial^3 w}{\partial y \partial t^2} \tag{2.3.97}$$

where ρ is the density of the plate. If we assume that the rotary inertia term on the right-hand side of (2.3.96) and (2.3.97) is negligible (thin plate), and further, we substitute (2.3.92), (2.3.93) and (2.3.94) for M_x, M_y and M_{xy} respectively, we obtain:

$$Q_x = -D\frac{\partial}{\partial x}\left[\frac{\partial^2 w}{\partial x^2} + \frac{\partial^2 w}{\partial y^2}\right] \tag{2.3.98}$$

$$Q_y = -D\frac{\partial}{\partial y}\left[\frac{\partial^2 w}{\partial x^2} + \frac{\partial^2 w}{\partial y^2}\right] \tag{2.3.99}$$

Q_x and Q_y have the units of force per unit length.

Returning to Fig. 2.21(b), and summing forces in the z-direction (remembering that for flexural waves, $N_x = N_y = N_{xy} = 0$), we obtain:

$$\frac{\partial Q_x}{\partial x} + \frac{\partial Q_y}{\partial y} + q = \rho h\frac{\partial^2 w}{\partial t^2} \tag{2.3.100}$$

where the external force q has the units of force per unit area.

Substituting (2.3.98) and (2.3.99) into (2.3.100) gives the classical plate equation for an isotropic plate:

$$D\nabla^4 w + \rho h\frac{\partial^2 w}{\partial t^2} = q \tag{2.3.101}$$

where $\nabla^4 = \nabla^2\nabla^2$. In cartesian coordinates

$$\nabla^2 = \frac{\partial^2}{\partial x^2} + \frac{\partial^2}{\partial y^2}$$

In polar coordinates

$$\nabla^2 = \frac{\partial^2 w}{\partial r^2} + \frac{1}{r}\frac{\partial w}{\partial r} + \frac{1}{r^2}\frac{\partial^2 w}{\partial \theta^2}$$

If the external force were replaced with external moments M_{ex} and M_{ey} (units of moment per unit area) about the x- and y-axes respectively, then (2.3.96) and (2.3.97) become:

$$Q_x - \frac{\partial M_x}{\partial x} - \frac{\partial M_{xy}}{\partial y} + M_{xe} = \frac{\rho h^3}{12}\frac{\partial^3 w}{\partial x\partial t^2} \tag{2.3.102}$$

$$Q_y - \frac{\partial M_{xy}}{\partial x} - \frac{\partial M_y}{\partial y} + M_{ye} = \frac{\rho h^3}{12}\frac{\partial^3 w}{\partial y\partial t^2} \tag{2.3.103}$$

and consequently, the wave equation becomes

$$D\nabla^4 w + \rho h\frac{\partial^2 w}{\partial t^2} = -\frac{\partial M_{ex}}{\partial y} + \frac{\partial M_{ey}}{\partial x} \tag{2.3.104}$$

Note that if the external force q were acting simultaneously, it would be simply added to the right-hand side of (2.3.104).

Effects of shear deformation and rotary inertia

In the same way as was found for beams, the effects of shear deformation and rotary inertia become important for thick plates and/or high frequencies (short wavelengths).

We can still make use of the three fundamental equations (2.3.96), (2.3.97) and (2.3.100) which we derived using classical theory. This time, however, we cannot neglect the rotary inertia terms on the right-hand side of (2.3.96) and (2.3.97). The flexural deflection w is now made up of a bending contribution and a shear deformation contribution. Thus, the angles of rotation of lines normal to the mid-plane before deformation cannot be directly related to the displacement w as was done in (2.3.84) and (2.3.85) in which shear deformation was neglected. Thus, the in-plane displacements must now be expressed in terms of the rotation angles ψ_x of a plane normal to the x-axis and ψ_y of a plane normal to the y-axis. Thus, (2.3.84) and (2.3.85) become

$$\xi_x = -z\psi_x \tag{2.3.105}$$

$$\xi_y = -z\psi_y \tag{2.3.106}$$

where only displacements associated with flexural wave propagation have been included.

Using the same procedure as for classical theory we obtain the following expressions for the bending moments:

$$M_x = -D\left[\frac{\partial\psi_x}{\partial x} + \nu\frac{\partial\psi_y}{\partial y}\right] \tag{2.3.107}$$

$$M_y = -D\left[\frac{\partial\psi_y}{\partial y} + \nu\frac{\partial\psi_x}{\partial x}\right] \tag{2.3.108}$$

$$M_{xy} = M_{yx} = -\frac{D(1-\nu)}{2}\left[\frac{\partial\psi_y}{\partial x} + \frac{\partial\psi_x}{\partial y}\right] \tag{2.3.109}$$

The transverse shearing forces Q_x and Q_y are now obtained by integrating the transverse shearing stresses over the plate thickness:

$$Q_x = \int_{-h/2}^{h/2} \sigma_{xz}\,\mathrm{d}z \tag{1.2.110}$$

$$Q_y = \int_{-h/2}^{h/2} \sigma_{yz}\,\mathrm{d}z \tag{2.3.111}$$

Substituting (2.3.2) into (2.3.1), then into (2.3.110) and (2.3.111), and then integrating results in the following:

$$Q_x = -\kappa^2 Gh \left[\psi_x - \frac{\partial w}{\partial x} \right] \tag{2.3.112}$$

$$Q_y = -\kappa^2 Gh \left[\psi_y - \frac{\partial w}{\partial y} \right] \tag{2.3.113}$$

where κ^2 is a constant introduced to account for the shear stresses not being constant over the thickness of the plate. Mindlin (1951) chose κ to make the dynamic theory predictions consistent with the known exact theory of elasticity prediction of the frequency of the fundamental 'thickness–shear' mode of vibration. Thus,

$$\kappa^2 \approx 0.76 + 0.3\nu \tag{2.3.114}$$

See Mindlin (1951) for a more detailed discussion of this constant.

Using (2.3.107) to (2.3.109), (2.3.112) and (2.3.113), equations (2.3.96), (2.3.97) and (2.3.100) can be expressed in terms of ψ_x, ψ_y and w as follows:

$$\frac{D}{2}\left[(1-\nu)\nabla^2\psi_x + (1+\nu)\frac{\partial\Phi}{\partial x}\right] - \kappa^2 Gh\left[\psi_x + \frac{\partial w}{\partial x}\right] = \frac{\partial h^3}{12}\frac{\partial^2\psi_x}{\partial t^2} \tag{2.3.115}$$

$$\frac{D}{2}\left[(1-\nu)\nabla^2\psi_y + (1+\nu)\frac{\partial\Phi}{\partial y}\right] - \kappa^2 Gh\left[\psi_y + \frac{\partial w}{\partial y}\right] = \frac{\partial h^3}{12}\frac{\partial^2\psi_y}{\partial t^2} \tag{2.3.116}$$

$$\kappa^2 Gh(\nabla^2 w + \Phi) + q = \rho h \frac{\partial^2 w}{\partial t^2} \tag{2.3.117}$$

where

$$\Phi = \frac{\partial\psi_x}{\partial x} + \frac{\partial\psi_y}{\partial y} \tag{2.3.118}$$

A single differential equation in w may be obtained by eliminating ψ_x and ψ_y from the preceding equations. Equations (2.3.115) and (2.3.116) are first differentiated with respect to x and y respectively and then added to give

$$\left[D\nabla^2 - G'h - \frac{\rho h^3}{12}\frac{\partial^2}{\partial t^2}\right]\Phi = G'h\nabla^2 w \tag{2.3.119}$$

where $G' = \kappa^2 G$. The quantity Φ is then eliminated between (2.3.119) and (2.3.117) to give

$$\left[\nabla^2 - \frac{\rho}{G'}\frac{\partial^2}{\partial t^2}\right]\left[D\nabla^2 - \frac{\rho h^3}{12}\frac{\partial^2}{\partial t^2}\right] w + \rho h \frac{\partial^2 w}{\partial t^2} = \left[1 - \frac{D\nabla^2}{G'h} + \frac{\rho h^2}{12G'}\frac{\partial^2}{\partial t^2}\right] q \tag{2.3.121}$$

The effect of ignoring shear and rotary inertia is shown in Fig. 2.24, where c_B is the flexural wave velocity and c_S is defined as $c_s = \sqrt{G/\rho}$. Note that shear deformation by itself accounts for almost all of the discrepancy between classical plate theory and three-dimensional elasticity theory. If rotary inertia terms are omitted from (2.3.117), we obtain:

$$D\left[\nabla^2 - \frac{\rho}{G'}\frac{\partial^2}{\partial t^2}\right]\nabla^2 w + \rho h \frac{\partial^2 w}{\partial t^2} = \left[1 - \frac{D\nabla^2}{G'h}\right] q \tag{2.3.122}$$

If transverse shear deformation only is neglected, then we obtain:

$$\left[d\nabla^2 - \frac{\rho h^3}{12}\frac{\partial^2}{\partial t^2}\right]\nabla^2 w + \rho h \frac{\partial^2 w}{\partial t^2} = q \tag{2.3.123}$$

If both rotary inertia and transverse shear are neglected, then (2.3.101) is obtained.

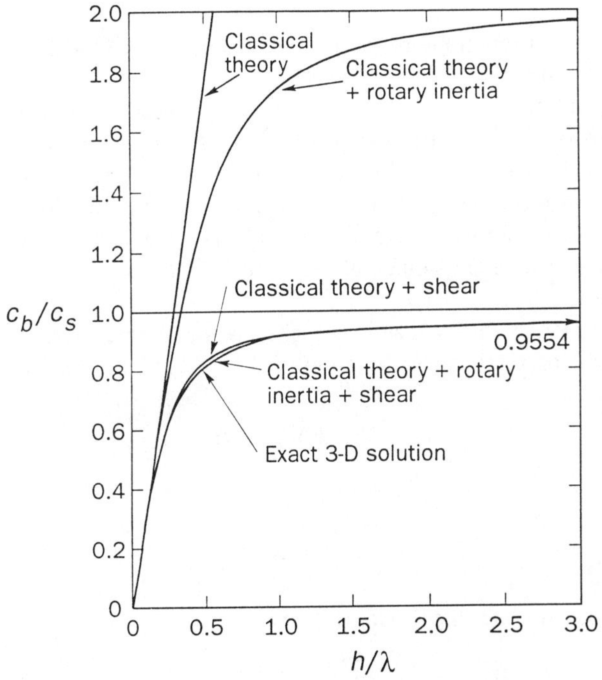

Fig. 2.24 Effect of rotary inertia and transverse shear on flexural wave speed predictions for a plate. The two curves, 'classical theory + rotary inertia + shear' and 'exact 3-D solution' lie on top of one another.

In all of the preceding equations q is an externally applied force per unit area. If this were replaced by a point force F, located at (x_0, y_0), then q could be written as

$$q = F\delta(x - x_0)\delta(y - y_0) \tag{2.3.124}$$

If the plate were subjected to externally applied moments, (2.3.96) and (2.3.97) would be replaced with (2.3.102) and (2.3.103), and (2.3.121) would become

$$\left[\nabla^2 - \frac{\rho}{G'}\frac{\partial^2}{\partial t^2}\right]\left[D\nabla^2 - \frac{\rho h^3}{12}\frac{\partial^2}{\partial t^2}\right] w + \rho h \frac{\partial^2 w}{\partial t^2} = -\frac{\partial M_{ex}}{\partial y} + \frac{\partial M_{ey}}{\partial x} \tag{2.3.125}$$

In a similar way as was done for a beam, an external point moment M_{xp} acting about the x-axis at $(x = x_0, y = y_0)$ may be included in (2.3.125) using

$$M_{ex}(x,y) = M_{xp}\,\delta(x - x_0)\,\delta(y - y_0) \tag{2.3.126}$$

Similarly, a line moment, M_{xL} of length b extending parallel to the x-axis at $x = x_0$ may be written as

$$M_{ex}(x,y) = \frac{M_{xL}}{b}\delta(x - x_0)\,[u(y - y_1) - u(y - y_2)] \tag{2.3.127}$$

where y_1 and y_2 represent the beginning and end of the line and $b = y_1 - y_2$; $u(\,)$ is the unit step function.

The equations of motion just derived can be used to accurately calculate plate resonance frequencies and mode shapes for a variety of boundary conditions and plate shapes by setting $q = M_{xe} = M_{ye} = 0$. Here only rectangular shaped plates will be considered. Other plate shapes have been considered by a number of researchers (including Mindlin and Deresiewicz (1954), who considered circular plates), much of the work is summarized by Leissa (1969).

Mindlin (1951) shows that in the absence of external surface loading the plate equation of motion can be simplified. Using the notation of Skudrzyk (1968) we obtain

$$(\nabla^2 + k_1^2)\, w_1 = 0 \tag{2.3.128}$$

$$(\nabla^2 + k_2^2) w_2 = 0 \tag{2.3.129}$$

$$(\nabla^2 + k_3^2) H = 0 \tag{2.3.130}$$

where the normal plate displacement is given by

$$w = w_1 + w_2 \tag{2.3.131}$$

The in-plane plate rotations may be written as

$$\psi_y = (\alpha_1 - 1)\frac{\partial w_1}{\partial x} + (\alpha_2 - 1)\frac{\partial w_2}{\partial x} + \frac{\partial H}{\partial y} \tag{2.3.132}$$

$$\psi_x = (\alpha_1 - 1)\frac{\partial w_1}{\partial y} + (\alpha_2 - 1)\frac{\partial w_2}{\partial y} - \frac{\partial H}{\partial x} \quad (2.3.133)$$

where

$$\alpha_1,\alpha_2 = \frac{2(k_1^2),(k_2^2)}{(1 - \nu)k_3^2} \quad (2.3.134)$$

The wavenumbers k_1, k_2 and k_3 are related to the classical plate bending wavenumber k_b found by substituting the solution

$$w = Ae^{j(\omega t - k_b x - k_b y)} \quad (2.3.135)$$

into the classical equation of motion for the unloaded plate. The wavenumbers are defined in the following equations:

$$k_1^2,\ k_2^2 = \frac{k_b^4}{2}\left\{F + I' \pm \left[(F-I')^2 + 4/k_b^4\right]^{1/2}\right\} \quad (2.3.136)$$

$$k_3^2 = \frac{2}{(1-\nu)}\left(I'k_b^4/h - 1/F\right) \quad (2.3.137)$$

where

$$k_b^4 = \left[\frac{\omega}{c_b}\right]^4 = \frac{\rho h\omega^2}{D} = \frac{12(1-\nu^2)\,\rho\omega^2}{Eh^2} \quad (2.3.138)$$

$$I' = \frac{h^3}{12} \quad (2.3.139)$$

$$F = 2h^2/(1-\nu)\pi^2 \quad (2.3.140)$$

and where h is the plate thickness. Note that A may be a complex quantity, depending upon the plate boundary conditions.

Equations (2.3.121), (2.3.122) and (2.3.123) (with the right-hand sides set equal to zero) can be solved for the plate resonance frequencies and mode shapes for any defined set of plate boundary conditions. These boundary conditions are in the form of specification of one of each of the pairs $(M_\nu,\ \psi_\nu)$ $(M_{\nu s},\ \psi_s)$ and (Q_ν, w) where ν is the normal and s is the tangent to the boundary edge.

For simply supported edges: $M_\nu = \psi_s = w = 0$ (2.3.141)

For clamped edges: $\psi_\nu = \psi_s = w = 0$ (2.3.142)

For free edges: $M_\nu = M_{\nu s} = Q_\nu = 0$ (2.3.143)

Along an edge parallel to the x-axis, $\nu = y$ and $s = x$.

Solutions to the equations of motion for rectangular plates are of the form:

$$w_1 = A_1 \sin(\alpha_1 x) \sin(\beta_1 y) \tag{2.3.144}$$

$$w_2 = A_2 \sin(\alpha_2 x) \sin(\beta_2 y) \tag{2.3.145}$$

$$H = A_3 \cos(\alpha_3 x) \sin(\beta_3 y) \tag{2.3.146}$$

where

$$\left.\begin{aligned} \alpha_1^2 + \beta_1^2 &= k_1^2 \\ \alpha_2^2 + \beta_2^2 &= k_2^2 \\ \alpha_3^2 + \beta_3^2 &= k_3^2 \end{aligned}\right\} \tag{2.3.147}$$

The coefficients α and β are determined by substitution of the appropriate boundary conditions into the preceding four equations.

For classical plate theory, only two boundary conditions are needed; thus, Q_ν and $M_{\nu s}$ combine together into a single boundary condition (see Leissa, 1973, p. 338). They usually take the form of specification of one in each of the following groups:

$$(w,\ F_\nu) \text{ and } \left[\frac{\partial w}{\partial \nu},\ M_\nu\right]$$

(where $F_\nu = Q_\nu + \dfrac{\partial M_{\nu s}}{\partial s}$).

For simply supported edges: $w = M_\nu = 0$ (2.3.148)

For clamped edges: $w = \dfrac{\partial w}{\partial \nu} = 0$ (2.3.149)

For free edges: $Q_\nu = M_\nu = 0$ (2.3.150)

where ν is the in-plane coordinate normal to the edge of the plate.

A solution to the classical wave equation for rectangular plates is:

$$w = A_1 X(x)\, Y(y) \tag{2.3.151}$$

For a plate simply supported at $x = 0$ and $x = L_x$,

$$X(x) = \sin\frac{m\pi x}{L_x} \qquad m = 1,\ 2,\ \ldots.. \tag{2.3.152}$$

where L_x is the plate dimension in the x-direction.

For a plate clamped at $x = 0$ and $x = L_x$,

$$X(x) = \cos\gamma_1\left[\frac{x}{L_x} - \frac{1}{2}\right] + \frac{\sin(\gamma_1/2)}{\sinh(\gamma_1/2)}\cosh\gamma_1\left[\frac{x}{L_x} - \frac{1}{2}\right] \tag{2.3.153}$$

$$m = 1,\ 3,\ 5\ \ldots..$$

γ_1 are solutions of

$$\tan(\gamma_1/2) + \tanh(\gamma_1/2) = 0 \tag{2.3.154}$$

and

$$X(x) = \sin\gamma_2\left[\frac{x}{L_x} - \frac{1}{2}\right] - \frac{\sin(\gamma_2/2)}{\sinh(\gamma_2/2)}\sinh\gamma_2\left[\frac{x}{L_x} - \frac{1}{2}\right], \quad m = 2,\ 4,\ 6 \tag{2.3.155}$$

γ_2 are solutions of

$$\tan(\gamma_2/2) - \tanh(\gamma_2/2) = 0 \tag{2.3.156}$$

Note that the integer m represents the number of nodal lines minus one. Similar expressions apply for $Y(y)$. Expressions for free edge plates and for plates with different boundary conditions at $x = 0$ and $x = L_x$ are given by Leissa (1969).

2.3.5 Waves in thin circular cylinders

The derivation of the wave equations or equations of motion for a thin circular cylindrical shell is a complex procedure and will only be outlined briefly here. For a more detailed treatment the reader is advised to consult the excellent publication by Leissa (1973).

Various researchers have derived these equations from first principles, making various simplifying assumptions along the way. The simplest theory is that due to Donnell and Mushtari (Leissa 1973) and this is considered sufficiently accurate for use in the complex analyses involving vibration isolation of a rigid body from a cylindrical support structure, as discussed in Chapter 12. However, the more complicated theory of Goldenveizer–Novozhilov has been shown (Pope, 1971) to give more accurate results for resonance frequencies and mode shapes of thin, simply supported (or shear diaphragm supported) circular cylinders which is an advantage for the analysis of sound transmission into cylindrical enclosures; this will be considered in Chapter 9. Both of these will be discussed here, although the derivations will not be included for the Donnell–Mushtari theory. On the other hand, some authors prefer to use Flügge's theory and for this reason the results derived using this approach will also be given, but the derivations will be omitted.

There are three different methods which have been used by various authors to derive the equations of motion for a curved shell (of which the circular cylinder is a special case). The method most widely used applies Newton's laws by summing the forces and moments which act on a shell element in much the same way as was done for a beam element in Section 2.5.1. The second method begins with the equations of motion of an infinitesimal element of the three-dimensional theory of elasticity and integrates these equations over the thickness to obtain the equations for a shell element. The third method is a class of variational methods, one of which involves the use of Hamilton's principle, which was discussed earlier. This latter method will be outlined briefly here.

Although the various wave types could be considered independently for beams and plates, it is necessary to consider the displacements of the surface in all three directions (radially, tangentially and axially) simultaneously. This is mainly because a radial displacement of the wall of a cylinder produces tensile or compressive tangential and axial membrane stresses, causing flexural waves to couple with longitudinal and shear waves.

The assumptions made in the following analysis are as follows:

1. The cylinder material is isotropic.
2. The cylinder wall thickness is uniform and small compared with the cylinder radius and length.
3. Strains and displacements are sufficiently small so that quantities of second and higher order in the strain displacement relations may be neglected in comparison with first order terms.
4. The transverse normal stress σ_{zz} is small and may be neglected in comparison with the other normal stress components.
5. Normals to the undeformed middle surface of the wall of the cylinder remain straight and normal to the deformed middle surface.

The stresses and their sign conventions for a cylindrical shell element are illustrated in Fig. 2.25.

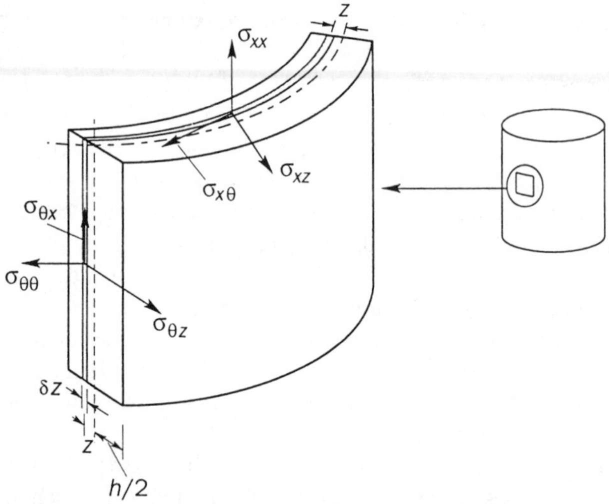

Fig. 2.25 Cylinder element showing normal and shear stresses.

For the element shown in Fig. 2.25, a is its radius of curvature, σ_{xz}, $\sigma_{\theta z}$, $\sigma_{x\theta}$ and $\sigma_{\theta x}$ are shear stresses, σ_{xx} and $\sigma_{\theta\theta}$ are normal stresses in the axial and tangential directions respectively and z is the distance of the infinitesimal segment δz from the central axis of the shell element. Note that z rather than r is used as the radial coordinate, as it has its origin at the centre of the shell element rather than the centre of curvature of the element.

The third assumption outlined above implies that the normal stress $\sigma_{zz} = 0$. The fourth assumption is known as Kirchoff's hypothesis and implies that

$$\sigma_{xz} = \sigma_{\theta z} = e_{xy} = e_{\theta z} = 0 \tag{2.3.157}$$

where e_{ik} represents the shear strain at element δz. In practice, σ_{xz} and $\sigma_{\theta z}$ are small but non-zero, as their integrals must supply the shearing forces needed for equilibrium.

The strain displacement equations for a circular cylinder may be derived from the general three-dimensional equations of motion (Leissa, 1973), and for the segment δz, they may be written as:

$$e_{xx} = \epsilon_{xx} + z\kappa_x \tag{2.3.158}$$

(for Donnell – Mushtari, Goldenveizer – Novozhilov and Flügge shell theories);

$$e_{\theta\theta} = \frac{1}{(1 + z/a)}(\epsilon_{\theta\theta} + z\kappa_\theta) \tag{2.3.159a}$$

(for Goldenveizer – Novozhilov and Flügge shell theories only);

$$e_{\theta\theta} = \epsilon_{\theta\theta} + z\kappa_\theta \tag{2.3.159b}$$

(for Donnell – Mushtari shell theory only);

$$e_{x\theta} = \frac{1}{(1 + z/a)}\left[\epsilon_{x\theta} + z\left(1 + \frac{z}{2a}\right)\tau\right] \tag{2.3.160a}$$

(for Goldenveiser – Novozhilov and Flügge shell theories only) and

$$e_{x\theta} = \epsilon_{x\theta} + z\tau \tag{2.3.160b}$$

(for Donnell – Mushtari shell theory only)

where e_{xx}, $e_{\theta\theta}$ and $e_{x\theta}$ are the normal and shear strains of the arbitrary segment δz, and ϵ_{xx}, $\epsilon_{\theta\theta}$ and $\epsilon_{x\theta}$ are the normal and shear strains of the surface in the middle of the wall thickness (mid-surface, $z = 0$). τ is the angular twist of this mid-surface and κ_x and κ_θ are the changes in curvature of the same surface. These six latter quantities are given by (for both Goldenveiser and Flügge shell theories):

$$\epsilon_{xx} = \frac{\partial \xi_x}{\partial x} \tag{2.3.161}$$

$$\epsilon_{\theta\theta} = \frac{1}{a}\frac{\partial \xi_\theta}{\partial \theta} + \frac{w}{a} \tag{2.3.162}$$

Note the presence of the additional strain caused by the radial displacement which is not taken into account in (2.3.1):

$$\epsilon_{x\theta} = \frac{1}{a}\frac{\partial \xi_x}{\partial \theta} + \frac{\partial \xi_\theta}{\partial x} \tag{2.3.163}$$

$$\tau = -\frac{2}{a}\frac{\partial^2 w}{\partial x \partial \theta} + \frac{2}{a}\frac{\partial \xi_\theta}{\partial x} \tag{2.3.164}$$

$$\kappa_x = -\frac{\partial^2 w}{\partial x^2} \tag{2.3.165}$$

$$\kappa_\theta = \frac{1}{a^2}\left[\frac{\partial \xi_\theta}{\partial \theta} - \frac{\partial^2 w}{\partial \theta^2}\right] \tag{2.3.166}$$

where ξ_x, ξ_θ and w are the axial, tangential and radial displacements respectively of the cylinder (see Fig. 2.26). Note that w is positive outwards.

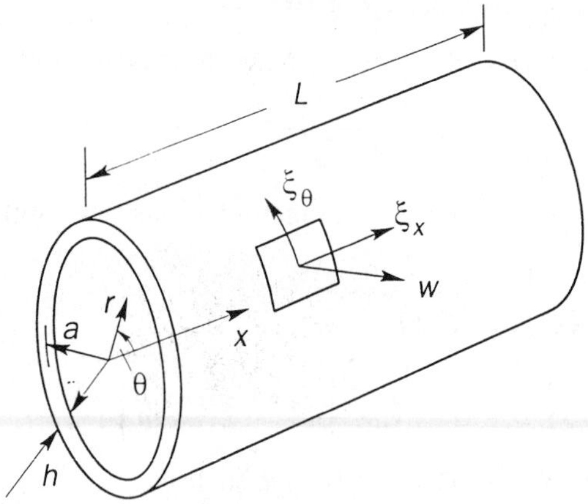

Fig. 2.26 Circular cylinder coordinates and displacements.

For Donnell–Mushtari theory, the same previous six equations apply except that the term involving ξ_θ in (2.2.164) and (2.3.166) is omitted.

The equations of motion of a thin cylindrical shell will now be derived using the Goldenveiser–Novozhilov method. Applying the Kirchoff hypothesis to the expression for strain energy derived from the theory of elasticity for a circular cylinder we obtain the following (Leissa, 1973):

$$V = \frac{1}{2}\int_{V'} (\sigma_{xx} e_{xx} + \sigma_{\theta\theta} e_{\theta\theta} + \sigma_{x\theta} e_{x\theta})\, \mathrm{d}V' \tag{2.3.167}$$

where $\mathrm{d}V'$ is an elemental volume, which when expressed in cylindrical shell coordinates is:

$$\mathrm{d}V' = (1 + z/a)\, a\, \mathrm{d}x\, \mathrm{d}\theta\, \mathrm{d}z \tag{2.3.168}$$

Before we can expand (2.3.167) further we need expressions for the stresses in terms of the strains. Using the assumptions outlined previously, the well known three-dimensional form of Hooke's law can be written for the segment δz as:

$$e_{xx} = \frac{1}{E}(\sigma_{xx} - \nu\sigma_{\theta\theta}) \tag{2.3.169}$$

$$e_{\theta\theta} = \frac{1}{E}(\sigma_{\theta\theta} - \nu\sigma_{xx}) \tag{2.3.170}$$

$$e_{x\theta} = \frac{2(1+\nu)}{E}\sigma_{x\theta} \tag{2.3.171}$$

where ν is Poisson's ratio.

Inverting the preceding three equations gives:

$$\sigma_{xx} = \frac{E}{(1-\nu^2)}(e_{xx} + \nu e_{\theta\theta}) \tag{2.3.172}$$

$$\sigma_{\theta\theta} = \frac{E}{(1-\nu^2)}(e_{\theta\theta} + \nu e_{xx}) \tag{2.3.173}$$

$$\sigma_{x\theta} = \frac{E}{2(1+\nu)}e_{x\theta} \tag{2.3.174}$$

Substituting (2.3.168) and (2.3.172) to (2.3.174) into (2.3.167) gives:

$$V = \frac{E}{2(1-\nu^2)}\int_{V'}\left[e_{xx}^2 + e_{\theta\theta}^2 + 2\nu e_{xx}e_{\theta\theta} + \frac{(1-\nu)}{2}e_{x\theta}^2\right]dV' \tag{2.3.175}$$

Substituting further the expressions (2.3.158) to (2.3.160) for e_{xx}, $e_{\theta\theta}$ and $e_{x\theta}$ into (2.3.175) gives

$$V = \frac{E}{2(1-\nu^2)}\int_{-h/2}^{h/2}\int_0^{2\pi}\int_0^{L}(Q_0 + zQ_1 + z^2Q_2)\,a\,\mathrm{d}x\,\mathrm{d}\theta\,\mathrm{d}z \tag{2.3.176}$$

where

$$Q_0 = (\epsilon_{xx} + \epsilon_{\theta\theta})^2 - 2(1-\nu)(\epsilon_{xx}\epsilon_{\theta\theta} - \epsilon_{x\theta}^2/4) \tag{2.3.177}$$

$$\begin{aligned} Q_1 = {} & 2(\epsilon_{xx}\kappa_x + \epsilon_{\theta\theta}\kappa_\theta) + 2\nu(\epsilon_{xx}\kappa_\theta + \epsilon_{\theta\theta}\kappa_x) \\ & + (1-\nu)\epsilon_{x\theta}\tau + \frac{1}{a}(\epsilon_{xx}^2 - \epsilon_{\theta\theta}^2) \end{aligned} \tag{2.3.178}$$

$$\begin{aligned} Q_2 = {} & (\kappa_x + \kappa_\theta)^2 - 2(1-\nu)(\kappa_x\kappa_\theta - \tau^2/4) \\ & + \frac{2}{a}(\epsilon_{xx}\kappa_x - \epsilon_{\theta\theta}\kappa_\theta) - \frac{(1-\nu)\epsilon_{x\theta}\tau}{2a} \\ & + \frac{\epsilon_{\theta\theta}^2}{a^2} + \frac{(1-\nu)\epsilon_{x\theta}^2}{2a^2} \end{aligned} \tag{2.3.179}$$

Integrating (2.3.176) with respect to z gives

$$V = \frac{Eh}{2(1-\nu^2)} \int_0^{2\pi} \int_0^L \left[Q_0 + \frac{h^2}{12} Q_2 \right] a\,\mathrm{d}x\,\mathrm{d}\theta \tag{2.3.180}$$

The equations of motion for the cylinder may now be derived by invoking Hamilton's variational principle. That is,

$$\delta \int_{t_1}^{t_2} (T-V)\,\mathrm{d}t = 0 \tag{2.3.181}$$

The kinetic energy T of the cylinder is

$$T = \frac{1}{2}\rho h \int_0^{2\pi} \int_0^L \left[\left(\frac{\partial \xi_x}{\partial t} \right)^2 + \left(\frac{\partial \xi_\theta}{\partial t} \right)^2 + \left(\frac{\partial w}{\partial t} \right)^2 \right] a\,\mathrm{d}x\,\mathrm{d}\theta \tag{2.3.182}$$

Substituting (2.3.161) to (2.3.166) into (2.3.177) and (2.3.179), then substituting the result into (2.3.180) gives the potential energy in terms of the displacements ξ_x, ξ_θ and w and their partial derivatives. Substituting this result and (2.3.182) into (2.3.181) gives the following

$$\delta \int_{t_1}^{t_2} \int_0^{2\pi} \int_0^L F \left[\xi_x, \xi_\theta, w, \frac{\partial \xi_x}{\partial x}, \frac{\partial \xi_x}{\partial \theta}, \frac{\partial \xi_x}{\partial t}, \frac{\partial \xi_\theta}{\partial x}, \frac{\partial \xi_\theta}{\partial \theta}, \frac{\partial \xi_\theta}{\partial t}, \frac{\partial w}{\partial x}, \frac{\partial w}{\partial \theta}, \frac{\partial w}{\partial t}, \frac{\partial^2 w}{\partial x^2}, \frac{\partial^2 w}{\partial \theta^2}, \frac{\partial^2 w}{\partial x \partial \theta} \right] \mathrm{d}x\,\mathrm{d}\theta\,\mathrm{d}t = 0 \tag{2.3.183}$$

where the functions ξ_x, ... $\dfrac{\partial^2 w}{\partial \theta^2}$ are functions of x, θ and t and the function F is $(T - V)/\mathrm{d}x\mathrm{d}\theta$.

From the calculus of variations discussed in Section 2.2, the conditions that (2.3.183) be satisfied are the Lagrange equations given by:

$$\frac{\partial F}{\partial \xi_x} - \frac{\partial}{\partial x}\left(\frac{\partial F}{\partial \xi_x^x} \right) - \frac{\partial}{\partial \theta}\left(\frac{\partial F}{\partial \xi_x^\theta} \right) - \frac{\partial}{\partial t}\left(\frac{\partial F}{\partial \xi_x^t} \right) = 0 \tag{2.3.184}$$

$$\frac{\partial F}{\partial \xi_\theta} - \frac{\partial}{\partial x}\left(\frac{\partial F}{\partial \xi_\theta^x} \right) - \frac{\partial}{\partial \theta}\left(\frac{\partial F}{\partial \xi_\theta^\theta} \right) - \frac{\partial}{\partial t}\left(\frac{\partial F}{\partial \xi_\theta^t} \right) = 0 \tag{2.3.185}$$

$$\frac{\partial F}{\partial w} - \frac{\partial}{\partial x}\left(\frac{\partial F}{\partial w^x} \right) - \frac{\partial}{\partial \theta}\left(\frac{\partial F}{\partial w^\theta} \right) - \frac{\partial}{\partial t}\left(\frac{\partial F}{\partial w^t} \right) + \frac{\partial^2}{\partial x^2}\left(\frac{\partial F}{\partial w^{xx}} \right) + \frac{\partial^2}{\partial x \partial \theta}\left(\frac{\partial F}{\partial w^{x\theta}} \right) + \frac{\partial^2}{\partial \theta^2}\left(\frac{\partial F}{\partial w^{\theta\theta}} \right) = 0 \tag{2.3.186}$$

where, for example, $\dfrac{\partial F}{\partial \xi_x^x}$ denotes the partial derivative of the function F with respect to the function $\dfrac{\partial \xi_x}{\partial x}$, and $w^{x\theta} = \dfrac{\partial^2 w}{\partial x \partial \theta}$.

Replacing F by $(T - V)/\mathrm{d}x\mathrm{d}\theta$ and substituting (2.3.180) and (2.3.182) for V and T respectively into equations (2.3.184) to (2.3.186) gives the required cylinder equations of motion (or the wave equations). These are then

$$a\frac{\partial^2 \xi_x}{\partial x^2} + \frac{(1-\nu)}{2a}\frac{\partial^2 \xi_x}{\partial \theta^2} + \frac{(1+\nu)}{2}\frac{\partial^2 \xi_\theta}{\partial x \partial \theta} + \nu\frac{\partial w}{\partial x} = \frac{\rho a(1-\nu^2)}{E}\frac{\partial^2 \xi_x}{\partial t^2} \tag{2.3.187}$$

$$\frac{(1+\nu)}{2}\frac{\partial^2 \xi_x}{\partial x \partial \theta} + \frac{a(1-\nu)}{2}\frac{\partial^2 \xi_\theta}{\partial x^2} + \frac{1}{a}\frac{\partial^2 \xi_\theta}{\partial \theta^2} + \frac{1}{a}\frac{\partial w}{\partial \theta}$$

$$+ \frac{h^2}{12a^2}\left[2(1-\nu)a\frac{\partial^2 \xi_\theta}{\partial x^2} + \frac{1}{a}\frac{\partial^2 \xi_\theta}{\partial \theta^2} - a(2-\nu)\frac{\partial^3 w}{\partial x^2 \partial \theta} - \frac{1}{a}\frac{\partial^3 w}{\partial \theta^3}\right] \tag{2.3.188}$$

$$= \frac{\rho a(1-\nu^2)}{E}\frac{\partial^2 \xi_\theta}{\partial t^2}$$

$$\nu\frac{\partial \xi_x}{\partial x} + \frac{1}{a}\frac{\partial \xi_\theta}{\partial \theta} + \frac{w}{a} + \frac{h^2}{12a^2}\left[-a(2-\nu)\frac{\partial^3 \xi_\theta}{\partial x^2 \partial \theta} - \frac{1}{a}\frac{\partial^3 \xi_\theta}{\partial \theta^3} + a^3\frac{\partial^4 w}{\partial x^4}\right. \tag{2.3.189}$$

$$\left. + 2a\frac{\partial^4 w}{\partial x^2 \partial \theta^2} + \frac{1}{a}\frac{\partial^4 w}{\partial \theta^4}\right] = \frac{-\rho a(1-\nu^2)}{E}\frac{\partial^2 w}{\partial t^2}$$

In matrix form, the equations of motion are

$$\begin{bmatrix} a_{11} & a_{12} & a_{13} \\ a_{21} & a_{22} & a_{23} \\ a_{31} & a_{32} & a_{33} \end{bmatrix}\begin{Bmatrix} \xi_x \\ \xi_\theta \\ w \end{Bmatrix} + \frac{h^2}{12a^2}\begin{bmatrix} b_{11} & b_{12} & b_{13} \\ b_{21} & b_{22} & b_{23} \\ b_{31} & b_{32} & b_{33} \end{bmatrix}\begin{Bmatrix} \xi_x \\ \xi_\theta \\ w \end{Bmatrix} = 0 \tag{2.3.190}$$

where

$$a_{11} = a\frac{\partial^2}{\partial x^2} + \frac{(1-\nu)}{2a}\frac{\partial^2}{\partial \theta^2} - \frac{\rho a(1-\nu^2)}{E}\frac{\partial^2}{\partial t^2} \tag{2.3.191}$$

$$a_{12} = \frac{(1+\nu)}{2}\frac{\partial^2}{\partial x \partial \theta} \tag{2.3.192}$$

$$a_{13} = \nu\frac{\partial}{\partial x} \tag{2.3.193}$$

$$a_{21} = \frac{(1+\nu)}{2}\frac{\partial^2}{\partial x \partial \theta} \tag{2.3.194}$$

$$a_{22} = \frac{a(1-\nu)}{2}\frac{\partial^2}{\partial x^2} + \frac{1}{a}\frac{\partial^2}{\partial \theta^2} - \frac{\rho a(1-\nu^2)}{E}\frac{\partial^2}{\partial t^2} \tag{2.3.195}$$

$$a_{23} = \frac{1}{a}\frac{\partial}{\partial \theta} \tag{2.3.196}$$

$$a_{31} = \nu\frac{\partial}{\partial x} \tag{2.3.197}$$

$$a_{32} = \frac{1}{a}\frac{\partial}{\partial \theta} \tag{2.3.198}$$

$$a_{33} = \frac{1}{a} + \frac{h^2}{12a^2}\left[a^3\frac{\partial^4}{\partial x^4} + 2a\frac{\partial^4}{\partial x^2\partial\theta^2} + \frac{1}{a}\frac{\partial^4}{\partial\theta^4}\right] + \frac{\rho a(1-\nu^2)}{E}\frac{\partial}{\partial t^2} \tag{2.3.199}$$

$$b_{11} = b_{12} = b_{13} = b_{21} = b_{31} = b_{33} = 0 \tag{2.3.200}$$

$$b_{22} = 2a(1-\nu)\frac{\partial^2}{\partial x^2} + \frac{1}{a}\frac{\partial^2}{\partial\theta^2} \tag{2.3.201}$$

$$b_{32} = b_{23} = -a(2-\nu)\frac{\partial^3}{\partial x^2\partial\theta} - \frac{1}{a}\frac{\partial^3}{\partial\theta^3} \tag{2.3.202}$$

If the Donnell–Mushtari theory had been used to derive the equations of motion, the coefficients a_{ij} would remain the same but all b_{ij} would be zero.

If Flügge's theory had been used, the coefficients a_{ij} would remain the same but the coefficients b_{ij} would be replaced with the following;

$$b_{12} = b_{21} = 0 \tag{2.3.203}$$

$$b_{11} = \frac{(1-\nu)}{2a}\frac{\partial^2}{\partial\theta^2} \tag{2.3.204}$$

$$b_{13} = b_{31} = -a^2\frac{\partial^3}{\partial x^3} + \frac{(1-\nu)}{2}\frac{\partial^3}{\partial x\partial\theta^2} \tag{2.3.205}$$

$$b_{22} = \frac{3a(1-\nu)}{2}\frac{\partial^2}{\partial x^2} \tag{2.3.206}$$

$$b_{23} = b_{32} = -\frac{a(3-\nu)}{2}\frac{\partial^3}{\partial x^2\partial\theta} \tag{2.3.207}$$

$$b_{31} = a^2\frac{\partial^3}{\partial x^3} + \frac{(1-\nu)}{2}\frac{\partial^3}{\partial x\partial\theta^2} \tag{2.3.208}$$

$$b_{33} = a + 2a\frac{\partial^2}{\partial\theta^2} \tag{2.3.209}$$

Equation (2.3.190) is used together with appropriate boundary conditions to determine the resonance frequencies and mode shapes for a particular cylinder. Generally only modes involving the radial displacement w are of interest and analyses are usually restricted accordingly.

The solution for harmonic vibration involves assuming a form of solution for ξ_x, ξ_θ and w and then substituting this back into (2.3.190). The determinant of the a_{ij} coefficient matrix is then set equal to zero which allows the eigen frequencies to be determined. These frequencies are then used together with the assumed solutions for ξ_x, ξ_θ and w in (2.3.190) to determine the corresponding mode shapes.

2.3.5.1 Boundary conditions

These are specified in terms of cylinder displacements at each end and the forces and moments acting on the cylinder at each end. Four boundary conditions must be specified for each end; one from each of the pairs listed below:

$$\xi_x = 0 \quad \text{or} \quad N_x = 0 \tag{2.3.210}$$

$$\xi_\theta = 0 \quad \text{or} \quad N_{x\theta} + \frac{M_{x\theta}}{a} = 0 \tag{2.3.211}$$

$$w = 0 \quad \text{or} \quad Q_x + \frac{1}{a}\frac{\partial M_{x\theta}}{\partial\theta} = 0 \tag{2.3.212}$$

$$\frac{\partial w}{\partial x} = 0 \quad \text{or} \quad M_x = 0 \tag{2.3.213}$$

The quantities N_x, $N_{x\theta}$, Q_x, $M_{x\theta}$ and M_x have not yet been defined. The first two are in-plane forces, the third is a force normal to the cylinder surface, the fourth is a twisting moment and the fifth term is a bending moment. They are defined in Fig. 2.27 for an element of a cylinder. The sign convention follows that of Leissa (1973).

Consider the faces of the element shown in Fig. 2.25. The resultant forces per unit length acting on each face can be calculated by integrating the stresses over the face thickness. For the vertical face we have,

$$N_\theta = \int_{-h/2}^{h/2} \sigma_{\theta\theta}\,dz \tag{2.3.214}$$

$$N_{\theta x} = \int_{-h/2}^{h/2} \sigma_{\theta x}\,dz \tag{2.3.215}$$

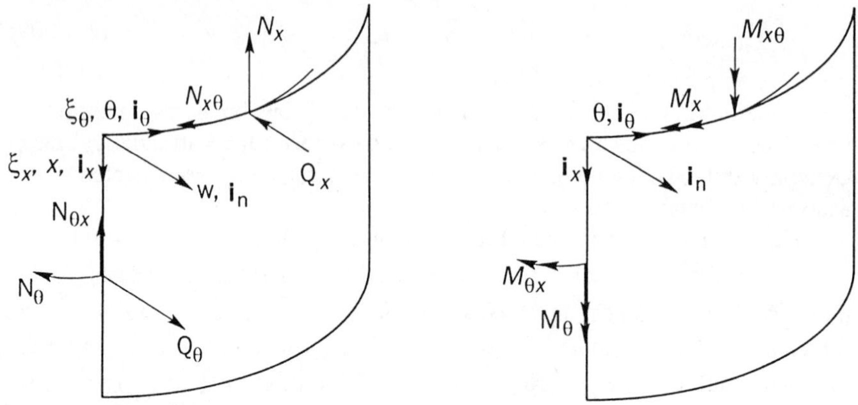

Fig. 2.27 Forces and moments acting on a cylindrical element.

$$Q_\theta = \int_{-h/2}^{h/2} \sigma_{\theta z}\, dz \tag{2.3.216}$$

For the top surface, the situation is a little different as the surface is curved. The arc length of the middle surface representing an incremental angle $d\theta$ is

$$ds = a\, d\theta \tag{2.3.217}$$

where a is the cylinder radius. The arc length for other surfaces parallel to middle surface and spaced a distance z from it is

$$ds_z = a(1 + z/a)\, d\theta \tag{2.3.218}$$

The difference between the two is $(1 + z/a)$. Thus, the net forces per unit length on the top horizontal face of Fig. 2.25 are given by:

$$N_x = \int_{-h/2}^{h/2} \sigma_{xx} (1 + z/a)\, dz \tag{2.3.219}$$

$$N_{x\theta} = \int_{-h/2}^{h/2} \sigma_{x\theta} (1 + z/a)\, dz \tag{2.3.220}$$

$$Q_x = \int_{-h/2}^{h/2} \sigma_{xz} (1 + z/a)\, dz \tag{2.3.221}$$

Similarly, the moment of the infinitesimal force $\sigma_{xx} ds\, dz$ about the θ line through the centre of the section is simply $z\sigma_{xx} ds\, dz$. The moment resultant M_x is obtained by integrating the moment over the thickness and dividing by $a d\theta$. Thus, we obtain:

$$M_x = \int_{-h/2}^{h/2} \sigma_{xx} (1 + z/a) z\, dz \tag{2.3.222}$$

$$M_{\theta x} = \int_{-h/2}^{h/2} \sigma_{\theta x} z \, dz \tag{2.3.223}$$

$$M_{\theta} = \int_{-h/2}^{h/2} \sigma_{\theta\theta} z \, dz \tag{2.3.224}$$

$$M_{x\theta} = \int_{-h/2}^{h/2} \sigma_{x\theta} (1 + z/a) z \, dz \tag{2.3.225}$$

To express the net forces and moments in terms of in-plane and normal cylinder displacements, we must first express them in terms of the strain components. To do this we return to the functional of (2.3.167) and taking its variation we obtain:

$$\delta V = \int_{V'} \left(\sigma_{xx} \, \delta e_{xx} + \sigma_{\theta\theta} \, \delta e_{\theta\theta} + \sigma_{x\theta} \, \delta e_{x\theta}\right) dV' \tag{2.3.226}$$

Substituting (2.3.158) to (2.3.160) for the total strains into (2.3.226) gives (for Goldenveiser–Novozhilov theory):

$$\delta V = \int_0^L \int_0^{2\pi} \int_{-h/2}^{h/2} \Big[\sigma_{xx} (1 + z/a) (\delta \epsilon_{xx} + z \delta \kappa_x) + \sigma_{\theta\theta} (\delta \epsilon_{\theta\theta} + z \delta \kappa_\theta) \tag{2.3.227}$$
$$+ \sigma_{x\theta} (\delta \epsilon_{x\theta} + z(1 + z/2a) \tau)\Big] a \, dz \, d\theta \, dx$$

Making use of the definitions for moments and forces of (2.3.214) to (2.3.225), equation (2.3.227) can be written as

$$\delta V = \int_0^L \int_0^{2\pi} \Big[N_x \delta \epsilon_{xx} + N_\theta \delta \epsilon_{\theta\theta} + N_{\theta x} \delta \epsilon_{x\theta} + M_x \delta \kappa_x + M_\theta \delta \kappa_\theta \tag{2.3.228}$$
$$+ \tfrac{1}{2} \left(M_{x\theta} + M_{\theta x}\right) \delta \tau \Big] a \, d\theta \, dx$$

Returning to (2.3.180), Leissa (1973) shows that the last four terms in Q_2 (see (2.3.179)) can be neglected and we obtain

$$V = \frac{Eh}{2(1 - \nu^2)} \int_0^L \int_0^{2\pi} \Big\{ \left(\epsilon_{xx} + \epsilon_{\theta\theta}\right)^2 - 2(1 - \nu) \left(\epsilon_{xx} \epsilon_{\theta\theta} - \epsilon_{x\theta}^2 /4\right) \tag{2.3.229}$$
$$+ \frac{h^2}{12} \Big[(\kappa_x + \kappa_\theta)^2 - 2(1 - \nu) (\kappa_x \kappa_\theta - \tau^2 /4) \Big] \Big\} a \, d\theta \, dx$$

Taking the variation of (2.3.229) yields

$$\delta V = \frac{Eh}{1 - \nu^2} \int_0^L \int_0^{2\pi} \Big\{ \left(\epsilon_{xx} + \nu \epsilon_{\theta\theta}\right) \delta \epsilon_{xx} + \left(\epsilon_{\theta\theta} + \nu \epsilon_{xx}\right) \delta \epsilon_{\theta\theta} + \frac{(1 - \nu)}{2} \epsilon_{x\theta} \delta \epsilon_{x\theta}$$
$$+ \frac{h^2}{12} \Big[\left(\kappa_x + \nu \kappa_\theta\right) \delta \kappa_x + (\kappa_\theta + \nu \kappa_x) \delta \kappa_\theta + \frac{(1 - \nu)}{2} \tau \delta \tau \Big] \Big\} a \, d\theta \, dx \tag{2.3.230}$$

Comparing (2.3.228) and (2.3.230) gives expressions (corresponding to the Goldenveiser–Novozhilov analysis method) for the moments and forces in terms of cylinder strains as follows:

$$N_x = \frac{Eh}{(1-\nu^2)}\left(\epsilon_{xx} + \nu\epsilon_{\theta\theta}\right) \tag{2.3.231}$$

$$N_\theta = \frac{Eh}{(1-\nu^2)}\left(\epsilon_{\theta\theta} + \nu\epsilon_{xx}\right) \tag{2.3.232}$$

$$N_{\theta x} = \frac{Eh}{2(1+\nu)}\epsilon_{x\theta} \tag{2.3.233}$$

$$M_x = \frac{Eh^3}{12(1-\nu^2)}\left(\kappa_x + \nu\kappa_\theta\right) \tag{2.3.234}$$

$$M_\theta = \frac{Eh^3}{12(1-\nu^2)}\left(\kappa_\theta + \nu\kappa_x\right) \tag{2.3.235}$$

$$\tfrac{1}{2}\left(M_{x\theta} + M_{\theta x}\right) = \frac{Eh^3\tau}{24(1+\nu)} \tag{2.3.236}$$

From symmetry of the stress tensor in Fig. 2.27, $\sigma_{x\theta} = \sigma_{\theta x}$. Thus, using (2.3.215), (2.3.220) and (2.3.225) we can show that

$$N_{x\theta} = N_{\theta x} + \frac{M_{\theta x}}{a} \tag{2.3.237}$$

Also, from (2.3.223) and (2.3.225) we obtain

$$M_{\theta x} = M_{x\theta} + \frac{1}{a}\int_{-h/2}^{h/2} \sigma_{x\theta} z^2 \,dz \tag{2.3.238}$$

It can be easily shown (Leissa, 1973) that the second term on the right-hand side of (2.3.238) is very small compared with the first. Thus, with the help of (2.3.236) we obtain

$$M_{x\theta} = M_{\theta x} = \frac{Eh^3\tau}{24(1+\nu)} \tag{2.3.239}$$

Substituting (2.3.233) and (2.3.239) into (2.3.237) we obtain

$$N_{x\theta} = \frac{Eh}{2(1+\nu)}\left[\epsilon_{x\theta} + \frac{h^2}{12a}\tau\right] \tag{2.3.240}$$

For Donnell–Mushtari theory, (2.3.231) to (2.3.236), (2.3.239) and (2.3.240) are identical except for (2.3.240), where the term containing τ is omitted.

If the Flügge analysis method is used, (2.3.231) to (2.3.235), (2.3.239) and (2.3.240) become:

$$N_x = \frac{Eh}{(1-\nu^2)}\left[\epsilon_{xx} + \nu\epsilon_{\theta\theta} + \frac{h^2\kappa_x}{12a}\right] \tag{2.3.241}$$

$$N_\theta = \frac{Eh}{(1-\nu^2)}\left[\epsilon_{\theta\theta} + \nu\epsilon_{xx} - \frac{h^2}{12a}\left(\kappa_\theta - \frac{\epsilon_{\theta\theta}}{a}\right)\right] \tag{2.3.242}$$

$$N_{x\theta} = \frac{Eh}{2(1+\nu)}\left[\epsilon_{x\theta} + \frac{h^2\tau}{24a}\right] \tag{2.3.243}$$

$$N_{\theta x} = \frac{Eh}{2(1+\nu)}\left[\epsilon_{x\theta} - \frac{h^2}{12a}\left(\frac{\tau}{2} - \frac{\epsilon_{x\theta}}{a}\right)\right] \tag{2.3.244}$$

$$M_x = \frac{Eh^3}{12(1-\nu^2)}\left[\kappa_x + \nu\kappa_\theta + \frac{\epsilon_{xx}}{a}\right] \tag{2.3.245}$$

$$M_\theta = \frac{Eh^3}{12(1-\nu^2)}\left[\kappa_\theta + \nu\kappa_x - \frac{\epsilon_{\theta\theta}}{a}\right] \tag{2.3.246}$$

$$M_{x\theta} = \frac{Eh^3\tau}{24(1+\nu)} \tag{2.3.247}$$

$$M_{\theta x} = \frac{Eh^3}{24(1+\nu)}\left(\tau - \frac{\epsilon_{x\theta}}{a}\right) \tag{2.3.248}$$

Expressions for Q_x and Q_θ in terms of the other forces and moments will be derived later by consideration of the equations of motion of a shell element (see (2.3.270) and (2.3.268)). However, for convenience, the results will be written down here. Thus,

$$Q_x = \frac{\partial M_x}{\partial x} + \frac{1}{a}\frac{\partial M_{\theta x}}{\partial \theta} \tag{2.3.249}$$

$$Q_\theta = \frac{1}{a}\frac{\partial M_\theta}{\partial \theta} + \frac{\partial M_{x\theta}}{\partial x} \tag{2.3.250}$$

Note that (2.3.241) and (2.3.242) only apply if there is no external loading acting on the cylinder, and they are valid for both Flügge and Goldenveiser–Novozhilov shell theories. However, the Donnell–Mushtari theory assumes that $Q_x = Q_\theta = 0$.

The quantities on the left-hand side of (2.3.231) to (2.3.235), (2.3.239) and (2.3.240) can be expressed in terms of the displacements ξ_x, ξ_θ and w by making use of the relationships expressed in (2.3.161) to (2.3.166) to give the following:

$$N_x = \frac{Eh}{a(1-\nu^2)}\left(\frac{\partial \xi_x}{\partial x} + \nu\frac{\partial \xi_\theta}{\partial \theta} + \nu w\right) \tag{2.3.251}$$

$$N_\theta = \frac{Eh}{a(1-\nu^2)}\left[\frac{\partial \xi_\theta}{\partial \theta} + w + \nu\frac{\partial \xi_x}{\partial x}\right] \tag{2.3.252}$$

$$N_{x\theta} = \frac{Eh}{2a(1+\nu)}\left[\frac{\partial \xi_x}{\partial \theta} + \frac{\partial \xi_\theta}{\partial x} + \frac{h^2}{6a^2}\left[-\frac{\partial^2 w}{\partial x \partial \theta} + \frac{\partial \xi_\theta}{\partial x}\right]\right] \tag{2.3.253}$$

$$N_{\theta x} = \frac{Eh}{2a(1+\nu)}\left[\frac{\partial \xi_x}{\partial \theta} + \frac{\partial \xi_\theta}{\partial x}\right] \tag{2.3.254}$$

$$M_{x\theta} = M_{\theta x} = \frac{Eh^3}{(1+\nu)}\left[-\frac{\partial^2 w}{\partial x \partial \theta} + \frac{\partial \xi_\theta}{\partial x}\right] \tag{2.3.255}$$

$$Q_x = -\frac{Eh}{a(1-\nu^2)}\left[\frac{\partial^2 \xi_x}{\partial x^2} + \frac{\nu \partial^2 \xi_\theta}{\partial x \partial \theta} + \nu\frac{\partial w}{\partial x}\right] - \frac{Eh}{2a(1+\nu)}\left[\frac{\partial^2 \xi_x}{\partial \theta^2} + \frac{\partial^2 \xi_\theta}{\partial x \partial \theta}\right] \tag{2.3.256}$$

$$Q_\theta = -\frac{Eh}{a(1-\nu^2)}\left[\frac{\partial^2 \xi_\theta}{\partial \theta^2} + \frac{\partial w}{\partial \theta} + \nu\frac{\partial^2 \xi_x}{\partial x \partial \theta}\right] - \frac{Eh}{2a(1+\nu)}\left[\frac{\partial^2 \xi_x}{\partial x \partial \theta} + \frac{\partial^2 \xi_\theta}{\partial x^2} + \frac{h^2}{6a^2}\left[-\frac{\partial^3 w}{\partial x^2 \partial \theta} + \frac{\partial^2 \xi_\theta}{\partial x^2}\right]\right] \tag{2.3.257}$$

The boundary condition which is closest to the equivalent of a simply supported plate boundary condition is referred to as the shear diaphragm (or SD) condition where

$$w = M_x = N_x = \xi_\theta = 0 \tag{2.3.258}$$

That is, the cylinder is closed at the end with a thin flat circular cover plate. The plate has considerable stiffness in its own plane, thus restraining the ν and w components of cylinder displacement. As the end plate is not very stiff in its transverse plane, it would generate very little bending moment M_x and very little longitudinal membrane force N_x, which explains the choice of boundary conditions for this case.

For a cylinder which has free ends, all of the boundary conditions on the left-hand side of (2.3.210) to (2.3.213) would be satisfied.

2.3.5.2 Cylinder equations of motion: alternative derivation

Referring to the cylinder element of thickness h shown in Fig. 2.25, and considering its equilibrium, under the influence of internal force and moment resultants and external applied forces and moments, we can write two equations of motion, one involving forces and the other involving moments, as follows (Leissa, 1973):

$$\frac{\partial \boldsymbol{F}_x}{\partial x}\mathrm{d}x + \frac{\partial \boldsymbol{F}_\theta}{\partial \theta}\mathrm{d}\theta + \boldsymbol{q}a\mathrm{d}x\mathrm{d}\theta = 0 \tag{2.3.259}$$

$$\begin{aligned}&\frac{\partial \boldsymbol{M}_{xT}}{\partial x}\mathrm{d}x + \frac{\partial \boldsymbol{M}_{\theta T}}{\partial \theta}\mathrm{d}\theta - (\boldsymbol{F}_x \times \boldsymbol{i}_\theta)\frac{a\mathrm{d}\theta}{2} - (\boldsymbol{F}_\theta \times \boldsymbol{i}_x)\frac{\mathrm{d}x}{2}\\ &+ \left[\boldsymbol{F}_x + \frac{\partial \boldsymbol{F}_x}{\partial x}\mathrm{d}x\right] \times \left[\mathrm{d}x\boldsymbol{i}_x + \frac{a\mathrm{d}\theta}{2}\boldsymbol{i}_\theta\right] + \left[\boldsymbol{F}_\theta + \frac{\partial \boldsymbol{F}_\theta}{\partial \theta}\mathrm{d}\theta\right]\left[a\mathrm{d}\theta\boldsymbol{i}_\theta + \mathrm{d}\frac{x}{2}\boldsymbol{i}_x\right]\\ &+ \boldsymbol{M}_e a\mathrm{d}x\mathrm{d}\theta = 0\end{aligned} \tag{2.3.260}$$

The external forces per unit area $\boldsymbol{q}$ and moments per unit area $\boldsymbol{M}_e$ are defined as

$$\boldsymbol{q} = q_x\boldsymbol{i}_x + q_\theta\boldsymbol{i}_\theta + q_n\boldsymbol{i}_n \tag{2.3.261}$$

$$\boldsymbol{M}_e = M_{ex}\boldsymbol{i}_x + M_{e\theta}\boldsymbol{i}_\theta + M_{en}\boldsymbol{i}_n \tag{2.3.262}$$

The total internal forces $\boldsymbol{F}_x$ and $\boldsymbol{F}_\theta$ are defined as

$$\boldsymbol{F}_x = (N_x\boldsymbol{i}_x + N_{x\theta}\boldsymbol{i}_\theta + Q_x\boldsymbol{i}_n)a\mathrm{d}\theta \tag{2.3.263}$$

$$\boldsymbol{F}_\theta = (N_{\theta x}\boldsymbol{i}_x + + N_\theta\boldsymbol{i}_\theta + Q_\theta\boldsymbol{i}_n)\mathrm{d}x \tag{2.3.264}$$

The total internal moments $\boldsymbol{M}_{xT}$ and $\boldsymbol{M}_{\theta T}$ are defined as

$$\boldsymbol{M}_{xT} = (-M_{x\theta}\boldsymbol{i}_x + M_x\boldsymbol{i}_\theta)a\mathrm{d}\theta \tag{2.3.265}$$

$$\boldsymbol{M}_{\theta T} = (-M_\theta\boldsymbol{i}_x + M_{\theta x}\boldsymbol{i}_\theta)\,\mathrm{d}x \tag{2.3.266}$$

Using (2.3.261), (2.3.263) and (2.3.264), (2.3.259) can be expanded into its three scalar components as follows (Leissa, 1973):

$$\frac{\partial N_x}{\partial x} + \frac{1}{a}\frac{\partial N_{\theta x}}{\partial \theta} + q_x = 0 \tag{2.3.267}$$

$$\frac{1}{a}\frac{\partial N_\theta}{\partial \theta} + \frac{\partial N_{x\theta}}{\partial x} + \frac{Q_\theta}{a} + q_\theta = 0 \tag{2.3.268}$$

$$-\frac{N_\theta}{a} + \frac{\partial Q_x}{\partial x} + \frac{1}{a}\frac{\partial Q_\theta}{\partial \theta} + q_n = 0 \tag{2.3.269}$$

Using (2.3.261) to (2.3.266), (2.3.260) can be expanded into its three scalar components (Leissa, 1973):

$$\frac{\partial M_x}{\partial x} + \frac{1}{a}\frac{\partial M_{\theta x}}{\partial \theta} - Q_x + M_{e\theta} = 0 \tag{2.3.270}$$

$$\frac{1}{a}\frac{\partial M_\theta}{\partial \theta} + \frac{\partial M_{x\theta}}{\partial x} - Q_\theta + M_{ex} = 0 \tag{2.3.271}$$

$$N_{x\theta} - N_{\theta x} - \frac{M_{\theta x}}{a} = 0 \tag{2.3.272}$$

Using (2.3.270) and (2.3.271) with $M_{e\theta} = M_{ex} = 0$, we obtain (2.3.241) and (2.3.242).

Eliminating Q_θ and Q_x from (2.3.268) and (2.3.269) by using (2.3.270) and (2.3.271), the number of equations of motion can be reduced to three. (Note that (2.3.272) is satisfied identically and is not a useful equation of motion.)

Substitution of expressions for N_x, N_θ, $N_{x\theta}$, M_x, M_θ and $M_{x\theta}$ in terms of displacements ξ_x, ξ_θ and w into the three equations of motion as described above will result in the equations previously derived using Hamilton's principle ((2.3.187) to (2.3.189)).

2.3.5.3 Solution of the equations of motion

Functions of the following forms are found to describe the motion of the cylinder for all types of boundary condition:

$$\xi_x = U_n e^{sx/a} \cos n\theta \cos\omega t \tag{2.3.273}$$

$$\xi_\theta = V_n e^{sx/a} \sin n\theta \cos\omega t \tag{2.3.274}$$

$$w = W_n e^{sx/a} \cos n\theta \cos\omega t \tag{2.3.275}$$

where U_n, V_n, W_n and s are undetermined constants and n is an integer describing the circumferential displacement distribution.

Note for shear diaphragm boundary conditions the solutions become

$$\xi_x = U_n \cos(m\pi x/L) \cos n\theta \cos\omega t \tag{2.3.276}$$

$$\xi_\theta = V_n \sin(m\pi x/L) \sin n\theta \cos\omega t \tag{2.3.277}$$

$$w = W_n \sin(m\pi x/L) \cos n\theta \cos\omega t \tag{2.3.278}$$

To find the resonance frequencies and mode shapes, the following steps are implemented (see Warburton, 1965):

1. Substitute the solutions (2.3.273) to (2.3.275) into the equations of motion (2.3.190). This will produce a quartic equation in w^2 with coefficients which are a function of ρ, a, ν, ω, E, h and n.

2. As there will be eight roots for s (for each value of n) from the quartic equation derived in 1. above, the quantity $W_n e^{sx/a}$ may be expressed in terms of eight real constants as

$$W_n e^{sx/a} = \sum_{r=1}^{8} B_r e^{s_r x/a} \tag{2.3.279}$$

where s_r is rth root of the quartic in s^2 and B_r is the rth constant.

3. Use the equations of motion to find the ratios (ξ_x/w) and (ξ_θ/w) and then the quantities $U_n e^{sx/a}$ and $V_n e^{sx/a}$ can also be written in terms of the constants B_r.
4. Substitute the solutions for u, v and w into the eight boundary condition equations (four for each end of the cylinder) in eight unknown coefficients. The characteristic frequency equation is then found by setting the determinant of the unknown coefficients equal to zero. The eigen frequencies corresponding to a specific value of n are then found by solving this characteristic frequency equation. There will be three roots for each value of n.

2.3.5.4 Effect of longitudinal and circumferential stiffeners

Provided that the stiffeners are relatively closely spaced and only low order long wavelength modes are considered, the resulting increased stiffness may be smeared out along and over the shell to produce an orthotropic shell constructed from isotropic materials (Mikulas and McElman, 1965). In this case, the equations of motion may be written as in (2.3.190), with different coefficients a_{ij} defined below:

$$a_{11} = a\frac{\partial^2}{\partial x^2} + \frac{C_{66}}{aC_{11}}\frac{\partial^2}{\partial \theta^2} - \frac{\rho h a}{C_{11}}\frac{\partial^2}{\partial t^2} \tag{2.3.280}$$

$$a_{12} = \frac{(C_{12} + C_{22})}{C_{11}}\frac{\partial^2}{\partial x \partial \theta} \tag{2.3.281}$$

$$a_{13} = \frac{C_{12}}{C_{11}}\frac{\partial}{\partial x} \tag{2.3.282}$$

$$a_{21} = \frac{(C_{12} + C_{22})}{C_{11}}\frac{\partial^2}{\partial x \partial \theta} \tag{2.3.283}$$

$$a_{22} = \frac{a(C_{66} + D_{66})}{C_{11}}\frac{\partial^2}{\partial x^2} + \frac{(C_{22} + D_{22})}{aC_{11}}\frac{\partial^2}{\partial \theta^2} - \frac{\rho h a}{C_{11}}\frac{\partial}{\partial t^2} \tag{2.3.284}$$

$$a_{23} = \frac{C_{22}}{aC_{11}}\frac{\partial}{\partial\theta} - \frac{D_{22}}{aC_{11}}\frac{\partial^3}{\partial\theta^3} - \frac{(D_{12}+D_{66})}{C_{11}}\frac{\partial^3}{\partial x\partial\theta^2} \tag{2.3.285}$$

$$a_{31} = \frac{C_{12}}{C_{11}}\frac{\partial}{\partial x} \tag{2.3.286}$$

$$a_{32} = \frac{C_{22}}{aC_{11}}\frac{\partial}{\partial\theta} - \frac{D_{22}}{aC_{11}}\frac{\partial^3}{\partial\theta^3} - \frac{a(D_{12}+D_{66})}{C_{11}}\frac{\partial^3}{\partial x^2\partial\theta} \tag{2.3.287}$$

$$a_{33} = \frac{C_{22}}{aC_{11}} + a^3\frac{D_{11}}{C_{11}}\frac{\partial^4}{\partial x^4} + \frac{D_{22}}{a}\frac{\partial^4}{\partial\theta^4} + \frac{a(2D_{12}+D_{66})}{C_{11}}\frac{\partial^4}{\partial x^2\partial\theta^2} + \frac{\rho h a}{C_{11}}\frac{\partial^2}{\partial t^2} \tag{2.3.288}$$

where C_{11}, C_{12}, C_{22} and C_{66} are the extensional stiffness constants and D_{11}, D_{12}, D_{22} and D_{66} are the flexural stiffness constants defined by

$$C_{11} = \frac{E_{LS}}{L_{R\theta}}\left[A_L + hL_{R\theta}/(1-\nu^2)\right] \tag{2.3.289}$$

$$C_{12} = \nu C_{11} \tag{2.3.290}$$

$$C_{22} = \frac{1}{L_{Rx}}\left[E_F A_F + E_{LS}hL_{Rx}/(1-\nu^2)\right] \tag{2.3.291}$$

$$C_{66} = (1-\nu)C_{11}/2 \tag{2.3.292}$$

$$D_{11} = \frac{E_{LS}}{L_{R\theta}}\left[I_{Lx} + I_{Sx}(1-\nu^2)\right] \tag{2.3.293}$$

$$D_{12} = \nu D_{11} \tag{2.3.294}$$

$$D_{22} = \frac{1}{L_{Rx}}\left[E_F I_{F\theta} + E_{LS}I_{SS}/(1-\nu^2)\right] \tag{2.3.295}$$

$$D_{66} = 2(1-\nu)D_{11} \tag{2.3.296}$$

and where

$$I_{F\theta} = I_F + A_F(y_F + h - r_\theta)^2 \tag{2.3.297}$$

$$I_{Lx} = I_L + A_L(y_L + h - r_x)^2 \tag{2.3.298}$$

$$I_{SS} = 3L_{Rx}h^3/48 + 3\beta L_{Rx}h(r_\theta - h/2)^2/4 \tag{2.3.299}$$

$$I_{Sx} = L_{R\theta}h^3/12 + L_{R\theta}h(r_x - h/2)^2 \tag{2.3.300}$$

$$r_\theta = \frac{A_F(y_F + h) + 3\beta L_{Rx}h^2/8}{A_F + 3\beta L_{Rx}h/4} \tag{2.3.301}$$

$$r_x = \frac{A_L(y_L + h) + L_{R\theta}h^2/2}{A_L + L_{R\theta}h} \tag{2.3.302}$$

where

A_F and A_L	=	cross-sectional areas of rings (frames) and stringers (longerons) respectively
I_F and I_L	=	second moments of area of frames and stringers about their own centroidal axes respectively
I_{Lx} and I_{Sx}	=	second moments of area of the stringers and skins respectively about the centroidal axis of the skin/stringer cross-section
I_{FS} and I_{SS}	=	second moments of area of the frames and skins respectively about the centroidal axis of the frame/skin cross-section
h	=	skin (wall) thickness
y_F and y_L	=	distances from the centroidal axes of the frames and stringers respectively to the underside of the skin.
E_F and E_{LS}	=	moduli of elasticity of the frames and stringers (and skins) respectively
L_{Rx} and $L_{R\theta}$	=	lengths of the repeating sections in the axial and circumferential directions respectively
β	=	0 if the skin is attached to the stringers but not to the frames
	=	1 if the skin is attached to the stringers and frames

2.3.5.5 Other complicating effects

Various complicating effects such as uniform preloading and inclusion of the effects of rotary inertia and shear deformation in the analysis are beyond the scope of this text but are considered elsewhere (Leissa, 1973).

Inclusion of the effects of rotary inertia and shear deformation results in a displacement vector containing five instead of three components and the equations of motion (2.3.190)' are now represented by a (5×5) instead of a (3×3) matrix which adds considerable algebraic complexity to the solution. The effect of including rotary inertia and shear deformation in the analysis is only significant for relatively thick ($a/h < 10$) cylinders or high order modes, neither of which are of concern in this book.

2.4 STRUCTURAL SOUND RADIATION, SOUND PROPAGATION AND GREEN'S FUNCTIONS

This section contains background definitions and information which will be essential for understanding the material in Chapters 7 – 11. The Green's function technique is a convenient approach to solving sound radiation problems, whether sound radiation by a structure into free space is considered or whether sound transmission through a structure into an enclosed space is of interest. Evaluating the Green's function which characterizes a particular physical system is an important step in determining the optimum control source and error sensor configuration and the maximum vibration or acoustic control which can be achieved by an optimal feedforward electronic controller applied to the particular physical system.

Physically, a Green's function is simply a transfer function which relates the response at one point in an acoustic medium or a structure to an excitation by a unit point source at another point. The value of the Green's function for a particular physical system is dependent on the location of the source and observation points and the frequency of excitation. Note that here neither the type of excitation source nor the type of response has been defined. This will be done when specific examples are considered.

Although Green's functions are not unique to acoustics and vibrations problems (see Morse and Feshbach, 1953, Chapter 7), the discussion here will be restricted to these types of problem in the interests of clarity and relevance. In particular, we will concern ourselves with problems involving sound radiation from vibrating structures, sound transmission through structures into enclosed spaces, vibration transmission through connected structures, and sound propagation in an acoustic medium.

A structure vibrating in contact with a compressible fluid such as air or water will generate pressure fluctuations in the fluid which, in turn, will react back on the structure and modify its vibration behaviour. This loading by the pressure waves in the fluid is known as radiation loading and generally for structures radiating into air it can be ignored. Consequently, the dynamic response of a structure can be evaluated as though it were vibrating in a vacuum and the pressure field generated by the vibrating structure can be evaluated independently by equating the velocity of the fluid to that of the structure at the structure/fluid interface.

However, for structures radiating into relatively dense fluids such as water or oil, the forces acting on the structure are significantly modified by the radiation loading, and since the acoustic pressure is dependent upon the structural response, a feedback coupling between the fluid and structure exists. Thus, the structural vibration and acoustic pressure responses must be evaluated simultaneously.

Another type of problem involves sound radiation into an enclosed space, where the response of the enclosed space is coupled to the response of the structure through which the sound is transmitted. In this instance, the response of the coupled system is derived from the mode shapes of the structure vibrating in a vacuum and the mode shapes of the enclosed space calculated with the assumption that the boundaries enclosing it are perfectly rigid. The two responses are then coupled together at the boundaries of the enclosure where the external to internal pressure difference across the boundary is related to the normal velocity of the structure by using the structural Green's function. Of course, this method is not mathematically rigorous (as the assumption of rigid enclosure boundaries for the acoustic mode shape calculations results in small errors, as does the assumption of a surrounding vacuum for the calculation of the structural mode shapes) but the results obtained are sufficiently accurate to justify its use.

For sound propagation in an acoustic medium, the response at a particular location in the medium due to sources acting at other locations can be calculated using the appropriate Green's function. Sound propagation in ducts, both plane wave and higher order mode propagation, and the implementation of active acoustic sources can be analysed by use of these Green's functions.

The solutions to the types of problems just mentioned are conveniently expressed in terms of acoustic Green's functions and structural Green's functions. Note that some authors prefer to refer to structural Green's functions as influence coefficients in an attempt to avoid confusion with acoustic Green's functions but this is done at the expense of additional complication of the terminology and will not be done here.

The underlying assumptions in the development of the solutions to the acoustic propagation, transmission and radiation problems discussed in this text are listed below:

1. Linearity: for a structure, each component of stress is a linear function of the corresponding strain component, and for a fluid, the acoustic pressure fluctuations about the mean are a linear fraction of the corresponding density fluctuations about the mean.
2. Dissipation: frictional dissipation of energy is assumed to take place in solid structures, as this is a necessary requirement if meaningful solutions are to be obtained for the structural response. The dissipation mechanism will be simulated here by using a small structural loss factor associated with the stress component proportional to the strain rate. That is, a complex modulus of elasticity $E(1+j\eta)$ will be assumed, where E is Young's modulus and η is the structural loss factor.
3. Homogeneity: the structure and fluid are both regarded as homogeneous. This assumption is not valid if sound propagation over large distances in the ocean or atmosphere is considered. However, it does provide good results for short distance propagation which can then be used with an appropriate model of the ocean or atmosphere to calculate long-range propagation without further consideration of the source.
4. Inviscid fluid: it is assumed that the acoustic fluid has no viscosity and therefore cannot support shear forces. Thus, the only component of structural displacement which contributes to the radiated sound field is that

which is normal to the surface of the structure. Similarly the acoustic medium can only apply normal loads to the structure.

It is possible to calculate the Green's function from classical analysis only for physically simple systems, and examples of some of these are discussed in the following sections. For more complex systems, finite element and boundary element methods can be used to numerically evaluate what constitutes an equivalent Green's function; this will be discussed in Chapters 8 and 9.

Once the Green's function has been determined, it can be used to calculate the total system response due to a finite size source by integrating over the boundary of the source. Similarly, if n point sources are considered, the total sound field at any point can be calculated by summing the product of Green's function corresponding to each source with the source strength of each source.

Both primary and control sources can be included in the analysis, and optimization techniques (see Chapter 8) can then be used to optimize the control force and error sensor locations to obtain maximum control of acoustic power radiation for sound radiation into one-, two- or three-dimensional free space, acoustic potential energy for sound radiation into enclosed spaces, and vibratory power transmission for vibrating structures.

At this point it will be valuable to consider a rigorous mathematical definition of the acoustic Green's function which may be defined as the solution to the inhomogeneous scalar Helmholtz equation (wave equation for a periodic disturbance with simple harmonic time dependence) for an acoustic medium containing a periodic driving source of unit strength. In other words, it is the solution of the inhomogeneous wave equation (or inhomogeneous scalar Helmholtz equation) with a singularity at the source point.

2.4.1 Acoustic Green's function: unbounded medium

The acoustic Green's function for an unbounded medium is defined as the solution of

$$\nabla^2 G(r, r_0, \omega) + k^2 G(r, r_0, \omega) = -\delta(r - r_0) \tag{2.4.1}$$

This wave equation is discussed in detail in Section 2.1 and by Morse and Feshbach (1953) and Pierce (1981).

The function on the right-hand side of (2.4.1) is the three-dimensional Dirac delta function, representing a unit point source at location r_0. The Dirac delta function allows a discontinuous point source to be described mathematically in terms of source strength per unit volume. In other words, it concentrates a uniformly distributed source onto a single point. The reason for doing this is that a uniformly distributed source is much more easily handled mathematically. Thus, if $q'(r)$ is the source strength per unit volume at any location r, a point source $q(r_0)$ at location r_0 may be expressed in terms of q' using the Dirac delta function as follows:

$$q'(r) = q(r)\delta(r - r_0) \tag{2.4.2}$$

Integrating q' over any enclosed volume gives the following:

$$\begin{aligned}\iiint_V q'(r)\,\mathrm{d}r = \iiint_V q(r)\,\delta(r - r_0)\,\mathrm{d}r &= q(r_0), && r_0 \text{ in } V \\ &= \frac{1}{2}q(r_0), && r_0 \text{ on boundary of } V \\ &= 0, && r_0 \text{ outside } V\end{aligned} \tag{2.4.3}$$

Note that q' has the units of T^{-1} and q the units of $\mathrm{L}^3\mathrm{T}^{-1}$.

It is clear from (2.4.3) that the function $\delta(r - r_0)$ has the dimensions L^{-3}. Thus, the dimensions of the Green's function in equation (2.4.1) must be L^{-1}. The Dirac delta function

$$\delta(r - r_0) = \delta(x - x_0)\,\delta(y - y_0)\,\delta(z - z_0) \tag{2.4.4}$$

is thus defined as a very high, very large and very narrow step function of source strength centred at r_0 and with an area of unity under each of the curves of force *vs* $(x - x_0)$, $(y - y_0)$ and $(z - z_0)$.

If the unit source represented by $\delta(r - r_0)$ in (2.4.1) has the units of volume velocity, then the function $G(r, r_0, \omega)$ may be interpreted as a velocity potential (units $\mathrm{L}^2\mathrm{T}^{-1}$).

A solution of (2.4.1) is found by application of Gauss's integral theorem (Pierce, 1981), and is:

$$G(r, r_0, \omega) = \frac{\mathrm{e}^{-jkR}}{4\pi R} \tag{2.4.5}$$

in which case, the unit volume flow of the source is defined as:

$$q(r_0) = \lim_{R \to 0}\left[-4\pi R^2 \frac{\partial G}{\partial R}\right] \tag{2.4.6}$$

$$\text{where } R = |\,r - r_0\,| \tag{2.4.7}$$

Note that in some textbooks the Green's function is defined without the 4π term in the denominator. In this case, the quantity 4π is included in the right-hand side of (2.4.1). Also, those texts which use negative ($\mathrm{e}^{-i\omega t}$) rather than positive ($\mathrm{e}^{j\omega t}$) time dependence show the Green's function as $\mathrm{e}^{ikR}/4\pi R$.

Equation (2.4.5) is known as the free field Green's function for an acoustic medium; that is, for any three-dimensional gas, liquid or solid supporting longitudinal wave propagation.

The Green's function (as well as being a solution of (2.4.1)) must also satisfy the Sommerfeld radiation condition, to ensure that only outward travelling waves are represented. That is:

$$\lim_{R\to 0} R\left[\frac{\partial G}{\partial R} + jkG\right] = 0 \tag{2.4.8}$$

The solution to (2.4.1) and (2.4.8) (given by (2.4.5)) is not subject to any boundary condition at finite range and thus is referred to as the free space Green's function. Remember that the Green's function represents the effect of a unit point source at any point in the system, on the response at any other point in the system.

As the units of the Green's function are L^{-1}, the pressure response at any location $\boldsymbol{r}$ in the acoustic medium due to a point source of strength $q(\boldsymbol{r}_0,\omega)\,\mathrm{m^3s^{-1}}$ is obtained by multiplying the product of the Green's function and the source strength by $\rho\omega$. Thus,

$$p(\boldsymbol{r},t) = p(\boldsymbol{r})\mathrm{e}^{j\omega t} = j\rho\omega G(\boldsymbol{r},\boldsymbol{r}_0,\omega)\,q(\boldsymbol{r}_0,\omega)\,\mathrm{e}^{j\omega t} \tag{2.4.9}$$

It may seem that we are discussing a fairly idealized case when we speak of a point source. However, the pressure response at any point in an acoustic medium due to a distributed source can be found by integrating the product of the source distribution $q(\boldsymbol{r}_0,\omega)$ with the Green's function over the space of the distributed source. Thus (with the time dependence $\mathrm{e}^{j\omega t}$ omitted),

$$p(\boldsymbol{r}) = j\rho\omega\int\int_V\int G(\boldsymbol{r},\boldsymbol{r}_0,\omega)\,q'(\boldsymbol{r}_0,\omega)\,\mathrm{d}\boldsymbol{r}_0 \tag{2.4.10}$$

where $q'(\boldsymbol{r}_0,\omega)$ is the volume velocity per unit volume at location $\boldsymbol{r}_0$.

For the acoustic sources considered in this book, the source distribution is usually over a defined surface, so the volume integral in (2.4.10) can usually be replaced with a surface integral.

2.4.2 Reciprocity of Green's functions

Before we can discuss reciprocity it is necessary to introduce Green's theorem which is a special case of Gauss's theorem and relates an area integral to a volume integral. Note that Gauss's theorem states that for an incompressible fluid, the fluid generated per unit time by all sources in a given volume is equal to the fluid that leaves the volume per unit time through its boundary. That is, for any two scalar functions $A(\boldsymbol{r})$, $B(\boldsymbol{r})$ of position $\boldsymbol{r}$ we may write

$$\int_S\int\left[A\nabla B - B\nabla A\right]\cdot\mathrm{d}\boldsymbol{S} = \int\int_V\int\left[A\nabla^2 B - B\nabla^2 A\right]\mathrm{d}V \tag{2.4.11}$$

As stated in the previous section the Green's function $G(\boldsymbol{r},\boldsymbol{r}_0,\omega)$ satisfies the equation

$$\nabla^2 G(r, r_0, \omega) + k^2 G(r, r_0, \omega) = -\delta(r - r_0) \tag{2.4.12}$$

However, the Green's function $G(r, r_1, \omega)$ satisfies the equation

$$\nabla^2 G(r, r_1, \omega) + k^2 G(r, r_1, \omega) = -\delta(r - r_1) \tag{2.4.13}$$

If we multiply the first equation by $G(r, r_1, \omega)$ and the second equation by $G(r, r_0, \omega)$ and integrate the difference over the volume V enclosed by an arbitrary boundary surface of area S we obtain

$$\begin{aligned} &-\iiint_V \left[G(r, r_0, \omega)\nabla^2 G(r, r_1, \omega) - G(r, r_1, \omega)\nabla^2 G(r, r_0, \omega)\right] \mathrm{d}V \\ &= \iiint_V G(r, r_0, \omega)\,\delta(r - r_1)\,\mathrm{d}V - \iiint_V G(r, r_1, \omega)\,\delta(r - r_0)\,\mathrm{d}V \end{aligned} \tag{2.4.14}$$

Using Green's theorem and the definition of the delta function the preceding equation can be written as

$$\begin{aligned} &-\iint_S \left[G(r, r_0, \omega)\nabla G(r, r_1, \omega) - G(r, r_1, \omega)\nabla G(r, r_0, \omega)\right] \mathrm{d}S \\ &= G(r_1, r_0, \omega) - G(r_0, r_1, \omega) \end{aligned} \tag{2.4.15}$$

However, as will be shown in the next section, the functions G by definition must satisfy one of the following types of boundary condition on the surface S:

$$G = 0, \quad \frac{\partial G}{\partial n} = 0 \quad \text{or} \quad \frac{\partial G}{\partial n}/G = \text{const}$$

where $\frac{\partial G}{\partial n}$ represents the normal gradient of G at the boundary surface S. Therefore, the integrand of the previous equation vanishes and we are left with

$$G(r_1, r_0, \omega) = G(r_0, r_1, \omega) \tag{2.4.16}$$

The physical interpretation of this relationship is that if a source at r_0 produces a certain response at r_1 it would produce the same response at r_0 if it were moved to r_1. This is known as reciprocity and it is fundamental to many acoustical analyses.

2.4.3 Acoustic Green's function for a three-dimensional bounded fluid

Equation (2.4.1), of which the Green's function is a solution, describes a medium which is homogeneous everywhere except at one point, the source point. When the point is on the boundary of a medium, the Green's function may be used to satisfy boundary conditions (for the homogeneous wave equation) that require neither the acoustic response nor the gradient of the acoustic response to be zero on the boundary. These are referred to as inhomogeneous boundary conditions. Conversely, when the point is a source point within the medium and not on the boundary, the Green's function is used to satisfy the inhomogeneous wave equation with homogeneous boundary conditions. It is implicit in the use of Green's functions solutions to the inhomogeneous wave equation with homogeneous boundary conditions or the homogeneous wave equation with inhomogeneous boundary conditions, that the two conditions of inhomogeneity do not coexist. If they do, then solutions must be obtained for only one inhomogeneity condition at once and the two solutions added together to give the solution corresponding to the coexistence of an inhomogeneous equation (where a source point is contained within the medium) and an inhomogeneous boundary (where a source point is on the boundary). Thus, it is implicit in the Green's function solution of the inhomogeneous wave equation representation of a point excitation source in an acoustic medium, that on the boundary of the medium at least one of the following conditions must be satisfied:

$$G = 0 \quad \frac{\partial G}{\partial n} = 0 \quad \text{or} \quad \frac{\partial G}{\partial n}/G = \text{const} \tag{2.4.17}$$

These conditions are referred to as homogeneous boundary conditions.

For the homogeneous wave equation with inhomogeneous boundary conditions, it is implicit that on the boundary surface of the medium the function $G(r_0, r, \omega)$ has specified values (not everywhere zero) or that $\dfrac{\partial G}{\partial n}$ has specified values (not everywhere zero) or that

$$aG + b(\partial G/\partial n) = F \tag{2.4.18}$$

Equation (2.4.10) represents the solution to the inhomogeneous wave equation with homogeneous boundary conditions, in terms of acoustic pressure. For a volume of fluid enclosed in within a bounding surface, containing volume velocity sources, the total solution for the pressure response at frequency ω is (2.4.10) plus the solution for the homogeneous wave equation with inhomogeneous boundary conditions; namely

$$\nabla^2 p(r, \omega) + k^2 p(r, \omega) = 0 \tag{2.4.19}$$

To solve this equation with boundary conditions at finite surfaces, we multiply (2.4.1) by $p(r, \omega)$ and (2.4.19) by $G(r, r_0, \omega)$ and subtract the first result from the second to obtain:

$$G(r,r_0,\omega)\,\nabla^2 p(r,\omega) - p(r,\omega)\,\nabla^2 G(r,r_0,\omega) = p(r,\omega)\,\delta(r - r_0) \tag{2.4.20}$$

where $r_0 = r_s$ is now a point on the boundary surface.

If we now interchange r and r_0 (r and r_0 are any points in the volume enclosed by the boundary), use reciprocity ($G(r,r_0,\omega) = G(r_0,r,\omega)$ and $\delta(r - r_0) = \delta(r_0 - r)$), and if we integrate over the volume defined by $x_0 y_0 z_0$ we obtain:

$$\int\int_V\int \left[G(r,r_0,\omega)\,\nabla^2 p(r_0,\omega) - p(r_0,\omega)\,\nabla^2 G(r,r_0,\omega)\right] \mathrm{d}r$$
$$= \int\int_V\int p(r_0,\omega)\,\delta(r-r_0)\,\mathrm{d}r_0 \tag{2.4.21}$$

Using Gauss's integral theorem and Green's identity (Junger and Feit, 1986, p. 81), the first integral in (2.4.21) can be written as a surface integral over the bounding surface and the second integral is (from 2.4.3) equal to $p(r,\omega)$. Thus, (2.4.21) can be written as:

$$p(r,\omega) = -\int_S\int \left[G(r,x,\omega)\frac{\partial}{\partial n_s}p(x,\omega) - p(x,\omega)\frac{\partial}{\partial n_s}G(r,x,\omega)\right]\mathrm{d}x \tag{2.4.22}$$

which is known as the Helmholtz integral equation. In this equation, S is the area of boundary surface, x is a vector location on the boundary surface, V is the volume enclosed by the boundary and the vector n_s is the normal to the local boundary surface, directed into the fluid.

Adding the solution for the inhomogeneous wave equation with homogeneous boundary conditions, we obtain the Kirkhoff–Helmholtz integral equation (in which the pressure gradient has been replaced with the particle velocity, u_n multiplied by $j\omega\rho_0$):

$$p(r,\omega) = \int_S\int \left[j\omega\rho_0 u_n(x,\omega)\,G(r,x,\omega) + p(x,\omega)\frac{\partial}{\partial n_s}G(r,x,\omega)\right]\mathrm{d}x$$
$$+ \int\int_V\int j\omega\rho_0\, q'(r_0,\omega)\,G(r,r_0)\,\mathrm{d}r_0 \tag{2.4.23}$$

The first integral is evaluated over all of the bounding surfaces and the second is evaluated over the bounded volume. Equation (2.4.23) is a special harmonic case of a more general integral equation, in which the time dependence is arbitrary and where the phase, $k\,|\,r\,|$, is replaced by a time difference $(t - |\,r\,|/c)$.

If the Green's function G can be chosen to satisfy one of the boundary conditions $G = 0$ or $\dfrac{\partial G}{\partial n_s} = 0$ over the entire boundary (as well as the wave equation and the Sommerfeld radiation condition), then one of the surface integral terms in (2.4.23) will disappear.

For the special case of an infinitely baffled, plane surface radiating into free space, choosing a Green's function consisting of (2.4.5) multiplied by two to account for an image source (reflection of the source in the plane surface), and ignoring the last term in (2.4.23) (as all sources are on the surface) allows (2.4.23) to be written in the form of Rayleigh's well-known integral equation as follows:

$$p(\boldsymbol{r},\omega) = \frac{j\omega\rho_0}{2\pi}\int_S\int \frac{\boldsymbol{u}_n(\boldsymbol{x},\omega)\,\mathrm{e}^{-jkR}}{R}\,\mathrm{d}\boldsymbol{x} \tag{2.4.24}$$

A solution to (2.4.19) subject to a rigid wall boundary condition ($\partial p/\partial \boldsymbol{n} = 0$) can be written as:

$$p(\boldsymbol{r})\mathrm{e}^{j\omega_n t} = A_n \psi_n(\boldsymbol{r})\mathrm{e}^{j\omega_n t} \tag{2.4.25}$$

where A_n is a complex constant and ψ_n is the pressure mode shape function for the nth acoustic mode in the rigid walled volume, with a resonance frequency of ω_n. Substituting (2.4.25) into (2.4.19) gives:

$$\nabla^2\psi_n(\boldsymbol{r}) + k_n^2\psi_n(\boldsymbol{r}) = 0 \tag{2.4.26}$$

where $k_n = \omega_n/c_0$. For a discrete set of values, k_n, the functions ψ_n satisfy the condition $\dfrac{\partial\psi_n}{\partial\boldsymbol{n}} = 0$ on the enclosure walls; thus, they can be incorporated into a Green's function which satisfies the same condition and which can be expressed as:

$$G(\boldsymbol{r},\boldsymbol{r}_0,\omega) = \sum_{n=0}^{\infty} B_n\psi_n(\boldsymbol{r}) \tag{2.4.27}$$

where B is a complex constant.

Equation (2.4.1) can now be written as:

$$-\sum_{n=0}^{\infty} k_n^2 B_n\psi_n(\boldsymbol{r}) + k^2\sum_{n=0}^{\infty} B_n\psi_n(\boldsymbol{r}) = -\delta(\boldsymbol{r}-\boldsymbol{r}_0) \tag{2.4.28}$$

where the following relations have been used (from (2.4.27) and (2.4.26)):

$$\nabla^2 G(\boldsymbol{r},\boldsymbol{r}_0,\omega) = \sum_{n=0}^{\infty} B_n\nabla^2\psi_n(\boldsymbol{r}) = -\sum_{n=0}^{\infty} k_n^2 B_n\psi_n(\boldsymbol{r}) \tag{2.4.29}$$

Using the condition that the natural modes of closed elastic systems are mutually orthogonal, we obtain (assuming a uniform mean fluid density):

$$\int_V \rho_0(\boldsymbol{r})\psi_\iota(\boldsymbol{r})\,\psi_n(\boldsymbol{r})\,dV = \begin{cases} 0 & \iota\neq n \\ \Lambda_n & \iota = n \end{cases} \tag{2.4.30}$$

where

$$\Lambda_n = \int_V \rho_0(\boldsymbol{r})\,\psi_n^2(\boldsymbol{r})\,\mathrm{d}V \tag{2.4.31}$$

If the medium has a uniform mean density then we may set $\rho_0(\boldsymbol{r}) = \rho_0$. Multiplying (2.4.28) by $\rho_0(\boldsymbol{r})\psi_\iota(\boldsymbol{r})$, integrating over the fluid volume, then setting $n = \iota$, we obtain

$$B_n\Lambda_n(k^2-k_n^2) = -\psi_n(\boldsymbol{r}_0)\,\rho_0 \tag{2.4.32}$$

or

$$B_n = \frac{\rho_0 \psi_n(r_0)}{\Lambda_n (k_n^2 - k^2)} \tag{2.4.33}$$

Substituting (2.4.33) into (2.4.27) gives the Green's function for an enclosed acoustic space with rigid boundaries as:

$$G(r, r_0, \omega) = \sum_{n=0}^{\infty} \frac{\rho_0 \psi_n(r) \psi_n(r_0)}{\Lambda_n (k_n^2 - k^2)} \tag{2.4.34}$$

The sound pressure at any point in the acoustic medium due to a point source of frequency ω and strength q, located at r_0 can be calculated by substituting (2.4.34) into (2.4.9) to give

$$p(r) = q(r_0)\rho_0 c_0^2 \omega \sum_{n=0}^{\infty} \frac{\psi_n(r)\psi_n(r_0)}{\Lambda_n (k_n^2 - k^2)} e^{j\omega t} = \sum_{n=0}^{\infty} A_n(\omega)\psi_n(r) e^{j\omega t} \tag{2.4.35a,b}$$

For more complicated distributed sources, (2.4.34) is substituted into (2.4.10).

The derivation of the Green's function for an enclosed space with viscous damping expressed in terms of a loss factor η is more complicated and will not be presented here. However, the result is that the wavenumber k_n becomes complex and equal to $k_n(1 + j\eta)$, and the Green's function becomes:

$$G(r, r_0, \omega) = \sum_{n=0}^{\infty} \frac{\psi_n(r) \psi_n(r_0)}{\Lambda_n (k_n^2 - k^2 + j\eta k k_n)} \tag{2.4.36}$$

Expressed in terms of frequency we have

$$G(r, r_0, \omega) = \rho_0 c_0^2 \sum_{n=0}^{\infty} \frac{\psi_n(r) \psi_n(r_0)}{\Lambda_n (\omega_n^2 - \omega^2 + j\eta \omega \omega_n)} \tag{2.4.37}$$

where ρ_0, c_0 are the density and speed of sound respectively in the acoustic medium, and η is a measure of the medium damping which is modelled as viscous and referred to as a loss factor (twice the critical damping ratio, ζ, see Chapter 4).

For a particular enclosed volume, the loss factor η is related to the enclosure reverberation time (time for a sound field to decay by 60 dB after the source is shut down) by:

$$\eta = \frac{2.2}{T_{60} f} \tag{2.4.38}$$

where f is the frequency of excitation in hertz, and T_{60} is the 60 dB decay time (seconds).

2.4.4 Acoustical Green's function for a source in a one-dimensional duct, of infinite length

The formulation is similar to that for a three-dimensional enclosed space, except that the mode shape functions are only defined in two dimensions, as no reflections occur in one of the coordinate directions.

To begin, consider a duct infinitely long with a unit point source of sound placed half way along it. To be consistent with the notation used by other authors, the plane of the duct cross section will be denoted the x,y-plane and the duct axis, the z-axis. Expressing the Dirac delta function of (2.4.1) in cartesian coordinates gives:

$$\nabla^2 G(r,r_0,\omega) + k^2 G(r,r_0,\omega) = -\delta(x-x_0)\,\delta(y-y_0)\,\delta(z-z_0) \quad (2.4.39)$$

where $r = (x,y,z)$ and $r_0 = (x_0,y_0,z_0)$.

As shown by Morse and Ingard (1968, p. 495), a solution to (2.4.19) for a duct is given by

$$p(r)\,e^{j\omega_n t} = A_n \psi_n(x,y)\,e^{-jk_{zn}z}\,e^{j\omega_n t} \quad (2.4.40)$$

Substituting this expression into (2.4.19) and separating out the transverse mode shape function $\psi_n(x,y)$ gives

$$\left(\frac{\partial^2}{\partial x^2} + \frac{\partial^2}{\partial y^2}\right)\psi_n + \kappa_n^2 \psi_n = 0 \quad (2.4.41)$$

and

$$k_{zn}^2 + \kappa_n^2 = k^2 \quad (2.4.42)$$

where $\psi_n \equiv \psi_n(x,y)$, κ_n is the wavenumber of the nth mode shape in the plane of the duct cross-section and k_{zn} is the wavenumber of the nth mode shape along the duct axis.

Solutions which fit an appropriate boundary condition at the duct walls occur only for a discrete set of values for the wavenumber κ_n, these values being called characteristic (or eigen) values and the corresponding solutions ψ_n being called eigen vectors. Note that in the case of rigid walls, the eigen vectors ψ_n satisfy the relation

$$\int_S\!\!\int \psi_n(x,y)\,\psi_\iota(x,y)\,\mathrm{d}x\,\mathrm{d}y = \begin{cases} 0 & \iota \neq n \\ \Lambda_n & \iota = n \end{cases} \quad (2.4.43)$$

The functions ψ_n satisfy the condition

$$\frac{\partial \psi_n}{\partial n} = 0 \quad (2.4.44)$$

around the duct perimeter for rigid duct walls. Thus, these functions can be incorporated into a Green's function, satisfying the same condition, which may be expressed as:

$$G(\boldsymbol{r},\boldsymbol{r}_0,\omega) = \sum_{n=0}^{\infty} B_n(z)\psi_n(x,y) \tag{2.4.45}$$

Thus,

$$\nabla^2 G(\boldsymbol{r},\boldsymbol{r}_0,\omega) = \sum_{n=0}^{\infty} \left\{ B_n \left[\frac{\partial^2\psi}{\partial x^2} + \frac{\partial^2\psi}{\partial y^2} \right] + \psi_n \frac{\partial^2 B_n}{\partial z^2} \right\} \tag{2.4.46}$$

Substituting (2.4.41) into (2.4.46) gives

$$\nabla^2 G(\boldsymbol{r},\boldsymbol{r}_0,\omega) = -\sum_{n=0}^{\infty} \left[B_n\psi_n\kappa_n^2 - \psi_n \frac{\partial^2 B_n}{\partial z^2} \right] \tag{2.4.47}$$

where $\psi_n = \psi_n(x,y)$ and $B_n = B_n(z)$. Substituting (2.4.47) into (2.4.39) gives

$$-\sum_{n=0}^{\infty} \left[B_n\psi_n\kappa_n^2 - \psi_n \frac{\partial^2 B_n}{\partial z^2} \right] + k^2 \sum_{n=0}^{\infty} B_n\psi_n = -\delta(\boldsymbol{r} - \boldsymbol{r}_0) \tag{2.4.48}$$

Multiplying (2.4.48) by ψ_m, integrating over the cross-sectional area of the duct, and making use of (2.4.43) gives

$$-\Lambda_n \left[B_n\kappa_n^2 - \frac{\partial B_n}{\partial z^2} \right] + k^2 B_n\Lambda_n = -\psi_n(x_0,y_0)\,\delta(z-z_0) \tag{2.4.49}$$

That is,

$$\left[\frac{\partial}{\partial z^2} + k_{zn}^2 \right] B_n = \frac{-\psi_n(x_0,y_0)\,\delta(z-z_0)}{\Lambda_n} \tag{2.4.50}$$

where

$$k_{nz}^2 = k^2 - \kappa_n^2 \tag{2.4.51}$$

The function $B_n(z)$ can be found by integrating (2.4.50) over z from $z_0 - \alpha$ to $z_0 + \alpha$ and then letting α go to zero. Thus,

$$\int_{z_0-\alpha}^{z_0+\alpha} \frac{\partial^2 B_n}{\partial z^2}\,\mathrm{d}z + k_{zn}^2 \int_{z_0-\alpha}^{z_0+\alpha} B_n\,\mathrm{d}z = -\int_{z_0-\alpha}^{z_0+\alpha} \frac{\psi_n(x_0,y_0)\delta(z-z_0)}{\Lambda_n}\,\mathrm{d}z \tag{2.4.52}$$

or

$$\left[\frac{\partial B_n}{\partial z} \right]_{z_0-\alpha}^{z_0+\alpha} + k_{zn}^2 \int_{z_0-\alpha}^{z_0+\alpha} B_n\,\mathrm{d}z = -\frac{\psi_n(x_0,y_0)}{\Lambda_n} \tag{2.4.53}$$

Before we can continue, we need to make an assumption regarding the form of the function $B_n(z)$, in particular the z dependence. For a duct extending infinitely in both directions from the source point z_0, the wave travelling on the positive size side of z_0 must be represented by a constant multiplied by $e^{j(\omega t - k_{zn} z)}$ and the wave travelling to the left of z_0 must be represented by a constant multiplied by $e^{j(\omega t + k_{zn} z)}$. However, as the value of $B(z)$ must be continuous across z_0, the constants must be adjusted so that we can write:

$$B_n(z) = D_n e^{-jk_{zn} | z - z_0 |} \tag{2.4.54}$$

$$\left.\begin{array}{l} \text{if } z > z_0, \text{ then } | z - z_0 | = z - z_0 \\ \text{if } z < z_0, \text{ then } | z - z_0 | = z_0 - z \end{array}\right\} \tag{2.4.55a,b}$$

$$\begin{aligned} \lim_{\alpha \to 0} \left[\frac{\partial B_n}{\partial z} \right]_{z_0 - \alpha}^{z_0 + \alpha} &= \lim_{\alpha \to 0} \left[\left(\frac{\partial B_n}{\partial z} \right)_{z_0 + \alpha} - \left(\frac{\partial B_n}{\partial z} \right)_{z_0 - \alpha} \right] \\ &= \lim_{\alpha \to 0} \left[-jk_{zn} D_n e^{-jk_{zn}\alpha} - jk_{zn} D_{ne}^{-jk_{zn}\alpha} \right] \\ &= -2jk_{zn} D_n \end{aligned} \tag{2.4.56a,b,c}$$

As α is a very small quantity, $B_n(z)$ may be considered a constant over the interval $z_0 + \alpha$ and $z_0 - \alpha$ (Morse and Ingard, 1968, p. 133). Thus,

$$k_{zn}^2 \int_{z_0 - \alpha}^{z_0 + \alpha} B_n \mathrm{d}z = 2k_{zn}^2 \alpha B_n \tag{2.4.57}$$

As $\alpha \to 0$, (2.4.57) $\to 0$.

Thus, in the limit as $\alpha \to 0$, (2.4.53) becomes

$$-2jk_{zn} D_n = \frac{-\psi_n(x_0, y_0)}{\Lambda_n} \tag{2.4.58}$$

Combining (2:4.54) and (2.4.58) gives

$$B_n(z) = \frac{-j\psi_n(x_0, y_0)}{2\Lambda_n k_{zn}} e^{-jk_{zn} | z - z_0 |} \tag{2.4.59}$$

and the Green's function is given by

$$G(r, r_0, \omega) = -\sum_{n=0}^{\infty} \frac{j\psi_n(x_0, y_0)\psi_n(x, y) e^{-jk_{zn} | z - z_0 |}}{2\Lambda_n k_{zn}} \tag{2.4.60}$$

If the duct is excited by a harmonic sound source of frequency ω, located at $z = 0$, the solution for the pressure at any location (x, y, z) in the duct can be written as:

$$p(x, y, z, \omega) = \sum_{n=0}^{\infty} A_n(\omega)\, \psi_n(x,y)\, e^{j(\omega t - k_{zn} z)} \tag{2.4.61}$$

where the coefficient $A_n(\omega)$ can be evaluated for any source type by substituting (2.4.60) into (2.4.9) or (2.4.10) and setting the result equal to (2.4.61). Alternatively, the coefficients could be found experimentally using the techniques outlined in Section 7.5.2.

For a rectangular section duct, we can replace the modal index n with a double index (m,n), where m is the number of horizontal nodal lines and n is the number of vertical nodal lines in a duct cross-section. The quantities in equation (2.4.61) are then defined as follows (Morse and Ingard, 1968):

$$\Psi_{mn}(x,y) = \cos\left[\frac{m\pi x}{b}\right] \cos\left[\frac{n\pi y}{d}\right] \tag{2.4.62}$$

$$\Lambda_{mn} = \frac{1}{S}\int_S \Psi_{mn}^2\, \mathrm{d}S \tag{2.4.63}$$

$$k_{mn} = \left(k^2 - \kappa_{mn}^2\right)^{1/2} = \sqrt{\left(\frac{\omega}{c}\right)^2 - \left(\frac{\pi m}{b}\right)^2 - \left(\frac{\pi n}{d}\right)^2} \tag{2.4.64}$$

where b and d are the duct cross-section dimensions, and S is the duct cross-sectional area.

2.4.5 Green's function for a vibrating surface

The general two-dimensional wave equation for a surface, vibrating at frequency ω, can be written as (Cremer *et al.*, 1973, p. 285):

$$L[w(\boldsymbol{x})] - m_s(\boldsymbol{x})\omega^2 w(\boldsymbol{x}) = 0 \tag{2.4.65}$$

where $L[\ ]$ represents a differential operator ($EI'\nabla^4$ for a homogenous thin plate, where I' is the second moment of area of the plate cross-section per unit width), m_s is the mass per unit area of the surface at location $\boldsymbol{x}$, $\boldsymbol{x} = (x, y)$ is the vector location of a point on the vibrating surface and $w(\boldsymbol{x})$ is the normal displacement of the surface at location $\boldsymbol{x}$.

As for the acoustic case discussed previously, a solution to (2.4.65) can be written as

$$w(\boldsymbol{x})e^{j\omega_n t} = A_n \psi_n(\boldsymbol{x})e^{j\omega_n t} \tag{2.4.66}$$

where A_n is a complex constant and ψ_n is the displacement mode shape function for the nth structural mode, assuming no fluid loading by the surrounding medium. Substituting (2.4.66) into (2.4.65) gives for each structural mode

$$L[\psi_n(x)] - m_s(x)\omega_n^2\psi_n(x) = 0 \tag{2.4.67}$$

If the boundary conditions are such that no energy can be conducted across the boundaries, then the functions $\psi_n(x)$ are orthogonal; that is,

$$\int_S\int m_s(x)\psi_n(x)\psi_\iota(x)\,\mathrm{d}x = 0 \quad \text{if } \iota \neq n \\ = m_n \quad \text{if } \iota = n \tag{2.4.68}$$

where m_n is known as the modal mass of the nth mode. The functions ψ_n satisfy the boundary condition expressed in (2.4.44) around the boundaries of the surface. Thus, they can be incorporated into a Green's function which satisfies the same condition and which can be expressed as,

$$G_s(x, x_0, \omega) = \sum_{n=0}^{\infty} B_n\psi_n(x) \tag{2.4.69}$$

where the Green's function is a solution of

$$L[G_s(x, x_0, \omega)] - m_s(x)\omega^2 G_s(x, x_0, \omega) = \delta(x - x_0) \tag{2.4.70}$$

and where B_n is a complex constant. Using (2.4.67) and (2.4.69), (2.4.70) can be written as:

$$\sum_{n=0}^{\infty} B_n m_s(x)\omega_n^2\psi_n(x) - m_s(x)\omega^2\sum_{n=0}^{\infty} B_n\psi_n(x) = \delta(x - x_0) \tag{2.4.71}$$

Multiplying (2.4.71) by $\psi_m(x)$ and integrating over the surface of the vibrating structure gives:

$$(B_n\omega_n^2 m_n - B_n\omega^2 m_n) = \psi(x_0) \tag{2.4.72}$$

Thus,

$$B_n = \frac{\psi_n(x_0)}{m_n(\omega_n^2 - \omega^2)} \tag{2.4.73}$$

and the Green's function is given by:

$$G_s(x, x_0, \omega) = \sum_{n=0}^{\infty} \frac{\psi_n(x)\psi_n(x_0)}{m_n(\omega_n^2 - \omega^2)} \tag{2.4.74}$$

where the modal mass, m_n is defined by (2.4.68). The displacement at any point, $x = (x,y)$ on the surface can be written as

$$w(x, \omega) = \sum_{n=0}^{\infty} A_n(\omega)\psi_n(x)e^{j\omega t} \tag{2.4.75}$$

where the coefficient $A_n(\omega)$ can be evaluated for any source type by substituting (2.4.74) into (2.4.9) or (2.4.10) and setting the result equal to (2.4.75). Alternatively, the coefficients could be found experimentally using modal analysis as described in Chapter 4.

If structural damping, characterized by a hysteretic loss factor η_n is included, the resonance frequency ω_n of mode n becomes complex (as a result of the surface bending stiffness EI' becoming complex and equal to $EI'(1+j\eta)$ – see Cremer *et al.* (1973, p. 290)), and equal to $\omega_n(1+j\eta)$. Thus, for a damped structure the Green's function is:

$$G_s(x,x_0,\omega) = \sum_{n=0}^{\infty} \frac{\psi_n(x)\psi_n(x_0)}{m_n(\omega_n^2 - \omega^2 + j\omega_n^2\eta)} \tag{2.4.76}$$

2.4.6 General application of Green's functions

Now that the acoustic and structural Green's functions of interest have been derived, we will show how they can be used to find the response of a structure or acoustic medium to point or distributed excitation forces. Only general results will be presented here; application to specific cases is left until later chapters.

To avoid confusion in the following discussion, the structural Green's function will be denoted with a subscript s, so we have $G_s(x,x_0,\omega)$.

2.4.6.1 Excitation of a structure by point forces

The structural displacement response $w(x)$ at location x on the structure and frequency ω to N point forces $F_i(x_i,\omega)$, $i = 1, N$ at locations x_i on the surface of a structure is given by

$$w(x,\omega) = \sum_{i=1}^{N} G_s(x,x_i,\omega)\,F(x_i,\omega) \tag{2.4.77}$$

In (2.4.77), the units of w are metres, the units of F are newtons and the units of G are metres per newton. Thus, in this case, the Green's function is the displacement of the structure at x due to a unit point force at x_i.

2.4.6.2 Excitation of a structure by a distributed force

For a distributed force such as an incident acoustic field, the displacement response of the structure at any location x is given by

$$w(x,\omega) = \int\!\!\int_S p(x_0,\omega)\,G_s(x,x_0,\omega)\,\mathrm{d}x_0 \tag{2.4.78}$$

where the integration is over the area of the source.

If the structure were a panel or shell subjected to a differential pressure $(p_0(\boldsymbol{x}_0, \omega) - p_i(\boldsymbol{x}_0, \omega))$ due to different acoustic pressures on the outside and inside surfaces, then (2.4.78) may be written as:

$$w(\boldsymbol{x}, \omega) = \int_S\!\!\int \left[p_0(\boldsymbol{x}_0, \omega) - p_i(\boldsymbol{x}_0, \omega)\right] G_s(\boldsymbol{x}, \boldsymbol{x}_0, \omega) \, \mathrm{d}\boldsymbol{x}_0 \tag{2.4.79}$$

where the units of the distributed force p are Nm^{-2}. Note that the direction of positive pressure is the same as that of positive structural displacement.

If both point excitation and distributed excitation exist simultaneously, then (2.4.77) and (2.4.78) may be added together to give the total structural response:

$$w(\boldsymbol{x}, \omega) = \int_S\!\!\int p(\boldsymbol{x}_0, \omega) \, G_s(\boldsymbol{x}, \boldsymbol{x}_0, \omega) \, \mathrm{d}\boldsymbol{x}_0 + \sum_{i=1}^{N} F(\boldsymbol{x}_i, \omega) \, G_s(\boldsymbol{x}, \boldsymbol{x}_i, \omega) \tag{2.4.80}$$

This equation is based on the assumption that the structural response is not affected by any sound field which it radiates and that excitation forces are normal to the structure.

2.4.6.3 Excitation of an acoustic medium by a number of point acoustic sources

When the source of excitation is acoustic, the acoustic pressure at any point $\boldsymbol{r}$ in the acoustic medium due to N point sources with a volume velocity of $\omega w(\boldsymbol{r}_i, \omega) \, \mathrm{d}S_i$ is given by

$$p(\boldsymbol{r}, \omega) = -\omega^2 \rho_0 \sum_{i=1}^{N} G(\boldsymbol{r}, \boldsymbol{r}_i, \omega) w(\boldsymbol{r}_i, \omega) \, \mathrm{d}S_i \tag{2.4.81}$$

where $\mathrm{d}S_i$ is the surface area of the ith point source boundary. For interest, the preceding equation may be compared with (2.4.9).

If each of the acoustic sources are distributed rather than point sources then each term in the sum would become a triple integral over the volume of each source, or if the sources were distributed over a boundary surface, then the integral would be a double integral over the boundary surface.

2.4.6.4 Excitation of an acoustic medium by a vibrating structure

When an acoustic medium is excited by a vibrating structure with displacement distribution $w(\boldsymbol{x}, \omega)$, the acoustic pressure at any point $\boldsymbol{r}$ in the acoustic medium is given by:

$$p(\boldsymbol{r}, \omega) = j\omega^2 \rho_0 \int_S\!\!\int w(\boldsymbol{x}, \omega) \, G(\boldsymbol{r}, \boldsymbol{x}, \omega) \, \mathrm{d}\boldsymbol{x} \tag{2.4.82}$$

In this case, the units of G are L^{-1}. Note that the points x lie on the surface S of the structure. For radiation from a plane surface surrounded by a large baffle, this equation reduces to the well known Rayleigh integral of (2.4.24), where $u_n(x, \omega) = j\omega w(x,\omega)$. The complex displacement $w(\mathbf{x}, \omega)$ at frequency ω may be determined by summing the contributions (amplitude and phase) due to each structural mode at location x.

It can be seen from (2.4.82) that the acoustical Green's function relates the acoustic pressure at some point r in the acoustic medium to the volume velocity of the source. Physically, this means that the total acoustic pressure at a point in space is determined by summing the complex contributions (amplitude and phase) from all points on the radiating surface. Although the vibration modes describing the motion of the surface are orthogonal in terms of structural vibration (see Chapter 4), they are not orthogonal on terms of describing their individual contributions to the radiated sound field. This means that the radiated sound pressure squared or radiated sound power cannot be calculated simply by adding together all of the contributions from each mode. This is because when the quantity $w(x, \omega)$ in (2.4.82) is squared to allow the pressure squared at $\boldsymbol{r}$ to be calculated, the result is made up of products of mode shapes squared as well as products of each mode shape with all of the others, the latter being referred to as cross-coupling terms. If the space averaged surface vibration amplitude squared were of interest, then the cross-coupling terms would integrate to zero as the modes are orthogonal. However, the cross-coupling term contributions to the squared radiated pressure field do not, in general, go to zero (except for the case of radiation from a uniform spherical shell) when integrated over an imaginary surface in space to give the radiated sound power because they are multiplied by the acoustical Green's function prior to integration. This is an extremely important concept from the viewpoint of active control of sound radiated by structures, as it illustrates that attempting to control one or more structural modes individually may not necessarily lead to a reduction in overall radiated sound power, even if the controlled modes would be the most efficient radiators if present in isolation. This concept and the associated design of shaped vibration sensors to sense structural sound radiation is discussed in detail in Chapter 8.

The idea that modal sound powers cannot be added to give the total power may be easier to understand if one imagines that the sound radiation efficiency of a surface is a function of the overall velocity distribution over it, and is independent of whether the velocity distribution is described in terms of modes or individual amplitudes and phases as a function of surface location. Generally, the more complicated the surface vibration pattern (or the greater the number of in-phase and out-of-phase areas), the less efficient will be the sound radiation. This is because adjacent areas of the surface which are out of phase effectively cancel the sound radiation from one another (provided that the separation of their mid-points is much less than a wavelength of sound in the adjacent acoustic medium). This results in a much less efficiently radiating surface at low frequencies, although at high frequencies where the separation between the mid-points of adjacent areas is much greater than a wavelength of sound in the acoustic medium, there will be no noticeable decrease in efficiency. From the active control viewpoint, this observation alone tells us qualitatively that many more control sources will be necessary to control high frequency radiation, as the areas of constant phase on the surface will need to be much smaller than necessary for control of low frequency radiation.

Another important concept involves the difference in sound fields radiated by a structure or panels excited by an incident acoustic wave and one excited by a mechanical localized force. In the former case the structure will be forced to respond in modes which are characterized by bending waves having wavelengths equal to the trace wavelengths of the incident acoustic field. Thus, at excitation frequencies below the structure critical frequency, the modes which are excited will not be resonant because the structural wavelength of the resonant modes will always be smaller than the wavelength in the acoustic medium. Thus, lower order modes will be excited at frequencies above their resonance frequencies. As these lower order modes are more efficient than the higher order modes which would have been resonant at the excitation frequencies, the radiated sound will be higher than it would be for a resonantly excited structure having the same mean square velocity levels at the same excitation frequencies. As excitation of a structure by a mechanical force results in resonant structural response, then it can be concluded that sound radiation from an acoustically excited structure will be greater than that radiated by a structure excited mechanically to the same vibration level (McGary, 1988). A useful item of information which follows from this conclusion is that structural damping will only be effective for controlling mechanically excited structures because it is only the resonant structural response which is significantly influenced by damping.

2.4.7 Structural sound radiation and wavenumber transforms

While on the topic of sound radiation, it is of interest to examine another technique for characterizing it. This is a method known as wavenumber transforms. This technique is useful because it provides some insight into the physical mechanisms involved in the active control of structural sound radiation which are discussed in depth in Chapter 8.

A wavenumber transform essentially transforms a quantity expressed as a function of spatial coordinates (for example, surface vibration velocity) into a quantity expressed as a function of wavenumber variables; the inverse transform does vice versa. It is essentially the same Fourier transform operation used in transforming a signal from the time domain to the frequency domain as will be discussed in Chapter 3. Thus, just as we obtain a discrete number of frequency bins in the frequency domain, we obtain a discrete number of wavenumber bins in the wavenumber domain. Remember that the wavenumber k is defined as ω/c_0, where c_0 is the speed of sound in the structure or acoustic medium of interest.

The wavenumber transform, as indicated by its name, essentially describes the response of a structure in terms of waves (rather than in terms of modes, as was done earlier in this section) and the acoustic radiation in terms of waves coupling with the acoustic medium rather than modes coupling with it. An important property of the wavenumber transform is that each of the wavenumbers so determined correspond to a particular angle of sound radiation which happens to coincide with the matching of the trace wavelength of the acoustic field with the wavelength of the structural vibration. The trace wavelength of the acoustic field radiated from a plane surface at some angle is defined as the wavelength that it would effectively 'project' on the surface as shown in Fig. 2.28.

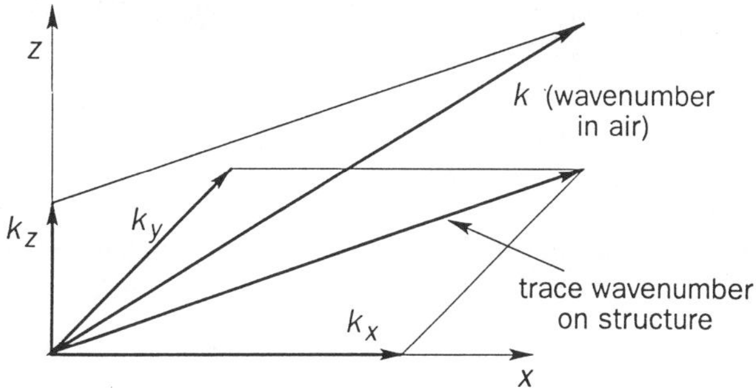

Fig. 2.28 Relationship between wavevectors, k, k_x, k_y and k_z for a radiating plane surface located in the $x-y$ plane.

Only wavenumbers corresponding to structural wavelengths greater than the wavelength in the acoustic medium surrounding the structure will radiate sound. These wavenumbers are referred to as 'supersonic', because they are characterized by a higher phase speed in the structure than the speed of sound in the acoustic medium at the same frequency. Except at structural boundaries and discontinuities, it is not possible for structural waves with wavelengths shorter than the corresponding waves in the acoustic medium to radiate sound because there is no angle of incidence at which the acoustic wavelength will be equal to the trace wavelength on the structure.

Thus, the purpose of using wavenumber transforms is to describe the structural vibration in terms of waves of different wavenumbers, which is really an alternative to describing it in terms of modes. The main advantage in using this method is that it is easy to identify those wavenumbers which are responsible for sound radiation, and then perhaps it may be possible to design shaped vibration sensors to detect only these wavenumbers (Fuller and Burdisso, 1991). Another advantage of wavenumber transforms which will be discussed in Chapter 8 is that they enable the acoustic radiation integral expression to be transformed to a partial differential equation which is usually more easily solved to obtain the structural radiation efficiency, for example.

The basic description of wavenumber transforms applied to waves in a general sense as well as to flexural waves on structures radiating sound has been discussed in detail by Fahy (1985). Here, a brief review of wavenumber transforms applied to flexural waves in structures will be given with the aim of formulating the general comments made above into quantitative equations which can be usefully applied to radiating structures.

As we shall see in Chapter 3, the Fourier transform, which is used to obtain the frequency spectrum of a time domain signal, is given as

$$X(f) = \int_{-\infty}^{\infty}\int_{-\infty}^{\infty} x(t) e^{-j2\pi ft} \, dt \tag{2.4.83}$$

where $x(t)$ is a time varying signal and f is the frequency in hertz of the Fourier component of the signal.

The two-dimensional spatial equivalent of this equation for a plane vibrating surface is:

$$U(k_x, k_y) = \int_{-\infty}^{\infty}\int_{-\infty}^{\infty} u(x, y) \, e^{-j(k_x x + k_y y)} \, dx \, dy \tag{2.4.84}$$

where $u(x, y)$ is the surface normal velocity distribution, and k_x, k_y are flexural bending wavenumbers on the surface in the x and y directions respectively, so that

$$k_x = \frac{\omega}{c_{bx}}, \qquad k_y = \omega / c_{by} \tag{2.4.85a,b}$$

where $\omega = 2\pi f$. The inverse transform is expressed in a similar way to the inverse Fourier transform discussed in Chapter 3 and may be written as follows:

$$u(x, y) = \frac{1}{(2\pi)^2} \int_{-\infty}^{\infty}\int_{-\infty}^{\infty} U(k_x, k_y) e^{j(k_x x + k_y y)} \, dk_x \, dk_y \tag{2.4.86}$$

The vibration velocity field $u(x, y)$ on the plane surface can be thought of as made up of an infinite number of sinusoidal travelling waves, each of which is described by

$$u_k(x, y) = U(k_x, k_y) \, e^{j(k_x x + k_y y)} \tag{2.4.87}$$

in a similar way to the transient time signal $x(t)$ of (2.4.83) being thought of as consisting of an infinite number of pure tones.

Although the surface velocity distribution $u(x, y)$ has been selected as the variable of interest because it is directly related to sound power radiation, the transform equations are equally valid for surface displacement, and indeed also for a two-dimensional acoustic wave in air with acoustic pressure used instead of surface velocity as the transform variable.

For a plane surface radiating sound into the surrounding medium, the normal velocity of the fluid at the surface is equal to the normal velocity of the surface. Thus, following a similar line of reasoning to that used in Section 2.1, the acoustic pressure gradient normal to the surface at the surface is related to the surface normal velocity by

$$j\omega\rho u_z(x, y) = -\frac{\partial p(x, y, 0)}{\partial z} \tag{2.4.88}$$

where z represents the axis normal to the surface lying in the $x-y$ plane. Equation (2.4.88) applies to either instantaneous values or amplitudes of pressure and velocity, provided that the same descriptor is used for each at any one time. The boundary condition of (2.4.88) can also be expressed in terms of the transformed variables P and U, as follows:

$$j\omega\rho U_z(k_x, k_y) = -.\frac{\partial P(k_x, k_y, z)}{\partial z}\Big|_{z=0} \tag{2.4.89}$$

The acoustic pressure field must also satisfy the transformed Helmholtz (wave) equation. That is,

$$\int_{-\infty}^{\infty}\int_{-\infty}^{\infty}\left[\frac{\partial^2}{\partial x^2} + \frac{\partial^2}{\partial y^2} + \frac{\partial^2}{\partial z^2} + k^2\right]p(x, y, z)\mathrm{e}^{-j(k_x x + k_y y)}\,\mathrm{d}x\,\mathrm{d}y = 0 \tag{2.4.90}$$

which can be written in terms of the pressure transform as (Junger and Feit, 1986)

$$\left[k^2 - k_x^2 - k_y^2 + \frac{\partial^2}{\partial z^2}\right]P(k_x, k_y, z) = 0 \tag{2.4.91}$$

It is clear that a solution to (2.4.91) is

$$P(k_x, k_y, z) = A\mathrm{e}^{-jk_z z} \tag{2.4.92}$$

where

$$k_z^2 = k^2 - k_x^2 - k_y^2 \tag{2.4.93}$$

Another solution would be the same as (2.4.92) with a positive exponent, but this would imply waves converging on the radiating surface from infinity, thus not satisfying the Sommerfeld radiation condition (Junger and Feit, 1986), and so it is not allowed.

Substitution of (2.4.92) into (2.4.89) gives

$$A = \frac{\omega\rho U_z(k_x, k_y)}{k_z} \tag{2.4.94}$$

Thus, substituting (2.4.94) into (2.4.92) we obtain

$$P(k_x, k_y, z) = \frac{\omega\rho U_z(k_x, k_y)\,\mathrm{e}^{-jk_z z}}{k_z} \tag{2.4.95}$$

Taking the inverse transform gives an expression for the sound pressure at any location in the near or far field of the vibrating surface as follows:

$$p(x, y, z) = \frac{\omega\rho}{(2\pi)^2}\int_{-\infty}^{\infty}\int_{-\infty}^{\infty}\frac{U_z(k_x, k_y)\mathrm{e}^{j(k_x x + k_y y - k_z z)}}{k_z}\,\mathrm{d}k_x\,\mathrm{d}k_y \tag{2.4.96}$$

This type of integral is almost always analytically intractable in the near field of vibrating surfaces, although it is generally possible to find an analytical solution for the far field sound pressure. However, in most cases (both near and far field), it is usually the simplest to use fast Fourier transform techniques to evaluate the integral.

It is of interest to examine the physical significance of the wavenumber, k_z, or wavevector as it is sometimes called (given its vector nature). We should hasten to point out that although wavenumbers may be thought of as vector quantities, the related quantity, wavelength, is always a scalar. The subscript z on k_z indicates that the wavevector quantity k_z represents the component of the wavevector in the z-direction. The actual radiated wave characterized by k does not necessarily radiate normally to the surface, and in fact it hardly ever does. The wavevectors k_x and k_y on the structure correspond to the x and y components of the acoustic wavevector k, which explains the physical significance of (2.4.93). Thus each discrete wavevector pair (k_x, k_y) corresponds to a wave radiating at a particular angle from a vibrating surface. On any particular surface where more than one vibration mode is excited, there will always be more than one direction in which waves will radiate from the surface.

From (2.4.93), it can be seen that if $k^2 < k_x^2 + k_y^2$, the wavevector k_z will be imaginary and will correspond to a wave which decays exponentially with distance from the surface and does not contribute to far field sound radiation. As wavenumber is inversely proportional to wave phase speed, this condition corresponds to waves in the structure with speeds less than the speed of the wave in the acoustic medium, and are thus referred to as subsonic structural waves.

The condition $k^2 > k_x^2 + k_y^2$, corresponds to supersonic structural waves which radiate far field sound well, as there is always some angle of radiation for which the trace wavelength on the radiating surface of the acoustic wave matches the structural wavelength (see Fig. 2.28 where the relationship between k, k_x, k_y and k_z is illustrated).

Wavenumber transforms are also useful for allowing the radiated acoustic power to be described in terms of quantities which can be measured on the surface of the radiating structure. In terms of applying active noise control, this means that correct measurement of these quantities will eliminate the need for acoustic error sensors (microphones) in an adaptive active noise control system designed to minimize radiated sound power.

The sound power radiated to the far field by a harmonically vibrating surface with a complex normal surface velocity distribution of $\boldsymbol{u}_z(x, y)$ is given by (see Section 2.5)

$$W = \frac{1}{2}\mathrm{Re}\left\{ \int\!\!\int_S p(x, y, 0)\, u_z^*(x, y)\, \mathrm{d}x\, \mathrm{d}y \right\} \tag{2.4.97}$$

where $p(x, y, 0)$ is the complex acoustic pressure in the fluid adjacent to the vibrating surface.

Following an argument similar to that used by Fahy (1985) for the one-dimensional problem, it can be shown that the equivalent expression in terms of transformed pressure and velocity is given by

$$W = \frac{1}{8\pi^2}\,\mathrm{Re}\left\{\int_{-\infty}^{\infty}\int_{-\infty}^{\infty} P(k_x, k_y)\, U_z^*(k_x, k_y)\, \mathrm{d}k_x\, \mathrm{d}k_y\right\} \tag{2.4.98}$$

Setting (2.4.97) and (2.4.98) equal is really a way of expressing Parseval's theorem in two-dimensional form (see Jenkins and Watts, 1968, for the equivalent one-dimensional expression for Fourier analysis).

Substituting (2.4.95) for P and then (2.4.93) for k_z into (2.4.98) gives

$$W = \frac{\omega\rho}{8\pi^2}\,\mathrm{Re}\left\{\int_{-\infty}^{\infty}\int_{-\infty}^{\infty} \frac{|\,U(k_x, k_y)\,|^2}{\sqrt{k^2 - k_x^2 - k_y^2}}\, \mathrm{d}k_x\, \mathrm{d}k_y\right\} \tag{2.4.99}$$

Only wavenumber components which satisfy $k \geq \sqrt{k_x^2 + k_y^2}$ contribute to the real part of (2.4.99); thus, the equation can be rewritten as

$$W = \frac{\omega\rho}{8\pi^2}\,\mathrm{Re}\left\{\iint_{k_x^2 + k_y^2 \leq k^2} \frac{|\,U(k_x, k_y)\,|^2}{\sqrt{k^2 - k_x^2 - k_y^2}}\, \mathrm{d}k_x\, \mathrm{d}k_y\right\} \tag{2.4.100}$$

The sound power can also be written in terms of the wavenumber transform of the acoustic pressure immediately adjacent to the radiating surface as

$$W = \frac{\omega\rho_0}{8\pi^2}\,\mathrm{Re}\left\{\iint_{k_x^2 + k_y^2 \leq k^2} |\,P(k_x, k_y, 0)\,|^2 \sqrt{k^2 - k_x^2 - k_y^2}\, \mathrm{d}k_x\, \mathrm{d}k_y\right\} \tag{2.4.101}$$

Using the principles of acoustic holography (Veronesi and Maynard, 1987), the transform of the surface acoustic pressure (or the pressure at any other plane away from the surface) can be derived from the transform of acoustic pressure measurements taken at an array of points on a plane parallel to the radiating surface and a distance z_m from it as follows:

$$P(k_x, k_y, z) = P(k_x, k_y, z_m)\, \mathrm{e}^{-jk_z(z - z_m)} \tag{2.4.102}$$

where $z = 0$ if the pressure transform on the surface is desired.

Before leaving the topic of sound radiation from vibrating surfaces, it is of interest to point out that structures excited by fluid-borne acoustic disturbances generally radiate more efficiently than do structures which are excited mechanically (McGary, 1988). This is, because structures excited acoustically are forced to vibrate in modes which are characterized by structural wavelengths equal to the trace acoustic wavelength of the incident sound field. These modes are generally of higher order than the modes which would be resonant at the excitation frequency and thus have higher radiation efficiencies. On the other hand, structures which are excited mechanically vibrate in modes which are resonant at the excitation frequency and as these modes are generally of lower order than those forced by acoustic waves with the same frequency content, the corresponding structural radiation efficiency is less.

2.4.8 Effect of fluid loading on structural sound radiation

The formulation discussed thus far is based on the assumption that the source strengths are independent of the response of the structure or acoustic medium. This assumption is often referred to as the 'uncoupled' assumption. For the case of a structure radiating into an acoustic medium this implies that the radiation field generated by the structure does not contribute significantly to the oscillatory forces driving the structure, and hence to the surface velocity distribution of the structure. For the case of external forces driving a structure, the 'uncoupled' assumption implies that the response of the structure does not influence the driving forces.

For structures radiating into dense fluids such as water, the effect of the radiated sound field on the structural response cannot be ignored. In this case the pressure field generated by the structure reacts on the structure and changes its response. Thus, the acoustic surface pressure generated by the structure contributes to the dynamic forces driving it.

Taking (2.4.80) for the total structural response due to acoustic and vibration sources, and substituting (2.4.82) for $p(\boldsymbol{x}_0,\omega)$ where $\boldsymbol{r} = \boldsymbol{x}_0$ (location $\boldsymbol{r}$ in the acoustic medium is adjacent to a structure surface location $\boldsymbol{x}$), we obtain

$$w(\boldsymbol{x},\omega) = \int_S\!\!\int G_s(\boldsymbol{x},\boldsymbol{x}_0,\omega)\left[j\omega^2\rho_0\int_S\!\!\int w(\boldsymbol{x}'',\omega)\,G(\boldsymbol{x}_0,\boldsymbol{x}'',\omega)\,\mathrm{d}\boldsymbol{x}''\right]\mathrm{d}\boldsymbol{x}_0 \tag{2.4.103}$$

$$+\sum_{i=1}^{N} F(\boldsymbol{x}_i,\omega)\,G_s(\boldsymbol{x},\boldsymbol{x}_i,\omega)$$

Since, $w(\boldsymbol{x}'',\omega)$ is unknown, the response $w(\boldsymbol{x},\omega)$ of the structure is the solution of an integral equation, which can be written as an inhomogeneous Friedholm equation of the second kind (Morse and Feshbach, 1953, p. 949):

$$w(\boldsymbol{x},\omega) = j\omega^2\rho_0\int_S\!\!\int \kappa(\boldsymbol{x},\boldsymbol{x}'',\omega)\,w(\boldsymbol{x}'',\omega)\,\mathrm{d}\boldsymbol{x}'' + f(\boldsymbol{x},\omega) \tag{2.4.104}$$

whose kernel is:

$$\kappa(\boldsymbol{x},\boldsymbol{x}'',\omega) = \int_S\!\!\int G_s(\boldsymbol{x},\boldsymbol{x}_0,\omega)\,G(\boldsymbol{x}_0,\boldsymbol{x}'',\omega)\,\mathrm{d}\boldsymbol{x}_0 \tag{2.4.105}$$

The inhomogeneous term is:

$$f(\boldsymbol{x},\omega) = \sum_{j=1}^{N} F(\boldsymbol{x}_i,\omega)\,G_s(\boldsymbol{x},\boldsymbol{x}_i,\omega) \tag{2.4.106}$$

For a structure excited by an incident sound wave rather than a number of forces, (2.4.106) is replaced by:

$$f(\boldsymbol{x}, \omega) = j\omega^2 \rho_0 \int_S \int p(\boldsymbol{x}', \omega) G_s(\boldsymbol{x}, \boldsymbol{x}', \omega) \mathrm{d}\boldsymbol{x}' \tag{2.4.107}$$

The preceding analysis can be performed in a similar way using wavenumber transforms, as described by Fahy (1985).

2.5 IMPEDANCE AND INTENSITY

Impedance and intensity will now be discussed in detail, as a thorough understanding of their physical meaning and measurement is assumed in the discussion in later chapters.

2.5.1 Acoustic impedances

There are four types of impedance commonly used in acoustics. Each type is directly related to the other three and can be derived from any one of the other three. Impedances are generally complex quantities (characterized by an amplitude and a phase) which are defined as a function of frequency. Thus, the quantities used in the equations to follow are complex amplitudes defined at specific frequencies, allowing the time dependent term $e^{j\omega t}$, to be omitted.

Impedances are generally associated with acoustic sources or acoustic propagation and a number of specific examples will be discussed later on in this section. However, before discussing any specific examples, we will differentiate between the different types of impedances.

2.5.1.1 Specific acoustic impedance, Z_s

The specific acoustic impedance is defined as the ratio of the acoustic pressure to particle velocity, u, in the direction of wave propagation, anywhere in an acoustic medium, including the surface of a noise source. Thus,

$$Z_s = \frac{p}{u} \tag{2.5.1}$$

2.5.1.2 Acoustic impedance, Z_A

The acoustic impedance is particularly useful for describing sound propagation in ducts, and for a plane wave in a duct of cross-sectional area S, it is defined as the ratio of the acoustic pressure to the volume velocity at a duct cross section. Thus

$$Z_A = \frac{p}{uS} = \frac{Z_s}{S} \tag{2.5.2a,b}$$

2.5.1.3 Mechanical impedance, Z_m

The mechanical impedance is defined as the ratio of the force F acting on a surface or system to the velocity of the system at the point of application of the force. If the system is a vibrating surface, then the surface velocity is equal to the acoustic particle velocity at an adjacent point in the surrounding fluid. If the vibrating surface of area S is subject to a uniform acoustic pressure p and is vibrating with a uniform normal velocity u, then the mechanical impedance is given by

$$Z_m = \frac{F}{u} = \frac{pS}{u} = Z_s S = Z_A S^2 \tag{2.5.3a,b,c,d}$$

For a more generally vibrating surface, the mechanical impedance is given by

$$Z_m = \frac{1}{<u^2>} \int_S \int p(\boldsymbol{x})\, \boldsymbol{u}^*(\boldsymbol{x})\,\mathrm{d}\boldsymbol{x} \tag{2.5.4}$$

where p is the acoustic pressure, $\boldsymbol{u}^*$ is the complex conjugate of the acoustic particle velocity adjacent to and normal to the vibrating surface at $\boldsymbol{x}$, $<u^2>$ is the mean square velocity averaged over the vibrating surface S and $\boldsymbol{x}$ is a vector location of a point on the surface.

Note that in (2.5.1) to (2.5.4) all quantities may be instantaneous, peak or root mean square, provided that only one type is used at any one time.

2.5.1.4 Radiation impedance and radiation efficiency

The radiation impedance of a surface or acoustic source is a measure of the reaction of the acoustic medium against the motion of the surface or source. It is really a special case of a mechanical impedance applied to a source of sound. Thus, (2.5.3) and (2.5.4) may also be used to define the radiation impedance.

When the radiation impedance is normalized by the characteristic impedance ($\rho_0 c_0$) of the acoustic medium and the surface area S of the noise source, the resulting quantity is known as the radiation efficiency σ (sometimes called radiation ratio, as it can sometimes exceed unity). Physically, the radiation efficiency is the ratio of the sound power radiated by a particular sound source of surface area S, to the power which would be carried in one direction by a plane wave of area S.

For a monopole source (pulsating sphere) of radius a and surface area S, the radiation efficiency σ_m is given by

$$\sigma_m = \frac{pS}{uS\rho_0 c_0} = \frac{p}{u\rho_0 c_0} = \frac{Z_s}{\rho_0 c_0} \tag{2.5.5a,b,c}$$

It can be shown (Bies and Hansen, 1996) that

$$\sigma_m = \frac{jka}{1+jka} = \cos\beta\, e^{j\beta} \tag{2.5.6a,b}$$

where

$$\beta = \tan^{-1}(1/ka) \tag{2.5.7}$$

and where $k = 2\pi/\lambda$ = wave number of the radiated sound.

The real and imaginary parts of σ_m are shown in Fig. 2.29 as a function of kr where

$$\sigma_M = \sigma_R + j\sigma_I \tag{2.5.8}$$

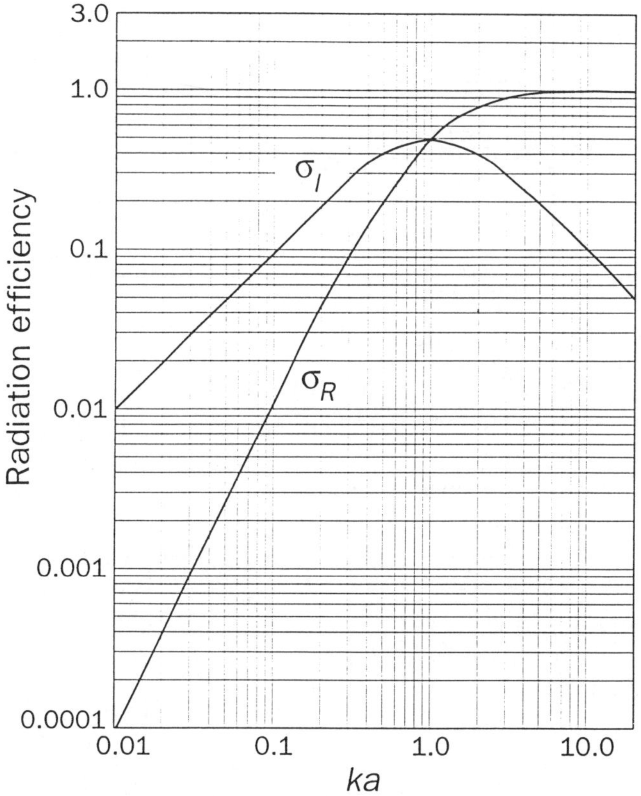

Fig 2.29 Real and imaginary parts of the normalized specific acoustic impedance $Z_s/\rho c$ of the air load on a pulsating sphere of radius a located in free space. Frequency is plotted on a normalized scale where $ka = 2\pi fa/c = 2\pi a/\lambda$. Note also that the ordinate is equal to $Z_M\rho cS$, where Z_M is the mechanical impedance; and to $Z_AS/\rho c$, where Z_A is the acoustic impedance. The quantity S is the area for which the impedance is being determined, and ρc is the characteristic impedance of the medium.

The real part of the radiation efficiency is associated with the radiation of acoustic energy to the far field, while the imaginary part is associated with energy storage in the near field of the vibrating surface (see Bies and Hansen, 1996, Chapter 6).

The real and imaginary parts of the radiation efficiency of a vibrating structure can be determined by measuring the space averaged complex acoustical intensity in the acoustic medium at the surface of the structure. Thus,

$$\sigma = \frac{\langle I \rangle}{\langle u^2 \rangle \rho_0 c_0} \tag{2.5.9}$$

where $\langle u^2 \rangle$ is the space and time averaged normal velocity of the radiating surface and $\langle I \rangle$ is the space averaged complex acoustic intensity in a direction normal to the surface.

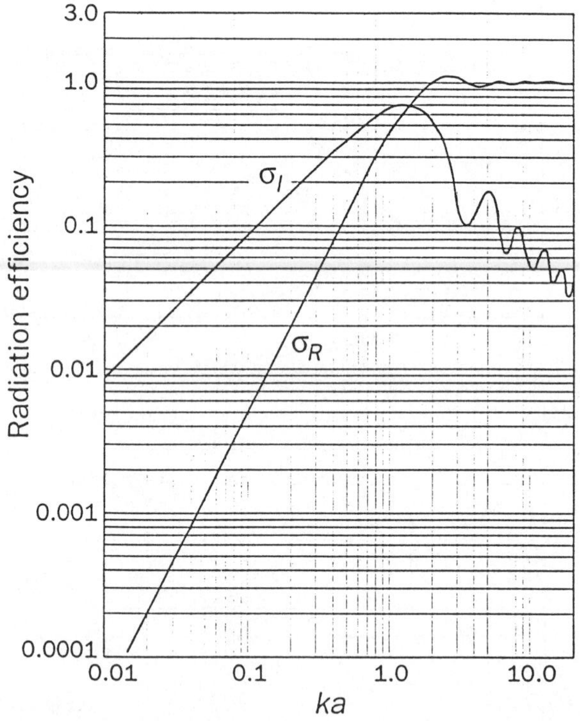

Fig. 2.30 Real and imaginary parts of the normalized mechanical impedance $(Z_M/\pi a^2 \rho c)$ of the air load on one side of a plane piston of radius a mounted in an infinite flat baffle. Frequency is plotted on a normalized scale, where $ka = 2\pi fa/c = 2\pi a/\lambda$. Note also that the ordinate is equal to $Z_A \pi a^2/\rho c$, where Z_A is the acoustic impedance.

If only the real part of the radiation efficiency (relevant to far field sound radiation) is needed, then the radiation efficiency can be determined by measuring the sound power W radiated by the surface or structure and using the following relation:

$$\sigma = \frac{W}{\langle u^2 \rangle \, S\rho c} \tag{2.5.10}$$

where S is the area of the radiating surface.

Another useful example is that of a plane circular piston, that is, a uniformly vibrating surface radiating into free space. Three cases will be considered: radiation from one side of the piston mounted in an infinite rigid baffle; radiation from the piston mounted in the end of a long tube; and radiation from both sides of a piston in free space.

For the piston mounted in the infinite rigid baffle, the expression for the complex radiation efficiency is (Kinsler *et al.*, 1982):

$$\sigma_B = \sigma_R + j\sigma_I = R(2ka) + jX(2ka) \tag{2.5.11a,b}$$

where

$$R(x) = \frac{x^2}{2\times 4} - \frac{x^4}{2\times 4^2\times 6} + \frac{x^6}{2\times 4^2\times 6^2\times 8} - \ldots . \tag{2.5.12}$$

and

$$X(x) = \frac{4}{\pi}\left[\frac{x}{3} - \frac{x^3}{3^2\times 5} + \frac{x^5}{3^2\times 5^2\times 7} - \ldots .\right] \tag{2.5.13}$$

where a is the radius of the circular piston and $k = 2\pi/\lambda =$ wavenumber. The real and imaginary parts of σ_B are plotted as a function of ka in Fig. 2.30.

The case of a plane circular piston in the end of a pipe has been analysed in detail by Levine and Schwinger (1948); the results are shown in Fig. 2.31.

The case of a piston radiating from both sides into free space was analysed in detail by Wiener (1951) and the results are given in Fig. 2.32. Note that at high frequencies, the real part of the radiation efficiency asymptotes to 2 rather than unity, as sound is radiated from both sides.

Another example which will now be considered is the radiation efficiency of a plane circular plate vibrating in one of its resonant modes (not necessarily at the resonance frequency), mounted in the plane of an infinite rigid baffle and radiating into free space. Three specific cases will be considered: sound radiation from a simply supported circular plate vibrating in one of its first seven low order modes; sound radiation for a clamped edge circular plate; and sound radiation from a simply supported rectangular plate, vibrating in its first few low order modes. For the latter case, only the real part of the radiation efficiency will be given.

The analysis used to derive the results for the circular plates is complicated, making use of the oblate spheroidal coordinate system, and is discussed in detail elsewhere (Hansen, 1980). Results are given for the clamped edge and simply supported edge circular plates, both real and imaginary components in Figs. 2.33 (a) to (h).

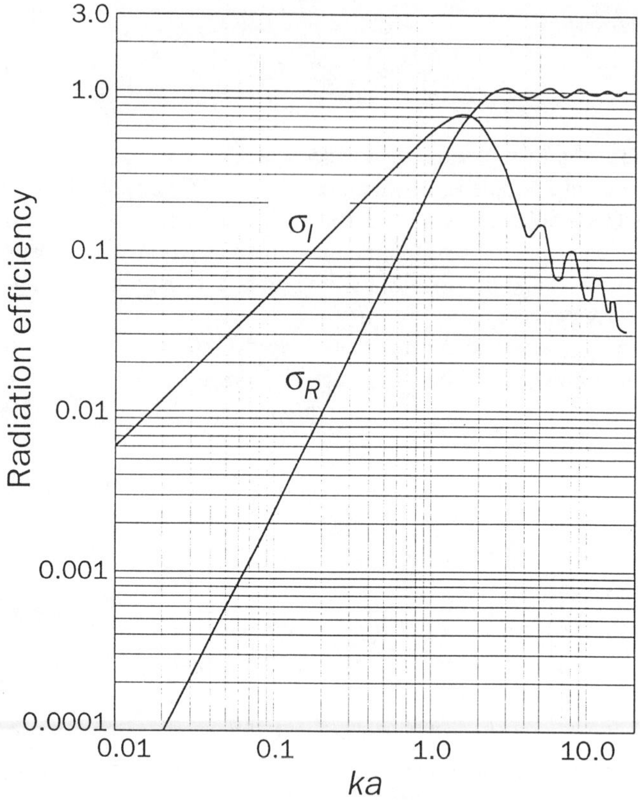

Fig. 2.31 Real and imaginary parts of the normalized mechanical impedance ($Z_M/\pi a^2 \rho c$) of the air load upon one side of a plane piston of radius a mounted in the end of a long tube. Frequency is plotted on a normalized scale, where $ka = 2\pi fa/c = 2\pi a/\lambda$. Note also the ordinate is equal to $Z_A \pi a^2/\rho c$, where Z_A is the acoustic impedance.

The concept of modes of vibration as a means of representing the motion of a vibrating surface is discussed more thoroughly in Chapter 4.

If only the real part of the radiation efficiency (associated with the radiation of energy to the far field) is of interest, then the following expression may be used to calculate just the real part:

$$\sigma = \frac{1}{<u^2> S\rho_0 c_0} \int_0^{\pi} \int_0^{2\pi} \frac{|\bar{p}^2(r)|}{2\rho_0 c_0} r^2 \mathrm{d}\theta \, \mathrm{d}\phi \tag{2.5.14}$$

where $|\bar{p}(r)|^2$ is the square of the modulus of the acoustic pressure amplitude at some location $\boldsymbol{r} = (r, \theta, \phi)$ in the far field of the vibrating surface and $<u^2>$ is the space and time averaged normal acoustic particle velocity, u, averaged over the surface (equal to the normal surface velocity) given by

$$<u^2> = \frac{1}{S} \int_S \left[\frac{1}{T} \int_0^T u^2(x, t) \, \mathrm{d}t \right] \mathrm{d}x \tag{2.5.15}$$

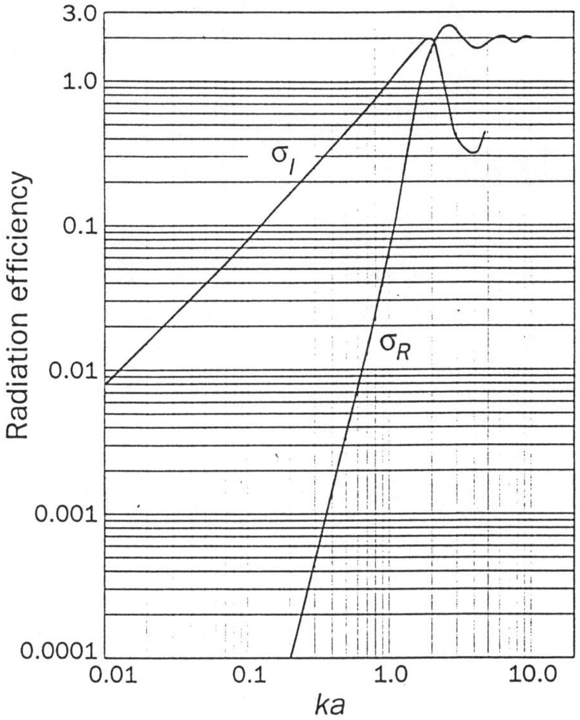

Fig. 2.32 Real and imaginary parts of the normalized mechanical impedance ($Z_M/\pi a^2\rho c$) of the air load upon *both* sides of a plane circular disk of radius a in free space. Frequency is plotted on a normalized scale, where $ka = 2\pi fa/c = 2\pi a/\lambda$. Note also the ordinate is equal to $Z_A\pi a^2/\rho c$, where Z_A is the normalized acoustic impedance.

where S is the area of plane vibrating surface, $\boldsymbol{x}$ is a vector location (x, y) on the surface and T is a suitable time period over which to estimate the mean square velocity of the surface. Note that the integration in (2.5.14) is over a hemisphere at a distance r from the centre of the vibrating surface. Note that (2.5.10) and (2.5.14) can be combined to give an expression for the radiated sound power in terms of the farfield radiated sound pressure.

The acoustic pressure $p(\boldsymbol{r}, t)$ at location $\boldsymbol{r}$ in space, radiated by a plane, infinitely baffled surface vibrating arbitrarily at angular frequency ω may be calculated using the following well known integral formulation first introduced by Lord Rayleigh in 1896 and discussed in Section 2.4:

$$p(\boldsymbol{r}, t) = \frac{j\omega\rho_0}{2\pi} \int_S\int \frac{u(\boldsymbol{x}, t)\,\mathrm{e}^{-jkr}}{r}\,\mathrm{d}\boldsymbol{x} \qquad (2.5.16)$$

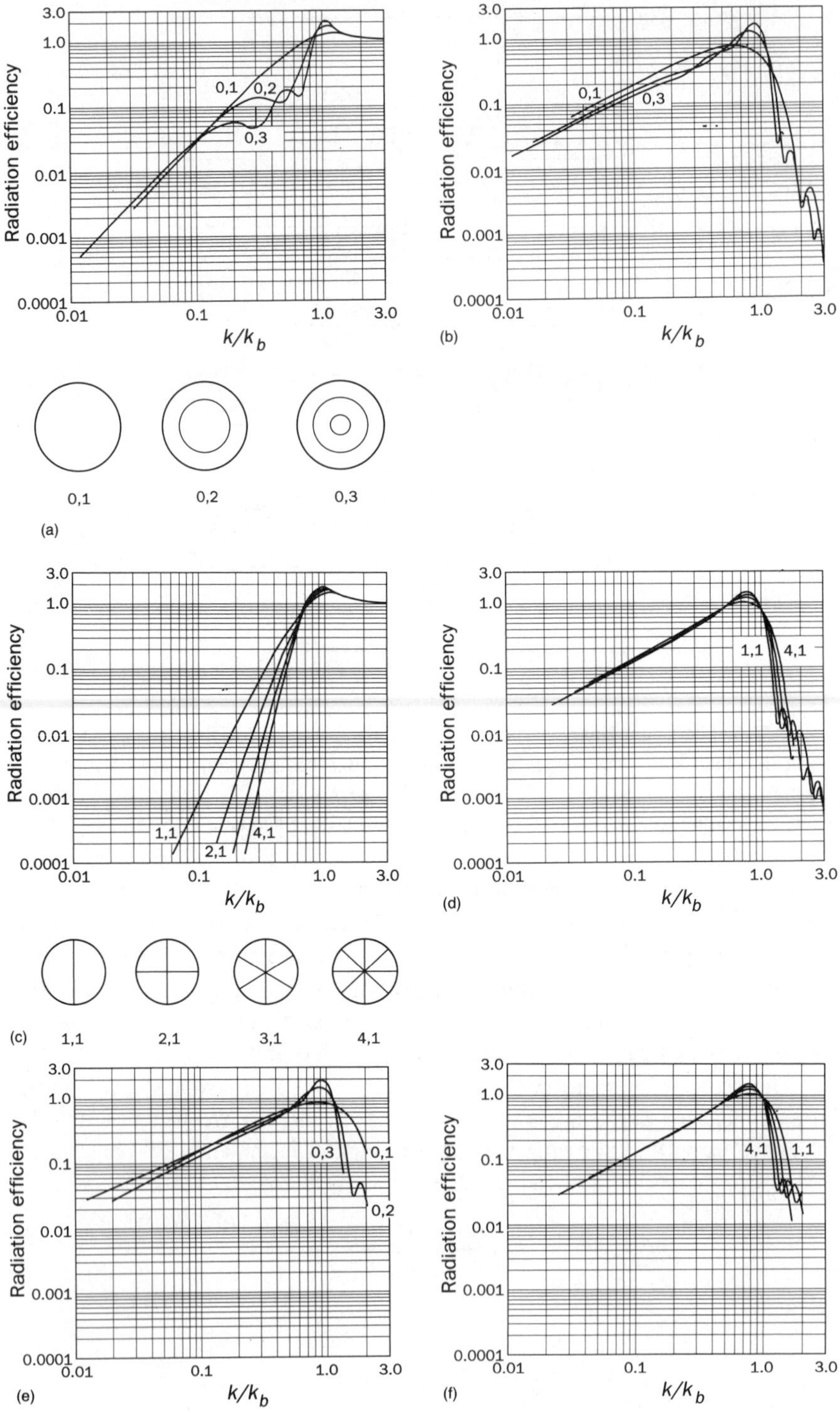

Radiation efficiency
3.0
1.0
0.1
0.01
0.001
0.0001
0.01
0.1
1.0
3.0
k/k_b
0,1
0,2
0,3
(a)
(b)
1,1
2,1
3,1
4,1
(c)
(d)
(e)
(f)

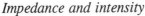

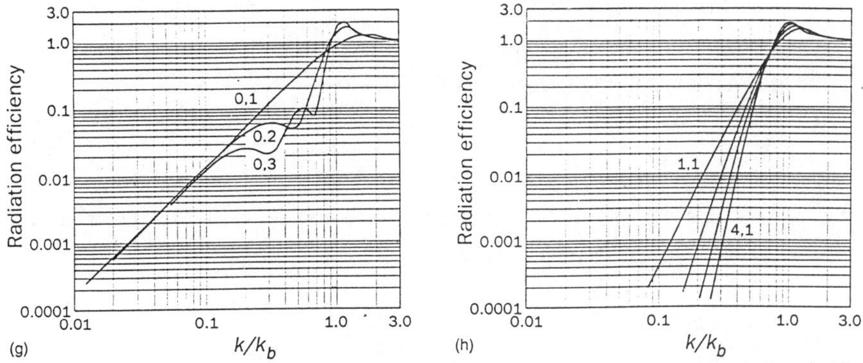

Fig. 2.33 (a) Resistive radiation efficiency for modes with circular nodes and plates with clamped edges calculated using classical plate theory. (b) Reactive radiation efficiency for modes with circular nodes and plates with clamped edges calculated using classical plate theory. (c) Resistive radiation efficiency for modes with diametral nodes and plates with clamped edges calculated using classical plate theory. (d) Reactive radiation efficiency for modes with diametral nodes and plates with clamped edges calculated using classical plate theory. (e) Reactive radiation efficiency for modes with circular nodes and plates with simply supported edges calculated using classical plate theory. (f) Reactive radiation efficiency for modes with diametral nodes and plates with simply supported edges calculated using classical plate theory. (g) Resistive radiation efficiency for modes with circular nodes and plates with simply supported edges calculated using classical plate theory. (h) Resistive radiation efficiency for modes with diametral nodes and plates with simply supported edges calculated using classical plate theory.

where $u(\boldsymbol{x}, t)$ is the normal surface velocity amplitude at location $\boldsymbol{x}$ and r is the distance from the location $\boldsymbol{x}$ on the surface to location $\boldsymbol{r}$ in space and for a fixed point in space r varies with location $\boldsymbol{x}$ on the surface. At first it may seem that an integral expression for calculating sound radiation from an infinitely baffled plane surface for a single frequency is a bit restrictive. In practice, the baffle need be only a few wavelengths in size, provided that radiation from the back of the surface is prevented from interfering with that radiated from the front (for example the wall of an enclosure. In fact, at low frequencies (radiating surface small compared with a wavelength of sound) no baffle is needed at all and very good predictions for the radiated field can be obtained if the 2 in the denominator of (2.5.16) is replaced with 4 to account for radiation into a spherical rather than a hemispherical space, (compare for example Figs. 2.30 and 2.31). Equation (2.5.16), in practice, can also be generalized to apply to a narrow band of noise with a centre frequency of ω. If multiple frequencies or frequency bands are considered, the sound field due to each can be combined by adding pressures squared to give the total sound pressure level.

Given the normal surface velocity amplitude distribution $\bar{u}(\boldsymbol{x})$ over a plane vibrating surface, the previous expressions may be used to calculate the real part of the surface radiation impedance or radiation efficiency. Wallace (1972) presented the analysis for a simply supported rectangular plate where he used the following modal velocity distribution for the m, nth mode (m, n are integers and for the lowest order mode $m = n = 1$). Note that the assumption of light structural damping is implicit in this formulation:

$$\bar{u}_{mn}(\boldsymbol{x}) = \bar{u}_{mn}(x,y) = A_{mn}\sin(m\pi x/L_x)\sin(n\pi y/L_y) \qquad (2.5.17a,b)$$

where A_{mn} is the modal velocity amplitude and L_x and L_y are the dimensions of the plate. Thus we obtain the following result for the farfield radiated sound pressure amplitude at frequency ω at a point (r,θ,ϑ) in space.

$$p(\boldsymbol{r}) = \frac{jA_{mn}k\rho_0 c_0}{2\pi r}\,\mathrm{e}^{-jkr}\,\frac{L_x L_y}{mn\pi^2}\left[\frac{(-1)^m\mathrm{e}^{-j\alpha}-1}{(\alpha/m\pi)^2-1}\right]\left[\frac{(-1)^n\mathrm{e}^{-j\beta}-1}{(\beta/n\pi)^2-1}\right] \qquad (2.5.18)$$

where

$$\alpha = kL_x\sin\theta\cos\vartheta \text{ and } \beta = kL_y\sin\theta\sin\vartheta \qquad (2.5.19a,b)$$

and where r is the distance of the point in space from the corner of the plate where $x = y = 0$.

The following result is then obtained for the radiation efficiency of the mnth mode:

$$\sigma_{mn} = \frac{64k^2L_xL_y}{\pi^6m^2n^2}\int_0^{\pi/2}\int_0^{\pi/2}\left[\frac{\begin{matrix}\cos\\ \sin\end{matrix}\left(\frac{\alpha}{2}\right)\begin{matrix}\cos\\ \sin\end{matrix}\left(\frac{\beta}{2}\right)}{[(\alpha/m\pi)^2-1]\,[(\beta/n\pi)^2-1]}\right]^2 \sin\theta\,\mathrm{d}\theta\,\mathrm{d}\vartheta \qquad (2.5.20)$$

where $\cos(\alpha/2)$ is used when m is odd and $\sin(\alpha/2)$ is used when m is even. Similarly $\cos(\beta/2)$ is used when n is odd and $\sin(\beta/2)$ is used when n is even.

The mode order m, n minus one, refers to the number of nodal lines across the plate in the vertical and horizontal directions respectively. That is, number of vertical nodal lines $= (n-1)$.

Results for the radiation efficiency of some low order vibration modes of a simply supported rectangular plate are given in Figs. 2.34 (a) to (d) (after Wallace, 1972).

When a plate is excited 'off-resonance', several modes are likely to contribute to the overall plate response and to the overall sound radiation. In this case, the far field sound pressure may be calculated in one of two ways as follows:

1. The amplitude and relative phase of the plate response can be calculated or measured at a large number of points on the plate and the resulting far field sound pressure calculated by replacing the integral in (2.5.16) by a sum over all of the calculated displacements (which are all complex, being represented by an amplitude and a phase).
2. The amplitudes and relative phases of each vibration mode on the plate can be determined by calculation (for a known forcing function) or by a modal decomposition using plate response data measured on the plate surface.

Next, the complex sound pressure radiated by each mode to any particular location in space can be calculated and the contributions from all modes added together (taking relative phases into account) to give the total sound pressure.

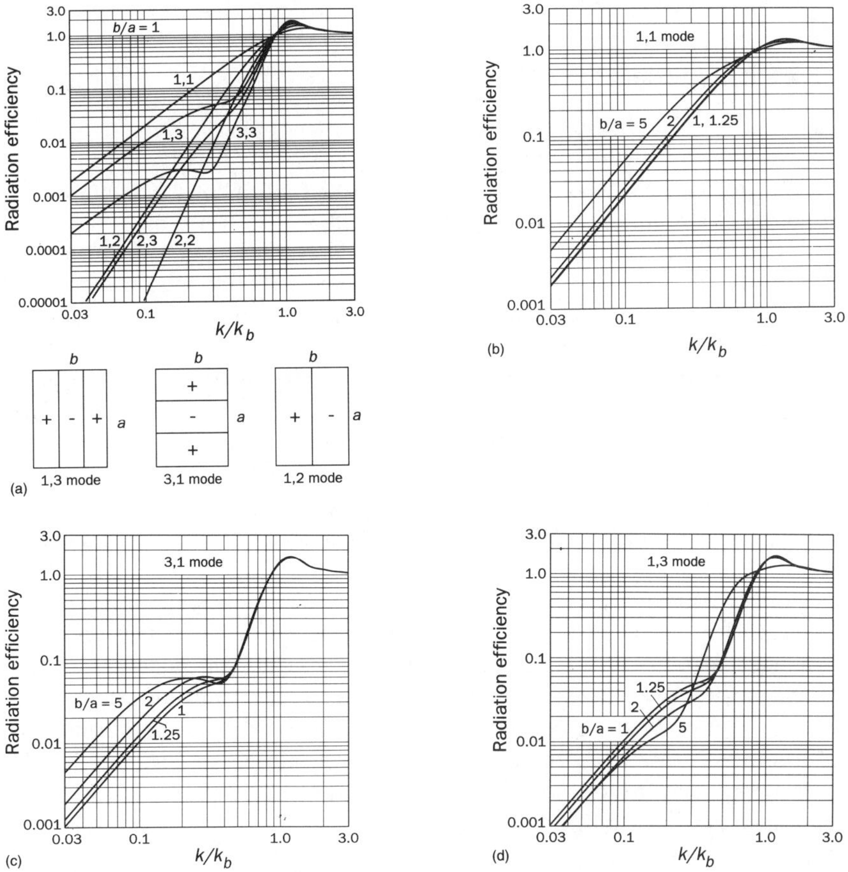

Fig. 2.34 Radiation efficiencies for various modes and dimensional aspect ratios for a simply supported rectangular plate (after Wallace, 1972).

The radiation efficiency can then be calculated as before by integrating the calculated total squared sound pressure over a hemispherical surface surrounding the plate, and then dividing the result by the plate surface area, S, the characteristic impedance, ρc, and the mean square surface velocity of the plate. Note that the relative amplitudes and phases of the modes excited will be dependent on the location and type of exciting force. This will become apparent in Chapter 8 in which active control of sound radiation is discussed.

Sometimes, the sound radiation efficiency is dominated by modes other than those which dominate the structural response, even if the response is close to resonance. This is because some very efficient modes could be excited at a much lower level than one or more modes which are not very efficient radiators. This phenomenon occurs more frequently when the structure is excited with an acoustic wave than when it is excited mechanically.

2.5.2 Structural input impedance

The input impedance (sometimes called the mechanical impedance) of a structure is a quantity which allows the vibrational input power or energy transfer to the structure to be calculated for a defined force or moment acting at a defined point or points. It also allows the energy transfer from one structure to another to be expressed in fairly simple terms, and is a function of excitation frequency and location on the structure. Structural input impedance essentially relates the motion of a structure to a disturbance applied at the same location. The reciprocal of structural input impedance is called the point mobility and is more commonly used in the discipline of structural dynamics. If the structural response is desired at a different location to the applied disturbance, then the related quantities are the transfer impedance or its reciprocal, transfer mobility.

Structural input impedance is an important concept used in the application of active vibration control to reduce vibratory power transmission in structures, as it enables the power input to the structure generated by the control sources to be calculated, and the type and necessary strength of the control source to be quantified. Although the expressions to follow refer to a point force or point moment, they can be applied in practice to forces or moments that act over a small area. The concept of structural input impedance also facilitates the analysis of vibratory power transmission through a complex structure which is an essential part of the design of active systems to control this power transmission. It is also very useful in the analysis of the effectiveness of active vibration isolation systems (as well as passive isolation systems).

There are three types of structural input impedance which are important; force impedance Z_F, moment impedance Z_M and wave impedance Z_W. The first two are defined as follows:

$$Z_F = F/u \tag{2.5.21}$$

$$Z_M = M/\alpha \tag{2.5.22}$$

where F is the excitation force, u is the velocity of the structure at the point of application of the force, M is the excitation moment and α is the angular velocity of the structure at the point of application of the moment. Thus for the concept of structural input impedance to be applied, it is assumed that the force or moment excitation is localized in a region which is small compared to a wavelength of sound. Impedance is usually defined as a function of frequency and is a complex quantity characterized by an amplitude and phase. Thus the quantities

F, u, M and α in the preceding equations are usually complex amplitudes at the frequency of interest, implying a phase shift between F and u, and M and α.

If a force and moment act together on a beam for example, the above results for independent force and moment impedances cannot be simply added together, as there will be coupling between the force and moment response. This arises because the point force will result in a rotation as well as a lateral displacement of the beam section and the moment will result in a displacement as well as a rotation of the beam section. This is discussed in more detail in Section 10.2. Similarly, if multiple forces or moments act at different locations, the impedance matrix will have coupling terms containing transfer impedances from one point to another.

The complex (real and imaginary) power transmission into a structure, generated by a harmonic (single frequency) point excitation force of complex amplitude F, is given by:

$$W = \frac{1}{2} F u_o^* \tag{2.5.23}$$

where the $*$ indicates the complex conjugate and u_o is the complex velocity amplitude of the structure at the point of application of the force F. In terms of impedance the complex power is given by:

$$W = \frac{FF^*}{2Z_F} = \frac{1}{2} u_o u_o^* Z_F \tag{2.5.24a,b}$$

The time averaged propagating part of the complex power is proportional to the product of the exciting force and the in-phase component of the structural velocity at the point of application of the force, and is given by:

$$\mathrm{Re}\{W\} = \frac{|F|^2}{2\mathrm{Re}\{Z_F\}} = \frac{1}{2} |u_o|^2 \mathrm{Re}\{Z_F\} \tag{2.5.25a,b}$$

The part which represents the amplitude of the non-propagating stored energy is proportional to the product of the excitation force and the in-quadrature component of the structural velocity at the point of application of the force, and is given by:

$$\mathrm{Im}\{W\} = \frac{|F|^2}{2\mathrm{Im}\{Z_F\}} = \frac{1}{2} |u_o|^2 \mathrm{Im}\{Z_F\} \tag{2.5.26a,b}$$

Similarly, the power injected by a moment excitation M is:

$$\mathrm{Re}\{W\} = \frac{|M|^2}{2\mathrm{Re}\{Z_M\}} = \frac{1}{2} |\alpha_o|^2 \mathrm{Re}\{Z_M\} \tag{2.5.27a,b}$$

where α_o is the angular velocity of the structure at the point of application of the moment M.

The reciprocal of the force impedance Z_F is referred to as the mobility of a structure and this latter quantity is often referred to in the literature. Force and moment impedances of simple structures can be calculated analytically using the wave equation; however, for more complex structures it is necessary to determine these quantities by measurement, or from the results of a modal analysis (see Chapter 3). Force impedance is usually measured by exciting the structure with a shaker (electrodynamic, hydraulic, piezoelectric, etc.), then measuring the force input (with a piezoelectric crystal) and the acceleration (with an accelerometer) at the point of interest on the structure. Sometimes the force and acceleration measurements are made using an impedance head, which is a single transducer containing two piezoelectric crystals, one for measuring force and one for measuring acceleration. When measuring the force impedance or mobility of a structure, it is important that the shaker is connected to the structure using a ball joint or a length of thin wire so that bending moments are not transmitted to the structure (see, for example, Ewins, 1984). It is also important to mount the force transducer at the structure end of the shaker attachment. When an impedance head is used, the end containing the force crystal should be attached to the structure. When using an impedance head it is also possible for the stiffness of the joint between the force crystal and the acceleration crystal to cause errors in the phase of the acceleration measurement at higher frequencies, so it is generally better to use separate force and acceleration transducers and mount both directly on to the structure (preferably with a screwed mounting stud).

The measurement of the moment impedance is more difficult and requires the use of two shakers exciting the structure at two locations as close together as possible. The phases and amplitudes of the input forces and accelerations at each location are measured and these measurements together with the distance between the excitation points are used to obtain the input moment and the angular acceleration of the structure midway between the two excitation points.

The derivation of theoretical force and moment impedances can be tedious, even for simple structures. However, an example derivation will be given here for the flexural wave point force impedance at the centre of a thin, infinitely long beam. In practice, a beam of finite length may be considered infinitely long if, either there is a large amount of damping at its ends, or if its internal loss factor η is sufficiently large that waves reflected from the ends have a sufficiently diminished amplitude by the time they arrive back at their source that they may be ignored. For a beam of length $2L$, excited in the centre and with negligible damping at the ends, the required value of the internal loss factor (for flexural waves) for the beam to be considered infinite is such that (Fahy, 1985)

$$\eta >> 2/k_b L \tag{2.5.28}$$

where the bending (or flexural) wave number k_b is given by

$$k_b = (\omega^2 m/EI)^{1/4} \tag{2.5.29}$$

ω is the excitation frequency (radians s^{-1}), m is the beam mass/unit length, E is Young's modulus of elasticity, and I is the second moment of area of the beam cross-section about an axis perpendicular to the beam axis and the flexural wave displacement.

2.5.2.1 Force impedance of an infinite beam (flexural waves)

In Section 2.3, the classical wave equation for flexural waves in a beam was found to be

$$EI\frac{\partial^4 w(x,t)}{\partial x^4} + m\frac{\partial^2 w(x,t)}{\partial t^2} = 0 \tag{2.5.30}$$

where w is the flexural wave displacement at axial location x along the beam. Note that this equation is only valid if the bending wavelength $\lambda = 2\pi/k_b$ is much larger than the thickness h of the beam (beam dimension in the direction of the flexural wave displacement). In general (2.5.30) is valid if $\lambda > 6h$. For beams which are wide with respect to their thickness, E must be replaced with $E(1 - \nu^2)$. Even with this adjustment, the beam equation will only hold at frequencies below which bending waves begin to propagate across the width. For a beam excited by a point force, F, at a location $x = a$, (2.5.30) becomes

$$\frac{\partial^4 w(x,t)}{\partial x^4} + m\frac{\partial^2 w(x,t)}{\partial t^2} = F\delta(x-a)\mathrm{e}^{j\omega t} \tag{2.5.31}$$

where $\delta(x-a)$ is the Dirac delta function discussed in detail in Section 2.4.

Equation (2.5.30) is equivalent to (2.5.31) at every point on the beam except where the force is applied. The solution to equation (2.5.30) was given in Section 2.3 as,

$$w(x,t) = \left[A_1\mathrm{e}^{-jk_bx} + A_2\mathrm{e}^{jk_bx} + A_3\mathrm{e}^{-k_bx} + A_4\mathrm{e}^{k_bx}\right]\mathrm{e}^{j\omega t} \tag{2.5.32}$$

Here we will assume that the centre of the beam is at $x = 0$ and the beam extends to infinity in each direction. Thus in the region $x < 0$, B and D must be zero (no reflected waves from the left end) and where $x > 0$, A and C are zero. The coefficients A, B, C and D may be found by satisfying equilibrium conditions immediately to the left ($x = 0^-$) and right ($x = 0^+$) of $x = 0$.

As shown in Section 2.3, the elastic shear stresses produce an upward force of magnitude

$$EI\frac{\partial^3 w(x)}{\partial x^3} \tag{2.5.33}$$

at the left end of an elemental beam section and a downward force of equal magnitude at the right-hand end of the section (see Fig. 2.35). As the applied force is considered positive in the positive w direction, then at $x = 0^+$

$$\frac{F}{2} - EI(jk_b^3 A_1 - k_b^3 A_3) = 0 \qquad (2.5.34)$$

and at $x = 0^-$

$$\frac{F}{2} + EI(-jk_b^3 A_2 + k_b^3 A_4) = 0 \qquad (2.5.35)$$

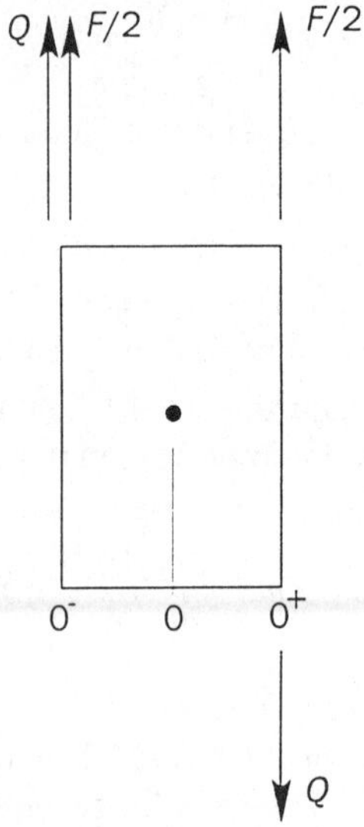

Fig. 2.35 External forces F and internal shear forces Q acting on an infinitesimally small cross-sectional element at the centre of a beam.

Because of symmetry, the slope of the beam at $x = 0$ is zero. Hence,

$$-jk_b A_1 - k_b A_3 = jk_b A_2 + k_b A_4 = 0 \qquad (2.5.36)$$

Equations (2.5.34) to (2.5.36) give

$$A_1 = A_2 = jA_3 = jA_4 \qquad (2.5.37a,b,c)$$

and

$$A_1 = -\frac{jF}{4EIk_b^3} \qquad (2.5.38)$$

Thus at $x = 0^+$, $x = 0^-$, the displacement w is given by (2.5.32) as

$$w(0^+) = w(0^-) = \left[\frac{-jF}{4EIk_b^3}\right](1 - j) \qquad (2.5.39a,b)$$

The point impedance is thus

$$Z_F = \frac{F}{\frac{\partial}{\partial t} w(0)} = (2EIk_b^3 \omega)(1 + j) = 2mc_b(1 + j) \qquad (2.5.40a,b,c)$$

where k_b is given by (2.5.29) and c_b is the wave speed of the flexural waves given by $c_b = \omega / k_b$. A more complicated expression applies for thick beams and can be derived using the wave equation for thick beams discussed in Section 2.3.

2.5.2.2 Summary of impedance formulae for infinite and semi-infinite isotropic beams and plates

The expressions given in Table 2.2 to follow for the point force and point moment impedances of thin beams and plates were derived using the appropriate wave equation and boundary conditions, and the method just illustrated for an infinite beam. Note that the bending wave equation for the angular velocity α has the same form as that for the normal velocity u. In all the formulae given, it is assumed that the source impedance Z_s is sufficiently small not to contribute to the total impedance of the beam or plate. Where this is not true, the impedance of the source can simply be added to that of the beam or plate provided that there are no reflected waves present (that is, the ends of the beam or edges of the plate not adjacent to the source must be an infinite distance from the source or absorptive). Where waves are reflected back to the driving point and Z_s cannot be neglected, the reflection coefficients of both the propagating and non-propagating waves at the driving point must be used to determine the point impedance of the driving source and beam or plate in combination.

In Table 2.2, S is the cross-sectional area of the beam, E is Young's modulus of elasticity of the beam or plate material, ρ is the density of the beam or plate material, m is the mass per unit area of the plate or mass per unit length of the beam, k_b is the bending wave number ($= \omega / c_b$), c_b is the bending wave speed, f ($= \omega / 2\pi$) is the excitation frequency in hertz, and a is the moment arm, or distance between the two forces used to generate the moment on the plate. The wavenumber, $k_b = \omega / c_b$, for a beam is defined by (2.5.29) and for an isotropic plate by

$$k_b = \left[\frac{12\omega^2 m(1 - \nu^2)}{Eh^3} \right]^{1/4} \qquad (2.5.41)$$

where m is the mass per unit area of the plate and ω is the angular frequency (rad s^{-1}).

Note that in Table 2.2 (a) is the distance between the two forces used to generate the moment.

Table 2.2 Summary of impedance formulae for isotropic thin beams and plates excited by a point force or a point moment (see Cremer *et al.*, 1973, p. 281) ($k_b = \omega/c_b$)

Wave type	Structural element	Z_f	Z_M
Longitudinal	Semi-infinite thin beam 	$S\sqrt{E\rho}$	-
Flexural	Infinite thin beam 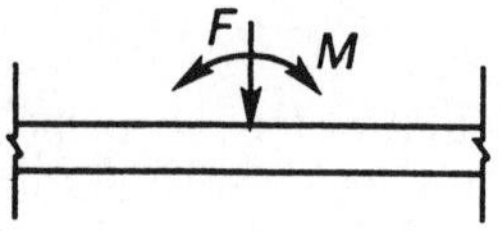	$2mc_b(1+j)$	$2mc_b(1-j)/k_b^2$
Flexural	Semi-infinite thin beam 	$\frac{1}{2}mc_b(1+j)$	$\frac{1}{2}mc_b(1-j)/k_b^2$
Flexural	Infinite, thin, isotropic plate 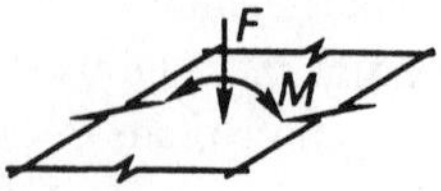	$8\omega m/k_b^2$	$\frac{16\omega m}{k_b^4}\left[1-\frac{4j}{\pi}\ln(0.9k_b a)\right]$
Flexural	Semi-infinite, thin isotropic plate simply supported on one edge 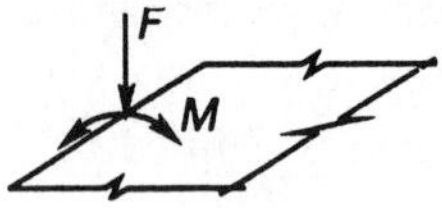	$2.3\omega m/k_b^2$	$\frac{5.3\omega m}{k_b^4}(1-1.46j\ln 0.9k_b a)^{-}$

It is of interest to note that the point force impedances of infinite systems such as those listed in Table 2.2 correspond to frequency averaged impedances of equivalent finite systems as illustrated in Fig. 2.36. The only difference is that the finiteness of the finite systems add the peaks and troughs which span the line for the infinite system (Skudrzyk, 1980). Thus the infinite system impedances provide a good estimate of the average impedance of finite systems with the modal peaks being smaller as structural damping is increased. An excellent treatment of this topic which includes tables of impedance formulae for many infinite and semi-infinite systems is given by Pinnington (1988).

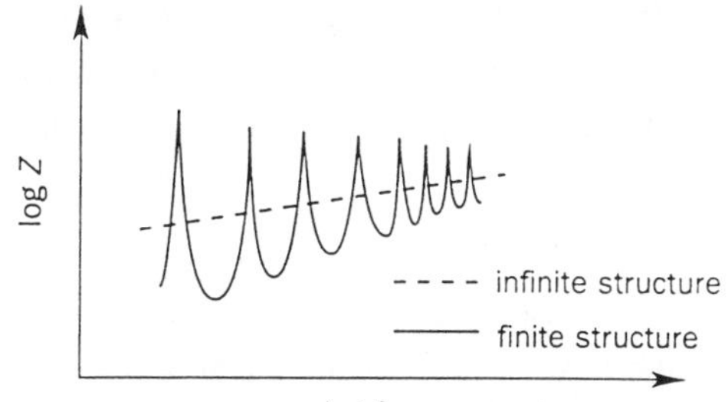

Fig. 2.36 Infinite system impedance approximation to the impedance of a similar finite system.

2.5.2.3 Point force impedance of finite systems

As will be discussed in Chapter 3, the motion of a finite vibrating structure can be characterized in terms of its modes of vibration, and the normal surface velocity $u(\mathbf{x})$ at location $\mathbf{x}$ due to an applied force $F(\mathbf{x}_F)$ at location $\mathbf{x}_F$ is given by

$$u(\mathbf{x}) = F(\mathbf{x}_F) \sum_{n=1}^{\infty} \frac{j\omega \psi_n(\mathbf{x}_F)\, \psi_n(\mathbf{x})}{m_n(\omega_n^2 - \omega^2 + j\eta\omega_n^2)} \tag{2.5.42}$$

from which the force impedance at location $\mathbf{x}_F$ is given by

$$Z_F = \frac{F(\mathbf{x}_F)}{u(\mathbf{x}_F)} = \left[\sum_{n=1}^{\infty} \frac{j\omega \psi_n^2(\mathbf{x}_F)}{m_n(\omega_n^2 - \omega^2 + j\eta_n\omega_n^2)} \right]^{-1} \tag{2.5.43}$$

where the modal mass m_n is defined as

$$m_n = \int\!\!\int_S \rho(\mathbf{x})\psi_n^2(\mathbf{x})\,\mathrm{d}\mathbf{x} \tag{2.5.44}$$

where $\rho(\mathbf{x})$ is the surface density (mass per unit area) at vector location $\mathbf{x}$ on the surface, $\psi_n(\mathbf{x}_F)$ is the value of the mode shape function at location $\mathbf{x}_F$, η_n is the structural loss factor for mode n and $\mathbf{x}_F$ is the vector location on the surface at which the point force F is applied.

To apply the preceding equations to a practical structure, the resonance frequencies, mode shapes and modal damping must be determined. Modal damping cannot be calculated, but it can be estimated from experience with similar structures, or it can be measured (if the structure exists) using the methods outlined in Chapter 3. Structural resonance frequencies and mode shapes can be calculated from first principles for simple structures and by using finite element analysis for more complex structures.

As an example, it will be shown briefly how the mode shapes and resonance frequencies may be derived for a beam of finite length and free ends. The solution to the wave equation for a thin beam is given by (2.5.32). One end of the beam is at $x = 0$ and the other is at $x = L$. Considering the boundary conditions at $x = 0$, as the beam is free at this end, the bending moment and shear force vanish. Thus,

$$\frac{\partial^2 w(x)}{\partial x^2} = 0 \quad \text{and} \quad \frac{\partial^3 w(x)}{\partial x^3} = 0 \tag{2.5.45a,b}$$

The solution to the beam wave equation may also be written in terms of hyperbolic and transcendental functions as follows:

$$w(x,t) = (A\cos k_b x + B\sin k_b x + C\cosh k_b x + D\sinh k_b x)\,e^{j\omega t} \tag{2.5.46}$$

and in fact this solution form is more commonly used to determine the beam resonance frequencies and mode shapes as it simplifies the analysis.

The boundary conditions of (2.5.45) applied to the end $x = 0$, lead to the following relations:

$$A = C \quad \text{and} \quad B = D \tag{2.5.47a,b}$$

Substituting (2.5.47) into (2.5.46) and ignoring the time dependence gives:

$$w(x) = A(\cos k_b x + \cosh k_b x) + B(\sin k_b x + \sinh k_b x) \tag{2.5.48}$$

Applying the first of the boundary conditions of (2.5.45) to equation (2.5.48) evaluated at $x = L$ gives:

$$B = -A\frac{\cosh k_b L - \cos k_b L}{\sinh k_b L - \sin k_b L} \tag{2.5.49}$$

Substituting (2.5.49) into (2.5.48) and eliminating the constant A gives the following result for the beam mode shape function:

$$\psi(x) = \left[\cosh k_b x + \cos k_b x - \frac{\cosh k_b L - \cos k_b L}{\sinh k_b L - \sin k_b L}(\sinh k_b x + \sin k_b x)\right] \tag{2.5.50}$$

Applying the second of the boundary conditions of (2.5.45) to (2.5.50) at $x = L$ allows an expression for the eigen solutions k_b to be obtained as follows

$$\cos(k_b L)\cosh(k_b L) = 1 \tag{2.5.51}$$

The solutions of (2.5.51) correspond to the resonance frequencies of the modes of vibration of the beam and when substituted into (2.5.50), they allow the corresponding mode shape functions to be calculated. Frequency equations and mode shapes corresponding to various beam end conditions and for a simply supported panel are given in Table 2.3. Note that for the beam, the resonance frequency is related to the solution k_b of the frequency equation, by (2.5.29). For the simply supported plate n and m are integer numbers which correspond to mode order (n, m).

2.5.2.4 Point force impedance of cylinders

Infinite cylinder

In this case, as there are no waves reflected from the ends of the cylinder, the modal displacements are independent of axial location and the mode shapes for radial, axial and circumferential vibration are:

$$\psi_{wn} = \cos n\theta \tag{2.5.52}$$

$$\psi_{\xi_\theta n} = \sin n\theta \tag{2.5.53}$$

$$\psi_{\xi_x n} = \cos n\theta \tag{2.5.54}$$

The resonance frequencies are given by

$$\Omega_n^2 = 0, 1 \qquad n = 0 \tag{2.5.55}$$

$$\Omega_n^2 = \frac{1}{2}(1-\nu)n^2 \quad \text{(axial mode)} \tag{2.5.56}$$

$$\Omega_n^2 = \frac{1}{2}\left[(1+n^2)(1+k_t n^2)\right] \mp \frac{1}{2}\left[(1+n^2)^2 - 2k_t n^2(1-6n^2+n^4)\right]^{1/2} \tag{2.5.57}$$

(Goldenveizer theory, radial and circumferential modes)

$$\text{or} \quad \Omega_n^2 = \frac{1}{2}\left[1+n^2+k_t n^4\right] \mp \frac{1}{2}\left[(1+n^2)^2 - 2k_t n^6\right]^{1/2} \tag{2.5.58}$$

(Flügge theory, radial and circumferential modes)

where

$$k_t = h^2/12a^2 \tag{2.5.59}$$

and where Ω_n is the non-dimensional frequency, defined as

$$\Omega_n = \omega_n a\sqrt{\frac{\rho(1-\nu^2)}{E}} \tag{2.5.60}$$

Table 2.3 Mode shape and resonance frequency equations for beams and plates

End Conditions	Frequency Equation	Mode Shape $\psi(x)$
	$\cos k_b L \cosh k_b L = 1$	$\psi(x) = (\sin k_b x - \sinh k_b x) + A(\cos k_b x - \cosh k_b x)$ $A = -\dfrac{\sin k_b L - \sinh k_b L}{\cos k_b L - \cosh k_b L}$
	$\tan k_b L = \tanh k_b L$	$\psi(x) = (\sin k_b x - \sinh k_b x) + A(\cos k_b x - \cosh k_b x)$ $A = -\dfrac{\sin k_b L + \sinh k_b L}{\cos k_b L + \cosh k_b L}$
	$\sin k_b L = 0$	$\psi(x) = \sin k_b x$
	$\cos k_b L \cosh k_b L = -1$	$\psi(x) = (\sin k_b x - \sinh k_b x) + A(\cos k_b x - \cosh k_b x)$ $A = -\dfrac{\sin k_b L + \sinh k_b L}{\sin k_b L - \sinh k_b L}$
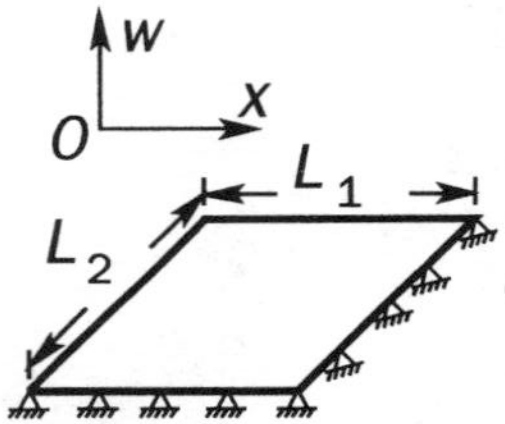 Simply supported thin, isotropic	$\omega_{n,m} = \left[\dfrac{Eh^3}{12m_s(1-\nu^2)} \left\{ \left(\dfrac{n\pi}{L_1}\right)^2 + \left(\dfrac{m\pi}{L_2}\right)^2 \right\} \right]$	$\psi(x,y) = \sin\left(\dfrac{n\pi x}{L_1}\right) \sin\left(\dfrac{m\pi y}{L_2}\right)$

Equations (2.5.52) to (2.5.60) may be substituted into (2.5.42) to (2.5.44) to give the point force impedance an infinite cylinder. In cases involving sound radiation, only the radial vibration component w mode shape (2.5.52) is of interest.

Finite cylinder – shear diaphragm ends

For this case, it is assumed that the ends of the cylinder are closed with a thin plate which is supported so that it cannot move normal to the cylinder surface. That is,

$$w = M_x = N_x = \xi_\theta = 0 \qquad (2.5.61a,b,c,d)$$

The mode shape functions are:

$$\psi_{w\iota,n} = \sin\frac{\iota\pi x}{L}\cos n\theta \qquad (2.5.62)$$

$$\psi_{\xi_\theta\iota,n} = \sin\frac{\iota\pi x}{L}\sin n\theta \qquad (2.5.63)$$

$$\psi_{\xi_x\iota,n} = \cos\frac{\iota\pi x}{L}\cos n\theta \qquad (2.5.64)$$

The non-dimensional resonance frequencies are solutions of:

$$\Omega^6 - (K_2 + k_t\Delta K_2)\Omega^4 + (K_1 + k_t\Delta K_1)\Omega^2 + (K_0 + \Delta K_0) = 0 \qquad (2.5.65)$$

where:

$$K_2 = 1 + \frac{1}{2}(3-\nu)(n^2+\lambda_x^2) + k_t(n^2+\lambda_x^2)^2 \qquad (2.5.66)$$

$$K_1 = \frac{1}{2}(1-\nu)\left[(3+2\nu)\lambda_x^2 + n^2 + (n^2+\lambda_x^2)^2 + \frac{(3-\nu)}{(1-\nu)}k_t(n^2+\lambda_x^2)\right] \qquad (2.5.67)$$

$$K_0 = \frac{1}{2}(1-\nu)\left[(1-\nu^2)\lambda_x^4 + k_t(n^2+\lambda^2)^4\right] \qquad (2.5.68)$$

where $\lambda_x = m\pi x/L$ (2.5.69)

For the Goldenveiser–Novozhilov theory:

$$\Delta K_2 = 2(1-\nu)\lambda_x^2 + n^2 \qquad (2.5.70)$$

$$\Delta K_1 = 2(1-\nu)\lambda_x^2 + n^2 + 2(1-\nu)\lambda_x^4 - (2-\nu)\lambda_x^2 n^2 - \frac{1}{2}(3+\nu)n^4 \qquad (2.5.71)$$

$$\Delta K_0 = \frac{1}{2}(1-\nu)[4(1-\nu^2)\lambda_x^4 + 4\lambda_x^2 n^2 + n^4 - 2(2-\nu)(2+\nu)\lambda_x^4 n^2 - 8\lambda_x^2 n^4 - 2n^6] \tag{2.5.72}$$

For the Flügge theory:

$$\Delta K_2 = \Delta K_1 = 0 \tag{2.5.73}$$

$$\Delta K_0 = \frac{1}{2}(1-\nu)[2(2-\nu)\lambda_x^2 n^2 + n^4 - 2\nu\lambda_x^6 - 6\lambda_x^4 n^2 - 2(4-\nu)\lambda_x^2 n^4 - 2n^6] \tag{2.5.74}$$

Mode shape functions and modal resonance frequencies equations for various other cylinder end conditions are given by Leissa (1973). The mode shape functions and solutions to the resonance frequency equations may be used together with (2.5.42) to (2.5.44) to obtain the point force impedance. Note that for $n \geq 2$ the mode shape functions and resonance frequencies are not very dependent upon the boundary conditions at the cylinder ends. Also, for all but the lowest order modes, the modal masses are equal to one-quarter of the total cylinder mass.

2.5.2.5 Wave impedance of finite structures

When a structure is excited by an incident acoustic field, its response is governed by its wave impedance, just as its response to a point force is governed by its point force input impedance. The wave impedance is defined as the ratio of the complex force per unit area, p (or pressure) to the complex velocity, u at a point on the structure (Fahy, 1985). Thus,

$$Z_{ws} = \frac{p_i}{u} \tag{2.5.75}$$

The wave impedance of a structure can be associated with a specific wavenumber, or frequency and phase speed combination, $k = \omega/c$. It is evaluated mathematically by applying a force to a structure in the form of a sinusoidal travelling wave and using the structure equation of motion to derive the structural response. It can be applied in practice to a random or multiple frequency sound field by using Fourier analysis to separate the signal into its frequency components and superposition to obtain the total structural response.

As an example, we will derive the wave impedance for an infinite undamped isotropic panel of thickness h subject to a transverse force in the form of a plane travelling wave characterized by a wavenumber, k. Note that only bending waves will be considered, as in practice when structures are excited by acoustic waves in a fluid medium, other waves cannot be generated because the fluid is not capable of supporting shear forces.

The equation of motion for bending waves in a plate subjected to an applied force of q/unit area is given by (2.3.101). This equation can be written for bending waves propagating in the x-direction only as:

$$D\frac{\partial^4 w(x,t)}{\partial x^4} + \rho h\frac{\partial^2 w(x,t)}{\partial t^2} = q\,e^{j(\omega t - kx)} \tag{2.5.76}$$

with a solution of the form:

$$w(x,t) = A\,e^{j(\omega t - kx)} \tag{2.5.77}$$

where A is the complex displacement amplitude, q is the complex amplitude of the applied force per unit area and

$$D = Eh^3/12(1-\nu^2) \tag{2.5.78}$$

Substituting (2.5.77) into (2.5.76) and making use of the complex modulus $E(1+j\eta)$ as a replacement for E in (2.5.78) (to allow the inclusion of damping), we obtain:

$$\left[\frac{E(1+j\eta)h^3}{12(1-\nu^2)}k^4 - \rho h\omega^2\right]A = q \tag{2.5.79}$$

The wave impedance is defined by (2.5.75), which is equivalent to:

$$Z_{ws} = \frac{q}{j\omega A} \tag{2.5.80}$$

Substituting for $\frac{q}{A}$ from (2.5.71) into (2.5.72) gives

$$Z_{ws} = \frac{Eh^3k^4\eta}{12(1-\nu^2)\omega} - j\left[\frac{Eh^3k^4}{12(1-\nu^2)\omega} - \rho h\omega\right] \tag{2.5.81}$$

For structures radiating into dense fluids such as water, the effect of fluid loading must also be taken into account using a fluid wave impedance as discussed by Fahy (1985). However, as the loading is usually negligible for radiation into air, it is not considered further here.

2.5.3 Sound intensity and sound power

Sound intensity is defined as a measure of the rate of local acoustic energy flow in an acoustic medium. It is a vector quantity characterized by a magnitude, a direction and a specific point location in the acoustic medium. The sound power being transmitted through an imaginary surface can be obtained by integrating the real component of sound intensity normal to the surface over the area of the surface. More specifically, sound intensity is commonly defined as the long time average rate of flow of sound energy through unit area of acoustic fluid. However, sound intensity is a complex vector, having both real and imaginary components. The real (or active) component is what is commonly used and has just been defined. The imaginary (or reactive) component is a measure of the energy stored in the sound field.

Note that the active component has a time averaged non-zero value, corresponding to a net transport of energy, while the reactive component has a zero time averaged value corresponding to local oscillatory transport of energy. As the reactive intensity is zero when averaged over time, it is generally expressed as an amplitude. It is associated with potential energy storage and does not propagate anywhere. At any single frequency, these active and reactive intensity components are associated with components of the acoustic particle velocity which are, respectively, in phase and in-quadrature with the local acoustic pressure. In-quadrature components occur in the near field (within half a wavelength) of sound sources or sound reflecting objects, or in any part of the field where sound waves are travelling in more than one direction simultaneously. Reactive intensity only has meaning for a single frequency or a narrow band sound field. When averaged over a wide frequency band the result is not the sum of the reactive intensity for the individual frequencies as it is for the active component. In fact, the reactive intensity is zero when averaged over a wide frequency band.

Sound intensity has an important application to active noise control systems where the aim is usually to minimize the sound power being transmitted in one or several directions. Often it may not be sufficient to minimize sound pressure at an error sensor or sensors to obtain optimal results. In instances where reflected waves are present in addition to the original primary disturbance, minimization of sound intensity will generally provide better results than minimization of sound pressure. Here, we will discuss both theoretical and practical aspects of sound intensity measurement as they apply to active noise control. A more in depth treatment is undertaken by Fahy (1995).

As sound intensity is an energy based quantity, its measurement or calculation requires the determination of two independent quantities, namely, the acoustic pressure and the acoustic particle velocity, together with the relative phase between the two quantities.

The instantaneous sound intensity at vector location $\boldsymbol{r}$ and at time t is the product of the instantaneous sound pressure $p(\boldsymbol{r},t)$ and acoustic particle velocity $\boldsymbol{u}(\boldsymbol{r},t)$, and is given by

$$\boldsymbol{I}_i(\boldsymbol{r},t) = p(\boldsymbol{r},t)\,\boldsymbol{u}(\boldsymbol{r},t) \tag{2.5.82}$$

The intensity normal to a particular surface, along the vector direction $\boldsymbol{n}$ is given by

$$I_n(\boldsymbol{r},t) = \boldsymbol{I}(\boldsymbol{r},t)\cdot\boldsymbol{n} \tag{2.5.83}$$

A general expression for the sound intensity $\boldsymbol{I}(\boldsymbol{r})$, is the time average of the instantaneous intensity given by (2.5.82), which may be written as follows:

$$\boldsymbol{I}(\boldsymbol{r}) = \langle p(\boldsymbol{r},t)\,\boldsymbol{u}(\boldsymbol{r},t)\rangle = \lim_{T\to\infty}\int_0^T p(\boldsymbol{r},t)\,\boldsymbol{u}(\boldsymbol{r},t)\,\mathrm{d}t \tag{2.5.84a,b}$$

For the special case of single frequency sound, complex notation may be introduced.

The sound pressure at a location in three-dimensional space may be represented as

$$p(\boldsymbol{r},t) = \bar{p}(\boldsymbol{r})\mathrm{e}^{j(\omega t + \theta_p(\boldsymbol{r}))} \tag{2.5.85}$$

where both the amplitude, $\bar{p}(\boldsymbol{r})$, and the phase, $\theta_p(\boldsymbol{r})$, are real, space dependent quantities. The phase term, $\theta_p(\boldsymbol{r})$, includes the term $-kr$. In the text to follow, the $\boldsymbol{r}$ dependence of $\bar{p}$ and θ_p will be omitted to reduce the complexity of the notation. The particle velocity may be obtained using Euler's equation (2.1.24) to give

$$\boldsymbol{u}(\boldsymbol{r},t) = \frac{j}{\omega\rho_0}\nabla p(\boldsymbol{r},t) = \frac{1}{\omega\rho_0}\left[-\bar{p}\,\nabla\theta_p + j\nabla\bar{p}\right]\mathrm{e}^{j(\omega t+\theta_p)} \tag{2.5.86a,b}$$

which may also be written as

$$\boldsymbol{u}(\boldsymbol{r},t) = \bar{\boldsymbol{u}}(\boldsymbol{r})\mathrm{e}^{j(\omega t + \theta_u(\boldsymbol{r}))} \tag{2.5.87}$$

The instantaneous intensity cannot be determined simply by multiplying (2.5.85) and (2.5.86) together. This is because the complex notation formulation $\mathrm{e}^{j(\omega t+\theta_p)}$ can only be used for linear quantities. Thus the product of two vector quantities represented in complex notation is given by the product of their real components only (Skudrzyk, 1971, p. 28). Thus,

$$\boldsymbol{I}_i(\boldsymbol{r},t) = \mathrm{Re}\{p(\boldsymbol{r},t)\}\mathrm{Re}\{\boldsymbol{u}(\boldsymbol{r},t)\} = -\frac{1}{\omega\rho_0}\left[\bar{p}^2\nabla\theta\right]\cos^2(\omega t + \theta_p)$$
$$- \frac{1}{2\omega\rho_0}\left[\bar{p}\,\nabla\bar{p}\,)\right]\sin 2(\omega t + \theta_p) \tag{2.5.88a,b}$$

The first term in (2.5.88b) is the product of the real part of the pressure with the real part of the velocity which is in phase with the pressure (active intensity), while the second term is the product of the real part of the pressure with the real part of the velocity which is in-quadrature with the pressure (reactive intensity); see Fig. 2.37.

The time average of the active intensity (first term in (2.5.88b) is thus given by

$$\boldsymbol{I}(\boldsymbol{r}) = -\frac{1}{2\rho_0\omega}\bar{p}^2\nabla\theta_p \tag{2.5.89}$$

Although the time averaged reactive intensity is zero (as it is an oscillating quantity), its amplitude is given by the amplitude of the second term in (2.5.88b) as:

$$\boldsymbol{I}_r(\boldsymbol{r}) = -\frac{1}{2\rho_0\omega}\bar{p}\,\nabla\bar{p} = -\frac{1}{4\rho_0\omega}\nabla\bar{p}^2 \tag{2.5.90a,b}$$

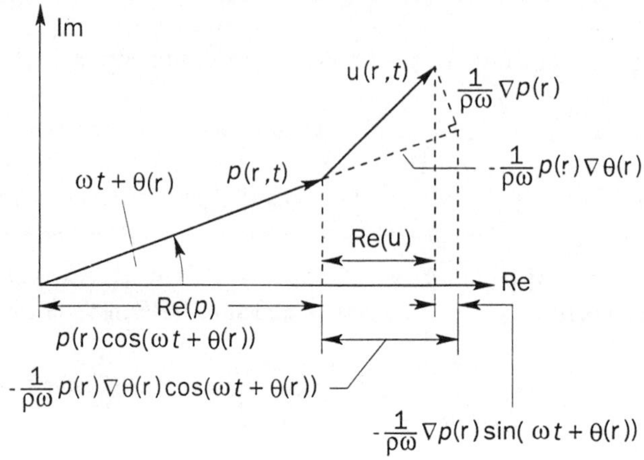

Fig. 2.37 Graphical representation of the calculation of the complex instantaneous intensity.

which in future will be referred to as the reactive intensity $\boldsymbol{Q}$.

Thus equation (2.5.88b) can be written as

$$\boldsymbol{I}_i(\boldsymbol{r},t) = \boldsymbol{I}(\boldsymbol{r})\left[1 + \cos 2(\omega t + \theta_p)\right] + \boldsymbol{Q}(\boldsymbol{r})\sin 2(\omega t + \theta_p) \qquad (2.5.91)$$

which may be rewritten as

$$\boldsymbol{I}_i(\boldsymbol{r},t) = \mathrm{Re}\left\{\left[\boldsymbol{I}(\boldsymbol{r}) + j\boldsymbol{Q}(\boldsymbol{r})\right]\left[1 + \mathrm{e}^{-2j(\omega t + \theta_p)}\right]\right\} \qquad (2.5.92)$$

Returning to (2.5.82), (2.5.85) and (2.5.87) the instantaneous intensity for a harmonic wave of frequency ω may also be written as:

$$\boldsymbol{I}_i(\boldsymbol{r},t) = \bar{p}\,\bar{\boldsymbol{u}}\cos(\omega t + \theta_p)\cos(\omega t + \theta_u) \qquad (2.5.93)$$

where $\bar{p}$ and $\bar{\boldsymbol{u}}$ are real amplitudes of the acoustic pressure and particle velocity respectively and θ_p and θ_u are the relative phase angles of the same quantities. Note that for simplicity in notation the dependence of $\bar{p}$, $\bar{\boldsymbol{u}}$, θ_p and θ_u on $\boldsymbol{r}$ has been omitted. Equation (2.5.93) may be rearranged (Fahy 1995) to give

$$\boldsymbol{I}_i(\boldsymbol{r},t) = \mathrm{Re}\left\{\frac{1}{2}\bar{p}\,\bar{\boldsymbol{u}}\,\mathrm{e}^{j\theta_r}\left[1 + \mathrm{e}^{-2j(\omega t + \theta_p)}\right]\right\} \qquad (2.5.94)$$

where

$$\theta_r = \theta_p - \theta_u \qquad (2.5.95)$$

Comparison of (2.5.92) and (2.5.94) shows that the active and reactive intensities can also be written as

$$\boldsymbol{I}(\boldsymbol{r}) = \frac{1}{2}\mathrm{Re}\{p(\boldsymbol{r})\boldsymbol{u}^*(\boldsymbol{r})\} = \frac{1}{2}\bar{p}\,\bar{\boldsymbol{u}}\cos\theta_r \qquad (2.5.96\text{a,b})$$

$$Q(r) = \frac{1}{2}\ \mathrm{Im}\{p(r)\boldsymbol{u}(r)\} = \frac{1}{2}\bar{p}\,\bar{\boldsymbol{u}}\sin\theta_r \tag{2.5.97a,b}$$

where the * denotes the complex conjugate, and where $p(\boldsymbol{r})$ and $\boldsymbol{u}(\boldsymbol{r})$ are complex amplitudes and $\bar{p}$ and $\bar{\boldsymbol{u}}$ are real amplitudes of the pressure and particle velocity respectively at location $\boldsymbol{r}$, so that

$$p(\boldsymbol{r}) = \bar{p}\mathrm{e}^{j\theta_p} \tag{2.5.98}$$

and

$$\boldsymbol{u}(\boldsymbol{r}) = \bar{\boldsymbol{u}}\mathrm{e}^{j\theta_u} \tag{2.5.99}$$

The sound power radiated through a surface of area S is given by

$$W = \int_S\!\!\int \left(\boldsymbol{I}_r(\boldsymbol{r}) + j\boldsymbol{Q}_r(\boldsymbol{r})\right).\boldsymbol{n}\,\mathrm{d}S \tag{2.5.100}$$

The real part of W corresponds to propagating energy while the imaginary part corresponds to energy stored in the sound field. The vector $\boldsymbol{n}$ is the normal to the surface through which the power transmission is to be calculated, at location $\boldsymbol{r}$.

2.5.3.1 Measurement of acoustic intensity

The measurement of sound intensity provides a means for directly determining the magnitude and direction of the acoustic power transmission at any location in space. Measuring and averaging the acoustic intensity over an imaginary surface surrounding a machine allows determination of the total acoustic power radiated by the machine.

Theoretically, measurements can be conducted in the near field of a machine, in the presence of reflecting surfaces and near other noisy machinery. However, if the reactive field associated with reflecting surfaces, or the near field of the sound source, is greater than the active field by 10 dB or more, or if the contributions of other nearby sound sources is 10 dB or more greater than the sound pressure level of the source under investigation, then in practice, reliable sound intensity measurements cannot be made.

The measurement of sound intensity requires the simultaneous determination of sound pressure and particle velocity. The determination of sound pressure is straightforward, but the determination of particle velocity presents some difficulties; thus, there are two principal techniques for the determination of particle velocity and consequently the measurement of sound intensity. Either the acoustic pressure and particle velocity are measured directly (p−u method) or the acoustic pressure is measured simultaneously at two closely spaced points and the mean pressure and particle velocity are calculated (p−p method). In either case the pressure is multiplied by the particle velocity to produce the instantaneous intensity and the time average intensity.

For the p−u method, the acoustic particle velocity is measured using two parallel ultrasonic beams travelling from source to receiver in opposite directions. Any particle movement in the direction of the beams will cause a phase difference in the received signals at the two receivers. The phase difference is related to the acoustic particle velocity in the space between the two receivers and may be used to estimate the particle velocity.

For the p−p method, the determinations of acoustic pressure and acoustic particle velocity are both made using a pair of high quality condenser microphones. The microphones are generally mounted side by side or facing one another and separated by a fixed distance (6 mm to 50 mm) depending upon the frequency range to be investigated. A signal proportional to the particle velocity at a point mid-way between the two microphones and along the line joining their acoustic centres is obtained using the finite difference in measured pressures to approximate the pressure gradient while the mean is taken as the pressure at the mid-point.

The assumed positive sense of the determined intensity is in the direction of the centre line from microphone 1 to microphone 2. For convenience, where appropriate in the following discussion, the positive direction of intensity will be indicated by unit vector $\boldsymbol{n}$.

Taking the gradient in direction $\boldsymbol{n}$ and using (2.1.26) gives the equation of motion relating the pressure gradient to the particle acceleration. That is,

$$\frac{\partial p}{\partial n} = -\rho_0 \frac{\partial u_n}{\partial t} \tag{2.5.101}$$

where p and u_n are both functions of the vector location $\boldsymbol{r}$ and time t. The component u_n of particle velocity is thus

$$u_n(t) = -\frac{1}{\rho_0} \int_{-\infty}^{t} \frac{\partial p(\tau)}{\partial n} \, \mathrm{d}\tau \tag{2.5.102}$$

where the assumption is implicit that the particle velocity is zero at time $t = -\infty$.

The integrand of (2.5.102) is approximated using the finite difference between the pressure signals p_1 and p_2 from microphones 1 and 2, respectively, which are separated by a distance of Δ:

$$u_n(t) = -\frac{1}{\rho_0 \Delta} \int_{-\infty}^{t} \left[p_2(\tau) - p_1(\tau) \right] \mathrm{d}\tau \tag{2.5.103}$$

The pressure mid-way between the two microphones is approximated as the mean:

$$p(t) = \frac{1}{2}\left(p_1(t) + p_2(t) \right) \tag{2.5.104}$$

Thus the instantaneous intensity in direction $\boldsymbol{n}$ at time t is approximated as:

$$I_n(t) = -\frac{1}{2\rho_0\Delta}\left[p_1(t) + p_2(t)\right]\int_{-\infty}^{t}\left[p_1(\tau) - p_2(\tau)\right]\mathrm{d}\tau \tag{2.5.105}$$

The mean active sound intensity in direction $\boldsymbol{n}$ (from microphone 1 to microphone 2) is:

$$I_n = -\frac{1}{2\rho_0\Delta}\left\langle\left(p_1(t) + p_2(t)\right)\int_{-\infty}^{t}\left(p_1(t) - p_2(t)\right)\mathrm{d}\tau\right\rangle \tag{2.5.106}$$

where $\langle\ \rangle$ denotes time average. That is, intensity in a direction from 1 to 2 is positive, as it is in the positive x-direction.

In a stationary field characterized by field variables x_1 and x_2 it is necessary that

$$x^2 = y^2 = xy = 0 \tag{2.5.107a-c}$$

from which it follows,

$$x_1\frac{\mathrm{d}x_1}{\mathrm{d}t} = x_2\frac{\mathrm{d}x_2}{\mathrm{d}t} = 0 \tag{2.5.108a,b}$$

and

$$\frac{\mathrm{d}x_1}{\mathrm{d}t}x_2 + \frac{\mathrm{d}x_2}{\mathrm{d}t}x_1 = 0 \tag{2.5.109}$$

Let the time derivatives of the field variables x_1 and x_2 be the acoustic pressures $p_1(t)$ and $p_2(t)$:

$$\frac{\mathrm{d}x_i}{\mathrm{d}t} = p_i(t) \qquad i = 1,2 \tag{2.5.110}$$

The field variables x_1 and x_2 are related to the integrals of the acoustic pressures $p_1(t)$ and $p_2(t)$ as follows,

$$x_i = \int_{-\infty}^{t} p_i(\tau)\mathrm{d}\tau \qquad i = 1,2 \tag{2.5.111}$$

Expansion of (2.5.106) into a series of four terms involving products of pressures and integrals of pressures, and use of the relations given by (2.5.108) to (2.5.111) shows that for stationary sound fields the instantaneous intensity can be obtained from the product of the signal from one microphone and the integrated signal from a second microphone in close proximity to the first (Fahy, 1995):

$$I_n(t) = \frac{1}{\rho_0\Delta}p_2(t)\int_{-\infty}^{t} p_1(\tau)\mathrm{d}\tau \tag{2.5.112}$$

The time average of (2.5.112) gives the following expression for the time average intensity in direction $\boldsymbol{n}$:

$$I_n = \lim_{T\to\infty} \frac{1}{\rho_0 \Delta T} \int_0^T \left[p_2(t) \int_{-\infty}^{t} p_1(\tau)\,\mathrm{d}\tau \right] \mathrm{d}t \qquad (2.5.113)$$

Thus, the mean intensity for a stationary sound field can be obtained from the product of the signal from one microphone and the integrated signal from another identical microphone in close proximity. Commercial instruments with digital filtering (e.g. one third octave and one octave) are available to implement (2.5.113).

As an example, consider the following two harmonic pressure signals: $p_1(t) = \bar{p}_1 \mathrm{e}^{j(\omega t + \theta_1)}$ and $p_2(t) = \bar{p}_2 \mathrm{e}^{j(\omega t + \theta_2)}$. Substitution of the real components of these quantities in (2.5.112) gives for the instantaneous intensity the following result:

$$I_n(t) = \frac{1}{4\rho_0 \Delta \omega} \left[\bar{p}_1^2 \sin(2\omega t + 2\theta_1) - \bar{p}_2^2 \sin(2\omega t + 2\theta_2) + 2\bar{p}_1 \bar{p}_2 \sin(\theta_1 - \theta_2) \right] \qquad (2.5.114)$$

Taking the time average of (2.5.114) gives the following expression for the (active) intensity:

$$I_n = \frac{\bar{p}_1 \bar{p}_2 \sin(\theta_1 - \theta_2)}{2\rho_0 \omega \Delta} \approx \frac{\bar{p}_1 \bar{p}_2 (\theta_1 - \theta_2)}{2\rho_0 \omega \Delta}; \qquad (\theta_1 - \theta_2) \ll 1 \qquad (2.5.115a,b)$$

This equation also follows directly from (2.5.89) where the finite difference approximation is used to replace $\nabla \theta_p$ with $(\theta_1 - \theta_2)/\Delta$ and $\bar{p}^2$ is approximated by $\bar{p}_1 \bar{p}_2$. The first two terms of the right-hand side of (2.5.114) describe the reactive part of the intensity. If the phase angles θ_1 and θ_2 are not greatly different, for example, the sound pressures $\bar{p}_1$ and $\bar{p}_2$ are measured at points which are closely spaced compared to a wavelength, the magnitude of the reactive component of the intensity is approximately:

$$Q_n = \frac{\left[\bar{p}_1^2 - \bar{p}_2^2 \right]}{4\omega \rho_0 \Delta} \qquad (2.5.116)$$

Equation (2.5.116) also follows directly from (2.5.90b) where $\bar{p}$ is replaced by $(\bar{p}_1 - \bar{p}_2)/2$ and ∇p is replaced with the finite difference approximation $(\bar{p}_1 - \bar{p}_2)/\Delta$.

Thus, measurements of the intensity in a stationary sound field could be made with only one microphone which in turn is located at two suitably spaced points, together with a phase meter to determine the phase of each signal relative to a stable reference signal. Indeed, the three-dimensional sound intensity vector field could be measured by automatically traversing a single microphone stepwise over the area of interest in two parallel planes spaced a short distance apart. Use of a single microphone for intensity measurements eliminates problems associated with microphone and amplifier/integrator phase mismatch and reduces the diffraction problem enormously. Although useful in the laboratory, this technique is difficult to implement in the field.

Note that for both p−u and p−p type measurements, the total instantaneous intensity vector $\boldsymbol{I}_i(t)$ can only be determined if three orthogonal components of particle velocity (or acoustic pressure gradient) can be measured simultaneously. However, if only the mean intensity vector in a steady sound field is needed, the results of the sequential measurement of the three orthogonal intensity components may be combined vectorially.

2.5.3.2 Frequency decomposition of the intensity

In the measurement and active control of acoustic power transmission, it is often necessary to decompose the intensity signal into its frequency components. This may be done either directly or indirectly.

Direct frequency decomposition

For a p−u probe, the frequency distribution of the mean intensity may be obtained by passing the two output signals (p and u) through identical band pass filters prior to performing the time averaging. With a p−p probe, the frequency distribution may be determined by passing the two signals through appropriate identical band pass filters, either before or after performing the sum, difference and integration operations of (2.5.106) and then time averaging the resulting outputs.

Indirect frequency decomposition

Determination of the intensity using this method is based upon Fourier analysis of the two probe signals (either the p−u signals or the p−p signals) Fahy (1995) shows that for a p−u probe, the intensity as a function of frequency is given by the single sided cross-spectrum G_{pu} (see Chapter 3 for a full explanation of this quantity) between the two signals:

$$I_n(\omega) = \mathrm{Re}\big[G_{pu}(\omega)\big] \tag{2.5.117}$$

$$Q_n(\omega) = -\mathrm{Im}\big[G_{pu}(\omega)\big] \tag{2.5.118}$$

As before, $I_n(\omega)$ represents the real (or active) time averaged intensity at frequency ω and $Q_n(\omega)$ represents the amplitude of the reactive component in direction $\boldsymbol{n}$ which is along the line joining the two microphones, positive from 1 to 2.

As will be discussed in Chapter 3, the cross-spectrum of two signals is defined as the product of the complex instantaneous spectrum of one signal with the complex conjugate of the complex instantaneous spectrum of the second signal. Thus, if $G_p(\omega)$ and $G_u(\omega)$ represent the complex single sided spectra of the pressure and velocity signals respectively, then the associated cross-spectrum is given by:

$$G_{pu}(\omega) = G_p^*(\omega)\, G_u(\omega) \tag{2.5.119}$$

where the $*$ represents the complex conjugate.

Note that for random noise, $G(\omega)$ represents a cross-spectral density function. Thus for random noise the expressions on the left-hand side of (2.5.117) and (2.5.118) represent intensity per hertz. For single frequency signals and harmonics, $G(\omega)$ is the cross-spectrum. Alternatively, the intensity corresponding to each harmonic may be obtained by multiplying the cross-spectral density function by the bandwidth of each FFT filter (or the frequency resolution). Most modern spectrum analysers have the capability to present the results in terms of either cross-spectrum or cross-spectral density. It is up to the user to ensure that the correct representation is used for the signal being analysed. This is discussed in more detail in Chapter 3.

For the case of the p−p probe, Fahy (1995) shows that the mean active intensity I_n and amplitude Q_n of the reactive intensity in direction $\boldsymbol{n}$ (from microphone 1 to microphone 2) at frequency ω are:

$$I_n(\omega) = -\frac{1}{\rho_0 \omega \Delta}\,\mathrm{Im}\left[G_{p_1 p_2}(\omega)\right] \tag{2.5.120}$$

$$Q_n(\omega) = -\mathrm{Im}\left[G_{pu}(\omega)\right] = \frac{1}{2\rho_0 \omega \Delta}\left[G_{p_1 p_1}(\omega) - G_{p_2 p_2}(\omega)\right] \tag{2.5.121a,b}$$

where $G_{p_1 p_2}$ is the cross-spectrum of the two pressure signals and $G_{p_1 p_1}$ and $G_{p_2 p_2}$ represent the auto spectral densities (see Chapter 3).

In the case of a stationary harmonic sound field, it is possible to determine the sound intensity by using a single microphone and the indirect frequency decomposition method just described by taking the cross-spectrum between the microphone signal and a stable reference signal (referred to as A) for two locations p_1 and p_2 of the microphone. Thus the effective cross-spectra $G_{p_1 p_2}$ for use in the preceding equations can be calculated as follows:

$$G_{p_1 p_2}(\omega) = \frac{G_{p_1}{}^* G_A\, G_{p_2}{}^* G_{p_2}}{G_{p_2}{}^* G_A} = \frac{G_{p_1 A}\, G_{p_2 p_2}}{G_{p_2} A} \tag{2.5.122a,b}$$

Equation (2.5.120) is an expression describing the intensity in the direction along the line joining the two microphone locations p_1 and p_2. If three pairs of microphone locations are used so that three orthogonal lines (one for each pair) are defined, then for a stationary sound field it will be possible to measure the intensity in three orthogonal directions, allowing the overall intensity vector to be determined. Thus the overall intensity vector can be mapped as a function of location in a stationary sound field by using a single microphone attached to a three-dimensional traversing system, and cross-spectral measurements between the microphone signal and a stable reference signal. A system such as this has been used by the authors to investigate the effect of active control of sound radiation from vibrating surfaces on both the reactive and active intensities in the vicinity of the surface.

2.5.3.3 Errors in the determination of acoustic intensity

Errors inherent in acoustic intensity measurements are associated with

1. finite difference approximation;
2. instrumentation phase mismatch;
3. nearfield effects.

Each of these effects will now be discussed briefly.

Finite difference approximation

Equations (2.5.103) and (2.5.104) rely on a finite difference approximation to determine the particle velocity and acoustic pressure at a point mid-way between the two microphones. This approximation is valid if the pressure gradient is small, which implies that it is more appropriate for low frequencies. The error due to the finite difference approximation for measurements made in a harmonic sound field is (Crocker 1993):

$$\frac{I_M}{I_T} = \frac{\sin k\Delta}{k\Delta} \tag{2.5.123}$$

where I_M and I_T are respectively the measured and true active intensities. Thus, the microphone spacing, Δ, places an upper bound on the frequency of the intensity that can be measured with an acceptable degree of accuracy.

Instrumentation phase mismatch

At low frequencies, the actual physical phase difference between the two microphone signals becomes very small and is thus affected by even small phase errors in the matching of the two channels of the instrumentation. This problem sets a lower bound on the frequency of the intensity which can be measured with a specified accuracy using instrumentation of a particular accuracy.

Thus any particular microphone spacing and associated instrumentation will be characterized by a frequency range over which the intensity error will be confined within certain limits. Suppliers of intensity measuring equipment usually provide graphs or bar charts indicating appropriate frequency ranges for particular microphone spacings and their instrumentation.

Nearfield effects

Close to a complicated source, the actual sound intensity may change significantly within the distance of the microphone spacing which makes the measured intensity a function of probe spacing and distance between the source and the probe.

Other errors

Other possible errors are associated with:

1. random errors in the cross-spectral estimate if the FFT method of (2.5.120) is used;
2. interference with the sound field by the microphones;
3. uncertainty in the effective separation distance between the microphones.

All of the errors mentioned are discussed in more detail by Fahy (1995).

2.5.4 Structural intensity and power transmission

Structural intensity allows quantification of the rate of local vibratory power transmission in a structure in a similar way that acoustic intensity allows the quantification of the acoustic power transmission in a fluid medium. The structural power transmission through an imaginary structural cross-section is obtained by integrating the component of structural intensity normal to the cross-section over the area of the cross-section. Structural intensity is commonly defined as the long time rate of vibratory energy flow through unit area of a solid structure. An alternative definition of structural intensity which is often used is that it is the vibrational power per unit width of structure in a given direction. However this alternative definition will not be used here. Like sound intensity, structural intensity is a complex vector, having both real and imaginary components, with similar definitions as given for sound intensity in Section 2.5.3. However, unlike sound intensity which only applies to longitudinal wave propagation in fluids, structural intensity can apply to longitudinal, shear (or torsional) and flexural waves in solid structures.

Like sound intensity, structural intensity has important applications to the active control of vibratory power transmission in structures, where the aim is often to minimize the transmission of vibratory power. Measuring the structural intensity allows the determination of the residual vibratory power transmission through a particular structural section, and can be used as a cost function for an active control system.

The measurement of structural intensity requires the determination of the local vibratory stress and particle velocity, and the relative phase between them. Real structures are characterized by both active and reactive structural intensity fields, and the instantaneous structural intensity is defined as:

$$\boldsymbol{I} = -\boldsymbol{u} \cdot \mathfrak{I} \tag{2.5.124}$$

where $\boldsymbol{u}$ is the particle velocity vector and $\mathfrak{S}$ is the stress tensor. For a three-dimensional coordinate system, (2.5.124) may be written as

$$\boldsymbol{I} = \begin{bmatrix} I_1 \\ I_2 \\ I_3 \end{bmatrix} = - \begin{bmatrix} \sigma_{11} & \sigma_{12} & \sigma_{13} \\ \sigma_{21} & \sigma_{22} & \sigma_{23} \\ \sigma_{31} & \sigma_{32} & \sigma_{33} \end{bmatrix} \begin{bmatrix} u_1 \\ u_2 \\ u_3 \end{bmatrix} \tag{2.5.125a,b}$$

where the σ_{ij}, $i, j = 1,3$ represent shear stresses, the σ_{ii} represent tensile or compressive stresses and $\boldsymbol{I}$ is the instantaneous intensity vector which is a time dependent quantity.

For a cartesian coordinate system the subscripts in (2.5.125) have the equivalent $x = 1$, $y = 2$, $z = 3$. For a cylindrical system, $r = 1$, $\theta = 2$, $x = 3$ where r is the radial coordinate, θ is the angular coordinate and x is the axial coordinate. The negative sign in (2.5.124) and (2.5.125) appears because the stress is directly related to the derivative of the displacement or the displacement gradient. If this gradient is negative, then the intensity vector must be orientated in the positive direction.

The intensity in direction 1 can be derived simply from (2.5.125) and is

$$I_1 = -(\sigma_{11}u_1 + \sigma_{12}u_2 + \sigma_{13}u_3) = -\sum_{k=1}^{3} \sigma_{1k}u_k \tag{2.5.126a,b}$$

The instantaneous power transmission per unit width through any beam, plate or shell cross-section can be found by averaging the intensity over the cross-sectional area as follows. For cartesian coordinates, the power transmission per unit width in the axial x direction for a shell of thickness h is given by

$$P_x = \int_{-h/2}^{h/2} I_x \, \mathrm{d}z \tag{2.5.127}$$

where z is the thickness coordinate having the value of zero at the centre of the cross-section. Power transmission in the y direction is described by a similar relationship. However, for a curved surface described by a cylindrical coordinate system, the axial power transmission (in the non-curved direction) is described by a slightly more complex expression which takes into account the curvature of the cross section through which the power is flowing (see Fig. 2.38), as follows (Romano *et al.*, 1990):

$$P_x = \int_{-h/2}^{h/2} I_x \left(1 + \frac{z}{a}\right) \mathrm{d}z \tag{2.5.128}$$

where a is the radius of curvature of the shell. The term $(1 + z/a)$ in the integral arises because the arc length of lines parallel to the $z = 0$ line shown in figure 2.38 differs from the arc length at $z = 0$ by this factor. Thus this must be included in the integral to provide the proper area weighting. The nomenclature and sign convention used here differ from that used by Pavic (1976) and Cremer *et al.* (1973). However, they are consistent with those used by Romano *et al.* (1990) and Leissa (1969, 1973).

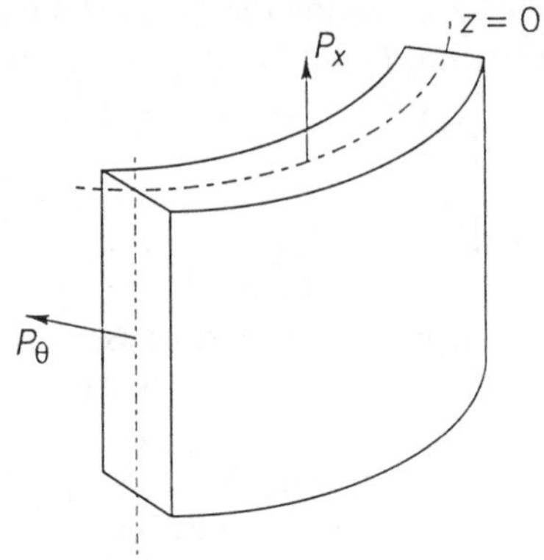

Fig. 2.38 Shell power transmission in axial and tangential directions.

For a flat plate, $a = \infty$ and (2.5.128) becomes the same as (2.5.127). As the section surface in the θ direction in Fig. 2.38 is the same shape as for a flat plate, the expression for power transmission in this direction is obtained from (2.5.127) by substituting θ for x.

In the equations to follow, the x and t dependencies which would normally appear in brackets following the symbols for displacements, rotations, moments and forces have been omitted to reduce the complexity of the notation; for example, $w(x,t)$ has been replaced simply by w.

Substituting (2.5.126) into (2.5.128), and using the cylindrical coordinate system, we obtain the following for the power transmission in the axial x direction per unit length of cylinder circumference:

$$P_x = -\int_{-h/2}^{h/2} \left(\sigma_{xz}\dot{w} + \sigma_{x\theta}\dot{\xi}_\theta + \sigma_{xx}\dot{\xi}_x\right)\left(1 + z/a\right)\mathrm{d}z \tag{2.5.129}$$

and for the power transmission in the circumferential θ direction per unit length of cylinder

$$P_\theta = -\int_{-h/2}^{h/2} \left(\sigma_{\theta\theta}\dot{\xi}_\theta + \sigma_{\theta x}\dot{\xi}_x + \sigma_{\theta z}\dot{w}\right)\mathrm{d}z \tag{2.5.130}$$

where the dot denotes differentiation with respect to time. Note that the displacements ξ_θ, ξ_x and w refer to the displacement of the elemental thickness $\mathrm{d}z$ (at location $a + z$) in the θ, x and z directions respectively and are not the same as the displacements of the section centre line. To express (2.5.129) and (2.5.130) in terms of centre-line displacements, we use a series expansion as follows (Romano *et al.*, 1990):

$$\begin{bmatrix} w(a+z) \\ \xi_x(a+z) \\ \xi_\theta(a+z) \end{bmatrix} = \begin{bmatrix} w(a) \\ \xi_x(a) \\ \xi_\theta(a) \end{bmatrix} + \begin{bmatrix} w^{(1)}(a) \\ \xi_x^{(1)}(a) \\ \xi_\theta^{(1)}(a) \end{bmatrix} \frac{z}{1!} + \begin{bmatrix} w^{(2)}(a) \\ \xi_x^{(2)}(a) \\ \xi_\theta^{(2)}(a) \end{bmatrix} \frac{z^2}{2!} + \ldots \tag{2.5.131}$$

where the superscript (1) denotes differentiation with respect to x and the superscript (2) denotes double differentiation with respect to x.

The Kirchhoff assumptions discussed in Section 2.3 for a cylinder allow (2.5.131) to be written as (see Romano *et al.* 1990):

$$w(a+z) = w(a) \tag{2.5.132}$$

$$\xi_x(a+z) = \xi_x(a) - z\frac{\partial w(a)}{\partial x} \tag{2.5.133}$$

$$\xi_\theta(a+z) = \xi_\theta(a) + \left[\frac{\xi_\theta(a)}{a} - \frac{1}{a}\frac{\partial w(a)}{\partial \theta}\right] z \tag{2.5.134}$$

If we substitute (2.5.132) to (2.5.134) into (2.5.129) and (2.5.130) and use (2.3.214) to (2.3.225) from Section 2.3 we obtain the following:

$$P_x = -\left[\dot{w}Q_x + \dot{\xi}_\theta N_{x\theta} + \left[\frac{\dot{\xi}_\theta}{a} - \frac{1}{a}\frac{\partial \dot{w}}{\partial \theta}\right] M_{x\theta} + \dot{\xi}_x N_x - \frac{\partial \dot{w}}{\partial x} M_x\right] \tag{2.5.135}$$

$$P_\theta = -\left[\dot{w}Q_\theta + \dot{\xi}_\theta N_\theta + \left[\frac{\dot{\xi}_\theta}{a} - \frac{1}{a}\frac{\partial \dot{w}}{\partial \theta}\right] M_\theta + \dot{\xi}_x N_{\theta x} - \frac{\partial \dot{w}}{\partial x} M_{\theta x}\right] \tag{2.5.136}$$

Note that the same coordinate system, sign conventions and definitions of moments and forces as used in Section 2.3 are used here.

In (2.5.135), shear waves are associated with the second term, longitudinal waves with the fourth term and flexural waves with the remaining terms.

The equation for power transmission per unit width in a plate in the x-direction can be obtained directly from (2.5.135) and (2.5.136) by replacing $\frac{1}{a}\frac{\partial}{\partial \theta}$ with $\frac{\partial}{\partial y}$, θ with y and ξ_θ/a with 0. Thus, we obtain for a plate

$$P_x = -\left[\dot{w}Q_x - \frac{\partial \dot{w}}{\partial y} M_{xy} - \frac{\partial \dot{w}}{\partial x} M_x + \dot{\xi}_y N_{xy} + \dot{\xi}_x N_x\right] \tag{2.5.137}$$

The expression for the y component of power transmission is obtained by interchanging the x and y subscripts in (2.5.137).

The first three terms in (2.5.137) correspond to power transmission associated with flexural wave propagation. The first of these is the shear force contribution, the second is the twisting contribution and the third term is the bending contribution. The fourth term corresponds to shear wave propagation while the fifth corresponds to longitudinal wave propagation.

For wave propagation in a beam, (2.5.137) becomes:

$$P_x = -\left[-\dot{w}Q_x - \dot{\theta}_x M_{xy} - \frac{\partial \dot{w}}{\partial x} M_x + \dot{\xi}_x N_x\right] \tag{2.5.138}$$

where the first and third terms represent the flexural wave power transmission, the second term represents the torsional wave power transmission and the last term represents the longitudinal wave power transmission. The quantity θ_x is the angle of rotation about the x-axis caused by a torsional wave. It is defined as:

$$\theta_x = \frac{\partial w}{\partial y} = -\frac{\partial w}{\partial z} \tag{2.5.139a,b}$$

The force and moment variables are defined and derived in Section 2.3. Note the difference in sign between the beam and plate for the $\dot{w}Q_x$ term. This is because the sign convention for positive Q_x on the plate is different to that for a beam (see Figs. 2.18 and 2.21(b)).

For a regular section beam, the flexural wave power transmission can be separated into two components corresponding to transverse displacement along two orthogonal axes. For the rectangular section beam shown in Fig. 2.13, these would be the y-and z-axes, and in (2.5.138) the displacement w would be replaced with w_y or w_z, depending upon the wave of interest. If both wave components were present simultaneously, then the total power transmission would simply be the arithmetic sum of the two components.

The quantities of (2.5.135) to (2.5.138) are functions of time as well as location on the cylinder, plate or beam. To enable us to express quantities in terms of net energy flow or time averaged power transmission, it is useful to find expressions for time averaged intensity. To do this we may use the same approach as that adopted for acoustic intensity. Thus, for a harmonic vibration on a beam at frequency ω we obtain

$$<P_x>_t = -\frac{1}{2}\mathrm{Re}\left\{\dot{w}^* Q_x + \dot{\theta}_x^* M_{xy} - \frac{\partial \dot{w}^*}{\partial x} M_x + \dot{\xi}_x^* N_x\right\} \tag{2.5.140}$$

where $<P_x>_t$ is the time averaged power transmission and the * denotes the complex conjugate.

Similar expressions can be obtained for plates and cylinders. In these latter two cases the left side of the equation will be time averaged power transmission per unit width. Also, as for acoustic intensity, the amplitude of the imaginary component of structural power transmission can be obtained by replacing Re in (2.5.140) with Im.

The measurement of structural intensity or structural power transmission is really only practical on simple structures such as beams, plates and shells. Even on these simple structures, it is extremely difficult to measure reactive intensity, due to the need for a minimum of four measurement locations and stringent requirements on the relative phase calibration between transducers.

In beams, plates and shells, the determination of structural intensity is possible from measurements on the surface of these elements, because relatively simple relationships have been derived between the variables which govern the energy flow through the structure and the vibration on the surface, thus enabling the determination of the flow of energy in a beam plate or shell by measurement of surface vibrations only. These simple relationships, however, only hold at lower frequencies where the motion of the interior of the structure is uniquely related to the motion of the surface. However, this is the frequency range in which active control is most useful; thus structural intensity seems a possible quantity to use as a cost function to be minimized in an active control system.

Here, expressions will be presented for the structural intensity in beams, plates and cylindrical shells in terms of the normal and in-plane surface displacements and their spatial derivatives. Experimental measurement of these quantities thus allows the determination of both active and reactive intensity components throughout a structural cross-section without the need to measure the stress directly. It will be shown how, in special cases, it is possible to use just two accelerometers to measure single wave intensities in beams and plates. The measurement of structural intensity on more general structures is still in its early development. At the time of writing, a technique has only been outlined for the measurement of surface intensity by using strain gauges together with vibration transducers (Pavic, 1987). However, it is envisaged that in the not too distant future, strain and vibration sensors will be incorporated into smart composite structures as they are manufactured, and measurement of the structural intensity at any location in the structure (and hence determination of the total structural vibratory power transmission) will be possible.

Before proceeding with the analysis for specific examples, some general observations about determining the product of two transducer signals will be summarised. Consider two time varying signals $x(t)$ and $y(t)$. The following notation will be used to denote the time averaged product in the time domain:

$$<xy>_t = <x(t)\,y(t)>_t \tag{2.5.141}$$

$$<x\int y>_t = <x(t)\int_0^t y(\tau)\,\mathrm{d}\tau>_t \tag{2.5.142}$$

For sinusoidal signals, the notation

$$x(t) = \bar{x}e^{j\omega t} \tag{2.5.143}$$

$$y(t) = \bar{y}e^{j\omega t} \tag{2.5.144}$$

where $\bar{x}$ and $\bar{y}$ may be complex, will be used. For broadband signals, the one sided cross-spectral density $G(x,y,\omega)$ will be used.

Table 2.4 contains some time and frequency domain relationships which will be used in the intensity measurement procedures to be described later.

2.5.4.1 Intensity measurement and power transmission in beams

The vibratory power propagating in a beam is given by the integral of the structural intensity due to all wave types over the beam cross section at the point of interest. Because vibratory power transmission rather than intensity at a point

Table 2.4 Signal processing relationships in time and frequency domains

Time Domain	Frequency Domain	
	Sinusoidal	Broadband
$\langle xy \rangle_t = \langle yx \rangle_t$ Amplitude of active intensity	$\frac{1}{2}\mathrm{Re}\{\bar{x}^*\bar{y}\}$	$\int_0^\infty \mathrm{Re}\{G_{x,y}(\omega)\}\mathrm{d}\omega$
Amplitude of reactive intensity	$\frac{1}{2}\mathrm{Im}\{\bar{x}^*\bar{y}\}$	$\int_0^\infty \mathrm{Im}\{G_{x,y}(\omega)\}\mathrm{d}\omega$
$\langle x\int y \rangle_t = -\langle y\int x \rangle_t$	$\frac{1}{2\omega}\mathrm{Im}\{\bar{x}^*\bar{y}\}$	$\int_0^\infty \frac{\mathrm{Im}\{G_{x,y}(\omega)\}\mathrm{d}\omega}{\omega}$
$\langle x\int\int y \rangle_t = \langle y\int\int x \rangle_t$	$-\frac{1}{2\omega^2}\mathrm{Re}\{\bar{x}^*\bar{y}\}$	$-\int_0^\infty \frac{\mathrm{Re}\{G_{x,y}(\omega)\}\mathrm{d}\omega}{\omega^2}$
$\langle \int x\int\int y \rangle_t = -\langle \int y\int\int x \rangle_t$	$\frac{1}{2\omega^3}\mathrm{Im}\{\bar{x}^*\bar{y}\}$	$\int_0^\infty \frac{\mathrm{Im}\{G_{x,y}(\omega)\}\mathrm{d}\omega}{\omega^3}$

is the preferred cost function for an active vibration control system, the following analysis for beams will express the results in terms of total power transmission. Some authors refer to this power transmission quantity as intensity, but as the units are actually power units, and to avoid confusion with the actual intensity defined in (2.5.125), the quantity will be referred to here as power. As accelerometers measure linear rather than angular accelerations, the cartesian coordinate system will be used, even for circular section beams. The beam displacements will be denoted w_y, w_z and ξ_x corresponding to lateral displacement in the y direction, lateral displacement in the z direction and axial displacement along the length of the beam (see Fig. 2.12). In the following analysis, classical (or Bernoulli–Euler) beam theory will be used.

To avoid confusion between beam power transmission and beam flexural displacement in the following analysis, the symbol used for structural power transmission will be P. (Note that the symbol W was used for power transmission in an acoustic medium.)

We will begin with (2.5.138) and derive power transmission expressions for each wave type (longitudinal, torsional and bending) in terms of the in-plane and normal beam displacements.

Longitudinal waves

From (2.5.138) we have for the power transmission

$$P_L(t) = -\dot{\xi}_x N_x = -\dot{\xi}_x \sigma_{xx} S = -\dot{\xi}_x E \epsilon_{xx} S = -SE \frac{\partial \xi_x}{\partial t}\frac{\partial \xi_x}{\partial x} \qquad (2.5.145a,b,c,d)$$

where S is the beam cross-sectional area and E is Young's modulus of elasticity. If other waves are present simultaneously, then ξ_x is the longitudinal displacement of the centre of the beam and will be written as ξ_{xo} to indicate this.

In the remainder of this section we will be discussing amplitudes of harmonically varying quantities as well as instantaneous values of these quantities. The instantaneous values are related to the amplitude as follows:

$$\xi_x = \bar{\xi}_x e^{j\omega t} \qquad (2.5.146)$$

Equation (2.5.145d) can be rewritten in terms of velocity and acceleration as follows:

$$P_L(t) = -SE \frac{\partial \xi_x}{\partial t}\frac{\partial \xi_x}{\partial x} = -SEu_x \int \frac{\partial u_x}{\partial x} = -SE \int a_x \iint \frac{\partial a_x}{\partial x} \qquad (2.5.147a\text{-}c)$$

where

$$\xi_x = \iint a_x(t)\, \mathrm{d}t\, \mathrm{d}t = \iint a_x = \int u_x \qquad (2.5.148a\text{-}c)$$

$$\frac{\partial \xi_x}{\partial t} = \int a_x(t)\, \mathrm{d}t = \int a_x = u_x \qquad (2.5.149a\text{-}c)$$

To determine the gradient, $\frac{\partial a_x}{\partial x}$ it is necessary to take simultaneous measurements at two points closely spaced a distance Δ apart, and then use a finite difference approximation. Thus the acceleration and acceleration gradient at a point midway between them are given by:

$$a_x = (a_{x1} + a_{x2})/2 \tag{2.5.150}$$

$$\frac{\partial a_x}{\partial x} = \frac{a_{x2} - a_{x1}}{\Delta} \tag{2.5.151}$$

where Δ is the spacing between the two accelerometers. A recommended value for Δ is between one-fifteenth and one-twentieth of a wavelength (Hayek *et al.*, 1990), although considerations outlined in Section 2.5.4.4 suggest that λ/10 may be more appropriate. The best value is dependent upon the structure characteristics, such as thickness and lateral dimensions, but is probably in the range stated above for most structures encountered in practice.

The accelerometers are numbered such that number 2 corresponds to a larger x-coordinate than number 1. Thus positive intensity (in the positive x-direction) is energy transmission from position 1 to position 2.

Substituting (2.5.150) and (2.5.151) into (2.5.147c) we obtain

$$P_L(t) = -\frac{SE}{2\Delta}\left[\int (a_{x1} + a_{x2}) \int\int (a_{x2} - a_{x1})\right] \tag{2.5.152}$$

The time average vibratory power transmission can thus be written as,

$$P_{La} = \langle P_L(t)\rangle_t = -\frac{SE}{2\Delta}\left\langle \int (a_{x1} + a_{x2}) \int\int (a_{x2} - a_{x1})\right\rangle_t \tag{2.5.153}$$

For harmonic excitation, $\int\int (a_{x2} - a_{x1}) = -(a_{x2} - a_{x1})/\omega^2$ and (2.5.152) can be rewritten as

$$P_L(t) = \frac{SE}{2\Delta\omega^2}\left[\left(\int (a_{x1} + a_{x2})\right)(a_{x2} - a_{x1})\right] \tag{2.5.154}$$

Thus, the corresponding time averaged power transmission can be written as

$$P_{La} = -\frac{SE}{2\omega^2\Delta}\left\langle (a_{x1} + a_{x2}) \int (a_{x2} - a_{x1})\, d\tau \right\rangle_t \tag{2.5.155}$$

or

$$P_{La} = \frac{SE}{2\Delta\omega^2}\left\langle a_{x2} \int a_{x1} \right\rangle_t \tag{2.5.156}$$

where the following properties of the two harmonic signals have been used:

$$\left\langle a_{x1} \int a_{x1} \right\rangle_t = \left\langle a_{x2} \int a_{x2} \right\rangle_t = 0 \tag{2.5.157}$$

$$\left\langle a_{x1} \int a_{x2} \right\rangle_t = -\left\langle a_{x2} \int a_{x1} \right\rangle_t \tag{2.5.158}$$

From the similarity between (2.5.155) and the corresponding (2.5.106) for acoustic intensity it may be deduced that an acoustic intensity analyser may be used to determine the structural intensity of a harmonic wave field on a beam in the far field of any sources or reflections, by replacing the pressure signals with accelerometer signals in (2.5.106) and (2.5.105) and using a different pre-multiplier.

If active control is to be used to minimize harmonic power transmission, then it can be seen from (2.5.156) that the quantity to be minimized is the time averaged product of the acceleration at the second accelerometer with the velocity at the first, where power transmission is positive in the direction from one to two. If instantaneous power transmission (both broadband and harmonic) is to be minimized, then it can be seen from (2.5.152) that the quantity to be minimized is the instantaneous product of the sum of the two velocities with the difference between the displacements.

For harmonic excitation, it is possible to derive a more convenient form of (2.5.156) by beginning with (2.5.145d) and immediately assuming harmonic excitation. Thus, for harmonic excitation

$$\bar{\xi}_x = A_1 e^{-jk_L x} + A_2 e^{jk_L x} = \bar{A}_1 e^{-j(k_L x - \theta_1)} + \bar{A}_2 e^{j(k_L x + \theta_2)} \tag{2.5.159}$$

where $\bar{A}_1$ and $\bar{A}_2$ are real amplitudes with the units of displacement and θ_1 and θ_2 represent the phases of the waves at $x = 0$. Equation (2.5.145d) can be rewritten as

$$P_{La} \doteq -\mathrm{Re}\left\{ \frac{SE}{2} (j\omega \xi_x)^* \frac{\partial \xi_x}{\partial x} \right\} = \mathrm{Re}\left\{ \frac{j\omega SE}{2}\, \bar{\xi}_x^* \frac{\partial \bar{\xi}_x}{\partial x} \right\} = \frac{\omega^2 SE}{2\, c_L} (\bar{A}_1^2 - \bar{A}_2^2)$$

(2.5.160a-c)

where $k_L = \omega / c_L$ is the wavenumber for the longitudinal wave and c_L is the wave speed. Note the absence of a nearfield component for longitudinal waves.

The amplitude of the fluctuating reactive power is:

$$P_{Lr} = \mathrm{Im}\left\{ \frac{j\omega SE}{2}\, \bar{\xi}_x^* \frac{\partial \bar{\xi}_x}{\partial x} \right\} = -\frac{\bar{A}_1 \bar{A}_2 \omega^2 SE}{c_L} \sin(2k_L x + \theta_2 - \theta_1)$$

(2.5.161a,b)

Because there is no nearfield wave component, the reactive power is associated solely with the interaction of waves travelling in opposite directions.

In terms of velocity (if a laser Doppler velocimeter is used to determine the beam response), (2.5.160b) may be written as:

$$P_{La} = \mathrm{Re}\left\{ \frac{jSE}{2\omega} \bar{u}_x^* \frac{\partial \bar{u}_x}{\partial x} \right\} \tag{2.5.162}$$

where u_x is the longitudinal velocity of the centre of the beam section (the time derivative of the displacement) at location x and the bar denotes the complex amplitude of the time varying signal.

If accelerometers were used and mounted to measure axial (or longitudinal) acceleration, then (2.5.160b) could be written as

$$P_{La} = \mathrm{Re}\left\{ \frac{jSE}{2\omega^3} \bar{a}_x^* \frac{\partial \bar{a}_x}{\partial x} \right\} \tag{2.5.163}$$

and (2.5.161a) could be written as

$$P_{Lr} = \mathrm{Im}\left\{ \frac{jSE}{2\omega^3} \bar{a}_x^* \frac{\partial \bar{a}_x}{\partial x} \right\} \tag{2.5.164}$$

where a_x is the longitudinal acceleration of the centre of the beam section.

Substituting (2.5.150) and (2.5.151) into (2.5.163) gives for the active power

$$P_{La} = -\mathrm{Im}\left\{ \frac{SE}{4\omega^3\Delta} (\bar{a}_{x1} + \bar{a}_{x2})^* (\bar{a}_{x2} - \bar{a}_{x1}) \right\} = \frac{SE}{2\omega^3\Delta} \, |\bar{a}_{x1}| \, |\bar{a}_{x2}| \sin(\theta_1 - \theta_2) \tag{2.5.165a,b}$$

and substituting (2.5.150) and (2.5.151) into (2.5.164) gives for the reactive power

$$P_{Lr} = -\mathrm{Re}\left\{ \frac{SE}{4\omega^3\Delta} (\bar{a}_{x1} + \bar{a}_{x2})^* (\bar{a}_{x2} - \bar{a}_{x1}) \right\} = \frac{SE}{4\omega^3\Delta} \left(|\bar{a}_{x2}|^2 - |\bar{a}_{x1}|^2 \right) \tag{2.5.166a,b}$$

Thus the time averaged longitudinal wave active power transmission for harmonic waves can be measured in practice by multiplying the amplitudes of two closely spaced accelerometers with the sine of the phase difference between the two. The direction of positive power transmission is from accelerometer 1 to 2. It can be shown easily that (2.5.165) is equivalent to (2.5.156); however, in many experimental situations it is probably easier to measure the amplitudes of and the phase difference between two sinusoidal signals than it is to accurately perform analogue integrations and multiplications.

Especially for broadband signals (but also for harmonic wave fields), often the most convenient way to determine the vibratory power is to use a measurement of the cross-spectrum between the two accelerometer signals, as will now be explained.

Taking the time average of (2.5.145d) gives for the active power transmission

$$<P_L(t)>_t = P_{La} = -SE\left\langle \frac{\partial \xi_x}{\partial t}\,\frac{\partial \xi_x}{\partial x} \right\rangle_t = -SE\left\langle \int a_x \int\int \frac{\partial a_x}{\partial x} \right\rangle_t \qquad (2.5.167a\text{-}c)$$

Using the relationships in Table 2.4, we can obtain the frequency dependent active power for a harmonic or broadband signal as

$$P_{La}(\omega) = \frac{SE}{\omega^3}\,\mathrm{Im}\left[G\left[\frac{\partial a_x}{\partial x}\,,\, a_x\,,\, \omega \right] \right] \qquad (2.5.168)$$

Note that $\mathrm{Im}[G(a_1, a_2, \omega)] = -\mathrm{Im}[G(a_2, a_1, \omega)]$, and the cross-spectrum is denoted as $G_{xy}(\omega)$ or $G(x, y, \omega)$. The cross-spectrum G can be determined by inputting the two acceleration signals into a spectrum analyser.

Substituting (2.5.150) and (2.5.151) into (2.5.168) gives:

$$P_{La}(\omega) = -\frac{SE}{\Delta\omega^3}\,\mathrm{Im}\left[G_{a_{x1},\, a_{x2}}(\omega) \right] \qquad (2.5.169)$$

In short hand notation, (2.5.169) may be written as:

$$P_{La}(\omega) = \frac{SE}{\Delta\omega^3}\,\mathrm{Im}\, G_{21}(\omega) \qquad (2.5.170)$$

If the accelerometer numbering convention is reversed, then G_{21} is replaced by G_{12}.

In a similar way as demonstrated by Fahy (1995) for acoustic intensity, the amplitude of the fluctuating reactive power can be shown to be

$$P_{Lr}(\omega) = -\frac{SE}{\Delta\omega^3}\left[G_{a_{x2},\, a_{x2}}(\omega) - G_{a_{x1},\, a_{x1}}(\omega) \right] = -\frac{SE}{\Delta\omega^3}\left[G_{22}(\omega) - G_{11}(\omega) \right]$$

(2.5.171a,b)

Note that for random noise, the functions $G(\omega)$ may be interpreted as cross-spectral density functions, so that the powers on the left of (2.5.169) to (2.5.171) are actually power per hertz. For single frequency or harmonic signals, the functions $G(\omega)$ are power spectrum functions, representing the total spectrum power. Alternatively, the power corresponding to each harmonic may be obtained by multiplying the cross-spectral density function by the bandwidth of each FFT filter.

For non-harmonic excitation, the concept of reactive power is meaningless because, unlike active intensity, the quantity determined over a wide frequency band will not simply be the sum of the values determined for any set of narrower frequency bands, which together make up the wide frequency band.

Torsional waves
From (2.5.138) we have

$$P_T(t) = \dot{\theta}_x M_{xy} = C\frac{\partial\theta_x}{\partial t}\frac{\partial\theta_x}{\partial x} \tag{2.5.172a,b}$$

where C is the torsional stiffness of the beam defined in Table 2.1 and (2.3.21). Equation (2.5.172) is in a similar form to (2.5.145). Thus, expressions for the active and reactive powers can be obtained from the previous section simply by replacing ξ_x with θ_x, SE with $-C$ and a_{x1}, a_{x2} with α_{x1}, α_{x2} as appropriate in all of the equations. The determination of α_x from accelerometer measurements will be discussed later.

Flexural waves
From (2.5.138) the power transmission expression for flexural waves characterized by deflections in the z-direction is given by

$$P_{Bz}(t) = -\left[-\dot{w}_z Q_x - \frac{\partial\dot{w}}{\partial x}M_x\right] \tag{2.5.173}$$

Using the definitions of Q_x and M_x from Section 2.3 for classical beam theory we obtain

$$P_{Bz}(t) = EI_{yy}\left[\frac{\partial w_z}{\partial t}\frac{\partial^3 w_z}{\partial x^3} - \frac{\partial^2 w_z}{\partial t\partial x}\frac{\partial^2 w_z}{\partial x^2}\right] \tag{2.5.174}$$

For flexural waves travelling in the x-direction and characterized by normal displacement in the y-direction, the power transmission equation is found by substituting w_y for w_z and I_{zz} for I_{yy} in (2.5.174). Note that I_{yy} is the second moment of area of the beam cross-section about the y-axis. If more than one wave type is present at one time, w_z would be replaced by w_{zo}, where the subscript o denotes displacement of the centre of the beam section.

Determination of the intensity or power transmission associated with flexural waves is more complicated than for torsional and longitudinal waves due to the higher order derivatives involved (see (2.5.174)). In fact it is necessary to use a minimum of four accelerometers to enable the third order derivative to be evaluated. Two types of accelerometer configuration, illustrated in Fig. 2.39, have been shown to give good results (Pavic, 1976; Hayek *et al.*, 1990).

For the configuration shown in Fig. 2.39(a), the derivatives are calculated using the following relations:

$$w = w_o = \frac{w_3 + w_2}{2} \tag{2.5.175a,b}$$

$$\frac{\partial w}{\partial x} = \frac{w_3 - w_2}{\Delta} \tag{2.5.176}$$

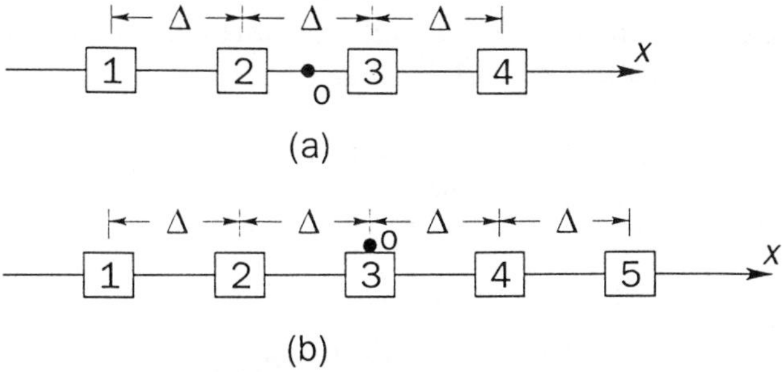

Fig. 2.39 Accelerometer configurations for determining higher order derivatives of the beam displacement at o: (a) Pavic (1976); (b) Hayek (1990).

$$\frac{\partial^2 w}{\partial x^2} = \frac{w_1 - w_2 - w_3 + w_4}{2\Delta^2} \tag{2.5.177}$$

$$\frac{\partial^3 w}{\partial x^3} = \frac{-w_1 + 3w_3 - 3w_2 + w_4}{\Delta^3} \tag{2.5.178}$$

Note that the accelerometer numbering convention adopted has been increasing number in the direction of increasing x. Although this is opposite to that adopted by Pavic (1976), it is the convention almost universally used by others. Also, as mentioned previously, we have maintained the moment and force convention which has been used for decades by those concerned with the dynamic analysis of plates. Unfortunately this convention has not been followed universally by those involved in the measurement of structural intensity. Thus some of the results presented here may look a little different to what appears in some of the published literature.

For the configuration in Fig. 2.39(b), the following expressions apply:

$$w = w_o \tag{2.5.179}$$

$$\frac{\partial w}{\partial x} = \frac{w_4 - w_2}{2\Delta} \tag{2.5.180}$$

$$\frac{\partial^3 w}{\partial x^2} = \frac{-w_1 + 2w_2 - 2w_4 + w_5}{2\Delta^3} \tag{2.5.181}$$

The latter equations (involving five accelerometers) give more accurate results than the equations for the four accelerometer case. However, Hayek *et al.* (1990) show that no further benefit is gained by using more measurement points, and that the optimum value for Δ is one-twentieth of a structural wavelength.

Taking the time average of (2.5.174) gives the active component of the power transmission. Thus,

$$P_{Bza} = \langle P_{Bz}(t) \rangle_t = EI_{yy} \left\langle \left(\frac{\partial w_z}{\partial t}\frac{\partial^3 w_z}{\partial x^3} - \frac{\partial}{\partial t}\left[\frac{\partial w_z}{\partial x}\right]\frac{\partial^2 w_z}{\partial x^2} \right) \right\rangle_t \tag{2.5.182a,b}$$

Using the notation introduced earlier, and assuming that the measurements are made with accelerometers, we have

$$w_z = \int\int a_z \tag{2.5.183}$$

$$\frac{\partial w_z}{\partial t} = \int a_z \tag{2.5.184}$$

Thus,

$$P_{Bza} = EI_{yy}\left\langle \int a_z \int\int \frac{\partial^3 a_z}{\partial x^3} - \int \frac{\partial a_z}{\partial x} \int\int \frac{\partial^2 a_z}{\partial x^2} \right\rangle_t \tag{2.5.185}$$

Using (2.5.175) to (2.5.178) (assuming four accelerometers) we obtain (Pavic, 1976):

$$P_{Bza} = \frac{EI_{yy}}{\Delta^3}\left\langle 4\int a_{z2}\int\int a_{z3} - \int a_{z2}\int\int a_{z4} - \int a_{z1}\int\int a_{z3}\right\rangle_t \tag{2.5.186}$$

Thus, using the relations in Table 2.4, the power is given in terms of the cross-spectrum by

$$P_{Bza}(\omega) = \frac{EI_{yy}}{\Delta^3\omega^3}\left[4G_{a_2,a_3}(\omega) - G_{a_2,a_4}(\omega) - G_{a_1,a_3}(\omega)\right] \tag{2.5.187}$$

If the measurements were performed using a laser Doppler velocimeter rather than accelerometers corresponding cross-spectral expressions for structural power transmission can be derived using the relationships of Table 2.4. If u_z represents the measured velocity, then the relationship corresponding to (2.5.187) for the active power transmission is

$$P_{Bza}(\omega) = \frac{EI_{yy}}{\Delta^3\omega}\left[4G_{u_2,u_3}(\omega) - G_{u_2,u_4}(\omega) - G_{u_1,u_3}(\omega)\right] \tag{2.5.188}$$

A simpler expression can be obtained if the accelerometers are located in the far field of all vibration sources and reflections. In this case, it may be shown that the shear force and bending moment contributions to the power transmission are the same (Noiseux, 1970) and the instantaneous power transmission may be written as

$$P_{Bz}(t) = -2EI_{yy}\left[\frac{\partial^2 w_z}{\partial t\,\partial x}\frac{\partial^2 w_z}{\partial x^2}\right] \tag{2.5.189}$$

The time averaged (or active) power may be written as

$$<P_{Bz}(t)> = P_{Bza} = -2EI_{yy}\left\langle \int \frac{\partial a}{\partial x}\int\int \frac{\partial^2 a}{\partial x^2}\right\rangle \tag{2.5.190a,b}$$

Thus, from Table 2.4, we obtain for the frequency domain:

$$P_{Bza}(\omega) = 2EI_{yy}\frac{\mathrm{Im}\left\{ G\left[\frac{\partial^2 a}{\partial x^2}, \frac{\partial a}{\partial x}, \omega\right]\right\}}{\omega^3} \tag{2.5.191}$$

In the absence of near fields, the relation between the Fourier components of $\frac{\partial^2 a}{\partial x^2}$ and a is

$$\frac{\partial^2 a}{\partial x^2} = -k_b^2 a \qquad (2.5.192)$$

Remembering from Section 2.3 that

$$\omega(m'/B)^{1/2} = k_b^2 \qquad (2.5.193)$$

where m' is the beam mass per unit length and $B = EI_{yy}$, and remembering the finite difference approximations

$$a = (a_{z1} + a_{z2})/2 \qquad (2.5.194)$$

$$\frac{\partial a}{\partial x} = (a_{z2} - a_{z1})/\Delta \qquad (2.5.195)$$

we obtain the following for the active power transmission:

$$P_{Bza}(\omega) = \frac{2(Bm')^{1/2}}{\omega^2\Delta} Im\left\{ G_{a_{z2},a_{z1}}(\omega)\right\} \qquad (2.5.196)$$

The corresponding expression for the reactive intensity is:

$$P_{Bzr}(\omega) = \frac{(Bm')^{1/2}}{\omega^2\Delta}\left[G_{a_{z1},a_{z1}} - G_{a_{z2},a_{z2}}\right] \qquad (2.5.197)$$

where $G_{a_{z1},a_{z1}}$ is the auto (power) spectral density of the accelerometer signal z_1.

Assuming a harmonic wave field and substituting (2.5.192) and (2.5.193) into (2.5.190b), gives for the time average active power

$$P_{Bza} = -\frac{2k_b^2 EI_{yy}}{\omega^2}\left\langle\left[\int\frac{\partial a}{\partial x}\right] a\right\rangle_t = -\frac{2(Bm')^{1/2}}{\omega}\left\langle\left[\int\frac{\partial a}{\partial x}\right] a\right\rangle_t \qquad (2.5.198a,b)$$

Using (2.5.194) and (2.5.195), (2.5.198) may be rewritten as

$$P_{Bza} = -\frac{(Bm')^{1/2}}{\omega\Delta}\left\langle(a_{z1} + a_{z2})\int(a_{z2} - a_{z1})\,d\tau\right\rangle_t \qquad (2.5.199)$$

Using (2.5.157) and (2.5.158), (2.5.199) can be rewritten as

$$P_{Bza} = \frac{2(Bm')^{1/2}}{\omega\Delta}\left\langle a_{z2}\int a_{z1}\right\rangle_t \qquad (2.5.200)$$

From the similarity between (2.5.199) and the corresponding equation (2.5.106) for acoustic intensity it may be deduced that an acoustic intensity analyser may be used to determine the structural intensity for a harmonic wave field on a beam in the far field of any sources or reflections, by replacing the pressure signals with accelerometer signals in (2.5.106). Alternatively, (2.5.200) and an analogue multiplying circuit may be used to multiply the signal from one accelerometer with the integrated signal from the other.

As was done for the calculation of longitudinal wave power, an alternative relationship can be derived by beginning with (2.5.174) and assuming harmonic excitation immediately. Thus, from (2.5.174), the time averaged active power transmission is

$$P_{Bza} = \mathrm{Re}\left\{\frac{EI_{yy}}{2}\left[\left(j\omega\bar{w}_z(x)\right)^* \frac{\partial^3\bar{w}_z(x)}{\partial x^3} - \frac{\partial\left(j\omega\bar{w}_z(x)\right)^*}{\partial x}\frac{\partial^2\bar{w}_z(x)}{\partial x^2}\right]\right\} \tag{2.5.201}$$

or

$$P_{Bza} = -\mathrm{Re}\left\{\frac{j\omega EI_{yy}}{2}\left[\bar{w}_z^*(x)\frac{\partial^3\bar{w}_z(x)}{\partial x^3} - \frac{\partial\bar{w}_z^*(x)}{\partial x}\frac{\partial^2\bar{w}_z(x)}{\partial x^2}\right]\right\} \tag{2.5.202}$$

and the amplitude of the fluctuating reactive component is

$$P_{Bzr} = -\mathrm{Im}\left\{\frac{j\omega EI_{yy}}{2}\left[\bar{w}_z^*(x)\frac{\partial^3\bar{w}_z(x)}{\partial x^3} - \frac{\partial\bar{w}_z^*(x)}{\partial x}\frac{\partial^2\bar{w}_z(x)}{\partial x^2}\right]\right\} \tag{2.5.203}$$

where the * denotes the complex conjugate.

As for the general case, simpler expressions can be obtained for P_{Bza} if the measurements are conducted at least one half of a wavelength away from any power source or sources of reflection; that is, in a region where near field effects may be neglected. Recalling the solution for a sinusoidal flexural wave travelling in a beam derived in Section 2.3, we have:

$$w_z(x,t) = \left[A_1 e^{-jk_b x} + A_2 e^{jk_b x} + A_3 e^{-k_b x} + A_4 e^{k_b x}\right]e^{j\omega t} \tag{2.5.204}$$

The first and second terms represent the propagating vibration field in the positive and negative x directions respectively, while the second two terms represent the decaying near field. Thus if we ignore the near field, we can set $A_3 = A_4 = 0$, and if we also omit the time dependent term $e^{j\omega t}$, (2.5.204) becomes

$$\bar{w}_z(x) = A_1 e^{-jk_b x} + A_2 e^{jk_b x} = \bar{A}_1 e^{-j(k_b x - \theta_1)} + \bar{A}_2 e^{j(k_b x + \theta_2)} \tag{2.5.205}$$

where A_1 and A_2 are complex numbers, $\bar{A}_1$ and $\bar{A}_2$ are real, and θ_1, θ_2 are the signal phases at $x = 0$. Substituting (2.5.205) into (2.5.202) gives for the active power:

$$P_{Bza} = EI_{yy}\,\omega\, k_b^3\,(\bar{A}_1^2 - \bar{A}_2^2) \tag{2.5.206}$$

However,

$$\mathrm{Re}\left\{ j\left[\bar{w}_z^* \frac{\partial \bar{w}_z}{\partial x} - \bar{w}_z \frac{\partial \bar{w}_z^*}{\partial x} \right] \right\} = 2\,k_b\,(\bar{A}_1^2 - \bar{A}_2^2) \qquad (2.5.207)$$

Thus,

$$P_{Bza} = \mathrm{Re}\left\{ \left[\frac{j\omega}{2} \right] EI_{yy}\, k_b^2 \left[\bar{w}_z^* \frac{\partial \bar{w}_z}{\partial x} - \bar{w}_z \frac{\partial \bar{w}_z^*}{\partial x} \right] \right\} = \omega EI_{yy} k_b^2 \,\mathrm{Im}\left\{ \bar{w}_z \frac{\partial \bar{w}_z^*}{\partial x} \right\}$$

(2.5.208a,b)

That is, the contribution due to the shear force is equal to the contribution due to the bending moment in the far field. This is in contrast to the situation in the near field of sources where it has been found (Pavic, 1990) that the shear force component dominates the bending wave component.

Substituting (2.5.205) into (2.5.203) gives for the reactive power

$$P_{Bzr} = 2EI_{yy}\,\omega k_b^3 \bar{A}_1 \bar{A}_2 \sin(2k_b x + \theta_2 - \theta_1) = \omega EI_{yy} k_b^2 \mathrm{Re}\left\{ \bar{w}_z \frac{\partial \bar{w}_z^*}{\partial x} \right\} \qquad (2.5.209a,b)$$

If accelerometers were used and mounted to measure lateral (or flexural) acceleration, then (2.5.208b) could be written as

$$P_{Bza} = \mathrm{Re}\left\{ \frac{jEI_{yy} k_b^2}{\omega^3} \bar{a}_z^* \frac{\partial \bar{a}_z}{\partial x} \right\} \qquad (2.5.210)$$

and (2.5.209b) could be written as

$$P_{Bzr} = \mathrm{Im}\left\{ \frac{jEI_{yy} k_b^2}{\omega^3} \bar{a}_z^* \frac{\partial \bar{a}_z}{\partial x} \right\} \qquad (2.5.211)$$

where a_z is the acceleration of the centre of the beam section in the z-direction.

Substituting (2.5.194) and (2.5.195) into (2.5.210) gives for the active power

$$P_{Bza} = -\mathrm{Im}\left\{ \frac{EI_{yy} k_b^2}{2\omega^3 \Delta} (\bar{a}_{z1} + \bar{a}_{z2})^* (\bar{a}_{z2} - \bar{a}_{z1}) \right\} = \frac{EI_{yy} k_b^2}{\omega^3 \Delta} \mid \bar{a}_{z1} \mid \; \mid \bar{a}_{z2} \mid \sin(\theta_1 - \theta_2)$$

(2.5.212a,b)

and substituting (2.5.194) and (2.5.195) into (2.5.211) gives for the reactive power

$$P_{Bzr} = -\mathrm{Re}\left\{ \frac{EI_{yy}k_b^2}{2\omega^3\Delta}(\bar{a}_{z1} + \bar{a}_{z2})^* (\bar{a}_{z2} - \bar{a}_{z1}) \right\} = \frac{EI_{yy}k_b^2}{\omega^3\Delta}\left(|\bar{a}_{z2}|^2 - |\bar{a}_{z1}|^2 \right)$$

(2.5.213a,b)

Thus the time averaged flexural wave active power transmission for harmonic waves can be measured in practice by multiplying the amplitudes of two closely spaced accelerometers with the sine of the phase difference between the two. The direction of positive power transmission is from accelerometer 1 to 2. It can be shown easily that (2.5.212) is equivalent to (2.5.199); however, in many experimental situations it is probably easier to measure the amplitudes of and the phase difference between two sinusoidal signals than it is to accurately perform analog integrations and multiplications.

Two element probes with the accelerometers mounted on a common base are commercially available and have been designed specifically for measuring flexural wave structural intensity in the far field away from structural discontinuities or vibration sources. However, at high frequencies large errors can occur using these probes due to the phase error introduced because of the flexibility of the base on which the accelerometers are mounted.

Total power transmission

Using (2.5.145), (2.5.172) and (2.5.174), the total instantaneous power transmission in the beam as a result of the propagation of all wave types (two orthogonal flexural waves, one longitudinal wave and one torsional wave) can be written in matrix form as:

$$P(t) = \frac{\partial}{\partial t} \mathbf{W}_0^{\mathrm{T}}(t)\, \mathbf{\Lambda}\, \mathbf{W}_0(t) \tag{2.5.214}$$

where

$$\mathbf{W}_0(t) = [\xi_{xo}(t),\ w_{yo}(t),\ w_{zo}(t),\ \theta_x(t),\ \theta_y(t),\ \theta_z(t)]^{\mathrm{T}} \tag{2.5.215}$$

where T denotes the transpose of a matrix or vector and where ξ_{xo}, w_{yo} and w_{zo} represent the displacements of the centre of the beam section in the x, y and z directions respectively and θ_x, θ_y, θ_z represent the rotations of the whole section about the x-, y- and z-axes respectively. The diagonal matrix Λ is given by:

$$\mathbf{\Lambda} = \begin{bmatrix} -ES\dfrac{\partial}{\partial x} &&&&& \\ & EI_{zz}\dfrac{\partial^3}{\partial x^3} &&&& \\ && EI_{yy}\dfrac{\partial^3}{\partial x^3} &&& \\ &&& -C\dfrac{\partial}{\partial x} && \\ &&&& -EI_{yy}\dfrac{\partial}{\partial x} & \\ &&&&& -EI_{zz}\dfrac{\partial}{\partial x} \end{bmatrix} \tag{2.5.216}$$

where I_{yy} and I_{zz} are the second moments of area of the cross-section about the y and z axes respectively, E is Young's modulus of elasticity, C is the torsional stiffness and S is the area of beam cross-section.

The angular rotations are defined as

$$\theta_x = \frac{\partial w_{zo}}{\partial y} = -\frac{\partial w_{yo}}{\partial z} \tag{2.5.217a,b}$$

$$\theta_y = -\frac{\partial w_{zo}}{\partial x} \tag{2.5.218}$$

$$\theta_z = \frac{\partial w_{yo}}{\partial x} \tag{2.5.219}$$

For single frequency beam excitation, the time averaged (or active) power transmission is given by:

$$P_a = -\frac{1}{2}\mathrm{Re}\left\{j\omega\,\bar{\mathbf{W}}_0^{\mathrm{H}}\,\mathbf{\Lambda}\,\bar{\mathbf{W}}_0\right\} \tag{2.5.220}$$

and the amplitude of the fluctuating imaginary component is:

$$P_r = -\frac{1}{2}\mathrm{Im}\left\{j\omega\bar{\mathbf{W}}_0^{\mathrm{H}}\,\mathbf{\Lambda}\,\bar{\mathbf{W}}_0\right\} \tag{2.5.221}$$

where the bar indicates the complex amplitude of a time varying quantity, and H is the transpose of the complex conjugate.

If the beam vibration is broadband as opposed to sinusoidal, the overall active intensity can be obtained by averaging the quantities in (2.5.220) in the time domain or by using the cross spectral representation in the frequency domain.

Measurement of beam accelerations

In the previous analysis it has been assumed that the overall displacements of the beam cross-section (w_{yo}, a_{zo}, a_{xo} and α_{xo}) and θ_x or corresponding accelerations a_{yo}, a_{zo}, a_{xo} and α_{xo}) can be determined by simple measurements on the beam surface. In practice, the measurements are not as straightforward as they seem at first, due to the dependence on more than one wave type of the surface displacement in any one direction. However this problem can be mostly overcome for symmetrical beams using an arrangement suggested by Verheij (1990) for measurements on circular pipes, and illustrated in Fig. 2.40. Means of extracting the amplitudes corresponding to each wave type are given in the figure caption. Note that the accelerometers are orientated and their outputs combined in such a way as to isolate each of the wave types from any influence from the others. For example, the Poisson contraction and expansion associated with longitudinal wave propagation acts in opposite directions on two opposite faces of the beam; thus subtracting one from the other of the outputs of the two accelerometers shown for measuring bending waves (w_y or w_z) will null this effect. Similarly, the longitudinal displacement of the beam surface as a result of section rotation due to bending wave propagation is nulled by adding the outputs of the two accelerometers shown for measuring longitudinal waves (or ξ_x). Thus, use of the measurement scheme shown in Fig. 2.40 theoretically allows measurement of the amplitude of each wave type regardless of the simultaneous presence or otherwise of other wave types. However, in practice problems can arise as a result of the non-zero cross-axis sensitivity of the accelerometers if all three wave types are present. Cross-axis sensitivity is discussed in more detail in Chapter 16, but simply put, it means that accelerometers are sensitive to motion in directions other than along their main axis as well as along their main axis. The cross-axis sensitivity is direction dependent with a maximum value usually of about 5% of the main axis sensitivity. Some manufacturers indicate the direction of least sensitivity with a mark on the accelerometer, and this can be two orders of magnitude less than the maximum value. Ignoring the effects of cross-axis sensitivity can lead to serious measurement errors, especially if one is attempting to measure the amplitude of a wave which is much smaller than the amplitude of the other types. For any measurement using the configuration shown in Fig. 2.40, it is possible to minimize the error resulting from cross-axis sensitivity by appropriate orientation of the accelerometers used for the measurements. Another problem associated with the scheme shown in Fig. 2.40 is that it only applies to beams which are rectangular, circular or ellipsoidal in cross-section.

An alternative scheme which allows the amplitude of each wave type to be determined accurately and which also allows the cross-axis sensitivity of the accelerometers to be taken into account, involves the measurement of the x-, y- and z-components of the displacement on the beam surface at three or more locations. In addition, the beam does not have to be rectangular, circular or ellipsoidal in cross section for this method to work, in contrast to the scheme shown in Fig. 2.36. However, the y- and z-axes must coincide with the two

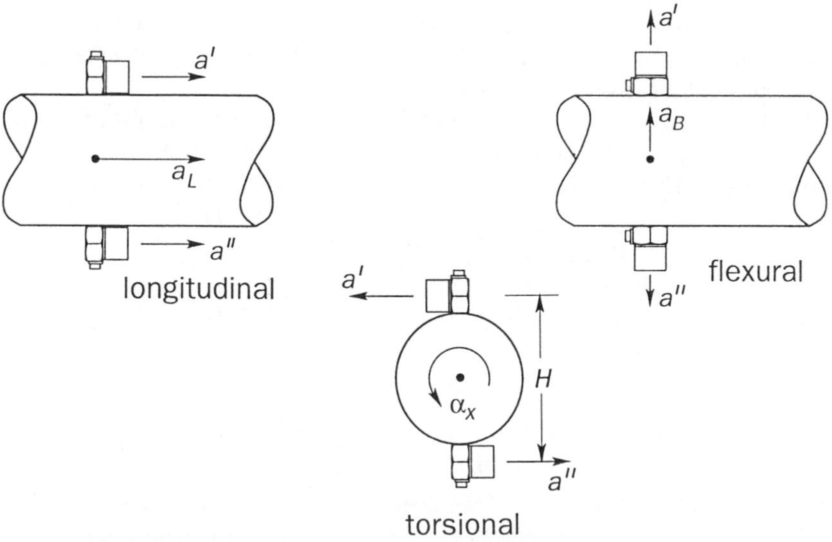

Fig. 2.40 Simplified scheme for determining accelerations at a point on the beam, assuming that each wave exists in isolation from the others. A similar scheme would also apply to rectangular cross-section beams: $a_L = (a' + a'')/2$, $a_B = (a' - a'')/2$, $a_x = (a' + a')/H$.

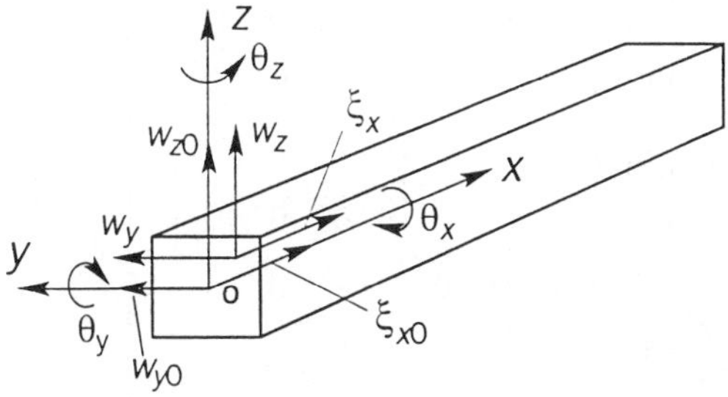

Fig. 2.41 Coordinate system for a rectangular beam.

principal orthogonal axes of the beam cross section, as shown in Fig. 2.41.

The displacement components w_z, w_y and ξ_x at any location on a beam cross-section are described by the following matrix equation which relates them to the displacements and rotations of the centre of the section (or section as a whole):

$$\begin{bmatrix} \xi_x \\ w_y \\ w_z \end{bmatrix} = \begin{bmatrix} 1 & 0 & 0 & 0 & z & -y \\ 0 & 1 & 0 & -z & 0 & 0 \\ 0 & 0 & 1 & y & 0 & 0 \end{bmatrix} \begin{bmatrix} \xi_{x0} \\ w_{y0} \\ w_{z0} \\ \theta_x \\ \theta_y \\ \theta_z \end{bmatrix} \tag{2.5.222}$$

and where y and z are coordinate locations indicating the distance of the measurement point from the x-axis of the beam. The subscript 0 indicates a measurement at the centre of the beam.

In theory at least, (2.5.222) can be used to determine the displacements of the beam section as a whole (or the displacements of the centre) from only six measurements. Thus,

$$\begin{bmatrix} \xi_{x1} \\ w_{y1} \\ w_{z1} \\ \xi_{x2} \\ w_{y2} \\ w_{z2} \end{bmatrix} = \begin{bmatrix} 1 & 0 & 0 & 0 & z_1 & -y_1 \\ 0 & 1 & 0 & -z_1 & 0 & 0 \\ 0 & 0 & 1 & y_1 & 0 & 0 \\ 1 & 0 & 0 & 0 & z_2 & -y_2 \\ 0 & 1 & 0 & -z_2 & 0 & 0 \\ 0 & 0 & 1 & y_2 & 0 & 0 \end{bmatrix} \begin{bmatrix} \xi_{x0} \\ w_{y0} \\ w_{z0} \\ \theta_x \\ \theta_y \\ \theta_z \end{bmatrix} \tag{2.5.223}$$

Unfortunately, the determinant of the coefficient matrix on the right-hand side of (2.5.223) is zero, which indicates that more measurements are needed. If a third measurement location is used, we have,

$$\begin{bmatrix} \xi_{x1} \\ w_{y1} \\ w_{z1} \\ \xi_{x2} \\ w_{y2} \\ w_{z2} \\ \xi_{x3} \\ w_{y3} \\ w_{z3} \end{bmatrix} = \begin{bmatrix} 1 & 0 & 0 & 0 & z_1 & -y_1 \\ 0 & 1 & 0 & -z_1 & 0 & 0 \\ 0 & 0 & 1 & y_1 & 0 & 0 \\ 1 & 0 & 0 & 0 & z_2 & -y_2 \\ 0 & 1 & 0 & -z_2 & 0 & 0 \\ 0 & 0 & 1 & y_2 & 0 & 0 \\ 1 & 0 & 0 & 0 & z_3 & -y_3 \\ 0 & 1 & 0 & -z_3 & 0 & 0 \\ 0 & 0 & 1 & y_3 & 0 & 0 \end{bmatrix} \begin{bmatrix} \xi_{x0} \\ w_{y0} \\ w_{z0} \\ \theta_x \\ \theta_y \\ \theta_z \end{bmatrix} \tag{2.5.224}$$

For future reference, the 9 × 6 location matrix in (2.5.224) will be denoted $\boldsymbol{A}$. A unique solution will exist for (2.5.224) only if the matrix $\boldsymbol{A}^{\mathrm{T}}\boldsymbol{A}$ is non-singular. If the accelerometer locations are so selected, $\boldsymbol{W}_0$ can be calculated by:

$$\boldsymbol{W}_0 = \left(\boldsymbol{A}^{\mathrm{T}}\boldsymbol{A}\right)^{-1} \left[\xi_{x1}, w_{y1}, w_{z1}, \xi_{x2}, w_{y2}, w_{z2}, \xi_{x3}, w_{y3}, w_{z3},\right]^{\mathrm{T}} \tag{2.5.225}$$

For a rectangular section beam, the matrix will be non-singular if the measurement locations are in the centre of any three of the four sides of a given section. For a circular beam, the measurement locations should be on the orthogonal y- and z-axes (these axes are defined for all but perfectly circular beams and in this latter case, the measurements should be on any two cross-sectional axes, y and z separated by 90°).

Note that (2.5.222) to (2.5.224) give complex amplitudes of the displacements of the centre of the beam cross-section which allows determination of the relative phases of the various propagating wave types.

One limitation of the method just described is that the Poisson contraction effect, resulting in strain in the transverse direction as a result of longitudinal waves travelling in the x-direction has been ignored. In most cases, this will not be a problem as it will be too small to be of importance. The magnitude of the lateral displacement at location z on a particular cross-section as a result of longitudinal wave propagation is:

$$w_z = z\epsilon_{zz} = -\nu z\epsilon_{xx} = -\nu z\frac{\partial \xi_x}{\partial x} \qquad (2.5.226a,b,c)$$

Substituting (2.5.148) into (2.5.226), assuming a wave travelling in only one direction and carrying out the differentiation it can be shown that for a material with a Poisson's ratio of approximately 0.3, the lateral displacement due to longitudinal wave propagation is approximately

$$w_z = \frac{h}{\lambda_L}\xi_x \qquad (2.5.227)$$

For a rectangular beam, the maximum value of z is half the beam thickness, and for many practical beams the longitudinal wavelength of longitudinal waves is much greater than the beam thickness, so the quantity in (2.5.227) is usually very small compared to the lateral displacement due to flexural waves. Nevertheless there will be cases where longitudinal waves will dominate the response and the Poisson contraction effect must be taken into account. For beams of regular section, this is best done using the arrangements shown in Fig. 2.40, where the Poisson effect is automatically cancelled by using two flexural wave accelerometers mounted on opposite sides of the beam.

It is interesting to note that in a thin wide beam, it is difficult to excite the torsional wave and the flexural wave characterized by displacement along the width of the beam, and the problem is reduced to one of identifying only one flexural wave and the longitudinal wave. In this case, the arrangement shown in Fig. 2.40 can lead to quite accurate results if the accelerometers for measuring the longitudinal waves are orientated so that their direction of minimum cross-axis sensitivity is in the direction of the flexural wave displacement and if the accelerometers for measuring the flexural wave are orientated so that their direction of minimum cross-axis sensitivity is in the direction of the longitudinal wave displacement. Only one accelerometer need be used for measuring the longitudinal wave if it is mounted on the thin edge of the beam.

Means of determining the amplitudes of waves travelling simultaneously in two different directions along the beam, using the acceleration measurements described in the preceding paragraphs, are discussed in Section 10.2.7. Briefly, at least four accelerometers are needed to resolve each pair of bending waves (provided that they are in the far field of any source or beam discontinuity), at least four accelerometers are needed to resolve the pair of longitudinal waves and at least four accelerometers are needed to resolve the pair of torsional waves. They must be mounted on at least two different beam cross-sections separated by no less than one-fifteenth of a wavelength and no more than one-third of a wavelength. Use of more accelerometers and more measurement cross-sections increases the accuracy of the results substantially. If more than two beam cross-sections are used, then the separation between the sections furthest apart should be no more than one-third of a wavelength. Note that the same cross-sectional locations may be different for each wave type (mainly because the wavelengths corresponding to each wave type are generally very different). For lightweight structures, the number of accelerometers needed may be sufficient to significantly affect the beam dynamics.

Effect of transverse sensitivity of accelerometers

If accelerometers are used for the measurements, they will invariably exhibit some degree of sensitivity (transverse sensitivity) along axes at right angles to their measurement axis, thus resulting in inaccuracies in the determination of the measured (ξ_x, w_z, w_y). However, if the cross-axis sensitivity as a fraction of the main axis sensitivity is known or measured beforehand, then the following relation may be used to determine the actual displacements (or accelerations) at any location from the measured ones. Assuming the actual displacements are ξ_x, w_y and w_z at a given measurement point, and the measured displacements are ξ_{mx}, w_{my} and w_{mz}, the two are related by the accelerometer cross-axis sensitivities as follows:

$$\begin{bmatrix} \xi_{mx} \\ w_{my} \\ w_{mz} \end{bmatrix} = \begin{bmatrix} 1 & \alpha_x & \alpha_x \\ \alpha_y & 1 & \alpha_y \\ \alpha_z & \alpha_z & 1 \end{bmatrix} \begin{bmatrix} \xi_x \\ w_y \\ w_z \end{bmatrix} \tag{2.5.228}$$

or

$$\begin{bmatrix} \xi_x \\ w_y \\ w_z \end{bmatrix} = \begin{bmatrix} 1 & \alpha_x & \alpha_x \\ \alpha_y & 1 & \alpha_y \\ \alpha_z & \alpha_z & 1 \end{bmatrix}^{-1} \begin{bmatrix} \xi_{mx} \\ w_{my} \\ w_{mz} \end{bmatrix} \tag{2.5.229}$$

where α_x, α_y and α_z are the cross-axis sensitivities of the accelerometers measuring ξ_{mx}, w_y and w_z respectively. Note that the cross-axis sensitivity of an accelerometer is usually strongly dependent upon the direction of the axis of interest (or angular orientation of the accelerometer) and this must also be taken into account during calibration and mounting of the accelerometers. Some accelerometers are available for which the direction of minimum cross-axis sensitivity is clearly marked, and in some cases this sensitivity is negligible, thus making the corrections outlined in this section unnecessary for situations involving the propagation of only two wave types.

2.5.4.2 Intensity measurement in plates

The instantaneous power transmission per unit width in a plate given by (2.5.137) may be expressed in terms of plate displacements using (2.3.92) to (2.3.94), (2.3.98) and (2.3.99) for classical plate theory and (2.3.107) to (2.3.109), (2.3.112) and (2.3.113) for Mindlin–Timoshenko plate theory which should be used for thick plates and/or high frequencies.

It can be seen from (2.5.137) that first, second and third order derivatives must be approximated to completely determine the sound intensity in a thin plate. However, it will be shown here that under certain conditions, the two accelerometer method (discussed for beams) may be used to determine the flexural wave component of the total power transmission. It can also be seen from (2.5.137) that in both the near and far field of sources it is necessary to measure the in-plane plate displacements and the second and third derivatives of the normal plate displacements to determine the structural power transmission in longitudinal and flexural waves.

As discussed by Pavic (1976) it is necessary to use eight accelerometers to evaluate the required derivatives in (2.5.137). The required finite difference equations may be formulated as for a beam. Here we will examine the simpler case of a harmonic sound field for which the following expressions for the active and reactive intensity can be derived from (2.5.137):

$$P_{xa} = \frac{1}{2}\mathrm{Re}\left\{ j\omega \left[\bar{w}^*\bar{Q}_x - \frac{\partial \bar{w}^*}{\partial x}\bar{M}_x - \frac{\partial \bar{w}}{\partial y}\bar{M}_{xy} + \bar{\xi}_x^*\bar{N}_x + \xi_y^*\bar{N}_{xy} \right] \right\} \tag{2.5.230}$$

$$P_{xr} = \frac{1}{2}\mathrm{Im}\left\{ j\omega \left[\bar{w}^*\bar{Q}_x - \frac{\partial \bar{w}^*}{\partial x}\bar{M}_x - \frac{\partial \bar{w}^*}{\partial y}\bar{M}_{xy} + \bar{\xi}_x^*\bar{N}_x + \bar{\xi}_y^*\bar{N}_{xy} \right] \right\} \tag{2.5.231}$$

where the asterisk denotes the complex conjugate and all quantities in the equations are complex (expressed as an amplitude and a relative phase). The first three terms in (2.5.230) and (2.5.231) are associated with bending waves, the fourth term is associated with longitudinal waves and the last term is associated with in-plane shear waves. Simpler expressions which can be implemented using two accelerometers will now be derived for each of these wave types.

Longitudinal waves

From (2.5.137), the power transmission per unit plate width associated with longitudinal wave propagation in the x-direction is given by

$$P_L(t) = -\dot{\xi}_x N_x \tag{2.5.232}$$

From equations (2.3.78), (2.3.84) to (2.3.87) and (2.3.89), the force component, N_x, may be written as,

$$N_x = \int_{-h/2}^{h/2} \sigma_{xx}\,\mathrm{d}z = \frac{Eh}{1-\nu^2}\left[\frac{\partial \xi_x}{\partial x} + \nu\frac{\partial \xi_y}{\partial y}\right] \tag{2.5.233a,b}$$

For longitudinal wave propagation in the x-direction only, ξ_y is negligible and we may write (2.5.232) as

$$P_L(t) = -\frac{Eh}{1-\nu^2}\frac{\partial \xi_x}{\partial t}\frac{\partial \xi_x}{\partial x} \tag{2.3.234}$$

which is equivalent to (2.5.147a) for a beam, where the cross-sectional area S, for a beam has been replaced with $h/(1-\nu^2)$, where h is the plate thickness and ν is Poisson's ratio.

Thus, all of the equations and measurement techniques derived previously for longitudinal wave power transmission in a beam are valid for longitudinal wave power transmission in a plate provided that S is replaced with $h/(1-\nu^2)$.

Similar arguments hold for longitudinal wave propagation in the y-direction, where the same equations may be used with the x and y subscripts interchanged.

Transverse shear waves

From (2.5.137), the power transmission per unit width associated with shear waves is

$$P_s(t) = -\dot{\xi}_y N_{xy} \tag{2.5.235}$$

From (2.3.72) to (2.3.75), the force component, N_{xy}, can be written as

$$N_{xy} = Gh\frac{\partial \xi_y}{\partial x} \tag{2.5.236}$$

For shear wave only propagation, (2.5.235) may be written as

$$P_s(t) = -Gh\frac{\partial \xi_y}{\partial t}\frac{\partial \xi_y}{\partial x} \tag{2.5.237}$$

which has the same form as (2.5.147a) for a beam, except that the longitudinal displacement is measured in the y-direction rather than the x-direction. Thus, all of the previously derived expressions for a beam can be used if the quantity SE is replaced with Gh and the measurement transducers are configured to measure longitudinal displacement in the y-direction (remember that only power transmission in the x-direction is being considered for now). Note that although the displacement in the y-direction is to be measured, it is the gradient of this in the x-direction which is required; thus, the two measurement transducers must be aligned with the x-axis.

Similar arguments hold for shear wave propagation in the y-direction, where the same equations may be used with the x and y subscripts interchanged.

Flexural waves

From (2.5.137), the power transmission per unit plate width associated with flexural wave propagation in the x-direction is given by

$$P_{Bx}(t) = -\dot{w}Q_x + \frac{\partial \dot{w}}{\partial y}M_{xy} + \frac{\partial \dot{w}}{\partial x}M_x \tag{2.5.238}$$

Substituting (2.3.92), (2.3.94) and (2.3.98) into (2.3.238) gives

$$P_{Bx}(t) = D\left[\frac{\partial w}{\partial t}\left(\frac{\partial^3 w}{\partial x^3} + \frac{\partial^3 w}{\partial x \partial y^2}\right)\right.$$
$$\left. - (1-\nu)\frac{\partial^2 w}{\partial t \partial y}\frac{\partial^2 w}{\partial x \partial y} - \frac{\partial^2 w}{\partial t \partial x}\left(\frac{\partial^2 w}{\partial x^2} + \nu\frac{\partial^2 w}{\partial y^2}\right)\right] \tag{2.5.239}$$

where $D = Eh^3/12(1-\nu^2)$. The expression for the y-component of power transmission is obtained simply by interchanging the x and y subscripts in (2.5.239).

Using eight accelerometers to obtain the gradients in (2.5.239), as demonstrated by Pavic (1976), and taking the time averaged result allows an expression similar in form to (2.5.186), but much more complex, to be obtained. Because of the many accelerometers and gradient estimates required, it is very difficult to obtain accurate results. The interested reader is referred to Pavic's article (Pavic, 1976) for more details.

For the special case of harmonic wave propagation, it is possible to simplify the expression for power transmission in the x-direction. For this case, the time average of (2.5.239) may be written as

$$P_{Bxa} = -\frac{1}{2}\mathrm{Re}\left\{ j\omega D\left[w\left(\frac{\partial^3 w^*}{\partial x^3} + \frac{\partial^3 w^*}{\partial x \partial y^2}\right)\right.\right.$$
$$\left.\left. - (1-\nu)\frac{\partial w}{\partial y}\frac{\partial^2 w^*}{\partial x \partial y} - \frac{\partial w}{\partial x}\left(\frac{\partial^2 w^*}{\partial x^2} + \nu\frac{\partial^2 w^*}{\partial y^2}\right)\right]\right\} \tag{2.5.240}$$

If a general solution is assumed for the plate equation of motion for any arbitrary plate edge boundary conditions, and if we further assume that the measurements will be made in the far field of any sources, the following may be written:

$$w = X(x)\,Y(y) = \left(A_1 e^{-jk_x x} + A_2 e^{jk_x x}\right)\left(B_1 e^{-jk_y y} + B_2 e^{jk_y y}\right) \tag{2.5.241a,b}$$

Then

$$\frac{\partial^2 w}{\partial x^2} = -k_x^2 w \tag{2.5.242}$$

$$\frac{\partial^2 w}{\partial y^2} = -k_y^2 w \tag{2.5.243}$$

and (2.5.240) becomes

$$\begin{aligned} P_{Bxa} = -\frac{1}{2}\mathrm{Im}\Bigg\{ \omega D \Bigg[w \left(-k_x^2 \frac{\partial w^*}{\partial x} - k_y^2 \frac{\partial w^*}{\partial x} \right) \\ - (1-\nu)\frac{\partial w}{\partial y}\frac{\partial^2 w^*}{\partial x \partial y} - \frac{\partial w}{\partial x}\left(-k_x^2 w^* - \nu k_y^2 w^* \right) \Bigg] \Bigg\} \end{aligned} \tag{2.5.244}$$

Using the relationship, $\mathrm{Im}(ab^*) = -\mathrm{Im}(a^*b)$, and assuming A_1, A_2, B_1, B_2, k_x and k_y are less than unity, then it can be shown that (2.5.244) can be written approximately as

$$P_{Bxa} = \omega D k_x^2 \mathrm{Im}\left\{ w \frac{\partial w^*}{\partial x} \right\} \tag{2.5.245}$$

which is similar to (2.5.208b) for flexural wave propagation in a beam.

A similar expression can be obtained for wave propagation in the *y*-direction. Thus, the intensity vector for flexural waves in a plate can be measured by measuring the *x*- and *y*-components, then calculating the vector magnitude and direction in the usual way.

As (2.5.245) is similar to (2.5.208b) for beams, except for the constant multiplier (Dk_x^2 instead of $EI_{yy}k_b^2$), all of the techniques for power transmission measurement embodied in (2.5.196), (2.5.200) and (2.5.210) to (2.5.213) are also valid for any particular direction on a plate. The quantities, k_x and k_y, are dependent upon the plate boundary conditions, but in many cases may be approximated as

$$k_x^2 = k_y^2 = \omega\sqrt{\frac{\rho h}{D}} \tag{2.5.246}$$

Indeed, this is the approximation which is implicitly assumed when the two accelerometer technique is used to determine the intensity vector in a plate. Although the approximation embodied in (2.5.245) and (2.5.246) gives good results in many cases, in general it is necessary to use eight accelerometers and evaluate all of the gradients in (2.4.240) directly (Pavic, 1976). This needs to be done for each of the *x*- and *y*-components of intensity to obtain the overall intensity vector.

The equivalence of (2.5.240) and (2.5.245) can be demonstrated numerically for particular cases (Pan and Hansen, 1994, 1996).

Intensity measurement in circular cylinders

The power transmission per unit width in a cylinder is given by (2.5.135) and (2.5.136). Again, the two accelerometer method will provide good results away from vibration sources and structural discontinuities, only if flexural waves are all that are present. Unfortunately, all wave types are coupled and exist on a cylinder surface, and thus the validity of the two accelerometer method in this case is open to question (see Pan and Hansen, 1997, for further discussion).

As for the plate, the derivatives to evaluate (2.5.135) and (2.5.136) can be determined using the finite difference technique and eight accelerometers. Note that additional accelerometers would be needed to measure the in-plane displacements ξ_x and ξ_θ.

Sources of error in the measurement of structural intensity

Because of the increased complexity of structural wave fields compared to acoustic fields, it is much more difficult to obtain accurate structural intensity measurements than it is to obtain accurate acoustic intensity measurements. Sources of error in structural intensity measurements are associated with mass loading effects of accelerometers (which may be avoided by using very small accelerometers or by using laser doppler velocimetry), the presence of wave types other than the one which is being measured, phase matching inaccuracies between the measurement channels, inaccuracies associated with the finite difference approximation and the presence of highly reactive fields associated with sources, sinks, discontinuities and boundaries, or with the simultaneous presence of reflected and incident wave fields.

When the reactive field is small compared to the active field and when only one wave type is present, errors associated with phase mismatch between instrumentation channels are only significant at low frequencies ($\Delta/\lambda < 0.1$, where Δ is the transducer separation and λ is the structural wavelength). For the same field situation, errors associated with the use of the finite difference approximation for the derivative of the displacement results in significant errors (approximately 20%) at frequencies above $\Delta/\lambda = 0.14$. Taylor (1990) showed that this error could be made insignificant for the four accelerometer method measurement of flexural wave power if the calculated power were multiplied by K_{fd} where K_{fd} is defined as

$$K_{fd} = \frac{1}{\sin k\delta}\,\frac{K^3\Delta^3}{2(\cos k\delta - \cos 3k\delta)} \tag{2.5.247}$$

where $\delta = \Delta/2$ and k is the flexural wavenumber.

In view of the preceding discussion, it is probably advisable to adjust Δ to be in the region of $\lambda/10$ when using either two or four accelerometers to measure the structural intensity of only one wave type in the absence of any reactive field.

In cases where the reactive field is significant, the effects of phase errors in the instrumentation are magnified. Such a situation can exist in the near field close to a vibration source, sink or structural discontinuity, or alternatively if incident and reflected waves exist simultaneously, causing the structure to have a partial or fully modal response. In the former cases, to minimize the effects of the near field, accelerometers must be at least half a structural wavelength away from structural discontinuities, boundaries or vibration sources. A structural wavelength for the various wave types which can propagate in beams and plates may be calculated using the relations in Table 2.5, where

$$\lambda = \frac{2\pi}{k} \quad \text{for all wave types} \tag{2.5.248}$$

and where k is the structural wavenumber for the wave of interest. Unfortunately, because of the finite size of structures, it is often not possible to make measurements in the absence of near field effects and many measurement situations are characterized by highly reactive fields, making phase matching between the two measurement channels of crucial importance.

Table 2.5 Expressions for the wavenumber for various wave types in beams and thin plates (k_L, k_T, k_b)

Wave type	Beam expression	Plate expression
Longitudinal	$\omega[\rho/E]^{1/2}$	$\omega[\rho(1-\nu^2)/E]$
Shear	$\omega[\rho/G]^{1/2}$ (torsional)	$\omega[\rho/G]^{1/2}$
Flexural	$[\omega^2 m'/EI]^{1/4}$	$\left[\left(\frac{m\pi}{L_1}\right)^2 + \left(\frac{m\pi}{L_2}\right)^2\right]^{1/2}$ (simply supported rectangular plates) $\left[\left(\frac{\pi\beta_{mn}}{r}\right)^2 + m^2/r^2\right]^{1/2}$ (circular plates)

Notes: k = wave number, ρ = density of beam or plate material, E, G = Young's modulus and shear modulus for the plate material, ν = Poisson's ratio for the plate material, ω = frequency of excitation, m' = mass per unit length of beam, m, n = modal indices, L_1, L_2 = dimensions of rectangular plate, r = radius of circular plate and β_{mn} = wave equation solutions for flexural waves on the circular plate (see Table 2.6).

When the reactive field is a result of the presence of reflected waves, the measurement accuracy is also influenced by the mode shapes of the structure at the measurement location, as measurements made near a vibration antinode will be more sensitive to instrumentation phase errors than those made near a node. This can be explained by considering the spatial variation of displacement between two measurement points separated by Δ. At an antinode this variation is very small and the phase difference between the two measurements will also be very small. Thus any relative phase error due to the instrumentation in this case would have a large effect on the measured results. This effect has been quantified by Taylor (1990) for measurements on beams in the presence of reflected waves. He showed that very large power transmission errors (over 200%) can result from instrumentation phase errors of less than 0.1° for $\Delta/\lambda \leq 0.05$. As the quantity Δ/λ (λ is the structural wavelength) increases, the error reduces but it is still 20% at $\Delta/\lambda = 0.12$. We could conclude that in the presence of reflected waves of half the amplitude or more of the incident waves, that structural power transmission measurements are so inaccurate as to be useless. In other words, structural intensity measurements will be questionable when made at frequencies corresponding to structural resonances or on structures which are lightly damped. Note also that if the vibration of a structure can be expressed in terms of a sum of normal modes (see Chapter 4), then there will be no active power transmission unless these modes are complex; that is, unless there is some structural damping within the structure or at the boundaries. Otherwise the power per unit width will be entirely reactive.

Errors also arise because of the presence of wave types other than the one which is to be measured for two reasons. First, a large portion of the total vibratory power may be missed and it may turn up in the other parts of the structure as the wave type being measured (due to energy conversion at structural discontinuities). Second, the presence of other wave types results in phase and amplitude errors in the data provided by the measurement transducers, the former becoming particularly critical in cases involving a significant reactive field.

These error considerations suggest that structural power transmission may not be a very suitable cost function for use in active structural vibration control.

It is interesting to note that flexural wave intensity measurements made in the far field using two accelerometers appear to be more accurate than measurements made using four accelerometers, probably because the four accelerometer technique is more sensitive to relative phase errors and the finite difference approximation implicit in the finite separation distance between the accelerometers.

2.5.5 Power transmission through vibration isolators into machine support structures, and power transmission into structures from an excitation source

The vibrational power transmission through a machine support point into a supporting structure can be measured using two distinct methods. The first method involves the measurement of the force input and acceleration (or velocity) at either the machine side of the mount or the structure side of it. In this case the instantaneous power transmission is given by:

$$P(t) = F(t)u(t) = F(t)\int a(t) \tag{2.5.249a,b}$$

For sinusoidal excitation, the active power transmission is given by

$$P_a = \frac{1}{2}\mathrm{Re}\left[\bar{F}\bar{u}^*\right] \tag{2.5.250}$$

where

$$F(t) = \bar{F}e^{j\omega t} \tag{2.5.251}$$

$$u(t) = \bar{u}e^{j\omega t} \tag{2.5.252}$$

The reactive component is given by

$$P_r = \frac{1}{2}\mathrm{Im}\left[\bar{F}\bar{u}^*\right] \tag{2.5.253}$$

The cross-spectral formulation for either single frequency or broadband signals is:

$$P_a(\omega) = \mathrm{Re}\left[G_{F,u}(\omega)\right] = \frac{1}{\omega}\mathrm{Im}\left[G_{F,a}(\omega)\right] \tag{2.5.254a,b}$$

The method just described is also suitable for measuring the power transmission into a structure from an external excitation source such as an electrodynamic shaker. Note that the right-hand side of (2.5.254) contains the cross-spectral density function G. Thus, for random signals, the power is actually the power per hertz. For single frequency signals or harmonics, the power associated with each harmonic is found by multiplying (2.5.254) by the bandwidth of each FFT filter (or the frequency resolution).

The reactive power (corresponding to stored energy) is given by

$$P_r(\omega) = \frac{1}{\omega}\mathrm{Im}\left[G_{F,u}(\omega)\right] = \frac{1}{\omega}\mathrm{Re}\left[G_{F,a}(\omega)\right] \tag{2.5.255a,b}$$

In (2.5.254) and (2.5.255), a is the measured acceleration, F is the measured force and u is the measured velocity.

The active power transmission can also be measured using an ordinary sound intensity analyser, placing the force transducer signal in one channel and the acceleration transducer in the other. Equation (2.5.249) can be used to express the active power transmission as

$$P_a = \left\langle F(t)\int a(t)\,\mathrm{d}t\right\rangle \tag{2.5.256}$$

which for a stationary signal can be shown to be equal to

$$P_a = \left\langle \frac{F+a}{2} \int_{-\infty}^{t} (F-a)\mathrm{d}\tau \right\rangle \tag{2.5.257}$$

which is directly proportional to the result obtained with a sound intensity analyser with the force signal input to one channel and the acceleration input to the other. The intensity analyser result would have to be multiplied by $\rho_0\Delta$ to give the correct absolute value for P_a. Here, ρ_0 is the density of air and Δ is the microphone spacing assumed by the sound intensity analyser. Equation (2.5.257) can be shown to be equal to (2.5.256) by multiplying terms in equation (2.5.256) and noting that $F\int a\mathrm{d}t = -a\int F\mathrm{d}t$ for stationary signals and $\left\langle F\int F\mathrm{d}t\right\rangle = \left\langle a\int a\mathrm{d}t\right\rangle = 0$.

The force and acceleration at the structure side or machine side of the mount may be measured using a piezoelectric force crystal and a piezoelectric accelerometer. In some cases, successful measurements can be taken using an impedance head transducer which has the force and accelerometer crystals mounted together in a single unit. However care must be taken to ensure that the accelerometer crystal is not between the force gauge and the mount (for measurements of power transmission into the mount) and not between the force gauge and the structure (for measurements of power transmission out of the mount). Otherwise the true force into the mount or into the structure will not be measured, due to the lack of stiffness of the accelerometer causing a small amplitude and phase shift. This problem is especially important when attempting to measure the power transmission into a structure from an excitation source (such as an electrodynamic shaker) during active control experiments (see Fig. 2.42). In many cases, however, it is preferable not to use an impedance head to measure the input power to a structure, but rather, a separate force transducer and acceleration transducer. Two suitable arrangements are shown in Fig. 2.43.

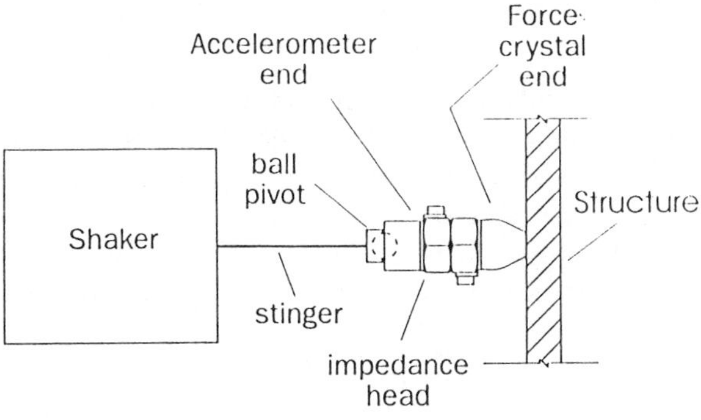

Fig. 2.42 Correct arrangement to measure power flow into a structure using an impedance head.

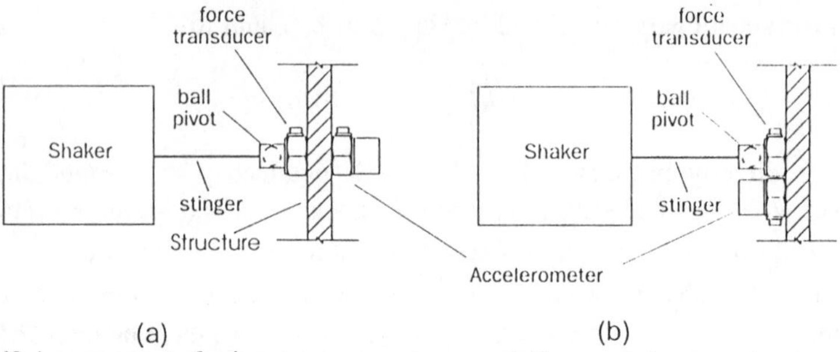

Fig. 2.43 Arrangements for input power measurements from a shaker to a structure using separate force and acceleration transducers.

These arrangements are only suitable provided that the structure thickness in Fig. 2.43(a) is much less than a structural wavelength and the distance between the centre of the force transducer and the centre of the acceleration transducer in Fig. 2.43(b) is much less than a structure wavelength.

The second method for power transmission measurement applies only to transmission from a machine support point through a supporting structure and not to the measurement of the power transmission into a structure from a shaker.

For a spring isolator, the axial force acting on the structure is given by the product of the difference in displacement between the two ends of the isolator and the spring constant. Thus, the real or active power transmitted through the isolator to the support structure will be given by

$$P_a = k\langle u_2(w_1 - w_2)\rangle \tag{2.5.258}$$

where k is the isolator axial stiffness, u_2 is the velocity of the base of the isolator and w_1 and w_2 are respectively the displacements of the top and base of the isolator. If the supporting structure is much more rigid than the isolator, then the difference ($w_1 - w_2$) can be approximated simply with the displacement, w_2, of the equipment side of the isolator (Pavic, 1977). Thus,

$$P_a = k\langle u_2 w_1\rangle \tag{2.5.259}$$

The preceding results only hold for frequencies below the first resonance frequency of the isolation system. If the vibration field can be considered stationary, this expression is equivalent to (2.5.84a) where acoustic pressure has been replaced by structural displacement. Thus, below the first resonance frequency where the power transmission into the isolator equals the power transmission out and the measurements are made using accelerometers, (2.5.259) can be written as

$$P_a(\omega) = \frac{k}{2\omega^2}\left\langle (a_2 + a_1) \int_{-\infty}^{t} (a_1 - a_2)\,\mathrm{d}\tau \right\rangle \tag{2.5.260}$$

which is directly proportional to the result obtained by placing the two accelerometer signals into the two channels of a sound intensity analyser. The calibration (or proportionality) constant is $\frac{\rho \Delta k}{\omega^2}$ where Δ is the microphone spacing assumed by the intensity analyser. However, if the purpose of the measurement is to evaluate the performance of an active element in an isolator or the effectiveness of the installation of an active vibration control system, then only the difference (in dB) in power transmission before and after active control is needed and thus absolute calibration is unnecessary.

Alternatively, if a spectrum analyser rather than an intensity analyser is available, the power transmission can be written as

$$P_a = \frac{k}{\omega^3} \mathrm{Im}\left[G_{a_1,a_2}(\omega)\right] \tag{2.5.261}$$

REFERENCES

Beranek, L.L. (1986). *Acoustics*. 2nd edn, American Institute of Physics, New York.

Bernhard, R.J. and Mickol, J.D. (1990). Probe mass effects on power transmission in lightweight beams and plates. In *Proceedings of the Third International Congress on Intensity Techniques*. Senlis, France, 307–314.

Bies, D.A. and Hansen, C.H. (1996). *Engineering Noise Control: Theory and Practice*, 2nd edn., E & FN Spon, London.

Cremer, L., Heckl, M. and Ungar, E.E. (1973). *Structure-borne Sound*. Springer Verlag, Berlin.

Crocker, M.J. (1993). The measurement of sound intensity and its application to acoustics and noise control. *Proceedings of Noise '93*, St Petersburg, Russia: Interpublish Ltd.

Ewins, D.J. (1984) *Modal Testing*. Research Studies Press, Letchworth.

Fahy, F.J. (1985) *Sound and Structural Vibration*. Academic Press, London.

Fahy, F.J. (1995). *Sound Intensity*. 2nd edn., E & FN Spon, London.

Flügge, W. (1962). *Stresses in Shells*. Springer-Verlag, Berlin.

Fuller, C.R. and Burdisso, R.A. (1991). A wavenumber domain approach to the active control of sound and vibration. *Journal of Sound and Vibration*, **148**, 355–360.

Graff, K.F. (1975). *Wave Motion in Elastic Solids*. Clarendon Press, Oxford.

Hansen, C.H. (1980). *A Study of Modal Sound Radiation*. PhD thesis, University of Adelaide, South Australia.

Hayek, S.I., Pechersky, M.J. and Sven, B.C. (1990). Measurement and analysis of near and far field structural intensity by scanning laser vibrometry. In *Proceedings of the Third International Congress on Intensity Techniques*. Senlis, France, 281–288.

Heckl, M. (1990). Waves, intensities and powers in structures. In *Proceedings of the Third International Congress on Intensity Techniques*. Senlis, France, 13–20.

Inman, D.J. (1989). *Vibration*. Prentice Hall, NJ.

Jenkins, G.M. and Watts, D.G. (1968). *Spectral Analysis and its Applications*. Holden-Day, San Francisco.

Junger, M.C. and Feit, D. (1986). *Sound Structures and their Interaction*. 2nd edn., MIT Press, Cambridge, MA.

Kinsler, L.E., Frey, A.R., Coppers, A.B. and Sanders, J.V. (1982). *Fundamentals of Acoustics*. 3rd edn., Wiley, New York.

Leissa, A.W. (1969). *Vibration of Plates*. NASA SP-160, reprinted by the Acoustical Society of America, 1993.

Leissa, A.W. (1973). *Vibration of Shells*. NASA SP-288, reprinted by the Acoustical Society of America, 1993.

Levine, H. and Schwinger, J. (1948). On the radiation of sound from an unflanged circular pipe. *Physics Review*, **73**, 383–406.

McCuskey, S.W. (1959). *An Introduction to Advanced Dynamics*. Addison Wesley, Reading, MA.

McGary, M.C. (1988). A new diagnostic method for separating airborne and structure borne noise radiated by plates with application to propeller aircraft. *Journal of the Acoustical Society of America*, **84**, 830–840.

Mierovitch, L. (1970). *Methods of Analytical Dynamics*. McGraw Hill, New York.

Mikulas, M.M. and McElman, J.A. (1965). *On Free Vibrations of Eccentrically Stiffened Cylindrical Shells and Flat Plates*. NASA TN D-3010.

Mindlin, R.D. (1951). Influence of rotary inertia and shear on the flexural motions of isotropic elastic plates. *Journal of Applied Mechanics*, **18**, 31–38.

Mindlin, R.D. and Deresiewicz, H. (1954). Thickness-shear and flexural vibrations of a circular disk. *Journal of Applied Physics*, **25**, 1329–1332.

Morse, P.M. and Feshbach, H. (1953). *Methods of Theoretical Physics*. McGraw Hill, New York.

Morse, P.M. and Ingard, K.U. (1968). *Theoretical Acoustics*. McGraw Hill, New York.

Noiseux, D.U. (1970). Measurement of power flow in uniform beams and plates. *Journal of the Acoustical Society of America*, **47**, 238–247.

Novozhilov, V.V. (1968). *The Theory of Thin Elastic Shells*. P. Noordhoff Lts, Groningen, Netherlands.

Pan, J. and Hansen, C.H. (1991). Active control of total vibratory power flow in a beam. I. physical system analysis. *Journal of the Acoustical Society of America*, **89**, 200–209.

Pan, X. and Hansen, C.H. (1994). Active control of vibratory power transmission along a semi-infinite plate. *Journal of Sound and Vibration*, **184**, 585–610.

Pan, X. and Hansen, C.H. (1996). Power transmission characteristics in an actively controlled semi-infinite plate. *Journal of Active Control*, **1**, (3).

Pan, X. and Hansen, C.H. (1997). Active control of vibration transmission in a cylindrical shell. *Journal of Sound and Vibration*.

Pavic, G. (1976). Measurement of structure borne wave intensity. Part 1: Formulation of the methods. *Journal of Sound and Vibration*, **49**, 221–230.

Pavic, G. (1977). Energy flow through elastic mountings. In *Proceedings of the 9th International Congress on Acoustics*. Madrid, Spain, 293.

Pavic, G. (1987). Structural surface intensity: an alternative approach in vibration analysis and diagnosis. *Journal of Sound and Vibration*, **115**, 405–422.

Pavic, G. (1988). Acoustical power flow in structures: a survey. In *Proceedings of Internoise '88*, 559–564.

Pavic, G. (1990). Energy flow induced by structural vibrations of elastic bodies. In *Proceedings of the Third International Congress on Intensity Techniques*. Senlis, France, 21–28.

Pierce, A.D. (1981). *Acoustics: An Introduction to its Physical Principles and Applications*. McGraw Hill, New York.

Pinnington, R.J. (1988). *Approximate Mobilities of Built up Structures*. ISVR Technical Report No. 162, Institute of Sound and Vibration Research, Southampton.

Pope, L.D. (1971). On the transmission of sound through finite closed shells: statistical energy analysis, modal coupling and non-resonant transmission. *Journal of the Acoustical Society of America*, **50**, 1004–1018.

Quinlan, D. (1985). *Measurement of Complex Intensity and Potential Energy Density in Thin Structures*. Masters Thesis, Pennsylvania State University.

Redman-White, W. (1983). The experimental measurement of flexural wave power flow in structures. In *Proceedings of the International Conference on Recent Advances in Structural Dynamics*. University of Southampton, 467–474.

Romano, A.J., Abraham, P.B. and Williams, E.G. (1990). A Poynting vector formulation for thin shells and plates, and its application to structural intensity analysis and source localisation. Part 1: Theory. *Journal of the Acoustical Society of America*, **87**, 1166–1175.

Skudrzyk, E. (1968) *Simple and Complex Vibratory Systems*. Pennsylvania State University Press, University Park, PA.

Skudrzyk, E. (1971). *The Foundations of Acoustics*. Springer-Verlag, New York.

Skudrzyk, E. (1980). The mean-value method of predicting the dynamic response of complex vibrators. *Journal of the Acoustical Society of America*, **67**, 1105–1135.

Soedel, W. (1993). *Vibration of Shells and Plates*. 2nd edn., Marcel Dekker.

Taylor, P.D. (1990). Nearfield structureborne power flow measurements. In *Proceedings of the International Congress on Recent Developments in Air and Structure-borne Sound and Vibration*. Auburn University, USA.

Tichy, J. (1989). Applications of intensity technique for noise and vibration analysis. In *Proceedings of Noise and Vibration '89*. Singapore, pp. 15–18.

Timoshenko, S. (1921). On the correction for shear of the differential equation for transverse vibrations of prismatic bars. *Philosophical Magazine*, **41**, 744–746.

Timoshenko, S. and Goodier, J.N. (1951). *Theory of Elasticity*. 2nd edn., McGraw-Hill, New York.

Timoshenko, S. and Woinowsky-Krieger, S. (1959). *Theory of Plates and Shells*. 2nd edn., McGraw-Hill, New York.

Tse, F.S., Morse, J.E. and Hinkle, R.T. (1978). *Mechanical Vibrations*. 2nd edn., Allyn and Bacon, Boston, MA.

Verheij, J.W. (1980). Cross spectral density methods for measuring structure borne power flow on beams and pipes. *Journal of Sound and Vibration*, **70**, 133–139.

Verheij, J.W. (1990). Measurements of structure-borne wave intensity on lightly damped pipes. *Noise Control Engineering Journal*, **35**, 69–76.

Veronesi, W.A. and Maynard, J.D. (1987). Nearfield acoustic holography II: Holographic reconstruction algorithms and computer implementation. *Journal of the Acoustical Society of America*, **81**, 1307–1321.

Wallace, C.E. (1972). Radiation resistance of a rectangular panel. *Journal of the Acoustical Society of America*, **51**, 946–952.

Warburton, G.B. (1965). Vibration of thin cylindrical shells. *Journal of Mechanical Engineering Science*, **7**, 399–407.

Wiener, F.M. (1951). On the relation between the sound fields radiated and diffracted by plane obstacles. *Journal of the Acoustical Society of America*, **23**, 697–700.

3

Spectral analysis

Spectral analysis (or frequency analysis) is the process of representing a time varying signal, such as the output from an acceleration transducer, in the frequency domain. This transformation of a signal from the time domain to the frequency domain has important implications for active noise and vibration control, both in analysis and implementation. In this chapter, the fundamental principles will be reviewed and some functions which will be of use later will be derived.

There are two common ways of transforming a signal from the time domain to the frequency domain. The first involves the use of band limited digital or analogue filters. The second involves the use of Fourier analysis, where the time domain signal is transformed using a Fourier series. This is implemented in practice digitally (referred to as the DFT – discrete Fourier transform) using a very efficient algorithm known as the FFT (fast Fourier transform). The use of digital filters will be discussed first.

3.1 DIGITAL FILTERING

The most common forms of analogue and digital filters are standardized octave, 1/3 octave, 1/12 and 1/24 octave bands. Such filters are referred to as constant percentage bandwidth filters, meaning that the filter bandwidth is a constant percentage of the band centre frequency. For example, the octave bandwidth is always about 70.1% of the band centre frequency, the 1/3 octave bandwidth is 23.2% of the band centre frequency and the 1/12 octave is 5.8% of the band centre frequency, where the band centre frequency is defined as the geometric mean of the upper and lower frequency bounds of the band.

The stated percentages are approximate, as a compromise has been adopted in defining the bands to simplify and to ensure repetition of the centre band

Table 3.1 Preferred frequency bands (Hz)

Band number	Octave band centre frequency	One-third octave band centre frequency	Band limits	
			Lower	Upper
14		25	22	28
15	31.5	31.5	28	35
16		40	35	44
17		50	44	57
18	63	63	57	71
19		80	71	88
20		100	88	113
21	125	125	113	141
22		160	141	176
23		200	176	225
24	250	250	225	283
25		315	283	353
26		400	353	440
27	500	500	440	565
28		630	565	707
29		800	707	880
30	1 000	1 000	880	1 130
31		1 250	1 130	1 414
32		1 600	1 414	1 760
33	2 000	2 000	1 760	2 250
34		2 500	2 250	2 825
35		3 150	2 825	3 530
36	4 000	4 000	3 530	4 400
37		5 000	4 400	5 650
38		6 300	5 650	7 070
39	8 000	8 000	7 070	8 800
40		10 000	8 800	11 300
41		12 500	11 300	14 140
42	16 000	16 000	14 140	17 600
43		20 000	17 600	22 500

frequencies. The compromise that has been adopted is that the logarithms to the base ten of the one-third octave centre band frequencies are tenth decade numbers such that the band number is $10\log_{10} f_c$, where f_c is the filter centre frequency. Standard octave and one-third octave bands are shown in Table 3.1. At this stage it is of interest to point out the difference between what is commonly referred to as white random noise and pink random noise. The former represents a signal which contains energy in any particular frequency band such that the energy per hertz is the same at all frequencies. The latter represents a signal containing an equal amount of energy in each octave band.

Besides constant percentage bandwidth filters, instruments with constant frequency bandwidth filters are also available. However, these instruments have largely been replaced by FFT analysers which give similar results in a fraction of the time and generally at a lower cost. When a time varying signal is filtered using either a constant percentage bandwidth or a constant frequency bandwidth filter a r.m.s. amplitude signal is obtained which is proportional to the sum of the total energy content of all frequencies included in the band.

When discussing digital filters and their use, an important consideration is the filter response time, T_R (s), which is the minimum time required for the filter output to reach steady state. The minimum time generally required is the inverse of the filter bandwidth, B (Hz). That is,

$$BT_R = \left[\frac{B}{f}\right] \cdot (fT_R) = bn_R \approx 1 \qquad (3.1.1a,b,c)$$

where b is the relative bandwidth of the filter (e.g. for a 1% filter, $b = 0.01$), f is the filter centre frequency and n_R is the number of cycles of the signal which it takes for the filter output to approach its final value. For example, for a one-third octave filter $b = 0.23$ and the number of cycles $n_R \approx 4.3$. A typical response of a one-third octave filter is shown in Fig. 3.1, where it will be noted that the actual response time is perhaps five cycles or more, depending on the desired accuracy.

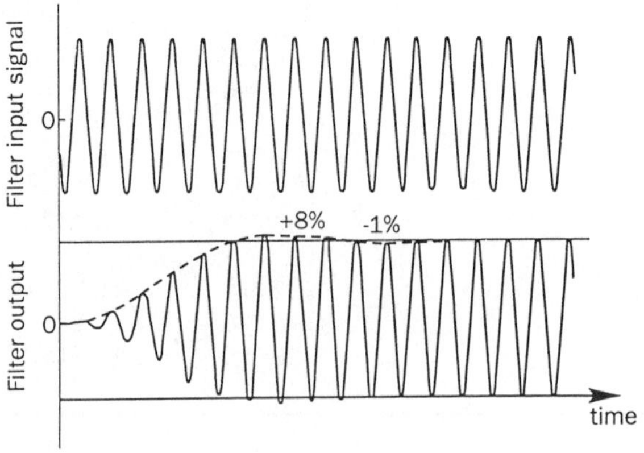

Fig. 3.1 Typical filter response from a one-third octave filter.

Where the r.m.s. value of a filtered signal is required it is necessary to determine the average value of the integrated squared output of the filtered signal over some prescribed period of time called the averaging time. The longer the averaging time the more nearly constant will be the r.m.s. value of the filtered output.

For a sinusoidal input of frequency f (Hz), or for several sinusoidal frequencies within the band where f (Hz) is the minimum separation between components, the variation in the average value will be less than 1/4 dB for an averaging time, $T_A \geq 3/f$ (s). For many sinusoidal components or for random noise and $BT_A \geq 1$, the error in the r.m.s. signal may be determined in terms of the statistical error, ϵ, calculated as follows:

$$\epsilon = 0.5(BT_A)^{-1/2} \tag{3.1.2}$$

For random noise the actual error has a 63.3% probability of being within the range $\pm\,\epsilon$ and a 95.5% probability of being within the range, $\pm\,2\epsilon$.

The calculated statistical error may be expressed in decibels as follows:

$$20\log_{10} e^{\epsilon} = 4.34(BT_A)^{-1/2} \quad \text{dB} \tag{3.1.3}$$

3.2 DIGITAL FOURIER ANALYSIS

The second method for transforming a signal from the time domain to the frequency domain is by Fourier analysis. This technique allows any time domain signal to be represented as the sum of a number of individual sinusoidal components, each characterized by a specific frequency, amplitude and relative phase angle. Whereas the filtering process described earlier provides information about the amplitudes of the frequency components, the information is insufficient to reconstruct the original time varying signal, and returning from the frequency domain to the time domain is not possible. By contrast, Fourier analysis provides sufficient information in the output to reconstruct the original time varying signal.

A general Fourier representation of a periodic time varying signal of period T and fundamental frequency $f_1 = 1/T$ such that $x(t) = x(t + NT)$ where $N = 1,2,\ldots$ takes the following form:

$$x(t) = \sum_{n=1}^{\infty} [A_n \cos(2\pi n f_1 t) + B_n \sin(2\pi n f_1 t)] \tag{3.2.1}$$

For example, the square wave illustrated in Fig. 3.2 is represented by (3.2.1) where for n odd $A_n = 4/(\pi n)$, for n even $A_n = 0$ and $B_n = 0$. Note that the component characterized by frequency, nf_1, is referred to as the nth harmonic of the fundamental frequency, f_1.

Use of Euler's well known equation (Abramowitz and Stegun, 1965) allows (3.2.1) to be rewritten in the following alternative form:

$$x(t) = \frac{1}{2}\sum_{n=0}^{\infty}\left[(A - jB)\,e^{j2\pi n f_1 t} + (A + jB)\,e^{-j2\pi n f_1 t}\right] \tag{3.2.2}$$

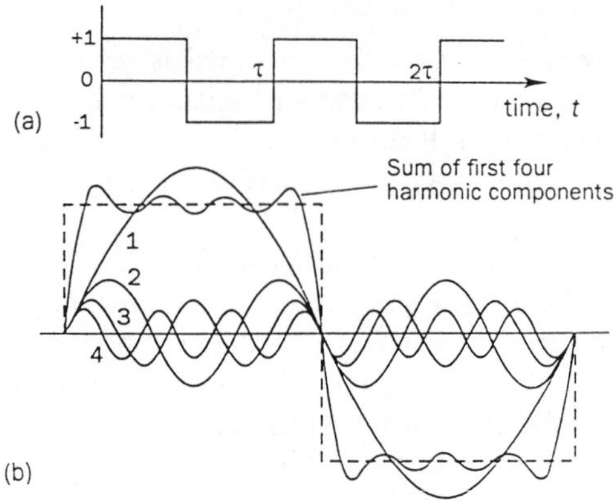

Fig. 3.2. Fourier analysis of a square wave: (a) periodic square wave; (b) harmonic components of a square wave.

A further reduction is possible by defining the complex spectral amplitude components $X_n = (A - jB)/2$ and $X_{-n} = (A + jB)/2$. Denoting by * the complex conjugate, the following relation may be written:

$$X_n = X^*_{-n} \tag{3.2.3}$$

Clearly (3.2.3) is satisfied, thus ensuring that the right-hand side of (3.2.2) is real. The introduction of (3.2.3) in equation (3.2.2) allows the following more compact expression to be written:

$$x(t) = \sum_{n=-\infty}^{\infty} X_n \, e^{j2\pi n f_1 t} \tag{3.2.4}$$

The spectrum of (3.2.4) now includes negative as well as positive values of n giving rise to components $-nf_1$. The spectrum is said to be two sided. The spectral amplitude components, X_n, may be calculated using the following expression:

$$X_n = X(f_n) = \frac{1}{T} \int_{-T/2}^{T/2} x(t) \, e^{-j2\pi n f_1 t} \, dt \tag{3.2.5}$$

It is of interest to examine the distribution with frequency of the power content of the signal. The instantaneous power of the time signal is equal to $[x(t)]^2$ and the mean power over any one period of length T is

$$W_{mean} = \frac{1}{T} \int_0^T [x(t)]^2 \, dt \tag{3.2.6}$$

Substitution of (3.2.1) in (3.2.6) and carrying out the integration gives the following result:

$$W_{mean} = \frac{1}{2} \sum_{n=1}^{\infty} [A_n^2 + B_n^2] \tag{3.2.7}$$

Thus, one result of Parseval's theorem has been demonstrated; the total power in a periodic signal may be determined by integrating the time domain signal over one period or by summing the squared amplitudes of all the components in the frequency spectrum. The spectrum of squared amplitudes is known as the power spectrum.

Figure 3.3 shows several possible interpretations of the Fourier transform. Figure 3.3(a) the two sided power spectrum; (b) the one sided spectrum obtained by adding the corresponding negative and positive frequency components; (c) the r.m.s. amplitude (square root of the power spectrum) and (d) the logarithm to the base 10 of the power spectrum in decibels, which is the form usually used in the analysis of noise and vibration problems.

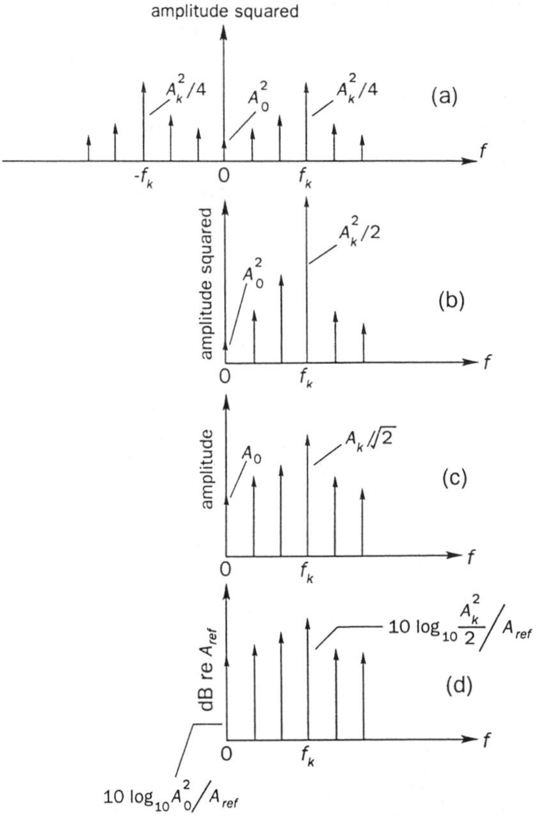

Fig. 3.3 Various spectrum representations: (a) two sided power spectrum; (b) one sided power spectrum; (c) r.m.s. amplitude spectrum; (d) dB spectrum (after Randall, 1988).

The previous analysis may be extended to the more general case of non-periodic (or random noise) signals by allowing the period T to become indefinitely large. In this case X_n becomes $X'(f)$, a continuous function of frequency, f. It is to be noted that whereas the units of X_n are the same as those of $x(t)$, the units of $X'(f)$ are those of $x(t)$ per hertz. With the proposed changes, (3.2.5) takes the following form:

$$X'(f) = \int_{-\infty}^{\infty} x(t)\mathrm{e}^{-j2\pi ft}\,\mathrm{d}t \tag{3.2.8}$$

The spectral density function $X'(f)$ is complex, characterized by a real and an imaginary part (or amplitude and phase).

Equation (3.2.4) becomes

$$x(t) = \int_{-\infty}^{\infty} X'(f)\mathrm{e}^{j2\pi ft}\,\mathrm{d}f \tag{3.2.9}$$

Equations (3.2.8) and (3.2.9) form a Fourier transform pair with the former referred to as the forward transform and the latter as the inverse transform.

In practice, a finite sample time T is always used to acquire data and the spectral representation of (3.2.5) is the result calculated by spectrum analysis equipment. This latter result is referred to as the spectrum and the spectral density is obtained by multiplying by the sample period, T.

Where a time function is represented as a sequence of samples taken at regular intervals, an alternative form of Fourier transform pair is as follows. The forward transform is

$$X(f) = \sum_{k=-\infty}^{\infty} x(t_k)\,\mathrm{e}^{-j2\pi ft_k} \tag{3.2.10}$$

where $t_k = k\Delta t$ is the time corresponding to the kth time sample, and the inverse transform is

$$x(t_k) = \frac{1}{f_s}\int_{-f_s/2}^{f_s/2} X(f)\,\mathrm{e}^{j2\pi ft_k}\,\mathrm{d}f \tag{3.2.11}$$

where f_s is the sampling frequency.

The form of Fourier transform pair used in spectrum analysis instrumentation is referred to as the discrete Fourier transform, for which the functions are sampled in both the time and frequency domains. Thus,

$$X(f_n) = \frac{1}{N}\sum_{k=0}^{N-1} x(t_k)\mathrm{e}^{-j2\pi nk/N} \qquad n = 1, 2, 3, \ldots N \tag{3.2.12}$$

$$x(t_k) = \sum_{n=0}^{N-1} X(f_n)\mathrm{e}^{j2\pi nk/N} \qquad k = 1, 2, 3, \ldots N \tag{3.2.13}$$

where k and n represent discrete sample numbers in the time and frequency domains respectively.

In (3.2.12), the spacing between frequency components in hertz is dependent on the time T to acquire the N samples of data in the time domain and is equal to $1/T$ or f_s/N. Thus the effective filter bandwidth B is equal to $1/T$.

The four Fourier transform pairs are shown graphically in Fig. 3.4. In (3.2.12) and (3.2.13), the functions have not been made symmetrical about the origin, but because of the periodicity of each, the second half of each sum also represents the negative half period to the left of the origin, as can be seen by inspection of the figure.

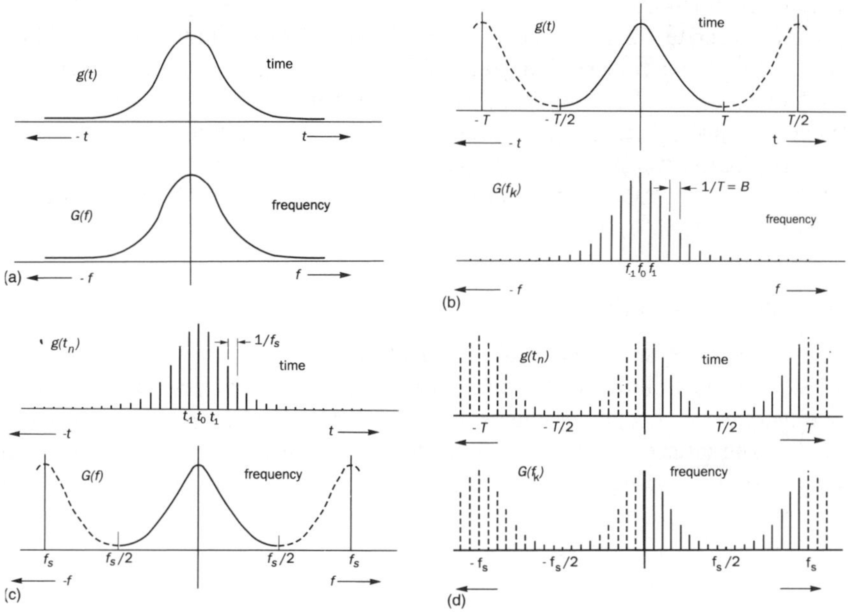

Fig. 3.4 Various Fourier transform pairs (after Randall, 1988):
(a) integral transform; signal infinite and continuous in both the time and frequency domains:

$$X'(f) = \int_{-\infty}^{\infty} x(t)\, e^{-j2\pi ft}\, dt \qquad \text{and} \qquad x(t) = \int_{-\infty}^{\infty} X'(f)\, e^{j2\pi ft}\, df$$

(b) Fourier series; signal periodic in the time domain and discrete in the frequency domain:

$$X(f_n) = \frac{1}{T} \int_{-T/2}^{T/2} g(t)\, e^{-j2\pi f_n t}\, dt \qquad \text{and} \qquad x(t) = \sum_{n=-\infty}^{\infty} X(f_n)\, e^{j2\pi f_n t}$$

(c) sampled function; signal discrete in the time domain and periodic in the frequency domain:

$$X(f) = \sum_{k=-\infty}^{\infty} x(t_k)\, e^{-j2\pi f t_k} \qquad \text{and} \qquad x(t_k) = \frac{1}{f_s} \int_{-f_s/2}^{f_s/2} X(f)\, e^{j2\pi f t_k}\, df$$

(d) discrete Fourier transform; signal discrete and periodic in both the time and frequency domains:

$$X(f_n) = \frac{1}{N} \sum_{k=0}^{N-1} x(t_k)\, e^{-j2\pi nk/N} \qquad \text{and} \qquad x(t_k) = \sum_{n=0}^{N-1} X(f_n)\, e^{j2\pi nk/N}$$

The frequency components above $f_s/2$ in Fig. 3.4 can be more easily visualized as negative frequency components and in practice, the frequency content of the final spectrum must be restricted to less than $f_s/2$. This is explained later when aliasing is discussed.

The discrete Fourier transform is well suited to the digital computations performed in instrumentation. Nevertheless, it can be seen by referring to (3.2.12) that to obtain N frequency components from N time samples, N^2 complex multiplications are required. Fortunately this is reduced by the use of the fast Fourier transform (FFT) algorithm to $N \log_2 N$ which for a typical case of $N = 1024$ speeds up computations by a factor of 100. This algorithm is discussed in detail by Randall (1988).

3.2.1 Power spectrum

The power spectrum is the most common form of spectral representation used in acoustics and vibration. For discussion assume that the measured variable is randomly distributed about a mean value (also called expected value or average value). The mean value of $x(k)$ is obtained by an appropriate limiting operation in which each value assumed by $x(k)$ is multiplied by its probability of occurrence, $p(x)$. This gives

$$E[x(k)] = \int_{-\infty}^{\infty} x p(x)\,\mathrm{d}x \tag{3.2.14}$$

where $E[\ ]$ represents the expected value over the index k of the term inside the bracket. Similarly, the expected value of any real single-valued continuous function $g(x)$ of the random variable $x(k)$ is given by (Bendat and Piersol, 1980):

$$E[g(x(k))] = \int_{-\infty}^{\infty} g(x) p(x)\,\mathrm{d}x \tag{3.2.15}$$

The two sided power spectrum is defined in terms of the amplitude spectrum $X(f)$ of (3.2.10) as:

$$S_{xx}(f) = \lim_{T \to \infty} E\left[X^*(f)\,X(f)\right] \tag{3.2.16}$$

and the power spectral density as

$$S'_{xx}(f) = \lim_{T \to \infty} T\,E\left[X^*(f)\,X(f)\right] = \lim_{T \to \infty} \frac{1}{T} E\left[X'^*(f)\,X'(f)\right] \tag{3.2.17a,b}$$

For a finite record length, T the two sided power spectrum may be estimated using

$$S_{xx}(f_n) \approx \frac{1}{q} \sum_{i=1}^{q} X_i^*(f_n)\,X_i(f_n) \tag{3.2.18}$$

where i is the spectrum number and q is the number of spectra over which the average is taken. The larger the value of q, the more closely will the estimate of $S_{xx}(f_n)$ approach its true value.

The power spectral density can be obtained from the power spectrum by dividing by the frequency spacing of the components in the frequency spectrum or by multiplying by the time T to acquire one record of data. Although it is often appropriate to express random noise spectra in terms of power spectral density, the same is not true for tonal components. Only the power spectrum will give the true energy content of a tonal component.

In practice, the two sided power spectrum $S_{xx}(f_n)$ is expressed in terms of the one sided power spectrum $G_{xx}(f_n)$ where

$$\begin{aligned} G_{xx}(f_n) &= 0 & f_n &< 0 \\ G_{xx}(f_n) &= S_{xx}(f_n) & f_n &= 0 \\ G_{xx}(f_n) &= 2S_{xx}(f_n) & f_n &> 0 \end{aligned} \tag{3.2.19}$$

Note that if successive spectra $X_i(f_n)$ are averaged, the result will be zero as the phases of each spectral component vary randomly from one record to the next. Thus, in practice, power spectra are more commonly used as they can be averaged together to give a more accurate result. This is because power spectra, $S_{xx}(f_n)$ are only represented by an amplitude; phase information is lost when the spectra are calculated (see (3.2.18)).

There remain a number of important concepts and possible pitfalls of frequency analysis which will now be discussed.

3.2.2 Uncertainty principle

The uncertainty principle states that the frequency resolution of a Fourier transformed signal is equal to the reciprocal of the time T to acquire the signal. Thus, for a single spectrum, $BT = 1$. An effectively higher BT product can be obtained by averaging several spectra together until an acceptable error is obtained according to (3.1.2) where B is the filter bandwidth or frequency resolution and T is the total sample time.

3.2.3 Sampling frequency and aliasing

The sampling frequency is the frequency at which the input signal is digitally sampled. If the signal contains frequencies greater than half the sampling frequency, then these will be 'folded back' and appear as frequencies less than half the sampling frequency. For example, if the sampling frequency were 20 000 Hz and the signal contained energy at 25 000 Hz, then it would appear as 5000 Hz. Similarly, if the signal contained energy at 15 000 Hz, it would appear as 5000 Hz. This phenomenon is known as 'aliasing' and in a spectrum analyser it is important to have analogue filters which have a sharp roll off for frequencies above about 0.4 of the sampling frequency. Aliasing is illustrated in Fig. 3.5 and discussed further in Chapter 13.

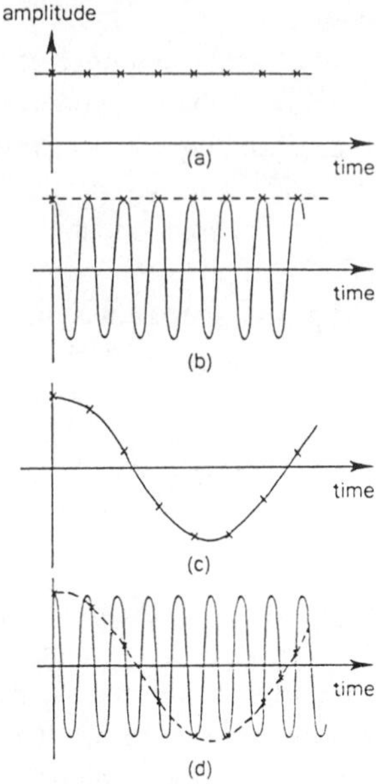

Fig. 3.5 Illustration of aliasing: (a) zero frequency or DC component; (b) spectrum component at sampling frequency f_s interpreted as d.c.; (c) spectrum component at $(1/N)f$; (d) spectrum component at $[(N+1)/N]f_s$ (after Randall, 1988).

3.2.4 Weighting functions

When sampling a signal in practice, a spectrum analyser must start and stop somewhere in time and this may cause a problem due to the effect of the discontinuity where the two ends of the record join in a loop. The solution is to apply a weighting function called a window which suppresses the effect of the discontinuity. The discontinuity, in the absence of weighting, causes side lobes to appear in the spectrum for a single frequency as shown by the solid curve in Fig. 3.6, which is effectively the same as applying a rectangular window function weighting. In this case, all signal values before sampling begins and after it ends are multiplied by zero and all values in between are multiplied by one.

A better choice of window is one which places less weight on the signal at either end of the window and maximum weight in the middle of the window such as the Hanning window, the effect of which is illustrated by the dashed curve in Fig. 3.6. Even though the main lobe is wider, implying coarser frequency resolution, the side lobe amplitudes fall away more rapidly.

The properties of various weighting functions are summarized in Table 3.2. The best for amplitude accuracy is the flat top (the name refers to the frequency domain effect, not the window shape). This is often used for calibration because the frequency of the calibration tone could lie anywhere between two lines of the analyser. However, this window provides the poorest frequency resolution. Maximum frequency resolution (and minimum amplitude accuracy) is achieved with the rectangular window, so this is sometimes used to separate two spectral peaks which have a similar amplitude and a small frequency spacing. A good compromise most commonly used is the Hanning window.

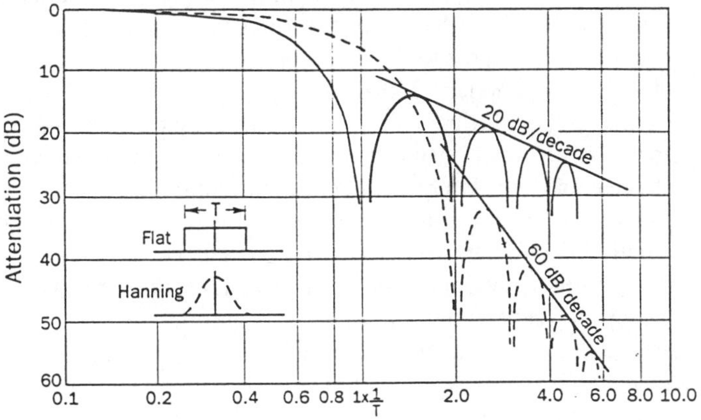

Fig. 3.6 Comparison of the filter characteristics of the rectangular and Hanning time weighting functions.

Table 3.2 Properties of the various time weighting functions (after Randall, 1988)

Window type	Highest sidelobe (dB)	Sidelobe falloff (dB/decade)	Noise bandwidth	Maximum amplitude error (dB)
Rectangular	-13	-20	1.00	3.9
Hanning	-32	-60	1.50	1.4
Hamming	-43	-20	1.36	1.8
Kaiser-Bessel	-69	-20	1.80	1.0
Truncated Gaussian	-69	-20	1.90	0.9
Flat-top	-93	0	3.77	<0.01

When a Fourier analysis is done in practice using a digitally implemented FFT algorithm, the resulting frequency spectrum is divided into a number of bands of finite width. Each band may be viewed as a filter, the shape of which is dependent upon the weighting function used. If the frequency of a signal falls in the middle of a band, its amplitude will be measured accurately. However, if it falls midway between two bands, the error in amplitude varies from 0.0 dB for the flat top window to 3.9 dB for the rectangular window. At the same time, the frequency bands obtained using the flat top window are 3.8 times wider, so the frequency resolution is 3.8 times poorer than for the rectangular window.

3.3 SIGNAL TYPES

Before designing an active noise control system, it is important to define the type of signal which is to be controlled. Figure 3.7 indicates the basic divisions into different signal types. The most fundamental division is into stationary and non-stationary signals. A stationary signal is essentially one for which average properties do not vary with time and is thus independent of the sample record used to define it. A non-stationary signal is defined as anything which does not satisfy the above definition. In practice, acoustic or vibration systems which are to be actively controlled are rarely absolutely stationary. However, in most cases, the variation is sufficiently slow over a short sample period for the system to be regarded as quasi-stationary, and treated as if it were stationary for the purpose of analysing control system stability and convergence properties. In addition to the continuous non-stationary signal type just discussed, there is a transient signal which begins and ends in one sample period. As these types of signal and truly non-stationary signals are not amenable to adaptive active control, they will not be discussed further here.

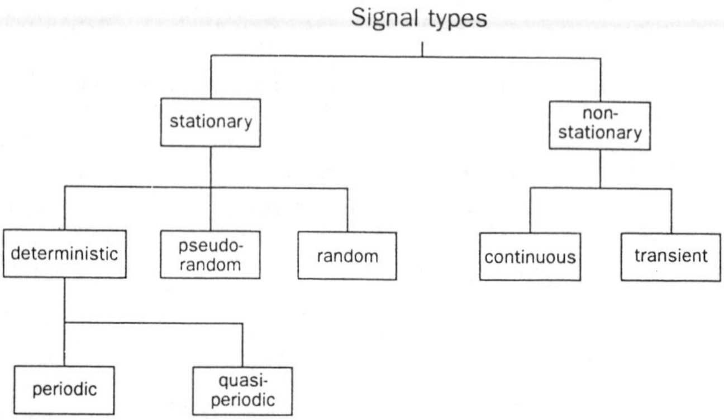

Fig. 3.7 Division into different signal types.

3.3.1 Stationary deterministic signals

Referring to Fig. 3.7, stationary deterministic signals may be divided into periodic and quasi-periodic types. In both cases, the frequency spectrum consists entirely of discrete sinusoidal components, and the value of the power spectrum will be independent of the filter bandwidth used, provided that only one component lies in each band. Periodic signals are those for which all of the discrete frequencies in the spectrum are multiples of some fundamental frequency (see Fig. 3.8(a)). For quasi-periodic signals, the discrete frequencies are unrelated to one another (see Fig. 3.8(b)). Systems excited by either periodic or quasi periodic signals are eminently suitable for active noise or vibration control.

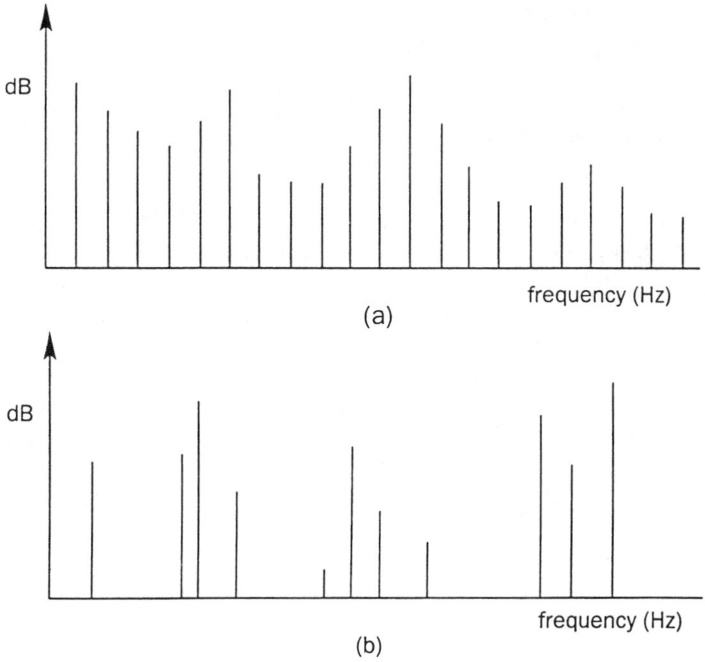

Fig. 3.8 Typical periodic and quasi-periodic spectra: (a) periodic; (b) quasi-periodic.

3.3.2 Stationary random signals

In contrast to deterministic signals, random signals are characterized by a spectrum which is continuously distributed with frequency, as shown in Fig. 3.9. Thus, the power transmitted by a filter will vary according to the filter bandwidth. The ordinate in Fig. 3.9 will be dependent upon the filter bandwidth Δf, and indirectly upon the upper frequency limit used in the Fourier analysis of the signal. The effect of filter bandwidth can be removed by dividing the power transmitted by each filter by the filter bandwidth to obtain the power spectral density.

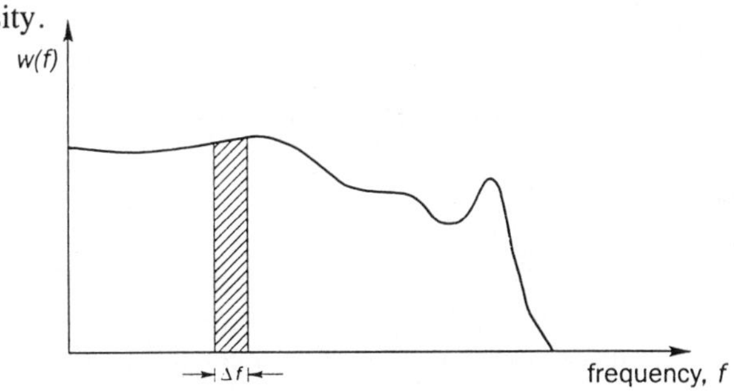

Fig. 3.9 Continuous spectrum of a stationary random signal.

The error in the spectrum for a stationary random signal is given by (3.1.3). For a single spectrum, the BT product is unity and the error ϵ is quite large. To obtain reasonably accurate results, a number of spectra must be averaged which effectively increases the BT product.

3.3.3 Pseudo-random signals

These signals are a particular type of periodic signal often used to simulate random signals. Even though they are periodic, the period T is sufficiently long for the spectral components in the frequency spectrum to be very close together. However, the phase relationship between adjacent spectral lines is random (see Fig. 3.10). When a pseudo-random signal is used to excite a physical system, the result is very similar to excitation by a random signal, provided that the bandwidth of any resonance peaks of the physical system spans a large number of spectral lines in the pseudo-random signal spectrum. Pseudo-random noise is often used as a cancellation path system identification signal in feedforward active control systems, as it is possible to use a level low enough not to significantly interfere with the signal being controlled. This is discussed in more detail in Chapter 6.

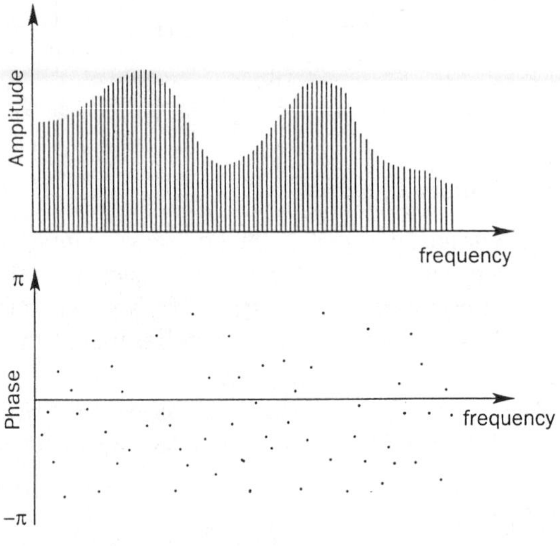

Fig. 3.10 Amplitude and phase spectra for a pseudo-random signal.

3.4 CONVOLUTION

The concept of the convolution of two signals in the time domain is equivalent to the multiplication of two signals in the frequency domain. The convolution of two time functions $f(t)$ and $h(t)$ is defined mathematically as

$$g(t) = \int_{-\infty}^{\infty} f(\tau) h(t-\tau) \mathrm{d}\tau \tag{3.4.1}$$

For convenience, (3.4.1) is often represented as:

$$g(t) = f(t) * h(t) \tag{3.4.2}$$

where the $*$ means 'convolved with'. One important application of this relationship is to the case where $f(t)$ represents the input to a physical system with an impulse response of $h(t)$. Then the output $g(t)$ is defined by (3.4.1).

The impulse response function (IRF) of a physical system is the response of the system (as a function of time) to an impulse of unit amplitude applied over an infinitesimally short period of time.

The convolution operation can best be explained with reference to Fig. 3.11. Fig. 3.11(a) represents a time signal $f(t)$ and Fig. 3.11(b) represents the impulse response $h(t)$ to which it is applied. The assumption is made that each point in $f(t)$ can be considered as an impulse weighted by the value of $f(t)$ at that point. Each such impulse excites an impulse response, the scaling of which is proportional to the level of $f(t)$ and whose time origin corresponds to the time of the impulse, as shown in Fig. 3.11(c). The solid line in Fig. 3.11(c) represents the response of the system to the input impulse at time t_n indicated in Fig. 3.11(a). The output signal at time t, $g(t)$, consists of the sum of these scaled impulse responses, each of which has been delayed by the time corresponding to the excitation of the impulse. The result is shown in Fig. 3.11(d).

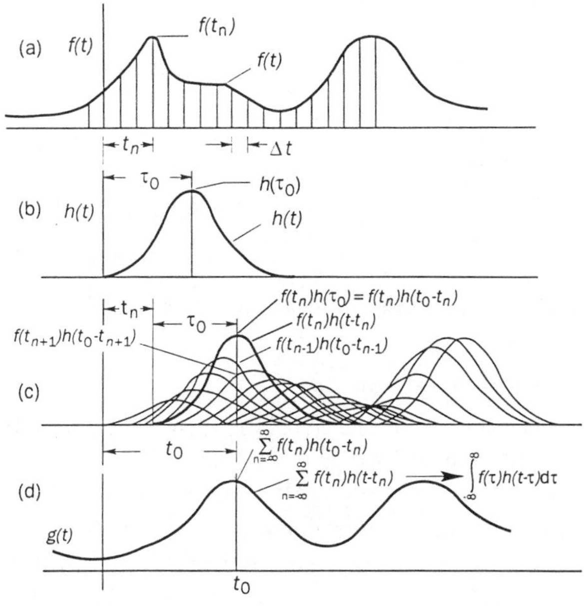

Fig. 3.11 Convolution of two time functions (after Randall, 1988).

Convolution of two digital signals at time sample k may be effected by using the relation

$$g(k) = f(k) * h(k) = \sum_{i=0}^{N-1} f(k-i)h(i) \tag{3.4.3}$$

where $h(i)$ is the ith element of the finite impulse response vector $h(i)$ of length N. This is discussed in more detail in Chapters 5 and 6.

3.4.1 Convolution with a delta function

A special case worthy of consideration involves convolution of a signal with a delta function, as illustrated in Fig. 3.12. In this case, $h(t)$ is a unit delta function $\delta(t-\tau_0)$ with a delay time of τ_0. The overall effect is to delay the signal $f(t)$ by τ_0 but otherwise to leave it unchanged. If the delta function is weighted with a scalar quantity, then the output $g(t)$ is weighted by the same amount. Thus, the effect of convolving a time signal with a delta function is to shift the origin of the time signal to the delta function delay, τ_0.

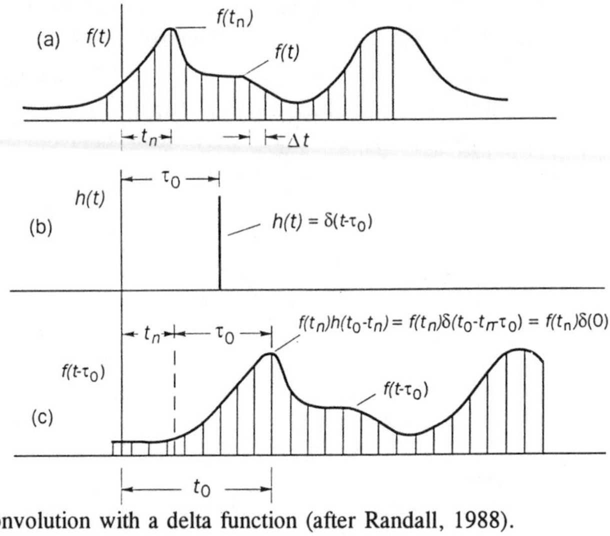

Fig. 3.12 Convolution with a delta function (after Randall, 1988).

3.4.2 Convolution theorem

This theorem states that the Fourier transform transforms a convolution between two time signals into a multiplication (complex between two frequency spectra), and the inverse Fourier transform transforms a multiplication in the frequency domain to a convolution in the time domain. Thus, if $G(f)$ is the Fourier transform of $g(t)$, $F(f)$ is the Fourier transform of $f(t)$ and $H(t)$ is the Fourier transform of $h(t)$, and if

$$g(t) = f(t) * h(t) \tag{3.4.4}$$

then

$$G(f) = F(f) \cdot H(f) \tag{3.4.5}$$

The benefits of this theorem are immediately apparent when interpreted in terms of the excitation and response of a physical system. The output spectrum $G(f)$ is obtained simply by multiplying the input spectrum with the frequency response function $H(f)$ at each frequency. As we have seen, the equivalent convolution in the time domain is a much more complicated procedure.

3.5 IMPORTANT FREQUENCY DOMAIN FUNCTIONS

In the discussion of the application of active control to reduce noise and vibration generated by physical systems, it is important that an understanding is developed of a number of functions which characterize the relationship between the input to and the output from the controller. Each of these important functions will now be discussed.

3.5.1 Cross-spectrum

The two sided cross-spectrum between two signals $x(t)$ and $y(t)$ in the frequency domain can be defined in a similar way as the power spectrum (or auto spectrum) of (3.2.16) and (3.2.18), using (3.2.15) as follows:

$$S_{xy}(f) = \lim_{T \to \infty} E\left[X^*(f)\,Y(f)\right] \tag{3.5.1}$$

or

$$S_{xy}(f_n) \approx \frac{1}{q} \sum_{i-1}^{q} X_i^*(f_n)\,Y_i(f_n) \tag{3.5.2}$$

where $X_i(f)$ and $Y_i(f)$ are instantaneous spectra and $S_{xy}(f_n)$ is estimated by averaging over a number, q, of instantaneous spectrum products obtained with finite time records of data. In contrast to the power spectrum which is real, the cross-spectrum is complex, characterized by an amplitude and a phase.

In practice, the amplitude of $S_{xy}(f_n)$ is the product of the two amplitudes $X(f_n)$ and $Y(f_n)$ and its phase is the difference in phase between $X(f_n)$ and $Y(f_n)$ $(= \phi_y - \phi_x)$. This function can be averaged as the relative phase between x and y is not random. Note that the cross-spectrum S_{yx} has the same amplitude but opposite phase to S_{xy}.

The cross-spectral density can also be defined in a similar way to the power spectral density by multiplying (3.5.2) by the time taken to acquire one record of data (which is the reciprocal of the spacing between the components in the frequency spectrum, or the reciprocal of the spectrum bandwidth).

As for power spectra, the two sided cross-spectrum can be expressed in single sided form as

$$
\begin{aligned}
G_{xy}(f_n) &= 0 & f_n &< 0 \\
G_{xy}(f_n) &= S_{xy}(f_n) & f_n &= 0 \\
G_{xy}(f_n) &= 2S_{xy}(f_n) & f_n &> 0
\end{aligned}
\tag{3.5.3}
$$

Note that $G_{xy}(f)$ is complex, with real and imaginary parts referred to as the co-spectrum and quad-spectrum respectively. As for power spectra, the accuracy of the estimate of the cross-spectrum improves as the number of records over which the averages are taken increases. The statistical error for a stationary Gaussian random signal is given (Randall, 1988) as:

$$\epsilon = \frac{1}{\sqrt{\gamma^2(f) q}} \tag{3.5.4}$$

where $\gamma^2(f)$ is the coherence function (see next section) and q is the number of averages.

The cross-spectrum is used in intensity analysis (as discussed in Section 2.5) and also for calculating the frequency response function discussed in Section 3.5.3.

The amplitude of $G_{xy}(f)$ gives a measure of how well the two functions X and Y correlate as a function of frequency and the phase angle of $G_{xy}(f)$ is a measure of the phase shift between the two signals as a function of frequency.

3.5.2 Coherence

The coherence function is a measure of the degree of linear dependence between two signals, as a function of frequency. It is calculated from the two autospectra (or power spectra) of the signals and their cross-spectrum as follows:

$$\gamma^2(f) = \frac{|\, G_{xy}(f) \,|^2}{G_{xx}(f) \cdot G_{yy}(f)} \tag{3.5.5}$$

By definition $\gamma^2(f)$ varies between 0 and 1, with 1 indicating a high degree of linear dependence between the two signals, X and Y. Thus, in a physical system where Y is the output and X is the input signal, the coherence is a measure of the degree to which Y is linearly related to X. If random noise is present in either X or Y then the value of the coherence will diminish. Other causes of diminished coherence are a non-linear relationship between $x(t)$ and $y(t)$; insufficient frequency resolution in the frequency spectrum; poor choice of window function; or a time delay, of the same order as the length of the record, between $x(t)$ and $y(t)$.

The main application of the coherence function is in checking the validity of frequency response measurements. Another more direct application is the to the calculation of signal S to noise N ratio as a function of frequency

$$S/N = \frac{\gamma^2(f)}{1-\gamma^2(f)} \tag{3.5.6}$$

3.5.3 Frequency response (or transfer) functions

The frequency response function $H(f)$ is the frequency domain equivalent of the impulse response function $h(t)$ in the time domain. For a stable, linear, time invariant system with an impulse response $h(t)$, the relationship between the input $x(t)$ and the output $y(t)$ is

$$y(t) = x(t) * h(t) \tag{3.5.7}$$

By the convolution theorem it follows that

$$Y(f) = X(f) \cdot H(f) \tag{3.5.8}$$

Thus, the frequency response function $H(f)$ is defined as

$$H(f) = \frac{Y(f)}{X(f)} \tag{3.5.9}$$

Note that the frequency response function $H(f)$ is the Fourier transform of the system impulse response function, $h(t)$.

In practice, it is desirable to average $H(f)$ over a number of spectra, but as $Y(f)$ and $X(f)$ are both instantaneous spectra, it is not possible to average either of these. For this reason it is convenient to modify (3.5.9). There are a number of possibilities, one of which is to multiply the numerator and denominator by the complex conjugate of the input spectrum. Thus,

$$H_1(f) = \frac{Y(f)\cdot X^*(f)}{X(f)\cdot X^*(f)} = \frac{G_{xy}(f)}{G_{xx}(f)} \tag{3.5.10a,b}$$

A second version is found by multiplying with $Y^*(f)$ instead of $X^*(f)$. Thus,

$$H_2(f) = \frac{Y(f)\cdot Y^*(f)}{X(f)\cdot Y^*(f)} = \frac{G_{yy}(f)}{G_{yx}(f)} \tag{3.5.11a,b}$$

Either of the above two forms of frequency response function are amenable to averaging, but $H_1(f)$ is the preferred version if the output signal $y(t)$ is more contaminated by noise than the input signal $x(t)$, whereas $H_2(f)$ is preferred if the input signal $x(t)$ is more contaminated by noise than the output (Randall, 1988).

3.5.4 Correlation functions

The cross-correlation function $R_{xy}(\tau)$ is a measure of how well two time domain signals correlate with one another as a function of the time shift, τ, between them. For stationary signals, the cross-correlation function is defined as

$$R_{xy}(\tau) = \lim_{T\to\infty} \frac{1}{T} \int_{-T/2}^{T/2} x(t)\, y(t+\tau)\, \mathrm{d}t \qquad (3.5.12)$$

Alternatively, the cross-correlation function can be derived by taking the inverse Fourier transform of the cross-spectrum $S_{xy}(f)$. Thus,

$$R_{xy}(\tau) = \mathscr{F}^{-1}\left\{S_{xy}(f)\right\} \qquad (3.5.13)$$

The major application of the cross-correlation function is to detect time delays between two signals and to extract a common signal from noise.

An example of the determination of time delays between two signals is illustrated by using a sound source in a reverberant room. One signal $x(t)$ is taken from a microphone close to the sound source and a second, $y(t)$, is taken from a microphone at some distance from the source. The cross-correlation function will show a peak at time τ_0 which corresponds to the time for an acoustic signal to travel from the first microphone to the second microphone and additional peaks at times corresponding to the time taken for acoustic waves to travel from the first microphone to the second by way of some reflected path. Note that the relative amplitudes of the peaks in the cross-correlation function are dependent on the strength of the signal $y(t)$ and also on the presence or otherwise of contaminating noise.

The auto correlation function is simply the cross-correlation function with $y(t + \tau)$ replaced by $x(t + \tau)$. The autocorrelation function can also be used to detect echoes, for example in a reverberant room. Thus, if $x(t)$ contains echoes (that is, scaled down versions of the main signal) it can be seen that when τ equals one of the echo delay times, the signal will correlate well with the delayed version of itself and result in a peak in the auto-correlation function.

REFERENCES

Abramowitz, M. and Stegun, I.A. (1965). *Handbook of Mathematical Functions with Formulas, Graphs, and Mathematical Tables.*

Bendat, J.S. and Piersol, A.G. (1980). *Engineering Applications of Correlation and Spectral Analysis*. Wiley, New York.

Braun, S. (ed.) (1986). *Mechanical Signature Analysis*. Academic Press, London.

Brigham, E.O. (1974). *The Fast Fourier Transform*. Prentice-Hall, NJ.

Broch, J.T. (1990). *Principles of Experimental Frequency Analysis*. Elsevier Applied Science, London.

Jenkins, G.M. and Watts, D.G. (1968). *Spectral Analysis and its Applications*. Holden-Day, San Francisco.

Oppenheim, A.V. and Schafer, R.W. (1975). *Digital Signal Processing*. Prentice-Hall, NJ.

Randall, R.B. (1988). *Frequency Analysis*. Bruel and Kjaer, Copenhagen.

4

Modal analysis

4.1 MODAL ANALYSIS: ANALYTICAL

A structure or a discrete system made of point masses connected by massless springs will vibrate in a way which can be described in terms of normal modes. Each normal mode of vibration is characterized in terms of a resonance frequency, mode shape and damping, much like the simple single-degree-of-freedom oscillator.

Modal analysis is the process of analytically determining the dynamic properties (resonance frequencies and mode shapes) of a system made up of a number of particles (or point masses) connected together in some way by a number of massless stiffnesses. Such a system may also be a continuous elastic structure which has been idealized as a number of point masses by use of some discretization procedure. Modal properties are determined by measuring the structural response to a known excitation. Once estimated, the modal model of a structure can be used, together with an estimate or measure of the system damping, to calculate the response of the system to any other applied forces, which may be point forces, distributed forces, point moments or distributed moments. When applied to an elastic structure, modal analysis results in the complete representation of the dynamic properties of a structure in terms of its modes of vibration. Modes of vibration are solutions to an eigenvalue problem as formulated from the differential equations which describe the motion of the structure or system of particles. The modal parameters (mode shape and resonance frequency) can be found either analytically or experimentally while the damping parameter can only be found experimentally.

It should be remembered that both analytical and experimental modal analyses are generally only practical for identifying the lowest order (first 20 or 30) vibration modes of a structure. However, the application of active control is generally only practical in this range, so the use of statistical energy analysis for the analysis of higher order modal response will not be considered here.

Very simple continuous structures such as beams, plates and cylinders are also amenable to analysis from first principles by using the appropriate wave equation together with suitable boundary conditions to determine resonance frequencies and mode shapes. Examples of this type of analysis were discussed in Section 2.3. In this section, however, the principles of analysis of discretized systems will be developed, and the results obtained for the system response at a particular location (or locations) to a particular force or forces as a function of system resonance frequencies, modal damping and mode shapes, will also be applicable to the simple systems analysed using the wave equation. As we will find several times later on in this book, mode shape functions of active noise and vibration control 'targets' make a convenient basis for physical system analysis and active control system design.

4.1.1 Single-degree-of-freedom system

It is assumed here that the reader has been exposed to basic texts on vibration. The purpose of this section is to develop the ideas and principles which will be of use in the analysis of multi-degree-of-freedom discrete systems.

An idealized single-degree-of-freedom system may be described as one which consists of a mass, not free to rotate and free to move in only one direction. The mass is attached to a rigid boundary through a massless spring and a massless viscous damper, as shown in Fig. 4.1.

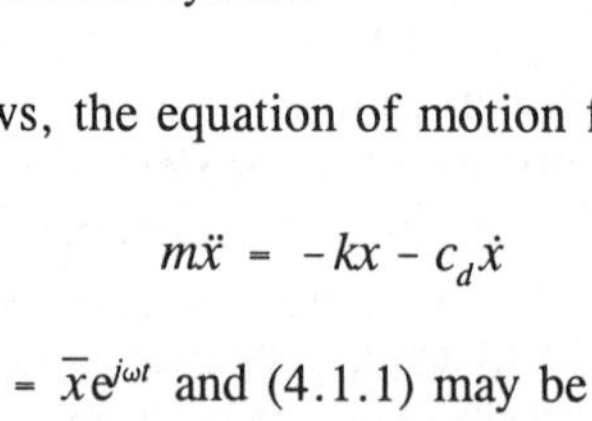

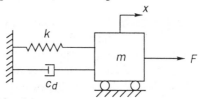

Fig. 4.1 Single-degree-of-freedom system.

Using Newton's laws, the equation of motion for this system with $f = 0$, may be written as

$$m\ddot{x} = -kx - c_d\dot{x} \tag{4.1.1}$$

For harmonic motion $x = \bar{x}e^{j\omega t}$ and (4.1.1) may be written as

$$-m\omega^2\bar{x} + j\omega c_d\bar{x} + k\bar{x} = 0 \tag{4.1.2}$$

where ω is the frequency of oscillation in rad s^{-1}.

If $c_d = 0$, (4.1.2) may be solved to give the undamped natural frequency of the system as

$$\omega_n = \sqrt{\frac{k}{m}} \quad \text{rad s}^{-1} \tag{4.1.3}$$

A single-degree-of-freedom system is critically damped if after removal of an exciting force, its motion decays to zero as rapidly as is possible with no oscillation about a mean value. In this case, the damping is

$$c_d = 2\sqrt{km} \tag{4.1.4}$$

The critical damping ratio ζ (or damping factor) is defined as the ratio of the damping c_d to the critical damping $2\sqrt{km}$. Thus,

$$\zeta = \frac{c_d}{2\sqrt{km}} \tag{4.1.5}$$

Substituting $s = j\omega$ into (4.1.2) using (4.1.3) and (4.1.5) and rearranging gives

$$s^2 + 2\zeta\omega_n s + \omega_n^2 = 0 \tag{4.1.6}$$

When $\zeta < 1$ (as in most practical systems) the solutions to (4.1.6) are two complex conjugates:

$$s = -\zeta\omega_n \pm j\omega_n\sqrt{1 - \zeta^2} \tag{4.1.7}$$

Thus, the natural frequency of vibration is complex, with an imaginary oscillating part of $\omega_n\sqrt{1-\zeta^2}$ and a real decaying part of $\zeta\omega_n$. The quantity, $\omega_n\sqrt{1 - \zeta^2}$, is often referred to as ω_d, the damped natural frequency.

The solution for the motion x is $x = \bar{x}e^{st}$, or

$$x = \bar{x}e^{-\zeta\omega_n t}\, e^{j[\omega_n\sqrt{(1-\zeta^2)}]t} \tag{4.1.8}$$

which represents a cyclic decaying motion, as shown in Fig. 4.2.

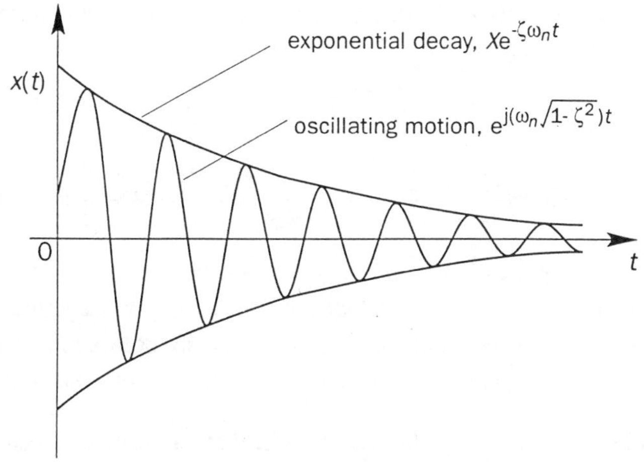

Fig. 4.2 Free vibration characteristics of a damped single-degree-of-freedom system.

If the forcing function f is now non-zero and equal to $\bar{f}e^{j\omega t}$ (that is, the right-hand side of (4.1.2) is equal to $\bar{f}e^{j\omega t}$), it can be shown that the response of the damped system is

$$x = \frac{\bar{f}e^{j\omega t}}{(k-\omega^2) + j\omega c_d} \tag{4.1.9}$$

The receptance function, defined as $\bar{x}/\bar{f}$, is then given by

$$\frac{\bar{x}}{\bar{f}} = \frac{1}{(k - \omega^2 m) + j\omega c_d} = \frac{1}{m\omega_n^2[1 - (\omega/\omega_n)^2 + 2j(\omega/\omega_n)\zeta]} \tag{4.1.10a,b}$$

The quantity $\bar{x}/\bar{f}$ is dependent upon the frequency ω of the excitation and is known as the receptance frequency-response function, because $\bar{x}$ represents displacement of the point mass. If the velocity $\bar{u}$ $(= \bar{x}\omega)$ were used instead of the displacement then $\bar{u}/\bar{f}$ would be the mobility frequency response function. If the acceleration $\bar{a}$ $(= \bar{x}\omega^2)$ were used, then $\bar{a}/\bar{f}$ would be the inertance frequency response function. It can be shown easily that the frequency of maximum displacement is given by $\omega_n[1 - 2\zeta^2]^{1/2}$, the frequency of maximum velocity given by ω_n and the frequency of maximum acceleration by $\omega_n[1 - 2\zeta^2]^{-1/2}$.

Up to now, the type of damping which has been considered has been viscous; that is, the damping force is proportional to the velocity of the point mass. However, this type of damping is not very representative of real structures, where it is often observed that damping varies at a rate approximately proportional to frequency. As the free vibration problem gives some analytical problems, only a forced response analysis will be considered for this type of damping. Assuming a damping of the form $h = c_d\omega$, the equation of motion for a single-degree-of-freedom (SDOF) system may be written as,

$$x(-\omega^2 m + k + jh)e^{j\omega t} = \bar{f}e^{j\omega t} \tag{4.1.11}$$

Thus,

$$\frac{\bar{x}}{\bar{f}} = \frac{1}{(k - \omega^2 m) + jh} = \frac{1}{m\omega_n^2[1 - (\omega/\omega_n)^2 + j\eta]} \tag{4.1.12a,b}$$

where $\eta = h/k$ is the structural loss factor.

From (4.1.12b), it may be concluded that the maximum displacement occurs at a frequency of $\omega = \omega_n$, and the maximum velocity occurs at a frequency of $\omega = \omega_n(1 + \eta^2)^{1/2}$ (assuming a forcing function which is constant with frequency).

It can be seen from (4.1.10) and (4.1.12) that the phase between the excitation force and displacement changes by 180° as the frequency of the excitation passes from below resonance to above resonance. For an undamped system, the phase changes instantaneously at the resonance frequency and as the damping becomes larger the phase change becomes less sudden.

4.1.2 Measures of damping

At this stage, it is worthwhile explaining the relationships between the different measures of damping in common use.

The structural loss factor η is related to the critical damping ratio ζ as follows:

$$\eta = 2\zeta/(1 - \zeta^2)^{1/2} \tag{4.1.13}$$

provided ζ is small. The quality factor Q is related to η by

$$Q = 1/\eta \tag{4.1.14}$$

The logarithmic decrement δ (or damping factor) is related to the critical damping ratio ζ as follows:

$$\delta = 2\pi\zeta/(1-\zeta^2)^{1/2} = \pi\eta \tag{4.1.15a,b}$$

There are a number of ways of measuring the loss factor of a system of masses or a structure, and two of these will be outlined here.

The first method requires a measurement of the time taken for the structural vibration to decay by 60 dB after cessation of an excitation force. The loss factor is then given by

$$\eta = \frac{2.2}{T_{60}f} \tag{4.1.16}$$

where T_{60} is the reverberation time in seconds and f is the frequency of excitation.

The second measurement method involves plotting the velocity response of a structure as a function of frequency, as shown in Fig. 4.3. If the response spans a resonance peak (corresponding to a vibration mode) as shown in the figure then the loss factor is given by

$$\eta = b/f_n \tag{4.1.17}$$

where f_n is the resonance frequency of the mode and b is the 3 dB bandwidth (Hz).

4.1.3 Multi-degree-of-freedom systems

In this section and in other parts of this text we will be concerned with structures having many degrees of freedom (or modes of vibration), and their analysis will be dependent upon the use and manipulation of matrices and vectors in a general way. To help visualize what analysis processes are being used, a two-degree-of-freedom system will be used as an example, although the general equations and solutions will apply to systems with any number of degrees of freedom.

Consider the damped two-degree-of-freedom system shown in Fig. 4.4. The type of damping has not been specified but may be viscous or structural. For the

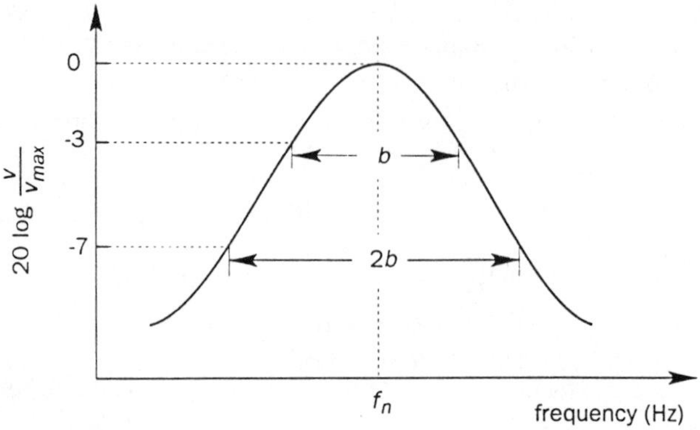

Fig. 4.3 Measurement of loss factor from resonance bandwidth data.

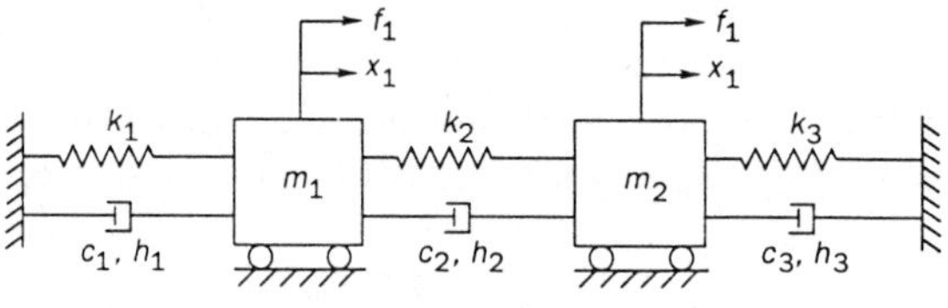

Fig. 4.4 Damped two-degree-of-freedom system.

case of viscous damping, the two equations of motion are

$$\left.\begin{aligned} m_1\ddot{x}_1 + (c_1+c_2)\dot{x}_1 + (k_1+k_2)x_1 - c_2\dot{x}_2 - k_2x_2 &= f_1(t) \\ -c_2\dot{x}_1 - k_2x_1 + m_2\ddot{x}_2 + (c_2+c_3)\dot{x}_2 + (k_2+k_3)x_2 &= f_2(t) \end{aligned}\right\} \quad (4.1.18a,b)$$

In matrix form, these equations may be written as:

$$\begin{bmatrix} m_1 & 0 \\ 0 & m_2 \end{bmatrix} \begin{bmatrix} \ddot{x}_1 \\ \ddot{x}_2 \end{bmatrix} + \begin{bmatrix} c_1+c_2 & -c_2 \\ -c_2 & c_1+c_2 \end{bmatrix} \begin{bmatrix} \dot{x}_1 \\ \dot{x}_2 \end{bmatrix} + \begin{bmatrix} k_1+k_2 & -k_2 \\ -k_2 & k_1+k_2 \end{bmatrix} \begin{bmatrix} x_1 \\ x_2 \end{bmatrix} = \begin{bmatrix} f_1(t) \\ f_2(t) \end{bmatrix} \quad (4.1.19)$$

In the absence of damping or any external excitation force, (4.1.19) becomes

$$\begin{bmatrix} m_1 & 0 \\ 0 & m_2 \end{bmatrix} \begin{bmatrix} \ddot{x}_1 \\ \ddot{x}_2 \end{bmatrix} = \begin{bmatrix} k_1+k_2 & -k_2 \\ -k_2 & k_1+k_2 \end{bmatrix} \begin{bmatrix} x_1 \\ x_2 \end{bmatrix} = 0 \quad (4.1.20a,b)$$

Assuming harmonic solutions of the form $x_1 = \bar{x}_1 e^{j\omega t}$ and $x_2 = \bar{x}_2 e^{j\omega t}$, (4.1.20) may be written in matrix form as

$$-\omega^2 \boldsymbol{M}\bar{\boldsymbol{x}} + \boldsymbol{K}\bar{\boldsymbol{x}} = 0 \quad (4.1.21)$$

for which the only non-trivial solution is

$$\det | \boldsymbol{K} - \omega^2 \boldsymbol{M} | = 0 \tag{4.1.22}$$

That is,

$$\det \begin{vmatrix} k_1 + k_2 - m_1\omega^2 & -k_2 \\ -k_2 & k_2 + k_3 - m_2\omega^2 \end{vmatrix} = 0 \tag{4.1.23}$$

$$\text{or } \omega^4 - \left[\frac{k_1 + k_2}{m_1} + \frac{k_2 + k_3}{m_2} \right] \omega^2 + \frac{k_1 k_3 + k_1 k_2 + k_2 k_3}{m_1 m_2} = 0 \tag{4.1.24}$$

This equation has two positive solutions, ω_1 and ω_2. Substituting these solutions into (4.1.21) and rearranging gives

$$\left[\frac{\bar{x}_1}{\bar{x}_2} \right]_1 = \frac{k_2}{k_1 + k_2 - \omega_1^2 m_1} \tag{4.1.25}$$

$$\left[\frac{\bar{x}_1}{\bar{x}_2} \right]_2 = \frac{k_2}{k_1 + k_2 - \omega_2^2 m_1} \tag{4.1.26}$$

Equations (4.1.25) and (4.1.26) represent the relative displacements of the masses m_1 and m_2 (or the mode shape) for modes 1 and 2 respectively. The modal matrix ψ is a 2×2 matrix and consists of one column for each mode. Thus,

$$\psi = \begin{bmatrix} a_{11} & a_{12} \\ a_{21} & a_{22} \end{bmatrix} = \begin{bmatrix} 1 & 1 \\ (\bar{x}_2/\bar{x}_1)_1 & (\bar{x}_2/\bar{x}_1)_2 \end{bmatrix} \tag{4.1.27a,b}$$

Note that the coordinates x_1 and x_2 are not unique generalized coordinates. We could just as easily have selected $q_1 = x_1$ and $q_2 = x_2 - x_1$ which results in the following equation for the undamped system

$$\begin{bmatrix} m_1 & 0 \\ m_2 & m_2 \end{bmatrix} \begin{bmatrix} \ddot{q}_1 \\ \ddot{q}_2 \end{bmatrix} + \begin{bmatrix} k_1 & -k_2 \\ k_3 & k_2 + k_3 \end{bmatrix} \begin{bmatrix} q_1 \\ q_2 \end{bmatrix} = 0 \tag{4.1.28}$$

Thus, the most general expression for (4.1.19) is

$$\begin{bmatrix} m_{11} & m_{12} \\ m_{21} & m_{22} \end{bmatrix} \begin{bmatrix} \ddot{q}_1 \\ \ddot{q}_2 \end{bmatrix} + \begin{bmatrix} c_{11} & c_{12} \\ c_{21} & c_{22} \end{bmatrix} \begin{bmatrix} \dot{q}_1 \\ \dot{q}_2 \end{bmatrix} + \begin{bmatrix} k_{11} & k_{12} \\ k_{21} & k_{22} \end{bmatrix} \begin{bmatrix} q_1 \\ q_2 \end{bmatrix} = \begin{bmatrix} Q_1(t) \\ Q_2(t) \end{bmatrix} \tag{4.1.29}$$

or

$$M\ddot{q} + C\dot{q} + Kq = Q(t) \tag{4.1.30}$$

which also applies to a system with n degrees of freedom, where n is any integer number. The matrices M, C and K are $n \times n$ matrices and $\ddot{q}$, $\dot{q}$, q and $Q(t)$ are $n \times 1$ vectors.

If the system is undamped and not subject to any external forces, the complete solution may be expressed in two $n \times n$ matrices as

$$\begin{bmatrix} \diagdown & & \\ & \omega_r^2 & \\ & & \diagdown \end{bmatrix} \quad \text{and} \quad \psi$$

where ω_r is the resonance frequency for the rth mode and the rth column in ψ is a vector which describes the rth mode shape. The matrix ψ is not unique; any column may be multiplied by a scaling factor which will be a function of the normalization used in the calculation procedure.

Vibration modes of a system are characterized by resonance frequencies and mode shapes, and possess orthogonal properties which may be stated mathematically as

$$\psi^{\mathrm{T}} M \psi = \begin{bmatrix} \diagdown & & \\ & m_r & \\ & & \diagdown \end{bmatrix} \tag{4.1.31}$$

and

$$\psi^{\mathrm{T}} K \psi = \begin{bmatrix} \diagdown & & \\ & k_r & \\ & & \diagdown \end{bmatrix} \tag{4.1.32}$$

where the resonance frequency for the rth mode is

$$\omega_r^2 = k_r / m_r \tag{4.1.33}$$

Although m_r and k_r are both affected by the scaling used for the mode shapes, the ratio k_r/m_r is unique for a given mode r. The quantities m_r and k_r are often referred to as the modal (or generalized) mass and modal (or generalized) stiffness respectively.

A result of the modal orthogonality is that the scalar product of any two column vectors in the mode shape matrix must be equal to zero. In other words, the eigen vectors obey the relation

$$\int_S\int \rho(x,y)\, h(x,y)\, \psi_m(x,y)\, \psi_n(x,y)\, \mathrm{d}x\mathrm{d}y = 0 \quad \text{for } m \neq n \tag{4.1.34}$$

where $\rho(x,y)$ is the structural density at (x,y) and $h(x,y)$ is the thickness of the surface at (x,y).

For a surface of uniform density and thickness, (4.1.34) implies that the space averaged product of any two mode shapes is zero. Equation (4.1.34) is derived by Cremer *et al.*, (1973, p. 284) who also show that the eigen vectors of systems which are not closed need not satisfy the orthogonality relation. That is, if energy can be removed from the system, orthogonality of modes is generally violated. A system is closed if its edges are free or completely restrained; added masses or non-conducting impedances also extract no energy. A system is 'open' if it is connected to other systems or if its edges can dissipate energy into its supports.

The equations of motion can be uncoupled by the correct choice p of the generalized coordinates so that for mode r we can write the following independent equation of motion:

$$m_r \ddot{p}_r + k_r p_r = 0 \tag{4.1.35}$$

where p_r is referred to as the rth principal coordinate. The principal coordinates are related to the generalized coordinates $\boldsymbol{x}$ by

$$\boldsymbol{p} = \boldsymbol{\psi}^{-1}\boldsymbol{x} \tag{4.1.36}$$

For a discrete system with the mass defined at n locations, the modal mass for the rth mode may be calculated from the mode shape vector for the rth mode using

$$m_r = \sum_{k=1}^{n} \sum_{i=1}^{n} m_{ki} \psi_{kr} \psi_{ir} \tag{4.1.37}$$

For a continuous non-discrete surface for which the mode shape is known, the modal mass is given by

$$m_r = \int_A \int m(\boldsymbol{r})\, \psi^2(\boldsymbol{r})\, \mathrm{d}\boldsymbol{r} \tag{4.1.38}$$

where $m(\boldsymbol{r})$ is the mass per unit area at location $\boldsymbol{r}$ on the surface.

4.1.3.1 Forced response of undamped systems

The general equation for a multi-degree-of-freedom system with no damping but with external forces acting at each point mass is

$$\boldsymbol{M}\ddot{\boldsymbol{x}}(t) + \boldsymbol{K}\boldsymbol{x}(t) = \boldsymbol{f}(t) \tag{4.1.39}$$

The components of the force vector $\boldsymbol{f}$ may have any amplitudes and phases, but we will assume here that all forces are of the same frequency. Forces at different frequencies may be treated separately and the results combined using superposition. Thus, the force vector is defined as

$$f(t) = \bar{f}e^{j\omega t} \tag{4.1.40}$$

and the solution is

$$\boldsymbol{x}(t) = \bar{\boldsymbol{x}}e^{j\omega t} \tag{4.1.41}$$

Note that $\bar{f}$ and $\bar{x}$ are complex vectors; that is, each element is characterized by an amplitude and a phase. Equation (4.1.39) may now be rewritten using (4.1.40) and (4.1.41) to give

$$(\boldsymbol{K} - \omega^2 \boldsymbol{M})\bar{\boldsymbol{x}}e^{j\omega t} = \bar{f}e^{j\omega t} \tag{4.1.42}$$

Equation (4.1.42) may be rearranged to solve for the unknown responses $\bar{x}$:

$$\bar{\boldsymbol{x}} = (\boldsymbol{K} - \omega^2 \boldsymbol{M})^{-1}\bar{f} \tag{4.1.43}$$

which can be written as

$$\bar{\boldsymbol{x}} = \alpha(\omega)\bar{f} \tag{4.1.44}$$

where $\alpha(\omega)$ is the $n \times n$ receptance matrix for the system. The general element $\alpha_{ik}(\omega)$ in the receptance matrix is defined as

$$\alpha_{ik}(\omega) = (\bar{x}_j/\bar{f}_k); \qquad \bar{f}_m = 0; \; m = 1, \ldots n \neq k \tag{4.1.45a,b}$$

Equation (4.1.45) represents the individual receptance function which is similar to that defined earlier for the SDOF system.

Values for the elements of $\alpha(\omega)$ can be determined by substituting appropriate values for the mass, stiffness and force matrices in (4.1.43). However, this involves inversion of the system matrix, which is impractical for systems with a large number of modes of vibration. Also, no insight is gained into the form of the various properties of the frequency response function.

Returning to (4.1.43), we can write

$$(\boldsymbol{K} - \omega^2 \boldsymbol{M}) = \alpha^{-1}(\omega) \tag{4.1.46}$$

Premultiplying both sides by ψ^{T} and postmultiplying both sides by ψ we obtain

$$\psi^{\mathrm{T}}(\boldsymbol{K} - \omega^2 \boldsymbol{M})\psi = \psi^{\mathrm{T}}\alpha^{-1}(\omega)\psi \tag{4.1.47}$$

or

$$\begin{bmatrix} \diagdown & & \\ & m_r & \\ & & \diagdown \end{bmatrix} \begin{bmatrix} \diagdown & & \\ & \omega_r^2 - \omega^2 & \\ & & \diagdown \end{bmatrix} = \psi^{\mathrm{T}}\alpha^{-1}(\omega)\psi \tag{4.1.48}$$

which gives

$$\alpha(\omega) = \psi \left\{ \begin{bmatrix} \diagdown & & \\ & m_r & \\ & & \diagdown \end{bmatrix} \begin{bmatrix} \diagdown & & \\ & \omega_r^2 - \omega^2 & \\ & & \diagdown \end{bmatrix} \right\}^{-1} \psi^{\mathrm{T}} \tag{4.1.49}$$

It is clear that the receptance matrix defined in (4.1.47) is symmetric and this is recognized as a principle of reciprocity which applies to many characteristics of practical systems. In this case, the implication is that

$$\alpha_{ik} = \bar{x}_i / \bar{f}_k = \alpha_{ki} = \bar{x}_k / \bar{f}_i \tag{4.1.50a,b,c}$$

which demonstrates the principle of reciprocity.

Any individual frequency response function can be expressed using (4.1.49) as

$$\alpha_{ik} = \sum_{r=1}^{n} \frac{\psi_{ir}\psi_{kr}}{m_r(\omega_r^2 - \omega^2)} \tag{4.1.51}$$

Any arbitrary normalization of the mode shape matrix ψ will be reflected in the modal mass matrix $\left[\diagdown\ m_r\ \diagdown\right]$ so that α_{ik} is independent of any normalization process.

Note that up to now, as damping has been excluded from the analysis, the mode shapes, modal masses and modal stiffnesses are all real quantities.

4.1.3.2 Damped MDOF systems: proportional damping

Proportional damping is a special type of damping which simplifies system analysis. The damping may be viscous or hysteretic but the damping matrix must be proportional to either or both of the mass and stiffness matrices. Proportional viscous damping is the type usually assumed in the analysis of acoustic waves in fluid media.

The advantage in using proportional damping is that the mode shapes for both the damped and undamped cases are the same and the modal resonance frequencies are also very similar. Thus, the properties of a proportionally damped system may be determined by analysing in full the undamped system and then making a small correction for the damping. Although this is done in many commercial software packages, it is only valid for this very special type of damping.

To begin, we will examine the case where the damping matrix is proportional to the stiffness matrix. As derived earlier, the general equation describing the motion of a damped multi-degree-of-freedom system subject to external forces is

$$\boldsymbol{M}\ddot{\boldsymbol{x}} + \boldsymbol{C}\dot{\boldsymbol{x}} + \boldsymbol{K}\boldsymbol{x} = \boldsymbol{f}(t) \tag{4.1.52}$$

In this case the damping matrix $\boldsymbol{C}$ is proportional to the mass matrix. Thus,

$$\boldsymbol{C} = \beta \boldsymbol{M} \tag{4.1.53}$$

If we pre- and postmultiply the damping matrix by the eigenvector (or mode shape) matrix for the undamped system, as was done previously for the mass and stiffness matrices, we find that

$$\boldsymbol{\psi}^{\mathrm{T}} \boldsymbol{C} \boldsymbol{\psi} = \beta \begin{bmatrix} \diagdown & & \\ & m_r & \\ & & \diagdown \end{bmatrix} = \begin{bmatrix} \diagdown & & \\ & c_r & \\ & & \diagdown \end{bmatrix} \tag{4.1.54a,b}$$

where the diagonal elements c_r represent the generalized (or modal) damping values.

Because this matrix is diagonal, the mode shapes of the damped system are identical to those of the undamped system. This can be shown by taking the general equation of motion (4.1.52) with no excitation forces $\boldsymbol{f}(t)$, and transforming it to principal coordinates using the undamped system mode shape matrix $\boldsymbol{\psi}$. We obtain

$$\begin{bmatrix} \diagdown & & \\ & m_r & \\ & & \diagdown \end{bmatrix} \ddot{\boldsymbol{P}} + \begin{bmatrix} \diagdown & & \\ & c_r & \\ & & \diagdown \end{bmatrix} \dot{\boldsymbol{P}} + \begin{bmatrix} \diagdown & & \\ & k_r & \\ & & \diagdown \end{bmatrix} \boldsymbol{P} = 0 \tag{4.1.55}$$

where $\boldsymbol{P} = \boldsymbol{\psi}^{-1} \boldsymbol{x}$. The rth individual equation is then

$$m_r \ddot{p}_r + c_r \dot{p}_r + k_r p_r = 0 \tag{4.1.56}$$

which is clearly the equation of motion for a single-degree-of-freedom system or for a single mode of a multi-degree-of-freedom system. This mode has a complex resonance frequency with an oscillatory part of

$$\omega_r' = \omega_r \sqrt{1 - \zeta^2} \tag{4.1.57}$$

and a decaying part of $\zeta_r \omega_r$ where

$$\omega_r^2 = k_r / m_r \text{ and } \zeta_r = \frac{c_r}{2\sqrt{k_r m_r}} = \frac{\beta}{2\omega_r} \tag{4.1.58a,b,c}$$

A more general form of proportional damping is where the damping matrix is related to the mass and stiffness matrices as follows:

$$\boldsymbol{C} = \beta \boldsymbol{M} + \gamma \boldsymbol{K} \tag{4.1.59}$$

In this case, the damped system will have the same mode shape vectors as the undamped system and the resonance frequencies will be

$$\omega_r' = \omega_r \sqrt{1 - \zeta_r^2} \; ; \; \zeta_r = \frac{\beta}{2\omega_r} + \frac{\gamma \omega_r}{2} \tag{4.1.60a,b}$$

Forced response analysis

For both types of proportional damping, the frequency response matrix for forced excitation is given by

$$\alpha(\omega) = \left[K + j\omega C - \omega^2 M\right]^{-1} \tag{4.1.61}$$

The ratio of the displacement $\bar{x}$ at location i to a force $\bar{f}$ of frequency ω at location k, represented by $\alpha_{ik}(\omega)$ is given by

$$\alpha_{ik}(\omega) = \sum_{r=1}^{n} \frac{\psi_{ir}\psi_{kr}}{m_r(\omega_r^2 - \omega^2) + j\omega c_r} \tag{4.1.62}$$

or

$$\alpha_{ik}(\omega) = \sum_{r=1}^{n} \frac{\psi_{ir}\psi_{kr}}{\omega_r^2 m_r\left[1 - (\omega/\omega_r)^2 + 2j(\omega/\omega_r)\zeta_r\right]} \tag{4.1.63}$$

where

$$\zeta_r = \frac{c_r}{2\sqrt{k_r m_r}}$$

The same procedure as outlined above can be followed for proportional hysteretic damping. The equations of motion are written as

$$M\ddot{x} + (K + jH)x = f \tag{4.1.64}$$

and the hysteretic damping matrix H is proportional to the mass and stiffness matrices as follows

$$H = \beta M + \gamma K \tag{4.1.65}$$

Again, we find that the mode shapes for the damped system are identical to those for the undamped system and the eigenvalues (or resonance frequencies) are complex of the following form

$$\omega_r' = \omega_r\sqrt{1 + j\eta_r}\ ; \quad \eta_r = \gamma + \beta/\omega_r^2; \quad \omega_r^2 = k_r/m_r \tag{4.1.66a,b,c}$$

The expression for an element of the general frequency response function matrix is written as

$$\alpha_{ik}(\omega) = \sum_{r=1}^{n} \frac{\psi_{ir}\psi_{kr}}{m_r(\omega_r^2 - \omega^2) + j\eta_r k_r} \tag{4.1.67}$$

In many practical structures, even though the damping may not be strictly proportional it is often sufficiently small that for the purposes of estimating the resonance frequencies, mode shapes and frequency response function, it is sufficiently accurate to assume proportional damping; this assumption is commonly used in the analysis of space structures which are discussed in Chapter 11. With this assumption it is possible to calculate the undamped resonance frequencies and mode shapes and then calculate the actual resonance frequencies using (4.1.66a) and the actual frequency response function using (4.1.67).

The output from most finite element software packages are the resonance frequencies, mode shapes, and modal masses for the undamped system. The forced response at any frequency ω can then be calculated using additional software which uses (4.1.67) and estimates of structural damping η.

4.1.3.3 Damped MDOF systems: general structural damping

In the practical analysis of structural vibrations, there are many cases where the assumption of proportional damping cannot be made. Thus, here we will consider the properties of a system with general structural (or hysteretic) damping.

The equation of motion may be written as before as

$$\boldsymbol{M}\ddot{\boldsymbol{x}} + (\boldsymbol{K} + j\boldsymbol{H})\boldsymbol{x} = \boldsymbol{f}(t) \tag{4.1.68}$$

Considering first the case where $\boldsymbol{f}(t) = 0$ we assume a solution

$$\boldsymbol{x} = \bar{\boldsymbol{x}}e^{j\lambda t} \tag{4.1.69}$$

Substituting this solution into (4.1.68) yields a solution consisting of complex eigen frequencies and mode shapes. The complex mode shape matrix consists of elements defined by a relative amplitude and phase so that the relative phase of the motion of the point masses can vary between 0 and 180° instead of being confined to one or the other as was the case for undamped or proportionally damped systems.

The rth eigenvalue may be written as

$$\lambda_r^2 = \omega_r^2(1 + i\eta_r) \tag{4.1.70}$$

where ω_r is a natural frequency and η_r is the loss factor for the rth mode. Note that ω_r is not exactly equal to the natural frequency for the undamped system, although it is close (Ewins, 1984).

The eigen vectors possess the same type of orthogonality properties as the undamped system and these may be defined by the following equations:

$$\boldsymbol{\psi}^{\mathrm{T}}\boldsymbol{M}\boldsymbol{\psi} = \begin{bmatrix} \ddots & & \\ & m_r & \\ & & \ddots \end{bmatrix} \tag{4.1.71}$$

The generalized (or modal) mass and stiffness parameters are now complex and the eigen solution for mode r is given by

$$\psi^{\mathrm{T}}(K+jH)\psi = \begin{bmatrix} \ddots & & \\ & k_r & \\ & & \ddots \end{bmatrix} \tag{4.1.72}$$

$$\lambda_r^2 = k_r/m_r \tag{4.1.73}$$

Forced response analysis

For the forced response analysis, a direct solution to (4.1.68) for a single frequency (harmonic) exciting force vector is:

$$\bar{x} = \left(K + jH - \omega^2 M\right)^{-1}\bar{f} = \alpha(\omega)\bar{f} \tag{4.1.74a,b}$$

where $f(t) = \bar{f}e^{j\omega t}$.

Following the same procedure as for the proportionally damped system we can show that:

$$\alpha_{ik}(\omega) = \sum_{r=1}^{n} \frac{\psi_{ir}\psi_{kr}}{m_r(\omega_r^2 - \omega^2 + j\eta_r\omega_r^2)} \tag{4.1.75}$$

The only difference between (4.1.75) and the equivalent (4.1.67) is that in (4.1.75) the mode shapes are complex numbers.

If it is desired to know the response of the system to a number of forces acting simultaneously, the following expression may be used:

$$\bar{x}(\omega) = \sum_{r=1}^{n} \frac{\psi_r \bar{f}(\omega)\psi_r}{m_r(\omega_r^2 - \omega^2 + j\eta_r\omega_r^2)} \tag{4.1.76}$$

where $\bar{f}(\omega)$ is the force amplitude vector at frequency ω.

4.1.3.4 Damped MDOF systems: general viscous damping

Although hysteretic damping is representative of the damping found in structures, it is not representative of the damping used in vehicle suspension systems. In this latter case the damping is closer to viscous. As active adaptive vehicle suspensions systems as well as active control of structural vibration and noise radiation will both be considered later in this book, it is worthwhile devoting a little space here to the analysis of a multi-degree-of-freedom system with general viscous damping (proportional viscous damping has been discussed already).

The analysis of the general viscous damping case is a complex problem which will only be briefly considered here. The general equation for forced excitation with viscous damping may be written as

$$M\ddot{x} + C\dot{x} + Kx = f(t) \tag{4.1.77}$$

As before we consider the case of zero excitation force and assume a solution of the form

$$x = \bar{x}e^{st} \tag{4.1.78}$$

Substituting this solution into the appropriate equation of motion gives

$$(s^2 M + sC + K)\bar{x} = 0 \tag{4.1.79}$$

This equation has $2n$ eigenvalue solutions which occur in complex conjugate pairs, of the following form:

$$s_r = \omega_r(-\zeta_r \pm j\sqrt{1-\zeta_r^2}\,) \tag{4.1.80}$$

where ω_r is the natural frequency (not exactly the same as the undamped natural frequency) and ζ_r is the critical damping ratio. The eigen vectors (or mode shapes) ψ_r and ψ_r^* which result from substitution of the solutions (4.1.80) into the equation of motion are also complex conjugates.

It can be shown that

$$\omega_r^2 = k_r / m_r \tag{4.1.81}$$

and

$$2\omega_r\zeta_r = c_r / m_r \tag{4.1.82}$$

where

$$\left.\begin{aligned} c_r &= \psi_r^* C \psi_r \\ m_r &= \psi_r^* M \psi_r \\ k_r &= \psi_r^* K \psi_r \end{aligned}\right\} \tag{4.1.83}$$

Forced response analysis

To derive expressions for the frequency response function equivalent to those derived for structural damping it is necessary to decouple the equations of motion using a new coordinate vector defined as

$$y = \begin{bmatrix} x \\ \dot{x} \end{bmatrix} \tag{4.1.84}$$

Substituting this in the equation of motion (with no force) gives

$$[C : M]\,\dot{y} + [K : 0]\,y = 0 \tag{4.1.85}$$

which represents n equations in $2n$ unknowns. Thus, we need an identity equation of the type

$$[M : 0]\,\dot{y} + [0 : -M]\,y = 0 \tag{4.1.86}$$

Combining (4.1.85) and (4.1.86) gives

$$\begin{bmatrix} \boldsymbol{C} & \boldsymbol{M} \\ \boldsymbol{M} & \boldsymbol{0} \end{bmatrix} \dot{y} + \begin{bmatrix} \boldsymbol{K} & \boldsymbol{0} \\ \boldsymbol{0} & -\boldsymbol{M} \end{bmatrix} y = 0 \tag{4.1.87}$$

which can be written as

$$\boldsymbol{A}\,\dot{y} + \boldsymbol{B}\,\dot{y} = \boldsymbol{0} \tag{4.1.88}$$

These $2n$ equations are now in the standard eigenvalue form and by assuming a solution of the form

$$y = \bar{y}e^{st} \tag{4.1.89}$$

we can obtain the $2n$ eigen values s_r and eigen vectors $\boldsymbol{\Theta}_r$ of the system which together satisfy the general equation,

$$(s_r\boldsymbol{A} + \boldsymbol{B})\,\boldsymbol{\Theta}_r = \boldsymbol{0} \qquad r = 1, \ldots 2n \tag{4.1.90}$$

These eigen values will be complex conjugate pairs and will possess orthogonality properties which may be stated as

$$\boldsymbol{\Theta}^{\mathrm{T}}\boldsymbol{A}\,\boldsymbol{\Theta} = \begin{bmatrix} \diagdown & & \\ & a_r & \\ & & \diagdown \end{bmatrix} \tag{4.1.91}$$

$$\boldsymbol{\Theta}^{\mathrm{T}}\boldsymbol{B}\,\boldsymbol{\Theta} = \begin{bmatrix} \diagdown & & \\ & b_r & \\ & & \diagdown \end{bmatrix} \tag{4.1.92}$$

and which have the characteristic that

$$s_r = -b_r/a_r \tag{4.1.93}$$

The forcing vector may now be expressed in terms of the new coordinate as

$$\boldsymbol{P} = \begin{bmatrix} f \\ \boldsymbol{0} \end{bmatrix} \tag{4.1.94}$$

Assuming a harmonic forcing function and response, and following a similar development as outlined in (4.1.42) to (4.1.51) we obtain

$$\begin{bmatrix} \bar{x} \\ \cdots \\ j\omega\bar{x} \end{bmatrix} = \sum_{r=1}^{2n} \frac{\boldsymbol{\Theta}_r^{\mathrm{T}}\boldsymbol{P}\boldsymbol{\Theta}_r}{a_r(j\omega - s_r)} \tag{4.1.95}$$

As the eigenvalues and eigen vectors appear in complex conjugate pairs, equation (4.1.95) may be written as

$$\begin{bmatrix} \bar{x} \\ \cdots \\ j\omega\bar{x} \end{bmatrix} = \sum_{r=1}^{n} \left[\frac{\boldsymbol{\Theta}_r^{\mathrm{T}}\boldsymbol{P}\boldsymbol{\Theta}_r}{a_r(j\omega - s_r)} + \frac{\boldsymbol{\Theta}_r^{*\mathrm{T}}\boldsymbol{P}\boldsymbol{\Theta}_r^{*}}{a_r^{*}\,(j\omega - s_r^{*})} \right] \tag{4.1.96}$$

Extracting a single frequency response function element we obtain

$$\alpha_{ik}(\omega) = \sum_{r=1}^{n} \frac{\Theta_{ir}\Theta_{kr}}{a_r[\omega_r\zeta_r + j(\omega - \omega_r\sqrt{1-\zeta_r^2}\,)]} + \frac{\Theta_{ir}^*\Theta_{kr}^*}{a_r^*\,[\omega_r\zeta_r + j(\omega + \omega_r\sqrt{1-\zeta_r^2}\,)]} \tag{4.1.97}$$

4.1.4 Summary

From the preceding sections it can be seen that the resonance frequencies and mode shapes for a vibroacoustic system may be determined by discretising it; that is, dividing it into a number of point masses connected by massless springs and dampers.

The equations of motion for such a system may be derived using Newton's laws or Lagrange's equations as outlined in Section 2.2. Once the equations of motion have been derived they may be expressed in matrix form in terms of mass, stiffness, damping and force matrices which can be solved for resonance frequencies and mode shapes as well as for the response of the structure at any specified location and for any excitation frequency. In practice, the use of commercial finite element software packages makes the derivation of the equations of motion, the corresponding mass, stiffness and damping matrices and the solution of the equations of motion transparent to the user. However, it is still necessary for the user to have a fundamental understanding of the physical principles involved (as outlined briefly in this section) so that the limitations of any analysis may be properly evaluated.

Of particular concern in the use of finite element analysis, or indeed in the use of any theoretical analysis for the determination of the response of a structure to one or more excitation forces, is the accurate estimation of the structural damping quantity. No analysis can provide a value for this; it is a required input to the analysis and the results of the response analysis are crucially dependent on its value. Damping values are usually determined from measurements on other similar structures and sometimes the analyst just uses values derived from past experience. Thus, in many cases the results from a response analysis will be approximate only; their accuracy almost solely depends on the accuracy in the estimation of the structural damping.

4.2 MODAL ANALYSIS: EXPERIMENTAL

It is not the intention of this section to cover the subject of experimental modal analysis in an exhaustive manner. The reader is referred to the excellent book by Ewins (1984) for this treatment. However, it is intended here to introduce the concepts necessary for the understanding of the use of experimental modal analysis as an aid for the design, optimization and performance evaluation of active noise and vibration control systems.

In this text, we will use experimental modal analysis in two ways; the first is the traditional use, involving the determination of the resonance frequencies and mode shapes for a complex structure which cannot be analysed from first principles. The second, less common use, involves determining the contributions of each vibration mode to the total vibration response of a structure at any particular excitation frequency for one or more excitation sources acting on a structure for which the theoretical mode shapes are known *a priori*. This latter technique is particularly useful for evaluating the effect of active control sources on each mode of vibration of a system and thus is a useful aid for both system design and performance evaluation.

Four assumptions are basic to the traditional experimental determination of modal resonance frequencies, mode shapes and damping of an elastic structure. They are listed as follows:

1. The structure is assumed to be linear. Associated with any displacement from equilibrium there will arise a restoring force of opposite sign proportional (to a first approximation) to the displacement. For example, a restraint which is of much greater stiffness in one direction than in the opposite direction is excluded from this analysis. Linearity has the consequence that the sum of the effects of two forces is the same as the effect of applying the sum of the two forces.
2. The structure's behaviour is time invariant. This is important, as repeated testing can be used to obtain statistical accuracy. Alternatively, if the system response varies with time, then the statistical approach implicit in the methods of modal analysis are obviously inappropriate.
3. The structural response is observable. That is, enough vibration modes or degrees of freedom can be measured to obtain an adequate behavioural model of the structure.
4. Maxwell's law of reciprocity is assumed to hold. This law, which follows from system linearity, may be stated as follows: 'The displacement at position A due to a unit force applied at position B is equal to the displacement at B due to a unit force applied at A.'

Traditional experimental modal analysis is often referred to as modal testing. Modal testing is used to verify theoretical models of structures and is also used to create accurate dynamic models of existing structures (modal resonance frequencies, damping and mode shapes) so that the effect of proposed structural modifications can be evaluated.

Experimental modal analysis (or modal testing) was begun back in the early days of the space programme in the USA, in the 1950s. There were no FFT analysers or laboratory based minicomputers then, so modal testing was done mostly with analogue instrumentation such as oscillators, amplifiers and oscilloscopes. With this type of testing, commonly referred to as sine testing or normal mode testing, a structure is excited one mode at a time with sinusoidal excitation. This excitation is provided by attaching one or more electrodynamic shakers to the structure and driving them with a sinusoidal signal. The frequency of the signal is adjusted to coincide with the natural frequency of one of the structure's modes. When this is done, the structure will readily absorb the energy and its predominant motion will be the mode shape of the mode being excited. Modal damping is measured by shutting off the shaker(s) and measuring the decay rate of the sinusoidal motion in the structure, as the mode is naturally damped out.

4.2.1 The transfer function method: traditional experimental modal analysis

With the discovery of the fast Fourier transform in 1965 and the advent of FFT analysers soon afterwards, a whole new approach to modal testing was born. This has become known as the transfer function method. This method has gained much in popularity in recent years because it is faster and easier to perform, and is much cheaper to implement than the normal mode method. Another real advantage is that a variety of different excitation signals (transient, random and swept sine) can be used to excite the structure. The structure is excited over a broad band of frequencies with these signals, and consequently many modes are excited at once. To identify modal parameters, a special type of FFT analyser must be used which can measure a so-called frequency response function (FRF) between two points (A and B) on the structure (see Chapter 3). An FRF measurement contains all of the information necessary to describe the dynamic response of the structure at point B due to an excitation force at point A.

To identify modal parameters, a whole set of these FRFs must be measured, typically between a single excitation point and many response points. (Alternatively, a set of measurements between many excitation points and a single response point can also be used. This latter set is typically measured when a small portable hammer is used to excite the structure.) A real advantage of this method is that these measurements can be made one at a time, thus requiring fewer transducers and less signal conditioning equipment than the normal mode method.

Once a set of FRF measurements is obtained with the FFT analyser and stored in a digital mass storage memory, they are then put through a 'curve fitting' process to identify the modal parameters of the structure. Curve fitting, also known as parameter estimation or identification, has undergone much development during the past decade, and a variety of methods are used today for obtaining modal parameters. These are discussed in detail by Ewins (1984).

4.2.1.1 Test procedure

A complete modal analysis test covers three phases: test preparation and set-up; transfer function (frequency response) measurements; and modal parameter identification. Best results are obtained if the input force as well as the structural response is measured, allowing the transfer function (response spectrum divided by the input force spectrum) to be determined. If the input force is assumed to be of uniform amplitude over the frequency range of interest, and is thus not measured, then by measuring the power spectrum of the structural response, it is still possible to obtain reasonable results for resonance frequencies and mode shapes, although damping results will have large errors.

The equipment required for experimental modal analysis consists of the following items:

1. swept frequency oscillator (or random noise generator) and power amplifier; or instrumented hammer (with piezoelectric load cell and charge amplifier);
2. accelerometer and charge amplifier;
3. two channel FFT spectrum analyser and some form of interface to a personal computer;
4. modal analysis software;
5. personal computer and plotter.

Test set-up

The first step in a modal analysis test is to support the structure so that it is unconstrained and is not affected significantly by its environment. To simulate this condition, the structure can be suspended from flexible cords, preferably attached to vibration nodes. Alternatively, a resilient support such as a foam mat may be used. The support should be designed so that the frequency of the highest frequency rigid body mode is less than 10% of the frequency of the lowest frequency flexural mode, to minimize flexural modal distortion. If a constrained support is not possible (for example, if the structure is very heavy), then it is sometimes acceptable to test them by resting them on a hard concrete slab. This is referred to as the 'grounded' condition, but it is less than ideal.

Rigid body modes are those associated with the body considered as rigid and mounted on a flexible suspension system. Essentially, these are the modes of the suspension system. The flexural modes, on the other hand, are associated with flexure of the body, treated as a system with distributed mass and stiffness. In the former case, energy absorption would be confined to losses in the suspension system, whereas in the latter case, energy absorption may occur throughout the suspended body as well.

The next step in a modal analysis test is to decide which type of excitation to use and how to apply it to the structure. There are three types of excitation which are in common use: step relaxation, shaker and impact. Each of these types will be discussed in detail in the following paragraphs.

Excitation by step relaxation

With step relaxation, the structure is preloaded (often in the shape of the mode of interest if a single mode is being studied) with a measured force and then released. The transient response thus generated propagates through and energizes the structure. This method is often used in modal analysis of piping systems.

Excitation by electrodynamic shaker

An electromagnetic shaker converts an electrical signal into a mechanical force applied to the structure. Shakers can be attached at different locations on the structure to ensure even distribution of the exciting force, but often only one shaker is used at one or two different positions. For each position of the shaker, accelerometer measurements are taken at many points on the structure, and the number of accelerometer locations required is generally a little more than twice the order of the highest mode to be investigated. The structure is usually excited at one or two locations and the vibration pickup (usually an accelerometer) is moved about to take measurements at a large number of locations (whose geometrical positions have been previously entered into the computer).

The shaker may be driven by a single sine wave, the frequency of which is incremented after each test, or it may be driven by a swept sine signal, by pseudo-random noise or by random noise. The main advantage of the sine or swept sine excitation methods is the higher signal to noise ratio and low crest factor (ratio of peak amplitude to r.m.s.), but this is at the expense of a longer measuring time. The sweep rate of the frequency sweep must be sufficiently slow not to distort the frequency response of the structure. This is especially important for lightly damped structures. One way of checking whether the sweep rate is acceptable is to compare the frequency response obtained by sweeping upwards in frequency with that obtained by sweeping downwards at the same rate. If the two results are not identical, the sweep rate is too fast.

The advantages of pseudo-random noise (random amplitude at a fixed set of frequencies) over random noise (random frequency and amplitude) include better signal to noise ratio and elimination of leakage. Leakage is defined here as movement of energy at one frequency in the spectrum to other frequencies. This leakage problem occurs with random noise because all of the excitation frequencies do not coincide with the finite set represented by the spectrum analyser. The FFT analysis of a finite sample length means that only certain frequencies are represented in the resulting spectrum whereas for random noise, energy exists at all frequencies. This results in a leakage of all the signal energy into the set of discrete frequencies which are represented. On the other hand, if pseudo-random noise is used which is represented by energy at the same discrete frequencies as those which characterize the analyser, no leakage occurs and the structural response signal will be much less 'noisy'.

Excitation by impact hammer

Impact excitation is generally implemented using an instrumented hammer, so that the impact force applied to the structure can be measured. For this method to be successful, the force applied with each impact should not vary too much between impacts. With impact excitation, either one of two techniques may be employed. The first involves exciting the structure at one or two locations and measuring the response with accelerometers at many other locations. Alternatively, to avoid shifting accelerometers around, equally good results are obtained by fixing one or two accelerometers to the structure and exciting it with the impact hammer at many other locations. This latter technique is generally faster to implement.

The impact hammer (see Fig. 4.5) is comprised of four parts:

1. handle;
2. head;
3. force transducer; and
4. tip.

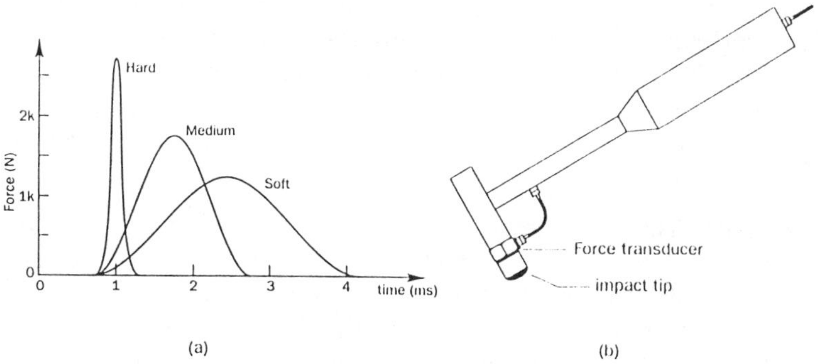

Fig. 4.5 Impact hammer and response: (a) hammer force as a function of time for tips of varying hardness (b) impact hammer.

The impact hammer excites the structure over a broad frequency range. Both the magnitude and the frequency content of the excitation applied to the structure can be controlled by the user. The magnitude of the excitation can be controlled by the mass of the hammer and the velocity of hammer impact.

The frequency range which is excited by the impact hammer is dependent on the duration of the transient spike generated in the time domain and is controlled by the stiffness of the contact surfaces and the mass of the impact hammer. Shorter transient durations result in higher frequency excitation.

There are various ways of controlling the magnitude of the input force for impact hammer excitation:

1. The input force to the structure is dependent upon the velocity of the impact hammer at the instant of impact between the tip and the structure. If consistency can be established between the contact velocity of each hit, the impact force magnitude can be accurately controlled. However, this is very hard to accomplish manually.
2. The magnitude of the input force can also be varied by changing the mass of the hammer head. By adding mass to the hammer (increasing the weight of the head), the input force magnitude will be increased. Changing the mass of the head is easily accomplished by attaching a different impact head.

The frequency content of the input force can be controlled by controlling the duration of the transient time domain signal. This can be accomplished in two ways:

1. Adding mass to the hammer will increase the time domain pulse width of the impact. This in turn will decrease the effective frequency range of excitation.
2. Decreasing the stiffness of the contact surface between the hammer and the test structure will also decrease the effective frequency range of excitation. Without altering the structure, the easiest means of altering the interface stiffness is by changing the tip of the impact hammer. Typically, three tips are supplied with a hammer kit. This gives the user the ability to excite different frequency ranges, depending on the requirements of the test. The softer an impact tip, the lower the interface stiffness, the longer the time domain duration of the pulse, and the lower the effective frequency range of excitation.

An impact hammer is best calibrated by using it to measure the dynamic behaviour of a simple structure with known dynamic behaviour. The calibration is usually conducted using a suspended mass (at least 100 times the accelerometer mass), as shown in Fig. 4.6. The accelerometer is attached to one side of the mass and the other side is impacted with the hammer. The spectrum analyser is used to measure the dimensionless frequency response function E_a/E_f (acceleration/force or inertance). Since the inertance frequency response of the pendulum mass is a constant $1/m$, any dynamic characteristics detected are those of the instrumentation.

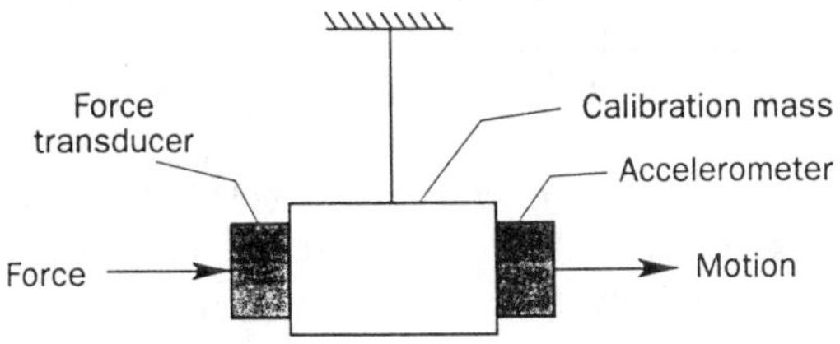

Fig. 4.6 Calibrating the measuring system (after Ewins, 1984).

From Newton's law, $(E_f/S_f)(1/m) = (E_a/S_a)$. A quantity, R, may be defined which can be used in subsequent measurements to convert voltages to ms^{-2} N^{-1}:

$$R = S_a/S_f = mE_a/E_f \qquad (4.2.1a,b)$$

where m is the calibration mass, E_a and E_f are voltages measured at calibration and S_a and S_f are the sensitivities of the accelerometer and force transducer respectively. Any subsequent frequency response measurements (inertances) made using this measurement system will be calibrated in both amplitude and phase when divided by R.

The main advantage of the impact hammer excitation of the structure is its short measurement time, but this is generally at the expense of a poor signal to noise ratio, and a relatively high crest factor which can exacerbate any non-linearity effects in the structural response. As with random noise excitation, leakage can be a problem. However, this problem may be minimised by correct windowing of the input force and response signals, as discussed a little later in this section.

Response transducers
The third step in setting up a modal test is to select and mount the structural response transducers. The transducers most commonly used to measure the structural response are accelerometers connected to charge amplifiers, so that a voltage proportional to the structural acceleration is generated.

In placing response sensors, for example accelerometers, care should be taken not to mass load the structure. Alternatively, if a finite element analysis is available, it may conveniently be altered to account for the accelerometer masses. Subsequently, their effects can be investigated analytically by letting masses tend to zero. If a finite element analysis is not available, tests may be repeated with a lighter (or heavier) accelerometer to experimentally investigate any loading effect. If mass loading is a problem, then the structure's vibration response should be measured using a non-contacting sensor such as a close mounted microphone or laser doppler velocimeter. Care must also be taken to ensure that accelerometers are securely fastened to the structure under test. However, this problem is generally only acute at high frequencies. Care should also be taken in the use of force transducers to measure the force input to the structure. The force transducer should be mounted as close as possible to the point of force input to the structure. Any mass or damping (such as a linear bearing in a shaker) between the transducer and the structure will lead to errors. In addition, piezoelectric force transducers are sensitive to bending moments, so care should be taken to avoid bending of the transducer.

FRF measurement points
The last step in setting up a modal test is to determine the structural geometry model. This is a set of points on the structure where forces will be input and/or response will be measured to provide the data necessary to model the overall structural response. Convenient local coordinates, for example, cartesian, cylindrical or spherical coordinate systems may be used for locating identification points. The computer can be used to put these into one global coordinate system. The emphasis should be on convenience for subsequent interpretation.

If results from a finite element analysis are available, they can be used as a guide to determine the most suitable points. Alternatively, one or two preliminary frequency sweeps and response measurements at a few locations may be used to obtain a rough idea of resonance frequencies and mode shapes prior to a full modal analysis.

4.2.1.2 Transfer function (or frequency response) measurements

Until now we have considered harmonic excitation forces of frequency ω. However, if a structure is excited by a band of frequencies simultaneously, it is desirable to obtain the frequency response function phase and amplitude for the band of frequencies simultaneously. The most convenient way to achieve this is by Fourier analysis which was discussed in detail in Chapter 3.

When taking transfer function measurements using either random noise or impact excitation it is necessary to take many averages to ensure that the system 'noise' is minimized. A way of checking that this is so is to look at the coherence function γ^2 (see Chapter 3) which is automatically calculated by most spectrum analysers at the same time as the transfer function. The coherence function gives an indication of how much of the output signal is caused by the input signal and varies between 0 and 1. A value of 1 represents a valid measurement, while a value of less than about 0.7 represents an invalid measurement.

However, great care should be exercised when interpreting coherence data, as its meaning is dependent on the number of averages taken. For example, a single average will give a coherence of unity which certainly does not imply an error free frequency response measurement. As a guide, the expected random error in a frequency response measurement as a function of the coherence value and number of averages taken is given in Fig. 4.7 which is adapted from the ISO standard on mobility measurement.

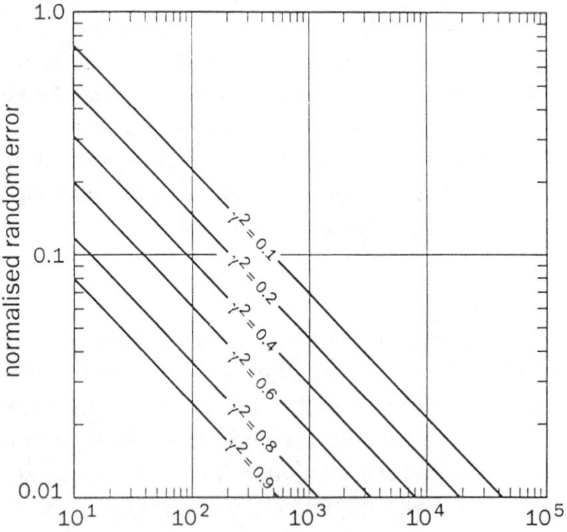

Fig. 4.7 Accuracy of FRF estimates vs. coherence between the input force and structural response (after Ewins, 1984).

In general, low coherence values can indicate one or more of five problems:

1. insufficient signal level (turn up gain on the analyser);
2. poor signal to noise ratio;
3. presence of other extraneous forcing functions;
4. insufficient averages taken;
5. leakage.

Insufficient signal level (case (1) above) is characterized by a rough plot of coherence vs. frequency, even though the average may be close to one (see Fig. 4.8). To overcome this, the gains of the transducer preamplifiers should be turned up or the attenuator setting on the spectrum analyser adjusted, or both.

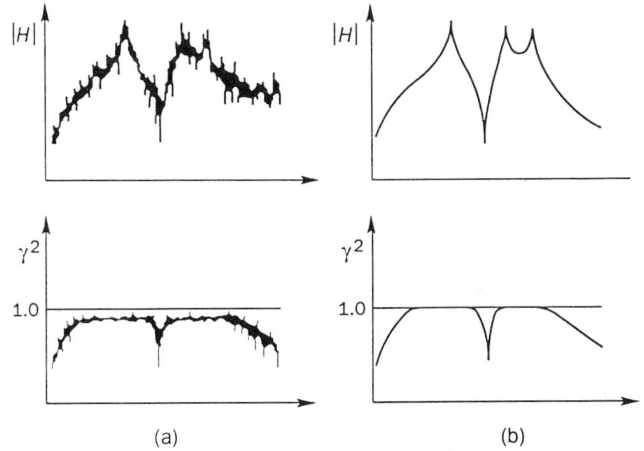

Fig. 4.8 Effect of insufficient signal level on the frequency response function H, and the coherence γ^2 : (a) measurement with noise due to incorrect attenuator setting (insufficient signal level); (b) same measurement with optimum attenuator setting (after Ewins, 1984).

For excitation of a structure by random noise via a shaker, poor coherence between the excitation and response signals is often measured at structural resonances. This is partly due to the leakage problem discussed previously and partly because the input force is very small (close to the instrumentation noise floor) at these frequencies.

Poor signal to noise ratio (case (2) above) can have two causes in the case of impact excitation. The first is due to the bandwidth of the input force being less than the frequency range of interest or the frequency range set on the spectrum analyser. This results in force zeros at higher frequencies, giving the false indication of many high frequency vibration modes. Conversely, it is not desirable for the bandwidth of the force to extend beyond the frequency range of interest to avoid the problem of exciting vibration modes above the frequency range of interest and thus contaminating the measurements with extraneous signals. A good compromise is for the input force auto power spectrum to be between 10 and 20 dB down from the peak value at the highest frequency of interest.

The second reason for poor signal to noise ratio during impact testing is the short duration of the force pulse in relation to the duration of the time domain data block. In many cases, the pulse may be defined by only a few sample points comprising only a small fraction of the total time window, the rest being noise. Thus, when averaged into the measurement, the noise becomes significant. This S/N problem can be minimized by using special force and response data windows which are specifically designed for impact testing. Such windows allow the force pulse to pass unattenuated and then selectively attenuate the following noise. An ideal window is illustrated in Fig. 4.9.

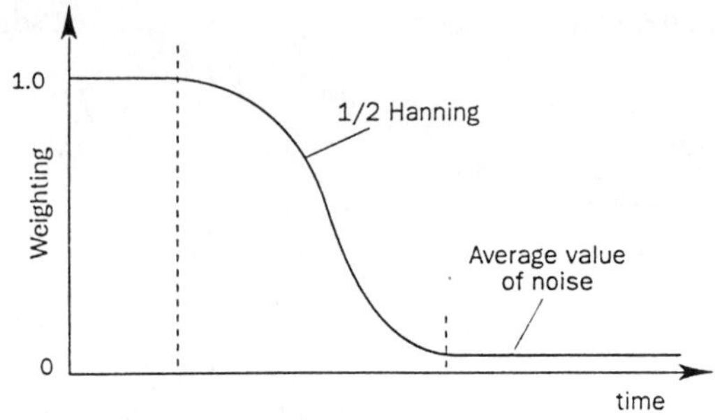

Fig. 4.9 Ideal window for impact testing (after Ewins, 1984).

The first portion is represented by a multiplier of one while the second portion is a half Hanning window. The results of using the window are shown in Fig. 4.10. In practice, it is sufficient to use an adjustable width rectangular window for the impact force signal and an adjustable length single sided exponential window for the response signal. This ensures that the force signal and the structural response caused by it are weighted strongly compared with any extraneous noise which may be present. These windows, which are available in many commercial spectrum analysers, are illustrated in Fig. 4.11.

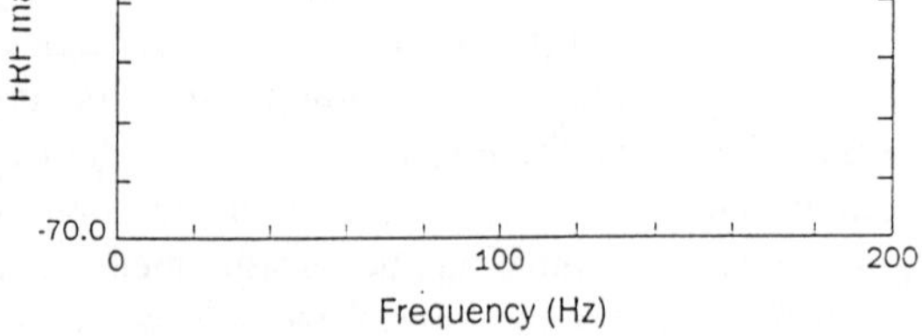

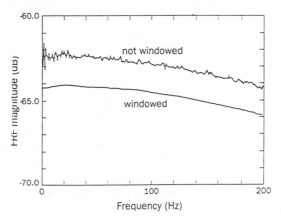

Fig. 4.10 Effect of correct windowing of the impact force input signal.

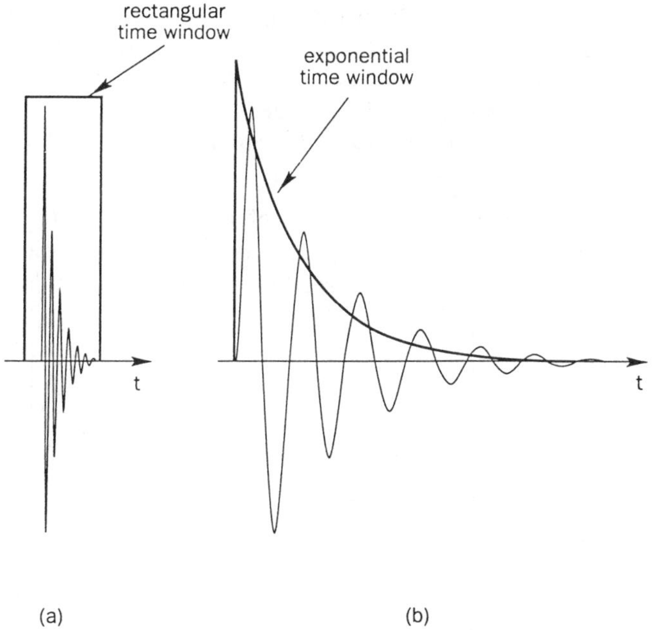

Fig. 4.11 Typical impact force and response windows available in many commercial FFT analysers: (a) rectangular force window; (b) exponential response window.

Another way (if the above mentioned alternative is not possible) to overcome the problem of short force duration is to use a number of randomly spaced impacts over the duration of the sample record and use a standard Hanning window. If it is possible to use only one impact, and the spectrum analyser does not have the optimum impact window mentioned above, then a rectangular window must be used, otherwise the force pulse will be effectively missed and the transfer function of noise will be measured.

Problems (3) and (4) above have obvious solutions and will not be discussed further. The leakage problem (case (5) above) can be caused by incorrect windowing of the data in the time domain. Leakage results in a broadening of the resonance peaks and always occurs when a signal is truncated by the measurement time window. Leakage is minimized for random noise excitation by using either a Hanning or Kaiser–Bessel time window function. For impact excitation, leakage will not be a problem in the windows suggested above to minimize the S/N problem are used. The only problem could occur when a rectangular window was used and if the force pulse or response pulse duration was so long that it had not decayed sufficiently by the end of the time window.

In some cases where the system is lightly damped, the resonance peaks may be so sharp that the frequency resolution of the analyser in the chosen baseband is insufficient. In this case, the response peak will follow the shape of the window function used and the calculated damping (using the modal analysis software) will be incorrect. In this case it is necessary to use zoom analysis (not simply expanding the displayed frequency response spectrum, but a new analysis with finer frequency resolution) to increase the frequency resolution. This can be done by some analysers on individual parts of the spectrum and the results combined to cover the entire frequency range.

In some cases, the resonance peak may appear inverted (see Fig. 4.12) due to the extremely small value of the cross-spectrum $G_{xy}(j\omega)$ being divided by an extremely small value of the power spectrum of the applied force $G_{xx}(j\omega)$. Note that in this latter case, the value of the coherence function between the applied force and the structural response is also small, indicating unreliable data.

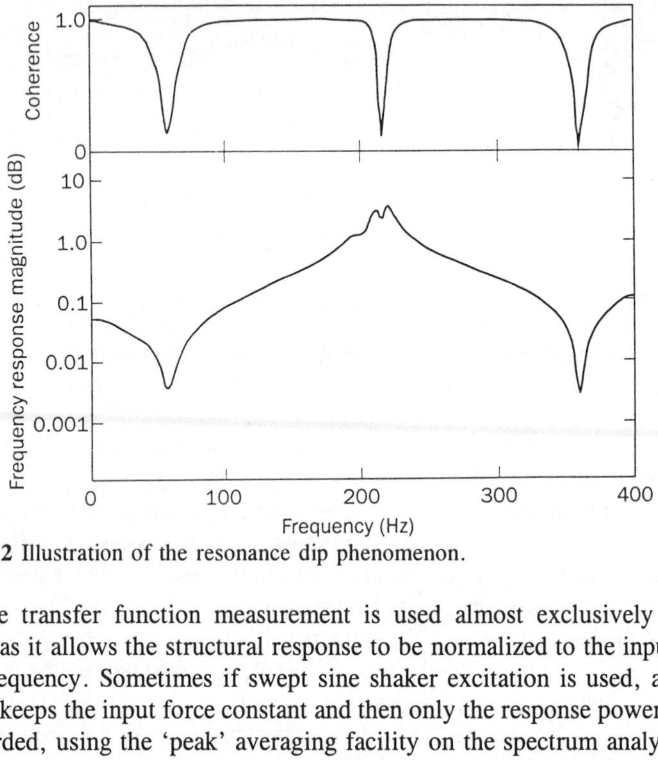

Fig. 4.12 Illustration of the resonance dip phenomenon.

The transfer function measurement is used almost exclusively in modal testing as it allows the structural response to be normalized to the input force at each frequency. Sometimes if swept sine shaker excitation is used, a feedback system keeps the input force constant and then only the response power spectrum is recorded, using the 'peak' averaging facility on the spectrum analyser which allows recording of the maximum response at each frequency during the sweep. Some analysers allow use of the 'peak' averaging facility when recording a transfer function. In this case, the transfer function recorded corresponds to the peak of the response at each frequency and there is no need for a feedback system to control the excitation force magnitude. Swept sine excitation is not in common use today, given the shorter testing times and higher accuracy available with random excitation. However, where structures are heavily damped, a slow sine sweep will often provide better results.

Overloading of transducer pre-amplifiers and the spectrum analyser amplifiers can also lead to serious errors in the transfer function measurements. This is best avoided by setting the spectrum analyser input attenuators low enough so that the spectrum analyser will overload before the transducer pre-amplifiers will. Some spectrum analysers can be set up to discard a data record if it contains any overload measurements, otherwise the overload light should be carefully monitored during testing.

4.2.1.3 Modal parameter identification

The transfer function data corresponding to the ratio of the response at many locations to the input force at another location or vice versa is used to produce modal resonance frequencies, modal damping and mode shape data for each vibration mode identified. This is generally done by downloading all the transfer function data and associated coordinate locations to a personal computer which contains one of the commercially available modal analysis software packages (which at the time of writing range in price from US$3000 to $25 000).

For structures that have several modes, most software has the option of employing specialized curve fitting procedures which determine modal resonance frequencies and damping for all modes at once (multimodal curve fitting). If modes are closely spaced in frequency and/or well damped then treating each one individually and curve fitting each one individually (single mode curve fitting) will lead to large errors; thus, the reason for treating all modes together. However, the multimodal curve fitting procedures are very time consuming and memory hungry, and thus suited to environments where accuracy is more important than speed.

Often memory limitations only allow parts of the frequency response curve to be curve fitted at any one time. However, once the frequency range of measurement has been covered, it is possible to regenerate the entire fitted frequency response function so the global goodness of fit can be evaluated by comparison with the overall measured frequency response function. Aspects of curve fitting frequency response data to determine modal parameters are discussed in detail by Braun and Ram (1987a,b) and will not be discussed further here.

Mode shapes

Evaluation of the mode shapes involves calculation of the relative vibration amplitude at each test point on the structure; however, almost all modal analysis software packages use only linear interpolation between adjacent geometric points on the structure resulting in a distorted view of the true mode shape if an insufficient number of points are used. Most software packages also allow an animated view of the mode shape for a particular mode.

Once the transfer functions and mode shapes have been calculated for a particular structure, most modal analysis software allows calculation of the response of the structure (see (4.1.75) and (4.1.76)) at each node to a defined excitation force which may be in the from a time history, random noise or a number of sinusoids. Some software also presents an animated total response shape as a result of a single frequency sinusoidal force applied at any number of nodes. However, a limitation of most modal analysis software is its inability to provide information regarding the contributions of each mode to the overall response when the structure is excited by a known force.

A further problem inherent in most modal analysis software is the inability to provide relative contributions of each mode to the overall structural response when the structure is excited by an unknown force. This could be done by measuring the transfer functions between the response at one reference point on the structure and the response at all other points on the structure, and performing a modal decomposition.

A capability offered by some modal analysis software is the calculation of the effect of adding or subtracting mass at various nodes, the effect of changing the stiffness between nodes or between the ground and particular nodes, and the effect of adding damping between nodes or between the ground and various nodes. This capability is invaluable when it is desired to modify the structural response to suit particular operating requirements.

Single-degree-of-freedom curve fitting of FRF data

The most useful single-degree-of-freedom curve fitting technique is known as the circle fit. A curve fitting operation is performed on data in the vicinity of a resonance peak and this allows extraction of resonance frequency and damping information for a single mode. If the FRF data for a single-degree-of-freedom system are presented in the form of a Nyquist plot (real vs. imaginary), then the data should fall on the circumference of a circle. In MDOF systems, the data around a particular resonance peak will fall on the arc of a circle; provided the peaks are reasonably well separated. This arc of data points is then fitted to a circle.

If viscous damping is present, then the mobility FRF function must be used and if structural damping is present, then the receptance FRF must be used as other FRF's do not result in data points forming the arc of a circle; rather, they will form part of an ellipse.

The circle fit method is a refinement of the very simple peak amplitude method which assumes that the system resonance frequencies correspond to the peaks in the FRF and that the damping associated with each peak may be calculated from the bandwidth corresponding to the two data points either side of the peak which are less than the peak amplitude by a factor of $\sqrt{2}$ (or 3 dB). The damping is given by (see Fig. 4.13):

$$\eta_r = (\omega_a^2 - \omega_b^2)/\omega_r^2 \approx \Delta\omega/\omega_r \qquad (4.2.2a,b)$$

In contrast to the simple amplitude method, the circle fit method includes the effects of other modes, provided that they are not too close in frequency to the mode under consideration. Thus, the circle fit method only assumes that in the vicinity of a resonance peak, the system response is dominated by a single mode and that in the small frequency range about resonance the effects of all other

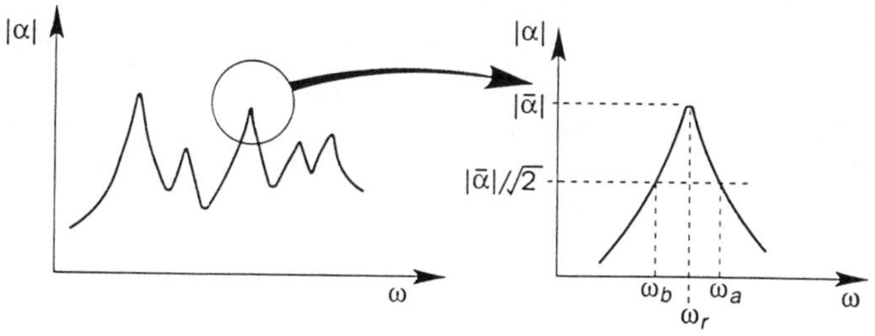

Fig. 4.13 Modal parameter estimation using the amplitude method.

modes are essentially constant and independent of frequency. This assumption can be expressed as follows. From Section 4.1, we have

$$\alpha_{i\iota}(\omega) = \sum_{s=1}^{N} \frac{\psi_{is}\,\psi_{\iota s}}{m_s(\omega_s^2 - \omega^2 + j\eta_s\,\omega_s^2)} \tag{4.2.3}$$

which can be written as

$$\alpha_{i\iota}(\omega) = \frac{\psi_{ir}\,\psi_{\iota r}}{m_r(\omega_r^2 - \omega^2 + j\eta_r\,\omega_r^2)} + \sum_{s=1\,\neq r}^{N} \frac{\psi_{is}\,\psi_{\iota s}}{m_s(\omega_s^2 - \omega^2 + j\eta_s\,\omega_s^2)} \tag{4.2.4}$$

It is assumed that the second term in (4.2.4) is independent of frequency (over the small frequency range under consideration) and (4.2.4) may be written as

$$\begin{aligned} \alpha_{i\iota}(\omega)_{\omega\approx\omega_r} &= \frac{\psi_{ir}\,\psi_{\iota r}}{m_r(\omega_r^2 - \omega^2 + j\eta_r\,\omega_r)} + B_{i\iota r} \\ &= \frac{A_{i\iota r}}{\omega_r^2 - \omega^2 + j\eta_r\,\omega_r} + B_{i\iota r} \end{aligned} \tag{4.2.5a,b}$$

where $A_{i\iota}$ is referred to as the modal constant. This can be illustrated by a specific example, shown in Fig. 4.13. Using a four DOF system, the receptance properties have been computed with (4.2.5) and each of the two terms has been plotted separately, in Figs. 4.14(a) and 4.14(b). Also shown in Fig. 4.14(c) is the corresponding plot of the total receptance over the same frequency range. What is clear in this example is the fact that the first term (that relating to the mode under examination) varies considerably through the resonance region, sweeping out the expected circular arc, while the second term, which includes the combined effects of all the other modes, is effectively constant through the narrow frequency range covered. Thus, we see from the total receptance plot in Fig. 4.14(c) that this may, in effect, be treated as a circle with the same properties as the modal circle for the specific mode in question but which is displaced from the origin of the Argand plane by an amount determined by the contribution of all the other modes. Note that this is not to say that the other modes are unimportant or negligible – quite the reverse, their influence can be considerable – but rather than their combined effect can be represented as a constant term around this resonance.

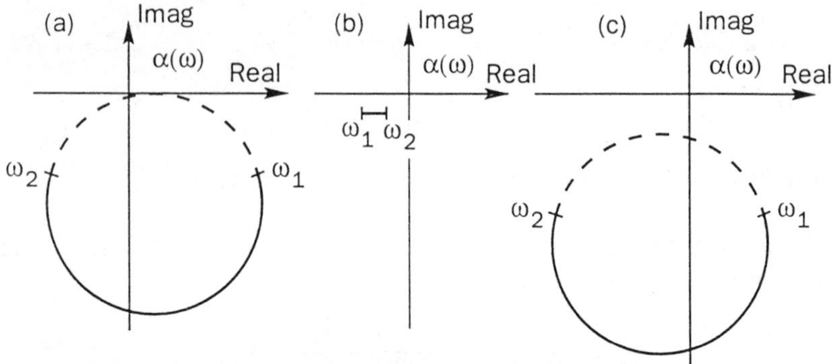

Fig. 4.14 Nyquist plot of four-degree-of-freedom receptance data: (a) first term; (b) second term; (c) total (after Ewins, 1984).

When undertaking a circle fit, the basic function of interest is:

$$\alpha = \frac{1}{\omega_r^2\,(1-(\omega/\omega_r)^2 + j\eta_r)} \tag{4.2.6}$$

since the effect of including the product $(\psi_{ir}\ \psi_{\iota r}/m_r)$ (referred to as the modal constant $A_{i\iota r}$) is to scale the size of the circle by $|\psi_{ir}\ \psi_{\iota r}/m_r|$ and to rotate it by the phase of $\psi_{ir}\ \psi_{\iota r}$. Equation (4.2.6) is plotted in Fig. 4.15.

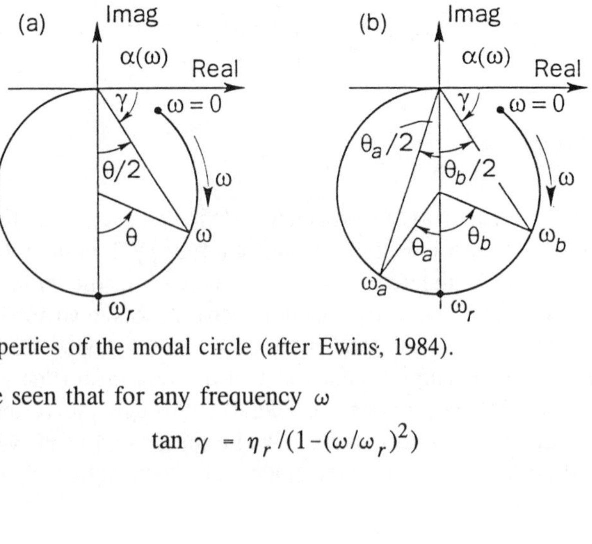

Fig. 4.15 Properties of the modal circle (after Ewins, 1984).

It can be seen that for any frequency ω

$$\tan\gamma = \eta_r/(1-(\omega/\omega_r)^2) \tag{4.2.7}$$

$$\tan\,(90° - \gamma) = \tan(\theta/2) = [1 - (\omega/\omega_r)^2]/\eta_r \tag{4.2.8a,b}$$

from which we obtain:

$$\omega^2 = \omega_r^2\,[1 - \eta_r \tan(\theta/2)] \tag{4.2.9}$$

If we differentiate (4.2.9) with respect to θ, we obtain:

$$\frac{d\omega^2}{d\theta} = (-\omega_r^2\eta_r/2)\left\{1 + [1 - (\omega/\omega_r)^2]^2/\eta_r^2\right\} \tag{4.2.10}$$

The reciprocal of this quantity is a measure of the rate at which the locus sweeps around the circular arc and reaches a maximum value (maximum sweep rate) when $\omega = \omega_r$, the natural frequency of the oscillator. This is shown by further differentiation, this time with respect to frequency, that:

$$\frac{d}{d\omega}\left[\frac{d\omega^2}{d\theta}\right] = 0 \quad \text{when} \quad (\omega_r^2 - \omega^2) = 0 \tag{4.2.11}$$

This result is useful for analysing MDOF system data since, in general, it is not known exactly where the natural frequency is. However, examination of the relative spacing of the measured data points around the circular arc near each resonance allows its value to be determined.

Another useful result can be obtained from this basic modal circle. Consider two specific points on the circle, one corresponding to a frequency, ω_b, below the natural frequency, and the other to one, ω_a, above the natural frequency. Referring to Fig. 4.14 we can write:

$$\tan(\theta_b/2) = (1 - (\omega_b/\omega_r)^2)/\eta_r \tag{4.2.12}$$

$$\tan(\theta_a/2) = ((\omega_a/\omega_r)^2 - 1)/\eta_r \tag{4.2.13}$$

and from these two equations we can obtain an expression for the damping of the mode:

$$\eta_r = \frac{\omega_a^2 - \omega_b^2}{\omega_r^2\,[\tan(\theta_a/2) + \tan(\theta_b/2)]} \tag{4.2.14}$$

and if we restrict our interest to the two points for which $\theta_a = \theta_b = 90°$ (the half-power points), we obtain the familiar formula:

$$\eta_r = (\omega_2 - \omega_1)/\omega_r \tag{4.2.15}$$

When scaled by the product $\psi_{ir}\ \psi_{\iota r}$, the circle diameter is given by

$$D_{i\iota r} = \frac{|\,\psi_{ir}\ \psi_{\iota r}\,|}{\omega_r^2\eta_r m_r} \tag{4.2.16}$$

The whole circle will be rotated so that the principle diameter which passes through ω_r is orientated at an angle to the imaginary axis equal to the phase angle of the $\psi_{ir}\ \psi_{\iota r}$ product.

Circle fit analysis procedure
The steps involved in a circle fit are:

1. select the points to be used;
2. fit the circle, calculate the quality of fit;
3. locate the natural frequency;
4. calculate multiple damping estimates, and scatter;
5. determine the modal constant, $A_{i\iota r} = \psi_{ir}\psi_{\iota r}/m_r$.

Step (1) can be made automatic by selecting a fixed number of points on either side of any identified maximum in the response modulus or it can be effected by the operator whose judgement may be better able to discern true modes from spurious perturbations on the plot and to reject certain suspect data points. The points chosen should not be influenced to any great extent by the neighbouring modes and, whenever possible without violating that first rule, should encompass some 270° of the circle. This is often not possible and a span of less than 180° is more usual, although care should be taken not to limit the range excessively as the result then becomes highly sensitive to the accuracy of the few points used. No fewer than six points should be used.

The second step, (2), can be performed by one of numerous curve-fitting routines and consists simply of finding a circle which gives a least-squares deviation for the points included. 'Errors' of the order of 1–2% are commonplace, and an example of the process is shown in Fig. 4.16(a). The quality of fit is related to the mean square error between the data points and the fitted circle.

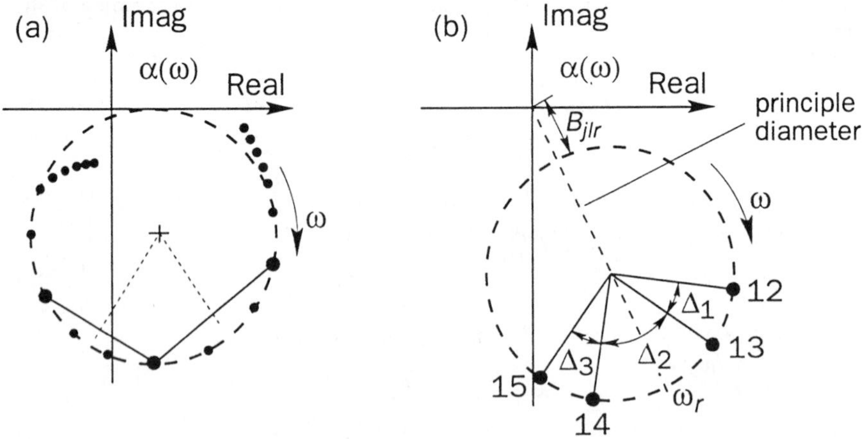

Fig. 4.16 Circle fit to FRF data: (a) circle-fit; (b) natural frequency location (after Ewins, 1984).

Step (3) can be implemented by numerically constructing radial lines from the circle centre to a succession of points equally spaced in the frequency domain around the resonance and by noting the angles they subtend with each other. Then, the rate of frequency sweep through the region can be estimated and the frequency at which it reaches a maximum can be deduced. If, as is usually the case, the frequencies of the points used in this analysis are spaced at regular intervals (i.e. a linear frequency increment), then this process can be effected using a finite difference method. Such a procedure enables one to pinpoint the natural frequency with a precision of about 10% of the frequency increments between the points. Fig. 4.16(b) shows the results from a typical calculation.

Next, for step (4), we are able to compute a set of damping estimates using every possible combination from our selected data points of one point below resonance with one above resonance using (4.2.14). With all these estimates we can either compute the mean value or we can choose to examine them individually to see whether there are any particular trends. Ideally, they should all be identical and so an indication not only of the mean but also of the deviation of the estimates is useful. If the deviation is less than 4–5%, then we have generally succeeded in making a good analysis. If, however, the scatter is 20 or 30%, there is something unsatisfactory. If the variations in damping estimate are random, then the scatter is probably due to measurement errors but if it is systematic, then it could be caused by various effects (such as poor experimental setup, interference from neighbouring modes, non-linear behaviour, etc.), none of which should, strictly, be averaged out.

Lastly, step (5) is a relatively simple one in that the magnitude and phase of the modal constant, $A_{i\iota r}$, can be determined (using (4.2.16)) from the circle diameter passing through the point ω_r, and from its orientation relative to the real and imaginary axes. This calculation is straightforward once the natural frequency has been located and the damping estimates obtained.

If it is desired to construct a theoretically regenerated FRF plot against which to compare the original measured data, it will be necessary to determine the contribution to this resonance of the other modes and this requires determining the $B_{i\iota r}$ of (4.2.5). This quantity is the distance from the top of the principal diameter (the one passing through the point ω_r) to the origin as shown in Fig. 4.16.

Residuals

At this point the concept of residual terms needs to be discussed. These terms take into account the effect of modes which we do not analyse directly but which exist nevertheless and have an influence of the FRF data. This influence must somehow be accounted for.

If we regenerate an FRF curve from the modal parameters we have extracted from the measured data, we would use the following equation:

$$y_{i\iota}(\omega) = \sum_{r=m_1}^{m_2} \frac{j\omega A_{i\iota r}}{\omega_r^2 - \omega^2 + j\eta_r\omega_r^2} \tag{4.2.17}$$

where $y_{i\iota}(\omega)$ is the mobility function. However, the equation which would fit the measured FRF more closely is

$$y_{i\iota}(\omega) = \sum_{r=1}^{N} \frac{j\omega A_{i\iota r}}{\omega_r^2 - \omega^2 + j\eta_r\omega_r^2} \tag{4.2.18}$$

which can be written as

$$y_{i\iota}(\omega) = \left[\underset{\substack{(\text{low} \\ \text{frequency} \\ \text{modes})}}{\sum_{r=1}^{m_1-1}} + \sum_{r=m_1}^{m_2} + \underset{\substack{(\text{high} \\ \text{frequency} \\ \text{modes})}}{\sum_{r=m_2+1}^{N}} \right] \frac{j\omega A_{i\iota r}}{\omega_r^2 - \omega^2 + j\eta_r\omega_r^2} \tag{4.2.19}$$

Figure 4.17 shows typical values of each of the three terms separately, and the middle one is all that is computed using modal data extracted from the modal analysis. To make the model accurate within the frequency range of the tests, it is necessary to correct the regenerated plot within the central frequency range to take account of the low frequency and high frequency modes.

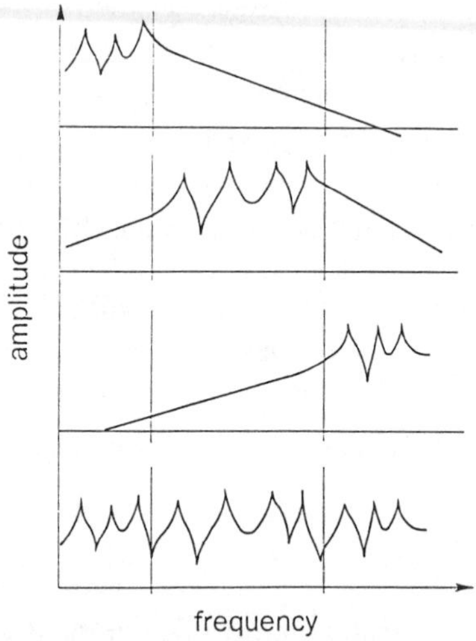

Fig. 4.17 Contribution of various terms to the total FRF: (a) low frequency modes; (b) identified modes; (c) high frequency modes; (d) all modes (after Ewins, 1984).

From the sketch it can be seen that in the frequency range of interest, the first term of (4.2.19) approximates to a mass-like behaviour while the third term approximates to a stiffness-like behaviour. On this basis we may quantify the residual terms, and rewrite (4.2.19) as

$$y_{i\iota}(\omega) = -\frac{j\omega}{\omega^2 M_{i\iota}^R} + \sum_{r=m_1}^{m_2} \left[\frac{j\omega A_{i\iota r}}{\omega_r^2 - \omega^2 + j\eta\omega_r^2} \right] + \frac{j\omega}{K_{i\iota}^R} \qquad (4.2.20)$$

where $M_{i\iota}^R$ and $K_{i\iota}^R$ are the residual mass and stiffness for that particular FRF. The residual terms are calculated using the difference between the actual FRF and the FRF constructed from the modal parameters at a few points each end of the FRF curve. The points at the low frequency end allow the required value of $K_{i\iota}^R$ to be calculated which will make the regenerated FRF values similar to the measured values. Similarly the points at the high frequency end allow the value of $M_{i\iota}^R$ to be calculated.

Multi-degree-of-freedom curve fitting FRF data

As mentioned previously, most commercially available modal analysis software packages utilize some form of MDOF curve fitting where a section of the FRF, containing several modes, is curve fitted to extract the modal parameters for all modes present in the frequency range selected. Users are usually asked how many modes they would like included in the fit. A sufficient number of modes will produce an almost perfect fit but will invariably produce computational modes which are included expressly to satisfy the curve-fitting requirement and which do not necessary represent any physical behaviour of the structure at all. Taken to extremes, computational modes will 'fit' small irregularities or errors in the measured data. Clearly, it is an important requirement that such computational modes can be distinguished from the genuine modes, and eliminated from the analysis before the results are presented to the user.

Computational mode elimination

One way of distinguishing computational modes from real modes is to specify more modes than really exist in the fitting procedure. Then 10 different fits to the FRF data can be made, each fit using every 10th data point of the original FRF data and each fit using a different subset of the original data. Thus, eventually all data points are used by the ten curve fits, but each curve fit uses a different 10% of the total data.

The result will be 10 estimates of frequency and damping for each of the four modes. The real modes will correspond to estimates of resonance frequency and damping which do not vary much from run to run; however, the computational modes will be characterized by large variances (greater than 5%) and can be rejected on that basis.

Note that a single analysis would produce the same average results for resonance frequency and damping but the information on variances which enables computational modes to be distinguished would not exist.

If the system being excited is non-linear, then different values of resonance frequency and damping will be obtained for different levels of force input. Other non-linearities, such as stiffness, being a non-linear function of frequency will produce distortions in the FRF curves which can lead to inconsistencies in damping estimates which will vary depending upon the calculation method. This topic is discussed in detail elsewhere (Ewins, 1984).

Global fitting FRF data

This is the process of analysing the curve fitted results of a number of different FRF curves, corresponding to force inputs and response measurements at various different parts of the structure, to yield the full modal model from which animated mode shapes are derived. As all FRF curves do not provide exactly the same resonance frequencies and damping values, these quantities are generally averaged.

Up to now, we have discussed how to obtain resonance frequency and damping data from a single frequency response function of a structure. We have also noted that the resonance frequency and damping values so obtained differ slightly depending upon the location of the excitation force and response measurement.

To obtain mode shape data, the barest minimum number of measurements required is one force measurement at one force impact location and acceleration measurements at this and all other measurement points on the structure. Thus, we need to measure a set of FRF curves, one curve for each point on the structure, with each curve sharing the same excitation point. Often it is prudent to repeat the measurements with a second force impact location to improve the likelihood of exciting modes which may have been excited poorly at the first force location and to replace poor FRF data which is often obtained when the force input and response measurement are a long way apart. To improve the accuracy even more, it is sometimes necessary to measure the point inertance at each point on the structure by exciting the structure and measuring its response at the same location, although in practice this additional work is generally not contemplated.

The resonance frequencies, damping values, and mode shapes are determined by averaging the overdetermined data.

4.2.1.4 Response models

These are needed to enable a prediction to be made of the response of the analysed structure to a given excitation force at any given location. Thus, the response model is an FRF matrix whose order ($n \times n$) is dictated by the number of coordinates n used in the test. The FRFs which were measured are regenerated using the resonance frequencies, damping values and modal coefficients calculated by curve fitting. The FRFs which were not measured are synthesized using the coefficients calculated from the measured FRFs. In principle, this presents no problem, as it is always possible to calculate the full response model from a modal model using

$$\alpha_{n \times n} = \psi_{n \times m} \begin{bmatrix} \diagdown & & \\ & m_r(\omega_r^2 + j\eta_r \omega_r^2 - \omega^2) & \\ & & \diagdown \end{bmatrix}^{-1} \psi^{\mathrm{T}}_{m \times n} \tag{4.2.21}$$

where n is the number of coordinates and m is the number of modes considered. However, this latter process is only successful if the effect of modes outside of the analysis range are taken into account; otherwise large errors can result. The effect mentioned above is taken into account by using a residue matrix $\boldsymbol{R}$ which must somehow be estimated. Thus, the correct response model is given by

$$\alpha = \psi \begin{bmatrix} \diagdown & & \\ & m_r(\omega_r^2 + j\eta_r\omega_r^2 - \omega^2) & \\ & & \diagdown \end{bmatrix}^{-1} \psi^{\mathrm{T}} + \boldsymbol{R} \tag{4.2.22}$$

where ψ, m_r, ω_r and η_r are derived from measured FRFs.

The most accurate means of estimating the individual elements of $\boldsymbol{R}$ is to measure all (or more than half at least) of the elements of the FRF matrix, but this would constitute a major escalation in the work involved. A second possibility, and a reasonably practical one, is to extend the frequency range of the modal test beyond that over which the model is eventually required. In this way, much of the content of the residual terms is included in separate modes and their actual magnitudes are reduced to relatively unimportant dimensions. The main problem with this approach is that one does not generally know when this last condition has been achieved, although a detailed examination of the regenerated curved using all the modes obtained and then again less the highest one(s) will give some indication in this direction.

A third possibility is to try to assess which of the many FRF elements are liable to need large residual terms and to make sure that these are included in the list of those which are measured and analysed. In practice, it is the point receptances which are expected to have the highest valued residuals and the remote transfer receptances which will have the smallest. Thus, the significant terms in the $\boldsymbol{R}$ matrix will generally be grouped close to the leading diagonal of the FRF matrix, and this suggests making measurements of most of the point mobility parameters. Such a procedure will seldom be practical unless analysis indicates that the response model is ineffective without such data, in which case it may be the only option.

Attempts to measure all elements or half of the elements (as the FRF matrix is symmetric) of the FRF matrix raises additional problems due to inconsistencies in estimated resonance frequencies and damping values. At the very least, natural frequencies and damping values should be averaged throughout the model and the mode shapes recalculated using the average damping values. In any case it is usually necessary to derive the response model by regenerating the FRF functions from the estimated average coefficients.

4.2.1.5 Structural response prediction

Another reason for deriving an accurate mathematical model for the dynamics of a structure is to provide the means to predict the response of that structure to more complicated and numerous excitations than can readily be measured directly in laboratory tests. Hence the idea that by performing a set of measurements under relatively simple excitation conditions, and analysing these data appropriately, we can then predict the structure's response to several excitations applied simultaneously.

The basis of this philosophy is itself quite simple and is summarized in the standard equation:

$$\boldsymbol{x}e^{j\omega t} = \alpha(\omega)\boldsymbol{f}e^{j\omega t} \tag{4.2.23}$$

where the required elements in the FRF matrix can be derived from the modal model by the familiar formula:

$$\alpha(\omega) = \psi \begin{bmatrix} \diagdown & & \\ & m_r(\omega_r^2 + j\eta_r\omega_r^2 - \omega^2) & \\ & & \diagdown \end{bmatrix}^{-1} \psi^{\mathrm{T}} \tag{4.2.24}$$

In general, this prediction method is capable of supplying good results provided sufficient modes are included in the modal model from which the FRF data used are derived.

4.3 MODAL AMPLITUDE DETERMINATION FROM SYSTEM RESPONSE MEASUREMENTS

A knowledge of the amplitudes of known mode shape functions can facilitate active noise and vibration control system design as will become apparent in Chapters 7, 8 and 11. Here we will briefly consider the problem of measuring the amplitudes of known modal shape functions. Further consideration to the topic is given in Section 11.3.2, in the discussion of 'modal filters'.

In extending the previous treatment in this chapter from general single- and multi-degree-of-freedom systems to vibroacoustic systems it will be convenient to redefine some variables. To avoid confusion which may occur when acoustic and structural systems are considered together, the variable ψ will be used for structural mode shape functions and ϕ will be used for acoustic mode shape functions. The variable p will be used to describe the acoustic pressure response and w will be used to describe the transverse structural response to a particular force input. The vector quantity $\boldsymbol{x}$ will be used to define a general location on a structural surface and $\boldsymbol{r}$ will be used to define a location in an acoustic space.

Using these variable definitions, the displacement of a structure at location $\boldsymbol{x}$ may be written in terms of the modal amplitudes and mode shape function at location $\boldsymbol{x}$ and time t as follows:

$$w(\boldsymbol{x},t) = \sum_{r=1}^{\infty} a_r(t)\psi_r(\boldsymbol{x}) \tag{4.3.1}$$

where $a_r(t)$ is the amplitude of mode r at time t and $\psi_r(\boldsymbol{x})$ is the mode shape function evaluated at $\boldsymbol{x}$.

If we multiply both sides of (4.3.1) by the mode shape function $\psi_r(\boldsymbol{x})$, integrate over the surface of the structure and use the orthogonality property of the mode shape functions, we obtain

$$a_r(t) = \int_S \psi_r(\boldsymbol{x}) w(\boldsymbol{x},t)\,\mathrm{d}\boldsymbol{x} \tag{4.3.2}$$

Measurements of quantities such as displacement or pressure are usually provided by point sensors, such as accelerometers or microphones. It is possible to extract modal amplitudes from such measurements by actually implementing (4.3.1); the discrete measurements can be interpolated to estimate the continuous function (such as displacement), and the integration performed numerically (see, for example, Meirovitch and Baruh (1982) for a vibration implementation, and Moore (1979) for an acoustic implementation). However, there is a second approach to resolving modal amplitudes which is better suited to real time implementation. If we return to (4.3.1) and use it to express the displacement at a number of measurement locations $\boldsymbol{x}_i$, then for each measurement location $\boldsymbol{x}_i$

$$w(\boldsymbol{x}_i,t) = \sum_{r=1}^{\infty} a_r(t)\psi_r(\boldsymbol{x}_i) \tag{4.3.3}$$

If we restrict the number of modes considered to n and the number of measurement locations to N_e, then we obtain a set of N_e simultaneous equations which can be written in matrix form as

$$\boldsymbol{w}(t) = \boldsymbol{\Psi A}(t) \tag{4.3.4}$$

where

$$\boldsymbol{w}(t) = \left[w(\boldsymbol{x}_1,t),\, w(\boldsymbol{x}_2,t) \ldots\ldots w(\boldsymbol{x}_{N_e},t)\right]^{\mathrm{T}} \tag{4.3.5}$$

$$\boldsymbol{\Psi} = \left[\boldsymbol{\psi}_1,\, \boldsymbol{\psi}_2, \ldots\ldots \boldsymbol{\psi}_n\right] \tag{4.3.6}$$

$$\boldsymbol{\psi}_r = \left[\psi_r(\boldsymbol{x}_1),\, \psi_r(\boldsymbol{x}_2), \ldots\ldots\ldots \psi_r(\boldsymbol{x}_{N_e})\right]^{\mathrm{T}} \tag{4.3.7}$$

$$\boldsymbol{A}(t) = \left[a_1(t),\, a_2(t), \ldots\ldots\ldots a_n(t)\right]^{\mathrm{T}} \tag{4.3.8}$$

The amplitude of the rth mode is then

$$a_r(t) = \sum_{i=1}^{N_e} f_{ri} w(\boldsymbol{x}_i,t) \tag{4.3.9}$$

Now we define

$$\boldsymbol{f}_r = \left[f_{r_1},\, f_{r_2} \ldots\ldots\ldots f_{r_{N_e}}\right]^{\mathrm{T}} \tag{4.3.10}$$

and

$$F = [f_1, f_2, \ldots\ldots f_n] = \Psi[\Psi^T\Psi]^{-1} \tag{4.3.11}$$

The latter extension is the generalized pseudo-inverse of the mode shape matrix Ψ, which is used in place of the inverse when the matrix to be inverted is not square. To obtain a well conditioned matrix for use during the pseudo-inversion process, it is important that there be many more measurement points than modes to be resolved. For an optimally conditioned matrix Ψ, the measurement locations should not be uniformly spaced, and at least $2.5-3$ times as many measurement points as the order of the highest order mode of interest are needed. Thus if the first 6 modes are of interest, the displacement must be measured at 15 locations, none of which should be on a modal node. In all cases, more measurement points will increase the accuracy of the modal amplitudes which are resolved. For cases where higher order modes exist which are not of interest, the amplitudes of these modes will be included (aliased) in the amplitudes of the lower order modes if the mode order is more than half the total number of measurement locations. This aliasing problem is very similar to the aliasing in the frequency domain when a time domain signal is digitized by sampling at a certain rate, and then the discrete Fourier transform used to convert it to the frequency domain. In this latter case, higher frequency components in the original signal with a frequency greater than half the sampling frequency will be aliased into the amplitudes of the lower frequencies.

For single frequency excitation, it is more convenient to express modal amplitudes as time independent complex quantities, defined by an amplitude and phase or by a real and imaginary part. In this case (4.3.9) becomes

$$\mathrm{Re}\{\bar{a}_r\} = \sum_{i=1}^{N_e} f_{ri}\mathrm{Re}\{\bar{w}(x_i)\} \tag{4.3.12}$$

$$\mathrm{Im}\{\bar{a}_r\} = \sum_{i=1}^{N_e} f_{ri}\,\mathrm{Im}\{\bar{w}(x_i)\} \tag{4.3.13}$$

As an example, we will demonstrate the application of this method to finding the complex modal amplitudes of a harmonically excited simply supported plate (Hansen *et al.*, 1989). The complex displacement amplitude at location (x,y) on a simply supported plate is given by

$$w(x,y) = \sum_{i=1}^{N}\sum_{q=1}^{M} a_{iq}\sin\frac{i\pi x}{L_x}\sin\frac{q\pi y}{L_y} \tag{4.3.14}$$

where a_{iq} is the complex modal amplitude for a mode having the indices i,q.

In matrix form this can be written as

$$\begin{bmatrix} w_1 \\ w_2 \\ \cdot \\ \cdot \\ \cdot \\ w_{N_e} \end{bmatrix} =$$

$$\begin{bmatrix} \sin\frac{\pi x_1}{L_x} & \sin\frac{\pi y_1}{L_y} & \sin\frac{2\pi x_1}{L_x} & \sin\frac{\pi y_1}{L_y} & \cdots & \sin\frac{N\pi x_1}{L_x} & \sin\frac{M\pi y_1}{L_y} \\ \cdot & \cdot & \cdot & \cdot & \cdots & \cdot & \cdot \\ \cdot & \cdot & \cdot & \cdot & \cdots & \cdot & \cdot \\ \cdot & \cdot & \cdot & \cdot & \cdots & \cdot & \cdot \\ \cdot & \cdot & \cdot & \cdot & \cdots & \cdot & \cdot \\ \sin\frac{\pi x_{N_e}}{L_x} & \sin\frac{\pi y_{N_e}}{L_y} & \sin\frac{2\pi x_{N_e}}{L_x} & \sin\frac{2\pi y_{N_e}}{L_y} & \cdots & \sin\frac{N\pi x_{N_e}}{L_x} & \sin\frac{M\pi y_{N_e}}{L_y} \end{bmatrix} \begin{bmatrix} a_{11} \\ a_{21} \\ \cdot \\ \cdot \\ \cdot \\ a_{NM} \end{bmatrix} \tag{4.3.15}$$

or in short form

$$\boldsymbol{w} = \Psi \boldsymbol{A} \tag{4.3.16}$$

Thus

$$\boldsymbol{A} = \Psi\left[\Psi^{\mathrm{T}}\Psi\right]^{-1}\boldsymbol{w} \tag{4.3.17}$$

In (4.3.17), the real part of $\boldsymbol{w}$ may be used to obtain the real part of $\boldsymbol{A}$ and similarly for the imaginary part. Of course in many cases, accelerometers or laser doppler velocimeters are used to measure acceleration or velocity respectively and the displacement is not necessarily the quantity of interest. In this case, the acceleration or velocity measurements may be used directly in place of $w(x,y)$ and the resulting modal amplitudes are acceleration or velocity amplitudes. If the displacement amplitude is desired, then it is better to replace the quantity $w(x,y)$ in (4.3.14) with $j\omega\, w(x,y)$ or $-\omega^2 w(x,y)$ (depending on whether velocity or acceleration is being measured) than it is to use an integrating circuit in the measurement instrumentation, because the latter is generally a significant source of error. However, when the structure is being excited by random noise and (4.3.3) is used to obtain time varying modal amplitudes an integrating circuit must be used to obtain displacement from acceleration or velocity measurements. Alternatively, if modal amplitudes are required as a function of frequency (rather than time), then the measured vibration signal can be Fourier transformed to the frequency domain as described in Chapter 3, and each frequency component treated as described above for single frequency excitation. Note that if it is desired to average over several spectra, the power spectrum must be used; thus phase data are lost and only modal amplitude data will be available.

As mentioned earlier, it is possible to approximate the integral of (4.3.2) directly for structures where the mode shape can be described in functional form, by using equally spaced displacement or acceleration measurement locations.

To illustrate how this may be done, two examples will be considered. The first will be a simply supported plate, the displacement of which has the form of (4.3.1) and can be written as

$$w(x,y,t) = \sum_{i=1}^{\infty}\sum_{q=1}^{\infty} a_{iq}(t)\cos\frac{i\pi x}{L_x}\cos\frac{q\pi y}{L_y} \tag{4.3.18}$$

Note that the origin of the coordinates has been chosen to be the centre of the plate. If the bottom left corner of the plate were chosen as the origin, then the cosine functions would be replaced with sine functions. The index r of (4.3.1) is the mode order of mode (i,q).

Multiplying both sides of (4.3.18) by $\cos\frac{m\pi x}{L_x}$, integrating in a line across the plate at a constant value of $y = y_c$ and using modal orthogonality gives

$$\frac{2}{L_x}\int_{-L_x/2}^{L_x/2} w(x,y_c,t)\cos\frac{m\pi x}{L_x}\,dx = \sum_{q=1}^{\infty} a_{mq}(t)\cos\frac{q\pi y}{L_y} \tag{4.3.19}$$

If we now multiply (4.3.19) by $\cos\frac{n\pi y}{L_y}$, integrate from 0 to L_y in a line down the plate at a constant value of $x = x_c$ and use modal orthogonality, we obtain

$$\frac{4}{L_xL_y}\int_{-L_x/2}^{L_x/2} w(x,y_c,t)\cos\frac{m\pi x}{L_x}\,dx\int_{-L_y/2}^{L_y/2} w(x_c,y,t)\cos\frac{n\pi y}{L_y}\,dy = a_{mn}(t) \tag{4.3.20}$$

In practice, (4.3.20) is implemented by replacing the integrals with finite sums, using a series of measurements in a line across the plate at $y = y_c$ to evaluate the first sum and using a second series of measurements in a line down the plate at $x = x_c$ to evaluate the second sum. It is important that the measurement points are in lines parallel to the x and y axes respectively so that the integrals of (4.3.20) can be separated. Thus we may write

$$a_{mn}(t) = \frac{4}{L_x L_y}\left[\sum_{i=1}^{N_x} w(x_i, y_c, t)\cos\frac{m\pi x_i}{L_x}\Delta x\right]\left[\sum_{i=1}^{N_y} w(x_c, y_i, t)\cos\frac{n\pi y_i}{L_y}\Delta y\right] \tag{4.3.21}$$

where Δx is the spacing of the measurement points in the line at $y = y_c$, Δy is the spacing of the measurement points in the line at $x = x_c$, x_i are the measurement point locations on the line at $y = y_c$ and y_i are the measurement point locations on the line at $x = x_c$. Note that x_c or y_c must not correspond to a nodal line of any of the modes which are to be resolved. In practice they should not even be close to nodal lines or the error in $a_{mn}(t)$ will be large. If the excitation is random and modal amplitudes as a function of frequency are desired, then a Fourier transform can be made of the data at each measurement point and the real and imaginary parts of the modal amplitudes obtained by using the corresponding real and imaginary parts of w in (4.3.21).

For single frequency excitation, the amplitudes and phases of the displacement (relative to some fixed reference) at each measurement point may be determined and converted to give the real and imaginary components which can then be used separately to find the real and imaginary component of the modal amplitude.

As a second example of approximation of the integral of (4.3.2) we will consider resolving circumferential modes from response measurements taken at points around the circumference of a circular cylinder at a constant axial location (Jones and Fuller, 1986).

The radial displacement of a cylinder at any given time can be represented as

$$w(\theta, t) = \sum_{n=0}^{\infty} [a_n \cos(n\theta) + b_n \sin(n\theta)]\, e^{j\omega t} \tag{4.3.22}$$

When a cylinder is excited, circumferential waves propagate in both directions around the cylinder combining to create an interference pattern or standing wave. To solve for the complex modal amplitudes a_n and b_n, (4.3.22) is multiplied by $\cos(m\theta)$ and $\sin(m\theta)$ respectively, and integrated from 0 to 2π. Thus,

$$\int_0^{2\pi} w(\theta)\cos(m\theta)\,d\theta = \sum_{n=0}^{\infty}\left[\int_0^{2\pi} a_n \cos(n\theta)\cos(m\theta)\,d\theta + \int_0^{2\pi} b_n \sin(n\theta)\cos(m\theta)\,d\theta\right] \tag{4.3.23}$$

$$\int_0^{2\pi} w(\theta)\sin(m\theta)\,d\theta = \sum_{n=0}^{\infty}\left[\int_0^{2\pi} a_n\cos(n\theta)\sin(m\theta)\,d\theta + \int_0^{2\pi} b_n\sin(n\theta)\sin(m\theta)\,d\theta\right] \tag{4.3.24}$$

where $m = 0, 1, 2, 3, \ldots \infty$ and the time dependence $e^{j\omega t}$ has been omitted. By utilizing the orthogonality characteristics of the Fourier series, (4.3.23) and (4.3.24) can be reduced and rearranged to solve explicitly for the modal amplitudes. The resulting equations are

$$a_n = \frac{1}{\epsilon\pi}\int_0^{2\pi} w(\theta)\cos(n\theta)\,d\theta \tag{4.3.25}$$

$$b_n = \frac{1}{\epsilon\pi}\int_0^{2\pi} w(\theta)\sin(n\theta)\,d\theta \tag{4.3.26}$$

where $\epsilon = 2$ for $n = 0$ and $\epsilon = 1$ for $n > 0$; $n = 0, 1, 2, 3, \ldots, \infty$.

If $w(\theta)$ is known completely as a function of θ, all of the modal amplitudes can be determined. In practice, however, $w(\theta)$ is known only at discrete points around the cylinder. Therefore, the integrals of (4.3.25) and (4.3.26) can be represented as Fourier summations of the form

$$a_n = \frac{1}{\epsilon\pi}\sum_{p=1}^{N_p} w(\theta_p)\cos(n\theta_p)\Delta\theta_p \tag{4.3.27}$$

$$b_n = \frac{1}{\epsilon\pi}\sum_{p=1}^{N_p} w(\theta_p)\sin(n\theta_p)\Delta\theta_p \tag{4.3.28}$$

where N_p is the number of circumferential positions where measurements are acquired and $\Delta\theta_p = 2\pi/N_p$ for equally spaced measuring points.

For both of the examples just given, the summations are really only approximations to integrals and rely on the contributions of modes of order greater than one half N_x (for m modes) and one half N_y (for n modes) being negligible. If this is not the case, aliasing will occur as explained previously. Also the higher the mode order, the less accurate will be the final result which implies that increasing the number of measurement points increases the accuracy of the results.

In summary, it may be said that the first modal decomposition method outlined in this section which used randomly located measurement points generally gives better results than the second method which uses evenly spaced measurement points.

REFERENCES

Braun, S.G. and Ram, Y.M. (1987a). Structural parameter identification in the frequency domain: the use of overdetermined systems. *Transactions of ASME, Journal of Dynamic Systems, Measurement and Control*, **109**, 120–123.

Braun, S.G. and Ram, Y.M. (1987b). Determination of structural modes via the Prony model. *Journal of the Acoustical Society of America*, **81**, 1447–1459.

Cremer, L., Heckl, M. and Ungar, E.E. (1973) *Structure-borne Sound*. Springer-Verlag, Berlin.

Ewins, D.J. (1984). *Modal Testing: Theory and Practice*. Research Studies Press, Letchworth.

Jones, J.D. and Fuller, C.R. (1986). Effects of an internal floor on low frequency sound transmission into aircraft cabins: an experimental investigation. In *Proceedings of the 10th Aeroacoustics Conference*. AIAA paper, AIAA-86-1939.

Meirovitch, L. and H. Baruh (1982). Control of self-adjoint distributed parameter systems. *Journal of Guidance, Control, and Dynamics*, **5**, 60–66.

Moore, C.J. (1979). Measurement of radial and circumferential modes in annular and circular fan ducts. *Journal of Sound and Vibration*, **62**, 235–256.

Morgan, D.R. (1991). An adaptive modal-based active control system. *Journal of the Acoustical Society of America*, **89**, 248–256.

5

Modern control review

5.1 INTRODUCTION

Active noise and vibration control can be viewed as a specialized section of the larger field of automatic, or feedback, control. Here we will consider some general background theory from this field which will be pertinent to the later specialized discussions. The work discussed here is relevant mainly to feedback noise and vibration control systems; most of the necessary background for feedforward systems is found in Chapter 6. The difference between the two, and the reasons for choosing one over the other, is discussed in this chapter, as well as Chapters 1 and 6.

Historically, feedback control systems have been used as solutions to a number of problems: stabilization of insufficiently stable systems, attainment of some desired system transfer function, tracking of a desired trajectory, reduction of system response to noise, and improvement of a system's robustness to changes in its open-loop (uncontrolled) dynamics are some of the more common reasons for applying a feedback control system. However, from the perspective of 'active noise and vibration control', it is the problem of reducing a system's response to noise or vibration, or unwanted disturbance, which is of primary interest.

Just as there are a variety of uses for feedback control, there are a variety of modelling techniques used as the basis for designing a system. Three of the more common are transfer function models, state variable models and matrix fraction models. Transfer function models are used in classical control techniques, which can be viewed as evolving principally before 1960. With these techniques, the relationship between the input and output signals to and from the system, or the system transfer function, was considered all important. Design of control systems was carried out based upon the system transfer function using a variety of principally graphical techniques, such as Nyquist, Bode and Root–Locus methods. The techniques used were conceptually simple and computationally inexpensive (developed without the convenience of modern digital computers they had to be!).

Classical control theory, however, has several drawbacks. Chief among these is that it is, generally, only applicable to single input, single output linear time invariant systems. These restrictions limit the utility of classical control methods when they are applied to the design of active noise and vibration control systems, which are usually multiple input, multiple output and often time varying. Further, classical techniques are not able to provide a control law which is optimal in any prespecified sense; that is, it is not possible to use classical control theory to optimize some desired performance criteria. In light of these drawbacks, classical control techniques will not be discussed in detail in this chapter.

'The trend in modern engineering has been to design complex systems which are the result of a compromise between a number of competing performance criteria, such as weight, size, accuracy, cost, etc. Modern control systems, and in particular for this book active noise and vibration control systems, must be able to accommodate, indeed complement, this philosophy. The design of such systems must also be able to efficiently utilize modern tools such as the digital computer. In order to fulfil these requirements, control theory since the 1960s has been largely developed around the concept of state rather than the concept of frequency response or transfer function. Control theory derived using this concept is loosely termed state space, or modern, control theory. State space control laws can be derived for multiple input, multiple output, linear or non-linear, time invariant or time varying systems. They can also be derived with the view to optimizing some specified set of performance criteria.

Recently, considerable research effort has been devoted to the development of control system design methodologies which use matrix fraction models of target systems. These methodologies can be viewed as multiple input, multiple output generalizations of classical control techniques, with results which are analogous to classical quantities such as phase and gain margins. These design techniques have not yet received widespread use in the active noise and vibration control community, and will be largely beyond the scope of this text; the interested reader is referred to Vidyasagar (1985), Francis (1987), Doyle *et al.* (1989), and MacFarlane and Glover (1990) as starting points for these techniques.

This chapter will be largely concerned with the design of control systems using state space, or modern, control methodologies. The aim is to provide a review of key points which will be pertinent to the understanding and design of active noise and vibration control systems. We will begin with a review of state space model derivation, then progress onto the design of control systems. For further information, there are a multitude of introductory texts for feedback control; two of our favourites are Franklin *et al.* (1986) and Ogata (1992). There are also a variety of texts dedicated to state space control methodologies and linear system theory. Some of our favourites are Chen (1970), Kwakernaak and Sivan (1972), Kailath (1978) and Friedland (1986).

5.2 SYSTEM ARRANGEMENTS

5.2.1 General system outlines

We will begin our overview of modern control with the definition of a few terms which will continually arise in the course of this book. A system is defined as a set of individual components acting together as a whole. The systems considered in this text fall into three broad categories; acoustic, structural and structural/acoustic. Any of these systems, when acted upon by some form of excitation, will exhibit a certain response. The excitations of interest to us here are classified as disturbance inputs, which are also referred to as primary excitation or primary disturbance in active control literature, and control inputs. Disturbance inputs are responsible for unwanted excitation of the system, and control inputs are purposely introduced into the system with the aim of obtaining some desired response (for us, this is usually attenuation of the response of the system to the disturbance inputs). An example of an acoustic system is the acoustic environment in an air handling duct, in which the disturbance input (usually from a fan noise source) is responsible for sound propagation. An example of a structural system is a beam, where the disturbance input (usually some dynamic mechanical force) is responsible for unwanted vibration. An example of a structural/acoustic system is an aircraft interior, where the disturbance input is the external acoustic field or the direct vibration excitation of the fuselage. The disturbance inputs excite the fuselage into vibration, which then generates the unwanted acoustic field in the cabin.

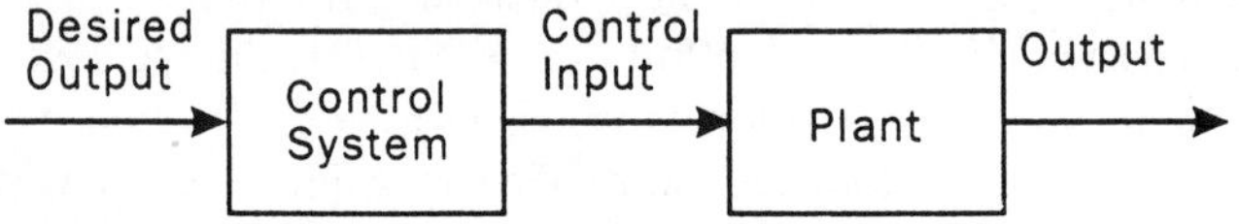

Fig. 5.1 Open loop control system.

In general, the most basic form of control system is one in which the system output has no effect upon the control input, and is referred to as an open loop control system. A block diagram of a typical open loop control system is shown in Fig. 5.1, where a desired output, also referred to as a reference input in control literature, is fed to the controller to produce a control input to the dynamic system, or plant. Such an arrangement can be found, for example, in a toaster, where the 'darkness', or timer, knob provides the desired output signal, and the control input determines the system output, which is the output of a certain darkness of toast. If the level of toast 'darkness' is unsatisfactory, due to fluctuations in bread quality or initial bread conditions, there is no way for the toaster to automatically alter the length of time heat is applied; the output of the system has no influence upon the control input.

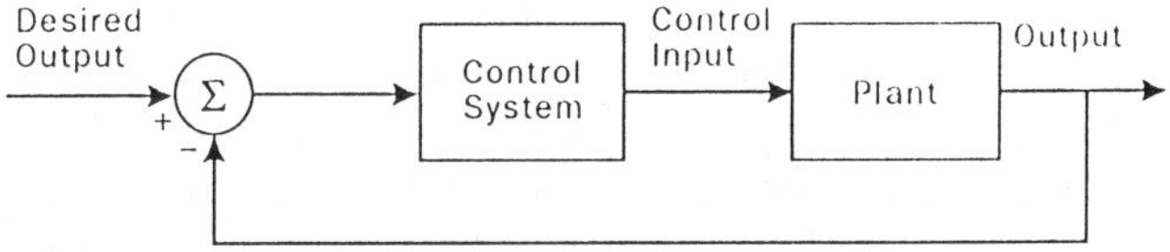

Fig. 5.2 Closed loop control system.

An arrangement which has the potential for improvement upon the open loop control system is one in which the system output does have an influence upon the control input, referred to as a closed loop, or feedback, control system. A block diagram of a typical closed loop control system is shown in Fig. 5.2. Here some output quantity is measured and compared to a desired value, and the resulting error used to correct the system's output. For example, if our toaster was fitted with a closed loop control system to ensure correct darkness, the measured darkness of the toast would be compared to the desired darkness and the control system action, to pop up or not to pop up, would be based upon the results of this comparison.

In our study of active control of sound and vibration we are not, in general, interested in toasting an object, or for that matter moving or altering the equilibrium state of a system. We are principally interested in disturbance attenuation. For us, then, the measured system output is an acoustic or vibrational disturbance, and the desired system output is normally zero. Therefore, the typical feedback control structure of interest here is as shown in Fig. 5.3, where the system output is used to derive the control input.

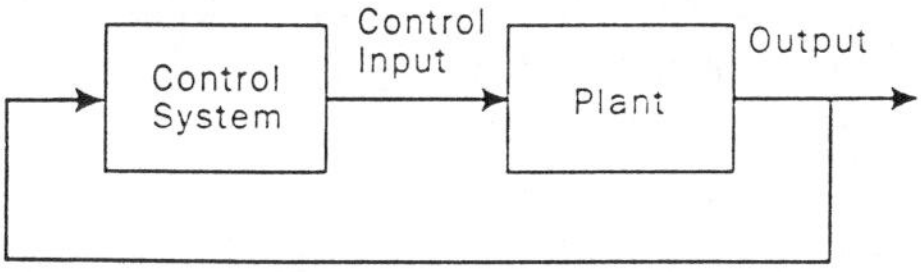

Fig. 5.3 Closed loop control arrangement as implemented in a typical feedback active control system.

In the implementation of active control systems it will in many instances be possible to obtain some *a priori* measure of the impending disturbance input, often referred to in active control literature as a reference signal. An example of this occurs when the disturbance propagates along a waveguide (such as an air handling duct), where it is possible to obtain an upstream measurement. A second example is where the source of the disturbance (the primary source) is rotating machinery, the disturbance is periodic, and a tachometer signal is available which is related to the disturbance. In these instances it is possible to feed-forward a measure of the disturbance to provide attenuation, producing a feedforward control system as shown in Fig. 5.4. Feedforward control systems, when they can be implemented, often offer the potential for greater disturbance attenuation than feedback control systems. Heuristically, the feedforward control system can be viewed as offering prevention of the disturbance, producing an output to counteract the disturbance upon its arrival, while feedback control systems must wait until the disturbance has occurred and been measured at the system output before they can act to attenuate it. Feedforward and feedback control systems can be implemented together to produce a control system which will both effectively attenuate the referenced disturbance to the degree maximally possible, and also provide some attenuation of the unreferenced component of the disturbance.

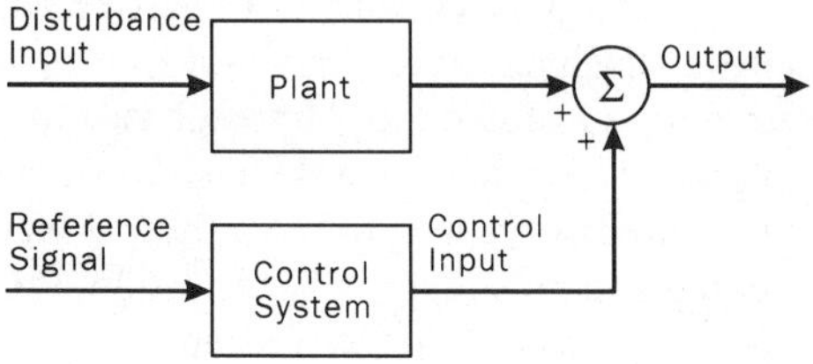

Fig. 5.4 Feedforward control system.

The feedforward control system shown in Fig. 5.4 is an open loop control system. This was the form of control system originally envisaged for active control by Paul Lueg in his patent of 1933, where the control system was set to produce a control input which is 180° out of phase with the primary disturbance at the point of application. However, such a control strategy is unable to cope with changes in the system, and attenuation would be greatly reduced after some period of time. The form of feedforward control system currently implemented in active control systems is an adaptive strategy, such as shown in Fig. 5.5. Here a measure of the system output is used to adjust the control system to provide maximum attenuation, which is effectively a closed loop implementation of a feedforward control strategy.

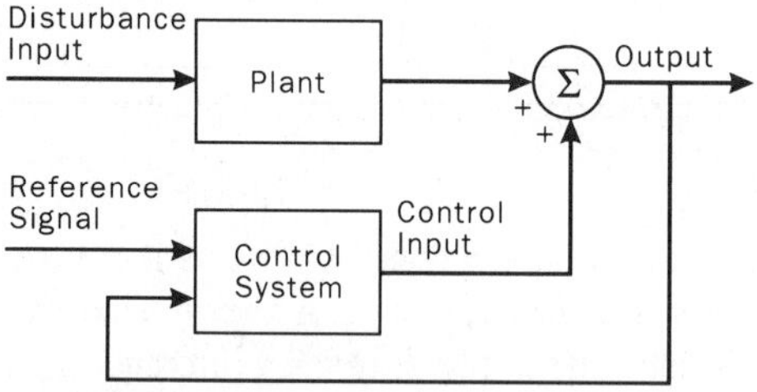

Fig. 5.5 Adaptive feedforward control system.

In this section we will be specifically interested in feedback control systems. Feedforward control systems will be considered in depth in Chapter 6.

5.2.2 Additions for digital implementation

The majority of active noise and vibration control systems will be implemented using digital controllers, which require some additional componentry to be added to the previous block diagrams. The generic feedback control system of Fig. 5.3

is shown modified for digital implementation in Fig. 5.6. The digital implementation involves the addition of an antialiasing filter, sample and hold circuitry, and an analogue to digital converter (ADC) on the input to the controller, and the addition of a digital to analogue converter (DAC), sample and hold circuitry, a reconstruction filter on the output of the controller, and the addition of a clock to synchronize events such as sample.

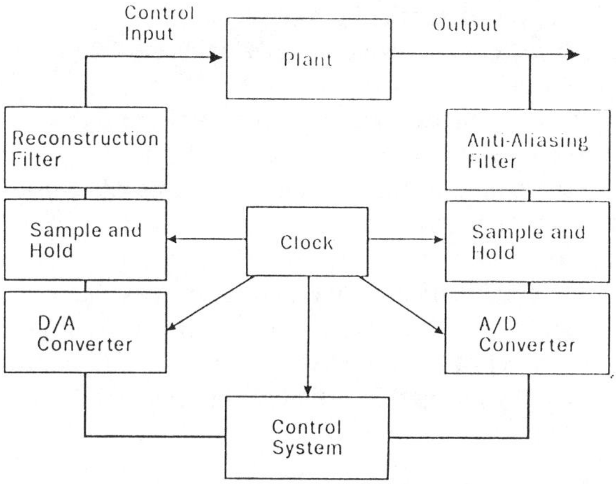

Fig. 5.6 Digital implementation of feedback control system of Fig. 5.3.

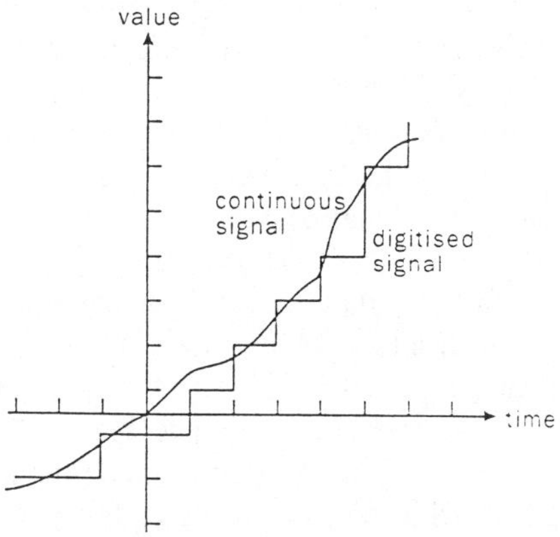

Fig. 5.7 Discrete representation of a continuous signal.

ADCs and DACs provide an interface between the real (continuous) world and the world of a digital system. ADCs take some physical variable, usually an electrical voltage, and convert it to a stream of numbers which are sent to the digital system. Referring to Fig. 5.7, these numbers usually arrive at increments of some fixed time period, or sample period. The numbers arriving from the

ADC are usually representative of the value of the signal at the start of the sample period, as the data input to the ADC is normally sampled then held constant during the conversion process to enable an accurate conversion, as will be discussed further below. Commonly, the sample period is implicitly referred to by a sample rate, which is the number of samples taken in one second. Thus, the ADC provides discrete time samples of a continuous (in time) physical variable. The entire system, consisting of both continuous and discrete time signals, is referred to as a sampled data system.

The digital signal coming from the ADC is quantized in level. This simply means that the stream of numbers sent to the digital control system has some finite number of digits, hence finite accuracy. Referring to Fig. 5.7, it can be seen that, as a result, while the value of the analogue signal fed into the ADC is increased in a continuous nature, the output is increased by discrete increments given by the quantum size. Normally, the digital signal has a binary representation, expressed as a number of bits, each with a state of 0 or 1. This leads to ADCs being classified by the number of bits they use to represent a quantity. For example, a 16-bit ADC converter will represent a sampled physical system variable as a set of 16 bits, each with a value of 0 or 1. It follows that the accuracy of the digital representation of the analogue (continuous) value is limited by the quantum size, given by

$$\text{quantum size} = \frac{\text{full scale range}}{2^n} \tag{5.2.1}$$

where n is the number of bits. For example, if the full scale range of the ADC is ± 10 volts, the accuracy of the 16-bit digital representation is limited to (20 volts)/(2^{16}) = 0.305 mV. The difference between the actual analogue value and its digital representation is referred to as the quantization error. The dynamic range of the ADC is also determined by the number of bits used to digitally represent the analogue value, and is usually expressed in decibels, or dB. For example, a 16-bit ADC has a dynamic range of (20 log (2^{16})), or 96.3, dB.

A signal is said to be digital if it is both discrete in time and quantized in level. The values of both the sample rate and quantum size, which define the digital system, have significant influences upon its performance. This will be noted several times throughout this book.

The DAC works in an opposite fashion to the ADC in the sense that it provides a continuous output signal in response to an input stream of numbers. This continuous output is achieved using the sample and hold circuit, normally incorporated 'on chip'. This circuit is designed to progressively extrapolate the output signal between successive samples in some prescribed manner. The most commonly used hold circuit is the zero order hold, which simply holds the output voltage constant between successive samples. With the zero order hold circuit, the output of the DAC/sample and hold circuitry is continuous in time, but quantised in level. To smooth out this pattern, a reconstruction filter is normally placed at the output of the DAC/sample and hold circuitry as shown in Fig. 5.7. This filter is low pass, which has the effect of removing the high frequency 'corners' from the stepped signal.

Further discussion of the components required to digitally implement a control system can be found in Chapter 13.

5.3 STATE SPACE SYSTEM MODELS FOR FEEDBACK CONTROL

Before we can consider implementing a feedback control system to actively attenuate some unwanted structural or acoustic disturbance, we must have a model of the targeted system. While classical control theory is based explicitly upon an input–output relationship, using a transfer function model of the system in the controller design, modern control theory is based upon a model of the system which is constructed from a set of first order differential equations, which can be combined into a first order matrix differential equation. The matrix notation greatly simplifies the mathematical representation of the system, and provides a form of problem expression which is readily amenable to computer solution. In this section we will review the derivation of continuous state space equations, with particular reference to systems commonly encountered in active noise and vibration control work. Extension of this work to discrete time systems will be undertaken in the section which follows.

As was outlined in Chapter 2, mathematical models of continuous, dynamic systems usually have the form of differential equations. The type of differential equation depends upon the type of system parameters used to develop the model. If the system components can be lumped (a lumped parameter system) in such a way that the parameters are not explicitly dependent upon spatial coordinates, the governing differential equations will be ordinary differential equations. The mass/spring/damper systems commonly considered in control texts generally fall into this category. So too do finite element models of systems. However, if the system components cannot be lumped (distributed parameter system), then the parameters are explicitly dependent upon spatial coordinates and the governing differential equations will be partial differential equations. The description of structures such as beams and plates typically falls into this category.

The input–output response of both distributed and lumped parameter systems is dependent upon time. That is not to say that the coefficients (parameters) which define the governing differential equations necessarily vary with time. If they are constant, the system is referred to as time invariant. If these coefficients vary with time, then the system is referred to as time-varying.

5.3.1 Development of state equations

Consider the generic dynamic (lumped parameter) system described by the (ordinary) differential equation

$$a_n \frac{\mathrm{d}^n y(t)}{\mathrm{d}t^n} + a_{n-1} \frac{\mathrm{d}^{n-1} y(t)}{\mathrm{d}t^{n-1}} + \ldots + a_1 \frac{\mathrm{d}y(t)}{\mathrm{d}t} + a_0 y(t) = x(t) \qquad (5.3.1)$$

The a terms (a_0, a_1,...) are the coefficients or system parameters. This system is said to be of order n, as the highest derivative is the nth derivative. As the dependent variable $y(t)$ and its derivatives are all first order variables, the system is also said to be linear. An important property of linear systems is that they obey the principal of superposition, which means that the response of the system to the combined action of several inputs can be found by determining the response of the system to each individual input, and summing. In other words, if the system has two inputs,

$$u(t) = c_1 u_1(t) + c_2 u_2(t) \tag{5.3.2}$$

the total response of the system is equal to the sum of the responses to the individual inputs:

$$y(t) = c_1 y_1(t) + c_2 y_2(t) \tag{5.3.3}$$

For a non-linear system,

$$y(t) \neq c_1 y_1(t) + c_2 y_2(t) \tag{5.3.4}$$

In many instances, the magnitude of the output $y(t)$ will influence the linearity of the system. This is true, for example, in acoustics, where sound pressures are assumed to be linear if the amplitudes are kept small.

State space modelling is based on the fact that a continuous, linear system can be characterized by a set of first order differential equations, which may be combined and expressed in matrix form. These characterizing equations are concerned with three types of variables; input variables, output variables and state variables. State variables are those comprising the smallest set of n variables, x_1,x_2 ... x_n, which are needed to completely describe the behaviour of the dynamic system of interest. Grouped together, the state variables form a state vector. State space is the n-dimensional vector space whose axes are described by the state vector.

So, in general, a linear, continuous system can be described by a set of n first order differential equations related to n state variables in the form

$$\dot{\boldsymbol{x}} = \boldsymbol{f}(\boldsymbol{x},\boldsymbol{u},t), \tag{5.3.5}$$

where the dot denotes differentiation with respect to time, $\boldsymbol{x}$ is the $(n \times 1)$ state vector, $\boldsymbol{u}$ is the $(r \times 1)$ vector of r system inputs, or input vector, and t is time (note here that bold lower case denotes a vector and bold upper case will denote a matrix). If the system under consideration is time invariant then the vector function $\boldsymbol{f}$ of (5.3.5) will not be explicitly dependent upon time. In a similar way to the input, the output of the system is related to the state variables in the form

$$\boldsymbol{y} = \boldsymbol{g}(\boldsymbol{x},\boldsymbol{u},t) \tag{5.3.6}$$

where $\boldsymbol{y}$ is the $(m \times 1)$ output vector.

From (5.3.5) and (5.3.6), the state equation and output equation can be written as

$$\dot{\boldsymbol{x}} = \boldsymbol{Ax} + \boldsymbol{Bu} \tag{5.3.7}$$

$$\boldsymbol{y} = \boldsymbol{Cx} + \boldsymbol{Du} \tag{5.3.8}$$

where $\boldsymbol{A}$ is the $(n \times n)$ state matrix, $\boldsymbol{B}$ is the $(n \times r)$ input matrix, $\boldsymbol{C}$ is the $(m \times n)$ output matrix, and $\boldsymbol{D}$ is the $(m \times r)$ direct transmission matrix. If the dynamic system is time invariant the matrices $\boldsymbol{A}$, $\boldsymbol{B}$, $\boldsymbol{C}$ and $\boldsymbol{D}$ will be constant coefficient (not explicitly dependent upon time), which is the case with which we will be largely concerned. For the majority of active noise and vibration control applications considered in this book $\boldsymbol{D} = 0$, and therefore $\boldsymbol{D}$ will usually be omitted in this section for simplification.

Example 5.1

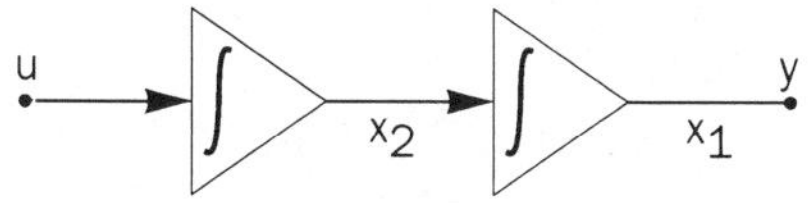

Fig. 5.8 Simple system.

Consider the simple single input system shown in Fig. 5.8. For this system, the generic state space model outlined in (5.3.7) and (5.3.8) can be specialized as

$$\dot{\boldsymbol{x}} = \boldsymbol{Ax} + \boldsymbol{bu} \tag{5.3.9}$$

$$\begin{bmatrix} \dot{x}_1 \\ \dot{x}_2 \end{bmatrix} = \begin{bmatrix} 0 & 1 \\ 0 & 0 \end{bmatrix} \begin{bmatrix} x_1 \\ x_2 \end{bmatrix} + \begin{bmatrix} 0 \\ 1 \end{bmatrix} u \tag{5.3.10}$$

$$y = \boldsymbol{cx} = \begin{bmatrix} 1 & 0 \end{bmatrix} \begin{bmatrix} x_1 \\ x_2 \end{bmatrix} \tag{5.3.11}$$

Example 5.2
Develop a state space model of a sound source in an enclosed, three-dimensional undamped acoustic space (see Dohner and Shoureshi, 1989).

Solution
The starting point for the development of the model is the inhomogeneous wave equation, derived in Section 2.1 and stated in (2.2.41). Using (2.2.37), this can be stated in terms of acoustic pressure $p(\boldsymbol{r},t)$ and the source volume velocity $q(\boldsymbol{r}_p,t)$ as

$$\left[\frac{1}{c^2}\frac{\partial^2}{\partial t^2} + \nabla^2\right] p(\boldsymbol{r},t) = \rho_o \left[\frac{\partial}{\partial t}\right] q(\boldsymbol{r}_p,t) \tag{5.3.12}$$

For an enclosed acoustic space, the acoustic pressure at any point $\boldsymbol{r}$ can be expressed as an infinite sum of contributions from acoustic modes ϕ,

$$p(\boldsymbol{r},t) = \sum_{n=1}^{\infty} a_n(t)\,\phi_n(\boldsymbol{r}) \tag{5.3.13}$$

The eigenvalue problem associated with the self-adjoint (orthogonal) acoustic modes is

$$\nabla^2\phi_n(\boldsymbol{r}) = \lambda_n\phi_n(\boldsymbol{r}) \tag{5.3.14}$$

where λ_n is the nth eigenvalue, and by orthogonality

$$\int_V \phi_m(\boldsymbol{r})\,\phi_n(\boldsymbol{r})\,\mathrm{d}\boldsymbol{r} = \delta(m-n) \tag{5.3.15}$$

where δ is a Kronecker delta function, equal to 1 if $m = n$ and 0 otherwise. Substituting the modal expansion of acoustic pressure into the wave equation produces

$$\frac{1}{c^2}\sum_{n=1}^{\infty}\frac{\partial^2 a_n}{\partial t^2}\phi_n(\boldsymbol{r}) + \sum_{n=1}^{\infty}\lambda_n a_n\phi_n(\boldsymbol{r}) = \rho_0\frac{\partial}{\partial t}q(\boldsymbol{r}_p,t) \tag{5.3.16}$$

If this equation is multiplied through by ϕ_m and integrated over the acoustic space, the result is (for the mth acoustic mode)

$$\frac{1}{c^2}\frac{\partial^2 a_m}{\partial t} + \lambda_m a_m = \rho_0\phi_m(\boldsymbol{r}_p)q(t) \tag{5.3.17}$$

This expression for the response of the mth mode can be written in state space form as

$$\dot{\boldsymbol{x}}_m(t) = \boldsymbol{A}_m\boldsymbol{x}_m(t) + \boldsymbol{b}_m u(t) \tag{5.3.18}$$

where the modal state vector is defined as

$$\boldsymbol{x}_m(t) = \left[x_m(t)\ \ \dot{x}_m(t)\right]^{\mathrm{T}} \tag{5.3.19}$$

and

$$\boldsymbol{A}_m = \begin{bmatrix} 0 & 1 \\ -c^2\lambda_m & 0 \end{bmatrix},\quad \boldsymbol{b}_m = \begin{bmatrix} 0 \\ \rho_0 c^2\phi_m(\boldsymbol{r}_p) \end{bmatrix},\quad u = q(t) \tag{5.3.20}$$

This can be expanded to include all modelled modes as

$$\dot{\boldsymbol{x}} = \boldsymbol{A}\boldsymbol{x} + \boldsymbol{b}u \tag{5.3.21}$$

where

$$
\boldsymbol{x} = \begin{bmatrix} a_1 \\ \dot{a}_1 \\ a_2 \\ \dot{a}_2 \\ \vdots \end{bmatrix}, \quad \boldsymbol{A} = \begin{bmatrix} 0 & 1 & 0 & 0 & \cdots \\ -c^2\lambda_1 & 0 & 0 & & \cdots \\ 0 & 0 & 0 & 1 & \cdots \\ 0 & 0 & -c^2\lambda_2 & 0 & \cdots \\ \cdots & \cdots & \cdots & \cdots & \cdots \end{bmatrix},
$$

$$
\boldsymbol{b} = \begin{bmatrix} 0 \\ \rho_0 c^2 \phi_1(\boldsymbol{r}_p) \\ 0 \\ \rho_0 c^2 \phi_2(\boldsymbol{r}_p) \\ \cdots \end{bmatrix}, \quad u = q(t) \tag{5.3.22}
$$

The output equation can be expressed as

$$p(\boldsymbol{r},t) = \boldsymbol{c}\boldsymbol{x} \tag{5.3.23}$$

where

$$\boldsymbol{C} = \left[\phi_1(\boldsymbol{r}) \;\; 0 \;\; \phi_2(\boldsymbol{r}) \;\; 0 \;\; \cdots\right] \tag{5.3.24}$$

Example 5.3

Develop a state space model for an input to a damped, distributed parameter system (discussed further in Chapter 11, as well as in Balas, 1978a, 1978b, 1982; Meirovitch and Baruh, 1982; Meirovitch *et al.*, 1983).

Solution

As discussed in Chapter 4, the motion of a damped, flexible (distributed parameter) system can be expressed as the partial differential equation

$$m(\boldsymbol{x})\ddot{w}(\boldsymbol{x},t) + c_d\dot{w}(\boldsymbol{x},t) + \kappa w(\boldsymbol{x},t) = u(\boldsymbol{x},t) \tag{5.3.25}$$

which represents the displacement of a point $\boldsymbol{x}$ to the applied force distribution u. Here m is the distributed mass, κ is a time-invariant, non-negative differential operator, and c_d is defined as

$$c_d = 2\zeta\sqrt{\kappa/m} \tag{5.3.26}$$

where ζ is the (non-negative) critical damping ratio. The eigenvalue problem associated with this differential equation is

$$\kappa\psi_n = \lambda_n m\psi_n \tag{5.3.27}$$

where λ_n is the nth eigenvalue, and ψ_n is the associated eigenfunction (mode shape function). The eigenvalues are related to the undamped natural frequencies by

$$\lambda_n = \omega_n^2 \tag{5.3.28}$$

The eigenfunctions are orthogonal, satisfying the expression

$$\int_S m(\boldsymbol{x})\psi_m(\boldsymbol{x})\psi_n(\boldsymbol{x})\, \mathrm{d}\boldsymbol{x} = \delta(m-n) \tag{5.3.29}$$

where $\delta(m - n)$ is a Kronecker delta function, described in the previous example. The displacement at any point $\boldsymbol{x}$ can be expressed as a sum of L modal contributions

$$w(\boldsymbol{x},t) = \sum_{i=1}^{L} w_i(t)\psi_i(\boldsymbol{x}) \tag{5.3.30}$$

In theory, $L = \infty$. However, in practice L is taken to be some finite number suitably large to accurately model the system dynamics. If this modal expansion is substituted into the equations of motion, by multiplying through by ψ_n and integrating over the surface of the structure it is found that, for the nth mode,

$$\ddot{w}_n(t) + 2\zeta\lambda_n^{1/2}\dot{w}_n(t) + \lambda_n w_n(t) = f_n(t) \tag{5.3.31}$$

where f_n is a modal generalized force

$$f_n(t) = \int_S \psi_n(\boldsymbol{x})u(\boldsymbol{x},t)\, \mathrm{d}\boldsymbol{x} \tag{5.3.32}$$

The control force is often taken to be applied by a discrete number, N, of point actuators. If this is the case, the nth modal generalized force can be expressed as

$$f_n(t) = \sum_{i=1}^{N} \psi_n(\boldsymbol{x}_i)f_i(t) \tag{5.3.33}$$

If the modal state vector is defined as

$$\boldsymbol{x}_n(t) = \left[w_n(t) \;\; \dot{w}_n(t)\right]^{\mathrm{T}} \tag{5.3.34}$$

the equation of motion of the nth mode becomes

$$\dot{\boldsymbol{x}}_n(t) = \boldsymbol{A}_n\boldsymbol{x}_n(t) + \boldsymbol{B}_n\boldsymbol{u}(t) \tag{5.3.35}$$

where

$$\boldsymbol{A}_n = \begin{bmatrix} 0 & 1 \\ -\lambda_n & -2\xi\lambda_n^{1/2} \end{bmatrix}, \quad \boldsymbol{B}_n = \begin{bmatrix} 0 & \cdots & 0 \\ \psi_n(\boldsymbol{x}_1) & \cdots & \psi_n(\boldsymbol{x}_m) \end{bmatrix}, \tag{5.3.36}$$

$$\boldsymbol{u} = \left[f_1(t) \;\cdots\; f_m(t)\right]^{\mathrm{T}}$$

This can be expanded to include all modelled modes as

$$\dot{\boldsymbol{x}} = \boldsymbol{A}\boldsymbol{x} + \boldsymbol{B}\boldsymbol{u} \tag{5.3.37}$$

where

$$\boldsymbol{x} = \begin{bmatrix} x_1 \\ x_2 \\ x_3 \\ \vdots \end{bmatrix}, \quad \boldsymbol{A} = \begin{bmatrix} A_1 & & & \\ & A_2 & & \\ & & A_3 & \\ & & & \ddots \end{bmatrix}, \quad \boldsymbol{B} = \begin{bmatrix} B_1 \\ B_2 \\ B_3 \\ \vdots \end{bmatrix} \tag{5.3.38}$$

If the system output is displacement at a point, the output equation can be expressed as

$$w(x,t) = \boldsymbol{c}\boldsymbol{x}(t) \tag{5.3.39}$$

where

$$\boldsymbol{c} = \begin{bmatrix} \psi_1(x) & 0 & \psi_2(x) & 0 & \cdots \end{bmatrix} \tag{5.3.40}$$

Equations (5.3.7) and (5.3.8) do not provide a unique definition of the system. Rather, any state vector χ which is a non-singular linear transformation of $\boldsymbol{x}$ is suitable,

$$\chi = \boldsymbol{T}\boldsymbol{x} \tag{5.3.41}$$

where $\boldsymbol{T}$ is some transformation matrix. By substituting (5.3.41) into (5.3.7) and (5.3.8), the set of transformed system equations is found to be

$$\begin{aligned} \dot{\chi} &= \boldsymbol{T}(\boldsymbol{A}\boldsymbol{x} + \boldsymbol{B}\boldsymbol{u}) = \boldsymbol{T}\boldsymbol{A}\boldsymbol{x} + \boldsymbol{T}\boldsymbol{B}\boldsymbol{u} = \boldsymbol{T}\boldsymbol{A}\boldsymbol{T}^{-1}\chi + \boldsymbol{T}\boldsymbol{B}\boldsymbol{u} \\ &= \boldsymbol{A}'\chi + \boldsymbol{B}'\boldsymbol{u} \end{aligned} \tag{5.3.42}$$

$$y = \boldsymbol{C}\boldsymbol{T}^{-1}\chi = \boldsymbol{C}'\chi \tag{5.3.43}$$

where

$$\boldsymbol{A}' = \boldsymbol{T}\boldsymbol{A}\boldsymbol{T}^{-1}, \quad \boldsymbol{B}' = \boldsymbol{T}\boldsymbol{B}, \quad \boldsymbol{C}' = \boldsymbol{C}\boldsymbol{T}^{-1} \tag{5.3.44}$$

Equations (5.3.42) and (5.3.43) are identical in form to (5.3.7) and (5.3.8), demonstrating that the transformation has no overall effect on the system input and output, only on its (transformed) internal state. In this regard, state representation of a dynamic system is advantageous because it standardizes the format of the required information into three matrices, $\boldsymbol{A}$, $\boldsymbol{B}$ and $\boldsymbol{C}$, regardless of the problem. Further, it may be advantageous, when confronting the problem of controller design, to change the state to bring the description matrices $\boldsymbol{A}$, $\boldsymbol{B}$ and $\boldsymbol{C}$ into canonical forms, as was done explicitly in developing the previous two examples.

At this point it is important to note, for future reference, that while the state equations do not provide a unique definition of the system, the eigenvalues of the state matrix $\boldsymbol{A}$ *are* unique. As we will discuss later in this chapter, the eigenvalues of the state matrix $\boldsymbol{A}$ are the roots of the characteristic equation

$$| \lambda \boldsymbol{I}-\boldsymbol{A} | = 0 \tag{5.3.45}$$

Consider the modification of the state matrix carried out in (5.3.44):

$$\boldsymbol{A}' = \boldsymbol{T}\boldsymbol{A}\boldsymbol{T}^{-1} \tag{5.3.46}$$

For the eigenvalues to be invariant the characteristic equations of the original and transformed state equations must be equivalent. Consider the latter of these,

$$| \lambda \boldsymbol{I}-\boldsymbol{A}' | = | \lambda \boldsymbol{I}-\boldsymbol{T}\boldsymbol{A}\boldsymbol{T}^{-1} | = | \lambda \boldsymbol{T}^{-1}\boldsymbol{T}-\boldsymbol{T}^{-1}\boldsymbol{A}\boldsymbol{T} | = | \boldsymbol{T}^{-1}(\lambda \boldsymbol{I}-\boldsymbol{A})\boldsymbol{T} |$$
$$= | \boldsymbol{T}^{-1} | \; | \lambda \boldsymbol{I}-\boldsymbol{A} | \; | \boldsymbol{T} | = | \boldsymbol{T}^{-1} | \; | \boldsymbol{T} | \; | \lambda \boldsymbol{I}-\boldsymbol{A} | \tag{5.3.47}$$

Note that the product of the determinants $|\boldsymbol{T}^{-1}|$ and $|\boldsymbol{T}|$ is the determinant of the product $|\boldsymbol{T}^{-1}\boldsymbol{T}|$,

$$| \lambda \boldsymbol{I}-\boldsymbol{T}^{-1}\boldsymbol{A}\boldsymbol{T} | = | \boldsymbol{T}^{-1}\boldsymbol{T} | \; | \lambda \boldsymbol{I}-\boldsymbol{A} | = | \lambda \boldsymbol{I}-\boldsymbol{A} | . \tag{5.3.48}$$

Therefore, the characteristic equations of the original and transformed state equations, and hence the eigenvalues of the system, will be invariant.

5.3.2 Solution of the state equation

To examine the characteristics of a system described by the state equations (5.3.7) and (5.3.8) it is often necessary to obtain the general solution of the linear time invariant state equation, a matrix differential equation. This is also necessary for deriving the form of equation required for digital implementation where, as we will see later in this chapter, the differential equation description of the continuous system response must be re-expressed in terms of difference equations. The general solution will be obtained in this section in two steps: first, derivation of the homogeneous solution (system response with zero control input), and second, derivation of a particular solution for a non-zero control input.

Let us first consider the homogeneous, or unforced, solution to the state equation (5.3.7), which amounts to solving the matrix differential equation

$$\dot{\boldsymbol{x}} = \boldsymbol{A}\boldsymbol{x} \tag{5.3.49}$$

To solve this equation, we will first assume that the solution will have the form of a vector power series in time t,

$$\boldsymbol{x}(t) = \boldsymbol{b}_0 + \boldsymbol{b}_1 t + \boldsymbol{b}_2 t^2 + \cdots + \boldsymbol{b}_i t^i \tag{5.3.50}$$

Substituting this assumed form into (5.3.49) yields

$$\boldsymbol{b}_1 + 2\boldsymbol{b}_2 t + 3\boldsymbol{b}_3 t^2 + \cdots + i\boldsymbol{b}_i t^{i-1} = \boldsymbol{A}(\boldsymbol{b}_0 + \boldsymbol{b}_1 t + \boldsymbol{b}_2 t^2 + \cdots + \boldsymbol{b}_i t^i) \tag{5.3.51}$$

For this solution to be valid it must hold for all time t. Therefore,

$$\begin{aligned} b_1 &= Ab_0 \\ b_2 &= \frac{1}{2}Ab_1 = \frac{1}{2}A^2 b_0 \\ &\cdots \\ b_i &= \frac{1}{i!}A^i b_0 \end{aligned} \tag{5.3.52}$$

By letting $x(0) = b_0$, the solution to (5.3.49) can be seen to be

$$x(t) = (I + At + \frac{1}{2!}A^2t^2 + \cdots + \frac{1}{i!}A^it^i)\, x(0) \tag{5.3.53}$$

or

$$x_h(t) = e^{A(t-t_0)}x_h(t_0) = \Phi(t)x_h(t_0) \tag{5.3.54}$$

where the subscript h denotes homogeneous. Here the matrix exponential is defined by

$$\begin{aligned} e^{A(t-t_0)} &= I + A(t-t_0) + A^2\frac{(t-t_0)^2}{2!} + A^3\frac{(t-t_0)^3}{3!} + \cdots \\ &= \sum_{i=0}^{\infty} A^i\frac{(t-t_0)^i}{i!} \end{aligned} \tag{5.3.55}$$

Note that, from (5.3.54), the solution to the homogeneous state equation (5.3.7) is simply a transformation of the initial state. The transformation is described by the state transition matrix $\Phi(t)$,

$$\Phi(t) = e^{A(t-t_0)} \tag{5.3.56}$$

This matrix is unique to any particular system, and contains all of the information required to describe the characteristics of the free (homogeneous) motion of the system.

It is useful to briefly consider two properties of the matrix exponential, and hence transition matrix. Consider two values of time, t_1 and t_2. Since the homogeneous solution of (5.3.56) is unique,

$$x_h(t_1) = e^{A(t_1-t_0)}x_h(t_0), \quad \text{and } x_h(t_2) = e^{A(t_2-t_0)}x_h(t_0) \tag{5.3.57}$$

As t_0 is an arbitrary starting time, $x_h(t_2)$ can be written as if $t_0 = t_1$,

$$x_h(t_2) = e^{A(t_2-t_1)}x_h(t_1) \tag{5.3.58}$$

Substituting (5.3.57) for $x_h(t_1)$ into (5.3.58) gives

$$x_h(t_2) = e^{A(t_2-t_1)}e^{A(t_1-t_0)}x_h(t_0) \tag{5.3.59}$$

Since the solution for $x_h(t)$ is unique,

$$x_h(t_2) = e^{A(t_2-t_0)}x_h(t_0) = e^{A(t_2-t_1)}e^{A(t_1-t_0)}x_h(t_0). \tag{5.3.60}$$

Therefore,

$$e^{A(t_2-t_0)} = e^{A(t_2-t_1)}e^{A(t_1-t_0)} \tag{5.3.61}$$

Now again since t_0 is arbitrary, if $t_0=t_2$,

$$e^{A0} = I = e^{-A(t_1-t_0)}e^{A(t_1-t_0)} \tag{5.3.62}$$

Therefore, the inverse of e^{At} is equal to e^{-At}. This property will be used in finding the particular solution of the state equation shortly.

The second property of the matrix exponential which should be remembered is

$$\begin{aligned} e^{(A+B)t} &= e^{At}e^{Bt} \qquad \text{iff } AB = BA \\ e^{(A+B)t} &\neq e^{At}e^{Bt} \qquad \text{if } AB \neq BA \end{aligned} \tag{5.3.63}$$

This is easily seen by comparing the expansions

$$e^{(A+B)t} = I + (A+B)t + \frac{(A+B)^2}{2!}t^2 + \frac{(A+B)^3}{3!}t^3 + \cdots \tag{5.3.64}$$

and

$$\begin{aligned} e^{At}e^{Bt} &= \left[I + At + \frac{A^2t^2}{2!} + \frac{A^3t^3}{3!} + \cdots\right]\left[I + Bt + \frac{B^2t^2}{2!} + \frac{B^3t^3}{3!} + \cdots\right] \\ &= I + (A+B)t + \frac{A^2t^2}{2!} + ABt^2 + \frac{B^2t^2}{2!} + \frac{A^3t^3}{3!} + \frac{A^2Bt^3}{2!} + \frac{AB^2t^3}{2!} + \frac{B^3t^3}{3!} + \cdots. \end{aligned} \tag{5.3.65}$$

The difference between (5.3.64) and (5.3.65) is

$$\begin{aligned} &e^{(A+B)t} - e^{At}e^{Bt} = \\ &\frac{BA-AB}{2!}t^2 + \frac{BA^2+ABA+B^2A+BAB-2A^2B-2AB^2}{3!}t^3 + \cdots. \end{aligned} \tag{5.3.66}$$

From (5.3.66) it is apparent that the difference is zero only if $\mathbf{AB} = \mathbf{BA}$.

A method of solving the homogeneous state equation which provides an alternative to the direct solution of (5.3.49) involves the use of Laplace transforms. Taking the Laplace transform of both sides of (5.3.49) produces

$$sx(s) - x(0) = Ax(s) \tag{5.3.67}$$

Rearranging this,

$$(sI-A)x(s) = x(0) \tag{5.3.68}$$

or

$$x(s) = (sI-A)^{-1}x(0) \tag{5.3.69}$$

Once the equation has been put in this form, the inverse Laplace transform can be used to solve it. Noting that the bracketed part of (5.3.69) can be expanded as

$$(s\boldsymbol{I}-\boldsymbol{A})^{-1} = \frac{\boldsymbol{I}}{s}+\frac{\boldsymbol{A}}{s^2}+\frac{\boldsymbol{A}^2}{s^3}+\cdots, \tag{5.3.70}$$

the inverse Laplace transform is

$$\mathcal{L}^{-1}\left((s\boldsymbol{I}-\boldsymbol{A})^{-1}\right) = \boldsymbol{I}+\boldsymbol{A}t+\frac{\boldsymbol{A}^2t^2}{2!}+\cdots = e^{\boldsymbol{A}t}. \tag{5.3.71}$$

Therefore, the solution to the homogeneous state equation is

$$\boldsymbol{x}(t) = e^{\boldsymbol{A}t}\boldsymbol{x}(0) \tag{5.3.72}$$

This is, of course, exactly the same solution which was obtained earlier in this section. The important point about this is not so much the actual solution, but rather the fact that (5.3.69) provides a means of obtaining a closed form solution for the matrix exponential. Further consideration of the problem of solving matrix exponentials can be found in Moler and van Loan (1978).

Example 5.4

Calculate the state transition matrix, or matrix exponential, for the system described by the equation

$$m\ddot{w} + kw = 0 \tag{5.3.73}$$

Solution

For this problem the first step is to re-express the equation in state variable form. This can be done in a manner similar to the previous examples in this section. Noting that the eigenvalue problem with this differential equation is

$$k\psi_n = \lambda_n m\psi_n = \omega_n^2 m\psi_n \tag{5.3.74}$$

and noting also that the displacement at any point $\boldsymbol{x}$ can be expressed as a sum of modal contributions

$$w(\boldsymbol{x},t) = \sum_{i=1}^{\infty} w_i(t)\psi_i(\boldsymbol{x}) \tag{5.3.75}$$

the original equation can be expressed as a sum of modal contributions

$$\sum_{i=1}^{\infty}\left(\ddot{w}_i+\omega_i^2 w_i\right) = 0 \tag{5.3.76}$$

Considering for simplicity only the nth mode, the state equation is

$$\dot{\boldsymbol{x}}_n = \boldsymbol{A}\boldsymbol{x}_n \tag{5.3.77}$$

where

$$\boldsymbol{x}_n = \begin{bmatrix} w_n \\ \dot{w}_n \end{bmatrix}, \quad \boldsymbol{A} = \begin{bmatrix} 0 & 1 \\ -\omega_n^2 & 0 \end{bmatrix} \tag{5.3.78}$$

Therefore,

$$(s\boldsymbol{I}-\boldsymbol{A}_n) = \begin{bmatrix} s & -1 \\ \omega_n^2 & s \end{bmatrix} \tag{5.3.79}$$

As this is only a (2 × 2) matrix, the inverse can be found readily as follows:

$$(s\boldsymbol{I}-\boldsymbol{A}_n)^{-1} = \frac{1}{s^2+\omega_n^2}\begin{bmatrix} s & 1 \\ -\omega_n^2 & s \end{bmatrix} \tag{5.3.80}$$

Using standard results which can be found in virtually any introductory controls text,

$$\mathcal{L}^{-1}\left((s\boldsymbol{I}-\boldsymbol{A}_n)^{-1}\right) = \mathrm{e}^{\mathrm{A}_n t} = \Phi_n(t) = \begin{bmatrix} \cos\omega_n t & \frac{1}{\omega_n}\sin\omega_n t \\ -\omega_n\sin\omega_n t & \cos\omega_n t \end{bmatrix} \tag{5.3.81}$$

Alternatively, the matrix exponential could be calculated directly from (5.3.55):

$$\Phi_n(t) = \mathrm{e}^{\boldsymbol{A}_n t} = \boldsymbol{I}+\boldsymbol{A}t+\frac{\boldsymbol{A}^2t^2}{2!}+\frac{\boldsymbol{A}^3t^3}{3!}\cdots \tag{5.3.82}$$

Expanding this gives

$$\Phi_n(t) = \mathrm{e}^{\boldsymbol{A}_n t} =$$

$$\begin{bmatrix} 1-\frac{1}{2!}(\omega_n t)^2+\frac{1}{4!}(\omega_n t)^4-\cdots & \frac{1}{\omega_n}\left[\omega_n t-\frac{1}{3!}(\omega_n t)^3+\frac{1}{5!}(\omega_n t)^5-\cdots\right] \\ -\omega_n\left[\omega_n t-\frac{1}{3!}(\omega_n t)^3+\frac{1}{5!}(\omega_n t)^5-\cdots\right] & 1-\frac{1}{2!}(\omega_n t)^2+\frac{1}{4!}(\omega_n t)^4-\cdots \end{bmatrix} \tag{5.3.83}$$

or

$$\Phi_n(t) = \mathrm{e}^{\boldsymbol{A}_n t} = \begin{bmatrix} \cos\omega_n t & \frac{1}{\omega_n}\sin\omega_n t \\ -\omega_n\sin\omega_n t & \cos\omega_n t \end{bmatrix} \tag{5.3.84}$$

Having solved the homogenous state equation, and examined a few of the properties of the matrix exponential, we are in a position to obtain the particular solution to the state equation. The particular solution to (5.3.7), when the control input no longer is equal to zero, is found using variation of parameters. With this method we first guess the form of the solution to be

$$\boldsymbol{x}_p(t) = \mathrm{e}^{\boldsymbol{A}(t-t_0)}\alpha(t) \tag{5.3.85}$$

where the subscript p denotes particular. Here, $\alpha(t)$ is a vector of variable parameters to be determined (this is in contrast to the vector of constant parameters, $\boldsymbol{x}_h(t)$, for the homogeneous case). Substituting the solution equation (5.3.85) back into (5.3.7) produces

$$\begin{aligned}\dot{\boldsymbol{x}} &= \boldsymbol{A}e^{\boldsymbol{A}(t-t_0)}\alpha(t) + e^{\boldsymbol{A}(t-t_0)}\dot{\alpha}(t) \\ &= \boldsymbol{A}e^{\boldsymbol{A}(t-t_0)}\alpha(t) + \boldsymbol{Bu}(t)\end{aligned} \tag{5.3.86}$$

Therefore,

$$\dot{\alpha}(t) = e^{-\boldsymbol{A}(t-t_0)}\boldsymbol{Bu}(t) \tag{5.3.87}$$

Assuming that the application of control begins at t_0, $\alpha(t)$ can be found from

$$\alpha(t) = \int_{t_0}^{t} e^{-\boldsymbol{A}(\tau-t_0)}\boldsymbol{Bu}(\tau)\ d\tau \tag{5.3.88}$$

Substituting this into (5.3.85) gives

$$\boldsymbol{x}_p(t) = e^{\boldsymbol{A}(t-t_0)}\int_{t_0}^{t} e^{-\boldsymbol{A}(\tau-t_0)}\boldsymbol{Bu}(\tau)\ d\tau \tag{5.3.89}$$

Using the property of the matrix exponential given in (5.3.61), this can be re-expressed as

$$\boldsymbol{x}_p(t) = \int_{t_0}^{t} e^{\boldsymbol{A}(t-\tau)}\boldsymbol{Bu}(\tau)\ d\tau \tag{5.3.90}$$

Combining the homogeneous solution of (5.3.54) with the particular solution of (5.3.90) produces the total solution for the case of non-zero system input:

$$\boldsymbol{x}(t) = e^{\boldsymbol{A}(t-t_0)}\boldsymbol{x}(t_0) + \int_{t_0}^{t} e^{\boldsymbol{A}(t-\tau)}\boldsymbol{Bu}(\tau)\ d\tau \tag{5.3.91}$$

We will now use the solutions to the state equation in developing state equations for discrete time (digital) systems.

5.4 DISCRETE TIME SYSTEM MODELS FOR FEEDBACK CONTROL

One way of designing a control system explicitly for digital implementation is to utilize linear difference equations to model the dynamics of the system of interest. An alternative way is to design the control system using continuous time representations of the problem, utilizing first order differential equations (as in the previous section), then to implement it digitally by transforming the control law into a discrete time format. In this section we will use the former approach, involving linear difference equations, because this will lead to a rationale for using digital filters to control a dynamic system. The conversion from continuous to discrete time control laws is almost trivial when using specialized software such as Matlab, and will not be covered here. The interested reader is referred to any of the modern control texts outlined in the introduction to this chapter. Here we will first develop linear difference equations which characterize the

Here we will first develop linear difference equations which characterize the dynamics of the system of interest, and then use the z-transform to solve the equations. Following this we will discuss realization of this form of solution in a digital filter, going on then to discuss the problem of system identification using the filters.

5.4.1 Development of difference equations

Consider the case where, at each sample period, the electronic control system receives an input sample u, and in response produces an output signal, y. The set of input samples up to and including the kth sample, $\boldsymbol{u}(k)$, define the set

$$\boldsymbol{u}(k) = [u(0) \;\; u(1) \;\; \cdots \;\; u(k)]^{\mathrm{T}} \tag{5.4.1}$$

The set of output signals leading up to this time defines the set

$$\boldsymbol{y}(k-1) = [y(0) \;\; y(1) \;\; \cdots \;\; y(k-1)]^{\mathrm{T}} \tag{5.4.2}$$

The kth output signal $y(k)$, given in response to receiving the kth input sample $u(k)$, is some function of the set of input samples $\boldsymbol{u}(k)$ and of the previous output signals $\boldsymbol{y}(k)$,

$$y(k) = f\ \{\boldsymbol{u}(k),\ \boldsymbol{y}(k-1)\} \tag{5.4.3}$$

Assuming that this function is linear, and depends only on m input samples and n output signals prior to the kth sample, enables the kth output signal to be expressed as

$$\begin{aligned} y(k) = {} & a_1 y(k-1) + a_2 y(k-2) + \cdots + a_n y(k-n) \\ & + b_0 u(k) + b_1 u(k-1) + \cdots + b_m u(k-m) \end{aligned} \tag{5.4.4}$$

Equations of the form of (5.4.4) are known as linear difference, or recurrence, equations. This name arises because the equation can be expressed using $y(k)$ plus a series of differences in $y(k)$, defined as

$$\begin{aligned} \Delta y(k) &= y(k) - y(k-1) && \text{(first difference)} \\ \Delta^2 y(k) &= \Delta y(k) - \Delta y(k-1) && \text{(second difference)} \\ \Delta^n y(k) &= \Delta^{n-1} y(k) - \Delta^{n-1} y(k-1) && (n\text{th difference}) \end{aligned} \tag{5.4.5}$$

As an example, consider a second order equation with coefficients a_1, a_2, and b_0 (let b_1, b_2 equal zero for simplicity). Then, using (5.4.5), $y(k)$, $y(k-1)$, and $y(k-2)$ can be re-expressed in terms of the difference equations

$$\begin{aligned} y(k) &= y(k) \\ y(k-1) &= y(k) - \nabla y(k) \\ y(k-2) &= y(k-1) - \nabla y(k-1) = y(k) - \nabla y(k) - \nabla(y(k) - \nabla y(k)) \\ &= y(k) - 2\nabla y(k) + \nabla^2 y(k). \end{aligned} \tag{5.4.6}$$

Therefore, the second order simplification of (5.4.4) being considered here can be written as

$$y(k) = a_1 y(k-1) + a_2 y(k-2) + b_0 u(k) = a_1(y(k) - \nabla y(k)) + a_2(y(k) - 2\nabla y(k) + \nabla^2 y(k)) + b_0 u(k) \tag{5.4.7}$$

In more conventional form,

$$-a_2 \nabla^2 y(k) + (a_1 + 2a_2)\nabla y(k) + (1 - a_1 - a_2) y(k) = b_0 u(k) \tag{5.4.8}$$

If the a and b coefficients in (5.4.4) are constant, the electronic control system provides an output signal $y(k)$ by solving a constant coefficient difference equation which models the system dynamics. By solving such an equation, the electronic control system can both control and emulate linear constant dynamic systems, and be implemented as a digital filter.

A well-known example of the use of discrete difference equations to approximate a continuous system is the discrete approximation of an integral. The simplest (and roughest) of these approximations is done using the trapezoidal rule. Suppose we have the continuous input signal shown in Fig. 5.9, and we wish the output of the system to be the integral

$$y(t) = \int_0^t u(t)\, dt \tag{5.4.9}$$

The solution to this integral must be approximated using the discrete input sample set $\boldsymbol{u}(k)$. Suppose an approximation to the integral at sample $k - 1$, $y(k - 1)$, exists, and we wish to use it to approximate the solution of the integral at the next input sample, $y(k)$. From Fig. 5.9, it can be seen that this problem reduces to finding the area under the curve defining the continuous function $u(k)$ for the sample period between samples $k - 1$ and k, and adding it to the previous approximation $y(k - 1)$. This area can be approximated by the trapezoid

$$\text{area} = (t(k) - t(k-1)) \frac{u(k) + u(k-1)}{2} = T\frac{u(k) + u(k-1)}{2} \tag{5.4.10}$$

where T is the sample period. Using this equation, the system output at time k, $y(k)$, is defined by the difference equation

$$y(k) = y(k-1) + \frac{T}{2}(u(k) + u(k-1)) \tag{5.4.11}$$

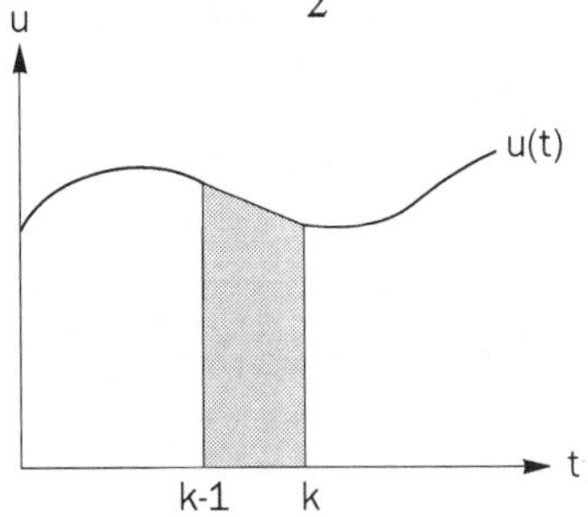

Fig. 5.9 Integration of continuous signal using a trapezoid.

5.4.2 State space equations for discrete time systems

In Section 5.3 we derived the total (homogeneous and particular) solution to continuous time state equations. This solution can be used to derive discrete time state space equations, in the process demonstrating the relationship between the continuous time (dynamic) system and its discrete time (digital) model. This derivation is undertaken by evaluating the previously obtained solution over one sample period to formulate the required difference equation for digital implementation.

Let the arbitrary starting point, t_0, be kT samples (where k is some arbitrary integer number), and let t be one sample period later, equal to $kT+T$. Substituting these discrete values into the total solution of (5.3.91) produces

$$\boldsymbol{x}(kT+T) = e^{AT}\boldsymbol{x}(kT) + \int_{kT}^{kT+T} e^{A(kT+T-\tau)}\boldsymbol{B}u(\tau)\,d\tau \tag{5.4.12}$$

Assume now that the value of $u(\tau)$ during this sample period is equal to $u(t_0) = u(kT)$ (that is, there is no delay on the sampled signal and a zero order hold is being used). (A discussion of the effect of delay can be found in Franklin *et al.* (1990)). If we define the variable κ as

$$\kappa = kT + T - \tau \tag{5.4.13}$$

(5.4.12) can be written as

$$\boldsymbol{x}(kT+T) = e^{AT}\boldsymbol{x}(kT) + \int_0^T e^{A\kappa}\,d\kappa\,\boldsymbol{B}u(kT) \tag{5.4.14}$$

Defining

$$\boldsymbol{\Phi} = e^{AT} \tag{5.4.15}$$

and

$$\boldsymbol{\Gamma} = \int_0^T e^{A\kappa}\,d\kappa\,\boldsymbol{B} \tag{5.4.16}$$

allows (5.4.14) to be written in the standard state space format (outlined in Section 5.3) as

$$\begin{aligned}\boldsymbol{x}(k+1) &= \boldsymbol{\Phi}\boldsymbol{x}(k) + \boldsymbol{\Gamma}u(k)\\ y(k) &= \boldsymbol{C}\boldsymbol{x}(k)\end{aligned} \tag{5.4.17}$$

To evaluate the system equations, $\boldsymbol{\Phi}$ can be expressed as

$$\boldsymbol{\Phi} = \boldsymbol{I} + T\boldsymbol{A}\boldsymbol{\Theta} \tag{5.4.18}$$

where

$$\boldsymbol{\Theta} = \boldsymbol{I} + \sum_{i=1}^{\infty} \frac{\boldsymbol{A}^i T^i}{(i+1)!} \tag{5.4.19}$$

Similarly, (5.4.16) can be evaluated and rewritten as

$$\mathbf{\Gamma} = T\mathbf{\Theta}\mathbf{B} \tag{5.4.20}$$

5.4.3 Discrete transfer functions

Difference equations form the basis of digital control. If the coefficients in (5.4.4) are linear, the relationship between the input sample $u(k)$ and the output signal $y(k)$ can be described by a transfer function. Transfer functions of linear discrete (digital) systems, based on difference equations, can be obtained using z-transform analysis.

The z-transform has the same use in discrete system analysis that the Laplace transform has in continuous system analysis, in that it enables the solution of the equation describing the system dynamics to become algebraic. If a signal has discrete values $x(0)$, $x(1)\ldots x(k)$, the z-transform of the signal is defined as

$$x(z) = z\{x(k)\} = \sum_{k=-\infty}^{\infty} x(k)\, z^{-k} \tag{5.4.21}$$

Here it is assumed that a range of values of the magnitude of the complex variable z can be found for which the series of (5.4.21) converges.

Before considering general use of the z-transform, it will be beneficial to first demonstrate its use in the simple integral problem considered previously in Section 5.4.1. Using the definition of (5.4.21), the z-transform of the output signal of (5.4.9) is

$$y(z) = \sum_{k=-\infty}^{\infty} y(k)\, z^{-k} \tag{5.4.22}$$

To analyse the integrating system, (5.4.11) must first be multiplied through by z^{-k}, and summed over all possible values of k, producing

$$\sum_{k=-\infty}^{\infty} y(k)\, z^{-k} = \sum_{k=-\infty}^{\infty} y(k-1)\, z^{-k} + \frac{T}{2}\left[\sum_{k=-\infty}^{\infty} u(k)\, z^{-k} + \sum_{k=-\infty}^{\infty} u(k-1)\, z^{-k}\right] \tag{5.4.23}$$

From (5.4.22), the left-hand side of (5.4.23) is simply the z-transform of the output signal. The first term on the right-hand side of (5.4.23) is

$$\sum_{k=-\infty}^{\infty} y(k-1)\, z^{-k} = \sum_{i=-\infty}^{\infty} y(i)\, z^{-i+1} = z^{-1}y(z) \tag{5.4.24}$$

The second term on the right-hand side of (5.4.23) is the z-transform of the input signal, while the final term on the right-hand side of the equation will have a form similar to (5.4.24). Using these simplifications, (5.4.23) can be written as

$$y(z) = z^{-1}y(z) + \frac{T}{2}\left(u(z) + z^{-1}u(z)\right) \tag{5.4.25}$$

The output function $y(z)$ can now be solved in terms of the input function $u(z)$ by simple algebra to yield

$$y(z) = \frac{T}{2}\frac{1+z^{-1}}{1-z^{-1}}u(z) \tag{5.4.26}$$

The transfer function of the discrete system, $G(z)$, is defined as the ratio of the z-transform of the output signal to the z-transform of the input signal:

$$G(z) = \frac{y(z)}{u(z)} \tag{5.4.27}$$

Therefore, the transfer function of the integrating system in the z-domain is

$$G(z) = \frac{T}{2}\frac{1+z^{-1}}{1-z^{-1}} \tag{5.4.28}$$

This same form of analysis can be used to derive the transfer function of the general difference (5.4.4), producing

$$G(z) = \frac{b_0 + b_1z^{-1} + b_2z^{-2} + \cdots + b_mz^{-m}}{1 - a_1z^{-1} - a_2z^{-2} - \cdots - a_nz^{-n}} \tag{5.4.29}$$

If the order of the denominator, n, is greater than or equal to the order of the numerator, m, then (5.4.29) can be expressed as a ratio of polynomials in z:

$$G(z) = \frac{b_0z^n + b_1z^{n-1} + \cdots + b_mz^{n-m}}{z^n - a_1z^{n-1} - \cdots - a_n} = \frac{b(z)}{a(z)} \tag{5.4.30}$$

For comparison, the discrete system state equations can be written in the discrete transfer function format of (5.4.30) by taking the z-transforms of (5.4.17):

$$\{z\boldsymbol{I} - \boldsymbol{\Phi}\}\boldsymbol{x}(z) = \boldsymbol{\Gamma}u(z) \tag{5.4.31}$$

$$y(z) = \boldsymbol{cx}(z) \tag{5.4.32}$$

This leads to

$$G(z) = \frac{y(z)}{u(z)} = \boldsymbol{c}\{z\boldsymbol{I} - \boldsymbol{\Phi}\}^{-1}\boldsymbol{\Gamma} \tag{5.4.33}$$

It will be useful at this point to give some physical meaning to the variable z. To arrive at this meaning, consider the case where all coefficients in (5.4.29) are equal to 0, with the exception of b_1, which will be set equal to 1. With this simplification,

$$G(z) = \frac{y(z)}{u(z)} = z^{-1} \tag{5.4.34}$$

However, $G(z)$ represents the transform of the difference equation (5.4.4). Substituting the coefficient values of 1 for b_1 and 0 for all others into (5.4.4) produces

$$y(k) = u(k-1) \tag{5.4.35}$$

Therefore, for this case, the output signal is equal to the input signal delayed by one sample period. From this it can be deduced that the transfer function z^{-1} is equal to a delay of one sample period, or a unit delay. This is shown in block form in Fig. 5.10. It follows that z^{-2} is a delay of two samples, z^{-3} is a delay of three samples, etc.

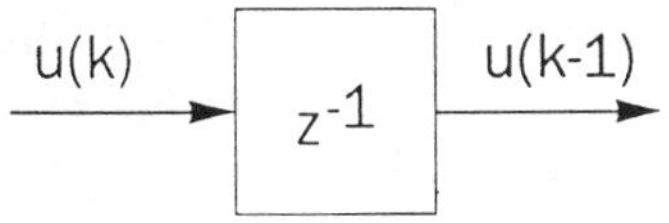

Fig. 5.10 z-operator, or unit delay.

It will also be helpful to give some further meaning as to what the transfer function $G(z)$ physically describes. The transfer function relates the input signal $u(z)$ to the output signal $y(z)$ by

$$y(z) = G(z)u(z) \tag{5.4.36}$$

Now, the z-transformed input signal is defined in (5.4.21) as

$$u(z) = \sum_{k=-\infty}^{\infty} u(k)\, z^{-k} \tag{5.4.37}$$

If the input signal $u(k)$ is a Kronecker delta function δ, defined as

$$\delta(k) = \begin{cases} 1 & for\ k=0 \\ 0 & for\ k \neq 0 \end{cases} \tag{5.4.38}$$

then it follows that the z-transform of the input signal is

$$u(z) = z\{\delta(k)\} = 1 \tag{5.4.39}$$

Substituting this into (5.4.36), the z-transformed output signal $y(z)$ related to this input signal is

$$y(z) = G(z) \tag{5.4.40}$$

Thus, the transfer function $G(z)$ is the transform of the response of the system to a unit pulse input, or the impulse response.

It should be pointed out here that the discussion of the use and properties of the z-transform given in this section is extremely limited. For a more thorough discussion, the reader is referred to any introductory digital control text such as Franklin *et al.* (1990), to a dedicated text such as Stearns (1975), or to an introductory digital signal processing text such as Oppenheim and Schafer (1975) or Rabiner and Gold (1975).

5.4.4 Transfer function realization in a digital filter

The transfer function representation of the system dynamics developed using linear difference equations can be practically realized in a digital filter. This can be done in either software, hardware (such as by using specialized signal processing chips with multipliers, accumulators and unit delays), or a combination of the two. This realization can be developed from a block diagram representation of the transfer function. We will describe two of the more common digital filter arrangements, which are classified by the duration of the impulse response; infinite impulse response filters, and finite impulse response filters.

Consider the transfer function representation of (5.4.30) which has n 'poles' in the denominator and m 'zeros' (the physical significance of poles and zeros will be discussed later in this chapter) in the numerator. Fig. 5.11 illustrates a block diagram directly realizing this. This type of realization is sometimes referred to as 'direct', in which the denominator and numerator of the transfer function have separate delay chains (series of delay elements). For the case considered here, this means that $(m+n)$ unit delays are employed.

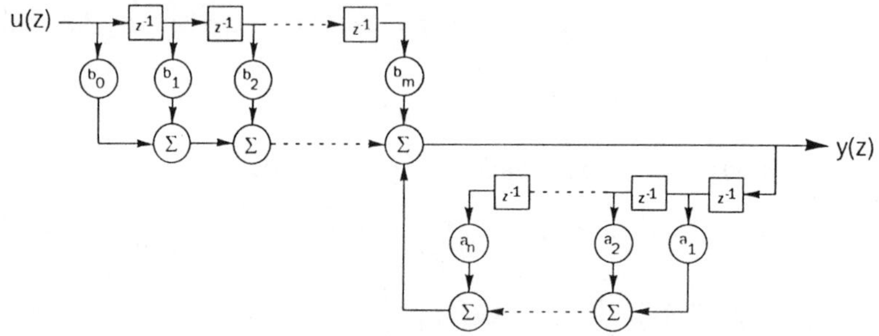

Fig. 5.11 Direct realization of a digital transfer function in a digital filter.

The number of unit delays utilized in the transfer function realization can be reduced from $(m + n)$ to simply m or n, whichever is greater. This reduction to the minimum possible number of delay elements is sometimes referred to as standard programming. To re-express the transfer function in a form amenable to this reduced architecture, (5.4.29) defining $G(z)$ must be rewritten as

$$G(z) = \frac{y(z)}{u(z)} = \frac{y(z)}{G(z)}\frac{G(z)}{u(z)}$$
$$= (b_0 + b_1 z^{-1} + \cdots + b_m z^{-m})\frac{1}{1 - a_1 z^{-1} - a_2 z^{-1} - \cdots - a_n} \tag{5.4.41}$$

where

$$\frac{y(z)}{G(z)} = b_0 + b_1 z^{-1} + \cdots + b_m z^{-m} \tag{5.4.42}$$

and

$$\frac{G(z)}{u(z)} = \frac{1}{1 - a_1 z^{-1} - \cdots - a_n z^{-n}} \tag{5.4.43}$$

Equation (5.4.42) can be rewritten as

$$y(z) = b_0 G(z) + b_1 z^{-1} G(z) + \cdots + b_m z^{-m} G(z) \tag{5.4.44}$$

and (5.4.43) as

$$G(z) = u(z) + a_1 z^{-1} G(z) + \cdots + a_n z^{-n} G(z) \tag{5.4.45}$$

Equations (5.4.44) and (5.4.45) can be realized in two block forms. One of these, shown in Fig. 5.12, is referred to as observer canonical form because the feedback loops all originate from the output or observed signal. It is a direct canonical form, as the coefficients used in the architecture are obtained directly from the transfer expression. The second canonical form, shown in Fig. 5.13, is referred to as the control canonical form, as all feedback loops return to the input of the control signal. As with the observer canonical architecture, it is a direct canonical form, with the coefficients being obtained directly from the transfer function expression.

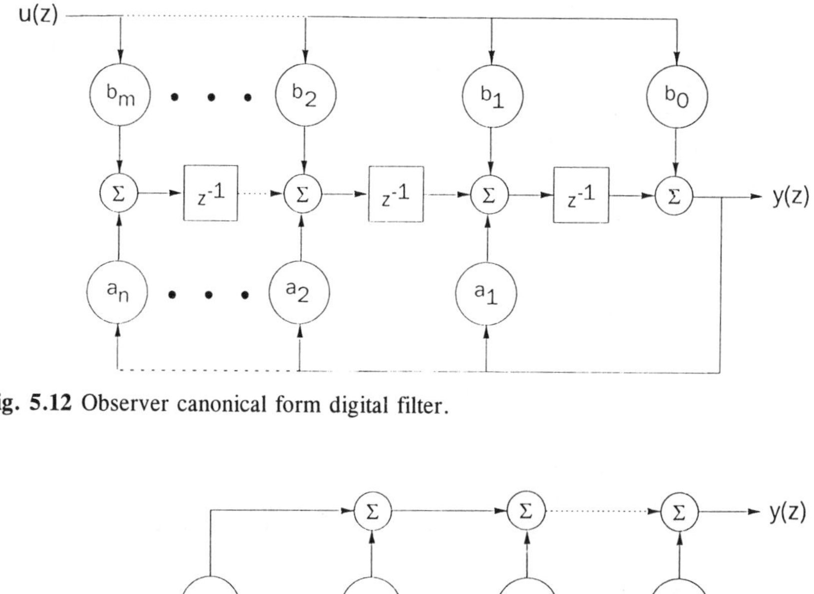

Fig. 5.12 Observer canonical form digital filter.

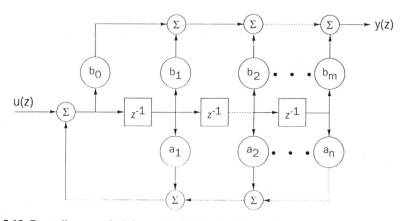

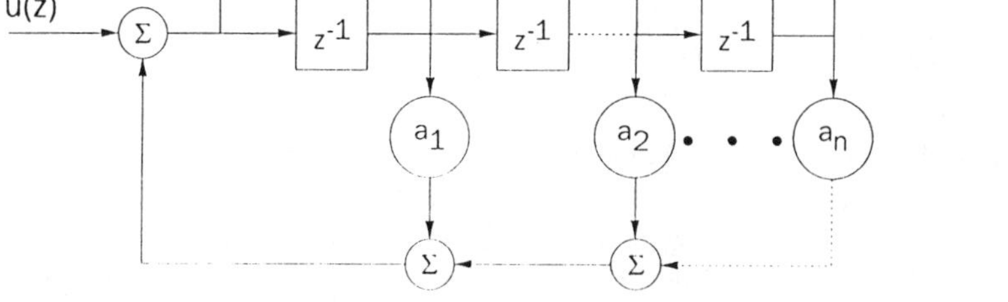

Fig. 5.13 Controller canonical form digital filter.

The impulse response of the filter architectures illustrated in Figs. 5.11–5.13, assuming that not all of the feedback loop coefficients (the a's) are zero, will be infinite in duration (although the magnitudes of the signal samples may become infinitely small as the time becomes infinitely large). As such, these filters are referred to as infinite impulse response (IIR) filters. They may also be called recursive filters, because the previous values of the output signal are used in the calculation of the current output signal. For the special case where all of the feedforward loop coefficients (the b's) are equal to zero, the digital filters of Figs. 5.11–5.13 are known as all-pole filters, and the transfer function is referred to as an all-pole model.

Consider now the case where all of the coefficients in the denominator of the transfer function are set equal to zero, or

$$G(z) = \frac{y(z)}{u(z)} = b_0 + b_1 z^{-1} + \cdots + b_m z^{-m} \tag{5.4.46}$$

This case may occur if there are no poles in the transfer function, or if the division of the transformed output by the transformed input is undertaken, and the resulting infinite series truncated. The realization of this transfer function representation is illustrated in block form in Fig. 5.14. The impulse response of this filter architecture will be finite in duration, hence the filter is referred to as a finite impulse response (FIR) filter. It may also be called a non-recursive filter, or an all-zero filter, as previous output samples are not used in calculating the value of the present output sample, or a moving average filter.

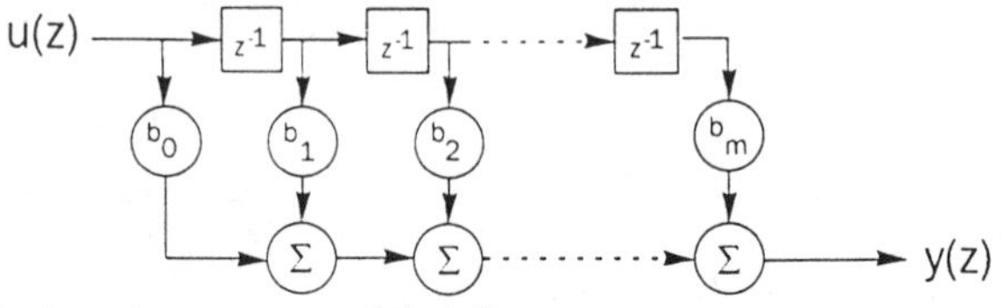

Fig. 5.14 Finite impulse response digital filter.

Infinite impulse response filters and finite impulse response filters are commonly used tools in digital signal processing, and are used in the majority of practically implemented active noise and vibration control systems. Characteristics of these architectures will be considered in more depth in the chapter on feedforward control, Chapter 6.

5.4.5 System identification using digital filters

We have now determined that dynamic, continuous time systems can be emulated using discrete time models which can be implemented readily as digital filters. For the remainder of this section we will expand upon this idea, looking at the mechanics of how such a model may be derived and looking at the problem of system identification (identification of the parameters of a model of the system). System identification can be divided into two categories: parametric and non-parametric. Parametric system identification is directed towards obtaining a mathematical model involving a set of parameters. These parameters could be, for example, the coefficients of a discrete transfer function. Non-parametric system identification is directed towards obtaining the frequency response of the system, such as would be used in classical control system analysis methods described in the next section. We will be concerned here with the first of these two categories.

We will examine two types of parametric system identification algorithms in this section: batch processing and recursive processing. Recursive system identification has the advantage of being suitable for implementation in real-time systems (systems which produce an output sample in response to each input sample). We will restrict ourselves to consideration of algorithms which are explicitly related to the adaptive signal processing algorithms which receive widespread use in active noise and vibration control, algorithms which are related to those outlined in Chapter 6.

The work outlined in the remainder of this section is, by necessity, representative of only a small portion of that available on the topic of system identification. The interested reader is referred to several texts dedicated to various aspects of the topic (Goodwin and Payne, 1977; Ljung and Söderström, 1983; Goodwin and Sin, 1984; Norton, 1986; Ljung, 1987; Söderström and Stoica, 1989).

5.4.5.1 Least squares prediction

As we derived previously in this section, the response of a dynamic system can be represented by a discrete transfer function as stated in (5.4.29), where the coefficients a_1, $a_2 \ldots a_n$ and b_0, $b_1 \ldots b_m$, are the parameters of the system. The aim of system identification is to derive coefficient values which will enable the output of the system model to most closely match the output of the actual dynamic system, for the same input sequence. The difference between the two, e, shown in Fig. 5.15, is referred to as the estimation or prediction error, or the residual.

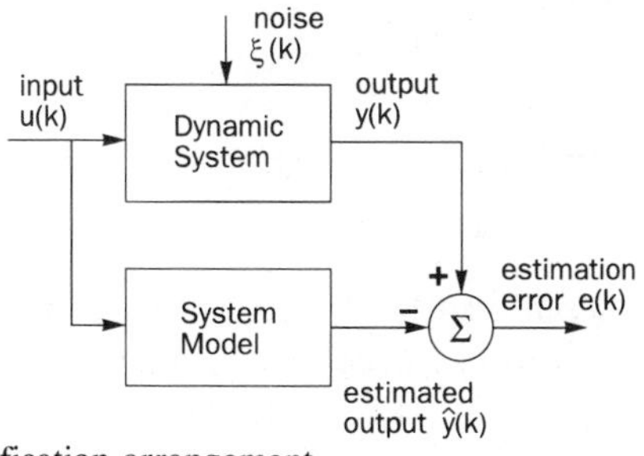

Fig. 5.15 System identification arrangement.

One important point to recognized about Fig. 5.15 is that if the system identification problem were arranged in such a way that the error criterion was minimization of the estimation error, then the solution would force the estimation error to assume some large negative value. This, of course, would be of little practical use. What is of interest to us is not that the actual (signed) value of the estimation error becomes small, but rather that the magnitude of the estimation error becomes small. Therefore, minimizing the square of the prediction error, rather than the prediction error itself, has much more utility as an error criterion. (This fact was probably first recognized by Karl Friedrich Gauss in the late 18th century, when he developed the basic concepts of least squares estimation, and applied it towards determining the orbits of the planets (Sorenson, 1970).) This minimization of the squared error leads to the name least squares prediction.

The (commonly used) system identification models which will be considered here are often referred to as ARMAX, or autoregressive moving average with an exogenous signal, models, a name arising from the statistics roots of the field. Two equivalent block diagrams of the ARMAX model are shown in Fig. 5.16, where $v(k)$ is an additional noise variable, and $c(z)/a(z)$ has the form of a discrete transfer function. In our initial development of the least squares (or linear regression) problem, we will neglect the noise term $v(k)$.

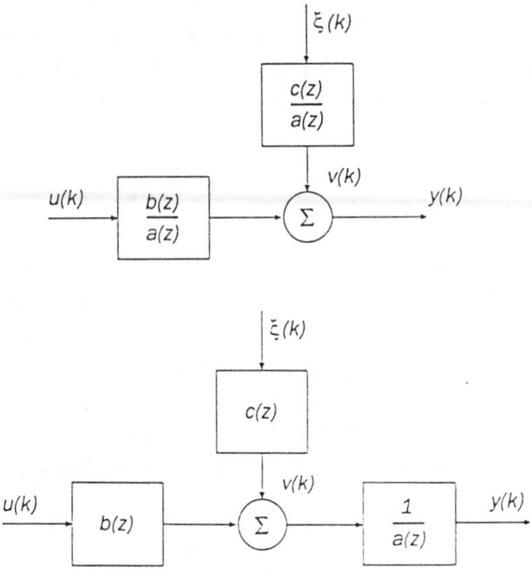

Fig. 5.16 ARMAX system indentification model.

Expanding the discrete transfer function of (5.4.4), provides the following expression for the output of the model which estimates the output of the actual dynamic system at time k, $\hat{y}(k)$, in response to the input $u(k)$:

$$\hat{y}(k) = a_1\hat{y}(k-1) + a_2\hat{y}(k-2) + \cdots + a_n\hat{y}(k-n) + b_0u(k) + b_1u(k-1) + \cdots + b_mu(k-m) \tag{5.4.47}$$

This can be expressed in matrix form as

$$\hat{y}(k) = \boldsymbol{\phi}^{\mathrm{T}}(k)\boldsymbol{\theta} \tag{5.4.48}$$

where $\boldsymbol{\phi}(k)$ is the vector of regressor variables, which are known, and $\boldsymbol{\theta}$ is the vector of (unknown) parameters:

$$\boldsymbol{\phi}^{\mathrm{T}}(k) = [\hat{y}(k-1) \quad \hat{y}(k-2) \cdots \hat{y}(k-n) \quad u(k) \quad u(k-1) \cdots u(k-m)] \tag{5.4.49}$$

$$\boldsymbol{\theta} = [a_1 \quad a_2 \cdots a_n \quad b_0 \quad b_1 \cdots b_m]^{\mathrm{T}} \tag{5.4.50}$$

The model of (5.4.48) is referred to as a regression model.

The estimation error $e(k)$ is defined as the difference between the actual dynamic system output $y(k)$ and the estimated system output $\hat{y}(k)$:

$$e(k) = y(k) - \hat{y}(k) = y(k) - \boldsymbol{\phi}^{\mathrm{T}}(k)\boldsymbol{\theta} \tag{5.4.51}$$

As the output of the system model depends upon past data up to n samples ago (assuming the order n of the denominator of the transfer function is greater than or equal to the order m of the numerator of the transfer function), the first 'legitimate' test of the model can only be conducted at time n. Hence, the first estimation error is $e(n)$. Now, suppose we have a known set of input and output data up to the period N. We can then formulate a vector of estimation errors, or a residual vector, $\boldsymbol{e}$, as follows:

$$\boldsymbol{e}(N) = [e(n) \quad e(n+1) \quad \cdots \quad e(N)]^{\mathrm{T}} \tag{5.4.52}$$

and the associated quantities $\boldsymbol{y}$, $\hat{\boldsymbol{y}}$, and $\boldsymbol{\Phi}$,

$$\boldsymbol{y}(N) = [y(n) \quad y(n+1) \cdots y(N)]^{\mathrm{T}} \tag{5.4.53}$$

$$\hat{\boldsymbol{y}}(N) = [\hat{y}(n) \quad \hat{y}(n+1) \cdots \hat{y}(N)]^{\mathrm{T}} \tag{5.4.54}$$

$$\boldsymbol{\Phi}(N) = \begin{bmatrix} \boldsymbol{\phi}^{\mathrm{T}}(n) \\ \boldsymbol{\phi}^{\mathrm{T}}(n+1) \\ \vdots \\ \boldsymbol{\phi}^{\mathrm{T}}(N) \end{bmatrix} \tag{5.4.55}$$

In terms of these quantities, the residual vector can be expressed as

$$\boldsymbol{e}(N) = \boldsymbol{y}(N) - \hat{\boldsymbol{y}}(N) = \boldsymbol{y}(N) - \boldsymbol{\Phi}(N)\boldsymbol{\theta} \tag{5.4.56}$$

The error criterion J, to be minimized, is the sum of the squares of the residuals:

$$J(\theta,N) = \sum_{i-n}^{N} e^2(i) \tag{5.4.57}$$

In terms of the quantities defined in (5.4.52) to (5.4.55),

$$J(\theta,N) = \boldsymbol{e}^{\mathrm{T}}(N)\boldsymbol{e}(N) \tag{5.4.58}$$

or

$$J(\boldsymbol{\theta},N) = [\boldsymbol{y}(N)-\boldsymbol{\Phi}(N)\boldsymbol{\theta}]^{\mathrm{T}} [\boldsymbol{y}(N) - \boldsymbol{\Phi}(N)\boldsymbol{\theta}]$$
$$= \boldsymbol{y}^{\mathrm{T}}(N)\boldsymbol{y}(N) - \boldsymbol{y}^{\mathrm{T}}(N)\boldsymbol{\Phi}(N)\boldsymbol{\theta} - \boldsymbol{\theta}^{\mathrm{T}}\boldsymbol{\Phi}^{\mathrm{T}}(N)\boldsymbol{y}(N) + \boldsymbol{\theta}^{\mathrm{T}}\boldsymbol{\Phi}^{\mathrm{T}}(N)\boldsymbol{\Phi}(N)\boldsymbol{\theta} \tag{5.4.59}$$

From (5.4.59) it can be seen that the error criterion, J, is a quadratic function of the unknown parameters $\boldsymbol{\theta}$. As the matrix $\boldsymbol{\Phi}^{\mathrm{T}}\boldsymbol{\Phi}$ is positive definite, the quadratic expression must have a minimum, corresponding to the optimum estimate of the vector of (unknown) parameters, $\hat{\boldsymbol{\theta}}$. This optimum vector estimate can be found in two ways. First, it can be found directly by rewriting (5.4.58) and (5.4.59) as

$$J(\boldsymbol{\theta},N) = \boldsymbol{e}^{\mathrm{T}}(N)\boldsymbol{e}(N) + \boldsymbol{y}^{\mathrm{T}}(N)\boldsymbol{\Phi}(N)[\boldsymbol{\Phi}^{\mathrm{T}}(N)\boldsymbol{\Phi}(N)]^{-1}\boldsymbol{\Phi}^{\mathrm{T}}(N)\boldsymbol{y}(N)$$
$$- \boldsymbol{y}^{\mathrm{T}}(N)\boldsymbol{\Phi}(N)[\boldsymbol{\Phi}^{\mathrm{T}}(N)\boldsymbol{\Phi}(N)]^{-1}\boldsymbol{\Phi}^{\mathrm{T}}(N)\boldsymbol{y}(N)$$
$$= \boldsymbol{y}^{\mathrm{T}}(N)[\boldsymbol{I} - \boldsymbol{\Phi}(N)[\boldsymbol{\Phi}^{\mathrm{T}}(N)\boldsymbol{\Phi}(N)]^{-1}\boldsymbol{\Phi}^{\mathrm{T}}(N)]\boldsymbol{y}(N)$$
$$+ [\boldsymbol{\theta} - [\boldsymbol{\Phi}^{\mathrm{T}}(N)\boldsymbol{\Phi}(N)]^{-1}\boldsymbol{\Phi}^{\mathrm{T}}(N)\boldsymbol{y}(N)]^{T}\boldsymbol{\Phi}^{\mathrm{T}}(N)\boldsymbol{\Phi}(N)[\boldsymbol{\theta} - [\boldsymbol{\Phi}^{\mathrm{T}}(N)\boldsymbol{\Phi}(N)]^{-1}\boldsymbol{\Phi}^{\mathrm{T}}(N)\boldsymbol{y}(N)] \tag{5.4.60}$$

As the first term on the right-hand side of (5.4.60) is not a function of $\boldsymbol{\theta}$, $\hat{\boldsymbol{\theta}}$ can be found by minimizing the second term on the right-hand side. Noting that the two bracketed parts of this term are identical, $\hat{\boldsymbol{\theta}}$ will cause these to be equal to zero (least square); thus,

$$\hat{\boldsymbol{\theta}} = [\boldsymbol{\Phi}^{\mathrm{T}}(N)\boldsymbol{\Phi}(N)]^{-1}\boldsymbol{\Phi}^{\mathrm{T}}(N)\boldsymbol{y}(N) \tag{5.4.61}$$

or

$$\boldsymbol{\Phi}^{\mathrm{T}}(N)\boldsymbol{\Phi}(N)\hat{\boldsymbol{\theta}} = \boldsymbol{\Phi}^{\mathrm{T}}(N)\boldsymbol{y}(N) \tag{5.4.62}$$

Equation (5.4.62) is referred to as the normal equation, and its solution provides the optimum parameter vector estimate $\hat{\boldsymbol{\theta}}$, which will minimize the magnitude of the estimation error.

The second method used to derive the equation for the optimum parameter vector is to differentiate (5.4.59) with respect to $\boldsymbol{\theta}$, and set the gradient equal to zero for the minimum. Differentiating produces

$$\frac{\partial J(\boldsymbol{\theta},N)}{\partial \boldsymbol{\theta}} = -2\boldsymbol{y}^{\mathrm{T}}(N)\boldsymbol{\Phi}(N) + 2\boldsymbol{\theta}^{\mathrm{T}}\boldsymbol{\Phi}^{\mathrm{T}}(N)\boldsymbol{\Phi}(N) \tag{5.4.63}$$

Setting this equal to zero gives

$$\boldsymbol{y}^{\mathrm{T}}(N)\boldsymbol{\Phi}(N) = \boldsymbol{\theta}^{\mathrm{T}}\boldsymbol{\Phi}^{\mathrm{T}}(N)\boldsymbol{\Phi}(N) \tag{5.4.64}$$

Transposing this again produces the normal equation

$$\boldsymbol{\Phi}^{\mathrm{T}}(N)\boldsymbol{\Phi}(N)\hat{\boldsymbol{\theta}} = \boldsymbol{\Phi}^{\mathrm{T}}(N)\boldsymbol{y}(N) \tag{5.4.65}$$

Calculation of the optimum parameter vector using (5.4.62) assumes that the matrix $[\Phi^T\Phi]$ is invertible. This condition can be referred to as an excitation condition (Astrom and Wittenmark, 1989), arising from the fact that for invertibility, the input sequence $\boldsymbol{u}$ must be sufficiently time varying (Astrom and Eykhoff, 1971).

5.4.5.2 Application to state/space modelling

System identification techniques aim to obtain a model of a given system, a model which can be used in the development of control systems. Therefore, having determined the parameters of the discrete transfer function which will model a given dynamic system, the question arises of how this can be applied to a state space model to enable a control system to be designed. In discrete state/space form, the state and output equations are given by (5.4.17). For an nth order system, assuming for simplicity that $m = n$, the dimensions of $\boldsymbol{\Psi}$ are $(n \times n)$, of $\boldsymbol{\Gamma}$ are $(n \times 1)$, and of $\boldsymbol{c}$ are $(n \times 1)$. Thus, there are a total of $(n^2 + 2n)$ parameters to be estimated. If this system were represented by a discrete transfer function, however, there would only be $(2n)$ parameters to estimate. As we have already shown that the two representations were equivalent, it can be surmised that in the state space representation there are (n^2) parameters that are in some sense redundant, and may be chosen somewhat arbitrarily. We can use this to our advantage in determining the 'arrangement' of the state/space model to accommodate simplified parameter estimation. The standard arrangements arising from this are canonical forms discussed previously in this section. For example, arranged in the observer canonical form shown in Fig. 5.12 (without a direct transmission term b_0),

$$\boldsymbol{\Psi} = \begin{bmatrix} a_1 & 1 & 0 & \cdots & 0 \\ a_2 & 0 & 1 & \cdots & 0 \\ & & & \vdots & \\ a_{n-1} & 0 & 0 & \cdots & 1 \\ a_n & 0 & 0 & \cdots & 0 \end{bmatrix} \tag{5.4.66}$$

$$\boldsymbol{\Gamma} = \begin{bmatrix} b_1 \\ b_2 \\ \vdots \\ b_{n-1} \\ b_n \end{bmatrix} \tag{5.4.67}$$

and

$$\boldsymbol{c} = [1 \quad 0 \quad 0 \cdots 0 \quad 0] \tag{5.4.68}$$

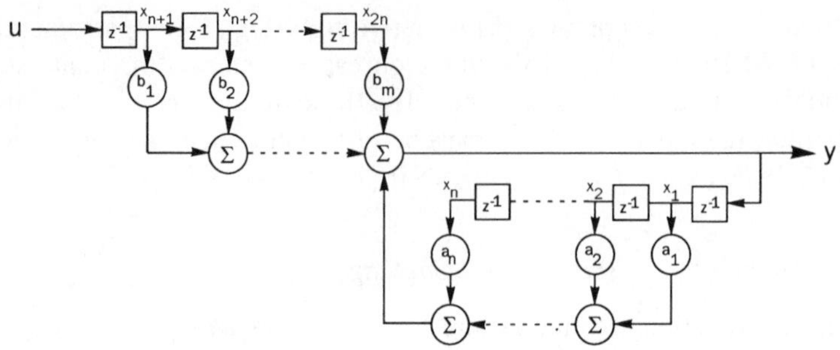

Fig. 5.17 ARMA system model.

Possibly the most commonly used canonical form, however, is the direct, or ARMA (autoregressive moving average) form, shown in Fig. 5.17. Again excluding the direct transmission term, the state matrices for this case are

$$\boldsymbol{\Psi} = \begin{bmatrix} a_1 & a_2 & \cdots & a_n & b_1 & b_2 & \cdots & b_n \\ 1 & 0 & \cdots & 0 & 0 & 0 & \cdots & 0 \\ 0 & 1 & \cdots & 0 & 0 & 0 & \cdots & 0 \\ & & & \vdots & & & & \\ 0 & 0 & \cdots & 0 & 0 & 0 & \cdots & 1 \end{bmatrix} \tag{5.4.69}$$

$$\boldsymbol{\Gamma} = [0 \ 0 \ \cdots \ 0 \ 1 \ 0 \ \cdots \ 0]^{\mathrm{T}} \tag{5.4.70}$$

and

$$\boldsymbol{c} = [a_1 \ \ a_2 \ \cdots \ a_n \ \ b_1 \ \ b_2 \ \cdots \ b_n] \tag{5.4.71}$$

The ARMA representation is not optimized in terms of the minimum number of states required for representation, having $2n$ states as opposed to n states in the observer and control canonical forms. However, it is the following form of the state vector that makes the ARMA representation so advantageous:

$$\boldsymbol{x}_{\mathrm{ARMA}} = [y(k-1) \ \ y(k-2) \ \cdots \ y(k-n) \ \ u(k-1) \ \ u(k-2) \ \cdots \ u(k-n)]^{\mathrm{T}} \tag{5.4.72}$$

Equation (5.4.72) shows that when the model is expressed in ARMA format, the state vector is composed solely of past inputs and outputs, which can be measured readily. The output equation is now simply:

$$\begin{aligned} y(k) &= \boldsymbol{c}\boldsymbol{x}(k) \\ &= a_1 y(k-1) + a_2 y(k-2) + \cdots + a_n y(k-n) \\ &\quad + b_1 u(k-1) + b_2 u(k-2) + \cdots + b_n u(k-n) \end{aligned} \tag{5.4.73}$$

This equation is identical to that of the discrete transfer function representation given in (5.4.4). Therefore, arranged in ARMA form, the problem of identifying the state/space model parameters becomes identical to the problem of identifying the discrete transfer function model parameters. It is worthwhile here to (again) mention two points. First, any state variable representation of a system is not unique. This enables re-expression of a state/space model in ARMA form. Second, the aim of the form of system identification being outlined here is to develop the simplest model which will predict the dynamic response of the system of interest. In arranging the state equations into ARMA form, we are mixing together a conglomerate of physical variables. This, however, is often perfectly acceptable for the task at hand. If physical variables need to be identified, a modal analysis approach (such as that outlined in Chapter 4) is a better option.

5.4.5.3 Problems with least squares prediction

We have thus far neglected the external noise term $v(k)$ in our development of the linear regression problem. It does, however, have a significant impact upon the use of what are linear regression techniques for system identification. By introducing $v(k)$, the output of the system $y(k)$, becomes

$$y(k) = \boldsymbol{\phi}^{\mathrm{T}}(k)\boldsymbol{\theta} + v(k) \tag{5.4.74}$$

If the estimate of the system parameter vector $\hat{\boldsymbol{\theta}}$ provides a reasonably accurate model of the dynamic system response, then the estimation error $\boldsymbol{e}$ can be expressed as

$$e(k) = \hat{\boldsymbol{\theta}} - \boldsymbol{\theta} =$$

$$\left[\frac{1}{N}\sum_{k=1}^{N}\boldsymbol{\phi}(k)\boldsymbol{\phi}^{\mathrm{T}}(k)\right]^{-1}\left[\frac{1}{N}\sum_{k=1}^{N}\boldsymbol{\phi}(k)y(k) - \frac{1}{N}\sum_{k=1}^{N}\boldsymbol{\phi}(k)\boldsymbol{\phi}^{\mathrm{T}}(k)\boldsymbol{\theta}\right] \tag{5.4.75}$$

$$= \left[\frac{1}{N}\sum_{k=1}^{N}\boldsymbol{\phi}(k)\boldsymbol{\phi}^{\mathrm{T}}(k)\right]^{-1}\left[\frac{1}{N}\sum_{k=1}^{N}\boldsymbol{\phi}(k)v(k)\right]$$

As the number of samples N becomes large, the bracketed quantities in (5.4.75) tend towards their expected values. Therefore, for the estimation error to tend towards zero as the number of samples becomes large (that is, for the estimated parameter vector to tend towards the actual parameter vector), two factors must be true:

$$E\{\boldsymbol{\phi}(k)\boldsymbol{\phi}^{\mathrm{T}}(k)\} \text{ is non-singular,} \qquad E\{\boldsymbol{\phi}(k)v(k)\} = 0 \tag{5.4.76}$$

where $E\{\ \}$ denotes the expected value of the bracketed term. The first of these conditions is normally true, with a few exceptions being:

1. The input is not persistently exciting (sufficiently time varying).
2. The data are completely noise free and the model order is chosen too high.
3. The input is generated by a linear low order feedback from the output.

The second condition of (5.4.76), however, is often *not* true, the notable exception being when $v(k)$ is white noise. As a result, linear regression (or least squares prediction) will often give a biased estimate of the system parameters. In some cases this bias will be acceptably small, as when the signal to noise ratio $u(k)/v(k)$ is large. In other cases, however, the estimation error will be unacceptably large, and the system identification method will have to be modified to decrease the effect of bias. These modifications, which will be discussed in the next section, lead to the development of more general prediction error (generalized least squares) methods based on 'more detailed' model structures.

5.4.5.4 Generalized least squares estimation

As mentioned above in relation to Fig. 5.15, system identification can be viewed as a prediction process; that is, a model is developed which will enable the prediction of the response of a physical system (the output) to a given disturbance (the input). The predicted output is based upon previous values of both system inputs and outputs. As the majority of systems are stochastic in nature (must be described in terms of statistics), this prediction will not be exact. Therefore, the aim of the system identification exercise is to derive a parameter vector, θ, such that the prediction error, defined in (5.4.51), is small.

Consider the ARMAX model of Fig. 5.16. The estimated output of this system is

$$a(z)y(k) = b(z)u(k) + v(k) \tag{5.4.77}$$

where $v(k)$ is described by the autocorrelation function

$$v(k) = c(z)\xi(k) \tag{5.4.78}$$

In (5.4.78) $c(z)$ is a discrete transfer function and $\xi(k)$ is a white noise input. Equation (5.4.77) can also be written in the form

$$a(z)y'(k) = b(z)u'(k) + \xi(k) \tag{5.4.79}$$

where

$$\begin{aligned} y'(k) &= \frac{1}{c(z)}y(k) \\ u'(k) &= \frac{1}{c(z)}u(k) \end{aligned} \tag{5.4.80}$$

The solution of (5.4.79) represents an ordinary least squares problem of the same type considered in the previous section. The problem is that $c(z)$ is usually not known, and must also be estimated. One iterative procedure to determine $a(z)$, $b(z)$ and $c(z)$ consists of the following steps:

1. Do an ordinary least squares fit of $a(z)$ and $b(z)$ of (5.4.30).
2. Analyse the residuals $v(k)$ and fit (using a least squares procedure similar to the one outlined previously):

$$\frac{1}{C(z)}v(k) = \xi(k) \tag{5.4.81}$$

3. Filter the inputs and outputs (through the estimate of $1/c(z)$) to get $y(k)$ and $u(k)$ from (5.4.33).
4. Repeat.

Greater detail of the steps involved for the multiple input, multiple output case can be found in Goodwin and Payne (1977) and Söderström and Stoica (1989).

5.4.5.5 Recursive least squares estimation

The least squares estimation procedures considered thus far for system identification assume that a set of data is available to perform the modelling algorithm off-line (it is not necessary to derive a model output in response to each new input sample). Often, however, it is necessary to undertake system identification on line, modifying the parameter estimates with each new sample (a procedure which will often be necessary for the adaptive feedforward control system implementations discussed in Chapter 6). This type of system identification is referred to as a recursive estimation.

Recursive system identification is usually implemented as part of an adaptive control system (see Fig. 5.18), feeding the system model (implicitly or explicitly) to the electronic controller. Recursive system identification is implemented in this role for a number of reasons:

1. The action of the controller is based upon the most recent model of the system.
2. Memory requirements for modelling are reduced, as not all input and output data are stored.
3. It is possible to track time varying system parameters.

To reformulate the least squares estimation problem for recursive implementation, let us define a variable $\boldsymbol{P}$:

$$\boldsymbol{P}(k) = (\boldsymbol{\Phi}^{\mathrm{T}}(k)\boldsymbol{\Phi}(k))^{-1} = \left[\sum_{i=1}^{k}\phi^{\mathrm{T}}(i)\phi(i)\right]^{-1} \tag{5.4.82}$$

From this, it follows that

$$\boldsymbol{P}^{-1}(k+1) = \boldsymbol{P}^{-1}(k) + \phi(k+1)\phi^{\mathrm{T}}(k+1) \tag{5.4.83}$$

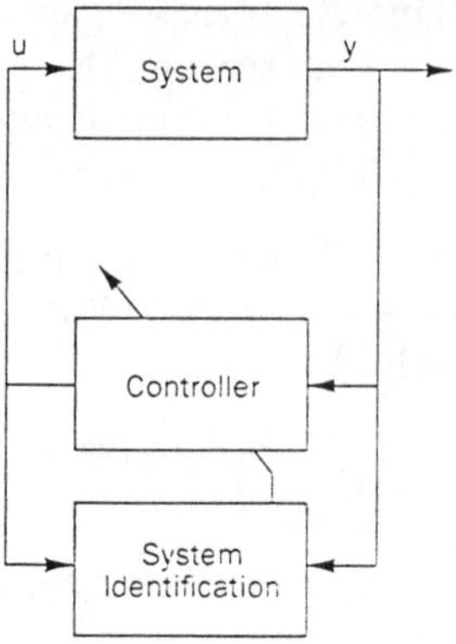

Fig. 5.18 Arrangement of system identification in an adaptive control system.

or

$$\boldsymbol{P}(k+1) = \left(\boldsymbol{P}^{-1}(k) + \phi(k+1)\phi(k)^{\mathrm{T}}\right)^{-1} \tag{5.4.84}$$

Note that by substituting $\boldsymbol{P}$ as defined in (5.4.82) into equation (5.4.62), the normal equation can be expressed as

$$\boldsymbol{P}^{-1}(k)\hat{\theta}(k) = \phi^{\mathrm{T}}(k)\boldsymbol{y}(k) \tag{5.4.85}$$

From (5.4.85) it follows that

$$\begin{aligned}\hat{\theta}(k+1) &\\ &= \hat{\theta}(k) - \boldsymbol{P}(k+1)\phi(k+1)\phi^{\mathrm{T}}(k+1)\hat{\theta}(k) + \boldsymbol{P}(k+1)\phi(k+1)y(k+1) \\ &= \hat{\theta}(k) + \boldsymbol{P}(k+1)\phi(k+1)[y(k+1) - \phi^{\mathrm{T}}(k+1)\hat{\theta}(k)]\end{aligned} \tag{5.4.86}$$

The bracketed part of the second term on the right-hand side of (5.4.86) is simply the estimation error at time $(k+1)$. Therefore,

$$\hat{\theta}(k+1) = \hat{\theta}(k) + \boldsymbol{P}(k+1)\phi(k+1)e(k+1) \tag{5.4.87}$$

or; defining

$$\kappa(k) = \boldsymbol{P}(k)\phi(k) \tag{5.4.88}$$

$$\hat{\theta}(k+1) = \hat{\theta}(k) + \kappa(k+1)e(k+1) \tag{5.4.89}$$

It is possible to derive a recursive relationship for $\boldsymbol{P}(k)$, using the matrix inversion lemma

$$[\boldsymbol{A} + \boldsymbol{BCD}]^{-1} = \boldsymbol{A}^{-1} - \boldsymbol{A}^{-1}\boldsymbol{B}[\boldsymbol{C}^{-1} + \boldsymbol{DA}^{-1}\boldsymbol{B}]^{-1}\boldsymbol{DA}^{-1} \tag{5.4.90}$$

and letting

$$\begin{aligned}\boldsymbol{A} &= \boldsymbol{P}^{-1}(k)\\ \boldsymbol{B} &= \phi(k+1)\\ \boldsymbol{C} &= \boldsymbol{I}\\ \boldsymbol{D} &= \phi^{\mathrm{T}}(k+1)\end{aligned} \tag{5.4.91}$$

From (5.4.86), the recursive relationship obtained is

$$P(k+1) = P(k) - P(k)\phi(k)[I + \phi^T(k+1)P(k)\phi(k+1)]^{-1}\phi^T(k+1)P(k) \tag{5.4.92}$$

From (5.4.88),

$$k(k+1) = P(k+1)\phi(k+1) = P(k)\phi(k+1)[I - \phi^T(k+1)P(k)\phi(k+1)]^{-1} \tag{5.4.93}$$

To formulate a recursive system identification procedure, all that is now required is the selection of some starting conditions. These are often taken as setting $\hat{\theta}(0) = 0$ and $P(0) = \alpha I$, where α is some large scalar value. A suggested suitably large value of α is (Franklin *et al.*, 1990)

$$\alpha = \frac{10}{N+1}\sum_{i=0}^{N} y^2(i) \tag{5.4.94}$$

Summarizing the operations of this section, a recursive system identification procedure consists of the following steps:

1. Select initial values of P, $\hat{\theta}$ as outlined above.
2. Collect input and output samples, $u(0)\ldots u(n)$, $y(0)\ldots y(n)$ to form ϕ^T.
3. Begin recursive loop. Calculate $k(k+1)$ using (5.4.93).
4. Collect input and output data, $u(k+1)$ and $y(k+1)$.
5. Calculate $\hat{\theta}(k+1)$ using (5.4.89).
6. Calculate $P(k+1)$ using (5.4.92).
7. Formulate $\phi(k+1)$.
8. Repeat from step (3).

There are several points about this system identification procedure which should be noted. First, note that the new estimate of the parameter vector, $\hat{\theta}(k+1)$, is made by adding a correction to the old estimate, $\hat{\theta}(k)$, based on the estimation error between the true dynamic system output and the previously predicted output. The matrix k determines the gain of this correction factor. Note also that no matrix inversion is required, as $\phi^T(k+1)P(k)\phi(k+1)$ is a scalar.

5.4.5.6 Inclusion of a forgetting factor

In systems targeted for active noise and vibration control, especially those implementing adaptive feedforward control (which, as outlined in Chapter 6, explicitly require system identification, usually of the form being outlined here), it is not uncommon to find system parameters which are time varying. In this case, it is often beneficial to include a forgetting factor into the algorithm, which will discount the value of the old data. With this addition, the error criterion becomes

$$J(\theta,N) = \sum_{i=n}^{N} \lambda^{N-i} e^2(i) \tag{5.4.95}$$

where λ is a forgetting factor, $0 < \lambda \leq 1$. Thus, older data is exponentially discounted. With this addition, the equations of steps three and six become (Astrom and Wittenmark, 1989):

$$\boldsymbol{k}(k+1) = \boldsymbol{P}(k)\boldsymbol{\phi}(k+1)[\lambda \boldsymbol{I} + \boldsymbol{\phi}^{\mathrm{T}}(k+1)\boldsymbol{P}(k)\boldsymbol{\phi}(k+1)]^{-1} \tag{5.4.96}$$

$$\boldsymbol{P}(k+1) = \frac{\boldsymbol{P}(k)[\boldsymbol{I} - \boldsymbol{k}(k+1)\boldsymbol{\phi}^{\mathrm{T}}(k+1)]}{\lambda} \tag{5.4.97}$$

5.4.5.7 Extended least squares algorithm

As with the batch processing least squares algorithm, the recursive least squares algorithm will give biased results when operating in the presence of correlated (with itself) noise. One way to help get around this problem is to implement an extended least squares algorithm.

Consider again the output of the ARMAX model of Fig. 5.16,

$$a(z)y(k) = b(z)u(k) + c(z)\xi(k) \tag{5.4.98}$$

This can be written as an extended (or pseudo-) linear regression problem

$$y(k) = \boldsymbol{\phi}^{\mathrm{T}}(k)\boldsymbol{\theta} + \xi(k) \tag{5.4.99}$$

where

$$\begin{aligned}\boldsymbol{\phi}^{\mathrm{T}}(k) = [&y(k-1) \;\; y(k-2) \cdots y(k-n) \;\; u(0) \;\; u(k-1) \;\; u(k-2) \\ &\cdots u(k-n) \;\; \xi(k-1) \;\; \xi(k-2) \cdots \xi(k-n)]\end{aligned} \tag{5.4.100}$$

$$\boldsymbol{\theta} = [a_1 \;\; a_2 \cdots a_n \;\; b_0 \;\; b_1 \;\; b_2 \cdots b_n \;\; c_1 \;\; c_2 \cdots c_n] \tag{5.4.101}$$

The problem with fitting a least squares estimation technique to this model is that the noise terms $\xi(k-1), \xi(k-2)\ldots\xi(k-n)$ are not known. They can, however, be approximated by the estimation errors, e. This leads to a recursive algorithm similar to the recursive least squares described in the start of this section, the terms of which are defined by

$$\begin{aligned}
\hat{\boldsymbol{\theta}}(k) &= \hat{\boldsymbol{\theta}}(k-1) + \boldsymbol{k}(k)e(k) \\
e(k) &= y(k) - \boldsymbol{\phi}^{\mathrm{T}}(k)\hat{\boldsymbol{\theta}}(k-1) \\
\boldsymbol{k}(k) &= \boldsymbol{P}(k)\boldsymbol{\phi}(k) = \boldsymbol{P}(k-1)\boldsymbol{\phi}(k)[\boldsymbol{I} + \boldsymbol{\phi}^{\mathrm{T}}(k)\boldsymbol{P}(k-1)\boldsymbol{\phi}(k)]^{-1} \\
\boldsymbol{P}(k) &= \boldsymbol{P}(k-1) - \boldsymbol{P}(k-1)\boldsymbol{\phi}(k)\boldsymbol{\phi}^{\mathrm{T}}(k)\boldsymbol{P}(k-1)[\boldsymbol{I} + \boldsymbol{\phi}^{\mathrm{T}}(k)\boldsymbol{P}(k-1)\boldsymbol{\phi}(k)] \\
\boldsymbol{\phi}(k) &= [y(k-1) \;\; y(k-2) \cdots y(k-n) \;\; u(k) \;\; u(k-1) \\
&\quad \cdots u(k-n) \;\; e(k-1) \;\; e(k-2) \cdots e(k-n)]^{\mathrm{T}}
\end{aligned} \tag{5.4.102}$$

5.4.5.8 Stochastic gradient algorithm

It is common for the recursive least squares algorithm to be shortened in practice for ease of implementation. This shortening leads to what is often referred to as the stochastic gradient algorithm.

Consider the recursive parameter estimation of (5.4.89). If $\boldsymbol{k}$ is now set equal to

$$\boldsymbol{k}(k) = \boldsymbol{\phi}(k)\frac{\boldsymbol{Q}}{r} \tag{5.4.103}$$

where $\boldsymbol{Q}$ is a (square) weighting matrix and r is a scalar, the general form of the stochastic gradient algorithm is produced.

Example 5.5

One of the most popular stochastic gradient algorithms is the least mean square or LMS algorithm, which will be considered in depth in Chapter 6. This algorithm is for use in a finite impulse response filter, where

$$\begin{aligned}\boldsymbol{\phi} &= [u(k) \quad u(k-1) \; \cdots \; u(k-n)]^{\mathrm{T}} \\ \hat{\boldsymbol{\theta}} &= [b_0 \quad b_1 \; \cdots \; b_n]^{\mathrm{T}}\end{aligned} \tag{5.4.104}$$

For the LMS algorithm, the weighting matrix $\boldsymbol{Q}$ is usually omitted and r is set equal to a constant, producing (effectively)

$$\hat{\boldsymbol{\theta}}(k+1) = \hat{\boldsymbol{\theta}}(k) + \frac{\boldsymbol{\phi}(k)e(k)}{r} \tag{5.4.105}$$

Example 5.6

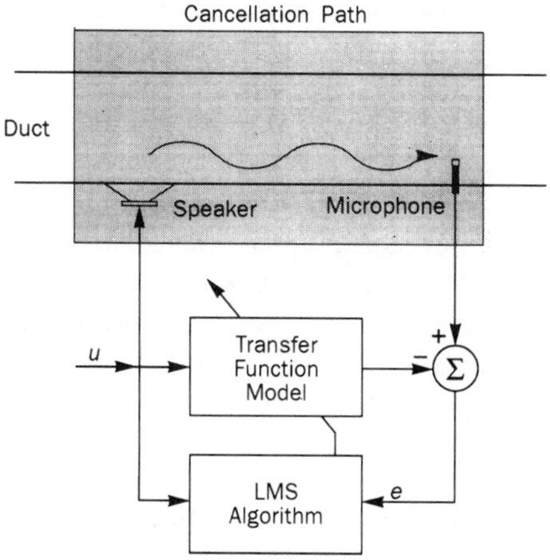

Fig. 5.19 Modelling of the cancellation path transfer function in an active control system.

As will be discussed in Chapter 6, for stable operation of adaptive feedforward controllers in active noise or vibration control systems, the transfer function of the 'cancellation path' must be identified (see Fig. 5.19). One of the first practical implementations of this requirement was achieved using the LMS algorithm in a single channel duct active noise control system (Poole *et al.*, 1984). With this system, an input signal was sent to both the control source and an FIR filter, used to model the system. The output of the model, $\hat{y}(k)$, was then compared to the actual system output measured by the error microphone, $y(k)$, with the difference being the estimation error. The weight coefficients of the FIR system model were then updated using the LMS algorithm, being held to fixed values when adaptation was complete.

There is, however, a problem which limits the practical viability of this method of error path system identification. This problem arises because the system modelling was conducted while the primary noise disturbance was switched off, with the final values of the system model weight coefficients being hard-wired into the system. The cancellation path, however, will have time varying parameters owing to changes in environmental variables such as airspeed and temperature, which will change the speed of sound in the system, and changes in mechanical variables, such as transducer response due to fungal growth! If the system model differs from the actual system by too great a margin (particularly, if the phase response is off by more than $\pm 90°$), then the adaptive control system will become unstable. To overcome this problem, it is necessary to conduct 'on-line' identification of the cancellation path (during operation of the adaptive control system), as in the next example.

Example 5.7
For the cancellation path system identification procedure of the previous section to be implemented continuously during the operation of the active control system, the input signal used for the identification must be uncorrelated with the other noise in the system (to avoid biasing both the system identification and the adaptive control system). The solution, therefore, is to introduce a random noise disturbance into the system and use it for system identification (Eriksson and Allie, 1988, 1989). (Note that this additional disturbance will add to the residual, uncontrolled disturbance, although normally the addition is small.) This random noise disturbance is fed into the control source with the control signal, and measured through the error sensor. The estimation error is then used in the LMS algorithm to adjust the model weights.

5.4.5.9 The projection algorithm

For the recursive least squares system identification procedure we have been considering, two state variable based matrices, $\hat{\boldsymbol{\theta}}$ and $\boldsymbol{P}$, must be updated at each iteration. The updating of $\boldsymbol{P}$ can become computationally intensive as the number of state variables used begins to increase. Therefore, several algorithms have been developed which avoid the need to update $\boldsymbol{P}$, at the expense of slower convergence time. One if these is the projection algorithm.

Before considering the projection algorithm, it will be useful to give some geometric significance to the problem of least squares estimation. Let us first re-express the equation defining the residuals vector, (5.4.51), as

$$\boldsymbol{e}(N) = \boldsymbol{y}(N) - \hat{\boldsymbol{y}}(N) = \boldsymbol{y}(N) - \boldsymbol{\Phi}(N)\boldsymbol{\theta} \tag{5.4.106}$$

or

$$\begin{bmatrix} e(n) \\ e(n+1) \\ \vdots \\ e(N) \end{bmatrix}$$

$$= \begin{bmatrix} y(n) \\ y(n+1) \\ \vdots \\ y(N) \end{bmatrix} - a_1 \begin{bmatrix} \hat{y}(n)-1 \\ \hat{y}(n+1)-1 \\ \vdots \\ \hat{y}(N)-1 \end{bmatrix} - a_2 \begin{bmatrix} \hat{y}(n)-2 \\ \hat{y}(n+1)-2 \\ \vdots \\ \hat{y}(N)-2 \end{bmatrix} \cdots - b_m \begin{bmatrix} u(n)-m \\ u(n+1)-m \\ \vdots \\ u(N)-m \end{bmatrix}$$

$$= \ddot{\boldsymbol{y}}(N) - a_1\boldsymbol{\phi}_{\text{col }1}(N) - a_2\boldsymbol{\phi}_{\text{col }2}(N) - \cdots - b_m\boldsymbol{\phi}_{\text{col }(m+n)}(N) \tag{5.4.107}$$

Equation (5.5.107) describes a geometric problem in $R^{(n+m+1)}$ vector space; that of finding the set of values of $\boldsymbol{\theta}(a_1,a_2,\ldots,a_n,b_0,b_1,\ldots,b_m)$, such that the vector $\boldsymbol{y}$ is best approximated by a linear combination of the column vectors of the matrix $\boldsymbol{\Phi}$, as shown in Fig. 5.20 (note that $(n+m+1)$ is equal to the number of elements in the parameter vector $\boldsymbol{\theta}$). Clearly, to uniquely determine $\boldsymbol{\theta}$, it will be necessary to make $(n+m+1)$ measurements of the system output.

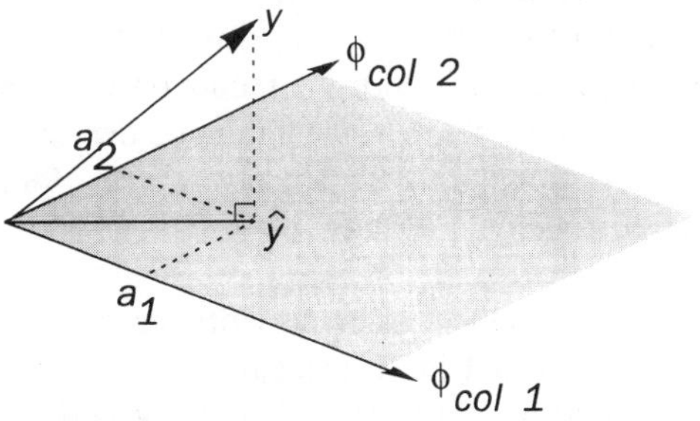

Fig. 5.20 Arrangement for derivation of the projection algorithm.

Consider now a single system output,

$$y(k) = \boldsymbol{\phi}^{\mathrm{T}}(k)\boldsymbol{\theta} \tag{5.4.108}$$

This single measurement describes the projection of the parameter vector, $\boldsymbol{\theta}$, on to the vector $\phi(k)$. If an estimate of the parameter vector, $\theta(k - 1)$, already exists, it can be used to improve upon the estimate

$$\hat{\theta}(k) = \hat{\theta}(k-1) + \mu\phi(k) \tag{5.4.109}$$

Substituting (5.5.109) into (5.5.108) produces

$$y(k) = \phi^{\mathrm{T}}(k)\hat{\theta}(k) = \phi^{\mathrm{T}}(k)\hat{\theta}(k-1) + \mu\phi^{\mathrm{T}}(k)\phi(k) \tag{5.4.110}$$

Rearranging this provides an expression for μ:

$$\mu = \frac{1}{\phi^{\mathrm{T}}(k)\phi(K)}(y(t) - \phi^{\mathrm{T}}(k)\hat{\theta}(k+1)) \tag{5.4.111}$$

Substituting this back into (5.4.109), we obtain

$$\hat{\theta}(k) = \hat{\theta}(k-1) + \frac{\phi(k)}{\phi^{\mathrm{T}}(k)\phi(k)}(y(k) - \phi^{\mathrm{T}}(k)\hat{\theta}(k-1)) \tag{5.4.112}$$

This is the projection algorithm (Goodwin and Sin, 1984), also referred to as Kaczmarz's algorithm (Astrom and Wittnemark, 1989). In practice, to avoid problems when ϕ contains all zero elements, the algorithm is implemented as

$$\hat{\theta}(k) = \hat{\theta}(k-1) + \frac{\phi(k)}{b + \phi^{\mathrm{T}}(k)\phi(k)}(y(k) - \phi^{\mathrm{T}}(k)\hat{\theta}(k-1)) \tag{5.4.113}$$

where b is some small scalar number. The projection algorithm is considered further in Chapter 6, in discussions on identification of the cancellation path transfer function, and in discussions of the normalized LMS algorithm.

5.4.5.10 A note on model order selection

Throughout this section it has been assumed that the order of the system (hence model) to be identified has been known. The question arises of how this order is determined. The usual procedure is basically trial and error, sometimes referred to as repeated least squares. This consists of identifying the system with increasing numbers of coefficients until the reduction in the error criterion, for adding more coefficients, dips below some minimum defined value, as shown in Fig. 5.21 (for a description, see Jategaonkar *et al.*, 1982). Other possible methods are to use correlation methods, described in Söderström and Stoica (1989).

Note also that in many active noise and vibration control systems there are explicit time delays which need to be identified. This can often best be done using a FIFO (first in, first out) buffer placed in front of the system model. The number of stages used in this buffer can also be estimated using the iterative procedure outlined above.

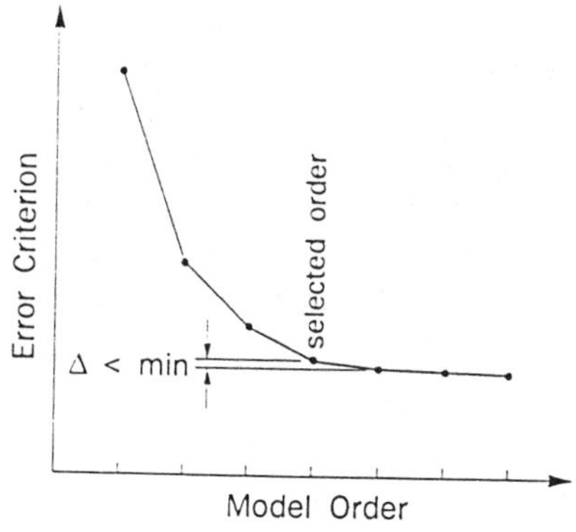

Fig. 5.21 Typical plot of minimum value of error criterion as a function of model order.

5.5 FREQUENCY DOMAIN ANALYSIS OF POLES, ZEROS AND SYSTEM RESPONSE

5.5.1 Introduction

The principal aim of this chapter is to review the concept of control system design and analysis using state space techniques. However, despite the level to which state space methods have been developed, and the natural way in which they apply to active control problems, there are a number of insights which arise from analysis in the frequency domain which are not readily apparent from state space analysis. In this section we will briefly consider some of these.

The fundamental concept used in frequency domain analysis of control systems is the transfer function, which as we have already noted defines the relationship between the Laplace transform $y(s)$ of the system output $y(t)$ and the Laplace transform $u(s)$ of the input signal $u(t)$

$$y(s) = G(s)u(s) \tag{5.5.1}$$

where $G(s)$ is the transfer function of the system. Equation (5.5.1) is valid for any linear time invariant (LTI) system, regardless of the order of the system. The Laplace transform variable is complex,

$$s = \sigma + j\omega \tag{5.5.2}$$

and is sometimes referred to as a complex frequency; frequency domain analysis techniques rely on the identification of s as such.

The frequency domain relationship of (5.5.1) can also be expressed in the time domain, as the convolution integral

$$y(t) = \int_0^t G(t-\tau)u(\tau)\, d\tau. \tag{5.5.3}$$

The convolution theorem can be used to go from (5.5.3) to (5.5.1) by enabling an expression for $G(s)$ to be derived. The convolution theorem states that the Laplace transform of a convolution of two functions is the product of the Laplace transforms of the two (individual) functions. Therefore

$$G(s) = \mathcal{L}\{G(t)\} = \int_0^\infty e^{-st} G(t)\, dt \tag{5.5.4}$$

To relate the concept of a transfer function to our previous work in this chapter, if the system of interest is expressed in standard state space format

$$\begin{aligned} \dot{x} &= \boldsymbol{A}x + \boldsymbol{B}u \\ y &= \boldsymbol{C}x + \boldsymbol{D}u \end{aligned} \tag{5.5.5}$$

the state equations can be written in a transfer function format by taking Laplace transforms

$$\begin{aligned} (s\boldsymbol{I}-\boldsymbol{A})x(s) &= \boldsymbol{B}u(s) \\ y(s) &= \boldsymbol{C}x(s) + \boldsymbol{D}u(s) \end{aligned} \tag{5.5.6}$$

This leads to

$$G(s) = \frac{y(s)}{u(s)} = \boldsymbol{C}(s\boldsymbol{I}-\boldsymbol{A})^{-1}\boldsymbol{B} + \boldsymbol{D} \tag{5.5.7}$$

which can be expanded as

$$G(s) = \boldsymbol{C}(s\boldsymbol{I}-\boldsymbol{A})^{-1}\boldsymbol{B} + \boldsymbol{D} = \frac{\boldsymbol{C}(\alpha_1 s^{k-1} + \alpha_2 s^{k-2} + \dots + \alpha_k)\boldsymbol{B}}{s^k + a_1 s^{k-1} + \dots + a_k} + \boldsymbol{D} \tag{5.5.8}$$

The denominator of the transfer function in (5.5.6) is the characteristic equation

$$|\, s\boldsymbol{I}-\boldsymbol{A}\,| = s^k + a_1 s^{k-1} + \dots + a_k \tag{5.5.9}$$

and the α's are the coefficient matrices of the adjoint matrix for the resolvent $(s\boldsymbol{I} - \boldsymbol{A})^{-1}$.

5.5.2 Block diagram manipulation

In many control systems, the system can be viewed as a set of non-interacting components, where the input of one is the output of another. One of the useful features of analysis in the frequency domain is that the overall transfer function of such an arrangement can be obtained by algebraic manipulation of the transfer functions of these 'components'. The three basic subsystem arrangements which are encountered are series transfer functions, parallel transfer functions, and feedback loops, as shown in Fig. 5.22 in the form of block diagrams. Block diagram representation is useful in that it leads to solution for the overall transfer function of the system by graphical simplification rather than algebraic manipulation, which is often simpler and more insightful.

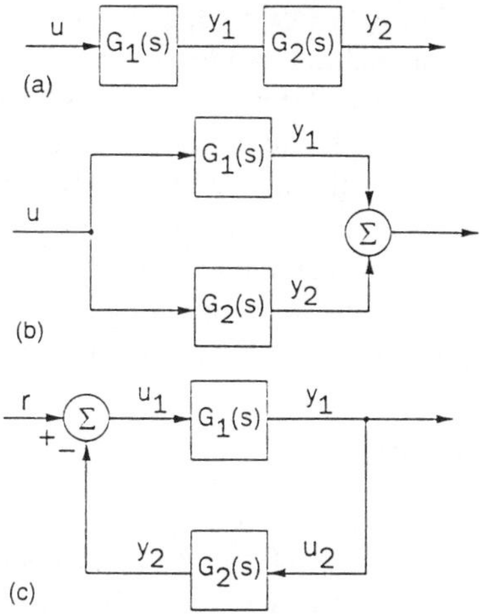

Fig. 5.22. Transfer functions: (a) in series; (b) in parallel; (c) in feedback loop.

For the arrangement shown in Fig. 5.22(a), where the transfer function G_1 is in series with the transfer function G_2, the overall transfer function of the system is given by the product G_1G_2. For the arrangement shown in Fig. 5.22(b), where the transfer function G_1 is in parallel with the transfer function G_2, the overall transfer function of the system is given by the sum of the two transfer functions G_1+G_2. These two results are easily obtained from consideration of the equations which describe the block diagrams. The feedback loop case is slightly more complicated, and will be of more interest in subsequent discussion, and so it is worthwhile deriving the result. To do this, observe that we have three simultaneous equations which describe the feedback arrangement shown in Fig. 5.22(c):

$$\begin{aligned} u_1(s) &= r(s) + y_2(s) \\ y_2(s) &= G_2(s)G_1(s)u_1(s) \\ y_1(s) &= G_1(s)u_1(s) \end{aligned} \tag{5.5.10}$$

It is straightforward to obtain the solution to these equations, as follows:

$$y_1(s) = \frac{G_1(s)}{1 - G_1(s)G_2(s)}r(s) \tag{5.5.11}$$

or, in the form of a transfer function relating output to input,

$$\frac{y_1(s)}{r(s)} = \frac{G_1(s)}{1 - G_1(s)G_2(s)} \tag{5.5.12}$$

Thus the transfer function of a single loop feedback arrangement is equal to the forward gain divided by 1 minus the loop gain, where the loop gain is the overall transfer function of the path that leads from a given variable back to the same variable, the loop path.

By repeated graphical simplification of the series, parallel and feedback loop subsystems which comprise an overall transfer function, it is often possible to derive the transfer function of a fairly complex system without performing a great deal of algebraic manipulation. An extended rule for the reduction of any block diagram was developed by Mason (1953, 1956), and can be found in many control texts. If we defined a path gain as the product of the component transfer functions which make up the path, and define that two paths touch if they have a common component, then for the special case where all forward paths and loop paths touch, Mason's rule states that the gain (transfer function) of the feedback system is given by the sum of the forward path gain (transfer function) divided by 1 minus the sum of the loop gains (transfer functions).

5.5.3 Control gain trade offs

The denominator of the transfer function of the system containing a feedback loop is of special interest, and is referred to as the return difference equation $F(s)$. Restricting ourselves to consideration of negative feedback gain, the return difference equation is

$$F(s) = 1 + G_1(s)G_2(s) \tag{5.5.13}$$

The zeros of the determinant of the return difference equation are the poles of the system. The return difference equation will become important in the next section when considering the stability of a system.

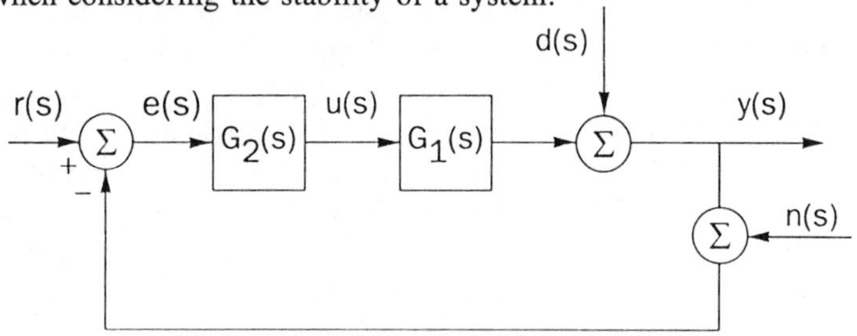

Fig. 5.23. Typical feedback control loop.

Before we begin our examination of system stability, it will be useful to discuss some of the trade offs associated with feedback gains. To do this, consider the control loop shown in Fig. 5.23, where r, e, u, y, d and n are the reference input, measured tracking error, plant input, plant output, disturbance input and measurement noise (as we mentioned in the introduction, in active control we are normally interested explicitly in disturbance attenuation, not in tracking a reference signal, so that r will usually be zero). In this diagram the transfer function G_2 represents the plant, or the system to be controlled, and the transfer function G_1 represents the controller. Using the block diagram manipulation rules we just discussed, it is straightforward to see that

$$y(s) = \frac{G_2(s)G_1(s)[r(s) - n(s)]}{1 + G_2(s)G_1(s)} + \frac{d(s)}{1 + G_2(s)G_1(s)} \tag{5.5.14}$$

$$e(s) = r(s) - y(s) - n(s) = \frac{r(s) - d(s)}{1 + G_2(s)G_1(s)} - \frac{n(s)}{1 + G_2(s)G_1(s)} \tag{5.5.15}$$

and

$$u(s) = \frac{G_1(s)[r(s)-n(s)-d(s)]}{1 + G_2(s)G_1(s)} \tag{5.5.16}$$

We will assume that the system is 'stable', the meaning of which we will discuss shortly.

Let us now define two quantities: the sensitivity function $S(s)$

$$S(s) = [1+G_2(s)G_1(s)]^{-1} \tag{5.5.17}$$

and the complementary sensitivity function $T(s)$

$$T(s) = \frac{G_2(s)G_1(s)}{1 + G_2(s)G_1(s)} \tag{5.5.18}$$

Note that the sensitivity function is the inverse of the return difference equation. Note also that $S(s)+T(s) = 1$. With these definitions we can make some observations which illustrate the trade offs associated with feedback gain selection:

1. From (5.5.15), for good tracking, such that $\|r\text{-}y\|$ is small when d and n are zero, the sensitivity function should be small (or the return difference equation large). This suggests that a large feedback loop gain is desirable.
2. From (5.5.14), for good suppression of the unwanted disturbance d, such that the resultant system output y is small, the sensitivity function should be small (or the return difference equation large). This again suggests a large feedback loop gain is desirable.
3. From (5.5.14), for good suppression of measurement noise n, such that the influence it has on the system output y is small, the complementary sensitivity function should be small (or the return difference equation small). Conversely from points (1) and (2), this suggests a small feedback loop gain is desirable. These observations demonstrate the need for a compromise between disturbance rejection and sensor noise rejection.

We can make a further observation with regard to the implications of having a large loop gain as suggested by points (1) and (2) above:

4. If the sensitivity function is small, then from (5.5.14)

$$u \approx G_2^{-1}(s)[r-n-d] \tag{5.5.19}$$

If this relationship holds outside of the bandwidth of the plant, where its frequency response magnitude is small, the control signal will become large and the control actuators could saturate.

These trade offs must be kept in mind when designing the control system.

5.5.4 Poles and zeros

Before considering the concepts of stability and various aspects of system response, it will be worthwhile to attach further meaning to the concepts of poles and zeros. A pole of the transfer function $G(s)$ is, by definition, a value of s such that the transfer function equation has a non-zero solution with a system input of zero. For a system described in transfer function notation of (5.5.1), the poles are defined as the values of s such that $u(s) = 0$ when $y(s) \neq 0$. For a system described in state space notation, (5.5.7) shows the poles to be defined by the eigenvalue expression

$$[s\boldsymbol{I}-\boldsymbol{A}]u(s) = 0 \tag{5.5.20}$$

or, in the standard form of an algebraic eigenvalue problem

$$|\, s\boldsymbol{I}-\boldsymbol{A} \,| = 0 \tag{5.5.21}$$

Example 5.8
To obtain a better picture of what poles physically are, consider again the simple second order system discussed previously in an example in Section 5.3. For this system, the homogeneous state space equation for the nth mode was derived as

$$\dot{\boldsymbol{x}}_n = \boldsymbol{A}_n \boldsymbol{x}_n \tag{5.5.22}$$

where

$$\boldsymbol{x}_n = \begin{bmatrix} x_n \\ \dot{x}_n \end{bmatrix}, \quad \boldsymbol{A}_n = \begin{bmatrix} 0 & 1 \\ -\omega_n^2 & 0 \end{bmatrix} \tag{5.5.23}$$

The eigenvalue expression for this system is

$$|\, s\boldsymbol{I}-\boldsymbol{A} \,| = s^2+\omega_n^2 = 0 \tag{5.5.24}$$

or

$$s = \pm j\omega_n \tag{5.5.25}$$

The poles of the system are the natural frequencies of the system plotted on the complex, or s, plane, and are therefore governed by physical system constraints, such as shape, material properties and boundary conditions.

Example 5.9
Consider the addition of damping in the second order system, such that its behaviour is now described by the expression

$$m\ddot{x}+2\zeta k^{1/2}m^{-1/2}\dot{x}+kx = 0 \tag{5.5.26}$$

The state space equation of the system for the nth mode can now be written directly as

$$\dot{\boldsymbol{x}}_n = \boldsymbol{A}_n\boldsymbol{x}_n \tag{5.5.27}$$

where

$$\boldsymbol{x}_n = \begin{bmatrix} x_n \\ \dot{x}_n \end{bmatrix}, \quad \boldsymbol{A}_n = \begin{bmatrix} 0 & 1 \\ -\omega_n^2 & -2\zeta\omega_n \end{bmatrix} \tag{5.5.28}$$

The poles of this system are defined by the characteristic equation

$$| s\boldsymbol{I}-\boldsymbol{A} | = s^2+2\zeta\omega_n s+\omega_n^2 = 0 \tag{5.5.29}$$

Solving this equation, the poles of the system are found to be

$$s = -\zeta\omega_n \pm j\omega_n\sqrt{1-\zeta^2} = -\zeta\omega_n \pm j\omega_d \tag{5.5.30}$$

where ω_d is the damped natural frequency. Thus, the poles are complex conjugates, the values of which are governed by the modal damping and the modal natural frequency.

The previous two examples show that the poles of the uncontrolled (homogeneous) system are directly related to the natural frequencies of the system. Poles have a significant effect upon the stability of the system. To quantify this effect, note again that the solution to the following homogeneous state equation

$$\dot{\boldsymbol{x}} = \boldsymbol{A}\boldsymbol{x} \tag{5.5.31}$$

has the form

$$\boldsymbol{x}(t) = \sum_{i=1}^{N} \boldsymbol{\Psi}_i e^{s_i t} \tag{5.5.32}$$

where s_i is the ith eigenvalue, or pole, as defined by (5.5.21), and ψ_i is its associated eigenvector. Both the eigenvalues and eigenvectors can be complex. Note from the form of (5.5.32), where the pole appears in an exponential term, that it is the real part of the eigenvalues, or poles, which will govern the stability of the system; the imaginary part of the system poles represents an oscillatory component of the response. Noting that a system is unstable if its response increases without bounds, a number of conclusions can be drawn. Specifically, a system is perturbed from its equilibrium position, and:

1. the real part of all system poles are negative, the solution to the state equations, governing the response of the system, asymptotically approaches

zero (the equilibrium point), and the system is asymptotically stable;
2. the real part of any pole is zero, the response of the system at the pole frequency neither increases nor decreases, and the system is said to be stable;
3. the real part of at least one pole is positive, the response of the system grows exponentially with no further input disturbance, so the system is unstable.

These three criteria can be envisaged most easily in terms of the complex plane, as shown on Fig. 5.24. Any poles which are on the left-hand side of the real axis will no cause the system to be unstable, and any poles on the right-hand side will cause the system to be unstable. The imaginary coordinate of the poles will determine the oscillatory component of the system response.

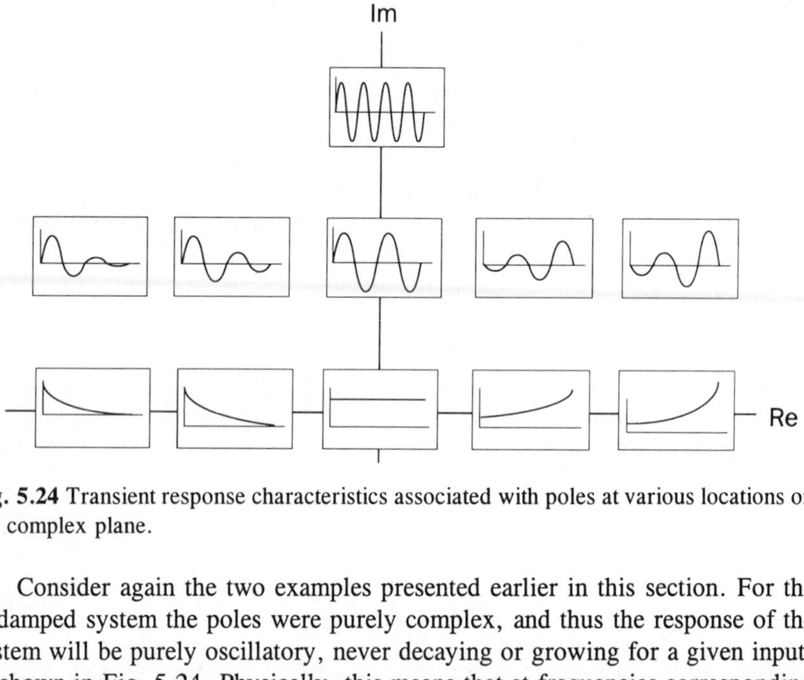

Fig. 5.24 Transient response characteristics associated with poles at various locations on the complex plane.

Consider again the two examples presented earlier in this section. For the undamped system the poles were purely complex, and thus the response of the system will be purely oscillatory, never decaying or growing for a given input, as shown in Fig. 5.24. Physically, this means that at frequencies corresponding to these poles energy can flow between the internal storage elements of compliance and inertia without loss. (The problem is, however, that because noise is present in practical systems, energy is continually entering this mode, and the response will continue to grow. Therefore, in practical terms the system will be unstable.) For the damped system, however, the poles are complex conjugates with negative real parts, the values of which are determined by the amount of damping and the resonance frequency. For small amounts of damping the response of the system will decay asymptotically towards zero, with an oscillatory component of frequency equal to the damped natural frequency of the system. However, if the value of damping increases to a value of 1.0 (critical damping) or more, the poles become purely negative real, and the response looses its oscillatory component, as shown in the figure.

System zeros are frequencies where a non-zero input produces a zero output, or in terms of the transfer function notation of (5.5.1) are values of s such that $y(s) = 0$ when $x(s) \neq 0$. In state space notation, they are defined by the equation

$$\begin{bmatrix} s\boldsymbol{I}-\boldsymbol{A} & -\boldsymbol{B} \\ \boldsymbol{C} & 0 \end{bmatrix} \begin{bmatrix} x(s) \\ C(s) \end{bmatrix} = \boldsymbol{0} \qquad (5.5.33)$$

Looking at this defining equation it can be concluded that, whereas system poles were dependent only upon physical system parameters, system zeros are dependent both upon system parameters and the location of the inputs and outputs. This conclusion can be drawn by looking at the content of the terms in the vectors $\boldsymbol{B}$ and $\boldsymbol{C}$, which are all source and sensor location dependent. In other words, zeros represent a 'dead zone' in the transfer of energy between one point and another.

The question then arises of what system zeros physically correspond to. One case where a system zero arises is where contributions from a set of modes at a sensor location sum to zero. Consider, for example, the cantilever beam shown in Fig. 5.25, where the response is assumed for simplicity to be due only to two modes. For the actuator/sensor arrangement shown in the figure, the sign of the mode shape functions at both the sensor and actuator location is the same, depicted as a negative number. Recalling that for an undamped system there is a phase reversal (change by 180°) in the input impedance of a mode after passing through resonance, it becomes apparent that at frequencies between their resonance frequencies these modes will be vibrating out of phase at the sensing point. It can therefore be envisaged that at some frequency between the two resonances the contributions from each mode to the response measured at the sensor will sum to zero. This frequency is a system zero for the transfer function between the input and output. One point to note concerning this mechanism for generating a transmission zero is that as more modes are included in a model, the zero frequencies will change (Lindner *et al.*, 1993). Therefore, care must be taken in constructing a model so as to include enough modes to adequately mimic the frequency response function (see Lindner *et al.*, 1993).

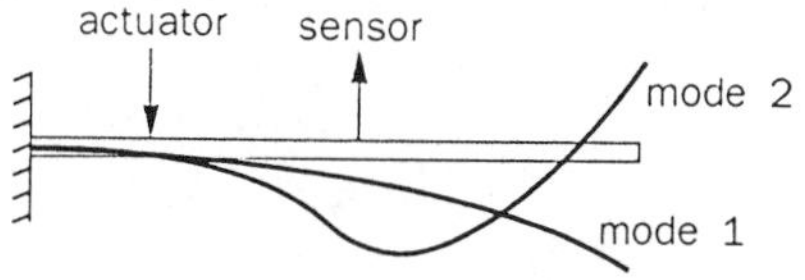

Fig. 5.25 Cantilever beam with sketches of the first two mode shape functions.

There is another physical meaning behind system zeros that is particularly applicable to large flexible systems; that is, system zeros correspond to resonances of a constrained substructure (Miu, 1991). This definition is perhaps best explained by means of an example.

Example 5.10

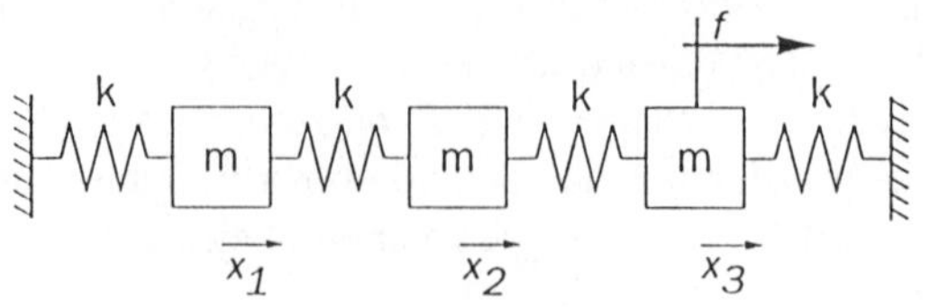

Fig. 5.26 Mass/spring system arrangement.

Let us define the poles and zeros of the spring/mass system shown in Fig. 5.26 (see Miu, 1991). For this system, the transfer function can be shown to be

$$G(s) = \frac{x_2(s)}{f(s)} = \frac{ms^2+2k}{(ms^2+k)(ms^2+3k)} \tag{5.5.34}$$

Thus the system has two sets of complex conjugate poles, $s = \pm j(k/m)^{1/2}$ and $s = \pm j(3k/m)^{1/2}$, and one set of complex conjugate zeros, $s = \pm j(2k/m)^{1/2}$. The poles correspond to the natural frequencies of the system. The zeros, however, correspond to the natural frequencies of the system constrained between the sensor and actuator location, which in this case is equivalent to a single mass, two spring system. This physical insight can be applied readily to more complex systems (Miu, 1991).

In terms of transient response characteristics, zeros primarily affect the overshoot and rise time of the response; as a zero moves towards the imaginary axis, both the rise time and overshoot increase.

One point to note here is that the effect of having a zero on the right-hand side of the complex plane is that the transient response of the dynamic system first goes negative, and the amplitude of this initial negative peak may be greater than one for zeros just outside the unit circle. Systems with this characteristic are termed non-minimum phase systems because, for the equivalent magnitude response, these zeros impart a greater phase shift than zeros on the left hand side of the complex plane.

Let us now turn our attention to poles and zeros of discrete time transfer functions. The definitions of poles and zeros given for a continuous time system hold for a sampled data system. For a system described in discrete time transfer function notation

$$G(z) = \frac{b(z)}{a(z)} \tag{5.5.35}$$

the poles are defined as the values of z such that $a(z) = 0$. For a system described in state space notation,

$$\begin{aligned} \boldsymbol{x}(k+1) &= \boldsymbol{\Phi x}(k) + \boldsymbol{\Gamma} u(k) \\ y(k) &= \boldsymbol{cx}(k) \end{aligned} \tag{5.5.36}$$

the poles are defined by the eigenvalue expression

$$[z\boldsymbol{I} - \boldsymbol{\Phi}]x(z) = 0 \tag{5.5.37}$$

Similarly, the zeros are defined by the expression

$$\begin{bmatrix} z\boldsymbol{I}-\boldsymbol{\Phi} & -\boldsymbol{\Gamma} \\ c & 0 \end{bmatrix} \begin{bmatrix} x(z) \\ c(z) \end{bmatrix} = 0 \tag{5.5.38}$$

We will examine system poles first. To do this, consider as an input signal the exponential function

$$u(k) = \begin{cases} r^k & (k \geq 0) \\ 0 & (k<0) \end{cases} \tag{5.5.39}$$

This function is useful because it can increase or decrease, and/or oscillate, with time, depending upon the value of k. From the definition of the z-transform given in (5.3.12), the z-transform of this input signal is

$$u(z) = z\{u(k)\} = \sum_{k=-\infty}^{\infty} r^k z^{-k} = \sum_{k=0}^{\infty} r^k z^{-k} = \sum_{k=0}^{\infty} (rz^{-1})^k \tag{5.5.40}$$

For $|rz^{-1}|<1$,

$$u(z) = \frac{1}{1 - rz^{-1}} = \frac{z}{z - r} \qquad (|z| > |r|) \tag{5.5.41}$$

Thus, the single pole for this system is located at $(z = r)$. From the definition of the input function of (5.3.39), several characteristics of pole location can be surmised:

1. If a pole is located at $|r|>1$, then from the definition of the input signal given in (5.3.39), the signal will grow in amplitude in an unbounded fashion with increasing time k. Hence, the system is deemed unstable with these poles.
2. If a pole is located at $|r|=1$, then the signal will remain constant in amplitude with increasing time k. This system is deemed marginally stable with these poles.
3. If a pole is located at $0<|r|<1$, then the signal will decay away exponentially, with time. The closer the magnitude pole is to 0, the faster the decay will be (fast poles), while the closer the magnitude of the pole is to 1, the slower the decay will be (slow poles).
4. If a pole is located at $r = 0$, then the signal will be a transient of finite duration (d.c. response).

These observations lead to the concept of the 'unit circle' in digital control, a circle centred about 0 with a radius of 1. All poles located inside the unit circle are stable, poles on the unit circle are marginally stable, and poles outside the unit circle are unstable. Figure 5.27 depicts the dynamic response for various locations of discrete system poles on a unit circle.

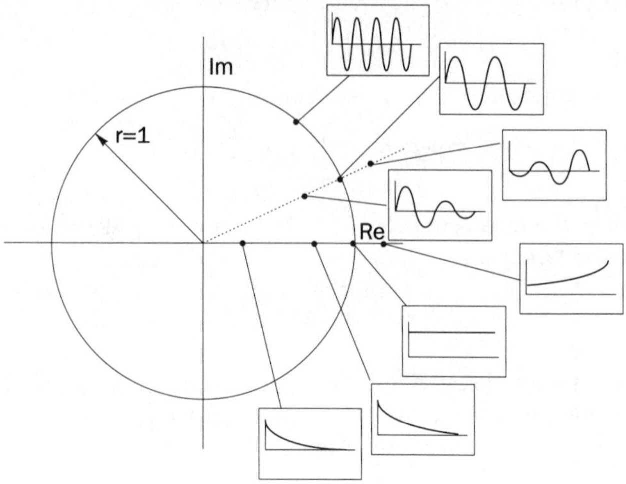

Fig. 5.27 Transient response characteristics associated with poles at various locations on the unit circle.

Consider now the addition of an oscillating component in the input signal:

$$u(k) = \begin{cases} 0 & (k<0) \\ r^k \cos(k\theta) & \text{otherwise} \end{cases} \tag{5.5.42}$$

This input signal can be re-expressed as the sum of two complex exponentials

$$u(k) = \frac{1}{2}(r^k e^{jk\theta} + r^k e^{-jk\theta}) \tag{5.5.43}$$

The z-transform of the first of the bracketed terms in (5.5.43) is

$$z\{r^k e^{jk\theta}\} = \sum_{k=0}^{\infty} r^k e^{jk\theta} z^{-k} = \sum_{k=0}^{\infty} (re^{j\theta}z^{-1})^k = \frac{1}{1-r^{j\theta}z^{-1}} = \frac{z}{z-re^{j\theta}} \tag{5.5.44}$$

$$(\text{for } |z| > r)$$

Similarly, the z-transform of the second bracketed term is

$$z\{r^k e^{-jk\theta}\} = \frac{z}{z-re^{-j\theta}} \tag{5.5.45}$$

Combining these (as the z-transform is a linear function), the z-transform of the input signal defined in (5.5.42) is

$$u(z) = \frac{1}{2}\left[\frac{z}{z-re^{j\theta}} + \frac{z}{z-re^{-j\theta}}\right] \tag{5.5.46}$$

Thus, this signal has two complex poles, located at $z = re^{\pm j\theta}$. The angle θ is defined by the sample rate of system

$$\theta = \frac{2\pi \text{ radians}}{N} = \frac{360°}{N} \tag{5.5.47}$$

where N is the number of samples per cycle. Heuristically, if the sample rate is low, and the angle θ large, then the system will be slow to respond to the input. Hence, the rise time will be long. Alternatively, if the sample rate is high, and the angle Θ small, then the system will respond quickly to the input. Hence, the rise time will be reduced. Figure 5.27 illustrates these concepts by showing the transient response characteristics of several pole locations.

System zeros have the same characteristics in discrete time transfer functions as they did for continuous time transfer functions. One point to note, however, is that a non-minimum phase system will have at least one zero outside of the unit circle, corresponding to the case of a zero on the right-hand side of the complex plane for the continuous case considered earlier in this section.

5.5.5 Stability

A system is said to be stable if it does not 'break down' in response to a given input disturbance. There are two categories of stability which are generally of interest. The first is concerned with the ability of the system to return to equilibrium after some arbitrary displacement, and the second is concerned with the ability of the system to produce a bounded output for every bounded input (BIBO stability). In non-linear or time varying systems these two categories are distinct; it is possible to be stable in one category and not the other. However, in linear time invariant (LTI) systems the criteria are practically equivalent.

5.5.5.1 BIBO stability

To derive a test for bounded input, bounded output stability, consider a system which has an input $u(t)$, an output $y(t)$, and impulse response $h(t)$. Then

$$y(t) = \int_0^t h(t-\tau)u(\tau)\, d\tau \tag{5.5.48}$$

If the output is bounded, then there is some constant c such that $|u| \leq c \leq \infty$, and

$$|y| = \left| \int_0^t h(t-\tau)u(\tau)\, d\tau \right| \leq \int_0^t |h|\,|u|\, d\tau \leq c\int_0^t |h(\tau)|\, d\tau \tag{5.5.49}$$

It is obvious that if the integral on the right-hand side of (5.5.49) is bounded, then the output of the system must be bounded. Therefore, a system has BIBO stability if

$$\int_0^t |h(\tau)| \, d\tau < \infty \tag{5.5.50}$$

Still restricting ourselves to LTI systems, recall now from our previous discussion in this section that the impulse response of the system is sum of time-weighted exponentials, defined by the decay time of each system 'mode', associated with each system pole. If the system is asymptotically stable, such that the response of all of the exponentials goes to zero over time, then the impulse response of the system must be bounded. Therefore, if the system is asymptotically stable it must also have BIBO stability. However, if the system is unstable it may still have BIBO stability for some inputs.

5.5.5.2 Routh–Hurwitz stability

We have seen that it is possible to assess the stability of a system by calculating the roots of the characteristic equation of a system, which define the system poles. It is, however, possible to assess the stability of a system without performing these tests, using algorithms developed by Routh and Hurwitz last century (both of which can be derived from the Lyapunov stability criterion to be considered shortly (Parks, 1962)). While calculating system poles (using computer based algorithms) is not the difficult task it was when the algorithms were developed, they are still useful tools. Developments of the criteria can be found in many linear systems books, and will not be repeated here. Rather, we will simply state the results. Also, we will only state results for the Routh algorithm, as the Hurwitz algorithm is essentially equivalent.

Consider the characteristic equation of an nth order system:

$$|s\boldsymbol{I}-\boldsymbol{A}| = s^n + a_1 s^{n-1} + a_2 s^{n-2} + \cdots + a_n \tag{5.5.51}$$

A necessary condition for system stability is that the roots of (5.5.51) all have negative real parts, so that all of the a coefficients must be positive. If any coefficients are zero or negative then the system will have poles outside the left-hand plane (LHP). A necessary and sufficient condition is that all of the elements in the first column of the Routh array be positive.

The Routh array is a triangular array constructed as shown in Table 5.1. Here the coefficients of the characteristic equation are arranged in two rows, beginning with the first and second coefficients and followed by the subsequent odd and even coefficient terms. The array is then completed as shown in the table. If all elements in the first column, with subscripts of 1, are positive, the system is stable.

Table 5.1 General form of a Routh array

	1	a_2	a_4	a_6 ...
	a_1	a_3	a_5	a_7 ...
$\alpha_1 = 1/a_1$	$b_1 = a_2\text{-}\alpha_1 a_3$	$b_2 = a_4\text{-}\alpha_1 a_5$	$b_3 = a_6$ - $\alpha_1 a_7$	...
$\alpha_2 = a_1/b_1$	$c_1 = a_3\text{-}\alpha_1 b_2$	$c_2 = a_5\text{-}\alpha_2 b_3$		
$\alpha_3 = b_1/c_1$	$d_1 = b_2\text{-}\alpha_3 c_2$	...	...	
$\alpha_4 = c_1/d_1$	...			
⋮				

5.6 CONTROLLABILITY AND OBSERVABILITY

5.6.1 Introduction

While analysis of system behaviour in the frequency domain may convey many insights in a relatively simple manner (graphically), there are a variety of characteristics which are not so easily explained. Controllability and observability fall into this category.

The concepts of controllability and observability deal with the ability of a control system to measure and control the states of a given system. A state is said to be controllable if there is some piecewise continuous (for continuous) or constant (for discrete time systems) control signal $\boldsymbol{u}(t)$ which is capable of changing the state from any given initial value to any desired final value in a finite length of time. If the state variable cannot be influenced by the control inputs $\boldsymbol{u}$, that state is said to be uncontrollable. If all states are controllable then the system is said to be controllable. Similarly, a system is said to be observable if every initial state $\boldsymbol{x}(0)$ can be determined by observing the system output $\boldsymbol{y}$ over some finite time period (for continuous time systems) or finite number of samples (for discrete time systems).

Frequency domain analysis techniques start off with the assumption that the response of the system can be completely determined by its transfer function, or, in other words, that the system is both controllable and observable. However, it was shown early in the development of state space analysis techniques that this is not always the case (Kalman, 1963). In fact, in general a system can be viewed as comprised of four subsystems: one which is both controllable and observable, one which is controllable but not observable, one which is observable but not controllable, and one which is neither controllable nor observable. It is possible, then, that the transfer function of a single input, single output system is of a lower degree that the state space dimension, and the system contains uncontrollable or unobservable states. These leads us to the conclusion that the transfer function of the system does not enable determination of the response (away from the sensing point) of the (generic) system as a whole.

The simple fact that a system has uncontrollable and/or unobservable states does not necessarily present a problem. If all system poles are in the left-hand side of the complex plane, then any initial conditions in the uncontrollable and unobservable states will decay to zero over time and the system will be stable.

The poles of the uncontrollable and/or unobservable states can, however, be in the right-hand side of the complex plane, making these states unstable. A system which has uncontrollable states with stable poles is referred to as stabilizable. A system which has unobservable states with stable poles is referred to as detectable. (A system which has uncontrollable or unobservable states with unstable poles is referred to as a disaster!)

Uncontrollable and/or unobservable states can arise from a variety of situations. Two not uncommon problems in active noise and vibration control experiments which will give rise to such states are the occurrence of a physically uncontrollable system due to poor control actuator placement, and too much symmetry in the system. Other situations such as the inclusion of redundant system variables can also give rise to uncontrollable and/or unobservable states. (For an introductory discussion, see Friedland, 1986.)

In this section we will develop the commonly used tests for system controllability and observability. We will restrict our consideration to LTI systems, although the controllability and observability tests we will develop can also applied (with minor modification) to time-varying systems (see, for example, Kailath, 1980). We will then use these tests to discuss the 'duality' of controllability and observability.

5.6.2 Controllability

Consider the generic continuous time system defined by the state equation

$$\dot{\boldsymbol{x}} = \boldsymbol{Ax} + \boldsymbol{Bu} \tag{5.6.1}$$

As was stated previously, the system is said to be controllable if it is possible to transfer any state with any set of initial conditions to any final state in some finite time period. Without loss of generality we will consider the case where the final state is zero.

In Section 5.3, equation (5.3.91), we derived the solution to the state equation (5.6.1). Specializing this to the case being considered here

$$\boldsymbol{x}(t_{\text{final}}) = \boldsymbol{0} = \mathrm{e}^{\boldsymbol{A}t_{\text{final}}}\boldsymbol{x}(0) + \int_0^{t_{\text{final}}} \mathrm{e}^{\boldsymbol{A}(t_{\text{final}}-\tau)}\boldsymbol{Bu}(\tau)\,\mathrm{d}\tau \tag{5.6.2}$$

or

$$\boldsymbol{x}(t_{\text{final}}) = \boldsymbol{0} = \boldsymbol{\Phi}(t_{\text{final}})\boldsymbol{x}(0) + \int_0^{t_{\text{final}}} \boldsymbol{\Phi}(t_{\text{final}}-\tau)\boldsymbol{Bu}(\tau)\,\mathrm{d}\tau \tag{5.6.3}$$

where $\boldsymbol{\Phi}$ is the state transition matrix. Let us define the matrix $\boldsymbol{P}(T-t)$

$$\boldsymbol{P}(T-t) = \int_t^T \boldsymbol{\Phi}(T-\tau)\boldsymbol{BB}^{\mathrm{T}}\boldsymbol{\Phi}^{\mathrm{T}}(T-\tau)\,\mathrm{d}\tau \tag{5.6.4}$$

In terms of this quantity, the control input which will take the states from $\boldsymbol{x}(0)$ to $\boldsymbol{x}(t_{\text{final}})$ is

$$\boldsymbol{u}(\tau) = \boldsymbol{B}^{\mathrm{T}}\boldsymbol{\Phi}^{\mathrm{T}}(t_{\text{final}}-\tau)\boldsymbol{P}^{-1}(t_{\text{final}}-0)\left[\boldsymbol{x}(t_{\text{final}})-\boldsymbol{\Phi}(t_{\text{final}}-0)\boldsymbol{x}(0)\right] \tag{5.6.5}$$

This can be verified by back-substitution of (5.6.5) into (5.6.3), to give

$$\boldsymbol{x}(t_{\text{final}}) = \boldsymbol{0} = \boldsymbol{\Phi}(t_{\text{final}})\boldsymbol{x}(0)$$
$$+ \left[\int_0^{t_{\text{final}}} \boldsymbol{\Phi}(t_{\text{final}}-\tau)\boldsymbol{B}\boldsymbol{B}^{\mathrm{T}}\boldsymbol{\Phi}^{\mathrm{T}}(t_{\text{final}}-\tau)\ \mathrm{d}\tau\right] \boldsymbol{P}^{-1}(t_{\text{final}})\left[\boldsymbol{x}(t_{final})-\boldsymbol{\Phi}(t_{\text{final}})\boldsymbol{x}(0)\right] \tag{5.6.6}$$

The integral part of (5.6.6) is equal to $\boldsymbol{P}$ as defined in (5.6.4), and so (5.6.6) can be simplified to

$$\boldsymbol{x}(t_{\text{final}}) = \boldsymbol{0} = \boldsymbol{\Phi}(t_{\text{final}})\boldsymbol{x}(0) + \boldsymbol{x}(t_{\text{final}})-\boldsymbol{\Phi}(t_{\text{final}})\boldsymbol{x}(0) \tag{5.6.7}$$

which is, of course, true.

The important point to note about the solution for the control input $\boldsymbol{u}(\tau)$ given in (5.6.5) is that for the solution to exist, the matrix $\boldsymbol{P}$ must be invertible. If $\boldsymbol{P}$ is singular for the initial time (taken as 0 here) and/or for all times greater than the initial, it is not possible to transfer all states from initial conditions defined at $t = 0$ to any given final conditions in some finite time period, and so the system is uncontrollable (Kalman *et al.*, 1963). Therefore, a system is controllable if and only if the matrix $\boldsymbol{P}$ defined in (5.6.4) is non-singular for t and for some $T>t$. The matrix $\boldsymbol{P}$ is referred to as the controllability grammian.

For LTI systems, an algebraic test is also available to assess the controllability of a system. To derive this, note that the state equation solution given in (5.6.2) can be written as

$$\boldsymbol{x}(0) = -\int_0^{t_{\text{final}}} \mathrm{e}^{\boldsymbol{A}(t_{\text{final}}-\tau)}\boldsymbol{B}\boldsymbol{u}(\tau)\ \mathrm{d}\tau \tag{5.6.8}$$

For an nth order system, the matrix exponential in (5.6.8) can be rewritten using the Cayley–Hamilton theorem as

$$\mathrm{e}^{-\boldsymbol{A}(\tau)} = \sum_{k=0}^{n-1} \alpha_k(\tau)\boldsymbol{A}^k \tag{5.6.9}$$

where the α terms are the coefficients of the characteristic polynomial of $\boldsymbol{A}$. This enables (5.6.8) to be expressed as

$$\boldsymbol{x}(0) = -\sum_{k=0}^{n-1} \boldsymbol{A}^k\boldsymbol{B}\int_0^{t_{\text{final}}} \alpha_k(\tau)\boldsymbol{u}(\tau)\ \mathrm{d}\tau \tag{5.6.10}$$

Letting

$$\beta_k = \int_0^{t_{\text{final}}} \alpha_k(\tau)\boldsymbol{u}(\tau)\ \mathrm{d}\tau \tag{5.6.11}$$

(5.6.10) becomes

$$x(0) = -\sum_{k=0}^{n-1} A^k B\beta_k = -\begin{bmatrix} B & AB & \cdots & A^{n-1}B \end{bmatrix} \begin{bmatrix} \beta_0 \\ \beta_1 \\ \vdots \\ \beta_{n-1} \end{bmatrix} \tag{5.6.12}$$

For the system to be controllable, and hence for (5.6.12) to be valid for any and all initial state vectors, the matrix on the right-hand side of (5.6.12) must be non-singular, or be of rank n. Defining this as the controllability matrix $\mathbb{C}$ for the continuous time system

$$\mathbb{C} = \begin{bmatrix} B & AB & \cdots & A^{n-1}B \end{bmatrix} \tag{5.6.13}$$

it can be said that if the system is controllable, the controllability matrix is full rank. This provides with a second test of controllability for an LTI system.

Note that from the definition of the controllability matrix given in (5.6.13), for it to be non-singular, the rows of the controllability matrix must be linearly independent. An often used test for the linear independence of functions is the grammian test, which relates directly to the use of the controllability grammian to assess controllability (Kailath, 1980).

A criterion similar to that outlined using the controllability matrix in (5.6.13) can be developed for discrete time systems. Consider the discrete time control system defined by the state equation

$$x(k+1) = Fx(k) + Gu(k) \tag{5.6.14}$$

By definition, if the system is controllable there is some piecewise constant control signal u which will transfer the initial state $x(0)$ to some (desired) final state, $x(\text{final})$ in n samples, or

$$\begin{aligned} x(1) &= Fx(0) + Gu(0) \\ x(2) &= Fx(1) + Gu(1) = F^2x(0) + FGu(0) + Gu(1) \\ &\vdots \\ x(\text{final}) &= x(n) = F^n x(0) + \sum_{i=0}^{n-1} F^{n-i-1}Gu(k) \end{aligned} \tag{5.6.15}$$

If the state is controllable, then the piecewise constant control sequence will be able to drive any state to zero, or

$$F^n x(0) + \sum_{i=0}^{n-1} F^{n-i-1}Gu(k) = \mathbf{0} \tag{5.6.16}$$

The piecewise constant control sequence which will bring the initial state to rest is defined by

$$x(n) - Fx(0) = \left[G \mid FG \mid F^2G \mid \cdots \mid F^{n-1}G\right] \begin{bmatrix} u(n-1) \\ u(n-2) \\ \vdots \\ u(0) \end{bmatrix} = \mathbf{0} \tag{5.6.17}$$

In viewing (5.6.14)–(5.6.16) it is evident that they define a set of n simultaneous algebraic equations. For there to be a solution to the set of simultaneous equations, the $(n \times n)$ matrix in equation must be non-singular, or be of rank n. Defining this $(n \times n)$ matrix as the controllability matrix $\mathbb{C}$ for discrete time systems

$$\mathbb{C} = \left[G \mid FG \mid F^2G \mid \cdots \mid F^{n-1}G\right] \tag{5.6.18}$$

it can be said that for the system to be controllable the controllability matrix must be non-singular, or the rank of the controllability matrix $\mathbb{C}$ must be n.

5.6.3 Observability

Let us turn our attention now to the problem of observability. Consider the state space equations

$$\begin{aligned} \dot{x} &= Ax + Bu \\ y &= Cx \end{aligned} \tag{5.6.19}$$

Starting from the initial state $x(0)$, the output y is defined by the relationship

$$y(\tau) = C\Phi(\tau - 0)x(0) \tag{5.6.20}$$

Multiplying both sides of this relation by $\Phi^{\mathrm{T}}(\tau)C^{\mathrm{T}}$ and integrating over the time period from the initial to the final states gives

$$\int_0^{t_{\text{final}}} \Phi^{\mathrm{T}}(\tau)C^{\mathrm{T}}y(\tau)\, \mathrm{d}\tau = \int_0^{t_{\text{final}}} \Phi^{\mathrm{T}}(\tau)C^{\mathrm{T}}C\Phi(\tau)\, \mathrm{d}\tau\; x(0) \tag{5.6.21}$$

We stated previously in this section that if a system is observable then it is possible to determine the initial state $x(0)$ from the final state $x(t_{\text{final}})$. Defining the matrix $M(T - t)$

$$M(T-t) = \int_t^T \Phi^{\mathrm{T}}(T-\tau)C^{\mathrm{T}}C\Phi(T-\tau)\, \mathrm{d}\tau \tag{5.6.22}$$

the solution for $x(0)$ is

$$x(0) = M^{-1}(t_{\text{final}}) \int_0^{t_{\text{final}}} \Phi^{\mathrm{T}}(t_{\text{final}}-\tau)C^{\mathrm{T}}y(t_{\text{final}}-\tau)\, \mathrm{d}\tau \tag{5.6.23}$$

This solution relies on the matrix $\boldsymbol{M}$, which is referred to as the observability grammian, being non-singular, or of full rank. Therefore, we can state that for some initial time t, if the observability grammian is non-singular for some $T>t$ the system is observable.

As with controllability, for LTI systems we can define a second, algebraic, observability criteria. To derive this, we can rewrite (5.6.20) using the expansion outlined in (5.6.9) as

$$\boldsymbol{y} = \sum_{k=0}^{n-1} \alpha_k \boldsymbol{C}\boldsymbol{A}^k \boldsymbol{x}(0) \tag{5.6.24}$$

This again describes a set of simultaneous algebraic equations, which can be expressed in matrix form as

$$\boldsymbol{y} = \begin{bmatrix} \alpha_1 \boldsymbol{I} & \alpha_2 \boldsymbol{I} & \cdots & \alpha_{n-1} \boldsymbol{I} \end{bmatrix} \begin{bmatrix} \boldsymbol{C} \\ \boldsymbol{CA} \\ \vdots \\ \boldsymbol{CA}^{n-1} \end{bmatrix} \boldsymbol{x}(0) \tag{5.6.25}$$

For this set of equations to be solved, the second matrix on the right-hand side of (5.6.25) must be non-singular. Defining this matrix as the observability matrix

$$\mathrm{O} = \begin{bmatrix} \boldsymbol{C} \\ \boldsymbol{CA} \\ \vdots \\ \boldsymbol{CA}^{n-1} \end{bmatrix} \tag{5.6.26}$$

it can be said that the system is observable if the observability matrix is non-singular, or of full rank, or of rank n.

The development of a criterion for discrete time system observability closely parallels that for the continuous time system. Consider the discrete time system described by the homogeneous state equations

$$\begin{aligned} \boldsymbol{x}(k+1) &= \boldsymbol{F}\boldsymbol{x}(k) \\ \boldsymbol{y}(k) &= \boldsymbol{C}\boldsymbol{x}(k) \end{aligned} \tag{5.6.27}$$

The system outputs are

$$\begin{aligned} y(0) &= \boldsymbol{C}\boldsymbol{x}(0) \\ y(1) &= \boldsymbol{C}\boldsymbol{x}(1) = \boldsymbol{CF}\boldsymbol{x}(0) \\ y(2) &= \boldsymbol{C}\boldsymbol{x}(2) = \boldsymbol{CF}\boldsymbol{x}(1) = \boldsymbol{CF}^2\boldsymbol{x}(0) \\ &\vdots \\ y(n-1) &= \boldsymbol{CF}^{n-1}\boldsymbol{x}(0) \end{aligned} \tag{5.6.28}$$

These outputs form a set of algebraic simultaneous equations, which can be expressed in matrix form as

$$y = \begin{bmatrix} C \\ CF \\ \vdots \\ CF^{n-1} \end{bmatrix} x(0) \tag{5.6.29}$$

For these equations to be solvable for the initial state vector $x(0)$, the matrix on the right-hand side of (5.6.29) must be non-singular, or be of rank n. Defining this matrix as the observability matrix for discrete time systems

$$\mathbf{O} = \begin{bmatrix} C \\ CF \\ \vdots \\ CF^{n-1} \end{bmatrix} \tag{5.6.30}$$

it can be said that the system is observable if the observability matrix is non-singular, or full rank, or of rank n.

5.6.4 A brief comment on joint relationships between controllability and observability

It may have appeared, when going through the last two sections, that the concepts of controllability and observability are closely related. This is, in fact, the case, and in this section we will briefly defined some of the relationships between the two concepts.

Consider some system S which is described by the state equations

$$\begin{aligned} \dot{x} &= Ax+Bu \\ y &= Cx \end{aligned} \tag{5.6.31}$$

and the 'dual' system S_d described by the state equations

$$\begin{aligned} \dot{z} &= A^{T}z+C^{T}v \\ w &= B^{T}z \end{aligned} \tag{5.6.32}$$

where x,z are state vectors, u,v are control vectors, and y,w are output vectors. The principal of duality states that the system S is controllable (observable) if the dual system S_d is observable (controllable). This is straightforward to verify, by simply writing down the controllability and observability equations.

The duality of controllability and observability means that the dimension of the 'controllable subspace' in the total state space is the same dimension of the 'observable subspace'. This means that if a system has unobservable modes, it also has uncontrollable modes. This is intuitively obvious, as a system control input is derived based upon a measurement of the current system response. If the response of some mode is not present in this measurement, then no control input is derived to attenuate the response. Hence, the mode is uncontrollable because it is unobservable.

Continuing with this line of thought, if a transfer function matrix $G(s)$ is related to the state space description of a system by

$$G(s) = C^{T}(sI-A)^{-1}B \tag{5.6.33}$$

then A has the minimal dimension if and only if the system described by $[A,B]$ is completely controllable, and (so) $[A,C]$ is completely observable (the term 'minimal dimension' refers to the smallest number of states which can completely describe the response of a system). If this is the case, the system described by the terms A, B and C is termed the minimal realization of G.

One final definition which will be of use later in this book when discussing controller reduction is that of a balanced realization. If a system is minimal, then the realization of the system in which the controllability and observability grammians are related by the relationship

$$P = Q = \text{diag}[\sigma_1,\sigma_2,\cdots,\sigma_n] \tag{5.6.34}$$

is referred to as a balanced realization (Moore, 1981). In (5.6.34) the sigma terms are the Hankel singular values of the system. Note also that if a system is minimal but not balanced, it is possible to perform a transformation which will change the coordinate system to a balanced one.

5.6.5 Lyapunov stability

Previously in this chapter we considered the problem of assessing system stability, finding that a system would be stable if the real parts of the eigenvalues were all negative. It was noted at that time, however, that it is often difficult to solve for these quantities, which lead us on to a discussion of simplified methods for systems expressed in transfer function notation. There is also a simplified method for assessing the stability of a system using the system state equations, which is suitable for application to both linear and non-linear systems. The reason for including the discussion of this technique in this section will become apparent at the very end of this section.

The method for examining system stability we will briefly discuss was developed by the Russian mathematician A.M. Lyapunov in the late nineteenth century. Lyapunov developed two methods for testing the stability of systems, particularly non-linear systems, governed by ordinary differential equations expressed in state-variable format. The second, or direct, method does not require explicit knowledge of the form of solution of the differential equations, and is therefore quite convenient for examining the stability of a dynamic system. It is also not restricted to LTI systems (although for simplicity it is these types of systems we will consider here). The following section provides only a brief overview of the second method of Lyapunov. Readers wanting to know more about Lyapunov stability, particularly for non-linear systems, are referred to Vidyasagar (1978). A translation of the original work can be found in (Lyapunov, 1992).

Consider the generic set of state equations

$$\dot{x} = f(x,t) \tag{5.6.35}$$

where x is a state vector. Limiting our discussion to LTI systems, (5.6.35) can be expressed in the (homogeneous) state variable format

$$\dot{x} = Ax \tag{5.6.36}$$

Note, however, if the system is non-linear, (5.6.35) could be rewritten as

$$\dot{x} = Ax + \alpha(x) \tag{5.6.37}$$

where $\alpha(x)$ contains the higher order terms of x, and is governed by

$$\lim_{\|x\|\to 0} \frac{\|\alpha(x)\|}{\|x\|} = 0 \tag{5.6.38}$$

For the systems being considered here there is, in general, a unique solution to the state equation which is dependent upon the initial conditions. The equilibrium state x_e of the system is defined by the expression

$$f(x_e,t) = 0 \qquad \text{for all time } t \tag{5.6.39}$$

For LTI systems there will be only one equilibrium state if the state matrix A is positive definite. If it is not already there, the origin of the state space can be moved to the equilibrium state location by a simple coordinate transformation, and it will be assumed that this has been done in the remainder of the discussion.

To provide a qualitative description of system stability which will lead to the quantitative criterion of Lyapunov's second method, it is useful for us to consider the state transformation described by (5.6.36) as tracing a path in time, or a trajectory, originating from the state space position $x(0)$. The location of this trajectory at any time t is defined by the solution of the state equation at time t, which we have derived in Section 5.3. If a second order system is being considered, this trajectory can easily be envisaged in the plane defined by the two states x_1 and x_2.

The distance between the equilibrium state x_e and any given state $x(t)$ is given by the Euclidean norm

$$\|x(t)\| = \|x(t)-x_e\| = \sqrt{(x_1(t)-x_{1,e})^2+(x_2(t)-x_{2,e})^2 + \cdots} \tag{5.6.40}$$

To assist in assessment of stability, let us define the radius r such that

$$\|x(0)-x_e\| = \|x(0)\| \leq r \tag{5.6.41}$$

and some larger radius R such that

$$\|x(t)-x_e\| = \|x(t)\| < R \tag{5.6.42}$$

The equilibrium state of the system, x_e, can be said to be stable if there is a radius r such that trajectories leaving from within r do not go beyond R for all time t (up to a infinite period). This idea is shown in Fig. 5.28. Further, the equilibrium state is said to be asymptotically stable if, in addition to there being

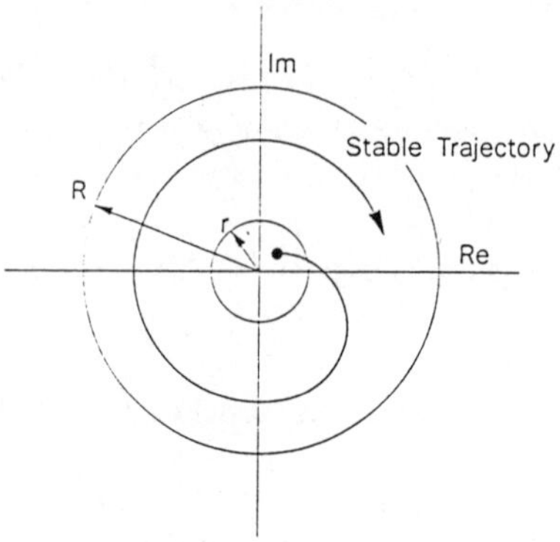

Fig. 5.28 Path of a stable trajectory.

an R for every r such that $\|x(0)\| < r$ and $\|x(t)\| < R$ for all time t, the trajectory converges to the equilibrium state as $t \to \infty$. If asymptotic stability holds for all initial positions $x(0)$ in the state space then the system is said to be asymptotically stable at large.

If no R can be found for r such that $\|x(0)\| < r$ and $\|x(t)\| < R$ for all time t, the system is said to be unstable. This idea is shown in Fig. 5.29. Using these definitions of stability and instability it would be possible to solve for a given trajectory and examine its characteristics to determine stability or instability. This, however, can be even more tedious than solving for the eigenvalues of the LTI systems to see if it is stable.

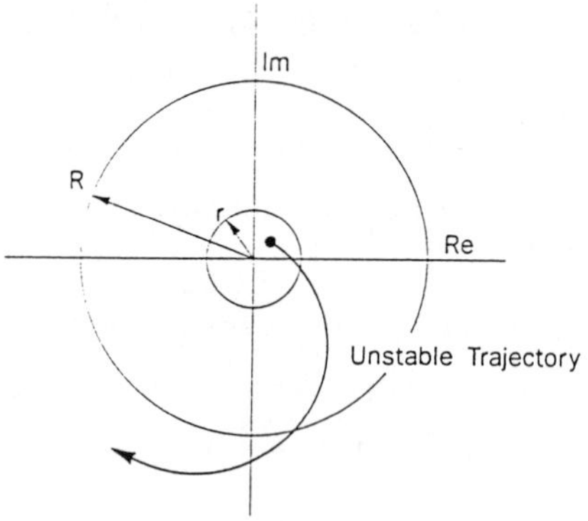

Fig. 5.29 Path of an unstable trajectory.

Lyapunov approached the problem from a more qualitative point of view. Extending from our description of stable systems, a vibrating system is stable if its total energy, which is a positive quantity, is continually decreasing in time, which is a negative derivative, until equilibrium is reached. This general concept can be applied to any system with an asymptotically stable equilibrium state; that is, the system will be stable if its total energy is continuously decreasing with time. The problem is mathematically how to define a generic 'total energy function' which can be used to assess the stability of any given system. To address this problem Lyapunov introduced a scalar function V (now referred to as a Lyapunov function), which has the same properties as total system energy, namely

1. $V(0) = 0$
2. $V(\boldsymbol{x}) > 0, \; \|\boldsymbol{x}\| \neq 0$
3. V is continuous, and has continuous derivatives with respect to all states in $\boldsymbol{x}$, and
4. $\dot{V} \leq 0$ along the trajectories.

The first two properties state that V is positive if any state is non-zero, and equal to zero otherwise. The third property defines the shape of V as smooth and bowl-like near the origin. The fourth property dictates that the trajectory will never climb up the bowl, away from the origin. If the fourth property is made stricter so that $\dot{V} < 0$, then the trajectories will converge to the origin, or equilibrium point.

Consider the continuous time system described in state space form by (5.6.36). A quantity fulfilling the definition of a Lyapunov function V is a quadratic measure of location:

$$V = \boldsymbol{x}^{\mathrm{T}}\boldsymbol{P}\boldsymbol{x} \tag{5.6.43}$$

where $\boldsymbol{P}$ is any symmetric positive-definite matrix. It will, in general, be possible to find a matrix $\boldsymbol{T}$ such that

$$\boldsymbol{P} = \boldsymbol{T}^{\mathrm{T}}\boldsymbol{T} \tag{5.6.44}$$

Using this, we can restate the Lyapunov function as

$$V = \sum_{i=1}^{n} \chi_i^2 \tag{5.6.45}$$

where

$$\chi = \boldsymbol{T}\boldsymbol{x} \tag{5.6.46}$$

For such a matrix $\boldsymbol{P}$, V satisfies the first three properties of a Lyapunov function, leaving only the fourth. The derivative of V is

$$\dot{V} = \frac{\mathrm{d}}{\mathrm{d}t}\boldsymbol{x}^{\mathrm{T}}\boldsymbol{P}\boldsymbol{x} = \dot{\boldsymbol{x}}^{\mathrm{T}}\boldsymbol{P}\boldsymbol{x} + \boldsymbol{x}^{\mathrm{T}}\boldsymbol{P}\dot{\boldsymbol{x}} = \boldsymbol{x}^{\mathrm{T}}(\boldsymbol{A}^{\mathrm{T}}\boldsymbol{P} + \boldsymbol{P}\boldsymbol{A})\boldsymbol{x} = -\boldsymbol{x}^{\mathrm{T}}\boldsymbol{Q}\boldsymbol{x} \tag{5.6.47}$$

where

$$\boldsymbol{A}^{\mathrm{T}}\boldsymbol{P} + \boldsymbol{P}\boldsymbol{A} = -\boldsymbol{Q} \tag{5.6.48}$$

and $\boldsymbol{A}$ is defined by the state equation (5.6.36).

Lyapunov showed that, for any positive definite matrix $\boldsymbol{Q}$, the solution $\boldsymbol{P}$ of (5.6.48), the Lyapunov equation, is positive if and only if all of the characteristic roots of the system matrix $\boldsymbol{A}$ have negative real parts. In other words, to test if a given system is stable (in the sense of Lyapunov), a positive definite matrix $\boldsymbol{Q}$ can be chosen (such as the identity matrix $\boldsymbol{I}$), and the Lyapunov equation can be solved using matrix $\boldsymbol{A}$ of the system of interest. The solution $\boldsymbol{P}$ can then be tested to see if it is positive definite (such as by seeing if the determinants of the n principal minors are positive definite). If it is, then the system exhibits Lyapunov stability.

Example 5.11

Consider a simple linear state variable system described by

$$\dot{\boldsymbol{x}} = \boldsymbol{A}\boldsymbol{x} \tag{5.6.49}$$

where

$$\boldsymbol{A} = \begin{bmatrix} -a_1 & a_2 \\ -a_2 & -a_1 \end{bmatrix}, \quad a_1 > 0 \tag{5.6.50}$$

To assess its stability, $\boldsymbol{Q}$ can be chosen to be any positive definite matrix, and therefore the identity matrix is a suitable choice. If this system is stable, then a positive definite matrix $\boldsymbol{P}$ must exist such that the Lyapunov equation (5.6.48) is satisfied. Letting

$$\boldsymbol{P} = \begin{bmatrix} p_{1,1} & p_{1,2} \\ p_{2,1} & p_{2,2} \end{bmatrix} \tag{5.6.51}$$

then

$$\boldsymbol{A}^{\mathrm{T}}\boldsymbol{P} + \boldsymbol{P}\boldsymbol{A} = -\boldsymbol{Q} \tag{5.6.52}$$

or

$$\begin{bmatrix} -a_1 & -a_2 \\ a_2 & -a_1 \end{bmatrix} \begin{bmatrix} p_{1,1} & p_{1,2} \\ p_{2,1} & p_{2,1} \end{bmatrix} + \begin{bmatrix} p_{1,1} & p_{1,2} \\ p_{2,1} & p_{2,2} \end{bmatrix} \begin{bmatrix} -a_1 & a_2 \\ -a_2 & -a_1 \end{bmatrix} = \begin{bmatrix} -1 & 0 \\ 0 & -1 \end{bmatrix} \tag{5.6.53}$$

Multiplying this out,

$$\begin{aligned} -2a_1p_{1,1} - 2a_2p_{1,2} &= -1 \\ -2a_1p_{1,2} + a_2p_{1,1} - a_2p_{2,2} &= 0 \\ -2a_1p_{2,1} + a_2p_{1,1} - a_2p_{2,2} &= 0 \\ -2a_1p_{2,2} + a_2p_{1,2} + a_2p_{2,1} &= -1 \end{aligned} \tag{5.6.54}$$

One solution to this set of equations is

$$p_{1,2} = p_{2,1} = 0, \quad p_{1,1} = p_{2,2} = \frac{1}{2a_1} \tag{5.6.55}$$

producing the matrix

$$P = \begin{bmatrix} \frac{1}{2a_1} & 0 \\ 0 & \frac{1}{2a_1} \end{bmatrix} \tag{5.6.56}$$

The determinants of the principal minors of $\boldsymbol{P}$ must be positive for the matrix to be positive definite, and hence the system to be stable. Therefore, the system described by (5.6.49) and (5.6.50) is stable if $a_1 > 0$.

Lyapunov stability can also be defined for discrete time control systems. However, rather than solving for the derivative of the function to establish criteria for system stability, we can use the difference between $V(k+1)$ and $V(k)$

$$\Delta V(\boldsymbol{x}(t)) = V(\boldsymbol{x}(k+1)) - V(\boldsymbol{x}(k) = \boldsymbol{x}^{\mathrm{T}}(k)\left(\boldsymbol{F}^{\mathrm{T}}\boldsymbol{P}\boldsymbol{F} - \boldsymbol{P}\right)\boldsymbol{x}(k) = -\boldsymbol{x}^{\mathrm{T}}(k)\boldsymbol{Q}\boldsymbol{x}(k) \tag{5.6.57}$$

where

$$(\boldsymbol{F}^{\mathrm{T}}\boldsymbol{P}\boldsymbol{G} - \boldsymbol{F}) = -\boldsymbol{Q} \tag{5.6.58}$$

and $\boldsymbol{F}$ is the discrete time system matrix, taking the place of $\boldsymbol{A}$ in continuous time systems. Therefore, if a positive definite matrix $\boldsymbol{P}$ can be found which satisfies the criterion of (5.6.58), where $\boldsymbol{Q}$ is any positive definite matrix, the discrete time system is stable.

The concepts of controllability and observability can be directly related to Lyapunov stability. The controllability and observability grammians $\boldsymbol{P}$ and $\boldsymbol{Q}$ are, in fact, also the unique, positive definite solutions to the following Lyapunov equations:

$$\boldsymbol{A}\boldsymbol{P} + \boldsymbol{P}\boldsymbol{A}^{\mathrm{T}} + \boldsymbol{B}\boldsymbol{B}^{\mathrm{T}} = 0 \tag{5.6.59}$$

and

$$\boldsymbol{A}^{\mathrm{T}}\boldsymbol{Q} + \boldsymbol{Q}\boldsymbol{A} + \boldsymbol{C}^{\mathrm{T}}\boldsymbol{C} = 0 \tag{5.6.60}$$

5.7 CONTROL LAW DESIGN VIA POLE PLACEMENT

Thus far in this chapter we have concentrated our efforts on developing methods to model systems and examine certain response characteristics, such as stability, frequency response, and controllability and observability. In this and the following sections we will begin to tackle the problem of designing a control system to modify the response of our 'plant' in some desired fashion. The method we will begin with what is referred to as pole placement.

5.7.1 Transformation into controller canonical form

We will begin with the problem of the design of a controller for a single input, single output system. It is common to refer to feedback controller as a regulator. Neglecting the possibility of feedforward control, the control law is the negative feedback of a linear combination of all states

$$u = -\boldsymbol{k}\boldsymbol{x} \tag{5.7.1}$$

where $\boldsymbol{k}$ a column vector of control coefficients:

$$\boldsymbol{k} = [k_1 \;\; k_2 \;\; \cdots \;\; k_n] \tag{5.7.2}$$

Defining the control input u in this way, the standard form of state equation becomes

$$\dot{\boldsymbol{x}} = (\boldsymbol{A}-\boldsymbol{b}\boldsymbol{k})\boldsymbol{x} \tag{5.7.3}$$

In stating the regulator equation (5.7.1), we have assumed that all state variables are available for use, or measurable in some way. This is often not the case, and some or all state variables must be estimated in some other way in order to implement the control law. We will consider this problem further in later sections.

To study the effect which incorporating a regulator has upon the system, we need to examine the influence it has upon the location of the (closed loop) poles. Taking the Laplace transform of equation (5.7.3):

$$\left(s\boldsymbol{I}-\boldsymbol{A}+\boldsymbol{b}\boldsymbol{k}\right)\boldsymbol{x}(s) = 0 \tag{5.7.4}$$

Therefore, the characteristic equation $\gamma_c(s)$ of the closed loop system is

$$\gamma_c(s) = \left| \; s\boldsymbol{I}-\boldsymbol{A}+\boldsymbol{b}\boldsymbol{k} \; \right| = 0 \tag{5.7.5}$$

Note that the regulator has the effect of changing the characteristic equation, and therefore changing the location of the system poles. In theory, if the system is controllable, the poles can be placed anywhere (we will prove this shortly). This means that, in theory, we can completely specify the closed loop performance of a system; we can speed up the response of a slow system at will, or add damping to a lightly damped system. In practice, however, we do not have such a free hand. Speeding the response of a slow system can require large control signals which cannot practically be generated because real amplifiers 'saturate'. Adding large amounts of damping to systems with poles near the imaginary axis can be somewhat risky, both because of the magnitudes of the control signal which are sometimes required, and because the gains are very sensitive to the location of the poles; a slight change in the open loop system away from its modelled behaviour can result in a serious change in the performance of the closed loop system. Therefore, in practice we must be 'moderate' in our goals.

Given that it is possible to move the system poles to some predefined location using a regulator, what remains now is to determine the regulator gains which produce the desired result. Given a set of desired pole locations ξ_1, ξ_2, ..., the desired characteristic equation is

$$\gamma_c(s) = (s-\xi_1)(s-\xi_2) \cdots = 0 \tag{5.7.6}$$

The required coefficients in $\boldsymbol{k}$ can be found by equating (5.7.5) and (5.7.6). In general, this can be a very tedious task if the order of the characteristic equation is greater than two or three. One exception to this is when the system equations are such that the control structure takes on the controller canonical form. This was discussed in Section 5.4 for discrete time systems, and is shown as a block diagram for continuous time systems in Fig. 5.30, where single sample time delays have been replaced by integrals for the continuous time system. For this system, the state equations are described by the terms

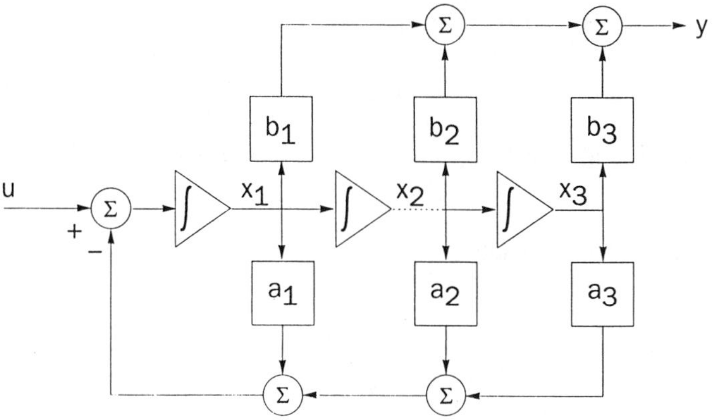

Fig. 5.30. Continuous system arranged in controller canonical form.

$$\boldsymbol{A} = \begin{bmatrix} -a_1 & -a_2 & \cdots & -a_{n-1} & -a_n \\ 1 & 0 & \cdots & 0 & 0 \\ 0 & 1 & \cdots & 0 & 0 \\ & & \vdots & & \\ 0 & 0 & \cdots & 1 & 0 \end{bmatrix}, \text{ and } \boldsymbol{b} = \begin{bmatrix} 1 \\ 0 \\ 0 \\ \vdots \\ 0 \end{bmatrix} \tag{5.7.7}$$

Substituting these values into (5.7.5), the characteristic equation of the system is

$$| s\boldsymbol{I}-\boldsymbol{A}+\boldsymbol{bk} | = \begin{vmatrix} s+a_1+k_1 & a_2+k_2 & \cdots & a_{n-1}+k_{n-1} & a_n+k_n \\ 1 & s & \cdots & 0 & 0 \\ 0 & 1 & \cdots & 0 & \\ & & \vdots & & \\ 0 & 0 & \cdots & 1 & s \end{vmatrix} \tag{5.7.8}$$

or

$$s^n + (a_1+k_1)s^{n-1} + \cdots + (a_{n-1}+k_{n-1})s + (a_n+k_n) = 0 \tag{5.7.9}$$

Therefore, if the desired root locations result in the characteristic equation

$$\gamma_c(s) = s^n + \alpha_1 s^{n-1} + \cdots + \alpha_{n-1}s + \alpha_n = 0 \tag{5.7.10}$$

the required control gains are defined by the expression

$$k_i = \alpha_i - a_i, \qquad i=1,2,\cdots,n \tag{5.7.11}$$

While most system equations are not originally expressed in control canonical form, it is often possible to transform the state equations to take advantage of this structure for regulator design (recall that the poles of the system are invariant under state transformation). Let us define a matrix $\boldsymbol{T}$ which will transform a given set of state equations into controller canonical form. This transformation matrix can be expressed as the product of two submatrices

$$\boldsymbol{T} = \boldsymbol{MN} \tag{5.7.12}$$

The first matrix, $\boldsymbol{M}$, is actually the controllability matrix

$$\boldsymbol{M} = \left[\boldsymbol{b} \;\; \boldsymbol{Ab} \;\; \cdots \;\; \boldsymbol{A}^{n-1}\boldsymbol{b}\right] \tag{5.7.13}$$

Application of this matrix will transform the state matrix into observer canonical form as follows:

$$\boldsymbol{M}^{-1}\boldsymbol{AM} = \begin{bmatrix} 0 & 0 & \cdots & -a_n \\ 1 & 0 & \cdots & -a_{n-1} \\ 0 & 1 & \cdots & -a_{k-1} \\ & & \vdots & \\ 0 & 0 & \cdots & -a_1 \end{bmatrix} \tag{5.7.14}$$

(Observer canonical form was discussed in Section 5.4 for discrete time systems, and is shown in Fig. 5.31 for a continuous time system, where single sample time delays have been replaced by integrals for the continuous time case.) The second matrix $\boldsymbol{N}$ is a triangular Toeplitz matrix

$$\boldsymbol{N} = \begin{bmatrix} 1 & a_1 & \cdots & a_{n-1} \\ 0 & 1 & \cdots & a_{n-2} \\ & & \vdots & \\ 0 & 0 & \cdots & 1 \end{bmatrix} \tag{5.7.15}$$

Applying the transformation to the state equation,

$$\dot{\boldsymbol{x}}' = \boldsymbol{A}'\boldsymbol{x}' + \boldsymbol{b}'\boldsymbol{k}'\boldsymbol{x}' \tag{5.7.16}$$

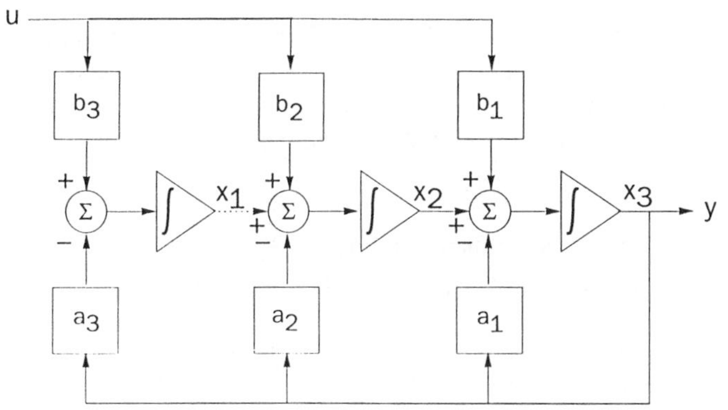

Fig. 5.31 Continuous system arranged in observer canonical form.

where

$$x' = T^{-1}x,\ A' = T^{-1}AT,\ b' = T^{-1}b,\ k' = kT \tag{5.7.17}$$

Once the gains have been calculated using the relationship of (5.7.11), they can be transformed back for use with the original states, using

$$k = k'T^{-1} \tag{5.7.18}$$

The form of the transformation matrix provides us with a criterion for when it is possible to undertake this transformation, and hence move the poles of a system: the system must be controllable, or else the controllability matrix M in the transformation matrix will not be invertible, and hence the transformation will not be possible. If the system is stabilizable the states can be split into controllable and uncontrollable groups, and the transformation conducted on the controllable subset; the pole locations of the uncontrollable subset will have to be accepted as is. The conclusion is only that part of the system which is controllable is open to pole placement.

This provides us with a procedure for determining the gains of a regulator which will move the open-loop poles to some desired location:

1. Check for controllability. If the system is not controllable but is stabilizable split the states into controllable and uncontrollable groups, and work with the controllable set.
2. Determine the desired pole locations (discussed further shortly), and construct the desired characteristic equation.
3. Transform the state equations into controller canonical form, using the transformation matrix outlined above.
4. Calculate the (transformed) control gains using (5.7.11).
5. Transform the result back into the original state space using (5.7.18).

5.7.2 Ackermann's formula

There are a number of other methods which can be used to determine the gains for pole placement, of which perhaps the best known is Ackermann's formula (originally, Ackermann, 1972; see also Kailath, 1980; Franklin *et al.*, 1991). To derive this, we will first define the variable

$$\tilde{A} = A - bk \tag{5.7.19}$$

The desired characteristic equation is then

$$| sI - A + bk | = | sI - \tilde{A} | = s^n + \alpha_1 s^{n-1} + \cdots + \alpha_{n-1} s + \alpha_n = 0 \tag{5.7.20}$$

The Cayley–Hamilton theorem states that $\tilde{A}$ satisfies its own characteristic equation, and so

$$\gamma(\tilde{A}) = \tilde{A}^n + \alpha_1 \tilde{A}^{n-1} + \cdots + \alpha_{n-1}\tilde{A} + \alpha_n I = 0 \tag{5.7.21}$$

To simplify the discussion, let us consider a specific system of third order; the final result can be generalized. If we expand (5.7.21) for this system, and use the definition of (5.7.19), we obtain

$$\begin{aligned}\gamma(\tilde{A}) = {} & A^3 + \alpha_1 A^2 + \alpha_2 A + \alpha_3 I - A^2 bk - bk\tilde{A}^2 \\ & - Abk\tilde{A} - \alpha_1 Abk - \alpha_1 bk\tilde{A} - \alpha_2 bk = 0\end{aligned} \tag{5.7.22}$$

Noting that

$$\gamma(A) = A^3 + \alpha_1 A^2 + \alpha_2 A + \alpha_3 I \neq 0 \tag{5.7.23}$$

we can re-express (5.7.22) as

$$\begin{aligned}\gamma(\tilde{A}) = {} & \gamma(A) - A^2 bk - bk\tilde{A}^2 \\ & - Abk\tilde{A} - \alpha_1 Abk - \alpha_1 bk\tilde{A} - \alpha_2 bk = 0\end{aligned} \tag{5.7.24}$$

Rearranging (5.7.24) gives

$$\begin{aligned}\gamma(A) &= b(\alpha_2 k + \alpha_1 k\tilde{A} + k\tilde{A}) + Ab(\alpha_1 k + k\tilde{A}) + A^2 bk \\ &= \begin{bmatrix} b & Ab & A^2 b \end{bmatrix} \begin{bmatrix} \alpha_2 k + \alpha_1 k\tilde{A} + k\tilde{A}^2 \\ \alpha_1 k + k\tilde{A} \\ k \end{bmatrix}\end{aligned} \tag{5.7.25}$$

The first of these matrices is simply the controllability matrix. If the system is controllable then the inverse of the controllability matrix exists. Assuming this, the control gains can be solved from (5.7.25) through the operations

$$[0\ \ 0\ \ 1]\left[b\ \ Ab\ \ A^2b\right]^{-1}\gamma(A) = [0\ \ 0\ \ 1]$$

$$\begin{bmatrix} \alpha_2 k + \alpha_1 k\tilde{A} + k\tilde{A}^2 \\ \alpha_1 k + k\tilde{A} \\ k \end{bmatrix} = k \tag{5.7.26}$$

Rearranging this into a general result

$$k = [0\ \ 0\ \ 0\ \ \cdots\ \ 0\ \ 1]M^{-1}\gamma(A) \tag{5.7.27}$$

where M is the controllability matrix. This is referred to as Ackermann's formula, used for determining the required control gains for moving the poles of a system to desired locations.

5.7.3 A note on gains for MIMO systems

Thus far, we have restricted our discussion to SISO systems. The methods we have developed can, however, be applied to MIMO systems. The only problem is that of redundancy.

Consider the system described by the state equation

$$\dot{x} = Ax + Bu \tag{5.7.28}$$

where there is now a vector of control inputs, defined by the expression

$$u = -Kx \tag{5.7.29}$$

where K is now a matrix of control gains. There are now more control gains than poles, by a factor equal to the number of control inputs into the system (recall that the number of poles is equal to the number of states). This means that there are more gains than (theoretically) required to move the poles to their desired locations. While this provides more flexibility, it is now up to the designer to determine what to do with the extra degrees of freedom in the control system. It is possible for the designer to simply set some gains equal to zero, to simplify the controller architecture. It is also possible to introduce some measure of symmetry into the system, splitting the requirements amongst the inputs. However, perhaps the best approach is to explicitly adopt a design strategy which allows the extra degrees of freedom to satisfy other constraints, such as control effort minimisation. This is possible through linear quadratic design methods (optimal control), which we will consider next.

5.8 OPTIMAL CONTROL

5.8.1 Introduction

In Section 5.7 we found that it is possible to design a control system which will place the poles of a (closed loop) system at desired locations in the complex plane. This means that it is possible to specify some amount of damping or bandwidth for a given system, and design a control system which will meet these requirements.

Why, then, would we ever want more? There are, in fact, a number of problems which are not readily solved by the pole placement approach to control system design. The first problem is the simple task of actually knowing what the optimum pole locations are. While there are a variety of methods for determining potential candidate locations, and while experience with a given system will often lead to an intuitive 'feel' for the best locations, it is not uncommon for the control system designer not to know exactly.

Another problem is that, for MIMO systems, the equation description of the problem is overdetermined and so the control and estimator gains which will produce the desired result are not unique. For example, if there are m control inputs and n system states, there are m times as many parameters as there are poles to place (recall that the number of poles is equal to the number of states). There are, in fact, an infinite number of gain combinations which can be used to place the poles. Which is best?

Another problem which can arise from the pole placement exercise is the specification of an overly large control input. In general, the further a control system has to push a pole from its original location, the greater the required control input, or 'effort'. If the control effort is too large, the control actuator will saturate. When this happens the actual system response may look nothing like the desired system response, and may even be unstable.

What we will discuss in this section is a method for overcoming all of these problems simultaneously, an approach to control system design called optimal control. Rather than beginning with a set of desired pole locations, optimal control begins with a quadratic performance index, or cost function, which is to be minimized through the application of feedback control. The form of this cost function enables the designer to explicitly take into account the control effort required to achieve the desired result with the control system, while its very use overcomes the problem of needing to know the optimal pole locations *a priori*.

Our discussion of optimal control in this section will be necessarily brief. We will not discuss any of the base 'theory' behind optimal control, such as the Pontryagin minimum principle, nor will we discuss finite time solutions to the optimal control problem. As we mentioned in the introduction, in active noise and vibration control we are typically interested in improving the steady state performance of a system in rejecting unwanted acoustic or vibratory disturbances. We will also limit our discussion to linear problems, for which we will derive linear control laws. For a detailed discussion of optimal control, the reader is referred to Athans and Falb (1966), Bryson and Ho (1969), Anderson and Moore (1979, 1990), and Lewis (1986).

5.8.2 Problem formulation

To formulate the optimal control problem, we again begin with a system, modelled in state space format as

$$\dot{x} = Ax + Bu \tag{5.8.1}$$

where we are considering the general, MIMO, case. We wish to modify the dynamic response of the system by introducing a control input u, derived from state feedback as follows:

$$u = -Kx \tag{5.8.2}$$

where the problem is to determine the gain matrix K which will facilitate our requirements.

To derive this gain matrix, the optimal control problem is formulated as one of choosing a control input which will minimize the performance index J:

$$J = \int_t^T [x^T(\tau)Qx(\tau) + u^T(\tau)Ru(\tau)]\, d\tau \tag{5.8.3}$$

where Q is a positive-definite (or positive semi-definite) Hermitian or real symmetric matrix, and R is a positive-definite Hermitian or real symmetric matrix. Note that in this equation, as well as those that follow, if the terms are complex then the transpose T would be replaced by the conjugate transpose H.

There are several points which should be made concerning the performance index in (5.8.3). The first of these is that any control input which will minimize J will also minimize a scalar number times J. This is mentioned because it is not uncommon to find the performance index in (5.8.3) with a constant of 1/2 in front.

The second item to point out is in regard to the limits on the integral in the performance index. As it is written in (5.8.3), the lower limit t is the present time, and the upper limit T is the final, or terminal, time. This means that the control input which minimizes the performance index in (5.8.3) will have some finite time duration, after which it will stop. This form of behaviour represents the general case, and is applied in practice to cases such as missile guidance systems. However, in our area of interest, active control of sound and vibration, we normally require a system which will provide attenuation of unwanted disturbances forever. This corresponds to the special case of (5.8.3) where the terminal time is infinity:

$$J = \int_0^\infty [x^T(\tau)Qx(\tau) + u^T(\tau)Ru(\tau)]\, d\tau \tag{5.8.4}$$

Equation (5.8.4) is referred to as a steady state optimal control problem, and is the form of the problem to which we will confine our discussion.

The final point to discuss in relation to (5.8.3) is in regard to the matrices Q and R. These are often referred to as the state weighting and control weighting matrices, respectively. As suggested by their name, these matrices weight the optimal control problem (weight the relative importance of attenuating the response of certain states and limiting the control effort), and in doing so have a great influence upon the control gain matrix $\boldsymbol{K}$ which is derived. Choosing the weighting matrices is not, unfortunately, an exact science, but more of an 'art'. Later in this chapter we will discuss some of the considerations surrounding selection of $\boldsymbol{Q}$ and $\boldsymbol{R}$, with a more detailed discussion in Chapter 11. However, for a thorough discussion the reader is referred to Anderson and Moore (1990).

There are a number methods which can be used to obtain a control gain matrix which will minimize the optimal control performance index of (5.8.3); two of the more common approaches cited in texts are 'dynamic programming' (Bellman and Dreyfus 1962), and application of the 'minimum principle' (Pontryagin *et al.*, 1962). The former of these is based upon the 'principal of optimality', which basically states that if a control is optimal over some time interval $(t_{(0)}, t_{(f)})$, then it is optimal over all sub-intervals $(t_{(k)}, t_{(f)})$. Using this idea, the optimal control sequence is calculated using backward recursion relations, starting from the desired final (optimum) point. The latter approach is based on calculus of variations. Both of these approaches have the ability not only to derive the control law, but also to prove that the linear control law is, in fact, optimal.

In our discussion we will take a slightly different approach, using the second method of Lyapunov to solve the form of problem in (5.8.4). We adopt this approach for two reasons; first, it builds upon our discussion of Lyapunov stability in Section 5.6, and secondly, if we design our system using a Lyapunov approach we can feel confident that the result will be stable. In fact, the relationships derived are the same regardless of which of the three mentioned methods of solution are used. For details of other methods, the interested reader should consult the specialized optimal control texts cited previously in this section.

5.8.3 A preview: evaluation of a performance index using Lyapunov's second method

Before we begin our discussion of the optimal control problem, it will prove worthwhile to briefly discuss the relationship between Lyapunov equations and performance indices. To do this, let us consider the problem of evaluating the performance index

$$J = \int_0^\infty \boldsymbol{x}^{\mathrm{T}} \boldsymbol{Q} \boldsymbol{x} \, dt \tag{5.8.5}$$

where there is no control input, so that the system is described by the homogeneous state equation

$$\dot{\boldsymbol{x}} = \boldsymbol{A}\boldsymbol{x} \tag{5.8.6}$$

A Lyapunov function can be used to evaluate the performance index of (5.8.5). To derive the expression, let us assume that

$$\boldsymbol{x}^{\mathrm{T}} \boldsymbol{Q} \boldsymbol{x} = -\frac{d}{dt}\left(\boldsymbol{x}^{\mathrm{T}} \boldsymbol{P} \boldsymbol{x}\right) \tag{5.8.7}$$

where $\boldsymbol{P}$ is some positive-definite Hermitian matrix. Evaluating (5.8.7), we obtain

$$\boldsymbol{x}^{\mathrm{T}}\boldsymbol{Q}\boldsymbol{x} = -\dot{\boldsymbol{x}}^{\mathrm{T}}\boldsymbol{P}\boldsymbol{x} - \boldsymbol{x}^{\mathrm{T}}\boldsymbol{P}\dot{\boldsymbol{x}} = -\boldsymbol{x}^{\mathrm{T}}\boldsymbol{A}^{\mathrm{T}}\boldsymbol{P}\boldsymbol{x} - \boldsymbol{x}^{\mathrm{T}}\boldsymbol{P}\boldsymbol{A}\boldsymbol{x} = -\boldsymbol{x}^{\mathrm{T}}(\boldsymbol{A}^{\mathrm{T}}\boldsymbol{P} + \boldsymbol{P}\boldsymbol{A})\boldsymbol{x} \tag{5.8.8}$$

From our discussion of Lyapunov's second method we know that if $\boldsymbol{A}$ is stable, for a given $\boldsymbol{Q}$ there exists a $\boldsymbol{P}$ such that

$$\boldsymbol{A}^{\mathrm{T}}\boldsymbol{P} + \boldsymbol{P}\boldsymbol{A} = -\boldsymbol{Q} \tag{5.8.9}$$

We can use the relationship of (5.8.9) to determine the elements of the matrix $\boldsymbol{P}$ in (5.8.8). Knowing these, we can evaluate the performance index as

$$J = \int_0^\infty \boldsymbol{x}^{\mathrm{T}}\boldsymbol{Q}\boldsymbol{x}\,\mathrm{d}t = -\boldsymbol{x}^{\mathrm{T}}\boldsymbol{P}\boldsymbol{x}\Big|_0^\infty = -\boldsymbol{x}^{\mathrm{T}}(\infty)\boldsymbol{P}\boldsymbol{x}(\infty) + \boldsymbol{x}^{\mathrm{T}}(0)\boldsymbol{P}\boldsymbol{x}(0) \tag{5.8.10}$$

As our system is stable, the values of the states will decay towards zero as the time becomes infinite. Therefore, the performance index is simply

$$J = \boldsymbol{x}^{\mathrm{T}}(0)\boldsymbol{P}\boldsymbol{x}(0) \tag{5.8.11}$$

The performance index can be determined by knowing the initial conditions of the states, $\boldsymbol{x}(0)$, and the value of $\boldsymbol{P}$, which is dependent upon the matrices $\boldsymbol{A}$ and $\boldsymbol{Q}$, as defined in (5.8.9). While this result is potentially useful unto itself as a means of assessing the influence of system parameter changes upon the error criterion, its real value is in the form of the solution, which we shall use in solving the optimal control problem.

5.8.4 Solution to the quadratic optimal control problem

Let us return now to the quadratic optimal control problem, where we wish to derive a gain matrix $\boldsymbol{K}$ which will minimize the steady-state performance index of (5.8.4). Substituting (5.8.2) into (5.8.1), including the control law in the state equation, produces

$$\dot{\boldsymbol{x}} = \boldsymbol{A}\boldsymbol{x} - \boldsymbol{B}\boldsymbol{K}\boldsymbol{x} = (\boldsymbol{A} - \boldsymbol{B}\boldsymbol{K})\boldsymbol{x} \tag{5.8.12}$$

In our derivation of the solution for the optimal control gain, we will assume that $(\boldsymbol{A} - \boldsymbol{B}\boldsymbol{K})$ is stable, so has eigenvalues with negative real parts.

Substituting (5.8.12) into (5.8.4) allows us to express the steady state performance index as

$$J = \int_0^\infty \boldsymbol{x}^{\mathrm{T}}(\boldsymbol{Q} + \boldsymbol{K}^{\mathrm{T}}\boldsymbol{R}\boldsymbol{K})\boldsymbol{x}\,\mathrm{d}t \tag{5.8.13}$$

Based upon the results of Section 5.8.3 and in particular (5.8.7), let

$$\boldsymbol{x}^{\mathrm{T}}(\boldsymbol{Q} + \boldsymbol{K}^{\mathrm{T}}\boldsymbol{R}\boldsymbol{K})\boldsymbol{x} = -\frac{\mathrm{d}}{\mathrm{d}t}(\boldsymbol{x}^{\mathrm{T}}\boldsymbol{P}\boldsymbol{x}) \tag{5.8.14}$$

where $\boldsymbol{P}$ is some positive-definite Hermitian matrix. We can rearrange (5.8.14) as

$$\begin{aligned} \boldsymbol{x}^{\mathrm{T}}(\boldsymbol{Q} + \boldsymbol{K}^{\mathrm{T}}\boldsymbol{R}\boldsymbol{K})x &= -\dot{\boldsymbol{x}}^{\mathrm{T}}\boldsymbol{P}\boldsymbol{x} - \boldsymbol{x}^{\mathrm{T}}\boldsymbol{P}\dot{\boldsymbol{x}} \\ &= -\boldsymbol{x}^{\mathrm{T}}\left[(\boldsymbol{A}-\boldsymbol{B}\boldsymbol{K})^{\mathrm{T}}\boldsymbol{P} + \boldsymbol{P}(\boldsymbol{A}-\boldsymbol{B}\boldsymbol{K})\right]x \end{aligned} \tag{5.8.15}$$

From this equation, we can deduce that

$$(\boldsymbol{A}-\boldsymbol{B}\boldsymbol{K})^{\mathrm{T}}\boldsymbol{P} + \boldsymbol{P}(\boldsymbol{A}-\boldsymbol{B}\boldsymbol{K}) = -(\boldsymbol{Q} + \boldsymbol{K}^{\mathrm{T}}\boldsymbol{R}\boldsymbol{K}) \tag{5.8.16}$$

Equation (5.8.16) is in the form of a Lyapunov equation; if $\boldsymbol{A} - \boldsymbol{B}\boldsymbol{K}$ is stable, then there exists a positive-definite matrix $\boldsymbol{P}$ that satisfies the expression. Based upon Section 5.8.3, the resulting value of the performance index will be

$$J = \boldsymbol{x}^{\mathrm{T}}(0)\boldsymbol{P}\boldsymbol{x}(0) \tag{5.8.17}$$

To solve the quadratic optimal control problem, we begin by factoring the positive-definite Hermitian matrix $\boldsymbol{R}$ as

$$\boldsymbol{R} = \boldsymbol{T}\boldsymbol{T}^{\mathrm{T}} \tag{5.8.18}$$

where $\boldsymbol{T}$ is a non-singular matrix. Using this relationship, (5.8.16) (the defining equation for $\boldsymbol{P}$) can be rewritten as

$$(\boldsymbol{A}^{\mathrm{T}}-\boldsymbol{K}^{\mathrm{T}}\boldsymbol{B}^{\mathrm{T}})\boldsymbol{P} + \boldsymbol{P}(\boldsymbol{A}-\boldsymbol{B}\boldsymbol{K}) = -\boldsymbol{Q} - \boldsymbol{K}^{\mathrm{T}}\boldsymbol{T}\boldsymbol{T}^{\mathrm{T}}\boldsymbol{K} \tag{5.8.19}$$

or, after some rearrangement, as

$$\begin{aligned} &\left[\boldsymbol{T}^{\mathrm{T}}\boldsymbol{K}-\boldsymbol{T}^{-1}\boldsymbol{B}^{\mathrm{T}}\boldsymbol{P}\right]^{\mathrm{T}}\left[\boldsymbol{T}^{\mathrm{T}}\boldsymbol{K}-\boldsymbol{T}^{-1}\boldsymbol{B}^{\mathrm{T}}\boldsymbol{P}\right] \\ &= \boldsymbol{P}\boldsymbol{B}\boldsymbol{R}^{-1}\boldsymbol{B}^{\mathrm{T}}\boldsymbol{P} - \boldsymbol{Q} - \boldsymbol{A}^{\mathrm{T}}\boldsymbol{P} - \boldsymbol{P}\boldsymbol{A} \end{aligned} \tag{5.8.20}$$

To minimize the error criterion, we wish to minimize both sides of this equation. As the left-hand side is non-negative, the minimum possible value it can have is zero, which occurs when

$$\boldsymbol{T}^{\mathrm{T}}\boldsymbol{K} = \boldsymbol{T}^{-1}\boldsymbol{B}^{\mathrm{T}}\boldsymbol{P} \tag{5.8.21}$$

From the result of (5.8.21), the optimum gain matrix is defined by the expression

$$\boldsymbol{K} = (\boldsymbol{T}^{\mathrm{T}})^{-1}\boldsymbol{T}^{-1}\boldsymbol{B}^{\mathrm{T}}\boldsymbol{P} = \boldsymbol{R}^{-1}\boldsymbol{B}^{\mathrm{T}}\boldsymbol{P} \tag{5.8.22}$$

under which conditions $\boldsymbol{P}$ will be defined by the expression

$$\boldsymbol{A}^{\mathrm{T}}\boldsymbol{P} + \boldsymbol{P}\boldsymbol{A} - \boldsymbol{P}\boldsymbol{B}\boldsymbol{R}^{-1}\boldsymbol{B}^{\mathrm{T}}\boldsymbol{P} + \boldsymbol{Q} = 0 \tag{5.8.23}$$

Equation (5.8.22) is known as the algebraic Ricatti equation, and its solution is required to determine the optimal control gain matrix of (5.8.21).

As we noted in the derivation of the optimum control gain, for a solution to exist the matrix $\boldsymbol{A} - \boldsymbol{B}\boldsymbol{K}$ must be stable. It is possible to show that this requirement is met if the matrix pair $[\boldsymbol{A},\boldsymbol{B}]$ is controllable, and the pair $[\boldsymbol{A},\boldsymbol{D}]$ is observable, where $\boldsymbol{D}$ is defined by the factorization $\boldsymbol{Q} = \boldsymbol{D}\boldsymbol{D}^{\mathrm{T}}$. If these requirements are met, then the algebraic Ricatti equation has a unique, positive definite solution $\boldsymbol{P}$ which minimizes the error criterion, or performance index, when the control law of (5.8.22) is implemented.

It should be noted that solving the matrix Ricatti equation is not a simple task, normally requiring an iterative approach (see, for example, Potter, 1966; Kleinman, 1968; Hitz and Anderson, 1972; Laub, 1979). With the advent of computer packages such as Matlab, which contain 'canned' algorithms, there is perhaps little point in the infrequent user considering programming his or her own routines.

5.8.5 Robustness characteristics

Having derived expressions for the optimum gain matrices, it is now of interest to assess the stability and robustness properties of the resulting systems using the classical control concepts of gain and phase margins discussed in Section 5.5. To do this, we must derive expressions related to the return difference equation, which was at the heart of our previous discussion.

Consider the steady state Ricatti equation (5.8.23). Using the expression for the optimum control gain matrix in (5.8.22), (5.8.23) can be re-expressed as

$$(-j\omega\boldsymbol{I}-\boldsymbol{A}^{\mathrm{T}})\boldsymbol{P} + \boldsymbol{P}(j\omega\boldsymbol{I}-\boldsymbol{A}) + \boldsymbol{K}^{\mathrm{T}}\boldsymbol{R}\boldsymbol{K} = \boldsymbol{Q} \tag{5.8.24}$$

If we now premultiply terms by $\boldsymbol{B}^{\mathrm{T}}(-j\omega\boldsymbol{I} - \boldsymbol{A}^{\mathrm{T}})^{-1}$, and postmultiply by $(j\omega\boldsymbol{I} - \boldsymbol{A})^{-1}\boldsymbol{B}$, and use the result of (5.8.22) for the optimum gain matrix (which states $\boldsymbol{B}^{\mathrm{T}}\boldsymbol{P} = \boldsymbol{R}\boldsymbol{K}$), we find that

$$\begin{aligned}\boldsymbol{B}^{\mathrm{T}}(-j\omega\boldsymbol{I}-\boldsymbol{A}^{\mathrm{T}})^{-1}\boldsymbol{K}^{\mathrm{T}}\boldsymbol{R} &+ \boldsymbol{R}\boldsymbol{K}(j\omega\boldsymbol{I}-\boldsymbol{A})^{-1}\boldsymbol{B}\\ &+ \boldsymbol{B}^{\mathrm{T}}(-j\omega\boldsymbol{I}-\boldsymbol{A}^{\mathrm{T}})^{-1}\boldsymbol{K}^{\mathrm{T}}\boldsymbol{R}\boldsymbol{K}(j\omega\boldsymbol{I}-\boldsymbol{A})^{-1}\boldsymbol{B}\\ &= \boldsymbol{B}^{\mathrm{T}}(-j\omega\boldsymbol{I}-\boldsymbol{A}^{\mathrm{T}})^{-1}\boldsymbol{Q}(j\omega\boldsymbol{I}-\boldsymbol{A})^{-1}\boldsymbol{B}\end{aligned} \tag{5.8.25}$$

From (5.8.25), it is straightforward to derive the equality

$$\begin{aligned}&\boldsymbol{R} + \boldsymbol{B}^{\mathrm{T}}(-j\omega\boldsymbol{I}-\boldsymbol{A}^{\mathrm{T}})^{-1}\ \boldsymbol{Q}\ (j\omega\boldsymbol{I}-\boldsymbol{A})^{-1}\boldsymbol{B} =\\ &[\boldsymbol{I}-\boldsymbol{B}^{\mathrm{T}}(-j\omega\boldsymbol{I}-\boldsymbol{A}^{\mathrm{T}})^{-1}\boldsymbol{K}^{\mathrm{T}}]\ \boldsymbol{R}\ [\boldsymbol{I}-\boldsymbol{K}(j\omega\boldsymbol{I}-\boldsymbol{A})^{-1}\boldsymbol{B}]\end{aligned} \tag{5.8.26}$$

Equation (5.8.26) is known as the return difference equality, where the expression $\boldsymbol{I} - \boldsymbol{K}(-j\omega\boldsymbol{I} - \boldsymbol{A})^{-1}\boldsymbol{B}$ is the equivalent of the return difference equation when the system is arranged as shown in Fig. 5.32.

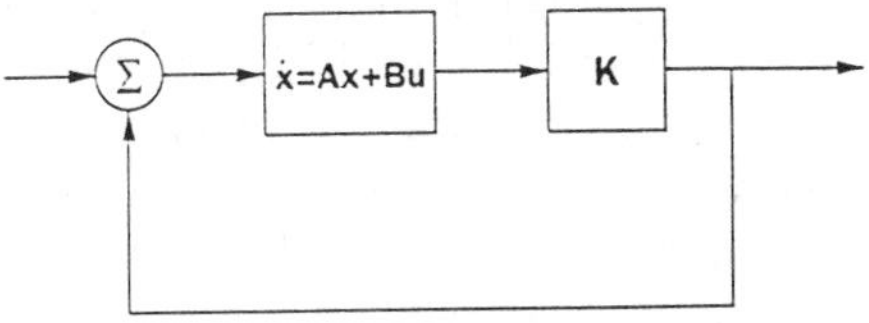

Fig. 5.32 State space closed loop control system.

Starting from Equation (5.8.16) and using similar steps, it is possible to derive the expression

$$\boldsymbol{R} - \boldsymbol{B}^{\mathrm{T}}(-j\omega\boldsymbol{I}-\boldsymbol{A}^{\mathrm{T}}-\boldsymbol{K}^{\mathrm{T}}\boldsymbol{B}^{\mathrm{T}})^{-1}\,\boldsymbol{Q}\,(j\omega\boldsymbol{I}-\boldsymbol{A}-\boldsymbol{B}\boldsymbol{K})^{-1}\boldsymbol{B} = [\boldsymbol{I}+\boldsymbol{B}^{\mathrm{T}}(-j\omega\boldsymbol{I}-\boldsymbol{A}^{\mathrm{T}}-\boldsymbol{K}^{\mathrm{T}}\boldsymbol{B}^{\mathrm{T}})^{-1}\boldsymbol{K}^{\mathrm{T}}]\;\boldsymbol{R}\;[\boldsymbol{I}-\boldsymbol{K}(j\omega\boldsymbol{I}-\boldsymbol{A}-\boldsymbol{B}\boldsymbol{K})^{-1}\boldsymbol{B}] \tag{5.8.27}$$

The relationships given in (5.8.26) and (5.8.27), while looking rather horrible, provide us with a means of assessing the robustness characteristics of a linear quadratic regulator (LQR) control system (a control system designed using linear quadratic (especially optimal control) methods). The SISO case is the simplest, and so will be used for our discussion. It can be shown that properties similar to those we will derive for the SISO case can be derived for the MIMO case (Safanov and Athans, 1977; Lehtomaki *et al.*, 1981; Anderson and Moore, 1990).

For an SISO system, (5.8.26) can be stated as

$$r + \boldsymbol{b}^{\mathrm{T}}(-j\omega\boldsymbol{I}-\boldsymbol{A}^{\mathrm{T}})^{-1}\,\boldsymbol{Q}\,(j\omega\boldsymbol{I}-\boldsymbol{A})^{-1}\boldsymbol{b} = r\,|\,1-\boldsymbol{k}(j\omega\boldsymbol{I}-\boldsymbol{A})^{-1}\boldsymbol{b}\,|^{2} \tag{5.8.28}$$

Factorizing $\boldsymbol{Q}$ as outlined previously produces

$$r + |\,\boldsymbol{d}^{\mathrm{T}}(j\omega\boldsymbol{I}-\boldsymbol{A})^{-1}\boldsymbol{b}\,|^{2} = r\,|\,1-\boldsymbol{k}(j\omega\boldsymbol{I}-\boldsymbol{A})^{-1}\boldsymbol{b}\,|^{2} \tag{5.8.29}$$

This relationship can also be expressed as

$$|\,1-\boldsymbol{k}(j\omega\boldsymbol{I}-\boldsymbol{A})^{-1}\boldsymbol{b}\,|^{2} \geq 1 \tag{5.8.30}$$

Thus the magnitude of the return difference equation $(1 - \boldsymbol{k}(j\omega\boldsymbol{I} - \boldsymbol{A})^{-1}\boldsymbol{b})$ is lower bounded by 1 for all frequencies (for the MIMO case, this translates into the fact that the singular values of the return differences are all lower bounded by 1).

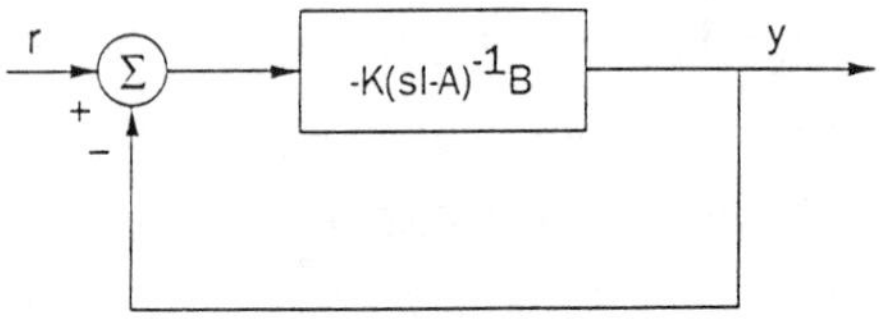

Fig. 5.33 System arrangement in terms of return difference equation.

We can relate this result to features on a Nyquist plot to obtain measures of gain and phase margin. Arranging the SISO system as shown in Fig. 5.33 the Nyquist plot will be the plot of real versus imaginary values of $-\boldsymbol{k}(j\omega\boldsymbol{I} - \boldsymbol{A})^{-1}\boldsymbol{b}$, for $-\infty \leq \omega \leq \infty$. From the inequality of (5.8.32), the plot of these values will always be at least unity distance from the point $(-1,0)$ (the plot will avoid a unit disk centred at (-1,0)). This concept is depicted in an example plot of Fig. 5.34. Further, if the system is detectable and therefore asymptotically stable, the number of counterclockwise encirclements of this disk must be equal to the number of poles in the transfer function $-\boldsymbol{k}(s\boldsymbol{I} - \boldsymbol{A})^{-1}\boldsymbol{b}$ with real parts greater than zero. An example for a system with one such pole is shown in Fig. 5.35.

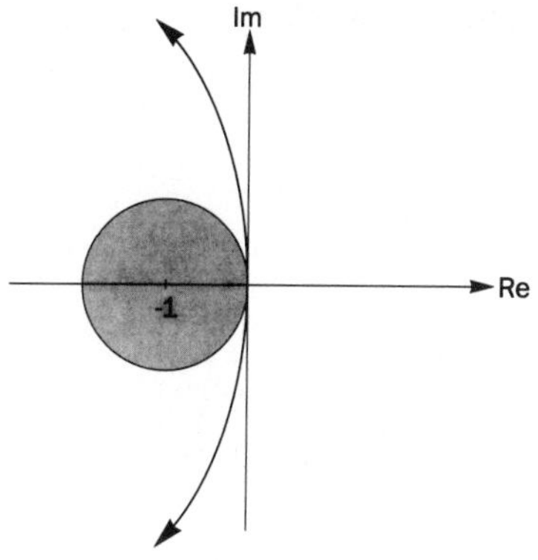

Fig. 5.34 Example of a Nyquist plot of return difference equation for an LQR controller.

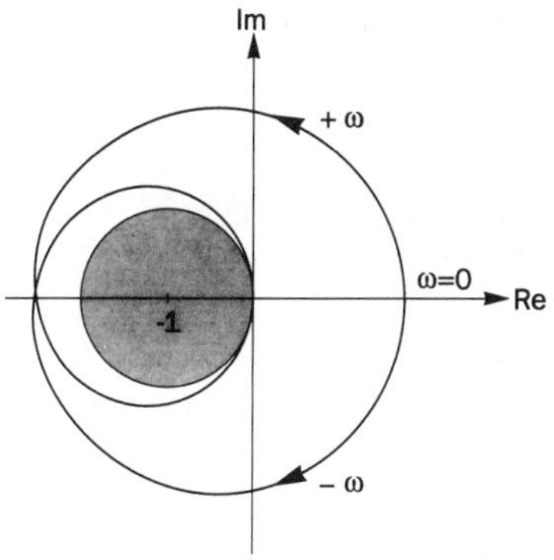

Fig. 5.35 A second example of a Nyquist plot of return difference equation for an LQR controller.

To assess the gain margin of the system, recall now that a closed loop system with a gain β will be asymptotically stable if it has a Nyquist plot which encircles the point $-1/\beta$ in a counterclockwise direction once for each pole in the transfer function $-\boldsymbol{k}(s\boldsymbol{I} - \boldsymbol{A})^{-1}\boldsymbol{b}$ with real parts greater than zero. From our discussion in the previous section, this means that for all $\beta > 1/2$ the system will be stable. In other words, the LQR system has (theoretically) an infinite gain margin (β can become infinitely large), and has a downsize margin (by which it can be reduced) of 1/2, or 6 dB.

Let us turn our attention now to assessing the phase margin of the system. Recall that phase margin is assessed from points on the Nyquist plot which are unit distance from the origin. Referring to Fig. 5.36, the smallest possible phase margin is therefore $\pm 60°$.

In conclusion, therefore, an SISO LQR regulator (theoretically) has an infinite gain margin, a downsize margin of 6 dB, and a phase margin of 60°

(originally derived by Anderson and Moore, 1971). (Note that this does not mean that the system will be stable if a phase change approaching 60° and a magnitude change approaching 1/2 are simultaneously introduced).

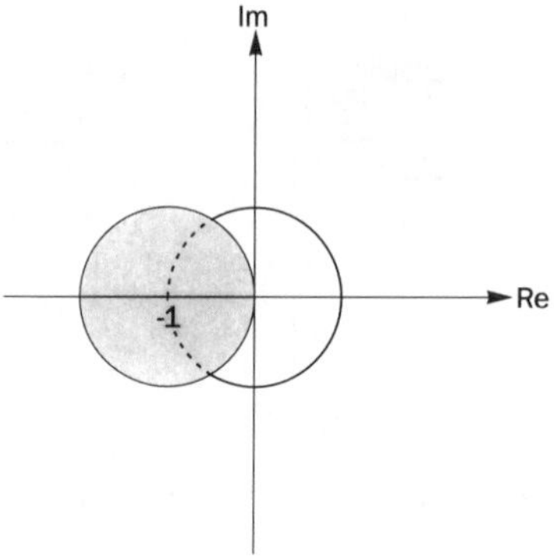

Fig. 5.36 Permissible locations (solid line) for a Nyquist plot at unit distance from the origin.

It would appear, then, that a control system designed using LQR methodology possesses some very desirable robustness properties. There are also, however, some undesirable features. First, time delays in the system will decrease the stability of the closed loop system. A time delay can be viewed as a frequency dependent phase shift, from which it is apparent that a delay $t < \pi/(3\omega)$ will cause instability unto itself. However, time delays will also reduce the (theoretically infinite) gain margin of the system, as the Nyquist plot will cross the real axis to the left of the origin.

Another undesirable feature relates to bandwidth. Note that for our system, the complementary sensitivity function is

$$T = 1-[1-\boldsymbol{k}(j\omega\boldsymbol{I}-\boldsymbol{A})^{-1}\boldsymbol{b}]^{-1} \tag{5.8.31}$$

As the frequency increases,

$$\lim_{\omega \to \infty} j\omega T = -\boldsymbol{kb} \tag{5.8.32}$$

This means that the complementary sensitivity function rolls off at a rate inversely proportion to the frequency, which is unattractively slow, especially for active noise and vibration control systems with uncertainties at high frequencies. Further, if the LQR design of $\boldsymbol{k}$ is conducted with little control effort weighting, the closed loop bandwidth, dictated by the eigenvalues of $\boldsymbol{A}+\boldsymbol{bk}$, can significantly exceed the open loop bandwidth, dictated by the eigenvalues of $\boldsymbol{A}$. It is possible (see Grimble and Owens, 1986) that the system will have very little robustness to variations in the entries of $\boldsymbol{b}$.

5.8.6 Frequency weighting

It is sometimes desirable to frequency-weight the design problem to improve the performance of the derived system. This situation occurs, for example, where a control system is being design to attenuate radiated acoustic power from a vibrating structure, which as an error criterion has implicit frequency dependence associated with it (this specific case is discussed further in Chapter 8). The most straightforward way to incorporate this weighting is to augment the plant model with frequency shaping filters, such that the performance index is additionally penalized in a frequency dependent manner (for a thorough discussion of this technique, the interested reader is referred to Gupta (1980), Anderson and Mingori (1985), Moore and Mingori (1987), and Anderson and Moore (1990)). In order to do this, we must first have a state space model of the desired frequency weighting

$$\dot{x}_f = Ax_f + B_f u_f$$
$$y_f = C_f x_f \tag{5.8.33}$$

where $u_f = y$. The augmented plant is therefore

$$\begin{bmatrix} \dot{x} \\ \dot{x}_f \end{bmatrix} = \begin{bmatrix} A & 0 \\ B_f C & A_f \end{bmatrix} \begin{bmatrix} x \\ x_f \end{bmatrix} + \begin{bmatrix} B \\ 0 \end{bmatrix} u \tag{5.8.34}$$

This augmented model can then be used in the standard LQR design approach.

5.9 OBSERVER DESIGN

In our discussions of LQR and pole placement design techniques in the previous sections, it was assumed that the entire state vector was available for feedback. Often, however, this is not the case; it is either impractical to measure some or all of the states, or the quality of the measurement, in terms of signal to noise ratio, is poor enough to reduce the utility of the compensator. In these instances we must estimate the unavailable state variables, using an observer (also referenced to an estimator). This approach will normally enable us to implement the compensator with acceptable performance.

There are two types of observers which we will discuss in this chapter. The first of these, discussed in this section, are linear observers (Luenberger, 1964, 1966, 1971). When implemented in an observable system, linear observers can be designed such that the difference between the actual system states and the states of the observer can be made to go to zero. The second type of observer, the Kalman filter, will be discussed in Section 5.11. This type of observer is designed with a knowledge of the process and observation noise, and can be viewed as an optimal observer.

If an observer is used to provide estimates of all state variables, it is referred to as a full order state observer. There are times, however, when it is possible to obtain a satisfactory measurement of some, but not all, states. In these instances a reduced order state observer may be implemented, targeting the estimation of only the unmeasurable states. In this section, we will consider the design of both full and reduced order linear state observers.

5.9.1 Full order observer design

Consider the case where we have a dynamic system, modelled in state variable format:

$$\dot{\boldsymbol{x}} = \boldsymbol{A}\boldsymbol{x} + \boldsymbol{B}\boldsymbol{u} \tag{5.9.1}$$

for which we have designed a compensator to attenuate some unwanted disturbance:

$$\boldsymbol{u} = -\boldsymbol{K}\boldsymbol{x} \tag{5.9.2}$$

Unfortunately, we do not have measures of the system states to implement the compensator. Rather, we are only able to measure the system output:

$$\boldsymbol{y} = \boldsymbol{C}\boldsymbol{x} \tag{5.9.3}$$

Such a case may exist, for example, where the vibration of a structure is targeted for attenuation, the model of the system is based on the system modes, but only a set of accelerometer measurements (the system output) is available, not a direct measurement of the system modes. In theory, if the observation matrix $\boldsymbol{C}$ is full rank, it would be possible to invert the matrix and obtain a measurement of the system states. This, however, is usually not a practical option.

To construct a better approach for estimating the states of a system, let us start with the idea that we want to model the plant dynamics

$$\frac{\mathrm{d}\hat{\boldsymbol{x}}}{\mathrm{d}t} = \boldsymbol{A}\hat{\boldsymbol{x}} + \boldsymbol{B}\boldsymbol{u} \tag{5.9.4}$$

where $\hat{\boldsymbol{x}}$ is an estimate of the actual system states. Here is assumed that we know the variables $\boldsymbol{A}$ and $\boldsymbol{B}$, as we used these in constructing the control law. If we define the error in the state estimate $\boldsymbol{e}$ as the difference between the true value of the states and the estimated value as follows:

$$\boldsymbol{e} = \boldsymbol{x} - \hat{\boldsymbol{x}} \tag{5.9.5}$$

then the dynamics of the error, found by subtracting (5.9.1) from (5.9.4), are governed by the expression

$$\dot{\boldsymbol{e}} = \boldsymbol{A}\boldsymbol{e} \tag{5.9.6}$$

This error converges at the same rate as the uncontrolled system, reaching the equilibrium value at the same time as the system. This is somewhat unacceptable, because if we were happy with the convergence of the dynamic system to equilibrium, we would not be contemplating the implementation of a control system!

To improve the convergence of the estimate of the state values to the actual values, we can employ a feedback-like scheme, feeding back some measure of the difference between what we have and what we want. An adequate measure can be provided by the system output. The difference between the actual system output and the system output we would expect based upon our estimated states is defined by the relationship

$$y - C\hat{x} = Cx - C\hat{x} = Ce \tag{5.9.7}$$

If we now define our (improved) estimate of the system dynamics by

$$\frac{\mathrm{d}\hat{x}}{\mathrm{d}t} = A\hat{x} + Bu + L(y - C\hat{x}) \tag{5.9.8}$$

where L is some gain matrix (the observer gains), the dynamics of the error, found by subtracting (5.9.1) from (5.9.8), will be defined by the relationship

$$\dot{e} = Ae + LCe \tag{5.9.9}$$

Equation (5.9.8) is the relationship for a full order linear observer, used for estimating all of the system states. A block diagram for the implementation of the observer is given in Fig. 5.37.

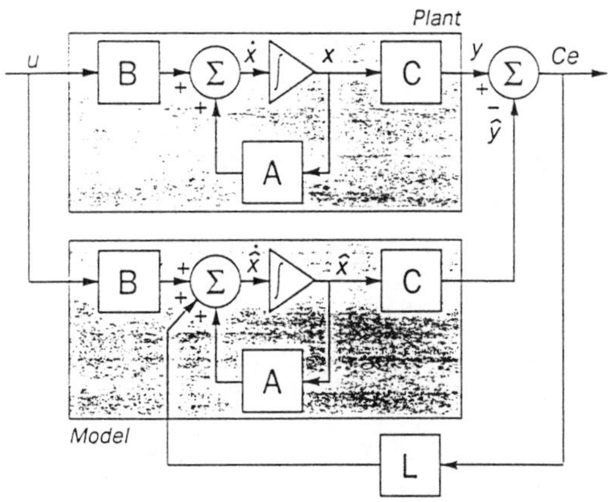

Fig. 5.37 Observer arrangement, where the observer models the plant.

There are several important points to emphasize concerning the relationship of (5.9.8). The first of these is that the estimation error is fed back to the model, and used directly to improve the estimate. This is the same sort of arrangement used in the system identification procedures discussed in Section 5.5, and has some intuitive appeal. The second point to note is that the impact which the error has upon the process of improving the result is governed by the observer gain L, a quantity we have not yet selected, and will discuss next.

The third point to note is that (5.9.8) is essentially the same form of equation we used in the pole placement exercise. Taking the Laplace transform of (5.9.8), the pole location of the errors in the state estimates are governed by the characteristic equation

$$|\, sI - A - LC \,| = s^n + \tilde{a}_1 s^{n-1} + \cdots + \tilde{a}_n \tag{5.9.10}$$

Therefore, selection of the observer gains will determine the location of the poles of the characteristic equation governing the convergence of the errors in our state estimates.

Observing that transposing a matrix does not change its eigenvalues, so that the eigenvalues of $\boldsymbol{A} - \boldsymbol{LC}$ are the same as the eigenvalues of $\boldsymbol{A}^{\mathrm{T}} - \boldsymbol{C}^{\mathrm{T}}\boldsymbol{L}^{\mathrm{T}}$, we can draw a direct analogy from our work in the previous section on pole placement for compensator gain determination. For a SISO system, the characteristic equation (5.9.10) is identical in form to the characteristic equation (5.7.5). Based upon this fact, if the desired 'response' of the estimation error is governed by a set of poles producing the desired characteristic equation

$$\gamma_e(s) = s^n + \alpha_1 s^{n-1} + \cdots + \alpha_{n-1}s + \alpha_n = 0 \tag{5.9.11}$$

then the required observer gain matrix is defined by the expression

$$\boldsymbol{l} = \boldsymbol{l}'\boldsymbol{T}^{-1} \tag{5.9.12}$$

where, analogous to (5.7.11)

$$l_i = \alpha_i - \tilde{a}_i, \qquad i = 1,2,\cdots,n \tag{5.9.13}$$

the $\tilde{a}$ terms are the poles of the original system (the eigenvalues of $\boldsymbol{A}$), and the transformation matrix $\boldsymbol{T}$ is the product of two submatrices,

$$\boldsymbol{T} = \boldsymbol{MN} \tag{5.9.14}$$

Whereas in the compensator gain pole placement exercise the first matrix $\boldsymbol{M}$ was the controllability matrix, in the observer gain problem it is the observability matrix:

$$\boldsymbol{M} = \left[\boldsymbol{c}^{\mathrm{T}}\boldsymbol{A}^{\mathrm{T}}\boldsymbol{c}^{\mathrm{T}} \;\cdots\; (\boldsymbol{A}^{\mathrm{funcn}}\boldsymbol{T})^{n-1}\boldsymbol{c}^{\mathrm{T}}\right] \tag{5.9.15}$$

The second matrix $\boldsymbol{N}$ is still the triangular Toeplitz matrix defined in (5.7.15). We can conclude from this that it is possible to arbitrarily place the estimated state poles only if the system is observable.

This provides us with a procedure for determining the gains of a linear full state estimator which will place the estimated state poles at some desired location:

1. Check for observability.
2. Determine the desired pole locations, and construct the desired characteristic equation.
3. Calculate the observer gains using (5.9.11).

As we may have guessed, it is also possible to use Ackermann's formula to derive the required observer gains. Going through the identical steps as the previous section, it is straightforward to shown that the observer gains are found by the relationship

$$\boldsymbol{L} = \gamma_e(\boldsymbol{A}^{\mathrm{T}})^{\mathrm{T}}\, \mathbf{O}^{-1}\, [0 \;\; 0 \;\; \cdots \;\; 0 \;\; 1]^{\mathrm{T}} \tag{5.9.16}$$

where **O** is the observability matrix.

In using pole placement to determine observer gains, it is common to place the observer poles well to the left of the controller poles on the complex plane. The rationale behind this is to make the observer poles relatively 'fast', so that the state estimates converge to (virtually) the correct values in a period of time which is relatively short.

5.9.2 Reduced order observers

The observer developed in the previous section is a full order observer, designed to reconstruct the entire state vector from the measured data. It is not uncommon, however, to have a system where it is possible to measure some, but not all, of the system states. In these instances it is possible to implement a reduced order observer (Luenberger, 1964, 1971; Gopinath, 1971), which will estimate only those state vectors which cannot be directly measured.

To develop a reduced order observer, we can partition the states into those $\boldsymbol{x}_m$ which can be measured and those $\boldsymbol{x}_u$ which are unmeasured and so must be estimated, so that the state equation is

$$\begin{bmatrix} \dot{\boldsymbol{x}}_m \\ \dot{\boldsymbol{x}}_u \end{bmatrix} = \begin{bmatrix} \boldsymbol{A}_{mm} & \boldsymbol{A}_{mu} \\ \boldsymbol{A}_{um} & \boldsymbol{A}_{uu} \end{bmatrix} \begin{bmatrix} \boldsymbol{x}_m \\ \boldsymbol{x}_u \end{bmatrix} + \begin{bmatrix} \boldsymbol{B}_m \\ \boldsymbol{B}_u \end{bmatrix} \boldsymbol{u} \tag{5.9.17}$$

with a system output

$$y = \begin{bmatrix} \boldsymbol{C}_m & \boldsymbol{0} \end{bmatrix} \begin{bmatrix} \boldsymbol{x}_m \\ \boldsymbol{x}_u \end{bmatrix} \tag{5.9.18}$$

We now have effectively two state equations, one for the measured variables

$$\dot{\boldsymbol{x}}_m = \boldsymbol{A}_{mm}\boldsymbol{x}_m + \boldsymbol{A}_{mu}\boldsymbol{x}_u + \boldsymbol{B}_m\boldsymbol{u} \tag{5.9.19}$$

and one for the unmeasured variables

$$\dot{\boldsymbol{x}}_u = \boldsymbol{A}_{uu}\boldsymbol{x}_u + (\boldsymbol{A}_{um}\boldsymbol{x}_m + \boldsymbol{B}_u\boldsymbol{u}) \tag{5.9.20}$$

It is useful now to compare the state equation for the unmeasured variables, (5.9.20), with the full state observer equation (5.9.8). For the full state observer equation the right-hand side terms are, in order, a term $\boldsymbol{A}$ describing the homogeneous system response, a constant input, $\boldsymbol{Bu}$, and a feedback term $\boldsymbol{LC}(\boldsymbol{x} - \hat{\boldsymbol{x}})$ based upon a measure of state estimation error. The right-hand side of (5.9.20) contains the first two of these; a term $\boldsymbol{A}_{uu}$ describing the homogeneous system response, and a constant input, $\boldsymbol{A}_{um}\boldsymbol{x}_m+\boldsymbol{b}_u\boldsymbol{u}$. To obtain the third term, which will allow us to fit the reduced order observer problem into the format we developed for the full order observer, note that the measured state (5.9.19) can be reordered as

$$\dot{\boldsymbol{x}}_m - \boldsymbol{A}_{mm}\boldsymbol{x}_m - \boldsymbol{B}_m\boldsymbol{u} = \boldsymbol{A}_{mu}\boldsymbol{x}_u \tag{5.9.21}$$

The left-hand side of (5.9.21) is a completely measurable, or known, quantity, while the right-hand side is unknown, and must be estimated. Therefore, this equation can be used to generate an error expression for the unmeasured states:

$$\text{error} = \dot{x}_m - A_{mm}x_m - B_m u - A_{mu}\hat{x}_u \qquad (5.9.22)$$

Using this, we can construct a reduced order observer, which will estimate only the unmeasured states, as

$$\frac{d\hat{x}_u}{dt} = A_{uu}\hat{x}_u + A_{um}x_m + B_u u + L(\dot{x}_m - A_{mm}x_m - B_m u - A_{mu}\hat{x}_u) \qquad (5.9.23)$$

If we defined our estimation error as the difference between the actual values of the estimated states and the estimated values as follows:

$$e_u = x_u - \hat{x}_u \qquad (5.9.24)$$

we find that the dynamics of the error are described by the relationship

$$\dot{e}_u = (A_{uu} - LA_{mu})\, e_u \qquad (5.9.25)$$

The characteristic equation of the estimation error is therefore

$$\gamma_e(s) = |\ sI-(A_{uu}-LA_{mu})\ | \qquad (5.9.26)$$

The gains of the reduced order observer can now be determined using the pole placement-like procedures we developed for the full order observer. If the procedure based upon transformation into controller canonical form is used, the submatrix M of (5.9.15) is now

$$M = \left[A_{mu}{}^{\mathrm{T}} \;\; A^{\mathrm{T}}A_{mu}{}^{\mathrm{T}} \;\; \cdots \;\; (A^{\mathrm{T}})^{n-1}A_{mu}{}^{\mathrm{T}}\right] \qquad (5.9.27)$$

If Ackermann's formula is used, the observability matrix is now equal to

$$\mathbf{O} = \begin{bmatrix} A_{mu} \\ A_{mu}A_{uu} \\ A_{mu}A_{uu}^2 \\ \vdots \\ A_{mu}A_{uu}^{n-2} \end{bmatrix} \qquad (5.9.28)$$

There is, however, one remaining problem with the reduced order observer of (5.9.23): it requires differentiation of a term. Differentiation in a controller or observer is inadvisable, as it tends to amplify high frequency noise in the measurements. To get around this, we can define a new state vector x_c

$$x_c = \hat{x}_u - Ly \qquad (5.9.29)$$

In terms of this new state vector, the reduced order observer is implemented as

$$\dot{x}_c = (A_{uu} - LA_{mu})\hat{x}_u + (A_{um} - LA_{mm})y + (B_u - LB_m)u \qquad (5.9.30)$$

5.10 RANDOM PROCESSES REVISITED

From the standpoint of active control, the principal purpose of using feedback is to decrease the response of a given system to disturbance inputs. Thus far, we have considered the disturbance input to be essentially some unspecified set of initial conditions, and have designed control systems to improve the response of the dynamic system in returning to equilibrium. While this approach is useful from the design perspective, it is somewhat unrealistic; in practice, the disturbance input is likely to be somewhat more complicated, and is often a random process. This is compounded by noise in measurements when using practical sensors, noise which is also random. If our control system is really 'optimal', as we claimed using the design methodology of the previous section, it must be able to accommodate these random disturbances inflicted upon our system.

We have discussed random processes previously in this book, in Chapter 3 from the standpoint of spectral analysis. In this section we will revisit the random process, adding more information which is directly relevant to both the discussion in the next section on optimal observers, and the discussion in the next chapter on the Weiner filtering problem. As with other topics in this 'review' chapter, our discussion of random processes will be necessarily brief. For a more comprehensive discussion, the reader is referred to Papoulis (1965), Jazwinski (1970), and Bendat and Piersol (1980).

5.10.1 Models and characteristics

From the standpoint of feedback control system design, for a system subject to random disturbances, our aim is to design a control law for a system modelled in state space format as

$$\begin{aligned} \dot{\boldsymbol{x}} &= \boldsymbol{A}\boldsymbol{x} + \boldsymbol{B}\boldsymbol{u} + \boldsymbol{v} \\ \boldsymbol{y} &= \boldsymbol{C}\boldsymbol{x} + \boldsymbol{w} \end{aligned} \tag{5.10.1}$$

where $\boldsymbol{v}$ and $\boldsymbol{w}$ are random processes. As the disturbance inputs and sensor noise are random processes, the response of our system at any instant in time will also be random. Therefore, we want to optimize our design not on the basis of response to a single disturbance, but rather on the basis of some statistical criterion, or some form of average behaviour. To formulate this criterion, we will need to examine the properties of random processes in a little more depth.

Unlike the deterministic processes we have considered thus far in this chapter, the state of a random process at any instant in the future *cannot* be predicted solely from a knowledge of the present state. Conceptually, a random process can be viewed as comprised of a set of individual processes which are statistically similar to each other, but because of some small variations, have outputs which are all different; the ensemble of outputs is the output of the random process. If we had access to each member of this ensemble of processes, and could take the ensemble average, or expected value, of their characteristics, we would have the characteristics of the random process. For example, if there were an ensemble of N individual processes making up a given random process, the mean value of the random process would be defined by the relationship

$$\bar{x}(t) = \frac{1}{N}\sum_{i=1}^{N} x_i(t) \tag{5.10.2}$$

the mean square value by

$$\overline{x^2(t)} = \frac{1}{N}\sum_{i=1}^{N} x_i^2(t) \tag{5.10.3}$$

the variance by

$$\sigma^2(t) = \overline{x^2(t)} - [\ \bar{x}(t)\ ]^2 \tag{5.10.4}$$

and the correlation by

$$r(t,\tau) = \frac{1}{N}\sum_{i=1}^{N} x_i(t)x_i(\tau) \tag{5.10.5}$$

There are two obvious problems which confront us at this stage. First, if we design a control system based upon some set of statistics derived from past and present measurements, what assurances do we have that the statistical model will continue to be valid in the future? Based upon our previous discussion, the answer is, in general, none. Second, how much data are required before we can make statistical inferences about a random process? The answer is, we don't know; it is system dependent. In fact, it is not uncommon to pick statistical parameters in response to the need for data (to facilitate our calculations), and not vice versa. In response to these somewhat cynical observations we could ask, why bother to design a control system based upon statistical measures? The answer is, what choice do we have? In fact, the whole point of this discussion is to bring home the fact that when we talk about 'optimal' control system designs in response to some form of random excitation, we must say it with some reservation; the design will only be optimal for the model of the excitation process we have chosen.

From a theoretical standpoint, a random process is characterized by an infinite series of probability density functions, which describe the probability of a given state having a given value at some point in time (probability density functions will be denoted here by $pdf\{x,t\}$). Obtaining a measure of these functions is normally not a practical option, and so their use is in mathematical development of problems. In terms of the probability density function, the statistical measures outlined in (5.10.2)–(5.10.5) are defined by the relationships

$$\bar{x}(t) = E\{x(t)\} = \int_{-\infty}^{\infty} x\ pdf\{x,t\}\ \mathrm{d}x \tag{5.10.6}$$

$$\overline{x^2(t)} = E\{x^2(t)\} = \int_{-\infty}^{\infty} x^2\ pdf\{x,t\}\ \mathrm{d}x \tag{5.10.7}$$

$$\sigma^2(t) = E\{[x(t)-\bar{x}(t)]^2\} = \int_{-\infty}^{\infty} [x-\bar{x}(t)]^2 \; pdf\{x,t\} \; \mathrm{d}t \tag{5.10.8}$$

and

$$r(t,\tau) = E\{x(t)x(\tau)\} = \int_{-\infty}^{\infty}\int_{-\infty}^{\infty} x_1 x_2 \; pdf\{x_1,x_2;t,\tau\} \; \mathrm{d}x_1 \mathrm{d}x_2 \tag{5.10.9}$$

In (5.10.6)−(5.10.9), $E\{\ \}$ denotes the expected value, or the ensemble average. The first three of these parameters are referred to as first order statistics, as they utilize a first order probability density function (concerned with the probability of the value of a single state). The final parameter, correlation, is a second order statistic, describing the probability of the values of two states.

The first order statistics can be applied to vector as well as scalar processes. For example, if

$$\boldsymbol{x}(t) = \begin{bmatrix} x_1(t) \\ x_2(t) \\ \vdots \\ x_n(t) \end{bmatrix} \tag{5.10.10}$$

then

$$\bar{\boldsymbol{x}}(t) = E\{\boldsymbol{x}(t)\} = \begin{bmatrix} \bar{x}_1(t) \\ \bar{x}_2(t) \\ \vdots \\ \bar{x}_n(t) \end{bmatrix} = \begin{bmatrix} E\{x_1(t)\} \\ E\{x_2(t)\} \\ \vdots \\ E\{x_n(t)\} \end{bmatrix} \tag{5.10.11}$$

For the second order statistic, the generalization of the correlation function is the correlation matrix

$$\boldsymbol{R}(t,\tau) = E\{\boldsymbol{x}(t)\boldsymbol{x}^{\mathrm{T}}(t)\} =$$

$$\begin{bmatrix} E\{x_1(t)x_1(\tau)\} & E\{x_1(t)x_2(\tau)\} & \cdots & E\{x_1(t)x_n(\tau)\} \\ E\{x_2(t)x_1(\tau)\} & E\{x_2(t)x_2(\tau)\} & \cdots & E\{x_2(t)x_n(\tau)\} \\ & & \vdots & \\ E\{x_n(t)x_1(\tau)\} & E\{x_n(t)x_2(\tau)\} & \cdots & E\{x_n(t)x_n(\tau)\} \end{bmatrix} \tag{5.10.12}$$

The infinite set of probability density functions which describe a random process are, in general, functions of time. If the probability density functions are time invariant, the random process is stationary 'in the strict sense'. If it is only known that the first and second order statistics are stationary (as higher order statistics are often hard to determine), the process is referred to as stationary 'in the wide sense'.

One final point to note in this discussion of models and characteristics is that, in general, there is no relationship between the ensemble average of a random process and the time average of a random process. However, in the special case of a stationary random process having an ensemble and time average which are equal, the process is referred to as 'ergodic'.

5.10.2 White noise

White noise is a random process with a mean (expected) value of zero, and a flat power spectrum. From our discussion of power spectra in Chapter 3, this latter property can be expressed as

$$S(\omega) = W \tag{5.10.13}$$

where W is some constant number. As the inverse Fourier transform of a constant is a unit impulse, the correlation function of white noise, defined by the inverse Fourier transform of the power spectrum, is

$$r(\tau) = W\delta(\tau) \tag{5.10.14}$$

where $\delta(\tau)$ is a unit impulse at the origin. For a vector process, the correlation function for a white noise process is

$$\boldsymbol{R}(\tau) = E\{\boldsymbol{x}(t)\boldsymbol{x}^{\mathrm{T}}(t+\tau)\} = \boldsymbol{W}\delta(\tau) \tag{5.10.15}$$

where $\boldsymbol{W}$ is a square matrix.

It is convenient to model a random disturbance exciting a linear system as white noise, because it leads to relatively simple expressions for the power spectrum (based upon the system transfer function) and correlation function (based upon the system impulse response matrix) of the system output. To derive these expressions, consider a linear system with an input $u(t)$ and output $y(t)$. We can relate the input and output through the superposition integral

$$\boldsymbol{y}(t) = \int_0^t \boldsymbol{H}(t,\gamma)\boldsymbol{u}(\gamma)\,\mathrm{d}\gamma \tag{5.10.16}$$

where $\boldsymbol{H}(t,\gamma)$ is the impulse response matrix (for an MIMO system, which becomes a single impulse response function for an SISO system). The are two aspects of equation (5.10.16) which should be noted at this stage. First, as the expected value (mean) of the input $\boldsymbol{u}$ is zero for white noise, the expected value (mean) of the output $\boldsymbol{y}$ is also zero. Second, note that the description in (5.10.16) is of a causal system, with the lower bounds on the integral equal to zero.

Beginning with the derivation of the correlation function, from the description in (5.10.12), the correlation matrix for this process is

$$\boldsymbol{R}_y(t,\tau) = E\{\boldsymbol{y}(t)\boldsymbol{y}^{\mathrm{T}}(\tau)\}$$

$$= E\left\{\int_0^T \boldsymbol{H}(t,\gamma)\boldsymbol{u}(\gamma)\,\mathrm{d}\gamma \times \int_0^\tau \boldsymbol{u}^{\mathrm{T}}(\lambda)\boldsymbol{H}^{\mathrm{T}}(\tau,\lambda)\,\mathrm{d}\gamma\right\} \qquad (5.10.17)$$

$$= E\left\{\int_0^t \int_0^\tau \boldsymbol{H}(t,\gamma)\boldsymbol{u}(\gamma)\boldsymbol{u}^{\mathrm{T}}(\lambda)\boldsymbol{H}^{\mathrm{T}}(\tau,\lambda)\,\mathrm{d}\gamma\,\mathrm{d}\lambda\right\}$$

Let us assume now that all random variation in the bracketed expression in (5.10.17) occurs at the system input, and that the input $\boldsymbol{u}$ is white noise. From the latter assumption, it follows that

$$E\{\boldsymbol{u}(\gamma)\boldsymbol{u}^{\mathrm{T}}(\lambda)\} = \boldsymbol{W}\delta(\gamma-\lambda) \qquad (5.10.18)$$

Using the former assumption and the result of (5.10.18), (5.10.17) can be simplified to

$$\boldsymbol{R}_y(t,\tau) = \int_0^t \int_0^\tau \boldsymbol{H}(t,\gamma)\ \boldsymbol{W}\delta(\gamma-\lambda)\ \boldsymbol{H}^{\mathrm{T}}(\tau,\lambda)\,\mathrm{d}\gamma\,\mathrm{d}\lambda \qquad (5.10.19)$$

Using the property of the delta function

$$\int_{t_1}^{t_2} f(\lambda)\delta(\gamma-\lambda)\,\mathrm{d}\lambda = f(\gamma), \qquad t_1 \le \gamma \le t_2 \qquad (5.10.20)$$

(5.10.19) can be further simplified to

$$\boldsymbol{R}_y(t-\tau) = \int_0^t \boldsymbol{H}(t,\gamma)\ \boldsymbol{W}\ \boldsymbol{H}^{\mathrm{T}}(\tau,\gamma)\,\mathrm{d}\gamma \qquad (5.10.21)$$

Let us now make the further assumption that our system is LTI, such that

$$\boldsymbol{H}(t,\tau) = \boldsymbol{H}(t-\tau) \qquad (5.10.22)$$

Substituting this into (5.10.21), we obtain

$$\boldsymbol{R}_y(t,\tau) = \int_0^t \boldsymbol{H}(t-\tau)\ \boldsymbol{W}\ \boldsymbol{H}^{\mathrm{T}}(\tau-\gamma)\,\mathrm{d}\gamma \qquad (5.10.23)$$

or, replacing τ by $t+\tau$

$$\boldsymbol{R}_y(t,t+\tau) = \int_0^t \boldsymbol{H}(t-\gamma)\ \boldsymbol{W}\ \boldsymbol{H}^{\mathrm{T}}(t-\gamma+\tau)\,\mathrm{d}\gamma$$

$$(5.10.24)$$

$$= \int_0^t \boldsymbol{H}(\eta)\ \boldsymbol{W}\ \boldsymbol{H}^{\mathrm{T}}(\eta+\tau)\,\mathrm{d}\eta$$

where $\eta = t - \gamma$. Assuming that our system us stable, as the process approaches steady state we find that the correlation matrix for a system excited by white noise is defined by the expression

$$\lim_{t \to \infty} \boldsymbol{R}_y(t,t+\tau) = \overline{\boldsymbol{R}}_y(\tau) = \int_0^{\infty} \boldsymbol{H}(\eta)\ \boldsymbol{W}\ \boldsymbol{H}^{\mathrm{T}}(\eta+\tau)\ d\eta \tag{5.10.25}$$

Note that (5.10.25) shows that the correlation matrix is defined entirely by the impulse response of the system, and the amplitude of the white noise input.

Let us turn our attention now to the output power spectrum. As the output power spectrum matrix is the Fourier transform of the correlation matrix, given for white noise excitation in (5.10.25), it is defined by the expression

$$\boldsymbol{S}(\omega) = \int_{-\infty}^{\infty} \int_0^{\infty} \boldsymbol{H}(\eta)\ \boldsymbol{W}\ \boldsymbol{H}^{\mathrm{T}}(\eta+\tau)\ \mathrm{d}\eta\ \mathrm{e}^{-j\omega\tau} \mathrm{d}\tau \tag{5.10.26}$$

To simplify (5.10.26), let us rewrite the integral expression as

$$\boldsymbol{S}(\omega) = \int_0^{\infty} \boldsymbol{H}(\eta)\ \boldsymbol{W} \left[\int_{-\infty}^{\infty} \boldsymbol{H}^{\mathrm{T}}(\eta+\tau)\ \mathrm{e}^{-j\omega\tau} \mathrm{d}\tau \right] \mathrm{d}\eta \tag{5.10.27}$$

The bracketed part of (5.10.27) can be re-expressed as

$$\begin{aligned} \int_{-\infty}^{\infty} \boldsymbol{H}^{\mathrm{T}}(\eta+\tau)\ \mathrm{e}^{-j\omega\tau}\ \mathrm{d}\tau &= \int_{-\infty}^{\infty} \boldsymbol{H}^{\mathrm{T}}(\gamma)\ \mathrm{e}^{-j\omega(\gamma-\eta)}\ \mathrm{d}\gamma \\ &= \boldsymbol{H}^{\mathrm{T}}(j\omega)\ \mathrm{e}^{j\omega\eta} \end{aligned} \tag{5.10.28}$$

where $\boldsymbol{H}(j\omega)$ is the transfer function

$$\boldsymbol{H}^{\mathrm{T}}(j\omega) = \int_{-\infty}^{\infty} \boldsymbol{H}^{\mathrm{T}}(\gamma)\ \mathrm{e}^{-j\omega\gamma}\ \mathrm{d}\gamma \tag{5.10.29}$$

Note that as our system is causal, there are no time components less than zero, so that

$$\boldsymbol{H}^{\mathrm{T}}(j\omega) = \boldsymbol{H}^{\mathrm{T}}(s) = \int_0^{\infty} \boldsymbol{H}^{\mathrm{T}}(\gamma)\ \mathrm{e}^{-j\omega\gamma}\ \mathrm{d}\gamma \tag{5.10.30}$$

Returning to (5.10.27), substituting (5.10.28) for the bracketed part of the expression produces

$$\boldsymbol{S}(\omega) = \int_0^{\infty} \boldsymbol{H}(\eta)\ \mathrm{e}^{j\omega\eta}\ \mathrm{d}\eta \times \boldsymbol{W}\ \boldsymbol{H}^{\mathrm{T}}(j\omega) \tag{5.10.31}$$

Noting again that our system is causal, the bounds on the integral in (5.10.31) can be expanded to $-\infty$ without altering the final result, enabling the expression to be simplified to

$$\boldsymbol{S}(\omega) = \boldsymbol{H}(-j\omega)\ \boldsymbol{W}\ \boldsymbol{H}^{\mathrm{T}}(j\omega) \tag{5.10.32}$$

Thus, the output power spectrum of a system excited by white noise is defined by the product of the system transfer function at negative frequencies, the white noise spectral density matrix, and the system transfer function at positive frequencies.

5.10.3 State space models

Let us now examine the problem of a system modelled using state space representation and subject to random excitation, described by (5.10.1). In this equation, $\boldsymbol{v}$ is the (white noise) disturbance input, responsible for excitation of the system, and $\boldsymbol{w}$ is observation noise at the measurement sensor. Our aim now is to characterize the response of the system to the white noise process. To do this, we will initially ignore both the observation noise and control input, and restrict our consideration to the excitation of the system by the disturbance input.

Ignoring the observation noise and control input terms, the system model of (5.10.1) becomes

$$\begin{aligned} \dot{\boldsymbol{x}} &= \boldsymbol{A}\boldsymbol{x} + \boldsymbol{v} \\ y &= \boldsymbol{C}\boldsymbol{x} \end{aligned} \tag{5.10.33}$$

To examine the response of the system to the white noise input, we begin with the solution to the state equations derived in Section 5.3, stated in (5.3.91). For the present system, the solution to the state equation (5.10.33) is

$$\boldsymbol{x}(t) = \boldsymbol{\Phi}(t,t_0)\boldsymbol{x}(t_0) + \int_{t_0}^{t} \boldsymbol{\Phi}(t,\tau)\boldsymbol{v}(\tau)\,\mathrm{d}\tau \tag{5.10.34}$$

where t_0 is some starting time and $\boldsymbol{\Phi}$ is the state transition matrix. Based upon our work in the previous section, a suitable way to characterize the system response is in terms of the correlation matrix. To derive a relationship describing this, we can begin by writing

$$\begin{aligned} \boldsymbol{x}(t)\boldsymbol{x}^{\mathrm{T}}(\tau) &= \boldsymbol{\Phi}(t,t_0)\boldsymbol{x}(t_0)\boldsymbol{x}^{\mathrm{T}}(t_0)\boldsymbol{\Phi}^{\mathrm{T}}(\tau,t_0) + \boldsymbol{\Phi}(t,t_0)\boldsymbol{x}(t_0) \\ &\times \left[\int_{t_0}^{t} \boldsymbol{\Phi}(t,\gamma)\boldsymbol{v}(\gamma)\,\mathrm{d}\gamma\right]^{\mathrm{T}} \int_{t_0}^{t} \boldsymbol{\Phi}(t,\gamma)\boldsymbol{v}(\gamma)\,\mathrm{d}\gamma \times \boldsymbol{x}^{\mathrm{T}}(t_0)\boldsymbol{\Phi}^{\mathrm{T}}(\tau,t_0) \\ &+ \int_{t_0}^{t}\int_{t_0}^{\tau} \boldsymbol{\Phi}(t,\gamma)\boldsymbol{v}(\gamma)\boldsymbol{v}^{\mathrm{T}}(\lambda)\boldsymbol{\Phi}^{\mathrm{T}}(\tau,\lambda)\,\mathrm{d}\lambda\,\mathrm{d}\gamma \end{aligned} \tag{5.10.35}$$

If we now take expected values of both sides of (5.10.35), and note that for white noise excitation

$$E\left\{\int_{t_0}^{t} \boldsymbol{\Phi}(t,\gamma)\boldsymbol{v}(\gamma)\,\mathrm{d}\gamma\right\} = \int_{t_0}^{t} \boldsymbol{\Phi}(t,\gamma)\,E\{\boldsymbol{v}(\gamma)\}\,\mathrm{d}\gamma = 0 \tag{5.10.36}$$

we obtain

$$R_x(t,\tau) = \Phi(t,t_0)\ E\{x(t_0)x^{\mathrm{T}}(t_0)\}\ \Phi^{\mathrm{T}}(t,t_0) + \int_{t_0}^{t}\int_{t_0}^{\tau} \Phi(t,\gamma)\ E\{v(\gamma)v^{\mathrm{T}}(\lambda)\}\ \Phi^{\mathrm{T}}(\tau,\lambda)\ \mathrm{d}\lambda\ \mathrm{d}\gamma \tag{5.10.37}$$

The first expected value expression in (5.10.37) is not precisely known for random excitation. Therefore, we will say

$$E\{x(t_0)x^{\mathrm{T}}(t_0)\} = P(t_0) \tag{5.10.38}$$

where $P(t_0)$ is the covariance matrix of $x(t_0)$.

The second expected value expression can be evaluated simply as

$$E\{v(\gamma)v^{\mathrm{T}}(\lambda)\} = W_v(\gamma)\delta(\gamma-\lambda) \tag{5.10.39}$$

Using this latter result, the expression in the double integral in (5.10.37) can be re-expressed as

$$\int_{t_0}^{t}\int_{t_0}^{\tau} \Phi(t,\gamma)\ E\{v(\gamma)v^{\mathrm{T}}(\lambda)\}\ \Phi^{\mathrm{T}}(\tau,\lambda)\ \mathrm{d}\lambda\ \mathrm{d}\gamma = \int_{t_0}^{t} \Phi(t,\gamma)W_v(\gamma)\left[\int_{t_0}^{\tau} \delta(\gamma-\lambda)\Phi^{\mathrm{T}}(t,\lambda)\ \mathrm{d}\lambda\right]\mathrm{d}\gamma \tag{5.10.40}$$

Note that the bracketed part of (5.10.40) can be evaluated as

$$\int_{t_0}^{\tau} \delta(\gamma-\lambda)\Phi^{\mathrm{T}}(t,\lambda)\ \mathrm{d}\lambda = \begin{cases} \Phi^{\mathrm{T}}(t,\gamma) & t_0<\gamma<t \\ 0 & \text{otherwise} \end{cases} \tag{5.10.41}$$

Using the results of (5.10.38)–(5.10.41), the expression defining the correlation matrix in (5.10.37) can be simplified to

$$R_x(t,\tau) = \Phi(t,t_0)P(t_0)\Phi^{\mathrm{T}}(t,t_0) + \int_{t_0}^{t_m} \Phi(t,\gamma)W_v(\gamma)\Phi^{\mathrm{T}}(\tau,\lambda)\ \mathrm{d}\lambda \tag{5.10.42}$$

where t_m is the minimum of t and τ.

Note that the transition matrices in (5.10.42) can be re-expressed as the products

$$\Phi(\tau,t_0) = \Phi(\tau,t)\Phi(t,t_0) \tag{5.10.43}$$

$$\Phi(\tau,\gamma) = \Phi(\tau,t)\Phi(t,\gamma) \tag{5.10.44}$$

Using these relationships, we can express the correlation matrix in terms of the covariance matrix as

$$R_x(t,\tau) = P(t)\Phi^{\mathrm{T}}(\tau,t) \quad \tau \geq t \tag{5.10.45}$$

where

$$P(t) = R_x(t,t) = \Phi(t,t_0)P(t_0)\Phi(t,t_0) + \int_{t_0}^{t} \Phi(t,\gamma)W_v(\gamma)\Phi^{\mathrm{T}}(t,\lambda)\ \mathrm{d}\lambda \tag{5.10.46}$$

Equation (5.10.45) and (5.10.46) describe the evolution of the correlation matrix with time, for $\tau \geq t$. To express this evolution in the form of a differential equation, we can write

$$\dot{P}(t) = \frac{\partial \Phi(t,t_0)}{\partial t} P(t_0)\Phi^{\mathrm{T}}(t,t_0) + \Phi(t,t_0)\frac{\partial \Phi^{\mathrm{T}}(t,t_0)}{\partial t} + \frac{\partial}{\partial t}\left[\int_{t_0}^{t} \Phi(t,\gamma)W_v(\gamma)\Phi^{\mathrm{T}}(t,\lambda)\ \mathrm{d}\lambda \right] \tag{5.10.47}$$

Noting that to differentiate an integral,

$$\frac{\partial}{\partial t}\int_0^t f(t,\lambda)\ \mathrm{d}\lambda = f(t,t) + \int_0^t \frac{\partial f(t,\lambda)}{\partial t}\ \mathrm{d}\lambda \tag{5.10.48}$$

and also the relationship

$$\frac{\partial \Phi(t,\tau)}{\partial t} = A(t)\Phi(t,\tau) \tag{5.10.49}$$

where A is state matrix, it is straightforward to simplify (5.10.48) to

$$\dot{P} = AP + PA^{\mathrm{T}} + W_v \tag{5.10.50}$$

where the initial conditions are

$$P(t)\Big|_{t=t_0} = P(t_0) \tag{5.10.51}$$

Equation (5.10.50) is referred to as the variance equation, which defines the evolution of the covariance matrix without the need to calculate the transition matrix. Note that if the white noise process is steady state, such that the derivative is zero, (5.10.50) can be simplified to

$$0 = AP + PA^{\mathrm{T}} + W_v \tag{5.10.52}$$

or in the more common form of a Lyapunov equation,

$$A^{\mathrm{T}}P + PA = -W_v \tag{5.10.53}$$

The correlation matrix for the output y is readily obtained from that derived for the state vector. Noting that

$$y(t)y^{T}(\tau) = C(t)x(t)x^{T}(\tau)C^{T}(\tau) \tag{5.10.54}$$

the correlation matrix for the output is simply

$$\begin{aligned} R_y(t,\tau) &= E\{y(t)y^{T}(\tau)\} = C(t)\ E\{x(t)x^{T}(\tau)\}\ C^{T}(\tau) \\ &= C(t)R_x(t,\tau)C^{T}(\tau) \end{aligned} \tag{5.10.55}$$

In terms of the covariance matrix derived previously,

$$P_y(t) = R_y(t,t) = C(t)P(t)C^{T}(t) \tag{5.10.56}$$

We will now make use of these results in deriving an optimal observer, a Kalman filter.

5.11 OPTIMAL OBSERVERS: THE KALMAN FILTER

In this section we will undertake a short discussion of the optimal filtering problem, the result of which is a derivation of the Kalman filter (Kalman, 1960; Kalman and Bucy, 1961). This instrument must rank as one of the most important contributions to control theory in the twentieth century, and can be applied to a variety of filtering, estimation, prediction and control problems.

Prior to the 1960s, the Weiner filter was the principal vehicle for optimal estimation problems. This derived a system model based upon the statistical properties of the input and output time histories, expressing the results in terms of transfer functions or impulse response functions (the Weiner filter is discussed further in the next chapter, as an introduction to adaptive signal processing). Unfortunately, the transfer function basis of the Weiner solution makes it difficult to implement in most systems, expensive to calculate, and limits its utility to stationary, linear systems.

The Kalman filter appeared in the 1960s, at a time when state space control methods were beginning to receive serious attention. Unlike the transfer function basis of its Weiner filter predecessor, the Kalman filter was expressed in the form of a differential equation, the form of a state space equation. Its gains were computationally cheap to derive, which makes it ideal for use with modest computers (the only type available in the 1960s). Its original derivation can also be extended to enable use in non-linear systems.

In our discussion we will limit ourselves to the use of Kalman filters as optimal state estimators. Our consideration will be necessarily brief, barely touching on the broad range of information available. For further information, the reader is referred to Anderson and Moore (1970).

5.11.1 Problem formulation

Consider the dynamic process described by the state equation

$$\dot{x} = Ax + Bu + v \tag{5.11.1}$$

where v is a white noise process, having a known spectral density matrix. The observations (outputs) of this process are governed by the expression

$$y = Cx + w \tag{5.11.2}$$

where w is also a white noise process with a known spectral density matrix. Our problem is to construct an optimal state observer for x.

Based on our previous work in Section 5.9, we can postulate that the desired observer has the form

$$\frac{\mathrm{d}\hat{x}}{\mathrm{d}t} = A\hat{x} + Bu + L(y - C\hat{x}) \tag{5.11.3}$$

Our problem is to find the optimal (under any reasonable error criterion) observer gain matrix L, which is the derivation of the Kalman filter. There are two points which should be noted here. First, the Kalman filter was derived several years before the concept of an observer was presented; the optimal observer came before the observer! Second, by assuming the form of the Kalman filter to be defined as in (5.11.3) the derivation is simplified. It is possible to derive the Kalman filter without any assumptions regarding the form of solution, but this is beyond the scope of this text.

The optimal state estimate we desire, x_e, is one which minimizes the error variance (a minimum variance estimate). In other words, given the measurement of $y(t)$, $t_i \le t \le t_f$, we want an estimate of the state vector $\hat{x}$ which minimizes the norm of the covariance matrix of the error:

$$P_e = E\{e(t_f)e^{\mathrm{T}}(t_f)\} = E\{[x(t_f) - \hat{x}(t_f)][x(t_f) - \hat{x}(t_f)]^{\mathrm{T}}\} \tag{5.11.4}$$

(recall that the estimation error $e(t_f)$ is defined as the difference between the actual and estimated state values, $x(t_f) - \hat{x}(t_f)$). This optimal estimate is easily shown to be the conditional mean of the state vector, or the expected value of $x(t)$ given the observation data $y(t)$,

$$x_e(t_f) = E\{x(t_f) \mid y(t),\ t_i \le t \le t_f\} \tag{5.11.5}$$

To derive the optimal observer gain, we begin with the differential equation describing the estimation error 'process':

$$\begin{aligned} \dot{e} = \dot{x} - \dot{x}_e &= Ax + v - Ax_e - L(Cx + w - Cx_e) \\ &= (A - LC)e + v - Lw \end{aligned} \tag{5.11.6}$$

As both v and w are white noise processes, the difference

$$\xi = v - Lw \tag{5.11.7}$$

is also a white noise process. Therefore, the error process is simply defined by the differential equation of a system excited by white noise,

$$\dot{e} = (A - LC)e + \xi \tag{5.11.8}$$

Note that the form of (5.11.8) is the same as that of (5.10.33). Therefore, we can describe the covariance of the estimation error using a variance equation of the form given in (5.10.50), derived for describing the covariance of the state vector. To do this, we must first calculate the covariance of the white noise process ξ. Using the definition given in (5.11.7), the covariance matrix of ξ is defined by the expression

$$\begin{aligned} E\{\xi(t)\xi^{T}(\tau)\} &= E\{v(t)v^{T}(\tau)\} - L(t)E\{w(t)v^{T}(\tau)\} \\ &\quad - E\{v(t)w^{T}(\tau)\}L^{T}(\tau) + L(t)E\{w(t)w^{T}(\tau)\}L^{T}(\tau) \end{aligned} \tag{5.11.9}$$

As all the processes are white noise, we can evaluate the following expected values:

$$E\{v(t)v^{T}(\tau)\} = V(t)\delta(t-\tau), \quad E\{v(t)\} = 0 \tag{5.11.10}$$

$$E\{v(t)w^{T}(\tau)\} = X(t)\delta(t-\tau) \tag{5.11.11}$$

$$E\{w(t)w^{T}(\tau)\} = W(t)\delta(t-\tau), \quad E\{w(t)\} = 0 \tag{5.11.12}$$

Therefore,

$$E\{\xi(t)\xi^{T}(\tau)\} = \Xi(t)\delta(t-\tau), \quad E\{\xi(t)\} = 0 \tag{5.11.13}$$

where

$$\Xi(t) = V(t) - L(t)X^{T}(t) - X(t)L^{T}(t) + L(t)W(t)L^{T}(t) \tag{5.11.14}$$

Using the result of (5.11.14) and (5.10.50), we can write an expression describing the covariance of the estimation error as

$$\begin{aligned} \dot{P}_e &= (A-LC)P_e + P_e(A-LC)^T \\ &\quad + V - LX^T - XL^T + LWL^T \end{aligned} \tag{5.11.15}$$

Let us now state our desired objective as one of minimizing the steady state covariance of the estimation error (where the derivative in (5.11.15) is equal to zero) by choosing the 'optimal' observer gain matrix L. Further, let us make the simplifying assumption that the excitation process and sensor noise are independent, such that the cross-correlation term X is equal to zero. With this assumption, the steady state covariance of the estimation error can be described by the expression

$$(A-LC)P_e + P_e(A-LC)^{T} = -(V+LWL^{T}) \tag{5.11.16}$$

This expression is in the form of a Lyapunov equation, the significance of which is the fact that gain matrix L which satisfies this relationship will produce a stable system. This equation is identical in form to (5.8.16), the equation from which we derived the optimal controller gain matrix. We can use the same series of steps here to derive the optimal observer gain matrix.

First, let us write the sensor noise power spectrum matrix W as

$$W = T^{\mathrm{T}}T \tag{5.11.17}$$

where T is a non-singular matrix. Note that we can express W in this way because it is a positive-definite Hermitian matrix. We can now write (5.11.16) as

$$(A-LC)P_e + P_e(A-LC)^{\mathrm{T}} + V + LT^{\mathrm{T}}TL^{\mathrm{T}} = 0 \tag{5.11.18}$$

or, expanding, as

$$\begin{aligned} &AP_e + P_eA^{\mathrm{T}} \\ &+ \left[TL^{\mathrm{T}}-(T^{\mathrm{T}})^{-1}CP_e\right]^{\mathrm{T}}\left[TL^{\mathrm{T}}-(T^{\mathrm{T}})^{-1}CP_e\right] \\ &- P_eC^{\mathrm{T}}W^{-1}CP_e + V \end{aligned} \tag{5.11.19}$$

Taking $\partial P_e/\partial L$ and setting the result equal to zero, the optimum observer gain is found to be

$$L = W^{-1}CP_e \tag{5.11.20}$$

where P_e is defined by the expression

$$AP_e + P_eA^{\mathrm{T}} - P_eC^{\mathrm{T}}W^{-1}CP_e + V = 0 \tag{5.11.21}$$

Equation (5.11.21) is again an algebraic (matrix) Riccati equation, and the optimal L is the Kalman filter gain matrix.

Therefore, the optimal (Kalman filter) observer gains can be calculated in the same manner as the optimum (LQR) controller gains discussed in Section 5.8. In our derivation, the state weighting matrix for the controller problem, Q, has been replaced by the power spectral matrix of the disturbance input, V, and the control effort weighting matrix, R, has been replaced by the power spectral matrix of the observation noise, W (as such, the Kalman filter can also be derived in a manner directly analogous to the LQR problem; refer to Anderson and Moore (1990)). This means, of course, that the derivation of the Kalman filter requires a knowledge of the noise processes in the system, which can often require significant approximations.

5.12 COMBINED CONTROL LAW/OBSERVER: COMPENSATOR DESIGN

Earlier in this chapter we considered the design of a feedback control system on the assumption that all system states were available for implementation. As this is often not a practical assumption, we have also discussed to construction of deterministic and optimal state observers to estimate the value of the system states. In this section we will put the two together, and look at the design of a regulator, or dynamic compensator. What we will be interested in is the influence which the observer has upon the performance and stability characteristics of the system, and ways of optimizing the implementation. Once again, our discussion will be brief. Interested readers will find a good discussion of various aspects of system design incorporating state observers in Anderson and Moore (1990).

5.12.1 Steady state relationships

Consider the system arrangement shown in Fig. 5.38, where the plant is defined by the standard state variable expressions

$$\dot{x} = Ax + Bu \qquad y = Cx \tag{5.12.1}$$

which we will assume to be both controllable and observable. In this arrangement an observer is used to estimate the system states, so that the control input to the system will be defined by the relationship

$$u = -K\hat{x} \tag{5.12.2}$$

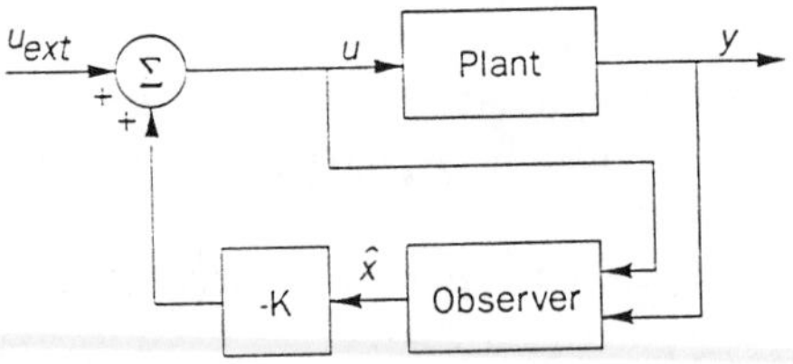

Fig. 5.38 Compensator arrangement, to open the loop at the plant input.

Substituting (5.12.2) into (5.12.1), the state equation becomes

$$\dot{x} = Ax - BK\hat{x} = (A - BK)x + BK(x - \hat{x}) \tag{5.12.3}$$

What we are interested in initially is assessing the steady state characteristics of the system, in terms of the closed loop transfer function and eigenvalue placement.

Defining again the estimation error (vector) as the difference between the actual and estimated state value,

$$e = x - \hat{x} \tag{5.12.4}$$

we can restate the state equation (5.12.3) in terms of this quantity:

$$\dot{x} = (A - BK)x + BKe \tag{5.12.5}$$

In the previous section we showed, for a full state observer, that the dynamics of the estimation error during operation of the state observer were governed by the relationship

$$\dot{e} = (A - LC)e \tag{5.12.6}$$

By combining (5.12.5) and (5.12.6) we can obtain a relationship describing the dynamics of the observed state feedback control system:

$$\frac{d}{dt}\begin{bmatrix} x \\ e \end{bmatrix} = \begin{bmatrix} A-BK & BK \\ 0 & A-LC \end{bmatrix} \begin{bmatrix} x \\ e \end{bmatrix} \tag{5.12.7}$$

$$y = [C \;\; 0] \begin{bmatrix} x \\ e \end{bmatrix} \tag{5.12.8}$$

The first point of interest is the characteristic equation of the system. This is defined by

$$\begin{vmatrix} sI-A+BK & -BK \\ 0 & sI-A+LC \end{vmatrix} = 0 \tag{5.12.9}$$

or

$$|\; sI-A+BK \;| \;\; |\; sI-A+LC \;| = 0 \tag{5.12.10}$$

This is simply the product of the characteristic equations defining the controller pole placements and the observer pole placements. It can be surmised that the poles of the estimated state controlled system are simply the sum of the control law defined poles and the observer defined poles, with the order of the system increasing by a factor of two.

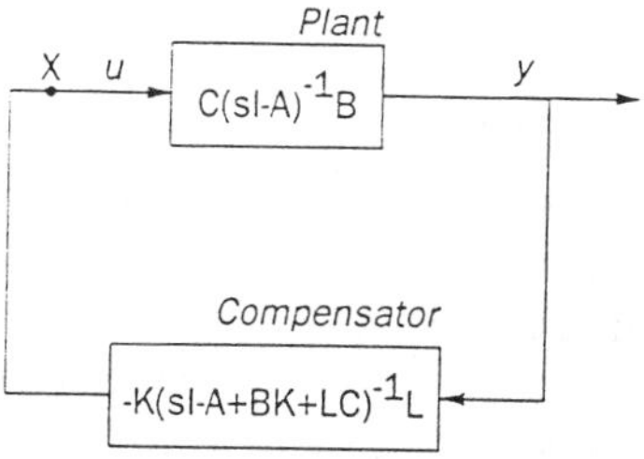

Fig. 5.39 Compensator structure.

The second point of interest is the closed loop system transfer function. To calculate this function, consider the introduction of some external input, as shown in Fig. 5.39. This input could, for example, be some form of reference trajectory in a tracking control system; for our purposes we will use it simply as a vehicle for transfer function calculations. With this addition, (5.12.7) becomes

$$\frac{d}{dt}\begin{bmatrix} x \\ e \end{bmatrix} = \begin{bmatrix} A-BK & BK \\ 0 & A+LC \end{bmatrix} \begin{bmatrix} x \\ e \end{bmatrix} + \begin{bmatrix} B \\ 0 \end{bmatrix} u_{ext} \tag{5.12.11}$$

The transfer function between the external input and system output can be calculated simply as

$$H(s) = C(sI-A+BK)^{-1}B \tag{5.12.12}$$

This is exactly the same transfer function which exists when using full state feedback. Therefore, in considering the steady state case, use of an observer has no influence upon the system. This makes intuitive sense, as when steady state has been reached the difference between the estimated and actual state values approaches zero.

From these results, it can be surmised that calculation of the control gains and observer gains are separate problems, and can be undertaken independently; a complete control system can be constructed by first calculating the feedback gains on the premise of full state feedback, then calculating the observer gains, and finally substituting the observed states for the actual states in the control law. This independence of the control and observation problems is referred to as the separation theorem (a proof of the separation theorem can be found in Anderson and Moore (1990)).

One final point to note is that while the steady state performance may be unchanged, the transient performance of the system employing an observer will differ from that using full state feedback. Note again that the system employing an observer is twice the dimension of the system using full state feedback, and although it will still be asymptotically stable, the convergence of the system to equilibrium will be governed by two sets of poles, those of the (full state feedback) controller plus those of the observer.

5.12.2 Robustness

To examine the robustness of the controller, we need to examine the open loop transfer functions of the system. The values of these are dependent upon where the loop is opened; to analyse input robustness, the loop is usually opened at the plant input. The loop gain can then be computed as the product of the plant and controller transfer functions, as depicted in Fig. 5.39. The transfer function of the plant is simply $\boldsymbol{C}(s\boldsymbol{I} - \boldsymbol{A})^{-1}\boldsymbol{B}$, as we have derived previously. To derive the transfer function of the controller, note that the full state observer is described by the relationship

$$\frac{d\hat{\boldsymbol{x}}}{dt} = \boldsymbol{A}\hat{\boldsymbol{x}} - \boldsymbol{BK}\hat{\boldsymbol{x}} - \boldsymbol{LC}\hat{\boldsymbol{x}} + \boldsymbol{Ly} = (\boldsymbol{A}-\boldsymbol{BK}-\boldsymbol{LC})\hat{\boldsymbol{x}} + \boldsymbol{Ly} \qquad (5.12.13)$$

Using this, the transfer function of the controller can be written as

$$\boldsymbol{H}(s) = -\boldsymbol{K}(s\boldsymbol{I}-\boldsymbol{A}+\boldsymbol{BK}+\boldsymbol{LC})^{-1}(\boldsymbol{L}) \qquad (5.12.14)$$

Therefore, the loop transfer function is

$$\boldsymbol{H}(s) = -\boldsymbol{K}(s\boldsymbol{I}-\boldsymbol{A}+\boldsymbol{BK}+\boldsymbol{LC})^{-1}\boldsymbol{L} \cdot \boldsymbol{C}(s\boldsymbol{I}-\boldsymbol{A})^{-1}\boldsymbol{B} \qquad (5.12.15)$$

This is certainly not the same loop gain $\boldsymbol{K}(s\boldsymbol{I} - \boldsymbol{A})^{-1}\boldsymbol{B}$ as for a full state feedback arrangement, and therefore the (return difference equations, and hence) robustness properties of the full state feedback implementation cannot be expected to hold for the observer-implemented system. Note also that the roll off the loop gain has increased (apparent when $j\omega$ is substituted for s), from 6 dB per octave to 12 dB per octave. Therefore, is appears that implementing the system with an observer has increased the high frequency robustness, and decreased the passband robustness. This is true regardless of whether the controller and observer were designed using linear quadratic or pole placement methods. This means that while an LQR system as outlined in Section 5.8 has desirable gain and phase margins, these may simply evaporate when the system is implemented with an observer. A simple example (taken from Doyle, 1978) will serve to highlight this.

Consider a system with two eigenvalues at +1, described by the state equations

$$\begin{bmatrix} \dot{x}_1 \\ \dot{x}_2 \end{bmatrix} = \begin{bmatrix} 1 & 1 \\ 0 & 1 \end{bmatrix} \begin{bmatrix} x_1 \\ x_2 \end{bmatrix} + \begin{bmatrix} 0 \\ 1 \end{bmatrix} u + \begin{bmatrix} 1 \\ 1 \end{bmatrix} v$$

$$y = [1 \quad 0] \begin{bmatrix} x_1 \\ x_2 \end{bmatrix} + w \tag{5.12.16}$$

Here the noise intensities are $V = \sigma > 0$, $W = 1$. If the control gains are derived using an LQR approach, and the observer gains calculated using a Kalman filter, with weighting matrices defined by

$$\boldsymbol{Q} = \rho \begin{bmatrix} 1 & 1 \\ 1 & 1 \end{bmatrix}, \quad R = 1 \tag{5.12.17}$$

then the controller and observer gain matrices are defined by

$$\boldsymbol{K} = -\alpha[1 \quad 1], \quad \alpha = 2+\sqrt{4+\rho} \tag{5.12.18}$$

and

$$\boldsymbol{L} = -\beta[1 \quad 1], \quad \beta = 2+\sqrt{4+\sigma} \tag{5.12.19}$$

(Note that if $\rho=\sigma$, then the solutions are identical.) The system matrix is found to be

$$\boldsymbol{A} = \begin{bmatrix} 1 & 1 & 0 & 0 \\ 0 & 1 & -m\alpha & -m\alpha \\ \beta & 0 & 1-\beta & 1 \\ \beta & 0 & -\beta-\alpha & 1-\alpha \end{bmatrix} \tag{5.12.20}$$

where m is the plant input gain, $m=1$ for the nominal case. The characteristic equation for this is

$$s^4 + f(s^3) + f(s^2) + [\beta+\alpha-4+2(m-1)\alpha\beta]s + 1 + (1-m)\alpha\beta = 0 \tag{5.12.21}$$

where $f(\ )$ denotes an expression not involving m. For the system to be stable, the last two coefficients must be positive. For large ρ,σ, producing large α,β, there will be instability for arbitrarily small perturbations from unity in the plant gain m. Therefore, this LQG controller demonstrates that arbitrarily small gain margins can exist when implementing a system with an observer, even when the observer is 'optimal'.

The question then arises, is there any way to recover the desirable robustness characteristics of the LQR design by adjustment of the observer design process? There are, in fact, a number of approaches which have been taken to improve the robustness of the observer-implemented system, referred to collectively as loop recovery techniques. The original concept can be traced back to Doyle and Stein (1979), who considered loop recovery for minimum phase plants. Their method generally consists of adding 'fictitious' noise to the plant input model, which can be viewed heuristically as representing plant variations or uncertainties. This 'noise' changes the observer gains. As the fictitious noise increases in amplitude, the robustness properties of the LQR design are recovered (along with the somewhat undesirable roll off characteristics of the controller). A thorough discussion of loop recovery techniques is beyond the scope of this 'review' section. The interested reader is referred to Anderson and Moore (1990) for a good overview of the techniques.

REFERENCES

Ackermann, J. (1972). Der entwurf linearer regelungssysteme im zustandsraum. *Regelungstechnik and Prozessdatenverarbeitung*, **7**, 297–300.

Anderson, B.D.O. and Mingori, D.L. (1985). Use of frequency dependence in linear quadratic control problems to frequency shape robustness. *Journal of Guidance, Control, and Dynamics*, **8**, 397–401.

Anderson, B.D.O. and Moore, J.B. (1979). *Optimal Filtering*. Prentice-Hall: Englewood Cliffs, NJ.

Anderson, B.D.O. and Moore, J.B. (1979). *Optimal Control, Linear Quadratic Methods*. Prentice-Hall: Englewood Cliffs, NJ.

Astrom, K.J. and Eykhoff, P. (1971). System identification: a survey. *Automatica*, **7**, 123–162.

Astrom, K.J. and Wittenmark, B. (1989). *Adaptive Control*. Addison-Wesley: Reading, MA.

Athans, M. and Falb, P.L. (1966). *Optimal Control*. McGraw-Hill: New York.

Balas, M.J. (1978a). Feedback control of flexible systems. *IEEE Transactions on Automatic Control*, **AC-23**, 673–679.

Balas, M.J. (1978b). Active control of flexible systems. *Journal of Optimization Theory and Applications*, **25**, 415–436.

Balas, M.J. (1982). Trends in large space structure control theory: fondest hopes, wildest dreams. *IEEE Transactions on Automatic Control*, **AC-27**, 522–535.

Bellman, R.E. and Dreyfus, S.E. (1962). *Applied Dynamic Programming*. Princeton University Press: Princeton, NJ.

Bendat, J.S. and Piersol, A.G. (1980). *Engineering Applications of Correlation and Spectral Analysis*. Wiley: New York.

Bryson, A.E. Jr. and Ho, Y.C. (1969). *Applied Optimal Control*. Blaisdell: Waltham, MA.

Chen, C.T. (1984). *Linear System Theory and Design*. Holt, Rinehart and Winston: New York.

Dohner, J.L. and Shoureshi, R. (1989). Modal control of acoustic plants. *Journal of Vibration, Acoustics, Stress, and Reliability in Design*, **111**, 326-330.

Doyle, J.C. (1978). Guaranteed margins in LQG regulators. *IEEE Transactions on Automatic Control*, **AC-23**, 664–665.

Doyle, J.C. and Stein, G. (1979). Robustness with observers. *IEEE Transactions on Automatic Control*, **AC-24**, 607–611.

Doyle, J.C., Glover, K., Khargonekar, P.P. and Francis, B.A. (1989). State-space solutions to standard H_2 and H_∞ control problems. *IEEE Transactions on Automatic Control*, **34**, 831–847.

Eriksson, L.J. and Allie, M.C. (1988). A practical system for active attenuation in ducts. *Sound and Vibration*, **22**, 30–34.

Eriksson, L.J. and Allie, M.C. (1989). Use of random noise for on-line transducer modelling in an adaptive active attenuation system. *Journal of the Acoustical Society of America*, **89**, 797–802.

Evans, W.R. (1948). Graphical analysis of control systems. *Transactions AIEE*, **67**, 547–551.

Francis, B.A. (1987) *A Course in H_∞ Control Theory*, Springer-Verlag: Berlin.

Franklin, G.F., Powell, J.D. and Emami-Naeini, A. (1991). *Feedback Control of Dynamic Systems*. Addison-Wesley: Reading, MA.

Franklin, G.F., Powell, J.D. and Workman, M.L. (1990). *Digital Control of Dynamic Systems*. Addison-Wesley: Reading, MA.

Friedland, B. (1986). *Control System Design: An Introduction to State-Space Methods*. McGraw-Hill: New York.

Goodwin, G.C. and Payne, R.L. (1977). *Dynamic System Identification: Experimental Design and Data Analysis*. Academic Press: New York.

Goodwin, G.C. and Sin, K.S. (1984). *Adaptive Filtering Prediction and Control*. Prentice-Hall: Englewood Cliffs, NJ.

Gopinath, B. (1971). On the control of linear multiple input-output systems. *Bell System Technical Journal*, **50**, 1063–1081.

Grimble, M.J. and Owens, T.J. (1986). On improving the robustness of LQ regulators. *IEEE Transactions on Automatic Control*, **AC-31**, 54–55.

Gupta, N.K. (1980). Frequency-shaped loop functionals: Extensions of linear-quadratic-gaussian design methods. *Journal of Guidance and Control*, **3**, 529–535.

Hitz, K.L. and Anderson, B.D.O. (1972). Iterative method of computing the limiting solution of the matrix Ricatti differential equation. *Proceedings of the IEEE*, **119**, 1402–1406.

Jategaonkar, R.V., Raol, J.R. and Balakrishna, S. (1982). Determination of model order for dynamical system. *IEEE Transactions on Systems, Man, and Cybernetics*, **SMC-12**, 56–62.

Jazwinski, A.H. (1970). *Stochastic Processes and Filtering Theory*. Academic Press: New York.

Kailath, T. (1980). *Linear Systems*. Prentice-Hall: Englewood Cliffs, NJ.

Kalman, R.E. (1960). A new approach to linear filtering and prediction problems. *ASME Journal of Basic Engineering*, **82D**, 35–45.

Kalman, R.E. (1963). Mathematical description of linear dynamic systems. *SIAM Journal on Control*, ser. A., **1**, 152–192.

Kalman, R.E. and Bucy, R.S. (1961). New results in linear filtering and prediction theory. *ASME Journal of Basic Engineering*, **83D**, 95–108.

Kalman, R.E., Ho, Y.C. and Narendra, K.S. (1963). Controllability of linear dynamic systems. *Contributions to Differential Equations*, **1**, 189–213.

Kleinman, D.L. (1968). On an iterative technique for Ricatti equation computations. *IEEE Transactions on Automatic Control*, **AC-13**, 114–115.

Kwakernaak, H. and Sivan, R. (1972). *Linear Optimal Control Systems*. Wiley-Interscience: New York.

Laub, A.J. (1979). A Schur method for solving algebraic matrix Ricatti equations. *IEEE Transactions on Automatic Control*, **AC-24**, 913–921.

Lehtomaki, N.A., Sandell, N.R., Jr., and Athans, M. (1981). Robustness results in linear-quadratic gaussian based multivariable control. *IEEE Transactions on Automatic Control*, **AC-26**, 75–92.

Lewis, F.L. (1986). *Optimal Control*. Wiley: New York.

Lindner, D.K., Reichard, K.M. and Tarkenton, L.M. (1993). Zeroes of modal models of flexible structures. *IEEE Transactions on Automatic Control*, **AC-38**, 1384–1388.

Ljung, L. (1987). *System Identification: Theory for the User*. Prentice-Hall: Englewood Cliffs, NJ.

Ljung, L. and Söderström, T. (1983). *Theory and Practice of Recursive Identication*. MIT Press: Cambridge, MA.

Luenberger, D.G. (1964). Observing the state of a linear system. *IEEE Transactions on Military Electronics*, **MIL-8**, 74–80.

Luenberger, D.G. (1966). Observers for multivariable systems. *IEEE Transactions on Automatic Control*, **AC-11**, 190–197.

Luenberger, D.G. (1971). An introduction to observers. *IEEE Transactions on Automatic Control*, **AC-16**, 596-602.

Lyapunov, A.M. (1992). *The General Problem of the Stability of Motion*, translation from Russian in *International Journal of Control*, **55**, 531–773.

MacFarlane, D.C. and Glover, K. (1990). *Robust Controller Design Using Normalized Coprime Factor Plant Descriptions*. Springer-Verlag: Berlin.

Mason, S.J. (1953). Feedback theory: some properties of signal flow graphs. *Proceedings IRE*, **41**, 1144–1156.

Mason, S.J. (1956). Feedback theory: further properties of signal flow graphs. *Proceedings IRE*, **44**, 920–926.

Meirovitch, L. and Baruh, H. (1982). Control of self-adjoint distributed parameter systems. *Journal of Guidance, Control, and Dynamics*, **5**, 60–66.

Meirovitch, L., Baruh, H. and Oz, H. (1983). A comparison of control techniques for large flexible systems. *Journal of Guidance, Control, and Dynamics*, **6**, 302–310.

Miu, D.K. (1991). Physical interpretation of transfer function zeroes for simple control systems with mechanical flexibilities. *Journal of Dynamic Systems, Measurement, and Control*, **113**, 419–424.

Moler, C. and van Loan, C. (1978). Nineteen dubious ways to compute the exponential of a matrix. *SIAM Review*, **20**, 810–836.

Moore, B.C. (1981). Principal component analysis in linear systems: Controllability, observability, and model reduction. *IEEE Transactions on Automatic Control*, **AC-26**, 17–32.

Moore, J.B. and Mingori, D.L. (1987). Robust frequency-shaped LQ control. *Automatica*, **23**, 641–646.

Norton, J.P. (1986). *An Introduction to Identication*. Academic Press: London.

Nyquist, H. (1932). Regeneration theory. *Bell Systems Technical Journal*, **11**, 126–147.

Ogata, K. (1992). *Modern Control Engineering*. Prentice Hall: Englewood Cliffs, NJ.

Oppenheim, A.V. and Schafer, R.W. (1975). *Digital Signal Processing*. Prentice-Hall: Englewood Cliffs, NJ.

Papoulis, A. (1965). *Probability, Random Variables, and Stochastic Processes*. Wiley: New York.

Parks, P.C. (1962). A new proof of the Routh–Hurwitz stability criterion using the second method of Lyapunov. *Proceedings of the Cambridge Philosophical Society*, **58**, 694–702.

Pontryagin, L.S., Boltysanskii, V.G., Gamkrelidze, R.V. and Mischenko, E.F. (1962). *The Mathematical Theory of Optimal Processes*, translated by K.N. Trirogoff. Interscience: New York.

Poole, L.A., Warnaka, G.E. and Cutter, R.C. (1984). The implementation of digital filters using a modified Widrow–Hoff Algorithm the the adaptive cancellation of acoustic noise. *Proceedings ICASSP '84*, **2**, 21.7.1–21.7.4.

Potter, J.E. (1966). Matrix quadratic solutions. *SIAM Journal of Applied Mathematics*, **14**, 496–501.

Rabiner, L.R. and Gold, B. (1975). *Theory and Application of Digital Signal Processing*. Prentice-Hall: Englewood Cliffs, NJ.

Safanov, M.G. and Athans, M. (1977). Gain and phase margins of multiloop LQG regulators. *IEEE Transactions on Automatic Control*, **AC-22**, 173–179.

Söderström, T. and Stoica, P. (1989). *System Identification*. Prentice-Hall: New York.

Sorenson, H.W. (1970). Least-squares estimation: from Gauss to Kalman. *IEEE Spectrum*, **7**, 63–68.

Stearns, S.D. (1975). *Digital Signal Analysis*. Hayden: Rochelle Park, NJ.

Vidyasagar, M. (1978). *Nonlinear Systems Analysis*. Prentice-Hall: Englewood Cliffs, NJ.

Vidyasagar, M. (1985). *Control System Synthesis: A Coprime Factorization Approach*. MIT Press: Cambridge, MA.

6

Feedforward control system design

6.1 INTRODUCTION

In the previous chapter we considered the design of feedback control systems, where the control input was derived using the error signal as a reference. While this is perhaps the most widely applicable control system arrangement, it does have a performance limitation: with persistent excitation it is not possible to drive the error signal to zero, as then there would be no control input. The smaller the error signal is driven, the higher the control gain must be, and the less stable will be the system.

There are some instances, however, where it is possible to predict the impending (primary) disturbance. In the context of active noise and vibration control, this commonly occurs where the primary source comprises rotating machinery, such that the disturbance is harmonic, or where the propagation path is down a waveguide, where the disturbance at any given point is a function of the disturbance at an 'upstream' point some time previously. In such cases the basis for the disturbance prediction can be used as a reference for control signal generation, thereby removing the non-zero restriction on the error signal. Such arrangements are referred to as feedforward control systems. The basic structure of a feedforward control system is shown in Fig. 6.1, in which a reference signal, in some way correlated with the impending primary disturbance, is used to derive the control input. If the correlation between the reference signal and the error signal is 'perfect', it is (theoretically) possible to drive the error signal to zero, which can be a very attractive feature.

Many of the electronic systems utilized in feedforward control schemes derive control inputs via modified adaptive signal processing algorithm/architecture combinations. Adaptive digital signal processing is a field born out of the requirements of modern telecommunications systems. In these systems, the need often arises to filter a signal, so that it can be extracted from contaminating noise. 'Conventional' signal processing systems employed to do this operate in an open loop fashion, using a filter with fixed characteristics. The underlying assumptions accompanying the use of fixed filters are that a description of the input signal is known, and that the system disturbance and response characteristics are time invariant. If this is the case, a satisfactory filter may be designed. It is often the case, however, that the characteristics of the input signal and system response are unknown, or may be slowly changing with time. In these instances the use of a filter with fixed characteristics may not give satisfactory performance. What is required is a filter which will 'learn' what characteristics are best at a given instant, and be able to 'relearn' to cope with slow changes in the signal structure. What is required is an adaptive filter.

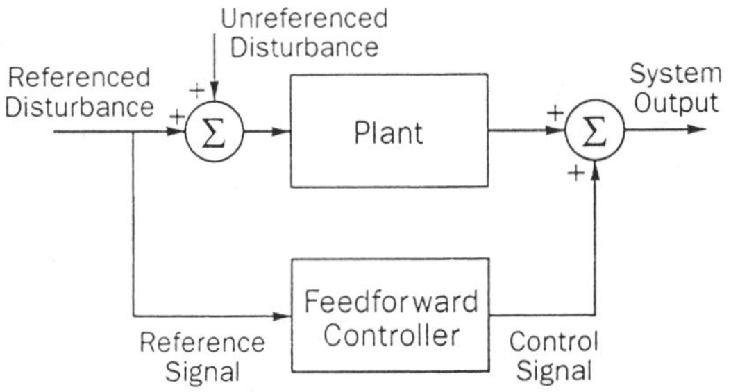

Fig. 6.1 Basic feedforward control system arrangement.

With the advances in digital technology over the past four decades, adaptive filters have become inexpensive and increasingly efficient. Indeed, adaptive digital signal processing has become a firmly established field, encompassing a wide range of applications, one of which is the active control of sound and vibration. However, as we will see, the application of adaptive digital signal processing techniques to active noise and vibration control problems requires several unique modifications to the 'standard' format of the algorithms.

In this chapter we will study the design of feedforward control systems. Because these control systems can (in theory) minimise the disturbances at a set of error sensors to whatever degree is physically possible, the control systems described in this chapter can be 'attached' to the sensor and actuator arrangements described in Chapters 7 – 10, and the levels of attenuation predicted from consideration of the physical characteristics of the target system achieved (for example, in an air handling duct). In this respect, the design of a feedforward active control system can be viewed as illustrated in Fig. 6.2, where the design of the electronic and physical systems are separate processes. While this idea is somewhat simplified, as certainly the transient performance of the electronic control system is dependent upon the characteristics of the physical system to which it is attached, and feedback of the control signal to the reference sensor can deteriorate performance, it is approximately true for the steady state case.

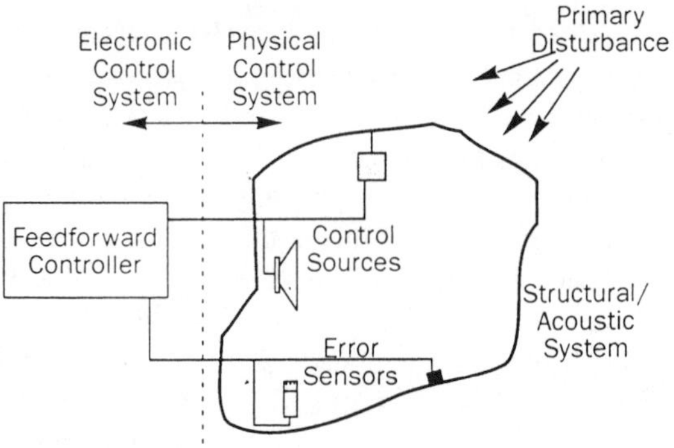

Fig. 6.2 Feedforward control system as comprised of two parts, the 'physical' control system (actuators and sensors) and the 'electronic' control system.

We will begin this chapter by studying feedforward control systems with fixed characteristics. This will provide some of the groundwork required to progress onto adaptive feedforward controllers. It will also be assumed that the reader has covered the material in Sections 5.2 and 5.4 concerning digital control systems. In studying adaptive control systems, we will generally consider the 'standard' adaptive signal processing problem first, before modifying the arrangement to accommodate the particular needs of an active control implementation. With adaptive systems, we will first study the most common (and simple) linear, non-recursive case before progressing on to the more complicated linear recursive case. Finally, we will consider non-linear controllers based upon artificial neural networks.

The adaptive digital signal processing material presented in this chapter is not intended to constitute a thorough treatment of the subject. Rather, it is meant to provide an overview of the algorithms commonly used in active noise and vibration control systems, to provide a basis for the discussion of their implementation. For a more general treatment of the topic, the reader is referred to several dedicated texts (Honig and Messerschmitt, 1984; Goodwin and Sin, 1984; Cowan and Grant, 1985; Widrow and Stearns, 1985; Alexander, 1986; Treichler *et al.*, 1987). Also, no treatment will be given to the problem of (fixed) digital filter design. There are several good books on this subject, including (Oppenheim and Schafer, 1975; Rabiner and Gold, 1975; and Lynn, 1982).

6.2 WHAT DOES FEEDFORWARD CONTROL DO?

Before considering the design of feedforward control systems, it is perhaps first appropriate to consider what they do and how they provide control in terms of modifications to the response characteristics of the system. In Chapter 5 we saw that feedback control systems provide attenuation of an unwanted disturbance by altering the poles, or natural frequencies, of the system. Could this also be the way in which feedforward control systems work?

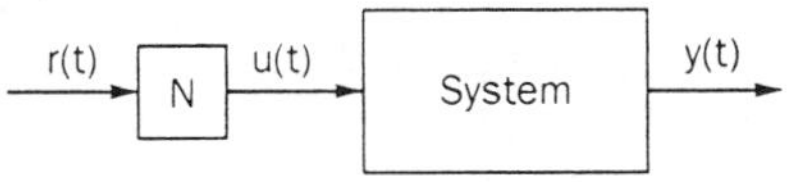

Fig. 6.3 Single input, single output system.

To examine this question, it is again useful to adopt the state space notation developed in Chapter 5, and it is easiest to consider the problem in the time domain. Considering the single input, single output (SISO) arrangement for controlling the output of the 'system' shown in Fig. 6.3, and neglecting any primary disturbance for the moment, the state space description of the equations of motion of some generic system with a single input and single output is as developed in Chapter 5:

$$\begin{aligned} \dot{\boldsymbol{x}}(t) &= \boldsymbol{A}\boldsymbol{x}(t)+\boldsymbol{b}u(t) \\ y(t) &= \boldsymbol{c}\boldsymbol{x}(t) \end{aligned} \tag{6.2.1}$$

where $\boldsymbol{x}$ is the $(n \times 1)$ state vector, $\boldsymbol{A}$ the $(n \times n)$ state matrix, $\boldsymbol{b}$ the $(n \times 1)$ input vector, u the control input, y the system output, and $\boldsymbol{c}$ the $(1 \times n)$ output vector. For simplicity, the control input we will consider here is a referenced sinusoid multiplied by some factor N,

$$u(t) = Nr(t) = Nr_0 e^{st} \tag{6.2.2}$$

where r_0 is the amplitude of the sinusoidal reference signal, and $s=j\omega$.

If (6.2.2) is substituted into (6.2.1), and the entire expression transformed into the frequency domain s by means of a Laplace transform, assuming zero initial conditions the equations of motion become

$$\begin{aligned} s\boldsymbol{x}(s) &= \boldsymbol{A}\boldsymbol{x}(s)+N\boldsymbol{b}r(s) \\ y(s) &= \boldsymbol{c}\boldsymbol{x}(s) \end{aligned} \tag{6.2.3}$$

The transformed state vector $\boldsymbol{x}(s)$ can be rewritten as

$$\boldsymbol{x}(s) = \frac{N\boldsymbol{b}r(s)}{s\boldsymbol{I}-\boldsymbol{A}} \tag{6.2.4}$$

where $\boldsymbol{I}$ is the identity matrix. If this is substituted into the transformed output expression in (6.2.3), the system transfer function is found to be

$$\frac{y(s)}{r(s)} = \boldsymbol{c}(s\boldsymbol{I}-\boldsymbol{A})^{-1}N\boldsymbol{b} \tag{6.2.5}$$

Note that the poles of the transfer function, defined by the denominator expression $(s\boldsymbol{I} - \boldsymbol{A})$, are unchanged from those without feedforward control. In fact, (6.2.5) shows that only the zeros of the transfer function, defined by the numerator expression, can be modified by the feedforward control input. This is intuitively sensible, as the poles of a system are defined by frequencies which have a non-zero output for zero input. Feedforward control signals are strictly externally generated signals, having no dependency on system states, and so by definition should not influence poles. For this reason feedforward control systems are inherently stable, which is a very different characteristic from feedback control systems. The only feedforward control system which can become unstable is an adaptive one, where the adaptive algorithm itself can become unstable.

Having seen that a feedforward input can only modify the zeroes of a system, let us consider the particular case where the control input is introduced to 'cancel' a primary disturbance, with which it is perfectly correlated, at the error sensor (system output) location. With this addition of a primary disturbance, the equations of motion become

$$\begin{aligned} \dot{x}(t) &= \boldsymbol{A}\boldsymbol{x}(t)+\boldsymbol{b}u(t)+\boldsymbol{w}(t) \\ y(t) &= \boldsymbol{c}\boldsymbol{x}(t) \end{aligned} \tag{6.2.6}$$

where $\boldsymbol{w}(t)$ is the $(n \times 1)$ primary disturbance vector, equal to a vector $\boldsymbol{m}$ (transfer function between the input sinusoid $r(t)$ and the primary disturbance $w(t)$ at the system output) multiplied by the sinusoid

$$\boldsymbol{w}(t) = \boldsymbol{m}r(t) \tag{6.2.7}$$

Note that if the feedforward control input is to cancel the primary disturbance at the error sensor, then following an analysis similar to that used to obtain (6.2.3) and (6.2.4), it can be shown that

$$\boldsymbol{c}\left[\boldsymbol{m}r(t)+N\boldsymbol{b}r(t)\right] = 0 \tag{6.2.8}$$

Substituting (6.2.7) into the equations of motion (6.2.6) and transforming the result into the frequency domain, it is straightforward to show that the transfer function between the output y and reference signal r is

$$\frac{y(s)}{r(s)} = \boldsymbol{c}\left(s\boldsymbol{I}-\boldsymbol{A}\right)^{-1}\left(N\boldsymbol{b}+\boldsymbol{m}\right)r(s) \tag{6.2.9}$$

which, from (6.2.8) is simply

$$\frac{y(s)}{r(s)} = 0 \tag{6.2.10}$$

That is, the feedforward controller places a zero of transmission at the reference signal frequency. An example of this is illustrated in Fig. 6.4, which depicts the (experimental) frequency response characteristics of a cantilever beam subject to primary point excitation of a 75 Hz tone in random noise, with a feedforward controller, driving a point control force, used to attenuate the tone. In this case, the frequency response is measured between the primary disturbance and the error sensor (a further description of the experimental arrangement can be found in Snyder and Tanaka, 1993a). Note that the characteristics of the beam response are unchanged with the exception of an added zero at the tonal frequency which is to be attenuated.

One final point which is worthwhile mentioning is that while it is strictly true that feedforward control systems modify the transmission zeros of a system, it is possible to interpret the effect as one of modifying the boundary conditions of the structure as seen by the feedforward controller referenced frequency band,

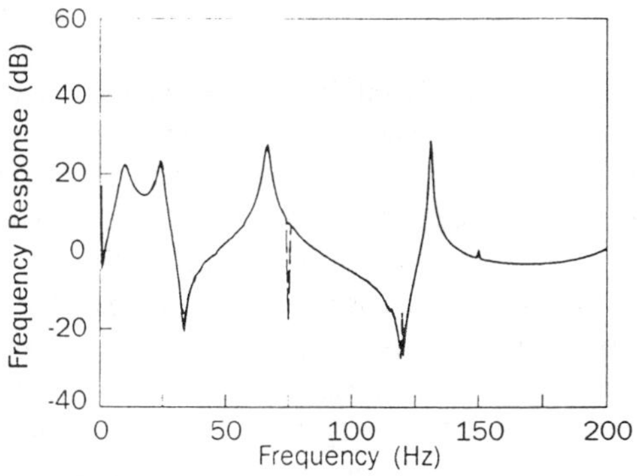

Fig. 6.4 Frequency response of a beam without (solid line) and with (dashed line) feedforward control at 75 Hz.

forcing a node at the error sensing location. In other words, the characteristic response, hence eigenfunctions, appear (to the frequency components in the reference signal band) to have changed (Burdisso and Fuller, 1992). With this point of view, it is possible to calculate new controlled 'resonant frequencies', which help to explain patterns of global response when minimizing a discrete error signal. These response characteristics, however, are only seen by frequency components in the reference signal provided to the feedforward control system, and any other frequency components will only cause the same response as they would have in the uncontrolled system.

6.3 FIXED CHARACTERISTIC FEEDFORWARD CONTROL SYSTEMS

To begin our discussion of feedforward control system design, we will consider the problem of designing controllers with fixed characteristics. For these controllers we will base the design on the transfer functions of components of the system, using them to derive an open loop control law which will minimise some desired error criterion. Only single input, single output (SISO) control arrangements will be considered here; that is, where there is a single control output and a single error sensor at which the acoustic or vibrational disturbance is to be minimized. Further, it will be assumed that all components in the systems are linear. While this is obviously an approximation for transducers such as speakers, it is in practice usually a reasonably valid one.

We will, in fact, implicitly consider the design of fixed characteristic feedforward controllers further in Chapters 7–10 for specific problems such as sound propagation in ducts, sound radiation into free space, and sound transmission into coupled enclosures. When we derive the optimum control source outputs for a given problem, the governing relationship defines the optimum controller characteristics (excluding the effects of the frequency response characteristics of the control source). In this section, feedforward controller design will be considered from a much more explicit standpoint, focusing our discussion not so much on characterizing the structural/acoustic physical phenomena which govern the system response to primary and control excitation, but rather on the relationship between 'generic' signals and transfer functions at and between components in the control system.

Open loop control systems such as those to be described in this section have performance limitations which arise from their fixed characteristics. This is especially true in active noise and vibration control systems, where the transfer functions of various parts of the system will be altered with changing environmental and operating conditions. From a practical standpoint, an adaptive feedforward control system, such as described later in this chapter, is often a more desirable option. It is, however, worth devoting some time to the study of open loop feedforward control systems, if for no other reason than to provide an optimum solution to a given fixed problem. Such a solution is a useful tool for examining the influence of a number of system parameters on feedforward control system performance, whether the controller is adaptive or open loop.

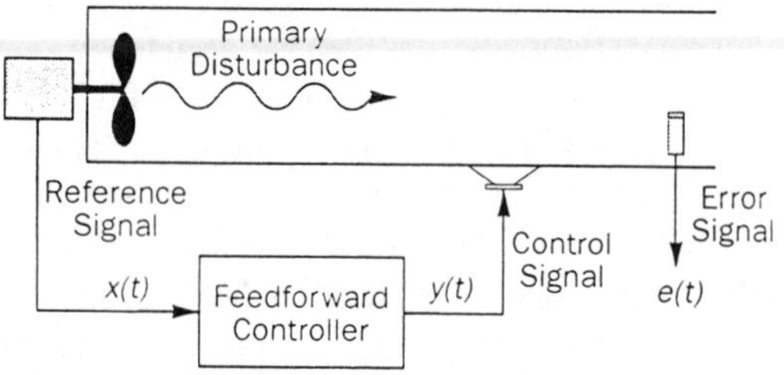

Fig. 6.5 Simple feedforward control arrangement for attenuating sound propagation in a duct, where the reference signal is taken from a measurement of the machinery rotation.

Let us first consider the controller design problem shown in Fig. 6.5. Here a piece of rotating machinery, such as a fan, is responsible for a disturbance propagating in an air handling duct. A signal $x(t)$, which is correlated with the disturbance, can be taken directly from the machine and provided to the feedforward controller as a reference signal, to be used in deriving a control signal $y(t)$. This arrangement is ideal as far as feedforward control implementation goes, because the reference signal will not be corrupted by the operation of the control system. This characteristic, which simplifies the problem, is valid for a wide range of practical problems where the unwanted primary disturbance originates from rotating machinery and the reference signal can be provided by some form of tachometer. The aim of this exercise is to derive the characteristics of the control system which will minimize the disturbance as measured by a downstream error sensor $e(t)$. This problem arrangement is by no means constrained to duct systems, but is common among all active noise and vibration control problems; a duct is used here simply to provide an intuitive model.

A block diagram of the problem is shown in Fig. 6.6, where all quantities have been restated in the frequency domain. The reference signal provided to the controller $X(\omega)$ is modified by the controller transfer function $G(s)$, to produce a control signal $Y(\omega)$,

$$Y(\omega) = G(s)X(\omega) \tag{6.3.1}$$

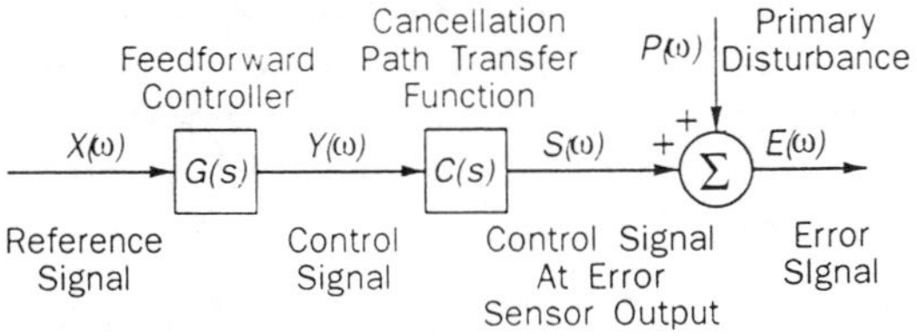

Fig. 6.6 Equivalent block diagram of duct feedforward control system in Fig. 6.5.

The control signal then propagates through the 'physical' system, being modified by the response characteristics of the control source, the duct section between the control source and error sensor, the response characteristics of the error sensor, and various electronic components such as filters. The influence of all of these can be lumped into a single 'cancellation path transfer function' $C(s)$, in such a way that the control signal measured at the output of the error sensor $S(\omega)$ is defined by the expression

$$S(\omega) = C(s)Y(\omega) \tag{6.3.2}$$

After this modification, it is added (superposition) with the primary disturbance $P(\omega)$ to produce an error signal. The problem is to derive the controller transfer function $G(s)$ that produces a control signal which will minimize the error signal.

This is a particularly easy problem in this instance. The error signal $E(\omega)$ is the superposition of the primary and control source components

$$E(\omega) = P(\omega)+S(\omega) = P(\omega)+C(s)Y(\omega) \tag{6.3.3}$$

Substituting (6.3.2) into (6.3.1),

$$E(\omega) = P(\omega)+C(s)G(s)X(\omega) \tag{6.3.4}$$

As the system is linear, this can be rewritten as

$$E(\omega) = P(\omega)+G(s)F(\omega) \tag{6.3.5}$$

where $F(\omega)$ is known as the filtered reference signal, which is the reference signal modified by the frequency response of the cancellation path transfer function,

$$F(\omega) = C(s)X(\omega) \tag{6.3.6}$$

From this it is immediately apparent that the error signal will be nullified if the controller transfer function is defined by the expression

$$G(s) = -\frac{P(\omega)}{F(\omega)} \tag{6.3.7}$$

While this transfer function is optimal, it is not constrained to be causal. If the disturbance is periodic, which is the case for rotating machinery, this does not present a problem, although the same cannot be said for random noise excitation.

The frequency response characteristics of the optimal controller can also be derived by considering signal power- and cross-spectral densities. Minimizing the error signal is equivalent to minimizing the power spectral density of the error signal at each frequency component

$$S_{ee}(\omega) = E\{E^*(\omega)E(\omega)\} \tag{6.3.8}$$

where $S_{ee}(\omega)$ is the error signal power spectral density. From (6.3.5) this can be expanded as

$$\begin{aligned} S_{ee}(\omega) = E\{&P^*(\omega)P(\omega) + P^*(\omega)F(\omega)G(s) \\ &+ G^*(s)F^*(\omega)P(\omega) + G^*(s)F^*(\omega)F(\omega)G(\omega)\} \end{aligned} \tag{6.3.9}$$

Equation (6.3.9) can be simplified by writing it in terms of signal power spectral densities and cross-spectral densities as follows:

$$\begin{aligned} S_{ee}(\omega) &= S_{pp}(\omega) + S_{fp}^*(\omega)G(s) \\ &+ G^*(s)S_{fp}(\omega) + G^*(s)S_{ff}(\omega)G(s) \end{aligned} \tag{6.3.10}$$

where $S_{pp}(\omega)$ is the power spectral density of the primary source signal, given by

$$S_{pp}(\omega) = E\{P^*(\omega)P(\omega)\} \tag{6.3.11}$$

$S_{fp}(\omega)$ is the cross-spectral density between the filtered reference signal and the primary source signal, given by

$$S_{fp}(\omega) = E\{F^*(\omega)P(\omega)\} \tag{6.3.12}$$

and $S_{ff}(\omega)$ is the power spectral density of the filtered reference signal

$$S_{ff}(\omega) = E\{F^*(\omega)F(\omega)\} \tag{6.3.13}$$

Expressed in this way, the error criterion (error signal power spectral density) is a simple complex scalar quadratic equation. Noting that the filtered reference signal power spectral density is real and positive, the optimum frequency response characteristics of the transfer function $G(s)$ are defined by the expression

$$G_{\text{opt}}(s) = -S_{ff}^{-1}(\omega)S_{pf}(\omega) \tag{6.3.14}$$

This relationship, which is purely feedforward (all zero), is a modified version of the solution to the Weiner–Hopf integral equation, where quantities which would normally pertain to the reference signal have been replaced by quantities pertaining to the filtered reference signal. As we will see later in this chapter, the problem can be restated in a discrete time format to defined the optimum weight coefficients for a FIR filter.

In the system shown in Fig. 6.5, a reference signal could be obtained which would not be corrupted by the operation of the feedforward control system. While this is often the case with harmonic disturbances, it is not true in general, especially when the disturbance is random noise. A more general feedforward control arrangement for the duct problem is shown in Fig. 6.7. In this case, the reference signal is obtained by an acoustic sensor placed in the duct upstream from the control system. The reference signal supplied by the sensor is now open to corruption by the upstream propagating component of the control output, as well as by measurement noise due to turbulence in the duct.

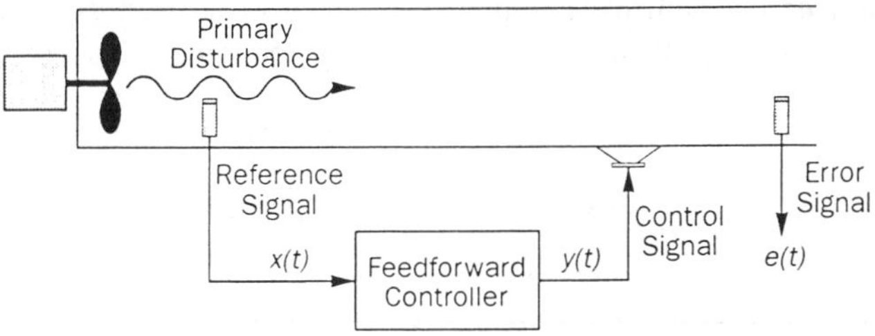

Fig. 6.7 Feedforward control arrangement for attenuating sound propagation in a duct, where the reference signal is provided by an upstream measurement of the disturbance.

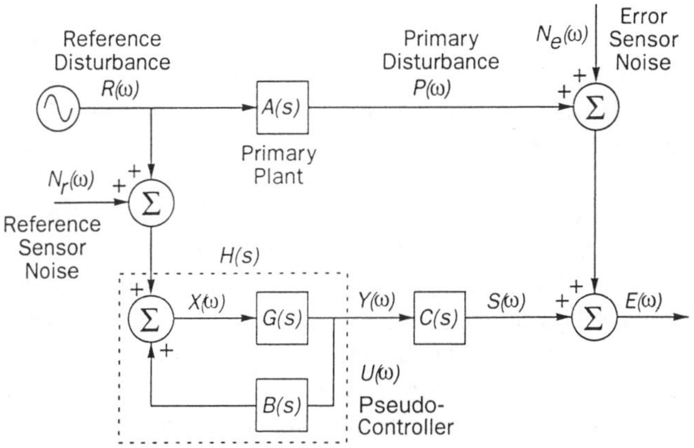

Fig. 6.8 Equivalent block diagram of duct feedforward control system in Fig. 6.7.

A block diagram of this more general arrangement is shown in Fig. 6.8. Here the primary source component of the error signal $P(\omega)$ is equal to the signal present at the reference sensor $R(\omega)$ modified by the frequency response of the primary 'plant', the latter quantity being the transfer function $A(s)$ between the reference sensor and the output of the error sensor,

$$P(\omega) = R(\omega)A(s) \tag{6.3.15}$$

The control source generated component $S(\omega)$ of the error signal is again equal to the control signal $Y(\omega)$ modified by the cancellation path transfer function $C(s)$. The relationship governing calculation of the control signal, however, is more complicated than it was in the previous model with an 'incorruptible' reference signal. The input to the control filter $X(\omega)$ can be viewed as comprised of three parts: the 'desired' reference signal $R(\omega)$, which is correlated with the primary noise disturbance, measurement noise at the reference sensor $N_r(\omega)$ such as due to turbulence, and feedback from the upstream propagating component of the control output. This latter quantity is equal to the control output modified by some transfer function $B(s)$. Referring to Fig. 6.8, it can be seen that by combining the first two of these quantities into a 'pseudo' reference signal $U(\omega)$, the combination of the controller output and the feedback can be modelled as a 'pseudo' controller transfer function, having the appearance of an infinite impulse response filter with a transfer function $H(s)$ given by (Eriksson *et al.*, 1987):

$$H(s) = \frac{Y(\omega)}{U(\omega)} = \frac{G(s)}{1-G(s)B(s)} \tag{6.3.16}$$

Finally, there is also measurement noise $N_e(\omega)$ associated with the error signal acquisition, such that the total error signal is the combination of three parts as follows:

$$E(\omega) = P(\omega)+S(\omega)+N_e(\omega) \tag{6.3.17}$$

The problem now is to determine a feedforward controller transfer function $G(s)$ which will minimize the acoustic or vibration disturbance at the error sensor. As we are again working in the frequency domain, this is equivalent to minimizing power spectral density S_{ee} of the error signal at each frequency component. The easiest way to do this is to first determine the optimum transfer function of the 'pseudo' controller $H(s)$, then use the solution and a knowledge of the feedback transfer function $B(s)$ to determine the optimum controller transfer function. Taking this approach, and combining the primary disturbance and error sensor measurement noise into a single 'desired' signal $D(\omega)$, the problem can be drawn as shown in Fig. 6.9. As the system is linear, we can reorder the problem as shown in Fig. 6.10, once again turning it into a simple least squares estimation problem in the frequency domain (similar to the non-feedback problem we just discussed). Note that by modelling the system as shown in Fig. 6.9, the product (in the frequency domain) of the pseudo reference signal $U(\omega)$ and the cancellation path transfer function is now equivalent to the filtered reference signal $F(\omega)$, where the pseudo reference signal $U(\omega)$ is actually a combination of the signal correlated with the primary disturbance $R(\omega)$ and the

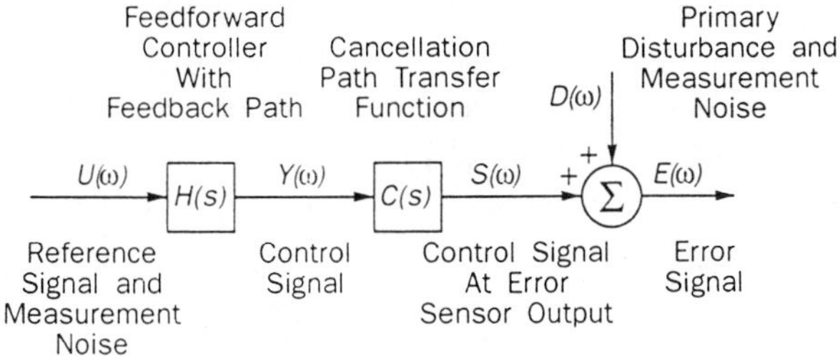

Fig. 6.9 Simplification of block diagram of Fig. 6.8.

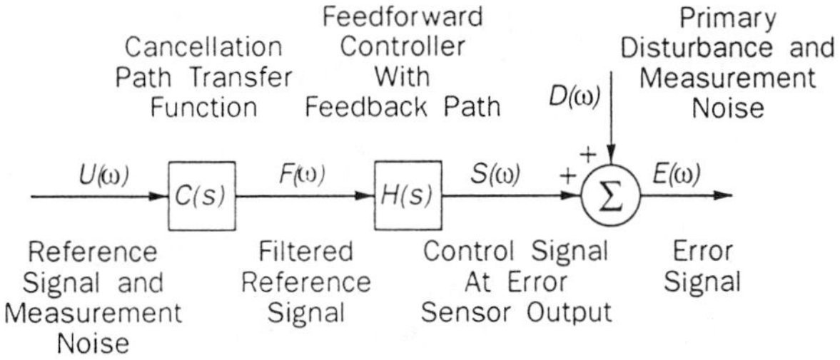

Fig. 6.10 Reordering of the block diagram of Fig. 6.9.

reference sensor measurement noise $N_r(\omega)$. The power spectral density in (6.3.10) can be expanded using the block diagram in Fig. 6.9 as

$$S_{ee}(\omega) = E\{D^*(\omega)D(\omega) + D^*(\omega)F(\omega)H(s) + H^*(s)F^*(\omega)D(\omega) + H^*(s)F^*(\omega)F(\omega)H(\omega)\} \tag{6.3.18}$$

This can be simplified by writing the equation in terms of signal power spectral densities and cross-spectral densities as follows:

$$S_{ee}(\omega) = S_{dd}(\omega) + S_{fd}^*(\omega)H(s) + H^*(s)S_{fd}(\omega) + H^*(s)S_{ff}(\omega)H(s) \tag{6.3.19}$$

where $S_{dd}(\omega)$ is the power spectral density of the desired signal $D(\omega)$

$$S_{dd}(\omega) = E\{D^*(\omega)D(\omega)\} \tag{6.3.20}$$

and $S_{fd}(\omega)$ is the cross-spectral density between the filtered reference signal and the desired signal:

$$S_{fd}(\omega) = E\{F^*(\omega)D(\omega)\} \tag{6.3.21}$$

Expressed in this way, the error criterion of power spectral density is again a simple complex scalar quadratic equation. The filtered reference signal power spectral density is a positive quantity, so that the optimum frequency response characteristics of the pseudo controller $H(s)$ are defined by the expression

$$H_{\text{opt}}(s) = -S_{ff}^{-1}(\omega)S_{fd}(\omega) \tag{6.3.22}$$

This is again in the form of a solution to the Weiner–Hopf integral expression, similar to the result obtained in the absence of feedback from the control source to the reference sensor. The 'residual' (error signal) power spectral density which exists after implementation of the optimum control law is found by substituting the result of (6.3.15) into (6.3.13), which yields the expression

$$S_{ee}(\omega)_{\min} = S_{dd}(\omega) - S_{ff}^{-1}(\omega) \mid S_{fd}(\omega) \mid^2 \tag{6.3.23}$$

Having determined the optimum characteristics of the transfer function $H(s)$, it is now straightforward to determine the optimum feedforward controller frequency response. From (6.3.16) this is simply

$$G_{\text{opt}}(s) = \frac{H_{\text{opt}}(s)}{1 + B(s)H_{\text{opt}}(s)} \tag{6.3.24}$$

It should be pointed out, however, that determination of the optimum characteristics does not necessarily produce a causal controller, one which can be implemented in the time domain.

It is interesting to compare the optimum controller characteristics with and without feedback from the control source to the reference signal. Without feedback, the optimum controller is purely feedforward, the characteristics being defined by the solution to a Weiner filtering problem. With feedback, however, the optimum controller has both zeros and poles. This provides some basis for selecting a filter architecture later in this chapter when the use of adaptive feedforward controllers is considered.

In addition to feedback, it is interesting also to consider the influence which measurement noise has upon the characteristics of the optimal feedforward controller, and upon the attenuation which it can achieve. To do this, we can first expand the power spectral density and cross-spectral density terms in (6.3.19). The cross-spectral density of the desired and filtered reference signals S_{fd} can be expanded as

$$S_{fd}(\omega) = E\left\{\left(C(s)[R(\omega) + N_r(\omega)]\right)^* \left(A(s)R(\omega) + N_e(\omega)\right)\right\} \tag{6.3.25}$$

Assuming that the measurement noise signals $N_r(\omega)$ and $N_e(\omega)$ are uncorrelated with each other and all other signals, then taking expected values produces

$$S_{fd}(\omega) = C^*(s)A(s)S_{rr}(\omega) \tag{6.3.26}$$

where S_{rr} is the power spectral density of the primary source-correlated component of the reference signal $R(\omega)$ (remember that the primary source signal is the signal on the output of the noiseless error sensor when only the source of unwanted noise is operating). The power spectral density of the filtered reference signal $S_{ff}(\omega)$ can be expanded as

$$S_{ff}(\omega) = E\left\{\left(C(s)[R(\omega) + N_r]\right)^* \left(C(s)[R(\omega) + N_r(\omega)]\right)\right\} \tag{6.3.27}$$

which can be simplified, using the assumption of uncorrelated measurement noise as

$$S_{ff}(\omega) = C^*(s)C(s)[S_{rr}(\omega)+S_{N_rN_r}(\omega)] \tag{6.3.28}$$

where $S_{N_rN_r}(\omega)$ is the power spectral density of the reference sensor measurement noise. Substituting (6.3.26) and (6.3.28) into the relation for the optimum characteristics of the transfer function $H_{\text{opt}}(s)$ yields the expression

$$H_{\text{opt}}(s) = -\frac{C^*(s)A(s)S_{rr}(\omega)}{C^*(s)C(s)[S_{rr}(\omega)+S_{N_rN_r}(\omega)]} \tag{6.3.29}$$

The signal to noise ratio (SNR) of the reference signal supplied to the controller is defined as the ratio of the power spectral density of the primary source-correlated component of the reference signal and the noise

$$SNR = \frac{S_{rr}(\omega)}{S_{N_rN_r}(\omega)} \tag{6.3.30}$$

In terms of this quantity, the optimum value of the transfer function $H(s)$, as defined in (6.3.29), can be written as (Roure, 1985; see also Widrow *et al.*, 1975a; Ffowcs-Williams *et al.*, 1985; Eriksson and Allie, 1988):

$$H_{\text{opt}}(s) = -\frac{SNR(\omega)}{1+SNR(\omega)}\frac{A(s)}{C(s)} \tag{6.3.31}$$

Thus as the noise level increases, and the *SNR* decreases, the gain of the transfer function $H_{\text{opt}}(s)$ and optimal controller transfer function $G_{\text{opt}}(s)$ decreases. Note that it is only noise at the reference sensor which is responsible for this result, not noise at the error sensor. To quantify the effect which this has upon the performance of the system, the power spectral density of the filtered reference signal $S_{ff}(\omega)$ and cross-spectral density between the filtered reference signal and the desired signal $S_{fd}(\omega)$ can be rewritten as

$$S_{ff}(\omega) = |C(s)|^2 S_{uu}(\omega) \tag{6.3.32}$$

$$S_{fd}(\omega) = C^*(s)S_{ud}(\omega) \tag{6.3.33}$$

where $S_{uu}(\omega)$ is the power spectral density of the reference signal $U(\omega)$, and $S_{ud}(\omega)$ is the cross-spectral density of the reference signal $U(\omega)$ and the desired signal $D(\omega)$. The attenuation of the power spectral density of the error signal which results from implementation of the optimum controller is therefore

$$\frac{S_{ee,\min}(\omega)}{S_{dd}(\omega)} = 1-\frac{|S_{ud}(\omega)|^2}{S_{uu}(\omega)S_{dd}(\omega)} = 1-\gamma_{ud}^2(\omega) \tag{6.3.34}$$

where $\gamma_{ud}^2(\omega)$ is the coherence function between the signals $U(\omega)$ and $D(\omega)$, which is the coherence function between the output of the reference sensor and the output of the error sensor without the active control system operating.

Equation (6.3.34) shows that the maximum levels of attenuation of the power spectral density of the error signal, which are limited by the correlation between the output of the reference sensor and the output of the error sensor at the frequency of interest in the absence of active control. Therefore, when contemplating a feedforward active control system design it is possible to predict the maximum performance by simply measuring the coherence of two signals (the reference sensor output and the error sensor output) without having to actually build a controller. If, for example, the coherence $\gamma_{ud}^2(\omega)$ at a given frequency is 0.90, then the maximum level of power attenuation at that frequency is 10 dB. If the coherence is 0.99, the maximum level of power attenuation is 20 dB, and so on.

It should be noted that these values are the *maximum* levels of attenuation. There are many factors which will reduce these levels in a practical implementation, such as:

1. The impulse response of the controller described by the transfer function $G_{opt}(s)$ may be non-causal, at least at some frequencies, and therefore it may not be possible to implement the transfer function exactly. Even if the impulse response is causal, the design of a filter to implement it exactly may be very complicated or impractical.
2. Simply minimizing the power spectral density of the signal at a single location does not necessarily mean that the attenuation is global. This is true even in the case of very low frequency excitation when the error sensor is in the near field of a sound source, such that significant quantities of the near field disturbance (components of the field which decay exponentially with distance) will be present in the signal.
3. As discussed in the introduction, a filter with fixed characteristics is designed based on the assumption that the system is stationary. In practice, many active noise and vibration control targets will be non-stationary systems, with transfer functions varying with temperature, air speed, passenger numbers, etc.
4. The attenuation predicted by (6.3.34) is for the referenced disturbance component of the sound field only. If there are other significant contributors to the noise field, such as boundary layer noise in aircraft, this will not be affected by the control source so that the net attenuation will be reduced.

One final point to note in this discussion concerns the validity of the assumption that the measurement noise at the reference sensor is uncorrelated with the measurement noise at the error sensor. While this will normally be the case, in the control of sound propagating in ducts measurement noise due to the boundary layer in the duct may, in fact, be correlated over a significant distance (Bull, 1968). This will be a particular problem for adaptive feedforward control systems, discussed in the remainder of this chapter, which will attempt to minimize both the primary disturbance and the correlated part of the measurement noise, thereby downgrading performance. Reduction of the turbulent flow measurement noise at the sensors can be achieved by using microphone arrangements which effectively average the acoustic pressure over an extended distance, as discussed in Chapter 14.

The final point to be considered here is how the transfer functions which define the frequency response characteristics of the optimum feedforward controller might be measured. The optimum controller is defined in equation (6.3.24) in terms of the optimum characteristics of the 'pseudo' controller $H(s)$, which incorporated both the feedforward control signal and the portion of the control output which is fed back through the reference sensor. The quantity $H_{\text{opt}}(s)$ can be re-expressed as

$$H_{\text{opt}}(s) = -\frac{S_{ud}(\omega)}{C(s)S_{uu}(\omega)} \tag{6.3.35}$$

By injecting white noise $v(\omega)$, uncorrelated with the primary disturbance, into the control source the cancellation path transfer function $C(s)$ which defines the frequency response between the signal input to the control source the signal output from the error sensor, and the feedback path transfer function $B(s)$ which defines the frequency response between the input signal to the control source and the output signal from the reference sensor, can both be measured using common two channel spectrum analysers. In terms of power and cross-spectral densities, the transfer functions are

$$C(s) = \frac{S_{ve}(\omega)}{S_{vv}(\omega)} \tag{6.3.36}$$

and

$$B(s) = \frac{S_{vx}(\omega)}{S_{vv}(\omega)} \tag{6.3.37}$$

where $S_{vv}(\omega)$ is the power spectral density of the white noise test signal, $S_{ve}(\omega)$ is the cross-spectral density between the white noise test signal and the error sensor signal, and S_{vx} is the cross-spectral density between the white noise test signal and the reference sensor signal.

What remains now is to determine the frequency response function between the reference sensor and the error sensor $P_n(s)$, where the n denotes that the estimate will be 'noisy', inherently incorporating the *SNR* terms in (6.3.30). This can again be measured easily as the frequency response function between the error and reference signals with both the white noise test signal and active control input turned off. In terms of spectral densities,

$$P_n(s) = \frac{S_{xe}(\omega)}{S_{xx}(\omega)} = \frac{S_{ud}(\omega)}{S_{uu}(\omega)} \tag{6.3.38}$$

With this measurement, $H_{\text{opt}}(s)$ is simply

$$H_{\text{opt}}(s) = -\frac{P_n(s)}{C(s)} \tag{6.3.39}$$

This can be substituted back into (6.3.24) with the measurement of $B(s)$ to obtain the characteristics of the optimum controller:

$$G_{opt}(s) = -\frac{P_n(s)}{C(s)-P_n(s)B(s)} \tag{6.3.40}$$

We have approached the work in this section principally from the standpoint of providing a set of 'base' characteristics, many of which will be used in the examination of adaptive feedforward control systems in the remainder of this section. It is, however, entirely possible to experimentally determine the characteristics of the optimum controller in (6.3.40) and implement the system directly (Roure, 1985). This will entail using an inverse Fourier transform to convert the desired frequency response characteristics into an impulse response function for time domain implementation. There are, however, a few practical difficulties which arise (Roure, 1985):

1. The estimates of the desired frequency response can be poor at low frequencies due to poor actuator response and measurement noise. The estimates also tend to have large values at high frequencies. Therefore, a window is used to attenuate the influence of the low and high frequency components to avoid windowing errors. The window used by Roure (1985) zeroed the response below 50 Hz, and smoothly attenuated the response above the cut-on frequency of the first higher order mode in the duct with which he was concerned.
2. The frequency response function will in general have non-causal components, which must be removed from the impulse response function after calculation.
3. The initial estimates of the frequency response functions used in defining the characteristics of the optimum controller will usually *not* produce the maximum levels of attenuation, owing to measurement errors and non-linearities in the actuators, and therefore must be iteratively updated (described by Elliott and Nelson, 1984; Roure, 1985).

For further information on fixed filter feedforward control systems, the reader is referred to Roure (1985) and Ross (1982).

6.4 WAVEFORM SYNTHESIS

As we noted in the beginning of the previous chapter, many of the problems targeted for active noise and vibration control have characteristics which are time varying, resulting in optimum controller characteristics which are also time varying. Because of this, an adaptive feedforward control system is often the most desirable option, as it has the ability to accommodate changes in the system, modifying itself to maintain the maximum levels of attenuation. One of the first such control systems was developed in the 1970s, and is referred to as 'waveform synthesis' (Chaplin, 1983). Although not as widely used as the modified adaptive signal processing approaches to be described in the sections which follow, and not as generally applicable, it is a practical technique which provides a useful entrance into the area of adaptive feedforward control system design.

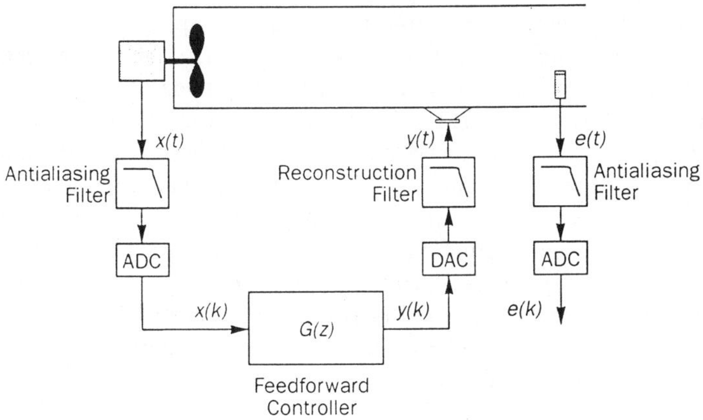

Fig. 6.11 Feedforward control arrangement for attenuating sound propagation in a duct, where the control system is explicitly digital.

Waveform synthesis specifically targets repetitive sound, such as sinusoidal disturbances. The basic idea revolves around the digitizing of a repetitive waveform, as is done by a common sample and hold operation required for analogue to digital signal conversion. Consider the system arrangement shown in Fig. 6.11, where the antialiasing and reconstruction filters used in a practical digital implementation are explicitly shown. The control filter is now explicitly digital, and the approach we will take to construct a control signal is waveform synthesis. As we have discussed previously, the output of the control filter is modified by a cancellation path transfer function before it appears in the error signal. It will be advantageous here to model this transfer function as an m-order (time invariant) finite impulse response function (vector) $\boldsymbol{c}$, given by

$$\boldsymbol{c} = \begin{bmatrix} c_0 & c_1 & c_2 & \cdots & c_{m-1} \end{bmatrix}^{\mathrm{T}} \tag{6.4.1}$$

such that the control source component of the error signal at time k, $s(k)$, is equal to the convolution of the output $y(k)$ of control filter at time k and the finite impulse response function c, as follows:

$$s(k) = y(k) * c = \boldsymbol{y}^{\mathrm{T}}(k)\boldsymbol{c} \tag{6.4.2}$$

where $\boldsymbol{y}(k)$ is the $(m \times 1)$ vector of most recent filter outputs:

$$\boldsymbol{y}(k) = \begin{bmatrix} y(k) & y(k-1) & y(k-2) & \cdots & y(k-m+1) \end{bmatrix} \tag{6.4.3}$$

The total error signal at time k, $e(k)$, is equal to the superposition of the primary $p(k)$ and control $s(k)$ components:

$$e(k) = p(k)+s(k) = p(k)+\boldsymbol{y}^{\mathrm{T}}(k)\boldsymbol{c} \tag{6.4.4}$$

From a qualitative perspective, for the SISO arrangement shown in Fig. 6.10 it will always be possible to generate some repetitive control signal, or 'waveform', which will completely cancel the repetitive primary source signal, regardless of the characteristics of the cancellation path (provided that the cancellation path does not contain a transfer function zero at the frequency component(s) of the signal, in which case the amplitude of the control signal would have to be 'infinite'). The object of the waveform synthesis is to find that waveform, and inject it into the system in response to each cycle of the primary disturbance. A relatively simple way of doing this is to have a reference sensor which is somehow sequenced with the primary disturbance, so that it delivers a single pulse at the same point in the primary waveform at each cycle. The sampling rate of the system is set so that there is an integer number of sampling periods between each reference pulse. If the total number of samples taken by the digital system during a single cycle of the primary disturbance is N, then the reference input to the control system is a Kronecker impulse train of period N,

$$x(k) = \sum_{i=-\infty}^{\infty} \delta(k-iN) \tag{6.4.5}$$

where δ is a Kronecker delta function, such that

$$\delta(k) = \begin{cases} 1 & k=0 \\ 0 & \text{otherwise} \end{cases} \tag{6.4.6}$$

A good example of such an arrangement is the synchronous sampling of a periodic waveform from a rotating machine (Elliott and Darlington, 1985), where the reference pulse is provided by a tachometer signal.

If there are N weights in the digital filter, the output $y(k)$ is simply equal to the value of the weight coefficient at the relevant point in the cycle, corresponding to time k:

$$y(k) = \sum_{j=0}^{N-1} x(k-j)w_j = \sum_{i=-\infty}^{i=\infty} \sum_{j=0}^{N-1} w_j \delta(k-iN-j) \tag{6.4.7}$$

The control filter output is therefore a repetitive display of weight coefficient values. If these weights are suitably adjusted, the periodic primary noise disturbance can be suppressed. One simple way of deriving the 'optimum' set of weight coefficients is to simply adjust, in turn, single weight coefficients on a 'trial and error' basis; if the amplitude of the error signal is reduced in response to the change the modification is retained, if the amplitude is increased the modification is discarded (Smith and Chaplin, 1983). This methodology can be quite slow, however, due both to the nature of the algorithm and the coupling between the weights in the error criterion (described more in the next section) and more efficient algorithms which adjust all weight coefficients simultaneously have been suggested (Smith and Chaplin, 1983). Such an adjustment could be conducted using the filtered-x LMS algorithm, which we will discuss later in this chapter.

6.5 THE NON-RECURSIVE (FIR) DETERMINISTIC GRADIENT DESCENT ALGORITHM

While the concept of waveform synthesis may provide a suitable basis from which to design an adaptive feedforward controller for repetitive disturbances where it is possible to synchronize the controller sampling rate with the disturbance frequency, it is not suitable for the more general case of random noise. It can also be slow in adaptation, and require a much larger number of weight coefficients than would be needed if the reference signal were continuous, as opposed to the impulse train used in waveform synthesis. The requirement of a sampling rate which provides an integer number of samples in each disturbance can also prove to be constraining in many instances.

In the remainder of this chapter, a more general approach to the design of an adaptive feedforward control system will be developed, based upon the concepts of adaptive signal processing. In the next four sections we will consider systems based upon the simplest filter architecture, the finite impulse response (FIR) filter. Initially (in the next two sections), the 'standard' implementation of the adaptive filters will be studied. This is the form of implementation which may exist, for example, in a telephone echo cancellation circuit. This will provide the groundwork for the active noise and vibration control implementation, which is an extension of the standard arrangement. Following our examination of the FIR filter, we will discuss the more complicated infinite impulse response (IIR) and artificial neural network implementations.

6.5.1 The FIR filter

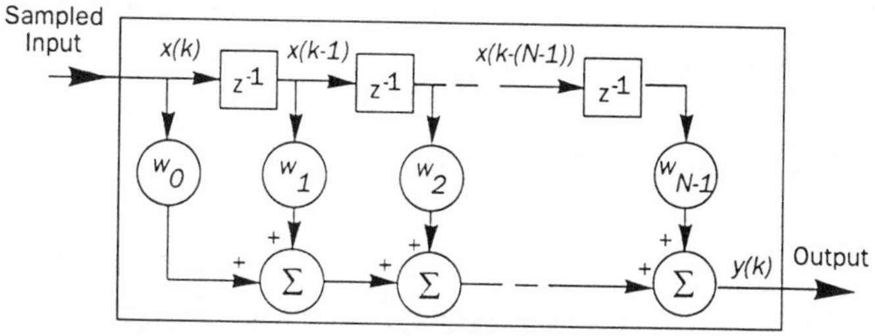

Fig. 6.12 The finite impulse response (FIR) filter.

The non-recursive, or finite impulse response (FIR), filter is probably the most utilised architecture in adaptive digital signal processing. The common direct realization of the FIR filter, known as the transversal filter or tapped delay line, is shown in Fig. 6.12. While for a given number of filter weights its performance may not be as good as its infinite impulse response filter cousin, its inherent stability and simple structure, which results in relatively simple algorithms being able to adaptively tune it, often make it the most practical choice. The next two sections will consider algorithms for the (standard) adaptive use of the FIR filter.

At some time (sample) k, the output $y(k)$ of the FIR filter is simply a weighted combination of past input samples

$$y(k) = \sum_{n=0}^{N-1} w_n(k)x(k-n) = \mathbf{w}^T(k)\mathbf{x}(k) = \mathbf{x}^T(k)\mathbf{w}(k) \tag{6.5.1}$$

where there are N stages in the filter, $\mathbf{w}$ is an $(N \times 1)$ vector of filter weight coefficients, given by

$$\mathbf{w}(k) = \left[w_0(k) \;\; w_1(k) \;\; \cdots \;\; w_{N-1}(k)\right]^T \tag{6.5.2}$$

$\mathbf{x}$ is an $(N \times 1)$ vector of input samples in the delay chain of the filter, given by

$$\mathbf{x}(k) = \left[x(k) \;\; x(k-1) \;\; \cdots \;\; x(k-(N-1))\right]^T \tag{6.5.3}$$

and T denotes transpose. The input delay chain of the finite impulse response filter is also called a tapped delay line (because the values in the delay chain, or line, are 'tapped off', multiplied and accumulated (added together) in the output derivation process), and the filter is also referred to as a tapped delay filter. As discussed in Chapter 5, the discrete transfer function of the FIR filter is all zero, having no poles (or terms in the denominator), and is written as

$$H(z) = \frac{Y(z)}{X(z)} = w_0 + w_1 z^{-1} + w_2 z^{-1} + \cdots + w_{N-1} z^{N-1} \tag{6.5.4}$$

Physically, this means that the filter output is only a function of present and past input samples, and not a function of past output samples. This lack of poles in the filter leads to its inherent stability. It also leads to its name, because when it is subject to a unit pulse input, it will produce an output for a finite period of time (a finite impulse response), the duration of which is determined by the number of delay stages in the filter.

6.5.2 Development of the error criterion

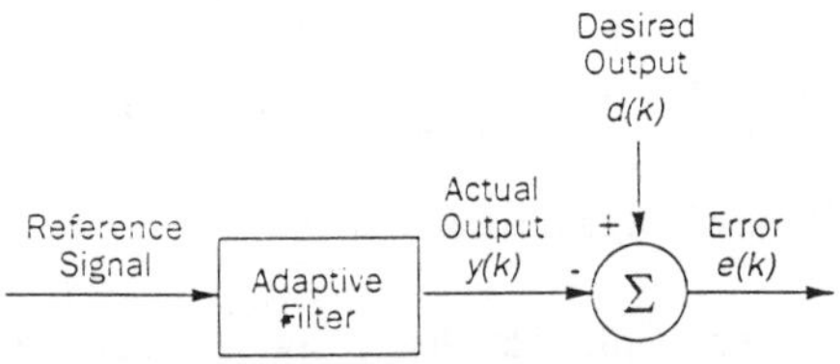

Fig. 6.13 Standard adaptive filtering problem: adjust the weights of the filter such that the output most closely matches some desired output.

Consider the implementation of an FIR filter in the adaptive filtering problem shown in Fig. 6.13, where a reference signal is input to the adaptive (FIR) filter, and the aim of the exercise is to have the filter output match some desired output as closely as possible (in an active control implementation, this desired signal will be the phase inverse of the primary signal as measured by the error sensor). The error signal $e(k)$ is defined as the difference between the desired output $d(k)$ and the actual output $y(k)$ of the filter at time k,

$$e(k) = d(k) - y(k) \tag{6.5.5}$$

The ideal error criterion for such a problem is the mean square error, ξ, defined as the ensemble average, or expected value, of the squared value of the error signal e

$$\xi(k) = E\{e^2(k)\} \tag{6.5.6}$$

As discussed is Chapter 5, mean square error is not the result of temporal averaging, but is rather the expected value, as denoted by the statistical expectation operator $E\{\}$, of the square of the error signal at any given instant in time.

With adaptive filtering problems it is often the case that the environment in which the problem is cast is stochastic in nature; that is, the signals utilized by the system are randomly varying and must be described by their statistics. This tends to complicate the analysis of adaptive signal processing algorithms operating in realistic environments. It is therefore common to invoke the assumption that all variables used are equal to their expected value at that instant in time, or $x(k) = E\{x(k)\}$. This is equivalent to converting the problem from a stochastic one to a deterministic one. While from a quantitative standpoint this tends to produce results which are 'overly optimistic', from a qualitative standpoint it greatly simplifies analysis of the adaptive filtering system and enables the derivation of a set of characteristics which clearly describe the nature of the influence which various parameters have upon the performance of the system. In this section we will likewise make this assumption. More exact results for stochastic implementation will be discussed in Section 6.6.

Returning to (6.5.6), equations (6.5.1) and (6.5.5) can be used to re-express the mean square error as

$$\begin{aligned} \xi(k) &= E\left\{ \left(d(k) - \boldsymbol{x}^{\mathrm{T}}(k)\boldsymbol{w}(k)\right)^2 \right\} \\ &= E\left\{d^2(k) - 2d(k)\boldsymbol{x}^{\mathrm{T}}(k)\boldsymbol{w}(k) + \boldsymbol{w}^{\mathrm{T}}(k)\boldsymbol{x}(k)\boldsymbol{x}^{\mathrm{T}}(k)\boldsymbol{w}(k)\right\} \end{aligned} \tag{6.5.7}$$

or

$$\begin{aligned} \xi(k) &= E\left\{d^2(k)\right\} - 2\, E\left\{d(k)\boldsymbol{x}^{\mathrm{T}}(k)\right\}\, \boldsymbol{w}(k) \\ &\quad + \boldsymbol{w}^{\mathrm{T}}(k)\, E\left\{\boldsymbol{x}(k)\boldsymbol{x}^{\mathrm{T}}(k)\right\}\, \boldsymbol{w}(k) \end{aligned} \tag{6.5.8}$$

(It should be noted that in the derivation of (6.5.8) an assumption is made that the weight coefficient vectors, $\boldsymbol{w}$, are uncorrelated with the input signal vector, $\boldsymbol{x}$. For (subjectively) 'slow' adaptation, where convergence of the filter weights to the final values may take hundreds or thousands of iterations, this assumption appears to be a valid one. For 'fast' adaptation, however, it is not strictly correct. Despite this, the qualitative algorithm behavioural characteristics to be examined are still valid, so the reader is simply asked to remember that for 'fast' adaptation the results may not be quantitatively exact.) The first term in (6.5.8) is equal to the mean square power σ_d^2 of the desired signal. The expected part of the second term is defined as the cross correlation $\boldsymbol{p}$ between the desired response and the input vector,

$$\boldsymbol{p} = E\{d(k)\boldsymbol{x}(k)\} \tag{6.5.9}$$

The expected part of the third term is defined as the input autocorrelation matrix $\boldsymbol{R}$,

$$\boldsymbol{R} = E\{\boldsymbol{x}(k)\boldsymbol{x}^{\mathrm{T}}(k)\} \tag{6.5.10}$$

In terms of these quantities, the mean square error is defined by the relationship

$$\xi(k) = \sigma_d^2 - 2\boldsymbol{p}^{\mathrm{T}}\boldsymbol{w}(k) + \boldsymbol{w}^{\mathrm{T}}(k)\boldsymbol{R}\boldsymbol{w}(k) \tag{6.5.11}$$

Observe that by employing the assumption that variables are equal to their expected values we have turned the form of the problem from stochastic into deterministic (in terms of signal cross-correlations and autocorrelations).

Equation (6.5.11) shows the mean square error to be a quadratic function of the filter weight coefficients. Therefore, the error, or performance, criterion of mean square error as a function of weight coefficient values describes a hyper-parabolic surface in $(N+1)$-dimensional space, where N is the number of weights in the filter (there are N principal, or independent, axes and one dependent axis of mean square error). Figure 6.14 illustrates a typical error surface for the case of two weight coefficients ($N = 2$), in three-dimensional space.

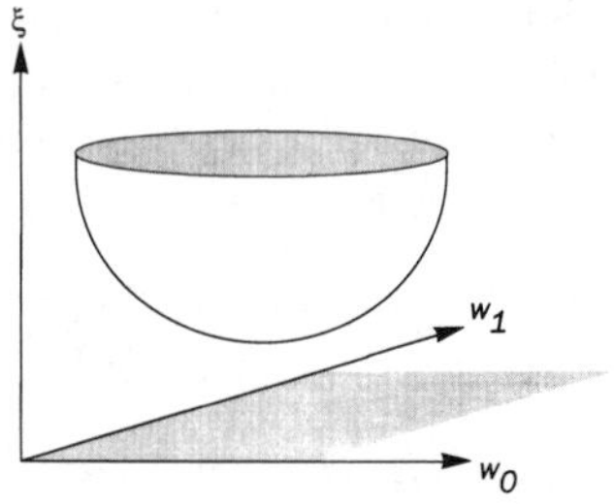

Fig. 6.14 Typical error surface ('bowl') for a two weight transversal filter.

6.5.3 Characterization of the error criterion

The stability/convergence characteristics of the adaptive algorithms considered later in this chapter are governed by the characteristics of the 'error surface', the topography of which is described by the error criterion of (6.5.11). For this reason, a closer examination of the error criterion is in order.

Probably the most important property of the mean square error criterion is the fact that it has only one extremum, and this extremum is a minimum. This characteristic is apparent in the plot of the error surface for a filter with two weight coefficients shown in Fig. 6.14. The aim of the adaptive filtering process is to derive an optimum set of weight coefficients, $\boldsymbol{w}_{\mathrm{opt}}$, such that the value of the mean square error is a minimum. Referring to Fig. 6.14, this is geometrically equivalent to finding the coordinates of the 'bottom of the bowl'. As the minimum is the only extremum, the optimum weight coefficient vector can be found by differentiating the mean square error ξ, as defined in (6.5.11), with respect to the weight coefficient vector $\boldsymbol{w}$ and setting the resulting gradient expression equal to zero. Doing this, we obtain

$$\frac{\partial \xi}{\partial \boldsymbol{w}} = 2\boldsymbol{R}\boldsymbol{w} - 2\boldsymbol{p} = 0 \tag{6.5.12}$$

Therefore, the optimum weight coefficient vector is defined by the relationship

$$\boldsymbol{w}_{\text{opt}} = \boldsymbol{R}^{-1}\boldsymbol{p} \tag{6.5.13}$$

The relationship of (6.5.13) is the discrete form of the solution to the Weiner–Hopf integral equation. This solution is the optimum weight coefficient vector for an FIR filter arranged as an estimator in the configuration shown in Fig. 6.15. With these weights the response of the FIR filter will match the (phase) inverse of the response of the system as closely as is possible.

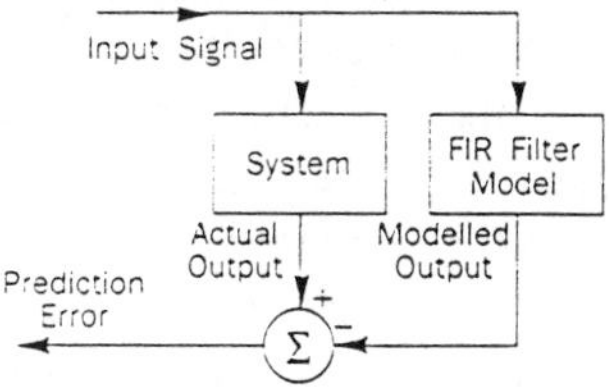

Fig. 6.15 Estimation problem, where the FIR filter weights are formulated so that the mean square value of the prediction error, the difference between the actual and modelled system outputs, is minimized.

If the optimum weight coefficient vector of (6.5.13) is substituted back into the defining equation of the mean square error, (6.5.11), an expression for the minimum mean square error ξ_{min} is produced:

$$\begin{aligned}
\xi_{\text{min}} &= \sigma_d^2 - 2\boldsymbol{p}^{\text{T}}\boldsymbol{w}_{\text{opt}} + \boldsymbol{w}_{\text{opt}}^{\text{T}}\boldsymbol{R}\boldsymbol{w}_{\text{opt}} \\
&= \sigma_d^2 - \boldsymbol{p}^{\text{T}}\boldsymbol{w}_{\text{opt}} - \boldsymbol{p}^{\text{T}}\boldsymbol{R}^{-1}\boldsymbol{p} + \boldsymbol{p}^{\text{T}}\boldsymbol{R}^{-1}\boldsymbol{R}\boldsymbol{R}^{-1}\boldsymbol{p} \\
&= \sigma_d^2 - \boldsymbol{p}^{\text{T}}\boldsymbol{w}_{\text{opt}} - \boldsymbol{p}^{\text{T}}\boldsymbol{R}^{-1}\boldsymbol{p} + \boldsymbol{p}^{\text{T}}\boldsymbol{R}^{-1}\boldsymbol{p} \\
&= \sigma_d^2 - \boldsymbol{p}^{\text{T}}\boldsymbol{w}_{\text{opt}}
\end{aligned} \tag{6.5.14}$$

The minimum mean square error defines the offset, or 'height' of the mean square error surface above the origin of the coordinate system defined by the weight coefficients, as depicted in Fig. 6.16. This value will often (usually) not be zero, because of 'noise' components in the desired signal which are uncorrelated with the reference signal and/or because the filter is of insufficient length. The minimum mean square error defines the absolute best possible result of the adaptive filtering problem; that is, how close the FIR filter output can come to matching the desired signal for a given filter size and given reference signal. The optimum weight coefficient vector $\boldsymbol{w}_{opt}$ of (6.5.13) and the minimum mean square error ξ_{min} of (6.5.14) define the coordinates of the base of the error surface.

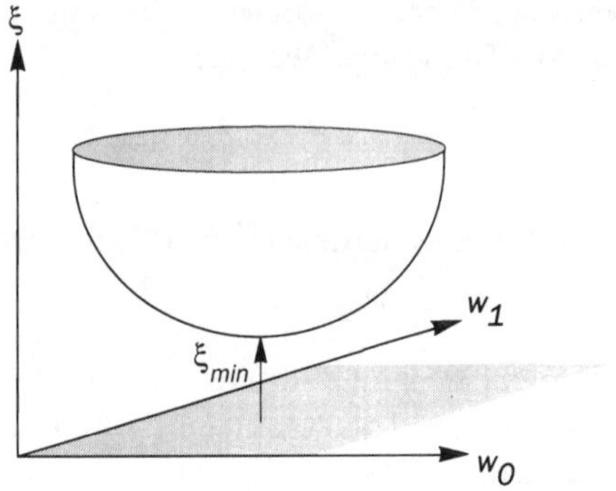

Fig. 6.16 Minimum mean square error is the offset of the 'bowl' from the origin.

As discussed in Section 6.3, the value of the minimum mean square error can be related to the correlation between the reference signal and the desired signal; for any reduction in mean square error to be had by inclusion of the output signal $y(k)$, there must be some correlation between the desired signal d and the input (reference) signal x. This follows from the definition of $\boldsymbol{p}$ given in (6.5.9). If these signals are *not* correlated to some degree then $\boldsymbol{p} = 0$. In this case, (6.5.13) dictates that the optimum weight coefficients are all equal to zero, and from (6.5.14) the minimum mean square error is equal to the mean square value of the desired response.

It is often useful to explicitly redefine the mean square error ξ in terms of the minimum mean square error component ξ_{min}, which cannot be reduced by the adaptive system, and the component of the mean square error in excess of this, the excess mean square error ξ_{ex},

$$\xi(k) = \xi_{min} + \xi_{ex}(k) \tag{6.5.15}$$

The utility of separating the mean square error into these two constituents is that we know the excess mean square error ξ_{ex} will be equal to zero when the filter weights are optimal, which cannot be said for the (total) mean square error ξ. This separation can be done by using (6.5.11) and (6.5.14) and the following steps:

$$\begin{aligned}
\xi(k) &= \xi_{\min}+\xi_{ex}(k)\\
&= \sigma_d^2-\boldsymbol{p}^{\mathrm{T}}\boldsymbol{w}_{\mathrm{opt}}+\boldsymbol{p}^{\mathrm{T}}\boldsymbol{w}_{\mathrm{opt}}(k)-2\boldsymbol{p}^{\mathrm{T}}\boldsymbol{w}(k)+\boldsymbol{w}^{\mathrm{T}}(k)\boldsymbol{R}\boldsymbol{w}(k)\\
&= \xi_{\min}+\boldsymbol{p}^{\mathrm{T}}\boldsymbol{R}^{-1}\boldsymbol{p}-2\boldsymbol{p}^{\mathrm{T}}\boldsymbol{w}(k)+\boldsymbol{w}^{\mathrm{T}}(k)\boldsymbol{R}\boldsymbol{w}(k)\\
&= \xi_{\min}+\boldsymbol{p}^{\mathrm{T}}\boldsymbol{R}^{-1}\boldsymbol{R}\boldsymbol{R}^{-1}\boldsymbol{p}-2\boldsymbol{p}^{\mathrm{T}}\boldsymbol{w}(k)+\boldsymbol{w}^{\mathrm{T}}(k)\boldsymbol{R}\boldsymbol{w}(k)\\
&= \xi_{\min}+\boldsymbol{w}_{\mathrm{opt}}^{\mathrm{T}}\boldsymbol{R}\boldsymbol{w}_{\mathrm{opt}}+\boldsymbol{w}^{\mathrm{T}}(k)\boldsymbol{R}\boldsymbol{w}(k)-\boldsymbol{w}_{\mathrm{opt}}^{\mathrm{T}}\boldsymbol{R}\boldsymbol{w}(k)-\boldsymbol{w}^{\mathrm{T}}(k)\boldsymbol{R}\boldsymbol{w}_{\mathrm{opt}}(k)\\
&= \xi_{\min}+(\boldsymbol{w}(k)-\boldsymbol{w}_{\mathrm{opt}})^{\mathrm{T}}\boldsymbol{R}(\boldsymbol{w}(k)-\boldsymbol{w}_{\mathrm{opt}})\\
&= \xi_{\min}+\boldsymbol{v}^{\mathrm{T}}(k)\boldsymbol{R}\boldsymbol{v}(k)
\end{aligned} \tag{6.5.16}$$

where $\boldsymbol{v}$ is the weight error vector, defined as the difference between the optimum and actual weight coefficient vectors,

$$\boldsymbol{v}(k) = \boldsymbol{w}(k)-\boldsymbol{w}_{\mathrm{opt}} \tag{6.5.17}$$

The excess mean square error is therefore equal to

$$\xi_{ex}(k) = \boldsymbol{v}^{\mathrm{T}}(k)\boldsymbol{R}\boldsymbol{v}(k) \tag{6.5.18}$$

Geometrically, restating the error criterion in terms of the weight error vector can be viewed as simply an axis translation, moving the origin of the coordinate system to the 'bottom of the bowl', as shown in Fig. 6.17.

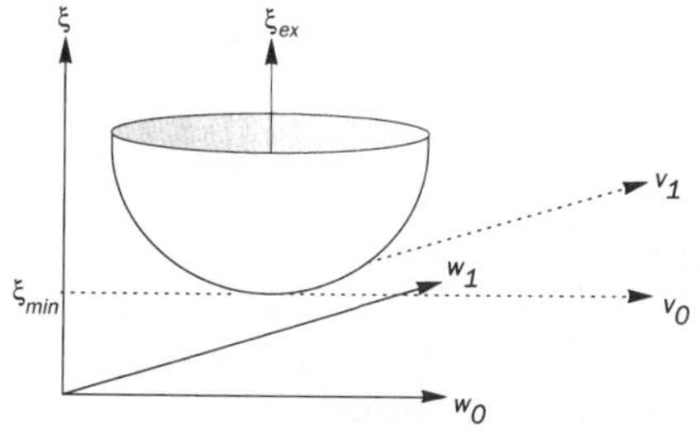

Fig. 6.17 The origin of the coordinate system can be moved to the base of the error surface by expressing the axis in terms of weight error.

It should be observed at this point that the transversal filter is not, in general, an orthogonal filtering structure. By this we mean that each weight in the filter does not make an independent contribution to the mean square error; rather, there is some 'cross-coupling' between the weights. Mathematically, this means that the change in mean square error which results from a change in one filter weight is dependent upon the present values of the other weights. Geometrically, this means that if one were looking down on the three-dimensional mean square error surface of Fig. 6.14 one would see a set of ellipses, or constant mean square error contours, as outlined in Fig. 6.18, and the principal axes of these ellipses would not be aligned with the coordinate system defined by the filter weights (had they been aligned, the filter weights would be orthogonal). As we will see shortly, however, from the standpoint of examining the stability of gradient descent algorithms, it is beneficial to consider the problem with the coordinate axes aligned with the principal axes. In this way the problem becomes 'decoupled', and the descent down each error surface 'slope' can be considered individually.

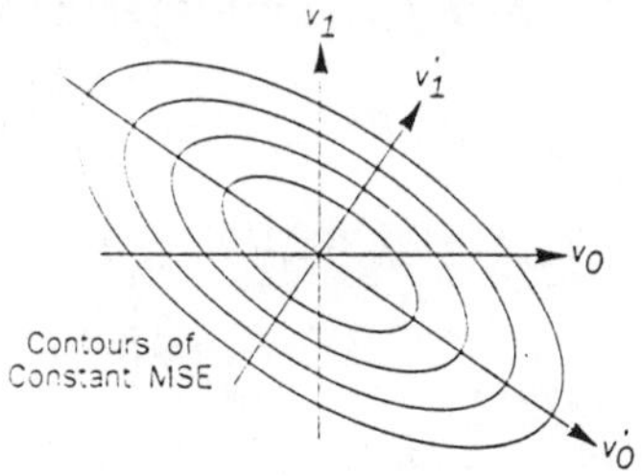

Fig. 6.18 Constant mean square error ellipses, the principal axis of which is defined by v'.

To decouple the error criterion, let us defined as follows a transformation matrix $\boldsymbol{Q}$ which will align the axes of the weight error with the principal axes of the error surface:

$$\boldsymbol{v}(k) = \boldsymbol{Q}\boldsymbol{v}'(k) \tag{6.5.19}$$

where $\boldsymbol{v}'$ is the transformed weight coefficient error vector, the elements of which define the principal axes of the error surface as shown in Fig. 6.18. Using this definition, the expression for the mean square error developed in (6.5.16) can be written in terms of the principal axes of the error surface as follows:

$$\xi(k) = \xi_{min} + \boldsymbol{v}'^{T}(k)\left(\boldsymbol{Q}^{T}\boldsymbol{R}\boldsymbol{Q}\right)\boldsymbol{v}'(k) \tag{6.5.20}$$

When the mean square error is decoupled by expressing the problem in terms of the principal axes of the error surface, then the problem becomes the summation of a set of scalar equations. Therefore, the bracketed part of (6.5.20) must be a diagonal matrix; hence the transformation of the input autocorrelation matrix $\boldsymbol{R}$ by the matrix $\boldsymbol{Q}$ must diagonalize $\boldsymbol{R}$. One matrix which can fulfil this role of $\boldsymbol{Q}$ is the orthonormal transformation matrix of $\boldsymbol{R}$, the columns of which are the eigenvectors of $\boldsymbol{R}$. With this definition of $\boldsymbol{Q}$,

$$\boldsymbol{Q}^{T}\boldsymbol{R}\boldsymbol{Q} = \Lambda = \boldsymbol{Q}^{-1}\boldsymbol{R}\boldsymbol{Q} \tag{6.5.21}$$

where Λ is a diagonal matrix, the elements of which are the eigenvalues of $\boldsymbol{R}$. (This transformation to the orthogonal principal axes can always be made as the input autocorrelation matrix $\boldsymbol{R}$ is a real symmetric matrix, for which real orthogonal eigenvectors will exist.) It can be surmised from this discussion that the eigenvectors of the input autocorrelation matrix define the principal axes of the error surface. Substituting (6.5.21) into (6.5.20) allows the mean square error to be expressed in terms of the principal axes of the error surface as

$$\xi(k) = \xi_{\min} + v'^{\mathrm{T}}(k)\Lambda v'(k) \tag{6.5.22}$$

To give some physical relevance to the eigenvalues of the error surface, (6.5.20) can be differentiated twice, yielding

$$\frac{\partial \xi}{\partial v'} = 2\Lambda v', \quad \frac{\partial^2 \xi}{\partial v'^2} = 2\Lambda \tag{6.5.23}$$

The first derivative in (6.5.23) is the vector of gradients of the error surface along the principal axes, and the second derivative is the vector of changes in the gradient for the principal axis, or 'acceleration down the slope'. It follows from this equation that for the error surface extremum to be a minimum, all eigenvalues must be positive (otherwise the acceleration would be away from the extremum). From the definition of the input autocorrelation matrix given in (6.5.10), this will always be the case for the systems considered in this section (we will find, however, that this will not always be the case for the active noise and vibration control implementation of these systems, where the autocorrelation matrix used by the algorithm may be subject to some 'phase error').

6.5.4 Development and characteristics of the deterministic gradient descent algorithm

The optimum weight coefficient vector w_{opt} was defined in (6.5.13) as the solution to the (discrete) Weiner–Hopf integral equation. However, it is usually impractical to directly solve (6.5.13) to obtain the set of optimum weight coefficients owing the problems such as difficulty in inverting the input autocorrelation matrix $\boldsymbol{R}$, changes in system variables which change the optimum weight coefficient values, and the averaging required to obtain good estimates of the expected values of the various terms. Rather, the optimum weight coefficient vector is normally found by using some numerical search routine. As the error surface is a hyper-paraboloid, with a single (global) minimum, a simple gradient descent type algorithm is often implemented.

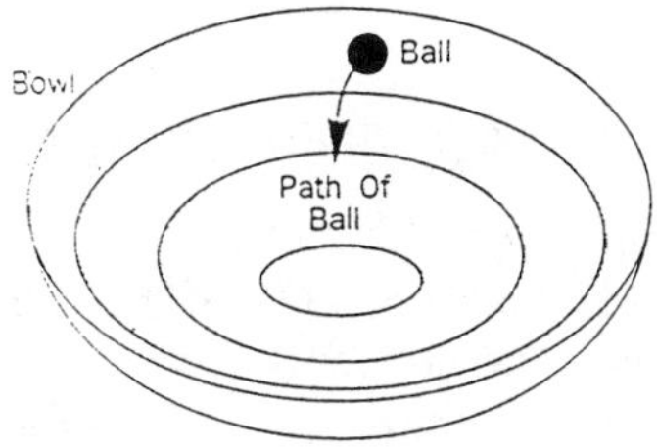

Fig. 6.19. Ball and bowl arrangement.

To obtain an intuitive derivation of a gradient descent algorithm for calculating the optimum weight coefficients of the FIR filter, consider what would happen if the error criterion 'bowl' was constructed and a ball was placed at some point on its edges, as shown in Fig. 6.19. When released, the ball would roll down the sides of the bowl, eventually coming to rest (after some oscillation) at the bottom. This is exactly what we would like our algorithm to do to find the optimum weight coefficient. When first released, the ball will roll in the direction of maximum (negative) change in the slope, or gradient, of the error surface. If we examine the position of the ball at discrete moments in time as it descends we would find that its new position is equal to its old position (one discrete moment ago) plus some distance down the negative gradient of the bowl.

These characteristics are somewhat formalized in a gradient descent algorithm. This type of algorithm attempts to arrive at a calculation of the optimum set of filter weights (at the bottom of the bowl) by adding to the present estimate of the optimum weight coefficient vector a portion of the negative gradient of the error surface at the location defined by this estimate. In this way the current value of the mean square error descends down the sides of the error 'bowl', eventually arriving at the bottom (the location corresponding to the optimum weight coefficients). Mathematically, a generic gradient descent algorithm can be expressed as

$$\boldsymbol{w}(k+1) = \boldsymbol{w}(k) - \mu\Delta\boldsymbol{w}(k) \tag{6.5.24}$$

where $\Delta\boldsymbol{w}$ is the gradient of the error surface at the location given by the current weight coefficient vector, and μ is the portion of the negative gradient to be added, referred to as the convergence coefficient. For the problem being considered here, this gradient was expressed previously in (6.5.12). Substituting this into (6.5.24) produces the deterministic gradient descent algorithm

$$\boldsymbol{w}(k+1) = \boldsymbol{w}(k) + 2\mu(\boldsymbol{p} - \boldsymbol{R}\boldsymbol{w}(k)) \tag{6.5.25}$$

From the standpoint of examining the characteristics of the gradient descent algorithm of (6.5.25), it is easier to first re-express the algorithm in terms of the weight error vector $\boldsymbol{v}$ defined in (6.5.17), as the elements in this vector must always converge towards zero if the algorithm is descending towards the 'bottom of the bowl'. Equation (6.5.25) can be expressed in this manner as

$$\boldsymbol{w}(k+1) - \boldsymbol{w}_{opt} = \boldsymbol{w}(k) - \boldsymbol{w}_{opt} + 2\mu(\boldsymbol{p} - \boldsymbol{R}\boldsymbol{w}_{opt} + \boldsymbol{R}\boldsymbol{w}_{opt} - \boldsymbol{R}\boldsymbol{w}(k)) \tag{6.5.26}$$

or

$$\boldsymbol{v}(k+1) = \boldsymbol{v}(k) + 2\mu(\boldsymbol{p} - \boldsymbol{p} - \boldsymbol{R}\boldsymbol{v}(k)) = \boldsymbol{v}(k) - 2\mu\boldsymbol{R}\boldsymbol{v}(k) \tag{6.5.27}$$

Equation (6.5.27) is a coupled equation, where the new value of weight error is based on a combination of both the old value of itself, plus the old values of the other weight errors. As discussed previously, this coupling arises geometrically because the coordinate system defined by the weight errors is not (in general) aligned with the principal axes of the error surface, and mathematically because the input autocorrelation matrix $\boldsymbol{R}$ is not diagonal. It is easier to examine the algorithm if it is first decoupled, so that the new value of the weight error is dependent only upon the old value of itself. This axes rotation can be accomplished by using the orthonormal transformation of (6.5.21).

Transforming (6.5.27) by multiplying through by the matrix $\boldsymbol{Q}$ yields

$$Q^{-1}v(k+1) = Q^{-1}v(k) - 2\mu Q^{-1}(RQQ^{-1}v(k)) \tag{6.5.28}$$

or

$$v'(k+1) = v'(k) - 2\mu\Lambda v'(k) = (I - 2\mu\Lambda)v'(k) \tag{6.5.29}$$

Decoupled in this manner, (6.5.29) is simply a set of scalar equations of the form

$$v_j'(k+1) = (1 - 2\mu\lambda_j)v_j'(k) \tag{6.5.30}$$

where λ_j is the eigenvalue associated with the jth eigenvector, or error surface principal axis. For the gradient descent algorithm to converge in a stable manner towards the optimum solution corresponding to the 'bottom of the bowl', each of the scalar equations of (6.5.29) must converge towards zero, or

$$\left| \frac{v_j'(k+1)}{v_j'(k)} \right| < 1 \tag{6.5.31}$$

for all j. Therefore,

$$-1 < (1 - 2\mu\lambda_j) < 1 \tag{6.5.32}$$

or

$$0 < \mu < \frac{1}{\lambda_j} \tag{6.5.33}$$

As it is the scalar equation with the maximum eigenvalue which will dictate the overall stability of the algorithm, the bounds placed on the convergence coefficient for stable (convergent) operation of the gradient descent algorithm are

$$0 < \mu < \frac{1}{\lambda_{\max}} \tag{6.5.34}$$

It is enlightening to assign some 'physical meaning' to the bounds dictated by (6.5.34). It was shown in (6.5.23) that the eigenvalue λ characterized the change in gradient, or acceleration, of the error surface associated with principal axes. Therefore, the stability of the gradient descent algorithm is limited by the acceleration down the steepest slope of the error surface. In terms of the bowl/ball analogy, the convergence coefficient μ can be regarded as the force with which we 'push' the ball (representing the weight coefficient calculations) down the sides of the bowl, and the strength of this push is limited by the amount of 'speed' this estimate will pick up as it slides down the slope as dictated by the acceleration, or eigenvalues of the input autocorrelation matrix. If we try to move the ball too quickly down this slope, the 'velocity' is has developed when it reaches the bottom will be such that the force of 'gravity', or the acceleration against its motion presented by the upwards slope of the opposite side of the bowl, will not be enough to contain its motion. Once the algorithm has gone past the 'critical velocity', it will diverge away from then optimum solution. This means that the ball will 'launch' itself out of the bowl!

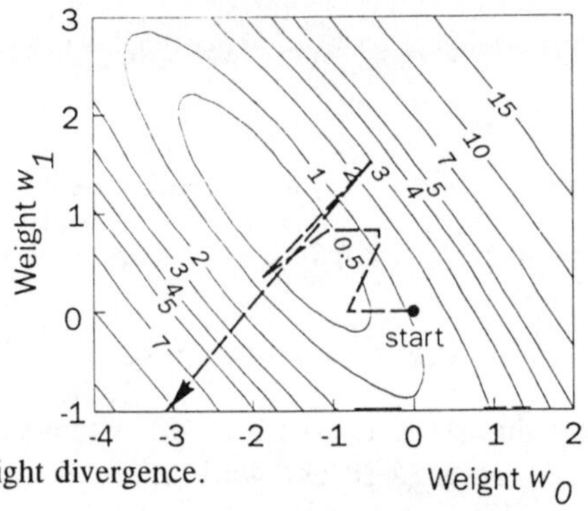

Fig. 6.20 Path of weight divergence.

While the bowl/ball analogy may seem somewhat simplistic, it does, in fact, provide quite an accurate qualitative picture of the algorithm characteristics. Figure 6.20 illustrates the divergence of the algorithm resulting from the choice of slightly too large a convergence coefficient. As can be seen, the algorithm lines up with the steepest slope, accelerates down it, and shoots off into the great unknown, never to return! (Another, although less accurate, analogy can be drawn between this and running down a hill. It is easy to run down a slight grade at high speed and still maintain 'personal' stability; however, if an attempt is made to run down a mountain, personal stability may be compromised!)

While the stability criterion of (6.5.34) is correct from an analytical point of view, it is not always easy to assess from a practical point of view. A more accessible, yet conservative, criterion can be formulated intuitively as follows: The value of the maximum eigenvalue of the input autocorrelation matrix cannot be greater than the trace (the sum of the diagonal elements) of the matrix, or

$$\lambda_{max} \leq tr[\boldsymbol{R}] \tag{6.5.35}$$

where $tr[\ ]$ denotes the trace of the matrix. For white noise, the trace of the matrix is equal to

$$tr[\boldsymbol{R}] = \sum_{k=0}^{N-1} E\{x^2(k)\} = N\, E\{x^2(k)\} \tag{6.5.36}$$

But $E\{x^2(k)\}$ is simply the mean square power of the input signal, σ_x^2. Therefore, alternative bounds on the convergence coefficient for stable operation of the deterministic gradient descent algorithm are

$$0 < \mu < \frac{1}{N\sigma_x^2} \tag{6.5.37}$$

Further insight into the convergence of the weight coefficients can be gained by considering the value of the bracketed term in (6.5.30), $(1 - 2\mu\lambda_j)$, in terms of different values of the convergence coefficient μ. Using the analogy of dynamic system transient response, if $\mu < ½\lambda_j$, then the algorithm is overdamped, leading to a long rise time and no overshoot. If $\mu = ½\lambda_j$, then the algorithm is critically damped, having the shortest rise time possible without any overshoot, and the shortest settling time (1 sample). If $\mu > \frac{1}{2}\lambda_j$, then the algorithm is underdamped, and has overshoot. These concepts are illustrated in Fig. 6.21.

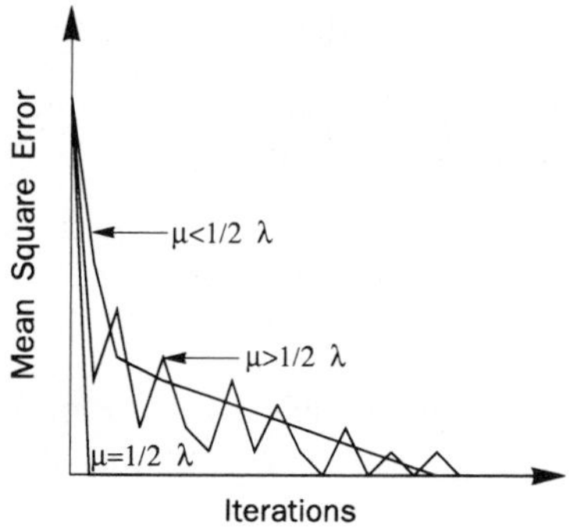

Fig. 6.21 Effect of convergence coefficient upon algorithm dynamic response.

Having discussed the convergence of the filter weights towards their optimum values, let us turn our attention to the problem of convergence of the mean square error, as defined in (6.5.22). As $\mathbf{\Lambda}$ is a diagonal matrix, this can be re-expressed as

$$\begin{aligned}\xi(k) &= \xi_{\min} + \boldsymbol{v}'^{\mathrm{T}}(k)\boldsymbol{\Lambda}\boldsymbol{v}'(k) = \xi_{\min} + tr\left[\boldsymbol{\Lambda}\boldsymbol{v}'(k)\boldsymbol{v}'^{\mathrm{T}}(k)\right] \\ &= \xi_{\min} + tr\left[\boldsymbol{\Lambda}\boldsymbol{C}'(k)\right]\end{aligned} \tag{6.5.38}$$

where $\boldsymbol{C}'$ is the covariance matrix of the weight errors, defined as

$$\boldsymbol{C}'(k) = \boldsymbol{v}'(k)\boldsymbol{v}'^{\mathrm{T}}(k) \tag{6.5.39}$$

Equation (6.5.38) shows that the convergence of the mean square error will be dictated by the convergence of the weight covariance matrix. From (6.5.29), this convergence is characterized by the expression

$$\boldsymbol{C}'(k+1) = (\boldsymbol{I} - 2\mu\boldsymbol{\Lambda})^2\boldsymbol{C}'(k) \tag{6.5.40}$$

In a similar way to the previous case of the transformed weight error vector, the diagonal elements of $\boldsymbol{C}'$ can be viewed as a set of scalar equations of the form

$$c_j'(k+1) = (1 - 2\mu\lambda_j)^2 c_j'(k) \tag{6.5.41}$$

From this viewpoint, it is apparent that the convergence of the mean square error is an exponential process, a point which can be seen more clearly by expressing (6.5.38) as

$$\xi(k) = \xi_{\min} + \sum_{j=0}^{N-1} (1 - 2\mu\lambda_j)^{2k} c_j'(0) = \xi_{\min} + \sum_{j=0}^{N-1} c_j'(0)e^{-2k/\tau_j} \tag{6.5.42}$$

where the time constant of the *j*th scalar equation, or 'mode', is

$$\tau_j = \left\{ \ln\left[\frac{1}{(1-\mu\lambda_j)^2} \right] \right\}^{-1} \tag{6.5.43}$$

and it is assumed that the process began at time $k = 0$.

To put the time constant of (6.5.43) into a more 'friendly' form, consider the expansion

$$e^{-1/\tau} = 1 - \frac{1}{\tau} + \frac{1}{2!\tau^2} - \dots \tag{6.5.44}$$

For the case of slow adaptation the higher order terms can be ignored. Letting $k = 1$ in (6.5.42) gives

$$(1-2\mu\lambda_j)^2 \approx \left[1 - \frac{1}{\tau_j} \right]^2 \tag{6.5.45}$$

Therefore, the commonly stated time constant of adaptation of the mean square error is (Widrow *et al.*, 1976)

$$\tau_j \approx \frac{1}{4\mu\lambda_j} \tag{6.5.46}$$

Equation (6.5.46) shows that the initial convergence of the algorithm will be dictated by the maximum eigenvalue, which has associated with it the shortest time constant. The overall time for convergence to the optimum value is dependent upon the minimum eigenvalue, which has associated with it the longest time constant. Further, it is interesting to consider the effect of a large eigenvalue spread (referring to the ratio between the maximum and minimum eigenvalues) on the speed of convergence by combining the time constant expression of (6.5.46) with the bounds placed on the convergence coefficient in (6.5.34) for stable operation. (A large eigenvalue spread is typical for a system with tonal peaks contained in low (or zero) levels of random noise, with significant oversampling of the tones (say, $> \sim 8$ samples per cycle).) If we re-express μ in terms of the bounds placed upon it by the maximum eigenvalue of the input autocorrelation matrix, the maximum time constant in the adaptation process (which, as outlined, is dependent upon the minimum eigenvalue of the input autocorrelation matrix) can be expressed as

$$\tau_{max} > \frac{\lambda_{max}}{4\lambda_{min}} \tag{6.5.47}$$

Therefore, the greater the eigenvalue spread, the slower the algorithm will be to reach steady state (Widrow *et al.*, 1976; Gardner, 1984; Freij and Cheetham, 1987).

Finally, it should be noted that, from (6.2.25), if the convergence coefficient is within the bounds set by (6.2.34),

$$\lim_{k\to\infty} c_j(k) = 0 \tag{6.5.48}$$

or, in other words, the excess mean square error converges towards zero, and the mean square error converges towards its minimum value. One quantity which will be examined later in this chapter is the algorithm misadjustment, M, which is defined as

$$M = \lim_{k\to\infty}\frac{\xi'(k)}{\xi_{min}} \tag{6.5.49}$$

For the deterministic gradient descent algorithm considered in this section, the misadjustment is equal to zero.

6.6 THE LMS ALGORITHM

6.6.1 Development of the LMS algorithm

In the previous section we considered the use of a gradient descent algorithm to calculate the optimum set of weights, which will minimise the mean square error criterion for a FIR filter. The reason for adopting this approach was to avoid explicitly solving the (discrete) Weiner–Hopf equation, which requires inversion of the input autocorrelation matrix. There is, however, still one major drawback to the deterministic gradient descent algorithm of the previous section, and that is the averaging required to obtain accurate values of the second order statistics of the system, specifically the terms in the input autocorrelation matrix and the cross-correlation vector. Also, systems are seldom perfectly stationary, and so the quantities must be recalculated over time. This can prove restrictive when implementing the algorithm. Therefore, some approximation to the previously derived deterministic gradient descent algorithm must be found, to avoid the limitations resulting from the need to use averaged quantities.

To derive this approximation, consider again the definition of the mean square error,

$$\begin{aligned}\xi(k) &= E\{e^2(k)\} = E\{(d(k)-y(k))^2\} \\ &= E\{(d(k)-\boldsymbol{w}^{\mathrm{T}}(k)\boldsymbol{x}(k))^2\}\end{aligned} \tag{6.6.1}$$

The gradient of the error surface defined by (6.6.1) was derived in (6.5.12), in terms of the input autocorrelation and cross-correlation matrices. These, however, are exactly the quantities it is desirable to avoid in a practical implementation. An alternative approach, however, is to approximate the mean square error at time k by the instantaneous error squared at time k:

$$\xi(k) \approx e^2(k) = \left(d(k)-\boldsymbol{w}^{\mathrm{T}}(k)\boldsymbol{x}(k)\right)^2 \tag{6.6.2}$$

Differentiating the value of instantaneous error squared with respect to the weight coefficient vector, the gradient estimate is

$$\Delta\boldsymbol{w}(k) \approx \frac{\partial e^2(k)}{\partial \boldsymbol{w}(k)} = -2e(k)\boldsymbol{x}(k) \tag{6.6.3}$$

Substituting this into the gradient descent algorithm format of (6.5.24) yields the expression

$$w(k+1) = w(k) + 2\mu e(k) x(k) \tag{6.6.4}$$

This is the least mean square, or LMS, algorithm, credited to Widrow and Hoff (1960) (see Widrow *et al.*, 1975a, for an early classic paper on the algorithm). It is also known as the stochastic gradient descent algorithm, as it is the stochastic approximation of the deterministic gradient descent algorithm of the previous section. Note that all that is required to implement the algorithm, to adjust the weights of the FIR filter, is a knowledge of the reference signal values in the delay chain, $x(k)$, and the resultant error from using the current weights to derive the output, $e(k)$. Therefore, a complete adaptive signal processing system could be formulated using the following steps:

1. Advance the values in the FIR filter delay chain one stage, and input a new reference sample.
2. Calculate a new FIR filter output, using (6.5.1).
3. Get the resultant error signal, the difference between the actual filter output and the desired output.
4. Calculate new weight coefficients using the LMS algorithm as outlined in (6.6.4).
5. Repeat.

Note also that the gradient estimate is only dependent upon what is in the filter at the time the output is calculated, so that weight coefficient updates do not have to be made with each new sample. This will not be so straightforward with the recursive filters considered later in the chapter.

It is to be expected that the stochastic approximation of the gradient of the error surface taken in (6.6.3) will have an influence upon the learning properties of the LMS algorithm. This is, in fact, the case, and a general formulation to describe these influences can be found in Gardner (1984). It is, however, very difficult to obtain a tractable analysis without making some assumptions about the statistics of the data being considered. One of the most common assumptions taken is that zero-mean Gaussian data is being used, as the characteristics of this data can be fully described. As an example of the 'trends' which result from the stochastic approximation, it can be shown (Horowitz and Senne, 1981; Tate and Goodyear, 1983) that for the Gaussian random data assumption, the bounds placed on the convergence coefficient for stable operation (comparable to (6.6.34)) are

$$0 < \mu < \frac{1}{3\lambda_{max}} \tag{6.6.5}$$

and

$$\eta(\mu) = \sum_{i=0}^{N-1} \frac{\mu\lambda_i}{1-2\mu\lambda_i} < 1 \tag{6.6.6}$$

where N is the number of stages in the filter. Even more stringent bounds derived for the same case using a slightly different approach are (Gholkar, 1990):

$$0 < \mu < \frac{1}{2\lambda_{max} + tr[\boldsymbol{R}]} \tag{6.6.7}$$

Using an intuitive formulation of the type used to obtain (6.5.37), a more practical, yet conservative, bounds (comparable to (6.5.37)) for all of these are

$$0 < \mu < \frac{1}{3N\sigma_x^2} \tag{6.6.8}$$

Comparing (6.6.5) and (6.6.8) to (6.5.34) and (6.5.37) shows the stability of the algorithm to be decreased by a factor of three as a result of the approximations made in its derivation.

Similarly, the misadjustment of the algorithm, defined in (6.2.50), of the LMS algorithm for the Gaussian random data assumption is

$$M = \frac{\eta(\mu)}{1-\eta(\mu)} \tag{6.6.9}$$

where $\eta(\mu)$ is by definition equal to the left-hand side of (6.6.6). For the assumption of a small convergence coefficient, defined by

$$\mu_{small} \sum_{i=0}^{N-1} \lambda_i \ll \frac{1}{2} \tag{6.6.10}$$

the misadjustment can be approximated by (Widrow *et al.*, 1976)

$$M \approx \mu \; tr[\boldsymbol{R}] \tag{6.6.11}$$

This can be compared to the misadjustment of the deterministic gradient descent algorithm, which is equal to zero.

6.6.2 Practical improvements for the LMS algorithm

There are several modifications to the LMS algorithm of (6.6.4) which can be made to improve its practical performance. Three of the more common of these are discussed in this section. The first addresses the long-term stable operation of the LMS algorithm in a quantized digital environment, enhanced by the introduction of tap leakage. The second concerns the selection of a convergence coefficient which will minimize the steady state mean square error. The third concerns the selection of a convergence coefficient which will optimize the tracking speed of the algorithm. As will be discussed later in this chapter, the first of these modifications can be applied directly to the active control implementation of the algorithm, while the second and third must be implemented with caution.

6.6.2.1 Introduction of tap leakage

In the implementation of an adaptive digital filter, there are two sources of quantization error: the quantization error which occurs in the analogue to digital (A/D) signal conversion (see Section 5.2.2 for a discussion of this), and the truncation error which occurs when multiplying two numbers in a finite precision environment. It may be tempting to ignore these errors in the implementation of the adaptive algorithm, as they would appear on the surface to random in sign, and of an order less than the least significant part of the system. As will be shown, however, such assumptions can lead to disastrous results.

Consider the adaptation gradient vector of the LMS algorithm, denoted here as $\boldsymbol{a}(k)$. From (6.6.4), this is defined by the expression

$$\boldsymbol{a}(k) = 2\mu e(k)\boldsymbol{x}(k) \tag{6.6.12}$$

In a practical digital implementation, this adaptation vector will be subject to both A/D quantization error and truncation error. Therefore, the estimated, or practical, adaptation vector is

$$\hat{\boldsymbol{a}}(k) = \big(2\mu(e(k)+\Delta e(k))+\Delta t_1\big)\big(\boldsymbol{x}(k)+\Delta\boldsymbol{x}(k)\big)+\Delta t_2 \tag{6.6.13}$$

where $\Delta e(k)$ and $\Delta\boldsymbol{x}(k)$ are the A/D quantization error of these quantities, Δt_1 and Δt_2 are the truncation errors associated with each multiplication, and it is assumed that the $2\mu e(k)$ term is calculated first, then multiplied by $\boldsymbol{x}(k)$. Lumping together these quantization errors, and defining

$$\tilde{\boldsymbol{a}}(k) = \hat{\boldsymbol{a}}(k)-\boldsymbol{a}(k) \tag{6.6.14}$$

the LMS algorithm can now be written as

$$\boldsymbol{w}(k+1) = \boldsymbol{w}(k)+\boldsymbol{a}(k)+\tilde{\boldsymbol{a}}(k) \tag{6.6.15}$$

The quantization errors are now explicit in the algorithm. Taking expected values of (6.6.15), an equation similar to (6.5.23) is obtained (Cioffi, 1987):

$$\boldsymbol{w}(k+1) = (\boldsymbol{I}-2\mu\boldsymbol{R})\boldsymbol{w}(k)+2\mu\boldsymbol{p}+E\{\tilde{\boldsymbol{a}}(k)\} \tag{6.6.16}$$

The steady state weight coefficient vector of (6.6.16), found by setting $\boldsymbol{w}(k+1) = \boldsymbol{w}(k)$ (such that the gradient is equal to zero), is (Cioffi, 1987):

$$\boldsymbol{W}(\infty) = \boldsymbol{R}^{-1}\boldsymbol{P}+\frac{1}{\mu}\boldsymbol{R}^{-1}\bar{\boldsymbol{a}} \tag{6.6.17}$$

where $\boldsymbol{a}$ is the vector mean of $\tilde{\boldsymbol{a}}(k)$. The first term in (6.6.17) is the optimum weight coefficient vector defined in (6.5.11). The second term is a deviation from this optimum caused by the previously outlined quantization errors. From (6.6.17) it is explicit that the quantization errors will bias the convergence of the LMS algorithm, so that the steady state weight coefficient vector is not the optimum one.

The deviation from the optimum in (6.6.17) can be decoupled using the orthonormal transformation of (6.5.21), which produces

$$\frac{1}{\mu}\boldsymbol{R}^{-1}\overline{\boldsymbol{a}} = \frac{1}{\mu}\sum_{i=0}^{N-1}\lambda_i^{-1}\boldsymbol{q}_i\boldsymbol{q}_i^{\mathrm{T}}\overline{\boldsymbol{a}} \tag{6.6.18}$$

where $\boldsymbol{q}_i$ is the eigenvector of the ith principal axes, or mode. The important point to note about (6.6.18) is that, as the deviation is proportional to the inverse of the eigenvalue of the autocorrelation matrix, when there are small eigenvalues the deviation will become very large. This is commonly the case in active noise and vibration control when discrete tones are targeted for control. The deviation will often lead, especially in the occurrences of small eigenvalues, to saturation (overflow) of some of the weight coefficients, seriously degrading the performance of the adaptive control system.

Heuristically, the result of (6.6.18) can be viewed as follows. Quantization errors tend to increase the 'energy level' of each weight coefficient in the adaptive filter. As was shown in (6.5.21), the eigenvalues of the input autocorrelation matrix are related to the slopes of the principal axes of the error surface. If there is a small eigenvalue, then the slope of the associated principal axes is also small. Therefore, for a given amount of additional 'energy' it is easy for the weight coefficient to wander a long way up this shallow grade, until it eventually overflows. The effects of this deviation from the optimum weight coefficient will usually not be apparent at first, as they build up from small quantization errors. As illustrated in Fig. 6.22, the weight coefficients will tend to follow a path of roughly equal mean square error, so that their travels are not apparent at the error sensor(s). However, after a finite operation time (sometimes only a few minutes at high sampling rates) one of the weights will saturate, and then performance will significantly diminish.

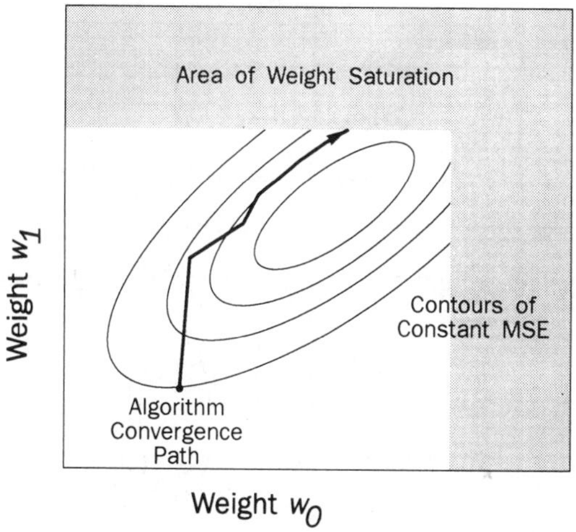

Fig. 6.22 Typical convergence path of two filter coefficients leading to overflow.

For the LMS algorithm, there is a relatively simple fix for this problem. In viewing Fig. 6.22, it is apparent that, besides minimizing the mean square error, what is also desired is the minimization of the magnitude of the weight coefficients. Thus the error criterion can be re-expressed in an optimal control-like format as (Cioffi, 1987):

$$\text{minimize } \left\{ e^2(k) + \frac{\alpha}{2} \| w(k) \|^2 \right\} \tag{6.6.19}$$

where $\| \ \|$ is the vector norm, or length of the vector, and α, referred to as the leakage coefficient, is some multiplying factor related to the importance of minimising the magnitude of the weight coefficients. With this revised error criterion, the gradient expression of (6.6.3) becomes

$$\nabla(k) = \frac{\partial(\text{error criterion})}{\partial \boldsymbol{w}} = -2e(k)\boldsymbol{x}(k) + \alpha \boldsymbol{w}(k) \tag{6.6.20}$$

Using this, the algorithm of (6.6.4) becomes

$$\boldsymbol{w}(k+1) = \boldsymbol{w}(k)(1-\alpha\mu) + 2\mu e(k)\boldsymbol{x}(k) \tag{6.6.21}$$

This is referred to as the 'tap leakage', or simply 'leaky', LMS algorithm (Ungerboeck, 1976; Gitlin *et al.*, 1982; Segalen and Demoment, 1982; Cioffi, 1987). In this discription, tap leakage refers to the continual removal, or leakage, of value from the weights. The addition of tap leakage to the LMS algorithm will bias the results of the algorithm, bounding the weight coefficients and hence preventing overflow. However, it will also increase the value of minimum mean square error. The optimum choice of leakage coefficient, α, must present a compromise between these two effects. Therefore, it is important to quantify both.

With tap leakage, the input auotcorrelation matrix essentially becomes

$$\boldsymbol{R}' = \boldsymbol{R} + \alpha \boldsymbol{I} \tag{6.6.22}$$

Therefore, the steady state weight coefficients are (comparable to (6.5.13)):

$$\boldsymbol{W}(\infty) = [\boldsymbol{R} + \alpha \boldsymbol{I}]^{-1} \boldsymbol{P} \tag{6.6.23}$$

Comparing this result with (6.5.13), it can be deduced that

$$\boldsymbol{W}(\infty) = [\boldsymbol{R} + \alpha \boldsymbol{I}]^{-1} \boldsymbol{R} \boldsymbol{w}_{\text{opt}} \tag{6.6.24}$$

With this steady state weight coefficient vector, the steady state value of mean square error, ξ_∞, will be

$$\begin{aligned} \xi_\infty &= E\left\{ \left(d(k) - \boldsymbol{w}(\infty)^{\mathrm{T}} \boldsymbol{x}(k) \right)^2 \right\} \\ &= E\left\{ \left(d(k) - [\boldsymbol{R} + \alpha \boldsymbol{I}]^{-1} \boldsymbol{R} \boldsymbol{w}_{\text{opt}}^{\mathrm{T}} \boldsymbol{X} \right)^2 \right\} \\ &= E\{d^2(k)\} - \boldsymbol{p}^{\mathrm{T}} \boldsymbol{w}_{\text{opt}} + \left(1 - [\boldsymbol{R} - \alpha \boldsymbol{I}]^{-1} \boldsymbol{R} \right) \boldsymbol{w}_{\text{opt}}^{\mathrm{T}} \boldsymbol{p} \\ &= \xi_{\min} + \left(1 - [\boldsymbol{R} + \alpha \boldsymbol{I}]^{-1} \boldsymbol{R} \right) \boldsymbol{w}_{\text{opt}}^{\mathrm{T}} \boldsymbol{p} \end{aligned} \tag{6.6.25}$$

Equations (6.6.23) and (6.6.25) could be used to determine a suitable value of leakage coefficient α for a given case. What is desired on one hand is that α be large enough such that the modification of the autocorrelation matrix in (6.6.23) produces a matrix which is not poorly conditioned for inversion (which is inherently what the LMS algorithm does). On the other hand, α must not be so large that the bracketed term in the final step in (6.6.25) is much greater than zero. In general, the choice of a 'small' leakage factor (such that $\mu\alpha$ is of the order of one to two bits of the algorithm calculation word length) is often suitable.

The importance of including a small amount of tap leakage in algorithms cannot be overestimated, especially in fixed-point processor implementations and in systems which will run for extended periods of time.

6.6.2.2 Selection of a convergence coefficient based on system error

It was stated earlier in this chapter, in (6.5.34) and (6.6.6), that there is an upper bound placed on the convergence coefficient for stable operation of the gradient descent algorithms (deterministic and LMS) which is inversely proportional to the maximum eigenvalue of the input autocorrelation matrix. It was also stated in (6.6.11) that the final misadjustment of the LMS algorithm is proportional to the size of the convergence coefficient. It would appear from these properties that the best choice of convergence coefficient is a very small one, which will both be stable and minimize the final misadjustment. Although this will mean that the speed of convergence is reduced (as the time constant of convergence of the mean square error given in (6.5.47) is inversely proportional to the value of convergence coefficient), this may be viewed as not terribly detrimental in many cases. This is because the timescale of active noise and vibration control systems is constrained to be longer than the timescale in most other digital filtering applications owing to significant propagation times between control sources and error sensors.

These properties, however, are based upon the 'analogue' characteristics of the LMS algorithm. In an analogue, or infinite precision, implementation of the LMS algorithm, reducing the convergence coefficient will reduce the residual mean square error ad infinitum. In fact, for this case a good balance between speed and accuracy can be attained by continuously decreasing the convergence coefficient during the adaptation process (Kesten, 1958; Sakrison, 1966). For the digital implementation of the algorithm, however, smaller is not always better. In fact, if the convergence coefficient is chosen to be too small, the final value of the mean square error will be increased (Gitlin *et al.*, 1973; Caraiscos and Liu, 1984; Cioffi, 1987). This comes about due to the quantization errors discussed in the previous section, and is explained as follows. First, note that the deviation from the optimum weight vector in (6.6.17) is inversely proportional to the convergence coefficient. Clearly, the convergence coefficient can only be reduced to the point where this term becomes important. If the input autocorrelation matrix $\boldsymbol{R}$ has small eigenvalues, a property which accompanies the eigenvalue disparity problem associated with adaptive filtering of spectra with large tonal peaks, then the restriction on the lower bounds of the convergence coefficient can become critical to satisfactory algorithm performance. Second, if the gradient estimate used in the LMS algorithm, stated in (6.6.3), is equal to less than half the value of the least significant bit of the digital control system, convergence will stop. Although this may seem an obvious point, it is one for which the implications cannot be overlooked; making the convergence coefficient too small will stop adaptation too soon. Increasing the convergence coefficient value will rectify this. It may be surprising to note that if the algorithm is initially adapted using a given value of convergence coefficient, and when steady state is reached the value is reduced with the aim of reducing the final misadjustment, the result may actually be an *increase* in the value of mean square error (Gitlin *et al.*, 1973).

So combining the analogue error characteristics associated with large values of convergence coefficient, with the digital error characteristics associated with small values of convergence coefficient, a typical plot of steady state mean square error as a function of convergence coefficient is shown in Fig. 6.23. The question to be asked now is, how can the optimum value of convergence coefficient be chosen? If the statistics of the environment in which the adaptive control system is operating are known, then it could be calculated, or at least bounded, using analysis such as those in Gitlin *et al.* (1973) and Caraiscos and Liu (1984). There are, however, more heuristic methods which can be incorporated in the adaptive algorithm.

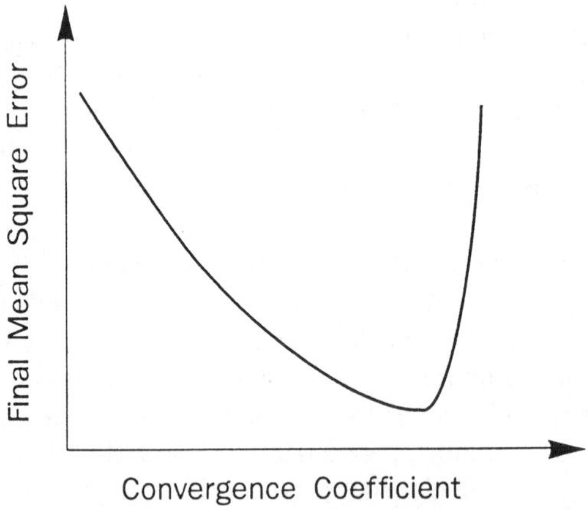

Fig. 6.23 Typical effect of convergence coefficient on steady state mean square error for a digital system.

One of these methods is implemented in the variable step (VS) version of the LMS algorithm (Harris *et al.*, 1986; also Kwong and Johnson, 1992). This algorithm assigns a separate convergence coefficient to each filter weight, so that the VS LMS algorithm is

$$w(k+1) = w(k)+2e(k)M(k)x(k) \tag{6.6.26}$$

where $M(k)$ is an $(N \times N)$ diagonal matrix of time varying convergence coefficients,

$$M(k) = \begin{bmatrix} \mu_0(k) & 0 & \cdots & 0 \\ 0 & \mu_1(k) & \cdots & 0 \\ 0 & 0 & \vdots & \\ 0 & 0 & \cdots & \mu_{N-1}(k) \end{bmatrix} \tag{6.6.27}$$

These convergence coefficients are all initialized at some value, then allowed to increase or decrease between limits depending on the 'quality' of the convergence. This quality is assessed by measuring the sign of the gradient estimate given in (6.6.3), for the associated filter weight. If the sign remains constant for m_0 successive calculations, the adaptation is too slow and the convergence coefficient should be increased by a multiplying factor of γ. If the sign alternates on m_1 successive calculations, then the convergence coefficient is too large, and should be reduced by a factor of γ. Typical values quoted for these quantities are m_0, $m_1 = 3-5$, and $\gamma = 2$ (Harris *et al.*, 1986). It should be noted that besides tending to minimize the final value of mean square error, this algor-ithm will also help speed convergence in cases with large eigenvalue disparities.

Another method of adjusting the value of the convergence coefficient is to actually use a second adaptive algorithm to adjust it (Kwong and Johnston, 1992). This second adaptive algorithm has a form similar to the LMS algorithm, and requires its own convergence coefficient, which must be assigned. One of the problems which the authors have had with this approach is the lack of predicability and, in some instances, stability which arises from implementing three adaptive algorithms simultaneously (one for the weight adaptation, one for modelling of the cancellation path transfer function, which will be described as part of the active control implementation of the algorithm, and a third for optimizing the convergence coefficient).

A further method of adjusting the algorithm convergence coefficient (one which will not, however, minimize the value of mean square error), is use of a normalized algorithm, as will now be described.

6.6.2.3 Normalized LMS algorithm

In the previous section, the choice of convergence coefficient was considered based upon error considerations; that is, what value of convergence coefficient will minimize the final value of mean square error. Tracking speed was not explicitly considered as a variable. There are times, however, when tracking speed *is* important. For these cases the algorithm can be modified using a convergence coefficient which is derived with explicit reference to convergence time.

Consider the measured error at time $k+1$, $e(k+1)$. This can be re-expressed as a Taylor series (Mikhael *et al.*, 1986):

$$e(k+1) = e(k) + \sum_{i=0}^{N-1} \frac{\partial e(k)}{\partial w_i(k)} \Delta w_i$$

$$+ \frac{1}{2!} \sum_{i=0}^{N-1} \sum_{j=0}^{N-1} \frac{\partial e^2(k)}{\partial w_i(k) \partial w_j(k)} \Delta w_i \Delta w_j + \cdots \tag{6.6.28}$$

where

$$\Delta w_i = w_i(k+1) - w_i(k) = -2\mu_i(k) e(k) \frac{\partial e(k)}{\partial w_i(k)} \tag{6.6.29}$$

Note here that a separate, time varying convergence coefficient has been assigned to each filter weight for generality. As the error term, given in (6.5.2), is linear, the higher order terms in (6.6.28) will be equal to zero. Truncating (6.6.28) accordingly, the instantaneous squared error can be expressed as

$$e^2(k+1) = e^2(k) \left[1 - 2\sum_{i=1}^{N-1} \mu_i(k) \left[\frac{\partial e(k)}{\partial w_i(k)} \right]^2 \right]^2 \tag{6.6.30}$$

Differentiating (6.6.30) with respect to the convergence coefficient produces the expression

$$\frac{\partial e^2(k+1)}{\partial \mu_i}$$

$$= -4e^2(k) \left[1 - 2\sum_{i=0}^{N-1} \mu_i(k) \left[\frac{\partial e(k)}{\partial w_i(k)} \right]^2 \right] \left[\frac{\partial e(k)}{\partial w_i(k)} \right]^2 \tag{6.6.31}$$

Setting this derivative equal to zero will produce the expression defining the 'optimum' convergence coefficient μ_i^0 with respect to minimizing the instantaneous error squared (in the fastest possible time):

$$\sum_{i=0}^{N-1} \mu_i^0 \left[\frac{\partial e(k)}{\partial w_i(k)} \right]^2 = \frac{1}{2} \tag{6.6.32}$$

The simplest case is when all of the convergence coefficients are equal in value,

$$\mu_0^0(k) = \mu_1^0(k) = \cdots = \mu^0(k) \tag{6.6.33}$$

For this case, (6.6.32) becomes

$$\mu^0(k) = \frac{1}{2\sum_{i=0}^{N-1} \left[\frac{\partial e(k)}{\partial w_i(k)} \right]^2} \tag{6.6.34}$$

Evaluating the partial derivative in (6.6.34),

$$\mu^0(k) = \frac{1}{2\sum_{i=0}^{N-1} x_i^2(k)} = \frac{1}{2\boldsymbol{x}^{\mathrm{T}}\boldsymbol{x}(k)} \tag{6.6.35}$$

Substituting this optimum convergence coefficient into the LMS algorithm of (6.6.4) yields

$$\boldsymbol{w}(k+1) = \boldsymbol{w}(k) + \frac{e(k)\boldsymbol{x}(k)}{\boldsymbol{x}^{\mathrm{T}}(k)\boldsymbol{x}(k)} \tag{6.6.36}$$

The algorithm of (6.3.36) is the normalized LMS algorithm (Bitmead and Anderson, 1980; Mikhael *et al.*, 1984; Bershad, 1986; Mikhael *et al.*, 1986; Tarrab and Feuer, 1988). Note also that it is the projection algorithm of (5.5.61), which was derived from geometric considerations of the error surface. This is the fastest version of the LMS algorithm.

It should be noted again that the normalized LMS algorithm is optimal in terms of convergence speed, or tracking capabilities, and will not necessarily be optimal in terms of final mean square error (Tarrab and Feuer, 1988).

6.6.2.4 A final note on convergence coefficients

A number of methods for optimizing the convergence coefficient of the standard LMS algorithm have been put forward in the literature. There is often a problem, however, in using these methodologies in an active control implementation. That problem arises because in an active noise or vibration control implementation the signal characteristics are not the only factor which influences the optimal and/or stable value of convergence coefficient, as is the case in the standard implementation. In an active noise or vibration control implementation, the time delay inherent between the generation of a control signal, and its appearance in the measured error signal, has a significant influence upon the choice of a convergence coefficient (this will be discussed in some depth later in this chapter). This time delay arises from the separation distance between control sources and error sensors, and the group delay through analogue anti-aliasing and reconstruction filters and transducers. Therefore, when implementing a methodology for adapting the convergence coefficient in an active noise or vibration control implemention of an adaptive filtering algorithm, the absolute values of the convergence coefficient derived by the methodology are often far too large for system stability; if each weight has its own convergence coefficient, then the relative values between individual convergence coefficients may be used, but the absolute values will not be suitable in practice.

6.7 ADAPTIVE FILTERING IN THE FREQUENCY DOMAIN

The adaptive filtering arrangements considered thus far in this chapter have all been implemented in the time domain. It is also possible to implement such an arrangement in the frequency domain, performing a Fourier transform on the data prior to use. In the field of adaptive signal processing, one common benefit of implementation in the frequency domain is a reduction in the computational complexity of large FIR filters which accompanies the use of 'block' updating strategies, where fast Fourier transform (FFT) algorithms efficiently perform the required convolution and correlation operations (Dentino *et al.*, 1978; Ferrara, 1980, 1985; Clark *et al.*, 1981, 1983; Mansour and Gray, 1982; Shynk, 1992). However, the applicability of such techniques to active noise and vibration control problems is somewhat dubious, as the filter output and new weight coefficients are calculated only after a block of data has been accumulated. One approach which can be used to overcome this problem is adaptation of the weight coefficients in the frequency domain, but implementation of the actual filtering process in the time domain (Reichard and Swanson, 1993), a technique which is similar to that described at the end of Section 6.3 (Roure, 1985). A second approach, which we will consider here, is to calculate the filter output at every sample, requiring a Fourier transform on the data after every new sample. Such an implementation of FFT is referred to as a sliding FFT. The advantages of adaptive filtering in the frequency domain in this instance are not reductions in computational complexity, but rather an improvement in the convergence rate by taking advantage of some properties of the Fourier transform. A third approach, not discussed here, is to replace the sliding FFT with a frequency sampling filter bank; refer to Shynk (1992) for details.

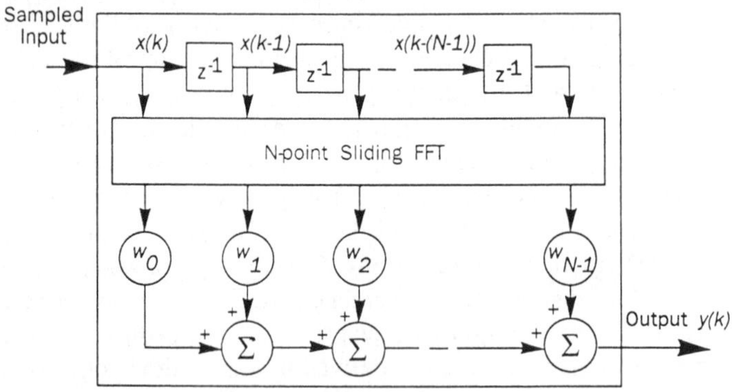

Fig. 6.24 Sliding FFT implementation of an FIR filter.

A block diagram of the system arrangement of interest is shown in Fig. 6.24. With this arrangement, a Fourier transform is performed on the sampled input data in the tapped delay, producing a transformed reference signal matrix, which is a diagonal matrix, the terms of which are the bin values X_j resulting from a Fourier transform of the (time domain) reference signal vector:

$$X(k) = \mathrm{diag}\left[\mathscr{F}\{x(k)\}\right] = \begin{bmatrix} X_0 & & & \\ & X_1 & & \\ & & \ddots & \\ & & & X_{N-1} \end{bmatrix} \tag{6.7.1}$$

where the are N bins being used in the calculation. The output of the frequency domain implemented FIR filter is derived in the same way as the time domain implementation. Defining a vector of complex weight coefficients,

$$\boldsymbol{w}(k) = \left[w_0 \;\; w_1 \;\; \cdots \;\; w_{N-1}\right]^{\mathrm{T}} \tag{6.7.2}$$

the frequency domain output vector $\boldsymbol{y}(k)$ is defined by the matrix expression

$$\boldsymbol{y}(k) = \boldsymbol{X}(k)\boldsymbol{w}(k) \tag{6.7.3}$$

This output can be converted back to a time domain signal by taking a (sliding) inverse Fourier transform of $\boldsymbol{y}$, and using only the first element of the resultant output vector. Alternatively, the elements of the frequency domain vector could simply be added together to form an output (Shynk, 1992) as the first component of an inverse Fourier transform is simply the sum of the input vector elements divided by N. This is the arrangement shown in Fig. 6.24.

The error signal is once again the difference between the desired and actual outputs, which can be expressed in terms of the frequency domain output vector as

$$e(k) = d(k) - \mathbf{1}^{\mathrm{T}}\boldsymbol{y}(k) \tag{6.7.4}$$

where $\mathbf{1}$ is a vector of ones. The error criterion to employ is minimization of the instantaneous error squared, used as an approximation of mean square error. In the frequency domain, with complex signals within the filter, this is

$$\xi \approx e^2(k) = e^*(k)e(k) \tag{6.7.5}$$

where * denotes complex conjugate. Here the complex conjugate of the error signal is

$$e^*(k) = d(k) - \mathbf{1}^{\mathrm{T}}\boldsymbol{y}^*(k) = d(k) - \mathbf{1}^{\mathrm{T}}\boldsymbol{X}^*(k)\boldsymbol{w}^*(k) \tag{6.7.6}$$

where it is assumed here that the desired signal $d(k)$ is real (actually it will make no difference to the algorithm if it too is complex). We must now determine the gradient of the error surface with respect to both the real and imaginary parts of the weight coefficient vector. For the real part

$$\frac{\partial e^2(k)}{\partial w_R} = -e(k)\boldsymbol{X}^*(k) - e^*(k)\boldsymbol{X}(k) \tag{6.7.7}$$

For the complex part,

$$\frac{\partial e^2(k)}{\partial w_I} = je(k)\boldsymbol{X}^*(k) - je^*(k)\boldsymbol{X}(k) \tag{6.7.8}$$

Putting these together, the gradient descent algorithm used for the frequency domain adaptive filtering arrangement shown in Fig. 6.24 is

$$\mathbf{w}(k+1) = \mathbf{w}(k) + 2\mu \mathbf{X}^{\mathrm{H}}(k)\mathbf{1}\mathbf{e}(k) \tag{6.7.9}$$

where ^{H} denotes the matrix Hermitian, or complex conjugate/transpose. This is referred to as the complex LMS algorithm (Widrow *et al.*, 1975b).

As mentioned previously, one advantage to adaptive filtering in the frequency domain is the potential for improved convergence rate. We have seen in the previous sections that the convergence rate and the limit placed upon the maximum value of convergence coefficient are dependent upon the eigenvalues of the input autocorrelation matrix. In the frequency domain, this corresponds to the signal power in each frequency bin. Therefore, convergence speed can be optimized by assigning to each frequency bin a separate, time varying convergence coefficient which is inversely proportional to the inverse of the signal power in that bin, or

$$\mathbf{w}(k+1) = \mathbf{w}(k) + 2\mu(k) \mathbf{X}^{\mathrm{H}}(k)\mathbf{1}\mathbf{e}(k) \tag{6.7.10}$$

where $\mu(k)$ is a diagonal matrix of time and frequency bin dependent convergence coefficients:

$$\mu(k) = \mu_{\mathrm{base}} \begin{bmatrix} P_0^{-1}(k) & & & \\ & P_1^{-1}(k) & & \\ & & \ddots & \\ & & & P_{N-1}^{-1}(k) \end{bmatrix} \tag{6.7.11}$$

Here μ_{base} is some 'base' value of convergence coefficient, and $P_n^{-1}(k)$ is the inverse of the current estimate of the signal power in the nth frequency bin, which is updated according to the relationship

$$P_n(k) = \alpha P_n(k-1) + \beta \mid X_n(k) \mid^2 \tag{6.7.12}$$

where $\alpha = 1 - \beta$ is a forgetting factor, and β is typically some small number. (We have encountered the forgetting factor previously in Chapter 5 when considering system identification. Its purpose is to bias the estimate in terms of the most recent data samples.) The algorithm in (6.7.10) is referred to as the transform domain LMS algorithm (Narayan *et al.*, 1983; Lee and Un, 1986). It is not confined to use with the Fourier transform, but can be used with various orthogonal transforms such as the discrete cosine transform, as outlined in the previous references.

One final point to note, as was mentioned in the previous section, is that one of the differences between a 'standard' adaptive filtering algorithm implementation and an active noise or vibration control implementation is that the selection of a convergence coefficient for an active noise implementation is dependent upon both signal characteristics and transfer functions between the control sources and error sensors (again, this will be discussed in depth later in this chapter). The choice of a convergence coefficient for a standard algorithm is dependent only upon signal characteristics. The principal advantage of implementing the adaptive algorithm in the frequency domain is the ease with which convergence coefficient values can derived, based upon signal characteristics using (6.7.11) and (6.7.12). When implementing such an algorithm in an active noise or vibration control system, it is important to have μ_{base} small enough to account for the destabilizing effects of time delays in the system.

6.8 SINGLE CHANNEL FILTERED-*X* LMS ALGORITHM

The combination of an FIR filter and gradient descent algorithm, particularly the LMS algorithm, form probably the most widely implemented adaptive filtering system. The system is simple and robust, yet effective and, for all practical purposes, usually reasonably quick to converge to a near-optimal solution. This combination is also the most common choice for active noise or vibration control system implementations. However, the form of the standard stochastic gradient descent (LMS) algorithm must be modified slightly.

In the following two sections we will develop an adaptive feedforward control system based upon the concepts of adaptive signal processing discussed in the previous three sections, using a FIR filter and gradient descent algorithm. In this section we will consider the simplest such system, an SISO (single channel) arrangement based upon a generalization of the LMS algorithm used to adaptive the weights of an FIR filter. In the section which follows we will expand the arrangement for MIMO (multichannel) implementation.

The single channel FIR system we will develop in this section is probably the simplest of the adaptive feedforward controllers based upon adaptive signal processing techniques. The resulting controller is best suited to implementation in active control systems targeting 'single mode' problems, such as control of plane wave sound propagation in air handling ducts, where significant levels of attenuation can be achieved using only a single control source and error sensor. The controller will be extended in the next section for implementation in multichannel systems. It is also best suited to implementation in systems where the reference signal will not be corrupted by the control signal, as this can lead to deterioration of control performance. If the reference signal will be corrupted, it is better to consider the use of an IIR filter based system, which will be discussed later in this chapter.

This section will begin by deriving an adaptive algorithm to adjust the weights of an FIR filter for the SISO active control implementation. The algorithm is a simple extension of the LMS algorithm derived previously in this chapter, in Section 6.6, so if the reader has not read that section, which details not only algorithm derivation but several practical modifications to the algorithm which can or should be implemented in an active noise or vibration control system, this should be done now. After deriving the algorithm, we will look at stability aspects which are important when implementing the algorithm in a feedforward active control system.

6.8.1 Derivation of the SISO filtered-x LMS algorithm

A block diagram of the SISO adaptive feedforward control arrangement which will be considered in this section is shown in Fig. 6.25. The control system functions can be divided effectively into two parts: control signal derivation, and filter weight adaptation. In this section we will limit discussion to control signal generation by an FIR filter and extend the analysis to an IIR filter later in the chapter. Our first task for this section is to derive an algorithm for the weight adaptation part of the control system such that the control signal derived by the FIR filter is in some sense optimal.

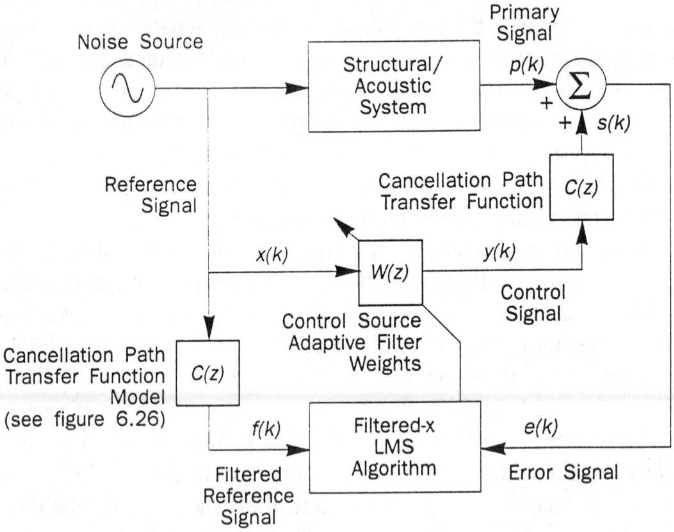

Fig. 6.25 Block diagram of adaptive SISO feedforward active control system.

To provide a basis for weight adaptation, the adaptive algorithm is provided with an 'error signal', a measure of the unwanted vibration or acoustic disturbance provided by some transducer in the system. The measurement provided by the error sensor $e(k)$ is again the sum of the primary source generated disturbance $p(k)$ and the control source generated disturbance $s(k)$,

$$e(k) = p(k) + s(k) \tag{6.8.1}$$

Referring back to Fig. 6.12, the output of the N-stage FIR control filter at time k, $y(k)$, is equal to the convolution operation

$$y(k) = \sum_{i=0}^{N-1} w_i(k)x(k-i) = \boldsymbol{w}^{\mathrm{T}}(k)\boldsymbol{x}(k) = \boldsymbol{x}^{\mathrm{T}}(k)\boldsymbol{w}(k) \tag{6.8.2}$$

where $\boldsymbol{x}(k)$ is the $(N \times 1)$ vector of reference samples x in the filter delay chain at time k,

$$\boldsymbol{x}(k) = \left[x(k)\ \ x(k-1)\ \ x(k-2)\ \cdots\ x(k-(N-1))\right]^{\mathrm{T}} \tag{6.8.3}$$

and $\boldsymbol{w}(k)$ is the $(N \times 1)$ vector of filter weight coefficients at time k

$$\boldsymbol{w}(k) = \left[w_0(k) \;\; w_1(k) \;\; w_2(k) \;\; \cdots \;\; w_{(N-1)}(k)\right]^{\mathrm{T}} \tag{6.8.4}$$

In an active noise or vibration control system, the feedforward control signal $y(k)$ derived by the FIR control filter will not be equal to the control source component $s(k)$ of the error signal. As we have noted in the previous sections, active noise and vibration control systems target 'physical' signals (sound and vibrational disturbances), as opposed to the electrical signals directly utilized by the adaptive electronic components, and so transducers are required to convert between the two regimes. These transducers (control sources and error sensors) will have characteristic frequency responses, or transfer functions. Also, there will be an acoustic, structural or structural/acoustic transfer function between the point of application of the control disturbance and the location of the error sensor. These transfer functions will often include a propagation delay, due to the separation distance between the control source and error sensor and the inherent time delay in analogue antialiasing filters, as well as a transfer function component due to the characteristic response of the system being controlled. As was mentioned in Section 6.3, in developing the adaptive algorithm all of the transfer functions between the control filter output and the error sensor output can be lumped into a single 'cancellation path transfer function'. This transfer function can be modelled in the time domain as the m-order finite impulse response function (vector) $\boldsymbol{c}$,

$$\boldsymbol{c} = \left[c_0 \;\; c_1 \;\; \cdots \;\; c_{m-1}\right]^{\mathrm{T}} \tag{6.8.5}$$

The control source generated component $s(k)$ of the error signal is then equal to the convolution of the filter output $y(k)$ and this finite impulse response function,

$$s(k) = y(k) * c = \sum_{i=0}^{m-1} y(k-i)c_i = \boldsymbol{y}^{\mathrm{T}}(k)\boldsymbol{c} \tag{6.8.6}$$

where $\boldsymbol{y}(k)$ is an $(m \times 1)$ vector of present and past control filter outputs,

$$\boldsymbol{y}(k) = \left[y(k) \;\; y(k-1) \;\; \cdots \;\; y(k-m+1)\right]^{\mathrm{T}} \tag{6.8.7}$$

The quantity $\boldsymbol{y}(k)$ can be obtained by the matrix multiplication

$$\boldsymbol{y}(k) = \boldsymbol{X}^{\mathrm{T}}(k)\boldsymbol{w} \tag{6.8.8}$$

where $\boldsymbol{X}(k)$ is an $(N \times m)$ matrix of present and past reference signal vectors, the columns of which are the m most recent reference signal vectors:

$$\boldsymbol{X}(k) = \left[\boldsymbol{x}(k) \;\; \boldsymbol{x}(k-1) \;\; \cdots \;\; \boldsymbol{x}(k-m+1)\right]$$

$$= \begin{bmatrix} x(k) & x(k-1) & \cdots & x(k-m+1) \\ x(k-1) & x(k-2) & \cdots & x(k-m) \\ & & \vdots & \\ x(k-N+1) & x(k-N) & \cdots & x(k-m-N+2) \end{bmatrix} \tag{6.8.9}$$

The control source component of the error signal can now be re-expressed by substituting (6.8.8) into (6.8.6):

$$s(k) = \left[X^T(k)w\right]^T c = w^T X^T(k) c = w^T f(k) = f^T(k) w \tag{6.8.10}$$

where $f(k)$ is the 'filtered' reference signal vector:

$$f(k) = X^T(k)c = \left[f(k)\ f(k-1)\ \cdots\ f(k-N+1)\right]^T \tag{6.8.11}$$

The ith term in $f(k)$, $f(k-i)$, is equal to the m most recent reference signal samples, $x(k-i)$ through $x(k-i-m+1)$, at the ith control filter stage (the ith position in the input delay chain) used in the generation of the control filter output, convolved with the impulse response function model of the cancellation path transfer function.

Ideally, the error criterion of the active control system is minimization of the mean square value of the error signal, $\xi(k)$,

$$\xi(k) = E\{e^2(k)\} \tag{6.8.12}$$

However, as discussed in Section 6.6 in relation to the LMS algorithm, in a practical implementation the stochastic approximation

$$\xi(k) \approx e^2(k) \tag{6.8.13}$$

is made, chiefly to avoid the need to average signal quantities over extended periods of time which will inherently accompany the use of an 'exact' mean square error criterion. The problem is now one of how to adapt the weights of the control source filter to minimize the error criterion. To this end we will use a series of steps identical to those employed in the derivation of the LMS algorithm in Section 6.6 (for (non-active control) implementations without a cancellation path transfer function) to derive a stochastic gradient descent algorithm for the purpose.

Recall from Section 6.5 that a gradient descent algorithm has the form

$$w(k+1) = w(k) - \mu \Delta w(k) \tag{6.8.14}$$

where $\Delta w(k)$ is the gradient of the error criterion with respect to the weights in the filter, and μ is the portion of the negative gradient to be added to the current weight coefficients with the aim of improving the performance of the system, known as the convergence coefficient. Using (6.8.1) and (6.8.10), the error criterion of (6.8.13) can be expanded as

$$e^2(k) = \left[p(k) + s(k)\right]^2 = \left[p(k) + w^T f(k)\right]^2 \tag{6.8.15}$$

Differentiating this with respect to the weight coefficient vector gives,

$$\Delta w(k) \approx \frac{\partial e^2(k)}{\partial w} = 2e(k)f(k) \tag{6.8.16}$$

Note that in (6.8.16) the true partial derivative $\Delta w(k)$ has be replaced by a functional derivative. This is a common approximation, and provided the convergence coefficient is small, such that weight changes are small with each iteration, the difference is minimal. Substituting (6.8.16) into (6.8.14), the gradient descent algorithm used for adapting the weights in the control source FIR filter is

$$w(k+1) = w(k) - 2\mu e(k)f(k) \tag{6.8.17}$$

This is known as the (SISO) filtered-*x* LMS algorithm, which has appeared in various forms in the literature at various times. The algorithm appears to have been first proposed by Morgan (1980), and then independently for feedforward control by Widrow *et al.* (1981) and specifically for active control by Burgess (1981). When the cancellation path is a pure delay the algorithm is simply the delayed LMS algorithm (Widrow, 1971; Qureshi and Newhall, 1973; Kabal, 1983; Long *et al.*, 1989). The name 'filtered-*x*' LMS algorithm was coined in Widrow and Stearns (1985, pp. 288–294).

It is straightforward to show, using the same steps taken in Section 6.6 for the LMS algorithm, that the complex number equivalent of this, which is suitable for implementation in the frequency domain, is

$$w(k+1) = w(k) - 2\mu e(k)f^*(k) \tag{6.8.18}$$

In this, the filter weights are complex numbers, as are the measured data (error signal and filtered reference signal), as would be the case following a Fourier transform. In (6.8.18) * denotes complex conjugate.

There are two characteristics of the algorithm given in (6.8.17) that should be noted in particular. The first is that the adaptation of the weight coefficients involves use of the 'filtered' reference signal vector f, as opposed to the reference signal vector x used in the derivation of the filter output and in the standard LMS algorithm (hence the name 'filtered-*x*' LMS algorithm). It is therefore intuitive that the characteristics of the cancellation path transfer function will have an influence upon the stability of the algorithm, an influence which will be investigated shortly. Second, observe that the right-hand side of (6.8.17) involves subtraction, rather than addition as for the 'standard' LMS algorithm. This is because acoustic and vibration signals must be physically 'added' (superposition) as opposed to electrical signals that can be 'subtracted'.

In the practical implementation of the algorithm in (6.8.17), the filtered reference signal term $f(k)$ at time k is derived by either convolving the reference signal vector $x(k)$ with an impulse response function model (estimate) of the cancellation path transfer function in the time domain before being used in the adaptive algorithm, as illustrated in Fig. 6.26 (if the complex number version of (6.8.18) is implemented, the reference signal could be multiplied by an estimate of the transfer function in the frequency domain). Thus, the calculations are based upon an estimate of the (actual) filtered reference signal vector

$$\hat{f}(k) = \left[\hat{f}(k) \;\; \hat{f}(k-1) \;\; \cdots \;\; \hat{f}(k-N+1)\right]^{\mathrm{T}} \tag{6.8.19}$$

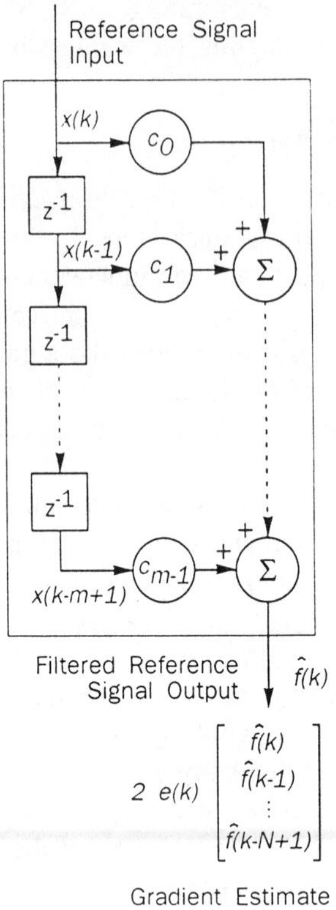

Fig. 6.26 Derivation of the filtered reference signal in the time domain by convolution of the reference signal vector with a model of the cancellation path transfer function.

where ^ denotes an estimated quantity. Therefore, the practical implementation of the algorithm is

$$\mathbf{w}(k+1) = \mathbf{w}(k) - 2\mu e(k)\hat{\mathbf{f}}(k) \tag{6.8.20}$$

or for the complex algorithm

$$\mathbf{w}(k+1) = \mathbf{w}(k) - 2\mu e(k)\hat{\mathbf{f}}^{*}(k) \tag{6.8.21}$$

The estimates of the filtered rêference signal vector used in (6.8.27) and (6.8.28) will, in general, be inaccurate to some degree. This can be expected to have an effect upon the stability of the filtered-*x* LMS algorithm in a practical system. For small inaccuracies, the effect will be minimal. However, for large inaccuracies the result can be disastrous. We will quantify these effects later in this section.

As a final note, a simple way to check that a system is functioning properly is to connect the controller output to the error signal input, start the system with a set of random weights, and use an input to the system similar to the expected reference signal. If the algorithm is correct the controller output should go towards zero. While the system output could be going towards zero for reasons other than a correct implementation, if it does *not* goes towards zero then it is safe to assume that the system is not working.

6.8.2 Solution for the optimum weight coefficients and examination of the error surface

Before considering the problem of algorithm stability we should first obtain some properties of the error surface in which the algorithm will operate, searching for the optimum set of weight coefficients which will minimise the error criterion. For this we will employ statistical expectation operators, making the error criterion of interest the mean square value of the error signal, the ideal error criterion which was approximated in (6.8.19) in the process of deriving the filtered-x LMS algorithm. Using (6.8.15), mean square error can be expressed as

$$\xi = E\{e^2(k)\} = E\{ \left(p(k) + \boldsymbol{w}^{\mathrm{T}}\boldsymbol{f}(k)\right)^2 \} \tag{6.8.22}$$

Note that in a stationary system, the mean square error is a function of the weight coefficients in the filter, not of time. Making the assumption of statistical independence between the weight coefficients and the signal quantities, the mean square error can be restated as

$$\xi = \sigma_p^2 + 2\boldsymbol{a}^{\mathrm{T}}\boldsymbol{w} + \boldsymbol{w}^{\mathrm{T}}\boldsymbol{B}\boldsymbol{w} \tag{6.8.23}$$

where σ_p^2 is the primary source signal power, given by

$$\sigma_p^2 = E\{p^2(k)\} \tag{6.8.24}$$

$\boldsymbol{a}$ is the cross-correlation vector between the filtered primary signal and filtered reference signal vector,

$$\boldsymbol{a} = E\{p(k)\boldsymbol{f}(k)\} \tag{6.8.25}$$

and $\boldsymbol{B}$ is the autocorrelation matrix of the filtered reference signal

$$\boldsymbol{B} = E\{\boldsymbol{f}(k)\boldsymbol{f}^{\mathrm{T}}(k)\} \tag{6.8.26}$$

The error criterion of (6.8.23) is a quadratic function of the control filter weight coefficients, as was the deterministic (purely electronic) algorithm considered in Section 6.5. As such it will have only one extremum, and it can easily be verified that the extremum is a minimum. The error surface described by the error criterion is therefore a hyper-paraboloid of dimension $(N+1)$, the same as that for the 'standard' FIR adaptive filtering problem, illustrated in Fig. 6.14 for the simple two weight case. The optimum set of weight coefficient vectors $\boldsymbol{w}_0$ can therefore be found by differentiating the error criterion of (6.8.23) with respect to this quantity and setting the resulting gradient expression equal to zero. Taking the derivative,

$$\frac{\partial \xi}{\partial \boldsymbol{w}} = 2\boldsymbol{B}\boldsymbol{w} + 2\boldsymbol{a} \tag{6.8.27}$$

Setting the derivative to zero yields the optimum weight coefficient vector

$$w_0 = -B^{-1}a \tag{6.8.28}$$

The optimum weight coefficient vector of (6.8.28) is a form of the solution to the discrete Weiner–Hopf integral equation which takes into account the transfer functions associated with the cancellation path in an active noise or vibration control system. It is a discrete time version of (6.3.22), which describes the optimum control filter transfer function characteristics in the frequency domain using power and cross-spectral densities. Equation (6.8.28) is comparable to the solution for the optimum set of weight coefficients for a standard (no cancellation path transfer function) adaptive FIR filter arrangement, stated in (6.5.11), where the autocorrelation matrix of the filtered reference signal $\boldsymbol{B}$ is replaced by the autocorrelation matrix of the (unfiltered) reference signal $\boldsymbol{R}$, and the cross-correlation vector between the primary signal and filtered reference signal $\boldsymbol{a}$ is replace by the cross correlation vector between the desired signal and the reference signal $\boldsymbol{p}$. The signs of the two solutions are different, owing to the superposition constraints of the physical system. Another way to accommodate the sign change is to consider that the desired signal d in the case of an active control implementation is actually the negative value of the primary signal, $d(k) = -p(k)$, because this will give the maximum level of attenuation.

To check that the extremum in the error surface is in fact a minimum, the second derivative of the error criterion must be a positive quantity, or

$$\frac{\partial^2 \xi}{\partial w^2} = 2B > 0 \tag{6.8.29}$$

In other words, the autocorrelation matrix of the filtered reference signal must be a positive definite quantity. From its definition in (6.8.26), it can be deduced that this will be true for the case considered here.

If this optimum weight coefficient vector is substituted back into the error criterion of (6.8.23), noting that the filtered autocorrelation matrix $\boldsymbol{B}$, and hence its inverse, is symmetric, the minimum (value of) mean square error $\xi_{\min}$ is found to be

$$\begin{aligned}\xi_{\min} &= \sigma_p^2 + 2a^{\mathrm{T}}w_o + w_o^{\mathrm{T}}Bw_o \\ &= \sigma_p^2 + a^{\mathrm{T}}w_o - a^{\mathrm{T}}B^{-1}a + a^{\mathrm{T}}B^{-1}BB^{-1}a \\ &= \sigma_p^2 + a^{\mathrm{T}}w_0\end{aligned} \tag{6.8.30}$$

As discussed in Section 6.5, the minimum mean square error defines the location of the error surface minimum on the dependent, mean square error, axis of the coordinate system. Using 'geometric intuition', this can be viewed as the 'height' of the error surface 'bowl' off the coordinate system defined by the filter weight coefficients (see Fig. 6.16, displaying this idea for the standard filtering problem). The examination of the properties of the error surface, and hence algorithm stability which will we consider shortly, will be simplified if we perform an axes translation such that the error surface minimum defines the origin of the coordinate system. This can be accomplished by re-expressing the error criterion in terms of excess mean square error ξ_{ex}, defined as the difference between the mean square error with the current weight coefficients and the minimum mean square error, below which it is impossible to go. Noting once again that $\boldsymbol{B}$ is symmetric, the excess mean square error can be expressed as

$$\begin{aligned}\xi_{ex} = \xi - \xi_{min} &= \sigma_p^2 + 2\boldsymbol{a}^T\boldsymbol{w} + \boldsymbol{w}^T\boldsymbol{B}\boldsymbol{w} - \sigma_p^2 - \boldsymbol{a}^T\boldsymbol{w}_0 \\ &= 2\boldsymbol{a}^T\boldsymbol{w} + \boldsymbol{w}^T\boldsymbol{B}\boldsymbol{w} - \boldsymbol{a}^T\boldsymbol{w}_0 \\ &= 2\boldsymbol{a}^T\boldsymbol{B}^{-1}\boldsymbol{B}\boldsymbol{w} + \boldsymbol{w}^T\boldsymbol{B}\boldsymbol{w} - \boldsymbol{a}^T\boldsymbol{B}^{-1}\boldsymbol{B}\boldsymbol{B}^{-1}\boldsymbol{a} \\ &= \boldsymbol{v}^T\boldsymbol{B}\boldsymbol{v}\end{aligned} \tag{6.8.31}$$

where $\boldsymbol{v}$ is the weight error vector, defined as the difference between the current values of the weight coefficients and those which are optimal:

$$\boldsymbol{v} = \boldsymbol{w} - \boldsymbol{w}_0 \tag{6.8.32}$$

The reason this re-expression will simplify the analysis is because we know that for the algorithm to be converging towards the optimum solution the value of excess mean square error, and hence the values in the weight error vector, must always be converging towards zero.

Expressed in terms of the weight error, (6.8.31) shows the characteristics of the error surface to be defined by the characteristics of the autocorrelation matrix of the filtered reference signal $\boldsymbol{B}$. Referring to the definition of (6.8.26), $\boldsymbol{B}$ will usually not be diagonal and as such (6.8.31) is a coupled representation of the error criterion. By this statement we mean that the contribution to the mean square error by a given weight coefficient error is dependent upon the values of other weight coefficient errors, the relationship being defined by the off-diagonal terms in $\boldsymbol{B}$. As $\boldsymbol{B}$ is symmetric, the problem can be decoupled by diagonalizing this matrix via an orthonormal transformation. We define an $(N \times N)$ orthonormal transformation matrix $\boldsymbol{Q}$, the columns of which are the eigenvectors of $\boldsymbol{B}$, as follows:

$$\boldsymbol{\Lambda} = \boldsymbol{Q}^{-1}\boldsymbol{B}\boldsymbol{Q} = \boldsymbol{Q}^T\boldsymbol{B}\boldsymbol{Q} \tag{6.8.33}$$

where $\boldsymbol{\Lambda}$ is the diagonal matrix of eigenvalues of $\boldsymbol{B}$. Using this transformation, the excess mean square error can be expressed as

$$\xi_{ex} = \boldsymbol{v}'^T\boldsymbol{\Lambda}\boldsymbol{v}' \tag{6.8.34}$$

where

$$\boldsymbol{v}' = \boldsymbol{Q}^T\boldsymbol{v} = \boldsymbol{Q}^{-1}\boldsymbol{v} \tag{6.8.35}$$

From the results of this transformation it is readily apparent that the eigenvectors of $\boldsymbol{B}$ define the principal axes of the error surface, because multiplying the weight error vector redefines the problem as the (eigenvalue) weighted sum of independent contributions to the error criterion, which corresponds to contributions along each principal axis.

We can summarize our findings thus far by stating that the characteristics of the error surface in the SISO active control problem employing the filtered-x LMS algorithm are identical in form to those found in the 'standard' adaptive filtering problem employing the LMS algorithm, with the *filtered* reference signal f replacing the reference signal x in all quantities. The optimum weight coefficient values are defined by the autocorrelation matrix of the *filtered* reference signal and the cross-correlation between the desired signal (equal to $-p(k)$) and the *filtered* reference signal, and the principal axes of the error surface are defined by the eigenvectors of the autocorrelation matrix of the *filtered* reference signal. It can also be anticipated, based upon our work in Section 6.5, that the eigenvalues of the autocorrelation matrix of the *filtered* reference signal will play a large part in determining the stability of the filtered-x LMS algorithm. It will therefore be of some interest to see if we can determine a qualitative relationship between the autocorrelation matrix of the filtered reference signal $\boldsymbol{B}$ and the autocorrelation matrix of the reference signal $\boldsymbol{R}$ to examine the influence of the cancellation path transfer function (through which the reference signal is 'filtered') upon the characteristics of the error surface.

To establish a relationship, we can begin by examining the terms which make up the filtered reference signal vector $\boldsymbol{f}$, as these are the basis of the autocorrelation matrix $\boldsymbol{B}$. This will be simplest if we consider a specific problem and then generalize. We will select perhaps the simplest problem of all: use of a two tap FIR filter, and a two stage finite impulse response representation of the cancellation path transfer function. For this system, the terms in the filtered reference signal vector are

$$f(k) = \begin{bmatrix} c_0x(k)+c_1x(k-1) \\ c_0x(k-1)+c_1x(k-2) \end{bmatrix} \tag{6.8.36}$$

where c_1 and c_2 are the coefficients in the cancellation path transfer function impulse response vector. Substituting this into (6.8.26), $\boldsymbol{B}$ is found to be given by

$$\boldsymbol{B} = \begin{bmatrix} (c_0^2+c_1^2)\sigma_x^2+2c_0c_1\gamma(1) & (c_0^2+c_1^2)\gamma(1)+c_0c_1(\sigma_x^2+\gamma(2)) \\ (c_0^2+c_1^2)\gamma(1)+c_0c_1(\sigma_x^2+\gamma(2)) & (c_0^2+c_1^2)\sigma_x^2+2c_0c_1\gamma(1) \end{bmatrix} \tag{6.8.37}$$

where σ_x^2 is the signal power of the reference signal, and $\gamma(n)$ is the correlation between reference signals n samples apart:

$$\gamma(n) = E\{x(k)x(k-n)\} \tag{6.8.38}$$

(note that $\gamma(n) = \gamma(-n)$, and that $\gamma(0) = \sigma_x^2$). If this result is generalized, it is found that the elements in the autocorrelation matrix of the filtered reference signal are defined by the expression

$$\boldsymbol{B}(\alpha,\beta) = \sum_{i=0}^{m-1}\sum_{j=0}^{m-1} c_ic_j\gamma(i-j+\alpha-\beta) \tag{6.8.39}$$

where α,β denote the row,column of the element of interest. This can be compared to the defining expression for the terms in $\boldsymbol{R}$,

$$\boldsymbol{R}(\alpha,\beta) = \gamma(\alpha-\beta) \tag{6.8.40}$$

A simple comparison of (6.8.39) and (6.8.40) shows that the values in autocorrelation matrix of the filtered and unfiltered reference signal will, in general, be different. Let us consider then under what circumstances the orientation, or principal axes, of the error surface are unaltered. For this to be true the eigenvectors of $\boldsymbol{B}$ and $\boldsymbol{R}$ must be the same, although the eigenvalues can be different. In terms of the weight error, this means that

$$\boldsymbol{v}'^{\mathrm{T}}\boldsymbol{\Lambda}_B\boldsymbol{v}' = \boldsymbol{v}'^{\mathrm{T}}\boldsymbol{\Lambda}_R\mathrm{T}\boldsymbol{v} \tag{6.8.41}$$

where $\boldsymbol{\Lambda}_B$ and $\boldsymbol{\Lambda}_R$ are the diagonal matrices of $\boldsymbol{B}$ and $\boldsymbol{R}$, respectively, and $\boldsymbol{T}$ is some diagonal matrix defining the change in eigenvalues. The simplest example of this is where all of the diagonal values in $\boldsymbol{T}$ are the same; that is,

$$\boldsymbol{T} = T\boldsymbol{I} \tag{6.8.42}$$

where T is some constant and $\boldsymbol{I}$ is the identity matrix. In this case

$$\boldsymbol{B} = T\boldsymbol{R} \tag{6.8.43}$$

From (6.8.39) and (6.8.40), this means that

$$T = \frac{1}{\gamma(\alpha-\beta)}\sum_{i=0}^{m-1}\sum_{j=0}^{m-1} c_i c_j \gamma(i-j+\alpha-\beta) \qquad \text{for all } \alpha,\beta \tag{6.8.44}$$

If the reference and filtered reference signals are harmonic at frequency ω, and the sampling rate of the system is ω_s, then

$$\gamma(n) = \cos n\phi \tag{6.8.45}$$

where $\phi = 2\pi\omega/\omega_s$ is the angular increment of each sample. Referring to (6.8.44), note that for each $(i,j),(j,i)$ pair

$$\begin{aligned} &\frac{c_i c_j \gamma\{(i-j)+(\alpha-\beta)\} + c_j c_i \gamma\{(j-i)+(\alpha-\beta)\}}{\gamma\{\alpha-\beta\}} \\ &= \frac{2c_i c_j \cos\{(i-j)\phi\}\cos\{(\alpha-\beta)\phi\}}{\cos\{(\alpha-\beta)\phi\}} \\ &= 2c_i c_j \cos\{(i-j)\phi\} \end{aligned} \tag{6.8.46}$$

The important point about the result of (6.8.46) is that it is independent of the element location (α,β) in the matrix. This means that for harmonic excitation,

$$\boldsymbol{B} = T\boldsymbol{R} \qquad T = \sum_{i=0}^{m-1}\sum_{j=0}^{m-1} c_i c_i \cos\{(i-j)\phi\} \tag{6.8.47}$$

Therefore, with harmonic excitation the eigenvectors of the error surface, which define the principal axes, are the same for the autocorrelation matrix of both the filtered and unfiltered reference signal vectors; the cancellation path transfer function simply translates and uniformly compresses or expands the error surface.

To contrast this, let us consider the case of random noise. Here the correlation between successive input samples will be zero, or

$$\gamma(n) = \begin{cases} \sigma_x^2 & i=j \\ 0 & i\neq j \end{cases} \tag{6.8.48}$$

The input autocorrelation matrix of the reference signal $\boldsymbol{R}$ is therefore a diagonal matrix of signal powers, given by

$$\boldsymbol{R} = \sigma_x^2 \boldsymbol{I} \tag{6.8.49}$$

The associated orthonormal matrix, the columns of which are the eigenvectors of $\boldsymbol{R}$, is simply the identity matrix. From (6.8.41) this means that for the orientation of the principal axes to be unchanged by the cancellation path transfer function, the autocorrelation matrix of the filtered reference signal must be diagonal. However, from the definition given in (6.8.26), the terms in $\boldsymbol{B}$ with a random noise input will be

$$\boldsymbol{B}(\alpha,\beta) = \sum_{i=0}^{m-i} \chi_i \sigma_x^2 \qquad \chi_i = \begin{cases} c_i c_{i-(\alpha-\beta)} & 0 \le i-(\alpha-\beta) \le m \\ 0 & \text{otherwise} \end{cases} \tag{6.8.50}$$

This shows that $\boldsymbol{B}$ will *not*, in general, be diagonal with a random input to the filter, although it will be symmetric. The exception to this occurs when there is a single coefficient in the impulse response model, so that the cancellation path transfer function evenly amplifies or attenuates the output. This means that the cancellation path transfer function can completely alter the shape of the error surface in this instance. This makes sense intuitively, as the cancellation path transfer function will effectively frequency weight the filtered reference signal, as some frequency components will respond more strongly than others. In response to this weighting, $\boldsymbol{B}$ loses its perfect symmetry, with the resulting shape of the error surface (defined by the eigenvectors of $\boldsymbol{B}$) dependent upon the form of the weighting. This has a correspondence with the harmonic excitation case, where we can see from (6.8.37) that the shape of the error surface is dependent upon frequency through the correlation function γ (harmonic excitation can be viewed as the limiting case of frequency weighting).

6.8.3 Stability analysis of the exact algorithm

From our earlier discussion it is clear that when analysing the stability of the filtered-x LMS algorithm we will be concerned with two categories of influence: influence of the structural/acoustic system on the stability of the exact algorithm, where the model of the cancellation path transfer function is exact, and the influence of cancellation path modelling errors upon the stability of the algorithm. We will now address the first of these categories.

For investigation of the stability of the algorithm we will employ statistical expectation operators to examine mean weight convergence, making the problem deterministic. This was the approach taken in Section 6.5 with the deterministic version of the LMS algorithm. As was outlined then, such an approach tends to produce results which are 'overly optimistic' in terms of stability, in the case of the LMS algorithm by as much as a factor of 3. The utility of such a simplified analysis is perhaps not so much in its quantitative results, but rather in its qualitative results which outlined the influence of various system parameters on stability. In this section we will be concerned with qualifying the influence of structural/acoustic system parameters on the algorithm so that we can, for example, predict the influence of moving sensors and actuators or of increasing amplifier gains. The approach we will take in analysing the algorithm is identical to that used in Section 6.5 for the deterministic gradient descent algorithm, so if the reader has not yet done so it will be useful to first read that section.

Our first task is to restate the filtered-*x* LMS algorithm to examine mean weight convergence. We can do this simply by replacing the stochastic gradient estimate of (6.8.16), based upon the instantaneous squared error, with the exact gradient estimate of (6.8.27), based upon mean square error. Doing this, we obtain

$$w(k+1) = w(k) - 2\mu\left[\boldsymbol{B}w(k) + \boldsymbol{a}\right] \tag{6.8.51}$$

To examine the stability of this deterministic version of the filtered-*x* LMS algorithm it will again prove useful to translate the origin of the coordinate system to the location of the optimum weight coefficient vector, restating the algorithm in terms of weight error which must be tending towards zero for the algorithm to be converging. Referring to (6.8.28), the algorithm can be restated in terms of the weight error vector as

$$w(k+1) - w_0 = w(k) - w_0 - 2\mu\left(\boldsymbol{B}w(k) - \boldsymbol{B}w_0 + \boldsymbol{a} + \boldsymbol{B}w_0\right) \tag{6.8.52}$$

or

$$v(k+1) = v(k) - 2\mu\boldsymbol{B}v(k) \tag{6.8.53}$$

The algorithm can now be diagonalized using the orthonormal transformation defined in (6.8.33), which will re-express it as a set of independent scalar equations. Multiplying (6.8.53) through by the orthonormal transformation matrix yields

$$\boldsymbol{Q}^{-1}v(k+1) = \boldsymbol{Q}^{-1}v(k) - 2\mu\boldsymbol{Q}^{-1}\boldsymbol{B}\boldsymbol{Q}\boldsymbol{Q}^{-1}v(k) \tag{6.8.54}$$

or

$$v'(k+1) = v'(k)\left[\boldsymbol{I} - 2\mu\boldsymbol{\Lambda}\right] \tag{6.8.55}$$

where

$$v'(k) = \boldsymbol{Q}^{-1}v(k) \tag{6.8.56}$$

As the algorithm is now decoupled, it can be examined as a set of scalar equations of the form

$$v_r'(k) = v_r'(k)\left[1-2\mu\lambda_r\right] \tag{6.8.57}$$

where λ_r is the eigenvalue of the rth scalar equation. For the filtered-x LMS algorithm to be stable over time, the weight errors must converge to some finite value for each scalar equation, or

$$\left| \frac{v_r'(k+1)}{v_r'(k)} \right| < 1 \tag{6.8.58}$$

for all r. From this criterion, it can be deduced that

$$0 < \mu < \frac{1}{\lambda_r} \tag{6.8.59}$$

Equation (6.8.59) shows that it will be the maximum eigenvalue of $\boldsymbol{B}$ which will dictate the overall stability requirements, so that the bound placed on the convergence coefficient μ for stable operation reduces to

$$0 < \mu < \frac{1}{\lambda_{max}} \tag{6.8.60}$$

where λ_{max} is the maximum eigenvalue of the diagonal matrix Λ.

The bounds on the convergence coefficient μ for stable operation of the (deterministic) filtered-x LMS algorithm are identical in form to the bounds placed on the convergence coefficient for stability in the implementation of the standard (deterministic) LMS algorithm, given in (6.2.28). The difference here, however, is that the eigenvalues of interest are those of the autocorrelation matrix of the filtered reference signal, $\boldsymbol{B}$, rather than the (standard) input autocorrelation matrix, $\boldsymbol{R}$. Thus, the cancellation path transfer function inherent in an active noise or vibration control system modifies the stability bound of the algorithm by modifying the characteristic eigenvalues of what would otherwise be a purely electronic system.

In light of this, we can easily assess the effect which a change in the amplitude of the cancellation path transfer function has upon the stability of the filtered-x LMS algorithm. Such a change in amplitude can, for example, be caused by changing the gain of control source or error sensor amplifiers or by moving the associated transducers. From the definition of the filtered input autocorrelation matrix given in (6.8.26), it is apparent that increasing the magnitude of the transfer functions will cause the magnitude of the eigenvalues of $\boldsymbol{B}$ to be increased proportionally squared. Equation (6.8.60) shows that the maximum stable value of convergence coefficient should be reduced proportionally squared in response to this. Figure 6.27 illustrates the effect of increasing the magnitude of the transfer function between the control source and error sensor, keeping the phase constant, for a two-weight filter being used to control a single sinusoid. The maximum allowable value of convergence coefficient for system stability, plotted against relative transfer function magnitude, is shown in this figure. In viewing this, an inverse proportional squared relationship clearly exists.

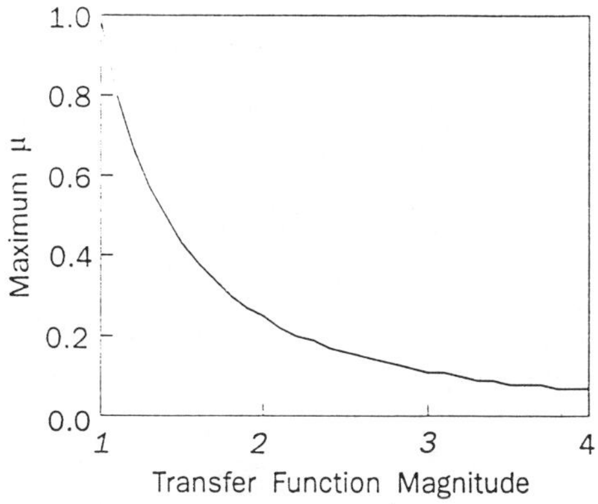

Fig. 6.27 Typical plot of maximum stable value of convergence coefficient as a function of cancellation path transfer function magnitude.

6.8.4 Effect of continuously updating the weight coefficients

The bounds placed on the convergence coefficient, μ, for algorithm stability given in (6.2.39) were effectively derived with the assumption that the weight coefficient vectors were adjusted only after the result of the previous modification was known. However, if the system is continuously adapting, the time delay associated with the cancellation path transfer function will have a significant effect upon the stability of the algorithm. This is because the present weight coefficient vector modification is based upon the results of a previous modification conducted n samples ago.

To examine this effect, we rewrite (6.8.55) for the case where the weight coefficient vector is updated at every sample, but with an explicit n sample time delay in the cancellation path transfer function, as follows:

$$v'(k+1) = v'(k) - 2\mu\Lambda v'(k-n) \tag{6.8.61}$$

Taking the z-transform of the ith scalar equation in (6.8.61) produces (Kabal, 1983):

$$V_i'(z) = \frac{z^{n+1}v_i'(0)}{z^{n+1}-z^n+2\mu\lambda_i} \tag{6.8.62}$$

For the algorithm to be stable, the poles of (6.8.62) (determined by the roots of the characteristic equation) must be within the unit circle.

The values of $2\mu\lambda_i$ for which the characteristic equation has roots on the unit circle can be found by substituting $e^{j\phi}$ for z, and setting the equation equal to zero, or

$$2\mu\lambda_i = e^{jn\phi} - e^{j(n+1)\phi} \tag{6.8.63}$$

As the matrix $\boldsymbol{B}$ is real symmetric, the eigenvalues will all be real. Therefore, equating real and imaginary parts of (6.8.63), we obtain

$$2\mu\lambda_i = \cos(n\phi) - \cos((n+1)\phi) \tag{6.8.64}$$

and

$$0 = \sin(n\phi) - \sin((n+1)\phi) \tag{6.8.65}$$

Therefore, from (6.8.65)

$$\phi = \frac{\pi}{2n+1} \tag{6.8.66}$$

where n is the acoustic time delay expressed in sample periods, as defined earlier. Substituting this value of ϕ into (6.8.64) yields the expression

$$2\mu\lambda_i = 2\sin\left[\frac{\pi}{2(2n+1)}\right] \tag{6.8.67}$$

Rearranging this result, the bound placed upon the convergence coefficient in a continuously adapting system with an explicit time delay is found to be (Kabal, 1983):

$$0 < \mu_i < \sin\left[\frac{\pi}{2(2n+1)}\right]\frac{1}{\lambda_i} \tag{6.8.68}$$

Comparing this result to (6.8.60), the bound placed upon the convergence coefficient in a non-continuously adapting system, it is found that the uncertainty which the inclusion of a delay introduces into the adaptive strategy reduces stability by a factor of $\sin(\pi/(2(2n+1))$. This effect is apparent in Fig. 6.28, which depicts the maximum allowable value of convergence coefficient for

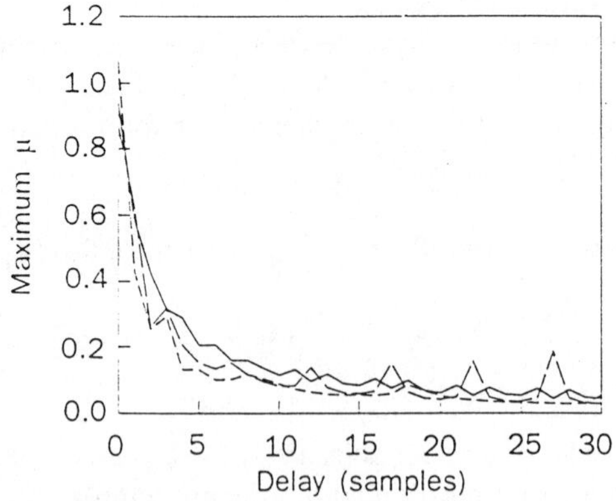

Fig. 6.28 Maximum stable value of convergence coefficient as a function of time delay in the cancellation path; solid line is a 10 sample delay, long dash is 25 sample delay, short dash is 50 sample delay.

system stability for a single control source, single error sensor system, plotted as a function of acoustic time delay (in samples). With this data, the primary disturbance is taken to be single frequency sinusoidal excitation. Three sets of data are shown, where the sampling rate is equal to 10 times, 25 times, and 50 times the excitation frequency.

It should be mentioned here that the delayed LMS algorithm (DLMS), which is a version of the LMS algorithm derived in Section 6.6 with a delay in the weight adaptation as described above, has received a significant amount of research activity in recent years. Further analysis of the algorithm, without the idealisations associated with the analysis above, can be found in Long *et al.* (1989).

An interesting point concerning the above result is that while the acoustic time delay reduces the maximum stable value of convergence coefficient when the weights are continuously updated, the choice of continuously updating the weight coefficient vectors, or updating only after the effect of the previous change is known (waiting for the delay) appears to have no significant effect upon the convergence speed of the algorithm. Figure 6.29 depicts the convergence of the mean square error for a single control source, single error sensor system where the acoustic time delay between the source and sensor is equal to 10 samples. Two curves are plotted, one for the continuously updating system and one where the weight coefficients are updated every 10 samples. The convergence coefficients for these are equivalent, scaled by a factor of $\sin(\pi/(2(2n+1)))$. Clearly, the difference in convergence speed is minimal.

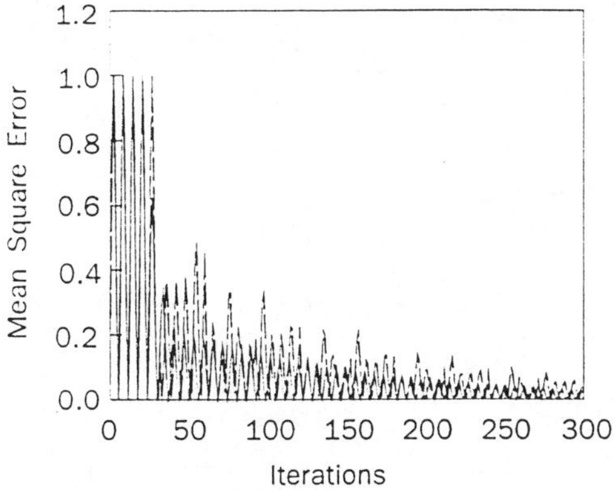

Fig. 6.29 Evolution of system error for a continuously adapting system (solid line), and one where the weights are adjusted only after the effect of the previous change is known (dashed line).

The influence which time delays in the system have upon algorithm stability is important when implementing some form of self-tuning convergence coefficient, such as outlined in Section 6.5. The majority of 'adaptive' convergence coefficient strategies are based upon assessing some characteristic of the reference signal, usually signal power. This is acceptable for a 'standard' electronic implementation of an adaptive algorithm, because algorithm stability is indirectly a function of signal power, because the eigenvalues of the input autocorrelation matrix are functions of signal power. However, for an active noise or vibration control system, system stability is dependent upon both the signal power of the filtered reference signal, and the time delay in the cancellation path transfer function, as outlined in (6.8.68). Furthermore, the influence of the time delay is often dominant, and not easily estimated 'on-line'. In practice, a system dependent multiplier can be used to modify the convergence coefficient value determined via some adaptive strategy (such as calculation via a normalized algorithm) to accommodate the time delay effects.

6.8.5 Effect of transfer function estimation errors: frequency domain algorithm, sine wave input

From the preceding development of the filtered-x LMS algorithm it is clear that estimates of the transfer function between the control source input and the error sensor output must be included in the implementation. However, these estimates will be prone to errors. The question arises of what influence these errors will have upon the stability of the system, or conversely, how close the estimates must be to the actual transfer function if the adaptive algorithm is to function satisfactorily. The next two sections will examine these questions for various system arrangements.

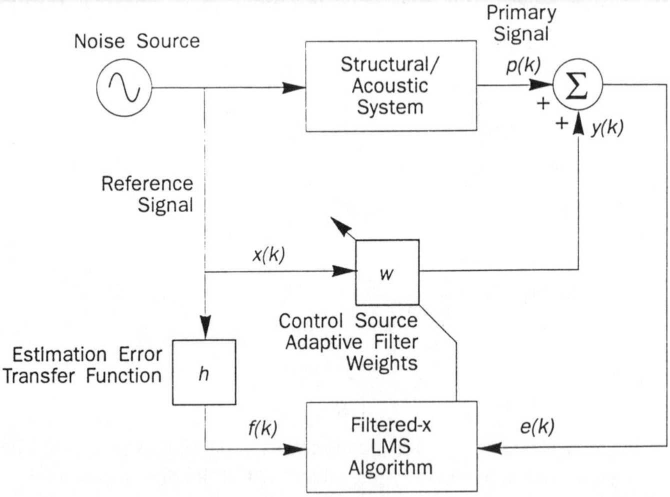

Fig. 6.30 Block diagram of adaptive SISO feedforward control system with an estimation error transfer function ***H*** inserted for analysis.

To examine the effect of these 'estimation errors' only, the cancellation path transfer function will be omitted from the system (equivalent to saying that the transfer function is unity gain and 0° phase), and a single 'estimation error transfer function', ***H***, will be inserted as shown in Fig. 6.30. In this section we will consider the simplest case, where the algorithm is operating in the frequency domain and the input signal is a sine wave. For this arrangement the estimation error transfer function can be represented by a single complex number, h, bringing about a simple gain and phase change.

For the arrangement shown in Fig. 6.30, the filtered-x LMS algorithm will operate according to

$$w(k+1) = w(k) - 2\mu e(k)\left(hx(k)\right)^{*} \tag{6.8.69}$$

where h is the complex estimation error transfer function

$$h = h_R + jh_I \tag{6.8.70}$$

and the error $e(k)$ is defined by the expression

$$e(k) = p(k) + \boldsymbol{w}^{\mathrm{T}}(k)\boldsymbol{x}(k) = p(k) + \boldsymbol{x}^{\mathrm{T}}(k)\boldsymbol{w}(k) \tag{6.8.71}$$

For the deterministic case being considered here, (6.8.69) can be rewritten as

$$w(k+1) = w(k) - 2\mu h^{*}\left[\boldsymbol{p} + \boldsymbol{R}\boldsymbol{w}\right] \tag{6.8.72}$$

where $\boldsymbol{p}$ is the cross-correlation vector between the primary signal and the input reference vector

$$\boldsymbol{p} = E\{p(k)\boldsymbol{x}^{*}(k)\} \tag{6.8.73}$$

and $\boldsymbol{R}$ is the reference signal autocorrelation matrix,

$$\boldsymbol{R} = E\{\boldsymbol{x}^{*}(k)\boldsymbol{x}^{\mathrm{T}}(k)\} \tag{6.8.74}$$

The reference signal autocorrelation matrix $\boldsymbol{R}$ is symmetric, and so can be diagonalized by an orthonormal transformation, as we have done several times before. Using the previously outlined notation, the orthonormal transformation applied to (6.8.72) yields

$$w'(k+1) = w'(k) - 2\mu h^{*}\left[p' + \Lambda w'(k)\right] \tag{6.8.75}$$

where Λ is now the diagonal matrix of eigenvalues of the autocorrelation matrix of the reference signal $\boldsymbol{R}$ (as opposed to the eigenvalues of the autocorrelation matrix of the filtered reference signal). In terms of the weight error vector $v(k)$ (6.8.75) can be restated as

$$v'(k+1) = v'(k) - 2\mu h^{*}\Lambda v'(k) = \left[\boldsymbol{I} - 2\mu h^{*}\Lambda\right]v'(k) \tag{6.8.76}$$

As (6.8.76) is decoupled, we can view it as a set of N scalar equations, each of which must converge towards zero as $k \to \infty$ for the algorithm to be stable. The ith scalar equation is

$$v'(k+1) = \left[1 - 2\mu h\lambda_i\right]v'(k) \tag{6.8.77}$$

For this to converge as $k \to \infty$,

$$\left|1 - 2\mu h^{*}\lambda_i\right| < 1 \tag{6.8.78}$$

As h is complex, we can rewrite (6.8.78) in terms of the real and imaginary parts as

$$\sqrt{(1-2\mu h_R \lambda_i)^2 + (2\mu h_I \lambda_i)^2} < 1 \tag{6.8.79}$$

or

$$(1-2\mu h_R \lambda_i)^2 + (2\mu h_I \lambda_i)^2 < 1 \tag{6.8.80}$$

Expanding this yields

$$1-4\mu h_R \lambda_i + 4\mu^2 h_R^2 \lambda_i^2 + 4\mu^2 h_I^2 \lambda_i^2 < 1 \tag{6.8.81}$$

We can re-express this as

$$\mu^2 \lambda_i^2 (h_R^2 + h_I^2) - \mu \lambda_i h_R < 0 \tag{6.8.82}$$

or

$$\mu \lambda_i \mid h \mid^2 - h_R < 0 \tag{6.8.83}$$

Equation (6.8.83) can be rewritten as

$$\mu < \frac{h_R}{\lambda_i \mid h \mid^2} \tag{6.8.84}$$

or

$$\mu < \frac{\cos(\phi_h)}{\lambda_i \mid h \mid} \tag{6.8.85}$$

where ϕ_h is the phase change caused by the (error in the) transfer function. Therefore, the bounds placed on the convergence coefficient μ for algorithm stability are (Snyder and Hansen, 1990):

$$0 < \mu < \frac{\cos(\phi_h)}{\lambda_{\max} \mid h \mid} \tag{6.8.86}$$

If we compare the result of (6.8.86) with the 'base' result of equation (6.8.60) we can deduce that for a sinusoidal input signal and the complex filtered-x LMS algorithm the effect of imperfections in the estimates of the error loop time delay and transfer functions can be considered in two parts. First, an error in the estimation of the phase of the error loop transfer function will reduce the maximum stable value of convergence coefficient by an amount proportional to $\cos(\phi_h)$ (Snyder and Hansen, 1990; Boucher *et al.*, 1991). It follows that if the estimate is in error by more than $\pm 90°$ the algorithm will become unstable regardless of the size of the convergence coefficient, a property well noted in the literature (Morgan, 1980; Burgess, 1981; Elliott and Darlington, 1985; Snyder and Hansen, 1990). Second, errors in the estimation of the magnitude of the transfer function will reduce the size of the maximum allowable convergence coefficient by an amount proportional to the inverse of that error. These trends are shown in Figs. 6.31 and 6.32, which illustrate the maximum stable value of convergence coefficient for a sinusoidal disturbance being controlled in the frequency domain, plotted as a function of the error in the estimation of the transfer function phase and amplitude respectively.

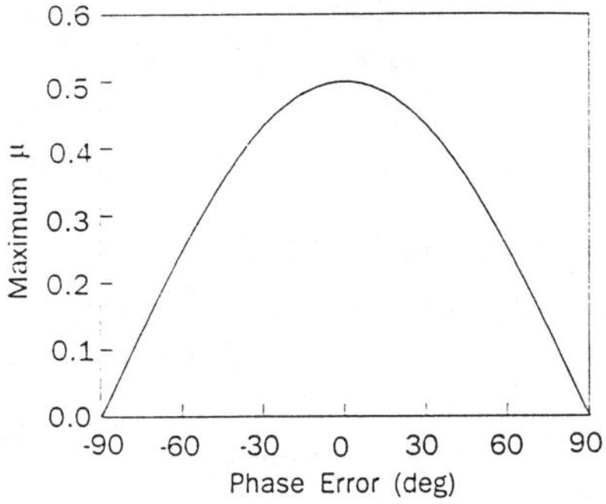

Fig. 6.31 Plot of maximum stable value of convergence coefficient as a function of an error in the estimate of the phase response of the cancellation path, frequency domain algorithm implementation.

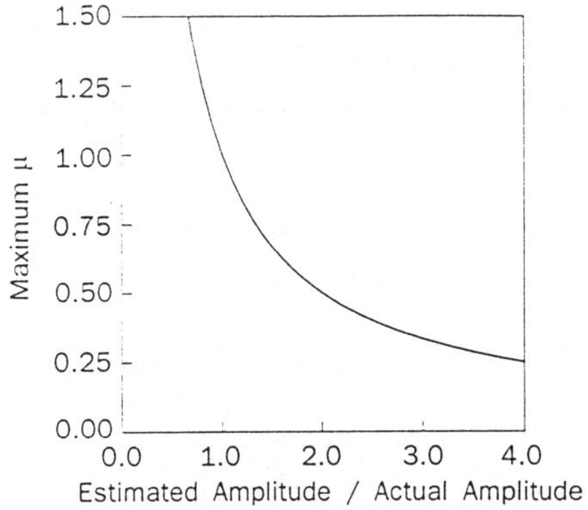

Fig. 6.32 Plot of maximum stable value of convergence coefficient as a function of an error in the estimate of the amplitude response of the cancellation path, frequency domain algorithm implementation.

6.8.6 Effect of transfer function estimation errors: time domain algorithm, sine wave input

It is reasonably straightforward to show that with the complex algorithm the maximum stable value of convergence coefficient will be altered by a multiplying factor proportional to the cosine of the phase error, and inversely proportional to the magnitude error. These results are not always appropriate, however, as many active noise and vibration control systems do not implement a complex version of the filtered-*x* LMS algorithm; rather they perform the adaptive filtering operation in the time domain. What we must therefore do is rederive the results for the real number problem.

The practical implementation of the time domain filtered-*x* LMS algorithm, where the filtered reference signal samples are derived using an estimate of the cancellation path transfer function, was stated previously in (6.8.20). Employing expectation operators to examine mean weight vector convergence, and using the assumption of statistical independence between the quantities, (6.8.20) can be written as:

$$E\{\boldsymbol{w}(k+1)\} = E\{\boldsymbol{w}(k)\} - 2\mu\left[\hat{\boldsymbol{c}}+\hat{\boldsymbol{B}}\, E\{\boldsymbol{w}(k)\}\right] \tag{6.8.87}$$

where $\hat{\boldsymbol{c}}$ is the cross-correlation vector between the primary source signal and the estimated filtered reference signal vector,

$$\hat{\boldsymbol{c}} = E\{p(k)\hat{\boldsymbol{f}}(k)\} \tag{6.8.88}$$

and $\hat{\boldsymbol{B}}$ is the cross-correlation matrix between the estimated and actual filtered reference signal vectors,

$$\hat{\boldsymbol{B}} = E\{\hat{\boldsymbol{f}}(k)\boldsymbol{f}(k)^{\mathrm{T}}\} \tag{6.8.89}$$

If the model of the cancellation path transfer function is exact, $\hat{\boldsymbol{B}}$ is the filtered reference signal autocorrelation matrix $\boldsymbol{B}$. The vector of weight coefficient values to which the algorithm will converge, which will produce a gradient term in (6.8.87) equal to zero (the bracketed part of the equation), is

$$\boldsymbol{w}_{\infty} = -\hat{\boldsymbol{B}}^{-1}\hat{\boldsymbol{c}} \tag{6.8.90}$$

Errors in the estimation of the cancellation path transfer function can be considered in two parts; errors in amplitude estimation, and errors in phase estimation. It is somewhat obvious from (6.8.89) that any error in the estimation of the magnitude of the transfer function will proportionally alter the magnitude of the autocorrelation matrix, and hence its eigenvalues, leading to an inverse proportional alteration in the maximum stable value of convergence coefficient (the same relationship we found for the complex implementation). The influence of a phase estimation error, however, is less intuitive than that of a magnitude estimation error, and is best considered with respect to a specific example. The example we will choose is the simple case of a two-tap filter being used to control a sinusoidal primary disturbance having an amplitude of one and a phase lag of 30° with respect to the reference signal. For simplicity, the actual cancellation path transfer function will be set to unity gain and 0° phase, and it will be assumed that there is no propagation time delay between the control signal output and the error signal input. For these conditions, at any given instant in time, $\boldsymbol{f}$ and $\hat{\boldsymbol{f}}$ are equal to

$$\boldsymbol{f}(k) = \left[\sin\left(\theta(k)+\gamma\right) \;\; \sin\,\theta(k)\right]^{\mathrm{T}} \tag{6.8.91}$$

and

$$\hat{\boldsymbol{f}}(k) = \left[\sin\left(\theta(k)+\gamma+\phi\right) \;\; \sin\left(\theta(k)+\phi\right)\right]^{\mathrm{T}} \tag{6.8.92}$$

where $\theta(k)$ is some (arbitrary) reference angle, ϕ is the phase error, and γ is the angular increment of each new sample. Using (6.8.91) and (6.8.92), $\hat{\boldsymbol{B}}$ for this particular case is equal to

$$\hat{\boldsymbol{B}} = \frac{1}{2}\begin{bmatrix} \cos\phi & \cos(\gamma+\phi) \\ \cos(\gamma-\phi) & \cos\phi \end{bmatrix} \tag{6.8.93}$$

The eigenvalues of the cross-correlation matrix $\hat{\boldsymbol{B}}$ are defined by the characteristic equation

$$|\,\hat{\lambda}_e \boldsymbol{I} - \hat{\boldsymbol{B}}\,| = 0 \tag{6.8.94}$$

where the subscript e denotes that the phase error ϕ is included. Solving this using (6.8.93),

$$\begin{aligned}\hat{\lambda}_e &= \frac{1}{2}\left(\cos\phi \pm\sqrt{\cos(\gamma+\phi)\cos(\gamma-\phi)}\right) \\ &= \frac{1}{2}\left(\cos\phi \pm\sqrt{\cos^2\phi - \sin^2\gamma}\right)\end{aligned} \tag{6.8.95}$$

The eigenvalues of (6.8.95) are those 'seen' by the filtered-x LMS algorithm based upon the phase errored filtered reference signal it is provided. If we consider the result of (6.8.95) in light of the bound placed upon the convergence coefficient in (6.8.60) for stable adaptation of the algorithm, it can be deduced that when the phase error is in excess of $\pm 90°$ the real part of at least one eigenvalue will become negative, and algorithm stability cannot be guaranteed. This was the same criterion which we derived for the complex implementation.

While it is useful to have bounds placed upon the tolerable levels of the cancellation path transfer function estimation error, it would be of enhanced utility if we knew the effect which estimation errors within this allowable range have upon algorithm stability. Calculating this effect, however, is impeded by the nature of the influence that the transfer function phase estimation errors have upon the error surface, the characteristics of which are defined by the autocorrelation matrix $\boldsymbol{B}$. To explain this, note that the eigenvectors $\boldsymbol{q}_e$ associated with the eigenvalues of (6.8.95) are defined by the expression

$$\hat{\boldsymbol{B}}\boldsymbol{q}_e = \hat{\lambda}_e \boldsymbol{q}_e \tag{6.8.96}$$

If (6.8.96) is expanded using (6.8.93) and (6.8.95) the relationship between the two elements of each eigenvector is found to be

$$q_{e,2} = \pm q_{e,1}\sqrt{\frac{\cos(\gamma-\phi)}{\cos(\gamma+\phi)}} \tag{6.8.97}$$

where the choice of $\pm$ in (6.8.97) is dependent upon, and the same as, the choice made in the eigenvalue expression (6.8.95). For the two eigenvectors to be orthogonal, their inner product must be equal to zero, which from (6.8.97) amounts to the criterion

$$|\cos(\gamma+\phi)| = |\cos(\gamma-\phi)| \tag{6.8.98}$$

Clearly, this is true only if the phase of estimation error is 0° or 180° (and if it is 180° the algorithm will, of course, be hopelessly unstable), or if $\gamma = \pi/2(\text{mod}\ \pi)$. Therefore, as the criterion of (6.8.98) is rarely satisified, it can be surmised that, in general, errors in the estimation of the phase of the cancellation path transfer function will result in non-orthogonal eigenvectors, or non-orthogonal error surface principal axes, being 'seen' by the adaptive algorithm. As such, the maximum stable value of convergence coefficient within the limiting phase estimation error bounds of ±90° cannot be determined by considering individual eigenvalues alone, greatly increasing the complexity of the analysis.

To examine this further, it will be convenient to rewrite the algorithm in (6.8.87) in terms of weight error as

$$\boldsymbol{v}(k+1) = \left[\boldsymbol{I}-2\mu\hat{\boldsymbol{B}}\right]\boldsymbol{v}(k) \tag{6.8.99}$$

If $\hat{\boldsymbol{B}}$ is separated into the sum of its 'correct', $\boldsymbol{B}$, and 'in error', $\tilde{\boldsymbol{B}}$, parts, then

$$\boldsymbol{v}(k+1) = \left[\boldsymbol{I}-2\mu\boldsymbol{B}\right]\boldsymbol{v}(k)-2\mu\tilde{\boldsymbol{B}}\boldsymbol{v}(k) \tag{6.8.100}$$

where

$$\boldsymbol{B} = \begin{bmatrix} 1 & \cos\gamma \\ \cos\gamma & 1 \end{bmatrix} \tag{6.8.101}$$

and

$$\tilde{\boldsymbol{B}} = \begin{bmatrix} \cos\phi - 1 & \cos(\gamma+\phi) - \cos\gamma \\ \cos(\gamma-\phi) - \cos\gamma & \cos\phi - 1 \end{bmatrix} \tag{6.8.102}$$

The filtered reference signal autocorrelation matrix $\boldsymbol{B}$ *is* symmetric, and so we can diagonalize it by use of an orthonormal transformation. For the case being considered here, the orthonormal transformation matrix, whose columns are the eigenvectors of $\boldsymbol{B}$, is

$$\boldsymbol{Q} = \frac{1}{\sqrt{2}}\begin{bmatrix} 1 & 1 \\ 1 & -1 \end{bmatrix} \tag{6.8.103}$$

Using the previously outlined notation, and multiplying (6.8.100) through by $\boldsymbol{Q}^{-1}$ produces

$$\begin{aligned} \boldsymbol{v}'(k+1) &= \left(\boldsymbol{I}-2\mu\Lambda\right)\boldsymbol{v}'(k)-2\mu\tilde{\boldsymbol{B}}'\boldsymbol{v}'(k) \\ &= \left(\boldsymbol{I}-2\mu[\Lambda+\tilde{\boldsymbol{B}}']\right)\boldsymbol{v}'(k) \end{aligned} \tag{6.8.104}$$

where

$$\tilde{\boldsymbol{B}}' = \boldsymbol{Q}^{-1}\tilde{\boldsymbol{B}}\boldsymbol{Q} \tag{6.8.105}$$

Using (6.8.102) and (6.8.103), we can expand $\tilde{\boldsymbol{B}}'$ as (Snyder and Hansen, 1994):

$$\tilde{\boldsymbol{B}}' = \begin{bmatrix} (\cos\phi - 1)(1 + \cos\gamma) & \sin\phi \sin\gamma \\ -\sin\phi\sin\gamma & (\cos\phi - 1)(1 - \cos\gamma) \end{bmatrix} \quad (6.8.106)$$

There are several interesting algorithm behaviour characteristics which can be explained in terms of (6.8.106). First, because of the appearance of sin ϕ terms, the effect of the cancellation path transfer function phase estimation error is not be expected to be symmetric about the 0° error position. The exception to this is when the sampling rate is equal to four times the input frequency (γ = 90°), such that the sin γ terms are equal to one, cos γ terms equal to zero, and the two eigenvalues in $\boldsymbol{\Lambda}$ are both equal to 0.5. With this condition, the 'inverse' +/- nature of the top and bottom rows produces identical coupling on both sides of the 0° phase error point. This is confirmed in Figs. 6.33 and 6.34, which show the maximum stable value of convergence coefficient plotted as a function of phase estimation error for sampling rates four (γ = 90°) and five (γ = 72°) times the input frequency, respectively. These results are very different from those which we derived for the complex algorithm implementation, where the reduction in the maximum stable value of convergence coefficient was found to be proportional to the cosine of the transfer function phase estimation error.

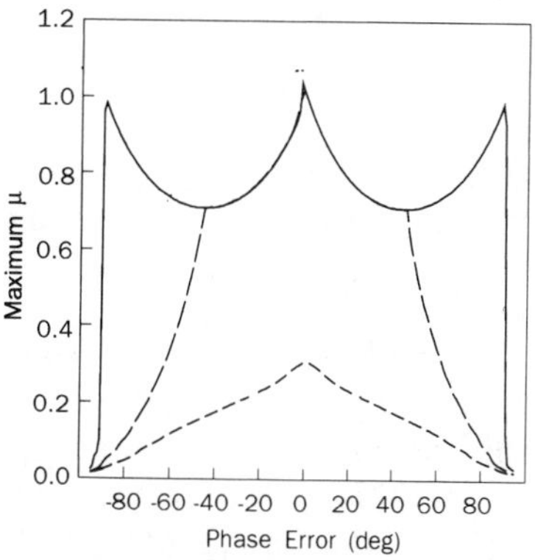

Figs. 6.33 Maximum stable value of convergence coefficient plotted as a function of transfer function phase estimation error for a sampling rate four times the input frequency (γ = 90°) and a time delay of zero (solid line), one (long dashes) and four (short dashes) samples, time domain algorithm.

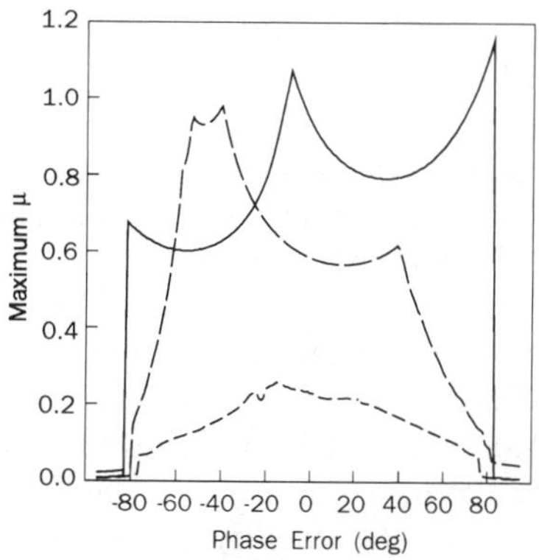

Fig. 6.34 Maximum stable value of convergence coefficient plotted as a function of transfer function phase estimation error for a sampling rate five times the input frequency (γ = 72°) and a time delay of zero (solid line), one (long dashes) and four (short dashes) samples, time domain algorithm.

The non-symmetry of the effect of phase error can become more pronounced at higher sampling rates. Figure 6.35 illustrates the convergence of the algorithm with a convergence coefficient μ = 0.5, where the sampling rate is equal to 12 times the input frequency ($\gamma = 30°$), for phase errors of $\pm 60°$. Here maximum stable value of convergence coefficient for the $+60°$ case is more than twice that of the $-60°$ case. The coupling of the error surface due to the phase estimation error may also in some instances actually increase the algorithm stability as compared to the case of a 'perfect' transfer function estimate ($\phi = 0°$); with a sampling rate equal to 12 times the input frequency, maximum stability is found to occur at a phase error equal to $-26°$.

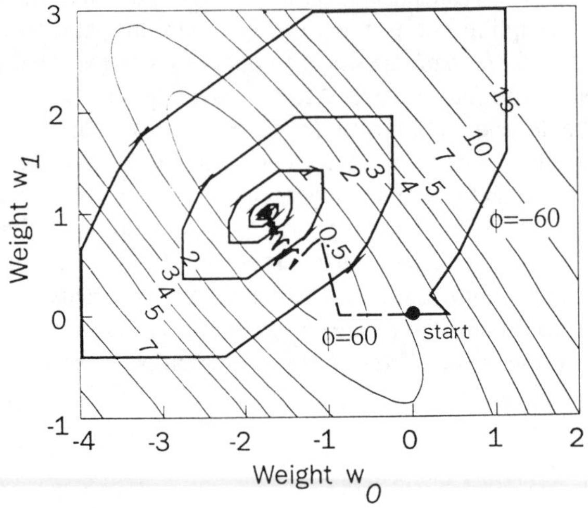

Fig. 6.35 Convergence path of algorithm for phase estimation errors of $\pm 60°$ plotted against contours of constant mean square error, $\mu = 0.5$; (sampling rate is 12 times the input frequency ($\gamma = 30°$)).

One other point to mention is that if the filtered-*x* LMS algorithm is used to update the weight coefficients with every new input sample, the inherent time delays present in the active control system will alter the values of maximum stable convergence coefficient, as we saw previously. This tends to significantly decrease the stability of the algorithm as the phase error approaches $\pm 90°$. Figures 6.33 and 6.34 also depict the maximum stable value of convergence coefficient for time delays of one sample and one cycle. Observe, especially in Fig. 6.34, that the characteristics of the relationship between the phase error and maximum stable convergence coefficient have changed. Also, with the time delays the peaks near the $\pm 90°$ bound are removed, and so the shape of the curve approximates the cosine function shape of the complex algorithm (Boucher *et al.*, 1991). However, the curve is still not, in general, symmetric about the $\phi = 0°$ point, the exception being at a sampling rate of four times the input frequency ($\gamma = 90°$).

The final interesting point to mention concerning the alteration of the error surface as seen by the algorithm due to the phase estimation error is that it may actually increase the algorithm stability. Figure 6.36 illustrates the convergence of the algorithm for a convergence coefficient of 1.1, where the sampling rate is 12 times the input frequency. Two cases are shown, one where the phase error is $-26°$ and one where there is no phase error. Here the algorithm becomes more stable with the error, the error helps the algorithm to avoid the 'rocking motion' between the two steep sides of the error surface. This again, however, is by no means a universal result.

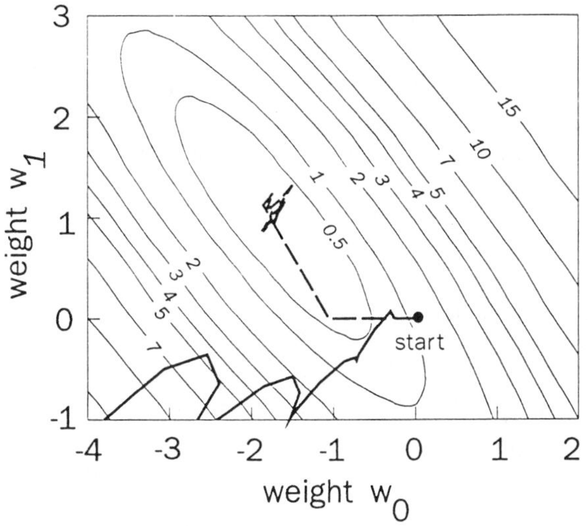

Fig. 6.36 Convergence path of algorithm for a phase estimation error of 0° and $-26°$ plotted against contours of constant mean square error, $\mu = 1.1$; sampling rate is 12 times the input frequency.

While these convergence characteristics of the time domain filtered-x LMS algorithm with a cancellation loop transfer function phase estimation error may be somewhat interesting, they are also somewhat discouraging from the viewpoint of deriving a simple relationship between transfer function phase error and algorithm stability. In fact, it is almost impossible to provide a more quantitative assessment of the effect of the cancellation path transfer function phase estimation error beyond stating that the tolerable bounds of this error are $|\phi| < 90°$.

6.8.7 Equivalent transfer function representation

When the disturbance to be actively controlled is periodic, it is possible for us to rearrange the adaptive feedforward control system to give it the appearance of a feedback control system, where the input to the system is the adaptive algorithm error signal. This arrangement, referred to as an equivalent transfer function representation, will have some benefits for stability analysis and in examining the out-of-(reference signal) band response of the system as modified by the adaptive feedforward controller. In deriving the equivalent transfer function representation we will refer to the error signal as ϵ, rather than e, to distinguish it from the exponential terms will arise.

With this change, the filtered-x LMS algorithm is represented as

$$w(k+1) = w(k) - 2\mu\epsilon(k)f(k) \tag{6.8.107}$$

For the case of interest here, where the referenced disturbance is harmonic at frequency ω_f, the ith reference signal sample in the controller FIR filter delay chain at time k is defined by the expression

$$x_i(k) = r\cos(\omega_f\, kT + \theta_i) \tag{6.8.108}$$

For sinusoidal excitation we can view the cancellation path transfer function as simply invoking an amplitude c and phase change ϕ at the referenced frequency. The ith filtered reference signal sample used in the filtered-x LMS algorithm is then

$$f_i(k) = cr\cos(\omega_f\, kT + \phi_i) \tag{6.8.109}$$

In (6.8.108) and (6.8.109) T is the sample period duration, θ_i is some arbitrary phase angle, and $\phi_i = (\theta_i + \phi)$.

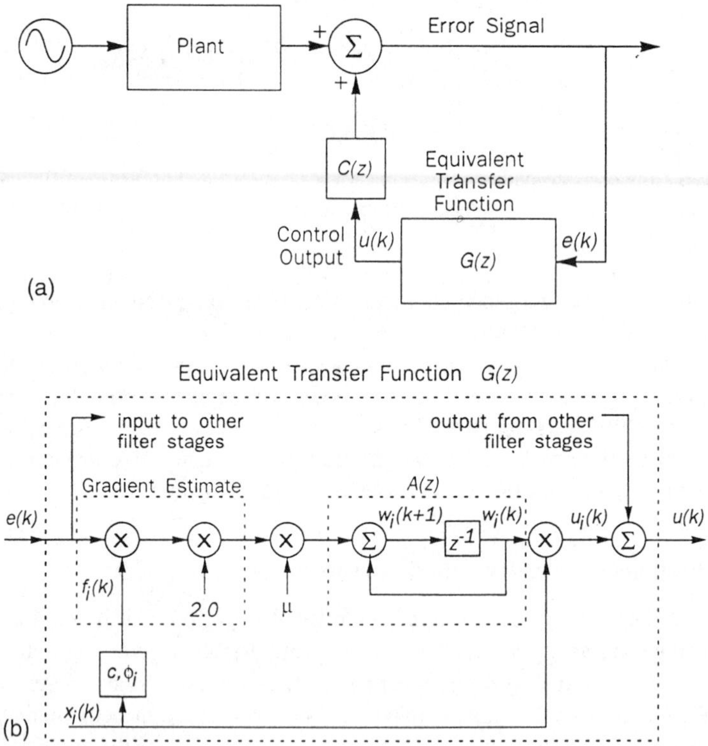

Fig. 6.37 (a) System arrangement with the adaptive feedforward control system replaced by an equivalent transfer function $G(z)$; (b) details of the equivalent transfer function $G(z)$.

Calculation of the 'equivalent transfer function' $G(z)$ between the error signal input to the filtered-x LMS algorithm and the output of the control FIR filter can be accomplished through a minor modification of the approach developed by Glover (1977) for analysing an FIR filter/LMS algorithm used as an adaptive notch filter. Using this approach, the filter/algorithm combination can be viewed as shown in Figs. 6.37(a) and 6.37(b), where the error signal is the input to the system and the output is the control signal. Referring to the detailed representation of $G(z)$ shown in Fig. 6.37(b), starting at the left-hand side, the z transform of the gradient estimate used by the filtered-x LMS algorithm in updating the ith control filter weight is

$$Z\{2e(k)f_i(k)\}$$

$$= cr e^{j\phi_i} Z\{\epsilon(k) e^{j\omega_f kT}\} + cr e^{-j\phi_i} Z\{\epsilon(k) e^{-j\omega_f kT}\} \tag{6.8.110}$$

$$= cr[e^{j\phi_i} E(ze^{-j\omega_f T}) + e^{-j\phi_i} E(ze^{j\omega_f T})]$$

where $E(ze^{-j\omega_f T})$ is the z transform of the error signal $E(z)$ rotated counter-clockwise around the unit circle through an angle $\omega_f T$, and $E(ze^{j\omega_f T})$ is the z transform of the error signal $E(z)$ rotated clockwise around the unit circle through an angle $\omega_f T$. The weight update calculation can be represented as a digital integration with a transfer function $A(z)$:

$$A(z) = \frac{1}{z-1} \tag{6.8.111}$$

Combining (6.8.110) and (6.8.111), the z transform of the ith weight can be expressed as

$$W_i(z) = \mu cr A(z)[e^{j\phi_i} E(ze^{-j\omega_f T}) + e^{-j\phi_i} E(ze^{j\omega_f T})] \tag{6.8.112}$$

The z transform of the ith filter stage output is

$$Y_i(z) = Z\{w_i(k) x_i(k)\}$$

$$= \frac{r}{2}[W_i(ze^{-j\omega_f T}) e^{j\theta_i} + W_i(ze^{j\omega_f T}) e^{-j\theta_i}] \tag{6.8.113}$$

Substituting (6.8.112) into (6.8.113) and summing over all N stages, the z transform of the control filter output at time k is

$$Y(z) = \sum_{i=0}^{N-1} Y_i(z) = \left\{ \frac{N\mu cr^2}{2} E(z)[e^{-j\phi} A(ze^{-j\omega_f}) + e^{j\phi} A(ze^{j\omega_f T})] \right\}$$

$$+ \left\{ \frac{\mu cr^2}{2} A(ze^{-j\omega_f T}) E(ze^{-j2\omega_f T}) \sum_{i=0}^{N-1} e^{j(2\theta_i + \phi)} + \frac{\mu cr^2}{2} A(ze^{j\omega_f T}) E(ze^{j2\omega_f T}) \sum_{i=0}^{N-1} e^{-j(2\theta_i + \phi)} \right\} \tag{6.8.114}$$

To simplify the summations in the second bracketed term in (6.8.114), we can note that as the reference samples are from a tapped delay line

$$\theta_i = \theta - \omega_f T(i-1) \tag{6.8.115}$$

where θ is some 'master' reference phase angle. Using this relationship, the summations can be evaluated as

$$\begin{aligned} &\sum_{i=0}^{N-1} e^{\pm j(2\theta_i + \phi)} \\ &= e^{\pm j[2\theta + \phi - \omega_f T(N-1)]} \frac{\sin N\omega_f T}{\sin \omega_f T} \\ &= \beta_{\pm}(\omega_f T, N) \end{aligned} \tag{6.8.116}$$

where the + or - subscript on β will be the same as that on the exponential.

We can now make the observation that the first bracketed term in (6.8.114) will be the basis of the linear, time invariant (LTI) part of the equivalent transfer function, while the second bracketed term is non-linear, introducing frequency shifted components into the response. Comparing the linear and non-linear terms it is apparent that under 'normal' operating conditions, where a filter with only a few weight coefficients is used to suppress a harmonic disturbance, the linear terms will be dominant but the non-linear terms will not, in general, be negligible. From (6.8.114) and (6.8.116) it is apparent that there are two instances when the equivalent transfer function can be viewed as purely LTI: first, where the number of taps N in the filter is large so that the first bracketed term in (6.8.114) is completely dominant, as discussed by Glover (1977). Second, under the synchronous sampling condition of $N\omega_f T = n180°$ and $\omega_f T \neq n180°$, where n is some integer, as discussed by Elliott and Darlington (1985) and Elliott *et al.* (1987). If one of these conditions is met, such that only the first, LTI, term in (6.8.114) needs to be considered, then

$$\begin{aligned} G(z) &= \frac{Y(z)}{E(z)} = \frac{N\mu c r^2}{2}\left[e^{-j\phi}A(ze^{-j\omega_f T}) + e^{j\phi}A(ze^{j\omega_f T})\right] \\ &= \frac{N\mu c r^2}{2}\left[\frac{e^{-j\phi}}{ze^{-j\omega_f T}-1} + \frac{e^{j\phi}}{ze^{j\omega_f T}-1}\right] \\ &= N\mu c r^2 \frac{z\cos(\omega_f T - \phi) - \cos\phi}{z^2 - 2z\cos\omega_f T + 1} \end{aligned} \tag{6.8.117}$$

In this instance, the equivalent transfer function takes on the form of a classical bandpass filter, with poles at the reference frequency, a point discussed further by both Glover (1977) and Sievers and von Flotow (1992). As an example, a plot of the amplitude response characteristics of $G(z)$ in (6.8.117) is illustrated in Fig. 6.38 for two values of convergence coefficient, μ = 0.05 and 0.01, N = 4 taps in the control FIR filter, a cancellation path transfer function

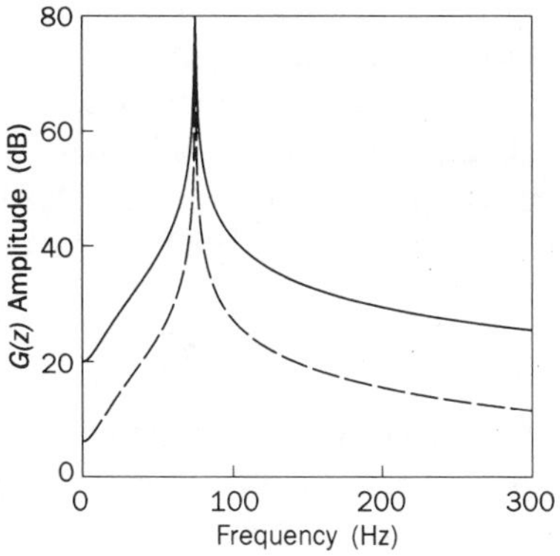

Fig. 6.38 Amplitude of equivalent transfer function $G(z)$ for $\mu = 0.05$ (solid line) and $\mu = 0.01$ (dashed line).

amplitude c of 1.0 and phase of 0°, a reference signal amplitude $r = 10.0$, and a feedforward reference frequency of 75 Hz. The important characteristic to note here, for future reference, is the reduction in out of (reference signal) band response which accompanies a reduction in convergence coefficient.

The equivalent transfer function $G(z)$ described the transfer function between the error signal input to the system and the control output. Expanding the error signal for this sinusoidal case, using

$$E(z) = P(z) + C(z)Y(z) \tag{6.8.118}$$

where $P(z)$ is the z transform of the primary source disturbance, the transfer function $H(z)$ between the error signal output and the primary disturbance input can be defined as

$$\frac{E(z)}{P(z)} = H(z) = \frac{1}{1 - C(z)G(z)} \tag{6.8.119}$$

If we limit consideration to only the LTI portion of the equivalent transfer function, and define the variable β, as

$$\beta = N\mu c r^2 \tag{6.8.120}$$

$H(z)$ can be expanded as (Elliott *et al.* 1987):

$$H(z) = \frac{z^2 - 2z\cos\omega_f T + 1}{z^2 - 2z\cos\omega_f T + 1 + \beta C(z)\left[z\cos(\omega_f T - \phi) - \cos\phi\right]} \tag{6.8.121}$$

This is a very useful relationship, because it enables the full behaviour of the system to be determined analytically. The only assumptions are that the disturbance is harmonic, and the non-linear part of the equivalent transfer function $G(z)$ is negligible.

As one example of the use of (6.8.121) in analysing system response, consider the case where the reference signal amplitude r is 1.0, the control filter has $N = 2$ taps, the system sampling rate is four times the excitation frequency such that $\omega_f T$ is $\pi/2$, and the cancellation path transfer function (estimate) amplitude c is 1.0 and phase ϕ is 0° but $C(z)$ is actually a one cycle (four sample) delay (considered in Elliott *et al.*, 1987). In this case,

$$H(z) = \frac{1 + z^{-2}}{1 + z^{-2} - 2\mu Z^{-6}} \tag{6.8.122}$$

The poles of this transfer function can be found analytically by solving the cubic equation in z^{-2} in the denominator. Doing this, it is found that the poles are real for a value of convergence coefficient μ less that 2/27, but imaginary for greater values. This corresponds to the transition between an 'overdamped' response, where there is no overshoot in the mean square error, and an 'underdamped' response, characterized by oscillations in the value of the mean square error during the convergence process, as discussed in Section 6.5 for a purely electronic system. If the convergence coefficient exceeds approximately 0.3, the poles move outside the unit circle and the system becomes unstable. Equivalent transfer function representation-based methodologies can also be used to examine the influence of factors such as cancellation path transfer function error (Elliott *et al.*, 1987; Boucher *et al.*, 1991) and transient response characteristics of the algorithm (Morgan and Sanford, 1992). It can also be used to examine the influence which an adaptive feedforward controller has upon the response characteristics of the system outside the referenced frequency band, as we will now discuss.

6.8.8 A note on implementing adaptive feedforward control systems with other control systems

In some cases it may be of interest to implement an adaptive feedforward controller in conjunction with a feedback control system. In such cases the feedforward controller would target periodic disturbances from specific sources such as rotating machinery on a flexible platform, while the feedback control system would be used to damped transient responses and excitation from non-periodic noise sources. To be of use in such an arrangement the adaptive feedforward controller must not (significantly) alter the poles of the system and/or deteriorate the system response away from its referenced frequency band. The combination of reported results from isolated implementations of adaptive feedforward control systems, which do not show deterioration in system response away from the referenced frequency band, and our knowledge that non-adaptive feedforward control affects only the zeros of a system, may lead us to the impression that both of these requirements are met.

With this in mind, consider the system arrangement shown in Fig. 6.39, which was used to generate the experimental data shown in Section 6.2 in the discussion of what a feedforward control system actually does. Here a steel cantilever beam of dimensions 1110 mm × 31 mm × 5.6 mm is subject to primary point force excitation 745 mm from the fixed end of the beam, and to feedforward control (point) excitation 1030 mm from the fixed end of the beam. In this arrangement, the first four resonance frequencies of the beam are found to be at 10 Hz, 24 Hz, 66.5 Hz and 131 Hz. The primary disturbance consists of a tone in low level random noise, with the amplitude of the tone approximately 40 dB above that of the noise. The adaptive feedforward control system, constructed from a four tap FIR filter and the filtered-x LMS algorithm (discussed in more detail in the next section), is provided a reference signal only for the tonal component of the primary disturbance. We may encounter a similar situation in practice, for example, when controlling vibratory power flow in a flexible platform where a piece of (feedforward referencable) rotating machinery is present. We can refer to the tonal component as the referenced disturbance, while the remainder of the primary disturbance, the low level random noise, can be referred to as the unreferenced disturbance. A velocity error signal is provided to the adaptive control system by an accelerometer mounted 1080 mm from the fixed end of the beam.

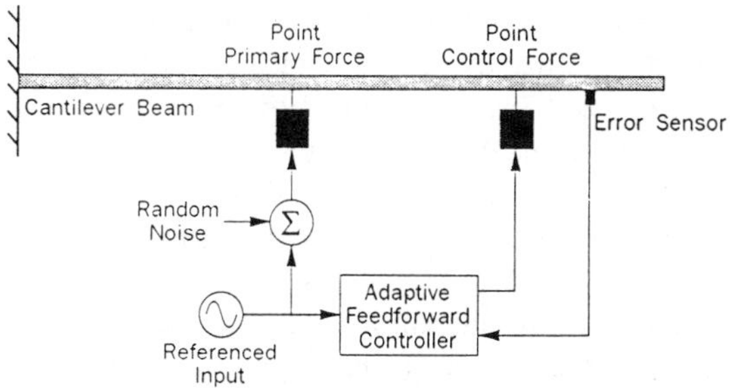

Fig. 6.39 Cantilever beam experimental arrangement.

When the tonal component of the primary disturbance is 75 Hz, the autospectrum of the error signal before and after implementation of the control system is as shown in Fig. 6.40. We observe that the adaptive feedforward control system has attenuated the referenced component of the spectrum by approximately 40 dB and remains stable. Its implementation has, however, also resulted in an increase in the response of the system at the 66.5 Hz resonance by over 20 dB, and the introduction of a new peak at 83 Hz; to a large extent the controlled response is inferior to the alternative with no control at all. This occurs in spite of the fact that the feedforward controller reference signal does not contain these frequency components. The increase in response would therefore appear to be in some way related to the presence of the frequency components in the error signal of the adaptive system, which inherently forms a sort of 'feedback loop', albeit a very indirect one through the weight adaptation process to the control signal output. This feedback loop appears to have modified the response characteristics of the beam; we would expect the zero at 75 Hz, as it is the target of the feedforward control system, but the changes at 66.5 Hz and 83 Hz are somewhat unexpected. It is therefore important to explain this phenomenon, and determine what can be done to overcome it.

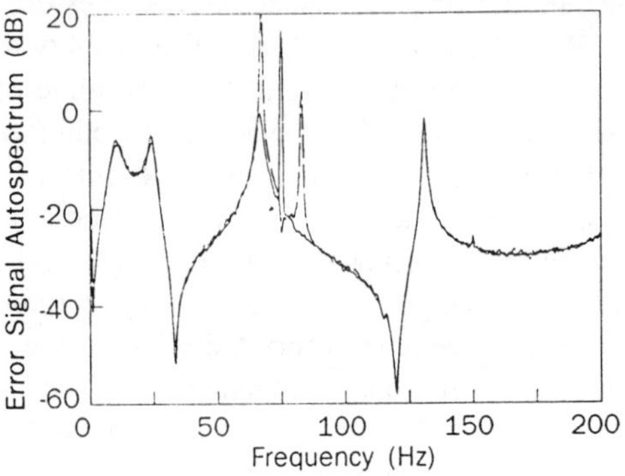

Fig. 6.40 Error signal autospectrum without feedforward control (solid line) and with feedforward control at 75 Hz (dashed line).

The first item which may be of interest is whether an adaptive feedforward control system can significantly alter the location of the open loop poles of the system being controlled. This idea is plausible because while the control signal derivation is itself purely feedforward, the control system is adaptive, with the filter weights being altered in response to an error signal which can contain all frequency components; we have already shown that it is possible to quantify the transfer function from the error signal input to the control signal output, highlighting the existence of this inherent feedback loop. To examine this possibility, we will use some of the rersults from the previous chapter and consider the effect of introducing a control signal derived via the adaptive feedforward control system into the discrete state equation of a single beam mode, with a single point control force and a single velocity system output. The state equations can be written as

$$\begin{aligned} \boldsymbol{x}_n(k+1) &= \boldsymbol{A}_n\boldsymbol{x}_n(k) + \boldsymbol{b}_n u(k) \\ y(k) &= \boldsymbol{c}_n\boldsymbol{x}_n(k) \end{aligned} \tag{6.8.123}$$

where

$$\boldsymbol{x}_n(k) = \begin{bmatrix} x_n(k) \\ \dot{x}_n(k) \end{bmatrix} \tag{6.8.124}$$

$$A_n = e^{-\xi_n \omega_n T} \times \begin{bmatrix} \cos\omega_{nd}T + \dfrac{\xi_n\omega_n}{\omega_{nd}}\sin\omega_{nd}T & \dfrac{1}{\omega_{nd}}\sin\omega_{nd}T \\ -\dfrac{\omega_n^2}{\omega_{nd}}\sin\omega_{nd}T & \cos\omega_{nd}T - \dfrac{\xi_n\omega_n}{\omega_{nd}}\sin\omega_{nd}T \end{bmatrix} \tag{6.8.125}$$

$$b_n = \phi(x_c)\, e^{-\xi_n\omega_n T} \times \begin{bmatrix} \left(\dfrac{e^{\xi_n\omega_n T} - \cos\omega_{nd}T - \dfrac{\xi_n\omega_n}{\omega_{nd}}\sin\omega_{nd}T}{\xi_n^2\omega_n^2 - \omega_{nd}^2} \right) \\ \dfrac{1}{\omega_{nd}}\sin\omega_{nd}T \end{bmatrix} \tag{6.8.126}$$

$$c_n = [0 \quad \phi_n(x_e)] \tag{6.8.127}$$

the quantity x_n is the displacement of the mode, ξ_n is the modal damping, the nth damped natural frequency is $\omega_{nd} = \omega_n\sqrt{1-\xi_n^2}$, $\phi_n(x_c)$ is the value of the nth mode shape function at the control location x_c, and $\phi_n(x_e)$ is the value of the nth mode shape function at the error sensor location x_e. In the adaptive feedforward control scheme the system output $y(k)$ will be the algorithm error signal $\epsilon(k)$, given by

$$\epsilon(k) = y(k) = c_n x(k) \tag{6.8.128}$$

Taking the z transform of (6.8.123) and (6.8.128), and limiting our consideration to only the linear time invariany (LTI) equivalent transfer function defined in (6.8.117) for simplicity, to relate the error signal of (6.8.128) to the control signal u in (6.8.123), the characteristic equation of the system is found to be

$$| zI - A_n - B_n' | = 0 \tag{6.8.129}$$

where

$$B_n' = N\mu c r^2 \frac{z\cos(\omega_f T - \phi) - \cos\phi}{z^2 - 2z\cos\omega_f T + 1} b_n c_n \tag{6.8.130}$$

Observe that the adaptive control system *will*, in fact, modify the poles of the mode, even though derivation of the control signal is purely feedforward. Use of the error signal in adaptive weight coefficient calculations is responsible for this, providing an inherent feedback path from the error sensor to the control output (if the adaptive algorithm is stopped, equivalent to setting the convergence coefficient $\mu = 0$, then (6.8.129) becomes the characteristic equation of the open loop system, which is the commonly stated result for the introduction of (non-adaptive) feedforward control) (Snyder and Tanaka, 1993a). The question we must answer now is, how significant is the change?

To answer this question, consider the change in the location of the 65.5 Hz pole during operation of the feedforward adaptive control system, calculated as a function of convergence coefficient μ with the cancellation path transfer function amplitude $c = 1.0$, phase $\phi = 0°$, reference signal amplitude $r = 10.0$ and frequency 75 Hz, modal damping $\xi_n = 0.01$, $N = 4$ taps in the FIR filter, a sampling rate of 1600 Hz, and with control source and error sensor locations outlined previously, to approximate the situation under which the experimental results were obtained. The locus of the pole originally at $(0.964 + j0.258)$ is shown in the (positive + j positive) section of the unit circle in Fig. 6.41 for a variation in μ from 0 to 0.150 (the congugate pole undergoes the mirror image translation). In this case the damping of the pole is increased in response to operation of the adaptive feedforward control system, with a slight initial increase in resonance frequency, reducing with larger values of convergence coefficient. Note that this behaviour is very similar to that observed with 'standard' velocity feedback using collocated sources and sensors, as might be expected for a velocity error signal with the control source and error sensor located in the same nodal area of the associated third beam mode. This result suggests that the change in pole location is *not* responsible for the increase in the resonant peak amplitude during operation of the adaptive feedforward control system (as the damping has increased). Also, in practice the adaptive algorithm in the feedforward control system would have become unstable long before the value of convergence coefficient μ reached 0.150 (in practice, at less than half this value) owing to the nearby out-of-band resonance (see Morgan and Sanford, 1992), limiting the actual movement of the pole. The adaptive feedforward control system is most likely responsible for the very subtle frequency increase in the 66.5 Hz resonant peak seen in Fig. 6.40, however further investigation is required to determine the cause of the significant increase in the existing resonant peak, and the introduction of a new peak.

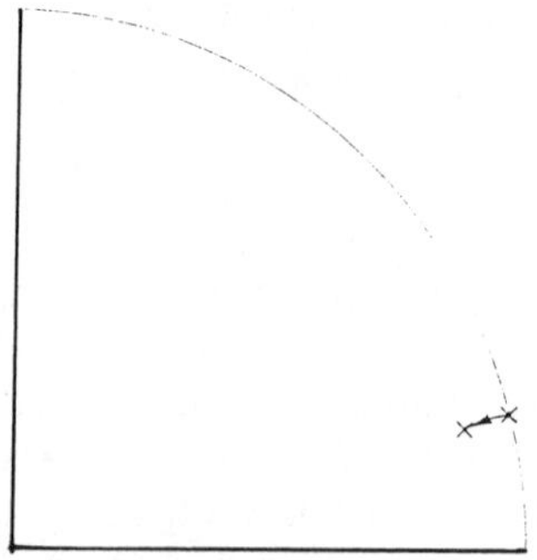

Fig. 6.41 Locus of the pole at $(0.964 + j0.258)$ with μ increasing from 0.0 to 0.15.

As we can effectively rule out a modification to the system poles as the cause of the results shown in the introduction, it may prove enlightening to consider further the idea of the equivalent transfer function as a bandpass filter. With this view it is plausible that if the feedforward control system is operating with a reference frequency close to either an observed resonant peak or an unreferenced periodic disturbance, then this frequency component in the error signal will pass through the adaptive feedforward controller and be present in its output. As we have shown, the open loop poles of the controlled system to be largely unaffected by the inherent error signal to control signal feedback loop with small values of convergence coefficient μ; thus, this process can be viewed as simply a feedforward output in response to the error signal, as opposed to the reference signal.

To investigate this idea, consider the case where the adaptive feedforward control system has effectively nulled the referenced disturbance at frequency ω_f, and the residual error signal is from a second order process:

$$E(z) = \frac{z(z - \cos\omega T)}{z^2 - 2z\cos\omega T + 1} = \frac{z}{2}\left[\frac{1}{z-e^{-j\omega T}} + \frac{1}{z-e^{j\omega T}}\right] \tag{6.8.131}$$

If consideration is initially limited to the LTI part of the equivalent transfer function $G(z)$, then substituting (6.8.131) into (6.8.117) yields the following expression for the controller output:

$$Y(z) = E(z)G(z) = N\mu cr^2 \frac{z(z - \cos\omega T)(z\cos(\omega_f T-\phi) - \cos\phi)}{(z-e^{-j\omega T})(z-e^{j\omega T})(z-e^{-j\omega_f T})(z-e^{j\omega_f T})} \tag{6.8.132}$$

The controller output is now defined by two pole pairs, the referenced pair at $e^{\pm j\omega_f T}$ and an additional unreferenced pair at $e^{\pm j\omega T}$. The output of the adaptive feedforward controller mirrors, to some extent, the unreferenced frequency response characteristics measured at the error sensor; the adaptive feedforward control system is effectively behaving as if it had two reference signals, from both the reference input and the error input to the adaptive algorithm. The relative influence of the reference and error signals on the control output is dependent upon the magnitude of the convergence coefficient, as will be discussed later in this section.

To explain the existence of new peaks in the spectrum we must expand consideration to account for the non-linear term in the equivalent transfer function $G(z)$ (the second bracketed expression in (6.8.114)). The first part of the non-linear term produces an output in response to the second order process error signal of (6.8.131) defined by the expression

$$\frac{\mu c r^2}{2} A(ze^{-j\omega_f T}) E(ze^{-j2\omega_f T}) \beta_+(\omega_f T, N) = \frac{\mu c r^2}{2} \beta_+(\omega_f T, N)$$

$$\times \frac{1}{ze^{-j\omega_f T} - 1} \frac{ze^{-j2\omega_f T}}{2} \left[\frac{1}{ze^{-j2\omega_f T} - e^{-j\omega}} + \frac{1}{ze^{-j2\omega_f T} - e^{j\omega}} \right] \qquad (6.8.133)$$

$$= \frac{\mu c r^2}{4} \beta_+(\omega_f T, N) \frac{ze^{j\omega_f T}\left(2z - e^{j(2\omega_f T - \omega T)} - e^{j(2\omega_f T + \omega T)}\right)}{\left(z - e^{j(2\omega_f T - \omega)}\right)\left(z - e^{j(2\omega_f T + \omega)}\right)\left(z - e^{j\omega_f T}\right)}$$

Similarly, the second part of the non-linear term in (6.8.114) produces an output in response to the second order process error signal defined by the expression

$$\frac{\mu c r^2}{2} A(ze^{j\omega_f T}) E(ze^{j2\omega_f T}) \beta_-(\omega_f T, N) = \frac{\mu c r^2}{2} \beta_-(\omega_f T, N)$$

$$= \frac{1}{ze^{j\omega_f T} - 1} \frac{ze^{j2\omega_f T}}{2} \left[\frac{1}{ze^{j2\omega_f T} - e^{-j\omega}} + \frac{1}{ze^{j2\omega_f T} - e^{j\omega}} \right] \qquad (6.8.134)$$

$$= \frac{\mu c r^2}{4} \beta_-(\omega_f T, N) \frac{ze^{-j\omega_f T}\left(2z - e^{-j(2\omega_f T - \omega T)} - e^{-j(2\omega_f T + \omega T)}\right)}{\left(z - e^{-j(2\omega_f T - \omega)}\right)\left(z - e^{-j(2\omega_f T + \omega)}\right)\left(z - e^{-j\omega_f T}\right)}$$

The results of (6.8.133) and (6.8.134) show that the non-linear part of the transfer function between the error signal input and the control output introduces pole pairs in the controller response at $e^{\pm j(2\omega_f T \pm \omega T)}$

In light of these results, let us consider again the data shown in Fig. 6.40. In response to the adaptive feedforward control system input, the 66.5 Hz resonant peak is amplified; the controller is driving it, as predicted by (6.8.132). The new peak is at approximately 83 Hz, which is equal to $(2\omega_f - \omega) = (150-66.5)$ Hz in response to the resonant peak as predicted by the result of (6.8.133) and (6.8.134). Therefore, the change in the response spectrum is *not* due to a change in the response characteristics of the beam, but rather is a result of the characteristics of the adaptive control system; that is, the adaptive control system is behaving like a non-linear bandpass filter, with the output both mirroring the response characteristics of the system in the absence of the referenced disturbance and introducing new peaks at $(2\omega_f \pm \omega)$ (Snyder and Tanaka, 1993a).

There are a few points which should be noted in relation to the results of the previous section. First, as the adaptive controller is behaving in a manner analogous to a bandpass filter, the problem of additional frequency components appearing in the control output from the error signal is most bothersome when the referenced disturbance is close in frequency to a second, unreferenced, response peak. Consider, for example, the effect of implementing the adaptive feedforward controller utilized in the introduction study to remove a tonal component at 140 Hz. In viewing the resulting response spectrum shown in Fig. 6.42 it can be seen that the amplitude of the fourth resonance peak at 131 Hz has increased significantly, with a 'mirrored' peak at $(2\omega_f - \omega) = (2 \times 140 \text{ Hz} - 131 \text{ Hz})$

= 150 Hz as predicted by the result of (6.8.133) and (6.8.134). How-ever, the increase in the 66.5 Hz peak, and the additional peak at 83 Hz observed in Fig. 6.40 which represents the results obtained with a 75 Hz reference signal, has subsided. It should also be noted that the effect which out-put of unreferenced frequency components has upon the response of the system is implementation dependent, varying with control source and error sensor position; in some instances the measured response will increase, in some instances it will actually be attenuated, depending upon the phasing of the signals.

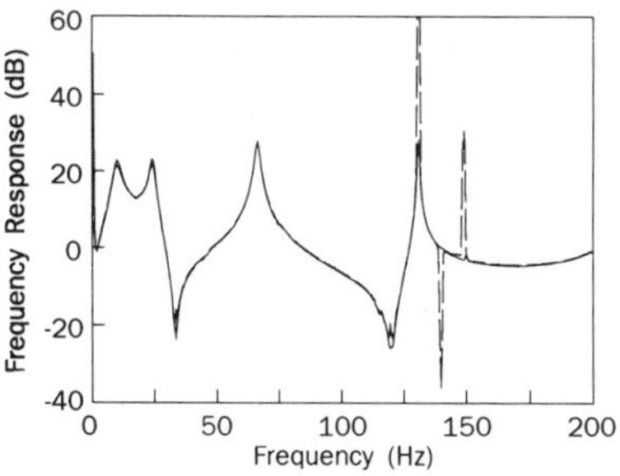

Fig. 6.42 Frequency response magnitude measured between the primary source and the error sensor, without feedforward control (solid line) and with feedforward control at 140 Hz (dashed line).

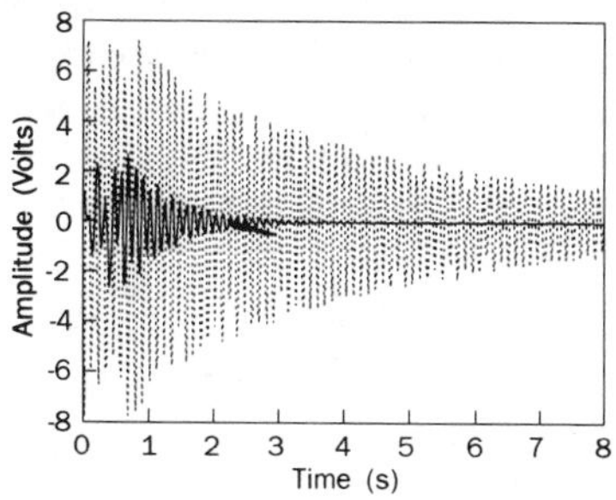

Fig. 6.43 Decay of beam response in the frequency band 115−165 Hz without feedforward control (solid line) and with feedforward control at 140 Hz (dashed line).

Second, the adaptive controller can have the effect of increasing the impulse response duration of beam modes in much the same way as a reduction in damping (an idea correlating with the increase in the amplitude of the resonance peaks). To demonstrate this, consider again the result described in the previous paragraph, shown in Fig. 6.42. If the random noise is turned off with and without the feedforward control system operating, the decay of the beam response in the frequency range from 115 Hz to 165 Hz is as shown in Fig. 6.43. Note that not only is the initial response of the beam increased, but the actual decay time is greatly lengthened.

The final, and perhaps most important, point to consider is how we can overcome these problems. As was noted, there are several ways to make the non-linear part of the equivalent transfer function negligible. This will have the effect of reducing the new response peaks at $(2\omega_f \pm \omega)$, but will have no influence upon the linear component of the equivalent transfer function which is responsible for increased response at the frequency ω. Virtually the only guaranteed way to remove the negative effects of signal components passing from the error signal to the control output is to reduce the value of the convergence coefficient to one much smaller than that dictated by stability constraints. This has the effect of reducing the out-of-band sensitivity of the controller, as shown in Fig. 6.38. From (6.8.110), this will reduce the amplitude of *all* terms in the transfer function between the error signal input to the adaptive algorithm and the control signal output from the controller. Figure 6.44 shows the same case as Fig. 6.40 with the convergence coefficient cut by a factor of 5, from 0.05 to 0.01. This was the value of convergence coefficient found experimentally to eliminate all traces of peaks for the particular arrangement described in the introduction, which gives us essentially the same result as the non-adaptive feedforward controller shown in Section 6.2. This value is less than 15% of that maximally stable, found experimentally to be just in excess of 0.07 ($\mu = 0.07$ is stable). This has the adverse effect of slowing tracking speed and slightly reducing the attenuation of the tone, but is what is required to achieve satisfactory results.

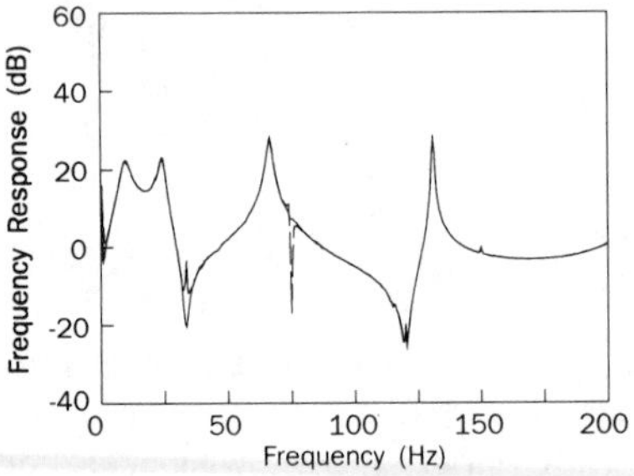

Fig. 6.44 Frequency response magnitude measured between the primary source and the error sensor, the same as Fig. 6.41 with the adaptive algorithm convergence coefficient reduced by 500%. Solid line is data measured without feedforward control, dashed line is data measured with feedforward control at 75 Hz.

6.9 THE MULTIPLE INPUT, MULTIPLE OUTPUT FILTERED-*X* LMS ALGORITHM

Perhaps the majority of the structural/acoustic systems which are targeted for active noise and vibration control exhibit complex, multimodal response to the primary excitation to which they are subjected. For an active control system to guarantee global attenuation of the offending disturbance, all of the principal offending structural/acoustic modes must be observable to, and controllable by, the system. It follows that for a single mode system, such as plane wave sound propagation in a duct, a relatively simple combination of a single control source and error sensor can achieve the desired result. Such a system could implement the SISO version of the filtered-*x* LMS algorithm considered in the previous section. For more complex systems, however, the required number of control sources and error sensors can increase dramatically. In these instances the algorithm must be extended to enable it to be used in multiple input, multiple output active control systems. In this section we will outline the multiple input, multiple output (MIMO) version of the filtered-*x* LMS algorithm, and extend the results of the previous section to consider some of the effects which system parameters have upon its stability and convergence characteristics.

6.9.1 Algorithm derivation

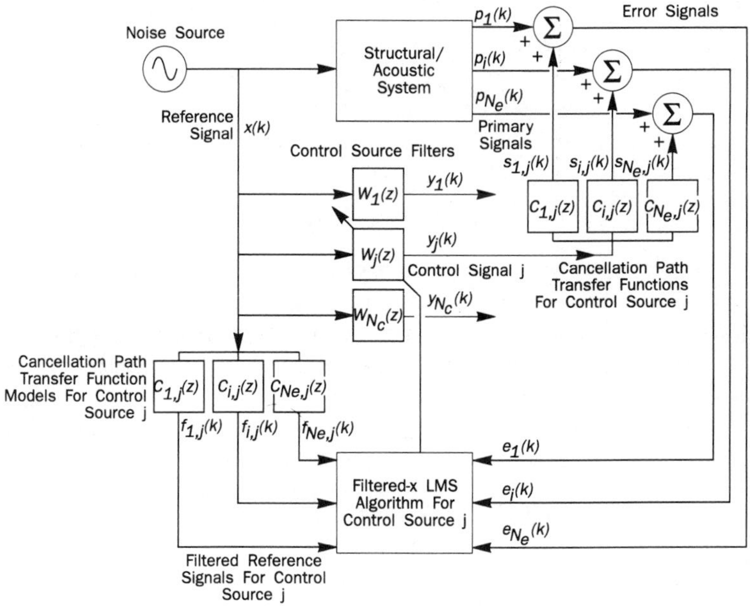

Fig. 6.45 Block diagram of an adaptive MIMO feedforward active control system.

The MIMO-generalization of the active control system arrangement discussed in the previous section is illustrated in Fig. 6.45. Here the signal supplied to each control source is generated by a separate FIR filter, and we will assume without loss of generality that each control source filter has N stages. The output of the ith error sensor at time k, $e_i(k)$, can be thought of as being comprised of two parts: that due to primary excitation $p_i(k)$, and that due to the sum of contributions from each of the N_c control sources $s_{i,j}(k)$,

$$e_i(k) = p_i(k) + \sum_{j=1}^{N_c} s_{i,j}(k) \tag{6.9.1}$$

As we discussed in the previous section for the SISO case, the jth control source component of the ith error signal is not, in general, equal to the output of the jth control filter $y_j(k)$, but is rather equal to a version of the control signal which has been modified by the cancellation path transfer function between the output of the jth control filter and the ith error sensor output. Modelling this transfer function as an m-stage finite impulse response function (vector) $\boldsymbol{c}_{i,j}$, and assuming that the system is time invariant, we can write

$$s_{i,j}(k) = y(k) * c_{i,j} = \boldsymbol{y}_j^{\mathrm{T}}(k)\boldsymbol{c}_{i,j} \tag{6.9.2}$$

where $y_j(k)$ is an $(m \times 1)$ vector of most recent outputs from the jth control filter:

$$\boldsymbol{y}_j(k) = \left[y_j(k) \;\; y_j(k-1) \;\; \cdots \;\; y_j(k-m+1)\right]^{\mathrm{T}} \tag{6.9.3}$$

If the output of the jth control filter is expanded, as

$$y_j(k) = \boldsymbol{x}^{\mathrm{T}}(k)\boldsymbol{w}_j \tag{6.9.4}$$

then $s_{i,j}$ can be expressed in the 'filtered reference signal' format used in the previous section, as

$$s_{i,j}(k) = \left[\boldsymbol{X}^{\mathrm{T}}(k)\boldsymbol{w}_j\right]^{\mathrm{T}}\boldsymbol{c}_{i,j} = \boldsymbol{w}_j^{\mathrm{T}}\left[\boldsymbol{X}(k)\boldsymbol{c}_{i,j}\right] = \boldsymbol{w}_j^{\mathrm{T}}\boldsymbol{f}_{i,j}(k) \tag{6.9.5}$$

where $\boldsymbol{X}(k)$ is an $(N \times m)$ matrix of m most recent reference signal vectors (the mth row of $\boldsymbol{X}(k)$ is the reference signal used in deriving the control filter output at time $k - m+1$) as defined previously in (6.8.9), and $\boldsymbol{f}_{i,j}$ is the i,jth filtered reference signal vector, 'filtered' by the cancellation path transfer function between the jth control filter output and ith error sensor output. Each element $f_{i,j}$ in this vector is equal to the reference signal x convolved with the finite impulse response model of the cancellation path transfer function:

$$f_{i,j}(k) = x(k) * c_{i,j} = \boldsymbol{x}^{\mathrm{T}}(k)\boldsymbol{c}_{i,j} \tag{6.9.6}$$

In the MIMO control arrangement the aim of the exercise, which we will make our error criterion, is to minimize the sum of the mean square values of the signal from each of the error sensors, Ξ; that is,

$$\Xi = \sum_{i=1}^{N_e} E\{e_i^2(k)\} = \sum_{i=1}^{N_e} \xi_i \tag{6.9.7}$$

where there are N_e error sensors in the system. We can again employ a gradient descent algorithm to search the error criterion for the optimum weight coefficients. However, as we have discussed several times, it is generally not practical to base a working feedforward active control system on this error criterion. Instead, we make the stochastic approximation

$$\Xi \approx \sum_{i=1}^{N_e} e_i^2(k) \tag{6.9.8}$$

and differentiate this expression with respect to the control filter weight coefficients to obtain an estimate of the gradient of the error criterion at the current filter settings (which is required to implement the gradient descent algorithm). To do this, (6.9.8) can be expanded, using (6.9.1) and (6.9.5), as

$$\sum_{i=1}^{N_e} e_i^2(k) = \sum_{i=1}^{N_e}\left[p_i(k) + \sum_{j=1}^{N_c} \boldsymbol{w}_j^{\mathrm{T}}\boldsymbol{f}_{i,j}(k)\right]^2 \tag{6.9.9}$$

Differentiating this with respect to the jth weight coefficient vector produces an expression for the gradient estimate

$$\Delta w_j(k) \approx \sum_{i=1}^{N_e} \frac{\partial e_i^2(k)}{\partial w_j} = 2\sum_{i=1}^{N_e} f_{i,j}(k) e_i(k) \tag{6.9.10}$$

where we are again approximating the partial derivative with a functional derivative. Using this gradient estimate in the standard gradient descent format produces the MIMO filtered-x LMS algorithm. For the jth control source this is expressed as (Elliott and Nelson, 1985; Elliott *et al.*, 1987):

$$w_j(k+1) = w_j(k) - 2\mu\sum_{i=1}^{N_e} f_{i,j}(k) e_i(k) \tag{6.9.11}$$

If all of the control sources are to adapted simultaneously, the algorithm can be expressed as

$$\left[w_1(k+1) \;|\; \cdots \;|\; w_{N_c}\right] = \left[w_1(k) \;|\; \cdots \;|\; w_{N_c}(k)\right] - 2\mu\sum_{i=1}^{N_e}\left[f_{i,1}(k) \;|\; \cdots \;|\; f_{i,N_c}(k)\right] e_i(k) \tag{6.9.12}$$

The algorithm of (6.9.11) can also be easily restated in a complex form for implementation in the frequency domain as

$$w_j(k+1) = w_j(k) - 2\mu\sum_{i=1}^{N_e} f_{i,j}^*(k) e_i(k) \tag{6.9.13}$$

where * denotes complex conjugate.

As outlined in the previous section for the single input, single output case, in a practical system estimates of the cancellation path transfer functions must be used in calculating the filtered reference signal. Therefore, for the jth weight coefficient vector, the practical implementation of the algorithm is

$$w_j(k+1) = w_j(k) - 2\mu\sum_{i=1}^{N_e} \hat{f}_{i,j}(k) e_i(k) \tag{6.9.14}$$

where ^ denotes estimated value.

6.9.2 Solution for the optimum set of weight coefficient vectors

Before considering the problem of algorithm stability, it will again be worthwhile to derive an expression for the set of optimum weight coefficient vectors. To do this we will need to restate the MIMO filtered-x LMS algorithm with explicit consideration of the mean square error criterion of (6.9.7), rederiving an expression for the gradient term in the algorithm with reference to this quantity. With this aim, (6.9.7) can be expanded using (6.9.1) and (6.9.5) as

$$\Xi = \sum_{i=1}^{N_e} E\left\{ \left[p_i(k) + \sum_{j=1}^{N_c} f_{i,j}^{\mathrm{T}}(k) w_j \right]^2 \right\} \tag{6.9.15}$$

or

$$\Xi = \sum_{i=1}^{N_e} \left[E\{p_i^2(k)\} + 2\sum_{j=1}^{N_c} a_{i,j}^{\mathrm{T}} w_j + \sum_{j=1}^{N_c} \sum_{\iota=1}^{N_c} w_j^{\mathrm{T}} \; E\{f_{i,j}(k) f_{i,\iota}^{\mathrm{T}}(k)\} \; w_\iota \right] \tag{6.9.16}$$

where $a_{i,j}$ is the cross-correlation vector between the primary source signal component of the ith error signal and the i, jth filtered reference signal:

$$a_{i,j} = E\{p_i(k) f_{i,j}(k)\} \tag{6.9.17}$$

The total mean square error, as defined in (6.9.7), can be written as

$$\Xi = E\{e^{\mathrm{T}}(k) e(k)\} \tag{6.9.18}$$

where $e(k)$ is an $(N_e \times 1)$ vector of error signals supplied to the control system at time k,

$$e(k) = \left[e_1(k) \;\; e_2(k) \;\; \cdots \;\; e_{N_e}(k) \right]^{\mathrm{T}} \tag{6.9.19}$$

This can be expressed as

$$\Xi = w_a^{\mathrm{T}} B_a w_a + a_a^{\mathrm{T}} w_a + w_a^{\mathrm{T}} a_a + d \tag{6.9.20}$$

where w_a is the vector of weight coefficients in all N_c control source filters:

$$w_a = \left[w_1^{\mathrm{T}} \mid w_2^{\mathrm{T}} \mid \cdots \mid w_{N_c}^{\mathrm{T}} \right]^{\mathrm{T}} \tag{6.9.21}$$

B_a is the autocorrelation matrix of the (total) filtered reference signal matrix F_a:

$$B_a = E\left\{ F_a(k) F_a^{\mathrm{T}}(k) \right\} \tag{6.9.22}$$

where F_a is the matrix of filtered reference signal vectors:

$$F_a(k) = \begin{bmatrix} f_{1,1}(k) & f_{1,2}(k) & \cdots & f_{1,N_c}(k) \\ f_{2,1}(k) & f_{2,2}(k) & \cdots & f_{2,N_c}(k) \\ & & \vdots & \\ f_{N_e,1}(k) & f_{N_e,2}(k) & \cdots & f_{N_e,N_c}(k) \end{bmatrix} \tag{6.9.23}$$

a_a is a cross-correlation vector, given by

$$a_a^{\mathrm{T}} = \left[\sum_{i=1}^{N_e} a_{i,1}^{\mathrm{T}} \;\Big|\; \sum_{i=1}^{N_e} a_{i,2}^{\mathrm{T}} \;\Big|\; \cdots \;\Big|\; \sum_{i=1}^{N_e} a_{i,N_c}^{\mathrm{T}} \right] \tag{6.9.24}$$

d is the total primary signal power, given by

$$d = E\{p^{\mathrm{T}}(k) p(k)\} \tag{6.9.25}$$

and $\boldsymbol{p}(k)$ is the vector of primary source signals at the output of the error sensors:

$$\boldsymbol{p}(k) = \left[p_1(k)\ p_2(k)\ \cdots\ p_{N_e}(k)\right]^{\mathrm{T}} \tag{6.9.26}$$

We can obtain a relationship defining the optimum set of weight coefficient vectors by differentiating Ξ as expressed in (6.9.20) with respect to the augmented weight coefficient vector, and setting the result equal to zero. Differentiating (6.9.20) yields

$$\frac{\partial \Xi}{\partial \boldsymbol{w}_a} = 2\boldsymbol{B}_a \boldsymbol{w}_a + 2\boldsymbol{a}_a \tag{6.9.27}$$

from which the optimum augmented weight coefficient vector is found to be

$$\boldsymbol{w}_{a,\mathrm{opt}} = -\boldsymbol{B}_a^{-1}\boldsymbol{a}_a \tag{6.9.28}$$

Note that this is simply the multichannel version of the result derived previously in (6.8.28), for the optimal frequency response characteristics of a set of control sources expressed in terms of signal power and cross–spectral densities. Substituting this optimum weight coefficient vector back into (6.9.20), the minimum mean square error is found to be

$$\Xi_{\mathrm{min}} = d + \boldsymbol{a}_a^{\mathrm{T}} \boldsymbol{w}_{a,\mathrm{opt}} \tag{6.9.29}$$

6.9.3 Solution for a single optimum weight coefficient vector

In addition to the 'complete' analysis of the previous section, it is informative to consider the optimum weight coefficients of a single control filter, where the effects which the control sources, 'coupled' through the structural/acoustic system, have upon each other becomes explicit. The optimum jth weight coefficient vector $\boldsymbol{w}_{j,\mathrm{opt}}$ is found by differentiating (6.9.9) with respect to the weight coefficient vector and setting the resulting gradient expression equal to zero. Before doing this, it will be advantageous to expand (6.9.9) as

$$\Xi = \sum_{i=1}^{N_e} \left[E\{p_i^2(k)\} + 2\sum_{j=1}^{N_c} \boldsymbol{a}_{i,j}^{\mathrm{T}} \boldsymbol{w}_j + \sum_{j=1}^{N_c}\sum_{q=1}^{N_c} \boldsymbol{w}_j^{\mathrm{T}}\, E\{\boldsymbol{X}(k)\boldsymbol{c}_{i,j}\boldsymbol{c}_{i,q}^{\mathrm{T}}\boldsymbol{X}^{\mathrm{T}}(k)\}\, \boldsymbol{w}_q \right] \tag{6.9.30}$$

This can be rewritten as

$$\Xi = \sum_{i=1}^{N_e} \left[E\{p_i^2(k)\} + 2\sum_{j=1}^{N_c} \boldsymbol{a}_{i,j}^{\mathrm{T}} \boldsymbol{w}_j + N\sum_{j=1}^{N_c}\sum_{q=1}^{N_c} \boldsymbol{w}_j^{\mathrm{T}} \boldsymbol{RT}_{i,j/q} \boldsymbol{w}_q \right] \tag{6.9.31}$$

where the subscript q refers to control source q, N is the number of taps in the filter, and

$$N\boldsymbol{R} = E\{\boldsymbol{X}(k)\boldsymbol{X}^{\mathrm{T}}(k)\} \tag{6.9.32}$$

where $\boldsymbol{R}$ is the autocorrelation matrix of the reference signal, defined previously as

$$\boldsymbol{R} = E\{\boldsymbol{x}(k)\boldsymbol{x}^{\mathrm{T}}(k)\} \tag{6.9.33}$$

Technically, the transformation matrix $\boldsymbol{T}_{i,j/q}$ in (6.9.31) is defined by the expression

$$\boldsymbol{T}_{i,j/q} = \boldsymbol{R}^{-1}\, E\{\boldsymbol{f}_{i,j}(k)\boldsymbol{f}_{i,q}{}^{\mathrm{T}}(k)\} \tag{6.9.34}$$

However, we can use some of the results from the SISO system examination to derive an expression for $\boldsymbol{T}_{i,j/q}$ for two common reference signals. If the reference signal is sinusoidal then a similar analysis as that leading to (6.8.47) shows

$$\boldsymbol{T}_{i,j/q} = T_{i,j/q}\boldsymbol{I} \tag{6.9.35}$$

where $T_{i,j/q}$ is a scalar number defined by the expression

$$T_{i,j/q} = \sum_{\iota=0}^{m-1}\sum_{n=0}^{m-1} c_{i,j,\iota}c_{i,q,n}\cos\{(i-j)\phi\} \tag{6.9.36}$$

where $c_{i,j,\iota}$ is the ιth coefficient in the finite impulse response model of the cancellation path transfer between the jth control filter and the ith error signal, and ϕ is the angular increment of the reference sinusoid during each sample period. If the reference signal is random noise, a similar analysis as that leading to (6.8.50) shows that the (α,β)th term of $\boldsymbol{T}_{i,j/q}$ is equal to

$$T_{i,j/q}(\alpha,\beta) = \begin{cases} \sum_{\iota=0}^{m-1} c_{i,j,\iota}c_{i,q,\iota-(\alpha-\beta)} & 0\le\iota-(\alpha-\beta)\le m \\ 0 & \text{otherwise}\end{cases} \tag{6.9.37}$$

Heuristically, $\boldsymbol{T}_{i,j/q}$ can be thought of as providing a measure of the orthogonality of the transfer functions between error sensor i and control sources j and q, as it can be written as

$$\boldsymbol{T}_{i,j/q} = \left[\boldsymbol{t}_{i,j}{}^{\mathrm{T}}\boldsymbol{t}_{i,q}\right]\boldsymbol{I} + \text{additional terms} \tag{6.9.38}$$

where $\boldsymbol{I}$ is the identity matrix. If the two transfer functions of interest are orthogonal then the first quantity in (6.9.38) will be equal to zero.

Differentiating (6.9.31), the gradient expression for the jth weight coefficient vector is

$$\frac{\partial \Xi}{\partial \boldsymbol{w}_j} = \sum_{i=1}^{N_e}\left[\sum_{q=1}^{N_c} 2\boldsymbol{R}\boldsymbol{T}_{i,j/q}\boldsymbol{w}_q + 2\boldsymbol{a}_{i,j}\right] \tag{6.9.39}$$

Setting (6.8.28) equal to zero produces the expression for the jth optimum weight coefficient vector (Snyder and Hansen, 1992)

$$\boldsymbol{w}_{j,\mathrm{opt}} = \sum_{i=1}^{N_e} -(\boldsymbol{R}\boldsymbol{T}_{i,j/j})^{-1}(\boldsymbol{a}_{i,j} + \sum_{q=1,\neq j}^{N_c} \boldsymbol{T}_{i,j/q}\boldsymbol{w}_q) \tag{6.9.40}$$

The influence of coupling in the system is readily apparent in (6.8.29): the first term in the expression is that which would be optimal in the absence of all other sources, and the second term is the modification due to the coexistence of the other control sources.

6.9.4 Stability and convergence of the MIMO filtered-x LMS algorithm

Having derived expressions for the optimum weight coefficient in an MIMO active control system employing FIR control filters, we are in a position to examine the convergence properties of the algorithm and how they are affected by the structural/acoustic system to which the control system is attached. As was done in examining the SISO implementation, we can view the total mean square error (criterion) Ξ as comprised of two parts; the minimum (total) mean square error Ξ_{min}, which is the smallest value of mean square error that an ideal controller is capable of attaining, and the excess (total) mean square error Ξ_{ex}, that part of the error criterion which the controller can reduce,

$$\Xi = \Xi_{min} + \Xi_{ex} \tag{6.9.41}$$

The behaviour of the algorithm during the convergence process, including whether the adaptation is stable or unstable, can be characterized by examining the change in the excess mean square error. In this section we will examine the convergence of the algorithm, neglecting the 'delayed adaptation' part of the problem and concentrating the analysis on the effects of other system variables such as control source and error sensor number. A discussion of the influence of time delays on a continuously adapting system can be found in the previous section for the SISO implementation of the system. In the analysis we will again concentrate on mean weight convergence, taking expected values of system variables. While this will produce overly optimistic absolute results for system stability, it will simplify the analysis to enable elucidation of the influence which many physical system parameters have upon the stability of the algorithm.

The first quantity we will study is the excess mean square error, which is equivalent to examining the stability of the MIMO filtered-x LMS algorithm. The expression for the excess mean square error, which is that part of the sum of the squared signals from the N_e error transducers which exceeds the minimum (obtainable) mean square error, can be found by subtracting the minimum mean square error of (6.9.29) from the total mean square error as defined in (6.9.20). By using the expression for the optimum weight coefficient vector, given in (6.9.28), this results in

$$\Xi_{ex} = \Xi - \Xi_{min} = \boldsymbol{v}_a^{\mathrm{T}} \boldsymbol{B}_a \boldsymbol{v}_a \tag{6.9.42}$$

where the augmented weight coefficient error vector $\boldsymbol{v}_a$ is defined as the difference between the (augmented) optimum weight coefficient vector and the current weight coefficient values:

$$\boldsymbol{v}_a = \boldsymbol{w}_a - \boldsymbol{w}_{a,\text{opt}} \tag{6.9.43}$$

As we have seen in previous section, excess mean square error is directly related to the values in the weight error vector. For the algorithm to be converging towards the optimum weight coefficient values the terms in the weight error vector, and hence the excess mean square error, must be converging towards zero.

To examine the convergence characteristics of the algorithm it will be advantageous to first decouple (6.9.42). As the augmented filtered input autocorrelation matrix $\boldsymbol{B}_a$ is symmetric, it may be diagonalized by the orthonormal transformation

$$\boldsymbol{\Lambda}_a = \boldsymbol{Q}^{-1}\boldsymbol{B}_a\boldsymbol{Q} \tag{6.9.44}$$

where $\boldsymbol{Q}$ is the orthonormal matrix of $\boldsymbol{B}_a$ (the columns of which are the eigenvectors of $\boldsymbol{B}_a$), and $\boldsymbol{\Lambda}_a$ is the diagonal matrix of eigenvalues of $\boldsymbol{B}_a$. Note again that the terms, and hence eigenvalues, of $\boldsymbol{B}_a$ are related to the filtered reference signals $\boldsymbol{f}$ rather than simply the reference signal $\boldsymbol{x}$ used in the derivation of the control signals. Premultiplying all vectors in (6.8.37) by the transpose of the orthonormal matrix will decouple the equation

$$\Xi_{\text{ex}} = \boldsymbol{v}_a'^{\text{T}}\boldsymbol{\Lambda}_a\boldsymbol{v}_a' \tag{6.9.45}$$

where

$$\boldsymbol{v}_a' = \boldsymbol{Q}^{-1}\boldsymbol{v}_a = \boldsymbol{Q}^{\text{T}}\boldsymbol{v}_a \tag{6.9.46}$$

As (6.9.45) is decoupled, it can be viewed as a set of scalar equations, each of which must converge to a set value (ideally zero) over time for the algorithm to be stable. As the MIMO filtered-x LMS algorithm adapts, the rth scalar equation (in the absence of any time delay) is

$$v_r'(k+1) = v_r'(k) - 2\mu\lambda_r v_r'(k) \tag{6.9.47}$$

or

$$v_r'(k+1) = v_r'(k)(1 - 2\mu\lambda_r) \tag{6.9.48}$$

For the excess mean square to converge to some finite value over time, the bracketed part of (6.9.48) must be of magnitude less than one, or

$$\left| \frac{v_r'(k+1)}{v_r'(k)} \right| < 1 \tag{6.9.49}$$

Therefore, for the excess mean square error to converge to some finite value over time, the convergence coefficient, μ, must be bounded by

$$0 < \mu < \frac{1}{\lambda_r} \tag{6.9.50}$$

Thus, as with the SISO system, the stability of the algorithm is dependent upon the eigenvalues of the autocorrelation matrix of the filtered reference signal. We can therefore qualitatively assess the influence of physical system parameters by examining the influence they have upon $\boldsymbol{B}_a$. The specific factors of interest in this section are the number of control sources and error sensors, and the magnitude of the cancellation path transfer functions.

Let us first consider the effect which adding additional error sensors into the system has upon the stability of the algorithm. To do this, it will prove advantageous to define an augmented filtered reference signal vector $\boldsymbol{f}_{ai}$ containing the filtered reference signal vectors derived from consideration of the transfer function between each control filter and the output of the ith error sensor:

$$\boldsymbol{f}_{ai}(k) = \left[\boldsymbol{f}_{i,1}^{\mathrm{T}}(k) \ | \ \boldsymbol{f}_{i,2}^{\mathrm{T}}(k) \ | \ \cdots \ | \ \boldsymbol{f}_{i,N_c}^{\mathrm{T}}(k) \right]^{\mathrm{T}} \tag{6.9.51}$$

such that the matrix of filtered reference signal vectors defined in (6.9.23) can be written as

$$\boldsymbol{F}_a(k) = \left[\boldsymbol{f}_{a1}(k) \ | \ \boldsymbol{f}_{a2}(k) \ | \ \cdots \ | \ \boldsymbol{f}_{aN_e}(k) \right]^{\mathrm{T}} \tag{6.9.52}$$

In terms of this vector, the (ι,j)th term of $\boldsymbol{B}_a$ is equal to

$$\boldsymbol{B}_a(\iota,j) = E\left\{ \sum_{i=1}^{N_e} f_{ai}(\iota,k) f_{ai}(j,k) \right\} \tag{6.9.53}$$

where $f_{ai}(\iota,k)$ and $f_{ai}(j,k)$ are the ιth and jth terms of the ith augmented filtered reference signal vector at time k. Written in this way it becomes clear that the addition of error sensors into the system will increase the value of the elements of $\boldsymbol{B}_a$ (reflected by an increase in the numbers in the summation of (6.9.53)) and hence its eigenvalues, thereby decreasing the maximum stable value of the convergence coefficient by an amount proportion to this increase. The increase in the value of the summation caused by the addition of an error sensor will be dependent upon the squared amplitude of the transfer function between the control sources and the new error sensor. A typical relationship is illustrated in Fig. 6.46, which depicts the maximum value of convergence coefficient for which stable behaviour is maintained on a simulated single control source system with a variable number of error sensors. In this simulation, the transfer functions between the control source and each error sensor were set to be the same phase and amplitude, as were the transfer functions between the primary source and each error sensor, to eliminate any additional effects. Observe that an inverse proportional relationship clearly exists.

The effect which changing the magnitude of the transfer functions between the control sources and error sensors has upon algorithm stability is now somewhat obvious. From (6.9.53), it can be deduced that as the cancellation path transfer function magnitudes are changed, such as by relocation of control sources or error sensors or by a change in the amplification of the signals, the magnitudes of the elements of $\boldsymbol{B}_a$ and hence its eigenvalues, are altered by a proportional squared relationship. It follows that this will cause an inverse proportional squared change in the stability of the algorithm. This effect is the same as was outlined for the single input, single output case in the previous section, and illustrated in Fig. 6.22. The only difference in the MIMO implementation is that the effect of a change in one cancellation path transfer function magnitude is moderated to some degree by the transfer functions of the other cancellation paths, as evidenced by the summation in the expression for the individual terms in $\boldsymbol{B}_a$ given in (6.8.48). An MIMO system has greater robustness to destabilising changes in individual components of the system because of this.

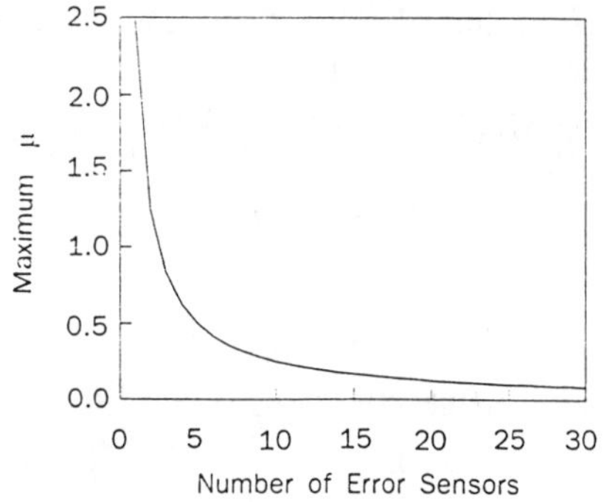

Fig. 6.46 Typical plot of maximum stable value of convergence coefficient as a function of the number of error sensors in the system.

To examine the effect which the control sources, 'coupled' together through the structural/acoustic system, have upon each other, let us consider now the adaptation of a single weight coefficient vector. With the assumption that the models of the cancellation path transfer functions are exact, expanding the sampled error term of (6.9.12) produces, for the jth control source,

$$\boldsymbol{w}_j(k+1) = \boldsymbol{w}_j(k) - 2\mu \sum_{i=1}^{N_e} \boldsymbol{f}_{i,j}(k) \left[p_i(k) + \sum_{q=1}^{N_c} \boldsymbol{f}_{i,q}^{\mathrm{T}}(k) \boldsymbol{w}_q(k) \right] \tag{6.9.54}$$

It is useful here to partition the primary source disturbance measurement $p_i(k)$ provided by the ith error sensor into two components: the 'excess', $p_{\mathrm{ex},i}(k)$, which will be 'cancelled' when all weight coefficients in the system are equal to their optimum value, and the 'minimum' component $p_{\mathrm{min},i}$ which will not be cancelled. The part of the primary source disturbance which will be cancelled can be further partitioned into components $p_{\mathrm{ex},i,j}(k)$ 'assigned' to each control source. For the purposes of analysis, the primary source component assigned to each control source is equivalent in amplitude, and opposite in phase, to the control signal sensed at each microphone if under final optimized conditions the primary source and all other control sources were switched off. Thus, the measured primary source disturbance provided by the ith error sensor can be represented as

$$p_i(k) = \sum_{j=1}^{N_c} p_{\mathrm{ex},i,j}(k) + p_{\mathrm{min},i}(k) \tag{6.9.55}$$

where

$$p_{\mathrm{ex},i,j}(k) = -\boldsymbol{f}_{i,j}^{\mathrm{T}}(k)\boldsymbol{w}_{j,\mathrm{opt}} \tag{6.9.56}$$

Substituting (6.9.55) and (6.9.56) into (6.9.11) yields the expression

$$\boldsymbol{w}_j(k+1) = \boldsymbol{w}_j(k) - 2\mu\sum_{i=1}^{N_e}\boldsymbol{f}_{i,j}(k)\left[p_{\min,i}(k) + \sum_{q=1}^{N_c}\left[p_{i,q}(k) + \boldsymbol{f}_{i,q}^{\mathrm{T}}(k)\boldsymbol{w}_q(k-n)\right]\right]$$

$$= \boldsymbol{w}_j(k) - 2\mu\sum_{i=1}^{N_e}\left[\boldsymbol{f}_{i,j}(k)\sum_{q=1}^{N_c}\boldsymbol{f}_{i,q}(k)(\boldsymbol{w}_q(k) - \boldsymbol{w}_{q,\mathrm{opt}}) + \boldsymbol{f}_{i,j}(k)p_{\min,i}(k)\right] \tag{6.9.57}$$

Equation (6.9.57) can be re-expressed in terms of the weight error vector as

$$\boldsymbol{v}_j(k+1) = \boldsymbol{v}_j(k) - 2\mu\sum_{i=1}^{N_e}\left[\boldsymbol{f}_{i,j}(k)\sum_{q=1}^{N_c}\boldsymbol{f}_{i,q}^{\mathrm{T}}(k)\boldsymbol{v}_q(k) + \boldsymbol{f}_{i,j}(k)p_{\min,i}(k)\right] \tag{6.9.58}$$

where

$$\boldsymbol{v}_j(k) = \boldsymbol{w}_j(k) - \boldsymbol{w}_{j,\mathrm{opt}} \tag{6.9.59}$$

Taking expected values to examine mean weight convergence, (6.9.58) becomes

$$\boldsymbol{v}_j(k+1) = \boldsymbol{v}_j(k) - 2\mu\sum_{i=1}^{N_e}\left[\sum_{q=1}^{N_c}E\{\boldsymbol{f}_{i,j}(k)\boldsymbol{f}_{i,q}^{\mathrm{T}}(k)\}\ \boldsymbol{v}_q(k)\right] \tag{6.9.60}$$

It should be noted here that the individual quantities $E\{\boldsymbol{f}_{i,j}(k)\ p_{\min,i}\}$ may be correlated for the MIMO case, as the residual signal from an single error sensor may be correlated with the reference signal (especially when the number of error sensors exceeds the number of control sources). However,

$$E\{\ \sum_{i=1}^{N_e}\boldsymbol{f}_{i,j}(k)p_{\min,i}(k)\ \} = 0 \tag{6.9.61}$$

Physically, this means that cancellation at any single error sensor in a multi-error system may not be complete, even when the control system is optimal.

In terms of the transformation $\boldsymbol{T}_{i,j/q}$ a transformed weight error vector can be defined as

$$\tilde{\boldsymbol{v}}_{j/q}(k) = \boldsymbol{T}_{i,j/q}\boldsymbol{v}_q(k) \tag{6.9.62}$$

enabling (6.9.60) to be expressed as

$$\boldsymbol{v}_j(k+1) = \boldsymbol{v}_j(k) - 2\mu\boldsymbol{R}\sum_{i=1}^{N_e}\sum_{q=1}^{N_c}\tilde{\boldsymbol{v}}_{j/q}(k) \tag{6.9.63}$$

To examine the convergence behaviour of (15.3.54), which will govern the convergence behaviour of the excess mean square error (hence algorithm stability), the equation must first be decoupled. As before, we can employ an orthonormal transformation matrix $\boldsymbol{Q}$ to accomplish this, where $\boldsymbol{Q}$ is now defined by the expression

$$\Lambda = \boldsymbol{Q}^{-1}\boldsymbol{R}\boldsymbol{Q} = \boldsymbol{Q}^{\mathrm{T}}\boldsymbol{R}\boldsymbol{Q} \tag{6.9.64}$$

where Λ is the diagonal matrix of eigenvalues of the reference signal autocorrelation matrix $\boldsymbol{R}$ (the columns of $\boldsymbol{Q}$ are the eigenvectors of $\boldsymbol{R}$). Premultiplying equation (6.8.58) by $\boldsymbol{Q}^{\mathrm{T}}$ produces

$$v'_{i,j}(k+1) = v'_{i,j}(k) - 2\mu\Lambda\sum_{i=1}^{N_e}\sum_{q=1}^{N_c}\tilde{v}'_{j/q}(k) \tag{6.9.65}$$

where

$$v' = \boldsymbol{Q}^{-1}v = \boldsymbol{Q}^{\mathrm{T}}v \tag{6.9.66}$$

As (6.9.65) is decoupled, it can be viewed as a set of independent scalar equations. For algorithm stability to be maintained, the sum of the N_e scalar equations for any given transformed weight error v_j' must converge to some finite value over time. Therefore, for the rth scalar equation

$$\left|\sum_{i=1}^{N_e}\frac{v'_{r,i,j}(k+1)}{v'_{r,i,j}(k)}\right| < 1 \tag{6.9.67}$$

From (6.9.65) this sets bounds on the convergence coefficient of (Snyder and Hansen, 1992)

$$0 < \mu < \left[\lambda_r\sum_{i=1}^{N_e}\sum_{q=1}^{N_c}\frac{\tilde{v}'_{r,j/q}(k)}{v'_{r,i,j}(k)}\right]^{-1} \tag{6.9.68}$$

where λ_r is the eigenvalue of the scalar equation of interest.

The influence attributable to the coupling of control sources through the structural/acoustic system can be deduced by considering the result of (6.9.68). The last term in this equation is related to the sum of the errors of the jth weight coefficient of interest to the other corresponding orthonormal transformed weight coefficients in the system. As described previously, the transformation of (6.9.66) can be viewed as providing a measure of the independence of these vector spaces. If, at one extreme, all of the transfer functions between each control source and error microphone were orthogonal, then the value of the summation would be relatively small, and so the system stability would be enhanced. At the other extreme, if all of the cancellation path transfer functions were the same, then the value of the summation would be relatively large, and the maximum stable value of the convergence coefficient would be reduced. This conclusion, that orthogonality of control source placement in terms of the structural/acoustic modal response has an effect on the maximum allowable convergence coefficient for system stability, would seem intuitively obvious; that it is predicted explicitly by the algorithm stability bounds is therefore not surprising.

These effects are illustrated in Fig. 6.47, which depicts the maximum stable value of convergence coefficient for a 10 error sensor system as a function of the number of control sources in the system. In this simulation, the magnitude of the transfer function between each control source and error sensor was taken as equal. As we would expect for this case, the maximum stable value of convergence coefficient is reduced by an amount proportional to the inverse of the number of control sources. If this same case were rerun, where each control source could be sensed at only one of the error sensors, it would be found that the stability of the system would be greatly enhanced, and that including additional control sources would have no effect. Viewing this case objectively, however, would indicate that it is almost trivial, being equivalent to 10 SISO systems running independently.

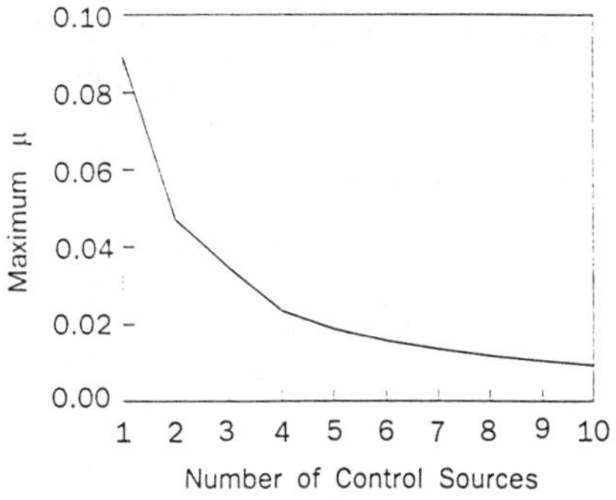

Fig. 6.47 Typical plot of maximum stable value of convergence coefficient as a function of the number of control sources in the system.

If we continue with this line of discussion, there is another aspect of the transformed weight error ratio terms which must be taken into account in system design. In a system with many control sources and error sensors, where convergence time is not of major concern, it may be tempting to update the weight coefficient vectors on a 'round robin' basis, one weight coefficient vector at a time, to save on hardware costs. If the control source placement is such that the active control problem is effectively underdetermined, there may not be a unique set of optimum weight coefficient vectors. Examples of where this situation may occur are where there are more control sources than error sensors, where there are more control sources and error sensors than structural/acoustic modes contributing (significantly) to the system response, and where there are more control sources than required to 'completely' attenuate the primary disturbance. For these cases if all of the weight coefficient vectors are adjusted simultaneously, the algorithm will inherently try to find a solution which requires the least amount of overall weight coefficient adjustment from the initial values. In doing this, it tends to divide up the overlapping parts of the primary source disturbance which can be 'cancelled' by a number of control sources. However, if one control source is adjusted at a time, the algorithm tries to adjust the first weight coefficient vector to control all of the primary source disturbance which it possibly can before beginning to adjust the next weight coefficient vector; that is, there is no division of the overlapping parts of the primary source components, but rather an 'all or nothing' solution.

This has two obvious implications for active control system design, the first of which is related to control effort. Updating a multichannel control system on a round robin basis could easily result in one control source being overdriven while others are hardly driven at all. Thus, in this case, control effort would need to be included in the criteria used to decide when a control source is adjusted sufficiently, and the algorithm begins to adjust the next control source in the round robin, so that no control sources are overdriven. These effects are evident in the data shown in Figs. 6.48(a) and 6.48(b), which depict the convergence path of the weight coefficients in a two control source, two error sensor system using both round robin and simultaneous weight coefficient vector adaptation. For these cases, the final value of the (minimum) mean square error was the same. In viewing these plots it is clear that, if unchecked, round robin adaptation can lead to control sources being overdriven.

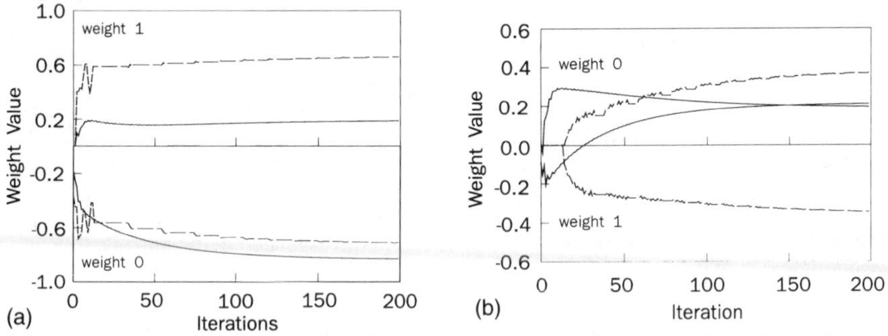

Fig. 6.48 Evolution of filter weights in a two source, two sensor system, with continuous (solid line) and round robin (dashed line) adaptation schemes: (a) filter 1; (b) filter 2.

A second implication of this 'division of labour' effect is related to algorithm stability. Considering the bounds placed on the convergence coefficient for algorithm stability given in (15.3.59), the weight error ratio can be written as (Snyder and Hansen, 1992):

$$\left[\sum_{j=1}^{N_c} \frac{\tilde{v}^{\prime}_{j/q}(k)}{v^{\prime}_j(k)}\right]^{-1} = \left[\frac{\tilde{v}^{\prime}_{j/j}(k-n)}{v^{\prime}_j(k)} + \sum_{j=1,\neq q}^{N_c} \frac{\tilde{v}^{\prime}_{j/q}(k)}{v^{\prime}_j(k)}\right]^{-1} \tag{6.9.69}$$

The first term on the right-hand side of (6.8.64) will be approximately the same, regardless of whether one or all weight coefficient vectors are adapted at the same time. The second term, however, will differ. This is because if the overlapping parts of the primary source components are not divided up, the final (optimized) output of the control sources which are updated later in the round robin procedure will have a reduced output (from the all or nothing primary source component division) compared to the output they would have had if all weight coefficient vectors are updated simultaneously (thus dividing up the overlapping primary source components). Therefore, if the initial weight coefficients are all set to a value of zero (as is commonly done), the errors in the weight coefficients adapted later in the round robin will be reduced (because their final values will be smaller due to their later position in the round robin procedure). As a result, the summation in (6.9.69) will be smaller than it would have been if all filters were adjusted simultaneously, and (6.9.68) shows that this will enhance system stability by an amount dependent upon the characteristics of the cancellation path transfer functions.

6.9.5 The effect of transfer function estimation errors upon algorithm stability

The effect of transfer function estimation errors upon the stability of the MIMO filtered-x LMS algorithm is more complex to examine than the SISO algorithm considered in the previous section, owing to the coupling of the control sources and error sensors through the structural/acoustic operating environment. This coupling, however, adds an element of 'forgiveness' to individual 'outlier' errors, arising from very poor transfer function estimates, not present for the SISO algorithm implementation. In fact, the trend of effects of transfer function estimation errors for the MIMO algorithm implementation can be viewed as generally the same as for the SISO system, but with the addition of 'averaging' between the channels.

Consider first the effect of transfer function magnitude estimation error. In viewing the analysis of (6.9.61)–(6.9.68), it can be deduced that an error in magnitude estimation will change the transformed weight error vector of (6.9.63), increasing (or decreasing) the size of its elements. From (6.9.68) it can be deduced that such an effect will reduce the maximum stable value of convergence coefficient. However, the reduction will not be proportional, as it will be 'averaged out' by the summation over all of the controller channels.

Consider the simplest case of a single control source, two error sensor system. Figure 6.49 depicts the effect which increasing the transfer function magnitude estimation error between the control source and one of the error sensors has upon the maximum stable value of convergence coefficient, for the case where the other error path has no transfer function estimation error, and has a fixed magnitude error of a factor of two (for this data the actual transfer function magnitude between the control source and both error sensors was taken to be equal). Clearly the magnitude estimation error reduces the stability of the system, but the effect is not directly proportional as it was for the SISO system; rather, it has been moderated by the other channel.

Now turn to the effects of phase estimation error, specifically for the case of a tonal (sine wave) disturbance and a time domain implementation of the control system. As with the magnitude estimation error it effects will be 'averaged out'. Consider again the simple case of a single control source, two error sensor system, where the is no structural/acoustic propagation time delay and no magnitude estimation error, and the control source, which has two weight coefficients, is being used to control a sinusoidal disturbance. Conducting an

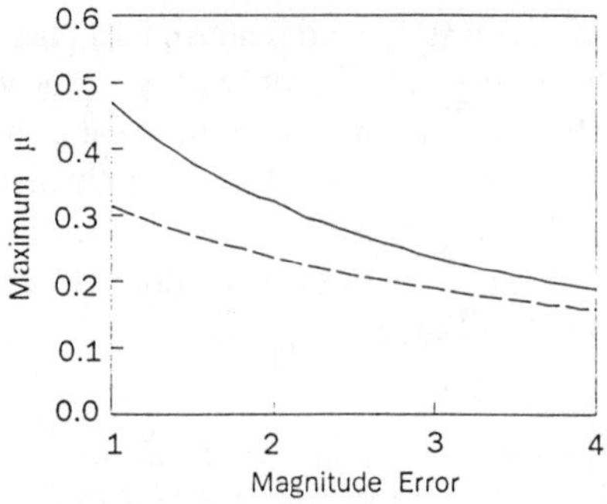

Fig. 6.49 Plot of maximum stable value of convergence coefficient as a function of an error in the estimate of the amplitude response of the cancellation path for a one source, two sensor system. Data shown in solid line is for a perfect estimate in one channel, axes value of error in the other. Data shown in dashed line are for an magnitude estimation error of two in one channel, axes value of error in the other.

analysis similar to that in Section 6.8.6 for an SISO arrangement, it can be shown that the eigenvalues of the phase errored filtered input autocorrelation matrix are

$$\lambda_e = \frac{1}{2}\left((\cos(\phi_1)+\cos(\phi_2)) \pm \sqrt{\cos^2\gamma(\cos(\phi_1)+\cos(\phi_2))^2 - \sin^2\gamma(\sin(\phi_1)+\sin(\phi_2))^2}\right) \tag{6.9.70}$$

where ϕ_1, ϕ_2 are the phase estimation errors of the transfer functions between the control source and the first and second error sensor, respectively, and γ is the angular increment occuring with each new sample. It is clear in (6.9.70) that the phase errors are 'averaged'. Figure 6.50 illustrates the maximum stable value of convergence coefficient for the system outlined above, plotted as a function of the phase estimation error of one of the transfer functions, where the phase estimation error of the other is fixed at 30° and the sampling rate is five times the excitation frequency (this is directly comparable to Fig. 6.34 for the SISO case). Here the system is stable in the region encompassed by a $\pm 150°$ phase error, a region which is found to expand or contract linearly with decreasing or increasing phase error in the other error path. This trend can also be found in frequency domain implementations of the algorithm, where the variation in maximum stable value of convergence coefficient is a smooth function of phase error, as was the case in the SISO system.

Therefore, to summarize conservatively, it can be said that the MIMO filtered-x LMS algorithm can be made stable if the errors in the estimation of the phase of the cancellation path transfer functions are all within the bounds of $\pm 90°$.

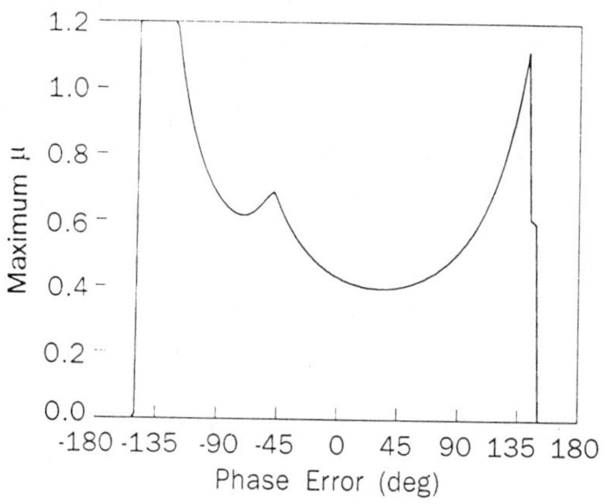

Fig. 6.50 Maximum stable value of convergence coefficient for a single source, two sensor system plotted as a function of phase estimation error in one change, with a phase estimation error of 30° in the other, time domain algorithm implementation.

6.9.6 Convergence properties of the control system

In this section we have examined the stability of the MIMO filtered-x LMS algorithm, and to some degree the effect which the physical system to which it is attached has upon stability. In addition to influencing algorithm stability, the physical system also influences the behaviour of the algorithm as it converges (in a stable fashion) to its steady state parameters. We will finish this section by quantifying this influence. Perhaps the simplest way to do this is to examine the convergence behaviour of a frequency domain version of the algorithm with tonal excitation.

Consider an adaptive feedforward control system, where the controller is adapted to minimise an error criterion which has an optimal control-like form, being a function of both the measured error signal and the control signal magnitude:

$$J = \boldsymbol{e}^{\mathrm{H}}\boldsymbol{e} + \alpha \boldsymbol{y}^{\mathrm{H}}\boldsymbol{y} \tag{6.9.71}$$

Here $\boldsymbol{e}$ is an $(N_e \times 1)$ vector of error signals provided by N_e error sensors, $\boldsymbol{y}$ is the $(N_c \times 1)$ vector of outputs from the N_c control filters, and α is a (small) scalar. We have used this form of error criterion previously in Section 6.6 to derive the leaked LMS algorithm.

As the analysis here is in the frequency domain, and the excitation is assumed to be tonal, the vector of error signals can be expressed as

$$\boldsymbol{e} = \boldsymbol{p} + \boldsymbol{C}\boldsymbol{y} \tag{6.9.72}$$

where $\boldsymbol{p}$ is the vector of primary source signals as measured by the error sensors, and $\boldsymbol{C}$ is an $(N_e \times N_c)$ matrix of (cancellation path) transfer functions between the control filter outputs and error signals. In this instance, this latter matrix contains complex numbers which describe the gain and phase changes (at the excitation frequency) between the control filter outputs and the error signals. Using (6.9.72), the system error criterion of (6.9.71) can be expanded as

$$J = \boldsymbol{y}^{\mathrm{H}}\left[\boldsymbol{C}^{\mathrm{H}}\boldsymbol{C}+\alpha\boldsymbol{I}\right]\boldsymbol{y} + \boldsymbol{y}^{\mathrm{H}}\boldsymbol{C}^{\mathrm{H}}\boldsymbol{p} + \boldsymbol{p}^{\mathrm{H}}\boldsymbol{C}\boldsymbol{y} + \boldsymbol{p}^{\mathrm{H}}\boldsymbol{p} \tag{6.9.73}$$

The form of the system error criterion in equation (6.9.73) will be encountered many times in this book, especially in Chapters 7–10. The important point to note at this stage is that if the matrix $[\boldsymbol{C}^{\mathrm{H}}\boldsymbol{C} + \alpha\boldsymbol{I}]$ is positive definite, there will be a unique (global) minimum in the quadratic error criterion, and hence a unique optimal set of control outputs $\boldsymbol{y}$. If the number of error sensors is greater than or equal to the number of control sources, this will often (usually) be the case.

It should be noted that the analysis in this section is being conducted with reference to the control filter outputs $\boldsymbol{y}$, rather than the filter weights as has been the case thus far in this section. This will facilitate assessment of the effect which the physical system has upon the convergence behaviour of the system as a whole, rather than on individual filter weights. Because of this change, the filtered-x LMS algorithm will need to be restated with explicit reference to the filter output. To do this, the gradient of the error criterion of (6.9.73) with respect to the filter outputs can be calculated as

$$\begin{aligned} \frac{\partial J}{\partial \boldsymbol{y}} &= \frac{\partial J}{\partial \boldsymbol{y}_R} + j\frac{\partial J}{\partial \boldsymbol{y}_I} \\ &= 2\boldsymbol{C}^{H}\boldsymbol{p} + 2\left[\boldsymbol{C}^{\mathrm{H}}\boldsymbol{C} + \alpha\boldsymbol{I}\right]\boldsymbol{y} = 2\boldsymbol{C}^{\mathrm{H}}\boldsymbol{e} + \alpha\boldsymbol{y} \end{aligned} \tag{6.9.74}$$

Therefore, the gradient-descent algorithm which is the frequency domain equivalent to the filtered-x LMS is

$$\boldsymbol{y}(k+1) = (1-\mu\alpha)\boldsymbol{y}(k) - 2\mu\boldsymbol{C}^{\mathrm{H}}\boldsymbol{e}(k) \tag{6.9.75}$$

and the optimum vector of control signals defined by the relationship

$$\boldsymbol{y}_{\mathrm{opt}} = -\left[\boldsymbol{C}^{\mathrm{H}}\boldsymbol{C}+\alpha\boldsymbol{I}\right]^{-1}\boldsymbol{C}^{\mathrm{H}}\boldsymbol{p} \tag{6.9.76}$$

To examine the convergence behaviour of the algorithm of (6.9.75), it will be useful to restate the algorithm in terms of an output error vector, analogous to the weight error vectors we have used previously in the chapter. It will also be advantageous to separate the error criterion into minimum and excess components, where the minimum component is that which is residual when the control outputs are equal to their optimal values. Defining the output error vector $\boldsymbol{v}$ by

$$\boldsymbol{v}(k) = \boldsymbol{y}(k) - \boldsymbol{y}_{\mathrm{opt}} \tag{6.9.77}$$

it is straightforward to write an expression for the evolution of the output error vector as

$$\boldsymbol{v}(k+1) = \left[\boldsymbol{I} - 2\mu(\boldsymbol{C}^{\mathrm{H}}\boldsymbol{C}+\alpha\boldsymbol{I})\right] \boldsymbol{v}(k) \tag{6.9.78}$$

Further, at any time k the value of the error criterion is related to the output error by

$$J(k) = J_{\min} + \boldsymbol{v}^{\mathrm{H}}(k)\left[\boldsymbol{C}^{\mathrm{H}}\boldsymbol{C}+\alpha\boldsymbol{I}\right]\boldsymbol{v}(k) \tag{6.9.79}$$

where $J_{\min}$ is the minimum error criterion value, which is the value of the error criterion when the output is optimal.

Because the matrix quantity in parentheses in (6.9.78) is positive definite and symmetric, it can be diagonalized by an orthonormal transformation as

$$\boldsymbol{C}^{\mathrm{H}}\boldsymbol{C} + \alpha\boldsymbol{I} = \boldsymbol{Q}\boldsymbol{\Lambda}\boldsymbol{Q}^{\mathrm{T}} \tag{6.9.80}$$

As we have discussed previously in this chapter, the orthonormal transformation of (6.9.80) can be used to decouple the algorithm to examine convergence along the principal axes of the error surface. Defining the transformed quantity

$$\boldsymbol{v}' = \boldsymbol{Q}^{-1}\boldsymbol{v} \tag{6.9.81}$$

evolution of the (transformed) output error vector (along the principal axes of the error surface) is seen to be defined by the relationship

$$\boldsymbol{v}'(k+1) = [\boldsymbol{I} - 2\mu\boldsymbol{\Lambda}]\boldsymbol{v}'(k) \tag{6.9.82}$$

or, in terms of the initial value of output error,

$$\boldsymbol{v}'(k) = \left[\boldsymbol{I}-2\mu\boldsymbol{\Lambda}\right]^{k}\boldsymbol{v}'(0) \tag{6.9.83}$$

Observe that this is simply a set of scalar equations, where evolution of the rth equation is

$$v_r'(k) = (1-2\mu\lambda_r)^k v_r'(0) \approx \mathrm{e}^{-2\mu\lambda_r k} v_r'(0) \tag{6.9.84}$$

Evolution of the error criterion is therefore governed by the relationship

$$J(k) = J_{\min} + \boldsymbol{v}'(0)\left[\boldsymbol{I}-2\mu\boldsymbol{\Lambda}\right]^{2k}\boldsymbol{\Lambda}\boldsymbol{v}'(0) \tag{6.9.85}$$

or, alternatively (Elliott *et al.*, 1992):

$$J(k) = J_{\min} + \sum_{i=1}^{N_e} \mid v_i(0) \mid {}^2\lambda_i(1-2\mu\lambda_i)^{2k} \tag{6.9.86}$$

The result of (6.9.86) shows that convergence of the (stable) control system towards its steady state values is governed by the eigenvalues of the matrix $[\boldsymbol{C}^{\mathrm{H}}\boldsymbol{C}+\alpha\boldsymbol{I}]$; the evolution of the error criterion has the form of a sum of a set of modal decays, where the time constant of each 'mode' is a function of the associated eigenvalue (see Elliott *et al.* (1992) for examples). Recall now that terms in the matrix $\boldsymbol{C}$ describe the gain and phase change between the control signal and error signal, where the individual terms are governed by a relationship of the form

$$C(i,j) = \sum_{\iota=1}^{N_m} \frac{\phi_\iota(r_i)\phi_\iota(r_j)}{m_\iota z_\iota} \tag{6.9.87}$$

where $\phi_\iota(r_i)$ is the value of the ιth mode shape function at the location of the ith error sensor, $\phi_\iota(r_j)$ is the value of the ιth mode shape function at the location of the jth control source, m_ι is the ιth modal mass and z_ι is the ιth modal input impedance (which is a function of the ιth resonance frequency). From this discussion, it can be deduced that the speed of convergence is a function of both the geometry of the control system and resonance frequencies of the modes of the physical system, where often the resonance frequencies determine the speed of the fastest modes, and the geometry of the control source and error sensor configuration determines both the speed of the slowest, and the disparity between the fastest and slowest modes. Observe also that the leakage factor α puts a limit on the convergence speed of the slowest mode, by putting a limit on the size of the smallest eigenvalue. (Further discussion on the calculation of the eigenvalues, eigenvectors and time constants can be found in Elliott *et al.* (1992).)

To conclude the examination, it is interesting to look at the evolution of the system control effort as the error criterion is reduced. Noting that the control signal vector can be written as

$$\boldsymbol{y}(k) = \boldsymbol{Q}\left[\boldsymbol{v}'(k)-\boldsymbol{v}'(0)\right] \tag{6.9.88}$$

control effort can be expressed as

$$\boldsymbol{y}^{\mathrm{H}}(k)\boldsymbol{y}(k) = \left[\boldsymbol{v}'(k)-\boldsymbol{v}'(0)\right]^{\mathrm{H}}\left[\boldsymbol{v}'(k)-\boldsymbol{v}'(0)\right] \tag{6.9.89}$$

or

$$\boldsymbol{y}^{\mathrm{H}}(k)\boldsymbol{y}(k) = \boldsymbol{v}'^{\mathrm{H}}(0)\left[\boldsymbol{I}-(\boldsymbol{I}-2\mu\boldsymbol{\Lambda})^k\right]^2\boldsymbol{v}'(0) \tag{6.9.90}$$

The important point to note here is that the modes associated with the smaller eigenvalues require a greater effort to control. Therefore, the algorithm leakage factor places a limit on the control effort which can be exerted by the algorithm, by limiting the size of the smallest eigenvalue (Elliott *et al.*, 1992). This result is to be expected by the very nature of the error criterion, where the leakage factor determines the contribution of control effort to the total value. It also demonstrates the importance of including algorithm leakage in any practical algorithm implementation, as outlined in Section 6.6.

6.10 CANCELLATION PATH TRANSFER FUNCTION ESTIMATION

In the previous two sections of this chapter the necessity of having a model of the cancellation path transfer function for implementing an adaptive feedforward control system has become very salient. We should therefore mention briefly how this model may be obtained. The simplest method is to obtain the model off-line prior to control system start up and with the primary source turned off. In this case the measurement of the transfer function is an adaptive estimation problem which we have discussed in some depth previously in both Section 5.4

on system identification and in Sections 6.5 and 6.6 on adaptive filtering. Here a 'modelling disturbance' can be injected into both the cancellation path transfer function 'system', the input of which is the control filter output and the output of which is the error sensor output, and into an adaptive filter. If a fixed frequency disturbance is the target of the active control system, such as generated by a rotating machine, the speed of which is synchronized with the mains frequency or by an oscillator in an experiment, then the modelling disturbance can be this frequency and the length of the adaptive filter can be short. Otherwise, broadband excitation should be used.

The problem with obtaining a fixed coefficient model of the cancellation path transfer function prior to start-up is that slight changes in the environment during system operation may lead to adaptive algorithm instability. Note that while a change in the transfer function phases in excess of 90° will force this result, smaller deviations still require reduced convergence coefficient values and so may also lead to instability. This problem is somewhat reduced in MIMO systems and in systems where little variation in the transfer functions can be expected, such as controlling low frequency sound fields in small enclosures (such as an automobile). However, in many practical implementations it is advisable to employ on-line cancellation path transfer function modelling. A variety of methods have been tried for this (see Eriksson, 1991a,b for a review). What we will discuss here are two of the more simple ones which follow on directly from previous work in Sections 5.4 and 6.6.

In Section 5.4 it was stated that system identification could be conducted with reasonable accuracy in the presence of other processes, provided the modelling disturbance was uncorrelated with the other external disturbances. This general concept holds true when the 'other' processes include the primary and control disturbance generation. Therefore, random noise can be injected into the system for cancellation path transfer function modelling during the operation of the active control system, with the control signal generation and identification processes separate as shown in Fig. 6.51. While this may seem to be a somewhat counter-productive exercise, injecting additional noise into a system targeted for active control, the level of the modelling disturbance can be very low (say, 30 dB below the unwanted disturbance) and still provide a model of suitable accuracy over a relatively long time.

One potential problem with this approach is model bias which may arise from correlation within the system, as discussed in Section 5.4. This can be particularly true in systems where the reference sensor is open to contamination from the control source output, such that some of the modelling disturbance can find its way into the system. With a low-level modelling disturbance this has been reported to be not a problem. However, it is possible that an extended least squares approach as outlined in Section 5.4 would improve the results if bias was a problem.

A second approach to modelling the cancellation path transfer function could be viewed as arising out of the question, if we are already injecting an external (control) signal into the system, why not use that signal to perform system

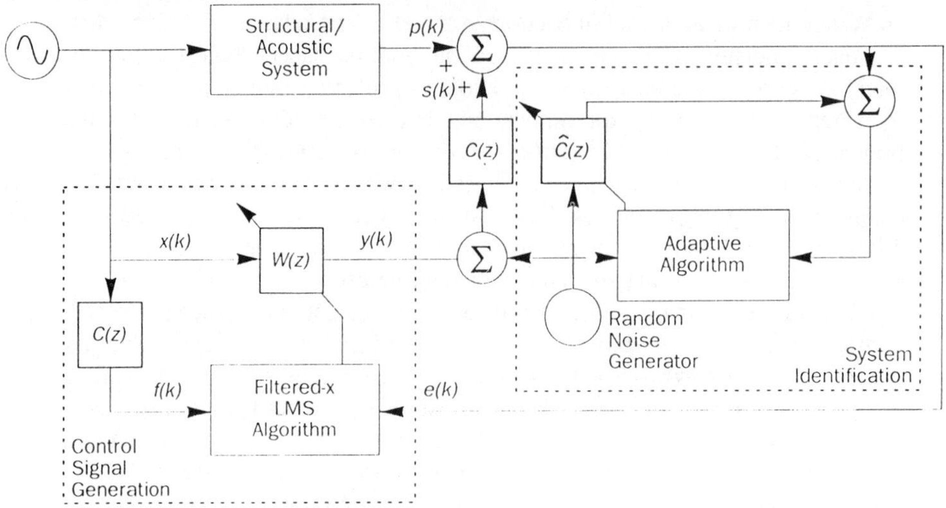

Fig. 6.51 Control signal derivation and cancellation path transfer function identification as separate processes in a practical controller.

identification? This is essentially the other extreme to injecting a random noise modelling disturbance, because the signal is injected into an environment which is (it is hoped) extremely rich in correlated signals, and so a model formulated using the control signal as a modelling disturbance is extremely prone to bias. The only possibility for obtaining a model of reasonable accuracy in this instance is to employ an extended least squares approach.

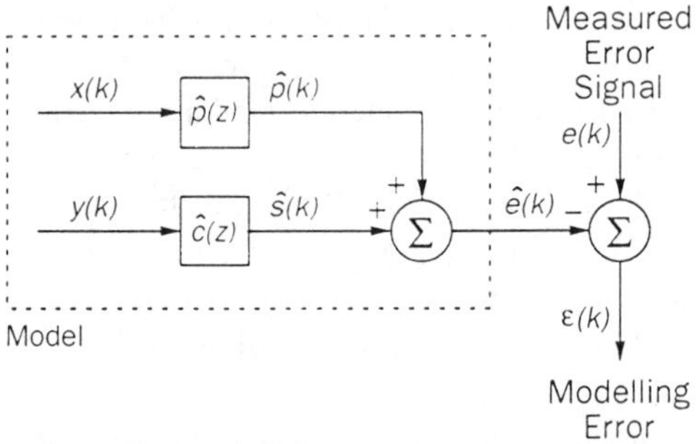

Fig. 6.52 Outline of modelling problem for development of an extended least squares modelling approach.

To do this, the problem can be viewed as shown in Fig. 6.52. Here the control signal $s(k)$ is again modelled as the output signal $y(k)$ convolved with the cancellation path transfer function c, and the primary disturbance is modelled as the reference signal $x(k)$ convolved with an n-stage finite impulse response model of some primary source transfer function p, such that

$$\hat{p}(k) = x(k) * p = \sum_{i=0}^{n} x(k-i)p_i = \mathbf{x}^{\mathrm{T}}(k)\mathbf{p} \tag{6.10.1}$$

where $\boldsymbol{p}$ is an $(n \times 1)$ vector of primary transfer function model coefficients. The error signal can therefore be estimated as the sum of the primary and control source model outputs. This can be written as an inner product using augmented matrices:

$$\hat{e}(k) = \boldsymbol{\phi}^{\mathrm{T}}(k)\boldsymbol{\theta} \tag{6.10.2}$$

where $\boldsymbol{\phi}$ is the augmented data vector

$$\boldsymbol{\phi}(k) = \begin{bmatrix} y(k) & y(k-1) & \cdots & y(k-m+1) & | & x(k) & x(k-1) & \cdots & x(k-n+1) \end{bmatrix}^{\mathrm{T}} \tag{6.10.3}$$

and $\boldsymbol{\theta}$ is the augmented parameter vector

$$\boldsymbol{\theta} = \begin{bmatrix} c_0 & c_1 & \cdots & c_{m-1} & | & p_0 & p_1 & \cdots & p_{n-1} \end{bmatrix}^{\mathrm{T}} \tag{6.10.4}$$

The difference between the measured system error $e(k)$ and its estimate is the modelling error $\epsilon(k)$:

$$\epsilon(k) = e(k) - \hat{e}(k) \tag{6.10.5}$$

What we now have is the makings of a simple adaptive filtering problem, the aim of which is to adjust the parameter vector $\boldsymbol{\theta}$ so as to minimise the mean square value of the modelling error $\epsilon(k)$. This could be undertaken with a simple gradient descent algorithm:

$$\boldsymbol{\theta}(k+1) = \boldsymbol{\theta}(k) - \mu\Delta\boldsymbol{\theta}(k) \tag{6.10.6}$$

It is straightforward to show, using the same steps taken in formulating the LMS algorithm, that

$$\Delta\boldsymbol{\theta}(k) \approx -2\epsilon(k)\boldsymbol{\phi}(k) \tag{6.10.7}$$

Therefore, the parameter vector can be updated according to

$$\boldsymbol{\theta}(k+1) = \boldsymbol{\theta}(k) + 2\mu\epsilon(k)\boldsymbol{\phi}(k) \tag{6.10.8}$$

To model the cancellation path transfer function using the algorithm of (6.10.8), the data vector must be updated with each control filter output derivation. The convolution of (6.10.2) can then be performed, and the result subtracted from the measured error signal to produce the modelling error of (6.10.5). This can then be used in (6.10.8) to update the parameter. The updated model of the cancellation path transfer function can be obtained by dividing the augmented matrix as shown in (6.10.4).

Speed in obtaining a suitably accurate estimate of the cancellation path transfer function is often more important than obtaining an extremely precise model. For this reason a normalized version of the algorithm in (6.10.8) is perhaps a better choice (effectively what is suggested in Sommerfeldt and Tichy, 1990). Based upon our discussion of the normalized LMS algorithm in Section 6.6, the normalised version of (6.10.8) can be written directly as

$$\theta(k+1) = \theta(k) - \frac{\epsilon(k)\phi(k)}{\phi^{T}(k)\phi(k)} \tag{6.10.9}$$

In practice, the gradient part of the expression, the second term on the right-hand side of (6.10.9), can be slightly reduced (by a multiplying factor) to enhance stability, although this occurs at the expense of algorithm speed. Note also that if a system is started up with all parameters 'zeroed', with zero values assigned to the control filter weights and the values in the parameter vector $\boldsymbol{\theta}$, then the values in the cancellation path transfer function model part of $\boldsymbol{\theta}$, and hence the filter weights, will never change. Some non-zero value must be inserted somewhere in the system prior to start up, such as in the first weight coefficient of the control filter.

The authors have found this approach to modelling the cancellation path transfer function to be effective with harmonic disturbances, where relatively short FIR models are used. There are a few points which seem to improve performance in such implementations. First, with harmonic disturbances the cancellation path transfer function and the primary source transfer function will be simple gain and phase changes. The models can therefore both be completely wrong yet still have minimal total estimation error. With harmonic disturbances it is found that a good approach to overcoming this problem is to form a model of the primary source transfer function before starting the control system, done simply by zeroing the control filter weights and cancellation path transfer function parameters. After an initial model of the primary source transfer function has been obtained then some initial weights can be loaded into the control filter to start the system. The coefficients in the parameter vector also have a tendency to 'wander' over time, especially if the active control error signal is very small. This problem can be overcome somewhat by setting a minimum value of estimation error below which no modification to the parameter vector values will be made.

Finally, with either modelling method, if the disturbance is sinusoidal and so the system identification has been restricted to only the frequency of interest, then if the frequency changes the system may become unstable. A 'race' then develops between the diverging control filter and the converging new estimate of the cancellation path transfer function, a race which should obviously be biased towards the new model. In such 'reduced systems' it is advisable to update the cancellation path transfer function model more often than the control system filter weights, say two or three times more often. The characteristics of the system response to a frequency change are then something like the data shown in Fig. 6.53, where the active control system is targeting a pure tone disturbance and the cancellation path transfer function model has only four stages and is adapted on line using the extended least squares approach outlined above. Clearly the model responds more quickly to the frequency change than the control filter algorithm, holding the system stable.

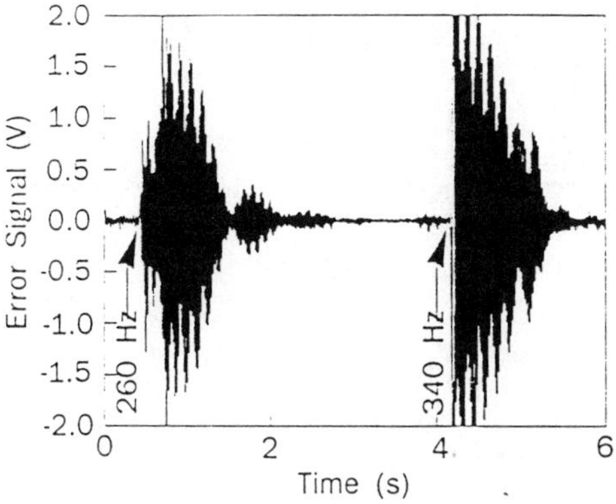

Fig. 6.53 Typical change in the error signal during a change in cancellation path transfer function when employing the extended least squares-like approach.

6.11 ADAPTIVE SIGNAL PROCESSING USING RECURSIVE (IIR) FILTERS

6.11.1 Why use an IIR Filter?

In the previous six sections we have considered adaptive signal processing using FIR filters, both in purely electronic implementations and in active control implementations. In fact, adaptive signal processing using FIR filters is a relatively easy process. The algorithms are simple, the structure is inherently stable, and provided care is taken in selecting the various algorithm parameters such as the convergence coefficient and leakage factor, convergence of the weight coefficients to near their optimum values will eventually take place. So why would there ever be any need to use a different filter structure?

There are, in fact, a few situations which may arise in active noise and vibration control where the use of a FIR filter may not provide the desired results. Consider the general block diagram of a feedforward active control system shown in Fig. 6.54. Active control is really a phase inverse modelling problem, where it is desired to introduce a (feedforward) controlling disturbance which will be 180° out of phase with the primary disturbance when it arrives at the error sensor. This control signal is derived by passing a reference input, which is correlated with the primary disturbance, through a model of the structural/acoustic system with 180° shifted phase characteristics. If the structural/acoustic system has many resonances in or near the frequency band of the reference signal, the structural/acoustic transfer function will have poles in it. Therefore, the length of the FIR filter required to accurately model the system will be very long, and the associated computational burden very large. In these instances, it would be desirable to employ a filter which contains poles in addition to the zeros of the FIR filter.

A second situation where it may not be advantageous to use a FIR filter in an active noise and vibration control system is where the control signal will

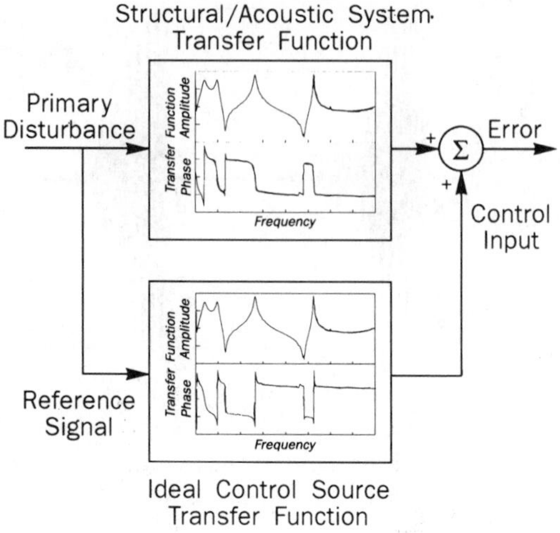

Fig. 6.54 Active control as an inverse phase modelling problem.

'feedback' into the reference sensor. We found in Section 6.3 that this feedback introduces poles into the optimal control source transfer function. If an FIR filter is used in the active control system, it must be very long to accommodate these poles.

In these instances, where the utility of a FIR filter is limited, it may be advantageous to consider the use of an infinite impulse response, or IIR, filter. The general arrangement of an IIR filter is shown in Fig. 6.55. Here we see that the filter output is derived from a weighted summation of past and present input samples, as was the FIR filter, plus past output samples. This latter 'feedback' contribution is responsible for the introduction of poles into the filter transfer function, which is

$$H(z) = \frac{y(z)}{x(z)} = \frac{b_0 + b_1 z^{-1} + b_2 z^{-2} + \cdots + b_{M-1} z^{-(M-1)}}{1 - a_1 z^{-1} - a_2 z^{-2} - \cdots - a_N z^{-N}} \tag{6.11.1}$$

where the a coefficients are the weights of the 'feedback' loop, and the b coefficients are the weights of the 'feedforward' loop. This is referred to as an infinite impulse response filter because in response to a unit pulse input the filter will (theoretically) produce an output forever, as the data samples in the feedback loop will never go to zero.

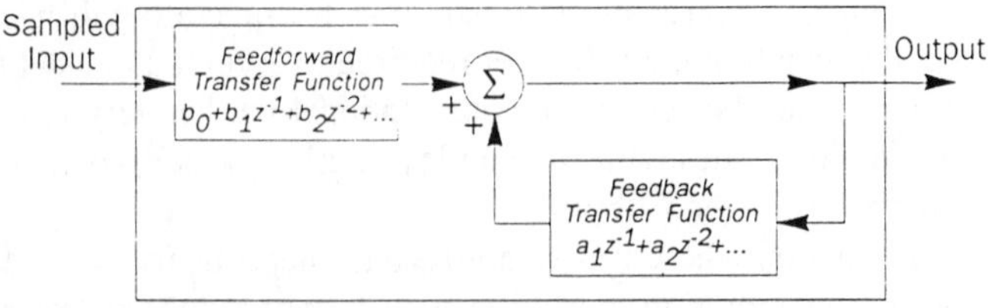

Fig. 6.55 Block diagram of an infinite impulse response (IIR) filter.

The use of IIR filters is not, however, without some drawbacks. The structure of IIR filters is more complex than FIR filters, and so too are the algorithms required to adapt them. IIR filters are not inherently stable, owing to the poles in the transfer function. Also, the error surface, and hence gradient, of an IIR filter can be much more complicated than for an FIR filter, as will be outlined. Combined with this is the problem that some of the filter poles may have very long time constants, making a convergence analysis of a IIR filter very complicated, if not impossible. Despite these problems, adaptive IIR filters can be used successfully in active noise and vibration control systems if care in implementation is taken.

This section will consider the use of gradient based algorithms with adaptive IIR filters, as implemented in purely electronic systems. In the section which follows we will extend the algorithm to active control implementation. Although this type of algorithm has some drawbacks, as will be pointed out, it is probably the most common and simple. Readers who are interested in adaptive algorithms which are faster, more computationally efficient, or are more amenable to a tractable convergence analysis are referred to the texts of Cowan and Grant (1985) and Treichler *et al.* (1987), as well as introductory papers by Johnson (1984) and Shynk (1989).

6.11.2 Error formulations

Before beginning the derivation of algorithms to adapt IIR filters, the problem of error formulation must be addressed. There are numerous error formulations for IIR-type filter arrangements, most of which were derived in the field of system identification (see Ljung and Söderström (1983) and Ljung (1987) for a discussion). There are, however, two error formulations which are predominantly used in adaptive signal processing, the equation error formulation and the output error formulation (Johnson, 1984; Shynk, 1989), and it is these which will be discussed here.

Consider first the equation error formulation, corresponding to the filter arrangement shown in Fig. 6.56. The output of the IIR filter is the sum of two (non-recursive) FIR filters:

$$y(k) = \sum_{i=1}^{N} a_i d(k-i) + \sum_{j=0}^{M-1} b_j x(k-j) \tag{6.11.2}$$

where $d(k)$ is the desired output signal of the filter at time k and there are M and N stages in the feedforward and feedback paths, respectively. The 'equation error' is defined as the difference between the desired filter output and the actual filter output,

$$e_{eq}(k) = d(k) - y(k) \tag{6.11.3}$$

The important aspect of the equation error arrangement of Fig. 6.56 is that old versions of the desired signal are used in the formulation of the current estimate of the desired signal (in the signal identification implementation, it is old values of the output of the system to be identified, which is the desired signal, which are used). The arrangement shown in Fig. 6.56 is, in fact, referred to as 'equation error' because the data samples in both the feedforward and feedback tapped delay lines are exact, and therefore only an error in the weights (hence the transfer function equation) will be responsible for the mean square error. As the IIR filter output is the sum of two independent FIR filter outputs the equation error is a linear function of the weight coefficients. Therefore, the mean square error will be a quadratic function with a single minimum (as it was for the FIR filter cases of the previous sections).

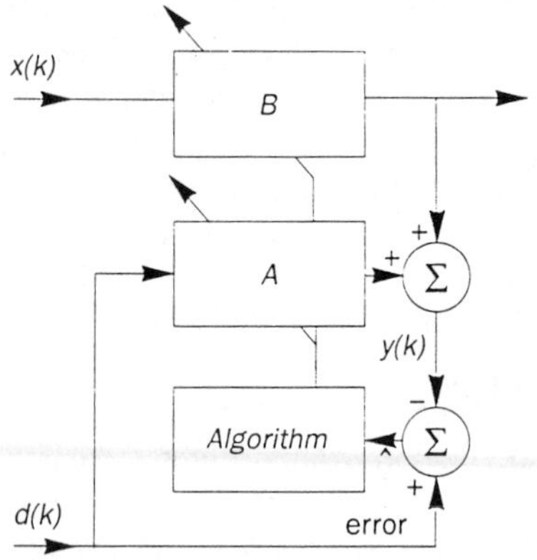

Fig. 6.56 Equation error formulation of an IIR filtering problem.

One drawback of the equation error arrangement of Fig. 6.56 is that if the desired signal is 'noisy', the results of the adaptive filtering operation will become biased (Johnson, 1984; Shynk, 1989). In fact, the bias of the weight coefficients may become so bad that the performance of the filter may be completely unsatisfactory. (As discussed in Section 5.4, this problem was addressed in the system identification context in the form of 'extended' least squares prediction.) Also, in most active noise and vibration control systems only the combined primary and control signals can be measured at the error sensor. Therefore, the desired signal (which is the phase inverse of the primary signal) is not explicitly known, making the implementation of equation error formulated filters difficult (although one possible implementation is outlined by Eriksson, 1991b). For these reasons equation error formulated filters will not be considered further here. Interested readers are referred to Ljung and Söderström (1983).

The other error formulation being considered is the output error formulation, shown in Fig. 6.57. With this arrangement, the output of the IIR filter is derived from the filter input and past outputs,

$$y(k) = \sum_{i=1}^{N} a_i y(k-i) + \sum_{j=0}^{M-1} b_j x(k-j) \tag{6.11.4}$$

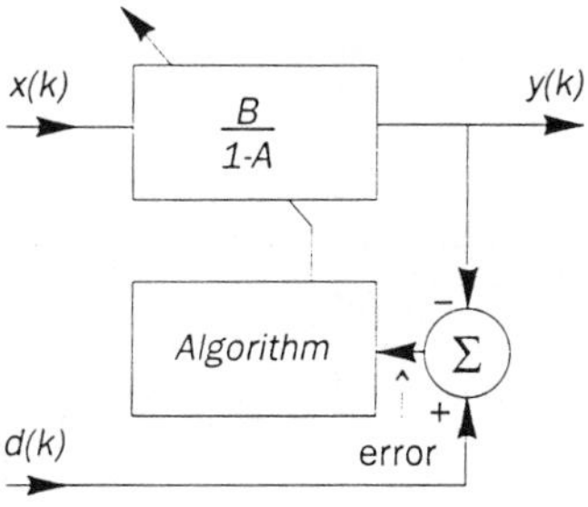

Fig. 6.57 Output error formulation of an IIR filtering problem.

The difference between the equation error and output error formulated filters is that the equation error filter uses old values of the desired output to derive the current filter output, while the output error filter uses old values of the actual filter output in the derivation. A result of using past values of the actual filter output in the derivation of the present filter output is that the output error, defined as

$$e_{\text{out}}(k) = d(k) - y(k) \tag{6.11.5}$$

is a non-linear function of the filter weight coefficients. The mean square error is therefore not a quadratic function of the weight coefficients, and may have multiple minima (Stearns, 1981; Söderström and Stoica, 1982; Fan and Nayeri, 1989). As such, gradient based algorithms may converge to a minimum which is not the global minimum, and become 'stuck'.

However, on a positive note, the weight coefficients are unbiased, unlike the equation error arrangement. Also, for active noise and vibration control systems, where the desired signal is not usually known explicitly, they are easier to implement, For these reasons, output error formulated filters will be considered in this section as candidates for implementation in active noise and vibration control systems (the actual implementation will be discussed in the next section).

6.11.3 Formulation of a gradient based algorithm

A general block diagram of the IIR filter problem (for an output error formulation) is shown in Fig. 6.58. The output of the filter is governed by (6.11.4), where a and b are the feedback and feedforward weight coefficients respectively. This can be restated in a form similar to that used in the section on system identification, Section 5.5, as

$$y(k) = \phi^{\mathrm{T}}(k)\theta = \theta^{\mathrm{T}}\phi(k) \tag{6.11.6}$$

where θ is the parameter vector of weight coefficients, given by

$$\theta = \left[a_1 \;\; a_2 \;\; \cdots \;\; a_N \;\; b_0 \;\; b_1 \;\; \cdots \;\; b_M\right]^{\mathrm{T}} \tag{6.11.7}$$

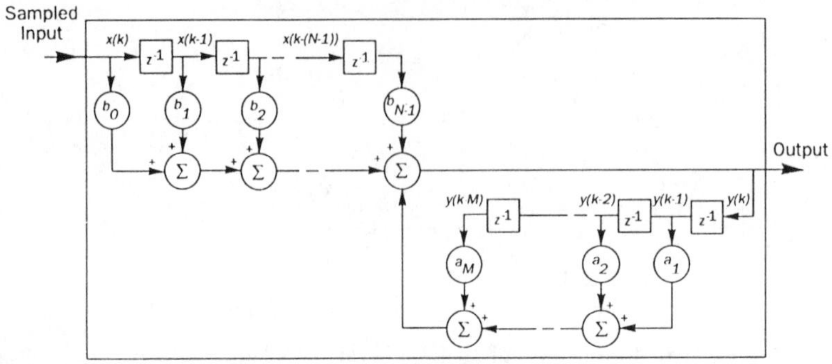

Fig. 6.58 Direct form implementation of an IIR filter.

and ϕ is the data vector, or regressor

$$\phi(k) = \left[y(k-1) \;\; y(k-2) \;\; \cdots \;\; y(k-N) \;\; x(k) \;\; x(k-1) \;\; \cdots \;\; x(k-M)\right]^{\mathrm{T}} \tag{6.11.8}$$

As was the case for the FIR filter based system, the aim of the adaptive algorithm is to arrive at a parameter vector, θ, such that the filter output most closely matches some desired output, $d(k)$. The (output) error signal, $e(k)$, is defined as the difference between the desired output and the actual output:

$$e(k) = d(k)-y(k) = d(k)-\theta^{\mathrm{T}}\phi(k) \tag{6.11.9}$$

The (ideal) error criterion, as before, is the minimization of the mean square error:

$$\begin{aligned}\xi(k) &= E\{e^2(k)\} = E\left\{ \left(d(k)-y(k)\right)^2 \right\} \\ &= E\left\{ \left(d(k)-\theta^{\mathrm{T}}\phi(k)\right)^2 \right\}\end{aligned} \tag{6.11.10}$$

As was outlined in Section 6.5, a gradient descent algorithm operates by adding to the present estimate of the optimal set of filter parameters a portion of the negative gradient of the error surface at the present coefficient values (location), in that way descending down the error surface towards a minimum. To formulate such an algorithm for the IIR filter for practical implementation we will again make the stochastic approximation of mean square error, $\xi(k) \approx e^2(k)$. Differentiating the squared value of instantaneous error with respect to the parameter vector produces

$$\frac{\partial e^2(k)}{\partial \theta(k)} = -2e(k)\frac{\partial y(k)}{\partial \theta(k)} \tag{6.11.11}$$

This can be separated into the feedback a and feedforward b parameters,

$$\frac{\partial e^2(k)}{\partial a_i(k)} = -2e(k)\frac{\partial y(k)}{\partial a_i(k)} \tag{6.11.12}$$

and

$$\frac{\partial e^2(k)}{\partial b_j(k)} = -2e(k)\frac{\partial y(k)}{\partial b_j(k)} \tag{6.11.13}$$

Evaluating the derivatives in (6.11.12) and (6.11.13)

$$\frac{\partial y(k)}{\partial a_i(k)} = y(k-i) + \sum_{j=1}^{N} a_j(k)\frac{\partial y(k-j)}{\partial a_i(k)} \tag{6.11.14}$$

and

$$\frac{\partial y(k)}{\partial b_j(k)} = x(k-j) + \sum_{i=1}^{N} a_i(k)\frac{\partial y(k-i)}{\partial b_j(k)} \tag{6.11.15}$$

The complexity of (6.11.14) and (6.11.15) makes the determination of closed form solutions extremely difficult, if not impossible. Therefore, some simplifying assumptions must be made. One common one (Parikh and Ahmed, 1978) is that the convergence coefficient is small such that the change in parameters at each iteration is also small, enabling the assumption

$$\frac{\partial y(k-j)}{\partial a_i(k)} \approx \frac{\partial y(k-j)}{\partial a_i(k-j)} \tag{6.11.16}$$

With this assumption, (6.11.14) can be simplified to

$$\frac{\partial y(k)}{\partial a_i(k)} \approx y(k-i) + \sum_{j=1}^{N} a_j(k)\frac{\partial y(k-j)}{\partial a_i(k-j)} \tag{6.11.17}$$

Similarly, for the derivative on the right-hand side of (6.11.15)

$$\frac{\partial y(k-i)}{\partial b_j(k)} \approx \frac{\partial y(k-i)}{\partial b_j(k-i)} \tag{6.11.18}$$

so that

$$\frac{\partial y(k)}{\partial b_j(k)} \approx x(k-j) + \sum_{i=1}^{N} a_i(k)\frac{\partial y(k-i)}{\partial b_j(k-i)} \tag{6.11.19}$$

The partial derivatives of (6.11.17) and (6.11.19) are now recursive in old versions of the partial derivative, which will significantly simplify their implementation in an adaptive algorithm. Combining (6.11.14) and (6.11.15), the result can be re-expressed in terms of the parameter and regressor vectors as

$$\frac{\partial y(k)}{\partial \theta(k)} \approx \phi(k) + \sum_{j=1}^{N} a_j(k)\frac{\partial y(k-j)}{\partial \theta(k-j)} \tag{6.11.20}$$

To simplify the presentation, define the variable

$$\psi(k) \approx \frac{\partial y(k)}{\partial \theta(k)} \tag{6.11.21}$$

so that

$$\psi(k) = \phi(k) + \sum_{j=1}^{N} a_j(k)\psi(k-j) \tag{6.11.22}$$

Substituting this back into (6.11.11), the gradient estimate at time k is

$$\frac{\partial e^2(k)}{\partial \theta(k)} = -2e(k)\psi(k) \tag{6.11.23}$$

Therefore, a gradient descent algorithm for the IIR filter is

$$\theta(k+1) = \theta(k) + 2\boldsymbol{M}\psi(k)e(k) \tag{6.11.24}$$

where M is a diagonal matrix of convergence coefficients,

$$\boldsymbol{M} = \begin{bmatrix} \mu_{a1} & & & & & \\ & \ddots & & & & \\ & & \mu_{aN} & & & \\ & & & \mu_{b0} & & \\ & & & & \ddots & \\ & & & & & \mu_{bM} \end{bmatrix} \tag{6.11.25}$$

6.11.4 Simplifications to the gradient algorithm

There are two simplifications which can be made to the algorithm of (6.11.24) to reduce its computation load. Define the 'filtered' versions of x and y as

$$x_f(k) = \frac{\partial y(k)}{\partial b_0(k)} \tag{6.11.26}$$

and

$$y_f(k) = \frac{\partial y(k)}{\partial a_1(k)} \tag{6.11.27}$$

From (6.11.17) and (6.11.19), these filtered variables evolve recursively in time as

$$x_f(k) = x(k) + \sum_{i=1}^{N} a_i(k)x_f(k-i) \tag{6.11.28}$$

and

$$y_f(k) = y(k) + \sum_{j=1}^{N} a_j(k)y_f(k-j) \tag{6.11.29}$$

By comparing (6.11.28) and (6.11.29) with (6.11.17) and (6.11.19) respectively, it can be seen that the gradient of each weight coefficient with respect to the appropriate datum, which from (6.11.26) and (6.11.27) is the 'filtered' signal value, is approximately equal to the gradient (filter signal value) of the weight preceding it, delayed by one sample. In other words

$$\frac{\partial x(k)}{\partial b_j(k)} \approx x_f(k-j) \tag{6.11.30}$$

and

$$\frac{\partial y(k)}{\partial a_i(k)} \approx y_f(k-i) \tag{6.11.31}$$

Therefore, the vector $\psi(k)$ of (6.11.22) can be approximated as (Söderström *et al.*, 1978; Horvath, 1980):

$$\psi(k) \approx \psi_f(k) = \left[y_f(k-1) \; \cdots \; y_f(k-N) \;\; x_f(k) \; \cdots \; x_f(k-M)\right] \tag{6.11.32}$$

The simplified result of (6.11.32) is much less computationally laborious than (6.11.22), and with small values of convergence coefficient essentially results in no degradation of performance, and as such is the form of algorithm normally used.

The second simplification is also directed at reducing the computational load associated with the vector ψ. Based on the results of the LMS algorithm, it was suggested that the recursion of past gradient estimates in (6.11.20), the final term on the right hand side, be ignored completely, and ψ be defined as (Feintuch, 1976):

$$\psi(k) \approx \phi(k) \tag{6.11.33}$$

In this form the adaptive algorithm for the IIR filter is a simple extension of the LMS algorithm. For this reason, the algorithm produced by the approximation given in (6.11.33) is known as the recursive LMS, or RLMS, algorithm.

The gradient estimate of the RLMS algorithm is significantly poorer than that yielded by using (6.11.22) or (6.11.32). As a result, it has been demonstrated that the RLMS algorithm can converge to a false minimum in the error surface (Johnson and Larimore, 1977) even when the algorithm without the approximation of equation (6.11.33) converges to the actual minima (Parikh and Ahmed, 1978). Despite this, the RLMS algorithm has been successfully implemented in many active noise control systems (Eriksson *et al.*, 1987), without any reports of significant suboptimal performance.

As mentioned in the introduction to this section, two of the drawbacks of IIR adaptive filtering are the possibility of instability arising from the recursive nature of the filter, where the poles are modified, and the difficulty in analytically studying the convergence behaviour of the algorithms. For these reasons it is strongly advisable to include some form of stability monitoring in the algorithm implementation to check for unstable poles, lying outside the unit circle. One very simple check to make sure all poles are contained within the unit circle is to monitor them:

$$\sum_{i-1}^{N} | a_i | < 1.0 \tag{6.11.34}$$

This, however, is a very conservative test, especially for filters with large numbers of coefficients. Jury's test (Jury, 1964, 1974) will provide a more sophisticated method of stability checking, but is more computationally expensive to implement.

If the possibility of an unstable pole has been detected, then it must be 'projected' back into the unit circle. This can be done simply by reducing the size of the convergence coefficient given in (6.11.25), or by ignoring the unstable weight vector update all together (Ljung and Söderström, 1982, pp.366–368). The possible drawback of this is that the algorithm may 'stall' indefinitely. More complex, and computationally intense, solutions to this problem have been suggested (Johnson, 1984), but so far none has been shown to be infallible. If the algorithm does stall, perhaps the best solution is to change the filter size and start again.

6.12 APPLICATION OF ADAPTIVE IIR FILTERS TO ACTIVE CONTROL SYSTEMS

In the previous section the implementation of adaptive IIR filters in 'standard', purely electronic systems was discussed. However, as with the implementation of adaptive FIR filters considered earlier in this chapter, the standard adaptive algorithms must be modified when implementing the adaptive filter in an active noise or vibration control system. These modifications are required to account for the cancellation path transfer function, the transfer function which exists between the output of the digital filter and the input of the error signal to the adaptive algorithm.

In this section we rederive the gradient descent algorithm for optimizing the weights of an IIR filter for active control implementation. After deriving the algorithm, we examine the influence which some of the commonly employed simplifying assumptions taken in the algorithm derivation have upon system performance, and upon the architecture of the active control system.

6.12.1 Basic algorithm development

To derive a gradient descent based adaptive algorithm for implementation of IIR filters in active sound or vibration control systems, consider the block diagram of a generic active control arrangement shown in Fig. 6.59. Here the (unwanted) primary disturbance $p(k)$ is modelled as derived from an ARMA process (discussed in the previous chapter). A reference signal $x(k)$, correlated with the primary disturbance, is provided to an IIR control filter to derive a control input $y(k)$. The control input propagates through a cancellation path transfer function, which incorporates the frequency-dependent response characteristics of the control source (for example, a speaker) and error sensor (for example, a microphone), as well as the response characteristics of the 'physical' system to which both are attached (for example, an air handling duct). The cancellation path transfer function is modelled in the time domain by an Mth order impulse response vector $\boldsymbol{c}$. The resulting control disturbance $s(k)$ combines with the primary disturbance to produce the error signal $e(k)$ which will be supplied to an adaptive algorithm for use in tuning the filter; the aim of this section is to develop an appropriate algorithm (the error smoothing filter will be discussed further shortly). We will develop the adaptive algorithm with reference to the single input, single output (SISO) problem for clarity of notation; extension to a multiple input, multiple output (MIMO) problem is relatively straightforward, with more complex algebra.

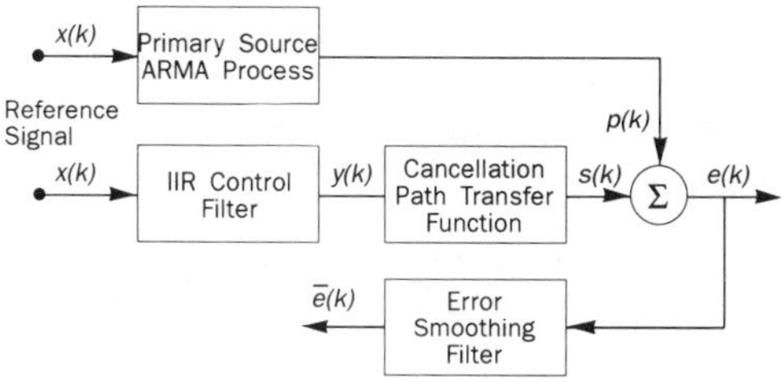

Fig. 6.59 Block diagram of active control problem for IIR filter-based system implementation.

The IIR filters of interest here are direct form architectures, discussed in the previous section and illustrated in Fig. 6.58. As stated, the output of this filter is defined by the expression

$$y(k) = \sum_{i=0}^{N_f} b_i x(k-i) + \sum_{j=1}^{N_b} a_j y(k-j) = \boldsymbol{x}^{\mathrm{T}}(k)\boldsymbol{b} + \boldsymbol{y}^{\mathrm{T}}(k-1)\boldsymbol{a} \tag{6.12.1}$$

where $\boldsymbol{x}(k)$ and $\boldsymbol{y}(k\text{-}1)$ are the vectors of most recent input samples and previous output samples respectively:

$$\begin{aligned} \boldsymbol{x}(k) &= \left[x(k) \;\; x(k-1) \;\; \cdots \;\; x(k-N_f)\right]^{\mathrm{T}} \\ \boldsymbol{y}(k-1) &= \left[y(k-1) \;\; y(k-2) \;\; \cdots \;\; y(k-N_b)\right]^{\mathrm{T}} \end{aligned} \tag{6.12.2}$$

and $\boldsymbol{a}$ and $\boldsymbol{b}$ are the vectors of feedforward and feedback filter weights respectively:

$$\boldsymbol{a} = \left[a_1 \;\; a_2 \;\; \cdots \;\; a_{N_b}\right]^{\mathrm{T}}, \quad \boldsymbol{b} = \left[b_o \;\; b_1 \;\; \cdots \;\; b_{N_f}\right]^{\mathrm{T}} \tag{6.12.3}$$

The transfer function of this filter is

$$G(z) = \frac{b_o + b_1 z^{-1} + \cdots + b_{N_f} z^{-N_f}}{1 - a_1 z^{-1} - \cdots - a_{N_b} z^{-N_b}} \tag{6.12.4}$$

The aim of the adaptive filtering operation is to minimize the performance criterion J of the mean square value of the error signal provided to the electronics by the transducer, or error sensor, defined as

$$J = E\{e^2(k)\} = E\{\,[p(k)+s(k)]^2\,\} \tag{6.12.5}$$

However, as with the adaptive algorithm developments undertaken previously in this chapter, the stochastic approximation of instantaneous squared error for mean square error will be made, such that

$$J = e^2(k) = [p(k)+s(k)]^2 \tag{6.12.6}$$

The gradient of the error criterion with respect to the filter weights is defined by the expression

$$\begin{aligned} \nabla J_{a_j}(k) &= \frac{\partial e^2(k)}{\partial a_j} = 2e(k)\frac{\partial s(k)}{\partial a_j(k)} \\ \nabla J_{b_i}(k) &= \frac{\partial e^2(k)}{\partial b_i} = 2e(k)\frac{\partial s(k)}{\partial b_i(k)} \end{aligned} \tag{6.12.7}$$

To obtain expressions for the partial derivatives in (6.12.7), the control disturbance can be expanded, including the time-dependence of the filter parameters during adaptation, as

$$\begin{aligned} s(k) &= \sum_{\iota=0}^{M} c_\iota y(k-\iota) \\ &= \sum_{\iota=0}^{M} c_\iota \left[\sum_{i=0}^{N_f} b_i(k-\iota)x(k-\iota-i) + \sum_{j=1}^{N_b} a_j(k-\iota)y(k-\iota-j) \right] \end{aligned} \tag{6.12.8}$$

Assuming 'slow' filter convergence, so that

$$a_j(k-\iota-j) \approx a_j(k), \quad b_i(k-\iota-i) \approx b_i(k) \tag{6.12.9}$$

the gradient of the error criterion with respect to the feedback weights is defined by the expression

$$\begin{aligned} \nabla J_{a_j}(k) &= 2e(k)\sum_{\iota=0}^{M} c_\iota \frac{\partial y(k-\iota)}{\partial a_j(k-\iota)} \\ &= 2e(k)\sum_{\iota=0}^{M} c_\iota \left[y(k-\iota-j) + \sum_{n=1}^{N_b} a_n(k)\frac{\partial y(k-\iota-n)}{\partial a_j(k)} \right] \end{aligned} \tag{6.12.10}$$

and the gradient of the error criterion with respect to the feedforward weights is defined by the expression

$$\nabla J_{b_i}(k) = 2e(k)\sum_{\iota=0}^{M} c_\iota \frac{\partial y(k-\iota)}{\partial b_i(k-\iota)}$$

$$= 2e(k)\sum_{\iota=0}^{M} c_\iota \left[x(k-\iota-i) + \sum_{n=1}^{N_b} a_n(k)\frac{\partial y(k-\iota-n)}{\partial b_i(k)} \right] \tag{6.12.11}$$

If the dependence of the current gradient (estimate) upon past gradient (estimates) is ignored (the summation term in the brackets in (6.12.10) and (6.12.11)), the resulting adaptive algorithm is referred to as the 'filtered-u' algorithm (Eriksson, 1991b). This has exactly the same form as the 'filtered-x' algorithm derived previously in the chapter for the use of adaptive FIR filter based active noise and vibration control systems. It is also the active noise and vibration control equivalent of the standard RLMS algorithm (without a transfer function in the cancellation path) discussed in the previous section.

To simplify the gradient estimate in a manner less inaccurate than simply ignoring terms, the variables x_f and y_f can be defined as

$$x_f(k) = \frac{\partial y(k)}{\partial b_o(k)} \tag{6.12.12}$$

$$y_f(k) = \frac{\partial y(k)}{\partial a_0(k)} \tag{6.12.13}$$

(note that a_0 is defined as equal to zero in the filter description; the utility of defining y_f in this manner will be apparent shortly). These variables evolve recursively in time as

$$x_f(k) = x(k) + \sum_{n=1}^{N_b} a_n(k)x_f(k-n) \tag{6.12.14}$$

$$y_f(k) = y(k) + \sum_{n=1}^{N_b} a_n(k)y_f(k-n) \tag{6.12.15}$$

Observe now that the gradient estimate of the error criterion at any given time with respect to each filter weight is approximately equal to the gradient estimate with respect to the weight preceding it, delayed by one sample. In other words, it can be approximated that

$$\frac{\partial x(k)}{\partial b_i(k)} \approx x_f(k-i) \tag{6.12.16}$$

and

$$\frac{\partial y(k)}{\partial a_j(k)} = y_f(k-j) \tag{6.12.17}$$

Therefore, the gradient of the error criterion with respect to the feedback weights is approximately equal to

$$\nabla J_{a_j}(k) = 2e(k)\sum_{\iota=0}^{M} c_\iota \frac{\partial y(k-\iota)}{\partial a_j(k-\iota)} = 2e(k)\sum_{\iota=0}^{M} c_\iota \, y_f(k-\iota-j) \tag{6.12.18}$$

and the gradient of the error criterion with respect to the feedforward weights is approximately equal to

$$\nabla J_{b_i}(k) = 2e(k)\sum_{\iota=0}^{M} c_\iota \frac{\partial y(k-\iota)}{\partial b_i(k-\iota)} = 2e(k)\sum_{\iota=0}^{M} c_\iota \, x_f(k-\iota-i) \tag{6.12.19}$$

If the 'filtered' reference sample f_x and filtered output sample f_y are now defined by

$$f_y(k) = \sum_{\iota=0}^{M} c_\iota \, y_f(k-\iota) \tag{6.12.20}$$

and

$$f_x(k) = \sum_{\iota=0}^{M} c_\iota \, x_f(k-\iota) \tag{6.12.21}$$

the adaptive algorithms for the updating the feedback and feedforward weights are defined by the expressions

$$a_j(k+1) = a_j(k) - 2\mu e(k) f_y(k-j) \tag{6.12.22}$$

and

$$b_i(k+1) = b_i(k) - 2\mu e(k) f_x(k-i) \tag{6.12.23}$$

where μ is the convergence coefficient, which may be different for each weight or weight section. The algorithm defined by (6.12.22) and (6.12.23) will be referred to here as the 'correct gradient' algorithm.

The algorithms of (6.12.22) and (6.12.23) are the same in appearance as those described by the filtered-*u* methodology. The difference, however, is the transfer function through which the data samples are 'filtered' prior to their use in the adaptive algorithm. With the correct gradient algorithm, the filtering is through both the recursive (or feedback) part of the control filter and a model of the cancellation path transfer function, as depicted in Fig. 6.60. With the filtered-*u* algorithm, filtering the data samples through the feedback part of the control filter is ignored.

As stated previously, extension of the adaptive algorithm to MIMO problems is relatively straighforward. If there are N_s control sources and N_e error sensors in the system, the objective is to minimize the sum of the squared error signals provided by the set of sensors. Defining the finite impulse response vector model of the cancellation path transfer function between the *u*th control source and *v*th error sensor as $\mathbf{c}_{u,v}$, the adaptive algorithms for the updating the feedback and feedforward weights in the *u*th control filter in the MIMO system are defined by

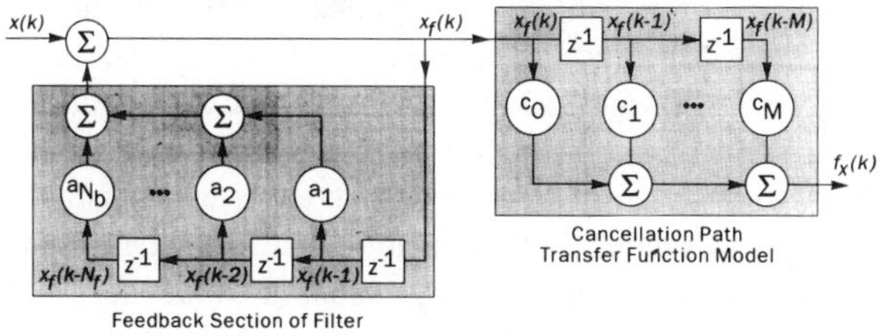

Fig. 6.60 Derivation of the filtered reference signal for the correct gradient implementation of an IIR filter-based system.

the expressions

$$a_{u,j}(k+1) = a_{u,j}(k) - 2\mu \sum_{n=1}^{N_e} e_n(k) f_{y,u,n}(k-j) \tag{6.12.24}$$

and

$$b_{u,i}(k+1) = b_{u,i}(k) - 2\mu \sum_{n=1}^{N_e} e_n(k) f_{x,u,n}(k-i) \tag{6.12.25}$$

where

$$f_{y,u,v}(k) = \sum_{\iota=0}^{M} c_{u,v,\iota}\, y_{f,u}(k-\iota) \tag{6.12.26}$$

and

$$f_{x,u,v}(k) = \sum_{\iota=0}^{M} c_{u,v,\iota}\, x_{f,u}(k-\iota) \tag{6.12.27}$$

with $x_{f,u}$ and $y_{f,u}$ defined by (6.12.16) and (6.12.17), respectively, and with the a coefficients being those in the uth control filter.

6.12.2 Simplification through system identification

As was outlined in Section 6.10, in a practical active sound or vibration control system, modelling of the cancellation path transfer function is typically done 'on-line', to account for changes in the structural/acoustic environment and transducer response over time. As with most adaptive control schemes requiring some measure of system identification, random or pseudo-random noise is normally injected into the system for this purpose. In conventional active control systems, the identification exercise is targeted solely at obtaining a model of the cancellation path transfer function; white noise is added to the control filter output for this purpose. The data samples are then filtered through this model prior to use in the adaptive algorithm. However, from the preceding development it is clear that when using direct-form IIR filters, the data samples should be filtered through both the feedback part of the control filter and the model of the cancellation path transfer function, as was outlined in Fig. 6.60. This process could be simplified by modelling the feedback part of the filter and the cancellation path transfer function in series. This could be accomplished by injecting random noise (for modelling) into the system at the point of summation of the feedforward and feedback sections of the filter, as illustrated in Fig. 6.61 (Snyder, 1994). The 'total transfer function model', which is assumed to be a finite impulse response model, is then adapted using standard techniques (such as the LMS algorithm) to minimize the squared value of estimation error, thereby providing a model of both the feedback loop and cancellation path transfer function; the data samples could be passed through the 'total' model prior to use in the adaptive algorithm. This approach to algorithm implementation, which will be referred to here as the 'estimated gradient algorithm', can provide algorithm performance which is superior to that of the (commonly used) filtered-*u* algorithm, which uses only a model of the cancellation path transfer function (Snyder, 1994).

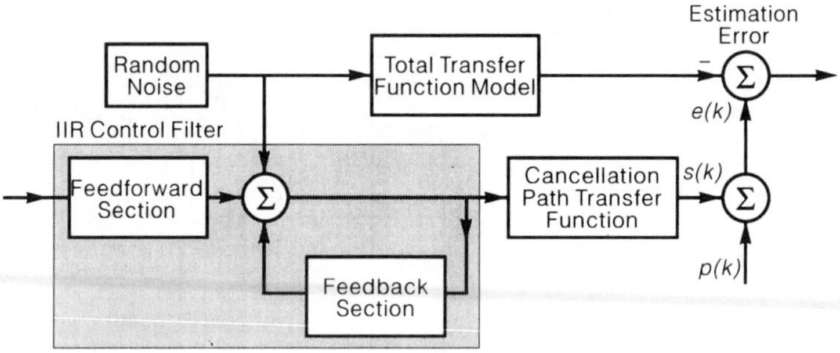

Fig. 6.61 Modelling of 'total transfer function' for gradient estimation.

We may now ask the question, why bother to model both the feedback part of the filter and the cancellation path transfer function in series? As the filter weights are known, these quantities can be used exactly in the feedback filtering part of the gradient formulation. Considering that the (FIR) total transfer function model is attempting to model poles, the approach suggested here will probably require a greater length filter than the total used in the combination of the feedback filter section and cancellation path transfer function model outlined in Fig. 6.60 to produce a model with adequate fidelity. The reason for adopting this approach centres around the way in which practical active control systems are often implemented. In many implementations, where the active control system is required to share microprocessor time with other system functions (such as an air conditioner noise control implementation, where the active noise control system is implemented on the same microprocessor as the motor control system), only the control signal derivation part of the adaptive control system needs to be implemented in real time, and it is therefore not uncommon to adapt the filter coefficients at a rate slower than real time; calculation of the control signal, which is interrupt controlled, requires the majority of available CPU time during the sample period, and so a single filter weight update, which includes calculation of filtered data samples, can take several sample periods to complete. If the feedback part of the gradient estimate is implemented in the literal sense of Fig. 6.60, both $x_f(k)$ and $y_f(k)$ must be calculated at each sample period. This can put a restrictive overhead on the size of the control filters which can be implemented. Alternatively, if an FIR filter is used to model both the feedback and the cancellation path transfer function, calculation of the filtered data samples can be done at a rate slower than real time, using a set of 'captured' data samples. Therefore, while the overall computational requirements of the 'total' system identification approach outlined here may be more than those of the arrangement shown in Fig. 6.60, the resulting calculation procedure actually enables the use of larger control filters, and therefore should produce a better result in terms of sound or vibration attenuation.

6.12.3 SHARF smoothing filter implementation

The simple hyperstable adaptive recursive filter, or SHARF, is a simplified version of an adaptive IIR filter algorithm derived from consideration of non-linear stability theory (Johnson, 1979; Larimore *et al.*, 1980; Johnson *et al.*, 1981). A thorough discussion of the algorithm is beyond the scope of this text. However, its functioning can be placed into the framework of a filtered gradient algorithm, such as the filtered-*x* LMS algorithm, as will be shown.

As implemented in an active control system, the SHARF algorithm is essentially a modified version of the filtered-*u* algorithm (or a modified version of the RLMS algorithm in the absence of the cancellation path transfer function). Returning to Fig. 6.59, the SHARF algorithm for active noise or vibration control implementation is described by

$$a_j(k+1) = a_j(k) - 2\mu e_s(k) f_y(k-j) \tag{6.12.28}$$

and

$$b_i(k+1) = b_i(k) - 2\mu e_s(k) f_x(k-i) \tag{6.12.29}$$

where f_x and f_y are the filtered signal values as described by the filtered-*u* algorithm (filtered only through a model of the cancellation path transfer function):

$$f_x(k) = \sum_{\iota=0}^{M} c_\iota x(k-\iota), \quad f_y(k) = \sum_{\iota=0}^{M} c_\iota y(k-\iota) \tag{6.12.30}$$

and $e_s(k)$ is a 'smoothed' error signal, which is an error signal passed through a (SHARF smoothing) FIR filter:

$$e_s(k) = e(k) + \sum_{i=1}^{N} v_i e(k-i) \tag{6.12.31}$$

The weights used in the smoothing filter are commonly equal to the negative values of the feedback weights in the control filter: $v_i = -a_i$.

To see how this arrangement relates to the more common filtered algorithm implementations, note that in the absence of the error smoothing filter (the standard filtered-u algorithm), by comparing (6.12.28) and (6.12.29) with (6.12.10) and (6.12.11), and the simplifications of these latter equations given in (6.12.14) and (6.12.15) and illustrated in Fig. 6.60, it is apparent that the gradient estimate used by the filtered-u algorithm is in error: there is an unaccounted-for transfer function in the 'overall' cancellation path (the feedback part of the cancellation path transfer function shown in Fig. 6.60). We have shown previously that this situation has the potential to cause adaptive algorithm instability. Based upon the previous results of this chapter, there are two ways to overcome this problem. The first of these is to filter the reference samples through this (previously unaccounted for) transfer function, which is the method employed by the correct and estimated gradient algorithms described in this section, and the approach taken in formulating the adaptive algorithms for active control systems previously described in this chapter. The second means of overcoming this problem is to filter the error signal through the inverse of the (previously unaccounted for) transfer function, which is the approach taken by the SHARF algorithm.

We have shown in relation to the 'standard' filtered algorithms, where the reference signal samples are passed through a model of the cancellation path transfer function, that the cancellation path transfer function model does not need to be exact; it can (theoretically) be within 90° of the correct (phase) value for all frequency components and still maintain system stability (recall that errors in the estimation of the magnitude of the cancellation path transfer function will only affect the size of the convergence coefficient which can be used). When adopting the approach of filtering the error signal through a model of the inverse of the transfer function, the equivalent condition is that the combination of the transfer function in the cancellation path to be compensated for, and transfer function through which the error signal is filtered:

$$H(z) = \frac{1 + \sum_{i=1}^{N} v_i z^{-i}}{1 - \sum_{i=1}^{M} a_i z^{-i}} \tag{6.12.32}$$

be strictly positive real (SPR) (Johnson *et al.*, 1981). In other words, the transfer function through which the error signal is filtered must be within 90° of the inverse of the transfer function in the cancellation path which is to be accounted for by the error filtering, which is the feedback part of the gradient estimate as illustrated in Fig. 6.60.

To lend further support to the view that the SHARF algorithm is simply a 'filtered' version of the RLMS (or filtered-u) algorithm implemented in an alternative fashion, recall from Section 6.8 that if the filtered-x LMS algorithm is implemented in the time domain, with the reference signal filtered through a model of the cancellation path transfer function which is within 90° of the correct value at all frequencies but is not exact, then the principal axes of the error surface may no longer be orthogonal. This leads to extremely complex convergence behaviour; at some values of cancellation path transfer function phase estimation error there may be a significant improvement in convergence behaviour, while at the mirror image value of cancellation path transfer function phase estimation error about the exact value (for example, the pair 30° and −30° transfer function phase estimation error), the convergence behaviour may be significantly degraded. This altering of the error surface has also been previously documented for the SHARF algorithm (Johnson *et al.*, 1981) as the transfer function through which the error signal is filter is varied from the inverse of the recursive part of the IIR filter transfer function (equivalent to having an error in the estimate of the phase of the transfer function). Interestingly, this phenomenon used to argue that the SHARF algorithm was *not* a gradient descent algorithm. The 'optimum' value of phase estimation error is rather difficult to predict *a priori*, and so in this paper no deliberate deviation from the 'exact' model will be taken.

There is one final point, which is only heuristic, which should be made in relation to the difference between filtering a reference signal and inverse filtering an error signal to account for a transfer function in the cancellation path. For a control strategy which is causal, such as is required when the disturbance is random noise, it is the error signal which is a function of previous reference samples, and not the reference signal which is a function of previous errors. Therefore, while either approach may be adequate from the standpoint of algorithm stability, it will not be too surprising if the approach of filtering the reference signal through a transfer function model (such that a given error signal is paired with a weighted set of past reference samples) produces, in some instances, a result which is superior to that obtained by inverse filtering the error signal.

6.12.4 Comparison of algorithms

Computer simulation studies are used here to compare the performance of the four adaptive algorithms for IIR filters outlined (the correct gradient algorithm, the estimated gradient algorithm, the filtered-*u* algorithm, and the filtered-*u* algorithm with a SHARF smoothing filter). The problem we will investigate is an extremely simple one, where the primary source path (from the reference sensor output to the error sensor output) is described by the second order transfer function:

$$H(z) = \frac{\beta_0 + \beta_1 z^{-1}}{1 - \alpha_1 z^{-1} - \alpha_2 z^{-2}} \tag{6.12.33}$$

The control filter is a first order system:

$$G(z) = \frac{b_0}{1 - a_1 z^{-1}} \tag{6.12.34}$$

and the excitation source (reference signal) is random noise with unity signal power. This problem structure was chosen for several reasons. First, it is reasonably straightforward to obtain relationships which describe to where the adaptive algorithms will converge (the final (mean) filter weight values for the problem), facilitating explanation of the simulation results. Second, it is generally not possible for the control filter to completely attenuate the primary disturbance, and so the efficiency of using the algorithms with a control filter having an order of less than the primary plant (often the case in active noise or vibration control when 'short' filter lengths are used) can be investigated. Finally, the problem format has been used previously in studies of the convergence behaviour of the RLMS algorithm (Johnson and Larimore, 1977), where some interesting results were obtained. As two of the algorithms under investigation are loosely based upon the RLMS algorithm, it is reasonable to expect that this problem format may also be useful in differentiating the performance characteristics of the algorithms of interest here.

Solutions for the mean weight values obtained when employing the various adaptive algorithms outlined here can be derived by finding when the expected value of the gradient estimate used in the algorithm is equal to zero. The general form of the result is

$$\boldsymbol{\theta} = -\boldsymbol{B}^{-1}\boldsymbol{d} \tag{6.12.35}$$

where θ is the vector of control filter weights,

$$\boldsymbol{\theta} = \begin{bmatrix} b_0 \\ a_1 \end{bmatrix} \tag{6.12.36}$$

$\boldsymbol{B}$ is a (2×2) matrix, and $\boldsymbol{d}$ is a (2×1) vector (for example, for the LMS algorithm $\boldsymbol{B}$ would be the reference signal autocorrelation matrix and $\boldsymbol{d}$ the negative value of the crosscorrelation (vector) between the desired signal and reference signal vector, negative because the error is superposition of two signals, rather than the difference between two signals). The general results for values of $\boldsymbol{B}$ and $\boldsymbol{d}$ for the various algorithms are given below. The important point to note at this stage is that the mean weight values for the algorithms are *not*, in general, the same.

For the correct gradient algorithm, the mean weight values are defined by:

$$
\boldsymbol{B} = \begin{bmatrix} \sum_{\iota=0}^{M}\sum_{j=0}^{\iota} c_\iota c_j a_1^{(\iota-j)} & \sum_{\iota=0}^{M}\sum_{j=0}^{M}\sum_{n=0}^{\infty} c_\iota c_j a_1^{(2n+j+1-\iota)} b_0 \\ \sum_{\iota=1}^{M}\sum_{j=0}^{\iota-1} c_\iota c_j (\iota-j) a_1^{(\iota-j-1)} b_0 & \sum_{\iota=0}^{M}\sum_{j=0}^{M}\sum_{n=0}^{\infty} c_\iota c_j (n+1) a_1^{(2n+|\iota-j|)} b_0^2 \end{bmatrix}
$$

$$
\boldsymbol{d} = \begin{bmatrix} \sum_{\iota=0}^{M}\sum_{n=0}^{\infty}\sum_{j=0}^{(\iota+n)\operatorname{div}2} c_\iota a_1^n \alpha_1^\gamma \alpha_2^j \beta_0 \begin{bmatrix}\gamma\\ j\end{bmatrix} + \sum_{\iota=0}^{M}\sum_{n=\eta}^{\infty}\sum_{j=0}^{(\iota+n-1)\operatorname{div}2} c_\iota a_1^n \alpha_1^{(\gamma-1)} \alpha_2^j \beta_1 \begin{bmatrix}\gamma-1\\ j\end{bmatrix} \\ \sum_{\iota=0}^{M}\sum_{n=0}^{\infty}\sum_{j=0}^{(\iota+n+1)\operatorname{div}2} (n+1) c_\iota a_1^n b_0 \alpha_1^{(\gamma+1)} \alpha_2^j \beta_0 \begin{bmatrix}\gamma+1\\ j\end{bmatrix} + \sum_{\iota=\iota}^{M}\sum_{n=0}^{\infty}\sum_{j=0}^{(\iota+n)\operatorname{div}2} (n+1) c_\iota a_1^n b_0 \alpha_1^\gamma \alpha_2^j \beta_1 \begin{bmatrix}\gamma\\ j\end{bmatrix} \end{bmatrix} \tag{6.12.37}
$$

where $\gamma = \iota+n-2j$, $\eta = 1$ if $\iota = 0$, and 0 otherwise, div denotes truncated division, and

$$
\begin{bmatrix}a\\ b\end{bmatrix} = \frac{(a+b)!}{a!b!} \tag{6.12.38}
$$

For the filtered-u algorithm, the mean weight values are defined by

$$
\boldsymbol{B} = \begin{bmatrix} \sum_{\iota=0}^{M} c_\iota^2 & \sum_{\iota=1}^{M}\sum_{j=0}^{\iota-1} c_\iota c_j a_1^{(\iota-j-1)} b_0 \\ \sum_{\iota=1}^{M}\sum_{j=0}^{\iota-1} c_\iota c_j a_1^{(\iota-j-1)} b_0 & \sum_{\iota=0}^{M}\sum_{j=0}^{M}\sum_{n=0}^{\infty} c_\iota c_j a_1^{(2n+|\iota-j|)} b_0^2 \end{bmatrix}
$$

$$
\boldsymbol{d} = \begin{bmatrix} \sum_{\iota=0}^{M}\sum_{j=0}^{\iota\operatorname{div}2} c_\iota \alpha_1^\gamma \alpha_2^j \beta_0 \begin{bmatrix}\gamma\\ j\end{bmatrix} + \sum_{\iota=1}^{M}\sum_{j=0}^{(\iota-1)\operatorname{div}2} c_\iota \alpha_1^{(\gamma-1)} \alpha_2^j \beta_1 \begin{bmatrix}\gamma-1\\ j\end{bmatrix} \\ \sum_{\iota=0}^{M}\sum_{n=0}^{\infty}\sum_{j=0}^{(\iota+n+1)\operatorname{div}2} (n+1) c_\iota a_1^n b_0 \alpha_1^{(\gamma+n+1)} \alpha_2^j \beta_0 \begin{bmatrix}\gamma+n+1\\ j\end{bmatrix} + \sum_{\iota=\iota}^{M}\sum_{n=0}^{\infty}\sum_{j=0}^{(\iota+n)\operatorname{div}2} c_\iota a_1^n b_0 \alpha_1^{(\gamma+n)} \alpha_2^j \beta_1 \begin{bmatrix}\gamma+n\\ j\end{bmatrix} \end{bmatrix} \tag{6.12.39}
$$

where $\gamma = \iota - 2j$.

For the filtered-*u* algorithm with a SHARF smoothing filter, the mean weight values are defined by

$$\boldsymbol{B} =$$

$$\begin{bmatrix} \sum_{\iota=0}^{M} c_\iota^2 + \sum_{\iota=1}^{M} c_\iota c_{\iota-1} a_1 & \sum_{\iota=1}^{M}\sum_{j=0}^{\iota-1} c_\iota c_j a_1^{(\iota-j-1)} b_0 + \sum_{\iota=2}^{M}\sum_{j=0}^{\iota-2} c_\iota c_j a_1^{(\iota-j-1)} b_0 \\ \sum_{\iota=1}^{M}\sum_{j=0}^{\iota-1} c_\iota c_j a_1^{(\iota-j-1)} b_0 + \sum_{\iota=0}^{M}\sum_{j=0}^{\iota} c_\iota c_j a_1^{(\iota-j-1)} b_0 & \sum_{\iota=0}^{M}\sum_{j=0}^{M}\sum_{n=0}^{\infty} c_\iota c_j a_1^{(2n+|\iota-j|)} b_0^2 + \sum_{\iota=0}^{M}\sum_{j=1}^{M}\sum_{n=0}^{\infty} c_\iota c_j a_1^{(2n+|\iota-j|} \end{bmatrix}$$

$$\boldsymbol{d} =$$

$$\begin{bmatrix} \sum_{\iota=0}^{M}\sum_{j=0}^{\iota \operatorname{div} 2} c_\iota \alpha_1^{\gamma} \alpha_2^{j} \beta_0 \begin{pmatrix} \gamma \\ j \end{pmatrix} \\ + \sum_{\iota=1}^{M}\sum_{j=0}^{(\iota-1)\operatorname{div} 2} c_\iota \alpha_1^{(\gamma-1)} \alpha_2^{j} (\beta_1 + a_1\beta_0) \begin{pmatrix} \gamma-1 \\ j \end{pmatrix} + \sum_{\iota=2}^{M}\sum_{j=0}^{(\iota-2)\operatorname{div} 2} c_\iota a_0 \alpha_1^{(\gamma-2)} \alpha_2^{j} \beta_1 \begin{pmatrix} \gamma-2 \\ j \end{pmatrix} \\ \sum_{\iota=0}^{M}\sum_{n=0}^{\infty}\sum_{j=0}^{(\iota+n+1)\operatorname{div} 2} c_\iota a_1^{n} b_0 \alpha_1^{(\gamma+n+1)} \alpha_2^{j} \beta_0 \begin{pmatrix} \gamma+n+1 \\ j \end{pmatrix} \\ + \sum_{\iota=0}^{M}\sum_{n=0}^{\infty}\sum_{j=0}^{(\iota+n)\operatorname{div} 2} c_\iota a_1^{n} b_0 \alpha_1^{(\gamma+n)} \alpha_2^{j} (\beta_1 + a_1\beta_0) \begin{pmatrix} \gamma+n \\ j \end{pmatrix} \\ + \\ \sum_{\iota=0}^{M}\sum_{n=\eta}^{\infty}\sum_{j=0}^{\iota+n-1} c_\iota a_1^{(n+1)} b_0 \alpha_1^{(\gamma+n-1)} \alpha_2^{j} \beta_1 \begin{pmatrix} \gamma+n-1 \\ j \end{pmatrix} \end{bmatrix} \quad (6.12.40)$$

For the first computer simulation, the cancellation path transfer function will be taken to be simply a phase inversion. This will simplify the analytical evaluation of algorithm convergence, while still making it necessary to compensate for the cancellation path transfer function is algorithm stability is to be maintained. With this cancellation path transfer function, the quantities defining the final mean weight values can be significantly shortened. For the correct gradient algorithm, the final mean weight values are defined by the quantities

$$\boldsymbol{B} = \begin{bmatrix} 1 & \sum_{n=0}^{\infty} a_1^{2n+1} b_0 \\ 0 & \sum_{n=0}^{\infty} (n+1) a_1^{2n} b_0^2 \end{bmatrix}$$

$$\boldsymbol{d} = \begin{bmatrix} -\sum_{n=0}^{\infty}\sum_{j=0}^{\infty} a_1^{n+2j} \alpha_1^n \alpha_2^j b_0 \begin{Bmatrix} i \\ j \end{Bmatrix} \\ -\sum_{n=0}^{\infty}\sum_{j=0}^{\infty} a_1^{(n+2j+1)} \alpha_1^n \alpha_2^j b_1 \begin{Bmatrix} n \\ j \end{Bmatrix} \\ -\sum_{n=0}^{\infty}\sum_{j=0}^{\infty} (n+2j) a_1^{(n+2j-1)} \alpha_1^n \alpha_2^j b_0 \beta_0 \begin{Bmatrix} n \\ j \end{Bmatrix} \\ -\sum_{n=0}^{\infty}\sum_{j=0}^{\infty} (n+2j+1) a_1^{(n+2j)} \alpha_1^n \alpha_2^j b_0 \beta_1 \begin{Bmatrix} n \\ j \end{Bmatrix} \end{bmatrix} \tag{6.12.41}$$

where

$$\begin{Bmatrix} a \\ b \end{Bmatrix} = \frac{(a+b)!}{a!b!} \tag{6.12.42}$$

The estimated gradient algorithm will converge to the same weight values.

For the filtered-u algorithm, the final mean weight values are defined by the quantities

$$\boldsymbol{B} = \begin{bmatrix} 1 & 0 \\ 0 & \sum_{n=0}^{\infty} a_1^{2n} b_0^2 \end{bmatrix}$$

$$\boldsymbol{d} = \begin{bmatrix} -\beta_0 \\ -\sum_{n=0}^{\infty}\sum_{j=\eta}^{\infty} a_1^{(n+2j-1)} \alpha_1^n \alpha_2^j b_0 \beta_0 \begin{Bmatrix} n \\ j \end{Bmatrix} \\ -\sum_{n=0}^{\infty}\sum_{j=0}^{\infty} a_1^{(n+2j)} \alpha_1^n \alpha_2^j b_0 \beta_1 \begin{Bmatrix} n \\ j \end{Bmatrix} \end{bmatrix} \tag{6.12.43}$$

where η = 1 for n = 0, and 0 otherwise. The filtered-u algorithm with the SHARF smoothing filter will converge to the same set of weight values as the the regular filtered-u algorithm.

Two problems will be considered for the system with a phase inversion cancellation path transfer function, both of which have been used previously to examine convergence behaviour of adaptive IIR filters (Johnson and Larimore, 1977). For the first of these, the primary disturbance process is governed by the transfer function

$$G(z) = \frac{0.05}{1-1.75z^{-1}+0.81z^{-2}} \tag{6.12.44}$$

Illustrated in Fig. 6.62 is the average (of 50 ensemble members which defined mean square error) convergence path for all four adaptive algorithms described in the previous section. The starting location of the convergence path is (a_1 = 0.5, b_0 = 0.4). For all but the estimated gradient algorithm, an exact model of the cancellation path transfer function was used. For the estimated gradient algorithm, random noise with an amplitude 30 dB below that of the reference signal was injected into the system in the location shown in Fig. 6.61 to model the total transfer function. When the adaptation process was started, the transfer function model was zeroed; the algorithm started without any *a priori* knowledge of its operating environment. The total transfer function was modelled using an eight tap FIR filter. The time frame shown for algorithm convergence was 4000 iterations for all but the estimated gradient algorithm, which was 8000 iterations. These convergence times could be improved upon in all cases, but with increased overshoot, which tended to mask the final weight locations.

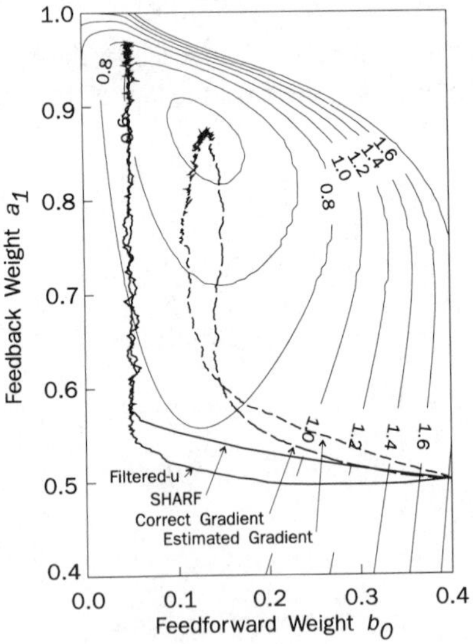

Fig. 6.62 Convergence path of the four adaptive algorithms, average of 50 ensemble members.

The main point of interest is that neither the filtered-u, nor the filtered-u algorithm with the SHARF smoothing filter, converge to the error surface minimum, while both the correct gradient and estimated gradient algorithms do. The former result is predicted by the analytical results of (6.12.43), which shows that the feedforward weight *must* converge to the corresponding (β_0) value of feedforward weight in the primary source transfer function (0.05); it is not a function of any other parameters. The converged feedback weight value is a function of all filter parameters, but does not converge to the correct solution either, presumably because of the error in the feedforward weight. This lack of convergence to the global minimum in a unimodal (one minimum) error surface is a result of using an incorrect gradient algorithm with a control system having an order less than that of the primary system which is the target of its actions. Had the control filter and primary plant been of the same dimension, or the control filter had a greater dimension, convergence would have been to a minimum. This begs the question, how often in an active control problem is the order of the primary plant known?

As mentioned earlier, the fidelity of the total transfer function model can be expected to have a bearing upon the convergence behaviour of the estimated gradient algorithm. Figure 6.63 illustrates the convergence of the estimated gradient algorithm using a one, four, and eight tap FIR model of the total transfer function. As the accuracy of the model (and hence tap point numbers) increases, so too does the accuracy of the algorithm. Observe, by comparing these results to those in Fig. 6.62, that even with a one tap model, the levels of attenuation achieved using the estimated gradient algorithm are superior to those achieved by either version of the filtered-u algorithm.

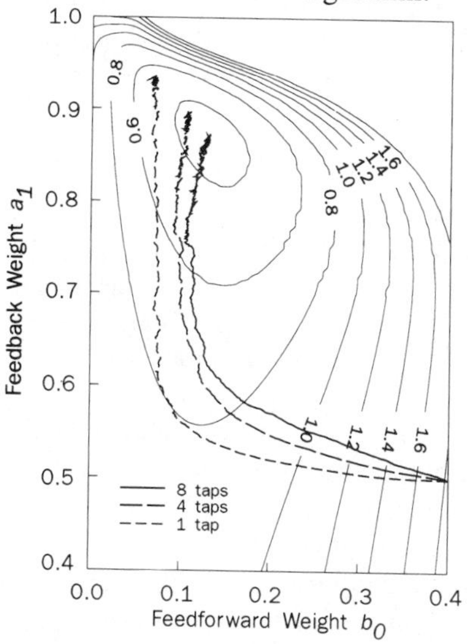

Fig. 6.63 Average convergence of adaptive algorithm with different cancellation path transfer function model lengths.

The second case simulated with a phase inversion cancellation path transfer function is also taken from Johnson and Larimore (1977), where the primary source transfer function is described by

$$H(z) = \frac{0.05 - 0.4z^{-1}}{1 - 1.1314z^{-1} + 0.25z^{-2}} \tag{6.12.45}$$

The error surface for this problem differs from that of the previous problem in that it is bimodal (has two error surface minima).

The average weight convergence path for the filtered-*u* algorithm, with and without the SHARF smoothing filter, for three different starting points is shown in Figs. 6.64 and 6.65 respectively. Note that in each instance the filter weights converge to the same value, which does *not* correspond to the global, or even a local, error surface minimum. As before, the feedforward weight converges to a value of = 0.05.

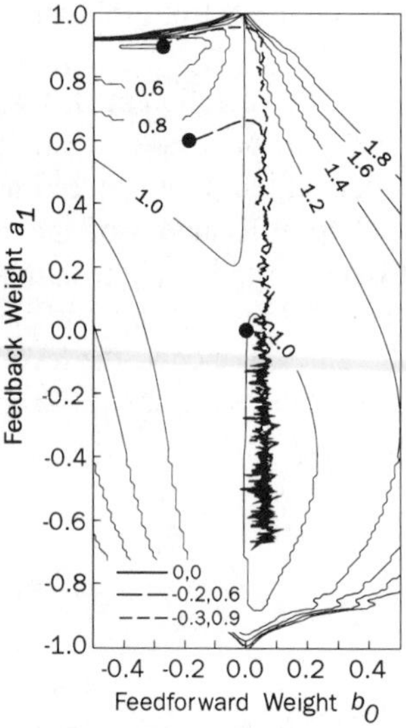

Fig. 6.64 Average convergence paths for three different starting points, filtered-*u* algorithm with SHARF smoothing filter.

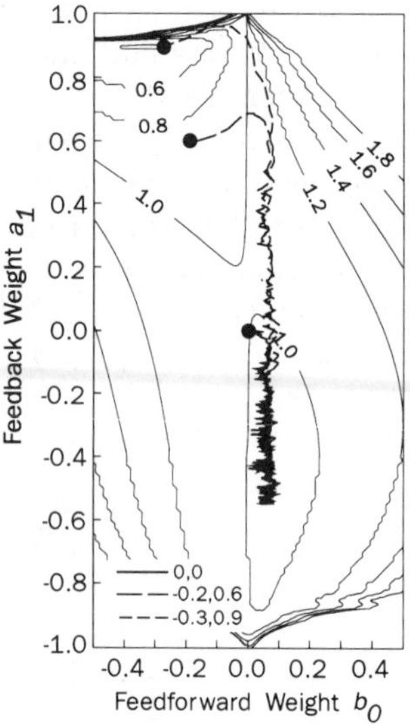

Fig. 6.65 Average convergence paths for three different starting points, filtered-*u* algorithm without SHARF smoothing filter.

The average convergence path for the filter weights calculated using the correct gradient algorithm starting from two different initial weight values is shown in Fig. 6.66. Here the weight values converge to the nearest minimum on the error surface, either the global or the weak local. This is the type of behaviour expected from a gradient descent algorithm.

The average convergence path for the filter weights calculated using the estimated gradient algorithm with three different initial values is depicted in Fig. 6.67. Observe that regardless of the starting location, even when starting with weight values corresponding to the error surface local minimum, the algorithm

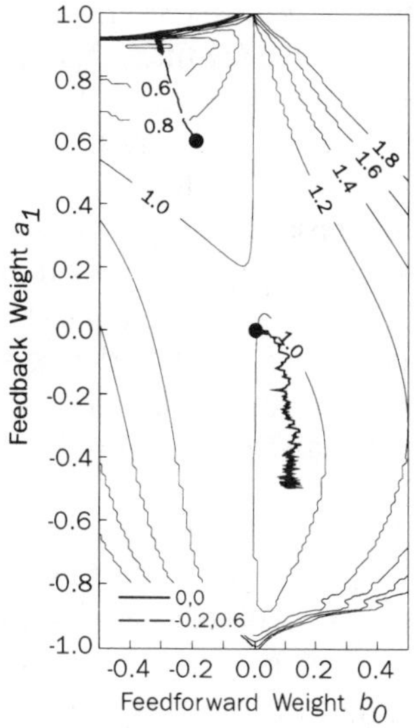

Fig. 6.66 Average convergence paths for two different starting points, correct gradient algorithm.

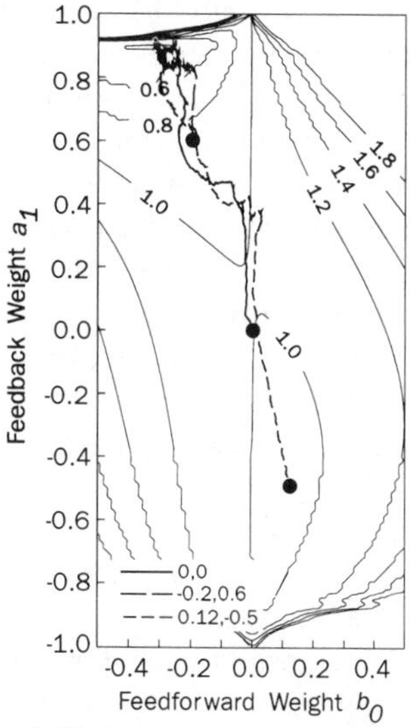

Fig. 6.67 Average convergence paths for three different starting points, estimate gradient algorithm.

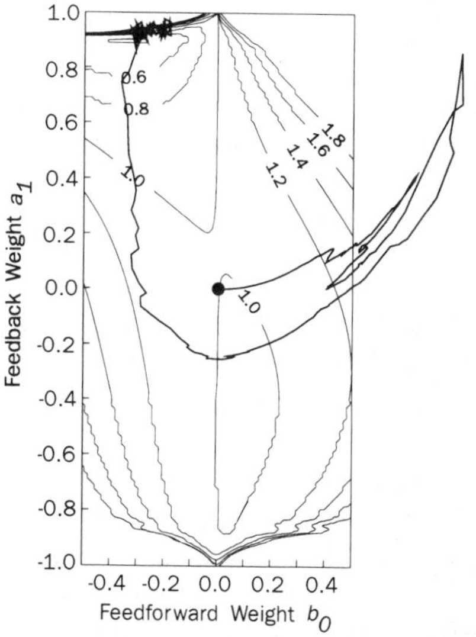

Fig. 6.68 Single convergence path for estimated gradient algorithm.

converges to the *global* minimum. Occasionally the path of convergence can appear quite 'tortuous', as illustrated by the single convergence path plotted in Fig. 6.68. The weights do, however, always converge to the same values. The reason for this improvement in convergence behaviour over that of the 'correct' algorithm is unknown at this stage. It was initially thought that the addition of random noise into the feedback part of the filter was responsible for the improved performance. However, adding random noise into the filter employing the correct gradient algorithm had no affect upon its convergence behaviour. Whatever the reason, the convergence behaviour is certainly desirable.

For the final simulation problem, the cancellation path transfer function was changed to one described by the finite impulse response vector (c_0, c_1, c_2) = (0.50, 0.35, 0.30). The primary plant was as described in (6.12.45). With this cancellation path transfer function, the error surface becomes unimodal. In addition, the filtered-*u* algorithm implementation employing the SHARF smoothing filter will converge to a different location than that of the standard filtered-*u* algorithm, a fact apparent from the relationships given in (6.12.39) and (6.12.40).

For this final simulation, Fig. 6.69 illustrates the average convergence path for the filter weights under the action of the four adaptive algorithms of interest. Note that, as with the previous unimodal problem, only the correct and estimated gradient algorithms converge to the error surface minimum. While the filtered-*u* algorithm employing the SHARF smoothing filter has a performance superior to that of the 'plain' version, it still fails to converge to the correct location.

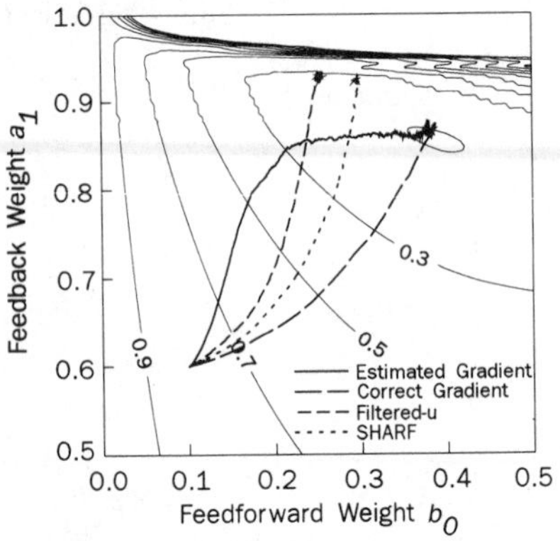

Fig. 6.69 Average convergence path of the four adaptive algorithms.

Based upon these simulation results, it is clear that if a small IIR filter is to be used for broadband active control, employment of the filtered-*u* algorithm, with or without the SHARF smoothing filter, does not guarantee an optimal result. Employment of either the correct or estimated gradient algorithm may provide some improvement. However, if the disturbance is tonal, or the filter long, past experience suggests that the filtered-*u* algorithm provides adequate performance.

6.13 ADAPTIVE FILTERING USING ARTIFICIAL NEURAL NETWORKS

The FIR and IIR adaptive filtering arrangements outlined thus far in this chapter have one common characteristic: they are both linear. That is, the output of each filter is a linear function of the present and past inputs and/or outputs. There will be instances, however, where the filtering problem of interest will have some form of non-linearity associated with it. In active noise and vibration control, examples of this arise where a sinusoidal reference signal is used to derive a signal to control a disturbance containing both the reference tone and several harmonics; systems where the control actuator has some non-linear performance characteristic, such as where it introduces harmonics into the system, which must be compensated for; and where a power based (intensity) error signal is used, which will be twice the frequency of the reference signal. In each of these cases a linear filter will produce suboptimal results at best. What is desired in these situations is a non-linear filter. One such filtering arrangement which has received increased attention in recent years is the artificial neural network, or simply neural net.

The architecture of artificial neural networks is, roughly speaking, based upon the current understanding of the human brain. As might be expected from these beginnings, there are a wide variety of proposed neural network structures and associated 'training' (adaptive) algorithms. All, however, are constructed using dense, parallel interconnections of simple computational units, or nodes. This parallelism provides a neural network with the 'side benefit' of a degree of fault tolerance not present in the sequential filtering arrangements discussed thus far. Typically, each node has associated with it some non-linearity, which provides the network with the capability of performing non-linear filtering operations. In addition to these characteristics, the roots from which neural networks have been spawned lend a romantic touch to the implementation of adaptive filters! Thus this nomenclature will be retained here, although a strong case can be made that neural network implementation is simply a form of multivariate statistical analysis.

Rather than describe the range of known neural network structures, only one will be considered here; the multilayer perceptron feedforward neural network. There are three reasons for this. First, this form of network has received the most attention from the control community as a potential non-linear filtering tool. Second, the network can be implemented easily with discrete signals, making it amenable to microprocessor implementation. Third, as will be shown shortly, this network and its associated training algorithm can be considered as generalizations of the linear FIR adaptive filtering arrangement considered previously in this chapter. The interested reader is directed to a number of books and review papers for information on other neural network topographies and algorithms (Rumelhart *et al.*, 1986; Lippman, 1987; Hertz *et al.*, 1991; Kosko, 1992a, 1992b).

In this section we will outline the 'basics' of adaptive filtering using artificial neural networks. In the section which follows we will adapt these systems for use in active noise and vibration control arrangements.

6.13.1 The perceptron

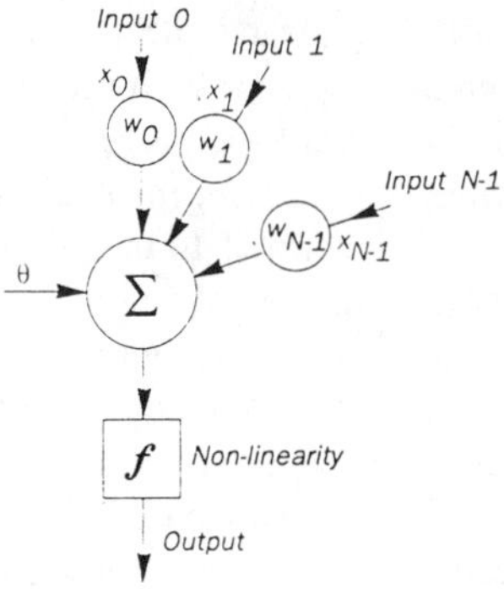

Fig. 6.70 A single perceptron.

As mentioned earlier, neural networks are constructed from dense interconnections of simple computational units, or nodes. For the networks considered here, these units are the perceptron, shown in Fig. 6.70. The output of the perceptron, y, is derived by summing a set of weighted input signals and a nodal bias, θ, and passing the result through some non-linear function, f:

$$y = f\left[\sum_{i=0}^{N-1} w_i x_i + \theta\right] \tag{6.13.1}$$

It should be noted that in practice the nodal bias, θ, is normally implemented using an additional (adaptive) weight with a constant unitary input signal, $x_{\text{bias}} = 1.0$ (the authors have found, however, that a smaller value, say $x_{\text{bias}} = 0.1$, will sometimes lead to a more stable implementation of the adaptive algorithm). The bias will not be explicitly stated in the remainder of this section, rather being simply considered as another input and weight.

The most commonly used non-linear functions for the type of neural network being considered here are the sigmoid function

$$f(x) = \frac{1}{1+e^{-x}} \tag{6.13.2}$$

and the hyperbolic tangent

$$f(x) = \frac{1-e^{-2x}}{1+e^{-2x}} \tag{6.13.3}$$

These functions are plotted in Fig. 6.71. An important characteristic of these functions, for reasons to become apparent shortly, is they are both differentiable. It should be pointed out that it is also possible to have linear nodes, without any non-linear function, such that the node is essentially an FIR filter with several inputs as opposed to a single input and a tapped delay line.

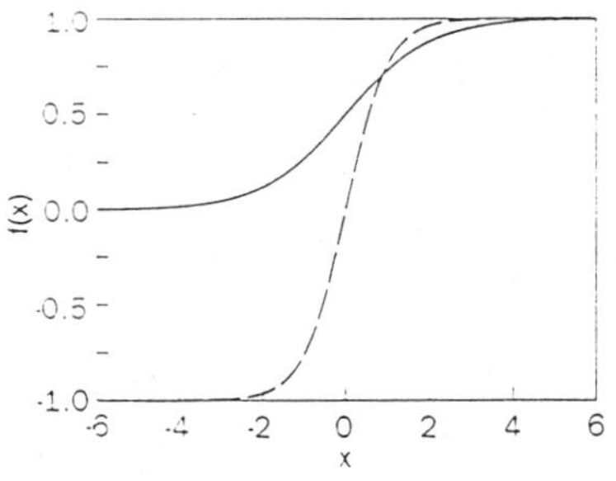

Fig. 6.71 Common nodal non-linearities: solid line is the sigmoid function, dashed line is a hyperbolic tangent.

In viewing the perceptron, the similarities between it and a standard FIR filter are obvious. In fact, a single perceptron, or a single 'layer' of perceptrons as illustrated in Fig. 6.72, has many of the characteristics of the FIR filter. The simplest way to examine these characteristics is to consider classification, or mapping, problems. In these problems a set of inputs is presented to the network, and the network is 'trained' to differentiate the sets into specific groups. It can be shown that a single perceptron is capable of distinguishing between the sets provided the groups are separated by a hyperplane, as shown in Fig. 6.73 (a hyperplane is simply the generalization of a straight line). This, however, is the limit of the single perceptron, which can prove to be a significant weakness (Minsky and Papert, 1969). This limitation can be overcome, however, by adding additional layers of perceptrons to form a 'network', as shown in Fig. 6.74.

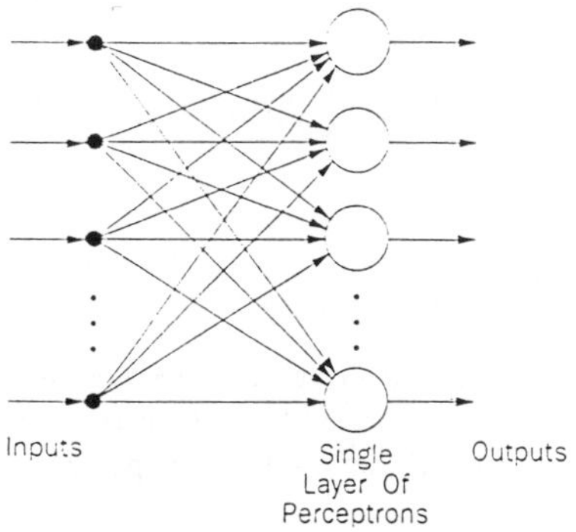

Fig. 6.72 Implementation of a single layer of perceptrons.

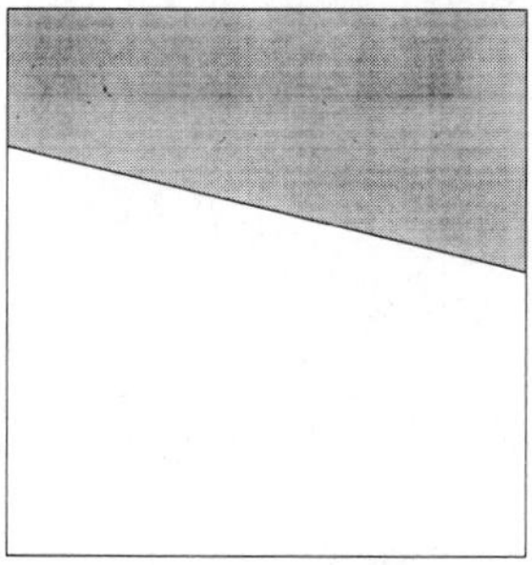

Fig. 6.73 Typical decision region which can be formed by a single layer of perceptrons, a 'hyperline' division.

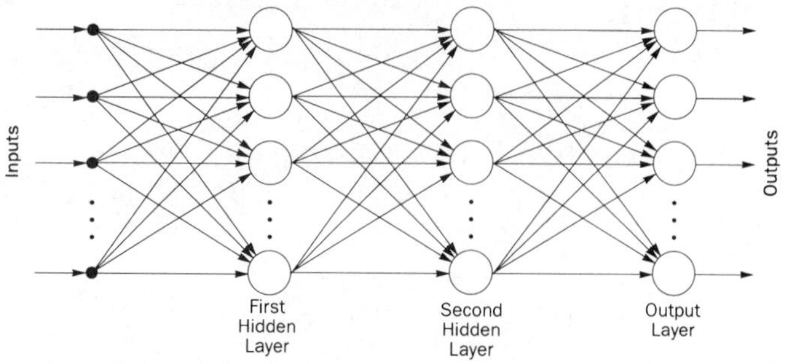

Fig. 6.74 Feedforward neural network constructed from several layers of perceptrons.

We may now ask the question, how many layers and nodes are required to solve a given problem? The question of layers is answered more easily than the question of nodes. As was stated earlier, each perceptron is capable of separating a given decision space by a single line. A layer of perceptrons is therefore capable of providing as many lines as there are perceptrons (nodes) in that layer. A second layer is able to combine these lines to form a single, enclosed boundary, such as that shown in Fig. 6.75. Continuing with this line of thought, a third layer is able to place a number of these enclosing boundaries, of differing shapes, at arbitrary locations in the decision space, as shown in Fig. 6.76. In fact, it can be shown mathematically that a three layer neural network can generate any arbitrary decision region (Lippmann, 1987).

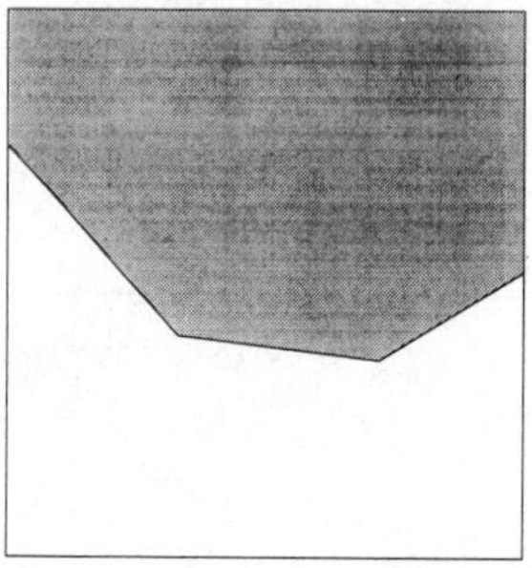

Fig. 6.75 Typical decision region which can be formed by a two layer neural network, a convex region.

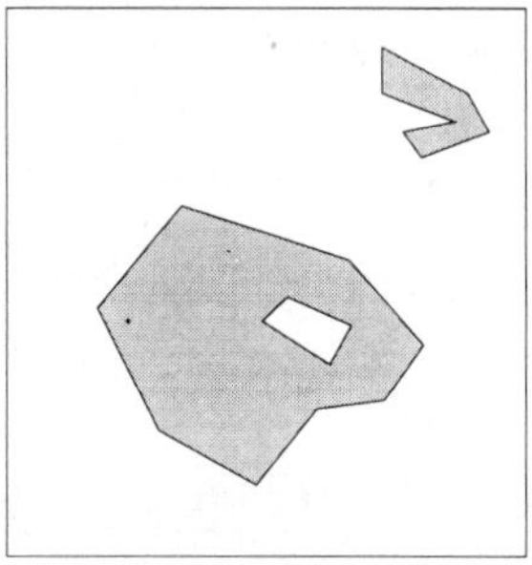

Fig. 6.76 Typical decision region which can be formed by a three layer neural network, an arbitrary shape.

The point of such an analysis is that no more than three layers are required in (feedforward) perceptron based networks because three layer networks can generate arbitrarily complex decision spaces (Lippmann, 1987). The only problem is that there is no such precise guidance at present concerning the number of nodes in each layer required to actually do it. It is possible, however, to arrive at some qualitative criteria in relation to the relative number of nodes in the layers based upon the discussion of the previous paragraph (following

Lippmann, 1987). The number of nodes in the second layer, which are responsible for combining the planar decision boundaries of the first layer into enclosed boundaries, must be greater than one if the decision regions are disconnected or cannot be constructed from a single convex-shaped area. In the worst case, the number of second layer nodes required will be equal to the number of disconnected regions in the input distributions. The number of nodes in the first layer should typically be sufficient to provide three or more edges for each convex area generated by every second layer node. Therefore, there should typically be three or more times as many nodes in the first layer as in the second.

The neural networks considered here are to be applied to a dynamic system, and as such, a simple static mapping operation will not be sufficient to produce the results we desire. For this reason, the neural networks will be given an input signal via a tapped delay line, as shown in Fig. 6.77. In this form, the feedforward neural network really does appear to be simply a generalization of the FIR filter.

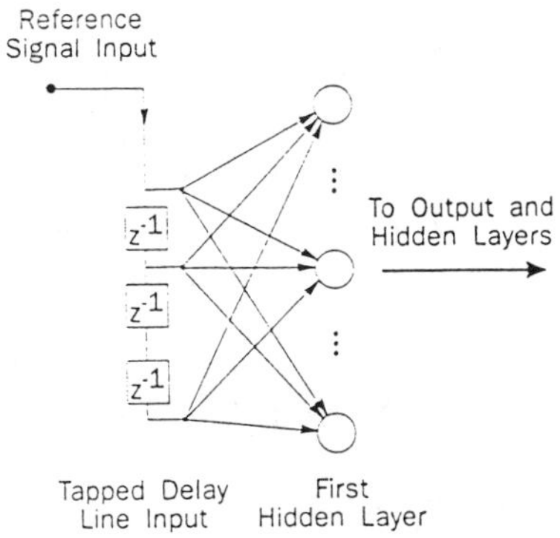

Fig. 6.77 Tapped delay line providing an input to a neural network.

One final point which should be mentioned here is the naming of the layers in a feedforward neural network. Referring back to Fig. 6.74, the final layer is referred to as the output layer, for obvious reasons. The internal layers are referred to as hidden layers, as their existence is not visible to the outside world.

6.13.2 The back propagation algorithm

The discussion of the previous section provides some indication of the capabilities of multilayer perceptron neural networks. However, some means of 'training' the network, or adapting the weight coefficients, must be developed to realise this potential. The most common form of algorithm utilized to conduct this training is the gradient descent algorithm, which we have used throughout this chapter. However, unlike the 'bowl' shaped error surface of the FIR filter, the error surface of the neural network may not be so well behaved; it may contain many local minima, capable of trapping any gradient descent based algorithm in a suboptimal solution. Despite this, the gradient descent-based algorithm 'appears' to converge to the desired solution in the majority of cases (Rumelhart *et al.*, 1986). More quantitative descriptions of the algorithm performance are the topic of on-going research.

As we have used throughout this chapter, the ideal error criterion for the adaptive filtering problem is minimization of the mean square value of the error signal:

$$\text{minimize } \xi(k) = E\left\{\sum_{i=1}^{N_{out}} e_i^2(k)\right\} \tag{6.13.4}$$

Note that the error criterion is based upon multiple, N_{out}, outputs from the neural network. The gradient descent algorithm acts to adjust the weights of the neural network to minimize this performance measure by adding to each weight a portion of the negative gradient of the error criterion with respect to the weight of interest,

$$w(k+1) = w(k) - \mu\Delta w(k) \tag{6.13.5}$$

where Δw is the gradient of the squared error as a function of the weight value, and μ is the convergence coefficient, or portion of the negative gradient to be added. What is required to implement the algorithm is an expression for the gradient with respect to each weight in the neural network. For practicality, the stochastic approximation

$$\xi(k) \approx \sum_{i=1}^{N_{out}} e_i^2(k) = \sum_{i=1}^{N_{out}} \left(d_i(k) - y_i(k)\right)^2 \tag{6.13.6}$$

is made, where $d_i(k)$ is the ith desired signal and $y_i(k)$ is the ith neural network output. As $d_i(k)$ is in no way a function of the weights of the neural network,

$$\Delta w \approx \frac{\partial e^2(k)}{\partial w} = 2e(k)\frac{\partial e(k)}{\partial w} = -2e(k)\frac{\partial y(k)}{\partial w} \tag{6.13.7}$$

The aim of the following analysis, based upon the preceding framework, is to find a solution to (6.13.7) for each weight in the controller neural network. The solution can then be substituted into the generic gradient descent algorithm of (6.13.5) to adapt each weight in the control system. Before attempting this for a generic feedforward neural network, it will prove useful to consider a simple 'single path' model as shown in Fig. 6.78, which traces the data through one possible route through a neural network with one hidden layer and an output layer.

Starting at the end of the single path, the output of the neural network is

$$y(k) = f_o\{x_h(k)\ w_o(k)\} \tag{6.13.8}$$

where $x_h(k)$ is the output from hidden layer node h at time k, and $w_o(k)$ is the weighting factor of this signal used as an input to the output layer node o.

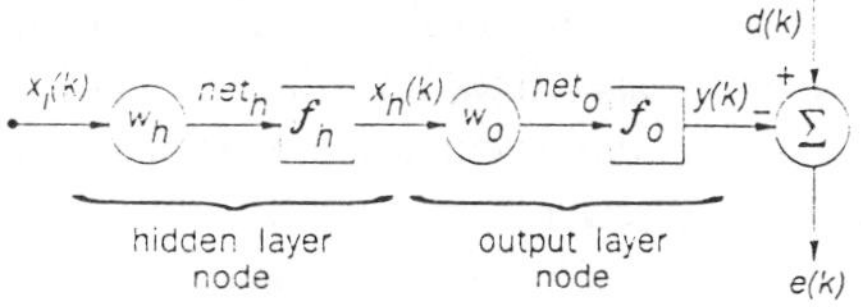

Fig. 6.78 Single data flow path through a two layer neural network.

Differentiating (6.13.8), the gradient of the error criterion given in (6.13.7) with respect to the weight w_o is found to be

$$\Delta w_o = 2e(k)\dot{f}_o\{x_h(k)\ w_o(k)\}x_h(k) = 2e(k)\dot{f}_o\{net_o(k)\}x_h(k) \quad (6.13.9)$$

where net_o is the result of the o nodal multiplication/accumulation operation (the same as an FIR filtering operation, although with only one weight in this instance), which is to be filtered through the nodal non-linearity

$$net_o(k) = x_h(k)w_o(k) \quad (6.13.10)$$

Equation (6.13.9) shows the importance of being able to differentiate the nodal non-linearity, as this derivative is required to calculate the gradient estimate.

To find the gradient with respect to the weights in the (hidden) layer preceding this, $x_h(k)$ can be rewritten as

$$x_h(k) = f_h\{x_i(k)\ w_h(k)\} \quad (6.13.11)$$

Substituting this expansion into (6.13.8) and differentiating, the gradient estimate for weight w_h is found to be

$$\Delta w_h = 2e(k)\dot{f}_o\{net_o(k)\}\ w_o(k)\ \dot{f}_h\{net_h\}\ x_i(k) \quad (6.13.12)$$

Note that the gradient estimates given in (6.13.9) and (6.13.12) can be expressed in the form

$$\Delta w_{\text{node},input} = \delta_{\text{node}}(k)x_{\text{input}}(k) \quad (6.13.13)$$

where x_{input} is the input to the node of interest (x_h for the output layer node and x_i for the hidden layer node), whose weighting function is $w_{\text{node,input}}$, and δ_{node} is the 'back propagated' error for the node of interest:

$$\delta_{node}(k) = \begin{cases} 2e(k)\dot{f}\{net_{\text{output}}(k)\} & \text{output layer node} \\ \dot{f}\{net_{node}(k)\} \displaystyle\sum_{j=1}^{N_{\text{nodes (layer+1)}}} \delta_j(k)\ w_{j,\text{node}}(k) & \text{other nodes} \end{cases} \quad (6.13.14)$$

where $N_{\text{nodes(layer+1)}}$ is the number of nodes in the layer immediately following ('downstream') the layer of interest, and $w_{j,\text{node}}(k)$ is the current value of the weighting coefficient for the output of the node of interest used as an input to node j in the next layer, layer+1.

The concepts of this simple example can be generalized to derive a (stochastic) gradient descent algorithm for a feedforward neural network of any size. Returning to (6.13.7), the gradient expression for the weights of the ith node of the jth layer can be rewritten as

$$\frac{\partial e^2}{\partial w_{i,j}} = \frac{\partial e^2}{\partial net_{j,i}} \frac{\partial net_{j,i}}{\partial w_{j,i}} \tag{6.13.15}$$

In (6.13.15) $net_{j,i}$ is the product of the multiply/accumulate (FIR filtering) operation

$$net_{j,i} = \boldsymbol{x}_{j-1}{}^{\mathrm{T}} \boldsymbol{w}_{j,i} \tag{6.13.16}$$

where $\boldsymbol{x}_{j-1}$ is the vector of outputs from the previous $j - 1$ layer, which are the inputs to the jth layer. Therefore, the second term on the right hand side of (6.13.15) is simply the input to the node,

$$\frac{\partial net_{j,i}}{\partial \boldsymbol{w}_{j,i}} = \boldsymbol{x}_{j-1} \tag{6.13.17}$$

To simplify the first term on the right-hand side of (6.13.15), let us rewrite the partial derivative as

$$\frac{\partial e^2}{\partial net_{j,i}} = \frac{\partial e^2}{\partial o_{j,i}} \frac{\partial o_{j,i}}{\partial net_{j,i}} \tag{6.13.18}$$

where o is the node output after passing through the non-linearity

$$o_{j,i} = f_{j,i}\{net_{j,i}\} \tag{6.13.19}$$

Therefore, the second term on the right-hand side of (6.13.18) is

$$\frac{\partial o_{j,i}}{\partial net_{j,i}} = \dot{f}_{j,i}\{net_{j,i}\} \tag{6.13.20}$$

Once again, note the importance of being able to differentiate the non-linear function.

The first term on the right-hand side of (6.13.18) will vary depending upon where the node and layer are placed in the network. If the node is located in the output layer, then $o_{j,i}=y_i$, and so

$$\frac{\partial e^2}{\partial o_{j,i}} = 2e \tag{6.13.21}$$

Therefore, for an output node, substituting (6.13.17), (6.13.19), and (6.13.21) into (6.13.15) produces the expression for the gradient

$$\Delta \boldsymbol{w}_{j,i} = 2\boldsymbol{x}_{j-1}\dot{f}_{j,i}\{net_{j,i}\}e_i = \boldsymbol{x}_{j-1}\delta_{j,i} \tag{6.13.22}$$

where for the output layer node j,i,

$$\delta_{j,i} = 2e_i\dot{f}\{net_{j,i}\} \tag{6.13.23}$$

Note that this equation is exactly the same as (6.13.14), derived for the simple single path problem.

If the node is located in the first hidden layer, immediately preceding the output layer, then we can rewrite (6.13.18) as

$$\frac{\partial e^2}{\partial net_{j,i}} = \frac{\partial e^2}{\partial o_{j+1,\iota}} \frac{\partial o_{j+1,\iota}}{\partial net_{j,i}} \tag{6.13.24}$$

where $o_{j+1,\iota}$ is the output from the output layer node ι. The first term on the left hand side of this expression was solved for in (6.13.21). It is straightforward to show that the second term is

$$\frac{\partial o_{j+1,\iota}}{\partial net_{j,i}} = \dot{f}_{j+1,\iota}\{net_{j+1,\iota}\}\ w_{j+1,\iota,i}\ \dot{f}_{j,i}\{net_{j,i}\} \tag{6.13.25}$$

where $w_{j+1,\iota,i}$ is the coefficient in the $(j+1)$ output layer node ι which weights the input signal taken from the output of node i in the previous (jth) layer. Considering all nodes in the $(j+1)$ output layer, the gradient estimate for the weights in a node of the first hidden layer is

$$\Delta \boldsymbol{w}_{j,i} = \boldsymbol{x}_{j-1}\ \dot{f}_{j,i}\{net_{j,i}\} \sum_{\iota=1}^{N_{\text{node } j+1}} w_{j+1,\iota,i}\ \dot{f}_{j+1,\iota}\{net_{j+1,\iota}\}\ 2e_\iota = \boldsymbol{x}_{j-1}\delta_{j,i} \tag{6.13.26}$$

where $N_{\text{node } j+1}$ is the number of node in layer $j+1$, the output layer in this instance, and

$$\begin{aligned} \delta_{j,i} &= \dot{f}_{j,i}\{net_{j,i}\} \sum_{\iota=1}^{N_{\text{node } j+1}} w_{j+1,\iota,i}\ \dot{f}_{j+1,\iota}\{net_{j+1,\iota}\}\ 2e_\iota \\ &= \dot{f}_{j,i}\{net_{j,i}\} \sum_{\iota=1}^{N_{\text{node } j+1}} w_{j+1,\iota,i}\delta_{j+1,\iota} \end{aligned} \tag{6.13.27}$$

It is straightforward to show that the form of (6.13.27) applies to all nodes and layers in the neural network. Therefore, the gradient descent algorithm for the feedforward neural network is as stated in (6.13.5), where the gradient estimate is simply a generalisation of (6.13.13) and (6.13.14), equal to

$$\Delta \boldsymbol{w}_{node\ ,\ input} = \delta_{\text{node}}(k)\boldsymbol{x}_{input}(k) \tag{6.13.28}$$

where $\boldsymbol{x}_{\text{input}}(k)$ is the vector of input values to the node of interest at time k, and δ_{node} is the 'back propagated' error for the node of interest,

$$\delta_{node}(k) = \begin{cases} 2e(k)\dot{f}\{net_{\text{output}}(k)\} & \text{output layer node} \\ \dot{f}\{net_{\text{node}}(k)\} \displaystyle\sum_{j=1}^{N_{\text{nodes (layer}+1)}} \delta_j(k)\ w_{j,\text{node}}(k) & \text{other nodes} \end{cases} \tag{6.13.29}$$

This is referred to as the back propagation algorithm, because the procedure used to update the weights in the network is to start at the output layer nodes, compute the gradient estimates, then 'back propagate' the error terms, the δ's, to the nodes in the preceding layer by multiplying them by the weight coefficients used in the data path between the two nodes, and to proceed in this way until the first layer in the network is reached. Therefore, when implementing the neural network one works from the top down to derive an output, then from the bottom up to adapt the weights.

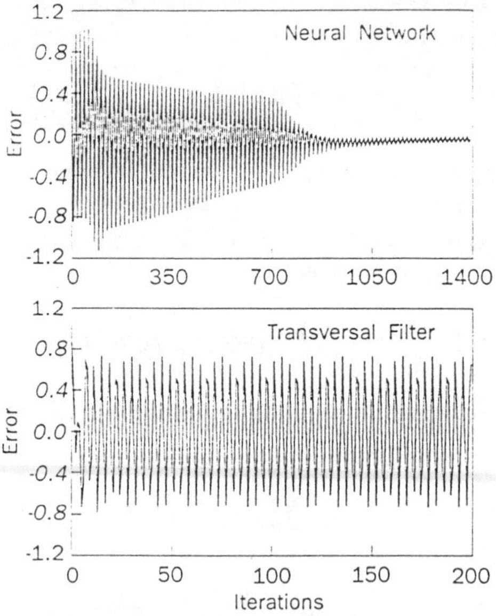

Fig. 6.79 Residual error for an 8 input node, 8 hidden layer node neural network, and an 8 tap FIR filter, where a tonal reference signal is used to derive a signal to suppress a primary disturbance consisting of the tone plus its first harmonic.

While the use of neural networks is still in its infancy, they are known to possess some characteristics which make their application to active noise and vibration control problems attractive. As an example of this, consider the problem of using a tonal reference signal to control a primary disturbance consisting of the tone and an equal magnitude portion of the first harmonic. Figure 6.79 illustrates the residual error signal for a neural network with 8 input nodes and 8 hidden layer nodes, as well as the residual error for control using a standard 8 tap transversal filter, adapted using the LMS algorithm. For both of these cases the sampling rate was taken as 7.5 times the reference signal tonal frequency. Comparing these it can be seen that the FIR filter quickly suppresses the tone (the change in error at the very beginning of the plot), but is incapable of providing any suppression of the harmonic (as is expected). The neural network, however, is capable of suppressing both the tone and the harmonic. The output signal from the neural network, shown unfiltered in Fig. 6.80, clearly illustrates the formation of the harmonic due to the non-linearity of the hidden layer. The use of neural networks to control harmonics of the feedforward controller reference signal will be considered further in the next section.

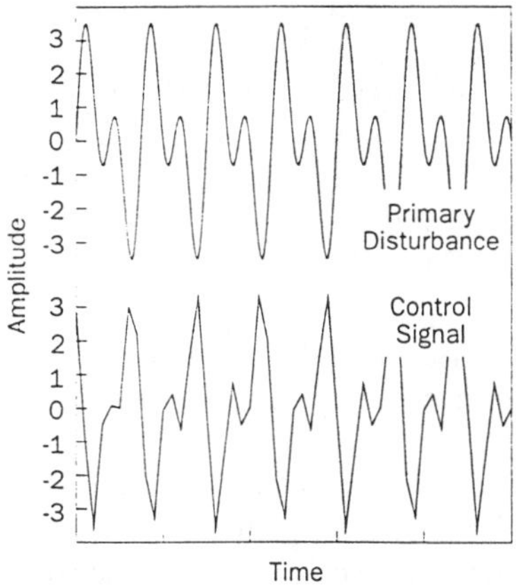

Fig. 6.80 Primary disturbance and neural network control signals after 1400 iterations, corresponding to the final error shown in Fig. 6.79.

6.14 NEURAL NETWORK BASED FEEDFORWARD ACTIVE CONTROL SYSTEMS

Having now studied the 'basics' of adaptive filtering using artificial neural networks, we are in a position to explore their use in feedforward active control systems. Basically, the feedforward neural network investigated in the previous section is simply a non-linear generalization of an FIR filter, and the (gradient descent) back propagation algorithm used to adapt it simply a generalization of the LMS algorithm. In light of this and our previous work in this chapter it is intuitively obvious that the algorithm must be modified when implementing the neural network in an active control system to account for the cancellation path transfer function(s) if algorithm stability is to be maintained. To derive an algorithm which will facilitate stable adaptation of a neural network based feedforward active control system, we can view the control problem as shown in Fig. 6.81. With this view, the cancellation path transfer function is modelled as a second neural network, the input of which is the control signal and the output of which is the feedforward control signal measured at the output of the error sensor. Using the idea of back propagation we discussed in the previous section, the measured error signal can then be simply back propagated through the transfer function model to be suitably conditioned for stable adaptation of the controller neural network.

Literature on the use of neural networks for non-linear feedforward control problems is widespread (see, for example, Narendra and Parthasarathy, 1990, 1991; Nguyen and Widrow, 1991; Hoskins *et al.*, 1991; Saint-Donat *et al.*, 1991; Snyder and Tanaka, 1992, 1993b; Tanaka *et al.*, 1993). What will be

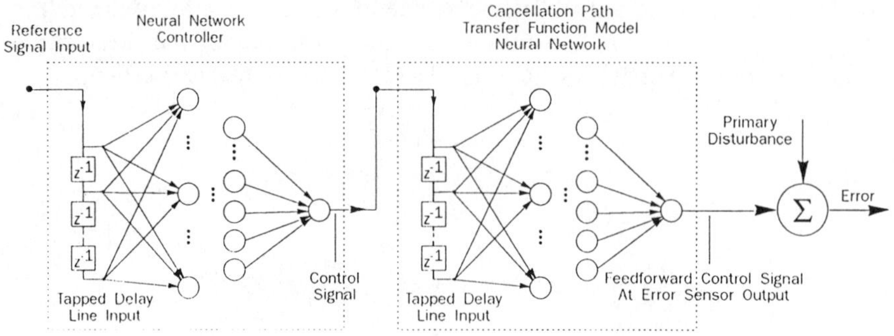

Fig. 6.81 Neural network feedforward controller with a neural network model of the cancellation path transfer function.

outlined here, however, will be details of a specific implementation targeted at supplanting the linear feedforward adaptive control arrangements studied thus far in this chapter with a non-linear controller for some applications. This means that the neural network must 'fit' into the physical and conceptual space defined by the linear controller, with no additional requirements. The adaptive algorithm to be formulated for such an arrangement will be shown later in the chapter to be an extension of the linear filtered-x LMS algorithm. While the algorithm falls within a recently described general framework (Narendra and Parthasarathy, 1991) it is perhaps better viewed as an extension of the use of the general feed-forward neural network for non-linear signal processing (Lippmann, 1987; Hertz *et al.*, 1991; Kosko, 1992a), in much the same way the linear arrangement which it aims to supplant is viewed as an extension of a linear adaptive signal processing arrangement.

6.14.1 Algorithm development: Simplified single path model

Development of the algorithm for adapting the neural network based controller will be best facilitated by first considering the simplified 'single path' model shown in Fig. 6.82. Here a reference input sample at time k, $x_{in}(k)$, which is in some way related to (but not necessarily linearly correlated with) the impending primary disturbance, $p(k)$, is used to derive a control signal $x_0(k)$ via the neural network controller. This control input to the system is modified by some system dependent cancellation loop transfer function, modelled as a second neural network, to produce the feedforward control signal measured at the output of the error sensor, $s(k)$. In deriving $s(k)$ from the control signal x_0 the transfer function model utilises both present and past control signals (via a tapped delay line), which enables the modelling of explicit system time delays to maintain causality within the control scheme. The system error signal is then the sum of the primary and control input signals (superposition of the signals in the structural/acoustic environment):

$$e(k) = p(k) + s(k) \tag{6.14.1}$$

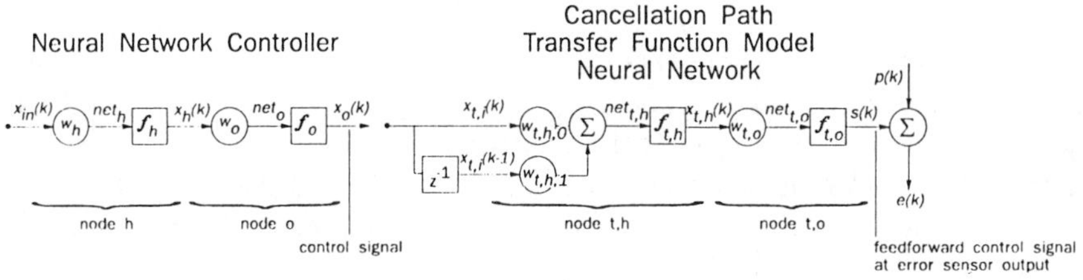

Fig. 6.82 Simplified 'single path' model.

We will assume that the neural network model of the cancellation path transfer function has already been formulated with some 'reasonable' accuracy (which, extrapolating from the previous linear filter results, will be taken to mean within 90° of the correct phase response), the model obtained by placing a neural network in the estimation arrangement described in the previous section and adapted using the standard back-propagation algorithm described in the previous section. The aim is to use the model to facilitate stable adaptation of the neural controller, via a gradient descent algorithm, to minimize the system error criterion, which is the mean square value of the error signal. For practicality, the stochastic approximation

$$\xi(k) \approx e^2(k) \tag{6.14.2}$$

is again made, and the functional derivative $\frac{\partial e^2(k)}{\partial w}$ used to approximate the desired quantity $\Delta w(k)$. Noting that the primary disturbance $p(k)$ is in no way a function of the weights of the control system or cancellation loop transfer function model:

$$\Delta w(k) \approx \frac{\partial e^2(k)}{\partial w} = 2e(k)\frac{\partial e(k)}{\partial w} = 2e(k)\frac{\partial s(k)}{\partial w} \tag{6.14.3}$$

Our task now is to find a solution to (6.14.3) for each weight in the controller neural network. The solution can then be substituted into a generic gradient descent algorithm format to adapt each weight in the control system. To obtain a solution to (6.14.3) for the weights in the controller neural network it is necessary to start at the output of the system as measured by the error sensor, $e(k)$, and work backwards (back propagate) through the transfer function model, producing an error signal which is suitably conditioned to enable derivation of the required gradient expressions. In conducting this back propagation it will prove useful, from the standpoint of developing an algorithm which is readily generalized, to first derive gradients of the error surface with respect to the weight coefficients in the cancellation path transfer function model neural network. These gradients are used only for the algorithm formulation, and will *not* be used to adapt the weights of the model during controller adaptation.

Starting at the end of the transfer function model, the system output is

$$s(k) = f_{t,o}\{x_{t,h}(k)\ w_{t,o}(k)\} \tag{6.14.4}$$

where the subscript t denotes that the quantities of interest are related to the cancellation loop transfer function model, $x_{t,h}(k)$ is the output from hidden layer node t,h at time k, and $w_{t,o}(k)$ is the weighting factor of this signal used as an input to the transfer function model output node t,o. Differentiating (6.14.4), the gradient of the error criterion given in (6.14.3) with respect to the weight $w_{t,o}$ is found to be

$$\begin{aligned}\Delta w_{t,o} &= 2e(k)f'_{t,o}\{x_{t,h}(k)\ w_{t,o}(k)\}x_{t,h}(k) \\ &= 2e(k)f'_{t,o}\{net_{t,o}(k)\}x_{t,h}(k)\end{aligned} \tag{6.14.5}$$

where $net_{t,o}$ is the result of the t,o nodal multiplication/accumulation operation, which is to be filtered through the nodal non-linearity,

$$net_{t,o}(k) = x_{t,h}(k)w_{t,o}(k) \tag{6.14.6}$$

To find the gradient with respect to the weights in the layer preceding this, $x_{t,h}(k)$ can be rewritten as

$$x_{t,h}(k) = f_{t,h}\{x_{t,i}(k)\ w_{t,h,0}(k) + x_{t,i}(k-1)\ w_{t,h,1}(k)\} \tag{6.14.7}$$

Substituting this expansion into (6.14.4) and differentiating, the gradient estimates for the weights $w_{t,h,0}$ and $w_{t,h,1}$ are, respectively,

$$\Delta w_{t,h,0} = 2e(k)f'_{t,o}\{net_{t,o}(k)\}\ w_{t,o}(k)\ f'_{t,h}\{net_{t,h}\}\ x_{t,i}(k) \tag{6.14.8}$$

and

$$\Delta w_{t,h,1} = 2e(k)f'_{t,o}\{net_{t,o}(k)\}\ w_{t,o}(k)\ f'_{t,h}\{net_{t,h}\}\ x_{t,i}(k-1) \tag{6.14.9}$$

where

$$net_{t,h} = \sum_{j=1}^{1} x_{t,i}(k-j)w_{t,h,j} \tag{6.14.10}$$

Note that the gradient estimates given in (6.14.5), (6.14.8) and (6.14.9) are simply those which would be used in the standard back-propagation algorithm we derived in the previous section, with the equations usually written in the form

$$\Delta w_{\text{node,input}} = \delta_{\text{node}}(k)x_{\text{input}}(k) \tag{6.14.11}$$

where x_{input} is the *input* to the *node* of interest, whose weighting function is $w_{\text{node,input}}$, and δ_{node} is the back propagated error for the node of interest,

$$\delta_{\text{node}}(k) = \begin{cases} 2e(k)f'\{net_{\text{output}}(k)\} & \text{output layer node} \\ f'\{net_{\text{node}}(k)\} \displaystyle\sum_{j=1}^{N_{\text{nodes}(layer+1)}} \delta_j(k)\ w_{j,\text{node}}(k) & \text{other nodes} \end{cases} \tag{6.14.12}$$

where $N_{\text{nodes(layer+1)}}$ is the number of nodes in the layer immediately following ('downstream') the layer of interest, and $w_{j,\text{node}}(k)$ is the current value of weighting coefficient for the output of the node of interest used as an input to node j in the next layer, layer+1.

Let us now consider the gradient of the error criterion with respect to the final weight in the control network, w_o. To calculate this, $x_{t,i}$ must be re-expressed as

$$x_{t,i}(k) = x_o(k) = f_o\{x_h(k)w_o(k)\} = f_o\{net_o(k)\} \tag{6.14.13}$$

and the result substituted into (6.14.7). Note, however, that both $x_{t,i}(k)$ and $x_{t,i}(k-1)$ appear in (6.14.7). Therefore,

$$\Delta w_o = \delta_{t,h}(k)\left(w_{t,h,0}(k)f_o'\{net_o(k)\}x_h(k) + w_{t,h,1}(k)f_o'\{net_o(k-1)\}x_h(k-1)\right) \tag{6.14.14}$$

or

$$\Delta w_o = \delta_{t,h}(k)\sum_{j=0}^{N_s-1} w_{t,h,j}(k)f_o'\{net_o(k-j)\}x_h(k-j) \tag{6.14.15}$$

where δth is the back propagated error signal at the first node of the cancellation loop transfer function model neural network:

$$\begin{aligned}\delta_{t,h}(k) &= 2e(k)f_{t,o}'\{net_{t,o}(k)\}\ w_{t,o}(k)\ f_{t,h}'\{net_{t,h}\} \\ &= f_{t,o}'\{net_{t,o}(k)\}\delta_{t,o}(k)w_{t,o}(k)\end{aligned} \tag{6.14.16}$$

$\delta_{t,o}$ is the back propagated error at the output node, N_s is the number of stages, or taps, in the cancellation loop transfer function model input delay chain, which is equal to 2 for the simplified 'single path' model being discussed here. If there were only a single input stage, hence no delay, (6.14.15) would again simply lead to the back propagation algorithm. However, since the error has been back propagated through a delay chain, the gradient estimate has been modified slightly, now taking into account contributions from both the present and past nodal outputs. Intuitively this is a logical result, as the feedforward control signal measured at the error sensor output, $s(k)$, is also a function of both past and present nodal outputs. To maintain consistency with our previous nomenclature, (6.14.15) can be written as

$$\Delta w_o = \sum_{j=0}^{N_s-1} \delta_o(k-j)x_h(k-j) \tag{6.14.17}$$

where

$$\delta_o(k-j) = f_o'\{net_o(k-j)\}\delta_{t,h}(k)w_{t,h,j}(k) \tag{6.14.18}$$

Finally, consider the gradient of the error criterion with respect to the weight w_h. This can be calculated by first noting that

$$x_h(k) = f_h\{x_i(k)w_h(k)\} = f_h\{net_h(k)\} \tag{6.14.19}$$

Substituting this into (6.14.17)

$$\Delta w_h = \sum_{j=0}^{N_s-1} \delta_h(k-j)x_i(k-j) \tag{6.14.20}$$

where

$$\delta_h(k-j) = \delta_o(k-j)\ w_o(k-j)\ f_h'\{net_h(k-j)\} \tag{6.14.21}$$

As with the output weight gradient estimate, defined in (6.14.17), the hidden layer weight gradient estimate given in (6.14.20) utilizes both the present and past nodal outputs in its derivation. This concept will, in fact, be shown to be a general one for all weights in the controller network, regardless of its size. One point to note briefly here in relation to (6.14.21) is that it is found in practice that if the adaptation is 'slow' it is sufficient to use the current value of weights $w_o(k)$, rather than the old values $w_o(k - j)$, thereby reducing the data storage requirements slightly.

6.14.2 Generalization of the algorithm

Having now studied the simplified model it is straightforward for us to derive a more general algorithm for the MIMO feedforward control problem. With this, ideally the error criterion to be minimized would be the sum of the mean square error signals from N_e error sensors:

$$\Xi = \sum_{n=0}^{N_e-1} \xi_n = \sum_{n=0}^{N_e-1} E\{e_n^2(k)\} \tag{6.14.22}$$

However, for practical considerations the stochastic approximation

$$\sum_{n=0}^{N_e-1} e_n^2(k) \tag{6.14.23}$$

will again be used. Minimization of this error criterion is to be carried out using N_c control inputs. Between this group of control outputs and the error signals there is a system dependent cancellation path transfer function. Assuming that the system itself is linear (as opposed to the control problem being linear, which is not assumed), the transfer functions can be considered individually for each control signal. This enables the system to be modelled as shown in Fig. 6.83, where a separate neural network is used to model the cancellation path transfer function between any given control signal and each of the N_e error sensors. Each neural network model has N_s inputs, from a delay chain, and N_e outputs, one for each error sensor, as shown in Fig. 6.84. Note that if the system is not linear then some simple modifications to the following analysis can be made, where a single cancellation path transfer function model neural network, which incorporates all control signals as inputs, is used. The advantage of separating the transfer functions is an implementation one; for several control signals and several error sensors, it is often found to be easier to obtain several small transfer function models as opposed to a large, all-inclusive one.

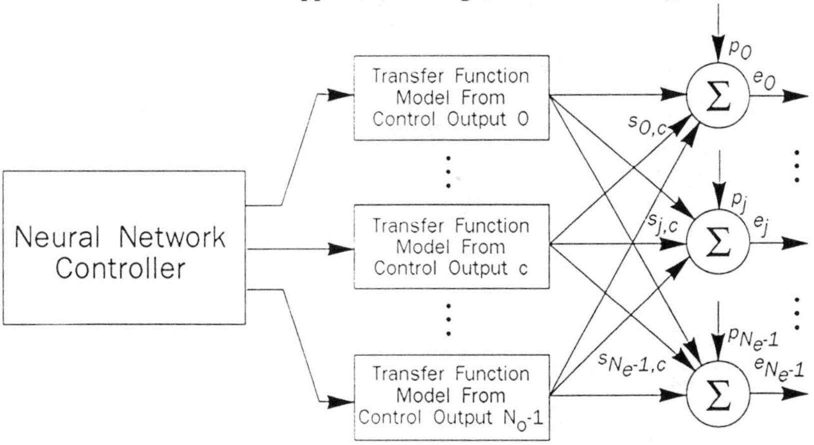

Fig. 6.83 Multiple input, multiple output neural network active control arrangement.

To derive the gradient descent algorithm for the neural network controller, we start again at the output layer of the cancellation path transfer function model. Considering the model for the cth control signal, as shown in Fig. 6.84, the jth output from the model is

$$s_{j,c}(k) = f_{tc,o,j}\left\{\boldsymbol{w}^{\mathrm{T}}_{tc,o,j}(k)\boldsymbol{x}_{tc,hf}(k)\right\} \tag{6.14.24}$$

where $\boldsymbol{x}_{tc,hf}(k)$ is the vector of output values from the layer preceding the output layer, which is the final hidden layer, hf, and $\boldsymbol{w}_{tc,o,j}(k)$ is the vector of weight coefficients for these values used as inputs to the jth node of the output layer, o,j, in the cancellation path transfer function model between the cth control signal and the error sensors, the quantities of which are denoted by the subscript tc. It was stated in (6.14.22) that the error criterion to be minimized is the sum of the squared error signals. However, only the jth error signal is a function of cancellation loop transfer function model's jth output. Therefore, the gradient estimate for the weight is

$$\begin{aligned}\Delta \boldsymbol{w}_{tc,o,j} &= \sum_{n=0}^{N_e-1} \frac{\partial e_n^2(k)}{\partial \boldsymbol{w}_{tc,o,j}} = 2e_j(k)f'_{tc,o,j}\{net_{tc,o,j}(k)\}\boldsymbol{x}_{tc,hf}(k) \\ &= \delta_{tc,o,j}(k)\boldsymbol{x}_{tc,hf}(k)\end{aligned} \tag{6.14.25}$$

where

$$\delta_{tc,o,j}(k) = 2e_j(k)f'_{tc,o,j}\{net_{tc,o,j}(k)\} \tag{6.14.26}$$

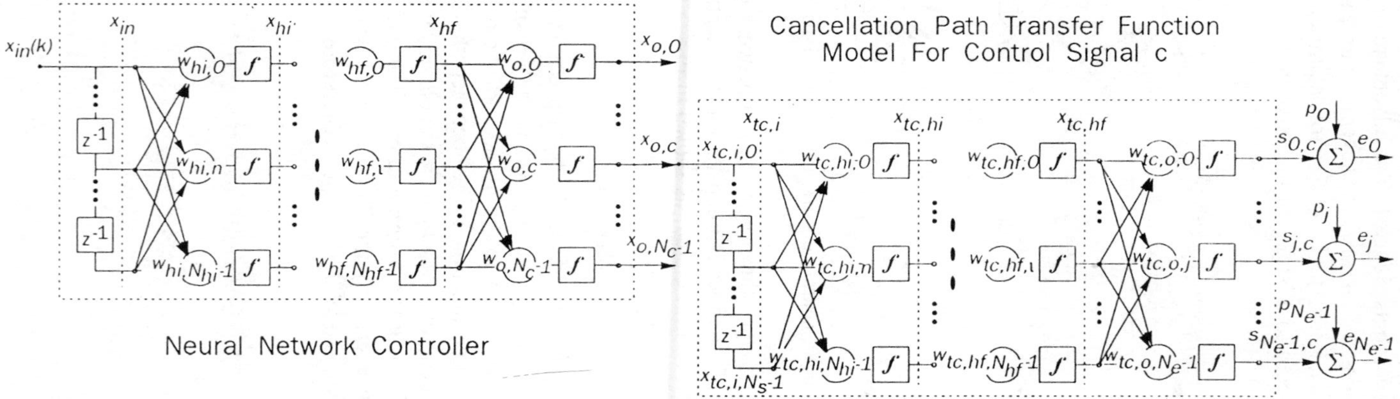

Fig. 6.84 Detail of controller and transfer function model neural networks.

Now consider the vector of weights associated with the ιth node in the layer immediately preceding the output layer, which is the final hidden layer, $\boldsymbol{w}_{tc,hf,\iota}$. To calculate the gradient of the error criterion with respect to these, note that the signal derived from this node, which is used as an input to the output layer of the cancellation path transfer function model neural network, is

$$x_{tc,hf,\iota}(k) = f_{tc,hf,\iota}\left(\boldsymbol{w}^{\mathrm{T}}_{tc,hf,\iota}(k)\boldsymbol{x}_{tc,hf-1}(k)\right) \tag{6.14.27}$$

where $\boldsymbol{x}_{tc,hf\text{-}1}(k)$ is the vector of output values from the layer preceding the final hidden layer, $hf - 1$, and $\boldsymbol{w}_{tc,hf,\iota}(k)$ is the vector of weight coefficients for these values used as inputs to the ιth final hidden layer node, hf,ι. Substituting this expansion into (6.14.24), and considering the entire set of N_e transfer function model outputs (corresponding to N_e error sensors), the gradient is found to be

$$\Delta\boldsymbol{w}_{tc,hf,\iota}$$

$$= \sum_{n=0}^{N_e-1} 2e_n(k) f'_{tc,o,n}\{net_{tc,o,n}(k)\} w_{tc,o,n,\iota}\, f'_{tc,hf,\iota}\{net_{tc,hf,\iota}(k)\}\boldsymbol{x}_{tc,hf-1}(k) \tag{6.14.28}$$

where $w_{tc,o,n,\iota}$ is the weighting factor for node n in the output layer associated with an input signal from node ι in the layer preceding it, the final hidden layer node. Written in more standard form,

$$\Delta\boldsymbol{w}_{tc,hf,\iota} = \delta_{tc,hf,\iota}(k)\boldsymbol{x}_{tc,hf-1}(k) \tag{6.14.29}$$

where

$$\delta_{tc,hf,\iota}(k)$$

$$= \sum_{n=0}^{N_e-1} 2e_n(k) f'_{tc,o,n}\{net_{tc,o,n}(k)\} w_{tc,o,n,\iota}\, f'_{tc,hf,\iota}\{net_{tc,hf,\iota}(k)\} \tag{6.14.30}$$

This gradient estimate is again simply the standard form used in the back propagation algorithm. We can surmise that, as with the simple 'single path' model, the standard form of the algorithm is used to back propagate the error signal through all layers in the cancellation path transfer function model(s) according to the relation given in (6.14.12). Once the back propagated error, $\delta(k)$, has reached the first hidden layer in the cancellation loop transfer function models then some modifications to the gradient estimate must be made, to enable back propagation of the error signal to continue through the delay chain, so that it can be used to modify the weights in the controller neural network.

Let us now consider the output layer of the controller neural network. To derive the gradient of the error criterion with respect to the weights of the cth output node, it must be noted that the samples in the delay chain input to the cancellation path transfer function model associated with it are related to the weights of this node by

$$x_{tc,i}(k) = x_{o,c}(k) = f_{o,c}\left\{\boldsymbol{w}^{\mathrm{T}}_{o,c}(k)\boldsymbol{x}_{hf}(k)\right\} \tag{6.14.31}$$

Taking into account all of the elements in the delay chain, the gradient $\Delta \boldsymbol{w}_{o,c}$ can be expressed

$$\Delta \boldsymbol{w}_{o,c} = \sum_{j=0}^{N_s - 1} \delta_{o,c}(k-j) \boldsymbol{x}_{hf}(k-j) \tag{6.14.32}$$

where

$$\delta_{o,c}(k-j) = f'_{o,c}\{net_{o,c}(k-j)\} \sum_{n=1}^{N_{tc,hi}} \delta_{tc,hi,n}(k) w_{tc,hi,n,j}(k) \tag{6.14.33}$$

$N_{tc,hi}$ is the number of nodes in the initial hidden layer (denoted by *hi*) of the transfer function model and $w_{tc,hi,n,j}$ is the weight of node n in the initial hidden layer of the neural network transfer function model associated with control signal c which weights the input from stage j in the tapped delay line. Comparing this to (6.14.17) for the simplified model, it can be seen that the form is the same. The only difference is the summation over the set of nodes in the initial hidden layer, $N_{tc,hi}$. If there was only a single node then the two equations would be identical.

Deriving the gradient estimate for the weight coefficients in any other layer of the controller neural network is now a relatively simple task. Consider the layer immediately preceding the output layer, which is the final hidden layer, denoted by the subscript *hf*. Viewing the form of (6.14.21), the gradient for the weight coefficient vector associated with node ι in this layer, $\boldsymbol{w}_{hf,\iota}$, can be written directly as

$$\Delta \boldsymbol{w}_{hf,\iota} = \sum_{j=0}^{N_s - 1} \delta_{hf,\iota}(k-j) \boldsymbol{x}_{hf-1}(k-j) \tag{6.14.34}$$

where

$$\delta_{hf,\iota}(k-j) = \sum_{\gamma=0}^{N_o - 1} \delta_{o,\gamma}(k-j) w_{o,\gamma,\iota}(k-j) f'_{hf,\iota}\{net_{hf,\iota}(k-j)\} \tag{6.14.35}$$

and N_o is the number of nodes in the controller neural network output layer. Thus the gradient estimate for any nodal weight coefficient vector in the controller neural network can be expressed as

$$\Delta \boldsymbol{w}_{\text{layer,node}} = \sum_{j=0}^{N_s - 1} \delta_{\text{layer,node}}(k-j) \boldsymbol{x}_{\text{layer}-1}(k-j) \tag{6.14.36}$$

where

$$\delta_{\text{layer,node}}(k-j) =$$

$$\begin{cases} f'_{\text{layer,node}}\{net_{\text{layer,node}}(k-j)\} \displaystyle\sum_{n=0}^{N_{\text{nodes}(t\ \text{node},hi)} - 1} \delta_{t\ \text{node},hi,n}(k) w_{t\ \text{node},hi,n,j}(k) & \text{output node} \\ f'_{\text{layer,node}}\{net_{\text{layer,node}}(k-j)\} \displaystyle\sum_{\gamma=0}^{N_{\text{nodes}(\text{layer}+1)} - 1} \delta_{\gamma}(k) w_{\text{layer}+1,\gamma,\text{node}}(k-j) & \text{other nodes} \end{cases} \tag{6.14.37}$$

and t node denotes the cancellation loop transfer function model associated with the control output node.

One point to note is that although it is the error signal which is thought of heuristically as being 'back propagated' through the delay chain input to the transfer function model, it is in fact past and present versions of the nodal outputs which are used in updating the controller network weights, and not past and present values of the error signal. This makes sense intuitively, as the error signal is a function of past and present nodal outputs, and not vice versa.

6.14.3 Comparison to the filtered-*x* LMS algorithm

It will be useful at this stage to compare the algorithm for adapting the neural network controller with the filtered-x LMS algorithm. Recall that the SISO version of the filtered-x LMS algorithm is

$$\boldsymbol{w}(k+1) = \boldsymbol{w}(k) - 2\mu e(k)\boldsymbol{f}(k) \tag{6.14.38}$$

where $\boldsymbol{w}(k)$ is the vector of weight coefficients in the filter at time k, μ is the convergence coefficient, $e(k)$ is the error signal at time k, and $\boldsymbol{f}(k)$ is the filtered reference signal, the elements of which are produced by convolving the reference signal samples with an n-stage finite impulse response model of the cancellation path transfer function. The neural network controller/cancellation path transfer function model combination considered here can be made equivalent architecturally to the FIR controller/cancellation loop transfer function model combination by representing both neural networks as having only a single (output) layer, with a single node, having a 'non-linear' output function f which is actually linear. If this is the case then the back propagated error signal at the output of the transfer function model is, from (6.14.26):

$$\delta_{t0,o,0}(k) = 2e_0(k)f'\{net_{t0,o,0}(k)\} = 2e(k) \tag{6.14.39}$$

where e_0 denotes error signal 0, and t_0 denotes the transfer function model associated with control output 0. If this is now 'back propagated' to the controller network, the gradient estimate, given in (6.14.36), becomes

$$\Delta \boldsymbol{w}_{o,0}(k) = \sum_{j=0}^{N_s-1} \delta_{o,0}(k-j)\boldsymbol{x}_i(k-j) \tag{6.14.40}$$

where

$$\delta_{o,0}(k-j) = \sum_{n=0}^{N_{t0,o,0}-1} e_0(k) w_{t0,o,0,n}(k) \tag{6.14.41}$$

Substituting this into (6.14.34), the resultant equation can be written in matrix form as

$$\Delta \boldsymbol{w}_{o,0} = \sum_{j=0}^{N_s-1} e(k)\boldsymbol{w}_{t0,o,0}^{\mathrm{T}}(k)\boldsymbol{x}(k-j) \tag{6.14.42}$$

which can be expressed as

$$\Delta w_{o,0} = e_0(k)f(k) \tag{6.14.43}$$

where $f(k)$ is the filtered reference signal sample vector. When this is substituted into the generic gradient descent algorithm format the result is simply the filtered-x LMS algorithm. It can therefore be surmised that the algorithm derived here is simply a generalisation of the filtered-x LMS algorithm, in the same way the standard back propagation algorithm is a generalization of the LMS algorithm.

6.14.4 Example

As a brief example of the capabilities of a neural network based feedforward control system, consider the problem of controlling harmonic excitation of a cantilever beam where the control actuator will introduce harmonics (rattle) if it is driven too hard (taken from Snyder and Tanaka, 1992). If the primary disturbance is 'light' such that no harmonics will be introduced and so the problem is linear, the autospectrum of the error signal under primary and control excitation using either a 6 tap FIR filter controller or a $6\times6\times1$ neural network controller (6 inputs, 6 (non-linear) hidden layer nodes, and 1 (linear) output node), combined with a 6×1 cancellation path transfer function model (6 inputs and 1 (linear) output node), is illustrated in Fig. 6.85. The error signal autospectra look identical for both systems. In fact, the output from the neural network controller is purely sinusoidal, as illustrated in Fig. 6.86.

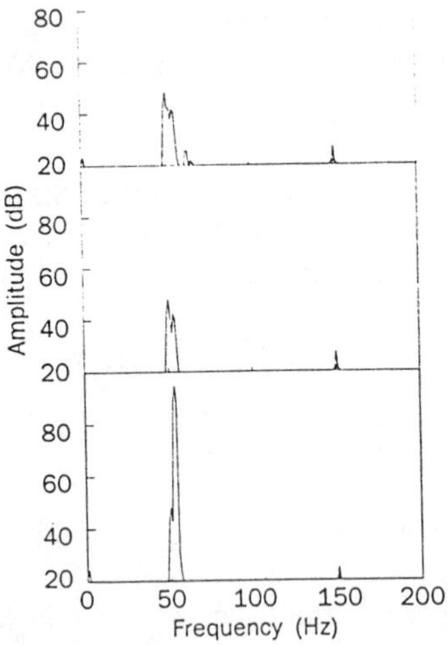

Fig. 6.85 Error signal autospectrum for the case of 53 Hz light primary excitation and no error filtering, from bottom to top under primary excitation, during neural network control, and during linear feedforward control.

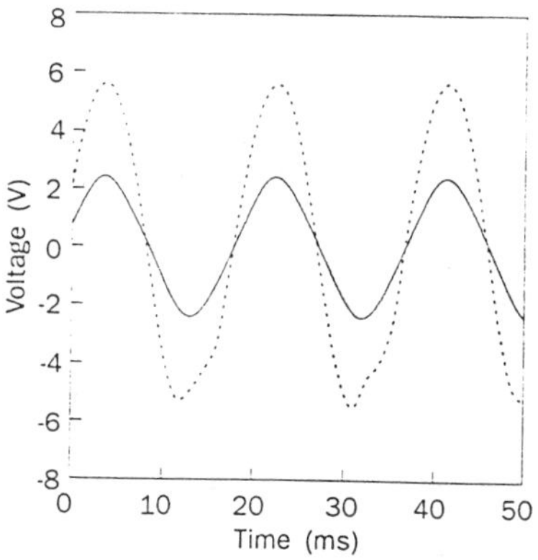

Fig. 6.86 Neural network controller output signals, solid line in response to a linear control problem (corresponding to the controlled spectrum of Fig. 6.85), dashed line in response to a non-linear control problem (corresponding to the controlled spectrum of Fig. 6.87).

If the primary disturbance is increased in amplitude, such that the control actuator begins to rattle, the autospectrum of the error signal during primary excitation and under control either by the 6-tap FIR filter system or by a $4 \times 6 \times 4 \times 1$ neural network (4 inputs, 6 nodes in the first hidden layer, 4 nodes in the second hidden layer, and a single output node), the latter adapted using a $4 \times 4 \times 1$ neural network cancellation path transfer function model is shown in Fig. 6.87. The attenuation of the primary tone provided by the neural network is not quite as large as was achieved with the linear FIR filter system. However, the linear system produced a 16 dB increase in the level of the first harmonic, largely offsetting the benefits obtained by suppressing the tone, while there was no increase in the first harmonic level with the neural network controller. The control signal produced by the neural network is shown by the dashed line in Fig. 6.86. We can see clearly that there is some slight distortion of the waveform, which combats the non-linearity of the control actuator. There is no mechanism in the linear filter/algorithm combination for forming this distortion, hence the inferior control performance in this instance.

If the harmonics problem is now increased, the error signal autospectrum during primary excitation and under control by a 6 tap FIR filter and a $4 \times 8 \times 4 \times 1$ neural network controller adapted using a $4 \times 4 \times 1$ neural network cancellation path transfer function model is illustrated in Fig. 6.88. Here the attenuation in the level of the primary 53 Hz tone produced by the neural network controller is less than that obtained using the linear controller, 9.3 dB versus 13.6 dB. This, however, is more than compensated for by the 24.4 dB reduction in the dominant 106 Hz harmonic achieved by the neural network system as compared to virtually no change with the linear controller. In terms of 'real' levels of control, as would be perceived by a human monitor, the neural network controller performance is vastly superior to that of the linear controller.

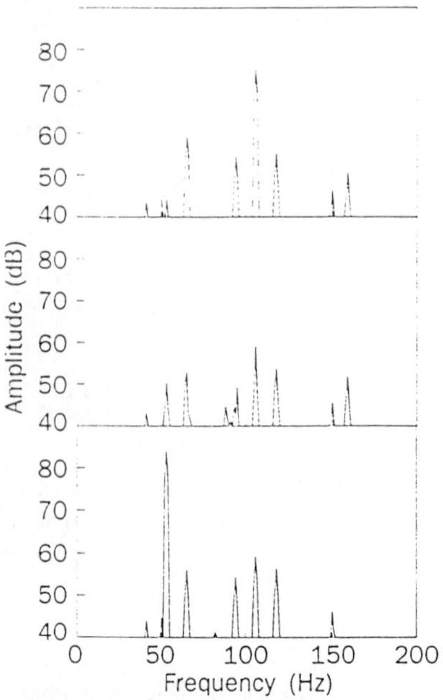

Fig. 6.87 Error signal autospectrum for the case of 53 Hz hard primary excitation and error filtering with the bandpass center frequency at 85 Hz, from bottom to top under primary excitation, during neural network control, and during linear feedforward control.

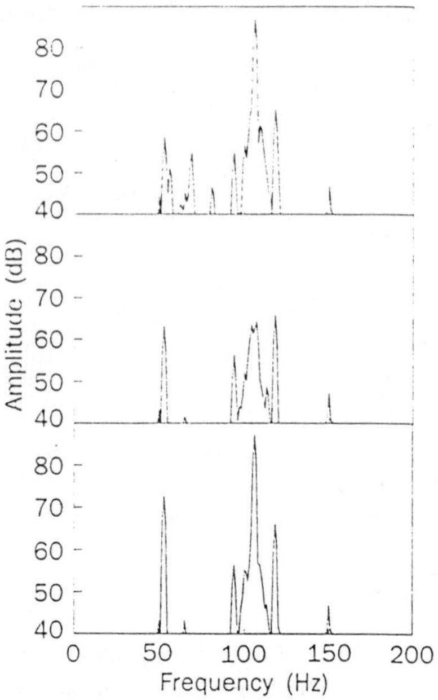

Fig. 6.88 Error signal autospectrum for the case of 53 Hz hard primary excitation and error filtering with the bandpass center frequency at 106 Hz, from bottom to top under primary excitation, during neural network control, and during linear feedforward control.

So then, we may ask ourselves why neural network based systems are not widely used in place of the linear systems discussed previously in this chapter. The principal problem, not evident from the experimental data, is a lack of predicability and consistency as to what a neural network controller will do. As we have seen in this chapter, the level of knowledge concerning what influence various structural/acoustic system parameters have upon the performance of the linear filter/algorithms is sufficiently advanced to enable the design of stable control systems. If a parameter in the system is changed (for example, additional error sensors being included in the system), we can compensate for it in the adaptive algorithm used in the linear feedforward control system to maintain stability and performance. Also, if a linear control system is effective under one set of conditions at a given instant in time, then it can be assumed that it will be effective under the same set of conditions at some other point in time. No such statements could be made regarding the neural controller used in the above experiments. While for a given set of conditions the neural network controller would be consistently stable, the levels of control obtained would not be consistent. Sometimes the controller would simply turn itself off, presumably because there was a local minimum in the error criterion for this result and the random initial weights were not sufficient to enable the algorithm to avoid getting trapped in it. Also, how control is achieved would be inconsistent. Will convergence be fast or slow? Would the control signal be clean or distorted in some unpredictable manner? These questions cannot always be answered based on previously acquired knowledge.

The most predictable cases are linear control problems, but even these can occasionally have unforeseen results. Consider for a moment the results of Figs. 6.85 and 6.86. When controlling a pure tone the neural controller performed very well, producing 52.4 dB of attenuation as a result of the 'beautiful' sine wave control signal shown in Fig. 6.86. It was said in describing the test that no bandpass filtering of the error signal was used in this instance. What if the previously described bandpass filter was used on the error signal, with the centre frequency set at 53 Hz, such that (practically) only the fundamental tonal component of the error signal was available to the adaptive controller? If the controller were linear this would have no detrimental effect; the control output is constrained to be some linear function of the 53 Hz reference signal provided to it. However, this constraint is not placed upon the neural network controller, and so the addition of filtering is responsible for the control signal depicted in Fig. 6.89. This control signal produces 48.0 dB attenuation of the 53 Hz tone, but from a subjective point of view the end result was disastrous (owing to excitation of out-of-band harmonics). The neural network controller does not know that, however.

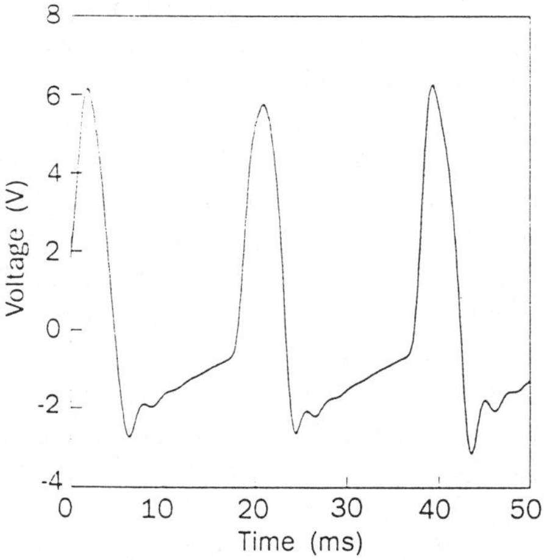

Fig. 6.89 Neural network control signal for the case of 53 Hz light primary excitation and bandpass error filtering at 53 Hz.

Therefore, based on the experimental work conducted and presented in this section, the following can be said: The neural network controller/algorithm scheme presented here for feedforward active control shows the potential to be equal in performance to a linear control scheme for a linear control problem, and to have (far) superior performance for non-linear problems. For this potential to be fully realized, however, a great deal more work needs to be done, work directed towards constraining the network not only to perform the desired task, but to perform it consistently and within some acceptable bounds of 'side effects', such as convergence speed and signal distortion.

6.15 ADAPTIVE FILTERING USING A GENETIC ALGORITHM

In the previous section we considered the use of neural networks for cases where the controller filter requirements are non-linear. It was pointed out that for application to active noise and vibration control systems, it was necessary to determine, to a reasonable degree of accuracy, the cancellation path transfer function. In this section, an alternative algorithm is considered for adaptation of non-linear filter weights. This algorithm is a modification of the genetic algorithm which is well known in the discipline of optimization. The principal advantage of the genetic algorithm is that it is stable with no knowledge of the cancellation path transfer function and is capable of optimally adapting the weights of any non-linear filter structure including a neural network. Thus, the error signal does not have to be linearly correlated with the reference signal.

Because the algorithm is virtually independent of the type of filter structure which is used, the best type of filter structure for a particular problem can be selected easily by using trial and error. The principal disadvantage of the genetic algorithm is that it is relatively slow, limiting its usefulness to relatively steady systems or systems which vary slowly with time. This is a result of the averaging time required for performance measurement which is at least half the period of the lowest frequency signal encountered in the error signals, which is the lowest frequency to be controlled if a suitable high pass filter is used.

The genetic algorithm is an optimization/search technique based on evolution, and is essentially a guided random search. It has been applied to many optimization problems, and in the field of active sound and vibration control it has also been used to optimise the placement of control sources (Wang, 1993; Katsikas *et al.*, 1993; Baek and Elliott, 1993; Tsahalis *et al.*, 1993 and Rao *et al.* 1991). Here, we discuss how the genetic algorithm may be used to adapt the coefficients of a digital filter (Wangler and Hansen, 1993, 1994). A general control system schematic arrangement for application of the genetic algorithm to filter weight adaptation is depicted in Fig. 6.90.

Use of the genetic algorithm enables any filter structure to be treated as a 'black box' which processes reference signals to produce control signals, based on different sets of filter weights. Basic genetic algorithm operation requires the testing of solutions (sets of filter weights), which involves loading the filter weights into the filter and subsequently evaluating the performance of the filter in minimizing a cost function based on the error sensor outputs. The genetic algorithm in essence combines high performance solutions while also including a random search component.

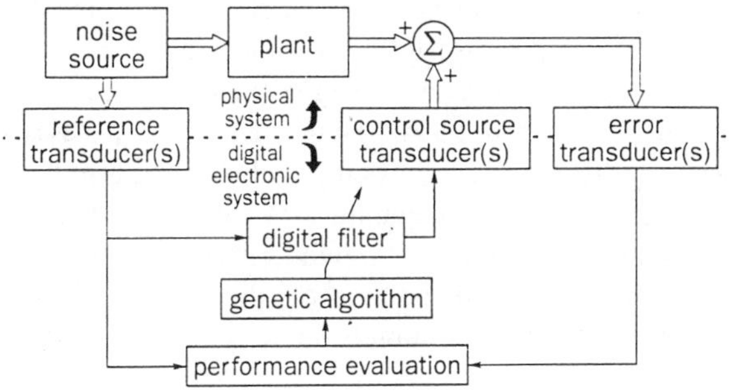

Fig. 6.90 Control system arrangement for genetic algorithm implementation.

6.15.1 Algorithm implementation

Implementation of the genetic algorithm described here has three basic stages: fitness evaluation, selection and breeding. Fitness evaluation requires the testing of the performance of all individuals in the population. Here an individual is considered to be a separate set of filter weights, with the fitness of the individual being a measure of the filter's performance when these weights are being used for filter output calculation. The population then consists of a collection of these individuals. Selection involves killing a given proportion of the population based on probabilistic 'survival of the fittest'. Killed individuals are replaced by children, which are created by breeding the remaining individuals in the population. Typically 70% of the population are killed, with the remaining 30% forming the mating pool for breeding. For each child produced, breeding first requires probabilistic selection of two (possibly the same) parent individuals, with fitter individuals being more likely to be chosen. The probability of selection is high for parents of 'good' fitness and low for parents of 'poor' fitness. For optimal results, it is best to vary the probability distribution depending upon the stage of convergence which the algorithm has reached. Typical probability distributions used at the beginning and at the end of convergence are illustrated in Fig. 6.91, where it can be seen that in the beginning, there is a heavy selection bias (or high selection pressure) towards the fitter individuals. Application of the crossover and mutation operators on the parent pair produces the new child. The crossover operator combines the information contained in two parent strings (or two sets of filter weights) by probabilistic copying of information from either parent to each corresponding string element (or single filter weight) of the child being produced. In this case, the probability of copying a particular weight value to the child from either of the two parents is the same. Mutation introduces random copying 'errors' during the information copying stage of crossover, and gives the algorithm a random search capability.

Mutation plays a minor role in the implementation of the genetic algorithm in standard optimization problems, in that it is used to replace lost bits in the binary encoding of the problem. As binary data have only two states, small mutation probabilities work well with the 'standard' implementation where data loss is minor. This is not the case in the implementation most suited to active noise and vibration control, where a weight string is used instead of binary encoding (Wangler and Hansen, 1994). Here, mutation is necessary to maintain population diversity (differences between individuals) and also to allow 'homing in' on optimal solutions, as the population data corresponding to one weight in the string will not fully represent the weight's entire data range. However, in practice it is necessary to place bounds on the allowed range of mutation (mutation amplitude), the optimal bounds being somewhat problem dependent.

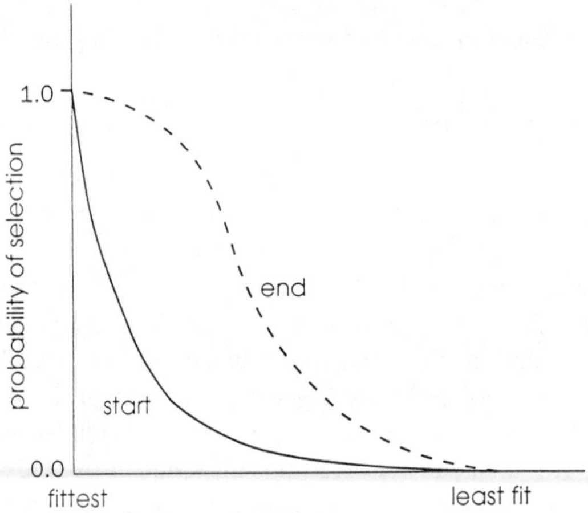

Fig. 6.91 Typical probability distributions used for parent selection for breeding.

Two selection processes are carried out during the operation of the genetic algorithm, namely the choice of individuals to be killed, and the choice of parents during breeding. Both selection processes have been implemented using a simulated roulette wheel, where each segment (or slot) on the roulette wheel is allocated a size proportional to the individual's probability of being chosen (selection probability) which is allocated according to Fig. 6.91. Each spin of the roulette wheel results in one 'winner' being selected. Selection probabilities are assigned such that low performance individuals are more likely to be killed, and such that high performance individuals are more likely to be chosen as parents for breeding. Selection without replacement is used for killing, where once an individual is chosen it is removed from the roulette wheel. For breeding, selection with replacement (no removal) is used for choosing the parents, hence the entire mating pool is used in the selection of each parent individual.

Many aspects of the genetic algorithm used in standard optimization implementations have been changed to give the desired on-line optimization performance required for active noise and vibration control (Wangler and Hansen, 1994), as discussed in the subsections to follow.

6.15.1.1 Killing selection instead of survivor selection

Choosing individuals to be killed rather than those to survive allows higher survival probabilities to be realised for the higher performing individuals. This enables greater selective pressure (bias towards survival and breeding of the higher performance individuals) to be applied, which can be used to give faster convergence when high levels of mutation are used to sustain population diversity. Use of killing selection also allows the best performing individual to be assigned a killing probability of zero to ensure its survival.

6.15.1.2 Weight string instead of binary encoding

The 'genetic code' of each individual is normally encoded as a binary string from the problem variables; in this case, it would imply that each weight would be coded as a binary string and the strings connected together to form a complete individual or set of weights, with the crossover and mutation operators working at the single bit level. Mutation of the upper bits of weight variables would result in large jumps in weight values when filter weights are encoded in this way, which significantly degrades on-line performance. To alleviate this problem in active noise and vibration control systems, a weight string is used, with the crossover and mutation operators applied using whole weight values as the smallest operational element.

6.15.1.3 Mutation probability and amplitude

Application of mutation to whole weight values enables a limit to be placed on the deviation of filter weight values about their current values, which gives control over the spread of the filter's performance. Mutation is applied to all child string variables at a given probability (mutation probability, typically 20 to 30%). The weights chosen to be mutated are modified by a random change in value, which is limited to a specified range (mutation amplitude). For best results the mutation amplitude should be relatively high at the start of convergence and low towards the end. Typical values range from 15% of the maximum possible weight value (at the start) to 0.01% of the maximum possible weight value (at the end of convergence).

6.15.1.4 Rank-based selection (killing and breeding)

Rank-based selection removes the scaling problems associated with fitness proportionate selection (assigning selection probabilities proportional to fitness values), and gives exceptional control over selective pressure (Whitley, 1989; Whitley and Hanson, 1989). Rank-based selection, used by Whitley and Hanson (1989) for breeding (parent selection) has been extended in the active noise and vibration control application to include killing selection. Selection probabilities, for both killing and breeding, are assigned based on the rank position of each individual's performance. This essentially means that the individuals are sorted into order from best to worst performance, then each allocated a fixed selection probability (probability of being chosen) based on their position in this list. The performance evaluation method used thus becomes irrelevant as long as the rank positions are the same (or similar). Separate (adjustable) probability distributions are used for killing and parent selection, with killing being more probable for lower ranked individuals and selection to be a parent being more probable for higher ranked individuals.

6.15.1.5 Uniform crossover

Uniform crossover nearly always combines the information of two parent strings more effectively than one or two point crossover (Syswerda, 1989). One point crossover is where a position along the string is selected at random, and information is copied (to the child being created) from one parent for the first part of the child string and from the other parent for the second part. Similarly two point crossover involves selecting two points along the string, and copying from one parent between these two points, and from the other parent for the rest of the child string. In uniform crossover each position along the child string is produced by randomly copying from either parent, with both parents being equally likely to be chosen as the information source. For active noise and vibration control problems it has been found that it is best to use a modified form of uniform crossover (Wangler and Hansen, 1994), for which the probability of copying information from the lower ranked parent is supplied, and whole weight values are the smallest elements that are copied (compared to single bits for binary encoded strings).

6.15.1.6 Genetic algorithm parameter adjustment

As suggested by De Jong (1985), adjustment of the operating parameters (probabilities, population size, etc.) can improve the performance of the algorithm. The adjustable parameters used in the active noise and vibration control implementation discussed here (population size, survival ratio, killing and breeding rank-probability distributions, crossover probability, mutation probability, and mutation amplitude) provide good control over the stages of adaptation needed when good on-line performance is required.

6.15.1.7 Performance measurement

To evaluate the fitness of an individual (set of filter weights) in minimizing the error signal it is necessary to average the mean square error signal over a period of time greater than the period of the lowest frequency signal present. There should also be a delay of twice this between each fitness (or performance) evaluation to allow any transient effects resulting from implementation of the previous individual (set of weights) to subside.

It is interesting to note that the genetic algorithm can handle any form of performance measure, including a measure of power or intensity, whereas previously discussed algorithms, because of the instantaneous nature of their cost functions, cannot use power or intensity error criteria easily.

In multichannel systems, the performance measure for the genetic algorithm would be the sum of the average square error measured by each error sensor. For applications involving 'more important' and 'less important' error sensors, it is easy to weight the signal from individual error sensors accordingly.

6.15.2 Implementation example

A single channel example similar to that discussed in Section 6.14.4, except that here the beam was fixed at both ends, will now be used to demonstrate the effectiveness of the genetic algorithm with three different types of filter structure. The first filter used was a linear FIR filter which was only capable of producing frequencies present in the reference signal, in this case a 133 Hz sinusoid.

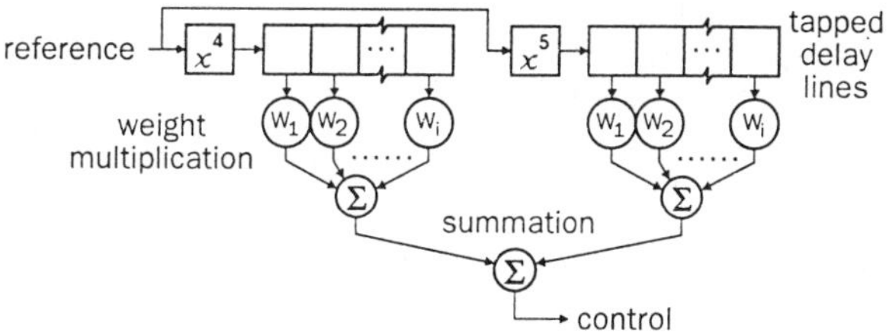

Fig. 6.92 Polynomial (P4P5) filter structure for genetic algorithm implementation.

The second filter structure used was a non-linear polynomial filter illustrated in Fig. 6.92. This filter consisted of two 50 tap FIR filters, the inputs of which were the reference signal raised to the fourth and fifth powers respectively, with the control signal obtained by adding the filter outputs. Raising a signal to the fourth power creates a signal consisting of the second and fourth order harmonics of the initial signal content. Similarly, raising a signal to the fifth power gives a signal with first (fundamental), third, and fifth order harmonic content. Hence this polynomial filter, referred to as a P4P5 filter, can only produce harmonics (including the fundamental) of the reference signal up to the fifth order.

The neural network based filter structure used is shown in Fig. 6.93, and has 50 taps, one hidden layer with 20 nodes, and one (linear) output layer node (designated $50\times20\times1$). Four different transfer functions were utilized simultaneously in the hidden layer, as shown in Fig. 6.93, with equal numbers (that is, five) of each type being used.

As for the example outlined in Section 6.14.4, the non-linearity in the control excitation was introduced by not attaching the control shaker properly to the beam. There was also considerable harmonic distortion in the primary excitation. Final converged vibration levels obtained using each of the three types of filter structure are shown in Figs. 6.94 to 6.96. The genetic algorithm adapted FIR filter gave a maximum of 12dB mean square error (MSE) reduction within 40 seconds. The P4P5 filter gave 12 dB at 50 seconds, and a maximum of 36 dB within 3 minutes. The 50×20×1 neural network filter gave 24 dB at 50 seconds, 30dB in 6 minutes, and a maximum of 32dB MSE reduction within 15 minutes.

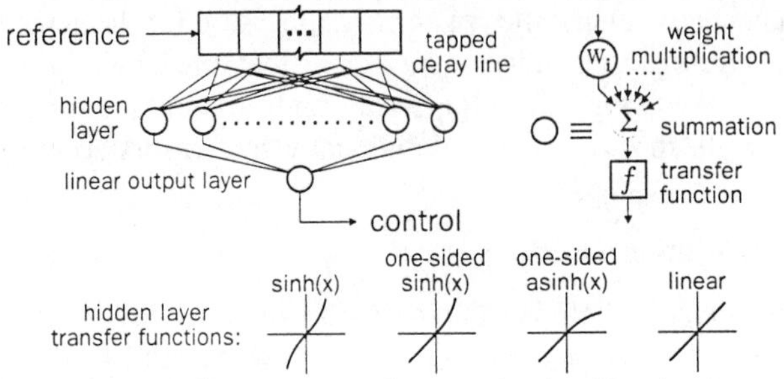

Fig. 6.93 Neural network filter structure for genetic algorithm implementation.

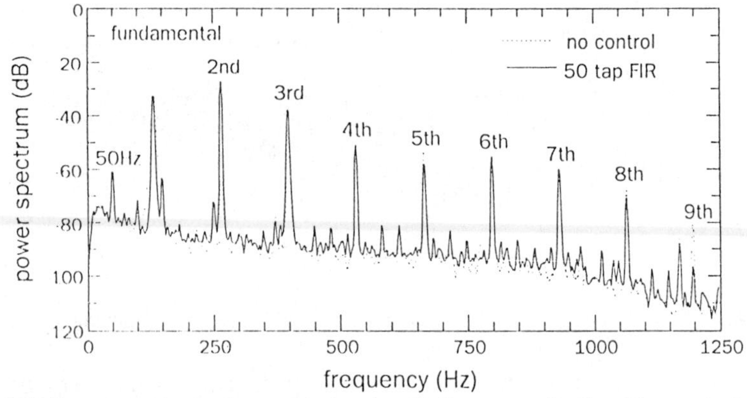

Fig. 6.94 Power spectrum of an error signal, with linear genetic algorithm control using an FIR filter structure.

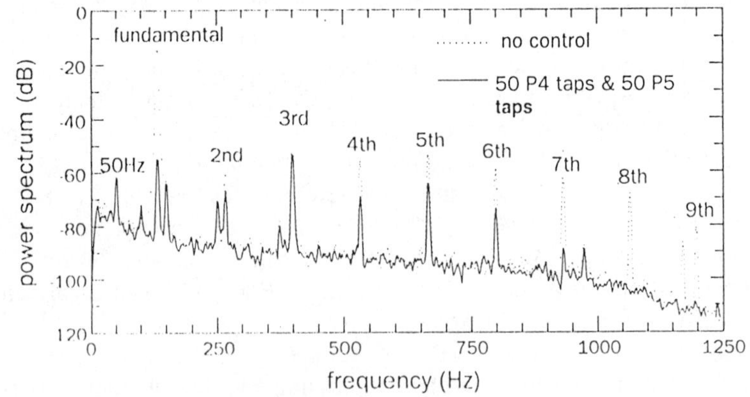

Fig. 6.95 Power spectrum of an error signal, with non-linear genetic algorithm control using the P4P5 filter structure.

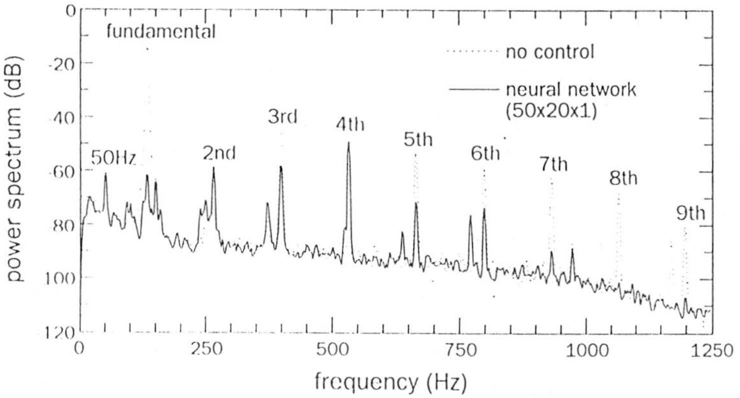

Fig. 6.96 Power spectrum of an error signal, with non-linear genetic algorithm control using the neural network filter structure.

Power spectra showing final converged vibration levels (measured at the error sensor) obtained using each of the three types of filter structure are shown in Figs. 6.94 to 6.96 respectively. The higher order harmonics present with no control applied are due to non-linear output from the primary electrodynamic shaker. A summary of the attenuation achieved at the harmonic peaks is given in Table 6.1.

Table 6.1 Error signal power spectrum attenuation (dB) at each harmonic for the FIR, P4P5 and Neural Network (NN) filters

	1st	2nd	3rd	4th	5th	6th	7th	8th	9th
FIR	18	-32	-8	-5	4	-4	-3	3	16
P4P5	40	8	7	13	10	15	26	37	29
NN	47	-2	12	-7	17	14	27	33	26

For the FIR filter case, attenuation of the fundamental peak at 133 Hz is limited due to the introduction of the higher order harmonics by the non-linear control source. The P4P5 filter achieved the best overall reduction, with all harmonic peaks being attenuated. In comparison, the neural network filter structure has given greater control of the first, third and fifth order components, but has caused the second and fourth order components to increase.

Note that the presence of small quantities of higher order harmonics in the reference signal seen by the controller (due to harmonic distortion in the signal generator) have allowed the attenuation of higher order harmonics that would not normally be possible for the FIR (eighth and ninth harmonics) and P4P5 (sixth to ninth harmonics) filter structures when given a purely sinusoidal reference.

REFERENCES

Alexander, S.T. (1986). *Adaptive Signal Processing: Theory and Applications*. Springer-Verlag: New York.

Baek, K.H. and Elliott, S.J. (1993). Genetic algorithms for choosing source locations in active control system. *Proceedings of the Institute of Acoustics*, **15**, 437–445.

Bershad, N.J. (1986). Analysis of the normalised LMS algorithm with Gaussian inputs. *IEEE Transactions on Acoustics, Speech, and Signal Processing*, **ASSP-34**, 793–806.

Bull, M.K. (1968). Boundary layer pressure fluctations. In *Noise and Acoustic Fatigue in Aeronautics*, E.J. Richards and D.J. Mead, ed. Wiley: New York.

Bitmead, R.R. and Anderson, B.D.O. (1980). Performance of adaptive estimation algorithms in dependent random environments. *IEEE Transactions on Automatic Control*, **AC-25**, 788–794.

Boucher, C.C., Elliott, S.J. and Nelson, P.A. (1991). Effect of errors in the plant model on the performance of algorithms for adaptive feedforward control. *IEE Proceedings, pt.F*, **138**, 313–319.

Burdisso, R.A. and Fuller, C.R. (1991). Eigenproperties of feedforward controlled flexible structures. *Journal of Intelligent Material Systems and Structures*, **2**, 494–507.

Burdisso, R.A. and Fuller, C.R. (1992). Theory of feedforward controlled system eigenproperties. *Journal of Sound and Vibration*, **153**, 437–451 (also, comments and reply, *Journal of Sound and Vibration*, **163**, 363–371).

Burgess, J.C. (1981). Active adaptive sound control in a duct: a computer simulation. *Journal of the Acoustical Society of America*, **70**, 715–726.

Caraiscos, C. and Liu, B. (1984) A roundoff error analysis of the LMS adaptive algorithm. *IEEE Transactions on Acoustics, Speech, and Signal Processing*, **ASSP-32**, 34–41.

Chaplin, G.B.B. (1983). Anti-sound: The Essex breakthrough. *Chartered Mechanical Engineer*, **30**, 41–47.

Cioffi, J.M. (1987). Limited precision effects in adaptive filtering. *IEEE Transactions on Circuits and Systems*, **CAS-34**, 821–833.

Clark, G.A., Mitra, S.K. and Parker, S.R. (1981) Block implementation of adaptive digital filters. *IEEE Transactions on Circuits and Systems*, **CAS-28**, 584–592.

Clark, G.A., Parker, S.R. and Mitra, S.K. (1983) A unified approach to time- and frequency-domain realization of FIR adaptive digital filters. *IEEE Transactions on Acoustics, Speech, and Signal Processing*, **ASSP-31**, 1073–1083.

Cowan, C.F.N. and Grant, P.M. ed. (1985). *Adaptive Filters*. Prentice Hall: Englewood Cliffs, NJ.

De Jong, K. (1985). Genetic algorithms: a 10 year perspective. In *Proceedings of the 1st International Conference on Genetic Algorithms and Their Applications*, 169–177.

Dentino, M., McCool, J.M. and Widrow, B. (1978). Adaptive filtering in the frequency domain. *Proceedings of the IEEE*, **66**, 1658–1659.

Elliott, S.J. and Nelson, P.A. (1984). *Models for Describing Active Noise Control in Ducts*. ISVR Technical Report 127.

Elliott, S.J. and Darlington, P. (1985). Adaptive cancellation of periodic, synchronously sampled interference. *IEEE Transactions on Acoustics, Speech, and Signal Processing*, **ASSP-33**, 715–717.

Elliott, S.J. and Nelson, P.A. (1985). Algorithm for multichannel LMS adaptive filtering. *Electronics Letters*, **21**, 979–981.

Elliott, S.J., Stothers, I.M. and Nelson, P.A. (1987). A multiple error LMS algorithm and its application to the active control of sound and vibration. *IEEE Transactions on Acoustics, Speech, and Signal Processing*, **ASSP-35**, 1423–1434.

Elliott, S.J., Boucher, C.C. and Nelson, P.A. (1992). The behaviour of a multiple channel active control system. *IEEE Transactions on Signal Processing*, **40**, 1041–1052.

Eriksson, L.J. (1991a). Recursive algorithms for active noise control. *Proc. International Symposium on Active Control of Sound and Vibration*, Tokyo, Japan, 9–11 April, 137–146.

Eriksson, L.J. (1991b). Development of the filtered-U algorithm for active noise control. *Journal of the Acoustical Society of America*, **89**, 257–265.

Eriksson, L.J., Allie, M.C. and Greiner, R.A. (1987). The selection and application of an IIR adaptive filter for use in active sound attenuation. *IEEE Transactions on Acoustics, Speech, and Signal Processing*, **ASSP-35**, 433–437.

Eriksson, L.J. and Allie, M.C. (1988). System considerations for adaptive modelling applied to active noise control. *Proceedings of ICASSP '88*, 2387–2390.

Fan, H. and Nayeri, M. (1989). On error surfaces of sufficient order adaptive IIR filters: Proofs and counterexamples to a unimodality conjecture. *IEEE Transactions on Acoustics, Speech, and Signal Processing*, **ASSP-37**, 1436–1442.

Feintuch, P.L. (1976). An adaptive recursive LMS filter. *Proceedings of the IEEE*, **64**, 1622–1624.

Ferrara, E.R. Jr. (1980). Fast implementation of LMS adaptive filters. *IEEE Transactions on Acoustics, Speech, and Signal Processing*, **ASSP-28**, 474–475.

Ferrara, E.R. Jr. (1985). Frequency domain adaptive filtering, in *Adaptive Filters*, C.F.N. Cowan and P.M. Grant, ed. Prentice Hall: Englewood Cliffs, NJ, Chapter 6, 145–179.

Feuer, A. and Weinstein, E. (1985). Convergence analysis of LMS filters with uncorrelated Gaussian data. *IEEE Transactions on Acoustics, Speech, and Signal Processing*, **ASSP-33**, 222–230.

Ffowcs-Williams, J.E., Roebuck, I. and Ross, C.F. (1985). Antiphase noise reduction. *Physics in Technology*, **6**, 19–24.

Foley, J.B. and Boland, F.M. (1988). A note on the convergence analysis of LMS adaptive filters with Guassian data. *IEEE Transactions on Acoustics, Speech, and Signal Processing*, **ASSP-36**, 1087–1089.

Freij, G.J. and Cheetham, B.M.G. (1987). Performance of LMS algorithm as a function of input signal pole loci. *Electronics Letters*, **26**, 1705–1706.

Gardner, W.A. (1984). Learning characteristics of stochastic-gradient-descent algorithms: a general study, analysis, and critique. *Signal Processing*, **6**, 113–133.

Gitlin, R.D., Mazo, J.E. and Taylor, M.G. (1973). On the design of gradient algorithms for digitally implemented adaptive filters. *IEEE Transactions on Circuit Theory*, **CT-20**, 125–136.

Gitlin, R.D., Meadors, H.C., Jr. and Weinstein, S.B. (1982) The tap-leakage algorithm: an algorithm for the stable operation of a digitally implemented, fractionaly spaced adaptive equaliser. *Bell System Technical Journal*, **61**, 1817–1839.

Glover, J.R. Jr. (1977). Adaptive noise canceling applied to sinusoidal interferences. *IEEE Transactions on Acoustics, Speech, and Signal Processing*, **ASSP-25**, 484–291.

Goodwin, G.C. and Sin, K.S. (1984). *Adaptive Filtering: Prediction and Control*. Prentice Hall: Englewood Cliffs, NJ.

Gholkar, V.A. (1990). Means square convergence analysis of LMS algorithm. *Electronics Letters*, **26**, 1705–1706.

Harris, R.W., Chabries, D.M. and Bishop, F.A. (1986). A variable step (VS) adaptive filter algorithm. *IEEE Transactions on Acoustics, Speech, and Signal Processing*, **ASSP-34**, 309–316.

Hertz, J., Krogh, A. and Palmer, R.G. (1991). *Introduction to the Theory of Neural Computation*. Addison-Wesley: Redwood City, CA.

Honig, M.L. and Messerschmitt, D.G. (1984). *Adaptive Filters: Structures, Algorithms, and Applications*. Kluwer Academic: Hingham, MA.

Horowitz, L.L. and Senne, K.D. (1981). Performance advantage of complex LMS for controlling narrow-band adaptive arrays. *IEEE Transactions on Acoustics, Speech, and Signal Processing*, **ASSP-29**, 722–736.

Horvath, S. Jr. (1980). A new adaptive recursive LMS filter, in *Digital Signal Processing*, V. Cappellini and A.G. Constantinides, ed. Academic Press: New York, 21–26.

Hoskins, D.A., Hwang, J.N. and Vagners, J. (1992). Iterative inversion of neural networks and its application to adaptive control. *IEEE Transactions on Neural Networks*, **3**, 292–301.

Johnson, C.R. Jr. (1979). A convergence proof for a hyperstable adaptive recursive filter. *IEEE Transactions on Information Theory*, **IT-25**, 745–749.

Johnson, C.R. Jr. (1984). Adaptive IIR filtering: current results and open issues. *IEEE Transactions on Information Theory*, **IT-30**, 237–250.

Johnson, C.R. Jr. and Larimore, M.G. (1977). Comments and additions to 'An adaptive recursive LMS filter'. *Proceedings of the IEEE*, **65**, 1399–1401.

Johnson, C.R. Jr., Larimore, M.G., Treichler, J.R. and Anderson, B.D.O. (1981). SHARF convergence properties. *IEEE Transactions on Acoustics, Speech and Signal Processing*, **ASSP-29**, 659–670.

Jury, E.I. (1964). *Theory and Applications of the z-Transform Method*. Wiley: New York.

Jury, E.I. (1974). *Inners and Stability of Dynamic Systems*. Wiley: New York.

Kabal, P. (1983). The stability of adaptive minimum mean square error equalizers using delayed adjustment. *IEEE Transactions on Communications*, **COM-31**, 430–432.

Katsikas, S.K., Tsahalis, D., Manolas, D. and Xanthakis, S. (1993). Genetic algorithms for active noise control. In *Proceedings of NOISE-93* St. Petersburg, Russia, May 31 – June 3, **2**, 167–171.

Kesten, H. (1958). Accelerated stochastic approximation. *Annals of Mathematical Statistics*, **29**, 41–59.

Kosko, B., ed. (1992a). *Neural Networks for Signal Processing*. Prentice Hall: Englewood Cliffs, NJ.

Kosko, B., ed. (1992b). *Neural Networks and Fuzzy Systems, A Dynamic Systems Approach to Machine Intelligence*. Prentice Hall: Englewood Cliffs, NJ.

Kwong, R.H. and Johnson, E.W. (1992). A variable step size LMS algorithm. *IEEE Transactions on Signal Processing*, **40**, 1633–1642.

Larimore, M.G., Treichler, J.R. and Johnson, C.R., Jr. (1980). SHARF: An algorithm for adapting IIR digital filters. *IEEE Transactions on Acoustics, Speech and Signal Processing*, **ASSP-28**, 428–440.

Lee, J.C. and Un, C.K. (1986). Performance of transform-domain LMS adaptive filters. *IEEE Transactions on Acoustics, Speech, and Signal Processing*, **ASSP-34**, 499–510.

Lippmann, R.P. (1987). An introduction to computing with neural networks. *IEEE ASSP Magazine*, **April**, 4–22.

Ljung, L. (1987). *System Identification: Theory for the User*. Prentice Hall: Englewood Cliffs, NJ.

Ljung, L. and Söderström, T. (1983). *Theory and Practice of Recursive Identification.* MIT Press: Cambridge, MA.

Long, G., Ling, F. and Proakis, J.G. (1989). The LMS algorithm with delayed coefficient adaption. *IEEE Transactions on Acoustics, Speech, and Signal Processing*, **ASSP-37**, 1397–1405.

Lynn, P.A. (1982). *An Introduction to the Analysis and Processing of Signals*. Macmillan: London.

Mansour, D. and Gray, A.H. Jr. (1982). Unconstrained frequency-domain adaptive filter. *IEEE Transactions on Acoustics, Speech, and Signal Processing*, **ASSP-30**, 726–734.

Mikhael, W.B., Wu, F., Kang, G. and Fransen, L. (1984). Optimum adaptive algorithms with applications to noise cancellation. *IEEE Transactions on Circuits and Systems*, **CAS-31**, 312–315.

Mikhael, W.B., Wu, F.H., Kazovsky, L.G., Kang, G.S. and Fransen, L.J. (1986). Adaptive filters with individual adaption of parameters. *IEEE Transactions on Circuits and Systems*, **CAS-33**, 677–685.

Minsky, M.L. and Papert, S.A. (1969). *Perceptrons*. MIT Press: Cambridge, MA.

Morgan, D.R. (1980). An analysis of multiple correlation cancellation loops with a filter in the auxiliary path. *IEEE Transaction on Acoustics, Speech, and Signal Processing*, **ASSP-28**, 454–467.

Morgan, D.R. and Sanford, C. (1992). A control theory approach to the stability and transient analysis of the filtered-x LMS adaptive notch filter. *IEEE Transactions on Signal Processing*, **40**, 2341–2346.

Narendra, K.S. and Parthasarathy, K. (1990). Identification and control of dynamical systems using neural networks. *IEEE Transactions on Neural Networks*, **1**, 4–27.

Narendra, K.S. and Parthasarathy, K. (1991). Gradient methods for the optimization of dynamical systems containing neural networks. *IEEE Transactions on Neural Networks*, **2**, 252–262.

Narayan, S.S., Peterson, A.M. and Narasimha, M.J. (1983). Transform domain LMS algorithm. *IEEE Transactions on Acoustics, Speech, and Signal Processing*, **ASSP-31**, 609–615.

Nguyen, D.H. and Widrow, B. (1991). Neural networks for self-learning control systems. *International Journal of Control*, **54**, 1439–1451.

Oppenheim, A.V. and Shafer, R.W. (1975). *Digital Signal Processing*. Prentice Hall: Englewood Cliffs, NJ.

Parikh, D. and Ahmed, N. (1978). On an adaptive algorithm for IIR filters.

Proceedings of the IEEE, **66**, 585–588.

Qureshi, S.K.H. and Newhall, E.E. (1973). An adaptive receiver for data transmission of time dispersive channels. *IEEE Transactions on Information Theory*, **IT-19**, 448–459.

Rabiner, L.R. and Gold, B. (1975). *Theory and Appplication of Digital Signal Processing*. Prentice Hall: Englewood Cliffs, NJ.

Rao, S.S, Pan, T-S and Venkayya, V.B. (1991). Optimal placement of actuators in actively controlled structures using genetic algorithms. *AIAA Journal*, **29**, 942–943.

Reichard, K.M. and Swanson, D.C. (1993). Frequency-domain implementation of the filtered-x algorithm with on-line system identification. *Proceedings of the Second Conference on Recent Advances in Active Control of Sound and Vibration*, 562–537.

Rosenblatt, F. (1962). *Principles of Neurodynamics*. Spartan: New York.

Ross, C.F. (1982). An algorithm for designing a broadband active sound control system. *Journal of Sound and Vibration*, **80**, 373–380.

Roure, A. (1985). Self-adaptive broadband active sound control system. *Journal of Sound and Vibration*, **101**, 429–441.

Rumelhart, D.E., McClelland, J.L. and the PDP Research Group (1986). *Parallel Distributed Processing: Explorations in the Microstructure of Cognition*. MIT Press: Cambridge, MA.

Sakrison, D.J. (1966). Stochastic approximation: a recursive method for solving regression problems. In *Advances in Communication Theory*, **2**, A.V. Balakhrishnan, ed. Academic Press: New York.

Saïnt-Donat, J., Bhat, N. and McAvoy, T.J. (1991). Neural network based model predictive control. *International Journal of Control*, **54**, 1453–1468.

Segalen, A. and Demoment, G. (1982). Constrained LMS adaptive algorithm. *Electronics Letters*, **18**, 226–227.

Seivers, L.A. and von Flotow, A.H. (1992). Comparison and extensions of control methods for narrow-band disturbance rejection. *IEEE Transactions on Signal Processing*, **40**, 2377–2391.

Shynk, J.J. (1989). Adaptive IIR filtering. *IEEE ASSP Magazine*, **April**, 4–21.

Shynk, J.J. (1992). Frequency-domain and multirate adaptive filtering. *IEEE Signal Processing Magazine*, **January**, 14–37.

Smith, R.A.. and Chaplin, G.B.B. (1983). A comparison of Essex algorithms for major industrial applications. *Proceedings of InterNoise '83*, 407–410.

Snyder, S.D. and Hansen, C.H. (1990). The influence of transducer transfer functions and acoustic time delays on the LMS algorithm in active noise control systems. *Journal of Sound and Vibration*, **140**, 409–424.

Snyder, S.D. and Hansen, C.H. (1992). Design considerations for active noise control systems implementing the multiple input, multiple output LMS algorithm. *Journal of Sound and Vibration*, **159**, 157–174.

Snyder, S.D. and Tanaka, N. (1992). Active vibration control using a neural network. *Proceedings First International Conference on Motion and Vibration Control (MOVIC)*, Yokohama, Japan, Sept. 7–11.

Snyder, S.D. and Tanaka, N. (1993a). Modification to overall system response when using narrowband adaptive feedforward control systems. *ASME Journal of Dynamic Systems, Measurement, and Control*, **115**, 621–626.

Snyder, S.D. and Tanaka, N. (1993b). A neural network for feedforward controlled smart structures. *Journal of Intelligent Material Systems and Structures*, **4**,

373–378.

Snyder, S.D. and Hansen, C.H. (1994). The effect of transfer function estimation errors on the filtered-*x* LMS algorithm. *IEEE Transactions on Signal Processing*, **42**, 950–953.

Söderström, T. and Stoica, P. (1982). Some properties of the output error method. *Automatica*, **18**, 93–99.

Söderström, T., Ljung, L. and Gustavsson, I. (1978). A theoretical analysis of recursive identification methods. *Automatica*, **14**, 231–244.

Sommerfeldt, S.D. and Tichy, J. (1990). Adaptive control of a two-stage vibration isolation mount. *Journal of the Acoustical Society of America*, **88**, 938–944.

Stearns, S.D. (1981). Error surfaces of recursive adaptive filters. *IEEE Transactions on Circuits and Systems*, **CAS-28**, 603–606.

Syswerda, G. (1989). Uniform crossover in genetic algorithms. In *Proceedings of the 3rd International Conference on Genetic Algorithms*. 2–9.

Tanaka, N., Snyder, S.D., Kikushima, Y. and Kuroda, M. (1993). Active control of non-linear vibration using a neural network (On the suppression of harmonics by neuro control and FIR filter). *Journal of the Japanese Society of Mechanical Engineers, pt. C*, **59**, 700–707.

Tarrab, M. and Feuer, A. (1988). Convergence and performance analysis of the normalised LMS algorithm with uncorrelated Gaussian data. *IEEE transactions of Information Theory*, **IT-34**, 680–691.

Tate, C.N. and Goodyear, C.C. (1983). Note on the convergence of linear predictive filters, adapted using the LMS algorithm. *IEE Proceedings Part G*, **130**, 61–64.

Treichler, J.R., Johnson, C.R., Jr., and Larimore, M.G. (1987). *Theory and Design of Adaptive Filters*. Wiley-Interscience: New York.

Ungerboeck, G. (1976). Fractional tap-spacing equaliser and consequences for closk recovery in data modems. *IEEE Transactions on Communication*, **COM-24**, 856–864.

Wang, B-T. (1993). Application of genetic algorithms to the optimum design of active control system. In *Proceedings of NOISE-93,* St Petersburg, Russia, May 31 – June 3, **2**, 231–236.

Wangler, C.T. and Hansen, C.H. (1993). Genetic algorithm adaptation of non-linear filter structures. In *Proceedings of Progress in Acoustics, Noise and Vibration Control*, Australian Acoustical Society Annual Conference, 9–10 November, 150–157.

Wangler, C.T. and Hansen, C.H. (1994). Genetic algorithm adaptation of non-linear filter structures for active sound and vibration control. In *Proceedings of 1994 IEEE International Conference on Acoustics, Speech and Signal Processing*, 19–22 April.

Whitley, D. (1989). The genitor algorithm and selection pressure: why rank based allocation of reproductive trials is best. In *Proceedings of the 3rd International Conference on Genetic Algorithms*, 116–121.

Whitley, D. and Hanson, T. (1989). Optimising neural networks using faster, more accurate genetic search. In *Proceedings of the 3rd International Conference on Genetic Algorithms*. 391–396.

Widrow, B. and Hoff, M. Jr. (1960). Adaptive switching circuits. *IRE WESCON Convention Record*, **Pt.4**, 96–104.

Widrow, B. (1971). Adaptive filters. In *Aspects of Network and System Theory*, R.E.

Kalman and M. DeClaris, ed. Holt, Rinehart and Winston: New York.

Widrow, B., Glover, J.R., Jr., McCool, J.M., Kaunitz, J., Williams, C.S., Hearn, R.H., Zeidler, J.R., Dong, E., Jr. and Goodlin, R.C. (1975a). Adaptive noise cancelling: principles and applications. *Proceedings of the IEEE*, **63**, 1692–1716.

Widrow, B., McCool, J.M. and Ball, M. (1975b). The complex LMS algorithm. *Proceeding of the IEEE*, **63**, 719–720.

Widrow, B., McCool, J.M., Larimore, M.G. and Johnson, C.R., Jr. (1976). Stationary and nonstationary learning characteristics of the LMS adaptive filter. *Proceedings of the IEEE*, **64**, 1151–1162.

Widrow, B., Shur, D. and Shaffer, S. (1981). On adaptive inverse control. *Proceedings of the 15th ASILOMAR Conference on Circuits, Systems, and Computers*, 185–195.

Widrow, B. and Stearns, S.D. (1985). *Adaptive Signal Processing*. Prentice Hall: Englewood Cliffs, NJ.

7

Active control of noise propagating in ducts

7.1 INTRODUCTION

The first known application of active noise control was concerned with sound propagating in a duct and was conceived by Paul Lueg in Germany in the early 1930s. He filed a patent application in Germany in 1933 and in the USA in 1934. Unfortunately, Lueg was never able to demonstrate his idea successfully, partly because the field of electronics was not sufficiently advanced to enable the required precision amplifiers to be constructed and partly because his proposed system was oversimplified.

Twenty years later, Olson and May (1953, 1956) proposed a slightly different system which reappears now and then with different names like 'virtual earth', 'near field' or 'tightly coupled monopole'. In essence, this is a feedback system which attempts to generate a pressure null at a microphone located in front of the control loudspeaker, and in so doing uses the signal from this error microphone to generate the required control signal. This may be contrasted to Lueg's system which was a feedforward system, making use of the signal from a microphone located in the duct between the primary source and control loudspeaker to provide the required control signal (after passing through an amplifier). Note that feedback systems aim to modify the transient response characteristics of the system, whereas feedforward systems aim to modify the impedance presented to the primary source or to modify the amplitude of the incoming disturbance by using the control source to absorb it. Feedback and feedforward controllers are discussed in more detail in Chapters 5 and 6 respectively.

It is Lueg's feedforward system rather than Olson's feedback system which forms the basis of modern day controllers used in most commercial duct active noise control systems. However, these controllers bear little resemblance to

Lueg's original system due to the many additions necessary to make them practical. These systems and their application to both periodic and random noise will be discussed in detail in the next section.

For the purposes of active control, noise propagating in ducts may be divided into plane waves and higher order modes. Plane waves propagate with the speed of sound in free space and are characterized by a uniform sound pressure distribution across the duct. On the other hand, higher order modes propagate at speeds dependent on the frequency of the sound and the mode order and are characterized by a non-uniform sound pressure distribution across the duct section. Thus, it is clear that higher order mode propagation is much more difficult to control with an active system than is plane wave propagation. Even sensing the sound pressure associated with a higher order mode becomes complicated. Sound sources and sound sensors must be placed at locations where they can drive and sense respectively the modes of interest.

Fortunately, higher order modes do not begin to propagate and assume importance in duct noise transmission until the frequency of the sound exceeds the cut-on frequency for the first higher order mode. Below this frequency, higher order modes, if generated, decay rapidly with distance from the noise source (a property referred to as evanescence) and thus are not a problem, provided that the sensing microphones are placed in the far field of the primary and control sources, so that they do not sense the pressure contribution due to these higher order, non-propagating modes. Thus, below the duct cut-on frequency, simple single channel plane wave controllers can achieve significant attenuations in duct noise propagation.

Ducts lined with porous acoustic material allow higher order mode propagation at any frequency; thus, where possible, the active control source should be placed in an unlined section of duct. If it is desired to extend the operating frequency range of a single channel controller to above the cut-on frequency of the first higher order mode, it may be possible to partition a section of duct into two to four segments, so that the cross-sectional size of each segment is sufficiently small to inhibit the propagation of higher order modes. Each segment can then be controlled using an independent control source driven by the same signal used to drive the control sources in the other segments. However, one error sensor would be needed in each segmented duct section to avoid the contribution of higher order modes to the error signal. The controller could be adapted to minimize the sum of the squared error signals. Note that because of the difficulty in obtaining and maintaining perfectly matched speakers, this type of system would probably only be practical if it were controlled by a multi-channel controller driving each control source independently.

In the optimization of a practical active system to minimize sound propagating in a duct, the electronic controller plays only a partial role. Of equal importance is the arrangement of the control sources, reference sensor and error sensors. To gain an appreciation of how this arrangement is optimized it is necessary to develop an understanding of the physical processes involved in active control, and this will be done by way of an analytical model for periodic noise control which will be discussed later. At this stage it is sufficient to say that control of periodic noise is achieved principally by a reduction in the sound power radiated by the primary source as a result of the impedance change caused by the control source. When a single source is used to control random noise it can be shown that the control mechanism (at optimal control) is reflection and dissipation of energy between the primary and control sources. If more than one closely spaced control source is used in this case, then at least one source may absorb power when optimally controlled. To gain a full understanding of the physical processes involved, it is necessary to include the primary source in any model concerned with harmonic noise; it is not sufficient to assume a wave arriving from some unspecified source somewhere upstream unless the sound source is emitting random noise only. Thus, the analytical model discussed in this chapter for harmonic noise is formulated in terms of the acoustic power radiated by the primary and control sources and the effect which each source has on the other's radiated power.

This chapter will begin with a discussion of the formulation and implementation of various types of controller which have been used in the past, and this will be followed by a detailed theoretical development, first for the control of plane waves and then for the control of higher order modes. The effects of primary source type, impedance termination at the primary source end of the duct, primary source location (in the plane of the duct cross-section or in the duct wall), and the effects of the duct termination at the other end from the primary source will be discussed. Acoustic control mechanisms will also be explained, a technique for measuring the acoustic power output of speaker control sources will be described, and acoustic error sensors will be discussed. Other topics directly related to the one-dimensional (or plane wave) duct problem which will be discussed briefly are sound radiation from engine exhaust exits, pulsations in liquid filled pipes and active headsets for hearing protection.

7.1.1 Active vs. passive control

Traditionally, sound propagating in ducts has been controlled using reactive or dissipative mufflers. The former type of muffler reduces the sound transmitted down the duct by changing the radiation impedance seen by the source, whereas the latter type of muffler dissipates the sound energy. Reactive mufflers usually consist of expansion chambers and perforated pipes whereas dissipative mufflers consist of baffles lined with sound absorbing material such as rockwool or glass fibre. The design of both types of muffler is covered in other texts (for example, Chapter 9 of Bies and Hansen, 1996) and will not be considered further here.

A commonly asked question is, 'What are the advantages of active control over reactive and dissipative mufflers?' The main advantages are the small size, low pressure drop (and associated energy savings in large air handling systems – Eghtesadi *et al.*, 1984, 1986) and good low frequency performance. However, there are many applications where passive control is preferred over active control. For example, for high frequency noise control, dissipative or combined dissipative/reactive mufflers are generally less expensive and have superior performance over active control systems. This is because high frequency noise is usually associated with the propagation of higher order modes in addition to plane waves in a duct. Active systems to control higher order mode propagation must be multi-channel and thus are much more complex than those for controlling plane waves. For a rectangular section duct, the cut-on frequency is given by $f_{cu} = c_0/2d$ where d is the largest cross-sectional dimension and c_0 is the speed of sound in free space. For circular section ducts, the cut-on frequency is given by $f_{cu} = 0.586c_0/d$ where d is the duct diameter. Thus, for a rectangular section duct with its largest cross-sectional dimension equal to 1 m, and carrying air at room temperature, the highest frequency which can be controlled with a simple single channel controller is 170 Hz.

On the other hand, passive systems to control low frequency sound are generally bulky and expensive, and often impractical to install. Thus, many air handling systems silenced by using passive mufflers suffer from a low frequency rumble noise problem which can be very annoying (Eriksson *et al.*, 1988). Generally, active systems are best suited to the control of low frequency harmonic noise for which their cost and performance advantages far outweigh the passive alternatives. In some instances, active control is the preferred alternative for broadband low frequency noise as well as tonal noise, particularly in cases where the muffler pressure drop must be minimized or for large exhaust stacks where the cost of passive low frequency mufflers can be several hundred thousand dollars per muffler, and installation results in a major construction project which is expensive and often disruptive to production schedules. Active systems combined with passive systems for both low and high frequency noise control have been discussed by Munjal and Eriksson (1989a).

Unfortunately, use of even simple active control systems for plane wave control in ducts is not without its implementation difficulties. Acoustic feedback from the control source to the noise detecting microphone can cause controller instability. Turbulent pressure fluctuations associated with flow in ducts contaminates the microphone signals and causes the controller to generate acoustic cancelling signals which add to the acoustic noise and do nothing to attenuate the turbulent pressure fluctuations which propagate at the same speed as the flow and not the speed of sound. The poor frequency response of loudspeakers at low frequencies and lack of uniform response at higher frequencies further complicates the controller design. Reflections from the loudspeakers, duct bends and duct ends also complicate the control problem. Contaminated flows cause special problems for microphones and loudspeakers. The life of loudspeakers in typical installations is short, varying from one to three years, because of the large cone excursions generally found to be necessary. Means of estimating cone displacements and loudspeaker power requirements for a particular installation are discussed in Chapter 14. A further problem, often overlooked, is the transmission of vibration along the duct wall which is radiated as sound downstream of the active or passive muffler, thus compromising the perceived performance. The wall vibration can also affect the error sensor output directly.

7.2 CONTROL SYSTEM IMPLEMENTATION

There are two fundamentally different types of controller; feedback and feedforward. Each type can be further categorized as non-adaptive or adaptive. Adaptive feedforward systems which may also be further categorized as periodic noise or random noise controllers are the only type of system in practical use today for the control of noise propagating in ducts. However, it is of interest to review the earlier system types as it allows the evolution of the technology from 1934 to the present day to be properly understood. In this section and the remainder of the chapter it is assumed that the system to be controlled is linear; that is, the principle of superposition holds whereby the response at a point in the duct resulting from the simultaneous action of a number of acoustic sources is equal to the sum of the responses due to each individual force.

7.2.1 Feedback control

To begin, the feedback controller first introduced by Olson (1953) will be discussed. This arrangement, shown in Fig. 7.1, is intended to drive the control loudspeaker to produce a pressure null at the microphone which acts to reflect sound waves propagating down the duct back towards the source.

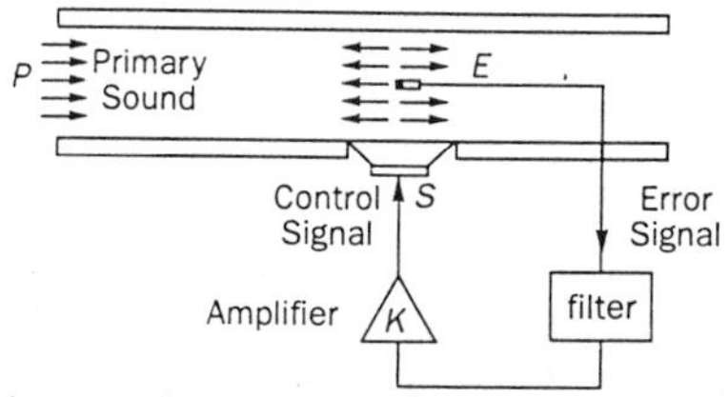

Fig. 7.1 Simple feedback control system for a duct (after Olson, 1956).

For the non-adaptive system proposed by Olson, the gain K of the amplifier is fixed. Note that feedback controllers, by definition, modify the transient response characteristics of the system being controlled by using a control signal derived from the error sensor. Thus, they work best at controlling longitudinal resonances in finite length ducts (Trinder and Nelson, 1983b) and are considered unsuitable for controlling random noise (Eriksson, 1991).

Unfortunately, in a feedback control system, the error signal can never go to zero (the desired level) or no control signal will be derived, and this limits the system performance. The larger the controller gain, the smaller the error signal can go; however, high gains result in the system being only marginally stable, capable of bursting into wild oscillations. Since duct acoustics influence the feedback loop, maintaining the desired amplifier gain to ensure good attenuation is difficult, and practical systems for active control of duct noise using this arrangement are non-existent, although similar principles are used in the design of controllers for active hearing protectors (see Section 7.8). Nevertheless, active feedback systems applied to duct noise control received a significant amount of attention in the 1980s (Eghtesadi and Leventhall, 1981; Wheeler and Parramore, 1981; Eghtesadi *et al.*, 1983; Trinder and Nelson, 1983a,b; Hong *et al.*, 1987). The performance of the system proposed by Hong *et al.* (1987) (Fig. 7.2(a)) to control noise from a large axial fan is shown in Fig. 7.2(b), where it can be seen that between 20 and 30 dB of attenuation of random noise was obtained over a 1.5 octave frequency band. However, the stability of system such as this cannot be guaranteed over a long period in a practical installation. Note that a system with a fixed gain feedback can only be effective in controlling a very narrow band of frequencies. However, the possibility also exists for using an adaptive algorithm to continually update the gain K which would effectively result in a system with much better performance and stability characteristics.

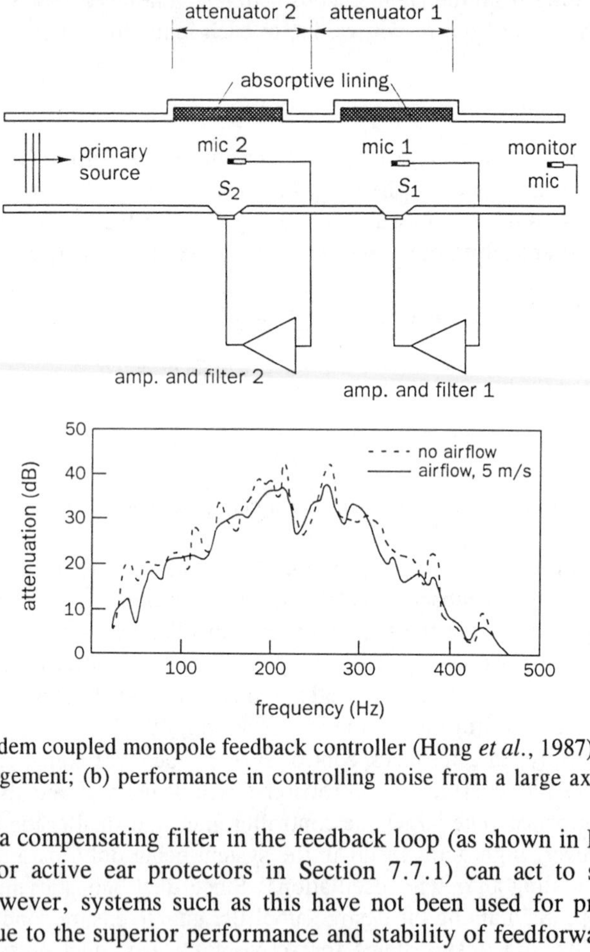

Fig. 7.2 Tandem coupled monopole feedback controller (Hong *et al.*, 1987): (a) physical system arrangement; (b) performance in controlling noise from a large axial fan.

Use of a compensating filter in the feedback loop (as shown in Fig. 7.1 and described for active ear protectors in Section 7.7.1) can act to stabilize the system. However, systems such as this have not been used for practical duct problems due to the superior performance and stability of feedforward systems, and the relative ease of sampling the incoming signal with a reference microphone in sufficient time to generate the required control signal at the control source located at some distance downstream from the reference microphone. The inherent difficulty associated with sampling the incoming signal to a hearing protector is the reason that feedback systems have been more fully developed for the latter application. More details on systems for hearing protectors are given in Section 7.8.

Another application in which a feedback system in a duct has met with some success is in the reduction of engine exhaust noise (Mori *et al.*, 1991), where the control source and error microphone were placed in an expansion chamber which was part of the exhaust system. Similar noise reductions were obtained as were obtained with a conventional passive silencer, but the resulting system pressure drop was much less.

7.2.2 Feedforward control

The simple feedforward control scheme mentioned in Section 7.1 and first proposed by Lueg (1933) is shown in Fig. 7.3. One problem which limits the effectiveness of this system is the acoustic feedback of sound from the control source to the reference (or detection) microphone. This results in the reference signal input to the controller not being a true representation of the primary noise which the control system is attempting to attenuate. However, the main problem, which took 50 years to solve, is the electronic implementation of the transfer function $G(j\omega)$ which acts on the microphone signal to produce the required signal to the loudspeaker. In the sections to follow, we will derive optimal expressions for $G(j\omega)$ for various control configurations and then discuss briefly how these may be implemented electronically. In all of the following analyses, the important assumption is made that all electrical, acoustic and electroacoustic elements of the system are linear, an approximation which normally can be regarded as reasonable.

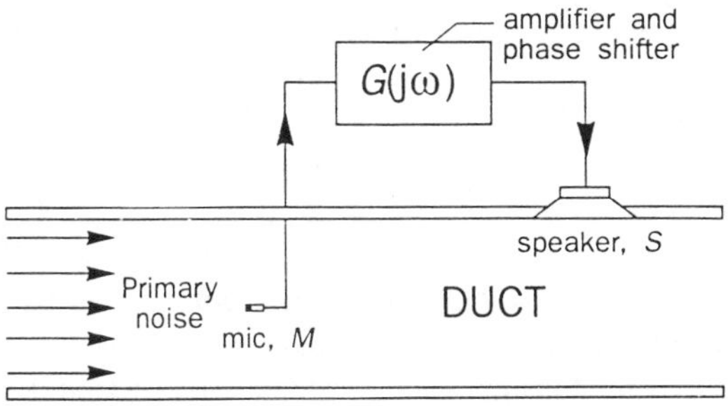

Fig. 7.3 Simple feedforward system for control of noise in a duct (after Lueg, 1936).

7.2.2.1 Independent reference signal

Use of an independent reference signal which cannot be contaminated by feedback from the control source eliminates one of the problems mentioned above (see Fig. 7.4). For periodic noise, this signal could be derived, for example, from a tachometer which measures the rotational speed of the machine making the noise. For random noise control only one example of a non-acoustic reference signal has been reported in the literature and that is, the use of intensity of light emitted by a flame to control its noise generation. However, the control of random noise in a duct invariably requires the use of an acoustically derived reference signal and thus is susceptible to acoustic feedback. This will be discussed in more detail below.

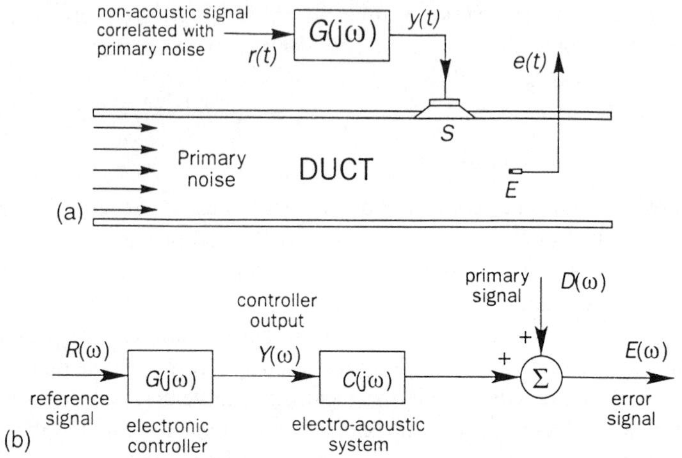

Fig. 7.4 Feedforward control system for a duct with no acoustic feedback from the control source to the reference signal. The signal correlated with the primary source could be derived from a shaft tachometer signal, for example, if the noise source were a fan in a duct. (a) Physical system; (b) Electrical block diagram.

For the system shown in Fig. 7.4, the cancellation path transfer function $C(j\omega)$ between the electrical input to the control loudspeaker and the electrical output from the error microphone at which the acoustic signal in the duct is to be minimised is given by

$$C(j\omega) = \frac{E(\omega)}{Y(\omega)} \tag{7.2.1}$$

where the upper case symbols are the frequency domain equivalents of the time domain signals denoted by lower case symbols. This transfer function could be measured in practice by turning off the primary noise source and injecting a signal into the control source.

From Fig. 7.4(b), the error signal may be written as

$$E(\omega) = D(\omega) + G(j\omega)\, C(j\omega)\, R(\omega) \tag{7.2.2}$$

and the optimal controller transfer function (for zero error) is then

$$G_o(j\omega) = \frac{-D(\omega)}{R(\omega)\, C(j\omega)} \tag{7.2.3}$$

The ratio $D(\omega)/R(\omega)$ can be measured by measuring the ratio of the error signal output to the controller input (after the signal conditioning electronics) with the controller turned off.

In practice, if the reference signal is generated using a tachometer, it is usually necessary to convert the tachometer output (which is normally a pulse train) to a periodic analogue signal constructed by adding together sinusoidal signals representing the fundamental and the harmonics which are to be controlled. This can be done by using appropriate electronic signal conditioning. Alternatively, the harmonics and the fundamental can all be kept separate and a controller designed to act on each one independently. This was first done by Conover (1956) for the control of noise radiated by an electrical transformer. In this case, however, the reference signal was the transformer supply voltage and the harmonics were generated by rectifying the signal and passing the result through narrow bandpass filters.

7.2.2.2 Waveform synthesis

In the late 1970s Chaplin (1980) developed a system which used a microphone to construct the waveform to drive the control source to minimize the error signal. The waveform consisted of a number of steps of constant voltage, the length and level of each step being adjusted iteratively to minimize the signal output from an error microphone placed in the duct downstream from the control source. The waveform repeated itself periodically with the period set by the tachometer input. Because of the need for a synchronizing signal and because of the relatively slow adaptation process, this type of system is only suitable for control of periodic noise which is synchronized to the tachometer signal. The slow adaptation process also results in a limited ability to track varying signals. This type of system is discussed in more detail by Nelson and Elliott (1992, Section 6.4).

7.2.2.3 Acoustic feedback

For random noise in ducts, and even for some situations involving just periodic noise or a combination of periodic and random noise, it is necessary to derive the reference signal from a microphone placed in the duct. This results in the reference signal being contaminated by the control source output, a phenomenon referred to as 'acoustic feedback'.

A physical representation of a system involving acoustic feedback is shown in Fig. 7.5(a) and the corresponding electrical block diagram is shown in Fig. 7.5(b). It is relatively straightforward (Nelson and Elliott, 1992, Section 6.6) to show that the transfer function $H_o(j\omega)$ required to drive the error signal to zero is given by

$$H_o(j\omega) = -\frac{P(j\omega)}{C(j\omega)} \tag{7.2.4}$$

where the subscript o refers to optimum.

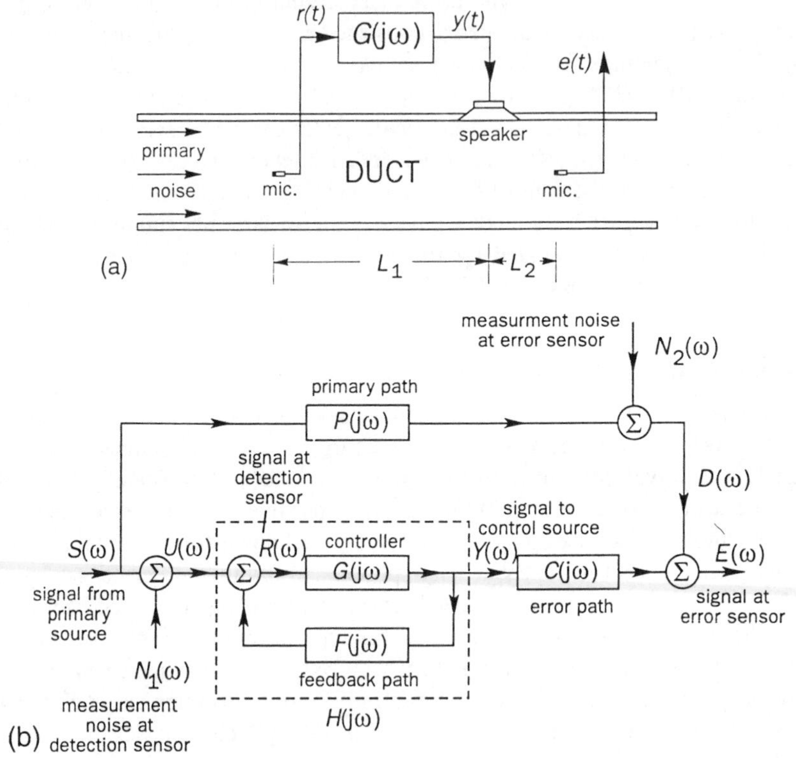

Fig. 7.5 Feedforward control system for a duct with acoustic feedback from the control source to the reference input: (a) physical system; (b) electrical block diagram.

From Fig. 7.5(b) it can be seen that $P(j\omega)$ is the ratio of the error microphone output signal to the controller input signal with the control source turned off. $C(j\omega)$ is the ratio of the error microphone output to the controller output signal with the primary source turned off, and can be measured by injecting a signal into the control source. Similarly, the feedback transfer function $F(j\omega)$ is the ratio of the reference microphone output to the signal injected into the control source, with both the controller output to the control source and primary source turned off. Obviously the signal to drive the control source must come from an external source to determine this transfer function.

From Fig. 7.5(b) it can be seen that $H(j\omega)$ is also defined as

$$H(j\omega) = \frac{G(j\omega)}{1 - G(j\omega)F(j\omega)} \tag{7.2.5}$$

where $G(j\omega)$ is the controller transfer function and $F(j\omega)$ represents the acoustic feedback path. Rearranging (7.2.5) gives

$$G(j\omega) = \frac{H(j\omega)}{1 + F(j\omega)H(j\omega)} \tag{7.2.6}$$

Substituting (7.2.4) into (7.2.6) gives, for the optimal controller (Roure, 1985):

$$G_o(j\omega) = \frac{-P(j\omega)}{C(j\omega) - F(j\omega)P(j\omega)} \tag{7.2.7}$$

All of the transfer functions on the right-hand side of (7.2.7) can be measured directly.

If measurement noise exists in the reference microphone, then $P(j\omega)$ in (7.2.7) is replaced by $P'(j\omega)$ (Nelson and Elliott, 1992), where

$$P'(j\omega) = \left[\frac{SNR(\omega)}{1 + SNR(\omega)}\right] P(j\omega) \tag{7.2.8}$$

where $SNR(\omega)$ is the power (or pressure squared) signal to noise ratio at the detection sensor at frequency ω. From (7.2.7) and (7.2.8) it can be seen that the optimal controller is balancing the cancellation of the acoustic noise against the amplification of the measurement noise. It is interesting to note that measurement noise in the error sensor signal has no effect on the optimal controller, $G_o(j\omega)$ (Nelson and Elliott, 1992).

Roure (1985) described one possible method of implementing the frequency domain design controller of (7.2.7) and (7.2.8). First $G_o(j\omega)$ was determined by measuring all of the required transfer functions. Next $G_o(j\omega)$ was transformed using an inverse discrete fast Fourier transform into a sampled impulse response which can be implemented as an FIR filter. A number of practical difficulties and means for overcoming them were discussed by Roure (1985) and are listed below.

1. At low frequencies, deficiencies in the loudspeaker response and turbulence noise result in large errors in the estimate of $G_o(j\omega)$. At high frequencies the estimate of $G_o(j\omega)$ is usually large so that if the entire response over the whole frequency range were inverse Fourier transformed, large windowing errors would result. Thus, Roure found it necessary to set values of $G_o(j\omega)$ equal to zero below 50 Hz and to smoothly attenuate $G_o(j\omega)$ above the first higher order mode cut-on frequency of the duct.
2. The calculated impulse response function to implement $G_o(j\omega)$ generally has non-causal components. Roure used a temporal window to set the non-causal part of the impulse response to zero and also smoothly attenuated the response for long time delays. (Note that the non-causal problem does not exist for periodic noise as for this type of signal, future and past have no meaning.)
3. As a result of errors in the estimation of $G_o(j\omega)$ and the windowing operations described above, the FIR filter using the calculated coefficients

does not result in the maximum achievable attenuation. Roure improved the attenuation by measuring the residual error signal and recalculating $G_o(j\omega)$ by multiplying the old estimate by a correction term calculated using an iterative algorithm. This was done on a continuous basis, although the time scale associated with each update was long compared to the sample rate. This is in contrast to an adaptive controller where the filter weight update time scale is of the order of the sample rate.

Results obtained using Roure's method to attenuate axial fan noise in a duct having cross-sectional dimensions of 0.4 m × 0.3 m are shown in Fig. 7.6.

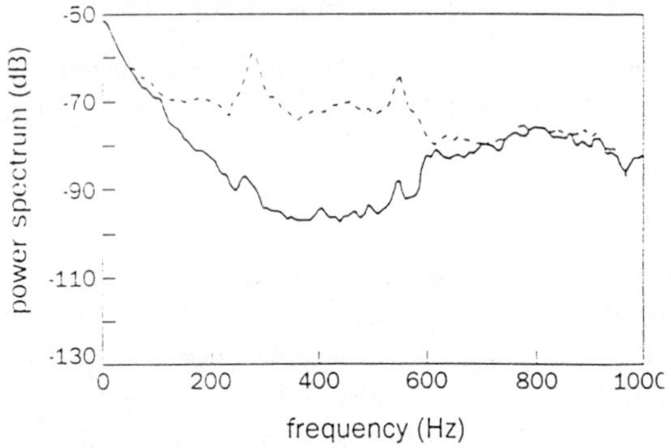

Fig. 7.6 Results of experiments on active control of axial fan noise in a duct with a mean air flow of 9 $m^{-1}s$ (after Roure, 1985): - - - - - - - - attenuator off; ———————— attenuator on. A 256 coefficient FIR filter and a sample rate of 4096 Hz were used. Below about 100 Hz turbulence noise was too great for appreciable attenuation to be obtained.

An alternative approach which involves the use of auto and cross correlation measurements to design the controller directly in the time domain has been used by Ross (1982a), who has also discussed iterative algorithms (1982b).

It is of interest to express the optimal controller of (7.2.7) in terms of acoustic parameters so that the effect of these parameters can be evaluated. Referring to (7.2.7), Nelson and Elliott (1992) show that for a duct terminated anechoically at each end (assuming that only plane waves are present), the transfer functions can be written in terms of physical variables as

$$P(j\omega) = \frac{M_e}{M_r} e^{-jk(L_1+L_2)} \tag{7.2.9}$$

$$C(j\omega) = L_c M_e e^{-jkL_2} \tag{7.2.10}$$

$$F(j\omega) = L_c D_c M_r D_r e^{-jkL_1} \tag{7.2.11}$$

For a finite duct with no anechoic terminations at either end, the corresponding expressions are (Nelson and Elliott, 1992)

$$P(j\omega) = \frac{M_e e^{-jk(L_1+L_2)}(1+D_e R_e)}{M_r\left(1+D_r R_e e^{-2jk(L_1+L_2)}\right)} \tag{7.2.12}$$

$$C(j\omega) = \frac{M_e L_c e^{-jkL_2}(1+D_e R_e)\left(1+D_r R_r e^{-2jkL_1}\right)}{1-R_r R_e e^{-2jk(L_1+L_2)}} \tag{7.2.13}$$

$$F(j\omega) = \frac{M_r L_c e^{-jkL_1}(D_r+R_r)(D_c+R_e e^{-2jkL_2})}{1-R_r R_e e^{-2jk(L+L_2)}} \tag{7.2.14}$$

In the above equations, M_r and M_e (volt/Pa) are respectively the complex frequency dependent responses of the reference microphone and error microphone, L_c (Pa/volt) is the complex frequency dependent response of the control source, D_r is the complex frequency dependent directivity of the reference microphone in the direction of the control source, D_c is the complex frequency dependent directivity of the control source in the direction of the reference microphone, L_1 is the distance between the reference microphone and control source (see Fig. 7.5(a)), L_2 is the distance between the control source and error sensor, k is the wavenumber of the acoustic wave, R_r is the complex frequency dependent reflection coefficient of the duct to the left of the reference microphone (measured at the reference microphone) and R_e is the complex frequency dependent reflection coefficient of the duct to the right of the error microphone (measured at the error microphone).

Substituting (7.2.9) to (7.2.11) into (7.2.7) gives

$$G_o(j\omega) = \frac{-e^{-jkL_1}}{L_c M_r\left(1-D_c D_r e^{-2jkL_1}\right)} \tag{7.2.15}$$

The same equation is obtained if (7.2.12) to (7.2.14) are substituted into (7.2.7). Thus, it is clear that the optimal controller transfer function is the same for infinite and finite length ducts. The quantities L_c, M_r, D_c and D_r are all complex, and can be represented by an amplitude and a phase. In the presence of a mean downstream flow of Mach number M (mean flow speed/speed of sound), Elliott and Nelson (1984) show that the optimal controller can be written as

$$G_o(j\omega) = \frac{-e^{-jkL_1/(1+M)}}{L_c M_r\left(1-D_c D_r e^{-2jkL_1/(1-M^2)}\right)} \tag{7.2.16}$$

From (7.2.15) it can be seen that $G_o(j\omega)$ is independent of any error sensor characteristics, as the pressure at this point is driven to zero. However, it is important to remember that it has been assumed that only plane waves are sensed by the error sensor. This means that the error sensor must be located sufficiently far downstream that it is not influenced by the near field of the control source. Similarly, the reference microphone has been assumed to only be detecting plane waves. Thus, it should be placed far enough from the primary noise source to be in its far field.

Returning to (7.2.15), it can be seen that the numerator represents the controller compensation needed to account for the time taken for a downstream travelling plane wave to travel from the reference microphone to the control source. The product L_cM_r compensates for the electroacoustic response of the reference microphone and control source and the quantity $(1 - D_c D_r e^{-2jkL_1})$ compensates for any direct feedback from the control source to the reference microphone. Note that a reflecting termination at the downstream end of the duct does not affect $G_o(j\omega)$ because the theoretically optimal controller cancels all downstream propagating waves, leaving nothing to be reflected. A reflecting termination upstream of the reference microphone will have some effect on $G_o(j\omega)$ because waves originating from the control source will be reflected, further contaminating the reference signal.

It is interesting to note from equation (7.2.15) that if either the reference microphone or control source were perfectly directional such that D_r or $D_c = 0$, then $G_o(j\omega)$ would have a very simple feedforward structure. If omnidirectional sources were used such that D_r or $D_c = 1$, then (7.2.15) becomes

$$G_o(j\omega) = \frac{-e^{-jkL_1}}{L_cM_r(1 - e^{-2jkL_1})} = \frac{j}{L_cM_r 2\sin(kL_1)} \qquad (7.2.17a,b)$$

This implies that in the absence of any damping in the duct, the controller response will have to be infinite at frequencies corresponding to the distance between the reference microphone and source being an integer multiple of half wavelengths.

Clearly then, it is desirable for either or both of the control source or error sensor to exhibit directional properties. A considerable amount of work was done in the 1970s and early 1980s (Swinbanks, 1973; Eghtesadi and Leventhall, 1983; Berengier and Roure, 1980b; La Fontaine and Shepherd, 1983, 1985) developing and using arrays of sources and sensors which were directional. The original intention of this work was to eliminate the feedback from the control source to the reference microphone. Equation (7.2.11) shows that for an infinitely long duct or a duct terminated anechoically at each end, the feedback transfer function will be zero if the directivity of either the control source or reference microphone is zero. However, for a finite length duct, it can be seen from (7.2.14) that the feedback transfer function will be non-zero even if both the directivity of the control source and the directivity of the reference microphone is zero. Nevertheless, it is still beneficial to examine how directional sources and sensors can be realized in practice as their use invariably improves the performance of practical duct active control systems.

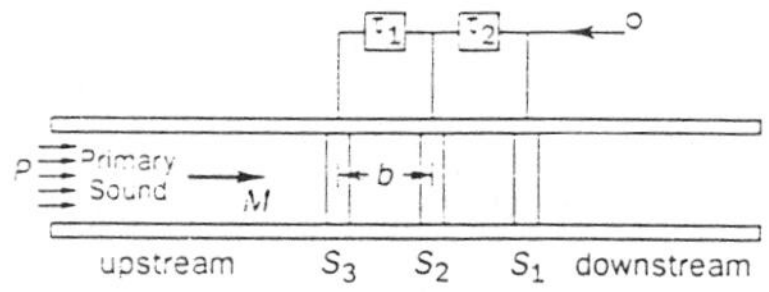

Fig. 7.7 Directional source (after Swinbanks, 1973).

Swinbanks (1973) proposed a directional control source system consisting of two or three ring monopole sources as shown in Fig. 7.7. Considering only two sources (s_3 and τ_2 assumed non-existent), Swinbanks (1973) showed that if the time τ_1 by which the upstream speaker driving signal is delayed with respect to the downstream speaker driving signal, is set equal to $b/[c_0(1 - M)]$ (where c_0 is the speed of sound in the duct and M is the flow Mach number) then there will be no energy propagated in the upstream direction. In practice, difficulty is encountered in exactly matching the speaker frequency responses resulting in imperfect directionality over a wide frequency range. Another severe disadvantage is the limited bandwidth over which a reasonable output can be obtained. If the efficiency η is defined as the ratio of the amplitude of the wave generated by the two source configuration to that generated by a single source (with the same volume velocity as the upstream source), then Swinbanks (1973) showed that this is given by

$$\eta = 2 \left| \sin\frac{\omega\tau_o}{2} \right| \tag{7.2.18}$$

where

$$\tau_o = \frac{2b}{c_0(1 - M^2)} \tag{7.2.19}$$

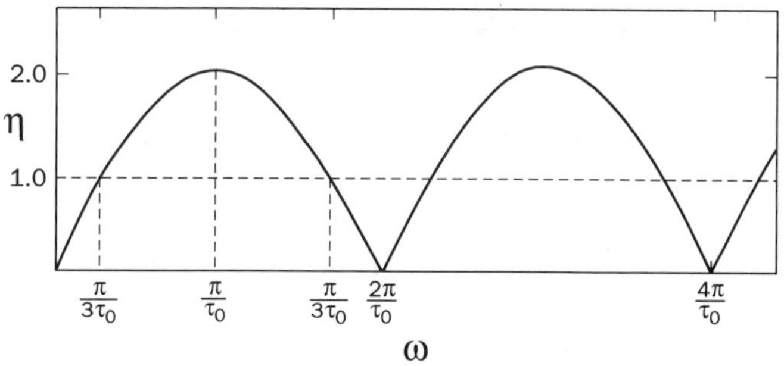

Fig. 7.8 Useful frequency range for a two element directional source (after Swinbanks, 1973).

The quantity η is plotted as a function of radian frequency in Fig. 7.8 where it can be seen that the efficiency is 1 or greater only over a bandwidth specified by

$$\frac{\pi}{3\tau_o} < \omega < \frac{5\pi}{3\tau_o} \tag{7.2.20}$$

with a centre frequency (in hertz) given by

$$f_o = \frac{1}{2\tau_o} \quad \text{(Hz)} \tag{7.2.21}$$

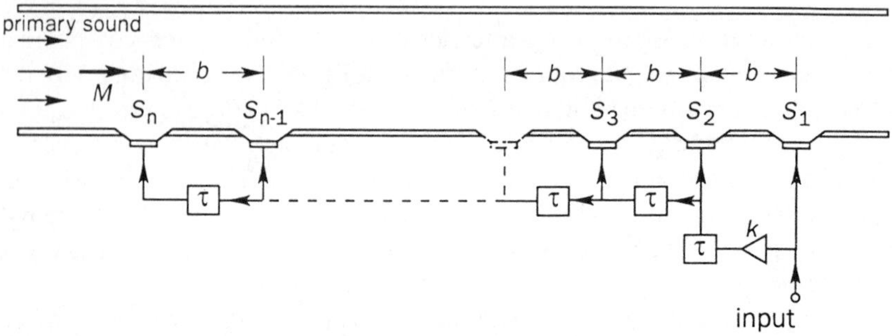

Fig. 7.9 *n*-element directional source (after Eghtesadi and Leventhall, 1983b). The amplifier gain K is given by $K = 1/(n - 1)$.

To increase the effective bandwidth of the directional source, Eghtesadi and Leventhall (1983b) suggested an array consisting of n monopole sources as shown in Fig. 7.9. To obtain zero output in the upstream direction, the following must be satisfied for each source output s_i

$$s_1(t) = -\frac{1}{n-1} \times$$

$$\times \left[s_2\left(t + \frac{b}{c_0(1-M)}\right) + s_3\left(t + \frac{2b}{c_0(1-M)}\right) + \ldots\ldots s_n\left(t + (n-1)\frac{b}{c_0(1-M)}\right)\right] \tag{7.2.22}$$

Thus, if all of the source separations are equal to b and all the time delays equal to τ which is defined as

$$\tau = \frac{b}{c_0(1-M)} \tag{7.2.23}$$

then the efficiency of downstream radiation can be written as (Eghtesadi and Leventhall, 1983b)

$$\eta^2 = 1 + \frac{\sin^2[(n-1)\omega\tau_o/2]}{(n-1)^2\sin^2(\omega\tau_o/2)} - \frac{2\sin[(n-1)\omega\tau_o/2]\cos(n\pi\tau_o/2)}{(n-1)\sin(\omega\tau_o/2)} \tag{7.2.24}$$

where τ_o is defined by (7.2.19).

If the number of sources n is even, then the maximum value of η (which occurs at f_o – see (7.2.21)) is given by

$$\eta_{max} = n/(n-1) \tag{7.2.25}$$

The effective bandwidth vs number of sources is listed in Table 7.1 and the efficiency is plotted as a function of frequency for $n = 3$, 4 and 5 in Fig. 7.10.

Table 7.1 Effective bandwidth vs. number of elements in the directional source (after Eghtesadi and Leventhall, 1983b)

Number of Elements	Bandwidth (octaves)
2	2.33
3	3.00
4	3.37
5	3.70
6	4.00
7	4.21
8	4.41

Exactly the same procedure can be used to design a directional microphone array, except in this case, the array sensitivity compared to the sensitivity of one microphone would be what was characterized by the efficiency η. Also the microphone array for the reference sensor would be orientated in the opposite sense, with the first microphone closest to the primary source.

Directional microphone arrays are generally easier to implement over a wide frequency range than are directional sound sources, as it is easier to obtain microphones with phase matched responses than it is to obtain phase matched loudspeakers (Elliott and Nelson, 1984). A system with two-element directional sources and sensors is illustrated in Fig. 7.11.

Loudspeakers and microphones acting as elements of these directional arrays are discussed in more detail in Chapter 14, as is the use of a probe tube as a more practical way of achieving a directional microphone. The probe tube microphone has the additional advantage that it also filters out a substantial amount of turbulence noise and indeed this is its primary purpose in currently available commercial systems.

In the derivation of the optimal controller equations in this section, it has been assumed that the control source internal impedance is infinite; that is, it is

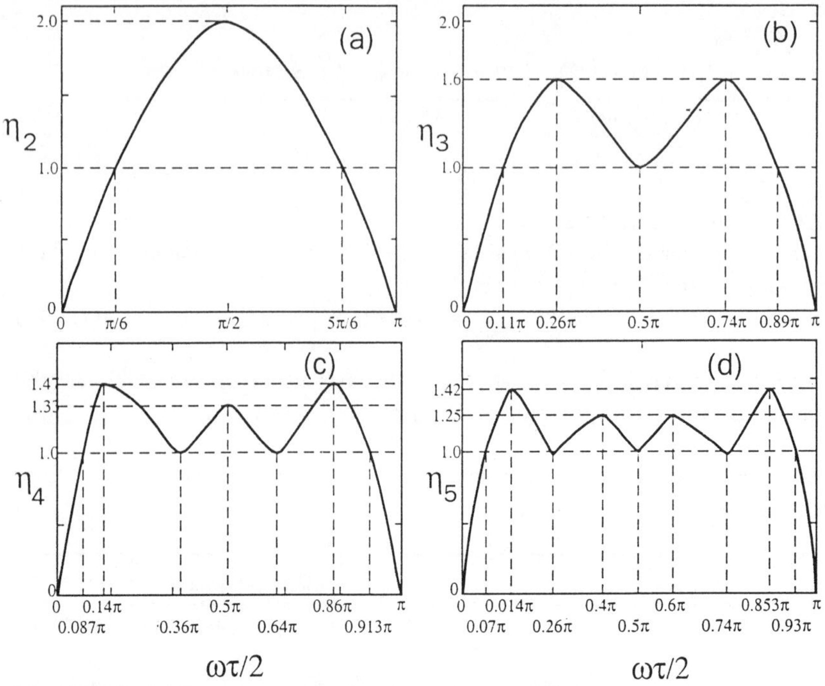

Fig. 7.10 Useful frequency ranges for three-, four- and five-element directional sources.

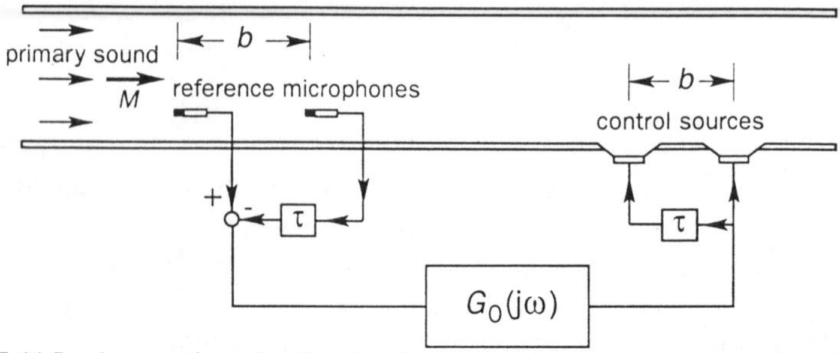

Fig. 7.11 Implementation of a directional sensor and source system in a duct. The delay $\tau = b/c_0(1 - M)$ where c_0 = speed of sound in the duct and b is the separation between the two elements in each array. M is the flow Mach number.

assumed to be a constant volume velocity source. If this assumption is relaxed, waves propagating in the duct will be reflected at the control source due to the impedance discontinuity as well as from the duct ends. Munjal and Eriksson (1988) showed that the effect on the optimal controller of relaxing this assumption is significant. However, loudspeakers using motional feedback of the cone, can be shown to be well approximated as constant volume velocity sources, even at frequencies close to the resonance associated with the suspension and backing cavity (Shepherd *et al.*, 1986a). However, for loudspeakers without motional feedback, the assumption of constant volume velocity is not very good, except at low frequencies when a small airtight cavity is used to enclose the back of the loudspeaker. The non-infinite internal impedance of the loudspeaker as well as other complicating effects such as measurement noise in the microphone signals and differences between the implemented controller and ideal controller frequency response mean that in general, the simple expressions of (7.2.15) and (7.2.16) for the optimal controller can only be used as a guide to the structure and approximate number of coefficients required for a practical controller.

In practice, it is always necessary to use either an iterative scheme or adaptive algorithm to update the controller coefficients, and the controller will never in practice correspond exactly to the optimal controller. Thus, although the optimal controller is theoretically capable of completely suppressing sound propagation downstream from the control source for both periodic and random primary noise, this result is never achieved in practice. For this reason it is of interest to develop acoustic models of systems so that the effect of using a non-optimal controller can be evaluated. Such models are discussed in detail later in this chapter. For example, it has been demonstrated that use of directional sources and microphones to reduce the effect of acoustic feedback does improve system performance (La Fontaine and Shepherd, 1983). Also control over a wider bandwidth and fewer dips in the attenuation curve over that bandwidth can be achieved if the control source response is made more uniform over the band, especially at low frequencies. This was demonstrated by La Fontaine and Shepherd (1983) who used feedback (effectively cone velocity feedback) from the back e.m.f. signal generated in the voice coil of a loudspeaker to achieve a more uniform loudspeaker response.

Before discussing the acoustic models, it is of interest to consider in a little more detail ways of implementing the optimal controller electronically. Use of a single FIR filter was discussed earlier in this chapter. An alternative architecture which has been suggested by Wanke (1976) and Davidson and Robinson (1977) is to implement two FIR filters, one to cancel the acoustic feedback path and the other to cancel the feedforward path (see Fig. 7.12).

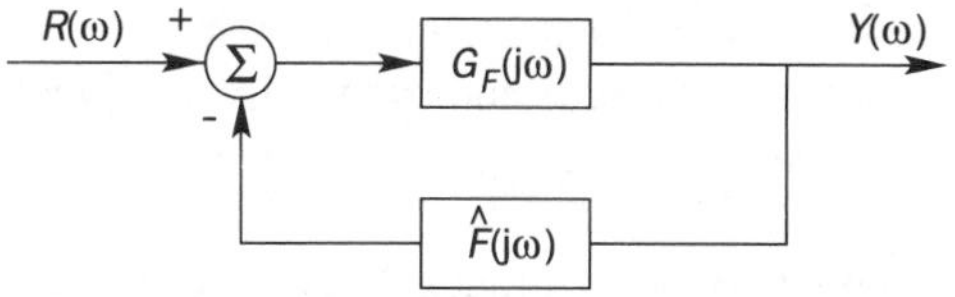

Fig. 7.12 Alternative controller architecture.

However, this type of design can be very inefficient to implement, as will be made clear by comparing the transfer function of this architecture with the optimal controller transfer function:

$$G(j\omega) = \frac{G_F(j\omega)}{1 + \hat{F}(j\omega)\,G_F(j\omega)} \tag{7.2.26a,b}$$

$$G_o(j\omega) = \frac{-P(j\omega)}{C(j\omega) - P(j\omega)\,F(j\omega)}$$

The ^ symbol over the F means that it is an estimate of F. Thus, assuming $\hat{F}(j\omega) = F(j\omega)$, then $G_F(j\omega) = -P(j\omega)/C(j\omega)$. For ducts with reflective ends, the impulse response of $F(j\omega)$ will be very long, thus requiring a long filter. Similarly, $G_F(j\omega)$ could also have a very long impulse response. Nelson and Elliott (1992) argue that the net response $G(j\omega)$ is much simpler than the two components $G_F(j\omega)$ and $(1 + \hat{F}(j\omega)\, G_F(j\omega))$ due to common terms in each which cancel one another, leading to the conclusion that the architecture of Fig. 7.12 is inefficient. An alternative is to use a recursive controller or IIR filter as implemented by Eriksson and his co-workers (1985, 1987 and 1991) and shown in Fig. 7.13.

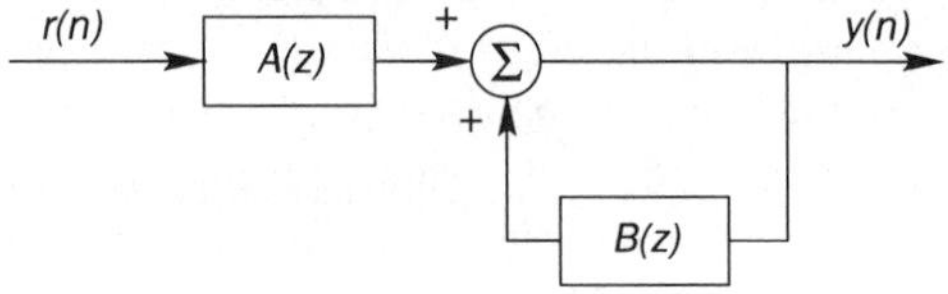

Fig. 7.13 IIR filter used by Eriksson and co-workers (1985, 1987 and 1991). The quantity (z) denotes a sampled data transfer function rather than a continuous transfer function.

In practice, the coefficients in $A(z)$ and $B(z)$ must be adjusted adaptively using the error microphone output and algorithms discussed in Chapter 6. Adaptive adjustment implies that the coefficients are changed on time scale equal to the time between consecutive samples. This is different to iterative adjustment discussed previously and used by Roure (1985) which makes changes based on some time averaged measurement of the error signal.

It is worth reiterating here that when adaptive algorithms are implemented, it is invariably necessary to include some leakage coefficient in the algorithm which subtracts a small amount from every filter coefficient at each update. If this is not done any small d.c. component present in the error sensor signal (due to d.c. offsets in the A/D converters for example) will cause a d.c. level to be fed to the control source in an attempt to cancel the d.c. component in the error signal. However, as the physical system will not respond to d.c., the d.c. output to the system will gradually increase and eventually saturate the D/A converters rendering the system useless, unless some form of leakage is adopted in the control algorithm.

Care must also be taken in any practical active control system to ensure that the error sensor is placed in the far field of the control source, otherwise it will detect the decaying evanescent field generated by the control source, thus limiting the controller performance. As the controller acts to minimize the sound pressure at the error microphone, the pressure of higher order modes will reduce the amount by which the plane wave is attenuated. This is discussed in more detail in Section 7.4.

One may be led to believe that a more robust system might use two error sensors to either act as a directional detector to downstream waves or to measure the acoustic intensity of downstream propagating waves, thus ignoring any waves reflected from the downstream end of the duct as a result of an imperfect control system. A directional microphone as an error sensor would emphasize the importance of some frequencies with respect to others due to large variations in its frequency response function. Two error sensors acting as an intensity sensor are unlikely to give good results due to the stringent microphone phase matching requirements and the presence of a significant reactive field in the duct due to reflections from the end. For these reasons, use of a single element error sensor is preferred for plane wave control.

In some cases, because of the hostile in-duct environment it may be desirable to place the error sensor outside the duct near the exit. However, great care must be taken in this case, as it is unlikely that minimizing the sound pressure at a single error sensor will be sufficient to achieve global noise reduction. It is likely that a number of error sensors with the controller minimizing the sum of the squared error signals will be necessary to achieve this.

It is important to remember that for the controller to be causal, the acoustic delay in the duct associated with propagation of the acoustic wave from the reference microphone to the control source must be greater than that associated with the analogue filters, A/D and D/A converters in the controller. Each pole of an analogue antialiasing filter has approximately ⅛ cycle of delay at the filter cut-off frequency which is typically set at one-third of the sampling frequency. Assuming one sample delay for the A/D and D/A converters, the total delay in the analogue path is (Elliott and Nelson, 1992):

$$\tau_A = T\left[1 + \frac{3n}{8}\right] \tag{7.2.27}$$

where T is the time in seconds between samples and n is the number of poles in the antialiasing filters. As an example, a sampling rate of 2000 Hz (giving a controller bandwidth of approximately 650 Hz) and a six pole antialiasing filter (assuming no reconstruction filter on the output of the controller) would result in $\tau_A = 1.6$ ms, which corresponds to a minimum separation of 0.6 m between the reference microphone and control source in a duct containing air at room temperature. Although some practitioners advocate the use of reconstruction filters to eliminate high frequency noise in the output to the control source (caused by the digital to analogue conversion), proper choice of control source response characteristics sometimes makes this unnecessary.

When using an adaptive algorithm to update the controller coefficients, there is a need to compensate for the electroacoustic transfer functions of the control source and error sensor and the time delay between them. This is referred to collectively as the cancellation path transfer function, and the need to compensate for this was first alluded to by Burgess (1981) and led to the development of the so-called filtered-x LMS algorithm (see Chapter 6). The derivation of the filtered-x LMS adaptive algorithm used to update the coefficients of the filter A and means for obtaining the estimate $\hat{C}$ of the cancellation path transfer function are discussed in Chapter 6. An adaptive scheme using this algorithm is illustrated in Fig. 7.14.

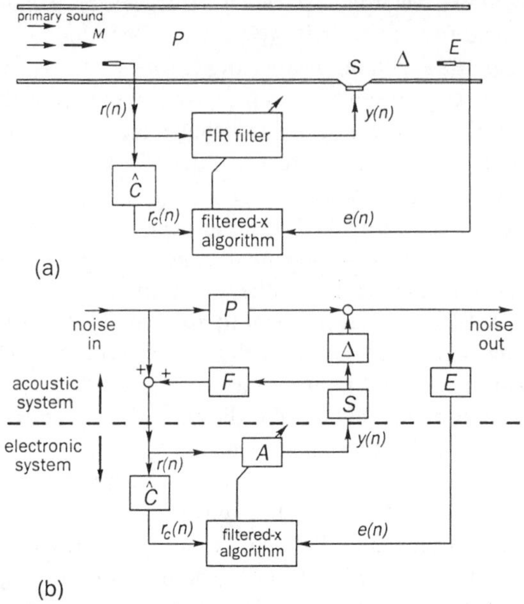

Fig. 7.14 Implementation of the adaptive filtered-x algorithm for active control of duct noise. $\hat{C}$ is an FIR filter which models the system cancellation path (between the electrical input to the control source and the electrical output from the error sensor): (a) physical system; (b) equivalent electrical block diagram.
A = FIR filter, F = acoustic feedback path, P = primary path from reference microphone to error microphone, Δ = propagation time delay between control source and error sensor, S = loudspeaker transfer function, E = error microphone transfer function, and $\hat{C}$ = estimate of the cancellation path transfer function (S,M,Δ).

Thus, using this approach, the controller FIR filter is updated each time the noise in the duct is sampled by the A/D converter. Thus, the algorithm can converge on a timescale comparable with the delay associated with the cancellation path, and so can rapidly track signal changes. The allowable maximum phase error in the estimate $\hat{C}(j\omega)$ is discussed in Chapter 6 and is generally close to 90°. This means that the system is fairly robust, but in systems where the flow speed or air temperature in the duct is likely to change, some means of updating the estimate $\hat{C}(j\omega)$ 'on-line' is needed. Means for making these estimates are also discussed in Chapter 6.

The most serious problem associated with the system shown in Fig. 7.14 is the effect of the feedback path F which can cause the LMS algorithm to become unstable. This is because it is possible for the adaptive filter to pass through a state in which there is unity gain around the feedback path/controller loop, resulting in saturation of the hardware used to implement the FIR filter. One way of minimizing this problem is to use a second FIR filter to compensate for the feedback path (Warnaka *et al.*, 1984; Poole *et al.*, 1984), as discussed previously and illustrated in Fig. 7.15. The filter $\hat{F}$ is designed by measuring the feedback transfer function using random noise injected into the control source with the primary source turned off.

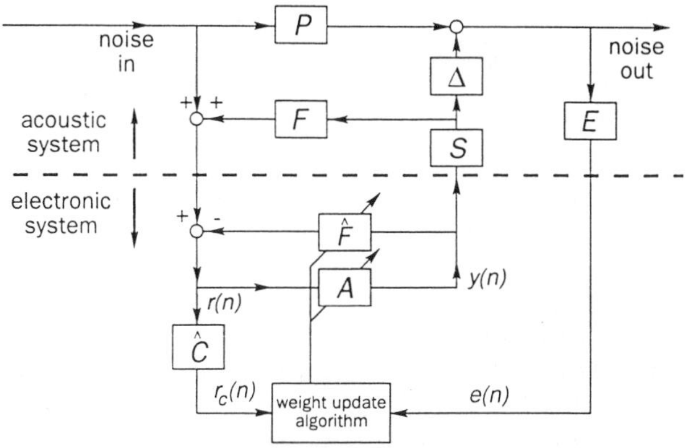

Fig. 7.15 Inefficient FIR controller arrangement with an additional compensator to account for acoustic feedback (after Warnaka *et al.*, 1984).

As mentioned previously, the arrangement involving a separate filter to compensate for the feedback path is very inefficient, and Eriksson and his co-workers (1987, 1989 and 1991) have thus developed an equivalent algorithm for use with a more efficient IIR filter. They have also developed a means of continuously measuring the cancellation path transfer function on-line using random noise input. The resulting arrangement is illustrated in Fig. 7.16, and the weight update equations for the FIR filters *A* and *B* making up the IIR controller are discussed in Chapter 6. Results obtained by Eriksson and Allie (1989) using the recursive controller illustrated in Fig. 7.16 are shown in Fig. 7.17.

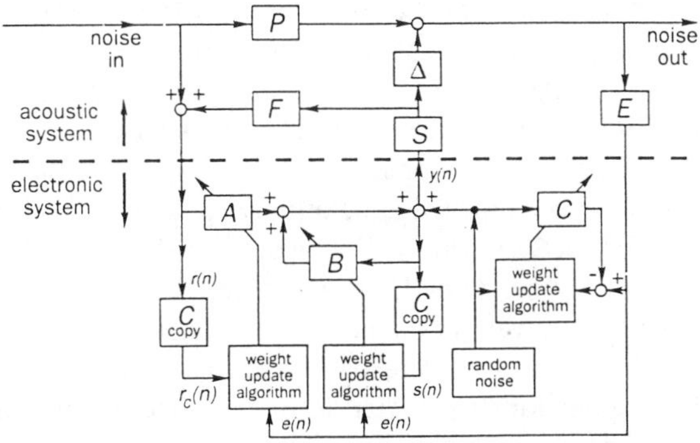

Fig. 7.16 Efficient adaptive recursive controller used by Eriksson including on-line identification of cancellation path using random noise.

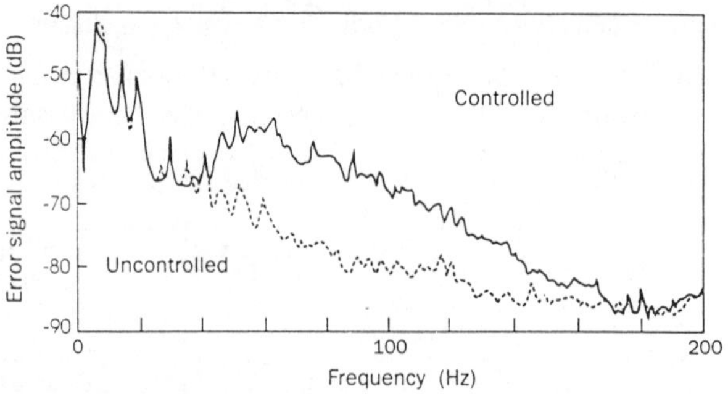

Fig. 7.17 Results obtained by Eriksson and Allie (1989) for control of random noise in an air conditioning duct with an air flow of approximately 14 m s^{-1}, using a recursive controller.

Note that use of the IIR filter has advantages other than compensating for acoustic feedback, especially when random noise is to be controlled. For random noise control an FIR filter needs many coefficients *N*, and for periodic noise control, two or three coefficients (or tap points) are needed for each harmonic to be controlled. On the other hand, because an IIR filter contains poles as well as zeros, far less filter coefficients are needed for effective control of random noise. However, the IIR filter also suffers from several disadvantages such as lack of inherent stability and the existence of bimodal error surfaces resulting in convergence to a non-optimum error signal. This is discussed more fully in Chapter 6.

7.3 HARMONIC (OR PERIODIC) PLANE WAVES

The purpose of the analysis outlined in this section is to develop an understanding of the physical mechanisms associated with the active control of harmonic noise propagating in a duct. Much of the work discussed here originates from investigations of the sound fields produced by both simple and finite size sound sources in hard-walled ducts of infinite and finite length (Doak, 1973a and b). Although this work was undertaken with no intention of applying it to active control, it forms the basis of the analysis of an active system, as it can be applied to both primary and control sources acting singly and in combination.

A number of authors have considered this problem in the past, although much of the work has been directed at determining an optimal controller transfer function (Elliott and Nelson, 1984). Berengier and Roure (1980), Snyder and Hansen (1989), and Snyder (1991), on the other hand, have investigated the physical mechanisms involved for both optimal and non-optimal controllers. The analysis presented here is an extension of the latter work, and involves the formulation of a model for the prediction of the effect of one or more active control sources on plane wave harmonic sound propagation in a rigid walled, semi-infinite duct, with the primary source located at the finite end in the plane of the duct cross-section. The primary source termination at z = 0 is modelled as either a rigid or uniform impedance surface. The effect of a finite impedance at the other end will also be discussed later in this section. The fluid contained within the duct is assumed to be at rest with constant and uniform ambient temperature and density. Excitation of the fluid by finite size, harmonic rectangular sources is assumed. The excitation frequency is constrained to be below the first higher order mode cut-on frequency. The model assumes that both the primary and control sources are constant volume velocity sources, which is equivalent to assuming infinite impedance sources. Some sources such as loudspeakers backed by a small, airtight cavity, and reciprocating compressors are best modelled in this way, although sources such as fans are best modelled as constant pressure sources. For this reason, a constant pressure primary source will also be considered.

In the following analysis, the positive time-dependent term $e^{j\omega t}$ is omitted with the assumption that the volume velocities, particle velocities and acoustic pressures used are amplitudes, unless otherwise stated. To begin, only a single control source will be considered. However, the analytical techniques can easily be extended to multiple control sources; the algebra just gets more complex.

The model will be used to calculate the acoustic power transmission associated with individual sources to determine the effect of source variables such as size, location, strength and relative phase. Control is assumed to be implemented using a feedforward control system.

The source arrangement and coordinate axes are illustrated in Fig. 7.18, where the finite size primary source is shown mounted in the termination plane of the duct cross section at $z = 0$, while the finite size control sources may be mounted in any of the four duct walls. Only positive values of z are allowed by this arrangement. Note that for plane wave control, the size of the primary source, the dimension of the control source across the duct and their location in the $x-y$ plane are not important; however, these parameters become important for the higher order mode case, which will be discussed later.

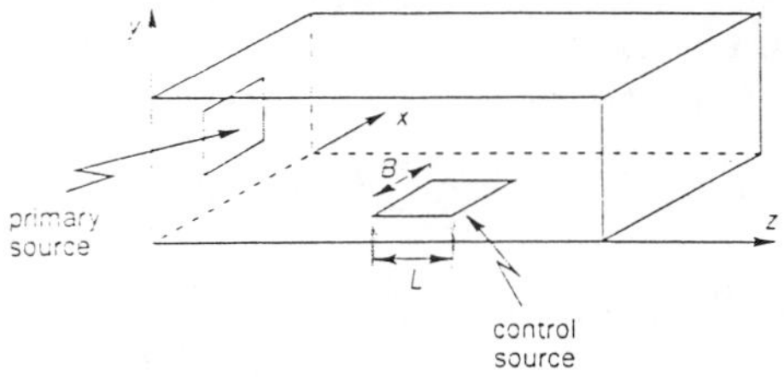

Fig. 7.18 Duct and source coordinates.

The total power transmission along the duct is the sum of the powers from the primary and control sources. The reduction due to active control is the difference in total power with and without the control source operating. If the sources are loudspeakers, the acoustic particle velocity across each source will be essentially uniform and the total radiated system acoustic power, for a single control source, will be given by:

$$W_{tot} = \frac{1}{2}\mathrm{Re}\left[\bar{Q}_p \bar{p}_p^*\right] + \frac{1}{2}\mathrm{Re}\left[\bar{Q}_c \bar{p}_c^*\right] \qquad (7.3.1)$$

where $\bar{Q}$ and $\bar{p}$ are, respectively the complex source volume velocity amplitude and complex acoustic pressure amplitude at the surface of the source. The subscripts p and c denote primary and control source respectively. If the acoustic particle velocity were not uniform across the face of the primary source, then the product of acoustic particle velocity and the complex conjugate of the acoustic pressure would have to be integrated over the source surface.

The purpose of the following analysis is to determine the optimum value of $\bar{Q}_c$ which will minimize W. This first requires that expressions for $\bar{p}_p$ and $\bar{p}_c$ be found for the case of both sources operating. We begin with the expression for the Green's function for an infinite duct (Morse and Ingard, 1968), which effectively relates the acoustic pressure and particle velocity at any location in the duct. This is

$$G(\boldsymbol{x}, \boldsymbol{x}_0, \omega) = \frac{-j}{S}\sum_m \sum_n \frac{\Psi_{mn}(x,y)\Psi_{mn}(x_0,y_0)}{\Lambda_{mn} k_{mn}} \mathrm{e}^{-jk_{mn}|z-z_0|} \qquad (7.3.2)$$

where ω is the angular frequency of the excitation, $\boldsymbol{x}_0 = (x_0, y_0, z_0)$ are the coordinates describing the location of the source and $\boldsymbol{x} = (x, y, z)$ is an arbitrary point within the duct. The quantity S $(= b \times d)$ is the cross-sectional area of the duct, k_{mn} is the modal wavenumber, Ψ_{mn} is the characteristic or mode shape function of the duct and Λ_{mn} is the modal normalisation factor. For a rectangular duct:

$$\Psi_{mn}(x,y) = \cos\left[\frac{m\pi x}{b}\right]\cos\left[\frac{n\pi y}{d}\right] \qquad (7.3.3)$$

where m, n are the modal indices and b and d the duct cross-section dimensions. Note that if plane waves only are considered $m = n = 0$. The modal normalisation factor for mode (m, n) is defined as,

$$\Lambda_{mn} = \frac{1}{S}\int_S \Psi_{mn}^2 \,\mathrm{d}S \qquad (7.3.4)$$

and the wavenumber for mode (m, n) is

$$k_{mn} = \left(k^2 - \kappa_{mn}^2\right)^{1/2} = \sqrt{\left[\frac{\omega}{c_0}\right]^2 - \left[\frac{\pi m}{b}\right]^2 - \left[\frac{\pi n}{d}\right]^2} \qquad (7.3.5a,b)$$

where c_0 is the speed of sound in free space.

As we are restricting the analysis to only plane waves for now, (7.3.2) can be written as:

$$G(z, z_0, \omega) = -\frac{j}{Sk}\mathrm{e}^{-jk|z-z_0|} \qquad (7.3.6)$$

where k is the free space acoustic wave number, ω/c_0.

In the analysis of this section and Section 7.4 to follow, the time dependent term $e^{j\omega t}$ in the expressions for acoustic pressure, acoustic particle velocity and source volume velocity is omitted to simplify the equations. Thus, strictly speaking, all of these quantities are complex amplitudes which, to be consistent with Chapter 2, should be represented with a bar over the symbol. However, this bar will be omitted from now on in this chapter to simplify the notation.

The simplified Green's function of (7.3.6) can be used to find the complex pressure amplitude, $p(z)$, at the axial location z in the duct as follows (Fahy, 1985):

$$p(z) = j\rho_0 c_0 k \int_S u(\boldsymbol{x}_0) G(\boldsymbol{x}, \boldsymbol{x}_0, \omega) \, \mathrm{d}\boldsymbol{x}_0 \tag{7.3.7}$$

where u is the complex particle velocity amplitude at a position $\boldsymbol{x}_0 = (x_0, y_0, z_0)$ on the face of the source.

If we assume that the centre of the primary source is located at $\boldsymbol{x}_p = (x_p, y_p, 0)$, and occupies the entire duct cross-section, an expression for the sound field generated by the primary source operating alone can be obtained by substituting (7.3.6) into (7.3.7), and writing the result in terms of primary source volume velocity amplitude, Q_p, as follows:

$$p_p(z) = \frac{\rho_0 c_0}{S} Q_p \mathrm{e}^{-jkz} \tag{7.3.8}$$

The preceding expression holds for a point source, or for any size uniform velocity source having a volume velocity of Q_p.

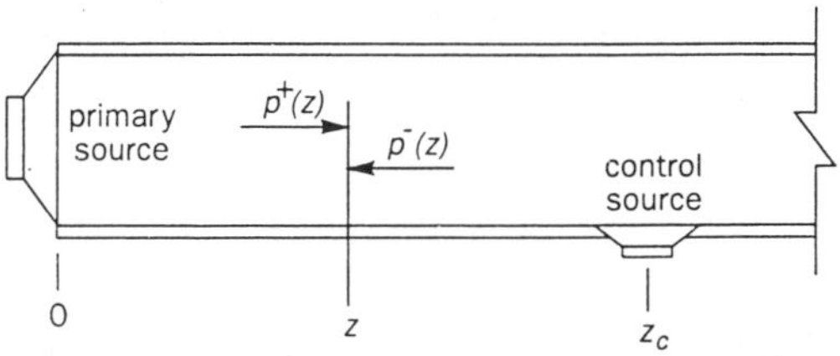

Fig. 7.19 Incident and reflected waves originating from the control source in a semi-infinite duct.

Consider the sound pressure field generated by a point control source operating with a volume velocity amplitude of Q_c. Referring to Fig. 7.19, the total sound pressure at some position z in the duct is the sum of a direct wave and a wave which has been reflected from the primary source end. If we represent the complex amplitude reflection coefficient at the primary source end of the duct as $\mathrm{e}^{-2\Phi}$, we obtain the following for the complex sound pressure amplitude $p_c(z)$ at any axial location z in the duct due to the point control source operating alone

$$p_c(z) = \frac{\rho_0 c_0 Q_c}{2S}\left[e^{-jk(z_c+z)} e^{-2\Phi} + e^{-jk|z-z_c|}\right] \tag{7.3.9}$$

where

$$\Phi = \pi\alpha + j\pi\beta \tag{7.3.10}$$

The scalar magnitude of the amplitude reflection coefficient at the primary source is $e^{-2\pi\alpha}$ and the phase angle between incident and reflected waves is $2\pi\beta$. The factor of 2 in the denominator of the term outside of the brackets in (7.3.9) arises because waves from the control source can travel in both directions.

If z is located between the control and primary sources, (7.3.9) may be written more simply as

$$p_c(z) = \frac{\rho_0 c_0 Q_c}{S} e^{-\Phi} e^{-jkz_c} \cosh(\Phi + jkz) \tag{7.3.11}$$

The variables α and β can be determined directly by measuring the standing wave in the section of the duct between the primary and control sources, in the same way that an impedance tube is used to measure the specific acoustic impedance of a sample placed at its end. The scalar magnitude of the sound pressure at any location in this region is

$$\begin{aligned} |p_c(z)| &= \sqrt{p_c(z) p_c^*(z)} \\ &= \frac{Q_c \rho_0 c_0}{S} e^{-\pi\alpha} \sqrt{\cosh^2(\pi\alpha) - \cos^2(\pi\beta + kz)} \end{aligned} \tag{7.3.12a,b}$$

This can be re-expressed as

$$|p_c(z)| = \frac{Q_c \rho_0 c_0}{S} e^{-\pi\alpha} \sqrt{\cosh^2(\pi\alpha) - \cos^2(\pi\beta')} \tag{7.3.13}$$

where

$$\beta' = \beta + \frac{z}{\lambda/2} \tag{7.3.14}$$

where λ is the wavelength of sound at the frequency of interest. From the preceding discussion, it can be seen that the pressure amplitude is a minimum when β' is an even multiple of ½, and is a maximum when β' is an odd multiple of ½. From (7.3.14) it is apparent that the distance between successive points of minimum or maximum sound pressure is ½ wavelength, and that the distance between the first acoustic pressure minimum and the terminated end, divided by ½ wavelength, is equal to $1 - \beta$. The ratio of the sound pressure amplitudes between the minimum and maximum pressure locations in the standing wave will be equal to

$$\frac{|p_c(z)|_{\min}}{|p_c(z)|_{\max}} = \frac{\sqrt{\cosh^2(\pi\alpha) - 1}}{\sqrt{\cosh^2(\pi\alpha)}} = \tanh(\pi\alpha) \tag{7.3.15a,b}$$

Equation (7.3.11) describes the sound field between the control source and the primary source for an idealized point source model of the control source; that is, one which has no physical size. However, as real acoustic sources are of finite size, some modification to (7.3.11) must be made to account for this. As shown in Fig. 7.18, the acoustic control source can be modelled roughly as a rectangular piston of width B and axial length D, whose centre line is located a distance $(z = z_c)$ from the primary source terminated end. To find the total sound pressure generated by the motion of this source, it is necessary to integrate the contributions from all points on the source. Considering only the plane wave mode, using (7.3.11) and assuming that the velocity distribution across the face of the piston is uniform, the acoustic pressure due to the control source at some location z between the control source and the primary source terminated end is

$$p_c(z) = B \int_{z_c - D/2}^{z_c + D/2} u_c \frac{\rho_0 c_0}{S} \mathrm{e}^{-\Phi} \mathrm{e}^{-jkz_c} \cosh(\Phi + jkz)\, \mathrm{d}z \tag{7.3.16}$$

where u_c is the particle velocity at any point on the source, taken to be uniform.

Evaluating this integral gives the expression for the pressure distribution produced by the control source alone between the control source and the terminated end as,

$$p_c(z) = Q_c \frac{\rho_0 c_0}{S} \gamma \mathrm{e}^{-\Phi} \mathrm{e}^{-jkz_c} \cosh(\Phi + jkz) \tag{7.3.17}$$

where the variable γ is defined as a control source size factor, and is given by

$$\gamma = \frac{2}{kD} \sin\left[\frac{kD}{2}\right] = \operatorname{sinc}\left[\frac{kD}{2}\right] \tag{7.3.18a,b}$$

7.3.1 Constant volume velocity primary source

Equations (7.3.8) and (7.3.17) can be used to determine the total acoustic pressure at the surface of both the primary and control sources, thereby enabling the calculation of the (real) acoustic power output of these sources, which is to be minimized by the active noise control system. Consider first the primary source, located at a position $(z = 0)$. Evaluating (7.3.8) and (7.3.17) at this location shows that for a constant volume velocity primary source of volume velocity Q_p before and after control, the total acoustic pressure at the primary source, due to both the control source and primary source operating together is

$$p(z = 0) = Q_p \frac{\rho_0 c_0}{S} + Q_c \frac{\rho_0 c_0}{S} \gamma \mathrm{e}^{-\Phi} \mathrm{e}^{-jkz_c} \cosh\Phi \tag{7.3.19}$$

Thus, the acoustic power output of the primary source under the influence of the sound pressure field produced by the control source is

$$W_p = \frac{1}{2}\mathrm{Re}\left\{ Q_p \left[Q_p \frac{\rho_0 c_0}{S} + Q_c \frac{\rho_0 c_0}{S} \gamma e^{-\Phi} e^{-jkz_c} \cosh\Phi \right]^* \right\} \tag{7.3.20}$$

Consider the acoustic power output of the control source. The sound pressure at any point z_s on the finite size source can again be calculated by evaluating (7.3.8) and (7.3.9) at this location (note that this is equivalent to integrating the local acoustic intensity over the surface of the control source, as the volume velocity is assumed contant over the face of the rectangular piston). The total acoustic pressure is then found by integrating over the surface of the source. Thus, the total acoustic pressure 'seen' by the control source is

$$p(z = z_c) = B \int_{z_c - D/2}^{z_c + D/2} Q_p \frac{\rho_0 c_0}{S} e^{-jkz_s}$$

$$+ Q_c \frac{\rho_0 c_0}{S} \gamma e^{-\Phi} e^{-jkz_c} \cosh(\Phi + jkz_s)\, dz_s \tag{7.3.21a,b}$$

$$= Q_p \frac{\rho_0 c_0}{S} \gamma e^{-jkz_c} + Q_c \frac{\rho_0 c_0}{S} \gamma^2 e^{-\phi} e^{-jkz_c} \cosh(\phi + jkz_c)$$

The acoustic power output of the control source, during operation of both the control and acoustic sources, is therefore

$$W_c = \frac{1}{2}\mathrm{Re}\left\{ Q_c \left[Q_p \frac{\rho_0 c_0}{S} \gamma e^{-jkz_c} + Q_c \frac{\rho_0 c_0}{S} \gamma^2 e^{-\Phi} e^{-jkz_c} \cosh(\Phi + jkz_c) \right]^* \right\} \tag{7.3.22}$$

The total acoustic power output of the system is given by the sum of (7.3.20) and (7.3.22). This sum can be used to determine the optimum control source volume velocity, which is that volume velocity which will minimize the acoustic power output of the total (primary and control) acoustic system, as will be outlined in the following sections.

7.3.1.1 Optimum control source volume velocity: idealized rigid primary source termination

Consider first the idealized case of the duct termination at the primary source end being perfectly rigid, corresponding to $\alpha = 0$ and $\beta = 0.5$. Substituting these values into (7.3.20), the primary source power output for this idealized case is:

$$W_p = \frac{1}{2} \mid Q_p \mid^2 \frac{\rho_0 c_0}{S} + \frac{1}{2}\frac{\rho_0 c_0}{S}\gamma \,\mathrm{Re}\{Q_p (Q_c \mathrm{e}^{-jkz_c})^*\} \tag{7.3.23}$$

For the control source, the acoustic power output is:

$$W_c = \frac{1}{2} \mid Q_c \mid^2 \frac{\rho_0 c_0}{S}\gamma^2 \cos^2(kz_c) + \frac{1}{2}\frac{\rho_0 c_0}{S}\gamma \,\mathrm{Re}\{ Q_c (Q_p \mathrm{e}3^{-jkz_c})^*\} \tag{7.3.24}$$

Noting that $Q_p Q_c^*$ in (7.3.23) is the complex conjugate of $Q_c Q_p^*$ in (7.3.24), the total (real) acoustic power output of the active controlled system can be expressed as a quadratic fuction of complex control source volume velocity amplitude

$$W_{tot} = Q_c^* a Q_c + Q_c^* b_1 + b_1^* Q_c + c \tag{7.3.25}$$

where

$$a = \frac{1}{2}\frac{\rho_0 c_0}{S}\gamma^2 \cos^2(kz_c) \tag{7.3.26}$$

$$b_1 = \frac{1}{2}\frac{\rho_0 c_0}{S}\gamma Q_p \cos(kz_c) \tag{7.3.27}$$

$$c = \frac{1}{2}\frac{\rho_0 c_0}{S} \mid Q_p \mid^2 \tag{7.3.28}$$

Equation (7.3.25) is identical in form to that describing the control of two free field monopole sources, investigated by Nelson *et al.* (1987). The optimum control source volume velocity is found by differentiating the equation with respect to this quantity, and setting the gradient equal to zero. Doing this, the optimum control source volume velocity is found to be,

$$Q_{c,\mathrm{opt}} = -\frac{b_1}{a} = \frac{Q_p}{\gamma \cos(kz_c)} \tag{7.3.29a,b}$$

Substituting (7.3.29) into (7.3.25) gives $W_{\mathrm{tot,\,min}} = 0$. That is, the total acoustic power transmission can be completely suppressed (theoretically) with a single control source, regardless of control source location. To simplify the notation, the minimized total power will be referred to in the future as W_{min}.

7.3.1.2 Optimum control source volume velocity: arbitrary uniform impedance termination at the primary source

Consider now the case of some arbitrary uniform primary source termination described by α and β. For this more general case, the total acoustic power output of the controlled system can again be written as a quadratic function of control source volume velocity:

$$W_{tot} = Q_c^* \mathrm{Re}\{a\} Q_c + \mathrm{Re}\{b_1 Q_c\} + \mathrm{Re}\{b_2 Q_c^*\} + c \tag{7.3.30}$$

where

$$a = \frac{1}{2}\frac{\rho_0 c_0}{S}\gamma^2\left(e^{-\Phi}e^{-jkz_c}\cosh(\Phi + jkz_c)\right)^* \tag{7.3.31}$$

$$b_1 = \frac{1}{2}\frac{\rho_0 c_0}{S}\gamma\left(Q_p e^{-jkz_c}\right)^* \tag{7.3.32}$$

$$b_2 = \frac{1}{2}\frac{\rho_0 c_0}{S}\gamma Q_p\left(e^{-\Phi}\, e^{-jkz_c}\cosh\Phi\right)^* \tag{7.3.33}$$

$$c = \frac{1}{2}\frac{\rho_0 c_0}{S}\mid Q_p \mid^2 \tag{7.3.34}$$

Note that (7.3.30) is non-symmetric, as b_1 and b_2 are not complex conjugates, as opposed to the 'symmetric' equation (7.3.25). The implications of this will be discussed later in this chapter.

The optimum control source volume velocity can again be determined by differentiating (7.3.30) with respect to the real and imaginary parts of Q_c, and setting the result equal to zero. This produces

$$Q_{c,\mathrm{opt}} = -\frac{1}{2}(b_1^* + b_2)/\mathrm{Re}\{a\} \tag{7.3.35}$$

Substituting this value back into (7.3.30), the expression for the minimum acoustic power output is found to be

$$W_{\min} = c - \frac{1}{4}(b_1^* + b_2)^*\ \mathrm{Re}\{a\}^{-1}(b_1^* + b_2) \tag{7.3.36}$$

An expression for the uncontrolled primary source power output may be derived using (7.3.8) and the relation:

$$W_{\mathrm{unc}} = \frac{1}{2}\mathrm{Re}\{p_p^*(0)\, Q_p\} \tag{7.3.37}$$

Thus,

$$W_{\mathrm{unc}} = \frac{1}{2}\frac{\rho_0 c_0}{S}\mid Q_p \mid^2 \tag{7.3.38}$$

The acoustic power attenuation in dB is

$$\Delta W = -10\,\log_{10}\left[\frac{W_{\min}}{W_{\mathrm{unc}}}\right] \tag{7.3.39}$$

For a rigid primary source termination ($\alpha = 0.0$ and $\beta = 0.5$), the right-hand side of (7.3.36) becomes zero and the right hand side of (7.3.39) becomes infinite, which agrees with the result found in the previous section.

7.3.1.3 Effect of control source location

It has been shown in Section 7.3.1.1 that for a rigid termination at the primary source end of the duct, it is theoretically possible to completely suppress the total system acoustic power output with the control source at any location. However, it can be seen by inspection of (7.3.35) that it is most efficient in terms of control source volume velocity requirements to place the control source at an integer multiple of half wavelengths from the noise source (note that if the control source is placed at an odd multiple of quarter wavelengths from the primary source, an infinite volume velocity would be required to achieve total control).

We will now examine the effect which α and β being different from the values of $\alpha = 0$, $\beta = 0.5$ (corresponding to a rigid termination at the primary source) has on the maximum achievable sound power reduction.

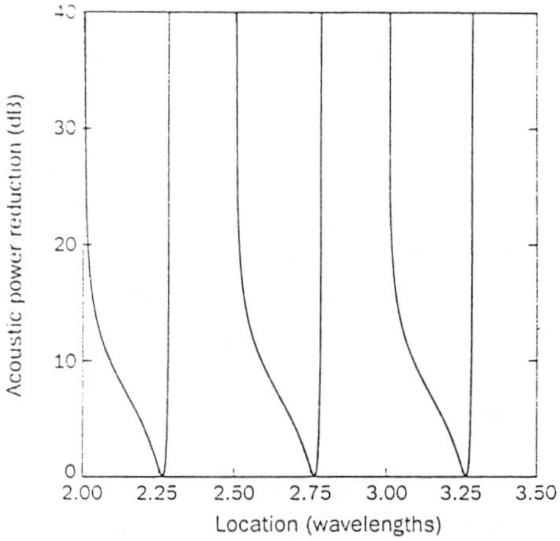

Fig. 7.20 Total acoustic power reduction as a function of primary source/control source separation, with the primary source defined by $\alpha = 0.002$, $\beta = 0.47$. Primary source frequency = 400 Hz. Constant volume velocity primary source.

For the case of α and β being slightly different from the values corresponding to a rigid termination (that is, $\alpha = 0.002$, $\beta = 0.47$), the power reduction which can be achieved is shown in Fig. 7.20 as a function of control source/primary source separation distance expressed in wavelengths. Plotted as a solid line in Fig. 7.21 is the associated volume velocity ratio (defined as the ratio of control source to primary source volume velocity magnitudes), with some experimentally measured data. The corresponding curve for a rigid termination at the primary source end is shown as a dashed line in the figure. In viewing these data, it is evident that significant levels of acoustic power attenuation can still be achieved with the control source at an integer multiple of a half-wavelength from the primary source, but that the attenuation that can be achieved at odd quarter wavelengths has been significantly reduced compared to that achievable with a rigid termination at the primary source end of the duct. The volume of the velocities required to achieve the maximum levels of control away from the optimum have also been reduced.

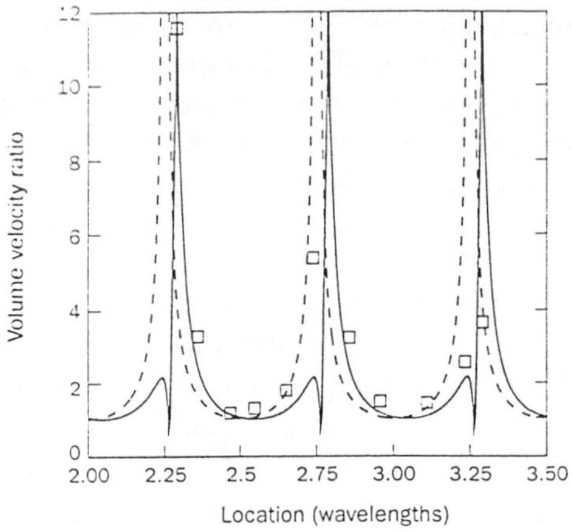

Fig. 7.21 Volume velocity ratio required to achieve maximum total acoustic power reduction as a function of primary source/control source separation. Primary source frequency = 400 Hz. Constant volume velocity primary source.
———————— = volume velocity ratio for $\alpha = 0.002$, $\beta = 0.47$
------------ = volume velocity ratio for rigid primary source termination ($\alpha = 0$, $\beta = 0.5$)
□ = experimental data ($\alpha = 0.002$, $\beta = 0.47$)

It is of interest to note that for a duct which is infinite in both directions (or terminated such that no waves are reflected from the ends), that the required control source volume velocity amplitude (but not phase) will be independent of its location and its amplitude will be equal to the primary source volume velocity amplitude.

7.3.1.4 Effect of control source size

From (7.3.29) the effect which size (inversely proportional to the parameter γ of (7.3.18)) has upon the volume velocity required to achieve maximum noise control can be deduced. While from (7.3.36) it can be deduced that the control source size (theoretically) has no influence upon the levels of control which can be achieved, it can be seen that as the size of the source begins to approach one half of a wavelength, the volume velocity required to achieve control increases dramatically.

7.3.1.5 Effect of error sensor location

It is quite apparent that in a semi-infinite duct, the acoustic power transmission can be adequately sensed by a single microphone placed in the duct in the far field of the primary and control sources. If the microphone is placed near enough to the control source that the measurement is influenced significantly by the source, then the performance of the control system will be compromised, as minimizing the sound pressure at the error microphone will not necessarily result in minimization of the far field propagating sound. Similar requirements hold for the proximity of the error microphone to duct discontinuities and terminations and also to the proximity of the reference microphone to sources and duct discontinuities.

Additional care must be taken with the error microphone location when the duct is characterized by a reflecting termination, resulting in a standing wave between the control source and the termination. In practice, it has been found that best results are obtained for a harmonic sound field when the error microphone is positioned at an antinode of the standing wave field. If the standing wave is large, then location of the error microphone at a minimum in the standing wave can produce very disappointing control results for harmonic waves.

When random noise propagating in a finite length duct with reflecting ends is considered, the location of the error sensor (provided it is in the far field of all sources of direct and reflected waves) is not important, as the reflected waves are uncorrelated with the incident waves.

7.3.2 Constant pressure primary source

In the previous section, we considered the acoustic power output of a constant volume velocity primary noise source which was being actively attenuated by the addition of a constant volume velocity control source. This represents a good approximation of a primary noise source such as a reciprocating compressor, although aerodynamic sources such as fans, are better modelled as constant pressure sources (Bies and Hansen, 1996). It will be useful to modify the previous analysis to consider the constant pressure aerodynamic type of primary source, and then use this to examine the differences in source acoustic power transmission between the constant volume velocity and constant pressure primary sources subjected to active control.

With a constant pressure source, the magnitude of the acoustic pressure at its face will remain constant before and after the application of active control. From (7.3.8), the initial acoustic pressure is

$$p_p(z = 0) = Q_p \frac{\rho_0 c_0}{S} \tag{7.3.40}$$

From (7.3.19), the acoustic pressure at the face of the primary source after the application of active control is

$$p(z=0) = Q_p'\frac{\rho_0 c_0}{S} + Q_c\frac{\rho_0 c_0}{S}\gamma e^{-\Phi} e^{-jkz_c} \cosh\Phi \tag{7.3.41}$$

where the primed primary source volume velocity under active control, Q_p', is different from the initial primary source volume velocity, Q_p (note that the control source is still considered to be a constant volume velocity source, approximating a loudspeaker with a small, airtight backing cavity).

Equating the primary source face pressure before the application of active control, given in (7.3.40), with that after the application of active control, given in (7.3.41), enables the determination of the controlled primary source volume velocity, Q_p'. Thus,

$$Q_p' = Q_p - Q_c \gamma e^{-\Phi} e^{-jkz_c} \cosh\Phi \tag{7.3.42}$$

Thus, the primary source acoustic power output under the action of active noise control is

$$W_p = \frac{1}{2}\mathrm{Re}\{Q_p' p_p^*\} \tag{7.3.43}$$

$$= \frac{1}{2}\mathrm{Re}\left\{\left(Q_p - Q_c \gamma e^{-\Phi} e^{-jkz_c} \cosh\Phi\right) Q_p^* \frac{\rho_0 c_0}{S}\right\} \tag{7.3.44}$$

Consider now the acoustic power output of the (constant volume velocity) control source. The pressure at the face of this source can be determined by substituting the final primary source volume velocity into (7.3.21b) to give,

$$p(z=z_c) = Q_p'\frac{\rho_0 c_0}{S}\gamma e^{-jkz_c} + Q_c\frac{\rho_0 c_0}{S}\gamma^2 e^{-\Phi} e^{-jkz_c} \cosh(\Phi + jkz_c) \tag{7.3.45}$$

Expanding Q_p' using (7.3.42), the acoustic pressure at the face of the control source is found to be

$$\begin{aligned} p(z=z_c) = {} & Q_p\frac{\rho_0 c_0}{S}\gamma e^{-jkz_c} \\ & + Q_c\frac{\rho_0 c_0}{S}\gamma^2 e^{-\Phi} e^{-jkz_c}\left(\cosh(\Phi + jkz_c) - e^{-jkz_c}\cosh\Phi\right) \end{aligned} \tag{7.3.46}$$

Thus, the control source acoustic power output is

$$W_c = \frac{1}{2}\text{Re}\left\{ Q_c \left[Q_p \frac{\rho_0 c_0}{S} \gamma e^{-jkz_c} \right.\right. \tag{7.3.47}$$

$$\left.\left. + Q_c \frac{\rho_0 c_0}{S} \gamma^2 e^{-\Phi} e^{-jkz_c} \left[\cosh(\Phi + jkz_c) - e^{-jkz_c} \cosh\Phi\right] \right]^* \right\}$$

Combining (7.3.44) and (7.3.47), the total system acoustic power output can be expressed as a quadratic function similar to (7.3.30):

$$W_{tot} = Q_c \text{Re}\{a\} Q_c^* + \text{Re}\{Q_c b_1\} + \text{Re}\{Q_c b_2\} + c \tag{7.3.48}$$

where

$$a = \frac{1}{2}\frac{\rho_0 c_0}{S}\gamma^2 \left(e^{-\Phi} e^{-jkz_c} \left(\cosh(\Phi + jkz_c) - e^{-jkz_c} \cosh\Phi \right) \right)^* \tag{7.3.49}$$

$$b_1 = \frac{1}{2}\frac{\rho_0 c_0}{S}\gamma \left(Q_p e^{-jkz_c} \right)^* \tag{7.3.50}$$

$$b_2 = -\frac{1}{2}\frac{\rho_0 c_0}{S}\gamma Q_p^* e^{-\Phi} e^{-jkz_c} \cosh\Phi \tag{7.3.51}$$

$$c = \frac{1}{2} | Q_p |^2 \frac{\rho_0 c_0}{S} \tag{7.3.52}$$

Differentiating (7.3.48) with respect to the control source volume velocity, and setting the gradient equal to zero, produces the optimum control source volume velocity equal to

$$Q_{c,\text{opt}} = -\frac{1}{2}(b_1^* + b_2^*)/\text{Re}\{a\} \tag{7.3.53}$$

Substituting (7.3.53) into (7.3.48), gives the following expression for the minimum acoustic power output:

$$W_{\min} = c - \frac{1}{4}(b_1 + b_2)\text{Re}\{a\}^{-1}(b_1 + b_2)^* \tag{7.3.54}$$

Substituting (7.3.49) to (7.3.52) into (7.3.54) gives a result for the minimum power which is independent of the control source size parameter, γ.

Consider first the case where the primary source termination is rigid, with $\alpha = 0.0$ and $\beta = 0.5$. Substituting these values into (7.3.49) to (7.3.51) and using (7.3.52) and (7.3.54), it is found that the minimum sound power is zero. Therefore, as with the constant volume velocity case, when the primary source termination is perfectly rigid, it is (theoretically) possible to completely suppress the total system acoustic power output with any control source location. Substituting the boundary conditions for the rigidly terminated primary source end into (7.3.49) to (7.3.51) and using (7.3.52) and (7.3.53), the optimum control source volume velocity for this simplified model is found to be

$$Q_c = \frac{jQ_p}{\gamma \sin(kz_c)} \tag{7.3.55}$$

which is dependent on the source size parameter, γ. Thus, the required source volume velocity to achieve optimal control is inversely proportional to the source size while the achievable sound power reduction is independent of the source size. This result is similar to that obtained for the constant volume velocity primary source.

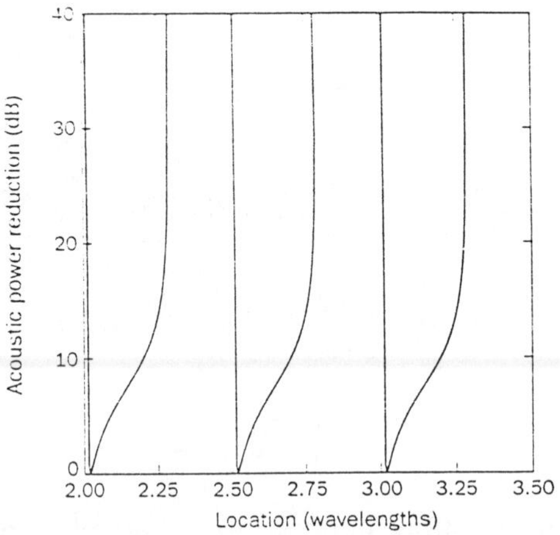

Fig. 7.22 Total acoustic power reduction as a function of primary source/control source separation, with the primary source defined by $\alpha = 0.002$, $\beta = 0.47$. Primary source frequency = 400 Hz. Constant pressure primary source.

It is interesting to contrast this optimum control source volume velocity result with that obtained for the similar constant volume velocity primary source case, given in (7.3.29). Whereas the most efficient control source placement for the constant volume velocity case was at half wavelength intervals, the most efficient control source placement for the constant pressure case is at odd quarter wavelength intervals. Also, the phase difference between the primary and control sources for the constant pressure case is modulated, as a function of the separation distance, between $\pm 90°$ by the presence of the imaginary term. Finally, note that as with the constant volume velocity case, the control source size does influence the optimum control source volume velocity; as the size of the source approaches one half of a wavelength, the required volume velocity increases dramatically. Consider now the effect of slightly relaxing the rigid termination boundary conditions. Figure 7.22 depicts the total acoustic power attenuation which can be achieved for a primary source termination defined by

$\alpha = 0.002$, $\beta = 0.47$ plotted as a function of control source/primary source separation distance expressed in wavelengths. Plotted in Fig. 7.23 is the control source volume velocity required to achieve control for a rigid primary source termination (solid line) and a termination characterized by $\alpha = 0.002$, $\beta = 0.47$ (dashed line). As can be seen from the figure, the optimum control source location occurs at separations between the primary and control sources of a little less than integer multiples of half a wavelength, where significant levels of acoustic power attenuation can still be achieved. At separations between the primary and control sources of odd quarter wavelengths, however, very little acoustic power attenuation is possible.

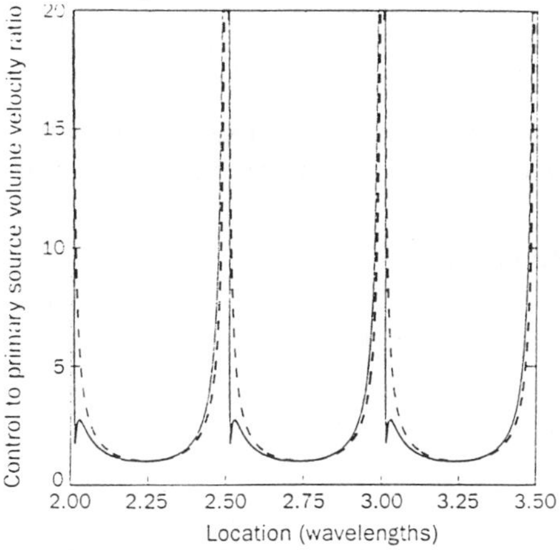

Fig. 7.23 Volume velocity ratio required to achieve maximum total acoustic power reduction as a function of primary source/control source separation. Primary source frequency = 400 Hz. Constant pressure primary source.
——————— = volume velocity ratio for $\alpha = 0.002$, $\beta = 0.47$
----------- = volume velocity ratio for rigid primary source termination ($\alpha = 0$, $\beta = 0.5$)

7.3.3 Primary source in the duct wall

The case of the primary source in the duct wall (Elliott and Nelson, 1984, 1986; Sha and Tian, 1987), as shown in Fig. 7.24, is different to the previously discussed case where the primary source was mounted in one of the duct termination planes, as now an incident wave will have a phase variation across the face of both the control and primary sources. Thus, this model will give a different indication to that given by the previously discussed model as to the potential performance of an active noise control system to control practical duct sound sources such as compressors or fans, as these sources are almost universally mounted in the plane of the duct cross section. It is useful to develop this model to allow an examination of the differences in predicted trends obtained as a result of making this small (and generally incorrect) simplification. Here, the analysis will be restricted to a constant volume velocity primary source.

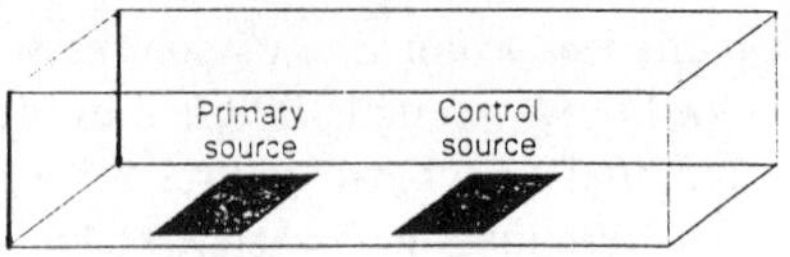

Fig. 7.24 Duct arrangement showing the primary source mounted in the duct wall.

Equation (7.3.8) describes the acoustic pressure field generated by the primary source mounted on the end of an infinite duct. For a point primary source, this can be modified for the doubly infinite geometry of Fig. 7.24 to give

$$p_p(z) = Q_p \frac{\rho_0 c_0}{2S} e^{-jkz} \tag{7.3.56}$$

For the case considered here, the primary source has a finite length along the duct axis. Therefore, the total acoustic pressure at any location z must be the sum of contributions integrated across the face of the source. Again, modelling the source as a rectangular piston of dimensions $B \times D$, assuming that the velocity distribution across the face of the source is uniform, and considering only the plane wave mode, this integral is

$$p_p(z) = B \int_{z_0-D/2}^{z_0+D/2} u_p \frac{\rho_0 c_0}{2S} e^{-jkz_s} \, dz_s \tag{7.3.57}$$

Evaluating this produces

$$p(z) = Q_p \frac{\rho_0 c_0}{2S} \gamma e^{-jkz} \tag{7.3.58}$$

where γ is defined in (7.3.18). Note that, as the primary source and control source are identically mounted, this expression can describe the sound pressure field of either the primary or control sources.

Consider now the acoustic pressure on the face of these sources operating in the presence of the other's sound field. For the primary source with its centre at $z = 0$, the pressure is

$$p(0) = \int_{-D/2}^{D/2} Q_p \frac{\rho_0 c_0}{2S} \gamma e^{-jkz} \, dz + \int_{z_c-D/2}^{z_c+D/2} Q_c \frac{\rho_0 c_0}{2S} \gamma e^{-jkz} \, dz$$

$$= Q_p \frac{\rho_0 c_0}{2S} \gamma^2 + Q_c \frac{\rho_0 c_0}{2S} \gamma^2 e^{-jkz_c} \tag{7.3.59a,b}$$

Thus, the acoustic power output of the primary source is

$$W_p = \frac{1}{2} Q_p \frac{\rho_0 c_0}{2S} \gamma^2 Q_p^* + \frac{1}{2} \mathrm{Re} \left\{ Q_p \left[Q_c \frac{\rho_0 c_0}{2S} \gamma^2 e^{-jkz_c} \right]^* \right\} \tag{7.3.60}$$

Using a similar analysis, the power output of the control source is found to be

$$W_c = \frac{1}{2} Q_c \frac{\rho_0 c_0}{2S} \gamma^2 Q_c^* + \frac{1}{2} \mathrm{Re} \left\{ Q_c \left[Q_p \frac{\rho_0 c_0}{2S} \gamma^2 e^{-jkz_c} \right]^* \right\} \tag{7.3.61}$$

Noting that $Q_p Q_c^*$ in (7.3.60) is the complex conjugate of $Q_p^* Q_c$ in (7.3.61), the total acoustic power transmission, found by combining (7.3.60) and (7.3.61), can be written as a quadratic function of the control source volume velocity as,

$$W_{tot} = Q_c^* a Q_c + Q_c^* b + b^* Q_c + c \tag{7.3.62}$$

where

$$a = \frac{1}{2} \frac{\rho_0 c_0}{2S} \gamma^2 \tag{7.3.63}$$

$$b = \frac{1}{2} Q_p \frac{\rho_0 c_0}{2S} \gamma^2 \cos(kz_c) \tag{7.3.64}$$

$$c = \frac{1}{2} Q_p Q_p^* \frac{\rho_0 c_0}{2S} \gamma^2 \tag{7.3.65}$$

Differentiating (7.3.62) with respect to the control source volume velocity, and setting the gradient equal to zero, produces the optimum control source volume velocity of

$$Q_{c,\mathrm{opt}} = -b/a = -Q_p \cos(kz_c) \tag{7.3.66a,b}$$

Substituting (7.3.66a) into (7.3.62), gives the expression for the minimum acoustic power output as follows:

$$W_{\min} = c - b^* a^{-1} b \tag{7.3.67}$$

Expanding this expression by using (7.3.63)−(7.3.65) and noting that the uncontrolled power is equal to c (as it has been for all cases considered), it is found that the ratio of controlled (residual) acoustic power transmission, $W_{\min}$, to the initial uncontrolled acoustic power transmission, W_{unc}, is

$$\frac{W_{\min}}{W_{\mathrm{unc}}} = 1 - \cos^2(kz_c) \tag{7.3.68}$$

It should be noted that the volume velocity of (7.3.66) is the one which will minimize the total acoustic power, propagating both upstream and downstream in the duct. It is sometimes desirable to simply stop the acoustic power transmission in one direction, irrespective of what changes occur in acoustic power transmission in the other direction. This is the concept employed by Trinder and Nelson (1983) in their acoustic virtual earth technique, which involves minimization of the acoustic pressure on the face of the control source using a feedback control system, thereby stopping any acoustic power transmission past the control source in the infinitely extending downstream duct section. This is equivalent to minimizing

$$p(z=z_c) = Q_p \frac{\rho_0 c_0}{2S} \gamma e^{-jkz_c} + Q_c \frac{\rho_0 c_0}{2S} \gamma \tag{7.3.69}$$

Here, the optimum control source volume velocity is

$$Q_{c,\text{opt}} = -Q_p e^{-jkz_c} \tag{7.3.70}$$

Substituting (7.3.69) into (7.3.62) gives the expression for the total acoustic power output corresponding to the minimum downstream power as

$$\frac{W_{\text{min}}}{W_{\text{unc}}} = 2(1 - \cos^2 kz_c) \tag{7.3.71}$$

which is twice that obtained when the total radiated power is minimized. Note that both the minimized power and the control source volume velocity required to achieve it are independent of the control source size parameter γ, for all cases with the primary source in the duct wall.

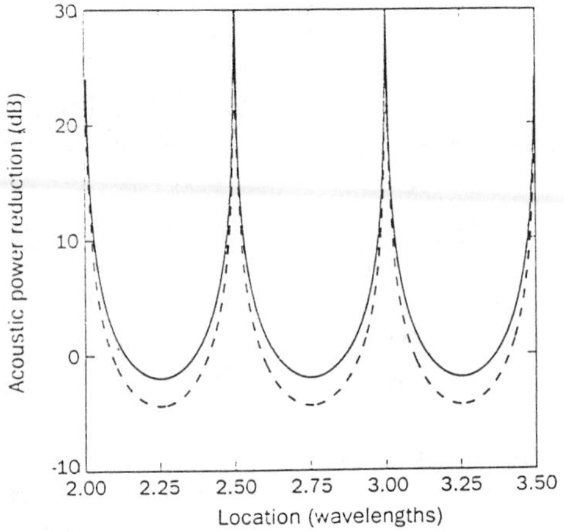

Fig. 7.25 Total acoustic power reduction at 400 Hz as a function of primary source/control source separation distance, for the primary source mounted in the duct wall.
———————— = minimization of total radiated acoustic power
---------------- = minimization of downstream radiated power.

Figure 7.25 illustrates the acoustic power attenuation that can be achieved for such a system plotted as a function of source separation distance expressed in wavelengths. Figure 7.26 depicts the associated control source/primary source volume velocity amplitude ratio. Comparing these plots to those of the similar, non-anechoically terminated duct, shown in Figs. 7.20 and 7.21, it shows that many of the trends are, in fact, the same; namely, the optimum source separation distance is at ½ wavelength intervals, with the worst results achieved at odd ¼ wavelength intervals. This system, however, will always be a symmetric one due

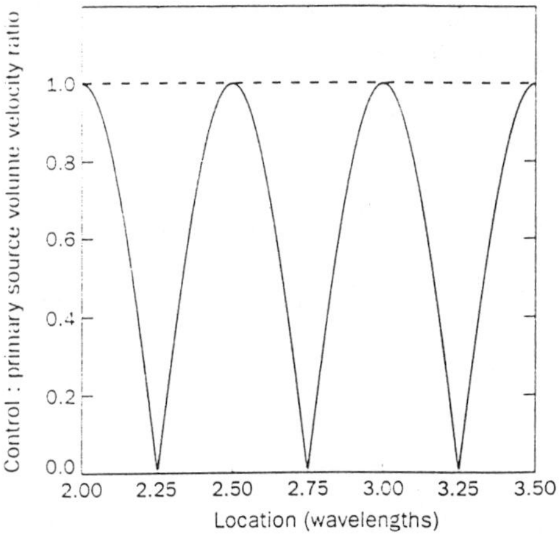

Fig. 7.26 Volume velocity ratio at 400 Hz required to achieve acoustic power reduction as a function of primary source/control source separation distance, for the primary source in the duct wall.
——————— = minimization of total radiated acoustic power
---------------- = minimization of downstream radiated power.

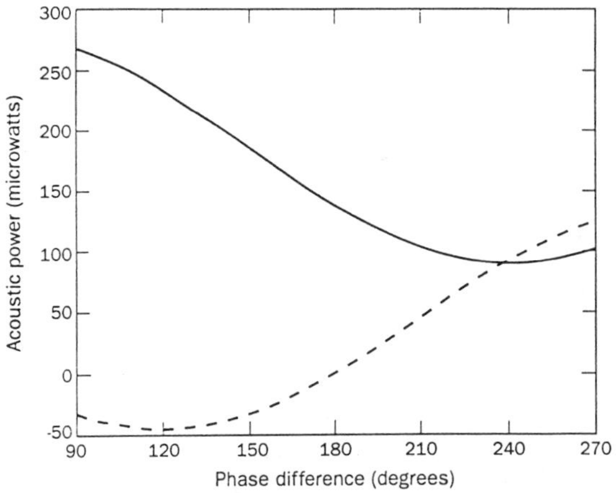

Fig. 7.27 Source acoustic power output at 400 Hz as a function of source phase difference for the primary source in the duct wall, with a source separation distance of 1 metre.
——————— = primary source
---------------- = control source

to the lack of a phase and amplitude modified reflected wave. This means that absorption of acoustic power will never be an optimal control mechanism, only suppression of the primary source power will be. Consider the data shown in Figs. 7.27 and 7.28, which illustrates the acoustic power output of the primary and control sources around the optimum phase difference of 180° for two constant volume velocity sources separated by 1 metre and operating at 400 Hz.

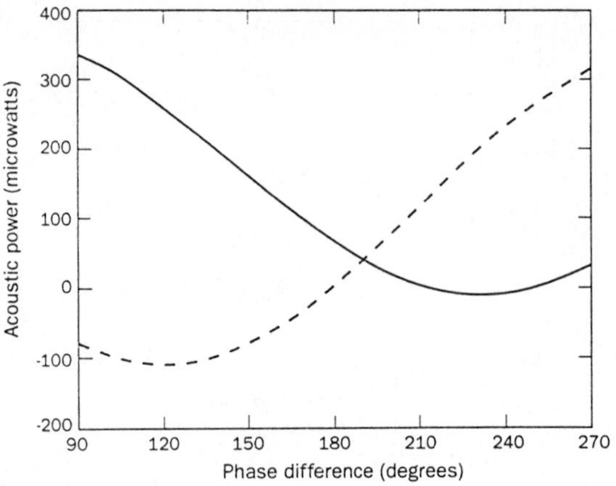

Fig. 7.28 Source acoustic power output at 400 Hz as a function of source phase difference for a primary source termination defined by $\alpha = 0.05$, $\beta = 0.45$, with a source separation distance of 1 metre.
———————— = primary source
---------------- = control source

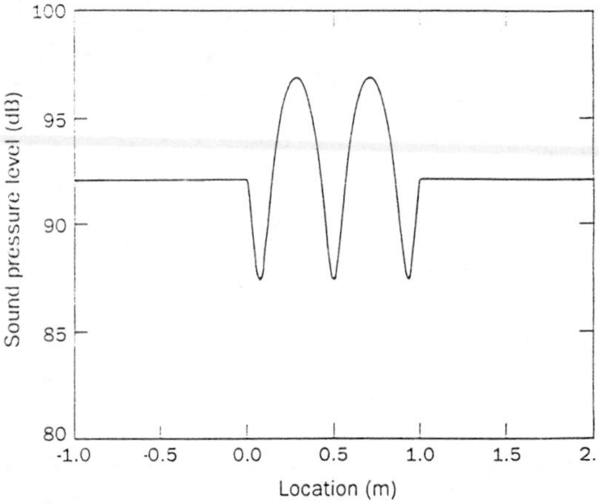

Fig. 7.29 Sound pressure distribution in the duct as a result of minimizing the total acoustic power output for the arrangement with the primary source in a wall of the duct. The primary source frequency is 400 Hz, its axial location is at 0.0 and the control source is located at 1.0 m.

The data shown in Fig. 7.27 are for the doubly infinite arrangement of Fig. 7.24 (where both the primary and control sources are mounted in the wall of an infinite duct), and the data in Fig. 7.28 are for the terminated arrangement of Fig. 7.18 with the termination conditions defined by $\alpha = 0.05$, $\beta = 0.45$. For the doubly infinite case, optimal control is achieved when the control source is producing zero acoustic power (at an operating phase difference of 180°). For the other case, however, optimal control is achieved when the control source is, in fact, absorbing 29 μW of the residual 96 μW acoustic power produced by the primary source (at an operating phase difference of 169.6°).

The difference between the optimum control source volume velocity for total sound power minimization, given in (7.3.66), and the optimum control source volume velocity for downstream power (or acoustic pressure on the face of the control source) minimization, given in (7.3.70), is that total power minimization minimizes only the component of the sound pressure in-phase with the source volume velocity, while the other minimizes the total sound pressure. This difference has a marked effect upon the final (controlled) sound pressure distribution. Figs. 7.29 and 7.30 show the final sound pressure distribution for two identical systems, where Fig. 7.29 shows the results of minimizing the total acoustic power output, and Fig. 7.30 shows the results of minimizing the acoustic pressure at the control source. The system shown here is operating at 400 Hz, and the sources are separated by 1 metre. Clearly, the downstream power transmission for the pressure minimized case is substantially less than for the power minimized case, but the upstream radiated acoustic power is increased as a result.

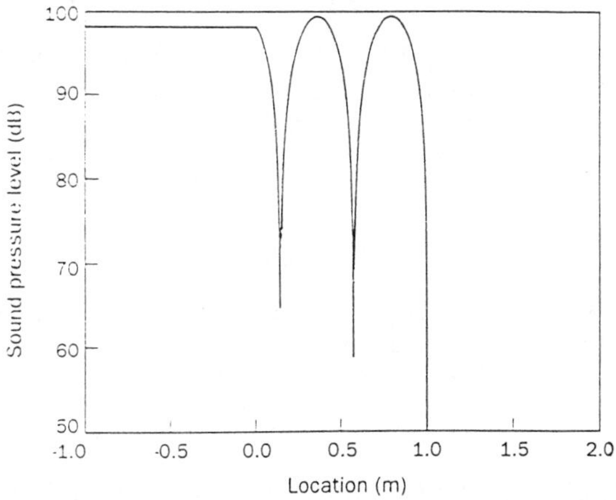

Fig. 7.30 Sound pressure distribution in the duct as a result of minimizing the downstream acoustic power output (or acoustic pressure on the face of the control source) for the primary source in a wall of the duct. The primary source frequency is 400 Hz, its axial location is at 0.0 and the control source is located at 1.0 m.

7.3.4 Finite length ducts

The preceding analysis has been concerned with a semi-infinite duct. When the other end of the duct is also terminated in an arbitrary impedance, the analysis becomes more complex. However, the results are of interest as they have a significant effect on the optimum locations for control sources and error sensors. The active control of harmonic sound fields in ducts of finite length has been considered in the past by a number of authors but the work generally has been directed towards the design of an optimal feed forward controller (Elliott and

Nelson, 1984; Billoud *et al.*, 1987; Ross and Yorke, 1987; Munjal and Eriksson, 1988; Curtis *et al.*, 1990). Other work has been concerned with feedback control system design (Trinder and Nelson 1983a,b; Eghtesadi and Leventhall, 1983a), and this work was discussed in detail in Section 7.2.1. The effect of finite length ducts on the control of random noise is discussed in Section 7.3.8. Here the effects of the finite length duct on the acoustic pressure in the duct for a harmonic source will be considered. It is then a relatively straightforward exercise to replace the expressions for a semi-infinite duct with these expressions to determine the effect of the finite duct length on the optimum location of the control source, assuming either a constant pressure or constant volume velocity primary source.

To do this we begin with (7.3.9) which gives the sound pressure in a semi-infinite duct as a result of a direct sound wave from a point control source and a wave once reflected from the primary source end. The sound pressure resulting from the wave reflected from the downstream end of the duct can be written as

$$p_1(z) = \frac{\rho_0 c_0 Q_c}{2S} e^{jk(z_c+z)} e^{-2\Phi_d} e^{-j2kL} \tag{7.3.72}$$

where z_c is the axial location of the control source, L is the length of the duct ($z = 0$ at the primary source) and Φ_d represents the reflection properties of the downstream duct end in much the same way as Φ of (7.3.9) represented the reflection properties of the primary source end.

As for Φ, Φ_d may be written as

$$\Phi_d = \pi\alpha_d + j\pi\beta_d \tag{7.3.73}$$

The sound pressure at location z resulting from a wave reflected from the downstream end and then from the primary source end is

$$p_2(z) = \frac{\rho_0 c_0 Q_c}{2S} e^{jk|z_c-z|} e^{-2\Phi} e^{-2\Phi_d} e^{-j2kL} \tag{7.3.74}$$

Adding (7.3.9), (7.3.72) and (7.3.74) plus pressure components as a result of additional reflections from the terminations (the number of significant contributions being dependent on the termination impedances) gives the following for the total pressure in the duct:

$$\begin{aligned} p_c(z) = \frac{\rho_0 c_0 Q_c}{2S} \Big[& e^{-jk|z_c-z|} + e^{-jk(z_c+z)} e^{-2\Phi} \\ & + e^{jk(z_c+z)} e^{-j2kL} e^{-2\Phi_d} \\ & + e^{jk|z_c-z|} e^{-2\Phi} e^{-2\Phi_d} e^{-j2kL} \Big] T_o \end{aligned} \tag{7.3.75}$$

where T_0 is the reverberation factor given by

$$T_0 = \sum_{i=0}^{\infty} \left[e^{-j2kL} e^{-2\Phi} e^{-2\Phi_d} \right]^i = \left[1 - e^{-j2kL} e^{-2\Phi} e^{-2\Phi_d} \right]^{-1} \tag{7.3.76a,b}$$

Similarly, an expression can be derived for the sound pressure in the duct due to the primary source located at $z_p = 0$. In this case, the second term in (7.3.75) will be omitted and we obtain

$$p_p(z) = \frac{\rho_0 c_0 Q_p}{S}\Big[e^{-jkz} + e^{jkz}\, e^{-j2kL}\, e^{-2\Phi_d} + e^{jkz}\, e^{-j2kL}\, e^{-2\Phi}\, e^{-2\Phi_d}\Big]T_o \qquad (7.3.77)$$

The effect of source size may be taken into account in the same way as it was for a semi-infinite duct, and a similar analysis used to derive an expression for the total power radiated down the duct with both sources operating. For example, (7.3.19) becomes

$$\begin{aligned} p_p(z=0) &= Q_p\frac{\rho_0 c_0}{S}\Big[1 + e^{-j2kL}\, e^{-2\Phi_d}\big(1 + e^{-2\Phi}\big)\Big]T_o \\ &+ \frac{Q_c\rho_0 c_0}{2S}\Big[e^{-jk\,|\,z_c - z\,|} + e^{-jk(z_c+z)}\, e^{-2\Phi} \\ &+ e^{jk(z_c+z)}\, e^{-j2kL}\, e^{-2\Phi_d} \\ &+ e^{jk\,|\,z_c - z\,|}\, e^{-2\Phi}\, e^{-2\Phi_d}\, e^{-j2kL}\Big]T_o \end{aligned} \qquad (7.3.78)$$

Similar modifications can be made to the other equations to give the desired results. Then the effect of control source location on the maximum achievable reduction in power transmission down the finite duct can be investigated as was done for the semi-infinite duct. Of course, the results will be dependent on the magnitude and phase of the reflection from the end of the duct. For a fixed frequency source it can be shown that for a fixed ratio of control to primary source volume velocity there will be some source locations in the duct where control will not be possible due to excessive volume velocity requirements. However, these locations are very much dependent on the duct termination impedance.

Measurement of the quantities Φ and Φ_d was discussed in Section 7.2.

7.3.5 Acoustic control mechanisms

Mechanisms will be examined for both constant pressure and constant volume velocity primary sources. In all cases, however, a constant volume velocity control source will be assumed.

7.3.5.1 Constant volume velocity primary source

In this section it will be shown by using both analytical and experimental results that the physical control mechanisms are a result of source power transmission control rather than wave cancellation. The experimental model used to obtain the results presented here was a 2 mm wall thickness duct, 215 mm × 215 mm square in cross section, terminated anechoically at one end and closed at the other end by the primary source (a 200 mm diameter circular speaker) mounted in the plane of the duct cross section. The anechoic wedge was 1.2 m long, constructed of rockwool, and was found to produce a standing wave ratio for primary excitation of less than 0.5 dB. The control source was located in one of the duct walls with its centre 1.25m from the primary source. The control source was a 100 mm diameter circular speaker, which was approximated in the theoretical analysis as a square speaker of equal area (89 mm × 89 mm).

Values of α and β which characterized the impedance of the end of the duct containing the primary source can be determined by examining the standing wave in the region between the two sources, as outlined in Section 7.3.1. When the primary source was a loudspeaker, it was found that these values were close to those expected for a rigid termination, typically $\alpha = 0.002$ (vs. 0.0 for a rigid termination), and $\beta = 0.48$ (vs. 0.5 for a rigid termination). This slight variation, however, was enough to alter the results substantially from what would be expected from the idealized assumption, as outlined earlier in this chapter.

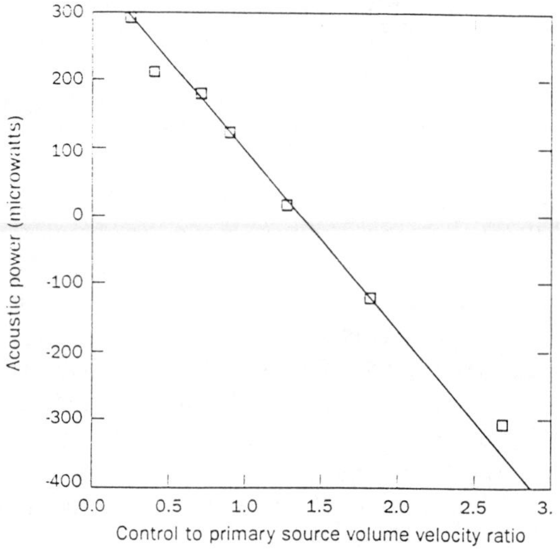

Fig. 7.31 Effect of varying the volume velocity ratio on the primary source power output, at 400 Hz, for a constant volume velocity primary source in the end of the duct, characterized by a termination impedance corresponding to $\alpha = 0.002$, $\beta = 0.48$.
———————— = theory
□ = experiment

We will examine first the effect which varying the volume velocity of the control source, for a fixed primary source volume velocity and drive source phase difference, has upon the source acoustic power transmission. Figure 7.31 illustrates both the theoretical and experimental primary source acoustic power transmission (for $\alpha = 0.002$, $\beta = 0.48$), plotted against the (scalar) ratio of the control source to primary source volume velocity magnitudes (the source phase difference for these points was 4.0°, measured as the phase difference between the acoustic pressure in the control source and primary source speaker enclosures). The figure indicates that, for this particular ambient temperature, source configuration and relative phase angles between the control and primary source driving signals, the primary source will begin to absorb sound power when the ratio of control source to primary source volume velocity amplitudes exceeds 1.2. For other source configurations and relative phase angles, the amplitude ratio will differ from 1.2.

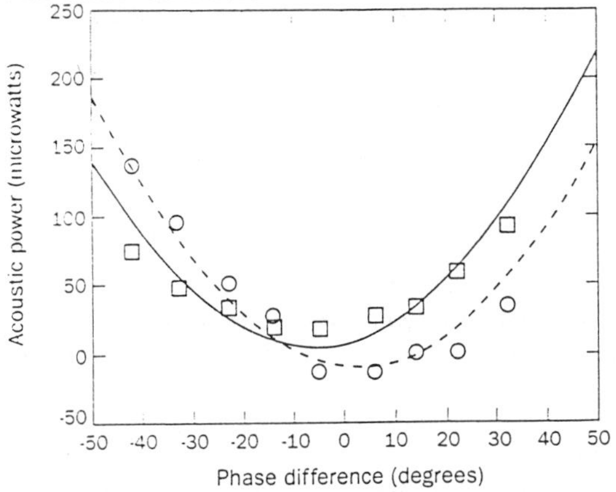

Fig. 7.32 Effect of varying source driving phase difference on acoustic power output at 400 Hz, for a constant volume velocity primary source in the end of the duct, characterized by a termination impedance corresponding to α = 0.002, β = 0.48.

———————	= primary source theory
□	= primary source experiment
----------------	= control source theory
O	= control source experiment

In Fig. 7.32, the variation in both primary source and control source acoustic power output is shown as a function of the phase difference ($\angle$control — $\angle$primary) between the primary and control source driving signals (measured as the phase difference of the acoustic pressures in the speaker enclosures), for a primary source volume velocity of 200 $\mu m^3 s^{-1}$, and a control source excited at 400 Hz with a volume velocity of 265 $\mu m^3 s^{-1}$. The total sound power radiated downstream is shown in Fig. 7.33 (with only the primary source operating the sound power output was 370 μW). In Fig. 7.32 it can be seen that for this particular arrangement, there is no phase angle where the primary source will absorb energy; however, the control source will absorb energy when the phase angle between the primary and control source driving signals is between $-3°$ and 13°. Measurements of control source power transmission made at phase angles close to and within this range were subject to error because the measured phase angle between the control source volume velocity and surface acoustic pressure was in the range of 269 to 271°, where small errors in phase angle measurement led to large errors in power transmission predictions (as the results are dependent upon the cosine of the phase angle).

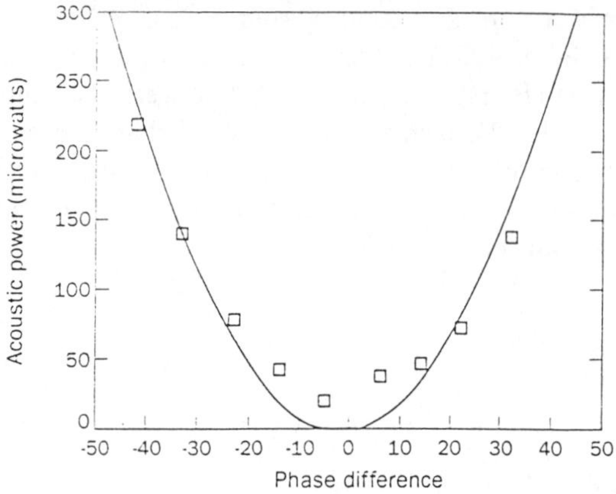

Fig. 7.33 Effect of varying phase difference on total system acoustic power output at 400 Hz, for a constant volume velocity primary source in the end of the duct, characterized by a termination impedance corresponding to $\alpha = 0.002$, $\beta = 0.48$.

——————— = theory

□ = experiment

The results just described indicate that the active (or real) acoustic power output of the primary source is affected by operation of the control source, and that near optimum control, the primary source power output is greatly reduced. Measurement of the total primary source impedance reveals little change in the magnitude of the reactive (or imaginary) component before and after operation of the control source, suggesting that the reduced active power is not re-routed into non propagating modes in the near field of the source (reactive power). Rather, it is simply not produced.

Although the primary source is unloaded by the control source, and is either producing very little (real) power or absorbing it, there is a large standing wave present between the primary and control sources. The stored energy represented by this standing wave is a result of the finite time it takes for the unloading to occur. This duct section is acting like an 'acoustic capacitance', storing the energy which is emitted during that time period.

7.3.5.2 Constant pressure primary source

The duct configuration used here is similar to that used for the previous section (Fig. 7.18), except that the primary source termination will be idealized here as rigid ($\alpha = 0$, $\beta = 0.5$) for convenience, with the constant pressure primary source generating sound at 400 Hz. A constant volume velocity control source is located in the wall of the duct 1 m downstream of the primary source. Figure 7.34 shows the level of acoustic power being transmitted out of each source when the control source is operating at the optimum volume velocity amplitude, as the phase difference between the two sources ($\angle$control $-$ $\angle$primary) is varied. In viewing the data of Fig. 7.34, it is evident that the acoustic power output of the primary source is greatly reduced, being equal to zero at the optimum phase difference. Thus, the physical control mechanism demonstrated here would again appear to be one of source unloading; that is, the radiation impedance of the primary noise source must be significantly altered.

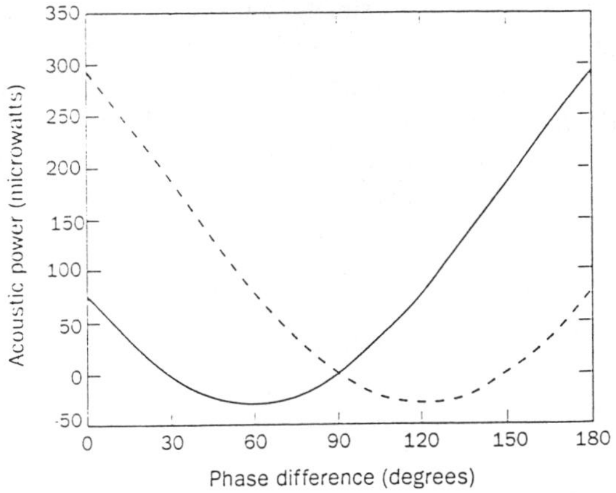

Fig. 7.34 Source acoustic power output at 400 Hz, for a constant pressure primary source in the end of the duct, characterized by a termination impedance corresponding to α = 0.002, β = 0.48, and for the control source 1 metre downstream.
———————— = primary source
---------------- = control source

As the primary source is a constant pressure source, the volume velocity must be reduced to achieve sound power attenuation. Alternatively, as the control source is constant volume velocity, the sound pressure at its face must be reduced when both sources are operational. Therefore, as radiation impedance is defined as the ratio of sound pressure to volume velocity, it can be surmised that for a constant pressure source, the acoustic power output is reduced by an increase in the radiation impedance, which causes a suppression of source volume velocity. Alternatively, if the noise source is a constant volume velocity source, fluid unloading causes a reduction in source radiation impedance, which in turn reduces the in-phase sound pressure on the face of the source, and the radiated sound power. It is interesting to note that these different mechanisms lead to different optimal phase differences between the primary and control sources when a constant volume velocity control source (such as a speaker) is used to control either a constant pressure, or constant volume velocity, primary source. Figure 7.35 illustrates the effect which phase difference has on the total power attenuation for the physical arrangement described previously, when the primary source is either a constant pressure or constant volume velocity type (the control source volume velocity amplitudes are fixed at the optimum value for these plots). Note that the optimum phase difference for the constant pressure source is 90°, while for the constant volume velocity source it is the more commonly cited 180°. This difference arises from the different ways the source impedance is changed. For a constant pressure source, control is achieved by a change in source volume velocity, while for a constant volume velocity source it is achieved by a reduction in source face sound pressure. As these quantities (pressure and velocity) are out of phase by 90° between the control and primary sources, for a constant volume velocity source to control a constant pressure source the phase difference should be ±90°. However, for two constant volume velocity sources, the phase difference should either be 0° or 180°. Note that these phase difference values only apply for a rigid termination (α = 0, β = 0.5) at the primary source end. For a non-rigid termination (or for different values of α and/or β) the relative phase values will differ substantially from these, as indicated in Figs. 7.32 and 7.33.

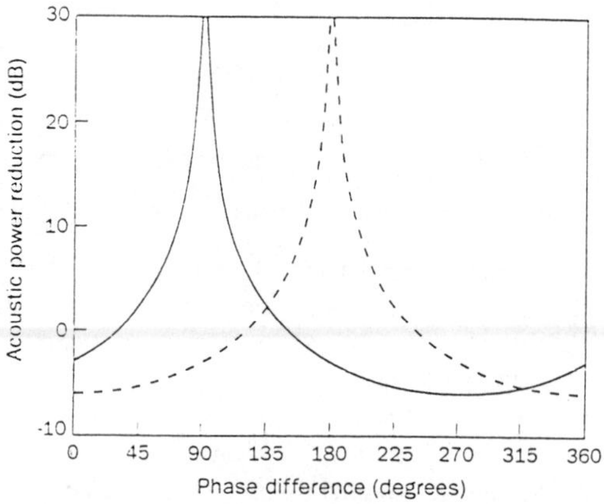

Fig. 7.35 Effect of phase difference on total acoustic power reduction at 400 Hz, with the control source 1 metre downstream of the primary source, which is mounted in the end of the duct and characterized by a termination impedance corresponding to α = 0.002, β = 0.48.

——————— = constant pressure primary source

---------------- = constant volume velocity primary source

7.3.6 Effect of mean flow

A mean airflow down the duct will result in an effective phase shift between positive and negative travelling waves, as the wavelength for waves travelling in the same direction as the mean flow will be increased, while the wavelength of waves travelling in the opposite direction will be reduced.

Thus, for a mean flow of mach number M (mean flow speed over speed of sound) from the primary source to the control source, (7.3.8) would become (Munjal, 1987):

$$p_p(z) = \frac{\rho_0 c_0}{S} Q_p \, e^{-jkz/(1+M)} \tag{7.3.79}$$

(7.3.9) would become:

$$p_c(z) = \frac{\rho_0 c_0 Q_c}{2S} \mathrm{e}^{-jk(z_c + z)/(1+M)} \mathrm{e}^{-2\Phi} + \mathrm{e}^{-jk\,|\,z - z_c\,|\,/(1-M)} \tag{7.3.80}$$

and (7.3.21b) would become:

$$p(z = z_c) = Q_p \frac{\rho_0 c_0}{S} \gamma \mathrm{e}^{-jkz_c/(1+M)}$$
$$+ Q_c \frac{\rho_0 c_0}{S} \gamma^2 \mathrm{e}^{-\Phi} \mathrm{e}^{-jkz_c/(1+M)} \left[\mathrm{e}^{-\Phi} \mathrm{e}^{-jkz_c/(1+M)} + \mathrm{e}^{\Phi} \mathrm{e}^{jkz_c/(1-M)} \right] \tag{7.3.81}$$

The remainder of the analysis to determine the minimum radiated power, the residual sound field and the corresponding control source volume velocity would closely follow that for a duct with no flow, with expressions containing e^{-jkz} or e^{jkz} being modified by dividing the exponent by $(1 + M)$ or $(1 - M)$ respectively. This is relatively straightforward and will not be done here. The trends shown in the results for no flow are similar to those with flow, except that specific phase angles between the control and primary sources to achieve similar results are different.

7.3.7 Multiple control sources

The volume velocity requirements for a single control source controlling plane waves in a finite length duct is dependent upon the source location and can approach very large values for some locations. As these optimal locations are frequency dependent, there will always be some frequencies which will require a large volume velocity output from a control source in a fixed location. As the volume velocity requirements can easily exceed the capability of practical sources, it may be preferable to use a two channel control system (two control sources and two error sensors) to control plane waves over a wide frequency band. As mentioned previously, this would also help eliminate problems caused by a single error sensor being located at a node in the standing wave caused by reflections from the duct terminations or discontinuities.

7.3.8 Random noise

It is of interest to analyse the case of random noise propagating down a finite length duct with a non-optimal controller. For random noise it can be assumed that the reflections from the ends of the duct are uncorrelated with the noise arriving directly from the sources. However, the primary source noise and the noise radiated by the control source are by definition well correlated. In the following analysis (which follows closely that of La Fontaine and Shepherd, 1985) an expression will be derived for the power spectral density of sound propagating past the control source for a non-ideal controller which has parameters similar to those which will be encountered in practice. The system which will be analysed is shown schematically in Fig. 7.36.

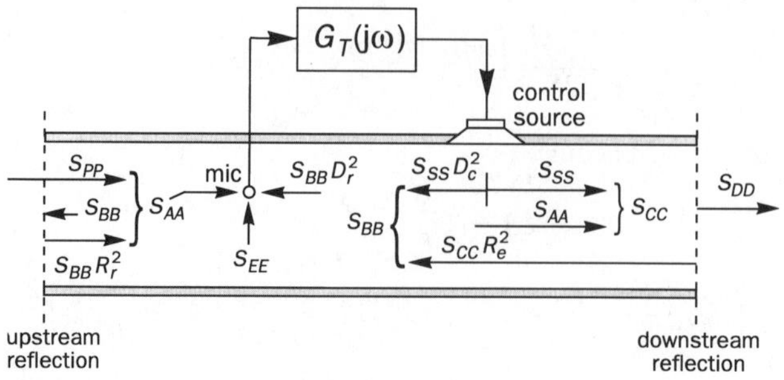

Fig. 7.36 Finite length duct with random noise.
S_{PP} = power spectrum of the primary random noise, S_{SS} = power spectrum of the random noise radiated by the loudspeaker, S_{CC} = residual noise in the duct, S_{DD} = noise radiated from the duct exit, S_{BB} = total downstream propagating noise, S_{AA} = total upstream propagating noise apart from that generated by the loudspeaker, S_{EE} = extraneous noise due to turbulence and instrumentation noise, R_x = upstream pressure reflection coefficient, R_e = downstream pressure reflection coefficient, D_x = amplitude ratio of sensitivity of microphone in upstream direction to that in downstream direction, D_s = amplitude ratio of noise radiated upstream to that radiated downstream by the control source and $G_T(j\omega)$ is the combined transfer function of the loudspeaker, controller and reference microphone (= $L_S M_x G(j\omega)$).

The transfer function G_T of the entire electroacoustic system can be characterised by a gain g and phase ϕ such that $G_T = -ge^{j\phi}$. In the following analysis the quantity ($j\omega$) which should qualify all spectral density symbols S and transfer functions will be omitted to simplify the resulting expressions. Also the reference microphone and control source direcivities, D_r and D_c respectively, represent amplitudes only, whereas previously they included a phase term.

The power spectrum, S_{DD}, of the noise escaping from the duct exit is given by

$$S_{DD} = S_{CC}(1 - R_e^2) \tag{7.3.82}$$

where,

$$S_{CC} = S_{AA} + S_{AS} + S_{SA} + S_{SS} \tag{7.3.83}$$

and where S_{CC}, measured near the duct exit, is the spectrum of the total sound travelling towards the duct exit, S_{AA}, measured near the reference microphone, is the spectrum of the total sound travelling downstream towards the reference microphone (including reflections), S_{SS} is the spectrum of the sound radiated downstream by the control source, S_{AS} is the cross-spectrum between S_{AA} and S_{SS} and R_e is the pressure amplitude reflection coefficient of the duct exit.

From Chapter 4 and Fig. 7.36 it is clear that

$$S_{AS} = S_{SA}^* = G_T S_{AA} \tag{7.3.84a,b}$$

and $$S_{SS} = D_r^2 \mid G_T \mid^2 S_{BB} + \mid G_T \mid^2 S_{AA} + \mid G_T \mid^2 S_{EE} \tag{7.3.85}$$

where S_{BB} is the spectrum of the total sound travelling upstream towards the primary source and S_{EE} is the spectrum of the electronic and turbulence noise. All of the spectral quantities and their associated measurement locations can be better visualised by inspecting Fig. 7.36.

Substituting (7.3.84) and (7.3.85) into (7.3.83) gives

$$S_{CC} = S_{AA} + (G_T + G_T^*)S_{AA} + D_r^2 \mid G_T \mid^2 S_{BB} + \mid G_T \mid^2 S_{AA} + \mid G_T \mid^2 S_{EE} \tag{7.3.86}$$

Substituting for G_T in (7.3.86) gives

$$S_{CC} = XS_{AA} + D_r^2 g^2 S_{BB} + g^2 S_{EE} \tag{7.3.87}$$

where

$$X = 1 - 2g\cos\phi + g^2 \tag{7.3.88}$$

The power spectrum S_{AA} is made up of the primary noise and reflections of S_{BB}. Thus,

$$S_{AA} = S_{PP} + (1 - R_r^2) S_{BB} \tag{7.3.89}$$

Similarly the power spectrum S_{BB} is made up of the upstream radiation from the control source and the reflections from the downstream end of the duct. Thus,

$$\begin{aligned} S_{BB} &= D_c^2 S_{SS} + R_e^2 S_{CC} \\ &= D_c^2 \left(D_r^2 g^2 S_{BB} + g^2 S_{AA} + g^2 S_{EE}\right) + R_e^2 S_{CC} \end{aligned} \tag{7.3.90 a,b}$$

Substituting (7.3.89) into (7.3.90b) and rearranging gives

$$S_{BB} = \frac{D_c^2 g^2 (S_{PP} + S_{EE}) + R_e^2 S_{CC}}{1 - D_c^2 g^2 (D_r^2 + R_r^2)} \tag{7.3.91}$$

Substituting (7.3.89) and (7.3.91) into (7.3.86), the result into (7.3.82) and rearranging gives

$$S_{DD} = \frac{(1-R_e^2)\left\{(XS_{PP} + g^2 S_{EE})\left[1 - D_c^2 g^2 (D_r^2 + R_r^2)\right] + D_c^2 g^2 (XR_r^2 + D_r^2 g^2)(S_{PP} + S_{EE})\right\}}{1 - D_c^2 g^2 (D_r^2 + R_r^2) - (XR_r^2 + D_r^2 g^2) R_e^2} \tag{7.3.92}$$

If the reference microphone and control source are both perfectly directional and if there are no reflections from the ends of the duct and if the turbulence and electronic noise are zero, then D_c, D_r, R_r, R_e and S_{EE} are all zero and equation (7.3.92) becomes

$$S_{DD} = XS_{PP} = (1 - 2g\cos\phi + g^2) S_{PP} \tag{7.3.93a,b}$$

It is clear that for a perfect controller ($g = 1$, $\phi = 0$) both (7.3.92) and (7.3.93) give $S_{CC} = 0$ and thus from (7.3.82), the sound S_{DD} radiated from the duct exit is also zero.

With the control system turned off,

$$S_{CC} = S_{AA} = \frac{S_{PP}}{1 - R_r^2 R_e^2} \tag{7.3.94a,b}$$

and,

$$S_{DD} = \frac{(1 - R_e^2) S_{PP}}{1 - R_r^2 R_e^2} \tag{7.3.95}$$

The control system performance (or insertion loss of the active attenuator) is calculated as $10\log_{10}$ of the ratio of (7.3.95) to (7.3.92).

Typical values of D_c and D_r for two element directional loudspeakers and microphones measured by La Fontaine and Shepherd were 0.18 and 0.1 respectively. Typical values of g and ϕ were 0.9 and 5 degrees respectively which represent a non-optimal controller. Equations (7.3.92) and (7.3.95) may be used to explore the effect of reflections from the ends of the duct on the performance of a non-optimal control system. La Fontaine and Shepherd (1985) show that for $g = 0.9°$ and $\phi = 5°$, the insertion loss of the active attenuator falls from 20 dB for a non reflecting upstream termination to 13 dB for a reflecting termination. They also show that for a system without phase error ($\phi = 0$), the effect of a reflecting termination downstream has only a small influence on the active attenuator insertion loss provided the upstream reflection coefficient remains small. However, as the upstream reflection coefficient approaches one, the downstream reflection coefficient becomes more important and the ratio of (7.3.92) and (7.3.95) becomes approximately

$$\frac{(S_{DD})_{\text{off}}}{(S_{DD})_{\text{on}}} = \left[\frac{(1+g)/(1-g) - R_e^2}{1 - R_e^2}\right]^{1/2} \tag{7.3.96}$$

which tends to infinity if the gain g approaches unity (perfect controller) or if R_e approaches unity. In the latter case an active attenuator would not be needed as sound would not be exiting from the duct.

La Fontaine and Shepherd (1985) also show that for a non-optimal controller ($\phi = 5°$) large upstream reflection coefficients increase the loudspeaker output power requirements by up to a factor of five, although downstream reflections have only a small effect. However, if phase error is present in the controller, the controller performance is strongly dependent on both the upstream and downstream reflection coefficients as shown in Fig. 7.37 (La Fontaine and Shepherd, 1985).

In practice, La Fontaine and Shepherd (1985) have found that reflection coefficients in the frequency range and duct sizes of interest for active control range from $R_r = 0.05$ to 0.6 and $R_e = 0.5$ to 1 (with higher values corresponding to lower frequencies).

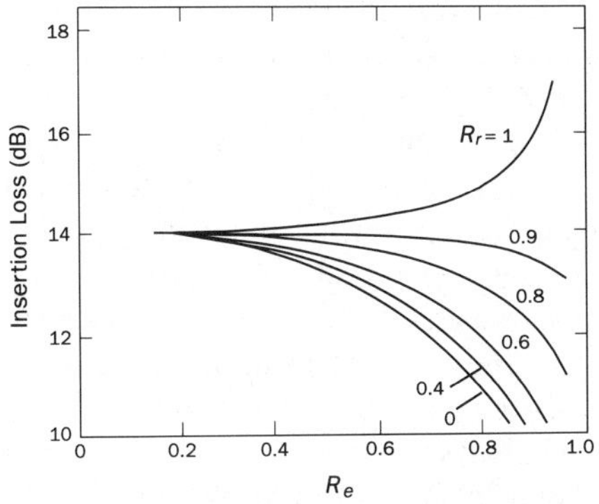

Fig. 7.37 Effect of duct reflection coefficients on the insertion loss of a non-ideal active attenuator for random noise ($g = 0.9$, $\phi = 5°$, $D_x = 1$, $D_s = 0.18$).

Note that the high value of insertion loss corresponding to large values of upstream reflection coefficient are a little misleading and do not mean that it is a good idea to increase the upstream reflection coefficient in practice as this will increase the exit noise levels before the active attenuator is switched on. The greater insertion loss of the active attenuator will not compensate for this initial increase in the exit noise due to the primary source.

Equations (7.3.92) and (7.3.95) can also be used to show that use of a directional control source and reference microphone as discussed in Section 7.2 results in an improved performance of the non-ideal controller. However, this is at the expense of reduced control source power output and microphone sensitivity (La Fontaine and Shepherd, 1985).

7.4 HIGHER ORDER MODES

The problem of controlling higher order modes in ducts has been addressed by a number of authors (Fedoryuk, 1975; Mazanikov *et al.*, 1977; Tichy and Warnaka, 1983; Maxwell *et al.*, 1989; Doelman, 1989; Eriksson *et al.*, 1989; Silcox and Elliott, 1990; Stell and Bernhard, 1990, 1991; Neise and Koopman, 1991; Zander and Hansen, 1992).

The three earlier papers essentially identified likely problems associated with the presence of propagating higher order modes. Eriksson *et al.* (1989) demonstrated control of a single higher order duct mode using a two channel controller and demonstrated that it was possible. Doelman (1989) was concerned primarily with control of multi-modal sound fields in an enclosure. Silcox and Elliott (1990) pointed out that the optimum controller transfer function derived in Section 7.2 becomes impractical to implement with one error microphone, as any changes in the modal structure in the duct will change the transfer functions involved and thus affect the controller performance even under ideal conditions. However, they demonstrated effective control of random noise at frequencies below the second higher order mode cut on frequency using a two channel system and a similar configuration to Eriksson *et al.* (1989). Neise and Koopman

(1991) showed that loudspeakers mounted on the cut-off of a centrifugal fan could significantly attenuate energy in higher order modes in the inlet and discharge ducts, provided the error sensors were arranged to adequately detect the modes to be controlled. As the authors did not have a multi-channel control system available, their results did not reflect the true potential of this control method. Their technique was also restricted to control of harmonic noise related to the blade passing frequency. Control of broadband noise is not possible with their technique, as the closeness of the primary and control sources meant that the reference signal must be obtained using a tachometer signal from the fan drive shaft. It is interesting to note that the same authors (Koopman *et al.*, 1988), using similar techniques reported attenuations of up to 23 dB for periodic noise associated with only plane wave propagation. This work is particularly interesting because it demonstrates active control working by direct modification of the impedance 'seen' by the primary noise source which in this case is the aerodynamic noise generated by the fan blade passing the fan cut-off.

Stell and Bernhard (1990) and Zander and Hansen (1992) addressed the theoretical analysis of active control of periodic noise propagating as higher order modes, with the view to defining the requirements of the physical system layout to achieve optimal control and possible limitations which might characterise systems for control of higher order modes. This analysis was later extended to include finite length ducts (Zander and Hansen, 1993). The latter work forms the basis of the analysis outlined in this section.

In 1991, Stell and Bernhard pointed out that for effective control of random noise propagating as higher order modes, an independent reference, error sensor and control source was required for each propagating mode, although they did not address how this may be implemented in practice. Zander and Hansen (1993) pointed out that the power propagating in each mode could be measured using a microphone array in one plane of the duct and that the signal provided in this way could be used together with one control source for each mode, to minimize noise propagating as higher order modes. As the noise was periodic in this case, it was sufficient to use a single microphone to provide the reference signal. However, for the control of random noise, the reference signal would have to be obtained using an array of microphones to decompose the noise in the duct into its constituent modes. This is discussed in more detail in Section 7.5.2.

In 1993, Laugesen showed that when multiple modes are to be controlled in a large rectangular section smoke stack with a limited number of error sensors, then if the error sensors have to be mounted on the walls, the best place is in the centre of each wall section. Ideally, however, the number of error sensors needed for optimal control is equal to one plus the number of propagating modes (including the plane wave mode) and the number of control sources needed is equal to the number of propagating modes. Laugesen (1993) states that the location of control sources is unimportant provided that they are not placed in the duct corners; however, one suspects that the control effort to achieve optimal control would be quite dependent on control source location.

Although the analysis in the following sections is based on a single frequency primary source, it is also useful for predicting the best performance which would be achievable for a random noise primary source. The difference between controlling the two types of signal lies in the practical implementation of a causal adaptive controller which is capable of generating the required optimal source volume velocities at each frequency, compared to the relative ease of implementation for a periodic signal for which causality constraints do not exist.

Active control of plane waves in ducts, which was discussed in Section 7.3 is a special case of active control of higher order duct modes. Thus, we begin the analysis here with (7.3.2)–(7.3.5) of Section 7.3. The duct arrangement is similar to that used for the plane wave analysis and shown in Fig. 7.18 except that the primary source has been generalized to be less than or equal to the duct cross-sectional area and located at (x_p, y_p, z_p); and the control source is specified by a length, width and (x_c, y_c, z_c) coordinate locations as shown in Fig. 7.38.

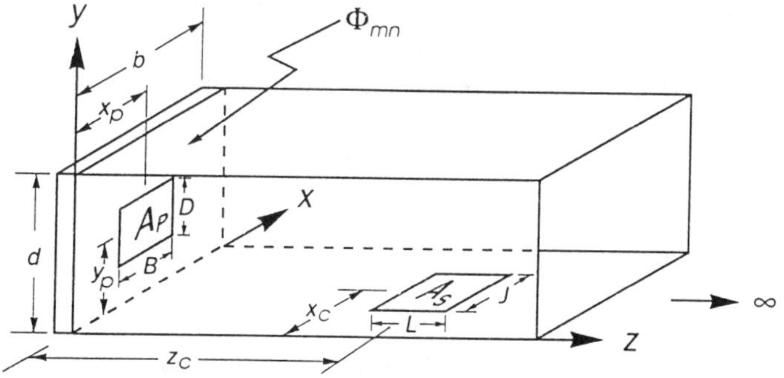

Fig. 7.38 Duct coordinate system and source arrangement for control of higher order modes in a semi-infinite duct.

The pressure at a point $\boldsymbol{x} = (x, y, z)$ in a semi-infinite duct due to a finite size, plane, periodic primary source radiating at a frequency of ω (rad s^{-1}) located with its centre at $\boldsymbol{x}_p = (x_p, y_p, 0)$ at the finite end of the duct in a plane normal to the duct axis is given by Fahy (1985) as:

$$p_p(x,y,z) = j\rho_0\omega \int_{A_p} u_p(x_0, y_0) G(\boldsymbol{x}, \boldsymbol{x}_0, \omega)\, \mathrm{d}x_0 \mathrm{d}y_0 \tag{7.4.1}$$

where $u_p(x_0, y_0)$ is the particle velocity at location (x_0, y_0) on the face of the primary source of width B and height D which has a surface area A_p. The duct cross-sectional dimensions are $b \times d$. Note that if p is a complex amplitude, then q must also be a complex amplitude. The pressure amplitude at an arbitrary point (x, y, z) due to a finite size primary source located at $(x_0, y_0, 0)$ is obtained from (7.3.2) and (7.4.1) and is:

$$p_p(x,y,z) = \frac{\rho_0\omega}{S} \sum_m \sum_n \int_{A_p} u_p \frac{\Psi_{mn}(x,y)\Psi_{mn}(x_0,y_0)}{\Lambda_{mn} k_{mn}} \mathrm{e}^{-jk_{mn}z} \mathrm{d}x_0 \mathrm{d}y_0 \tag{7.4.2}$$

Solving the integral for the rectangular duct case and a rectangular primary source of volume velocity Q_p mounted in the plane of the duct cross-section, as illustrated in Fig. 7.38, yields:

$$p_p(x,y,z) = \frac{\rho_0 \omega Q_p}{S A_p} \sum_n \frac{\Psi_{mn}(x,y)\Psi_{mn}(x_p,y_p)\gamma_{pmn}}{\Lambda_{mn} k_{mn}} e^{-jk_{mn}z} \tag{7.4.3}$$

where (x_p, y_p) is the location of the centre of the source which is in the $z = 0$ plane. γ_{pmn} is the primary source finite source factor given by:

$$\gamma_{pmn} = \begin{cases} \dfrac{2b}{m\pi}\sin\left[\dfrac{m\pi B}{2b}\right]\dfrac{2d}{n\pi}\sin\left[\dfrac{n\pi D}{2d}\right] & m,\ n \neq 0 \\ BD & m = n = 0 \end{cases} \tag{7.4.4}$$

If only plane wave propagation is considered $(m = n = 0)$, $\Psi_{00} = 1$, $\Lambda_{00} = 1$ and $k_0 = \omega/c_0$; thus, (7.4.3) becomes identical to (7.3.8).

If the primary source were a point source, then (7.4.3) would become:

$$p_p(x,y,z) = \frac{\rho_0 \omega Q_p}{S} \sum_{m=0}^{\infty} \sum_{n=0}^{\infty} \frac{\Psi_{mn}(x,y)\Psi_{mn}(x_0,y_0)}{\Lambda_{mn} k_{mn}} e^{-jk_{mn}z} \tag{7.4.5}$$

where $\Psi_{mn}(x, y)$, Λ_{mn} and k_{mn} are defined by (7.3.3), (7.3.4) and (7.3.5) respectively.

The mean pressure over the face of the finite size primary source is:

$$p_{p/p} = \frac{1}{A_p} \int_{A_p} p_p(x_0,y_0,z_p)\,dx_0\,dy_0 \tag{7.4.6}$$

The subscript p/p refers to a quantity at the primary source due to the primary source. Similarly p/c refers to a quantity at the primary source due to the control source. This notation will be used throughout the analysis. Solving for the case considered yields:

$$p_{p/p} = \frac{\rho_0 \omega Q_p}{A_p^2 S} \sum_m \sum_n \frac{\gamma_{pmn}^2 \Psi_{mn}^2(x_p,y_p)}{\Lambda_{mn} k_{mn}} \tag{7.4.7}$$

For plane wave propagation, this expression becomes equal to (7.3.8) with $z = 0$.

Equation (7.4.6) can be rewritten for the mean pressure at the control source located in the z, x plane of a duct wall at $y = 0$ due to the primary source as:

$$p_{c/p} = \frac{1}{A_c} \int_{A_c} p_p(x,y,z)\,dx\,dz \tag{7.4.8}$$

where A_c is the surface area of the control source of length L in the axial direction and lateral length J. Substituting the expression for the pressure at a point due to the primary source, given in (7.4.3), into (7.4.8) yields the mean pressure over the face of the finite size control source due to the primary source, namely:

$$p_{c/p} = \frac{\rho_0 \omega Q_p}{A_p A_c S} \sum_m \sum_n \frac{\gamma_{pmn}\,\gamma_{cmn}\,\Psi_{mn}(x_p, y_p)\,\Psi_{mn}(x_c, y_c)}{\Lambda_{mn} k_{mn}} e^{-jk_{mn} z_c} \qquad (7.4.9)$$

z_c is the axial distance from the primary source face at $z = 0$ to the centre of the control source which is located at (x_c, y_c, z_c). γ_{cmn} is the control source finite source factor, given by:

$$\gamma_{cmn} = \begin{cases} (-1)^n \dfrac{2b}{m\pi} \sin\left[\dfrac{m\pi J}{2b}\right] \dfrac{2}{k_{mn}} \sin\left[\dfrac{k_{mn} L}{2}\right] & m,n \neq 0 \\ JL & m = n = 0 \end{cases} \qquad (7.4.10)$$

Equations (7.4.9) and (7.4.10) are unchanged for a source mounted in the $y = d$ duct wall. For a source mounted in the $x = b$ or $x = 0$ walls, (7.4.9) remains unchanged and in part (a) of (7.4.10), b is replaced with d.

The acoustic pressure at any point in the semi-infinite duct due to a source mounted part way along the duct in one of the duct walls consists of the sum of the direct pressure from the source and the pressure reflected from the finite end.

The contribution to the pressure coming directly from a simple point control source at location (x_c, y_c, z_c) is given by:

$$p_{c_d}(x,y,z) = \frac{\rho_0 \omega Q_c}{2S} \sum_m \sum_n \frac{\Psi_{mn}(x,y)\Psi_{mn}(x_c,y_c)}{\Lambda_{mn} k_{mn}} e^{-jk_{mn}|z - z_c|} \qquad (7.4.11)$$

To simplify the analysis to begin with, it will be assumed that there is a rigid termination at the primary source end of the duct. For this case, the pressure at point (x,y,z) due to the component reflected from the primary source end is given by:

$$p_{c_r}(x,y,z) = \frac{\rho_0 \omega Q_c}{2S} \sum_m \sum_n \frac{\Psi_{mn}(x,y)\Psi_{mn}(x_c,y_c)}{\Lambda_{mn} k_{mn}} e^{-jk_{mn}(z + z_c)} \qquad (7.4.12)$$

where the subscript d refers to the pressure component coming directly from the control source, and the subscript r denotes the pressure component due to the reflection from the primary source end of the duct.

The total pressure at a point is equal to the sum of the direct and reflected pressure components, namely:

$$p_c(x,y,z) = p_{c_d}(x,y,z) + p_{c_r}(x,y,z) \qquad (7.4.13)$$

For a point (x,y,z) downstream of the control source, that is, for $z \geq z_c$:

$$p_c(x,y,z) = \frac{\rho_0 \omega Q_c}{S} \sum_m \sum_n \frac{\Psi_{mn}(x,y)\Psi_{mn}(x_c,y_c)}{\Lambda_{mn} k_{mn}} \mathrm{e}^{-jk_{mn}z} \left[\frac{\mathrm{e}^{jk_{mn}z_c} + \mathrm{e}^{-jk_{mn}z_c}}{2} \right] \tag{7.4.14}$$

or:

$$p_c(x,y,z) = \frac{\rho_0 \omega Q_c}{S} \sum_m \sum_n \frac{\Psi_{mn}(x,y)\Psi_{mn}(x_c,y_c)}{\Lambda_{mn} k_{mn}} \mathrm{e}^{-jk_{mn}z} \cosh(jk_{mn}z_c) \tag{7.4.15}$$

Thus,

$$p_c(x,y,z) = \frac{\rho_0 \omega Q_c}{S} \sum_m \sum_n \left[\frac{\Psi_{mn}(x,y)\Psi_{mn}(x_c,y_c)}{\Lambda_{mn} k_{mn}} \mathrm{e}^{-jk_{mn}z} \cos(k_{mn}z_c) \right] \quad \textit{for } z \geq z_c \tag{7.4.16}$$

For plane waves only (7.4.16) is identical to (7.3.9).

A similar approach can be used to derive the total pressure, due to the control source only, at a point *(x,y,z)* located between the primary source and the control source. Hence the total pressure at point *(x,y,z)* for a simple point control source is:

$$p_c(x,y,z) = \frac{\rho_0 \omega Q_c}{S} \sum_m \sum_n \left[\frac{\Psi_{mn}(x,y)\Psi_{mn}(x_c,y_c)}{\Lambda_{mn} k_{mn}} \mathrm{e}^{-j(k_{mn}z_c + \beta)} \cos(k_{mn}z) \right] \quad \textit{for } 0 \leq z \leq z_c \tag{7.4.17}$$

where (x_c, y_c, z_c) is the coordinate of the control source. For plane waves only, (7.4.17) is identical to (7.3.11) with $\Phi = 0$ (rigid termination at $z = 0$).

Using a similar approach to that used for the primary source, the mean pressure over the face of a finite size rectangular control source of dimensions $L \times J$, due to the sound field produced by the control source, is found to be:

$$p_{c/c} = \frac{\rho_0\omega Q_c}{A_c^2 S}\sum_m\sum_n \frac{\gamma_{cmn}^2\Psi_{mn}^2(x_c,y_c)}{\Lambda_{mn}k_{mn}} e^{-jk_{mn}z_c}\cos(k_{mn}z_c) \tag{7.4.18}$$

The pressure over the face of the primary source due to the control source is:

$$p_{p/c} = \frac{\rho_0\omega Q_c}{A_c A_p S}\sum_m\sum_n \frac{\gamma_{cmn}\gamma_{pmn}\Psi_{mn}(x_p,y_p)\Psi_{mn}(x_c,y_c)}{\Lambda_{mn}k_{mn}} e^{-jk_{mn}z_c} \tag{7.4.19}$$

Alteration of the termination plane at $z = 0$ from a rigid termination to a uniform impedance surface allows greater flexibility in the model, and enables a more accurate prediction of a real system's performance. Again the pressure at a point in the duct will be equal to the sum of the contribution directly from the source and the contribution of waves reflected from the termination plane at $z = 0$. A plane wall of uniform impedance will not couple modes or reflect a multiple number of modes from a single incident mode. Hence an *mn*th order mode incident upon the termination will be reflected from the surface as an *mn*th order mode. The alteration of the incident wave due to the impedance surface can be expressed in exponential form for the *mn*th mode as:

$$e^{-2\Phi_{mn}} \tag{7.4.20}$$

where:

$$\Phi_{mn} = \pi\alpha_{mn} + j\pi\beta_{mn} \tag{7.4.21}$$

The relationship between these quantities, the complex reflection coefficient and the specific acoustic impedance of the duct termination is discussed in Section 7.5.

Using a similar approach to that used in Section 7.3 for plane waves, it can be shown that the pressure, due to the control source only, at a point (x,y,z) positioned between the primary and control source (that is, for $0 \le z \le z_c$), is given for a simple point control source by:

$$p_c(x,y,z) =$$

$$\frac{\rho_0\omega Q_c}{S}\sum_m\sum_n\left[\frac{\Psi_{mn}(x,y)\Psi_{mn}(x_c,y_c)}{\Lambda_{mn}k_{mn}} e^{(-\Phi_{mn}-jk_{mn}z_c)}\cosh(\Phi_{mn}+jk_{mn}z)\right] \tag{7.4.22}$$

Similarly for an axial location $z \ge z_c$:

$$p_c(x,y,z) =$$

$$\frac{\rho_0\omega Q_c}{S}\sum_m\sum_n\left[\frac{\Psi_{mn}(x,y)\Psi_{mn}(x_c,y_c)}{\Lambda_{mn}k_{mn}} e^{(-\Phi_{mn}-jk_{mn}z)}\cosh(\Phi_{mn}+jk_{mn}z_c)\right] \tag{7.4.23}$$

For plane waves only, (7.4.22) is identical to (7.3.11). Expressions for p_p, $p_{p/p}$ and $p_{c/p}$ corresponding to finite size sources are identical to expressions (7.4.3), (7.4.7) and (7.4.9) respectively for a rigid termination at the primary source. However, expressions for $p_{p/c}$ and $p_{c/c}$ are different to those for a rigid termination and may be written as follows:

$$p_{p/c} = \frac{\rho_0 \omega Q_c}{A_c A_p S} \sum_m \sum_n \left[\frac{\gamma_{pmn} \gamma_{cmn} \Psi_{mn}(x_p, y_p) \Psi_{mn}(x_c, y_c)}{\Lambda_{mn} k_{mn}} \mathrm{e}^{-(jk_{mn} z_c + \Phi_{mn})} \cosh(\Phi_{mn}) \right] \tag{7.4.24}$$

$$p_{c/c} = \frac{\rho_0 \omega Q_c}{A_c^2 S} \sum_m \sum_n \left[\frac{\gamma_{cmn}^2 \Psi_{mn}^2(x_c, y_c)}{\Lambda_{mn} k_{mn}} \mathrm{e}^{-(jk_{mn} z_c + \Phi_{mn})} \cosh(\Phi_{mn} + jk_{mn} z_c) \right] \tag{7.4.25}$$

Until now, the acoustic pressure at a point has been expressed as the sum of n modal contributions, with n extending from zero to infinity. The acoustic pressure may be otherwise expressed in terms of its real and imaginary components by considering that for an excitation frequency ω there are a finite number of modes which are 'cut on', and hence contribute to the propagating part of the pressure, while those modes with a cut-on frequency greater than the excitation frequency contribute only to the non-propagating component of the pressure. This may be written as:

$$\begin{aligned} \omega &\geq \kappa_{mn} c_0 \quad \text{mode cut on} \\ \omega &< \kappa_{mn} c_0 \quad \text{mode cut off} \end{aligned} \tag{7.4.26}$$

where κ_{mn} is the modal eigenvalue given by:

$$\kappa_{mn} = \sqrt{\left[\frac{\pi m}{b}\right]^2 + \left[\frac{\pi n}{d}\right]^2} \tag{7.4.27}$$

Hence the propagating part of the pressure may be written as the sum of contributions from the $N+1$ cut-on modes extending from zero to N, as the plane wave or fundamental mode is cut-on for all frequencies of excitation, where:

$$\kappa_N < \frac{\omega}{c_0} < \kappa_{N+1} \tag{7.4.28}$$

Acoustic modes having modal eigenvalues greater than ω/c_0 do not propagate along the duct as waves, but decay in an exponential manner with axial distance from the source, in contrast to the cut-on modes which, for the rigid walled case considered here, are not attenuated and propagate along the duct. For these cut-off modes the pressure and axial particle velocity are exactly out of phase, and hence the modes transport no mean energy.

7.4.1 Constant volume velocity primary source, single control source

The total acoustic pressure amplitude at the face of the primary source is given by:

$$p_p = p_{p/p} + p_{p/c} \tag{7.4.29}$$

which for a constant volume velocity source is found by adding together (7.4.7) and (7.4.19).

The total pressure at the face of the control source is:

$$p_c = p_{c/p} + p_{c/c} \tag{7.4.30}$$

which is found by adding together (7.4.9) and (7.4.18). The total power radiated down the duct is then found by substituting (7.4.29) and (7.4.30) into (7.3.1).

7.4.1.1 Optimum control source volume velocity: idealized rigid primary source termination

For the case of a rigid duct termination plane at $z = 0$, it is possible to express the total acoustic power output of the system as a quadratic function of the complex control source volume velocity amplitude Q_c, where:

$$Q_c = \bar{A}_c \mathrm{e}^{-j\beta_c} \tag{7.4.31}$$

and A_c is the scalar volume velocity amplitude of the control source, and β_c is its phase with respect to the primary source volume velocity. For a system consisting of a finite size primary source and a single control source this is given by:

$$W = Q_c^* a Q_c + Q_c b + b^* Q_c + c \tag{7.4.32}$$

where

$$a = \frac{1}{2}\frac{\rho_0 \omega}{A_c^2 S} \sum_m \sum_n \frac{\gamma_{cmn}^2 \Psi_{mn}^2(x_c, y_c)}{\Lambda_{mn} k_{mn}} \cos^2(k_{mn} z_c) \tag{7.4.33}$$

$$b = \frac{1}{2}\frac{\rho_0 \omega Q_p}{A_c A_p S} \sum_m \sum_n \frac{\gamma_{pmn}\gamma_{cmn}\Psi_{mn}(x_p, y_p)\Psi_{mn}(x_c, y_c)}{\Lambda_{mn} k_{mn}} \cos(k_{mn} z_c) \tag{7.4.34}$$

$$c = \frac{1}{2}\frac{\rho_0 \omega}{A_p^2 S} |Q_p|^2 \sum_m \sum_n \frac{\gamma_{pmn}^2 \Psi_{mn}^2(x_p, y_p)}{\Lambda_{mn} k_{mn}} \tag{7.4.35}$$

In the preceding equations, the solution for plane waves is found by setting $m = n = 0$, and $\Lambda_{00} = 1$.

Although the preceding quadratic equation is not differentiable with respect to the complex volume velocity Q_c, it is possible to express the total acoustic power W in terms of real and imaginary components, and to differentiate these components with respect to the real and imaginary parts of the control source volume velocity. Equating these gradients to zero yields the complex control source volume velocity which minimizes the total acoustic power. This optimum volume velocity amplitude is given by:

$$Q_{c_{opt}} = -\frac{b}{a} \tag{7.4.36}$$

Note that complex notation is used to express the required phase of the control source volume velocity with respect to the primary source velocity.

7.4.1.2 Optimum control source volume velocity: arbitrary uniform impedance termination at the primary source

For a duct with a uniform impedance surface termination plane, the total radiated power for a system comprising one primary source and a single control source may be calculated by using (7.3.1), (7.4.29), (7.4.30), (7.4.7), (7.4.9), (7.4.24) and (7.4.25).

That is,

$$W = Q_c^* \mathrm{Re}\{a\} Q_c + \mathrm{Re}\{b_1 Q_c\} + \mathrm{Re}\{b_2 Q_c^*\} + c \tag{7.4.37}$$

where

$$a = \frac{1}{2}\mathrm{Re}\left\{\frac{\rho_0 \omega}{A_c^2 S} \sum_m \sum_n \frac{\gamma_{cmn}^2 \Psi_{mn}^2(x_c, y_c)}{\Lambda_{mn} k_{mn}} e^{-jk_{mn} z_c} e^{-\Phi_{mn}} \cosh(jk_{mn} z_c + \Phi_{mn})\right\} \tag{7.4.38}$$

$$b_1 = \frac{1}{2}\frac{\rho_0 \omega Q_p^*}{A_c A_p S} \mathrm{Re}\left\{\sum_m \sum_n \frac{\gamma_{pmn}\gamma_{cmn}\Psi_{mn}(x_p, y_p)\Psi_{mn}(x_c, y_c)}{\Lambda_{mn} k_{mn}} e^{-jk_{mn} z_c}\right\} \tag{7.4.39}$$

$$b_2 = \frac{1}{2}\frac{\rho_0 \omega Q_p}{A_c A_p S} \mathrm{Re}\left\{ \sum_m \sum_n \frac{\gamma_{pmn}\gamma_{cmn}\Psi_{mn}(x_p, y_p)\Psi_{mn}(x_c, y_c)}{\Lambda_{mn} k_{mn}} \mathrm{e}^{-jk_{mn}z_c}\, \mathrm{e}^{-\Phi_{mn}} \cosh\Phi_{mn} \right\} \tag{7.4.40}$$

$$c = \frac{1}{2}\frac{\rho_0 \omega}{A_p^2 S} \mid Q_p \mid^2 \sum_m \sum_n \frac{\gamma_{pmn}^2 \Psi_{mn}^2(x_p, y_p)}{\Lambda_{mn} k_{mn}} \tag{7.4.41}$$

Differentiating the total power expression with respect to the real and imaginary components of the complex control source volume velocity yields the optimum control source volume velocity which is given by:

$$Q_{c_{opt}} = -\frac{1}{2}\mathrm{Re}\{a\}^{-1}(b_1^* + b_2) \tag{7.4.42}$$

7.4.1.3 Dual control sources

For the following analysis the subscript 1 will refer to the control source c_1, and the subscript 2 will refer to the control source c_2. A source arrangement for multiple sources is shown in Fig. 7.39. Using a procedure similar to that used for the single control source case the additional acoustic pressure terms are derived in the following manner.

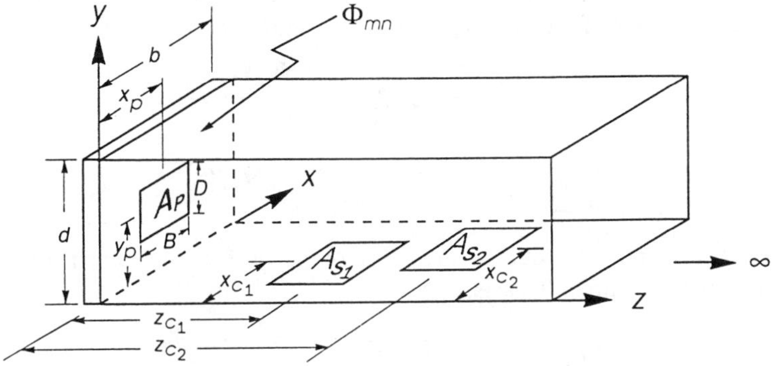

Fig. 7.39 Dual control source duct model.

Assuming the control source c_2 to be located at a greater axial distance from the primary source than control source c_1, that is, for $z_{c2} \geq z_{c1}$, the mean pressure over the face of c_2 due to c_1 is:

$$p_{c_2/c_1} = \frac{\rho_0 \omega Q_{c_1}}{A_{c_1} A_{c_2} S} \sum_m \sum_n \frac{\gamma_{c_1 mn}\gamma_{c_2 mn}\Psi_{mn}(x_{c_1}, y_{c_1})\Psi_{mn}(x_{c_2}, y_{c_2})}{\Lambda_{mn} k_{mn}} \cos(k_{mn} z_{c_1})\, \mathrm{e}^{-jk_{mn}z_{c_2}} \tag{7.4.43}$$

For plane waves only, (7.4.43) becomes:

$$p_{c2/c1} = \frac{\rho_0 c_0}{S} Q_{c1} e^{-jk_o z_{c2}} \cos(kz_{c1}) \operatorname{sinc}\left[\frac{kL_1}{2}\right] \operatorname{sinc}\left[\frac{kL_2}{2}\right] \tag{7.4.44}$$

Similar expressions hold for $p_{c1/c2}$ and the total pressure at source c_1 is now:

$$p_{c1} = p_{c1/p} + p_{c1/c1} + p_{c1/c2} \tag{7.4.45}$$

where $p_{c1/p}$ is equivalent to $p_{c/p}$ of (7.4.9) and $p_{c1/c1}$ is equivalent to $p_{c/c}$ of (7.4.18). Similar expressions hold for p_{c2} and p_p.

The total acoustic power radiated down the duct for the dual control source system is:

$$W = \frac{1}{2} \operatorname{Re}\left[Q_p p_p^* + Q_{c1} p_{c1}^* + Q_{c2} p_{c2}^*\right] \tag{7.4.46}$$

7.4.2 Constant pressure primary source

The procedure for calculating the total power transmission and the optimum control source volume velocity for a constant pressure primary source is similar to that used for plane waves and outlined in Section 7.3. Essentially, the acoustic pressure on the face of the primary source is the same before and after introduction of the control source; however, the primary source volume velocity changes. Thus, using (7.4.9) and (7.4.29) we can write the following for the pressure at the face of the primary source after introduction of the control source

$$p_p = \frac{\rho_0 \omega Q_p'}{A_p^2 S} \sum_m \sum_n \frac{\gamma_{pmn}^2 \Psi_{mn}^2(x_p, y_p)}{\Lambda_{mn} k_{mn}} + \frac{\rho_0 \omega Q_c}{A_c A_p S} \sum_m \sum_n \frac{\gamma_{pmn} \gamma_{cmn} \Psi_{mn}(x_p, y_p) \Psi_{mn}(x_c, y_c)}{\Lambda_{mn} k_{mn}} e^{-jk_{mn} z_c} \tag{7.4.47}$$

However, this must be equal to the pressure on the face of the primary source with no control source. That is,

$$P_p = \frac{\rho_0 \omega Q_p}{A_p^2 S} \sum_m \sum_n \frac{\gamma_{pmn}^2 \Psi_{mn}^2(x_p, y_p)}{\Lambda_{mn} k_{mn}} \tag{7.4.48}$$

Setting the right-hand sides of (7.4.47) and (7.4.48) equal allows an expression for Q_p' to be obtained in terms of Q_p. This can be substituted for Q_p in (7.4.7), (7.4.9), (7.4.18) and (7.4.19) to obtain expressions for the total pressure at the face of the primary and control sources after introduction of the control source. A similar procedure can be followed for a non-rigid termination at the primary source using the constant volume source equations already derived for that termination.

7.4.3 Finite length duct

If the downstream end of the duct is terminated non-anechoically waves will be reflected back upstream, if the controller is non-optimal. If higher order modes are propagating, each mode will be characterized by a different amplitude reflection coefficient, $e^{-2\Phi_{2mn}}$.

An expression will now be derived for the sound pressure in a finite length duct at location $\boldsymbol{x} = (x,y,z)$ as a result of a point control source located at an arbitrary location $\boldsymbol{x}_c = (x_c,y_c,z_c)$ in the duct. The arrangement is illustrated in Fig. 7.40.

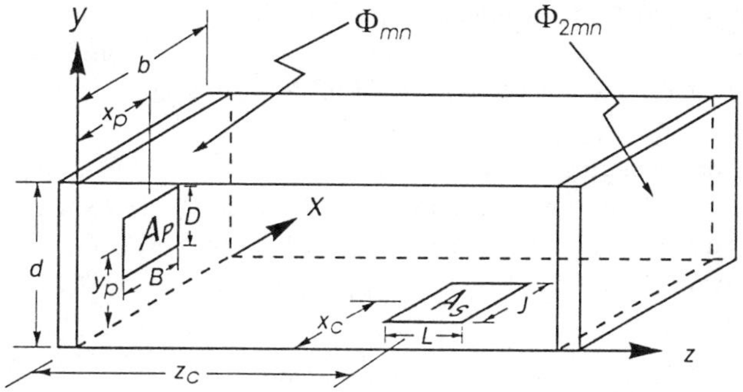

Fig. 7.40 Finite duct model.

The sound pressure due to the direct wave is

$$p_1(\boldsymbol{x}) = \sum_m \sum_n A_{mn}\, e^{-jk_{mn}|z-z_c|} \tag{7.4.49}$$

The sound pressure due to the wave reflected from the left end of the duct is

$$p_2(\boldsymbol{x}) = \sum_m \sum_n A_{mn}\, e^{-jk_{mn}(z_c+z)}\, e^{-2\Phi_{mn}} \tag{7.4.50}$$

The sound pressure due to the wave reflected from the right end of the duct is

$$p_3(\boldsymbol{x}) = \sum_m \sum_n A_{mn}\, e^{jk_{mn}(z_c+z)}\, e^{-2\Phi_{2mn}}\, e^{-j2k_{mn}L} \tag{7.4.51}$$

The sound pressure due to a wave reflected from both ends is

$$p_4(\boldsymbol{x}) = \sum_m \sum_n A_{mn}\, e^{jk_{mn}|z_c-z|}\, e^{-2\Phi_{2mn}}\, e^{-2\Phi_{mn}}\, e^{-j2k_{mn}L} \tag{7.4.52}$$

where
$$A_{mn} = \frac{\rho_0 \omega Q_c}{2S} \frac{\Psi_{mn}(\boldsymbol{x})\,\Psi_{mn}(\boldsymbol{x}_c)}{\Lambda_{mn} k_{mn}} \tag{7.4.53}$$

$e^{-2\Phi_{mn}}$ is the impedance function characterizing the left end of the duct for mode m,n and $e^{-2\Phi_{2mn}}$ is the impedance function characterizing the right end of the duct for mode m,n.

Adding (7.4.49) to (7.4.52) together with additional significant reflections, the number of which are dependent on the values of Φ_{mn} and Φ_{2mn}, gives the total sound pressure at $\boldsymbol{x}$ due to the point source at $\boldsymbol{x}_c$, as follows:

$$p(x) = \sum_m \sum_n A_{mn} \Big[e^{-jk_{mn} |z_c - z|} + e^{-jk_{mn}(z_c + z)} e^{-2\Phi_{mn}}$$

$$+ e^{jk_{mn}(z_c + z)} e^{-j2k_{mn}L} e^{-2\Phi_{2mn}} + e^{jk_{mn} |z_c - z|} e^{-2\Phi_{2mn}} e^{-2\Phi_{mn}} e^{-j2k_{mn}L} \Big] T_n \tag{7.4.54}$$

where T_n is the modal reverberation factor given by

$$T_n = \sum_{i=0}^{\infty} \left[e^{-j2k_{mn}L} e^{-2\Phi_{mn}} e^{-2\Phi_{2mn}} \right]^i = \left[1 - e^{-j2k_{mn}L} e^{-2\Phi_{mn}} e^{-2\Phi_{2mn}} \right]^{-1} \tag{7.4.55a,b}$$

For a source mounted at the left end of the duct in the plane of the duct cross section, the second term in brackets in (7.4.54) is omitted. For a finite size control source mounted in a duct wall (in the x-z plane) the factor γ_{cmn} of (7.4.10) may be used to correct the preceding equations for the source size. That is, A_{mn} becomes $A_{mn}\gamma_{cmn}$.

For a primary source mounted in the plane of the duct cross section, at $z = 0$, the result for a point source is multiplied by the finite size source factor, γ_{pmn}, of (7.4.4). That is, A_{mn} becomes $A_{mn}\ \gamma_{pmn}$. For both primary and control sources of finite size, γ_{pmn} is replaced with $A_{mn}\gamma_{pmn}\gamma_{cmn}$.

7.4.4 Effect of control source location and size

When higher order modes are to be controlled in addition to the plane wave mode, the optimum control source volume velocity and phase for minimum total power transmission down the duct will not result in minimizing the power in each mode. Rather, a compromise will be reached, as the required optimal control source volume velocity will be different for each mode. Also, the optimum control source location for each mode (both axially as well as in the duct cross section) will also be different for each mode.

In general, at least one control source should be used for each mode to be controlled, although the arrangement of the control sources should be optimized to minimize the total power transmission, probably resulting in each control source affecting more than one mode.

The previously described theory can be used to calculate the total power reduction for periodic noise propagating in a duct for a specified duct size and control source arrangement. In Section 7.4.3, this analysis was extended to include the effect of reflection of sound from the duct exit, which somewhat complicates the algebra, but the general conclusions regarding the effects of control source size and arrangement are essentially unchanged.

Note that for higher order mode control, the optimum control source location will be a function of each coordinate direction for a rectangular duct, in contrast to the case of plane wave control where it is just a function of control source axial location. For a semi-infinite duct, the optimum axial separation distance between the control source and primary source for control of a particular mode is equal to an integer multiple of half wavelengths for that particular mode, where the wavelength of the *mn*th mode is given by:

$$\lambda_{mn} = 2\pi \left[\left(\frac{\omega}{c_0} \right)^2 - \left(\frac{\pi m}{b} \right)^2 - \left(\frac{\pi n}{d} \right)^2 \right]^{-\frac{1}{2}} \quad \text{metres} \tag{7.4.56}$$

where m and n correspond to the numbers of nodal lines along the x and y directions respectively, c_0 is the speed of sound in the free space (m s^{-1}), ω is the angular frequency (rad s^{-1}), and b and d are the duct cross-section dimensions.

Thus, if more than one mode is present, optimum attenuation will occur when the control source is located so that it is an integer multiple of half wavelengths from the primary source for all modes which are present. Thus, it is clear that if more than two or three modes are present it will be difficult to find a control source location which is an integer multiple of half wavelengths from the primary source for all modes. Even if such a location is found, it will be frequency dependent for higher order modes.

Even if the correct source location is found it is difficult to achieve high levels of control, as the optimum volume velocity to control one mode will not be the same as that to control another mode. This suggests that for control of higher order modes to be effective, a multichannel controller with at least one channel per mode to be controlled is needed. For some modes, one controller channel driving two control sources will produce optimal results, by minimizing spill-over of controller energy into other propagating modes.

As for the plane wave case, control sources should be as small as possible along the length of the duct to minimize the volume velocity requirement for control.

7.4.5 Effect of error sensor type and location

For similar reasons as for plane waves, implementation of a higher order mode power transmission (or intensity) sensor would not be feasible in practice. Indeed, for higher order modes, the intensity varies across the duct so even an accurate point measurement of sound intensity would not provide a measure of the power transmission. Thus, one would need to ensure that at least as many error sensors as modes to be controlled were used and that each error sensor is placed so that it can detect at least one of the modes to be controlled. Unfortunately, these locations will vary with frequency and with duct flow and temperature conditions so it appears that an excess of sensors should be used if a robust control system is to be realized.

For finite length ducts, the presence of reflected waves will always make optimal location of the error sensors more difficult, as they should not be placed at nodes in any standing wave field produced as a result of the reflected waves. As these nodal locations are frequency dependent this is again a difficult problem and points to the desirability of many error sensors feeding a multichannel control system if higher order modes are to be controlled effectively.

Zander and Hansen (1993) investigated analytically the effectiveness of various error sensor strategies to control a propagating multi-mode sound field in a duct. These strategies were: minimization of the squared pressure amplitude at a point; minimization of the sum of the squared pressure amplitudes at a number of locations throughout the duct; minimization of the sum of the squared pressure amplitudes at a number of locations downstream of the control source; minimization of the total real acoustic power output of the primary and control sources; and minimization of the acoustic power transmission downstream of the control source, as determined by modal decomposition of the duct sound field. From the results obtained using the five different error sensor strategies the most appropriate strategy for minimizing the sound field downstream of the control source was found to be minimization of the downstream power transmission. Each of the other strategies were found to yield poorer levels of downstream power transmission reduction for at least one of the test cases, implying that the estimate of downstream power transmission obtained from modal decomposition of the duct sound field was the most robust technique in terms of varying excitation frequency and varying termination conditions. The downstream power transmission estimate technique yielded levels of power transmission reduction equal to or greater than the other error sensor strategies for all of the tests conducted. Nevertheless, for completeness, each of the above-mentioned control strategies will now be discussed. In the following paragraphs the quantity to be minimized (squared pressure or sound power as discussed above) will be denoted F so that a general solution for the optimum control forces and minimum achievable value of F can be formulated.

The value of the function F, to be minimized under the influence of the primary and control sources can be expressed as a quadratic function of the control source volume velocities, $\boldsymbol{Q}_c = \begin{bmatrix} Q_{c_1} & Q_{c_2} & \cdots & Q_{c_M} \end{bmatrix}^{\mathrm{T}}$, for M control sources, such that

$$F = \boldsymbol{Q}_c^{\mathrm{H}} \boldsymbol{a} \boldsymbol{Q}_c + \boldsymbol{b}_1 \boldsymbol{Q}_c + \boldsymbol{Q}_c^{\mathrm{H}} \boldsymbol{b}_2 + c \tag{7.4.57}$$

The composition of the matrices $\boldsymbol{a}$, $\boldsymbol{b}_1$, $\boldsymbol{b}_2$, and the value of the variable c are all dependent upon the function F to be minimized. Note that the matrix form of the preceding equation allows for any number of primary and control sources. Differentiating (7.4.57) with respect to the real and imaginary components of the control source volume velocity, $\boldsymbol{Q}_c$, and equating the result to zero, yields the optimum control source volume velocity as

$$\boldsymbol{Q}_{c_{\mathrm{opt}}} = -\boldsymbol{a}^{-1} \boldsymbol{b} \tag{7.4.58}$$

where

$$\boldsymbol{b} = \frac{1}{2}\left\{\boldsymbol{b}_1^{\mathrm{H}} + \boldsymbol{b}_2\right\} \tag{7.4.59}$$

The specific form of the matrices $\boldsymbol{a}$, $\boldsymbol{b}_1$, $\boldsymbol{b}_2$, and the variable c will now be outlined for each error sensor strategy.

For an error criterion of minimization of the pressure amplitude at a point, $|p(\boldsymbol{x})|^2$, such that the error function F in (7.4.57) is equal to $|p(\boldsymbol{x})|^2$, the matrices take the form

$$\boldsymbol{a} = \boldsymbol{Z}_c^{\mathrm{H}} \boldsymbol{Z}_c \tag{7.4.60}$$

$$\boldsymbol{b}_1 = \boldsymbol{b}_2^{\mathrm{H}} \tag{7.4.61}$$

$$\boldsymbol{b}_2 = \boldsymbol{b} = \boldsymbol{Z}_c^{\mathrm{H}} \boldsymbol{Z}_p \boldsymbol{Q}_p \tag{7.4.62}$$

$$c = \boldsymbol{Q}_p^{\mathrm{H}} \boldsymbol{Z}_p^{\mathrm{H}} \boldsymbol{Z}_p \boldsymbol{Q}_p \tag{7.4.63}$$

where $\boldsymbol{Z}_p$ relates the pressure at the point $\boldsymbol{x}$ to the primary source volume velocity $\boldsymbol{Q}_p = \left[Q_{p_1} \; Q_{p_2} \; \cdots \; Q_{p_N}\right]^{\mathrm{T}}$, for N primary sources, by

$$\boldsymbol{Z}_p = \left[\frac{p_{p_1}(\boldsymbol{x})}{Q_{p_1}}, \; \cdots, \; \frac{p_{p_N}(\boldsymbol{x})}{Q_{p_N}}\right] \tag{7.4.64}$$

and $\boldsymbol{Z}_c$ similarly relates the pressure at $\boldsymbol{x}$ to the control source volume velocity $\boldsymbol{Q}_c$, such that

$$\boldsymbol{Z}_c = \left[\frac{p_{c_1}(\boldsymbol{x})}{Q_{c_1}}, \; \cdots, \; \frac{p_{c_M}(\boldsymbol{x})}{Q_{c_M}}\right] \tag{7.4.65}$$

where $p_{p_1}(\boldsymbol{x})$ is the acoustic pressure at location $\boldsymbol{x}$ due only to primary source, p_1 and similarly for $p_{c_1}(\boldsymbol{x})$.

A practically achievable estimate of the acoustical potential energy, E_p, in a region of the duct is given by the sum of the squares of the sound pressures at a large number of locations, l, distributed throughout the region (Curtis *et al.*, 1987). The minimization of the pressure at a single point is a subset of this error strategy for the case of $l = 1$. The quadratic function for the minimization of the acoustic potential energy estimate, J_p, is equal to

$$F = J_p = \frac{1}{4\rho c_o^2 l} \sum_{i=1}^{l} |p(\boldsymbol{x}_i)|^2 \tag{7.4.66}$$

where the constant factor $1/4\rho c_o^2 l$ is introduced such that the estimate, J_p, is compatible with the actual acoustic potential energy, E_p (Curtis *et al.*, 1987). The matrices in the quadratic expression take the same form as those in (7.4.61), (7.4.62) and (7.4.63).

Minimization of the total real acoustic power output of the primary and control sources, W, gives (Nelson *et al.*, 1987) (7.4.57) where

$$a = \mathrm{Re}\left\{ Z_c(x_c)^{\mathrm{H}} \right\} \tag{7.4.67}$$

$$b_1 = Q_p^{\mathrm{H}} \,\mathrm{Re}\left\{ Z_p(x_c)^{\mathrm{H}} \right\} \tag{7.4.68}$$

$$b_2 = \mathrm{Re}\left\{ Z_c(x_p)^{\mathrm{H}} \right\} Q_p \tag{7.4.69}$$

$$c = Q_p^{\mathrm{H}} \,\mathrm{Re}\left\{ Z_p(x_p)^{\mathrm{H}} \right\} Q_p \tag{7.4.70}$$

where $Z_p(x_c)$, an $M \times N$ matrix for M control sources and N primary sources, relates the pressure at the control source location x_c due to the primary source volume velocity Q_p, such that

$$Z_p(x_c) = \begin{bmatrix} \dfrac{p_{p_1}(x_{c_1})}{Q_{p_1}} & \cdots & \dfrac{p_{p_N}(x_{c_1})}{Q_{p_N}} \\ \vdots & \ddots & \vdots \\ \dfrac{p_{p_1}(x_{c_M})}{Q_{p_1}} & \cdots & \dfrac{p_{p_N}(x_{c_M})}{Q_{p_N}} \end{bmatrix} \tag{7.4.71}$$

and similarly for $Z_c(x_p)$ $(N \times M)$, $Z_c(x_c)$ $(M \times M)$, and $Z_p(x_p)$ $(N \times N)$. It should be noted that a direct measurement of W is difficult in a practical context, and hence is treated chiefly as a theoretical strategy.

Some earlier investigations into minimization of downstream power as an error sensor strategy have expressed the power as the area integral of the acoustic intensity over the duct cross section, namely

$$W = \frac{1}{2} \int_S \mathrm{Re}\{ p u^* \} \mathrm{d}S \tag{7.4.72}$$

which has in the past led researchers to remark that it is probably impractical to implement such a control strategy due to the difficulty of monitoring power (Stell and Bernhard, 1990).

A way around this difficulty is to formulate the propagating acoustic power in terms of modal amplitudes which, in turn are formulated in terms of the total pressure at a discrete number of points in the duct. As will be seen, this allows a realizable measure of the propagating acoustic power, and is hence suitable for use as an error sensor strategy.

Modal amplitudes A_{mni} of modes mn in the sound field propagating towards the duct opening and the amplitude A_{mnr} of the same modes travelling in the opposite direction, away from the opening may be determined by taking sound pressure measurements on two cross-sectional planes in the duct as discussed in detail in Section 7.5. To resolve N modes, N measurements are needed on each plane. For now, it will be assumed that these measurements have allowed the determination of the amplitudes of the modes propagating in both directions. Consequently, the total real acoustic power W_i transmitted along the duct towards one end and the power W_r transmitted away from the opening, and back down the duct can be written as:

$$W_i = \sum_m \sum_n \frac{bdk_{mn} \mid A_{mni} \mid^2}{\rho_0 \omega} \tag{7.4.73}$$

and

$$W_r = \sum_m \sum_n \frac{bdk_{mn} \mid A_{mnr} \mid^2}{\rho_0 \omega} \tag{7.4.74}$$

where W_i and W_r represent the total acoustic power propagating towards, and away from, the termination, respectively.

The modal amplitudes of the incident and reflected modes may be placed for convenience in a $2N \times 1$ matrix, with the elements in the diagonal alternately representing the amplitude of an incident mode and the amplitude of the same mode reflected from the end. Thus, the first element of $\boldsymbol{A}$ corresponds to the amplitude of the first mode propagating towards the duct exit, the second element in the diagonal corresponds to the amplitude of the first mode propagating away from the exit, etc. A $2N \times 2N$ diagonal selection matrix, $\boldsymbol{S}$, can be constructed to make the components of the $2N \times 1$ modal amplitude matrix, $\boldsymbol{A}$, in a specific direction, equal to the total acoustic power transmission in that direction. Hence, multiplication of $\boldsymbol{A}$ by the selection matrix $\boldsymbol{S}$ enables the calculation of the acoustic power transmission in one direction, in this case towards the duct exit. If a total of N modes (including the plane wave mode) are considered, we may define $\boldsymbol{S}$ as:

$$\boldsymbol{S} = \begin{bmatrix} s_0 & & & \\ & 0 & 0 & \\ & & \ddots & \\ & 0 & s_{N-1} & \\ & & & 0 \end{bmatrix} \tag{7.4.75}$$

where, for propagating modes,

$$s_{mn} = \sqrt{\frac{bdk_{mn}}{\rho_0 \omega}} \tag{7.4.76}$$

and for evanescent modes (or modes that are not 'cut on') and for modes that are propagating away from the duct exit, $s_{mn} = 0$. Thus every second element on the diagonal of the matrix $\boldsymbol{S}$ is zero, such that the resulting modal amplitudes are representative of the downstream acoustic power transmission, W_d, which is given by

$$W_d = \boldsymbol{A}^{\mathrm{H}} \boldsymbol{S}^{\mathrm{H}} \boldsymbol{S} \boldsymbol{A} \tag{7.4.77}$$

For minimization of the downstream power W_d, the error function F of (7.4.57) is equal to W_d, where

$$a = Z_c^{\mathrm{H}} \left(\Omega^{-1}\right)^{\mathrm{H}} S^{\mathrm{H}} S \, \Omega^{-1} Z_c \tag{7.4.78}$$

$$b_1 = Q_p^{\mathrm{H}} Z_p^{\mathrm{H}} \left(\Omega^{-1}\right)^{\mathrm{H}} S^{\mathrm{H}} S \, \Omega^{-1} Z_c \tag{7.4.79}$$

$$b_2 = Z_c^{\mathrm{H}} \left(\Omega^{-1}\right)^{\mathrm{H}} S^{\mathrm{H}} S \, \Omega^{-1} Z_p Q_p \tag{7.4.80}$$

$$c = Q_p^{\mathrm{H}} Z_p^{\mathrm{H}} \left(\Omega^{-1}\right)^{\mathrm{H}} S^{\mathrm{H}} S \, \Omega^{-1} Z_p Q_p \tag{7.4.81}$$

where Z_p relates the pressure at the measurement point $\boldsymbol{x} = (x, y, z)$ to the primary source volume velocity $\boldsymbol{Q}_p = [Q_{p1}, Q_{p2}, \cdots, Q_{pN}]^{\mathrm{T}}$ for N primary sources, by

$$Z_p = \left[\frac{p(\boldsymbol{x}, \boldsymbol{x}_{p_1})}{Q_{p_1}} \quad \cdots \quad \frac{p(\boldsymbol{x}, \boldsymbol{x}_{p_N})}{Q_{p_N}} \right] \tag{7.4.82}$$

and Z_c similarly relates the pressure at $\boldsymbol{x}$ to the control source volume velocity, $\boldsymbol{Q}_c = [Q_{c1}, Q_{c2}, \cdots, Q_{cM}]^{\mathrm{T}}$, for M control sources, such that

$$Z_c = \left[\frac{p(\boldsymbol{x}, \boldsymbol{x}_{c_1})}{Q_{c_1}} \quad \cdots \quad \frac{p(\boldsymbol{x}, \boldsymbol{x}_{c_M})}{Q_{c_M}} \right] \tag{7.4.83}$$

The power is minimized when the control source matrix, $\boldsymbol{Q}_c$ is as defined by (7.3.35). The quantity Ω relates the modal amplitudes to the acoustic pressure measurements in the duct and is defined in Section 7.5.2. In the preceding equations for any number of primary sources and M control sources, $\boldsymbol{a}$ is an M $\times$ M matrix, b_1 is 1 $\times$ M, b_2 is $M \times 1$ and c is a scalar representing the uncontrolled power.

If the duct exit is characterized by a uniform impedance, energy in a particular mode will not be converted into other modes on reflection. In this case, minimization of the power in a particular mode will not be converted into other modes on reflection. In this case, only one error sensing plane (rather than two) and correspondingly only half the number of error sensors are required. This will enable the total modal amplitude (incident + reflected) to be determined without allowing the two individual components to be resolved, giving the total power propagating both ways in the duct as

$$W_d = \sum_m \sum_n \frac{bdk_{mn} \mid A_{mn} \mid^2}{\rho_0 \omega} \tag{7.4.84}$$

The modal amplitude matrix $\boldsymbol{A}$ representing the amplitudes for N modes, is now an $N \times 1$ matrix rather than a $2N \times 1$ matrix, and the selection matrix $\boldsymbol{S}$ for converting the modal amplitudes to modal powers is an $N \times N$ matrix, with elements defined by (7.4.76). The optimum control source volume velocities required to minimize the total modal power may be calculated using (7.4.78) to (7.4.83) and equation (7.3.35).

In a practical system, the signals from the microphone array used for the modal decomposition would need to be processed at some stage to obtain an error signal proportional to the downstream acoustic power transmission. This may be performed digitally within the adaptive feedforward controller, or prior to the controller input by a separate circuit.

7.5 ACOUSTIC MEASUREMENTS IN DUCTS

In the evaluation of the feasibility of active control for a specific application and the estimation of hardware requirements, it is necessary to be able to determine the level of the sound field associated with sound propagating from the noise source to the site of the proposed control source, the level of sound reflected from the duct terminations or discontinuities and the contribution of turbulent pressure fluctuations to any measurements made using a microphone. For the analysis outlined in Sections 7.3 and 7.4, it is useful to be able to determine the reflection coefficients of duct ends or discontinuities, the total sound power propagating down the duct and the sound power radiated by both primary and control sources. All of these measurement aspects will be considered in the following subsections.

7.5.1 Duct termination impedance

The specific acoustic impedance of a duct termination is both frequency and mode dependent. That is, each mode experiences a different impedance which depends on frequency. As the modal specific impedance is directly related to the duct complex modal reflection coefficient, it is of interest to derive an expression for this quantity in terms of measurable duct parameters.

The normal acoustic particle velocity at a duct termination plane may be written as

$$u_{mn}(\boldsymbol{x}_s) = \bar{u}_{mn}\psi_{mn}(\boldsymbol{x}_s) \tag{7.5.1}$$

where $\bar{u}_{mn}$ is the velocity amplitude of the acoustic mode at the exit plane. The pressure at a point $\boldsymbol{x} = (x,y)$ on the exit plane due to mode (m, n), distribution to a first approximation is given by (Fahy, 1985):

$$p_{mn}(\boldsymbol{x}) = \frac{j\rho_0\omega}{2\pi}\int_S \frac{\bar{u}_{mn}\Psi_{mn}(\boldsymbol{x}_s)\,e^{-jkr}}{r}\,\mathrm{d}\boldsymbol{x}_s \tag{7.5.2}$$

where $r = |\boldsymbol{x}_s - \boldsymbol{x}|$ and k is the wavenumber. Note that the effect of the radiation load on the modal velocity at the duct exit has been ignored, which means that the preceding expression is only approximately true.

The modal specific acoustic impedance at the exit plane, $Z_{s_{mn}}$, is defined as

$$Z_{s_{mn}} = \frac{1}{S}\int_S \frac{p_{mn}(\boldsymbol{x})}{u_{mn}(\boldsymbol{x})}\,\mathrm{d}\boldsymbol{x} \tag{7.5.3}$$

which becomes

$$Z_{s_{mn}} = \frac{j\rho_0\omega}{2\pi S}\int_S \Psi_{mn}^{-1}(x)\left[\int_S \frac{\Psi_{mn}(x_s)\,e^{-jkr}}{r}\,dx_s\right]dx \tag{7.5.4}$$

and reduces to

$$Z_{s_0} = \frac{j\rho_0\omega}{2\pi S}\int_S\int_S \frac{e^{-jkr}}{r}\,dx_s\,dx \tag{7.5.5}$$

for the case of plane wave propagation.

In a similar problem, Morse (1948) calculated the plane wave (0,0) mode radiation impedance for a circular opening, and resolved the pressure distribution over the opening into higher order modes to determine the direct and coupling impedances for the (0,0) mode. The graphs shown indicate that the coupling impedances are small except near the cut-on frequencies of the higher order modes. Hence, in this analysis it will be assumed that the coupling impedances are negligible, which implies that a mode incident upon the duct termination will be reflected, as the same mode and will not be coupled into other modes.

Morse and Ingard (1968) show that the modal specific impedance at the termination is related to the modal impedance function by

$$Z_{s_{mn}} = \rho_0 c_0 \coth\Phi_{mn} = \rho_0 c_0 \tanh[\pi\alpha_n - j\pi(\beta_n - 1/2)] \tag{7.5.6a,b}$$

As the complex amplitude reflection coefficient R_{mn} for the mnth mode is related to the impedance function Φ_{mn} by

$$R_{mn} = e^{-2\Phi_{mn}} \tag{7.5.7}$$

The specific acoustic impedance for mode mn at the termination is related to the complex amplitude reflection coefficient as

$$Z_{s_{mn}} = \rho_0 c_0\left[\frac{1+R_{mn}}{1-R_{mn}}\right] \tag{7.5.8}$$

where R_{mn} is the complex ratio of the reflected to the incident mnth modal amplitudes.

If the duct termination is radiating into a space the sound power radiated may be calculated using the complex radiation efficiency defined by

$$\sigma_{mn} = \frac{W_{mn}}{<u_{mn}^2>_{st} S\rho_0 c_0} = \frac{Z_{s_{mn}}}{\rho_0 c_0} = \frac{1+R_{mn}}{1-R_{mn}} \tag{7.5.9a,b,c}$$

where W_{mn} is the complex power at the opening due to mode mn radiating by itself (the real part of which propagates away from the opening) and $<u_{mn}^2>_{st}$ is the mean square velocity at the duct termination, averaged in time and over the area S of the termination. Note that this relationship is only meaningful if a single mode only exists at the duct exit, as power radiated by individual modes cannot be added to give the total power radiated by several modes radiating simultaneously; only the complex far field sound pressures due to each mode can be added, and then these used to determine the total radiated sound power.

From (7.5.7) it is clear that the impedance function Φ_{mn} characterizing the duct termination impedance can be calculated if the amplitudes of the reflected and incident waves at the duct termination can be measured. In the next section, means of measuring the amplitude of these waves at some location within the duct will be discussed. Once this is done, the reflection coefficient of the duct termination is given by

$$R_{mn} = e^{-2\Phi_{mn}} = \frac{A_{mn_r}}{A_{mn_i}} e^{j2k_{mn}(L - z_1)} \qquad (7.5.10a,b)$$

where $(L - z_1)$ is the distance from the duct termination to the measurement plane, and A_{mni} and A_{mnr} are respectively, the complex modal amplitudes of the incident and reflected waves determined at the measurement plane z_1.

7.5.2 Sound pressure associated with waves propagating in one direction

For tonal noise and only plane waves, the sound pressure associated with the wave propagating towards the duct exit can be determined by measuring the maximum and minimum sound pressure associated with the standing wave in the duct. The mean square sound pressure associated with the wave propagating towards the duct exit is then

$$p^2 = p_{min} \times p_{max} \qquad (7.5.11)$$

where p_{min} and p_{max} are rms quantities at the frequency of interest.

For broadband random noise, the sound pressure associated with the wave propagating towards the duct exit can be calculated by measuring the cross correlation function $R_{12}(\tau)$ between two microphones well separated axially in the duct (Shepherd *et al.*, 1986b). Two peaks will appear in the function $R_{12}(\tau)$ at delays τ equal to the sound propagation times between points $\pm z/c$ where z is the microphone separation distance. All parts of the function $R_{12}(\tau)$ except the peak of interest are then edited out and the result is Fourier transformed to give the power spectral density of the noise propagating away from the source.

When higher order modes are propagating in the frequency range of interest, it is desirable to be able to determine the contributions of each mode to the propagating wave. This will also allow the determination of the total pressure wave amplitude in each duct segment when a duct is divided into sections which are small enough to only allow plane wave propagation in the frequency range of interest. If a multichannel controller is contemplated to control the higher order modes directly, then it is important to know which modes require the most attenuation.

One method of determining the contribution from each higher order mode involves measurement of the cross spectrum between two microphones located in the same duct cross section at various different positions (Bolleter and

Crocker, 1972; Shepherd *et al.*, 1986b).

Alternatively, transfer function measurements may be made between a single reference microphone and another microphone (or number of microphones) which is moved from point to point over a particular duct cross section (Åbom, 1989). As will be seen later, the effect of turbulent pressure fluctuations is limited by restricting both the reference and scanning microphone locations to a single duct cross section. However, if reflected waves are present, measurements over two cross sections are necessary to resolve the amplitudes of the direct and reflected waves. As discussed in Section 7.4.5, it is sometimes unnecessary to resolve the direct and reflected amplitudes, and only a single amplitude is needed for each mode. In this case, measurements may be restricted to a single duct cross section (with only half the total number of sensors as used over two cross sections), and the result will be the sum of the incident and reflected amplitudes for each mode. Note that to simplify the process and avoid the influence of evanescent waves, measurements should be made in the far field of any noise sources or duct discontinuities.

The sound pressure at any location (x,y,z) in the duct may be written as

$$p(x,y,z,t) = \sum_m \sum_n \left[A^+_{mn}\, e^{-jk^+_{mn}z} + A^-_{mn}\, e^{-jk^-_{mn}z}\right]\psi_{mn}(x,y)e^{j\omega t} \tag{7.5.12}$$

where, if no flow is present, k_{mn} is defined by (7.3.5), and in the presence of a mean flow of Mach number M, k_{mn} is given by

$$k_{mn} = \frac{[(k^2 - \kappa^2_{mn})(1 - M^2)]^{1/2} - kM}{1 - M^2} \tag{7.5.13}$$

The quantity M is defined as positive in the direction of k^+ wave propagation and negative in the direction of k^- wave propagation. κ_{mn} is defined in (7.3.5). A^+_{mn} and A^-_{mn} are the complex modal amplitudes characterizing the waves propagating to the right and left respectively, and ψ_{mn} is the mode shape function.

In the frequency domain (taking the Fourier transform of (7.5.12)) we can write for the acoustic pressure amplitude at frequency ω,

$$p(\boldsymbol{x},\omega) = \sum_m \sum_n \left[A^+_{mn}(\omega)\, e^{-jk^+_{mn}z} + A^-_{mn}(\omega)\, e^{-jk^-_{mn}z}\right]\psi_{mn}(x,y) \tag{7.5.14}$$

where $\boldsymbol{x} = (x,y,z)$. In the preceding equation, all modes above cut-on as well as any modes below cut-on which have significant levels at the measurement positions should be included. If N modes are included, then a minimum of $2N$ transfer function measurements will be needed to determine the modal amplitudes of waves propagating in both directions along the duct. Note that the measurements should be independent. Using a matrix formulation, we can write (for a single frequency ω)

$$\bar{p} = \Omega A \tag{7.5.15}$$

where $\bar{p}$ is a $2N \times 1$ matrix containing the complex acoustic pressure measurements derived from the transfer function measurements using

$$\bar{p} = \bar{p}_1 \boldsymbol{H} \tag{7.5.16}$$

where $\bar{p}_1$ is the acoustic pressure amplitude at the reference location and H is the $2N \times 1$ transfer function matrix representing the transfer function between $\bar{p}_1$ and the sound pressure at the measurement locations. For convenience, a minimum of two cross-sectional planes in the duct can be used with the first N measurements taken in the first plane and the measurements $N + 1$ to $2N$ taken in the second plane.

In (7.5.15), Ω is a $2N \times 2N$ matrix which represents the transfer function between two measurement planes. Each row corresponds to the modal contributions at the ith microphone location, with each element in the row representing the contribution from one modal component. Thus, row i of the matrix has the form

$$\left[\psi_{oo}(x_i, y_i)e^{-jk_{oo}z_i} \; \psi_{oo}(x_i, y_i)e^{jk_{oo}z_i} \; \ldots\ldots \; \psi_{mn}(x_i, y_i)e^{-jk_{mn}z_i} \; \psi_{mn}(x_i, y_i)e^{jk_{mn}z_i}\right] \tag{7.5.17}$$

where the oo subscript corresponds to the plane wave and where z_i is the axial distance between the first measurement plane and the measurement point.

The quantity, $\boldsymbol{A}$ is a $2N\times1$ matrix containing the modal amplitudes of the incident and reflected waves at the first measurement plane. Thus,

$$\boldsymbol{A} = \left[A_{oo_i} A_{oo_r} \ldots\ldots\ldots A_{mn_i} A_{mn_r}\right]^{\mathrm{T}} \tag{7.5.18}$$

Rearranging (7.5.15) provides a solution for the modal amplitudes as follows:

$$\boldsymbol{A} = \boldsymbol{\Omega}^{-1}\bar{p} \tag{7.5.19}$$

If all of the measurements are not independent at the frequency of interest, then the matrix $\boldsymbol{\Omega}$ will be singular and will not be invertible to give $\boldsymbol{A}$. One way around this is to take more measurements M than modes to be resolved. In this case, $\bar{p}$ will be an $M\times1$ and Ω will be an $M\times N$ matrix. Equation (7.5.19) can then be written as,

$$\boldsymbol{A} = \left[\boldsymbol{\Omega}^{T}\boldsymbol{\Omega}\right]^{-1}\bar{p} \tag{7.5.20}$$

The preceding analysis applies equally well to circular or rectangular section ducts, provided the correct mode shape functions ψ_{mn} are used. For rectangular section ducts this function is defined by (7.3.3).

If N pressure measurements are taken on only one plane (as discussed earlier), so that a single amplitude is obtained for each mode (representing the sum of amplitudes corresponding to both directions of propagation), then $\boldsymbol{A}$ becomes an $N \times 1$ matrix, given by

$$A = [A_{00}, \cdots, A_{mn}]^{\mathrm{T}} \tag{7.5.21}$$

and the ith row of the $N \times N$ matrix Ω (representing the ith measurement location) becomes

$$\Omega = [\psi_0(x_i, y_i), \cdots, \psi_{mn}(x_i, y_i)]^{\mathrm{T}} \tag{7.5.22}$$

Equation (7.5.19) is still used to find the N modal amplitudes represented by the matrix, A, but the matrices are of order N, rather than $2N$.

7.5.3 Turbulence measurement

It is important to be able to determine the contribution of turbulent pressure fluctuations to acoustic signals from in-duct microphones, to be able to evaluate the potential noise reduction which can be achieved with an active controller. At best, the controller will only be able to reduce the acoustic noise to 3 dB below the turbulent pressure fluctuations sensed by the reference microphone (Shepherd *et al.*, 1986b), as the acoustic signal output by the control loudspeakers to cancel the turbulent pressure fluctuations will instead add to the residual acoustic noise at the error sensor.

The cross-spectrum measured with two microphones is given by

$$S_{12}(j\omega) = S_1^*(j\omega)S_2(j\omega) = \left[S_{M_1}(j\omega) + S_{T_1}(j\omega)\right]^* \left[S_{M_2}(j\omega) + S_{T_2}(j\omega)\right] \tag{7.5.23a,b}$$

where $S_{M_1}(j\omega)$ and $S_{M_2}(j\omega)$ are the complex spectra of the acoustic components and $S_{T_1}(j\omega)$ and $S_{T_2}(j\omega)$ are the complex spectra of the turbulence components. If the microphones are sufficiently far apart that the turbulent signals do not correlate, then (7.5.23) can be written as

$$S_{12}(j\omega) = S_{M_1}^*(j\omega)\, S_{M_2}(j\omega) \tag{7.5.24}$$

If the microphone locations are chosen (for example in a single plane in a duct containing only plane waves) such that

$$|S_{M_1}| = |S_{M_2}|$$

then

$$|S_{12}| = |S_{M_1}|^2 = |S_{M_2}|^2 = |S_M|^2 \tag{7.5.25a,b,c,d}$$

where $j\omega$ has been omitted for brevity. The turbulent component of the total spectrum is then

$$|S_{T_1}| = |S_1| - |S_{12}| \tag{7.5.26}$$

where S_{T_1} S_M and S_1, are respectively the turbulence spectrum component, acoustic spectrum component and total spectrum measured at microphone location 1.

Equation (7.5.25) is always satisfied for two points on the same cross-section with only plane waves propagating. If higher order modes are present, then (7.5.25) is still satisfied for symmetric locations in a cross section when there is no cross-modal correlation (a reasonable assumption according to Bolletter and Crocker, 1972).

If one microphone can be positioned outside the duct, then the proportion of the in duct signal which is attributable to acoustics is given by the coherence function

$$\gamma^2 = \frac{|S_{12}|^2}{S_{11} S_{22}} \tag{7.5.27}$$

where S_{11} and S_{22} are the auto (or power) spectra at locations 1 (in the duct) and 2 (outside the duct) respectively (see Chapter 4). In computing γ^2 care must be taken to ensure that the time window is large enough to allow for the propagation delay between the two microphones. Alternatively the signal from the microphone in the duct could be delayed by an appropriate amount.

7.5.4 Total power transmission measurements

For plane waves only, the total acoustic power W_a propagating down the duct is obtained using (7.5.11) as follows:

$$W_a = \frac{\rho_0 c_0}{S} p_{max} p_{min} \tag{7.5.28}$$

where S is the duct cross-sectional area and $\rho_0 c_0$ is the characteristic impedance of the gas in the duct.

7.5.5 Measurement of control source power output

Measuring the power contributions from each source when all sources are operating requires a more sophisticated approach. Attempting to measure (the changes in) acoustic power radiation by measuring the electrical power supplied to the source is extremely difficult and has yet to be done successfully, as the electrical power (in watts) is several orders of magnitude larger than the acoustical power (in microwatts). What follows is a description of a method for directly measuring the active acoustic power output, which is equal to the product of the cone volume velocity and the in-phase part of the acoustic pressure adjacent to the cone. The acoustic pressure in the duct adjacent to the cone can be measured by using a suitably located microphone. The volume velocity of the speaker cone can be determined by enclosing the back of the speaker in a small box, measuring the pressure p_i in the box, and using the following expression for the acoustic impedance of a small volume (Bies and Hansen, 1996):

$$Z_v = \frac{p_i}{Q} = -j\frac{\rho_0 c_0^2}{V\omega} \tag{7.5.29a,b}$$

where p_i is the acoustic pressure measured inside the box, V is the volume of the box, and ω is the angular frequency. The phase between the cone volume velocity and the acoustic pressure in the duct at the cone face is 270° greater than the measured phase between the acoustic pressure in the speaker box and the acoustic pressure in the duct (as there is a 90° phase difference between acoustic pressure and acoustic volume velocity in the box, and a 180° phase difference between the acoustic volume velocity on the top and bottom of the speaker cone).

Accurate measurement of the phase difference between the acoustic volume velocity and pressure at the speaker cone face is essential when measuring acoustic power output. In practice, it has been found that the phase is essentially uniform throughout the small enclosure. However, phase varies quite dramatically as the microphone in the duct is moved away from the front of the speaker cone. It is crucial that the microphone be positioned as close to the cone as possible for accurate measurements. Pressure also varies across the face of the speaker (on the duct side), even at frequencies at which wavelengths are much greater than the speaker diameter.

Thus, it is necessary to position the measuring microphone at the correct lateral position across the speaker face to correctly determine the sound power. This can be achieved by adjusting the microphone location on a trial and error basis, with just one sound source operating, measuring the resulting pressure distribution in the duct, and comparing the power transmission determined from (7.5.28) to that determined from the measured volume velocity and acoustic pressure at the speaker cone face (Snyder and Hansen, 1989).

Figures 7.41(a) and (b) show the amplitude and phase variation across one diameter of a control source (200 mm speaker) mounted in the wall of a 0.2 m × 0.2 m cross-section duct anechoically terminated at one end and terminated with another speaker at the other end. In viewing these it can be seen that there is a significant phase and amplitude variation across the speakers, especially near the edges. This is not surprising, as the cone radius is actually approximately 10 mm shorter than the overall speaker radius, with the outer 10 mm being a flexible rubber strip. In the centre region, however, the phase and amplitude are reasonably constant. Note that for a source mounted in a duct wall, both upstream and downstream measurements are required to find the total power transmission.

The acoustic impedance Z 'seen' by the control source before and during control can be found simply by replacing p_i in (7.5.29) with the pressure p_c measured on the centre of the face of the duct side of the control source. Thus,

$$Z = \frac{p_c}{Q} = \frac{-j\rho_0 c_0^2 p_c}{V\omega p_i} \tag{7.5.30a,b}$$

Clearly this is only applicable for plane wave propagation.

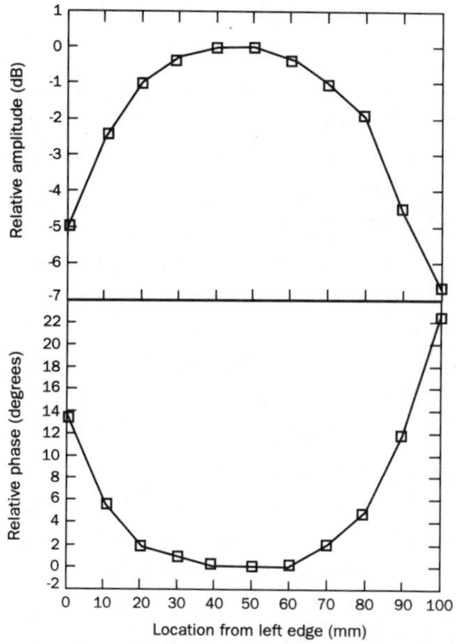

Fig. 7.41 Variation in pressure amplitude measured across a 200 mm diameter loudspeaker, (with its back enclosed in a small cavity) at a distance of 5 mm in front of the cone, for a driving frequency of 400 Hz, with the speaker mounted in the wall of a 0.2 m × 0.2 m cross-section duct: (a) pressure amplitude variation; (b) pressure phase variation.

7.6 SOUND RADIATED FROM EXHAUST OUTLETS

The active control of sound radiated from exhaust outlets is similar to the control of noise propagating in ducts except that the control sources and error sensors are located outside of the duct. A feedforward system for controlling diesel engine harmonic noise was first proposed by Chaplin (1980) and involved a single loudspeaker mounted at the exit of the exhaust system as shown in Fig. 7.42. The reference signal was obtained from a tachometer directed at the engine cam shaft.

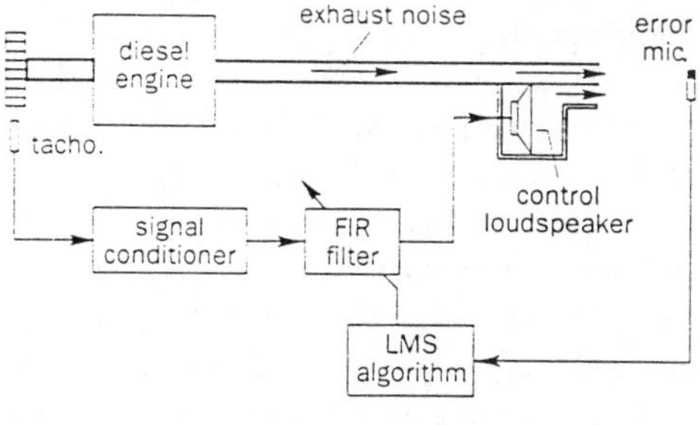

Fig. 7.42 Active control of exhaust noise (after Chaplin, 1980).

In practice, better results are obtained if more than a single control source is used. For example use of another speaker above the exhaust pipe shown in Fig. 7.42 would result in a much less efficient longitudinal quadrupole rather than the dipole formed with just a single sound source. This type of system was demonstrated by Trinder *et al.* (1986) for control of motor vehicle noise.

Kido *et al.* (1987, 1989) have investigated the control of random noise using a feedforward arrangement similar to that illustrated in Fig. 7.42. The only difference is that the reference signal is taken from a microphone mounted in the duct. The authors use bends in the duct between the reference microphone and duct exit to minimize acoustic feedback, although the usefulness of the bends in preventing feedback is questionable for low frequency noise.

The physical mechanism associated with the use of a control source adjacent to a duct exit is that the control source acts to change the simple monopole source to a less efficient dipole. The use of two control sources on opposite sides of the duct exit results in an even less efficient quadrupole. Clearly this mechanism will only be effective if the duct exit size is small compared to a wavelength of sound. As shown by Bies and Hansen (1996, Chapter 5), the maximum theoretical sound power attenuation Δ which can be achieved by changing a point monopole source to a dipole is given by

$$\Delta = 10\log_{10}\left[2/(2kd)^2\right] \tag{7.6.1}$$

where $k = 2\pi/\lambda$ is the wavenumber of sound at the frequency of interest and d is the distance between the two point sources. In practice, the sources are of finite size and d is the distance between the centre of the duct exit and the centre of the control sources. The effect of the finite size sources is to limit the maximum achievable power reduction to a little less than the theoretical optimum for point sources (see Chapter 8 for a more detailed discussion of monopole source control).

Changing a monopole to a longitudinal quadrupole results in a maximum theoretical reduction Δ in sound power given by

$$\Delta = 10\log_{10}\left[5/(2kd)^4\right] \tag{7.6.2}$$

The resulting radiation patterns are typical of those exhibited by dipoles and quadrupoles respectively, with the minimum sound pressure being directly along the axis of the exhaust pipe for the configuration shown in Fig. 7.42 (Bies and Hansen 1996, Chapter 5). Larger noise reductions than the theoretical maximum reductions outlined above have been achieved by placing the control source so that it faces the duct as shown in Fig. 7.43(a) (Hall *et al.*, 1990).

The reason for the configuration shown in Fig. 7.43(a) being more effective is because the control source increases the effective reflection coefficient of the end of the duct, as well as changing the source from a monopole to a dipole. However, this configuration is not very practical, as it adds considerably to pressure losses of air flowing out of the duct; thus, the slightly less efficient configuration shown in Fig. 7.43(c) is generally preferred.

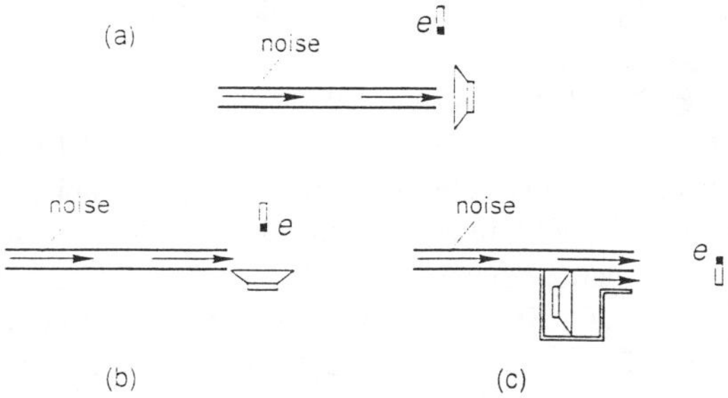

Fig. 7.43 Configurations for active control of noise radiated from the exit of a duct: (a) configuration for maximum attenuation; (b) configuration for least attenuation; (c) most practical configuration. The optimal location for the error sensor is shown in each case.

For all configurations, the best location for a single error microphone was found to be along the line of expected pressure minimum of the resulting dipole radiation. This location is shown in each of the figures. Generally, sound power reduction results are independent of the error microphone distance from the sources, although the results are less sensitive to errors in the angular location of the microphone as distance from the sources is increased.

For very large duct exit sections, where the product kd exceeds 0.5 (k is the wavenumber and d is the duct exit diameter), it is unlikely that significant global noise reductions will be achieved using this technique, although significant local reductions in directions corresponding to error sensor locations will still be achievable provided that enough control sources are used to surround the duct exit.

It is interesting to note that the upper frequency limit imposed by the requirement that $kd < 0.5$ is lower than the cut on frequency for higher order mode propagation in the duct. Thus, it appears that where possible, the control source should be placed inside the duct as invariably better results will be obtained. As mentioned earlier, if the error sensor is then located outside of the duct, more than one may be needed to ensure that global noise reduction is achieved, rather than just reduction in the direction of the error sensor location. When more than one error sensor is used, the control system acts to minimize the sum of the instantaneous squared pressures detected at all of the error sensors.

Unfortunately, when attempting to actively control internal combustion engine noise, it is generally impractical to install the sound sources inside the exhaust pipe or in the pipe wall, because it is difficult to find sound sources capable of withstanding the hot and dirty environment. It is also difficult for conventional sound sources to generate the required high sound levels (typically 170–180 dB) needed in the exhaust pipe for control (in fact, about 8 kW of electrical power is needed in a typical installation). If air modulated valves are used as sound sources (with no attention paid to optimizing the acoustical efficiency), experience has shown that the large amount of compressed air needed (equivalent to the output from two large portable compressors as would be used to drive jackhammers), makes this option impractical. Thus, for engine exhaust silencing, it is more practical to locate the control sound sources outside the exhaust exit. Holt (1993) reported that for a typical production car exhaust, the volume displacement required of a control source at the exhaust exit to achieve significant noise reductions at significant engine loads was equivalent to that produced by a 400 mm diameter loudspeaker. Holt also reported on the use of a Ling electropneumatic valve which resulted in significant noise attenuations at frequencies above 100 Hz, but very little at lower frequencies. He also reported on the use of a rotating valve compressed air source which was limited in that it could only control a single frequency at any one time.

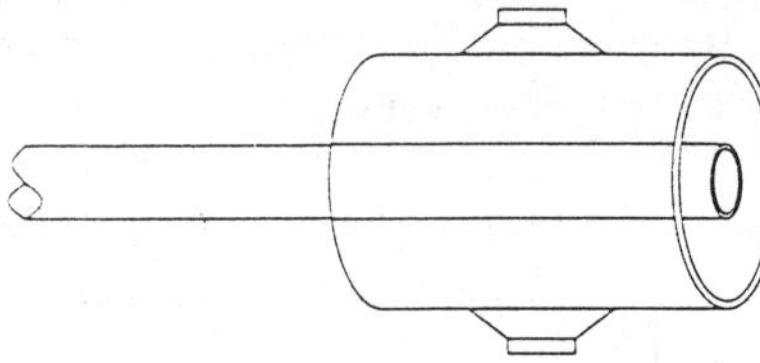

Fig. 7.44 Vehicle exhaust muffler with concentric cylinder to contain and direct the control sound.

An alternative control source arrangement to those outlined above and one which is especially suited to engine exhaust silencing using feedforward control, is illustrated in Fig. 7.44 (after Foller, 1992). It consists of a second larger diameter circular duct which is concentric with the one carrying the primary noise. Controlling sound is introduced into the second duct using horn drivers or loudspeakers as shown in the figure. This controlling sound interacts with the primary sound at the duct exit. For $kd > 0.5$, (where d is the diameter of the exit of the outer pipe), the performance of the active silencer will be severely reduced, as the exiting sound can no longer be assumed to be a simple spherical wave. Replacing the outer concentric pipe with a horn such as a catenoidal horn will make the controlling sound generation more efficient, thus reducing the power requirements of the loudspeakers or horn drivers.

Overall, it seems likely that the advantages (such as less back pressure and reduced fuel consumption) associated with using active silencing in engine exhausts will only outweigh the disadvantages (increased cost and complexity) for trucks and buses and not for standard passenger cars. The performance of active mufflers is similar to or better than passive mufflers at low frequencies (10 to 12 dB attenuation), but not as good at higher fequencies (greater than about 800 Hz). Thus, a practical installation would probably consist of a mixture of both; active silencing for low frequencies and a straight through, low pressure drop, dissipative, passive silencer for high frequencies.

Feedback control of motor vehicle noise was investigated by Mori *et al.* (1991) who used a feedback control structure which they described as a tight-coupled monopole using a similar control source arrangement to that illustrated in Fig. 7.44. The feedback error microphone is located adjacent to the control loudspeakers and as this is a feedback system, no reference microphone is used. The performance of this muffler in an actual exhaust system is compared to that of a conventional muffler in Fig. 7.45, where it can be seen that the exhaust noise is similar for the two types of muffler. The advantage of the active muffler is reduced back pressure which translates into reduced fuel consumption. The authors estimate a reduction in fuel consumption of between 5 and 10%, but this is probably a little optimistic; 2 to 3% would be more realistic. This type of muffler is still not available in a production vehicle probably because of the difficulty in protecting the components from the high temperatures present over a life of many years.

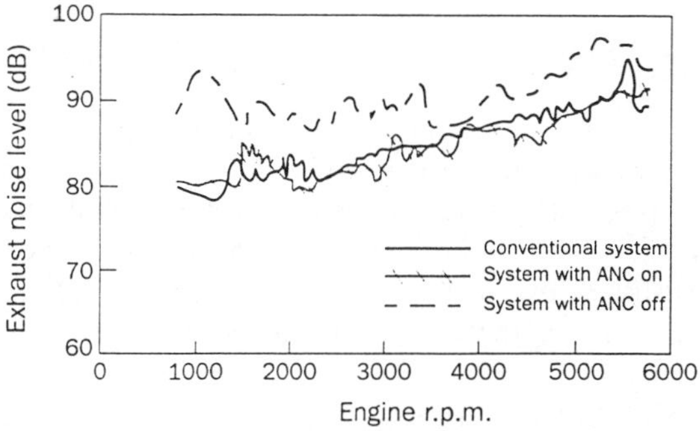

Fig. 7.45 Performance of an active car exhaust muffler compared to the performance of a passive muffler (after Mori *et al.* 1991).

Semi-active exhaust mufflers have also been designed and used in automobiles (Holt, 1993). They consist essentially of switched valves which allow the passive exhaust system arrangement or muffler internal arrangement to be changed depending on the engine operating conditions (speed, load, exhaust back pressure). Although they are much less expensive than active mufflers to manufacture and simpler to install, and their performance is similar, they exhibit similar pressure drops to passive mufflers.

7.7 CONTROL OF PRESSURE PULSATIONS IN LIQUID FILLED DUCTS

The principles of active control of noise propagating in liquid filled ducts are much the same as those for air ducts (Culbreth *et al.*, 1988). However, the very much higher speeds of sound in liquids means that higher order modes do not become significant until the frequency or duct size is increased by approximately five-fold over the corresponding limits for an air duct. For ducts with flexible walls, this difference is much less; also, considerable care must be exercised to minimize the contribution of pressure waves transmitted directly to the sound sensors through a mechanical connection or by being reradiated from the walls into the liquid in the duct.

The other significant difference between control systems for liquid filled ducts and air ducts is associated with the types of control sources and sound sensors which are used. Microphones, which are used in air ducts are replaced by hydrophones for liquid filled ducts. Control sources which are commercially available include sonar sources (for water filled ducts). Alternatively, sources could be constructed by using mechanical shakers, piezoelectric actuators or magnetostrictive actuators to drive a thin diaphragm located in a tube attached to the pipe wall. The diaphragm material would obviously need to be chemically inert to the liquid in the pipe.

One potential industrial application of this type of system would be in reducing fluid borne pulsations in pipework attached to reciprocating compressors.

7.8 ACTIVE HEADSETS AND HEARING PROTECTORS

It is well known that conventional passive hearing protectors are not very effective in protecting the wearer from low frequency noise, and that communication using standard headsets in noisy areas is extremely difficult. Both active headsets and active hearing protectors enhance hearing protection at low frequencies (usually below 1500 Hz). Active hearing protectors differ from active headsets in that the former include passive elements to further attenuate high frequency sound (above 250 Hz), and the latter allow radio communication to be heard clearly. As the principles of operation of active headsets and active hearing protectors are similar, the two devices will be treated together here.

Active headsets and hearing protectors are included in this chapter on active control of noise propagating in ducts, as plane waves in ducts and the sound field in a hearing protector are both one-dimensional problems, thus enabling good results to be achieved with a single channel control system.

The need for increased performance of passive hearing protectors in a number of applications is well established, especially in the low frequency range where the performance is particularly poor. In addition to the needs in noisy industries such as sheet metal and forging, better hearing protectors are needed for occupants of tracked military vehicles and military aircraft, as is demonstrated in Fig. 7.46.

From Fig. 7.46, it can be seen that for the two applications shown, an active system is required to enhance the performance of the passive protectors by a maximum of 19 dB in the 63 Hz octave band to a minimum of 8 dB in the 1000 Hz octave band. Enhancement at higher frequencies is not necessary for these applications.

Although a number of researchers have been working on the development of active hearing protectors for some time, almost all of the work reported has been in the form of brief conference papers and patents rather than detailed journal papers. This lack of detailed reporting of particular designs probably stems from the commercial sensitivity of the work. However, two detailed studies have been recorded in the doctoral theses of Wheeler (1986) and Carme (1987), and these along with the paper by Salloway and Twiney (1985) provide some useful insights. In this section, various designs of hearing protectors and

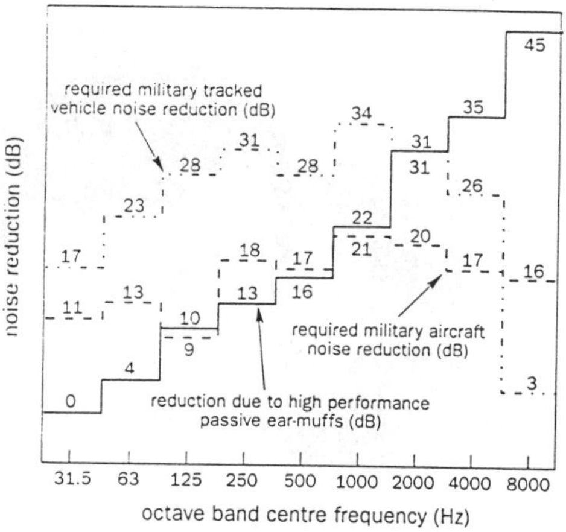

Fig. 7.46 Required performance and actual performance of passive hearing protectors for two applications (data extracted from Tichy *et al.* (1987))

head sets will be discussed with reference to their principles of operation, practical implementation problems and potential performance.

There are two main types of control system for active hearing protectors and headsets; feedback and feedforward. Each will be described in detail in the following sections and the associated advantages and disadvantages will be discussed.

7.8.1 Feedback systems

A typical active feedback system for hearing protectors is illustrated in Fig. 7.47. In practice, it is necessary to place the microphone as close as possible to the ear canal as this is where the sound pressure will be minimized.

The design challenge is to develop a compensation filter *C* which allows the gain *K* to be large without causing the system to become unstable. Before discussing how this may be done, the system in Fig. 7.47 will be analysed so that the sound pressure at the microphone after control can be expressed in terms of the compensating filter transfer function *C*.

The total complex sound pressure $p_t(\omega)$ at the microphone at frequency ω with the control system in operation is given by

$$p_t(\omega) = p_p(\omega) + Kp_t(\omega)\,C(j\omega)H(j\omega) \tag{7.8.1}$$

where $p_p(\omega)$ is the sound pressure without the control system operating, *K* is the gain of the amplifier, $C(j\omega)$ is the transfer function of the filter at frequency ω and $H(j\omega)$ is the combined transfer function of the speaker, hearing protector cavity and microphone at frequency ω. Note that all terms in (7.8.1) are complex; that is, they are characterized by an amplitude and a relative phase.

Equation (7.8.1) may be rearranged to give

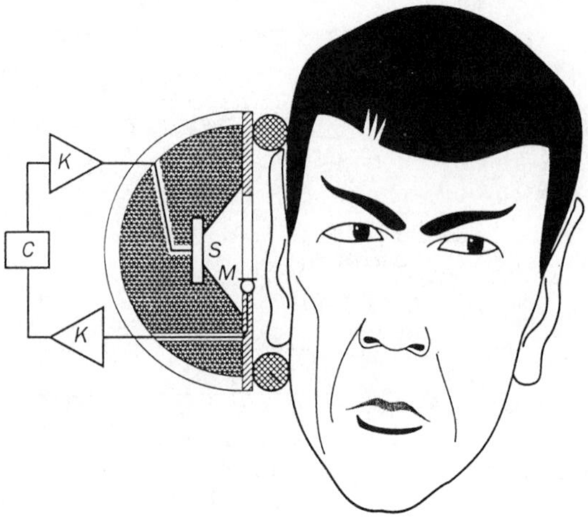

Fig. 7.47 Feedback control system for an active hearing protector; K = amplifier gain, S = speaker, M = microphone, C = compensation filter.

$$p_t(\omega) = p_p(\omega)/[1 - KC(j\omega)H(j\omega)] \qquad (7.8.2)$$

Thus $| p_t(\omega) |$ approaches zero (maximum control) when $| 1\text{-}KC(j\omega)H(j\omega) |$ becomes very large. This is usually achieved by making K large.

It is well known that a control system such as the one described by (7.8.2) will become unstable if the Nyquist plot of the function $KC(j\omega)H(j\omega)$ encloses the point $(1, j0)$. (Note that the Nyquist plot is a plot of the real part of $KC(j\omega)H(j\omega)$ vs. the imaginary part of this function for increasing values of ω, as is discussed in many basic control books.) Also, the system will be unstable if the overall loop gain is greater than unity when the phase shift is $-360°$. A more detailed discussion of these stability criteria is provided by Nelson and Elliott (1992). As shown in Fig. 7.48, the phase of the transfer function $H(j\omega)$ decreases steadily as the frequency increases. Thus, to avoid violating the stability criteria, it is necessary to introduce a compensating filter which will amplify the low frequency loop gain and attenuate the high frequency gain; that is, a low pass filter. However, the filter must not add too much to the loop phase shift which would result in $-360°$ being reached at a lower frequency.

A feedback system such as the one shown in Fig. 7.47 was first suggested by Dorey *et al*. (1975), although problems with instability were reported and the feasibility of attenuation of random noise was not established. The work was directed at developing a headset suitable for aircrew, and for this reason it also involved the introduction of a communications signal between the compensating filter and the amplifier. Equation (7.8.2) for this system then becomes

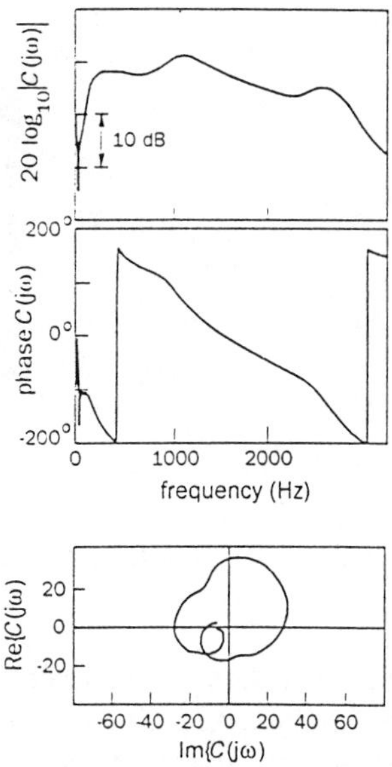

Fig. 7.48 Measurements of the uncompensated electro-acoustic frequency response function made by Carme (1987) on a prototype active ear defender.

$$p_t(\omega) = \frac{p_p(\omega)}{[1 - KC(j\omega)H(j\omega)]} + \frac{S_p(\omega)C(j\omega)L(j\omega)}{[1 - KC(j\omega)H(j\omega)]} \tag{7.8.3}$$

where $S_p(\omega)$ is the introduced communication signal and $L(\omega)$ is the loudspeaker transfer function. As the communication signal is clearly affected by the feedback loop, it must be conditioned prior to being introduced, so that this effect may be minimized.

An alternative location for injection of the communications signal is between the microphone and the compensation filter (Carme, 1988). In this case, (7.8.2) becomes

$$p_t(\omega) = \frac{p_p(\omega) + KS_p(\omega)\,C(j\omega)\,H(j\omega)}{1 - KC(j\omega)\,H(j\omega)} \tag{7.8.4}$$

A third alternative arrangement is illustrated in Fig. 7.49 which is similar to the first case cited above except that the radio signal is removed from the microphone signal prior to it going to the compensation filter.

For large values of the gain K, (7.8.4) can be written as

$$|\,p_t(\omega)\,| = \epsilon + S_p(\omega)\,, \text{ where } \epsilon << 1 \tag{7.8.5}$$

Thus, not only is the communications signal $S_p(\omega)$ not affected by the compensation filter, it is also free from the distortion usually caused by the transfer function of the loudspeaker and cavity, thus resulting in a much clearer and more easily heard signal.

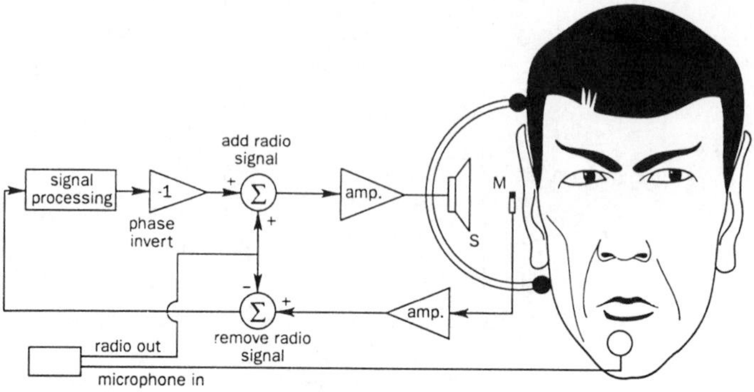

Fig. 7.49 Alternative arrangement for a feedback controlled headset intended for radio communication.

In 1978, Wheeler *et al.* reported on an improvement of the previous system which was effective for random noise and which demonstrated the feasibility of introducing the communication signal in the way described above. Neither of the above papers provided any details of the design of the compensation filter C, although in his thesis of 1986, Wheeler described and used a compensating filter which minimized the added phase shift, attenuated the high frequencies and had a transfer function (or strictly speaking, the frequency response) given by

$$C(j\omega) = \frac{(j\omega - z)}{(j\omega - p)} \tag{7.8.6}$$

where z and p are the corresponding real zero and real pole of the first order filter. A plot of the response of this filter (together with the gain k of the amplifier) used by Wheeler is shown in Fig. 7.50.

Placing z in the negative or left plane of a pole zero diagram will minimize the phase of $C(j\omega)$ while providing the required reduction in gain at higher frequencies. The pole is also placed in the left plane to ensure stability. Note that in the high frequency limit this filter has a zero phase effect which has the desired effect of not lowering the frequency at which the total loop phase shift is $-360°$.

Realization of this type of filter in practice can be achieved simply by using an analogue circuit, such as a standard 'transient lag network' where appropriate values are chosen for the two resistors and capacitor making up the circuit.

Carme (1987) concluded that for practical purposes, the most appropriate filter had the frequency response function given by

$$C(j\omega) = \frac{(j\omega - z)(j\omega - z^*)}{(j\omega - p)(j\omega - p^*)} \tag{7.8.7}$$

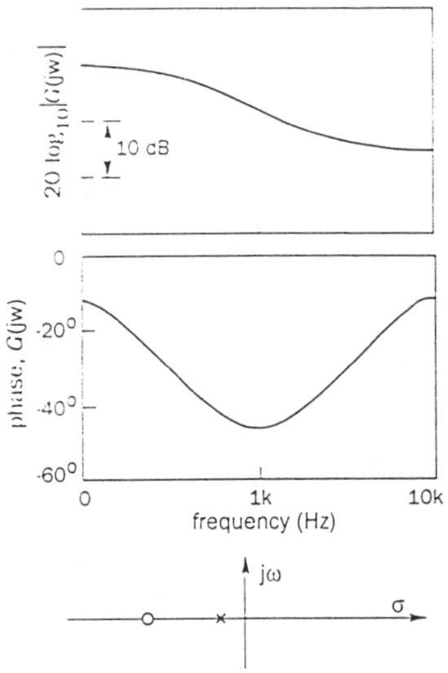

Fig. 7.50 Frequency response function of a first order compensator used by Wheeler (1986).

Note that the * denotes the complex conjugate. Carme adopted the standard 'bi-quad' realization of this frequency response function where again the pole and zero locations are determined by the appropriate selection of simple electronic components. The amplitude and phase of the function defined by (7.8.7) are shown in Fig. 7.51. Typical reductions in noise obtained by Carme with the compensating filter described by (7.8.4) are shown in Fig. 7.52.

In 1988, Carme introduced an optimal filter design which could be optimized for a particular input noise spectrum. The cost function which he minimized is given by:

$$E = \left[\sum_{i=1}^{I} \left| 1 - K C(j\omega_i) H(j\omega_i) \right|^2 \right] e^{\alpha(R_{max} - \beta)} \tag{7.8.8}$$

where the exponential term is used to prevent continued parameter optimisation if the process is heading towards an unstable solution. The quantity R_{max} defines a stability constraint and the constants α and β are used to adjust its influence on the optimization process. The modulus term in (7.8.8) is optimized by adjusting the compensation function $C(j\omega)$ which is defined by

$$C(j\omega) = \frac{\sum_{n=1}^{N} A_n (j\omega)^{n-1}}{\sum_{m=1}^{M} B_m (j\omega)^{m-1}} \tag{7.8.9}$$

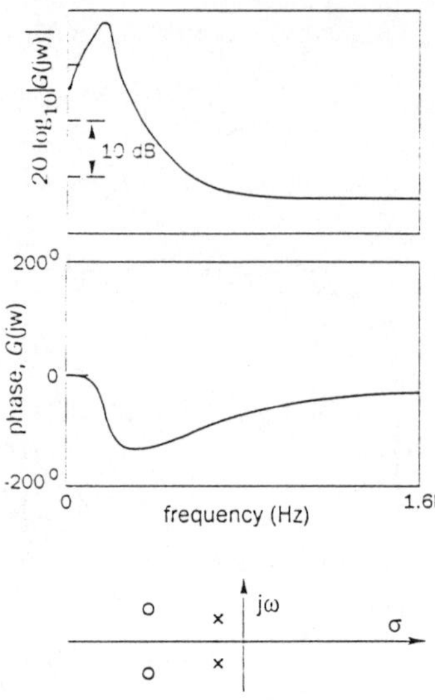

Fig. 7.51 Frequency response function and pole zero map of the second order filter used by Carme (1987).

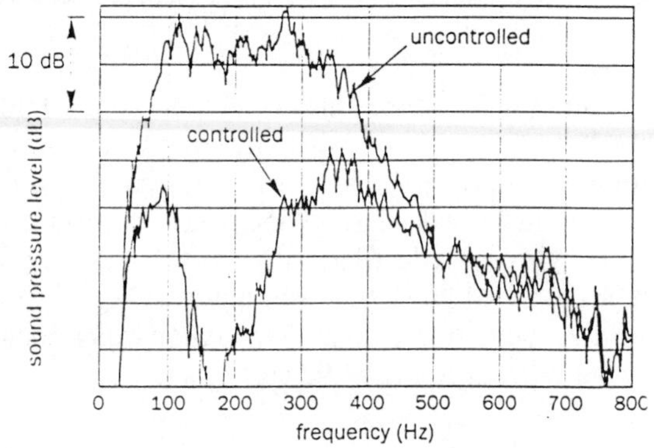

Fig. 7.52 Reduction in broadband noise power spectrum at the entrance to the ear canal achieved using Carme's second order filter.

The coefficients A_n and B_m are optimized using a gradient descent algorithm such that the cost function E is minimized. In practice Carme used $N = M = 2$; that is, a second order filter. However, he did not make clear how many frequency intervals, N, he divided the frequency spectrum into to evaluate the cost function. Nor did he provide any values for α, β or $R_{\max}$.

Results obtained by Carme using this optimized filter are much better than those shown in the previous figure, extending the range of significant noise reduction over almost two decades in frequency. The frequency response of the optimised filter used and the noise reduction results obtained are shown in Fig. 7.53.

One problem associated with these feedback control systems for active head sets is the large variability in the transfer functions of the cavity depending upon who is wearing the device and upon the quality of the acoustic seal between the device and the head. This can result in the onset of instability if an ear protector optimally adjusted for one person is worn by another or if the device slips a little on the wearer's head. Trinder and Jones (1987) claim to have developed a filter which minimizes the effect of the acoustic seal on the results and is also effective for an open backed headset, for which they obtained more than 15 dB of reduction at the ear over a decade in frequency (60 to 600 Hz). However, no details of the filter design are provided by the authors.

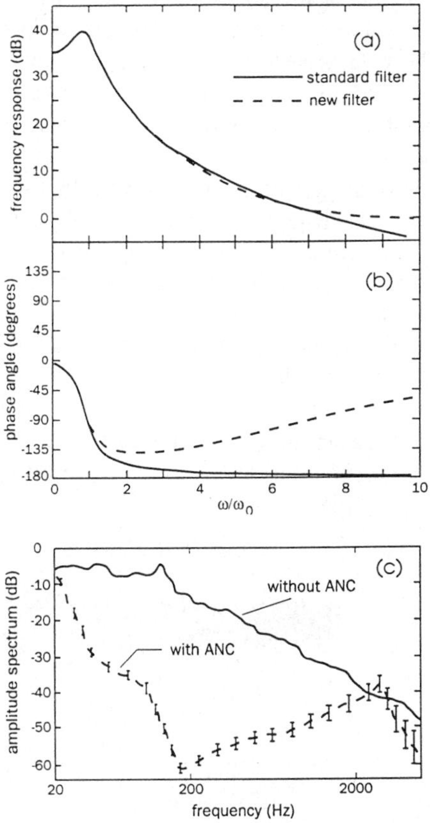

Fig. 7.53 Results obtained using the optimized filter of Carme (1988): (a) amplitude of the filter frequency response; (b) phase of the filter frequency response; (c) attenuation achieved at the ear.

It is clear that feedback control systems are effective in significantly reducing broadband noise in ear protectors and their practical application is the subject of much current attention.

Open back headsets are preferable to fully enclosed hearing protectors for pilots and other industrial workers who are required to wear them continuously for long periods of time, as heat build up in the enclosed cavity and the pressure to maintain the acoustic seal causes considerable discomfort to the wearer. Another study by Veit (1988) reported on the effectiveness of a feedback active control system using an open backed headset with the microphone mounted externally. This microphone location was justified because of the low frequency nature of the noise and had the effect of minimizing the effect of the compensation network on the radio communication signals incorporated as part of the system. Veit reported maximum noise reductions of 20 dB in one ⅓ octave band and 10 dB over two octave bands (from 250 Hz to 1000 Hz). The frequency band corresponding to maximum attenuation was adjustable.

Although a considerable effort has been devoted to developing commercial active headsets and hearing protectors using feedback controllers, their use to date has been very limited, possible due to potential system instabilities caused by high pressure impulsive noise or low frequency pressure pulsations which can overdrive the loudspeaker.

Removing the low frequency pressure pulsations by use of a high pass filter is described in a patent by Twiney and Salloway (1990). However, high level impulsive noise in the control frequency range can still result in system instability if the control speaker is overdriven as a result.

An interesting recent development concerns low cost active headsets which have been conceived using analogue electronics to implement what is basically a feedback system. These open backed headsets are being trialled for use by passengers in commercial airliners to improve the audibility of in-flight entertainment and are expected to have a unit cost to the airline companies of less than US$10. The measured noise reduction of a prototype unit is shown as a function of frequency in Fig. 7.54.

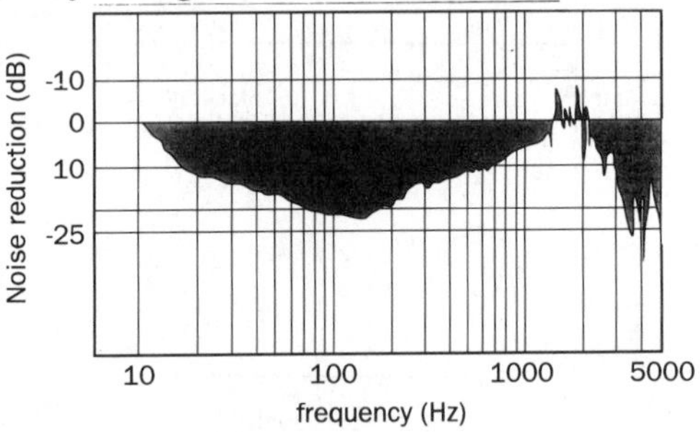

Fig. 7.54 Performance of a low cost open backed headset for airline passengers.

More sophisticated active headsets with a retail price of less than US$200 are also being manufactured by at least one company in the USA.

7.8.2 Feedforward systems

As discussed in earlier chapters, feedforward control systems rely on the availability of a reference signal which contains information on the frequency content of the noise signal to be attenuated. For an adaptive feedforward system an error microphone provides a measure of the remaining acoustic signal after action of the control loudspeaker and this signal is used to update the filter which operates on the reference signal prior to feeding it to the control source. Two types of adaptive feedforward system are shown in Fig. 7.55.

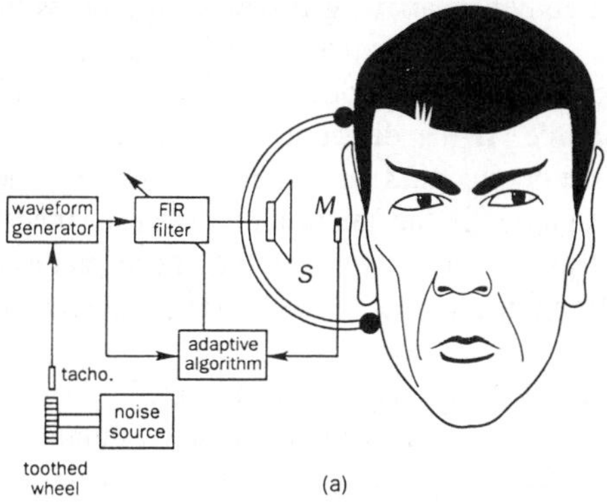

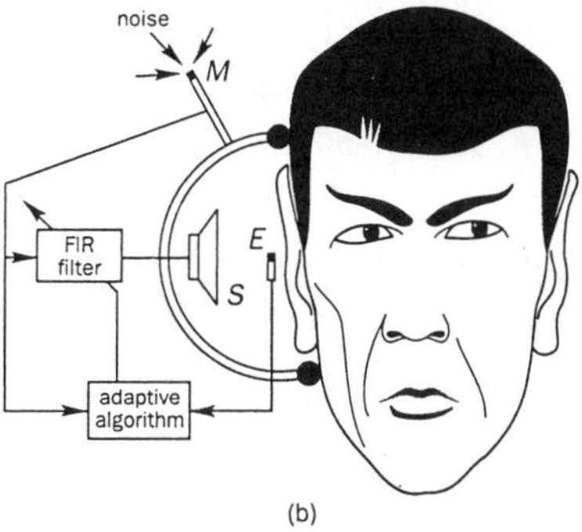

Fig. 7.55 Adaptive feedforward control systems for active head sets: (a) tacho reference signal; (b) microphone reference signal. E = error microphone, M = reference microphone and S = loudspeaker.

The system shown in Fig. 7.55(a) was first reported by Jones and Smith (1983) and is the subject of a patent awarded to Chaplin *et al.* (1987). It will only control periodic noise originating from the noise source attached to the toothed wheel sampled with the tachometer. In practice, this is limited to the fundamental rotational frequency and the first few harmonics. Any random or periodic noise originating from other noise sources will not be attenuated. However, in certain applications this may be an advantage rather than a disadvantage. For example, in the mining industry it is desirable to attenuate rotational equipment noise but not the random noise associated with 'roof talk' which gives miners some warning of an impending cave in. It is also feasible to use radio transmitted signals to transmit the reference signal from a transducer on the rotating equipment to the headset electronics.

In other situations, it is desirable to control the random noise component as well and a system which is capable of controlling both periodic and random noise is illustrated in Fig. 7.55(b). The main problem associated with controlling random noise is the need to ensure that the noise signals arrive at the microphone sufficiently long enough before arriving at the ear to process the signal and send it to the loudspeaker. Satisfaction of this requirement means that for random noise control, the reference microphone must be mounted on a boom pointed towards the noise source. If the direction from which the noise originates is unknown or varies, the boom could be designed so that the microphone location is adjustable. For example, if the microphone were connected to the hearing protector with a ball-joint or if it were on the end of a flexible rod, the user could adjust its location with respect to the hearing protector until the noise was minimized.

Feedforward controllers are advantageous in that the control signals they generate do not attempt to interfere with any radio communication signal which is not sampled by the reference microphone. In Fig. 7.55, the communication signal would be introduced immediately prior to the loudspeaker. The same reference microphone or tachometer signal could also be used to drive a second adaptive noise cancelling system to minimize noise in radio communication signals initiated by the wearer of the active hearing protector, using the circuit shown in Fig. 7.56.

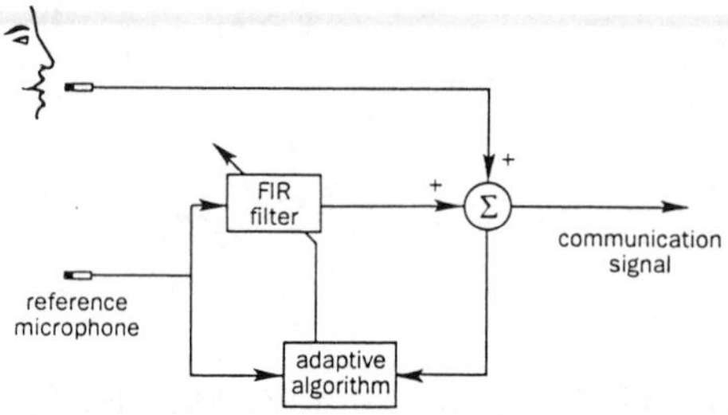

Fig. 7.56 Active control of background noise in a radio communication signal.

7.8.3 Transducer considerations

One of the most difficult tasks in the design of an active hearing protector is the design of an appropriate loudspeaker which is capable of producing the required cancelling signals. The loudspeaker must be small enough to fit in a standard hearing protector and yet perform well at low frequencies. It must also have the required 'headroom' to support undistorted speech communication signals at the required level. In an intense noise environment such as found in a tracked vehicle, achieving this performance can be difficult.

Fortunately, procuring a microphone suitable for measuring the unwanted noise is relatively easy, as there are several commercially available miniature electret microphones which are more than adequate for the task.

REFERENCES

Åbom, M. (1989). Modal decomposition in ducts based on transfer function measurements between microphone pairs. *Journal of Sound and Vibration*, **135**, 95–114.

Berengier, M. and Roure, A. (1980a). Radiation impedance of one or several real sources mounted in a hard-walled rectangular waveguide. *Journal of Sound and Vibration*, **71**, 389–398.

Berengier, M. and Roure, A. (1980b). Broad-band active sound absorption in a duct carrying uniformly flowing fluid. *Journal of Sound and Vibration*, **68**, 437–449.

Bies, D.A. and Hansen, C.H. (1996). *Engineering noise control: Theory and Practice*. 2nd edn. E & FN Spon, London.

Billoud, G., Galland, M.A. and Sunyach, M. (1987). Anti sound systems in short ducts: an experimental investigation of causality and stability effects. *Journal of Theoretical and Applied Mechanics*, **6** suppl., 111–124.

Bolleter, U. and Crocker, M.J. (1972) Theory and movement of modal spectra in hard-walled cylindrical ducts. *Journal of the Acoucistical Society of America*, **51**, 1439–1447.

Burgess, J.C. (1981). Active adaptive sound control in a duct: a computer simulation. *Journal of the Acoustical Society of America*, **70**, 715–726.

Carme, C. (1987). *Absorption acoustique active dans les cavities*. Doctoral thesis, Faculte des Sciences de Luminy, Universite D'Aix-Marseille II, France.

Carme, C. (1988). A new filtering method by feedback for A.N.C. at the ear. In *Proceedings of Internoise '88*. Institute of Noise Control Engineering, pp. 1083–1086.

Chaplin, G.B.B. (1980). The cancellation of repetitive noise and vibration. In *Proceedings of Inter-Noise 1980*. Institute of Noise Control Engineering, 699–702.

Chaplin, G.B.B. (1983). Anti-noise: the Essex breakthrough. *Chartered Mechanical Engineer*, **30**, 41–47.

Chaplin, G.B.B. and Smith, R.A. (1978). The sound of silence. *Engineering*, July.

Chaplin, G.B.B., Smith, R.A. and Bramer, T.P.C. (1987). Method and apparatus for reducing repetitive noise entering the ear. US Patent 4654,871.

Conover, W.B. (1956). Fighting noise with noise. *Noise Control*, **2**, 78–82.

Culbreth, W.G., Hendricks, E.W. and Hansen, R.J. (1988). Active cancellation of noise in a liquid-filled pipe using an adaptive filter. *Journal of the Acoustical Society of America*, **83**, 1306–1310.

Curtis, A.R.D., Nelson, P.A. and Elliott, S.J. (1990). Active reduction of a one-dimensional enclosed sound field: an experimental investigation of three control strategies. *Journal of the Acoustical Society of America*, **88**, 2265–2268.

Curtis, A.R.D., Nelson, P.A., Elliott, S.J., and Bullmore, A.J. (1987). Active suppression of acoustic resonance. *Journal of the Acoustical Society of America*, **81**, 624–631.

Davidson, A.R. and Robinson, T.G.F. (1977). Noise cancellation apparatus. US Patent No. 4025 724.

Davis, M.R. (1989). Reduction of noise radiated from open pipe terminations. *Journal of Sound and Vibration*, **132**, 213–225.

Doak, P.E. (1973a). Excitation, transmission and radiation of sound from source distributions in hard-walled ducts of finite length (I): the effects of duct cross-section geometry and source distribution space–time pattern. *Journal of Sound and*

Vibration, **31**, 1–72.

Doak, P.E. (1973b). Excitation, transmission and radiation of sound from source distributions in hard-walled ducts of finite length (II): the effects of duct length. *Journal of Sound and Vibration*, **31**, 137–174.

Doelman, N.J. (1989). Active control of sound fields in an enclosure of low modal density. In *Proceedings of Internoise '89*. Institute of Noise Control Engineering, 451-454.

Dorey, A.P. Pelc, S.F. and Watson P.R. (1975). An active noise reduction system for use with ear defenders. In *Proceedings of the 8th International Aerospace Symposium*. Cranfield, 24–27.

Eghtesadi K. and Leventhall, H.G. (1981a). Active control of noise and vibration. *Proceedings of the Institute of Acoustics*.

Eghtesadi, K. and Leventhall, H.G. (1981b). Active attenuation of noise: the Chelsea dipole. *Journal of Sound and Vibration*, **75**, 127–134.

Eghtesadi, K. and Leventhall, H.G. (1983a). The effects of non-ideal elements and geometry on the performance of the Chelsea dipole active attenuator. *Journal of Sound and Vibration*, **91**, 1–10.

Eghtesadi, K. and Leventhall, H.G. (1983b). A study of n-source active attenuator arrays for noise in ducts. *Journal of Sound and Vibration*, **91**, 11–19.

Eghtesadi, K., Hong, W.K.W. and Leventhall, H.G. (1983). The tight-coupled monopole active attenuator in a duct. *Noise Control Engineering*, **20**, 16–20.

Eghtesadi, K., Hong, W.K.W. and Leventhall, H.G. (1984). Economics of active attenuation of noise in ducts. In *Proceedings of Internoise '84*. Institute of Noise Control Engineering, Honolulu, 447–452.

Eghtesadi, K., Hong, W.K.W. and Leventhall, H.G. (1986). Energy conservation by active noise attenuation in ducts. *Noise Control Engineering Journal*, **27**, 90–94.

Eghtesadi, K., Hong, W.K.W. and Leventhall, H.G. (1987). Active attenuation of noise: the Chelsea N-source attenuator arrays. In *Proceedings of Inter-Noise 1987*. Institute of Noise Control Engineering, 509–512.

Elliott, S.J. and Darlington, P. (1985). Adaptive cancellation of periodic, synchronously sampled interference. *IEEE Transactions on Circuits and Systems*, **33**, 715–717.

Elliott, S.J. and Nelson, P.A. (1984). *Models for Describing Active Noise Control in Ducts*. Institute of Sound and Vibration Research Technical Report No. 127.

Elliott, S.J., Stothers, I.M. and Nelson, P.A. (1987). A multiple error LMS algorithm and its application to the active control of sound and vibration. *IEEE Transactions on Acoustics, Speech and Signal Processing*, **ASSP-35**, 1423–1434.

Eriksson, L.J. (1985). *Active Sound Attenuation using Adaptive Digital Signal Processing Techniques*. PhD Thesis, University of Wisconsin-Madison, Wisconsin.

Eriksson, L.J. (1991a). The development of the filtered – U algorithm for active noise control. *Journal of the Acoustical Society of America*, **89**, 257–265.

Eriksson, L.J. (1991b). Recursive algorithms for active noise control. In *International Symposium on Active Control of Sound and Vibration*. Tec, Japan.

Eriksson, L.J. and Allie, M.C. (1989a). Use of random noise for on-line transducer modelling in an adaptive active attenuation system. *Journal of the Acoustical Society of America*, **85**, 797–802.

Eriksson, L.J. and Allie, M.C. (1989b). Use of random noise for on-line transducer modelling in an adaptive active attenuation system. *Journal of the Acoustical Society of America*, **85**, 797–802.

Eriksson, L.J., Allie, M.C. and Greiner, R.A. (1987). The selection and application of an IIR adaptive filter for use in active sound attenuation. *IEEE transactions on Acoustics, Speech and Signal Processing*, **ASSP−35**, 433−437.

Eriksson, L.J., Allie, M.C., Bremigan, C.D. and Gilbert, J.A. (1988). Active noise control and specifications for fan noise problems. In *Proceedings of Noise-Con '88*. Institute of Noise Control Engineering.

Eriksson, L.J., Allie, M.C., Hoops, R.H. and Warner, J.V. (1989). Higher order mode cancellation in ducts using active noise control. In *Proceedings of Inter-Noise '89*. Institute of Noise Control Engineering, 495−500.

Fahy, F.J. (1985). *Sound and Structural Vibration: Radiation, Transmission and Response*. Academic Press, London.

Fedoryuk, M.V. (1975). The suppression of sound in acoustic waveguides. *Soviet Physics Acoustics*, **21**, 174−176.

Foller, D. (1992). Antischall – chancen und grenzen. *Automobiltechnische Zeitschrift*, **94**, 88−93.

Ford, R.D. (1984). Power requirements for active noise control in ducts. *Journal of Sound and Vibration*, **92**, 411−417.

Hall, H.R., Ferren, W.B. and Bernhard, R.J. (1990). Active control of radiated sound from ducts. In *Active Noise and Vibration Control*. ASME Publication NCA Vol. 8, 143-152.

Holt, D.J. (1993). Advanced exhaust silencing. *Automotive Engineering*, **101**, 13−16.

Hong, W.K.W., Eghtesadi, K. and Leventhall, H.G. (1987). The tight-coupled monopole and tight-coupled tandem attenuators: theoretical aspects and experimental attenuation in an air duct. *Journal of the Acoustical Society of America*, **81**, 376−388.

Jessel, M.J.M. and Mangiante, G.A. (1972). Active sound absorbers in an air duct. *Journal of Sound and Vibration*, **23**, 383−390.

Jones, O. and Smith, R.A. (1983). The selective anti-noise ear defender. In *Proceedings of Internoise '83*. Institute of Noise Control Engineering, 375−378.

Kido, K., Morikawa, S. and Abe, M. (1987). Stable method for active cancellation fo duct noise by synthesised sound. *Journal of Vibration, Acoustics, Stress and Reliability in Design*, **109**, 37−42.

Kido, K., Kanai, H. and Abe, M. (1989). Active reduction of noise by additional noise source and its limit. *Journal of Vibration, Acoustics, Stress and Reliability in Design*, **111**, 480−485.

Koopmann, G.H., Fox, D.J. and Neise, W. (1988). Active source cancellation of the blade tone fundamental and harmonics in centrifugal fans. *Journal of Sound and Vibration*, **126**, 209−220.

La Fontaine, R.F. and Shepherd, I.C. (1983) An experimental study of a broad band active attenuator for cancellation of random noise in ducts. *Journal of Sound and Vibration*, **90**, 351−362.

La Fontaine, R,F. and Shepherd, I.C. (1985). The influence of waveguide reflections and system configuration on the performance of an active noise attenuator. *Journal of Sound and Vibration*, **100**, 569−579.

Laugesen, S. (1993). Active control of tonal noise in a large chimney stack. In *Proceedings of Noise '93*. St Petersburg, 2, 179−184.

Leventhall, H.G. and Eghtesadi, K. (1979). Active attenuation of noise: dipole and monopole systems. In *Proceedings of Inter-Noise '79*. Institute of Noise Control Engineering, 175−180.

Lueg, P. (1936). Process of silencing sound oscillations. US Patent No. 2043 416 Filed 1934, granted June 6, 1936.

Maxwell, R.G., Sjosten, P. and Lindqvist, E.A. (1989). Active noise control of pure tones in ducts for nonplane waves: a case study. In *Proceedings of Inter-Noise '89*. Institute of Noise Control Engineering, 447-450.

Mazanikov, A.A., Tyutekin, V.V. and Ukolov, A.T. (1977). An active system for the suppression of sound fields in a multimode waveguide. *Soviet Physics-Acoustics*, **23**, 276–277.

Mori, K., Nishiwaki, N., Takemori, Y., Saeki, N., Taki, M., and Morishita, T. (1991). Application of AAC silencer to reduce automobile exhaust noise. In *Proceedings of Inter-Noise '91*. Institute of Noise Control Engineering, 529–532.

Morse, P.M. (1948) *Vibration and Sound*. McGraw-Hill: New York (reprinted by the Acoustical Society of America, 1981).

Morse, P.M. and Ingard, K.U. (1968). *Theoretical Acoustics*. McGraw-Hill, New York, Chapter 9.

Munjal, M.L. (1987) *Acoustics of Ducts and Mufflers with Application to Exhaust and Ventilation System Design*. John Wiley, New York.

Munjal, M.L. and Eriksson, L.J. (1988). An analytical one-dimensional standing wave model of a linear active noise control system in a duct. *Journal of the Acoustical Society of America*, **84**, 1086–1093.

Munjal, M.L. and Eriksson, L.J. (1989a). Analysis of a hybrid noise control system for a duct. *Journal of the Acoustical Society of America*, **86**, 832–834.

Munjal, M.C. and Eriksson, L.J. (1989b). An exact one-dimensional analysis of the acoustic sensitivity of the antiturbulence probe tube in a duct. *Journal of the Acoustical Society of America*, **85**, 582–587.

Neise, W. and Koopman, G.H. (1991). Active sources in the cut-off of centrifugal fans to reduce blade tones at higher-order duct mode frequencies. *Journal of Vibration and Acoustics*, **113**, 123–131.

Nelson, P.A. and Elliott, S.J. (1992). *Active Control of Sound*. Academic Press, London, Chapter 7.

Nelson, P.A., Curtis, A.R.D., Elliott, S.J., and Bullmore, A.J. (1987). The active minimization of harmonic enclosed sound fields, part I: theory. *Journal of Sound and Vibration*, **117**, 1–13.

Olson, H.F. (1956). Electronic control and noise, vibration and reverberation. *Journal of the Acoustical Society of America*, **28**, 966–972.

Olson, H.F. and May, E.G. (1953). Electronic sound absorber. *Journal of the Acoustical Society of America*, **25**, 1130–1136.

Poole, J.H.B. and Leventhall, H.G. (1976). An experimental study of Swinbanks' method of active attenuation of sound in ducts. *Journal of Sound and Vibration*, **49**, 257–266.

Poole, J.H.B. and Leventhall, H.G. (1978). Active attenuation of noise in ducts. *Journal of Sound and Vibration*, **57**, 308–309.

Poole, L.A., Warnaka, G.E. and Cutter, R.C. (1984). The implementation of digital filters using a modified Widrow-Hoff algorithm for the adaptive cancellation of acoustic noise. In *Proceedings of the International Conference on Acoustics, Speech and Signal Processing*, **2**, 21.7.1–21.7.4.

Ross, C.F. (1982a). An algorithm for designing a broadband active sound control system. *Journal of Sound and Vibration*, **80**, 373–380.

Ross, C.F. (1982b). An adaptive digital filter for a broadband active sound control system. *Journal of Sound and Vibration*, **80**, 381–388.

Ross, C.F. and Yorke, A.V. (1987). Energy flow in active control systems. *Journal of*

Theoretical and Applied Mechanics, **6** suppl., 99–110.

Roure, A. (1985). Self adaptive broadband active sound control systems. *Journal of Sound and Vibration*, **101**, 429–441.

Salloway, A.J. and Twiney, R.C. (1985). Earphone active noise reduction systems. In *Proceedings of the Institute of Acoustics*, **7**, 95.

Sha, J. and Tian, J. (1987). Acousticdal mechanism of active noise attenuator in a duct. In Proceedings of Internoise '87.

Shepherd, I.C., La Fontaine, R.F. and Cabelli, A. (1984). Active attenuation in turbulent flow ducts. In *Proceedings of Inter-Noise '84*. Institute of Noise Control Engineering, 497–502.

Shepherd, I.C., La Fontaine, R.F. and Cabelli, A. (1985). A bi-directional microphone for the measurement of duct noise. *Journal of Sound and Vibration*, **101**, 563–573.

Shepherd, I.C., La Fontaine, R.F. and Cabelli, A. (1986a). *Attenuation in Flow Ducts: Assessment of Prospective Applications*. ASME Paper 86-WA/NCA-26.

Shepherd, I.C., Cabelli, A. and La Fontaine, R.F. (1986b). Characteristics of loudspeakers operating in an active noise attenuator. *Journal of Sound and Vibration*, **110**, 471–481.

Shepherd, I.C., La Fontaine, R.F. and Cabelli, A. (1989). The influence of turbulent pressure fluctuations on an active attenuator in a flow duct. *Journal of Sound and Vibration*, **130**, 125–135.

Silcox, R.J. and Elliott, S.J. (1990). *Active Control of Multi-dimensional Random Sound in Ducts*. NASA Technical Memorandum, 102653.

Snyder, S.D. and Hansen, C.H. (1989). Active noise control in ducts: some physical insights. *Journal of the Acoustical Society of America*, **86**, 184–194.

Snyder, S.D. (1991). *A Fundamental Study of Active Noise Control System Design*. PhD Thesis, University of Adelaide, Department of Mechanical Engineering, Adelaide, South Australia, 5005.

Sommerfeldt, S.D. and Tichy, J. (1990). Adaptive control of a two-stage vibration isolation mount. *Journal of the Acoustical Society of America*, **88**, 938–944.

Stell, J.D. and Bernhard, R.J. (1990). Active control of high order acoustical modes in a semi-infinite waveguide. In *Proceedings of the ASME ANVC Winter Annual Meeting*. ASME, New York, **8**, 131–142.

Swinbanks, M.A. (1973). The active control of sound propagation in long ducts. *Journal of Sound and Vibration*, **27**, 411–436.

Tichy, J. and Warnaka, G.E. (1983). Effect of evanescent waves on the active attenuation of sound in ducts. In *Proceedings of Internoise '83*. Institute of Noise Control Engineering, 435–438.

Tichy, J., Poole, L.A. and Warnaka, G.E. (1987). Requirements for active and passive noise control in hearing protectors. In *Proceedings of Noise-Con '87*. Institute of Noise Control Engineering, 389–392.

Trinder, M.C.J. and Jones, O. (1987). Active noise control at the ear. In *Proceedings of Noise-Con '87*. Institute of Noise Control Engineering, 393–398.

Trinder, M.C.J. and Nelson, P.A. (1983a). The acoustical virtual earth and its application to ducts with reflecting terminations. In *Proceedings of Inter-Noise '83*. Institute of Noise Control Engineering, 447–450.

Trinder, M.C.J. and Nelson, P.A. (1983b). Active noise control in finite length ducts. *Journal of Sound and Vibration*, **89**, 95–105.

Trinder, M.C.J., Chaplin, G.B.B. and Nelson, P.M. (1986). Active control of commercial vehicle exhaust noise. In *Proceedings of Internoise '86*. Institute of

Noise Control Engineering, 611–616.

Twiney, R.C. and Salloway, A.J. (1990). Active noise reduction systems reducing unwanted signal enhancement. US Patent No. 4953,217.

Veit, I. (1988). A lightweight headset with an active noise compensation. In *Proceedings of Internoise '88*. Institute of Noise Control Engineering, 1087–1090.

Wanke, R.L. (1976). Acoustic abatement method and apparatus. US Patent No. 3936,606.

Warnaka, G.E., Poole, L.A. and Tichy, J. (1984). Active acoustic attenuator. US Patent number 4473,906.

Wheeler, P.D. (1986). *Voice Communications in the Cockpit Noise Environment: The Role of Active Noise Reduction*. PhD thesis, University of Southampton, England.

Wheeler, P.D., Rawlinson, R.D., Pelc, S.F. and Dorey, A.P. (1978). The development and testing of an active noise reduction system for use in ear defenders. In *Proceedings of Internoise '78*. Institute of Noise Control Engineering, 977–982.

Wheeler, P.D. and Parramore, T.S. (1981). Active control of noise and vibration. In *Proceedings of the Institute of Acoustics*.

Widrow, B. and Stearns, S.D. (1985). *Adaptive Signal Processing*. Prentice Hall, New Jersey.

Zander, A.C. and Hansen, C.H. (1992). Active control of higher order acoustic modes in ducts. *Journal of the Acoustical Society of America*, **92**, 244–257.

Zander, A.C. and Hansen, C.H. (1993). A comparison of error sensor strategies for the active control of duct noise. *Journal of the Acoustical Society of America*, **94**, 841–848.

8

Active control of free field sound radiation

8.1 INTRODUCTION

In this chapter we will examine the second of our general classes of active noise control problems, the active control of sound radiation into free space. As with the previous chapter on controlling sound propagation in ducts, principal consideration will be given to the 'physical' part of the active noise control system; models will be derived which enable prediction of the effects of applying active noise control, both on the acoustic field and on the acoustic radiation characteristics of the noise sources. These models can then be used in analysis of the physical mechanisms of active control, and in system design exercises, where the aim is to find an arrangement of sources and sensors which is capable of providing the maximum level of disturbance attenuation. In practice, this arrangement would be driven by the feedforward control systems described in Chapter 6, or the feedback arrangements described in Chapter 5.

Some of the first attempts to control sound radiation into free space can be found in the work directed at controlling tonal noise radiated by large electrical transformers (Conover, 1956; Ross, 1978; Hesselman, 1978; Angevine, 1981, 1992, 1993; Berge *et al.*, 1987, 1988). This work involved the use of loudspeakers placed near the transformer tank to control the noise radiated to one or more community locations. It was found that if global control was to be achieved in all directions, then it was necessary to use a large array of speakers, almost as large as the transformer tank itself. Use of one or two speakers resulted in reduced noise levels in some directions at the expense of increased noise levels in others. In addition the angular spread of the directions of reduced levels were generally quite narrow. Although not stated by the authors, the physical reason for this behaviour is that to achieve global noise reduction using acoustic control sources, it is necessary for the sources to change the radiation impedance 'seen' by the transformer tank. This can only be done by using a large array of control sources. If only one or two small sources are used, areas of reduced noise level are achieved solely by local destructive interference effects at the expense of other areas of increased level where constructive interference takes place. Thus in this latter case, the radiation impedance 'seen' by the transformer (and hence its radiated sound power) is barely changed, and as a result the overall radiated sound power of the transformer plus control sources is generally larger than the sound power radiated by the transformer itself when only one or two control sources are used, even though there will be some locations (particularly error sensor locations) where the sound level will be reduced.

It is only relatively recently that there has been a concentrated effort to develop practical control systems to control the sound power radiated by vibrating surfaces. Knyazev and Tartakovskii (1967) were the first to investigate the control of sound radiation by using control forces on the vibrating structure. In 1988, Deffayet and Nelson analysed the active control of sound radiated by a finite rectangular panel using acoustic control sources. Hansen *et al.* (1989) compared the relative effectiveness of control forces and acoustic sources for controlling sound radiated by a rectangular panel. Analytical modelling of the active control of sound radiated by a rectangular panels has been undertaken by Walker (1976), Fuller (1990), Pan *et al.* (1992) and Wang and Fuller (1992).

More recently, research efforts have focused upon the control of orthotropic panels (Meirovitch and Thangjitham, 1990a), the control of fluid loaded panels (Meirovitch and Thangjitham, 1990b; Gu and Fuller, 1993), combined active and passive control (Koshigoe and Murdock, 1993), broadband disturbances (Baumann *et al.*, 1992), the use of piezoceramic crystals to provide the control forces (Hansen *et al.*, 1989; Wang *et al.*, 1990; Fuller *et al.*, 1991a, 1991b; Hansen *et al.*, 1991; Metcalf *et al.*, 1992) and shaped PVDF (poly vinyl diflouride) sensors instead of microphones to provide the required controller error signal (Lee and Moon, 1990; Clark and Fuller, 1992a,b,c; Elliott and Johnson, 1993; Naghshineh and Koopmann, 1993; Snyder and Tanaka, 1993b, 1994a,b). Use of vibration error sensors to minimize sound radiation is much more convenient than the use of microphones mounted in the far field of a noise radiating surface and the problem of shaping vibration sensors to provide such a signal is an interesting and complex problem which is why so much effort is currently being devoted to research in this area.

Investigation of the physical mechanisms (Hansen *et al.*, 1989; Snyder and Hansen, 1991; Snyder, 1991; Elliott *et al.*, 1991a; Elliott and Johnson, 1992) involved in controlling sound radiation from a simple vibrating surface has provided an understanding of the complexity of the problem and has also resulted in the determination of the influence of geometric and structural/acoustic variables on the maximum achievable reduction in sound power. This work has led to the formulation of strategies for the optimum design of multi-channel systems for the simultaneous control of a number of sources and error sensors (Hansen *et al.*, 1990; Thi *et al.*, 1991; Snyder and Hansen, 1991).

This chapter will begin with an examination of the active control of harmonic sound radiation from a set of monopole sources in free space by the introduction of a second set of monopole sources. While this problem is somewhat idealized, it will provide a variety of physical and methodological insights which will prove valuable in examining 'more realistic' problems. Problems of controlling sound radiation from vibrating structures will then be discussed. In doing this, we will have our first look at the use of vibration sources for attenuating acoustic radiation by modifying the velocity distribution of the source. It will be shown that while the control of sound radation using acoustic and vibration control sources may appear very different on the surface, the physical mechanisms behind the attenuation are actually closely related.

Once control of harmonic sound radiation from idealized monopole sources, and more realistic structure model, has been studied, attention will be given to the control of non-harmonic acoustic radiation, random noise and impulse excitation. It will be shown that these problems are related to the harmonic excitation cases studied earlier in the chapter, both in control mechanism and optimal performance considerations. The chapter will conclude with a brief discussion of the relationship between feedback control of a vibrating structure, discussed in Chapter 11, and feedback control of acoustic radiation from that structure.

8.2 CONTROL OF HARMONIC SOUND PRESSURE AT A POINT

Possibly the most basic active noise control problem which could be envisaged is the minimization, or 'cancellation', at some point of harmonic sound, radiated by one acoustic monopole (primary) source, through the introduction of a second (control) monopole source, the geometry of which is shown in Fig. 8.1. As such, an examination of this problem provides an ideal introduction to the problem of free field active noise control.

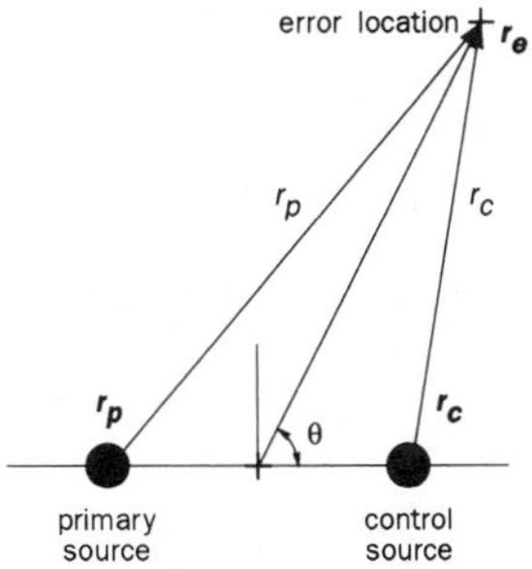

Fig. 8.1 Single monopole primary source / single monopole control source system arrangement.

It was shown in Chapter 2 that the sound pressure at some location $\boldsymbol{r_e}$ in space due to the harmonic operation of an acoustic monopole source located at a position $\boldsymbol{r_q}$ is

$$p(\boldsymbol{r_e}) = \frac{j\omega\rho_0 q e^{-jkr}}{4\pi r} \tag{8.2.1}$$

where $r = |\mathbf{r}_e - \mathbf{r}_q|$, q is the volume velocity, or strength, of the monopole source, k is the acoustic wavenumber at the frequency ω of interest, and positive harmonic time dependence of the form $e^{j\omega t}$ is implicit in the equation (this will be the case throughout this chapter). It will be assumed throughout this chapter that the acoustic sources are all constant volume velocity sources (the volume velocity of the source does not change, regardless of other additions to the environment in which it is placed). Physically, this translates into the ideal that the internal impedance of the source is infinite. However, many practical sources and environments, such as speakers (with a small airtight backing enclosure) operating in air, exhibit characteristics which are sufficiently close to this (where the mechanical impedance of the sound source is several orders of magnitude greater than the radiation impedance it sees looking into the acoustic environment), so that the results obtained theoretically using this idealization accurately represent real-world phenomena.

As the system being considered is linear, the concept of superposition is valid, so that the sound pressure at some location $\mathbf{r}_e$ during the operation of both the primary and control sources is simply the sum of the individual pressures,

$$p(\mathbf{r}_e) = p_p(\mathbf{r}_e) + p_c(\mathbf{r}_e) \tag{8.2.2}$$

where the subscripts p and c denote primary and control sources respectively. Using (8.2.1), this can be rewritten as

$$p(\mathbf{r}_e) = \frac{j\omega\rho_0 q_p e^{-jkr_p}}{4\pi r_p} + \frac{j\omega\rho_0 q_c e^{-jkr_c}}{4\pi r_c} \tag{8.2.3}$$

where r_p and r_c are the distances between the primary and control sources and the point in space of interest, $r_p = |\mathbf{r}_e - \mathbf{r}_p|$, $r_c = |\mathbf{r}_e - \mathbf{r}_c|$.

The object of the exercise here is, given some primary source volume velocity q_p, to find the volume velocity q_c which will minimise the acoustic pressure at some location $\mathbf{r}_e$. For a single control source and single 'error' location, the minimised pressure will, in fact, be zero. This objective is easily accomplished using (8.2.3). For $p(\mathbf{r}_e) = 0$,

$$\frac{j\omega\rho_0 q_p e^{-jkr_p}}{4\pi r_p} = -\frac{j\omega\rho_0 q_c e^{-jkr_c}}{4\pi r_c} \tag{8.2.4}$$

or

$$q_{c,p} = -\frac{r_c}{r_p} q_p e^{-jk(r_p - r_c)} \tag{8.2.5}$$

where the subscript c,p will be used to denote a control source volume velocity which has been derived with the aim of minimizing acoustic pressure at some discrete location. Equation (8.2.5) describes the relationship between the primary and control source volume velocities which will result in the minimization of harmonic sound pressure at some point in space. It states that the control source volume velocity, relative to the primary source volume velocity, must be proportional in amplitude to the relative distances to the error location of interest

(r_c/r_p), and produce a pressure signal which is 180° out of phase with the primary source generated pressure when it arrives at the error location $(-e^{-jk(r_p-r_c)})$.

For this extremely simple system it is quite straightforward to derive the relationship between the primary and control sources which will result in the minimization of acoustic pressure at some point in space. It will not, however, always be so easy to simply write down the solution to the problem, and so it is useful to rederive the result of (8.2.5) in a way which will be more easily generalized later in this chapter. Therefore, consider again the problem of minimizing the acoustic pressure amplitude at some point in space. This is equivalent to minimizing the squared pressure amplitude:

$$|p(r)|^2 = p^*(r)p(r) \tag{8.2.6}$$

where * denotes the complex conjugate. During the operation of both the primary and control sources, the pressure at any point in space is as defined in (8.2.3). Substituting this into (8.2.6) gives

$$|p(r_e)|^2 = \left[\frac{j\omega\rho_0 q_p e^{-jkr_p}}{4\pi r_p} + \frac{j\omega\rho_0 q_c e^{-jkr_c}}{4\pi r_c}\right]^* \left[\frac{j\omega\rho_0 q_p e^{-jkr_p}}{4\pi r_p} + \frac{j\omega\rho_0 q_c e^{-jkr_c}}{4\pi r_c}\right]$$

$$= q_c^* a q_c + q_c^* b + b^* q_c + c \tag{8.2.7}$$

where

$$a = \left[\frac{\omega\rho_0}{4\pi r_c}\right]^2, \quad b = \left[\frac{\omega\rho_0}{4\pi}\right]^2 \frac{1}{r_c r_p} q_p e^{-jk(r_p-r_c)}, \quad c = \left[\frac{\omega\rho_0}{4\pi r_p}\right]^2 \tag{8.2.8}$$

Equation (8.2.7) shows that the squared acoustic pressure amplitude at any location in space is a real quadratic function of the complex control source volume velocity, q_c. This is perhaps more easily seen by rewriting the equation in terms of its real and imaginary parts as follows,

$$|p(r)|^2 = aq_{cR}^2 + 2b_R q_{cR} + aq_{cI}^2 + 2b_I q_{cI} + c \tag{8.2.9}$$

where the subscripts I and R denote the real and imaginary components, respectively. Thus a plot of the squared acoustic pressure amplitude as a function of the real and imaginary components of the control source volume velocity forms a 'bowl', as shown in Fig. 8.2. The bottom of the bowl defines the optimum control source volume velocity, which will minimize the error criterion, the squared acoustic pressure amplitude. This value of volume velocity can be found by differentiating (8.2.9) with respect to its real and imaginary components, and setting the result equal to zero. Doing this gives

$$\frac{\partial |p(r)|^2}{\partial q_{cR}} = 2aq_{cR} + 2b_R = 0 \tag{8.2.10}$$

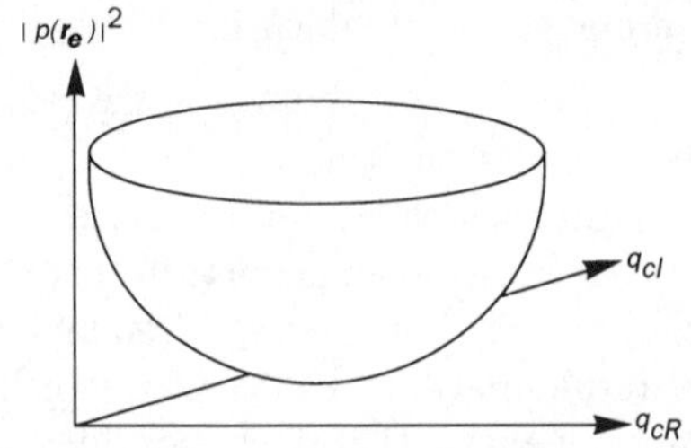

Fig. 8.2 Typical plot of squared acoustic pressure amplitude as a function of the real and imaginary parts of control source volume velocity.

and

$$\frac{\partial\,|\,p(r)\,|^{\,2}}{\partial q_{cI}} = 2aq_{cI} + 2b_I = 0 \tag{8.2.11}$$

Multiplying the result of (8.2.11) by the imaginary number j, and putting it together with the result of (8.2.10), produces the following expression for the optimum control source volume velocity (Nelson *et al.*, 1985, 1987),

$$q_{c,p} = -a^{-1}b \tag{8.2.12}$$

For the particular case of interest here, if the definitions of (8.2.8) are substituted into (8.2.12) the optimum control source volume velocity is found to be

$$q_{c,p} = -\frac{r_c}{r_p}q_p e^{-jk(r_p - r_c)} \tag{8.2.13}$$

This result is, of course, identical to that derived in (8.2.5). The difference is that in this instance it was derived using quadratic optimization techniques, which will be the methodology adopted throughout this chapter when considering the minimization of acoustic pressure and power for more complex problems.

There are two points which should be briefly noted here while on the topic of quadratic optimization. First, the result of (8.2.12) is unique only if the value of a is non-zero, and will define a minimum only if the value of a is positive. In other words, for quadratic optimization to produce the desired result for an optimum control source volume velocity, a must be positive definite. From the definition of a given in (8.2.8), it is obvious that this property is true in this instance. Second, if the result of (8.2.12) is substituted back into (8.2.7), the minimum squared acoustic pressure amplitude at the error sensing location is found to be

$$|\,p(r)\,|^{\,2}_{\min} = c - b^* a^{-1} b = c + b^* q_{c,p} \tag{8.2.14}$$

If the terms in (8.2.14) are expanded using the definitions of a, b, and c given in (8.2.7) for this problem, the minimum pressure will be found to be equal to zero.

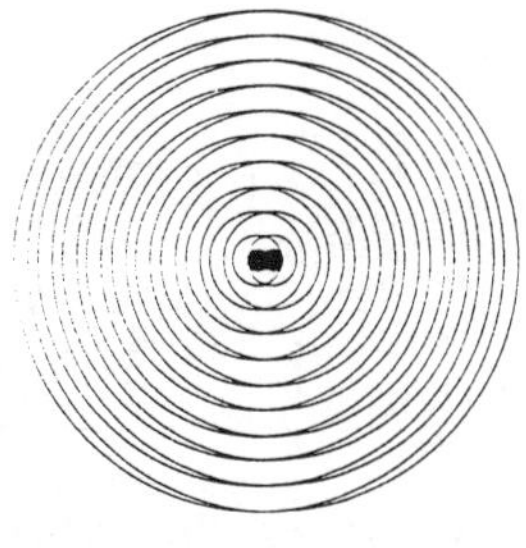

Fig. 8.3 Interference pattern of two coherent sources.

It does not automatically follow that minimizing the sound pressure at a single point in space will minimize, or even reduce, the overall levels of sound pressure in the combined radiated field. It is well known that two coherent sound sources will interfere to produce (in a linear system) a pattern of constructive and destructive interference 'fringes', as shown in Fig. 8.3. The question arises, how close must the control source be to the primary source if, when minimizing the acoustic pressure at some point in space, no fringes are to appear? In other words, how close must the control source be to the primary source before minimizing the sound pressure at some particular point in space will guarantee a pressure reduction at every point in the farfield? The answer to this question can be found by making some farfield approximations to the result of (8.2.5).

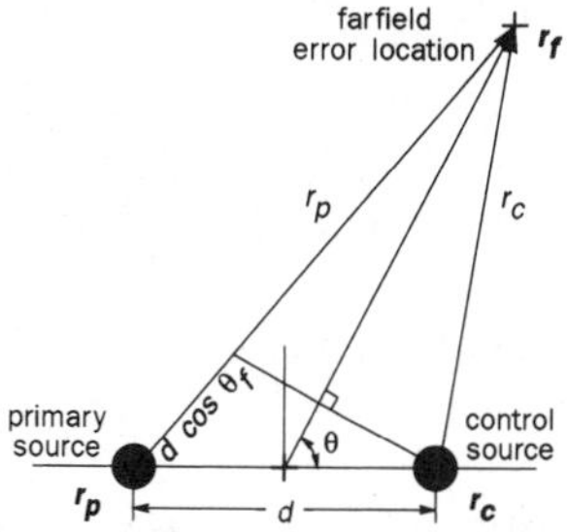

Fig. 8.4 Single monopole primary source/single monopole control source system arrangement with farfield approximations.

Consider the primary/control source arrangement shown in Fig. 8.4, where r_f is some location in the farfield such that $r_c/r_p \approx 1.0$. If the separation distance between the primary and control sources is d, then the optimum relationship between the control source and primary source volume velocities for minimizing the acoustic pressure amplitude at this point, stated in its general form in (8.2.5), can be approximated as

$$q_{c,p} \approx -q_p e^{-jk(d\cos\theta_f)} \tag{8.2.15}$$

Substituting this relationship back into (8.2.3), gives the resultant sound pressure at any point $\boldsymbol{r}$ in space during the operation of both the primary and control sources as

$$p(\boldsymbol{r}) = \frac{j\omega\rho_0}{4\pi}q_p\left[\frac{e^{-jkr_p}}{r_p} - \frac{e^{-jk(r_c+d\cos\theta_f)}}{r_c}\right] \tag{8.2.16}$$

or

$$p(\boldsymbol{r}) = \frac{j\omega\rho_0}{4\pi}q_p e^{-jkr_p}\left[\frac{1}{r_p} - \frac{e^{-jk(r_c-r_p+d\cos\theta_f)}}{r_c}\right] \tag{8.2.17}$$

Confining consideration to any point in the farfield, and again invoking the farfield assumptions already outlined, this can be written as

$$p(\boldsymbol{r}) \approx \frac{j\omega\rho_0}{4\pi r_f}q_p e^{-jkr_p}\left(1 - e^{-jkd(\cos\theta_f - \cos\theta)}\right) \tag{8.2.18}$$

If acoustic pressure amplitude is reduced, then

$$\frac{|\,p(\boldsymbol{r})\,|^2}{|\,p_p(\boldsymbol{r})\,|^2} < 1 \tag{8.2.19}$$

Substituting (8.2.1) and (8.2.18) into (8.2.19) yields the criterion

$$2\left(1-\cos\left[kd\,\{\cos\theta_f - \cos\theta\}\right]\right) < 1 \tag{8.2.20}$$

As the maximum value of the expression in the curly brackets is equal to 2.0, this criterion reduces to (Nelson and Elliott, 1992)

$$1 - \cos 2kd < 1/2 \tag{8.2.21}$$

Therefore, for the sound pressure to be reduced at all locations in the farfield as a result of minimizing the sound pressure at some arbitrary error location, $kd < \pi/6$, or, in terms of wavelength of the frequency of sound of interest, $d < \lambda/12$.

The extremely strict separation criterion of (8.2.21) corresponds to a worst case situation, where $(\cos\theta_f - \cos\theta) = 2$, or $\theta_f = 0$ and $\theta = \pi$. The criterion can be relaxed a bit with some *a priori* knowledge of a more judicious positioning of the error location. Consider, for example, the case where $\theta_f = \pi/2$, reducing the maximum value of the expression in the curly brackets in (8.2.20) to 1.0, at $\theta = \pi$. For this case, therefore, the maximum allowable separation distance for attenuation of the sound field in all directions is increased to $d = \lambda/6$. This fact is borne out in the plots of the residual sound field for the minimization of acoustic pressure at $\theta_f = 0$ and π, shown in Fig. 8.5. The lessons of this are, that the control source must be located in close proximity of the primary source for overall sound attenuation to be achieved, and that the error sensor location has a significant effect upon the overall levels of sound attenuation which are achieved by an active noise control system. The problem of optimizing the error sensor location will be considered three sections from now, after the problem of establishing a basis for evaluating how much attenuation of the total radiated acoustic power is physically possible.

8.3 THE MINIMUM ACOUSTIC POWER OUTPUT OF TWO FREE FIELD MONOPOLE SOURCES

In the previous section the problem of minimizing the acoustic pressure at some point in space, radiated from one harmonically oscillating monopole (primary) source, by the introduction of a second (control) source radiating at the same frequency was considered. It was seen, however, that the result of such an exercise is not necessarily a reduction in the amplitude of the total radiated sound field, or total radiated acoustic power. To be able to assess the quality of the control effect provided by minimizing the sound pressure at a discrete error location, two questions arise. How much attenuation in the total radiated acoustic power is actually possible, and what is the relationship between the primary and control sources which will achieve this result?

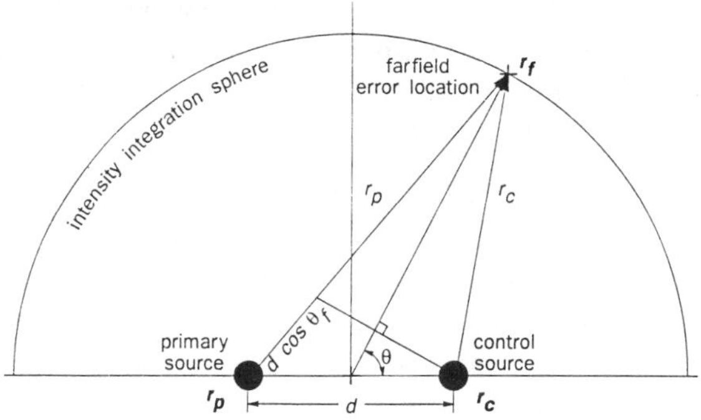

Fig. 8.6 Geometry used to determine sound power by farfield integration of acoustic intensity.

To answer these questions, consider again the problem of using one harmonically radiating point source to control the output of another, but this time using as an error criterion, minimization of the total radiated acoustic power. It was seen in Chapter 2 that the total radiated acoustic power is equal to the integration of the real acoustic intensity out of a sphere encompassing the sound sources. Using the geometry of Fig. 8.6, the acoustic power output W of the sources is

$$W = \int_0^{2\pi}\int_0^{\pi/2} \frac{|p(r)|^2}{2\rho_0 c_0} r^2 \sin\theta \, d\theta \, d\phi = 2\pi r^2 \int_0^{\pi} \frac{|p(r)|^2}{2\rho_0 c_0} \sin\theta \, d\theta \tag{8.3.1}$$

where r is the radius of the enclosing sphere, which is in the farfield of the sources ($r >> d$). The total acoustic pressure amplitude squared at any point in the space during the operation of both the primary and control sources was stated in the previous section, in (8.2.7). Substituting this into (8.3.1), the total acoustic power output of the primary source/control source combination can be written as

$$W_{tot} = q_c^* a_w q_c + q_c^* b_w + b_w^* q_c + c_w \tag{8.3.2}$$

where

$$a_w = \frac{\pi r^2}{\rho_0 c_0}\int_0^{\pi}\left[\frac{\omega\rho_0}{4\pi r_c}\right]^2 \sin\theta \ d\theta \tag{8.3.3}$$

$$b_w = \frac{\pi r^2}{\rho_0 c_0}\int_0^{\pi}\left[\frac{\omega\rho_0}{4\pi}\right]^2 \frac{1}{r_c r_p} q_p e^{-jk(r_p - r_c)} \sin\theta \ d\theta \tag{8.3.4}$$

$$c_w = \frac{\pi r^2}{\rho_0 c_0}\int_0^{\pi}\left[\frac{\omega\rho_0}{4\pi r_p}\right]^2 \sin\theta \ d\theta \tag{8.3.5}$$

Thus, as with the squared acoustic pressure amplitude at a point, the total radiated acoustic power is a real quadratic function of the control source volume velocity. Using the result of (8.2.12), the optimum control source volume velocity can be written directly as

$$q_{c,w} = -a_w^{-1} b_w \tag{8.3.6}$$

where the subscript c, w will be used to denote a control source volume velocity which has been derived with the aim of minimizing the total radiated acoustic power. Thus, to evaluate the optimum control source volume velocity with regard to minimizing the total acoustic power output of the system, it is necessary to solve two integrals, and a further third if the minimum power output is required (see (8.3.15)). For the case of two monopole acoustic sources being considered here, the integrals of (8.3.3) can be solved analytically, thereby greatly simplifying the final result. To do this, note that b_w can be rewritten using the farfield approximation of (8.2.15) as

$$b_w \approx \frac{\pi r^2}{\rho_0 c_0}\int_0^{\pi}\left[\frac{\omega\rho_0}{4\pi}\right]^2 \frac{1}{r^2} q_p e^{-jk(d\cos\theta)} \sin\theta \ d\theta \tag{8.3.7}$$

and that

$$\int_0^{\pi} \cos(kd\cos\theta)\sin\theta \ d\theta = \frac{2\sin(kd)}{kd} = 2 \ \text{sinc} \ kd$$
$$\int_0^{\pi} \sin(kd\cos\theta)\sin\theta \ d\theta = 0 \tag{8.3.8}$$

Using these results, the terms in (8.3.3)–(8.3.5) can be evaluated as

$$a_w = \frac{\omega k \rho_0}{8\pi} = \frac{\omega^2 \rho_0}{8\pi c_0} \tag{8.3.9}$$

$$b_w = \frac{\omega k \rho_0}{8\pi} q_p \text{ sinc } kd \tag{8.3.10}$$

$$c_w = \frac{\omega k \rho_0}{8\pi} \tag{8.3.11}$$

resulting in the optimum control source volume velocity of (8.3.6) being expressed as (Nelson and Elliott, 1986):

$$q_{c,w} = -q_p \text{ sinc } kd \tag{8.3.12}$$

The minimum acoustic power output of the primary source/control source pair can be found by substituting the result of (8.3.12) back into (8.3.2), producing

$$W_{\min} = c_w - b_w^* a_w^{-1} b_w = c_w + b_w^* q_{c,w} = \frac{\omega^2 \rho_0}{8\pi c_0} \mid q_p \mid^2 (1-\text{sinc}^2 kd) \tag{8.3.13}$$

It will be shown shortly (in (8.3.24)) that the acoustic power radiated by the primary source operating alone is

$$W_{p,unc} = \frac{\omega^2 \rho_0}{8\pi c_0} \mid q_p \mid^2 \tag{8.3.14}$$

therefore, the reduction in total radiated acoustic power which is possible by introducing the control source is (Nelson and Elliott, 1986)

$$\frac{W_{\min}}{W_p} = 1-\text{sinc}^2 kd \tag{8.3.15}$$

Before discussing the significance of the results obtained so far it will be worthwhile to note that the control source volume velocity definition derived in (8.3.12) using farfield intensity integration, can be arrived at in a slightly different manner using a 'nearfield' analysis technique (Levine, 1980a, 1980b, 1980c; Nelson *et al.*, 1987). It will be worthwhile rederiving the previously obtained results using this methodology, as it will provide a basis for the work on minimizing the acoustic power output from more than one primary source by using more than one control source, which is considered later on in this chapter. This methodology is more akin to that used in the previous chapter to formulate the optimum control source volume velocity characteristics for minimizing the acoustic power propagation in air handling ducts, in that the acoustic power output at the source location is used as the basis of examination, rather than integrating to obtain the acoustic power in the farfield.

It was stated in Chapter 2 that the real acoustic power output of a source can be calculated by integrating the real acoustic intensity travelling out of a surface enclosing the source,

$$W = \int_S \boldsymbol{I} \cdot \boldsymbol{n} \, \mathrm{d}S \tag{8.3.16}$$

where $\boldsymbol{n}$ is the outward normal vector from the surface S enclosing the source. Performing this integration over a sphere which is enclosing the monopole source

$$W = \int_S \frac{1}{2}\mathrm{Re}\{p^*(\boldsymbol{r})\boldsymbol{u}(\boldsymbol{r})\}\cdot\boldsymbol{n}\ \mathrm{d}S \tag{8.3.17}$$

or

$$W = \frac{1}{2}\mathrm{Re}\{p^*(\boldsymbol{r})q\} \tag{8.3.18}$$

It should be noted that p and q in (8.3.18) are the acoustic pressure and volume velocity amplitudes respectively.

From (8.3.18) it can be deduced that, with a knowledge of the source volume velocity, calculation of the acoustic pressure at the monopole source location will enable the calculation of the acoustic power radiated by the source. The acoustic pressure due to the operation of the source at some location $\boldsymbol{r}$ relative to it, defined in (8.2.1), can be expressed in the form of an impedance as follows

$$p(\boldsymbol{r}) = q\ z(\boldsymbol{r}) \tag{8.3.19}$$

where z is a complex impedance given by

$$z(\boldsymbol{r}) = j\omega\rho_0\frac{\mathrm{e}^{-jkr}}{4\pi r} \tag{8.3.20}$$

where r is the distance between the source location $\boldsymbol{r}_q$ and the point of interest $\boldsymbol{r}$

$$r = |\boldsymbol{r}-\boldsymbol{r}_q| \tag{8.3.21}$$

The impedance can be expressed in terms of real and imaginary parts as

$$z(\boldsymbol{r}) = \frac{\omega k\rho_0}{4\pi}\left[\frac{\sin kr}{kr}+j\frac{\cos kr}{kr}\right] \tag{8.3.22}$$

Note that as $r\to 0$ the impedance $z(\boldsymbol{r})$ becomes infinite. However, on closer inspection it can be seen that this is due only to the imaginary part becoming infinite, as

$$\lim_{r\to 0}\frac{\sin kr}{kr} = 1 \tag{8.3.23}$$

Therefore, substituting (8.3.23) into (8.3.22) and then into (8.3.18) produces the expression for the acoustic power radiated by the source:

$$W = \frac{1}{2}|q|^2\ \mathrm{Re}\{z(\boldsymbol{r})\} = \frac{1}{2}|q|^2 z_0 \tag{8.3.24}$$

where

$$z_0 = \frac{\omega k\rho_0}{4\pi} \tag{8.3.25}$$

Note that this is the result which was 'anticipated' prior to (8.3.14).

Having now obtained a set of expressions for calculating the power output of a monopole source based upon the characteristics of the acoustic environment at its location, we are in a position to calculate the power output of the monopole primary and control sources when they are operating together. Consider first the primary monopole source. From (8.3.18), the sound power radiated by the primary monopole source in the presence of the sound field produced by the control monopole source will be

$$W_p = \frac{1}{2}\,\mathrm{Re}\left\{\left(p_p(\boldsymbol{r}_p)+p_c(\boldsymbol{r}_p)\right)^* q_p\right\} \tag{8.3.26}$$

where $p_p(\boldsymbol{r}_p)$ and $p_c(\boldsymbol{r}_p)$ are the acoustic pressures at the location of the primary monopole source, $\boldsymbol{r}_p$, due to the operation of the primary and control sources, respectively. Using (8.3.19) this can be re-expressed in terms of impedance

$$W_p = \frac{1}{2}\,\mathrm{Re}\left\{\left(q_p z_{p/p}+q_p z_{p/c}\right)^* q_p\right\} \tag{8.3.27}$$

where $z_{p/p}$ is the component of the primary source impedance due to self-operation,

$$z_{p/p} = \frac{p_p(\boldsymbol{r}_p)}{q_p} \tag{8.3.28}$$

and $z_{p/c}$ is the component of the primary source impedance due to the operation of the control source

$$z_{p/c} = \frac{p_c(\boldsymbol{r}_p)}{q_p} \tag{8.3.29}$$

Using the expression for the acoustic pressure at some location $\boldsymbol{r}$ relative to a monopole source (8.2.1), and the definition of z_0 given in (8.3.25), (8.3.27) can be simplified to Using a similar analysis, the acoustic power output of the

$$W_p = \frac{1}{2}\,|\,q_p\,|^2\,\mathrm{Re}\{\,(z_{p/p}+z_{p/c})^*\,\}$$

$$= \frac{1}{2}\,|\,q_p\,|^2 z_0\,\mathrm{Re}\left\{1+\frac{q_c^*}{q_p^*}\,\frac{\sin kd}{kd}\right\} = \frac{1}{2}\,|\,q_p\,|^2 z_0\,\mathrm{Re}\left\{1+\frac{q_c^*}{q_p^*}\,\mathrm{sinc}\,kd\right\} \tag{8.3.30}$$

control source operating in the presence of the primary source sound field can be evaluated as

$$W_c = \frac{1}{2}\,|\,q_c\,|^2\mathrm{Re}\left\{\,(z_{c/c}+z_{c/p})^*\,\right\} = \frac{1}{2}\,|\,q_c\,|^2 z_0\,\mathrm{Re}\left\{1+\frac{q_p^*}{q_c^*}\,\mathrm{sinc}\,kd\right\} \tag{8.3.31}$$

Combining (8.3.30) and (8.3.31), the total sound power output of the actively controlled system is

$$W_{tot} = W_p + W_c = \frac{1}{2}\Big(|q_c|^2 \operatorname{Re}\{z^*_{c/c}\} + q_c^* \operatorname{Re}\{z^*_{c/p}\} q_p \qquad (8.3.32)$$

$$+ q_p^* \operatorname{Re}\{z^*_{p/c}\} q_c + |q_p|^2 \operatorname{Re}\{z^*_{p/p}\} \Big)$$

Noting that from reciprocity $\operatorname{Re}\{z_{p/c}\} = \operatorname{Re}\{z_{c/p}\}$, (8.3.32) can be re-expressed in a form more amenable to solution by the quadratic optimization techniques used earlier:

$$W_{tot} = q_c^* a_w q_c + q_c^* b_w + b_w^* q_c + c_w \qquad (8.3.33)$$

where

$$a_w = \frac{1}{2}\operatorname{Re}\{z^*_{c/c}\} = \frac{\omega k \rho_0}{8\pi} \qquad (8.3.34)$$

$$b_w = \frac{1}{2}\operatorname{Re}\{z^*_{c/p}\} q_p = \frac{\omega k \rho_0}{8\pi} q_p \operatorname{sinc} kd \qquad (8.3.35)$$

$$c_w = \frac{1}{2} |q_p|^2 \operatorname{Re}\{z^*_{p/p}\} = |q_p|^2 \frac{\omega k \rho_0}{8\pi} \qquad (8.3.36)$$

These terms, however, show that the expression for the total radiated acoustic power given in (8.3.33), formulated using the nearfield technique, is exactly the same as that obtained using farfield integration, stated in (8.3.2) using the terms defined in (8.3.9)–(8.3.11) (as well it should be!). Using quadratic optimization techniques on (8.3.33) to determine the optimum control source volume velocity which will minimize the total radiated acoustic power will produce a result identical to that stated in (8.3.12):

$$q_{c,w} = -a_w^{-1} b_w \qquad (8.3.37)$$

Therefore, whether considering the minimization of real source power output (nearfield analysis) or total radiated acoustic power (farfield intensity integration), the optimum control source volume velocity relation is the same.

This idea is a general one: there are two ways to derive the necessary quadratic equation for acoustic power output, by considering some farfield measure of acoustic power output (integration of acoustic intensity in the farfield), or by directly considering the real acoustic power output at the source location. The merits of using either method of formulation will vary from case to case; however, the end result should be the same no matter what method is used.

There are several points of interest to note concerning the optimum control source volume velocity, stated in (8.3.12), and the resultant minimum acoustic power output, stated in (8.3.13). The first is that the amplitude of the optimum control source volume velocity is always *less* than the primary source volume velocity. This can be contrasted to the problem of nulling the acoustic pressure at a single point, considered in the previous section, where is was seen that the control source volume velocity amplitude was greater than the primary source volume velocity amplitude for points closer to the primary source than the control source, and less than the primary source volume velocity amplitude for points closer to the control source. Also, note that the control source volume velocity is constrained to be exactly 180° out of phase (for separation distances of odd intervals of half-wavelengths) or in phase (for separation distances of even intervals of half-wavelengths). This is in contrast to the problem of nulling the acoustic pressure at a single point, where the only constraint on phase is that the primary and control source pressures must be out of phase at the error sensing location. (Both of these facts have obvious implications for optimum error sensor placement, discussed later in this chapter.) These characteristics of the control source volume velocity can be seen in Fig. 8.7, which illustrates the optimum control source volume velocity as a function of separation distance *d*.

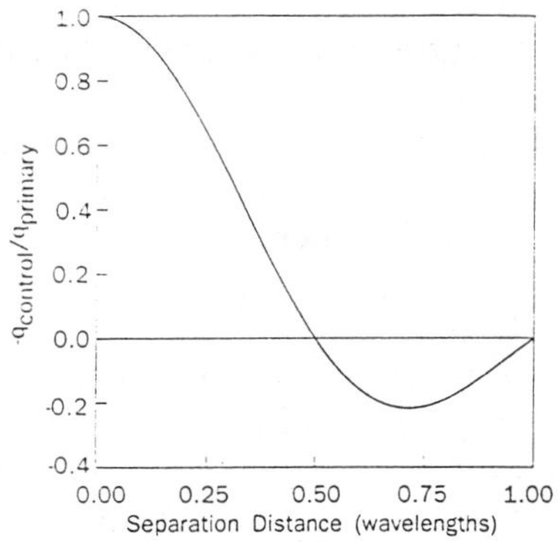

Fig. 8.7 Optimum control source volume velocity as a function of primary monopole source/control monopole source separation distance.

The next point to note is that the maximum achievable levels of acoustic power attenuation drop off rapidly as the separation distance between the control source and primary source increases, as shown in Fig. 8.8. These levels represent the absolute best result which can be achieved, no matter how good the electronics or how large the number of error sensors. (The appearance of the integrations in (8.3.3)−(8.3.5) used in formulating the result is equivalent to using an infinite number of error sensors, although the maximum levels of acoustic power attenuation can be achieved using a number which is greatly reduced from this in practice, in fact only 1 for this problem, with some judicious placement. This point will be considered later in the chapter.) So, for example, if 10 dB of acoustic power attenuation is desired, the control source must be placed within approximately 1/10 wavelength from the primary source as a prerequisite, before considering the practical problems of error sensor placement and electronic control system design.

One question which may arise out of the previously derived results is whether a reduction in the total radiated sound power via the control source

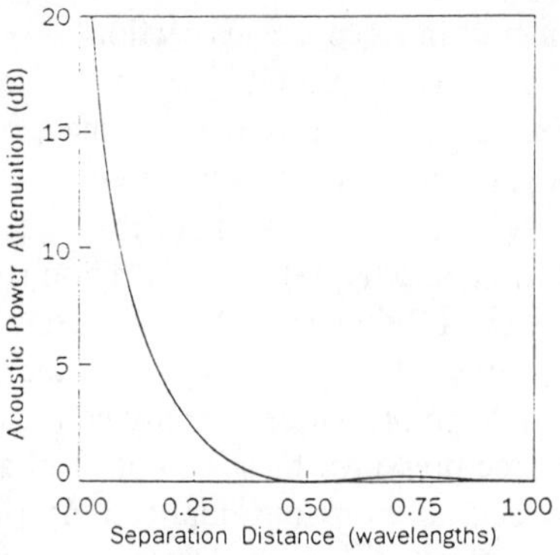

Fig. 8.8 Maximum acoustic power attenuation as a function of primary monopole source/control monopole source separation distance.

volume velocity of (8.3.12) necessarily corresponds to a reduction in the acoustic pressure amplitude at all locations in the farfield of the control source/primary source location. To answer this question, the control source volume velocity defined in (8.3.12) can be substituted into the expression for farfield acoustic pressure derived in the previous section, (8.2.18), producing

$$p(r) = p_p(r)\left(1 - \text{sinc}\ kd\ e^{jk(d\cos\theta)}\right) \tag{8.3.38}$$

where $p_p(r)$ is the acoustic pressure at some point $\boldsymbol{r}$ in the farfield when the primary source is operating alone. Therefore, the ratio of the squared acoustic pressure amplitudes at any point in the farfield before and after the application of active control is

$$\frac{|\,p(r)\,|^2}{|\,p_p(r)\,|^2} = 1 + \text{sinc}^2 kd - 2\ \text{sinc}\ kd\ \cos\,(kd\ \cos\theta) \tag{8.3.39}$$

For this to be less than one for all θ,

$$\text{sinc}^2 kd - 2\ \text{sinc}\ kd\ \cos kd < 0 \tag{8.3.40}$$

or

$$\text{sinc}\ kd < 2\cos kd \tag{8.3.41}$$

This criterion is valid for $\approx d < \lambda/6$, which is more stringent than the criterion for achieving a reduction in the total radiated acoustic power. Therefore, a reduction in the total radiated acoustic power is not necessarily accompanied by a reduction in the acoustic pressure at all points in the farfield. This fact is borne out in Fig. 8.9, which illustrates the residual farfield sound pressure levels when minimizing total radiated acoustic power at various separation distances.

In viewing the farfield pressure distributions of Fig. 8.9, it is readily apparent that when the separation distance is small, the residual sound field looks virtually identical to that of a dipole. However, as the separation distance increases, this characteristic becomes less exact. It is interesting to compare the

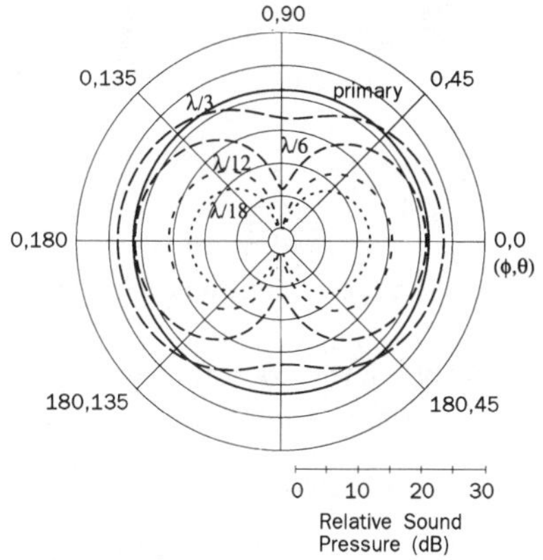

Fig. 8.9 Residual acoustic pressure distribution for various separation distances (labelled as a fraction of the wavelength of the acoustic disturbance) during maximum power attenuation.

optimal control source volume velocity with that of a dipole. With a dipole, the control source volume velocity $q_c = -q_p$, which is the limiting case of the optimum control source volume velocity of (8.3.12) as the separation distance tends towards zero. If this control source volume velocity is substituted into (8.3.32), the total acoustic power output of the primary source/control source pair is found to be

$$W_d = \frac{1}{2} \mid q_p \mid ^2 z_o \{2(1-\text{sinc } kd)\} \tag{8.3.42}$$

resulting in a change over the sole operation of the primary source of (Ffowcs Williams, 1984)

$$\frac{W_d}{W_p} = 2(1-\text{sinc } kd) \tag{8.3.43}$$

This result is directly comparable to that given in (8.3.15) for the change obtained when the optimum control source volume velocity is used. Observe that for the dipole arrangement, as the separation distance of the primary source and control source increases, the acoustic power output of the dipole becomes greater than that of the primary source operating alone, tending towards double the original power output as the two sources effectively become completely independent. The trend is different than what we found when using the optimum control source volume velocity, where as the separation distance increases the power reduction simply tends towards zero, as the optimum volume velocity 'law' dictates that the control source actually turns itself off when there are large separation distances involved.

It will be interesting to consider the time domain equivalent of the optimum control source volume velocity relationship stated in (8.3.12). To transform this frequency domain result into the time domain, the derivation of the control

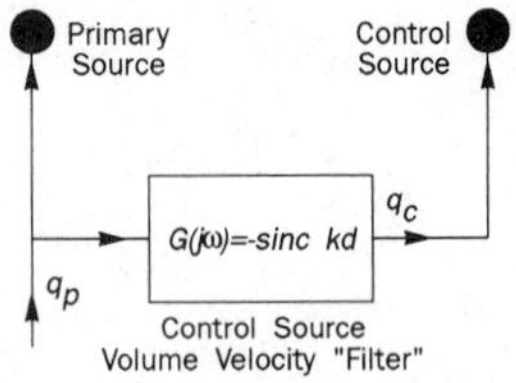

Fig. 8.10 Control source volume velocity derivation viewed as a filtering operation.

source volume velocity can be viewed as a filtering operation, as illustrated in Fig. 8.10. The frequency response of the filter, from (8.3.12), is:

$$G(\omega) = -\text{sinc } kd = -\frac{\sin kd}{kd} \tag{8.3.44}$$

The impulse response of this filter, found by taking the inverse Fourier transform of (8.3.44), is (Elliott and Nelson, 1986; Nelson and Elliott, 1986):

$$g(t) = -\frac{\left\{ H\left[t+\dfrac{d}{c_0}\right] - H\left[t-\dfrac{d}{c_0}\right] \right\}}{\dfrac{2d}{c_0}} \tag{8.3.45}$$

where $H(\alpha)$ is the Heaviside unit step function,

$$H(\alpha) = \begin{cases} 0 & \alpha<0 \\ 1 & \alpha>0 \end{cases} \tag{8.3.46}$$

The impulse response of (8.3.45) is shown on Fig. 8.11. Evaluation of this impulse response function requires a knowledge of what happened a time period (d/c_0) ago (the time of propagation between the two sources), as well as a knowledge of what will happen at a time (d/c_0) in the future. Hence, the use of the control source volume velocity of equation (8.3.12) results in a non-causal control system. This situation is quite satisfactory for harmonic sound problems, which are what is being considered here (and is possibly what constitutes the majority of noise problems targeted for active control). However, for non-periodic excitation, this control strategy is not appropriate. This point will be considered in more depth later in this chapter.

The final point to discuss is perhaps the most fundamental. What is the physical mechanism by which a reduction in the total radiated acoustic power is achieved? To answer this question, it is informative to calculate the acoustic

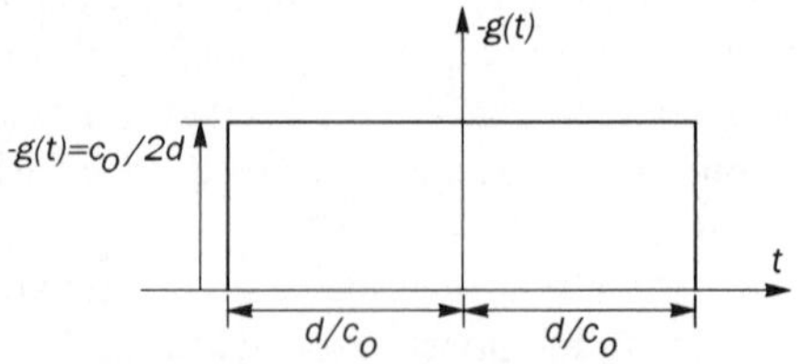

Fig. 8.11 Impulse response of the optimum control source volume velocity relationship.

power output of the primary and control sources during the operation of the active control system, where the control source volume velocity is as defined by (8.3.12). Consider first the control source. If the optimum volume velocity of equation (8.3.12) is substituted into the equation defining the control source power output derived via the nearfield technique (8.3.31), the result is

$$W_c = \frac{1}{2} \mid q_c \mid^2 z_0 \ \mathrm{Re}\left\{ 1 + \frac{q_p^*}{-q_p^* \operatorname{sinc} kd} \operatorname{sinc} kd \right\} = 0 \tag{8.3.47}$$

In other words, when the control source is driven to produce the optimal volume velocity, its acoustic power output is zero. Turning our attention to the primary source, if a similar substitution is done into (8.3.30), the actively controlled primary source acoustic output is found to be

$$W_p = \frac{1}{2} \mid q_p \mid^2 z_0 \ \mathrm{Re}\left\{ 1 - \frac{-q_p^* \operatorname{sinc} kd}{q_p^*} \operatorname{sinc} kd \right\} \tag{8.3.48}$$
$$= \frac{1}{2} \mid q_p \mid^2 z_0 \ (1 - \operatorname{sinc}^2 kd)$$

Comparing this to the original (uncontrolled) primary source power output, using (8.3.24),

$$W_{p,\mathrm{unc}} = \frac{1}{2} \mid q_p \mid^2 z_0 \tag{8.3.49}$$

it can be deduced that the introduction of the control source has reduced the power output of the primary source by a factor of $(1 - \operatorname{sinc}^2 kd)$. But how has this reduction been accomplished? It was stated in the beginning of Section 8.2 that the acoustic sources being considered here are all assumed to have a constant volume velocity, so that a change in this quantity can be ruled out as a mechanism for acoustic power reduction. Referring back to the defining equation for the primary source acoustic power, (8.3.18), it can therefore be deduced that a reduction in acoustic power output could only have come about by a reduction in the component of pressure at the source which is in-phase with the velocity. This is perhaps better stated in terms of the real, or active, radiation impedance presented to the source. For the primary monopole source, the active radiation impedance 'seen' under optimally controlled conditions is, from (8.3.30),

$$\mathrm{Re}\{z_{p/p} + z_{p/c}\} = z_0 \mathrm{Re}\left\{ 1 + \frac{q_c^*}{q_p^*} \operatorname{sinc} kd \right\} = z_0 \mathrm{Re}\{1 - \operatorname{sinc}^2 kd\} \tag{8.3.50}$$

Equation (8.3.50) states during the action of active control the real radiation impedance seen by the primary source is reduced. In other words, the source is unloaded by the introduction of the control source into the system. For the control source, under optimum controlled conditions the impedance is, from (8.3.31),

$$\mathrm{Re}\{z_{c/c}+Z_{c/p}\} = z_0\mathrm{Re}\left\{ 1+\frac{q_p^*}{q_{cw}^*}\mathrm{sinc}\; kd \right\} = z_0\mathrm{Re}\{1-1\} = 0 \tag{8.3.51}$$

This shows that while in the action of applying active control, the control source is itself unloaded, in fact perfectly so (the real radiation impedance is equal to zero). It can therefore be concluded that the physical mechanism by which attenuation of the total radiated acoustic power is obtained is a mutual unloading, or a mutual reduction in the real radiation impedance of the sources. The sources continue to displace the fluid medium (they are constant volume velocity sources), but this action is no longer efficiently converted into propagating acoustic power. This mechanism can also be referred to as one of acoustic power *suppression*. In fact, it has been known for many years that sound sources operating in close proximity to each other will mutually alter each other's radiation impedance, and hence acoustic power output; active control simply aims to optimize this phenomenon to minimize the total acoustic power output.

The reader may be wondering at this stage why the optimal control mechanism for free field radiation is exclusively one of suppression, and not a combination of suppression and absorption as was the case for two acoustic sources interacting in an air handling duct, considered in the previous chapter. The difference in mechanism is a direct result of the validity of acoustic reciprocity in free space, a concept which may not hold true in a non-rigidly terminated duct. This will be considered in greater depth later in this chapter.

It should be pointed out that the result of (8.3.51) does not imply that the control source cannot be *forced* to absorb sound; rather, in this instance it is simply not the optimal thing to do. To demonstrate this, consider the case where there are two monopole sources (control and primary) in close proximity, and constrained to have identical volume velocities while allowed a variable phase relationship, such that

$$q_c = q_p e^{j\phi} \tag{8.3.52}$$

With this scenario, the primary source, control source, and total system acoustic power output are, from (8.3.30)–(8.3.32):

$$\begin{aligned} W_p &= \frac{1}{2}\,|\,q_p\,|^2 z_0\ \mathrm{Re}\left\{1+\frac{q_c^*}{q_p^*}\ \mathrm{sinc}\ kd\right\} \\ &= \frac{1}{2}\,|\,q_p\,|^2 z_0 \left[1+\frac{\sin(kd-\phi)}{kd}\right] \end{aligned} \tag{8.3.53}$$

$$W_c = \frac{1}{2} \mid q_c \mid^2 z_0 \ \mathrm{Re}\left\{ 1 + \frac{q_p^*}{q_c^*} \ \mathrm{sinc}\ kd \right\} = \frac{1}{2} \mid q_p \mid^2 z_0 \left[1 + \frac{\sin(kd+\phi)}{kd} \right] \tag{8.3.54}$$

$$W_{tot} = W_p + W_c = \frac{1}{2} \mid q_p \mid^2 z_0 \left\{ \left[1 + \frac{\sin(kd-\phi)}{kd} \right] + \left[1 + \frac{\sin(kd+\phi)}{kd} \right] \right\} \tag{8.3.55}$$

Using these three equations, Fig. 8.12 illustrates the source and total acoustic power output for this scenario. It can be seen that at some source phase differences, especially 90° and 270°, one of the sources is absorbing acoustic power, which 'induces' the other source to substantially increase its acoustic power output. The absorption does not completely compensate for the increased output from the other source, however, and the result is a non-optimal change in the total acoustic power flow.

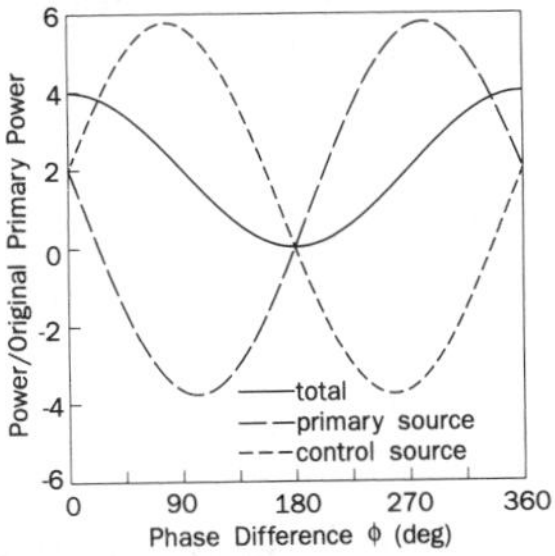

Fig. 8.12 Typical plot of primary source, control source, and total acoustic power output as a function of phase difference between the sources, where the sources are of equal volume velocity.

To continue with this line of thought, it will be interesting to consider what would happen if, rather than trying to minimize the total radiated acoustic power, the error criterion was changed to one of maximizing the power absorption of the control source. Such an arrangement could be viewed as equivalent to having a 'passive' control element being acted upon by the sound field (Elliott *et al.*, 1991). To derive the optimum control source volume velocity with the aim of producing such a system, the control source power output as defined in (8.3.31) can be re-expressed as

$$W_c = q_c^* a_w q_c + \frac{1}{2}\left(q_c^* z_{c/p} q_p + q_p^* z_{c/p}^* q_p \right) \tag{8.3.56}$$

where a_w is as defined in (8.3.34). Differentiating this equation with respect to the control source volume velocity, and setting the resultant gradient expression equal to zero, produces an expression for the control source volume velocity which will minimize the control source power output (i.e. maximize absorption):

$$q_{ca} = -\frac{1}{2} a_w^{-1} z_{c/p} q_p = -\frac{1}{2} q_p \left[\frac{\sin kd}{kd} + j \frac{\cos kd}{kd} \right] \tag{8.3.57}$$

where the subscript c,a denotes that the control source volume velocity has been derived with respect to maximizing absorption. The control source volume velocity is now based upon the total acoustic pressure from the primary source presented to it, rather than just the real component, as was used when minimizing the total acoustic power output. If this control source volume velocity is substituted into the defining equation for the primary source power output, (8.3.30), the controlled power output, written in terms of the original (uncontrolled) primary source power, is found to be

$$W_p = W_{p,\text{unc}} \left\{ 1 + \frac{1}{2}\left[\left(\frac{\cos kd}{kd}\right)^2 - \left(\frac{\sin kd}{kd}\right)^2 \right] \right\} \tag{8.3.58}$$

$$= W_{p,\text{unc}} \left\{ 1 + \frac{\cos 2kd}{2(kd)^2} \right\}$$

Making a similar substitution into (8.3.30) for the control source, its power output while being configured to maximally absorb, expressed in terms of the uncontrolled primary source power, is

$$W_c = -W_{p,\text{unc}} \left\{ \frac{1}{4}\left[\left(\frac{\cos kd}{kd}\right)^2 + \left(\frac{\sin kd}{kd}\right)^2 \right] \right\} \tag{8.3.59}$$

$$= -W_{p,\text{unc}} \left\{ \frac{1}{4(kd)^2} \right\}$$

where the negative sign indicates that the power flow is into the control source (power absorption). Using these expressions, the primary source, control source, and total power flow is shown in Fig. 8.13, plotted as a function of source separation distance. It can be seen that within a small band, between approximately $\lambda/12$ and $5\lambda/12$, some attenuation of the total system power output is achieved. However, as the separation distance becomes small, such that $kd \to 0$, the power output of the primary source increases dramatically. The absorption of the control source also increases, but not sufficiently to compensate for it; hence the total system power output also increases dramatically. This behaviour is predicted by (8.3.58) and (8.3.59), which show the limiting total power flow as the separation distance approaches zero to be (Elliott *et al.*, 1991)

$$\lim_{kd \to 0} W_{tot} = W_{p,\text{unc}} \frac{1}{4(kd)^2} \tag{8.3.60}$$

Thus, the net effect of an active control system configured to have the control source maximally absorb sound is to produce an enormous *increase* in the total system acoustic power output as the control source comes within close proximity of the primary source (Elliott *et al.*, 1991).

In concluding this discussion of the minimum power output of two free field monopole sources, it is useful to summarize a few important results. First, it is

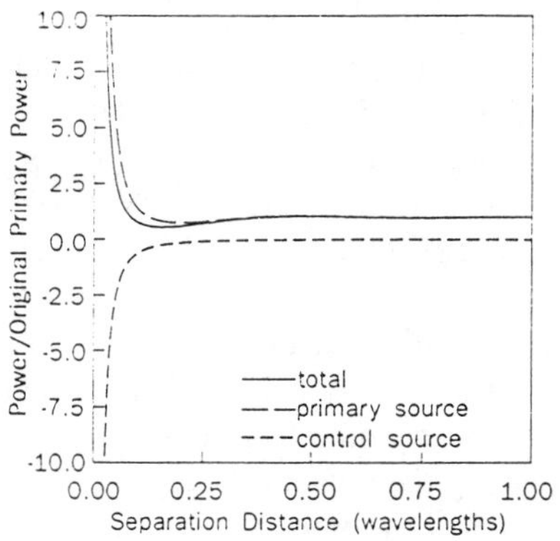

Fig. 8.13 Primary source, control source, and total acoustic power output when the control source is maximally absorbing.

possible to formulate analytically an expression for a control source volume velocity which will minimize the total acoustic power output of the system, using quadratic optimization techniques. This volume velocity was shown in (8.3.12) to be $q_c = -q_p$ sinc *kd*, which will reduce the total acoustic power output by a factor $\Delta W = 1 - \text{sinc}^2 kd$. From these expressions it is clear that the control source and primary source must be in close proximity to each other (say, within $\lambda/10$) for any appreciable reduction in radiated acoustic power to be obtained. The optimum control source volume velocity is less than that of the primary source, decreasing as the separation distance increases. In the limiting case as the separation distance $kd \rightarrow 0$ the control source and primary source form a dipole, but only in the limiting case. If the control source is constrained to be a dipole the results become significantly degraded as the separation distance increases. Also, the optimal control source volume velocity produces a system which acts non-causally with respect to the primary source. Finally, the mechanism of control for maximum acoustic power attenuation is exclusively one of power suppression, owing to the fact that the impedances seen by the primary and control sources when operating alone are equal. Any forcing of the control source to absorb the acoustic power output of the primary source will degrade the control effect; this effect will become more significant as the separation distance decreases and the rate of absorption increases.

8.4 ACTIVE CONTROL OF ACOUSTIC RADIATION FROM MULTIPLE PRIMARY MONOPOLE SOURCES USING MULTIPLE CONTROL MONOPOLE SOURCES

The examination of the use of a single harmonically oscillating monopole control source to attenuate the output from a similarly oscillating single primary monopole source undertaken in the previous two sections can be extended to include the use of multiple primary and control sources, as well as multiple error sensing locations. The general methodology used to examine these enlarged systems, to determine the optimum control source volume velocities for minimizing pressure or acoustic power, is the same as that used to examine the simple two source arrangement. However, the single complex quantities of pressure and volume velocity will need to be replaced with vectors of the acoustic pressures at a set of error sensing locations, or volume velocities from a set of sources. This extension will provide an introduction to the problem of controlling sound radiation from a vibrating structure, considered later in this chapter, where we will consider sound radiation from a set of structural modes of vibration. Both the problems of minimizing sound pressure at a set of discrete error sensing locations, and of minimizing the total acoustic power output, will be considered in this section. The former of these problems will be discussed first.

One related point worth mentioning here is that it is (theoretically) possible to group a set of monopole sources in a compact region and arrange the amplitude and phasing such that an arbitrary farfield acoustic pressure distribution is achieved. Therefore, given an unwanted sound field, a 'cancelling' field could be derived using a (possibly infinite) set of monopole sources. This concept, suggested in (Kempton, 1976), will not be discussed here, owing to our perceived lack of practicality. For further reading, refer to Kempton (1976) or Nelson and Elliott (1992).

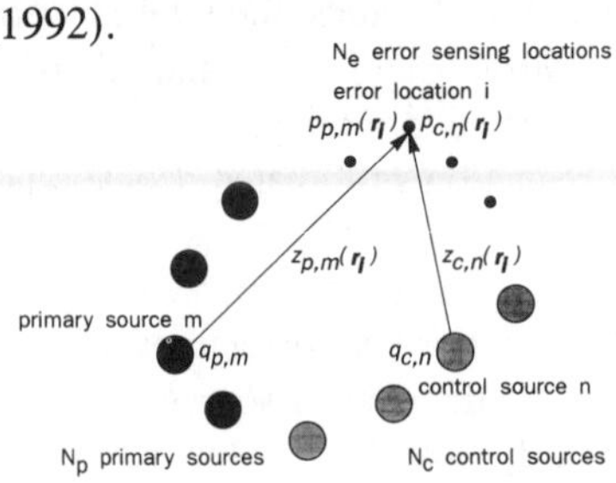

Fig. 8.14 Multiple monopole primary sources/multiple monopole control sources system arrangement.

Consider the system arrangement shown in Fig. 8.14, where there are N_p harmonically oscillating primary monopole sources, N_c harmonically oscillating control monopole sources, and N_e discrete error sensing locations where it is desired to minimise the acoustic pressure amplitude. The acoustic pressure at the ith error sensor location is, by superposition, the sum of contributions from the N_p primary sources and N_c control sources:

$$p(\boldsymbol{r}_i) = \sum_{m=1}^{N_p} p_{p,m}(\boldsymbol{r}_i) + \sum_{n=1}^{N_c} p_{c,n}(\boldsymbol{r}_i) \tag{8.4.1}$$

Using (8.2.1), which defines the acoustic pressure at some location $\boldsymbol{r}$ relative to a monopole source, (8.4.1) can be rewritten in terms of volume velocities:

$$p(\boldsymbol{r}_i) = \sum_{m=1}^{N_p} q_{p,m} z_{p,m}(\boldsymbol{r}_i) + \sum_{n=1}^{N_c} q_{c,n} z_{c,n}(\boldsymbol{r}_i) \tag{8.4.2}$$

where for the primary sources

$$z_{p,m}(\boldsymbol{r}_i) = \frac{j\omega\rho_0 e^{-jk|\boldsymbol{r}_i - \boldsymbol{r}_{p,m}|}}{4\pi |\boldsymbol{r}_i - \boldsymbol{r}_{p,m}|} \tag{8.4.3}$$

and for the control sources

$$z_{c,n}(\boldsymbol{r}_i) = \frac{j\omega\rho_0 e^{-jk|\boldsymbol{r}_i - \boldsymbol{r}_{c,n}|}}{4\pi |\boldsymbol{r}_i - \boldsymbol{r}_{c,n}|} \tag{8.4.4}$$

Equation (8.4.2) can be re-expressed in matrix form as

$$p(\boldsymbol{r}_i) = \boldsymbol{z}_p(\boldsymbol{r}_i)^{\mathrm{T}} \boldsymbol{q}_p + \boldsymbol{z}_c(\boldsymbol{r}_i)^{\mathrm{T}} \boldsymbol{q}_c \tag{8.4.5}$$

where $\boldsymbol{q}_p$ and $\boldsymbol{q}_c$ are the $(N_p \times 1)$ vector of primary monopole source volume velocities and $(N_c \times 1)$ vector of control monopole source volume velocities, respectively,

$$\boldsymbol{q}_p = \begin{bmatrix} q_{p,1} \\ q_{p,2} \\ \vdots \\ q_{p,N_p} \end{bmatrix}, \quad \boldsymbol{q}_c = \begin{bmatrix} q_{c,1} \\ q_{c,2} \\ \vdots \\ q_{c,N_c} \end{bmatrix} \tag{8.4.6}$$

and $\boldsymbol{z}_p(\boldsymbol{r}_i)$ and $\boldsymbol{z}_c(\boldsymbol{r}_i)$ are the $(N_p \times 1)$ vector of primary source radiation transfer functions to the error location $\boldsymbol{r}_i$ and the $(N_c \times 1)$ vector of control source radiation transfer functions to the error location $\boldsymbol{r}_i$, respectively,

$$\boldsymbol{z}_p(\boldsymbol{r}_i) = \begin{bmatrix} z_{p,1}(\boldsymbol{r}_i) \\ z_{p,2}(\boldsymbol{r}_i) \\ \vdots \\ z_{p,N_p}(\boldsymbol{r}_i) \end{bmatrix}, \quad \boldsymbol{z}_c(\boldsymbol{r}_i) = \begin{bmatrix} z_{c,1}(\boldsymbol{r}_i) \\ z_{c,2}(\boldsymbol{r}_i) \\ \vdots \\ z_{c,N_c}(\boldsymbol{r}_i) \end{bmatrix} \tag{8.4.7}$$

Equation (8.4.5) can be expanded to include consideration of the entire set of N_e error sensing locations:

$$\boldsymbol{p}_e = \boldsymbol{Z}_p \boldsymbol{q}_p + \boldsymbol{Z}_c \boldsymbol{q}_c \tag{8.4.8}$$

where $\boldsymbol{p}_e$ is the $(N_e \times 1)$ vector of pressures at the error locations,

$$\boldsymbol{p}_e = \begin{bmatrix} p(\boldsymbol{r}_1) \\ p(\boldsymbol{r}_2) \\ \vdots \\ p(\boldsymbol{r}_{N_e}) \end{bmatrix} \tag{8.4.9}$$

and Z_p and Z_c are the matrices of radiation transfer functions between the error locations and the primary and control source volume velocities respectively:

$$Z_p = \begin{bmatrix} z_p(r_1)^T \\ z_p(r_2)^T \\ \vdots \\ z_p(r_{N_e})^T \end{bmatrix} \qquad Z_c = \begin{bmatrix} z_c(r_1)^T \\ z_c(r_2)^T \\ \vdots \\ z_c(r_{N_e})^T \end{bmatrix} \tag{8.4.10}$$

The aim of the exercise in the first instance is to minimize the sound pressure at a set of error sensing locations, or to minimize the sum of the squared acoustic pressure amplitudes:

$$\sum_{i=1}^{N_e} | p(r_i) |^2 = \sum_{i=1}^{N_e} p^*(r_i)p(r_i) \tag{8.4.11}$$

Equation (8.4.11) can be written in matrix form:

$$\sum_{i=1}^{N_e} | p(r_i) |^2 = p_e^H p_e \tag{8.4.12}$$

where H is the Hermitian, or complex conjugate and transpose, of the matrix. Using (8.4.8), this can be expanded as

$$\begin{aligned} p_e^H p_e &= [Z_p q_p + Z_c q_c]^H [Z_p q_p + Z_c q_c] \\ &= q_c^H Z_c^H Z_c q_c + q_c^H Z_c^H Z_p q_p + q_p^H Z_p^H Z_c q_c + q_p^H Z_p^H Z_p q_p \end{aligned} \tag{8.4.13}$$

Equation (8.4.13) can be written more conveniently as

$$J_{prs} = p_e^H p_e = q_c^H A_{prs} q_c + q_c^H b_{prs} + b_{prs}^H q_c + c_{prs} \tag{8.4.14}$$

where

$$A_{prs} = Z_c^H Z_c \tag{8.4.15}$$

$$b_{prs} = Z_c^H Z_p q_p \tag{8.4.16}$$

and

$$c_{prs} = q_p^H Z_p^H Z_p q_p \tag{8.4.17}$$

Here the subscript *prs* is being used to denote pressure minimization. Note that c_{prs} is the sum of the squared pressure amplitudes at the error sensing locations during primary excitation only, prior to the commencement of active control.

Equation (8.4.14) shows the error criterion J_{prs}, the minimization of the squared acoustic pressure amplitudes at a set of discrete error locations, to be a real quadratic function of the *vector* of control source volume velocities. Hence the error surface, which is the error criterion plotted as a function of the real and imaginary components of the control source volume velocities, will be a $(2N_c+1)$-dimensional hyper-paraboloid whose axes are defined by the real and imaginary components of each complex volume velocity. This is a generalization of the three-dimensional 'bowl' shown in Fig. 8.2 for the case of $N_c = 1$. Being a quadratic function of q_c, the error criterion of (8.4.14) can be differentiated with respect to this quantity, and the resultant gradient expression set equal to zero to derive the vector of optimum control source volume velocities. Performing these operations on the real and imaginary parts separately gives

$$\frac{\partial J_{prs}}{\partial q_{c_R}} = 2A_{prs_R}q_{c_R} + 2b_{prs_R} = 0 \tag{8.4.18}$$

and

$$\frac{\partial J_{prs}}{\partial q_{c_I}} = 2A_{prs_I}q_{c_I} + 2b_{prs_I} = 0 \tag{8.4.19}$$

Multiplying the imaginary part by j and adding it to the real part produces the following expression for the optimum vector of control source volume velocities:

$$q_{c,p} = -A_p^{-1}b_p \tag{8.4.20}$$

The result of (8.4.20) is directly comparable to equation (8.2.12) for a system arrangement consisting of a single primary and control source, and a single error sensor. The difference is that (8.4.20) uses matrix quantities, and produces as a result the vector of optimum control source volume velocities.

Substitution of the result of (8.4.20) back into (8.4.14) yields the expression for the minimum residual (controlled) sum of the squared sound pressures at the error locations:

$$\left\{p_e^H p_e\right\}_{\min} = c_{prs} - b_{prs}^H A_{prs}^{-1} b_{prs} = c_{prs} + b_{prs}^H q_{c,p} \tag{8.4.21}$$

It should be noted that if the operation of (8.4.20) is to define a unique vector of control source volume velocities which will collectively minimize the sum of the squared acoustic pressure amplitudes at the error sensing locations, the matrix A_{prs} must be positive definite. For the case being considered here, that translates into the requirement that there be at least as many error locations as there are control sources. If there are fewer, then the matrix A_{prs} will be singular, resulting in an infinite number of 'optimum' control source volume velocity vectors. If there are an equal number of control sources and error locations then the vector of optimum control source volume velocities will completely null the acoustic pressure at the error locations. If there are more error locations than control sources, then the residual acoustic pressure at the error locations will probably not be equal to zero, although this may not have a negative influence upon the overall levels of acoustic power attenuation attained.

Before leaving this discussion of minimisation of the acoustic pressure at a set of discrete error locations, it is informative to have a brief qualitative look at the form of the matrix A_{prs}. If we examine the definition of this matrix given in (8.4.15) it can be deduced that its formulation can be written heuristically as

$$A_{prs} = \begin{bmatrix} \text{control source 1 transfer functions} \\ \text{control source 2 transfer functions} \\ \cdots \\ \text{control source } N_c \text{ transfer functions} \end{bmatrix}^* \begin{bmatrix} \text{control source 1 transfer functions} \\ \text{control source 2 transfer functions} \\ \cdots \\ \text{control source } N_c \text{ transfer functions} \end{bmatrix}^T \tag{8.4.22}$$

Viewed in this way, it can be deduced that the diagonal elements in $\boldsymbol{A}_{prs}$ are those terms which would be used in deriving the optimal control source volume velocities if the associated control source were the sole one operating. The off-diagonal terms represent the required modification to, or influence upon, this solution owing to the presence of other operating control sources. In other words, the off-diagonal terms are responsible for coupling the individually-optimal control source volume velocities. Consider, for example, a 2 primary source, 2 control source, 2 error sensor system. For this system the matrix $\boldsymbol{A}$ will be

$$\boldsymbol{A} = \frac{j\omega\rho_0}{4\pi} \begin{bmatrix} \dfrac{r_{1,1}+r_{1,2}}{r_{1,1}r_{1,2}} & \dfrac{e^{-jk(r_{2,1}-r_{1,1})}}{r_{1,1}r_{2,1}} + \dfrac{e^{-jk(r_{2,2}-r_{1,2})}}{r_{2,2}r_{1,2}} \\ \dfrac{e^{-jk(r_{1,1}-r_{2,1})}}{r_{1,1}r_{2,1}} + \dfrac{e^{-jk(r_{1,2}-r_{2,2})}}{r_{2,2}r_{1,2}} & \dfrac{r_{2,1}+r_{2,2}}{r_{2,1}r_{2,2}} \end{bmatrix} \tag{8.4.23}$$

where the quantity $r_{a,b}$ is the distance between control source a and error sensor b. The diagonal terms are only concerned with a single source, while the off-diagonal terms couple the solution by considering the transfer functions between both sources and error sensors. If an attempt were made to obtain the set of optimal control source volume velocities by considering each control source individually, deriving the volume velocity of one source which on its own would minimize the pressure at the error sensing locations, then deriving the next, and so on, the final result would be suboptimal; it would be equivalent to forcing all of the off-diagonal terms in $\boldsymbol{A}_p$ to be equal to zero. For the optimal set of control source volume velocities to be derived, the problem must be considered holistically, taking into account the inherent coupling of the control sources which is a result of mutual operation in the acoustic space, rather than individually. The quotation 'no man is an island' is never more true than when applied to multiple control sources in an active control system.

Having now considered the minimization of acoustic power at a number of discrete error locations, let us now turn our attention to the second of the two error criteria being studied here; minimization of total radiated acoustic power.

It was stated in the previous section that acoustic power was equal to the integration of the real acoustic intensity travelling out of a sphere enclosing the noise sources as follows:

$$W = 2\pi r^2 \int_0^\pi \frac{|\,p(r)\,|^2}{2\rho_0 c_0} \sin\theta \; d\theta \tag{8.4.24}$$

In a similar way to the single control source case, the pressure modulus squared term in (8.4.24) can be expanded to re-express the total radiated power as a quadratic function of the vector of control source volume velocities. Equation (8.4.5) provides an expression for the acoustic pressure at some location in space during the operation of the N_p primary sources and N_c control sources. Using this, the pressure modulus squared at any location is

$$\begin{aligned} |\,p(r)\,|^2 &= p(r)^* p(r) \\ &= \boldsymbol{q}_c^{\mathrm{H}} \boldsymbol{z}_c^* \boldsymbol{z}_c^{\mathrm{T}} \boldsymbol{q}_c + \boldsymbol{q}_c^{\mathrm{H}} \boldsymbol{z}_c^* \boldsymbol{z}_p^{\mathrm{T}} \boldsymbol{q}_p + \boldsymbol{q}_p^{\mathrm{H}} \boldsymbol{z}_p^* \boldsymbol{z}_c^{\mathrm{T}} \boldsymbol{q}_c + \boldsymbol{q}_p^{\mathrm{H}} \boldsymbol{z}_p^* \boldsymbol{z}_p^{\mathrm{T}} \boldsymbol{q}_p \end{aligned} \tag{8.4.25}$$

If (8.4.25) is substituted into (8.4.24), it is found that the total radiated acoustic power can be expressed as

$$W = \boldsymbol{q}_c^{\mathrm{H}} \boldsymbol{A}_w \boldsymbol{q}_c + \boldsymbol{q}_c^{\mathrm{H}} \boldsymbol{b}_w + \boldsymbol{b}_w^{\mathrm{H}} \boldsymbol{q}_c + c \tag{8.4.26}$$

where

$$\boldsymbol{A}_w = \frac{\pi r^2}{\rho_0 c_0} \left\{ \int_0^\pi \boldsymbol{z}_c^* \boldsymbol{z}_c^{\mathrm{T}} \sin\theta \; d\theta \right\} \tag{8.4.27}$$

$$\boldsymbol{b}_w = \frac{\pi r^2}{\rho_0 c_0} \left\{ \int_0^\pi \boldsymbol{z}_c^* \boldsymbol{z}_p^{\mathrm{T}} \sin\theta \; d\theta \right\} \boldsymbol{q}_p \tag{8.4.28}$$

and

$$c_w = \frac{\pi r^2}{\rho_0 c_0} \boldsymbol{q}_p^{\mathrm{H}} \left\{ \int_0^\pi \boldsymbol{z}_p^* \boldsymbol{z}_p^{\mathrm{T}} \sin\theta \; d\theta \right\} \boldsymbol{q}_p \tag{8.4.29}$$

From the form of (8.4.26), where power is expressed as a quadratic function of the control source volume velocities, $\boldsymbol{q}_c$, it is immediately apparent that the expression for the vector of control source volume velocities which will minimize the total radiated acoustic power will have the same form as (8.4.20); that is,

$$\boldsymbol{q}_{c,w} = -\boldsymbol{A}_w^{-1} \boldsymbol{b}_w \tag{8.4.30}$$

which will result in a residual total acoustic power output, found by substituting (8.4.29) into (8.4.26), of

$$W_{\min} = c_w - \boldsymbol{b}_w^{\mathrm{H}} \boldsymbol{A}_w^{-1} \boldsymbol{b}_w = c_w + \boldsymbol{b}_w^{\mathrm{H}} \boldsymbol{q}_{c,w} \tag{8.4.31}$$

Note again that the term c_w is equal to the value of the error criterion during the operation of the primary sources only, in this case the primary source radiated acoustic power prior to the onset of active control. From (8.4.31), the reduction in total acoustic power output is

$$\Delta W = \frac{W_{\min}}{W_{p,\mathrm{unc}}} = 1 - c_w^{-1} \boldsymbol{b}_w^H \boldsymbol{A}_w^{-1} \boldsymbol{b}_w = 1 + c_w^{-1} \boldsymbol{b}_w^H \boldsymbol{q}_{c,w} \tag{8.4.32}$$

For the problem being considered here, the form of the terms $\boldsymbol{A}_w$, $\boldsymbol{b}_w$, and c_w can be simplified by solving analytically the integral expressions in equation (8.4.27)−(8.4.29). Consider first $\boldsymbol{A}_w$. The integration required for the m,nth term is, from (8.4.4) and (8.4.27):

$$\boldsymbol{A}_w(m,n) = \frac{\pi r^2}{\rho_0 c_0} \int_0^{\pi} \left[\frac{j\omega\rho_0 e^{-jk|\boldsymbol{r}-\boldsymbol{r}_{c,m}|}}{4\pi |\boldsymbol{r}-\boldsymbol{r}_{c,m}|} \right]^* \left[\frac{j\omega\rho_0 e^{-jk|\boldsymbol{r}-\boldsymbol{r}_{c,n}|}}{4\pi |\boldsymbol{r}-\boldsymbol{r}_{c,n}|} \right] \sin\theta \, d\theta \tag{8.4.33}$$

As the integration is taking place in the farfield of the sources, we can approximate the integration sphere as having its origin midway along a line connecting the mth and nth control sources. By doing this we have arranged a geometry which is the same as that shown in Fig. 8.4, where $|\boldsymbol{r} - \boldsymbol{r}_{c,n}| \approx |\boldsymbol{r} - \boldsymbol{r}_{c,m}|$, and $|\boldsymbol{r} - \boldsymbol{r}_{c,m}| - |\boldsymbol{r} - \boldsymbol{r}_{c,n}| \approx d_{cm,cn}\cos\theta$, where $d_{cm,cn}$ is the separation distance between the mth and nth control sources. Making these approximations, (8.4.33) can be written as

$$\boldsymbol{A}_w(m,n) = \frac{\omega^2\rho_0}{16\pi c_0} \int_0^{\pi} e^{-jk(d\cos\theta)} \sin\theta \, d\theta \tag{8.4.34}$$

Using the standard results stated in (8.3.8), $\boldsymbol{A}_w(m,n)$ can now be evaluated as

$$\boldsymbol{A}_w(m,n) = \frac{\omega^2\rho_0}{8\pi c_0} \operatorname{sinc} kd_{cm,cn} \tag{8.4.35}$$

Note again that for the diagonal terms, where $d_{cm,cn} = 0$, sinc $d_{cm,cn} = 1.0$.

Using a similar procedure, $\boldsymbol{b}_w$ in (8.4.27) can be simplified to

$$\boldsymbol{b}_w = \boldsymbol{B}_w \boldsymbol{q}_p \tag{8.4.36}$$

where $\boldsymbol{B}_w$ is an $(N_c \times N_p)$ matrix, whose terms are defined by

$$\boldsymbol{B}_w(m,n) = \frac{\omega^2\rho_0}{8\pi c_0} \operatorname{sinc} kd_{cm,pn} \tag{8.4.37}$$

where $d_{cm,pn}$ is the separation distance between the mth control source and nth primary sources. Finally, c_w can also be simplified to

$$c_w = \boldsymbol{q}_p^{\mathrm{H}} \boldsymbol{C}_w \boldsymbol{q}_p \tag{8.4.38}$$

where $\boldsymbol{C}_w$ is an $(N_p \times N_p)$ matrix, the terms or which are defined by

$$\boldsymbol{C}_w(m,n) = \frac{\omega^2\rho_0}{8\pi c_0} \operatorname{sinc} kd_{pm,pn} \tag{8.4.39}$$

where $d_{pm,pn}$ is the separation distance between the mth and nth primary sources.

Before considering the results obtained thus far for minimizing the total radiated acoustic power by the set of N_p primary sources and N_c control sources, it will again be worthwhile rederiving the results using a nearfield, rather than farfield, methodology (as was done in the previous section for the single primary source, single control source case; see also Nelson *et al.* (1987)). As was stated in the previous section, while it is not the most generally applicable methodology for analysing free field radiation problems, particularly when a radiating structure is involved, the nearfield methodology does provide a 'neat' means of solution for the monopole radiation problem.

It was stated in the previous section, in (8.3.20), that the acoustic power output from a monopole source is equal to the real part of the product of the source volume velocity and the complex conjugate of the acoustic pressure at the source location:

$$W = \frac{1}{2}\,\mathrm{Re}\{p(r_q)^* q\} \tag{8.4.40}$$

If we consider the acoustic power output of the primary monopole sources, the total power output of the group is

$$W = \frac{1}{2}\sum_{i=1}^{N_p}\mathrm{Re}\{p(r_{pi})^* q_{pi}\} = \frac{1}{2}\,\mathrm{Re}\{\boldsymbol{p}_p^{\mathrm{H}}\boldsymbol{q}_p\} \tag{8.4.41}$$

where $\boldsymbol{p}_p$ is the ($N_p \times 1$) vector of complex acoustic pressures at the primary source locations,

$$\boldsymbol{p}_p = \begin{bmatrix} p(r_{p1}) \\ p(r_{p2}) \\ \vdots \\ p(r_{pN_p}) \end{bmatrix} \tag{8.4.42}$$

The acoustic pressures at the primary source locations, during the operation of the active control system, are the sum (superposition) of the pressures at that location attributable to the entire set of N_p primary sources and N_c control sources. Thus $\boldsymbol{p}_p$ can be re-expressed in terms of the volume velocities of the entire set of sources, in a manner similar to that of (8.4.8):

$$\boldsymbol{p}_p = \boldsymbol{Z}_{p/p}\boldsymbol{q}_p + \boldsymbol{Z}_{p/c}\boldsymbol{q}_c \tag{8.4.43}$$

where

$$\boldsymbol{Z}_{p/p} = \begin{bmatrix} z_p(r_{p1})^{\mathrm{T}} \\ z_p(r_{p2})^{\mathrm{T}} \\ \vdots \\ z_p(r_{pN_p})^{\mathrm{T}} \end{bmatrix} \qquad \boldsymbol{Z}_{p/c} = \begin{bmatrix} z_c(r_{p1})^{\mathrm{T}} \\ z_c(r_{p2})^{\mathrm{T}} \\ \vdots \\ z_c(r_{pN_p})^{\mathrm{T}} \end{bmatrix} \tag{8.4.44}$$

and $z_p(r_i)$, $z_c(r_i)$ are as defined in (8.4.7). Substituting this expression into (8.4.40), the total power output of the primary sources operating in the presence of the control sources is

$$W_p = \frac{1}{2}\,\mathrm{Re}\left\{q_p^{\mathrm{H}} Z_{p/p}^{\mathrm{H}} q_p + q_p^{\mathrm{H}} Z_{p/c}^{\mathrm{H}} q_c\right\} \tag{8.4.45}$$

An identical set of steps can be undertaken to derive a matrix expression for the total control source acoustic power output when operating in the presence of the primary sources. The result of doing this is

$$W_c = \frac{1}{2}\,\mathrm{Re}\left\{q_c^{\mathrm{H}} Z_{c/c}^{\mathrm{H}} q_c + q_c^{\mathrm{H}} Z_{c/p}^{\mathrm{H}} q_p\right\} \tag{8.4.46}$$

where

$$Z_{c/c} = \begin{bmatrix} z_c(r_{c1})^{\mathrm{T}} \\ z_c(r_{c2})^{\mathrm{T}} \\ \vdots \\ z_c(r_{cN_c})^{\mathrm{T}} \end{bmatrix} \qquad Z_{c/p} = \begin{bmatrix} z_p(r_{c1})^{\mathrm{T}} \\ z_p(r_{c2})^{\mathrm{T}} \\ \vdots \\ z_p(r_{cN_c})^{\mathrm{T}} \end{bmatrix} \tag{8.4.47}$$

and $z_p(r_i)$, $z_c(r_i)$ are again as defined in (8.4.7). Combining equations (8.4.44) and (8.4.45), the total acoustic power output of the system is

$$\begin{aligned} W_{tot} &= W_p + W_c \\ &= \frac{1}{2}\,\mathrm{Re}\left\{q_c^{\mathrm{H}} Z_{c/c}^{\mathrm{H}} q_c + q_c^{\mathrm{H}} Z_{c/p}^{\mathrm{H}} q_p + q_p^{\mathrm{H}} Z_{p/c}^{\mathrm{H}} q_c + q_p^{\mathrm{H}} Z_{p/p}^{\mathrm{H}} q_p\right\} \end{aligned} \tag{8.4.48}$$

Equation (8.4.48) can be re-expressed in a format more amenable to minimization using quadratic optimization techniques, which by now are well known to the reader, as

$$W_{tot} = q_c^{\mathrm{H}} A_w q_c + q_c^{\mathrm{H}} b + b^{\mathrm{H}} q_c + c \tag{8.4.49}$$

where

$$A_w = \frac{1}{2}\,\mathrm{Re}\left\{Z_{c/c}\right\} \tag{8.4.50}$$

$$b_w = \frac{1}{2}\,\mathrm{Re}\left\{Z_{p/c}^{\mathrm{H}} q_p\right\} \tag{8.4.51}$$

and

$$c_w = \frac{1}{2} q_p^{\mathrm{H}}\,\mathrm{Re}\left\{Z_{p/p}^{\mathrm{H}}\right\} q_p \tag{8.4.52}$$

However, the terms in these equations are exactly the same as those given in (8.4.33), (8.4.35), and (8.4.37) respectively, derived using the farfield measure of acoustic power as a basis. Therefore, once again the quadratic expression is the same, whether a farfield or nearfield measure of power is employed in its derivation.

We can now use the multiple source formulation developed in this section to examine the influence which control source and error sensor placement has upon overall system performance.

8.5 THE EFFECT OF TRANSDUCER LOCATION

In the previous sections expressions defining control source volume velocities which will minimize the total radiated acoustic power from a set of acoustic monopole sources, or which will minimize the acoustic pressure amplitude at a number of discrete error locations, have been derived. Being able to calculate these quantities is a first step in the design of an active control system. The next step is to use these quantities to optimally place the control sources and error sensors. The aim of this section is to use some of the results of the previous sections to demonstrate the effects which transducer placement has upon the performance of the monopole-based systems considered thus far. This will provide a 'primer' for the structural radiation discussion undertaken later in this chapter.

As mentioned in the introduction to this chapter and elsewhere in this book, the design of a feedforward active control system can be viewed as a hierarchical procedure aimed at optimizing the performance of the four principal components of the system; introduction of the control signal, extraction of an error signal, acquisition of a primary disturbance-correlated reference signal for the controller, and design of the controller itself. If any one of these facets is improperly designed then the system is doomed, as it is only as good as its weakest link. The first two of these facets are of principal interest here.

While it is true that all components must be optimized for the system to function, the placement of the control sources, to introduce the controlling disturbance into the system, is arguably the most important stage in designing the active control system. This is because the location of the control sources will place the absolute bound upon the level of acoustic power attenuation which can be achieved, and regardless of the number of error sensors used to sense the residual sound field, regardless of the quality and expense of the electronics used to drive the system. Unfortunately, optimizing the control source locations is also one of the least straightforward aspects of designing an active control system. This is because acoustic power attenuation is not a linear function of control source location. Thus while it is possible to determine directly the maximum levels of acoustic power attenuation which are possible with a given control source arrangement using the quadratic optimization approach of the previous sections, it is not possible to use such an approach to determine the control source locations which will maximize these levels. Usually, to place the control sources, some form of (numerical) 'search' routine must be undertaken, using the calculation of acoustic power attenuation for a given control source arrangement (calculated using the quadratic optimization procedure of the previous sections) to provide the error criterion. There are a few extremely simple monopole arrangements where an equation for power reduction in terms of control source location can be derived, and it is these which will be studied here.

For the most simple system of all, where a single monopole control source is used to attenuate acoustic radiation from a single monopole primary source, the only location consideration is separation distance. It was shown in Section 8.3 that the effect which separation distance has upon the levels of acoustic power attenuation which can be achieved is quantified in the expression

$$\frac{W_{\text{controlled}}}{W_{\text{uncontrolled}}} = 1 - \text{sinc}^2 kd \tag{8.5.1}$$

It is worthwhile emphasizing again the extremely rapid reduction in acoustic power attenuation which accompanies an increase in the distance between the primary and control sources. Using the approximate measure of 10 dB reduction at a separation distance of 1/10 wavelength, consider that at 344 Hz the control source must be with 100 mm of the primary source to provide this rather moderate amount of attenuation. If the separation distance is increased to 500 mm no attenuation of the total radiated acoustic power is possible. Close source proximity is a must in active control of sound power.

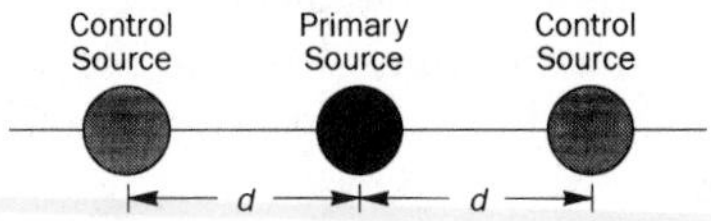

Fig. 8.15 Single monopole primary source/dual monopole control source system arrangement.

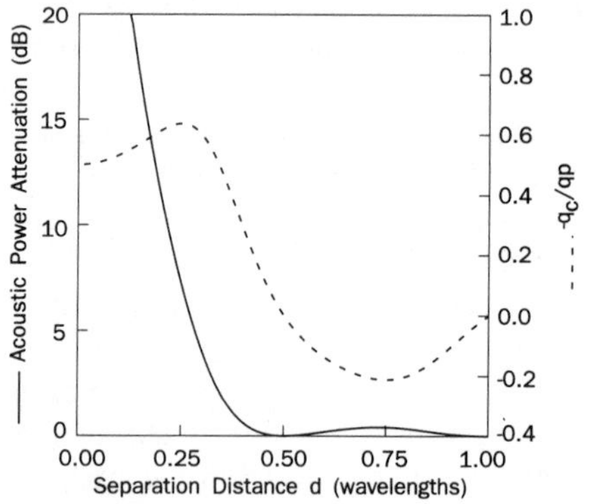

Fig. 8.16 Optimum control source volume velocities and maximum acoustic power attenuation as a function of separation distance *d* for the single primary/dual control monopole source system.

The addition of a second control source can improve the acoustic power attenuation when the sources are in close proximity, but can do little to improve the result when the source separation distance is greater than one half of an acoustic wavelength. Consider the arrangement depicted in Fig. 8.15, where two monopole control sources are used to attenuate the radiated acoustic power from a single monopole primary source. All sources are on a common axis, and the control sources are an equal distance from the primary source. The maximum levels of acoustic power attenuation for this arrangement are plotted as a function of separation distance in Fig. 8.16. Note that while the result is better than that obtained using a single control source for 'small' separation distances, the improvement deteriorates to a negligible level when the separation distance exceeds one half acoustic wavelength.

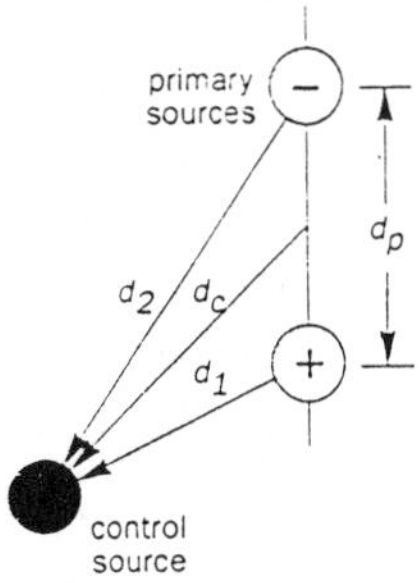

Fig. 8.17 Dual primary source/monopole control source system arrangement.

A third case where the effect of control source location on acoustic power attenuation can be relatively easily calculated is that of a single monopole control source being used to attenuated acoustic radiation from a dipole-like pair of primary sources. The optimum control source volume velocity relationship can be calculated using the quadratic optimization approach outlined in the previous section. For the arrangement shown in Fig. 8.17, it can easily be deduced that

$$a = \frac{\omega^2 \rho_0}{8\pi c_0}$$

$$b = \frac{\omega^2 \rho_0}{8\pi c_0} q_p (\operatorname{sinc} kd_1 - \operatorname{sinc} kd_2) \tag{8.5.2}$$

$$c = \frac{\omega^2 \rho_0}{4\pi c_0} \mid q_p \mid^2 (1 - \operatorname{sinc} kd_p)$$

where q_p is the primary source volume velocity, taken to be equal in amplitude but opposite in sign for the two sources, d_1 and d_2 are the separation distances between the control source and the two primary sources, and d_p is the separation distance between the two primary sources. Substituting these definitions into (8.4.31), for the case being considered here, the minimum achievable radiated acoustic power is:

$$W_{\min} = \frac{\omega^2 \rho_0}{8\pi c_0} \mid q_p \mid^2 \left\{ 2(1 - \operatorname{sinc} kd_p) - (\operatorname{sinc} kd_1 - \operatorname{sinc} kd_2)^2 \right\} \tag{8.5.3}$$

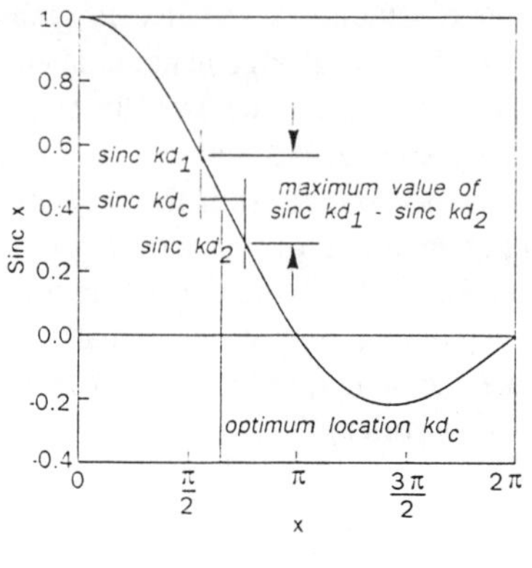

Fig. 8.18 Plot of sinc function used to determine optimum control source location.

From (8.5.3) it is clear that the optimum control source volume velocity is one which maximizes the quantity (sinc kd_1 − sinc kd_2)^2, or maximizes the difference between sinc kd_1 and sinc kd_2. The problem of determining the optimum control source location can therefore be solved heuristically, referring to the shape of the sinc function curve shown in Fig. 8.18. Optimally, the control source should be placed on the primary source axis (where $(d_1 - d_2)$ will be maximized), situated at a distance from the mid-point between the two sources such that the slope (sinc kd_1 − sinc kd_2) is maximized, which for this separation distance and frequency is approximately $kd_c = 2\pi/3$. This trend is displayed clearly in Fig. 8.19, which depicts the levels of acoustic power attenuation possible for a given control source location, where the primary source separation distance is 0.1 m and the excitation frequency is $\omega = 2\pi c_0$. Note that the optimum control source placements are at mirror image locations of $kd_c \approx \pm 2\pi/3$, and are not at the location of either primary source.

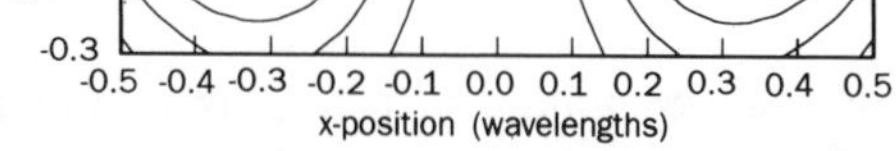

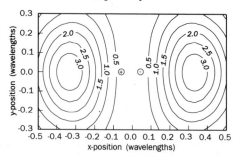

Fig. 8.19 Maximum acoustic power attenuation as a function of control source location for a dipole primary source, primary source separation distance is 1/10 of a wavelength.

If the strengths of the two primary sources are unequal, the control source placement characteristics outlined above change. Consider again the arrangement depicted in Fig. 8.17, but where the volume velocities of primary source 1 and 2 are αq_p and $-q_p$, respectively, where α can be viewed as a 'dipole modifying factor'. Working through the same analysis referenced previously, the terms a, b, and c are now found to be equal to

$$a = \frac{\omega^2 \rho_0}{8\pi c_0}$$

$$b = \frac{\omega^2 \rho_0}{8\pi c_0} q_p (\alpha \text{ sinc } kd_1 - \text{sinc } kd_2) \tag{8.5.4}$$

$$c = \frac{\omega^2 \rho_0}{4\pi c_0} | q_p |^2 (1 + \alpha^2 + (1-\alpha) \text{ sinc } kd_p)$$

This leads to a minimum total acoustic power output of:

$$W_{\min} = \frac{\omega^2 \rho_0}{8\pi c_0} | q_p |^2 \left\{ (1+\alpha^2+(1-\alpha)\text{sinc } kd_p) - (\alpha \text{ sinc } kd_1 - \text{sinc } kd_2)^2 \right\} \tag{8.5.5}$$

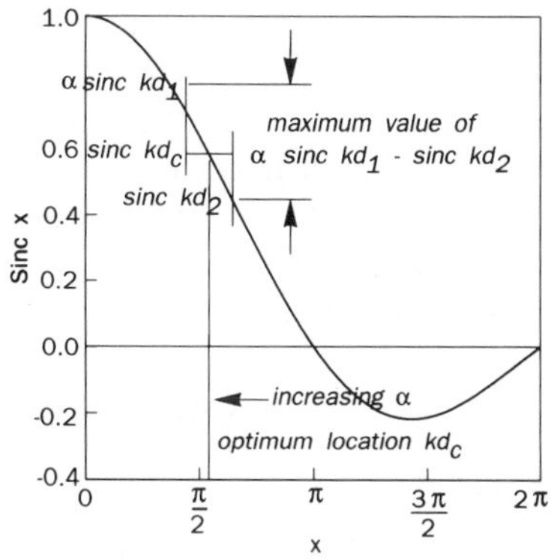

Fig. 8.20 Evaluation of optimum control source location when the primary source monopoles are of unequal strengths.

The influence of the different primary monopole source strengths can be qualified by considering the fact, from (8.5.5), that to achieve the lowest minimum acoustic power output the term $(\alpha \text{ sinc } kd_1 - \text{sinc } kd_2)^2$ must be maximized. In light of the characteristics of the sinc function, illustrated in Fig. 8.20, it can be deduced that as α increases from a value of 1.0 the optimum control source location moves closer to the stronger source, so that the difference between (α sinc kd_1) and (sinc kd_2) is maximized. This difference also grows in size, meaning that larger values of attenuation are possible. This intuitively makes sense, as the monopole component of the sound field has increased, and the single control source will chiefly attenuate this component. Further, as α departs from a unity value so too does the phenomenon of mirror image optimum control source locations.

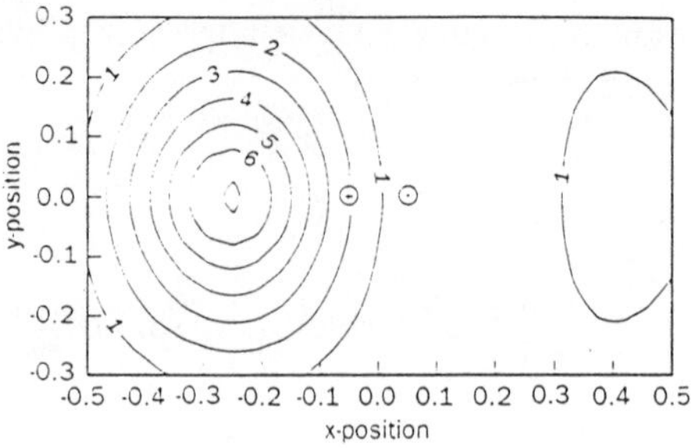

Fig. 8.21 Maximum acoustic power attenuation as a function of control source location for a dipole-like primary source, where one primary source is 20% stronger than the other, and the primary source separation distance is 1/10 of a wavelength.

These qualitative views are mirrored in Fig. 8.21, which illustrates the maximum levels of acoustic power attenuation as a function of control source location for a single sinusoidal primary source output with only a small deviation from equal source strengths, $\alpha = 1.2$ (the change in the harmonic excitation case is identical, and so will not be shown). Here the optimum control source location has moved from a value $kd_c = 2\pi/3$ to a position $kd_c = \pi/2$. Note also that the maximum levels of sound power attenuation have increased by a factor of two.

Once the control source location has been selected, the next step in the design of the active control system is to optimally locate the error sensor(s). For sinusoidal excitation, the control source output which will minimize the acoustic pressure amplitude at a location in space for a single control source and error sensor was given in (8.2.5). At optimum error microphone positions this volume–velocity relationship will be equal to the optimum for minimizing the total radiated acoustic power, given in (8.3.12). For primary source/control source arrangements which are closely spaced, therefore, the nearest optimum error sensor placements will run on a line perpendicular to the control source/primary source axis, centred at a position between the primary and control sources such that r_c/r_p=sinc kd, which will always be closer to the control source. Figure 8.22 illustrates the resultant power attenuation from minimizing the acoustic pressure amplitude at a single location for harmonic excitation, with a primary source/control source separation distance of 0.1λ and an exciting frequency $\omega = 2\pi c_0$. The outlined characteristics of optimal error sensor locations are clearly evident.

There are two important points which are evident from this result. First, placing an error sensor in the near field of the sources, especially the primary source, is a 'risky' pursuit; the acoustic power attenuation achieved when minimizing the acoustic pressure at the sensor(s) varies rapidly with location. Second, the optimum error sensor location(s) are always at the locations of greatest acoustic pressure attenuation when the control source is generating the optimum volume–velocity relationship. Both of these results are general for the active control of free field sound radiation.

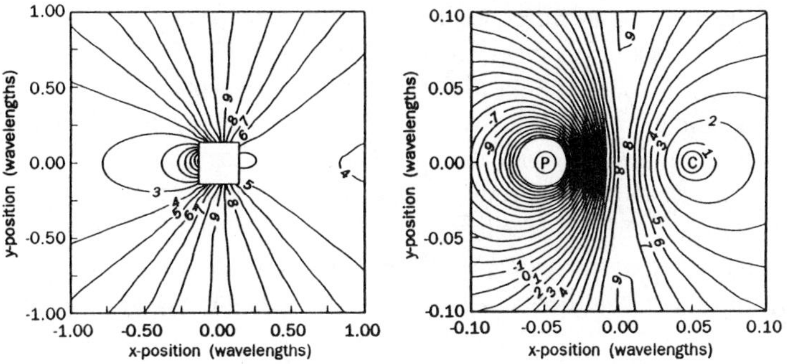

Fig. 8.22 Acoustic power attenuation as a function of error sensor placement, single monopole primary and control sources separated by 1/10 wavelength.

8.6 REFERENCE SENSOR LOCATION CONSIDERATIONS

Thus far in this section we have considered the problem of calculating the attenuation of a radiated sound field for a given 'physical system', control source(s) and error sensor(s), arrangement. In doing this we have essentially calculated an optimum open loop control law for minimizing the radiated acoustic power or acoustic pressure at the error sensor location, and applied the resulting signal to the control source(s). The assumption behind this approach is that when a feedforward control system is attached to the system through the control sources and error sensors, it will do the same thing. However, feedforward control systems require a reference signal which is correlated with the impending primary disturbance from which to derive the control input. When the unwanted (primary) disturbance is periodic it is often possible to acquire this reference signal directly from the source of excitation, such as from a tachometer attached to an item of rotating machinery, although in general a sensor such as a microphone must be used to measure the primary and disturbance, providing the reference signal to the feedforward controller.

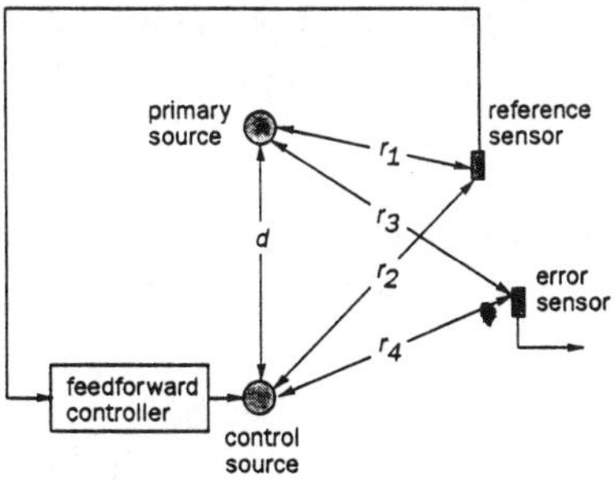

Fig. 8.23 Single primary source, control source, and error sensor system with feedforward control and reference sensor (single channel feedforward active control system).

This, however, can introduce the new problems into the system design. Referring to the system arrangement shown in Fig. 8.23, which depicts a single monopole control source being used to attenuate the acoustic radiation from a single monopole primary source with a single reference sensor and error sensor (microphones), it can be seen that feedback from the control source to the reference sensor can occur. From the equivalent block diagram in Fig. 8.24, it is readily apparent that this feedback loop has the potential to drive the system unstable, purely from the geometry of the system. In this section we will investigate this problem, and derive some criteria for maintaining system stability.

8.6.1 Problem formulation

For the control system to null the acoustic pressure at the error sensor, the control and primary signals at its location must be equal in amplitude and opposite in phase:

$$S(s) = -P(s) \tag{8.6.1}$$

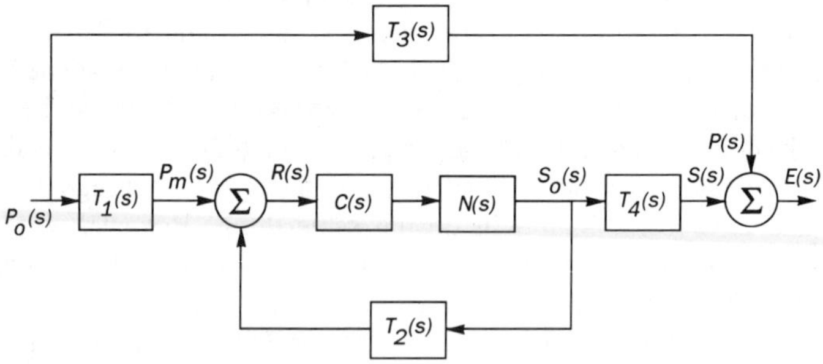

Fig. 8.24 Block diagram of single channel feedforward active control system shown in Fig. 8.23.

From Fig. 8.24, it may be seen that the primary source signal at the error sensor is equal to the primary source signal at the origin of the disturbance, $P_0(s)$, modified by the transfer function between this point and the error sensor, $T_3(s)$,

$$P(s) = T_3(s)P_0(s) \tag{8.6.2}$$

Similarly, the control source signal at the error sensor is equal to the control source signal $S_0(s)$ modified by the transfer function between this location and the error sensor, $T_4(s)$,

$$S(s) = T_4(s)S_0(s) \tag{8.6.3}$$

The control source signal can also be expressed in terms of the primary source signal as

$$S_0(s) = \frac{C(s)N(s)T_1(s)}{1 - C(s)N(s)X_2(s)}P_0(s) \tag{8.6.4}$$

where $C(s)$ is the Laplace transform of the controller, $N(s)$ is the Laplace transform of the other 'peripheral' components of the control system, including the reference sensor and control source frequency response characteristics, as well the frequency response characteristics of any filters, amplifiers, and other electronics, $T_1(s)$ is the transfer function between the primary disturbance origin and the reference sensor location, and $T_2(s)$ is the transfer function between the control source location and the reference sensor location. Substituting (8.6.2), (8.6.3), and (8.6.4) into (8.6.1), the frequency response characteristics of the optimum controller, which will null the acoustic pressure at the error sensor, are defined by the relationship

$$C(s) = \frac{T_3(s)}{N(s)\left[T_2(s)T_3(s) - T_1(s)T_4(s)\right]} \tag{8.6.5}$$

Observe from (8.6.5) that feedback of the control signal to the reference sensor has placed poles in the optimum control source transfer function, the same result which we encountered in Chapter 6 when considering feedback to the reference sensor for a duct noise problem. Thus feedback alters the result from the simple gain and phase changes which described the optimum control source characteristics without feedback to the reference sensor, which we have derived thus far in this chapter. There are several important conclusions which can be drawn from this result:

1. An adaptive feedforward control system based upon infinite impulse response (IIR) filters will be better suited to this problem than a system based upon finite impulse response (FIR) filters, owing to the IIR filter being a pole-zero architecture, and the FIR filter being an all-zero architecture.
2. It is possible that the feedback to the reference sensor, which forms a closed-loop system, will cause the control system to go unstable when an attempt to null the acoustic pressure at the error sensor is made. It is this last point which will be of interest to us here.

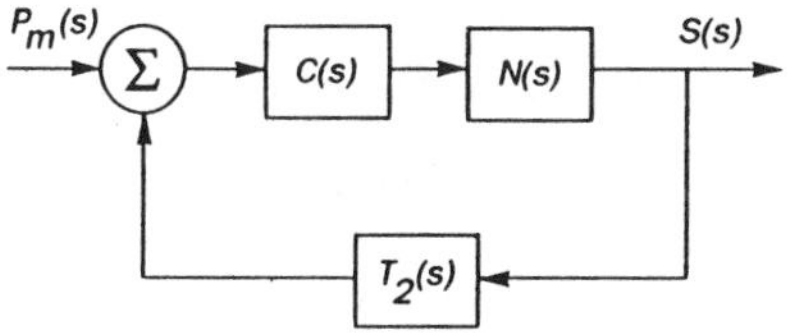

Fig. 8.25 Block diagram of feedback part of control system.

To examine the possibility of instability arising from feedback of the control signal to the reference sensor, we need only to consider the closed loop part of the system shown in Fig. 8.25. Using this arrangement, the control signal is now defined by the expression

$$S(s) = C(s)N(s)[P_m(s)+T_2(s)S(s)] \tag{8.6.6}$$

where $P_m(s)$ is the primary source disturbance as measured by the reference sensor. We can express this in the more standard form of a closed loop transfer function as

$$\frac{S(s)}{P_m(s)} = \frac{C(s)N(s)}{1+G(s)} \tag{8.6.7}$$

where $G(s)$ is defined by

$$G(s) = -C(s)N(s)T_2(s) \tag{8.6.8}$$

From (8.6.7), the roots of the characteristic equation $(1 + G(s))$ should all lie in the left-hand side of the imaginary axis for the system to be stable.

To examine the bounds of stability and instability, we can derive relationships between the gain and phase margins of the system and the locations of the sources and sensors. Recall that the gain margin is the factor whereby the gain is less than the critical value (above which the system becomes unstable). If we separate $G(s)$ into explicit gain and phase components,

$$G(j\omega) = B(\omega)e^{j\theta(\omega)} \tag{8.6.9}$$

then the gain margin K_g is defined by the expression

$$K_g = \frac{1}{B(\omega)} \quad \text{when } \theta(\omega) = -\pi \tag{8.6.10}$$

Phase margin is the additional phase K_θ required to make the system unstable, which is the amount by which the phase exceeds $-180°$ when the gain is equal to 1. For our problem here, phase margin is defined by

$$K_\theta = \theta(\omega) + \pi \quad \text{when } B(\omega) = 1 \tag{8.6.11}$$

To calculate the gain and phase margins of the optimal system, we can substitute the optimum controller transfer function of (8.6.5) into (8.6.8). After some minor manipulation we find that (Tokhi and Leitch, 1991, 1992)

$$G(s) = \frac{1}{\dfrac{T_1(s)T_4(s)}{T_2(s)T_3(s)}-1} \tag{8.6.12}$$

We can simplify this as

$$G(j\omega) = \frac{1}{H(\omega)e^{j\phi(\omega)}-1} \tag{8.6.13}$$

where $H(\omega)$ and $e^{j\phi(\omega)}$ are gain and phase terms,

$$H(\omega)e^{j\phi(\omega)} = \frac{T_1(j\omega)T_4(j\omega)}{T_2(j\omega)T_3(j\omega)} \tag{8.6.14}$$

Substituting (8.6.14) back into (8.6.13) and the result into (8.6.9), the gain and phase of $G(s)$ respectively are found to be defined by

$$B(\omega) = \left[H^2(\omega) - 2H(\omega)\cos\phi(\omega) + 1\right]^{-1/2} \tag{8.6.15}$$

and

$$\theta(\omega) = \tan^{-1}\frac{H(\omega)\sin\phi(\omega)}{1 - H(\omega)\cos\phi(\omega)} + 2m\pi, \quad m=0,\pm1,\ldots \tag{8.6.16}$$

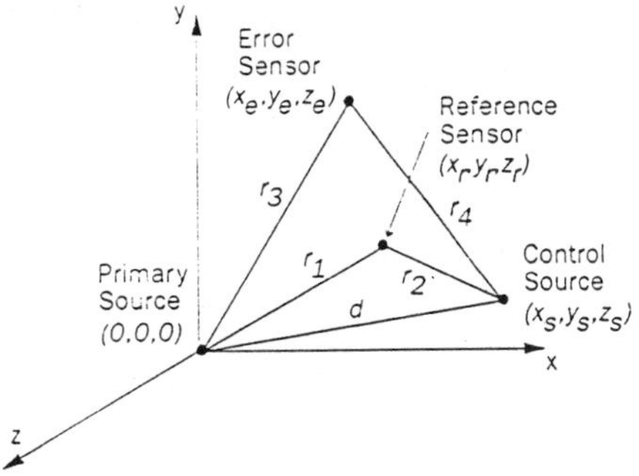

Fig. 8.26 Source and sensor geometry for single channel system.

Our aim here is to examine system stability based upon the physical arrangement of the sources and sensors. To do this, we can place the primary source at a location (0,0,0), control source at (x_s,y_s,z_s), reference sensor at (x_r,y_r,z_r), and error sensor at (x_e,y_e,z_e) as shown in Fig. 8.26. The various separation distances shown in the figure are now defined by the expressions

$$r_1 = \left[x_r^2+y_r^2+z_r^2\right]^{1/2} \tag{8.6.17}$$

$$r_2 = \left[(x_r-x_s)^2+(y_r-y_s)^2+(z_r-z_s)^2\right]^{1/2} \tag{8.6.18}$$

$$r_3 = \left[x_e^2+y_e^2+z_e^2\right]^{1/2} \tag{8.6.19}$$

$$r_4 = \left[(x_e-x_s)^2+(y_e-y_s)^2+(z_e-z_s)^2\right]^{1/2} \tag{8.6.20}$$

and

$$d = \left[x_s^2+y_s^2+z_s^2\right]^{1/2} \tag{8.6.21}$$

Noting that in the acoustic medium the transfer function between any two points is defined by

$$T_i = \frac{e^{-jkr_i}}{r_i} \tag{8.6.22}$$

the gain and phase terms in (8.6.14) respectively become (Tokhi and Leitch, 1992)

$$H(\omega) = \frac{r_2 r_3}{r_1 r_4} = \frac{\frac{r_3}{r_4}}{\frac{r_1}{r_2}} \tag{8.6.23}$$

and

$$\phi(\omega) = k\left[(r_3 - r_4) - (r_1 - r_2)\right] = k(r_{34} - r_{12}) \tag{8.6.24}$$

where $r_{34} = r_3 - r_4$ and $r_{12} = r_1 - r_2$. Using these relationships we can examine the effect which geometry has upon system stability.

8.6.2 Gain margin

We will begin by examining gain margin. To calculate gain margin we must first derive where the phase of the transfer function is 180°. Rewriting the phase relationship in (8.6.16) as

$$\frac{H(\omega)\sin\phi(\omega)}{H(\omega)\cos\phi(\omega) - 1} = \tan(2m+1)\pi \tag{8.6.25}$$

it is apparent that for the phase of the transfer function to be 180°

$$H(\omega)\sin\phi(\omega) = 1 \quad \text{and} \quad H(\omega)\cos\phi(\omega) - 1 < 0 \tag{8.6.26}$$

As $H(\omega)$ is a real, positive number, this criterion becomes

$$\begin{aligned} \sin\phi(\omega) &= 0, \quad \cos\phi(\omega) = 1 \text{ and } H(\omega) < 1 \\ \sin\phi(\omega) &= 0, \quad \cos\phi(\omega) = -1 \text{ and } H(\omega) > 0 \end{aligned} \tag{8.6.27}$$

Therefore,

$$\phi(\omega) = 2m\pi \quad \text{for } H(\omega) < 1, \quad (2m+1)\pi \quad \text{for } H(\omega) > 0 \tag{8.6.28}$$

or

$$\phi(\omega) = (2m+1)\pi \text{ for } H(\omega) \geq 1 \quad m\pi \text{ for } 0 < H(\omega) < 1 \tag{8.6.29}$$

Substituting (8.6.24) into (8.6.29), the conditions under which the phase of $G(s)$ is $-180°$, where the gain margin can be evaluated, are found to be (Tokhi and Leitch, 1992)

$$r_{34} - r_{12} = \begin{cases} (2m+1)\dfrac{\lambda}{2} & \text{for } H(\omega) \geq 1 \\ m\dfrac{\lambda}{2} & \text{for } 0 \leq H(\omega) < 1 \end{cases} \tag{8.6.30}$$

Consider first the case where $\phi(\omega)$ is an even multiple of π. Substituting $\phi(\omega) = 2m\pi$ into (8.6.15) produces the expression

$$K_g = \left[H^2(\omega) - 2H(\omega) + 1\right]^{1/2} = |\, H(\omega) - 1 \,| \tag{8.6.31}$$

As $H(\omega)$ is less than 1 in this case (from (8.6.29)),

$$K_g = 1-H(\omega) \quad \text{for } 0<H(\omega)<1 \quad \text{and} \quad \phi(\omega) = 2m\pi \tag{8.6.32}$$

If $\phi(\omega)$ is now an odd multiple of π, substituting $\phi(\omega) = (2m+1)\pi$ into (8.6.15) produces

$$K_g = \left[H^2(\omega) + 2H(\omega)+1\right]^{1/2} = |\ H(\omega) + 1\ | \tag{8.6.33}$$

As $H(\omega)$ is greater than zero,

$$K_g = 1+H(\omega) \quad \text{for } H(\omega)>0 \quad \text{and} \quad \phi(\omega) = (2m+1)\pi \tag{8.6.34}$$

Combining (8.6.32) and (8.6.34), the gain margin of the system is found to be

$$K_g = \begin{cases} 1-H(\omega) & \text{for } \phi(\omega) = 2m\pi \text{ and } 0<H(\omega)<1 \\ 1+H(\omega) & \text{for } \phi(\omega)=(2m+1)\pi \text{ and } H(\omega)>0 \end{cases} \tag{8.6.35}$$

As mentioned previously, for the system to be stable the gain margin must be greater than 1. In viewing the criteria in (8.6.35), it is clear that this can only happen if $H(\omega)>1$, in which case the gain margin is greater than 2 (Tohki and Leitch, 1992). Equation (8.6.23) shows $H(\omega)$ to be entirely defined by the location of the sources and sensors. Therefore, we can use the requirement of $H(\omega)>1$ to examine what system arrangements will be stable, and what arrangements will be unstable.

Referring to (8.6.23) and Fig. 8.23, for $H(\omega)$ to be greater than 1,

$$\frac{r_3}{r_4} > \frac{r_1}{r_2} \tag{8.6.36}$$

In words, for the system to be stable, the ratio of the distance between the primary source and error sensor to the distance between the control source and error sensor must be greater than the ratio of the distance between the primary source and reference sensor and the distance between the control source and reference sensor. Further geometric consideration of this requirement can be found in Tohki and Leitch (1992).

8.6.3 Phase margin

What remains for us now is to assess the phase margin. Referring to (8.6.15), the transfer function gain $B(\omega)$ has a unity value when

$$\left[H^2(\omega) - 2H(\omega)\cos\phi(\omega) + 1\right]^{-1/2} = 1 \tag{8.6.37}$$

or

$$\cos\phi(\omega) = \frac{H(\omega)}{2} \tag{8.6.38}$$

From our analysis of gain margin, we found that the amplitude term $H(\omega)$ must be greater than 1 for the system to be stable. Based upon (8.6.38), the limit on $H(\omega)$ is therefore

$$1 \leq H(\omega) \leq 2 \tag{8.6.39}$$

Therefore, the range of $\phi(\omega)$ is

$$2m\pi - \frac{\pi}{3} \leq \phi(\omega) \leq 2m\pi + \frac{\pi}{3} \quad \text{for } 1 \leq H(\omega) \leq 2 \tag{8.6.40}$$

Substituting (8.6.24) into (8.6.40) and simplifying yields the allowable distance difference $r_{34} - r_{12}$ as

$$(6m-1)\frac{\lambda}{6} \leq r_{34} - r_{12} \leq (6m+1)\frac{\lambda}{6} \quad 1 \leq H(\omega) \leq 2 \tag{8.6.41}$$

Equation (8.6.41) provides the conditions under which $B(\omega) = 1$. To evaluate the resultant phase margin, note first that

$$\sin\phi(\omega) = \begin{cases} \frac{1}{2}\sqrt{4-H^2(\omega)} & \text{for } 0 \leq \phi(\omega) \leq \frac{(6m+1)\pi}{3} \\ -\frac{1}{2}\sqrt{4-H^2(\omega)} & \text{for } \frac{(6m-1)\pi}{3} \leq \phi(\omega) < 0 \end{cases} \tag{8.6.42}$$

Substituting (8.6.38) and (8.4.42) into (8.6.16), the phase margin is found to be (Tohki and Leitch, 1992)

$$K_\theta = \tan^{-1}\frac{H(\omega)\sqrt{4-H^2(\omega)}}{2-H^2(\omega)} + 2m\pi \quad \text{for } 0 \leq \phi(\omega) \leq \frac{(6n+1)\pi}{3}$$

$$-\tan^{-1}\frac{H(\omega)\sqrt{4-H^2(\omega)}}{2-H^2(\omega)} + 2m\pi \quad \text{for } \frac{(6m-1)\pi}{3} \leq \phi(\omega) < 0$$

$$\tag{8.6.43}$$

Note that for values of $H(\omega)$ greater than 1 the phase margin is positive, indicating that the system is stable. Above a value of 2, however, the system may become unstable. Therefore, the conclusion that can be drawn is that for a feedforward system to be stable, the value of $H(\omega)$, defined in (8.6.23) as purely a function of the location of the sources and sensors, must be between 1 and 2.

8.7 THE ACTIVE CONTROL OF HARMONIC SOUND RADIATION FROM PLANAR STRUCTURES: GENERAL PROBLEM FORMULATION

The work presented in the preceeding sections of this chapter has been directed at the use of a set of idealized acoustic monopole control sources to attenuate the sound radiation from a second set of idealized monopole primary acoustic sources. Such a series of studies has two principal aims. First, there are many practical noise problems which can be viewed to a first approximation as originating from one or more point sources. However, possibly a more important reason for studying the relatively simple cases of the previous sections is to obtain insight, concerning both physical mechanisms and theoretical methodology, which can be used in the design of active control systems for other, possibly more practical, free field radiation problems. It is the philosophy behind this second reason which will be of use in the next sections, where we will examine the active control of harmonic sound radiation from planar, vibrating structures situated in an infinite baffle. In the course of this study we will use the knowledge gained in the examination of the monopole problems to develop a theoretical framework for obtaining optimum control source volume velocities for the minimization of acoustic pressure at a set of discrete error locations, and for minimizing the total radiated acoustic power. We will then repeat the exercise for the use of vibration sources, directly attached to the structure, to minimize the acoustic radiation. These theoretical formulations will be 'put to task' in the next section by applying them to the problem of controlling sound radiation from a vibrating panel. Finally, problems associated with controlling structural vibration with the aim of controlling the radiated sound field will be discussed.

The cynical reader might protest at this point, 'What will we actually gain in this extension to the previous analysis? How many planar, vibrating structures which are situated in an infinitely extending baffle actually exist?' In fact, we stand gain a considerable amount from this extension. First, it provides us with our first opportunity to analytically study something resembling a real structural radiation problem, where the forced response of the structural modes is what is responsible for generating the sound field. Second, there are many structures which can, in first approximation, be represented by finite planar radiators in an infinite baffle. The results of the idealized study conducted in the next three sections will provide at least a qualitative feeling for what can be accomplished by active control, the means by which attenuation is attained, and the importance of various system parameters. Third, consideration of a vibrating structure offers the opportunity to assess the potential of a different control arrangement than considered previously; the use of vibration control sources attached directly to structure, which will manipulate the velocity distribution of the structure in such a way as to attenuate the acoustic radiation.

To begin with, in this section we will develop the theoretical framework required for examining the control of harmonic sound radiation from planar, vibrating structures using either acoustic or vibration control sources. The development undertaken here is a generalisation of work published in recent years; see, for example, Deffayet and Nelson (1988) or Pan *et al.* (1992) for acoustic control source applications, or Fuller (1988, 1990) or Pan *et al.* (1992) for vibration source applications. As with the monopole cases discussed previously, there are two possible error criteria which can be used: minimization of the acoustic pressure at a set of error locations, and minimization of the total radiated acoustic power. The former of these will be examined first.

8.7.1 Minimization of acoustic pressure at discrete locations using acoustic monopole sources

Consider an active control problem where a vibrating structure is responsible for the primary noise disturbance, and we wish to use a set of N_c acoustic monopole sources to attenuate the acoustic pressure at a set of N_e error locations. The error criterion, J_{prs}, is the sum of the squared acoustic pressure amplitudes at the error locations

$$J_{prs} = \sum_{i=1}^{N_e} | p(r_i) |^2 = \sum_{i=1}^{N_e} p^*(r_i) p(r_i) \tag{8.7.1}$$

For the equivalent monopole problem of Section 8.3, the optimum control source volume velocity for minimizing this error criterion was formulated by restating the criterion as a quadratic function of the vector of control source volume velocities, and applying quadratic optimization techniques to obtain the optimal control source volume velocity vector. This would seem again to be the logical approach to take. To take this path, however, we must first formulate a set of expressions for predicting the acoustic pressure at some location in space due to the vibration of the structure, which is the primary source.

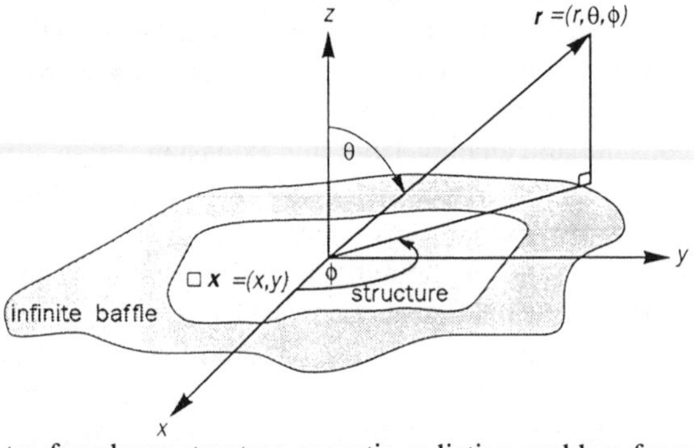

Fig. 8.27 Geometry for planar structure acoustic radiation problem formulation.

The geometry which will be used for formulating this required expression is shown in Fig. 8.27. A planar structure, situated in an infinite baffle, is subject to a harmonic exciting pressure field, p_0, of the form $e^{j\omega t}$, at its interface with the fluid half-space. The fluid will be confined to being of relatively low density, which will allow us to neglect any fluid loading of the structure as it radiates. The velocity $v(x)$ of the structure at some point $x = (x,y)$ due to this exciting force can be determined from the Green's function response equation (see Section 2.4):

$$v(x) = j\omega \int_S G_s(x \mid x')\, p_o(x')\, \mathrm{d}x' \tag{8.7.2}$$

where $G_s(x|x')$ is the structural Green's function, defined as

$$G_s(x \mid x') = \sum_{\iota=1}^{\infty} \frac{\psi_\iota(x)\ \psi_\iota(x')}{M_\iota Z_\iota} \tag{8.7.3}$$

and the integration is over the surface of the structure S. Recall that in the Green's function $\psi_\iota(x)$ is the ιth structural mode shape function evaluated at the location x, M_ι is the modal mass of the ιth mode,

$$M_\iota = \int_S \rho_s(x)h(x)\psi_\iota^2(x)\ \mathrm{d}x \tag{8.7.4}$$

ρ_s is the density of the structure material, h is the material thickness, and Z_ι is the input impedance of the ιth mode:

$$Z_\iota = \omega_\iota^2(1+j\eta_\iota)-\omega^2 \tag{8.7.5}$$

The quantity ω_ι is the resonance frequency, in radians per second, of the ιth structural mode, and ω is the frequency of excitation, also in radians per second. The exciting pressure field p_0 is comprised of two components: the blocked pressure, and the radiation reaction pressure caused by the vibration of the structure. When the fluid medium surrounding the structure is of low density, so that the radiation reaction can pressure can be ignored, the exciting pressure field can be thought of as entirely due to the blocked pressure p_{bl}, which is the pressure distribution from the primary disturbance:

$$p_0(x') = p_{bl}(x') \tag{8.7.6}$$

The velocity of the structure can be expanded in terms of an infinite set of *in vacuo* structural mode shape functions,

$$v(x) = \sum_{i=1}^{\infty} v_i\psi_i(x) \tag{8.7.7}$$

where v_i is the complex velocity amplitude of the ith structural mode. Substituting this expansion into (8.7.2) enables the response of the structure to the primary exciting force to be expressed as:

$$\sum_{i=1}^{\infty} v_i\psi_i(x) = j\omega \int_S \left[\sum_{\iota=1}^{\infty} \frac{\psi_\iota(x)\psi_\iota(x')}{M_\iota Z_\iota}\right] p_0(x')\ \mathrm{d}x' \tag{8.7.8}$$

From (8.7.8), the response of a single, ιth, mode can be written more compactly as

$$(Z_\iota M_\iota)v_\iota = \gamma_\iota \tag{8.7.9}$$

where γ_ι is the modal generalized force of the ιth structural mode, which is based upon the blocked pressure only, and defined as

$$\gamma_\iota = \int_S p_{bl}(x')\psi_\iota(x')\ \mathrm{d}x' = \int_S p_0(x')\psi_\iota(x')\ \mathrm{d}x' \tag{8.7.10}$$

From the form of (8.7.9), it is apparent that when the forced response of the structure is expressed in terms of the normal modes of vibration, the problem is decoupled; that is, it can be expressed as the sum of an infinite set of (complex) scalar equations. This comes about as a direct result of neglecting fluid loading on the structure, assuming that it is radiating sound into air or some other low density fluid medium. If the fluid medium is of greater density, such as water, the assumption cannot be made, and the set of modal equations becomes coupled, complicating the analysis. This coupling is described, for example, by Davies (1971), Pope and Leibowitz (1974), Lomas and Hayek (1977), and Sandman (1977), as well as by Fahy (1985) and Junger and Feit (1986); active control of infinite fluid-loaded plates is described by Gu and Fuller (1991), and for finite plates by Meirovitch and Thangjitham (1990a) and Gu and Fuller (1993).

The response of the structure, expressed in (8.7.7) as an infinite summation of *in vacuo* structural modes, can be estimated by summing over m structural modes only. Making this truncation, (8.7.9) can be written in matrix form for the entire set of m structural modes as

$$\boldsymbol{Z}_I \boldsymbol{v} = \boldsymbol{\gamma} \tag{8.7.11}$$

where $\boldsymbol{v}$ is the $(m \times 1)$ vector of complex modal amplitudes, $\boldsymbol{Z}_I$ is the $(m \times m)$ diagonal matrix of modal input transfer functions, the (i,i)th element of which is

$$z_I(i,i) = M_i\, Z_i \tag{8.7.12}$$

where Z_i is defined by (8.7.5) and γ is the $(m \times 1)$ vector of (blocked pressure) modal generalized forces. Thus, the amplitude of the structural modes under primary excitation can be expressed as

$$\boldsymbol{v}_p = \boldsymbol{Z}_I^{-1} \boldsymbol{\gamma}_p \tag{8.7.13}$$

where the subscript p denotes primary excitation.

Once the planar, infinitely baffled structure has been set into motion the resultant acoustic pressure field can be calculated using the Rayleigh integral as follows:

$$\begin{aligned} p(\boldsymbol{r}) &= j\omega\rho_0 \int_S \frac{v(\boldsymbol{x})\mathrm{e}^{-jkr}}{2\pi r}\, \mathrm{d}\boldsymbol{x} = j\omega\rho_0 \int_S G_f(\boldsymbol{x} \mid \boldsymbol{r}) v(\boldsymbol{x})\, \mathrm{d}\boldsymbol{x} \\ &= j\omega\rho_0 \sum_{i=1}^{m} v_i \int_S G_f(\boldsymbol{x} \mid \boldsymbol{r}) \psi_i(\boldsymbol{x})\, \mathrm{d}\boldsymbol{x} \end{aligned} \tag{8.7.14}$$

where $G_f(\boldsymbol{x} \mid \boldsymbol{r})$ is a Green's function, defined as

$$G_f(\boldsymbol{x} \mid \boldsymbol{r}) = \frac{\mathrm{e}^{-jkr}}{2\pi r} \tag{8.7.15}$$

r being the separation distance between the point on the structure at $\boldsymbol{x}$ and the point in the acoustic space at $\boldsymbol{r}$. (Note here that (8.7.14) is identical in form to (8.2.1), which described the sound pressure at some location due to a monopole acoustic source, with the source volume velocity equal to

$$q = 2 \int_S v(x)\ \mathrm{d}x \tag{8.7.16}$$

The constant 2 is included because the sound field is generated using only ½ of the total fluid displacement of the structure, owing to the infinite baffle which blocks any contribution from the back of the panel.)

Equation (8.7.14) shows that the total acoustic pressure at any location $\boldsymbol{r}$ in space is constructed from the sum of contributions from the m structural modes. The radiated sound pressure can therefore be expressed in matrix form as

$$p(\boldsymbol{r}) = \boldsymbol{z}_{rad}^{\mathrm{T}}\boldsymbol{v} \tag{8.7.17}$$

where $\boldsymbol{z}_{rad}$ is the $(m \times 1)$ vector of modal radiation transfer functions, the ith element of which is

$$z_{rad}(i,\boldsymbol{r}) = j\omega\rho_0 \int_S G_f(\boldsymbol{x} \mid \boldsymbol{r})\psi_i(\boldsymbol{x})\ \mathrm{d}\boldsymbol{x} = \frac{j\rho_0\omega}{2\pi r}\int_S \mathrm{e}^{-jkr}\psi_i(\boldsymbol{x})\ \mathrm{d}\boldsymbol{x} \tag{8.7.18}$$

Combining (8.7.13) and (8.7.18) enables the radiated acoustic field under primary excitation to be evaluated as a function of the primary (pressure) forcing function

$$p_p(\boldsymbol{r}) = \boldsymbol{z}_{rad}^{\mathrm{T}}\boldsymbol{v}_p = \boldsymbol{z}_{rad}^{\mathrm{T}}\boldsymbol{Z}_I^{-1}\boldsymbol{\gamma}_p \tag{8.7.19}$$

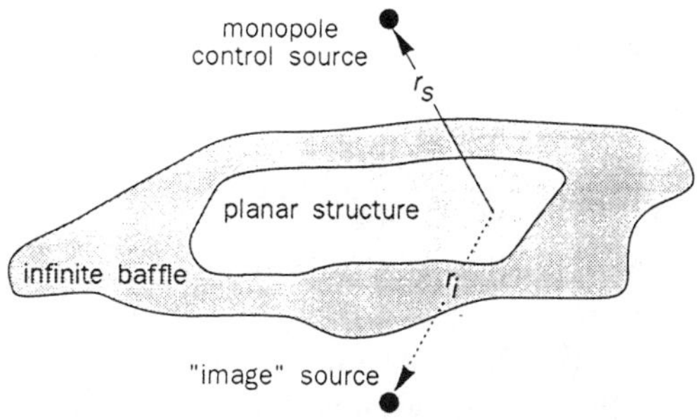

Fig. 8.28 Monopole control source/image source arrangement.

Let us now turn our attention to the problem of formulating a description of the sound field radiated by the control monopole acoustic sources. The sound field radiated by the monopole sources in the arrangement being considered here will be somewhat different from that considered previously in this chapter owing to the structure/infinite baffle arrangement, which will cause a reflection of the sound field. Making the assumption that the velocity distribution of the structure will be unaffected by the inclusion of the control sources in the system (another result of assuming a non-dense fluid medium), the total acoustic pressure at any location $\boldsymbol{r}$ in space during the operation of the control source can in this instance be modelled as coming from both the control source and its mirror image, as shown in Fig. 8.28; the sound pressure radiated by each monopole control source can be determined by considering the source and its mirror image as two separate entities. Using the expression for the sound field at location $\mathbf{r}$ produced by a monopole source given in (8.2.1), the radiated sound field produced by any singular control source in the presence of the reflecting surface is

$$p_c(\mathbf{r}) = \frac{j\omega\rho_0 q_c \mathrm{e}^{-jkr_s}}{4\pi r_s} + \frac{j\omega\rho_0 q_c \mathrm{e}^{-jkr_i}}{4\pi r_i} \tag{8.7.20}$$

where the subscripts s and i denote the distance from the location of the source and the mirror image, respectively. This can be rewritten more compactly as

$$p_c(\mathbf{r}) = \frac{j\rho_0\omega q_c}{4\pi}\left[\frac{\mathrm{e}^{-jkr_s}}{r_s} + \frac{\mathrm{e}^{-jkr_i}}{r_i}\right] \tag{8.7.21}$$

If N_c monopole control sources are being used, the total control source-generated acoustic field will be the superposition of the sound pressures radiated by each of the sources

$$p_c(\mathbf{r}) = \sum_{n=1}^{N_c} \frac{j\rho_0\omega q_{c,n}}{4\pi}\left[\frac{\mathrm{e}^{-jkr_{n,s}}}{r_{n,s}} + \frac{\mathrm{e}^{-jkr_{n,i}}}{r_{n,i}}\right] \tag{8.7.22}$$

This can be expressed in matrix form as

$$p_c(\mathbf{r}) = \mathbf{z}_{mono}^{\mathrm{T}}(\mathbf{r})\,\mathbf{q}_c \tag{8.7.23}$$

where $\mathbf{z}_{mono}$ is the $(N_c \times 1)$ vector of control source radiation transfer functions, which define the relationship between source volume velocity and pressure, and $\mathbf{q}_c$ is the $(N_c \times 1)$ vector of control source volume velocities. The nth element of $\mathbf{z}_{mono}$ is

$$z_{mono}(\mathbf{r},n) = \frac{j\rho_0\omega}{4\pi}\left[\frac{\mathrm{e}^{-jkr_{n,s}}}{r_{n,s}} + \frac{\mathrm{e}^{-jkr_{n,i}}}{r_{n,i}}\right] \tag{8.7.24}$$

where $r_{n,s}$ and $r_{n,i}$ are the distances from the nth actual and image source locations respectively.

Having derived matrix expressions for the sound pressure generated at any point in space due to the primary noise source (the radiating structure) and the monopole control sources, it is possible to express the error criterion, the sum of the squared acoustic pressure amplitudes at a discrete set of error sensing locations, as a quadratic function of control source volume velocities. The optimum set of volume velocities can then be calculated by means of the quadratic optimization techniques already used in this chapter. The sum of the squared acoustic pressure amplitudes at the set of N_e error sensing locations can be expressed as

$$J_p = \sum_{i=1}^{N_e} |\, p(\mathbf{r}_i)\,|^2 = \sum_{i=1}^{N_e} p^*(\mathbf{r}_i)p(\mathbf{r}_i) = \mathbf{p}_e^{\mathrm{H}}\mathbf{p}_e \tag{8.7.25}$$

where $\mathbf{p}_e$ is the $(N_e \times 1)$ vector of acoustic pressures at the error locations. In our linear system the acoustic pressures are the superposition of the primary and control source generated pressures,

$$\boldsymbol{p}_e = \boldsymbol{p}_p + \boldsymbol{p}_c \tag{8.7.26}$$

where $\boldsymbol{p}_p$ and $\boldsymbol{p}_c$ are the $(N_e \times 1)$ vectors of primary and control source generated pressures at the error locations, respectively, so that

$$J_p = \boldsymbol{p}_e^{\mathrm{H}}\boldsymbol{p}_e = \boldsymbol{p}_c^{\mathrm{H}}\boldsymbol{p}_c + \boldsymbol{p}_c^{\mathrm{H}}\boldsymbol{p}_p + \boldsymbol{p}_p^{\mathrm{H}}\boldsymbol{p}_c + \boldsymbol{p}_p^{\mathrm{H}}\boldsymbol{p}_p \tag{8.7.27}$$

The vector of primary source pressures can be re-expressed using (8.7.17) as

$$\boldsymbol{p}_p = \boldsymbol{Z}_r\boldsymbol{v}_p \tag{8.7.28}$$

where $\boldsymbol{Z}_r$ is the $(N_e \times m)$ matrix of modal radiation transfer functions between the m structural modes being used in the calculation and the N_e error locations, the rows of which are the transposed vectors of structural modal radiation transfer functions, defined in (8.7.18), to each error location

$$\boldsymbol{Z}_r = \begin{bmatrix} \boldsymbol{z}_{rad}^{\mathrm{T}}(\boldsymbol{r}_1) \\ \boldsymbol{z}_{rad}^{\mathrm{T}}(\boldsymbol{r}_2) \\ \vdots \\ \boldsymbol{z}_{rad}^{\mathrm{T}}(\boldsymbol{r}_{N_e}) \end{bmatrix} \tag{8.7.29}$$

The vector of control source pressures can be re-expressed in terms of (8.7.23) as

$$\boldsymbol{p}_c = \boldsymbol{Z}_m\boldsymbol{q}_c \tag{8.7.30}$$

where $\boldsymbol{Z}_m$ is the $(N_e \times N_c)$ matrix of radiation transfer functions between the N_c monopole control sources and the N_e error locations, the rows of which are the transposed vectors of control source radiation transfer functions, defined in (8.7.24), to each error location

$$\boldsymbol{Z}_m = \begin{bmatrix} \boldsymbol{z}_{mono}^{\mathrm{T}}(\boldsymbol{r}_1) \\ \boldsymbol{z}_{mono}^{\mathrm{T}}(\boldsymbol{r}_2) \\ \vdots \\ \boldsymbol{z}_{mono}^{\mathrm{T}}(\boldsymbol{r}_{N_e}) \end{bmatrix} \tag{8.7.31}$$

Substituting (8.7.28) and (8.7.30) into the error criterion as stated in (8.7.27) enables the latter to be written as

$$J_{prs} = \boldsymbol{q}_c^{\mathrm{H}}\boldsymbol{Z}_m^{\mathrm{H}}\boldsymbol{Z}_m\boldsymbol{q}_c + \boldsymbol{q}_c^{\mathrm{H}}\boldsymbol{Z}_m^{\mathrm{H}}\boldsymbol{Z}_r\boldsymbol{v}_p + \boldsymbol{v}_p^{\mathrm{H}}\boldsymbol{Z}_r^{\mathrm{H}}\boldsymbol{Z}_m\boldsymbol{q}_c + \boldsymbol{v}_p^{\mathrm{H}}\boldsymbol{Z}_r^{\mathrm{H}}\boldsymbol{Z}_r\boldsymbol{v}_p \tag{8.7.32}$$

or

$$J_{prs} = \boldsymbol{q}_c^{\mathrm{H}}\boldsymbol{A}\boldsymbol{q}_c + \boldsymbol{q}_c^{\mathrm{H}}\boldsymbol{b} + \boldsymbol{b}^{\mathrm{H}}\boldsymbol{q}_c + c \tag{8.7.33}$$

where

$$\boldsymbol{A} = \boldsymbol{Z}_m^{\mathrm{H}}\boldsymbol{Z}_m \tag{8.7.34}$$

$$b = Z_m^H Z_r v_p \tag{8.7.35}$$

and

$$c = v_p^H Z_r^H Z_r v_p \tag{8.7.36}$$

The form of the error criterion given in (8.7.33) is exactly that which we were aiming for, where it is expressed as a quadratic function of the vector of control source volume velocities $\boldsymbol{q}_c$. The optimum set of volume velocities for minimising the sum of the squared sound pressure levels at N_e error locations can be found directly, by differentiating (8.7.33) with respect to $\boldsymbol{q}_c$ and setting the gradient equal to zero. The result of this exercise, as outlined in Section 8.4, is

$$q_{c,prs} = -A^{-1}b \tag{8.7.37}$$

where the subscripts *c,prs* on the control source volume velocity vector indicate that it is optimal with respect to minimizing the acoustic pressure amplitude at a set of discrete error locations. (As outlined in Section 8.4, the matrix $\boldsymbol{A}$ must be positive definite for a unique solution to exist, and for the solution to be the optimum one.) Substituting this result back into (8.7.33) produces the expression for the minimum sum of the squared acoustic pressures

$$J_{prs,\min} = \left\{ \sum_{i=1}^{N_e} |p(r_i)|^2 \right\}_{\min} = c - b^H A^{-1} b = c + b^H q_{c,p} \tag{8.7.38}$$

Note here again that the quantity c is equal to the value of the error criterion, the sum of the squared pressures, during operation of the primary noise source alone.

8.7.2 Minimization of total radiated acoustic power using acoustic monopole sources

The procedure for calculating the control source volume velocity or force matrix that minimizes the total radiated acoustic power is much the same as that outlined for minimizing the sound pressure at a discrete point or points. The difference is that the matrices $\boldsymbol{A}$ and $\boldsymbol{b}$, and the quantity c, must be modified to include the 'spatial integration' necessary in sound power calculations. There are two commonly employed approaches to calculating the control source volume velocities which will minimize the total radiated acoustic power of the system, both of which are aimed at expressing this error criterion as a quadratic function of control source output. The derived expression can then be differentiated with respect to control source volume velocities, and the resultant gradient equation set equal to zero to yield the optimum value. One approach to obtaining the desired quadratic expression is to construct the problem based upon a farfield measure of acoustic power, where the total radiated acoustic power is evaluated by integrating the farfield acoustic intensity over a hemisphere enclosing the radiating structure and control sources. Using the geometry shown in Fig. 8.27, this can be written as

$$W = \int_0^{2\pi}\int_0^{\pi/2} \frac{|p(r)|^2}{2\rho_0 c_0} |r|^2 \sin\theta \, d\theta \, d\phi \tag{8.7.39}$$

where the location $\boldsymbol{r}$ is defined by the spherical coordinates $\boldsymbol{r} = (r,\theta,\phi)$. The squared acoustic pressure amplitude at any point is

$$|p(r)|^2 = p^*(r)p(r) \tag{8.7.40}$$

which, written explicitly in terms of the primary and control source contributions, is

$$\begin{aligned} |p(r)|^2 &= \left(p_c(r)+p_p(r)\right)^* \left(p_c(r)+p_p(r)\right) \\ &= p_c^*(r)p_c(r)+p_c^*(r)p_p(r)+p_p^*(r)p_c(r)+p_p^*(r)p_p(r) \end{aligned} \tag{8.7.41}$$

Using (8.7.17) and (8.7.23) this can be re-expressed as

$$\begin{aligned} |p(r)|^2 = \; & \boldsymbol{q}_c^{\mathrm{H}} \boldsymbol{z}_{mono}^* \boldsymbol{z}_{mono}^{\mathrm{T}} \boldsymbol{q}_c + \boldsymbol{q}_c^{\mathrm{H}} \boldsymbol{z}_{mono}^* \boldsymbol{z}_{rad}^{\mathrm{T}} \boldsymbol{v}_p \\ & + \boldsymbol{v}_p^{\mathrm{H}} \boldsymbol{z}_{rad}^* \boldsymbol{z}_{mono}^{\mathrm{T}} \boldsymbol{q}_c + \boldsymbol{v}_p^{\mathrm{H}} \boldsymbol{z}_{rad}^* \boldsymbol{z}_{rad}^{\mathrm{T}} \boldsymbol{v}_p \end{aligned} \tag{8.7.42}$$

If (8.7.42) is now substituted into the defining (8.7.39), the total radiated acoustic power can be written as a quadratic function of control source volume velocity

$$W = \boldsymbol{q}_c^{\mathrm{H}} \boldsymbol{A} \boldsymbol{q}_c + \boldsymbol{q}_c^{\mathrm{H}} \boldsymbol{b} + \boldsymbol{b}^{\mathrm{H}} \boldsymbol{q}_c + c \tag{8.7.43}$$

where

$$\boldsymbol{A} = \int_0^{2\pi}\int_0^{\pi/2} \frac{\boldsymbol{z}_{mono}^* \boldsymbol{z}_{mono}^{\mathrm{T}}}{2\rho_0 c_0} |r|^2 \sin\theta \, d\theta \, d\phi \tag{8.7.44}$$

$$\boldsymbol{b} = \int_0^{2\pi}\int_0^{\pi/2} \frac{\boldsymbol{z}_{mono}^* \boldsymbol{z}_{rad}^{\mathrm{T}} \boldsymbol{v}_p}{2\rho_0 c_0} |r|^2 \sin\theta \, d\theta \, d\phi \tag{8.7.45}$$

and

$$c = \int_0^{2\pi}\int_0^{\pi/2} \frac{\boldsymbol{v}_p^{\mathrm{H}} \boldsymbol{z}_{rad}^* \boldsymbol{z}_{rad}^{\mathrm{T}} \boldsymbol{v}_p}{2\rho_0 c_0} |r|^2 \sin\theta \, d\theta \, d\phi \tag{8.7.46}$$

In this case, the quantity c is the acoustic power radiated by the primary source operating alone.

As before, differentiating (8.7.43) with respect to $\boldsymbol{q}_c$ and setting the gradient equal to zero will produce an expression which defines the optimum control source volume velocities,

$$\boldsymbol{q}_{c,w} = -\boldsymbol{A}^{-1}\boldsymbol{b} \tag{8.7.47}$$

where the subscripts c,w on the control source volume velocity vector indicate that it is optimal with respect to minimizing the total radiated acoustic power. Substituting this result back into (8.7.43) produces the expression for the minimum radiated acoustic power

$$W_{\min} = c - \boldsymbol{b}^{\mathrm{H}}\boldsymbol{A}^{-1}\boldsymbol{b} = c + \boldsymbol{b}^{\mathrm{H}}\boldsymbol{q}_{c,w} \tag{8.7.48}$$

In Sections 8.3 and 8.4 similar expressions were derived for the problem of controlling the total acoustic power output of a set monopole sources. In that exercise it was found that the integrations which are the equivalent of those stated in (8.7.44)–(8.7.46) could be solved analytically, leading to a much more compact statement of the problem. Unfortunately, this will not normally be possible for the problem of controlling structural radiation. Consider the three terms in (8.7.43) which require integration, and the specialised case of only one acoustic control source. Referring to (8.7.24), the term $\boldsymbol{A}$ can be specialized in this instance as

$$\boldsymbol{A} = \int_0^{2\pi} \frac{\left[\dfrac{j\rho_0\omega}{4\pi}\left\{\dfrac{\mathrm{e}^{-jkr_s}}{r_s} + \dfrac{\mathrm{e}^{-jkr_i}}{r_i}\right\}\right]^* \left[\dfrac{j\rho_0\omega}{4\pi}\left\{\dfrac{\mathrm{e}^{-jkr_s}}{r_s} + \dfrac{\mathrm{e}^{-jkr_i}}{r_i}\right\}\right]}{2\rho_0 c_0} \,|\,\boldsymbol{r}\,|^2 \sin\theta \,\mathrm{d}\theta \,\mathrm{d}\phi \tag{8.7.49}$$

or

$$\boldsymbol{A} = \frac{\omega^2 r^2}{16\pi} \int_0^{2\pi} \left\{\frac{\mathrm{e}^{-jkr_s}}{r_s} + \frac{\mathrm{e}^{-jkr_i}}{r_i}\right\}^* \left\{\frac{\mathrm{e}^{-jkr_s}}{r_s} + \frac{\mathrm{e}^{-jkr_i}}{r_i}\right\} \sin\theta \,\mathrm{d}\theta \,\mathrm{d}\phi \tag{8.7.50}$$

Following from our monopole work in Section 8.4, as the integration is being calculated in the farfield the approximations $r_s \approx r_i \approx r$ and $r_i - r_s = z \cos\theta$ can be made. This allows (8.7.50) to be restated, after some simple algebra, as

$$\boldsymbol{A} = \frac{\omega^2 \rho_0}{8\pi c_0} \int_0^{2\pi} \{1 + \cos(2kz \cos\theta)\} \sin\theta \,\mathrm{d}\theta \tag{8.7.51}$$

Noting again that

$$\int_0^{\pi/2} \cos(x \cos\theta) \sin\theta \,\mathrm{d}\theta = \operatorname{sinc} x \tag{8.7.52}$$

enables the term in $\boldsymbol{A}$ to be evaluated as

$$\boldsymbol{A} = \frac{\omega^2 \rho_0}{8\pi c_0}(1 + \operatorname{sinc} 2kz) \tag{8.7.53}$$

From this, it can be seen that the problem is not in evaluating the $\boldsymbol{A}$ term. In fact, the problems are encountered are in the evaluation of the $\boldsymbol{b}$ and c terms. In each of these terms the vector z_{rad} appears, the elements of which, from the definition of (8.7.18), must themselves be evaluated through integration (recall that the terms in z_{rad} define the transfer function between the velocity of a given structural mode and the acoustic pressure at some location in space). It will rarely be possible to do this analytically, even with the consideration limited to farfield radiation and the most regular structure geometries. If such a result can be obtained, it will almost surely be a function of the angular coordinates θ, ϕ (and of course radius r, which will be constant for the integration). To evaluate the terms in $\boldsymbol{b}$ and c, the result obtained for z_{rad} must be multiplied by other spatially dependent terms and integrated. It can easily be deduced from this description of events that numerical techniques are usually required. At low frequencies, however, it is sometimes possible to formulate approximate solutions, which we will discuss shortly.

The second approach taken to obtain the desired quadratic equation for calculating the optimum set of control source volume velocities is to consider source power output directly, this quantity being equal to

$$W = \frac{1}{2} \operatorname{Re} \left\{ \int_S v(\boldsymbol{x})\, p^*(\boldsymbol{x})\, \mathrm{d}\boldsymbol{x} \right\} \tag{8.7.54}$$

where Re denotes the real part of the expression, $v(\boldsymbol{x})$ is the velocity at some location $\boldsymbol{x}$ on the surface of the source, $p^*(\boldsymbol{x})$ is the complex conjugate of the acoustic pressure at this same location, and the integration is conducted over the surface of the source. Of course, both the farfield and nearfield developments should lead to the same solution for the optimum control source output.

Consider again an active control system where a set of N_c acoustic monopole control sources is being used to attenuate the sound field radiated by a vibrating structure. The total acoustic power output W_t of this arrangement can be written in the form

$$W_t = W_c + W_p \tag{8.7.55}$$

where the subscript t is used to denote a total (combined primary and control source) quantity. In (8.7.55), W_c is the total acoustic power output of the control source distribution, equal to the sum of contributions from the N_c monopole control sources, the integral expression (8.7.54) corresponding to the monopole control sources being replaced by multiplication of volume velocity and the complex conjugate of acoustic pressure at the monopole source location, to give

$$W_c = \frac{1}{2} \operatorname{Re} \left\{ \sum_{i=1}^{N_c} q_{ci}\, p_t^*(\boldsymbol{r}_{ci}) \right\} \tag{8.7.56}$$

where q_{ci} is the volume velocity of the ith control source, located at a position $\boldsymbol{r}_{ci}$ in space. W_p in (8.7.55) is the acoustic power output of the primary source (the radiating structure), given by

$$W_p = \frac{1}{2} \operatorname{Re} \left\{ \int_S v_p(\boldsymbol{x})\, p_t^*(\boldsymbol{x})\, \mathrm{d}\boldsymbol{x} \right\} \tag{8.7.57}$$

As the total pressure at any location $\boldsymbol{r}$ can be considered to be the superposition of the primary and control source pressures at that location, (8.7.55) can be expanded to

$$W_t = \frac{1}{2}\,\mathrm{Re}\left\{ \sum_{i=1}^{N_c} q_{ci}\left[p_c(\boldsymbol{r}_{ci})+p_p(\boldsymbol{r}_{ci})\right]^* + \int_S v(\boldsymbol{x})\left[p_c(\boldsymbol{x})+p_p(\boldsymbol{x})\right]^* \,\mathrm{d}\boldsymbol{x} \right\} \tag{8.7.58}$$

Using (8.7.19) and (8.7.23), the total acoustic power output of the system, defined in (8.7.58), can be written in matrix form as:

$$W_t = \boldsymbol{q}_c^{\mathrm{H}}\boldsymbol{A}\boldsymbol{q}_c + \boldsymbol{b}^{\mathrm{H}}\boldsymbol{q}_c + \boldsymbol{q}_c^{\mathrm{H}}\boldsymbol{b} + c \tag{8.7.59}$$

where

$$\boldsymbol{A} = \frac{1}{2}\mathrm{Re}\left\{ \begin{bmatrix} \boldsymbol{z}_{mono}^{\mathrm{T}}(\boldsymbol{r}_{c1}) \\ \boldsymbol{z}_{mono}^{\mathrm{T}}(\boldsymbol{r}_{c2}) \\ \vdots \\ \boldsymbol{z}_{mono}^{\mathrm{T}}(\boldsymbol{r}_{cN_c}) \end{bmatrix} \right\} \tag{8.7.60}$$

$$\boldsymbol{b} = \frac{1}{2}\mathrm{Re}\left\{ \begin{bmatrix} \boldsymbol{z}_{rad}^{\mathrm{T}}(\boldsymbol{r}_{c1}) \\ \boldsymbol{z}_{rad}^{\mathrm{T}}(\boldsymbol{r}_{c2}) \\ \vdots \\ \boldsymbol{z}_{rad}^{\mathrm{T}}(\boldsymbol{r}_{cN_c}) \end{bmatrix} \right\} \boldsymbol{v}_p = \frac{1}{2}\sum_{i=1}^{N_m} \left[\int_S \mathrm{Re}\left\{\boldsymbol{z}_{mono}^{*}(\boldsymbol{x})\right\}\psi_i(\boldsymbol{x})\,\mathrm{d}\boldsymbol{x}\cdot v_{p,i} \right] \tag{8.7.61}$$

$$c = \boldsymbol{v}_p^{\mathrm{H}}\boldsymbol{Z}_w\boldsymbol{v}_p \tag{8.7.62}$$

and the i,ιth term of the matrix $\boldsymbol{Z}_w$, which will be referred to as the acoustic power transfer matrix in more in-depth studies later in this chapter, is defined by

$$Z_w(i,\iota) = \frac{\omega\rho_0}{4\pi}\int_S\int_S \psi_i^*(\boldsymbol{x}')\frac{\sin kr}{r}\psi_\iota(\boldsymbol{x})\;\mathrm{d}\boldsymbol{x}\;\mathrm{d}\boldsymbol{x}' \tag{8.7.63}$$

Expressed in this 'standard' form, the vector of optimum control source volume velocities, calculated by differentiating (8.7.59) with respect to $\boldsymbol{q}_c$ and setting the resultant gradient expression equal to zero, is found to be

$$\boldsymbol{q}_{c_{\mathrm{opt}}} = -\boldsymbol{A}^{-1}\boldsymbol{b} \tag{8.7.64}$$

the use of which will result in a minimum acoustic power output from the controlled system of

$$W_{\min} = c - \boldsymbol{b}^{\mathrm{H}}\boldsymbol{A}^{-1}\boldsymbol{b} \tag{8.7.65}$$

where c, defined in (8.7.62), is the acoustic power output of the system in the absence of active control.

As we mentioned in the previous farfield based minimum acoustic power formulation, it would greatly simplify the exercise if we could solve the integrations in (8.7.61) and (8.7.62) analytically. We found in the farfield acoustic power formulation that this was not generally possible. This is also a problem in a nearfield formulation of the problem. Consider first the matrix $\boldsymbol{A}$. The m,nth term in this real, symmetric matrix is the real part of the radiation transfer function between the mth and nth monopole control sources. From the definition given in (8.7.60) this is

$$A_{m,n} = \frac{\omega^2 \rho_0}{8\pi c_0}\left(\text{sinc}\; kr_{mn,s} + \text{sinc}\; kr_{mn,i}\right) \tag{8.7.66}$$

where $r_{mn,s}$ and $r_{mn,i}$ are respectively the distance between the m and nth real sources (the same distance which exists between the two image sources) and the distance between an image and real source,

$$\begin{aligned} r_{mn,s} &= \sqrt{(x_m - x_n)^2 + (y_m - y_n)^2 + (z_m - z_n)^2}\;, \\ r_{mn,i} &= \sqrt{(x_m - x_n)^2 + (y_m - y_n)^2 + (z_m + z_n)^2} \end{aligned} \tag{8.7.67}$$

Note that if a diagonal term of $\boldsymbol{A}$ is being evaluated, then the result is the radiation resistance of a monopole source near an infinite plane surface,

$$A_{m,m} = \frac{\omega^2 \rho_0}{8\pi c_0}\left(1 + \text{sinc}\; 2kz_m\right) \tag{8.7.68}$$

This is the same expression as was derived in (8.7.53) for the farfield formulation, as is to be expected.

The terms $\boldsymbol{b}$ and c are not as straightforward to evaluate, owing to the integrals contained within them. However, if the equations are reconsidered in the wavenumber domain the problem will lend itself to low-frequency approximations. Considering the former of these terms, in viewing (8.7.61) it can be deduced that each term in the vector $\boldsymbol{b}$ describes the change in acoustic power output one source (a single control source or the vibrating structure) elicits in the other source (structure or control) per unit control source volume velocity (as reciprocity holds in free space the influence one source has upon the other will be mutual). Thus each term in $\boldsymbol{b}$ is the result of the modal summation

$$b_n = \sum_{i=1}^{m} \left\{ \left[\frac{\omega \rho_0}{4\pi} \int_S \psi_i(x) \frac{\sin kr}{r}\, \mathrm{d}x \right] \cdot v_i \right\} \tag{8.7.69}$$

where r is the distance between a point on the vibrating structure and the acoustic control source, $\psi_i(\boldsymbol{x})$ is the value of the ith mode shape function at location $\boldsymbol{x}$ on the structure, and v_i is the velocity amplitude of the ith mode. The integral expression in the square brackets in (8.7.69) can be approximated in the low frequency regime as follows. First, the $(\sin kr)/r$ term can be re-expressed using a MacLaurin series as

$$\frac{\sin kr}{r} = \sum_{s=0}^{\infty} \frac{(-1)^s (kr)^{2s+1}}{r\,(2s+1)!} = \sum_{s=0}^{\infty} \frac{(-1)^s k^{2s+1} r^{2s}}{(2s+1)!} \tag{8.7.70}$$

The separation distance r, defined in (8.7.67), can be rewritten using a binomial expansion as

$$r^{2s} = \left[(x-x_{cn})^2 + (y-y_{cn})^2 + z_{cn}^2\right]^s =$$

$$\sum_{m=0}^{s}\sum_{l=0}^{m}\sum_{p=0}^{2m-2l}\sum_{q=0}^{2l} \begin{bmatrix} s \\ m \end{bmatrix} \begin{bmatrix} m \\ l \end{bmatrix} \begin{bmatrix} 2m-2l \\ p \end{bmatrix} \begin{bmatrix} 2l \\ q \end{bmatrix} x_{cn}^{2m-2l-p} x^p y_{cn}^{2l-q} y^q z_{cn}^{2s-2m} \tag{8.7.71}$$

where

$$\begin{bmatrix} a \\ b \end{bmatrix} = \frac{a!}{b!(a-b)!} \tag{8.7.72}$$

and the monopole control source is at a location (x_{cn}, y_{cn}, z_{cn}). Therefore, the bracketed part of (8.7.69) can be written as

$$v_i \cdot \frac{\rho_0 c_0}{4\pi} \sum_{s=0}^{\infty} \frac{(-1)^s k^{2s+2}}{(2s+1)!} \sum_{m=0}^{s}\sum_{l=0}^{m}\sum_{p=0}^{2m-2l}\sum_{q=0}^{2l} \begin{bmatrix} s \\ m \end{bmatrix} \begin{bmatrix} m \\ l \end{bmatrix} \begin{bmatrix} 2m-2l \\ p \end{bmatrix} \begin{bmatrix} 2l \\ q \end{bmatrix}$$

$$\times\, x_{cn}^{2m-2l-p} y_{cn}^{2l-q} z_{cn}^{2s-2m} \int_S \phi_i(\mathbf{x}) x^p y^q \, \mathrm{d}\mathbf{x} \tag{8.7.73}$$

The integral in (8.7.73) can be re-expressed as a partial derivative by considering the problem in k-space (see Section 2.4). Doing this requires calculation of the Fourier transform of the mode shape function, defined by the expression

$$\tilde{\psi}(k_x, k_y) = \int_S \psi(\mathbf{x}) \mathrm{e}^{-\mathrm{j}k_x x} \mathrm{e}^{-\mathrm{j}k_y y} \mathrm{d}\mathbf{x} \tag{8.7.74}$$

The moments of the mode shape function (the integral) in (8.7.73) (in the plane $z = 0$) can be expressed as derivatives in k-space evaluated at $k_x = k_y = 0$, so that the integral in (8.7.73) can be written in the form

$$\int_S \psi(\mathbf{x}) x^a y^b \, \mathrm{d}\mathbf{x} = j^{a+b} \frac{\partial^{a+b} \tilde{\psi}(k_x, k_y)}{\partial^a k_x \, \partial^b k_y} \bigg|_{k_x = k_y = 0} \tag{8.7.75}$$

Thus the bracketed part of (8.7.69) can be approximated as

$$v_i \cdot \frac{\rho_o c_o}{4\pi} \sum_{s=0}^{\infty} \frac{(-1)^s k^{2s+2}}{(2s+1)!} \sum_{m=0}^{s}\sum_{l=0}^{m}\sum_{p=0}^{2m-2l}\sum_{q=0}^{2l} \begin{bmatrix} s \\ m \end{bmatrix} \begin{bmatrix} m \\ l \end{bmatrix} \begin{bmatrix} 2m-2l \\ p \end{bmatrix} \begin{bmatrix} 2l \\ q \end{bmatrix}$$

$$\times\, x_{cn}^{2m-2l-p} y_{cn}^{2l-q} z_{cn}^{2s-2m} \, j^{p+q} \left\{ \left[\frac{\partial}{\partial k_x}\right]^p \left[\frac{\partial}{\partial k_y}\right]^q \tilde{\psi}_i \right\} \bigg|_{k_x = k_y = 0} \tag{8.7.76}$$

This expression, while looking rather foreboding, in fact often provides a straightforward means of approximating the terms in the vector $\boldsymbol{b}$, as the transformed mode shape functions of many commonly studied structures can be approximated by an infinite series and the derivatives easily evaluated (see Williams (1983) for a related discussion with examples).

The terms in the matrix $\boldsymbol{Z}_w$, upon which the term c is based, can be evaluated using the same approach. Doing this, it is straightforward to show that (8.7.63) can be expressed as

$$Z_w(i,\iota) = \frac{\rho_0 c}{4\pi} \sum_{m=0}^{\infty} \frac{k^{2m+2}}{(2m+1)!} \sum_{l=0}^{m} \sum_{p=0}^{2m-2l} \sum_{q=0}^{2l} \binom{m}{l} \binom{2m-2l}{p} \binom{2l}{q}$$
$$\times \left\{ \left[\frac{\partial}{\partial k_x}\right]^{2m-2l-p} \left[\frac{\partial}{\partial k_y}\right]^{2l-q} \tilde{\psi}_i^* \right\} \left\{ \left[\frac{\partial}{\partial k_x}\right]^{p} \left[\frac{\partial}{\partial k_y}\right]^{q} \tilde{\psi}_\iota \right\} \Bigg|_{k_x=k_y=0} \tag{8.7.77}$$

This approach has been used in calculating the radiation efficiencies for a variety of structural mode shape functions (Williams, 1983), as well as for calculating total acoustic power output of a structure (Snyder and Tanaka, 1992). Examples of the use of these equations will be given in the next section.

8.7.3 Minimization of acoustic pressure at discrete locations using vibration sources

Having derived a set of expressions which enable the determination of the optimal volume velocities for a set of monopole control sources used to attenuate radiation from an idealized vibrating structure, it is time to divert our attention to an alternative means of achieving control. When the sound field to be attenuated is produced by a vibrating structure, vibration control sources may be effective in providing the desired reduction in total radiated sound power, or acoustic pressure amplitude at a set of error sensing locations. Vibration control sources can provide this attenuation by altering the velocity distribution of the structure. To assess this capability, what is required is some means of calculating the optimum control source output with respect to the error criterion of interest, acoustic power or acoustic pressure.

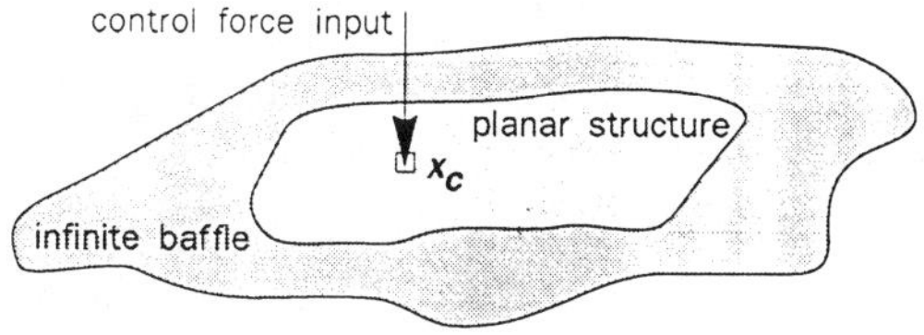

Fig. 8.29 Vibration control source model.

Consider the system illustrated in Fig. 8.29, where a force input is used to reduce the acoustic radiation from a vibrating structure. In practice the force input will be provided by a vibration source which is modelled as a point force for two reasons: first, it approximates the characteristics of a shaker attached to the structure at a point, such as through a stinger, and second, analytically it will prove to make to required calculations reasonably straightforward. (More complex piezoelectric ceramic vibration actuators are discussed later in Chapter 15; see also Fuller *et al.* (1991b), Clark *et al.* (1991), Clark and Fuller (1992c) and Koshigoe and Murdock (1993)). With a point source input to the structure, a quantity which can be manipulated in a manner analogous to our previous treatment of acoustic control source volume velocity is input force. Based on our previous experience, a logical approach to this problem would therefore seem to be to attempt to express the error criterion as a quadratic function of the control source input forces, differentiate the expression with respect to this quantity, and set the resultant gradient expression equal to zero to obtain the optimum set of complex force amplitudes. This necessitates formulating expressions for both the primary and control source radiated acoustic pressure distribution, the latter in terms of input force. The required expressions for the primary source pressure distribution have already been derived, and stated in (8.7.19), so we are half way there.

To obtain an expression for the acoustic pressure radiated in response to the control point source input, note that the problem is virtually identical to that of determining the primary source radiated sound pressure. The only difference is that in the primary source case the forcing function, the modal generalized force γ defined in (8.7.10), is assumed to be known either directly or implicitly in a knowledge of the existing vibration of the structure through (8.7.13). If an expression for the modal generalized force in terms of the control source input force can be found, then the previous primary source directed derivation can be used to determine the control source pressure distribution.

Referring to (8.7.10), to evaluate the modal generalized force an expression for the force input per unit area, or blocked pressure, from the forcing function must be obtained. For a point force, this expression is

$$p_{bl,c} = f_c \delta(\boldsymbol{x}-\boldsymbol{x}_c) \tag{8.7.78}$$

where f_c is the complex amplitude of the control force, $\delta(\boldsymbol{x} - \boldsymbol{x}_c)$ is a Dirac delta function, defined by

$$\delta(\boldsymbol{x}-\boldsymbol{x}_c) = \begin{cases} 1 & \boldsymbol{x}=\boldsymbol{x}_c \\ 0 & \text{otherwise} \end{cases} \tag{8.7.79}$$

and $\boldsymbol{x}_c$ is the attachment location of the point control force. Substituting this into (8.7.10), the ιth modal generalized force due to the control force input is

$$\gamma_\iota = \int_S f_c\, \psi_\iota(\boldsymbol{x})\, \delta(\boldsymbol{x}-\boldsymbol{x}_c)\, \mathrm{d}\boldsymbol{x} = f_c \psi_\iota(\boldsymbol{x}_c) \tag{8.7.80}$$

In words, (8.7.80) states that the modal generalized force resulting from the control point force input is equal to the amplitude of the control force multiplied by the value of the ιth mode shape function at the application point (note again that harmonic time dependence is implicit in this relationship). If N_c control forces are to be used, then the total value of the ιth modal generalised force is the superposition of the N_c individual contributions,

$$\gamma_\iota = \sum_{i=1}^{N_c} f_{ci}\ \phi_\iota(\boldsymbol{x}_i) \tag{8.7.81}$$

which can be written in matrix form as

$$\gamma_\iota = \boldsymbol{\psi}_{c\iota}^{\mathrm{T}} \boldsymbol{f}_c \tag{8.7.82}$$

where $\boldsymbol{\psi}_{c\iota}$ is the $(N_c \times 1)$ vector of values of the ιth mode shape function at the control force input positions,

$$\boldsymbol{\psi}_{c\iota} = \begin{bmatrix} \psi_\iota(\boldsymbol{x}_1) \\ \psi_\iota(\boldsymbol{x}_2) \\ \vdots \\ \psi_\iota(\boldsymbol{x}_{N_c}) \end{bmatrix} \tag{8.7.83}$$

and $\boldsymbol{f}_c$ is the $(N_c \times 1)$ vector of complex control force amplitudes

$$\boldsymbol{f}_c = \begin{bmatrix} f_{c1} \\ f_{c2} \\ \cdots \\ f_{cN_c} \end{bmatrix} \tag{8.7.84}$$

If m structural modes are being considered in the calculations, then the vector of modal generalized forces resulting from the control force inputs, $\boldsymbol{\gamma}_c$, can be expressed as a product of two matrices,

$$\boldsymbol{\gamma}_c = \boldsymbol{\Psi}_c \boldsymbol{f}_c \tag{8.7.85}$$

where $\boldsymbol{\Psi}_c$ is the $(m \times N_c)$ matrix of structural mode shape functions evaluated at the control source input locations, the rows of which are the transposed vectors of the mode shape functions evaluated at the control source locations defined in (8.7.83),

$$\boldsymbol{\Psi}_c = \begin{bmatrix} \boldsymbol{\psi}_{c1}^{\mathrm{T}} \\ \boldsymbol{\psi}_{c2}^{\mathrm{T}} \\ \vdots \\ \boldsymbol{\psi}_{cm}^{\mathrm{T}} \end{bmatrix} \tag{8.7.86}$$

Equation (8.7.86) can now simply be substituted into the form of (8.7.19) to produce an expression for the acoustic sound field at location $\boldsymbol{r}$ resulting from the control force input:

$$p_c(\boldsymbol{r}) = \boldsymbol{z}_{rad}^{\mathrm{T}} \boldsymbol{Z}_I^{-1} \boldsymbol{\Psi}_c \boldsymbol{f}_c \tag{8.7.87}$$

Now armed with an expression for the acoustic pressure which results from a given control force input, it is possible to consider expressing the error criterion as a quadratic function of the control force input, and using this expression to derive the optimum set of control forces. Consider first the minimisation of the sum of the squared acoustic pressure amplitudes at a discrete set of error sensing locations. This criterion was expressed in (8.7.27) in terms of the vectors of sound pressures at the points of interest due to the control and primary sources operating alone. Using (8.7.87), the $(N_e \times 1)$ vector of acoustic pressures $\boldsymbol{p}_c$ due to the control force inputs will be

$$\boldsymbol{p}_c = \boldsymbol{Z}_r \boldsymbol{Z}_I^{-1} \boldsymbol{\Phi}_c \boldsymbol{f}_c \tag{8.7.88}$$

where $\boldsymbol{Z}_r$ is the $(N_e \times m)$ matrix of modal radiation transfer functions between the m structural modes being used in the calculation and the N_e error locations, as defined in (8.7.29). Using this expression, in conjunction with that for the vector of acoustic pressures generated during primary excitation, expressed in (8.7.28), the error criterion when vibration control sources are being used can be expanded as

$$J_p = \boldsymbol{q}_c^{\mathrm{H}} \boldsymbol{A} \boldsymbol{q}_c + \boldsymbol{q}_c^{\mathrm{H}} \boldsymbol{b} + \boldsymbol{b}^{\mathrm{H}} \boldsymbol{q}_c + c \tag{8.7.89}$$

where

$$\boldsymbol{A} = \boldsymbol{\Psi}_c^{\mathrm{H}} \{\boldsymbol{Z}_I^{-1}\}^{\mathrm{H}} \boldsymbol{Z}_r^{\mathrm{H}} \boldsymbol{Z}_r \boldsymbol{Z}_I^{-1} \boldsymbol{\Psi}_c \tag{8.7.90}$$

$$\boldsymbol{b} = \boldsymbol{\Psi}_c^{\mathrm{H}} \{\boldsymbol{Z}_I^{-1}\}^{\mathrm{H}} \boldsymbol{Z}_r^{\mathrm{H}} \boldsymbol{Z}_r \boldsymbol{v}_p \tag{8.7.91}$$

and

$$c = \boldsymbol{v}_p^{\mathrm{H}} \boldsymbol{Z}_r^{\mathrm{H}} \boldsymbol{Z}_r \boldsymbol{v}_p \tag{8.7.92}$$

The form of the error criterion given in (8.7.89) is identical to that of (8.7.33) developed for the use of acoustic control sources. Differentiating it and setting the resultant gradient expression equal to zero produces an expression for the vector of optimum control forces $\boldsymbol{f}_c$

$$\boldsymbol{f}_{c,prs} = -\boldsymbol{A}^{-1} \boldsymbol{b} \tag{8.7.93}$$

where the subscripts on the control force vector c,prs indicate that it is optimal with respect to minimizing the acoustic pressure amplitude at a set of discrete error locations. Substituting this result back into (8.7.89) produces the expression for the minimum sum of the squared pressures:

$$J_{prs,\min} = \left\{ \sum_{i=1}^{N_e} | p(r_i) |^2 \right\}_{\min} = c - \boldsymbol{b}^{\mathrm{H}} \boldsymbol{A}^{-1} \boldsymbol{b} = c + \boldsymbol{b}^{\mathrm{H}} \boldsymbol{f}_{c,prs} \tag{8.7.94}$$

8.7.4 Minimization of total radiated acoustic power using vibration sources

The procedure for calculating the control source force inputs which minimize the total radiated acoustic power is much the same as that outlined previously for the use of acoustic control sources, where we need to express total radiated acoustic power as a quadratic function of the control force inputs. It is possible to develop this expression based upon either a nearfield or farfield measure of acoustic power. We will do both here, beginning with the farfield case. Using (8.7.87), it is straightforward to show, using the same steps taken in deriving (8.7.43), that the total radiated acoustic power can be expressed as a quadratic function of the control force inputs as

$$W = \boldsymbol{f}_c^{\mathrm{H}}\boldsymbol{A}\boldsymbol{f}_c + \boldsymbol{f}_c^{\mathrm{H}}\boldsymbol{b} + \boldsymbol{b}^{\mathrm{H}}\boldsymbol{f}_c + c \tag{8.7.95}$$

where

$$\boldsymbol{A} = \boldsymbol{\Psi}_c^{\mathrm{H}}\{\boldsymbol{Z}_I^{-1}\}^{\mathrm{H}}\left\{\int_0^{2\pi}\int_0^{\pi/2}\frac{\boldsymbol{z}_{rad}^{*}\boldsymbol{z}_{rad}^{\mathrm{T}}}{2\rho_0 c_0}\,|\,\boldsymbol{r}\,|^2\sin\theta\;\mathrm{d}\theta\;\mathrm{d}\phi\right\}\boldsymbol{Z}_I^{-1}\boldsymbol{\Psi}_c \tag{8.7.96}$$

$$\boldsymbol{b} = \boldsymbol{\Psi}_c^{\mathrm{H}}\{\boldsymbol{Z}_I^{-1}\}^{\mathrm{H}}\left\{\int_0^{2\pi}\int_0^{\pi/2}\frac{\boldsymbol{z}_{rad}^{*}\boldsymbol{z}_{rad}^{\mathrm{T}}}{2\rho_0 c_0}\,|\,\boldsymbol{r}\,|^2\sin\theta\;\mathrm{d}\theta\;\mathrm{d}\phi\right\}\boldsymbol{v}_p \tag{8.7.97}$$

and

$$c = \boldsymbol{v}_p^{\mathrm{H}}\left\{\int_0^{2\pi}\int_0^{\pi/2}\frac{\boldsymbol{z}_{rad}^{*}\boldsymbol{z}_{rad}^{\mathrm{T}}}{2\rho_0 c_0}\,|\,\boldsymbol{r}\,|^2\sin\theta\;\mathrm{d}\theta\;\mathrm{d}\phi\right\}\boldsymbol{v}_p \tag{8.7.98}$$

As before, differentiating (8.8.95) with respect to $\boldsymbol{f}_c$ and setting the gradient expression equal to zero will produce an expression which defines the vector of optimum control force inputs:

$$\boldsymbol{f}_{c,w} = -\boldsymbol{A}^{-1}\boldsymbol{b} \tag{8.7.99}$$

where the subscripts c,w indicate that it is optimal with respect to minimising the total radiated acoustic power. Substituting this result back into (8.8.95) produces the expression for the minimum radiated acoustic power

$$W_{\min} = c - \boldsymbol{b}^{\mathrm{H}}\boldsymbol{A}^{-1}\boldsymbol{b} = c + \boldsymbol{b}^{\mathrm{H}}\boldsymbol{f}_{c,w} \tag{8.7.100}$$

As with the acoustic control source problem, obtaining an analytical solution for the terms in $\boldsymbol{A}$, $\boldsymbol{b}$ and c in (8.7.96)–(8.7.98) is not, in general, a viable option. Rather, the spatial integrations must be performed numerically. However, as we will see shortly, at low frequencies it may be possible to make some approximations.

The other (nearfield) approach to deriving the necessary quadratic equation is to consider source power output directly (at the plate surface), this quantity being expressed previously in (8.7.54). The total acoustic power output W_t of this linear structural radiation problem in response to the combination of primary and control excitation can be written in the form

$$W_t = \frac{1}{2}\,\mathrm{Re}\left\{ \int_S v_t(x) p_t^*(x) \mathrm{d}x \right\} \tag{8.7.101}$$

where t denotes a total (combined primary and control) quantity. Expanding the total acoustic pressure as the superposition of primary and control source generated components, (8.7.101) becomes

$$W_t = \frac{1}{2}\,\mathrm{Re}\left\{ \int_S v_c(x) p_c^*(x)\,\mathrm{d}x + \int_S v_c(x) p_p^*(x)\,\mathrm{d}x + \int_S v_p(x) p_c^*(x)\,\mathrm{d}x + \int_S v_p(x) p_p^*(x) \mathrm{d}x \right\} \tag{8.7.102}$$

Observe that the expansion produces a set of four expressions, from left to right describing the acoustic power output of the structure due to excitation by the control source distribution operating alone, the modification to this quantity due to the coexisting primary excited acoustic pressure field, the modification to the primary excited acoustic power output due to the coexisting control source generated acoustic pressure field, and the acoustic power output of the structure due to the sole operation of the primary forcing function.

For the planar structure being considered here, the acoustic pressure $p(x)$ can be expressed in terms of surface velocity using the Rayleigh integral as outlined in (8.7.14). If (8.7.14) is used to expand (8.7.102) the result is a set of four expressions

$$W_t = W_{cc} + W_{cp} + W_{pc} + W_{pp} \tag{8.7.103}$$

each with the form of the surface integral for acoustic power as described, for example, in Williams (1983). In the terms on the right-hand side of (8.7.103), the first subscript refers to the surface velocity generating source distribution, and the second subscript the acoustic field generating source distribution. For example, W_{cp} is the component of total radiated acoustic power attributable to the combination of the surface velocity induced by the control source distribution (operating alone) and the acoustic field induced by the primary source distribution (also operating alone)

$$W_{cp} = \frac{\omega\rho_0}{4\pi}\,\mathrm{Re}\left\{ \int_S \int_S v_p^*(x') \frac{\sin kr}{r} v_c(x)\,\mathrm{d}x'\,\mathrm{d}x \right\} \tag{8.7.104}$$

with similar relationships for all other terms. If the velocity at any point on the structure is expanded in terms of modal contributions, then (8.7.104) can be expanded as

$$W_{cp} = \frac{\omega\rho_0}{4\pi} \int_S \int_S \left[\sum_{i=1}^{m} v_{pi}^*(x') \frac{\sin kr}{r} \sum_{\iota=1}^{m} v_{c\iota}(x)\,\mathrm{d}x'\,\mathrm{d}x \right] \tag{8.7.105}$$

again with similar relationships for the other terms. Using (8.7.13) and (8.7.85) it is straightforward to re-express (8.7.103) in our standard form (as a quadratic function of the control force inputs) as

$$W_t = f_c^H A f_c + b^H f_c + f_c^H b + c \tag{8.7.106}$$

where

$$A = \Psi_c^H \{Z_I^{-1}\}^H Z_w Z_I^{-1} \Psi_c \tag{8.7.107}$$

$$b = \Psi_c^H \{Z_I^{-1}\}^H Z_w v_p \tag{8.7.108}$$

$$c = v_p^H Z_w v_p \tag{8.7.109}$$

and Z_w is the acoustic power transfer matrix as defined previously in (8.7.63). Expressed in this form, the vector of optimal control forces is by now well known to be equal to

$$f_c = -A^{-1} b \tag{8.7.110}$$

In viewing (8.7.107)–(8.7.109), it is apparent that to obtain an analytical solution for the optimum set of control force inputs, it will be necessary to obtain an analytical solution for the terms in the acoustic power transfer matrix Z_w. While this is not, in general, a viable option, it is often possible to obtain an approximate solution if the excitation frequency is 'low', as outlined in (8.7.77). The next section will specialize this expression for a simply supported, baffled panel, and will shown that the off-diagonal terms in the power transfer matrix can be calculated by some simple algebraic manipulation of the diagonal terms.

We now have all of the tools required to examine the effectiveness, or otherwise, of applying active control to planar, vibrating structures situated in an infinite baffle. If acoustic control sources are of interest then, for a given arrangement of sources and a given frequency, use of equations (8.7.34)–(8.7.37) will enable us to derive a set of volume velocities which will minimize the acoustic pressure at a set of potential error sensing locations, or use of equations (8.7.44)–(8.7.47) or (8.7.60)–(8.7.64) will produce a set of volume velocities which will minimize the total radiated acoustic power. If vibration control sources are of interest, then for a given arrangement of source and a given frequency use of (8.7.90)–(8.7.93) will enable us to derive a set of input forces which will minimize the acoustic pressure at a set of potential error sensing locations, or use of (8.7.96)–(8.7.99) or (8.7.107)–(8.7.110) will produce a set of input forces which will minimize the total radiated acoustic power. In the next section the use of this methodology will be demonstrated by applying it to the specific problem of controlling sound radiation from a infinitely baffled, simply supported rectangular panel.

8.8 AN EXAMPLE: CONTROL OF SOUND RADIATION FROM A RECTANGULAR PANEL

In the previous section, generalized analytical models were developed that could be used to determine the optimum control source volume velocities and forces for minimizing periodic sound radiation from baffled, planar vibrating structures. In this section a specific example will be considered to demonstrate how to specialize the general models for a specific problem. For this purpose, the case where the radiating 'structure' is a simply supported rectangular panel will be used. The reason for choosing this arrangement for a closer inspection of the active control problem is the relative simplicity of analytically modelling the system, which arises from its regular geometry. This simplicity correlates well with its use as a model structure in published literature. This specific example will be used several times in the sections which follow to demonstrate certain characteristics of the active control problem.

8.8.1 Specialization for examining minimization of acoustic pressure at discrete locations

We will begin our example study by considering the problem of minimizing the acoustic pressure amplitude at discrete locations in space. A quick critique of the main equations of interest for calculating the optimum control source volume velocities and forces which minimize acoustic pressure, (8.7.34)–(8.7.37) and (8.7.90)–(8.7.93), shows that the matrices which must be specialised are $\boldsymbol{\Psi}_c$, $\boldsymbol{Z}_l$ and $\boldsymbol{Z}_r$ (note that the terms in $\boldsymbol{Z}_m$, defined in (8.7.31), are the same for all baffled, planar structure problems and can be traced back to equation (8.7.21)). Starting with the first of these, the mode shape function of the rectangular panel, with the origin at the lower left corner, is

$$\psi(\boldsymbol{x}) = \sin\frac{M\pi x}{L_x}\sin\frac{N\pi y}{L_y} \tag{8.8.1}$$

where M and N are the modal indices in the x and y directions respectively, and L_x, L_y are the panel dimensions. Therefore, the terms in the vector ψ_{cl}, defined in equation (8.7.83), which is used in constructing the rows of $\boldsymbol{\Psi}_c$ as defined in (8.7.86), are

$$\psi_{c\iota} = \begin{bmatrix} \psi_\iota(\boldsymbol{x}_1) \\ \psi_\iota(\boldsymbol{x}_2) \\ \vdots \\ \psi_\iota(\boldsymbol{x}_{N_c}) \end{bmatrix} = \begin{bmatrix} \sin\dfrac{M_\iota \pi x_1}{L_x}\sin\dfrac{N_\iota \pi y_1}{L_y} \\ \sin\dfrac{M_\iota \pi x_2}{L_x}\sin\dfrac{N_\iota \pi y_2}{L_y} \\ \vdots \\ \sin\dfrac{M_\iota \pi x_{N_c}}{L_x}\sin\dfrac{N_\iota \pi y_{N_c}}{L_y} \end{bmatrix} \tag{8.8.2}$$

In words, the terms in the $(N_c \times 1)$ vector $\Psi_{c\iota}$ are the values of the ιth structural mode shape function evaluated at the application point of each of the N_c vibration control sources. The $(m \times N_c)$ matrix Ψ_c is then

$$\Psi_c = \begin{bmatrix} \sin\frac{M_1 \pi x_1}{L_x}\sin\frac{N_1 \pi y_1}{L_y} & \cdots & \sin\frac{M_1 \pi x_{N_c}}{L_x}\sin\frac{N_1 \pi y_{N_c}}{L_y} \\ & \vdots & \\ \sin\frac{M_m \pi x_1}{L_x}\sin\frac{N_m \pi y_1}{L_y} & \cdots & \sin\frac{M_m \pi x_{N_c}}{L_x}\sin\frac{N_m \pi y_{N_c}}{L_y} \end{bmatrix} \tag{8.8.3}$$

where there are N_c control sources and m structural modes being modelled in the calculations.

To specialise the second of these matrices, $\mathbf{Z}_I$, what is required (from the definition in (8.7.12)) are the modal input impedances and modal masses. The former of these requires calculation of the resonance frequencies associated with the panel modes. These are found from the expression (Junger and Feit, 1986)

$$\omega_i = \left[\frac{D}{\rho_s h}\right]^{1/2} \left\{ \left[\frac{M\pi}{L_x}\right]^2 + \left[\frac{N\pi}{L_y}\right]^2 \right\} \tag{8.8.4}$$

where M,N are the modal indices of the ith mode, D is the panel bending stiffness, given by

$$D = EI = \frac{Eh^3}{12(1-\nu)^2} \tag{8.8.5}$$

I is the moment of inertia of the panel, ρ_s is the structure material density, h is the panel thickness, E the modulus of elasticity, and ν is Poisson's ratio. The modal mass can be calculated by substituting the mode shape function of (8.8.1) into the defining equation of modal mass, (8.7.4). Assuming a panel which is homogeneous in both thickness and material density, this produces

$$\begin{aligned} M_\iota &= \int_S \rho_s(\mathbf{x}) h(\mathbf{x}) \psi_\iota^2(\mathbf{x}) \, \mathrm{d}\mathbf{x} \\ &= \int_0^\pi \int_0^\pi \rho_s h \left[\sin\frac{M_\iota \pi x}{L_x} \sin\frac{N_\iota \pi y}{L_y} \right]^2 \mathrm{d}x \, \mathrm{d}y = \frac{\rho_s h A}{4} \end{aligned} \tag{8.8.6}$$

where A is the panel area, $A = xy$. Therefore, the (i,i)th term in the diagonal matrix $\mathbf{Z}_I$, as defined in (8.7.12), is

$$z_I(i,i) = M_i Z_i = \frac{\rho_s h A}{4} \left\{ \omega_i^2 (1 + j\eta) - \omega^2 \right\} \tag{8.8.7}$$

where Z_i is defined in (8.7.5).

The final matrix to specialize is $\mathbf{Z}_r$, which requires calculation of the terms in the vector z_{rad}. From the definition given in (8.7.18), this requires evaluation of the integral expression

$$z_{rad}(i,\boldsymbol{r}) = 2j\omega\rho_0 \int_S G_f(\boldsymbol{x} \mid \boldsymbol{r})\psi_i(\boldsymbol{x})\,\mathrm{d}\boldsymbol{x} \tag{8.8.8}$$

using the modal shape function of (8.8.1). It is not possible to evaluate this integral analytically for any arbitrary location $\boldsymbol{r}$ in space. However, if the problem is restricted to locations in the farfield of the structure, such that $r >> L_x$ and $r >> L_y$, an approximate solution can be found (Wallace, 1972):

$$z_{rad}(i,\boldsymbol{r}) \approx \frac{j\omega\rho_0}{2\pi r}\,\mathrm{e}^{-jkr}\,\frac{L_x L_y}{M_i N_i \pi^2}\left[\frac{-1^{M_i}\mathrm{e}^{-j\alpha}-1}{\left[\dfrac{\alpha}{M_i\pi}\right]^2-1}\right]\left[\frac{-1^{N_i}\mathrm{e}^{-j\beta}-1}{\left[\dfrac{\beta}{N_i\pi}\right]^2-1}\right] \tag{8.8.9}$$

where

$$\alpha = kL_x \sin\theta\,\cos\phi,\ \beta = kL_y \sin\theta\,\sin\phi \tag{8.8.10}$$

The terms in the $(N_e \times m)$ matrix $\mathbf{Z}_r$ are simply the evaluation of this expression for the combination of each of the m structural modes at the N_e error sensing locations.

With these specializations we can now quantify the terms in (8.7.34) and (8.7.35), and (8.7.90) and (8.7.91), and using the relationships in (8.7.37) and (8.7.90) calculate the control inputs which will minimize the acoustic pressure amplitude at discrete locations. It is then straightforward to calculate the residual acoustic field at any location by adding together (superposition) the primary and control source pressures; for the primary source, the acoustic pressure is defined by (8.7.19), for acoustic control sources by (8.7.23), and for vibration control sources by (8.7.87). This enables us to paint a complete picture of what will happen during the application of an active control system which minimizes acoustic pressure at discrete error sensing locations in space.

Before using our model to examine some of the phenomena associated with the active control of acoustic radiation from our rectangular panel, it is of interest to compare the results predicted by our theoretical model with some obtained experimentally. There have been a number of theoretical and experimental investigations of this problem, as well as other related panel radiation problems, published, including Fuller *et al.* (1989, 1991a, 1991b), Thomas *et al.* (1990), Hansen and Snyder (1991), and Metcalf *et al.* (1992); here we will use results related to those of Pan *et al.* (1992). With this study a simply supported rectangular steel panel of dimensions was situated in a baffle in an anechoic room, with a primary noise disturbance from a non-contacting electro-magnetic shaker mounted above the panel center. The steel panel was 380 mm × 300 mm in size, and either 2 mm or 9.5 mm thick. The important resonance frequencies are listed in Table 8.1. The results presented here will be the rad-iated sound field measured across a horizontal arc 1.8 metres from the panel center, a distance which can be considered as just in the farfield (Beranek, 1986, p. 100), from $\boldsymbol{r}$ = (1.8 m, 90°, 180°) to (1.8 m, 90°, 0°). The primary

Table 8.1 Experimental and theoretical panel resonances

Mode	1.9 mm thick panel		9.5 mm thick panel	
	Theoretical resonance (Hz)	Experimental resonance (Hz)	Theoretical resonance (Hz)	Experimental resonance (Hz)
(1,1)	86.3	88	418.5	444
(2,1)	185.8	187	900.6	920
(1,2)	245.9	244	1191.9	1196
(2,2)	345.4	343	1674.0	1688
(3,1)	351.6	349	1703.9	1692
(3,2)	511.1	—	2477.4	—
(1,3)	511.9	501	2481.0	2456
(4,1)	583.6	581	2828.8	2796
(2,3)	611.3	595	2963.0	2928
(4,2)	743.0	732	3602.2	—

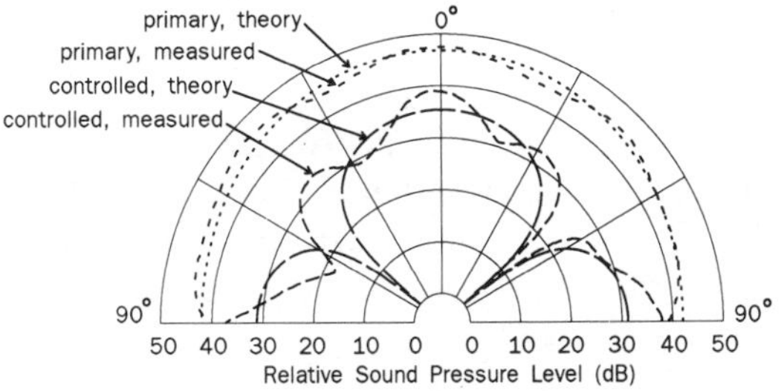

Fig. 8.30 Theoretical and experimental acoustic pressure distributions for a single acoustic control source in the panel center, error sensor at (1.8 m, 50°, 180°), 338 Hz.

disturbance will be modelled as a point force (a thorough discussion of the experimental conditions can be found in Pan *et al.* (1992)).

Consider first the use of an acoustic control source to attenuate radiation from the 2 mm thick panel at 338 Hz. The acoustic source used here is a horn driver located 20 mm in front of the center of the panel, with a horn diameter of 50 mm at the exit (the acoustic control source will be modelled as a monopole source 20 mm from the front of the panel). Figure 8.30 illustrates the effect of minimizing the sound pressure at r = (1.8 m, 50°, 180°). The comparison between theory and experiment shows that the general agreement is good, although diffraction around the horn driver has slightly altered the acoustic field, introducing an interference pattern into the result. This sort of deviation from an idealised model is to be expected, where practical transducers are not infinitely small. However, the ability of the model to predict the major characteristics of the residual sound field is clearly evident.

It would be expected that this diffraction effect would become more pronounced as both the frequency, and number of control sources (placed in front of the panel), are increased. To examine this, we will increase the panel thickness to 9.5 mm, the frequency of excitation to 1707 Hz, and the number of control sources to three. The two additional horn drivers are placed 100 mm in the x direction on either side of the center of the panel. To simply the experiment, the magnitudes of each horn driver output are constrained to be equal, with the relative phases varying 0°/180°/0° across the panel. With this arrangement, a single error sensor will be adequate to implement the control system; had the three control sources been completely independent a single error sensor would result in an underdetermined system, and the overall experimental result somewhat unpredictable. To model this arrangement analytically, the term in the vector z_{mono} defined in (8.7.24) can be modified to

$$z_{mono} = \frac{j\rho_0\omega}{4\pi}\left[\frac{e^{-jkr_{s1}}}{r_{s1}} + \frac{e^{-jkr_{i1}}}{r_{i1}} + \frac{e^{-jkr_{s2}}}{r_{s2}} + \frac{e^{-jkr_{i2}}}{r_{i2}} + \frac{e^{-jkr_{s3}}}{r_{s3}} + \frac{e^{-jkr_{i3}}}{r_{i3}}\right] \tag{8.8.11}$$

The theoretical and experimental sound fields under primary excitation are illustrated in Fig. 8.31. Observe that the diffraction has become worse, although the general characteristics still match quite well. The theoretical and experimental residual sound fields obtained when minimizing the sound pressure at r = (1.8 m, 0°, 0°) are shown in Fig. 8.32. Here the interference pattern has become markedly worse, owing to multiple sound sources diffracting around multiple objects (horns). However, the average amplitudes of the predicted and measured residual sound fields, as well as the general characteristics of the directivity pattern, match quite well.

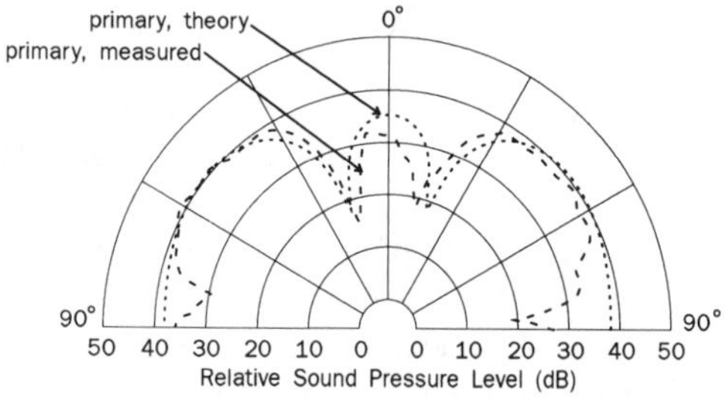

Fig. 8.31 Theoretical and experimental primary source acoustic pressure distributions at 1707 Hz.

Let us turn our attention now to the use of vibration control sources to attenuate the radiated acoustic field. For this purpose, vibration control will be applied using an electrodynamic shaker through a stinger attached to the panel 150 mm to the left of the panel centre. A plot of the theoretical and measured primary radiated and controlled residual sound pressure levels for excitation at 338 Hz and sound pressure minimization at r = (1.8 m ,0°, 0°) are given in

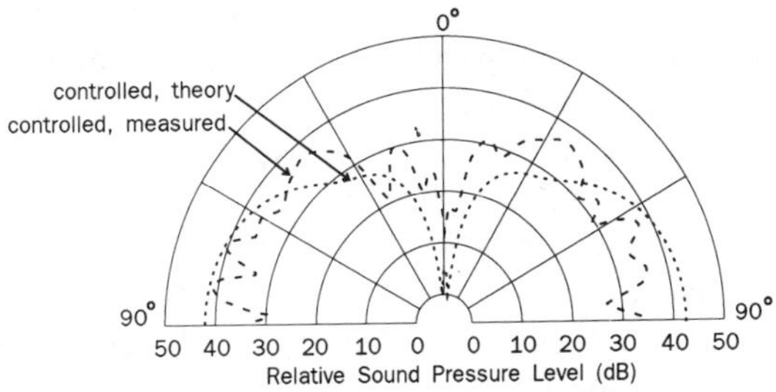

Fig. 8.32 Theoretical and experimental residual pressure distributions for a three acoustic control sources, error sensor at (1.8 m, 0°, 0°), 1707 Hz.

Fig. 8.33. Here the agreement between theory and experiment is very good, as the diffraction phenomena associated with having the acoustic sources in front of the panel is removed.

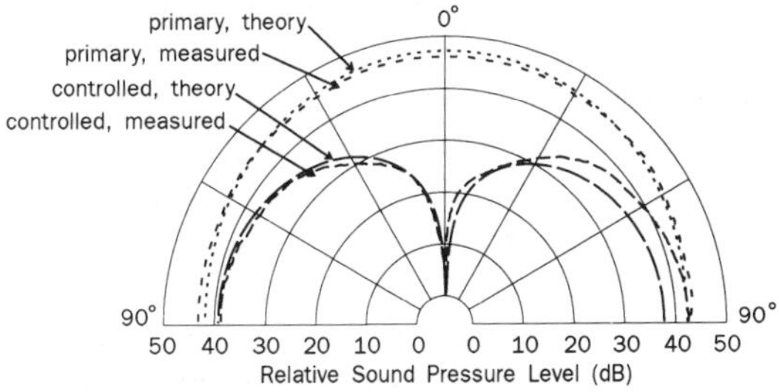

Fig. 8.33 Theoretical and experimental pressure distributions for a vibration control source, error sensor at (1.8 m, 0°, 0°), 338 Hz.

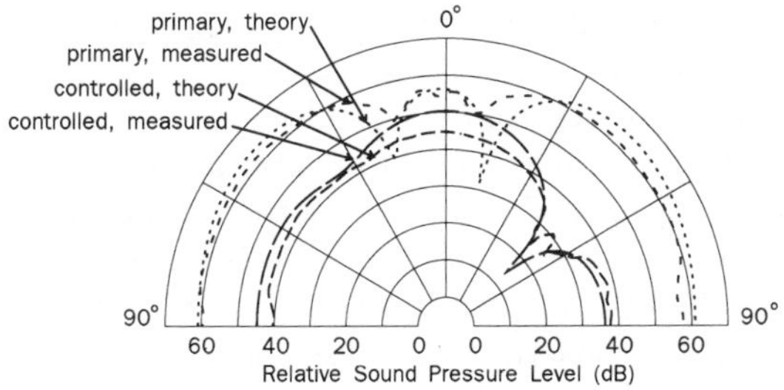

Fig. 8.34 Theoretical and experimental pressure distributions for a piezoelectric ceramic path vibration control source, error sensor at (1.8 m , 50°, 180°), 1707 Hz.

One of the interesting vibration actuators to become available in recent years is the piezoelectric ceramic patch, described in Chapter 15. These actuators are relatively small, and can be bonded directly to the panel surface to help create a 'smart structure', a structure with embedded actuators, sensors and a control system to modify its response characteristics. As discussed in Chapter 15, these actuators can be used as vibration sources to attenuate structural acoustic radiation. For our final experimental result, we will bond a piezoelectric crystal actuator in the centre of the 9.5 mm panel, and use it to control the sound field at 1707 Hz. Modelling the crystal as a simple point force, the theoretical and experimental results are as plotted in Fig. 8.34. Observe that while there is some evidence of diffraction (from the experimental baffle and microphone boom) which is a result of the high frequency, the correspondence between the theoretical and experimental results is very good. The conclusion we can draw from this and our other tests is that our relatively simple theoretical model is quite accurate in predicting the effect of applying active control using either acoustic or vibration control sources.

8.8.2 Minimization of radiated acoustic power

Let us now turn our attention to the problem of minimizing the acoustic power output of the system. In general, to investigate this topic we need to use the system specific quantities of the previous section (mode shape function, resonance frequencies and radiation transfer function) in (8.7.44)–(8.7.47) and/or (8.7.60)–(8.7.64) for acoustic control sources, or (8.7.96)–(8.7.99) and/or (8.7.107)–(8.7.110) for vibration control sources, and perform the requried integrations numerically to calculate the optimum control source outputs. From there it is straightforward to find the acoustic power attenuation using the minimum power expressions of (8.7.48), (8.7.65), (8.7.100) or (8.7.110), noting that the term c in each of these is the uncontrolled acoustic power output of the system. Later in this chapter, when we briefly discuss some aspects of designing an active control system, this is the approach we will take. However, at low frequencies it is often possible to obtain approximate expression for the acoustic power attenuation which do not require numerical integration, expressions which can be quite useful when trying to paint a qualitative picture of the influence which certain parameters have upon system performance. The techniques required to obtain these approximate solutions were outlined in the previous section, where the problem was transformed into the wavenumber domain to change the integrals into partial derivatives. In this section we will concentrate on obtaining these approximate solutions for our rectangular panel system, demonstrating how to fill out the terms in (8.7.60)–(8.7.64) and (8.7.107)–(8.7.110). These solutions will be used in subsequent sections.

A general form for low frequency approximate solutions for the elements in the vector $\boldsymbol{b}$ defined in equation (8.7.61) for the problem of minimizing radiated acoustic power using an acoustic control source was derived in (8.7.76), and for the acoustic power transfer matrix $\boldsymbol{Z}_w$ in (8.7.77). To use these expressions we must first obtain the Fourier transform of the mode shape function of our simply supported rectangular panel system. The mode shape function of the simply supported panel is separable in the x and y directions, as follows

$$\phi(\boldsymbol{x}) = \phi_x(x)\phi_y(y) = \sin\frac{M\pi x}{L_x}\,\sin\frac{N\pi y}{L_y} \tag{8.8.12}$$

Taking the Fourier transform of the x component of the mode shape function as outlined in (8.7.74), if the modal index is an odd number

$$\tilde{\phi}_x(x) = \frac{2\dfrac{M\pi}{L_x}\cos\dfrac{k_xL_x}{2}}{\left[\dfrac{M\pi}{L_x}\right]^2 - k_x^2} \tag{8.8.13}$$

and if the modal index is an even number

$$\tilde{\phi}_x(x) = j\frac{2\dfrac{M\pi}{L_x}\sin\dfrac{k_xL_x}{2}}{\left[\dfrac{M\pi}{L_x}\right]^2 - k_x^2} \tag{8.8.14}$$

with similar expressions for the y component of the mode shape function. In many cases, solution for the required partial derivatives of the Fourier transformed mode shape functions is simplified if the transformed mode shape functions are first expressed as a series expansion (although this is of debatable value for our simple example). Working towards this goal, a general expression for the Fourier transformed, simply supported rectangular panel mode shape function is

$$\tilde{\phi}(k_x,k_y) = \frac{4L_xL_y}{MN\pi^2} f(k_x)\, f(k_y) \tag{8.8.15}$$

where $k_x = k \sin\theta \cos\phi$, $k_y = k \sin\theta \sin\phi$, and for odd modes

$$f(k_x) = \frac{\left[\dfrac{M\pi}{L_x}\right]^2 \cos\dfrac{k_xL_x}{2}}{k_x^2 - \left[\dfrac{M\pi}{L_x}\right]^2} \tag{8.8.16}$$

while for even modes

$$f(k_x) = j\frac{\left[\dfrac{M\pi}{L_x}\right]^2 \sin\dfrac{k_xL_x}{2}}{k_x^2 - \left[\dfrac{M\pi}{L_x}\right]^2} \tag{8.8.17}$$

with similar expressions for k_y where the y dimension is used in place of the x dimension, and the y modal index in place of the x modal index. A MacLaurin expansion of (8.8.16) and (8.8.17) can be easily taken. For odd modes, (8.8.16) is expanded as

$$f(k_x) = -\left[1 - \frac{(k_xL_x)^2}{8} + \cdots\right]\left[1 + \left[\frac{k_xL_x}{M\pi}\right]^2 + \cdots\right] \tag{8.8.18}$$

and for even modes (8.8.17) is expanded as

$$f(k_x) = -j\left[\frac{k_xL_x}{2} - \frac{(k_xL_x)^3}{48} + \cdots\right]\left[1 + \left[\frac{k_xL_x}{M\pi}\right]^2 + \cdots\right] \tag{8.8.19}$$

The Fourier transformed mode shape functions of (8.8.15), used in conjunction with the odd/even expansions of (8.8.18) and (8.8.19), can be substituted into (8.7.77) to calculate the terms in the acoustic power transfer matrix $\mathbf{Z}_w$ and into (8.7.76) to calculate the term $\boldsymbol{b}$ used in the acoustic control source minimum acoustic power formulation for our simply supported rectangular panel system. These can then be used in conjunction with (8.7.60)–(8.7.64) and (8.7.107)–(8.7.110) to provide an approximate calculation of the optimum control source outputs and (8.7.65) and (8.7.111) for the residual radiated acoustic power at low frequencies.

Let us first consider calculation of the terms in the acoustic power transfer matrix $\mathbf{Z}_w$. One point to mention before undertaking this calculation is that only similar index-type modal combinations will produce a non-zero result ($\mathbf{Z}_w(i,\iota)$ will be non-zero if i and ι are both (odd,odd), (odd,even), (even,odd) or (even,even) modes). While this result is to be expected, it is not immediately obvious from the form of (8.7.77). Upon substituting values into the equation, however, it is found that the inherent 'squareness' of the equation (the sum of the orders of the partial derivatives in each term must be an even number, equal to $2m$), combined with the differences in the form of (8.7.18) and (8.7.19) (8.7.18) is an even power expansion, (8.7.19) is an odd power expansion) always combine to enforce this result. For this reason we will discuss only similar index-type modal combinations.

Consider first the calculation of the term $\mathbf{Z}_w(i,\iota)$ when i and ι are both odd,odd index modes. To match the accuracy of the commonly stated results for radiation efficiency, also derived using low-frequency approximations, the infinite summation in (8.7.77) can be truncated at $m = 1$ (this places a low frequency constraint placed upon the solution, as for such a premature truncation of the infinite series to yield an accurate result, $(ka/M\pi) << 1$, $(kb/N\pi) << 1$). The term $\mathbf{Z}_w(i,\iota)$ for odd,odd modal combinations is

$$\mathbf{Z}_{w_{o,o}}(i,\iota) = \frac{4\rho_0c_0k^2L_x^2L_x^2}{M_iN_iM_\iota N_\iota\pi^5}\left\{1 - \frac{k^2L_xL_y}{24}\left[\left[1-\frac{8}{(M_i\pi)^2}\right]\frac{L_x}{L_y} + \left[1-\frac{8}{(N_i\pi)^2}\right]\frac{L_y}{L_x} + \left[1-\frac{8}{(M_\iota\pi)^2}\right]\frac{L_x}{L_y} + \left[1-\frac{8}{(N_\iota\pi)^2}\right]\frac{L_y}{L_x}\right]\right\} \tag{8.8.20}$$

For even,odd mode types, in order to obtain the same accuracy it will be necessary to consider terms in equation (8.7.77) up to $m = 2$. Doing this, the terms in $\boldsymbol{Z}_w$ for even,odd mode combinations is

$$Z_{w_{elo}}(i,\iota) = \frac{\rho_0 c_0 k^4 L_x^4 L_y^2}{3M_i N_i M_\iota N_\iota \pi^5}\left\{1 - \frac{k^2 L_x L_y}{40}\left[\left(1-\frac{24}{(M_i\pi)^2}\right)\frac{L_x}{L_y} + \left(1-\frac{8}{(N_i\pi)^2}\right)\frac{L_y}{L_x} + \left(1-\frac{24}{(M_\iota\pi)^2}\right)\frac{L_x}{L_y} + \left(1-\frac{8}{(N_\iota\pi)^2}\right)\frac{L_y}{L_x}\right]\right\} \tag{8.8.21}$$

Note also that this expression will hold for odd,even modes, as the power transfer matrix term $\boldsymbol{Z}_w(i,\iota)$ will be the same as (8.8.21), with the x and y modal indices and dimensions swapped. Finally, for even,even mode types, to obtain the desired accuracy it will be necessary to include terms in the infinite series of equation (8.7.77) up to m=3. Doing this, the term $\boldsymbol{Z}_w(i,\iota)$ for even,even modal combinations is found to be

$$Z_{w_{ele}}(i,\iota) = \frac{\rho_0 c_0 k^6 L_x^4 L_y^4}{60M_i N_i M_\iota N_\iota \pi^5}\left\{1 - \frac{k^2 L_x L_y}{28}\left[\left(1-\frac{24}{(M_i\pi)^2}\right)\frac{L_x}{L_y} + \left(1-\frac{24}{(N_i\pi)^2}\right)\frac{L_y}{L_x} + \left(1-\frac{24}{(M_\iota\pi)^2}\right)\frac{L_x}{L_y} + \left(1-\frac{24}{(N_\iota\pi)^2}\right)\frac{L_y}{L_x}\right]\right\} \tag{8.8.22}$$

Let us now turn our attention to the task of obtaining approximate solutions for the terms in the $\boldsymbol{b}$ vector used in the acoustic control source problem, defined in (8.7.61), by filling out the terms in the expansion of (8.7.76). Here it is found that the number of terms in the expansion required to obtain the same accuracy as for the expressions above is the same for the same mode types. For odd,odd modes, the low frequency approximate solution is

$$b_i = v_i\frac{\rho_0 c_0 k^4 L_x L_y}{6M_i N_i \pi^3}\left\{1 - \frac{k^2 L_x L_y}{40}\left[\frac{|\,2\boldsymbol{r}\,|^2}{L_x L_y} + \left(1-\frac{24}{(M_i\pi)^2}\right)\frac{L_x}{L_y} + \left(1-\frac{8}{(N_i\pi)^2}\right)\frac{L_y}{L_x}\right]\right\} \tag{8.8.23}$$

where

$$|\,2\boldsymbol{r}\,|^2 = 4\left[(x-\frac{L_x}{2})^2+(y-\frac{L_y}{2})^2+z^2\right] \tag{8.8.24}$$

For even,odd and odd,even modes,

$$b_i = v_i \frac{\rho_0 c_0 k^4 L_x^2 L_y \left[x - \frac{L_x}{2}\right]}{6 M_i N_i \pi^3}$$

$$\left\{1 - \frac{k^2 L_x L_y}{40}\left[\frac{|2r|^2}{L_x L_y} + \left[1 - \frac{24}{(M_i \pi)^2}\right]\frac{L_x}{L_y} + \left[1 - \frac{8}{(N_i \pi)^2}\right]\frac{L_y}{L_x}\right]\right\} \tag{8.8.25}$$

Finally, for even,even modes,

$$b_i = v_i \frac{\rho_0 c_0 k^6 L_x^2 L_y^2 \left[x - \frac{L_x}{2}\right]\left[y - \frac{L_y}{2}\right]}{60 M_i N_i \pi^3}$$

$$\left\{1 - \frac{k^2 L_x L_y}{28}\left[\frac{|2r|^2}{L_x L_y} + \left[1 - \frac{24}{(M_i \pi)^2}\right]\frac{L_x}{L_y} + \left[1 - \frac{24}{(N_i \pi)^2}\right]\frac{L_y}{L_x}\right]\right\} \tag{8.8.26}$$

As we mentioned, one of the purposes of deriving low frequency approximate solutions is to obtain a qualitative picture of the effects of system parameters on the ability of the system to provide acoustic power attenuation. For example, using the above solutions in (8.7.60)–(8.7.64), it is found that with a single acoustic control source and a single structural mode the leading terms in the acoustic power attenuation series expansion are, for an odd/odd mode (after Deffayet and Nelson, 1988):

$$\frac{W_{\min}}{W_{pri}} \approx \frac{k^2 |r|^2}{3} \tag{8.8.27}$$

for an even,odd or odd,even mode:

$$\frac{W_{\min}}{W_{pri}} \approx 1 - \frac{k^2 \left[x - \frac{L_x}{2}\right]^2}{3}\left[1 - \frac{k^2 |r|^2}{5}\right] \tag{8.8.28}$$

and for an even,even

$$\frac{W_{\min}}{W_{pri}} \approx 1 - \frac{k^4 \left[x - \frac{L_x}{2}\right]^2 \left[y - \frac{L_y}{2}\right]^2}{120}\left[1 - \frac{4k^2 |r|^2}{14}\right] \tag{8.8.29}$$

As a side note, one immediately obvious parameter having the greatest influence over the system performance is control source location. Observe from

(8.8.27)−(8.8.29) that with an odd,odd mode the optimum control source location (for low frequency excitation) is in the centre of the panel, while a control source a this location for the other mode types will be ineffective (Deffayet and Nelson, 1988).

8.9 ELECTRICAL TRANSFORMER NOISE CONTROL

Noise from large electrical transformers is characterized by single frequency components at twice, four times, six times and eight times the a.c. line frequency (50 Hz in Europe, Japan and Australia and 60 Hz in North America), although in some cases higher harmonics also contribute. When transformers are located close to residential communities, the characteristic low frequency humming noise is often a cause of widespread complaints. This noise is caused by vibrations of the core (caused principally by magnetostriction) which are transmitted through the oil bath to the outer tank which then vibrates and radiates noise. As it is extremely difficult to reduce the amplitude of the magnetostriction and thus control the problem at its source, traditional noise control involves the construction of a massive enclosure around the transformer. This enclosure must cooled with forced air ventilation drawn into the enclosure and exhausted through silenced ductwork. This is extremely expensive and in many cases very inconvenient for maintenance and inspection purposes. For this reason, the prospect of active control has become increasingly attractive as practical automatic control systems have become more easily realized.

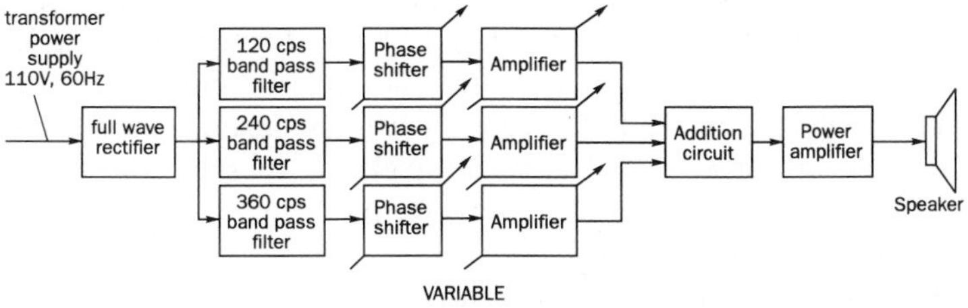

Fig. 8.35 Conover's system for the control of electrical transformer noise.

The first published attempt to control electrical transformer noise was reported by Conover in 1956. He used the voltage signal, from the low voltage side of the transformer, and a full wave rectifier to generate a reference signal. This reference signal was passed through narrow bandpass filters to isolate periodic noise at 120 Hz, 240 Hz and 360 Hz (see Fig. 8.35). Each of these three signals was then passed through a variable gain amplifier and variable phase shifter before being recombined and fed to a single loudspeaker placed next to the transformer. The amplitudes and phases of the three individual signals were adjusted manually using the variable gain amplifier and phase shifter until the sound at an error microphone was minimized. Conover found that although he could achieve noise reductions of up to 25 dB at the error microphone, this reduction was restricted to a very small angle from the line joining the error microphone to the centre of the tank wall against which the loudspeaker was placed. In other directions, the overall noise level invariably increased with the control loudspeaker turned on. He also found that as the transformer noise varied considerably from day to day, the system had to be adjusted regularly. He suggested that this might be done automatically but did not follow this up.

More than 20 years later, Ross (1978) used an almost identical system to demonstrate control of transformer noise transmitted to an office adjacent to two transformers in a courtyard. It appears that Ross developed this independently as no mention of Conover's work was made. At about the same time Hesselmann (1978) and then Angevine (1981) demonstrated that global control could be achieved for a transformer in an anechoic room provided that the transformer was completely surrounded by loudspeakers. Angevine used independent single channel controllers to minimize the sound level in the vicinity of each control source. He found that the attenuation was dependent on the number of control sources and that this dependence was stronger at lower frequencies. For example, increasing the number of control sources from 10 to 30 increased the attenuation at 100 Hz from 9 dB to 19 dB.

Berge *et al.* (1987, 1988) reported on an attempt to actively control electrical transformer noise using an approach similar to that of Conover (1956) except that the gains and phases of the individual frequency components (100 Hz and 200 Hz) were adjusted automatically to minimize an r.m.s. signal at a microphone, using an iterative algorithm which updated the gains and phases every one to four minutes. Rapid updating (every 12 seconds) seemed to degrade the performance for some unexplained reason. Using only one loudspeaker control source, they found that significant noise attenuation could only be achieved over a very narrow angle either side of the error microphone. In addition, expected levels in other directions invariably increased when the active control system was turned on. Clearly, the single control source had a negligible effect on the radiation impedance 'seen' by the transformer tank and served only to provide an interfering sound field which must increase noise levels in some locations if reductions are to be achieved at other locations.

More recently Angevine (1992, 1993, 1994, 1995) has reported some success using multiple loudspeaker sources in front of a transformer tank to achieve significant noise reductions over a wide area (15 to 20 dB over 35° to 40°).

Another possible means of controlling electrical transformer noise is to use active force actuators on the tank to suppress the vibration of modes which contribute most to sound radiation (Mcloughlin *et al.*, 1994). These force actuators could be driven by a multichannel controller to minimize the noise at a number of error microphone locations. It is highly likely that a system such as this would require a controller with considerably less channels than would be required if loudspeakers were used. However, it may be difficult to generate sufficient forces with practical vibration actuators as the high internal impedance of the transformer tank filled with oil would make it difficult to suppress the vibration of the radiating modes. On the other hand, if the modal rearrangement mechanism discussed earlier in this chapter proves effective, then the control forces required by the actuators may be considerably less than at first thought. Possible vibration control actuators could be piezo electric, magnetostrictive or hydraulic (see Chapter 12). Means of mounting these actuators which require no backing mass are discussed at the end of Chapter 12.

A third possible means for actively reducing electrical transformer noise is to surround it with an enclosure made from thin, perforated sheet metal of about 30% open area. Control actuators could be used to drive the perforated sheet metal enclosure (which would have a low internal impedance) to minimize the sound at the error microphone locations. The enclosure could be driven so that the solid part was out of phase with the fluid in the holes, thus forming a low efficiency multiple element sound source. The perforations would still allow natural cooling of the transformer tank without the need for forced convection. The feasibility of this approach has been investigated for sound radiation from a rectangular panel (Burgemeister and Hansen, 1993).

The idea of using vibration actuators to control enclosure wall vibration may be a generally efficient means of improving passive enclosure transmission loss at low tonal frequencies generated by rotating equipment inside.

8.10 A CLOSER LOOK AT CONTROL MECHANISMS AND A COMMON LINK AMONG ALL ACTIVE CONTROL SYSTEMS

In the previous two sections we have spent some time developing analytical models which can be used to assess the outcome of applying active control with a given system arrangement. What we have yet to do is really examine, in physical terms, how this outcome (acoustic power attenuation) is achieved; that is, what the actual control mechanisms are. In this section we will direct our efforts towards this task.

Although the principal aim of the previous sections was to develop and give examples of general theoretical models which could be used in an analytical study or design exercise, the few results presented tend to leave one with the feeling that acoustic- and vibration-control-source based systems are orthogonal entities. Certainly the acoustic-control-source efforts could be viewed as an extension to the theoretical work undertaken earlier in the chapter on monopole source arrangements, with qualitative characteristics likely to be common. The efforts directed at the use of vibration control sources to suppress structural radiation are, however, unique to the structural radiation problem, with an apparently unique set of qualitative system characteristics (such as the influence of source location). It may seem unlikely that any important similarities between acoustic-source based control systems and vibration-source based control systems would exist. In this section we will see that this is not the case. In fact, there is a common link between all active control systems, found at the fundamental level of control mechanisms and the 'control source' generated value of the global error criterion of interest. Here active noise control systems will be considered specifically, although the conclusions drawn are not confined to the use of an acoustic power error criterion, but will exist with other global error criteria such as kinetic energy in vibrating structures.

We will begin this section by deriving a ‘common link’, then progress on to a discussion of how this applies to acoustic- and vibration-source based systems. In doing this we will undertake a closer examination of how active control systems provide attenuation of the radiated acoustic field, especially the more complicated question of how vibration-source based systems provide attenuation of the radiated acoustic field by altering the velocity distribution of the structure to which they are attached.

8.10.1 A common link

Consider the active noise control arrangement shown in Fig. 8.36, where some generic control source distribution is being used to attenuate the acoustic field originating from some generic primary source distribution. The primary source is subject to harmonic excitation of the form $e^{j\omega t}$, which will again be left implicit in the equations, with no specifications yet made concerning the characteristics of the acoustic environment. In the development of a ‘common link’ it will prove insightful to examine the acoustic power output of the control source during application of active control. To do this, what is first required is calculation of the optimum control source output (volume velocity for acoustic control sources, or input force for vibration control sources), which will minimize the total radiated acoustic power of the system. We have done this previously in Sections 8.4 and 8.7 using two different approaches (nearfield and farfield), both of which were aimed at expressing the error criterion (total radiated acoustic power) as a quadratic function of control source output. This expression was then differentiated with respect to control source output, and the resultant gradient equation set equal to zero to yield the optimum value. The process will be repeated here in a much more general way, using only the nearfield formulation of the error criterion.

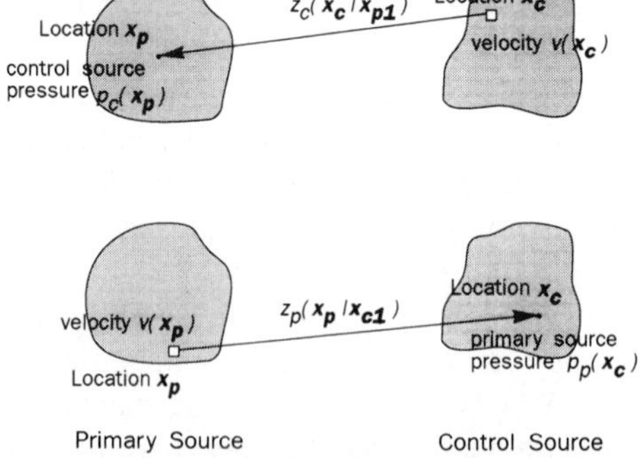

Fig. 8.36 Geometry for a generic control source distribution being used to attenuation acoustic radiation from a generic primary source distribution.

Recall that with a nearfield formulation of the error criterion, source acoustic power output is considered directly, this quantity being equal to

$$W = \frac{1}{2}\,\mathrm{Re}\left\{\int_S v(\boldsymbol{x})\,p^*(\boldsymbol{x})\,\mathrm{d}\boldsymbol{x}\right\} \tag{8.10.1}$$

where Re denotes the real part of the expression, $v(\boldsymbol{x})$ is the velocity at some location $\boldsymbol{x}$ on the surface of the source, $p^*(\boldsymbol{x})$ is the complex conjugate of the acoustic pressure at this same location, and the integration is conducted over the surface of the source. Returning to Fig. 8.36, we can rewrite the total acoustic power output W_t of this arrangement in the form

$$W_t = W_c + W_p = \frac{1}{2}\,\mathrm{Re}\left\{\int_{S_c} v(\boldsymbol{x}_c)p_t^*(\boldsymbol{x}_c)\mathrm{d}\boldsymbol{x}_c + \int_{S_p} v(\boldsymbol{x}_p)p_t^*(\boldsymbol{x}_p)\mathrm{d}\boldsymbol{x}_p\right\} \tag{8.10.2}$$

where W_c is the acoustic power output of the control source

$$W_c = \frac{1}{2}\,\mathrm{Re}\left\{\int_{S_c} v(\boldsymbol{x}_c)p_t^*(\boldsymbol{x}_c)\,\mathrm{d}\boldsymbol{x}_c\right\} \tag{8.10.3}$$

W_p is the acoustic power output of the primary source

$$W_p = \frac{1}{2}\,\mathrm{Re}\left\{\int_{S_p} v(\boldsymbol{x}_p)\,p_t^*(\boldsymbol{x}_p)\,\mathrm{d}\boldsymbol{x}_p\right\} \tag{8.10.4}$$

and the subscripts p, c, and t are used to denote primary source, control source, and total (combined primary and control source) quantities. As the total pressure at any location can be considered to be the superposition of the primary and control source pressures at that location, (8.10.2) can be expanded to

$$W_t = \frac{1}{2}\,\mathrm{Re}\left\{\int_{S_c} v(\boldsymbol{x}_c)\big(p_c(\boldsymbol{x}_c)+p_p(\boldsymbol{x}_c)\big)^*\,\mathrm{d}\boldsymbol{x}_c + \int_{S_p} v(\boldsymbol{x}_p)\big(p_c(\boldsymbol{x}_p)+p_p(\boldsymbol{x}_p)\big)^*\,\mathrm{d}\boldsymbol{x}_p\right\} \tag{8.10.5}$$

The control source generated acoustic pressure at a location $\boldsymbol{r}$ will be related to the control source velocity distribution through an equation of the form

$$p_c(\boldsymbol{r}) = \int_{S_c} v(\boldsymbol{x}_c)z_c(\boldsymbol{x}_c\mid\boldsymbol{r})\,\mathrm{d}\boldsymbol{x}_c = q_c\int_{S_c} z_v(\boldsymbol{x}_c)z_c(\boldsymbol{x}_c\mid\boldsymbol{r})\,\mathrm{d}\boldsymbol{x}_c \tag{8.10.6}$$

where $z_c(\boldsymbol{x}_c|\boldsymbol{r})$ is the radiation transfer function between the control source velocity at location $\boldsymbol{x}_c$ and the acoustic pressure at location $\boldsymbol{r}$, q_c is the complex amplitude of the control source output, and $z_v(\boldsymbol{x}_c)$ is the transfer function between this overall amplitude and the specific velocity level on the control source at location $\boldsymbol{x}_c$. Note that q_c is completely general, and fills the role of either volume velocity or input force amplitude, considered separately in the derivations of Section 8.7. It could even represent a more 'unusual' quantity, such as the amplitude of a structural mode or the amplitude of some section of the vibrating structure, if it was of interest to find out what the optimum value of these quantities, with respect to minimizing the error criterion, would be.

Substituting (8.10.6) into (8.10.5), the total radiated acoustic power can be written as

$$W_t = \frac{1}{2}\,\mathrm{Re}\left\{ q_c \int_{S_c} z_v(\mathbf{x}'_c) \left[q_c \int_{S_c} z_v(\mathbf{x}_c) z_c(\mathbf{x}_c \mid \mathbf{x}'_c)\,\mathrm{d}\mathbf{x}_c + p_p(\mathbf{x}'_c) \right]^* \mathrm{d}\mathbf{x}'_c \right. \tag{8.10.7}$$

$$\left. + \int_{S_p} v(\mathbf{x}_p) \left[q_c \int_{S_c} z_v(\mathbf{x}_c) z_c(\mathbf{x}_c \mid \mathbf{x}_p)\,\mathrm{d}\mathbf{x}_c + p_p(\mathbf{x}_p) \right]^* \mathrm{d}\mathbf{x}_p \right\}$$

or

$$W_t = q_c^* a q_c + \mathrm{Re}\left\{ q_c b_1^* \right\} + \mathrm{Re}\left\{ b_2 q_c^* \right\} + c \tag{8.10.8}$$

where

$$a = \frac{1}{2}\,\mathrm{Re}\left\{ \int_{S_c} z_v(\mathbf{x}'_c) \int_{S_c} z_v^*(\mathbf{x}_c) z_c^*(\mathbf{x}_c \mid \mathbf{x}'_c)\,\mathrm{d}\mathbf{x}_c\,\mathrm{d}\mathbf{x}'_c \right\} \tag{8.10.9}$$

$$b_1^* = \frac{1}{2} \int_{S_c} z_v(\mathbf{x}'_c) p_p^*(\mathbf{x}'_c)\,\mathrm{d}\mathbf{x}'_c \tag{8.10.10}$$

$$b_2 = \frac{1}{2} \int_{S_p} v(\mathbf{x}_p) \int_{S_c} z_v^*(\mathbf{x}_c) z_c^*(\mathbf{x}_c \mid \mathbf{x}_p)\,\mathrm{d}\mathbf{x}_c\,\mathrm{d}\mathbf{x}_p \tag{8.10.11}$$

and

$$c = \frac{1}{2}\,\mathrm{Re}\left\{ \int_{S_p} v(\mathbf{x}_p) p_p^*(\mathbf{x}_p)\,\mathrm{d}\mathbf{x}_p \right\} \tag{8.10.12}$$

Equation (8.10.8) is the desired quadratic form defining the total acoustic power output of the system, which can be used to derive the optimum value of q_c, the complex control source amplitude. It should again be noted that no specifications concerning the characteristics of the primary source distribution, control source distribution, or the acoustic environment have yet been made. In fact, it is possible to derive a quadratic expression of the form of (8.10.8) for the majority of error criteria used in the development of active control systems for harmonic excitation problems.

To use (8.10.8) to obtain the optimum control source output, it must first be rewritten in terms of its real and imaginary parts:

$$W_t = a q_{c\mathrm{R}}^2 + a q_{c\mathrm{I}}^2 + b_{1\mathrm{R}} q_{c\mathrm{R}} + b_{1\mathrm{I}} q_{c\mathrm{I}} + b_{2\mathrm{R}} q_{c\mathrm{R}} + b_{2\mathrm{I}} q_{c\mathrm{I}} + c \tag{8.10.13}$$

where the quantities a and c are defined in (8.10.9) and (8.10.12) as being strictly real. This can then be differentiated with respect to the real and imaginary parts of the control source output, and the resultant gradient expressions set equal to zero to define the optimum values:

$$\frac{\partial W_t}{\partial q_{c\mathrm{R}}} = 2aq_{c\mathrm{R}} + b_{1\mathrm{R}} + b_{2\mathrm{R}} = 0$$
$$\frac{\partial W_t}{\partial q_{c\mathrm{I}}} = 2aq_{c\mathrm{I}} + b_{1\mathrm{I}} + b_{2\mathrm{I}} = 0 \qquad (8.10.14)$$

Recombining the real and imaginary parts, the optimum control source output is found to be defined by the relationship

$$q_{c,\mathrm{opt}} = -\frac{1}{2}a^{-1}(b_1+b_2) \qquad (8.10.15)$$

Note that for the relationship of (8.10.15) to actually define a unique value of control source output which will minimize the total radiated acoustic power, the term a must be positive, which from consideration of the physical meaning of the terms in (8.10.8) (which we will discuss next), will surely be the case.

It is worthwhile here to assign some physical meaning to the terms defined in (8.10.9)−(8.10.12), specifically in terms of the acoustic power error criterion which is of most interest to us in this chapter. The simplest of these is the term c, which is the value of the error criterion (acoustic power output) of the primary source in the absence of active control. The term a can be viewed as the radiation resistance of the control source, the quantity which defines the acoustic power produced by the source, operating alone, for a given source output. From this it is apparent that a must always be positive, because if it were zero, the control source would be incapable of radiating any (real) acoustic power, regardless of the amplitude of its output, and if it were negative it would simply absorb energy in response to operating (an acoustic black hole!). It is the quantities b_1 and b_2, however, which will prove to be the most important for the topic we are considering here. The terms in (8.10.8) in which b_1 and b_2 appear are the modifications to the control source and primary source acoustic power outputs respectively, due to the addition of the other sound source distribution into the acoustic environment. From a mathematical perspective, b_1 is simply equal to the integration of half the product of the primary source generated acoustic pressure and the control source velocity transfer function over the surface area of the control source, which when multiplied by the control source velocity amplitude will define the change in control source acoustic power output. Heuristically, it can be viewed as representing the desire of the control system to absorb the primary source radiated power. Similarly, from a mathematical perspective the term b_2 is equal to the integration over the primary source surface of half the value of the product of the velocity at some point on the surface of the primary source and the complex conjugate of the transfer function between the control source amplitude and the acoustic radiation to that point, which when multiplied by the control source velocity amplitude will define the change in primary source acoustic power output. However, the heuristic interpretation of b_2 is that it represents the desire of the control system to suppress the primary source power output. Therefore, the bracketed part of (8.10.15) can be viewed as the balance between wanting to absorb the primary source power output and wanting to suppress it. The final acoustic power output of the control source will be a result of this balance. In line with these interpretations, the a term can be viewed heuristically as simply a magnitude modulating term.

If the optimum control source output of (8.10.15) is substituted into (8.10.3) and (8.10.4), in terms of the quantities defined in (8.10.9)–(8.10.12) the controlled power outputs of the primary and control sources are found to be

$$W_c = \frac{1}{4}a^{-1}(b_1+b_2)^*(b_1+b_2) - \frac{1}{2}a^{-1}\mathrm{Re}\left\{(b_1+b_2)b_1^*\right\} \tag{8.10.16}$$

and

$$W_p = -\frac{1}{2}a^{-1}\mathrm{Re}\left\{(b_1+b_2)^*b_2\right\}+c \tag{8.10.17}$$

After some minor algebraic manipulation we can express the former of these, the acoustic power output of the control source, as

$$W_c = \frac{1}{4}a^{-1}\left(|b_2|^2-|b_1|^2\right) \tag{8.10.18}$$

From this rearrangement, it can be deduced that the control source power output will be zero if

$$|b_1| = |b_2| \tag{8.10.19}$$

or heuristically if the desire to suppress the primary source radiation is equal to the desire to absorb it. On the surface this would appear an unlikely occurrence. It is, however, very common, as we will see.

To give more credence to this line of heuristic thought, we can consider what happens if b_1 and b_2 are related by some multiplying factor, so that $b_1 = \gamma b_2$. Substituting this into (8.10.18) the power output of the control source then becomes

$$W_c = \frac{1}{4}a^{-1}\left\{|b_2|^2(1-\gamma^2)\right\} \tag{8.10.20}$$

From this relationship, if γ is greater than 1, interpreted as the desire to absorb, b_1, being greater than the desire to suppress, b_2, the final power output of the control source will be negative and absorption will occur. Similarly, if γ is less than 1, interpreted as the desire to absorb, b_1, being less than the desire to suppress, b_2, the final power output of the control source will be positive.

To quantify the circumstances under which the desires to suppress and absorb are perfectly balanced, and hence the control source acoustic power output is equal to zero, further consideration of the terms b_1 and b_2 is necessary. The primary source acoustic pressure at the control source location, $p_p(\boldsymbol{x}_c)$, which appears in the term b_1 as defined in (8.10.10), can be rewritten in terms of the velocity distribution of the primary source as follows

$$p_p(\boldsymbol{x}_c) = \int_{S_p} v(\boldsymbol{x}_p) z_p(\boldsymbol{x}_p \mid \boldsymbol{x}_c) \, \mathrm{d}\boldsymbol{x}_p \tag{8.10.21}$$

where $z_p(\boldsymbol{x}_p|\boldsymbol{x}_c)$ is the radiation transfer function between the surface velocity at the location $\boldsymbol{x}_p$ on the primary source and control source location $\boldsymbol{x}_c$. Substituting this into (8.10.10), the term b_1^* can be rewritten as

$$b_1^* = \frac{1}{2} \int_{S_c} z_v(\boldsymbol{x}_c') \int_{S_p} v^*(\boldsymbol{x}_p) z_p^*(\boldsymbol{x}_p \mid \boldsymbol{x}_c') \, \mathrm{d}\boldsymbol{x}_p \, \mathrm{d}\boldsymbol{x}_c' \tag{8.10.22}$$

If we expand the zero control source power output criterion of (8.10.19) using this result in conjunction with (8.10.11), for zero control source acoustic power output, we obtain

$$\begin{aligned} &\left| \frac{1}{2} \int_{S_c} z_v(\boldsymbol{x}_c')^* \int_{S_p} v(\boldsymbol{x}_p) z_p(\boldsymbol{x}_p \mid \boldsymbol{x}_c') \, \mathrm{d}\boldsymbol{x}_p \, \mathrm{d}\boldsymbol{x}_c' \right| \\ &= \left| \frac{1}{2} \int_{S_p} v(\boldsymbol{x}_p) \int_{S_c} z_v(\boldsymbol{x}_c)^* z_c(\boldsymbol{x}_c \mid \boldsymbol{x}_p)^* \, \mathrm{d}\boldsymbol{x}_c \, \mathrm{d}\boldsymbol{x}_p \right| \end{aligned} \tag{8.10.23}$$

The criterion of (8.10.23) will only be satisfied if z_p and z_c are equal, real numbers. While it is unlikely that this will occur directly, there is a common situation that can occur which will enable the neglect of the imaginary components of these terms, and in so doing enable fulfilment of the criterion. To examine this, it is first necessary to combine the two middle terms of the acoustic power (8.10.8), producing

$$W_t = q_c^* a q_c + \mathrm{Re}\left\{q_c b_1^* + b_2 q_c^*\right\} + c \tag{8.10.24}$$

This newly combined term can be expressed, using (8.10.11) and (8.10.22), as

$$\begin{aligned} &\mathrm{Re}\left\{q_c b_1^* + b_2 q_c^*\right\} \\ &= \frac{1}{2} \int_{S_c} \int_{S_p} \mathrm{Re}\Big\{ \left(q_c z_v(\boldsymbol{x}_c)\right) v^*(\boldsymbol{x}_p) z_p^*(\boldsymbol{x}_p \mid \boldsymbol{x}_c) \\ &\quad + \left(q_c z_v(\boldsymbol{x}_c)\right)^* v(\boldsymbol{x}_p) z_c^*(\boldsymbol{x}_c \mid \boldsymbol{x}_p)\Big\} \, \mathrm{d}\boldsymbol{x}_p \, \mathrm{d}\boldsymbol{x}_c \end{aligned} \tag{8.10.25}$$

Suppose for a moment that $z_p(\boldsymbol{x}_p|\boldsymbol{x}_c) = z_c(\boldsymbol{x}_c|\boldsymbol{x}_p)$. If this is the case, then the terms in the integral in (8.10.25) have the form

$$\mathrm{Re}\{de^* f^* + d^* e f^*\} = de^* \, \mathrm{Re}\{f\} + d^* e \, \mathrm{Re}\{f\} \tag{8.10.26}$$

Here the term f is the transfer function $z_p(\boldsymbol{x}_p|\boldsymbol{x}_c) = z_c(\boldsymbol{x}_c|\boldsymbol{x}_p)$. This result means that if $z_p(\boldsymbol{x}_p|\boldsymbol{x}_c) = z_c(\boldsymbol{x}_c|\boldsymbol{x}_p)$, only the real part of these terms contributes to the acoustic power calculation, and so the imaginary components of these terms can be neglected without any alteration to the final result of the equation. If that is the case then the criterion for zero control source acoustic power output is satisfied. However, stating that $z_p(\boldsymbol{x}_p|\boldsymbol{x}_c) = z_c(\boldsymbol{x}_c|\boldsymbol{x}_p)$ is simply stating that acoustic reciprocity exists. Therefore, the principal result of this section is that if acoustic reciprocity exists between each point on the primary source and each point on the control source, the acoustic power output of the control source during optimal operating conditions will be zero (Snyder and Tanaka, 1993a). (Note here again that we have considered only harmonic excitation in this section, and hence in the preceding statement. Later in this chapter we will consider random noise and find a rather different result.) The systems we are considering in this chapter all radiate into free space, where reciprocity exists, neglecting wind effects. Therefore, we can state that the acoustic power output of the control source under optimal conditions must be zero.

The question now arises, if there are multiple control sources what is the acoustic power output of each source? Simply having a total acoustic power output of zero from the control source array does not in itself mean that contributions from the constituent elements will be zero. To answer this, we can start by generalizing (8.10.8) for N_c multiple, separate control sources as

$$\ddot{W}_t = \mathrm{Re}\{\boldsymbol{q}_c^{\mathrm{H}}\boldsymbol{A}\boldsymbol{q}_c\} + \mathrm{Re}\{\boldsymbol{b}_1^{\mathrm{H}}\boldsymbol{q}_c\} + \mathrm{Re}\{\boldsymbol{q}_c^{\mathrm{H}}\boldsymbol{b}_2\} + c \tag{8.10.27}$$

where $\boldsymbol{q}_c$ is an N_c length vector of control source outputs, $\boldsymbol{A}$ is an $(N_c \times N_c)$ matrix of transfer functions, whose terms are defined by the expression

$$A(i,\iota) = \frac{1}{2}\,\mathrm{Re}\left\{\int_{S_{ci}} z_v(\boldsymbol{x}'_{ci}) \int_{S_{c\iota}} z_v^*(\boldsymbol{x}_{c\iota}) z_c^*(\boldsymbol{x}_{c\iota} \mid \boldsymbol{x}'_{ci})\,\mathrm{d}\boldsymbol{x}_{c\iota}\,\mathrm{d}\boldsymbol{x}'_{ci}\right\} \tag{8.10.28}$$

$\boldsymbol{b}_1$ is an N_c length vector whose terms are defined by

$$b_1(i) = \frac{1}{2}\int_{S_{ci}} z_v^*(\boldsymbol{x}'_{ci}) p_p(\boldsymbol{x}'_{ci})\mathrm{d}\boldsymbol{x}'_{ci} \tag{8.10.29}$$

and $\boldsymbol{b}_2$ is an N_c length vector whose terms are defined by

$$b_2(i) = \frac{1}{2}\int_{S_p} v(\boldsymbol{x}_p) \int_{S_{ci}} z_v^*(\boldsymbol{x}_{ci}) z_c^*(\boldsymbol{x}_{ci} \mid \boldsymbol{x}_p)\,\mathrm{d}\boldsymbol{x}_{ci}\,\mathrm{d}\boldsymbol{x}_p \tag{8.10.30}$$

Let us suppose again that acoustic reciprocity exists in the system. If then we define a variable

$$\boldsymbol{b}_1 = \boldsymbol{b}_2^* = \boldsymbol{b} \tag{8.10.31}$$

the total acoustic power output can be expressed as

$$W_t = \boldsymbol{q}_c^{\mathrm{H}}\boldsymbol{A}\boldsymbol{q}_c + \boldsymbol{q}_c^{\mathrm{H}}\boldsymbol{b} + \boldsymbol{b}^{\mathrm{H}}\boldsymbol{q}_c + c \tag{8.10.32}$$

This is the standard expression of an error criterion as a quadratic function of control source output, which we have used several times already in this chapter. Based on our previous work it is straightforward to write an expression defining the vector of optimum control source amplitudes

$$\boldsymbol{q}_c = -\boldsymbol{A}^{-1}\boldsymbol{b} \tag{8.10.33}$$

Now, the total acoustic power output of the control source array is defined by the first two terms in (8.10.27) as follows:

$$W_c = \mathrm{Re}\{\boldsymbol{q}_c^{\mathrm{H}}\boldsymbol{A}\boldsymbol{q}_c\} + \mathrm{Re}\{\boldsymbol{b}_1^{\mathrm{H}}\boldsymbol{q}_c\} \tag{8.10.34}$$

From this total, the acoustic power output of a single, ith, control source is equal to

$$W_{ci} = \mathrm{Re}\{\boldsymbol{q}_c^{\mathrm{H}}\boldsymbol{A}(\mathrm{col}\ i)q_c(i)\} + \mathrm{Re}\{b^*(i)q_c(i)\} \tag{8.10.35}$$

where $\boldsymbol{A}(\mathrm{col}\ i)$ is the ith column of the matrix $\boldsymbol{A}$, $b(i)$ is the ith element in the vector $\boldsymbol{b}$, and $q_c(i)$ is the complex amplitude of the ith control source, the ith element of the vector $\boldsymbol{q}_c$. If the vector of control source outputs is set equal to the optimum value defined in (8.10.33), the acoustic power output of the ith control source will be

$$W_{ci} = \mathrm{Re}\{[-\boldsymbol{A}^{-1}\boldsymbol{b}]^{\mathrm{H}}\boldsymbol{A}(\mathrm{col}\ i)q_c(i)\} + \mathrm{Re}\{b(i)^*q_c(i)\} = 0 \tag{8.10.36}$$

Therefore, the acoustic power output of each element in the control source array will also be zero.

It should be stressed that we have derived this result from a completely general basis, without any specification about the type of primary or control source. In fact, this result could have been derived for any global error criterion which can be expressed as a quadratic function of the control source output, such as minimization of structural kinetic energy or acoustic potential energy in an enclosed space. This result, therefore, provides a common link among all active control systems. The problem now is how to apply it to attain a maximum understanding of the physical mechanisms responsible for providing attenuation in an active system. As we will see, the concept is straightforward in an acoustic-source based system, but will require some 'lateral thinking' in a vibration-source based system.

8.10.2 Acoustic control source mechanisms and the common link

In our examination of this common link and the relationship it has to control mechanisms, we will begin with consideration of acoustic-control-source based systems. With these systems, it is straightforward to apply the above result, as the control source is an obviously separate vibrating entity which will have zero acoustic power output under ideal operating conditions. We have already seen the result in Section 8.4 in relation to sources with monopole primary and control sources. It will, however, be worth rederiving the result in terms of the theory of this section before progressing on to consideration of the control of sound radiation from an infinitely baffled, planar structure using a monopole control source.

Consider first the two monopole arrangement shown in Fig. 8.37. As we stated in Section 8.2, the sound pressure distribution of a monopole source is defined as

$$p(r) = \frac{j\omega\rho q}{4\pi r}e^{-jkr} = j\omega\rho q G_f(r_q \mid r) \tag{8.10.37}$$

where r_q is the location of the source, and G_f is the free space Green's function,

$$G_f(x \mid r) = \frac{e^{-jkr}}{4\pi r} \tag{8.10.38}$$

The control source output which needs to be optimized in this instance, q_c in (8.10.15), is its volume velocity.

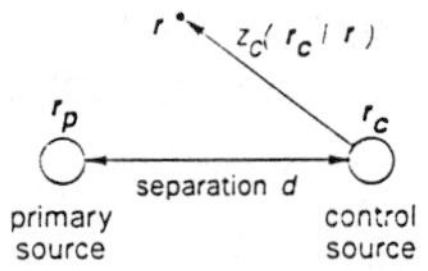

Fig. 8.37 Single primary source/single control source arrangement.

We can use the relationship of (8.10.38) to specialize the terms in a, b_1, and b_2 in (8.10.9)–(8.10.11) to examine the power output of the control source. To do this, we set the term q_c in (8.10.8) equal to the control source volume velocity amplitude, with a velocity transfer function z_v equal to 1.0. The radiation transfer function z_c is, from (8.10.30),

$$z_c(x_c \mid r) = j\omega\rho\frac{e^{-jkr}}{4\pi r} = j\omega\rho G_f(x_c \mid r) \tag{8.10.39}$$

with an identical expression for the primary source radiation transfer function z_p. For monopole sources, as the velocity is isolated to a single location r_q in space, the integrations can be replaced by multiplications of the values at the source locations. Taking these factors into account, the term a, which is the radiation resistance of a monopole source operating alone, and the terms b_1, and b_2 become

$$a = \frac{1}{2}\mathrm{Re}\{j\omega\rho G_f(r_c \mid r_c)\} = \frac{\omega k\rho}{8\pi} \tag{8.10.40}$$

$$b_1 = \frac{1}{2}z_p(r_p \mid r_c)q_p = \frac{1}{2}j\omega\rho G_f(r_p \mid r_c)q_p = \frac{j\omega\rho e^{-jk\,|\,r_c-r_p\,|}}{8\pi\,|\,r_c-r_p\,|}q_p \tag{8.10.41}$$

and

$$b_2 = \frac{1}{2} z_c^*(r_c \mid r_p) q_p = -\frac{1}{2} j\omega\rho G_f^*(r_c \mid r_p) q_p = \frac{-j\omega\rho e^{jk\mid r_p - r_c \mid}}{8\pi \mid r_p - r_c \mid} q_p \tag{8.10.42}$$

By comparing (8.10.41) and (8.10.42) it is apparent that $z_p(r_p|r_c) = z_c(r_c|r_p)$, so that the imaginary parts of these transfer functions can be neglected in the overall acoustic power equation without altering the result; thus

$$\begin{aligned} W_t &= q_c^* a q_c + \mathrm{Re}\{q_c b_1^*\} + \mathrm{Re}\{b_2 q_c^*\} + c \\ &= q_c^* a q_c + q_c \mathrm{Re}\{z_p(r_p \mid r_c)\} q_p^* + q_c^* \mathrm{Re}\{z_c(r_c \mid r_p)\} q_p + c \end{aligned} \tag{8.10.43}$$

As we found in the theoretical development of this section, neglect of the imaginary components of these transfer functions in conjunction with equality of their real parts results in the control source acoustic power output being equal to zero. This can be shown by substituting a as defined in equation (8.10.40), and b_1 and b_2 as defined in equations (8.10.41) and (8.10.42) with only the real parts of the radiation transfer functions into (8.10.18), producing the following result for the control source acoustic power output:

$$\begin{aligned} W_c &= \frac{1}{4} a^{-1} \left(\mid b_2 \mid^2 - \mid b_1 \mid^2 \right) \\ &= \frac{1}{4} \frac{8\pi}{\omega k \rho} \left[\mid \frac{\omega\rho \sin kd}{8\pi d} \mid^2 - \mid \frac{\omega\rho \sin kd}{8\pi d} \mid^2 \right] = 0 \end{aligned} \tag{8.10.44}$$

This is the same result as we obtained in Section 8.3.

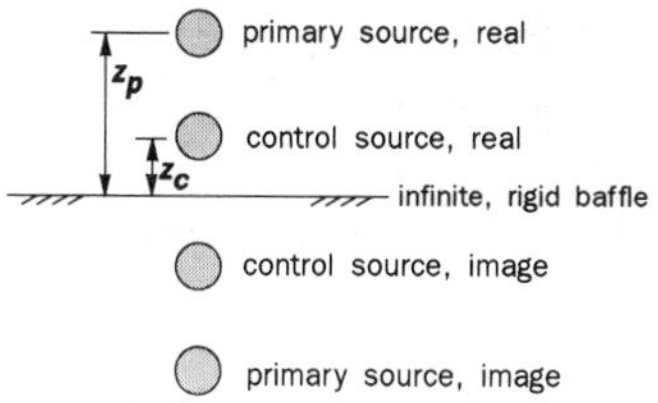

Fig. 8.38 Single primary source/single control source in the presence of an infinite rigid baffle.

An important point to note is that it is the radiation transfer functions between the acoustic sources which are responsible for the final acoustic power output of the control source, and not the real radiation impedance (radiation resistance) presented to the sources when they are operating alone. This quantity, which for the control source was shown to be equal to the term a in the total acoustic power equation (8.10.8), can best be viewed as an amplitude modulating factor. To demonstrate this, we can consider the arrangement depicted in Fig. 8.38, where the control and primary monopole sources are operating in the presence of an infinite, rigid baffle (a problem considered in Snyder and Tanaka (1993a) and Cunefare and Shepard (1993)). The sources are arranged coaxially in a line perpendicular to the baffle, with the control source, at a distance z_c, closer to the baffle than the primary source, at a distance z_p. As we have already discussed, in the presence of an infinite rigid baffle, the sound field produced by a monopole source can be modelled as the superposition of sound fields radiated by the real source and a mirror image source in the absence of the baffle; thus the acoustic pressure at location $\boldsymbol{r}$ is given by

$$p(r) = \frac{j\omega\rho q}{4\pi}\left[\frac{e^{-jkr_s}}{r_s} + \frac{e^{-jkr_i}}{r_i}\right] \tag{8.10.45}$$

where the subscripts s and i are used to denote the real source and image source. The acoustic power output of a single source, operating in the presence of the baffle, can be calculated using the acoustic pressure distribution of (8.10.45) in the defining (8.10.1), specialized for the monopole case as follows:

$$\begin{aligned} W &= \frac{1}{2}\mathrm{Re}\{q\, p_t(r_q)^*\} = |q|^2\frac{\omega\rho}{8\pi}\mathrm{Re}\left\{\frac{-je^{jkr_s}}{r_s} + \frac{-je^{jkr_i}}{r_i}\right\} \\ &= |q|^2\frac{\omega k\rho}{8\pi}(1 + \mathrm{sinc}\ 2kz) \end{aligned} \tag{8.10.46}$$

Comparing this result to (8.10.40), which is the radiation resistance of a monopole source in free space, it can be deduced that the radiation resistance has been increased by the introduction of the baffle by a factor of (1 + sinc $2kz$). As such, the primary and control sources in the system illustrated in Fig. 8.38 will have different radiation resistances, due to their different distances from the baffle. According to our theory, this should not make any difference to the final value of control source acoustic power output; it should still be equal to zero, as the radiation transfer functions between any two sources (real or image) will be the same. Only the amplitude of the control source output should change.

To test these predictions, we must calculate the terms b_1 and b_2, as well as the term a. From (8.10.46), the latter is

$$a = \frac{\omega k\rho}{8\pi}(1 + \mathrm{sinc}\ 2kz) \tag{8.10.47}$$

Using the description of acoustic pressure given in (8.10.45), the terms b_1 and b_2 can be calculated as

$$\begin{aligned} b_1 &= \frac{1}{2}z_p(r_p\,|\,r_c)q_p = \frac{1}{2}j\omega\rho\{G_f(r_{p,r}\,|\,r_c) + G_f(r_{p,i}\,|\,r_c)\}q_p \\ &= \frac{j\omega\rho}{8\pi}\left\{\frac{e^{-jk(z_p-z_c)}}{z_p-z_c} + \frac{e^{-jk(z_p+z_c)}}{z_p+z_c}\right\}q_p \end{aligned} \tag{8.10.48}$$

and

$$b_2 = \frac{1}{2}z_c(r_c \mid r_p)^* q_p = -\frac{1}{2}j\omega\rho\{G_f(r_{p,r} \mid r_c) + G_f(r_{p,i} \mid r_c)\}^* q_p$$

$$= \frac{-j\omega\rho}{8\pi}\left\{\frac{e^{jk(z_p - z_c)}}{z_p - z_c} + \frac{e^{jk(z_p + z_c)}}{z_p + z_c}\right\} q_p \tag{8.10.49}$$

From (8.10.48) and (8.10.49) it is apparent that $z_p(r_p|r_c) = z_c(r_c|r_p)$, which again leads to a zero control source power output, the same result for the previous system without the baffle. Thus the change in source (primary or control) radiation resistance has had no effect upon the final acoustic energy state of the control source. It has, however, altered the optimum value of the control source volume velocity from the unbaffled case, which was shown in Section 8.4 to be defined by the relationship $q_c = -q_p$ sinc kd, where d is the separation distance between the two sources. Substitution of (8.10.47)–(8.10.49) into (8.10.15) shows the optimal relationship in the presence of the baffle to be

$$q_{c,\text{opt}} = -q_p \left\{\text{sinc}(kz_p + kz_c) + \text{sinc}(kz_p - kz_c)\right\} \tag{8.10.50}$$

Both the zero acoustic power output and the altered optimum control source volume velocity were predicted qualitatively by the general theory of the previous section.

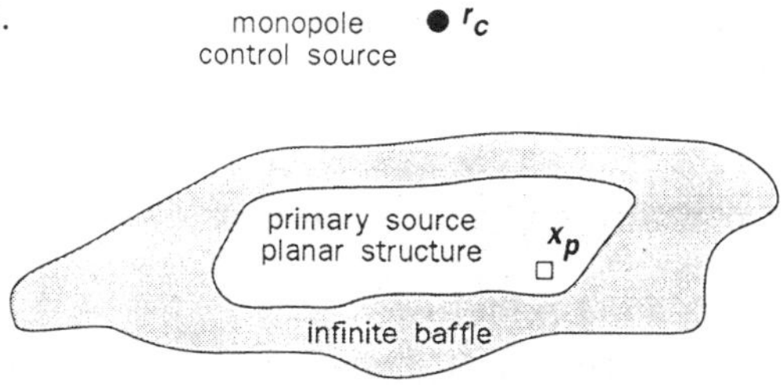

Fig. 8.39 Monopole control source being used to attenuate acoustic radiation from a vibrating panel.

Having seen how the 'common link' theory is applied to simple monopole source problems, let us turn our attention to attenuating acoustic radiation from a vibrating structure, the system arrangement of which is shown in Fig. 8.39. The description of the sound field radiated by the control source is identical to that of the previous case, for the monopole sources operating in the presence of the baffle, stated in (8.10.45). The sound field radiated by the structure is defined by the Rayleigh integral, in terms of surface velocity

$$p(r) = \frac{j\omega\rho}{2\pi}\int_{S_p} v(\boldsymbol{x}_p)\frac{e^{-jkr}}{r}\,d\boldsymbol{x}_p = j2\omega\rho\int_{S_p} v(\boldsymbol{x}_p)G_f(\boldsymbol{x}_p \mid r)\,d\boldsymbol{x}_p \tag{8.10.51}$$

where $\boldsymbol{r}$ is the distance between the location $\boldsymbol{x}_p$ on the structure and the location in space of interest. To calculate the control source volume velocity from the standpoint of mapping the residual acoustic pressure field, or calculating the precise value of acoustic power attenuation, (8.10.51) is often specialized for the structure of interest and separated in terms of modal contributions to the sound field. This was the approach taken in Section 8.7, which was later specialized for the rectangular panel example in Section 8.8. This approach, however, is not necessary to demonstrate the point of interest here. Rather, (8.10.47) can be used directly in the formulation of the terms b_1 and b_2, producing

$$b_1 = \frac{1}{2}\int_{S_p} v(\boldsymbol{x}_p) z_p(\boldsymbol{x}_p \mid \boldsymbol{r}_c)\, \mathrm{d}\boldsymbol{x}_p = \frac{j\omega\rho}{4\pi}\int_{S_p} v(\boldsymbol{x}_p)\frac{\mathrm{e}^{-jk\,|\boldsymbol{r}_c - \boldsymbol{x}_p|}}{|\boldsymbol{r}_c - \boldsymbol{x}_p|}\, \mathrm{d}\boldsymbol{x}_p \tag{8.10.52}$$

and

$$\begin{aligned} b_2 &= \frac{1}{2}\int_{S_p} v(\boldsymbol{x}_p) z_c(\boldsymbol{r}_c \mid \boldsymbol{x}_p)^*\, \mathrm{d}\boldsymbol{x}_p \\ &= -\frac{1}{2} j\omega\rho \int_{S_p} v(\boldsymbol{x}_p)\left\{G_f(\boldsymbol{r}_{p,r} \mid \boldsymbol{r}_c) + G_f(\boldsymbol{r}_{p,i} \mid \boldsymbol{r}_c)\right\}^*\, \mathrm{d}\boldsymbol{x}_p \\ &= \frac{-j\omega\rho}{4\pi}\int_{S_p} v(\boldsymbol{x}_p)\frac{\mathrm{e}^{jk\,|\boldsymbol{x}_p - \boldsymbol{r}_c|}}{|\boldsymbol{x}_p - \boldsymbol{r}_c|}\, \mathrm{d}\boldsymbol{x}_p \end{aligned} \tag{8.10.53}$$

In viewing (8.10.52) and (8.10.53) it is again apparent that, despite the fact that the primary source is a radiating structure and the control source is a monopole source, $z_p(\boldsymbol{x}_p|\boldsymbol{r}_c) = z_c(\boldsymbol{r}_c|\boldsymbol{x}_p)$, which is a direct result of acoustic reciprocity. As such the control source acoustic power output will again be zero under optimal operating conditions.

The final item of business in this discussion is to equate the above results with the control mechanisms discussed previously. For acoustic control sources this is a very simple point; as we have assumed that the primary and control sources have a constant volume velocity, the control mechanism must be one of pressure reduction at the face of both sources, or a reduction in radiation impedance seen by both the control source (to zero) and primary source. This control mechanism has been referred to previously in this book as one of acoustic power suppression, or source unloading.

8.10.3 Mechanism prelude: a vibration source example

While the determination of control mechanisms and application of our common link theory for systems using acoustic control sources is rather straightforward, the use of vibration control sources represents a much more complicated problem. Vibration control sources provide attenuation of the radiated acoustic field by altering the velocity distribution of the vibrating structure. Before examining this effect in terms of mechanisms and the common link theory it will prove informative to briefly consider an example.

Using the quantities defined in the previous section, the specific case we will consider is a simply supported rectangular panel of dimension 0.38 m × 0.30 m × 2 mm, excited in the four corners by point forces, at x = 0.019 m and 0.361 m, and y = 0.015 m and 0.285 m, with all forces equal in amplitude and phase. This is similar to the form of excitation that may be found on a rotating machinery cover plate. The excitation frequency is 350 Hz, which lies slightly below the fourth and fifth panel modes, the (2,2) and (3,1) modes, as shown in Table 8.2. This frequency was chosen for its ability to readily demonstrate all control mechanisms, as will be shown. The control source is a single point force that can be applied to any location on the panel.

Table 8.2 Theoretical panel resonances

Mode	Theoretical resonance (Hz)
(1,1)	88.1
(2,1)	189.6
(1,2)	250.9
(2,2)	352.4
(3,1)	358.7
(3,2)	521.6
(1,3)	522.3
(4,1)	595.5
(2,3)	623.8
(4,2)	758.4

Figure 8.40 illustrates the maximum achievable sound power attenuation as a function of control source location on the panel, showing three optimum control source locations, one on the horizontal centreline, on at each of the two edges and one in the centre of the panel. It will prove interesting to examine the changes in the panel vibration characteristics which accompany the application of acoustic power-optimal control with a control source at the centre and then at one edge.

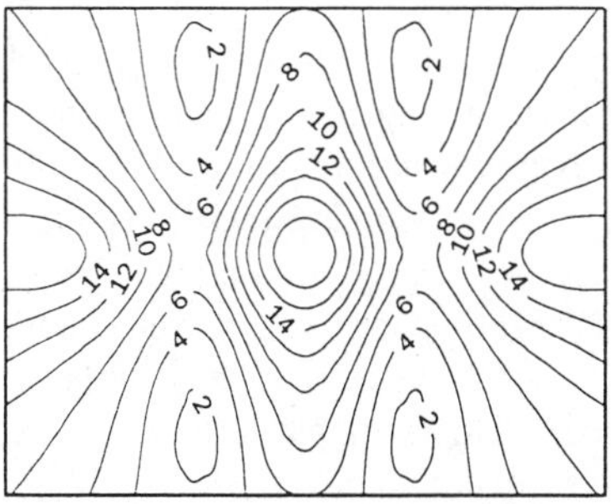

Fig. 8.40 Maximum acoustic power attenuation as a function of vibration control source location on the panel, 350 Hz.

The effect of controlling the panel radiation at the optimum location of (0.36 m, 0.15 m) upon the amplitudes and relative phases of the first 10 panel modes is evident in Fig. 8.41. Here it can be seen that the amplitudes of the

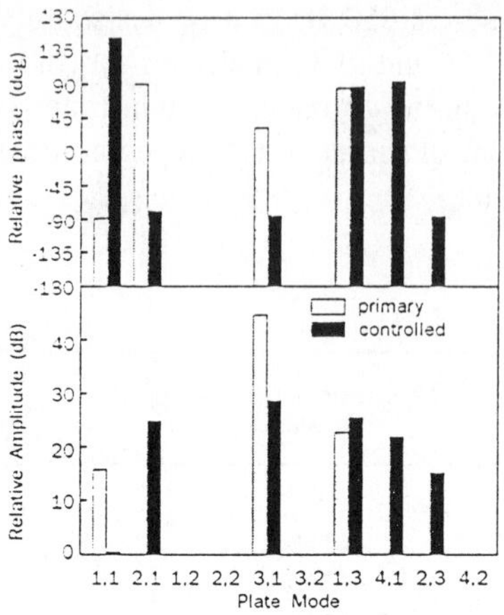

Fig. 8.41 Panel modal amplitudes and phases before and after control, vibration control source near the panel edge.

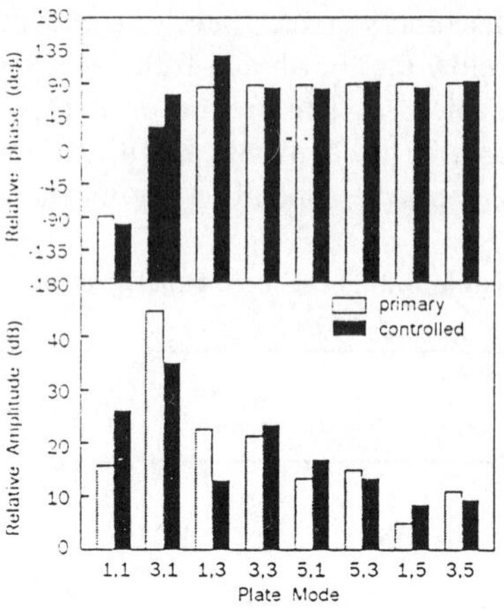

Fig. 8.42 Panel modal amplitudes and phases before and after control, vibration control source in panel centre.

nearly resonant (3,1) mode and the (1,1) mode are substantially reduced. These two modes, based on their velocity levels and radiation efficiencies at this frequency, have the greatest potential for sound power generation. Their reduction causes a significant reduction in the total radiated sound power, which is the primary control mechanism producing the 19.9 dB of attenuation achieved here. (One additional point to note concerning these results is the lack of excitation of any even numbered modes under the primary forcing function. This is due solely to the symmetric nature of the primary forcing function.)

The effect which using a vibration source at (0.19 m, 0.15 m) to control sound power radiation has upon the first 8 non-zero amplitude panel modes is shown in Fig. 8.42. Here the nearly resonant (3,1) mode, as well as the (1,3) mode, both have a reduction in amplitude of approximately 10 dB. This would appear to be offset, however, by an increase in the amplitude of the (1,1) mode of approximately 10 dB. Just in viewing these results, it would seem unlikely that the 22.7 dB reduction in radiated sound power would be achieved only by a reduction in the modal vibration amplitudes of the radiating modes. However, we can observe in the figure that the relative amplitudes and phases of the modes have changed significantly. This will prove to be what is responsible for the extent of the sound power reduction achieved.

For the purposes of our discussion here, it can be viewed there are two ways in which an alteration of the panel velocity distribution can provide a reduction in the radiated acoustic power. The first is simply a reduction in the velocity levels of the principal offending modes, as was the case in the first example. This type of control is implicitly modal control, and is an approach commonly taken in systems explicitly targeting a reduction in vibration level. The other way in which an alteration of the panel velocity distribution can provide a reduction in the radiated acoustic field is by reducing the radiation efficiency of the structure, without necessarily causing a reduction in the modal velocity levels (Snyder and Hansen, 1991a). This is possible in our example problem because the panel modes are not orthogonal radiators, so that the total acoustic power output depends not only upon the velocity of individual modes, but also upon the relative phases of the modal velocities (cross terms in the expression of acoustic power in terms of structural modes). Changing the amplitude and phase relationship between modes will therefore change the acoustic power output of the system. Minimizing acoustic power by altering the relative complex velocities of the structural modes to minimize radiation efficiency has been termed modal rearrangement, or modal restructuring, as opposed to modal control. (The details of the relationship between the modes of our rectangular panel system will be discussed in more detail in the next section.)

It should be noted at this point that these two control mechanisms, reduction of vibration level or reduction in overall radiation efficiency, have appeared under a variety of labels, the pros and cons of which can be debated. It can be viewed (Burdisso and Fuller, 1991, 1992) that during the application of active control the 'controlled eigenfunctions', or controlled mode shape functions, of the structure are different from those of the uncontrolled structure, and that these different mode shape functions are less efficient radiators. This is one way of describing the control mechanism because when the active control system is engaged, all control-system referenced disturbances see a different set of boundary conditions, made up of the original system boundary conditions plus those forced by the active controller (this is discussed further in Chapter 6 on feedforward control.

The application of active control can also be related to the wavenumber spectrum of the vibrating structure (Fuller and Burdisso, 1991; Clark and Fuller, 1992a; Wang and Fuller, 1992), where the radiating components of the spectrum are reduced (wavenumber domain considerations are discussed further in Chapter 2).

Still another way to view what happens under the action of active control is that the amplitude of orthogonally radiating sets of structural modes decrease when the active control system is implemented, which may or may not reduce the overall vibration levels on the structure (this concept will be discussed further in the next section). However, from the standpoint of developing a physical understanding of what leads to a reduction in the radiated acoustic field upon application of vibration-source based active control, we feel that the simple concepts of a reduction in vibration levels, or a reduction in the overall radiation efficiency of the panel, is best.

Figure 8.43 illustrates the mean square velocity levels of the panel before and after the application of active control for our second example. Note that an overall reduction in amplitude of approximately 6 dB has occurred. This is because the reduction in amplitude of the (3,1) and (1,3) modes is greater than the increase in amplitude of the (1,1) mode. This, however, is not enough to

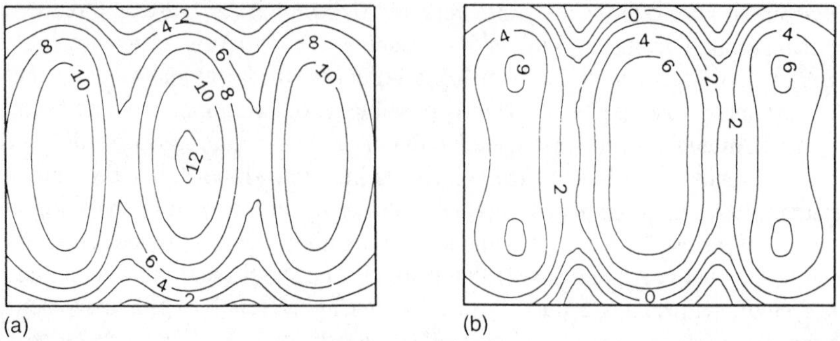

Fig. 8.43 Panel velocity distribution: (a) during primary excitation; (b) during application of control input.

account for the 22.7 dB of radiated sound power attenuation. The remainder of the sound attenuation must therefore be due to modal rearrangement.

8.10.4 Control sources and sources of control

Before relating the vibration control source results to our common link, it is worthwhile briefly considering the idea of a 'control source' in the light of physical 'sources of control'. Referring to the defining equation of acoustic power in (8.10.1), there are two classes of physical mechanism, or sources of control, which will attenuate the radiated acoustic power of the system. The first of these are acoustic control mechanisms, where the in-phase pressure, and hence radiation impedance, on the surface of the control source and/or other radiators in the system is reduced (note that acoustic absorption falls within this definition). The second are velocity control mechanisms, where the velocity levels of the radiating sources are reduced, or the structural input impedance to the primary excitation increased. We have already seen that an acoustic control source can be a source of acoustic control. It is also obvious that a vibration control source can be a source of velocity control. What is less obvious is the fact that a vibration control source can also be a source of acoustic control, which is possible through the modal rearrangement control mechanism because the vibration source is attached to a structure. Structural vibration will often form inherent sources of acoustic control, areas of the structure whose acoustic radiation is responsible for reducing the acoustic power output of other areas on the same structure. One common example of this is seen in the areas of negative acoustic intensity which frequently appear on the surface of vibrating structures, areas which obviously reduce the total acoustic power output of the structure. Vibration control sources are potential vehicles for optimising the attenuating effect of these inherent phenomena via modal rearrangement. Indeed, it is possible to actually *increase* in the velocity levels of a structure while providing attenuation of the radiated field. Clearly, in this case the vibration control source has greatly reduced the overall radiation efficiency of the structure, an effect which can only come about by enhancing the effectiveness of the physical phenomena responsible for limiting the acoustic power radiated from the structure, the inherent sources of acoustic control. Thus when the term 'control source' is used in our discussion here it will be rather generic; vibration control sources attached to a structure operating as sources of velocity control, or vibration control sources acting via the structure as sources of acoustic control. With this idea in mind it is possible to apply the common link theory to vibration-source based systems.

8.10.5 Vibration-source control mechanisms and the common link

Let us now turn our attention to vibration-source based systems, where a vibration control (point) force, attached directly to the vibrating structure, is used to modify the velocity distribution of the structure in such a way as to attenuate the radiated acoustic field. This situation represents a departure from our previous study in that the control source is no longer strictly acoustic nor distinct; both the primary and control radiation is from the same structure. It can, however, still be adequately represented by the previously outlined theory. To investigate the control source power output we must again specialize the terms a, b_1 and b_2 for the active control arrangement, where the term q_c is now the control force amplitude, rather than volume velocity. While it is specifically the radiation transfer functions z_p and z_c which require calculation to demonstrate zero acoustic power output, it is useful also to specialize the control source velocity transfer function z_v, to demonstrate how the control arrangement fits into the previously outlined theory.

The velocity transfer function $z_v(\boldsymbol{x}_c)$ is the transfer function between the control force output and the velocity at a location $\boldsymbol{x}_c$ on the structure. Expressing the velocity at this location in terms of the set of structural mode shape functions, the velocity transfer function can be defined using the following Green's function relationship for the velocity response

$$v(\boldsymbol{x}_c) = j\omega \int_{S_c} G_s(\boldsymbol{x}' \mid \boldsymbol{x}_c) f_c \delta(\boldsymbol{x}' - \boldsymbol{x}_f)\, \mathrm{d}\boldsymbol{x}' = f_c\, j\omega G_s(\boldsymbol{x}_c \mid \boldsymbol{x}_f) \qquad (8.10.54)$$

In (8.10.54) $\boldsymbol{x}_f$ is the application point of the control force, $\delta(\boldsymbol{x}' - \boldsymbol{x}_f)$ is a Dirac delta function, and $G_s(\boldsymbol{x}_c|\boldsymbol{x}_f)$ is the structural Green's function. Therefore,

$$z_v(\boldsymbol{x}_c) = j\omega G_s(\boldsymbol{x}_c \mid \boldsymbol{x}_f) \qquad (8.10.55)$$

The radiated sound field of the planar structure is governed by the Rayleigh integral. In viewing this, it is clear that z_p and z_c both have the form

$$z_c(\boldsymbol{x}_c \mid \boldsymbol{x}) = j2\omega\rho G_f(\boldsymbol{x}_c \mid \boldsymbol{x}) = \frac{j\omega\rho e^{-jk\,|\,\boldsymbol{x}-\boldsymbol{x}_c\,|}}{2\pi\,|\,\boldsymbol{x}-\boldsymbol{x}_c\,|} \qquad (8.10.56)$$

Therefore, terms b_1 and b_2 are

$$b_1 = \frac{1}{2}\int_{S_c} z_v^*(x_c') \int_{S_p} v(x_p) z_p(x_p \mid x_c)\, \mathrm{d}x_p\, \mathrm{d}x_c'$$

$$= \frac{1}{2}\int_{S_c} z_v^*(x_c') \int_{S_p} v(x_p) \left\{ \frac{j\omega\rho e^{-jk\mid x_p - x_c' \mid}}{2\pi \mid x_p - x_c' \mid} \right\} \mathrm{d}x_p\, \mathrm{d}x_c' \tag{8.10.57}$$

and

$$b_2 = \frac{1}{2}\int_{S_p} v(x_p) \int_{S_c} z_v^*(x_c) z_c^*(x_c \mid x_p)\, \mathrm{d}x_c\, \mathrm{d}x_p$$

$$= \frac{1}{2}\int_{S_p} v(x_p) \int_{S_c} z_v(x_c)^* \left\{ \frac{j\omega\rho e^{-jk\mid x_c - x_p \mid}}{2\pi \mid x_c - x_p \mid} \right\}^* \mathrm{d}x_c\, \mathrm{d}x_p \tag{8.10.58}$$

In viewing these expressions for b_1 and b_2, it is clear that the radiation transfer functions, the bracketed part of the expressions, are equal, which lead to a control source power output equal to zero. This result, however, could easily have been anticipated from the fact that the control source (radiation) and primary source (radiation) are (from) the same object, the vibrating structure.

For vibration control sources, the control source acoustic power output which will be zero, the term W_c in (8.10.3), is defined mathematically as the acoustic power output of the structure with a control source-induced velocity distribution operating in the total (combined primary and control) acoustic pressure field,

$$W_c = \frac{1}{2}\,\mathrm{Re}\left\{ q_c \int_{S_c} z_v(x_c)\, p_t^*(x_c)\, \mathrm{d}x_c \right\} \tag{8.9.59}$$

However, it is informative to relate this result to 'real world' phenomena, which would be witnessed upon implementation of a power-optimal feedforward active control system, using the potential control mechanisms outlined previously in this section. For vibration sources operating as sources of velocity control the concept is straightforward: the overall velocity levels of the structure will be minimized. For vibration sources operating as sources of acoustic control the meaning is more obscure, and is best elucidated by means of another example.

Consider a specific case where the vibrating 'structure' is again a simply supported, lightly damped, infinitely baffled rectangular panel. The concept of intercellular cancellation can be used to explain the small value of radiation efficiency of this panel at low frequencies. Referring to the sketch of the cross-section of a three-index mode shown in Fig. 8.44, at low frequencies the acoustic radiation from the centre nodal area can be viewed as 'cancelling' half of the acoustic radiation from each of the end nodal areas, a phenomena which leads to an area of negative acoustic intensity in the centre of the panel (an inherent source of acoustic control). With this idea in mind, it is possible to determine

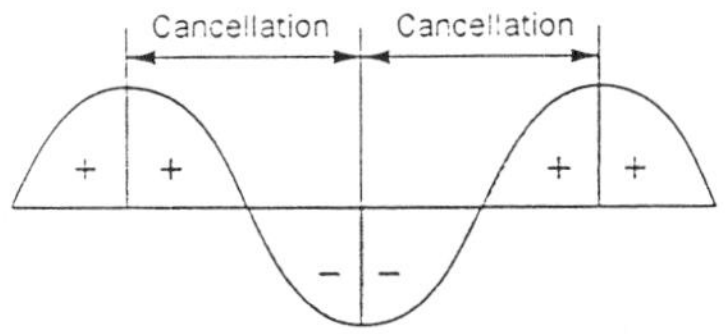

Fig. 8.44 Intercellular cancellation in a simply supported rectangular panel three-index mode.

analytically what centre nodal area amplitude, relative to the edge amplitudes, would minimize the total acoustic power output of the plate. This can be accomplished using our previously derived common link theory, where the control source velocity distribution is over the centre nodal area and the primary source velocity distribution is over the edge nodal areas. Working through the theory, the complex control source amplitude q_c introduced prior to (8.9.3) simply becomes the complex amplitude of the centre nodal area, for which an optimum value can be derived once the remaining terms in (8.9.11) are calculated. For example, for a 0.88 m × 1.80 m × 9 mm steel panel oscillating at 89 Hz, the optimum centre nodal area is found to be 1.66 times that of the edge nodal areas. More importantly for the discussion here, however, is that it follows that the inherent source of acoustic control, the area of negative acoustic intensity, will have an acoustic power output of zero when optimally tuned.

Figure 8.45 depicts the acoustic power output of the inherent source of acoustic control, the areas of negative acoustic intensity (negative acoustic power, or absorption), the 'primary source', the areas of positive acoustic intensity, and the total acoustic power output of the previously dimensioned panel excited at 89 Hz as the ratio of the amplitudes of the centre and edge nodal areas is varied. Figure 8.46 depicts the acoustic intensity distribution on the panel corresponding to several specific cases. As illustrated in Fig. 8.46, under normal conditions (where the amplitude ratio = 1) there is acoustic power flow out of the plate edges and into the centre; both positive and negative contributions to the total acoustic power output exist. As described previously, the centre nodal area can be viewed as attenuating the acoustic radiation from the outer nodal areas, performing the function of a source of acoustic control. Altering the relative amplitudes of the nodal areas alters this balance. If the inherent source of acoustic control, the area of negative intensity, is removed (amplitude ratio = 0) the acoustic power output of the plate increases, as expected. Conversely, if the magnitude of the amplitude ratio is substantially increased, the outer nodal areas become the acoustic power sinks. However, the most interesting result occurs at a value of amplitude ratio equal to 1.66, where the amplitude of the centre nodal area has increased relative to the edge nodal areas. At this point the minimum acoustic power output from the plate occurs. However, more importantly for the topic of discussion here is the fact that at this point the centre nodal area, the inherent source of acoustic control, stops absorbing and actually has an acoustic power output of zero, with the radiation from the outer nodal

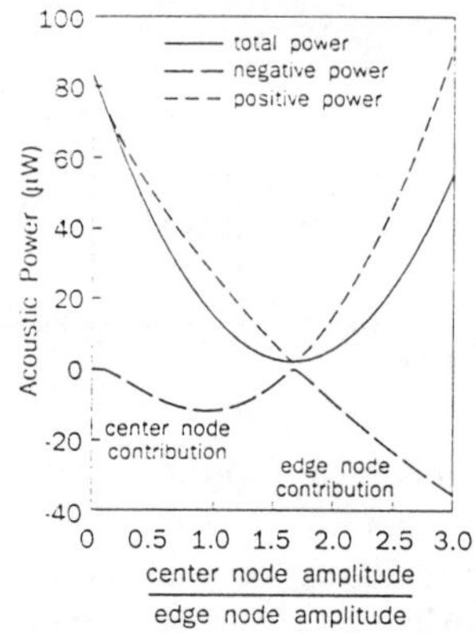

Fig. 8.45 Acoustic power output of negative and positive intensity regions (labelled negative and positive power) on the panel, as a function of the ratio of the edge nodal area amplitudes to the centre nodal area amplitude.

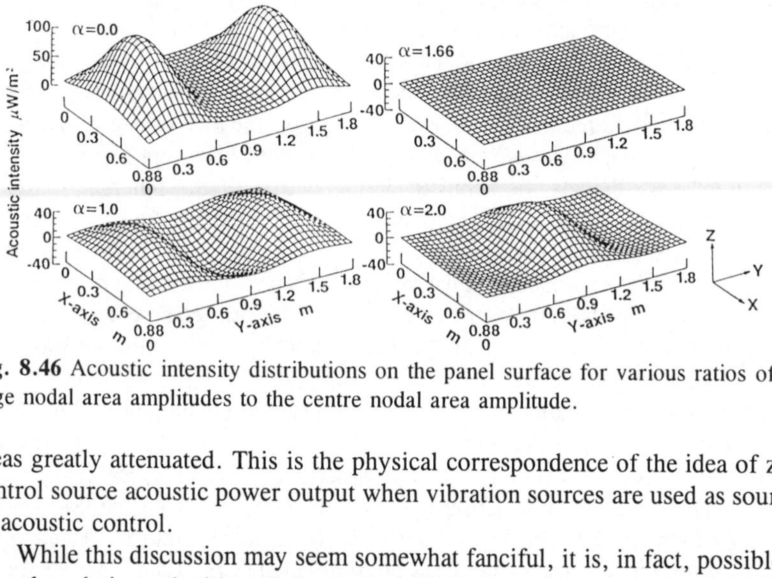

Fig. 8.46 Acoustic intensity distributions on the panel surface for various ratios of the edge nodal area amplitudes to the centre nodal area amplitude.

areas greatly attenuated. This is the physical correspondence of the idea of zero control source acoustic power output when vibration sources are used as sources of acoustic control.

While this discussion may seem somewhat fanciful, it is, in fact, possible to alter the relative velocities of the centre and edge nodal areas by superimposing a second odd-index mode, such as a 1-index mode, which will in turn alter the effectiveness of the intercellular cancellation phenomena. Vibration sources can be used to alter the relative amplitude and phasing of the two modes (this is the modal rearrangement control mechanism), thereby becoming vehicles to 'tune' this inherent source of acoustic control. Because there is a single global minimum in the acoustic power error criterion, when the problem is formulated in terms of vibration control source force inputs the resulting final state of the vibrating panel must be qualitatively the same as that found when the problem is formulated in terms of nodal areas in (8.9.13): the amplitude of the centre

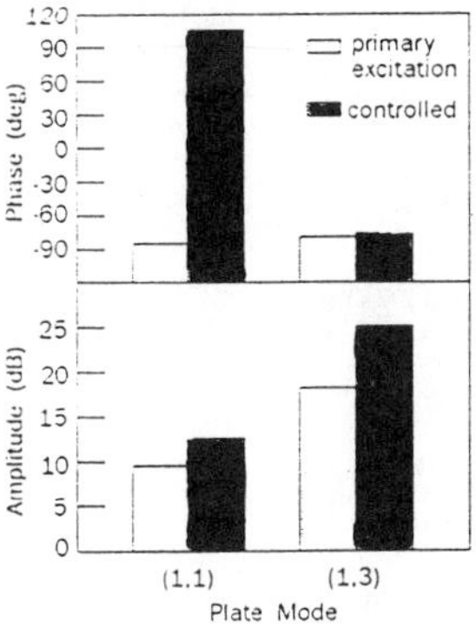

Fig. 8.47 Amplitudes and phases of the (1,1) and (3,1) modes before and after application of active control.

nodal area will have increased relative to the amplitude of the edge nodal areas, and the acoustic power output of this source of acoustic control will be zero; the area of negative intensity will disappear.

To demonstrate this, consider again the 0.88 m × 1.80 m × 9 mm panel, subject to harmonic primary excitation at 89 Hz by a point force located at a position $x = 0.44$ m, $y = 0.30$ m, with the origin of the coordinate system at the lower left corner of the panel. A vibration control source, attached at the panel centre, $x = 0.44$m, $y = 0.90$ m, is used to attenuate the radiated acoustic field. For clarity of explanation, only two modes, the (1,1) and (1,3), will be used in the calculations presented here, although qualitatively the phenomena are the same as we saw in the prelude example of Section 8.10.3. The resonance frequency of the (1,1) mode is 35.2 Hz, and for the (1,3) is 89.5 Hz. The velocities of the two modes in response to both primary excitation and optimum controlled conditions are plotted in Fig. 8.47. The acoustic power attenuation obtained by the application of the optimum vibration control force is 10.8 dB. However, the change in modal velocities alone does not reflect this attenuation. In fact, the velocities of both modes have increased substantially. Rather than being due to a reduction in plate velocity, attenuation is the result of an acoustic mechanism-based phenomena, arising from the change in the relative complex amplitudes of the modes. Figures 8.48(a) and 8.48(b) depict the plate velocity distribution before and after the application of active control. It can be seen that while the distribution pattern is the same, the relative amplitude of the centre nodal area has increased over the outside nodes, which is our hypothesized result.

The final question for us to answer here is why the vibration control source has employed an acoustic-type rather than a velocity-type control mechanism. The reason is that with the primary source towards the edge of the plate and the control source in the centre it is not possible to simultaneously reduce the amplitude of both modes, or to provide velocity (modal) control. However, the 'modal rearrangement' mechanism can be invoked, reducing the radiation

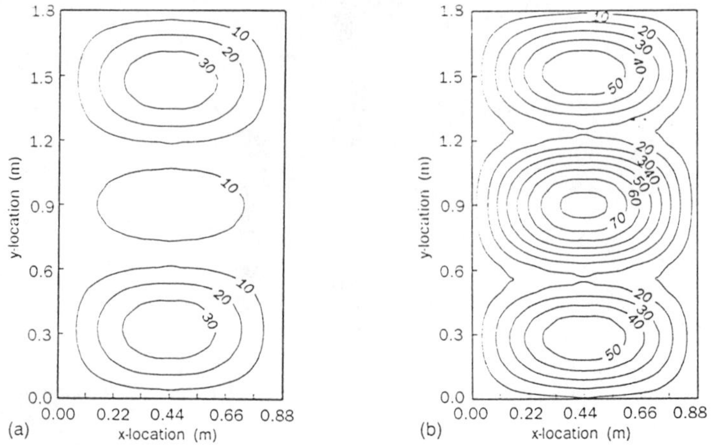

Fig. 8.48 Panel velocity distribution: (a) during primary excitation; (b) after application of active control.

efficiency of the panel. If the panel is very lightly damped and vibrating at resonance, the control effect would be a combination of modal control and modal rearrangement, a reduction in the amplitude of the (3,1) mode and subsequent increase in the amplitude of the (1,1) mode until the centre node to edge node amplitude ratio reached a value of 1.66. If the control source was moved to the panel edge this would change as the (3,1) and (1,1) modes can be reduced in amplitude simultaneously. Either way, one result is consistent; the acoustic power output of the 'control source', as defined in Section 8.9.4, is zero under optimally controlled conditions.

8.11 SENSING VIBRATION TO MINIMIZE ACOUSTIC RADIATION

In this chapter we have spent some time considering the use of vibration control sources to attenuate acoustic radiation from a vibrating structure. In structural/acoustic problems, the use of vibration control sources offers several potential advantages over the use of more conventional acoustic control sources, one of these being system compactness: somewhat bulky speakers and cabinets can be replaced by compact control sources such as surface mounted or embedded piezoelectric ceramic actuators (which are discussed further in Chapter 12), creating a 'smart structure' when driven by an appropriate control system. A potential hindrance to realizing this advantage is the use of acoustic sensors, microphones, to provide an error signal to the control system. As we saw in our monopole source studies in the beginning of this chapter, microphones should typically be placed in the farfield of the acoustic system to ensure that global sound attenuation accompanies the reduction in acoustic pressure at the sensor. It can be envisaged that an improvement would be had if vibration error sensors could be used instead of acoustic sensors.

The use of vibration error sensors, such as accelerometers, in systems targeting a reduction in structural vibration is commonplace. With these systems the control system can be designed to explicitly reduce structural modal amplitudes, using what is termed a 'modal filter' which will resolve a set of measured vibration signals into a set of modal signals (modal velocity, displacement, etc.). A more novel approach which we will discuss in Chapter 14 is the use of shaped piezoelectric polymer film sensors which discriminate between structural modes, thus eliminating the need for explicit error signal prefiltering. However, as we saw in the previous section, it does not always happen that minimizing the radiated acoustic power reduces the modal velocity levels of the principal offending modes; in fact, we saw an example in which an *increase* in structural velocity accompanied a reduction in radiated acoustic power. Clearly a control system which is constrained to reducing structural velocity is inappropriate for control of structural sound radiation.

In this section we will consider the problem of sensing some measure of structural vibration to provide a measure of the error criterion of interest, such as acoustic power or structural kinetic energy. Such a signal could be used as an error signal for either an adaptive feedforward active control system, or a feedback system, to reduce structural or acoustic disturbances. The development of the problem will be general, specifically targeting the questions of what should be sensed in terms of structural modes to maximize the levels of global attenuation of the error criterion of interest, and how should an adaptive feedforward active control system using vibration-based error signals be implemented? Application of the theory developed in this section to feedback control systems will be undertaken later in Sections 8.13 and 8.14.

The analysis which will be undertaken here is based upon the idea of finding orthogonal groupings of structural modes which radiate as a set, and is based principally on the work published by Snyder and Tanaka (1993b) and Snyder *et al.* (1994a). For further discussion relating to active control, the reader is referred to Baumann *et al.* (1991, 1992), Elliott and Johnson (1992, 1993) and Naghshineh and Koopmann (1992, 1993); related work was published by Keltie and Peng (1987), Borgiotti (1990), Cunefare (1991) and Naghshineh *et al.* (1992).

8.11.1 General theory

Consider some generic structure, subject to harmonic excitation by an unspecified primary forcing function. The aim of active control is to globally attenuate some measure of the system response, which can normally be expressed as a quadratic function of the structural modal velocities. This measure may be vibrational, such as the kinetic energy of the structure, or acoustic, such as radiated acoustic power or acoustic potential energy in a coupled enclosure problem. Each of these quantities, or global error criteria, can be expressed in the form

$$J = v^{H}Av \tag{8.11.1}$$

where J is the global error criterion of interest, $\boldsymbol{v}$ is the vector of complex modal velocity amplitudes, H is the matrix Hermitian (conjugate transpose), and $\boldsymbol{A}$ is a symmetric transfer matrix (symmetric provided reciprocity holds). Note that to write this relationship we have assumed that the infinite modal summation required to exactly decompose structural velocity has been truncated at some finite number of modes. The diagonal terms of the matrix $\boldsymbol{A}$ define the value of the error criterion which would result from the vibration of a single, isolated structural mode, and the off-diagonal terms define the modification to this value caused by the coexistence of the other structural modes.

As $\boldsymbol{A}$ is a symmetric matrix it can be diagonalized by the orthonormal transformation

$$A = Q\Lambda Q^{-1} \tag{8.11.2}$$

where $\boldsymbol{Q}$ is the (square) orthonormal transform matrix, the columns of which are the eigenvectors of the transfer matrix $\boldsymbol{A}$, and $\boldsymbol{\Lambda}$ is the diagonal matrix of associated eigenvalues. The symmetric form of $\boldsymbol{A}$ dictates that the eigenvectors in $\boldsymbol{Q}$ will be orthogonal. If the transformation of (8.11.2) is substituted into (8.11.1) the generic global error criterion expression becomes

$$J = v^{H}Av = v^{H}Q\Lambda Q^{-1}v = v'^{H}\Lambda v' \tag{8.11.3}$$

where the vector of transformed modal velocities $\boldsymbol{v}'$ is defined by the expression

$$v' = Q^{-1}v \tag{8.11.4}$$

and use has been made of the property of the orthonormal transform matrix

$$Q^{-1} = Q^{T} \tag{8.11.5}$$

Observe that the orthonormal transformation has decoupled the global error criterion expression, enabling us to express it as a weighted summation of the form

$$J = \sum_{i=1}^{m} \lambda_i \, | v_i' |^2 \tag{8.11.6}$$

where λ_i is the ith eigenvalue of $\boldsymbol{A}$, v'_i is the ith transformed modal velocity of the vector $\boldsymbol{v}'$, and the summation is over the m structural modes being modelled.

The important point to note about the error criterion as expressed in (8.11.6) is that v'_i is not necessarily equal to the velocity amplitude of the ith structural mode v_i, but is rather equal to some orthogonal grouping of modal contributions, defining the ith principle axis of the error 'surface' (the plot of error criterion as a function of modal velocity) defined by the quadratic criterion of (8.11.1) (the concept of error surfaces is discussed in more depth in Chapter 6 in relation to adaptive signal processing, a problem which will have the same qualitative characteristics). Consider, for example, some hypothetical problem where only two structural modes are being modelled and the error criterion is defined by the transfer matrix

$$A = \begin{bmatrix} 1.0 & 0.5 \\ 0.5 & 1.0 \end{bmatrix} \tag{8.11.7}$$

In this instance the orthonormal transform matrix Q and the eigenvalue matrix Λ are equal to

$$Q = \frac{1}{\sqrt{2}} \begin{bmatrix} 1.0 & 1.0 \\ 1.0 & -1.0 \end{bmatrix}, \quad \Lambda = \begin{bmatrix} 1.5 & 0.0 \\ 0.0 & 0.5 \end{bmatrix} \tag{8.11.8}$$

Therefore, the transformed velocities, as defined by (8.11.4), are

$$v_1' = \frac{1}{\sqrt{2}}(v_1 + v_2), \; v_2' = \frac{1}{\sqrt{2}}(v_1 - v_2) \tag{8.11.9}$$

These transformed velocities defined the m principle axes of the $(m + 1 = 3)$-dimensional hyper-parabolic error surface associated with the quadratic (in modal velocity) error criterion of (8.11.1).

Let us consider now the problem of implementing an adaptive feedforward active control system to attenuate the global error criterion of (8.11.1), where a vibration control source is being used. The desired error signal, to be minimized by our control system, is some measure of structural velocity which we can be decomposed into a set of orthogonal constituents by some electronic preprocessing operation such as modal filtering, or by the use of a custom sensor such as constructed from shaped piezoelectric polymer film. (We will discuss how to shape a piezoelectric polymer film sensor to obtain a measurement of some desired quantity in more depth in Chapter 14. The aim of the exercise in this chapter is to derive the quantity to be measured.) One approach we could take in decomposing the error signal is to base it upon the *in vacuo* mode shape functions of the structure. However, using this data in its 'raw' form may not produce the desired result, as strictly speaking the value of the global error criterion is based not solely upon (squared) individual *in vacuo* (untransformed) modal contributions but rather upon these terms in conjunction with modifications resulting from other coexisting modes (in other words, the weighting matrix A is not diagonal). Simplifying this holistic view by aiming to minimize only individual contributions can prove detrimental to performance, as when controlling acoustic radiation from a vibrating structure. A better approach is to resolve the orthogonal transformed modal velocities in v' and simply weight the contributions by the associated eigenvalues. Not only does this implicitly account for the off-diagonal, or cross-coupling, terms in the transfer matrix A, but we may also find that it reduces the number of error signal inputs to the controller required to obtain a satisfactory result, as some eigenvalues may be so small as to make contributions to the error criterion by the associated transformed modal velocities negligible. This latter point is becoming evident even in the previous simple two-mode example, where one eigenvalue is three times the value of the other, indicating that measurement of the associated eigenvector v'_1 is three times as important.

8.11.2 Minimizing vibration vs. minimizing acoustic power

To obtain some insight into differences that minimizing structural vibration has with minimizing radiated acoustic power, it will prove useful to first consider the problem of minimizing structural kinetic energy, which provides a measure of the vibration of a finite structure. Structural kinetic energy is defined here as

$$E_k = \frac{1}{2}\int_S \rho_s(\boldsymbol{x})h(\boldsymbol{x})v^2(\boldsymbol{x})\,\mathrm{d}\boldsymbol{x} \tag{8.11.10}$$

where $\rho_s(\boldsymbol{x})$, $h(\boldsymbol{x})$ and $v^2(\boldsymbol{x})$ are the material density, material thickness, and velocity at location $\boldsymbol{x}$ on the structure, and the integration is conducted over the surface area S of the structure. The velocity term in (8.11.10) can be expanded in terms of modal contributions as follows:

$$v(\boldsymbol{x}) = \sum_{i=1}^{N_m} v_i\psi_i(\boldsymbol{x}) \tag{8.11.11}$$

where v_i is the complex velocity of the ith mode, $\psi_i(\boldsymbol{x})$ is the value of the associated mode shape function at location $\boldsymbol{x}$, and there are N_m modes being modelled in the calculations. Substituting (8.11.11) into (8.11.10), and taking advantage of modal orthogonality enables structural kinetic energy to be expressed as

$$E_k = \frac{1}{2}\int_S \rho_s(\boldsymbol{x})h(\boldsymbol{x})\left[\sum_{i=1}^{N_m} v_i\psi_i(\boldsymbol{x})\right]^*\left[\sum_{i=1}^{N_m} v_i\psi_i(\boldsymbol{x})\right]\mathrm{d}\boldsymbol{x} = \frac{1}{2}\sum_{i=1}^{N_m} M_i\,|\,v_i\,|^2 \tag{8.11.12}$$

where M_i is the ith modal mass, given by

$$M_i = \int_S \rho_s(\boldsymbol{x})h(\boldsymbol{x})\psi_i^2(\boldsymbol{x})\,\mathrm{d}\boldsymbol{x} \tag{8.11.13}$$

and * denotes complex conjugate. Structural kinetic energy can therefore be written in the form of the global error criterion expression (8.11.1)

$$J = E_k = \boldsymbol{v}^{\mathrm{H}}\boldsymbol{A}\boldsymbol{v} \tag{8.11.14}$$

where $\boldsymbol{A}$ is a diagonal matrix of modal masses, with the elements defined as $A(i,i) = M_i/2$.

From the description of (8.11.14) it is apparent that in this problem the orthonormal transform matrix $\boldsymbol{Q}$ of (8.11.2) will simply be the identity matrix, and that the eigenvalues in $\boldsymbol{\Lambda}$ in (8.11.2) are equal to the modal masses. The transformed modal velocities v'_i of (8.11.4) are therefore simply equal to the *in vacuo* modal velocities v_i. In structures where all of the modal masses are equal, such as simply supported rectangular plates, the eigenvalue weighting can be neglected in a practical implementation (because the weighting given to each input will be the same), so that the optimal error signals to provide to the adaptive control system for minimization of the kinetic energy of the structure are simply the *in vacuo* structural modal velocities (this simple result will not, however, occur in general for structural acoustic problems, as will be seen in the next two sections).

A feedforward control system based on this idea for a general structure could look something like the arrangement shown in Fig. 8.49, where a common adaptive controller, such as based upon an FIR filter and filtered-x LMS algorithm described in Chapter 6, is being employed to 'tune' the control force inputs to minimize the weighted set of structural modal velocity error signals, as measured by shaped piezoelectric polymer film sensors (used, for example, in Lee and Moon (1990) and Collins *et al.* (1992); see also Clark and Fuller (1991, 1992b); Snyder *et al.* (1993, 1994a, 1994b). For the kinetic energy case the error signal weighting is frequency independent (defined by the eigenvalues), simply equal to the modal masses (note again that with some structures, such as rectangular simply supported panel, all modal masses are the same and so the weighting can be neglected).

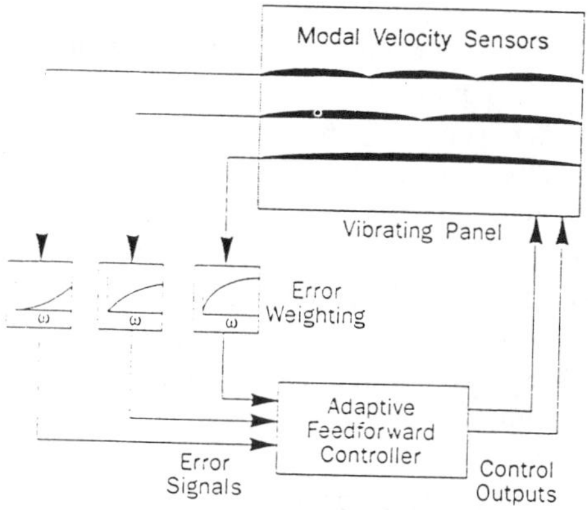

Fig. 8.49 Feedforward control arrangement using modal velocity sensors.

This result can be contrasted to that obtained when considering minimization of radiated acoustic power. With this latter problem the transfer matrix $\boldsymbol{A}$ of (8.11.2) is neither constrained to be diagonal nor to have equal magnitude terms, and as such both couples and weights the problem (couples because the contribution of one *in vacuo* mode the the error criterion is dependent upon the amplitudes of other *in vacuo* modes). Strictly speaking, to explicitly minimize radiated acoustic power it is not sufficient to consider individual structural modal amplitudes; rather, cross-coupling between the modes (defined by the off-diagonal terms of $\boldsymbol{A}$) must also be taken into account. This difference between minimizing structural velocity and acoustic power infers that any control approach aimed at minimizing the sum of (possibly weighted) individual structural modal velocities with a view to minimizing radiated acoustic power is fundamentally in error. Rather, acoustic power attenuation will be achieved by minimizing the weighted sum of the squared amplitudes of a set of coupled modal velocity groups, or by *altering* the velocity distribution. This alteration may result in reduced velocity levels for some or all structural modes, but this result is by no means guaranteed, even for resonant or nearly resonant modes.

8.11.3 An example: minimizing radiated acoustic power from a rectangular panel

To consider this idea further it is best to pick a specific example. The one chosen here is again a baffled, simply supported rectangular panel. With this structure and an acoustic power error criterion, the transfer matrix $\boldsymbol{A}$ is equal to the acoustic power transfer matrix $\mathbf{Z}_w$ defined initially in (8.7.63). As the acoustic power transfer matrix is real symmetric it can be diagonalized by an orthonormal transformation as defined in (8.11.2), enabling acoustic power to be expressed in the form of (8.11.3) and (8.11.6), as follows:

$$W = \boldsymbol{v}'^{H}\boldsymbol{\Lambda}\boldsymbol{v}' = \sum_{i=1}^{m} \lambda_i \,|\, v_i' \,|^2 \tag{8.11.15}$$

where the m transformed structural modes used in the summation are as defined in (8.11.4), with the orthonormal matrix $\boldsymbol{Q}$ containing the eigenvectors of the acoustic power transfer matrix $\mathbf{Z}_w$ (or $\boldsymbol{A}$). If the transformed modal velocities could be sensed directly and the signals weighted by the associated eigenvalue (which is effectively the radiation efficiency of the transformed mode) then the radiated acoustic power could be minimized by devising a control system which would minimize the set of weighted signals. In this case, a feedforward control system could again look something like the arrangement shown in Fig. 8.49, where transformed modal velocities are measured with shaped piezoelectric film sensors, and the weighting factors are the eigenvalues associated with each transformed mode.

While the orthonormal transformation described in the previous section enables the problem of controlling sound transmission from a vibrating panel to be described as a problem of minimizing a weighted set of amplitudes of orthogonal contributors to the error criterion, thereby simplifying a potential feedforward control arrangement, there is still a (commonly encountered) problem which must be discussed: the power transfer matrix $\mathbf{Z}_w$, and hence its eigenvectors used in calculating the transformed modes and eigenvalues used in weighting the problem, is frequency dependent. This means that the constituents of a given transformed mode, which could be measured using a shaped piezoelectric polymer film sensor, will be frequency dependent. At first, this would appear to be a serious impediment to the implementation of optimally shaped sensors for control of sound radiation over anything but the narrowest of frequency bands. However, it is insightful to qualitatively consider the form of the frequency dependence of the constituent terms of the acoustic power transfer matrix $\mathbf{Z}_w$. At low frequencies (such that $(kL_x/m\pi) << 1$ and $(kL_y/n\pi) << 1$, where k is the acoustic wavenumber at the frequency of interest) the (α,β)th term in the matrix $\mathbf{Z}_w$ can be expressed in terms of modal radiation efficiencies as

$$\mathbf{Z}_w(\alpha,\beta) = \frac{\rho_0 c_0 L_x L_y}{8} \frac{\left(1+(-1)^{M_\alpha+M_\beta}\right)\left(1+(-1)^{N_\alpha+N_\beta}\right)}{4} \times \frac{\frac{M_\alpha N_\alpha}{M_\beta N_\beta} S_\alpha + \frac{M_\beta N_\beta}{M_\alpha N_\alpha} S_\beta}{2} \tag{8.11.16}$$

where (M_α,N_α), (M_β,N_β) and S_α, S_β are the (x,y) modal indices and radiation efficiencies for modes α and β, respectively, the latter found in work published by Wallace (1972). The important point to note about this result is that it shows at low frequencies the relative values of the terms in the acoustic power transfer matrix $\mathbf{Z}_w$ will be related to each other by the relative values of the individual modal radiation efficiencies, while the absolute values will be related to the radiation efficiencies themselves. This is important because it means that while the eigenvectors of $\mathbf{Z}_w$, which define the constituents of the transformed modes, will vary with frequency. In the low frequency regime this variation can often be taken as insignificant when viewed in light of the attenuation of radiated acoustic power which can actually be obtained, because the ratio of the radiation efficiencies of any two modes may not vary greatly with frequency (see the graphs in Section 2.5). However, (8.11.16) also shows that the eigenvalues of $\mathbf{Z}_w$ will vary markedly with frequency, as these will be related to the individual values of radiation efficiency (as stated, the eigenvalues are equivalent to the radiation efficiencies of the transformed modal velocities).

Table 8.3 Theoretical panel resonances below 200 Hz

Mode	Theoretical resonance frequency (Hz)
1,1	39.5
2,1	62.3
3,1	100.4
1,2	135.0
4,1	153.7
2,2	157.9
3,2	195.9

To investigate this further, we will briefly study a steel panel of dimensions 1.8 m × 0.88 m × 10.1 mm, subject to harmonic excitation within the frequency band 30–200 Hz. In this range there are seven modal resonances, as outlined in Table 8.3. If the acoustic power transfer matrix $\mathbf{Z}_w$ is calculated at any given frequency in this band using 30 structural modes and then decomposed as outlined in (8.11.4) it is found that at low frequencies only three transformed

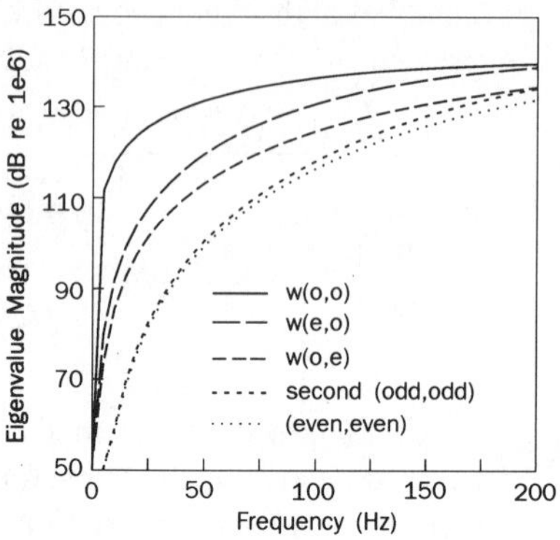

Fig. 8.50 Amplitude of the five largest eigenvalues of the acoustic power transfer matrix as a function of frequency.

acoustic power modes may be considered as 'significant', significance being defined by the associated eigenvalue which will weight the influence which the transformed mode has upon the calculation. These transformed modes are, in order of importance, based upon orthogonal groupings of (odd,odd), (even,odd), and (odd,even) structural modes, and will be referred to as transformed modes *w(o,o)*, *w(e,o)* and *w(o,e)*, respectively. As the frequency increases, two more modes, based upon a second grouping of (odd,odd) modes and a grouping of (even,even) modes, become important. A plot of the variation in eigenvalues of these transformed modes as a function of frequency is given in Fig. 8.50.

The interesting point to note here is that while the eigenvalues of these transformed modes vary greatly with frequency, the proportion of each *in vacuo* structural mode contributing to the transformed mode defined by the eigenvectors varies by a relatively small degree. Consider for example the w(o,o) mode, which is shown in Fig. 8.50 to have the greatest radiation potential. Table 8.4 lists the values of the non-zero contributions from the first 60 structural modes to this transformed mode at 30 Hz and 100 Hz. This mode is dominated by the (1,1), (3,1) and (1,3) structural modes, with the proportional contributions changing only slightly.

Table 8.4 Constituents of the first transformed mode *w(o,o)* as defined at 30 Hz and 100 Hz.

Mode	30 Hz value	100 Hz value
1,1	0.913	0.919
3,1	0.229	0.199
5,1	0.138	0.117
1,3	0.233	0.241
1,5	0.141	0.143

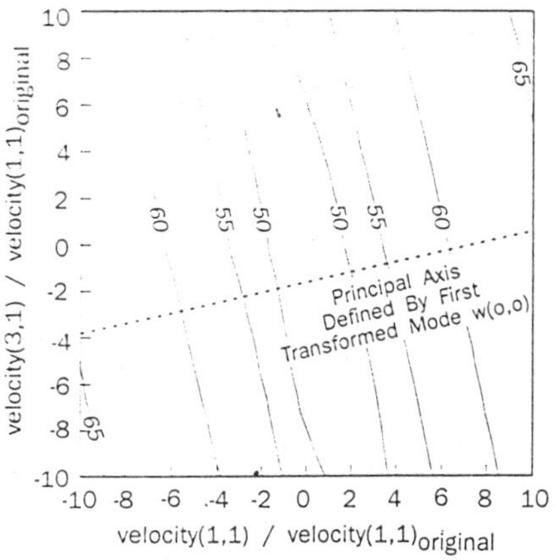

Fig. 8.51 Acoustic power output as a function of the amplitudes of the (1,1) and (3,1) structure modes.

To give further insight into the physical significance of transformed modes, it is useful to consider the variation in acoustic power output as a function of the relative amplitudes of the (1,1) and (3,1) structural modes. Figure 8.51 depicts the acoustic power output of the described panel at 100 Hz in response to primary point excitation at (0.35 m, 0.64 m), where the amplitude of the (1,1) and (3,1) modes are varied relative to the original amplitude of the (1,1) mode, and the amplitude of all other structural modes are kept constant. Observe that the principle axes of the quadratic surface are *not* parallel to the axes defined by the structural modal velocities, but rather are skewed. The first transformed mode *w(o,o)* describes the steepest principle axis of the function (the axis along which the greatest variation in acoustic power is seen). Clearly, minimizing the acoustic power amplitude along this axis (equivalent to minimizing the *w(o,o)* mode) will go a long way towards minimizing the total acoustic power output of the panel. For interest, the other principle axis of the figure is described by the second orthogonal grouping of (odd,odd) structural modes mentioned previously.

Note also that by considering the structural radiation in terms of orthogonal transformed modes it is perfectly clear that radiated acoustic power can be reduced while the vibration amplitude of the structure is increased. For example, consider two situations, where the amplitude of the (1,1) structural mode is 1.0 and the amplitude of the (3,1) structural mode is -1.0 and -5.0, respectively, while the amplitude of all other modes is 0.0. The kinetic energy of the structure is less in the first of these two instances. However, in viewing the constituents of the dominant transformed mode given in Table 8.4, it can be seen that the radiated acoustic power is lower for the second of these two cases (the sum of the products of the amplitudes and their proportions in the transformed mode is less), even though the kinetic energy is significantly increased. This phenomenon occurs because structural modes are *not* orthogonal radiators; transformed

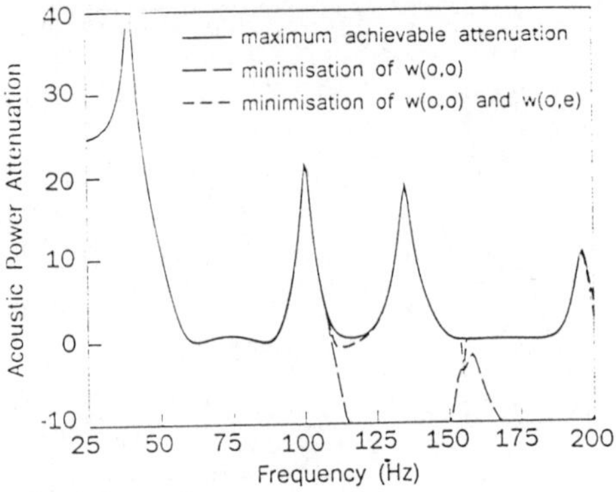

Fig. 8.52 Maximum acoustic power attenuation and attenuation achieved by minimizing the first, and first and third, transformed modes.

modes, as we have defined in this section, are. Clearly, we should aim to minimize the amplitudes of the weighted set of transformed modes in the active control of structural radiation.

There are two questions which arise from the results of Table 8.4 which need to be addressed. First, from the standpoint of using measurements of the amplitudes of transformed modes to formulate a control strategy, how important are the relatively small variations in the proportional contributions of the constituents as a function of frequency? Is it possible to base a feedforward control arrangement on minimizing transformed modes which are strictly only correct at a single frequency (calculated using the eigenvectors of the matrix $\mathbf{Z}_w$ at that frequency) and vary only the weighting factor (eigenvalue) when the frequency is changed? Second, what real advantage does the minimization of transformed modal velocities have over the minimization of ordinary structural modal velocities? Illustrated in Fig. 8.52 are the theoretical maximum levels of attenuation of acoustic power radiated from the previously described plate, as well as the attenuation achieved using the $w(o,o)$, and both the $w(o,o)$ and $w(o,e)$, transformed modal velocities as defined at 30 Hz (strictly only optimal at this frequency) and weighted by the frequency-correct eigenvalue (the form of implementation shown in Fig. 8.49), with a primary point force at (0.44 m,0.35 m), a control point force at (0.24 m,0.9m), and damping $\eta = 0.02$ for all modes. Note that at low frequencies minimization of a single transformed modal velocity achieves practically the optimal result, while the introduction of the second transformed modal velocity extends the region of control over the entire bandwidth. These results suggest that the slight variations in the relative amplitudes of the constituents of the transformed structure mode are essentially negligible in light of the acoustic power attenuation which is achievable with an ideal feedforward control system.

Figure 8.53 illustrates the levels of acoustic power attenuation which can be achieved by minimizing the kinetic energy of the structure (calculated using all

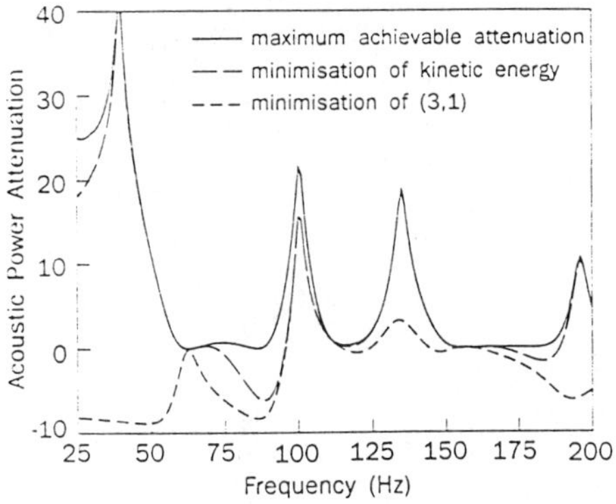

Fig. 8.53 Maximum acoustic power attenuation and attenuation achieved by minimizing the structural kinetic energy, and the amplitude of the (3,1) structural mode.

30 modelled modes), and which can be obtained by minimizing the amplitude of the (3,1) mode. Observe that the minimization of the kinetic energy of the structure achieves almost the same result as the minimization of the transformed modal velocities (a result which would not necessarily be found for a general structure). It falls short, however, in the vicinity of the (3,1) mode resonance frequency. This is a result of there being two ways to reduce radiated acoustic power, as we have discussed: reduction in velocity, and an alteration of the relative amplitudes and temporal phasing of the vibration modes to enhance the inherent acoustic pressure-control mechanisms, thereby reducing the radiation efficiency of the structure. The additional control in this vicinity is achieved by minimizing the amplitude of the transformed $w(o,o)$ mode which effectively balances the decrease in the amplitude of the (3,1) structural mode against the resulting increase in the amplitude of the (1,1) mode, such that the centre nodal area of the (3,1) mode is essentially increased in amplitude (superposition of the (1,1) and (3,1) modes) while the edge nodal areas are decreased in amplitude. This enhances the effect of intercellular cancellation, the inherent acoustic control mechanism, as we described in the previous section. It is not possible to achieve this result by simply minimizing the velocity distribution of the structure, or by minimizing the mode which contributes most to the radiated sound. Note also that minimizing the amplitude of the (3,1) mode only guarantees attenuation at the (3,1) resonance, while Fig. 8.52 shows that minimization of the single $w(o,o)$ transformed mode will provide attenuation at both the (1,1) and (3,1) resonances. Finally, it should again be mentioned that the relative quality of results obtained by minimizing the kinetic energy of the structure (as compared to the maximum possible levels of acoustic power attenuation) are source location dependent, while those obtained by minimizing the weighted set of transformed modal velocities are not.

Another point to consider is the importance of the eigenvalue weighting. Figure 8.54 illustrates the levels of acoustic power attenuation achieved by

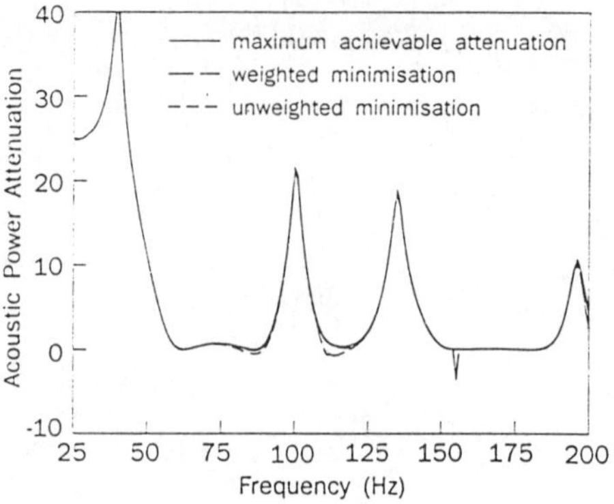

Fig. 8.54 Maximum acoustic power attenuation and attenuation achieved by minimizing the first three transformed modes with and without eigenvalue weighting.

minimizing the unweighted set and also the weighted set of transformed modal velocities. In comparing the data shown in the two curves, the results corresponding to the unweighted case are seen to be only slightly suboptimal. This is because the resonances of the structure inherently weight the problem to some degree. At low frequencies, where the $w(o,o)$ transformed mode is dominant, the (1,1) and (3,1) modes are resonant. As the frequency increases, such that the $w(e,o)$ and $w(o,e)$ transformed modes become important, the (1,2) and (2,1) resonances occur. This characteristic is not globally applicable, however, and neglecting the eigenvalue weighting with other source placements and in other problems can seriously degrade results. In the next chapter we will see that this is especially true in coupled structural/acoustic problems.

Active control systems targeting attenuation of acoustic radiation from vibrating panels have been implemented experimentally using shaped piezoelectric polymer film sensors to measure the transformed modes outlined in this section. For both broadband and tonal excitation, the attenuation in radiated acoustic power achieved was found to be near optimal. For further details, the reader is referred to Snyder *et al.* (1993, 1994a).

8.12 SOME NOTES ON APPROACHING THE DESIGN OF AN ACTIVE CONTROL SYSTEM FOR SOUND RADIATION FROM A VIBRATING SURFACE

Having developed analytical models which enable us to examine free space active noise control systems, it is a good time to briefly discuss the problem of designing applicable feedforward active control systems. We will assume here that an 'incorruptible' reference signal is available for implementation of the control system (reference signal corruption was discussed in Section 8.6), and concentrate on the problem of optimizing the control source/error sensor arrangement.

Although we have been successful in deriving equations which facilitate calculation of the optimum control source outputs for a given control source/error sensor arrangement, a direct analytical means of determining the optimum control source and error sensor location has not been found. This is because the error criterion of interest, radiated acoustic power or the squared acoustic pressure amplitude at a set of discrete locations, is not a simple quadratic function of source and sensor location. Therefore, system design tends to have the appearance of being largely a 'trial and error' process. It is possible, however, to give some structure to the task, and point out some potential pitfalls and shortcuts, which will be the aim of our discussion in this section.

8.12.1 Stepping through the design of a system

Before beginning the design process, the system, and its response under primary excitation, must be characterized. The minimum requirement here is that the primary sound field (amplitude and relative phase) on a test surface surrounding the structure in the farfield be measured at the frequency of interest using an appropriate measurement location distribution similar to that used when sound intensity is used to estimate sound power. If vibration control sources are to be used, the structural response must also be measured. If the structural response can be decomposed into its individual contributing modes this often proves helpful in painting a qualitative picture of the influence of system variables on performance (see the example in Section 8.8.2). It is not absolutely necessary, however, as the optimum control source outputs and reduction in the global error criterion for a given control source arrangement can be calculated using the 'spatial', rather than 'modal', theoretical model developed in Section 8.10 for our common link study.

Once the system to be controlled has been characterized, the task is to choose the type of control source to be used. As demonstrated in the previous sections, the use of either acoustic or vibration control sources can produce significant levels of reduction in the overall sound power radiated by a vibrating structure into free space. The mechanisms by which they achieve this attenuation, and the influence which various geometric and structural/acoustic parameters have upon the magnitudes of these reductions, are different for the two control source types. In choosing the control source type for a particular application, it is sometimes helpful to think of the following points:

1. Acoustic control sources are the easier of the two types with which to design a system. No information about the structural response is required, only measurements of the primary radiated sound field at the frequency of interest are necessary, and this is a very distinct advantage for the free space radiation problem. One disadvantage is that the system is constrained to using an acoustic error sensor(s) (such as a microphone) to provide feedback to the electronic control system. Another disadvantage is that the levels of sound attenuation that can be attained per source may be less than those

attained with vibration control source(s), especially if the panel dimensions are approximately equal to, or larger than, the acoustic wavelength at the frequency of interest, or if the panel is near resonance. Acoustic control sources can also be more obtrusive than some vibration sources, such as peizoelectric actuators.

2. Vibration control sources are more difficult to use than acoustic control sources for system design. A detailed description of the structural response, and also the resulting sound radiation field, must be known if an optimized design is to be undertaken. However, if this can be done, vibration control sources may exhibit an increased level of performance (on a per source basis) over acoustic control sources. The final system may also be more compact and unobtrusive for vibration control sources, especially if piezoelectric actuators and vibration error sensors can be used.

Once the system has been characterized, and the type and number of control sources chosen, the next step is to optimize the location of the control sources. This can be a particularly difficult exercise, as it is, in general, impossible to directly determine the optimum source locations due to it not being a linear function of sound power attenuation. A numerical search routine is therefore required to optimize the control source locations. What is required for this is a means to estimate of the maximum possible sound power attenuation for a given control source arrangement, which can be accomplished using the theoretical models already developed in this chapter, or estimated using the multiple regression approach outlined at the end of this section.

One problem with the use of a numerical search routine is that there may be local optima (minima) in the error surface. In this case, the starting point(s) for the search algorithm will influence the final result. The designer must be aware of this, restarting the procedure at several locations or by using a random search technique to determine the optimal starting location if it is not possible to choose a starting point based on 'common sense'. Such common sense guidelines would include initial placement of the control sources as close as possible to the antinodes of the modes contributing most to the sound radiation, and also placement of the vibration control sources in the least stiff parts of the structure so that control effort is minimized.

Once the control source location has been selected, the final step is to select the type and location of the error sensor(s). Regardless of the control source type, microphones sampling the far field will always be effective (provided there are no background noise problems from other sources). With the control source placed, there is a straightforward rule of thumb for selecting the optimum error sensor location: error sensors should be placed at the location of greatest pressure difference between the primary radiated and optimally controlled (control source output optimized with respect to minimizing acoustic power output) residual sound fields. A simple example of this is shown in Fig. 8.55, which illustrates the primary and controlled acoustic fields in the panel midsection 1.8 m away, and the acoustic power attenuation achieved when

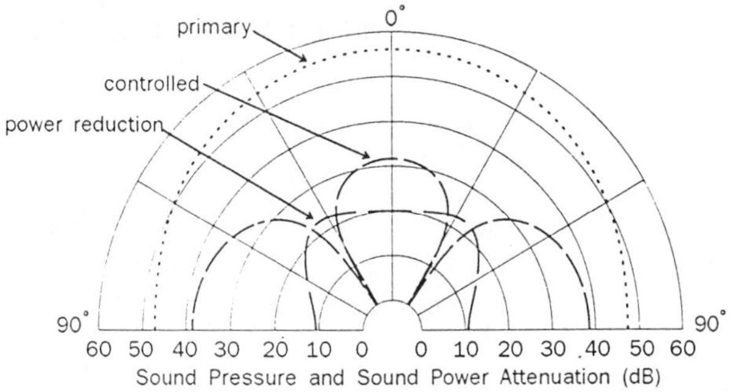

Fig. 8.55 Primary and residual acoustic pressure distributions, plotted with acoustic power attenuation as a function of error microphone position.

minimizing the acoustic pressure at a single point in this arc, for the 0.38 m × 0.30 m × 2 mm panel system described in Section 8.11.3 with a single point control force in the panel centre. Note that the maximum levels of acoustic power attenuation are achieved with the error sensors at the locations of the greatest pressure difference between the primary and optimally controlled (minimum acoustic power) residual sound fields. This concept also applies to multiple control sources (Snyder and Hansen, 1991).

If vibration control sources are being used, then vibration error sensors can also be used. The problem of using a measure of structural vibration to minimize radiated acoustic power was considered in some depth in the previous section.

8.12.2 A shortcut: determination of the optimum control source amplitudes and phases using multiple regression

For some systems targeted for active control it is not feasible to perform a detailed analysis in a reasonable length of time, owing to the complexity of the system. With others, perhaps all that is desired is some way of ranking the quality of potential control source/error sensor arrangements. One alternative to the detailed theoretical development of the previous section, with is readily amenable to the design of systems based upon measured data, is the use of multiple regression to determine the optimum control source amplitudes (Snyder and Hansen, 1991b). Multiple regression is a generalized linear least-squares technique, where several independent variables are used to predict the dependent variable of interest. Here the dependent variable of interest is the 180° inverse of the primary sound field (which will provide the greatest level of acoustic power flow attenuation), while the independent variables are the control source transfer functions and volume velocities or forces.

From our farfield based developments it can be concluded that minimization of the acoustic radiated sound power error criterion, used for determining the maximum possible attenuation with a given control source arrangement, is equivalent to minimizing the average squared sound pressure over a hemisphere enclosing the source for free field radiation. This can be estimated as the minimization of the sound pressure squared at a finite number of points, given by

$$\sum_{i=1}^{N} (p_{p,i}+p_{c,i})^2 \tag{8.12.1}$$

where the number of points N should be chosen so that (8.12.1) is representative of the radiated acoustic power. For vibration control sources, the control generated sound pressure at any point i can be expressed as

$$p_{c,i} = z_{vt,i}^{\mathrm{T}} f \tag{8.12.2}$$

where theoretically

$$z_{vt,i}^{\mathrm{T}} = z_{rad}^{\mathrm{T}} Z_I^{-1} \Psi_c \tag{8.12.3}$$

For acoustic control sources, $p_{c,i}$ is defined by

$$p_{c,i} = z_{mono}^{\mathrm{T}}(r_i) q_c \tag{8.12.4}$$

Note, however, that measured data can be used for $p_{p,i}$ and the terms in $z_{vt,i}$ and $z_{mono,i}$. As both of these quantities are complex they must be measured as transfer functions, so for pressure they must be measured relative to some reference signal. In doing this, note that the absolute value of the reference signal is unimportant in determining the 'quality' of a given control source placement, measured by the global attenuation levels.

Substituting these relations into (8.12.1), the optimization criterion for vibration control sources is that the following expression should be minimized:

$$\sum_{i=1}^{N} \left[p_{p,i} + \sum_{j=1}^{L} z_{vt\,j,i}\, f_j \right]^2 \tag{8.12.5}$$

and for acoustic control sources, the following should be minimized:

$$\sum_{i=1}^{N} \left[p_{p,i} + \sum_{j=1}^{L} z_{mono\,j,i}\, q_j \right]^2 \tag{8.12.6}$$

where the subscripts j,i denote the transfer function between the jth control source and the ith measurement point. Expressions (8.12.5) and (8.12.6) can be written in matrix form. For vibration control sources, (8.12.5) becomes

$$| Z_v f_c - \{-p_p\} | \tag{8.12.7}$$

and for acoustic control sources, (8.12.6) becomes

$$| Z_m q_c - \{-p_p\} | \tag{8.12.8}$$

where

$$p_p = \begin{bmatrix} p_{p1} & p_{p2} & \cdots & p_{pN} \end{bmatrix}^{\mathrm{T}} \tag{8.12.9}$$

$$\mathbf{Z}_v = \begin{bmatrix} z_{vt,1} & z_{vt,2} & \cdots & z_{vt,N} \end{bmatrix}^{\mathrm{T}} \tag{8.12.10}$$

and

$$\mathbf{Z}_m = \begin{bmatrix} z_{mono,1} & z_{mono,2} & \cdots & z_{mono,N} \end{bmatrix}^{\mathrm{T}} \tag{8.12.11}$$

The minimization problems of (8.12.7) and (8.12.8) can be solved by various methods, such as by the use of singular value decomposition, or by using one of many commercially available multiple regression software packages. The control forces and/or control volume velocities that result are those which are optimal (those which minimize the power flow) for the given control source positions.

The level of power attenuation achieved using active noise control can be estimated as

$$\Delta W = -10 \log_{10} \left[\sum_{i-1}^{N} \frac{(p_{p,i}+p_{c,i})^2}{(p_{p,i})^2} \right] \tag{8.12.12}$$

From (8.12.7) and (8.12.8) it can be seen that the control source sound pressure $p_{c,i}$ desired is actually the estimated inverse of the primary source sound pressure, $-p_{p,i}$. Therefore, (8.12.5) can be written as

$$\Delta W + -10 \log_{10} \left[\sum_{i-1}^{N} \frac{(p_{p,i}-p'_{p,i})^2}{(p_{p,i})^2} \right] \tag{8.12.13}$$

where $'$ denotes estimate. The denominator of (8.12.13) is equivalent to the sum of the squares of the measured dependent variable, SS_p, while the numerator is equivalent to the sum of squares of the residuals, SS_{res}. Using these equivalences, the estimated acoustic power reduction for the given control source arrangement can be written as

$$\Delta W = -10 \log_{10} \left[\frac{SS_{res}}{SS_p} \right] \tag{8.12.14}$$

(8.12.15)

$$= -10 \log_{10}(1-R^2)$$

where R is the multiple correlation coefficient.

Therefore, the multiple correlation coefficient can be used to estimate the acoustic power reduction under optimum control for a given control source arrangement. As the number of measurement points increases, so does the accuracy of the estimate. For a very large number of points, this method becomes equivalent to the integration methods of the previous section for determining the optimum control source volume velocities or forces for a given control source and error sensor arrangement. The main advantages of this method are the speed of calculation, and the ability to easily incorporate combined vibration and acoustic control sources. This makes it well suited to practical implementation in a multidimensional optimization routine.

It should be noted that this technique is also suitable for determining the control source volume velocities or forces which minimise the sum of the squared sound pressure at a specific point or points. In this case, only the error sensing locations would be used as measurement points in the equations.

There are further advantages to the use of multiple regression in active control system design. One of these concerns the fact, mentioned earlier in this section, that the optimum error microphone locations are at the points of maximum difference between the primary and controlled sound fields. These points can be determined directly using a commercial multiple regression package. These packages usually produce, as part of their output data, a vector of residuals. These residuals are the difference between the measured quantity (pressure here), and the value predicted by the regression equation. The point will be the point with the smallest residual value.

8.13 ACTIVE CONTROL OF FREE FIELD RANDOM NOISE

So far in this chapter we have concentrated our efforts on examining feedforward active control of harmonic excitation. Analysis of this form of problem is easily facilitated by considering it in the frequency domain, which produces an optimal control source output 'schedule', or control source output as a function of time, which is not constrained to be causal with respect to the primary disturbance. For a large number of problems targeted for active noise control, such an analytical approach and result is completely acceptable. There are instances, however, where a non-causal control source schedule is not practically realizable. One of these instances occurs when considering the active control of free field random noise. To analyse such a problem, and constrain the solution for the optimum control source output to be causal, the problem must be developed using a time domain, rather than frequency domain, approach. In this section we will undertake such an analysis for the problem of controlling the free field radiation from a point acoustic (primary) monopole source by the introduction of a second point acoustic (control) monopole source, and in doing so develop some qualitative results which can be applied to the range of causally constrained problems.

8.13.1 Analytical basis

Let us again consider the active control of sound radiation from one (primary) monopole source by the introduction of a second (control) monopole source, but in this instance let the primary monopole oscillation be random, rather than harmonic. As with all free field radiation problems, there are two possible error criteria which immediately come to mind; control of acoustic pressure at a point or points, and control of total radiated acoustic power. It will prove to be more straightforward in this instance if the control of acoustic pressure at a single error sensing point is considered first. The starting point for the analysis is the inhomogeneous wave equation, which was originally outlined in Section 2.1:

$$\left[\nabla^2 - \frac{1}{c_0^2}\frac{\partial^2}{\partial t^2}\right] p(r,t) = -\rho_0 \frac{\partial Q(r,t)}{\partial t} \tag{8.13.1}$$

As monopole acoustic sources are being considered, the source strength density, Q, can be taken as concentrated at a single location, r_q, such that $Q(r,t) = q(t)$ $\delta(r - r_q)$. With this assumption, the solution for the acoustic pressure fluctuation at some location r_e due to the operation of the monopole source located at r_q is

$$p(r_e,t) = \frac{\rho_0}{4\pi r_{q/e}} \frac{\partial q\left[t - \frac{r_{q/e}}{c_0}\right]}{\partial t} = \frac{\rho_0}{4\pi r_{q/e}} \dot{q}\left[t - \frac{r_{q/e}}{c_0}\right] \tag{8.13.2}$$

where $r_{q/e} = |r_e - r_q|$. This expression is simply the time domain version of (8.2.1), which describes the acoustic pressure radiated from a monopole source in the frequency domain, where the explicit time derivative in the time domain is equivalent to $j\omega$ in the frequency domain.

During operation of the active control system the residual acoustic pressure at any location r_e (e denoting error sensor location) can be viewed as the superposition of the primary source and control source generated acoustic pressure at that location as follows:

$$p(r_e,t) = \frac{\rho_0}{4\pi}\left[\frac{\dot{q}_p\left[t - \frac{r_{p/e}}{c_0}\right]}{r_{p/e}} + \frac{\dot{q}_c\left[t - \frac{r_{c/e}}{c_0}\right]}{r_{c/e}}\right] \tag{8.13.3}$$

where $r_{p/e} = |r_e - r_p|$ and $r_{c/e} = |r_e - r_c|$, with r_p and r_c being the locations of the primary and control monopole sources, respectively. It is readily apparent from (8.13.3) that the control source output which will minimize the acoustic pressure at the error sensing location is

$$\dot{q}_{c,\mathrm{opt}}(t) = \frac{r_{c/e}}{r_{p/e}} \dot{q}_p\left[t - \frac{r_{p/e} - r_{c/e}}{c_0}\right] \tag{8.13.4}$$

This control output is, in theory, realizable in a causally constrained system if the error sensor location is closer to the control source than the primary source, such that $t - (r_{p/e} - r_{c/e})/c_0$ is always a point in time prior to the control output time t (in practice, the control source must be closer than the primary source to the error sensor by an amount greater than or equal to the time it takes for a signal to propagate through the electronic control, as dicussed in Chapter 13). Conversely, if the error sensor is located at a point closer to the primary source than the control source, the constraints of causality make the realisation of the control source output described by (8.13.4) a practical impossibility. However, it is possible to use a knowledge of the statistical properties of the primary source output to *estimate* the primary source output at some time in the future based upon its past outputs. The form of the problem then becomes one of

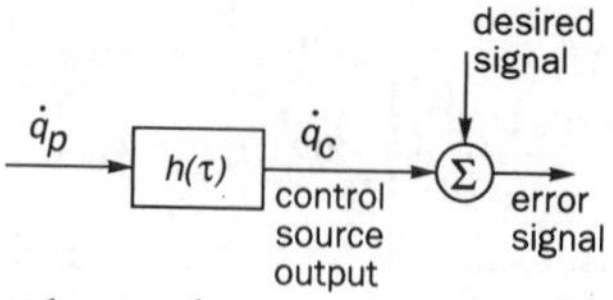

Fig. 8.56 Derivation of optimal control source output, and hence filter impulse response function.

optimally estimating the primary source output at time $t - (r_{p/e} - r_{c/e})/c_0$, which can viewed as a Weiner filtering problem.

Weiner filter theory has been encountered previously in Chapter 6 for the purpose of deriving an optimal finite impulse response filter, the output of which most closely matches some desired output based solely on previous samples. This is identical to the problem we have here. Consider the system arrangement shown in Fig. 8.56. Here the desired control source output, which will completely minimize the acoustic pressure amplitude at the error sensor, is as stated in (8.13.4). The actual control source output is defined by the relationship

$$\dot{q}_c(t) = \int_0^\infty h(\tau)\, \dot{q}_p(t-\tau)\, d\tau \tag{8.13.5}$$

where $h(\tau)$ is the impulse response function of the control source, and the lower limit on the convolution integral has been set to zero for physical realisability. Therefore, what is desired is some optimal control source impulse function, h_{opt}, which satisfies the relationship

$$\dot{q}_{c\ \text{opt}}(t) = \frac{r_{c/e}}{r_{p/e}}\, \dot{q}_p\left[t - \frac{r_{p/e} - r_{c/e}}{c_0}\right] = \int_0^\infty h_{\text{opt}}(\tau)\, \dot{q}_p(t-\tau)\, d\tau \tag{8.13.6}$$

The product of this desired output and the primary source volume velocity time derivative is

$$\dot{q}_p(t)\left[\frac{r_{c/e}}{r_{p/e}}\right] \dot{q}_p\left[t - \frac{r_{p/e} - r_{c/e}}{c_0}\right] = \int_0^\infty h_{\text{opt}}(\gamma)\, \dot{q}_p(t)\, \dot{q}_p(t-\tau-\gamma)\, d\gamma \tag{8.13.7}$$

Taking expected values, this can be written as a cross-correlation relationship

$$\left[\frac{r_{c/e}}{r_{p/e}}\right] R_{\dot{p}\dot{p}}\left[\tau - \frac{r_{p/e} - r_{c/e}}{c_0}\right] = \int_0^\infty h_{\text{opt}}(\gamma)\, R_{\dot{p}\dot{p}}(\tau-\gamma)\, d\gamma \tag{8.13.8}$$

or

$$\left[\frac{r_{c/e}}{r_{p/e}}\right] R_{\dot{p}\dot{p}}(\tau) = \int_0^\infty h_{\text{opt}}(\gamma)\, R_{\dot{p}\dot{p}}\left[\tau + \frac{r_{p/e} - r_{c/e}}{c_0} - \gamma\right] d\gamma \tag{8.13.9}$$

where $R_{\dot{p}\dot{p}}$ is the autocorrelation of the primary source time derivative. Equation (8.13.6) is simply the Weiner–Hopf integral equation, the solution of which defines the optimum control source impulse response function, h_{opt}.

The solution to the Weiner–Hopf integral equation was formulated in Chapter 6 in terms of spectral density functions, which are based upon the Fourier transforms of the quantities in the equation. It will be advantageous here,

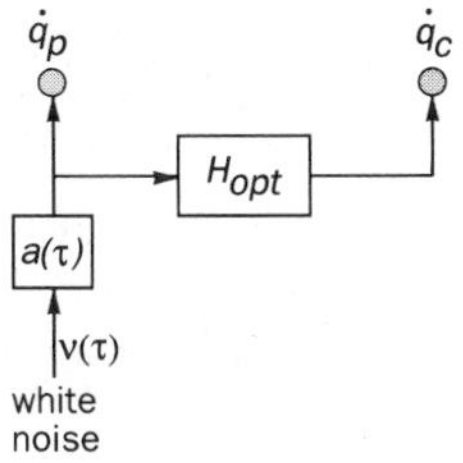

Fig. 8.57 Single channel active control system arrangement as a time domain filtering exercise.

however, to state the result in terms of Laplace transformed quantities, for ease of demonstrating an example. Referring to Fig. 8.57, modelling the primary source output as a white noise sequence ν passed through a filter with an impulse response function a such that

$$\dot{q}_p(t) = \int_0^\infty a(\tau)\ \nu(t-\tau)\ \mathrm{d}\tau \tag{8.13.10}$$

the optimum control source transfer function $H_{\text{opt}}(s)$, which is found by solving the Weiner–Hopf integral equation using the primary source characteristics given in (8.13.10), is

$$H_{\text{opt}}(s) = \int_0^\infty h_{\text{opt}}(t)\mathrm{e}^{-st}\ \mathrm{d}t = -\frac{r_{c/e}}{r_{p/e}}\ A^{-1}(s)\ B^{-1}(s) \tag{8.13.11}$$

where

$$A(s) = \int_0^\infty a(t)\mathrm{e}^{-st}\ \mathrm{d}t \tag{8.13.12}$$

$$B(s) = \int_0^\infty a\left[t+\frac{r_{p/e}-r_{c/e}}{c_0}\right]\ \mathrm{e}^{-st}\ \mathrm{d}t \tag{8.13.13}$$

The use of (8.13.11) in deriving a control source output schedule is best illustrated by example.

As an example, let us consider the case where the primary source volume velocity is modelled as passing white noise through a second order filter, the characteristics of which are defined by

$$\alpha(s) = \frac{\omega_n^2}{s^2+2\zeta\omega_n s+\omega_n^2} \tag{8.13.14}$$

where ω_n and ζ are the natural frequency and damping ratio of the system respectively. (This response from this model is similar to the response of a structure excited by a broadband source where there are resonances in the excitation frequency band.) The time derivative of the primary source volume velocity is therefore modelled as white noise passed through a filter whose characteristics are the derivative of $\alpha(s)$,

$$A(s) = \frac{s\omega_n}{s^2+2\zeta\omega_n s+\omega_n^2} \tag{8.13.15}$$

The impulse response of the above transfer function is a standard table result,

$$a(t) = -\frac{\omega_n^3}{\omega_d} e^{-\zeta\omega_n t} \sin(\omega_d t-\phi) \tag{8.13.16}$$

where ω_d is the damped natural frequency, $\omega_d = \omega_n\surd(1 - \zeta^2)$, and $\phi = \tan^{-1}[\surd(1 - \zeta^2)/\zeta]$. The impulse response $b(t)$ is found to be

$$b(t) = a(t+\Delta t) = -\frac{\omega_n^3}{\omega_d} e^{-\zeta\omega_n (t+\Delta t)}\sin(\omega_d t-\phi+\Delta t) \tag{8.13.17}$$

where $\Delta t = (r_{p/e} - r_{c/e})/c_0$. Taking the Laplace transform of this produces (Nelson *et al.*, 1988, 1990)

$$B(s) = -\frac{\omega_n^3}{\omega_d} e^{-\zeta\omega_n\Delta t} \left\{ \frac{s\left(\zeta\omega_n \sin\omega_d\Delta t - \omega_d\cos\omega_d\Delta t\right) + \omega_n^2\sin\omega_d\Delta t}{s^2+2\zeta\omega_n s+\omega_n^2} \right\} \tag{8.13.18}$$

Therefore, using the result of (8.13.11), the optimum control source transfer function is

$$H_{\text{opt}}(s) = \frac{r_{c/e}}{r_{p/e}} \frac{e^{-\zeta\omega_n\Delta t}}{\omega_d} \left[\left(\zeta\omega_d \sin\omega_d\Delta t\right) + \frac{\omega_n^2}{s} \sin\omega_d\Delta t \right] \tag{8.13.19}$$

Taking the inverse Laplace transform of this, the optimum control source output is found to be

$$\dot{q}_{c\,\text{opt}}(t) = \frac{r_{c/e}}{r_{p/e}} \left\{ q_p(t) \left[\frac{e^{-\zeta\omega_n\Delta t}}{\omega_d} \omega_n^2 \sin\omega_d\Delta t \right] - \dot{q}_p(t) \left[\frac{e^{-\zeta\omega_n\Delta t}}{\omega_d} \left[\omega_d \cos\omega_d\Delta t - \zeta\omega_n \sin\omega_d\Delta t\right] \right] \right\} \tag{8.13.20}$$

8.13.2 Minimum sound pressure amplitude at the error sensor

Having now derived an expression for determining the optimal control source output characteristics as a function of the primary noise statistics, the next step is to determine the sound pressure reduction which will be achieved at the error sensor when using this control signal. If the error sensor is closer to the control source than the primary source, such that the control schedule of (8.13.4) can be implemented exactly, the residual sound pressure will in theory be zero. However, if the error sensor is closer to the primary source than the control source, such that the control source characteristics must be based on the optimal estimator derived in (8.13.11), the residual sound pressure will probably not be equal zero. The actual value can be easily quantified by considering the analysis of Weiner filtering given in Chapter 6. In this analysis it was shown that the minimum mean square error was equal to

$$\xi_{\min} = E\{y^2(t)\} - E\{y(t)\hat{y}(t)\} \tag{8.13.21}$$

where y is the desired output, and $\hat{y}$ is the optimal estimate of the desired output. Therefore, by substituting the obtained (from (8.13.11)) optimal estimate of the desired control source output into (8.13.21), the residual sound pressure amplitude can be calculated. This again is best demonstrated by means of example. Consider the problem of calculating the minimum sound pressure amplitude at the error sensor for the control source output derived in the previous problem (from Nelson *et al.*, 1988, 1990). To calculate the residual sound pressure amplitude at the error sensor location, note that the first term in (8.13.21) is simply the squared sound pressure amplitude of the primary disturbance before the addition of active control, which can be denoted simply as J_p. Substituting the final result of the previous example into (8.13.21), the minimum mean square pressure becomes

$$J_{\min} = J_p - E\left\{ \frac{\rho_0 \dot{q}_p(t - r_{p/e}/c_0)}{4\pi r_{p/e}} \left[\frac{M\rho_0 \dot{q}_p\left[t - \dfrac{r_{c/e}}{c_0}\right]}{4\pi r_{c/e}} + \frac{N\rho_0 q_p\left[t - \dfrac{r_{c/e}}{c_0}\right]}{4\pi r_{c/e}} \right] \right\} \tag{8.13.22}$$

where

$$M = \frac{e^{-\zeta\omega_n \Delta t}}{\omega_d}\left(\zeta\omega_n \sin\omega_d \Delta t - \omega_d \cos\omega_d \Delta t\right) \tag{8.13.23}$$

$$N = \frac{e^{-\zeta\omega_n \Delta t}}{\omega_d}\,\omega_n^2 \sin\omega_d \Delta t \tag{8.13.24}$$

This can be simplified to

$$J_{\min} = J_p - \left[\frac{\rho_0}{4\pi r_{p/e}}\right]^2 \left(MR_{\dot{p}\dot{p}}(\Delta t) + NR_{p\dot{p}}(\Delta t)\right) \tag{8.13.25}$$

where

$$R_{\dot{p}\dot{p}}(\Delta t) = E\{\dot{q}_p(t)\dot{q}_p(t+\Delta t)\} \tag{8.13.26}$$

$$R_{p\dot{p}}(\Delta t) = E\{q_p(t)\dot{q}_p(t+\Delta t)\} \tag{8.13.27}$$

The autocorrelation and cross-correlation functions given above can be solved for the description of the primary noise source characteristics given in the beginning of the previous example, giving (Bendat and Piersol, 1986, Chapter 6; Nelson *et al.*, 1988, 1990)

$$R_{\dot{p}\dot{p}}(\Delta t) = \frac{e^{-\zeta\omega_n\Delta t}}{\omega_d}\frac{\omega_n^3}{8\zeta}\left(\omega_d \cos\omega_d\Delta t - \zeta\omega_n \sin\omega_d\Delta t\right) \quad (8.13.28)$$

$$R_{p\dot{p}}(\Delta t) = -\frac{e^{-\zeta\omega_n\Delta t}}{\omega_d}\frac{\omega_n^3}{8\zeta}\sin\omega_d\Delta t \quad (8.13.29)$$

The reduction in squared sound pressure amplitude at the error sensing location can be expressed as

$$\Delta p = \frac{J_{\min}}{J_p} \quad (8.13.30)$$

Using the previous equations, with some algebraic manipulation (Nelson, 1988), this can be expressed as

$$\Delta p = 1 - \frac{e^{-2\zeta\omega_n\Delta t}}{\omega_d^2}\left(\omega_n^2 \sin^2\omega_d\Delta t + (\omega_d \cos\omega_d\Delta t - \zeta\omega_n \sin\omega_d\Delta t)^2\right) \quad (8.13.31)$$

In viewing the description of the pressure reduction, it is apparent that the reduction which can be achieved is dependent upon two parameters, the damping of the primary source characteristics (determining how flat the response is), and the parameter $\omega\Delta t$. Note that this term can be expressed as

$$\omega_n\Delta t = \frac{\omega_n(r_{ple}-r_{cle})}{c_0} = \frac{2\pi(r_{ple}-r_{cle})}{\lambda_n} \quad (8.13.32)$$

which provides a direct representation of the effect of the difference in path length on the level of attenuation which can be achieved. As would be expected, as the difference in path length becomes small, or the damping of the primary source response decreases (such that there is a sharper peak, or the output becomes more 'tonal'), the levels of attenuation which can be achieved increase.

8.13.3 Minimization of total radiated acoustic power

Having considered the problem of minimizing acoustic pressure amplitude at a single point in space, the next case to consider is that of minimizing the radiated acoustic power from the primary source/control source pair. For the time domain analysis, the appropriate error criterion is instantaneous acoustic power, defined by the farfield integration of instantaneous acoustic intensity over a surface enclosing the source pair:

$$W = \int_S I(r,t)\, dS \quad (8.13.33)$$

where the farfield instantaneous acoustic intensity is directly related to the pressure amplitude squared:

$$I(r,t) = \frac{|\,p(r,t)\,|^2}{2\rho_0 c_0} \tag{8.13.34}$$

As such, minimizing total radiated power is equivalent to minimizing the spatially averaged squared sound pressure level. For the control of random noise, this means that the causally constrained control source output can be considered as a combination of exact 'signal delay' control, for the points on the intensity integration surface closer to the control source than the primary source, and optimal prediction control, for points on the integration surface closer to the primary source than the control source. However, rather than directly use the spatial integration to derive the optimum control source operating 'schedule', the nearfield approach used in analysing the power radiated by a harmonic monopole source (in the beginning of this chapter) will be used again to derive the desired control source characteristics.

Expressed in the time domain, the time averaged acoustic power output of a monopole source is

$$W = \int_S E\{p(r,t)v(r,t)\}\cdot \boldsymbol{n}\ \mathrm{dS} = E\{p(r_q,t)q(t)\} \tag{8.13.35}$$

However, as was mentioned in Section 2.4, the pressure at the source location r_q is a singular quantity. This problem can be overcome by rewriting (8.13.35) using (8.13.2) as follows:

$$W = \lim_{r/c_o \to 0} E\left\{\frac{\rho_0}{4\pi r}\,\dot{q}\left[t-\frac{r}{c_0}\right]\,q(t)\right\} \tag{8.13.36}$$

where r is the distance from the source. Substituting the expansion

$$q(t) = q(t-\frac{r}{c_0}) + (\frac{r}{c_0})\,\dot{q}(t-\frac{r}{c_0})\cdots \tag{8.13.37}$$

into (8.13.36), and noting that for a stationary random process

$$E\{\dot{x}(t)x(t)\} = 0 \tag{8.13.38}$$

(8.13.36) can be evaluated to give an expression for the time averaged power output from a monopole source, stated in the time domain, as

$$W = \frac{\rho_0}{4\pi c_0}\,E\,\{\dot{q}^2(t)\} \tag{8.13.39}$$

Having now formulated an expression to overcome the problem of acoustic pressure singularity when evaluating power output at a monopole source, we are now in a position to consider minimisation of the total radiated acoustic power. Using (8.13.35), the total radiated power from the primary source/control source pair is

$$W = E\left\{\,(p_p(r_p,t)+p_c(r_p,t))q_p(t) + (p_p(r_c,t)+p_c(r_c,t))q_c(t)\,\right\} \tag{8.13.40}$$

where $p_p(\boldsymbol{r},t)$, $p_c(\boldsymbol{r},t)$ denote primary and control source generated sound pressures at location $\boldsymbol{r}$ and time t, respectively. Using (8.13.39), (8.13.40) can be re-expressed as

$$W = E\left\{\frac{\rho_0}{4\pi c_0}\dot{q}_p^2(t) + p_c(r_p,t)q_p(t) + p_p(r_c,t)q_c(t) + \frac{\rho_0}{4\pi c_0}\dot{q}_c^2(t)\right\} \tag{8.13.41}$$

The second and third terms and (8.13.41) can be further restated using (8.13.2):

$$E\{p_c(r_p,t)q_p(t)\} = \frac{\rho_0}{4\pi d}\,E\left\{\dot{q}_c(t-\frac{d}{c_0})q_p(t)\right\} = \frac{\rho_0}{4\pi d}\,E\left\{\dot{q}_c(t)q_p(t+\frac{d}{c_0})\right\} \tag{8.13.42}$$

where d is the primary source/control source separation distance, and

$$E\{p_p(r_c,t)q_c(t)\}$$

$$= \frac{\rho_0}{4\pi d}\,E\left\{\dot{q}_p(t-\frac{d}{c_0})q_c(t)\right\} = -\frac{\rho_0}{4\pi d}\,E\left\{q_p(t-\frac{d}{c_0})\dot{q}_c(t)\right\} \tag{8.13.43}$$

Substituting (8.13.42) and (8.13.43) into (8.13.41), and noting that the first term in (8.13.41) is the acoustic power W_p radiated by the primary source in the absence of active control produces the following expression for the total radiated acoustic power:

$$W = W_p + \frac{\rho_0}{4\pi c_0}\,E\left\{\dot{q}_c^2(t) + \frac{c_0}{d}\dot{q}_c(t)\left[q_p(t+\frac{d}{c_0}) - q_p(t-\frac{d}{c_0})\right]\right\} \tag{8.13.44}$$

This expression can be used to formulate a set of control source characteristics which will minimise the total radiated acoustic power.

To derive the optimum control source characteristics, it will be useful to assume that the time derivative of the control source volume velocity is related to the primary source volume velocity, rather than its time derivative as was assumed when minimising sound pressure at a point. With this assumption, the control source output can be expressed as:

$$\dot{q}_c(t) = \int_0^\infty h_w(\tau)q_p(t-\tau)\,\mathrm{d}\tau \tag{8.13.45}$$

Deriving the required Weiner–Hopf integral equation is not quite as simple as the 'intuitive' approach used when minimizing sound pressure at a point, as the optimum control source output is not immediately apparent, and so variational methods will be employed for the task (Laning and Battin, 1956; Nelson *et al.*, 1988). With these methods, it is assumed that the impulse response h_w can be written as

$$h_w(t) = h_{w,\mathrm{opt}}(t) + \epsilon h_w'(t) \tag{8.13.46}$$

where $h_{w,\text{opt}}$ is the desired optimal impulse response function, ϵ is a real number, and h_w' is some arbitrary function of t. The approach taken to solving the problem is to re-express the power output in terms of ϵ, and then, noting that for the optimum impulse response function $\epsilon = 0$, differentiate the expression with respect to ϵ and set the result of the differentiation equal to zero. Substituting (8.13.46) into (8.13.45) produces

$$W = W_p + \frac{\rho_0}{4\pi c_0}\left\{\int_0^\infty\int_0^\infty \left(h_{w,\text{opt}}(\tau)+\epsilon h_w'(\tau)\right)\left(h_{w,\text{opt}}(\gamma)+\epsilon h_w'(\gamma)\right) R_{pp}(\tau-\gamma)\, \mathrm{d}_\tau \mathrm{d}_\gamma \right.$$

$$\left. + \frac{c_0}{d}\int_0^\infty \left(h_{w,\text{opt}}(\tau)+\epsilon h_w'(\tau)\right) R_{pp}(\tau+\frac{d}{c_0})\, \mathrm{d}\tau - \frac{c_0}{d}\int_0^\infty \left(h_{w,\text{opt}}(\tau)+\epsilon h_w'(\tau)\right) R_{pp}(\tau-\frac{d}{c_0})\, \mathrm{d}_\tau \right\} \tag{8.13.47}$$

Expanding (8.13.47) in terms of powers of ϵ, differentiating with respect to ϵ, and setting $\left(\frac{\partial W}{\partial \epsilon}\right)_{\epsilon=0} = 0$ produces the expression

$$\int_0^\infty\int_0^\infty \left(h_{w,\text{opt}}(\tau)h_w'(\tau)h_{w,\text{opt}}(\gamma)\right) R_{pp}(\tau-\gamma)\, \mathrm{d}\tau \mathrm{d}\gamma$$

$$+ \frac{c_0}{d}\int_0^\infty h_w'(\tau)\left[R_{pp}\left[\tau+\frac{d}{c_0}\right] - R_{pp}\left[\tau-\frac{d}{c_0}\right]\right] \mathrm{d}\tau = 0 \tag{8.13.48}$$

Noting that $R_{pp}(\tau - \gamma) = R_{pp}(\gamma-\tau)$, so that the first term in (8.13.48) can be expressed as

$$\int_0^\infty\int_0^\infty 2h_w'(\tau)h_{w,\text{opt}}(\gamma)\, R_{pp}(\tau-\gamma)\, \mathrm{d}\tau \mathrm{d}\gamma \tag{8.13.49}$$

(8.13.48) can be rewritten as

$$\int_0^\infty h_w'(\tau)\left\{\int_0^\infty 2h_{w,\text{opt}}(\tau)\, R_{pp}(\tau-\gamma)\, \mathrm{d}\gamma \right.$$

$$\left. + \frac{c_0}{d}\left[R_{pp}(\tau+\frac{d}{c_0}) - R_{pp}(\tau-\frac{d}{c_0})\right]\right\} \mathrm{d}\tau = 0 \tag{8.13.50}$$

As equation (8.13.50) must hold for any arbitrary function h_w', it can be deduced that the expression inside the curly brackets must be equal to zero. Therefore,

$$\int_0^\infty h_{w,\text{opt}}(\tau)\, R_{pp}(\tau-\gamma)\; \mathrm{d}\gamma \;+\; \frac{c_0}{2d}\left[R_{pp}\left[\tau+\frac{d}{c_0}\right] - R_{pp}\left[\tau-\frac{d}{c_0}\right]\right] = 0 \qquad (8.13.51)$$

This is the Weiner–Hopf integral equation which defines the optimum control source characteristics for minimizing the total radiated acoustic power (Nelson *et al.*, 1988, 1990).

Now that the desired form of the Weiner–Hopf integral equation has been derived, what remains is to solve it for the optimum control source impulse function. To do this, we will use the idea put forward in the beginning of this analysis, that the optimum impulse response function must be a combination of a 'delay' control, and optimal estimation, resulting in the following form for the solution:

$$h_{w,\text{opt}}(t) = \frac{c_0}{2d}\left[\delta\left[t-\frac{d}{c_0}\right] - h_{w,e}(t)\right] \qquad (8.13.52)$$

where $\delta(t-d/c_o)$ is the pure time delay and $h_{w,e}$ is the impulse response for the estimation part of the impulse response function. Substituting this assumed form of solution into (8.13.51) shows that $h_{w,e}$ must satisfy the expression

$$\int_0^\infty h_{w,e}(\gamma) R_{pp}(\tau-\gamma)\; \mathrm{d}\gamma - R_{pp}\left[\tau+\frac{d}{c_0}\right] = 0 \qquad \tau \geq 0 \qquad (8.13.53)$$

because

$$\frac{c_0}{2d}\left[\int_0^\infty \delta\left(\gamma-\frac{d}{c_0}\right) R_{pp}(\tau-\gamma)\; \mathrm{d}\gamma - R_{pp}\left[t-\frac{d}{c_0}\right]\right] = 0 \qquad (8.13.54)$$

From (8.13.53) it is clear that the impulse response function $h_{w,e}$ is the solution to the problem which requires optimal prediction of the primary source output at a time in the future corresponding to the propagation time delay due to the primary source/control source separation distance. Further, the pure delay component of the impulse response has corresponds physically to one of the control mechanisms (absorption) seen previously. Substituting (8.13.52) into (8.13.45) shows that the optimum control source output characteristics are defined by (Nelson *et al.*, 1988, 1990):

$$\dot{q}_{c,\text{opt}}(t) = \frac{c_0}{2d}\left[q_p\left[t-\frac{d}{c_0}\right] - \int_0^\infty h_{w,e}(\tau) q_p(t-\tau)\; \mathrm{d}\tau\right] \qquad (8.13.55)$$

The first term in (8.13.55) describes a control source output which maximizes the absorption of acoustic power, while the second maximizes suppression.

8.13.4 Calculation of the minimum power output

Now that an expression has been derived for the optimum control source output with respect to minimizing total radiated acoustic power, what remains is to find out what levels of attenuation it will provide. Substituting the optimum control source volume velocity time derivative of (8.13.55) into (8.13.41), which describes the total acoustic power output of the system, gives

$$W_{\min} = \frac{\rho_0}{4\pi c_0} E\left\{ \dot{q}_p^2(t) + \dot{q}_{c,\mathrm{opt}}^2(t) + \frac{c_0}{d}\dot{q}_{c,\mathrm{opt}}(t)\left[q_p\left[t+\frac{d}{c_0}\right] - q_p\left[t-\frac{d}{c_0}\right]\right]\right\} \tag{8.13.56}$$

Equation (8.13.56) can be re-expressed as

$$W_{\min} = \frac{\rho_0}{4\pi c_0} E\left\{ \dot{q}_p^2(t) + \frac{c_0}{2d}\dot{q}_{c,\mathrm{opt}}(t)\left[q_p(t+\frac{d}{c_0}) - q_p(t-\frac{d}{c_0})\right] \right. \tag{8.13.57}$$

$$\left. + \left[\dot{q}_{c,\mathrm{opt}}^2(t) + \frac{c_0}{2d}\dot{q}_{c,\mathrm{opt}}(t)\left[q_p(t+\frac{d}{c_0}) - q_p(t-\frac{d}{c_0})\right]\right]\right\}$$

It can be shown that the term in the square brackets in (8.13.57) is equal to zero by using the substitution

$$\dot{q}_{c,\mathrm{opt}}(t) = \int_0^\infty h_{w,\mathrm{opt}}(\tau) q_p(t-\tau)\,\mathrm{d}\tau \tag{8.13.58}$$

and noting that

$$E\left\{\left[\dot{q}_{c,\mathrm{opt}}^2(t) + \frac{c_0}{2d}\dot{q}_{c,\mathrm{opt}}(t)\left[q_p(t+\frac{d}{c_0}) - q_p(t-\frac{d}{c_0})\right]\right]\right\} = \tag{8.13.59}$$

$$\int_0^\infty h_{w,\mathrm{opt}}(\tau)\left[\int_0^\infty h_{w,\mathrm{opt}}(\gamma) R_{pp}(\tau-\gamma)\,\mathrm{d}\gamma + \frac{c_0}{2d}\left[R_{pp}(\tau+\frac{d}{c_0}) - R_{pp}(\tau-\frac{d}{c_0})\right]\right]$$

an equality which is a result of the defining Weiner–Hopf integral equation. Therefore, the minimum radiated acoustic power is

$$W_{\min} = \frac{\rho_0}{4\pi c_0} E\left\{ \dot{q}_p^2(t) + \frac{c_0}{2d}\dot{q}_{c,\mathrm{opt}}(t)\left[q_p(t+\frac{d}{c_0}) - q_p(t-\frac{d}{c_0})\right]\right\} \tag{8.13.60}$$

or

$$W_{\min} = W_p + \frac{\rho_0}{4\pi c_0} E\left\{ \frac{c_0}{2d}\dot{q}_{c,\mathrm{opt}}(t)\left[q_p(t+\frac{d}{c_0}) - q_p(t-\frac{d}{c_0})\right]\right\} \tag{8.13.61}$$

where W_p is the power radiated by the primary source in the absence of active control.

Therefore, the minimum acoustic power output of the system, as well as the control source operating schedule, is a function of both a desire to absorb the primary acoustic radiation which has already occurred, and the desire to suppress the acoustic radiation which will occur in the future. We will consider this compromise further in the next section when examining the active control of impulses.

8.14 ACTIVE CONTROL OF IMPACT ACCELERATION NOISE

Many of the prominent sources of industrial noise pollution in factories are machines concerned with the reshaping of a piece of material. Such machines normally perform their allotted task via some impulsive process, such as stamping, forging or riveting. The noise signature from such an operation is compiled from two separate generating mechanisms; the rapid acceleration or deceleration of some part of the workpiece and/or machine, and the pseudo-steady state vibration of the machine and/or workpiece following the impact, as it dissipates its vibrational energy (Osman *et al.*, 1974; Akay, 1978; Richards *et al.*, 1979). The result of this is often a signature which is characterized by two separate parts; a short duration, high amplitude initial pulse due to the acceleration noise, and a much longer duration, lower amplitude 'ringing' following this. While it is the residual structural vibrations which are the primary contributors to the overall noise level (Akay, 1978), it is perhaps the initial acceleration noise pulse which holds the greatest potential for hearing damage. This is due to its short duration, which limits the ability of the middle ear mechanism to stiffen and thus partially protect the inner ear from the pressure pulse (Akay, 1978), resulting in the pressure pulses passing directly into the inner ear with their full amplitude and hearing damage potential.

In this section we will examine the potential for applying active control to attenuate the sound generated by an impulsive process. We will specifically consider the initial acceleration impulse noise portion of the noise signature, because we will discuss feedback control of radiated acoustic power (probably the best approach for the acoustic radiation from the residual vibration) later in this chapter.

One of the characteristics of the acoustic pressure fluctuation caused by the initial acceleration is that it is essentially deterministic. Further, for elastic collisions in simple systems it can be shown that its amplitude is linearly correlated with the impact force (Koss and Alfredson, 1973; Koss, 1974), which can be measured easily with a load cell. These features enable us to contemplate the use of a feedforward active control system, where the impact force, or machine acceleration, is provided to the control system as a reference signal. Alternatively, if the characteristics of successive impulses are constant, a 'preview' signal, such as the acceleration of the machine, can be used as a reference signal. Therefore, the collisions we will be concerned with here will be limited to elastic collisions between two rigid bodies, such as the collision between the hammer and anvil in a forge. For these cases, the acoustic waveform generated by the initial acceleration is constructed from what is essentially a single sinusoidal pulse from each of the impacting bodies. We have already studied the active control of what is basically sinusoidal (harmonic) radiation from rigid bodies in this chapter, in regard to control of acoustic radiation from monopole sources. However, the impact noise problems being considered here have a distinguishing characteristic: they require control inputs which are constrained to be causal with respect to the primary disturbance. This rules out the use of the frequency domain analysis techniques employed in the examination of the performance potential of feedforward control for the previously mentioned harmonic excitation problems, as these methods were shown to produce control output schedules which are not constrained to be causal with respect to the primary noise disturbance. Causally constrained systems were studied in the previous section using an analysis akin to Weiner filter theory, employing statistical expectation operators to develop a control source whose characteristics minimize the mean square value of the acoustic quantity of interest (power or pressure). However, as the problem being considered here is essentially deterministic this approach is more complex than what is really needed.

Therefore, we will therefore begin this section by developing a methodology for calculating the optimum control source output 'schedule' for controlling both the overall (spatially averaged) levels of the initial impact pulse, as well as minimizing the pressure amplitude at prospective (discrete) error sensing locations. Following this, we will examine a very simple problem, the control of a single sinusoidal pulse from a monopole source.

8.14.1 Method for obtaining optimum control source pressure output schedules

Let us begin our study of the problem of actively attenuating impact acceleration noise by developing a methodology to facilitate calculation of the optimum control source output schedule, or output as a function of time, which will minimize the particular error criterion of interest, either the total radiated acoustic power or the acoustic pressure at a given set of potential error sensing locations. The method we will develop to calculate the desired control source output schedules is similar to both the previously developed frequency and time domain techniques in that it is least-squares based, but is best viewed as a linear regression approach to solving the problem.

Consider the active control problem illustrated in Fig. 8.58, where some generic control source distribution is being used to actively attenuate the acoustic radiation from some generic primary source distribution. The error criterion we will examine first in our time domain analysis of the problem is instantaneous radiated acoustic power, calculated as the integration of instantaneous acoustic intensity I over the surface of a volume enclosing both the primary and control source distribution. Instantaneous acoustic intensity can be expressed as

$$I(r,t) = \frac{|\,p(r,t)\,|^2}{\rho_0 c_0} \tag{8.14.1}$$

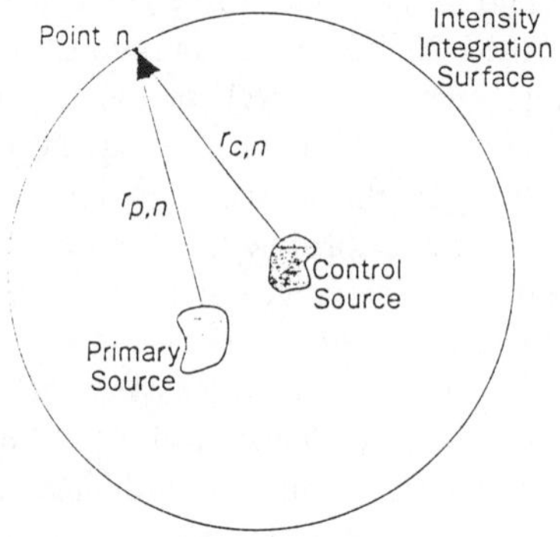

Fig. 8.58 Geometry for impulse problem formulation.

where $p(r,t)$ is the acoustic pressure at r at time t. This can be approximated at any given time by minimizing the squared acoustic pressure amplitude at a discrete number of points, N_n, distributed evenly over the surface of the enclosing space,

$$\text{minimizing } W \approx \text{minimizing} \sum_{n=1}^{N_n} |\, p(r_n,t) \,|^2 \tag{8.14.2}$$

The object now is to derive an operating schedule for the control source distribution which will do this during the time period when the impact acceleration noise is present.

For the impact acceleration noise problems we are considering here, the operating time of the primary source is bounded, beginning when the impact occurs and finishing when the impacted body returns to the point of equilibrium. The duration of the non-zero contribution to the acoustic pressure at all points on the integration surface from the impact is therefore also bounded, beginning at time $t_{p,i}$ when the pressure pulse initially contacts the closest point on the enclosing surface and finishing at time $t_{p,f}$ when the trailing edge of the pulse is exiting from the most distant point. The object of applying active control is therefore to minimize the sum of the squared acoustic pressure amplitudes at all points on the enclosing surface throughout the period $(t_{p,i},\ t_{p,f})$. As with the discretization of the continuous spatial integration, we can approximate the minimization of pressure squared during this time as the problem of minimizing the pressure squared at discrete time intervals during the period $(t_{p,i},\ t_{p,f})$, such that the error criterion used to analytically determine the optimum control source operating schedule is

$$\text{minimize} \sum_{\gamma=0}^{N_t-1} \sum_{n=1}^{N_n} |\, p(r_n, t=t_{p,i}+\gamma\Delta t_p) \,|^2 \tag{8.14.3}$$

N_t being the total number of points in time between $t_{p,i}$ and $t_{p,f}$ which are considered, and each point is incremented by Δt_p from the one before,

$$\Delta t_p = \frac{t_{p,i} - t_{p,f}}{N_t - 1} \tag{8.14.4}$$

During operation of the active noise control system, the acoustic pressure $p(\boldsymbol{r}_n,t)$ at any location $\boldsymbol{r}_n$ and time t can be considered as the superposition of the primary and control source contributions as follows:

$$p(r_n,t) = p_p(r_n,t) + p_c(r_n,t) \tag{8.14.5}$$

where the subscripts p, c denote primary and control respectively. The control sources of interest here will in practice be speakers, but will be modelled as acoustic monopole sources. The time domain description of the pressure fluctuation at some distance r from an acoustic monopole source due to its operation can be found from the inhomogenous wave equation

$$\left[\nabla^2 - \frac{1}{c_0^2}\frac{\partial^2}{\partial t^2}\right] p(r,t) = -\rho_0 \frac{\partial Q(r,t)}{\partial t} \tag{8.14.6}$$

Concentrating the source strength density Q at a single location $\boldsymbol{r}_q$, such that $Q(r,t) = q(t)\delta(r - r_q)$, enables us to write the solution for the acoustic pressure at some point $\boldsymbol{r}$ resulting from the monopole source as (see (8.13.2))

$$p(r,t) = \frac{\rho_0}{4\pi r} \frac{\partial q(t - \frac{r}{c_0})}{\partial t} = \frac{\rho_0}{4\pi r} \dot{q}(t - \frac{r}{c_0}) \tag{8.14.7}$$

From (8.14.7) we can surmise that the control source operating schedule, which will define the contribution of the control source to the acoustic pressure on the intensity integration surface, will in fact be a record of the time derivative of the control source volume velocity, $\dot{q}$, as a function of time.

If there are N_c multiple, separate control sources, then (8.14.5) can be further expanded as

$$p(r_n,t) = p_p(r_p,t) + \sum_{j=1}^{N_c} p_{cj}(r_n,t) \tag{8.14.8}$$

The acoustic pressure at the set of N_n points on the intensity integration surface at time t can be written as the sum of vectors:

$$\boldsymbol{p}(t) = \boldsymbol{p}_p(t) + \sum_{j=1}^{N_c} \boldsymbol{p}_{cj}(t) \tag{8.14.9}$$

where $\boldsymbol{p}$, $\boldsymbol{p}_p$, and $\boldsymbol{p}_{cj}$ are respectively the ($N_n \times 1$) vectors of the total acoustic pressure, the primary source contributed acoustic pressure, and the jth control source contributed acoustic pressure, defined as follows:

$$p(t) = \begin{bmatrix} p(r_1,t) \\ p(r_2,t) \\ \vdots \\ p(r_n,t) \end{bmatrix}, \quad p_p(t) = \begin{bmatrix} p_p(r_1,t) \\ p_p(r_2,t) \\ \vdots \\ p_p(r_n,t) \end{bmatrix}, \quad p_{cj} = \begin{bmatrix} p_{cj}(r_1,t) \\ p_{cj}(r_2,t) \\ \vdots \\ p_{cj}(r_n,t) \end{bmatrix} \tag{8.14.10}$$

Considered over the entire time period $(t_{p,i}, t_{p,f})$, the set of total, primary, and control sound pressures can therefore be expressed in matrix form as

$$\boldsymbol{P} = \boldsymbol{P}_p + \sum_{j=1}^{N_c} \boldsymbol{P}_{cj} = \boldsymbol{P}_p + \boldsymbol{P}_c \tag{8.14.11}$$

where

$$\boldsymbol{P} = \begin{bmatrix} p(t_0) \\ p(t_1) \\ \vdots \\ p(t_{N_t-1}) \end{bmatrix}, \quad \boldsymbol{P}_p = \begin{bmatrix} p_p(t_0) \\ p_p(t_1) \\ \vdots \\ p_p(t_{N_t-1}) \end{bmatrix}, \quad \boldsymbol{P}_{cj} = \begin{bmatrix} p_{cj}(t_0) \\ p_{cj}(t_1) \\ \vdots \\ p_{cj}(t_{N_t-1}) \end{bmatrix}, \quad \boldsymbol{P}_c = \sum_{j=1}^{N_c} \boldsymbol{P}_{cj} \tag{8.14.12}$$

Using (8.14.11) we can now write the error criterion of (8.14.3) simply in matrix form as,

$$\text{minimize } \boldsymbol{P}^{\mathrm{T}}\boldsymbol{P} = \boldsymbol{P}_c^{\mathrm{T}}\boldsymbol{P}_c + \boldsymbol{P}_c^{\mathrm{T}}\boldsymbol{P}_p + \boldsymbol{P}_p^{\mathrm{T}}\boldsymbol{P}_c + \boldsymbol{P}_p^{\mathrm{T}}\boldsymbol{P}_p \tag{8.14.13}$$

To use this form of the error criterion to derive the optimum control source operating schedules, note that the (non-zero) control source operating time, like the primary source operating time, will be finite, confined within the period $(t_{c,i}, t_{c,f})$ (also note that the control source operating period $(t_{c,i}, t_{c,f})$ is not constrained to be equivalent to the primary source operating period $(t_{p,i}, t_{p,f})$ in either starting or finishing times, or in total duration). By discretising this duration into N_{tc} separate periods, the matrix $\boldsymbol{P}_c$ can be written as

$$\boldsymbol{P}_c = \boldsymbol{ZS} \tag{8.14.14}$$

where $\boldsymbol{S}$ is the $(N_c N_{tc} \times 1)$ vector of control source output schedules,

$$\boldsymbol{S} = \left[\text{control source 1 schedule} \;\vdots\; \cdots \;\vdots\; \text{control source } N_c \text{ schedule}\right]^{\mathrm{T}}$$

$$= \left[\dot{q}_{c1}(\tau_0)\;\; \dot{q}_{c1}(\tau_1)\;\; \cdots\;\; \dot{q}_{c1}(\tau_{N_{tc}-1}) \;\vdots\; \cdots \;\vdots\; \dot{q}_{cN_c}(\tau_0)\;\; \dot{q}_{cN_c}(\tau_1)\;\; \cdots\;\; \dot{q}_{cN_c}(\tau_{N_{tc}-1})\right]^{\mathrm{T}} \tag{8.14.15}$$

The quantity $\dot{q}_{cj}(\tau)$ is the time derivative of the volume velocity generated by control source j at control source operating time τ, and $\boldsymbol{Z}$ is the $(N_n N_t \times N_c N_{tc})$

matrix of space and time dependent transfer functions between these control source volume velocity derivatives, and the point locations and times on the intensity integration surface, defined as follows:

$$Z = \left[z_c 1(\tau_0) \;\; z_c 1(\tau_1) \;\cdots\; z_c 1(\tau_{N_{tc}-1}) \;\vdots\; \cdots \;\vdots\; z_{cN_c}(\tau_0) \;\; z_{cN_c}(\tau_1) \;\cdots\; z_{cN_c}(\tau_{N_{tc}-1}) \right] \tag{8.14.16}$$

where

$$z_{cj}(\tau) = \left[z_{cj,\tau}(r_1,t_0) \;\cdots\; z_{cj,\tau}(r_{N_n},t_0) \;\vdots\; \cdots \;\vdots\; z_{cj,\tau}(r_1,t_{N_t-1}) \;\cdots\; z_{cj,\tau}(r_{N_n},t_{N_t-1}) \right]^{\mathrm{T}} \tag{8.14.17}$$

and, from (8.14.7), for monopole sources,

$$z_{cj,\tau}(r_1,t) = \begin{cases} \dfrac{\rho_0}{4\pi r} & \tau = t - \dfrac{r_1}{c_0} \\ 0 & \text{otherwise} \end{cases} \tag{8.14.18}$$

Note that for each point location $\boldsymbol{r}_n$ on the intensity integration surface, there will be only one non-zero transfer function between that point at a given time t, and the jth control source output schedule in the matrix $\boldsymbol{S}$, $\dot{q}_j(\tau)$, corresponding to time $\tau = t - r/c_0$, and hence only one non-zero entry in the columns of the transfer function matrix $\boldsymbol{Z}$.

Using (8.14.14), (8.14.13) can be written in the standard quadratic form as

$$\boldsymbol{P}^{\mathrm{T}}\boldsymbol{P} = \boldsymbol{S}^{\mathrm{T}}\boldsymbol{A}\boldsymbol{S} + \boldsymbol{S}^{\mathrm{T}}\boldsymbol{b} + \boldsymbol{b}^{\mathrm{T}}\boldsymbol{S} + c \tag{8.14.19}$$

where

$$\boldsymbol{A} = \boldsymbol{Z}^{\mathrm{T}}\boldsymbol{Z}, \; \boldsymbol{b} = \boldsymbol{Z}^{\mathrm{T}}\boldsymbol{P}_p, \; c = \boldsymbol{P}_p^{\mathrm{T}}\boldsymbol{P}_p \tag{8.14.20}$$

Expressed in this form, the vector of optimum control source operating schedules, $\boldsymbol{S}_{\mathrm{opt}}$, is defined by the expression

$$\boldsymbol{S}_{\mathrm{opt}} = -\boldsymbol{A}^{-1}\boldsymbol{b} \tag{8.14.21}$$

The minimum value of the error criterion, the sum of the squared acoustic pressure amplitudes over all points and times is, therefore

$$\boldsymbol{P}^{\mathrm{T}}\boldsymbol{P} = c - \boldsymbol{b}^{\mathrm{T}}\boldsymbol{A}^{-1}\boldsymbol{b} \tag{8.14.22}$$

Note that, from (8.14.20), the term c is the value of the error criterion during primary excitation only.

This procedure for deriving the optimum control source output pressure schedules is analogous to the frequency domain methods we used previously in this chapter, and is identical to the procedures used in a multiple regression problem. The dependent variable in this case is $-\boldsymbol{P}_p$, which is what is desired to achieve the maximum levels of attenuation of the radiated acoustic power, and the independent variables are the terms in the matrix of transfer functions, **Z**. Hence the optimum control source operating schedules can be easily calculated using commercially available multiple regression software.

Note finally that the methodology developed here for deriving the optimum control source operating schedules with respect to minimizing the total radiated acoustic power can also be used to derive similar schedules for minimizing the acoustic pressure amplitude at discrete error sensing locations. To do this the set of N_n distributed measuring points on the intensity integration sphere should simply be replaced with the error sensing locations of interest.

8.14.2 Example: Control of a sinusoidal pulse from a single source

The model problem we will consider here is the control of a single sinusoidal pulse from one acoustic (primary) monopole source by the introduction of a second acoustic (control) monopole source. This model is a good starting point for a study of feedforward active control of impact acceleration noise for three reasons. First, it is a very simple case, making its implementation in the theoretical framework presented in the previous section a straightforward exercise. Second, it can be viewed as half of the acceleration noise problem, which in first approximation is the emission of single sinusoidal pulses from two impacting bodies. Third, we have already studied the active control of harmonic sound radiation from one monopole acoustic source by a second monopole acoustic source, so that our previously obtained results provide a basis against which to compare results derived here.

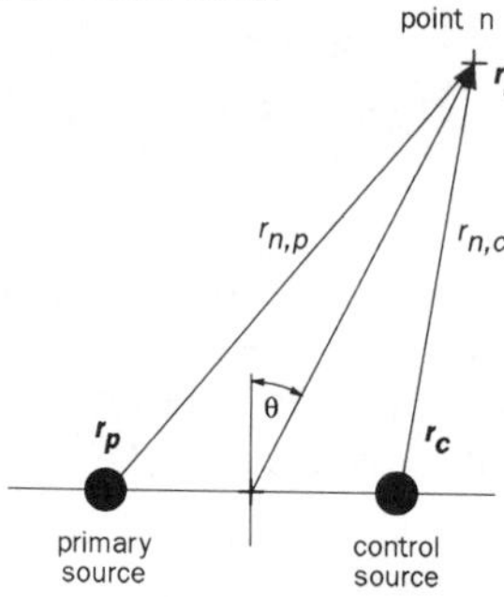

Fig. 8.59 Geometry for single primary source/single control source system.

The system to being studied in this section is illustrated in Fig. 8.59. Here a single monopole control source is being used to control the acoustic radiation from a single monopole primary source. The primary source generated acoustic pressure which will be arriving at location r_n at time t' is defined using (8.14.7) as follows:

$$p_p(r_n,t') = \frac{\rho_0}{4\pi r_{n,p}} \dot{q}_p(t-\frac{r_{n,p}}{c_0}) \tag{8.14.23}$$

where for the single sinusoidal pulse problem being considered in this section $\dot{q}_p$ is defined by

$$\dot{q}_p(t) = \begin{cases} \dot{q}_p \sin \omega t & 0<t<\dfrac{2\pi}{\omega} \\ 0 & \text{otherwise} \end{cases} \tag{8.14.24}$$

Without loss of generality $\dot{q}_p$ will be set equal to 1.0, so that our results can be viewed as normalized.

For comparison, recall from the work in Section 8.3 that when the arrangement shown in Fig. 8.59 is subject to harmonic excitation, the minimum acoustic power output is produced when the control source volume velocity relationship is

$$q_{c,\mathrm{opt}} = -q_p \ \mathrm{sinc}\ kd \tag{8.14.25}$$

where k is the acoustic wavenumber and d is the source separation distance. With this optimal value of control source volume velocity, the reduction in total radiated acoustic power from the initial state of primary source operation only was found to be

$$\Delta W = 1-\mathrm{sinc}^2 kd \tag{8.14.26}$$

The first, and possibly most important, question which must be answered is how much attenuation of the total radiated acoustic power of this single pulse can actually be achieved? Figure 8.60 illustrates the maximum levels of acoustic power reduction which can be achieved as a function of primary source/control source separation distance, plotted with the equivalent curve for the harmonic excitation case. The first point to note is that the levels of attenuation of the single pulse acoustic power fall off rapidly as the separation distance increases, conservatively following the $(1 - \mathrm{sinc}^2 kd)$ relationship of the harmonic excitation case and crossing the 10 dB level at approximately 1/10 wavelength separation. However, a perhaps more surprising result is that it is always possible to achieve slightly greater levels of acoustic power attenuation controlling a single sinusoidal pulse, using a causally constrained control signal, than would be similarly attained when controlling harmonic excitation with an unconstrained control signal.

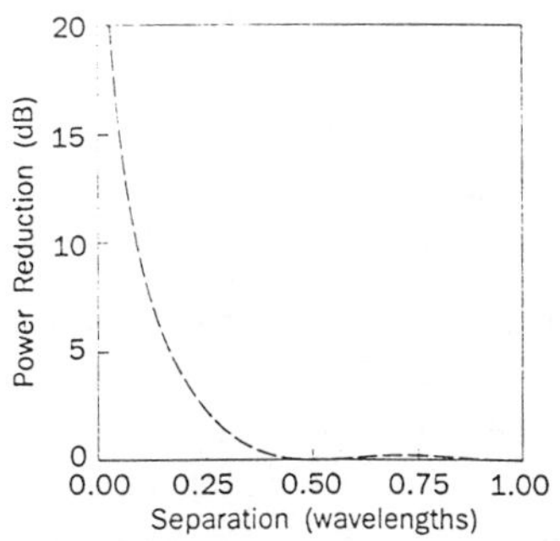

Fig. 8.60 Acoustic power attenuation as a function of source separation distance.

To explain why the performance potential has been improved, note that as with the previously studied frequency domain cases, the total acoustic power output of the system arrangement shown in Fig. 8.59 is the sum of four terms (relating to the four terms in (8.14.13)), which can be expressed as

$$W_{total} = W_c + \Delta W_c + \Delta W_p + W_p \tag{8.14.27}$$

where W_c, W_p are the acoustic power outputs of the control and primary sources respectively, which would be measured if they were operating on their own, and ΔW_c, ΔW_p are the modifications to these power outputs arising from the existence of the other source in the acoustic domain. For significant (or, even any) acoustic power attenuation to be had, the sum of these modification terms must be a negative value, as the acoustic power outputs of the sources operating on their own will always be positive values. For the arrangement being considered here, applying the analysis of the previous section (see (8.13.44)), the sum of ΔW_c and ΔW_p is equal to

$$\Delta W_c + \Delta W_p = E\left\{ \dot{q}_s(t) \frac{\rho_0}{4\pi d} q_p(t+\frac{d}{c_0}) \right\} + E\left\{ q_s(t)\, \dot{q}_p(t-\frac{d}{c_0}) \frac{\rho_0}{4\pi d} \right\} \tag{8.14.28}$$

where $E\{\}$ is the statistical expectation operator. In viewing this equation, it can be deduced that the optimum control source output at time t will be based upon two criteria: producing a pressure which will arrive at the primary source at time $(t + d/c_0)$ and be out of phase with its volume velocity, hence reducing its power output (primary source suppression), and producing a volume velocity which will be out of phase with the impinging primary source pressure, generated a time $(t - d/c_0)$ ago (self-suppression or absorption). Similar to the conclusion we derived in the common link theory derivation of Section 8.9, these two criteria represent a compromise, rather than a coalition. This can be seen more clearly by restating the second term on the right-hand side of (8.14.28) in terms of $\dot{q}_s$ to produce

$$\Delta W_c + \Delta W_p = E\left\{ \dot{q}_s(t) \frac{\rho_0}{4\pi d} q_p(t+\frac{d}{c_0}) \right\} - E\left\{ \dot{q}_s(t)\, q_p(t-\frac{d}{c_0}) \frac{\rho_0}{4\pi d} \right\} \tag{8.14.29}$$

With harmonic excitation we found in Sections 8.4 and 8.9 that this compromise leads, under optimal conditions, to a control source whose characteristic is zero acoustic power output, and a primary source with a reduced acoustic power output level. As the separation distance approaches $\lambda/2$ this compromise results in an ever decreasing control source volume velocity output, until at a separation distance of $\lambda/2$ the difference between the primary source output at $(\tau + d/c_0)$ and $(\tau - d/c_0)$ becomes zero; hence the best control output is no control output. With the impact acceleration noise problem, however, the control source is *not*

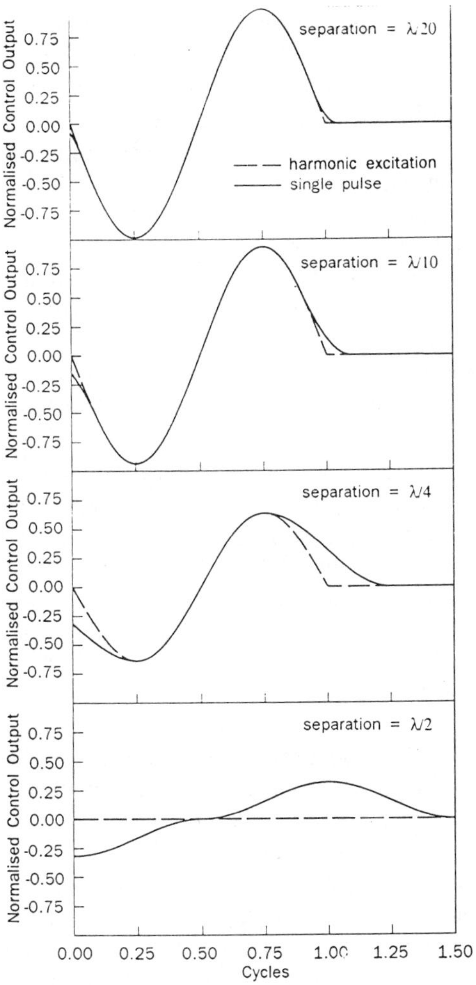

Fig. 8.61 Control source operating schedules as a function of source separation distance; solid line is for a single pulse, dashed line is for harmonic excitation.

constrained to provide a control source output which is a compromise between primary source acoustic power suppression and self-acoustic power suppression during the time periods $(0<t<d/c_0)$ and $(2\pi/\omega - d/c_0<t)$, owing to the finite operating time of the primary source; the control source is able to concentrate solely on initially suppressing the future primary source power output, and is similarly free to absorb some of the acoustic energy of trailing part of the pulse. Therefore, the improved performance is a direct result of the finite operating period of the primary noise disturbance.

This trend manifests itself in different control source schedules for the single sinusoidal pulse and harmonic excitation problems, as illustrated in Fig. 8.61. As we would expect, the two control source schedules are identical for the time period ($d/c_0 < t < 2\pi/\omega - d/c_0$). However, before and after this period the volume velocity outputs are different; they are based solely on the primary source output at time ($\tau + d/c_0$) and ($\tau - d/c_0$), respectively. It is in these periods that an increased control effect is facilitated. This is particularly apparent at a separation distance is $\lambda/2$, when the optimum control source output for the harmonic excitation problem is zero, but non-zero for the single sinusoidal pulse case. If, for example, the problem was extended to a pulse consisting of two sinusoidal cycles, the control output would follow the zero output schedule of the harmonic excitation case for the period ($d/c_0 < t < 4\pi/\omega - d/c_0$) and be non-zero outside this period.

One other point to notice in regard to the control schedules of Fig. 8.61 is the non-zero starting point of the single pulse schedules. This is clearly a practical impossibility, the result of which will be a slightly reduced level of acoustic power attenuation actually achieved. It should also be noted here that when practically implementing a system, the time delays inherent in the analogue antialiasing and reconstruction filters in a digital control system, as well as the digital (sampling) delay, further reduces the ability of the system to output a pulse 'quickly' (it takes a finite time for a reference signal to pass through the controller). Therefore, if a preview control system is not used, the practical results will be further deteriorated from the theoretical maximum levels of attenuation. We can therefore qualitatively surmise that the levels of acoustic power attenuation which can be achieved when controlling a single sinusoidal pulse from an acoustic point source are at least as high, probably slightly higher, than can be obtained when controlling harmonic excitation.

One additional point to mention here is that the above result may not hold when the primary source is essentially multipole, and a single control source is used. In these instances, the attenuation of the single pulse may be significantly less than the attenuation achieved during harmonic excitation.

Having now seen how much acoustic power attenuation is actually possible for a given control source/primary source arrangement, the next step is to consider the inclusion of an error microphone in the system to make it practically realizable, and to optimize its placement. For comparative purposes, recall that in Section 8.5 we found that for primary source/control source arrangements which are closely spaced and subject to harmonic excitation the nearest optimum error sensor placements will run on a line perpendicular to the control source/primary source axis, centred at a position between the primary and control sources such that $r_c/r_p = \text{sinc}\, kd$, which will always be closer to the control source.

8.15 FEEDBACK CONTROL OF SOUND RADIATION FROM VIBRATING STRUCTURES

Thus far in this chapter, we have considered the design of the 'physical' part of active control systems, the idea being to develop models which enable us to examine how much attenuation of the radiated acoustic field is possible for a given control source/error sensor placement. These models can then be used to optimize sensor and actuator placement for a particular problem, and to study the mechanisms behind the active attenuation. To implement an active control system using the desired transducer arrangement it has been assumed that a feedforward control system, like that outlined in Chapter 6, will be used. There are a range of active control problems where this approach to implementation is not desirable, cases where a reference signal is not (readily) available for feedforward control implementation. In these instances a feedback control strategy can be used.

Design of feedback control systems was reviewed in Chapter 5, and will receive further attention in Chapter 11 where feedback control of flexible structures will be studied. Many of the design requirements outlined in these chapters are relevant to feedback control of structural/acoustic radiation into free space. In fact, this topic can be viewed as a specific case of feedback control of flexible structures, where the optimal control state weighting matrix is chosen to reflect radiated acoustic power, or acoustic pressure in a particular direction. It may also be advantageous to transform the system states to better reflect the characteristics of structural/acoustic radiation.

In this section we will briefly consider feedback control of structural/acoustic radiation. As we will see, once developed it will be possible to apply most of the discussion in Chapter 11 to this active control problem with specific state weighting and transformation matrices. We will concern ourselves here with systems which explicitly target a reduction in radiated acoustic power, such as considered by Baumann *et al.* (1991, 1992) and Snyder *et al.* (1994b). Formulating the problem to explicitly reduce acoustic pressure at certain locations in space requires only minor modifications to the procedure to be outlined; for this problem, the reader is referred to Meirovitch and Thangjitham (1990a, 1990b) and Meirovitch (1993).

8.15.1 Derivation of structure state equations

Derivation of the state equations required for constructing a control law aimed at attenuating radiated acoustic power from a vibrating structure will be undertaken in two phases; derivation of the state equations for a damped, distributed parameter system, and modification of these equations to reflect the characteristics of acoustic power output. Derivation of the first of these, the state equations describing the motion of the distributed parameter system, will be discussed more fully in Chapter 11, but will be repeated here in brief for completeness.

The motion of a damped, flexible (distributed parameter) system can be expressed as the partial differential equation

$$m(\boldsymbol{x})\ddot{w}(\boldsymbol{x},t)+c_d\dot{w}(\boldsymbol{x},t)+\kappa w(\boldsymbol{x},t) = u(\boldsymbol{x},t) \tag{8.15.1}$$

where $w(\boldsymbol{x},t)$ is the displacement at location $\boldsymbol{x}$ on the structure at time t in response to the applied force distribution u (having units of force per unit area, or pressure). Here m is the surface density (mass per unit area), κ is a time-invariant, non-negative differential operator, and c_d is defined as

$$c_d = 2\zeta\kappa^{1/2} \tag{8.15.2}$$

where ζ is the (non-negative) damping coefficient. The eigenvalue problem associated with this differential equation is

$$\kappa\psi_n = \lambda_n m\psi_n \tag{8.15.3}$$

where λ_n is the nth eigenvalue, and ψ_n is the associated eigenfunction (mode shape function). The eigenvalues are related to the undamped natural frequencies by

$$\lambda_n = \omega_n^2 \tag{8.15.4}$$

The eigenfunctions are orthogonal, satisfying the relation

$$\int_S m(\boldsymbol{x})\psi_m(\boldsymbol{x})\psi_n(\boldsymbol{x})\ \mathrm{d}\boldsymbol{x} = \delta(m-n) \tag{8.15.5}$$

where $\delta(m - n)$ is a Kronecker delta function.

The displacement at any location $\boldsymbol{x}$ on the structure can be expressed as a sum of modal contributions as follows:

$$w(\boldsymbol{x},t) = \sum_{i=1}^{\infty} x_i(t)\psi_i(\boldsymbol{x}) \tag{8.15.6}$$

where $x_i(t)$ is the amplitude of the ith mode at time t, and $\Psi_i(\boldsymbol{x})$ is the value of the associated mode shape function at location $\boldsymbol{x}$ on the structure. If this modal expansion is substituted into the equation of motion (8.15.1), multiplying through by Ψ_n and integrating over the surface of the structure produces the following equation describing the behaviour of the nth mode:

$$\ddot{x}_n(t)+2\zeta\omega_n\dot{x}(t)+\omega_n^2 x_n(t) = f_n(t) \tag{8.15.7}$$

where $f_n(t)$ is the nth modal generalized force given by

$$f_n(t) = \int_S \psi_n(\boldsymbol{x})u(\boldsymbol{x},t) \tag{8.15.8}$$

Equation (8.15.7) can be expressed in state variable form as

$$\dot{\boldsymbol{x}}_n(t) = \boldsymbol{A}_n\boldsymbol{x}_n(t)+\boldsymbol{f}_n(t) \tag{8.15.9}$$

where

$$\boldsymbol{x}_n(t) = \begin{bmatrix} x_n(t) \\ \dot{x}_n(t) \end{bmatrix}, \quad \boldsymbol{A}_n = \begin{bmatrix} 0 & 1 \\ -\omega_n^2 & -2\zeta\omega_n \end{bmatrix}, \quad \gamma_n(t) = \begin{bmatrix} 0 \\ f_n(t) \end{bmatrix} \tag{8.15.10}$$

To calculate control forces which will minimize the radiated acoustic power, it is necessary to divide the modal generalized force into two components, one due to the primary disturbance f_p and one due to the control force input f_c:

$$f_n = f_{n,c} + f_{n,p} \tag{8.15.11}$$

where n is the mode of interest. For the control input modal generalized force it is useful to express the force distribution $u(\boldsymbol{x},t)$ as the product of some force amplitude $f(t)$ and a transfer function $z_f(\boldsymbol{x})$ to the location $\boldsymbol{x}$ (having units of force per unit area per unit force):

$$u(\boldsymbol{x},t) = f(t)z_f(\boldsymbol{x}) \tag{8.15.12}$$

such that the control modal generalized force is equal to

$$f_{n,c} = f(t)\int_S \psi_n(\boldsymbol{x})z_f(\boldsymbol{x})\ \mathrm{d}\boldsymbol{x} = f(t)\psi_{g,n} \tag{8.15.13}$$

where $\Psi_{g,n}$ is the nth modal generalized force transfer function:

$$\psi_{g,n} = \int_S \psi_n(\boldsymbol{x})z_f(\boldsymbol{x})\ \mathrm{d}\boldsymbol{x} \tag{8.15.14}$$

For the simplest control force, a point force, the modal generalized force is described by

$$f_{c,n} = f(t)\int_S \psi_n(\boldsymbol{x})\ \delta(\boldsymbol{x}-\boldsymbol{x}_f)\ \mathrm{d}\boldsymbol{x} = f(t)\psi_n(\boldsymbol{x}_f) \tag{8.15.15}$$

where $\boldsymbol{x}_f$ is the location of the point force input of the structure and $\delta(\boldsymbol{x} - \boldsymbol{x}_f)$ is a Dirac delta function. Therefore, the modal generalized force transfer function is simply equal to the value of the mode shape function at the control force application point. If there are N_c control sources then the control source modal generalized force can be written as the matrix expression

$$f_{c,n} = \boldsymbol{B}_n\boldsymbol{u}(t) \tag{8.15.16}$$

where

$$\boldsymbol{B}_n = \begin{bmatrix} 0 & 0 & \cdots & 0 \\ \psi_{gn,1} & \psi_{gn,2} & \cdots & \psi_{gn,N_c} \end{bmatrix}, \quad \boldsymbol{u}(t) = \begin{bmatrix} f_1(t) \\ f_2(t) \\ \vdots \\ f_{N_c}(t) \end{bmatrix} \tag{8.15.17}$$

Therefore, the state space equation of motion of mode n in (8.15.9) can be expressed as

$$\dot{\boldsymbol{x}}_n(t) = \boldsymbol{A}_n\boldsymbol{x}_n(t) + \boldsymbol{B}_n\boldsymbol{u}(t) + \boldsymbol{f}_{p,n} \tag{8.15.18}$$

Returning to consideration of the entire system, if there are N_m structural modes being modelled, the system equations of motion of the modelled modes can be written in state variable form as

$$\dot{x}(t) = Ax(t)+Bu(t)+f_p(t) \tag{8.15.19}$$

where

$$x(t) = \begin{bmatrix} x_1(t) \\ x_2(t) \\ \vdots \\ x_{N_m}(t) \end{bmatrix}, \quad A = \begin{bmatrix} A_1 & & & \\ & A_2 & & \\ & & \ddots & \\ & & & A_{N_m} \end{bmatrix}, \quad B = \begin{bmatrix} B_1 \\ B_2 \\ \vdots \\ B_{N_m} \end{bmatrix}, \quad f_p(t) = \begin{bmatrix} f_{p,1}(t) \\ f_{p,2}(t) \\ \vdots \\ f_{p,N_m}(t) \end{bmatrix} \tag{8.15.20}$$

Note that A is not truly a diagonal matrix, but rather a diagonal arrangement of modal matrices A_i, defined in (8.15.10).

The state equation describing the system response, (8.15.18), is the same equation we will encounter in Chapter 11 when considering the control of vibrating structures.

8.15.2 Modification of problem to consider acoustic radiation

The modal equation of motion (8.15.19) can easily be used to derive control laws aimed at attenuating the velocity levels of the structure. However, as we have seen already in this chapter this is not necessarily the best approach to take when considering the minimization of acoustic radiation. To restate the control problem in a form more amenable to calculation of a control law for this problem it is necessary to first consider the characteristics of the radiation of acoustic power from the vibrating structure. Acoustic power output for harmonic excitation can be calculated by integrating the real part of the acoustic intensity I over the surface of the vibrating structure,

$$W = \int_S \mathrm{Re}\{I(x)\}\ \mathrm{d}x = \int_S \mathrm{Re}\left\{\frac{1}{2}p(x)\dot{w}(x)^*\right\}\ \mathrm{d}x \tag{8.15.21}$$

where $\dot{w}(x)$ is the complex surface velocity at a location x on the structure, $p(x)$ is the complex acoustic pressure at this same location, and S denotes that integration is over the surface of the vibrating structure. For the time domain formulation being undertaken here, however, it is more appropriate to consider the integration of instantaneous acoustic intensity I_i, defined as

$$I_i(x,t) = \mathrm{Re}\{p(x,t)\}\ \mathrm{Re}\{\dot{w}(x,t)\} \tag{8.15.22}$$

Note that acoustic intensity I is equal to the temporal average of instantaneous acoustic intensity,

$$I(x) = \frac{1}{T}\int_0^T I_i(x,t)\ \mathrm{d}t \tag{8.15.23}$$

From this relationship it is clear that minimization of the spatial integral (over the surface of the vibrating structure) of instantaneous acoustic intensity at all points in time t will have the effect of minimizing radiated acoustic power. Therefore, the acoustic power-related error criterion J_w to be used here is this spatial integral

$$J_w(t) = \int_S \mathrm{Re}\{p(\mathbf{x},t)\}\ \mathrm{Re}\{\dot{w}(\mathbf{x},t)\}\ \mathrm{d}\mathbf{x} \tag{8.15.24}$$

To simplify the following discussion, consideration will be limited here to infinitely baffled, planar structures. The methodology used in deriving the expressions, however, can be adapted to other geometries.

If we restrict ourselves to planar structures for simplicity, the acoustic pressure $p(\mathbf{x},t)$ can be expressed in terms of surface velocity using the Rayleigh integral:

$$p(\mathbf{x},t) = \frac{j\omega\rho_0}{2\pi}\int_S \dot{w}(\mathbf{x}',t)\frac{\mathrm{e}^{-\mathrm{j}k\,|\,\mathbf{x}-\mathbf{x}'\,|}}{|\,\mathbf{x}-\mathbf{x}'\,|}\ \mathrm{d}\mathbf{x}' = \frac{j\omega\rho_0}{2\pi}\int_S \dot{w}(\mathbf{x}',t)\frac{\mathrm{e}^{-\mathrm{j}kr}}{r}\ \mathrm{d}\mathbf{x}' \tag{8.15.25}$$

where k is the acoustic wavenumber and r is the distance between the source location $\mathbf{x}'$ and the receiver location $\mathbf{x}$:

$$r = |\,\mathbf{x}-\mathbf{x}'\,| = \sqrt{(x-x')^2 + (y-y')^2} \tag{8.15.26}$$

Substituting (8.15.25) into (8.15.24) produces the following integral expression for the acoustic power error criterion:

$$J_w(t) = \frac{\rho_0\omega}{2\pi}\int_S\int_S \dot{w}(\mathbf{x}',t)\ \frac{\sin kr}{r}\ \dot{w}(\mathbf{x},t)\ \mathrm{d}\mathbf{x}'\ \mathrm{d}\mathbf{x} \tag{8.15.27}$$

Expanding the velocity in terms of modal contributions enables (8.15.27) to be written as

$$J_w(t) = \frac{\rho_0\omega}{2\pi}\int_S\int_S \left[\sum_{i=1}^{N_m}\dot{x}_i(t)\psi_i(\mathbf{x}')\right]\ \frac{\sin kr}{r}\ \left[\sum_{\iota=1}^{N_m}\dot{x}_\iota(t)\psi_\iota(\mathbf{x})\right]\ \mathrm{d}\mathbf{x}'\ \mathrm{d}\mathbf{x} \tag{8.15.28}$$

This notation can be simplified to the matrix expression

$$J_w(t) = \mathbf{v}(t)^{\mathrm{T}}\mathbf{Z}_w\mathbf{v}(t) \tag{8.15.29}$$

where $\mathbf{v}(t)$ is the $(N_m \times 1)$ vector of modal velocity amplitudes

$$\mathbf{v}(t) = \begin{bmatrix}\dot{x}_1(t)\\ \dot{x}_2(t)\\ \vdots\\ \dot{x}_{N_m}(t)\end{bmatrix} \tag{8.15.30}$$

and $\mathbf{Z}_w$ is a 'power transfer matrix', which we have discussed previously in this chapter, the terms of which are defined by

$$Z_w(i,\iota) = \frac{\rho_0\omega}{2\pi}\int_S\int_S \psi_i(\mathbf{x}')\ \frac{\sin kr}{r}\ \psi_\iota(\mathbf{x})^{\mathrm{T}}\ \mathrm{d}\mathbf{x}'\ \mathrm{d}\mathbf{x} \tag{8.15.31}$$

One point to note here is that the same analysis can be undertaken using modal displacements in place of modal velocities, the end result having the same form as (8.15.29) with the displacement power transfer matrix related to the velocity power transfer matrix by

$$\mathbf{Z}_{w,\text{ displacement}} = \omega^2 \mathbf{Z}_{w,\text{velocity}} \tag{8.15.32}$$

As we observed in Section 8.11 in our discussion of the differences between the control of structural vibration, and the resultant acoustic radiation, the power transfer matrix $\mathbf{Z}_w$ will not, in general, be diagonal, as structural modes do not radiate independently from each other. It may therefore not be appropriate to design a control law which aims to explicitly minimize modal amplitudes. This conclusion covers both unweighted and weighted summations, such as where the structural modes with the greatest radiation efficiency are preferentially treated.

The relationship in (8.15.29), as well as the similar relationship relating displacement to acoustic radiation through (8.15.32), will enable us to state an optimal control problem where the error criterion is explicitly a measure of radiated acoustic power. It should be noted that it is possible to develop the same form of equations to describe acoustic radiation to particular points in space, and to develop control laws to minimise this measure of acoustic radiation (Meirovitch and Thangjitham, 1990a, 1990b; Meirovitch, 1993). This problem is a straightforward modification of the one above and will not be explicitly studied here.

8.15.3 Problem statement in terms of transformed modes

To aid in the formulation of the control law it will be useful to decouple equation (8.15.29), so that the acoustic power error criterion can be expressed as the (weighted) sum of squared contributions. To do this we adopt the approach outlined in Section 8.11, noting that the power transfer matrix $\mathbf{Z}_w$ is real symmetric and so can be diagonalized by the orthonormal transformation

$$\mathbf{Z}_w = \boldsymbol{T}\boldsymbol{\Lambda}_w\boldsymbol{T}^{-1} \tag{8.15.33}$$

where $\boldsymbol{T}$ is the orthonormal transform matrix, the columns of which are the eigenvectors of the power transfer matrix $\mathbf{Z}_w$, and $\boldsymbol{\Lambda}_w$ is the diagonal matrix of associated eigenvalues, where the subscript w is used to denote eigenvalues associated with the acoustic power problem as opposed to eigenvalues associated with the vibration problem. Note that the transformation in (8.15.33) will produce the same eigenvectors for both the displacement and velocity formulation of the power transfer matrix, but different eigenvalues related by

$$\boldsymbol{\Lambda}_{w,\text{displacement}} = \omega^2 \boldsymbol{\Lambda}_{w,\text{velocity}} \tag{8.15.34}$$

The form of $\mathbf{Z}_w$ dictates that the eigenvectors in $\boldsymbol{T}$ will be orthogonal. If the transformation of (8.15.33) is substituted into (8.15.29) the acoustic power error criterion expression becomes

$$J_w = \boldsymbol{v}(t)^{\mathrm{T}}\mathbf{Z}_w\boldsymbol{v}(t) = \boldsymbol{v}(t)^{\mathrm{T}}\boldsymbol{T}\boldsymbol{\Lambda}_w\boldsymbol{T}^{-1}\boldsymbol{v}(t) = \boldsymbol{v}'(t)^{\mathrm{T}}\boldsymbol{\Lambda}_w\boldsymbol{v}'(t) \tag{8.15.35}$$

where the vector of transformed modal velocities $\boldsymbol{v}'$ is defined by the expression

$$\boldsymbol{v}'(t) = \boldsymbol{T}^{-1}\boldsymbol{v}(t) \tag{8.15.36}$$

and use has been made of the property of the orthonormal transform matrix

$$\boldsymbol{T}^{-1} = \boldsymbol{T}^{\mathrm{T}} \tag{8.15.37}$$

Observe that the orthonormal transformation has decoupled the global error criterion expression, so that it is now equal to a weighted summation of the form

$$J_w(t) = \sum_{i=1}^{N_m} \lambda_{wi}\left(\dot{x}_i'(t)\right)^2 \tag{8.15.38}$$

where λ_{wi} is the ith eigenvalue of $\boldsymbol{Z}_w$, $v'_i(t)$ is the ith transformed modal velocity of the vector $\boldsymbol{v}'(t)$, and the summation is over the N_m structural modes being modelled. It is also straightforward to show that a similar relationship can be developed using transformed modal displacements, with the related error criterion defined by the expression

$$J_{w,\text{displacement}}(t) = \sum_{i=1}^{N_m} \omega^2\lambda_{wi}\left(x_i'(t)\right)^2 \tag{8.15.39}$$

with the transformed modal displacement related to modal displacement in the same way that the transformed modal velocity is related to modal velocity, as defined in (8.15.36).

The orthonormal transformation matrix of (8.15.33) can be used to cast the state space modal equations of motion (8.15.19) into a form which will facilitate calculation of the control law for minimizing radiated acoustic power. The only problem is that the dimension of $\boldsymbol{x}$ is $(2N_m \times 1)$, as opposed the $(N_m \times 1)$ dimension of the orthonormal transformation matrix $\boldsymbol{T}$. Observe, however, that an expanded orthonormal transformation matrix $\boldsymbol{T}_e$ can be constructed as

$$\boldsymbol{T}_e = \begin{bmatrix} T(1,1) & 0 & T(1,2) & 0 & T(1,3) & 0 & \cdots \\ 0 & T(1,1) & 0 & T(1,2) & 0 & T(1,3) & \cdots \\ T(2,1) & 0 & T(2,2) & 0 & T(2,3) & 0 & \cdots \\ 0 & T(2,1) & 0 & T(2,2) & 0 & T(2,3) & \cdots \\ T(3,1) & 0 & T(3,2) & 0 & T(3,3) & 0 & \cdots \\ 0 & T(3,1) & 0 & T(3,2) & 0 & T(3,3) & \cdots \\ & & & \vdots & & & \end{bmatrix} \tag{8.15.40}$$

which will still have the orthonormal property $\boldsymbol{T}_e^{-1} = \boldsymbol{T}_e^{\mathrm{T}}$. This expanded orthonormal matrix can be applied to (8.15.19) as

$$\boldsymbol{T}_e^{-1}\dot{\boldsymbol{x}}(t) = \boldsymbol{T}_e^{-1}\boldsymbol{A}\boldsymbol{T}_e\boldsymbol{T}_e^{-1}\boldsymbol{x}(t) + \boldsymbol{T}_e^{-1}\boldsymbol{B}\boldsymbol{u}(t) + \boldsymbol{T}_e^{-1}\boldsymbol{f}_p(t) \tag{8.15.41}$$

or

$$\dot{\boldsymbol{x}}'(t) = \boldsymbol{A}'\boldsymbol{x}'(t) + \boldsymbol{B}'\boldsymbol{u}(t) + \boldsymbol{f}_p'(t) \tag{8.15.42}$$

where the transformed quantities, denoted by $'$, are defined by the relationships

$$x'(t) = T_e^{-1}x(t), \quad A' = T_e^{-1}AT_e, \quad B' = T_e^{-1}B, \quad f_p' = T_e^{-1}f_p \tag{8.15.43}$$

Observe now that with transformed modal velocities and displacements as the system states, the acoustic power output of the system will be minimized by minimizing the error criterion

$$J_w = x'^{\mathrm{T}}(t)Q_w(\omega)x'(t) \tag{8.15.44}$$

where $Q_w(\omega)$ is the $(2N_m \times 2N_m)$ diagonal matrix whose terms are defined by

$$Q_w(\omega) = \begin{bmatrix} \omega^2\lambda_1 & 0 & 0 & 0 & \cdots \\ 0 & \lambda_1 & 0 & 0 & \cdots \\ 0 & 0 & \omega^2\lambda_2 & 0 & \cdots \\ 0 & 0 & 0 & \lambda_2 & \cdots \\ & & \vdots & & \end{bmatrix} \tag{8.15.45}$$

The problem of suppressing radiated acoustic power from the vibrating structure can now be expressed as the optimal control problem

$$J = \int_0^\infty \left(x'^{\mathrm{T}}(t)Q_w(\omega)x'(t) + u^{\mathrm{T}}(t)Ru(t)\right) \mathrm{d}t \tag{8.15.46}$$

Note here that the weighting matrix $Q_w(\omega)$ in the optimal control error criterion is frequency dependent.

As we saw in the previous section, the state transformations of this section are not strictly required to formulate the optimal control problem in terms of minimising radiated acoustic power. From (8.15.29) and (8.15.32) it is easily shown that the optimal control error criterion could be re-expressed in terms of untransformed states, structural modal displacements and velocities, as

$$J = \int_0^\infty \left(x^{\mathrm{T}}(t)Q(\omega)x(t) + u^{\mathrm{T}}(t)Ru(t)\right) \mathrm{d}t \tag{8.15.47}$$

where $Q(\omega)$ is defined as

$$Q = \begin{bmatrix} \omega^2 Z_w(1,1) & 0 & \omega^2 Z_w(1,2) & 0 & \omega^2 Z_w(1,3) & 0 & \cdots \\ 0 & Z_w(1,1) & 0 & Z_w(1,2) & 0 & Z_w(1,3) & \cdots \\ \omega^2 Z_w(2,1) & 0 & \omega^2 Z_w(2,2) & 0 & \omega^2 Z_w(2,3) & 0 & \cdots \\ 0 & Z_w(2,1) & 0 & Z_w(2,2) & 0 & Z_w(2,3) & \cdots \\ \omega^2 Z_w(3,1) & 0 & \omega^2 Z_w(3,2) & 0 & \omega^2 Z_w(3,3) & 0 & \cdots \\ 0 & Z_w(3,1) & 0 & Z_w(3,2) & 0 & Z_w(3,3) & \cdots \\ & & & \vdots & & & \end{bmatrix} \tag{8.15.48}$$

The question then arises of what advantage there is in expressing the problem in terms of transformed states. In addition to the computational benefits which arise from problem decoupling, the chief advantage is a simplified means of model reduction. As we discussed in Section 8.11, at low frequencies only a few transformed modes need to be viewed as 'important', enabling a significantly reduced order model of the problem to be used. The transformed modes are described by orthogonal combinations of structural modes, and could therefore be measured using a set of modal filters (Meirovitch and Baruh, 1983; Morgan, 1992; see also Chapter 11) to resolve the structural modes, with these values combined in accordance with the eigenvectors of the power transfer matrix $\mathbf{Z}_w$ to produce the transformed modes. This approach only partially takes advantage of the benefit of formulating the problem in terms of transformed modes, as while the number of transformed modes required to model the problem is low, the number of structural modes required to accurately construct the transformed mode is somewhat higher. If custom sensors, such as piezoelectric polymer films, were used to measure the transformed modes, however, only a few measurements would be required. Experimental implementation of this approach can be found in Snyder *et al.* (1994b).

Finally, as we have noted, the state weighting matrix which is required for the optimal control problem to explicitly consider acoustic power minimization is frequency dependent. Perhaps the most straightforward way to account for this frequency dependence is to augment the state vector with additional variables the response of which is related to the frequency dependence of the error criterion. This is described in more detail by Baumann *et al.* (1991, 1992). A general discussion of the problem of frequency-weighted error criteria in optimal control problems is discussed by Gupta (1980), Anderson and Mingori (1985), Moore and Mingori (1987), and Anderson and Moore (1990).

REFERENCES

Akay, A. (1978). A review of impact noise. *Journal of the Acoustical Society of America*, **64**, 977–987.

Anderson, B.D.O. and Mingori, D.L. (1985). Use of frequency dependence in linear quadratic control problems to frequency shape robustness. *Journal of Guidance, Control, and Dynamics*, **8**, 397–401.

Anderson, B.D.O. and Moore, J.B. (1990). *Optimal Control, Linear Quadratic Methods*. Prentice Hall: Englewood Cliffs, NJ.

Angevine, O.L. (1981). Active acoustic attenuation of electric transformer noise. *Proceedings of InterNoise '81*, 303–306.

Angevine, O.L. (1992). Active cancellation of the hum of large electric transformers. In *Proceedings of Internoise '92*, 313–316.

Angevine, O.L. (1993). Active control of hum from large power transformers: the real world. In *Proceedings of the Second conference on Recent Advances in Active Control of Sound and Vibration*, Virginia Tech., Blacksburg, VA, 279–290.

Angevine, O.L. (1995). The prediction of transformer noise. *Sound and Vibration*, October, 16–18.

Angevine, O.L. (1995). Active systems for attenuation of noise. *International Journal of Active Control*, **1**, 65–78.

Baumann, W.T., Saunders, W.R. and Robertshaw, H.H. (1991). Active suppression of acoustic radiation from impulsively excited structures. *Journal of the Acoustical Society of America*, **90**, 3202–3208.

Baumann, W.T., Ho, F.S. and Robertshaw, H.H. (1992). Active structural acoustic control of broadband disturbances. *Journal of the Acoustical Society of America*, **92**, 1998–2005.

Bendat, J.S. and Piersol, A.G. (1986). *Random Data*. Wiley: New York.

Beranek, L.L. (1986). *Acoustics*. Acoustical Society of America: New York.

Berge, T., Pettersen, O.K.O. and Sorsdal, S. (1987). Active noise cancellation of transformer noise. *Proceedings of InterNoise '87*, 537–540.

Berge, T., Pettersen, O.K.O. and Sorsdal, S. (1988). Active cancellation of transformer noise: field measurements. *Applied Acoustics*, **23**, 309–320.

Borgiotti, G.V. (1990). The power radiated by a vibrating body in an acoustic fluid and its determination from boundary measurements. *Journal of the Acoustical Society of America*, **88**, 1884–1893.

Burdisso, R.A. and Fuller, C.R. (1991). Eigenproperties of feedforward controlled flexible structures. *Journal of Intelligent Material Systems and Structures*, **2**, 494–507.

Burdisso, R.A. and Fuller, C.R. (1992). Theory of feedforward controlled system eigenproperties. *Journal of Sound and Vibration*, **153**, 437–451 (also, comments and reply, *Journal of Sound and Vibration*, **163**, 363–371).

Burgemeister, K.A. and Hansen, C.H. (1993). Use of a secondary perferated panel to actively control the sound radiated by a heavy structure. *Proceedings of ASME, Winter Annual Meeting*, 93–WA/NCA–2, New Orleans, LA.

Clark, R.L. and Fuller, C.R. (1991). Control of sound radiation with adaptive structures. *Journal of Intelligent Material Systems and Structures*, **2**, 431–452.

Clark, R.L. and Fuller, C.R. (1992a). Active structural acoustic control with adaptive structures including wavenumber considerations. *Journal of Intelligent Material Systems and Structures*, **3**, 296–315.

Clark, R.L. and Fuller, C.R. (1992b). Modal sensing of efficient acoustic radiators with PVDF distributed sensors in active structural acoustic approaches. *Journal of the Acoustical Society of America*, **91**, 3321–3329.

Clark. R.L. and Fuller, C.R. (1992c). Experiments on active control of structurally radiated sound using multiple piezoceramic actuators. *Journal of the Acoustical Society of America*, **91**, 3313–3320.

Clark, R.L., Fuller, C.R. and Wicks, A.L. (1991). Characterization of multiple piezoceramic actuators for structural excitation. *Journal of the Acoustical Society of America*, **90**, 346–357.

Collins, S.A., Padilla, C.E., Notestine, R.J., von Flotow, A.H., Schmitz, E. and Ramey, M. (1992). Design, manufacture, and application to space robots of distributed piezoelectric film sensors. *Journal of Guidance, Control, and Dynamics*, **15**, 396–403.

Conover, W.B. (1956). Fighting noise with noise. *Noise Control*, **2**, 78–82.

Cunefare, K.A. (1991). The minimum multi-modal radiation efficiency of baffled finite beams. *Journal of the Acoustical Society of America*, **90**, 2521–2529.

Cunefare, K.A. and Shepard, S. (1993). The active control of point acoustic sources in a half-space. *Journal of the Acoustical Society of America*, **93**, 2732–2739.

Davies, H.G. (1971). Low frequency random excitation of water-loaded rectangular plates. *Journal of Sound and Vibration*, **15**, 107–126.

Deffayet, C. and Nelson, P.A. (1988). Active control of low-frequency harmonic sound radiated by a finite panel. *Journal of the Acoustical Society of America*, **84**, 2192–2199.

Elliott, S.J. and Johnson, M.E. (1992). *A Note on the Minimisation of Radiated Sound Power*. ISVR Tecnical Note 1992.

Elliott, S.J. and Johnson, M.E. (1993). Radiation modes and the active control of sound power. *Journal of the Acoustical Society of America*, **94**, 2194–2204.

Elliott, S.J. and Nelson, P.A. (1986). The implications of causality in active control. *Proceedings of InterNoise '86*, 583–588.

Elliott, S.J., Joseph, P., Nelson, P.A. and Johnson, M.E. (1991). Power output minimisation and power absorption in the active control of sound. *Journal of the Acoustical Society of America*, **90**, 2501–2512.

Fahy, F. (1985). *Sound and Structural Vibration: Radiation, Transmission, and Response*. Academic Press: London.

Ffowcs Williams, J.E. (1984). Anti-sound. *Proceedings of the Royal Society of London, Series A*, **395**, 63–88.

Fuller, C.R. (1988). Analysis of active control of sound radiation from elastic plates by force inputs. *Proceedings of InterNoise '88*, 1061–1064.

Fuller, C.R. (1990). Active control of sound transmission/radiation from elastic plates by vibration inputs. I. Analysis. *Journal of Sound and Vibration*, **136**, 1–15.

Fuller, C.R. and Burdisso, R.A. (1991). A wavenumber domain approach to the active control of sound and vibration. *Journal of Sound and Vibration*, **148**, 355–360.

Fuller, C.R., Hansen, C.H. and Snyder, S.D. (1989). Active control of structurally radiated noise using piezoceramic actuators. *Proceedings of InterNoise '89*, 509–511.

Fuller, C.R., Hansen, C.H. and Snyder, S.D. (1991a). Experiments on active control of sound radiation from a panel using a piezo ceramic actuator. *Journal of Sound and Vibration*, **150**, 179–190.

Fuller, C.R., Hansen, C.H. and Snyder, S.D. (1991b). Active control of sound radiation from a vibrating panel by sound sources and vibration inputs: An experimental comparison. *Journal of Sound and Vibration*, **145**, 195–216.

Gu, Y. and Fuller, C.R. (1991). Active control of sound radiation due to subsonic wave scattering from discontinuities on fluid loaded plates. I. Far-field pressure. *Journal of the Acoustical Society of America*, **90**, 2020–2026.

Gu, Y. and Fuller, C.R. (1993). Active control of sound radiation from a fluid-loaded rectangular uniform plate. *Journal of the Acoustical Society of America*, **93**, 337–345.

Gupta, N.K. (1980). Frequency-shaped loop functionals: Extensions of linear-quadratic-gaussian design methods. *Journal of Guidance and Control*, **3**, 529–535.

Hansen, C.H. and Snyder, S.D. (1991). Effect of geometric and structural/acoustic variables on the active control of sound radiation from a vibrating surface. *Proceedings of Recent Advances in Active Control of Sound and Vibration*, 487–506.

Hansen, C.H., Snyder, S.D. and Fuller, C.R. (1989). Reduction of noise radiated by a vibrating rectangular panel by active sound sources and active vibration sources: a comparison. *Proceedings of Noise and Vibration '89*, Singapore, 16–18 August, E50–E51.

Hesselmann, N. (1978). Investigation of noise reduction on a 100 kVA transformer tank by means of active methods. *Applied Acoustics*, **11**, 27–34.

Junger, M.C. and Feit, D. (1986). *Sound, Structures, and Their Interaction*. MIT Press: Cambridge, MA.

Keltie, R.F. and Peng, H. (1987). Acoustic power radiated from point-forced thin elastic plates. *Journal of Sound and Vibration*, **112**, 45–52.

Kempton, A.J. (1976). The ambiguity of acoustic sources: a possibility for active control? *Journal of Sound and Vibration*, **48**, 475–483.

Knyazev, A.S. and Tartakovskii, B.D. (1967). Abatement of radiation from flexurally vibrating plates by means of active local vibration dampers. *Soviet Physics Acoustics*, **13**, 115–116.

Koshigoe, S. and Murdock, J.W. (1993). A unified analysis of both active and passive damping for a plate with piezoelectric transducers. *Journal of the Acoustical Society of America*, **93**, 346–355.

Koss, L.L. and Alfredson, R.J. (1973). Transient sound radiated by spheres undergoing an elastic collision. *Journal of Sound and Vibration*, **27**, 59–75.

Koss, L.L. (1974). Transient sound from colliding spheres: normalised results. *Journal of Sound and Vibration*, **36**, 541–553.

Laning, J.H. and Battin, R.H. (1956). *Random Processes in Automatic Control*. McGraw–Hill: New York.

Lee, C.K. and Moon, F.C. (1990). Modal sensors/actuators. *Journal of Applied Mechanics*, **57**, 396–403.

Levine, H. (1980a). On source radiation. *Journal of the Acoustical Society of America*, **68**, 1199–1205.

Levine, H. (1980b). A note on sound radiation from distributed sources. *Journal of Sound and Vibration*, **68**, 203–207.

Levine, H. (1980c). Output of acoustical sources. *Journal of the Acoustical Society of America*, **67**, 1935–1946.

Lomas, N.S. and Hayek, S.I. (1977). Vibration and acoustic radiation of elastically supported rectangular plates. *Journal of Sound and Vibration*, **52**, 1–25.

McLoughlin, M., Hildebrand, S. and Hu, Z. (1994). A novel active transformer quietening system. In *Proceedings of Internoise '94.*

Meirovitch, L. and Baruh, H. (1983). On the problem of observation spillover in self-adjoint distributed-parameter systems. *Journal of Optimization Theory and Applications*, **39**, 269–291.

Meirovitch, L. and Thangjitham, S. (1990a). Control of sound radiation from submerged plates. *Journal of the Acoustical Society of America*, **88**, 402–407.

Meirovitch, L. and Thangjitham, S. (1990b). Active control of sound radiation pressure. *Journal of Vibration and Acoustics*, **112**, 237–244.

Metcalf, V.L., Fuller, C.R., Silcox, R.J. and Brown, D.E. (1992). Active control of sound transmission/radiation from elastic plates by vibrational inputs. II. Experiments. *Journal of Sound and Vibration*, **153**, 387–402.

Moore, J.B. and Mingori, D.L. (1987). Robust frequency-shaped LQ control. *Automatica*, **23**, 641–646.

Morgan, D.R. (1991). An adaptive modal-based active control system. *Journal of the Acoustical Society of America*, **89**, 248–256.

Naghshineh, K. and Koopmann, G.H. (1992). A design method for achieving weak radiator structures using active vibration control. *Journal of the Acoustical Society of America*, **92**, 856–870.

Naghshineh, K. and Koopmann, G.H. (1993). Active control of sound power using acoustic basis functions as surface velocity filters. *Journal of the Acoustical Society of America*, **93**, 2740–2752.

Naghshineh, K., Koopmann, G.H. and Belegundu, A.D. (1992). Material tailoring of structures to achieve a minimum radiation condition. *Journal of the Acoustical Society of America*, **92**, 841–855.

Nelson, P.A. and Elliott, S.J. (1986). The minimum power output of a pair of free field monopole sources. *Journal of Sound and Vibration*, **105**, 173–178.

Nelson, P.A. and Elliott, S.J. (1992). *Active Control of Sound*. Academic Press: London.

Nelson, P.A., Curtis, A.R.D. and Elliott, S.J. (1985). Quadratic optimization problems in the active control of free and enclosed sound fields. *Proceedings of the Institute of Acoustics*, **7**, 45–53.

Nelson, P.A., Curtis, A.R.D. and Elliott, S.J. (1986). On the active absorption of sound. *Proceedings of InterNoise '86*, 601–606.

Nelson, P.A., Curtis, A.R.D., Elliott, S.J. and Bullmore, A.J. (1987). The minimum power output of free field point sources and the active control of sound. *Journal of Sound and Vibration*, **116**, 397–414.

Nelson, P.A., Hammond, J.K., Joseph, P. and Elliott, S.J. (1988). *The Calculation of Causally Constrained Optima in the Active Control of Sound*. ISVR Technical Report 147.

Nelson, P.A., Hammond, J.K., Joseph, P. and Elliott, S.J. (1990). Active control of stationary random sound fields. *Journal of the Acoustical Society of America*, **87**, 963–975.

Osman, A., Sadek, M. and Knight, W. (1974). Noise and vibration analysis of an impact forming machine. *Journal of Engineering in Industry*, **February**, 233–240.

Pan, J., Snyder, S.D., Hansen, C.H. and Fuller, C.R. (1992). Active control of far field sound radiated by a rectangular panel: a general analysis. *Journal of the Acoustical Society of America*, **91**, 2056–2066.

Pope, L.D. and Leibowitz, R.C. (1974). Intermodal coupling coefficients for a fluid-loaded rectangular plate. *Journal of the Acoustical Society of America*, **56**, 408–415.

Richards, E.J., Westcott, M.E. and Jeyapalan, R.K. (1979). Prediction of impact noise, I. Acceleration noise. *Journal of Sound and Vibration*, **62**, 547–575.

Ross, C.F. (1978). Experiments on the active control of transformer noise. *Journal of Sound and Vibration*, **61**, 473–480.

Sandman, B.E. (1977). Fluid-loaded vibration of an elastic plate carrying a concentrated mass. *Journal of the Acoustical Society of America*, **61**, 1502–1510.

Snyder, S.D. (1991). *A Fundamental Study of Active Noise Control System Design*. PhD Thesis, University of Adelaide.

Snyder, S.D and Hansen, C.H. (1991a). Mechanisms of active noise control using vibration sources. *Journal of Sound and Vibration*, **147**, 519–525.

Snyder, S.D. and Hansen, C.H. (1991b). Using multiple regression to optimise active noise control system design. *Journal of Sound and Vibration*, **148**, 537–532.

Snyder, S.D. and Tanaka, N. (1993a). To absorb or not to absorb: Control source power output in active noise control systems. *Journal of the Acoustical Society of America*, **94**, 185–195.

Snyder, S.D. and Tanaka, N. (1993b). On feedforward active control of sound and vibration using vibration error signals. *Journal of the Acoustical Society of America*, **94**, 2181–2193.

Snyder, S.D. and Tanaka, N. (1995). Calculating total acoustic power output using modal radiation efficiencies. *Journal of the Acoustical Society of America*, **97**, 1702–1709.

Snyder, S.D., Tanaka, N. and Hansen, C.H. (1993). Shaped vibration sensors for feedforward control of structural radiation. *Proceedings of Recent Advances in Active Control of Sound and Vibration*, 177–188.

Snyder, S.D., Tanaka, N. and Kikushima, Y. (1995). The use of optimally shaped piezo-electric film sensors in the active control of free field structural radiation, part 2: Feedback control. To be published in *ASME Journal of Vibration and Acoustics*.

Snyder, S.D., Tanaka, N. and Kikushima, Y. (1995). The use of optimally shaped piezo-electric film sensors in the active control of free field structural radiation, part 1: Feedforward control. *ASME Journal of Vibration and Acoustics*, **17**, 311–322.

Thomas, D.R., Nelson, P.A. and Elliott, S.J. (1990). Experiments on the active control of the transmission of sound through a clamped rectangular plate. *Journal of Sound and Vibration*, **139**, 351–355.

Thi, J., Unver, E. and Zuniga, M. (1991). Comparison of design approaches in sound radiation suppression. *Proceedings of Recent Advances in Active Control of Sound and Vibration*, Virginia Polytechnic Institute and State University, Blacksburg, VA, 534–551.

Tokhi, M.O. and Leitch, R.R. (1991a). Design of active noise control systems operating in three-dimensional non-dispersive propagation medium. *Noise Control Engineering Journal*, **36**, 41–53.

Tokhi, M.O. and Leitch, R.R. (1991b). The robust design of active noise control systems based on relative stability measures. *Journal of the Acoustical Society of America*, **90**, 334–345.

Walker, L.A. (1976). Characteristics of an active feedback system for the control of plate vibrations. *Journal of Sound and Vibration*, **46**, 157–176.

Wallace, C.F. (1972). Radiation resistance of a rectangular plate. *Journal of the Acoustical Society of America*, **51**, 946–952.

Wang, B.T. and Fuller, C.R. (1992). Near-field pressure, intensity, and wave-number distributions for active structural acoustic control of plate radiation: Theoretical analysis. *Journal of the Acoustical Society of America*, **92**, 1489–1498.

Wang, B.T. Dimitradis, E.K., and Fuller, C.R. (1990). Active control of structurally radiated noise using multiple piezo electric actuators. *Proceedings of AIAA SDM Conference AIAA paper* **90–1172–CP**, 2409–2416.

Williams, E.G. (1983). A series expansion of the acoustic power radiated from planar sources. *Journal of the Acoustical Society of America*, **73**, 1520–1524.

9

Active control of enclosed sound fields

9.1 INTRODUCTION

In this chapter we will examine a third 'general group' of active noise control problems, that of controlling sound fields in enclosed spaces. This group of problems includes many commonly encountered systems, such as automobile interiors, aircraft cabins, and rooms in houses. Initial consideration of active control of enclosed sound fields was undertaken in the 1950s, with the aim being to achieve both global and local sound attenuation using an 'electronic sound absorber' (Olson and May, 1953; Olson, 1956). This device was basically a single speaker and microphone placed in close proximity of each other, with a feedback control system which could drive the speaker to either reduce the pressure at the microphone or absorb part of the incident acoustic field. Little became of this device at its time of inception, possibly due to limitations in electronics technology.

Over the next 30 years there appears to have been little work done on the active attenuation of enclosed sound fields. There was, however, research conducted on the related area of 'assisted resonance' in concert halls (Parkin and Morgan, 1965, 1970, 1971). This is basically active noise control where the aim is not to null the sound field, but rather to modify the reverberation characteristics of the enclosed space. Once again, this technology does not seem to have found widespread use.

Since the early 1980s interest in the active control of enclosed sound fields has increased dramatically, owing in part to parallel advances in microprocessor technology which makes possible the implementation of such systems. Passenger vehicles in particular have become targets of active noise control research.

Advances in engineering materials have led to an increase in the strength to weight ratio, particularly in aircraft and automobiles. It has also led to the development of more fuel efficient, yet louder (in the case of modern turbo-prop engines (Magliozzi, 1984)), propulsion systems. Passengers do not, however, expect modern advances to result in any decrease in creature comfort; on the contrary, we expect to be pampered with ever-increasing comfort. This has led to a problem for structural acousticians, as their old ally, mass, is being eliminated before their very eyes. Active noise control is viewed by many as a possible means to control low frequency noise without substantially increasing the mass of these sleek, modern carriers. Recently, there have been several experimental studies on implementations in aircraft (Zalas and Tichy, 1984; Dorling *et al.*, 1989; Elliott *et al.*, 1989a, 1990; Salikuddin and Ahuja, 1989; Simpson *et al.*, 1989, 1991), automobiles (Oswald, 1984; Elliott *et al.*, 1988a), and in tractor cabins (Nadim and Smith, 1983). In Japan it is possible to now purchase an automobile with an active control system installed.

Simply introducing a few transducers connected to an electronic controller into an enclosed sound field will not, however, guarantee satisfactory results, even if it is physically possible to substantially attenuate a given enclosed sound field. The work presented in this section is aimed at developing methods of analysis which will assist us in optimally implementing active control systems for this class of problem. We will concentrate on developing methods of calculating the maximum levels of attenuation which are possible for a given control source arrangement, predicting the levels of attenuation which will be achieved by minimizing the disturbance at discrete error sensing locations, examining the mechanisms by which active control provides attenuation, and developing a qualitative feeling for how effective active noise control will be for a given problem. The first part of the chapter will concentrate on feedforward control of low frequency harmonic excitation, first for sound sources in rigid enclosures and then for sound transmission into coupled enclosures. As in the previous section, the analysis will be directed at obtaining a set of governing equations in the form of matrix expressions, which can be manipulated simply to provide optimum control source outputs for a variety of error criteria using quadratic optimization techniques (see, for example, Piraux and Nayroles, 1980; Nelson *et al.*, 1985, 1987). These models will then be used to examine control mechanisms and the effect of source/sensor arrangements for feedforward control implementations. We will then consider control at higher frequencies, where the modal density in the enclosed space is high. Finally, there will be a limited discussion relating to the application of feedback control to the problem.

9.2 CONTROL OF HARMONIC SOUND FIELDS IN RIGID ENCLOSURES AT DISCRETE LOCATIONS

A good starting point for study of the control of enclosed sound fields is the simplest problem, the active control of sound fields in rigid walled enclosures at discrete locations. By specifying ‘rigid’ walls, all of the enclosure boundaries are assumed to be locally reacting, meaning that wave motion in the enclosing structure is not possible and therefore the enclosed sound field is entirely the result of sound sources contained within it or on its boundaries. (It should be noted that even in very thick walled enclosures such as reverberation rooms this assumption may not be strictly valid, especially at low frequencies (Pan and Bies, 1990a, 1990b), but is still commonly employed, yielding reasonably accurate results.) The aim of this section is to develop relationships for the volume velocities of acoustic control sources which will result in the minimization of acoustic pressure at discrete locations in the enclosed space, locations which may be potential error sensing points. As in the previous two chapters, the approach we will take is to formulate the governing equations which describe the enclosed sound field as matrix expressions, then describe the error criterion in terms of these matrices, which will greatly simplify the task of calculating the control source output. In the section following this the global minimization of the rigidly enclosed sound field, and means of obtaining the best possible result for a given control source arrangement, will be discussed.

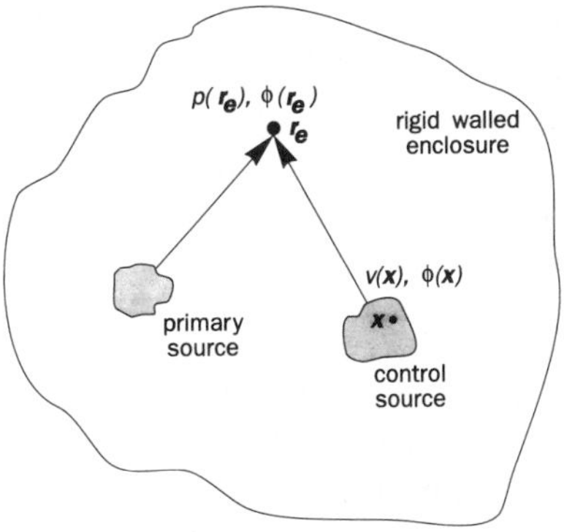

Fig. 9.1 Active control arrangement in a rigid walled enclosure.

As was outlined earlier in this book, the sound field within an enclosed space is composed of an infinite set of acoustic modes, ϕ. Referring to Fig. 9.1, the sound pressure $p_\iota(r_e)$ at some location r_e in the enclosure due to the ιth mode, excited by some acoustic source distribution contained within the enclosed space, is defined by the relationship

$$p_\iota(r_e) = j\rho_0\omega \int_V \frac{\phi_\iota(r)\phi_\iota(r_e)}{\Lambda_\iota(\kappa_\iota^2 - k^2)} s(r) \, \mathrm{d}r \tag{9.2.1}$$

where ρ_0 is the density of the acoustic medium, ω is the excitation frequency, $s(r)$ is the acoustic source strength in units of volume velocity per unit volume, $\phi_\iota(r)$ is the value of acoustic mode shape ι at location r in the enclosed space, $\phi_\iota(r_e)$ is the value of acoustic mode shape ι at location r_e in the enclosed space, Λ_ι is the volume normalization of acoustic mode ι,

$$\Lambda_\iota = \int_V \phi_\iota^2(r)\, dr \tag{9.2.2}$$

The quantity κ_ι is the eigenvalue associated with acoustic mode ι, k is the acoustic wavenumber at the excitation frequency, and the integrations in (9.2.1) and (9.2.2) are over the volume of the enclosed space V. Acoustic damping will be included in the system models developed here, so that κ_ι is a complex eigenvalue of the cavity, defined as $\kappa_\iota = \omega_\iota/[c_0(1 - j2\zeta_\iota)]$, where the imaginary component defines the acoustic modal damping, ζ_ι being the viscous damping ratio of the mode.

Consider, for example, a system where there is a single primary monopole source and a single control monopole source operating in the enclosed space, and the aim of the active control system is to minimize the ι*th* modal sound pressure at the error sensing location $\boldsymbol{r}_e$. From (9.2.1), the modal sound pressure at location $\boldsymbol{r}_e$ resulting from the operation of a monopole sound source at location $\boldsymbol{r}_q$ will be

$$p_\iota(\boldsymbol{r}_e) = j\rho_0\omega \int_V \frac{\phi_\iota(\boldsymbol{r}_q)\phi_\iota(\boldsymbol{r}_e)}{\Lambda_\iota(\kappa_\iota^2 - k^2)}\, q\, \delta(\boldsymbol{r}-\boldsymbol{r}_q)\, d\boldsymbol{r} \tag{9.2.3}$$

where the source strength $s(\boldsymbol{r})$ is equal to the product of the monopole source volume velocity amplitude q and a Dirac delta function $\delta(\boldsymbol{r} - \mathbf{r}_q)$, where the delta function has units of L^{-3}, or reciprocal volume. Evaluating the integral in (9.2.3), the modal sound pressure is found to be

$$p_\iota(\boldsymbol{r}_e) = j\rho_0\omega q \frac{\phi_\iota(\boldsymbol{r}_q)\phi_\iota(\boldsymbol{r}_e)}{\Lambda_\iota(\kappa_\iota^2 - k^2)} \tag{9.2.4}$$

As the system considered here is linear, the sound pressure at location $\boldsymbol{r}_e$ in space during operation of both the primary and control sources will be equal to the superposition of the two components,

$$p_\iota(\boldsymbol{r}_e) = p_{p,\iota}(\boldsymbol{r}_e) + p_{c,\iota}(\boldsymbol{r}_e) \tag{9.2.5}$$

Expanding (9.2.5) using (9.2.4) it is easily shown that the modal sound pressure at $\boldsymbol{r}_e$ can be completely nullified by the choice of control source volume velocity

$$q_c = -q_p \frac{\phi_\iota(\boldsymbol{r}_p)}{\phi_\iota(\boldsymbol{r}_c)} \tag{9.2.6}$$

where $\boldsymbol{r}_p$ and $\boldsymbol{r}_c$ are the locations of the primary and control source monopoles, respectively. Note that in the relationship given in (9.2.6) there is no dependency upon error location $\boldsymbol{r}_e$, because for a single mode, nullifying the acoustic pressure at one location is equivalent to nullifying the acoustic pressure at all locations.

The utility of the result of (9.2.6) is somewhat dubious, as it is unlikely that consideration of a single mode in an enclosed space will be adequate to represent any physical system. Theoretically, exact modelling of an enclosed sound field requires consideration of an infinite number of acoustic modes:

$$p(r_e) = j\rho_0\omega \int_V \sum_{\iota=1}^{\infty} \frac{\phi_\iota(r)\phi_\iota(r_e)}{\Lambda_\iota(\kappa_\iota^2 - k^2)} s(r)\, \mathrm{d}r = j\rho_0\omega \int_V G_a(r \mid r_e) s(r)\, \mathrm{d}r \quad (9.2.7)$$

where $G_a(r|r_e)$ is the Green's function of the interior space, defined in Chapter 2 as

$$G_a(r \mid r_e) = \sum_{\iota=1}^{\infty} \frac{\phi_\iota(r)\phi_\iota(r_e)}{\Lambda_\iota(\kappa_\iota^2 - k^2)} \quad (9.2.8)$$

In enclosures of low modal density (low frequency excitation) sufficiently accurate results can be obtained by truncating the infinite summation at a relatively low number of modes, N_n. We will restrict discussion in this section to systems of low modal density, explicitly truncating the acoustic modal summation at N_n modes, and delay discussion of the effects of higher modal density on the qualitative results Section 9.9.

Observe now that the integral expression in (9.2.7) can be rewritten, using the truncated summation, as

$$p(r_e) = j\rho_0\omega \sum_{\iota=1}^{N_n} \frac{\phi(r_e)}{\Lambda_\iota(\kappa_\iota^2 - k^2)} \int_V \phi_\iota(r) s(r)\, \mathrm{d}r \quad (9.2.9)$$

Expressed in this way the integral in (9.2.9) can be viewed as a modal generalized volume velocity, analogous to the structural modal generalized forces encountered in the previous chapter. It will be useful to further re-express this integral by writing the acoustic source strength distribution $s(r)$ as the product of a complex volume velocity amplitude, q, and a transfer function $z_v(r)$ to that point,

$$\int_V \phi_\iota(r) s(r)\, \mathrm{d}r = q \int_V \phi_\iota(r) z_v(r)\, \mathrm{d}r = q\phi_{g,\iota} \quad (9.2.10)$$

where $\phi_{g,\iota}$ is the modal generalized volume velocity transfer function for the ιth acoustic mode, which describes the volume velocity excitation of the mode for a unit amplitude source volume velocity:

$$\phi_{g,\iota} = \int_V \phi_\iota(r) z_v(r)\, \mathrm{d}r \quad (9.2.11)$$

If, for example, the source of interest was sufficiently small to be modelled as an acoustic monopole, the modal generalized volume velocity would be

$$q\, \phi_{g,\iota} = \int_V \phi_\iota(r)\, q\, \delta(r - r_q)\, \mathrm{d}r = q\, \phi_\iota(r_q) \quad (9.2.12)$$

so that the modal generalized volume velocity transfer function, $\phi_{g,\iota}$ is simply the value of the mode shape at the monopole source location:

$$\phi_{g,\iota} = \int_V \phi_\iota(r)\, \delta(r - r_q)\, \mathrm{d}r = \phi_\iota(r_q) \quad (9.2.13)$$

It is also straightforward to simplify (9.2.11) for other sources of regular geometry, such as rectangular pistons (as were considered in Chapter 7 for plane wave sound propagation in air handling ducts). This enables explicit modelling of the effects of control source size.

Returning to (9.2.9), by truncating the modal summation it is possible to express the total acoustic pressure at some location $\boldsymbol{r_e}$ in the enclosed space as the product of a set of matrices. If we define an $(N_n \times 1)$ vector of modal generalized volume velocity transfer functions, $\boldsymbol{\phi}_g$,

$$\boldsymbol{\phi}_g = \begin{bmatrix} \phi_{g,1} \\ \phi_{g,2} \\ \vdots \\ \phi_{g,N_n} \end{bmatrix} \tag{9.2.14}$$

an $(N_n \times N_n)$ diagonal matrix of acoustic modal transfer functions, $\boldsymbol{Z}_a$, the terms of which describe the complex amplitude of the acoustic modes for a unit modal generalized volume velocity,

$$Z_a(i,i) = \frac{j\rho_0\omega}{\Lambda_i(\kappa_i^2 - k^2)} \tag{9.2.15}$$

and an $(N_n \times 1)$ vector of acoustic mode shape values at the location $\boldsymbol{r_e}$ in the enclosed space, $\boldsymbol{\phi}(\boldsymbol{r_e})$:

$$\boldsymbol{\phi}(\boldsymbol{r_e}) = \begin{bmatrix} \phi_1(\boldsymbol{r_e}) \\ \phi_2(\boldsymbol{r_e}) \\ \vdots \\ \phi_{N_n}(\boldsymbol{r_e}) \end{bmatrix} \tag{9.2.16}$$

the acoustic pressure at $\boldsymbol{r_e}$ can be written as

$$p(\boldsymbol{r_e}) = \boldsymbol{\phi}^{\mathrm{T}}(\boldsymbol{r_e})\boldsymbol{Z}_a\boldsymbol{\phi}_g q \tag{9.2.17}$$

As we have seen in the previous two chapters, expressing the acoustic pressure in matrix form enables straightforward calculation of the control source volume velocity which will minimize the acoustic pressure amplitude, the sum of contributions from all N_n acoustic modes being modelled, at the error location of interest in the enclosed space.

As we are dealing with linear systems, the total acoustic pressure at any location in the enclosed space during operation of the active control system will be equal to the sum of the primary and control components:

$$p(\boldsymbol{r_e}) = p_c(\boldsymbol{r_e}) + p_p(\boldsymbol{r_e}) \tag{9.2.18}$$

Therefore, the squared amplitude of the acoustic pressure at $\boldsymbol{r_e}$ during operation of the control and primary source distributions is equal to

$$|p(\boldsymbol{r_e})|^2 = p(\boldsymbol{r_e})^* p(\boldsymbol{r_e}) = [p_c(\boldsymbol{r_e}) + p_p(\boldsymbol{r_e})]^* [p_c(\boldsymbol{r_e}) + p_p(\boldsymbol{r_e})] \tag{9.2.19}$$

Using (9.2.17), this can be rewritten as

$$|p(r_e)|^2 = q_c^* a q_c + q_c^* b + b^* q_c + c \tag{9.2.20}$$

where

$$a = \phi_{gc}^{\mathrm{H}} \mathbf{Z}_a^{\mathrm{H}} \mathbf{Z}_p \mathbf{Z}_a \phi_{gc} \tag{9.9.21}$$

$$b = \phi_{gc}^{\mathrm{H}} \mathbf{Z}_a^{\mathrm{H}} \mathbf{Z}_p \mathbf{Z}_a \phi_{gp} q_p \tag{9.9.22}$$

$$c = q_p^* \phi_{gp}^{\mathrm{H}} \mathbf{Z}_a^{\mathrm{H}} \mathbf{Z}_p \mathbf{Z}_a \phi_{gp} q_p \tag{9.9.23}$$

where

$$\mathbf{Z}_p = \phi(r_e)\phi^{\mathrm{T}}(r_e) \tag{9.2.24}$$

In (9.2.21)–(9.2.23) the subscripts p and c denote primary and control source related quantities. One point to note briefly here is that $\mathbf{Z}_p$ is a real, symmetric matrix, effectively a weighting matrix for minimizing the acoustic modal amplitudes. This concept will be discussed further in Section 9.7 in relation to optimum error sensor location.

Equation (9.2.20) shows the squared acoustic pressure amplitude at any location $\boldsymbol{r}_e$ in the enclosed space to be a real quadratic function of the complex control source volume velocity, q_c. This is seen more easily by rewriting the equation in terms of its real and imaginary parts:

$$|p(r_e)|^2 = aq_{c\mathrm{R}}^2 + 2b_{\mathrm{R}}q_{c\mathrm{R}} + aq_{c\mathrm{I}}^2 + 2b_{\mathrm{I}}q_{c\mathrm{I}} + c \tag{9.2.25}$$

where the subscripts I and R denote real and imaginary components respectively. As we have discussed previously this means that a plot of the squared acoustic pressure amplitude as a function of the real and imaginary components of the control source volume velocity forms a 'bowl', as was shown in Fig. 8.2. The bottom of the bowl defines the optimum control source volume velocity, which will minimize the error criterion, the squared acoustic pressure amplitude. This value of volume velocity can be found by differentiating (9.2.20) with respect to its real and imaginary components, and setting the resultant gradient expression equal to zero. Doing this,

$$\frac{\partial |p(r_e)|^2}{\partial q_{c\mathrm{R}}} = 2aq_{c\mathrm{R}} + 2b_{\mathrm{R}} = 0 \tag{9.2.26}$$

$$\frac{\partial |p(r_e)|^2}{\partial q_{c\mathrm{I}}} = 2aq_{c\mathrm{I}} + 2b_{\mathrm{I}} = 0 \tag{9.2.27}$$

Putting together the real, (9.2.26), and imaginary, (9.2.27), components of the criteria, through multiplying the latter by j and adding it to the former, produces the expression for the optimum control source volume velocity,

$$q_{c,p} = -a^{-1}b \tag{9.2.28}$$

where the subscript c,p denotes control source volume velocity derived with respect to minimizing a pressure error criterion. If this control source volume velocity is substituted back into (9.2.20), the minimum squared pressure amplitude is found to be

$$|p(r_e)|^2_{\min} = c - b^* a^{-1} b = c + b^* q_{c,p} \tag{9.2.29}$$

As we have seen before, the term c is the value of the error criterion, the squared acoustic pressure amplitude at $\boldsymbol{r}_e$, during primary excitation only.

This formulation for the optimum single control source output q_c which will minimize the acoustic pressure at a single error sensing location $\boldsymbol{r}_e$ in the enclosed space can easily be extended to include multiple control sources and multiple error sensors. The $(N_e \times 1)$ vector of sound pressures at the set of N_e discrete error sensing locations, $\boldsymbol{p}_e$, resulting from the operation of the set of N_s acoustic sources in the enclosed space can be expressed as

$$\boldsymbol{p}_e = \boldsymbol{\Phi}_e \boldsymbol{Z}_a \boldsymbol{\Phi}_g \boldsymbol{q} \tag{9.2.30}$$

where $\boldsymbol{\Phi}_e$ is the $(N_e \times N_n)$ matrix of acoustic mode shape functions evaluated at the set of error sensing locations,

$$\boldsymbol{\Phi}_e = \begin{bmatrix} \phi(r_1)^{\mathrm{T}} \\ \phi(r_2)^{\mathrm{T}} \\ \vdots \\ \phi(r_{N_e})^{\mathrm{T}} \end{bmatrix} \tag{9.2.31}$$

$\boldsymbol{\Phi}_g$ is the $(N_n \times N_s)$ matrix of modal generalized volume velocity transfer functions for the N_s sources,

$$\boldsymbol{\Phi}_g = \begin{bmatrix} \phi_{g1} & \phi_{g2} & \cdots & \phi_{gN_s} \end{bmatrix} \tag{9.2.32}$$

and $\boldsymbol{q}$ is the $(N_s \times 1)$ vector of complex source velocity amplitudes

$$\boldsymbol{q} = \begin{bmatrix} q_1 \\ q_2 \\ \vdots \\ q_{N_s} \end{bmatrix} \tag{9.2.33}$$

Using these quantities, the vector of acoustic pressure amplitudes at the N_e error sensing locations during combined operation of N_p primary and N_c control sources is

$$\boldsymbol{p}_e = \boldsymbol{p}_{ce} + \boldsymbol{p}_{pe} = \boldsymbol{\Phi}_e \boldsymbol{Z}_a \boldsymbol{\Phi}_{gc} \boldsymbol{q}_c + \boldsymbol{\Phi}_e \boldsymbol{Z}_a \boldsymbol{\Phi}_{gp} \boldsymbol{q}_p \tag{9.2.34}$$

where $\boldsymbol{p}_{pe}$ and $\boldsymbol{p}_{ce}$ are the $(N_e \times 1)$ vectors of acoustic pressures at the error sensing locations during primary and control excitation only respectively. The sum of the squared acoustic pressure amplitudes at the error sensing locations is therefore

$$\sum_{i=1}^{N_e} |p(r_{ei})|^2 = p_e^{\mathrm{H}} p_e = [p_{ce}+p_{pe}]^{\mathrm{H}}[p_{ce}+p_{pe}] \tag{9.2.35}$$

which can be written more compactly as

$$\sum_{i=1}^{N_e} |p(r_{ei})|^2 = q_c^{\mathrm{H}} A q_c + q_c^{\mathrm{H}} b + b^{\mathrm{H}} q_c + c \tag{9.2.36}$$

where

$$A = \Phi_{gc}^{\mathrm{H}} Z_a^{\mathrm{H}} Z_p Z_a \Phi_{gc} \tag{9.2.37}$$

$$b = \Phi_{gc}^{\mathrm{H}} Z_a^{\mathrm{H}} Z_p Z_a \Phi_{gp} q_p \tag{9.2.38}$$

$$c = q_p^{\mathrm{H}} \Phi_{gp}^{\mathrm{H}} Z_a^{\mathrm{H}} Z_p Z_a \Phi_{gp} q_p \tag{9.2.39}$$

and where Z_p is now based upon multiple error sensor locations,

$$Z_p = \Phi_e^{\mathrm{T}} \Phi_e \tag{9.2.40}$$

Equation (9.2.36) shows this expanded error criterion, the minimization of the squared acoustic pressure amplitudes at a set of discrete error locations, to be a real quadratic function of the *vector* of control source volume velocities. As discussed in Section 8.4, this means that the error surface, or error criterion plotted as a function of the real and imaginary components of the control source volume velocities, will be a $(2N_c+1)$-dimensional hyper-paraboloid, which is a generalization of the three-dimensional 'bowl' shown in Fig. 8.2 for the case of $N_c = 1$. Being a quadratic function of q_c the error criterion of (9.2.36) can again be differentiated with respect to this quantity, and the resultant gradient expression set equal to zero to derive the vector of optimum control source volume velocities. Performing these operations on the real and imaginary parts separately,

$$\frac{\partial p_e^{\mathrm{H}} p_e}{\partial q_{c_{\mathrm{R}}}} = 2A_{\mathrm{R}} q_{c_{\mathrm{R}}} + 2b_{\mathrm{R}} = 0 \tag{9.2.41}$$

$$\frac{\partial p_e^{\mathrm{H}} p_e}{\partial q_{c_{\mathrm{I}}}} = 2A_{\mathrm{I}} q_{c_{\mathrm{I}}} + 2b_{\mathrm{I}} = 0 \tag{9.2.42}$$

multiplying the imaginary part by j and adding it to the real part produces the expression for the optimum vector of control source volume velocities

$$q_{c,p} = -A^{-1} b \tag{9.2.43}$$

The result of (9.2.43) is simply an extension of (9.2.29), derived for a system arrangement consisting of a single primary and control source and a single error sensor. Substituting the result of (9.2.43) back into (9.2.36) produces the expression for the minimum residual (controlled) sum of the squared sound pressures at the error locations

$$[\boldsymbol{p}_e^{\mathrm{H}}\boldsymbol{p}_e]_{\min} = c - \boldsymbol{b}^{\mathrm{H}}\boldsymbol{A}^{-1}\boldsymbol{b} = c + \boldsymbol{b}^{\mathrm{H}}\boldsymbol{q}_{c,p} \tag{9.2.44}$$

It should again be pointed out that if the relationship of (9.2.44) is to define a unique vector of control source volume velocities which will collectively minimize the sum of the squared acoustic pressure amplitudes at the error sensing locations, the matrix $\boldsymbol{A}$ must be positive definite. For the case being considered here, this often translates into the requirement that there be at least as many error locations as there are control sources. If there are less, then the matrix $\boldsymbol{A}$ may be singular, and hence there will be an infinite number of optimum control source volume velocity vectors. If there are an equal number of control sources and error locations then the vector of optimum control source volume velocities will often completely null the acoustic pressure at the error locations. If there are more error locations than control sources, then the residual acoustic pressure at the error locations will probably not be equal to zero, although this may not have a negative influence upon the overall levels of acoustic power attenuation attained.

A specific example of minimizing acoustic pressure in a rigid walled enclosure will be presented at the end of the next section, after we derive expressions for determining the best possible result for the minimization of acoustic potential energy in the enclosed space for a given control source arrangement.

9.3 GLOBAL CONTROL OF SOUND FIELDS IN RIGID ENCLOSURES

As we have seen throughout this book, simply minimizing the sound field at discrete locations does not guarantee the best result in terms of global attenuation of the sound field throughout the enclosure. To determine the maximum levels of global sound attenuation which are possible for a given source arrangement the problem must be reformulated in terms of some global error criterion.

In the previous two chapters the global error criterion which we have used is real acoustic power, as it is this quantity which will propagate away from the sound source and become a nuisance to the surrounding population. When considering enclosed sound fields, however, acoustic power is not a particularly useful basis for examination, as for a given volume velocity source the only real acoustic power output will be a byproduct of acoustic damping. This is evident in the defining equation for acoustic pressure (9.2.1), where the acoustic pressure in the absence of damping is seen to be 90° out of phase with the velocity (recall that acoustic power is defined by the product of the in-phase components of pressure and velocity). A more suitable global measure of acoustic pressure when considering enclosed sound fields is acoustic potential energy, E_p,

$$E_p = \frac{1}{4\rho_0 c_0^2}\int_V |\, p(\boldsymbol{r})\,|^2 \,\mathrm{d}\boldsymbol{r} \tag{9.3.1}$$

which is explicitly based upon the volume integration of the squared acoustic pressure amplitudes in the enclosed space. To give some further physical meaning to acoustic potential energy, note that the acoustic pressure term in (9.3.1) can be expanded as the sum of modal contributions

$$p(r) = \sum_{\iota=1}^{\infty} p_\iota \phi_\iota(r) \tag{9.3.2}$$

where p_ι is the complex pressure amplitude of the ιth acoustic mode. Substituting this expansion into (9.3.1),

$$E_p = \frac{1}{4\rho_0 c_0^2} \int_V \left[\sum_{\iota=1}^{\infty} p_\iota \phi_\iota(r) \right]^* \left[\sum_{\iota=1}^{\infty} p_\iota \phi_\iota(r) \right] \mathrm{d}V \tag{9.3.3}$$

Acoustic modes in a rigid enclosure are orthogonal, so that

$$\int_V \phi_i(r)\phi_\iota(r)\,\mathrm{d}r = \begin{cases} \Lambda_\iota & i=\iota \\ 0 & i\neq\iota \end{cases} \tag{9.3.4}$$

where Λ_ι is the volume normalization of the ιth acoustic mode. Using this property the expression for acoustic potential energy given in (9.3.3) can be simplified to

$$E_p = \frac{1}{4\rho_0 c_0^2} \sum_{\iota=1}^{\infty} \Lambda_\iota \,|\, p_\iota \,|^2 \tag{9.3.5}$$

Equation (9.3.5) shows that minimizing acoustic potential energy in the enclosed space is equivalent to minimizing the weighted (by mode normalization) sum of the squares of the acoustic modal amplitudes. Therefore, to formulate expressions for the control source volume velocities which will minimize the enclosed sound field we must consider the acoustic modal amplitudes themselves, rather than their values at specific error sensing locations.

From (9.2.1) in the previous section, it is apparent that the amplitude of the ιth acoustic mode resulting from the operation of some acoustic source distribution located within the enclosed space is defined by the equation

$$p_\iota = j\rho_0\omega \int_V \frac{\phi_\iota(r)}{\Lambda_\iota(\kappa_\iota^2 - k^2)} s(r)\,\mathrm{d}r \tag{9.3.6}$$

Considering a truncated set of acoustic modes, the $(N_n \times 1)$ vector of acoustic modal amplitudes $\boldsymbol{p}$ resultant from the operation of a single velocity source can be expressed in matrix form using the quantities defined in the previous section

$$\boldsymbol{p} = \boldsymbol{Z}_a \,\boldsymbol{\phi}_g \, q \tag{9.3.7}$$

The sum of the squared acoustic modal amplitudes during operation of single control and primary source distributions is therefore equal to

$$\sum_{\iota=1}^{N_n} |\, p_\iota \,|^2 = \sum_{\iota=1}^{N_n} p_\iota^* p_\iota = [\boldsymbol{p}_c + \boldsymbol{p}_p]^{\mathrm{H}} [\boldsymbol{p}_c + \boldsymbol{p}_p] \tag{9.3.8}$$

Using (9.3.8), the acoustic potential energy for this single control and primary source arrangement can therefore be expressed as

$$E_p = \frac{1}{4\rho_0 c_0^2} \sum_{\iota=1}^{N_n} |p_\iota|^2 = q_c^* a q_c + q_c^* b + b^* q_c + c \tag{9.3.9}$$

where

$$a = \boldsymbol{\phi}_{gc}^{\mathrm{H}} \mathbf{Z}_a^{\mathrm{H}} \mathbf{Z}_E \mathbf{Z}_a \boldsymbol{\phi}_{gc} \tag{9.3.10}$$

$$b = \boldsymbol{\phi}_{gc}^{\mathrm{H}} \mathbf{Z}_a^{\mathrm{H}} \mathbf{Z}_E \mathbf{Z}_a \boldsymbol{\phi}_{gp} q_p \tag{9.3.11}$$

and

$$c = q_p^* \boldsymbol{\phi}_{gp}^{\mathrm{H}} \mathbf{Z}_a^{\mathrm{H}} \mathbf{Z}_E \mathbf{Z}_a \boldsymbol{\phi}_{gp} q_p \tag{9.3.12}$$

where the terms of the acoustic potential energy-based diagonal weighting matrix $\mathbf{Z}_E$ are defined by

$$Z_E(i,i) = \frac{\Lambda_i}{4\rho_0 c_0} \tag{9.3.13}$$

Observing that (9.3.9) is identical in form to (9.2.20) of the previous section, the optimum control source volume velocity with respect to providing global reduction of the enclosed sound field is defined by the same relationship as given in (9.2.28):

$$q_{c,E} = -a^{-1}b \tag{9.3.14}$$

where the subscript c,E denotes a control source output which is optimal with respect to minimizing acoustic potential energy. Substituting this relationship back into (9.3.9) the minimum value of acoustic potential energy is found to be

$$E_{p,\min} = c - b^* a^{-1} b = c + b^* q_{c,E} \tag{9.3.15}$$

where the term c is again the value of the error criterion, acoustic potential energy in the enclosed space, under primary excitation only.

There is an interesting observation which can be made here in regard to the phasing of the control source volume velocity relative to the primary source when minimizing acoustic potential energy. If the modal generalized volume velocity transfer functions for the primary and control sources are both real, which will usually be the case (for example when both sources are modelled as acoustic monopoles), then as the quantity a in (9.3.14) is real (as defined in (9.3.10)) and the quantity b will be equal to some real number multiplied by the primary source amplitude, the control source will always be either in-phase or 180° out of phase with the primary source. We have seen this rather generic trait throughout this book, where when using one source to control the output of a second equivalent source the optimum result will be obtained when the sources are precisely in or out of phase.

The previous single and primary control source formulation can easily be expanded to incorporate multiple control and primary sources. Again using quantities defined in the previous section, the acoustic potential energy in the enclosure with this expanded system arrangement is

$$E_p = \frac{1}{4\rho_0 c_0^2}\sum_{\iota=1}^{N_n} |p_\iota|^2 = q_c^{\mathrm{H}} A q_c + q_c^{\mathrm{H}} b + b^{\mathrm{H}} q_c + c \tag{9.3.16}$$

where

$$A = \Phi_{gc}^{\mathrm{H}}\, Z_a^{\mathrm{H}}\, Z_E\, Z_a\, \Phi_{gc} \tag{9.3.17}$$

$$b = \Phi_{gc}^{\mathrm{H}}\, Z_a^{\mathrm{H}}\, Z_E\, Z_a\, \Phi_{gp}\, q_p \tag{9.3.18}$$

and

$$c = q_p^{\mathrm{H}}\, \Phi_{gp}^{\mathrm{H}}\, Z_a^{\mathrm{H}}\, Z_E\, Z_a\, \Phi_{gp}\, q_p \tag{9.3.19}$$

As (9.3.16) is identical in form to (9.2.36) we can use the previous result of (9.2.43) to simply write the expression for the vector of optimum control source outputs as

$$q_{c,E} = -A^{-1} b \tag{9.3.20}$$

Substituting this result into (9.3.16) produces the expression for the minimum acoustic potential energy,

$$E_{p,\min} = c - b^{\mathrm{H}} A^{-1} b = c + b^{\mathrm{H}} q_{c,E} \tag{9.3.21}$$

Before considering a simple example of control of a rigidly enclosed sound field there is one final point which should be discussed, that of physical mechanisms. We have seen in previous chapters that when (constant volume velocity) acoustic sources are used in an active control system two control mechanisms are possible: acoustic power suppression, where the control source(s) mutually unload the primary source such that the total acoustic power output is reduced, and acoustic power absorption, where the control source absorbs some of the primary source acoustic power output. However, we have already pointed out that when considering the active control of sound fields in an enclosed space, acoustic power is not a particularly useful basis for examination, as for a given velocity source the only real acoustic power output will be a by-product of acoustic damping. This is the reason that acoustic potential energy, based upon volume-averaged acoustic pressure levels, as opposed to acoustic power, was used as a global error criterion. Therefore, when examining control mechanisms in a rigid enclosure, it is more appropriate to examine the individual source contributions to acoustic potential energy.

It was shown in (9.3.5) that acoustic potential energy is equal to the weighted sum of squared acoustic modal pressure amplitudes. This sum can be expressed in a manner analogous to total acoustic power as the sum of primary and control source contributions,

$$E_p = \frac{1}{4\rho_0 c_0^2}\sum_{\iota=1}^{n} \Lambda_\iota |p_\iota|^2 = E_{pp} + E_{pc} \tag{9.3.22}$$

where E_{pp} and E_{pc} are the contributions to acoustic potential energy by the primary source and control source distributions, respectively, which are equal to

$$E_{p(p/c)} = \frac{1}{4\rho_0 c_0^2} \sum_{\iota=1}^{n} \Lambda_\iota \, \mathrm{Re}\left\{p_{\iota,p/c} \, p_{\iota,tot}^*\right\} \tag{9.3.23}$$

Here p/c denotes primary or control source, *tot* denotes total (sum of primary and control source contributions), and Re denotes the real part of the expression. The acoustic modal amplitude resulting from some source distribution is defined in (9.3.7) for single sources, and for multiple sources it can easily be extended to

$$\mathbf{p} = \mathbf{Z}_a \mathbf{\Phi}_g \mathbf{q} \tag{9.3.24}$$

(where the terms must be specialized for either control sources or primary sources). Thus for a single control source, single primary source system it is straightforward to show that the control source contribution to the acoustic potential energy can be expressed as

$$E_{pc} = q_c^* a q_c + \mathrm{Re}\{q_c b^*\} \tag{9.3.25}$$

Similarly, the primary source contribution to the potential energy of the enclosed space is

$$E_{pp} = \mathrm{Re}\{q_c^* b\} + c \tag{9.3.26}$$

If the optimum value of control source output, as defined in (9.3.14), is substituted into (9.3.24), which defines the control source acoustic potential energy, it is found that under optimum conditions

$$\begin{aligned} E_{pc} &= q_{c,E}^* a q_{c,E} + \left\{q_{c,E} b^*\right\} \\ &= (-a^{-1}b)^* a(-a^{-1}b) + \mathrm{Re}\left\{(-a^{-1}b)b^*\right\} = 0 \end{aligned} \tag{9.3.27}$$

That is, the contribution made by the control source to the acoustic potential energy in the enclosed space is zero; it makes neither a negative nor positive contribution. The control mechanism which is optimally employed must the be one of pure suppression, where a reduction in acoustic potential energy is achieved by a mutual unloading of the primary and control source distribution, resulting in a zero contribution by the control source and a reduced contribution by the primary source to the acoustic potential energy. Note that if the optimum control source output of (9.3.14) is substituted into (9.3.25), which defines the primary source contribution to acoustic potential energy, the result is

$$E_{pp} = \mathrm{Re}\left\{q_{c,E}^* b + c\right\} = c - b^* a^{-1} b \tag{9.3.28}$$

which is the same as the minimum acoustic potential energy of the system, as outlined in (9.3.15). The residual acoustic potential energy can be entirely traced to the primary source distribution.

This result can be easily expanded to included consideration of multiple sources. In terms of the quantities defined in (9.3.17) and (9.3.18), (9.3.22) can be specialized for a set of control sources as

$$E_{pc} = q_c^{\mathrm{H}} A q_c + \mathrm{Re}\{b^{\mathrm{H}} q_c\} \tag{9.3.29}$$

If the vector of optimal control source outputs as defined in (9.3.20) is now inserted into this relation, noting that A is a Hermitian matrix, the acoustic potential energy attributable to the control source array is

$$E_{p,c} = [-A^{-1}b]^{\mathrm{H}} A[-A^{-1}b] + \mathrm{Re}\{b^{\mathrm{H}}[-A^{-1}b]\} = 0 \tag{9.3.30}$$

That is, the acoustic potential energy attributable to the control source array is zero, as may have been expected. Once again the control mechanism is one of pure suppression.

Simply having a total contribution to the acoustic potential energy of zero from the control source array does not in itself mean that contributions from the constituent elements will necessarily be zero. However, note that the acoustic potential energy attributable to the operation of a single, ith, control source can be expressed as

$$E_{pc}(i) = \mathrm{Re}\{q_{c,E}^{\mathrm{H}} A(\mathrm{col}\ i) q_{c,E}(i)\} + \mathrm{Re}\{b^*(i) q_{c,E}(i)\} \tag{9.3.31}$$

where $A(\mathrm{col}\ i)$ is the ith column of the matrix A, $b(i)$ is the ith element in the vector b, and $q_{c,E}(i)$ is the optimum volume velocity of the ith control source, the ith element of the vector $q_{c,E}$. If the vector of optimum control source volume velocities is again expanded then the result is

$$E_{pc}(i) = \mathrm{Re}\{[-A^{-1}b]^{\mathrm{H}} A(\mathrm{col}\ i) q_{c,E}(i)\} + \mathrm{Re}\{b(i)^* q_{c,E}(i)\} = 0 \tag{9.3.32}$$

It has therefore just been shown that the acoustic potential energy contribution of each element in the control source array will also be zero. Note that this is analogous to the free space result of the previous section, where it was shown that if reciprocity exists between each point of the primary source distribution (in this instance the vibrating structure) and each point on the control source array then the acoustic power output of the control sources, under optimal conditions, will be zero.

Example 9.1

Application of the above methodology to analysis the active control of sound fields in rigid walled enclosures has been undertaken for a number of problems (Bullmore, 1988; Bullmore *et al.*, 1987, 1990; Elliott *et al.*, 1987). As a simple example of the active control of a rigidly enclosed sound field we will choose the problem of low frequency sound attenuation in a rigid walled rectangular enclosure of dimensions 3.43 m × 0.1 m × 0.1 m, essentially the problem considered at length by Curtis *et al.* (1985, 1987, 1990). To fit this problem into the theoretical models we have developed thus far we require only a knowledge of the acoustic mode shape functions and resonance frequencies. For a rectangular enclosure, the acoustic mode shape function is

$$\phi_{l,m,n}(x,y,z) = \cos\frac{l\pi x}{L_x}\cos\frac{m\pi y}{L_y}\cos\frac{n\pi z}{L_z} \tag{9.3.33}$$

where l, m, n and L_x, L_y, and L_z are the modal indices and enclosure dimensions in the x, y, and z directions, respectively, with the origin of the coordinate system in the lower left corner of the enclosure. The resonance frequencies associated with these acoustic modes are defined by the expression

$$\omega_{l,m,n} = c_0\pi\sqrt{\left(\frac{l}{L_x}\right)^2 + \left(\frac{m}{L_y}\right)^2 + \left(\frac{n}{L_z}\right)^2} \tag{9.3.34}$$

The acoustic mode normalisation associated with this mode shape function is

$$\Lambda_{l,m,n} = \int_V \phi^2_{l,m,n}(\boldsymbol{r})\,\mathrm{d}\boldsymbol{r} = \frac{V}{8}\epsilon_l\epsilon_m\epsilon_n, \qquad \epsilon_{l,m,n} = \begin{cases} 1 & l,m,n \neq 0 \\ 2 & l,m,n = 0 \end{cases} \tag{9.3.35}$$

In considering the dimensions of the example enclosure in light of the defining expression for resonance frequency it is apparent that at low frequencies the problem is essentially one-dimensional, with the modal indices varying only in the x-direction ((0,0,0), (1,0,0), (2,0,0), etc.). A sketch of the enclosure geometry and the x-direction variation for several low-frequency resonant modes is shown in Fig. 9.2. Also, for an enclosure of these dimensions, the low order resonance frequencies will be spaced in 50 Hz intervals, the (0,0,0) at 0 Hz, (1,0,0) at 50 Hz, (2,0,0) at 100 Hz, etc. For this example we will consider the use of a single monopole primary source, located at $x = 0$ (note that for the low order modes the y and z locations are irrelevant), a single control source and a single error sensor.

The first point of interest is how much attenuation of acoustic potential energy is possible, and why. Figure 9.3 illustrates the levels of acoustic potential energy reduction possible for a single control source located at the end of the enclosure $x = L_x$ as a function of frequency for two levels of acoustic damping, 0.005 (light damping) and 0.05 (moderate damping). The data display a number of attenuation peaks corresponding to the resonance frequencies of the modes. To interpret these results there are two points which must be noted. First, the mode shapes are ordered such that with this source arrangement it is impossible to simultaneously reduce the amplitude of two sequential (in resonance frequency) modes with this primary source/control source arrangement; a reduction in the amplitude of one mode will always cause an increase in the amplitude of the neighbouring (in resonance frequency) modes. Second, it was shown previously in this section that acoustic potential energy is equal to the weighted sum of squared acoustic modal amplitudes, values of which are illustrated in Figs. 9.4 and 9.5 for the low frequency modes as a function of frequency for the two damping rates. At and near the resonance frequencies a

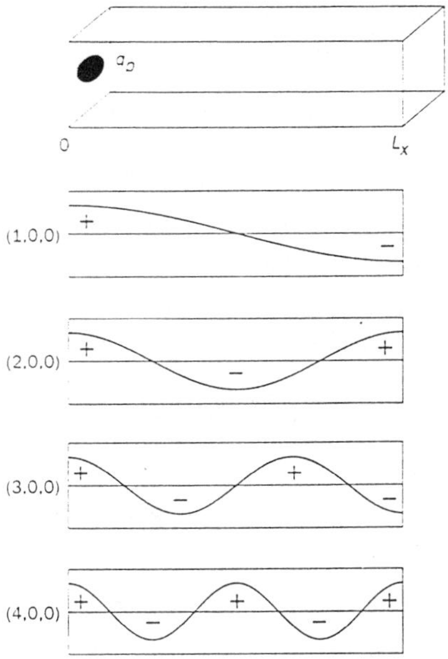

Fig. 9.2 One-dimensional rectangular enclosure with several low-frequency mode shape functions.

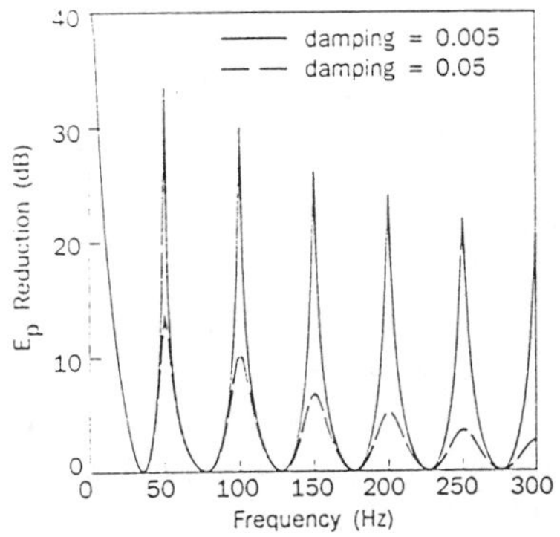

Fig. 9.3 Maximum possible acoustic potential energy reduction as a function of frequency with a single control source at $x = L_x$ with acoustic damping of 0.005 and 0.05.

single mode is clearly dominant in this summation, a mode which will have a very small input impedance, defined by the quantity $(\kappa_i^2 - k^2)$ in the denominator of the governing equation (9.2.1). This mode can be significantly decreased in amplitude with only a relatively small spillover effect, the increase in the amplitude of the two neighbouring modes, so that a large attenuation of acoustic potential energy is possible. As the damping of the enclosure increases the amplitude of the resonant mode decreases (due to the increase in the modal input

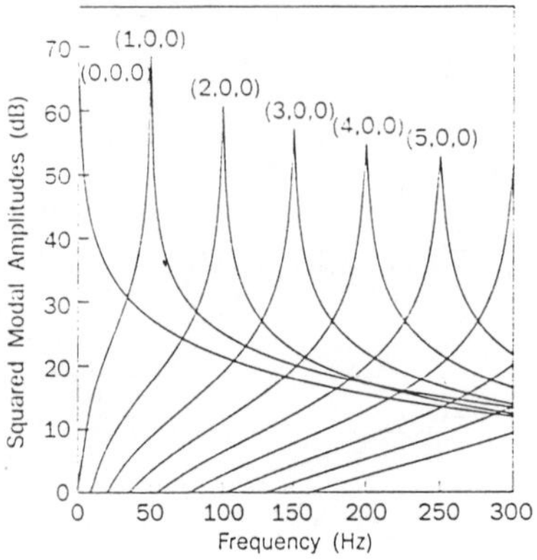

Fig. 9.4 Squared modal amplitudes as a function of frequency, damping = 0.005.

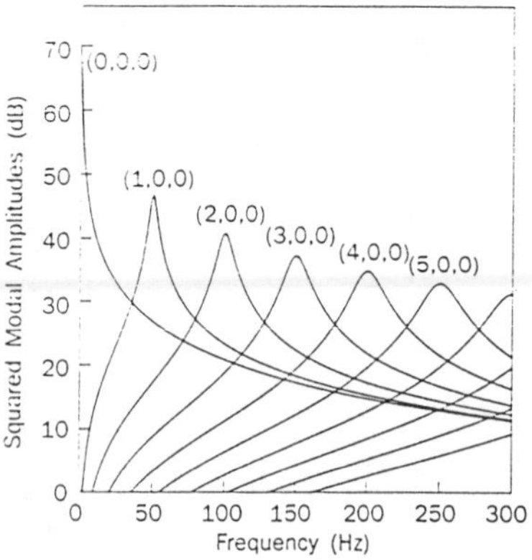

Fig. 9.5 Squared modal amplitudes as a function of frequency, damping = 0.05.

impedance), and so too does the reduction in acoustic potential energy. This trend is apparent when comparing the data in Figs. 9.3, 9.4 and 9.5. Away from resonance frequencies, however, the summation is not dominated by a single mode, but rather is very 'multi-modal'. Decreasing the amplitude of the most dominant mode will increase the amplitude of the next most important in the summation by a relatively significant amount. The end result is a very reduced potential for attenuation. Clearly a single control source is only effective at controlling a resonant response where a single acoustic mode is largely responsible for the sound field. As the number of modes increases so too does the number of control sources required to control them.

The location of the control source plays an important role in the level of attenuation which can be obtained. Figure 9.6 depicts the maximum levels of acoustic potential energy attenuation as a function of control source position for two frequencies, 100 Hz where the resonant (2,0,0) mode dominates the acoustic

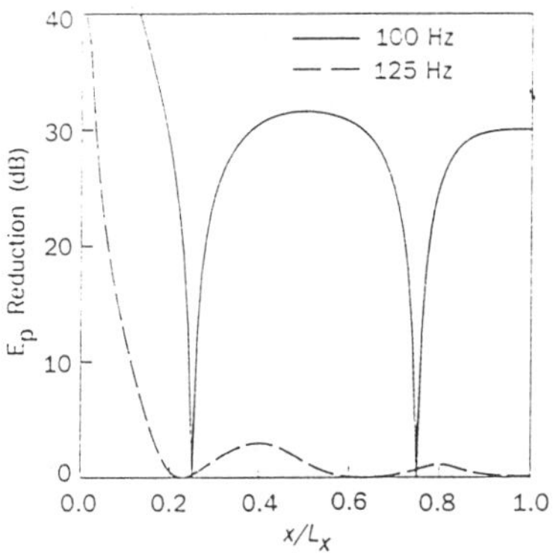

Fig. 9.6 Maximum possible acoustic potential energy reduction as a function of location of a single control source for 100 Hz and 125 Hz excitation.

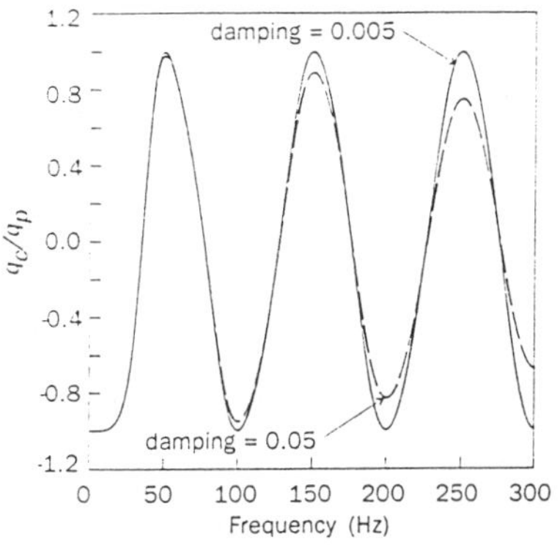

Fig. 9.7 Ratio of optimal control source to primary source volume velocity for a control source at $x = L_x$ with acoustic damping of 0.005 and 0.05.

potential energy summation, and 125 Hz, where two modes, the (2,0,0) and (3,0,0), have equal importance. For both of these cases the maximum levels of attenuation are achieved with the control sources near the primary source position, as at this position it will be possible to attenuate the amplitude of all modes simultaneously. At 100 Hz, significant levels of attenuation are possible with the control source well away from the primary source, as a single mode dominates the response. The troughs in the attenuation curve are at the nodal points (zero amplitude) of the resonant mode. The 125 Hz case does not fare so well for control source locations away from the primary source, however, as two modes must be reduced in amplitude simultaneously to obtain a satisfactory result.

It was pointed out previously that when a single control source is used to control an identical primary source it will either be exactly in or out of phase with it. Figure 9.7 depicts the ratio of control source to primary source volume velocity when the control source is at a position $x = L_x$, as a function of frequency. With light damping, the control source volume velocity amplitude is practically equal to the primary source amplitude at the resonance frequencies, once again indicating the dominance of a single mode (which is essentially being nulled using the relationship given in (9.2.6)) in the response. As the damping is increased, however, the amplitude required of the control source for maximum potential energy reduction decays with increasing frequency. This is due to a reduction in the amplitude of the resonant peaks in the response, and hence a reduction in the dominance of a single acoustic mode in the result. For both levels of damping, the optimum control source output is zero at some frequency between the resonant peaks. At this point, a decrease in one modal amplitude will be exactly offset by an increase in its neighbour; hence it is best to do nothing at all.

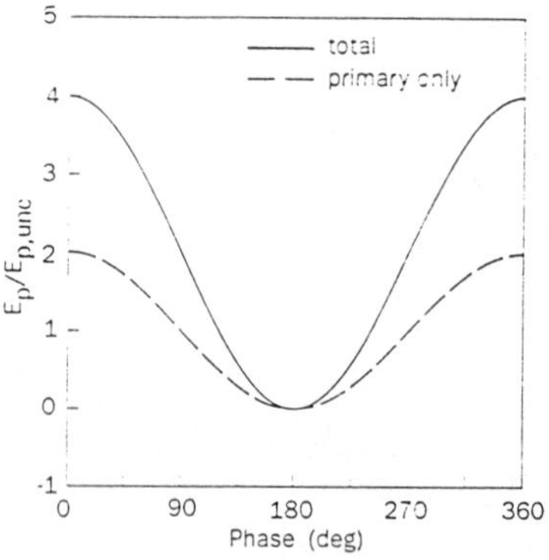

Fig. 9.8 Total normalized acoustic potential energy with equal amplitude primary and control inputs as a function of phase difference, plotted with the individual source contribution; control source at $x = L_x$, 100 Hz excitation.

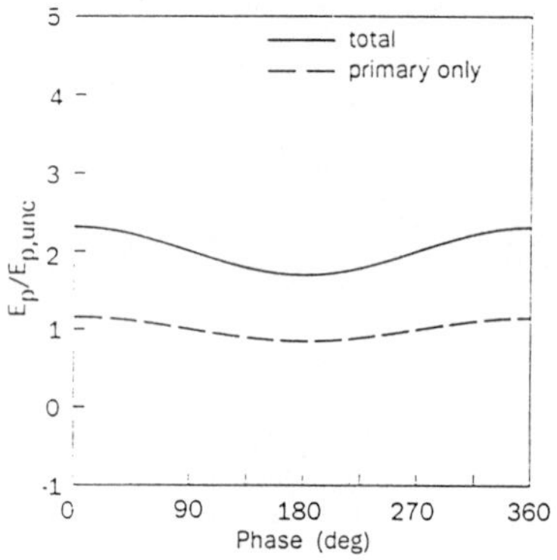

Fig. 9.9 Total normalized acoustic potential energy with equal amplitude primary and control inputs as a function of phase difference, plotted with the individual source contribution; control source at $x = L_x$, 125 Hz excitation.

It was shown theoretically that under optimum conditions the contribution of the acoustic control source(s) to the acoustic potential energy in the enclosure will be zero. Away from the optimum this will not be the case. Figures 9.8 and 9.9 illustrate the acoustic potential energy, and the primary and control source contributions, as a function of phase when the volume velocity amplitudes are fixed to be equal. In Fig. 9.8 data from a frequency of 100 Hz is plotted. In this figure, the primary and control source contributions are identical, and are shown plotted as one line. This is at the resonance frequency of a lightly damped mode, so that ideally the volume velocity amplitudes should practically be equal. Observe that in this instance there are only positive contributions to acoustic potential energy, and no negative. This is also the case at 125 Hz, the data for which are shown in Fig. 9.9. Here there is a net increase in acoustic potential energy for all phase differences, as the decrease in amplitude of one mode is offset by an increase in another. This result is different from that seen in the previous chapter for free field radiation, where it was seen that altering the phase difference between two closely spaced monopole sources of equal amplitude will cause one source to begin absorbing acoustic power.

To force one source to make a negative contribution to the acoustic potential energy total it is necessary to change the relative volume velocity amplitudes. Figure 9.10 illustrates the total acoustic potential energy, and the primary and control source contributions, for a control source mounted at $x = L_x$, operating out of phase with the primary source at 100 Hz, as a function of control source amplitude. Note that at low amplitudes, the control source is making a negative contribution to the total, or 'absorbing' acoustic potential energy (in fact, in a real source it is simply using the pressure field to help 'push' the source,

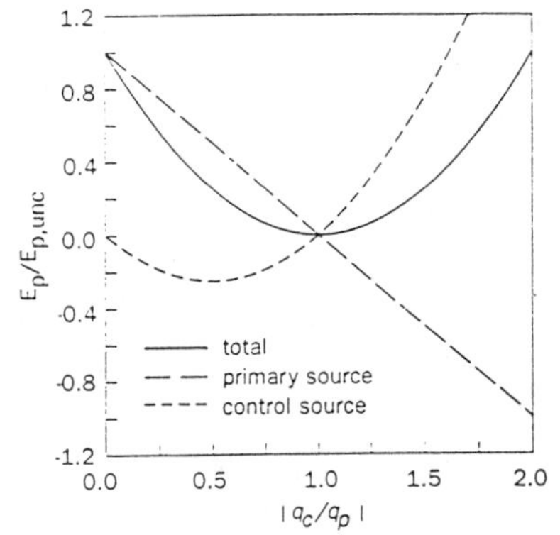

Fig. 9.10 Total and individual source normalized acoustic potential energy with out of phase primary and control inputs as a function of volume velocity amplitude ratio, control source at $x = L_x$, 100 Hz.

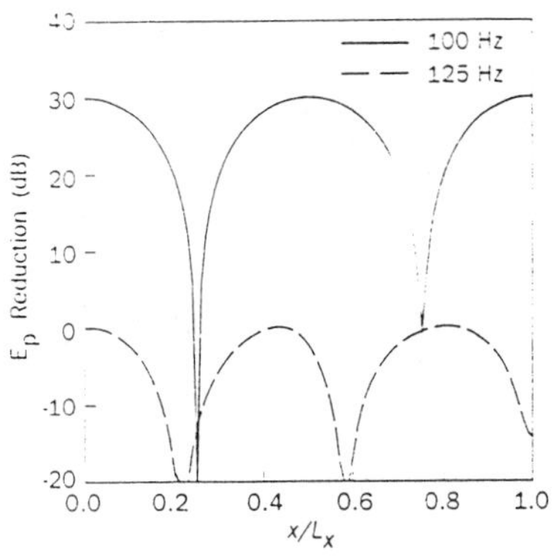

Fig. 9.11 Acoustic potential energy reduction resulting from minimizing the output of a single error sensor as a function of location of error sensor location for a single point control source at $x = L_x$ for 100 Hz and 125 Hz excitation.

reducing its mechanical impedance). At high control source amplitudes these roles have reversed. Only at one point, the optimum amplitude of approximately 1.0 for this frequency, is the control source neither producing or absorbing, and it is at this point the maximum attenuation of acoustic potential energy occurs.

Having now seen how much attenuation is maximally possible for a given control source arrangement, it is time to consider including a microphone into the system to make it practically realisable. Figure 9.11 illustrates the levels of acoustic potential energy reduction with a control source at $x = L_x$ as function of error microphone location for two frequencies, 100 Hz, which is dominated by the resonant response of the (2,0,0) mode, and 125 Hz, the response at which is not dominated by a single mode. For the 100 Hz case, significant levels of attenuation can be achieved at any error microphone location away from the nodes of the resonant mode, as the resonant mode dominates the pressure field as well as the acoustic potential energy. For the 125 Hz case, however, at most locations there is a significant increase in the sound field when minimizing acoustic pressure at a single point. These results are reflected in the sound field plots shown in Figs. 9.12 and 9.13. Observe that while there is a reduction in acoustic pressure at the error sensor for the 125 Hz case there is an overall increase in the sound pressure level or acoustic potential energy.

While considering residual sound fields, it is insightful to compare the residual sound fields under the action of an optimally adjusted control source, shown in Fig. 9.14 for the 100 Hz case and Fig. 9.13 for the 125 Hz case, with the plot of acoustic potential energy reduction as a function of error sensor position given in Fig. 9.11. Note that the locations of maximum pressure

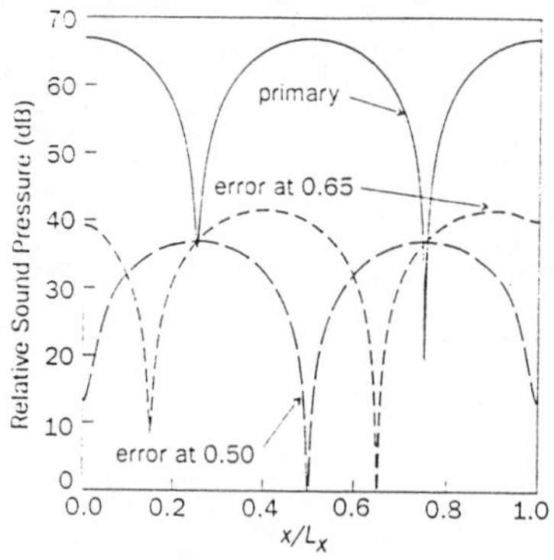

Fig. 9.12 Primary and controlled sound fields resulting from minimizing acoustic pressure at a single location, control source at $x = L_x$, 100 Hz. Illustrated are the primary sound field, the controlled sound field with an error sensor at $x/L_x = 0.50$, and the controlled sound field with an error at $x/L_x = 0.65$.

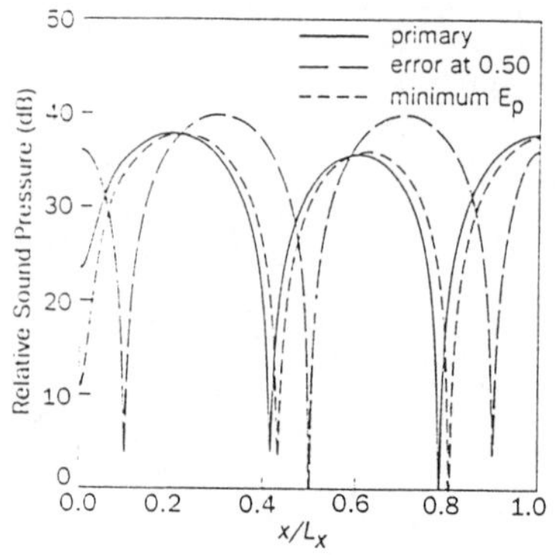

Fig. 9.13 Primary and controlled sound fields resulting from minimizing acoustic pressure at a single location, control source at $x = L_x$, 125 Hz. Illustrated are the primary sound field, the controlled sound field with an error sensor at $x/L_x = 0.50$, and the controlled sound field with an error at $x/L_x = 0.65$.

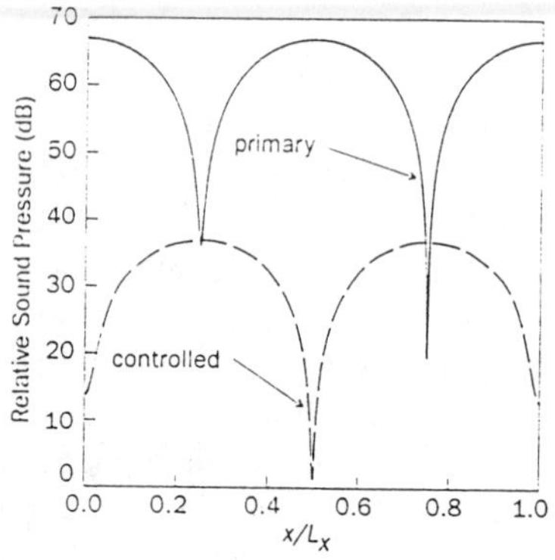

Fig. 9.14 Primary and controlled sound fields resultant from minimizing acoustic potential energy, control source at $x = L_x$, 100 Hz.

difference between the primary and optimal residual fields (determined that the acoustic potential energy can be accurately measured) are also the locations at which a single error sensor should be placed to obtain the maximum possible reduction in acoustic potential energy as a result of minimizing the acoustic pressure at one point. This is a general result, to be discussed further in Section 9.7, which can be used to optimally place acoustic error sensors once a control source arrangement has been decided upon.

A conclusion which can be drawn from this short example is that active control of sound fields in lightly damped enclosed spaces is most effective at resonances of the acoustic space. In these instances the problem is essentially the control of a single mode, which increases the potential for attenuation as well as making the placement of control sources and error sensors a more forgiving task. For multiple modes to be controlled the number of control sources and error sensors can be increased; however, the potential for attenuation is never as great as at a resonance frequency.

9.4 CONTROL OF SOUND FIELDS IN COUPLED ENCLOSURES AT DISCRETE LOCATIONS

In the previous two sections we have concentrated on the active control of harmonic sound fields in rigid walled enclosures, where discrete sound sources either contained within the enclosed space or on its boundary were solely responsible for the sound field. Many practical systems, however, have enclosed sound fields which actually result from the vibration of part or all of the enclosing structure. A common example of this is an aircraft, where the interior (cabin) noise is generated by the vibration of the fuselage. While the analysis of the previous two sections will provide useful qualitative results for coupled structural/acoustic systems such as this, in many ways it falls short. An obvious example of this is the fact that it is not possible to examine the use of vibration sources for controlling sound transmission and vibration sensors for providing an error signal unless the vibration of the enclosing structure is included in the model. The aim of the next two sections is to extend the work of the previous two sections by considering the active control of harmonic sound fields in coupled enclosures, where the vibration of the surrounding structure is responsible for the noise disturbance.

Several methods of modelling coupled enclosures for active control have been utilized in the past (Bullmore *et al.*, 1986; Fuller, 1986; Pan *et al.*, 1990). The one to be developed here is based upon modal coupling theory (Pope, 1971; Fahy, 1985; Pan and Bies, 1990a; Pan *et al.*, 1990; Snyder and Hansen, 1994a). Modal coupling theory uses the premise that there is 'weak' modal coupling between the vibrating structural and acoustic modes, such that the *in vacuo* mode shapes of the structure and the mode shapes of a rigidly enclosed space can be used to determine the coupled system response. If the fluid medium is non-dense, such as air, and the structure is not 'small', weak coupling can normally be assumed and this approach will provide an accurate model of the system response. The advantage of using modal coupling theory is that theoretically any weakly coupled system can be considered, the only requirement being a knowledge of the structural and acoustic resonance frequencies, the associated mode shape functions, and a calculation of the coupling between them. This section will develop the basic equations required for modelling the coupled system response, then go on to apply them to the problem of minimizing the acoustic field at discrete error sensing locations. The section which follows will

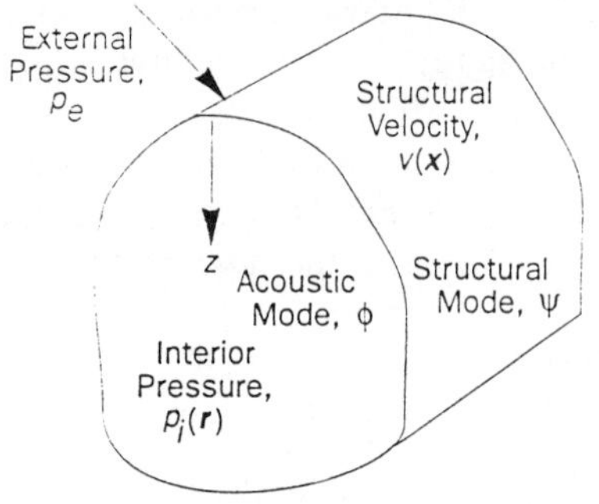

Fig. 9.15 Sound transmission into a coupled enclosure.

extend this analysis to consider minimization of acoustic potential energy in the enclosed space, as well as several 'hybrid' error criteria. Specific examples using the theoretical models developed here and in the next section will be given in Section 9.6 during a discussion of the control mechanisms, and in Section 9.7 during a discussion of the influence of control source and error sensor location.

As in the previous studies of rigid walled enclosures the aim of the analysis outlined here is to derive an equation defining acoustic pressure at a point as a matrix expression. From this equation it will be relatively easy to derive a relationship for the control source outputs which will minimize the pressure. The first step in the analysis is to derive an expression for the structural velocity levels which result from a given external forcing function. Once these are known we can then go on to examine the resulting interior sound field. Referring to Fig. 9.15, the velocity at any location $\boldsymbol{x}$ on the vibrating structure is defined by the relationship

$$v(\boldsymbol{x}) = j\omega \int_S G_s(\boldsymbol{x}' \mid \boldsymbol{x})(p_{ext}(\boldsymbol{x}') - p(\boldsymbol{x}'))\, \mathrm{d}\boldsymbol{x}' \tag{9.4.1}$$

where ω is the frequency of interest in rads^{-1}, p_{ext} and p are the external and internal acoustic pressures, respectively, and $\boldsymbol{x}$ is a location on the vibrating structure. $G_s(\boldsymbol{x}'|\boldsymbol{x})$ is the structural Green's function, defined as

$$G_s(\boldsymbol{x}' \mid \boldsymbol{x}) = \sum_{i=1}^{\infty} \frac{\psi_i(\boldsymbol{x}')\psi_i(\boldsymbol{x})}{M_i Z_i} \tag{9.4.2}$$

Recall that in the Green's function of (9.4.2) $\psi_i(\boldsymbol{x})$ is the ith mode shape of the structure evaluated at location $\boldsymbol{x}$ on the structure and M_i and Z_i are the modal mass and *in vacuo* structural input impedance of the ith mode, defined as

$$M_i = \int_S m(\boldsymbol{x})\psi_i^2(\boldsymbol{x})\, \mathrm{d}\boldsymbol{x}, \quad Z_i = (\omega_i^2 + j\eta_i\omega_i^2 - \omega^2) \tag{9.4.3}$$

where $m(\boldsymbol{x})$ is the surface density at location $\boldsymbol{x}$, $m(\boldsymbol{x}) = \rho_s(\boldsymbol{x})h(\boldsymbol{x})$, where $\rho_s(\boldsymbol{x})$ and $h(\boldsymbol{x})$ are respectively the density and thickness of the structure at $\boldsymbol{x}$, ω_i is the resonance frequency of the ith structural mode, and η_i is its associated hysterotic loss factor. For the analysis outlined here, a more useful set of quantities than surface velocity at certain locations on the structure are modal velocity amplitudes, v_i. Equation (9.4.1) can be re-expressed in terms of these by expanding the surface velocity term in terms of modes as

$$\sum_{i=1}^{\infty} v_i \psi_i(\boldsymbol{x}) = j\omega \int_S G_s(\boldsymbol{x}' \mid \boldsymbol{x})(p_{ext}(\boldsymbol{x}') - p(\boldsymbol{x}'))\, \mathrm{d}\boldsymbol{x}' \tag{9.4.4}$$

To simplify the notation we will consider a single structural mode, mode r, the velocity of which at location $\boldsymbol{x}$ is

$$v_r \psi_r(\boldsymbol{x}) = j\omega \left\{ \int_S \frac{\psi_r(\boldsymbol{x}')\psi_r(\boldsymbol{x})}{M_r Z_r} (p_{ext}(\boldsymbol{x}') - p(\boldsymbol{x}'))\, \mathrm{d}\boldsymbol{x}' \right\} \tag{9.4.5}$$

The analysis will then be extended at the end to included consideration of the entire set of modes to be modelled.

To derive a matrix-based expression for structural velocity we must simplify the bracketed integral expression in (9.4.5). For the first term, this is done simply by rewriting it as

$$\int_S \frac{\psi_r(\boldsymbol{x}')\psi_r(\boldsymbol{x})}{M_r Z_r} p_{ext}(\boldsymbol{x}')\, \mathrm{d}\boldsymbol{x}' = \frac{\gamma_r \psi_r(\boldsymbol{x})}{M_r Z_r} \tag{9.4.6}$$

where γ_r is the rth modal generalized force,

$$\gamma_r = \int_S p_{ext}(\boldsymbol{x}')\psi_r(\boldsymbol{x}')\, \mathrm{d}\boldsymbol{x}' \tag{9.4.7}$$

The simplification of the second term in the brackets, however, will require a bit more work.

Equation (9.2.1), defining acoustic pressure in the enclosed space resulting from some volume velocity source distribution, can be specialized for the case where the source distribution is the vibrating enclosing structure by replacing acoustic mode shape functions with structural mode shape functions as the pressure of interest is adjacent to the structure surface, as

$$p(\boldsymbol{x}') = j\rho_0\omega \int_S \sum_{\iota=1}^{\infty} \frac{\psi_\iota(\boldsymbol{x}'')\psi_\iota(\boldsymbol{x}')}{\Lambda_\iota(\kappa_\iota^2 - k^2)} v(\boldsymbol{x}'')\, \mathrm{d}\boldsymbol{x}'' \tag{9.4.8}$$

If the velocity term $v(\boldsymbol{x}'')$ is expanded in terms of structural modes, this can be written as

$$p(\boldsymbol{x}') = j\rho_0\omega \int_S \sum_{\iota=1}^{\infty} \frac{\psi_\iota(\boldsymbol{x}'')\psi_\iota(\boldsymbol{x}')}{\Lambda_\iota(\kappa_\iota^2 - k^2)} \sum_{i=1}^{\infty} v_i \psi_i(\boldsymbol{x}'')\, \mathrm{d}\boldsymbol{x}'' \tag{9.4.9}$$

or

$$p(\boldsymbol{x}') = j\rho_0\omega S \sum_{\iota=1}^{\infty} \frac{\psi_\iota(\boldsymbol{x}')}{\Lambda_\iota(\kappa_\iota^2 - k^2)} \sum_{i=1}^{\infty} v_i B_{\iota,i} \tag{9.4.10}$$

where S is the surface area of the structure and $B_{\iota,i}$ is the non-dimensional modal coupling coefficient between the ιth acoustic mode and ith structural mode:

$$B_{\iota,i} = \frac{1}{S}\int_S \psi_i(x'')\phi_\iota(x'')\,\mathrm{d}x'' \tag{9.4.11}$$

Equation (9.4.10) can be used to expand the second term in the bracketed integral expression in (9.4.5) as follows:

$$\begin{aligned}&\int_S \frac{\psi_r(x')\psi_r(x)}{M_r Z_r} p(x')\,\mathrm{d}x \\ &= j\rho_0\omega S\int_S \frac{\psi_r(x')\psi_r(x)}{M_r Z_r}\sum_{\iota=1}^{\infty}\frac{\psi_\iota(x')}{\Lambda_\iota(\kappa_\iota^2-k^2)}\sum_{i=1}^{\infty} v_i B_{\iota,i}\,\mathrm{d}x\end{aligned} \tag{9.4.12}$$

Equation (9.4.12) can be further simplified by again using the modal coupling coefficient defined in (9.4.11):

$$\int_S \frac{\psi_r(x')\psi_r(x)}{M_r Z_r} p(x')\,\mathrm{d}x = j\rho_0\omega S^2\frac{\psi_r(x)}{M_r Z_r}\sum_{\iota=1}^{\infty}\sum_{i=1}^{\infty}\frac{v_i B_{\iota,r} B_{\iota,i}}{\Lambda_\iota(\kappa_\iota^2-k^2)} \tag{9.4.13}$$

Substituting the simplified expressions of (9.4.6) and (9.4.13) into (9.4.5), and dividing through by the mode shape term $\psi(x)$, produces the defining expression for the rth modal velocity amplitude in response to an external exciting force:

$$v_r = j\frac{\omega\gamma_r}{M_r Z_r} + \frac{\rho_0\omega^2 S^2}{M_r Z_r}\sum_{\iota=1}^{\infty}\sum_{i=1}^{\infty}\frac{v_i B_{\iota,r} B_{\iota,i}}{\Lambda_\iota(\kappa_\iota^2-k^2)} \tag{9.4.14}$$

This can be restated as

$$\begin{aligned}&\left[j\rho_0 S^2\omega\left[\sum_{\iota=1}^{\infty}\frac{B_{\iota,r}B_{\iota,r}}{\Lambda_\iota(\kappa_\iota^2-k^2)}\right] - \frac{jM_r Z_r}{\omega}\right] v_r \\ &+ \sum_{i=1,\neq r}^{\infty} j\rho_0 S^2\omega\left[\sum_{\iota=1}^{\infty}\frac{B_{\iota,r}B_{\iota,i}}{\Lambda_\iota(\kappa_\iota^2-k^2)}\right] v_i = \gamma_{r,p}\end{aligned} \tag{9.4.15}$$

To write (9.4.15) as a matrix expression it will be necessary to truncate the infinite summation of structural modes to N_m modes, and the infinite summation of acoustic modes to N_n modes. As was discussed in Section 9.2, such a truncation effectively limits the utility of the analysis to low frequencies, where a 'manageable' number of modes can be used in the calculations and a sufficiently accurate answer still obtained. With these truncations, the (velocity) response of the structure to the external forcing function can be written as

$$\boldsymbol{v} = \boldsymbol{Z}_I^{-1}\boldsymbol{\gamma} \tag{9.4.16}$$

where $\boldsymbol{v}$ is the ($N_m \times 1$) vector of structural modal velocities resulting from the external forcing function, $\boldsymbol{Z}_I$ is the ($N_m \times N_m$) structural modal input impedance matrix, the terms of which are defined by the expressions

$$Z_I(u,u) = j\rho_0 S^2 \omega \sum_{\iota=1}^{n} \frac{B_{\iota,u} B_{\iota,u}}{\Lambda_\iota(\kappa_\iota^2 - k^2)} - \frac{jM_u Z_u}{\omega} = \text{diagonal terms} \tag{9.4.17}$$

$$Z_I(u,v) = j\rho_0 S^2 \omega \sum_{\iota=1}^{n} \frac{B_{\iota,u} B_{\iota,v}}{\Lambda_\iota(\kappa_\iota^2 - k^2)} = \text{off diagonal terms} \tag{9.4.18}$$

where u,v refer to the uth, vth structural modes, $Z_I(u,u)$ represents the velocity response of mode u to a unit excitation force, $Z_I(u,v)$ represents the response of mode v to a unit response of mode u, and γ is the $(N_m \times 1)$ vector of modal generalized forces, the ith element of which is

$$\gamma(i) = \gamma_i = \int_S \psi_i(\mathbf{x}) p_{ext}(\mathbf{x}) \, \mathrm{d}\mathbf{x} \tag{9.4.19}$$

Having derived expressions describing the structural velocity levels for a given external forcing function, we must turn our attention back to the acoustic space to derive expressions describing the resulting sound field. If the structural velocity term in (9.4.8) is expanded using the modelled set of N_m modes, the internal pressure distribution can be expressed as

$$\begin{aligned} p(\mathbf{r}) &= j\rho_0 \omega \int_S \frac{\phi_\iota(\mathbf{x})\phi_\iota(\mathbf{r})}{\Lambda_\iota(\kappa_\iota^2 - k^2)} v(\mathbf{x}) \, \mathrm{d}\mathbf{x} \\ &= j\rho_0 \omega \sum_{\iota=1}^{N_n} \frac{\phi_\iota(\mathbf{r})}{\Lambda_\iota(\kappa_\iota^2 - k^2)} \sum_{i=1}^{N_m} v_i \int_S \psi_i(\mathbf{x}) \phi_\iota(\mathbf{x}) \, \mathrm{d}\mathbf{x} \end{aligned} \tag{9.4.20}$$

This can be written more compactly using the modal coupling coefficient defined in (9.4.11) as

$$p(\mathbf{r}) = j\rho_0 \omega S \sum_{\iota=1}^{N_n} \frac{\phi_\iota(\mathbf{r})}{\Lambda_\iota(\kappa_\iota^2 - k^2)} \sum_{i=1}^{N_m} v_i B_{\iota,i} \tag{9.4.21}$$

Expressed in this way, (9.4.21) can also be written as a matrix expression,

$$p(\mathbf{r}) = \boldsymbol{\phi}^{\mathrm{T}}(\mathbf{r})\boldsymbol{p} = \boldsymbol{\phi}^{\mathrm{T}}(\mathbf{r}) \, \mathbf{Z}_a \, \mathbf{B} \, \mathbf{v} = \boldsymbol{\phi}^{\mathrm{T}}(\mathbf{r}) \, \mathbf{Z}_a \, \mathbf{B} \, \mathbf{Z}_I^{-1} \, \boldsymbol{\gamma} \tag{9.4.22}$$

Equation (9.4.22) is the target of this analysis, expressing acoustic pressure $p(\mathbf{r})$ at any location in the enclosed space resultant from vibration of the structure. In this $\boldsymbol{p}$ is the $(N_n \times 1)$ vector of complex acoustic modal amplitudes, and $\mathbf{Z}_a$ is the $(N_n \times N_n)$ diagonal matrix of acoustic modal radiation transfer functions, the terms of which were originally defined in (9.2.15) by the relation

$$Z_a(\iota,\iota) = \frac{j\rho_0 \omega}{\Lambda_\iota(\kappa_\iota^2 - k^2)} \tag{9.4.23}$$

and $\mathbf{B}$ is the $(N_m \times N_n)$ matrix of modal coupling coefficients, the elements of which are defined by

$$B(\iota,i) = SB_{\iota,i} \tag{9.4.24}$$

Before using (9.4.22) to derive control source relationships for minimizing the acoustic pressure at discrete error sensing locations, it is interesting to compare this expression with (9.2.30), the equivalent relationship for the use of multiple discrete velocity sources in rigid walled enclosures. For a single error sensing location the first two matrix terms, $\boldsymbol{\phi}^{\mathrm{T}}(\boldsymbol{r}_e)$ and $\boldsymbol{Z}_a$, are the same. If we consider the terms in the matrix $\boldsymbol{B}$ in light of the definition of the modal coupling coefficient given in (9.4.11) we find they are equivalent to the terms in the matrix of modal generalized volume velocity transfer functions $\boldsymbol{\Phi}_g$ used in the discrete source formulation, the terms of which are defined in (9.2.11). That is, the terms in $\boldsymbol{B}$ define the volume velocity excitation of the acoustic modes in the enclosed space in response to a unit velocity amplitude for each structural mode. In (9.4.22), each structural mode is being treated in a manner analogous to a single discrete source in a rigid walled enclosure, one which can be viewed as a 'modal' approach to the problem. Later in this chapter we will touch on a different approach, one which can be viewed as a 'spatial' approach to the problem using boundary element methods, where each node on a grid of locations placed over the structure is viewed as a discrete sound source. Each of these methods has pros and cons. If mode shapes can be found the modal approach often lends itself to an enhanced physical understanding of the physical phenomena employed by active control systems. However, for complicated systems modelled using finite element methods, a spatial approach may make more sense computationally.

One point which should be mentioned briefly here is that structural/acoustic modal coupling is far more selective in terms of what acoustic modes will be excited by a given structural mode than the modal generalized volume velocity transfer functions were in determining what acoustic modes will be excited by, for example, a monopole source. In the latter case a source must be located on a nodal line of an acoustic mode for it not to be excited. Modal coupling can be selective to the point of 'allowing' a structural mode to effectively excite only a single acoustic mode in a given frequency band, as is the case for a finite cylinder (considered in more depth in Section 9.7). Figure 9.16 illustrates a coupled and uncoupled structural/acoustic modal pair. In fact, for systems of regular geometry often a necessary but not sufficient requirement for coupling is that both structural and acoustic modes be either symmetric or anti-symmetric about some line of symmetry in the structure. In Fig. 9.16, only the anti-symmetric acoustic mode couples with the anti-symmetric structural mode. This characteristic of structural/acoustic modal coupling will become important when assessing control mechanisms, as will be discussed in Section 9.7.

Consider now the problem of minimizing the sound pressure at discrete error sensing locations in the enclosed space. We have two types of control source at our disposal to achieve this objective: discrete acoustic control sources, as were considered in Section 9.2 with rigid walled enclosures, and vibration control sources attached directly to the vibrating structure. The former acoustic control sources will be considered first.

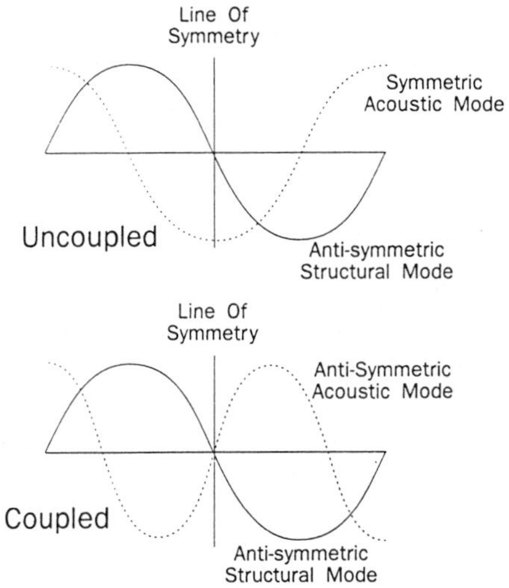

Fig. 9.16 Example of structural/acoustic modal coupling.

The squared amplitude of the acoustic pressure at some error sensing location r_e during operation of the control and primary source distributions is equal to

$$|p(r_e)|^2 = p(r_e)^* p(r_e) = [p_c(r_e)+p_p(r_e)]^* [p_c(r_e)+p_p(r_e)] \tag{9.4.25}$$

where the subscripts p and c denote primary and control source related quantities. Here the primary noise source is the vibrating structure, for which the governing acoustic pressure field relationship is given in (9.4.22). For acoustic control sources, in many weakly coupled structural/acoustic systems sufficient accuracy is attained by considering the enclosure to be rigid walled, neglecting the structural/acoustic interaction when formulating the sound field present in the enclosure during operation of the control sources alone. The exception to this is at a structural resonance, where neglecting this interaction may result in some discrepancy between the theoretical and experimental results. The interaction will be neglected here for brevity; those interested in including it can refer to the work of Pan (1992). Neglecting any structural/acoustic interaction enables us to use the previously derived relationship for the sound field in a rigid walled enclosure, (9.2.17), to describe the pressure field generated by the control source. Using (9.4.22) and (9.2.17) the squared acoustic pressure amplitude at r_e can be re-expressed as

$$|p(r_e)|^2 = q_c^* a q_c + q_c^* b + b^* q_c + c \tag{9.4.26}$$

where

$$a = \phi_{gc}^{\mathrm{H}} \mathbf{Z}_a^{\mathrm{H}} \mathbf{Z}_p \; \mathbf{Z}_a \phi_{gc} \tag{9.4.27}$$

$$b = \phi_{gc}^{H} Z_a^{H} Z_p Z_a B v_p \tag{9.4.28}$$

$$c = v_p^{H} B^{T} Z_a^{H} Z_p Z_a B v_p \tag{9.4.29}$$

where

$$Z_p = \phi(r)\phi^{T}(r) \tag{9.4.30}$$

In (9.4.28) and (9.4.29) the structural modal velocities resulting from the primary excitation v_p have been explicitly used in the equations. Alternatively, the primary acoustic modal amplitudes or structural input impedance and modal generalized force could be used, as shown in (9.4.22). Note also that when confronted with a real problem, measured data could easily be used in these primary source related matrix quantities.

Equation (9.4.26) is in what is now a fairly standard form, so that we can write the relationship for the optimum control source output directly as

$$q_{c,p} = -a^{-1}b \tag{9.4.31}$$

which will result in the minimized squared pressure amplitude at the error sensing location of

$$| p(r_e) |^2_{\min} = c - b^{*}a^{-1}b \tag{9.4.32}$$

For a single error sensor and single control source this minimum value will be theoretically equal to zero.

As with the rigid walled enclosure, this result can easily be extended to include consideration of multiple control sources and error sensing positions. Using quantities defined in Section 9.2, the sum of squared acoustic pressure amplitudes at a set of error sensing locations can be expressed as

$$\sum_{i=1}^{N_e} | p(r_e) |^2 = q_c^{H} A q_c + q_c^{H} b + b^{H} q_c + c \tag{9.4.33}$$

where

$$A = \Phi_{gc}^{H} Z_a^{H} Z_p Z_a \Phi_{gc} \tag{9.4.34}$$

$$b = \Phi_{gc}^{H} Z_a^{H} Z_p Z_a B v_p \tag{9.9.35}$$

$$c = v_p^{H} B^{T} Z_a^{H} Z_p \phi_e Z_a B v_p \tag{9.4.36}$$

where Z_p is now based upon multiple error sensing locations and is defined as

$$Z_p = \Phi_e^{T} \Phi_e \tag{9.4.37}$$

Equation (9.4.33) is again in standard form for multiple source problems, with the vector of optimum control source outputs defined by the relationship

$$q_{c,p} = c - b^{H} A^{-1} b \tag{9.4.38}$$

The sum of the residual squared acoustic pressure amplitudes under optimal conditions, found by substituting (9.4.38) into (9.4.33), is

$$\left[\sum_{i=1}^{N_e} |p(r_e)|^2\right]_{\min} = c - b^{\mathrm{H}} A^{-1} b \tag{9.4.39}$$

Let us now turn our attention to the second control source option, vibration sources. As the acoustic pressure at the error sensor(s) due to operation of these sources will be a result of structural radiation, the previous analysis, culminating in (9.4.22), will be used in formulating both the control and primary source contributions. As the systems which are of interest here are linear, it is possible to simply add the two in the analysis. What is required first is to re-express the modal generalized force term γ to directly reflect the control source input. To do this, note that the rth modal generalized force, as defined in (9.4.7), can be rewritten for a single vibration source as

$$\gamma_r = \int_S p_{ext}(x)\psi_r(x)\,\mathrm{d}x = f\int_S z_f(x)\psi_r(x)\,\mathrm{d}x \tag{9.4.40}$$

where f is a complex force amplitude, and $z_f(x)$ is a transfer function between this force and the pressure at location x. This can be simplified to

$$\gamma_r = f\psi_{g,r} \tag{9.4.41}$$

where $\psi_{g,r}$ is a modal generalized force transfer function, defined as

$$\psi_{g,r} = \int_S z_f(x)\psi_r(x)\,\mathrm{d}x \tag{9.4.42}$$

If, for example, the control source were a point force, the modal generalized force would be

$$f\,\psi_{g,r} = \int_S \psi_r(x)\,f\,\delta(x-x_f)\,\mathrm{d}x = f\,\psi(r_f) \tag{9.4.43}$$

where f is the complex amplitude of the force, input at a location x_f, and $\delta(x-x_f)$ is a Dirac delta function. In this case the modal generalized force transfer function, $\psi_{g,r}$, is simply the value of the mode shape at the point force input location, as follows

$$\psi_{g,r} = \int_S \psi_r(x)\,\delta(x-x_f)\,\mathrm{d}x = \psi_r(x_f) \tag{9.4.44}$$

Substituting this expansion in (9.4.22) gives the acoustic field generated by the vibration control source, which is defined by the relationship

$$p(r) = \phi^{\mathrm{T}}(r)\,Z_a\,B\,Z_I^{-1}\,\psi_{gc}\,f_c \tag{9.4.45}$$

where ψ_{gc} is the $(N_m \times 1)$ vector of control source modal generalized force transfer functions. Using (9.4.22) for the primary source and (9.4.45) for the control source, the squared acoustic pressure amplitude at $\boldsymbol{r}_e$ as defined in (9.4.25) can be expressed as

$$|p(\boldsymbol{r}_e)|^2 = f_c^* a f_c + f_c^* b + b^* f_c + c \tag{9.4.46}$$

where

$$a = \boldsymbol{\psi}_{gc}^{\mathrm{H}} \left\{\boldsymbol{Z}_I^{-1}\right\}^{\mathrm{H}} \boldsymbol{B}^{\mathrm{T}} \boldsymbol{Z}_a^{\mathrm{H}} \boldsymbol{Z}_p \boldsymbol{Z}_a \boldsymbol{B} \boldsymbol{Z}_I^{-1} \boldsymbol{\psi}_{gc} \tag{9.4.47}$$

$$b = \boldsymbol{\psi}_{gc}^{\mathrm{T}} \left\{\boldsymbol{Z}_I^{-1}\right\}^{\mathrm{H}} \boldsymbol{B}^{\mathrm{T}} \boldsymbol{Z}_a^{\mathrm{H}} \boldsymbol{Z}_p \boldsymbol{Z}_a \boldsymbol{B} \boldsymbol{Z}_I^{-1} \boldsymbol{v}_p \tag{9.4.48}$$

and

$$c = \boldsymbol{v}_p^{\mathrm{H}} \boldsymbol{B}^{\mathrm{T}} \boldsymbol{Z}_a^{\mathrm{H}} \boldsymbol{Z}_p \boldsymbol{Z}_a \boldsymbol{B} \boldsymbol{v}_p \tag{9.4.49}$$

where $\boldsymbol{Z}_p$ is as defined in (9.4.30). Equation (9.4.46) is written in standard form, this time as a quadratic function of vibration control source amplitude. This enables us simply to write the relationship for the optimum control force as

$$f_{c,p} = -a^{-1} b \tag{9.4.50}$$

which results in the minimized squared pressure amplitude at the error sensing location of

$$|p(r_e)|^2_{\min} = c - b^* a^{-1} b \tag{9.4.51}$$

Again, the subscripts p and c denote primary and control source, respectively, and for a single error sensor and single control source this minimum value will be theoretically equal to zero.

This analysis can also be easily extended to include consideration of multiple sources and error sensing positions. Defining the $(N_c \times 1)$ vector of control forces $\boldsymbol{f}_c$

$$\boldsymbol{f}_c = \begin{bmatrix} f_1 \\ f_2 \\ \vdots \\ f_{N_c} \end{bmatrix} \tag{9.4.52}$$

and the $(N_m \times N_c)$ matrix whose columns are the control source modal generalised force transfer function vectors $\boldsymbol{\Psi}_{gc}$

$$\boldsymbol{\Psi}_{gc} = \begin{bmatrix} \boldsymbol{\psi}_{g1} & \boldsymbol{\psi}_{g2} & \cdots & \boldsymbol{\psi}_{gN_c} \end{bmatrix} \tag{9.4.53}$$

the sum of squared acoustic pressure amplitudes at a set of error sensing locations can be expressed as

$$\sum_{i=1}^{N_e} |p(\boldsymbol{r}_e)|^2 = \boldsymbol{f}_c^{\mathrm{H}} \boldsymbol{A} \boldsymbol{f}_c + \boldsymbol{f}_c^{\mathrm{H}} \boldsymbol{b} + \boldsymbol{b}^{\mathrm{H}} \boldsymbol{f}_c + c \tag{9.4.54}$$

where

$$A = \Psi_{gc}^{\mathrm{H}} \left\{Z_I^{-1}\right\}^{\mathrm{H}} B^{\mathrm{T}} Z_a^{\mathrm{H}} Z_p Z_a B Z_I^{-1} \Psi_{gc} \tag{9.4.55}$$

$$b = \Psi_{gc}^{\mathrm{H}} \left\{Z_I^{-1}\right\}^{\mathrm{H}} B^{\mathrm{T}} Z_a^{\mathrm{H}} Z_p Z_a B \; v_p \tag{9.4.56}$$

and

$$c = v_p^{\mathrm{H}} B^{\mathrm{T}} Z_a^{\mathrm{H}} Z_p Z_a B v_p \tag{9.4.57}$$

where Z_p is now based upon multiple error sensors, as defined in (9.4.37). Equation (9.4.54) is in our standard form for multiple source problems, expressed as a quadratic function of vibration control source output. The vector of optimum control source outputs therefore is defined by the relationship

$$f_{c,p} = -A^{-1}b \tag{9.4.58}$$

The sum of the residual squared acoustic pressure amplitudes under optimal conditions, found by substituting (9.4.58) into (9.4.54), is

$$\left[\sum_{i=1}^{N_e} \left| p(r_e) \right|^2\right]_{\min} = c - b^{\mathrm{H}} A^{-1} b \tag{9.4.59}$$

9.5 MINIMIZATION OF ACOUSTIC POTENTIAL ENERGY IN COUPLED ENCLOSURES

The relationships developed in the previous section will facilitate calculation of the sound field in a coupled enclosure, as well as the acoustic or vibration control source outputs which will minimize the acoustic pressure at discrete locations in the field. What remains to be done is to extend these analyses to include consideration of the best possible result, minimization of acoustic potential energy in the enclosed space. In addition, this section will consider a few 'hybrid' control source and error sensor arrangements, such as the use of both acoustic and vibration control sources and the minimization of the velocity levels of the enclosing structure. Our first task, however, is the minimization of acoustic potential energy.

In Section 9.3, (9.3.5), we showed that acoustic potential energy is equal to the weight sum of squared acoustic modal amplitudes. We then derived a vector relationship describing the acoustic modal amplitudes resulting from the operation of a discrete acoustic source (see (9.3.7)). What we require here then is a matrix expression for the vector of acoustic modal amplitudes p resulting from excitation by the surrounding structure. In light of (9.4.21) and (9.4.22) this can be written directly as

$$p = Z_a B v = Z_a B Z_I^{-1} \gamma \tag{9.5.1}$$

Considering first the use of acoustic control sources, the acoustic potential energy in the enclosure can be written in matrix form as

$$E_p = \frac{1}{4\rho_o c_o^2} \sum_{\iota=1}^{N_n} \Lambda_\iota \left| p_\iota \right|^2 = q_c^{\mathrm{H}} A q_c + q_c^{\mathrm{H}} b + b^{\mathrm{H}} q_c + c \tag{9.5.2}$$

where

$$A = \Phi_{gc}^{H} Z_a^{H} Z_E Z_a \Phi_{gc} \tag{9.5.3}$$

$$b = \Phi_{gc}^{H} Z_a^{H} Z_E Z_a B v_p \tag{9.5.4}$$

and

$$c = v_p^{H} B^{T} Z_a^{H} Z_E Z_a B v_p \tag{9.5.5}$$

where the elements of the diagonal weighting matrix Z_E are defined by

$$Z_E(i,i) = \frac{\Lambda_i}{4\rho_0 c_0^2} \tag{9.5.6}$$

Equation (9.5.2) is in standard form for multiple source problems, with the vector of optimum control source outputs defined by the relationship

$$q_{c,E} = -A^{-1}b \tag{9.5.7}$$

The minimum value of acoustic potential energy under optimal conditions, found by substituting (9.5.7) into (9.5.2), is

$$E_{p,\min} = c - b^{H}A^{-1}b \tag{9.5.8}$$

A virtually identical set of equations can be written for the use of vibration control sources to minimize acoustic potential energy in the coupled enclosure. Using (9.5.1) in conjunction with the expansion of the modal generalized force described in (9.4.41) the acoustic potential energy in the enclosure can be written as a quadratic function of the vector of vibration control source outputs as

$$E_p = \frac{1}{4\rho_o c_o^2} \sum_{\iota=1}^{N_n} \Lambda_\iota \, | p_\iota |^2 = f_c^{H} A f_c + f_c^{H} b + b^{H} f_c + c \tag{9.5.9}$$

where

$$A = \Psi_{gc}^{T} \left\{ Z_I^{-1} \right\}^{H} B^{T} Z_a^{H} Z_E Z_a B Z_I^{-1} \Psi_{gc} \tag{9.5.10}$$

$$b = \Psi_{gc}^{T} \left\{ Z_I^{-1} \right\}^{H} B^{T} Z_a^{H} Z_E Z_a B v_p \tag{9.5.11}$$

and c and Z_E are as defined in (9.5.5) and (9.5.6) respectively. Written in this form the relationship defining the vector of optimum control source outputs can be written directly as

$$f_{c,E} = -A^{-1}b \tag{9.5.12}$$

leading to a minimum value of acoustic potential energy equal to

$$E_{p,\min} = c - b^{H}A^{-1}b \tag{9.5.13}$$

In addition to the basic system arrangements involving the use of either acoustic or vibration control sources to minimize the sound pressure at discrete points or throughout the enclosed space, there are a number of 'hybrid' control source and error sensor arrangements which may be worthwhile investigating in some situations. The first of these is the use of both acoustic and vibration control sources to minimize acoustic pressure or potential energy. The analytical model required to describe this is a simple extension of those presented in the previous two sections, where the control source generated sound pressure is now the summation of that generated by the vibration control source(s), $p_{cv}(\boldsymbol{r}_e)$, and the acoustic control source(s), $p_{ca}(\boldsymbol{r}_e)$:

$$p_c(\boldsymbol{r}_e) = p_{cv}(\boldsymbol{r}_e)+p_{ca}(\boldsymbol{r}_e) \tag{9.5.14}$$

Using the matrices outlined previously in Section 9.4, the vector of acoustic pressures at a set of error sensor locations in the enclosure can be expressed as

$$\boldsymbol{p}_{ce} = \left[\boldsymbol{\Phi}_e \boldsymbol{Z}_a \boldsymbol{B} \boldsymbol{Z}_I^{-1} \boldsymbol{\Psi}_{gc} \quad | \quad \boldsymbol{\Phi}_e \boldsymbol{Z}_a \boldsymbol{\Phi}_{gc}\right] \begin{bmatrix} \boldsymbol{f}_c \\ \boldsymbol{q}_c \end{bmatrix} \tag{9.5.15}$$

Using this relationship, the sum of the squared sound pressures at a set of error sensing locations is

$$\sum_{i=1}^{l} | p(\boldsymbol{r}_i) |^2 = \begin{bmatrix} \boldsymbol{f}_c \\ \boldsymbol{q}_c \end{bmatrix}^{\mathrm{H}} \boldsymbol{A} \begin{bmatrix} \boldsymbol{f}_c \\ \boldsymbol{q}_c \end{bmatrix} + \begin{bmatrix} \boldsymbol{f}_c \\ \boldsymbol{q}_c \end{bmatrix}^{\mathrm{H}} \boldsymbol{b} + \boldsymbol{b}^{\mathrm{H}} \begin{bmatrix} \boldsymbol{f}_c \\ \boldsymbol{q}_c \end{bmatrix} + c \tag{9.5.16}$$

where

$$\boldsymbol{A} = \left[\boldsymbol{\Phi}_e \boldsymbol{Z}_a \boldsymbol{B} \boldsymbol{Z}_I^{-1} \boldsymbol{\Psi}_{gc} \quad | \quad \boldsymbol{\Phi}_e \boldsymbol{Z}_a \boldsymbol{\Phi}_{gc}\right]^{\mathrm{H}} \left[\boldsymbol{\Phi}_e \boldsymbol{Z}_a \boldsymbol{B} \boldsymbol{Z}_I^{-1} \boldsymbol{\Psi}_{gc} \quad | \quad \boldsymbol{\Phi}_e \boldsymbol{Z}_a \boldsymbol{\Phi}_{gc}\right] \tag{9.5.17}$$

$$\boldsymbol{b} = \left[\boldsymbol{\Phi}_e \boldsymbol{Z}_a \boldsymbol{B} \boldsymbol{Z}_I^{-1} \boldsymbol{\Psi}_{gc} \quad | \quad \boldsymbol{\Phi}_e \boldsymbol{Z}_a \boldsymbol{\Phi}_{gc}\right]^{\mathrm{H}} \boldsymbol{\Phi}_e^{\mathrm{T}} \boldsymbol{Z}_a \boldsymbol{B} \boldsymbol{v}_p \tag{9.5.18}$$

and c is as defined in (9.4.36). The optimum set of control forces and volume velocities has the same form as in the previous sections, and is

$$\begin{bmatrix} \boldsymbol{f}_c \\ \boldsymbol{q}_c \end{bmatrix}_{opt} = -\boldsymbol{A}^{-1} \boldsymbol{b} \tag{9.5.19}$$

If minimization of acoustic potential energy is the error criterion of interest, the same form of equation as stated in (9.5.16) is used, with

$$\boldsymbol{A} = \begin{bmatrix} \boldsymbol{A}_v & \boldsymbol{A}_{vq}^{\mathrm{H}} \\ \boldsymbol{A}_{vq} & \boldsymbol{A}_q \end{bmatrix} \tag{9.5.20}$$

$$\boldsymbol{b} = \left[\boldsymbol{Z}_a \boldsymbol{B} \boldsymbol{Z}_I^{-1} \boldsymbol{\Psi}_{gc} \quad | \quad \boldsymbol{Z}_a \boldsymbol{\Phi}_{gc}\right]^{\mathrm{H}} \boldsymbol{Z}_a \boldsymbol{B} \boldsymbol{v}_p \tag{9.5.21}$$

and with c as defined in (9.5.5). Here $\boldsymbol{A}_v$ is the matrix $\boldsymbol{A}$ for vibration control sources operating alone, as defined in (9.5.10), $\boldsymbol{A}_q$ is the matrix $\boldsymbol{A}$ for acoustic control source operating alone, as defined in (9.5.3), and $\boldsymbol{A}_{vq}$ is

$$\boldsymbol{A}_{vq} = \boldsymbol{\Phi}_{gc}^{\mathrm{T}} \boldsymbol{Z}_a^{\mathrm{H}} \boldsymbol{Z}_E \boldsymbol{Z}_a \boldsymbol{B} \boldsymbol{Z}_I^{-1} \boldsymbol{\Psi}_{gc} \tag{9.5.22}$$

The next error criterion which may be of interest is the minimization of tructural velocity levels at discrete points, or the global error criterion of kinetic energy of the structure, when vibration control sources are used. As with sound pressure, the surface velocity at any location can be considered as the sum of the primary generated and control source generated velocities:

$$v(\boldsymbol{x}) = v_p(\boldsymbol{x}) + v_c(\boldsymbol{x}) \tag{9.5.23}$$

In terms of the matrices defined previously, the velocity levels at any point on the structure resulting from the primary and control sources can be expressed as

$$v_p(\boldsymbol{x}) = \boldsymbol{\psi}^{\mathrm{T}}(\boldsymbol{x})\boldsymbol{v}_p \tag{9.5.24}$$

and

$$v_c(\boldsymbol{x}) = \boldsymbol{\psi}^{\mathrm{T}}(\boldsymbol{x})\boldsymbol{Z}_I^{-1}\boldsymbol{\Psi}_{gc}\,\boldsymbol{f}_c \tag{9.5.25}$$

where $\boldsymbol{\psi}(\boldsymbol{x})$ is the $(N_m \times 1)$ vector of structural mode shape functions evaluated at location $\boldsymbol{x}$. Using these expressions, the sum of squared velocity levels at a set of error sensing points can be expressed as

$$\sum_{i=1}^{\iota} |\, v(\boldsymbol{x}_i)\,|^2 = \boldsymbol{f}_c^{\mathrm{H}}\boldsymbol{A}\boldsymbol{f}_c + \boldsymbol{f}_c^{\mathrm{H}}\boldsymbol{b} + \boldsymbol{b}^{\mathrm{H}}\boldsymbol{f}_c + c \tag{9.5.26}$$

where

$$\boldsymbol{A} = \boldsymbol{\Psi}_{gc}^{\mathrm{H}}\left\{\boldsymbol{Z}_I^{-1}\right\}^{\mathrm{H}}\boldsymbol{Z}_v\,\boldsymbol{Z}_I^{-1}\,\boldsymbol{\Psi}_{gc} \tag{9.5.27}$$

$$\boldsymbol{b} = \boldsymbol{\Psi}_{gc}^{\mathrm{H}}\left\{\boldsymbol{Z}_I^{-1}\right\}^{\mathrm{H}}\boldsymbol{Z}_v\,\boldsymbol{v}_p \tag{9.5.28}$$

$$c = \boldsymbol{v}_p^{\mathrm{H}}\,\boldsymbol{Z}_v\,\boldsymbol{v}_p \tag{9.5.29}$$

where $\boldsymbol{Z}_v$ is the weighting matrix based upon minimizing the velocity levels at discrete locations, defined as

$$\boldsymbol{Z}_v = \boldsymbol{\Psi}_e^{\mathrm{T}}\boldsymbol{\Psi}_e \tag{9.5.30}$$

and $\boldsymbol{\Psi}_e$ is the $(N_m \times \iota)$ matrix whose columns are the $(N_m \times 1)$ structural mode shape function vectors $\boldsymbol{\psi}^{\mathrm{T}}(\boldsymbol{x})$ evaluated at the N_e error sensing locations. By inspection it can be deduced that the optimum set of control forces, to minimize structural velocity levels, are found by using the expression

$$\boldsymbol{f}_{c,v} = -\boldsymbol{A}^{-1}\boldsymbol{b} \tag{9.5.31}$$

If minimization of the kinetic energy E_k of the structure

$$E_k = \int_S \rho_s(\boldsymbol{x})h(\boldsymbol{x})v^2(\boldsymbol{x})\,\mathrm{d}\boldsymbol{x} \tag{9.5.32}$$

is desired, the problem can be examined in the same way as the acoustic potential energy problem. Taking advantage of modal orthogonality, it can be shown that to minimize this quantity, the matrix $\boldsymbol{Z}_v$ in (9.5.30) must be replaced with a diagonal weighting matrix $\boldsymbol{Z}_k$, the terms of which are defined by

$$Z_k(i,i) = M_i \tag{9.5.33}$$

where M_i is the modal mass of the ith structural mode.

As with acoustic potential energy in Section 9.3, it will be useful here to give some further physical meaning to the concept of minimizing the kinetic energy of the structure. The velocity term in the definition of structure kinetic energy given in (9.5.32) can be expanded as the sum of modal contributions as follows:

$$v(\boldsymbol{x}) = \sum_{i=1}^{N_m} v_i \psi_i(\boldsymbol{x}) \tag{9.5.34}$$

where v_i is the complex amplitude of the ith structure mode. Therefore,

$$E_k = \int_S \rho_s(\boldsymbol{x}) h(\boldsymbol{x}) \left[\sum_{i=1}^{N_m} v_i \psi_i(\boldsymbol{x})\right]^* \left[\sum_{i=1}^{N_m} v_i \psi_i(\boldsymbol{x})\right] \mathrm{d}\boldsymbol{x} \tag{9.5.35}$$

Observe, however, that by using modal orthogonality this can be simplified to

$$E_k = \sum_{i=1}^{N_m} M_i \mid v_i \mid^2 \tag{9.5.36}$$

Thus minimizing the structure kinetic energy is equivalent to minimizing the weighted (by modal mass) sum of the squares of the structural modal amplitudes.

An extension to the use of vibration sources to control surface velocity levels is the use of vibration sources to control both surface velocity levels and interior acoustic levels, as may be encountered when both accelerometers and microphones are used as error sensors. The set of acoustic pressures and surface velocities under both primary and control excitation can be expressed in matrix form as

$$\begin{bmatrix} \boldsymbol{p} \\ \boldsymbol{v} \end{bmatrix}_p = \begin{bmatrix} \boldsymbol{\Phi}_e^{\mathrm{T}} \boldsymbol{Z}_a \boldsymbol{B} \boldsymbol{v}_p \\ \boldsymbol{\Psi}_e^{\mathrm{T}} \boldsymbol{v}_p \end{bmatrix} = \begin{bmatrix} \boldsymbol{\Phi}_e^{\mathrm{T}} \boldsymbol{Z}_a \boldsymbol{B} \\ \boldsymbol{\Psi}_e^{\mathrm{T}} \end{bmatrix} \boldsymbol{v}_p \tag{9.5.37}$$

and

$$\begin{bmatrix} \boldsymbol{p} \\ \boldsymbol{v} \end{bmatrix}_c = \begin{bmatrix} \boldsymbol{\Phi}_e^{\mathrm{T}} \boldsymbol{Z}_a \boldsymbol{B} \boldsymbol{Z}_I^{-1} \boldsymbol{\Psi}_{gc} \\ \boldsymbol{\Psi}_e^{\mathrm{T}} \boldsymbol{Z}_I^{-1} \boldsymbol{\Psi}_{gc} \end{bmatrix} \boldsymbol{f}_c = \begin{bmatrix} \boldsymbol{\Phi}_e^{\mathrm{T}} \boldsymbol{Z}_a \boldsymbol{B} \\ \boldsymbol{\Psi}_e^{\mathrm{T}} \end{bmatrix} \boldsymbol{Z}_I^{-1} \boldsymbol{\Psi}_{gc} \boldsymbol{f}_c \tag{9.5.38}$$

Therefore, the sum of the squared sound pressures and surface velocities at the set of error sensing points is

$$\begin{bmatrix} \boldsymbol{p} \\ \boldsymbol{v} \end{bmatrix}^{\mathrm{H}} \begin{bmatrix} \boldsymbol{p} \\ \boldsymbol{v} \end{bmatrix} = \boldsymbol{f}_c^{\mathrm{H}} \boldsymbol{A} \boldsymbol{f}_c + \boldsymbol{f}_c^{\mathrm{H}} \boldsymbol{b} + \boldsymbol{b}^{\mathrm{H}} \boldsymbol{f}_c + c \tag{9.5.39}$$

where $\boldsymbol{A}$, $\boldsymbol{b}$ and c are as given in (9.5.27)–(9.5.29), with $\boldsymbol{Z}_v$ now equal to

$$Z_v = \begin{bmatrix} \Phi_e^T Z_a B \\ \Psi_e^T \end{bmatrix}^H \begin{bmatrix} \Phi_e^T Z_a B \\ \Psi_e^T \end{bmatrix} \tag{9.5.40}$$

One point to note when modelling systems utilizing 'mixed medium' error sensors is that in a practical implementation these transducers will have different amplifications, which will inherently introduce some weighting into the problem. This may have an influence upon the overall control obtained by the system, and must be considered in the analytical modelling stage. This can be done by incorporating scalar weighting factors into the Z_v matrix of (9.5.36), producing

$$Z_w = \begin{bmatrix} w_a \Phi_e^T Z_a B \\ w_v \Psi_e^T \end{bmatrix}^H \begin{bmatrix} w_a \Phi_e^T Z_a B \\ w_v \Psi_e^T \end{bmatrix} \tag{9.5.41}$$

where w_a, w_v are scalar weighting factors for the minimization of the acoustic and vibration error sensors respectively. The scalar weighting factors may be replaced with vector quantities to allow more importance to be placed on some error sensors than others, enabling the sound levels to be reduced further at critical locations.

9.5.1 Multiple regression as a short cut

To implement the previous analytical models in a numerical search routine to optimize the placement of control sources/error sensors can be a lengthy process, and may not be practical in some instances due to the complexity of the structural/acoustic system, nor desired if the aim of the analysis is simply to discriminate between possible design options. The process can, however, be simplified using multiple regression (Snyder and Hansen, 1991, 1994a). This is a generalized linear least-squares technique, where several independent variables are used to predict the dependent variable of interest. Here the dependent variable of interest is the 180° inverse of the primary sound field (which will provide the greatest level of acoustic attenuation), while the independent variables are the control source transfer functions (dependent only upon the position of the sources and sensors) and volume velocities or forces. As will be seen shortly, one advantage to this approach is the ease with which measured data can be included in the design process.

Referring to (9.3.1), it can be concluded that minimization of the acoustic potential energy is equivalent to minimizing the average squared sound pressure in the enclosure or, using a finite number of points is equivalent to minimizing

$$\sum_{i=1}^{N} | p_p(r_i) + p_c(r_i) |^2 \tag{9.5.42}$$

where the number of points N at which the pressure is to be minimized should ideally be chosen so that (9.5.42) is representative of the acoustic potential energy; if the aim is simply to discriminate between design options, the number of points considered can be substantially reduced. For vibration control sources, the control generated sound pressure at any point i can be expressed as

$$p_c(r_i) = z_{vt}^{\mathrm{T}}(r_i)\, f_c \tag{9.5.43}$$

where the terms in $z_{vt}(r_i)$ could be calculated analytically as

$$z_{vt}^{\mathrm{T}}(r_i) = \boldsymbol{\phi}^{\mathrm{T}}(r_i)\boldsymbol{Z}_a\boldsymbol{B}\boldsymbol{Z}_I^{-1}\boldsymbol{\Psi}_{gc} \tag{9.5.44}$$

or measured data could be used. For acoustic control sources, $p_c(r_i)$ is defined by the expression

$$p_c(r_i) = z_{at}^{\mathrm{T}} q_c \tag{9.5.45}$$

where analytically

$$z_{at}^{\mathrm{T}} = \boldsymbol{\phi}^{\mathrm{T}}(r)\boldsymbol{Z}_a\boldsymbol{\Phi}_{gc} \tag{9.5.46}$$

Alternatively, measured data could again be used. Note also that the primary generated sound pressure, as well as the transfer functions between the response at the control source locations and any point on the structure or in the acoustic space, can either be calculated using the previously outlined analytical methods, or measured *in situ*.

Substituting these relations into (9.5.42) gives the optimization criterion for L vibration control sources; that is, the following expression should be minimized:

$$\sum_{i=1}^{N} \left| \, p_p(r_i) + \sum_{\iota=1}^{L} z_{vt}(r_i,\iota) f_\iota \, \right|^2 \tag{9.5.47}$$

and for L acoustic control sources, the following should be minimized:

$$\sum_{i=1}^{N} \left| \, p_p(r_i) + \sum_{\iota=1}^{L} z_{at}(r_i,\iota) q_\iota \, \right|^2 \tag{9.5.48}$$

Equations (9.5.47) and (9.5.48) can be written in matrix form. For vibration control sources, equation (9.5.47) becomes

$$\left| \, \boldsymbol{Z}_{vt} \boldsymbol{f}_c - (-\boldsymbol{p}_p) \, \right|^2 \tag{9.5.49}$$

where $\boldsymbol{Z}_{vt}$ is the $(N \times N_c)$ vector of transposed transfer function vectors $z_{vt}^{\mathrm{T}}(r_i)$ between the control force inputs and the error sensing locations. For acoustic control sources, (9.5.48) becomes:

$$\left| \, \boldsymbol{Z}_{at} \boldsymbol{q}_c - (-\boldsymbol{p}_p) \, \right|^2 \tag{9.5.50}$$

where $\boldsymbol{Z}_{at}$ is the $(N \times N_c)$ vector of transposed transfer function vectors $z_{at}^{\mathrm{T}}(r_i)$ between the control volume velocities and the error sensing locations.

Equations (9.5.49) and (9.5.50) can be solved by various methods, such as by the use of singular value decomposition, or by using one of many commercially available multiple regression software packages. The control forces and/or control volume velocities that result are those which are optimal for the given control source/error sensor positions. (It should be noted that a combination of acoustic and vibration control sources can be optimized in the same manner.)

The level of acoustic potential energy attenuation achieved using active noise control can be estimated as

$$\Delta W = -10 \log_{10} \left[\frac{\sum_{i=1}^{N} | p_p(r_i) + p_c(r_i) |^2}{\sum_{i=1}^{N} | p_p(r_i) |^2} \right] \tag{9.5.51}$$

From (9.5.47) and (9.5.48), it can be seen that the desired control source sound pressure, $p_c(r_i)$, is actually the estimated inverse of the primary source sound pressure, $-p_p(r_i)$. Therefore, (9.5.51) can be written as

$$\Delta W = -10 \log_{10} \left[\frac{\sum_{i=1}^{N} | p_p(r_i) - \hat{p}_p(r_i) |^2}{\sum_{i=1}^{N} | p_p(r_i) |^2} \right] \tag{9.5.52}$$

where ^ denotes estimate. For periodic sound, the mean (complex) sound pressure is zero (for both the real and imaginary parts). Thus, (9.5.52) can be expressed as

$$\Delta W = -10 \log_{10} \left[\frac{\sum_{i=1}^{N} | p_p(r_i) - \hat{p}_p(r_i) |^2}{\sum_{i=1}^{N} | p_p(r_i) - \tilde{p}_p(r_i) |^2} \right] \tag{9.5.53}$$

where ~ denotes the mean value. Expressed in this form the denominator of (9.5.53) is equivalent to the sum of the squares of the measured dependent variable, SS_p, while the numerator is equivalent to the sum of squares of the residuals, SS_{res}. Using this notation, the estimated acoustic potential energy reduction for the given control source arrangement can be written as

$$\Delta W = -10 \log_{10} \left[\frac{SS_{res}}{SS_p} \right] \tag{9.5.54}$$

or

$$\Delta W = -10 \log_{10} \left(1 - R^2 \right) \tag{9.5.55}$$

where R is the multiple correlation coefficient (Tabachnick and Fidell, 1989). Thus the multiple correlation coefficient can be used to estimate the acoustic power reduction under optimum control for a given control source arrangement. As the number of data points increases, so does the accuracy of the estimate. For a very large number of points, this method becomes equivalent to the quadratic optimization methods of the previous section for determining the optimum control source volume velocities or forces for a given control source and error sensor arrangement.

The main advantages to be had in using a multiple regression approach are the speed of calculation, and the ease with which practical transfer function measurements can be incorporated into the design process. An active control system could be optimally designed using all measured data and a multiple regression routine as follows:

1. A grid of prospective control source and error sensor locations can be laid out, and the transfer functions between each of these measured at the frequency(s) of interest. These measurements will comprise the terms in (9.5.44) and (9.5.46). If the aim of the analysis is to discriminate between potential design choices the grid can be coarse.
2. Next, the transfer function between some arbitrary reference point and each potential error sensor location can be measured under primary excitation to make up the terms in $\boldsymbol{p}_p$.
3. Different control source locations, or combinations of locations, can be inserted into the multiple regression routine. This could be done either manually or by using one of the 'stepwise' regression procedures commonly available. The multiple convergence coefficient R^2 can be used to estimate the acoustic potential energy reduction, as outlined in (9.5.55).

One point to note concerning the implementation of the outlined multiple regression routine is that the majority of terms (acoustic pressures, forces, volume velocities and impedances) are complex, and therefore cannot be directly incorporated into a commercial package. Rather, the real and imaginary components of the acoustic pressure at each location must be considered separately, doubling the size of the problem. Consider, for example, (9.5.49). For one pressure point and one control source, this can be written as

$$(z_{vtR} + jz_{vtI})(f_{cR} + jf_{cI}) = (-p_{pR} - jp_{pI}) \tag{9.5.56}$$

where the subscripts R and I refer to real and imaginary components respectively. Equation (9.5.56) can be written in a matrix form suitable for implementation as

$$\begin{bmatrix} z_{vtR} & -z_{vtI} \\ z_{vtI} & z_{vtR} \end{bmatrix} \begin{bmatrix} f_{cR} \\ f_{cI} \end{bmatrix} = \begin{bmatrix} -p_{pR} \\ -p_{pI} \end{bmatrix} \tag{9.5.57}$$

Besides those already mentioned, there is one further advantage to using multiple regression. As we have already seen in Section 9.3 and will discuss further in Section 9.7, the optimum error sensor locations are at the points of maximum sound pressure difference between the primary and optimally controlled residual sound fields. These points can be determined directly using a commercial multiple regression package, and the error sensors located accordingly. These packages, as part of their output data, generally produce a vector of residuals, which are the differences between the measured quantities (pressure here), and the values predicted by the regression equation. The point of minimum acoustic pressure in the residual sound field will be the point associated with the smallest residual value.

9.6 CALCULATION OF OPTIMAL CONTROL SOURCE VOLUME VELOCITIES USING BOUNDARY ELEMENT METHODS

In Sections 9.4 and 9.5 we developed equations that enabled calculation of the sound field in a coupled enclosure, as well as the control source outputs which would minimize this field. At that time we noted that this was essentially a modal approach to the problem, where each structural mode was being treated as a discrete velocity source. In this section we will briefly consider another approach to the problem, a spatial approach, where individual points on the enclosure boundary are explicitly considered in the calculation. This approach is based upon boundary element methods, and may be the only viable analytical methodology when the enclosure of interest has a complex geometry. There are two commonly utilized boundary element formulations for predicting acoustic response: direct boundary element methods, based directly upon the Helmholtz integral equations, and indirect boundary element methods, based upon Huygen's principle. The one we will consider here is the latter indirect boundary element, such as developed in (Mollo and Bernhard, 1987, 1989, 1990). Those interested in formulations based upon the direct boundary element method are directed to (Cunefare and Koopmann, 1991).

As stated, the indirect boundary element method is essentially a numerical implementation of Huygen's principle, where a fictitious distribution of sources on the boundary is used in the calculations. Details on the development of this methodology can be found in (Brebbia and Walker, 1980; Banerjee and Butterfield, 1981; Bernhard *et al.*, 1986). The indirect boundary element method produces a matrix equation for the acoustic pressure at a number of points:

$$\boldsymbol{p}_f = \boldsymbol{S\alpha} + \boldsymbol{Tq} \tag{9.6.1}$$

where $\boldsymbol{p}_f$ is the ($N_p \times 1$) vector of acoustic pressures at the N_p field points, $\boldsymbol{\alpha}$ is the vector of boundary conditions associated with each element, $\boldsymbol{q}$ is the vector of volume velocities of any acoustic point sources present in the domain, and $\boldsymbol{S}$ and $\boldsymbol{T}$ are boundary element matrices.

We have shown several times in this chapter that the acoustic potential energy in an enclosed space is equal to the weighted sum of squared acoustic modal amplitudes. However, in the boundary element formulation of the problem of minimising the enclosed sound field we are not dealing explicitly with modes but rather acoustic pressure at discrete points. Therefore, we must approximate acoustic potential energy by considering the squared acoustic pressure amplitude at discrete points, similar to the approach taken in the multiple regression formulation outlined in Section 9.5. In this instance we will included a weighting term for generalization, such that the error criterion to be minimized, J_p, is the weighted sum of squared acoustic pressure amplitudes at the field points,

$$J_p = \sum_{i=1}^{N_p} w_i \, | \, p_i \, |^2 \tag{9.6.2}$$

where w_i is the weighting factor associated with point p_i. Note that inclusion of the weighting term enables either local or global control to be investigated, as some of the weighting factors can be set equal to zero. Equation (9.6.2) can be written more compactly in matrix form as

$$J_p = p_f^H W p_f \tag{9.6.3}$$

where W is a diagonal matrix of weighting terms.

To investigate active noise control, the last term in (9.6.1) can be partitioned into primary and control source components, producing

$$p_f = S\alpha + T_p q_p + T_c q_c \tag{9.6.4}$$

where the subscripts p and c denote primary and control source related quantities, respectively. Using (9.6.4) we can express the error criterion as

$$J_p = q_c^H A q_c + b^H q_c + q_c^H b + c \tag{9.6.5}$$

where

$$A = T_c^H W T_c \tag{9.6.6}$$

$$b = T_c^H W \{S\alpha + T_p q_p\} \tag{9.6.7}$$

and

$$c = \{S\alpha + T_p q_p\}^H \; W \; \{S\alpha + T_p q_p\} \tag{9.6.8}$$

Equation (9.6.5) is in exactly the same form as the error criterion expressions we developed previously for multiple control source problems. As the matrix A is positive definite we can write the solution for the optimum control source volume velocities directly as

$$q_{c,B} = -A^{-1} b \tag{9.6.9}$$

where the subscript c,B is used to denote the optimum control source volume velocities with respect to a boundary element problem. If (9.6.9) is expanded using (9.6.6) and (9.6.7) the vector of optimum control source volume velocities is found to be

$$q_{c,B} = -\{T_c^H W T_c\}^{-1} T_c^H W \{S\alpha + T_p q_p\} \tag{9.6.10}$$

Substituting (9.6.9) back into (9.6.5), the minimum value of the error criterion is found to be

$$J_{p,\min} = c - b^H A^{-1} b \tag{9.6.11}$$

Note again that the approach taken in calculating the vector of optimum control source volume velocities is the same whether a spatial or modal approach to the problem formulation is taken.

Examples of the implementation of this boundary element approach for examining active control in enclosed spaces can be found in (Mollo and Bernhard, 1987, 1989, 1990).

9.7 CONTROL MECHANISMS

The analytical models developed in Sections 9.4 and 9.5 enable calculation of a number of important parameters in the design of active noise control systems, such as how much attenuation of acoustic potential energy is possible for a given control source arrangement, and how much attenuation will result by minimising the disturbance (vibrational or acoustic) at the error sensors. These models will be used in this section to consider a more fundamental point, that of how attenuation of the enclosed sound field is achieved physically. What are the physical mechanisms? From what we know so far, attenuation can be the result of two possible classes of physical mechanism, or sources of control: acoustic control mechanisms, where the acoustic pressure is reduced without necessarily a reduction in the velocity levels of the coupled structural modes, and velocity control mechanisms, where the velocity amplitudes of the coupled structural modes are explicitly reduced. It is intuitively obvious that acoustic control sources can be sources of acoustic control, and vibration control sources can be sources of velocity control. Perhaps not so obvious is the fact that vibration sources can also be sources of acoustic control, altering but not necessarily reducing the velocity levels of the coupled structural modes while decreasing the acoustic pressure in the enclosed space. What we wish to do in this section is expand on these general concepts. As the physical mechanisms by which acoustic control sources achieve active sound attenuation are less complicated than those of vibration control sources, they will be considered first.

9.7.1 Acoustic control source mechanisms

As we discussed in Section 9.3 in relation to active control in rigid enclosures, while acoustic power is a suitable basis quantity for examination of global sound attenuation in free field structural radiation problems, it is not particularly useful in the examination of enclosed sound fields. This is because any real acoustic power flow will only be as a result of acoustic damping, as evident in (9.2.1), and will not in general reflect the global sound pressure levels in the enclosed space. This is why acoustic potential energy was chosen as a global error criterion in Sections 9.2 and 9.4, and it is this latter quantity which will be used as the basis of examination of acoustic control mechanisms.

We have seen in previous chapters that when acoustic control mechanisms are employed, two outcomes are possible: suppression, where the control source(s) mutually unload the primary source such that the acoustic radiation is reduced, and absorption, where the control source absorbs some of the primary source acoustic radiation. Therefore, to examine acoustic control mechanisms employed in transmission problems we must examine the contribution made by the noise sources to the total acoustic potential energy under optimal operating conditions. It was outlined in Section 9.3 that acoustic potential energy is equal to the weighted sum of squared acoustic modal pressure amplitudes, quantities which themselves can be expressed as the sum of the primary source, E_{pp}, and control source, E_{pc}, contributions. These contributions are defined by the relationship

$$E_{p(p/c)} = \frac{1}{4\rho_0 c_0^2} \sum_{\iota=1}^{N_n} \Lambda_\iota \operatorname{Re}\left\{ p_{\iota,p/c} \, p_{\iota,tot}^* \right\} \tag{9.7.1}$$

where p/c denotes primary or control source, *tot* denotes total (sum of primary and control source contributions), and Re denotes the real part of the expression. The vector of acoustic modal amplitudes resulting from primary excitation generated by vibration of the surrounding structure were defined previously in (9.5.1) as

$$\boldsymbol{p}_p = \boldsymbol{Z}_a \, \boldsymbol{B} \, \boldsymbol{v}_p \tag{9.7.2}$$

Once again neglecting any structural/acoustic interaction from operation of the acoustic control sources alone, the resulting acoustic modal amplitudes are as given in (9.3.7),

$$\boldsymbol{p}_c = \boldsymbol{Z}_a \, \boldsymbol{\Phi}_{gc} \, \boldsymbol{q}_c \tag{9.7.3}$$

If (9.7.2) and (9,7.3) are substituted into (9.7.1), it is straightforward to show that the control source contribution to the acoustic potential energy can be expressed as

$$E_{pc} = \boldsymbol{q}_c^{\mathrm{H}} \boldsymbol{A} \boldsymbol{q}_c + \operatorname{Re}\left\{ \boldsymbol{b}^{\mathrm{H}} \boldsymbol{q}_c \right\} \tag{9.7.4}$$

where the terms $\boldsymbol{A}$ and $\boldsymbol{b}$ are those used in the formulation of the vector of optimum acoustic control source outputs, defined in (9.5.3) and (9.5.4), respectively. Noting that $\boldsymbol{A}$ is a symmetric matrix, if the vector of optimal acoustic control source outputs (as defined in (9.5.7)) is now inserted into (9.7.4) the acoustic potential energy attributable to the control source array is

$$E_{pc} = [-\boldsymbol{A}^{-1}\boldsymbol{b}]^{\mathrm{H}} \boldsymbol{A} [-\boldsymbol{A}^{-1}\boldsymbol{b}] + \operatorname{Re}\left\{ \boldsymbol{b}^{\mathrm{H}} [-\boldsymbol{A}^{-1}\boldsymbol{b}] \right\} = 0 \tag{9.7.5}$$

That is, the acoustic potential energy attributable to the control source array is zero. This is exactly the same result as derived for the use of acoustic control sources in rigid walled enclosures, in Section 9.3. This means that the acoustic control mechanism under ideal conditions is one of pure suppression, where the control source itself is neither providing a positive contribution to the acoustic potential energy total nor absorbing any of the primary source acoustic radiation. This fact is confirmed by considering the primary source acoustic potential energy contribution under optimally controlled conditions. If (9.7.2) and (9.7.3) are again substituted into (9.7.1) to calculated the primary source acoustic potential energy, it is straightforward to show that the result can be expressed as

$$E_{pp} = \operatorname{Re}\left\{ \boldsymbol{q}_c^{\mathrm{H}} \boldsymbol{b} \right\} + c = \operatorname{Re}\left\{ \boldsymbol{b}^{\mathrm{H}} \boldsymbol{q}_c \right\} + c \tag{9.7.6}$$

where c is as defined in (9.5.5). Substituting the vector of optimum control source outputs into (9.7.6) produces

$$E_{pp} = \operatorname{Re}\left\{ -\boldsymbol{b}^{\mathrm{H}} \boldsymbol{A}^{-1} \boldsymbol{b} \right\} + c = c - \boldsymbol{b}^{\mathrm{H}} \boldsymbol{A}^{-1} \boldsymbol{b} \tag{9.7.7}$$

which is identical to the expression for the minimum acoustic potential energy of the system, as defined in (9.5.8). The primary source is the only non-zero contributor to this total.

As we noted in Section 9.3, simply having a total contribution of zero to the acoustic potential energy from the control source array does not necessarily mean that contributions from the constituent elements will be zero. The acoustic potential energy attributable to a single acoustic control source in the array can be expressed as

$$E_{pc}(i) = \mathrm{Re}\{\boldsymbol{q}_c^{\mathrm{H}}\boldsymbol{A}(\mathrm{col}\ i)q_c(i)\} + \mathrm{Re}\{b^*(i)q_c(i)\} \tag{9.7.8}$$

where $\boldsymbol{A}(\mathrm{col}\ i)$ is the ith column of the matrix $\boldsymbol{A}$ defined in (9.5.3), and $b(i)$ is the ith element in the vector $\boldsymbol{b}$ as defined in (9.5.4). If the vector of optimum control source volume velocities is again expanded then the result is

$$E_{pc}(i) = \mathrm{Re}\{[-\boldsymbol{A}^{-1}\boldsymbol{b}]^{\mathrm{H}}\boldsymbol{A}(\mathrm{col}\ i)q_c(i)\} + \mathrm{Re}\{b(i)^*q_c(i)\} = 0 \tag{9.7.9}$$

Therefore the acoustic potential energy contribution of each element in the control source array must also be zero, as was the case for the rigid walled enclosure problem.

The fact that the control mechanism employed when using optimally tuned acoustic control sources is one of suppression infers under what general conditions the maximum levels of acoustic potential energy attenuation per control source will be obtained. In most coupled structural/acoustic systems the coupling characteristics for low frequency excitation will be 'multi-modal'; that is, any given acoustic mode will couple to several structural modes which are significantly excited and vice versa. At an acoustic resonance a single acoustic mode will dominate the acoustic pressure distribution in the enclosure, and hence the levels of acoustic potential energy. A single acoustic control source can easily excite this same single mode, and hence quite easily provide a significant reduction in pressure over the entire surface of the structure (unloading over the surface of the structure). At a structural resonance, however, several acoustic modes may be significantly excited. For the control source distribution to similarly (in relative phase and amplitude) excite the acoustic modes, control source placement and number can become critical. Therefore, generally speaking, acoustic control sources are recommended for controlling acoustic resonances. The effect of an acoustic control source is shown in Fig. 9.17, which illustrates the maximum levels of acoustic potential energy attenuation as a function of frequency for the case where a normally incident plane wave primary disturbance impinges upon a rectangular panel/cavity system (described in more detail shortly). The single acoustic control source is located in the bottom and centre of the enclosure, at a position (0.434,0.575,0.0) m, where the box interior dimensions are (0.868 × 1.15 × 1.0) m and the top panel is steel, 6 mm thick. Not surprisingly, the greatest levels of attenuation are achieved at acoustic resonances.

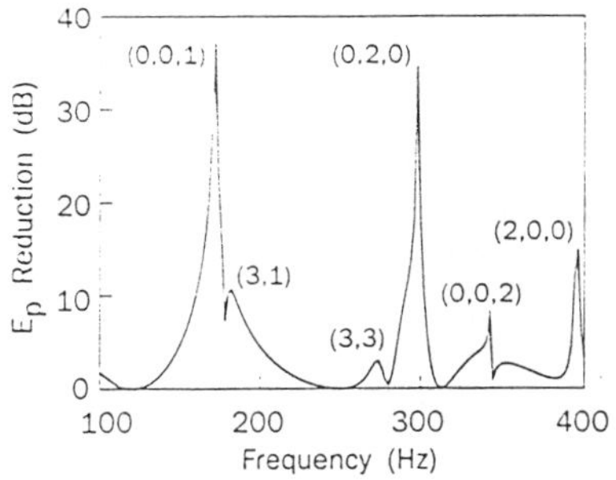

Fig. 9.17 Maximum levels of acoustic potential energy reduction for rectangular enclosure system as a function of frequency. Locations of acoustic modal resonances are represented by three integers in brackets and structural modes are represented by two integers.

9.7.2 Vibration control source mechanisms

Let us now turn our attention to the mechanisms employed by vibration control sources in providing attenuation of the acoustic potential energy in the coupled enclosure. Vibration control sources provide a reduction in the levels of acoustic potential energy in the enclosed space by altering the velocity distribution of the structure. From the viewpoint of active control, the alteration manifests itself in two different ways; a reduction in the amplitudes of the dominant coupled structural modes, and an alteration in the relative amplitudes and phases of the coupled structural modes. Each of these effects has the potential to provide attenuation, and may do so individually or in partnership with the other. For this reason the active control of sound transmission into enclosed spaces is more complicated (in terms of control mechanisms) when vibration control sources are used than when acoustic control sources are used, and a closer examination of the response of the coupled structural/acoustic system is required to obtain a qualitative overview of what is physically happening. As has been noted several times, a reduction in acoustic potential energy is equivalent to a reduction in the weighted (by mode normalization) sum of the squared acoustic modal amplitudes. Therefore, to obtain this overview we will examine the excitation of a single acoustic mode by the enclosing structure.

It was shown in Section 9.4, (9.4.21), that the amplitude of the ιth acoustic mode, excited by the vibration of the enclosing structure, is defined by the relationship

$$p_\iota = \frac{j\rho_o \omega S}{\Lambda_\iota(\kappa_\iota^2 - k^2)} \sum_{i=1}^{\infty} \beta_{\iota,i} v_i \tag{9.7.10}$$

If the infinite summation in this relationship is truncated at N_m structural modes, the expression defining the complex amplitude of the ιth acoustic mode can be written as the product of two vectors,

$$p_\iota = z_{a,\iota}^{\mathrm{T}} v \tag{9.7.11}$$

where v is the $(N_m \times 1)$ vector of structure modal velocity amplitudes and $z_{a,\iota}$ is the $(N_m \times 1)$ vector of radiation transfer functions between the structure modes and the ιth acoustic mode, the terms of which are defined by

$$z_{a,\iota}(i) = \frac{j\rho_0 \omega S}{\Lambda_\iota(\kappa_\iota^2 - k^2)} \beta_{\iota,i} \tag{9.7.12}$$

From this, the squared amplitude of the ιth acoustic mode is defined by the matrix expression

$$| p_\iota |^2 = v^{\mathrm{H}} z_{a,\iota}^{*} z_{a,\iota}^{\mathrm{T}} v \tag{9.7.13}$$

Minimizing the acoustic potential energy in the enclosed space is equivalent to minimizing a weighted set of equations of the form of (9.7.13).

Consider the problem of minimizing the amplitude of the ιth acoustic mode in light of the characteristics of its excitation by the structure as defined in (9.7.12). If the modal coupling characteristics and excitation frequency of the system are such that only one structural mode is essentially responsible for excitation of the acoustic mode, then attenuation will be provided by reducing the amplitude of that single coupled structural mode, hence by velocity, or modal, control. This control characteristic will apply to systems with extremely selective modal coupling characteristics and to systems excited near the resonance frequency of one of the (lightly damped) coupled structural modes. If, however, there are two or more structural modes which have the potential to significantly excite the acoustic mode of interest, there are two ways in which attenuation can be provided. The first is simply to reduce the amplitude of this set of structural modes, the mechanism employed again being one of velocity, or modal, control. The second way in which attenuation can be provided is to alter the relative amplitudes and phases of these structural modes such that the overall excitation of the acoustic mode, the sum of contributions from the structural modes, is reduced. The mechanism employed in this instance is one of acoustic control, where the acoustic pressure is reduced without necessarily a reduction in the velocity of the structure. In contrast to modal control this mechanism was labelled as 'modal rearrangement' in the previous chapter, where the relative complex amplitudes of the structural modes are altered, not necessarily reduced.

When there are two or more structural modes capable of significant excitation of the acoustic mode, there is then some element of choice in the control mechanism employed. The choice taken is dependent upon the characteristics of the primary forcing function and the control source arrangement. If there are as many or more control sources as there are 'significant' structural modes, then it is usually possible to provide modal control. If there are less, then the control mechanism employed is highly location dependent. If it is possible to reduce the amplitude of all structural modes simultaneously, the control mechanism employed will be velocity, or modal, control. If it is possible to simply reduce the total energy transfer from the structural modes into the acoustic mode by altering the relative complex amplitudes of the structural modes then acoustic control, or modal rearrangement, will be employed. Often the final result will be a combination of modal control and modal rearrangement. It should be emphasized that the mechanism employed is a function of the physical arrangement of the control source and error sensors; the attached electronic control control system simply aims to attenuate the unwanted disturbance at the error sensor(s), and has no effect upon the mechanism employed to do it.

To study these qualitative characteristics further it will useful to conduct a brief analytical investigation of several sound transmission problems. For this study, three separate coupled structural/acoustic systems will be considered; a rectangular panel/cavity system, a finite circular cylinder, and a finite circular cylinder with an integral longitudinal partition (floor). This combination of systems is selected for two reasons; first, solutions for the eigenvalues (natural frequencies) and eigenvectors (mode shapes) of these systems are obtained relatively easily, the rectangular and plain cylindrical systems presenting an analytically tractable problem, and the cylinder with floor requiring only a relatively simple numerical approach. Second, these three systems represent a range of structural/acoustic modal coupling characteristics, the nature and importance of which will become apparent. The first task in our study, however, is to specialize the theory of the previous two sections to the three outlined systems.

9.7.3 Specialization of theory for the rectangular enclosure case

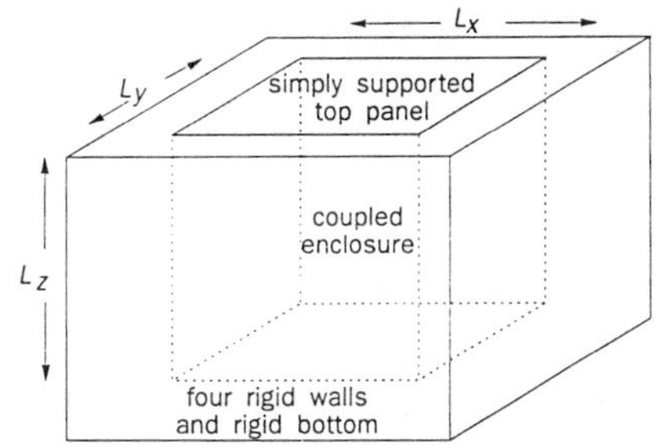

Fig. 9.18 Rectangular enclosure system.

The first system of interest is the rectangular enclosure (often referred to as a panel/cavity system) depicted in Fig. 9.18, having four rigid walls and a rigid bottom, with a flexible top comprising a simply supported rectangular panel. This arrangement has been considered extensively in published literature, both in fundamental studies of sound transmission into enclosures involving structural/acoustic modal coupling (Lyon, 1963; Pretlove, 1966; Bhattacharya and Crocker, 1969; Pan and Bies, 1990a, 1990b), and in fundamental studies of active control of sound transmission using vibration sources (Pan *et al.*, 1990; Pan and Hansen, 1991a, 1991b; Snyder and Hansen, 1994b). The reason for this is its simple geometry, which facilitates a clear study of physical phenomena. For our purposes, the simple geometry will be doubly beneficial, as it leads to what can be viewed as 'multi-modal' structural acoustic coupling at low frequencies.

As was outlined in Section 9.4 in the beginning of the development of the coupled structural/acoustic equations, the assumption of weak modal coupling implies that the response of the coupled structural/acoustic system can be described by the *in vacuo* response of the structural component, the rigid walled response of the acoustic component, and the coupling between the two. For the system shown in Fig. 9.18, the *in vacuo* structural mode shape functions of the simply supported rectangular top panel, and the acoustic mode shape functions of the rigid walled rectangular enclosure are respectively

$$\psi_{u,v}(x,y) = \sin\frac{u\pi x}{L_x}\,\sin\frac{v\pi y}{L_y} \tag{9.7.14}$$

and

$$\phi_{l,m,n}(x,y,z) = \cos\frac{l\pi x}{L_x}\,\cos\frac{m\pi y}{L_y}\,\cos\frac{n\pi z}{L_z} \tag{9.7.15}$$

where u and v are the structural modal indices, l, m, and n are the acoustic modal indices, and L_x, L_y and L_z are the (inside) dimensions of the enclosure. The origin of the coordinate system is the lower left corner of the panel shown in the figure. The structural and acoustic resonance frequencies associated with these modes are respectively

$$\omega_{u,v} = \left[\frac{D}{\rho_s h}\right]^{1/2}\left[\left[\frac{m\pi}{L_x}\right]^2 + \left[\frac{n\pi}{L_y}\right]^2\right] \tag{9.7.16}$$

and

$$\omega_{l,m,n} = c_o\pi\sqrt{\left[\frac{l}{L_x}\right]^2 + \left[\frac{m}{L_y}\right]^2 + \left[\frac{n}{L_z}\right]^2} \tag{9.7.17}$$

where D is the panel bending stiffness, given by

$$D = EI = \frac{Eh^3}{12(1-\nu^2)} \tag{9.7.18}$$

E is the modulus of elasticity, I is the moment of inertia, h is the panel thickness and ν is Poisson's ratio. The modal mass of the top panel and the acoustic mode normalisation are respectively

$$M_i = \int_S \rho_s(\boldsymbol{x})h(\boldsymbol{x})\psi_i^2(\boldsymbol{x})\,\mathrm{d}\boldsymbol{x} = \frac{\rho_s hS}{4} \tag{9.7.19}$$

and

$$\Lambda_{l,m,n} = \int_V \phi_{l,m,n}^2(\boldsymbol{r})\,\mathrm{d}\boldsymbol{r} = \frac{V}{8}\epsilon_l\epsilon_m\epsilon_n, \quad \epsilon_{l,m,n} = \begin{cases} 1 & l,m,n \neq 0 \\ 2 & l,m,n = 0 \end{cases} \tag{9.7.20}$$

where S and V are the surface area of the top panel and enclosed volume of the cavity respectively.

Once the characteristic eigenvalues and eigenvectors of the structural and acoustic subsystems are known, all that remains to be done to enable calculation of the system response is to define the coupling between them. Substituting the structural and acoustic mode shape functions outlined in (9.7.14) and (9.7.15) into the modal coupling relation, (9.4.11), shows the coupling between acoustic mode (l,m,n) and structural mode (u,v) to be defined by (Pan and Bies, 1990a)

$$B_{(u,v),(l,m,n)} = \frac{1}{S}\int_S \psi_{u,v}(\boldsymbol{x})\phi_{l,m,n}(\boldsymbol{x})\,\mathrm{d}\boldsymbol{x}$$

$$= \begin{cases} -1^n \dfrac{uv(-1^{l+u}-1)(-1^{m+v}-1)}{\pi^2(l^2-u^2)(m^2-v^2)} & l \neq u,\ m \neq v \\ 0 & \text{otherwise} \end{cases} \tag{9.7.21}$$

Note that, from (9.7.21), all odd index panel modes couple with even index acoustic modes, and vice versa. Even at low frequencies this means that several significantly excited or excitable panel modes (such as the (1,1), (1,3) and (3,1)) will be coupled to the same set of acoustic modes, each transferring energy from the structure into the acoustic space. This is what was meant by 'multi-modal' coupling at low frequencies, a characteristic which will be important in determining what physical mechanism is responsible for attenuation of the acoustic potential energy in the enclosed space.

Equations (9.7.14)–(9.7.21) can be used in the previously outlined theory to determine the acoustic pressure at a point, the acoustic potential energy in the enclosure, or the optimum control forces/volume velocities for a given error criterion. For example, if it is desired to know the acoustic pressure at some point $\boldsymbol{r}$ in the enclosure under primary excitation, where the primary forcing function and hence the modal generalised force matrix γ_p (the terms of which are defined in (9.4.7)) are known, (9.4.22) can be used, where $\phi(\boldsymbol{r})$ is now the vector of acoustic mode shape functions defined in (9.7.15) and evaluated at the point $\boldsymbol{r}$ of interest. The terms in $\boldsymbol{Z}_a$, as outlined in (9.4.23), are evaluated using the acoustic mode resonance frequencies determined from (9.7.17), and $\boldsymbol{Z}_I$, whose terms are defined in (9.4.17) and (9.4.18), is evaluated using equations (9.7.14)–(9.7.21).

9.7.4 Specialization of theory for the finite length circular cylinder case

The second system of interest, shown in Fig. 9.19, is a finite length circular cylinder, the ends of which are assumed to be simply supported and sufficiently capped such that the sound transmission through them can be ignored. This arrangement is also popular in published literature, owing to its resemblance of an aircraft fuselage. It has been examined previously in some depth using a modal coupling theoretical approach in the context of describing general sound transmission (Pope *et al.*, 1980, 1982), in which the methodology was found to

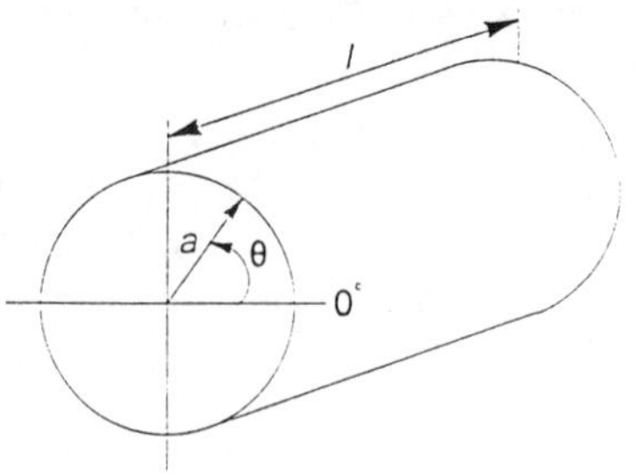

Fig. 9.19 Plain cylindrical enclosure system.

be quite accurate in its ability to predict the sound transmission. It has also been considered by a number of researchers in active control, using both acoustic and vibration control sources (for vibration control source studies, see for example (Fuller and Jones, 1987; Jones and Fuller, 1987; Snyder and Hansen, 1994b); for acoustic control source studies, see for example (Bullmore *et al.*, 1986; Silcox *et al.*, 1987)).

As was done in the previous section for the rectangular enclosure system, the quantities which are required to specialise the general theory to the circular cylinder case are the *in vacuo* response of the structure, the rigid walled response of the enclosed acoustic space, and the coupling between the two. The structural and acoustic mode shape functions for these are respectively

$$\psi_{u,v}(z,\theta) = \sin\frac{u\pi z}{L}\left(\cos(v\theta) + \sin(v\theta)\right) \tag{9.7.22}$$

and

$$\phi_{q,n,s}(r,\theta,z) = \cos\frac{q\pi z}{L}\, J_n\!\left(\frac{\gamma_{ns} r}{a}\right)\left(\cos(n\theta) + \sin(n\theta)\right) \tag{9.7.23}$$

where u,v are the axial and circumferential structural modal indices, L is the cylinder length, q,n,s, are the axial, circumferential, and radial acoustic modal indices, respectively, a is the cylinder radius, J_n is a Bessel function of the first kind of order n, and γ_{ns} is the value of the sth zero of the derivative of the Bessel function of order n:

$$J_n'(\gamma_{ns}) = 0 \tag{9.7.24}$$

Note that both sine and cosine functions are required to describe the angular distribution of the structural and acoustic modes on the cylinder.

The resonance frequencies associated with the structural mode shape functions can be found from the characteristic equation (Leissa, 1973):

$$(\Omega^2)^3 - (K_2+\tilde{h}\,\Delta K_2)(\Omega^2)^2 + (K_1+\tilde{h}\,\Delta K_1)(\Omega^2) - (K_0+\tilde{h}\,\Delta K_0) = 0 \tag{9.7.25}$$

where

$$\Omega^2 = \frac{\rho_s(1-\nu^2)}{E}a^2\omega_{u,v}^2 = \frac{a^2\omega_{u,v}^2}{c_s} \tag{9.7.26}$$

$$\tilde{h} = \frac{h^2}{12a^2} \tag{9.7.27}$$

and c_s is the speed of sound in the material. The constants K_0, K_1 and K_2 in (9.7.25) are from the Donnell–Mustari shell theory, and are defined as

$$\begin{aligned} K_0 &= \frac{1-\nu}{2}\left((1-\nu^2)\lambda^4 + \tilde{h}(v^2+\lambda^2)^4\right) \\ K_1 &= \frac{1-\nu}{2}\left[(3+2\nu)\lambda^2 + v^2 + (v^2+\lambda^2)^2 + \frac{3-\nu}{1-\nu}\tilde{h}(v^2+\lambda^2)^3\right] \\ K_2 &= 1 + \frac{3-\nu}{2}(v^2+\lambda^2) + \tilde{h}(v^2+\lambda^2)^2 \end{aligned} \tag{9.7.28}$$

where

$$\lambda = \frac{u\pi a}{L} \tag{9.7.29}$$

ΔK_0, ΔK_1, and ΔK_2 are the modifying constants of the Goldenveizer–Novozhilov/Arnold–Warburton shell theory, defined as

$$\begin{aligned} \Delta K_0 &= \frac{1-\nu}{2}\left(4(1-\nu^2)\lambda^4 + 4\lambda^2v^2 + v^4 - 2(2-\nu)(2+\nu)\lambda^4 - 8\lambda^2v^4 - 2v^6\right) \\ \Delta K_1 &= 2(1-\nu)\lambda^2 + v^2 + 2(1-\nu)\lambda^4 - (2-\nu)\lambda^2v^2 - \frac{3+\nu}{2}v^4 \\ \Delta K_2 &= 2(1-\nu)\lambda^2 + v^2 \end{aligned} \tag{9.7.30}$$

Goldenveizer–Novozhilov/Arnold–Warburton shell theory will be used in our study because of its previously demonstrated ability to accurately predict the modal response of a thin, circular finite length shell of the type we will consider (Pope *et al.*, 1980, 1982).

The cubic characteristic equation (9.7.25) has three roots for each set of structural modal indices, corresponding to three resonance frequencies, associated respectively with motion predominantly in each of the radial, tangential, and axial directions. The first root is the one associated with the radial response dominated resonance (see the discussion in Section 2.3), and will be the only one used in the calculations presented here, as it is the flexure of the cylinder which is responsible for the interior sound generation.

The modal mass associated with each structural mode (assuming that the structure has uniform material properties) is found from

$$M_i = \int_S \rho_s(x)h(x)\psi_{u,v}^2(x)\,\mathrm{d}S = \frac{\rho_s hS}{4} \tag{9.7.31}$$

where S is the surface area of the cylinder (excluding the ends) and h is the wall thickness. The resonance frequencies of the acoustic modes are found from

$$\omega_{q,n,s} = c_o \sqrt{\gamma_{ns}^2 + \left(\frac{q\pi}{L}\right)^2} \tag{9.7.32}$$

The associated acoustic mode normalization term is

$$\Lambda_{q,n,s} = \int_V \phi_{q,n,s}^2(r)\, dr = \frac{\gamma_{ns}^2 - n^2}{2\gamma_{ns}^2} \frac{J_n^2(\gamma_{ns})}{2} \pi a^2 L \epsilon_q \epsilon_n \tag{9.7.33}$$

where

$$\epsilon_q,\ \epsilon_n = \begin{cases} 2 & q,n = 0 \\ 1 & q,n > 0 \end{cases} \tag{9.7.34}$$

The final term to evaluate is the coupling factor between the structural and acoustic modes. Substituting the acoustic and structural mode shape functions of (9.7.22) and (9.7.23) into the coupling relation yields

$$B_{(u,v)(q,n,s)} = \begin{cases} \dfrac{J_n(\gamma_{ns})}{2L} \dfrac{\epsilon_n \kappa_u (1-(-1)^{u+q})}{\kappa_u^2 - \kappa_q^2} & n=v,\ \dfrac{u+q}{2} \neq \text{integer} \\ 0 & \text{otherwise} \end{cases} \tag{9.7.35}$$

where

$$\kappa_u = \frac{u\pi}{L}, \quad \kappa_q = \frac{q\pi}{L} \tag{9.7.36}$$

Note that the modal coupling characteristics of the circular cylinder, defined by (9.7.35), are much more selective than those of the rectangular enclosure given in (9.7.21). For the circular enclosure the structural and acoustic modes will only couple if the circumferential modal indices match, and the axial modal indices are an odd/even combination. For example, if the acoustic mode of interest is the (1,2,1) mode, only the (2,2), (4,2), ... structural modes will couple to it, modes whose resonance frequencies are far apart. From a practical standpoint this means that effectively only one structural mode is coupled to one acoustic mode when the excitation frequency is 'low', which is what will be considered here. This has implications for the physical control mechanism, to be considered shortly.

9.7.5 Specialization of general model for the cylinder with floor system

Use of the previously developed analytical models is not confined to systems with analytically tractable characteristic equations, with readily describable mode shape functions and resonance frequencies; rather, the analytical models are valid for any weakly coupled structural/acoustic system. One example of a system which is of interest theoretically and experimentally, but which does not have analytically tractable solutions for the mode shape functions and resonance

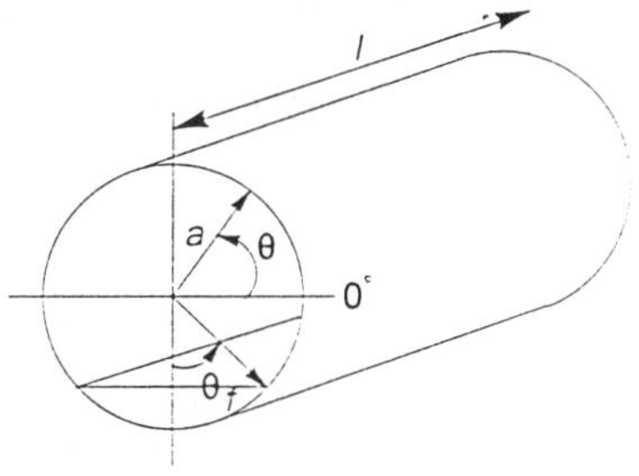

Fig. 9.20 Cylindrical enclosure with floor.

frequencies, is the finite length circular cylinder with an integral longitudinal partition, or floor, shown in Fig. 9.20. This arrangement is a more accurate representation of an aircraft fuselage than the plain circular cylinder considered in the previous section, and will be shown to be slightly different with respect to the physical control mechanisms available to reduce the acoustic potential energy in the enclosed space.

As with the two previous cases, what is required to study the active control of sound transmission into this system is a description of the response of the structural and acoustic subsystems, and the coupling between them. The response of the structural subsystem can be determined using component mode synthesis (Peterson and Boyd, 1978). Using this method, the basis function for the total structural mode shape is a combination of the mode shape basis functions for the plain cylinder and the floor (plate) structure components. Simply supported boundary conditions are assumed for the longitudinal end of both the cylinder and floor structure, so that the axial mode shape function of the structure is

$$\phi_u = \sin\frac{u\pi z}{L} \tag{9.7.37}$$

The acoustic mode shape functions can be found by using a finite difference implementation of the Helmholtz equation (Pope and Wilby, 1982; Pope *et al.*, 1983). The acoustic mode shape functions are then represented as eigenvectors of a set of values at a discrete set (grid) of points. Linear interpolation can be used to assess the value of the mode shape at any off-grid point. The axial mode shape function is equal to that of a one-dimensional enclosure,

$$\phi_q = \cos\frac{q\pi z}{L} \tag{9.7.38}$$

Coupling between the structural and acoustic modes can be evaluated numerically. When doing this evaluation it is found that the coupling characteristics become much more 'multi-modal' at low frequencies, similar to a rectangular enclosure, as the floor angle increases. By this we mean that most horizontally symmetric and anti-symmetric structural modes will couple, to some extent, to most horizontally symmetric and anti-symmetric acoustic modes, respectively, subject to the axial constraint of an odd/even combination of structural and acoustic axial modal indices. As a result, vibration control sources may have an expanded scope for employing both control mechanisms, as we will now see.

9.7.6 Examination of mechanisms

As was outlined in Section 9.4, a weakly coupled enclosure has the majority of the overall system energy associated with the individual responses of the structural and acoustic systems, with relatively little associated with the interaction between the two. Within this definition, therefore, it its possible to define two regimes of coupled system response; cavity controlled response, where the majority of the total system energy is associated with the response of the enclosed acoustic space (such as near an acoustic resonance) and structure controlled response, where the majority of the total system energy is associated with the response of the structure (such as near a structural resonance). The distinction between these two regimes becomes important in qualifying the control mechanisms associated with vibration control sources.

Consider first the case of a structure controlled response. If it is predominantly the response of the structure which is creating the noise problems, it would appear intuitively obvious that the best way to overcome the problem is to eliminate the principal offending structural mode or modes. This is, in fact, what can occur. Consider again the rectangular panel system mentioned at the end of the discussion of acoustic control mechanisms. Figure 9.21 illustrates the (theoretical) response of the dominant modes of the top panel of the rectangular enclosure system for an excitation frequency of 120 Hz, close to the (1,3) resonance of the panel, before and after the application of optimum vibration control at a panel location ($x = 0.434$, $y = 0.9$) m to attenuate the sound transmission of a normally incident plane wave. Clearly, the amplitudes of the dominant modes have been reduced, which is reflected in a 23.1 dB reduction in acoustic potential energy. This reduction in the amplitudes of the dominant structural modes will not, however, always be the only control mechanism with a structure-controlled response. Figure 9.22 illustrates the change in the amplitudes and phases of these same modes when the control source has been moved to the centre of the top panel. As before, the nearly resonant (1,3) panel mode has been reduced in amplitude. However, this reduction is much less than that achieved when the control source was located in its previous position, and is offset to some degree by the increase in amplitude of the (1,1) mode. Despite this, there is still a reduction of 22.6 dB in the enclosure acoustic potential energy.

From the standpoint of assessing the physical control mechanism, an important aspect of the structural modal amplitudes and phases of Fig. 9.22 is the alteration in the relative amplitudes and phases of (1,1), (1,3), and, to a lesser extent, (3,1) modes. As was outlined earlier in this section, the modal coupling characteristics of the rectangular enclosure are such that odd index structural modes couple with even index acoustic modes, and vice versa so that

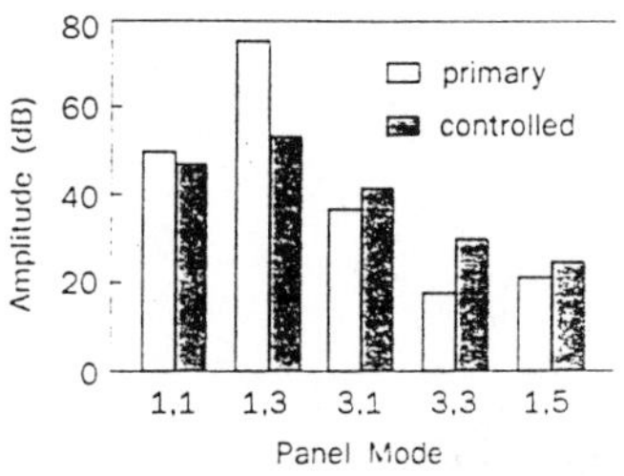

Fig. 9.21 Change in structural modal amplitudes of the top panel of the rectangular enclosure system with a point control force applied off centre at 120 Hz.

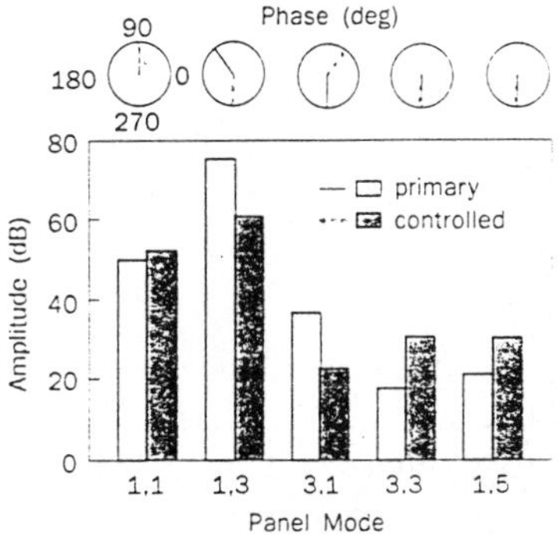

Fig. 9.22 Change in structural modal amplitudes of the top panel of the rectangular enclosure system with a point control force applied on centre at 120 Hz.

all three of these structural modes couple to the same set of acoustic modes. By altering the relative amplitudes and phases of these modes it is possible to reduce the total excitation of the (coupled) acoustic modes, which is the summation of the contributions from each (coupled) structural mode. This modal rearrangement control mechanism occurs when the vibration source is placed at the panel centre because the transfer function between the control force input and the two dominant structural modes, the (1,1) and (1,3) modes, differs in phase by a significant amount (greater than 90°), and so the amplitudes of both of these structural modes could not be reduced concurrently.

With this rectangular enclosure system the modal rearrangement control mechanism can be even more evident with an acoustic cavity controlled response. Figures 9.23 and 9.24 illustrate the phase and amplitudes of the dominant structural and acoustic modes for the active control of sound transmission from a normally incident plane wave at 171 Hz, near the (0,0,1) acoustic cavity resonance frequency, using a control force at the panel centre. Here there is no significant decrease in the amplitude of any of the structural modes, while the nearly resonant (0,0,1) acoustic mode has decreased in amplitude leading to a reduction in acoustic potential energy of 14.9 dB.

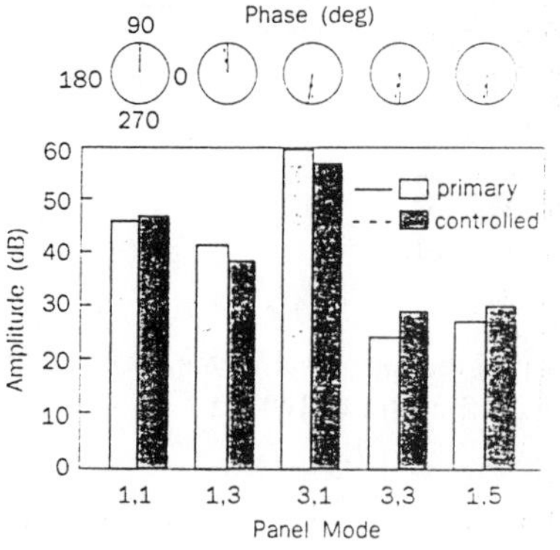

Fig. 9.23 Change in structural modal amplitudes of the top panel of the rectangular enclosure system with a point control force applied on centre at 171 Hz.

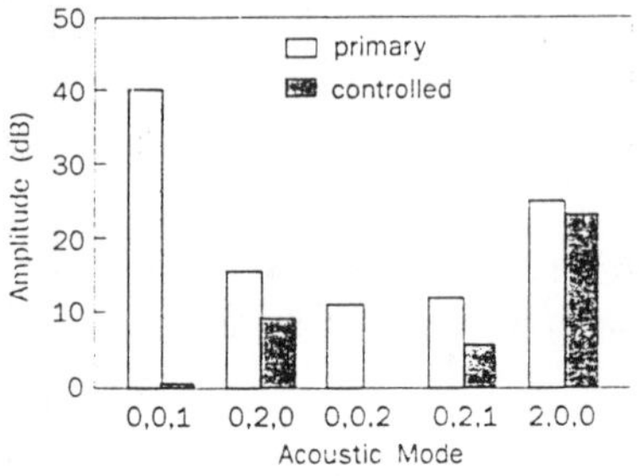

Fig. 9.24 Change in acoustic modal amplitudes of the rectangular enclosure system with a point control force applied on centre at 171 Hz.

It is important to note that the reason modal rearrangement is able to provide global sound attenuation in this system is that there are several structural modes with 'similar' magnitude input impedances coupled to a single acoustic mode; in other words, modal rearrangement works because there are several modes available to 'rearrange'. This seemingly obvious point has implications for the active control of coupled structural/acoustic systems other than the rectangular enclosure; systems such as a cylindrical enclosure.

Consider a simply supported aluminium cylindrical enclosure of 1.2 m length, 0.254 m diameter and 1.6 mm wall thickness, with a normally incident plane wave primary forcing function, and a control force located at (z = 0.6 m, θ = 0°). Figures 9.25 and 9.26 show the amplitudes of the dominant structural and acoustic modes for an excitation frequency of 272 Hz, near the (1,2) structural resonance. Here the amplitude of this nearly-resonant structural mode has been reduced, with a corresponding reduction in the (0,2,1) acoustic mode which is coupled to it, and an overall reduction of 17.7 dB in the acoustic potential energy. Figures 9.27 and 9.28 show the same modal amplitudes for control at a frequency of 656 Hz, near the (0,2,1) acoustic resonance. Again, modal amplitude control is the physical mechanism which provides the 11.1 dB reduction in acoustic potential energy. In fact, modal amplitude control is the *only* physical mechanism (for vibration control sources) which will provide global sound attenuation for a circular cylinder. This is because of its structural/acoustic modal coupling characteristics, which at low frequencies essentially limits coupling to single structural mode/acoustic mode pairs. With only a single structural mode exciting a given acoustic mode, it is not possible to use modal rearrangement.

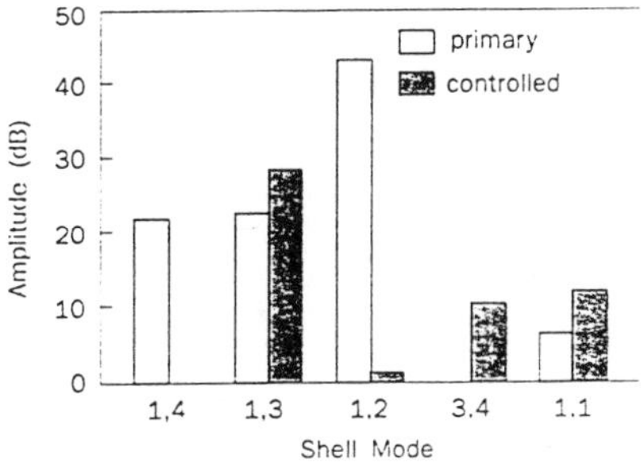

Fig. 9.25 Change in structural modal amplitudes of the plain cylindrical enclosure system with a point control force applied at the midsection, 272 Hz.

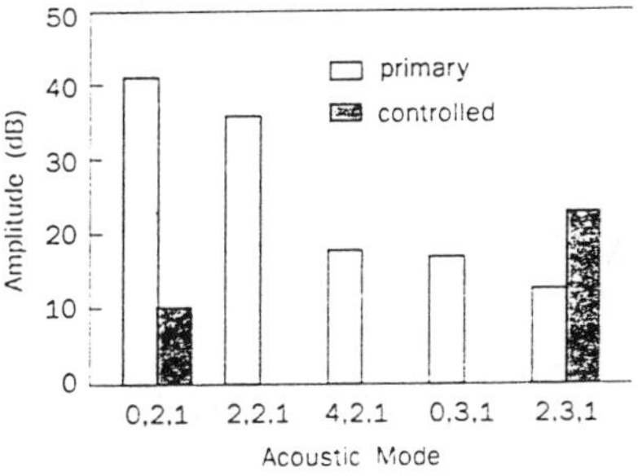

Fig. 9.26 Change in acoustic modal amplitudes of the plain cylindrical enclosure system with a point control force applied at the midsection, 272 Hz.

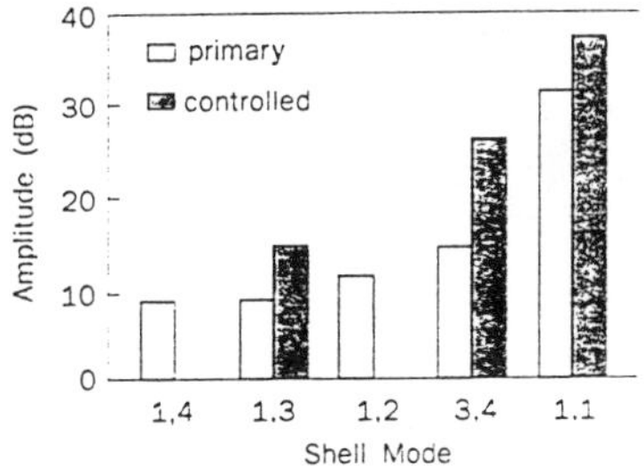

Fig. 9.27 Change in structural modal amplitudes of the plain cylindrical enclosure system with a point control force applied at the midsection, 656 Hz.

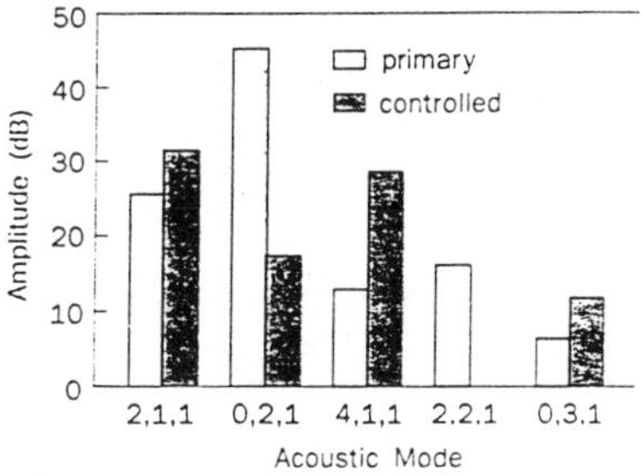

Fig. 9.28 Change in acoustic modal amplitudes of the plain cylindrical enclosure system with a point control force applied at the midsection, 656 Hz.

The final point to note here is that while a reduction in structural modal amplitudes is the physical mechanism which provides global sound attenuation in a circular cylinder when vibration sources are used, it does not necessarily follow that the overall levels of shell vibration will decrease when point forces are used. This fact is illustrated in Figs. 9.29 and 9.30, which plot the maximum possible levels of acoustic potential energy reduction which can be achieved for a normally incident plane wave primary forcing function, controlled by a single vibration (point) control force at (z = 0.6, θ = 0°), as a function of frequency, and the corresponding change in mean square velocity levels. Clearly the overall velocity levels increase in many cases. This spillover effect results from the (point) control force inadvertently exciting higher order structural modes, which do not drive the interior sound field efficiently, while controlling the lower order structural modes which are largely responsible for the interior sound field. Hence, this effect has little impact upon the controlled sound field.

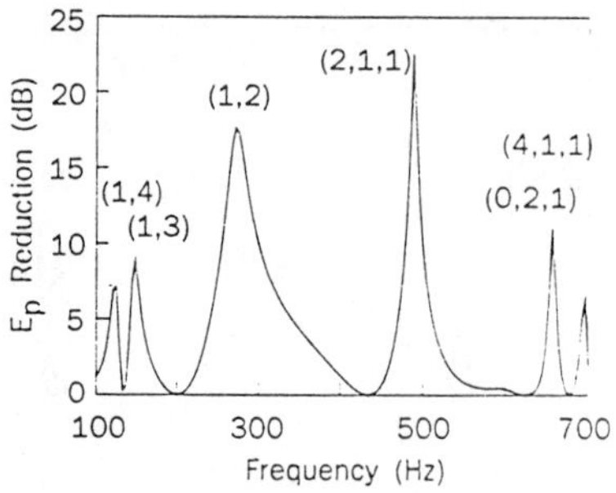

Fig. 9.29 Maximum achievable levels of acoustic potential energy reduction for the plain cylindrical enclosure system as a function of frequency.

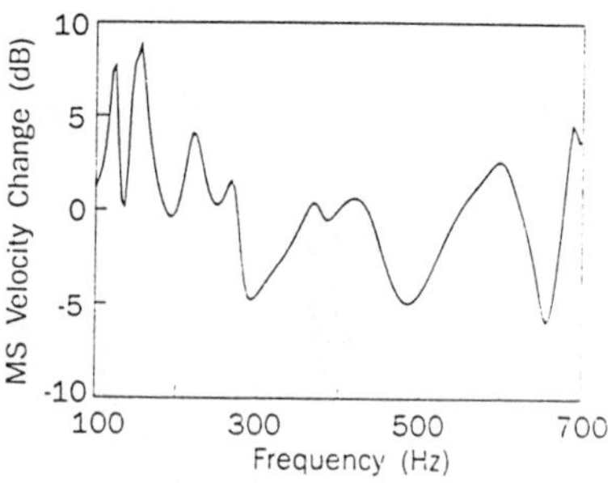

Fig. 9.30 Change in overall structure velocity levels corresponding to the maximum acoustic potential energy reduction as a function of frequency.

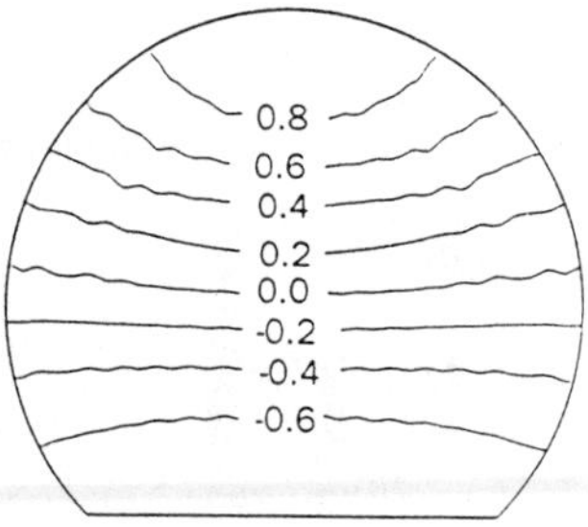

Fig. 9.31 Cross-section of the acoustic mode shape resonant at 465 Hz for the cylinder with floor system.

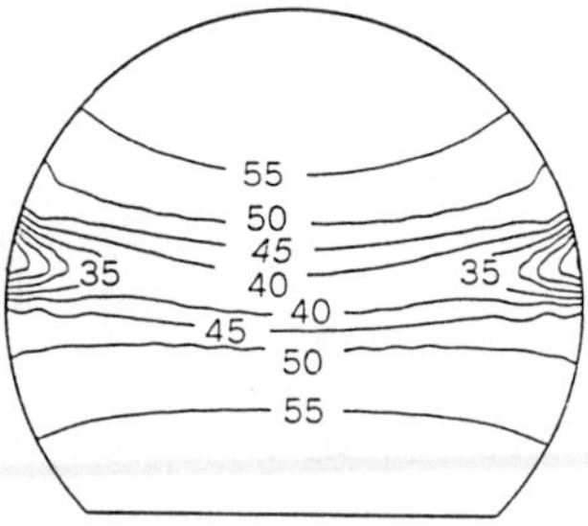

Fig. 9.32 Primary sound field at the midsection of the cylinder with floor under primary excitation, 465 Hz.

In a practical system such as an aircraft or automobile, then, will modal rearrangement be a mechanism which can provide global sound attenuation? The answer is that it probably will, as most practical systems do not exhibit such ideal single mode coupling as found in the cylindrical enclosure system. In fact, the characteristics of the circular cylinder can be altered by simply including a longitudinal 'floor' in the enclosure.

To illustrate this, consider the introduction, into the previously considered cylindrical enclosure, of a 1.6 mm longitudinal partition, mounted so that the two longitudinal edges subtend an angle ($2\theta_f$ in Fig. 9.20) of 90° with the central axis of the cylinder. Let the excitation frequency now be 465 Hz, which is near the cut-on (axial modal index $q = 0$) of the circumferential acoustic mode shape shown in Fig. 9.31. Figures 9.32 and 9.33 illustrate the sound fields in a cross-section midway along the cylinder length before and after control with a point force located in the same position ($z = 0.6$m, $\theta = 0°$) as the previous (plain cylinder) example. The reduction in sound pressure levels is readily apparent, and is matched by the 23.1 dB reduction in acoustic potential energy. This

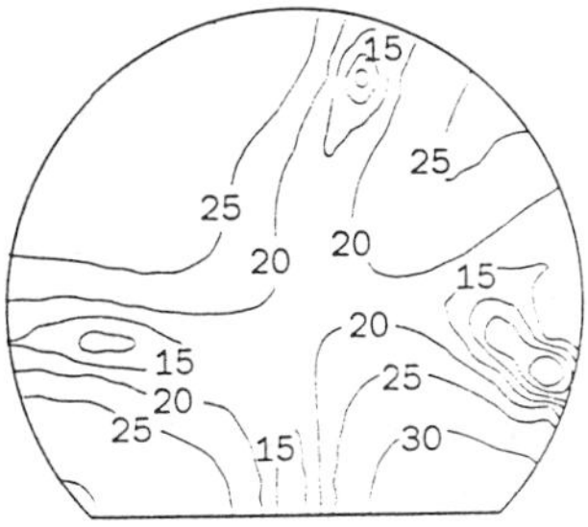

Fig. 9.33 Controlled sound field at the midsection of the cylinder with floor under primary excitation, 465 Hz.

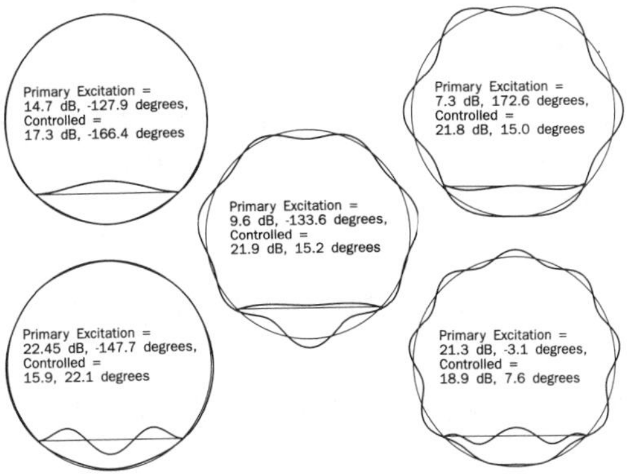

Fig. 9.34 Change in amplitude and phase of the dominant structural modes under control, 465 Hz. For all the modes shown in cross-section the axial modal indice $m = 1$.

reduction, however, is not matched by a similar reduction in the principal coupled structural modes, which are illustrated in Fig. 9.34. Many of the amplitudes of the dominant coupled structural modes do, in fact, increase, which is accompanied by an increase in the overall structural velocity levels, demonstrating that modal rearrangement is the control mechanism at work here.

Therefore, we can conclude that there are two mechanisms which are generally employable by vibration control sources: velocity control mechanisms, where the velocity of the coupled structural modes is explicitly reduced, and acoustic control mechanisms, where the relative complex amplitudes of the coupled structural modes are altered in relative phase and magnitude such that the total excitation of the acoustic mode is reduced, without necessarily a reduction in any of the individual structural modes. The choice of mechanism is dependent upon the modal coupling characteristics of the system and the placement of the primary and control sources.

9.8 INFLUENCE OF CONTROL SOURCE AND ERROR SENSOR ARRANGEMENT

As we have discussed in previous chapters, it has proved thus far not possible to directly (analytically) optimize the design active control systems, including those targeting sound transmission problems. This is because the maximum level of reduction in acoustic potential energy is not a linear function of control source location, and because the optimum error sensor locations are dependent upon the location of the control sources. Therefore, analytical models which have been developed thus far in this chapter, models which enable the prediction of the maximum levels of reduction in acoustic potential energy possible for a given control source arrangement, and the control source volume velocities/forces required to minimize the sound pressure/structural velocity at potential error sensing locations, must be implemented in some form of a numerical search routine to optimize the physical arrangement of the system. This section will examine some of the implications of control source and error sensor placement, the aim being to obtain some qualitative results which will aid in the optimization process.

9.8.1 Control source/error sensor type

When approaching the design of an active system to control sound transmission into a coupled structural/acoustic system, the first variables to consider are the types of control source and error sensor. In terms of control 'potential', a good general rule is that acoustic control sources are best at controlling acoustic resonances, while vibration control sources are best at controlling structural resonances. The reason is simply that in each of these cases the control source will be concerned principally with a single mode, which makes for a more forgiving design exercise. With most coupled structural/acoustic systems a structural resonance will drive several acoustic modes, while at an acoustic resonance the acoustic mode will be driven by several structural modes. Therefore when a control source of one medium is used to attenuate a resonance of another medium it must contend with several modes, complicating the design problem. Thus the best choice of control source is dependent upon which type of coupled modes (structurally dominated or acoustically dominated) contribute most to the interior sound field in the frequency range of interest. There are, however, other factors which need to be considered. From the standpoint of peace of mind regarding structural integrity (a point often considered important by aviation authorities!), and ease of installation and modification, acoustic control sources are often a better choice. On the other hand, piezoelectric vibration sources are much lighter than acoustic sources (speakers), and require less power to generate the required control signal.

In general with structural/acoustic problems, the choice of error sensor type is influenced by the choice of control source type. Obviously, acoustic control sources require acoustic error sensors. With vibration control sources, however, there is a choice, but caution must be exercised when using discrete vibration error sensors, as it has been shown that the maximum attenuation of acoustic potential energy can result in an increase in the vibration levels of the structure (during the employment of control mechanisms involving modal rearrangement). As we will see in the next section, the use of distributed vibration sensors to provide an error signal is also subject to some difficulty, leading to the conclusion that acoustic error sensors may, in most cases, be the better option.

9.8.2 Effect of control source arrangement/numbers

The next questions to consider are how critical the placement of control sources is in achieving the maximum levels of acoustic potential energy reduction, and whether there is only one (global) optimum arrangement or are several (local

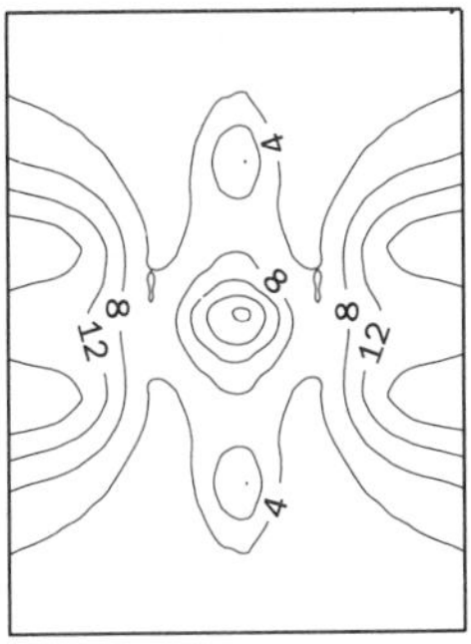

Fig. 9.35 Maximum levels of acoustic potential energy reduction as a function of vibration control source location on the top of the rectangular enclosure system, 190 Hz.

optima, or 'minima'). The answer to the first of these questions is 'very', and the answer to the second is 'there usually are several' (Pan and Hansen, 1991b; Snyder and Hansen, 1994b). Figure 9.35 shows the maximum levels of acoustic potential energy reduction which can be obtained as a function of the location of a single control force on the top panel of the rectangular enclosure studied in the previous section, for a plane wave primary disturbance at 190 Hz. This figure shows several 'optimal' control source placements, which are actually dominated by different control mechanisms (the edge optima achieve control by employment of velocity (modal control) mechanisms, while the center optimum relies on acoustic (modal rearrangement) mechanisms). The general locations of these optima are not surprising; they are at antinodes of the (1,3) structural mode, which dominates the acoustic transmission at this frequency.

It is important to note here is that if control sources are simply placed in 'convenient' locations in the system (plane, automobile, etc.), the level of global sound attenuation achieved may not be as good as that which is possible, and may in fact be discouraging. Ideally, the control sources should be placed using a numerical search routine, with the error criterion being the minimization of the acoustic potential energy or the minimization of the sound pressure level at a number of critical locations or a combination of both, the latter effect achieved by using the potential energy minimization procedure and applying a higher weighting to error sensors at critical locations. However, if this is not possible potential design arrangements can be discriminated by using a multiple regression approach with limited measured data, as was outlined in Section 9.4.

The next point to consider is how many control sources to use. As with location, this is not an easy question to answer. Heuristically, the best control will be achieved if the control sources are capable of 'duplicating' the response of the system to the primary disturbance. Determining the number of control sources required to do this is analogous to determining the length of a digital filter required to model a dynamic system, which we discussed briefly in Chapter 6. The basic way of determining the number of variables (sources/filter stages) in both of these is simply to include an additional one, reoptimize the arrangement, and see what the improvement in the result there is. If the improvement is not 'significant', the number of control sources can be considered adequate. It may also be tempting to simply include a very large number of control sources, with the aim of compensating for non-ideal placement. This temptation should be avoided, as increasing the number of control sources (and error sensors) will simply make it hard for the electronic control system to keep up with changes in the operating environment, as the increased computational load and the required decrease in the adaptive algorithm convergence coefficient (discussed in Chapter 13) will decrease its reaction time.

9.8.3 Effect of error sensor location

The data illustrated in Fig. 9.35 clearly demonstrate that the locations of the active control source(s) has a very significant influence upon the levels of reduction in acoustic potential energy which can be achieved. This is also true for the placement of the error sensor(s). Figure 9.36 illustrates the levels in acoustic potential energy reduction achieved by minimizing the sound pressure at a single point in the previously studies rectangular enclosure, in the $z = 0$ plane, with a vibration control source at (0.4,0.5) and a normally incident plane wave primary disturbance at 190 Hz. As can be seen, there is a single optimum error sensor location in this plane.

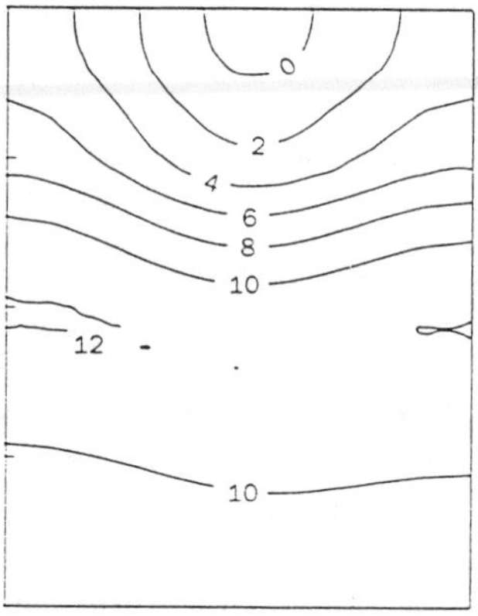

Fig. 9.36 Acoustic potential energy reduction achieved by minimizing the sound pressure level at one point on the bottom of the rectangular enclosure, vibration control source at (0.4,0.5), 190 Hz.

A quantitative assessment of the characteristics of an optimum error sensor arrangement can be obtained by comparing the governing equations of acoustic potential energy and the sum of squared acoustic pressure amplitudes at a number of discrete locations. For a rigid walled enclosure, comparing (9.2.36) and (9.3.16) shows that for the problem of minimizing acoustic pressure to be equivalent to the problem of minimizing acoustic potential energy

$$\mathbf{Z}_p = \mathbf{Z}_E \tag{9.8.1}$$

where these can be viewed as weighting matrices for the two problems, respectively. The same conclusion can be arrived at for the problem of controlling sound transmission into a coupled enclosure, by comparing (9.4.33) and (9.5.2) for an acoustic control source, and (9.4.54) and (9.5.9) for a vibration control source. If we consider the terms in these matrices, it can be concluded that for an optimal set of N_e error sensors,

$$\sum_{i=1}^{N_e} \begin{bmatrix} \phi_1^2(r_i) & \phi_i(r_i)\phi_2(r_i) & \cdots & \phi_1(r_i)\phi_{N_n}(r_i) \\ \phi_2(r_i)\phi_1(r_i) & \phi_2^2(r_i) & \cdots & \phi_2(r_i)\phi_{N_n}(r_i) \\ & & \vdots & \\ \phi_{N_n}(r_i)\phi_1(r_i) & \phi_{N_n}(r_i)\phi_2(r_i) & \cdots & \phi_{N_n}^2(r_i) \end{bmatrix} \rightarrow \begin{bmatrix} \Lambda_1 & 0 & \cdots & 0 \\ 0 & \Lambda_2 & \cdots & 0 \\ & & \vdots & \\ 0 & 0 & \cdots & \Lambda_{N_n} \end{bmatrix} \tag{9.8.2}$$

It is possible to use this as a criterion for assessing the effect of error sensor placement for a number of simple systems (Bullmore, 1988).

There is, however, a simpler way to optimally place error sensors. Considering the data shown in Fig. 9.36, it is possible to make a prediction of the optimal error sensor location by looking at the difference between the primary and optimally controlled sound fields for the given control source arrangement as shown in Fig. 9.37 for the $z = 0$ plane. Comparing this with the optimal error sensor location plot, it is apparent that the optimal error sensor location is at the point of greatest acoustic pressure reduction under optimum (in terms of acoustic potential energy reduction) control. This intuitively sensible result was seen previously in Section 9.3 when considering a simple one-dimensional example.

This characteristic is especially advantageous when using multiple regression to optimize the control source location. The points of maximum difference between the primary and optimally controlled residual sound fields can be determined directly using the residuals vector produced as an output by commercial multiple regression packages. These residuals are the difference between the measured quantity (pressure here), and the value predicted by the regression equation. The point of minimum acoustic pressure in the residual sound field will be the point with the smallest residual value. Illustrated in Fig. 9.38 is a plot of the 'normalized' residuals,

$$\text{normalized residual} = \frac{p_{\text{residual}}}{|p_{pri}|} \tag{9.8.3}$$

for the arrangement under discussion, where the control force has been optimized using multiple regression. When this result is compared to the plot of acoustic

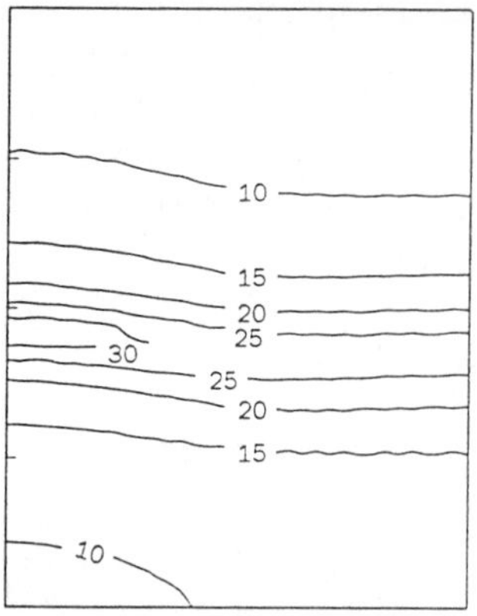

Fig. 9.37 Attenuation of the sound field under optimal vibration control applied at (0.4,0.5) in the bottom of the rectangular enclosure, 190 Hz.

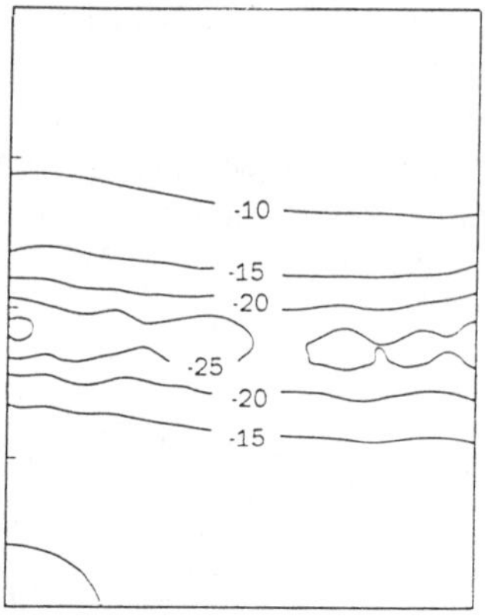

Fig. 9.38 Multiple regression residuals under optimal vibration control applied at (0.4,0.5) in the bottom of the rectangular enclosure, 190 Hz.

potential energy reduction as a function of error sensor location given in Fig. 9.36 it can be seen that the optimum error sensor location is at the location of the minimum normalized residual.

9.9 CONTROLLING VIBRATION TO CONTROL SOUND TRANSMISSION

In the previous section we saw that the levels of acoustic potential energy attenuation which are achieved by a practical system are highly dependent upon the location of discrete error sensors used in the system. In this section we will consider the implementation of a system with another form of error signal, where sets of structural modes are used as error signals in vibration control-source based systems. Sets of modes could be resolved using 'modal filters' (discussed in Chapter 11), where a modal decomposition is performed on a set of discrete error signals prior to their use by the electronic control system. Alternatively, shaped piezoelectric polymer film sensors could be used to spatially decompose the structural vibration into the desired sets of structural modes.

The problem of what function of structural vibration to measure and minimize to reduce the associated acoustic field was considered in chapter 8 for controlling sound radiation into free space. It was outlined there that most global error criteria can be expressed in the form

$$J = \boldsymbol{v}^{\mathrm{H}}\boldsymbol{A}\boldsymbol{v} \tag{9.9.1}$$

where J is the global error criterion of interest and $\boldsymbol{A}$ is some real, symmetric, positive definite transfer matrix. As such, $\boldsymbol{A}$ can be diagonalized by the orthonormal transformation

$$\boldsymbol{A} = \boldsymbol{Q}\boldsymbol{\Lambda}\boldsymbol{Q}^{-1} \tag{9.9.2}$$

where $\boldsymbol{Q}$ is the orthonormal transform matrix, the (square) column vector of eigenvectors of the transfer matrix $\boldsymbol{A}$, and $\boldsymbol{\Lambda}$ is the diagonal matrix of associated eigenvalues. The form of $\boldsymbol{A}$ dictates that the eigenvectors in $\boldsymbol{Q}$ will be orthogonal. If the transformation of (9.9.2) is substituted into (9.9.1) the global error criterion expression becomes

$$J = \boldsymbol{v}^{\mathrm{H}}\boldsymbol{A}\boldsymbol{v} = \boldsymbol{v}^{\mathrm{H}}\boldsymbol{Q}\boldsymbol{\Lambda}\boldsymbol{Q}^{-1}\boldsymbol{v} = \boldsymbol{v}'^{\mathrm{H}}\boldsymbol{\Lambda}\boldsymbol{v} \tag{9.9.3}$$

where the vector of transformed modal velocities $\boldsymbol{v}'$ is defined by the expression

$$\boldsymbol{v}' = \boldsymbol{Q}^{-1}\boldsymbol{v} \tag{9.9.4}$$

and use has been made of the property of the orthonormal transform matrix

$$\boldsymbol{Q}^{-1} = \boldsymbol{Q}^{\mathrm{T}} \tag{9.9.5}$$

The orthonormal transformation has decoupled the global error criterion expression, so that it is now equal to a weighted summation of the form

$$J = \sum_{i=1}^{N_m} \lambda_i \mid v_i' \mid^2 \tag{9.9.6}$$

where λ_i is the ith eigenvalue of $\boldsymbol{A}$, v'_i is the ith transformed modal velocity. As was discussed, it is the transformed modal velocities which ideally should be measured and used as error signals, where the measured signals are each weighted by the associated eigenvalues prior to use by the electronic control system (Snyder and Tanaka, 1992). The utility of this approach for the problem of controlling sound transmission into coupled enclosures is what will be investigated here.

With the control of sound transmission into coupled enclosures, the global error criterion we have been using is acoustic potential energy in the enclosure, E_p, shown in (9.3.5) to be equal to the weighted sum of squared acoustic modal amplitudes. From the result of (9.4.20) and (9.4.21) it is straightforward to show that the amplitude of the ith acoustic mode can be expressed as

$$p_i = \frac{j\rho_0\omega}{\Lambda_i(\kappa_i^2-k^2)} \int_S \sum_{\iota=1}^{N_m} \psi_\iota(x)\phi_i(x)v_\iota \, \mathrm{d}x = \frac{j\rho_0\omega S}{\Lambda_i(\kappa_\iota^2-k^2)} \sum_{\iota=1}^{N_m} \beta_{i,\iota} v_\iota \tag{9.9.7}$$

where $\beta_{i,\iota}$ is the non-dimensional modal coupling coefficient for the ith acoustic mode and the ιth structural mode, and is defined in (9.4.11). Equation (9.9.7) can be written as a matrix expression, where the complex amplitude of the ith acoustic mode is equal to

$$p_i = \boldsymbol{z}_{a,i}^{\mathrm{T}}\boldsymbol{v} \tag{9.9.8}$$

where $\boldsymbol{v}$ is the $(N_m \times 1)$ vector of structure modal velocity amplitudes and $z_{a,i}$ is the $(N_m \times 1)$ vector of radiation transfer functions between the structure modes and the ith acoustic mode, the terms of which are defined by

$$z_{a,i}(\iota) = \frac{j\rho_0 \omega S}{\Lambda_i(\kappa_i^2 - k^2)} \beta_{i,\iota} \tag{9.9.9}$$

Therefore, the vector of N_n acoustic modal amplitudes $\boldsymbol{p}$ is defined by the expression

$$\boldsymbol{p} = \boldsymbol{Z}_a \boldsymbol{v} \tag{9.9.10}$$

where $\boldsymbol{Z}_a$ is the $(N_n \times N_m)$ matrix of radiation transfer vectors

$$\boldsymbol{Z}_a = \begin{bmatrix} z_{a1}^{\mathrm{T}} \\ z_{a2}^{\mathrm{T}} \\ \vdots \\ z_{aN_m}^{\mathrm{T}} \end{bmatrix} \tag{9.9.11}$$

Equation (9.9.10) can be substituted into (9.3.5), defining acoustic potential energy as the weighted sum of squared acoustic modal amplitudes, to express the acoustic potential energy in the form of (9.9.1), as

$$J = E_p = \boldsymbol{v}^{\mathrm{H}} \boldsymbol{A} \boldsymbol{v} \tag{9.9.12}$$

where the transfer matrix $\boldsymbol{A}$ is defined by the relationship

$$\boldsymbol{A} = \boldsymbol{Z}_a^{*} \boldsymbol{W} \boldsymbol{Z}_a^{\mathrm{T}} \tag{9.9.13}$$

such that the (m,n)th term is equal to

$$A(m,n) = \sum_{i=1}^{N_n} \frac{\rho_0 k^2 S^2}{4\Lambda_i \, | \, (\kappa_i^2 - k^2) \, |^2} \beta_{i,m} \beta_{i,n} \tag{9.9.14}$$

As with radiated acoustic power, the defining relationship for the transfer matrix $\boldsymbol{A}$ given in (9.9.15) shows that it is not constrained to be diagonal. The amount of modal cross-coupling, the importance of the off-diagonal terms, will be dictated by the modal coupling characteristics of the system. If the coupling is very specific, to the point where a given acoustic mode is coupled to a single structural mode, then the transfer matrix $\boldsymbol{A}$ will be diagonal. The orthonormal transform matrix $\boldsymbol{Q}$ will then be the identity matrix, and the eigenvalues will be equal to

$$\lambda_m = \sum_{i=1}^{N_n} \frac{\rho_0 k^2 S^2}{4\Lambda_i \mid (\kappa_i^2 - k^2) \mid^2} \beta_{i,m}^2 \tag{9.9.15}$$

We have already seen that this is essentially the case for low frequency excitation of a simply supported circular cylinder, where for a structural mode to couple to an acoustic mode the circumferential modal indices must match and the axial modal indices must form and odd/even pair. This means that while a given acoustic mode, for example with circumferential index n and an odd axial index, will couple to the infinite set of structural modes $(2,n)$, $(4,n)$, ..., the separation (in frequency) of the resonance frequencies of the structural modes dictate that at low frequencies acoustic modes will often be driven by essentially a single structural mode. Most practical systems, however, will have a significant degree of cross-coupling, where several structural modes are coupled to a single acoustic mode, so that even at low frequencies (with a very limited set of modes being considered) the transfer matrix $\boldsymbol{A}$ will not be diagonal.

The values of the terms in the transfer matrix are frequency dependent, governed by both the actual frequency of excitation, and the admittance $(\kappa_i^2 - k^2)^{-1}$ of the acoustic modes at the excitation frequency. The eigenvalues of the transfer matrix will therefore peak at resonances of the acoustic modes to which the transformed modal velocities are coupled. This will have significant implications for the practicality of implementing the type of control system developed in Chapter 8 for minimizing acoustic radiation into free space using measurements of structural vibration, as will be discussed shortly.

To further investigate the problem of sensing vibration to control acoustic potential energy in a coupled enclosure, it will again be useful to specialize the problem. The system chosen is a rectangular cavity/panel system shown in Fig. 9.18, with a description of the governing equations given in Section 9.7. Here a simply supported rectangular panel of dimensions 1.8 m × 0.88 m × 9 mm is placed on top of a rigid box of 1.0 m depth, with vibration of the panel responsible for excitation of the enclosed sound field. The first ten structural and acoustic modal resonance frequencies are listed in Table 9.1. Recall that for this arrangement, there is a significant degree of cross-coupling in this system, as the only requirement for a structural and acoustic mode to couple is that the modal indices in both x- and y-directions form an odd/even (structural/acoustic) pair (the modal coupling relationship is defined in (9.7.21)). This means that the transfer matrix $\boldsymbol{A}$ will not be diagonal, even with a low number modal truncation at low frequencies.

For the calculations outlined here, 30 acoustic modes and 30 structural modes and assuming acoustic and structural loss factors of 0.01 are used in formulating the transfer matrix $\boldsymbol{A}$. Like the case in the Chapter 8, when decomposing the transfer matrix in the frequency range up to 250 Hz, it is found

Table 9.1 Resonance frequencies of first 10 structural and acoustic modes

Structural mode	Resonance frequency (Hz)	Acoustic mode	Resonance frequency (Hz)
(1,1)	35.17	(0,0,0)	0.00
(2,1)	55.52	(1,0,0)	95.28
(3,1)	89.44	(0,0,1)	171.50
(1,2)	120.33	(2,0,0)	190.56
(4,1)	136.94	(0,1,0)	194.89
(2,2)	140.68	(1,0,1)	196.19
(3,2)	174.60	(1,1,0)	216.93
(5,1)	198.00	(2,0,1)	256.37
(4,2)	222.09	(0,1,1)	259.60
(1,3)	262.25	(2,1,0)	272.57

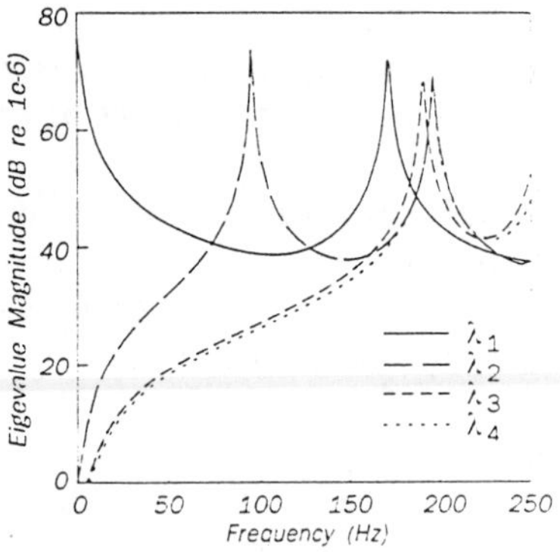

Fig. 9.39 Eigenvalues of the first four (most important) transformed modal velocities.

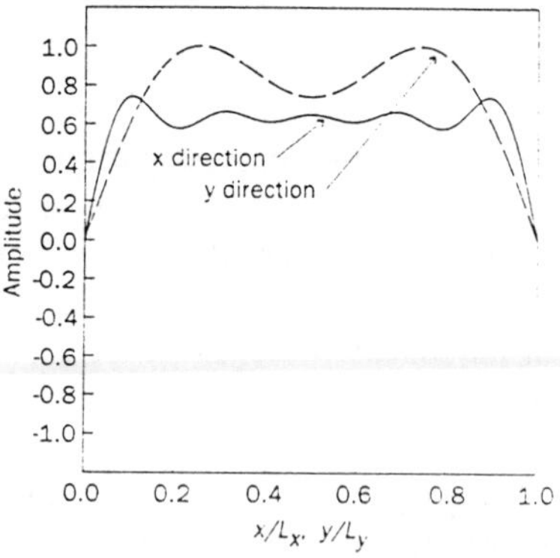

Fig. 9.40 First transformed mode shape.

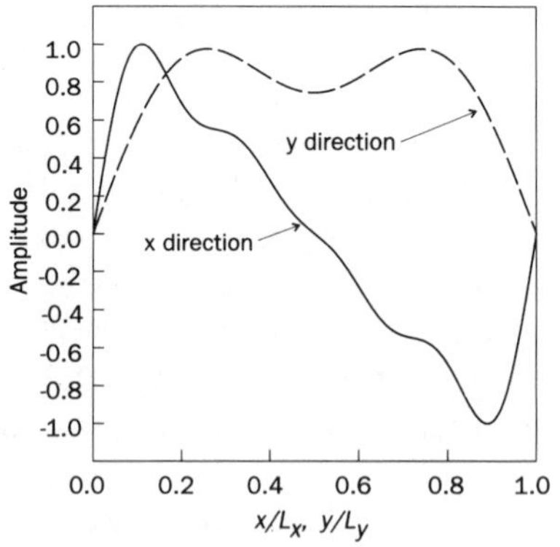

Fig. 9.41 Second transformed mode shape.

only a limited number of eigenvalues (four) are of significant amplitude. These eigenvalues are plotted as a function of frequency in Fig. 9.39. The transformed modal velocities associated with these are based predominantly upon the (1,1), (2,1), (3,1) and (1,2) structural modes, respectively. The first two of these, normalized to an amplitude of 1.0, are plotted in Figs. 9.40 and 9.41. Note the similarity between these transformed modal velocities and those obtained for the problem of radiated acoustic power in Chapter 8. Further, it is found that while both the eigenvalues and eigenvectors are frequency dependent, the variation in the eigenvectors with frequency is relatively small, once again opening the possibility of implementing a system using as error signals transformed modal velocities which are based upon eigenvectors which are strictly correct at only one frequency, such as could be measured using a shaped piezoelectric polymer film sensor, weighting these measurements by the frequency correct eigenvalue to provide an error signal and achieving near-optimum results (as was done for free field radiation in the previous chapter).

It is interesting to note that while the eigenvalues shown in Fig. 9.39 do peak at the resonance frequencies of the acoustic modes to which the transformed modal velocities are coupled, the coupling is extremely selective. For example, the first eigenvector, based predominantly upon the (1,1) structural mode, peaks at the $(0,0,n)$ acoustic resonances *only*, where $n = 0,1,...$ It does *not* peak at $(2,0,n)$ acoustic resonances, even though all of the structural modes in the eigenvector are (odd,odd) index modes, which will individually couple to this acoustic mode. Similarly, the eigenvalues associated with the third eigenvector, based predominantly upon the (3,1) structural mode, peak at resonances of $(2,0,n)$ acoustic modes *only*, and not at other (even,even,n) acoustic resonances. Decomposition of the transfer matrix $\boldsymbol{A}$ produces eigenvectors defining transformed modal velocities which drive a single x,y index set of acoustic modes. Therefore, when the modal density in the enclosed space is low, such as at low frequencies in relatively small enclosures, only a few transformed modal velocities need to be considered. As the modal density increases, however, more transformed modal velocities become important and the size of the control problem begins to increase.

Illustrated in Fig. 9.42 are the maximum levels of acoustic potential energy reduction, as a function of frequency, with a primary point force at a location ($x = 0.35$ m, $y = 0.64$ m) relative to the lower left corner, and a control point force placed symmetrically at ($x = 1.45$ m, $y = 0.24$ m). Figure 9.43 shows the acoustic potential energy reduction which results from minimizing the first and first to fourth transformed modal velocities. In this calculation the transformed modal velocities were based upon eigenvectors calculated at 30 Hz, and weighted in the calculation using the frequency correct eigenvalues shown in Fig. 9.39. As may have been expected, minimizing the amplitude of the first transformed modal velocity produces good results at resonances of its constituent structural modes, such as the (1,1) and (3,1), and at acoustic resonances of the form $(0,0,n)$. Minimizing the weighted set of four transformed modal velocities achieves practically maximal attenuation over the frequency range of interest.

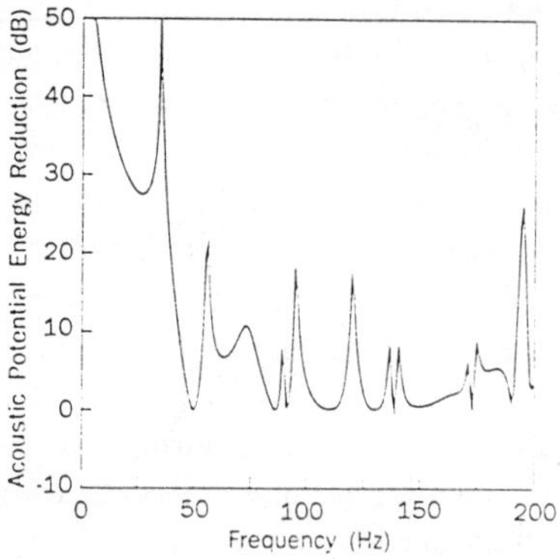

Fig. 9.42 Maximum levels of acoustic potential energy reduction, primary source at (0.35 m, 0.64 m) and control source at (1.45 m, 0.24 m).

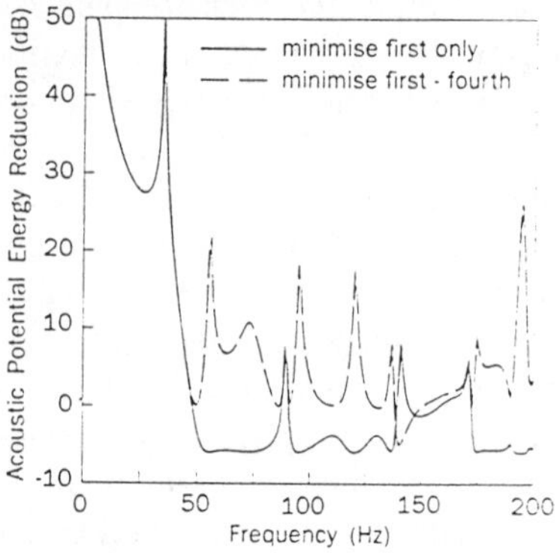

Fig. 9.43 Acoustic potential energy reduction achieved by minimizing the weighted sum of transformed modal velocities, primary source at (0.35 m, 0.64 m) and control source at (1.45 m, 0.24 m). Shown are the results from minimization of first transformed modal velocity, and minimization of first to fourth transformed modal velocities.

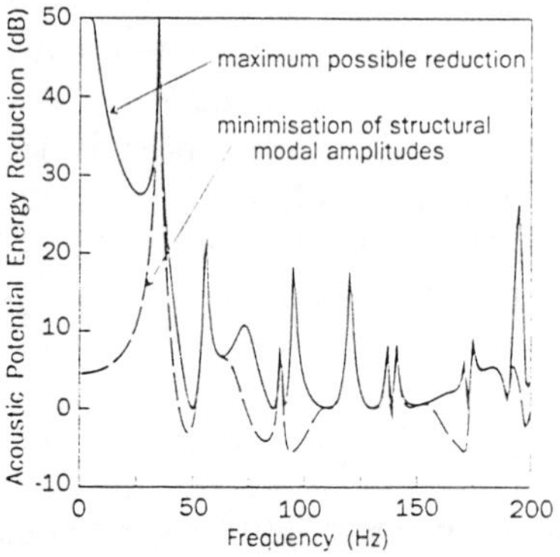

Fig. 9.44 Acoustic potential energy reduction achieved by minimizing structural modal amplitudes, primary source at (0.35 m, 0.64 m) and control source at (1.45 m, 0.24 m). Shown are the maximum possible reduction, and the result obtained from minimization of structural modal amplitudes.

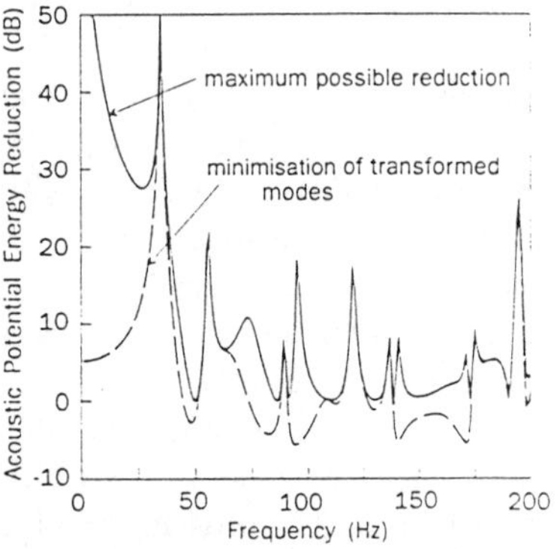

Fig. 9.45 Acoustic potential energy reduction achieved by minimizing the unweighted set of four transformed modal velocities, primary source at (0.35 m, 0.64 m) and control source at (1.45 m, 0.24 m). Shown are the maximum possible reduction, and the result obtained from minimization of transformed modal velocities.

Figure 9.44 illustrates the levels of attenuation which result from minimizing the kinetic energy of the vibrating panel, or the structural velocity amplitudes. While this approach produces good results near structural resonances, it fails to provide adequate levels of attenuation at acoustic resonances. At acoustic resonances the sound field will not reflect (in magnitude) the structural velocity distribution, owing to the disproportionate values of acoustic modal admittance $(\kappa_i^2 - k^2)^{-1}$. Clearly, minimizing structural velocity with the aim of attenuating the enclosed acoustic field is not, in general, a sound approach.

Having now seen that the approach of minimizing the sum of transformed modal velocity amplitudes, calculated using the eigenvectors of the transfer matrix $\boldsymbol{A}$ which are strictly only correct for a single frequency (as could be measured by a shaped sensor), and weighted by the frequency correct eigenvalue, is theoretically capable of producing the desired control effect, the question of practicality again arises. Unfortunately, for control of sound transmission into coupled enclosures this may present a problem. Figure 9.45 illustrates the levels of acoustic potential energy attenuation which are achieved using the set of four transformed modal velocities, but with the eigenvalue weighting neglected. While the attenuation is still satisfactory at many frequencies, the overall result is far less impressive than that obtained with the weighting included. One issue of practicality is therefore the ability to construct eigenvalue 'filters' with the frequency characteristics shown in Fig. 9.39, equivalent to the problem of constructing a set of filters which will mirror the response spectrum of a specific set of acoustic modes. While this is possible, it is not as straightforward as the construction of what are essentially high-pass (eigenvalue) filters for the control of acoustic power radiation into free space, as outlined in the previous chapter. There is also the problem of increasing modal density requiring an increasing number of transformed modal velocities to be measured. Therefore, while it is feasible to use structural velocity measurements as error signals in adaptive feedforward control systems targeting the attenuation of sound transmission into coupled enclosures using vibration control sources, practicality must be assessed very much on a per problem (system/frequency range) basis.

9.10 THE INFLUENCE OF MODAL DENSITY

Thus far in this chapter we have explicitly dealt with systems subject to low frequency excitation, where the response can be accurately modelled using a relatively small number of acoustic modes. It has been shown that in such systems active control is most effective at suppressing system resonances, where the response of the dominant mode can be reduced without having a significant spillover effect (non-resonant modes are inadvertently increased in amplitude in the control process). In this section, the applicability of these results to higher frequency excitation will be considered, where the response of the system is unlikely to be dominated by a single mode, and where the reduction in amplitude of one set of modes is likely to be offset by a similar increase in the amplitude of another set of modes. Specifically, a criterion for predicting the level of global attenuation which is maximally possible based upon the nature of the system response will be developed.

A parameter which is useful in developing such a criterion is modal overlap, $M(\omega)$, which quantifies the likely number of resonance frequencies of other modes lying within the 3 dB bandwidth of a given modal resonance. This parameter is equal to product of the average modal 3 dB bandwidth $\Delta\omega$ and modal density $n(\omega)$, which is the average number of resonance frequencies per unit frequency,

$$M(\omega) = \Delta\omega \; n(\omega) \tag{9.10.1}$$

Assuming viscous damping, the 3 dB bandwidth of the ιth mode is

$$\Delta\omega = 2\zeta_\iota\omega_\iota \tag{9.10.2}$$

where ζ_ι and ω_ι are the critical damping ratio and resonance frequency of the ιth acoustic mode, respectively. Therefore, modal density can be re-expressed as

$$M(\omega_\iota) = 2\zeta_\iota\omega_\iota n(\omega_\iota) \tag{9.10.3}$$

It is possible (Tohyama and Suzuki, 1987; Elliott, 1989) to derive an expression for the average reduction in acoustic potential energy for a rigid walled system containing a single primary source and a single control source remote from it as an explicit function of modal density. This will provide a qualitative picture of how well active noise control will work (globally) at higher frequencies, away from somewhat isolated modal resonances. The related problem of local control is considered in the next section.

As has been discussed several times, acoustic potential energy is equal to the weighted sum of squared acoustic modal amplitudes:

$$E_p = \frac{1}{4\rho_0 c_0^2}\sum_{\iota=1}^{\infty}\Lambda_\iota \,|\, p_\iota \,|^2 \tag{9.10.4}$$

A defining relationship for acoustic modal amplitudes resulting from excitation by a volume velocity source can be written, using (9.3.6) and (9.2.12), as

$$p_\iota = \frac{j\rho_0\omega}{\Lambda_\iota(\kappa_\iota^2 - k^2)}\phi_{g,\iota}q \tag{9.10.5}$$

where q is the complex volume velocity amplitude of the source and $\phi_{g,\iota}$ is the modal generalized volume velocity transfer function for the ιth mode. For the examination here it will be useful to state explicitly the complex eigenvalue κ_ι in terms of both resonance frequency and damping:

$$p_\iota = \frac{j\rho_0 c_0^2\omega}{\Lambda_\iota\left(\omega_\iota^2(1-j2\zeta) - \omega^2\right)}\phi_{g,\iota}q \tag{9.10.6}$$

Expanded in this way, it is possible to write the relation for acoustic potential energy:

$$E_p \geq \sum_{\iota=1}^{\infty} \frac{\rho_0 c_0 \omega^2}{4\Lambda_\iota\left(4\zeta_\iota \omega_\iota^2 \omega^2 + (\omega^2 - \omega_\iota^2)\right)^2} \mid \phi_{g,\iota} q_\iota \mid^2 \tag{9.10.7}$$

Consider now a system where there is a single primary source and single control source contained within a rigid walled enclosure, such that the total volume velocity excitation of any mode is equal to the sum of the primary and control source components. If it is now assumed that at the excitation frequency ω there is a single dominant acoustic mode, mode d, we can re-express the acoustic potential energy as the sum of dominant E_{pd} and residual E_{pr} modal components as

$$E_p = E_{pd} + E_{pr} \approx D \mid \phi_{gp,d} q_p + \phi_{gc,d} q_c \mid^2 + \epsilon_p \mid q_p \mid^2 + \epsilon_c \mid q_c \mid^2 \tag{9.10.8}$$

where

$$D = \frac{\rho_0 c_0 \omega^2}{4\Lambda_d\left(4\zeta_d^2 \omega_d^2 \omega^2 + (\omega^2 - \omega_d^2)^2\right)} \tag{9.10.9}$$

$$\epsilon_p = \sum_{\iota=1, \neq d}^{\infty} \frac{\rho_0 c_0 \omega^2}{4\Lambda_\iota\left(4\zeta_\iota^2 \omega_\iota^2 \omega^2 + (\omega^2 - \omega_\iota^2)^2\right)} \mid \phi_{gp,\iota} \mid^2 \tag{9.10.10}$$

and

$$\epsilon_c = \sum_{\iota=1, \neq d}^{\infty} \frac{\rho_0 c_0 \omega^2}{4\Lambda_\iota\left(4\zeta_\iota^2 \omega_\iota^2 \omega^2 + (\omega^2 - \omega_\iota^2)^2\right)} \mid \phi_{gc,\iota} \mid^2 \tag{9.10.11}$$

Here it has been assumed that, as the primary and control sources are remote from each other, modal cross terms of the form $\phi_{gp,\iota}\, \phi_{gc,\iota}$ will have random signs and so their sum over all modes will be small and can be ignored. Using (9.10.8)–(9.10.11), acoustic potential energy can be expressed in terms of the dominant and residual modal components as

$$E_p = q_c^* a q_c + q_c^* b + b^* q_c + c \tag{9.10.12}$$

where

$$a = D\phi_{gc,d}^2 + \epsilon_c \tag{9.10.13}$$

$$b = D\phi_{gc,d}\phi_{gp,d} q_p \tag{9.10.14}$$

and

$$c = (D\phi_{gp,d}^2 + \epsilon_p) \mid q_p \mid^2 \tag{9.10.15}$$

Based on previous work, the optimum control source volume velocity derived from a quadratic expression of the standard form used in (9.10.12) is defined by the relationship

$$q_{c,E} = -a^{-1}b = -\frac{D\phi_{gc,d}\phi_{gp,d}}{D\phi_{gc,d}^2 + \epsilon_c} q_p \tag{9.10.16}$$

Observe that if only the dominant mode existed (9.10.16) would be simply

$$q_{c,E} = -a^{-1}b = -\frac{\phi_{gp,d}}{\phi_{gc,d}} q_p \tag{9.10.17}$$

which is the same as (9.2.7), minimizing the amplitude of a single mode.

Similarly based on previous work, the reduction in acoustic potential energy is defined by the expression

$$\frac{E_{p,\min}}{E_{orig}} = \frac{c - b^{*} a^{-1} b}{c} \tag{9.10.18}$$

This can expanded using (9.10.13)–(9.10.15) as

$$\frac{E_{p,\min}}{E_{p,orig}} = 1 - \frac{D^2 \phi_{gc,d}^2 \phi_{gp,d}^2}{(D\phi_{gc,d}^2 + \epsilon_c)(D\phi_{gp,d}^2 + \epsilon_p)} \tag{9.10.19}$$

If we now make the simplifying assumptions that $\phi_{gc,d} \approx \phi_{gp,d} \approx 1$ and $\epsilon_p \approx \epsilon_c = \epsilon$, (9.10.19) can be written as

$$\frac{E_{p,\min}}{E_{p,orig}} \approx 1 - \frac{D^2}{(D+\epsilon)^2} = 1 - \left[1 + \frac{\epsilon}{D}\right]^{-2} \tag{9.10.20}$$

If $\epsilon/D << 1$, such as at an isolated acoustic resonance, then

$$\frac{E_{p,\min}}{E_{p,orig}} = \frac{2\epsilon}{D} \tag{9.10.21}$$

The result of (9.10.21) defines an explicit, albeit approximate, relationship between acoustic potential energy reduction and the excitation characteristics of the enclosure in terms of dominant and residual modes. What remains now is to re-express this relationship in terms of modal density, which will require us to consider a specific type of enclosure. The one we will choose is a rigid walled rectangular enclosure.

In a rectangular enclosure we can divide the infinite set of acoustic modes into three categories: axial modes, which are essentially one-dimensional (such as the (1,0,0) mode), tangential, which are essentially two-dimensional (such as the (1,1,0) mode), and oblique, which are essentially three-dimensional (such as the (1,1,1) mode). The modal density of axial modes is a rectangular enclosure is (Morse, 1948):

$$n(\omega) = \frac{l}{\pi c_0} \tag{9.10.22}$$

where l is the length of the enclosure over which the modal indices are varying. The modal density of tangential modes is (Morse, 1948):

$$n(\omega) = \frac{\omega S}{2\pi c_0^2} \tag{9.10.23}$$

where S is the area of the two-dimensional plane over which the modal indices are varying. The modal density of oblique modes is (Morse, 1948):

$$n(\omega) = \frac{\omega^2 V}{2\pi^2 c_0^3} \tag{9.10.24}$$

where V is the volume of the enclosure. Using (9.10.3), the modal overlap of these modal categories is then

$$M(\omega) = \frac{2\zeta\omega l}{\pi c_0} \tag{9.10.25}$$

for axial modes,

$$M(\omega) = \frac{\zeta\omega^2 S}{\pi c_0^2} \tag{9.10.26}$$

for tangential modes, and

$$M(\omega) = \frac{\zeta\omega^3 V}{\pi c_0^3} \tag{9.10.27}$$

for oblique modes. Responses which are dominated by oblique (from Tohyama and Suzuki, 1987) and axial (from Elliott, 1989) modes will be considered here.

Consider first the oblique mode dominated response. If the excitation frequency is set equal to the resonance frequency of the dominant mode d, then from (9.10.9)

$$D \geq \frac{\rho_0 c_0^2}{2\zeta_d^2\omega^2 V} \tag{9.10.28}$$

where use is made of the fact that for oblique modes the normalization term Λ is equal to $V/8$, as defined in (9.3.35). It will be assumed that the level of excitation of the other, residual, acoustic modes is equal to the diffuse field value. Assuming the $\zeta = \zeta_i$ for each mode, this is equal to (after Nelson *et al.*, 1987)

$$\epsilon = \frac{\omega\rho_0}{4\pi c_0\zeta} \tag{9.10.29}$$

Assuming that ζ_d is also equal to ζ, the residual to dominant mode ratio in (9.10.20) is equal to

$$\frac{\epsilon}{D} \approx \frac{\zeta\omega^3 V}{2\pi c_0^3} = \frac{\pi}{2}M(\omega) \tag{9.10.30}$$

where $M(\omega)$ is the oblique field modal overlap as given in (9.10.27). Substituting this relationship into (9.10.20), the reduction in acoustic potential energy resulting from suppression of the dominant acoustic mode is

$$\frac{E_{p,\min}}{E_{p,orig}} = 1 - \left[1 + \frac{\pi}{2}M(\omega)\right]^{-2} \tag{9.10.31}$$

If the modal density $M(\omega) << 1$, such as at low frequencies, this can be further simplified to

$$\frac{E_{p,\min}}{E_{p.orig}} \approx \pi M(\omega) \tag{9.10.32}$$

In an enclosure dominated by axial modes the result will be slightly different. As we noted in the simple example in Section 9.3, axial modes are uniformly spaced in natural frequency, the frequencies following the relationship

$$\omega_{\iota} = \frac{\iota c_0 \pi}{l} \tag{9.10.33}$$

where l is the length of the enclosure over which the modal indices are varying. The term ϵ governing residual mode acoustic potential energy can therefore be approximated by examining the acoustic potential energy in the enclosure contributed by the two modes which are immediately preceding and following a resonant mode, such that $\omega_{\iota} = \omega \pm c_0\pi/2l$. Assuming the $c_0\pi/2l << \omega$ and $2\zeta\omega << c_0\pi/l$, the residual mode acoustic potential energy is governed by

$$\epsilon = \frac{\rho_0 l^2}{2\pi^2 V} \tag{9.10.34}$$

With the excitation frequency equal to the resonance frequency of the dominant mode, the term D governing the acoustic potential energy contributed by the dominant mode is as stated in (9.10.28). The ratio of residual mode to dominant mode acoustic potential energy for the an axial mode dominated system is therefore

$$\frac{\epsilon}{D} = \frac{8\zeta^2\omega^2 l^2}{\pi^2 c_0^2} = 2M(\omega)^2 \tag{9.10.35}$$

where the modal overlap $M(\omega)$ is now that of an axial mode system, as outlined in (9.10.25). Substituting this into (9.10.20), the reduction in acoustic potential energy is

$$\frac{E_{p,\min}}{E_{p,orig}} \approx 1-\left(1+2M(\omega)^2\right)^{-2} \tag{9.10.36}$$

If the modal overlap $M(\omega) << 1$ this can be simplified to

$$\frac{E_{p,\min}}{E_{p,orig}} \approx 4M(\omega)^2 \tag{9.10.37}$$

In comparing (9.10.31) and (9.10.32) for oblique modes with (9.10.36) and (9.10.37) for axial modes it can be seen that if the modal overlap is small there is a significant difference in the effectiveness of active control in the two systems. However, as the modal overlap increases, the systems tend toward the same asymptotic result:

$$\frac{E_{p,\min}}{E_{p,orig}} \rightarrow 1 \text{ for } M(\omega) \geq 1 \tag{9.10.38}$$

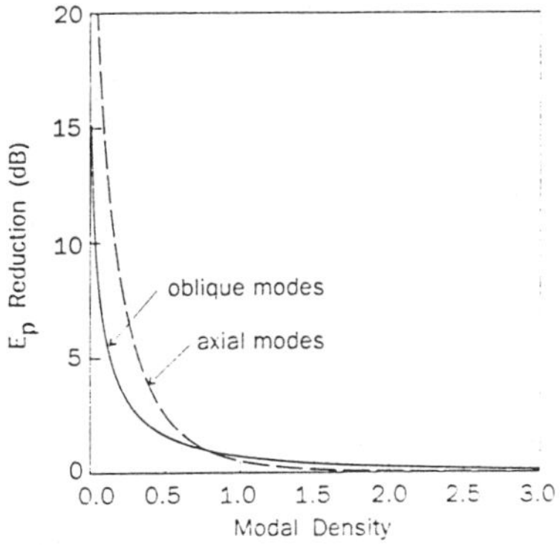

Fig. 9.46 Maximum attenuation of acoustic potential energy as a function of modal density for a sound field composed of oblique modes and a sound field composed of axial modes.

These results are evident in Fig. 9.46, which depict acoustic potential energy attenuation as a function of modal overlap for both mode types.

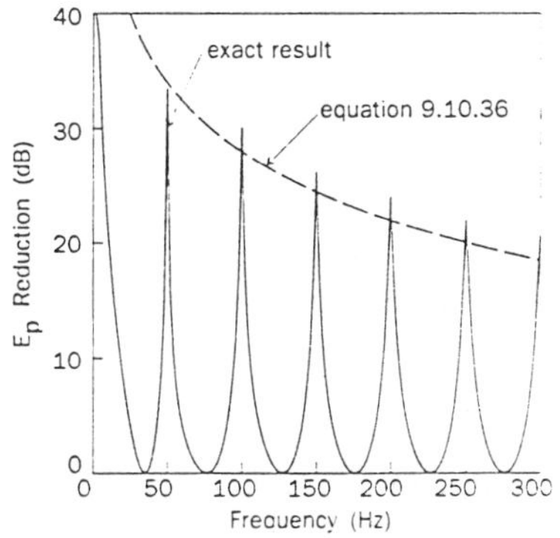

Fig. 9.47 Maximum acoustic potential energy reduction as a function of frequency with a single control source at $x = L_x$, damping of 0.005. Shown are the exact result, and the estimated result using (9.10.36).

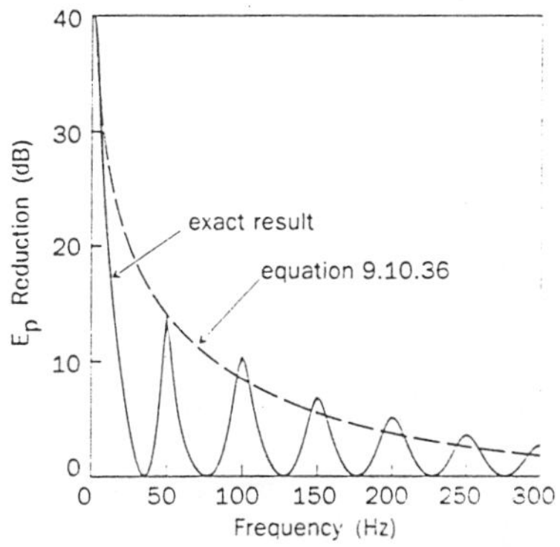

Fig. 9.48 Maximum acoustic potential energy reduction as a function of frequency with a single control source at $x = L_x$, damping of 0.05. Shown are the exact result, and the estimated result using (9.10.36).

To demonstrate the effectiveness of these predictive formulae, it is interesting to apply them to the simple one-dimensional example discussed in Section 9.3. Figures 9.47 and 9.48 illustrate the predicted attenuation using (9.10.36) with the previously calculated (exact) results for damping ratios of 0.005 and 0.05. The prediction is seen to envelope the maximum results, occurring at the resonance frequencies of the acoustic modes (the assumption made in deriving the criteria) very well.

These results demonstrate that the usefulness of (single source) active noise control is limited to essentially isolated resonance problems. The exception to this is when there is a single compact primary noise source, and the control source is placed in close proximity to it. To arrive at this latter conclusion, note that the expression for the control source volume velocity which is optimal with respect to minimising acoustic potential energy, stated in (9.3.14), can be rewritten as

$$q_{c,E} = -q_p \frac{\sum_{\iota=1}^{\infty} \frac{\omega^2 \phi_\iota(r_c)\phi_\iota(r_p)}{(\omega_\iota^2-\omega^2)^2+(2\zeta_\iota\omega_\iota\omega)^2}}{\sum_{\iota=1}^{\infty} \frac{\omega^2 \phi_\iota^2(r_p)}{(\omega_\iota^2-\omega^2)^2+(2\zeta\omega_\iota\omega)^2}} \tag{9.10.39}$$

$$= -q_p \frac{\sum_{\iota=1}^{N_m} |\, Y_\iota(\omega) \,|^2 \phi_\iota(r_c)\phi_\iota(r_p)}{\sum_{\iota=1}^{\infty} |\, Y_\iota(\omega) \,|^2 \phi_\iota^2(r_p)}$$

where

$$|\, Y_\iota(\omega) \,|^2 = \frac{\omega^2}{(\omega_\iota^2-\omega^2)^2+(2\zeta\omega_\iota\omega)^2} \tag{9.10.40}$$

In writing this expression consideration is again restricted to oblique modes. At high values of modal overlap, where the excitation is above the Schroeder frequency, the infinite summations in (9.10.39) can be approximated by an integral expression. With this approximation the numerator of (9.10.39) becomes

$$\sum_{\iota=1}^{\infty} |\, Y_\iota(\omega) \,|^2 \phi_\iota(r_p)\phi_\iota(r_c) \approx \int_0^{\infty} \langle |\, Y_\iota \,|^2 \rangle_u \langle \phi_\iota(r_p)\phi_\iota(r_c) \rangle_u n(u) \, du \tag{9.10.41}$$

where the continuous variable u has replaced the discrete variable ω_ι, the symbol $< >_u$ means that the value in the angular brackets is averaged over a small bandwidth centred on u, and $n(u)$ is the modal density at u. The first of the terms in the integral expression of (9.10.41) is evaluated as

$$\langle |\, Y_\iota(\omega) \,|^2 \rangle_u = \frac{\omega^2}{(u^2-\omega^2)^2+(2\zeta u\omega)^2} \tag{9.10.42}$$

where the critical damping ratio is now assumed to be constant for all modes. The second term in the integral expression is evaluated as (Nelson *et al.*, 1987)

$$\langle \phi_\iota(r_p)\phi_\iota(r_c) \rangle_u = \text{sinc}\frac{ud}{c_0} \tag{9.10.43}$$

where d is the separation distance between the sources. As only oblique modes are being considered, the modal density term in (9.10.39) can be expanded using (9.10.24), so that the integral expression can be written as

$$\sum_{\iota=1}^{N_m} | Y_\iota(\omega) |^2 \phi_\iota(r_p)\phi_\iota(r_c) = \frac{\omega^2 V}{2\pi^2 c_0^3} \int_0^\infty \frac{\omega^2}{(u^2-\omega^2)^2+(2\zeta u\omega)^2} \operatorname{sinc}\frac{ud}{c_0}\, du \tag{9.10.44}$$

With the further assumption of light damping, this integral can be evaluated as (Nelson *et al.*, 1987)

$$\sum_{\iota=1}^{N_m} | Y_\iota(\omega) |^2 \phi_\iota(r_p)\phi_\iota(r_c) = \frac{\omega^2 V}{8\pi c_0^3 \zeta} \operatorname{sinc} kd \tag{9.10.45}$$

Using the same series of steps on the denominator as were just used for the numerator of (9.10.39), it is found that

$$\sum_{\iota=1}^{N_m} | Y_\iota(\omega) |^2 \phi_\iota^2(r_p) = \frac{\omega^2 V}{8\pi c_0^3 \zeta} \tag{9.10.46}$$

Therefore, the relationship defining the control source volume velocity which is optimal with respect to minimizing the acoustic potential energy in the enclosure is

$$q_{c,E} = -q_p \operatorname{sinc} kd \tag{9.10.47}$$

which is exactly the same as the expression obtained for minimizing the radiated acoustic power for the same system arrangement operating in free space. At high modal density only control of the direct field of the source is possible, as is the case in free space, which is why the two solutions are the same. It follows that the maximum attenuation of acoustic potential energy is defined by the relationship

$$\frac{E_{p,\min}}{E_{p,orig}} = 1 - \operatorname{sinc}^2 kd \tag{9.10.48}$$

Thus, as the control source becomes remote from the primary source, such that $kd \geq \pi$, global attenuation of the sound field becomes impossible. This provides an explicit analytical demonstration that global control of enclosed sound fields of high modal density is *only* possible with closely spaced compact noise sources, and is virtually impossible in the general sense.

9.11 CONTROL OF SOUND AT A POINT IN ENCLOSURES WITH HIGH MODAL DENSITIES

While it may not always be possible to obtain significant levels of global sound attenuation in an enclosed space using active control, especially when the modal density of the sound field is high, it is usually theoretically possible to reduce the acoustic pressure at a single point. This problem has been considered previously in Sections 9.2 and 9.4 for systems operating in low modal density environments. In this section we will consider the problem for a system operating in a high modal density environment, examining both local and global effects of the control strategy. Rather than consider a discrete set of modal contributions, as we did in the low modal density studies of Sections 9.2 and 9.4, we will employ statistical operators to examine the sound field. For simplicity, we will restrict discussion to a system with a single monopole primary source, single monopole control source, and single (acoustic) error sensing location.

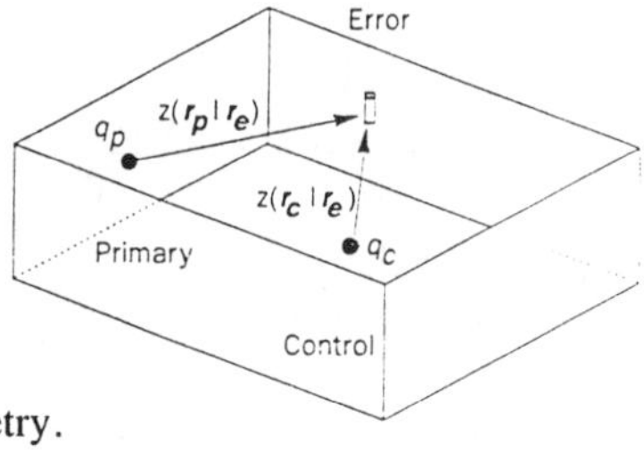

Fig. 9.49 System geometry.

Consider the problem as shown in Fig. 9.49, where the output characteristics of the control source are to be adjusted to minimize the acoustic pressure amplitude at the error sensing location. The acoustic pressure at the error sensing location $p(r_e)$ is equal to the sum of primary and control contributions,

$$p(r_e) = p_c(r_e)+p_p(r_e) = q_c z(r_c \mid r_e)+q_p z(r_p \mid r_e) \tag{9.11.1}$$

where q_c and q_p are the volume velocities of the control and primary sources respectively, and $z(r_c|r_e)$ and $z(r_p|r_e)$ are the transfer functions between the control and primary source locations and the error sensing location. It is obvious that the acoustic pressure at the error sensing location can be driven to zero by the simple relationship

$$q_c = -q_p \frac{z(r_p \mid r_e)}{z(r_c \mid r_e)} \tag{9.11.2}$$

What is of interest here, however, is the development of expressions which will facilitate examination of the overall effect of applying active control.

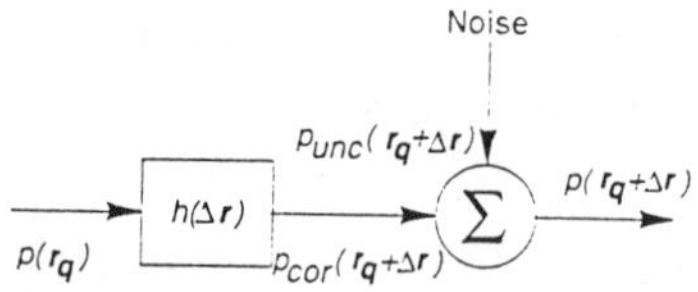

Fig. 9.50 Relationship between acoustic pressures $p(r_q)$ and $p(r_q+\Delta r)$.

Consider the acoustic pressure at some location r_q, and its relationship to the acoustic pressure at a point some distance Δr away, as shown in Fig. 9.50. It is possible to view the sound field at $(r_q+\Delta r)$ as composed of two parts, one perfectly correlated with the acoustic pressure at r_q, the other perfectly uncorrelated (Elliott *et al.*, 1989):

$$p(\boldsymbol{r}_q+\Delta\boldsymbol{r}) = p_{\mathrm{cor}}(\boldsymbol{r}_q+\Delta\boldsymbol{r})+p_{\mathrm{unc}}(\boldsymbol{r}_q+\Delta\boldsymbol{r}) \tag{9.11.3}$$

where the subscripts cor and unc denote the part of the acoustic pressure which is perfectly correlated and perfectly uncorrelated with the sound field at $\boldsymbol{r}_q$, respectively. Note that by this definition the cross-correlation between $p(\boldsymbol{r}_q)$ and $p_{\mathrm{unc}}(\boldsymbol{r}_q+\Delta\boldsymbol{r})$ is equal to

$$E\{p(\boldsymbol{r}_q)\ p_{\mathrm{unc}}(\boldsymbol{r}_q+\Delta\boldsymbol{r})\} = 0 \tag{9.11.4}$$

where $E\{\}$ denotes expected value of the variable in the brackets. In our linear system, the correlated part of $p(\boldsymbol{r}_q+\Delta\boldsymbol{r})$ is related to the pressure at $p(\boldsymbol{r}_q)$ by a simple linear relationship

$$p_{\mathrm{cor}}(\boldsymbol{r}_q+\Delta\boldsymbol{r}) = h(\Delta\boldsymbol{r})p(\boldsymbol{r}_q) \tag{9.11.5}$$

where $h(\Delta\boldsymbol{r})$ is a linear transformation operator, in this case a scalar quantity. Therefore, the cross-correlation between the acoustic pressure $p(\boldsymbol{r}_q)$ and $p_{\mathrm{cor}}(\boldsymbol{r}_q+\Delta\boldsymbol{r})$ is equal to

$$E\{p(\boldsymbol{r}_q)p_{\mathrm{cor}}(\boldsymbol{r}_q+\Delta\boldsymbol{r})\} = h(\Delta\boldsymbol{r})\ E\{p^2(\boldsymbol{r}_q)\} \tag{9.11.6}$$

It can be shown (Cook *et al.*, 1955; Morrow, 1971) that in a diffuse field

$$h(\Delta\boldsymbol{r}) = \operatorname{sinc} k\Delta r \tag{9.11.7}$$

where Δr is the length of $\Delta\boldsymbol{r}$. Noting that

$$E\{p_{\mathrm{cor}}(\boldsymbol{r}_q+\Delta\boldsymbol{r})\ p_{\mathrm{unc}}(\boldsymbol{r}_q+\Delta\boldsymbol{r})\} = 0 \tag{9.11.8}$$

it is possible to write the relationship

$$E\{p^2(\boldsymbol{r}_q+\Delta\boldsymbol{r})\} = E\{p^2_{\mathrm{cor}}(\boldsymbol{r}_q+\Delta\boldsymbol{r})\} + E\{p^2_{\mathrm{unc}}(\boldsymbol{r}_q+\Delta\boldsymbol{r})\} \tag{9.11.9}$$

As

$$\{p_{\mathrm{cor}}(\boldsymbol{r}_q+\Delta\boldsymbol{r})\} = E\{p(\boldsymbol{r}_q)\}\ \operatorname{sinc} k\Delta r \tag{9.11.10}$$

it is possible to express the expected pressure at any location as

$$E\{p^2\} = E\{p^2\}\operatorname{sinc}^2 k\Delta r + E\{p^2_{\mathrm{unc}}(\boldsymbol{r}_q+\Delta\boldsymbol{r})\} \tag{9.11.11}$$

so that

$$\{p^2_{\mathrm{unc}}(\boldsymbol{r}_q+\Delta\boldsymbol{r})\} = E\{p^2(\boldsymbol{r}_q)\}\ (1 - \operatorname{sinc}^2 k\Delta r) \tag{9.11.12}$$

Let us consider now what happens to the sound field when the acoustic pressure at the error sensing location $\boldsymbol{r}_e$ is minimized. The acoustic pressure at some point $\Delta\boldsymbol{r}$ away from the error sensing location is equal to the superposition of the primary and control source contributions:

$$p(\boldsymbol{r}_e+\Delta\boldsymbol{r}) = p_p(\boldsymbol{r}_e+\Delta\boldsymbol{r})+p_c(\boldsymbol{r}_e+\Delta\boldsymbol{r}) \tag{9.11.13}$$

The primary and control source acoustic pressures can each be divided into two components, one perfectly correlated with the acoustic pressure at the error sensing location r_e and one perfectly uncorrelated, so that

$$p_p(r_e+\Delta r) = p_{p,\mathrm{cor}}(r_e+\Delta r)+p_{p,\mathrm{unc}}(r_e+\Delta r) \tag{9.11.14}$$

where

$$\begin{aligned} p_{p,\mathrm{cor}}(r_e+\Delta r) &= p_p(r_e)\ \mathrm{sinc}\ k\Delta r \\ E\{p^2_{p,\mathrm{unc}}(r_e+\Delta r)\} &= E\{p^2_p(r_e)\}(1 - \mathrm{sinc}^2 k\Delta r) \end{aligned} \tag{9.11.15}$$

and

$$p_c(r_e+\Delta r) = p_{c,\mathrm{cor}}(r_e+\Delta r)+p_{c,\mathrm{unc}}(r_e+\Delta r) \tag{9.11.16}$$

where

$$\begin{aligned} p_{c,\mathrm{cor}}(r_e+\Delta r) &= p_c(r_e)\ \mathrm{sinc}\ k\Delta r \\ E\{p^2_{c,\mathrm{unc}}(r_e+\Delta r)\} &= E\{p^2_c(r_e)\}(1 - \mathrm{sinc}^2 k\Delta r) \end{aligned} \tag{9.11.17}$$

If we now assume that the primary and control sources are remote from each other, compared to the wavelength of sound, then

$$E\{p_{p,\mathrm{unc}}(r_e+\Delta r)\ p_{c,\mathrm{unc}}(r_e+\Delta r)\} = 0 \tag{9.11.18}$$

If the acoustic pressure at the error sensing location is driven to zero, then

$$p_c(r_e) = -p_p(r_e) \tag{9.11.19}$$

and so

$$p_{c,\mathrm{cor}}(r_e+\Delta r) = -p_{p,\mathrm{cor}}(r_e+\Delta r) \tag{9.11.20}$$

The squared acoustic pressure level at $(r_e+\Delta r)$is equal to

$$\begin{aligned} &E\{p^2(r_e+\Delta r)\} \\ &= E\{ (p_{c,\mathrm{cor}}(r_e+\Delta r)+p_{c,\mathrm{unc}}r_e+\Delta r)+p_{p,\mathrm{cor}}(r_e+\Delta r)+p_{p,\mathrm{unc}}(r_e+\Delta r))^2 \} \end{aligned} \tag{9.11.21}$$

Using the previous outlined relationships, this can be simplified to (Elliott *et al.*, 1989)

$$E\{p^2(r_e+\Delta r)\} = \left(\{p_p^2\}+\{p_c^2\} \right) (1 - \mathrm{sinc}^2 k\Delta r) \tag{9.11.22}$$

where $E\{p^2\}$ is the expected squared pressure amplitude when the source is operating alone.

Equation (9.11.22) shows that the squared acoustic pressure amplitude rapidly increases away from the error sensing location, tending towards the sum of the individual primary and control source quantities. For example, if a reduction of 10 dB is required at a certain point it must be within a distance of approximately 1/10 wavelength from the error sensor (cancellation point). Generally speaking, the sound field away from the error sensor increases in amplitude when the acoustic pressure at a single point in a diffuse sound field (with all sources and sensors remote from each other). This qualitative idea is evident in Fig. 9.51, the result of a computer simulation examining the pressure field around a single error sensor for a single primary and control source placed randomly in an enclosure containing a diffuse sound field (after Elliott *et al.*, 1988). The data shown is the average for 200 simulations, and clearly shows the extreme localization of the 'zone of quiet'. Observe that away from the error sensor the sound field has *increased* in amplitude.

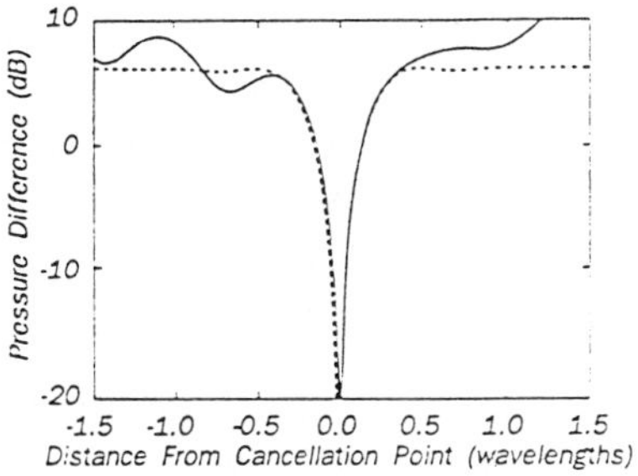

Fig. 9.51 Average value of mean square pressure difference as a function of distance from the cancellation point (after Elliott *et al.*, 1988). Shown are theoretical and computer simulated results.

It is interesting to note that this result is not exactly repeatable, although qualitatively the same characteristics could always be expected. To explain this, note that (9.11.22) shows the mean square acoustic pressure away from the error sensing location to be equal to the sum of the mean square values of primary and control source generated pressure. Thus the increase in level of the mean square value of the sound field is dependent upon the ratio of the control to primary source volume velocities under optimum (cancelling) conditions. Equation (9.11.2) shows this ratio to be

$$\frac{|q_c|^2}{|q_p|^2} = \frac{|z(r_p\,|\,r_e)|^2}{|z(r_c\,|\,r_p)|^2} \tag{9.11.23}$$

If the amplitude of transfer function between the control source and error sensor $z(r_c|r_e)$ is very small, such that the control source is not well coupled to the error sensor, then the amplitude of the control source volume velocity will be very large compared to that of the primary source, and the increase in the amplitude of the pressure field will be correspondingly large.

The probability density function of the real and imaginary components of acoustic pressure at any point in a diffuse sound field will be Gaussian. The probability density function of the ratio of (9.11.23) will therefore be chi-squared with two degrees of freedom, described by the relationship

$$f(x) = \frac{1}{(1+x)^2} \tag{9.11.24}$$

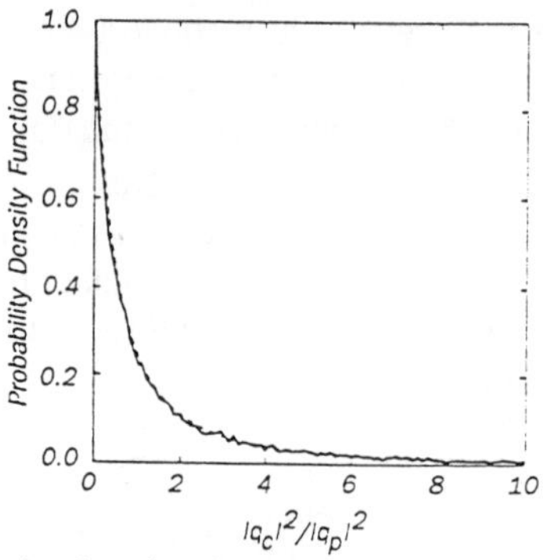

Fig. 9.52 Probability density function for the ratio of control to primary mean square source volume velocities (after Elliott *et al.*, 1988). Shown are the results of computer simulation, and a chi-squared distribution with two degrees of freedom.

where x is the variable of interest, in this case the ratio of the mean square volume velocities in (9.11.23). Figure 9.52 illustrates the probability density function associated with the simulation result of Fig. 9.51, along with the chi-squared probability density function with two degrees of freedom (after Elliott *et al.*, 1989). The correspondence is obvious. The point of this discussion, however, is that large increases in the mean square acoustic pressure away from the cancellation point can occur, with single frequency excitation, due to poor control source placement. As the quality of the control source placement will not be so readily apparent in sound fields of high modal density as in sound fields of low modal density when the sources and sensors are widely dispersed, and for broadband control the source placement is likely to be poor at some frequencies, there is no utility in implementing such an active control arrangement in systems of high modal density with the aim of sound attenuation over anything other than a very small spatial region.

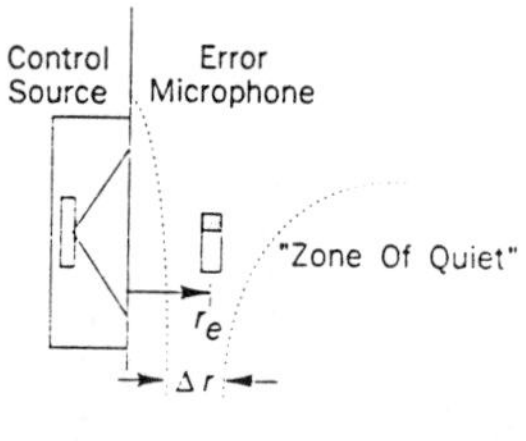

Fig. 9.53 Nearfield local control arrangement.

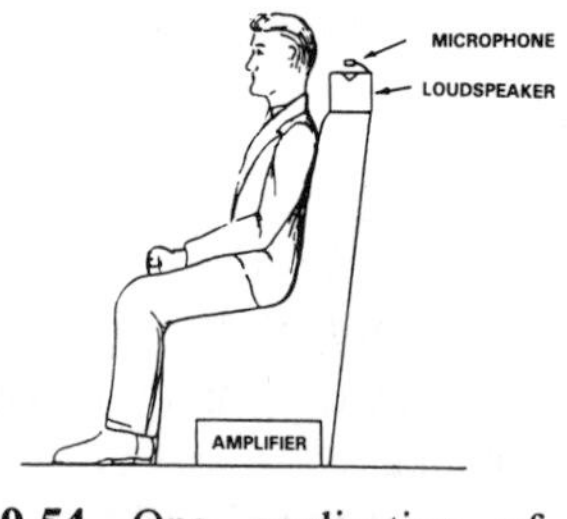

Fig. 9.54 One application of an 'electronic sound reducer' (after Olson and May, 1953).

One approach which can be taken to improve these results, at least from the standpoint of guarding against substantial increases in acoustic potential energy in the enclosed space when minimizing the acoustic pressure at a single location, is to place the error microphone directly in front of the control source as illustrated in Fig. 9.53. With this arrangement the control source volume velocity required to minimize the acoustic pressure at the error sensor is greatly reduced, and is largely independent of the statistical properties of the sound field. This, in fact, is one of the modes of operation of an 'electronic sound reducer' proposed in 1953 (Olson and May, 1953), although the control approach taken was feedback rather than feedforward. Part of the utility of this device was envisaged as being able to provide local acoustic control for a variety of arrangements, such as near an occupant's head in an automobile or aircraft, shown in the sketch in Fig. 9.54. We can extend the previous analysis to examine how effective such an arrangement could be.

To do this, first express the transfer function between the control source and error sensor as the sum of two components, that due to direct radiation and that due to the reverberant field:

$$z(\mathbf{r}_c \mid \mathbf{r}_e) = z_d(\mathbf{r}_c \mid \mathbf{r}_e) + z_r(\mathbf{r}_c \mid \mathbf{r}_e) \tag{9.11.25}$$

where the subscripts d and r are used to denote the direct and reverberant field components. Substituting this into (9.11.2), the optimum control source volume velocity becomes

$$q_c = -q_p \frac{z(\mathbf{r}_p \mid \mathbf{r}_e)}{z_d(\mathbf{r}_c \mid \mathbf{r}_e) + z_r(\mathbf{r}_c \mid \mathbf{r}_e)} \tag{9.11.26}$$

With the error microphone placed in the vicinity of the control source, the direct field transfer function will dominate (9.11.26), being much greater than either the primary source transfer function or the reverberant field component of the control source transfer function. This is why the control source volume velocity required to minimize the acoustic pressure at the error microphone will be greatly reduced, which will reduce the problem of greatly increased sound pressure levels away from the error sensor.

As the direct component of the control source transfer function will be dominant in its immediate vicinity we can, to a first approximation, neglect the control source generated reverberant field component when examining the sound field in this area. The acoustic pressure a small distance $\Delta\mathbf{r}$ away from the error sensor can therefore be approximated as

$$p(\mathbf{r}_e + \Delta\mathbf{r}) \approx p_p(\mathbf{r}_e + \Delta\mathbf{r}) + z_d(\mathbf{r}_c \mid \mathbf{r}_e + \Delta\mathbf{r}) q_c \tag{9.11.27}$$

This can be rewritten as

$$p(\mathbf{r}_e + \Delta\mathbf{r}) \approx p_p(\mathbf{r}_e + \Delta\mathbf{r}) - \frac{z_d(\mathbf{r}_c \mid \mathbf{r}_e + \Delta\mathbf{r})}{z_d(\mathbf{r}_c \mid \mathbf{r}_e)} p_p(\mathbf{r}_e) \tag{9.11.28}$$

The change in mean square pressure amplitude at the point $\Delta\mathbf{r}$ away from the error sensor is therefore

$$\frac{E\{\,|\,p(r_e+\Delta r)\,|\,^2\}}{E\{\,|\,p_p(r_e)\,|\,^2\}} \tag{9.11.29}$$

$$= 1 - 2\,\mathrm{Re}\left\{\frac{z_d(r_c\,|\,r_e+\Delta r)}{z_d(r_c\,|\,r_e)}\right\}\mathrm{sinc}\;\Delta r + \left|\frac{z_d(r_c\,|\,r_e+\Delta r)}{z_d(r_c\,|\,r_e)}\right|^2$$

As consideration is restricted to small distances Δr, the sinc Δr term in (9.11.29) can be approximated as equal to 1.0. The transfer function term $z(r_c|r_e+\Delta r)$ can be expanded as a Taylor series about r_e, retaining only the first term because the distance Δr is small, as

$$z_d(r_c\,|\,r_e+\Delta r) \approx z_d(r_c\,|\,r_e)+\nabla z_d(r_c\,|\,r_e)\cdot\Delta r \tag{9.11.30}$$

Substituting this expansion into (9.11.29) and using the approximation for small distances

$$E\{\,|\,p_p(r_e)\,|\,^2\} \approx E\{\,|\,p_p(r_e+\Delta r)\,|\,^2\} \tag{9.11.31}$$

produces the expression (after Joseph *et al.*, 1989)

$$\frac{E\{\,|\,p(r_e+\Delta r)\,|\,^2\}}{E\{\,|\,p_p(r_e+\Delta r)\,|\,^2\}} \approx \left|\frac{\nabla z_d(r_c\,|\,r_e)\cdot\Delta r}{z_d(r_c\,|\,r_e)}\right|^2 \tag{9.11.32}$$

This result shows that, to a first approximation, the change (hopefully reduction) in the mean square acoustic pressure amplitude at a location in the vicinity of the error sensor is dependent upon the ratio of the pressure gradient to the absolute value of the pressure in the near field of the control source, as

$$\left|\frac{\nabla z_d(r_c\,|\,r_e)\cdot\Delta r}{z_d(r_c\,|\,r_e)}\right|^2 = \left|\frac{\nabla p_d(r_c)\cdot\Delta r}{p_d(r_c)}\right|^2 \tag{9.11.33}$$

This result can be further simplified by re-expressing the pressure gradient in terms of particle velocity (for harmonic pressure) as

$$\nabla p_d(r_c) = -j\omega\rho_0 u_d(r_c) \tag{9.11.34}$$

where $u_d(r_c)$ is the particle velocity associated with the direct acoustic pressure. Assuming that the particle velocity is evaluated in the same direction as Δr, substituting (9.11.34) into (9.11.33), and expressing the result in terms of specific acoustic impedance,

$$z_d(r_c) = \frac{u_d(r_c)}{p_d(r_c)} \tag{9.11.35}$$

produces the expression

$$\frac{E\{\,|\,p(r_e+\Delta r)\,|\,^2\}}{E\{\,|\,p_p(r_e+\Delta r)\,|\,^2\}} \approx \left|\frac{\rho_0 c_0}{z_d(r_c)}\right|^2 k^2\Delta r^2 \tag{9.11.36}$$

This formula can be used to predict the change in the pressure field away from the error sensing location.

As an example, consider the simple case of a monopole control source, for which the specific acoustic impedance some distance Δr away is

$$z(\Delta r) = \rho_0 c_0 \frac{jk\Delta r}{1+jk\Delta r} \tag{9.11.37}$$

Substituting this into (9.11.36) produces the following expression for the change in acoustic pressure in the vicinity of the error microphone:

$$\frac{E\{\,|\,p(r_e+\Delta r)\,|\,^2\}}{E\{\,|\,p_p(r_e+\Delta r)\,|\,^2\}} \approx (1+k^2r_e^2)\left[\frac{\Delta r}{r_e}\right]^2 \tag{9.11.38}$$

where r_e is the distance between the monopole control source and the error sensor. For small separations between the control source and error sensor this can be approximated as

$$\frac{E\{\,|\,p(r_e+\Delta r)\,|\,^2\}}{E\{\,|\,p_p(r_e+\Delta r)\,|\,^2\}} \approx \left[\frac{\Delta r}{r_e}\right]^2 < \epsilon \tag{9.11.39}$$

where ϵ is the mean square pressure reduction. Therefore, a 10 dB zone of quiet will encircle the control source for a radial distance equal to twice the value of (9.11.39), or

$$\Delta r = 2r_e\sqrt{\epsilon} \approx 0.6r_e \tag{9.11.40}$$

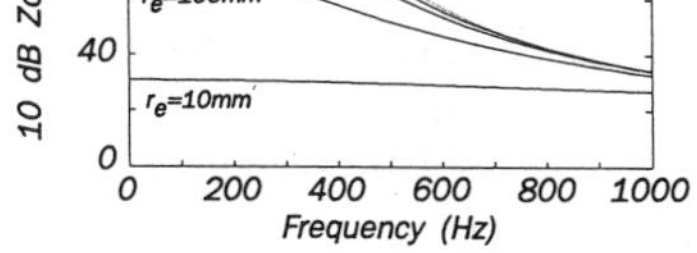

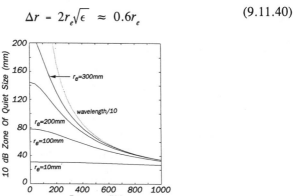

Fig. 9.55 The axial extent of a 10 dB minimum zone of quiet generated by a 0.05 m radius piston control source as a function of frequency for various control source/error sensor separation distances (after Elliott and David, 1992).

As with the case of a separated control source and error sensor, the area of pressure reduction is extremely localized; it is, in fact, even smaller than before. Recall from Fig. 9.51 that with the separated arrangement, a 10 dB zone of quiet existed at a radial distance of approximately 1/10 wavelength from the error microphone. Figure 9.55 depicts the size of the 10 dB zone of quiet around the error sensor as a function of frequency for various separations of the error sensor from the 0.05 m piston control source (see Elliott and David, 1992). These are, in fact, bounded by the 1/10 wavelength criterion, being much less than this amount for small separation distances. Clearly, the advantage of locating the error microphone close the control source is the reduction in the problem of significant increases in acoustic pressure away from the error sensor, and not an enhancement in the size of the 'zone of quiet'.

9.12 STATE SPACE MODELS OF ACOUSTIC SYSTEMS

While there has been a substantial amount of work conducted on the application of feedforward active control systems to attenuate enclosed sound fields, the application of feedback control systems to the problem has received relatively little attention. In this section we will consider this latter problem to a limited degree by simply formulating state space models of enclosed sound field which can be used to formulate control laws. The formulation is very similar to that used to examine distributed parameter (structural) systems, which will be considered later in Chapter 11.

The starting point for the development of state space model is the inhomogeneous wave equation, derived in Chapter 2. This can be stated in terms of acoustic pressure as

$$\left[\frac{1}{c^2}\frac{\partial^2}{\partial t^2} + \nabla^2\right] p(r,t) = \rho_0 \frac{\partial}{\partial t} s(r_s,t) \tag{9.12.1}$$

where s is the volume velocity distribution in the enclosure. For an enclosed acoustic space, the acoustic pressure at any location $\boldsymbol{r}$ can be expressed as contributions from an infinite set of acoustic modes, ϕ,

$$p(r,t) = \sum_{n=1}^{\infty} p_n(t)\ \phi_n(r) \tag{9.12.2}$$

where p_n is the complex amplitude of the nth acoustic mode. The eigenvalue problem associated with the (self-adjoint) acoustic modes is

$$\nabla^2\phi_n(r) = \lambda_n\phi_n(r) \tag{9.12.3}$$

where λ_n is the nth eigenvalue, and the acoustic mode shape functions will be orthonormalized in such a way that

$$\int_V \phi_m(r)\ \phi_n(r)\ \mathrm{d}r = \delta(m-n) \tag{9.12.4}$$

where δ is a Kronecker delta function,

$$\delta(m-n) = \begin{matrix} 1 & m=n \\ 0 & m\neq n \end{matrix} \tag{9.12.5}$$

Note that the relationship between the eigenvalue λ_n and the natural frequency ω_n of the nth mode is

$$\lambda_n = \omega_n^2 \tag{9.12.6}$$

Substituting the modal expansion of acoustic pressure into the wave equation produces

$$\frac{1}{c^2}\sum_{n=1}^{\infty}\frac{\partial^2 p_n(t)}{\partial t^2}\phi_n(r) + \sum_{n=1}^{\infty}\lambda_n p_n(t)\phi_n(r) = \rho_0\frac{\partial}{\partial t}s(r_s,t) \tag{9.12.7}$$

Multiplying this through by the mth acoustic mode shape, ϕ_m, and integrating over the enclosed space produces the equation governing the response of the mth mode

$$\frac{1}{c_0^2}\frac{\partial^2 p_m(t)}{\partial t} + \lambda_m p_m(t) = \rho_0\alpha_m(t) \tag{9.12.8}$$

or

$$\frac{\partial^2 p_m(t)}{\partial t} + c_0^2\omega_m^2 p_m(t) = \rho_0 c_0^2\alpha_m(t) \tag{9.12.9}$$

where $\alpha_m(t)$ in the modal generalized volume velocity of the mth acoustic mode:

$$\alpha_m(t) = \int_V \phi_m(r)s(r,t)\ \mathrm{d}r \tag{9.12.10}$$

Using this, the response of the mth acoustic mode can be written in state space form as

$$\dot{x}_m(t) = A_m x_m(t) + \alpha_m(t) \tag{9.12.11}$$

where

$$x_m(t) = \begin{bmatrix} p_m(t) \\ \dot{p}_m(t) \end{bmatrix}, \quad A_m = \begin{bmatrix} 0 & 1 \\ -c_0^2\omega_m^2 & 0 \end{bmatrix}$$
$$\alpha_m(t) = \begin{bmatrix} 0 \\ \rho_0 c_0^2\alpha_m(t) \end{bmatrix} \tag{9.12.12}$$

To formulate a control law for attenuating the enclosed sound field it is necessary to explicitly state the matrix $\boldsymbol{\alpha}_m$ in terms of a control input and a transfer function. To do this, the volume velocity distribution is expressed as the product of some amplitude q and a transfer function to the location of interest z_q,

$$s(r,t) = q(t)z_q(r) \tag{9.12.13}$$

Substituting this product into (9.12.10), the modal generalized volume velocity can be written as

$$\alpha_m(t) = q(t)\int_V \phi_m(r)z_q(r)\ \mathrm{d}r = q(t)\phi_{g,m} \tag{9.12.14}$$

where $\phi_{g,m}$ is the mth modal generalized volume velocity transfer function:

$$\phi_{g,m} = \int_V \phi_m(\boldsymbol{r}) z_q(\boldsymbol{r})\, \mathrm{d}\boldsymbol{r} \tag{9.12.15}$$

This term was encountered previously in Section 9.2, where it was shown in (9.2.13) that for the simplest source arrangement, a monopole acoustic source, the modal generalized volume velocity transfer function was simply equal to the value of the acoustic mode shape function at the monopole location. Using the expansion of (9.12.14) the state space equation (9.12.11) can be rewritten as

$$\dot{\boldsymbol{x}}_m(t) = \boldsymbol{A}_m \boldsymbol{x}_m(t) + \boldsymbol{B}_m \boldsymbol{u}(t) \tag{9.12.16}$$

where

$$\boldsymbol{B} = \begin{bmatrix} 0 & 0 & \cdots & 0 \\ \phi_{g,1} & \phi_{g,2} & \cdots & \phi_{g,N_c} \end{bmatrix}, \quad \boldsymbol{u}(t) = \begin{bmatrix} q_1(t) \\ q_2(t) \\ \vdots \\ q_{N_c}(t) \end{bmatrix} \tag{9.12.17}$$

where N_c sources are being used in the system. If the pressure field is being modelled using N_n acoustic modes then the state space description of the system can be written as

$$\dot{\boldsymbol{x}}(t) = \boldsymbol{A}\boldsymbol{x}(t) + \boldsymbol{B}\boldsymbol{u}(t) \tag{9.12.18}$$

where

$$\boldsymbol{x}(t) = \begin{bmatrix} \boldsymbol{x}_1(t) \\ \boldsymbol{x}_2(t) \\ \vdots \\ \boldsymbol{x}_{N_n}(t) \end{bmatrix}, \quad \boldsymbol{A} = \begin{bmatrix} \boldsymbol{A}_1 & & & \\ & \boldsymbol{A}_2 & & \\ & & \ddots & \\ & & & \boldsymbol{A}_{N_n} \end{bmatrix}, \quad \boldsymbol{B} = \begin{bmatrix} \boldsymbol{B}_1 \\ \boldsymbol{B}_2 \\ \vdots \\ \boldsymbol{B}_{N_n} \end{bmatrix} \tag{9.12.19}$$

Note that $\boldsymbol{A}$ is not a diagonal matrix, but rather a diagonal arrangement of modal state vectors, defined in (9.12.12).

Acoustic damping can be added to this model by rewriting the wave equation as

$$\left[\frac{1}{c^2}\frac{\partial^2}{\partial t^2} + \zeta \nabla^2 \frac{\partial}{\partial t} + \nabla^2\right] p(\boldsymbol{r},t) = \rho_0 \frac{\partial}{\partial t} s(\boldsymbol{r}_s,t) \tag{9.12.20}$$

where ζ is the critical damping ratio. Going through the same steps as before, the damped modal equation of motion is found to be

$$\frac{\partial^2 p_m(t)}{\partial t} + \zeta c_0^2 \omega_m^2 \frac{\partial p_m(t)}{\partial t} + c_0^2 \omega_m^2 p_m(t) = \rho_0 c_0^2 \alpha_m(t) \tag{9.12.21}$$

which can be written in the state space form of (9.12.11) with the modified state matrix

$$A_m = \begin{bmatrix} 0 & 1 \\ -c_{0^2}\omega_m^2 & -\zeta c_{0^2}\omega_m^2 \end{bmatrix} \tag{9.12.22}$$

This damped state matrix can then be used in place of the undamped state matrices in $\boldsymbol{A}$, as defined in (9.12.19).

If microphones are used to measure the sound field at discrete locations, the acoustic pressure at each location can be expressed as the sum of modal contributions,

$$p(r,t) = \sum_{i=1}^{N_n} p_i(t)\phi_i(r) \tag{9.12.23}$$

If N_e error sensors are being used, the vector of acoustic pressures at their locations, the output equation, can be expressed as

$$y(t) = \boldsymbol{C}x(t) \tag{9.12.24}$$

where $y(t)$ is the vector of pressures

$$y(t) = \begin{bmatrix} p(r_1,t) \\ p(r_2,t) \\ \vdots \\ p(r_{N_e},t) \end{bmatrix} \tag{9.12.25}$$

and $\boldsymbol{C}$ is the output matrix

$$\boldsymbol{C} = \begin{bmatrix} \phi_1(r_1) & 0 & \phi_2(r_1) & 0 & \cdots & \phi_{N_n}(r_1) & 0 \\ \phi_1(r_2) & 0 & \phi_2(r_2) & 0 & \cdots & \phi_{N_n}(r_2) & 0 \\ & & & & \vdots & & \\ \phi_1(r_{N_e}) & 0 & \phi_2(r_{N_e}) & 0 & \cdots & \phi_{N_n}(r_{N_e}) & 0 \end{bmatrix} \tag{9.12.26}$$

An example of the application of the state equations (9.12.18) and (9.12.24) in formulating a control law for attenuating the sound field in a rigid walled rectangular enclosure was given by Dohner and Shoureshi (1989), where an LQG approach was taken in formulating the control law.

9.13 AIRCRAFT INTERIOR NOISE

9.13.1 Introduction

It is well known and generally accepted that interior noise levels in general and commercial aviation propeller driven aircraft are unacceptable. Unfortunately, the weight penalty resulting from the use of passive control measures to reduce the noise to acceptable levels is even more unacceptable. Passive control is usually associated with the attachment to the fuselage interior surface of multi-layer trim panels, an example of which might be made up as follows: a layer of lightweight damping material to the fuselage, a central weighted layer separated from the fuselage by a fibreglass blanket and a stiff decorative inner trim panel separated from the central weighted layer by another fibreglass blanket (Silence, 1991).

Another form of passive control which is also characterized by a significant weight penalty is the use of tuned vibration dampers acting on the fuselage to reduce vibration at the fundamental and first harmonic of the propeller blade passing frequency (Emborg and Halvorsen, 1988). Experimental tests on a SAAB340 aircraft demonstrated in-flight noise reductions of up to 6 dB(A) at seat locations close to the plane of the propellers. Clearly, this type of control must be used together with some form of trim panel control as well.

Existing interior noise problems in propeller driven aircraft together with the resurgence of interest in the development of more fuel efficient, high speed turbo propeller aircraft has generated considerable interest in the application of active control to reduce interior noise levels with a small weight penalty. Available data (Wilby *et al.*, 1985; Patrick, 1986; Wilby and Wilby, 1987) indicate that 20−30 dB of noise reduction is required for the first three or four propeller blade passage harmonics and 10−20 dB for the next five or six to reduce them to levels resulting from turbulent boundary layer excitation of the fuselage.

As might be concluded from the preceding discussion, active control is particularly suited to controlling tonal noise such as generated by a propeller. Researchers are also working on reducing jet engine noise by placing acoustic control sources in the periphery of an extension to the shroud (or cowling) at the inlet and exit of the engine. However, control of noise generated by action of the turbulent boundary layer on the fuselage is considerably more difficult, although progress is likely to be made by applying adaptive feedback control to vibration actuators on the fuselage to minimize a cost function proportional to interior noise levels. Such a cost function could be derived from appropriately shaped PVDF film which could be configured to detect only those structural modes which contribute significantly to interior noise levels.

In recent years, there has been a major thrust from an European consortium (Advanced Study of Active Noise Control in Aircraft, ASANCA) (Borchers *et al.*, 1992; Van der Auweraer, 1993; Emborg and Ross, 1993) to implement active noise control systems in several medium sized passenger aircraft. In these studies, testing and theoretical work has been undertaken for four different partner aircraft (that is, Dornier 228, Saab 340, ATR 42, and Fokker 100). They were selected to identify general noise control information and demonstrate that active noise control was feasible in aircraft that have large differences in acoustic character.

The ASANCA consortium implemented active noise control using a fixed array of acoustic control sources, located in the interior trim and ceiling of the aircraft (Borchers *et al.*, 1992; Van der Auweraer, 1993; Emborg and Ross, 1993). A fuselage test cell of a Dornier 228 was subsequently used to experimentally verify the analytical results (Hackstein *et al.*, 1992). Reductions of 15 and 16 dB were achieved for the blade passage tone and first harmonic respectively.

Additional flight tests (Emborg and Ross, 1993) were performed using a similar system in a Saab 340 aircraft. A total of 48 error sensors and 24 loudspeakers were installed. Reductions in this aircraft were 10 dB and 3 dB for the blade passage tone and first harmonic respectively. These results were not as good as predicted for the Dornier 228 aircraft, but this is probably as a result of differences in the fuselage structure and the presence of cabin furniture and passengers. It is expected that a production system will be available in this aircraft by late 1994.

In an earlier series of flight tests, Elliott *et al.* (1990) used 26 configurations involving up to 16 loudspeakers and 32 microphones to actively reduce in-flight interior noise levels in a B.Ae.748, propeller driven aircraft. They reported significant reductions (of the order of 10 dB) in noise levels at the fundamental, second and third harmonics (88 Hz, 176 Hz and 264 Hz respectively) of the propeller blade passing frequency, which resulted in a 7 dB (A) reduction at forward seat locations near the plane of the propellers. However, reductions further from the propeller plane were much less.

Simpson *et al.* (1991) reported similar reductions using active vibration control sources acting on the fuselage structure of the aft section of a McDonnell Douglas DC-9. In this case, the aircraft section was on the ground and engine noise was simulated using shakers attached to engine pylons.

Other experimental tests on actual aircraft have involved in-flight tests on the ATR-42 (Paonessa *et al.*, 1993), and laboratory tests on a Handley Page 137 Jetstream III fuselage (Warner and Bernhard, 1987). Noise reductions for the ATR-42 of up to 12 dB and for the Handley Page fuselage of between 12 and 22 dB were obtained.

These systems have either been single-channel systems or not very sophisticated multichannel systems. It is clear that a considerable amount of analytical modelling and testing with multichannel controllers, as outlined in the remainder of this section, remains to be done.

At very low frequencies characterized by the first or second harmonic of a typical propeller blade passing frequency (below about 160 Hz) it is feasible to attempt noise reductions using local cancellation which does not result in global noise reductions but only reductions in specific areas, usually at the expense of increases in other areas (Elliott *et al.*, 1990). A single-channel feedback system may be used for example to produce a zone of quiet around a headrest in a seat. Elliott (1991) showed by computer simulation that it is possible to produce a spherical zone of quiet of diameter equal to one-tenth of a wavelength around a single microphone with a control source mounted in close proximity (for example, in a headrest). In this zone, noise levels can be expected to be 10 dB quieter on average than they were with just the primary field. However, this technique is only suitable for very low frequencies (at 200 Hz, $\lambda/10 \simeq 170$ mm) and is less preferable than the option of using feedforward control to achieve global noise reductions.

It is of interest to note in passing that of the many patents for active noise control devices, at least two are directed particularly at aircraft interior noise.

One (Fuller, 1987) addresses the use of vibration actuators on the fuselage to reduce interior noise levels and the other (Warnaka and Zalas, 1985) addresses the use of acoustic sources within the cabin.

9.13.2 Analytical modelling

Although a significant proportion of the energy exciting the fuselage and thus generating interior noise in an aircraft is structure borne from the engine mounts and the wings to the fuselage, most of the analytical modelling work has been directed at predicting the effects of excitation of the fuselage by an external pressure field. However, as will be shown, it is relatively straightforward to include the structure borne energy transmission in the acoustic model, provided the forcing function at the attachment points of the wing to the fuselage can be estimated. This forcing function is made up of a combination of mechanical energy at the engine mounts and aerodynamic pressures imposed by the propeller wake on the wing, the latter being much more important for high speed turboprop aircraft (Wilby, 1990).

As pointed out by Wilby (1989), large errors in predictions are obtained if the analytical model is too greatly simplified; for example, if the fuselage is modelled as a simple cylinder as was done by Bullmore *et al.* (1990). This is because the presence of the floor destroys the axi-symmetric nature of the interior acoustic modes, thus allowing more than one structural mode to couple with each acoustic mode. resulting in a second mechanism available for acoustic control, as described in Section 9.7. If the fuselage is modelled as a simple cylinder, then this second control mechanism is not analytically possible, and large prediction errors can occur at some frequencies.

As was made clear earlier in this chapter, the first step in developing an analytical model which can be used for the design of an active noise control system, is to develop an accurate model capable of predicting local as well as average fuselage vibration levels and interior noise levels. The cylinder with floor model discussed earlier in this chapter is adequate for investigating mechanisms of active control and likely achievable noise reductions, but it is inadequate for designing an active control system for an actual aircraft or for accurately predicting expected noise reductions due to active control. The same may be said of the detailed analytical model described by Wilby (1989) and Pope (1990), called PAIN (propeller aircraft interior noise) which is a combination of closed form and numerical approaches. In PAIN, the structural model is a cylinder with a floor and its resonance frequencies and mode shapes are characterized in closed form by smearing the effects of the stiffeners over the entire structural skin. This is done by calculating an equivalent average EI (where E is Young's modulus of elasticity and I is the second moment of area (or area moment of inertia)) for the skin and stiffeners and also an effective structural mass per unit area of the structural cross-section averaged over the skin and stiffeners. Finite difference methods are used to calculate the modal characteristics of the interior space (cylinder with a floor). To simplify the model with little loss of accuracy, the modal parameters of the structure are calculated for the fuselage in a vacuum, those for the interior space are calculated assuming a rigid structural boundary, and the results coupled together using a similar approach to that outlined earlier in this chapter.

The propeller noise field in the PAIN model is obtained either analytically or from test data, and involves the determination of amplitude and phase data describing the acoustic pressure associated with the exterior acoustic field at all exterior locations on the fuselage. Interior acoustic treatments (trim panels) have been included as an integral part of the PAIN model.

Perhaps the most promising modelling approach for practical fuselage active noise control system design is to use separate finite element analyses to determine the modal characteristics of both the in-vacuo structure and the rigidly enclosed acoustic space (Unruh and Dobosz, 1988; Martin, 1993). Once found, these can be coupled, and the effect of active control predicted using the same procedure as used for the cylinder with a floor earlier in this chapter. Use of finite element analysis enables discrete modelling of important structural and acoustic characteristics that influence the overall fuselage response. In some previous studies (Borchers *et al.*, 1992 – Dornier 228 aircraft; Van der Auweraer 1993 – Dornier 228 aircraft; Emborg and Ross 1993 – SAAB 340 aircraft; Paxton *et al.*, 1993 – MD-80 jet aircraft; Green 1992 – SAAB 2000 aircraft), the structure was modelled as a small length of fuselage adjacent to the propeller, where the sound transmission is at its highest. No consideration was given to the effect of wing attachments and the potential for structure borne excitation via that path, and the excitation input into the structure was primarily represented as a propeller blade passage pressure distribution on the fuselage surface. Modelling a small length of fuselage is computationally expedient, but the results can be of limited use in evaluating the effectiveness of global active noise control systems.

The structural modal parameters required for the analysis approach outlined above can be found easily by using most commercially available software packages. Although most commercial software packages do not have acoustic elements, solutions for acoustical modal properties may be obtained with structural FEA packages by using an appropriate displacement–pressure or stress–pressure analogy as described by Lamancusa (1988). In summary, a structural finite element analysis package can be 'fooled' into solving an acoustic problem using the displacement–pressure analogy by taking the following steps.

1. Set Poisson's ratio $\nu = 0$.
2. Set density $\rho = 1.206$ kg/m^3.
3. Set elastic modulus $E = \rho c^2$ (1.42×10^5 Pa).
4. Set modulus of rigidity $G = E$.
5. Set structural displacements equal to zero in all but one direction. Displacement in the remaining non-zero direction is acoustic pressure. Any direction is acceptable (as pressure is a scalar) except in an axi-symmetric problem it must be along the axis of symmetry.

6. At a free surface, the pressure is set equal to zero.
7. Along rigid walls no boundary conditions are needed.
8. Where the applied acoustic pressure is known, give the selected degree-of-freedom a forced displacement equal to the applied pressure.
9. At a surface where the normal component of displacement is known, apply an external force, $F = -S\ddot{a}_n(t)$ where S is the area of the moving surface and $\ddot{a}_n(t)$ is the normal component of surface acceleration.

One of the important steps in the development of an analytical model is to accurately define the forcing function acting on the fuselage, which is a combination of the exterior acoustic pressure field and the structure-borne forces at the wing attachment points. The exterior acoustic field can be determined to a reasonable accuracy by on-ground measurements, or to a lesser accuracy by approximating the propeller as a pair of dipoles whose axes intersect at right angles (Mahan and Fuller 1986), and which are 90° out of phase with one another. If a counter-clockwise rotating source is to be simulated, then each of the four monopoles making up the two dipoles must lag its clockwise neighbour by 90°. While the monopoles themselves remain motionless, they will produce a combined dipole type directivity pattern which rotates in the counter-clockwise direction with an angular frequency equal to the angular frequency of oscillation of the dipoles. Note that the rotation of the directivity pattern through one cycle does not represent one rotation of the propeller; rather, it represents one cycle of the fundamental, or a harmonic, of the complex acoustic pressure field generated by the motion of an individual propeller blade past the fuselage. The centre of the dipoles is between the propeller hub and the fuselage, at about 60% of the distance from the propeller hub to the propeller tip, in the plane of the propeller. In one example, Mahan and Fuller found that good results were obtained for a spacing of 10% of the fuselage radius between individual monopoles making up each of the two dipoles.

In summary, an analytical model to predict interior noise levels would be constructed using the following steps.

1. Development of a finite element model to determine *in vacuo* structural modal parameters.
2. Development of a finite element model to determine rigid boundary acoustic mode shapes.
3. Generation of the acoustic field forcing function and the structural forcing function at the wing attachment points.
4. Determination of the structural response to the forcing functions using experimentally determined or estimated structural loss factors.
5. Determination of the interior acoustic field using the coupled structural/acoustic modal analysis outlined in Section 9.4.
6. Optimization of control source types, locations and strengths as well as error sensor types, locations and weighting factors using the procedure outlined in Section 9.7.

9.13.3 Control sources and error sensors

There are a number of practical alternatives which can be used for control sources and error sensors. Acoustic control sources can be either loudspeakers or horn drivers placed at appropriate locations in the aircraft cabin. It is also possible to use ductwork to direct the control sound to locations where it is needed and where there may not necessarily be room for a loudspeaker. A special purpose lightweight speaker for aircraft has been developed (Warnaka *et al.*, 1992) using a rare earth permanent magnet encapsulated by steel to reduce stray magnetic fields. Generally, loudspeakers should be capable of generating approximately 105 – 110 dB at 1 m at frequencies to be controlled. In practice, the effectiveness of the loudspeakers is limited by the available cone excursion, rather than the required input power, which is usually between 1 and 4 W.

Vibration control sources, directed at controlling the fuselage vibration in a way which will minimize interior noise, can be thin piezoceramic crystals bonded to the fuselage, or piezoceramic stack actuators or magnetostrictive rod actuators, the latter two being mounted between the fuselage wall and the interior trim panel. Each of these actuator types is discussed in detail in Chapters 15 and 16. One possible disadvantage associated with using vibration actuators on the fuselage is that some energy will spill over into structural modes which are not excited by the primary source. Also, minimization of the interior acoustic field does not necessarily result in lower structural vibration levels and in some cases these levels may increase, thus increasing the potential for acoustic fatigue failure. Nevertheless, vibration control sources are more effective than acoustic sources in reducing interior noise levels dominated by structural controlled modes, although acoustic sources are more effective than vibration sources in reducing sound levels dominated by acoustic controlled modes. Thus a practical system would probably include both types of source with the fuselage vibration level as one of the error inputs which would be assigned a weighting to allow a balance to be obtained between interior noise level reductions and fuselage vibration level increases. Another way of minimizing fuselage vibration levels would be to use a 'control effort' error signal made up of the sum of the squares of all of the signals driving the control actuators, thus ensuring that the vibrational energy into the fuselage is minimized. This would also have the effect of ensuring that one control actuator was not driven significantly harder than the others; it would assist in the control effort being divided more equally between control actuators. If necessary, a similar error signal could be derived for acoustic sources.

Practical error sensors include interior mounted microphones and PVDF film (see Chapter 16) bonded to the fuselage. There is even a possibility that appropriate shaping of the PVDF sensor may allow a single vibration error signal to be obtained which is proportional to the average radiated interior noise levels. Signals from both vibration and acoustic error sensors could be appropriately weighted to place more or less importance on interior noise levels in specific locations.

9.14 AUTOMOBILE INTERIOR NOISE

With the ever increasing drive for greater fuel efficiency and lower manufacturing costs, there is considerable interest in reducing the weight of passenger cars. Thus there is a need for the development of appropriate technology to control engine noise, exhaust noise and road noise, so that noise levels in the passenger compartment will not be increased as the vehicle weight is reduced. One approach with great potential for low frequency noise reduction is active control. The use of active mufflers to control exhaust noise was discussed in Chapter 7 and the use of active vibration isolation to control low frequency engine noise will be discussed in Chapter 12. Here the discussion will be restricted to the active control of noise generated by the tyre road interaction. This noise is random in nature and physical constraints limit the extent of control which is achievable to 5 – 10 dB, depending on the particular vehicle being tested. The active systems discussed in this section have a very different purpose from the low bandwidth active suspension systems discussed in Chapter 12. Here the aim is to minimize noise levels in the passenger compartment in the audio frequency range, whereas the purpose of an active suspension is to improve the ride and reduce vibration levels at sub-audio frequencies.

Considerable effort in a number of countries has been focussed on the application of active control of road noise in automobile passenger compartments. Guicking *et al.* (1991) used measured road noise data and simulation techniques to estimate the likely benefit of active control. They found that it was important to limit control effort to avoid non-linear distortion in the loudspeakers. Since 1988, researchers at the Institute of Sound and Vibration Research have been publishing the results of their joint efforts with Lotus Engineering on active control of road noise (Elliott *et al.*, 1988a; Sutton *et al.*, 1990; Elliott *et al.*, 1990b; Sutton *et al.*, 1991, 1994). Their early simulation work using measured data showed that a relatively short processing delay in the controller could be tolerated, but after that further delays resulted in a significant degradation in performance. In later work, the research team demonstrated a reduction of 7 dB(A) in the frequency range 100 – 200Hz on a Citroen AX sedan travelling at 60 km h^{-1}, and published guidelines on optimal locations for reference sensors and control loudspeakers (Sutton *et al.*, 1994). In the same time period, Bernhard and his co-workers working on a similar and parallel program to that of the ISVR team have published a number of papers (summarized by Bernhard, 1995) outlining their work and have demonstrated results which are similar to those achieved by the ISVR team. Perhaps the most important contribution of Bernhard's team has been the development of design techniques for the optimal location of reference sensors.

Road noise is transmitted into the passenger compartment through the wheel spindle and then through various suspension components to the body panels which vibrate and radiate sound. Although the spectrum is generally broadband, it usually also includes predictable components due to the repeating nature of the tread vibration pattern which is even apparent on rough roads. As all of the energy is transmitted through the wheel spindle, this seems the obvious location for a reference sensor to be used with a feedforward active control system. However, as discussed by Bernhard (1995), often more than one reference sensor per wheel is needed to obtain good results. Bernhard (1995) also concluded that there are sufficient delays in the energy propagation path for feedforward control (with its inherent performance and stability advantages) to be effective for many applications.

In addition to the number and locations of reference sensors, other system parameters which must be considered in the active control system design are the number and optimal locations of control loudspeakers in the passenger compartment, the number and optimal locations of microphone error sensors, the length and type of adaptive control filters and algorithm convergence speed to maintain stability.

Bernhard (1995) summarized the optimal characteristics of many of the parameters just mentioned and these results will be summarized here. First he indicated the importance of the coherence between the reference and error sensors, as the maximum theoretically achievable noise reduction assuming an ideal controller is,

$$NR = 10\log_{10}(1 - \gamma^2) \tag{9.14.1}$$

where γ is the coherence for a single input–single output system and the multiple coherence for a multiple input–single output system. The optimum number of reference sensors was found to be either two or three, depending on the vehicle and the optimum locations were very vehicle specific and were chosen to maximize the coherence between the reference and error sensors (Kompella, 1992), while at the same time minimizing the coherence between one reference sensor and another. Good coherence between two or more reference sensors results in very slow convergence of the adaptive filters and poor tracking ability of the active noise control system. The reference sensors must also be located such that it is possible for the adaptive system to realize the optimal filter with the number of filter taps available.

As shown earlier in this chapter, one control source is required for each significant mode in the passenger compartment which must be controlled and the best locations are at the antinodes of the enclosure response. The passenger compartment response of most automobiles is dominated at low frequencies (up to 150 Hz for small cars and up to 200 Hz for large cars) by four modes (Bernhard, 1995) with antinodes at the corners of the compartment. Thus it is a practical alternative to use the four loudspeakers already present for the vehicle sound system as control sources, even if this requires some small change to the location of the speakers. With four control sources, four error microphones are needed and the best location for these is likely to be in the headrests of the seats, one for each seat.

Bernhard (1995) found that FIR filters perform as well as IIR filters with the same number of coefficients and that the optimal filter length varied between 300 and 600 taps for each control channel.

In conclusion, between 5 and 7 dB reduction of road noise in many automobile passenger compartments in the frequency range 50 Hz to 150 Hz may be achieved using a four-channel feedforward active control system. However, future reductions in excess of this are unlikely due to the complexity of the physical problem. For similar reasons, it is unlikely that the frequency range of optimal control will be extended upwards, except for very small vehicles.

REFERENCES

Banerjee, P.K. and Butterfield, R. (1981). *Boundary Element Methods in Engineering Science*. McGraw-Hill: London.

Bernhard, R.J. (1995). Active control of road noise inside automobiles. In *Proceedings of Active 95*, Institute of Noise Control Engineering, pp. 21–32.

Bhattacharya, M.C. and Crocker, M.J. (1969). Forced vibration of a panel and radiation of sound into a room. *Acustica*, **22**, 275–294.

Borchers, I.U., Emborg, U., Sollo, A., Waterman, E.H., Palliard,J., Larsen, P.N., Venet, G., Göransson, P. and Martin, V. (1992). Advanced study for active noise control in aircraft. In *Proceedings of the 4th NASA/SAE/DLR Aircraft Interior Noise Workshop*. Friedrichshafen, Germany, 19–20 May.

Brebbia, C.A. and Walker, S. (1980). *Boundary Element Techniques in Engineering*. Newnes-Butterworths: London.

Bernhard, R.J., Gardner, B.K., Mollo, C.G. and Kipp, C.R. (1986). Prediction of sound fields in cavities using boundary element methods. *AIAA Paper*, **AIAA-86-1864**.

Bullmore, A.J. (1988). *The Active Minimisation of Harmonic Enclosed Sound Fields with Particular Application to Propeller Induced Cabin Noise*. PhD thesis, University of Southampton.

Bullmore, A.J., Nelson, P.A., Curtis, A.R.D. and Elliott, S.J. (1987). The active minimisation of harmonic enclosed sound fields, Part II: A computer simulation. *Journal of Sound and Vibration*, **117**, 15–33.

Bullmore, A.J., Nelson, P.A. and Elliott, S.J. (1990). Theoretical studies of the active control of propeller-induced cabin noise. *Journal of Sound and Vibration*, **140**, 191–217.

Cook, R.K., Waterhouse, R.V., Berendt, R.D., Edelman, S. and Thompson, M.C. (1955). Measurement of correlation coefficients in reverberant sound fields. *Journal of the Acoustical Society of America*, **27**, 1072–1077.

Cunefare, K.A. and Koopmann, G.H. (1991). Global optimum active noise control: surface and far-field effects. *Journal of the Acoustical Society of America*, **90**, 365–373.

Curtis, A.R.D., Elliott, S.J. and Nelson, P.A. (1985). Active control of one-dimensional enclosed sound fields. *Proceedings of Internoise 85*, 579–582.

Curtis, A.R.D., Nelson, P.A., Elliott, S.J. and Bullmore, A.J. (1987). The active suppression of acoustic resonances. *Journal of the Acoustical Society of America*, **81**, 624–631.

Curtis, A.R.D., Nelson, P.A. and Elliott, S.J. (1990). Active reduction of a one-dimensional enclosed sound field: An experimental investigation of three control strategies. *Journal of the Acoustical Society of America*, **88**, 2265–2268.

Dohner, J.L. and Shoureshi, R. (1989). Modal control of acoustic plants. *Journal of Vibration, Acoustics, Stress, and Reliability in Design*, **111**, 326–330.

Dorling, C.M., Eatwell, G.P., Hutchins, S.M., Ross, C.F. and Sutcliffe, S.G. (1989). A demonstration of active noise reduction in an aircraft cabin. *Journal of Sound and Vibration*, **128**, 358–360.

Elliott, S.J. (1989). *The Influence of Modal Overlap in the Active Control of Sound and Vibration*. ISVR Memorandum, **695**.

Elliott, S.J. (1991). Active control of enclosed sound fields. In *Notes for a Short Course on Active Control of Noise and Vibration*, ed. C.H. Hansen. Department of Mechanical Engineering, University of Adelaide, GPO Box 498 Adelaide, South Australia, 9.1–9.38.

Elliott, S.J. and David, A. (1992). A virtual microphone arrangement for local active sound control. *Proceedings of the First International Conference on Motion and Vibration Control (MOVIC)*, 1027–1031.

Elliott, S.J., Curtis, A.R.D., Bullmore, A.J. and Nelson, P.A. (1987). The active minimisation of harmonic enclosed sound fields, Part III: Experimental verification. *Journal of Sound and Vibration*, **117**, 35–58.

Elliott, S.J., Stothers, I.M., Nelson, P.A., McDonald, A.M., Quinn, D.C. and Saunders, T. (1988a). The active control of engine noise inside cars. *Proceedings of Internoise 88*, 987–990.

Elliott, S.J., Joseph, P., Bullmore, A.J. and Nelson, P.A. (1988b). Active cancellation at a point in a pure tone diffuse sound field. *Journal of Sound and Vibration*, **120**, 183–189.

Elliott, S.J., Nelson, P.A., Stothers, I.M. and Boucher, C.C. (1989). Preliminary results of in-flight experiments on the active control of propeller-induced cabin noise. *Journal of Sound and Vibration*, **128**, 355–357.

Elliott, S.J., Nelson, P.A., Stothers, I.M. and Boucher, C.C. (1990a). In-flight experiments on the active control of propeller-induced cabin noise. *Journal of Sound and Vibration*, **140**, 219–238.

Elliott, S.J., Nelson, P.A. and Sutton, P.J. (1990b). The active control of low frequency engine and road noise inside automotive interiors. *Active Noise and Vibration Control,* ASME publication NCA-8, 125–129.

Emborg, U. and Halvorsen, W.C. (1988). Development of tuned dampers for control of bladepass tones in cabin of SAAB 340. In *Proceedings of the 3rd SAE/NASA Aircraft Interior Noise Workshop*. Hampton, Virginia, April, 11–12.

Emborg, U., and Ross, C.F., (1993) Active control in the Saab 340. In *Proceedings of the Recent Advances in Active Control of Sound and Vibration*. Blacksburg, VA, April, 100–109.

Fuller, C.R. (1987). Apparatus and method for global noise reduction. US Patent No. 4715 559.

Fahy, F. (1985). *Sound and Structural Vibration: Radiation, Transmission, and Response*. Academic Press: London.

Fuller, C.R. (1986). Analytical model for investigation of interior noise characteristics in aircraft with multiple propellers including synchrophasing. *Journal of Sound and Vibration*, **109**, 141–156.

Fuller, C.R. and Jones, J.D. (1987). Experiments on the reduction of propeller induced interior noise by active control of cylinder vibration. *Journal of Sound and Vibration*, **112**, 389–395.

Green, I.S. (1992). Vibro-acoustic FE analyses of the Saab 2000 aircraft. In *Proceedings of the 4th NASA/SAE/DLR Aircraft Interior Noise Workshop*. Friedrichshafen, Germany, 44–69.

Guicking, D., Bronzel, M. and Bohm, W. (1991). Active adaptive noise control in cars. In *Proceedings of Recent Advances in Active Control of Sound and Vibration*. Technomic Publishing Co., Inc., 657–670.

Guy, R.W. (1979). The response of a cavity backed panel to external airborne excitation: a general analysis. *Journal of the Acoustical Society of America*, **65**, 719–731.

Hackstein H.J., Borchers, I.U., Renger, K. and Vogt, K., (1992). The Dornier 328 acoustic test cell (ATC) for interior noise tests and selected test results. In *Proceedings of the Recent Advances in Active Control of Sound and Vibration*. Blacksburg, VA, April, pp. 35-43.

Jones, J.D. and Fuller, C.R. (1987). Active control of sound fields in elastic cylinders by multi-control forces. *AIAA Paper*, **AIAA–87–2707**.

Junger, M.C. and Feit, D. (1986). *Sound, Structures, and Their Interaction*. MIT Press: Cambridge, MA.

Kompella, M.S. (1992). *Improved Multiple Input/Multiple Output Modeling Procedures with Consideration of Statistical Information*. PhD thesis, Purdue University, USA.

Lamancusa, J.S. (1988). Acoustic finite element modelling using commercial structural analysis programs. *Noise Control Engineering Journal*, **30**, 65–71.

Leissa, A.W. (1973). *Vibration of Shells*. NASA SP-288.

Lyon, R.H. (1963). Noise reduction of rectangular enclosures with one flexible wall. *Journal of the Acoustical Society of America*, **35**, 1791–1797.

Magliozzi, B. (1984). Advanced turboprop noise: a historical review. *AIAA paper*, **AIAA 84-226**.

Mahan, J.R. and Fuller, C.R. (1986). A propeller model for studying trace velocity effects on interior noise. *Journal of Aircraft*, 23, 142–147.

Martin, V. (1993). Small-scale vibro-acoustic modelling for active noise control in aircraft. In *Proceedings of Noise-93*, St Petersburg, Russia, Vol. 2, 195–200.

Mollo, C.G. and Bernhard, R.J. (1987). *A Generalised Method for Optimization of Active Noise Controllers in Three-dimensional Spaces*. AIAA paper, **AIAA-87-2705**.

Mollo, C.G. and Bernhard, R.J. (1989). Generalized method of predicting optimal performance of active noise controllers. *AIAA Journal*, **27**, 1473–1478.

Mollo, C.G. and Bernhard, R.J. (1990). Numerical evaluation of the performance of active noise control systems. *Journal of Vibration and Acoustics*, **112**, 230–236.

Morrow, C.T. (1971). Point to point correlation of sound pressures in reverberant chambers. *Journal of Sound and Vibration*, **16**, 29–42.

Morse, P.M. (1948). *Vibration and Sound*. McGraw-Hill: New York (reprinted by the Acoustical Society of America, 1981).

Nadim, M. and Smith, R.A. (1983). Synchronous adaptive cancellation in vehicle cabins. *Proceedings of Internoise 83*, 461–464.

Nelson, P.A., Curtis, A.R.D. and Elliott, S.J. (1985). Quadratic optimisation problems in the active control of sound. *Proceedings of the Institute of Acoustics*, **7**, 45–53.

Nelson, P.A., Curtis, A.R.D., Elliott, S.J. and Bullmore, A.J. (1987). The active minimisation of harmonic enclosed sound fields, Part I: Theory. *Journal of Sound and Vibration*, **117**, 1–13.

Olson, H.F. (1956). Electronic control of noise, vibration, and reverberation. *Journal of the Acoustical Society of America*, **28**, 966–972.

Olson, H.F. and May, E.G. (1953). Electronic sound absorber. *Journal of the Acoustical Society of America*, **25**, 1130–1136.

Oswald, L.J. (1984). Reduction of diesel noise inside passenger compartments using active, adaptive noise control. *Proceedings of Internoise 84*, 483–488.

Pan, J. (1992). The forced response of an acoustic-structural coupled system. *Journal of the Acoustical Society of America*, **91**, 949–956.

Pan, J. and Bies, D.A. (1990a). The effect of fluid-structural coupling on sound waves in an enclosure – theoretical part. *Journal of the Acoustical Society of America*, **87**, 691–707.

Pan, J. and Bies, D.A. (1990b). The effect of fluid-structural coupling on sound waves in an enclosure – experimental part, *Journal of the Acoustical Society of America*, **87**, 708–717.

Pan, J., Hansen, C.H. and Bies, D.A. (1990). Active control of noise transmission through a panel into a cavity: I. Analytical study. *Journal of the Acoustical Society of America*, **87**, 2098–2108.

Pan, J. and Hansen, C.H. (1991a). Active control of noise transmission through a panel into a cavity: II. Experimental study. *Journal of the Acoustical Society of America*, **90**, 1488–1492.

Pan, J. and Hansen, C.H. (1991b). Active control of noise transmission through a panel into a cavity: III. Effect of actuator location, *Journal of the Acoustical Society of America*, **90**, 1493–1501.

Paonessa, A., Sollo, A., Paxton, M. Purver, M. and Ross, C.F. (1993). Experimental active control of sound in the ATR 42. In *Proceedings of Noise-Con '93*, 225–230.

Parkin, P.H. and Morgan, K. (1965). Assisted resonance. *Journal of Sound and Vibration*, **2**, 464.

Parkin, P.H. and Morgan, K. (1970). Assisted resonance in the Royal Festival Hall, London: 1965–1969, *Journal of the Acoustical Society of America*.

Parkin, P.H. and Morgan, K. (1971). Assisted resonance in the Royal Festival Hall, London: 1965–1969. *Journal of Sound and Vibration*, **15**, 127–141.

Patrick, H.V.L. (1986). Cabin noise characteristics of a small propeller powered aircraft. AIAA paper. In *Proceedings of the 10th Aeroacoustics Conference*, Seattle, 9–11 July, 86–1096.

Paxton, M., Purver, M., Ross, C.F., Baptist, M., Lang, M.A., May, D.N. and Simpson, M.A. (1993). Active control of sound in a MD-80. In *Proceedings of the Recent Advances in Active Control of Sound And Vibration*. Blacksburg, VA, April, 67-73.

Peterson, M.R. and Boyd, D.E. (1978). Free vibrations of circular cylinders with longitudinal, interior partitions. *Journal of Sound and Vibration*, **60**, 45–62.

Piraux, J. and Nayroles, B. (1980). A theoretical model for active noise attenuation in three-dimensional space. *Proceedings of Internoise 80*, 703–706.

Pope, L.D. (1971). On the transmission of sound through finite closed shells: Statistical energy analysis, modal coupling, and non-resonant transmission. *Journal of the Acoustical Society of America*, **50**, 1004–1018.

Pope, L.D. (1990). On the prediction of propeller tone sound levels and pressure gradients in an aeroplane cabin. *Journal of the Acoustical Society of America*, 88, 2755–2765.

Pope, L.D. and Wilby, E.G. (1982). *Analytical Prediction of the Interior Noise for Cylindrical Models of Aircraft Fuselages for Prescribed Exterior Noise Fields, Phase II*. NASA CR–165869.

Pope, L.D., Rennison, D.C. and Wilby, E.G. (1980). *Analytical Prediction of the Interior Noise for Cylindrical Models of Aircraft Fuselages for Prescribed Exterior Noise Fields*. NASA CR–159363.

Pope, L.D., Rennison, D.C., Willis, C.M. and Mayes, M.H. (1982). Development and validation of preliminary analytical models for aircraft noise prediction. *Journal of Sound and Vibration*, **82**, 541–575.

Pope, L.D., Wilby, E.G., Willis, C.M. and Mayes, W.H. (1983). Aircraft interior noise models: Sidewall trim, stiffened structures, and cabin acoustics with floor partition. *Journal of Sound and Vibration*, **89**, 371–417.

Pretlove, A.J. (1966). Forced vibrations of a rectangular panel backed by a closed rectangular cavity. *Journal of Sound and Vibration*, **3**, 252–261.

Salikuddin, M. and Ahuja, K.K. (1989). Application of localised active control to reduce propeller noise transmitted through fuselage surface. *Journal of Sound and Vibration*, **133**, 467–481.

Silcox, R.J., Fuller, C.R. and Lester, H.C. (1987). *Mechanisms of Active Control in Cylindrical Fuselage Structures*. AIAA Paper, **AIAA-87-2703**.

Silence, J.L. (1991). Meeting the noise control needs of today's aircraft. *Noise and Vibration Worldwide*, July, 14–16.

Simpson, M.A., Luong, T.M., Swinbanks, M.A., Russell, M.A. and Leventhall, H.G. (1989). Full scale demonstration tests of cabin noise reduction using active noise control. *Proceedings of Internoise 89*, 459–462.

Simpson, M.A., Luong, T.M., Fuller, C.R. and Jones, J.D. (1991). Full-scale demonstration tests of cabin noise reduction using active vibration control. *Journal of Aircraft*, **28**, 208–215.

Snyder, S.D. and Hansen, C.H. (1991). Using multiple regression to optimize active noise control system design. *Journal of Sound and Vibration*, **148**, 537–542.

Snyder, S.D. and Hansen, C.H. (1994a). The design of systems to actively control periodic sound transmission into enclosed spaces, Part 1. Analytical models. *Journal of Sound and Vibration*, **170**, 433–449.

Snyder, S.D. and Hansen, C.H. (1994b). The design of systems to actively control periodic sound transmission into enclosed spaces, Part 2. Mechanisms and trends, *Journal of Sound and Vibration*, **170**, 451–472.

Snyder, S.D. and Tanaka, N. (1992). On feedforward active control of sound and vibration using vibration error signals. *Journal of the Acoustical Society of America*, submitted.

Sutton, T.J., Elliott, S.J. and Nelson, P.A. (1990). The active control of road noise inside automobiles. In *Proceedings of Internoise '90*, 689–695.

Sutton, T.J., Elliott, S.J. and Moore, I. (1991). Use of non-linear controllers in the active attenuation of road noise inside cars. In *Proceedings of Recent Advances in Active Control of Sound and Vibration*. Technomic Publishing Co. Inc., 682–690.

Sutton, T.J., Elliott, S.J. McDonald, A.M. and Saunders, T.J. (1994). Active control of road noise inside vehicles. *Noise Control Engineering Journal*, **42**, 137–147.

Tabachnick, B.G. and Fidell, L.S. (1989). *Using Multivariante Statistics*. Harper and Row: New York.

Tohyama, M. and Suzuki, A. (1987). Active power minimisation of a sound source in a closed space. *Journal of Sound and Vibration*, **119**, 562–564.

Unruh, J.F. and Dobosz, S.A. (1988). Fuselage structural-acoustic modelling for structure-borne interior noise transmission. *Journal of Vibration, Acoustics, Stress and Reliability in Design*, **110**, 226–233.

Van der Auweraer, H., Otte, D., Venet, G. and Catalifaud, J. (1993). Aircraft interior sound field analysis in view of active control: results from the ASANCA project. In *Proceedings of Noise-Con '93*, 219–224.

Warnaka, G.E. and Zalas, J.M. (1985). Active attenuation of noise in a closed structure.

US Patent No. 4562 589.

Warnaka, G.E., Kleinle, M., Tsangaris, P., Oslac, M.J. and Moskow, H.J. (1992). A lightweight loudspeaker for aircraft communications and active noise control. In *Proceedings of the 4th MASA/SAE/DLR Aircraft Interior Noise Workshop*. Friedrichshafen, Germany, 19–20 May.

Warner, J.V. and Bernhard, R.J. (1987). Digital control of sound fields in three-dimensional enclosures. AIAA Paper 87–2706. In *Proceedings of the 11th Aeroacoustics Conference*. Palo Alto, CA, 19–21 October.

Wilby, J.F. (1989). Noise transmission into propeller-driven aeroplanes. *The Shock and Vibration Digest*, **21**, 3–10.

Wilby, J.F. and Wilby, E.G. (1987). Measurements of propeller noise in a light turboprop aeroplane. AIAA paper 87–2737, In *Proceedings of the 11th Aeroacoustics conference*. Palo Alto, CA, 19–21 October.

Wilby, J.F., McDaniel, C.D. and Wilby, E.G. (1985). In-flight Acoustic Measurements on a Light Twin-engined Turboprop Aeroplane. NASA-CR 178004.

Zalas, J.M. and Tichy, J. (1984). *Active Attenuation of Propeller Blade Passage Noise*. NASA CR-172386.

10

Feedforward control of vibration in beams and plates

The vibration control of structures to minimize their noise radiation was discussed in Chapters 8 and 9, and the control of global structural vibration levels will be discussed in Chapter 11. In this chapter, the control of vibration transmission along structural elements is discussed. Such control may be for the purpose of reducing the sound radiated from another structure attached to the structural element or it may be for the purpose of reducing vibration levels in an attached structure. Examples of the former case include the control of vibration transmission along a strut connecting the gearbox to the cabin in a helicopter (Elliott, 1993) or the control of vibratory energy transmitted along a submarine hull excited by the propeller pressure field. An example of the latter case is the control of vibratory power transmitted along a support arm to a space telescope.

Structural vibration may be described in terms of waves of different types travelling in different directions, or in terms of vibration modes. If the transmission of vibratory energy is to be controlled, then a wave description is more appropriate, whereas if overall structural vibration levels are to be controlled, a modal description gives better results. The control of modal vibration using feedforward control has not been discussed extensively in the literature, although Clark (1994) described a methodology for approaching such problems. However, as this chapter is concerned with the control of transmission of vibratory energy rather than global vibration control, a wave description is used in the analysis. All possible wave types (flexural, longitudinal and torsional (or shear)) are considered, because although longitudinal and torsional waves do not generally in themselves contribute significantly to sound radiation or feelable structural vibration, they are easily scattered into flexural waves at structural junctions and discontinuities, and these waves are generally associated with significant sound radiation.

As we are concerned here with vibration transmission, it is assumed that it is possible to obtain a measure of the incoming disturbance. For periodic signals it is not important when the measurement is obtained relative to the controller output; it is only important that it reflect the frequency content of the signal to be controlled. However, for random disturbances, the measurement must be made sufficiently far ahead of the control actuator for the wave propagation time from the measurement point to the control actuator to be greater than the delay through the control system. Otherwise the controller will be non-causal. Achieving causality can be very difficult for longitudinal waves which, in common structural materials, propagate at a speed which is commonly an order of magnitude faster than flexural waves. For flexural waves, the propagation speed is much less, but even for these waves, the non-propagating, evanescent field associated with vibration sources can result in error signals not being proportional to the propagating power transmission. This problem will be discussed in more detail later on in this chapter.

If it is not possible to obtain a measure of the incoming disturbance in sufficient time, then it is necessary to use feedback control and this is discussed in Chapter 11. Feedback control effectively adds damping, stiffness or mass to a structure (at the force input location), thus effectively changing system resonance frequencies. It can be very effective in reducing overall vibration levels in a reverberant structure, and also effective in controlling random propagating disturbances. However, if the disturbance is periodic, the performance achieved even for a reverberant structure is not as good as that obtained using feedforward control, because in the feedback system the error signal (which is diminished by the action of the controller) is used directly to obtain the control signal; thus, large gains in the feedback loop are needed to obtain a substantial reduction in the error signal and this comes at some cost to system stability margins. On the other hand, in the feedforward system, the error signal is used as the quantity to be minimized in a control algorithm and is not directly related to the control signal. When feedback control is used with a single force actuator to control the propagation of travelling waves in a beam, large feedback gains effectively result in a pinned or simply supported condition at the force location. This results in control of the component of the bending wave associated with transverse displacement, but provides no control for the component associated with rotation. As the two components are associated with equal energies, the maximum reduction in total propagating power which can be obtained is 3 dB. However, when displacement feedback (as opposed to acceleration, velocity or force feedback) is used, there is one frequency dependent value of the feedback gain which will result in total suppression of the travelling wave. Interestingly, this value is the same as the optimal feedforward gain with the error sensor co-located with the control force. This is discussed in more detail in Chapter 11.

Both feedforward and feedback systems are governed by laws of controllability and observability, and this is especially important in reverberant structures, where the vibration may be characterized by modes. A particular mode or wave type is controllable only if the control actuator is located so that it can drive that mode or configured so that it can drive that particular wave type. The same comments can be made about the law of observability and its applicability to the sensor used by the controller to derive the control signal.

It is interesting to note the differences in control mechanisms associated with the different types of control and input disturbance. For feedback control, the mechanism, as mentioned previously, effectively involves artificially increasing the structural stiffness, mass and/or damping, thus resulting in a smaller response to or a reflection of an incoming disturbance. On the other hand, with feedforward control the mechanism can be any one or all of three; reflection, absorption or suppression. For periodic excitation, the likely mechanisms are absorption of the propagating wave power by the control actuator or suppression of the primary source power generation by effectively changing the impedance it sees. For random noise, the power is either absorbed by the control actuator or reflected back from whence it came. In this case causality effects prevent the control source from affecting the impedance of the primary source and thus suppressing the power generation unless, of course, the control source is located in the near field of the primary force.

The active control of beam vibration is the simplest form of active control of structural vibration, being a similar type of problem to active control of plane waves in ducts, with the added complication that the flexural wavespeed is frequency dependent, thus making it more difficult to implement control for random noise disturbances. Active control of beam vibration was first discussed by Scheuren (1985) who considered the feedforward control of flexural waves in an infinite beam using a single control force. To control both the propagating flexural wave component and the evanescent near field, two control forces have been used (Mace, 1987; Redman-White *et al.*, 1987; Scheuren 1988; McKinnell, 1988, 1989).

In addition to control of flexural waves in infinite beams, effort has also been directed at the control of longitudinal and torsional waves (Pan and Hansen, 1991), as these wave types can be converted to the more efficiently radiating and more easily feelable flexural waves at structural junctions and discontinuities. The effect of error sensor location and type of cost function (power or vibration level) on the maximum achievable control of flexural waves has also been examined (Pan and Hansen, 1993a), as has the control of flexural wave propagation in finite beams with various end conditions (Pan and Hansen, 1993b).

Although the preceding analyses were mostly undertaken using harmonic primary excitation, McKinnell (1989) showed that the results are equally applicable to a random noise disturbance provided that the control source is in the far field of the primary source and provided that the time required to process the reference signal to obtain the desired control source signal is less than the time taken for the wave to propagate from the reference signal transducer to the control source. This effectively imposes an upper limit on the frequency of the waves which can be controlled, at least for dispersive type waves such as flexural waves. If an accelerometer or other form of vibration sensor is used as the error sensor, a low frequency limit will also exist which is governed by the frequency at which the wavelength of the travelling waves is sufficiently long that the error sensor encroaches on the near field of the control source. When this happens, the maximum achievable control is substantially reduced (Pan and Hansen, 1993a). In practice, for random noise excitation, a higher low frequency limit results from lack of coherence between the reference (or detection) and error signals (Elliott *et al.*, 1990).

If the overall structural vibration level rather than vibratory power transmission is to be suppressed, then a modal description of the structure is the more appropriate one, whether feedback or feedforward control is used. Feedback control for this case is discussed in Chapter 11 and feedforward control follows a similar approach to that used for minimizing sound radiation as discussed in Chapter 8 or for minimizing the potential energy in an enclosure as discussed in Chapter 9. The main difference is that the structural case involves the minimization of structural kinetic energy approximated using a number of vibration transducers whilst the acoustic case involves minimization of the acoustic potential energy approximated using a number of microphones. Thus, this aspect will not be considered further here.

The remainder of this chapter begins with a discussion of the feedforward control of vibratory power transmission in an infinitely long beam, with flexural, longitudinal and torsional waves all being taken into account. This discussion is followed by an analysis of the feedforward control of flexural waves in beams with various end conditions, which is followed by an analysis of the feedforward control of vibratory power transmission in a semi-infinite plate. The minimization of the vibratory power transmission at the error sensor is compared with the minimization of flexural wave vibration amplitude in terms of effectiveness in controlling vibration downstream of the error sensor. The effect of control source/primary source separation and error sensor/control source separation on vibration levels both upstream and downstream of the primary source is also considered. Note that the downstream direction is defined as the direction from the primary source to the control source. Means for measuring power transmission of all three wave types in a beam simultaneously are also discussed.

An advantage of feedforward systems is that the physical system can be analysed separately from the electronic control system. Thus, for any given control source and error sensor arrangement it is possible to determine the maximum achievable reduction in vibration level or vibratory power transmission assuming an ideal controller. Therefore, this chapter will only be concerned with the analysis of the physical system. Design of the feedforward controller is discussed in Chapters 6 and 13, and feedback control of structural vibration is discussed in Chapter 11.

10.1 INFINITE BEAM

Using feedforward control to actively control any one particular wave type in a beam is analogous to controlling plane waves propagating in an air duct as discussed in Chapter 7. For longitudinal and torsional waves on the beam there is no nearfield effect, with the result that the error sensor may be located close to a vibration source. Another difference arises for the case of flexural waves which are characterized as dispersive; that is, their propagation speed increases with increasing frequency.

A typical feedforward control system arrangement for controlling a single wave type on an infinite beam using accelerometer reference and error sensors is illustrated in Fig. 10.1(a) and the equivalent block diagram is shown in Fig. 10.1(b) (Elliott *et al.*, 1990). The quantity $\ddot{w}(x)$ is the acceleration measured by an accelerometer at location x. Note that the feedback path $F(j\omega)$ shown in the block diagram is part of the physical system, not the electronic control system, and represents the effect of the control source on the reference signal input into the electronic controller. The transfer function or frequency response $P(j\omega)$ represents the transfer function of the path from the reference (or detection) sensor along the beam to the error sensor and includes the electromechanical transfer functions of the reference and error transducers. Similarly, $C(j\omega)$ represents the transfer function from the controller electrical output through the control source, along the beam to the error sensor and through the error sensor to the error sensor output, also taking into account the electro-mechanical transfer functions of the control source and error transducer (referred to as the cancellation path transfer function). The electronic control system is represented by $T(j\omega)$ and the adaptive algorithm adjusts $T(j\omega)$ to minimize the error signal. Note that $R(\omega)$, $Y(\omega)$ and $E(\omega)$ are Fourier transforms of the signals $r(t)$, $y(t)$ and $e(t)$. The contribution of the primary source to the reference (or detection) sensor is $I(\omega)$, while the contribution to the error sensor is $D(\omega)$. Using the frequency domain for system analysis is purely a convenience, as the time domain equivalents (impulse responses) of frequency domain transfer functions cannot simply be multiplied together; they must be convolved as explained in Chapters 3 and 6, thus making algebraic manipulations more complex.

For an infinite beam, where the dynamic effects of the actuators and sensors are ignored, the transfer functions discussed above are defined formally as

$$\begin{aligned} C(j\omega) &= \frac{E(\omega)}{Y(\omega)}\bigg|_{I(\omega)=0} = \frac{\ddot{w}(x_e)}{F_c}\bigg|_{\ddot{w}(x_p)=0} \\ &= \frac{j\omega^2}{4EIk_b^3}\left[\mathrm{e}^{-jk_b(x_e-x_c)} - j\mathrm{e}^{-k(x_e-x_c)}\right] \end{aligned} \tag{10.1.1}$$

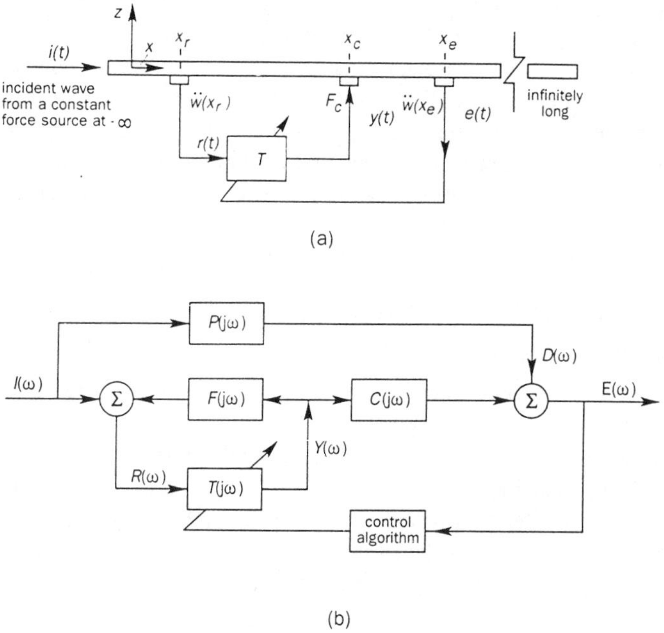

Fig. 10.1 Feedforward control of wave propagation in an infinite beam: (a) physical system arrangement; (b) equivalent electrical block diagram.

$$
\begin{aligned}
F(j\omega) &= \frac{R(\omega)}{Y(\omega)} \Big|_{I(\omega)=0} = \frac{\ddot{w}(x_r)}{F_c} \Big|_{\ddot{w}(x_p)=0} \\
&= \frac{j\omega^2}{4EIk_b^3}\left[e^{-jk_b(x_e - x_r)} - je^{-k(x_e - x_r)}\right]
\end{aligned}
\tag{10.1.2}
$$

$$
P(j\omega) = \frac{E(\omega)}{R(\omega)} \Big|_{T(j\omega)=0} = \frac{\ddot{w}(x_e)}{\ddot{w}(x_r)} \Big|_{F_c=0} = e^{-jk_b(x_e - x_r)} \tag{10.1.3}
$$

From Fig. 10.1(b), the total error signal is

$$
E(\omega) = I(\omega)\left[P(j\omega) + \frac{C(j\omega)\,T(j\omega)}{1 - T(j\omega)\,F(j\omega)}\right] \tag{10.1.4}
$$

If there is no measurement noise, the error signal $E(\omega)$ can be made zero if

$$
T(j\omega) = \frac{-P(j\omega)}{C(j\omega) - P(j\omega)\,F(j\omega)} \tag{10.1.5}
$$

For control of single frequency noise, $T(j\omega)$ can be modelled in the time domain using a transversal filter with two coefficients or stages. These two stages, however, may have very large weight values, thus presenting numerical difficulties during implementation which can be overcome using a few more stages. For control of broadband noise, two orders of magnitude more stages were found to be necessary (Elliott *et al.*, 1990). In practice, the filter must be made adaptive to account for changes in the physical system transfer functions which usually occur relatively slowly with time. When controlling broadband noise, the maximum attenuation in vibration level at the error sensor at frequency ω has been shown to be (Elliott *et al.*, 1990) $10\log_{10}\left[\frac{1}{1-\gamma_{xe}^2(\omega)}\right]$ where γ_{xe}^2 is the coherence between the controller reference input signal and error input signal at frequency ω.

In practice, the transfer functions $P(j\omega)$, $F(j\omega)$ and $C(j\omega)$ of (10.1.5) may be measured by driving the control actuators with random noise, with all other input forces deactivated. Changes in $C(j\omega)$ over the lifetime of the controller can cause it to go unstable and changes in the other transfer functions result in decreases in controller performance. Ways of overcoming this problem include using an adaptive filter and algorithm which includes an on-line estimate of $C(j\omega)$ as discussed in Chapter 6. Note that the inclusion of the feedback path $F(j\omega)$ only affects the desired controller transfer function for optimal control. It does not affect the maximum achievable controllability with an assumed ideal controller as discussed in the remainder of this chapter.

The transfer functions also clearly depend on the wave type being controlled; thus, it is clear that this single channel controller cannot be used to control more than one wave type simultaneously. Use of a two-channel feedforward controller to control a single frequency flexural and longitudinal wave was reported by Fuller *et al.* (1990) and the simultaneous control of two single frequency flexural waves and one longitudinal wave using a three channel controller was reported by Clark *et al.* (1992).

10.1.1 Flexural wave control: minimizing vibration

In this section an analysis of flexural waves propagating along an infinite beam will be undertaken with the idea of examining the effect of error sensor location on the maximum achievable reduction in vibration levels downstream from the error sensor. The coordinate system used for the analysis is shown in figure 10.2. It will be assumed that the excitation forces act normal to the $x-y$ plane, in the z-direction. A harmonic primary point force of magnitude F_p and zero relative phase acts at $x = 0$. It is assumed that the system to be controlled is linear; that is, the principle of superposition holds whereby the response at a point on the beam under the simultaneous action of a number of forces is equal to the sum of the responses due to each individual force. Here and in subsequent sections, Euler beam theory will be used; that is, the effects of shear and rotary inertia will be neglected. The consequences of this were discussed in Section 2.3.

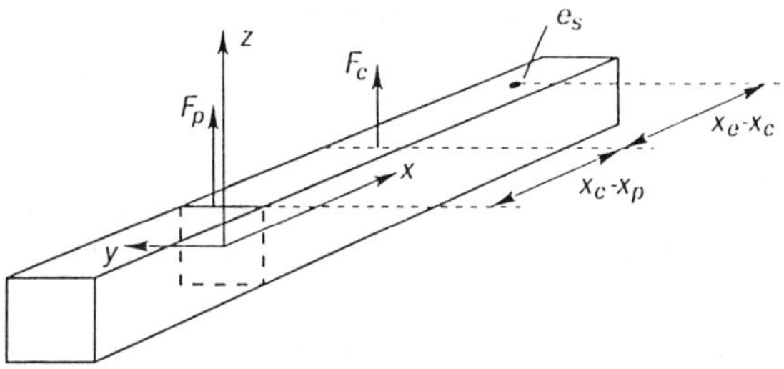

Fig. 10.2 Coordinate system for flexural wave propagation in a beam.

As discussed in Chapter 2, the general solution for flexural wave propagation in a beam is (omitting time dependence for convenience)

$$w(x) = A_1 e^{-jk_b x} + A_2 e^{jk_b x} + A_3 e^{-k_b x} + A_4 e^{k_b x} \tag{10.1.6}$$

Considering only waves propagating in the positive x direction we obtain $A_2 = A_4 = 0$.

If a point force F_p is applied at $x = 0$, the symmetry constrains the slope to be zero at that point. That is,

$$\frac{\partial w(x)}{\partial x} = 0 \quad \text{at } x = 0 \tag{10.1.7}$$

Thus,

$$-jk_b A_1 e^{-jk_b x} - k_b A_3 e^{-k_b x}\Big|_{x=0} = 0 \tag{10.1.8}$$

or

$$A_1 = jA_3 \tag{10.1.9}$$

Thus, (10.1.6) becomes

$$w(x) = A_1\left(e^{-jk_b x} - je^{-k_b x}\right) \tag{10.1.10}$$

Following the argument used in Section 2.5 to derive the point impedance for a beam, A_1 is given by

$$A_1 = \frac{-jF_p}{4EIk_b^3} \tag{10.1.11}$$

Thus, (10.1.10) can be written as

$$w(x) = \frac{-F_p}{4EIk_b^3}\left(je^{-jk_b x} + e^{-k_b x}\right) \tag{10.1.12}$$

If instead of being applied at the origin, the primary force is applied at x_p, and if an additional (control) force is applied at x_c, the equation for the displacement at any location x can be derived from (10.1.12) and written as

$$w(x) = F_p\beta_p + F_c\beta_c \tag{10.1.13}$$

where the coefficients β_p and β_c are defined as

$$\beta_{p,c} = -\frac{1}{4EIk_b^3}\left(je^{-jk_b|x-x_{p,c}|} + e^{-k_b|x-x_{p,c}|}\right) \tag{10.1.14}$$

Remember that the first term in (10.1.14) is the propagating component and the second term is the non-propagating or evanescent component. The control force required to make $w(x)$ equal to zero is then

$$F_c = -F_p\frac{\beta_p(x_e)}{\beta_c(x_e)} \tag{10.1.15}$$

The ratio of controlled to uncontrolled displacement amplitude downstream of the control source is

$$R = \left|\frac{(F_p\beta_p + F_c\beta_c)}{F_p\beta_p}\right| = \left|1 + \frac{F_c\beta_c(x)}{F_p\beta_p(x)}\right| \tag{10.1.16a,b}$$

By substituting (10.1.15) into this, we obtain

$$R = \left|1 - \frac{\beta_p(x_e)\beta_s(x)}{\beta_s(x_e)\beta_p(x)}\right| \tag{10.1.17}$$

The β_p and β_c factors can then be replaced by their definitions and the limit as x approaches infinity is taken so that the expression for the reduction in far field residual vibration is

$$\lim_{x\to\infty} R = \frac{\left|1 - e^{-k_b(x_c - x_p)(1-j)}\right|}{\left|je^{k_b(x_e - x_c)(1+j)} + 1\right|} \tag{10.1.18}$$

From the equation, it is clear that the reduction in farfield acceleration level depends on the separation between the control force and the primary force as well as the separation between the control source and the error sensor (at which the acceleration is minimized). If the distance between the control and primary source is large compared with a wavelength; that is, if $k_b(x_c - x_p) >> 1$, the numerator of (10.1.18) approaches unity. If in addition, the distance between the error sensor and the control source is large compared with a wavelength; that is, if $k_b(x_e - x_c) >> 1$, the final expression for the downstream residual vibration becomes

$$\lim_{x\to\infty} R \approx e^{-k_b(x_e - x_c)} \tag{10.1.19}$$

For harmonic excitation, the net vibrational power transmission in the beam is proportional to the square of the farfield acceleration level, and so the power transmission reduction is proportional to $\frac{20}{\log_e 10} k_b(x_e - x_c)$ dB.

The acceleration distribution along the beam may be calculated from (10.1.13) for both the controlled condition where the control force is given by (10.1.15) and the uncontrolled condition. In the example shown in Fig. 10.3(a),

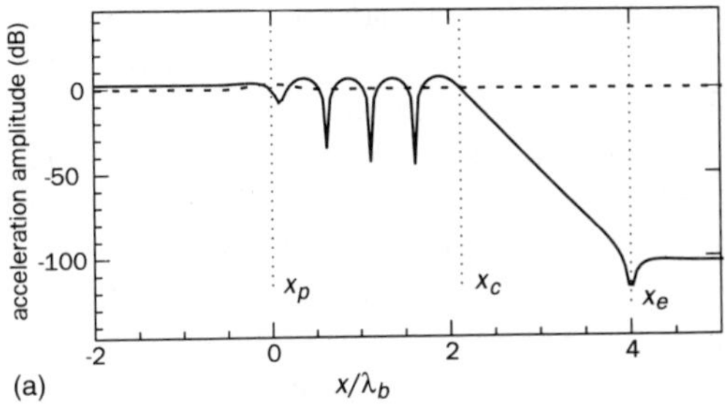

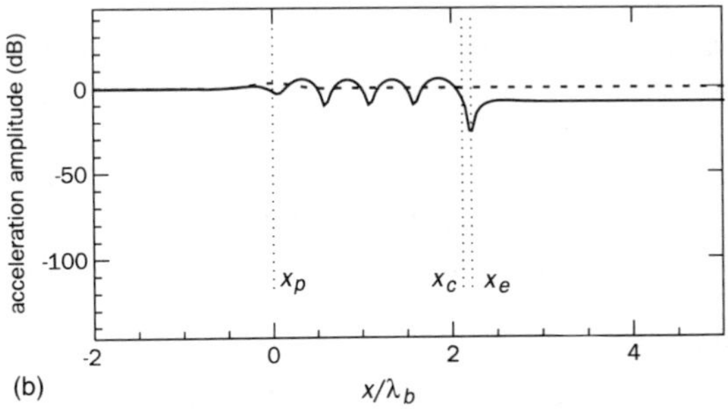

Fig. 10.3 Distribution of acceleration amplitude in an anechoically terminated beam of 25 mm × 50 mm section with the primary source at $x_p/\lambda_b = 0$ and the control source at $x_s/\lambda_b = 2.12$.

------------------ primary source only ———————— controlled

(a) Error sensor at $x_e/\lambda_b = 4.0$; (b) Error sensor at $x_e/\lambda_b = 2.22$.

unit primary force is applied at $x_p = 0$, the error sensor is placed in the farfield of the control force and the control force is also in the far field of the primary force. The dotted line represents the effect of the primary force acting alone, while the solid line shows the effect of adding at x_c the optimal control force defined by (10.1.15). Clearly, the (constant) residual acceleration level on the right of the error sensor at x_e is very much smaller than the uncontrolled level. For this particular case, with the primary source at $x_p = 0$, and the control source at $x_c/\lambda_b = 2.12$, ($\lambda_b = 0.4823$ m is the flexural wavelength) the reduction in vibration level is about 110 dB. Of course, it would not be feasible to achieve this with a practical control system. Nevertheless, the calculation does indicate the maximum theoretically possible reduction. Between the control source location and the error sensor, the vibration level decreases exponentially from the uncontrolled level to the residual level. As required in the analysis, the vibration amplitude at the error sensor ($x_e/\lambda_b = 4.0$) is zero ($-\infty$ dB). When the control force is placed in the near field of the primary force, the resulting reduction in vibration levels downstream of the error sensor is slightly greater, as expected by inspection of (10.1.18).

For the results shown in Fig. 10.3(b), the error sensor is in the near field of the control force. The residual acceleration level on the right of the error sensor is only 9 dB lower than the uncontrolled level. This is much less than that shown in Fig. 10.3(a) and indicates the importance of placing a vibration error sensor in the far field of the control force if the power transmission is to be controlled adequately using an acceleration sensor.

Between the primary and control sources there is a standing wave pattern in both figures. Because the applied forces represent structural discontinuities with finite rather than infinite or zero impedance, the primary and control sources appear at neither the nodes nor the antinodes of these standing wave patterns. It was shown by Brennan *et al.* (1992) that these standing waves can be eliminated by using two rather than one control force, as they can be configured to suppress the wave propagating towards the primary source as well as the downstream propagating wave. Two control forces can also be used to suppress the near (or evanescent) field downstream of the control source as well as the downstream propagating wave. If three control forces are used, it is possible to suppress all of the waves mentioned above simultaneously (Brennan *et al.* 1992) and if four closely spaced control forces are used, then it is possible to suppress the evanescent field on the upstream side of the control forces as well. Note that as the number of closely spaced (0.1 λ) control forces increases, the total control force effort increases exponentially (Brennan *et al.*, 1992). Also the required control force becomes prohibitively large as the control source separation approaches zero or an integer multiple of half wavelengths.

10.1.2 Flexural wave control: minimizing power transmission

It is of interest to discover what control improvements are possible if the error sensor is capable of measuring structural power transmission in the beam and providing the control algorithm with a measure of this quantity rather than just vibration amplitude.

The flexural wave power being transmitted along the beam through any one cross-sectional location is made up of an active component and a reactive component as discussed in Section 2.5. As shown by (2.5.202), the active or time averaged power transmission is

$$P_{Ba} = -\mathrm{Re}\left\{\frac{j\omega EI_{yy}}{2}\left[w^*(x)\frac{\partial^3 w(x)}{\partial x^3} - \frac{\partial w^*(x)}{\partial x}\frac{\partial^2 w(x)}{\partial x^2}\right]\right\} \tag{10.1.20}$$

where $\bar{w}_z$ in (2.5.202) has been replaced by $w(x)$ and the amplitude of the reactive component is

$$P_{Br} = -\mathrm{Im}\left\{\frac{j\omega EI_{yy}}{2}\left[w^*(x)\frac{\partial^3 w(x)}{\partial x^3} - \frac{\partial w^*(x)}{\partial x}\frac{\partial^2 w(x)}{\partial x^2}\right]\right\} \tag{10.1.21}$$

Using (10.1.13), (10.1.20) can be written in quadratic form in terms of the control force F_c as

$$P_{Ba} = R_e\left\{F_c^* a_B F_c + b_{1B} F_c + F_c^* b_{2B} + c_B\right\} \tag{10.1.22}$$

where c_B is the power due to the primary force only and is found by setting $F_c = 0$ in (10.1.13) and substituting the result into (10.1.20) to give

$$c_B = \frac{j\omega EI_{yy}}{2} F_p^2 \left[\beta_p^* \frac{\partial^3 \beta_p}{\partial x^3} - \frac{\partial \beta_p^*}{\partial x}\frac{\partial^2 \beta_p}{\partial x^2}\right] \tag{10.1.23}$$

where β_p is defined by (10.1.14).

The quantities a_B, b_{1B} and b_{2B} in (10.1.22) are defined as follows:

$$a_B = \frac{j\omega EI_{yy}}{2}\left[\beta_c^* \frac{\partial^3 \beta_c}{\partial x^3} - \frac{\partial \beta_c^*}{\partial x}\frac{\partial^2 \beta_c}{\partial x^2}\right] \tag{10.1.24}$$

$$b_{1B} = \frac{j\omega EI_{yy} F_p}{2}\left[\beta_p^* \frac{\partial^3 \beta_c}{\partial x^3} - \frac{\partial \beta_p^*}{\partial x}\frac{\partial^2 \beta_c}{\partial x^2}\right] \tag{10.1.25}$$

$$b_{2B} = \frac{j\omega EI_{yy} F_p}{2}\left[\beta_c^* \frac{\partial^3 \beta_p}{\partial x^3} - \frac{\partial \beta_c^*}{\partial x}\frac{\partial^2 \beta_p}{\partial x^2}\right] \tag{10.1.26}$$

In the preceding four equations, care must be exercised when differentiating β as the results are different for $x > x_p$ or $x > x_c$ than they are for $x < x_p$ or $x < x_c$.

Equation (10.1.22) is the familiar quadratic minimization problem which in this case is solved by differentiating the real part of the power (10.1.22) with respect to both the real and imaginary parts of the control force and setting each result to zero. The result is the optimal control force F_c, given by (Pan and Hansen, 1991)

$$F_c = -\frac{1}{2\mathrm{Re}\{a_B\}}(b_{1B}^* + b_{2B}) \tag{10.1.27}$$

If the power is minimized in the far field of the control and primary forces, then $b_{1B}^* = b_{2B}$. Substituting (10.1.23) to (10.1.26) into (10.1.27) gives (for minimizing farfield power transmission)

$$F_c = -F_p e^{-jk_b(x_c - x_p)} \tag{10.1.28}$$

Substituting (10.1.14) and (10.1.23) into (10.1.22) (with $F_c = 0$) and taking the limit as $x \to \infty$ gives the following result for the farfield flexural wave power transmission with no control:

$$P_{Bau} = \frac{F_p^2 \omega}{16 EI_{yy} k_b^3} \tag{10.1.29}$$

The controlled power transmission to the right of the error sensor location, x_e, is found by substituting (10.1.23) to (10.1.27) into (10.1.22) and is equal to zero. This is in contrast to the result obtained in (10.1.18) for acceleration control where it was shown that the power transmission downstream of the error sensor was dependent on both the separation between the control source and primary source (weakly) and the separation between the error sensor and the control source (strongly).

Between the primary and control sources, the reduction in power transmission due to the control force is found to be

$$\frac{P_{Ba}}{P_{Bau}} = 2\sin\left[k_b(x_c - x_p)\right]e^{-k_b(x_c - x_p)} \tag{10.1.30}$$

To the left of the primary source the reduction in power transmission is

$$\frac{P_{Ba}}{P_{Bau}} = 2\cos\left[2k_b(x_c - x_p)\right] - 2 \tag{10.1.31}$$

Equations (10.1.30) and (10.1.31) are zero if the following condition is satisfied:

$$x_c - x_p = n\frac{\pi}{k_b} = n\frac{\lambda_b}{2} \tag{10.1.32a,b}$$

Thus, best results are obtained for power being transmitted on both sides of the primary force and between the control force and primary force if the separation between the latter two forces is an integer multiple of half the structural wavelength. When this occurs, the primary source power is completely suppressed at the primary source with no absorption at the control source. Similar conclusions were drawn in Chapter 7 in the discussion of waves propagating in air ducts.

The relative effectiveness of a vibration error sensor compared with a power transmission (or intensity) sensor for the active control of power transmission in a beam is indicated in Fig. 10.4. In part (a), the maximum achievable power transmission control is plotted as a function of frequency for an aluminium beam of cross-sectional dimensions 50 mm × 25 mm, terminated anechoically, with the error sensor located 1 m from the control source and the control source located 1 m from the primary source.

It was shown by Pan and Hansen (1993a) that the reactive (or near) field is 20 dB below the active or propagating field at distances greater than $0.73\lambda_b$ from a point force source. This corresponds to $k_b(x_e - x_c) = 4.6$. From Fig. 10.4(b) it can be seen that the influence of the near field at this point is negligible. As mentioned earlier, if the error sensor is an accelerometer or velocity or displacement sensor, and is placed in the near field of a control source, then it will be much less effective in controlling the downstream power transmission than if it were in the far field of the control source. This places a lower frequency limit on the vibration which can be controlled (as the near field grows in size as the frequency of flexural waves decreases). Elliott and Billet

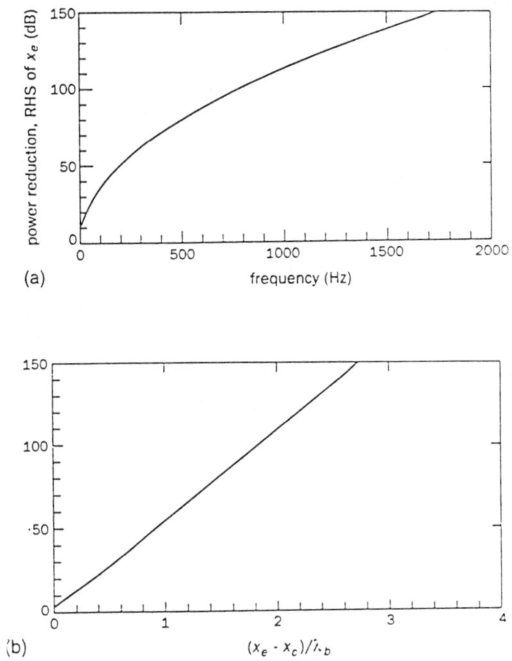

Fig. 10.4 Maximum achievable reduction in real power flow (and for field vibration level) for an infinitely long beam excited by a point force with an error sensor at which the acceleration is minimized located at $x = x_e$: (a) as a function of frequency for $(x_e - x_c) = 1$ m, and $(x_c - x_p) = 1$ m; (b) as a function of the ratio of the separation between the error sensor and control source, and the wavelength, for $(x_c - x_p)/\lambda_b = 1$ m.

(1993) showed that for practically sized beams this low frequency limit is below the limit imposed by other constraints such as obtaining a reference signal which is coherent with the vibration to be controlled. Elliott and Billet (1993) also pointed out a high frequency limitation which is associated with the increasing flexural wavespeed as the disturbance frequency is increased and the fixed, frequency independent, finite time taken by the controller to produce a control signal, once it receives a reference signal. This is clearly not a real limit for periodic vibration, but it is very important for random vibration which requires the controller to be causal. In the same article it was also shown that the bandwidth of controllability increases exponentially as the controller processing time is reduced. The group delay τ_g associated with a flexural wave travelling a distance L, along a beam is obtained using (2.3.58) and is given by

$$\tau_g = L / c_g = \frac{L}{2\omega^{1/2}} \left[\frac{m}{EI} \right]^{1/4} \tag{10.1.33}$$

10.1.3 Simultaneous control of all wave types: power transmission

In this section, expressions for the power being transmitted along an infinite beam in the form of flexural, longitudinal and torsional waves will be investigated for the purpose of determining the optimal control forces to control these wave types and the maximum control achievable assuming an ideal controller. Another reason for pursuing this analysis is to investigate the spillover of energy into other wave types when attempts are made to actively control flexural wave power with a control force which does not act exactly normal to the beam surface. Pan and Hansen (1991) have analysed the active control of all four wave types (two flexural, one torsional and one longitudinal) in a thick beam, and some experimental results have been presented by Clark *et al.* (1992). Brennan *et al.* (1992) reported a means of controlling the three wave types by using three magnetostrictive actuators mounted within a hollow tube used to model a helicopter strut. The analysis to follow is based principally on the work reported by Pan and Hansen (1991).

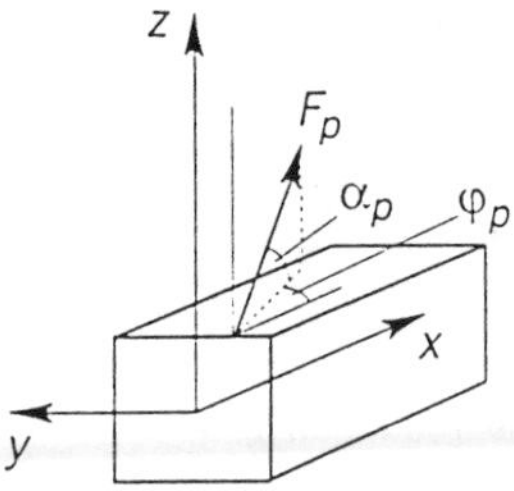

Fig. 10.5 Beam excited by a force at an arbitrary location and orientation.

Consider the rectangular section beam shown in Fig. 10.5 excited by a primary force at some angle α_p from the x-axis in the $x-z$ plane and angle ϕ_p from the x-axis in the $x-y$ plane. The force is located at (x_p, y_p, z_p) and can be represented in terms of unit vectors along the x, y and z axes. Thus,

$$F_{pL} = F_p\delta(x-x_p) = \left(F_{px}\boldsymbol{i} + F_{py}\boldsymbol{j} + F_{pz}\boldsymbol{k}\right)\delta(x-x_p) \qquad (10.1.34a,b)$$

where the Dirac delta function is used as a mathematical convenience so that the force can be represented as a force per unit length along the x-direction. Equation (10.1.34) indicates that the force is zero everywhere except at $x = x_p$. The line moment per unit length about the origin of the $x = x_p$ plane resulting from the action of F_{pL} is given by

$$\boldsymbol{M}_{pL} = \boldsymbol{\sigma}_p \times \boldsymbol{F}_{pL} = M_{px}\boldsymbol{i} + M_{py}\boldsymbol{j} + M_{pz}\boldsymbol{k} \qquad (10.1.35a,b)$$

where

$$\boldsymbol{\sigma}_p = x_p\boldsymbol{i} + y_p\boldsymbol{j} + z_p\boldsymbol{k} \qquad (10.1.36)$$

Each wave type will now be considered separately.

10.1.3.1 Longitudinal waves

Using the wave equation derived in Section 2.3 for free vibration and including the external force we may write

$$\frac{\partial^2 \xi_x}{\partial x^2} - \frac{1}{c_L^2}\frac{\partial^2 \xi_x}{\partial t^2} = -\frac{1}{ES}F_{pL}\boldsymbol{i} \tag{10.1.37}$$

10.1.3.2 Torsional waves

Using the analysis of Section 2.3 we obtain for the torsional wave equation:

$$\frac{\partial^2 \theta_x}{\partial x^2} - \frac{1}{c_T^2}\frac{\partial^2 \theta_x}{\partial t^2} = -\frac{1}{GJ}M_{pL}\boldsymbol{i} \tag{10.1.38}$$

10.3.1.3 Flexural waves

The forced wave equation for flexural waves corresponding to displacements in the y-direction is

$$\frac{\partial^4 w_y}{\partial x^4} + \frac{1}{c_{by}^2}\frac{\partial^2 w_y}{\partial t^2} = \frac{1}{EI_{zz}}\left[F_{pL}\boldsymbol{j} - \frac{\partial M_{pL}}{\partial x}\boldsymbol{k}\right] \tag{10.1.39}$$

where

$$c_{by}^2 = \frac{EI_{zz}}{\rho S} \tag{10.1.40}$$

The forced wave equation corresponding to displacements in the z-direction is

$$\frac{\partial^4 w_z}{\partial x^4} + \frac{1}{c_{bz}^2}\frac{\partial^2 w_z}{\partial t^2} = \frac{1}{EI_{yy}}\left[F_{pL}\boldsymbol{k} + \frac{\partial M_{pL}}{\partial x}\boldsymbol{j}\right] \tag{10.1.41}$$

where

$$c_{bz}^2 = \frac{EI_{yy}}{\rho S} \tag{10.1.42}$$

If a control force F_c is now introduced at location (x_c, y_c, z_c) the combined effect of both the control force and primary force can be taken into account in the preceding equations by replacing $F_{p\ell}$ with $F_{cL} + F_{pL}$, M_{pL} with $M_{pL} + M_{cL}$ and $\frac{\partial M_{pL}}{\partial x}$ with $\frac{\partial M_{pL}}{\partial x} + \frac{\partial M_{cL}}{\partial x}$.

With both the primary force and the control force included, the solutions to the preceding equations of motion are (Pan and Hansen, 1991)

$$w_{xo}(x,\omega,t) = \left(F_{px}\beta_{pL} + F_{cx}\beta_{cL}\right)e^{j\omega t} \tag{10.1.43}$$

$$w_{yo}(x,\omega,t) = \left(F_{py}\beta_{pby} + F_{cy}\beta_{cby} + M_{pz}\beta_{pbyM} + M_{cz}\beta_{cbyM}\right)e^{j\omega t} \tag{10.1.44}$$

$$w_{zo}(x,\omega,t) = \left(F_{pz}\beta_{pbz} + F_{c}\beta_{cbz} + M_{py}\beta_{pbzM} + M_{cy}\beta_{cbzM}\right)e^{j\omega t} \tag{10.1.45}$$

$$\theta_x(x,\omega,t) = \left(M_{px}\beta_{pT} + M_{cx}\beta_{cT}\right)e^{j\omega t} \tag{10.1.46}$$

The β coefficients corresponding to the primary excitation force are defined as follows. The subscript o indicates that the displacements are those of the centre of the beam cross-section at any particular location x.

$$\beta_{pL} = \frac{-je^{-jk_L|x-x_p|}}{2ESk_L} \tag{10.1.47}$$

$$\beta_{pby} = \frac{-je^{-jk_{by}|x-x_p|} - e^{-k_{by}|x-x_p|}}{4EI_{zz}k_{by}^3} \tag{10.1.48}$$

$$\beta_{pbz} = \frac{-je^{-jk_{bz}|x-x_p|} - e^{-k_{bz}|x-x_p|}}{4EI_{yy}k_{bz}^3} \tag{10.1.49}$$

$$\beta_{pT} = \frac{-je^{-jk_T|x-x_p|}}{2GJk_T} \tag{10.1.50}$$

$$\beta_{pbyM} = \frac{(x-x_p)\left(e^{-jk_{by}|x-x_p|} - e^{-k_{by}|x-x_p|}\right)}{4EI_{zz}k_{by}^3\,|x-x_p|} \tag{10.1.51}$$

$$\beta_{pbzM} = \frac{(x-x_p)\left(e^{-jk_{bz}|x-x_p|} - e^{-k_{bz}|x-x_p|}\right)}{4EI_{yy}k_{bz}^3\,|x-x_p|} \tag{10.1.52}$$

Similar expressions hold for the β coefficients with the c subscript which refers to the control forces and moments. In this latter case, x_p is replaced by x_c.

The wavenumbers k_{bz}, k_{by}, k_L and k_T correspond respectively to bending waves with displacements along the z-axis and the y-axis, longitudinal waves and torsional waves. They are defined as follows:

$$k_{bz}^4 = \frac{\omega^2 m}{EI_{yy}} = \left[\frac{\omega}{c_{bz}}\right]^4 \tag{10.1.53a,b}$$

$$k_{by}^4 = \frac{\omega^2 m}{EI_{zz}} = \left[\frac{\omega}{c_{by}}\right]^4 \tag{10.1.54a,b}$$

$$k_L = \omega\sqrt{\frac{\rho}{E}} = \frac{\omega}{c_L} \tag{10.1.55a,b}$$

$$k_T = \omega\sqrt{\frac{\rho}{G}} = \frac{\omega}{c_T} \tag{10.1.56a,b}$$

J is the polar second moment of area of the beam cross-section, S is the cross-sectional area, G is the shear modulus or modulus of rigidity of the beam material and E is its modulus of elasticity.

The general displacement vector for the centre of the beam can be written as

$$\boldsymbol{W}_0 = \left(\xi_{xo},\ w_{yo},\ w_{zo},\ \theta_x,\ \theta_y,\ \theta_z\right)^{\mathrm{T}} \tag{10.1.57}$$

Note that θ_y and θ_z are defined as

$$\theta_y = -\frac{\partial w_{zo}}{\partial x} \tag{10.1.58}$$

$$\theta_z = \frac{\partial w_{yo}}{\partial x} \tag{10.1.59}$$

The time independent part of the general displacement vector can be written, by combining (10.1.43) to (10.1.55), as

$$\boldsymbol{W}_0 = \boldsymbol{a}_p\boldsymbol{Q}_p + \boldsymbol{a}_c\boldsymbol{Q}_c \tag{10.1.60}$$

where the generalized forces are defined as

$$\boldsymbol{Q}_{p,c} = \left[F_x,\ F_y,\ F_z,\ M_x,\ M_y,\ M_z\right]^{\mathrm{T}}_{p,c} = \boldsymbol{R}_{p,c}F_{p,c} \tag{10.1.61a,b}$$

$$\boldsymbol{R}_{c,p} = [\cos\alpha_{p,c}\cos\phi_{p,c},\ \cos\alpha_{p,c}\sin\phi_{p,c},\ \sin\alpha_{p,c},\ y_{p,c}\sin\alpha_{p,c} - z_{p,c}\cos\alpha_{p,c}\sin\phi_{p,c},$$
$$z_{p,c}\cos\alpha_{p,c}\cos\phi_{p,c},\ -y_{p,c}\cos\alpha_{p,c}\sin\phi_{p,c}]^{\mathrm{T}} \tag{10.1.62}$$

(the subscript p corresponds to the primary force and $\boldsymbol{R}_p$, while the subscript c corresponds to the control source and $\boldsymbol{R}_c$).

The influence coefficient matrix $\boldsymbol{a}_p$ is defined as

$$\boldsymbol{a}_p = \begin{bmatrix} \beta_{pL} & 0 & 0 & 0 & 0 & 0 \\ 0 & \beta_{pby} & 0 & 0 & 0 & \beta_{pbyM} \\ 0 & 0 & \beta_{pbz} & 0 & \beta_{pbzM} & 0 \\ 0 & 0 & 0 & \beta_T & 0 & 0 \\ 0 & 0 & \beta'_{pbz} & 0 & \beta'_{pbzM} & 0 \\ 0 & \beta'_{pby} & 0 & 0 & 0 & \beta'_{pbyM} \end{bmatrix} \tag{10.1.63}$$

where the prime denotes differentiation with respect to x. $\boldsymbol{a}_s$ is defined by replacing the subscript p with the subscript s in (10.1.63).

Although the preceding equations have been derived for a single point primary force and a single point control force, they can be extended easily to the case of multiple discrete point forces or distributed forces. In the former case, the right hand side of (10.1.60) becomes a summation over all the point forces and in the latter case it becomes an integration over the region of the distributed force.

Equations (2.5.160), (2.5.172) and (2.5.102) give the following for the time averaged power transmission in the beam:

$$P_a = \tfrac{1}{2}\,\mathrm{Re}\left\{j\omega\,\bar{\boldsymbol{W}}_0^{\mathrm{H}}\,\boldsymbol{\Lambda}\,\bar{\boldsymbol{W}}_0\right\} \tag{10.1.64}$$

where $\bar{\boldsymbol{W}}_0$ is defined in (10.1.60) and where Λ is a diagonal matrix defined as

$$\Lambda = \begin{bmatrix} ES\dfrac{\partial}{\partial x} & 0 & 0 & 0 & 0 & 0 \\ 0 & -EI_{zz}\dfrac{\partial^3}{\partial x^3} & 0 & 0 & 0 & 0 \\ 0 & 0 & -EI_{yy}\dfrac{\partial^3}{\partial x^3} & 0 & 0 & 0 \\ 0 & 0 & 0 & -GJ\dfrac{\partial}{\partial x} & 0 & 0 \\ 0 & 0 & 0 & 0 & EI_{yy}\dfrac{\partial}{\partial x} & 0 \\ 0 & 0 & 0 & 0 & 0 & EI_{zz}\dfrac{\partial}{\partial x} \end{bmatrix} \tag{10.1.65}$$

Using (10.1.60), (10.1.61), (10.1.62) and (10.1.64), it is possible to write the equation for the time averaged power transmission in terms of the control force amplitude F_c. That is,

$$P_a = F_c^*\, a_B F_c + b_{1B} F_c + F_c^*\, b_{2B} + c_B \tag{10.1.66}$$

where

$$a_B = \boldsymbol{R}_c^{\mathrm{T}}\boldsymbol{a}_c^{\mathrm{H}}\boldsymbol{\Lambda}\boldsymbol{a}_c\boldsymbol{R}_c \tag{10.1.67}$$

$$b_{1B} = \boldsymbol{Q}_p\boldsymbol{a}_p^{\mathrm{H}}\boldsymbol{\Lambda}\boldsymbol{a}_c\boldsymbol{R}_c \tag{10.1.68}$$

$$b_{2B} = \boldsymbol{R}_c^{\mathrm{T}}\boldsymbol{a}_c^{\mathrm{H}}\boldsymbol{\Lambda}\boldsymbol{a}_p\boldsymbol{Q}_p \tag{10.1.69}$$

$$c_B = \boldsymbol{Q}_p\boldsymbol{a}_p^{\mathrm{H}}\boldsymbol{\Lambda}\boldsymbol{a}_p\boldsymbol{Q}_p \tag{10.1.70}$$

Differentiating (10.1.66) with respect to the real and imaginary parts of the control force and setting both results to zero gives the following for the optimum control force:

$$F_c^{\mathrm{opt}} = -\frac{1}{2\mathrm{Re}\{a_B\}}\left(b_{1B}^* + b_{2B}\right) \tag{10.1.71}$$

Substituting (10.1.71) into (10.1.66) provides an expression for the minimum achievable real power transmission, assuming an ideal electronic controller and some form of ideal power transmission (or intensity) sensor. In practice, it is difficult to construct a power transmission sensor of sufficient accuracy and vibration sensors are often used instead. These give good results if placed in the far field of the control source where power transmission is proportional to vibration velocity squared.

Pan and Hansen (1991) showed that where a single control force is used, its angular orientation (α_c, ϕ_c) is extremely important. For example, if it is desired to control flexural waves, effective control of total power transmission without a significant amount of spillover into other wave types can only be achieved if the control force orientation is within 1° or 2° of the normal to the beam surface. The net result is that one control force is needed for each wave type that is to be controlled. It is not possible to control successfully two or more wave types with only one control force as the optimum orientation and relative phase is very much dependent on the particular wave type to be controlled.

Control of three waves (two flexural and one longitudinal) simultaneously using piezoceramic crystal actuators bonded to a beam was demonstrated by Clark *et al.* (1992). The error sensor configuration consisted of six accelerometers which were configured to allow the vibration level corresponding to each wave type to be determined.

Flexural wave vibration amplitudes were measured as shown in Fig. 2.37, using two accelerometers located on the beam surfaces at $y = \pm L_y/2$. The outputs from these accelerometers were preamplified and subtracted using an analog computer to obtain a voltage proportional to the flexural wave component characterised by displacement in the y-direction. Two additional accelerometers located on the beam surface at $z = \pm L_z/2$ provided a signal proportional to the flexural wave component characterized by displacement in the z-direction. To measure the longitudinal wave component, two accelerometers were located on the beam at $z = 0$, $y = \pm L_y/2$. The accelerometers were mounted on their side as shown in Fig. 2.37 and their outputs were summed in an analog computer to give a voltage proportional to the amplitude of the longitudinal waves.

The accelerometer cross-axis sensitivity can be a problem in cases where the flexural wave amplitude is large compared to the longitudinal wave amplitude, as this results in the longitudinal wave amplitude measurement being contaminated significantly by the flexural wave.

The controller configuration used for the simultaneous minimization of the three wave types is discussed in detail by Clark *et al.* (1992). Means of determining the amplitudes of each wave type at various locations along the beam and the power transmission corresponding to each wave type before and after control is discussed in Section 2.5.5.2.

10.1.4 Effect of damping

As the beam is terminated anechoically and no reverberant field exists, the results obtained for the maximum achievable reductions in vibration level with active control will not be affected by the structural damping characteristics of the beam.

10.2 FINITE BEAMS

As mentioned previously, it is often important to control the transmission of vibratory power in structural elements such as beams and plates to minimize vibration in, or sound radiation from, attached structures. In the previous section, an analysis was presented which enabled the effect of active control on the reduction of power transmission in an infinite (or anechoically terminated) beam to be calculated.

Here, the more practical problem of active control of vibratory power and vibration level in finite length beams with arbitrary terminations, as shown in Fig. 10.6, will be examined, following closely the work of Hansen *et al.* (1993), Pan and Hansen (1993b) and Young and Hansen (1994a, 1994b). Only control of flexural waves will be discussed, as the analysis including all wave types is not easily managed. Support conditions at the end of the beam will be defined in terms of force and moment impedances, and referred to as boundary impedances, which may be defined in terms of force and moment impedances and are analogous to the concept of point impedance discussed in Section 2.5. In that section, the concept of impedance to a point force was discussed assuming that no applied moment was acting on the beam. However, in the presence of an applied moment, there will exist coupling between the moment and the force impedance, as the moment will generate a lateral displacement as well as a rotation. Similarly a pure applied force will generate rotation as well as lateral displacement. Thus, it is not possible to write the boundary impedance for a beam in terms of simple force and moment impedances. Rather, a fully coupled impedance matrix formulation must be used. The meaning of this will be made clearer by the following analysis.

The local lateral beam velocity and angular rotational velocity generated as a result of an applied moment M_e and an applied force F_e may be written as

$$\dot{w} = \frac{M_e}{Z_{mf}} + \frac{F_e}{Z_f} \tag{10.2.1}$$

$$\dot{\theta} = \frac{M_e}{Z_m} + \frac{F_e}{Z_{fm}} \tag{10.2.2}$$

where Z_{mf} and Z_{fm} ($Z_{mf} \neq Z_{fm}$) are the coupling impedances and Z_f and Z_m are the force and moment impedances respectively.

At the beam boundary, the moment and force applied to the beam by the supports is equal to the internal moment M and shear force Q respectively.

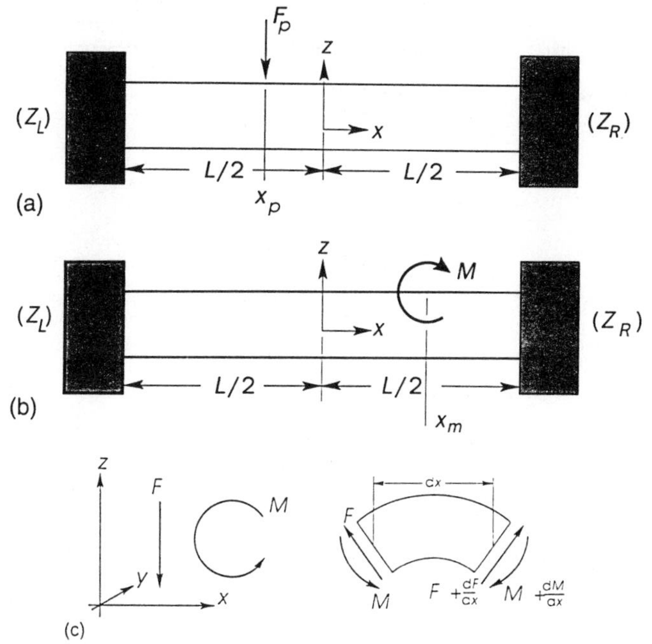

Fig. 10.6 Finite beam model: (a) point force excitation; (b) line moment excitation; (c) sign conventions.

Making these substitutions and rearranging allows (10.2.1) and (10.2.2) to be written as

$$\begin{bmatrix} Q \\ M \end{bmatrix} = \begin{bmatrix} Z_{f\dot{w}} & Z_{f\dot{\theta}} \\ Z_{m\dot{w}} & Z_{m\dot{\theta}} \end{bmatrix} \begin{bmatrix} \dot{w} \\ \dot{\theta} \end{bmatrix} \tag{10.2.3}$$

where

$$\begin{bmatrix} Z_{f\dot{w}} & Z_{f\dot{\theta}} \\ Z_{m\dot{w}} & Z_{m\dot{\theta}} \end{bmatrix} = \begin{bmatrix} 1/Z_m & 1/Z_{fm} \\ 1/Z_{mf} & 1/Z_f \end{bmatrix}^{-1} \tag{10.2.4}$$

Coefficients of the impedance matrix of (10.2.3) corresponding to various beam end conditions are listed in Table 10.1. Note that for all of the conditions listed, the cross-coupling terms from the matrix of (10.2.3) are zero. In the analysis to follow, dissipation of energy within the beam may be taken into account by replacing Young's modulus of elasticity E by $E(1+j\eta)$ where η is the beam loss factor at the frequency of interest.

10.2.1 Equivalent boundary impedance for an infinite beam

A finite beam can be made to behave as an infinite beam by applying appropriate boundary impedances to its ends. The appropriate impedance matrix is derived

Table 10.1 Impedances corresponding to standard terminations

End Condition	Representation	Boundary Condition	Impedance
Simply Supported		$w = 0$ $\frac{\partial^2 w}{\partial x^2} = 0$	$Z_{f\dot{w}} = \infty$ $Z_{m\theta} = 0$
Fixed		$w = 0$ $\frac{\partial w}{\partial x} = 0$	$Z_{f\dot{w}} = \infty$ $Z_{m\theta} = \infty$
Free		$\frac{\partial^2 w}{\partial x^2} = 0$ $\frac{\partial^3 w}{\partial x^3} = 0$	$Z_{f\dot{w}} = 0$ $Z_{m\theta} = 0$
Deflected Spring		$\frac{\partial^2 w}{\partial x^2} = 0$ $EI_{yy}\frac{\partial^3 w}{\partial x^3} = -K_D w$	$Z_{f\dot{w}} = j\frac{K_D}{\omega}$ $Z_{m\dot{\theta}} = 0$
Torsion Spring	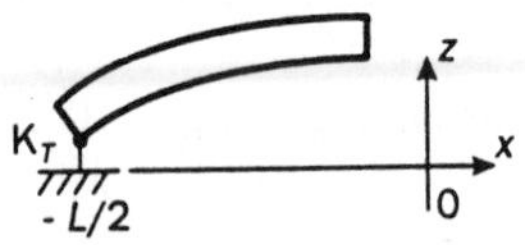	$w = 0$ $EI_{yy}\frac{\partial^2 w}{\partial x^2} = K_T\frac{\partial w}{\partial x}$	$Z_{f\dot{w}} = \infty$ $Z_{m\dot{\theta}} = -j\frac{K_T}{\omega}$
Mass		$\frac{\partial^2 w}{\partial x^2} = 0$ $EI_{yy}\frac{\partial^3 w}{\partial x^3} = -m\frac{\partial^2 w}{\partial t^2}$	$Z_{f\dot{w}} = -j\omega m$ $Z_{m\dot{\theta}} = 0$
Dashpot	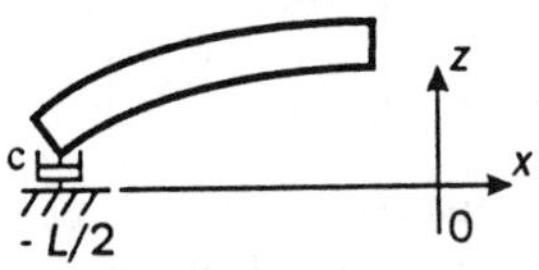	$\frac{\partial^2 w}{\partial x^2} = 0$ $EI_{yy}\frac{\partial^3 w}{\partial x^3} = -c\frac{\partial w}{\partial t}$	$Z_{f\dot{w}} = -c$ $Z_{m\theta} = 0$

as follows. In an infinite beam with a flexural wave characterized by displacement in the z-direction, travelling toward the right and generated by a sole point force excitation, the displacement amplitude in the z-direction can be written as

$$w(x) = B_1 e^{-jk_B x} + B_3 e^{-k_B x} \tag{10.2.5}$$

so that

$$\dot{w}(x) = j\omega B_1 e^{-jk_B x} + j\omega B_3 e^{-k_B x} \tag{10.2.6}$$

and

$$\dot{\theta}(x) = -\frac{\partial \dot{w}}{\partial x} = -\omega k_b B_1 e^{-jk_B x} + j\omega k_b B_3 e^{-k_B x} \tag{10.2.7a,b}$$

where B_1 and B_3 are arbitrary constants.

Equations (10.2.6) and (10.2.7) can be written in a matrix form as

$$\begin{bmatrix} \dot{w}(x) \\ \dot{\theta}(x) \end{bmatrix} = \begin{bmatrix} j\omega & j\omega \\ -\omega k_b & j\omega k_b \end{bmatrix} \begin{bmatrix} B_1 e^{-jk_B x} \\ B_3 e^{-k_B x} \end{bmatrix} \tag{10.2.8}$$

which can be inverted to give

$$\begin{bmatrix} B_1 e^{-jk_B x} \\ B_3 e^{-k_B x} \end{bmatrix} = \begin{bmatrix} \dfrac{(1-j)}{2\omega} & -\dfrac{(1-j)}{2\omega k_b} \\ -\dfrac{(1+j)}{2\omega} & \dfrac{(1-j)}{2\omega k_b} \end{bmatrix} \begin{bmatrix} \dot{w}(x) \\ \dot{\theta}(x) \end{bmatrix} \tag{10.2.9}$$

where the dot denotes differentiation with respect to time.

Equation (10.2.5) can be differentiated two more times to give the bending moment M and the shear force Q, and the result can be written in matrix form as:

$$\begin{bmatrix} Q(x) \\ M(x) \end{bmatrix} = -EI_{yy} \begin{bmatrix} -jk_b^3 & k_b^3 \\ -k_b^2 & k_b^2 \end{bmatrix} \begin{bmatrix} B_1 e^{-jk_B x} \\ B_3 e^{-k_B x} \end{bmatrix} \tag{10.2.10}$$

where E is Young's modulus of elasticity and I_{yy} is the second moment of area of the beam cross section about the y-axis.

The column vector on the right-hand side of (10.2.10) can be replaced with the right-hand side of (10.2.9). Thus,

$$\begin{bmatrix} Q(x) \\ M(x) \end{bmatrix} = \boldsymbol{Z_R} \begin{bmatrix} \dot{w}(x) \\ \dot{\theta}(x) \end{bmatrix} \tag{10.2.11}$$

where

$$\boldsymbol{Z_R} = \begin{bmatrix} (1+j)EI_{yy}k_b^3/\omega & -EI_{yy}k_b^2/\omega \\ EI_{yy}k_b^2/\omega & -(1-j)EI_{yy}k_b/\omega \end{bmatrix} \tag{10.2.12}$$

so that the boundary bending moment and shear force are expressed as the product of the impedance matrix and a column vector containing the velocity and angular velocity of the beam. The quantity, $\mathbf{Z}_R$ is known as the wave impedance matrix. By a similar process of considering a flexural wave travelling from a source toward the left in an infinite beam the left wave impedance matrix is

$$\mathbf{Z}_L = \begin{bmatrix} -(1+j)EI_{yy}k_b^3/\omega & -EI_{yy}k_b^2/\omega \\ EI_{yy}k_b^2/\omega & (1-j)EI_{yy}k_b/\omega \end{bmatrix} \qquad (10.2.13)$$

Note that $\mathbf{Z}_R \neq \mathbf{Z}_L$ and $\mathbf{Z}_R \neq -\mathbf{Z}_L$. By using the wave impedance matrix $\mathbf{Z}_R$ (or $\mathbf{Z}_L$) as the right (or left) boundary impedance matrix in the finite beam model, the boundary is effectively removed and the beam becomes semi-infinite. An infinite beam can be modelled by using both the left and right wave impedance matrices as boundary impedances at opposite ends of a finite length beam. This use of wave impedance matrices to model the termination of a finite length beam produces numerical results identical to those obtained by evaluating the expression derived from the analysis of an infinite beam.

10.2.2 Response to a point force

The purpose of this section is to determine the response to a simple harmonic point force excitation applied at $x = x_0$ of a finite beam with left and right boundary conditions specified as impedance matrices $\mathbf{Z}_L$ and $\mathbf{Z}_R$ as shown in Fig. 10.6(a). The applied point force produces a discontinuity in the shear force function $Q(x)$ and so it is necessary to calculate two sets of constant coefficients A_1, A_2, A_3 and A_4 on the left, and B_1, B_2, B_3 and B_4 on the right of the applied force, so that

$$w_1(x) = A_1 e^{-jk_b x} + A_2 e^{jk_b x} + A_3 e^{-k_b x} + A_4 e^{k_b x} \qquad (10.2.14)$$

and

$$w_2(x) = B_1 e^{-jk_b x} + B_2 e^{jk_b x} + B_3 e^{-k_b x} + B_4 e^{k_b x} \qquad (10.2.15)$$

At the common boundary $x = x_0$,

$$w_1(x_0) = w_2(x_0) \qquad (10.2.16)$$

$$w_1'(x_0) = w_2'(x_0) \qquad (10.2.17)$$

$$w_1''(x_0) = w_2''(x_0) \qquad (10.2.18)$$

$$w_1'''(x_0) = w_2'''(x_0) - \frac{F_e}{EI_{yy}} \qquad (10.2.19)$$

where the prime denotes differentiation with respect to x.

From (10.2.3) the left-hand boundary condition of the beam at $x = x_L$ can be written as

$$\begin{bmatrix} Q(x_L) \\ M(x_L) \end{bmatrix} = \begin{bmatrix} Z_{Lf\dot{w}} & Z_{Lf\dot{\theta}} \\ Z_{Lm\dot{w}} & Z_{Lm\dot{\theta}} \end{bmatrix} \begin{bmatrix} \dot{w}_1(x_L) \\ \dot{\theta}_1(x_L) \end{bmatrix} \tag{10.2.20}$$

By using (2.3.35) and (2.3.36) to replace the bending moment and shear force with a derivative of the displacement function, the following is obtained:

$$\begin{bmatrix} Z_{Lf\dot{w}} & Z_{Lf\dot{\theta}} \\ Z_{Lm\dot{w}} & Z_{Lm\dot{\theta}} \end{bmatrix} \begin{bmatrix} \dot{w}_1(x_L) \\ \dot{\theta}_1(x_L) \end{bmatrix} + EI_{yy} \begin{bmatrix} -w_1'''(x_L) \\ w_1''(x_L) \end{bmatrix} = 0 \tag{10.2.21}$$

Similarly, for the right-hand boundary of the beam at $x = x_R$:

$$\begin{bmatrix} Z_{Rf\dot{w}} & Z_{Rf\dot{\theta}} \\ Z_{Rm\dot{w}} & Z_{Rm\dot{\theta}} \end{bmatrix} \begin{bmatrix} \dot{w}_2(x_R) \\ \dot{\theta}_2(x_R) \end{bmatrix} + EI_{yy} \begin{bmatrix} -w_2'''(x_R) \\ w_2''(x_R) \end{bmatrix} = 0 \tag{10.2.22}$$

where

$$\theta_1(x) = -\frac{\partial w_1(x)}{\partial x}, \qquad \theta_2(x) = -\frac{\partial w_2(x)}{\partial x} \tag{10.2.23a,b}$$

Equations (10.2.14) and (10.2.15) are then differentiated to produce expressions for w_1, w_2, $\dot{\theta}_1$, $\dot{\theta}_2$, w_1', w_1'', w_1''', w_2', w_2'' and w_2''' which contain the unknown coefficients A_1, A_2, A_3, A_4, B_1, B_2, B_3 and B_4. These expressions can be substituted into (10.2.16), (10.2.17), (10.2.18), (10.2.19), (10.2.21) and (10.2.22) and the equations can be combined into a single system of linear equations, to obtain

$$\alpha X = F \tag{10.2.24}$$

where

$$X = [A_4, A_3, A_2, A_1, B_4, B_3, B_2, B_1]^{\mathrm{T}} \tag{10.2.25}$$

$$F = [0, 0, 0, 0, 0, 0, 0, \frac{F_e}{EI_{yy}k_b^3}]^{\mathrm{T}} \tag{10.2.26}$$

and

$$\alpha = \begin{bmatrix} j\omega\beta_L(Z_{Lm\dot{w}} - k_b Z_{Lm\dot{\theta}} + H_2) & \frac{j\omega(Z_{Lm\dot{w}} + k_b Z_{Lm\dot{\theta}} + H_2)}{\beta_L} & j\omega\beta_L^j(Z_{Lm\dot{w}} - jk_b Z_{Lm\dot{\theta}} - H_2) & \frac{j\omega(Z_{Lm\dot{w}} + jk_b Z_{Lm\dot{\theta}} - H_2)}{\beta_L^j} & 0 & 0 & 0 & 0 \\ j\omega\beta_L(Z_{Lf\dot{w}} - k_b Z_{Lf\dot{\theta}} - H_1) & \frac{j\omega(Z_{Lf\dot{w}} + k_b Z_{Lf\dot{\theta}} + H_1)}{\beta_L} & j\omega\beta_L^j(Z_{Lf\dot{w}} - jk_b Z_{Lf\dot{\theta}} + jH_1) & \frac{j\omega(Z_{Lf\dot{w}} + jk_b Z_{Lf\dot{\theta}} - jH_1)}{\beta_L^j} & 0 & 0 & 0 & 0 \\ 0 & 0 & 0 & 0 & j\omega\beta_R(Z_{Rm\dot{w}} - k_b Z_{Rm\dot{\theta}} + H_2) & \frac{j\omega(Z_{Rm\dot{w}} + k_b Z_{Rm\dot{\theta}} + H_2)}{\beta_R} & j\omega\beta_R^j(Z_{Rm\dot{w}} - jk_b Z_{Rm\dot{\theta}} - H_2) & \frac{j\omega(Z_{Rm\dot{w}} + jk_b Z_{Rm\dot{\theta}} - H_2)}{\beta_R^j} \\ 0 & 0 & 0 & 0 & j\omega\beta_R(Z_{Rf\dot{w}} - k_b Z_{Rf\dot{\theta}} - H_1) & \frac{j\omega(Z_{Rf\dot{w}} + k_b Z_{Rf\dot{\theta}} + H_1)}{\beta_R} & j\omega\beta_R^j(Z_{Rf\dot{w}} - jk_b Z_{Rf\dot{\theta}} + jH_1) & \frac{j\omega(Z_{Rf\dot{w}} + jk_b Z_{Rf\dot{\theta}} - jH_1)}{\beta_R^j} \\ \beta_0 & 1/\beta_0 & \beta_0^j & 1/\beta_0^j & -\beta_0 & -1/\beta_0 & -\beta_0^j & -1/\beta_0^j \\ \beta_0 & -1/\beta_0 & j\beta_0^j & -j/\beta_0^j & -\beta_0 & 1/\beta_0 & -j\beta_0^j & j/\beta_0^j \\ \beta_0 & 1/\beta_0 & -\beta_0^j & -1/\beta_0^j & -\beta_0 & -1/\beta_0 & \beta_0^j & 1/\beta_0^j \\ -\beta_0 & 1/\beta_0 & j\beta_0^j & -j/\beta_0^j & \beta_0 & -1/\beta_0 & -j\beta_0^j & j/\beta_0^j \end{bmatrix} \tag{10.2.27}$$

where $\beta_L = e^{k_b x_L}$, $\beta_0 = e^{k_b x_0}$, $\beta_R = e^{k_b x_R}$, $\beta_L^j = e^{jk_b x_L}$, $\beta_0^j = e^{jk_b x_0}$, $\beta_R^j = e^{jk_b x_R}$ $H_1 = EI_{yy}k_b^3/(j\omega)$ and $H_2 = EI_{yy}k_b^2/(j\omega)$. The solution vector $\boldsymbol{X}$ characterizes the response of a finite length beam to a point force simple harmonic excitation.

10.2.3 Response to a concentrated line moment

Here, we consider the response to a simple harmonic concentrated line moment across the beam applied at $x = x_0$ as shown in Fig. 10.6(b), with left and right boundary conditions specified as impedance matrices $\boldsymbol{Z}_L$ and $\boldsymbol{Z}_R$. Following the sign conventions shown in Fig. 10.6(c) (which are consistent with those used in Section 2.3), the equation of motion for the flexural vibration of the beam shown in Fig. 10.6(b) is

$$EI_{yy}\frac{\partial^4 w}{\partial x^4} + \rho S\frac{\partial^2 w}{\partial t^2} = \frac{dM_e}{dx}\delta(x-x_0)\, e^{j\omega t} \tag{10.2.28}$$

where M_e is the amplitude of a harmonic point moment applied at $x = x_0$ and $\delta(x-x_0)$ is the Dirac delta function. The equations describing the displacement on the left and right side of the applied moment are identical to (10.2.14) and (10.2.15) for a point force excitation.

The first two boundary conditions at the common boundary between the left and right sides of the beam are the same as (10.2.16) and (10.2.17) for a point force. The other two equations are slightly different and are

$$w_1''(x_0) = w_2''(x_0) - \frac{M_e}{EI_{yy}} \tag{10.2.29}$$

$$w_1'''(x_0) = w_2'''(x_0) \tag{10.2.30}$$

The solution procedure is the same as for a point force excitation; in fact, (10.2.20) to (10.2.27) for a point force also describe the solution for an applied moment except that the force vector of (10.2.26) is replaced by a moment vector defined as

$$\boldsymbol{M} = [0, 0, 0, 0, 0, 0, -\frac{M_e}{EI_{yy}k_b^2}, 0]^{\mathrm{T}} \tag{10.2.31}$$

10.2.4 Active vibration control with a point force

In the following analysis, a primary point force acts at $x = x_p$, a control point force acts at $x = x_c$ and an error sensor is located at $x = x_e$ as shown in Fig. 10.7. The cost function to be minimized is the beam flexural displacement amplitude squared at the location of the error sensor. The total beam response may be considered as the sum of the responses due to the primary and control forces, each of which may be calculated separately.

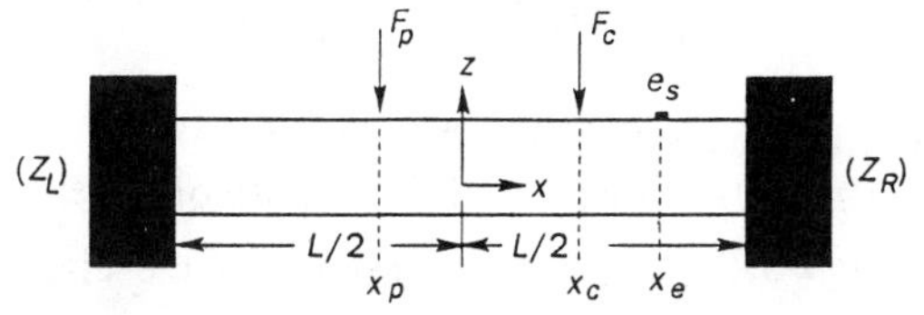

Fig. 10.7 Control force and error sensor configuration for point force control of a beam with arbitrary impedance terminations.

The boundary condition equation for the primary force is

$$\alpha_p \boldsymbol{X}_p = \boldsymbol{F}_p \tag{10.2.32}$$

or

$$\boldsymbol{X}_p = \alpha_p^{-1} \boldsymbol{F}_p \tag{10.2.33}$$

where the primary force vector $\boldsymbol{F}_p = [0, 0, 0, 0, 0, 0, 0, F_p/(EI_{yy}k_b^3)]^{\mathrm{T}}$, $\boldsymbol{X}_p$ is the boundary eigenvector for the primary force and $\boldsymbol{\alpha}_p$ is the matrix of boundary condition coefficients for the primary force. The boundary condition equation for the control force $\boldsymbol{F}_c$ is

$$\boldsymbol{\alpha}_c \boldsymbol{X}_c = \boldsymbol{F}_c \tag{10.2.34}$$

or

$$\boldsymbol{X}_c = \boldsymbol{\alpha}_c^{-1} \boldsymbol{F}_c \tag{10.2.35}$$

where the control force vector $\boldsymbol{F}_c = [0, 0, 0, 0, 0, 0, 0, F_c/(EI_{yy}k_b^3)]^{\mathrm{T}}$, $\boldsymbol{X}_c$ is the boundary eigenvector for the control force and $\boldsymbol{\alpha}_c$ is the matrix of boundary condition coefficients for the control force.

At the error sensor location x_e, the displacement due to the primary force is,

$$w_p = \boldsymbol{X}_p^{\mathrm{T}} \boldsymbol{E} \tag{10.2.36}$$

and the displacement due to the control force is

$$w_c = \boldsymbol{X}_c^{\mathrm{T}} \boldsymbol{E} \tag{10.2.37}$$

where

$$\boldsymbol{E} = \begin{bmatrix} 0 & 0 & 0 & 0 & \mathrm{e}^{k_b x_e} & \mathrm{e}^{-k_b x_e} & \mathrm{e}^{jk_b x_e} & \mathrm{e}^{-jk_b x_e} \end{bmatrix}^{\mathrm{T}} \tag{10.2.38}$$

By adding (10.2.36) and (10.2.37), the total displacement can be written as

$$w = w_p + w_c = \boldsymbol{X}_p^{\mathrm{T}} \boldsymbol{E} + \boldsymbol{X}_c^{\mathrm{T}} \boldsymbol{E} \tag{10.2.39a,b}$$

and by substituting (10.2.33) and (10.2.35) into (10.2.39), we obtain

$$\begin{aligned} w &= [\boldsymbol{\alpha}_p^{-1} \boldsymbol{F}_p]^{\mathrm{T}} \boldsymbol{E} + [\boldsymbol{\alpha}_c^{-1} \boldsymbol{F}_c]^{\mathrm{T}} \boldsymbol{E} \\ &= \frac{F_p}{EI_{yy}k_b^3} [\boldsymbol{\alpha}_p^{-1}]_{i,8}^{\mathrm{T}} \boldsymbol{E} + \frac{F_c}{EI_{yy}k_b^3} [\boldsymbol{\alpha}_c^{-1}]_{i,8}^{\mathrm{T}} \boldsymbol{E} \end{aligned} \tag{10.2.40a,b}$$

The optimal control force $\boldsymbol{F}_c$ for the primary force $\boldsymbol{F}_p$ may be found by letting $w = 0$ in (10.2.40). By writing the transpose of the eighth column of the inverse of $\boldsymbol{\alpha}_p$ as $\boldsymbol{P} = (\boldsymbol{\alpha}_p^{-1})_{i,8}^{\mathrm{T}}$ and the transpose of the eighth column of the inverse of $\boldsymbol{\alpha}_c$ as $\boldsymbol{C} = (\boldsymbol{\alpha}_c^{-1})_{i,8}^{\mathrm{T}}$, the optimal control force can be written as

$$F_c = -\frac{\boldsymbol{PE}}{\boldsymbol{CE}} F_p \tag{10.2.41}$$

for an error sensor in either the near field or the far field of the control force. If the wave impedances from (10.2.12) and (10.2.13) are substituted into (10.2.27), the numerical result is the same as that obtained by using the expression derived in Section 10.1 for an infinite beam, which is

$$F_c = -\frac{\beta_p}{\beta_c} F_p \tag{10.2.42}$$

Equation (10.2.41) is clearly more complicated than (10.2.42) because it takes into account not only the boundary conditions (boundary impedances), but also the relative locations of the primary and control sources, and error sensor in relation to the ends of the beam.

10.2.4.1 Effect of boundary impedance

Acceleration distributions along the beam for both optimally controlled and uncontrolled cases corresponding to a harmonic primary forcing frequency of 1000 Hz (which does not correspond to a beam resonance) applied at $x_p/\lambda_b = 0$, a point control force at $x_c/\lambda_b = 2.24$ and an error sensor at $x_e/\lambda_b = 4.48$, (where $\lambda_b = 0.48$ m is the wavelength of the flexural wave in the beam), for a range of boundary impedances are shown in Fig. 10.8. For the dB ordinate scale in these and all the following figures, the reference level is the far field uncontrolled infinite beam acceleration produced by the primary source. By optimally controlled, it is meant that the single control source has been driven in such a way as to obtain the maximum achievable vibration reduction at the error sensor. Because the error sensor is not near the point of application of a force or a discontinuity, near field effects on the beam response at the error sensor are numerically negligible.

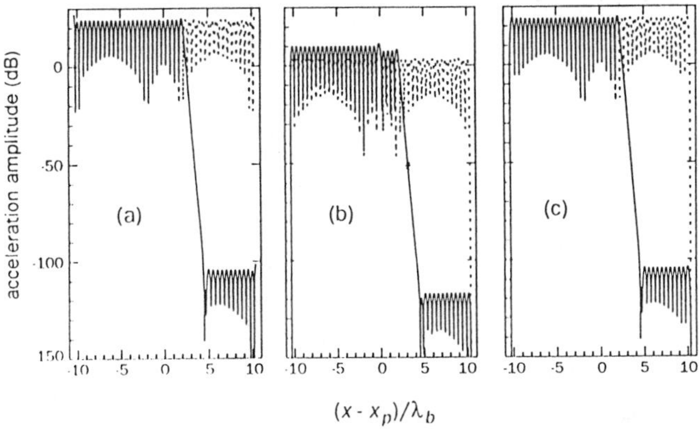

Fig. 10.8 Controlled and uncontrolled distribution of beam acceleration amplitude as a function of boundary impedances. Acceleration level minimized at the error sensor $(x_e - x_p)/\lambda_b = 4.48$, with $x_p/\lambda_b = 0$ and $(x_c - x_p)/\lambda_b = 2.24$. The beam extends from $(x - x_p)/\lambda_b = -10.0$ to $(x - x_p)/\lambda_b = 10.0$. The forcing frequency is 1000 Hz.
———————— controlled,
------------------ uncontrolled.

Three standard boundary conditions are: (a) $Z_{f\dot{\omega}} = 0$ and $Z_{m\dot{\theta}} = 0$ (free–free); (b) $Z_{f\dot{\omega}} = \infty$ and $Z_{m\dot{\theta}} = 0$ (pinned–pinned); (c) $Z_{f\dot{\omega}} = \infty$ and $Z_{m\dot{\theta}} = \infty$ (fixed–fixed).

For convenience, we start with a number of simple boundary conditions. Figure 10.8(a) shows the forced response of a free-free beam, Fig. 10.8(b) shows the forced response of a pinned–pinned beam and (simply supported) Fig. 10.8(c) shows the forced response of a fixed–fixed beam. From these figures, it can be seen that the free–free beam has the same acceleration distribution and potential for vibration reduction as the fixed-fixed beam except within about half a wavelength of the ends of the beam which are at $(x - x_p)/\lambda_b = \pm 10.32$. The pinned–pinned beam has a different acceleration distribution from that of the free–free and fixed–fixed beam; in particular, between each end and the adjacent source, the vibration level is about 15 dB lower. Also, the vibration level of the controlled beam between the primary and control sources is about 5 dB lower than on the left of the primary source. In practice, the extent of the reductions shown in the figures will not be realized. However, it is useful to present them to gain insight into the effect of the termination impedances on the beam controllability.

10.2.4.2 *Effect of control force location*

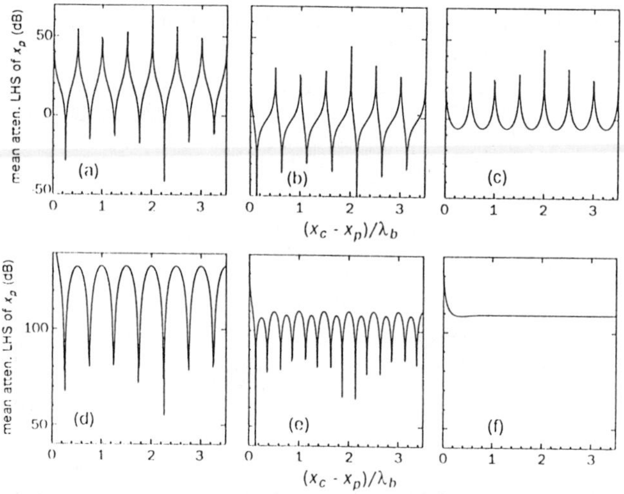

Fig. 10.9 Effect of control force location x_c/λ_b on the mean attenuation of acceleration. The location of the primary force and the excitation frequency are the same as in Fig. 10.9, with fixed $(x_e - x_c)/\lambda_b = 2.0$: (a) Mean attenuation of acceleration upstream of the primary force (free–free); (b) mean attenuation of acceleration upstream of the primary force (pinned–pinned); (c) mean attenuation of acceleration upstream of the primary force (infinite–infinite); (d) mean attenuation of acceleration downstream of the error sensor (free–free); (e) mean attenuation of acceleration downstream of the error sensor (pinned–pinned); (f) mean attenuation of acceleration downstream of the error sensor (infinite–infinite).

Figures 10.9(a), 10.9(c) and 10.9(e) show the maximum achievable mean attenuation of acceleration level in the far field upstream of the primary source for free–free, pinned–pinned (simply supported) and infinite–infinite beams respectively. Control of the fixed–fixed beam produces similar attenuation as the free–free beam, so the results for the fixed–fixed beam are not shown in this figure. In these examples, the mean attenuation of acceleration is shown as a function of the distance $(x_c - x_p)$ between the primary and control sources. The forcing frequency is 1000 Hz (which is non-resonant) and the error sensor is two wavelengths downstream of the control source such that $(x_e - x_c)/\lambda_b = 2.0$. From these figures, it can be seen that the acceleration upstream of the primary source is maximally reduced if the separation between the primary and control sources is an integer multiple of half of a wavelength. The attenuation minima for the free beam occur at odd multiples of a quarter wavelength separation between the control and primary sources. The pinned beam has a different acceleration distribution to that of the free–free beam; the vibration level is about 20 dB lower and its minima are not located at control and primary separations of multiples of a quarter wavelength. For the infinite beam, the peaks in attenuation also occur at half wavelength intervals and are at about the same level as those for the pinned–pinned beam.

Figures 10.9(b),(d) and (f) show the corresponding maximum achievable mean attenuation downstream of the error sensor, assuming an ideal feedforward controller. For the free beam (Fig. 10.9(b)), the maxima are at integer multiples of a half-wavelength and the minima are at odd integer multiples of a quarter-wavelength. The attenuation for the pinned beam (Fig. 10.9(d)) is greatest at integer multiples of a quarter wavelength and is minimal at odd integer multiples of one-eighth of a wavelength. The infinite beam (Fig. 10.9(f)) produces a constant maximum achievable mean attenuation of about 110 dB for the particular location which was used for the error sensor. The maximum attenuation would approach $-\infty$ as the separation between the control source and error sensor approached ∞. For the infinite beam, the independence of the separation between the primary and control sources ensures that good control is possible. In contrast, the extreme sensitivity of the finite beams to control source location indicates that it would be far more difficult to achieve satisfactory control if the primary source location or excitation frequency are not fixed. This implies that good control of a broadband signal, using an ideal feedforward controller, would only occur over narrow bands separated by very narrow bands where control is poor, unless multiple control sources and error sensors are used.

10.2.4.3 Effect of error sensor location

Figure 10.10(a) shows the maximum achievable mean attenuation of acceleration level in the far field upstream of the primary source as a function of both control source and error sensor locations. Each curve in the figure indicates the attenuation for a fixed but different value of control source location. The forcing

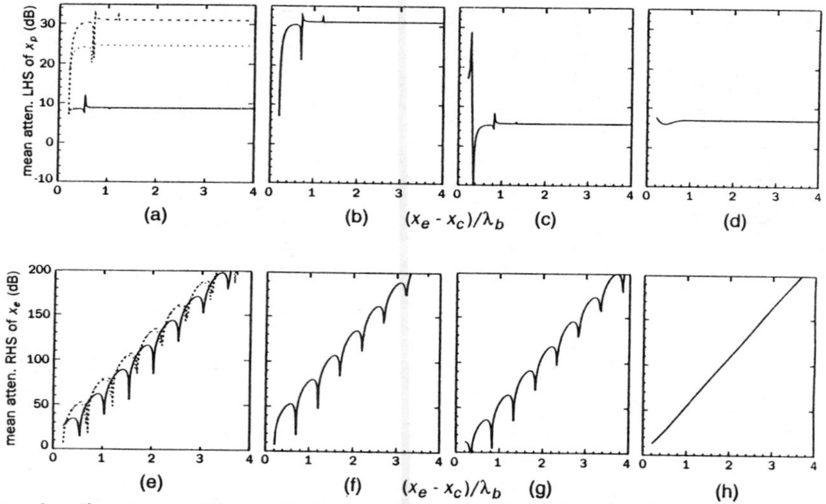

Fig. 10.10 Effect of error sensor location $(x_e - x_c)/\lambda_b$ on the mean attenuation of acceleration for various separations between the primary and control sources. The location of primary force and the excitation frequency are the same as in Fig. 10.9: (a) Mean attenuation of acceleration upstream of the primary force with varying $(x_c - x_p)/\lambda_b$ (free–free).
(b) mean attenuation of acceleration upstream of the primary force with $(x_c - x_p)/\lambda_b = 1.0$ (free–free); (c) mean attenuation of acceleration upstream of the primary force with $(x_c - x_p)/\lambda_b = 1.0$ (pinned–pinned); (d) mean attenuation of acceleration upstream of the primary force with $(x_c - x_p)/\lambda_b = 1.0$ (infinite–infinite); (e) mean attenuation of acceleration downstream of the error sensor with varying $(x_c - x_p)/\lambda_b$ (free–free); (f) mean attenuation of acceleration downstream of the error sensor with $(x_c - x_p)/\lambda_b = 1.0$ (free–free); (g) mean attenuation of acceleration downstream of the error sensor with $(x_c - x_p)/\lambda_b = 1.0$ (pinned–pinned); (h) mean attenuation of acceleration downstream of the error sensor with $(x_c - x_p)/\lambda_b = 1.0$ (infinite–infinite).

-------------------- $(x_c - x_p)/\lambda_b = 1.0$
.... $(x_c - x_p)/\lambda_b = 2.0$
———————— $(x_c - x_p)/\lambda_b = 0.2$

parts (a) and (e) only

frequency is 1000 Hz and the ends of the beam are free. This figure shows that the attenuation depends on the separation $(x_c - x_p)$, but not on the error sensor location, provided that the error sensor is in the far field of the control and primary sources. However, the nature of the dependence on $(x_c - x_p)$ is not clear.

Figure 10.10(b) shows the corresponding maximum achievable mean attenuation downstream of the error sensor. The average reduction in dB is clearly a linearly increasing function of $(x_e - x_c)/\lambda_b$. For each curve in this figure it is possible to draw an 'upper bound' straight line, which is a tangent to every lobe of the curve. The vertical distance between these straight lines is the same as the distance between the horizontal parts of the corresponding curves in Fig. 10.10(a).

Figures 10.10(c), 10.10(e) and 10.10(g) show the maximum achievable mean attenuation of acceleration level on the left hand side of the primary source for free–free, pinned–pinned and infinite–infinite beams. The control source is located at $(x_c - x_p)/\lambda_b = 1.0$. These figures show that the mean attenuation is different for each of the different boundary conditions. As in Fig. 10.10(a), the attenuation upstream of primary source in all three beams does not depend on the error sensor location if the error sensor is in the far field of the control and primary sources. Figures 10.10(d), 10.10(f) and 10.10(h) show the corresponding mean attenuation downstream of the error sensor. For the free beam (Fig. 10.10(d)), minima in the curve correspond to separations between the control source and error sensor which are odd integer multiples of one-quarter of a wavelength. As the error sensor is moved further downstream, local maxima in attenuation are encountered at intervals of half a wavelength separ-ation between the control source and error sensor. In the case of the pinned beam (Fig. 10.10(f)), although the distance between successive extrema is the same, their actual locations are different from those of the free beam. In the control source far field of the infinite beam (Fig. 10.10(h)), the attenuation is simply proportional to the distance between the error sensor and the control source.

10.2.4.4 *Effect of forcing frequency*

Figure 10.11(a) shows the maximum achievable mean attenuation of acceleration level in the far field upstream of the primary source as a function of the forcing frequency. Figure 10.11(b) shows the corresponding maximum achievable mean attenuation downstream of the error sensor. In these examples the boundary force and moment impedances are zero. In both cases, the acceleration is reduced maximally at frequencies for which the separation between the primary and control sources is an integer multiple of half of a wavelength. In Fig. 10.11(b), the average attenuation clearly increases with increasing frequency. The performance is poor at low frequencies because the error sensor is in the near field of the control source.

Figures 10.11(c) and 10.11(d) show the maximum achievable mean attenuation of vibration in an infinite beam as a function of frequency. There are

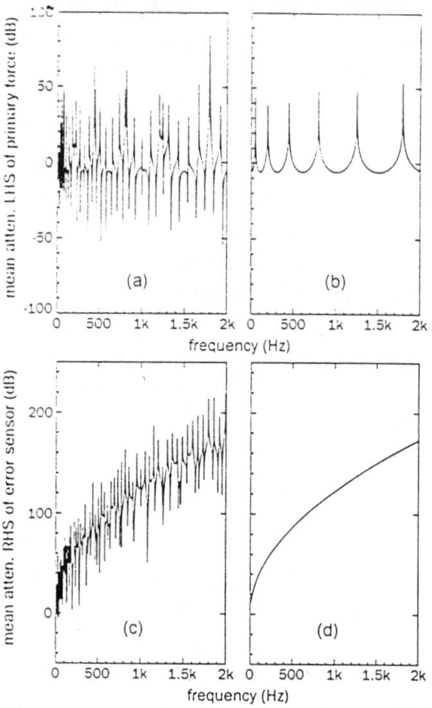

Fig. 10.11 Mean attenuation of acceleration as a function of forcing frequency. The location of primary and control sources and location of the error sensor are the same as in Fig. 10.9: (a) mean attenuation of acceleration upstream of the primary force (free–free); (b) mean attenuation of acceleration upstream of the primary force (infinite–infinite); (c) mean attenuation of acceleration downstream of the error sensor (free–free); (d) mean attenuation of acceleration downstream of the error sensor (infinite–infinite).

six sharp peaks in Fig. 10.11(c). At each of these peaks, the distance between control and primary sources is an integer multiple of half of a wavelength. The number of peaks is limited to six because there are no reflections from boundaries (i.e. no resonances). In Fig. 10.11(d), the attenuation is proportional to the square root of frequency because the attenuation is proportional to $(x_e - x_c)/\lambda_b$ and $\lambda_b \propto f^{-1/2}$. More detailed results are discussed by Pan and Hansen (1993b).

10.2.4.5 Summary of control results using a single control force

From the preceding results and those of Pan and Hansen (1993b), it can be concluded that although reflective boundaries reduce the maximum control achievable, it is still possible to achieve high levels of vibration reduction over a range of termination impedances and harmonic excitation frequencies with a single error sensor and a single control source. The extent of achievable control with a feedforward controller is strongly dependent upon excitation frequency, control source location and error sensor location.

10.2.5 Minimizing vibration using a piezoceramic actuator and an angle stiffener

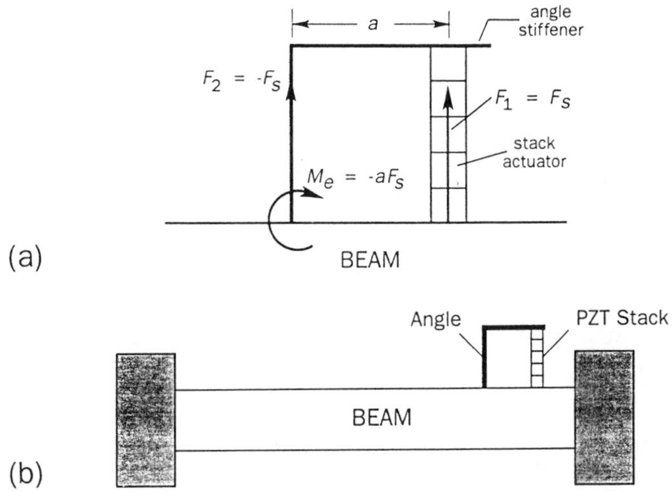

Fig. 10.12 Model for analysing the response of a finite length beam to excitation with a piezoceramic stack: (a) stack and angle model; (b) beam model with stack actuator.

Figure 10.12(b) shows the resultant forces and moments applied to the beam by the angle stiffener and piezoceramic stack (shown in Fig. 10.12(a)) with a primary force F_p at $x = x_p$. Control forces F_1 and F_2 act at $x = x_1$ and $x = x_2$ respectively, with the concentrated moment M_1 also acting at $x = x_1$. An error sensor is located at axial location $x = x_e$. The boundary condition equation for the primary (excitation) point force was shown in Section 10.2.2 to be

$$\boldsymbol{\alpha}_p \boldsymbol{X}_p = \boldsymbol{F}_p \tag{10.2.43}$$

or

$$\boldsymbol{X}_p = \boldsymbol{\alpha}_p^{-1} \boldsymbol{F}_p \tag{10.2.44}$$

where $\boldsymbol{F}_p = [0, 0, 0, 0, 0, 0, 0, \frac{F_p}{k_b^3 EI_{yy}}]^{\mathrm{T}}$, $\boldsymbol{\alpha}_p$ is the matrix of boundary condition coefficients for the primary force and $\boldsymbol{X}_p$ is the boundary eigenvector for the primary force. Similarly

$$\boldsymbol{X}_1 = \boldsymbol{\alpha}_1^{-1} \boldsymbol{F}_1 \tag{10.2.45}$$

$$X_2 = \alpha_2^{-1} F_2 \tag{10.2.46}$$

where X_1 and X_2 are the boundary eigenvectors for the two control forces. In addition

$$X_m = \alpha_m^{-1} M \tag{10.2.47}$$

where X_m is the boundary eigenvector for the control moment and M is defined by (10.2.31). At the error sensor ($x = x_e$), the displacement due to each force and moment is given by, for $z = p,\ 1,\ 2$ and m,

$$w_z = X_z^{\mathrm{T}} E \tag{10.2.48}$$

where E is defined by (10.2.38).

Using superposition to sum the displacements defined in (10.2.48), the total displacement is:

$$\begin{aligned} w &= w_p + w_1 + w_2 + w_m \\ &= X_p^{\mathrm{T}} E + X_1^{T} E + X_2^{\mathrm{T}} E + X_m^{\mathrm{T}} E \end{aligned} \tag{10.2.49a,b}$$

By substituting (10.2.44) to (10.2.47) into (10.2.49), we obtain

$$\begin{aligned} w &= [\alpha_p^{-1} F_p]^{\mathrm{T}} E + [\alpha_1^{-1} F_1]^{\mathrm{T}} E + [\alpha_2^{-1} F_2]^{\mathrm{T}} E + [\alpha_m^{-1} M]^{\mathrm{T}} E \\ &= \frac{F_p}{k_b^3 EI_{yy}} [\alpha_p^{-1}]_{i,8}^{\mathrm{T}} E + \frac{F_1}{k_b^3 EI_{yy}} [\alpha_1^{-1}]_{i,8}^{\mathrm{T}} E + \frac{F_2}{k_b^3 EI_{yy}} [\alpha_2^{-1}]_{i,8}^{\mathrm{T}} E + \frac{M_e}{k_b^2 EI_{yy}} [\alpha_m^{-1}]_{i,7}^{\mathrm{T}} E \end{aligned} \tag{10.2.50a,b}$$

Setting $w = 0$ to find the optimal control force, defining the transpose of the eighth column of the inverse of α_p as $P = [\alpha_p^{-1}]_{i,8}^{\mathrm{T}}$ (and similarly for $A = [\alpha_1^{-1}]_{i,8}^{\mathrm{T}}$, and $B = [\alpha_2^{-1}]_{i,8}^{\mathrm{T}}$), and defining the transpose of the seventh column of the inverse of α_m as $C = [\alpha_m^{-1}]_{i,7}^{\mathrm{T}}$, (10.2.50) can be rewritten as:

$$AEF_1 + BEF_2 + k_b CEM_e = -PEF_p \tag{10.2.51}$$

Analysis of the forces applied by the stack and angle stiffener gives $F_1 = -F_2 = F_c$ and $M_e = -aF_c$, where F_c is the compressive force applied by the piezoceramic stack (see Fig. 10.12(b)) and is positive as shown. The optimal control force F_c can be written as

$$F_c = -\frac{PE}{AE - BE - k_b aCE} F_p \tag{10.2.52}$$

The discussion in the following sections examines the effect of varying the forcing frequency, control source location; error sensor location and stiffener flange length, a (see Fig. 10.12) on the active control of flexural harmonic vibration in beams with two sets of end conditions, infinite and pinned. Although these two end conditions yield substantially different results, the trends observed for the pinned beam are similar for free or clamped ends or any end conditions resulting in a substantial reflected wave amplitude. End conditions which result in absorption of a substantial part of the incoming flexural wave would exhibit trends between those shown by the infinite and pinned beams.

In the following discussion, the control force amplitude is expressed as a fraction or multiple of the primary force amplitude and the control force phase is relative to the primary force phase. For convenience the primary excitation force is always located at $x = 0$. It is also assumed that in all cases the control force has been optimised to minimize the acceleration level at the error sensor. The acceleration reference level is the far field acceleration level with just the primary excitation force acting.

10.2.5.1 Effect of variations in forcing frequency, stiffener flange length and control location on the control force

The discussion that follows examines the effect of varying forcing frequency, control source location (which is defined here as the location of the angle/beam joint), error sensor location, and stiffener flange length on the active control of vibration in beams with simply supported (pinned) ends. For interest, results are compared to those obtained for beams of infinite length. The beam parameters (including location of the control source, primary source and error sensor) are listed in Table 10.2.

Table 10.2 Beam parameters used in the analysis

Parameter	Value
Beam length L_x	5.0 m
Beam width L_y	0.05 m
Beam height L_z	0.025 m
Young's modulus E	71.1 GPa
Excitation force location x_0	0.0 m
Frequency f	1000 Hz
Wavelength λ	0.4824 m

Control force amplitudes are expressed as multiples of the primary force amplitude, and the acceleration amplitude dB scale reference level is the far field uncontrolled infinite beam acceleration produced by the primary force acting alone. In all cases, the control force is assumed to be optimally adjusted to minimize the acceleration at the error sensor location.

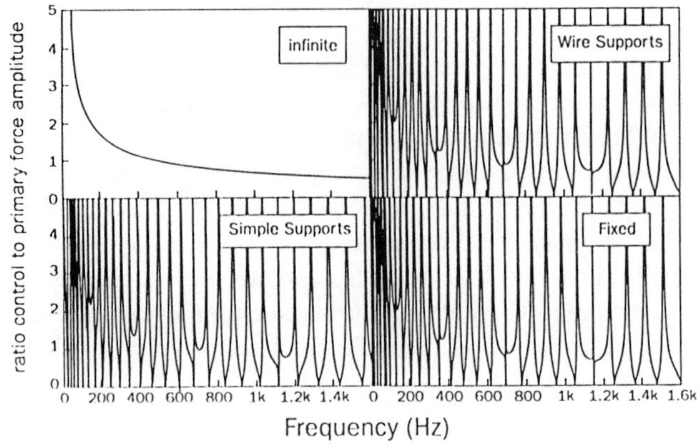

Fig. 10.13 Control force amplitude for optimal feedforward control as a function of frequency for various beam end conditions: minima occur at beam resonances; maxima occur at $(x_c - x_p) = (c + n/2)\lambda$, where c is the frequency dependent distance between x_p and the first node in the standing wave.

Figure 10.13 shows the effect of varying the forcing frequency on the magnitude of the control force required to minimize the beam vibration at the error sensor location for beams with four different end conditions; simply supported, free, fixed and infinite. In each case, the two ends of the beam have the same end conditions. The control force is located 1 m from the primary source and the error sensor 2 m from the primary source, as indicated by Table 10.2. The minima on the curves for the simply supported beam occur at resonance frequencies, when control is easier. At these frequencies the control force amplitude is small but non-zero. The difference in control effort required to control resonant and non-resonant response is associated with the variation from mode to mode in the phase of the optimal control force input required, with the result that decreasing the response of one mode can increase the response of another. At resonance only one mode dominates the response with the result that only a small force is required to achieve control. However off-resonance the response is a result of the combined effect of more than one mode, generally having widely differing optimal control force phases.

The maxima in Fig. 10.13 occur when the relative spacing between primary and control forces is given by $x = (c + n/2)\lambda$ for integer n and constant c. This is verified in Fig. 10.14 which shows the control force magnitude as a function of separation between the primary and control sources, with a constant error sensor location–control source location separation of 1 m (2.07λ) and an excitation frequency of 1000 Hz. The maxima occur because of the difficulty in controlling the flexural vibration when the controller is placed at a node of the standing wave caused by reflection from the terminations. The constant c represents the distance (in wavelengths) between the primary source and the first

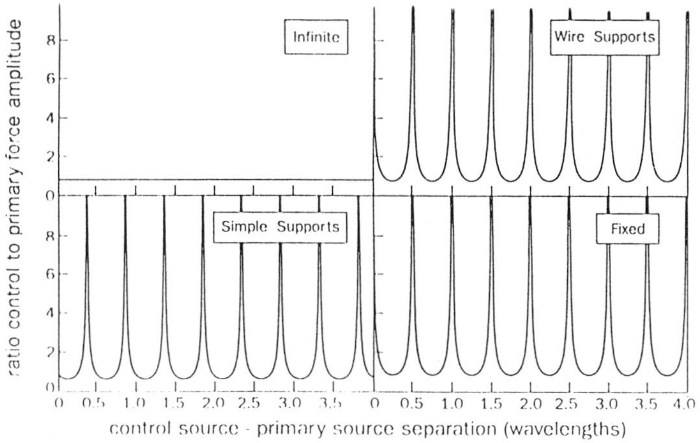

Fig. 10.14 Control force amplitudes for optimal feedforward control as a function of the separation between the control and primary forces $(x_c - x_p)/\lambda_b$.

node in the standing wave in the direction of the control source. This constant changes with frequency, and for $f = 1000$ Hz it is approximately zero.

Interestingly, the phase of the optimal control force relative to the primary force is zero (or 180°) for the simply supported beam and varies between 0° and 180° as a function of frequency for the infinite beam. This is because the response of a lightly damped beam can be modelled as the sum of many modal responses for which the control source location is either 0 or 180° out of phase with the primary source location. This also makes control of a finite beam difficult with a single control force off-resonance, because decreasing the response of one mode requires the oppositely phased signal to that required by the two modes nearest in resonance frequency to the first. Thus the resulting control only incrementally increases as the control force magnitude increases drastically. Conversely at resonance frequencies, where the beam response is dominated by a single resonant mode, control is much easier.

Figure 10.15 shows the control force magnitude plotted as a function of increasing stiffener flange length in wavelengths a/λ (see Fig. 10.12(b) for a definition of the stiffener flange length). The exponentially shaped decrease in control force magnitude with increasing stiffener flange length can be attributed to the increasing size of the angle relative to the flexural wavelength. When the wavelength is large compared to the stiffener flange length, the two control forces operating in opposite directions tend to cancel. This effect can also be seen in Fig. 10.13(a) (the infinite beam case) where the relative control force amplitude is plotted as a function of frequency, and to a lesser extent in Fig. 10.13(b).

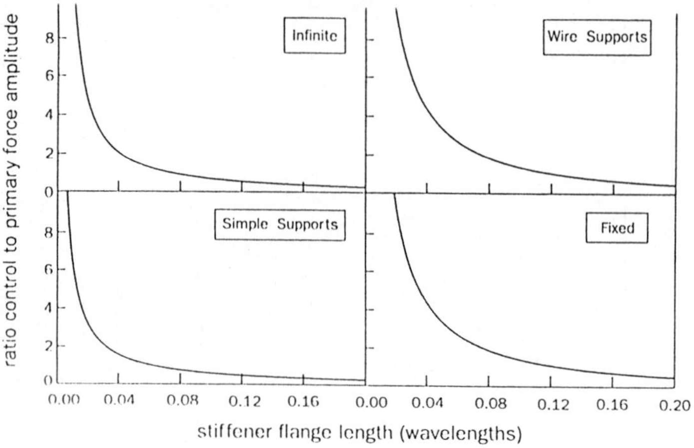

Fig. 10.15 Control force amplitudes for optimal feedforward control as a function of the stiffener flange length.

10.2.5.2 Acceleration distribution for controlled and uncontrolled cases

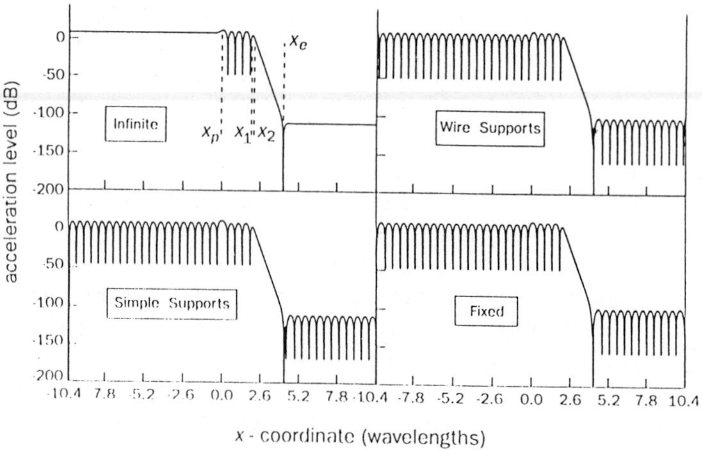

Fig. 10.16 Acceleration distribution along the optimally controlled beam.

Figure 10.16 shows the acceleration amplitude distribution for a controlled beam, excited at 1000 Hz using the control, primary and error sensor locations given in Table 10.2. The curves dip to a minimum at the error sensor location ($x = 4.14\lambda$) where acceleration has been minimized. In both cases, the theoretically achievable reduction in acceleration amplitude downstream of the error sensor, using an ideal feedforward controller, is over 100 dB. For the simply supported beam, the acceleration amplitude is also reduced upstream of the primary force. However, the amount of this reduction or increase in acceleration amplitude upstream of the primary force depends on the control source location, with the maximum attenuation upstream of the primary source being achieved when the control source–primary source separation is $(0.23 + n/2)\lambda$. This value is independent of the beam length and the excitation frequency, and noting also that the maxima occur for the infinite beam as well as the finite beams, we conclude that the maxima are a result of a modal beam response between the primary and control sources only. Note that the constant of 0.23 arises as the control force location has been defined at the point of attachment of the angle stiffener to the beam. However, the effective point of action of the control force is at a location between the point of attachment of the angle stiffener to the beam and the piezoceramic actuator stack, such that the constant would be 0.25.

10.2.5.3 Effect of control location on attenuation of acceleration level

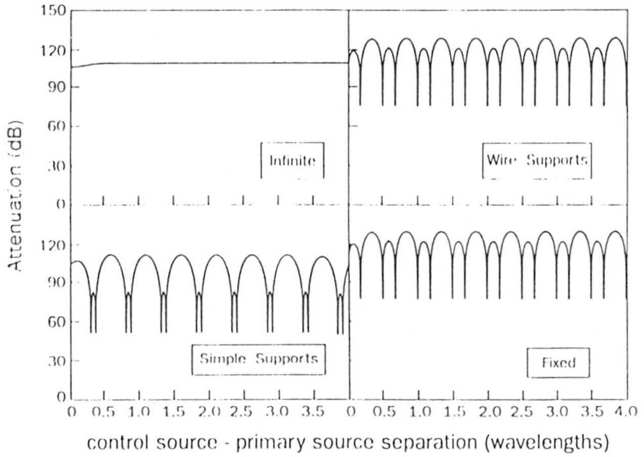

Fig. 10.17 Mean, maximum achievable attenuation of acceleration level for an optimally controlled beam, downstream of the error sensor as a function of separation between the primary and control forces. Each even minimum occurs at a minimum in the control force amplitude (see Fig. 10.14). Each odd minimum occurs at $(d + n/2)\lambda$, $n = 1, 2, \ldots$

Figure 10.17 shows the mean attenuation of acceleration level downstream of the error sensor as a function of separation between primary and control sources. The control force locations giving the best results upstream of the primary force also give high attenuation downstream of the error sensor. Every second minimum occurs at a location corresponding to a maximum in the control force magnitude (see Fig. 10.14), and again these are separated by half a wavelength. The odd numbered minima occur at control source–primary source separations of $(d + n/2)\lambda$, where d is a constant dependent on frequency.

10.2.5.4 Effect of error sensor location on attenuation of acceleration level

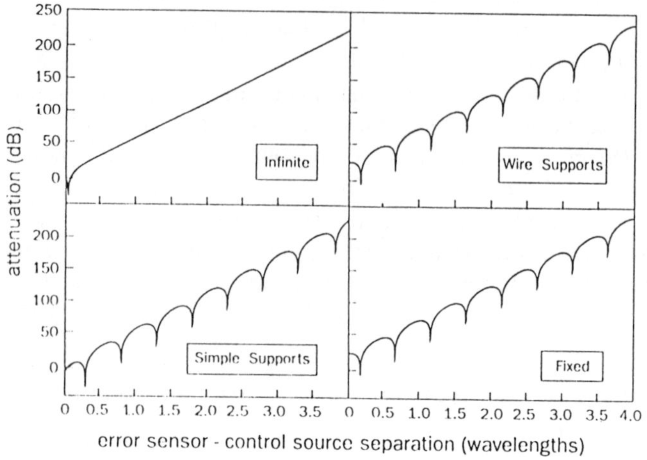

Fig. 10.18 Mean, maximum achievable attenuation of acceleration level for an optimally controlled beam, downstream of the error sensor as a function of separation between the control force and error sensor. Minima occur at $(d + n/2)\lambda$, $n = 1, 2, \ldots$

Figure 10.18 shows the mean attenuation downstream of the error sensor as a function of the separation between the control force and error sensor. It can be seen that downstream of the error sensor, mean attenuation increases with increasing separation between the error sensor and control location at the rate of around 50 dB per wavelength separation. The minima in the curves for the simply supported beam correspond to separations in the error sensor and control location of $(d + n/2)\lambda$, where d is the constant dependent on the frequency previously defined.

10.2.6 Determination of beam end impedances

In practice, rarely will beams be characterized by the ideal end impedances listed in Table 10.1. Thus, to be able to predict the effect of active control on a beam terminated with an arbitrary impedance it is necessary to determine the end impedances by some sort of measurement. Fortunately, it is not necessary to determine the fully coupled beam end impedance matrix directly; rather, an equivalent impedance matrix containing only two uncoupled terms instead of four terms can be shown to give the same beam displacements and derivatives as the actual four element impedance matrix, at all locations on the beam. In the following paragraphs it will be shown how this two element impedance matrix characterizing each end of the beam can be determined using data from four accelerometers mounted on the beam. It is not necessary for the two ends of the beam to be terminated in the same way.

Using (10.2.4), beam end conditions may be characterized by impedance matrices Z_L and Z_R corresponding respectively to the left and right ends of the beam as follows:

$$Z_L = \begin{bmatrix} Z_{Lf\dot{w}} & Z_{Lf\dot{\theta}} \\ Z_{Lm\dot{w}} & Z_{Lm\dot{\theta}} \end{bmatrix} = \begin{bmatrix} \dfrac{1}{Z_{Lm}} & \dfrac{1}{Z_{Lfm}} \\ \dfrac{1}{Z_{Lmf}} & \dfrac{1}{Z_{Lf}} \end{bmatrix}^{-1} \qquad (10.2.53a,b)$$

$$Z_R = \begin{bmatrix} Z_{Rf\dot{w}} & Z_{Rf\dot{\theta}} \\ Z_{Rm\dot{w}} & Z_{Rm\dot{\theta}} \end{bmatrix} = \begin{bmatrix} \dfrac{1}{Z_{Rm}} & \dfrac{1}{Z_{Rfm}} \\ \dfrac{1}{Z_{Rmf}} & \dfrac{1}{Z_{Rf}} \end{bmatrix}^{-1} \qquad (10.2.54a,b)$$

Equation (10.2.24) may be generalized for force or moment excitation by replacing the vector $\boldsymbol{F}$ by $\boldsymbol{B}$, where for moment excitation $\boldsymbol{B} = \boldsymbol{M}$ and for force excitation $\boldsymbol{B} = \boldsymbol{F}$, where $\boldsymbol{M}$ is defined by (10.2.31) and $\boldsymbol{F}$ is defined by (10.2.26). Substituting $\boldsymbol{B}$ for $\boldsymbol{F}$ in (10.2.24) and inverting the result gives the following:

$$\boldsymbol{X} = \alpha^{-1}\boldsymbol{B} \qquad (10.2.55)$$

Because the vector $\boldsymbol{B}$ consists of a non-zero element in either the seventh row or eighth row only, the only columns of importance in the inverse matrix α^{-1} are the seventh and eight columns, as all other columns will be multiplied by a zero element on the vector $\boldsymbol{B}$. The practical implication of this result is that only the larger elements of the first four rows of the α matrix will affect the solution vector $\boldsymbol{X}$. It is proposed that the accuracy of the solution vector $\boldsymbol{X}$ can be maintained with the diagonal elements $Z_{m\dot{w}}$ and $Z_{f\dot{\theta}}$ of the impedance matrix set to zero. This simplification is justified by examples rather than by formal proof, as inverting the complex matrix α symbolically is not practicable.

Once the impedance matrix has been approximated by the equivalent matrix with just two unknowns, determining the unknown equivalent impedance of a given beam termination from experimental data is possible. Beginning with the beam shown in Fig. 10.6(a), the unknown termination at the left-hand end may be described by the equivalent impedance matrix

$$Z_L = \begin{bmatrix} Z_{L1} & 0 \\ 0 & Z_{L2} \end{bmatrix} \sim \begin{bmatrix} Z_{Lf\dot{w}} & Z_{Lf\dot{\theta}} \\ Z_{Lm\dot{w}} & Z_{Lm\dot{\theta}} \end{bmatrix} \qquad (10.2.56a,b)$$

The right hand termination may be such that the impedance values are known, or it may be the same unknown termination used on the left hand end, in which case (following the sign conventions given in Fig. 10.6(c)) the equivalent impedance matrix Z_R is given by

$$Z_R = \begin{bmatrix} Z_{R1} & 0 \\ 0 & Z_{R2} \end{bmatrix} = \begin{bmatrix} -Z_{L1} & 0 \\ 0 & -Z_{L2} \end{bmatrix} \qquad (10.2.57a,b)$$

The method that follows will be the same regardless of whether a known or unknown termination is used at the right hand end.

Setting the coupling terms in (10.2.56(b)) equal to zero for our equivalent case, and using (10.2.3) and (10.2.54), we obtain

$$Z_{L1} = \frac{Q}{\dot{w}} \qquad (10.2.58)$$

and

$$Z_{L2} = \frac{M}{\dot{\theta}} \qquad (10.2.59)$$

For harmonic signals $\dot{w} = j\omega w$ and $\dot{\theta} = -j\omega w'$. Replacing the bending moment and shear force with derivatives of the displacement function (see (2.3.35) and (2.3.36)) gives

$$Z_{L1} = \frac{EI_{yy} w_1'''(x_L)}{j\omega w_1(x_L)} \qquad (10.2.60)$$

$$Z_{L2} = \frac{EI_{yy} w_1''(x_L)}{j\omega w_1'(x_L)} \qquad (10.2.61)$$

All that remains is to find the displacement and derivatives required in (10.2.60) and (10.2.61). We measure the accelerations a_1, a_2, a_3, ..., a_n (both amplitude and phase relative to an arbitrary, fixed reference signal) at n locations x_1, x_2, x_3, ..., x_n such that $x_L < x_i < x_0$, where x_L is the location of the left end of the beam and x_0 is the location of the excitation force. For $i = 1 - n$,

$$a_{ie}(x_i) = -\omega^2 w_{ie}(x_i) = -\omega^2\left(A_{1e}\,e^{-jk_b x_i} + A_{2e}e^{jk_b x_i} + A_{3e}e^{-k_b x_i} + A_{4e}e^{k_b x_i}\right) \qquad (10.2.62a,b)$$

where the subscript e denotes experimentally obtained values. Note that the acceleration $a_{ie}(x_i)$ at location (x_i) is complex, being defined in terms of a magnitude and a phase. In matrix form,

$$\begin{bmatrix} -\frac{a_{1e}}{\omega^2} \\ -\frac{a_{2e}}{\omega^2} \\ -\frac{a_{3e}}{\omega^2} \\ \vdots \\ -\frac{a_{ne}}{\omega^2} \end{bmatrix} = \begin{bmatrix} \beta_1 & \beta_1^{-1} & \beta_1^{j} & \beta_1^{-j} \\ \beta_2 & \beta_2^{-1} & \beta_2^{j} & \beta_2^{-j} \\ \beta_3 & \beta_3^{-1} & \beta_3^{j} & \beta_3^{-j} \\ \vdots & \vdots & \vdots & \vdots \\ \beta_n & \beta_n^{-1} & \beta_n^{j} & \beta_n^{-j} \end{bmatrix} \begin{bmatrix} A_{4e} \\ A_{3e} \\ A_{2e} \\ A_{1e} \end{bmatrix} \tag{10.2.63}$$

where $\beta_i = e^{k_b x_i}$, $\beta_i^{-1} = e^{-k_b x_i}$, $\beta_i^{j} = e^{jk_b x_i}$ and $\beta_i^{-j} = e^{-jk_b x_i}$. Rearranging gives

$$\begin{bmatrix} A_{4e} \\ A_{3e} \\ A_{2e} \\ A_{1e} \end{bmatrix} = \begin{bmatrix} \beta_1 & \beta_1^{-1} & \beta_1^{j} & \beta_1^{-j} \\ \beta_2 & \beta_2^{-1} & \beta_2^{j} & \beta_2^{-j} \\ \beta_3 & \beta_3^{-1} & \beta_3^{j} & \beta_3^{-j} \\ \vdots & \vdots & \vdots & \vdots \\ \beta_n & \beta_n^{-1} & \beta_n^{j} & \beta_n^{-j} \end{bmatrix}^{-1} \begin{bmatrix} -\frac{a_{1e}}{\omega^2} \\ -\frac{a_{2e}}{\omega^2} \\ -\frac{a_{3e}}{\omega^2} \\ \vdots \\ -\frac{a_{ne}}{\omega^2} \end{bmatrix} \tag{10.2.64}$$

where the superscript '-1' represents the generalized inverse or pseudo-inverse of the matrix. The matrix is better conditioned if the accelerometer spacings are random rather than uniform. Equation (10.2.64) represents a system of n equations in four unknowns. If $n = 4$, the system is determined, but the solution $[A_{4e}, A_{3e}, A_{2e}, A_{1e}]$ is extremely sensitive to errors in the measured accelerations. However, if an overdetermined system ($n > 4$) is used, the error is dramatically reduced, as will now be shown.

Let w_e be the displacement calculated from the constants A_{1e}, A_{2e}, A_{3e} and A_{4e}. For the beam described by the parameters of Table 10.2 with end conditions modelled as infinite, the error induced in the displacement w_e given an initial error in the accelerations (a_i, $i = 1 - n$) of 10% is plotted as a function of the number of accelerometers n in Fig. 10.19. For this case, $n \geq 10$ provides a satisfactory accuracy.

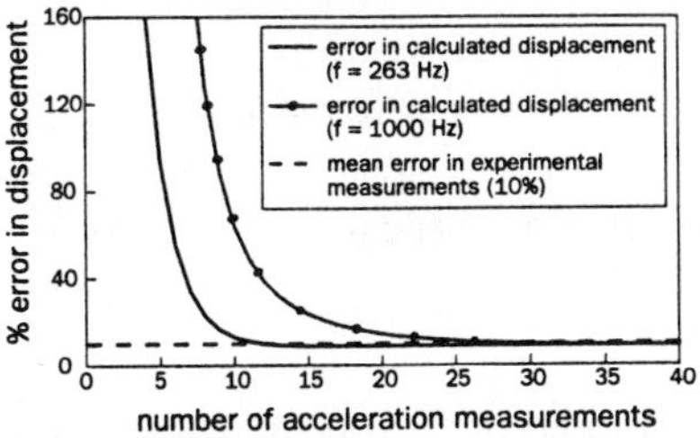

Fig. 10.19 Error induced in the calculated controlled beam response for an accelerometer measurement error under primary excitation of 10% (in both the real and imaginary parts).

Once calculated, A_{1e}, A_{2e}, A_{3e} and A_{4e} are substituted into equation (10.2.14) and differentiation carried out to find $w_{1e}(x_L)$, $w'_{1e}(x_L)$, $w''_{1e}(x_L)$ and $w'''_{1e}(x_L)$. Equations (10.2.60) and (10.2.61) can then be used to find the equivalent impedances Z_{L1} and Z_{L2}, which may be substituted into the α matrix to find the solution vector at any other location along the beam with and without control.

10.2.6.1 Accuracy of the approximation

It should be noted that use of the 'equivalent' impedance matrix obtained by eliminating the diagonal elements from the impedance matrix is only valid for analysis similar to that followed in this paper. It is not claimed that the resulting impedance matrix closely approximates the real impedance values of the termination in general circumstances. However, the numerical answers for all derivatives of displacement (and hence acceleration, etc.), calculated at any point along the beam, are correct to eight significant figures or are insignificant (less than 10^{-16}) when compared to the corresponding derivatives obtained by using the 'exact' impedance matrix. This is clearly well within the accuracy available from the experimental measurements!

The accuracy of this method has been tested with a variety of cases. The 'exact' impedance matrices and the corresponding approximations calculated using the method described are given in Table 10.3. All of these examples utilise a right hand impedance corresponding to an ideally infinite beam, and approximations are made for the various test cases at the left hand end of the beam. The parameters characterizing the beam are those given in Table 10.2. For all the examples, all derivatives of displacement calculated using the 'exact' and approximate impedance matrices are either identical to at least eight significant figures, or are insignificant (smaller than the reference by a factor of 10^{15} or more). It would be expected that the simplification might fail when the original matrix had large elements on the diagonal, but this is not the case, as shown by the first two examples. The third example shows an approximation for an impedance matrix with four complex elements. Examples 4, 5 and 6 show the exact and approximate matrices for the ideal free, fixed and infinite beam impedances respectively.

Note that if the measurements were taken at a large distance from the beam termination such that the terms A_{3e} and A_{4e} were very small, the matrix in (10.2.64) would be ill-conditioned and the steps outlined in Section 10.2.7 below would have to be undertaken.

10.2.7 Measuring amplitudes of waves travelling simultaneously in opposite directions

Examination of (10.2.64) reveals that the analysis of the previous section can

Table 10.3 Impedance matrices and corresponding approximations

Example No.	'Exact' matrix $[Z_L]$	Corresponding approximation (with zero diagonal elements)
1	$\begin{bmatrix} 10^{100}+10^{100}j & 0 \\ 0 & -10^{100}-10^{100}j \end{bmatrix}$	$\begin{bmatrix} 0 & 2.44\times10^{15}-5.00\times10^{16}j \\ -1.61\times10^{19}-9.30\times10^{18}j & 0 \end{bmatrix}$
2	$\begin{bmatrix} 10^{100} & 0 \\ 0 & -10^{100} \end{bmatrix}$	$\begin{bmatrix} 0 & 1.01\times10^{17}-4.28\times10^{16}j \\ -2.07\times10^{19}-1.27\times10^{18}j & 0 \end{bmatrix}$
3	$\begin{bmatrix} 0 & 0 \\ 0 & 0 \end{bmatrix}$	$\begin{bmatrix} 0 & -1.5\times10^{-15}-2.8\times10^{-16}j \\ 5.6\times10^{-14}-1.3\times10^{-13}j & 0 \end{bmatrix}$
4	$\begin{bmatrix} 0 & 10^{100} \\ 10^{100} & 0 \end{bmatrix}$	$\begin{bmatrix} 0 & 5.00\times10^{15}-8.13\times10^{16}j \\ -1.38\times10^{19}-8.48\times10^{17}j & 0 \end{bmatrix}$
5	$\begin{bmatrix} 124.99 & 9.596-9.596j \\ -1628.0-1628.0j & -124.99 \end{bmatrix}$	$\begin{bmatrix} 0 & 9.596 \\ -1628.0+1.35\times10^{-13}j & 0 \end{bmatrix}$
6	$\begin{bmatrix} 1+2j & 3+4j \\ 5+6j & 7+8j \end{bmatrix}$	$\begin{bmatrix} 0 & 4.963+6.101j \\ 222.396+10.889j & 0 \end{bmatrix}$

also be applied to determining the amplitudes of the two flexural waves (A_{1e} and A_{2e}) propagating in opposite directions, and the amplitudes of the non-propagating near field components (A_{3e} and A_{4e}). If the measurements are made in the far field of any vibration sources, beam discontinuities or boundaries such that A_{3e} and A_{4e} are very small, then the matrix requiring inversion in (10.2.64) will be ill-conditioned and the results for A_{1e} and A_{2e} will be inaccurate. In this situation better results are obtained by setting A_{3e} and A_{4e} equal to zero, removing A_{3e} and A_{4e} from the left side vector and removing the terms β_i and β_i^{-1} from the matrix.

Longitudinal waves may be measured in a similar way to flexural waves, provided the guidelines outlined in Chapter 2, Section 2.5.4.1 for obtaining accurate measurements are followed closely, especially if more than one wave type is present. In fact, determining the amplitudes of two longitudinal waves propagating in opposite directions is less prone to error than doing the same for flexural waves, as there are no near field effects to contend with and the waves may be described using

$$\xi_x = A_1 e^{-jkx} + A_2 e^{jkx}$$

where the complex wave amplitudes, A_1 and A_2 are determined as for flexural waves. Similar procedures can be followed to determine the amplitudes of two torsional waves travelling in opposite directions. As mentioned elsewhere in this book, the interest in longitudinal and torsional waves is not so much due to their direct contribution to feelable (or potentially damaging) vibration or sound radiation, but rather because their energy can be converted to the more problematic flexural waves at structural junctions and discontinuities, and especially at locations where there is effectively an eccentric mass.

10.3 ACTIVE CONTROL OF VIBRATION IN A SEMI-INFINITE PLATE

Here, we consider the active control of vibratory power transmission along a semi-infinite plate. This is of interest from the point of view of reducing vibration levels in, or sound radiation from, the plate or reducing the same quantities in an attached structure. To make the analytical problem tractable, only the more important flexural waves will be considered here. This does not mean that in a practical situation longitudinal waves could not be important; indeed, conversion of longitudinal wave energy into the more 'feelable' and more efficiently radiating flexural waves is a common phenomenon at structural junctions and discontinuities.

It is interesting to note that the problem of flexural waves in a plate is similar in some respects to higher order mode propagation in a two dimensional air duct. The model considered here is a semi-infinite plate, free at one end, anechoically terminated at the other and simply supported along the other two sides and so the analysis describes the vibration in terms of bending waves travelling from the free end to the anechoically terminated end. These waves consist of the fundamental wave and various higher order waves involving reflections at the simply supported edges in much the same way as plane wave and higher order mode propagation occur in an air duct.

10.3.1 Response of a semi-infinite plate to a line of point forces driven in phase

The plate considered here lies in the x-y plane with a free edge at $x = 0$ and simply supported edges at $y = 0$ and $y = L_y$ as shown in Fig. 10.20.

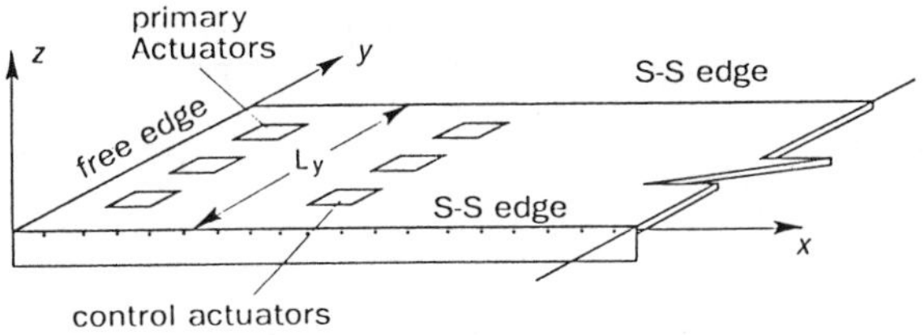

Fig. 10.20 Semi-infinite plate model. The origin is at the mid-point of the plate thickness.

The equation of motion for the plate (classical plate theory) is given by (2.3.101). The external force per unit area excitation represented by q in that equation is an array of n equally spaced point forces spanning the plate along a line parallel to the y-axis. At this stage, all of these forces at locations (x_0, y_i, $i = 1, n$), will be assumed to be driven in phase and with the same complex amplitude F_0 so that from (2.3.124)

$$q = F_0 \sum_{i=1}^{n} \delta(x - x_0)\delta(y - y_i) \tag{10.3.1}$$

The two sides ($y = 0$ and $y = L_y$) are simply supported, and so the following harmonic series solution in y can be assumed for the vibrational displacement:

$$w(x, y, t) = \sum_{m=1}^{\infty} w_m(x) \sin\frac{m\pi x}{L_y} e^{j\omega t} \tag{10.3.2}$$

where m is the mode number. Each eigen function $w_m(x)$ can be expressed in terms of unknown constants A_1, A_2, A_3 and A_4 and modal wavenumbers k_{im} as follows:

$$w_m(x) = A_1 e^{k_{1m}x} + A_2 e^{k_{2m}x} + A_3 e^{k_{3m}x} + A_4 e^{k_{4m}x} \tag{10.3.3}$$

To find the modal wavenumbers (i.e. eigenvalues), the homogeneous form of (2.3.101) is multiplied by $\sin\frac{m\pi y}{L_y}$ and integrated with respect to y over the width the plate (that is, from 0 to L_y), to give:

$$\frac{d^4 w_m(x)}{dx^4} - 2\left[\frac{m\pi}{L_y}\right]^2 \frac{d^2 w_m(x)}{dx^2} + \left[\left[\frac{m\pi}{L_y}\right]^4 - \frac{\rho h\omega^2}{D_h}\right] w_m(x) = 0 \tag{10.3.4}$$

where $D_h = Eh/[12(1 - \nu^2)]$ is the flexural rigidity of the plate. Solutions to this ordinary differential equation are assumed to be of the form $e^{k_m x}$, and its characteristic equation is:

$$k_m^4 - 2\left[\frac{m\pi}{L_y}\right]^2 k_m^2 + \left[\frac{m\pi}{L_y}\right]^4 - \frac{\rho h\omega^2}{D_h} = 0 \tag{10.3.5}$$

which has the roots

$$k_{1m,2m} = \pm\sqrt{\left[\frac{m\pi}{L_y}\right]^2 + \sqrt{\frac{\rho h\omega^2}{D_h}}} \tag{10.3.6}$$

$$k_{3m,4m} = \pm\sqrt{\left[\frac{m\pi}{L_y}\right]^2 - \sqrt{\frac{\rho h\omega^2}{D_h}}} \tag{10.3.7}$$

On each side of the applied force, the eigen function $w_m(x)$ is then a different linear combination of the terms $e^{k_{im} x}$ (with $i = 1, 2, 3, 4$). For $x < x_0$,

$$w_{1m}(x) = A_1 e^{k_{1m}x} + A_2 e^{k_{2m}x} + A_3 e^{k_{3m}x} + A_4 e^{k_{4m}x} \tag{10.3.8}$$

and for $x > x_0$,

$$w_{2m}(x) = B_2 e^{k_{2m}x} + B_4 e^{k_{4m}x} \tag{10.3.9}$$

The phenomenon of wave propagation in this plate is very similar to the propagation of longitudinal waves in an air duct. Below a certain cut-on frequency, the flexural wave propagation will be one-dimensional as in a beam or as longitudinal plane waves in an air duct. Above the cut-on frequency, higher order modes, characterized by wave reflections from the two simply supported plate edges, begin to propagate resulting in a very non-uniform vibration field on the plate. As the frequency becomes higher, more modes cut-on and the vibration field on the plate becomes even more complex. The cut-on frequencies of the higher order modes can be obtained from (10.3.8) by determining the frequency when the wavenumber first becomes imaginary. Thus, the mth higher order mode will cut-on when the excitation frequency is given by

$$f = \frac{\pi}{2}\left[\frac{m}{L_y}\right]^2 \sqrt{\frac{D_h}{\rho h}} \tag{10.3.10}$$

Note that in (10.3.9), the coefficients B_1 and B_3 have been omitted because, for a semi-infinite plate, there is no boundary to produce reflected waves with a negative horizontal velocity component. At the junction $x = x_0$, the following boundary conditions must be satisfied:

$$w_{1m} = w_{2m} \tag{10.3.11}$$

$$\frac{\partial w_{1m}}{\partial x} = \frac{\partial w_{2m}}{\partial x} \tag{10.3.12}$$

$$\frac{\partial^2 w_{1m}}{\partial x^2} = \frac{\partial^2 w_{2m}}{\partial x^2} \tag{10.3.13}$$

and

$$\frac{\partial^3 w_{1m}}{\partial x^3} - \frac{\partial^3 w_{2m}}{\partial x^3} = -\frac{2F_0}{L_y D_h} \sum_{i-1}^{n} \sin \frac{m\pi y_i}{L_y} \tag{10.3.14}$$

For the free edge at $x = 0$, the expression

$$-D_h \left(\frac{\partial^2 w}{\partial x^2} + \nu \frac{\partial^2 w}{\partial y^2} \right) \Big|_{x-0} = 0 \tag{10.3.15}$$

given by Leissa (1960) can be used to express the bending moment boundary condition:

$$M_x(0, y) = 0 \tag{10.3.16}$$

in terms of displacement, with the following result:

$$\sum_{m-1}^{\infty} \left[k_{1m}^2 A_1 + k_{2m}^2 A_2 + k_{3m}^2 A_3 + k_{4m}^2 A_4 - \nu \left(\frac{m\pi}{L_y} \right)^2 (A_1 + A_2 + A_3 + A_4) \right] \sin \frac{m\pi y}{L_y} = 0 \tag{10.3.17}$$

By again multiplying both sides by $\sin \frac{m\pi y}{L_y}$ and integrating from $y = 0$ to L_y, the following is obtained:

$$\left[k_{1m}^2 - \nu \left(\frac{m\pi}{L_y} \right)^2 \right] A_1 + \left[k_{2m}^2 - \nu \left(\frac{m\pi}{L_y} \right)^2 \right] A_2 + \left[k_{3m}^2 - \nu \left(\frac{m\pi}{L_y} \right)^2 \right] A_3 + \left[k_{4m}^2 - \nu \left(\frac{m\pi}{L_y} \right)^2 \right] A_4 = 0 \tag{10.3.18}$$

The free edge condition also requires that the net vertical force at $x = 0$ be zero (Leissa, 1960). Thus,

$$V_x = Q_x + \frac{\partial M_{xy}}{\partial y} = 0 \tag{10.3.19a,b}$$

which can be expressed in terms of displacement as follows:

$$V_x(x, y) \mid_{x=0} = -D_h \left[\frac{\partial^3 w}{\partial x^3} + (2 - \nu) \frac{\partial^3 w}{\partial x \partial y^2} \right] \mid_{x=0} = 0 \qquad (10.3.20a,b)$$

Thus,

$$\begin{aligned} &\left[k_{1m}^3 - (2 - \nu) \left[\frac{m\pi}{L_y} \right]^2 k_{1m} \right] A_1 \\ &+ \left[k_{2m}^3 - (2 - \nu) \left[\frac{m\pi}{L_y} \right]^2 k_{2m} \right] A_2 \\ &+ \left[k_{3m}^3 - (2 - \nu) \left[\frac{m\pi}{L_y} \right]^2 k_{3m} \right] A_3 \\ &+ \left[k_{4m}^3 - (2 - \nu) \left[\frac{m\pi}{L_y} \right]^2 k_{4m} \right] A_4 = 0 \end{aligned} \qquad (10.3.21)$$

Equations (10.3.11) to (10.3.14), (10.3.18) and (10.3.21) can be written as a 6 × 6 matrix equation as follows:

$$\begin{bmatrix} k_{1m}^2 - \nu H & k_{2m}^2 - \nu H & k_{3m}^2 - \nu H & k_{4m}^2 - \nu H & 0 & 0 \\ k_{1m}^3 - (2-\nu) H k_{1m} & k_{2m}^3 - (2-\nu) H k_{2m} & k_{3m}^3 - (2-\nu) H k_{3m} & k_{4m}^3 - (2-\nu) H k_{4m} & 0 & 0 \\ e^{k_{1m}x_0} & e^{k_{2m}x_0} & e^{k_{3m}x_0} & e^{k_{4m}x_0} & -e^{k_{2m}x_0} & -e^{k_{4m}x_0} \\ k_{1m} e^{k_{1m}x_0} & k_{2m} e^{k_{2m}x_0} & k_{3m} e^{k_{3m}x_0} & k_{4m} e^{k_{4m}x_0} & -k_{2m} e^{k_{2m}x_0} & -k_{4m} e^{k_{4m}x_0} \\ k_{1m}^2 e^{k_{1m}x_0} & k_{2m}^2 e^{k_{2m}x_0} & k_{3m}^2 e^{k_{3m}x_0} & k_{4m}^2 e^{k_{4m}x_0} & -k_{2m}^2 e^{k_{2m}x_0} & -k_{4m}^2 e^{k_{4m}x_0} \\ k_{1m}^3 e^{k_{1m}x_0} & k_{2m}^3 e^{k_{2m}x_0} & k_{3m}^3 e^{k_{3m}x_0} & k_{4m}^3 e^{k_{4m}x_0} & -k_{2m}^3 e^{k_{2m}x_0} & -k_{4m}^3 e^{k_{4m}x_0} \end{bmatrix} \begin{bmatrix} A_1 \\ A_2 \\ A_3 \\ A_4 \\ B_2 \\ B_4 \end{bmatrix} = \begin{bmatrix} 0 \\ 0 \\ 0 \\ 0 \\ 0 \\ -\dfrac{2F_0}{L_y D_h} \sum_{i=1}^{n} \sin \dfrac{m\pi y_i}{L_y} \end{bmatrix} \qquad (10.3.22)$$

which can be written as $\alpha X = F$. For each value of m, the solution X of the 6×6 system of equations is an eigenvector which describes a travelling wave mode shape and amplitude. The modal wavenumbers k_{1m}, k_{2m}, k_{3m} and k_{4m} required by (10.3 22) are calculated from (10.3.6) and (10.3.7), and the quantity $H = (m\pi/L_y)^2$. The plate response at any location (x, y) due to the row of in-phase point forces is:

$$w = F_0 w_{0\text{-}f} \tag{10.3.23}$$

where $w_{0\text{-}f}$ is the response to unit force excitation which is obtained by solving (10.3.22) and substituting the results for A_1, A_2, A_3, A_4, B_2 and B_4 into (10.3.8), (10.3.3) and (10.3.2) or (10.3.9) and (10.3.2), depending upon the location at which the plate response is to be evaluated. Finally, we require the response of the plate to unit primary and unit secondary force excitation. By expressing the solution for the eigenvectors as:

$$X = \alpha^{-1} F \tag{10.3.24}$$

the response at (x, y) is:

$$w_{0\text{-}f}(x, y) = \sum_{m=1}^{\infty} \frac{2}{L_y D_h} \left[\sum_{i=1}^{n} \sin\frac{m\pi y_i}{L_y} \right] \left[(\alpha)^{-1} \right]_{i,6}^{\mathrm{T}} E_m \sin\frac{m\pi y}{L_y} \tag{10.3.25}$$

where n is the number of forces in the array, m is the mode number and $\left[(\alpha)^{-1} \right]_{i,6}^{\mathrm{T}}$ is the sixth column of the inverse of the coefficient matrix α from (10.3.22).

For $x < x_0$,

$$E_m = \left[e^{k_{1m}x} \; e^{k_{2m}x} \; e^{k_{3m}x} \; e^{k_{4m}x} \; 0 \; 0 \right]^{\mathrm{T}} \tag{10.3.26}$$

and for $x > x_0$,

$$E_m = \left[0 \; 0 \; 0 \; 0 \; e^{k_{2m}x} \; e^{k_{4m}x} \right]^{\mathrm{T}} \tag{10.3.27}$$

10.3.2 Minimization of acceleration with a line of in-phase control forces

If the plate is excited by a line of in-phase primary point forces of complex amplitude F_p located at $x = x_p$, the plate response $w_p(x, y)$ at any location (x, y) is found by using (10.3.23) and (10.3.25) and is:

$$w_p(x, y) = F_p w_{p\text{-}f}(x, y) = F_p \sum_{m=1}^{\infty} \frac{2}{L_y D_h} \left[\sum_{i=1}^{n_p} \sin\frac{m\pi y_i}{L_y} \right] \left[(\alpha_p)^{-1} \right]_{i,6}^{\mathrm{T}} E_m \sin\frac{m\pi y}{L_y}$$

(10.3.28a,b)

where $w_{p-f}(x,y)$ is the plate response to unit primary force excitation, n_p is the number of primary forces, and α_p is similar to α with F_0 replaced by F_p and x_0

replaced by x_p. Similarly, if a line of control forces of complex amplitude F_c is placed at $x = x_c$, the plate response due to these acting alone is:

$$w_c(x, y) = F_c w_{c-f}(x, y) = F_c \sum_{m=1}^{\infty} \frac{2}{L_y D_h} \left[\sum_{i=1}^{n_c} \sin \frac{m\pi y_i}{L_y} \right] \left[(\alpha_c)^{-1} \right]_{i,6}^{T} E_m \sin \frac{m\pi y}{L_y} \tag{10.3.29a,b}$$

where n_c is the number of control actuators, and $w_{c-f}(x,y)$ is the plate response to unit control force excitation. The total plate response due to primary and secondary forces acting together is then:

$$w = w_p + w_c = F_p w_{p-f} + F_c w_{c-f} \tag{10.3.30a,b}$$

The optimal control force F_c for minimizing the acceleration across the width of the plate at a constant x may be found by integrating the mean square of the displacement defined in (10.3.30), and setting the partial derivatives of the integration with respect to the real and imaginary components of the control force equal to zero. The result is:

$$F_c = -F_p \frac{\int_0^{L_y} w_{p-f} w_{c-f}^{*} \, dy}{\int_0^{L_y} | w_{c-f} |^2 \, dy} \tag{10.3.31}$$

10.3.3 Minimization of acceleration with a line of *n* independently driven control forces

If the plate is driven by an array of in-phase primary point forces in a line at $x = x_p$ and n independent control point forces in a line at $x = x_c$, the total plate response may be written as:

$$w = w_p + w_c = F_p w_{p-f} + F_{c1} w_{c-f1} + F_{c2} w_{c-f2} + \cdots + F_{cn} w_{c-fn} \tag{10.3.32a,b}$$

The quantities w_{c-f1}, w_{c-f2}, $\cdots$, w_{c-fn} are each calculated in a similar way to w_{c-f} in the previous section, except that n_c is set equal to one in each case.

The optimal control forces for minimizing acceleration in a line across the plate may be found by integrating across the plate (x = const) the mean square of the displacement defined in (10.3.32) and setting the partial derivatives of the integration with respect to each of the real and imaginary components of the control forces equal to zero. The result is an optimal set of control forces as follows:

$$\begin{bmatrix} F_{c1} \\ F_{c2} \\ \vdots \\ F_{cn} \end{bmatrix} =$$

$$- \begin{bmatrix} \int_0^{L_y} |w_{c-f1}|^2 \, dy & \int_0^{L_y} w_{c-f1}^* w_{c-f2} \, dy & \cdots & \int_0^{L_y} w_{c-f1}^* w_{c-fn} \, dy \\ \int_0^{L_y} w_{c-f1} w_{c-f2}^* \, dy & \int_0^{L_y} |w_{c-f2}|^2 \, dy & \cdots & \int_0^{L_y} w_{c-f2}^* w_{c-fn} \, dy \\ \vdots & \vdots & \vdots & \vdots \\ \int_0^{L_y} w_{c-f1} w_{c-fn}^* \, dy & \int_0^{L_y} w_{c-f2} w_{c-fn}^* \, dy & \cdots & \int_0^{L_y} |w_{c-fn}|^2 \, dy \end{bmatrix}^{-1} \begin{bmatrix} \int_0^{L_y} w_{p-f} w_{c-f1}^* \, dy \\ \int_0^{L_y} w_{p-f} w_{c-f2}^* \, dy \\ \vdots \\ \int_0^{L_y} w_{p-f} w_{c-fn}^* \, dy \end{bmatrix} F_p \tag{10.3.33}$$

10.3.4 Power transmission

From Section 2.5, the expression for the x component of flexural wave vibration intensity (or power transmission per unit width) in a plate is:

$$P_x(t) = - \left[\dot{w} Q_x - \frac{\partial \dot{w}}{\partial x} M_x - \frac{\partial \dot{w}}{\partial y} M_{xy} \right] \tag{10.3.34}$$

Note that here the contribution from longitudinal and shear waves is assumed to be zero. The total instantaneous flexural wave power transmission through a section at constant x is then given by:

$$P_x(t) = - \int_0^{L_y} \left[\dot{w} Q_x - \frac{\partial \dot{w}}{\partial x} M_x - \frac{\partial \dot{w}}{\partial y} M_{xy} \right] dy \tag{10.3.35}$$

or, for a single frequency:

$$P_x(t) = - \int_0^{L_y} \left[j\omega w Q_x - j\omega \frac{\partial w}{\partial x} M_x - j\omega \frac{\partial w}{\partial y} M_{xy} \right] dy \tag{10.3.36}$$

In (10.3.35) and (10.3.36) all displacements, forces and moments are time dependent quantities and also dependent on the coordinate location (x, y) on the plate.

As discussed in Section 2.5, for single frequency excitation the real (or active) part of the power transmission along the plate is the product of the real part of the force term with the real part of the velocity term for each pair of terms in (10.3.36) and the result is time averaged. Thus, the active power transmission for harmonic excitation is given by:

$$P_{xa} = -\frac{1}{2}\int_0^{L_y} \mathrm{Re}\left\{ \left[j\omega w\right]^* Q_x - \left[j\omega\frac{\partial w}{\partial x}\right]^* M_x - \left[j\omega\frac{\partial w}{\partial y}\right]^* M_{xy} \right\} \mathrm{d}y \tag{10.3.37}$$

where the quantities Q_x, M_x and M_{xy} have been defined in Section 2.3.

For one line of primary actuators and a second line of control actuators parallel to the y-axis, the resulting total power transmission can be expressed in terms of the primary and control forces, using superposition.

10.3.5 Minimization of power transmission with a line of in-phase point control forces

The total power transmission resulting from a line of in-phase point primary forces and a line of in-phase point control forces acting together can be found by substituting (10.3.28) and (10.3.29) into (10.3.30) and the result into (10.3.37). The shear forces and moments may also be expressed in terms of the displacements associated with each excitation force using equations from Section 2.3 and (10.3.28), (10.3.29) and (10.3.30) to give:

$$M_x = \frac{Eh^3}{12(1-\nu^2)}\left[F_p\frac{\partial^2 w_{p\text{-}f}}{\partial x^2} + F_c\frac{\partial^2 w_{c\text{-}f}}{\partial x^2} + \nu\left(F_p\frac{\partial^2 w_{p\text{-}f}}{\partial y^2} + F_c\frac{\partial^2 w_{c\text{-}f}}{\partial y^2}\right)\right] \tag{10.3.38}$$

$$M_{xy} = -\frac{Eh^3}{12(1+\nu)}\left[F_p\frac{\partial^2 w_{p\text{-}f}}{\partial x\,\partial y} + F_c\frac{\partial^2 w_{c\text{-}f}}{\partial x\,\partial y}\right] \tag{10.3.39}$$

$$Q_x = \frac{\partial M_x}{\partial x} - \frac{\partial M_{xy}}{\partial y}$$

$$= -\frac{Eh^3}{12(1-\nu^2)}\left[F_p\frac{\partial^3 w_{p\text{-}f}}{\partial x^3} + F_c\frac{\partial^3 w_{c\text{-}f}}{\partial x^3} + F_p\frac{\partial^3 w_{p\text{-}f}}{\partial x\,\partial y^2} + F_c\frac{\partial^3 w_{c\text{-}f}}{\partial x\,\partial y^2}\right] \tag{10.3.40a,b}$$

where ν is Poisson's ratio and E is Young's modulus of elasticity. The preceding three equations and (10.3.30) can be substituted into the expression (10.3.37) for the power transmission through any plate cross section at axial location x to give:

$$P_{xa} = \frac{1}{2}\int_0^{L_y} \mathrm{Re}\left\{F_c F_c^* A + F_c B F_p^* + F_c^* C F_p + F_p F_p^* D\right\} \mathrm{d}y \tag{10.3.41}$$

where

$$A = \frac{j\omega Eh^3}{12(1-\nu^2)}\left[-\left(\frac{\partial^3 w_{c\text{-}f}}{\partial x^3} + \frac{\partial^3 w_{c\text{-}f}}{\partial x \partial y^2}\right) w_{c\text{-}f}^* \right. \tag{10.3.42}$$

$$\left. + \left(\frac{\partial^2 w_{c\text{-}f}}{\partial x^2} + \nu \frac{\partial^2 w_{c\text{-}f}}{\partial y^2}\right)\frac{\partial w_{c\text{-}f}^*}{\partial x} + (1-\nu)\frac{\partial^2 w_{c\text{-}f}}{\partial x \partial y}\frac{\partial w_{c\text{-}f}^*}{\partial y}\right]$$

$$B = \frac{j\omega Eh^3}{12(1-\nu^2)}\left[-\left(\frac{\partial^3 w_{c\text{-}f}}{\partial x^3} + \frac{\partial^3 w_{c\text{-}f}}{\partial x \partial y^2}\right) w_{p\text{-}f}^* \right. \tag{10.3.43}$$

$$\left. + \left(\frac{\partial^2 w_{c\text{-}f}}{\partial x^2} + \nu \frac{\partial^2 w_{c\text{-}f}}{\partial y^2}\right)\frac{\partial w_{p\text{-}f}^*}{\partial x} + (1-\nu)\frac{\partial^2 w_{c\text{-}f}}{\partial x \partial y}\frac{\partial w_{p\text{-}f}^*}{\partial y}\right]$$

$$C = \frac{j\omega Eh^3}{12(1-\nu^2)}\left[-\left(\frac{\partial^3 w_{p\text{-}f}}{\partial x^3} + \frac{\partial^3 w_{p\text{-}f}}{\partial x \partial y^2}\right) w_{c\text{-}f}^* \right. \tag{10.3.44}$$

$$\left. + \left(\frac{\partial^2 w_{p\text{-}f}}{\partial x^2} + \nu \frac{\partial^2 w_{p\text{-}f}}{\partial y^2}\right)\frac{\partial w_{c\text{-}f}^*}{\partial x} + (1-\nu)\frac{\partial^2 w_{p\text{-}f}}{\partial x \partial y}\frac{\partial w_{c\text{-}f}^*}{\partial y}\right]$$

The partial derivatives of the active (real) power transmission with respect to the real and imaginary components of the control force are:

$$\frac{\partial P_{xa}}{\partial F_r} = \frac{1}{2}\int_0^{L_y} \mathrm{Re}\left\{F_c^* A + F_c A + F_p^* B + F_p C\right\} \mathrm{d}y \tag{10.3.45}$$

$$\frac{\partial P_{xa}}{\partial F_j} = \frac{1}{2}\int_0^{L_y} \mathrm{Re}\left\{jF_c^* A - jF_c A + jF_p^* B - jF_p C\right\} \mathrm{d}y \tag{10.3.46}$$

respectively, where $F_c = F_r + jF_j$. The optimal force is found by requiring that both of these derivatives are zero. That is:

$$\int_0^{L_y} \mathrm{Re}\left\{F_c^* A + F_c A + F_p^* B + F_p C\right\} \mathrm{d}y = 0 \tag{10.3.47}$$

$$\int_0^{L_y} \mathrm{Re}\left\{jF_c^* A - jF_c A + jF_p^* B - jF_p C\right\} \mathrm{d}y = 0 \tag{10.3.48}$$

The result is:

$$F_r^{\text{opt}} = -\frac{\int_0^{L_y} \text{Re}\left\{F_p^* B + F_p C\right\} \text{dy}}{2\int_0^{L_y} \text{Re}\{A\} \text{dy}} = -\frac{\int_0^{L_y} \text{Re}\left\{F_p^* B\right\} \text{dy} + \int_0^{L_y} \text{Re}\{F_p C\} \text{dy}}{2\int_0^{L_y} \text{Re}\{A\} \text{dy}}$$

(10.3.49a,b)

$$F_j^{\text{opt}} = -\frac{\int_0^{L_y} \text{Re}\left\{jF_p^* B - jF_p C\right\} \text{dy}}{2\int_0^{L_y} \text{Re}\{A\} \text{dy}} = -\frac{-\int_0^{L_y} \text{Im}\left\{F_p^* B\right\} \text{dy} + \int_0^{L_y} \text{Im}\{F_p C\} \text{dy}}{2\int_0^{L_y} \text{Re}\{A\} \text{dy}}$$

(10.3.50a,b)

or

$$F_c^{\text{opt}} = F_r^{\text{opt}} + jF_j^{\text{opt}} = -\frac{\int_0^{L_y} B^* \,\text{dy} + \int_0^{L_y} C \,\text{dy}}{2\int_0^{L_y} \text{Re}\{A\} \text{dy}} F_p \qquad (10.3.51\text{a,b})$$

Equation (10.3.51) is an expression for the control force which minimizes the power transmission along the plate, and is of the same form (apart from the integration over y) as the expression for the optimal force required to control the power transmission in a beam (see Section 10.1). The optimum plate power transmission is given by:

$$P_{xa}^{\text{opt}} =$$

$$-\frac{1}{2}\left[\frac{\left|\int_0^{L_y} B \,\text{dy}\right|^2 + 2\text{Re}\left\{\int_0^{L_y} B \,\text{dy} \int_0^{L_y} C \,\text{dy}\right\} + \left|\int_0^{L_y} C \,\text{dy}\right|^2}{4\int_0^{L_y} \text{Re}\{A\} \text{dy}} - \int_0^{L_y} \text{Re}\{D\} dy\right] |F_p|^2$$

(10.3.52)

where the power transmission has been obtained by integrating the intensity over a plate cross-section. The differential of (10.3.52) with respect to x and y, which is the intensity, is a strongly varying function of location, but by conservation of energy, the total power transmission through a particular plate cross-section must be independent of the location and shape of the cross-section surface. It is assumed here that it is possible in practice to actually measure the total power transmission through a particular plate cross-section. In practice, this may be approximated by making a number of point measurements across the width of the plate and averaging the results. Pan and Hansen (1995) show that the required number of measurements to achieve an acceptable accuracy can be very small (as low as three in some cases).

10.3.6 Minimization of power transmission with a line of n independently driven point control forces

For this case, the plate response is given by (10.3.32) which can be substituted into (10.3.37) to give an expression for the total power being transmitted past a line across the plate at an axial location x, which can be written in matrix form as:

$$P_{xa} = \frac{1}{2}\int_0^{L_y} \mathrm{Re}\{\boldsymbol{F}^{\mathrm{H}}\boldsymbol{A}\boldsymbol{F}\}\,\mathrm{d}y \tag{10.3.53}$$

where

$$\boldsymbol{F} = [F_p \;\; F_{c1} \;\; F_{c2} \;\; \cdots \;\; F_{cn}]^{\mathrm{T}} \tag{10.3.54}$$

and

$$\boldsymbol{A} = \begin{bmatrix} A(1,1) & A(1,2) & A(1,3) & \cdots & A(1,n+1) \\ A(2,1) & A(2,2) & A(2,3) & \cdots & A(2,n+1) \\ A(3,1) & A(3,2) & A(3,3) & \cdots & A(3,n+1) \\ \vdots & \vdots & \vdots & \vdots & \vdots \\ A(n+1,1) & A(n+1,2) & A(n+1,3) & \cdots & A(n+1,n+1) \end{bmatrix} \tag{10.3.55}$$

The coefficients $A(i,\, j)$ $(i = 1,\, n+1;\, j = 1,\, n+1)$ of matrix $\boldsymbol{A}$ result from the product of terms in (10.3.37), each of which contains contributions from the $n+1$ different force elements of (10.3.54). Thus, $A(i,j)$ is the product of the contribution to the first part of each term in (10.3.37) due to the ith element of $\boldsymbol{F}$ with the contribution to the second part of each term due to the jth element of $\boldsymbol{F}$. For example,

$$A(1,1) = \frac{j\omega Eh^3}{12(1-\nu^2)}\left[-\left(\frac{\partial^3 w_{p\text{-}f}}{\partial x^3} + \frac{\partial^3 w_{p\text{-}f}}{\partial x\,\partial y^2}\right) w_{p\text{-}f}^{*} + \left(\frac{\partial^2 w_{p\text{-}f}}{\partial x^2} + \nu\frac{\partial^2 w_{p\text{-}f}}{\partial y^2}\right)\frac{\partial w_{p\text{-}f}^{*}}{\partial x} + (1-\nu)\frac{\partial^2 w_{p\text{-}f}}{\partial x\,\partial y}\frac{\partial w_{p\text{-}f}^{*}}{\partial y}\right] \tag{10.3.56}$$

$$A(2,1) = \frac{j\omega Eh^3}{12(1-\nu^2)}\left[-\left(\frac{\partial^3 w_{p-f}}{\partial x^3} + \frac{\partial^3 w_{p-f}}{\partial x\,\partial y^2}\right) w^*_{c-f1}\right.$$
$$\left. + \left(\frac{\partial^2 w_{p-f}}{\partial x^2} + \nu\frac{\partial^2 w_{p-f}}{\partial y^2}\right)\frac{\partial w^*_{c-f1}}{\partial x} + (1-\nu)\frac{\partial^2 w_{p-f}}{\partial x\,\partial y}\frac{\partial w^*_{c-f1}}{\partial y}\right] \tag{10.3.57}$$

$$A(n+1,1) = \frac{j\omega Eh^3}{12(1-\nu^2)}\left[-\left(\frac{\partial^3 w_{p-f}}{\partial x^3} + \frac{\partial^3 w_{p-f}}{\partial x\,\partial y^2}\right) w^*_{c-fn}\right.$$
$$\left. + \left(\frac{\partial^2 w_{p-f}}{\partial x^2} + \nu\frac{\partial^2 w_{p-f}}{\partial y^2}\right)\frac{\partial w^*_{c-fn}}{\partial x} + (1-\nu)\frac{\partial^2 w_{p-f}}{\partial x\,\partial y}\frac{\partial w^*_{c-fn}}{\partial y}\right] \tag{10.3.58}$$

By defining $F_{c1} = F_{r1} + jF_{j1}$, $F_{c2} = F_{r2} + jF_{j2}$, and $F_{cn} = F_{rn} + jF_{jn}$, the partial derivatives of the active (real) power transmission with respect to each real and imaginary control force may be written as:

$$\frac{\partial P_{xa}}{\partial F_{r1}} = \frac{1}{2}\int_0^{L_y} \mathrm{Re}\Big\{F_p^* A(1,2) + F_p A(2,1) + F_{c1} A(2,2) + F_{c1}^* A(2,2)$$
$$+ F_{c2}^* A(3,2) + F_{c2} A(2,3) + \cdots + F_{cn}^* A(n+1,2) + F_{cn} A(2,n+1)\Big\}\,\mathrm{d}y \tag{10.3.59}$$

$$\frac{\partial P_{xa}}{\partial F_{j1}} = \frac{1}{2}\int_0^{L_y} \mathrm{Re}\Big\{jF_p^* A(1,2) - jF_p A(2,1) + jF_{c1}^* A(2,2) - jF_{c1} A(2,2)$$
$$+ jF_{c2}^* A(3,2) - jF_{c2} A(2,3) + \cdots + jF_{cn}^* A(n+1,2) - jF_{cn} A(2,n+1)\Big\}\,\mathrm{d}y \tag{10.3.60}$$

$$\frac{\partial P_{xa}}{\partial F_{r2}} = \frac{1}{2}\int_0^{L_y} \mathrm{Re}\Big\{F_p^* A(1,3) + F_p A(3,1) + F_{c1}^* A(2,3) + F_{c1} A(3,2)$$
$$+ F_{c2}^* A(3,3) + F_{c2}^* A(3,3) + F_{cn}^* A(n+1,3) + F_{cn} A(3,n+1)\Big\}\,\mathrm{d}y \tag{10.3.61}$$

$$\frac{\partial P_{xa}}{\partial F_{j2}} = \frac{1}{2}\int_0^{L_y} \mathrm{Re}\Big\{jF_p^* A(1,3) - jF_p A(3,1) + jF_{c1}^* A(2,3) - jF_{c1} A(3,2)$$
$$+ jF_{c2}^* A(3,3) - jF_{c2} A(3,3) + jF_{cn}^* A(n+1,3) - jF_{cn} A(3,n+1)\Big\}\,\mathrm{d}y \tag{10.3.62}$$

$$\frac{\partial P_{xa}}{\partial F_{rn}} = \frac{1}{2}\int_0^{L_y} \mathrm{Re}\Big\{F_p^* A(1,n+1) + F_p A(n+1,1) + F_{c1}^* A(2,n+1) + F_{c1} A(n+1,2)$$

$$+ F_{c2}^* A(3,n+1) + F_{c2} A(n+1,3) + F_{cn}^* A(n+1,n+1) + F_{cn} A(n+1,n+1)\Big\}\,\mathrm{d}y \tag{10.3.63}$$

$$\frac{\partial P_{xa}}{\partial F_{jn}} = \frac{1}{2}\int_0^{L_y} \mathrm{Re}\Big\{jF_p^* A(1,n+1) - jF_p A(n+1,1) + jF_{c1}^* A(2,n+1) - jF_{c1} A(n+1,2)$$

$$+ jF_{c2}^* A(3,n+1) - jF_{c2} A(n+1,3) + jF_{cn}^* A(n+1,n+1) - jF_{cn}^* A(n+1,n+1)\Big\}\,\mathrm{d}y \tag{10.3.64}$$

An optimum set of control forces corresponding to a minimum power transmission is achieved when each of the derivatives is zero. The matrix form of the system of equations (10.3.59) to (10.3.64) is:

$$\begin{bmatrix} F_{c1} \\ F_{c2} \\ \vdots \\ F_{cn} \end{bmatrix} = - \begin{bmatrix} \int_0^{L_y}[A^*(2,2) + A(2,2)]\,\mathrm{d}y & \int_0^{L_y}[A^*(3,2) + A(2,3)]\,\mathrm{d}y & \cdots & \int_0^{L_y}[A^*(n+1,2) + A(2,n+1)]\,\mathrm{d}y \\ \int_0^{L_y}[A^*(2,3) + A(3,2)]\,\mathrm{d}y & \int_0^{L_y}[A^*(3,3) + A(3,3)]\,\mathrm{d}y & \cdots & \int_0^{L_y}[A^*(n+1,3) + A(3,n+1)]\,\mathrm{d}y \\ \vdots & \vdots & \vdots & \vdots \\ \int_0^{L_y}[A^*(2,n+1) + A(n+1,2)]\,\mathrm{d}y & \int_0^{L_y}[A^*(3,n+1) + A(n+1,3)]\,\mathrm{d}y & \cdots & \int_0^{L_y}[A^*(n+1,n+1) + A(n+1,n+1)]\,\mathrm{d}y \end{bmatrix}^{-1} \left(\begin{bmatrix} \int_0^{L_y} A^*(1,2)\,\mathrm{d}y \\ \int_0^{L_y} A^*(1,3)\,dy \\ \vdots \\ \int_0^{L_y} A^*(1,n+1)\,\mathrm{d}y \end{bmatrix} + \begin{bmatrix} \int_0^{L_y} A(2,1)\,\mathrm{d}y \\ \int_0^{L_y} A(3,1)\,\mathrm{d}y \\ \vdots \\ \int_0^{L_y} A(n+1,1)\,\mathrm{d}y \end{bmatrix} \right) F_p \tag{10.3.65}$$

By comparing (10.3.41) with (10.3.53) and (10.3.51) with (10.3.65), it can be seen that the single force changes to a force vector when in-phase force control changes to independent force control. The expression for power transmission given by (10.3.53) not only includes each force term, but also includes coupling force terms, which makes independent force control much more complex than in-phase force control to analyse. As has been shown in the analysis, the procedure can be used for any number n of independently controlled forces.

The matrix which must be inverted in (10.3.65) is ill-conditioned for the case where the location of power transmission minimization is in the near field of the control force array; thus, more reliable results are obtained if the point of minimization is always in the far field of the control source.

10.3.7 Numerical results

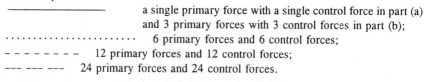

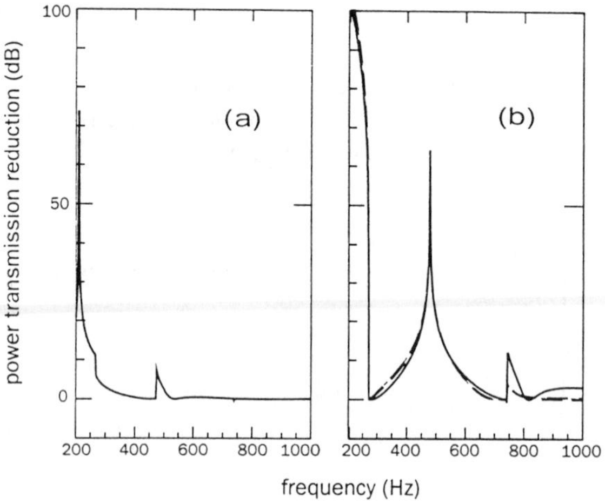

Fig. 10.21 Power flow reduction as a function of frequency with the control sources located at the optimum location for 210 Hz (from Pan and Hansen, 1995): (a) a single off-centre primary and control source; (b) equal numbers of primary and control sources driven in phase.

———————— a single primary force with a single control force in part (a) and 3 primary forces with 3 control forces in part (b);

.... 6 primary forces and 6 control forces;

- - - - - - - - - 12 primary forces and 12 control forces;

--- --- --- 24 primary forces and 24 control forces.

Numerical results obtained using the preceding analysis for a semi-infinite plate indicate that it is extremely difficult to achieve a significant degree of vibratory power transmission control using either a single control force or a row of in-phase control sources (Pan and Hansen, 1995). Although control force locations which result in significant (greater than 15–20 dB) power transmission reductions do exist for any particular frequency, the locations are extremely frequency dependent and for higher frequencies (above the cut-on frequency of the third cross mode) the locations are very small in area. As the frequency dependency of the location can also be viewed as speed of sound (or temperature) sensitivity, it may be concluded that the use of a single control force or line of in-phase control forces to control vibratory power transmission in a semi-infinite plate is impractical. This is demonstrated by the plots in Figs. 10.21(a) and (b), for which the control source location is optimum at 210 Hz. The figures show how the maximum achievable reduction in power transmission varies with frequency for a single primary and control force (Fig. 10.21(a)) and for a row of equal number of primary and control forces (Fig. 10.21(b)). As expected, good control is achieved only over very narrow frequency ranges in both cases, although the row of control sources performs somewhat better.

In contrast to the row of in-phase control sources, a row of three independently driven control sources produces large reductions in power transmission over a wide frequency range as indicated in Fig. 10.22.

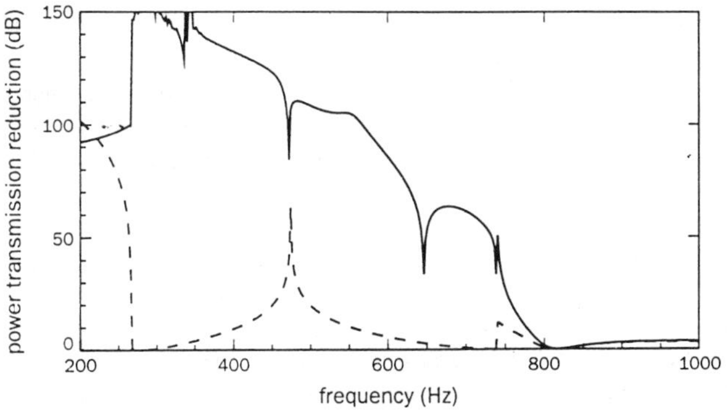

Fig. 10.22 Power flow reduction for three in-phase, uniform-amplitude primary forces and three independent control forces, as a function of frequency with the control sources located at the optimum location for 210 Hz.
·········· 3 primary forces and 3 control forces driven in phase;
——— 3 primary forces driven in phase and 3 control forces driven independently.

As an aside, it is of interest to note that the far field vibration generated by a pair of piezoceramic crystals placed on opposite sides of a thin plate or beam can be simulated by a point force of a specified amplitude (Pan and Hansen, 1994).

REFERENCES

Brennan, M.J., Elliott, S.J. and Pinnington, R.J. (1992). Active control of vibrations transmitted through struts. In *Proceedings of First International MOVIC Conference*. Japan, 605–609.

Clark, R.L. (1994). Adaptive feedforward modal space control. *Journal of the Acoustical Society of America*, **95**, 2989.

Clark, R.L., Pan, J and Hansen, C.H. (1992). An experimental study of the active control of multiple wave types in a beam. *Journal of the Acoustical Society of America*, **92**, 871–876.

Elliott, S.J. (1993). Active control of structure-borne noise. *Proceedings of the Institute of Acoustics*, **15**, 93–119.

Elliott, S.J., Stothers, I.M. and Billet, L. (1990). The use of adaptive filters in the active control of flexural waves in structures. In *Active Noise and Vibration Control.* ASME publication NCA. Vol. 8, 161–166.

Elliott, S.J. and Billet, L. (1993). Adaptive control of flexural waves propagating in a beam. *Journal of Sound and Vibration*, **163**, 295–310.

Fuller, C.R., Gibbs, C.P. and Silcox, R.J. (1990). Simultaneous control of flexural and extensional waves in beams. *Journal of Intelligent Material Systems and Structures*, **1**, 235–247.

Hansen, C.H., Young, A.J. and Pan, X. (1993). Active control of harmonic vibration in beams with arbitrary end conditions. In *Proceedings of the Second Conference on Recent Advances in Active Control of Sound and Vibration*. Technomic Publishing Co., Blacksburg, VA, 487–506.

Mace, R.B. (1987). Active control of flexural vibrations. *Journal of Sound and Vibration*, **114**, 253–270.

McKinnell, R.J. (1988). Active vibration isolation by cancellation of bending waves. *Proceedings of the Institute of Acoustics*, **10**, 581–588.

McKinnell, R.J. (1989). Active vibration isolation by cancelling bending waves. *Proceedings of the Royal Society of London*, **A421**, 357–393.

Pan, J. and Hansen, C.H. (1991). Active control of total vibratory power flow in a beam. 1. Physical system analysis. *Journal of the Acoustical Society of America*, **89**, 200–209.

Pan, X. and Hansen, C.H. (1993a). Effect of error sensor type and location on the active control of beam vibration. *Journal of Sound and Vibration*, **165**, 497–510.

Pan, X. and Hansen, C.H. (1993b). Effect of end conditions on the active control of beam vibration. *Journal of Sound and Vibration*, **168**, 429–448.

Pan, X. and Hansen, C.H. (1994). Piezo-electric crystal vs point force excitation of beams and plates. *Journal of Intelligent Material Systems and Structures*, **5**, 363–370.

Pan, X. and Hansen, C.H. (1995). Active control of vibratory power transmission along a semi-infinite plate. *Journal of Sound and Vibration*, **184**, 585–610.

Redman-White, W., Nelson, P.A. and Curtis, A.R.D. (1987). Experiments on the active control of flexural wave power flow. *Journal of Sound and Vibration*, **112**, 187–191.

Scheuren, J. (1985). Active control of bending waves in beams. In *Proceedings of Internoise '85*. Institute of Noise Control Engineering, 591–595.

Scheuren, J. (1988). Non-reflecting termination for bending waves in beams by active means. In *Proceedings of Internoise '88*. Institute of Noise Control Engineering, 1065–1068.

Scheuren, J. (1989). Iterative design of bandlimited FIR filters with gain constraints for active control of wave propagation. In *Proceedings of the International Conference on Acoustics Speech and Signal Processing*. IEEE, paper A2.3.

Young, A.J. and Hansen, C.H. (1994). Control of flexural vibration in a beam using a piezoceramic actuator and an angle stiffener. *Journal of Intelligent Material Systems and Structures*, **5**, 536–549.

11

Feedback control of flexible structures described in terms of modes

11.1 INTRODUCTION

One of the problems which is encountered in assembling a book such as this one is space; there is too much information for too few pages. As a result, some topics which are relevant to the subject matter of the text must be truncated, or even deleted altogether. While there is an associated risk that the reader will become 'unstable' in his or her application of the technology, trying to apply control techniques to problems which are outside their intended scope, the authors of the book are charged with the task of reducing the risk of this 'technique spillover'. This chapter, perhaps more than others in this book, has been limited in its scope, specifically to consideration of structures which are not too large or flexible.

Active vibration control of flexible structures (using feedback techniques) is a topic which has been vastly researched and published in recent years. Most of this effort has been directed at the problem of vibration attenuation in large flexible structures, specifically large space structures. The dynamic response of such structures is typically characterized by fundamental modal resonance frequencies in the order of single hertz, and a (very) large number of lightly damped, poorly modelled modes in the bandwidth of most ('practically' formulated) controlled systems. Therefore, there is a limit to which the response of such a structure, and hence the controller design for the problem, can be based upon system modes (von Flotow, 1988).

There are several good starting points for the study of large space structure control. Texts edited by Atluri and Amos (1988), and Junkins (1990) are two such points. The series of texts edited by Leondes are also valuable sources of information. Review papers by Balas (1982), Nurre *et al.* (1984) and Hallauer (1990) are also useful.

In this chapter we will consider vibration control of structures described in terms of system modes. This subject description limits our targets to low frequency controllers for 'small' structures and/or 'stiff' structures, the responses of which are largely characterized by well spaced resonances in the controller bandwidth. These response characteristics are similar to the response characteristics of the dynamic systems we have considered thus far in this book, such as the low frequency response of coupled structural/acoustic enclosures.

In the following sections we will consider the development of feedback control system based upon a knowledge of the modes of the target structure. We will begin by developing feedback laws for this 'modal control', and then go on to consider issues such as stability, model reduction, and sensor and actuator placement. Much of the work outlined in this chapter follows on directly from the discussion of Chapter 5, the review of modern control theory, which should be referred to for more fundamental discussion.

11.2 MODAL CONTROL

We will begin this chapter with the development of an analytical framework useful for examining various aspects of feedback control of flexible structures, one based upon system modes. This mathematical modelling approach has been used for many years, with early references found in, for example, Balas (1978a, 1978b, 1982), Meirovitch and Baruh (1982) and Meirovitch and Oz (1980a,b,c). Early related works on modal control include Simon and Mitter (1968), and Porter and Crossley (1972). Once developed, we will use the model to develop suitable control laws, and examine some aspects of 'modal control'.

As we mentioned in the introduction, there is a practical limit to any control approach formulated on the basis of modal analysis. This limit is approached as the maximum frequency in the controller bandwidth increases. Increasing frequency results in increasing modal density, including higher order modes in the controller bandwidth. These higher order modes are most susceptible to model errors, especially when finite element models are used, and the increasing number of poorly known modes is a potential disaster for the control system designer. Analysis and feedback control techniques based upon wave propagation have been put forward (for example, MacMartin and Hall, 1991; Miller and von Flotow, 1989; Miller *et al.*, 1990; von Flotow, 1986; von Flotow and Schafer, 1986), but are outside the scope of this text (however, wave propagation analysis was used in Chapter 10 for analysis of feedforward control systems for structures, and the interested reader is referred back to that chapter).

In formulating the control laws in this section, there are several practical constraints which we will impose upon ourselves (Athans, 1970): (1) we have a finite (small) number of sensors and actuators; (2) our controller must be finite (small) dimensional; and (3) there is some degree (usually small) of damping in the system. These constraints will influence our design, and implicitly the first two constraints will be responsible for a few problems (spillover), while the last has the potential to enhance the stability of the controller.

11.2.1 Development of governing equations

The systems which are of interest to us in this chapter, mechanically flexible structures, are distributed parameter systems, where the system parameters are spatially dependent. Such systems are, in theory, infinite dimensional, which means that in practice for a model to have good fidelity is must be large. As we will see later in this chapter, some of the fundamental problems associated with the active control of flexible structures arise from the use of a (relatively) small-dimension controller to control a large-dimension system.

The response of the distributed parameter systems of interest here can be described by the partial differential equation

$$m(\boldsymbol{r})\frac{\partial^2 w(\boldsymbol{r},t)}{\partial t^2} + d\frac{\partial w(\boldsymbol{r},t)}{\partial t} + kw(\boldsymbol{r},t) = f(\boldsymbol{r},t) \tag{11.2.1}$$

where $w(\boldsymbol{r},t)$ is the displacement of the structure at some location $\boldsymbol{r}$ at time t, $m(\boldsymbol{r})$ is the mass density at $\boldsymbol{r}$, d and k are time-invariant, non-negative differential operators describing the damping and stiffness of the structure respectively, and $f(\boldsymbol{r},t)$ is the applied force at location $\boldsymbol{r}$ at time t. The damping of a flexible structure, especially one which is large, is often small and not well known, so it is common to assume that $d = 0$. When the effects of damping are to be included in the model, d will be represented here by proportional damping, a class of viscous damping where the damping operator is a linear combination of the stiffness operator and mass distribution (as discussed in Chapter 4),

$$d = \alpha_1 m + \alpha_2 k \tag{11.2.2}$$

For our discussion here, it will be assumed that

$$d = 2\zeta\sqrt{k/m} \tag{11.2.3}$$

where ζ is a small (positive) damping coefficient.

For the analytical development being undertaken in this section, the control sources will be modelled as point forces, so that the applied force distribution can be written as the sum of a disturbance input and N_s control inputs u,

$$f(\boldsymbol{r},t) = \sum_{i=1}^{N_s} b_i(\boldsymbol{r})u_i(t) + \text{disturbances} \tag{11.2.4}$$

where $b_i(\boldsymbol{r})$ is an influence function value at $\boldsymbol{r}$. For point force excitation, the influence function is commonly taken to be a Dirac delta function.

The response of the structure to the applied force distribution is measured by a set of linearly independent sensors, the output of each of which is defined by the expression

$$y(t) = \int_S \left[c(\boldsymbol{r})w(\boldsymbol{r},t) + e(\boldsymbol{r})\frac{\partial w(\boldsymbol{r},t)}{\partial t} \right] \mathrm{d}x \tag{11.2.5}$$

where c is an influence function defining the measurement of the structure displacement, e is an influence function defining the velocity measurement, and integration is over the surface of the structure S. If the measurement is from a point displacement sensor, where the influence function $c(\mathbf{r})$ is defined by the product of a gain and a Dirac delta function, the output from the sensor is defined by the expression

$$y(t) = w(\mathbf{r},t)c \tag{11.2.6}$$

where c is the gain of the displacement sensing system. Similarly, if the measurement is from a point velocity sensor, where the influence function $e(\mathbf{r})$ is also defined by the product of a gain and a Dirac delta function, the output from the sensor is defined by the expression

$$y(t) = \dot{w}(\mathbf{r},t)e \tag{11.2.7}$$

where e is the gain of the velocity sensing system.

The domain of the stiffness operator k is assumed to contain all smooth functions satisfying the boundary conditions of the structure, and have the usual inner product (*,*) and norm $|| \ ||$. For the structures of interest to us, k is assumed to have a discrete spectrum, with isolated resonance frequencies defined by the eigenvalue problem associated with the governing partial differential (11.2.1)

$$k\psi_i = \lambda_i m\psi_i \tag{11.2.8}$$

where λ_i is the ith eigenvalue, and ψ_i is the associated eigenvector. The resonance frequencies ω are related to the eigenvalues of (11.2.8) by

$$\omega_i = \lambda_i^{1/2} \tag{11.2.9}$$

while the mode shape functions are defined by the eigenvectors. It is usual to assume that the stiffness operator is self-adjoint, such that all eigenvectors (mode shape functions) are orthogonal. In this chapter it will also be assumed that the mode shape functions are mass normalized, to be consistant will literature on this topic. This means that the mass and stiffness terms can be normalized as

$$\int_S m(\mathbf{r})\psi_i(\mathbf{r})\psi_\iota(\mathbf{r})\ \mathrm{d}S = \delta_{i,\iota} \tag{11.2.10}$$

and

$$\int_S \psi_i(\mathbf{r})k\psi_\iota(\mathbf{r})\ \mathrm{d}S = \lambda_i\delta_{i,\iota} = \omega_i^2\delta_{i,\iota} \tag{11.2.11}$$

where $\delta_{i,\iota}$ is the Kronecker delta function, defined by

$$\delta_{i,\iota} = \begin{cases} 0 & i \neq \iota \\ 1 & i = \iota \end{cases} \tag{11.2.12}$$

It is also usual to assume that the stiffness operator is positive definite, such that all resonance frequencies are greater than zero. For our discussion in this chapter, it will be assumed that the resonance frequencies (and hence eigenvalues) are ordered $0 \leq \omega_1 \leq \omega_2 \leq \ldots$

Having defined several properties of the systems of interest to us, the response of which are governed by the partial differential (11.2.1), it is straightforward to re-express (11.2.1) as a set of modal equations, where each equation defines the response of a particular structural mode. The structure displacement amplitude at time t and location r, $w(r,t)$, can be expressed as an infinite sum of modal contributions,

$$w(r,t) = \sum_{i=1}^{\infty} z_i(t)\psi_i(r) \tag{11.2.13}$$

where $\phi_i(\mathbf{r})$ is the value of the ith mode shape function at location $\mathbf{r}$, and $z_i(t)$ is the modal amplitude at time t. By substituting this expansion into (11.2.1), multiplying both sides by ϕ_i, and integrating over the surface of the structure S, the governing equation of motion can be re-expressed as an infinite sum of modal contributions,

$$\sum_{i=1}^{\infty} \left[\ddot{z}_i(t) + 2\zeta\omega_i \dot{z}_i(t) + \omega_i^2 z_i(t) = f_i(t) \right] \tag{11.2.14}$$

where $f_i(t)$ is a modal generalized force,

$$f_i(t) = \int_S f(r,t)\psi_i(r)\ \mathrm{d}r \tag{11.2.15}$$

Observe that each term in the infinite summation has the same form as the equation of motion for a single-degree-of-freedom system discussed in Chapter 4. Thus individual modes may be considered as behaving as single degree of freedom systems characterized by a resonance frequency and damping coefficient. Also, as the control inputs are assumed to be coming from point sources, the control source contribution to the total modal generalized force can be written as:

$$f_{c,i}(t) = \sum_{\iota=1}^{N_s} u_\iota(t)\psi_i(r_\iota) \tag{11.2.16}$$

where r_ι is the location at which force u_ι is applied.

The system outputs are now also defined in terms of structural modes. With the assumption that a measurement is done by point displacement sensor, the output of the ith mode is defined by

$$y_i(t) = z_i(t)\psi_i(r)c \tag{11.2.17}$$

Similarly, with the assumption that a measurement is done by point velocity sensor, the output of the ith mode is defined by

$$y_i(t) = \dot{z}_i(t)\psi_i(r)e \tag{11.2.18}$$

11.2.2. Discrete element model development

For large structures, it is common to use finite element methods (FEM), or some other form of discretization method, to model the response. A typical approximation of the governing partial differential (11.2.1) can be obtained from the relationship

$$w(\boldsymbol{r},t) \approx \sum_{i=1}^{N_\theta} q_i(t)\theta_i(\boldsymbol{r}) \tag{11.2.19}$$

where $\theta_i(\boldsymbol{r})$ is a (known) local approximation of structure, and $q_i(t)$ is the amplitude of the ith discrete element (note that if finite elements were being used, then the value of the local approximation of the structure at $\boldsymbol{r}$ would be zero for all elements bar the element actually located at $\boldsymbol{r}$). Using (11.2.19), a well known approximation of (11.2.1) is (Balas, 1990):

$$\boldsymbol{M}_0\ddot{\boldsymbol{q}} + \boldsymbol{D}_0\dot{\boldsymbol{q}} + \boldsymbol{K}_0\boldsymbol{q} = \boldsymbol{f}_0 \tag{11.2.20}$$

where $\boldsymbol{q}$ is the vector of element displacements,

$$\boldsymbol{q} = \begin{bmatrix} q_1 & q_2 & \cdots & q_{N_\theta} \end{bmatrix}^{\mathrm{T}} \tag{11.2.21}$$

$\boldsymbol{M}_0$ is a mass matrix, the i,ι element of which is defined by the relationship

$$\boldsymbol{M}_0(i,\iota) = \int_S m(\boldsymbol{r})\theta_i(\boldsymbol{r})\theta_\iota(\boldsymbol{r})\,\mathrm{d}\boldsymbol{r} \tag{11.2.22}$$

where $m(\boldsymbol{r})$ is the mass per unit area at $\boldsymbol{r}$, D_0 is a damping matrix, the i,ι element of which is defined by the relationship

$$\boldsymbol{D}_0(i,\iota) = \int_S \theta_i(\boldsymbol{r})c_d\theta_\iota(\boldsymbol{r})\,\mathrm{d}\boldsymbol{r} \tag{11.2.23}$$

and $\boldsymbol{K}_0$ is a stiffness matrix, the i,ι element of which is defined by the relationship

$$\boldsymbol{K}_0(i,\iota) = \int_S \theta_i(\boldsymbol{r})k_0\theta_\iota(\boldsymbol{r})\,\mathrm{d}\boldsymbol{r} \tag{11.2.24}$$

The control force input is defined by

$$\boldsymbol{f}_0 = \boldsymbol{B}_0\boldsymbol{u} \tag{11.2.25}$$

where the i,ι element of the control input matrix $\boldsymbol{B}_0$ is defined by the expression

$$\boldsymbol{B}_0(i,\iota) = \int_S \theta_i(\boldsymbol{r})b_\iota(\boldsymbol{r})\,\mathrm{d}\boldsymbol{r} \tag{11.2.26}$$

where b_ι is the influence function for the ιth control source.

The system output associated with (11.2.20), which is the finite element approximation of (11.2.5), is given by the expression

$$\boldsymbol{y} = \boldsymbol{C}_0\boldsymbol{q} + \boldsymbol{E}_0\dot{\boldsymbol{q}} \tag{11.2.27}$$

where C_0 and E_0 are displacement and velocity output matrices, defined by the relationships

$$C_0(i,\iota) = \int_S \theta_i(r)c_\iota(r)\ \mathrm{d}r \tag{11.2.28}$$

$$E_0(i,\iota) = \int_S \theta_i(r)e_\iota(r)\ \mathrm{d}r \tag{11.2.29}$$

This system model is sometimes referred to as a configuration space model, with the displacement vector referred to as the configuration vector. For complicated structures, it is more likely that the system model will be formulated in this lumped' parameter form than in the (exact) distributed parameter form of (11.2.1).

The finite element model of system response can also be restated in terms of system modes, using the generalized eigenvalue problem associated with (11.2.20):

$$K_0\psi_i = \lambda_i M_0\psi_i \tag{11.2.30}$$

where λ_i is the ith eigenvalue, and ψ_i is the associated eigenvector. Transformation from element coordinates to modal coordinates z is accomplished using the transformation matrix Ψ,

$$q = \Psi z \tag{11.2.31}$$

where the columns of Ψ are the eigenvectors defined by equation (11.2.30),

$$\Psi = \begin{bmatrix}\psi_1 & \psi_2 & \cdots & \psi_{N_\theta}\end{bmatrix} \tag{11.2.32}$$

and N_θ is the number of modes considered in the model. Observe that, as both the mass matrix M_0 and stiffness matrix K_0 are non-negative and symmetric, the transforms

$$\Psi^{\mathrm{T}}M_0\Psi = I \tag{11.2.33}$$

and

$$\Psi^{\mathrm{T}}K_0\Psi = \Lambda = \mathrm{diag}[\lambda_1\ \lambda_2\ \cdots\ \lambda_{N_\theta}] \tag{11.2.34}$$

exist, and so (11.2.20) can be transformed into 'approximate' (discrete models of continuous) modal coordinates as

$$\ddot{z} + 2\xi\omega\dot{z} + \omega^2 z = \tilde{B}_0 u \tag{11.2.35}$$

where

$$\tilde{B}_0 = \Psi^{\mathrm{T}}B_0 \tag{11.2.36}$$

The output equation is similarly transformed to

$$y = \tilde{C}_0 z + \tilde{E}_0\dot{z} \tag{11.2.37}$$

where

$$\tilde{C}_0 = C_0\Psi,\ \tilde{E}_0 = E_0\Psi \tag{11.2.38}$$

It should be noted that the accuracy of the mode shape functions derived from a finite element model decreases as the mode number increases. In general, the number of useable modes from the calculation is an order of magnitude less than the number of finite elements used in the calculation.

11.2.3. Transformation into state-space form

It is straightforward to re-express the modal equations of motion in a state-space format. To do so, let us define the state vector

$$x_i = \begin{bmatrix} z_i \\ \dot{z}_i \end{bmatrix} \tag{11.2.39}$$

Assuming point forces are used for control, and ignoring the (primary source) disturbance inputs in this development (which, as discussed in Chapter 5, is commonly done in development of state space models), the response of a single mode in (11.2.14) can be written as

$$\dot{x}_i(t) = A_i x_i(t) + B_i u(t) \tag{11.2.40}$$

where

$$A_i = \begin{bmatrix} 0 & 1 \\ -\omega_i^2 & -2\zeta\omega_i \end{bmatrix},\quad B_i = \begin{bmatrix} 0 & \cdots & 0 \\ \psi_i(r_1) & \cdots & \psi_i(r_{N_c}) \end{bmatrix} \tag{11.2.41}$$

and N_c is the number of control inputs, situated at locations $r_1,\ldots,r_{N}c$. The output of this mode, as measured by point displacement and velocity sensors, is

$$y = Cx \tag{11.2.42}$$

where y is the vector of measured outputs, and

$$C = \begin{bmatrix} \psi_1(r_1) & 0 \\ \vdots & \\ \psi_1(r_{N_d}) & 0 \\ 0 & \psi_1(r_1) \\ \vdots & \\ 0 & \psi_1(r_{N_v}) \end{bmatrix} \tag{11.2.43}$$

where there are N_d displacement sensors and N_v velocity sensors being used.

Equation (11.2.40) describes the response of a single mode of vibration. This can be expanded to extend consideration to the set of N_m modelled modes as

$$\dot{x} = Ax + Bu \tag{11.2.44}$$

where the state vector is now defined as

$$x = \begin{bmatrix} z_1 & z_2 & \cdots & z_{N_m} & \dot{z}_1 & \dot{z}_2 & \cdots & \dot{z}_{N_m} \end{bmatrix}^{\mathrm{T}} \tag{11.2.45}$$

and

$$A = \begin{bmatrix} 0 & I \\ -\Lambda & -2\zeta\Lambda^{1/2} \end{bmatrix}, \quad B = \begin{bmatrix} 0 \\ B^* \end{bmatrix} = \begin{bmatrix} 0 & \cdots & 0 \\ & \vdots & \\ 0 & \cdots & 0 \\ \psi_1(r_1) & \cdots & \psi_1(r_{N_c}) \\ & \vdots & \\ \psi_{N_m}(r_1) & \cdots & \psi_{N_m}(r_{N_c}) \end{bmatrix} \tag{11.2.46}$$

In (11.2.46), I is an $(N_m \times N_m)$ identity matrix, Λ is the $(N_m \times N_m)$ diagonal matrix of system eigenvalues, and B^* is the lower, non-zero part of B. Also, the damping coefficient is shown as the same for all modes, but this need not be the case. The output equation is now

$$y = Cx \tag{11.2.47}$$

where, if it is again assumed that displacement and velocity sensors are being used,

$$C = \begin{bmatrix} C_d & 0 \\ 0 & C_v \end{bmatrix} = \begin{bmatrix} \psi_1(r_1) & \cdots & \psi_{N_m}(r_1) & 0 & \cdots & 0 \\ \vdots & & & & & \\ \psi_1(r_{N_d}) & \cdots & \psi_{N_m}(r_{N_d}) & 0 & \cdots & 0 \\ 0 & \cdots & 0 & \psi_1(r_1) & \cdots & \psi_{N_m}(r_1) \\ \vdots & & & & & \\ 0 & \cdots & 0 & \psi_1(r_{N_v}) & \cdots & \psi_{N_m}(r_{N_v}) \end{bmatrix} \tag{11.2.48}$$

One final point to note is that in the absence of damping the eigenvalues of the system are purely real; any complex component to the eigenvalues arises from the damping. From our discussion in Chapter 5, where we saw that the response of a system mode is described by an exponential with a time constant proportional to the associated eigenvalue, we can conclude that the reverberation time of many of these systems with which we are interested is extremely long. Also, the stability margins of the systems are small. The former provides an impetus for applying active control to flexible structures, the latter dictates that care in system design is required.

11.2.4. Model reduction

The number of modes required to model the response of a structure with good fidelity may be quite large, and so it is generally impractical to consider all modelled modes in the development of a control strategy. Therefore, we need to formulate a reduced order model (ROM) of the system to derive a control strategy. A discussion of some of the ways to derive the 'best' reduced order model for a given system will be undertaken later in this chapter. In this section we will simply assume that some criteria have been applied to the set of modelled modes to facilitate partitioning into controlled (modelled in controller design) and residual or uncontrolled (unmodelled in controller design) mode subsets, and then proceed to formulate the control strategy solely from consideration of the controlled modes.

If we divide the set of modelled modes into N controlled modes and R uncontrolled modes, and define the state vector containing the displacements and velocities of the controlled modes as $\boldsymbol{x}_N$, and the state vector containing the displacements and velocities of the uncontrolled modes as $\boldsymbol{x}_R$, we have the following relationships:

$$\dot{\boldsymbol{x}}_N = \boldsymbol{A}_N \boldsymbol{x}_N + \boldsymbol{B}_N \boldsymbol{u} \tag{11.2.49}$$

and

$$\dot{\boldsymbol{x}}_R = \boldsymbol{A}_R \boldsymbol{x}_R + \boldsymbol{B}_R \boldsymbol{u} \tag{11.2.50}$$

where

$$\boldsymbol{A}_N = \begin{bmatrix} \boldsymbol{0} & \boldsymbol{I}_N \\ -\boldsymbol{\Lambda}_N & -2\zeta\boldsymbol{\Lambda}_N^{1/2} \end{bmatrix}, \quad \boldsymbol{B}_N = \begin{bmatrix} \boldsymbol{0} \\ \boldsymbol{B}_N^* \end{bmatrix} = \begin{bmatrix} 0 & \cdots & 0 \\ & \vdots & \\ 0 & \cdots & 0 \\ \psi_1(r_1) & \cdots & \psi_1(r_{N_c}) \\ & \vdots & \\ \psi_N(r_1) & \cdots & \psi_N(r_{N_c}) \end{bmatrix} \tag{11.2.51}$$

and

$$\boldsymbol{A}_R = \begin{bmatrix} \boldsymbol{0} & \boldsymbol{I}_R \\ -\boldsymbol{\Lambda}_R & -2\zeta\boldsymbol{\Lambda}_R^{1/2} \end{bmatrix}, \quad \boldsymbol{B}_R = \begin{bmatrix} \boldsymbol{0} \\ \boldsymbol{B}_R^* \end{bmatrix} = \begin{bmatrix} 0 & \cdots & 0 \\ & \vdots & \\ 0 & \cdots & 0 \\ \psi_1(r_1) & \cdots & \psi_1(r_{N_c}) \\ & \vdots & \\ \psi_R(r_1) & \cdots & \psi_R(r_{N_c}) \end{bmatrix} \tag{11.2.52}$$

Similarly, the system output is partitioned as

$$y = y_N + y_R = C_N x_N + C_R x_R \tag{11.2.53}$$

where

$$C_N = \begin{bmatrix} C_{Nd} & 0 \\ 0 & C_{Nv} \end{bmatrix} = \begin{bmatrix} \phi_1(r_1) & \cdots & \phi_N(r_1) & 0 & \cdots & 0 \\ \vdots & & & & & \\ \psi_1(r_{N_d}) & \cdots & \psi_N(r_{N_d}) & 0 & \cdots & 0 \\ 0 & \cdots & 0 & \psi_1(r_1) & \cdots & \psi_N(r_1) \\ \vdots & & & & & \\ 0 & \cdots & 0 & \psi_1(r_{N_v}) & \cdots & \psi_N(r_{N_v}) \end{bmatrix} \tag{11.2.54}$$

and

$$C_R = \begin{bmatrix} C_{Rd} & 0 \\ 0 & C_{Rv} \end{bmatrix} = \begin{bmatrix} \psi_1(r_1) & \cdots & \psi_R(r_1) & 0 & \cdots & 0 \\ \vdots & & & & & \\ \psi_1(r_{N_d}) & \cdots & \psi_R(r_{N_d}) & 0 & \cdots & 0 \\ 0 & \cdots & 0 & \psi_1(r_1) & \cdots & \psi_R(r_1) \\ \vdots & & & & & \\ 0 & \cdots & 0 & \psi_1(r_{N_v}) & \cdots & \psi_R(r_{N_v}) \end{bmatrix} \tag{11.2.55}$$

11.2.5 Modal control

Having developed a state space model of the flexible structures of interest to us, we can begin to develop feedback control laws to change the dynamic response of the systems in some desired way. In this section we will develop control laws based upon the reduced order model, and in the process look at the requirements for controllability and observability. In the section which follows we will examine the effect of considering only part of the structure modes in development of the control law.

Design of a feedback control system for flexible structures using state space methods follows the methodology reviewed in Chapter 5. Here the active control system consists of two parts: a state observer or estimator, which estimates the values of the system states at any time t, and a control law, or regulator, which filters the state estimates to produce control input signals. For our discussion here, the control law will consist of static gains, so that the control inputs are derived from linear combinations of the system states.

If the output signal to noise ratio is high, the state observer can be deterministic, as discussed in Section 5.9. However, if the signal to noise ratio is not high the state observer will need to be a Kalman filter, discussed in Section 5.11. Either way, the state observer has the form

$$\frac{d\hat{x}_N}{dt} = A_N \hat{x}_N + B_N u + L_N(y - C\hat{x}_N) \tag{11.2.56}$$

where ˆ denotes an estimated (or observed) quantity, N denotes quantities associated with the controlled modes in the reduced order model, and the time dependence of the state, control and output vectors has been dropped for clarity of notation.

The estimation error $\boldsymbol{e}_N$, defined as the difference between the actual state value and estimated state value,

$$\boldsymbol{e}_N = \hat{\boldsymbol{x}}_N - \boldsymbol{x}_N \tag{11.2.57}$$

is defined by the relationship

$$\dot{\boldsymbol{e}}_N = (\boldsymbol{A}_N - \boldsymbol{L}_N\boldsymbol{C}_N)\boldsymbol{e}_N + \boldsymbol{L}_N\boldsymbol{C}_R\boldsymbol{x}_R \tag{11.2.58}$$

If we ignore the residual system contribution, this becomes

$$\dot{\boldsymbol{e}}_N = (\boldsymbol{A}_N - \boldsymbol{L}_N\boldsymbol{C}_N)\boldsymbol{e}_N \tag{11.2.59}$$

The result of (11.2.59) is applicable to systems employing a deterministic observer; for those employing a Kalman filter or optimal observer, there is an additional white noise term, as outlined in Section 5.11. Thus the equations governing the observer for the reduced order system are the same as those governing a fully modelled system.

The control law of interest to us is (negative) linear state variable feedback, defined by the relationship

$$\boldsymbol{u} = -\boldsymbol{K}_N\boldsymbol{x}_N \tag{11.2.60}$$

As we discussed in Chapter 5, if the system is controllable and deterministic then the control gains can be calculated using a pole placement approach. If the system is controllable, the eigenvalues can (ignoring practicalities) be arbitrarily placed, giving the system any desired damping and stiffness characteristics.

In practice, the control law actually implemented uses estimated state values rather than the actual ones,

$$\boldsymbol{u} = -\boldsymbol{K}_N\hat{\boldsymbol{x}}_n \tag{11.2.61}$$

To quantify the effect which this has upon the system, the state equation can be augmented with the estimator error (11.2.59)

$$z_N = \begin{bmatrix} \boldsymbol{x}_N \\ \boldsymbol{e}_N \end{bmatrix} \tag{11.2.62}$$

it is straightforward to show, using the same series of steps taken in Section 5.12, that the overall response of the controller (which uses only the reduced-model modes) is governed by the expression

$$\dot{z}_N = \begin{bmatrix} \boldsymbol{A}_N - \boldsymbol{B}_N\boldsymbol{K}_N & \boldsymbol{B}_N\boldsymbol{K}_N \\ \boldsymbol{0} & \boldsymbol{A}_N - \boldsymbol{L}_N\boldsymbol{C}_N \end{bmatrix} z_N \tag{11.2.63}$$

As we discussed in Chapter 5, the form of (11.2.63) shows that the eigenvalues of the system are the combination of those of the controlled system with full (of the reduced order model) state feedback and those of the observer; the use of an observer does not change the location of the poles of the controlled system, but rather adds its own poles. This is the deterministic separation principal. A similar result exists for systems implementing an optimal (Kalman filter) observer, known as the stochastic separation principal.

The control gains could also be calculated using a steady state optimal control approach, where the control law is chosen to minimize the performance index

$$J_N = \int_0^\infty \left[\boldsymbol{x}_N^{\mathrm{T}}(t)\boldsymbol{Q}_n\boldsymbol{x}_N(t) + \boldsymbol{u}^{\mathrm{T}}(t)\boldsymbol{R}_N\boldsymbol{u}(t) \right] \mathrm{d}t \tag{11.2.64}$$

As the total energy in the controlled modes is given by the sum of the kinetic and potential energies of the (normalized) system:

$$E_N(t) = \frac{1}{2}\sum_{i=1}^{N} \left[\dot{z}_i^2 + \omega_i^2 z_i^2(t) \right] \tag{11.2.65}$$

where N is the number of controlled modes, a common choice for the state weighting matrix is

$$\boldsymbol{Q}_N = \frac{1}{2}\begin{bmatrix} \boldsymbol{\Lambda}_N & \boldsymbol{0} \\ \boldsymbol{0} & \boldsymbol{I}_N \end{bmatrix} \tag{11.2.66}$$

If the optimal control law is implemented with the estimated state, the controller is, in general, suboptimal. The increase in the performance index which results is (Bongiorno and Youla, 1968)

$$\Delta J = \int_0^\infty \boldsymbol{e}_N^{\mathrm{T}}(t)\boldsymbol{K}_N^{\mathrm{T}}\boldsymbol{R}_N\boldsymbol{K}_N\boldsymbol{e}_N(t)\ \mathrm{d}t \tag{11.2.67}$$

As we discussed in Section 5.8, deriving the optimal control gains normally involves solving a matrix Ricatti equation which is the same order as the state equation, which for the systems of interest to us is ($2N \times 2N$). This order can be reduced by taking advantage of the form of the state equations, as we will see shortly.

When the system response is described in terms of 'modal' state equations, it is straightforward to assess controllability and observability. From our discussion in Chapter 5, for an LTI system to be controllable, the controllability matrix for the pair ($\boldsymbol{A}_N$, $\boldsymbol{B}_N$), defined as

$$\left[\boldsymbol{B}_N \ \ \boldsymbol{A}_N\boldsymbol{B}_N \ \cdots \ \boldsymbol{A}_N^{2N-1}\boldsymbol{B}_N \right] \tag{11.2.68}$$

must be full rank, which in this case is rank $2N$. Referring to the definition of $\boldsymbol{A}_N$ given in (11.2.51), and ignoring the damping term, this is equivalent to the controllability matrix for the pair ($\boldsymbol{\Lambda}_N$, $\boldsymbol{B}_N$), defined as

$$\left[\boldsymbol{B}_N^* \ \ \boldsymbol{\Lambda}_N\boldsymbol{B}_N^* \ \cdots \ \boldsymbol{\Lambda}_N^{N-1}\boldsymbol{B}_N^* \right] \tag{11.2.69}$$

being rank N (Balas, 1978a,b). As $\mathbf{\Lambda}$ is diagonal, this means that each row of $\mathbf{B}_N^*$ must have a non-zero entry. In other words, for the system to be controllable, at least one control actuator must be at a non-zero location of each mode shape function. This criterion is a useful one for guiding control source placement, which we will discuss further later in this chapter.

Similar to the criterion for controllability, for our system to be observable, the observability matrix for the pair $(\mathbf{A}_N, \mathbf{C}_N)$, defined as

$$\begin{bmatrix} \mathbf{C}_N \\ \mathbf{C}_N\mathbf{A}_N \\ \vdots \\ \mathbf{C}_N\mathbf{A}_N^{\;2N-1} \end{bmatrix} \tag{11.2.70}$$

must be full rank, which is $2N$. When only displacement sensors are used, from the definition of $\mathbf{A}_N$ given in (11.2.51), this means that the observability matrix for the pair $(\mathbf{\Lambda}_N, \mathbf{C}_{Nd})$, defined as

$$\begin{bmatrix} \mathbf{C}_{Nd} \\ \mathbf{C}_{Nd}\mathbf{\Lambda}_N \\ \vdots \\ \mathbf{C}_{Nd}\mathbf{\Lambda}_N^{\;N-1} \end{bmatrix} \tag{11.2.71}$$

must be of rank N (Balas, 1978a,b). As $\mathbf{\Lambda}$ is diagonal, this means that each column of $\mathbf{C}_{Nd}$ must have a non-zero entry. In other words, for the system to be observable, at least one displacement sensor must be at a non-zero location of each mode shape function. This criterion is a useful one for guiding sensor placement, which we will discuss further later in this chapter.

When velocity sensors are used, the observability result differs slightly from that obtained with displacement sensors. Here the requirement is that the observability matrix for the pair $(\mathbf{\Lambda}_N, \mathbf{C}_{Nv})$, defined as

$$\begin{bmatrix} \mathbf{C}_{Nv} \\ \mathbf{C}_{Nv}\mathbf{\Lambda}_N \\ \vdots \\ \mathbf{C}_{Nv}\mathbf{\Lambda}_N^{\;N-1} \end{bmatrix} \tag{11.2.72}$$

be of rank N, and also that the matrix $\mathbf{\Lambda}_N$ be of rank N (Balas, 1978a,b). This follows from the fact that, with velocity sensors, for $(\mathbf{A}_N, \mathbf{C}_N)$ to be observable then both $(\mathbf{\Lambda}_N, \mathbf{C}_{Nv})$ and $(\mathbf{\Lambda}_N, \mathbf{C}_{Nv}\mathbf{\Lambda}_N)$ must be rank N. The second criterion is equivalent to $\mathbf{\Lambda}_N$ being of rank N when $(\mathbf{\Lambda}_N, \mathbf{C}_{Nv})$ is observable. The requirement that $\mathbf{\Lambda}_N$ be of rank N means that if there are rigid body modes (zero eigenvalues) a system which has only velocity sensors will not be observable; displacement sensors are (additionally or solely) required.

The above controllability and observability results hold for systems which have non-repeated eigenvalues. Therefore, if each eigenvalue of the modelled system has unit multiplicity, the system can be made controllable and observable with a single control actuator and sensor, provided they are located away from a zero-value location of any of the mode shape functions included in the modelled system. If, however, there are repeated eigenvalues (as may be the case when rigid body modes are present) then the diagonal block of $\boldsymbol{\Lambda}_N$ associated with the repeated eigenvalue, and the associated blocks of $\boldsymbol{B}_N$ and $\boldsymbol{C}_N$ must form an observable and controllable subsystem. This will only be true if the rank of the associated blocks in $\boldsymbol{B}_N$ and $\boldsymbol{C}_N$ is equal to the eigenvalue multiplicity (Chen, 1970, p. 191). Therefore, a single (point) actuator and sensor cannot produce an observable and controllable system when there are repeated eigenvalues. In fact, the number of sensors required is equal to, or greater than, the eigenvalue multiplicity (Balas, 1978a,b).

11.2.6. Spillover

Thus far, we have considered only the controlled part of the system, ignoring the influence which the residual (uncontrolled) part of the system has upon controller performance. This would be acceptable if both the residual input matrix $\boldsymbol{B}_R$ and residual output matrix $\boldsymbol{C}_R$ were zero. However, in general they are not. Referring to Fig. 11.1, the residual modes are usually excited to some degree by the active control system, and the sensor signal usually includes some measure of the response of the residual (uncontrolled) modes. These two effects, referred to as control spillover and observation spillover, respectively, can have a detrimental effect upon system performance.

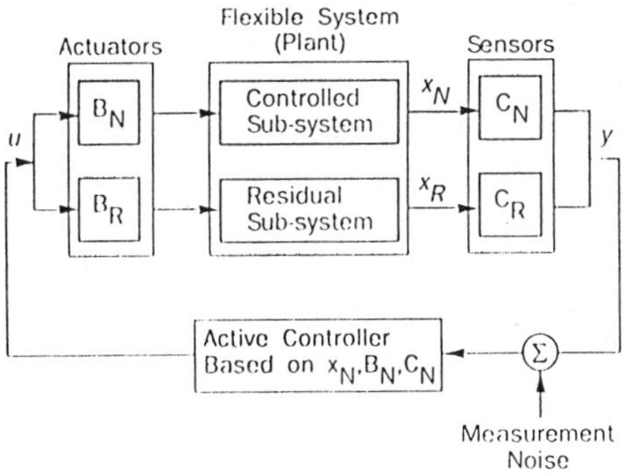

Fig. 11.1 Block diagram of active control system, illustrating where control and observation spillover arise (after Balas, 1978a, 1978b).

To examine this, let us define the augmented state vector χ,

$$\chi = \begin{bmatrix} z_N \\ x_R \end{bmatrix} \tag{11.2.73}$$

where z_N is as defined in Equation (11.2.62). Using (11.2.49), (11.2.50), (11.2.58) and (11.2.61), it is straightforward to show that the closed loop system equations satisfy the relationship

$$\dot{\chi} = \begin{bmatrix} H_{11} & H_{12} \\ H_{21} & H_{22} \end{bmatrix} \chi \tag{11.2.74}$$

where

$$H_{11} = \begin{bmatrix} A_N - B_N K_N & -B_N K_N \\ 0 & A_N - L_N C_N \end{bmatrix} \tag{11.2.75}$$

$$H_{12} = \begin{bmatrix} 0 \\ L_N C_R \end{bmatrix} \tag{11.2.76}$$

$$H_{21} = \begin{bmatrix} -B_R K_N & -B_R K_N \end{bmatrix} \tag{11.2.77}$$

and $H_{22} = A_R$.

If there is no observation spillover (in other words, the sensors measure only the controlled system), then $C_R = 0$ and the system response is identical to that considered in our previous discussion where the effects of the residual (uncontrolled) system are neglected. In this case the poles of the closed loop system are those of the modelled system under full state feedback control $A_N - B_N K_N$, the modelled system observer $A_N - L_N C_N$, and those of the residual system A_R.

If we now examine the response of the residual system when there is control spillover but not observation spillover, we find that it is defined by the expression

$$\dot{x}_R = A_R x_R + B_R K_N [x_N + e_N] \tag{11.2.78}$$

It is because the control input is purely 'external', not a function of the residual system state, that the poles of the residual system are unchanged. The conclusion we can draw is that when observation spillover is not present, while control spillover may degrade the overall response characteristics of the controlled system, it cannot destabilize the system (Balas, 1978a,b).

Let us consider now what happens when observation spillover is present. To do this, we can first divide the state matrix H as

$$H = H_c + H_0 \tag{11.2.79}$$

where

$$H_c = \begin{bmatrix} H_{11} & 0 \\ H_{21} & H_{22} \end{bmatrix}, \quad H_0 = \begin{bmatrix} 0 & H_{12} \\ 0 & 0 \end{bmatrix} \tag{11.2.80}$$

Observe that $H = H_c$ in the absence of observation spillover (control spillover only), and that H_0 describes the observation spillover. From this we can conclude that the poles of H are (continuous) functions of the (real) parameters of H_0, and hence C_R, and hence the degree of observation spillover. The conclusion which can be drawn then is that observation spillover can destabilise the controlled system (Balas, 1978a,b). Usually we can expect the poles of the residual system to be most susceptible to this effect, especially when the damping in the system is low (or negligible), because the controller and observer will have some degree of stability margin (recall that we have shown that the controlled system poles are the combination of the poles of the controller, observer, and residual system).

The question is then, how to avoid observation spillover. One way would be to locate the sensors on the zero locations of the residual mode shape functions. This, however, is normally impractical, especially when one considers that the residual modes usually include high frequency resonances with rapid spatial variations in mode shape. A common way to avoid observation spillover is to filter the sensor signal, to remove the resonances of the uncontrolled modes (or only to measure the resonances of the controlled modes). This can be done by using bandpass filters, or even phase locked loops (Balas, 1978a,b). Another way to avoid the destabilising effects of observation spillover is to using collocated sensors and actuators, which we will discuss further later in this chapter.

Example 11.1

As a numerical example of modal control, we will consider the active control of vibration in a simply supported beam (from Balas, 1978a,b). In this example we will assume that the beam has no damping, so that the response of the beam is governed by the Euler–Bernoulli partial differential equation

$$m\frac{\partial^2 w(x,t)}{\partial t^2} + EI\frac{\partial^4 w(x,t)}{\partial x^4} = f(x,t) \tag{11.2.81}$$

where $w(x,t)$ is the displacement of the (one-dimensional) beam at x at time t, $f(x,t)$ is the applied force distribution, E and I are the modulus of elasticity and moment of inertia of the beam, respectively, and $0 \le x \le L$, where L is the beam length. The simply supported boundary conditions are described by

$$w(0,t) = w(1,t) = 0 \tag{11.2.82}$$

and

$$\frac{\partial^2 w(x,t)}{\partial x^2} = \frac{\partial^2 w(x,t)}{\partial x^2} = 0 \tag{11.2.83}$$

The mode shape functions are the beam are described by

$$\phi_i(x) = \sin\frac{i\pi x}{L} \tag{11.2.84}$$

For simplicity, we will set E, I, m and L equal to one. The associated resonance frequencies are described by

$$\omega_i = \left(\frac{i\pi}{L}\right)^2 \tag{11.2.85}$$

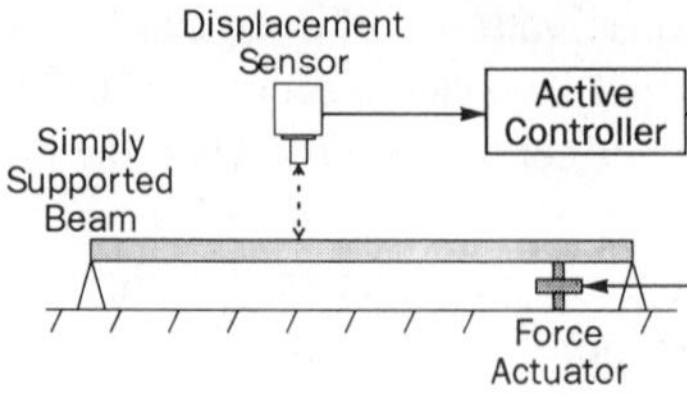

Fig. 11.2 Arrangement for beam simulation (after Balas, 1978a,b).

The beam/active controller arrangement is shown in Fig. 11.2. The beam is controlled by a single point control force at $x_c = L/6$, so that the applied force distribution is modelled by $f(x,t) = f(t)\,\delta(x_c - x)$. The beam output is measured by a single (unity gain) point displacement sensor at x_e, so that the beam output is $y = w(x_e,t)$. For this simple example, the transducer dynamics and system noise will be ignored.

The active controller in this example is designed to control the first three beam modes. With three controlled modes, the state observer must be six-dimensional, and, since noise in the system is ignored, it is deterministic. The control gains are calculated from an optimal control law which minimizes the (unweighted) energy in the first three modes. Therefore, with the state matrix

$$\boldsymbol{x} = \begin{bmatrix} z_1 & z_2 & z_3 & \dot{z}_1 & \dot{z}_2 & \dot{z}_3 \end{bmatrix}^{\mathrm{T}} \tag{11.2.86}$$

the weighting matrices in the optimal control law are

$$\boldsymbol{Q} = \begin{bmatrix} \boldsymbol{\Lambda}_N & \boldsymbol{0} \\ \boldsymbol{0} & \boldsymbol{I} \end{bmatrix}, \quad \boldsymbol{R} = 0.1\boldsymbol{I} \tag{11.2.87}$$

where $\boldsymbol{I}$ is the identity matrix. The resulting control gains are (Balas, 1978a,b):

$$\boldsymbol{k} = [0.5448 \;\; 5.018 \;\; 18.303 \;\; 3.162 \;\; 3.1597 \;\; 3.156] \tag{11.2.88}$$

The observer gains are chosen to place the observer poles to the left of the controlled system poles, using a pole placement routine. The resulting observer poles are (Balas, 1978a,b):

$$\boldsymbol{l} = [31.7821 \;\; -95.626 \;\; 160.9996 \;\; -118.304 \;\; 837.444 \;\; 11388.99]^{\mathrm{T}} \tag{11.2.89}$$

In the calculations the beam is given an initial displacement such that the first three modes have unity displacement, while the other modes are unperturbed. The sensor output is shown in Fig. 11.3, the control input in Fig. 11.4, and the energy in the controlled modes in Fig. 11.5. Observe that the response of the controlled beam modes decays at a rate which is slightly faster than if each mode had one per cent critical damping. The energy in the fourth (residual) mode is shown in Fig. 11.6. Observe that the energy in this mode is increasing over time. In fact, this mode is now unstable. This is a result of observation spillover. The observation spillover-induced instability could be eliminated by either filtering the sensor signal, or in this case by adding a small amount of passive damping which would shift the residual pole slightly to the

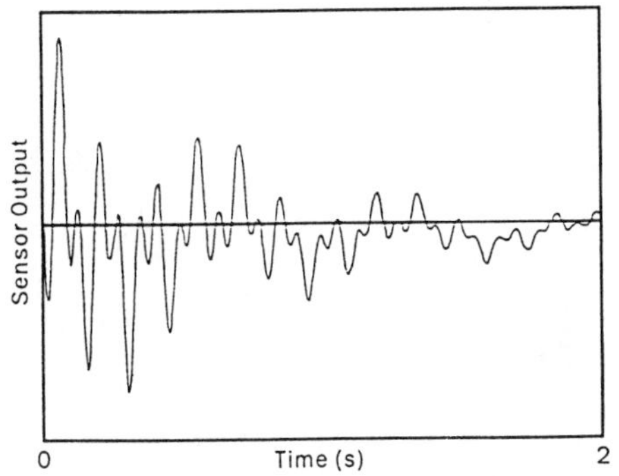

Fig. 11.3 Sensor output (after Balas, 1978a,b).

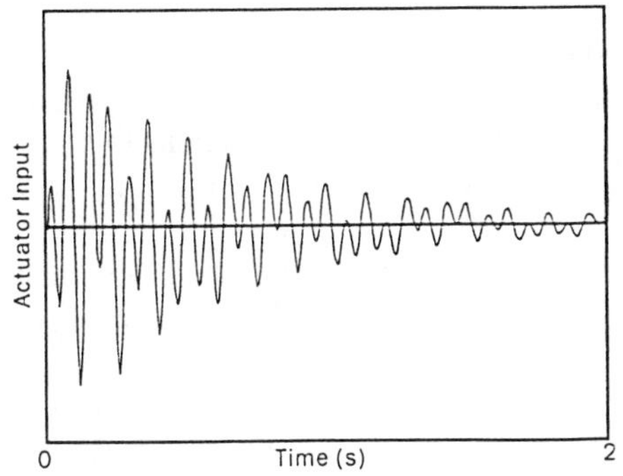

Fig. 11.4 Actuator input (after Balas, 1978a,b).

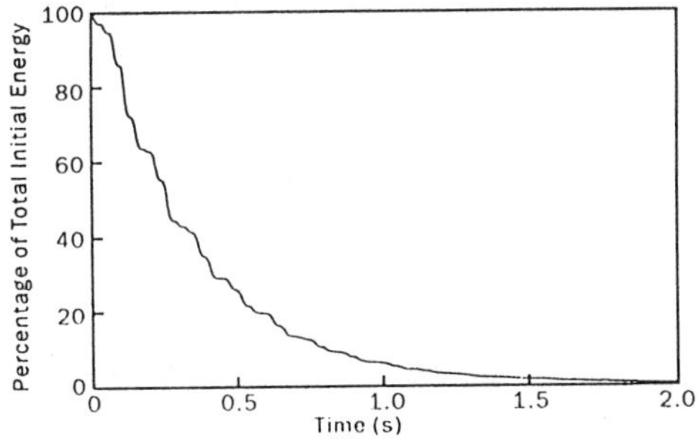

Fig. 11.5 Energy in controlled modes during application of active control (after Balas, 1978a,b).

left, increasing its stability margin and hence providing some tolerance to spillover effects.

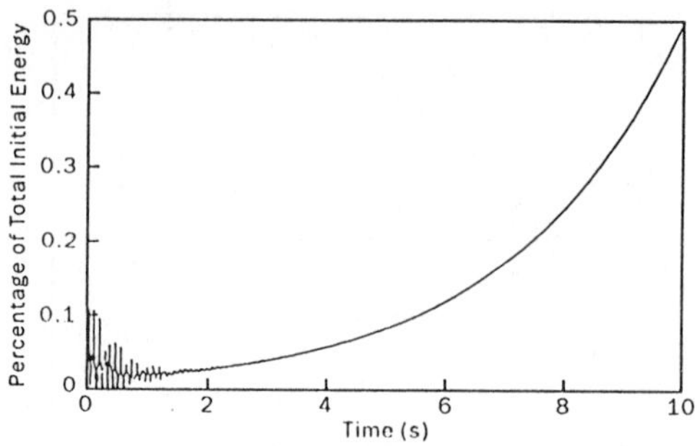

Fig. 11.6 Energy in residual mode during application of active control (after Balas, 1978a,b).

11.2.7. Optimal control gains for second order matrix equations

Thus far in this section we have considered the 'basics' of modal control, basing our analysis on state equations which describe the response of our system in terms of system modes. We can use these equations to calculate optimal control gains for attenuating the system response to the unwanted disturbance. Calculation of these gains can be carried out using the methods outlined in Chapter 5, which involves solving a matrix Ricatti equation of dimension $(2N \times 2N)$, where there are N modes in the reduced order system model.

In this section, we will consider further the calculation of optimal control gains for systems described by second order matrix equations, which is a generalization of the form of the modal state equations which we have considered thus far. Specifically, we will find that there are characteristics of the coefficient matrices which can be used to reduce the problem to one of solving three $(N \times N)$ matrix equations, which may significantly reduce the computational burden.

Recall from Chapter 5 that the steady state optimal control input for a system described in state space notation has the form

$$\boldsymbol{u} = -\boldsymbol{K}\boldsymbol{x} \tag{11.2.90}$$

where the gain matrix $\boldsymbol{K}$ is defined by the relationship

$$\boldsymbol{K} = \boldsymbol{R}^{-1}\boldsymbol{B}^{\mathrm{T}}\boldsymbol{P} \tag{11.2.91}$$

where $\boldsymbol{R}$ is the control effort weighting matrix, $\boldsymbol{B}$ is the input matrix, and the matrix $\boldsymbol{P}$ satisfies the steady state matrix Ricatti equation

$$\boldsymbol{A}^{\mathrm{T}}\boldsymbol{P} + \boldsymbol{P}\boldsymbol{A} - \boldsymbol{P}\boldsymbol{B}\boldsymbol{R}^{-1}\boldsymbol{B}^{\mathrm{T}}\boldsymbol{P} + \boldsymbol{Q} = 0 \tag{11.2.92}$$

where $\boldsymbol{A}$ is the state matrix.

The (full) state and input matrices for the systems of interest to us here are defined in (11.2.46), with the reduced order matrices defined in (11.2.51). Note that modal state equations provide a 'lumped' approximation of the equations governing the response of the distributed parameter system, and are a (truncated) set of second order matrix equations. The method of solution for the optimal control gains to be outlined here is valid for any such lumped parameter system, for which the equations of motion can be written as a set of simultaneous second order differential equations,

$$M\ddot{q} + D\dot{q} + Kq = Bu \tag{11.2.93}$$

where M, D, and K are $(n \times n)$ mass, damping, and stiffness matrices, respectively, q is an n-dimensional displacement vector and u is an n-dimensional control vector (there are $n/2$ modelled modes). When expressed in state variable format, the state vector, state matrix and input matrix have the form

$$x = \begin{bmatrix} q \\ \dot{q} \end{bmatrix}, \quad A = \begin{bmatrix} 0 & I \\ -M^{-1}K & -M^{-1}D \end{bmatrix}, \quad B = \begin{bmatrix} 0 \\ M^{-1}B' \end{bmatrix} \tag{11.2.94}$$

If we compare the terms in (11.2.94) with those of the (normalized) modal state equations given in (11.2.46), we find that $M^{-1}K = \Lambda$, $M^{-1}D = 2\zeta\Lambda^{1/2}$, and $M^{-1}B' = B^*$.

If the state weighting matrix used in the optimal control problem is diagonal,

$$Q = \begin{bmatrix} Q_1 & 0 \\ 0 & Q_2 \end{bmatrix} \tag{11.2.95}$$

we can take advantage of the form of the state and input matrices to put the matrix Ricatti equation in a convenient form for solution. As we discussed previously in this section, this is commonly the case in active noise and vibration control problems, where the weighting matrix Q is chosen such that $x^{\mathrm{T}}Qx$ represents the total system energy (in this generalized case, by setting $Q_1 = K$, $Q_2 = M$).

To simplify the optimal control problem, we can first divide the matrix P in the Ricatti equation into four $(n \times n)$ submatrices,

$$P = \begin{bmatrix} P_{11} & P_{12} \\ P_{12}^{\mathrm{T}} & P_{22} \end{bmatrix} \tag{11.2.96}$$

Using the notation of (11.2.60) and (11.2.62), the $(2n \times 2n)$ Ricatti equation (11.2.58) can be expressed as the following set of $(n \times n)$ matrix equations (Kwak and Meirovitch, 1993):

$$-Q_1 + KM^{-1}P_{12}^{\mathrm{T}} + P_{12}M^{-1}K + P_{12}M^{-1}B^*R^{-1}B^{*\mathrm{T}}M^{-1}P_{12}^{\mathrm{T}} = 0 \tag{11.2.97}$$

$$-P_{11} + KM^{-1}P_{22} + P_{12}M^{-1}D + P_{12}M^{-1}B^*R^{-1}B^{*\mathrm{T}}M^{-1}P_{22} = 0 \tag{11.2.98}$$

$$-P_{11} + P_{22}M^{-1}K + DM^{-1}P_{12}^{\mathrm{T}} + P_{22}M^{-1}B^*R^{-1}B^{*\mathrm{T}}M^{-1}P_{12}^{\mathrm{T}} = 0 \tag{11.2.99}$$

$$\begin{gathered} -Q_2 - P_{12} - P_{12}^{\mathrm{T}} + DM^{-1}P_{22} + P_{22}M^{-1}D + \\ P_{22}M^{-1}B^*R^{-1}B^{*\mathrm{T}}M^{-1}P_{22} = 0 \end{gathered} \tag{11.2.100}$$

Note again that these equations can be restated using the modal state space equation notation of (11.2.46) by setting $\boldsymbol{M}^{-1}\boldsymbol{K} = \Lambda$, $\boldsymbol{M}^{-1}\boldsymbol{D} = 2\xi\Lambda^{1/2}$ and $\boldsymbol{M}^{-1}\boldsymbol{B}' = \boldsymbol{B}^{*}$.

Owing to the form of the input matrix $\boldsymbol{B}$, not all of the submatrices of $\boldsymbol{P}$ are needed to derive the optimum control gains. If we expand (11.2.91), we find that

$$\boldsymbol{K} = \boldsymbol{R}^{-1}\begin{bmatrix}0 & \boldsymbol{B}'^{T}\boldsymbol{M}^{-1}\end{bmatrix}\begin{bmatrix}\boldsymbol{P}_{11} & \boldsymbol{P}_{12}\\ \boldsymbol{P}_{12}^{T} & \boldsymbol{P}_{22}\end{bmatrix} = \boldsymbol{R}^{-1}\boldsymbol{B}'^{T}\boldsymbol{M}^{-1}\begin{bmatrix}\boldsymbol{P}_{12}^{T} & \boldsymbol{P}_{22}\end{bmatrix} \tag{11.2.101}$$

From this it is apparent that the submatrix P_{11} is not used in calculation of the control gains, although it would be calculated if the Ricatti equation (11.2.92) were simply solved as written. If we eliminate this term from equations (11.2.98) and (11.2.99), we obtain the matrix equation (Kwak and Meirovitch, 1993):

$$\begin{aligned}&\boldsymbol{P}_{22}\boldsymbol{M}^{-1}\boldsymbol{K} - \boldsymbol{K}\boldsymbol{M}^{-1}\boldsymbol{P}_{22} + \boldsymbol{D}\boldsymbol{M}^{-1}\boldsymbol{P}_{12}^{T} - \boldsymbol{P}_{12}\boldsymbol{M}^{-1}\boldsymbol{D}\\ &+ \boldsymbol{P}_{22}\boldsymbol{M}^{-1}\boldsymbol{B}^{*}\boldsymbol{R}^{-1}\boldsymbol{B}^{*T}\boldsymbol{M}^{-1}\boldsymbol{P}_{12}^{T} - \boldsymbol{P}_{12}\boldsymbol{M}^{-1}\boldsymbol{B}^{*}\boldsymbol{R}^{-1}\boldsymbol{B}^{*T}\boldsymbol{M}^{-1}\boldsymbol{P}_{22} = 0\end{aligned} \tag{11.2.102}$$

The $(n \times n)$ matrix equations (11.2.97), (11.2.100) and (11.2.102) can be solved for $\boldsymbol{P}_{12}$ and $\boldsymbol{P}_{22}$ to facilitate calculation of the optimum control gains as an alternative to solving the full $(2n \times 2n)$ matrix Ricatti equation. In designs which involve a large number of modelled modes, this may present a (significant) computational saving.

On the surface, it may appear from (11.2.97), which contains only $\boldsymbol{P}_{12}$ and has the appearance of an $(n \times n)$ Ricatti equation, that it is possible to solve for $\boldsymbol{P}_{12}$ independently of $\boldsymbol{P}_{22}$. This is not, however, the case, as P_{12} is not in general a symmetric matrix. However, an iterative approach can be used to solve for the two variables (Kwak and Meirovitch, 1993).

To derive an iterative procedure, note first that the matrix $\boldsymbol{P}_{12}$ can be expressed as the sum of a symmetric and skew-symmetric matrix (Noble and Daniel, 1977),

$$\boldsymbol{P}_{12} = \bar{\boldsymbol{P}}_{12} + \tilde{\boldsymbol{P}}_{12} \tag{11.2.103}$$

where $\bar{}$ denotes the symmetric matrix, and $\tilde{}$ denotes the skew-symmetric matrix. Substituting the expansion of (11.2.103) into (11.2.97) yields the relationship

$$\begin{aligned}&-\boldsymbol{Q}_{1} + \boldsymbol{K}\boldsymbol{M}^{-1}\bar{\boldsymbol{P}}_{12} + \bar{\boldsymbol{P}}_{12}\boldsymbol{M}^{-1}\boldsymbol{K} + \bar{\boldsymbol{P}}_{12}\boldsymbol{M}^{-1}\boldsymbol{B}^{*}\boldsymbol{R}^{-1}\boldsymbol{B}^{*T}\boldsymbol{M}^{-1}\bar{\boldsymbol{P}}_{12}\\ &- \boldsymbol{K}\boldsymbol{M}^{-1}\tilde{\boldsymbol{P}}_{12} + \tilde{\boldsymbol{P}}_{12}\boldsymbol{M}^{-1}\boldsymbol{K} + \tilde{\boldsymbol{P}}_{12}\boldsymbol{M}^{-1}\boldsymbol{B}^{*}\boldsymbol{R}^{-1}\boldsymbol{B}^{*T}\boldsymbol{M}^{-1}\bar{\boldsymbol{P}}_{12}\\ &- \bar{\boldsymbol{P}}_{12}\boldsymbol{M}^{-1}\boldsymbol{B}^{*}\boldsymbol{R}^{-1}\boldsymbol{B}^{*T}\boldsymbol{M}^{-1}\tilde{\boldsymbol{P}}_{12} - \tilde{\boldsymbol{P}}_{12}\boldsymbol{M}^{-1}\boldsymbol{B}^{*}\boldsymbol{R}^{-1}\boldsymbol{B}^{*T}\boldsymbol{M}^{-1}\tilde{\boldsymbol{P}}_{12} = 0\end{aligned} \tag{11.2.104}$$

With a view to solving a Ricatti equation for the symmetric part of $\boldsymbol{P}_{12}$, we can rewrite (11.2.104) as

$$\begin{aligned}&-(\boldsymbol{Q}_{1}+\boldsymbol{S}) + \boldsymbol{K}\boldsymbol{M}^{-1}\bar{\boldsymbol{P}}_{12} + \bar{\boldsymbol{P}}_{12}\boldsymbol{M}^{-1}\boldsymbol{K}\\ &+ \bar{\boldsymbol{P}}_{12}\boldsymbol{M}^{-1}\boldsymbol{B}^{*}\boldsymbol{R}^{-1}\boldsymbol{B}^{*T}\boldsymbol{M}^{-1}\bar{\boldsymbol{P}}_{12} = 0\end{aligned} \tag{11.2.105}$$

where S is a symmetric matrix defined by the expression

$$\begin{aligned} S = &-\tilde{P}_{12}\left(M^{-1}K + M^{-1}B^{*}R^{-1}B^{*\mathrm{T}}M^{-1}\bar{P}_{12}\right) \\ &+ \left(KM^{-1} + \bar{P}_{12}M^{-1}B^{*}R^{-1}B^{*\mathrm{T}}M^{-1}\right)\tilde{P}_{12} \\ &+ \tilde{P}_{12}M^{-1}B^{*}R^{-1}B^{*\mathrm{T}}M^{-1}\tilde{P}_{12} \end{aligned} \qquad (11.2.106)$$

Observe that, for a given symmetric matrix S, solution of (11.2.105) is solution of an $(n \times n)$ matrix Ricatti equation.

Continuing, if the expansion of (11.2.103) is substituted in (11.2.100), we obtain a second Ricatti equation:

$$\begin{aligned} &-(Q_2+2\bar{P}_{12}) + DM^{-1}P_{22} + P_{22}M^{-1}D \\ &+ P_{22}M^{-1}B^{*}R^{-1}B^{*\mathrm{T}}M^{-1}P_{22} = 0 \end{aligned} \qquad (11.2.107)$$

Finally, if (11.2.103) is substituted into (11.2.102), we obtain the Lyapunov equation

$$\begin{aligned} &-\tilde{P}_{12}\left(M^{-1}D + M^{-1}B^{*}R^{-1}B^{*\mathrm{T}}M^{-1}P_{22}\right) \\ &- \left(DM^{-1} + P_{22}M^{-1}B^{*}R^{-1}B^{*\mathrm{T}}M^{-1}\right)\tilde{P}_{12} \\ &+ P_{22}\left(M^{-1}K + M^{-1}B^{*}R^{-1}B^{*\mathrm{T}}M^{-1}\bar{P}_{12}\right) \\ &- \left(KM^{-1} + \bar{P}_{12}M^{-1}B^{*}R^{-1}B^{*\mathrm{T}}M^{-1}\right)P_{22} \\ &+ DM^{-1}\bar{P}_{12} - \bar{P}_{12}M^{-1}D = 0 \end{aligned} \qquad (11.2.108)$$

An iterative procedure for solving for P_{12} and P_{22} can now be devised as follows (Kwak and Meirovitch, 1993): starting with some initial guess of S:

1. Insert matrix S into (11.2.105) and solve for the symmetric part of P_{12}.
2. Insert the just-computed value of P_{12} into (11.2.107) and solve this Ricatti equation for P_{22}.
3. Insert both the symmetric part of P_{12} and P_{22} into (11.2.108) and solve for the skew-symmetric matrix $\tilde{P}_{12}$.
4. Insert both the symmetric and skew-symmetric parts of P_{12} into (11.2.106) and calculate a new value of S.
5. Repeat steps (1)–(4) until the matrices have converged.

A numerical example of this procedure can be found in Kwak and Meirovitch (1993).

11.2.8. A brief note on passive damping

One of the constraints we outlined in the beginning of this section was that at least a small amount of passive damping is present in the structure to be controlled. The last point we wish to make in this section is that the passive damping is desirable, improving the stability of our controllers (Hughes and Abdel-Rahman, 1979; Spanos, 1989; Grandhi, 1990; Rao *et al.*, 1990; Gueler *et al.*, 1993). In some instances, it can also reduce the number of sensors and actuators required to establish controllability and observability (Hughes and Skelton, 1980a).

Consider the desired controller response outlined in Fig. 11.7. Here we have two requirements for the controller, the first of which is gain stabilization of the response of the unmodelled, or poorly modelled, modes outside the controller bandwidth (Gueler *et al.*, 1993). The loop gain of a flexible structure peaks at structural resonances, where the structural response is controlled only by its damping. As there is an inverse relationship between damping and response at resonance, it is obvious that any structure without damping cannot be gain stabilized. As the resonance frequency increases, the effectiveness of (viscous) passive damping also increases, which leads to the small high frequency damping requirement shown in Fig. 11.8.

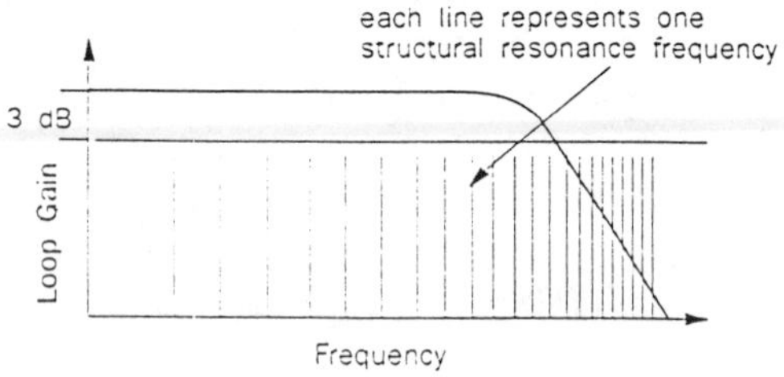

Fig. 11.7 Figurative depiction of problem statement for bandwidth to include many poorly modelled, lightly damped, closely spaced modes (after Gueler *et al.*, 1993).

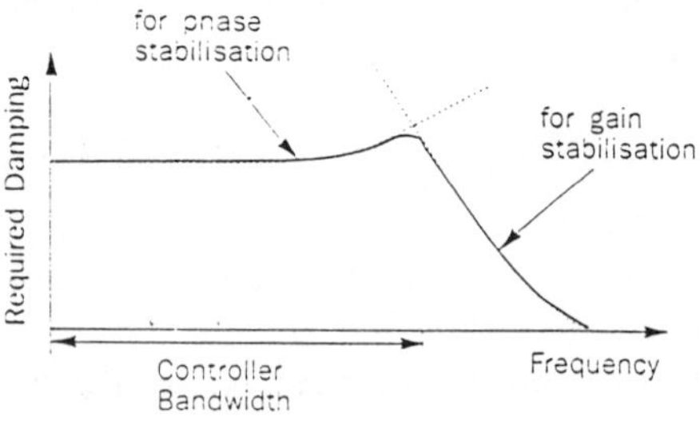

Fig. 11.8 Required level of passive damping to meet problem specification (after Gueler *et al.*, 1993).

The second requirement of the controller is phase stabilization of the system response within the controller bandwidth. This requirement follows from the fact that to achieve high bandwidth control of a flexible structure, the structure dynamics must be compensated, particularly at resonances. Often, the controllers implemented to do this effectively notch filter the response at structural resonances, cancelling the structure's poles and zeroes with the controllers zeroes and poles, respectively. For this to be achieved, an accurate model of the response of the target structure is needed. In an undamped system, uncertainty in this model can lead to instability.

To demonstrate this, consider Fig. 11.9, which illustrates the departure of the root locus of a system with a pole above or below a zero, a situation which may exist in a compensated, undamped system where there was uncertainty in the system model. If we assume that the remainder of the plant dynamics is responsible for a phase lag of $-90°$, the departure angle will be at $180°$ when the pole is above the zero, but $0°$ when the zero is above the pole. This means that if the dynamic system has no (or very little) damping, in the second instance the system will go unstable as the compensator gain is increased.

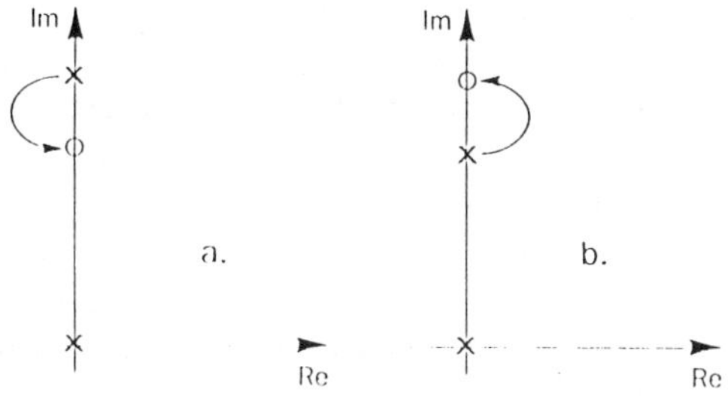

Fig. 11.9 Departure angles of root locus of a single oscillatory system as a result of uncertainty in pole location (after Gueler *et al.*, 1993): (a) pole above zero; (b) zero above pole.

If we assume that the poles migrate to the zeroes in a pattern which approximates a semi-circle, we can estimate the amount of damping which is required to avoid instability. As the radius of the semi-circle is dependent upon the pole-zero separation, the level of damping required to assure stability robustness is (Gueler *et al.*, 1993):

$$\zeta = \frac{|\omega_z - \omega_n|}{\omega_z + \omega_n} \tag{11.2.109}$$

where ω_z is the zero frequency, and ω_n is the resonance (pole) frequency. This level of passive damping will ensure that the pole does not migrate across the imaginary axis for any gain value.

To derive a relationship for the required degree of passive damping which is perhaps more readily applicable, let us consider the system response in the vicinity of the resonance of an isolated structure mode, which is governed by the relationship

$$G(s) = \frac{1}{s^2 + 2\zeta\omega_n s + \omega_n^2} \tag{11.2.110}$$

The phase angle of this transfer function follows the relationship

$$\theta(\omega) = -\tan^{-1}\frac{2\zeta\omega_n\omega}{\omega_n^2 - \omega^2} \tag{11.2.111}$$

Therefore, at resonance, the change in phase with respect to frequency is given by

$$\frac{d\theta}{d\omega} = -\frac{1}{\zeta\omega_n} \tag{11.2.112}$$

Therefore, the phase change as the system passes through resonance will be sharp if the damping is low. If we define the uncertainty in resonance frequency as $\delta\omega = \omega_n - \omega_{\text{actual}}$, then a first order approximation of the uncertainty in phase angle near resonance is given by (Gueler *et al.*, 1993):

$$\delta\theta = -\frac{\delta\omega}{\zeta\omega_n} \tag{11.2.113}$$

This relationship states that given an uncertainty in resonance frequency, the uncertainty in phase is inversely proportional to the damping.

It is possible to use the relationship of (11.2.113) to quantify the degradation in closed-loop stability resulting from imperfect pole-zero cancellation for a specific problem (see Gueler *et al.*, 1993). The important point to note, however, which is evident from the sketches in Fig. 11.9 is that if there is no passive damping in the system, the phase excursion can be 180°, passive damping reduces this value.

For a (local) phase margin of θ_{pm}, which from our definitions in Chapter 5 is the amount of additional phase lag required to make the system unstable at a given frequency, the permissible amount of uncertainty in plant natural frequency is given by

$$\delta\omega \le \theta_{pm}\zeta\omega_n \tag{11.2.114}$$

Therefore, the amount of passive damping needed, given a pole-zero mismatch of $\delta\omega$ and a phase margin θ_{pm}, is (Geuler *et al.*, 1993):

$$\zeta \ge \frac{\delta\omega}{\theta_{pm}\omega_n} \tag{11.2.115}$$

This expression leads to the damping requirement for phase stabilization in the controller bandwidth sketched in Fig. 11.8.

Further examples of the need for passive damping to avoid controller instability can be found in Gueler *et al.*, 1993).

11.3 INDEPENDENT MODAL SPACE CONTROL

In the previous section, we considered the design of active control systems for flexible structures using state space design methods, where the system states are modal displacements and velocities. The general problem framework fits into that reviewed in Chapter 5, where the control and observer gains can be calculated using either pole placement or optimal (linear quadratic) methodologies. There can, however, be computational problems with this approach when a large number of modes are included in the control calculations; even using the iterative approach outlined in the previous section, where solution of the matrix Ricatti equation was reduced from solution of a ($2N \times 2N$) matrix equation to solution of three ($N \times N$) matrix equations, may not reduce the computational burden sufficiently.

Computational problems can arise with a large system design because the control system is 'coupled': because the actuators and sensors excite and measure multiple structure modes (for point devices, theoretically all modes), a control input derived from consideration of one mode will excite additional modes (controlled and residual) upon application (thus feedback control destroys the independence of the open-loop modal equations). Because of this coupling, the various controller design equations (such as the Ricatti equation) will have cross-terms describing the influence which a control input derived from one state has upon other states, which limits the extent to which 'shortcuts' can be taken in the design process.

One approach to simplifying control design, then, would be to somehow eliminate the extensive cross-coupling between the states (or modes). If the independence of the modal equations could be maintained during implementation of a feedback control system (closed loop), then the control law could be derived as a set of independent control laws, one for each mode. This basic idea is behind a form of control referred to as independent modal-space control (IMSC) (Bennighof and Meirovitch, 1988; Meirovitch and Baruh, 1982; Meirovitch and Bennighof, 1986; Meirovitch and Oz, 1980a, 1980b, 1980c; Meirovitch *et al.*, 1977, 1979, 1983).

11.3.1 Control law development

To develop the independent modal space control (IMSC) methodology, we should begin by considering the control input to the structure, which is the root cause of control system coupling. If a set of point actuators is used in the control system, then the control input from each can be modelled using a Dirac delta function,

$$f(\boldsymbol{r},t) = u(t)\delta(\boldsymbol{r}-\boldsymbol{r}_c) \tag{11.3.1}$$

where $u(t)$ is the control input at time t, and $\boldsymbol{r}_c$ is the location of the control actuator on the structure. If there are N_s control actuators in the system, then the *i*th modal generalized force, defined in the previous section in (11.2.15), is simply

$$f_i(t) = \sum_{\iota=1}^{N_s} u_\iota(t)\psi_i(\boldsymbol{r}_\iota) \tag{11.3.2}$$

The relationship in (11.3.2) implies that all control inputs may excite all modelled structural modes (provided they are not located on a zero of the mode shape function). This is where the system coupling arises upon implementation of closed-loop control.

To simplify the development, let us write the controlled system equations in a slightly different manner than was done in the previous section. The response of the controlled system can still be written as the state space equation

$$\dot{\boldsymbol{x}}_N = \boldsymbol{A}_N\boldsymbol{x}_N + \boldsymbol{B}_N\boldsymbol{u} \tag{11.3.3}$$

However, this time the system states, which are still modal displacements and velocities, are arranged as

$$\boldsymbol{x}_N = \begin{bmatrix} z_1 & \dot{z}_1 & \cdots & z_N & \dot{z}_N \end{bmatrix}^{\mathrm{T}} \tag{11.3.4}$$

which puts the state matrix in a block diagonal form,

$$\boldsymbol{A}_N = \begin{bmatrix} \boldsymbol{A}_{N1} & 0 & \cdots & 0 \\ 0 & \boldsymbol{A}_{N2} & \cdots & 0 \\ & & \ddots & \\ 0 & 0 & \cdots & \boldsymbol{A}_{NN} \end{bmatrix} \tag{11.3.5}$$

where, for a lightly damped system:

$$\boldsymbol{A}_{Ni} = \begin{bmatrix} 0 & 1 \\ -\omega_{Ni}^2 & -2\zeta\omega_{Ni} \end{bmatrix} \tag{11.3.6}$$

and where N is the number of modelled modes, which in this instance is the same as the number of modes to be controlled. For N_s discrete control (point) inputs,

$$\boldsymbol{Bu} = \boldsymbol{f} = \begin{bmatrix} 0 & 0 & \cdots & 0 \\ \psi_1(\boldsymbol{r}_1) & \psi_1(\boldsymbol{r}_2) & \cdots & \psi_1(\boldsymbol{r}_{N_s}) \\ & & \vdots & \\ 0 & 0 & \cdots & 0 \\ \psi_N(\boldsymbol{r}_1) & \psi_N(\boldsymbol{r}_2) & \cdots & \psi_N(\boldsymbol{r}_{N_s}) \end{bmatrix} \begin{bmatrix} u_1 \\ u_2 \\ \vdots \\ u_{N_s} \end{bmatrix} = \begin{bmatrix} 0 \\ f_1 \\ \vdots \\ 0 \\ f_N \end{bmatrix} \tag{11.3.7}$$

where $\boldsymbol{f}$ is the vector of modal generalized forces.

The basic idea behind IMSC is to restate the problem as a set of independent modal equations, thereby eliminating coupling, which will simplify the controller design exercise. To do this, we can design the controller in terms of the modal generalized forces in the vector $\boldsymbol{f}$, then transform the result to obtain the actual control inputs $\boldsymbol{u}$,

$$u = B_N^{-1} f \tag{11.3.8}$$

where B_N^{-1} is a pseudo inverse in general, and an actual inverse if the control matrix B_N is square (B_N will be square if the number of control inputs is equal to the number of controlled modes).

To see how designing for a single mode will simplify the controller design exercise, consider first a pole placement exercise. If we are considering a single mode, the control input will be defined by the expression

$$bu = -bk_i x_i = -\begin{bmatrix} 0 \\ 1 \end{bmatrix} \begin{bmatrix} k_{i1} & k_{i2} \end{bmatrix} \begin{bmatrix} z_i \\ \dot{z}_i \end{bmatrix} \tag{11.3.9}$$

Using the state matrix outlined in (11.3.4) specialized for a single mode, the characteristic equation for this mode is

$$\left| sI - A_{Ni} + bk_i x_i \right| = \begin{vmatrix} s & -1 \\ \omega_i^2 + k_{i1} & s + 2\zeta\omega_i + k_{i2} \end{vmatrix} \tag{11.3.10}$$

or

$$s^2 + (k_{i2} + 2\zeta\omega_i)s + (\omega_i^2 + k_{i1}) = 0 \tag{11.3.11}$$

If the desired location of the poles for this mode defines the desired characteristic equation

$$\gamma = s^2 + \alpha_1 s + \alpha_2 \tag{11.3.12}$$

then the control gains are simply defined by the relationships

$$\begin{aligned} k_{i1} &= \alpha_2 - \omega_i^2 \\ k_{i2} &= \alpha_1 - 2\zeta\omega_i \end{aligned} \tag{11.3.13}$$

Optimal control problems are also more straightforward when designing for a single modal state equation, and deriving a modal generalized force. When the modal equations are independent, the cost function associated with the optimal control problem is simply the sum of the cost functions of the individual modes:

$$J = \sum_{i=1}^{N} J_i \tag{11.3.14}$$

With steady state optimal control, these modal cost functions are defined by the relationship

$$J_i = \int_0^\infty [x_i^T(\tau) Q x_i(\tau) + u_i^T(\tau) R u_i(\tau)]\, d\tau \tag{11.3.15}$$

where Q is the state weighting matrix, and R is the control effort weighting matrix.

For steady state optimal control, the control signal for the ith modal 'system' is defined by the relationship

$$u_i = -Kx_i \tag{11.3.16}$$

where the gain matrix K_i is defined by the relationship

$$K_i = R^{-1}B_i^{\mathrm{T}}P_i \tag{11.3.17}$$

where R is the control effort weighting matrix, B_i is the input matrix for the ith mode, and P_i satisfies the matrix Ricatti equation

$$A_i^{\mathrm{T}}P_i + P_iA_i - P_iB_iR^{-1}B_i^{\mathrm{T}}P_i + Q = 0 \tag{11.3.18}$$

where Q is the state weighting matrix.

When considering a modal generalized force we have the relationship

$$f_i = \begin{bmatrix} 0 \\ f_i \end{bmatrix} = -K_ix_i = -\begin{bmatrix} k_{11} & k_{12} \\ k_{21} & k_{22} \end{bmatrix} \begin{bmatrix} z_i \\ \dot{z}_i \end{bmatrix} \tag{11.3.19}$$

From this relationship it is apparent the $k_{11} = k_{12} = 0$. For this relationship to hold, the control effort weighting matrix must have the form (Oz and Meirovitch, 1980a)

$$R = \begin{bmatrix} \infty & 0 \\ 0 & R_{22} \end{bmatrix} \tag{11.3.20}$$

If we are interested in minimizing system energy, such that the state weighting matrix is

$$Q = \begin{bmatrix} \omega^2 & 0 \\ 0 & 1 \end{bmatrix} \tag{11.3.21}$$

(so that $x^{\mathrm{T}}Qx$ reflects the system kinetic energy), substituting values into (11.3.17) and (11.3.18) we find that the gain matrix is (Meirovitch and Baruh, 1982)

$$K_i = \begin{bmatrix} 0 & 0 \\ \omega_i^2\left(\sqrt{1+R_{22}^{-1}} - 1\right) & \left[2\omega_i\left(\sqrt{1+R_{22}^{-1}} - 1\right) + \omega_i^2 R_{22}^{-1}\right]^{1/2} \end{bmatrix} \tag{11.3.22}$$

Solving a set of these equations can be considerably faster than solving a single large Ricatti equation (see Meirovitch *et al.*, 1983). When implementing this control law, the closed loop poles are (Meirovitch and Baruh, 1982)

$$\lambda_i = -\frac{\omega_i}{2}\left(2\sqrt{1+R_{22}^{-1}} - 1 + R_{22}^{-1}\right)^{1/2} \pm \frac{\omega_i}{2}\left(2\sqrt{1+R_{22}^{-1}} - 1 - R_{22}^{-1}\right)^{1/2} \tag{11.3.23}$$

Once the control gains have been derived for each modelled mode, the result must be transformed back into the form of the original state equation. For a control input defined by

$$u = -Kx \tag{11.3.24}$$

where $\boldsymbol{K}$ is, in general, a $(2n \times m)$ matrix of gains for $2n$ states (n modes) and m control inputs, the transformation requires solving the relationship

$$-\boldsymbol{B}\boldsymbol{K}\boldsymbol{x} = \begin{bmatrix} 0 & 0 & \cdots & 0 \\ \psi_1(r_1) & \psi_1(r_2) & \cdots & \psi_1(r_m) \\ & & \vdots & \\ 0 & 0 & \cdots & 0 \\ \psi_n(r_1) & \psi_n(r_2) & \cdots & \psi_n(r_m) \end{bmatrix} \begin{bmatrix} k_{1,1} & k_{1,2} & \cdots & k_{1,2n} \\ & & \vdots & \\ k_{m,1} & k_{m,2} & \cdots & k_{m,2n} \end{bmatrix} \begin{bmatrix} x_1 \\ \dot{x}_1 \\ \vdots \\ x_n \\ \dot{x}_n \end{bmatrix}$$

$$= \begin{bmatrix} 0 \\ f_1 \\ \vdots \\ 0 \\ f_n \end{bmatrix} = \begin{bmatrix} 0 & 0 & \cdots & 0 & 0 \\ k_{11} & k_{12} & \cdots & 0 & 0 \\ & & \vdots & & \\ 0 & 0 & \cdots & k_{n1} & k_{n2} \end{bmatrix} \begin{bmatrix} x_1 \\ \dot{x}_1 \\ \vdots \\ x_n \\ \dot{x}_n \end{bmatrix} \tag{11.3.25}$$

Defining the terms

$$\boldsymbol{B}' = \begin{bmatrix} \psi_1(r_1) & \psi_1(r_2) & \cdots & \psi_1(r_m) \\ & & \vdots & \\ \psi_n(r_1) & \psi_n(r_2) & \cdots & \psi_n(r_m) \end{bmatrix}, \quad \boldsymbol{K}' = \begin{bmatrix} k_{11} & k_{12} & \cdots & 0 & 0 \\ & & \ddots & & \\ 0 & 0 & \cdots & k_{n1} & k_{n2} \end{bmatrix} \tag{11.3.26}$$

solving (11.3.25) is equivalent to solving the matrix expression

$$\boldsymbol{B}'\boldsymbol{K} = \boldsymbol{K}' \tag{11.3.27}$$

The control gain matrix $\boldsymbol{K}$ is therefore defined by the expression

$$\boldsymbol{K} = \boldsymbol{B}^{-1}\boldsymbol{K}' \tag{11.3.28}$$

where $\boldsymbol{B}^{-1}$ is the left inverse of the matrix $\boldsymbol{B}'$ (simply an inverse if the number of control inputs is equal to the number of modelled modes).

Numerical examples of the implementation of independent modal space control can be found in the references outlined in the beginning of this chapter.

It should be pointed out here that while independent modal space control has the advantage of simplifying the control law calculation, it is not without some drawbacks. Possibly chief among these arises from the operation in (11.3.28); the method is dependent upon calculation of inverses of matrices which can be large and ill-conditioned. The ill-conditioning problem is particularly true when the structure and transducer arrangement has some form of geometric symmetry.

Finally, as derivation of feedback control gains using the independent modal-space control methodology requires a matrix inversion, implementation is eased if the number of control actuators is set equal to the number of modes to be controlled (otherwise a pseudo-inverse is required). Alternatively, if a small number of actuators is used to control a larger number of modes, some form of time-sharing (controlling a smaller number of modes at any given time) can be used (Baz *et al.*, 1989; Baz and Poh, 1990). This is referred to as the modified independent modal space control (MIMSC) method.

11.3.2 Modal filters

To implement the independent modal-space control methodology, we require a measure of the displacement and velocity of the system modes. We could use the observers outlined in the previous section for this purpose (which provided estimates of modal displacements and velocities). In doing so, it would be important again to take precautions to avoid the destabilizing effects of observation spillover. A second method of isolating the (resonant) response of a particular structural mode would be to use a bandpass filter, with the passband center frequency set equal to the resonance frequency of the mode of interest (Hallauer *et al.*, 1982). However, this technique can be difficult to implement when the resonance mode of interest is not sufficiently isolated from the resonance frequencies of other modes, particularly a problem with higher order modes. A further alternative is to use a 'modal filters'. These could be implemented using shaped piezoelectric polymer film sensors (Lee and Moon, 1990), as we discussed in Chapter 8 for controlling structural/acoustic radiation into free space. They could also be implemented as spatial filters which resolve modal displacements and velocities from a set of discrete sensor measurements (Meirovitch and Baruh, 1982, 1985; Morgan, 1991). It is this last technique which will be of interest to us here.

The concepts underlying the construction of modal filters are the same as those we used to formulate the modal control problem. As we have discussed, if a set of modes are orthogonal (self-adjoint), then the displacement of the ith mode can be extracted from a known structural displacement distribution through the relationship

$$z_i(t) = \int_S m(r)\psi_i(r)w(r,t)\ \mathrm{d}r \tag{11.3.29}$$

where integration is over the surface of the structure. Similarly, the velocity of the ith mode can be extracted from a known structural velocity distribution through the relationship

$$\dot{z}_i(t) = \int_S m(r)\psi_i(r)\dot{w}(r,t)\ \mathrm{d}r \tag{11.3.30}$$

Usually, structural displacement and velocity sensors are discrete transducers. Therefore, (11.3.29) and (11.3.30) could be implemented by interpolating the discrete measurements to estimate the continuous displacement and velocity distributions, and the integrations performed numerically

(Meirovitch and Oz, 1982). However, there is a second approach to modal filtering which is better suited to real time implementation. Considering only displacement (development of the problem for velocity sensing would be identical), the estimated displacement distribution of the structure, based upon a measurement of displacement at N_e discrete locations, is defined by the expression

$$\hat{w}(r,t) = \sum_{j=1}^{N_e} G(r,r_j)w(r_j,t) \tag{11.3.31}$$

where $G(r,r_j)$ is the interpolation function between the measurement position r_j and location r. If we multiply (11.3.31) through by $m(r)\psi_i(r)$ and integrate over the surface of the structure, we obtain an estimate of the displacement of mode i based upon a set of discrete displacement measurements,

$$z_i = \sum_{j=1}^{N_e} f_{ij}w(r,t) \tag{11.3.32}$$

The important point to note about this equation is that the estimate of modal displacements from a set of discrete displacement measurements has become a simple input–output problem, a simple (modal) filtering problem. The coefficients f_{ij} are fixed for a given set of structural mode shape functions and sensor locations, and can be calculated prior to implementing the control system.

It is possible to derive an expression for the filter coefficients f_{ij} by first solving for the interpolation functions $G(r,r_j)$, a task which can be accomplished using either a Rayleigh–Ritz or finite element approach. For our purposes, however, we will tackle the problem from a least squares approach. What we desire is to form linear combinations of discrete sensor signals to extract the displacement response of a given structural mode, which is a spatial filtering operation. The coefficients in the ith spatial filter f_i can be expressed as a vector,

$$f_i = \begin{bmatrix} f_{i,1} & f_{i,2} & \cdots & f_{i,N_e} \end{bmatrix}^{\mathrm{T}} \tag{11.3.33}$$

We will assume for the moment that the number of modes to be resolved, N, is less than or equal to the number of sensors N_e. If the spatial filter f_i is to resolve the ith structural mode, then the desired response for the filter is defined by the relationship

$$f_i^{\mathrm{T}}\psi_j = \begin{cases} 1 & i=j \\ 0 & i \neq j \end{cases} \tag{11.3.34}$$

where ψ_j is the vector of mass-normalized mode shape values (for mode j) at the sensing locations,

$$\psi_j = \begin{bmatrix} \psi_j(r_1) & \psi_j(r_2) & \cdots & \psi(r_{N_e}) \end{bmatrix}^{\mathrm{T}} \tag{11.3.35}$$

If we assemble the filter and mode shape vectors into two matrices

$$\boldsymbol{\Psi} = [\psi_1 \ \psi_2 \ \cdots \ \psi_N] , \quad \boldsymbol{F} = [f_1 \ f_2 \ \cdots \ f_N] \tag{11.3.36}$$

the desired set of modal filters satisfies the relationship (Morgan, 1991)

$$\boldsymbol{F} = \boldsymbol{\Psi}[\boldsymbol{\Psi}^{\mathrm{T}}\boldsymbol{\Psi}]^{-1} \tag{11.3.37}$$

Note that $\boldsymbol{F}^{\mathrm{T}}$ is the generalized, or Moore–Penrose, inverse of the mode shape value matrix $\boldsymbol{\Psi}$.

In most real-world situations, the total number of modes excited in a structure is greater than the number of sensors, as well as the number of modes in the reduced order model. As we discussed previously in this chapter, this can lead to problems of observation spillover. One approach to overcoming this problem would be to consider more modes than simply those in the reduced order model when designing the modal filters, and design the filters to minimize the cross-response between the modes. Therefore, for the ith modal filter we wish to minimize

$$\sum_{j-1}^{N} \left(f_i^{\mathrm{T}}\psi_j\right)^2 \tag{11.3.38}$$

subject to the constraint

$$f_i^{\mathrm{T}}\psi_j = 1 \tag{11.3.39}$$

Using vector differentiation with the method of Lagrange multipliers, we find the solution to the problem posed in (11.3.38) and (11.3.39) to be (Morgan, 1991)

$$\boldsymbol{F} = [\boldsymbol{\Psi}\boldsymbol{\Psi}^{\mathrm{T}}]^{-1}\boldsymbol{\Psi}\boldsymbol{\Lambda} \tag{11.3.40}$$

where $\boldsymbol{\Lambda}$ is a diagonal scaling matrix calculated to satisfy the constraint (11.3.39). Note that apart from the scaling matrix, the solution in (11.3.40) is the generalized inverse of the transposed mode shape value matrix $\boldsymbol{\Psi}^{\mathrm{T}}$.

There are several points worth mentioning here. First, it is possible to weight the above least squares problem to take into account differences in the excitation of individual modes, if they are known (Morgan, 1991). Second, it is possible to estimate modal velocities using only modal displacement measurements, and vice versa, using a 'modal' observer. To see this, consider the ith modal state equation (as it would appear in an independent modal space control problem)

$$\dot{\boldsymbol{x}}_i(t) = \boldsymbol{A}_i\boldsymbol{x}_i + \boldsymbol{b}f_i(t) \tag{11.3.41}$$

where

$$\boldsymbol{x}_i = \begin{bmatrix} z_i \\ \dot{z}_i \end{bmatrix} , \quad \boldsymbol{A}_i = \begin{bmatrix} 0 & 1 \\ -\omega_i^2 & -2\zeta\omega_i \end{bmatrix} , \quad \boldsymbol{b} = \begin{bmatrix} 0 \\ 1 \end{bmatrix} \tag{11.3.42}$$

Consider the case where only modal displacement measurements are available (from modal filters using discrete displacement sensors). The modal output equation is then

$$y_i(t) = \boldsymbol{c}\boldsymbol{x}_i(t) \tag{11.3.43}$$

where

$$\boldsymbol{c} = \begin{bmatrix} 1 & 0 \end{bmatrix} \tag{11.3.44}$$

The question is now, how can we recover the modal velocities. To this end, consider a modal observer, described by (Meirovitch and Baruh, 1985)

$$\frac{\mathrm{d}\hat{\boldsymbol{x}}_i(t)}{\mathrm{d}t} = \boldsymbol{A}_i\hat{\boldsymbol{x}}_i(t) + \boldsymbol{b}f_i(t) + \boldsymbol{l}_i[y_i(t) - \hat{y}_i(t)] \tag{11.3.45}$$

where $\hat{x}$ is the estimated state, and $\boldsymbol{l}_i$ is the observer gain vector,

$$\boldsymbol{l}_i = \begin{bmatrix} l_{i,1} & l_{i,2} \end{bmatrix} \tag{11.3.46}$$

To assist in assigning the modal observer gains, we can define a modal error vector $\boldsymbol{e}_i$ as

$$\boldsymbol{e}_i(t) = \hat{\boldsymbol{x}}_i(t) - \boldsymbol{x}_i(t) \tag{11.3.47}$$

The evolution of the error vector is governed by the expression

$$\dot{\boldsymbol{e}}_i(t) = \frac{\mathrm{d}\hat{\boldsymbol{x}}_i(t)}{\mathrm{d}t} - \dot{\boldsymbol{x}}_i = (\boldsymbol{A}_i - \boldsymbol{l}_i\boldsymbol{c})\boldsymbol{e}_i(t) \tag{11.3.48}$$

Therefore, the observer gains can be assigned such that the poles of $(\boldsymbol{A}_i - \boldsymbol{l}_i\boldsymbol{c})$ have negative real parts that allow the observer to converge as quickly as possible. While the observer outlined above is deterministic, in a noisy environment a Kalman filter could also be used.

The final point to make here concerns the locations of discrete sensors for modal filtering applications. If we are attempting to numerically solve the integral expression (11.3.29) directly, then it seems logical to locate the discrete sensors in a regular manner. However, when attempting to implement modal filters in the manner described above, regular spacing of discrete sensors, especially on a structure with regular geometry, can lead to matrix conditioning problems. Usually, a more randomized placement, perhaps within certain bounds, is desirable.

Example 11.2

As a simple example of modal filter design (taken from Morgan (1991)), consider the problem where there are three modes and two sensors. This situation could arise where there are two modes in the reduced order model (that is, two controlled modes), but three modelled modes. Suppose that the (2 sensor $\times$ 3 mode) mode shape value matrix is equal to

$$\boldsymbol{\Phi} = \begin{bmatrix} 0 & \sqrt{3}/2 & 1 \\ 1 & 1/2 & 0 \end{bmatrix} \tag{11.3.49}$$

Therefore, the modal filter matrix is

$$\boldsymbol{F} = \left[\boldsymbol{\Phi}\boldsymbol{\Phi}^{\mathrm{T}}\right]^{-1}\boldsymbol{\Phi}\boldsymbol{\Lambda} = 1/8\begin{bmatrix} -\sqrt{3} & 2\sqrt{3} & 5 \\ 7 & 2 & -\sqrt{3} \end{bmatrix}\boldsymbol{\Lambda} \tag{11.3.50}$$

To satisfy the constraint (11.3.39), the diagonal scaling matrix $\boldsymbol{\Lambda}$ is

$$\boldsymbol{\Lambda} = \begin{bmatrix} 8/7 & 0 & 0 \\ 0 & 2 & 0 \\ 0 & 0 & 8/5 \end{bmatrix} \tag{11.3.51}$$

so that the modal filter matrix, the columns of which define the three modal filters, is

$$\boldsymbol{F} = \begin{bmatrix} -\sqrt{3}/7 & \sqrt{3}/2 & 1 \\ 1 & 1/2 & -\sqrt{3}/5 \end{bmatrix} \tag{11.3.52}$$

The measurements of modal displacement and/or velocity required to implement a modal control system could now be obtained by taking the measurements (displacement and/or velocity) from the two sensors and multiplying them by the coefficients in the columns of $\boldsymbol{F}$.

11.4 CO-LOCATED CONTROLLERS

Thus far in this chapter, we have been concerned with control strategies which are essentially suitable for controlling a low number of critical structural modes, where the risk of performance-reducing control spillover and destabilizing observation spillover, resulting from the presence of untargeted modes in the structure, exists. There is, however, another control approach where it can be guaranteed that all vibration modes will remain stable when the active control system is in operation, at the expense of control performance. That technique is referred to as low authority control, and is commonly implemented using velocity feedback (Balas, 1979; Aubrun, 1980; Joshi, 1981, 1986, 1989; Schulz and Heimbold, 1983; Martinovic *et al.*, 1990; Creamer and Junkins, 1991; Hanagud *et al.*, 1992). Low authority control specifically aims to augment the damping of a structure, and by using co-located (usually written 'collocated' in the literature) sensors and actuators can do so in a stable fashion. In this section we will briefly describe this result; for a thorough study of collocated (dissipative) controllers, refer to Joshi (1989).

To examine this control approach, let us begin with the open loop modal equation,

$$\ddot{z}(t) + 2\zeta\mathbf{\Lambda}^{1/2}\dot{z}(t) + \mathbf{\Lambda}z(t) + \boldsymbol{B}\boldsymbol{f}(t) \tag{11.4.1}$$

where z is a $(n \times 1)$ vector of modal displacement amplitudes, n is the number of modelled modes, $\mathbf{\Lambda}$ is a diagonal matrix of open-loop eigenvalues, such that $\mathbf{\Lambda}^{1/2}$ is a diagonal matrix of resonance frequencies, $\boldsymbol{f}$ is an $(N_c \times 1)$ vector of control (point) force inputs, and $\boldsymbol{B}$ is $(n \times N_c)$ matrix of influence functions, where the i,jth term is the value of the ith mode shape function at the location of the jth control force. The output equation associated with this is

$$y(t) = \boldsymbol{C}\boldsymbol{\Psi}z(t) + \boldsymbol{E}\boldsymbol{\Psi}\dot{z}(t) \tag{11.4.2}$$

where $\boldsymbol{y}$ is an $(N_e \times 1)$ vector of N_e sensor outputs, $\boldsymbol{C}$ and $\boldsymbol{E}$ are $(N_e \times N_e)$ diagonal gain matrices, and ψ is an $(N_e \times n)$ matrix whose i,jth term is the value of the jth mode shape function at the ith sensor location.

Let us now make several assumptions. First, let the number of sensors N_e be equal to the number of control actuators N_c. Second, let only point velocity sensors be used in the system. Third, make the control source and error sensors be collocated. With these assumptions, $\boldsymbol{C} = \boldsymbol{0}$, $\boldsymbol{E} = \boldsymbol{I}$, and $\boldsymbol{\Psi} = \boldsymbol{B}^{\mathrm{T}}$. Therefore, the output (11.4.2) becomes

$$y(t) = \boldsymbol{B}^{\mathrm{T}}\dot{z}(t) \tag{11.4.3}$$

With direct velocity feedback, the control law has the form

$$f(t) = -\boldsymbol{Q}y(t) \tag{11.4.4}$$

where $\boldsymbol{Q}$ is an $(N_e \times N_e)$ symmetric, non-negative gain matrix.

One of the interesting, and attractive, features of this control approach outlined in (11.4.4) is that it is energy dissipative (Balas, 1979), and therefore always stable (similar to the Lyapunov stability criterion discussed in chapter 5). Further, if there are no zero resonance frequencies in $\mathbf{\Lambda}$ (no rigid body modes, so that $\mathbf{\Lambda}$ is positive definite), then the closed loop system is stable for any damping coefficient $\zeta \geq 0$, and asymptotically stable if either $\zeta > 0$, or $\zeta \geq 0$ and $\boldsymbol{B}\boldsymbol{Q}^{1/2}$ is non-singular (Balas, 1979). To prove this, let us defined the total system energy $E(t)$ as

$$E(t) = 1/2\left(\dot{z}^{\mathrm{T}}\dot{z} + z^{\mathrm{T}}\mathbf{\Lambda}z\right) \tag{11.4.5}$$

Using (11.4.1), we can also write

$$\dot{E}(t) = \dot{z}^{\mathrm{T}}\ddot{z} + \dot{z}^{\mathrm{T}}\mathbf{\Lambda}z = -2\zeta\dot{z}^{\mathrm{T}}\mathbf{\Lambda}^{1/2}\dot{z} + \dot{z}^{\mathrm{T}}\boldsymbol{B}\boldsymbol{f} \tag{11.4.6}$$

From (11.4.3) and (11.4.4),

$$\boldsymbol{f} = -\boldsymbol{Q}\boldsymbol{B}^{\mathrm{T}}z \tag{11.4.7}$$

Therefore,

$$\dot{E}(t) = -\dot{z}^{\mathrm{T}}\left(2\zeta\mathbf{\Lambda}^{1/2} + \boldsymbol{B}\boldsymbol{Q}\boldsymbol{B}^{\mathrm{T}}\right)\dot{z} \tag{11.4.8}$$

Note, however, that $\Lambda^{1/2} \geq \mathbf{0}$, and $\boldsymbol{BQB}^{\mathrm{T}} \geq \mathbf{0}$, and so the derivative of the total system energy, defined in (11.4.8), must be negative. Therefore, the value of the total system energy must be decreasing, so the system is dissipative. If $\Lambda>0$, then the total system energy is positive while its first derivative is negative for all non-zero states $(z,\dot{z})$, then the total system energy $E(t)$ is a Lyapunov function. Therefore, from (11.4.8) the closed-loop system is stable for any $\zeta\geq 0$, and asymptotically stable if the matrix $\boldsymbol{W}=2\zeta\Lambda^{1/2}+\boldsymbol{BQB}^{\mathrm{T}}$ is positive definite. This is true if either $\zeta>0$, or $\xi\geq 0$ and $\boldsymbol{BQ}^{1/2}$ is non-singular, as stated above (Balas, 1979).

The important significance of this result is that for collocated sensors and actuators, system stability is maintained regardless of the number of modes in the model (to be controlled), regardless of the inaccuracy of the knowledge of parameters. The spillover problem is completely avoided.

If the system response does included contributions from rigid body modes, such that Λ contains zero eigenvalues, then a sufficient condition for stability is that the energy in the zero resonance frequency modes remain constant (that is, the control actuators do no excite the rigid body modes). If the zero resonance frequency modes are damped, then the system is asymptotically stable (Balas, 1979).

When velocity feedback is used with collocated sensors and actuators, the resulting closed-loop transfer function has alternating poles and zeroes on the imaginary axis (Martin and Bryson, 1980; Lee and Speyer, 1993). With this arrangement, the system can always be stabilized (with lead compensation) (Martin and Bryson, 1980). Unfortunately, sensor and actuator dynamics, if unmodelled, can alter this pole/zero arrangement, and lead to system instability (Goh and Caughey, 1985; Inman, 1990). As we discussed in Section 11.2, a small amount of passive damping can add stability robustness, and help to alleviate this problem.

The final point to mention here is that wave-absorbing controllers (von Flotow and Schafer, 1986), developed for large space structure control for which description of the system response in terms of modes is undesirable, bear a distinct resemblance to the collocated controllers described here. In fact, the wave-absorbing compensators designed by von Flotow and Schafer (1986) can be viewed as velocity feedback modified to have a slightly different phase, with the gain increasing with frequency.

11.5 A BRIEF NOTE ON MODEL REDUCTION

The controllers we have considered in this chapter are the same order as the systems themselves (feedback from each state). Given this fact, it is then worthwhile considering whether a smaller (lower-order) controller could be designed which would achieve approximately the same performance. Lower-order controllers are usually preferable to higher order controls for reasons of reduced complexity, which leads to simplified implementation and enhanced physical insight.

There are three basic ways in which the reduction of the order of a controller could be approached: approximation of the dynamic system (plant) by a reduced order model prior to the controller design, such that the latter process is based entirely upon the reduced order model; reduction of the (full order) controller size after it has been designed from consideration of the whole of the dynamic system; and formulation of the controller design problem with order constraints imposed from the start. The first and third of these methods are referred to in various texts as direct methods, while the second is referred to as an indirect method. In this section we will consider only the reduction of the model prior to control system design. For a thorough discussion of techniques for reducing the size of the controller after its formulation, refer to Anderson and Moore (1990); a comparison of several techniques can also be found in Liu and Anderson (1986). For a thorough discussion of the controllers with order constraints imposed upon them, refer to Bernstein and Hyland (1990). Also, for a thorough discussion of the model reduction problem we will briefly discuss here, the interested reader is referred to Gawronski and Juang (1990).

The approach of reducing the size of a system model prior to design of the feedback control system is one which we have used in the previous sections in this chapter. It is also one which general practitioners of control try to avoid, for a number of reasons. One objection is that contradicts the (heuristic, but generally correct) notion that if there is any approximation in the design process, it should be postponed as long as possible. In fact, in early work on control of distributed parameter systems, a practical constraint placed upon controller design was that 'the distributed nature of the system should be retained until numerical results are required' (Athans, 1970). A second objection arises from the idea (Anderson and Moore, 1990) that

> what constitutes satisfactory approximation of the plant necessarily involves the controller: it is the closed-loop behaviour that one is ultimately interested in, and it is clear that a controller design could yield a situation in which very big variations in the open loop plant in a limited frequency range had little effect on the closed-loop performance, while rather small variations in another frequency range could dramatically affect the closed-loop behaviour. Now since the definition of a good plant approximation involves the controller, and since the controller is not known at the time of approximation, one is caught in a logical loop.

Despite these objections, it is often the case that the system model must be reduced prior to controller design simply because its size is simply unmanageable in the calculation procedure. For systems described by large order finite element models, common techniques for model reduction are the Guyan-Irons reduction method (Craig, 1981), or application of component mode synthesis techniques (Craig, 1987). When the dynamic system is described in terms of system modes, a common technique for model reduction is modal cost analysis (Skelton and Hughes, 1980; Skelton and Yousuff, 1982,1983; Skelton *et al.*, 1990; Yousuff *et al.*, 1985; Yousuff and Breida, 1992).

The concept behind modal cost analysis can be summarized as follows. Given a dynamic system, describe in terms of its normal modes such that the system output is

$$y = \sum_{i=1}^{n} \left(c_i z_i + e_i \dot{z}_i\right) \tag{11.5.1}$$

where

$$\ddot{z}_i + 2\zeta_i \omega_i \dot{z}_i + \omega_i^2 z_i = f_i \tag{11.5.2}$$

find a subset of r system modes, the output of which is

$$y_r = \sum_{i=1}^{r} \left(c_i z_i + e_i \dot{z}_i\right) \tag{11.5.3}$$

such that the error between the full order model and reduced order model, defined as the difference between the output of the two,

$$e(t) = y(t) - y_r(t) \tag{11.5.4}$$

is minimized. For a given number of modes in the subset r, the optimal set of constituents would minimize the model error δJ, defined as

$$\delta J = \lim_{t\to\infty} E\left\{\frac{1}{t}\int_0^t e^{\mathrm{T}}(\tau)e(\tau)\,\mathrm{d}\tau\right\} \tag{11.5.5}$$

For a given system, the model error is usually computed after the reduced order model is specified. The model reduction process then takes on the form of an iterative procedure, where different combinations of systems modes are trialled, and the 'best' set retained as the reduced order model. The simplify this process, component cost analysis (referred to as modal cost analysis when the system is described in terms of its normal modes) can be employed. Modal cost analysis uses a modified version of (11.5.5) for the criterion for model reduction; modal cost analysis aims to minimize the cost error ΔJ, defined as

$$\Delta J = J - J_r \tag{11.5.6}$$

where

$$J = \lim_{t\to\infty} E\left\{\frac{1}{t}\int_0^t y^{\mathrm{T}}(\tau)y(\tau)\,\mathrm{d}\tau\right\} \tag{11.5.7}$$

and

$$J_r = \lim_{t\to\infty} E\left\{\frac{1}{t}\int_0^t y_r^{\mathrm{T}}(\tau)y_r(\tau)\,\mathrm{d}\tau\right\} \tag{11.5.8}$$

If the reduced order model is optimal, it will satisfy the orthogonality condition (Wilson, 1974):

$$\langle e,y_r\rangle = \lim_{t\to\infty} E\left\{\frac{1}{t}\int_0^t e^{\mathrm{T}}(\tau)y_r(\tau)\,\mathrm{d}\tau\right\} = 0 \tag{11.5.9}$$

where < > is an inner product. The model error in (11.5.5) can be expressed in terms of the cost error of (11.5.6) and the inner product in (11.5.9)

$$\delta J = \Delta J - 2\langle e,y_r\rangle \tag{11.5.10}$$

It can be surmised that minimizing the cost error of (11.5.9) is a satisfactory approach provided that the reduced order model is known to be at least near optimal. However, we still have the problem that we require the reduced order model to be known before we can calculate the error criterion. To overcome this problem, modal cost analysis uses a predicted cost error, defined as

$$\Delta J' = J - J_r' \tag{11.5.11}$$

where $'$ denotes a predicted value. J_r' is calculated from

$$J_r' = \sum_{i\in r} J_i' \tag{11.5.12}$$

where J_i' is the ith modal cost, defined as

$$J_i' = \frac{1}{2}\lim_{t\to\infty} E\left\{\frac{1}{t}\int_0^t \frac{\partial[y^{\mathrm{T}}(\tau)y(\tau)]}{\partial z_i(\tau)}z_i(\tau)\,\mathrm{d}\tau\right\} \tag{11.5.13}$$

(In the more general component cost analysis (Skelton and Yousuff, 1983), the modal state z_i is replaced by the state of some component on the structure.) As the modal costs can be calculated prior to calculation of the error criterion, the predicted cost error can be calculated before the formulation of the reduced order model. Therefore, modal cost analysis aims to derive an 'optimal' reduced order model by minimizing the predicted cost error defined in (11.5.11).

One problem with modal cost analysis is that, except in special circumstances, there is no guarantee that the predicted cost error is equal to the model error, and therefore the result may not be the optimal reduced order model. It is, however, possible to calculate the model error directly in much the same manner as the predicted cost error (Yousuff and Breida, 1992). Let us define a modal output as

$$y_i = c_i z_i + e_i\dot{z}_i \tag{11.5.14}$$

such that the system output is simply defined by the expression

$$y = \sum_{i\in N_n} y_i \tag{11.5.15}$$

If we then define the residual modes as the difference between the total modal set and the reduced order subset,

$$N_{res} = N_n - N_r \tag{11.5.16}$$

we find that

$$e = \sum_{i \in N_{res}} y_i \tag{11.5.17}$$

From this, the model error is

$$\delta J = \sum_{i \in N_{res}} \sum_{j \in N_{res}} y_{ij} \tag{11.5.18}$$

where

$$y_{ij} = \lim_{t \to \infty} E \left\{ \frac{1}{t} \int_0^t y_i^{\mathrm{T}}(\tau) y_j(\tau) \, \mathrm{d}\tau \right\} \tag{11.5.19}$$

This relationship can be used in the modal cost analysis procedure. Improved results obtained using this approach are found in Yousuff and Breida (1992).

The relationship between the 'maximum possible' disturbance attenuation and that provided by a control system designed using a reduced order model is directly related to the cost function J. If the cost is 'small', then the performance of the control system designed using a reduced order model can be expected to be good.

One final point to mention is that even if a reduced order model is used in the control law formulation, it may be necessary to reduce its order further for implementation. In the case, the various procedures outlined in Anderson and Moore (1990) for controller reduction are necessary.

11.6 SENSOR AND ACTUATOR PLACEMENT CONSIDERATIONS

In the control law development outlined in this chapter the approach taken has basically been one of assuming that the location of sensors and actuators on a structure is given, and the problem has been one of formulating an optimal control law (with reference to some performance objectives) given these location constraints. It is intuitively obvious, however, that sensor and actuator placement will have a large influence upon the ultimate performance of the control system. For example, in a poor system design an actuator could be placed on or near the nodal line of mode which is to be controlled, the result being that either an excessively large force is required for control at best, or that the mode is uncontrollable at worst. Similarly, a sensor could be placed on or near a nodal line of a mode to be controlled, meaning that the signal to noise ratio is poor at best, or the mode is unobservable (and hence uncontrollable) at worst.

A large number of strategies for optimising the placement of sensors and actuators have been developed in recent years, largely based upon the idea of minimising some performance index associated with transducer placement. For (probably) the majority of these, the performance index is based upon some measure of controllability and observability, derived from minimum energy considerations (see, for example, Hughes and Skelton, 1980b; Arbel, 1981;

Vander Velde and Carignan, 1984; Hac and Liu, 1992; Maghami and Joshi, 1993). In this section we will briefly outline some of the concepts associated with sensor and actuator placement based upon energy considerations, which will be seen to be intimately connected to ideas of controllability and observability. A more complete tutorial-like discussion of actuator and sensor placement issues can be found in an article by Baruh (1992).

11.6.1 Actuator placement

In placing actuators on a structure, our aim will be to excite the structure with the minimum control effort for the various operating conditions. The operating conditions we will be particularly interested in here are transient response recovery, and attenuation of persistent excitation.

Heuristically, the criterion of minimum control effort for actuator placement can be seen to take into account the metric of controllability, because if a mode is uncontrollable the control effort required is infinite. However, simply assessing controllability in itself is not necessarily a good criterion for actuator placement; an actuator can be extremely close to a nodal line of a mode of interest and the system will still be (theoretically) controllable, although the control effort required for the control may be in excess of that practically achievable.

The response of the systems we are interested in here are again described in terms of system modes, where the governing equation for each mode is

$$\ddot{z}_i(t) + 2\zeta\omega_i\dot{z}_i(t) + \omega_i^2 z_i(t) = f_i(t) \tag{11.6.1}$$

This can be expressed in state space form as

$$\dot{\boldsymbol{x}} = \boldsymbol{A}\boldsymbol{x} + \boldsymbol{B}\boldsymbol{u} \tag{11.6.2}$$

In this instance, however, we will write the state vector and matrix in a slightly different form. Defining the state vector (for N modelled modes) as

$$\boldsymbol{x} = \begin{bmatrix} x_1 & \omega_1\dot{x}_1 & \cdots & x_N & \omega_N\dot{x}_N \end{bmatrix}^{\mathrm{T}} \tag{11.6.3}$$

the state matrix is in block diagonal form, given by

$$\boldsymbol{A} = \begin{bmatrix} \boldsymbol{A}_1 & & \\ & \ddots & \\ & & \boldsymbol{A}_N \end{bmatrix}, \quad \boldsymbol{A}_i = \begin{bmatrix} 0 & \omega_i \\ -\omega_i & -2\xi\omega_i \end{bmatrix} \tag{11.6.4}$$

Accordingly, the control input matrix for N_c control forces is now defined as

$$\boldsymbol{B} = \begin{bmatrix} 0 & \cdots & 0 \\ \psi_1(\boldsymbol{r}_1) & \cdots & \psi_1(\boldsymbol{r}_{N_c}) \\ & \vdots & \\ 0 & \cdots & 0 \\ \psi_N(\boldsymbol{r}_1) & \cdots & \psi_N(\boldsymbol{r}_{N_c}) \end{bmatrix} \tag{11.6.5}$$

11.6.1.1 Transient excitation

Let us first consider the transient excitation case, where the system is subject to some perturbation and the control system is to return it to its original state. To guide the actuator placement, consider the problem of returning the perturbed system to equilibrium at some time t_f, subject to minimizing the energy criterion

$$J = \int_0^{t_f} \boldsymbol{u}^{\mathrm{T}}(\tau)\boldsymbol{u}(\tau)\,\mathrm{d}\tau \tag{11.6.6}$$

As we discussed in Chapter 5, the optimal solution to this problem is given by (Kalman *et al.*, 1962):

$$\boldsymbol{u}_{opt}(t) = -\boldsymbol{B}^{\mathrm{T}}\mathrm{e}^{\boldsymbol{A}(t_f-t)}\boldsymbol{P}^{-1}(t_f)(\mathrm{e}^{\boldsymbol{A}\mathrm{T}}\boldsymbol{x}(0)-\boldsymbol{x}(t_f)) \tag{11.6.7}$$

where $\boldsymbol{P}$ is the controllability grammian, defined originally in Chapter 5 as

$$\boldsymbol{P}(t) = \int_0^{t} \mathrm{e}^{\boldsymbol{A}(\tau)}\boldsymbol{B}\boldsymbol{B}^{\mathrm{T}}\mathrm{e}^{\boldsymbol{A}^{\mathrm{T}}(\tau)}\,\mathrm{d}\tau \tag{11.6.8}$$

The minimum energy associated with this optimal result is (Kalman *et al.*, 1962):

$$J_{\min} = \left[\mathrm{e}^{\boldsymbol{A}(t_f)}\boldsymbol{x}(0) - \boldsymbol{x}(t_f)\right]^{\mathrm{T}} \boldsymbol{P}^{-1}(t_f) \left[\mathrm{e}^{\boldsymbol{A}(t_f)}\boldsymbol{x}(0) - \boldsymbol{x}(t_f)\right] \tag{11.6.9}$$

Observe that, from (11.6.7), if the controllability grammian is small (such that the inverse is large), the optimum control forces will be large. Associated with this, the minimum value of the energy criterion, defined in (11.6.9), will be large. Therefore, it is desirable to make the controllability grammian as large as possible (this defines a relationship between controllability and energy considerations). Referring to the definition of the controllability grammian given in (11.6.8), we can see that it is dependent upon the control input matrix $\boldsymbol{B}$, which, from (11.6.5), is dependent upon actuator location. This dependency will enable us to define some degree of quality for a given actuator arrangement: a 'good' actuator arrangement is one which maximizes the norm of the controllability grammian.

As an example of how controllability can vary with actuator position, consider Fig. 11.10, which illustrates the degree of controllability of a free–free beam with three modelled modes. Observe that at the nodal location for each mode the controllability goes to zero, which is to be expected. The maximum controllability it at the ends of the beam, where all modes have an antinode.

Our last problem with using controllability to guide actuator placement arises because the controllability grammian is dependent upon the time designated for return of the system to its equilibrium state following perturbation, t_f. To overcome this problem, we will consider the steady-state solution to the controllability grammian relationship, where t_f is infinite. In this case, for asymptotically stable systems the controllability grammian satisfies the Lyapunov equation

$$\boldsymbol{A}\boldsymbol{P} + \boldsymbol{P}\boldsymbol{A}^{\mathrm{T}} + \boldsymbol{B}\boldsymbol{B}^{\mathrm{T}} = 0 \tag{11.6.10}$$

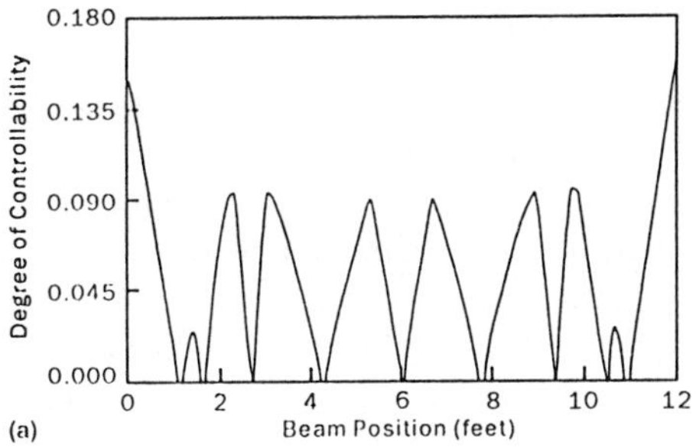

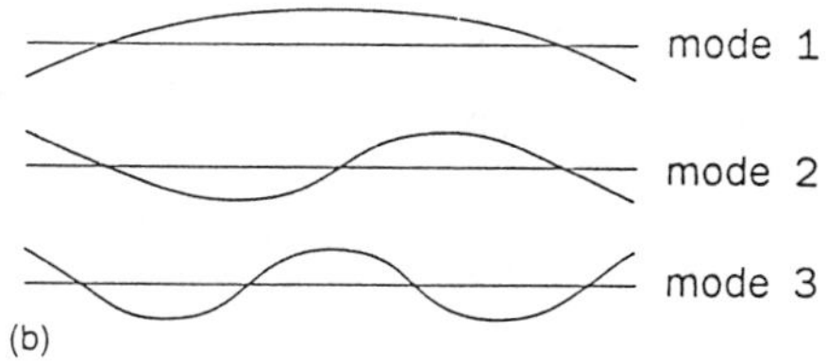

Fig. 11.10 (a) Degree of controllability for a free-free beam (after Vander Velde and Carignan, 1984); (b) sketch of first three mode shapes for a free–free beam.

Because of the special form of the state matrix $\boldsymbol{A}$, a closed form solution for the Lyapunov equation (11.6.10) can be found for system models of any order N (Gawronski and Juang, 1990). Further, if the damping is small, the controllability grammian is essentially block diagonal (Gawronski and Juang, 1990),

$$\boldsymbol{P} \approx \begin{bmatrix} \boldsymbol{P}_1 & & \\ & \ddots & \\ & & \boldsymbol{P}_N \end{bmatrix} \tag{11.6.11}$$

where

$$\boldsymbol{P}_i = \begin{bmatrix} \dfrac{\sum_{\iota=1}^{N_c} \psi_i^2(\boldsymbol{r}_\iota)}{4\zeta\omega_i} & 0 \\ 0 & \dfrac{\sum_{\iota=1}^{N_c} \psi_i^2(\boldsymbol{r}_\iota)}{4\zeta\omega_i} \end{bmatrix} \tag{11.6.12}$$

11.6.1.2 Persistent excitation

For persistent excitation, the objective of actuator placement is to arrive at an arrangement which maximizes the influence which the control system has upon the steady state behaviour. We can attempt to quantify this idea by stating that the energy transmitted to the structure as a whole, as well as to the individual modes, from the actuators should be maximized for a given control effort limitation. To quantify this, suppose the actuator signals are mutually uncorrelated white noise processes of unit intensity, and have a covariance matrix defined by

$$E\{\boldsymbol{u}(t)\boldsymbol{u}^{\mathrm{T}}(t)\} = \boldsymbol{I}\delta(t-\tau) \tag{11.6.13}$$

Under these conditions, the covariance matrix of the system state

$$E\{\boldsymbol{x}(t)\boldsymbol{x}^{\mathrm{T}}(t)\} = \boldsymbol{X}(t) \tag{11.6.14}$$

is a steady state solution of the Lyapunov equation (11.6.10) (Hac and Liu, 1992). Further, the mean value of the total energy of the modelled modes is (Hac and Liu, 1992)

$$E\{E(t)\} = \sum_{i=1}^{N} \frac{\sum_{i=1}^{N_c} \psi_i(\boldsymbol{r}_i)}{4\zeta\omega_i} \tag{11.6.15}$$

Therefore, the total energy is the sum of the diagonal elements of the controllability grammian divided by two.

From consideration of the transient and persistent excitation cases we have two different performance criteria. The transient response case performance criteria is dependent upon the product of the diagonal elements of the controllability grammian, while the persistent excitation case is based upon the sum of the diagonal elements. Therefore, in placing control actuators it is possible to use these criteria separately, or combine them. One proposed form of combination is (Hac and Liu, 1992)

$$J = \left[\sum_{i=1}^{2N} \lambda_j\right] \times \sqrt[2N]{\prod_{i=1}^{2N} \lambda_i} \tag{11.6.16}$$

where λ_i is the ith eigenvalue of the controllability grammian, which is a diagonal element when the controllability grammian is diagonal.

11.6.2 Sensor placement

Design of the sensing system can be viewed as the dual of the design of the actuating system. Rather than being interested in maximizing the response of the structure as a whole, as well as its individual modes, to a given control input, for the sensing problem we are interested in maximizing the (sensor) signal power for a given excitation of the structure. As before, we will be interested in both the transient excitation and persistent excitation conditions.

11.6.2.1 Transient excitation

Consider first the transient excitation problem, and for simplicity assume that only displacement sensors are being used in the sensing system (a similar development is possible with velocity sensors). In this case the system output is

$$y(t) = Cx(t) \tag{11.6.17}$$

where the output matrix C is defined by

$$C = \begin{bmatrix} \psi_1(r_1) & \cdots & \psi_1(r_{N_e}) \\ 0 & \cdots & 0 \\ & \vdots & \\ \psi_N(r_1) & \cdots & \psi_N(r_{N_e}) \\ 0 & \cdots & 0 \end{bmatrix} \tag{11.6.18}$$

where there are N_e sensors in the system. If the system is perturbed at time 0, starting with state $x(0)$, the output is defined by the relationship

$$y(\tau) = Ce^{A(\tau)}x(0) \tag{11.6.19}$$

The system output energy, which we wish to maximize with our sensor placement, is defined by

$$J = \int_0^{t_f} y^{\mathrm{T}}(\tau)y(\tau)\,\mathrm{d}\tau \tag{11.6.20}$$

Substituting (11.6.19) into (11.6.20), the system output energy can be written as

$$J = x^{\mathrm{T}}(0)M(t_f)x(0) \tag{11.6.21}$$

where M is the observability grammian which, from our definition in Chapter 5, can be written as

$$M(t_f) = \int_0^{t_f} \mathrm{e}^{A^{\mathrm{T}}(t_f)}C^{\mathrm{T}}C\mathrm{e}^{A(t_f)} \tag{11.6.22}$$

As an example of how the observability grammian can vary with sensor position, consider Fig. 11.11, which illustrates the degree of observability for the same free-free beam used in Fig. 11.10. Observe that again at the nodal location for each mode the observability goes to zero. The maximum observability is again at the edge of the beam, where all modes have an antinode.

As with the controllability grammian in the previous section, to overcome the time dependency in (11.6.21) (as t_f is usually not defined *a priori*), we can work with the steady state case. Here, for asymptotically stable systems, the controllability grammian satisfies the Lyapunov equation

$$A^{\mathrm{T}}M + MA + C^{\mathrm{T}}C = 0 \tag{11.6.23}$$

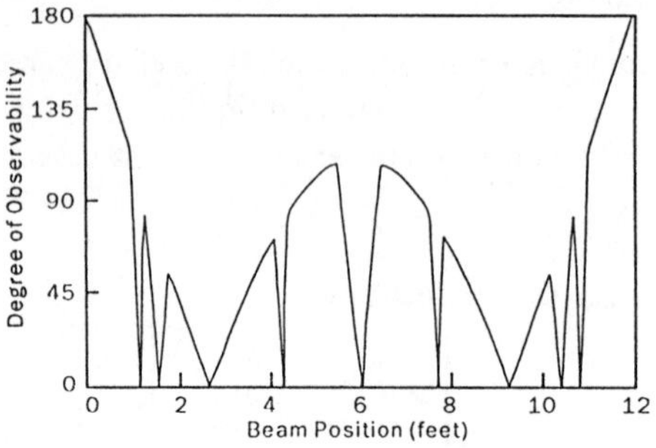

Fig. 11.11 Degree of observability for a free–free beam (after Vander Velde and Carignan, 1984).

Because of the special form of the state matrix $\boldsymbol{A}$, we can again find a closed form solution for the Lyapunov equation (11.6.23). Further, if the damping is small, the observability grammian is essentially block diagonal,

$$\boldsymbol{M} \approx \begin{bmatrix} \boldsymbol{M}_1 & & \\ & \ddots & \\ & & \boldsymbol{M}_N \end{bmatrix} \tag{11.6.24}$$

where (Hac and Liu, 1992)

$$\boldsymbol{M}_i = \begin{bmatrix} \dfrac{\sum_{\iota=1}^{N_e} \psi_i^2(\boldsymbol{r}_\iota)}{4\zeta\omega_i^3} & 0 \\ 0 & \dfrac{\sum_{\iota=1}^{N_e} \psi_i^2(\boldsymbol{r}_\iota)}{4\zeta\omega_i^3} \end{bmatrix} \tag{11.6.25}$$

In this case, the system output energy is (Hac and Liu, 1992)

$$\boldsymbol{x}^{\mathrm{T}}(0)\boldsymbol{M}\boldsymbol{x}(0) = \sum_{i=1}^{N} \frac{\sum_{\iota=1}^{N_e} \psi_i^2(\boldsymbol{r}_\iota)}{4\zeta\omega_i^3}[x_{2i-1}^2(0) + x_{2i}^2(0)] \tag{11.6.26}$$

Clearly, to maximize the system output energy, both the individual diagonal terms, as well as their sum, must be maximized. This can be accomplished by maximizing the norm of the observability grammian. This result is the dual of the actuator problem, where we found that the norm of the controllability grammian must be maximized.

11.6.2.2 Persistent excitation

Let us now consider the persistent excitation problem, where we will assume that a white noise process excites all modes with equal strength. In this case, the mean square value of the system output is

$$J = E\{y^{\mathrm{T}}(t)y(t)\} = tr[(C^{\mathrm{T}}C)X(t)] \tag{11.6.27}$$

where $X(t)$ is the previously defined covariance matrix of the state vector. If the eigenvalues of the state matrix are well spaced, so that the spectrum contains isolated resonances, and the damping is small then the convariance matrix is approximately diagonal. As a result, the mean square value of the system output is (Hac and Liu, 1992)

$$J = E\{y^{\mathrm{T}}(t)y(t)\} = \sum_{i=1}^{N} \frac{\sum_{\iota=1}^{N_e} \psi_i^2(r_\iota)}{4\zeta\omega_i^3} \tag{11.6.28}$$

Observe that this result is the dual of the actuator placement result for persistent excitation given in (11.6.15).

From consideration of the transient and persistent excitation cases we again have two different performance criteria; the transient response case performance criteria is dependent upon the product of the diagonal elements of the controllability grammian, while the persistent excitation case is based upon the sum of the diagonal elements. Therefore, as in placing control actuators, when placing sensors it is possible to use these criteria separately, or combine them. This combination can be the same as outlined in (11.6.16), but where λ_i is now the ith eigenvalue of the observability grammian. Because the placement criteria are duals, when velocity sensors and force actuators are used, their optimal placements coincide.

11.6.3 A few additional comments

There are a few additional points which need to be made in respect to the above developments. First, as the sensor and actuator placement exercise is dependent upon both the sum and product of the eigenvalues of the controllability and observability grammians, it is important that the eigenvalues be properly scaled (of similar value) to avoid any bias in the result. By using $[z_i, \omega_i\dot{z}_i]$ as system states, rather than $[z_i, \dot{z}_i]$, both eigenvalues associated with a given mode are equal. If different states are used, different eigenvalues result.

The development above was only concerned with sensing and actuating modes in the reduced order system model. The development could be expanded to account for the problem of control and observation spillover. To do this, the criterion of (11.6.16) could be formulated for both controlled modes and residual modes. The criterion used in actuator and sensor placement could then be equal to the controlled mode value minus some fraction of the residual mode value, the factor weighting the importance of overcoming spillover problems. Further consideration of actuator placement and spillover can be found in Lindberg and Longman (1982).

Finally, one other consideration in sensor and actuator placement which has been studied is the ability to recognise component failures, and minimise the reduction in controllability and observability which will arise (Vander Velde and Carignan, 1984; Baruh and Choe, 1990a, 1990b; Baruh, 1992). One approach to doing this is to assign a probability density function to state of failures of components, and use this to derive a measure of the expected value of the degree of controllability and/or observability over a given operating period. Further description of this can be found in Vander Velde and Carignan (1984).

REFERENCES

Anderson, B.D.O. and Moore, J.B. (1990). *Optimal Control, Linear Quadratic Methods*. Prentice-Hall: Englewood Cliffs, NJ.

Arbel, A. (1981). Controllability measures and actuator placement in oscillatory systems. *International Journal of Control*, **33**, 565–574.

Athans, M. (1970). Toward a practical theory of distributed parameter systems. *IEEE Transactions on Automatic Control*, **AC-15**, 245–247.

Atluri, S.N. and Amos, A.K. (1988). *Large Space Structures: Dynamics and Control*. Springer-Verlag: Berlin.

Aubrun, J.N. (1980). Theory of control of structures by low-authority controllers. *Journal of Guidance and Control*, **3**, 444–451.

Balas, M.J. (1978a). Feedback control of flexible systems. *IEEE Transactions on Automatic Control*, **AC-23**, 673–679.

Balas, M.J. (1978b). Active control of flexible systems. *Journal of Optimization Theory and Applications*, **25**, 415–436.

Balas, M.J. (1979). Direct velocity feedback control of large space structures. *Journal of Guidance and Control*, **2**, 252–253.

Balas, M.J. (1982). Trends in large space structure control theory: fondest hopes, wildest dreams. *IEEE Transactions on Automatic Control*, **AC-27**, 522–535.

Balas, M.J. (1990). Low order control of linear finite-element models of large flexible structures using second-order parallel architectures, in *Mechanics and Control of Large Flexible Structures*, J.L. Junkins, ed. AIAA Press: Washington, DC.

Baruh, H. (1992). Placement of sensors and actuators in structural control, in *Control and Dynamic Systems*, Vol. 52, C.T. Leondes, ed. Academic Press: San Diego.

Baruh, H. and Choe, K. (1990a). Sensor placement in structural control. *Journal of Guidance, Control, and Dynamics*, **13**, 524–533.

Baruh, H. and Choe, K. (1990b). Reliability issues in structural control, in *Control and Dynamic Systems*, Vol. 32, C.T. Leondes, ed. Academic Press: San Diego.

Baz, A., Poh, S. and Studer, P. (1989). Modified independent modal space control method for active control of flexible systems. *Proceedings of the Institution of Mechanical Engineers*, **203**, 103–112.

Baz, A. and Poh, S. (1990). Experimental implementation of the modified independent modal space control method. *Journal of Sound and Vibration*, **139**, 133–149.

Benhabibm R.J., Iwens, R.P. and Jackson, R.L. (1981). Stability of large space structure control systems using positivity concepts. *Journal of Guidance and Control*, **4**,

487–494.

Bennighof, J.K. and Meirovitch, L. (1988). Active vibration control of a distributed system with moving support. *Journal of Vibration, Acoustics, Stress, and Reliability in Design*, **110**, 246–253.

Bernstein, D.S. and Hyland, D.C. (1990). Optimal projection approach to robust fixed-structure control design, in *Mechanics and Control of Large Flexible Structures*, J.L. Junkins, ed. AIAA Press; Washington, DC.

Bongiorno, J. and Youla, D. (1968). On observers in multi-variable control systems. *International Journal of Control*, **8**, 221–243.

Chen, C. (1970). *Introduction to Linear System Theory*. Holt, Rinehart and Winston: New York.

Choe, K. and Baruh, H. (1992). Actuator placement in structural control. *Journal of Guidance, Control, and Dynamics*, **15**, 40–48.

Craig, R.R., Jr. (1981). *Structural Dynamics: An Introduction to Computer Methods*. Wiley: New York.

Craig, R.R., Jr. (1987). A review of time-domain and frequency-domain component-mode synthesis techniques. *International Journal of Analytical and Experimental Modal Analysis*, **2**, 59–72.

Creamer, N.G. and Junkins, J.L. (1991). Low-authority eigenvalue placement for second-order structural systems. *Journal of Guidance, Control, and Dynamics*, **14**, 698–701.

Gawronski, W. and Juang, J. (1990). Model reduction for flexible structures", in *Control and Dynamics*, Vol. 36, C.T. Leondes, ed. Academic Press: San Diego, 143–222.

Goh, C.J. and Caughey, T.K. (1985). On the problem caused by finite actuator dynamics in the collocated control of large space structures. *International Journal of Control*, **41**, 787–802.

Grandhi, R.V. (1990). Optimum design of space structures with active and passive damping. *Engineering With Computers*, **6**, 117–183.

Gueler, R., von Flotow, A.H. and Vos, D.W. (1993). Passive damping for robust feedback control of flexible structures. *Journal of Guidance, Control and Dynamics*, **16**, 662–667.

Hac, A. and Liu, L. (1992). Sensor and actuator location in motion control of flexible structures. *Proceedings of First International Conference on Motion and Vibration Control (MOVIC)*, Yokohama, 86–91.

Hallauer, W.L., Jr. (1990). Recent literature on experimental structural dynamics and control research, in *Mechanics and Control of Large Flexible Structures*, J.L. Junkins, ed. AIAA Press; Washington, DC.

Hallauer, W.L., Jr., Skidmore, G.R. and Mesquita, L.C. (1982). Experimental-theoretical study of active vibration control. *Proceedings of the International Modal Analysis Conference*, 39–45.

Hanagud, S., Obal, M.W. and Calise, A.J. (1992). Optimal vibration control by the use of piezoceramic sensors and actuators. *Journal of Guidance, Control, and Dynamics*, **15**, 1199–1206.

Hughes, P.C. and Skelton, R.E. (1980a). Controllability and observability of linear matrix-second-order systems. *Journal of Applied Mechanics*, **47**, 415–420.

Hughes, P.C. and Abdel-Rahman, T.M. (1979). Stability of proportional-plus-derivative-plus-integral control of flexible spacecraft. *Journal of Guidance and Control*, **2**, 499–503.

Hughes, P.C. and Skelton, R.E. (1980b). Controllability and observability for flexible spacecraft. *Journal of Guidance and Control*, **3**, 452–459.

Inman, D.J. (1990). Control/structure interaction: effects of actuator dynamics, in *Mechanics and Control of Large Flexible Structures*, J.L. Junkins, ed. *AIAA Progress in Astronautics and Aeronautics*, 507–533.

Joshi, S.M. (1981). *A Controller Design Approach for Large Flexible Space Structures. NASA Contractor Report*, CR–165717.

Joshi, S.M. (1986). Robustness properties of collocated controllers for flexible spacecraft. *Journal of Guidance, Control, and Dynamics*, **9**, 85–91.

Joshi, S.M. (1989). *Control of Large Flexible Space Structures*. Springer-Verlag: Berlin.

Junkins, J.L. ed. (1990). *Mechanics and Control of Large Flexible Structures*. AIAA Press: Washington, DC.

Kwak, M.K. and Meirovitch, L. (1993). An algorithm for the computation of optimal control gains for second order matrix equations. *Journal of Sound and Vibration*, **166**, 45–54.

Lee, C.-K. and Moon, F.C. (1990). Modal sensors/actuators. *Journal of Applied Mechanics*, **57**, 434–441.

Lee, Y.J. and Speyer, J.L. (1993). Zero locus of a beam with varying sensor and actuator locations. *Journal of Guidance, Control, and Dynamics*, **16**, 21–25.

Leondes, C.T. (1990). *Control and Dynamic Systems*. Academic Press: San Diego.

Lindberg, R.E. and Longman, R.W. (1982). *Optimization of Actuator Placement via Degree of Controllability Criteria Including Spillover Considerations. AIAA Paper*, **82-1435**.

Liu, Y. and Anderson, B.D.O. (1986). Controller reduction via stable factorization and balancing. *International Journal of Control*, **44**, 507–531.

MacMartin, D.G. and Hall, S.R. (1991). Structural control experiments using an H_∞ power flow approach. *Journal of Sound and Vibration*, **148**, 223–241.

Maghami, P.G. and Joshi, S.M. (1993). Sensor-actuator placement for flexible structures with actuator dynamics. *Journal of Guidance, Control, and Dynamics*, **16**, 301–307.

Martin, G.D. and Bryson, A.E., Jr. (1980). Attitude control of a flexible spacecraft. *Journal of Guidance and Control*, **3**, 37–41.

Martinovic, Z.N., Schamel, G.C., Hafta, R.T. and Hallauer, W.L. (1990). Analytical and experimental investigation of output feedback vs linear quadratic regulator. *Journal of Guidance, Control, and Dynamics*, **13**, 160–167.

Meirovitch, L. and Baruh, H. (1982). Control of self-adjoint distributed parameter systems. *Journal of Guidance, Control, and Dynamics*, **5**, 60–66.

Meirovitch, L. and Baruh, H. (1985). The implementation of modal filters for control of structures. *Journal of Guidance, Control, and Dynamics*, **8**, 707–716.

Meirovitch, L., Baruh, H. and Oz, H. (1983). A comparison of control techniques for large flexible systems. *Journal of Guidance, Control, and Dynamics*, **6**, 302–310.

Meirovitch, L. and Bennighof, J.K. (1986). Modal control of travelling waves in flexible structures. *Journal of Sound and Vibration*, **111**, 131–144.

Meirovitch, L. and Oz, H. (1980a). Modal-space control of large flexible spacecraft possessing ignorable coordinates. *Journal of Guidance and Control*, **3**, 569–577.

Meirovitch, L. and Oz, H. (1980b). Modal-space control of distributed gyroscopic systems. *Journal of Guidance and Control*, **3**, 140–150.

Meirovitch, L. and Oz, H. (1980c). Optimal modal-space control of flexible gyroscopic systems. *Journal of Guidance and Control*, **3**, 218–226.

Meirovitch, L., van Landingham, H.F. and Oz, H. (1977). Control of spinning flexible spacecraft by modal synthesis. *Acta Astronautica*, **4**, 985–1010.

Meirovitch, L., van Landingham, H.F. and Oz, H. (1979). Distributed control of spinning flexible spacecraft. *Journal of Guidance and Control*, **2**, 407–415.

Miller, D.W. and von Flotow, A.H. (1989). Power flow in structural networks. *Journal of Sound and Vibration*, **128**, 145–162.

Miller, D.W., Hall, S.R. and von Flotow, A.H. (1990). Optimal control of power flow at structural junctions. *Journal of Sound and Vibration*, **140**, 475–497.

Morgan, D.R. (1991). An adaptive modal-based active control system. *Journal of the Acoustical Society of America*, **89**, 248–256.

Nurre, G.S., Ryan, R.S., Scofield, H.N. and Sims, J.L. (1984). Dynamics and control of large space structures. *Journal of Guidance, Control, and Dynamics*, **7**, 514–526.

Noble, B. and Daniel, J.W. (1977). *Applied Linear Algebra*, 2nd edn. Prentice-Hall: Englewood Cliffs, NJ.

Porter, B. and Crossley, T.R. (1972). *Modal Control: Theory and Applications*. Taylor and Francis: London.

Rao, S.S., Pan, T. and Venkayya, V.B. (1990). Robustness improvements of actively controlled structures through structural modification. *AIAA Journal*, **28**, 353–361.

Schulz, G. and Heimbold, G. (1983). Dislocated actuator/sensor positioning and feedback design for flexible structures. *Journal of Guidance, Control, and Dynamics*, **6**, 361–367.

Simon, J.D. and Mitter, K. (1968). A theory of modal control. *Information and Control*, **13**, 316–353.

Skelton, R.E. and Hughes, P.C. (1980). Modal cost analysis of linear matrix second order systems. *Journal of Dynamic Systems, Measurement, and Control*, **102**, 151–158.

Skelton, R.E. and Yousuff, A. (1982). Component cost analysis of large-scale systems, in *Control and Dynamic Systems*, Vol. 18, C.T. Leondes, ed. Academic Press: San Diego.

Skelton, R.E. and Yousuff, A. (1983). Component cost analysis of large-scale systems. *International Journal of Control*, **37**, 285–304.

Skelton, R.E., Singh, R. and Ramakrishnan, J. (1990). Component model reduction by component cost analysis, in *Mechanics and Control of Large Flexible Structures*, J.L. Junkins, ed. AIAA Press: Washington, DC.

Spanos, J.T. (1989). Control-structure interaction in precision pointing servo loops. *Journal of Guidance, Control, and Dynamics*, **12**, 256–263.

Vander Velde, W.E. and Carignan, C.R. (1984). Number and placement of control system components considering possible failures. *Journal of Guidance, Control, and Dynamics*, **6**, 703–709.

von Flotow, A.H. (1986). Disturbance propagation in structural networks. *Journal of Sound and Vibration*, **106**, 433–450.

von Flotow, A.H. (1988). The acoustic limit of control of structural dynamics, in *Large Space Structures: Dynamics and Control*, S.N. Atluri and A.K. Amos, ed. Springer-Verlag: Berlin.

von Flotow, A.H. and Schafer, B. (1986). Wave-absorbing controllers for a flexible beam. *Journal of Guidance, Control, and Dynamics*, **9**, 673–680.

Wilson, D.A. (1974). Model reduction for multivariable systems. *International Journal of Control*, **20**, 57–64.

Yousuff, A. and Breida, M. (1992). Model reduction of mechanical systems. *Journal of Guidance, Control, and Dynamics*, **16**, 408–410.

Yousuff, A., Wagie, D.A. and Skelton, R.E. (1985). Linear systems approximations via covariance equivalent realizations. *Journal of Mathematical Analysis and Applications*, **106**, 91–114.

12

Vibration isolation

12.1 INTRODUCTION

Active vibration isolation involves the use of an active system to reduce the transmission of vibration from one body or structure to another. A broader definition would also include the reduction of vibration of a machine or structure by an active vibration absorber. Passive vibration isolation is covered adequately in many textbooks (see, for example Bies and Hansen, 1996) and will not be discussed here. In the analyses discussed in this chapter a constant force (or infinite impedance) source is assumed. That is, it is assumed that the driving force is independent of the structure and does not change significantly if the dynamics of the structure change. Although this idealized case is not often found in practice, the constant force assumption simplifies complex analyses and the results obtained are indicative of what can be achieved in many practical cases.

Active vibration isolation systems are usually much more complex and expensive than their passive counterparts which consist of steel or rubber springs and dashpots (illustrated diagrammatically in Fig. 12.1(a)) and which have been in use for many years. So one may well ask what advantages are offered by active systems which justify their increased cost and complexity. Of course, the main advantages are better static stability of the supported equipment and better performance, especially at low frequencies, which in many cases makes an active system the only feasible choice. Active systems can also be used to minimize vibration at critical locations on a flexible support structure, at some distance from the isolator attachment point and for some applications this is a distinct advantage. Active systems also have the capability of adjusting to changes in machine operating conditions (and thus vibration excitation frequencies) without any outside intervention. Another important advantage of active systems is that they can dissipate energy as well as supply it, although a major disadvantage in

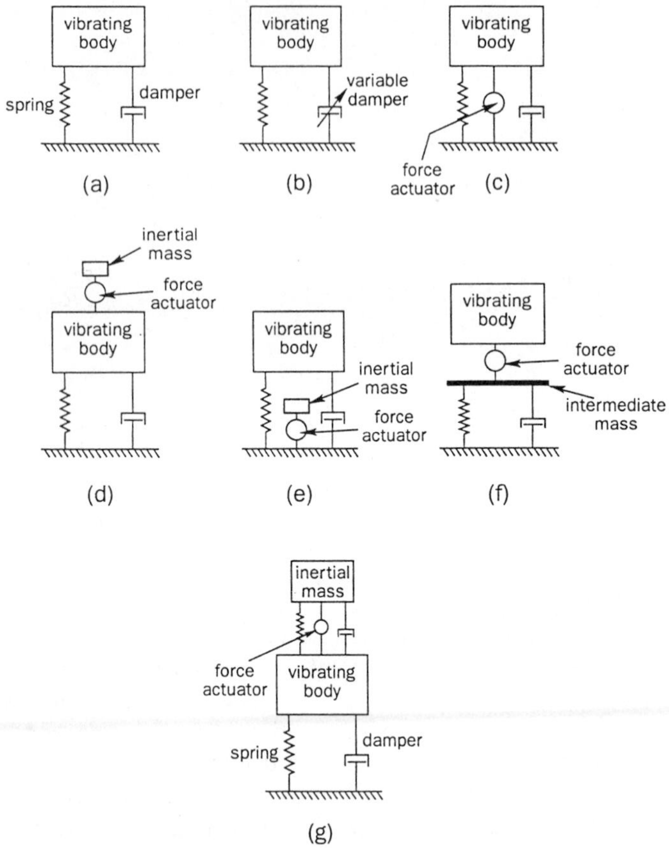

Fig. 12.1 Passive and semi-active vibration isolation: (a) passive system showing a conventional spring and damper; (b) semi-active system with a variable damper; (c) semi-active system with a force on both the vibrating body and the support structure; (d) semi-active system with a control force applied only to the vibrating body; (e) semi-active system with a control force applied only to the structure to be isolated from the vibration source; (f) semi-active system with the force actuator in series with the passive elements; (g) semi-active vibration absorber.

addition to cost and complexity is the requirement of an external power source and in many cases the need for numerous sensors as well as actuators which could have durability problems.

Active systems have been used in the past to isolate optical systems from support structure vibrations, vehicle cabins from tyre vibrations generated by an uneven road surface, space telescopes from vibrations generated by driving equipment, vehicles from engine induced vibrations, helicopter cabins from rotor gearbox vibration and the ground from vibrations generated by heavy machinery. Thus, in some cases it is desirable to isolate vibrations of an item of equipment from a support structure and in other cases it is desirable to isolate equipment from a vibrating support, the latter often being referred to as base excitation.

Sometimes active systems are used either in parallel with or in series with passive isolators. Such systems are often referred to as semi-active and consist of two main types. The first type, often used for suspension systems of luxury vehicles, involves control of the system damping, generally by varying the orifice size in a hydraulic damper. It is shown diagrammatically in Fig. 12.1(b) for a single-degree-of-freedom system. The second type of semi-active system involves the use of a force actuator driven by a control system. There are four ways of implementing this type of control, as illustrated in Fig. 12.1. As shown in the figure, the control force may be either in series with or parallel with the passive elements. Alternatively, the control force may act only on the vibrating body or only on the support structure. Each of the semi-active systems shown in Fig. 12.1 require a control system to drive it. However, this has been omitted from the figures for clarity.

Each of the systems illustrated in Fig. 12.1 are characterized by disadvantages and advantages. The advantage of the variable damper shown in (b) is its relative simplicity and low cost. In some cases its performance is comparable to systems containing a force actuator, but in many other cases the performance of a system such as this is not as good. The improved performance of the system shown in (c) at low frequencies is sometimes offset by the reduced performance at high frequencies due to high frequency force transmission through the actuator as a result of the limited actuator bandwidth, especially if hydraulic actuators are used. In practice, low frequency performance is also often limited by the high displacement requirement of the actuators which often precludes the use of magnetostrictive or piezoceramic materials. Instead, pneumatic, hydraulic or electromagnetic actuators with their associated weight problems and in the case of pneumatic and hydraulic actuators, their associated fluid supply inconvenience, must be used. With the force actuator acting either on the vibrating body or support structure, the loss in high frequency performance is no longer a problem as the system will be equivalent to a passive system at worst. However, for large machines a large actuator with a large inertial mass will be needed to provide the required control force and in some cases this may be impractical. However, if the frequency range over which control is desired is limited to very low frequencies or to a very narrow band of frequencies, the actuator force requirements can be significantly reduced (Tanaka and Kikushima, 1988), and as many practical actuators are characterized by relatively high force capability and low displacement capability, this may be the preferred configuration in many cases.

The configuration shown in Fig. 12.1(f) also does not suffer from the loss in high frequency performance as a result of limited actuator bandwidth. However, the actuator must be large enough to support the weight of the vibrating machine and passive suspension system. The main advantage of this configuration is that the active system is isolated from the dynamics of the supporting structure, which is an important advantage in terms of control system stability if the supporting structure is non-rigid.

In cases where it is desired to limit the vibration of a specific machine or structure at a single frequency, an active vibration absorber can be attached to it as shown in Fig. 12.1(g). The advantage of the active vibration absorber over the traditional passive one is that it can be adjusted to track variations in the exciting frequency resulting from speed variations of the machine causing the vibration problem.

Fully active systems contain no passive elements and are usually implemented in one of two ways as shown in Fig. 12.2. Again, the control system used to drive the actuator and the required vibration sensors have been omitted for clarity. These will be discussed later in this chapter.

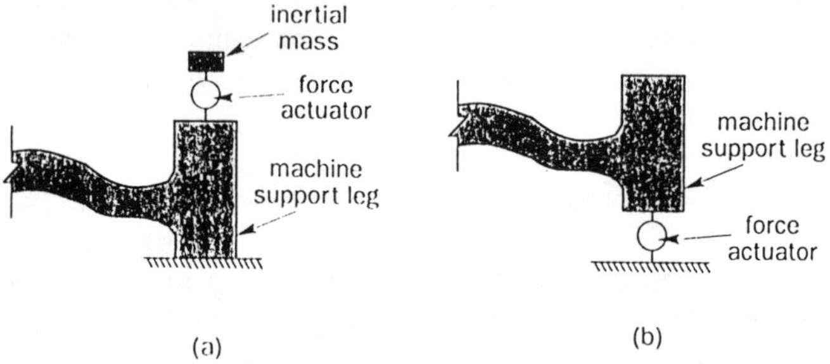

Fig. 12.2 Active vibration isolation: (a) force actuator acting only on the vibrating machine support leg; (b) actuator between the vibrating machine and the support structure and supporting the full weight of the machine.

The two configurations shown in Fig. 12.2 may be used to minimize vibration transmission from the machine to its supporting structure or to minimize vibration transmission from the structure to the equipment mounted on it, which may be an optical device such as a telescope or microscope.

The advantage of the system shown in part (a) of the figure is that the machine can be supported very rigidly so that there is no risk of movement and it will not vibrate. However, for large machines, a large actuator and seismic mass will be needed as the force actuator needs to exert sufficiently large dynamic forces on the support point to counteract the forces produced by the machine. For the configuration shown in part (b) of the figure, the force actuator must be capable of supporting the weight of the machine in a stable manner. A further disadvantage of this arrangement is that sometimes the force actuator will cause excessive vibrations in the machine as it attempts to minimize the vibration in the supporting structure.

In implementing active vibration isolation systems, great care must be taken to ensure that all vibration transmission paths are accounted for, because in some cases, control of one transmission path and not others could result in an increase in vibration level at locations where it should be reduced (Ross and Yorke, 1987). This is because in some instances, vibration energy arriving at a location by way of more than one transmission path can destructively interfere; removal of one of the transmission paths can reduce the destructive interference and result in an increase in vibration level. This is especially true for periodic vibration signals such as those generated by rotating or reciprocating machinery. Other likely transmission paths (or flanking paths) which may also need to be controlled are horizontal and rotational vibration transmission through the mounts, airborne acoustic waves exciting the support structure and vibration transmission through pipes and other fittings directly connecting the engine to the support structure. This is a very good reason for including all vibration types (vertical, horizontal and rotational) in the analyses of Sections 12.4 to 12.7.

Identification of all flanking paths is especially important when vibration isolation or structural vibration control is used to reduce structurally radiated noise into free space or into an enclosure, such as a car interior for example. For this example, the main vibration transmission path would be through the engine and gearbox isolation mounts, but flanking paths also exist via other mechanical attachments to the engine and via acoustic paths from the air intake and exhaust. An example of interior noise control in a car was presented by Quinn (1992) who demonstrated the effectiveness of controlling the main vibration transmission path with an active engine isolator, and using two interior loudspeakers to further reduce the interior noise. His system also used an accelerometer error sensor, four interior mounted error microphones, and a feedforward multi-channel control system, and was directed mainly at controlling periodic noise at the engine firing frequency.

Although Quinlan (1992) showed that it was feasible to substantially isolate low frequency periodic engine vibration from the vehicle body using a single isolator with an active element acting in only one direction, generally three translational and three rotational vibration transmission components must be included. This is especially true for isolation of higher frequency noise which is a problem when isolating submarine equipment platforms from the hull.

The design of an isolation mount is often a compromise between good vibration isolation and acceptable static rigidity. Use of an active element in the system helps to overcome this trade-off. To reduce the complexity of the mount, passive elements can be designed to provide good isolation in all directions but the one along which the transmission of vibration is the greatest, and this is the direction addressed by the active element. Jenkins *et al.* (1990, 1993) demonstrated such a mount which involved the use of an intermediate air bag to remove the shear and rotational components of the vibration. This is discussed in more detail in Section 12.4.

12.1.1 Feedforward vs. feedback control

As discussed earlier in this book, there are two fundamentally different approaches which have been used in the past for implementing active noise and vibration control systems; feedforward and feedback control. Feedforward control involves feeding a signal related to the disturbance input into the controller which then generates a signal to drive a control actuator in such a way as to cancel the disturbance. On the other hand, feedback control uses a signal derived from the system response to a disturbance which is amplified, passed through a compensator circuit and used to drive the control actuator to cancel the residual effects occurring after the initial disturbance has passed.

Feedforward control has been shown to provide better results than feedback control, provided that a signal well correlated with the disturbance input is available to the controller. Obtaining such a signal is especially simple for periodic excitation such as that generated by a rotating machine, or in cases where the active isolator is sufficiently far from the vibration source that the control system has time to respond to a measure of the incoming disturbance before it arrives at the actuator. For this latter case, it is possible to isolate random as well as periodic noise. However, in many cases for which active vibration isolation is applicable, it is not possible to obtain the required measure of the disturbance input, and a feedback control system must be used such as for active vehicle suspensions and for active isolation of sensitive equipment from a vibrating support structure.

As described in Chapter 6, practical implementation of a feedforward system for noise or vibration control usually requires the system to be adaptive so that it can adapt to changes in system parameters such as speed of sound in the structure which changes with temperature. An adaptive feedforward isolation system is best implemented using a digital filter, the weights of which are updated by an algorithm which uses as its inputs the primary disturbance signal and the signal from one or more error sensors which measure residual vibratory power transmission into or vibration on the structure to be controlled. The measured disturbance signal is passed through the filter into the control actuator to produce the force required to minimize the residual vibration in the structure.

An inherent disadvantage of feedback control systems is their tendency to go unstable if the feedback gain is set too high, but a high feedback gain is needed for good performance of active noise and vibration control systems. Another problem is that as the controller takes effect, the feedback signal is reduced in magnitude until it is no longer useful, thus limiting the potential performance of a feedback system. Thus, feedforward systems are preferred whenever it is possible to obtain a signal related to the disturbance input as, unlike feedback systems, they do not modify the dynamic response of the system being controlled, and are inherently much more stable. However, in many cases, a feedback system is the only feasible type, and care must be taken to limit the feedback gain so the system remains stable over the whole range of possible system inputs and variations in the dynamics of the system being controlled (due to ambient temperature changes, for example).

12.1.2 Flexible vs. stiff support structures

When the support structure is flexible, it is often characterized by vibration modes in the frequency range for which vibration isolation is required. In this case, a passive isolation system can be quite ineffective, especially at frequencies corresponding to resonance frequencies of the supporting structure. Even with an active isolation system (which will typically involve more than one force actuator) it may not be possible to drive the structural response to zero at each mount point. Power can be transmitted to the structure by moments as well as by forces along axes at 90° to the isolator axis. Thus, for a machine supported on a number of isolators, the control forces used to drive the actuators are not independent and must be derived using a multi-channel control system. Horizontal control forces may also sometimes be required. Thus, in many cases, the control system to drive each active isolator cannot be designed independently; usually, a multi-channel controller must be designed. Of course, this is not necessarily so for the case of a vehicular active suspension system where it is quite possible for each of the four wheel suspensions to act independently.

In the remainder of this chapter, simple feedback active isolators are discussed first of all, beginning with a single-degree-of-freedom system and covering base excited systems as well as vibration absorbers. Isolation of a machine from a flexible sub-structure is also discussed. Following a discussion of feedback control, feedforward control of single-degree-of-freedom systems and multi-degree-of-freedom systems involving flexible subsystems is considered.

12.2 FEEDBACK CONTROL

As discussed in Section 12.1, feedback control is the best control approach in situations where it is not possible to sample the incoming disturbance soon enough for a feedforward control system to be effective. In this section we will begin by discussing the vibration response of a single-degree-of-freedom system consisting of a mass supported on a spring and damper connected to a rigid foundation. The system will be excited by a single force acting on the mass and may be modelled as a second order system (by using a second order differential equation). The effect on the response of the mass of applying various types of feedback to drive a control force will then be examined. Next we will discuss the case where the foundation is free to move and we will examine the transmission of vibration from the foundation to the mass. This will be followed by an examination of the control of vibration of the mass by using a vibration absorber tuned by feedback control. Finally, a special case involving the active isolation of vibration from both a rigid and flexible substructure through a single mount will be discussed. The use of feedback control for situations involving multiple mounts will not be discussed here, as it is still at the research stage. However, some cases involving active vibration isolation of rigid bodies from flexible substructures using feedforward control will be discussed in Section 12.4.

12.2.1 Single-degree-of-freedom passive system

A single-degree-of-freedom passive isolation system is shown in Fig. 12.3. For now we will restrict our attention to the motion of the mass. The transmission of force into the foundation will be considered in Section 12.2.5.

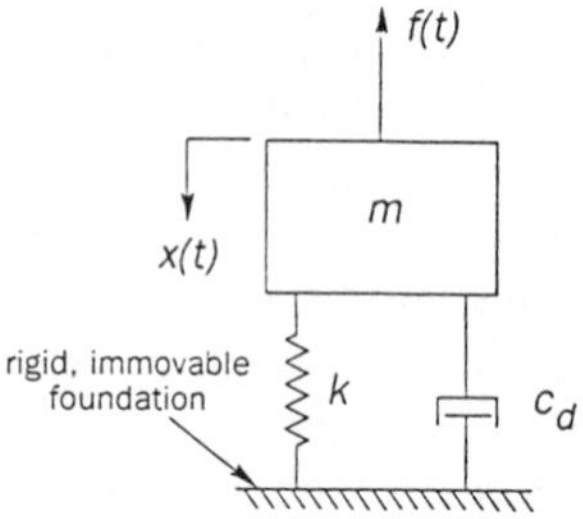

Fig. 12.3 Single-degree-of-freedom passive isolation system.

The second order differential equation which describes the motion of mass m is

$$m\ddot{x}(t) + c_d\dot{x}(t) + kx(t) = f(t) \tag{12.2.1}$$

where the $\cdot$ denotes derivative with respect to time and $\cdot\cdot$ denotes the double derivative with respect to time.

Taking the Laplace transform gives

$$m\left[s^2X(s) - sx(0) - \dot{x}(0)\right] + c_d\left[sX(s) - x(0)\right] + kX(s) = F(s) \tag{12.2.2}$$

Assuming zero initial conditions ($x(0) = \dot{x}(0) = 0$), equation (12.2.2) becomes

$$\left[ms^2 + c_d s + k\right]X(s) = F(s) \tag{12.2.3}$$

and the transfer function, relating displacement to force is

$$H(s) = \frac{X(s)}{F(s)} = \frac{1}{ms^2 + c_d s + k} \tag{12.2.4a,b}$$

The system natural frequency may be defined as:

$$\omega_0 = \sqrt{k/m} \tag{12.2.5}$$

and the viscous damping ratio as

$$\zeta = c_d/2m\omega_0 \tag{12.2.6}$$

Using these definitions, (12.2.4) can be written as

$$H(s) = \frac{1}{m}\frac{1}{s^2 + 2\zeta\omega_0 s + \omega_0^2} \tag{12.2.7}$$

To find the response of the system to a unit impulse we expand the transfer function into partial fractions as follows

$$H(s) = \frac{A}{(s+p_1)} - \frac{A}{(s+p_2)} \tag{12.2.8}$$

where

$$A = \frac{1}{2m\omega_0\sqrt{\zeta^2 - 1}} \tag{12.2.9}$$

$$p_1 = \omega_0\left(\zeta - \sqrt{\zeta^2 - 1}\right) \tag{12.2.10}$$

$$p_2 = \omega_0\left(\zeta + \sqrt{\zeta^2 - 1}\right) \tag{12.2.11}$$

Transforming back into the time domain we obtain

$$h(t) = Ae^{-\left(\zeta - \sqrt{\zeta^2 - 1}\right)\omega_0 t} - Ae^{-\left(\zeta + \sqrt{\zeta^2 - 1}\right)\omega_0 t} \tag{12.2.12}$$

The magnitude of the quantity $h(t)/A$ is shown in Fig. 12.4 as a function of time for various values of the viscous damping ratio ζ.

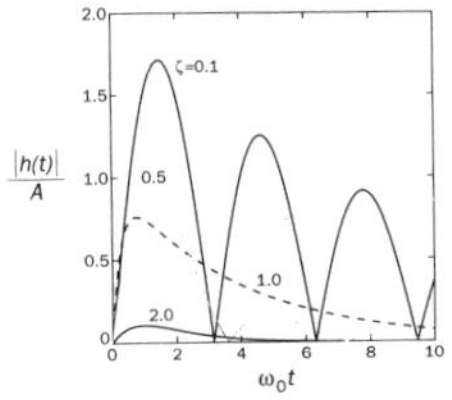

Fig. 12.4 Single-degree-of-freedom system impulse response for various values of damping ratio.

Inspection of (12.2.8) shows that if the poles p_1 and p_2 have positive real parts, they will lie in the left half of the s-plane plot (where Re$\{s\}$ is the horizontal axis and Im$\{s\}$ is the vertical axis) and thus the system will be stable. This will be true provided the damping ratio, $\zeta > 0$, which is a property of all actual passive isolation systems.

If we substitute $j\omega = s$ in (12.2.7), we obtain the frequency response function, $H(j\omega)$, given by

$$H(j\omega) = \frac{1}{m}\frac{1}{\omega_0^2 - \omega^2 + 2j\omega_0\omega\zeta} \tag{12.2.13}$$

or in dimensionless form,

$$m\omega_0^2 H(j\omega) = kH(j\omega) = \frac{1}{1 - (\omega/\omega_0)^2 + 2j\zeta\omega/\omega_0} \tag{12.2.14a,b}$$

The transfer function can also be written as

$$H(j\omega) = |\, H(j\omega)\,|\; e^{j\theta} \tag{12.2.15}$$

where

$$|\, H(j\omega)\,| = \frac{1/k}{\left[1 - (\omega/\omega_0)^2\right]^2 + \left[(2\zeta\omega/\omega_0)^2\right]^{1/2}} \tag{12.2.16}$$

and the relative phase angle between the force and displacement is given by

$$\theta = \tan^{-1}\left[\frac{-2\zeta\omega/\omega_0}{1-(\omega/\omega_0)^2}\right] \tag{12.2.17}$$

The modulus and phase angle of $H(j\omega)$ for various values of the critical damping ratio are plotted in Fig. 12.5.

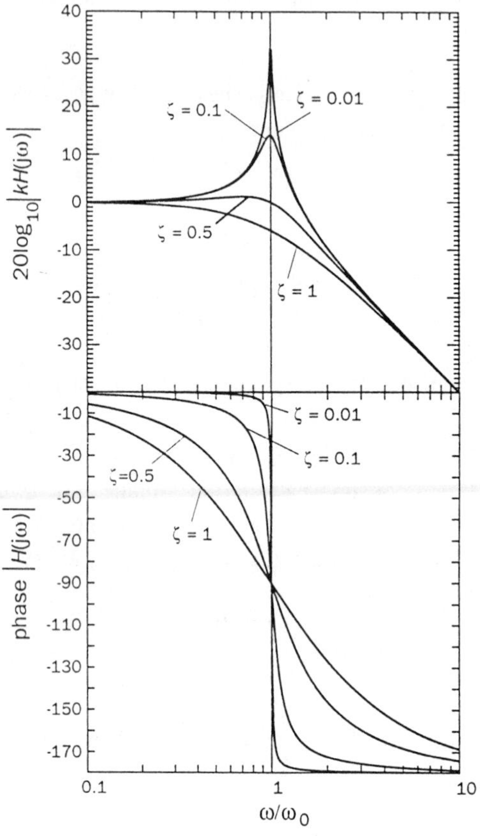

Fig. 12.5 Frequency response of a single-degree-of-freedom passive isolation system.

12.2.2 Feedback control of single-degree-of-freedom system

The dynamics of the system discussed in the previous section can be modified by adding a control force proportional to the displacement, velocity or acceleration (or a combination of these) to the vibrating mass. This is called feedback control.

With a control force $f_c(t)$ acting on the system, the equation of motion can be written as

$$m\ddot{x}(t) + c_d\dot{x}(t) + kx(t) = f(t) + f_c(t) \tag{12.2.18}$$

If we detect the acceleration, velocity and displacement of the mass and feed these quantities back through gains K_a, K_v and K_d to obtain $f_c(t)$, we may write

$$f_c(t) = -\left[K_a\ddot{x}(t) + K_v\dot{x}(t) + K_d x(t)\right] \tag{12.2.19}$$

A block diagram illustrating this feedback control arrangement is shown in Fig. 12.6(b) and the physical system is shown in Fig. 12.6(a).

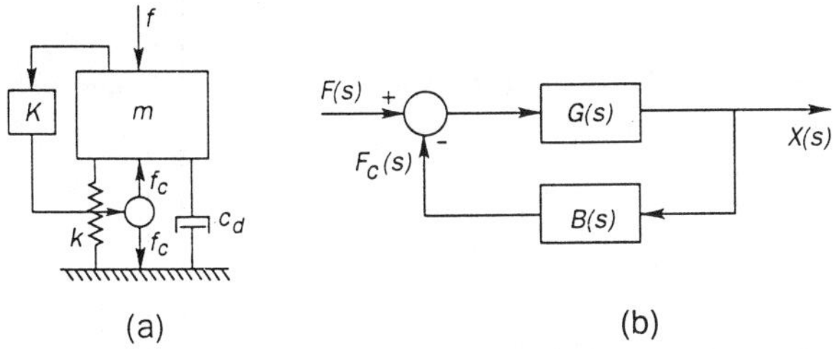

Fig. 12.6 Feedback control of a single degree of freedom isolation system: (a) physical system; (b) block diagram.

In practice, typical feedback control systems use some combination of acceleration, velocity, displacement or force feedback. In Fig. 12.6, the dynamics of the suspended mass on its own is represented by

$$G(s) = \frac{1}{ms^2 + c_d s + k} \tag{12.2.20}$$

and the feedback control force in the s domain is found by taking the Laplace transform of (12.2.19) with zero initial conditions. Thus,

$$F_c(s) = (K_a s^2 + K_v s + K_d)X(s) = B(s)X(s) \tag{12.2.21a,b}$$

The frequency response is given by

$$H(s) = \frac{X(s)}{F(s)} = \frac{G(s)}{1 + G(s)B(s)}$$

$$= \frac{1}{(m + K_a)s^2 + (c_d + K_v)s + k + K_d} \tag{12.2.22a,b,c}$$

The time domain equivalent of this equation which could also have been derived by substituting (12.2.19) into (12.2.18), is

$$(m + K_a)\ddot{x}(t) + (c_d + K_v)\dot{x}(t) + (k + K_d)x(t) = f(t) \tag{12.2.23}$$

In real physical systems, there is a finite time delay between receiving the signal from the vibration sensor, processing it, feeding it to the vibration actuator and propagating again to the vibration sensor. This affects the system stability and it can be shown that because of this phenomenon, velocity feedback systems are generally the most inherently stable (Fuller *et al.*, 1994).

12.2.2.1 Displacement feedback

The new resonance frequency and damping ratio for displacement feedback only (K_v, K_a = 0) are

$$\omega_0' = \sqrt{\frac{k+K_d}{m}} = \omega_0\sqrt{1+K_d/k} \quad (12.2.24a,b)$$

$$\zeta' = \frac{c_d}{2\sqrt{(k+K_d)m}} \quad (12.2.25)$$

If we let $K_a = K_v = 0$, use (12.2.5) and (12.2.6) and substitute $s = j\omega$ in (12.2.22c), we can write the following for the frequency response with displacement feedback:

$$kH(j\omega) = \frac{1}{1 + K_d/k - (\omega/\omega_0)^2 + 2j\zeta(\omega/\omega_0)} \quad (12.2.26)$$

where ω_0 and ζ are the same quantities as used in equation (12.2.14). Alternatively, ω_0' and ζ' may be substituted for ω_0 and ζ respectively in (12.2.14).

The normalized modulus of the closed loop frequency response function $|H(j\omega)|$ is plotted in Fig. 12.7 showing the effect of varying the displacement feedback gain K_d for various values of K_d/k, with $\zeta = 0.05$. It can be seen from the figure that increasing the displacement feedback increases low frequency isolation but decreases high frequency isolation and that the crossover from low to high frequency behaviour is dependent on the stiffness of the passive isolation system.

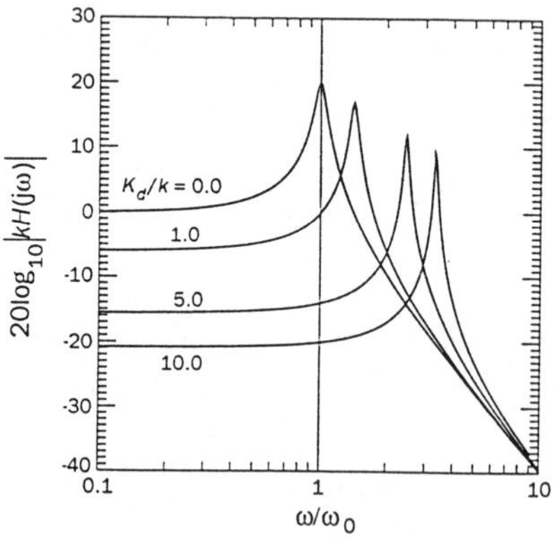

Fig. 12.7 Effect of displacement feedback on the response of a single-degree-of-freedom system for $\zeta = 0.05$.

12.2.2.2 Velocity feedback

In this case, the new resonance frequency and damping are given by

$$\omega_0' = \sqrt{\frac{k}{m}} = \omega_0 \tag{12.2.27a,b}$$

$$\zeta' = \frac{c_d + K_v}{2\sqrt{km}} = \zeta(1 + K_v/c_d) \tag{12.2.28a,b}$$

If we let $K_v = K_d = 0$, use (12.2.5) and (12.2.6) and substitute $s = j\omega$ in (12.2.22c), we can write the following for the frequency response with velocity feedback:

$$kH(j\omega) = \frac{1}{1 - (\omega/\omega_0)^2 + 2j\zeta(1 + K_v/c_d)(\omega/\omega_0)} \tag{12.2.29}$$

Alternatively, ω_0' and ζ' may be substituted for ω_o and ζ respectively in (12.2.14).

The normalized modulus of the closed loop frequency response function of (12.2.29) is plotted in Fig. 12.8, and shows the effect of varying the velocity feedback gain for various values of K_v/c_d with $\zeta = 0.05$. It can be seen from the figure that increasing velocity feedback effectively increases the system damping; thus, increasing the isolation in the region of the sytem resonance with little effect at low and high frequencies.

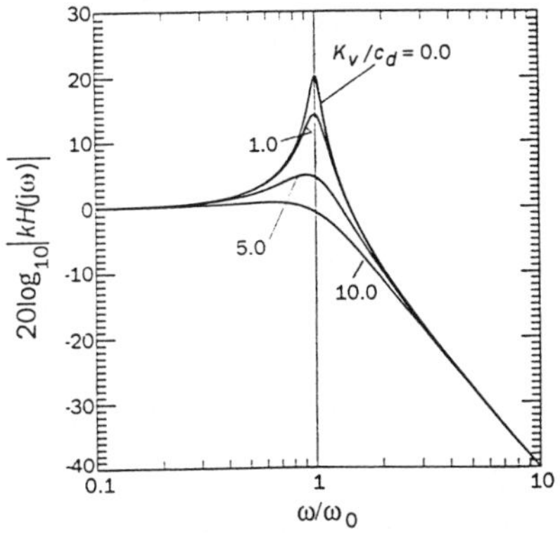

Fig. 12.8 Effect of velocity feedback on the response of a single-degree-of-freedom system for $\zeta = 0.05$.

12.2.2.3 Acceleration feedback

In this case, the new system resonance frequency and damping are given by

$$\omega_0' = \sqrt{\frac{k}{m + K_a}} \tag{12.2.30}$$

$$\zeta' = \frac{c_d}{2\sqrt{k(m+K_a)}} \tag{12.2.31}$$

If we let $K_a = K_d = 0$, use (12.2.5) and (12.2.6) and substitute $s = j\omega$ in (12.2.22c), we can write the following for the frequency response with acceleration feedback:

$$kH(j\omega) = \frac{1}{1 - (\omega/\omega_0)^2(1 + K_a/m) + 2j\zeta(\omega/\omega_0)} \tag{12.2.32}$$

Alternatively, ω_0' and ζ' may be substituted for ω_0 and ζ respectively in (12.2.14).

The normalized modulus of the closed loop frequency response function of equation (12.2.32) is plotted in Fig. 12.9, and shows the effect of varying the acceleration feedback gain for various values of K_a/m with $\zeta = 0.05$. It can be seen from the figure that increasing acceleration feedback decreases low frequency isolation but increases high frequency isolation.

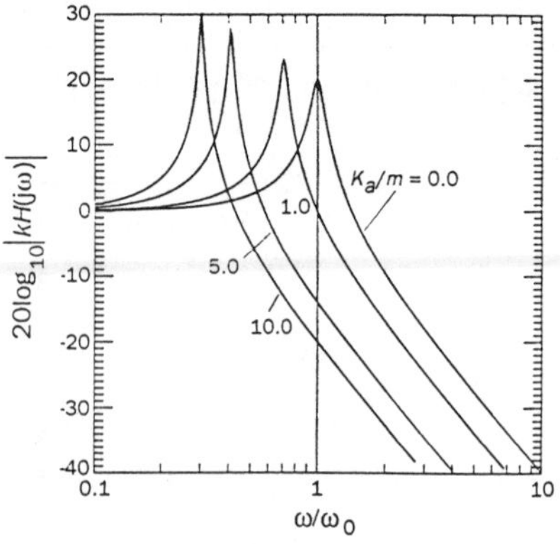

Fig. 12.9 Effect of acceleration feedback on the response of a single-degree-of-freedom system for $\zeta = 0.05$.

Thus, it can be seen that adding feedback to the single-degree-of-freedom system actually changes the system parameters (resonance frequency and damping). It will be shown in Section 12.4 that feedforward control does not do this. This characteristic of the two different types of control was discussed in detail in Chapter 6.

It can also be seen that all types of feedback control can either reduce the frequency response over the entire frequency range of interest or reduce it at some desired frequency by moving the system resonance frequency and increasing the effective damping ratio.

12.2.2.4 Closed loop stability

It is of interest to determine the bounds on the gains K_d, K_v and K_a for stability considerations. The new pole locations with the addition of all three types of feedback can be expressed in terms of the new resonance frequency and damping ratio. That is, poles occur when

$$s = -\omega_0\left(\zeta \pm \sqrt{\zeta^2 - 1}\right) \tag{12.2.33}$$

where

$$\omega_0 = \frac{\sqrt{k + K_d}}{m} \tag{12.3.34}$$

and

$$\zeta = \frac{c_d + K_v}{2\sqrt{(k+K_d)(m+K_a)}} \tag{12.2.35}$$

Thus, the poles will lie in the negative half plane (that is, the system will be stable) provided all of the following conditions are satisfied.

$$K_a + m > 0, \quad K_v + c_d > 0, \quad K_d + k > 0 \quad (12.2.36a,b,c)$$

12.2.3 Base excited second order system

For this system, the support is no longer considered fixed; it is free to move, although it is constrained to move vertically. It is of interest to investigate how the various types of feedback can be used to control vibration transmission from the support to the mass for the system shown in Fig. 12.10.

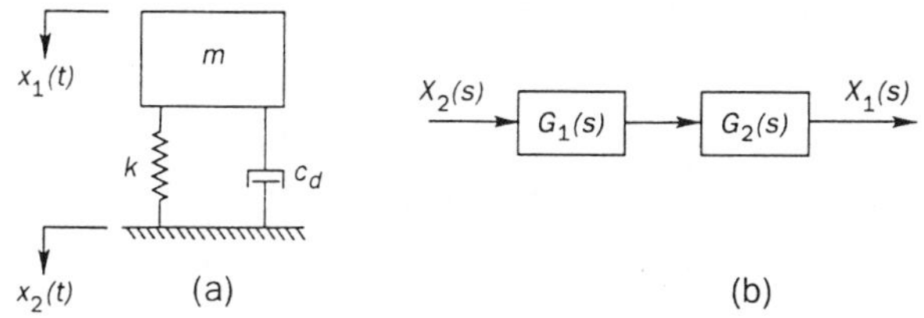

Fig. 12.10 Base excited system: (a) physical system; (b) block diagram in s-domain.

The equation of motion for the mass can be written as

$$m\ddot{x}_1(t) + c_d\dot{x}_1(t) + kx_1(t) - c_d\dot{x}_2(t) - kx_2(t) = 0 \tag{12.2.37}$$

Taking Laplace transforms and assuming zero initial conditions gives

$$\left[ms^2 + c_d s + k\right]X_1(s) - [c_d s + k]\,X_2(s) = 0 \tag{12.2.38}$$

or in terms of the block diagram shown in Fig. 12.10(b),

$$X_1(s) - G_1(s)\,G_2(s)X_2(s) = 0 \tag{12.2.39}$$

and thus the transfer function $H(s)$ relating the displacement of the mass to the displacement of the base can be written as

$$H(s) = \frac{X_1(s)}{X_2(s)} = G_1(s)G_2(s) = \frac{c_d s + k}{ms^2 + c_d s + k} \qquad (12.2.40a,b)$$

or in terms of the system resonance frequency and damping

$$H(s) = \frac{2\zeta\omega_0 s + \omega_0^2}{s^2 + 2\zeta\omega_0 s + \omega_0^2} \qquad (12.2.41)$$

where ω_0 and ζ are defined by (12.2.5) and (12.2.6) respectively.

Using $s = j\theta$, (12.2.41) can be written as

$$H(j\omega) = \frac{2j\zeta(\omega/\omega_0) + 1}{1 + 2j\zeta(\omega/\omega_0) - (\omega/\omega_0)^2} = |H(j\omega)| e^{j\theta} \qquad (12.2.42a,b)$$

where

$$|H(j\omega)| = \sqrt{\frac{1 + (2\zeta\omega/\omega_0)^2}{\left[1 - (\omega/\omega_0)^2\right]^2 + [2\zeta\omega/\omega_0]^2}} \qquad (12.2.43)$$

and

$$\theta = \tan^{-1}\left[\frac{2\zeta\omega/\omega_0}{(\omega_0/\omega)^2 - 1 + 4\zeta^2}\right] \qquad (12.2.44)$$

The normalized magnitude of $H(j\omega)$ is shown in Fig. 12.11 as a function of ω/ω_0 for various values of the damping ratio ζ. Note the reduced high frequency performance as a result of damping.

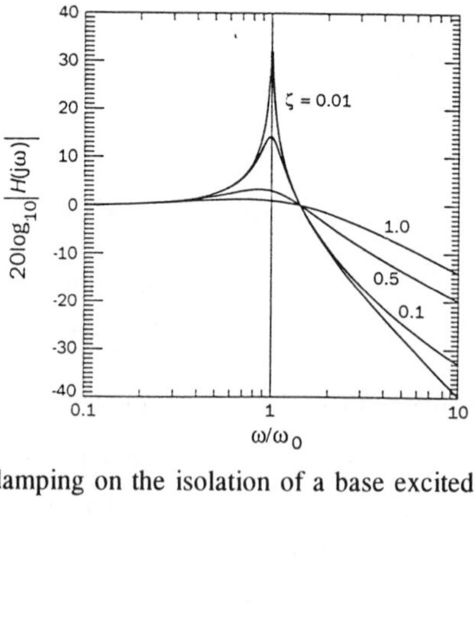

Fig. 12.11 Effect of damping on the isolation of a base excited system.

12.2.3.1 Relative displacement feedback

The block diagram shown in Fig. 12.12(b) represents the physical system shown in Fig. 12.12(a).

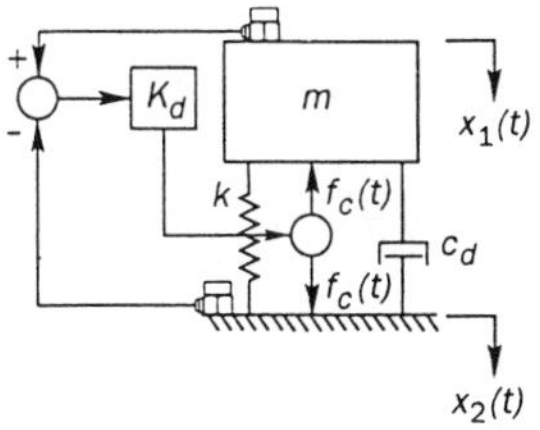

(a)

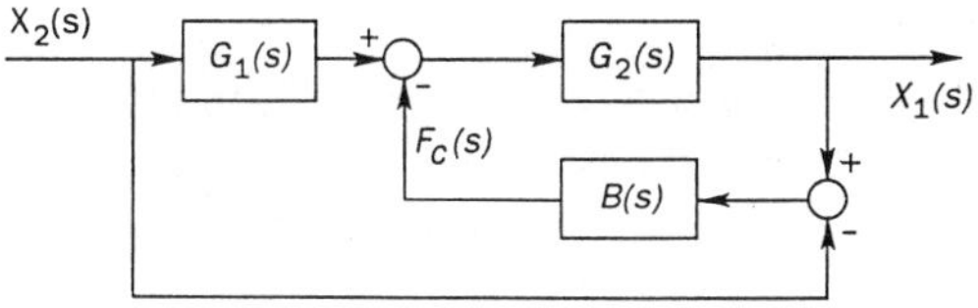

(b)

Fig. 12.12 Relative displacement feedback for base excited system: (a) physical system; (b) block diagram.

Here,

$$G_1(s) = c_d + k \qquad (12.2.45)$$

$$G_2(s) = \frac{1}{ms^2 + c_d s + k} \qquad (12.2.46)$$

$$B(s) = K_d \qquad (12.2.47)$$

$$F_c(s) = B(s) X_1(s) \qquad (12.2.48)$$

and

$$H(s) = \frac{X_1(s)}{X_2(s)} = \frac{G_1(s)\,G_2(s) + B(s)\,G_2(s)}{1 + G_2(s)\,B(s)} = \frac{G_1(s) + B(s)}{\dfrac{1}{G_2(s)} + B(s)} \qquad (12.2.49a,b,c)$$

Substituting (12.2.45) to (12.2.47) into (12.2.49) gives

$$H(s) = \frac{c_d s + k + K_d}{ms^2 + c_d s + k + K_d} \qquad (12.2.50)$$

In the time domain, the control force is

$$f_c(t) = -K_d(x_1(t) - x_2(t)) \quad (12.2.51)$$

and the system equation of motion is

$$m\ddot{x}_1(t) + c_d\dot{x}_1(t) + (k+K_d)x_1(t) - c_d\dot{x}_2(t) - (k+K_d)x_2(t) = 0 \quad (12.2.52)$$

Note that the same results are obtained for the motion of the mass (but not the support) if the control reacts against the support or if an inertial mass is used for it to react against. The new system resonance frequency and damping ratio are:

$$\omega_0' = \sqrt{\frac{k+K_d}{m}} = \omega_0\sqrt{1 + K_d/k} \quad (12.2.53a,b)$$

$$\zeta' = \frac{c_d}{2\sqrt{(k+K_d)m}} = \frac{\zeta}{\sqrt{1 + K_d/k}} \quad (12.2.54a,b)$$

Using (12.2.5) and (12.2.6) and substituting $s = j\omega$ in (12.2.50), we can write the following for the frequency response with relative displacement feedback:

$$H(j\omega) = \frac{1 + K_d/k + 2j\zeta(\omega/\omega_0)}{1 + K_d/k - (\omega/\omega_0)^2 + 2j\zeta(\omega/\omega_0)} \quad (12.2.55)$$

Alternatively, ω_0' and ζ' may be substituted for ω_0 and ζ respectively in (12.2.42).

The modulus of the closed loop frequency response function of (12.2.55) is plotted in Fig. 12.13, and shows the effect of varying the relative displacement feedback gain for various values of K_d/k, with $\zeta = 0.05$. It can be seen from the figure that increasing relative displacement feedback increases the effective system resonance frequency, but has little effect on the system damping, resulting in increased low frequency isolation but decreased high frequency isolation.

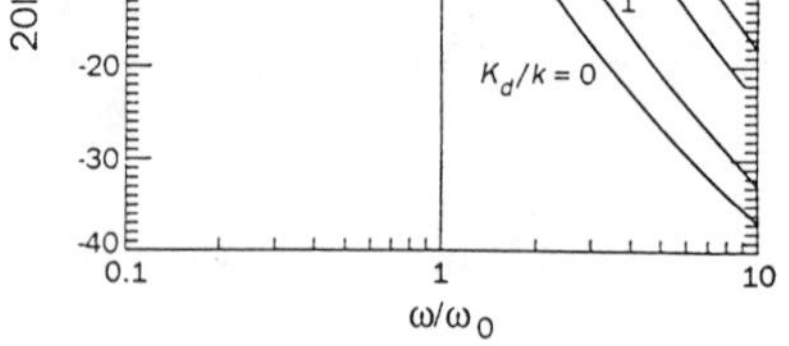

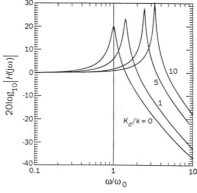

Fig. 12.13 Effect of relative displacement feedback gain on the frequency response of a base excited system for $\zeta = 0.05$.

12.2.3.2 Absolute displacement feedback

Again, the block diagram shown in Fig. 12.14(b) is an s-plane representation of the physical system shown in Fig. 12.14(a), with all of the quantities defined as in (12.2.45) to (12.2.48).

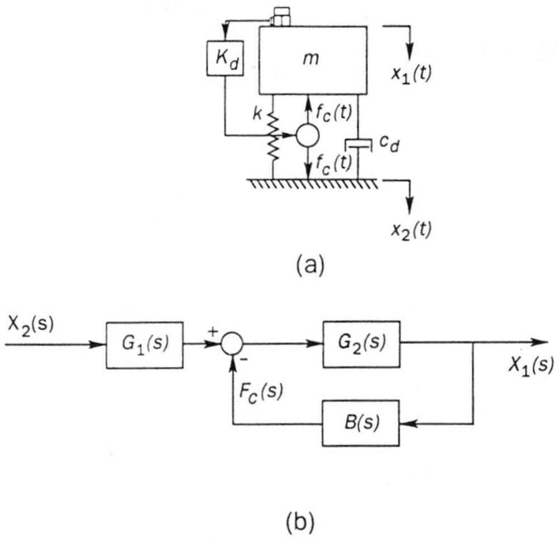

Fig. 12.14 Absolute displacement feedback.

In this case,

$$H(s) = \frac{G_1(s)\,G_2(s)}{1 + G_2(s)B(s)} = \frac{G_1(s)}{\dfrac{1}{G_2(s)} + B(s)} \qquad (12.2.56a,b)$$

or

$$H(s) = \frac{c_d s + k}{ms^2 + c_d s + k + K_d} \qquad (12.2.57)$$

In the time domain, the control force $f_c(t)$ and equation of motion of the mass are

$$f_c(t) = -K_d x_1(t) \qquad (12.2.58)$$

and

$$m\ddot{x}_1(t) + c_d\dot{x}_1(t) + (k + K_d)x_1(t) - c_d\dot{x}_2(t) - kx_2(t) = 0 \qquad (12.2.59)$$

It is not possible to define an equivalent damping ratio and resonance frequency as was done for relative displacement feedback. However, using (12.2.5) and (12.2.6) and substituting $s = j\omega$ in (12.2.57), we can write the following for the frequency response with absolute displacement feedback:

$$H(j\omega) = \frac{1 + 2j\zeta(\omega/\omega_0)}{1 + K_d/k - (\omega/\omega_0)^2 + 2j\zeta(\omega/\omega_0)} \qquad (12.2.60)$$

The modulus of the closed loop frequency response function of (12.2.60) is plotted in Fig. 12.15, and shows the effect of varying the absolute displacement feedback gain for various values of K_d/k, with $\zeta = 0.05$. It can be seen from the figure that increasing the feedback gain increases the system natural frequency, as it effectively increases the stiffness of the system. This results in increased low frequency isolation as K_d is increased, and is much more effective than relative displacement feedback.

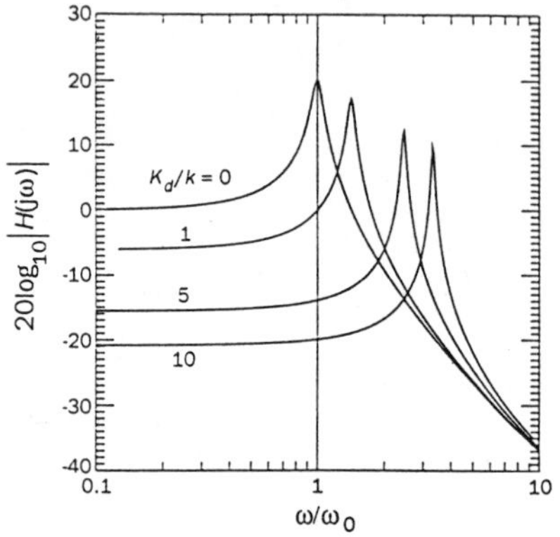

Fig 12.15 Effect of absolute displacement feedback gain on the frequency response of a base excited system.

12.2.3.3 Relative velocity feedback

The arrangement for this is similar to that shown in Fig. 12.12 for relative displacement feedback, except that K_d is replaced by $K_v s$. Using (12.2.49), we obtain for the transfer function

$$H(s) = \frac{X_1(s)}{X_2(s)} = \frac{(c_d + K_v)s + k}{ms^2 + (c_d + K_v)s + k} \qquad (12.2.61\text{a,b})$$

In this case, the control force in the time domain is given by

$$f_c(t) = -K_v\left[\dot{x}_1(t) - \dot{x}_2(t)\right] \qquad (12.2.62)$$

and the system equation of motion is

$$m\ddot{x}_1(t) + (c_d + K_v)\dot{x}_1(t) + kx_1(t) - (c_d + K_v)\dot{x}_2(t) - kx_2(t) = 0 \qquad (12.2.63)$$

Of course, we could have derived these equations in the reverse order by writing down the equation of motion from inspection of the physical system and then taken Laplace transforms to obtain (12.2.61).

The new resonance frequency and damping ratio are

$$\omega_0' = \omega_0 \qquad (12.2.64)$$

$$\zeta' = \frac{c_d + K_v}{2\sqrt{km}} \tag{12.2.65}$$

Thus, (12.2.61) can be written in terms of the system resonance frequency and critical damping ratio by replacing ω_0 and ζ in (12.2.42) with ω_0' and ζ'.

Alternatively, using (12.2.5) and (12.2.6) and substituting $s = j\omega$ in (12.2.61), we can write the following for the frequency response with relative velocity feedback:

$$H(j\omega) = \frac{1 + 2j\zeta(\omega/\omega_0)(1 + K_v/c_d)}{1 - (\omega/\omega_0)^2 + 2j\zeta(\omega/\omega_0)(1 + K_v/c_d)} \tag{12.2.66}$$

The modulus of the closed loop frequency response function of (12.2.66) is plotted in Fig. 12.16, and shows the effect of varying the relative velocity feedback gain for various values of K_v/c_d, with $\zeta = 0.05$. It can be seen from the figure that increasing the feedback gain increases isolation in the region of the system natural frequency, but reduces high frequency isolation.

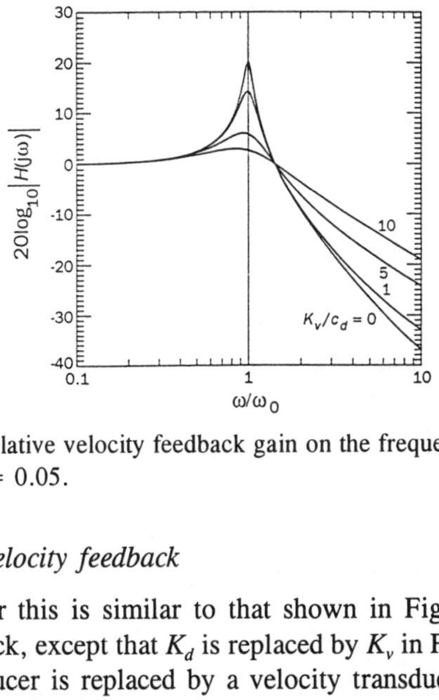

Fig. 12.16 Effect of relative velocity feedback gain on the frequency response of a base excited system for $\zeta = 0.05$.

12.2.3.4 Absolute velocity feedback

The arrangement for this is similar to that shown in Fig. 12.14 for absolute displacement feedback, except that K_d is replaced by K_v in Fig. 12.14(a) (and the displacement transducer is replaced by a velocity transducer) and $B(s)$ of Fig. 12.14(b) is now equal to $K_v s$. Using these substitutions in (12.2.56) we may write for the transfer function

$$H(s) = \frac{X_1(s)}{X_2(s)} = \frac{c_d s + k}{ms^2 + (c_d + K_v)s + k} \tag{12.2.67a,b}$$

In the time domain, the control force is

$$f_c = -K_v \dot{x}_1(t) \tag{12.2.68}$$

and the equation of motion of the mass is

$$m\ddot{x}_1(t) + (c_d + K_v)\dot{x}_1(t) + kx_1(t) - c_d\dot{x}_2(t) - kx_2(t) = 0 \qquad (12.2.69)$$

Using (12.2.5) and (12.2.6) and substituting $s = j\omega$ in (12.2.67), we can write the following for the frequency response with absolute velocity feedback:

$$H(j\omega) = \frac{1 + 2j\zeta(\omega/\omega_0)}{1 - (\omega/\omega_0)^2 + 2j\zeta(\omega/\omega_0)(1 + K_v/c_d)} \qquad (12.2.70)$$

The modulus of the closed loop frequency response function of (12.2.70) is plotted in Fig. 12.17, and shows the effect of varying the absolute velocity feedback gain for various values of K_v/c_d, with $\zeta = 0.05$. It can be seen from the figure that increasing the feedback gain increases the isolation in the vicinity of the system resonance frequency without significantly affecting the isolation at other frequencies.

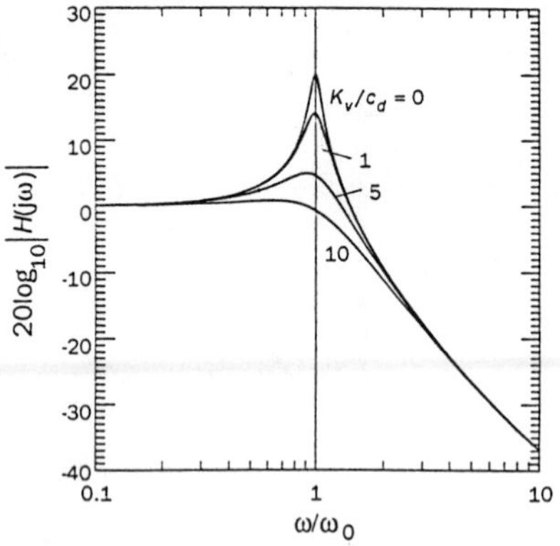

Fig. 12.17 Effect of absolute velocity feedback gain on the frequency response of a base excited system for $\zeta = 0.05$.

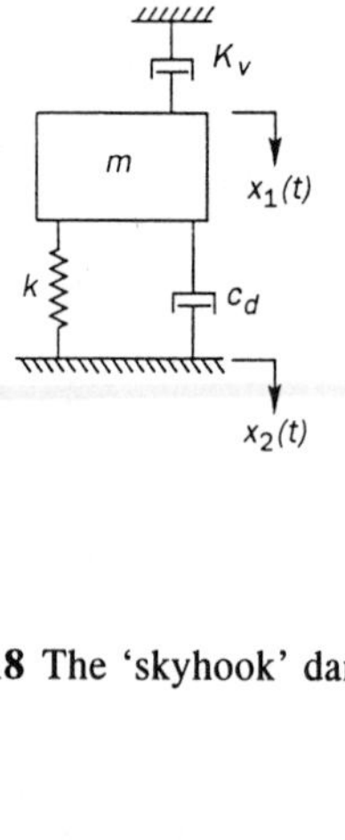

Fig. 12.18 The 'skyhook' damper.

The additional damping provided by feedback of absolute velocity can be considered equivalent to applying damping to the mass only by way of a sky hook (see Fig. 12.18).

12.2.3.5 *Relative acceleration feedback*

This arrangement is similar to that shown in Fig. 12.12 except that in part (a) K_d is replaced by K_a (and the displacement transducers are replaced with acceleration transducers) and in part (b), $B(s)$ is equal to $K_a s^2$.

Then, using these substitutions and (12.2.49), we obtain

$$H(s) = \frac{X_1(s)}{X_2(s)} = \frac{K_a s^2 + c_d s + k}{(K_a + m)s^2 + c_d s + k} \qquad (12.2.71a,b)$$

In this case, the control force in the time domain is given by

$$f_c(t) = -K_a[\ddot{x}_1(t) - \ddot{x}_2(t)] \tag{12.2.72}$$

and the equation of motion of the mass is

$$(m+K_a)\ddot{x}_1(t) + c_d\dot{x}_1(t) + kx_1(t) - K_a\ddot{x}_2(t) - c_d\dot{x}_2(t) - kx_2(t) = 0 \tag{12.2.73}$$

The new system resonance frequency and critical damping ratio are

$$\omega_o' = \sqrt{\frac{k}{K_a+m}} \tag{12.2.74}$$

$$\zeta' = \frac{c_d}{2\sqrt{k(K_a+m)}} \tag{12.2.75}$$

Using (12.2.5) and (12.2.6) and substituting $s = j\omega$ in equation (12.2.71), we can write the following for the frequency response with relative acceleration feedback

$$H(j\omega) = \frac{1 + 2j\zeta(\omega/\omega_0) - (K_a/m)(\omega/\omega_0)^2}{1 - (\omega/\omega_0)^2(1 + K_a/m) + 2j\zeta(\omega/\omega_0)} \tag{12.2.76}$$

Alternatively, ω_0' and ζ' may be substituted for ω_0 and ζ respectively in (12.2.42).

The modulus of the closed loop frequency response function of equation (12.2.76) is plotted in Fig. 12.19, and shows the effect of varying the relative acceleration feedback gain for various values of K_a/m, with $\zeta = 0.05$. It can be seen from the figure that increasing the feedback gain increases the isolation in the vicinity of the system resonance frequency but decreases the isolation achieved at other frequencies.

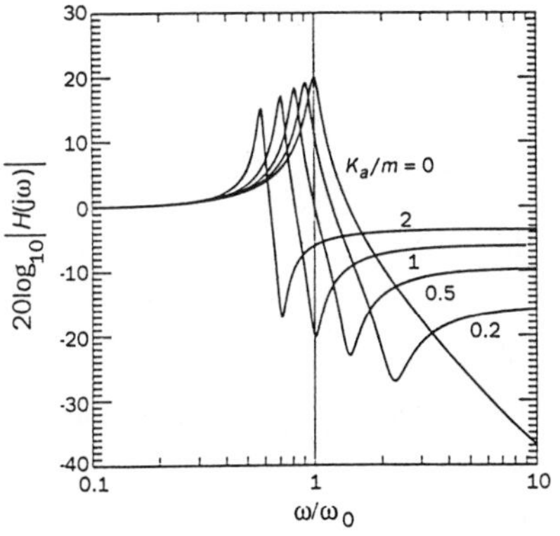

Fig. 12.19 Effect of relative acceleration feedback gain on the frequency response of a base excited system for $\zeta = 0.05$.

12.2.3.6 Absolute acceleration feedback

The arrangement for this is similar to that shown in Fig. 12.14 for absolute displacement feedback except that K_d is replaced by K_a (and the velocity transducer is replaced by an accelerometer) in Fig. 12.14(a) and $B(s)$ of Fig. 12.14(b) is now equal to $K_a s^2$. Using these substitutions in (12.2.49) we may write for the transfer function

$$H(s) = \frac{X_1(s)}{X_2(s)} = \frac{c_d s + k}{(K_a + m)s^2 + c_d s + k} \qquad (12.2.77a,b)$$

The system equation of motion for this case is

$$(m + K_a)\ddot{x}_1(t) + c_d\dot{x}_1(t) + kx_1(t) - c_d\dot{x}_2(t) - kx_2(t) = 0 \qquad (12.2.78)$$

Using (12.2.5) and (12.2.6) and substituting $s = j\omega$ in (12.2.77), we can write the following for the frequency response with absolute acceleration feedback:

$$H(j\omega) = \frac{1 + 2j\zeta(\omega/\omega_0)}{1 - (\omega/\omega_0)^2(1 + K_a/m) + 2j\zeta(\omega/\omega_0)} \qquad (12.2.79)$$

The modulus of the closed loop frequency response function of (12.2.79) is plotted in Fig. 12.20, and shows the effect of varying the absolute acceleration feedback gain for various values of K_a/m, with $\zeta = 0.05$. It can be seen from the figure that increasing the feedback gain is equivalent to adding mass to the system, resulting in a movement of the resonance frequency to lower frequencies, thus producing higher isolation at higher frequencies at the expense of lower isolation at low frequencies.

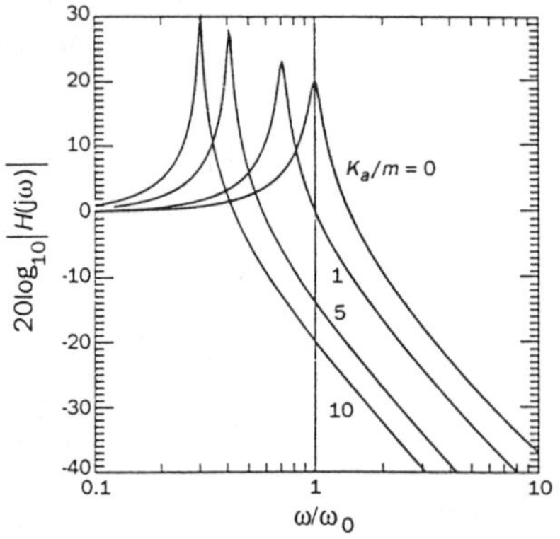

Fig. 12.20 Effect of absolute acceleration feedback gain on the frequency response of a base excited system for $\zeta = 0.05$.

12.2.3.7 Force feedback

Here we will consider the case where the total force transmitted by the base into the suspension system will be used as the feedback variable (see Fig. 12.21).

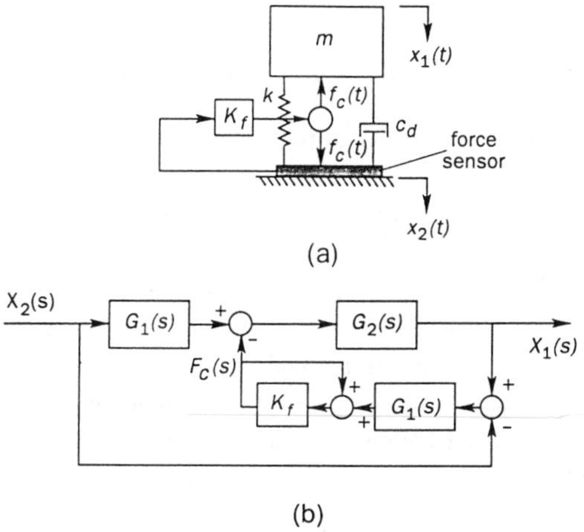

Fig. 12.21 Force feedback for a base excited system.

From Fig. 12.21, the transfer function in the frequency domain is given by

$$H(s) = \frac{X_1(s)}{X_2(s)} = \frac{G_1(s) + \dfrac{G_1(s)K_f}{1-K_f}}{\dfrac{1}{G_2(s)} + \dfrac{G_1(s)K_f}{1-K_f}} = \frac{c_d s + k}{(1-K_f)ms^2 + c_d s + k} \qquad (12.2.80a,b,c)$$

where $G_1(s)$ and $G_2(s)$ are defined by (12.2.45) and (12.2.46).

The dynamic equations describing the system in Fig. 12.21 are

$$m\ddot{x}_1 + c_d(\dot{x}_1 - \dot{x}_2) + k(x_1 - x_2) + K_f f_T(t) = 0 \qquad (12.2.81)$$

$$c_d(\dot{x}_1 - \dot{x}_2) + k(x_1 - x_2) + K_f f_T(t) = f_T \qquad (12.2.82)$$

where the transmitted force f_T and the control force f_c are related by $f_c(t) = K_f f_T(t)$.

The new system resonance frequency and critical damping ratio are given by:

$$\omega_0' = \sqrt{\frac{k}{m(1-K_f)}} \qquad (12.2.83)$$

$$\zeta' = \frac{c_d}{2\sqrt{km(1-K_f)}} \qquad (12.2.84)$$

It can be seen from the above equations that force feedback effectively increases the suspension critical damping ratio.

Using (12.2.5) and (12.2.6) and substituting $s = j\omega$ in (12.2.80c), we can write the following for the frequency response with force feedback (force acting on the base and suspended mass simultaneously):

$$H(j\omega) = \frac{1 + 2j\zeta(\omega/\omega_0)}{1 - (\omega/\omega_0)^2(1 - K_f) + 2j\zeta(\omega/\omega_0)} \tag{12.2.85}$$

Alternatively, ω_0' and ζ' may be substituted for ω_0 and ζ respectively in (12.2.42).

The modulus of the closed loop frequency response function of (12.2.85) is plotted in Fig. 12.22, and shows the effect of varying the force feedback gain for various values of K_f/m, with $\zeta = 0.05$. It can be seen from the figure that increasing the feedback gain effectively increases the damping and stiffness of the system. This results in an improved low frequency performance which extends also to high frequencies for high values of K_f/m. As will be discussed later, this extension to high frequencies will not occur in practice because the system becomes unstable if K_f/m is greater than one.

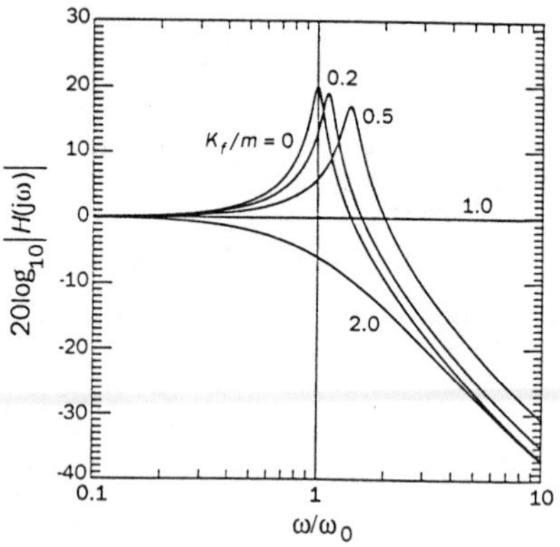

Fig. 12.22 Effect of force feedback gain on the frequency response of a base excited system for $\zeta = 0.05$.

If the system were configured so that the control force acted only on the suspended mass and not the base, then the innermost feedback loop of Fig. 12.21(b) would be eliminated and (12.2.80) would become

$$H(s) = \frac{c_d s + k}{\frac{m}{1 + K_f} s^2 + c_d s + k} \tag{12.2.86}$$

The modulus of the frequency response function would be the same as before except that ω_0' and ζ' would be defined as

$$\omega_0' = \sqrt{\frac{k(1 + K_f)}{m}} \tag{12.2.87}$$

$$\zeta' = \frac{c_d}{2\sqrt{\dfrac{km}{1+K_f}}} \tag{12.2.88}$$

Using (12.2.5) and (12.2.6) and substituting $s = j\omega$ in (12.2.86), we can write the following for the frequency response with force feedback (force acting only on the suspended mass):

$$H(j\omega) = \frac{1 + 2j\zeta(\omega/\omega_0)}{1 - (\omega/\omega_0)^2\dfrac{1}{(1+K_f)} + 2j\zeta(\omega/\omega_0)} \tag{12.2.89}$$

Alternatively, ω_0' and ζ' may be substituted for ω_o and ζ respectively in (12.2.42).

The modulus of the closed loop frequency response function of (12.2.89) is plotted in Fig. 12.23, and shows the effect of varying the force (on only the suspended mass) feedback gain for various values of K_f/m, with $\zeta = 0.05$. It can be seen from the figure that increasing the feedback gain effectively increases the damping and stiffness of the system. This results in an improved low frequency performance which, in contrast to the previous case, does not extend to high frequencies for high values of K_f/m.

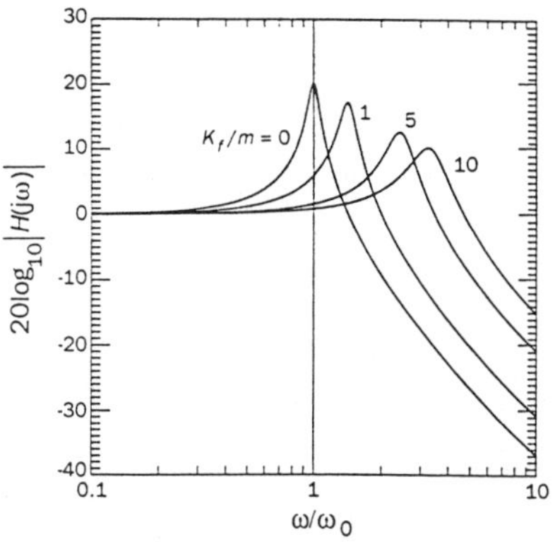

Fig. 12.23 Effect of force feedback gain on the frequency response of a base excited system with the control force only acting on the suspended mass for $\zeta = 0.05$.

12.2.3.7 Closed loop stability of the base excited system

As discussed in Section 12.2.1, the system will be stable provided that all of its poles are on the left of the $j\omega$ axis in the s plane. For the cases discussed, this condition is met provided that all of the conditions expressed in (12.2.36) are satisfied. Additionally, for the force feedback system, K_f/m must be less than one for the case of the force acting on the suspended mass and the base.

All of the simple feedback compensation circuits *(B(s))* (consisting only of a simple gain) discussed so far are the simplest form of feedback control possible. Better performance over a wider frequency band can be achieved in practice using more complicated compensator circuits ($B(s)$), depending upon the application, and this is where a large research effort is currently concentrated. Some examples of more complex controllers have been discussed in Chapter 11.

12.2.4 Dynamic vibration absorber

A dynamic vibration absorber is used to reduce the vibration amplitude of a structure or machine vibrating at a resonance frequency. Construction of the device involves suspending a mass from the machine or structure using one or more springs and dampers (see Fig. 12.24). If no damping is used, the effect of the absorber is to change the system from one characterized by a single resonance to one characterized by two resonances, the latter two appearing on either side (in the frequency domain) of the original. Although the system response at the original resonance frequency is substantially reduced, the response at the two new resonance frequencies is generally much larger than it was before. As the amount of damping is increased, the response at the two new resonance frequencies decreases and the total system response at the old original resonance frequency increases until a point is reached at which they are essentially equal, thus defining what is known as the optimum critical damping ratio ζ_2 for the suspended mass.

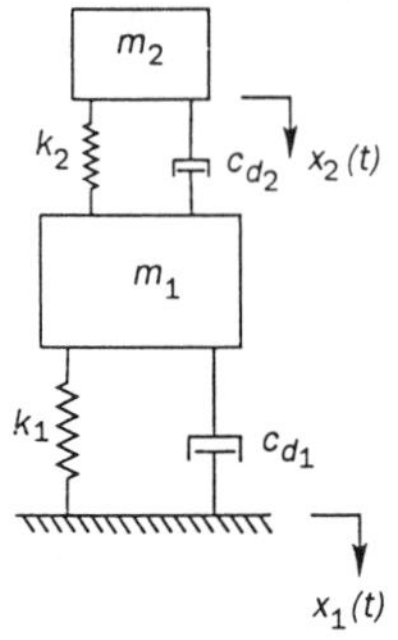

Fig. 12.24 Vibration absorber, $\zeta_2 = c_{d_2}/[2(k_2 m_2)^{1/2}]$.

Note that the response at the two new resonance frequencies is not necessarily the same; in fact, it will only be the same if the ratio of the stiffness k_2 to stiffness k_1 is optimized.

Values of the stiffness ratio and critical damping ratio for 'optimum' performance of the absorber are (Den Hartog, 1956; Davies, 1965):

$$\frac{k_2}{k_1} = \frac{m_1 m_2}{(m_1 + m_2)^2} \tag{12.2.90}$$

$$\zeta_2^2 = \frac{3(m_2/m_1)}{8(1+m_2/m_1)^3} \tag{12.2.91}$$

and the predicted amplitude of response of the mass m_1 in the frequency range including and between the two system resonances is

$$x_1/d = \sqrt{1+2m_1/m_2} \tag{12.2.92}$$

where d is the static deflection of the original mass m_1.

Note that in some cases the so called 'optimum' solution may not be the best. For example, if a machine is being excited by a single frequency which is very close to or coincides with the resonance frequency of the machine on its mounts, it is desirable to reduce the system response by as much as possible at the exciting frequency; the response at other frequencies is not so important. In this case, it will be better to use a value of ζ_2 smaller than the optimum, which will have the effect of reducing the system response at the original resonance frequency at the expense of increasing it at the two new resonance frequencies.

Note that regardless of the stiffness ratio or critical damping ratio ζ_2, the separation between the two new resonance frequencies and the original resonance frequency is dependent solely on the mass ratio m_2/m_1. The larger this ratio, the greater will be the frequency separation. Determination of the actual frequencies can be accomplished using published charts (Bies and Hansen, 1996).

The equation of motion for the mass m_2 in the system shown in Fig. 12.24 can be written as

$$\begin{aligned} m_2\left[\ddot{x}_2(t) - \ddot{x}_1(t)\right] + c_{d_2}\left[\dot{x}_2(t) - \dot{x}_1(t)\right] \\ + k_2\left[x_2(t) - x_1(t)\right] = -m_2\ddot{x}_1(t) \end{aligned} \tag{12.2.93}$$

Putting $z(t) = x_2(t) - x_1(t)$, and dividing through by m_2 we obtain

$$\ddot{z}(t) + 2\zeta_2\omega_{02}\dot{z}(t) + \omega_{02}^2 z(t) = -\ddot{x}_1(t) \tag{12.2.94}$$

where

$$\omega_{02} = \sqrt{k_2/m_2} \tag{12.2.95}$$

and

$$\zeta_2 = \frac{c_{d_2}}{2\sqrt{k_2 m_2}} \tag{12.2.96}$$

Equation (12.2.94) describes a second order system having the resonance frequency and critical damping ratio of the supplementary absorber system, but whose output is the relative displacement of the two masses and whose input is the acceleration of the main mass.

Equation (12.2.94) can be solved to give the required output $z(t) = x_2(t) - x_1(t)$ for a given input acceleration, damping ratio and resonance frequency of the secondary system. If the optimum values for damping ratio and stiffness ratio are required, then these can be used in the equation to find the corresponding

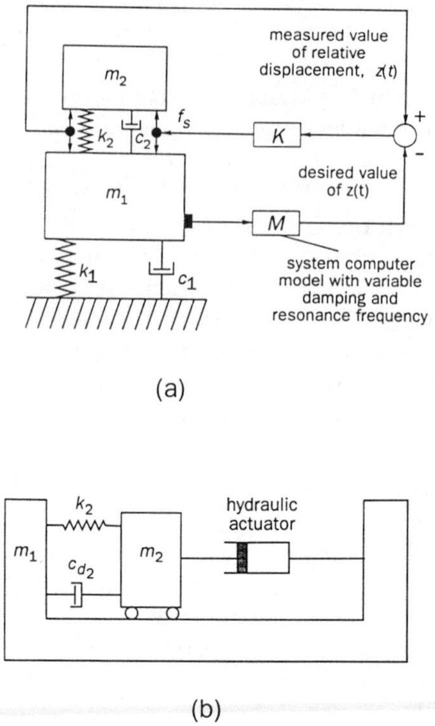

Fig. 12.25 Active tuning of a vibration absorber: (a) for a vertically mounted machine; (b) to control building vibrations.

optimum value of $z(t)$. The relevance of this process to the active tuning of an absorber will now be made clear (see Fig. 12.25).

If, for example, it were difficult to obtain accurate estimates of m_1 and k_1 of the original system, then it would be difficult to optimise values of k_2 and ζ_2 of the absorber at the design stage. However, this could be overcome by using values of k_2 and ζ_2 which were close to the desired values and tuning the absorber with a feedback control system. In one implementation reported by Lund (1980), (12.2.94) was implemented on an analogue computer such that the desired output $z(t)$ was calculated for a given input acceleration of the main mass (see Fig. 12.25).

Values of ω_{02} and ζ_2 can be made adjustable in the computer model so that they can be manually adjusted to optimize the response of the main mass. The control force $f_c(t)$ can then be determined so that it modifies the stiffness and damping of the attached absorber to the values used in the computer model. This is accomplished by driving the control force (as shown in Fig. 12.25) with a control signal proportional to the difference between the actual measured value of $z(t)$ and that desired according to the computer model solution of (12.2.94). A system such as the one just described has been used to reduce wind induced sway in tall buildings; an example of a specific installation is described by Petersen (1980).

In another implementation, a simple tuned vibration absorber system (using multiple tuned masses) has been used by Kaneda and Seto (1983), for vibration reduction of portable tools such as pavement breakers and impact hammers.

12.2.5 Vibration isolation of equipment from a rigid or flexible support structure

Until now we have been concerned with the reduction in the response of a mass suspended on a spring and damper system attached to a rigid foundation, which may or may not move. In practice such systems include optical equipment mounted on vibration isolators and vibration absorbers attached to structures, flexibly mounted equipment or buildings. However, there still remains a large class of vibration isolation problems which we have not yet considered, and which are characterized by the common requirement that vibration of a support structure caused by vibrating equipment attached to it is to be reduced by interposing an isolation system between the vibration source and the structure.

An ideal vibration isolation mount would provide high stiffness below some limiting frequency so that low frequency loads (including the weight) can be transmitted. Above the limiting frequency, the isolator should have zero stiffness so that vibrations at these higher frequencies are not transmitted. It is very difficult to approximate the ideal mount with a passive system; thus the attractiveness of an active system. Here we will only be concerned with a feedback system; isolation of a machine from a support structure using feedforward control is considered in Sections 12.4 onwards.

12.2.5.1 Rigid support structure

Returning to the single-degree-of-freedom system discussed in Section 12.2.1, we will examine the force transmitted through the isolation system to the support and use this as a measure of the vibratory energy transmitted.

The equation of motion for the mass is given by (12.2.1) and from Fig. 12.3 it can be seen clearly that the force transmitted into the structure is given by

$$f_T(t) = c_d\dot{x}(t) + kx(t) \tag{12.2.97}$$

Using (12.2.1) and (12.2.97) we obtain the following for the ratio of transmitted force to the force acting on the mass:

$$\frac{f_T(t)}{f(t)} = \frac{c_d\dot{x}(t) + kx(t)}{m\ddot{x}(t) + c_d\dot{x}(t) + kx(t)} \tag{12.2.98}$$

Taking Laplace transforms and zero initial conditions gives

$$H(s) = \frac{F_T(s)}{F(s)} = \frac{c_d s + k}{ms^2 + c_d s + k} \qquad (12.2.99a,b)$$

which is identical to (12.2.40) for the displacement transmission for a base excited system.

We will now examine the effect of various types of feedback on the force transmission: displacement, velocity, acceleration and force feedback.

Displacement feedback

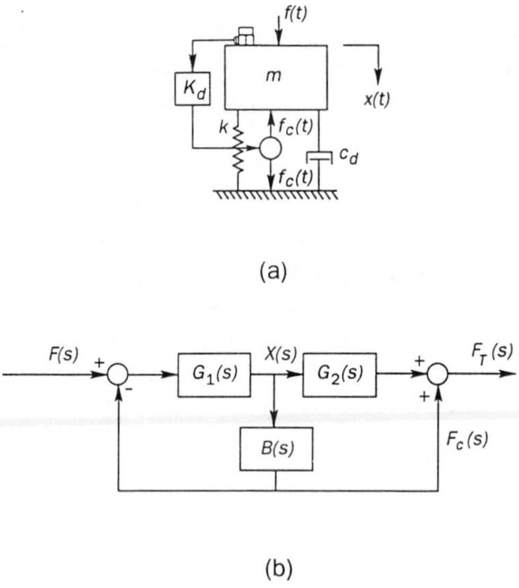

Fig. 12.26 Absolute displacement feedback for control of transmitted force $f_T(t)$: (a) physical system; (b) frequency domain block diagram.

The equations of motion for the system in Fig. 12.26(a) are

$$m\ddot{x}(t) + c_d\dot{x}(t) + (k + K_d)x = f(t) \qquad (12.2.100)$$

$$c_d\dot{x}(t) + (k + K_d)x = f_T(t) \qquad (12.2.101)$$

For the equivalent s-plane system shown in Fig. 12.26(b), we have

$$X(s) = \frac{F(s)\,G_1(s)}{1 + B(s)\,G_s(s)} \qquad (12.2.102)$$

Also,

$$F_T(s) = X(s)\left[G_2(s) + B(s)\right] = \frac{F(s)\left[G_2(s) + B(s)\right]}{\dfrac{1}{G_1(s)} + B(s)} \qquad (12.2.103a,b)$$

Thus,

$$H(s) = \frac{F_T(s)}{F(s)} = \frac{G_2(s) + B(s)}{\dfrac{1}{G_1(s)} + B(s)} = \frac{c_d s + k + K_d}{ms^2 + c_d s + k + K_d} \qquad (12.2.104a,b,c)$$

where $G_1(s)$ and $G_2(s)$ are defined by (12.2.45) and (12.2.46) respectively, and B(s) = K_d as defined in equation (12.2.47) for the relative displacement feedback of a base excited system.

If the control force only acted on the base then we would have

$$H(s) = \frac{c_d s + k + K_d}{ms^2 + c_d s + k} \qquad (12.2.105)$$

Using (12.2.5) and (12.2.6) and substituting $s = j\omega$ in (12.2.105), we can write the following for the frequency response with absolute displacement feedback (force acting only on the base):

$$H(j\omega) = \frac{1 + K_d/k + 2j\zeta(\omega/\omega_0)}{1 - (\omega/\omega_0)^2 + 2j\zeta(\omega/\omega_0)} \qquad (12.2.106)$$

The modulus of the closed loop frequency response function of (12.2.106) is plotted in Fig. 12.27, and shows the effect of varying the displacement feedback gain for various values of K_d/k, with $\zeta = 0.05$. It can be seen from the figure that increasing the negative feedback gain effectively increases the isolation at all frequencies without affecting the system resonance frequency. However, it is unlikely that this would result in a stable system.

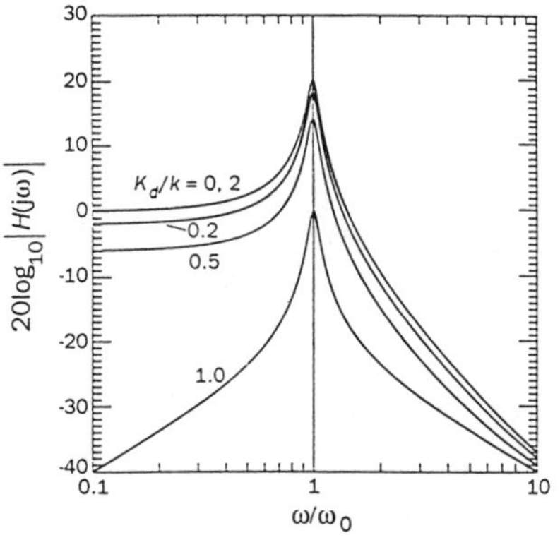

Fig. 12.27 Modulus of the frequency response for force transmission into a rigid base with displacement feedback, and with the control force acting only on the base for $\zeta = 0.05$.

The case of the control force acting only on the mass is the same as for absolute displacement feedback of the base excited system plotted in Fig. 12.14.

The case of the control force acting on the mass and reacting on the base is the same as for relative displacement feedback of the base excited system shown in Fig. 12.12.

Velocity feedback

Again, similar results are obtained as were obtained for displacement transmission of the base excited system discussed earlier, except for the case where the control force acted only on the base. In this case, the force transmission frequency response function is

$$H(s) = \frac{(c_d + K_v)s + k}{ms^2 + c_d s + k} \tag{12.2.107}$$

and $B(s)$ of Fig. 12.21(b) is

$$B(s) = K_v s \tag{12.2.108}$$

Using (12.2.5) and (12.2.6) and substituting $s = j\omega$ in (12.2.107), we can write the following for the force transmission frequency response with velocity feedback (force acting only on the base):

$$H(j\omega) = \frac{1 + 2j\zeta(\omega/\omega_0)(1 + K_v/c_d)}{1 - (\omega/\omega_0)^2 + 2j\zeta(\omega/\omega_0)} \tag{12.2.109}$$

The modulus of the closed loop frequency response function of (12.2.109) is plotted in Fig. 12.28, and shows the effect of varying the velocity feedback gain for various values of K_v/c_d, with $\zeta = 0.05$. It can be seen from the figure that increasing the negative feedback gain increases the isolation only slightly at high frequencies (provided the gain K_v/c_d is less than one).

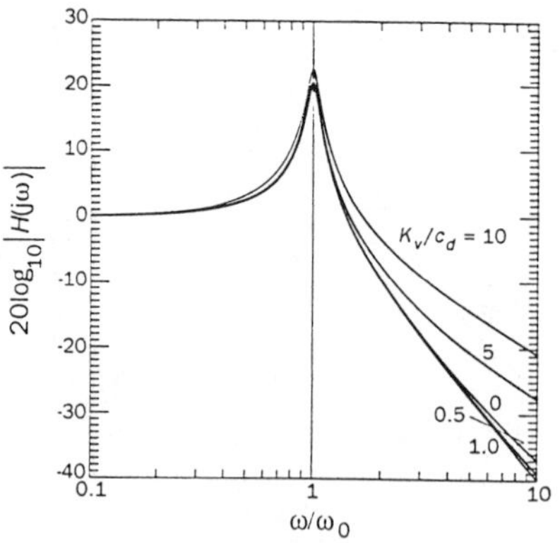

Fig. 12.28 Modulus of the frequency response for force transmission into a rigid base with velocity feedback and with the control force acting only on the base for $\zeta = 0.05$.

Acceleration feedback

Again similar results are obtained as were obtained for the base excited system, except where the control force acts only on the base, the force transmission is given by

$$H(s) = \frac{K_a s^2 + c_d s + k}{ms^2 + c_d s + k} \tag{12.2.110}$$

$B(s)$ of Fig. 12.26(b) is given by

$$B(s) = K_a s^2 \tag{12.2.111}$$

Using (12.2.5) and (12.2.6) and substituting $s = j\omega$ in (12.2.110), we can write the following for the force transmission frequency response with acceleration feedback (force acting only on the base):

$$H(j\omega) = \frac{1 + 2j\zeta(\omega/\omega_0) - (K_a/m)(\omega/\omega_0)^2}{1 - (\omega/\omega_0)^2 + 2j\zeta(\omega/\omega_0)} \tag{12.2.112}$$

The modulus of the closed loop frequency response function of (12.2.112) is plotted in Fig. 12.29, and shows the effect of varying the acceleration feedback gain for various values of K_a/m, with $\zeta = 0.05$. It can be seen from the figure that increasing the feedback gain increases the isolation at frequencies immediately above and immediately below the system resonance frequency with little effect at this frequency. Isolation at high frequencies is also markedly reduced.

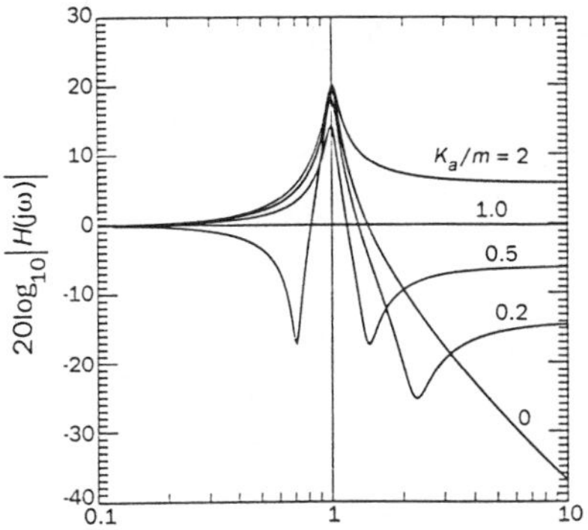

Fig. 12.29 Modulus of the frequency response for force transmission into a rigid base with acceleration feedback and with the force acting only on the base for $\zeta = 0.05$.

Force feedback

For this case, $G_1(s)$ and $G_2(s)$ are as defined in (12.2.45) and (12.2.46), and $B(s) = K_f$.

The equations of motion for the system shown in Fig. 12.30 are

$$m\ddot{x}(t) + c_d\dot{x}(t) + kx(t) + K_f f_T(t) = f(t) \tag{12.2.113}$$

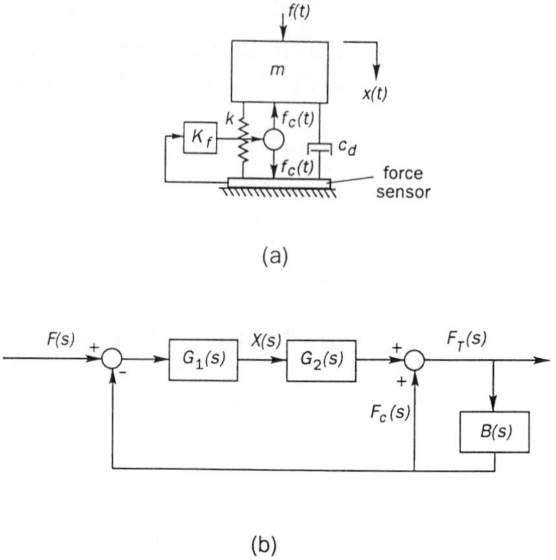

Fig. 12.30 Force feedback for control of transmitted force: (a) physical system; (b) frequency domain block diagram.

$$c_d \dot{x}(t) + kx(t) + K_f f_T(t) = f_T(t) \tag{12.2.114}$$

For the equivalent s-plane model shown in Fig. 12.30 we can write

$$F_T(s) = F(s)\,G_1(s)\,G_2(s) - F_T(s)\,B(s)\,G_1(s)\,G_2(s) + F_T(s)\,B(s) \tag{12.2.115}$$

or

$$H(s) = \frac{F_T(s)}{F(s)} = \frac{G_s(s)}{\dfrac{1}{G_1(s)} + B(s)\,G_2(s) - \dfrac{B(s)}{G_1(s)}} = \frac{c_d s + k}{ms^2(1 - K_f) + c_d s + k}$$

(12.2.116a,b,c)

Using (12.2.5) and (12.2.6) and substituting $s = j\omega$ in (12.2.116c), we can write the following for the force transmission frequency response with force feedback (force acting on the base and the mass):

$$H(j\omega) = \frac{1 + 2j\zeta(\omega/\omega_0)}{1 - (\omega/\omega_0)^2(1 - K_f) + 2j\zeta(\omega/\omega_0)} \tag{12.2.117}$$

The modulus of the closed loop frequency response function of (12.2.117) is plotted in Fig. 12.31, and shows the effect of varying the force feedback gain for various values of K_f/m, with $\zeta = 0.05$. It can be seen from the figure that

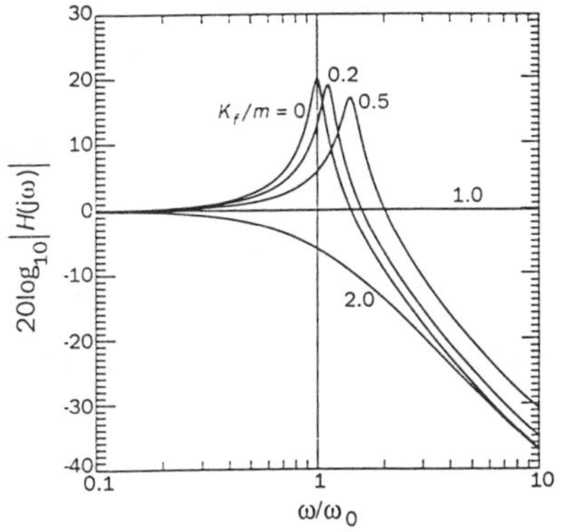

Fig. 12.31 Modulus of the frequency response for force transmission into a rigid base with force feedback and the control force acting on the vibrating body and the rigid base for $\zeta = 0.05$.

increasing the feedback gain effectively increases the system resonance frequency and critical damping ratio which results in increases in the isolation achieved at low frequencies at the expense of reduced isolation at high frequencies. It is likely that the system would be unstable for values of the gain K_f/m greater than one. Note that K_f represents the amplitude of the control force as a fraction of the transmitted force in the absence of control.

Force feedback: control force applied only to mass

If the control force were only applied to the mass, we would have the situation shown in Fig. 12.32. For this case, $G_1(s)$ and $G_2(s)$ are as defined in (12.2.45) and (12.2.46), and $B(s) = K_f$.

Figure 12.32(a) shows the force actuator located between the vibrating mass and the passive isolation system. This is usually referred to as a 'series' system as the active and passive elements are in series between the vibrating body and the support. It is assumed that the deformation of the force actuator is very small compared to the movement of the mass. For this case, it is clear that the force actuator must be capable of supporting the weight of the vibrating mass. This is in contrast to the situation shown in Fig. 12.30(a) where it can be seen that the force actuator does not have to support the mass of the machine but it must be strong enough to overcome the stiffness of the stiffness element k so that it can develop an actuation force. This latter system is referred to as a parallel system as the active and passive elements are mounted in parallel between the vibrating body and the support structure.

In the configuration shown in Fig. 12.32(b) the force is applied only to the vibrating rigid body. In dynamic terms this is identical to the configuration shown in Fig. 12.32(a), provided the force actuator is very rigid compared to the passive suspension. Such a transducer might be a piezoceramic stack or a magnetostrictive rod (see Chapter 16).

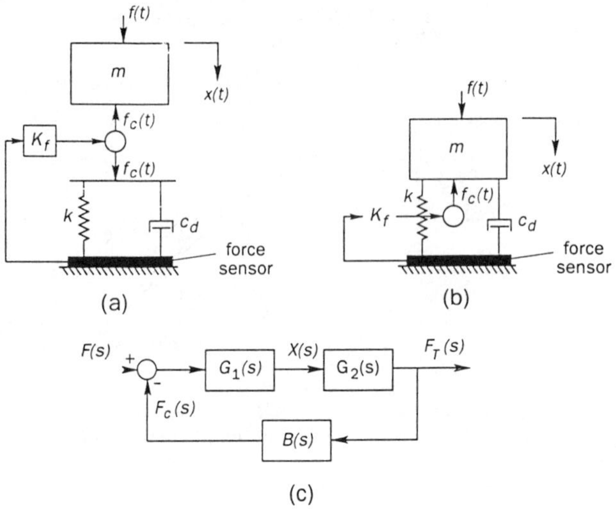

Fig. 12.32 Force feedback on vibrating body for control of force transmission to a rigid foundation (sometimes a rigid mass is placed between the actuator and the passive system): (a) physical system 1; (b) physical system 2; (c) frequency domain block diagram.

In practice, the control force shown in Fig. 12.32(b) could be applied by a shaker mounted on the vibrating body *m* and driving a second inertial mass attached only to the shaker.

The equations of motion for the system shown in Fig. 12.32 are

$$m\ddot{x}(t) + c_d\dot{x}(t) + kx(t) + K_f f_T(t) = f(t) \tag{12.2.118}$$

$$c_d\dot{x}(t) + kx(t) = f_T(t) \tag{12.2.119}$$

For the equivalent *s*-plane block diagram shown in Fig. 12.32(c), we can write

$$F_T(s) = F(s)G_1(s)G_2(s) - F_T(s)B(s)G_1(s)G_2(s) \tag{12.2.120}$$

$$H(s) = \frac{F_T(s)}{F_s} = \frac{G_1(s)G_2(s)}{1 + G_1(s)G_2(s)B(s)} = \frac{c_d s + k}{ms^2 + (1 + K_f)(c_d s + k)}$$

(12.2.121a,b,c)

Using (12.2.5) and (12.2.6) and substituting $s = j\omega$ in (12.2.121c), we can write the following for the force transmission frequency response with force feedback (force acting only on the mass):

$$H(j\omega) = \frac{1 + 2j\zeta(\omega/\omega_0)}{1 - (\omega/\omega_0)^2(1 + K_f) + 2j\zeta(\omega/\omega_0)(1 + K_f)} \tag{12.2.122}$$

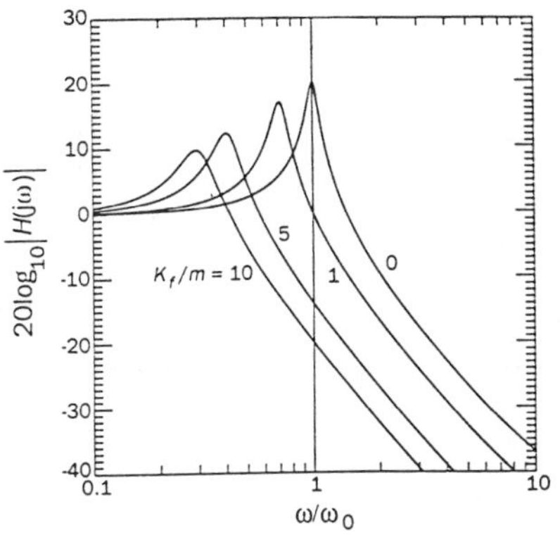

Fig. 12.33 Modulus of the frequency response for force transmission into a rigid base with force feedback and the control force acting on the rigid body only.

The modulus of the closed loop frequency response function of (12.2.122) is plotted in Fig. 12.33, and shows the effect of varying the force feedback gain for various values of K_f/m, with $\zeta = 0.05$. It can be seen from the figure that increasing the feedback gain effectively decreases the system resonance frequency and increases the critical damping ratio which results in increases in the isolation achieved at high frequencies at the expense of reduced isolation at low frequencies. It is likely that the system would be unstable for values of the gain K_f/m greater than one.

Force feedback: control force applied only to support

The series system shown in Fig. 12.34(a) is similar to that shown in figure 12.32(a), except that the control force is now acting on the structure rather than the vibrating body. For this case, $G_1(s)$ and $G_2(s)$ are as defined in (12.2.45) and (12.2.46), and $B(s) = K_f$.

The equations of motion for the system shown in Fig. 12.34 are

$$m\ddot{x}(t) + c_d\dot{x}(t) + kx(t) = f(t) \tag{12.2.123}$$

$$c_d\dot{x}(t) + kx(t) + K_f f_T(t) = f_T(t) \tag{12.2.124}$$

In the s-domain we have

$$\frac{F_T(s)}{F(s)} = \frac{G_1(s)G_2(s)}{1+K_f} = \frac{c_d s + k}{(ms^2 + c_d s + k)(1+K_f)} \qquad \text{(12.2.125a,b)}$$

Using (12.2.5) and (12.2.6) and substituting $s = j\omega$ in (12.2.125b), we can write the following for the force transmission frequency response with force feedback (force acting only on the base):

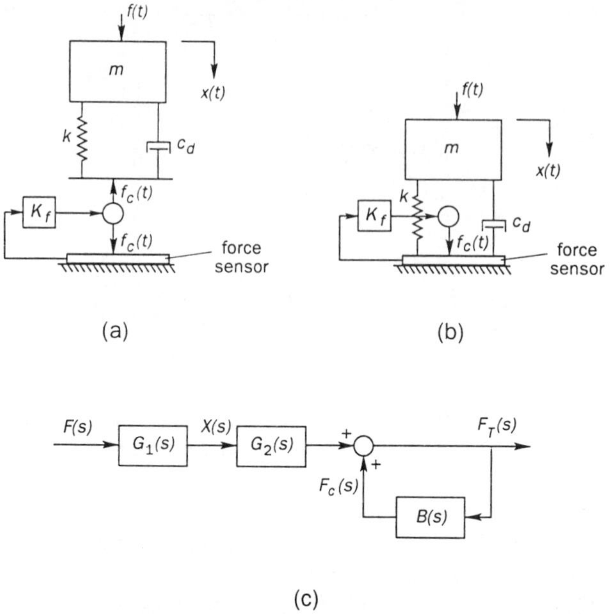

Fig. 12.34 Force feedback on the support structure for control of force transmission from a vibrating rigid body into the structure: (a) physical system 1; (b) physical system 2; (c) frequency domain block diagram.

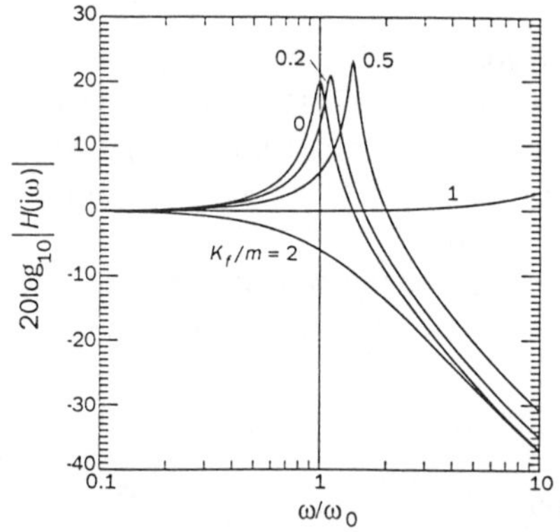

Fig. 12.35 Modulus of the frequency response for force transmission into a rigid base with force feedback and the control force acting only on the rigid base for $\zeta = 0.05$.

$$H(j\omega) = \frac{1 + 2j\zeta(\omega/\omega_0)}{1 - (\omega/\omega_0)^2(1 - K_f) + 2j\zeta(\omega/\omega_0)(1 - K_f)} \tag{12.2.126}$$

The modulus of the closed loop frequency response function of (12.2.126) is plotted in Fig. 12.35, and shows the effect of varying the force feedback gain for various values of K_f/m, with $\zeta = 0.05$. It can be seen from the figure that increasing the feedback gain effectively increases the system resonance frequency and decreases the critical damping ratio which results in an increase in the isolation achieved at low frequencies at the expense of reduced isolation at high frequencies. It is likely that the system would be unstable for values of the gain K_f/m greater than one.

12.2.5.2 Flexible support structure

Consider the simple system shown in Fig. 12.36. The equation of motion for the mass m is

$$m\ddot{x}_1(t) + c_d\dot{x}_1(t) + kx_1(t) - c_d\dot{x}_2(t) - kx_2(t) = f(t) \tag{12.2.127}$$

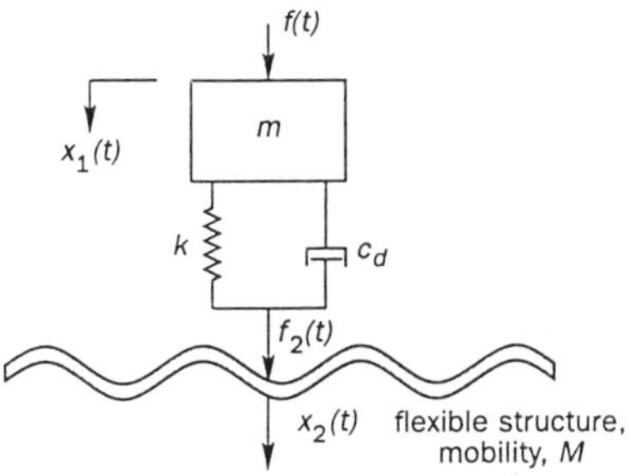

Fig. 12.36 Vibration isolation of a vibrating mass from a flexible structure.

Taking Laplace transforms and zero initial conditions gives

$$(ms^2 + c_d s + k)X_1(s) - (c_d s + k)X_2(s) = F(s) \tag{12.2.128}$$

What we desire is the transfer function between the input force $F(s)$, acting on the isolated mass, and the structural response $X_2(s)$. Thus, we need to eliminate $X_1(s)$ from the equation. We do this by defining the structural point mobility $M(s)$ as

$$M(s) = \frac{sX_2(s)}{F_2(s)} \tag{12.2.129}$$

where $F_2(s)$ is the force input to the structure, defined as

$$F_2(s) = (c_d s + k)X_1(s) - (c_d s + k)X_2(s) \tag{12.2.130}$$

Substituting (12.2.130) into (12.2.129) and rearranging gives

$$X_1(s) = X_2(s)\left[1 + \frac{s}{M(s)(c_d s + k)}\right] \tag{12.2.131}$$

Substituting (12.2.131) into (12.2.128) gives

$$(ms^2 + c_d s + k)\left[1 + \frac{s}{M(s)(c_d s + k)}\right] X_2(s) - (c_d s + k)\,X_2(s) = F(s) \qquad (12.2.132)$$

Rearranging gives

$$\frac{X_2(s)}{F(s)} = \left[ms^2 + \frac{ms^3 + c_d s^2 + ks}{M(s)(c_d s + k)}\right]^{-1} \qquad (12.2.133)$$

or

$$\frac{X_2(s)}{F(s)} = \frac{M(s)(c_d s + k)}{(s)(mc_d s^3 + mks^2) + ms^3 + c_d s^2 + ks} \qquad (12.2.134)$$

The point mobility of a flexible structure may be modelled as (Scribner *et al.*, 1990)

$$M(s) = \frac{sA \prod_{i=1}^{\infty}\left(1 - \dfrac{s}{z_i}\right)}{\prod_{i=1}^{\infty}\left(1 - \dfrac{s}{p_i}\right)} \qquad (12.2.135)$$

where z_i and p_i are complex zeros and poles and A is a real constant. The function in (12.2.135) is guaranteed to have alternating poles and zeros on the imaginary axis in the s-plane for the case of an undamped structure (Gevarter, 1970). On the other hand, for hysteretic structural damping, the poles and zeros will be to the left of the imaginary axis. However, Scribner *et al.* (1990) show that provided the structure is lightly damped, it is not very important from the point of view of controller design what the damping mechanism is, whether hysterectic, viscous or proportional viscous.

As (12.2.134) is very complex, the control of the structural vibration by using a force driven by simple velocity acceleration or displacement feedback is inadequate from either the performance or stability point of view. In this case, it is necessary to feed back either structural velocity, acceleration displacement or driving force through a compensation circuit, which is more complicated than a simple gain, to the control force transducer. The number of terms used in (12.2.135) to derive the optimal controller is dependent upon the upper frequency limit desired for control. In practice, it is extremely difficult to accurately determine the poles and zeros of (12.2.135), thus making it extremely difficult to design a robust controller with good performance over a wide frequency range, which is what would be desired in an ideal vibration isolation system. Such a system, which requires a detailed knowledge of the structural dynamics for its design, is referred to as a 'broadband' controller. However, as shown by Scribner *et al.* (1990) a narrow band controller can be designed without the need to know the details of the structural dynamics. Such a controller results in a mount with low stiffness over narrow frequency ranges which are harmonically related. Thus, using this type of feedback controller, it is feasible to achieve good performance and stability for control of vibration transmission at a machine operating frequency and its harmonics, for example. With a suitable designed compensator circuit, it is possible for the feedback controller to track changes in the machine operating frequency (Scribner *et al.*, 1990). The same authors also state that it is better to use force transmitted into the structure as the control variable as better performance is possible and also it has the effect of controlling vibration transmission from the mounted machine to the structure and also from the structure to the mounted machine. On the other hand, velocity, acceleration or displacement feedback will only control vibration transmission to the part of the system on which the transducer is mounted.

It is of interest to examine the isolation system and compensator circuit used by Scribner *et al.* 1990. The physical system is shown in Fig. 12.37(a) and the control block diagram in Fig. 12.37(b).

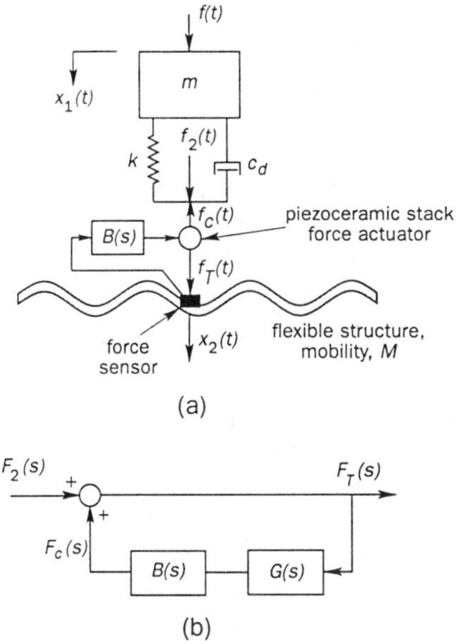

Fig. 12.37 Vibration isolation of a rigid body from a flexible support structure: (a) physical arrangement; (b) control block diagram.

The quantity $G(s)$ represents the transfer function between the voltage into the piezo force generator to the voltage out of the force sensor, $F_2(s)$ is the disturbance input from the machine to the structure and $B(s)$ is the compensator transfer function.

From Fig. 12.37(a) we can see that a series system (see Section 12.1) is being considered and from Fig. 12.37(b) we can derive the relationship between the disturbance input force $F_2(s)$ (with the actuator turned off) and the structural vibration output force $F_T(s)$ as

$$F_T(s) = F_2(s)\frac{1}{1+B(s)\,G(s)} \tag{12.2.136}$$

which shows that the product $B(s) \times G(s)$ must be large for $F_2(s)$ to have an insignificant effect on $F_T(s)$. The loop must also be stable. A compensator which satisfies these requirements is given by

$$B(s) = \frac{1}{s^2 + 2\zeta_c \omega_c s + \omega_c^2} \tag{12.2.137}$$

where the compensator damping ζ_c must be low for high performance and the compensator resonance frequency ω_c should be set to the fundamental machine operating frequency.

However, Scribner *et al.*, 1990 report a more clever compensation circuit (see Fig. 12.38) which also allows the compensator to track the machine operating frequency ω.

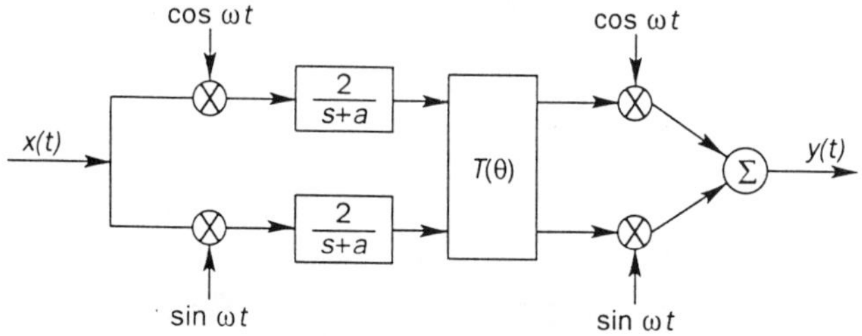

Fig. 12.38 Frequency tracking analogue compensator circuit (after Scribner *et al.*, 1990).

The signals $\sin\omega t$ and $\cos\omega t$ are derived from the fundamental machine excitation ω (usually the machine rotational speed). These signals are multiplied with the input $x(t)$ and the result is low pass filtered with a pole at $s = -a$. (Note that a should be about two orders of magnitude smaller than the machine operating frequency.) If $a = 0$, the multiplication process is equivalent to a Fourier transform of the time signal $x(t)$. As the pole a is so small in magnitude, the outputs from the low pass filters are very close to d.c. and these two outputs are then multiplied by an orthonormal rotation matrix $T(\theta)$ given by

$$T(\theta) = \begin{bmatrix} \cos\theta & -\sin\theta \\ \sin\theta & \cos\theta \end{bmatrix} \tag{12.2.138}$$

After the two signals have been rotated in phase, they are converted back to the time domain by again multiplying by $\cos\omega t$ and $\sin\omega t$ respectively. The two results are then summed together to give the compensator output. The frequency response of the compensator in the s-domain is

$$C(s) = \frac{Y(s)}{X(s)} = \frac{(s+a)\cos\theta + \omega\sin\theta}{(s+a)^2 + \omega^2} \tag{12.2.139a,b}$$

which has a zero at $s = -a - \omega\tan\theta$ and poles at $s = -a \pm j\omega$.

Note that as the system performance is best at the poles and worst at the zeros, θ should be chosen so that the zero occurs between the fundamental and first harmonic of the machine operational frequency and a should be made very small (100 times less than the machine operating frequency).

Although feedback control has been shown to be robust and effective over very narrow frequency ranges for controlling vibration transmission through an isolator to a flexible structure, it is very difficult to implement it over a wide frequency band. To do this, a very accurate identification of the structural dynamics is needed and as this usually changes over time, the identification should be carried out 'on-line'. This complicates the control problem to such an extent that a considerable amount of research is necessary before broadband feedback control will be practically implemented in vibration isolation systems. As feedforward systems are more effective than feedback systems for controlling periodic disturbances which can be converted to an electrical signal, the use of feedback controllers, even for narrow band control of vibration isolation systems is limited to those cases where it is not possible to measure the disturbance signal adequately. Feedback controllers are also the only option when the disturbance signal is a transient or random noise which cannot be measured sufficiently far enough ahead in time to allow a feedforward controller to produce a compensation signal.

Although feedback control over a narrow frequency range has been demonstrated for the control of vibration transmission along a single axis and at a single point into a structure, a considerable amount of research effort is needed to develop these ideas to control vibration translations along and rotations around multiple axes simultaneously and the control of vibration into several points on the structure simultaneously. Provided that all translations and rotations can be controlled at each structural excitation point, it may be possible to control each point with an independent control system, although control of all points with a single multichannel controller may be more effective as cross-coupling between mounts is likely to be important.

12.2.5.3 Use of an intermediate mass

To improve the isolation efficiency of passive systems, an intermediate rigid, massive plate can be placed between the vibrating body and the structure from which it is to be isolated. The mass of the plate needs to be sufficiently large for the additional resonance frequency introduced into the system by its presence to be well below the frequency range in which isolation is required. However, the mass should be no larger than necessary, so as to minimize the actuation force required. Use of an active system to control the vibration of the intermediate mass can reduce the amount of mass required and can improve the low frequency performance of the existing system (Ross *et al.*, 1988). As pointed out by Ross, it is important to control the rotations and horizontal translations, as well as the vertical translations of the mass to ensure that no vibration is transmitted to the support structure. Implementation of this idea in practice requires a minimum of six inertial shakers and six transducers which can measure the six possible degrees of freedom (rotational and translational) of the rigid mass. In many cases the degree of freedom involving rotation about the vertical axis can be ignored.

As the intermediate rigid mass is poorly coupled to the vibration source and the rigid body, Ross *et al*. claim that each mount can be controlled independently of the others in a multimount system which is isolating a mass from a flexible structure. Whether or not this is true in practice remains to be seen. Also the use of an intermediate mass decoupling the force actuator from the dynamics of the structure allows a more stable feedback control system to be realised. In practice, better performance has been achieved using feedback of the motion of the intermediate mass rather than the difference in motion between the mass and the structure as would be sensed by a gap sensor (Blackwood and von Flotow, 1993).

Although Ross *et al*. claim that the arrangement shown in Fig. 12.39 can control rotational as well as translational motion, it is difficult to imagine how this may be implemented in practice, as it is clear that the shakers cannot move the mass to the left horizontally at the same time as they rotate it clockwise about its centre. In fact, horizontal motion to the left is also associated with anti-clockwise rotation. However, the addition of a third shaker shown as a dashed outline in Fig. 12.39 will allow the simultaneous control of horizontal, vertical and rotational motion. A schematic layout of how a feedback controller may be implemented to control one horizontal translational, one rotational and the vertical translational degree of freedom is shown in Fig. 12.40.

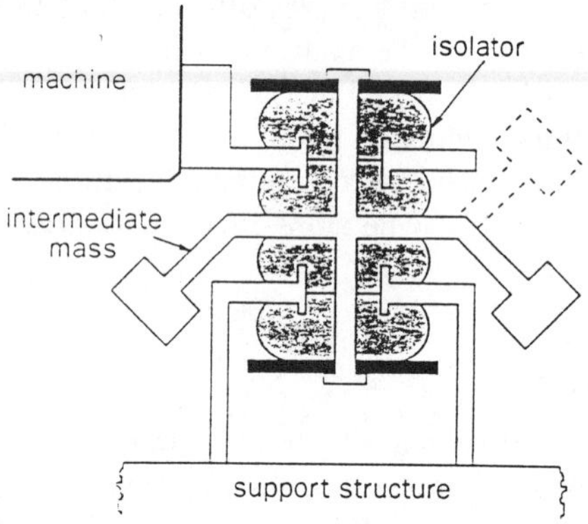

Fig. 12.39 Active vibration isolation mount with an intermediate mass (after Ross *et al.*, 1988) (The shaker shown as a dashed outline is not shown by Ross *et al.* and the additional damper at the bottom of the mount shown by Ross *et al.* has been omitted here for clarity.)

Control of five degrees of freedom with five inertial shakers would also be possible, although the arrangement would be even more complex than that shown in Fig. 12.40, as the output from each sensor would be used to drive four of the five shakers, with a different set of four being driven by each transducer signal.

Stability problems associated with the arrangement shown in Fig. 12.40 would be very difficult to overcome unless the interaction of the three transducer

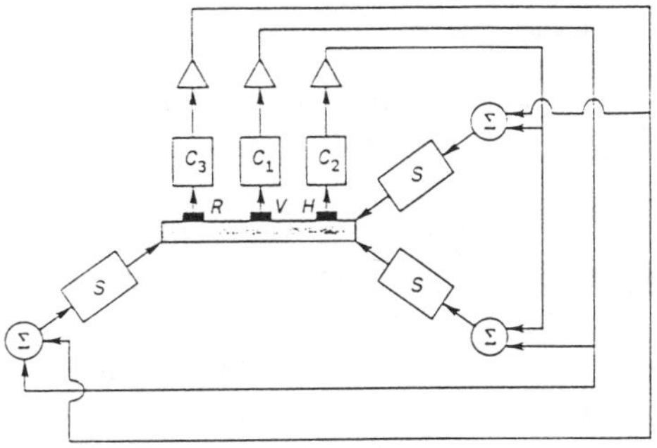

Fig. 12.40 Possible implementation of a feedback system to control three degrees of freedom of the rigid mass of Fig. 12.39. Σ = sum, C_1, C_2, C_3 = compensation filters, ▷ = amplifier, S = inertial shaker, and V, R and H are the vertical, rotational and horizontal motion transducers respectively.

signals could be made almost negligible. This would require horizontal motion transducers which were insensitive to vertical and rotational motion. Similar requirements would be necessary for the vertical and rotational motion sensors. In addition, the frequency response of the control shakers would have to be well matched so that the vertical motion feedback did not generate horizontal or rotational motion, and vice versa for horizontal and rotational motion feedback. These problems have so far prevented practical implementation of the system to isolation of motion other than vertical translation.

12.3 APPLICATIONS OF FEEDBACK CONTROL

In this section,various applications of feedback vibration isolation will be discussed, including the following topics:

1. vehicle suspension systems (active and semi-active);
2. transmission of vibration from rigidly mounted machinery to a stiff support structure;
3. transmission of vibration from a support structure to an item of sensitive equipment mounted on it;
4. vibration of tall buildings;
5. isolation of machinery from flexible structures (including engine mounts);
6. isolation of drive-train and rotor vibration from a helicopter cabin.

Many of these applications are also amenable to feedforward control, provided that a reference signal correlated with the disturbance can be obtained sufficiently far in advance for the controller to calculate the control signals so that they arrive at the error sensors at the same time as the primary disturbance used to obtain the reference signal. For periodic disturbances, feedforward control may indeed be preferable as the reference signal need not be obtained in advance of the controller output signal to satisfy causality constraints. Feedforward control will be discussed in Sections 12.4 to 12.8.

12.3.1 Vehicle suspension systems

As mentioned in Section 12.1, vehicle suspension systems can be split into four categories: passive, semi-active, slow-active and fully active. Only the latter three types will be considered here. The main difference between fully active and semi-active systems is that a fully active system can put power into the suspension as well as dissipate it, whereas a semi-active system can only dissipate power, usually by varying the system damping.

Semi-active systems may be further subdivided into the continuous type and the on/off type. In the former system, the damping force is continuously variable during the time when it is being applied and in the latter case, it is constant during this time. Both types spend some of the time in the 'off' state, when no damping force is applied. The criterion determining when the damping force will be applied and when it will not be applied is known as the control law.

Similar control laws describing the control damping force are used for fully active and continuously variable semi-active systems, except that when the fully active system is supplying power, the semi-active system switches off.

Fully active systems generally use hydraulic actuators to generate the required forces in the suspension system. In practical systems, the actuator is usually placed in parallel with a passive spring so that it does not have to support the weight of the vehicle, thus greatly reducing the actuator static force requirements. The actuator needs to have a large bandwidth to minimize the transmission of high frequency as well as low frequency loads to the vehicle. This translates into an expensive actuator. One way around this is to use a low bandwidth actuator in series with a passive spring (slow active system). In this system, the passive spring provides the required isolation at high frequencies while the actuator provides vibration control at low frequencies (usually below 3 Hz).

A disadvantage of actuators in active suspensions is the high power consumption (typically 10 kW for motor vehicles), although the power requirements for low bandwidth actuators are less (typically 2–3 kW) (Wendel and Stecklein, 1991). It is possible to reduce the power requirements of an active suspension by using one hydraulic pump for each suspension unit and driving all pumps with a common shaft. Thus if an individual suspension unit must release fluid at a particular time its pump will be displaced in the opposite direction, thus acting as a hydraulic motor which can help drive another pump which is required to supply fluid (as they are on the same drive shaft) (Wendel and Stecklein, 1991). However, the power requirements for semi-active systems are considerably smaller than either fully active system, as power is only needed to drive the sensors, damper valves and the controller, and is not needed to provide the control force. Because they only dissipate energy, semi-active systems are more failsafe and considerably less expensive than fully active systems. This is because semi-active systems do not require hydraulic pumps, accumulators, high performance filters, pipework actuators and servo-valves, but just a rapidly adjustable damper (Hine and Pearce, 1988). Also to be considered are the running costs of a fully active system which typically consumes a few kilowatts of power.

Although many journal papers concentrate on specifying the performance of an active suspension system in terms of ride quality (or the transmission of vibration from the wheel to the body), there are a number of other important parameters including those outlined below, which must also be considered.

1. Handling: this is pitch and roll of the vehicle body as a result of cornering and braking manoeuvres.
2. Road holding: this is the contact force between the tyres and the road. Clearly, this force should be as large as possible at all times.
3. Suspension travel: the allowable limit of suspension travel in any vehicle design will affect the performance achievable from the suspension systems.
4. Static deflection resulting from variable payload.

Although passive suspensions can theoretically perform almost as well as active systems in terms of ride quality, they would do so at the expense of unacceptable performance for the four parameters discussed above. Active suspensions on the other hand, do a much better job of optimizing all performance parameters, although it is not possible to achieve the actual optimum for each parameter simultaneously. The optimization ability of semi-active systems lies between that of passive and fully active systems.

Note that with the exception of very small damping ratios ζ, when more damping is added to a passive or on–off active suspension system, the ride quality is degraded but handling, road holding and suspension travel is improved. This is in contrast to the continuous variable semi-active or fully active systems where increasing the damping improves the ride quality (Miller, 1988a). This is caused because the damping force used in the latter systems is proportional to absolute velocity of the car body rather than relative velocity between the car body and the wheel. Thus the trade-off between ride comfort and handling which exists for the passive and on–off semi-active systems does not exist for the continuously variable semi-active and fully active systems. Note also that fully active systems can be designed to be self-levelling; that is, they can keep the vehicle on an even keel during cornering, etc. (Sharp and Hassan, 1986). In a practical system, however, some provision must be made for the wheel to body relative displacement feedback to be dominant at low frequencies, to ensure that the limit of the actuator stroke is not reached.

The great virtue of a fully active system is its ability to adapt to variable operating conditions, and to employ the full suspension working space (allowable suspension travel) to satisfy the ride comfort requirements and the four parameters listed previously. On the other hand, the major weakness of a suspension system which contains non-adjustable control forces is that it can be ideal for only one operating condition (or road type), and will probably be far from ideal in conditions differing from the one for which it was designed. Note that only feedback (and not feedforward) control systems are generally used for vehicle systems because of the difficulty in measuring the incoming disturbance.

However, some feedback systems have been designed using a preview of the incoming disturbance to improve performance (Foag, 1988).

Over the past 40 years, many papers have been published in the vehicle literature on 'intelligent' suspension systems. The idea of automatic (or active) suspensions for vehicles was first suggested by Federspeil-Labrosse in 1954. Since that time servo-mechanisms have been introduced to provide the control forces (Rockwell and Lawther, 1964; Kimica, 1965), and the simple one-degree-of-freedom analytical model has been extended to a more complicated two-degree-of-freedom model which has been analysed using state space techniques and optimal linear control theory by Thompson (1979). Semi-active suspensions were introduced in the early 1970s (Crosby and Karnopp, 1973) and since that time there have been numerous papers published comparing the relative merits of active and semi-active suspension systems and describing a number of variations aimed to improve performance.

It is not our intention to devote the considerable space needed for a comprehensive literative review here. However, the work done prior to 1987 has been covered in an excellent review by Sharp and Crolla (1987), and some of the more significant work since that time will be discussed here.

For a long time, controller design and performance predictions were based on a quarter car model which treated each wheel suspension as independent from all the others. Indeed, Sharp and Crolla (1987) showed that this simple model is capable of adequately representing the vehicle for calculations of ride comfort. To include vehicle handling as well as ride comfort, it has been assumed that it is sufficient to use suspension travel in the performance function to be minimized. However, to properly model the handling aspects it is necessary to use more complex half or full car models (depending on whether pitch, roll or both is to be modelled) and degrees of freedom in addition to simple vertical movement at each suspension point (Todd, 1990; Shannan and Vanderploeg, 1989; Malek and Hedrick, 1985; El-Madany *et al.*, 1987; Crolla and Abdel-Hady, 1991). Half or full car models have also been used for those cases involving a feedforward (or preview) part in the control system (Hrovat, 1991).

In almost all cases, whether they be full, half or quarter car models, each wheel suspension is modelled as a two-degree-of-freedom system, as shown in Fig. 12.41(b), where the upper mass, spring and damper represent the wheel mass and tyre stiffness. This allows the wheel dynamics to be included in the model and introduces an additional constraint of maximum tyre deflection which is not included in the single degree-of-freedom model shown in Fig. 12.41(a), in which the wheel dynamics are excluded. Reducing tyre deflection reduces the potential for wheel hop and results in improved vehicle handling. Since single degree-of-freedom models do not include the wheel hop constraint, the results obtained using these models give an upper limit to the possible suspension ride performance which can be achieved with the two-degree-of-freedom model (Tseng and Hrovat, 1990; Karnopp, 1989; Thompson, 1976). However, the single-degree-of-freedom system is sometimes used to demonstrate trends in cases where only the maximum achievable ride performance is of interest; for

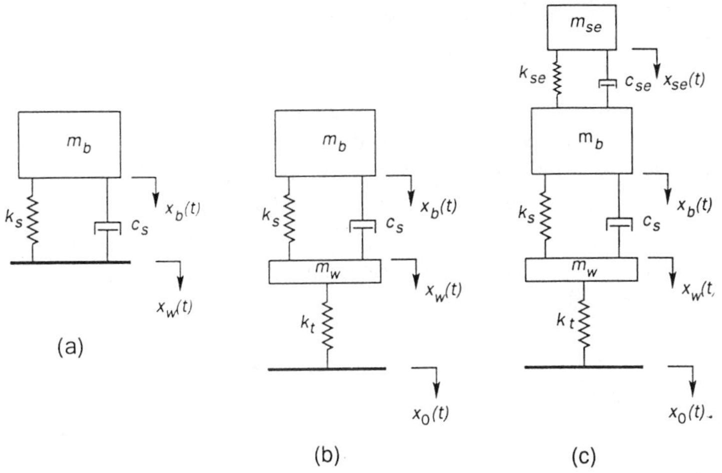

Fig. 12.41 Quarter car suspension model: (a) single-degree-of-freedom model (subscript *b* refers to the vehicle body, *s* refers to the suspension, *w* to the wheel and *t* to the tyre); (b) two degree-of-freedom model (includes wheel dynamics); (c) three degree-of-freedom model (includes seat dynamics, where the subscript *se* refers to the seat).

example, when different control strategies are being compared (Alanoly and Sankar, 1987), or where the two-degree-of-freedom problem becomes too complex to be manageable (Karnopp, 1989; Narayanan and Raju, 1990).

To obtain a truly accurate representation of passenger comfort, it is also necessary to include the seat dynamics to produce a three degree-of-freedom model as shown in Fig. 12.41(c). However, this is rarely done in the literature; indeed, it appears to have been considered only twice (Hennecke and Zieglmeier, 1988; Craighead, 1988). In the latter case, the seat was replaced by an ambulance stretcher.

Similar models to those shown in Fig. 12.41(c) have also been used to design active suspensions for track cabs (El-Madany and El-Razaz, 1988).

Using another approach, Cheok *et al.* (1989) developed a system for maximising passenger comfort using active control of the seat suspension only. Junker and Seewald (1984) also investigated an active seat suspension system, but for a tractor.

In virtually all suspension models discussed in the literature, the tyre damping is assumed to be zero as indicated by the absence of a damper in parallel with k_w in Fig. 12.41(b). Levitt and Zorka (1991) showed that the effect of including a small non-zero ($\approx$ 0.02) value for the tyre damping ratio, ζ_w, is to reduce the calculated vehicle body acceleration by up to 30%. Thus it is clear that if tyre damping is excluded from the model, the calculated car body accelerations (vertical direction) will be overestimates of actual values, resulting in a conservative suspension design.

Active suspension systems actually built and working in operational vehicles have been reported by Hine and Pearce (1988), Satoh *et al.* (1990), Aburaya *et al.* (1990) and Yashuda *et al.* (1991). However, the number of publications in this area is very limited, and gives no indication of the high level of effort being expended by automotive manufacturers, for obvious commercial reasons.

Applications of active suspensions to mobile vehicles other than automobiles have also been reported in the literature. Active suspensions for trains have been considered by a number of authors (Williams, 1985; Pollard and Simons, 1984, Okamoto *et al.*, 1987; Gimenez *et al.*, 1988; Goodall and Kortum, 1990). An active suspension for a magnetically levitated vehicle was discussed by Nagai *et al.* (1990). Wheeled robot active suspension systems have been considered by Tani *et al.* (1990), and various aspects of tractor active suspension systems have been considered by Stayner (1988).

According to Crolla (1988), the procedure used to model vehicle suspensions generally involves the following stages:

1. development of the equations of motion for the ¼ car model together with equations describing the (active or passive) force producing elements;
2. derivation of appropriate (sometimes optimal) control laws;
3. use of a random disturbance input at the tyre to calculate performance;
4. repeating the calculations for a range of road surfaces and vehicle speeds. As the suspension for most vehicles must be optimised to operate within a certain maximum vertical movement of the suspension, the performance of the various types is generally compared for a fixed maximum allowed vertical movement, usually $\pm 8-10$ cm.

In the sections to follow, the first two steps of the above procedure will be outlined for active and semi-active systems using the principles of feedback control system analysis discussed in Chapter 5. However, the detailed results for various disturbance inputs will not be discussed; these are covered adequately in the various references mentioned above so only general results will be discussed here.

12.3.1.1 Fully active suspensions

Fully active suspensions may be divided into two types; fast active (often referred to just as fully active) where the control actuator may be mounted in parallel with a passive spring (Fig. 12.42(a)) (or in some cases there may be no passive spring) and slow active, where the control actuator is mounted in series with a passive spring (see Fig. 12.42(b)). Slow active systems will be discussed in Section 12.3.1.2. The system shown in Fig. 12.42(a) can be represented schematically as shown in Fig. 12.43.

In many technical papers, positive x is considered upwards, but similar results are obtained regardless of the convention used. The equation of motion for the vehicle body is

$$m_b\ddot{x}_b + k_s(x_b - x_w) + f_d - f = 0 \qquad (12.3.1)$$

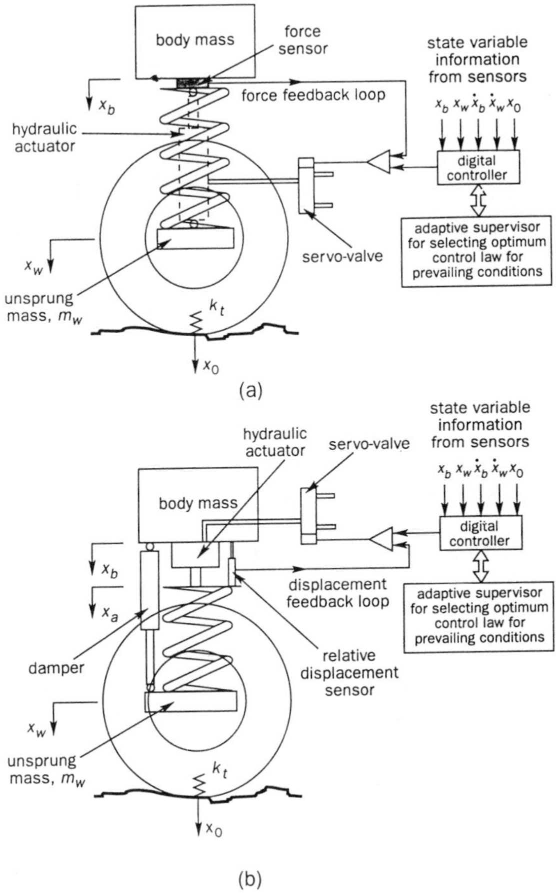

Fig. 12.42 Fully active suspension system configurations: (a) fast active; (b) slow active.

If there is no spring in parallel with the actuator, f_d, then $k_s = 0$.

In (12.3.1) and the equations to follow, the time varying nature of the quantities has been omitted to simplify the notation. The quantity x_b is the displacement of the car body, x_w is the displacement of the wheel, x_0 is the absolute displacement of the tyre surface, m_b is the mass of the quarter car body, m_w is the mass of the wheel, f is the disturbance force as a result of cornering or braking and f_d is the active control force.

The equation of motion for the wheel mass shown in Fig. 12.43(a) is

$$m_w\ddot{x}_w - c_s(\dot{x}_b - \dot{x}_w) - k_s(x_b - x_w) - f_d + c_t(\dot{x}_w - \dot{x}_0) + k_t(x_w - x_0) = 0 \tag{12.3.2}$$

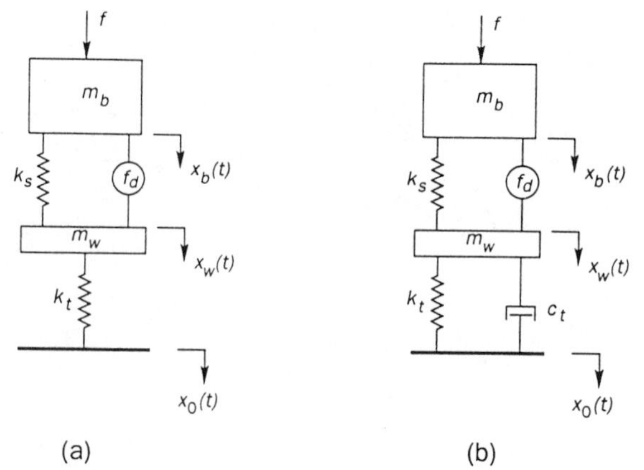

Fig. 12.43 Schematic representation of a two-degree-of-freedom quarter car model active suspension: (a) excluding tyre damping; (b) including tyre damping.

In its full feedback form, the control force demand is given by Crolla (1988) as

$$f_d = a_1 (x_b - x_0) + a_2 (x_w - x_0) + a_3 \dot{x}_b + a_4 \dot{x}_w \tag{12.3.3}$$

Due to the difficulty in measuring the relative displacement between the car body and the road, and the wheel and the road, the damping force is often calculated using only partial state feedback of the form

$$f_d = a_3 \dot{x}_b + a_4 \dot{x}_w \tag{12.3.4}$$

with results obtained being almost as good as those obtained using full state feedback.

An alternative control law which is very similar to (12.3.4) is given by Miller (1988a) as

$$f_d = \frac{c_{\text{on}}}{m_b} \dot{x}_b + \frac{c_{\text{off}}}{m_b} (\dot{x}_b - \dot{x}_w) \tag{12.3.5}$$

where c_{on}, c_{off}, a_3 and a_4 are determined using optimal control theory as discussed later. The reason for expressing the control law as shown in (12.3.5) is to demonstrate the similarity between fully active and semi-active systems which will be discussed in Section 12.3.1.3. The effective damping ratios corresponding to c_{on} and c_{off} are given by Miller (1988a):

$$\zeta_{\text{on}} = \frac{c_{\text{on}}}{2\sqrt{k_s m_b}} \quad ; \quad \zeta_{\text{off}} = \frac{c_{\text{off}}}{2\sqrt{k_s m_b}} \tag{12.3.6a,b}$$

Thus, (12.3.5) can be written as

$$f_d = 2\zeta_{\text{on}} \omega_b \dot{x}_b + 2\zeta_{\text{off}} \omega_b (\dot{x}_b - \dot{x}_w) \tag{12.3.7}$$

where

$$\omega_b = \sqrt{\frac{k_s}{m_b}} \tag{12.3.8}$$

Equation (12.3.7) is identical to (12.3.4), where

$$a_3 = 2\omega_b(\zeta_{\text{on}} + \zeta_{\text{off}}) \tag{12.3.9}$$

and

$$a_4 = -2\omega_b\zeta_{\text{off}} \tag{12.3.10}$$

The second term in (12.3.7) corresponds to a passive damping element in the system having a damping ratio ζ_{off}. The first term in (12.3.7) corresponds to feedback of absolute velocity (the 'skyhook' damper discussed in section 12.3.4) with a feedback gain of $2\zeta_{\text{on}}\,\omega_b$, which results in a continuously variable damping force c_{on}.

Another control law which is similar to that given in (12.3.3) is given by Tseng and Hrovat (1990) and El-Madany (1990) as

$$f_d = a_1(x_b - x_w) + a_2(x_w - x_0) + a_3\dot{x}_b + a_4\dot{x}_w \tag{12.3.11}$$

If we then substitute the state variables $x_1 = (x_b - x_w)$, $x_2 = (x_w - x_0)$, $x_3 = \dot{x}_b$ and $x_4 = \dot{x}_w$, we can write (12.3.11) as

$$f_d = a_1x_1 + a_2x_2 + a_3x_3 + a_4x_4 \tag{12.3.12}$$

In terms of these state variables, we can rewrite the equations of motion, for the system shown in Fig. 12.43, in matrix form as

$$\dot{\boldsymbol{x}} = \boldsymbol{A}\boldsymbol{x} + \boldsymbol{B}f_d + \boldsymbol{D}\dot{x}_0 \tag{12.3.13}$$

where the state vector is

$$\boldsymbol{x} = [x_1\ x_2\ x_3\ x_4]^{\mathrm{T}} \tag{12.3.14}$$

and

$$\boldsymbol{A} = \begin{bmatrix} 0 & 0 & 1 & -1 \\ 0 & 0 & 0 & 1 \\ -k_s/m_b & 0 & 0 & 0 \\ k_s/m_w & -k_t/m_w & 0 & 0 \end{bmatrix} \tag{12.3.15}$$

$$\boldsymbol{B} = \begin{bmatrix} 0 & 0 & -1/m_b & 1/m_w \end{bmatrix}^{\mathrm{T}} \tag{12.3.16}$$

$$\boldsymbol{D} = \begin{bmatrix} 0 & -1 & 0 & 0 \end{bmatrix}^{\mathrm{T}} \tag{12.3.17}$$

If tyre damping were included, as shown in Fig. 12.43(b), the matrices could be written as

$$\boldsymbol{A} = \begin{bmatrix} 0 & 0 & 1 & -1 \\ 0 & 0 & 0 & 1 \\ -\omega_b^2 & 0 & 0 & 0 \\ \omega_b^2/\rho & -\omega_w^2 & 0 & -2\zeta_t\,\omega_w \end{bmatrix} \tag{12.3.18}$$

$$\boldsymbol{B} = \begin{bmatrix} 0 & 0 & -1/m_b & 1/m_w \end{bmatrix}^{\mathrm{T}} \tag{12.3.19}$$

$$\boldsymbol{D}_1 = \begin{bmatrix} 0 & -1 & 0 & 2\zeta_t\omega_t \end{bmatrix}^{\mathrm{T}} \tag{12.3.20}$$

The variables in (12.3.18) to (12.3.20) are defined as

$$\omega_w^2 = k_t\,/m_w \tag{12.3.21}$$

$$\zeta_t = \frac{c_t}{2\sqrt{k_t m_w}} \tag{12.3.22}$$

$$\omega_b^2 = k_s\,/m_b \tag{12.3.23}$$

and $\rho = m_w/m_b$ is the ratio of wheel mass to body mass. Note that the inertia force f (shown in Fig. 12.43) caused by cornering or braking has been excluded from the matrix equations.

Values of m_b, m_w, k_s, ζ_t and k_t typical of automobiles are 250 kg, 25 kg, 20 000 Nm^{-1}, 0.02 and 120 000 Nm^{-1} respectively, and the controller bandwidth typically needed is 15 Hz. The control actuator is usually required to be powerful enough to produce damping ratios given by Miller (1988a):

$$\frac{c_s}{2\sqrt{k_s\, m_b}} \geq 1 \tag{12.3.24}$$

The vehicle excitation, x_0, is the apparent vertical roadway movement as a result of the vehicle's forward motion along an uneven road surface. It is generally accepted that the input excitation due to surface irregularities may be considered as a Gaussian random process with zero mean and single sided power spectral density (given by El-Madany and Abduljabbar 1989):

$$G_{x0}(\omega) = \frac{2\alpha V\sigma^2}{\pi}\,\frac{1}{\alpha^2V^2 + \omega^2} \tag{12.3.25}$$

where σ^2 is the variance of road irregularities, α is a coefficient depending on the shape of these irregularities and V is the vehicle forward speed. For a concrete road, $\alpha = 0.15$ m^{-1} and $\sigma = 0.87 \times 10^{-2}$ m (El-Madany and Abduljabbar, 1989).

The spectrum of (12.3.25) can be generated from a white noise process by using a filter of the form

$$\dot{x}_0 = -\alpha V x_0 + \xi \tag{12.3.26}$$

where ξ is a zero mean Gaussian white noise process with the expected value given by

$$E\left[\xi(t)\,\xi(t-\tau)\right] = 2\alpha V\sigma^2\delta(\tau) = q_f\,\delta(\tau) \qquad (12.3.27a,b)$$

Equations (12.3.13) and (12.3.26) can be combined into a single matrix equation describing the vehicle motion in state vector form as follows:

$$\dot{\boldsymbol{x}} = \boldsymbol{A}\boldsymbol{x} + \boldsymbol{B}f_d + \boldsymbol{D}\xi \qquad (12.3.28)$$

where the new state vector is

$$\boldsymbol{x} = [x_1\ x_2\ x_3\ x_4\ x_0]^{\mathrm{T}} \qquad (12.3.29)$$

and

$$\boldsymbol{A} = \begin{bmatrix} 0 & 0 & 1 & -1 & 0 \\ 0 & 0 & 0 & 1 & -1 \\ -\omega_b^2 & 0 & 0 & 0 & 0 \\ 0 & -\omega_w^2 & \omega_b^2/\rho & -2\zeta_t\omega_w & 2\zeta_t\omega_w \\ 0 & 0 & 0 & 0 & -\alpha V \end{bmatrix} \qquad (12.3.30)$$

$$\boldsymbol{D} = [0\ \ 0\ \ 0\ \ 0\ \ 1]^{\mathrm{T}} \qquad (12.3.31)$$

$$\boldsymbol{B} = [0\ \ 0\ \ 1/m_b\ \ 1/m_w\ \ 0]^{\mathrm{T}} \qquad (12.3.32)$$

The optimal control force is that force which will result in the minimisation of a quadratic performance index. One performance index which considers the magnitude of the control force and all state variables, except the road surface profile x_0, is (Yue *et al.*, 1989)

$$J = E\left[q_1x_1^2 + q_2x_2^2 + q_3x_3^2 + q_4x_4^2 + q_5x_5^2 + rf_d^2\right] \qquad (12.3.33)$$

where $x_5 = \dot{x}_3$ and where E is the expectation operator, q_1 is the performance index which represents the importance of suspension deflection, q_2 represents tyre deflection, q_3 represents car body velocity, q_4 represents wheel velocity, q_5 which is an additional state variable, represents car body acceleration which is related to passenger comfort and r represents the importance of minimizing the control force.

The performance index J may also be written as (Yoshimura *et al.*, 1986; Yoshimura and Ananthanarayana, 1987):

$$J = \lim_{T\to\infty}\frac{1}{T}\int_0^T E\left[q_1x_1^2 + q_2x_2^2 + q_3x_3^2 + q_4x_4^2 + q_5x_5^2 + rf_d^2\right]\mathrm{d}t \qquad (12.3.34)$$

$$= \lim_{T\to\infty}\frac{1}{T}\int_0^T E\left[\boldsymbol{x}^{\mathrm{T}}\boldsymbol{Q}\boldsymbol{x} + \boldsymbol{F}^{\mathrm{T}}\boldsymbol{R}\boldsymbol{F}\right]\mathrm{d}t \qquad (12.3.35)$$

where

$$Q = \begin{bmatrix} q_1 & 0 & 0 & 0 & 0 \\ 0 & q_2 & 0 & 0 & 0 \\ 0 & 0 & q_3 & 0 & 0 \\ 0 & 0 & 0 & q_4 & 0 \\ 0 & 0 & 0 & 0 & q_5 \\ 0 & 0 & 0 & 0 & 0 \end{bmatrix} \tag{12.3.36}$$

$$x = [x_1,\ x_2,\ x_3,\ x_4,\ x_5\ x_0]^{\mathrm{T}} \tag{12.3.37}$$

$$F = f_d = a_1x_1 + a_2x_2 + a_3x_3 + a_4x_4 + a_5x_5 + a_0x_0 \tag{12.3.38a,b}$$

$$R = r \tag{12.3.39}$$

The single element matrices $\boldsymbol{F}$ and $\boldsymbol{R}$ would have more elements if more control actuators were used (Yoshimura and Ananthananyana, 1987). Note that the control force f_d must be a function of all state variables included in the state vector $\boldsymbol{x}$.

A simpler and more commonly used performance index can be derived by combining the last two terms of (12.3.33) into one. This is justifiable as the control force input can be used as a measure of ride comfort. Also, the velocities of the wheel and car body are not of much interest in terms of suspension performance. Thus (12.2.33) can be simplified to (El-Madany and Abduljabbar, 1989; Tseng and Hrovat, 1990)

$$J = E\left[q_1x_1^2 + q_2x_2^2 + rf_d^2\right] \tag{12.3.40}$$

The problem of minimizing J is often referred to as the linear quadratic regulator problem (or LQ control), and the required values of a_1, a_2, a_3, a_4 and a_0 to minimize J can be obtained by using optimal control theory which involves solving the associated Ricatti equation (see Chapter 5). Following the standard procedure (Tseng and Hrovat, 1990), we can write the following for the solution for a_1, a_2, a_3, a_4 and a_0:

$$[a_1\ a_2\ a_3\ a_4\ a_0]^{\mathrm{T}} = \boldsymbol{R}^{-1}\boldsymbol{B}^{\mathrm{T}}\boldsymbol{P} \tag{12.3.41}$$

giving the optimal control force, $f_d^* = a_1^*x_1 + a_2^*x_2 + a_3^*x_3 + a_4^*x_4 + a_0^*x_0$, as

$$f_d^*(t) = \boldsymbol{R}^{-1}\boldsymbol{B}^{\mathrm{T}}\boldsymbol{P}\boldsymbol{x}(t) \tag{12.3.42}$$

where the asterisk denotes an optimal quantity and where the vector $\boldsymbol{x}$ is defined by (12.3.29), the vector $\boldsymbol{B}$ is defined by (12.3.32), and where $\boldsymbol{P} = \boldsymbol{P}^{\mathrm{T}}$ is the symmetric positive definite solution of the algebraic Ricatti equation:

$$\boldsymbol{PA} + \boldsymbol{A}^{\mathrm{T}}\boldsymbol{P} - \boldsymbol{PBR}^{-1}\boldsymbol{B}^{\mathrm{T}}\boldsymbol{P} + \boldsymbol{Q} = 0 \tag{12.3.43}$$

where the matrix $\boldsymbol{A}$ is defined by (12.3.30), and

$$\boldsymbol{Q} = \begin{bmatrix} q_1 & 0 & 0 & 0 & 0 \\ 0 & q_2 & 0 & 0 & 0 \\ 0 & 0 & 0 & 0 & 0 \\ 0 & 0 & 0 & 0 & 0 \\ 0 & 0 & 0 & 0 & 0 \end{bmatrix} \tag{12.3.44}$$

Sometimes it is not feasible to directly measure all of the state feedback variables, needed to evaluate the performance function J. Assuming that it is possible to measure at least one variable, it is possible to estimate the others using a Kalman filter (see Chapter 5), as follows. Equation (12.3.28) may be rewritten as

$$\frac{\partial \hat{\boldsymbol{x}}}{\partial t} = \boldsymbol{A}\hat{\boldsymbol{x}} + \boldsymbol{B}f_d + \boldsymbol{L}(\boldsymbol{x}_m - \boldsymbol{C}_m\hat{\boldsymbol{x}}) \tag{12.3.45}$$

where the ^ indicates an estimated quantity and where $\boldsymbol{x}_m$ is the vector of measured variables. Usually, this vector contains only one variable which is the suspension deflection. Thus,

$$\boldsymbol{x}_m = \left[(x_1 + v_1)\ 0\ 0\ 0\ 0\right]^{\mathrm{T}} = \boldsymbol{C}_m\boldsymbol{x} + v_1 \tag{12.3.46a,b}$$

where

$$\boldsymbol{C}_m = \left[1,\ 0,\ 0,\ 0\ 0\right] \tag{12.3.47}$$

and where v_1 represents measurement noise which is assumed white and which has an expected value of

$$E\left[v_1(t)\, v_1(t-\tau)\right] = \mu\delta(\tau) \tag{12.3.48}$$

where δ is the Dirac delta function and μ is a constant.

The optimal control problem is to choose the control force gains a_1, a_2, a_3, a_4 and a_0 (see (12.3.38)) and the Kalman filter gains $\boldsymbol{L} = [h_1, h_2, h_3, h_4, h_0]^{\mathrm{T}}$ to minimize the performance index J. This is referred to as linear quadratic Gaussian control (or LQG control) and the overall compensator transfer function (in the frequency domain) between the active force $f_d(j\omega)$ and the sensor output $x_m(j\omega)$ is

$$\frac{f_d(j\omega)}{x_m(j\omega)} = -\boldsymbol{G}\left(\boldsymbol{I} - \boldsymbol{A} + \boldsymbol{B}\,\boldsymbol{G} + \boldsymbol{L}\,\boldsymbol{C}_m\right)^{-1}\boldsymbol{L} \tag{12.3.49}$$

where $\boldsymbol{G} = (a_1, a_2, a_3, a_4, a_0)^{-1}$ and I is the identity matrix.

This LQG design procedure is guaranteed to produce a stable closed loop controller design if the system response is observable by x_m and controllable by f_d. The LQG system applied to a typical vehicle suspension has a performance characteristic approaching that for full feedback control.

The optimal control force is found by replacing $\boldsymbol{x}(t)$ in (12.3.42) with $\hat{\boldsymbol{x}}(t)$.

Thus,

$$f_d^*(t) = R^{-1} B^{\mathrm{T}} P \hat{x}(t) \tag{12.3.50}$$

where P is the solution of (12.3.43).

The Kalman filter gains L are found from

$$L = P_f C_m^{\mathrm{T}} \mu^{-1} \tag{12.3.51}$$

and where P_f is the solution of the Ricatti equation

$$P_f A + A^{\mathrm{T}} P_f - P_f C_m^{\mathrm{T}} \mu^{-1} C_m P_f + D Q_f D^{\mathrm{T}} = 0 \tag{12.3.52}$$

where D is defined by (12.3.31), A by (12.3.30), C_m by (12.3.47), and μ by (12.3.48).

The quantity Q_f is a diagonal matrix containing a single non-zero element in the fifth row. That is,

$$Q_f = \begin{bmatrix} 0 & 0 & 0 & 0 & 0 \\ 0 & 0 & 0 & 0 & 0 \\ 0 & 0 & 0 & 0 & 0 \\ 0 & 0 & 0 & 0 & 0 \\ 0 & 0 & 0 & 0 & q_f \end{bmatrix} \tag{12.3.53}$$

where q_f is defined by (12.3.26) and (12.3.27).

The layout of the optimal controller and Kalman filter is shown in block diagram form in Fig. 12.44. As discussed in Chapter 5, this type of approach is more prone to stability problems than if a state estimator is not used and less state variables are included in the system model.

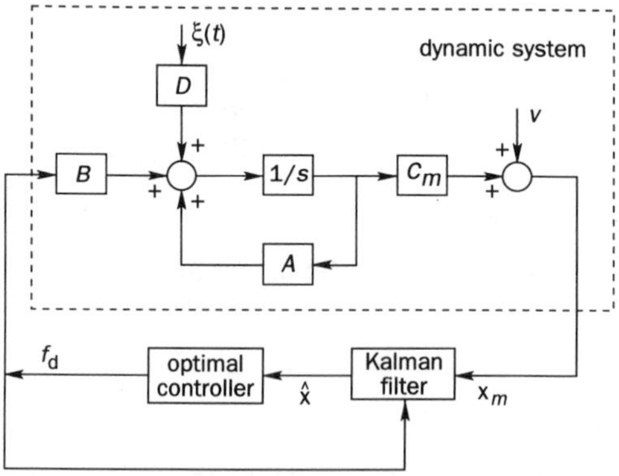

Fig. 12.44 Block diagram of an active suspension with partial state feedback.

Another technique used to determine the optimum control force demand is referred to as model reference adaptive control (MRAC), in which the disturbance and vibration of the vehicle is reduced to a level which is determined by an ideal conceptual suspension system – a reference model. This is done by comparing the actual suspension response with the model response to the same impact, and adjusting the force signal to the control actuator to minimize the difference between the two. Sunwoo *et al.* (1991) used this technique with a 'skyhook' damper (see Section 12.2.3.4) as the reference model. The two advantages of using MRAC over the standard active control method are that it adapts well to variations in suspension characteristics and the sprung load, and the characteristics of the reference model can be changed quickly to suit the desired ride condition by simply changing the computer algorithm used to determine the required actuator input.

Yet another very innovative approach to a fully active system is the use of MRAC concepts with a neural network controller which can establish its own control laws without requiring a complete knowledge of the system dynamics (Cheok and Huang, 1989). Although this approach is more difficult to implement, it promises superior performance over other active systems.

The stability of control systems with various linear quadratic (LQ) control laws determined using optimal control theory is discussed by Ulsoy and Hrovat (1990) and Yue *et al.* (1989), who compare the relative performance of various control laws including full state feedback, feedback of absolute velocity of the vehicle body ('skyhook' damper as discussed in Section 12.2.3.4.) and a linear quadratic gaussian (LQG) regulator using suspension deflection $(x_b - x_w)$ as the control variable to be minimized. They found that of the latter three types, the LQG regulator gave the best trade-off between ride quality, suspension travel and road holding constraints.

Preview control

The improved performance as a result of including a feedforward part (or preview control) with the feedback controller has been investigated by a number of authors. It was first proposed by Bender in 1968, who used a sensor to detect the road profile ahead of the vehicle to provide a signal to a controller designed to minimize the effect of the profile (or disturbance) input at the tyre/road contact point. Bender used Weiner filter theory to find the optimum controller, employing similar techniques to those outlined in Sections 5.11 and 6.5. Similar techniques were used by Sasaki *et al.* (1976) and Iwata and Nakano (1976) on a two-dimen-sional model. The main difficulty with this approach is the implementation of the optimal controller which is not guaranteed to be causal. This type of problem was discussed in relation to controlling random noise in ducts in Section 7.3.9.

Tomizuka (1975 and 1976) derived an optimal feedback controller which had a feedforward component consisting of a weighted sum of future road elevations taken over a preview distance; however, the application of this theory in practice was hindered by the complicated mathematics, requiring solution of a large set of regressive matrix equations to determine the optimal controller. Thompson *et al.* (1980) approached the problem as a simple, linear quadratic regulator with a vector disturbance input. The preview sensor only senses the distance to the road at a point ahead of the wheel; thus to determine the disturbance input, an estimate of the body displacement is necessary. This can be obtained by double integration of the signal from an accelerometer mounted on the body using integrators with feedback compensation to minimize low frequency noise problems. (Thompson *et al.*, 1980). The filtering of the preview sensor to provide the delayed signal for use in the controller could be achieved using an FIR filter with an adaptive algorithm for updating the weights, although this technique has not yet been reported in the literature. Thompson *et al.* (1980) used a multi-stage fixed delay line, and relied on changing the point at which the input is fed to the filter from the first stage to a later stage as the vehicle speed increases (to account for the shorter required delay).

More recent work (Foag, 1988) has shown that the use of preview significantly reduces the required control forces and actuator bandwidth compared to active systems without preview. In practice, the preview signal can be derived from an ultrasonic, infrared or radar device. However, one of the main difficulties associated with the practical application of preview control is locating a sensor sufficiently far in front of the vehicle to provide sufficient delay at high vehicle speeds to enable the controller to develop the low frequency actuator signals it needs. One way round this problem is to use preview only for the back wheels, with the disturbance input to the front wheels acting as the preview input to the controller for the back wheels (Hrovat, 1991). Analysis of this system requires a more complex two-dimensional, four-degree-of-freedom half vehicle model compared to the one-dimensional two-degree-of-freedom quarter vehicle model used by Thompson *et al.* (1980). Of course, providing preview for only the back wheels is not as effective as providing it to both front and back wheels but it offers significant performance advantages over active suspensions with no preview.

Yoshimura and Ananthanarayana (1991) approached the preview problem slightly differently by including it in an extended Kalman filter, thus allowing a controller to be developed which will function for an irregular road surface. The dynamic single wheel model with preview control is illustrated in Fig. 12.45.

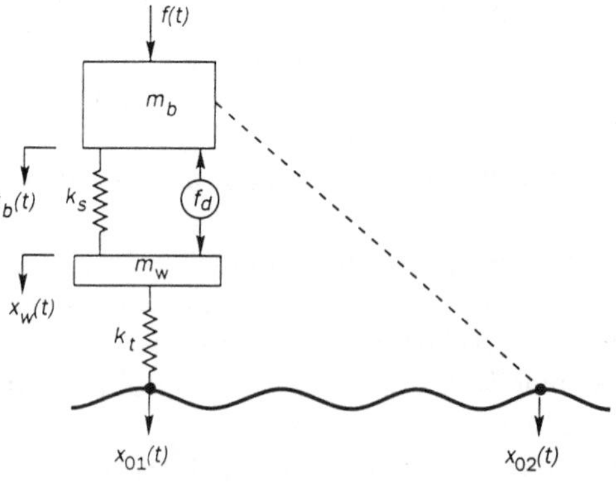

Fig. 12.45 Single wheel model with preview.

The analysis is similar to that outlined previously for the Kalman filter, except for the presence of an additional measured input in the measured state vector $\boldsymbol{x}_m$, which is the quantity $x_b - x_{02}$. Equations (12.3.45) and (12.3.46) still describe this system except that the variables are more complex, and are defined as follows:

$$\hat{\boldsymbol{x}} = [\hat{x}_1 \ \ \hat{x}_2 \ \ \hat{x}_3 \ \ \hat{x}_4 \ \ \hat{x}_{01} \ \ \hat{x}_{02}]^{\mathrm{T}} \tag{12.3.54}$$

$$\boldsymbol{x}_m = [x_{m_1} \ \ x_{m_2}]^{\mathrm{T}} \tag{12.3.55}$$

$$\boldsymbol{C}_m = \begin{bmatrix} 1 & 0 & 0 & 0 & 0 & 0 \\ 1 & 1 & 0 & 0 & 1 & -1 \end{bmatrix} \tag{12.3.56}$$

$$\boldsymbol{A} = \begin{bmatrix} 0 & 0 & 1 & -1 & 0 \\ 0 & 0 & 0 & 1 & -1 \\ -\omega_b^2 & 0 & 0 & 0 & 0 \\ 0 & -\omega_w^2 & -\omega_b^2/\rho & -2\zeta_t\omega_w & 2\zeta_t\omega_w \\ 0 & 0 & 0 & 0 & \alpha V \\ 0 & 0 & 0 & 0 & 0 \end{bmatrix} \tag{12.3.57}$$

$$\boldsymbol{B} = [0 \ \ 0 \ \ 1/m_b \ \ 1/m_w \ \ 0 \ \ 0]^{\mathrm{T}} \tag{12.3.58}$$

$$\boldsymbol{L} = \begin{bmatrix} l_{11} & l_{12} \\ l_{21} & l_{22} \\ l_{31} & l_{32} \\ l_{41} & l_{42} \\ l_{011} & l_{012} \\ l_{021} & l_{022} \end{bmatrix} \tag{12.3.59}$$

$$\boldsymbol{v} = \begin{bmatrix} v_1 \\ v_2 \end{bmatrix} \tag{12.3.60}$$

The elements of the measurement noise vector $\boldsymbol{v}$ (which replaces v_1 in (12.3.46b) have expected values of

$$\left.\begin{aligned} E\left[v_1(t)\, v_1(t-\tau)\right] &= \mu_1 \delta(\tau) \\ E\left[v_2(t)\, v_2(t-\tau)\right] &= \mu_2 \delta(\tau) \\ E\left[v_1(t)\, v_2(t-\tau)\right] &= 0 \end{aligned}\right\} \tag{12.3.61a,b,c}$$

The surface profile at locations 1 and 2 can be written as follows

$$\dot{x}_{01} = -\alpha V x_{01} + \xi_1 \tag{12.3.62}$$

$$\dot{x}_{02} = -\alpha V x_{02} + \xi_2 \tag{12.3.63}$$

where the expected values of ξ are

$$\left.\begin{aligned} E\left[\xi_1(t)\xi_1(t-\tau)\right] &= 2\alpha V\sigma^2\delta(\tau) = q_f\,\delta(\tau) \\ E\left[\xi_2(t)\xi_2(t-\tau)\right] &= 2\alpha V\sigma^2\delta(\tau) = q_f\,\delta(\tau) \\ E\left[\xi_1(t)\xi_2(t-\tau)\right] &= 2\alpha V\sigma^2\delta(\tau-p) = q_f\,\delta(\tau-p) \end{aligned}\right\} \tag{12.3.64a,b,c}$$

where p is the preview time or the time taken for the vehicle to travel from x_{01} to x_{02}.

The optimal control force coefficients a_1, a_2, a_3, a_4, a_{01} and a_{02} are found as before using (12.3.42) with $\boldsymbol{R}$ defined by (12.3.39) and $\boldsymbol{B}$ defined by (12.3.58). The symmetric matrix $\boldsymbol{P}$ is the solution of (12.3.43) with the matrix $\boldsymbol{A}$ given by (12.3.57) and the matrix $\boldsymbol{Q}$ given by:

$$\boldsymbol{Q} = \begin{bmatrix} q_1 & 0 & 0 & 0 & 0 & 0 \\ 0 & q_2 & 0 & 0 & 0 & 0 \\ 0 & 0 & 0 & 0 & 0 & 0 \\ 0 & 0 & 0 & 0 & 0 & 0 \\ 0 & 0 & 0 & 0 & 0 & 0 \\ 0 & 0 & 0 & 0 & 0 & 0 \end{bmatrix} \tag{12.3.65}$$

The Kalman filter gains are found by using (12.3.51) where (12.3.56) is used for $\boldsymbol{C}_m$ and the following is used for $\boldsymbol{\mu}$:

$$\boldsymbol{\mu} = [\mu_1 \; \mu_2]^{\mathrm{T}} \tag{12.3.66}$$

$\boldsymbol{P}_f$ is a solution of equation (12.3.52) with $\boldsymbol{A}$, $\boldsymbol{C}_m$ and $\boldsymbol{\mu}$ defined by equations (12.3.57), (12.3.56) and (12.3.62) respectively. The vector $\boldsymbol{D}$ in equation (12.3.52) is given by

$$\boldsymbol{D} = [0 \; 0 \; 0 \; 0 \; 1 \; 1]^{\mathrm{T}} \tag{12.3.67}$$

The matrix $\boldsymbol{Q}_f$ in (12.3.52) is defined for this case as:

$$\boldsymbol{Q}_f = \begin{bmatrix} 0 & 0 & 0 & 0 & 0 & 0 \\ 0 & 0 & 0 & 0 & 0 & 0 \\ 0 & 0 & 0 & 0 & 0 & 0 \\ 0 & 0 & 0 & 0 & 0 & 0 \\ 0 & 0 & 0 & 0 & q_f & 0 \\ 0 & 0 & 0 & 0 & 0 & q_f \end{bmatrix} \tag{12.3.68}$$

where q_f is defined by (12.3.62), (12.3.63) and (12.3.64).

A large part of the effort in making active vehicle suspension systems practically realizable has been expended in the development of suitable low cost sensors and actuators (Moore, 1988; Parker and Lau, 1988; Decker *et al.*, 1988; Shiozaki *et al.*, 1991; Hattori *et al.*, 1990; Asano *et al.*, 1991; Satoh and Osada, 1991; Akatsu *et al.*, 1990) and a considerable effort has also been expended in developing ways of monitoring the health of these sensors and actuators and maintaining system performance in the event of component failures (de Benito and Eckert, 1990; Litkouhi and Boustany, 1988). Sensors which have been developed include piezoceramic force and acceleration transducers, potentiometric position transducers and inductance position transducers, while actuator development has mainly focused on various hydraulic or air actuator valve designs, (for semi-active suspensions), piezoceramic actuators and low cost hydraulic actuators and associated servo-valves (for fully active suspensions) and electrorheological fluid. This latter actuator is a fluid formed by doping a dielectric base fluid with semiconductor particles. On application of an electric field of sufficient strength, the particles will form chains which link across the electrodes. The resistance to shear of these particles can be varied by controlling the strength of the electric field, thus forming a damping element suitable for a semi-active suspension (Stanway *et al.*, 1989).

12.3.1.2 Slow active systems

As mentioned in the previous section, the actuator bandwidth required for a fully active suspension is typically 15 Hz, and this is expensive. Considerable interest has developed in ways of reducing this bandwidth requirement without reducing performance. One idea is to use the actuator in series with passive springs so that the actuator controls low frequency (less than 3 Hz) motions while the passive spring controls the higher frequency motions in the range where the actuator is essentially very stiff. Such a system is illustrated in Fig. 12.42(b) and shown schematically in Fig. 12.46 (see also Sharp and Hassan, 1987b).

For the system shown in Fig. 12.46, the control law is now based on displacement demand for the control actuator rather than force demand as was used for the fully active system. In this case the equations of motion are

$$m_b\ddot{x}_b + k_s(x_s - x_w) + c_s(\dot{x}_b - \dot{x}_w) = 0 \qquad (12.3.69)$$

$$m_w\ddot{x}_w - k_s(x_s - x_w) - c_s(\dot{x}_b - \dot{x}_w) + k_t(x_w - x_0) = 0 \qquad (12.3.70)$$

and the control law is (Crolla, 1988)

$$(x_b - x_s) = a_1 x_b + a_2 x_w + a_3\dot{x}_b + a_4\dot{x}_w \qquad (12.3.71)$$

If fewer state measurements are available, then simpler control laws have been shown to work well. For example (Wilson *et al.*, 1986)

$$(x_b - x_s) = a_1(\dot{x}_b - \dot{x}_s) \qquad (12.3.72)$$

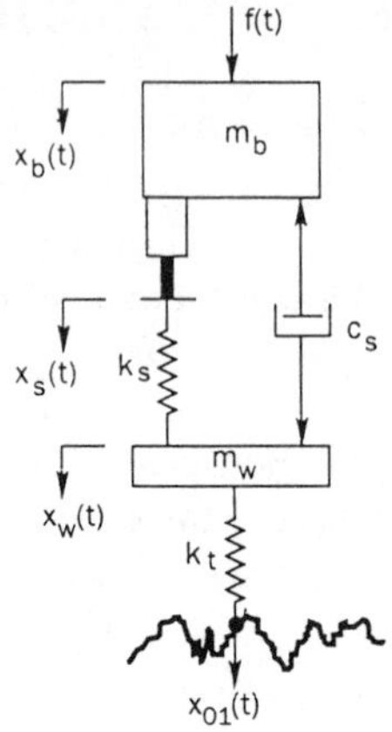

Fig. 12.46 Schematic of a slow active suspension system.

or

$$(x_b - x_s) = a_1\dot{x}_b - a_2\dot{x}_w \qquad (12.3.73)$$

Optimal control theory can then be used to evaluate the unknown constants a_1 a_2, etc., using the same approach as was taken in the last section for fully active systems. The performance obtained with these control laws and the arrangement shown in Fig. 12.46 is very close to that obtained for a fully active system, provided that the slow actuator has a bandwidth of at least 3 Hz (Sharp and Hassan, 1987).

The main disadvantage of the slow active system just described is the high levels of static force demanded from the actuator unless a means can be found to support the vehicle weight by a spring in parallel with the actuator. This is simple to model but impractical to implement with metallic spring elements (due to excessive space demands) but it is possible to implement at some cost premium with air or hydropneumatic suspension systems.

12.3.1.3 Semi-active suspension systems

These systems follow similar control laws to a fully active system, shown schematically in Fig. 12.43, except during times when the latter are required to supply power, the semi-active system damping is set to zero or a very small value. Thus semi-active systems supply no power but simply adjust the damping coefficient of an otherwise passive damper. During the period when an equivalent fully active damper would be absorbing power, the semi-active damper is set to it 'on' state and during the time when the equivalent fully active damper would be supplying power to the suspension, the semi-active damper is set to its 'off' state. Although the ideal damping in the 'off' state is zero if only ride quality is to be maximized, it should be a little greater than zero to optimize the trade-off between ride quality and road holding (Miller, 1988a). Thus in the following discussion, a finite value of passive damping at all times will be assumed.

Semi-active dampers may be divided essentially into two basic types: the simple on/off type where the damping in the 'on' state is constant but considerably greater than the damping in the 'off' state; and the continuously variable type where the damping in the 'on' state is continuously variable and in the 'off' state it is constant and much smaller than in the 'on' state. Note that a constant damping coefficient results in a damping force proportional to the relative velocity between the vehicle wheel and body – a property which is characteristic of a passive damper where the damping force f_d is

$$f_d = c_{\text{off}}(\dot{x}_b - \dot{x}_w) = 2\zeta_{\text{off}}\omega_b(\dot{x}_b - \dot{x}_w) \qquad (12.3.74\text{a,b})$$

The continuously variable damper commonly discussed in the literature will be referred to as a type 1 damper and is characterized by the following control law which comes directly from (12.3.7) for a fully active damper with partial state feedback:

$$\text{Type 1} \quad f_d = \begin{cases} 2\zeta_{\text{on}}\omega_b\dot{x}_b + 2\zeta_{\text{off}}\omega_b(\dot{x}_b - \dot{x}_w); & \dot{x}_b(\dot{x}_b - \dot{x}_w) > 0 \\ 2\zeta_{\text{off}}\omega_b(\dot{x}_b - \dot{x}_w); & \dot{x}_b(\dot{x}_b - \dot{x}_w) < 0 \end{cases} \qquad (12.3.75\text{a,b})$$

The simple on/off damper will be referred to as the type 2, and has the following control law:

$$\text{Type 2} \quad f_d = \begin{cases} 2\zeta_{\text{on}}\omega_b(\dot{x}_b - \dot{x}_w); & \dot{x}_b(\dot{x}_b - \dot{x}_w) > 0 \\ 2\zeta_{\text{off}}\omega_b(\dot{x}_b - \dot{x}_w); & \dot{x}_b(\dot{x}_b - \dot{x}_w) < 0 \end{cases} \qquad (12.3.76\text{a,b})$$

It can be seen that, for type 2, the semi-active damping coefficient takes on two discrete and fixed values, depending on the value of the condition function $\dot{x}_b(\dot{x}_b - \dot{x}_w)$. The disadvantage of the first type is that a servo-valve of high bandwidth is required to provide the continuously variable damping coefficient (usually done by varying the size of the orifice in a passive damper on a continuous basis). For the second type, the damping coefficient remains constant in each state but equal to a different value for each state, thus avoiding the need for a high bandwidth servo-valve.

However, both types 1 and 2 discussed above suffer from the disadvantage that they require a measurement of the absolute velocity of the vehicle which can only be obtained by integrating the signal from an accelerometer mounted on the vehicle. For the low frequency signals of interest, this integration is difficult and hampered by hardware limitations; thus, this type of control has yet to be implemented on an actual vehicle. In an attempt to overcome this latter problem, Rakheja and Sankar (1985) developed on alternative to the simple on/off damper (type 2) described above. This will be referred to as a type 3 control law and is based on the following logic.

When the spring force and damping force act in the same direction, the damping force tends to increase the acceleration of the suspended mass; thus to optimize ride comfort, a very small or zero damping force is preferred for this part of the cycle. When the spring force and damping force act in opposite directions, the force on the mass can be made zero if the magnitude of the damping force is made the same as the spring force. This logic is reflected in the condition function for the type 3 damper, the control law for which is:

$$\text{Type 3} \quad f_d = \begin{cases} 2\zeta_{\text{on}}\omega_b(\dot{x}_b - \dot{x}_w)\;|\;\dot{x}_b - \dot{x}_w\;|\;; & (x_b - x_w)(\dot{x}_b - \dot{x}_w) > 0 \\ 2\zeta_{\text{off}}\omega_b(\dot{x}_b - \dot{x}_w)\;|\;\dot{x}_b - \dot{x}_w\;|\;; & (x_b - x_w)(\dot{x}_b - \dot{x}_w) < 0 \end{cases} \tag{12.3.77a,b}$$

It can be seen from (12.3.77) that the only measurements which are required are the relative velocity and relative displacement between the vehicle body and the wheel.

Using the same reasoning as used to derive the type 3 on/off system, Alanoly and Sankar (1987) developed a continuously variable semi-active system which we shall refer to as a type 4 system. The control law for this system is

$$\text{Type 4} \quad f_d = \begin{cases} \alpha_{\text{on}}\omega_b^2(x_b - x_w); & (x_b - x_w)(\dot{x}_b - \dot{x}_w) > 0 \\ 0; & (x_b - x_w)(\dot{x}_b - \dot{x}_w) < 0 \end{cases} \tag{12.3.78a,b}$$

where the coefficient $\alpha_{\text{on}} > 0$ is the gain of the semi-active controller. Although not mentioned by the authors, it is likely that a better trade-off between ride quality and road holding would be achieved if the damping force in the 'off' state were made equal to $\alpha_{\text{off}}\,\omega_b^2(x_b - x_w)$ where α_{off} is much smaller than α_{on}.

Although the type 4 system requires a similar performing servo-valve to the type 1 system, it does not require a measurement of the absolute vertical velocity of the vehicle body, and has a similar overall performance.

Elimination of the need for a high bandwidth (>15 Hz) servo-valve and its associated cost penalty by using either the type 2 or type 3 on/off semi-active system comes at a considerable performance disadvantage, resulting in a suspension with a performance about midway between a passive and fully active system. On the other hand, use of either a type 1 or type 3 continuously variable semi-active system results in a performance very close to that achieved by a fully active system (Alanoly and Sankar, 1987; El-Madanay and Abduljabbar, 1989). Another fundamental difference between the passive and on/off semi-active systems on the one hand and the continuously variable semi-active and fully active on the other, is that in the former case increasing the system damping beyond some very small value reduces ride quality, whereas in the latter case increasing the damping increases ride quality. Note that high damping is desirable to maximize handling and vehicle stability. Thus the trade-off between ride quality and vehicle handling which characterizes passive and on/off type semi-active suspensions does not occur for continuously variable semi-active or fully active systems. Similarly, it has been shown (Miller, 1988a) that increasing the damping for the passive and on/off semi-active systems reduces road holding performance (as the body and wheel bounce together on the tyre stiffness resulting in large variations in tyre contact force). On the other hand, increasing the damping in fully active or continuously variable semi active systems improves road holding. However, it is difficult to match the road holding performance of a passive system with an active or semi-active system, although the performance of a semi-active system comes close with the addition of damping in the off-state (which results in a small reduction in ride quality).

Suspension travel in a passive system is reduced by increasing the damping, but in all types of active system the travel is increased as the damping increases beyond some very small value. Thus the performance of the active and semi-active suspensions is limited by the allowable space in which they can work (usually $\pm$ 10 cm).

In summary, the best design trade-offs for a vehicle result in ζ for a passive system equal to 0.3, ζ_{on} and ζ_{off} for an on/off semi active system equal to 0.4 to 0.6 and 0.1 to 0.2 respectively and for a continuously variable semi-active system maximum values of ζ_{on} and ζ_{off} should be equal to 1.0 or greater and 0.1 to 0.2 respectively. The desired values for a fully active suspension are similar to those for a continuously variable semi-active system.

Note that the theoretical performance of active and semi-active suspensions is somewhat reduced in practice due to inaccuracies in the hardware which converts velocities and displacements to electrical signals and electrical signals to damping coefficients. These problems are discussed in detail by Miller (1988b).

In summary, it can be said that all suspensions provide trade-offs between ride comfort, handling, road holding and suspension travel. However, active and continuously variable semi-active computer controlled suspensions can provide better compromises between the various requirements. As the performance of a continuously variable semi-active suspension can be made almost as good as a fully active system (and much better than a passive system), there is some doubt whether the slightly improved performance of the fully active or slow active systems can be justified in terms of the increased cost, number of components and complexity involved.

12.3.1.4 Switchable damper

A switchable damper is different from an on–off type semi-active damper in that the switching of the damper force from one value to another occurs over a relatively long time period, whereas the semi-active damper is switched every hundred milliseconds or so.

Even though the switchable damper may stay in one setting for a relatively long time, it is capable of changing from one setting to another very quickly (15–20 ms). This changeover speed is necessary for safety considerations, as an excessively soft suspension can be disastrous during severe cornering or sudden vehicle direction changes.

Switchable dampers currently being considered have three settings; soft, medium and hard. The setting which is automatically selected depends upon the roughness of the road and the severity of the driver's steering actions (which are related to the vehicle speed as well as the steering wheel movement). In some cases, it is possible to switch between settings manually, but in this case, the soft setting cannot be made so soft as in the automation-only switching case, for obvious safety reasons. One particular switchable damper design is discussed by Meller and Frühauf (1988).

12.3.2 Rigid mount active isolation

In cases where low frequency machine movement cannot be tolerated, it is sometimes necessary to be able to connect a machine rigidly to a support foundation and at the same time reduce the forces acting at the support point due to the machine vibration. In these cases, a reduction in the vibratory force acting on the mount can be achieved by using a shaker attached to the support point and driving an inertial mass.

A system similar to that shown in Fig. 12.47 was developed using a hydraulic actuator (and the slightly more complex control system needed) by Tanaka and Kikushima in 1988, who demonstrated a working system on an actual forge hammer with spetacular results. The purpose of the LVDT (linear variable differential transformer – see Chapter 15) was to sense the relative displacement of the actuator rod in the cylinder, so that it did not gradually drift to one extreme or another. Various forms of the optimal compensator B are discussed by Tanaka and Kikushima (1988).

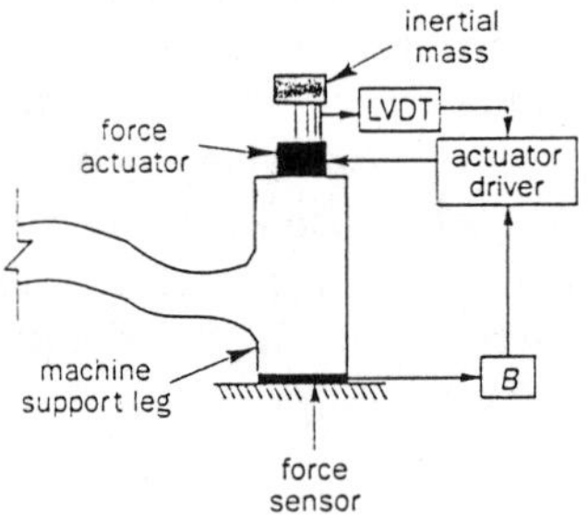

Fig. 12.47 An active isolator for a rigid machine mount.

12.3.3 Vibration isolation of optical equipment

The problem of isolating sensitive optical equipment from vibrations in its support structure is similar to the vehicle suspension system problem, except that the frequency range of interest usually extends a little higher.

Successful vibration isolation of a space telescope optical system from its structural support has been reported by Kaplow and Velman in 1980, who used a simple form of velocity feedback control to achieve the desired result.

More recently Fenn *et al.* (1990) have developed an integrated six-degree-of-freedom magnetic isolator, controlled with a non-linear feedback system to isolate microgravity experiments from space station movements and vibration. Stampleman and von Flotow (1990) have developed a mount intended for a similar purpose using PVDF film controlled with a simple feedback system incorporating acceleration feedback. A third group (Jones *et al.*, 1990) also working on the isolation of microgravity experiments from support structures in orbit have developed a six-degree-of-freedom isolator using simple displacement feedback, However, they point out that there are some problems associated with the initial experiment platform release when the spacecraft reaches orbit (as the experiments are anchored firmly during launch).

12.3.4 Vibration reduction in tall buildings

Excessive vibration (or low frequency sway) can be induced in tall buildings by either seismic activity or wind. Control can be effected either by using a tuned mass damper (or tuned vibration absorber, as discussed in Section 12.2.4), or by using tensioned cables in the wall of each level of the building as shown in Fig. 12.48. The tension in the cables (or active tendons as they are sometimes called) are controlled by hydraulic actuators, thus reducing building sway. Means of controlling these tensions using a feedback control system are discussed by Abdel-Rohman (1987) and Samali *et al.* (1985).

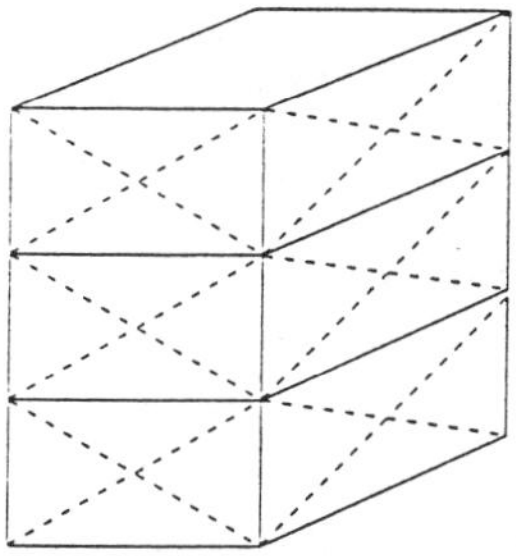

Fig. 12.48 Use of tensioned cables to control building sway.

12.3.5 Active isolation of machinery from flexible structures: engine mounts

This topic was discussed in detail in Section 12.2.5. However, one special case which has not been discussed previously is that of semi-active isolation of an engine from a flexible support structure (car) as discussed by Graf and Shoureshi (1987). These engine mounts are similar in principle to a semi-active suspension in that the mount damping is made adjustable by using an adjustable orifice through which hydraulic fluid passes, as shown in Fig. 12.49

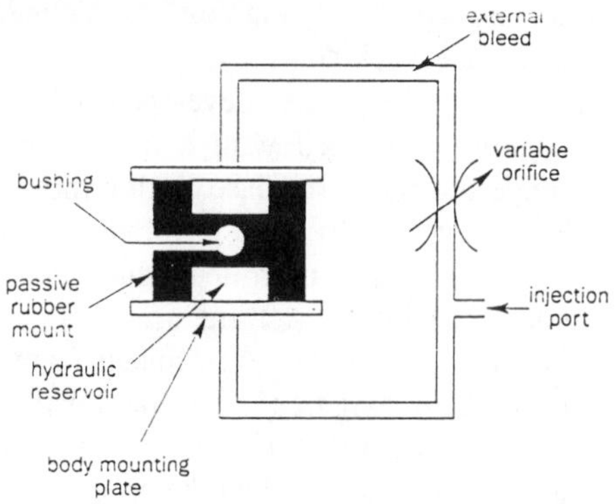

Fig. 12.49 Semi-active engine mount.

The bushing set in the mount is used to connect the mount to the engine and the vehicle body is attached to the mounting plates. The two fluid resevoirs are filled with fluid of density ρ, viscosity μ. The compliance of the reservoir volume is c_2, the length of the external bleed tube is l and the bleed tube diameter is d. The effective damping ratio of the mount is then given by (Graf and Shoureshi, 1987) as

$$\zeta = \frac{32\mu l}{d^3\sqrt{2\pi\rho / c_2}} \tag{12.3.79}$$

Another type of engine mount was described by Quinlan (1992) and is shown in Fig. 12.50, where the main rubber element supports the weight of the engine. At frequencies below about 20 Hz, it behaves as a conventional hydromount with vibration energy being dissipated by pumping fluid from the upper chamber to the lower chamber through an annular channel. At high frequencies the inertia of the fluid prevents significant flow through the orifice and the electromagnetic coil, driven by an electronic controller, acts on the metal diaphragm to compensate the forces produced by the engine.

Yet another type of engine mount which has been tested at the prototype stage is one which uses an electrorheological fluid to provide the damping forces (Sproston and Stanway, 1992). This type of fluid is discussed in Section 15.13.

12.3.6 Helicopter vibration control

Active control of rotor and gearbox induced vibrations in a helicopter fuselage has been a subject of much interest since first investigated by Bies (1968). Unfortunately, much of the work remains confidential and unpublished as it has been undertaken by private companies and Defence Departments.

One method of control which has been tried with some success is referred to as 'higher harmonic control' or HHC and involves using a feedback control system to oscillate the rotor blades in pitch at harmonic frequencies greater than

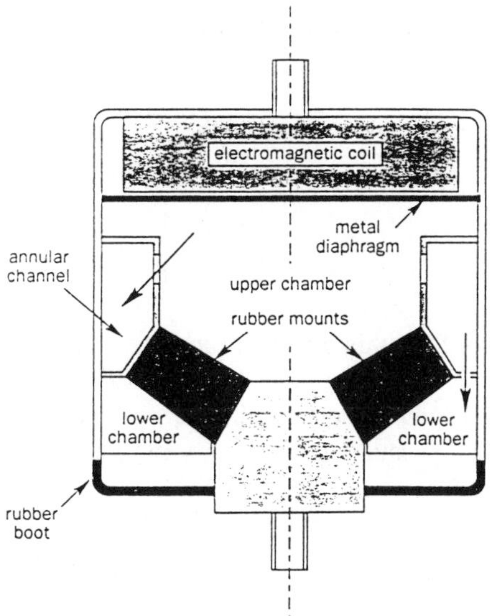

Fig. 12.50 Alternative semi-active engine mount (after Quinlan, 1992).

the fundamental rotational frequency of the rotor. The method was first suggested by Shaw (1967) and subsequent wind tunnel and flight tests were reported (McHugh and Shaw, 1978; Shaw and Albion, 1980). Some recent work in this area has been reported by Nguyen and Chopra (1990), Prasad (1991) and Hall and Wereley (1992). These authors base their work on the use of a feedback control system driving an actuator to minimize vibration levels at one or more locations on the airframe.

Shaw's control algorithm may be classified essentially as a classical narrowband disturbance rejection compensator. Its integral action rather than a detailed knowledge of helicopter dynamics is responsible for the harmonic disturbance rejection achieved. The algorithm requires a knowledge of the transfer function between the control force input on the rotor swashplate and the locations on the fuselage where vibration is to be minimized. At each step in the adaptation process, the vibration signal measured at the minimization points is harmonically analysed. The resulting vector of vibration amplitudes for each harmonic to be controlled is then multiplied by the inverse of the transfer function matrix (which is divided into sine and cosine terms) to give a vector of sine and cosine components of the harmonic control force input required for the next step of the adaptation process. Thus, the sine and cosine components of the nth harmonic at the measurement locations on the fuselage are

$$\boldsymbol{w} = \boldsymbol{T}\boldsymbol{F} + \boldsymbol{w}_0 \tag{12.3.80}$$

where $\boldsymbol{w}$ is the vector of vibration amplitudes, $\boldsymbol{T}$ is the (assumed constant with time) transfer function matrix, $\boldsymbol{F}$ is the control inut vector and $\boldsymbol{w}_0$ is the vector of vibration amplitudes with no control input. The rèsulting control law is

$$w_{n+1} = w_n - T^{-1} w_n \tag{12.3.81}$$

For a single input, single output system, the transfer function matrix may be written as

$$T = \begin{bmatrix} T_{cc} & T_{cs} \\ T_{sc} & T_{ss} \end{bmatrix} \tag{12.3.82}$$

Details describing how this algorithm may be implemented in discrete- or continuous-time are given by Hall and Wereley (1992). These authors indicate that implementation of the algorithm in the continuous-time domain offers significant performance and stability advantages over the discrete-time implementation.

A different method of implementing HHC was suggested by Gupta and Du Val (1982) and Du Val *et al.* (1984) and was essentially a linear quadratic regulator with frequency shaped cost functions (see Chapter 5). This approach provides similar performance results to that discussed above and is eloquently summarised by Hall and Wereley (1992).

Unfortunately the HHC method of fuselage vibration control has problems associated with reduced rotor performance and airworthiness implications.

A second and probably more practical method of reducing helicopter fuselage vibration is to use an active isolation system to isolate the gearbox and rotor from the fuselage. King (1988) reported on such a system used on a Westland Helicopter in the UK achieving fuselage vibration reductions of 15 to 25 dB. The active isolation system consisted of four hydraulic actuators (mounted in parallel with passive rubber elements so that the passive system supported the entire static load) located between the airframe and a raft structure which carried the gearbox and rotor. Ten vibration sensors were used on the fuselage to provide the feedback signal to the electronic controller which was used to drive the four actuators. Two types of controller were investigated. The first was a time domain controller designed to minimize fuselage modal vibration using the IMSC method discussed in Chapter 11. The second and preferred system was a frequency domain controller which acted to suppress vibration transmission at the dominant rotor blade passing frequency only. Note that although the frequency domain controller was more robust, it had a very slow response time ($\approx$ 1 second) limiting the vibration reduction capability during fast manoeuvres. The frequency domain approach involved optimizing a performance function J (see Chapter 5) which includes control effort as well as fuselage vibration response, as follows:

$$J = Y^{\mathrm{T}} C^{\mathrm{T}} Y + X^{\mathrm{T}} D X \tag{12.3.83}$$

where Y is the complex vector of fuselage vibration measurement, X is the complex vector of actuator forces, C is a weighting matrix which allows for the possibility of certain fuselage locations being more important than others in terms of the need for vibration reduction. The second term in (12.3.83) involving the weighting matrix, $\boldsymbol{D}$, is included to allow limiting the actuator commands within their practical restraints. This type of term in a performance function is often referred to as the control effort term and is commonly used to share the control effort between actuators as evenly as possible, as well as keeping the required effort within practical bounds.

The equation of motion relating the fuselage vibration to the control actuator inputs is

$$\boldsymbol{Y} = \boldsymbol{T}\boldsymbol{X} + \boldsymbol{B} \tag{12.3.84}$$

where $\boldsymbol{T}$ is a transfer matrix relating the actuator loads to the fuselage vibration and $\boldsymbol{B}$ is a vector representing the uncontrolled vibration at the measurement locations represented in $\boldsymbol{Y}$.

Minimization of the performance function of (12.3.83) gives for the optimal complex actuator forces:

$$\boldsymbol{X} = -\left[\boldsymbol{T}^{\mathrm{T}}\boldsymbol{C}\boldsymbol{T} + \boldsymbol{D}\right]^{-1}\boldsymbol{T}^{\mathrm{T}}\ \boldsymbol{C}\ \boldsymbol{B} \tag{12.3.85}$$

Note that all of the terms in the matrices discussed above are complex; that is, they are described by an amplitude and phase.

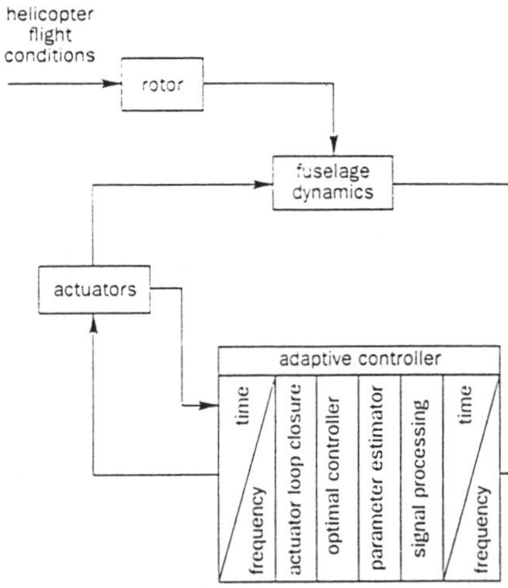

Fig. 12.51 Schematic of the system for active control of helicopter vibrations (after King, 1988).

A schematic representation of the control system used by King (1988), which incorporated the performance function just discussed is shown in Fig. 12.51. In this figure the incoming signals from the vibration sensors on the fuselage are first converted to the frequency domain using a fast Fourier transform (FFT) and then the blade passing frequency component is extracted in the signal processing section. The transfer matrix, $\boldsymbol{T}$, is estimated by the parameter estimator and then the optimal actuator forces needed to minimize the performance function are calculated by the optimal controller. Force feedback from the actuators is used to compensate for the actuator dynamics as indicated by the actuator loop closure block in the figure. The required actuator signals, calculated in the frequency domain, are then converted to time domain signals using an inverse DFT and then applied to the control actuators.

12.4 FEEDFORWARD CONTROL: BASIC SDOF SYSTEM

Feedforward control is the best approach for vibration isolation when the disturbance signal can be measured before the control actuator needs to act or when it is periodic (such as that generated by a rotating machine). Here, we will begin by examining feedforward control of a single-degree-of-freedom system which involves control of vibration transmission from a rigid mass to a stiff support. This will be followed in subsequent sections by a discussion of feedforward control of periodic vibration from a rigid body into a flexible beam, from a rigid body into a flexible panel and from a rigid body into a flexible cylinder.

For the single-degree-of-freedom system illustrated in Fig. 12.52, three different control force cases will be considered; the control force acting on the rigid body, the control force acting on the support structure and the control force acting on the rigid body but reacting against the support structure (Nelson *et al.*, 1987; Nelson, 1991).

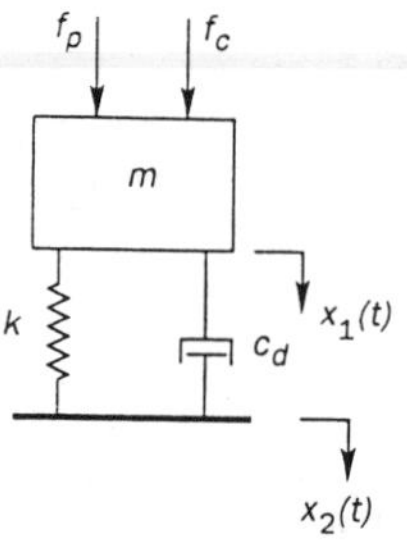

Fig. 12.52 Single-degree-of-freedom system with the control force acting on the rigid body.

12.4.1 Control force acting on the rigid body

For the system shown in Fig. 12.52, the equation of motion is

$$m\ddot{x}_1 + c_d(\dot{x}_1 - \dot{x}_2) + k(x_1 - x_2) = f_p + f_c \qquad (12.4.1)$$

For a periodic disturbing force f_p of frequency ω, we obtain

$$(-\omega^2 m + j\omega c_d + k)x_1 - (j\omega c_d + k)x_2 = f_p + f_c \qquad (12.4.2)$$

It can be seen from (12.4.2) (and intuitively) that the control force f_c which ensures that both x_1 and x_2 are zero is given by $f_c = -f_p$.

12.4.2 Control force acting on the support structure

For the system shown in Fig. 12.53, let the mobility of the support structure at the excitation frequency be M, such that

$$j\omega x_2 = Mf \tag{12.4.3}$$

where f is the force acting on the support structure. The total force acting on the support structure is

$$f = f_c + k(x_1 - x_2) + c_d(\dot{x}_1 - \dot{x}_2) \tag{12.4.4}$$

Thus, (12.4.3) can be written as

$$j\omega x_2 = M\left(f_c + k(x_1 - x_2) + c_d(\dot{x}_1 - \dot{x}_2)\right) \tag{12.4.5}$$

The equation of motion for the rigid mass, m is

$$(-\omega^2 m + j\omega c_d + k)x_1 - (j\omega c_d + k)x_2 = f_p \tag{12.4.6}$$

Setting $x_p = 0$ in (12.4.5) and (12.4.6) gives

$$f_c = f_p \frac{-(j\omega c_d + k)}{-\omega^2 m + j\omega c_d + k} \tag{12.4.7}$$

Note that the control force to produce zero displacement of the support structure is independent of the support structure mobility.

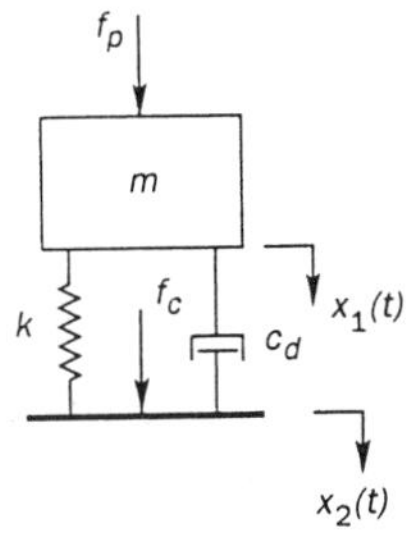

Fig. 12.53 Single-degree-of-freedom system with the control force acting on the support structure.

Expressed in terms of the system resonance frequency ω_0 and damping ratio ζ, equation (12.4.7) can be written as

$$f_c = f_p \left[\frac{2j\zeta\omega_0\omega + \omega_0^2}{\omega^2 - 2j\zeta\omega_0\omega - \omega_0^2} \right] \tag{12.4.8}$$

and the magnitude is given by

$$\left| \frac{f_c}{f_p} \right| = \sqrt{\frac{1 + (2\zeta\omega/\omega_0)^2}{\left[(\omega/\omega_0)^2 - 1\right]^2 + (2\zeta\omega/\omega_0)^2}} \tag{12.4.9}$$

which is identical to (12.2.43) which describes the magnitude of (x_1/x_2) for a base excited system.

12.4.3 Control force acting on the rigid body and reacting on the support structure

For the system shown in Fig. 12.54, the equation for the support structure velocity is identical to (12.4.5), but the equation of motion for the rigid body is,

$$(-\omega^2 m + j\omega c_d + k)x_1 - (j\omega c_d + k)x_2 = f_p - f_c \tag{12.4.10}$$

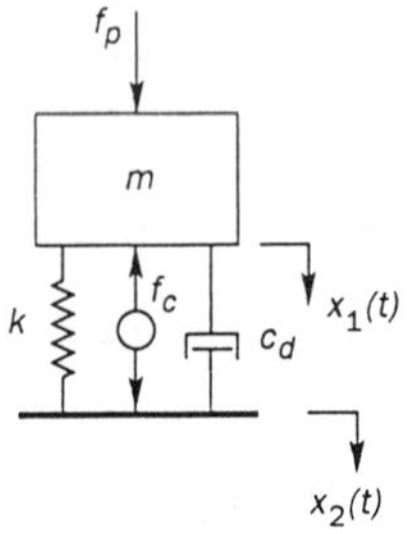

Fig. 12.54 Single-degree-of-freedom system with the control force acting on the rigid body and reacting on the support structure.

Setting $x_2 = 0$ in (12.4.5) and (12.4.6) gives

$$f_c = f_p \frac{j\omega c_d + k}{\omega^2 m} \tag{12.4.11}$$

Expressed in terms of the system resonance frequency ω_0 and damping ratio, ζ, (12.4.11) is

$$f_c = f_p \frac{2j\zeta\omega\omega_0 + \omega_0^2}{\omega^2} \tag{12.4.12}$$

and the amplitude ratio is

$$\left| \frac{f_c}{f_p} \right| = \sqrt{\frac{1 + (2\zeta\omega/\omega_0)^2}{(\omega/\omega_0)^2}} \tag{12.4.13}$$

12.4.4 Summary

It is important to note that in all three of the cases discussed above, the displacement of the support structure is reduced to zero by the control force. However, for a system consisting of multiple connections between the rigid body and support structure, with a control force at each mount, it is generally not possible to obtain a unique solution for the control forces required to reduce the amplitude to zero at each support point on the structure. In fact, because of the interaction or cross-coupling between forces acting at the base of each mount, it is possible for the control forces to 'fight' one another so that zero net displacement or force at the base of a mount may not be possible. Thus, where multiple mounts are used it is preferable to use a multichannel control system and total vibratory power transmission into the support structure through all support points as the cost function to be minimised, as it results in smaller required control forces and sometimes better overall reductions in structural vibration levels. Note also that feedforward control does not change the system resonance frequency and damping as did feedback control which was discussed in Section 12.2.

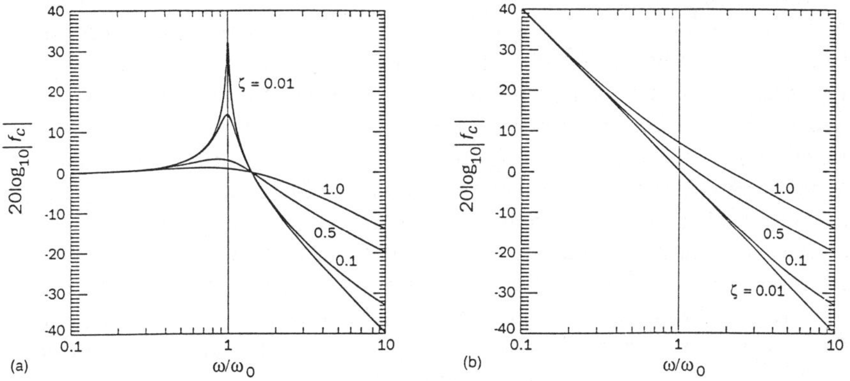

Fig. 12.55 Effect of damping ratio and excitation frequency on the magnitude of the required control force for a single-degree-of-freedom system: (a) control force acting only on the support structure (Fig. 12.52); (b) control force acting on the rigid body and reacting on the support structure (Fig. 12.54).

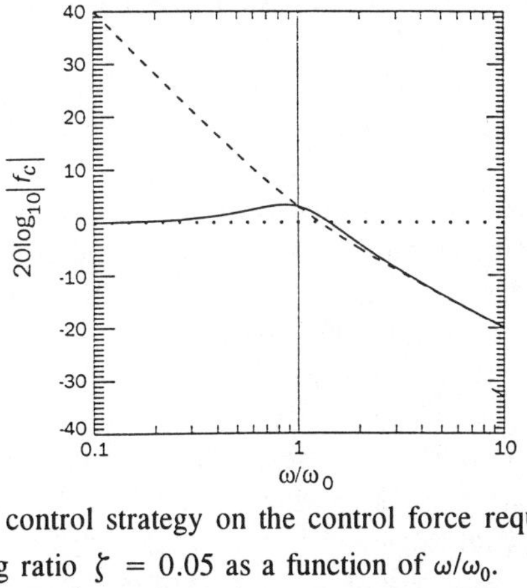

Fig. 12.56 Effect of control strategy on the control force required to achieve optimal control for a damping ratio $\zeta = 0.05$ as a function of ω/ω_0.
——— control force acting only on the support structure.
- - - control force acting on the rigid body and reacting on the support structure.
• • • control force acting on the rigid body.

It is interesting to plot (12.4.9) and (12.4.13) as a function of ω/ω_o for various values of ζ to examine what sort of control force is necessary in each case. This is done in Fig. 12.55, and in Fig. 12.56, the control forces required using the latter two control strategies discussed above are compared as a function of ω/ω_0 for $\zeta = 0.05$. For the first strategy the control force is simply equal to the inverse of the primary excitation force in all cases.

12.5 FEEDFORWARD CONTROL: SINGLE ISOLATOR BETWEEN A RIGID BODY AND A FLEXIBLE BEAM

In this section, a theoretical model will be developed to calculate the optimal feedforward control force required to minimize the vibratory power transmission from a rigid body through an isolator into a flexible beam as a function of excitation frequency. As there is only one attachment point, the minimum power transmission will be zero, as will be the force and displacement at the base of the isolator when the control force is optimal. However, in practice the controller is never able to achieve the exact optimal control force; thus there will always be some residual power transmitted into the supporting beam. Thus the analysis outlined here for deriving the optimal control force is useful for investigating the effects of the control forces varying slightly from the theoretical optimum.

We begin by developing equations of motion for each of the three parts of the total system: the rigid body, the isolator and the beam. These are then combined into a single matrix equation for the total system which allows the vibratory power transmission into the beam to be calculated as a function of control force phase and magnitude. Quadratic optimization is then used, as described in Chapter 8, to find the optimal control force. Once the optimal control force for a particular system configuration is found, the maximum achievable reduction in power transmission into the beam assuming an ideal feedforward controller is calculated.

In the following analysis the superscripts and subscripts 0 refer to the centre of gravity of the rigid body, t refers to the top of the isolators, b refers to the bottom of the isolators, J refers to the Jth isolator, I refers to the Ith beam vibration mode and c refers to the control force.

In this case, the primary excitation force is not simply a vertical force, but a combination of all possible translations and rotations (that is, it is arbitrary). Thus, the primary excitation force may be written as

$$Q_0 = \left[F_x F_y F_z M_x M_y M_z\right]^{\mathrm{T}} \tag{12.5.1}$$

To be consistent with published literature the vertical direction is now the z-direction, the coordinate x is in the direction of the beam axis and y is the coordinate across the beam as shown in Fig. 12.57.

The quantities labelled F and M in (12.5.1) refer to force and moment inputs respectively on the rigid body. Note that another rigid body is shown between the beam and the lower end of the suspension in Fig. 12.57. This more closely

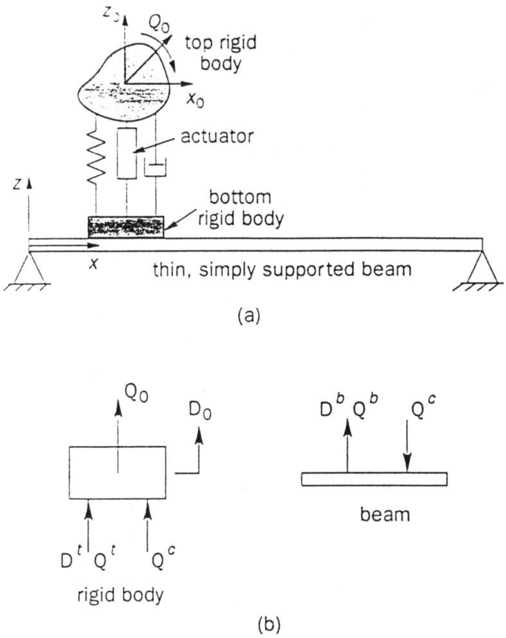

Fig. 12.57 Theoretical model for a single isolator between a rigid body and a flexible beam: (a) beam – isolator model; (b) sign conventions

simulates the practical implementation of a suspension system such as this, where the lower part of the actuator and spring support are well represented by a rigid mass. In this case, the rigid body is modelled as a point mass loading on the beam and included in the beam equation of motion.

12.5.1 Rigid body equation of motion

Referring to Fig. 12.57, the equation of motion of the rigid body can be written in matrix form as (Pan *et al.*, 1993):

$$\boldsymbol{Z}\boldsymbol{D}_0 = \boldsymbol{Q}_0 + \boldsymbol{R}^t\boldsymbol{Q}^t \tag{12.5.2}$$

where $\boldsymbol{Z} = -\omega^2\boldsymbol{m}_0$, ω is the primary force excitation frequency, $\boldsymbol{m}_0$ is the diagonal inertia matrix given by

$$\boldsymbol{m}_0 = \begin{bmatrix} m & & & & & \\ & m & & & & \\ & & m & & 0 & \\ & & & I_x & & \\ & 0 & & & I_y & \\ & & & & & I_z \end{bmatrix} \tag{12.5.3}$$

where m is the mass of the body and I_x, I_y and I_z are the moments of inertia of the body around its x-, y- and z-axes respectively. $\boldsymbol{D}_0$ in (12.5.2) is the complex displacement matrix of the centre of gravity of the rigid body and is given in terms of the three linear displacements x_0, y_0, z_0 and the three angular rotational displacements θ_{0x}, θ_{0y} and θ_{0z} as follows:

$$\boldsymbol{D}_0 = \left[x_0\ y_0\ z_0\ \theta_{x0}\ \theta_{y0}\ \theta_{z0}\right]^{\mathrm{T}} \tag{12.5.4}$$

The quantity $\boldsymbol{Q}^t = [F_x^t\, F_y^t\, F_z^t\, M_x^t\, M_y^t\, M_z^t]^{\mathrm{T}}$ is the elastic force matrix for the passive mount which acts at the point $(x_1^t\ y_1^t\ z_1^t)$, where the top of the mount is connected to the rigid body. The quantity $\boldsymbol{R}^t$ is the matrix which accounts for coupling between moments and translational displacements, and is defined as follows:

$$\boldsymbol{R}^t = \begin{bmatrix} 1 & 0 & 0 & 0 & 0 & 0 \\ 0 & 1 & 0 & 0 & 0 & 0 \\ 0 & 0 & 1 & 0 & 0 & 0 \\ 0 & -z^t & y^t & 1 & 0 & 0 \\ z^t & 0 & -x^t & 0 & 1 & 0 \\ -y^t & x^t & 0 & 0 & 0 & 1 \end{bmatrix} \tag{12.5.5}$$

The complex stiffness matrix of the mount (which will be used later) is defined as:

$$\boldsymbol{K} = \begin{bmatrix} k_x & 0 & 0 & 0 & 0 & 0 \\ 0 & k_y & 0 & 0 & 0 & 0 \\ 0 & 0 & k_z & 0 & 0 & 0 \\ 0 & 0 & 0 & k_{\theta x} & 0 & 0 \\ 0 & 0 & 0 & 0 & k_{\theta y} & 0 \\ 0 & 0 & 0 & 0 & 0 & k_{\theta z} \end{bmatrix} \tag{12.5.6}$$

where $k_x \cdots k_{\theta z}$ are complex stiffness coefficients of the mount corresponding to each degree of freedom:

$$k_x = k_x^0(1 + j\eta_x); \quad k_y = k_y^0(1 + j\eta_y); \quad \cdots \quad k_{\theta z} = k_{\theta z}^0(1 + j\eta_{\theta z}) \tag{12.5.7a,b,c}$$

where $k_x^0, \cdots k_{\theta z}^0$ are the stiffness coefficients and $\eta_x, \cdots, \eta_{\theta z}$ are the structural loss factors. For a cylindrical mount, the stiffness coefficients can be calculated according to Table 12.1, where E, ν, L and a are respectively Young's modulus of elasticity and Poisson's ratio for the material, the length of the mount and the section radius of the mount.

Table 12.1 Stiffness coefficients of a cylindrical isolator

k_x^0	k_y^0	k_z^0	$k_{\theta x}^0$	$k_{\theta y}^0$	$k_{\theta z}^0$
$\dfrac{3\pi E a^4}{4L^3}$	$\dfrac{3\pi E a^4}{4L^3}$	$\dfrac{\pi E a^2}{L}$	$\dfrac{\pi E a^4}{4L}$	$\dfrac{\pi E a^4}{4L}$	$\dfrac{\pi E a^4}{4(1+\nu)L}$

12.5.2 Supporting beam equation of motion

For the thin beam considered here, with its thickness far smaller than its width and its width far smaller than its length, the transverse vibration in the z-direction dominates its motion. Errors arising from ignoring other motions such as lateral vibration in the y-direction, rotational vibration around the x-direction and twisting vibration around the z-direction are expected to be insignificant, as the beam is sufficiently thin and the base of the mount is attached at $y_b = 0$.

Force and moment components in the force vector $Q^b = [F_{bx}, F_{by}, F_{bz}, M_{bx}, M_{by}, M_{bz}]^T$ for the mount acting on the beam are assumed to be point actions, so that Dirac delta functions and their differentials may be used for describing the forces and moments in terms of quantities per unit area. The point actions which generate the lateral vibration of the beam only include three such items:

$$F_{bz}\delta(x - x_b), \ \frac{h}{2}F_{bx}\frac{\partial\delta(x - x_b)}{\partial x} \text{ and } M_{by}\frac{\partial\delta(x - x_b)}{\partial x}$$

Thus, the transverse displacement $w(x, t: x_b)$ of the beam can be determined by using the following equation (see Section 2.3):

$$\rho S\frac{\partial^2 w}{\partial t^2} + EI_b\frac{\partial^4 w}{\partial x^4} = F_{bz}\delta(x - x_b) + (\frac{h}{2}F_{bx} + M_{by})\frac{\partial\delta(x - x_b)}{\partial x} + m^b\omega^2 w\,\delta(x - x_b) \tag{12.5.8}$$

where m^b is the mass of the isolator base.

Boundary conditions at both simply supported ends of the beam are:

$$w = \frac{\partial^2 w}{\partial x^2} = 0 \tag{12.5.9a,b}$$

where $F_{bx}, \cdots, M_{bz}$ are the elements of the force matrix, Q^b; ρ, S and I_b are respectively, the density of the material, the beam cross-sectional area and the second moment of area of the beam section about the neutral plane parallel with the z-axis. The location of the beam excitation point, or elastic mount support point is $(x_b, 0, h/2)$. Damping is included on a modal basis.

For harmonic force excitation, the lateral displacement w in the z-direction for the simply supported beam can be written as (see Chapter 2):

$$w = \left[\sin\frac{\pi x}{L_b},\ \sin\frac{2\pi x}{L_b},\ \cdots,\ \sin\frac{p\pi x}{L_b} \right] \boldsymbol{w}_p \tag{12.5.10}$$

where P is the number of modes included in the analysis, $\sin(I\pi x/L_b)$, $(I = 1, \cdots, P)$ are the mode shape functions, and $\boldsymbol{w}_p = [w_1, w_2, \cdots, w_P]^T$ is the modal complex amplitude coefficient matrix which depends upon x_b and ω. Substituting (12.5.10) into (12.5.8), and using the othogonal property of the mode shape functions, gives:

$$m_I(\omega_I^2 - \omega^2 + j\eta_I\omega_I^2)w_I$$

$$= \left[-\frac{I\pi h}{2L_b}\cos\frac{I\pi x_b}{L_b}, 0, \sin\frac{I\pi x_b}{L_b}, 0, -\frac{I\pi}{L_b}\cos\frac{I\pi x_b}{L_b}, 0 \right] \boldsymbol{Q}^b \tag{12.5.11}$$

$$+ \sum_{I'=1}^{P} C_{I'I} w_{I'} \qquad (I = 1, \cdots, P)$$

where L_b and h are the length and thickness of the beam respectively, and ω_I, $(I = 1, \cdots, P)$ are the resonance frequencies of the beam, given by:

$$\omega_I = \left(\frac{I^2\pi^2}{L_b^2} \right) \sqrt{\frac{EI_b}{\rho S}} \tag{12.5.12}$$

The quantity E is Young's modulus of elasticity, m_I is the modal mass for mode I which is equal to half of the beam mass for each mode:

$$m_I = \frac{1}{2}\rho A L_b \qquad (I = 1, \cdots, P) \tag{12.5.13}$$

The quantity η_I is the modal loss factor for mode I and is related to the 60 dB modal decay time T_I by:

$$\eta_I = \frac{4.4\pi}{T_I\omega_I} \tag{12.5.14}$$

$C_{I',I}$ $(I', I = 1, P)$ is the concentrated mass contribution given by:

$$C_{I',I} = m^b\omega^2 \sin\frac{I\pi x_b}{L_b} \sin\frac{I'\pi x_b}{L_b} \tag{12.5.15}$$

where m^b is the mass of the isolator base.

Equation (12.5.11) can be expressed compactly by making use of matrix notation. Thus,

$$\boldsymbol{Z}_p \boldsymbol{w}_p = \boldsymbol{R}^b \boldsymbol{Q}^b \tag{12.5.16}$$

where

$$Z_p = \begin{bmatrix} \chi_1 - C_{11} & -C_{12} & \cdots & -C_{1P} \\ -C_{21} & \chi_2 - C_{22} & & \vdots \\ \vdots & & \ddots & \\ -C_{P1} & \cdots & & \chi_P - C_{PP} \end{bmatrix} \tag{12.5.17}$$

where

$$\chi_I = m_I(\omega_I^2 + j\eta_I\omega_I^2 - \omega^2) \tag{12.5.18}$$

and

$$R^b = \begin{bmatrix} \frac{-h\pi}{2L_b}\cos\frac{\pi x_b}{L_b} & 0 & \sin\frac{\pi x_b}{L_b} & 0 & -\frac{\pi}{L_b}\cos\frac{\pi x_b}{L_b} & 0 \\ \frac{-2h\pi}{2L_b}\cos\frac{2\pi x_b}{L_b} & 0 & \sin\frac{2\pi x_b}{L_b} & 0 & -\frac{2\pi}{L_b}\cos\frac{2\pi x_b}{L_b} & 0 \\ \vdots & \vdots & \vdots & \vdots & \vdots & \vdots \\ \frac{-Ph\pi}{2L_b}\cos\frac{P\pi x_b}{L_b} & 0 & \sin\frac{P\pi x_b}{L_b} & 0 & -\frac{P\pi}{L_b}\cos\frac{P\pi x_b}{L_b} & 0 \end{bmatrix} \tag{12.5.19}$$

Using elastic thin plate theory (as the beam is much wider than it is thick), the elements of the displacement matrix $D^b = [\xi_x, \xi_y, w, \theta_x, \theta_y, \theta_z]^{\mathrm{T}}$ at the elastic mount support point can be described in terms of the lateral displacement w as follows:

$$\xi_x = -\frac{h}{2}\frac{\partial w}{\partial x},\ \xi_y = -\frac{h}{2}\frac{\partial w}{\partial y} = 0,\ \theta_x = \frac{\partial w}{\partial y} = 0,\ \theta_y = -\frac{\partial w}{\partial x},\ \theta_z = 0$$

(12.5.20a,b,c,d,e,f,g)

Substituting (12.5.10) into (12.5.20), we obtain,

$$D^b = (R^b)^{\mathrm{T}} w_p \tag{12.5.21}$$

12.5.3 System equation and power transmission

The matrices describing the forces and moments acting on the rigid body $Q^t = [F_{tx}, F_{ty}, F_{tz}, M_{tx}, M_{ty}, M_{tz}]^{\mathrm{T}}$ and on the support beam $Q^b = [F_{bx}, F_{by}, F_{bz}, M_{bx}, M_{by}, M_{bz}]^{\mathrm{T}}$ are both proportional to the mount's linear and angular deformations which are described by the relative displacement of its top and bottom ends. Thus,

$$Q^t = -Q^b = K(D^b - D^t) \tag{12.5.22a,b}$$

where D^t and D^b are respectively the displacement matrices for the top and bottom ends of the mount. It is easily shown that

$$D^t = (R^t)^{\mathrm{T}} D_0 \tag{12.5.23}$$

Selecting D_0 (defined in (12.5.4)) and w_p (defined following 12.5.10)) as the unknowns and substituting (12.5.22) into (12.5.2) and (12.5.16), the system can finally be expressed by the following complex linear equations with a symmetrical coefficient matrix:

$$\begin{bmatrix} A_{11} & A_{12} \\ A_{21} & A_{22} \end{bmatrix} \begin{bmatrix} D_0 \\ w_p \end{bmatrix} = \begin{bmatrix} Q_0 \\ \mathbf{0} \end{bmatrix} \tag{12.5.24}$$

where $\mathbf{0}$ is a $(P \times 1)$ zero vector and

$$A_{11} = Z + R^t K (R^t)^{\mathrm{T}} \tag{12.5.25}$$

$$A_{12} = (A_{21})^{\mathrm{T}} = -R^t K (R^b)^{\mathrm{T}} \tag{12.5.26a,b}$$

$$A_{22} = Z_p + R^b K (R^b)^{\mathrm{T}} \tag{12.5.27}$$

When $Q_0 = 0$, and all loss factors are assumed to be zero, (12.5.24) represents the free vibration of the coupled system. Solving the eigenvalue problem of the following coefficient matrix which, in this case, is a real symmetrical matrix:

$$A = \begin{bmatrix} A_{11} & A_{12} \\ A_{21} & A_{22} \end{bmatrix} = A_1 - \omega^2 A_2 \tag{12.5.28a,b}$$

gives $P + 6$ resonance frequencies and corresponding mode shape vectors for the coupled system. When a force actuator is connected in parallel with the passive mount, the resulting active control force vector Q_c acts on both the upper rigid body and the support beam simultaneously, with the same amplitudes but opposite phases, and (12.5.24) becomes:

$$\begin{bmatrix} A_{11} & A_{12} \\ A_{21} & A_{22} \end{bmatrix} \begin{bmatrix} D_0 \\ w_p \end{bmatrix} = \begin{bmatrix} Q_0 + R^t Q^c \\ -R^b Q^c \end{bmatrix} \tag{12.5.29}$$

A positive control force is defined here as one which acts at the bottom of the isolator in the direction of positive displacement (and in the opposite direction at the top of the isolator).

The time average power transmission P_0 into the support beam, which is the quantity which must be controlled, can be calculated by using the following equation:

$$P_0 = \mathrm{Re}\left\{ -\frac{1}{2} j\omega (Q^b - Q^c)^{\mathrm{H}} D^b \right\} = \frac{\omega}{2} \mathrm{Im}\{(Q^b - Q^c)^{\mathrm{H}} D^b\}$$

$$= -\frac{\omega}{2} \mathrm{Im}\{(D^b)^{\mathrm{H}} (Q^b - Q^c)\} \tag{12.5.30a,b,c}$$

where the symbol H represents the transpose and conjugate of a matrix. Note that removing j from the right-hand side of (12.5.30a) and taking the imaginary part causes a sign change, as does changing the transpose conjugate from the force matrix to the displacement matrix. Positive power transmission is defined as in the direction of positive force and displacement and is actually out of the beam. This is the reason for the minus sign in (12.5.30a). Because only one mount is being considered, the power transmission P_0 corresponding to the optimal control force vector should be zero.

12.5.4 Optimum control force and minimum power transmission

If the driving force matrix remains unchanged, each element of the matrices $\boldsymbol{D}^t$, $\boldsymbol{D}^b$ and $\boldsymbol{Q}^b$ will generally be a linear function of the control force matrix elements. Therefore, the output power P_0 into the supporting beam is generally a quadratic function of these control variables and depends upon the driving frequency ω. For a specified driving frequency, the magnitude of the power transmission reduction will depend upon system parameters. Analytically, it is possible to select these parameters so that the power transmission or power reduction reaches an optimal value in a given frequency range. Substituting (12.5.21) for $\boldsymbol{D}^b$ and (12.5.25) for $\boldsymbol{D}^t$, (12.5.22) can be used to write $(\boldsymbol{D}^b)^{\mathrm{H}}(\boldsymbol{Q}^b - \boldsymbol{Q}^c)$ as

$$(\boldsymbol{D}^b)^{\mathrm{H}}(\boldsymbol{Q}^b - \boldsymbol{Q}^c) = -(\boldsymbol{w}_p)^{\mathrm{H}}\boldsymbol{R}^b\left[\boldsymbol{K}(\boldsymbol{R}^b)^{\mathrm{T}}\boldsymbol{w}_p - \boldsymbol{K}(\boldsymbol{R}^t)^{\mathrm{T}}\boldsymbol{D}_0 + \boldsymbol{Q}^c\right] \quad (12.5.31)$$

Using (12.5.29), we can write

$$\boldsymbol{D}_0 = \boldsymbol{B}_{11}\left[\boldsymbol{Q}_0 + \boldsymbol{R}^t\boldsymbol{Q}^c\right] - \boldsymbol{B}_{12}\boldsymbol{R}^b\boldsymbol{Q}^c \quad (12.5.32)$$

$$\boldsymbol{w}_p = \boldsymbol{B}_{21}\left[\boldsymbol{Q}_0 + \boldsymbol{R}^t\boldsymbol{Q}^c\right] - \boldsymbol{B}_{22}\boldsymbol{R}^b\boldsymbol{Q}^c \quad (12.5.33)$$

where the matrices $\boldsymbol{B}_{11}$, $\boldsymbol{B}_{12}$, $\boldsymbol{B}_{21}$, $\boldsymbol{B}_{22}$ are the submatrix elements of the inverse of system matrix $\boldsymbol{A}$, as follows:

$$\begin{bmatrix} \boldsymbol{B}_{11} & \boldsymbol{B}_{12} \\ \boldsymbol{B}_{21} & \boldsymbol{B}_{22} \end{bmatrix} = \begin{bmatrix} \boldsymbol{A}_{11} & \boldsymbol{A}_{12} \\ \boldsymbol{A}_{21} & \boldsymbol{A}_{22} \end{bmatrix}^{-1} \quad (12.5.34)$$

$$\boldsymbol{B}_{11} = \left[\boldsymbol{A}_{11} - \boldsymbol{A}_{12}\boldsymbol{A}_{22}^{-1}\boldsymbol{A}_{21}\right]^{-1} \quad (12.5.35)$$

$$\boldsymbol{B}_{22} = \left[\boldsymbol{A}_{22} - \boldsymbol{A}_{21}\boldsymbol{A}_{11}^{-1}\boldsymbol{A}_{12}\right]^{-1} \quad (12.5.36)$$

$$\boldsymbol{B}_{12} = -\boldsymbol{A}_{11}^{-1}\boldsymbol{A}_{12}\left[\boldsymbol{A}_{22} - \boldsymbol{A}_{21}\boldsymbol{A}_{11}^{-1}\boldsymbol{A}_{12}\right]^{-1} \quad (12.5.37)$$

$$\boldsymbol{B}_{21} = -\boldsymbol{A}_{22}^{-1}\boldsymbol{A}_{21}\left[\boldsymbol{A}_{11} - \boldsymbol{A}_{12}\boldsymbol{A}_{22}^{-1}\boldsymbol{A}_{21}\right]^{-1} \quad (12.5.38)$$

Substituting (12.5.32) and (12.5.33) into (12.5.31) gives the following quadratic form of $(\boldsymbol{D}^b)^{\mathrm{H}}(\boldsymbol{Q}^b - \boldsymbol{Q}_c)$:

$$(\boldsymbol{D}^b)^{\mathrm{H}}(\boldsymbol{Q}^b - \boldsymbol{Q}^c) = (\boldsymbol{Q}^c)^{\mathrm{H}}\boldsymbol{a}\boldsymbol{Q}^c + (\boldsymbol{Q}^c)^{\mathrm{H}}\boldsymbol{b}_1 + \boldsymbol{b}_2\boldsymbol{Q}^c + c \quad (12.5.39)$$

where

$$a = G_1 K G_2 - G_1 \tag{12.5.40}$$

$$b_1 = G_1 K G_3 \tag{12.5.41}$$

$$b_2 = G_4 K G_2 - G_4 \tag{12.5.42}$$

$$c = G_4 K G_3 \tag{12.5.43}$$

where the matrices G_1, G_2, G_3 and G_4 are defined as:

$$G_1 = (R^t)^{\mathrm{T}} B_{21}^{\mathrm{H}} R^b - (R^b)^{\mathrm{T}} B_{22}^{\mathrm{H}} R^b \tag{12.5.44}$$

$$G_2 = (R^t)^{\mathrm{T}} B_{11} R^t - (R^t)^{\mathrm{T}} B_{12} R^b - (R^b)^{\mathrm{T}} B_{21} R^t + (R^b)^{\mathrm{T}} B_{22} R^b \tag{12.5.45}$$

$$G_3 = \left[(R^t)^{\mathrm{T}} B_{11} - (R^b)^{\mathrm{T}} B_{21}\right] Q_0 \tag{12.5.46}$$

$$G_4 = Q_0^{\mathrm{T}} B_{21}^{\mathrm{H}} R^b \tag{12.5.47}$$

Thus, the output power P_0 in (12.5.28) can be expressed as the following real quadratic function:

$$P_0 = -\frac{\omega}{2}\mathrm{Im}\{(D^b)^{\mathrm{H}}(Q^b - Q^c)\}$$
$$= -\frac{\omega}{2}\left\{(q^c)^{\mathrm{T}}\begin{bmatrix} a^i & a^r \\ -a^r & a^i \end{bmatrix} q^c + (q^c)^{\mathrm{T}}\begin{bmatrix} b_1^i \\ -b_1^r \end{bmatrix} + \begin{bmatrix} b_2^i & b_2^r \end{bmatrix} q^c + c^i\right\} \tag{12.5.48a,b}$$

or in terms of the following equivalent expression, with a symmetrical coefficient matrix for the quadratic term:

$$P_0 = -\frac{\omega}{2}\{(q^c)^{\mathrm{T}}\alpha q^c + (q^c)^{\mathrm{T}}\beta + \beta^{\mathrm{T}} q^c + c^i\} \tag{12.5.49}$$

where

$$\alpha = \alpha^{\mathrm{T}} = \frac{1}{2}\begin{bmatrix} a^i + (a^i)^{\mathrm{T}} & a^r - (a^r)^{\mathrm{T}} \\ -a^r + (a^r)^{\mathrm{T}} & a^i + (a^i)^{\mathrm{T}} \end{bmatrix} \tag{12.5.50a,b}$$

and

$$\beta = \frac{1}{2}\begin{bmatrix} (b_2^i)^{\mathrm{T}} + b_1^i \\ (b_2^r)^{\mathrm{T}} - b_1^r \end{bmatrix} \tag{12.5.51}$$

and the real matrices a^r, a^i, b_1^r, ··· , represent, respectively, the real and imaginary parts of the complex matrices a, b_1, b_2 and the constant c. Clearly, the real output power for the uncontrolled case where $q^c = 0$ is given by $P_0' = -\frac{1}{2}\omega c^i$. The real control 'force' vector q^c of 12 elements consists of real Q^{cr} and imaginary Q^{ci} parts of the control force vector Q^c; thus,

$$q^c = \begin{bmatrix} Q^{cr} \\ Q^{ci} \end{bmatrix} \tag{12.5.52}$$

A reduction in output power as a result of the action of the control forces will occur when the relationship $\Delta P = P_0' - P_0 < 0$ is satisfied (as positive power flow is defined upwards, out of the beam). In other words, control force vectors satisfying the following inequality can result in a reduction of the output power:

$$(q^c)^{\mathrm{T}}\alpha q^c + (q^c)^{\mathrm{T}}\beta + \beta^{\mathrm{T}} q^c < 0 \tag{12.5.53}$$

From Chapter 8, we know that the quadratic function of (12.5.49) has a minimum (which should be equal to zero) given by

$$P_{\min} = \frac{\omega}{2}\{\beta^{\mathrm{T}}\alpha^{-1}\beta - c^i\} \tag{12.5.54}$$

corresponding to an optimum control force vector of

$$q^c_{\mathrm{opt}} = -\alpha^{-1}\beta \tag{12.5.55}$$

where the coefficient matrix α is a positive definite matrix.

The quantity $P_{\min}$ is theoretically equal to zero for an ideal controller, but the preceding analysis allows the characteristics of a non-ideal controller to be taken into account.

12.6 FEEDFORWARD CONTROL: MULTIPLE ISOLATORS BETWEEN A RIGID BODY AND A FLEXIBLE PLATE

In many cases, for example in submarines, equipment is isolated from its final support structure using a two stage isolation system which consists of an intermediate structure inserted between the machine and the support structure, which is shown in its simplest form in Fig. 12.58.

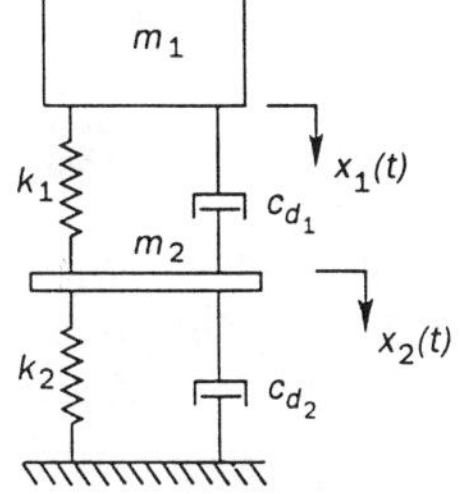

Fig. 12.58 Two stage vibration isolation system.

In a submarine, for example, the intermediate mass m_2 is the machinery support platform and the base support shown in Fig. 12.58 is the hull. However in this case, the intermediate structure cannot be modelled simply as a mass as it is itself a multi-modal structure, with many resonances in the frequency range of interest which serve to reduce the effectiveness of the double mounting arrangement (which otherwise would be much more effective than the single-degree-of-freedom system with no intermediate mass). Thus there is considerable interest in devising means of reducing the transmission of vibratory energy into the intermediate structure. One possibility is to use active control, and the purpose of this section is to investigate the performance improvement which can be achieved in terms of reduction in the response of the intermediate structure by the use of active control. Here it will be assumed that the final support structure is rigid. The use of a flexible cylinder as the final support structure has been considered by Pan and Hansen (1994).

In the work discussed in this section a periodic excitation force will be assumed, as the analysis is directed towards the use of a feedforward active control system. To begin, we will examine the transmission of only vertical excitation forces into a clamped edge flexible plate. This will be extended later to include horizontal forces and moments and simply supported and corner supported plates.

12.6.1 Vertical excitation forces only

This problem was first investigated by Jenkins *et al.* (1988, 1993) who examined the problem of four active isolators between a rigid body and an edge clamped flexible rectangular plate using finite element analysis. The active isolators consisted of an electromagnetic active element in parallel with a cylindrical flexible passive element with some internal damping, as shown in Fig. 12.59.

Only vertical input forces from the rigid body were assumed, and the purpose of the active isolators was to minimize the resulting flexural vibration levels in the flexible support plate, so the cost function J to be minimized was

$$J = \sum_{n=1}^{N} |w_n|^2 = \boldsymbol{W}^{\mathrm{H}}\boldsymbol{W} \qquad (12.6.1\text{a,b})$$

The total displacement at any point on the plate is a linear superposition of the displacements $\boldsymbol{W}_p$ due to the primary excitation force and those due to the control forces $\boldsymbol{f}_c$ so that

$$\boldsymbol{W} = \boldsymbol{W}_p + \boldsymbol{R}\boldsymbol{f}_c \qquad (12.6.2)$$

where $\boldsymbol{R}$ is the complex receptance matrix relating the displacement at any point on the plate to an input force at any other point.

Thus the cost function to be minimized can be written as

$$J = (\boldsymbol{W}_p + \boldsymbol{R}\boldsymbol{f}_c)^{\mathrm{H}}(\boldsymbol{W}_p + \boldsymbol{R}\boldsymbol{f}_c) \qquad (12.6.3)$$

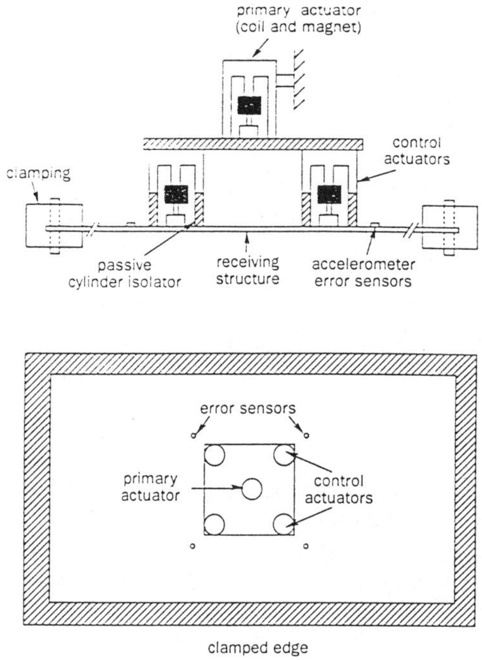

Fig. 12.59 Active isolation of a rigid machine from a flexible clamped plate.

Rearranging gives

$$J = \boldsymbol{f}_c^{\mathrm{H}} \boldsymbol{R}^{\mathrm{H}} \boldsymbol{R} \boldsymbol{f}_c + \boldsymbol{f}_c^{\mathrm{H}} \boldsymbol{R}^{\mathrm{H}} \boldsymbol{W}_p + \boldsymbol{W}_p^{\mathrm{H}} \boldsymbol{R} \boldsymbol{f}_c + \boldsymbol{W}_p^{\mathrm{H}} \boldsymbol{W}_p \tag{12.6.4}$$

which is a Hermitian quadratic form with a unique minimum (provided $\boldsymbol{R}^{\mathrm{H}}\boldsymbol{R}$ is positive). Finding the minimum of (12.6.4) results in the following optimal control force vector (the length of which is equal to the number of control forces which in this case is four):

$$\boldsymbol{f}_c = -(\boldsymbol{R}^{\mathrm{H}} \boldsymbol{R})^{-1} \boldsymbol{R}^{\mathrm{H}} \boldsymbol{W}_p \tag{12.6.5}$$

which is of the well known form

$$\boldsymbol{f}_c = -\boldsymbol{a}^{-1} \boldsymbol{b} \tag{12.6.6}$$

which was discussed in Chapter 8.

In practice, these optimal control forces can be determined by using a feedforward controller containing an adaptive filter, the weights of which are updated using a gradient descent algorithm as discussed in Chapter 6.

The reduction in vibration levels (averaged over 81 locations on the panel) for the system illustrated in Fig. 12.59 are shown in Fig. 12.60. Experimental data were obtained using a controller which minimized the acceleration levels at eight locations on the plate and the theoretical curves were obtained using finite element analysis.

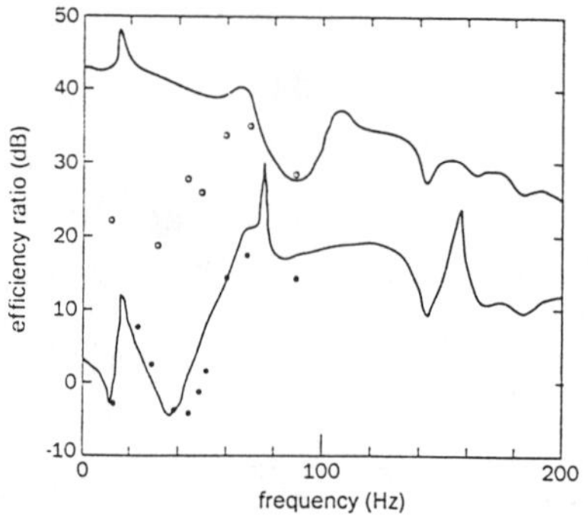

Fig. 12.60 Reduction in average plate acceleration levels for the four isolator system shown in Fig. 12.59 (after Jenkins *et al.*, 1993).
——— finite element analysis prediction for passive isolators only
- - - finite element analysis prediction for active plus passive isolation
• • • measured passive isolation
o o o measured active plus passive isolation

12.6.2 Generalized excitation forces

Unfortunately, the ideal situation discussed in the previous section in which only vertical forces are transmitted to the flexible support structure is difficult to realize in practice, although in some cases replacing the passive isolators with an air bag system to prevent transmission of horizontal forces and moments may be feasible (Jenkins *et al.*, 1990).

In this section, a complete analysis of the transmission of vibratory power from a rigid machine to a flexible, simply supported rectangular plate which takes into account all possible force and moment components is undertaken. We regard this as important, as in most practical systems, power transmission through an isolator can be very complicated, because equipment supported on isolators has many degrees of freedom. Thus, at the point of connection of each mount to the support structure various forces and moments exist which excite various types of waves in the support structure. In fact, the vibration of the machine and the support structure is coupled through the mounts and the modal response of the coupled system is directly associated with the magnitude of the power injected by each of the exciting forces and moments. In support of the preceding argument, White (1990) showed that the power transmission due to moment excitation induces more flexural wave power transmission into a flexible support panel than does excitation by normal forces at higher frequencies. Also, it is well known that longitudinal and shear waves (which do not radiate much noise or contribute significantly to feelable vibration levels on structures), can have large amounts of their energy converted to flexural waves at structural discontinuities; thus it is important to control the injection of power into these wave types as well.

Experimental work undertaken by Pinnington and White (1981), Pinnington (1987) and Pavic and Oreskovic (1977) indicates that vibratory power transmission through each isolator is a suitable cost function for minimizing the overall vibration response of the flexible support plate.

In the analysis to be outlined here, all of the power transmission components into the flexible support plate, associated with all the force and moment components from each elastic mount are taken into account. The total power transmission into the support plate from all of the mounts is analysed by identifying the vibration modes corresponding to the coupled suspension system. The machine is modelled as a six-degree-of-freedom rigid body which is supported by a flexible rectangular plate through multiple elastic mounts, as shown in Fig. 12.61.

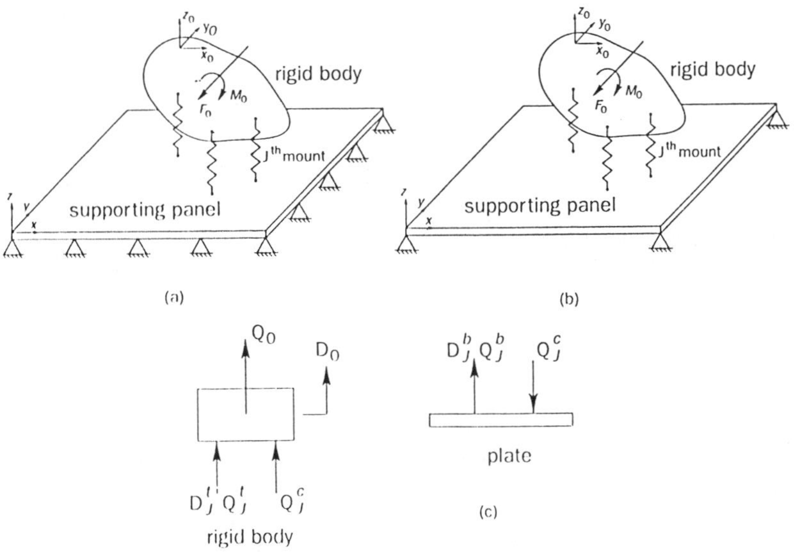

Fig. 12.61 Rigid body isolated from a flexible plate by multiple isolation mounts: (a) simply supported plate; (b) corner supported, free edge plate; (c) sign conventions showing directions of positive forces and displacements.

As shown in the figure, the plate is modelled with both simply supported and corner supported boundary conditions. The former edge condition is realized in practice by supporting the edge of the plate with a strip of thin shim spring steel attached to a rigid foundation, as illustrated in Fig. 12.62. The latter edge condition is realized in practice by supporting the plate at a point under each corner. The type of plate edge support only affects the plate mode shape functions as will become apparent in the analysis below.

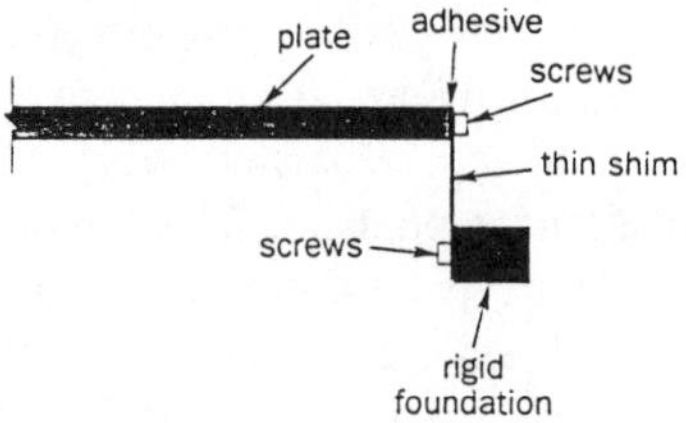

Fig. 12.62 Simply supported edge as realized in practice.

12.6.2.1 Rigid body equation of motion

To begin the analysis, we assume that the rigid body is excited by forces and moments which can be represented by a generalized force vector given by (12.5.1). An equation describing the motion of the rigid body can be derived using the same technique as used to derive the equation of motion (12.5.2) for the rigid body acted upon by a single isolator in Section 12.5. The only difference in this case is that multiple isolators are acting. Thus we obtain

$$\boldsymbol{Z}_0 \boldsymbol{D}_0 = \boldsymbol{Q}_0 + \sum_{J=1}^{L_1} \boldsymbol{R}_J^t \boldsymbol{Q}_J^t \tag{12.6.7}$$

where the variables without the subscript J were defined in Section 12.5, and L_I is the number of isolators. Note that the subscript J also needs to be added to the elements of $\boldsymbol{R}_J^t$.

12.6.2.2 Supporting plate equations of motion

At the bottom of each elastic mount, the supporting plate is driven by forces and moments, which for the Jth mount can be written as:

$$\boldsymbol{Q}_J^b = -\boldsymbol{Q}_J^t = \left[F_{xJ}^b,\ F_{yJ}^b,\ F_{zJ}^b,\ M_{xJ}^b,\ M_{yJ}^b,\ M_{zJ}^b\right]^{\mathrm{T}} \tag{12.6.8a,b}$$

Here, the force and moment components in $\boldsymbol{Q}_J^b$ are assumed to be concentrated at one point $\boldsymbol{\sigma}_J$ on the plate, so that Dirac delta functions can be used to describe the external forces and moments per unit area. The effect of the two shear forces at each mount on the plate vibration can be modelled as two moments around the x- and y-axes. The drilling degree of freedom of the thin plate is suppressed; therefore, the twisting moments M_{zJ}^b ($J = 1, \cdots, L_1$) on the plate surface do not generate any flexural waves in the plate. By considering the force (in the z-direction) and moments (around the x- and y-axes) equilibria, by including the influence of the external forces and moments and by modelling the mass loading m_J^b of each mount on the supporting plate at frequency ω as a concentrated inertial force $m_J^b \omega^2 w(\boldsymbol{\sigma}, \omega)\delta(\boldsymbol{\sigma} - \boldsymbol{\sigma}_J)$, the plate displacement, $w = w(\boldsymbol{\sigma},\omega)$ can be described by the following partial differential equation:

$$\rho h \frac{\partial^2 w}{\partial t^2} + \frac{Eh^3}{12(1-\nu^2)}\nabla^4 w = \sum_{J=1}^{L_1}\left[F_{zJ}^b \delta(\boldsymbol{\sigma} - \boldsymbol{\sigma}_J) + \frac{hF_{xJ}^b}{2}\frac{\partial\delta(\boldsymbol{\sigma} - \boldsymbol{\sigma}_J)}{\partial x} + \frac{hF_{yJ}^b}{2}\frac{\partial\delta(\boldsymbol{\sigma} - \boldsymbol{\sigma}_J)}{\partial x} - M_{xJ}^b\frac{\partial\delta(\boldsymbol{\sigma} - \boldsymbol{\sigma}_J)}{\partial y} + M_{yJ}^b\frac{\partial\delta(\boldsymbol{\sigma} - \boldsymbol{\sigma}_J)}{\partial x} + m_J^b\omega^2 w\,\delta(\boldsymbol{\sigma} - \boldsymbol{\sigma}_J)\right] \tag{12.6.9}$$

where ρ, E, ν and h are the density, Young's modulus of elasticity, Poisson's ratio and thickness respectively of the plate. The quantity $\boldsymbol{\sigma}_J = (x_J, y_J)$ is the location vector of the Jth mount on the supporting plate surface, and $\delta(\boldsymbol{\sigma} - \boldsymbol{\sigma}_J)$ is the Dirac delta function.

Using modal analysis, the plate displacement $w(\boldsymbol{\sigma}, \omega)$ can be expressed in terms of a mode shape matrix $\boldsymbol{\psi}_p(\boldsymbol{\sigma}) = [\psi_1(\boldsymbol{\sigma}), \psi_2(\boldsymbol{\sigma}), \cdots, \psi_P(\boldsymbol{\sigma})]^{\mathrm{T}}$ and a modal amplitude coefficient matrix $\boldsymbol{w}_p = [w_1, w_2, \cdots, w_p]^{\mathrm{T}}$ as:

$$w = w(\boldsymbol{\sigma}, \omega) = \boldsymbol{\psi}_p^{\mathrm{T}}(\boldsymbol{\sigma})\boldsymbol{w}_p \tag{12.6.10a,b}$$

where $\psi_I(\boldsymbol{\sigma})$ is the Ith plate mode shape function, which for a simply supported rectangular plate of dimensions L_x and L_y, is (Wallace 1972):

$$\psi_I(\boldsymbol{\sigma}) = \psi_{m,n}(x,y) = \sin\frac{m\pi x}{L_x}\sin\frac{n\pi y}{L_y} \tag{12.6.11}$$

where m and n are integers corresponding to the Ith mode order.

For a corner supported plate (that is, one which is point supported at each of its four corners), the mode shape functions may be expressed as (Reed, 1965):

$$\psi_I(\boldsymbol{\sigma}) = \sum_{m=0}\sum_{n=1}\left[a_{mn}\cos\frac{m\pi x}{L_x}\sin\frac{n\pi y}{L_y} + b_{mn}\cos\frac{m\pi y}{L_y}\sin\frac{n\pi x}{L_x}\right] \tag{12.6.12}$$

where in this case the coefficients a_{mn} and b_{mn} are different for each mode and dependent on the ratio L_x/L_y.

Substituting (12.6.10) into (12.6.9) and using the orthogonal property of the mode shape functions and Dirac delta functions, the coefficient w_I for the Ith plate mode becomes:

$$m_I(\omega_I^2 + j\eta_I\omega_I^2 - \omega^2)w_I = \sum_{J=1}^{L_1}\left[\frac{h}{2}\psi_{Ix}(\sigma_J), -\frac{h}{2}\psi_{Iy}(\sigma_J), \psi_I(\sigma_J), -\psi_{Iy}(\sigma_J), -\psi_{Ix}(\sigma_J), 0\right]\boldsymbol{Q}_J^t$$

$$+ \sum_{I'=1}^{P} C_{I',I} w_{I'} \qquad (I = 1, \cdots, P) \tag{12.6.13}$$

where the modal masses m_I are defined as:

$$m_I = \rho h \int_{A_f} \psi_I^2(\sigma)\,\mathrm{d}\sigma \tag{12.6.14a,b}$$

For a simply rectangular plate, the right side of (12.6.14b) is $\rho L_x L_y h/4$ and the angular resonance frequency, ω_I, of the Ith plate mode (with modal indices (m, n)) is given by

$$\omega_I = h\left[\left(\frac{m\pi}{L_x}\right)^2 + \left(\frac{n\pi}{L_y}\right)^2\right]\left[\frac{E}{12\rho(1-\nu^2)}\right]^{1/2} \tag{12.6.15}$$

For a corner supported plate, the resonance frequencies must be calculated using the Rayleigh–Ritz method as described by Reed (1965).

In the preceding equations, ρ, L_x, L_y and h are respectively the density, length dimensions and thickness of a rectangular plate of area A_f, and η_I is the loss factor of the Ith plate mode, defined by (12.5.14). $C_{I',I}$ $(I', I = 1, \cdots, P)$ is the concentrated mass contribution given by:

$$C_{I',I} = \sum_{J=1}^{L_1} m_J^b \omega^2 \psi_{I'}(\sigma_J)\,\psi_I(\sigma_J) \tag{12.6.16}$$

When I in (12.6.13) increments from 1 to P, P simultaneous equations can be obtained for the coefficients $w_1, w_2, \cdots, w_P$, and the matrix equation for $\boldsymbol{w}_p$ is obtained as:

$$\boldsymbol{Z}_p \boldsymbol{w}_p = \sum_{J=1}^{L_1} \boldsymbol{R}_J^b \boldsymbol{Q}_J^b \tag{12.6.17}$$

where $\boldsymbol{Z}_p$ is the uncoupled plate characteristic matrix including the influence of the concentrated masses of the mounts and is given by:

$$\boldsymbol{Z}_p = \begin{bmatrix} \chi_1 - C_{1,1} & -C_{1,2} & -C_{1,3} & \cdots & -C_{1,P} \\ -C_{2,1} & \chi_2 - C_{2,2} & -C_{2,3} & \cdots & -C_{2,P} \\ \vdots & \vdots & \vdots & \vdots & \vdots \\ -C_{P,1} & -C_{P,2} & -C_{P,3} & \cdots & \chi_P - C_{P,P} \end{bmatrix} \tag{12.6.18}$$

where χ_I $(I = 1, \cdots, P)$ is defined as:

$$\chi_I = m_I(\omega_I^2 + j\eta_I\omega_I^2 - \omega^2) \tag{12.6.19}$$

The quantity $\boldsymbol{R}_J^b$ is the force coupling matrix (on the supporting plate) for the Jth mount and is defined as follows:

$$\boldsymbol{R}_J^b = \begin{bmatrix} -\frac{h}{2}\psi_{1x}(\boldsymbol{\sigma}_J) & -\frac{h}{2}\psi_{1y}(\boldsymbol{\sigma}_J) & \psi_1(\boldsymbol{\sigma}_J) & \psi_{1y}(\boldsymbol{\sigma}_J) & -\psi_{1x}(\boldsymbol{\sigma}_J) & 0 \\ -\frac{h}{2}\psi_{2x}(\boldsymbol{\sigma}_J) & -\frac{h}{2}\psi_{2y}(\boldsymbol{\sigma}_J) & \psi_2(\boldsymbol{\sigma}_J) & \psi_{2y}(\boldsymbol{\sigma}_J) & -\psi_{2x}(\boldsymbol{\sigma}_J) & 0 \\ \vdots & \vdots & \vdots & \vdots & \vdots & \vdots \\ -\frac{h}{2}\psi_{Px}(\boldsymbol{\sigma}_J) & -\frac{h}{2}\psi_{Py}(\boldsymbol{\sigma}_J) & \psi_P(\boldsymbol{\sigma}_J) & \psi_{Py}(\boldsymbol{\sigma}_J) & -\psi_{Px}(\boldsymbol{\sigma}_J) & 0 \end{bmatrix} \tag{12.6.20}$$

For the simply supported rectangular plate and at $\boldsymbol{\sigma}_J = (x_{bJ}, y_{bJ})$, the functions $\psi_{Ix}(\boldsymbol{\sigma}_J)$ and $\psi_{Iy}(\boldsymbol{\sigma}_J)$ ($I = 1, \cdots, P$) are respectively:

$$\psi_{Ix}(\boldsymbol{\sigma}_J) = \frac{\partial \psi_I(\boldsymbol{\sigma}_J)}{\partial x} = \frac{m\pi}{L_x}\cos\frac{m\pi x_{bJ}}{L_x}\sin\frac{n\pi y_{bJ}}{L_y} \tag{12.6.21a,b}$$

$$\psi_{Iy}(\boldsymbol{\sigma}_J) = \frac{\partial \psi_I(\boldsymbol{\sigma}_J)}{\partial y} = \frac{n\pi}{L_y}\sin\frac{m\pi x_{bJ}}{L_x}\cos\frac{n\pi y_{bJ}}{L_y} \tag{12.6.22a,b}$$

For the corner supported panel, the functions are:

$$\psi_{Ix}(\boldsymbol{\sigma}_J) = \frac{\pi}{L_x}\sum_{m=0}^{\infty}\sum_{n=1}^{\infty} \times \left[-ma_{mn}\sin\frac{m\pi x_{bJ}}{L_x}\sin\frac{n\pi y_{bJ}}{L_y} + nb_{mn}\cos\frac{m\pi y_{bJ}}{L_y}\cos\frac{n\pi x_{bJ}}{L_x}\right] \tag{12.6.23}$$

$$\psi_{Iy}(\boldsymbol{\sigma}_J) = \frac{\pi}{L_y}\sum_{m=0}^{\infty}\sum_{n=1}^{\infty} \times \left[na_{mn}\cos\frac{m\pi x_{bJ}}{L_x}\cos\frac{n\pi y_{bJ}}{L_y} - mb_{mn}\sin\frac{m\pi y_{bJ}}{L_y}\sin\frac{n\pi x_{bJ}}{L_x}\right] \tag{12.6.24}$$

12.6.2.3 System equations of motion

Equations (12.6.7) and (12.6.17) are coupled through the elastic force vectors $\boldsymbol{Q}_J^b$ and $\boldsymbol{Q}_J^t$, ($J = 1, \cdots, L_1$) at the mounts. These force vectors are related to the relative displacements between the top and bottom ends of the mounts as described by (12.5.22) with the subscript J added to all quantities. The local displacements of the top and bottom surfaces of each mount can be related to the displacement vector $\boldsymbol{D}_0$ of the rigid body and $\boldsymbol{w}_p$ of the plate, because the local top displacements of the Jth mount are related to the rigid body displacement vectors $\boldsymbol{s}_0$ and $\boldsymbol{\theta}_0$ by:

$$\left[\xi_{xJ}^t, \xi_{yJ}^t, w_J^t\right]^{\mathrm{T}} = s_0 + \theta_0 \times r_{0J}. \tag{12.6.25}$$

$$\left[\theta_{xJ}^t, \theta_{yJ}^t, \theta_{zJ}^t\right]^{\mathrm{T}} = \theta_0 \tag{12.6.26}$$

where $s_0 = [x^0, y^0, z^0]^{\mathrm{T}}$ and $\theta_0 = [\theta_x^0, \theta_y^0, \theta_z^0]^{\mathrm{T}}$.

The corresponding matrix form for displacements at the top of the mount is:

$$D_J^t = (R_J^t)^{\mathrm{T}} D_0 \tag{12.6.27}$$

where R^t is defined in (12.5.5).

The local bottom displacements D_J^b of the Jth mount are related to the plate displacement vector $w(\sigma_J)$ at σ_J by:

$$D_J^b = \left[\xi_{xJ}^b, \xi_{yJ}^b, w_J^b, \theta_{xJ}^b, \theta_{yJ}^b, \theta_{zJ}^b\right]^{\mathrm{T}} = \left[-\frac{h}{2}\frac{\partial w}{\partial x}, -\frac{h}{2}\frac{\partial w}{\partial y}, w, \frac{\partial w}{\partial y}, -\frac{\partial w}{\partial x}, 0\right]^{\mathrm{T}} \tag{12.6.28a,b}$$

Substituting (12.6.10) into (12.6.28) allows the following matrix relation for displacements at the bottom of the mount to be obtained:

$$D_J^b = (R_J^b)^{\mathrm{T}} w_p \tag{12.6.29}$$

where R_J^b is defined in (12.6.20).

Consideration of the difference in displacement between the top and bottom of the isolators allows the following expression to be written:

$$Q_J^t = -Q_J^b = K_J (D_J^b - D_J^t) \tag{12.6.30}$$

where

$$K_J = \begin{bmatrix} k_{J1}(1 + j\eta_{J1}) & & & \\ & k_{J2}(1 + j\eta_{J2}) & & \\ & & \ddots & \\ & & & k_{J6}(1 + j\eta_{J6}) \end{bmatrix} \tag{12.6.31}$$

Substituting (12.5.5) (with the subscript J added so that it applies for each mount), (12.6.8), (12.6.20), (12.6.27) and (12.6.29) into (12.6.7) and (12.6.17), gives a matrix equation describing the response of the coupled system as follows:

$$\begin{bmatrix} A_{11} & A_{12} \\ A_{21} & A_{22} \end{bmatrix} \begin{bmatrix} D_0 \\ w_p \end{bmatrix} = \begin{bmatrix} Q_0 \\ 0 \end{bmatrix} \tag{12.6.32}$$

where $\mathbf{0}$ is a $(P \times 1)$ zero vector and

$$A_{11} = Z_0 + \sum_{J=1}^{L_1} R_J^t K_J (R_J^t)^{\mathrm{T}} \tag{12.6.33}$$

$$A_{12} = -\sum_{j=1}^{L_1} R_J^t K_J (R_J^b)^{\mathrm{T}} \tag{12.6.34}$$

$$A_{21} = -\sum_{j=1}^{L_1} R_J^b K_J (R_J^t)^{\mathrm{T}} \tag{12.6.35}$$

$$A_{22} = Z_p + \sum_{J=1}^{L_1} R_J^b K_J (R_J^b)^{\mathrm{T}} \tag{12.6.36}$$

If $Q_0 = 0$, (12.6.32) describes the free vibration behaviour of the coupled system, the mode shapes and system resonance frequencies of which can be determined by solving this free vibration problem. For a given external forcing function acting on the rigid body, Q_0, the displacement response at any location can be calculated from (12.6.32).

A passive isolator can be made active by introducing a force actuator (with a frequency response extending above the frequency range of the excitation to be isolated) connected in parallel with it. When used with a suitable feedforward control system, the actuator exerts a control force on the rigid body and the support point on the panel simultaneously. For active isolators acting at the same locations as the passive isolators, the right part of (12.6.32) is replaced by a combined force vector, and the equation of motion becomes:

$$\begin{bmatrix} A_{11} & A_{12} \\ A_{21} & A_{22} \end{bmatrix} \begin{bmatrix} D_0 \\ w_p \end{bmatrix} = \begin{bmatrix} Q_0 + \sum_{J=1}^{L_1} R_J^t Q_J^c \\ -\sum_{J=1}^{L_1} R_J^b Q_J^c \end{bmatrix} \tag{12.6.37}$$

where Q_J^c are the control force vectors generated by each actuator acting on the support panel.

12.6.2.4 Optimal control forces and minimum power transmission

As before, the cost function chosen for minimization is the time average power transmission into the supporting plate through the isolators, and can be expressed as follows:

$$P_0 = \mathrm{Re}\left\{ -\frac{1}{2} j\omega \sum_{J=1}^{L_1} (Q_J^b - Q_J^c)^{\mathrm{H}} D_J^b \right\} \tag{12.6.38a}$$

$$= \frac{\omega}{2} \mathrm{Im}\left\{ \sum_{J=1}^{L_1} (Q_J^b - Q_J^c)^{\mathrm{H}} D_J^b \right\} \tag{12.6.38b}$$

$$= -\frac{\omega}{2}\,\mathrm{Im}\left\{\sum_{J=1}^{L_1}(\boldsymbol{D}_J^b)^{\mathrm{H}}(\boldsymbol{Q}_J^b - \boldsymbol{Q}_J^c)\right\} \tag{12.6.38c}$$

Sign conventions are the same as those used for the beam problem of Section 12.5.

The time average power transmission from the rigid body into the isolators can be described as:

$$P_t = \mathrm{Re}\left\{-\frac{1}{2}j\omega\sum_{J=1}^{L_1}(\boldsymbol{Q}_J^t + \boldsymbol{Q}_J^c)^{\mathrm{H}}\boldsymbol{D}_J^t\right\} \tag{12.6.39}$$

and the input power due to the external driving force is given by:

$$P_i = \mathrm{Re}\left\{-\frac{1}{2}j\omega\boldsymbol{Q}_0^{\mathrm{T}}\boldsymbol{D}_0\right\} \tag{12.6.40}$$

It is possible to express the total output power P_0 by using an explicit quadratic function. Thus, the term $\sum_{J=1}^{L_1}(\boldsymbol{D}_J^b)^{\mathrm{H}}(\boldsymbol{Q}_J^b - \boldsymbol{Q}_J^c)$ in (12.6.38) can be expressed as:

$$\sum_{J=1}^{L_1}(\boldsymbol{D}_J^b)^{\mathrm{H}}(\boldsymbol{Q}_J^b - \boldsymbol{Q}_J^c) = (\boldsymbol{Q}^c)^{\mathrm{H}}\boldsymbol{a}\boldsymbol{Q}^c + (\boldsymbol{Q}^c)^{\mathrm{H}}\boldsymbol{b}_1 + \boldsymbol{b}_2\boldsymbol{Q}^c + \boldsymbol{c} \tag{12.6.41}$$

where $\boldsymbol{Q}^c = [\boldsymbol{Q}_1^c, \boldsymbol{Q}_2^c, \cdots, \boldsymbol{Q}_J^c]^{\mathrm{T}}$.

The form of the matrices $\boldsymbol{a}$, $\boldsymbol{b}_1$, $\boldsymbol{b}_2$ and $\boldsymbol{c}$ is dependent on the number of isolators used and to make the solution of this problem tractable, we will consider only four isolators between the rigid body and the support plate; that is, $L_1 = 4$. In this case, the matrices $\boldsymbol{a}$, $\boldsymbol{b}_1$, $\boldsymbol{b}_2$, and $\boldsymbol{c}$ may be defined as follows:

$$\sum_{J=1}^{4}(\boldsymbol{D}_J^b)^{\mathrm{H}}(\boldsymbol{Q}_J^b - \boldsymbol{Q}_J^c) = (\boldsymbol{Q}^c)^{\mathrm{H}}\boldsymbol{a}\boldsymbol{Q}^c + (\boldsymbol{Q}^c)^{\mathrm{H}}\boldsymbol{b}_1 + \boldsymbol{b}_2\boldsymbol{Q}^c + \boldsymbol{c} \tag{12.6.42}$$

where

$$\boldsymbol{a} = \boldsymbol{G}_1\boldsymbol{K}\boldsymbol{G}_2 - \boldsymbol{G}_1 \tag{12.6.43}$$

$$\boldsymbol{b}_1 = \boldsymbol{G}_1\boldsymbol{K}\boldsymbol{G}_3 \tag{12.6.44}$$

$$\boldsymbol{b}_2 = \boldsymbol{G}_4\boldsymbol{K}\boldsymbol{G}_2 - \boldsymbol{G}_4 \tag{12.6.45}$$

$$\boldsymbol{c} = \boldsymbol{G}_4\boldsymbol{K}\boldsymbol{G}_3 \tag{12.6.46}$$

and

$$G_1 = \begin{bmatrix} \boldsymbol{H}_1^{\mathrm{H}}\boldsymbol{R}_1^b & \boldsymbol{H}_1^{\mathrm{H}}\boldsymbol{R}_2^b & \boldsymbol{H}_1^{\mathrm{H}}\boldsymbol{R}_3^b & \boldsymbol{H}_1^{\mathrm{H}}\boldsymbol{R}_4^b \\ \boldsymbol{H}_2^{\mathrm{H}}\boldsymbol{R}_1^b & \boldsymbol{H}_2^{\mathrm{H}}\boldsymbol{R}_2^b & \boldsymbol{H}_2^{\mathrm{H}}\boldsymbol{R}_3^b & \boldsymbol{H}_2^{\mathrm{H}}\boldsymbol{R}_4^b \\ \boldsymbol{H}_3^{\mathrm{H}}\boldsymbol{R}_1^b & \boldsymbol{H}_3^{\mathrm{H}}\boldsymbol{R}_2^b & \boldsymbol{H}_3^{\mathrm{H}}\boldsymbol{R}_3^b & \boldsymbol{H}_3^{\mathrm{H}}\boldsymbol{R}_4^b \\ \boldsymbol{H}_4^{\mathrm{H}}\boldsymbol{R}_1^b & \boldsymbol{H}_4^{\mathrm{H}}\boldsymbol{R}_2^b & \boldsymbol{H}_4^{\mathrm{H}}\boldsymbol{R}_3^b & \boldsymbol{H}_4^{\mathrm{H}}\boldsymbol{R}_4^b \end{bmatrix} \tag{12.6.47}$$

$$G_2 =$$

$$\begin{bmatrix} (\boldsymbol{R}_1^t)^{\mathrm{T}}\boldsymbol{H}_5 - (\boldsymbol{R}_1^b)^{\mathrm{T}}\boldsymbol{H}_1 & (\boldsymbol{R}_1^t)^{\mathrm{T}}\boldsymbol{H}_6 - (\boldsymbol{R}_1^b)^{\mathrm{T}}\boldsymbol{H}_2 & (\boldsymbol{R}_1^t)^{\mathrm{T}}\boldsymbol{H}_7 - (\boldsymbol{R}_1^b)^{\mathrm{T}}\boldsymbol{H}_3 & (\boldsymbol{R}_1^t)^{\mathrm{T}}\boldsymbol{H}_8 - (\boldsymbol{R}_1^b)^{\mathrm{T}}\boldsymbol{H}_4 \\ (\boldsymbol{R}_2^t)^{\mathrm{T}}\boldsymbol{H}_5 - (\boldsymbol{R}_2^b)^{\mathrm{T}}\boldsymbol{H}_1 & (\boldsymbol{R}_2^t)^{\mathrm{T}}\boldsymbol{H}_6 - (\boldsymbol{R}_2^b)^{\mathrm{T}}\boldsymbol{H}_2 & (\boldsymbol{R}_2^t)^{\mathrm{T}}\boldsymbol{H}_7 - (\boldsymbol{R}_2^b)^{\mathrm{T}}\boldsymbol{H}_3 & (\boldsymbol{R}_2^t)^{\mathrm{T}}\boldsymbol{H}_8 - (\boldsymbol{R}_2^b)^{\mathrm{T}}\boldsymbol{H}_4 \\ (\boldsymbol{R}_3^t)^{\mathrm{T}}\boldsymbol{H}_5 - (\boldsymbol{R}_3^b)^{\mathrm{T}}\boldsymbol{H}_1 & (\boldsymbol{R}_3^t)^{\mathrm{T}}\boldsymbol{H}_6 - (\boldsymbol{R}_3^b)^{\mathrm{T}}\boldsymbol{H}_2 & (\boldsymbol{R}_3^t)^{\mathrm{T}}\boldsymbol{H}_7 - (\boldsymbol{R}_3^b)^{\mathrm{T}}\boldsymbol{H}_3 & (\boldsymbol{R}_3^t)^{\mathrm{T}}\boldsymbol{H}_8 - (\boldsymbol{R}_3^b)^{\mathrm{T}}\boldsymbol{H}_4 \\ (\boldsymbol{R}_4^t)^{\mathrm{T}}\boldsymbol{H}_5 - (\boldsymbol{R}_4^b)^{\mathrm{T}}\boldsymbol{H}_1 & (\boldsymbol{R}_4^t)^{\mathrm{T}}\boldsymbol{H}_6 - (\boldsymbol{R}_4^b)^{\mathrm{T}}\boldsymbol{H}_2 & (\boldsymbol{R}_4^t)^{\mathrm{T}}\boldsymbol{H}_7 - (\boldsymbol{R}_4^b)^{\mathrm{T}}\boldsymbol{H}_3 & (\boldsymbol{R}_4^t)^{\mathrm{T}}\boldsymbol{H}_8 - (\boldsymbol{R}_4^b)^{\mathrm{T}}\boldsymbol{H}_4 \end{bmatrix} \tag{12.6.48}$$

$$G_3 = \begin{bmatrix} (\boldsymbol{R}_1^t)^{\mathrm{T}}\boldsymbol{B}_{11} - (\boldsymbol{R}_1^b)^{\mathrm{T}}\boldsymbol{B}_{21} \\ (\boldsymbol{R}_2^t)^{\mathrm{T}}\boldsymbol{B}_{11} - (\boldsymbol{R}_2^b)^{\mathrm{T}}\boldsymbol{B}_{21} \\ (\boldsymbol{R}_3^t)^{\mathrm{T}}\boldsymbol{B}_{11} - (\boldsymbol{R}_3^b)^{\mathrm{T}}\boldsymbol{B}_{21} \\ (\boldsymbol{R}_4^t)^{\mathrm{T}}\boldsymbol{B}_{11} - (\boldsymbol{R}_4^b)^{\mathrm{T}}\boldsymbol{B}_{21} \end{bmatrix} \boldsymbol{Q}_0 \tag{12.6.49}$$

$$G_4 = \boldsymbol{Q}_0^{\mathrm{T}}\begin{bmatrix} \boldsymbol{B}_{21}^{\mathrm{H}}\boldsymbol{R}_1^b & \boldsymbol{B}_{21}^{\mathrm{H}}\boldsymbol{R}_2^b & \boldsymbol{B}_{21}^{\mathrm{H}}\boldsymbol{R}_3^b & \boldsymbol{B}_{21}^{\mathrm{H}}\boldsymbol{R}_4^b \end{bmatrix} \tag{12.6.50}$$

$$\boldsymbol{K} = \begin{bmatrix} \boldsymbol{K}_1 & & & \\ & \boldsymbol{K}_2 & & \\ & & \boldsymbol{K}_3 & \\ & & & \boldsymbol{K}_4 \end{bmatrix} \tag{12.6.51}$$

where matrices $H_1 - H_8$ are defined as follows:

$$\boldsymbol{H}_1 = \boldsymbol{B}_{21}\boldsymbol{R}_1^t - \boldsymbol{B}_{22}\boldsymbol{R}_1^b \tag{12.6.52}$$

$$\boldsymbol{H}_2 = \boldsymbol{B}_{21}\boldsymbol{R}_2^t - \boldsymbol{B}_{22}\boldsymbol{R}_2^b \tag{12.6.53}$$

$$\boldsymbol{H}_3 = \boldsymbol{B}_{21}\boldsymbol{R}_3^t - \boldsymbol{B}_{22}\boldsymbol{R}_3^b \tag{12.6.54}$$

$$\boldsymbol{H}_4 = \boldsymbol{B}_{21}\boldsymbol{R}_4^t - \boldsymbol{B}_{22}\boldsymbol{R}_4^b \tag{12.6.55}$$

$$\boldsymbol{H}_5 = \boldsymbol{B}_{11}\boldsymbol{R}_1^t - \boldsymbol{B}_{12}\boldsymbol{R}_1^b \tag{12.6.56}$$

$$H_6 = B_{11} R_2^t - B_{12} R_2^b \tag{12.6.57}$$

$$H_7 = B_{11} R_3^t - B_{12} R_3^b \tag{12.6.58}$$

$$H_8 = B_{11} R_4^t - B_{12} R_4^b \tag{12.6.59}$$

and the matrices $\boldsymbol{B}_{11}$, $\boldsymbol{B}_{12}$, $\boldsymbol{B}_{21}$, $\boldsymbol{B}_{22}$ in the above expressions are submatrix elements of the inverse of the system matrix $\boldsymbol{A}$, and are defined in (12.5.35) to (12.5.39).

The control force vector $\boldsymbol{Q}^c$ of dimension (24 × 1) in (12.6.42) is a combined vector of control forces as follows:

$$\boldsymbol{Q}^c = \left[\boldsymbol{Q}_1^c \;\; \boldsymbol{Q}_2^c \;\; \boldsymbol{Q}_3^c \;\; \boldsymbol{Q}_4^c\right]^{\mathrm{T}} \tag{12.6.60}$$

Thus, the total time average output power P_0 can be expressed as the following real quadratic function:

$$P_0 = \mathrm{Re}\left\{-\frac{1}{2} j\omega \sum_{J=1}^{4} (\boldsymbol{Q}_J^b - \boldsymbol{Q}_J^c)^{\mathrm{H}} \boldsymbol{D}_J^b\right\} = -\frac{\omega}{2} \mathrm{Im}\left\{\sum_{J=1}^{4} (\boldsymbol{D}_J^b)^{\mathrm{H}} (\boldsymbol{Q}_J^b - \boldsymbol{Q}_J^c)\right\}$$

$$= -\frac{\omega}{2}\left\{(\boldsymbol{q}^c)^{\mathrm{T}} \begin{bmatrix} \boldsymbol{a}^i & \boldsymbol{a}^r \\ -\boldsymbol{a}^r & \boldsymbol{a}^i \end{bmatrix} \boldsymbol{q}^c + (\boldsymbol{q}^c)^{\mathrm{T}} \begin{bmatrix} \boldsymbol{b}_1^i \\ -\boldsymbol{b}_1^r \end{bmatrix} + \left[\boldsymbol{b}_2^i \;\; \boldsymbol{b}_2^r\right] \boldsymbol{q}^c + c^i\right\} \tag{12.6.61a,b,c}$$

Following the same procedure as outlined in Section 12.5, (12.5.49) to (12.5.55) (noting that the control force now consists of four vectors as shown in (12.6.60), we can obtain an expression for the optimal control forces (identical to (12.5.55)) and the maximum achievable reduction in power transmission into the plate as a result of active control, the latter quantity being given by,

$$(\Delta P)_{\max} = \frac{\omega}{2} \beta^{\mathrm{T}} \alpha^{-1} \beta \tag{12.6.62}$$

From the control force vector $\boldsymbol{q}_{\mathrm{opt}}^c$ which results in the maximum power reduction, the amplitudes and phases of a set of optimal active control forces $(\boldsymbol{Q}_1^c)_{\mathrm{opt}}$, $(\boldsymbol{Q}_2^c)_{\mathrm{opt}}$, $(\boldsymbol{Q}_3^c)_{\mathrm{opt}}$, and $(\boldsymbol{Q}_4^c)_{\mathrm{opt}}$, can be obtained.

For a two-isolator system, the preceding analysis can also be used, except that the matrices need to be reduced in size to exclude the contributions from the terms represented by $J = 3, 4$.

The minimum power transmission for multiple mounts is theoretically zero if the control forces act in the same axes as the excitation forces. The preceding analysis, however, allows evaluation of cases where the controller is non-ideal or where the control force along a particular axis is limited or unavailable.

12.6.3 Rigid mass as the intermediate structure

The use of an intermediate rigid mass between a vibrating rigid body and a flexible support structure as a means of improving the vibration isolation of a passive system was mentioned in Section 12.2 where feedback control of the motion of the rigid mass was considered. For periodic excitation, the vibration of the rigid mass could be controlled using a feedforward system, the idea being that control of the rigid mass vibration will minimize the vibration transmitted to the support structure.

Jenkins *et al.* (1990) reported on the use of four isolators involving an intermediate rigid mass for the isolation of vibration of a diesel engine from a ribbed steel plate. The lower elastic mount beneath the rigid intermediate mass was an air bag spring which had little shear stiffness, resulting in no measurable transmission of rotational or horizontal translational forces. Thus the active system was only required to control the transmission of vertical forces. A schematic representation of the actuator used is shown in Fig. 12.63.

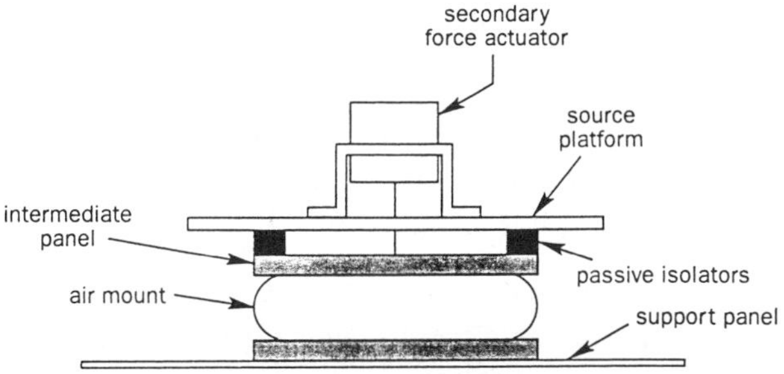

Fig. 12.63 Active isolator with an air bag suspension (after Jenkins *et al.*, 1990).

Good isolation (12 dB) was achieved by Jenkins *et al.* (1990) for the sixth harmonic of the diesel engine rotational speed. However, results obtained for the other harmonics were disappointing as the electrodynamic shaker used did not have a high enough force capability. An electrodynamic shaker of sufficient size would have weighed half as much as the diesel engine. Thus it has been suggested that the upper resilient mount should be redesigned to reduce vertical stiffness while retaining the shear stiffness which is necessary to maintain alignment of the shaker and its attachment point on the intermediate mass. This would result in a smaller force requirement of the shaker and potentially better performance in regard to isolation of the more important low order harmonics of the engine rotational speed.

12.7 FEEDFORWARD CONTROL: MULTIPLE ISOLATORS BETWEEN A RIGID BODY AND A FLEXIBLE CYLINDER

The particular case which will now be analysed is illustrated in Fig. 12.64. More details of the analysis and some numerical results are given by Pan, Howard and Hansen (1996).

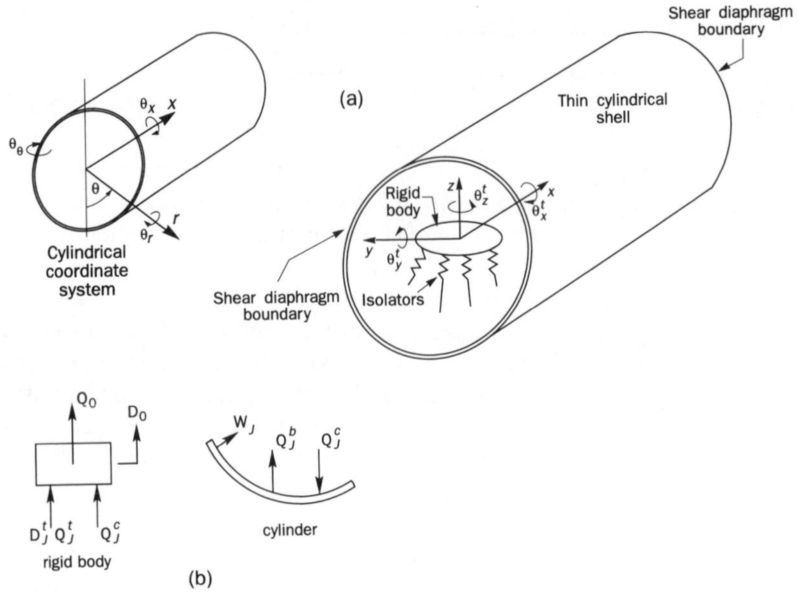

Fig. 12.64 Rigid body isolated from a flexible cylinder using multiple mounts: (a) theoretical model; (b) sign conventions showing directions of positive forces and displacements.

As in the previous section, the cost function to be minimized will be the total vibratory power transmission into the flexible cylinder. The active isolators act on the rigid body and react on the support cylinder and each acts in parallel with a passive element consisting of a stiffness and a viscous damper.

12.7.1 Rigid body equation of motion

The same equation of motion as used previously for the rigid body with a flexible plate support (12.6.7) also describes the rigid body motion in this case.

12.7.2 Supporting thin cylinder equations of motion

The supporting cylinder is driven by L_1 force vectors, defined as $\boldsymbol{Q}_J^b = [F_{xJ}^b, F_{yJ}^b, F_{zJ}^b, M_{xJ}^b, M_{yJ}^b, M_{zJ}^b]^{\mathrm{T}}$ in the cartesian coordinate system or $\boldsymbol{Q}_J = [F_{xJ}, F_{\theta J}, F_{rJ}, M_{xJ}, M_{\theta J}, M_{rJ}]^{\mathrm{T}}$ in the cylindrical coordinate system as shown in Fig. 12.64, where $J = 1, \cdots, L_1$. The force vectors acting on the top of each isolator are related to those acting on the bottom as follows:

$$\boldsymbol{Q}_J^b = -\boldsymbol{Q}_J^t \tag{12.7.1}$$

and the forces expressed in the cylindrical coordinate system are related to those expressed in cartesian coordinate system as follows:

$$Q_J = T_J Q_J^b \tag{12.7.2}$$

where T_J is a coordinate transformation matrix between the cartesian and cylindrical coordinates of the support point σ_J on the internal surface of the cylinder at $r = R-h/2$, and is defined as follows:

$$T_J = \begin{bmatrix} T_{0J} & \mathbf{0} \\ \mathbf{0} & T_{0J} \end{bmatrix} \tag{12.7.3}$$

where $\mathbf{0}$ is a 3 × 3 zero-order matrix and

$$T_{0J} = \begin{bmatrix} 1 & 0 & 0 \\ 0 & \cos\theta_J & -\sin\theta_J \\ 0 & \sin\theta_J & \cos\theta_J \end{bmatrix} \tag{12.7.4}$$

It can be shown that $(T_J)^{-1} = (T_J)^{T}$. The force and moment components in Q_J^b or Q_J are also assumed to be concentrated point actions at support points σ_J on the thin shell, so that Dirac delta functions and their partial derivatives can be used to describe the external force distribution on the inner surface of the cylinder. Compared with the radius R, the thickness h of the cylinder is sufficiently small for the vibration of the cylinder to be primarily radial, with the axial x and tangential θ displacements being small enough to allow the corresponding inertia terms in the axial and tangential directions in the equation of motion of the cylindrical shell to be neglected. Forces acting in the axial and tangential directions excite vibrational displacements in these directions which in turn give rise to some radial motion as a result of coupling between the different directions of motion. However, the radial vibration produced in this way may be considered small compared to that which is produced directly by moments and forces acting in the radial direction. The assumption of small indirect radial vibration is important because it simplifies the analysis enormously (allowing the RHS of the first two of the following three equations of motion to be set equal to zero); nevertheless it is possible that including the effect of indirect radial vibration could lead to more accurate results. Note that the axial and tangential forces produced on the inside of the cylinder at the base of the mounts will produce some direct radial displacement due to the induced moment about the centre of the cylinder thickness. This is taken into account in the following analysis.

The bending displacement $w(\theta,s:t)$ of the shell and resulting tangential displacements $\xi_s(s,\theta;t)$ in the x direction and $\xi_\theta(s,\theta;t)$ in the θ direction (where $s = x/R$) can be then described by the following equations (Donnell–Mushtari shell theory):

$$\frac{\partial^2\xi_s}{\partial s^2} + \frac{(1-\nu)}{2}\frac{\partial^2\xi_s}{\partial\theta^2} + \frac{(1+\nu)}{2}\frac{\partial^2\xi_\theta}{\partial s\,\partial\theta} + \nu\frac{\partial w}{\partial s} = 0 \tag{12.7.5}$$

$$\frac{(1+\nu)}{2}\frac{\partial^2 \xi_s}{\partial s \partial \theta} + \frac{(1-\nu)}{2}\frac{\partial^2 \xi_\theta}{\partial s^2} + \frac{\partial^2 \xi_\theta}{\partial \theta^2} + \frac{\partial w}{\partial \theta} = 0 \tag{12.7.6}$$

$$\nu\frac{\partial \xi_s}{\partial s} + \frac{\partial \xi_\theta}{\partial \theta} + w + \kappa \nabla^4 w + \frac{\rho(1-\nu^2)R^2}{E}\frac{\partial^2 w}{\partial t^2} = \frac{(1-\nu^2)}{Eh}\left[q_1(x,\theta) + q_2(x,\theta)\right]e^{j\omega t} \tag{12.7.7}$$

where the gradient operator is defined as

$$\nabla^4 = \nabla^2\nabla^2 = \left\{\frac{\partial^2}{\partial s^2} + \frac{\partial^2}{\partial \theta^2}\right\}^2$$

The quantities $\xi_s(s,\theta;t)$, $\xi_\theta(s,\theta;t)$ and w are the orthogonal components of the displacement in the axial x, circumferential θ and radial w directions and the force distribution functions q_1 , q_2 in the right side of (12.7.7) are:

$$q_1 = \sum_{J-1}^{L_1} m_J^b \omega^2 w(\sigma_J)\,\delta(\sigma - \sigma_J) \tag{12.7.8}$$

$$q_2 = p(s,\theta) \tag{12.7.9}$$

where L_1 is the number of shell modes considered and where ρ, E and ν are respectively the density, Young's modulus and Poisson's ratio of the shell material, R and h are respectively the radius and thickness of the cylinder, κ is a dimensionless shell thickness parameter defined as $\kappa = h^2/(12R^2)$, m_J^b is a concentrated mass modelling the base at the bottom of the Jth mount, and the forcing function, $p(x,\theta)$ can be expressed as follows:

$$p(s,\theta) = \sum_{J-1}^{L_1}\left[-\frac{h}{2}\frac{\partial\delta(\sigma-\sigma_J)}{\partial x}, -\frac{h}{2R}\frac{\partial\delta(\sigma-\sigma_J)}{\partial\theta}, \delta(\sigma-\sigma_J), -\frac{1}{R}\frac{\partial\delta(\sigma-\sigma_J)}{\partial\theta}, \frac{\partial\delta(\sigma-\sigma_J)}{\partial x}, 0\right]\boldsymbol{Q}_J$$

$$= \sum_{J-1}^{L_1}\left[-\frac{h}{2}\frac{\partial\delta(\sigma-\sigma_J)}{R\,\partial s}, -\frac{h}{2R}\frac{\partial\delta(\sigma-\sigma_J)}{\partial\theta}, \delta(\sigma-\sigma_J), -\frac{1}{R}\frac{\partial\delta(\sigma-\sigma_J)}{\partial\theta}, \frac{\partial\delta(\sigma-\sigma_J)}{R\,\partial s}, 0\right]\boldsymbol{Q}_J \tag{12.7.10a,b}$$

For a simply supported circular cylindrical shell, the following harmonic solutions can be employed:

$$\xi_s(s,\theta;t) = \xi_s(s,\theta)e^{j\omega t},\ \xi_\theta(s,\theta;t) = \xi_\theta(s,\theta)e^{j\omega t},\ w(s,\theta;t) = w(s,\theta)e^{j\omega t} \tag{12.7.11}$$

and

$$\xi_s(s,\theta) = \sum_{m=1}^{\infty}\sum_{n=1}^{\infty} a_{mn}\cos\lambda s \sin n\theta \tag{12.7.12}$$

$$\xi_\theta(s,\theta) = \sum_{m=1}^{\infty}\sum_{n=1}^{\infty} b_{mn}\sin\lambda s \cos n\theta \tag{12.7.13}$$

$$w(s,\theta) = \sum_{m=1}^{\infty}\sum_{n=1}^{\infty} c_{mn}\sin\lambda s \sin n\theta \tag{12.7.14}$$

where $\lambda = m\pi R/L_0$, a_{mn}, b_{mn} and c_{mn} are modal amplitude coefficients for mode mn, and L_0 is the length of the cylinder. Substituting (12.7.10) into (12.7.5) and (12.7.6) gives:

$$a_{mn} = \lambda\alpha_{mn}c_{mn} = \frac{\lambda(\nu\lambda^2 - n^2)}{(\lambda^2 + n^2)^2}c_{mn} \tag{12.7.15a,b}$$

$$b_{mn} = n\beta_{mn}c_{mn} = \frac{n[(2+\nu)\lambda^2 + n^2]}{(\lambda^2 + n^2)^2}c_{mn} \tag{12.7.16a,b}$$

where $\lambda = m\pi R/L_0$ and $m,n = 1, 2, \cdots$; $\alpha_{mn} = (\nu\lambda^2 - n^2)(\lambda^2 + n^2)^{-2}$ and $\beta_{mn} = [(2+\nu)\lambda^2 + n^2](\lambda^2 + n^2)^{-2}$.

For convenience, the modes will be identified by a single index k, rather than the double index mn and arranged in ascending order of resonance frequency. Thus, c_{mn} will be represented as c_k from now on. Substituting (12.7.11)–(12.7.16) into (12.7.7), multiplying each side by $\psi_k(\sigma_J)$ and integrating over the cylinder length L_0 (orthogonal property of the mode shape functions) and using the results $\int_0^{L_0/R} \sin\lambda_1 s \sin\lambda_2 s\, ds = \frac{L_0}{2R}$ if $\lambda_1 = \lambda_2$ and 0 if $\lambda_1 \neq \lambda_2$ and $\int_0^{2\pi} \sin n_1\theta \sin n_2\theta\, d\theta = \pi$ if $n_1 = n_2$ and 0 $n_1 \neq n_2$, and adding loss (damping) factors to every mode gives the following equation for the radial modal amplitude coefficients c_k, ($k = 1,2,\cdots$) for all shell modes:

$$\frac{m_s}{4}(\omega_k^2 + j\eta_k\omega_k^2 - \omega^2)c_k =$$

$$\sum_{J=1}^{L_1}\left\{\left[\frac{h}{2R}\psi_{ks}(\sigma_J),\ \frac{h}{2R}\psi_{k\theta}(\sigma_J),\ \psi_k(\sigma_J),\ \frac{1}{2R}\psi_{k\theta}(\sigma_J),\ -\frac{1}{R}\psi_{ks}(\sigma_J),\ 0\right]Q_J + \sum_{i=1}^{\infty}C_{ik}c_i\right\} \tag{12.7.17}$$

where c_i and c_k are the coefficients for the ith shell mode and kth shell mode respectively, The quantity $m_s = 2\pi\rho RhL_0$ is the mass of the cylindrical shell, ω_k

and η_k are, respectively, the shell mode resonance angular frequencies and loss factors arranged in ascending order, and C_{ik} is the concentrated mass contribution given by:

$$C_{ik} = \sum_{J=1}^{L_1} m_J^b \omega^2 \psi_i(\sigma_J) \psi_k(\sigma_J) \tag{12.7.18}$$

The three functions $\psi_k(\sigma_J)$, $\psi_{ks}(\sigma_J)$, and $\psi_{k\theta}(\sigma_J)$ are dimensionless and are evaluated at the point $\sigma_J(s_J, \theta_J)$ on the shell surface and defined as follows:

$$\psi_k(\sigma_J) = \sin\lambda s_J \sin n\theta_J \tag{12.7.19}$$

$$\psi_{ks}(\sigma_J) = \lambda \cos\lambda s_J \sin n\theta_J \tag{12.7.20}$$

$$\psi_{k\theta}(\sigma_J) = n \sin\lambda s_J \cos n\theta_J \tag{12.7.21}$$

where n is the circumferential order of the kth mode.

The shell mode resonance angular frequencies ω_k, $(k = 1, 2, \cdots)$ can be calculated by using the following relation:

$$\omega_k^2 = \left[\frac{(1-\nu^2)\lambda^4}{(\lambda^2+n^2)^2} + \kappa(\lambda^2+n^2)^2 \right] \frac{E}{\rho(1-\nu^2)R^2} \tag{12.7.22}$$

When only the first P modes are taken into account, (12.7.17) can be written in the following matrix form:

$$\boldsymbol{Z}_s \boldsymbol{c}^s = \sum_{J=1}^{L_1} \boldsymbol{R}_J^b \boldsymbol{Q}_J \tag{12.7.23}$$

where $\boldsymbol{Z}_s$ is the uncoupled shell characteristic matrix, including the influence of the concentrated masses of the isolating mounts:

$$\boldsymbol{Z}_s = \begin{bmatrix} \Omega_1 - C_{1,1} & -C_{1,2} & -C_{1,3} & \cdots & -C_{1,P} \\ -C_{2,1} & \Omega_2 - C_{2,2} & -C_{2,3} & \cdots & -C_{2,P} \\ \vdots & \vdots & \vdots & \vdots & \vdots \\ -C_{P,1} & -C_{P,2} & -C_{P,3} & \cdots & \Omega_P - C_{P,P} \end{bmatrix} \tag{12.7.24}$$

where $C_{ik} = C_{ki}$ is defined by (12.7.18), Ω_k $(k = 1, \cdots, P)$ is defined as:

$$\Omega_k = \frac{1}{4} m_s (\omega_k^2 + j\eta_k \omega_k^2 - \omega^2) \tag{12.7.25}$$

and $\boldsymbol{c}^s = [c_1, c_2, \cdots, c_P]^{\mathrm{T}}$.

The quantity $\boldsymbol{R}_J^b$ is a force coupling matrix corresponding to force vector $\boldsymbol{Q}_J$ acting on the Jth support point on the cylindrical shell and is defined as follows:

$$\boldsymbol{R}_J^b = \begin{bmatrix} \frac{h}{2R}\psi_{1s}(\sigma_J) & \frac{h}{2R}\psi_{1\theta}(\sigma_J) & \psi_1(\sigma_J) & \frac{1}{R}\psi_{1\theta}(\sigma_J) & -\frac{1}{R}\psi_{1s}(\sigma_J) & 0 \\ \frac{h}{2R}\psi_{2s}(\sigma_J) & \frac{h}{2R}\psi_{2\theta}(\sigma_J) & \psi_2(\sigma_J) & \frac{1}{R}\psi_{2\theta}(\sigma_J) & -\frac{1}{R}\psi_{2s}(\sigma_J) & 0 \\ \vdots & \vdots & \vdots & \vdots & \vdots & \vdots \\ \frac{h}{2R}\psi_{Ps}(\sigma_J) & \frac{h}{2R}\psi_{P\theta}(\sigma_J) & \psi_P(\sigma_J) & \frac{1}{R}\psi_{P\theta}(\sigma_J) & -\frac{1}{R}\psi_{Ps}(\sigma_J) & 0 \end{bmatrix} \tag{12.7.26}$$

In the cylindrical coordinate system, the rotational displacements of the elastic elements at a point on the shell surface can be calculated by using the following equations:

$$\theta_x = \frac{1}{R}\frac{\partial w}{\partial \theta}, \qquad \theta_\theta = -\frac{\partial w}{\partial x} \tag{12.7.27a,b}$$

Note that θ_r is zero. Using (12.7.12)–(12.7.16) and (12.7.26) and the preceding definition of $\boldsymbol{c}^s$, the following matrix expression can be obtained:

$$\boldsymbol{W}_J = \boldsymbol{\Psi}_J \boldsymbol{c}^s \qquad (J = 1, \cdots, L_1) \tag{12.7.28}$$

where $W_J = [\xi_s, \xi_\theta, w, \theta_x, \theta_\theta, \theta_r]^{\mathrm{T}}$ is the six-dimensional displacement vector of the Jth support point $\boldsymbol{\sigma}_J$ in the cylindrical coordinate system, and the matrix $\boldsymbol{\Psi}_J$ is given as follows:

$$\boldsymbol{\Psi}_J = [R_J^b]^{\mathrm{T}} = \begin{bmatrix} \frac{h}{2R}\psi_{1s}(\sigma_J) & \frac{h}{2R}\psi_{2s}(\sigma_J) & \cdots & \frac{h}{2R}\psi_{Ps}(\sigma_J) \\ \frac{h}{2R}\psi_{1\theta}(\sigma_J) & \frac{h}{2R}\psi_{2\theta}(\sigma_J) & \cdots & \frac{h}{2R}\psi_{P\theta}(\sigma_J) \\ \psi_1(\sigma_J) & \psi_2(\sigma_J) & \cdots & \psi_P(\sigma_J) \\ \frac{1}{R}\psi_{1\theta}(\sigma_J) & \frac{1}{R}\psi_{2\theta}(\sigma_J) & \cdots & \frac{1}{R}\psi_{P\theta}(\sigma_J) \\ -\frac{1}{R}\psi_{1s}(\sigma_J) & -\frac{1}{R}\psi_{2s}(\sigma_J) & \cdots & -\frac{1}{R}\psi_{Ps}(\sigma_J) \\ 0 & 0 & \cdots & 0 \end{bmatrix} \tag{12.7.29}$$

12.7.3 System equation of motion

Adapting (12.5.22) to multiple mounts, we obtain for each mount the relationship between the top and bottom displacements as:

$$\boldsymbol{Q}_J^t = -\boldsymbol{Q}_J^b = \boldsymbol{K}_J(\boldsymbol{D}_J^b - \boldsymbol{D}_J^t) \tag{12.7.30a,b}$$

where

$$D_J^t = (R_J^t)^{\mathrm{T}} D_0 \tag{12.7.31}$$

Sign conventions are the same as those used for the beam problem in Section 12.5. The quantity D_0 is defined by (12.5.4), R_J^t is defined by (12.5.5) and K_J is defined by (12.5.6) with a subscript J added.

The displacement vector D_J^b of the shell support point σ_J in the cartesian coordinate system can be expressed in terms of W_J in the cylindrical coordinate system by using the following relationship:

$$W_J = T_J D_J^b \tag{12.7.32}$$

Synthesising (12.6.7), (12.6.8), (12.6.27), (12.6.29), (12.7.1), (12.7.2), (12.7.23), (12.7.28) and (12.7.32) gives the following equation of motion for the coupled system:

$$\begin{bmatrix} A_{11} & A_{12} \\ A_{21} & A_{22} \end{bmatrix} \begin{bmatrix} D_0 \\ c^s \end{bmatrix} = \begin{bmatrix} Q_0 \\ 0 \end{bmatrix} \tag{12.7.33}$$

where the element matrices A_{11}, ···, A_{22} are given by the following expressions:

$$A_{11} = Z_0 + \sum_{J=1}^{L_1} R_J^t K_J (R_J^t)^{\mathrm{T}} \tag{12.7.34}$$

$$A_{12} = -\sum_{j=1}^{L_1} R_J^t K_J T_J^{\mathrm{T}} \Psi_J \tag{12.7.35}$$

$$A_{21} = -\sum_{j=1}^{L_1} R_J^b T_J K_J (R_J^t)^{\mathrm{T}} \tag{12.7.36}$$

$$A_{22} = Z_s + \sum_{J=1}^{L_1} R_J^b T_J K_J T_J^{\mathrm{T}} \Psi_J \tag{12.7.37}$$

The resonance frequencies and mode shapes of the coupled system can be obtained by solving the eigenvalue problem of the coefficient matrix containing the elements A_{ij}, $(i,j = 1,4)$ when $Q_0 = 0$.

The passive isolators can be made active by using force actuators connected in parallel with each of them. When used with a suitable feedforward control system, these actuators exert control forces on the rigid body and the shell support points simultaneously, to change the system response to some optimal condition. In this case, the right hand side of (12.7.33) is replaced by a combined force vector, and the equation becomes:

$$\begin{bmatrix} A_{11} & A_{12} \\ A_{21} & A_{22} \end{bmatrix} \begin{bmatrix} D_0 \\ c^s \end{bmatrix} = \begin{bmatrix} Q_0 + \sum_{J=1}^{L_1} R_J^t Q_J^c \\ -\sum_{J=1}^{L_1} R_J^b T_J Q_J^c \end{bmatrix} \tag{12.7.38}$$

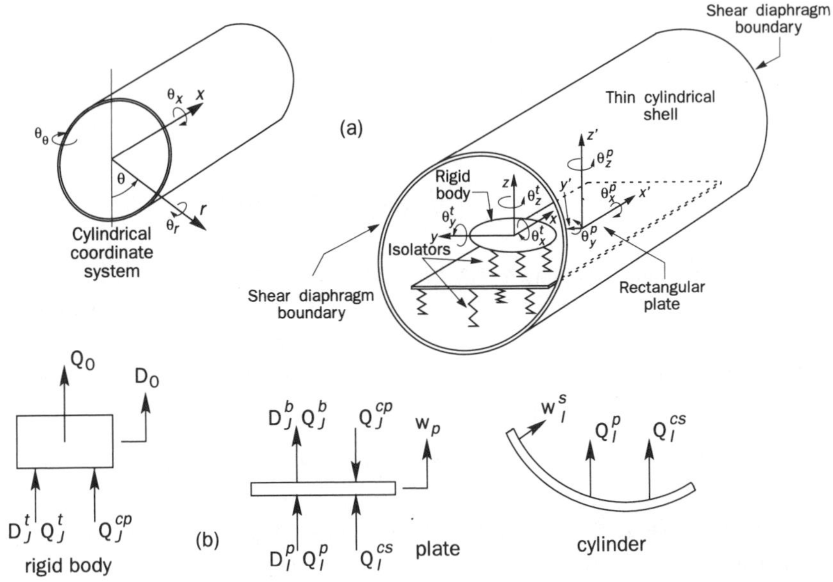

Fig. 12.65 Rigid body isolated through an intermediate flexible panel from a flexible cylinder using multiple mounts: (a) theoretical model; (b) sign conventions showing directions of positive forces and displacements.

where $\boldsymbol{Q}_J^c$ is the control force vector acting on the rigid body from the Jth actuator connected in parallel with the Jth mount attaching the rigid body to the flexible shell, and is defined for the Jth actuator in terms of forces and moments in the cartesian coordinate system as $\boldsymbol{Q}_J^c = \left[F_x^c,\ F_y^c,\ F_z^c,\ M_x^c,\ M_y^c,\ M_z^c\right]_J^{\mathrm{T}}$. It is defined as positive when acting in the direction of negative cylinder displacement at the bottom of the isolator which results in a control force acting in the direction of positive rigid body displacement at the top of the isolator.

The next stage of the analysis would be to include a flexible panel between the rigid body and the flexible cylinder (a good model of submarine equipment mounting). This is illustrated in Fig. 12.65 and the analysis is discussed by Pan and Hansen (1994).

12.7.4 Minimization of power transmission into the support cylinder

The total power transmission P_0 into the cylinder through the isolating mounts can be calculated as follows:

$$P_0 = -\mathrm{Re}\left\{\frac{1}{2}j\omega\sum_{J=1}^{L_1}(\boldsymbol{Q}_J - \boldsymbol{T}_J\boldsymbol{Q}_J^c)^{\mathrm{H}}\boldsymbol{W}_J\right\}$$
$$= -\frac{\omega}{2}\mathrm{Im}\left\{\sum_{J=1}^{L_1}\boldsymbol{W}_J^{\mathrm{H}}(\boldsymbol{Q}_J - \boldsymbol{T}_J\boldsymbol{Q}_J^c)\right\} \qquad (12.7.39a,b)$$

This is the quantity which may be used as the active control cost function to be minimized. It can be seen that the output power transmission P_0 depends upon the parameters of the passive system as well as on the control forces. For a given system, it is possible to express the output power by using an explicit quadratic function as follows:

$$\sum_{J=1}^{L_1}\boldsymbol{W}_J^{\mathrm{H}}(\boldsymbol{Q}_J - \boldsymbol{T}_J\boldsymbol{Q}_J^c) = (\boldsymbol{Q}^c)^{\mathrm{H}}\boldsymbol{a}\boldsymbol{Q}^c + (\boldsymbol{Q}^c)^{\mathrm{H}}\boldsymbol{b}_1 + \boldsymbol{b}_2\boldsymbol{Q}^c + c \qquad (12.7.40)$$

The matrices $\boldsymbol{a}$, $\boldsymbol{b}_1$, $\boldsymbol{b}_2$ and the constant c are dependent upon the numbers of isolators used. For a four-isolator system, the analysis is as outlined in Section 12.6.2.4, (12.6.43) to (12.6.62), except that the quantities R_1^b, R_2^b, R_3^b, R_4^b and D_J^b are replaced by $\boldsymbol{\Psi}_1^{\mathrm{T}}\boldsymbol{T}_1$, $\boldsymbol{\Psi}_2^{\mathrm{T}}\boldsymbol{T}_2$, $\boldsymbol{\Psi}_3^{\mathrm{T}}\boldsymbol{T}_3$, $\boldsymbol{\Psi}_4^{\mathrm{T}}\boldsymbol{T}_4$ and $\boldsymbol{W}_J$ respectively.

The comments made in Section 12.6.2.4 regarding zero power transmission are equally valid for the isolators considered here.

12.8 FEEDFORWARD CONTROL: SUMMARY

As mentioned at the beginning of Section 12.4, feedforward control is appropriate in situations where periodic machinery vibration is to be isolated from rigid or flexible support structures. The analyses outlined in Sections 12.5 to 12.7 can be used to estimate the maximum achievable reduction in vibratory power transmission from a machine into a support structure assuming the availability of an ideal feedforward electronic controller. Aspects to be considered in the design of such a controller are discussed in Chapters 6 and 13.

REFERENCES

Abdel-Rohman, M. (1987). Feasibility of active control of tall buildings against wind. *Journal of Structural Engineering*, **113**, 349–362.

Aburaya, T., Kawanishi, M., Kondo, H. and Hamada, T. (1990). Development of an electronic control system for active suspension. In *Proceedings of the 29th IEEE Conference on Decision and Control*, **4**, 2220–2225.

Akatsu, Y., Fukushima, N., Tukahashi, K. Satoh, M. and Kawarazaki, Y. (1990). Active suspension employing an electrohydraulic pressure control system. In *Proceedings of the 18th Fisita Congress: The Promise of New Technology in the Automotive Industry*. SAE, Warrendale, PA, 949–959.

Alanoly, J. and Sankar, S. (1987). A new concept in semi-active vibration isolation.

Journal of Mechanics, Transmissions and Automation in Design, **109**, 242–247.

Asano, S., Takahashi, J., Saito, T. and Matsumoto, H. (1991). Development of acceleration sensor and acceleration evaluation system for super low range frequency. In *Proceedings of the SAE International Congress on Sensors and Actuators*. SAE Publication, p. 242, 37–49.

Bender, E.K. (1968). Optimum linear preview control with application to vehicle suspension. *Journal of Basic Engineering*, **90**, 213–221.

Bies, D.A. (1968). *Feasibility Study of a Hybrid Vibration Isolation System*. Bolt Beranek and Newman Inc., Report 1620.

Bies, D.A. and Hansen, C.H. (1995). *Engineering Noise Control*, 2nd edn., E&FN Spon, London, Ch. 10.

Blackwood, G.H. and von Flotow, A.H. (1993). Active control for vibration isolation despite resonant structural dynamics: a trade study of sensors, actuators and configurations. In *Proceedings of the Second Conference on Recent Advances in Active Control of Sound and Vibration*, ed. R.A. Burdisso. Technomic Publishing Co., Blacksburg, VA, 482–494.

Cheok, K.C. and Huang, N.J. (1989). Lyapunov stability analysis for self-learning neural model with application to semi-active suspension control system. In *Proceedings of the IEEE International Symposium on Intelligent Control*, pp. 326–331.

Cheok, K.C., Hu, H.X. and Loh, N.K. (1989). Discrete-time frequency-shaping parametric control with application to active seat suspension control. *IEEE Transactions Industrial Electronics*, **36**, 383–390.

Craighead, I.A. (1988). An active suspension system for an ambulance stretcher. In *Proceedings of the International Conference on Advanced Suspensions*. Institution of Mechanical Engineers, London.

Crolla, D.A. (1988). Theoretical comparisons of various active suspension systems in terms of performance and power requirements. In *Proceedings of the Institution of Mechanical Engineers International Conference on Advanced Suspensions*. Mechanical Engineering Publications, Edmonds.

Crolla, D.A. and Abdel-Hady, M.B.A. (1991). Active suspension control; performance comparisons using control laws applied to a full vehicle model. *Vehicle System Dynamics*, **20**, 107–120.

Davies P.O.A.L. (1965). *Introduction to Dynamic Analysis and Automatic Control*. John Wiley.

de Benito, C.D. and Eckert, S.J. (1990). Control of an active suspension system subject to random component failures. *Transactions of ASME, Journal of Dynamic Systems, Measurement and Control*, **112**, 94–99.

Den Hartog J.P. (1956). *Mechanical Vibrations*, 4th edn. McGraw Hill, New York.

Decker, H., Schramm, W. and Kallenback, R. (1988). A practical approach towards advanced semi-active suspension systems. In *Proceedings of the International Conference on Advanced Suspensions*. Institution of Mechanical Engineers, London, 93–100.

Du Val, R.W., Gregory, C.Z. Jr. and Gupta, N.K. (1984). Design and evaluation of a state-feedback vibration controller. *Journal of the American Helicopter Society*, **29**, 30-37.

El-Madany, M.M. (1990). Ride performance potential of active fast load levelling systems. *Vehicle System Dynamics*, **19**, 19–47.

El-Madany, M.M. and Abduljabbar, Z. (1989). On the statistical performance of active and semi-active car suspension systems. *Computers and Structures*, **33**, 785–790.

El-Madany, M.M. and El-Razaz, Z.S. (1988). Performance of actively suspended cabs in highway trucks: evaluation and optimization. *Journal of Sound and Vibration*, **126**, 423–435.

El-Madany, M.M., El-Tamimi, A. and Al-Swailam, S.I. (1987). On some aspects of active control of road vehicles. *Modelling Simulation and Control*, **10** 13–23.

Federspiel-Labrosse, G.M. (1954). Contribution a l'etude et au perfectionement de la suspension des vehicles. *Journal de la Society Ing. Auto.*, December, 427–436.

Fenn, R.C., Downer, J.R., Gondhalekar, V. and Johnson B.G. (1990). An active magnetic suspension for space-based microgravity isolation. In *ASME Publication NCA*, Vol. 8, 49–56.

Foag, W. (1988). A practical control concept for passenger car active suspensions with preview. In *Proceedings of the International Conference on Advanced Suspensions*. Institution of Mechanical Engineers, London, 43–50.

Fuller, C.R. Elliott, S.J. and Nelson, P.A. (1996). *Active Control of Vibration* (forthcoming).

Gevarter, W.B. (1970). Basic relations for the control of flexible vehicles. *AIAA Journal*, **8**, 666–672.

Gimenez, J.G., Busturia, J.M., Abete, J.M. and Vinyolas, J. (1988). Theoretical-experimental modelling of active suspensions for railway vehicles. *ASME Applied Mechanics Division*, AMD, **96**, 149–156.

Goodall, R.M. and Kortum, W. (1990). Active suspensions for railway vehicles: an affordable luxury or an inevitable consequence? In *Proceedings of the 11th Triennial World Congress of the International Federation of Automatic Control*, **5**, Pergamon Press, New York.

Graf, P.L. and Shoureshi, R. (1987). *Modelling and Implementation of Semi-active Hydraulic Engine Mounts*. ASME paper 87-WA/DSC-28.

Gupta, N.K. and Du Val, R.W. (1982). A new approach for active control of rotorcraft vibration. *Journal of Guidance and Control*, **5**, 143–150.

Hall, S.R. and Wereby, N.M. (1989). Linear control issues in the higher control of helicopter vibrations. In *Proceedings of the 45th Annual Forum of the American Helicopter Society*, Boston, MA.

Hall, S.R. and Wereley, N.M. (1992). Performance of higher harmonic control algorithms for helicopter vibration reduction. *Journal of Guidance*, **16**, 793–797.

Hattori, K., Kizu, R. Yokoyo, Y. and Ohno, H. (1990). Linear pressure control valve for active suspension. In *Proceedings of the ASME International Computers in Engineering Conference*, 583–588.

Hennecke, D. and Zieglmeier, F.J. (1988). Frequency dependent variable suspension damping: theoretical background and practical success. In *Proceedings of the International Conference on Advanced Suspensions*. Institution of Mechanical Engineers, London, 101–105.

Hine, P.J. and Pearce, P.T. (1988). A practical intelligent damping system. In *Proceedings of the International Conference on Advanced Suspensions*. Institution of Mechanical Engineers, London, 141–148.

Hrovat D. (1991). Optimal suspension performance for 2-D vehicle models. *Journal of Sound and Vibration*, **146**, 93–110.

Iwata, Y and Nakano, M. (1976). Optimum preview control of vehicle air suspensions. *Bulletin of the Japanese Society of Mechanical Engineers*, **19**, 1485–1489.

Jenkins, M.D., Nelson, P.A. and Elliott, S.J. (1988). A finite element model for the prediction of the performance of an active vibration isolation system. In *Proceedings of Internoise '88*. Institute of Noise Control Engineering, 1057–1060.

Jenkins, M.D., Nelson, P.A. and Elliott, S.J. (1990). Active isolation of periodic machinery vibration from resonant substructures in *Active Noise and Vibration Control – 1990*. ASME Publication NCA, Vol. 8.

Jenkins, M.D., Nelson, P.A. Pinnington,R.J. and Elliott, S.J. (1993). Active isolation of periodic machinery vibrations. *Journal of Sound and Vibration*, **166**, 117–140.

Jones, D.I., Ownes, A.R., and Owen, R.G. (1990). A microgravity facility for in-orbit experiments. In *ASME Publication NCA*, Vol. 8, 67–74.

Junker, H. and Seewald, A. (1984). Theoretical investigation of an active suspension system for wheeled tractors. In *Proceedings of the 8th International Conference of the International Society for Terrain Vehicle Systems: The performance of Off-Road Vehicles and Machines*, 185–214.

Kaneda, K. and Seto, K. (1983). Vibration isolation using feedback control and compound dynamic absorber for portable vibrating tools. *Bulletin of the Japanese Society of Mechanical Engineers*, **26**, 1219–1225.

Kaplow, C.E. and Velman, J.R. (1980). Active local vibration isolation applied to a flexible space telescope. *Journal of Guidance and Control*, **3**, 227–233.

Karnopp, D. (1989). Analytical results for optimum actively damped suspensions under random excitation. *Transactions of the ASME, Journal of Vibration, Acoustics Stress and Reliability in Design*, **111**, 278–282.

Kimica, S. (1965). *Servo-controlled Pneumatic Isolators: Their Properties and Applications*. ASME publication 65-WA/MD-12.

King, S.P. (1988). Minimisation of helicopter vibration through blade design. *Aeronautical Journal*, **92**, 247–263.

Levitt, J.A. and Zorka, N.G. (1991). Influence of tire damping in quarter car active suspension models. *Journal of Dynamic Systems, Measurement and Control: Transactions of the ASME*, **113**, 134–137.

Litkouhi, B. and Boustany, N.M. (1988). On board sensor failure detection of an active suspension system using the generalized likelihood ratio approach. In *Proceedings of the IEEE 27th Conference on Decision and Control*.

Lund, R.A. (1980). Active damping of large structures in wind. In *Structural Control*, ed. H.H.E. Leipholtz. North Holland Publishing, Amsterdam.

Malek, K.M. and Hedrick, J.K. (1985). Decoupled active suspension design for improved automotive ride quality/handling performance. *Vehicle System Dynamics*, **14**, 78–81.

McHugh, F.J. and Shaw, J. (1978). Helicopter vibration reduction with higher harmonic blade pitch. *Journal of the American Helicopter Society*, **23**, 26–35.

Meller, T. and Frühauf, F. (1988). Variable damping: philosophy and experiences of a preferred system. In *Proceedings of the Institution of Mechanical Engineers International Conference on Advanced Suspensions*. Mechanical Engineering Publications, Edmunds.

Miller L.R. (1988a). Tuning passive, semi-active and fully active suspension systems. *Proceedings of the IEEE 27th Conference on Decision and Control*. Austin, TX, 2047–2053.

Miller, L.R. (1988b). The effect of hardware limitations on an on/off semi-active suspension. In *Proceedings of the International Conference on Advanced Suspensions*. Institution of Mechanical Engineers, London, 199–206.

Moore, J.H. (1988). Linear variable inductance position transducer for suspension systems. In *Proceedings of the International Conference on Advanced Suspensions*. Institution of Mechanical Engineers, London, 75–82.

Nagai, M. Moran, A. and Tanaka, S. (1990). *IFAC Symposia Series: Proceedings of a Triennial World Congress*, **4**, Pergamon Press, New York, 419–424.

Narayanan, S. and Raju G.V. (1990). Stochastic control of non-stationary response of a single-degree-of-freedom vehicle model. *Journal of Sound and Vibration*, **141**, 449–463.

Nelson, P.A. (1991). Active Vibration Isolation. In *Active Control of Noise and Vibration Course Notes*, ed. C.H. Hansen. University of Adelaide, Adelaide, South Australia.

Nelson, P.A., Jenkins, M.S. and Elliott, S.J. (1987). Active isolation of periodic vibrations. In *Proceedings of Noise Con '87.* Institute of Noise Control Engineering, 425–430.

Nguyen, K. and Chopra, I. (1990). Application of higher harmonic control to rotors operating at high speed and thrust. *Journal of the American Helicopter Society*, **35**, 78–89.

Pan, J.-Q. and Hansen, C.H. (1993). Active control of power flow from a vibrating rigid body to a flexible panel through two active isolators. *Journal of the Acoustical Society of America*, **93**, 1947–1953.

Pan, J.-Q. and Hansen, C.H. (1994). Power flow from a vibration source through an intermediate flexible panel to a flexible cylinder. *ASME Journal of Vibration and Acoustics*, **116**, 496–505.

Pan, J.-Q., Hansen, C.H. and Pan, J. (1993). Active isolation of a vibration source from a thin beam using a single active mount. *Journal of the Acoustical Society of America*, **94**, 1425–1434.

Pan, J.-Q., Howard, C.Q. and Hansen, C.H. (1996). Power transmission from a vibrating body to a circular cylindrical shell through active elastic isolators. *Journal of the Acoustical Society of America*, **100** (1).

Pan, J., Pan, J.-Q. and Hansen, C.H. (1992). Total power flow from a vibrating rigid body to a thin panel through multiple elastic mounts. *Journal of the Acoustical Society of America*, **92**, 895–907.

Pavic, G. and Oreskovic, G. (1977). Energy flow through elastic mountings. In *Proceedings of the 9th International Congress on Acoustics*, Paper G3.

Petersen, N.R. (1980). Design of large scale tuned mass dampers. In *Structural Control*, ed. H.H.E. Leipholtz. North Holland Publishing, Amsterdam.

Pinnington, R.J. (1987). Vibrational power transmission to a seating of a vibration isolated motor. *Journal of Sound and Vibration*, **118**, 515.

Pinnington, R.J. and White, R.G. (1981). Power flow through machine isolators to resonant and non-resonant beams. *Journal of Sound and Vibration*, **75**, 179–197.

Prasad, J.V.R. (1991). Active vibration control using fixed order dynamic compensation with frequency shaped cost functionals. *IEEE Control Systems Magazine*, **11**, 71–78.

Okamoto, I., Koyanagi,S., Higaki, H., Terada, K., Sebata, M. and Takai, H. (1987). Active suspension system for railroad passenger cars. In *Proceedings of the Joint ASME/IEEE/AAR Railroad Conference*. IEEE Service Centre, Piscataway, NJ, 141–146.

Parker, G.A. and Lau, K.S. (1988). A novel valve for semi-active vehicle suspension systems. In *Proceedings of the International Conference on Advanced Suspensions*. Institution of Mechanical Engineers, London, 69–74.

Pollard, N.G., and Simons, N.J.A. (1984). Passenger comfort: the role of active suspensions. *Railway Engineering*, **1**, 17–31.

Quinlan, D.C. (1992). Automotive active engine mount systems. *AVL Conference on engine and environment* '92.

Rakheja, S. and Sankar, S. (1985). *Vibration and a Shock Isolation Performance of a Semi-active On-off Damper*. ASME paper 85-DET-15.

Reed, R.E. (1965). *Comparison of Methods in Calculating Resonance Frequencies for Corner Supported Rectangular Plates*. NASA TN D-3030.

Rockwell, T.H. and Lawther, J.M. (1964). Theoretical and experimental results on active vibration dampers. *Journal of the Acoustical Society of America*, **36**, 1507–1515.

Ross, C.F., Scott, J.F. and Sutcliffe, S.G. (1988). Active control of vibration. *International Patent Application No. PCT/GB87/00902*.

Samali, B., Yang, J.N. and Liu, S.C. (1985). Active control of seismic excited buildings. *Journal of Structural Engineering*, **111**, 2165–2180.

Sasaki, M., Kamiya, J. and Shimogo, T. (1976). Optimal preview control of vehicle suspension. *Bulletin of the Japanese Society of Mechanical Engineers*, **19**, 265–273.

Satoh, M. and Osada, K. (1991). Long life potentiometric position sensor. Its material and application. In *Proceedings of the SAE International Congress on Sensors and Actuators*. SAE Publication P-242, 1–12.

Satoh, M., Fukushima, N., Akatsu, Y., Fujimura, I. and Fukuyama, K. (1990). An active suspension employing an electro-hydraulic pressure control system. In *Proceedings of the 29th IEEE Conference on Decision and Control*, **4**, 2226–2231.

Scribner, K.B., Sievers, L.A. and von Flotow, A.H. (1990). *Active Narrow Band Vibration Isolation of Machinery Noise from Resonant Substructures*. ASME publication NCA-8, 101–111.

Shannan, J.E. and Vanderploeg, M.J. (1989). Vehicle handling model with active suspensions. *Journal of Mech. Transm. Autom. Des.*, **111**, 375–381.

Sharp, R.S. and Crolla, D.A. (1987). Road vehicle suspension system design: a review. *Vehicle System Dynamics*, **16**, 167–192.

Sharp, R.S. and Hassan, S.A. (1986). The relative performance capabilities of passive, active and semi-active car suspension systems. In *Proceedings of the Institution of Mechanical Engineers*, **200**, part D3, 219–228.

Sharp, R.S. and Hassan, S.A. (1987a). Performance and design considerations for dissipative semi-active suspension systems for automobiles. In *Proceedings of the Institution of Mechanical Engineers*, **201,** Part D2, 149–153.

Sharp, R.S. and Hassan, S.A. (1987b). On the performance capabilities of active automobile suspension systems of limited bandwidth. *Vehicle System Dynamics*, **16**, 213–225

Shaw, J. (1967). *A Feasibility Study of Helicopter Vibration Reduction by Self-optimising Higher Harmonic Blade Pitch Control*. MS thesis, Department of Aeronautics and Astronautics, MIT, Cambridge, MA.

Shaw, J. and Albion, N. (1980). Active control of rotor blade pitch for vibration reduction: a wind tunnel demonstration. *Vertica*, **4**, 3–11.

Shaw, J. and Albion, N. (1981). Active control of the helicopter rotor for vibration reduction. *Journal of the American Helicopter Society*, **26**.

Shaw, J., Albion, N., Hanker, E. Jr., and Teal, R. (1989). Higher harmonic control: wind tunnel demonstration of fully effective vibratory hub force suppression. *Journal of the American Helicopter Society*, **34**.

Shiozaki, M., Kamiya, S., Kuroyanagi, M. and Matsui, K. (1991). *High Speed Control of Damping Force Using Piezo-electric Elements*. SAE: Warrendale, PA, 149–154.

Sommerfeldt, S.D. and Tichy, J. (1990). Adaptive control of a two-stage vibration isolation mount. *Journal of the Acoustical Society of America*, **88**, 938–944.

Sproston, J.L. and Stanway, R. (1992). Electrorheological fluids in vibration isolation.

In *Proceedings of Actuator '92, the 3rd International Conference on New Actuators*. VDI/VDE-Technogiezentrum Informationstechnik GmbH, Bremen, Germany, 116–117.

Stampleman, D.S. and von Flotow, A.H. (1990). Microgravity isolation mounts based upon piezo-electric film. In *ASME Publication NCA*, Vol. 8, 57–66.

Stanway, R., Sproston, J. and Firoozian, R. (1989). Identification of the damping law of an electro-rheological fluid: a sequential filtering approach. *Journal of Dynamic Systems Measurement and Control*, **111**, 91–96.

Stayner, R.M. (1988). Suspensions for agricultural vehicles. In *Proceedings of the International Conference on Advanced Suspensions*. Institution of Mechanical Engineers, London, 133–140.

Sunwoo, M., Cheok, K.C., and Huang, N.J. (1991). Application of model reference adaptive control to active suspension systems. *IEEE Transactions, Industrial Electronics*, **38**, 217–222.

Sussman, N.E. (1974). Statistical ground excitation models for high speed vehicle dynamic analysis. *High Speed Ground Transportation Journal*, **8**, 145–154.

Tanaka, N. and Kikushima, Y. (1988). Rigid support active vibration isolation. Journal of *Sound and Vibration*, **125**, 539–559.

Tani, K. Usui, S. Horiuchi, E., Shirai, N. and Hirobe, S. (1990). Computer controlled active suspension for a wheeled terrain robot. *International Journal of Comput. Appl. Technol.*, **3**, 100–104.

Thompson, A.G. (1976). An active suspension with optimal linear state feedback. *Vehicle System Dynamics* **5**, 187–203.

Thompson, A.G. and Pearce, C.E.M. (1979). *An Optimal Suspension for an Automobile on a Random Road*. SAE paper 790478.

Thompson, A.G., Davis B.R. and Pearce C.E.M. (1980). *An Optimal Linear Active Suspension with Finite Road Preview. Society of Automotive Engineers*, paper 800520.

Todd, K.B. (1990). Handling performance of road vehicles with different active suspensions. *ASME Applied Mechanics Division Transportation Systems,* AMD **108**, 19-26.

Tomizuka, M. (1975). Optimal continuous finite preview problems. *IEEE Transactions on Automatic Control*, **AC 20**, 362–365.

Tomizuka, M. (1976). Optimal linear preview control with application to vehicle suspension: revisited. *ASME Journal of Dynamic Systems, Measurement and Control*, **September**, 309–315.

Tseng, T. and Hrovat, D. (1990). Some characteristics of optimal vehicle suspensions based on quarter-car models. In *Proceedings of the IEEE 29th Conference on Decision and Control*, 2232–2237.

Ulsoy, A.G., Hrovat, D. (1990). Stability robustness of LQG active suspensions. In *Proceedings of the 1990 American Control Conference*, 1347–1356.

Wallace, C.F. (1972). Radiation resistance of a rectangular plate. *Journal of the Acoustical Society of America*, **51**, 946–952.

Wendel, G.R. and Stecklein, G.L. (1991). A regenerative active suspension system. *Vehicle Dynamics and Electronic Controlled Suspensions*. SAE Special Publication Number 861, 129–135.

White, A.D. and Cooper, D.G. (1984). An adaptive controller for multivariable active noise control. *Applied Acoustics*, **17**, 99–109.

Williams, R.A. (1985). Active suspensions: classical or optimal? *Vehicle System*

Dynamics, **14**, 127–132.

Wilson, D.A., Sharp R.S,. and Hassan, S.A. (1986). Application of linear optimal control theory to the design of automobile suspensions. *Vehicle System Dynamics*, **15**, 103–118.

Yashuda, E., Doi, S. Hattori, K., Suzuki, H. and Hayashi, Y. (1991). In *Vehicle Dynamics and Electronic Controlled Suspensions. SAE Special Publication,* **861**, SAE: Warrendale, PA, 155–162.

Yoshimura, T., Ananthanarayana, N. and Deepak, D. (1986). An active vertical suspension for track/vehicle systems. *Journal of Sound and Vibration*, **106**, 217–225.

Yoshimura, T. and Ananthanarayana, N. (1987). An active lateral suspension to a track/vehicle system using stochastic optimal control. *Journal of sound and Vibration*, **115**, 473–482.

Yoshimura, T. and Ananthanarayana, N. (1991). Stochastic optimal control of vehicle suspension with preview on an irregular surface. *International Journal of Systems Science*, **22**, 1599–1611.

Yue, C., Butsuen, T. and Hedrick, J.K. (1989). Alternative control laws for automotive active suspensions. *Transactions of ASME, Journal of Dynamic Systems, Measurement and Control*, **111**, 286–291.

13

A few electronic implementation issues

In this book a good deal of effort has been directed at describing what is required to attain active attenuation of some unwanted disturbance, which includes deriving a number of algorithms for control system design. In this chapter we will consider some of the issues associated with implementation of some of these algorithms in an electronic (digital) controller. This consideration will cover both what is physically required in terms of hardware, and performance-related issues in hardware selection. The discussion will be largely qualitative, as there are no fixed rules for optimizing hardware arrangements. The discussion will also be brief, concentrating on a few issues which are commonly encountered in active control system implementation. A detailed discussion of hardware issues is beyond the scope of this book, and would probably be out of date by the time it is read.

The discussion here will be largely directed at the implementation of feedforward control systems, as it is these systems which have been of principal interest in this book. The hardware requirements for feedforward and feedback implementations are, for the most part, the same. What will be lacking is discussion of strictly feedback implementation issues, such as integrator windup. For discussion of strictly feedback control issues, the reader is directed towards any good digital control text, such as Middleton and Goodwin (1990) and Franklin *et al.* (1990).

Illustrated in Fig. 13.1 is a block diagram of the main components of a digital active control system. As was outlined in Chapter 5, digital implementation of control systems involves the inclusion of a number of components to the standard 'textbook control arrangement. These additions include an anti-

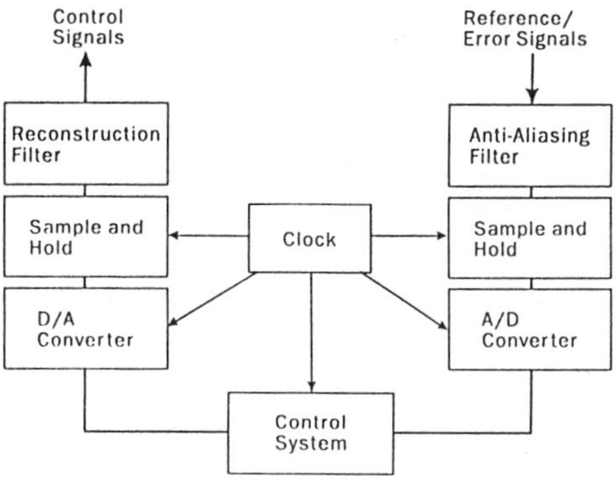

Fig. 13.1 Block diagram of the main components of a digital control system.

aliasing filter, sample and hold circuitry, and an analogue to digital converter (ADC) on the input to the controller, the inclusion of a digital to analogue conver-ter (DAC), sample and hold circuitry, a reconstruction filter on the output of the controller, and the addition of a clock to synchronize events such as sampling. Referring to Fig. 13.2, for the purposes of this chapter, this arrangement can be separated into two parts: the analogue/digital interface, and the main controller 'body', on which the control algorithms will run. This body is typically a micro-processor. Discussion in this chapter will also be separated along the lines of Fig. 13.2: design of the analogue to digital interface, and implementation of active control software on a microprocessor.

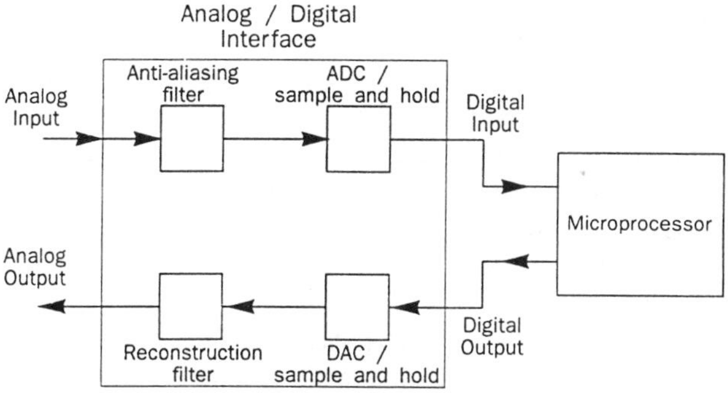

Fig. 13.2 Digital control system separated into two parts: an analogue/digital interface, and the main controller body.

13.1 THE ANALOGUE/DIGITIAL INTERFACE

Recall from Chapter 5 that the ADCs and DACs provide an interface between the real (continuous) world and the world of a digital system. ADCs take some physical variable, usually an electrical voltage, and convert it to a stream of numbers which are sent to controller for use in the control algorithm. These numbers usually arrive at increments of some fixed time period, or sampling period. The numbers arriving from the ADC are usually representative of the value of the signal at the start of the sampling period, as the data input to the ADC is normally sampled and then held constant during the conversion process to enable an accurate conversion, as will be discussed shortly. Commonly, the sampling period is implicitly referred to by a sampling rate, which is the number of samples taken in one second.

The digital signal coming from the ADC is quantized in level. This simply means that the stream of numbers sent to the digital control system has some finite number of digits, hence finite accuracy. This accuracy is normally quantified by the number of bits used by the ADC to represent the measured signal. For example, a 16-bit ADC converter will represent a sampled physical system variable as a set of 16 bits, each with a value of 0 or 1. It follows that the accuracy of the digital representation of the analogue value is limited by the 'quantum size', given by

$$\text{quantum size} = \frac{\text{full scale range}}{2^n} \tag{13.1.1}$$

where n is the number of bits. For example, if the full scale range of the ADC is ± 10 volts, the accuracy of the 16 bit digital representation is limited to better than (20 volts)/(2^{16}) = 0.305 millivolts. The difference between the actual analogue value and its digital representation is referred to as the quantization error. The dynamic range of the ADC is also determined by the number of bits used to digitally represent the analogue value, and is usually expressed in decibels, or dB. For example, a 16 bit ADC has a dynamic range of (20 log (2^{16})), or 96.3, dB.

(One point which should be mentioned here is that the 'ideal' dynamic range of a converter, as defined above, is often significantly less than the effective dynamic range which is achieved in implementation. This is partly because it is unlikely that the high end of the voltage range is consistently used, rendering some of the more significant bits useless. Measurement noise also often corrupts the signal, effectively rendering one or more of the less significant bits ineffective.)

The DAC works in an opposite fashion to the ADC in the sense that it provides a continuous output signal in response to an input stream of numbers. This continuous output is achieved using the sample and hold circuit, normally incorporated 'on chip'. This circuit is designed to progressively extrapolate the output signal between successive samples in some prescribed manner. The most commonly used hold circuit is the zero order hold, which simply holds the output voltage constant between successive samples. With the zero order hold circuit the output of the DAC/sample and hold circuitry is continuous in time, but quantised in level. To smooth out this pattern, a reconstruction filter is normally placed at the output of the DAC/sample and hold circuitry. This filter is low pass, which has the effect of removing the high frequency 'corners' from the stepped signal.

There are a wide range of variables associated with the design of the analogue/digital interface. Discussion here will concentrate on the most general parameters, such as sample rate selection, and avoid package-specific issues such as how the data is made available to the microprocessor (parallel versus serial chips). Further discussion of these issues can be found in application notes available from most component manufacturers, such as analogue devices.

13.1.1. Sample rate selection

The first issue of interest here is selection of a suitable sample rate. An absolute limit on the lower value of the sampling rate can be derived by studying the effect of sampling on the (continuous) system transfer function. To do this, it is useful to consider the simplified ADC/DAC arrangement shown in Fig. 13.3, comprising two parts: the sampler, and the hold. The signal coming from the sampler can be viewed as a set of impulses, or an impulse train,

$$x^*(t) = \sum_{k=-\infty}^{\infty} x(t)\ \delta(t-kT) \tag{13.1.2}$$

where T is the sampling period, taken here to be fixed. The Laplace transform of this is

$$\mathscr{L}\{x^*(t)\} = x^*(s) = \sum_{k=-\infty}^{\infty} x(kT)\mathrm{e}^{-skT} \tag{13.1.3}$$

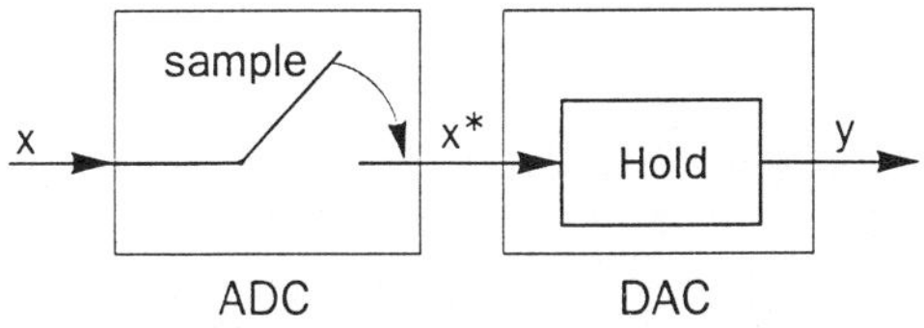

Fig. 13.3 Simple ADC/DAC arrangement viewed as two parts: a sampler and a hold.

For the zero-order hold considered here the output at time t is equal to

$$x_n(t) = x(kT) \qquad kT \le t < kT+T \tag{13.1.4}$$

This can be expressed in terms of a unit step function as

$$x_n(t) = x(kT)(1(t)-1(t-T)) \tag{13.1.5}$$

which enables the Laplace transform to be calculated:

$$\mathscr{L}\{x_n(t)\} = x_n(s) = \mathscr{L}\{x(kT)\}\frac{1-\mathrm{e}^{-sT}}{s} \tag{13.1.6}$$

Combining the sample of (13.1.3) and the hold of (13.1.6), the continuous time transfer function of the sample and hold is

$$y(s) = \sum_{k=-\infty}^{\infty} x(kT)e^{-skT}\frac{1-\mathrm{e}^{-sT}}{s} \tag{13.1.7}$$

Therefore, the digital control system can be modelled as in block diagram form as shown in Fig. 13.4.

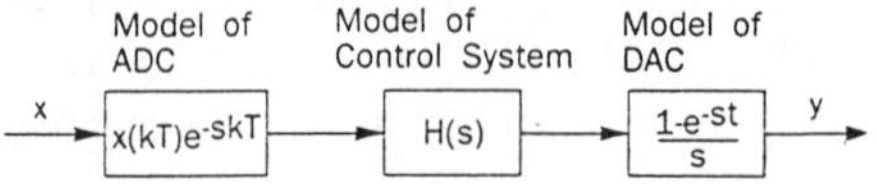

Fig. 13.4 Block diagram of the simple ADC/DAC arrangement.

Consider now the frequency domain representation of (13.1.2). The impulse train can be re-expressed as a Fourier series:

$$\sum_{k=-\infty}^{\infty} \delta(t-kT) = \sum_{n=-\infty}^{\infty} h_n e^{j\left(\frac{2\pi n}{T}\right)t} \tag{13.1.8}$$

where the coefficients H_n are found by integrating over a single sampling period,

$$h_n = \frac{1}{T}\int_{-T/2}^{T/2} \delta(t) e^{-j\omega_s t}\, dt \tag{13.1.9}$$

Here ω_s is the sampling frequency in rad s^{-1}, defined by

$$\omega_s = \frac{2\pi}{T} \tag{13.1.10}$$

Evaluation of this integral results in

$$h_n = \frac{1}{T} \tag{13.1.11}$$

Therefore, the impulse train can be expressed in frequency domain format as

$$\sum_{k=-\infty}^{\infty} \delta(t-kT) = \frac{1}{T}\sum_{n=-\infty}^{\infty} x(s-jn\omega_s) \tag{13.1.12}$$

Substituting this result back into (13.1.2), and taking the Laplace transform, it is found that

$$x^*(s) = \frac{1}{T}\sum_{n=-\infty}^{\infty} x(s-jn\omega_s) \tag{13.1.13}$$

The significance of this result of (13.1.13) is that it states that images of the true value of the sampled spectrum repeat themselves at infinite numbers of intervals of ω_s. This phenomenon is termed aliasing.

Practically, the phenomenon of aliasing means that it is impossible to tell the difference between two (or more) sinusoids based upon the sampled signal. This effect is illustrated in Fig. 13.5, where two sinusoids have exactly the same sampled values, and can therefore not be distinguished from one another based upon the sampled data. Aliasing can have a significant detrimental effect upon control system performance if there are substantial levels of 'high' frequency data, $\omega > \omega_s/2$, which are allowed to alias onto the 'low' frequency data, $\omega < \omega_s/2$. To combat this problem, antialiasing filters placed in front of the input

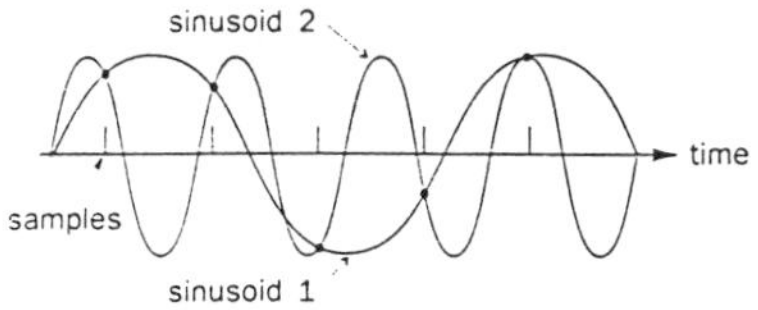

Fig. 13.5 A demonstration of aliasing, where two sinusoids have the same sampled values.

to the sample and hold/ADC arrangement. These are analogue low-pass filters which remove frequency components greater than half the sampling frequency, $\omega > \omega_s/2$, from the input spectrum.

From the above discussion, it can be surmised that the absolute lower value of sampling rate for a given problem is twice the highest frequency of interest. However, actually implementing a system with this sampling rate is not advisable. First, while it is theoretically possible to reconstruct a harmonic signal sampled at twice its frequency, the filter required to do so is of infinite length, and not BIBO stable (see the discussion of Shannon's reconstruction theory in Middleton and Goodwin (1990)). Second, there is no margin for error in the upper frequency limit; any slight change in upper frequency results in aliasing.

So what is a good sample rate? It was noted in Chapter 6 that for tonal excitation, synchronizing the sampling to be at four times the excitation has some beneficial effects from the standpoint of adaptive algorithm performance (the possibility of a uniform error surface, or equal eigenvalues, which leads to the 'best' compromise between algorithm speed and stability). In general, however, we cannot expect to be able to synchronize the sample rate of the controller with the unwanted disturbance, even if it is harmonic.

To paint a qualitative picture of a 'good' sample rate, consider the problem of sampling a step response of a system. If, for ease of computation, we assume the system has a bandwidth of 1 Hz, then the continuous signal will be defined by

$$y(t) = 1 - e^{-2\pi t} \tag{13.1.14}$$

Referring to Fig. 13.6, if the step response is sampled at 2 Hz, the step is indistinguishable to the viewer. Sampled at 5 Hz, the characteristics begin to appear. At 10 Hz, the step is apparent. In fact, if the samples are connected with straight lines, the reconstruction of the step is in error by less than 4% (Middleton and Goodwin, 1990). Intuitively then we can postulate that the filtering exercise, which is analogous the reconstruction of a signal, becomes 'easier' as the sampling rate increases.

There is, however, a limit to this process. At high sample rates, tens or even hundreds of times the disturbance frequency, there are numerical problems. In Chapter 6 it was shown that as the level of oversampling increased the disparity in the eigenvalues of the autocorrelation matrix of the input samples increased. These eigenvalues to a large extent govern the convergence behaviour of the algorithm. A large disparity in eigenvalues generally translates into slow algorithm convergence, and reduced algorithm stability. Fast sampling also

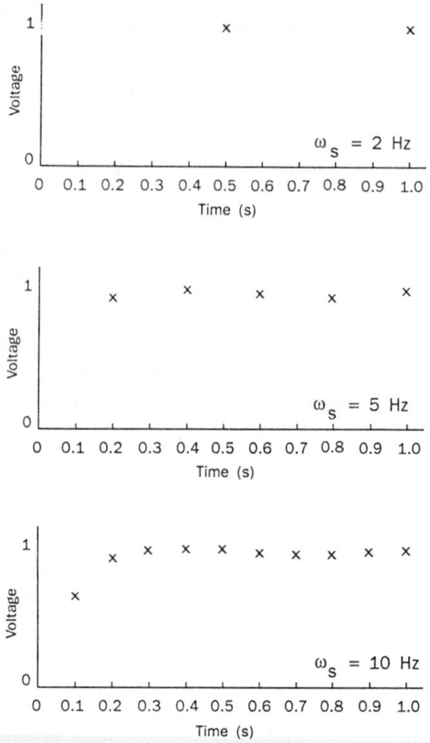

Fig. 13.6 Step response of a system with a 1 Hz bandwidth, sampled at 2 Hz, 5 Hz and 10 Hz.

enhances the numerical inaccuracies associated with finite word length (integer) calculations (Middleton and Goodwin, 1990).

Based on the above discussion, it can be surmised that the optimum sample rate is a compromise between fast and slow; both of these extremes lead to problems with adaptive algorithm convergence and, especially with fast sampling, stability. The 'optimum' sample rate compromise is often sited as ten times the frequency of interest. In practice, this sample rate provides for rapid convergence of the adaptive algorithm and reasonable levels of stability. In implementing active control systems we often find that for a given sampling rate ω_s the system will work reasonably well from frequencies approaching $\omega_s/100$ to frequencies up to $\omega_s/4$. On the low end of the scale, adaptation of the controller with frequencies below $\omega_s/100$ is often (extremely) slow, and not a particularly stable operation. While this is sometimes improved by increasing the length of the digital (control) filter, the only real solution is a reduction in sampling rate. On the high end of the scale, the adaptive algorithm appears ineffective with excitation frequencies above $\omega_s/4$.

13.1.2 Converter type and group delay considerations

A second important variable in the design of analogue/digital interface is the converter type, a selection which is influenced by group delay considerations. For our discussion here, group delay is a measure of the time it takes for a signal to pass from the input of the analogue/digital interface to the output. For analogue to digital conversion, what is of interest is the group delay through the anti-alias filters and the ADC. For digital to analogue conversion, what is of interest is the group delay through the DAC and the reconstruction filters.

Before being a discussion of factors related to group delay, it is necessary to understand why group delay is important. Group delay has a significant influence upon (at least) three variables in active control system design: physical size, level of control, and stability. Physical size is most obviously influenced by group delay in a causal feedforward active control system, such as a system designed to control (broadband) sound propagation in an air handling duct. If an upstream microphone is being used to supply a reference signal to the active control system, the group delay will limit the proximity of the reference sensor and control source, and hence the physical size of the system. If the group delay through each converter and filter is a few milliseconds, then for sound travelling at 343 m s^{-1} to be controlled the active control system must be at least two meters in length (once sensed, the sound will travel two metres by the time the controlling disturbance is introduced into the system). For compact systems, group delay minimization is desirable.

If a feedback control system is used, the effect of group delay is slightly different. First, group delay through the analogue/digital interface will be responsible for a phase shift in the signal provided to the controller; the phase of the signal provided to the control at time t will not be the same as the signal at the sensor at time t. If at all significant (say, $>10°$) this phase shift must be taken into account at the time of controller design. Even if it is taken into account, group delay often reduces the 'average' level of control which is achieved. This is because feedback systems are often implemented to attenuate the reverberant response of a system. If it takes 'longer' to sense the disturbance, it takes longer to control it. Averaged over the operating cycle of the system, this results in reduced levels of attenuation with longer group delays.

Group delay can also influence the stability of adaptive algorithms used in feedforward active control systems. As is a finite time delay between the derivation of a control signal and its 'appearance' in the error signal, there is a delay between a change in filter weights and its effect being measured. As was discussed in Chapter 6, when the weight adaptation is done at time intervals shorter than this delay the stability of the adaptive algorithm is reduced. With long group delays this is often the case.

In active control system implementation, perhaps the two most common types of ADCs employed are successive approximation converters and sigma delta converters. Successive approximation converters, which are probably the most common used at the time of writing this book, function by a sequence of comparisons between the measured signal and known voltage levels. These converters often require a separate sample and hold amplifier to 'capture' the signal for conversion, and external antialiasing filters. They range widely in price, but can be relatively expensive for highly accurate devices. The group delay through the converter (governed by the conversion time) can vary, but is often significantly less than the group delay through the anti-alias filter which precedes it.

The group delay through the anti-alias filter is influenced by a number of variables. The two most significant are the number of poles in the filter and the cut-off frequency. Increasing the number of poles, and decreasing the cut-off frequency, both increase the group delay in an approximately linear fashion.

Sigma delta converters are functionally different from successive approximation converters, and are often significantly cheaper. These converters sample internally at a rate significantly faster than the 'observed' sampling rate, such as 128 times per delivered sample. An internal conversion with an accuracy of only 1 bit is followed by a 'decimation' filtering process to derive the final sampled signal. Because of the high internal sampling rates, the antialiasing filter cut-off can be an extremely high frequency. This means that a single pole filter can be used, which is usually on-chip. The high sampling rate also means that the sample and hold amplifier is not required. The overall result is often a (significantly) cheaper converter with no additional (sample and hold and antialiasing filter) circuitry required.

Sigma delta converters also have some negative features. The most significant of these is group delay. The filtering process inherent in the sigma delta conversion technique requires time. At the time of writing this book, the time was typically 30 samples at the nominal (no internal) sampling rate. Therefore, if a sampling rate of 5000 Hz was used, the group delay on the input to the controller is in the order of 5−6 ms. This is significantly longer than the group delay which could be expected using a successive approximation converter and an antialiasing filter with a cut-off at 2000 Hz. For harmonic excitation this is not a problem, as the controller can be non-causal (if an adaptive algorithm is used stability will be reduced, but this can often be compensated for). However, for a causal system this may be unacceptable.

13.2 MICROPROCESSOR SELECTION

In a digital active control system implementation, the actual control algorithms, including the control filters, will run on a microprocessor of some type. Advances in microprocessor technology are coming at such a rapid pace that it is somewhat pointless to outlined specific 'chips', as by the time this book is published it is likely that the components will be out of date. It is no coincidence that interest in active noise control has paralleled advances in microprocessor technology; it is these advances which has made active noise and vibration control practically realisable. As microprocessor technology continues to improve, the applicability of active control will expand.

In this section we will briefly describe the types of microprocessors which are most commonly used at this stage, and the differences between them. Specific brands will not be discussed, only broad categories. The three types of microprocessors of interest here are 'general' microprocessors, digital signal processors and microcontrollers. In the future it is quite possible that the differences which separate these types of microprocessors will begin to evaporate, and the chip 'classes' described here will begin to merge as the technology advances.

'General' microprocessors, for the purposes of the discussion here, refer to microprocessors which have large instruction sets and can do a wide range of tasks. Such a microprocessor can be viewed as a 'jack of all trades but master of none'. There are several classes of chips which fall into this category. Perhaps the most common is a complex instruction set computer, or CISC chip. These are general purpose microcomputers which have assembly language instruction sets with complex instructions which take several microprocessor clock cycles to implement. These are typically the microprocessors found in personal computers.

A more advanced version of this type of chip is the reduced instruction set computer, or RISC chip. RISC chips have only a simple instruction set, where each instruction can be implemented in one cycle of the microprocessor clock. While they are typically faster than their CISC cousins, RISC chips tend to be more expensive and more complex. Their compilers are also more complex, as all complex routines must be broken down into a series of simple instructions.

In active control implementations, general purpose microprocessors often have the disadvantage that they are relatively slow to do an important task: multiply. It is not uncommon for a general purpose microprocessor to take tens of clock cycles to do a single multiplication. When implementing digital filters this can be a serious problem, greatly limiting the length of filters which can be implemented. They do, in general, have greater flexibility than the other two types of microprocessors being discussed here. This alone, however, is usually not enough to make them the most attractive option.

Digital signal processors (DSPs) are microprocessors with architectures which are optimized for digital signal processing and numeric computation. They can typically perform a multiply and accumulate (add) operation in a single microprocessor clock cycle. This makes them extremely attractive for active control system implementation, which relies heavily on multiplications and additions (the functions of a digital filter). These microprocessors tend not to support 'other' operations, such as division, as well as general purpose microprocessors, but are often still the microprocessor of choice.

Microcontrollers are essentially complete computer systems on one chip. They come with a wide variety of on-board options, including ADCs and DACs. They are also very cheap, commonly finding use in mass-produced products such as white goods. They are, however, lacking in some of the functions important in active control. They generally has less accuracy than the other devices discussed here, being, for example, 8-bit devices instead of 16- or 32-bit devices (which the other devices typically are). The ADCs and DACs are also typically 8 bits. Microcontrollers are also slow to multiply, as they lack the specialized architecture of the DSP. Their main asset is low cost, although at the time of writing this book DSPs were rapidly approaching the cost of microcontrollers in quantity.

13.3 SOFTWARE CONSIDERATIONS

Once a microprocessor and analogue/digital interface has been chosen, some though must be given to the layout of the software. The issues of interest here are division of the control code, language use and the choice of fixed or floating point implementation.

In a typical adaptive feedforward active control system, the controller software can be divided into two parts: that which must be done in real time (at each sample period), and that which can be done at a rate slower than real time (over several sample periods). Typically, calculation of the control filter output(s) must be done in real time, while all other operations can be done 'off line'. Real time code is typically 'interrupt driven'. With this, when a data sample is ready, a signal is sent to the chip which causes it to stop what it is doing and go perform a specified operation (in this instance, calculation of the filter output). Splitting controller code into an interrupt-driven real time part and an off-line background part often leads to the most efficient use of the microprocessor capabilities, especially for large implementations.

With most microprocessors, a number of programming languages can be used to generate controller code. These usually include a low-level assembly language and a variety of high-level languages, such as C. While it is generally possible to program tasks in any of the available languages, the speed of execution of the compiled code can vary significantly. In general, code written using the microprocessor assembly language executes faster (in other words, is more efficient) than code written using a higher level language. This means that it is advantageous to write the real time portion of the control code in assembly language, to give it maximum speed. The off-line code may then be written in a higher level language, where speed is not as important as the speed in with which a program can be written and debugged (higher level languages tend to be far simpler to program in that assembly language).

Finally, there is the choice of fixed point and floating point code. Basically, fixed point code is implemented using integer number, while floating point code is implemented using real numbers. Microprocessors, especially DSPs, can be purchased which explicit support fixed point and floating point calculations. In general, fixed point processors are (significantly) less expensive than floating point processors. However, fixed point calculations are more prone to a range of numerical problems than floating point calculations. As most fixed point devices have floating point 'libraries' which allow them to emulate the functions of fixed point processors, it is not uncommon to implement a controller on a fixed point chip with fixed point real time filtering code and floating point weight adaptation.

REFERENCES

Franklin, G.F., Powell, J.D. and Workman, M.L. (1990). *Digital Control of Dynamic Systems*. Addison-Wesley: Reading, MA.

Middleton, R.H. and Goodwin, G.C. (1990). *Digital Control and Estimation: A Unified Approach*. Prentice Hall: Englewood Cliffs, NJ.

14

Sound sources and sound sensors

Many of the active control systems discussed in previous chapters require the use of transducers to either generate or measure sound. To choose the most appropriate transducer for a particular application, and to predict the maximum controllability of a particular physical system, it is useful to have a basic understanding of the physical principles governing the operation of the various transducers.

14.1 LOUDSPEAKERS

When choosing a loudspeaker for an active noise control system, the important parameter for specification is cone volume displacement capability rather than power handling capacity. The electrical power needed to generate the required cone displacement depends on the volume of the enclosure backing the loudspeaker. At low frequencies (in the range generally of interest for active control) the effective stiffness associated with the loudspeaker will be inversely proportional to the enclosure volume. Thus, the electrical input power to the speaker will also be approximately inversely proportional to the square of the effective stiffness, as is shown by the following analysis.

At low frequencies where the speaker cone can be assumed to behave like a rigid piston, the thrust T applied by the voice coil is

$$T = BLI \tag{14.1.1}$$

where B is the magnetic flux density, L is the length of conductor and I is the current flowing through it. This thrust is balanced by the acoustic load and the mechanical impedance of the loudspeaker suspension and cone. Equating the two gives

$$BLI = Z_m v/S \tag{14.2.2}$$

where v is the cone volume velocity, S is the effective cone area and Z_m is the total mechanical impedance which is the sum of that due to the acoustic load (Z_{ma}) and that due to the loudspeaker (Z_{me}). At frequency, ω, this latter quantity is given by

$$Z_{me} = j\omega m + C + k/j\omega \tag{14.1.3}$$

where m is the effective mass of the cone and armature, C is the suspension damping and k is the suspension stiffness (including the effect of the enclosure backing the loudspeaker). Note that at low frequencies and for small enclosures, the enclosure stiffness (which is inversely proportional to the volume) will generally be much larger than the suspension stiffness and thus will dominate the term k.

The terminal voltage on the voice coil is given by

$$E = Z_E I + BLu \tag{14.1.4}$$

where Z_E is the blocked electrical impedance given by

$$Z_E = j\omega L + R + C_c/j\omega \tag{14.1.5}$$

where L, R and C_c are respectively the inductance, resistance and capacitance of the voice coil. The quantity, u, is the voice coil velocity which is equal to v/A if we assume that the cone is moving as a piston. Note that BLU is the back EMF due to motion of the coil. Shepherd *et al.* (1986) reported that using this back EMF as a feedback signal to the coil driving signal improves the smoothness of the speaker frequency response and makes the speaker behave more like a constant volume velocity source (infinite internal impedance).

The terminal voltage can be related to the cone movement by combining (14.1.2) and (14.1.4) to give

$$E = u\left[\frac{Z_E Z_m}{BL} + BL\right] \tag{14.1.6}$$

The electrical power consumed by the voice coil is given by

$$W = \frac{1}{2}I^2 R + \frac{1}{2}u^2 \mathrm{Re}\{Z_m\} \tag{14.1.7}$$

which can be written in terms of the volume velocity of the cone (of area S) using (14.1.2) to give

$$W = \frac{v^2}{2S^2}\left[\frac{|Z_m|^2 R}{(BL)^2} + \text{Re}\{Z_m\}\right] \tag{14.1.8}$$

The first term in brackets in (14.1.8) represents the electrical losses and the second term the mechanical losses.

For small backing enclosures and low frequencies, the stiffness term of Z_m will dominate and it can be seen from (14.1.8) that the power will then be approximately proportional to this quantity squared. It is also of interest to note that the required electrical power to the speaker is also proportional to the square of the volume velocity. The required control source volume velocity to achieve optimal control in a particular physical system is determined using the physical system analyses discussed elsewhere in this book.

In some cases it will be found that multiple loudspeakers are needed to obtain the required cone volume velocity. For sound propagating in a rectangular section duct, the required cone volume velocity can be shown to be related to the propagating pressure contribution of each loudspeaker, which for plane waves can be written as (Shepherd *et al.*, 1986):

$$u = \frac{2pS(\omega d/2c_0)\ \sin(\omega\tau)}{\rho_0 c_0 S_p \sin(\omega d/2c_0)} \tag{14.1.9}$$

where p and u represent the acoustic pressure amplitude and acoustic velocity amplitude at frequency ω, d is the side length of the square piston representing the speaker cone, c_0 is the speed of sound, S is the duct cross-sectional area and S_p is the speaker cone area. For a circular cone, d may be replaced by $\pi d/4$, where d is the diameter of the cone, to give an approximate result. If p in (14.1.9) is an r.m.s. quantity then v will be also.

In (14.1.9), the term sin $(\omega\tau)$ is included to account for a speaker source consisting of two speakers to make it directional (see Section 7.2 for a full discussion). The quantity τ is the time the signal into one speaker is delayed before being imput to the other. If only one speaker is used, the term $\sin(\omega\tau)$ is set equal to unity. If the loudspeaker is small compared to a wavelength of sound, the term, $(\omega d/2c_0)/\sin(\omega d/2c_0)$ from (14.1.9) is approximately unity and the required speaker response is much easier to realize in practice, thus minimizing signal distortion. Thus, use of a number of smaller speakers is always preferred over one large loudspeaker if it is practical. For example, in a duct system it would be preferable to use a number of smaller speakers placed in the duct walls at the same axial location along the duct, rather than a single large loudspeaker. For random noise, the peak cone velocity requirements for active control are likely to be four or five times the estimated r.m.s. velocity requirement (Shepherd *et al.*, 1986). Note also that loudspeakers should not be driven at maximum power or maximum cone deflection, as this results in substantial shortening of the speaker life.

Shepherd *et al.* (1986) and Ford (1984) show that the speaker electrical power requirement can vary over three orders of magnitude for the same acoustic power output. They also show that the electrical power requirements are minimized at the frequency corresponding to the mechanical resonance of the loudspeaker. Thus, it is suggested that the loudspeaker be designed for each application so that its mechanical resonance lies in the centre of the frequency range of interest. Unfortunately, this is not usually a practical alternative and doing this will require the use of motional feedback of the speaker cone to smooth the resulting frequency response. Most large loudspeakers have a fundamental mechanical resonance in free air of between 25 and 50 Hz. This can be varied to suit a particular application by either adding mass to the cone (to reduce it) or by increasing the effective suspension stiffness by adding a backing enclosure to the speaker (to increase the resonance frequency).

In cases where the required volume velocity exceeds the capability of available speakers, (usually 200–400 W of electrical power), five alternative options are available. As a first alternative, the number of speakers at each control source location can be increased (for example, by placing them around the perimeter of a duct cross section at the desired axial location).

As a second alternative, the loudspeaker coil and magnet could be replaced by a low-inertia, high-speed d.c. servo-motor similar to those used in computer tape drives. A belt arrangement is used to convert the motor rotary motion to linear motion. Because of its low inertia and ability to rotate in either direction, the motor is capable of responding to audio signals up to about 130 Hz. Peak to peak deflections achievable with this arrangement are about 30 mm, compared to 10 mm for a typical bass speaker (Leventhall, 1988). Speakers are commercially available which can produce 130 dB at 1 m, between 30 Hz and 130 Hz and 114 dB at 16 Hz with less than 2% harmonic distorton and 400 W of electrical power (Intersonics Inc).

As a third alternative, the loudspeaker could be replaced with an air modulated source which consists of a device which directs large volumes of air through slots which are adjusted in size using a vibration generator or servo-motor. Other techniques involve the continuous modulation of the size of an orifice through which high pressure air is passing by using a plug or valve attached to a vibration generator or servo-motor (Glendinning *et al.*, 1988). Because of the low acoustic efficiency of these types of sources, they need to be used in conjunction with a suitable horn, otherwise the compressed air requirements are very large; for example, it was found that two industrial air compressors typically used on building sites were needed to provide sufficient sound to actively control noise in an automotive exhaust system.

A fourth alternative is to use a horn to improve the coupling efficiency of the speaker in the frequency range of interest. The design of horns is discussed in the next section.

A fifth alternative is to use a well designed, tuned (and probably ported) backing enclosure optimized for the frequency range of interest. Software is commercially available to do this, using as input, speaker electrical and mechanical specifications and the characteristics of the space into which the speaker will be radiating. If the speaker is to be used in a duct, one of dimensions of the radiating space is set very large and the other two are the duct cross-sectional dimensions.

Use of conventional loudspeakers as active control sources in industrial environments requires considerable forethought and planning, even supposing that the speakers have been selected so that they will be subject to only a fraction of the maximum deflection capability. Special precautions must be observed in the speaker design to take account of the operating environment. For example, high temperature, water resistant adhesives must be used for the cone/coil connection in applications involving high humidity and/or high temperature. Special care is also needed with the speaker cone design; either the paper cone must be sprayed with a protective coating or substituted with some inert plastic, depending on the nature of the environment to be encountered. In cases of extreme heat, the loudspeaker cone must be protected with a heat shield which in some cases may require the addition of cooling air flowing past the shield. Alternatively, the loudspeaker could be removed from extreme conditions by placing it at the end of a short tube, although the space for this may not always be available. High temperature loudspeakers designed especially for active noise control applications are commercially available (Harmon Inc.) but they are relatively expensive.

Loudspeaker enclosures generally need drain holes to prevent water build-up caused by cyclic heating and cooling of enclosures, especially those located outdoors. The drain holes must be sufficiently small not to affect the stiffness of the enclosure at the frequencies of interest or alternatively must be taken into account in the design of the enclosure. Calculations of the effect of vents on the enclosure impedance can be included as part of the overall enclosure design.

A loudspeaker backed by a small enclosure can be approximated as a constant volume velocity source for the purpose of physical system analysis in the low frequency range generally of interest for active noise control (e.g. below 400 Hz). Larger enclosures could result in the suspension resonance lying in the frequency range of interest, and unless cone motional feedback is used (Shepherd *et al.*, 1986), the speaker cannot be considered a constant volume velocity source at frequencies near this resonance. However, at higher frequencies or for speakers with no backing enclosure, the constant pressure source offers a closer approximation to actual behaviour.

Manufacturers have been slow to produce loudspeakers designed specifically for active noise control applications, although some manufacturers are currently working in this area. Generally, loudspeakers used in clean industrial installations have been found to have a life of approximately three years, although this is likely to increase as speaker designs more appropriate to active control requirements (relatively low continuous power, high transient power and large cone excursions) are developed.

14.2 HORNS

Horns are used simply as a means of improving the electrical to acoustic efficiency of a loudspeaker. Use of an exponential or conical horn in a conventional installation will result in a very uneven frequency response, both in amplitude and phase, thus making control over abroad frequency range more difficult (Ford, 1984). This is because the horn impedance controls the acoustic response in regions of the horn resonance and the driver resistance controls the acoustic response in regions of anti-resonance. Ford concludes that the use of horns is likely to introduce more practical difficulties than any improvement in efficiency could justify.

Nevertheless, the authors have found that horn drivers fitted with a catenoidal horn can provide large volume velocities over a relative wide frequency range with a small horn exit diameter. Catenoidal horns have an advantage over exponential or conical shapes because they have a much higher efficiency, especially just above the cut-off frequency of the horn. The catenoidal horn cross-sectional shape is (Morse 1948)

$$S = S_0 \cosh^2(x/h) \tag{14.2.1}$$

where S is the cross-sectional area at axial location x from the throat of the horn and S_0 is the cross-sectional area of the throat. The quantity h is a scaling factor, set equal to 0.5 m for the abovementioned design.

The efficiency of any horn is reduced by reflections from the mouth back towards the throat which occur when the mouth diameter is small compared to a wavelength of sound. At these low frequencies Morse (1948) shows that the specific acoustic impedance presented by the throat of the horn to the driver is given by (assuming positive time dependence)

$$\frac{Z_0}{\rho_0 c_0} = \frac{\tau/2(a\omega/c_0)^2 + j\tan(\omega\tau L_e/c_0)}{\tau + (j\tau^2/2)(a\omega/c_0)^2 \tan(\omega\tau L_e/c_0)} \tag{14.2.2}$$

where ω is the driving frequency, c_0 is the speed of sound in free space, L_e is the effective length of the horn ($= L + 8a/3\pi$), L is the actual length of the horn, a is the radius of the horn mouth and τ is defined by

$$\tau^2 = 1 - (\omega_0/\omega)^2 \tag{14.2.3}$$

where ω_0 is the horn cut-off frequency given by

$$\omega_0 = c_0/h \tag{14.2.4}$$

The power radiated from the horn (assuming negligible losses in the horn and assuming a constant volume velocity driver) is given by

$$W = \frac{1}{2}\bar{u}_0^2 S_0 \,\mathrm{Re}\{Z_0\} \tag{14.2.5}$$

where $\bar{u}_0$ is the velocity amplitude of the driver.

The derivation of (14.2.2) is based on the assumption that the horn mouth is surrounded by an infinitely large baffle and radiates into free space. Radiation into a duct is a considerably more complex problem, and the resulting expression would depend on the impedance presented by the duct (or the location of the horn along the duct axis). Nevertheless, some general conclusions can be made as follows.

1. The acoustic volume velocity generated at the mouth of the horn will be strongly frequency dependent, as it depends on the impedance presented by the duct to the mouth as well as the horn transfer impedance from the throat to the mouth.
2. Although the horn driver at the throat may behave like a constant volume velocity source, the mouth of the horn will behave more like a constant pressure source. Thus, optimum locations for horns will be different to those for enclosure backed loudspeakers for the same physical system to be actively controlled.
3. At certain frequencies, it will be possible to achieve much larger volume velocities for a certain diameter source than would be possible using a loudspeaker. Thus, horn sources may prove advantageous in some situations requiring large reductions in tonal noise.

14.3 OMNI-DIRECTIONAL MICROPHONES

In modern digitally based active control systems, the frequency response of the microphone used is not very critical, as any lack of flatness in amplitude or phase is taken into account in the system identification algorithms. For this reason, it is common to find relatively inexpensive microphones used in active control systems. The two most common types are the the prepolarized condenser (or electret) microphone (see Fig. 14.1) and the piezoelectric microphone (see Fig. 14.2). The electret microphone is sensitive to dust and moisture on its diaphragm, but it is capable of reliable operation at elevated temperatures and is relatively insensitive to vibration. By contrast, the piezoelectric microphone is less sensitive to dust and moisture but it can be damaged by exposure to elevated temperatures and, in general, it tends to be quite microphonic; that is, a piezoelectric microphone may respond about equally well to vibration and sound whereas a condenser microphone will respond well to sound and effectively not at all to vibration.

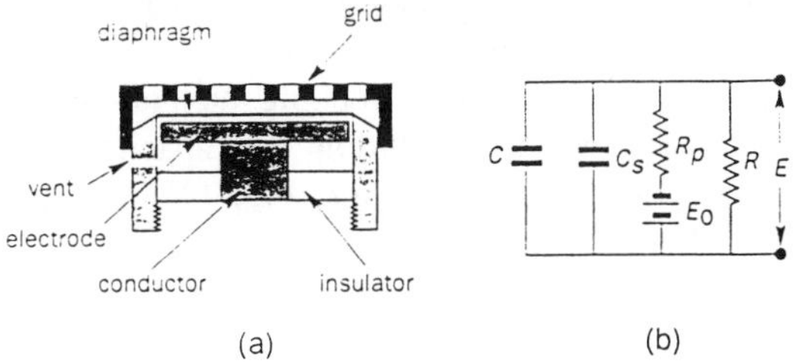

Fig. 14.1 Condenser microphone schematic and electrical circuit.

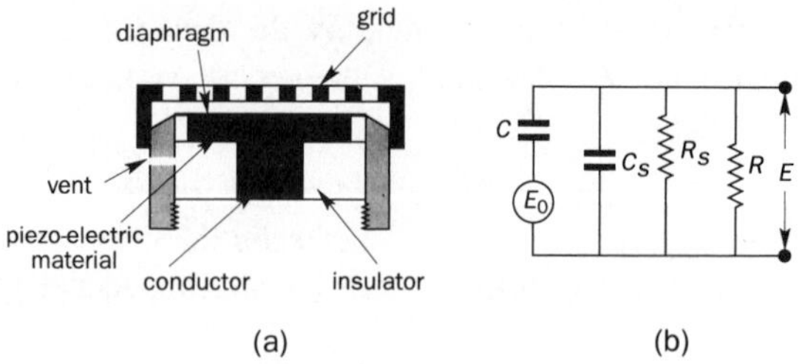

Fig. 14.2 Piezoelectric microphone schematic and electrical circuit.

14.3.1 Condenser microphone

A condenser microphone consists of a diaphragm which serves as one electrode of a condenser, and a polarized backing plate, parallel to the diaphragm and separated from it by a very narrow air gap, which serves as the other electrode. The condenser is polarized by means of a bound charge, so that small variations in the air gap due to pressure-induced displacement of the diaphragm result in corresponding variations in the voltage on the condenser.

The bound charge on the backing plate may be provided either by means of an externally supplied biasing voltage of the order of 200 V, or by use of an electret which forms either part of the diaphragm or the backing plate. It is possible to purchase very inexpensive electret microphones for less than US$3; however, a manual selection from a large batch is necessary to select those with a sufficient sensitivity, stability and frequency bandwidth. Details of the electret construction and its use are discussed in the literature (Fredericksen *et al.*, 1979). For the purpose of the present discussion, however, the details of the latter construction are unimportant. The essential features of a condenser microphone and a sufficient representation of its electrical circuit for the present purpose are provided in Fig. 14.1.

Referring to Fig. 14.1, the bound charge Q may be supplied by a d.c. power supply of voltage E_0 through a very large resistor R_p. Alternatively, the branch containing the d.c. supply and resistor R_p may be thought of as a schematic representation of the electret. The microphone response voltage is detected across the load resistor R. A good signal can be obtained at the input to a high internal impedance detector, even though the motion of the diaphragm is only a small fraction of the wavelength of light.

It is of interest to derive the equation relating the microphone output voltage to the diaphragm displacement. By definition, the capacitance of a device is its charge divided by the voltage across it. From Fig. 14.1 we can see that for the diaphragm at rest with a d.c. bias voltage of E_0 we have

$$\frac{Q}{C + C_s} = E_0 \tag{14.3.1}$$

The microphone capacitance is inversely proportional to the spacing at rest, h, between the diaphragm and the backing electrode. If the microphone diaphragm moves a distance x inward (negative displacement, positive pressure) so that the spacing becomes $h - x$, the microphone capacitance will increase from C to $C + \delta C$ and the voltage across the capacitors will decrease in response to the change in capacitance by an amount E to $E_0 - E$. Thus,

$$E = -\frac{Q}{C + \delta C + C_s} + E_0 \tag{14.3.2}$$

The microphone capacitance is inversely proportional to the spacing between the diaphragm and the backing electrode; thus,

$$\frac{C + \delta C}{C} = \frac{h}{h - x} \tag{14.3.3}$$

Equation (14.3.3) may be rewritten as

$$\delta C = C\left(\frac{1}{1 - x/h} - 1\right) \tag{14.3.4}$$

Substitution of (14.3.4) into (14.3.2) and use of (14.3.1) gives the following relation:

$$E = -\frac{Q}{C}\left[\frac{1 - x/h}{1 + (C_s/C)(1 - x/h)} - \frac{1}{1 + C_s/C}\right] \tag{14.3.5}$$

Equation (14.3.5) may be rewritten in the following form:

$$E = -\frac{Q}{C}\left[\frac{1 - x/h}{1 + C_s/C}\left(1 + \frac{(x/h)(C_s/C)}{1 + C_s/C} + \dots\right) - \frac{1}{1 + C_s/C}\right] \tag{14.3.6}$$

By design, $C_s/C << 1$ and $x/h << 1$; thus in a well designed microphone the higher order terms in (14.3.6) may be omitted and by defining an empirical constant $K_1 = 1/(Ch)$, (14.3.6) takes the following approximate form:

$$E \approx \frac{K_1 Q x}{1 + C_s/C} - \frac{K_1 Q x^2}{h(1 + C_s/C)^2} \tag{14.3.7}$$

As can be seen by the preceding equations, the constant K_1, depends upon the spacing, at rest, h, between the microphone diaphragm and the backing electrode and the capacitance of the device, C and must be determined by calibration. The quantity C_s is the stray capacitance in the electrical circuit. For good linear response, the capacitance ratio C_s/C must be kept as small as possible and similarly, the microphone displacement relative to the condenser spacing x/h should be very small. In a well designed microphone, the second term in (14.3.7) is thus negligible.

14.3.2 Piezoelectric microphone

A sketch of the essential features of a typical piezoelectric microphone and a schematic representation of its electrical circuit are shown in Fig. 14.2. In this case, sound incident upon the diaphragm tends to stress or unstress the piezoelectric element which, in response, induces a bound charge across its capacitance. The effect of the variable charge is like that of a voltage generator, as shown in the circuit.

For a piezoelectric microphone, the equivalent circuit of which is shown in Fig. 14.2, the microphone output voltage is given by

$$E = \frac{Q}{C + C_s} = \frac{Q/C}{1 + C_s/C} \tag{14.3.8a,b}$$

As $E_0 = Q/C$,

$$E = \frac{E_0}{1 + C_s/C} \tag{14.3.9}$$

The voltage, E_0, generated by the piezo crystal is proportional to its displacement. Thus,

$$E_0 = K_2 x \tag{14.3.10}$$

Substituting (14.3.10) into (14.3.9) gives

$$E = \frac{K_2 x}{(1 + C_s/C)} \tag{14.3.11}$$

Equation (14.3.11) is essentially the same as the linear term of (14.3.7). In either case, an acoustic wave of pressure p incident upon the surface of the microphone diaphragm of area S will induce a mean displacement x determined by the compliance K_3 of the diaphragm. Thus, the relation between acoustic pressure and mean displacement is

$$x = -K_3 S p \tag{14.3.12}$$

Substitution of (14.3.12) into either (14.3.7) or (14.3.11) results in a relation, formally the same in either case, between the induced microphone voltage and the incident acoustic pressure. Thus, for either microphone,

$$E = KSp/(1 + C_s/C) \tag{14.3.13}$$

In (14.3.13) the constant K must be determined by calibration.

14.3.3 Microphone sensitivity

It is customary to express the sensitivity of microphones in decibels relative to a reference level. Following accepted practice, we let the reference voltage E_{ref} be 1 V and the reference pressure p_0 be 1 Pa. We may then write, for the sensitivity S_m of a microphone,

$$S_m = 20\log_{10}[Ep_0/(E_{ref}p)] \tag{14.3.14}$$

Equation (14.3.14) may, in turn, be rewritten in terms of sound pressure level L_p:

$$S_m = 20\log_{10}E - L_p + 94 \qquad \text{(dB)} \tag{14.3.15}$$

Typical condenser microphone sensitivities range between −25 and −60 dB re 1 V Pa^{-1}. Piezoelectric microphone sensitivities lie at the lower end of this range. As an example, if the incident sound pressure level is 74 dB, then a microphone of −30 dB re 1 V Pa^{-1} sensitivity will produce a voltage which is down from 1 V by 50 dB. The voltage thus produced is $E = 10^{-50/20} = 3.15$ mV.

Reference to (14.3.11) shows that the output voltage from a microphone is directly proportional to the area of the diaphragm. Thus, the smaller a microphone of a given type, the smaller will be its sensitivity. As the example shows, the output voltage may be rather small, especially if very low sound pressure levels are to be measured, and the magnitude of the gain which is possible in practice is limited by the internal noise of the amplification devices. These considerations call for a microphone with a large diaphragm which will produce a corresponding relatively large output voltage.

In applications where very low level error signals must be detected, microphones with a low self-noise floor must be used and these are more expensive. For example, electrets with a reasonable frequency response from 20 Hz to 6 kHz and with a self-noise floor of 50 dB can be purchased for US$5−10. However, electrets with a flat frequency response from 20 Hz to 20 kHz and a self noise floor of −20dB can cost up to 100 times as much. Piezoelectric microphones are generally less expensive and less sensitive than electrets of the same diameter and are often adequate for use in active noise control systems.

14.4 DIRECTIONAL MICROPHONES

In cases where the direction of propagation of the acoustic disturbance is known, such as in a duct, it is desirable for the reference sensor to be directional so that it is insensitive to sound emanating from the control sources and only sensitive to sound arriving from the direction of the disturbance. In practice, the desired result can be achieved in any one or combination of three ways: using a directional control source (which consists of two or more sound sources, with appropriate relative phasing), using a directional microphone or by using an IIR filter to include the effect of the acoustic feedback signal in the electronic control system design. In this section we will be discussing directional microphones only. However, it should be pointed out that in many cases it is not possible to obtain a high performance stable active noise control system in a duct by just using an IIR filter and the reference sensor often needs some directional properties as well so that acoustic feedback from the control speakers to the reference microphone is minimized.

14.4.1 Tube microphones

A very simple directional microphone consists of a long tube with a small axial slit and with a standard condenser or piezoelectric microphone mounted at one end, as first suggested by Tamm and Kurtz in 1954. The principle of operation of this device relies on the pressure fluctuations generated by the external acoustic field travelling down the tube towards the microphone at the speed of sound. If the acoustic wave generating the pressure fluctuations in the tube is also travelling from the far end of the tube towards the microphone, then the pressure fluctuations generated inside the tube will all be in phase on arrival at the microphone and will reinforce one another. However, if the external acoustic wave is incident on the tube at some angle to the tube axis, the pressure fluctuations generated by it inside the tube will not be in-phase on arrival at the microphone, as the trace velocity of the external acoustic wave outside of the tube along the tube axis will be less than the velocity of the pressure fluctuations in the tube. Thus, the sensitivity of the device will decrease as the angle of the incident external sound field with respect to the tube axis increases from 0 to 180°, with 0° corresponding to a sound wave incident first on the end of the tube opposite the microphone. Thus, if this type of microphone is mounted in a duct, it should always point towards the sound source, regardless of the direction of any flow which is present.

The type of microphone just described is often referred to as a 'shotgun' microphone for obvious reasons. If the end opposite the one containing the microphone can be made totally absorptive, then the frequency response (or sensitivity as a function of frequency) of the device will be relatively flat. The best way to achieve an anechoic termination to a probe tube is to attach a long flexible, coiled tube to each end (Davy and Dunn, 1993). In this case the microphone is attached to the probe tube by way of a 'T-piece', which for the low frequency range generally of interest is insignificantly different to having it at one end.

If an anechoic termination is not achieved, the result is a wavy frequency response (± 1.5 dB for a 100% reflecting end, tube diameter 13 mm and tube length 500 mm). This frequency response characteristic can be made smoother by adding a hollow tube (with open ends), filled with porous acoustic material, to the inside of the tube. The hollow tube diameter would be about half the probe tube diameter as shown in Fig. 14.3. Also increasing the tube length and reducing its diameter will improve the flatness of frequency response (Munjal and Eriksson, 1989).

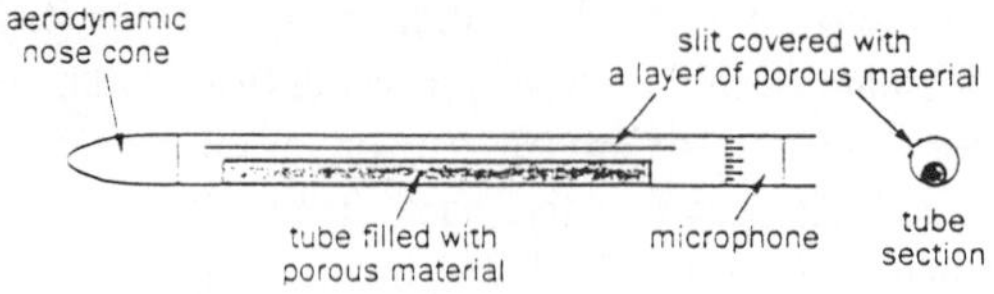

Fig. 14.3 Directional microphone with internal absorption tube.

In practice, it is necessary to cover the slit in the probe tube with a porous cloth to reduce the damping of the pressure fluctuations travelling through the inside of the tube. If this is not done, the pressure fluctuations generated at the far end of the tube away from the microphone will differ in amplitude to those produced close to the microphone, thus reducing the effect of the phase cancelling mechanism at the microphone location, which minimizes the contribution of acoustic signals resulting from waves not parallel to the tube axis. Thus, addition of the porous cloth enhances the directional properties of the device. However, if the cloth is too impervious, the overall sensitivity of the device will be substantially reduced. A good compromise, in practice, is to use a cloth with a flow resistance of about $2\rho c$ MKS rayls, where ρc is the characteristic impedance of the fluid in the duct. This results in a device with a sensitivity slightly higher (1 or 2 dB) than the microphone with no tube (Neise, 1975). Compare this to a reduction in sensitivity of 15 dB if a flow resistance of $20\rho c$ is used. The preceding values correspond to a slit width to tube circumference ratio of 0.025 and sound field frequency ω(rad s^{-1}) characterized by $\omega L/c > 2$ (Munjal and Eriksson, 1989), where L is the tube length (m) and c is the speed of sound in the fluid external to the tube (m s^{-1}).

The directional properties of the microphone and tube become more marked for longer tubes and higher frequencies. As a guide, the sensitivity difference between a sound wave arriving on-axis from the most sensitive direction and one arriving on axis from the opposite direction is 0 dB for $\omega L/c \leq 1$, 20 dB for $\omega L/c = 10$ and 40 dB for $\omega L/c = 100$. That is, the difference in dB varies in proportion to the log of the increase in the product of frequency and length, provided $\omega L/c \geq 1$. This assumes that the end of the tube opposite the microphone is 100% absorptive. If this is not so, the difference between the 0° and 180° response is much less, reducing to a constant 5 to 10 dB in the limit of a perfectly reflecting end for $\omega L/c > 2$.

As mentioned earlier, if the tube has a reflecting end opposite the microphone, the frequency response of the device will be wavy. This waviness increases if the flow resistance of the cloth over the slit increases, so this is another reason for keeping the flow resistance to a minimum.

14.4.2 Microphone arrays

Microphone arrays have also been used in the past to achieve directional sensing. The principle of operation is similar to a tube containing a microphone at one end, except that the array is made up of a number of discrete sensing points arranged in a line, rather than the continuous sensor represented by the tube. The signal from each microphone is phase shifted to account for the time taken for an acoustic signal to travel directly from one to the other, resulting in suppression of acoustic waves arriving off-axis, or from the wrong on-axis direction. The array is effective below frequencies for which the microphone separation distance is less than half a wavelength. Increasing the number of microphones increases the effective directivity, but very good results can be obtained with just two microphones. At very low frequencies, microphone arrays are much more effective than probe tubes and in a duct containing flow and low frequency noise, a combination of two or more microphones in a probe tube is likely to be necessary (at least for the reference sensor) to attack the dual problem of turbulence noise reduction and lack of microphone directivity. A disadvantage of the discrete microphones is the variability as a function of signal frequency required in the phase delay applied to the signal from the microphone closest to the primary noise source. This means that a fixed phase delay will result in a probe with good directivity properties over a narrow frequency range. This may be adequate for some problems and if not, the working frequency range can be increased by using more microphones or by using circuitry capable of providing a time delay independent of frequency. The authors have achieved a directivity of better than 20 dB between 30 and 40 Hz using just two microphones spaced 1m apart and a constant phase delay. The analysis of two and three speaker arrays (which is similar to that required for microphone arrays) was discussed in Section 7.2.2.3. Essentially, for a fixed time delay of τ_0 (which must be equal to the acoustic delay experienced by a signal travelling between the two microphones), the frequencies (in hertz) of maximum array sensitivity (twice that of a single microphone) are given by $f = 0.5n/\tau_0$, where n is an odd integer (see Fig. 7.8). At frequencies below $0.17\tau_0$, the microphone sensitivity is less than that of a single microphone. As the number of (equally spaced) microphones in the array increases, the frequency below which the sensitivity is less than that of a single microphone decreases ($0.07\tau_0$) for five microphones (see Fig. 7.10).

14.4.3 Gradient microphones

As the pressure gradient is greatest in the direction normal to the wave front of an acoustic disturbance, measurement of this quantity will provide a signal related to the direction of propagation of the wave. For a plane progressive wave of amplitude A, the first and second order pressure gradients along the x-axis are, respectively,

$$\frac{\partial p}{\partial x} = -jk_x A e^{-jk_x x} e^{-j\omega t} \tag{14.4.1}$$

$$\frac{\partial^2 p}{\partial x^2} = -k_x^2 A e^{-jk_x x} e^{-j\omega t} \tag{14.4.2}$$

where

$$k_x = \frac{\omega \cos\theta}{c_0} \tag{14.4.3}$$

Note the dependence of the sensitivity on frequency ω for the first order gradient and ω^2 for the second order gradient. The first order gradient microphone sensitivity is also dependent on cos θ where θ is the angle between the direction of propagation of the wave and the x-axis (which is normal to the surface of a gradient microphone). Clearly, a second order gradient device will have a $\cos^2\theta$ directivity. In the near field of a sound source, the pressure gradients will be larger than those given by (14.4.1) and (14.4.2) due to the decaying near field. However, this is in addition to the $\cos\theta$ or $\cos^2\theta$ dependence just discussed and will rarely be of importance in active noise control systems.

A first order gradient microphone may be realized by exposing both sides of an electret diaphragm (see section 14.3) to the sound field. A second order gradient microphone can be realized by placing two first order gradient microphones in small baffles spaced a short distance apart (Sessler *et al.*, 1989) as shown in Fig. 14.4.

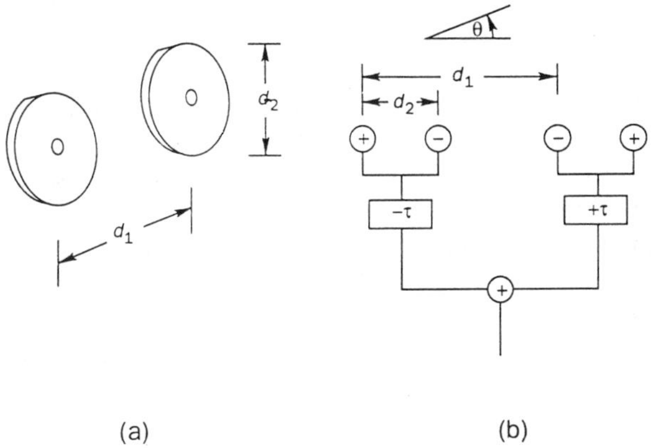

Fig. 14.4 Second order gradient microphone arrangement: (a) physical arrangement; (b) schematic representation.

The baffles in the second order gradient serve to delay the signal between each side of each of the two first order gradient microphones. The signal from the first order microphone on the left is delayed by $-\tau$ and the one on the right is delayed by $+\tau$. In practice, this is achieved by delaying the output of the right hand gradient microphone by 2τ and the left one by zero. Note that the outputs of the two first order gradient microphones are poled in anti-phase as shown in the figure.

The sensitivity of the second order gradient microphone shown in Fig. 14.4 at low frequencies is given by Sessler *et al.* 1989:

$$M_s = M_{os} k^2 d_1 d_2 \left[\frac{2c_0\tau}{d_1} + \cos\theta \right] \cos\theta \tag{14.4.4}$$

where M_{os} is the sensitivity of each of the first order gradient microphones, k is the wavenumber at the frequency of interest and the other quantities are defined in the figure.

Although gradient microphones are probably unsuitable for use as active control system sensors in ducts containing flow, they may be very useful as directional sensors for systems used to control sound radiation into free or enclosed spaces. A disadvantage of the simple first order gradient microphone is that it is bidirectional, i.e. it cannot distinguish the difference between waves arriving from in front of it or behind it. On the other hand, the second order gradient microphone just described is unidirectional and has a very strong null corresponding to sound waves arriving from the $\theta = 180°$ direction.

A further disadvantage of gradient microphones in general is their relatively high noise floor at low frequencies, although this may not be a limitation in situations where very high level sound fields are being controlled.

14.5 TURBULENCE FILTERING SENSORS

When active noise control systems are applied to duct systems containing flow, the performance is often compromised by contamination of the reference and error microphone signals with signals generated by turbulent pressure fluctuations which travel at the flow speed rather than the speed of sound. The turbulent pressure fluctuations contribute very little to the noise radiated from the end of the duct; thus, it is essential that their influence is removed from both the reference and error microphones used in an active noise control system. If this is not done, the performance of the active control system in reducing sound propagation will be severely impaired.

14.5.1 Probe tube microphones

The most common and simplest way of reducing the influence of turbulent pressure fluctuations is to use a probe tube microphone, which consists of a standard condenser or piezoelectric microphone placed at one end of a long tube of diameter similar to the microphone diameter. The walls of the tube are either porous, or contain holes or an axial slit. For best results, the microphone must be located at the end of the tube furthest from the sound source (the tube is oriented so that its longitudinal axis is parallel to the direction of sound propagation – see Fig. 14.5).

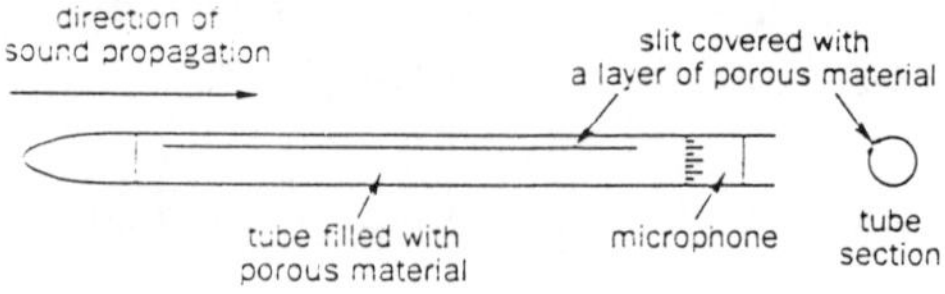

Fig. 14.5 Probe tube microphone in a duct.

The principle of operation of the probe tube microphone in minimizing the effect of the turbulent pressure fluctuations on the microphone signal relies on the difference in speed of propagation between acoustic waves and turbulent pressure fluctuations. Any fluctuating pressure field (whether acoustic or turbulent flow) acting on the outside of the tube will generate pressure disturbances inside the tube which will propagate with the speed of sound to either end of the tube. To simplify the explanation we will assume that the tube end opposite the microphone completely absorbs acoustic waves incident on it from inside the tube. Then the pressure at the microphone is only determined by the waves directly propagating toward it from the tube wall. Pressure fluctuations along the inside of the tube excited by an external acoustic field will propagate towards the microphone at the same speed as the acoustic wave in the external duct. Thus, the pressure fluctuations excited inside the tube at various locations along the length will all arrive in phase at the microphone. On the other hand, pressure fluctuations excited inside the tube by an external turbulent pressure field will not arrive in phase at the microphone, as the turbulent pressure field external to the tube propagates at the speed of the flow. Thus, the pressure fluctuations generated on the inside of the tube by the turbulent flow will cancel one another out at the microphone, due to their random differences in relative phase. As a further illustration of the principle of operation, Shepherd *et al.* (1989) show mathematically that to an observer moving at the speed of sound, the turbulent pressure fluctuations in a duct average to zero while the acoustic component appears as a steady signal. It is also clear that the faster the convection speed of the turbulent pressure fluctuations the longer the probe tube will need to be to adequately suppress their influence at the microphone. Of course, to completely suppress the turbulent pressure contribution at the microphone, the probe tube would need to be infinitely long. On the other hand, if the flow speed is too great, there will be a significant difference between the speed of sound in the duct and the speed of sound in the probe tube. This will result in a reduction in the sensitivity of the probe tube microphone to the acoustic signal. This is discussed in more detail in Section 14.5.1.2.

For reasons outlined in Section 14.4, the probe tube microphone also has a significant directional response, being much more sensitive to sound incident from one end than from the other end. Thus, in an active noise control system it serves the dual purpose of minimizing the effect of turbulent pressure fluctuations and also the acoustic feedback arriving at the reference microphone from the control source (in a duct).

The sensitivity of the microphone in the probe tube to acoustic pressure fluctuations and to turbulent pressure fluctuations is dependent upon a number of parameters, each of which were investigated by Neise (1975) and which will be discussed in the following sections. As pointed out by Munjal and Eriksson (1989), the analysis used by Neise only considers the effect of the external sound field in the duct on the internal field in the probe tube and neglects to consider the effect of the sound field in the probe tube (generated by the external field) on the sound field in the duct. This has the effect of introducing errors in both the acoustic and turbulent pressure sensitivity of the microphone, which become significant at frequencies ω (rad s^{-1}) characterized by $\omega L/c < 0.5$, where L is the probe tube length and c is the speed of sound in the fluid external to the probe tube. Specific values quoted in the sections to follow apply to the probe tube discussed by Neise (1975) and Munjal and Eriksson (1989). This tube was 13 mm in diameter and had a slit width to tube circumference ratio of 0.025. The tube length dependence is included in the parameters quoted.

14.5.1.1 Effect of slit flow resistance on acoustic sensitivity

To optimize the performance of the probe tube microphone, it is necessary to cover the slit with acoustically porous cloth, which serves to minimize the damping effect of the slit on acoustic waves propagating in the tube. Without the cloth cover, the acoustic waves generated in the tube at the far end away from the microphone would be significantly reduced in amplitude on reaching the microphone, and cancellation at the microphone of the various pressure fluctuations generated by the turbulent pressure field would not be as effective. However, if the cloth is insufficiently porous, the sensitivity of the microphone to the external acoustic field will be substantially reduced. Although the turbulent pressure sensitivity is also reduced by a similar amount, it is clearly undesirable to reduce the acoustic sensitivity of the microphone unnecessarily as the microphone signal will then be more susceptible to contamination by electronic noise.

A convenient compromise is to use fibreglass cloth having a flow resistance of approximately $2\rho_0c_0$ MKS rayls. This results in an acoustic sensitivity of the probe tube plus microphone of approximately 2 dB over that of the microphone with no tube, for frequencies above $\omega L/c_0 = 5$ and -3dB at frequencies below $\omega L/c_0 = 0.5$. In between, the sensitivity may be reasonably approximated as a linear function of $\omega L/c_0$, where L is the probe tube length.

For a cloth having a flow resistance of $2\rho c_0$, the sensitivity of the microphone plus tube is -15 dB at a frequency corresponding to $\omega L/U = 10$ (where U is the mean flow speed in the duct) and this decreases by 10 dB each time $\omega L/U$ increases by a factor of 10). For example, the sensitivity corresponding to $\omega L/U = 100$ is -25 dB. At lower frequencies, the sensitivity probably approaches 0 dB close to $\omega L/U = 1$, but this is uncertain, as the only analysis available is that by Neise (1975) which suffers from the one way acoustic interaction limitation mentioned earlier..

Provided that the acoustic sensitivity of the probe tube and microphone were independent of flow speed in the duct, the amplification of the acoustic signal over the turbulent pressure fluctuation signal would be given by the difference in the acoustic and turbulent pressure sensitivities which can be calculated as just outlined for a specified flow speed and frequency. Although the acoustic sensitivity does depend on flow speed at higher values of $\omega L/c_0$, it will be shown in the next section that this occurs outside the frequency range and flow speed range of general concern in active noise control systems.

The previous discussion has been based on the assumption that the end of the tube opposite the microphone is 100% absorbing. In practice, achieving this condition is difficult and best done as described by Davy and Dunn (1993). The effect of the reflecting end on the acoustic sensitivity corresponding to a slit covered by a cloth of flow resistance equal to $2\rho c$ is to cause it to vary between 0 and +3 dB as a function of frequency for frequencies above $\omega L/c = 2$ and to be equal to zero at lower frequencies. The effect of the reflecting end on the turbulent pressure sensitivity is to increase the average sensitivity by about 2 dB and introduce a small fluctuation (± 1.5 dB) as a function of frequency (Neise, 1975). As the effect of the reflecting end on the difference between the acoustic and turbulent pressure sensitivity is small (of the order of 2–3 dB), it is generally not considered worthwhile to attempt to make the probe tube end opposite the microphone absorptive. The ± 1.5 dB waviness in the frequency response could be reduced by adding into the probe tube a hollow smaller diameter tube open at the ends and filled with porous material as illustrated in Section 15.4 and described by Neise (1975), but this is usually unnecessary for digital active noise control systems.

14.5.1.2 Effect of flow speed on acoustic sensitivity

The effect of the mean flow speed in the duct on the turbulent pressure sensitivity of the probe tube microphone was discussed in the previous section. Here we will discuss the effect of flow speed U on the acoustic sensitivity. Neise showed that for the tube he investigated, the flow speed begins to have an effect on the acoustic sensitivity at frequencies corresponding to $\dfrac{\omega L}{c_0} > \dfrac{2}{M}$ where M is the Mach number of the mean flow. It appears that the sensitivity is reduced by about 20 dB for each factor of 10 increase in the product ωL, although the actual sensitivity curve is a wavy line as a function of frequency, varying ± 3 dB about the mean. The reason for this high frequency roll-off in microphone acoustic sensitivity is that as the flow speed increases the phase speed of the acoustic wave in the duct becomes more different to that in the probe tube (in which there is no mean flow). Thus, the same cancellation mechanism begins to occur as for the turbulent pressure fluctuations. Fortunately, at reasonable flow speeds and probe tube lengths, this high frequency roll off occurs beyond the highest frequency usually of interest for active control. In general, the difference between the acoustic pressure and turbulent pressure sensitivities of the microphone increases by 3–4 dB for each halving of the flow speed (Neise, 1975), at least in the frequency range below the frequency at which the acoustic sensitivity begins to roll off. Note that flow speeds characterized by $M = 0.05$ to 0.1 are typical of commercial air conditioning and industrial air handling systems.

14.5.1.3 Effect of probe tube orientation

It is clear from the discussion in Sections 14.4 and 14.5 so far, that the probe tube should be oriented so that the microphone end is furthest from the acoustic source generating the sound field to be measured. The direction of mean flow is not important. Orientation of the probe tube the opposite way round will result in a much smaller difference between the acoustic and turbulent pressure sensitivities, resulting in a much poorer turbulent pressure fluctuation filter.

14.5.1.4 Effect of probe tube diameter

Munjal and Eriksson (1989) show that increasing the probe tube diameter (for a fixed duct cross-sectional area) for a tube with completely reflecting ends results in reduced acoustic sensitivity and increased waviness in the acoustic sensitivity frequency response. They conclude that the maximum permissible probe tube diameter should be less than 0.1 of the effective duct diameter. Note that the waviness in the frequency response increases from ± 1.5 dB for a diameter ratio of 0.085 to ± 6 dB for a ratio of 0.34 (Munjal and Eriksson 1989).

14.5.1.5 Effect of a reflective duct termination

If the probe tube is being used to sample the acoustic field in a duct in which waves are reflected from the exit, then the waviness in the acoustic pressure sensitivity frequency response curve can be as much as ± 10 dB. This could make the electronic controller design more difficult as stability margins could decrease.

14.5.1.6 Probe tube design guidelines

1. Probe tube diameter should be less than 0.1 of the effective duct diameter.
2. Sound absorbing material in the probe tube or at the end opposite the microphone is unnecessary.
3. The flow resistance of the porous material covering the slit should be approximately $2\rho_0 c_0$.
4. The probe tube should be oriented so that its microphone is furthest from the sound source generating the sound field to be measured, irrespective of the direction of mean flow.
5. The ratio of slit width to probe tube circumference should be about 0.025.
6. The probe tube length should be at least 500 mm and longer if possible.
7. The cut-off frequency below which turbulent pressure fluctuations will not be attenuated is given by (Shepherd *et al.*, 1989).

$$f_c = \frac{c_0 M}{L(1-M)} \quad \text{(Hz)} \tag{14.5.1}$$

where M is the flow Mach number, c_0 the speed of sound and L the probe tube length.

At a flow speed of Mach number $M = 0.05$, the efficiency of a 13 mm probe tube, 500 mm long, designed using the preceding guidelines varies from about 10 dB at 50 Hz to about 25 dB at 500 Hz. (By efficiency we mean the relative sensitivity to acoustic and turbulent pressure fields.) Above about 500 Hz, the efficiency does not increase significantly.

Shepherd *et al.* (1989) point out that the effect of the probe tube on the microphone phase response as a function of frequency constitutes a complication for active noise control systems. However, probe tube microphones are the preferred means of sampling the sound field in the majority of commercial systems used to active control noise in ducts containing air flow. Shepherd *et al.* also show that in some cases, active control using a microphone with a standard, commercially available nose cone only allows the acoustic field to be attenuated to the level of the turbulent pressure fluctuations which may be a negligible amount in some cases. The probe tube microphone design described above can result in a further 10 to 25 dB reduction in acoustic levels and thus it is considered an essential part of many active control installations in air ducts.

14.5.1.7 Other probe tube designs

Other probe tube designs which have been used with some success include a tube with small holes covering its surface (Noiseux and Horwath, 1970; Bolleter *et al.*, 1970), a porous tube made of glass fibre (Nakamura *et al.*, 1971) and a porous rigid ceramic, plastic or sintered metal tube (Hoops and Eriksson, 1991) with an average cell dimension less than 100μm. The porous tube is now the preferred device in commercial active control systems for noise in ducts because of its low cost, simplicity of manufacture and simplicity of installation. It is difficult to judge from the literature, the relative performance of the various designs in terms of turbulence rejection capability. However, it appears that the tube with holes in its surface (and a cloth covering) does not perform as well as the tube with a slit (and cloth covering) or the porous ceramic tube.

14.5.2 Microphone arrays

Another method of suppressing the turbulent pressure signal is to use a microphone array consisting of a number of discrete microphones arranged in a line directed at the noise source. If the signal from each microphone is electronically delayed by the acoustic propagation time from it to the furthest microphone from the source and the signals all combined, then the slower propagating turbulent pressure signal will be suppressed, the amount of suppression being dependent on the number of microphones used. This system is rather like a discrete version of the probe tube, although it is much less practical, more expensive and no more effective. If the discrete microphones are separated sufficiently so that the turbulent pressure fluctuations at each microphone are uncorrelated, then Shepherd *et al.* show that the suppression of the turbulent pressure signal will be 10 log N dB, where N is the number of microphones.

The discrete microphone array also has a cut-off frequency below which there will be no attenuation of the turbulent pressure fluctuations. This is given by

$$f_c = \frac{c_0 M}{N L_s (1 - M)} \quad \text{(Hz)} \tag{14.5.2}$$

where c_0 is the speed of sound, M is the flow Mach number, N is the number of microphones in the array and L_s the separation between adjacent microphones. Note the similarity of this equation and (14.5.1).

14.5.3 Use of two microphones and a recursive linear optimal filter

A more sophisticated means of filtering out the turbulent pressure contribution from a single microphone signal was proposed by Bouc and Felix (1987). They used a microprocessor to construct a Kalman filter (see Chapter 5) to provide an optimal estimate of the acoustic and turbulent pressure fluctuations. This allowed single frequency acoustic signals to be extracted successfully. The same authors also addressed the problem of extracting a broadband acoustic signal (or suppressing the turbulent pressure contribution in a broadband acoustic signal). This was done by using a linear autoregressive moving average (ARMA) stochastic model, using an optimisation technique to estimate the parameters, and using the output signals from two microphones separated by one duct diameter and at the same axial location along the duct. Although they demonstrated a successful application of their technique, it has not received wide acceptance, probably due to the complexity of its implementation.

REFERENCES

Bolleter, U., Crocker, M.J. and Baade, P. (1970). Tubular microphone windscreen for in-duct fan sound power measurements. In *Proceedings of the 80th Meeting of the Acoustical Society of America*.

Bouc, R. and Felix, D. (1987). A real-time procedure for acoustic turbulence filtering in ducts. *Journal of Sound and Vibration*, **118**, 1–10.

Davy, J.L. and Dunn, I.P. (1993). The development of a flush mounted microphone turbulence screen for use in a power station chimney flue. *Noise Control Engineering Journal*, **41**, 313–322.

Flanagan, J.L., Berkley, D.A., Elko, G.W., West, J.E. and Sondhi, M.M. (1991). Autodirective microphone systems. *Acustica*, **73**, 58–71.

Ford, R.D. (1984). Power requirements for active noise control in ducts. *Journal of Sound and Vibration*, **92**, 411–417.

Frederiksen, E., Eirby, N. and Mathiasen, H. (1979). *Technical Review*, No. 4, Bruel and Kjaer: Denmark.

Glendinning, A.G., Elliott, S.J. and Nelson, P.A. (1988). *A High Intensity Acoustic Source for Active Attenuation of Exhaust Noise*. ISVR Technical Report No. 156.

Hoops, R.H. and Eriksson, L.J. (1991). Rigid foraminous microphone probe for acoustic measurement in turbulent flow. US Patent #4903 249.

Leventhall, H.G. (1988). Problems of transducers for active attenuation of noise. In *Proceedings of Internoise '88*. Institute of Noise Control Engineering, 1091–1094.

Morse, P.M. (1948) *Vibration and Sound*. McGraw-Hill: New York (reprinted by the Acoustical Society of America, 1981).

Morse, P.M. (1976). *Vibration and Sound*. American Institute of Physics, New York, 281–285.

Munjal, M.L. and Eriksson, L.J. (1989). An exact one-dimensional analysis of the acoustic sensitivity of the antiturbulence probe in a duct. *Journal of the Acoustical Society of America*, **85**, 582–587.

Nakamura, A. Sugiyama, A., Tanaka, T. and Matsumoto, R. (1971). Experimental investigation for detection of sound pressure level by a microphone in an airstream. *Journal of the Acoustical Society of America*, **50**, 40–46.

Neise, W. (1975). Theoretical and experimental investigations of microphone probes for sound measurements in turbulent flow. *Journal of Sound and Vibration*, **39**, 371–400.

Noiseux, D.U. and Horwath, T.G. (1970). Design of a porous pipe microphone for the rejection of axial flow noise. In *Proceedings of the 79th meeting of the Acoustical Society of America*.

Sessler, G.M., West, J.E. and Kubli, R.A. (1989). Uni-directional, second order gradient microphone. *Journal of the Acoustical Society of America*, **86**, 2063–2066.

Shepherd, I.C., Cabelli, A. and La Fontaine, R.F. (1986). Characteristics of loudspeakers operating in an active noise attenuator. *Journal of Sound and Vibration*, **110**, 471–481.

Shepherd, I.C., La Fontaine, R.F. and Cabelli, A. (1989). The influence of turbulent pressure fluctuations on an active attenuator in a flow duct. *Journal of Sound and Vibration*, **130**, 125–135.

Tamm, K. and Kurtze, G. (1954). Ein neuartiges mikrofon gro*β*er Richtungs-selektivität. *Acustica*, **4**, 469–470.

15

Vibration sensors and vibration sources

In this chapter, vibration sensors and actuators which have been used in the past in active noise and vibration control systems are discussed. Semi-active dampers which are of major importance in vehicle suspension design are strictly neither sensors nor actuators. As they are discussed in detail in Section 12.3 they will not be discussed further here. We begin this chapter with a discussion of vibration sensors, followed by a description of various vibration actuators.

Prior to discussing the various sensor types, it is useful to point out the relationships between the acceleration, velocity and displacement of a vibrating object. For single frequencies or narrow bands of noise, the displacement, d, velocity, v, and acceleration, a, are related by the frequency, ω (rad s^{-1}), as $d/\omega^2 = v/\omega = a$. In terms of phase angle, velocity leads displacement by 90° and acceleration leads velocity by 90°. For narrowband or broadband signals, velocity can also be derived from acceleration measurements using electronic integrating circuits. On the other hand, deriving velocity and acceleration signals by differentiating displacement signals is generally not practical due primarily to the limited dynamic range of displacement transducers and secondarily to the cost of differentiating electronics.

15.1 ACCELEROMETERS

Vibratory motion is often sensed by using an accelerometer attached to the vibrating object. These transducers may be either piezoresistive or piezoelectric.

Piezoelectric accelerometers consist of a small mass attached to a piezoelectric crystal. The inertia force due to acceleration of the mass causes stress in the crystal which in turn produces a voltage proportional to the stress (or acceleration of the mass). The mass may be mounted to produce either compressive/tensile stress or alternatively, shear stress in the crystal. The latter arrangement allows a smaller (and lighter) accelerometer for the same sensitivity. Three piezoelectric accelerometer types are illustrated in Fig. 15.1(a), (b) and (c) and a schematic of the delta shear type is shown in Fig. 15.1(d).

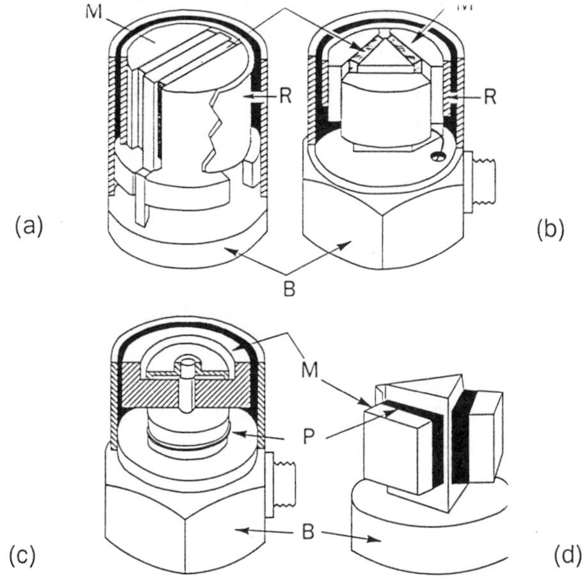

Fig. 15.1 Piezoelectric accelerometer configurations: (a) planar shear type; (b) delta shear type; (c) compression type; (d) schematic of delta shear type. (Brüel and Kjaer.) M = seismic mass, P = piezoelectric element, R = clamping ring and B = base.

Another type of piezoelectric accelerometer which is much less expensive is one made using piezoelectric polymer film (polyvinylidene fluoride or PVDF), in place of the piezoelectric crystal. As it is not possible to use the shear arrangement shown in Fig. 15.1(a) for PVDF film material, the resulting accelerometer sensitivity is usually about ten times less than that achieved with a piezo crystal mounted in a shear arrangement. However, the flexibility of the PVDF film allows it to be used in a beam-type arrangement (see Fig. 15.2) for greater sensitivity, at the expense of a reduced upper limiting operating frequency.

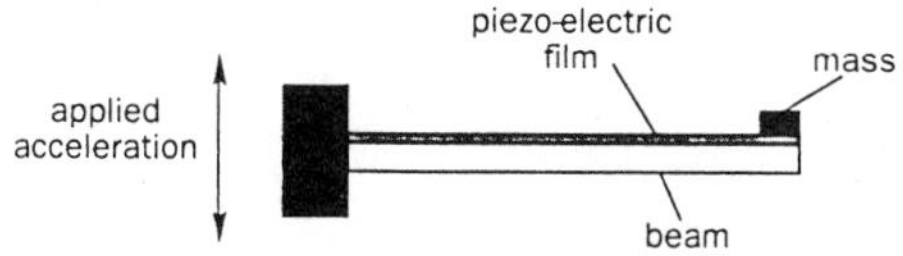

Fig. 15.2 Beam-type accelerometer.

PVDF film has also been used to make rotational acceleration transducers which are now commercially available. Both linear and rotational PVDF film accelerometers are likely to be preferred in future due to their high performance to cost ratio, especially when ordered in large quantities.

Piezoresistive accelerometers rely on the measurement of resistance change in a piezoresistive element subjected to stress. The element is generally mounted on a beam as in Fig. 15.2. They are less common than piezoelectric crystal accelerometers and generally less sensitive by an order of magnitude for the same size and frequency response. They also require a stable d.c. power supply to excite the piezoresistive element (or elements). However, piezoresistive accelerometers are capable of measuring down to d.c. (or zero frequency) are easily calibrated, and can be used effectively with low impedance voltage amplifiers.

When choosing an accelerometer, some compromise must always be made between weight and sensitivity. Small accelerometers are more convenient to use, can measure higher frequencies and are less likely to affect the vibration characteristics of the structure by mass loading it. However, they have low sensitivity, which puts a lower limit on the acceleration amplitude which can be measured. Accelerometers range in weight from miniature 0.65 g for high level vibration amplitude (up to a frequency of 18 kHz) on light weight structures, to 500 g for low level vibration measurement on heavy structures (up to a frequency of 1 kHz). Thus, prior to choosing an accelerometer, it is necessary to know approximately the range of vibration amplitudes and frequencies to be expected as well as detailed accelerometer characteristics, including the effect of various amplifier types. The latter information should be readily available from the accelerometer manufacturer.

Unlike a piezoresistive accelerometer which may only be treated as a voltage source, a piezoelectric accelerometer may be treated as either a charge or voltage source. Thus, its sensitivity can be expressed in terms of charge or voltage per unit of acceleration (pC ms^{-2} or millivolts ms^{-2}). The piezoelectric element acts as a capacitor C_a in parallel with a very high internal leakage resistance R_a which for practical purposes can be ignored. The element may be treated either as an ideal charge source in parallel with C_a and the cable capacitance C_c or as a voltage source V_a in series with C_a and loaded by C_c as shown in Fig. 15.3.

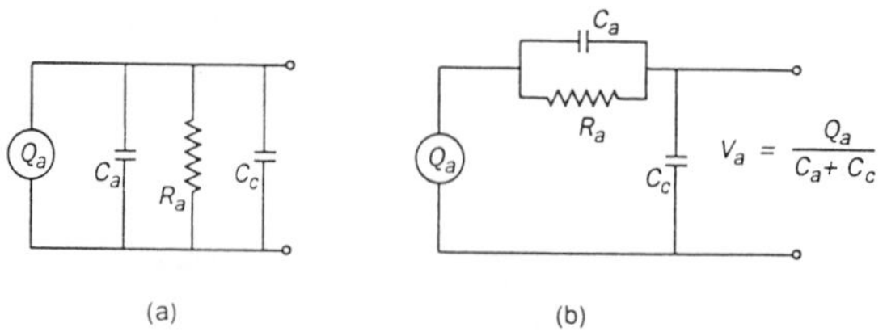

Fig. 15.3 Equivalent circuits for a piezoelectric accelerometer.

Where voltage amplification is used, the sensitivity is dependent on the length of cable between the accelerometer and amplifier. Any motion of the connecting cable can also result in spurious acceleration signals. Because the voltage amplifier decreases the electrical time constant of the accelerometer, the amplifier must have a very high input impedance to measure low frequency vibration and to not significantly load the accelerometer electrically, which would effectively reduce its sensitivity. Commercially available high impedance voltage amplifiers allow accurate measurement down to about 20 Hz.

Alternatively, charge amplifiers (which unfortunately are more expensive, although more commonly used due to the higher accuracy obtained for the measurements) are preferred as they have a very high input impedance and thus do not load the accelerometer output; they allow measurement of acceleration down to frequencies of 0.2 Hz; they are insensitive to cable lengths up to 500 m and they are relatively insensitive to cable movement. Many charge amplifiers also have the capability of integrating acceleration signals to produce signals proportional to velocity or displacement. Particularly at low frequencies, this facility should be used with care, as phase errors and high levels of electronic noise will be present, especially if double integration is used to obtain a displacement signal. Some larger accelerometers have inbuilt preamplifiers which avoids many of the cable induced noise problems associated with separate amplifiers.

The minimum vibration level which can be measured by an accelerometer is dependent upon its sensitivity and can be as low as 10^{-4} ms^{-2}. The maximum level is dependent upon size and can be as high as 10^{6} ms^{-2} for small shock accelerometers. Most commercially available accelerometers at least cover the range 10^{-2} to 5×10^{4} ms^{-2}. This range is then extended at one end or the other, depending upon accelerometer type.

The transverse sensitivity of an accelerometer is its maximum sensitivity in a direction at right angles to its main axis. The maximum value is usually quoted on calibration charts and should be less than 5% of the axial sensitivity. Clearly, this can affect acceleration readings significantly if the transverse vibration amplitude at the measurement location is an order of magnitude larger than the axial amplitude. Note that the transverse sensitivity is not the same in all directions; it can vary from zero up to the maximum, depending on the direction of interest. Thus, it is possible to virtually eliminate the transverse vibration effect if the transverse vibration only occurs in one known direction.

Accelerometer base strain, due to the structure on which it is mounted undergoing strain variations, will generate vibration signals. These effects are reduced by using a shear type accelerometer and are virtually negligible for piezoresistive accelerometers.

Magnetic fields have a negligible effect on an accelerometer. The effect of intense electric fields can be minimized by using a differential pre-amplifier with two outputs from the same accelerometer such that voltages common to the two outputs are cancelled out. This arrangement is generally necessary when using accelerometers near large generators or alternators.

Accelerometers are generally lightly damped and have a single-degree-of-freedom resonance (characterized by the seismic mass and piezoelectric crystal stiffness) well above the operating frequency range. Care should be taken to ensure that high frequency vibrations do not excite the accelerometer resonance and thus produce preamplifier or amplifier overloading or errors in measurements, especially when no frequency analysis is used. The effect of this resonance can be minimized by inserting a mechanical filter between the accelerometer and its mounting point. This results in loss of accuracy at lower frequencies, effectively shifting the ± 3 dB error point down in frequency by a factor of five (see Fig. 15.4). However, the transverse sensitivity at higher frequencies is also much reduced.

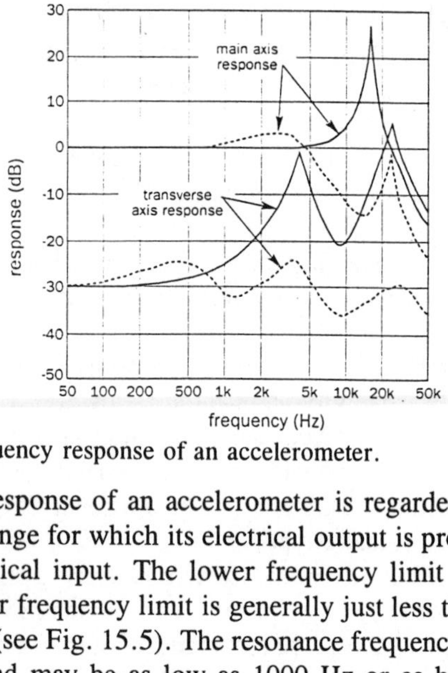

Fig. 15.4 Typical frequency response of an accelerometer.

The frequency response of an accelerometer is regarded as essentially flat over the frequency range for which its electrical output is proportional to within ± 5% of its mechanical input. The lower frequency limit has been discussed previously. The upper frequency limit is generally just less than one third of the resonance frequency (see Fig. 15.5). The resonance frequency is dependent upon accelerometer size and may be as low as 1000 Hz or as high as 180 kHz. In general, accelerometers with higher resonance frequencies are smaller in size and less sensitive.

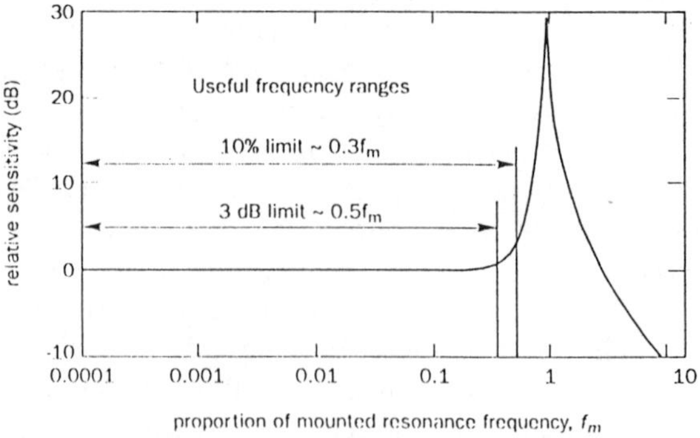

Fig. 15.5 Useful frequency range of a piezoelectric accelerometer.

15.1.1 Accelerometer mounting

There are numerous methods of attaching an accelerometer to a vibrating surface. Some are illustrated in Fig. 15.6. The method used will determine the upper frequency for which measurements will be valid.

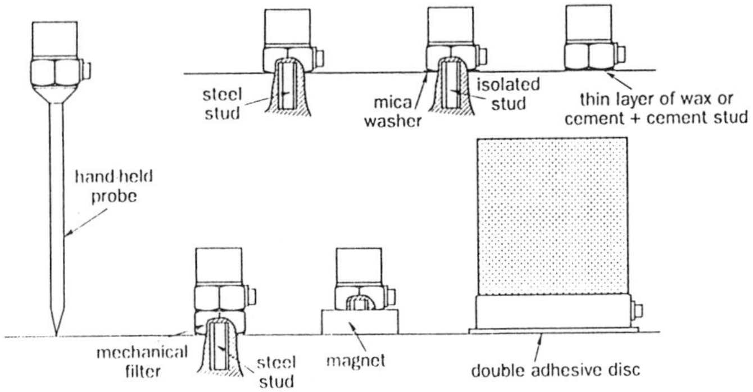

Fig. 15.6 Accelerometer mounting techniques (after Brüel and Kjaer).

The effect of various mounting techniques on the frequency response of a Brüel and Kjaer general purpose accelerometer is shown in Fig. 15.7. Mounting the accelerometer by bolting it to the surface to be measured with a steel stud is by far the best method and results in the resonance frequency quoted in the calibration chart. Best results are obtained if a thin layer of silicon grease is used between the accelerometer base and the structure, especially if the mounting surface is a little rough. If electrical isolation is desired, or if drilling holes in the surface is undesirable, a cementing stud may be used with little loss in performance. This stud is fixed to the surface using epoxy or cyanoacrylate adhesive. If electrical isolation is desired a *thin* layer of epoxy can be spread on the surface and allowed to dry, and the stud then stuck to the dried layer. Alternatively, the accelerometer could be fixed to the surface directly using adhesive but this can damage it superficially. An alternative method of electrical isolation if holes are allowed in the test surface is to use an isolated stud with a mica washer separating the surface from the accelerometer base.

Beeswax can be used for fixing the accelerometer to the vibrating surface with little loss in performance up to temperatures of 40°C provided only a thin layer is used. Thin double sided adhesive tape can also be used at the expense of reducing the useable upper frequency limit of the accelerometer by approximately 25%. Thick double sided tape (0.8 mm) reduces the upper frequency limit by up to 80%. The use of a permanent magnet to mount the accelerometer also affects its performance, reducing its mounted resonance frequency by approximately 50%. This mounting method is restricted to ferromagnetic materials.

For piezoelectric accelerometers with no integral preamplifier, the accelerometer cable should be fixed with tape to the vibrating surface to prevent

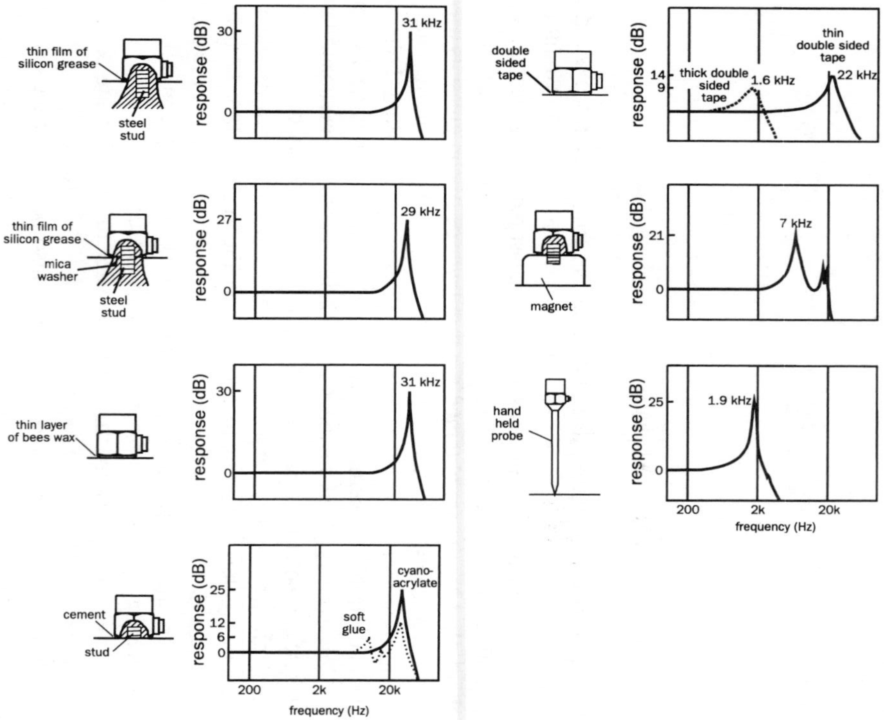

Fig. 15.7 Effect of mounting technique on the frequency response of an accelerometer with a solid mounted resonance frequency of 32 kHz.

the cable from excessive movement. This will minimize the effect of triboelectric noise which results from the cable screen being separated from the insulation around the inner core of the cable. This separation creates a varying electric field which results in a minute current flowing into the screen which will be superimposed on the accelerometer signal as a noise signal. For this reason low noise accelerometer cable should always be used.

15.1.2 Phase response

Accelerometers generally have a flat zero degree phase between the mechanical input and electrical output signals at frequencies below about half the mounted resonance frequency as shown in Fig. 15.8. If accelerometers are heavily damped to minimize the resonance effect, then the phase error between the mechanical input and electrical output will be significant, even at low frequencies.

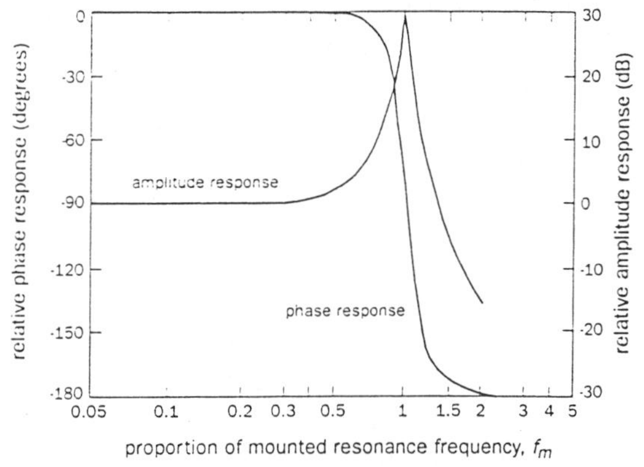

Fig. 15.8 Typical piezoelectric accelerometer phase response.

15.1.3 Temperature effects

Temperatures above 100°C can result in small reversible changes in accelerometer sensitivity of up to 12% at 200°C. If the accelerometer base temperature is kept down using a heat sink and mica washer with forced air cooling, then the sensitivity will change by less than 12% up to 400°C. Accelerometers cannot generally be used at temperatures in excess of 400°C.

15.1.4 Earth loops

If the test object is connected to ground, the accelerometer must be electrically isolated from it or an earth loop may result, producing a high level hum in the resulting acceleration signal at the mains power supply frequency (50 or 60 Hz).

15.1.5 Handling

Accelerometers should not be dropped on hard surfaces or subjected to temperatures in excess of 400°C. Otherwise permanent damage (such as cracking of the brittle piezoelectric element), indicated by a significant change in accelerometer sensitivity will result.

15.2 VELOCITY TRANSDUCERS

Velocity transducers are generally one of two types. The least common type is the non-contacting magnetic type consisting of a cylindrical permanent magnetic on which is wound an insulated coil. A voltage is produced by the varying reluctance between the transducer and the vibrating surface. This voltage is proportional to the surface velocity and the mean distance between the transducer and the surface. When non-ferrous vibrating surfaces are to be measured, a high permeability disc may be attached to the surface. This type of transducer is generally unsuitable for absolute measurements, but is very useful for relative vibration velocity measurements such as needed for vehicle active suspension systems. Another device suitable for vehicle active suspension systems consists of a magnet fixed to the vehicle body and a coil fixed to the suspension. Relative motion between the two produces a coil voltage proportional to the velocity (Wolfe and Jolly, 1990). This type of device is a subset of the most common general type of velocity transducer consisting of a moving coil surrounding a permanent magnet. Inductive EMF, which is proportional to the velocity of the coil with respect to the permanent magnet, is set up in the coil when it is vibrated. In the 10 Hz to 1 kHz frequency range, for which the transducers are suitable, the permanent magnet remains virtually stationary and the resulting voltage is directly proportional to the velocity of the surface on which it is mounted. Outside this frequency range, the velocity transducer electrical output is not proportional to its velocity. This type of velocity transducer is designed to have a low natural frequency (below its useful frequency range); thus, it is generally quite heavy and can significantly mass load light structures. Some care is needed in mounting but this is not as critical as for accelerometers, due to the relatively small upper frequency limit characterizing the basic transducer.

Velocity transducers generally cover the dynamic range of 1 to 100 mm s^{-1}. Some allow measurements down to 0.1 mm s^{-1} while others extend them to 250 mm s^{-1}. Sensitivities are generally of the order of 20 mV mm s^{-1}. Due to their limited dynamic range they are not as useful as accelerometers. Low impedance, inexpensive voltage amplifiers are suitable for amplifying the signal. Temperatures during operation or storage should not exceed 120°C.

Another type of velocity transducer is the laser Doppler velocimeter which allows vibration velocity measurement without the need to fix anything (except reflective tape) to the vibrating surface. One type of commercially available system is shown schematically in Fig. 15.9. The laser light (frequency f) is split by the beam-splitter into two separate beams – the target beam and the reference beam.

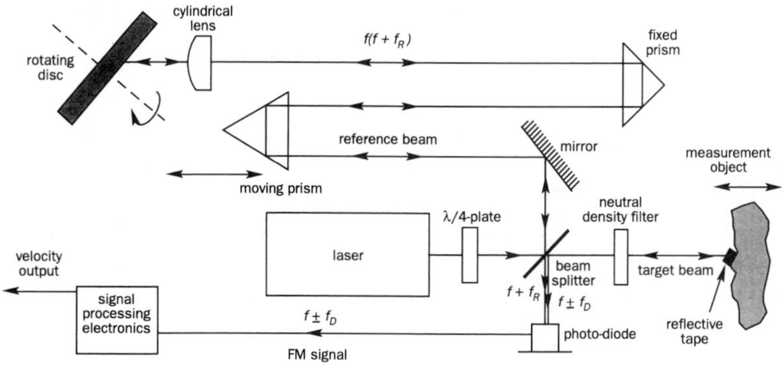

Fig. 15.9 Laser doppler velocimeter, schematic representation.

The target beam is directed on to the surface of a vibrating target, and is reflected back along its incident path by a small patch of retroreflective tape. The returned light is frequency-modulated by a variable Doppler shift due to the velocity of the target. The frequency range of the modulation is $f \pm f_D$, where f_D is the Doppler shift due to the maximum velocity of the target. This frequency modulation (FM) contains the required measurement but, since the frequencies are of the order of 10^{15} Hz, it is not practical to demodulate the signal directly. The returned target beam has first to be heterodyned with a reference beam so that the high frequency component can be reduced. The reference beam is directed, by an optical system, onto a rotating reference disc set at an angle to the direction of incidence. The surface of the disc is covered with retroreflective tape, and its speed of rotation is controlled by a crystal oscillator. The frequency of the returned reference beam ($f + f_R$) is increased by a constant Doppler shift due to the constant speed of the reference disc.

The two returned beams are recombined at the beam-splitter and directed onto the surface of a photodiode where they heterodyne. The output from the photodiode is an FM signal with a frequency range $f_R \pm f_D$. It is necessary to frequency-shift the reference beam to prevent loss of signal as the target velocity passes through zero, which also allows the direction of the velocity to be determined as either positive or negative.

The output from the photodiode is a frequency modulated signal centred on the reference frequency. Appropriate electronic signal processing is used to convert this to a voltage output proportional to the instantaneous velocity of the vibrating surface.

The commercial system represented in Fig. 15.9 is capable of measuring velocities over the frequency range from 0 to 20 kHz with a maximum amplitude of 1 m s^{-1}. The minimum measurable level is approximately 1 mm s^{-1} for broadband signals and 0.1 mm s^{-1} for pure tone signals. Note that velocity signals are also obtained in practice by integrating accelerometer signals, although this sometimes causes low frequency electronic noise problems.

15.3 DISPLACEMENT TRANSDUCERS

Although the dynamic range of displacement transducers is typically much smaller than it is for accelerometers, displacement transducers are often more practical at very low frequencies (0 – 10Hz), where vibration amplitudes are measured in terms of tenths or hundredths of a millimetre, and where corresponding accelerations are small. Also, in some active systems, displacement is the preferred control variable and attempts to derive this from an acceleration signal by double integration usually lead to excessive electronic noise, especially at very low frequencies (less than 10 Hz).

15.3.1 Proximity probe

The most common type of displacement transducer is the proximity probe. Two types of proximity probe are available, the capacitance probe and the eddy current probe. The first relies on the measurement of the change in electrical capacitance between the vibrating machine surface and the stationary probe. The eddy current probe relies on the generation of a magnetic field at the probe tip by a high frequency (>500 kHz) voltage applied to a coil of fine wire. This magnetic field induces eddy currents proportional to the size of the gap between the probe tip and machine surface. These currents oppose the high frequency voltage and reduce its amplitude in proportion to the size of the gap. Typical gap ranges in which the amplitude is a linear function of gap size vary from between 1 mm and 4 mm for smaller diameter probes to between 2 mm and 20 mm for larger probes. The carrier amplitude signal is demodulated to give a low impedance voltage output proportional to gap size over the linear range of the transducer.

Eddy current probes are more common and easier to use than capacitative type pickups so further discussion will be restricted to the former type. When mounting an eddy current proximity probe to measure the vibration of a rotating shaft, the surface or shaft to be measured must be free of all irregularities such as scratches, corrosion, out-of-roundness, chain marks, etc. Any irregularity will cause a change in probe gap which is not a shaft position change, resulting in signal errors. This is called mechanical run out.

The shaft material must be of uniform composition all the way round its surface so that the resistivity does not vary as the shaft rotates, resulting in an unwanted electronic noise signal. This is called electrical runout noise. Care should also be taken with the use of plated shafts, as thin plating can allow the eddy currents to penetrate to the main shaft material, resulting in two different material resistivities being detected as well as the rough interface surface between shaft and surface treatment. As probes are generally matched to a particular shaft material, this can lead to calibration problems, as well as electronic noise problems due to the rough interface.

Proximity probes require a power supply and low impedance voltage amplifier, both of which are generally supplied in the same module. Note that the length of cable between the power supply and probe significantly affects the probe sensitivity.

The dynamic range of a proximity probe is typically 100:1, although some have a range of 150:1 and others 60:1. The resolution to which the probe can measure varies from 0.02 mm to 0.4 mm, depending upon the absolute range (2 mm to 25 mm). The limited dynamic range restricts its practical application to frequencies of less than 200 Hz.

15.3.2 Linear variable differential transformer (LVDT)

This type of transducer consists of a single primary and two secondary coils wound around a cylindrical bobbin. A moveable nickel iron core is positioned inside the windings and it is the movement of this core which is measured (see Fig. 15.10).

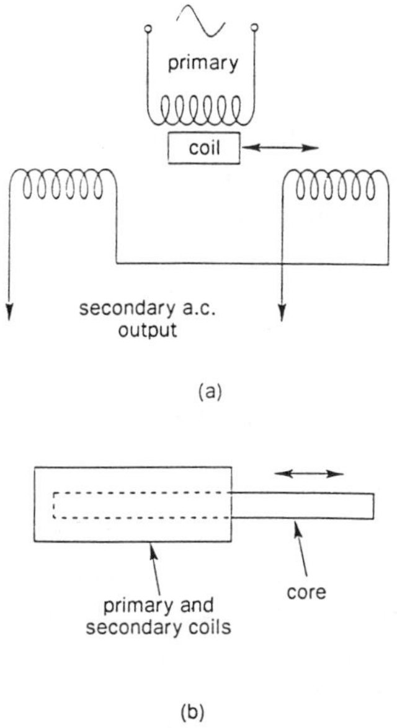

Fig. 15.10 (a) schematic of an LVDT transducer; (b) LVDT transducer.

To operate the transducer, it is necessary to drive the primary coil with a sine wave, usually at a frequency of between 1 and 15 kHz and amplitude between 1 and 10 volts (r.m.s.). The output from the secondary coil is a sine wave which contains positional information in terms of its amplitude and phase. The output with the core at the centre of the stroke is zero rising to a maximum amplitude with the core at either end of its stroke. The output is in phase with the primary drive with the core on one side of the centre and 180° out of phase when the core is on the other side of the centre.

Commercially available transducers are usually supplied with an inbuilt oscillator to generate the a.c. signal and appropriate demodulating electronics to convert the a.c. output into a d.c. level proportional to displacement. Alternatively, the LVDT, oscillator and demodulating electronics can be purchased separately, allowing greater flexibility. A typical dynamic range for this type of transducer is about 100:1, with maximum displacements measurable ranging from 1 mm to 100 mm, depending on the transducer selected. Note that because of the limited dynamic range, it is important to select the transducer with the appropriate upper limit for a particular task.

The frequency range characterizing most commercially available LVDT transducers is d.c. to 100 Hz. For long stroke ($\pm$15 to $\pm$100 mm) transducers, The diameter is typically between 12 and 20 mm, with a body length of about three times the maximum stroke (or displacement from centre). Short stroke transducers usually have a fixed length not less than 30–50 mm. More detail on the principles of operation of both of these devices are available in publications from commercial suppliers.

15.3.3 Linear variable inductance transducers (LVIT)

This type of transducer is particularly suited to measuring relative displacements in vehicle suspension systems, and makes use of a cylindrical coil, the inductance of which is arranged to vary with movement of a spoiler over it. The coil is excited at approximately 100 kHz and the variation in its inductance is caused by the introduction of a highly conductive, non-ferrous spoiler into the resulting electromagnetic field. The spoiler takes the form of a coaxial rod sliding inside the coil or a coaxial cylinder sliding over the outside of the coil.

The most effective way of determining variations in inductance, and one which is relatively insensitive to variations in inductor coil resistance due to temperature effects or manufacturing irregularities, is to use a resonant circuit. The inductor is used as one component of a simple LC oscillator, the frequency of which changes with inductance changes, resulting in a frequency modulated carrier signal which can be converted to a voltage proportional to relative axial displacement between the coil and spoiler by using a commercially available frequency demodulation device.

Transducer sizes vary from diameters of a few millimetres to tens of millimetres, and lengths depend on the displacement amplitude to be measured.

Improvements in transducer performance can be made by using two coils and detecting the frequency differential between two resonant circuits, each of which contains one coil. This is described in more detail by Moore (1988).

15.4 STRAIN SENSORS

When a structure vibrates either flexurally or longitudinally, its surface is subjected to cyclic strains which can be detected with a strain sensor. Three types of strain sensor will be considered here; conventional resistive strain gauges, PVDF film and optical fibres. Piezoceramic crystals can also be used but because of their brittle nature, relatively high cost and lower sensitivity, they cannot compare with PVDF film for this purpose and thus will not be considered here.

15.4.1 Resistive strain gauges

Strain gauges are constructed of a thin, electrical-conducting wire, foil or semiconductor sandwiched between two plastic sheets as shown in Fig. 15.11. Larger areas at the ends of the grid facilitate connection of cables. The plastic sheets which are bonded to the active element simplify handling and protect the element from mechanical damage.

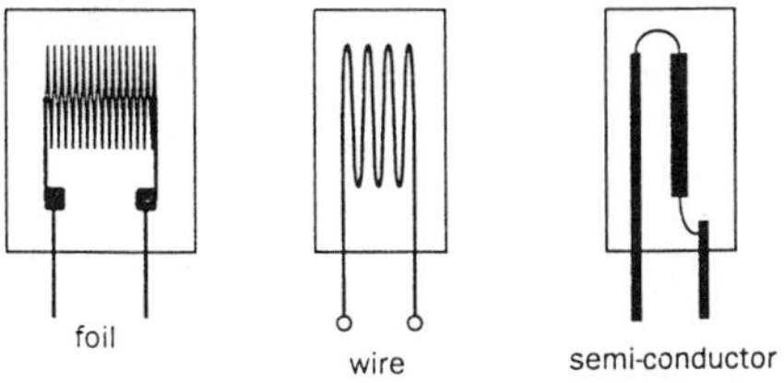

Fig. 15.11 Types of strain gauge.

Foil strain gauges are produced by photo-etching a metallic foil (3 to 5 μm thick). The material used for the metallic element is usually an alloy of nickel and copper (constantan), the exact composition of which is dependent upon the material on which the strain gauge is to be used. Some effort is made to match the temperature coefficient of expansion of the strain gauge with the material on which it is to be used.

Wire strain gauges are produced using metallic wire, 15 to 25 μm in diameter. Foil gauges are easier to make if the gauge length is approximately 6 mm or less whereas wire gauges are easier for lengths greater than approximately 6 mm. Wire gauges are also more suitable for high temperature applications, as more suitable materials are available and better mounting techniques can be used.

Semiconductor strain gauges contain a semiconductor element a few hundred micrometres wide and 20–30 μm thick. This type of gauge is the most expensive and the least preferred type for most applications.

The operation of metallic strain gauges is based on the principle that the electrical resistance of a metallic conductor changes if the conductor is subjected to an applied strain. The change in resistance of the strain gauge is partly due to the change in the geometry of the conductor and partly due to the change in the specific conductivity ρ_s of the conductor material due to changes in the material structure. The change in resistance ΔR as a fraction of the nominal resistance R_0 is given by

$$\frac{\Delta R}{R_0} = \epsilon \left[1 + 2\nu + \frac{1}{\rho_s} \frac{\partial \rho_s}{\partial \epsilon} \right] \tag{15.4.1}$$

where ν is Poisson's ratio, ϵ is the strain imposed on the strain gauge and the first two terms in brackets represent the geometrical component. The quantity in brackets in (15.4.1) is a constant over a wide range of values of ϵ for materials such as constantan which are used in strain gauges. Thus, (15.4.1) can be written as

$$\frac{\Delta R}{R_0} = K\epsilon \tag{15.4.2}$$

where K is the gauge factor (usually in the vicinity of 2).

The operation of semiconductor strain gauges is based on the piezoelectric effect; that is, mechanical stress applied to a semiconductor will result in a change in resistance as a result of a change in electron mobility.

The type of doping material implanted in the semiconductor will determine the conductivity (or gauge factor). It is also possible to obtain both negative or positive gauge factors.

The strain gauge relationship for semiconductor gauges is

$$\frac{\Delta R}{R_0} = \frac{T_0}{T} K_1 \epsilon + \left[\frac{T_0}{T}\right]^2 K_2 \epsilon^2 + \ldots . \tag{15.4.3}$$

It is this non-linearity and high unit cost which limit the use of this type of gauge.

As the resistance change ΔR is usually a very small value, it cannot be accurately measured directly. The most common means of determining ΔR is to insert the strain gauge as one leg in a Wheatstone bridge (see Fig. 15.12). The bridge is balanced by varying R_1 to produce zero output, with the vibrating structure at rest. When the structure begins to vibrate a voltage proportional to the strain (and thus structural vibration bending moment) will appear at the bridge output. Note that the bending moment is the double spatial derivative of the structural displacement.

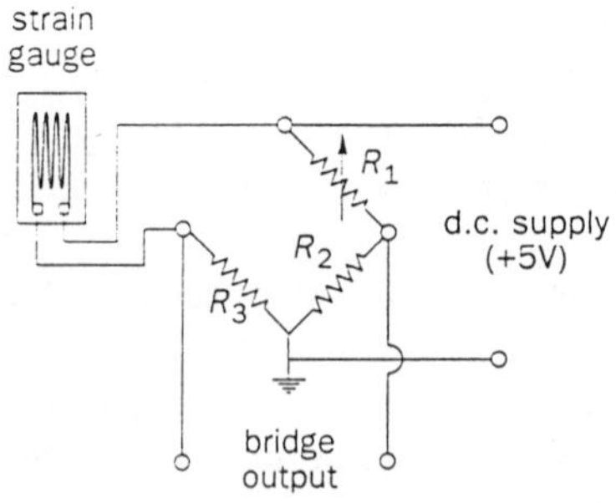

Fig. 15.12 Wheatstone bridge circuit.

The bridge output voltage V_0 is related to the change in resistance ΔR of the strain gauge, the bridge d.c. supply voltage V_s and the structural strain as

$$\frac{V_0}{V_s} = \frac{\Delta R}{R} = K\epsilon \tag{15.4.4a,b}$$

As V_s is normally about 5 V and K is about 2 V, the output voltage V_0 is about ten times the strain level. Thus, 1μ strain would produce 10 μV. Clearly, a high gain, low noise operational amplifier is needed to amplify the voltage to a useable level if strains as low as a few μ strain are to be measured. Also, the d.c. power supply to the bridge must be characterized by very low noise levels; often batteries rather than regulated mains power are necessary. The dynamic range of a strain gauge and its associated electrical system is from $2-5\mu$ strain to 10000μ strain over a frequency range limited only by the amplifying electronics. The frequency range of general interest for active noise and vibration control (0 to 500 Hz) is easily covered with standard circuitry.

In some cases, the sensitivity of the measurement system can be increased by replacing resistances R_3 or R_2 in the wheatstone bridge with a second strain gauge. If resistance R_2 is replaced with a strain gauge, the two strain gauges in the bridge must be tensioned and compressed out of phase with one another for the two effects to add together. This is achieved, for example, by using two strain gauges on opposite sides of a beam to detect flexural vibration. This configuration has the added advantage of minimising the d.c. drift in the output signal as a result of temperature changes occurring in the system being measured. If only one strain gauge is used (or if the second one replaces R_3 in the Wheatstone bridge circuit), then a temperature change in the system being measured will cause the structure to expand differently to the strain gauge (as the coefficients of thermal expansion of the test structure and the strain gauge are not exactly matched in practice), thus resulting in a strain gauge output not related to stress in the structure. However, this is rarely a problem when strain gauges are used just to sense vibration, as the temperature fluctuations generally occur well below the frequency range of interest.

When installing strain gauges on a surface for vibration measurement, many factors need to be considered such as adhesive selection, protection of the installed gauge from the environment and selection of a gauge size appropriate for the task. Larger gauges average the vibration over a larger surface area and are usually easier to install. Suitable adhesives include cyano-acrylate (super glue) provided the completed installation is covered with a waterproof compound. Other adhesives and protective compounds are available from strain gauge manufacturers.

15.4.2 PVDF film

When bonded to a surface, polyvinylidine diflouride film (PVDF film) produces either a charge or a voltage proportional to the strain of the surface. The charge may be amplified by a high impedance charge amplifier to produce a voltage proportional to the strain. The circuits used to describe the electrical characteristics of PVDF film are similar to those discussed in Section 15.1 for accelerometers. The advantage offered by PVDF film when compared to strain gauges is its ability to act as a distributed sensor and to be shaped so that it senses particular vibration modes, or combinations of modes, if so desired. For example, if it is desired to control sound radiation, then the sensor could be shaped so that it sensed the surface vibration distribution which contributed most to the sound radiation. This is discussed in more detail in Chapter 8. When acting as a distributed sensor, PVDF film provides an output charge or voltage proportional to the surface strain integrated over the area covered by the sensor.

Before discussing the use of shaped sensors in more detail, the general properties of PVDF film will be outlined, so that some understanding may be gained of how it works and the meaning of the coefficients used to define its properties.

PVDF film is an extremely flexible polymer which can be polarised across its thickness (usually between 9 and 110 μm) by a strong electric field. It can act either as a sensor or an actuator. In this section, only its function as a sensor will be discussed; its function as an actuator will be discussed in Section 15.11. Once polarised, PVDF film will provide a charge or voltage when subjected to an applied tensile force or strain. Conversely, it will produce a strain when subjected to an applied voltage, but this must be well below its original polarization voltage. The maximum operating voltage is 30 V μm^{-1} thickness and the breakdown voltage at which polarization is lost is 100 V μm^{-1} thickness. The polarization axis for PVDF film is usually normal to the surface, across the thickness (shown as the τ-axis in Fig. 15.13).

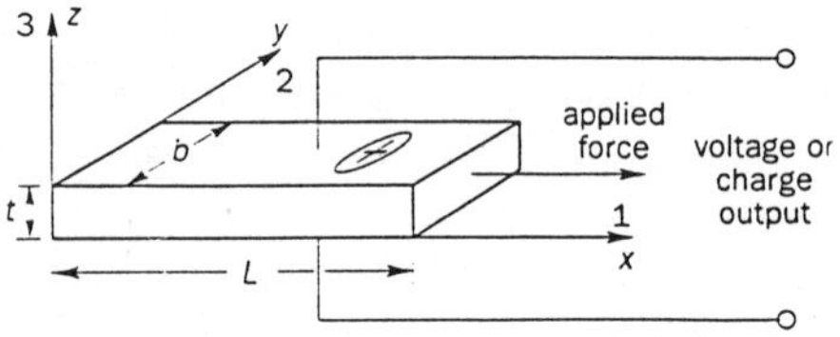

Fig. 15.13 PVDF film.

For a compression or tension force F applied in the x-direction, the voltage, V, and charge, Q, outputs are given by,

$$\frac{Q}{Lb} = \frac{F}{tb} d_{31} = \sigma d_{31} \quad \mathrm{C\,m^{-2}} \tag{15.4.5 a,b}$$

$$\frac{V}{t} = \frac{F}{tb} g_{31} = \sigma g_{31} \quad \mathrm{V\,m^{-2}} \tag{15.4.6 a,b}$$

where σ is the stress in the PVDF film, and x, y, b, t and L are defined in Fig. 15.13.

For a compression or tension force F applied in the y-direction, the voltage and charge outputs are found by substituting d_{32} and g_{32} for d_{31} and g_{31} in (15.4.5) and (15.4.6). The d_{ij} coefficients are referred to as charge (or strain) coefficients and are constants which characterize a particular PVDF film. Similarly, the g_{ij} coefficients are referred to as voltage (or stress) coefficients.

The charge coefficients are the ratios of electric charge generated per unit area to force applied per unit area and the voltage coefficients are the ratios of the electric field produced (volts/metre) to the force applied per unit area. The subscripts 1 and 2 refer to force applied in the x- and y-directions respectively and the subscript 3 refers to the axis of polarization, which is usually the z-axis. Typical values of the charge and voltage coefficients are:

$$g_{32} = g_{31} = 216 \times 10^{-3} \ \frac{\mathrm{V\,m^{-1}}}{\mathrm{N\,m^{-2}}}$$

$$d_{31} = 23 \times 10^{-12} \ \frac{\mathrm{C\,m^{-2}}}{\mathrm{N\,m^{-2}}}$$

$$d_{32} = 3 \times 10^{-12} \ \frac{\mathrm{C\,m^{-2}}}{\mathrm{N\,m^{-2}}}$$

The charge output due to an applied strain is given by

$$\frac{Q}{Lb} = \epsilon e_{31} \tag{15.4.7}$$

where ϵ is the applied strain and the coefficients e_{ij} are related to d_{ij} as (Lee and Moon, 1990a):

$$\begin{bmatrix} e_{31} \\ e_{32} \\ e_{36} \end{bmatrix} = \begin{bmatrix} E_p/(1-\nu_p^2) & \nu_p E_p/(1-\nu_p^2) & 0 \\ \nu_p E_p/(1-\nu_p^2) & E_p/(1-\nu_p^2) & 0 \\ 0 & 0 & E_p/2(1+\nu) \end{bmatrix} \begin{bmatrix} d_{31} \\ d_{32} \\ d_{36} \end{bmatrix} \tag{15.4.8}$$

The subscript 6 takes into account the possibility of the PVDF film principal axes not being coincident with the principal axes of the structure to which it is bonded. The quantity, d_{36}, is zero if the axes are coincident. E_p is the PVDF film elastic modulus (typically 2×10^9 Pa) and ν is Poisson's ratio (typically 0.35).

Typical thicknesses of PVDF film range from 9 μm to 110 μm. However, it can be seen from equation (15.4.7) that if the film is attached to a vibrating surface, the charge produced is independent of the film thickness, and only dependent upon the strain induced by the vibrating surface and the area of film used. On the other hand, (15.4.6) shows that the voltage produced is dependent on the thickness but not the area of film. However, the same disadvantages associated with using a piezoelectric accelerometer as a voltage source (discussed in Section 15.1) apply to the PVDF film as well.

As a guide to the sensitivity of PVDF film compared to a strain gauge for measuring dynamic strain, equation (15.4.7) shows that a piece 10 mm $\times$ 10 mm will produce a charge of approximately 5.5 pC when subjected to 1μ strain. A conventional charge amplifier noise floor is approximately 0.003 pC, so the strain detected can be as low as 10^{-9}. This compares very favourably with a strain gauge which is limited to about 10^{-6}. The upper limit of the dynamic strain which can be measured is governed by the tensile and compressive strength and is approximately 0.015, which is similar to a strain gauge.

It is of interest to examine how PVDF sensors may be shaped to respond to certain modes on a simply supported rectangular thin panel, illustrated in Fig. 15.14.

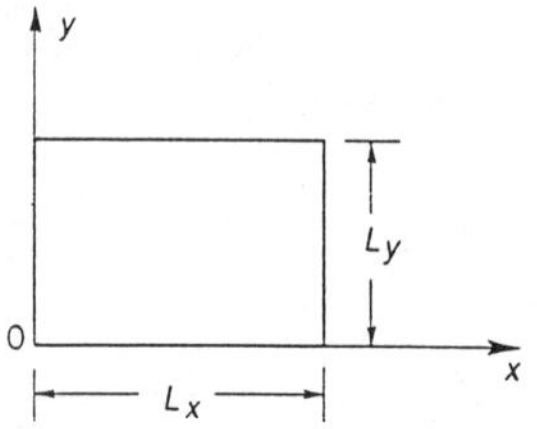

Fig. 15.14 Coordinate systems for a simply supported thin panel.

Lee and Moon (1990b) give the equation describing the closed circuit charge signal $q(t)$ measured at the electrode of a piezoelectric film as

$$q(t) = \frac{h_p + h_s}{2} \int_0^{L_x} \int_0^{L_y} f P_0 \left[e_{31} \frac{\partial^2 w}{\partial x^2} + e_{32} \frac{\partial^2 w}{\partial y^2} + 2e_{36} \frac{\partial^2 w}{\partial x \partial y} \right] \mathrm{d}y\mathrm{d}x \quad (15.4.9)$$

The quantities h_p and h_s are the thicknesses of the piezo film and plate respectively. The polarization profile is P_0 which for the case considered here is either $+1$ or -1. The quantity f represents the shape function of the piezo-film, and w is the flexural displacement of the panel.

15.4.2.1 One-dimensional sensor

If the sensor is narrow and long $(x >> y)$, then (15.4.9) can be written as

$$q_x(t) = \frac{h_p + h_s}{2} \int_0^{L_x} f P_0 e_{31} \frac{\partial^2 w}{\partial x^2} \mathrm{d}x \quad (15.4.10)$$

where $q(t)$ is the charge generated by a PVDF sensor which is relatively thin in the y-direction.

Neglecting time dependence, the flexural displacement of a simply supported panel can be expressed as

$$\bar{w}(x,y) = \sum_{m=1}^{\infty} \sum_{n=1}^{\infty} A_{mn} \sin\left[\frac{m\pi x}{L_x}\right] \sin\left[\frac{n\pi y}{L_y}\right] \quad (15.4.11)$$

where m and n represent the modal indices of the panel modes and A_{mn} represents the complex modal amplitude for mode (m,n).

Substituting (15.4.11) into (15.4.10) gives for the charge amplitude

$$q_x = -\frac{(h_p + h_s)}{2} \int_0^{L_x} f P_0 e_{31} \sum_{m=1}^{\infty} \sum_{n=1}^{\infty} A_{mn} \left[\frac{m\pi}{L_x}\right]^2 \sin\left[\frac{m\pi x}{L_x}\right] \sin\left[\frac{n\pi y_s}{L_y}\right] dx \tag{15.4.12}$$

The quantity y_s is the y-coordinate of the sensor extending in the x-direction.

We will now consider two different shapes for the PVDF film. For the first, a narrow strip of constant width will be examined (see Fig. 15.15(a)). For a strip such as this, $f = 1$ and P_0 = constant:

$$q_x = -\frac{(h_p + h_s)}{2} e_{31} P_0 \sum_{m=1}^{\infty} \sum_{n=1}^{\infty} A_{mn} \left[\frac{m\pi}{L_x}\right]^2 \sin\frac{n\pi y_s}{L_y} \int_0^{L_x} \sin\left[\frac{m\pi x}{L_x}\right] dx \tag{15.4.13}$$

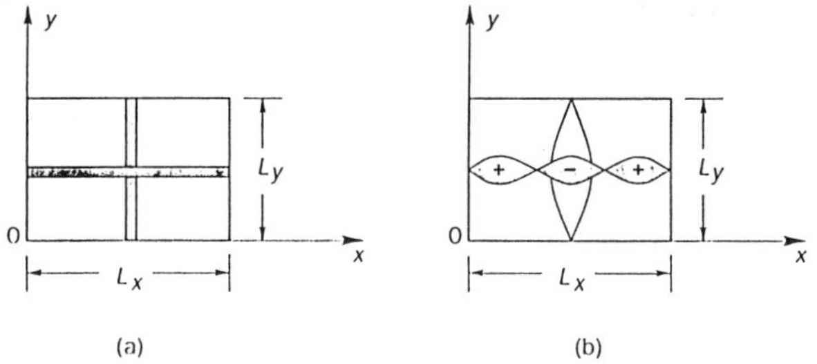

Fig. 15.15 Strip sensors for detection of desired modes: (a) linear sensors; (b) shaped sensors $m' = 3$, $n' = 1$.

Thus, if y_s is set equal to $L_y/2$; that is, if the strip is placed across the centre of the plate it can be seen that the contribution from all even (m even, n even) and odd even (m odd, n even) modes (that is, all modes with a nodal line along the sensor) will not contribute to the sensor output.

Now consider a sensor shape and pole connection such that

$$f P_0 = \mu_{m'} \sin\left[\frac{m'\pi x}{L_x}\right] \tag{15.4.14}$$

where $\mu_{m'}$ is the sensor scaling factor.

That is, the sensor shape and polarity is made proportional to the modal strain distribution across the panel for modes with the coefficient m' in the x-direction. The quantity m' can also be considered the modal index of the sensor. A sensor with $m' = 3$ in the x direction and another with $n' = 1$ in the y-direction are shown in Fig. 15.15(b).

Substituting (15.4.10) into (15.4.8) and using the orthogonality property of the modes we obtain

$$q_x = -\left[\frac{h_p + h_s}{2}\right] e_{31}\,\mu_{m'} \left[\frac{m\pi}{L_x}\right]^2 \frac{L_x}{2}\,\delta_{m'm} \sum_{n=1}^{\infty} A_{mn} \sin\left[\frac{n\pi y_s}{L_y}\right] \tag{15.4.15}$$

where

$$\delta_{m'm} = 1 \text{ if } m' = m$$
$$= 0 \text{ if } m' \neq m$$

It can be seen from (15.4.15) that the sensor will not detect modes where $m \neq m'$. If it is placed along the centre line of the panel so that $y_s = L_y/2$, it will only detect modes with $m = m'$ and with n odd.

A similar analysis procedure to that described above can be done for a sensor parallel to the y-axis with similar results.

15.4.2.3 Two-dimensional sensor

For a two-dimensional sensor configuration with zero skew, (15.4.9) can be written as

$$q(t) = \left[\frac{h_p + h_s}{2}\right] \int_{x_1}^{x_2} \int_{y_1}^{y_2} f P_0 \left[e_{31}\frac{\partial^2 w}{\partial x^2} + e_{32}\frac{\partial^2 w}{\partial y^2}\right] \mathrm{d}x\,\mathrm{d}y \tag{15.4.16}$$

where

$$x_1 = (L_x - L_x')/2, \quad x_2 = x_1 + L_x'/2$$
$$y_1 = (L_y - L_y')/2, \quad y_2 = y_1 + L_y'/2$$

L_x' and L_y' are the sensor dimensions (see Fig. 15.16).

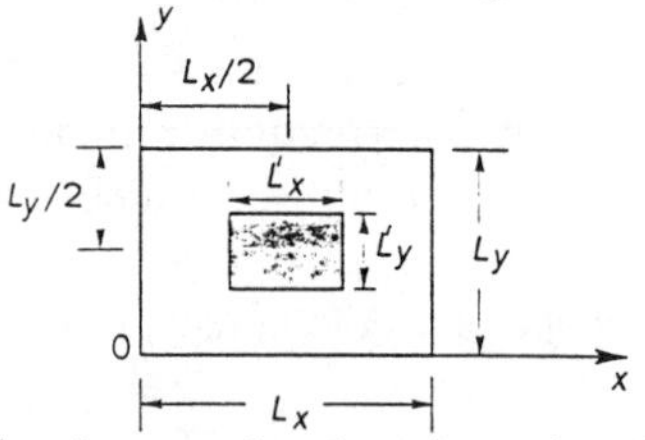

Fig. 15.16 Two-dimensional sensor for detecting only odd/odd panel modes.

For a two-dimensional rectangular sensor centred on a simply supported rectangular panel, (15.4.16) becomes:

$$q_{xy} =$$

$$-\left[\frac{h_p + h_s}{2}\right] \int_{x_1}^{x_2} \int_{y_1}^{y_2} f P_0 \sum_{m=1}^{\infty} \sum_{n=1}^{\infty} A_{mn} \left[e_{31}\left[\frac{m\pi}{L_x}\right]^2 + e_{32}\left[\frac{n\pi}{L_y}\right]^2\right] \sin\left[\frac{m\pi}{L_x}\right] \sin\left[\frac{n\pi y}{L_y}\right] \mathrm{d}y\,\mathrm{d}x$$

(15.4.17)

Setting the sensor polarity P_0 constant over the PVDF film and solving the integral gives $q_{xy} = 0$ for all modes except those for which m and n are both odd. In the latter case,

$$q_{xy} = -\left[\frac{h_p + h_s}{2}\right] P_0 \sum_{m=1}^{\infty} A_{mn} \left[e_{31} \left[\frac{m\pi}{L_x}\right]^2 + e_{32} \left[\frac{n\pi}{L_y}\right]^2 \right] \times \frac{4L_xL_y}{mn\pi^2} \cos\left[\frac{m\pi x_1}{L_x}\right] \cos\frac{n\pi y_1}{L_y} \tag{15.4.18}$$

Note that for a lightly damped panel, A_{mn} may be considered real.

Thus, a rectangular PVDF film sensor placed in the centre of a rectangular panel will only detect odd/odd modes. Shaped sensors for the detection of acoustic power radiation are discussed in Chapter 8.

15.4.3 Optical fibres

The use of optical fibres as strain sensors was first discussed by Butter and Hocker (1978) and they have since been applied as sensors for active control of beam vibration by Cox and Lindner (1991). An optical fibre sensor consists of a small diameter single strand of glass which can be embedded in or attached to a dynamic structure producing negligible dynamic loading.

Optical fibre sensing is expensive, requiring a laser and complex signal processing electronics, but for some applications it is the only feasible choice. Such applications include structures subjected to high level electromagnetic fields, high temperatures and pressures, and/or harsh chemical environments.

Modal domain (MD) optical fibre sensors appear to be the most practical (Cox and Lindner, 1991). They were first investigated by Layton and Bucaro (1979) and utilize the interference phenomenon between two or more modes propagating in a multimode optical fibre to sense the axial strain in that fibre. This configuration, which does not require the traditional interferometer reference arm, simplifies sensor mounting and makes it more stable. Power distributions in the fibres were stabilised by the introduction of elliptic core fibres (Kim *et al.*, 1987) which eliminated power coupling between spatially degenerate modes in the fibre as the fibre was strained. Another recent development was the attachment of insensitive lead-in and lead-out optical fibres to the sensing element.

An MD optical fibre sensor is characterized by an intensity pattern at its output which is affected by the integrated strain over the length of the fibre. The effect of strain on the physical characteristics of the fibre is threefold: first, the axial length is changed; second, the Poisson effect will result in a change in the fibre diameter; and third, the photoelastic effect results in the indices of refraction being dependent on strain. Thus, the functional dependence of the light intensity output on strain is complex. A typical optical fibre is constructed of an inner core and an outer cladding as shown in Fig. 15.17. Usually the index of refraction n_1 of the inner core is similar to that of the outer cladding n_2 so that $(n_1 - n_2) << 1$. This assumption leads to a simplified description of the electromagnetic waves in the fibre and will be used in the analysis to follow.

To determine the nature of the functional dependence of the light output intensity on the fibre strain, it is of interest to examine the solutions to Maxwell's electromagnetic wave equation, which resembles the acoustic wave equation discussed at some length in Chapter 2. Although the analysis referred to here assumes a circular section fibre core, the results are applicable in general terms to the elliptical section core which is used in practice for reasons mentioned previously Cox and Lindner 1991. The electric field $\boldsymbol{E}$ for a propagating electromagnetic field is described by the following wave equation

$$\nabla^2 \boldsymbol{E} - \frac{\partial^2}{\partial t^2}\boldsymbol{E} = 0 \qquad (15.4.19)$$

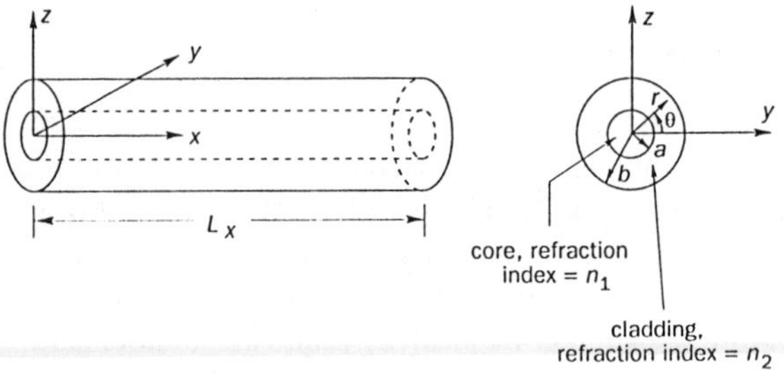

Fig. 15.17 Optical fibre coordinate system.

For single frequency laser light propagating in one direction (the x or axial direction down the fibre, as in Fig. 15.17) a solution to (15.4.19) in cylindrical coordinates is

$$\bar{\boldsymbol{E}}(r, \theta, t) = \boldsymbol{E}(r, \theta)e^{j(\omega t - \beta x - \psi)} \qquad (15.4.20)$$

where β is a propagation constant equivalent in meaning to an acoustic wave number, and ψ is a relative phase angle.

The electric field $E(r,\theta)$ can be decomposed into its vector components. Thus,

$$\boldsymbol{E}(r, \theta) = E_x(r, \theta)\boldsymbol{i} + E_y(r, \theta)\boldsymbol{j} + E_z(r, \theta)\boldsymbol{k} \qquad (15.4.21)$$

where $\boldsymbol{i}, \boldsymbol{j}, \boldsymbol{k}$ are unit vectors in the x-, y- and z-directions.

If $\boldsymbol{E}(r,\theta) \bullet \boldsymbol{k}$ is a constant, then the wave is said to be linearly polarized (LP). Solutions are found for $\boldsymbol{E}(r,\theta)$ in both the central core and outer cladding of the fibre by applying the boundary conditions imposed by the shape of the core and the cladding.

As laser light is usually polarized in a direction corresponding to the z- or y-axis (linearly polarized), it is convenient to decompose the field in the $y-z$ plane into two linearly independent solutions:

$$E(r,\theta) = \left[E_{xy}\boldsymbol{i} + E_{yy}\boldsymbol{j}\right]\left[E_{xz}\boldsymbol{i} + E_{zz}\boldsymbol{k}\right] \tag{15.4.22}$$

The first subscript corresponds to the field and the second to the polarization direction. Thus, the first two terms in (15.4.22) correspond to light polarized in the y-direction, while the last two terms correspond to light polarized in the z-direction.

For laser light polarized in only one direction it is sufficient to consider only the first solution of (15.4.22). This is obtained in terms of β satisfying the boundary conditions in both the central core and the outer cladding of the fibre.

Solutions for the propagation constant β are determined by solving the characteristic equation determined by matching the tangential components of the electric field across the boundary between the core and cladding. Each value of β corresponds to a particular mode of electromagnetic propagation along the x-axis. Values of ψ corresponding to each mode are determined by the conditions of entry of the electromagnetic wave into the sensitive section. The lowest frequency at which a particular mode can carry power is referred to as the cut-off frequency for that mode. As the light is linearly polarized, the modes are referred to as LP_{ij} modes where the subscript i refers to the electric field distribution in the y-z plane for fixed x while the subscript j refers to the distribution along the x-axis.

The LP_{01} mode has a zero cut-off frequency; it always carries power. For frequencies characterized by $0 < V < 2.405$ it is the only mode which carries power and optical fibres designed to operate in this regime are called single mode fibres. Note that V is the normalized frequency defined as

$$V = \frac{\omega}{c} a (n_1^2 - n_2^2)^{1/2} \tag{15.4.23}$$

where ω is the radian frequency of the light, c is the speed of light, a is the radius of the inner core and n_1, n_2 are, respectively, the indices of refraction of the inner core and outer cladding.

For frequencies defined by $2.405 < V < 3.832$, only the LP_{01} and LP_{11} modes carry power. Thus, for an optical fibre designed to operate in this frequency range, the intensity output is given by

$$I_f(r,\theta,z) = \left| E_{01}(r,\theta,z) + E_{11}(r,\theta,z) \right|^2 \tag{15.4.24}$$

At any cross-section of the fibre optic sensor analysed above, the two propagating modes interfere to generate a well defined intensity pattern. As each mode is characterized by a different propagation constant β_{ij}, the interference pattern is a function of axial location x.

When the fibre is subjected to axial strain, the optical path length for each mode is changed by a different amount. This change is caused by changes in the axial length of the fibre, in the fibre diameter (as a result of the Poisson effect) and in the fibre index of refraction (as a result of the photoelastic effect). As a result, the interference pattern at the fibre end face changes in a predictable way as a function of the applied strain. Thus, the sensor is essentially an interferometer with both legs contained in the optical fibre.

If it is assumed that a photodetector can be configured to measure just a small part of the intensity at the output of the sensitive part of the senor then the output of the photodetector can be shown to be (Cox and Lindner, 1991)

$$y_{PD} = K_e(I_0 + I_1 \cos\Gamma) \tag{15.4.25}$$

where K_e is the gain of the photodetector and Γ is the phase of the intensity given by

$$\Gamma = \Delta\beta L_x + \Delta\psi \tag{15.4.26}$$

where $\Delta\beta = \beta_{01} - \beta_{11}$ and $\Delta\psi = \psi_{01} - \psi_{11}$ and L_x is the length of sensitive fibre. Note that the total intensity emerging from the end of the fibre is unaffected by the strain in the fibre; only the intensity distribution is affected. Thus, it is important that the photodetector be configured to observe only part of the end face of the fibre. Means of doing this are discussed later.

Following the approach of Cox and Lindner (1991), the first order effect of strain is on the phase of the intensity and is given by

$$\Gamma(\epsilon) = \Delta\beta(\epsilon)L_x(\epsilon) + \Delta\psi \tag{15.4.27}$$

By considering the phase change due to axial strain on each fibre segment dx, it can be shown that

$$\Gamma(\epsilon) = \Gamma_0 + \Delta\hat{\beta}\int_0^{L_x} \epsilon(x,t)\,dx \tag{15.4.28}$$

where $\Delta\hat{\beta}$ represents the amplitude of the time varying quantity $\Delta\beta$. Thus, the photodetector output may be expressed as

$$y_{PD}(t) = K_e I_1\left[1 + \cos\left[\Gamma_0 + \Delta\hat{\beta}\int_0^{L_x} \epsilon(x,t)\,dx\right]\right] \tag{15.4.29}$$

where, for convenience $I_1 = I_0$ has been assumed.

The resulting photodetector output is a sinusoid with a d.c. offset. This d.c. offset can removed using a high pass filter and Γ_0 can be set equal to $2\pi m + 3\pi/2$ (m an integer) by applying some pre-strain to the fibre. This pre-strain is necessary to ensure that the sensor will operate in its linear range. After these operations, the output is given by

$$y_{MD}(t) = K_{MD}\sin\left[\Delta\hat{\beta}\int_0^{L_x} \epsilon(x,t)\,dx\right] \tag{15.4.30}$$

where K_{MD} is a constant. Note that the sinusoidal non-linearity can be ignored for small amplitudes of strain but must be taken into account when large amplitudes are to be measured.

As mentioned previously, when the modal domain (MD) sensor just described is implemented in practice, it is necessary to use on elliptical core to eliminate the problem of unpredictable coupling between four spatially degenerate LP_{11} modes which are all characterised by the same propagation constant β. It is also necessary to use a low pass (as well as the high pass) filter on the output to eliminate photodetector noise. To ensure that $\Delta\psi$ remains constant, it is essential to use an insensitive 'lead in' optical fibre which allows propagation of only the LP_{01} mode.

To measure light arriving from only part of the output face of the sensor it is necessary to splice it with an offset (as shown in Fig. 15.18) to an insensitive 'lead-out' fibre which only allows propagation of the LP_{01} mode. This has the effect of transforming the phase modulation (due to a time varying strain) at the output of the sensitive fibre to amplitude modulation of the light at the output of the lead-out fibre.

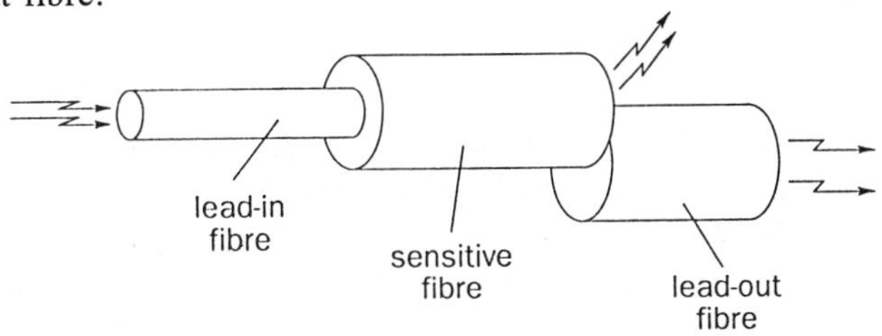

Fig. 15.18 Optically spliced modal domain strain sensor.

In practice, the fusion of the lead-out fibre to the sensitive fibre is performed while the sensor is actually sensing cyclic strain so that the misalignment can be adjusted for optimal modulation.

15.5 HYDRAULIC ACTUATORS

Hydraulic actuators generally consist of a piston in a hydraulic cylinder which has hydraulic fluid openings at each end. Introduction of high pressure fluid into one end of the cylinder causes the piston to move to the other end, expelling fluid from the opening at that end. Switching the supply of high pressure fluid to the other end of the cylinder causes the piston to move in the opposite direction. The most common method to alternate the hydraulic supply from one end of the cylinder to the other is to use a servo-valve which consists of a moveable spool connected to a solenoid.

Applying voltage to the solenoid coil causes the spool to move, which in turn results in the opening and closing of valves which are responsible for directing hydraulic fluid to either end of the hydraulic cylinder (see Fig. 15.19). Note that the servo-valve spool is usually maintained in its inactive position with springs at either end.

Servo-valves can be quite expensive, ranging from US$100 to US$3000 depending on the upper frequency at which they are required to operate, which can be 3 to 20 Hz for an active vehicle suspension or 150 Hz for structural vibration control.

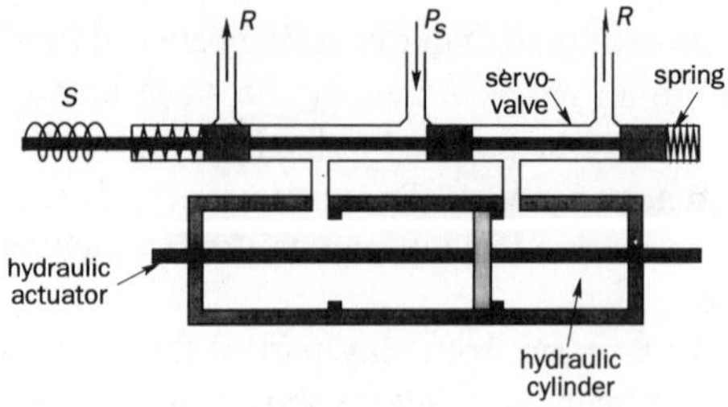

Fig. 15.19 Schematic of a hydraulic actuator driven by a servo-valve shown with the actuator driven to the right.
R = hydraulic return line
P_s = hydraulic supply pressure
S = solenoid

One advantage of hydraulic actuators is their large displacement and high force generating capability for a relatively small size. Disadvantages include the need for a hydraulic power supply which can be inherently noisy and non-linearities between the servo-valve input voltage and the hydraulic actuator force or displacement output. It is extremely important that the servo-valve be mounted as closely as possible to the hydraulic cylinder to minimise loss of performance, especially at high frequencies (above about 20 Hz). For best results at high frequencies, the servo-valve should be incorporated in the actuator assembly as was done by King (1988).

In an attempt to make the actuator motion correspond more closely to the solenoid drive current, some manufacturers have developed two stage servo-valves, in which torque motor is used to operate a valve which in turn directs hydraulic fluid to one side or the other of the servo-valve spool.

Note that the hydraulic supply is generated by a hydraulic pump with an accumulator (or large pressurized reservoir) between the pump and the servo-valve to smooth out any pressure pulsations.

Hydraulic actuators have been used in the past in active vehicle suspensions (Crolla, 1988; Stayner, 1988) and active control of helicopter cabin vibration (King, 1988).

15.6 PNEUMATIC ACTUATORS

Pneumatic actuators are very similar in operation to hydraulic actuators, except that the hydraulic fluid is replaced by air. One advantage of active pneumatic actuators is that they can use the same air supply as passive air springs which may be mounted in parallel with them. Pneumatic actuators may also be the preferred option in cases where an existing compressed air supply is available (such as in rail vehicle active suspension applications; see Buzan and Hedrick, 1983 and Cho and Hedrick, 1985).

The major disadvantage of pneumatic actuators is their relatively low bandwidth (less than 10 Hz) due to the compressibility of the air. Nevertheless, some success has been reported (Cho and Hedrick, 1985) in using pneumatic actuators in an active suspension for a rail vehicle.

15.7 PROOF MASS ACTUATOR

This type of actuator consists of a mass which is free to slide along a track. The mass is accelerated using an electromagnetic field (see Fig. 15.20a) so in one sense this type of actuator is similar to an electrodynamic shaker which will be described in the next section. The mass can also be accelerated by other means; for example, by connecting it to a ball screw (see Fig. 15.20(b); Clark *et al.*, 1990). The proof mass results in a reaction in the ball screw or electromagnetic stator which is attached to the structure to be controlled.

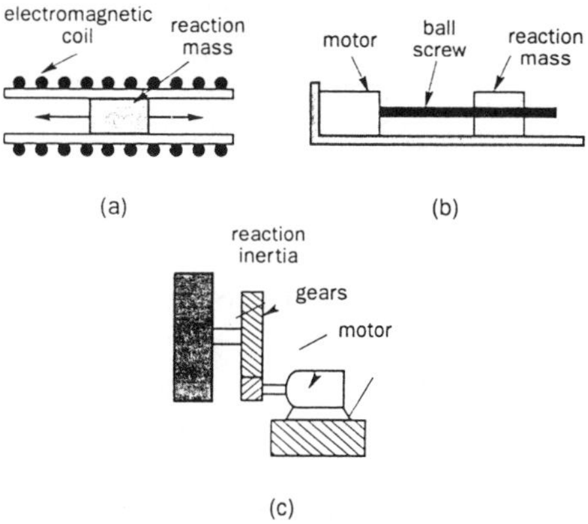

Fig. 15.20 Proof mass actuators: (a) linear, electromagnetic drives; (b) linear, ball screw drive; (c) rotational.

Another type of proof mass actuator is one which imparts moments rather than forces to a structure by control of a rotating mass which results in reaction moments on the support structure (see Fig. 15.20(c)).

One problem with linear proof mass actuators is the limited motion of the reaction mass which becomes more serious when low frequency vibration modes in a structure are to be controlled. However, the problem is much less for the ball screw type actuator as high forces can be produced on the reaction mass without producing large motions.

In cases where it is not possible to prevent the proof mass from encountering the limits of its travel, Politansky and Pilkey (1989) showed that a suboptimal feedback control system yielded good control results.

15.8 ELECTRODYNAMIC AND ELECTROMAGNETIC ACTUATORS

Electrodynamic actuators or shakers consist of a moveable core (or armature) to which is fixed a cylindrical electrical coil. The core and coil move back and forth in a magnetic field which is generated by a permanent magnet (smaller shakers) or by a d.c. current flowing through a second coil fixed to the stationary portion of the shaker (stator), as shown in Fig. 15.21. The moveable core is supported on mounts which are very flexible in the axial direction in which it is desired that the core move and rigid in the radial direction to prevent the core contacting the outer armature. To maximize lightness and stiffness, the core is usually constructed of high strength aluminium alloy and is usually hollow. Mechanical stops are also usually provided to prevent the core from being over driven. In some cases, the driving coil is air cooled.

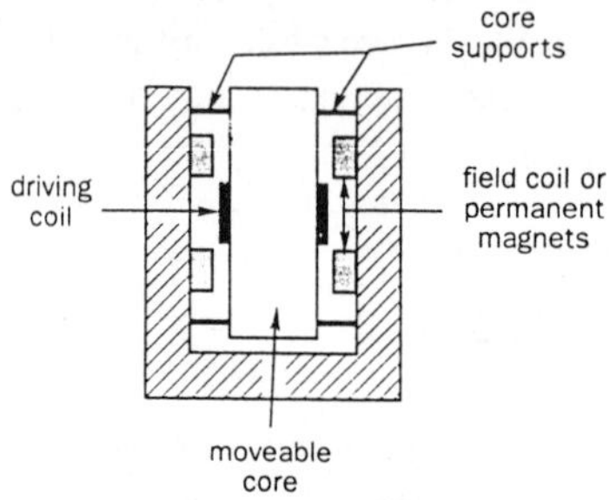

Fig. 15.21 Electrodynamic shaker schematic diagram.

When a sinusoidal voltage is applied to the drive coil, the polarity and strength of the drive coil magnetic field changes in phase with the applied voltage and so then does the force of attraction between the field coil (or permanent magnets) and the driving coil, resulting in axial movement of the core (or armature).

Electromagnetic actuators are similar in construction to electrodynamic shakers except that the inner core as well as the armature is fixed (see Fig. 15.22). It is constructed by surrounding an electrical coil with a permanent magnet. Supplying a sinusoidal voltage to the coil results in the production of a sinusoidally varying magnetic field which can be used to shake ferro-magnetic structures, or structures made from other materials, if a thin piece of shim steel is bonded to them. Clearly the electromagnetic actuator must be mounted on a rigid fixture, preferably not connected to the structure to be excited.

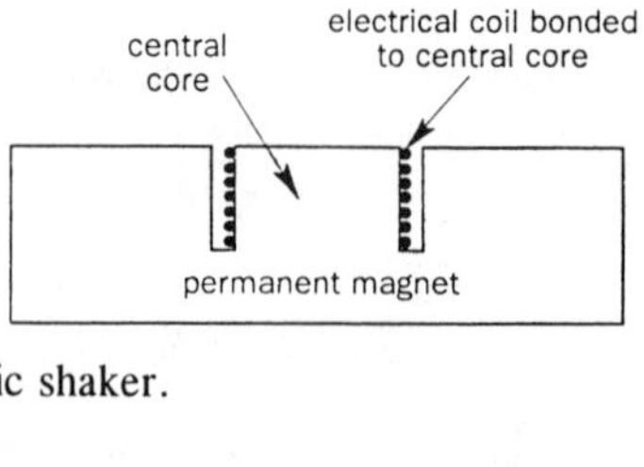

Fig. 15.22 Electromagnetic shaker.

If the permanent magnet were not used, the attraction force between the coil and the structure being shaken would not vary sinsoidally but would vary as

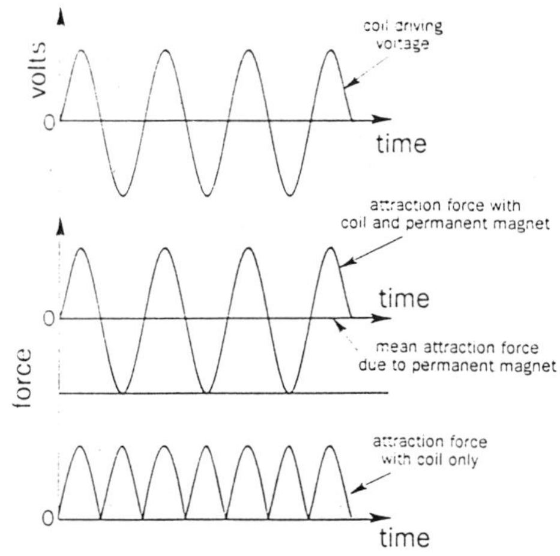

Fig. 15.23 Electromagnetic attraction force vs. coil driving voltage: (a) sinusoidal coil driving voltage; (b) corresponding attraction force for transducer with permanent magnet; (c) corresponding attraction force for transducer without permanent magnet.

shown in Fig. 15.23(b), resulting in most of the excitation energy being at twice the coil driving frequency.

Electromagnetic drivers such as those just described are easily and cheaply constructed by bonding the coil from a loudspeaker driver into the core of the permanent magnet where it is normally located in its at rest condition.

15.9 MAGNETOSTRICTIVE ACTUATORS

The phenomenon of giant magnetostriction in rare earth, Fe_2 compounds was first reported by Clark and Delson in 1972. The most effective alloy is referred to as terfenol, with the composition $Tb_{0.3}$ $Dy_{0.7}$ $Fe_{1.93}$. The subscripts refer to the atomic proportions of each element (terbium, dysprosium and iron). Multiplying these by the atomic weights gives the weight proportions (18%, 42% and 40% respectively). The notable properties of this material are its high strain ability (25 times that of nickel, the only other commonly used magnetostrictive material and 10 times that of piezoelectric ceramics – see next section) and high energy densities (100 times that of piezoceramics). Thus, terfenol-D can produce high force levels and high strains relative to other expanding materials. The maximum possible theoretical strain is 2440 μstrain, and 1400 μstrain is achievable in practice. The properties of terfenol are discussed in detail in the literature Hiller *et al.* (1989); Moffet *et al.* (1991,a,b). A typical commercially available terfenol actuator is illustrated in Fig. 15.24 (Edge Technologies, 1990).

The alternating magnetic field is applied to the terfenol rod by a coil which can produce a strain of 1000 μstrain in the rod for an applied magnetic field H of 700 Oe. This field strength is a function of the current flowing in the coil and can be calculated using

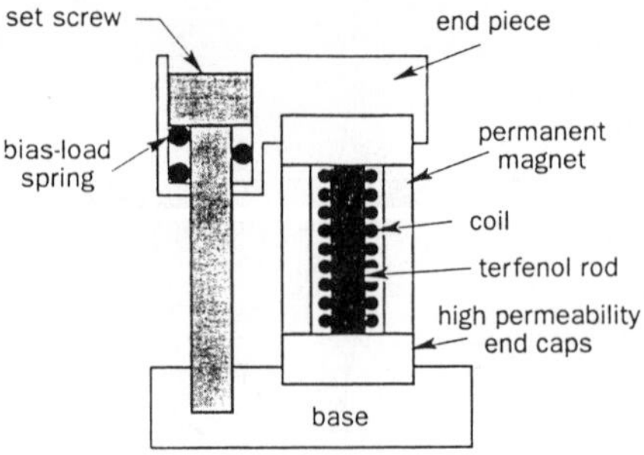

Fig. 15.24 Cylindrical magnetostrictive actuator cross-section.

$$H = \frac{f 2\pi NI}{d_i(\alpha - 1)} \sqrt{\frac{(\alpha^2 - 1)}{2\pi\beta}} \tag{15.9.1}$$

$$f = 0.2 \sqrt{\frac{2\pi\beta}{(\alpha^2 - 1)}} \log_e \left[\frac{\alpha + \sqrt{\alpha^2 + \beta^2}}{1 + \sqrt{1 + \beta^2}} \right] \tag{15.9.2}$$

where $\alpha = d_0/d_i$, I is the coil current (amps), N is the number of turns in coil, d_0 is the coil outer diameter (m), d_i is the coil inner diameter (m), $\beta = L/d_i$ and L is the coil length (m). The geometry factor f gives some indication of the efficiency of the particular coil design, and has a maximum theoretical value of 0.179.

Thus, for an actuator containing a 50 mm long terfenol rod, the maximum possible displacement will be $50 \times 10^{-3} \times 1400 \times 10^{-6}$m or 70 μm. In practice, actuators are generally limited to 1000 μstrain (or 50 μm for a 50 mm long terfenol rod) to minimize non-linearity.

Examination of the actuator illustrated in Fig. 15.24 shows the use of high permeability iron end caps to help reduce fringing of the magnetic flux lines at the rod ends. Flux density is also improved by the coil extending past the ends of the terfenol rod.

15.9.1 Magnetic bias

As application of a magnetic field of any orientation will cause the terfenol rod to expand, a d.c. magnetic bias is necessary if the terfenol rod expansion and contraction is to follow an a.c. input signal into the coil. Thus, a d.c. magnetic bias supplied either with a d.c. voltage signal applied to the coil or with a permanent magnet, is required so that the terfenol rod expands and contracts about a mean strain (usually about 500 μstrain). Use of d.c. bias current is characterized by problems associated with drive amplifier design and transducer overheating; thus, it is more common to use a permanent magnet bias. The cylindrical permanent magnet bias shown in Fig. 15.24 also acts as the flux return path for the exciter coil.

15.9.2 Mechanical bias or prestress

To optimize performance (maximize the strain ability of the actuator) it is necessary to apply a longitudinal prestress of between 7 and 10 MPa. The prestress ensures that the magnetically biased actuator will contract as much as it expands when a sinusoidal voltage is applied to the drive coil. It is essential that the prestress arrangement also be highly compliant to avoid damping the drive element. The mechanical bias also helps to mate components and allows transfer of forces with low loss.

15.9.3 Frequency response, displacement and force

Terfenol actuators are usually characterized by a resonance frequency in the few kilohertz range and generally have a reasonably flat frequency response at frequencies below about half the resonance frequency. The actual resonance frequency depends on the mass driven and the stiffness of the terfenol rod. Longer, thinner drives and larger masses result in lower resonance frequencies.

The force output for a clamped actuator may be calculated using $F = EA\Delta L/L$, where E is the elastic modulus (3.5×10^{10} N m^{-2}), A is the cross-sectional area of the terfenol rod (m^2) and $\Delta L/L$ is the strain. Thus, a 6 mm diameter rod operating in a magnetic field capable of producing ± 500 μstrain about the bias value of +500 can produce a force of ± 500 N. Specifications for one such actuator consisting of a 50 mm long $\times$ 6 mm terfenol rod are listed in Table 15.1.

Table 15.1 Typical terfenol actuator specifications

Actuator	
Useable displacement	± 25 μm
Current input for normal operation	± 1 amp
Force capability	± 500 N
Overall length	75 mm
Overall diameter	60 mm
Terfenol length	50 mm
Terfenol diameter	6 mm
Overall weight	0.6 kg
Coil	
Rating	300 Oe amp^{-1}, 1181 amp turns
Coil length	53 mm
Coil inner diameter	6.7 mm
Coil outer diameter	16.5 mm
Wire gauge	26 AWG

Note: See *Edge Technologies* manual for more information (Buttler, 1988).

15.9.4 Disadvantages of terfenol actuators

Terfenol is very brittle and must be carefully handled. Its tensile strength is low (100 MPa) although its compressive strength is reasonably high (780 MPa). Another important disadvantage is the low displacement capability, which would be a problem in low frequency vibration control (below 100 Hz). One way around this limitation is to use displacement expansion devices which rely on the lever principle to magnify the displacement at the expense of reduced force output. The main disadvantage associated with magnetostrictive actuators, however, is the hysteresis inherent in terfenol, which results in a non-linear actuation force for a linear voltage input. This results in excitation frequencies being generated which are not related to the voltage input frequencies. Although this is a serious problem for active vibration control systems at the present time, it is expected that non-linear controllers currently being developed will be able to cope adequately with such non-linear actuators such as these.

15.9.5 Advantages of terfenol actuators

The main advantage of terfenol is its high force capability for a relatively low cost. Electrodynamic shakers capable of the same force capability cost and weigh between 25 and 100 times as much and require a much larger driving amplifier. Thus, another advantage is small size and light weight, which makes the actuator ideal for use in situations where no reaction mass is necessary. Actuator arrangements for active vibration control with no need for a reaction mass are discussed in Section 15.12.1.

15.10 SHAPE MEMORY ALLOY ACTUATORS

A shape memory alloy is a material which when plastically deformed in its low temperature (or martensitic) condition, and the external stresses removed, will regain its original (memory) shape when heated. The process of regaining the original shape is associated with a phase change of the solid alloy from a martensite to an austenite crystalline structure.

The most commonly used shape memory alloy in active control systems is nitinol (an alloy of approximately equi-atomic composition of nickel and tin). Room temperature plastic strains of typically 6 to 8% can be completely recovered by heating the material beyond its transition temperature (which ranges from 45°C to 55°C). Restraining the material from regaining its memory shape can yield stresses of up to 500 MPa (for an 8% plastic strain and a temperature of 180°C – see Rogers, 1990). On being transformed from its martensite to its austenite phase, the elastic modulus of nitonol increases by a factor of three (from 25 to 75 GPa) and its yield strength increases by a factor of eight (from 80 to 600 MPa).

The attractiveness of nitinol over other shape memory alloys for active control is its high resistivity which makes resistive heating feasible. This means that nitinol can be conveniently heated by passing an electrical current through it.

Nitinol is commercially available in the form of thin wire or film which can be used for the active control of structural vibration by either embedding the material in the structure or fixing it to the surface of the structure. Before being embedded or fixed to the surface of the structure, the nitinol is plastically elongated. When used as fibres in composite structures, the nitinol fibres must be constrained so that they cannot contract to their normal length during the high temperature curing of the composite.

Active control of structural vibration using nitinol can be achieved either directly or indirectly. When nitinol wire of film is heated it tries to contract to its normal size. If it is bonded to the surface of a structure it will impose a bending moment on the structure as it tries to contract. If the structure were a vibrating thin panel or a beam, then the vibration could be controlled by using a shape memory alloy element on each side of the panel or beam. Applying current to the element on one side will cause a controlling moment acting in one direction and applying current to the element on the other side will cause a controlling moment in the opposite direction. Thus, alternating the supply of current from one side to the other will induce an alternating moment in the structure which can be used to control the structural vibration. This is referred to as direct control and application of this method to the control of the vibration of a cantilevered beam has been demonstrated successfully by Baz *et al.*, 1990. The upper frequency limit for control is about 10 Hz, due to the difficulty in cooling the nitinol between cycles. If the nitinol is embedded in the structure (on each side of the neutral axis) the upper frequency limit for control would be much lower due to the inherent cooling problems.

A second active control technique involves the use of shape memory alloy elements embedded in a structure, which when activated, change the dynamic characteristics of the structure by changing its effective stiffness. For this technique, shape memory fibres are usually embedded on the neutral axis of a composite structure. Applying a current to the fibres causes them to apply a strain to the structure which can increase its resonance frequencies. This technique has been demonstrated for a beam (Rogers, 1990) and a plate (Rogers *et al.*, 1991). In both cases, it was found that heating the nitinol resistively to about 120°C changed the resonance frequencies of the first few modes by up to 200%. Lower temperatures resulted in smaller changes. It follows that this technique can be used to optimize the sound radiation characteristics of a structure for a particular excitation. If the excitation frequency spectrum slowly changes, the structural stiffness could be readjusted by adjusting the current flow through the nitinol elements to maintain optimal sound radiation characteristics (Saunders *et al.*, 1991). This has obvious applications to composite submarine hulls, aircraft fuselages and other aerospace structures.

15.11 PIEZOELECTRIC (ELECTROSTRICTIVE) ACTUATORS

This type of actuator may be divided into two distinct categories; the thin plate or film type and the stack (or thick) type. The thin type may be further categorized into piezoceramic plates (PZT) or piezoelectric film (PVDF). The thin type is usually bonded to the structure it is to excite and generates excitation by imparting a bending moment to the structure. The stack type is used in a similar way to an electrodynamic shaker or magnetostrictive actuator, generally applying to the structure a force distributed over a small area. The thin type and stack type actuators will be considered separately in the following discussion. Both types of actuator generate forces and displacements when subjected to an electric field – hence their alternative description, electrostrictive.

15.11.1 Thin actuators

A detailed review of work on thin piezo actuators was done by Smits *et al.* (1991) and the reader is referred to this for a detailed survey of the uses of this material.

The behaviour of piezoceramic plate and PVDF film actuators is similar (although their properties differ significantly); thus, the two types will be considered together. Note that these actuator types can also act as sensors and PVDF film acting in this capacity was discussed in detail in Section 15.4.

For thin actuators, the extension as a result of an applied voltage is in a direction normal to the direction of polarization as shown in Fig. 15.25. For an applied voltage V, the free extension ΔL *(m)* is given in terms of strain by

$$\frac{\Delta L}{L} = d_{31} V/t = \Lambda \qquad (15.11.1a,b)$$

where t is the piezo thickness and d_{31} is its charge or strain constant, representing the strain produced by an applied electric field.

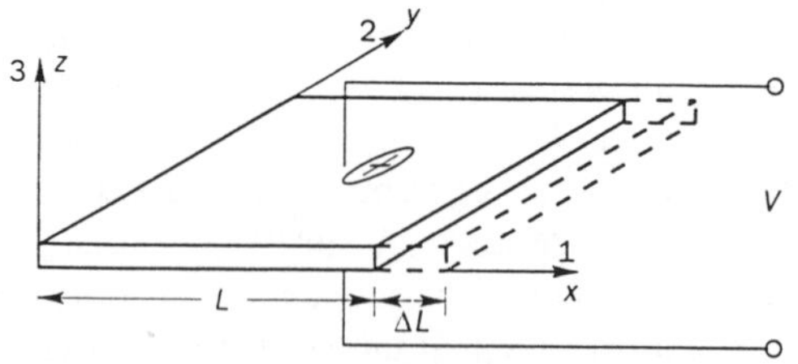

Fig. 15.25 Extension of a thin piezoelectric actuator as a result of an applied voltage.

If the piezo element is bonded to a structure, the increase in length will be less than this and a resulting bidirectional longitudinal force will be applied to the structure surface which will result in a moment induced in the structure.

For PZT (lead zirconate titanite) piezo ceramic, d_{31} for one commercially available material is 166 × 10^{-12}. On the other hand, d_{31} for commercially available PVDF film is 23 × 10^{-12}. Also, Young's modulus of elasticity for PZT is 63 × 10^9 N m^{-2}, whereas for PVDF film it is only 2 × 10^9 N m^{-2}. It will be shown in the following section that the strain induced in a simple structural element such as a beam (which is not too thin) is proportional to the product of d_{31} and E for a given applied voltage. Thus, it is clear that for the same thickness material and same applied voltage, PZT will produce about 150 times the strain in a beam than will PVDF film. Note, however, that the maximum operating voltage for PVDF film is about 25 times that for PZT (the latter being about 1 V r.m.s. μm^{-1} material thickness).

A PZT layer or PVDF film bonded to a structure such as a beam and subjected to an applied voltage will effectively generate a moment approximately equivalent to a line moment at each free edge, as shown in Fig. 15.26(a). If a second actuator is bonded to the opposite side of the beam and subjected to an applied voltage equal and opposite to that applied to the first actuator, then the beam will bend about its neutral axis. The moment distribution generated in the beam by the two assumed line moments is shown as the dashed line in Fig. 15.26(b). The actual distribution is more like that shown by the solid line (Crawley and de Luis, 1987). However, the effect of this deviation from the assumed ideal case is insignificant from the viewpoint of calculating the beam response.

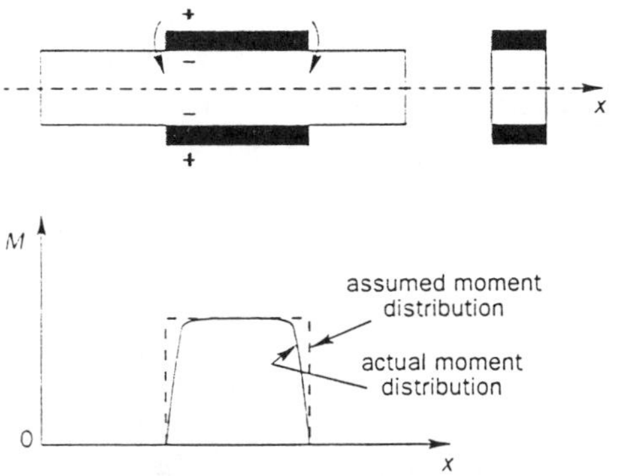

Fig. 15.26 Beam bending moment distribution due to a PZT layer.

Although PZT actuators can be bonded directly on to structures using epoxy adhesive, their brittleness makes this a task requiring considerable skill, especially on curved surfaces: Usually, a thin layer of epoxy is first bonded to the surface under a layer of glass which has release agent sprayed on it. When dry, the glass is removed, leaving a smooth surface on to which the PZT is bonded. Some researchers prefer to order the PZT from manufacturers already bonded to a thin brass shim. This shim is then bonded (with a lot less trouble) directly to the structure to be controlled.

Thin PZT actuators for active control were first used by Forward and

Swigert in 1981 for the control of vibration of a tall mast. In 1987 Crawley and de Luis presented a detailed analysis for the case of a beam excited by two PZT crystals bonded to it on opposite sides and driven 180° out of phase. Unfortunately this work suffers from the incorrect assumption of uniform strain across the thickness of the actuators. However, many of the results obtained remain very useful. In 1988, Baz and Poh developed a finite element model of a beam excited by a singe PZT crystal bonded to one side. In 1991, four important papers were published which explained how to accurately model PZT excitation of beams (Crawley and Anderson 1991; Clark *et al.*, 1991) and plates (Dimitriadis and Fuller, 1991; Kim and Jones, 1991). Unfortunately, the assumption of the slope of the stress distribution in the actuator being the same as that in the beam, which was made by Clark *et al.* (1991) and Dimitriadis *et al.* (1991) is not strictly correct, as the elastic modulus of the actuator material is not the same as that of the beam. The implications of this will be made clear in the next section.

15.11.1.1 One-dimensional actuator model: effective moment

It is useful to be able to derive the bending moment generated in a one dimensional structure such as a beam by a pair of piezoelectric actuators placed on opposite sides of it. Once derived, the bending moment can be used with the equations of Chapter 10 to calculate the response of a beam with any desired end conditions.

The configuration being analysed is shown in Fig. 15.27, which also shows the assumed strain distributions and corresponding stress distribution for a positive beam displacement in the z-direction when the actuators are energised and produce pure positive bending in the beam.

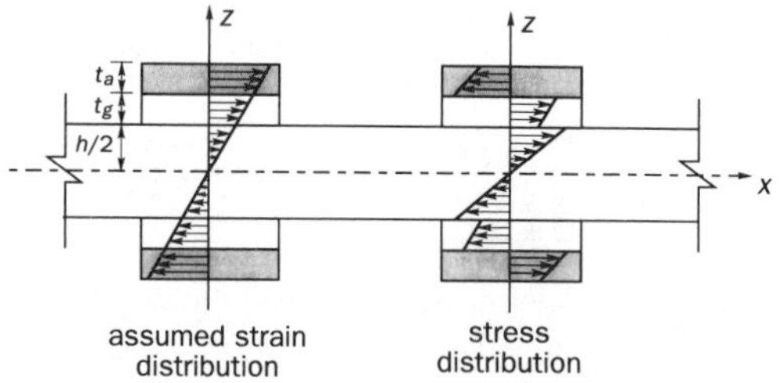

Fig. 15.27 One dimensional actuator model.

The linear strain distribution implies that the strains are continuous across each layer interface and the centres of the radii of curvature for each layer are concurrent. That is, the glue layer is not deformed during bending such that its thickness at the edges varies differently to its thickness in the centre.

Note that a glue layer has been included, although if this is very thin, it has a negligible effect on the results (Kim and Jones, 1991). Note also that the stress in the top actuator is compressive even though that in the glue layer and beam beneath it is in tension. This is because the strain in the actuator is less than it would be if it were not bonded to the beam (due to the applied electric field – see (15.11.1)).

The bending moment generated in the beam by the actuators per unit length on a plane perpendicular to the x-axis is given by

$$M_x = \int_{-h/2}^{h/2} \sigma_x z \mathrm{d}z \tag{15.11.2}$$

where h is the thickness of the beam, z is the distance from the centre of the beam to where σ_x is acting and σ_x is the axial normal stress acting along the x-axis, varying from zero at the neutral axis to a maximum at the top or bottom of the beam. The moment M_x exists in the beam over the full length of the actuator and may be generated by two external line moments acting in opposite directions (see Fig. 15.26(a)) applied to the beam at the free edges of the actuators, where each line moment is equal in magnitude to bM_x (b is the width of the beam).

The stress σ_x may be written in terms of the stress in the beam at the interface between the bonding (or glue) layer and the beam (σ_{ib}) as follows

$$\sigma_x = \frac{2\sigma_{ib} z}{h} \tag{15.11.3}$$

To obtain an expression for σ_{ib} in terms of known quantities, we use the following moment equilibrium condition:

$$\int_0^{h/2} \sigma_x z \mathrm{d}z + \int_{h/2}^{h/2+t_g} \sigma_g z \mathrm{d}z + \int_{h/2+t_g}^{h/2+t_g+t_a} \sigma_a z \mathrm{d}z = 0 \tag{15.11.4}$$

Before solving (15.11.4) for σ_{ib} it is necessary to express σ_x, σ_g and σ_a in terms of σ_{ib}.

The necessary equivalence for σ_x is given by (15.11.3). The expressions relating σ_g and σ_a to σ_{ib} can be derived from the assumption of uniform strain slope across the beam, glue and actuator as shown in Fig. 15.27. That is

$$\epsilon(z) = \epsilon_b = \epsilon_g = \epsilon_a = \mu z \tag{15.11.5a,b,c,d}$$

where μ is the strain slope which can be found by developing an expression for ϵ_b in terms of the beam stress σ_x and the beam modulus of elasticity, E_b. For a beam whose thickness is similar to its width we can use the plane stress assumption which gives

$$\epsilon_b = \frac{\sigma_x}{E_b} \tag{15.11.6}$$

Substituting (15.11.3) into (15.11.6) and using (15.11.5) gives

$$\mu = \frac{2\sigma_{ib}}{E_b h} \tag{15.11.7}$$

Clearly, the stress in the actuator is related to the free strain Λ due to the electric field (see (15.11.1)), Young's modulus of elasticity for the actuator and the actuator strain ϵ_a as a result of bending as follows:

$$\sigma_a = E_a(\epsilon_a - \Lambda) \tag{15.11.8}$$

As Λ is always larger than ϵ_a, the stress in the actuator will be of opposite sign to that in the beam.

Substituting (15.11.5) and (15.11.7) into (15.11.8) gives

$$\sigma_a = E_a\left[\frac{2\sigma_{ib}z}{E_b h} - \Lambda\right] \tag{15.11.9}$$

The stress in the bonding layer may be written as

$$\sigma_g = E_g\epsilon_g = 2\frac{E_g}{E_b}\sigma_{ib}\frac{z}{h} \tag{15.11.10a,b}$$

Substituting (15.11.3), (15.11.9) and (15.11.10) into (15.11.4) and integrating gives

$$\sigma_{ib} = \frac{3hE_a t_a(h + 2t_g + t_a)\Lambda}{2\left[\frac{h^3}{4} + \frac{E_g}{E_b}t_g\left[\frac{3}{2}h^2 + 3ht_g + 2t_g^2\right] + \frac{E_a}{E_b}t_a\left[3\frac{h^2}{2} + 6t_g^2 + 2t_a^2 + 6ht_g + 3ht_a + 6t_g t_a\right]\right]} \tag{15.11.11}$$

Substituting (15.11.11) into (15.11.3), then into (15.11.2) and integrating gives

$$M_x = \frac{h^3 E_a t_a(h + 2t_g + t_a)\Lambda}{\frac{h^3}{4} + \frac{E_g}{E_b}t_g\left[\frac{3}{2}h^2 + 3ht_g + 2t_g^2\right] + \frac{E_a}{E_b}t_a\left[\frac{3h^2}{2} + 6t_g^2 + 2t_a^2 + 6ht_g + 3ht_a + 6t_g t_a\right]} \tag{15.11.12}$$

If the bonding layer is very thin, its effect can be neglected (Kim and Jones, 1991) and we obtain

$$M_x = \frac{\rho_a(2 + \rho_a)}{4\left[1 + \frac{E_a}{E_b}\rho_a\left(3 + \rho_a^2 + 3\rho_a\right)\right]}h^2 E_a\Lambda \tag{15.11.13}$$

where $\rho_a = \dfrac{2t_a}{h}$ (15.11.14)

The preceding formulation is for a beam completely covered with the actuator. In cases where the actuators only extend partially along the beam, free edges will exist where the equilibrium condition requires that the normal stress at the actuator boundaries be zero, effectively invalidating the relationships just derived. However, it is generally accepted (Kim and Jones, 1991; Dimitriadis *et al.*, 1991) that the actuator stress field for a distributed actuator is unaffected by the free edge except within about four actuator thicknesses of the boundary. This has the effect of slightly altering the moment distribution in the beam (to the solid curve shown in Fig. 15.26(b)), which has a negligible effect on the resulting beam response.

Using the results of Chapter 2.3, the classical beam equation of motion for excitation by a piezoelectric actuator pair is

$$\frac{\partial^4 w}{\partial x^4} + \frac{\rho S}{E_b I_b}\frac{\partial^2 w}{\partial t^2} = -\frac{bM_x}{E_b I_b}\left[\delta'(x-x_1) - \delta'(x-x_2)\right] \qquad (15.11.15)$$

where the prime denotes differentiation with respect to x, x_1 and x_2 are the axial coordinates of the two edges of the piezo actuator and M_L of (2.3.50) is equal to bM_x. The quantity I_b is the second moment of area of the beam cross section $(bh^3/12)$, and E_b is the elastic modulus for the beam material. Note that the effects of transverse shear in the beam and rotary inertia are neglected in this equation which makes it valid only for frequencies where the wavelength is much larger than the beam cross-sectional dimensions. If this is not so, then the more complex equation of motion derived in Section 2.3 must be used.

If the actuators do not cover the full width of the beam, the right hand side of (15.11.15) must be multiplied by b_a/b where b_a is the actuator width and b is the beam width.

Note that the preceding analysis only applies to a beam whose thickness is similar to its width. For thin beams, the analysis and corresponding expression for the moment M_y acting on a thin plate, as derived in the next section must be used. Furthermore, the elastic modulus E_b for the beam in (15.11.15) must be replaced by $E_b/(1 - \nu^2)$.

15.11.1.2 Two-dimensional actuator analysis

The analysis for a two-dimensional actuator is similar to that just done for the one-dimensional case with the main difference being the assumption of plane strain (due to the thinness of the plate with respect to its dimensions).

For the plate the following stress strain relationships can be written (Timoshenko and Woinowsky-Krieger, 1959):

$$\sigma_{xp} = \frac{E_p}{1-\nu_p^2}\left(\epsilon_{xp} + \nu_p \epsilon_{yp}\right) \qquad (15.11.16)$$

$$\sigma_{yp} = \frac{E_p}{1-\nu_p^2}\left(\epsilon_{yp} + \nu_p \epsilon_{xp}\right) \qquad (15.11.17)$$

For an actuator which is bonded to a homogeneous plate and which gives equal free strains in the x- and y-directions (that is, $d_{31} = d_{32}$), then $\epsilon_{xp} = \epsilon_{yp}$ (Dimitriadis and Fuller, 1991); thus, the stresses in the x- and y-directions will also be equal. If this assumption is not made (and it may not be valid for some PVDF film), the analysis is still possible but the algebra becomes more complex.

Using the above mentioned equality, the subscripts x and y (denoting the x and y directions across the plate – see Fig. 15.28) may be dropped and (15.11.16) and (15.11.17) become

$$\sigma_p = \frac{E_p}{1-\nu_p}\epsilon_p \tag{15.11.18}$$

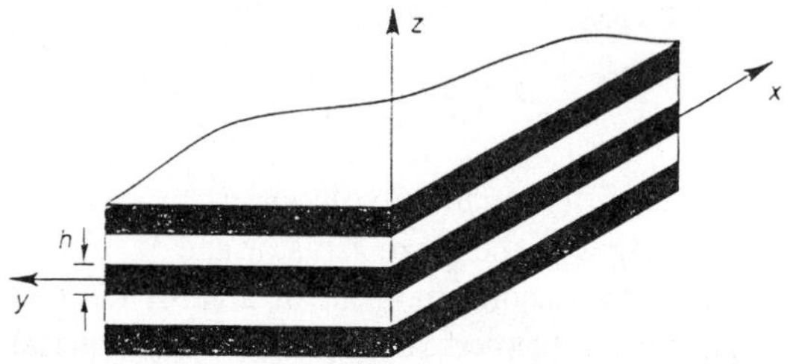

Fig. 15.28 Plate model for two-dimensional actuator analysis.

In terms of the stress, σ_{ip}, in the plate at the interface between the plate and the glue layer, the stress in the plate may be written as

$$\sigma_p = \frac{\sigma_{ip} z}{h} \tag{15.11.19}$$

The slope μ of the strain curve is now given by

$$\mu = \frac{2\sigma_{ip}(1-\nu_p)}{E_p h} \tag{15.11.20}$$

The stress distribution in the bonding layer may be written as

$$\sigma_g = \frac{E_g \epsilon_g}{(1-\nu_g)} = \frac{E_g z \mu}{(1-\nu_g)} = \frac{2E_g(1-\nu_p)}{E_p(1-\nu_g)}\sigma_{ip}\frac{z}{h} \tag{15.11.21a,b,c}$$

Modifying (15.11.8) to account for the Poisson effect we obtain the following for the actuator stress:

$$\sigma_a = \frac{E_a}{1-\nu_a}(\epsilon_a - \Lambda) \tag{15.11.22}$$

Substituting (15.11.5) and (15.11.20) into (15.11.22) gives

$$\sigma_a = \frac{2E_a(1-\nu_p)\sigma_{ip}}{E_p(1-\nu_a)h} - \frac{E_a\Lambda}{(1-\nu_a)} \tag{15.11.23}$$

Substituting (15.11.19), (15.11.21) and (15.11.23) into (15.11.4) (where σ_x has been replaced by σ_p) gives

$$\sigma_{ip} = \frac{3\gamma h t_a (h + 2t_g + t_a)\Lambda}{2\left[\frac{h^3}{4} + \alpha t_g\left[\frac{3}{2}h^2 + 3ht_g + 2t_g^2\right] + \beta t_a\left[\frac{3}{2}h^2 + 6t_g^2 + 2t_a^2 + 6ht_g + 3ht_a + 6t_g t_a\right]\right]} \quad (15.11.24)$$

where

$$\alpha = \frac{(1-\nu_p)E_g}{(1-\nu_g)E_p} \quad (15.11.25)$$

$$\beta = \frac{(1-\nu_p)E_a}{(1-\nu_a)E_p} \quad (15.11.26)$$

$$\gamma = \frac{E_a}{1-\nu_a} \quad (15.11.27)$$

Values of E_a, ν_a, and d_{31} for typical PZT and PVDF materials are given in Table 15.2.

The bending moments in the plate per unit length about planes perpendicular to the x and y axes are equal and are given by

$$M_x = M_y = \int_{-h}^{h} \sigma_p z dz = \frac{1}{h}\int_{-h}^{h} \sigma_{ip} z^2 dz = \frac{2}{3}h^2\sigma_{ip} \quad (15.11.28\text{a-d})$$

and the equivalent external line moments are $M_{xL} = M_y(y_2 - y_1)$ and $M_{yL} = M_x(x_2 - x_1)$. Substituting (15.11.24) into (15.11.28d) gives the required expressions for M_x and M_y.

For the case of zero glue layer thickness (which can usually be assumed) we obtain

Table 15.2 Properties of some commercially available thin piezoelectric actuators

	PZT			
Material	G-1195	G-1512	G-1278	PVDF Film
E_a (GPa)	63	63	60	2
ν_a	0.35	0.35	0.35	0.35
d_{31} (m V^{-1} or C N^{-1})	1.66×10^{-10}	2.3×10^{-10}	2.7×10^{-10}	0.23×10^{-10}
d_{33}(m V^{-1})	-3.6×10^{-10}	-4.8×10^{-10}	-5.85×10^{-10}	-0.33×10^{-10}

$$M_x = M_y = \frac{\rho_a(2+\rho_a)}{4\left(1+\beta\rho_a\left(3+\rho_a^2+3\rho_a\right)\right)} h^2\gamma\Lambda \qquad (15.11.29a,b)$$

where ρ_a is defined by (15.11.14).

Equations (15.11.24) to (15.11.29) can be used to calculate the applied moment as a function of the ratio of actuator thickness to plate thickness. When this is done it is found that for a constant applied electric field strength (volts per mm of actuator thickness), there is an optimum actuator thickness for producing the maximum applied moment (Kim and Jones, 1991). It can be shown that for commercially available piezoceramics, the optimum actuator thickness is about half of the plate thickness for a steel substructure and about one quarter of the plate thickness for an aluminium substructure. A similar analysis to find the optimum actuator thickness for a beam could be done using (15.11.12).

The moments generated in the plate by the piezoactuator can be approximated by external line moments acting on the plate at the edges of the piezoelectric layer. Using the results of Section 2.3, (2.3.127) the plate equation of motion (classical plate theory) for excitation by a piezoceramic actuator bonded to the plate surface as shown in Fig. 15.29 can be written as

$$\begin{aligned} D\nabla^4 w + \rho h\ddot{w} = M_x\Big[&-\left(\delta'(x-x_1) - \delta'(x-x_2)\right)\left(u(y-y_1)-u(y-y_2)\right) \\ &+\left(\delta^\dagger(y-y_1) - \delta^\dagger(y-y_2)\right)\left(u(x-x_1) - u(x-x_2)\right)\Big] \end{aligned} \qquad (15.11.30)$$

where M_x is defined by (15.11.29), $u(\,)$ is the unit step function (introduced as the piezoelectric layer does not extend to the plate edges), $\delta(\,)$ is the Dirac delta function, and D is the flexural rigidity of the plate given by

$$D = \frac{Eh^3}{12(1-\nu^2)} \qquad (15.11.31)$$

The prime $'$ denotes differentiation with respect to x, $\dagger$ denotes differentiation with respect to y, $\bullet$ denotes differentiation with respect to time, ρh is the mass per unit area of the plate and h is the plate half thickness. It is assumed that the piezoelectric actuators do not add significantly to the mass or stiffness of the plate.

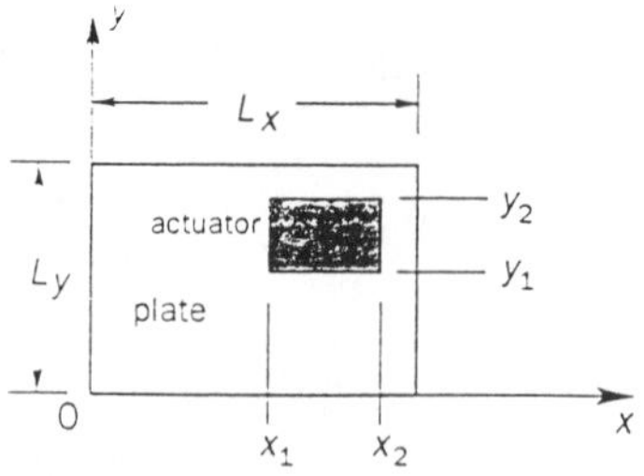

Fig. 15.29 Rectangular plate excited by a rectangular actuator (note that a second identical actuator is located in the same position on the other side of the plate).

The plate forcing function represented by the right-hand side of (15.11.30) is a result of external line moments acting around the edges of the actuators which act to generate the uniformly distributed moment in the plate.

It is of interest to examine the solution w to (15.11.30) for a simply supported rectangular plate for a harmonically varying piezoactuator exciting voltage of frequency ω. Using the results of chapter 4 we can write the following for the plate displacement $w(x,y,t)$ at any location (x,y) and time t in terms of its vibration modes (m,n):

$$w(x,y,t) = e^{j\omega t} \sum_{m=1}^{\infty} \sum_{n=1}^{\infty} A_{mn} \psi_{mn}(x,y) \tag{15.11.32}$$

where A_{mn} is the amplitude of mode m,n and $\psi_{mn}(x,y)$ is the mode shape at location x,y.

Using the results of Section 2.3, we have the following for the mode shape of a simply supported panel:

$$\psi_{mn}(x,y) = \sin\frac{m\pi x}{L_x} \sin\frac{n\pi y}{L_y} \tag{15.11.33}$$

The unknown modal amplitudes A_{mn} are found by substituting (15.11.33) into (15.11.32) and the result into (15.11.30). Both sides of equation (15.11.30) are then multiplied by $\psi_{m',n'}(x, y)$ and the result integrated over the surface of the plate. When this is done, the left side of the equation becomes

$$\int_0^{L_y}\int_0^{L_x} \sin(\gamma_{m'} x)\sin(\gamma_{n'} y) \sum_{m=1}^{\infty}\sum_{n=1}^{\infty} A_{mn} \sin(\gamma_m x)\sin(\gamma_n y)\left[D(\gamma_m^2 + \gamma_n^2) - \rho h\omega^2\right] \mathrm{d}x\,\mathrm{d}y \tag{15.11.34}$$

where $\gamma_m = m\pi/L_x$ and $\gamma_n = n\pi/L_y$.

As the modes are orthogonal, the product of any two different mode shape functions when integrated over the surface of the plate is zero. Thus, expression (15.11.34) becomes

$$\frac{S}{2} A_{mn}\left[D(\gamma_m^2 + \gamma_n^2) - \rho h\omega^2\right] \tag{15.11.35}$$

where $m' = m$, $n' = n$ and $S = L_x L_y$ is the area of the plate.

The plate resonance frequencies ω_{mn} are defined as the values of ω required to make (15.11.35) equal to zero. Thus,

$$\omega_{mn}^2 = \frac{D}{\rho h}\left(\gamma_m^2 + \gamma_n^2\right) \tag{15.11.36}$$

and (15.11.35) can be written as

$$\frac{\rho h S}{2} A_{mn}\left[\omega_{mn}^2 - \omega^2\right] \tag{15.11.37}$$

Multiplying the right side of (15.11.30) by $\psi_{m',n'}$ and integrating over the area S of the plate gives

$$M_x \int_0^{L_x} \int_0^{L_y} \sin(\gamma_{m'} x) \sin(\gamma_{n'} y)$$

$$\times \left[\frac{\partial}{\partial x} (\delta(x-x_1) - \delta(x-x_2)) \; (u(y-y_1) - u(y-y_2)) \right. \tag{15.11.38}$$

$$\left. + \frac{\partial}{\partial y} (\delta(y-y_1) - \delta(y-y_2)) \; (u(x-x_1) - u(x-x_2)) \right] \mathrm{d}y \mathrm{d}x$$

The above expression can be evaluated to give

$$\frac{-2M_x(\gamma_m^2 + \gamma_n^2)}{\gamma_m \gamma_n} \left(\cos\gamma_m x_1 - \cos\gamma_m x_2\right) \left(\cos\gamma_n y_1 - \cos\gamma_n y_2\right) \tag{15.11.39}$$

where $m' = m$ and $n' = n$.

An expression for the modal amplitude A_{mn} is obtained by equating expressions (15.11.35) and (15.11.39) to give

$$A_{mn} = \frac{4M_x(\gamma_m^2 + \gamma_n^2)}{\rho h S(\omega_{mn}^2 - \omega^2)\, \gamma_m \gamma_n} \left(\cos\gamma_m x_2 - \cos\gamma_m x_1\right) \left(\cos\gamma_n y_1 - \cos\gamma_n y_2\right) \tag{15.11.40}$$

where M_x is defined by (15.11.28).

15.11.2 Thick actuators

These actuators are generally thick compared to their other dimensions and act by expanding across their thickness in the direction of polarization. The voltage is applied to the polarized surfaces as for the thin actuators, but in this case, the expansion of the thickness rather than expansion in the other dimensions is used to perform useful work. Stack actuators (thick actuators) are made by bonding many thin actuators together as shown in Fig. 15.30. Commercially available stack actuators usually consist of up to 100 thin actuators (0.3 to 1 mm thick) stacked together, with diameters ranging from 15 to 50 mm.

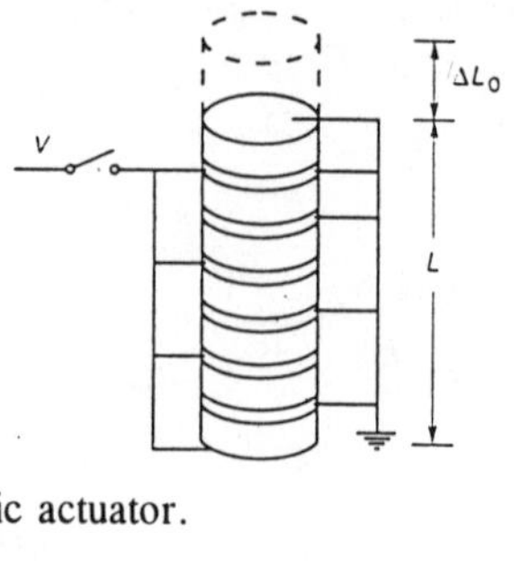

Fig. 15.30 Stack piezoelectric actuator.

When unloaded, the free expansion of this type of actuator is given by

$$\Delta L_0 = d_{33}\, n\, V \tag{15.11.41}$$

where n is the number of actuators in the stack, V is the applied voltage and d_{33} is the charge or strain constant for expansion in the thickness direction.

If the stacked actuator is to be used as a shaker, it should be preloaded with a compressive force greater than the maximum required dynamic force, as the actuators work much better at pushing than pulling.

The extension of a stack actuator when driving a stiffness k_s is given by

$$\Delta L = \Delta L_0 \frac{k_a}{k_a + k_s} \tag{15.11.42}$$

where k_a and k_s represent the stiffnesses of the actuator and structure being driven respectively.

The effective force that the actuator can generate is given by

$$F = k_a \Delta L_0 \left[1 - \frac{k_a}{k_a + k_s} \right] \tag{15.11.43}$$

The resonance frequency of an unloaded actuator is given by

$$f_0 = \frac{1}{2\pi} \sqrt{\frac{2k_a}{m}} \tag{15.11.44}$$

where m is the mass of the actuator and ΔL_0 is its extension.

When the actuator is used to drive an additional mass M or a structure with a modal mass M, the effective resonance frequency f_0' is given by

$$f_0' = f_0 \sqrt{\frac{m}{m + 2M}} \tag{15.11.45}$$

The upper limiting frequency of operation is given by

$$f_{max} = \frac{i_{max}}{2cU_0} \tag{15.11.46}$$

The quantity C is the total capacitance of the actuator (farads), i_{max} (amps) is the available current from the driving amplifier and U_o is the nominal operating voltage (usually 1000 V).

One problem with piezoelectric actuators (both the thin sheet type and stack type), which is also experienced by magnetostrictive actuators, is their hysteresis property which means that the expansion as a function of a given applied electric field (or voltage) is dependent upon whether the field level is rising or falling. It is also a non-linear function of the applied electric field in either direction. This results in non-linear excitation of a structure attached to one of these actuator types. Thus, if the exciting voltage used to drive the actuator is sinusoidal, the resulting structure motion will contain additional frequencies. In practice, a strong signal at the first harmonic of the exciting frequency is usually found. This poses obvious problems when these actuators are used in active control systems with linear controllers. Clearly, to be effective, these actuators need to be driven with non-linear controllers, an example of which could contain a tapped delay line whose weights are adjusted using a genetic algorithm.

15.12 SMART STRUCTURES

Smart structures are those which contain sensors and/or actuators as an integral part. Passive smart structures are those which contain only sensing elements which allow their state at any particular time to be determined. Such sensors could include piezoelectric film (PVDF) or optical fibres. Active smart structures contain in-built actuators as well as sensors which enable them to respond automatically to correct some undesirable state detected by the sensors.

Actuators which have been used in laboratory experiments in the past include piezoelectric film, piezoceramic crystal or shape memory alloy. Magnetostrictive terfenol may also be useful for this purpose, but is more likely to be bonded to the external surface of structures rather than embedded in them.

Structures that particularly lend themselves to the integration of sensors and actuators are carbon fibre and glass fibre composite structures which are made by laminating the glass or carbon cloth with a suitable epoxy resin. However, other non-composite structures can be made into smart structures by bonding actuators and sensors to their surface.

15.12.1 Novel actuator configurations

One problem with applying piezoceramic stacks, magnetostrictive rods, hydraulic or electrodynamic actuators to a structure in the traditional way is the need for a reaction mass and support for the actuator. The need for a reaction mass makes the application of active vibration control using these actuators impractical in many situations. However, the need for this reaction mass can be eliminated by using a more imaginative actuator configuration which applies control bending moments rather than control forces to the structure. For stiffened structures such as aircraft fuselages or submarine hulls, implementation of the actuator configurations is even more convenient.

Two possible actuator configurations which need no reaction mass are illustrated in Figs. 15.31(a) and (b). The first figure illustrates the arrangement where stiffeners already existed on the panel, and the second figure shows an alternative arrangement which might be used if no prior stiffeners existed. The configurations shown for panel vibration control can easily be extended to cylindrical structures such as aircraft fuselages and submarine hulls.

15.13 ELECTRORHEOLOGICAL FLUIDS

No chapter on actuators for active noise and vibration control would be complete

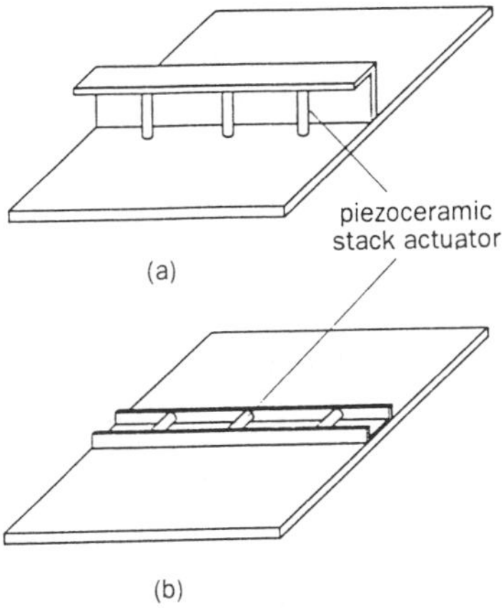

Fig. 15.31 Actuator configurations for applying a bending moment to a panel, with no need for a reaction mass.

without a discussion of electrorheological fluids. The most common type of electrorheological fluid being investigated commercially is the class of dielectric oils doped with semiconductor particle suspensions. On application of an electric field of sufficient strength, the particles form chains which link across the electrodes, resulting in an apparent change in viscosity (or resistance to flow). Perhaps the most obvious application of these types of fluid in active vibration control is to provide a variable damping force such as discussed in Section 12.3 for a semi-active vehicle suspension or for an active engine mount. Although it is possible to switch the fluids from their inactive to active state in 3–5 ms, it takes somewhat longer to reverse the process, thus severely restricting the useful frequency range of active dampers and isolators constructed from these materials. This seems to be the current problem which must be surmounted before these fluids will achieve widespread use for active vibration control. Other problems are associated with the relatively high voltage requirements (2–10 kV) and separation of the fluid/solid particles when the fluid is in its inactive state (Brennan, 1993, personal communication). More details on the current state of the art in the development of these fluids for active vibration control is discussed in the literature (Morishita and Mitsui, 1992; Sproston and Stanway, 1992; Weiss *et al.*, 1992; Austin, 1993).

REFERENCES

Austin, S.A. (1993). The vibration damping effect of an electrorheological fluid. *Journal of Vibration and Acoustics*, **115**, 136–140.

Baz, A. and Poh, S. (1988). Performance of an active control system with piezoelectric actuators. *Journal of Sound and Vibration*, **126**, 327–343.

Baz, A., Imam, K. and McCoy, J. (1990). Active vibration control of flexible beams using shape memory actuators. *Journal of Sound and Vibration*, **140**, 437–456.

Butter, C.D. and Hocker, G.B. (1978). Fibre optics strain gauge. *Applied Optics*, **17**, 2867–2869.

Buttler, J.L. (1988). *Application Manual for the Design of Etrema Terfenol-D Magnetostrictive Transducers*. Etrema Products Inc, 306 South 16th St, Ames, IA, USA.

Buzan, F.T. and Hedrick, J.K. (1983). Lateral active pneumatic suspensions for rail vehicle control. In *Proceedings of the 1983 American Control Conference*. San Francisco, 263–269.

Cho, D. and Hedrick, J.K. (1985). Pneumatic actuators for vehicle active suspension applications. *ASME Journal of Dynamic Systems Measurement and Control*, **107**, 67–72.

Clark, W.W., Robertshaw, H.H. and Warrington, T.J. (1990). A comparison of actuators for vibration control of the planar vibrations of a flexible cantilevered beam. *Journal of Intelligent Material Systems and Structures*, **1**, 289–308.

Clark, R.L., Fuller, C.R. and Wicks, A. (1991). Characterisation of multiple piezoelectric actuators for structural excitation. *Journal of the Acoustical Society of America*, **90**, 346–357.

Columbia Research Laboratories. *Theory and application of linear variable differential transformers*. Columbia Technical Publications Woodlyn, PA.

Cox, D.E. and Lindner, D.K. (1991). Active control for vibration suppression in a flexible beam using a modal domain optical fibre sensor. *ASME Journal of Vibration and Acoustics*, **113**, 369–382.

Crawley, E.F. and de Luis, J. (1987). Use of piezoelectric actuators as elements of intelligent structures. *AIAA Journal*, **25**, 1373–1385.

Crawley, E.F. and Anderson, E.H. (1990). Detailed models of piezoceramic actuation of beams. *Journal of Intelligent Material Systems and Structures*, **1**, 4–25.

Crolla, D.A. (1988). Theoretical comparisons of various active suspension systems in terms of performance and power requirements. In *Proceedings of the Institution of Mechanical Engineers International Conference on Advanced Suspensions*. London, 1-9.

Dimitriadis, E.K., Fuller, C.R. and Rogers, C.A. (1991). Piezoelectric actuators for distributed vibration excitation of thin plates. *ASME Journal of Vibration and Acoustics*, **113**, 100–107.

Edge Technologies (1990). RA101 *Actuator Handbook*.

Forward, R.L. and Swigert, C.J. (1981). Electronic damping of orthogonal bending modes in a cylindrical mast. *Journal of Spacecraft and Rockets*, **18**, 5–17.

Hiller, M.W, Bryant, M.D. and Umega, J. (1989). Attenuation and transformation of vibration through active control of magnetostrictive terfenol. *Journal of Sound and Vibration*, **134**, 507–519.

Kim, B.Y., Blake, J.H., Huong, S.Y. and Shaw, H.J. (1987). Use of highly elliptical core fibres in two-mode fibre devices. *Optical Letters*, **12**, 739.

Kim, S.J. and Jones, J.D. (1991). Optimal design of piezoactuators for active noise and vibration control. *AIAA Journal*, **29**, 2047–2053.

King, S.P. (1988). The minimisation of helicopter vibration through blade design and active control. *Aeronautical Journal*, **92**, 247–263.

Layton, M.R. and Bucaro, J.A. (1979). Optical fibre acoustic sensor utilising mode-mode interference. *Applied Optics*, **18**, 666.

Lee, C.K. and Moon, F.C. (1990a). *Transactions of ASME: Journal of Applied Mechanics*, **57**, 434–441.

Lee, C.K. and Moon F.C. (1990b). Laminated piezoelectric plates for torsion and bending sensors and actuators. *Journal of the Acoustical Society of America*, **85**, 2432–2439.

Moffett, M.B., Clark, A.E., Wun-Fogle, M., Linberg, J., Teter, J.P. and McLaughlin, E.A. (1991a). Characterisation of terfenol-D for magnetostrictive transducers. *Journal of the Acoustical Society of America*, **89**, 1448–1455.

Moffett, M.B., Powers, J.M. and Clark, A.E. (1991b). Comparison of terfenol-D and PZT-4 power limitations. *Journal of the Acoustical Society of America*, **90**, 1184–1185.

Moore, J.H. (1988). Linear variable inductance position transducer for suspension systems. In *Proceedings of the Institution of Mechanical Engineers International Conference on Advanced Suspensions*. London, 75–82.

Morishita, S and Mitsui, J. (1992). Controllable squeeze film damper (an application of electrorheological fluid). *Journal of Vibration and Acoustics*, **114**, 354–357.

Politansky, H. and Pilkey, W.D. (1989). Suboptimal feedback vibration control of a beam with a proof-mass actuator. *Journal of Guidance and Control*, **12**, 691–697.

Rogers, C.A. (1990). Active vibration and structural acoustic control of shape memory alloy hybrid composites: experimental results. *Journal of the Acoustical Society of America*, **88**, 2803–2811.

Rogers, C.A., Liang, C. and Fuller, C.R. (1991). Modelling of shape memory alloy hybrid composites for structural acoustic control. *Journal of the Acoustical Society of America*, **89**, 210–220.

Saunders, W.R., Robertshaw, H.H. and Rogers, C.A. (1991). Structural acoustic control of a shape memory alloy composite beam. *Journal of Intelligent Material Systems and Structures*, **2**, 508–527.

Serridge, M. and Torben, R.L. (1987). *Piezoelectric accelerometer and vibration preamplifier handbook*. Brüel and Kjaer, Copenhagen.

Smits, J.G., Dalke, S.I. and Cooney, T.K. (1991). The constituent equations of piezoelectric bimorphs. *Sensors and Actuators A*, **28**, 41–61.

Sproston, J.L. and Stanway, R. (1992). Electrorheological fluids in vibration isolation. In *Proceedings of Actuator '92, the 3rd International Conference on New Actuators*. VDI/VDE-Technogiezentrum Informationstechnik GmbH, Bremen, Germany, 116–117.

Stayner, R.M. (1988). Suspensions for agricultural vehicles. In *Proceedings of the Institution of Mechanical Engineers International Conference on Advanced Suspensions*. London, 133–137.

Timoshenko, S. and Woinowsky-Krieger (1959). *Theory of plates and shells*, 2nd edn. McGraw-Hill, New York, Chapters 1, 2 and 4.

Wallace, C.E. (1972). Radiation resistance of a rectangular panel. *Journal of the Acoustical Society of America*, **51**, 946–952.

Weiss, K.D., Coulter, J.P. and Carlson, J.D. (1992). Electrorheological materials and their usage in intelligent material systems and structures, Part 1: mechanisms, formulations and properties. In *Recent Advances in Sensory and Adaptive Materials and their Applications*. Techonomic Publishing Co., Lancaster PA, USA.

Wolfe, P.T. and Jolly, M.R. (1990). Velocity transducer for vehicle suspension system. US Patent No. 4979 573.

Appendix. A brief review of some results of linear algebra

The purpose of this appendix is to provide a brief review of many of the particular results of linear algebra used in this book. For more extensive treatments, the reader should consult any of the standard textbooks on linear algebra, such as Bellman (1970), Noble (1969) and Noble and Daniél (1977). There are also a number of specialized software packages which deal explicitly with linear algebra manipulations, such as Matlab. These packages are useful for solving the matrix equations encountered in active noise and vibration control research.

A1 MATRICES AND VECTORS

An $(m \times n)$ matrix is a collection of mn numbers (complex or real), a_{ij} $(i=1,2,\ldots,m,\ j=1,2,\ldots,n)$, written in an array of m row and n columns:

$$\boldsymbol{A} = \begin{bmatrix} a_{11} & a_{12} & \cdots & a_{1n} \\ a_{21} & a_{22} & \cdots & a_{2n} \\ \vdots & \vdots & \vdots & \vdots \\ a_{ml} & a_{m2} & \cdots & a_{mn} \end{bmatrix} \tag{A1}$$

The term a_{ij} appears in the ith row and jth column of the array. If the number of rows is equal to the number of columns, the matrix is said to be square.

An m vector, also referred to as an $(m \times 1)$ vector or column m vector, is a matrix with 1 column and m rows:

$$\boldsymbol{a} = \begin{bmatrix} a_1 \\ a_2 \\ \vdots \\ a_m \end{bmatrix} \tag{A2}$$

Throughout this book, matrices are denoted by bold capital letters, such as $\boldsymbol{A}$, while vectors are denoted by bold lower case letters, such as $\boldsymbol{a}$.

A2 ADDITION, SUBTRACTION AND MULTIPLICATION BY A SCALAR

If two matrices have the same number of rows and columns, they can be added or subtracted. When adding matrices, the individual corresponding terms are added. For example, if

$$\boldsymbol{A} = \begin{bmatrix} a_{11} & a_{12} & \cdots & a_{1n} \\ a_{21} & a_{22} & \cdots & a_{2n} \\ \vdots & \vdots & \vdots & \vdots \\ a_{ml} & a_{m2} & \cdots & a_{mn} \end{bmatrix} \tag{A3}$$

and

$$\boldsymbol{B} = \begin{bmatrix} b_{11} & b_{12} & \cdots & b_{1n} \\ b_{21} & b_{22} & \cdots & b_{2n} \\ \vdots & \vdots & \vdots & \vdots \\ b_{ml} & b_{m2} & \cdots & b_{mn} \end{bmatrix} \tag{A4}$$

then

$$\boldsymbol{A} + \boldsymbol{B} = \begin{bmatrix} a_{11}+b_{11} & a_{12}+b_{12} & \cdots & a_{1n}+b_{1n} \\ a_{21}+b_{21} & a_{22}+b_{22} & \cdots & a_{2n}+b_{2n} \\ \vdots & \vdots & \vdots & \vdots \\ a_{ml}+b_{ml} & a_{m2}+b_{m2} & \cdots & a_{mn}+b_{mn} \end{bmatrix} \tag{A5}$$

When subtracting matrices, the individual terms are subtracted. Note that matrix addition is commutative, as

$$\boldsymbol{A} + \boldsymbol{B} = \boldsymbol{B} + \boldsymbol{A} \tag{A6}$$

It is also associative, as

$$(\boldsymbol{A} + \boldsymbol{B}) + \boldsymbol{C} = \boldsymbol{A} + (\boldsymbol{B} + \boldsymbol{C}) \tag{A7}$$

Matrices can also be multiplied by a scalar. Here the individual terms are each multiplied by a scalar. For example, if k is a scalar,

$$kA = \begin{bmatrix} ka_{11} & ka_{12} & \cdots & ka_{1n} \\ ka_{21} & ka_{22} & \cdots & ka_{2n} \\ \vdots & \vdots & \vdots & \vdots \\ ka_{m1} & ka_{m2} & \cdots & ka_{mn} \end{bmatrix} \tag{A8}$$

A3 MULTIPLICATION OF MATRICES

Two matrices $\boldsymbol{A}$ and $\boldsymbol{B}$ can be multiplied together to form the product $\boldsymbol{AB}$ if the number of columns in $\boldsymbol{A}$ is equal to the number of rows in $\boldsymbol{B}$. If, for example, $\boldsymbol{A}$ is an $(m \times p)$ matrix, and $\boldsymbol{B}$ is a $(p \times n)$ matrix, then the product $\boldsymbol{AB}$ is defined by

$$\boldsymbol{C} = \boldsymbol{AB} \tag{A9}$$

where $\boldsymbol{C}$ is an $(m \times n)$ matrix, the terms of which are defined by

$$c_{ij} = \sum_{k=1}^{p} a_{ik} b_{kj} \tag{A10}$$

Matrix multiplication is associative, with the product of three (or more) matrices defined by

$$\boldsymbol{ABC} = (\boldsymbol{AB})\boldsymbol{C} = \boldsymbol{A}(\boldsymbol{BC}) \tag{A11}$$

Matrix multiplication is also distributive, where

$$\boldsymbol{A}(\boldsymbol{B} + \boldsymbol{C}) = \boldsymbol{AB} + \boldsymbol{AC} \tag{A12}$$

However, matrix multiplication is not commutative, as, in general,

$$\boldsymbol{AB} \neq \boldsymbol{BA} \tag{A13}$$

In fact, while the product $\boldsymbol{AB}$ may be formed, it may not be possible to form the product $\boldsymbol{BA}$.

The identity matrix $\boldsymbol{I}$ is defined as the $(p \times p)$ matrix with all principal diagonal elements equal to 1, and all other terms equal to zero:

$$\boldsymbol{I} = \begin{bmatrix} 1 & 0 & \cdots & 0 \\ 0 & 1 & \cdots & 0 \\ & & \ddots & \\ 0 & 0 & \cdots & 1 \end{bmatrix} \tag{A14}$$

For any $(m \times p)$ matrix $\boldsymbol{A}$, the identity matrix has the property

$$\boldsymbol{AI} = \boldsymbol{A} \tag{A15}$$

Similarly, if the identity matrix is $(m \times m)$,

$$\boldsymbol{IA} = \boldsymbol{A} \tag{A16}$$

A4 TRANSPOSITION

If a matrix is transposed, the rows and columns are interchanged. For example, the transpose of the $(m \times n)$ matrix $\boldsymbol{A}$, denoted by $\boldsymbol{A}^{\mathrm{T}}$, is defined as the $(n \times m)$ matrix $\boldsymbol{B}$,

$$\boldsymbol{A}^{\mathrm{T}} = \boldsymbol{B} \tag{A17}$$

where

$$b_{ij} = a_{ji} \tag{A18}$$

The transpose of a matrix product is defined by

$$(\boldsymbol{AB})^{\mathrm{T}} = \boldsymbol{B}^{\mathrm{T}}\boldsymbol{A}^{\mathrm{T}} \tag{A19}$$

This result can be extended to products of more than two matrices, such as

$$(\boldsymbol{ABC})^{\mathrm{T}} = \boldsymbol{C}^{\mathrm{T}}\boldsymbol{B}^{\mathrm{T}}\boldsymbol{A}^{\mathrm{T}} \tag{A20}$$

If

$$\boldsymbol{A} = \boldsymbol{A}^{\mathrm{T}} \tag{A21}$$

then the matrix $\boldsymbol{A}$ is said to be symmetric.

The Hermitian transpose of a matrix is defined as the complex conjugate of the transposed matrix (when taking the complex conjugate of a matrix, each term in the matrix is conjugated). Therefore, the Hermitian transpose of the $(m \times n)$ matrix $\boldsymbol{A}$, denoted by $\boldsymbol{A}^{\mathrm{H}}$, is defined as the $(n \times m)$ matrix $\boldsymbol{B}$

$$\boldsymbol{A}^{\mathrm{H}} = \boldsymbol{B} \tag{A22}$$

where

$$b_{ij} = a_{ji}^{*} \tag{A23}$$

If $\boldsymbol{A} = \boldsymbol{A}^{\mathrm{H}}$, then $\boldsymbol{A}$ is said to be a Hermitian matrix.

A5 DETERMINANTS

The determinant of the (2×2) matrix $\boldsymbol{A}$, denoted $|\boldsymbol{A}|$, is defined as

$$|\boldsymbol{A}| = \begin{vmatrix} a_{11} & a_{12} \\ a_{21} & a_{22} \end{vmatrix} = a_{11}a_{22} - a_{12}a_{21} \tag{A24}$$

The minor M_{ij} of the element a_{ij} of the square matrix A is the determinant of the matrix formed by deleting the ith row and jth column from A. For example, if A is a (3 × 3) matrix,

$$A = \begin{bmatrix} a_{11} & a_{12} & a_{13} \\ a_{21} & a_{22} & a_{23} \\ a_{31} & a_{32} & a_{33} \end{bmatrix} \tag{A25}$$

the minor M_{11} is found by taking the determinant of A with the first column and first row of numbers deleted,

$$M_{11} = \begin{vmatrix} a_{22} & a_{23} \\ a_{32} & a_{33} \end{vmatrix} \tag{A26}$$

The cofactor C_{ij} of the element a_{ij} of the matrix A is defined by

$$C_{ij} = (-1)^{i+j} M_{ij} \tag{A27}$$

The determinant of a square matrix of arbitrary size is equal to the sum of the products of the elements and their cofactors along any column or row. For example, the determinant of the (3 × 3) matrix A above can be found by adding the products of the elements and their cofactors along the first row:

$$|A| = a_{11}C_{11} + a_{12}C_{12} + a_{13}C_{13} \tag{A28}$$

Therefore, the determinant of a large square matrix can be broken up into a problem of calculating the determinants of a number of smaller square matrices.

If two matrices A and B are square, then

$$|AB| = |A|\,|B| \tag{A29}$$

A matrix is said to be singular if its determinant is equal to zero.

A6 MATRIX INVERSES

The inverse A^{-1} of the matrix A is defined by

$$AA^{-1} = A^{-1}A = I \tag{A30}$$

The matrix A must be square and, as we will see, be non-singular for the inverse to be defined.

The inverse of a matrix A can be derived by first calculating the adjoint $\hat{A}$ of the matrix. The adjoint $\hat{A}$ is defined as the transpose of the matrix of cofactors of A,

$$\hat{A} = \begin{bmatrix} C_{11} & C_{12} & \cdots & C_{1m} \\ C_{21} & C_{22} & \cdots & C_{2m} \\ \vdots & \vdots & \vdots & \vdots \\ C_{m1} & C_{m2} & \cdots & C_{mm} \end{bmatrix} \tag{A31}$$

The inverse A^{-1} of the matrix A is equal to the adjoint of A multiplied by the reciprocal of the determinant of A,

$$A^{-1} = \frac{1}{|A|}\hat{A} \tag{A32}$$

Note that if the matrix A is singular, such that the determinant is zero, the inverse is not defined.

While the definition given in (A32) is correct, using it to calculate a matrix inverse is inefficient for all by the smallest matrices (as the order of operations increases with the size m of the matrix by $m!$). There are a number of algorithms which require in the order of m^3 operations to compute the inverse of an arbitrary square matrix (outlined in many of the standard texts and in numerical methods books such as Press *et al.* (1986)). For matrices of special form, such as Toeplitz matrices, which are symmetric matrices in which the elements along any diagonal are equal (often encountered in adaptive signal processing work), the order of operations can be further reduced (to m^2 operations for Toeplitz matrices).

If we have an over determined (or under determined) system of equations written as $Ax = B$ where A is an $n \times m$ matrix ($n \neq m$), and we wish to solve for x, then

$$x = (A^{T}A)^{-1} A^{T}B \tag{A33}$$

where $(A^{T}A)^{-1} A^{T}$ is the pseudo-inverse (or generalized inverse) of A.

A7 RANK OF A MATRIX

The rank of the $(m \times n)$ matrix A is the maximum number of linearly independent rows of A and the maximum number of linearly independent columns of A. Alternatively, the rank of A is a positive integer r such that some $(r \times r)$ submatrix of A, formed by deleting $(m-r)$ rows and $(n-r)$ columns, is non-singular, whereas no $((r+1) \times (r+1))$ submatrix is non-singular. If rank A is equal to the number of columns or the number of rows of A, then A is said to have full rank.

A8 POSITIVE AND NON-NEGATIVE DEFINITE MATRICES

A matrix A is said to be positive definite if the quantity $x^{H}Ax$ is positive for all non-zero vectors x; if the quantity is simply non-negative, then A is said to be non-negative definite.

For A to be positive definite, all of the leading minors must be positive; that is,

$$a_{11} > 0, \quad \begin{vmatrix} a_{11} & a_{12} \\ a_{21} & a_{22} \end{vmatrix} > 0, \quad \begin{vmatrix} a_{11} & a_{12} & a_{13} \\ a_{21} & a_{22} & a_{23} \\ a_{31} & a_{32} & a_{33} \end{vmatrix} > 0, \ \ldots \text{ etc} \qquad \text{(A34)}$$

For A to be non-negative definite, all of the leading minors must be non-negative.

A9 EIGENVALUES AND EIGENVECTORS

Let A be a (square) $(n \times n)$ matrix. The polynomial $|\lambda I - A|$ is referred to as the characteristic equation of A. The solutions to the characteristic equation are the eigenvalues of A. If λ_i is an eigenvalue of A, then there exists at least one vector q_i which satisfies the relationship

$$Aq_i = \lambda_i q_i \qquad \text{(A35)}$$

The vector q_i is an eigenvector of A. If the eigenvalue λ_i is not repeated then the eigenvector q_i is unique. If an eigenvector λ_i is real, then the entries in the associated eigenvector q_i are real; if λ_i is complex, then so too are the entries in q_i.

The eigenvalues of a Hermitian matrix are all real, and if the matrix is also positive definite the eigenvalues are also all positive. If a matrix is symmetric, then the eigenvalues are also all real. Further, it is true that

$$|A| = \prod_{i=1}^{n} \lambda_i \qquad \text{(A36)}$$

If A is singular, then there is at least one eigenvalue equal to zero.

A10 ORTHOGONALITY

If a square matrix A has the property $A^H A = AA^H = I$, then the matrix A is said to be orthogonal. The eigenvalues of A then have a magnitude of unity. If q_i is an eigenvector associated with λ_i, and q_j is an eigenvector associated with λ_j, and if $\lambda_i \neq \lambda_j$ and $q_i^H q_j = 0$, then the vectors q_i and q_j are said to be orthogonal.

The eigenvectors of a Hermitian matrix are all orthogonal. Further, it is common to normalise the eigenvectors such that $q_i^H q_i = 1$, in which case the eigenvectors are said to be orthonormal. A set of orthonormal eigenvectors can be expressed as columns of a unitary matrix Q

$$Q = (q_1, q_2, \cdots, q_n) \qquad \text{(A37)}$$

which means that

$$Q^{H}Q = QQ^{H} = I \tag{A38}$$

The set of equations which define the eigenvectors, expressed for a single eigenvector in (A35), can now be written in matrix form as

$$AQ = Q\Lambda \tag{A39}$$

where Λ is the diagonal matrix of eigenvalues,

$$\Lambda = \begin{bmatrix} \lambda_1 & 0 & \cdots & 0 \\ 0 & \lambda_2 & \cdots & 0 \\ & & \ddots & \\ 0 & 0 & \cdots & \lambda_n \end{bmatrix} \tag{A40}$$

Post-multiplying both sides of (A39) by Q^{H} yields

$$A = Q\Lambda Q^{H} \tag{A41}$$

or

$$Q^{H}AQ = \Lambda \tag{A42}$$

Equations (A.41) and (A.42) define the orthonormal decomposition of A, where A is re-expressed in terms of its eigenvectors and eigenvalues.

A11 VECTOR NORMS

The norm of a vector A, expressed as $\|x\|$, is the length or size of the vector x. The most common norm is the Euclidean norm, defined for the vector $x = (x_1, x_2, \ldots, x_n)$ as

$$\|x\| = \left(\sum_{i=1}^{n} x_i^2 \right)^{1/2} \tag{A43}$$

Three properties of vector norms are:

1. $\|x\| \geq 0$ for all x, where the norm is equal to zero only if $x = \mathbf{0}$.
2. $\|ax\| = |a| \, \|x\|$ for any scalar a and all x.
3. $\|x+y\| \leq \|x\| + \|y\|$ for all x and y.

REFERENCES

Press, W.H., Flannery, B.P., Tenkolsky, S.A. and Vettering, W.T. (1986) *Numerical Recipes: The Art of Scientific Computing*. Cambridge University Press, Cambridge.

Index